T0234382

TRANSACTIONS

OF THE

INTERNATIONAL

ASTRONOMICAL UNION

VOLUME XXB – PROCEEDINGS

INTERNATIONAL ASTRONOMICAL UNION

UNION ASTRONOMIQUE INTERNATIONALE

TRANSACTIONS

OF THE

INTERNATIONAL ASTRONOMICAL UNION

VOLUME XXB

PROCEEDINGS OF THE TWENTIETH

GENERAL ASSEMBLY

BALTIMORE 1988

Edited by

DEREK McNALLY

General Secretary of the Union

KLUWER ACADEMIC PUBLISHERS

DORDRECHT / BOSTON / LONDON

Library of Congress Cataloging in Publication Data

International Astronomical Union. General Assembly (20th : 1988 :
Baltimore, Md.)
Proceedings of the Twentieth General Assembly, Baltimore, 1988 /
edited by Derek McNally.
p. cm. -- (Transactions of the International Astronomical
Union ; v. 20B)
At head of title: International Astronomical Union, Union
astronomique internationale.
ISBN 978-0-7923-0582-8 ISBN 978-94-009-0497-2 (eBook)
DOI 10.1007/978-94-009-0497-2

1. International Astronomical Union. General Assembly-
-Congresses. 2. Astronomy--Congresses. I. McNally, Derek.
II. Title. III. Series.
QB1.I6 vol. 20B
520 s--dc20
[520'.6] 89-24633

ISBN 978-0-7923-0582-8

Published on behalf of
the International Astronomical Union
by
Kluwer Academic Publishers, P.O. Box 17, 3300 AA Dordrecht, The Netherlands.

Kluwer Academic Publishers incorporates
the publishing programmes of
D. Reidel, Martinus Nijhoff, Dr W. Junk and MTP Press.

Sold and distributed in the U.S.A. and Canada
by Kluwer Academic Publishers,
101 Philip Drive, Norwell, MA 02061, U.S.A.

In all other countries, sold and distributed
by Kluwer Academic Publishers Group,
P.O. Box 322, 3300 AH Dordrecht, The Netherlands.

PRESIDENT DE L'UNION ASTRONOMIQUE INTERNATIONALE

JORGE SAHADE

PRESIDENT OF THE INTERNATIONAL ASTRONOMICAL UNION

1985–1988

CONTENTS

Preface

The XXth General Assembly of the International Astronomical Union was held in Baltimore, Maryland USA from August 02 to 11, 1988. The Inaugural Ceremony on August 02 was held in the presence of representatives of the United States Government, the State of Maryland, the City of Baltimore and the host institution -the Johns Hopkins University- as well as of the National and Local Organising Committees.

The scientific programme maintained the high standards of the Union and the scientific proceedings may be found either in this volume or in volume 8 of Highlights of Astronomy. The scientific programme was organised by the 40 Commission Presidents and coordinated by the General Secretary (1985-1988), Dr. J.-P. Swings. The local arrangements were effectively made through the National Organising Committee under the Chairmanship of Prof. F. Drake and the Local Organising Committee under the co-Chairmanship of Prof. A. Davidsen and Dr. R. Giacconi. The smooth day to day operation of the meeting resulted from the incomparable dedication of Karen Weinstock and Harold Screen.

This volume summarises the work of the XXth General Assembly. Chapter I contains the discourses at the Inaugural Ceremony and the music for the specially composed Fanfare for Brass Ensemble. Chapter II contains the proceedings of the Extraordinary General Assembly of August 02, 1988. Chapter III contains the proceedings of both sessions of the General Assembly, the report of the Finance Committee and the Accounts, the Resolutions adopted and other matters of Union's business. Chapter IV is the Report of the Executive Committee 1985-1988. Chapters II, III and IV record the business of the Union. Chapter V records the business and scientific sessions of the Commissions during the General Assembly. Chapters VI, VII and VIII contain the Statutes, By-Laws and Working Rules of the Union, the membership list (individual, geographical and by Commission) and the IAU Style book respectively. In order to make the sections of this volume of the Transactions more easily accessible, the Functions and Statutes of the Union, Chapter VI, have been printed on pink paper and the Membership details, Chapter VII, on green paper. A list of commonly used abbreviations in this volume appears immediately after the list of contents.

In the preparation of the volume I am deeply grateful to my predecessor, J.-P. Swings for his unstinted help, advice and translation of the Statute revisions into French; to Roger Cayrel for his work in translating the Resolutions into French; to George Wilkins for undertaking and bringing to fruition the preparation of the current IAU Style Book -and to the editors of the principal astronomical journals for their valuable suggestions; in particular to Monique Orine for her work in preparing the text of this volume and to both Monique Orine and Huguette Gigan of the Paris Secretariat for the preparation of the final camera ready copy during a period of considerable difficulty occasioned by a postal strike and the move of the Secretariat.

IAU Secretariat
98bis, bd Arago
75014 Paris
France

Derek McNally
General Secretary
April 24, 1989

Abbreviations

BIPM	Bureau International des Poids et Mesures
CCDS	Consultative Committee for the Definition of the Second
CCIR	International Radio Consultative Committee
CODATA	Committee on Data for Science & Technology
COSPAR	Committee on Space Research
COSTED	Committee on Science & Technology in Developing Countries
CTS	Committee on the Teaching of Science
EPS	European Physical Society
ESA	European Space Agency
ESO	European Southern Observatory
FAGS	Federation of Astronomical & Geophysical Services
IAF	International Astronautical Federation
IAG	International Association of Geodesy
ICSU	International Council of Scientific Unions
IERS	International Earth Rotation Service
IRAS	Infrared Astronomy Satelite
IUCAF	Inter-Union Commission on Frequency Allocations for Radio Astronomy & Space Science
IUGG	International Union of Geodesy and Geophysics
IUPAP	International Union of Pure & Applied Physics
IUWDS	International Ursigram & World Day Service
NASA	National Aeronautics & Space Administration
NSF	National Science Foundation
OECD	Organisation of Economic Cooperation & Development
RAS	Radio Astronomy Service
RDSS	Radio-Determination Satellite Service
SCOPE	Scientific Committee on Problems of the Environment
SCOSTEP	Scientific Committee on Solar-Terrestrical Physics
UNESCO	United Nations Educational, Scientific & Cultural Organization
URSI	Union Radio Scientifique Internationale
WARC	World Administrative Radio Conference

CHAPTER I

INAUGURAL CEREMONY

2 August 1988

The Inaugural Ceremony was held at the Convention Center, Baltimore, in the presence of distinguished representatives from the Government of the United States, the State of Maryland, the City of Baltimore, the Johns Hopkins University, the National Aeronautics & Space Administration, the and the US National Committee for the IAU.

The chair was taken by Dr. A. Davidsen, co-Chairman of the Local Organizing Committee, from the Johns Hopkins University.

A rousing aubade, specially composed for the occasion by Elam Ray Sprenkle, was beautifully performed by the Annapolis Brass Quintet under the direction of David Cren.

An Aubade for Brass Quartet,

composed by Elam Ray Sprenkle,

Commissioned for the XXth General Assembly

An Aubade

for Brass Quartet

Commissioned for the Astro-Physics Convention

Baltimore, Md., Summer, 1988

Sprenkle

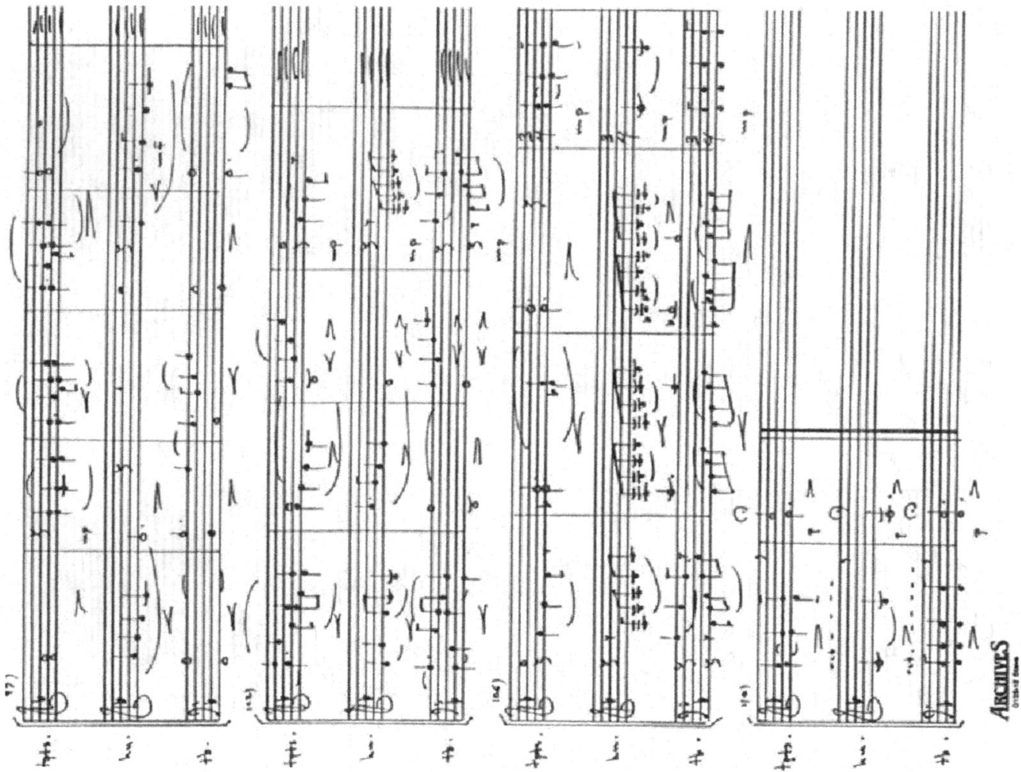

Address by Dr. A. Davidsen, co-Chairman of the Local Organizing Committee

"Dr Graham, Lt. Gov Steinberg, Mayor Schmoke, Members of the IAU, Distinguished Guests, Ladies and Gentlemen,

My name is Arthur Davidsen and I am co-Chairman of the Local Organizing Committee of the Twentieth General Assembly of the IAU. I have the honor of being the first to welcome you to Baltimore and to bring you greetings and good wishes from a veritable army of local astronomers and staff, who have labored extremely hard to arrange what we hope will be a very memorable meeting for each of you.

When we began planning this event in 1982, it was with the belief that Baltimore was on the verge of becoming one of the world capitals of astronomical research. This expectation was of course based on the establishment of the Space Telescope Science Institute on the Johns Hopkins Homewood Campus. In spite of the tragic event which has delayed the launch of the Hubble Space Telescope, I believe we have come a long way toward achieving our goal.

When this decade began there were only five astronomers in Baltimore; now there are over one hundred working here. From a narrow base of studies, primarily in ultraviolet astronomy, we have expanded to include a broad range of research in experimental, observational and theoretical astrophysics. In addition to the Hubble Space Telescope, Baltimore astronomers are involved in several other space missions, such as the Astro Observatory, also planned for launch aboard the space shuttle next year. Back on the ground, we are collaborating in the Magellan Project with the Carnegie Institution of Washington and the University of Arizona in the development of an eight meter telescope for Los Campanas, in Chile. If you take time to visit the Johns Hopkins campus this week or next, as I hope you will, you'll see that a major expansion of the Space Telescope Science Institute is nearing completion and that a magnificent new home for the Department of Physics and Astronomy and the Center for Astrophysical Sciences is being built directly across the street from the Institute. The successful sponsorship of this important meeting is yet another hallmark of the growth of astronomy here in Baltimore and will, we hope, set the stage for us to play an even more significant role in the decades to come.

On behalf of all the astronomers and staff of the Local Organizing Committee, I offer you a very warm welcome to the Twentieth General Assembly of the IAU and I have arranged the weather specifically to emphasize that point. And now I would like to introduce the co-Chairman of the Local Organizing Committee, Dr. Ricardo Giacconi, who is of course the Director of the Space Telescope Science Institute."

Address by Dr. R. Giacconi, co-Chairman of the Local Organizing Committee

"Ladies and Gentlemen, Distinguished Guests,

It was entirely appropriate that Arthur Davidsen should speak first, because although we are co-Chairmen, he did all the work, and I wanted to take this opportunity to make that clear.

We at Space Telescope Science Institute have been involved in helping to prepare this meeting, and we had the hope and fond expectation that the intellectual highlight of this meeting would be the fact that we would paper the walls with data from the now-unlaunched Hubble Space Telescope. As you know, the tragic accident that befell Challenger and the long period of recovery which is being required to put the NASA fleet back in operation has prevented us from doing that. Notwithstanding which, I hope you will take this opportunity to visit the Institute, to acquaint yourself with the instrumentation and with the details of how you may wish to operate and use the Space Telescope, those of you who do intend to use it. We have arranged, in a practical manner, buses to bring you up there with two visits a day, and to do that you just sign up at the Space Telescope desk.

The Association of Universities for Research in Astronomy (AURA) has asked me to transmit to you their warmest wishes, and the welcome that I am extending to you is not only from AURA, not only from the European and American staff of the Space Telescope Science Institute, but also from the literally thousands of people in the NASA/ESA family of institutions, its centers, its contractors, the academic and research institutions which have been involved now for so many years in the preparation of this very sophisticated, fantastic instrument. Their work is what is going to make it happen and they are extremely happy to share it with you. So once again, welcome to Baltimore."

Address by Dr. F. Drake, Chairman of the National Organizing Committe.

"I am here to welcome all of you on behalf of the American astronomical community.

Twenty-seven years have passed since the General Assembly last met in the United States. Twenty-seven eventful years in the history of astronomy, and I am sure there are many people here who remember those exciting days long ago in Berkeley and even perhaps the wonderful trip to the Napa Valley, when it was even hotter than it is in Baltimore. But so much has happened in that time.

When last we met in the United States, the words quasar and pulsar had never, not once, been uttered by a human voice in the entire history of the planet. The words "microwave background" would have meant nothing to any astronomer, and aperture synthesis was a hazy but fascinating concept understood by only a few. In those twenty-seven years, the words "Viking", "Venera", "Voyager", have taken on entirely new meanings, and indeed it is amazing to realize that in such a short time, we have explored almost all of the planets of the solar system.

It has been said that there have been two golden ages of astronomy; the first started at the time of Galileo and lasted for the next hundred years as the basic structure of the solar system and its laws were discovered. The second started but a few decades ago and is still rich with progress.

We astronomers have been privileged to live in it. The greatest discoveries continue: the events surrounding Supernova 1987A; the gravitational lenses; the eclipsing pulsar, destroying its companion before our very eyes. The potential for future discoveries is actually growing as we await the launch of the Hubble Space Telescope, follow with excitement the flight of two spacecraft to Phobos and watch the construction of a new generation of giant optical and radio telescopes on the Earth.

Astronomy continues as a prime cutting edge in the development of science. We have seen the way to understand the earliest moments of the history of the Universe. Astronomy may even serve as a source for major developments in physics, revealing in an indirect but profound way the nature of elementary particles. It may even tell us, someday, of the variety of biologies which exist in the Universe.

American astronomers have been fortunate and proud to be an important part of the recent golden age of astronomy. Our science is highly respected in our country and we enjoy generous support, not only from federal and state governments, but from private citizens and corporations. We have had a long tradition of excellence, at least long by American standards. This year, for example, we are celebrating the one-hundredth anniversary of the Lick Observatory, the first experiment in placing large telescopes on mountain tops and in fact a successful experiment, which has pointed the way to other new observatories ever since. We also celebrate, here in America, the one-hundredth anniversary of the founding of the Astronomical Society of the Pacific. Americans, both scientists and the general public, have long been admirers of astronomy. It is an important part of our culture, and serves as an intellectual, philosophic, and aesthetic magnet to attract our citizens to interests in science.

Now for more than three years, as Arthur Davidsen just mentioned, indeed from before the General Assembly in New Delhi, the National Organizing Committee and the very much harder working Local Organizing Committee have been preparing for this General Assembly. It has been a privilege to create the setting for a General Assembly which so fosters international cooperation and progress in astronomy. We welcome this opportunity to acquaint you more extensively with America and its astronomy and so, on behalf of the National Organizing Committee and indeed on behalf of all of the American astronomers, I welcome you to the General Assembly of the International Astronomical Union. We stand ready to do all that is possible to make your stay here a pleasant and productive one."

Address by Dr. S. Muller, President of Johns Hopkins University

"Ladies and Gentlemen, good afternoon,

It is a pleasure to add my welcome to Baltimore and to the Twentieth General Assembly of the International Astronomical Union on behalf of all of the Johns Hopkins University.

It strikes me as fitting that this University is serving again as host for this major international gathering. You may know that Johns Hopkins has had an international character since its founding in 1876. The Johns Hopkins University, in fact, owes its very existence to the German statesman and educator, Wilhelm von Humboldt who was the founding spirit of the modern German university; because Johns Hopkins, modeled explicitly on von Humboldt's precepts, became America's first true, modern research university, granting the doctorate, committed to freedom of teaching and research, dedicated to the unity of research and teaching.

As you already know, the Department of the University that is most deeply involved with IAU is, of course, the Department of Physics and Astronomy. We had great good luck in that in the initial, original faculty of the Johns Hopkins University, the first Professor of physics was Henry A. Rowland, perhaps now best known for the Rowland grating. Since Dr. Rowland founded the department in 1876, the university's faculty in physics and astronomy have attained great distinction at the university and of course in the past few years the department has been enhanced by the establishment of the Center for Astrophysical Sciences and very much so by the presence of Space Telescope Science Institute at Homewood, now in immediate geographic adjacency to the new building of the department.

The Johns Hopkins University is not resting on its laurels in astronomy with the Hopkins Ultraviolet Telescope and with the Space Telescope Science Institute. Recently, we've reached an agreement with the Carnegie Institution of Washington and the University of Arizona to collaborate in the design, construction and operation of an eight meter optical telescope at Los Campanas in Chile. The Magellan Telescope Project will give astrophysicists another major tool for the study of the Universe, one which will significantly complement the Hopkins Ultraviolet Telescope and the Hubble telescope. Formally, so far, we are committed only to the design phase of this project, but we have every hope that the Magellan Telescope will prove to be feasible and that it will serve to reinforce our commitment to the astrophysical sciences, which is already so substantial.

It is an honor for the Johns Hopkins University to serve as host institution for this General Assembly in cooperation with the National Academy of Sciences. We are deeply grateful to the National Aeronautics and Space Administration and the National Science Foundation for their magnificent support of this conference, and we are especially indebted to the corporate sponsors, whose generosity has made possible the special events that we hope will make this Assembly truly a memorable one for you.

I wish you productive and rewarding meetings, and as I close, it is my special privilege and pleasure to introduce to you the Mayor of the City of Baltimore, the Honorable Kurt L. Schmoke."

Address by K.L. Schmoke, Mayor of the City of Baltimore

"Thank you very much, President Muller. Good afternoon,

Every so often, cities, like people, are treated to something special. It's usually not hard to tell when it happens. If you've ever seen a child's face light up during a show at the planetarium, you know exactly what I mean. Well, for Baltimore, hosting the International Astronomical Union's Twentieth General Assembly is indeed something special. So on behalf of all of our citizens I want to warmly welcome you to Baltimore and to welcome Dr. William Graham, Science Advisor to President Reagan and Director, Office of Science and Technology Policy, and all of the Members and Friends of the IAU.

Baltimore, as many of you know, has a strong interest in astronomical discovery. We are very proud of the one hundred and six year old Baltimore Astronomical Society, and of the fact that Johns Hopkins University is the home of the Space Telescope Science Institute and the Center for Astrophysical Sciences. We are equally proud to have been chosen as the meeting place for this Assembly. I say that in particular because those things that will most concentrate the collective mind of the IAU over the next ten days: quasars, the Hubble Space Telescope, the Space Station, joint ventures to Mars, to name just a few, are, for all their complexity, the stuff of dreams and youthful fascination.

So I think that there is a message for the children of our city and children everywhere in the work of this Assembly: make your future one of hope, human adventure, and exploration. I should add that those themes also characterize the kind of city that Baltimore has become and I want to thank the IAU for helping to identify our city with the ideals that you represent.

Now I know that many of you are from out of town and from out of this country and I invite you to see Baltimore and to meet our citizens, especially our young people. You'll discover that they, like you, are curious, adventurous and open to new ideas. So again I want to welcome all of you here today, and I wish you much success with both this assembly and your ongoing scientific endeavors. Thank you all very much."

Address by M.A. Steinberg, Lieutenant Governor of the State of Maryland

"Thank you very much and good afternoon, everyone,

It is a privilege for me to represent the Governor today to issue a proclamation. But I would like to just take a short moment or two to express some personal comments. I have tremendous respect for the scientific community because as an elected official, and I've had the privilege of serving in elected office for twenty-two years, I've always respected the ability of the scientific community to remove the geographic barriers of countries and jurisdictions, to look at the Planet as a single entity and to utilize the collective minds of everyone on this Planet so that this quality of life may be enriched for all of us.

I'm very proud to be here today, for a number of reasons. Astronomy is one of the oldest sciences. But it's an old science, and man always was cognizant of the fact of the importance of the sun: that it gave the heat, that it provided the light; farmers always utilized the skies to see the change of the seasons; and how the science of astronomy has advanced in a relatively short time is just amazing to a lay individual.

As the Lieutenant Governor of this State, I am extremely proud that you chose the State of Maryland and the city of Baltimore, which is a jewel in the State of Maryland, to hold this conference. Why is it special? I mean, in your sixty-seventh year, this is only the third time that the United States has served as a host community. We are very proud of the Johns Hopkins University. We have an outstanding higher educational system in the State of Maryland, but I have always been very sensitive to the fact that higher education was important both in the private sector as well as in the public schools, because it's a higher education system to our citizens.

You know, before I read the proclamation, one of the earlier speakers indicated two things about which I want to make some personal observations. First, I think Dr. Davidsen alluded to the fact that he arranged a bright, sunny day today. It shows you it's a relative statement to make. As most of you know, we've been experiencing a terrible drought. Several weeks ago when the rains finally broke through, in July, I had the occasion to come before a group that was hosting its convention in this facility and as I came to the podium there was a downpour outside and I was very jubilant; and I said, "Ladies and gentlemen, I am extremely pleased to be here. I think your organization broke the drought, and we're very pleased to have a nice rainy day." So it's a relative situation that you're talking about.

The second point, I forget which speaker alluded to it, is that astronomy has experienced two basic golden ages: the time of Galileo and then the last two decades. I would hope and I sincerely pray that, perhaps twenty years from today, someone will be standing in a podium before one of your conferences and will say that the last two decades were diamond ages in the field of astronomy. Because as an elected official, I see the future with tremendous problems and concerns: the source of energy, the ability to have enough food to feed the people on this planet, our resources that are diminishing; and I believe your science, your discipline, holds the key to unlock some of the mysteries that we need to know for these answers. So before reading the proclamation, I say to you: I wish you well in the next ten days of your conference; I certainly hope that your intellectual discussions will lead to more progress and I wish you the greatest success. I look forward that in the next two decades, if I'm fortunate enough to be here, you will be able to continue the progress that you've made in the last two decades.

And finally, where Mayor Schmoke invited you to see Baltimore, for those of you from out of State and out of the country, Maryland is a microcosm of the United States, and so you can save a lot of money by just visiting the various sections of the State of Maryland and you will walk away from the State of Maryland seeing the United States. My home base, Annapolis, is a beautiful place as are the west of Maryland and the Eastern Shore. I also hope that you will avail yourself of the opportunity of looking at many of our historic sites and very beautiful facilities. At this time I would like to read the proclamation issued by Governor Schaefer. It reads as follows:

> **Whereas** it has been said that the future of mankind as we know it depends directly upon how far our dreams will take us; and over the years astronomers and other dedicated members of our international scientific community have properly utilized their skills and visions to further this truth; and **whereas**, since its founding in 1919, the IAU has proudly remained at the forefront in providing forums through which astronomers cooperate internationally while exchanging scientific information and boasts a membership of over 6000 active members hailing from more than 50 countries; and **whereas,** Maryland, home to The Johns Hopkins University and the renowned Space Telescope Science Institute, is extremely proud of the various scientific and educational institutions located throughout our great State, facilities which are consistently on the cutting edge of scientific exploration and technological breakthroughs; and **whereas,** Maryland is pleased to join in welcoming distinguished members of the International Astronomical Union to our State as they attend their organization's prestigious Twentieth General Assembly, as they nobly share in discussion, cameraderie, and friendship for the ultimate betterment of all mankind; **now I, William Donald Schaeffer, Governor of the Great State of Maryland, do hereby proclaim August 2-11, 1988, as International Astronomical Union Days in Maryland and do commend this observance to all of our citizens.**

And it's signed under today's date by Governor William Donald Schaefer. Thank you very much."

Address by Dr. F. Drake, on behalf of
Dr. B. Shakhashiri, Assistant Director for Science and Engineering Education
National Science Foundation

"The National Science Foundation was to have been represented by Dr. Bassam Shakhashiri, Assistant Director for Science and Engineering Education. Dr. Shakhashiri has unfortunately been detained in a remote airport, and does send his regrets, and I think it will not surprise any of the astronomers here to recognize that these things happen and that the event which happened this morning is not usual. Dr. Shakhashiri found himself marooned in a remote airport in a city known as Indianapolis. Our apologies to those of you from Indiana. He has given me the essence of the remarks he wished to make to you and he is truly disappointed he could not be here because he felt this was an extremely important occasion and he was looking forward to being able to participate in it. In any case, his wishes and his remarks were as follows:

> He does wish that this General Assembly be extremely successful, and on behalf of the National Science Foundation he sends greetings and welcome to all the participants in the General Assembly.
>
> As one of the two lead agencies supporting astronomy in the United States, he is proud of the role the National Science Foundation has taken in supporting university and other astronomical research and in particular in developing the national astronomy centers, which are so prominent in American astronomy and in supporting their operations now over several decades. These have become one of the jewels in the crown, not only of American astronomy, but of worldwide astronomy. Many of you have used them, whether you be Americans or people from other countries. He is also pleased that the National Science Foundation was able to join NASA in providing the funding which has made this General Assembly possible. Or, I might add parenthetically, made it possible for your registration fees to be less than astronomical on this occasion.
>
> He notes that the appeal of astronomy to people of all ages, and especially to youth, is extremely great. To those in education, astronomy is particularly important because it can serve as an excellent vehicle to stimulate excitement in science and to cultivate the senses to recognize and pursue intellectual activities. He notes that in our present world of growing population, of new aspirations to higher qualities of life, better standards of living, we need a much broader knowledge of science and appreciation of it and its limitations. He recognizes that scientists know this and calls upon us to be diligent in remembering that in our work.
>
> He notes that because of the increasing impact, both positive and negative, on the welfare of the world: its climate; again, standards of living; economic viability and welfare and wealth; that there is a great importance to communicate science, not only to other scientists, as will be done in an excellent fashion at this General Assembly, but also to non scientists. We need educated citizens in all countries that can think and behave in a rational way. We need educated citizens who can distinguish between astronomy and astrology. We, as scientists, have a responsibility to promote rational behavior on the part of other members of our society, who not only depend on us for leadership, but also for their own welfare and who of course provide the support for our activities. The intellectual excitement in making discoveries is something which must be shared outside the scientific community.

Indeed, scientists and engineers must excell in research in their subspecialties, but they should beware that that is not enough. A dangerous situation will develop if a gap widens between the specialists and the general population. He knows that the General Assemblies play a role in reducing that danger and allowing that communication to occur and thus he congratulates us on this, our General Assembly in the United States, and wishes great success in our undertaking."

Address by Dr. N.W. Hinners,
Associate Deputy Administrator, NASA Headquarters

"Honored Guests, Members of the IAU, Ladies and Gentlemen,

It is indeed a pleasure on behalf of NASA to be here today to welcome you to the State of Maryland and to Báltimore, home of the Space Telescope Science Institute. I'd like to take my brief time here today to talk about the basic state of the US space science programme, its relationship to the overall civil space programme, and the special place of astronomy, with note on the particular role of international cooperation, so well represented today by the IAU.

First, I think we're all aware that the top priority of NASA is indeed to return to safe flight with the shuttle. Despite annoying, but not unusual recent delays in this test or that, the launch of the shuttle is now measured in terms of weeks, not months or years. While we recognize the need for due caution, we're also cognizant of the large number of payloads awaiting launch, some of which have unique launch windows, such as Magellan and Galileo. Others, such as the Tracking and Data Relay Satellite, must precede the satellites they are to service. Add to this a priority for certain Department of Defense satellites, and one finds that the first available opportunity to launch the Hubble Space Telescope is the summer of 1989. Let me assure you that NASA is doing all it can to maintain that launch opportunity, recognizing the cost of delay both in dollars and to a waiting astronomy community and the public. While the delay is regrettable, the time has, in part, been well spent in tuning the very complex ground system of the Space Telescope.

NASA, in addition to the Department of Defense, realizes the extreme penalty we have paid for becoming reliant upon the shuttle for all our launches. It is one that we do not want to repeat, and we are committed to the concept of a mixed launch fleet. Missions that do not uniquely require the shuttle, either for launch or servicing, will be flown on expendable rockets. Conversion of the Cosmic Background Explorer to a Delta launch and conversion of the Extreme Ultraviolet Explorer and ROSAT to Deltas are the first steps in the implementation phase of that policy. The Gamma Ray Observatory, essentially complete, is tailored to the shuttle and will remain as a shuttle launch.

NASA's second priority has been to initiate the development of the Space Station, a key ingredient to the eventual fulfilment of the US goal to expand human presence and activity beyond earth orbit into the solar system. International in scope, the Space Station is going to enable certain science which requires either construction in orbit or frequent servicing or both. In the world of astronomy, such candidate missions as the Large Deployable Reflector, Astromag, and the Long Baseline Optical Space Interferometer immediately come to mind. The choice, however, of which missions to do in conjunction with the Space Station must be based primarily on the scientific requirements and only secondarily on using the Space Station because it is there. Initiation of Space Station has obvious major budget implications, especially in light of the serious US federal budget deficit. Indeed, it was recognition of the essential need to have a strong, healthy space science, technology and aeronautics programme that led NASA to propose, and the President to accept, a major increase in the NASA funding. The fate of the entire NASA budget is now in the hands of Congress, but I'm confident that out of it we'll survive a vigorous space science programme, highlighted by the beginnings of the Advanced X-Ray Astrophysics Facility. It's with only a little bit of chagrin that I recall telling Dr. Giacconi in 1978, that the AXAF new start would be delayed until 1982. I was only seven years off, Riccardo!

The Space Science Programme will, in addition to continuing the great observatory line, reemphasize Explorers and Scout class missions, enable faster response times and meet unique requirements. Similarly, sounding rockets and balloons, so important to Supernova 1987A observations, will be enhanced. No longer, I hope, will you hear of the potential phase-out of sounding rockets and balloons.

Now to the role of international cooperation. Clearly, history has a lesson for us. I am, as are many here, impressed with the extremely fruitful collaborations in the International Ultraviolet Explorer, the IUE, now in its tenth year of observatory operations; and in the Infrared Astronomy Satellite, IRAS, which has made so many fundamental discoveries. A host of additional collaborations exists, with ESA on the Hubble Space Telescope, Germany on ROSAT, the Gamma Ray Observatory. Many potentials exist for the future; the follow-on to IUE, called Lyman; the Space Station related missions; other free-flyers. There's no doubt that in our minds the international cooperation in astronomy is a healthy endeavor. There's simply so much to be done that it is beyond the financial ability of any one nation to go it alone. Repeated calls in our Congress for international participation make that abundantly clear. Science, by its nature, thrives on collaboration, and the talent is spread across national boundaries. Thus NASA remains committed to the concept and the reality of international cooperation.

We view international cooperation as something which must be approached systematically in the context of our overall national policy and goals. Clearly, the US and other nations aspire to pre-eminence in some scientific or technical arena, where they will want to go alone or have minor participation of others. Other arenas, by dint of where the talent is, or for reasons of priority or finance, will be amenable to major sharing. This applies not only to space hardware development, but more and more to the long term operations and data systems, including global data networking.

The key to successful international cooperation in the past has been the bottom up approach, whereby mutual interest generated at the scientist to scientist level has grown into country to country commitments. This is as it should be. Cooperation mandated from above is usually fraught with difficulty and simply suffers from well intentioned but frustrating bureaucratic impedimenta. The IAU is a major facilitator in generating scientific, technical and programmatic exchanges. Out of your activities in the next ten days can grow the right kind of international ventures.

Clearly, astronomy is the forefront space science, one which is on the cutting edge of technology, discovery and comprehension of that most marvelous invention of nature, the Universe.

Thank you."

Address by Dr. William R. Graham, Science Advisor to the President and Director of the Office of Science and Technology Policy

"On behalf of President Ronald Reagan, I would like to welcome you to the Twentieth General Assembly of the International Astronomical Union and for those of you from other countries, I'd like to welcome you to the United States of America as well. This meeting is a milestone in astronomy and also a milestone, I think, in international cooperation in astronomy, a field of scientific endeavor that has, over the centuries, led the way for international cooperation in research and discovery about the fundamental properties of the Universe.

President Reagan and the people of the United States strongly support scientific research in a wide range of areas and that certainly carries into the astronomical sciences as well. In fact, that support has been growing over the last decade and we have every expectation that it will continue to grow in the future. As a part of this, the United States strongly encourages international cooperation in scientific research. Many of the astronomical programmes that we're going to discuss here this week and next and many of the instruments have become so unique that they in all likelihood won't be duplicated, at least not exactly, but will be joined and used by scientists from many nations, and that is certainly as it should be. But even more basically, it is our experience that scientific research benefits all people. In terms of game theory, if you like that vernacular, it's truly a positive sum enterprise.

We recognize that scientific research draws heavily on the world's most valuable resources and those that are its scarcest: the human minds and the human skills of this Planet. But it also enriches and expands those human resources through the training of new generations of scientists, through the strengthening of our institutions and through enriching our cultures. And in recognition of the growing importance of international cooperation and the value of human ingenuity, last year, at the Organization of Economic Cooperation and Development, which is a group of twenty-four industrialized nations, the United States proposed a general framework of principles for international cooperation in science and technology. That proposal was carefully designed to make several explicit, fundamental principles known in the international area of science and technology cooperation.

Among the OECD recommendations, the following were key: equitable contributions from all industrialized nations in support of basic research and maintaining up-to-date research facilities; equitable contributions from all industrialized nations to the education and advanced training of the next generation of scientist and engineers; open access for researchers, scientists and engineers to basic research facilities and activities supported by member governments; dissemination of basic research information and results through publication in the generally available scientific literature and other customary practices -certainly this General Assembly is a highpoint of those customary practices- and, of course, acknowledgement and respect for the intellectual accomplishments of scientists worldwide.

These principles strongly apply to astronomy and have been growing in strength for many years. Examples of US support for international cooperation in astronomy abound and you've already heard of many this afternoon. I would add to the list Mauna Kea, possibly a unique and certainly a prime site for ground based astronomical observation, and one where the US has been privileged to provide site and assistance to France, Canada and the UK, as well as to the US astronomical facilities and a welcome reception to astronomers from around the world. Similarly, we are received at the Cerro Tololo Inter-American Observatory in Chile and at many other sites around the world. The names that you have heard: IRAS, the IRAS programme, its great data bank still being digested; the Very Long Baseline Radio Astronomy programmes; the Hubble Telescope and the Gamma Ray Observatory, about to be launched; the Microwave Observation Programme, moving now into the immediate future, with its million channel data acquisition capability, and its challenge is not only to radio astronomy, but to providing the artificial intelligence and the expert systems we'll need to reduce the data that that's able to collect; and then to the AXAF and to many more space based experiments.

Even more interesting, perhaps, and something hardly even imagined at the last General Assembly in the United States: we're moving beyond the photon to other forms of energy detection, including the deep underground detectors for neutrinos, detectors which recently identified the neutrino flux from Supernova 1987A; and on beyond that to gravity wave measurements, to more advanced planetary measurements and perhaps to forms of energy that we haven't yet detected or decided could be of astronomical use.

So the United States is proud to host this meeting and happy to welcome every qualified astronomer, of whatever origin, to our country. I hope, when the history of this meeting is recorded, that it will be recognized as a testament to the realization that combining our resources, human and physical, we can further accelerate the pace of our long adventure in scientific discovery, the most challenging and exciting course of discovery that this world has ever known. Thank you."

Address by Prof. J. Sahade,
President of the International Astronomical Union

"Ladies and Gentlemen, Distinguished Guests,

After such a remarkable performance of the Annapolis Brass Quintet and our listening to the inspired composition especially prepared for the occasion, I feel in the awkward situation of having to say a few words on behalf of the IAU when I would think that everyone of us would prefer to be left with the magic of the melodious piece we were just presented with ringing in our ears.

The International Astronomical Union is very happy to hold a General Assembly in the United States for a third time and very grateful to the US National Academy of Sciences for the warm invitation that was extended to us.

At this opening ceremony, where I enthusiastically welcome all of you, I would also like to express our gratefulness to the Johns Hopkins University for serving as the gracious host of it, the twentieth of our triennial gatherings.

I would like to thank President Muller for his words of welcome and I would also like to praise him very highly for the great support he gives to the development of Astronomy at Johns Hopkins and for the enthusiasm and the generosity of the contributions he has made as our host, towards the success of the General Assembly. We feel most honoured for your being with us at this ceremony, President Muller!

In a way, the Johns Hopkins University is living up to a long tradition. Those of us who are engaged in spectroscopic work do remember the father of the concave diffraction grating, Henry A. Rowland, the first Professor of Physics at Johns Hopkins University and the first President of the American Physical Society. I do not need to stress the importance of Rowland's discovery, or invention if you prefer, which he announced in 1882. For me, who started using in Cordoba, Argentina, a grating spectropgraph, the name of Johns Hopkins was always present in my mind: the grating was a replica made by Professor Robert W. Wood at Johns Hopkins University.

We are most pleased indeed that the setting for our meetings is provided by the Monument City, by the two hundred and forty-nine year old City of Baltimore, and we are pleased indeed that the City's Mayor could be with us today. In Baltimore we find the past, the present and the future blended together in a wonderful, constructive, inspiring way. Baltimore is one of the particularly important places in the early history of the United States, during the American Revolution; it is now a busy and booming industrial center world-wide famous for the academic and research levels of our host, the Johns Hopkins University, for its glorious Baltimore Symphony Orchestra that will delight us next Thursday and for the many expressions of art and culture that characterize the City. And projecting itself in a future which is already here, Baltimore houses, on the Johns Hopkins University premises, the American Space Telescope Science Institute that is connected to the forthcoming activities of the Hubble Space Telescope, that we astronomers the world over are so eagerly and impatiently awaiting -and in particular to receiving its invaluable information that has long been overdue.

I said that we feel very happy to meet once again in the United States. And this is something natural because this country is one of the first ever National Members that adhered to the IAU, upon its creation almost 70 years ago. Moreover, about 27 percent of our individual Members are US astronomers, and the category to which the country subscribes is the highest category that is listed in our By-Laws.

But there are even more profound reasons for our feelings, as you all know. The United States are noteworthy for the outstanding contributions to our astronomical knowledge that have always been made. The important optical astronomical centers started on the East Coast, Harvard, spread to the middle-west, Yerkes, and then to the West Coast, to California, also to Texas, and to Arizona, and even furthest west, to Hawaii, in search for the best skies available to place the large reflectors that were to be built.

The many contributions that have come from Lick, from Mount Wilson, from Mount Palomar, from McDonald, from Kitt Peak, from Mauna Kea are landmarks in our quest for learning more and more about the Universe and also a tribute to the people and institutions who provided the funding and to the hard work of the astronomers of this country, who, in many cases, carried out their research with the cooperation of astronomers from other lands.

The New Astronomies that started popping up practically immediately after the Second World War, and particularly since the advent of the space age, the new technologies that are being placed at our disposal, have so drastically changed our classical set up for observing and for analyzing the information that is obtained, that old-timers like myself are having an exciting time trying to adapt themselves to the new situation. And in all these changes the United States -its astronomers, its physicists, its engineers- have played and are playing an outstanding role.

As time goes by, Astronomy is becoming more and more exciting and to add to the excitement we are sometimes offered the possibility of attempting new approaches like in the case of Halley's Comet or of learning through the observations, about the characteristics of the actual progenitor of a supernova like the naked-eye one that appeared last year in the Large Magellanic Could, an object that is the subject of one of our Joint Discussions this week.

In the present day level of the astronomical activities in the United States there are two organizations that play a major role, the National Science Foundation and the National Aeronautics and Space Administration. We greatly appreciate that the two agencies are represented in this opening ceremony and I would like to take the opportunity to thank NSF and NASA for the very significant contributions that they have made to the Local Organizing Committee.

We feel also very honoured with the proclamation of the Governor of the State of Maryland, Terra Mariae, as it was called on the Latin Charter in 1632. Maryland is the home of several institutions where Astronomy is an important activity. I should particularly mention the University of Maryland and NASA's Goddard Space Flight Center, the seat of many outstanding astronomical space activities.

Unfortunately for our science, not everything is bright and rosy. There are, or there appear to be, some dark clouds coming up in our horizon that threaten to degrade drastically and increasingly the Earth's environement, thus adversely affecting "astronomical observations from the ground and from space".

The contamination of space could reach levels that are both ridiculous and frightening. We are very much concerned about it and equally concerned also with the people of this region in regard to the pollution of the Chesapeake Bay area which affects the quality of life both human and wild.

One of these frightening projects that will dramatically spoil our prospects of continuing contributing to our knowledge and understanding of the Universe originates in this country. I am referring to the so-called Celestis-Space Services paylaod launch proposal that aims at launching "cremated human remains into Earth orbit, using highly reflective containers". It seems incredible that business interests could attempt degrade in such a radical way the interests of mankind, of the human beings who aim at increasing knowledge and at living in a purer environment!

The international astronomical community is so concerned with Celestis-type projects that at our last General Assembly in Delhi, a resolution was unanimously passed that had the character of a general appeal to countries and space agencies. The reaction to such an appeal has not been as positive as we would have expected and, as a consequence, the relevant Commission of the Union is going to hold a press conference during this General Assembly and on August 13 to 16 will be holding, in Washington DC, a colloquium on "Light Pollution, Radio Interference and Space Debris". Let us hope that what will be said at the present General Assembly and the activities of our Commission 50 on "Protection of Existing and Potential Observatory Sites" will succeed in creating an awareness in decision-makers and in the general public on the important problem of the environmental pollution that goes beyond the immediate surroundings.

Let me finish my words of welcome by expressing our thanks to the Science Adviser to President Reagan for being with us and for speaking to us this afternoon, and the appreciation and congratulations of the Union to the wonderful work of the National and the Local Organizing Committees that has led to the perfect and detailed organization of this Assembly.

I am sure we will all enjoy, during the next few days, not only a good, carefully prepared scientific meeting but also the warm hospitality of Baltimore and a taste of your famous blue crabs.

And now, I declare the XXth General Assembly open!"

CHAPTER II

EXTRAORDINARY GENERAL ASSEMBLY

Held in the Baltimore Convention Center

2 August 1988, 15.00

Prof. J. Sahade, President in the Chair

An Extraordinary General Assembly was held immediately preceding the first session of the General Assembly on 1988 August 2 to amend the Statutes of the Union to permit the election of a President-Elect to the Executive Committee and to institute a new category of adherence to the Union -Associate National Member status for countries in the process of developing astronomical activity.

1. Extension of the membership of the Executive Committee by the election of a President-Elect

The President proposed to the Extraordinary General Assembly that the membership of the Executive Committee be extended by the election of a President-Elect in order to ensure continuity in the presidential activities of the Union. It would be the expectation that the President-Elect would normally become the President of the Union at the next General Assembly following election as President-Elect.

This proposal was accepted unanimously by the Extraordinary General Assembly.

2. Establishment of a new category of adherence to the Union, Associate National Member

The President proposed to the Extraordinary General Assembly that a new category of adherence -Associate National Member- be established, allowing countries in which astronomical activities are developing to adhere to the Union. Associate National Members will be accepted into the Union for a maximum of nine years, at the end of which time they either become a full Member of, or resign from, the Union.

This proposal was accepted unanimously by the Extraordinary General Assembly.

The Extraordinary General Assembly gave its assent to the renumbering and rewording of affected Statutes in order to accommodate the above agreed provisions.

CHAPTER III

TWENTIETH GENERAL ASSEMBLY

FIRST SESSION

Held in the Baltimore Convention Center

2nd August 1988, 16.00

Prof. J. Sahade, President in the Chair

1. Formal Opening by the President

"Mesdames, Messieurs,

Je voudrais commencer cette première séance de la Vingtième Assemblée Générale de l'Union Astronomique Internationale en vous souhaitant d'une part la bienvenue et, d'autre part, un séjour non seulement très agréable mais surtout très profitable au point de vue scientifique.

A Delhi, j'ai partagé avec vous les réflexions que j'avais à l'esprit au moment d'être élu Président de l'Union. Mon adresse de cette après-midi sera en quelque sorte une continuation de l'allocution de Delhi avec, en plus, l'expérience de presque trois années.

Comme à Delhi, je n'essaierai pas d'utiliser alternativement des paragraphes en anglais et en français tout au long de mon allocution, parce que, vraiment, je trouve que c'est impossible pour moi de faire cela d'une manière à la fois équitable et efficace ; d'autre part, il me semble qu'il serait même gênant de vous lire quelques parties de l'adresse deux fois, une fois en anglais et l'autre fois en français. Alors, aujourd'hui encore, je m'adresserai à vous en utilisant seulement la langue du pays où nous sommes réunis.

We start this General Assembly after having held an Extraordinary General Assembly that consisted in changes in the Statutes that incorporate a new position in the Executive Committee, that of President-Elect, and a new category of membership, that of Associate National Member.

I am convinced that both additions will contribute to the improvement of our Union and to the development of Astronomy in countries that have as yet a very low level of activity in the realm of our science.

At this first session of the General Assembly we are supposed, I believe, to report on what has happened in the Union during the time elapsed since Delhi, if anything has happened at all, and set the stage for the forthcoming term in our traditional sequence. The curve of growth of individual membership in our Union has a definite trend since about the time of our insertion in the space age. When I first noticed the tendency I became worried and also became convinced that it was time to start thinking as to whether we should continue holding General Assemblies in the way we have been doing it so far, at the risk of the meetings becoming unmanageable in size. But there has been no increase, perhaps a decrease, in attendance since Brighton (1970) and Grenoble (1976). So, for the time being, there seems to be no reason to worry in regard to the size of the General Assemblies. But, we should, however, ask ourselves: why is it that while the number of individual Members is steadily increasing in the Union, the General Assembly attendance is not going the same way?

The reasons may be several. Of course, there may be, at each time, particular circumstances that we would have to consider for a thorough evaluation. In addition, there is, on one side, a proliferation of scientific meetings and, on the other, a shortage of the funds generally available for research. Moreover, a good portion of the community of nations is facing serious economic problems, and for their scientists the level reached by travel expenses, subsistence expenses and registration fees in terms of their own salaries, makes attendance at meetings somewhat prohibitive, much more drastically so than it is now for almost every scientist. It might be a good exercise to look more deeply into the problem and see what can be done about it. A few days ago, I brought the question of the registration fees to the attention of the Third World Academy of Sciences in the hope that the Academy together with ICSU could consider it and try to find at least a partial solution.

The shortage of research funds has either led to the actual closing down or to threats of shutting off or of curtailing the activities of important, long established observatories. Fortunately, in most cases, good sense has prevailed. This suggests that it might perhaps be necessary to remind decision-makers, from time to time, in an appropriate way, of the importance of Astronomy, the oldest of scientific human endeavours, a pure, basic science, with a strong social component, that generates progress in other branches of science and technology and even gives rise to products of massive consumption, like pyrex ware or quartz watches, to give only two examples.

In Delhi I expressed my concern in regard to the gap that is increasing between the groups that have readily access to large telescopes, to VLBI installations, to space instrumentation... and those that cannot aim at having similar possibilities within their reach and feel that they are being left hopelessly behind. In order to try to contribute to the improvement of such a situation and to create a better general feeling among concerned astronomers, we have established a Working Group of the Executive Committee for the Promotion and Development of Astronomy. In addition, the General Secretary has been contacting UNESCO for the signing of a possible cooperative contract which would be coherent with the objectives of the Working Group.

One of the tasks that is assigned to the Group relates to the fundraising exercises that were strongly suggested by the Finance Committee of the Delhi General Assembly. We are starting cautiously as is advisable and I hope that in a short time the Working Group would have acquired important dimensions and will be leading the IAU into activities that would help bridge, at least partly, an undesirable gap and will be bringing to the forefront of modern research, sometimes perhaps in the framework of some kind of consortia arrangements, intelligent people who may otherwise have little chance to find satisfaction for their aspirations. I feel confident that this move we have made will work towards a healthier developement of Astronomy and towards a strengthening of our Union. On the other hand, we would just be fulfilling one of the objectives of the Union, as read in the Statutes.

During the interval between Delhi and Baltimore -and here, I ought to place a complaint because such interval is much less than three years, quite a discrimination, I would think!- there have been quite a number of highlights in astronomical research, of the calibre, for instance, the outstandingly successful multiple space experiments with Halley's Comet. We were also rewarded with the appearance of the first naked-eye supernova since Kepler's and with the discovery of its progenitor star.

The expectations for the present General Assembly were to learn of the first results coming from the Hubble Space Telescope. Unfortunately, the saddest tragedy of the Challenger, to whose crew we pay once again our sorrowful homage, delayed the launch, now scheduled for some time next year.

Also next year, ESA's Hipparcos is due to be in orbit and other astronomical satellites that will certainly yield important results are scheduled later in time.

Of course, we are eagerly waiting for the information that shortly will be coming from the Phobos missions of the Soviet Union.

So, the promise is for more exciting times for us, astronomers. I think that we are fortunate people, we enjoy what we do and we live excitement after excitement!

I am happy to extend once more a warm welcome to all of you, Members of the Union, invited participants and guests. I particularly welcome those Members that have belonged to the Union for fifty years or more, and are present at this General Assembly, namely,

Prof. W.J. Luyten
Prof. K.A. Strand

I would now propose that we send a telegram to past Presidents and past General Secretaries of the Union who are unable to be with us at this gathering,

Prof. J.H. Oort
Prof. V.A. Ambartsumian
Prof. L. Perek
Prof. C. de Jager
Prof. G. Contopoulos.

I am happy to extend a warm welcome to members who in the past have served on the Executive Committee of this Union, and who are attending the General Assembly:

Prof. A.A. Blaauw
Prof. W. Iwanowska
Prof. E. Müller
Prof. J.-C. Pecker
Prof. S. van den Bergh

and to the official representatives of the Adhering Organizations which support this Union.

Furthermore, I extend a hearty welcome to the official representatives of sister Unions, namely:

UNESCO	S. Raither
ESA	R. Bonnet
BIPM	B. Guinot
FAGS	E.A. Tandberg Hanssen
NASA	N. Hinners
IAF	D.M. Papagiannis
IUPAP	K. Larkin
IAG	I.I. Mueller
CCIR	A.R. Thompson
INTERCOSMOS	N.S. Kardashev
SCOSTEP	S.T. Wu
CODATA	D.R. Lide, Jr
COSPAR	H. Friedman
IUCAF	B. Robinson

I now ask those present to stand while the General Secretary reads the names of the members who have died since the XIXth General Assembly.

The General Secretary then read the following list:

AARONSON M.
ABRAHAM H.
ALLEN C.W.
BECKER F.
BOBROVNIKOFF N.
BONOV A.
BRAZ M.A.
CALAMAI G.
DE MOTTONI Y. P.
DOUGLAS A.V.
DUBOSHIN G.N.
ELYASBERG P.E.
FEDOROV E.P.
FISCHER P.L.
FRACASSINI M.
FRICKE W.
GOKMEN T.
GOLDBERG L.
HARO G.
HOPPE J.A.
HORSKY Z.
HUMPHREYS C.J.
HYNEK J.A.
INGRAO H.C.

IRIARTE B.
KABURAKI M.
KANNO M.
KEMP J.C.
KHOLOPOV P.N.
KOTSAKIS D.
KROOK M.
KULIKOV K.A.
KUPPERIAN J.E.
KUZMIN G.G.
LARINK J.
LOCHTE-HOLTGREVEN W.
MCGEE J.D.
MCVITTIE G.
MEURERS J.
MIYADI M.
MOFFETT A.T.
MOHLER O.C.
MUSTEL E.R.
NEVEN L.
NEVIN T.
NEWKIRK G.A.
NIKONOV V.B.
ORLOV A.A.
OVENDEN M.W.
PELSENEER J.K.
PELTIER L.
PRZYBYLSKI A.
ROBERTSON W.H.
SADLER D.H.
SANCHEZ M.C.
SCHWERDTFEGER H.M.
SERVAJEAN R.
SEVERNY A.B.
STEPANOV V.E.
STROBEL W.
STROMGREN B.
SUZUKI H.
TELEKI G.
TEMESVARY S.
THERNOE K.A.
TUZI K.
VASILEVSKI S.
VELGHE A.G.
WATERFIELD R.
WELIACHEW L.
WOOLEY R. v.d. R.
ZEL'DOVICH Ya
ZHANG YU-ZHE
ZIEBA A.

2. Appointment of Official Interpreters

The General Secretary appointed J. Lesh as official interpreter from French to English and R. Cayrel from English to French.

3. Report of the Executive Committee 1985-1987

The President invited discussion on the Report of the Executive Committee for the 3-year interval 1985-1988, as presented in IAU Information Bulletin No. 60.

The report has been given consideration by the Official Representatives of the Adhering Organisations. The financial section of the report will be scrutinised by the Finance Committee who will present their report in the second session of the General Assembly. (The Report of the Executive Committee follows in Chapter IV, pp. 79-101).

There being no points raised by Members of the Union from the floor, the General Assembly unanimously approved the report of the Executive Committee (1985-1987) subject to receiving the Report of the Finance Committee (see pp. 37-39).

4. Report of the General Secretary (January-August 1988)

Nominations for IAU Membership

The President asked the General Secretary to inform the General Assembly of the nominations for IAU Membership.

The General Secretary indicated that he had received almost 700 nominations for IAU membership. A further 53 nominations had been received from Commissions Presidents and Members of the Executive Committee.

Report by the General Secretary

First of all, I wish to thank most heartily the President of the IAU, Jorge Sahade and the Assistant General Secretary, Derek McNally, who were my partners in the team that attempted to run the affairs of the Union from the end of November 1985 until this date: it was a short term, but the amount of work was essentially the same as for the regular triennia, and, as many of you know, we had to take some non trivial decisions concerning the personnel in the Paris Secretariat. The team work is really appreciated when the situation is a bit difficult.

In the last two IAU Information Bulletins I indicated the changes that occurred in the Secretariat in Paris and do not plan to say more about those past events. I simply wish to state that the support of a very efficient Secretariat (containing a minimum of persons) is absolutely mandatory in order to successfully reduce the workload of the General Secretary, and to help him with all the activities (and deadlines!) inherent in his duty: this is something I definitely appreciated during a large fraction of my term, and I trust that my successor will be very well assisted by Monique Orine and Huguette Gigan. The secretarial support I received in Liège from Denise Fraipont since 1982 is gratefully acknowledged as well; it is furthermore a pleasure to mention that Viviana Soler in La Plata and Valerie Peerless in London were of great help to the other two Officers.

I also wish to thank the members of the Executive Committee and the Presidents of Commissions, Working Groups, and Committees for their collaboration.

In addition, I would like to express my gratitude to the Local Organizing Committee of the XXth General Assembly, in particular to Professor Arthur Davidsen, to the two "chevilles ouvrières", Karen Weinstock and Harold Screen, and to their collaborators, for the most impressive amount of efficient work they performed in the last years. It was really a pleasure being part of the team organizing this General Assembly.

On behalf of the Executive Committe I wish to say that we are thankful for the extensive financial support which has been given to this General Assembly and which has allowed the IAU together with the Local Organizers to support no less than 250 scientists, i.e. young astronomers, most of them from countries where our science is in a developing phase, and a few more senior researchers from countries with currency problems. I do hope that these grantees, as well as their colleagues from all over the world who come for the first time to an IAU General Assembly, and, in fact, that all the participants here in Baltimore will find this large gathering interesting and scientifically rewarding.

The programme in front of us is very full (250 sessions in 6.5 days; I take this opportunity to thank the Invited Speakers and all the organizers of the Joint Discussions, Joint Commission Meetings, Scientific sessions, etc.): the IAU members have received the programme a few weeks ago since it was printed in IB 60 and some of you have hopefully already had a look at it. In any case, all participants should be able to keep track of what is going on through the daily issues of the newspaper "IAU Today", and by regularly checking the notice boards. The boards, some of them beautifully decorated, are located at a strategic point near the mailboxes and the large and quite spectacular exhibit area.

As you realize, the coverage of this XXth General Assembly by the different types of news media is larger than on previous occasions: it is my hope that there will be an excellent collaboration between the scientists and the press, and that the numerous press conferences, interviews, etc., will lead to showing how exciting our science is, and how interesting this particular meeting is. Your best possible collaboration will be greatly appreciated.

Coming back to the Executive Committee report per se, I remind you that a draft report for the period November 85-December 87 has appeared in IAU IB 60 (pp. 20-26), and that the highlights of the activities of the Commissions have now been published in the "Reports on Astronomy" (IAU Transactions XXA, 708 pages). In the first 7 months of 1988 the largest part of our activities has been devoted to the preparation of this General Assembly; let me simply mention a few items that were not included in the report given in IB 60:

- one negative point first: because of the Secretariat changes, we were not able to keep the promise we had made earlier, i.e. to regularly publish updates of the IAU membership list. A new list will nevertheless appear as soon as possible after this General Assembly, and will include all the new members to be accepted here (around 790);

- the move of the IAU secretariat to its new offices will take place this coming November;

- we continued to provide some advice and help to Mr. R. Pansard-Besson who is preparing an extensive series of TV broadcasts on "Les Palais de la Découverte : de Stonehenge au Télescope Spatial": a very preliminary preview of the film will be shown here on August 11th, just before the second session of the General Assembly. It will be a reward for those who stay until the end of the Assembly... and who get up reasonably early after the closing banquet;

- work has continued on the IAU Style Manual, and a meeting with editors of the main astronomical journals has been organized. The results seem very encouraging; a quasi agreement exists on such difficult matters as abbreviations... and I hope this General Assembly will be in a position to endorse a resolution on this;

- more contacts have been made with several countries, which should lead to quite a number of new adherents to the Union;

- we are pleased to report that we received tremendous support from many countries concerning anti-space junk and anti-pollution activities; resolutions on these matters will be proposed here, and an IAU Colloquium on these topics will take place in Washington immediately after the General Assembly;

- the new International Earth Rotation Service (a joint venture of Commissions 19 and 31) is now a reality: birth took place on January 1st of this year;

- closer contacts with UNESCO exist, which may hopefully lead to joint programmes with our Union, i.e. to more activities in the frame of the promotion and development of astronomy, which is one important objective of the IAU.

There have of course been many other activities, and you will hear about most of them during the General Assembly. I will finish here by wishing you a most interesting and enjoyable meeting and by looking forward to the opportunity of seeing you in person during the coming days... even if the waiting list in front of my office may be a little long on some occasions."

The General Secretary's report was received with acclamation.

5. Report on the Work of the Special Nominating Committee

The President informed the Assembly that the Special Nominating Committee had selected the following IAU members for proposal as members of the Executive Committee from August 11, 1988.

As President

Professor Yoshihide Kozai (Japan)

As President-Elect

Professor Alexander A. Boyarchuk (USSR)

As Vice-Presidents continuing

Dr. Alan H. Batten (Canada)
Professor Per Olof Lindblad (Sweden)
Professor R. Kippenhahn (Germany, FR)

As Vice-Presidents

Professor V. Radhakrishnan (India)
Professor Morton Roberts (USA)
Professor Ye Shu Hua (China)

As General Secretary

Dr. D. McNally (UK)

As Assistant General Secretary

Dr. Jacqueline Bergeron (France)

As Advisers to the Executive Committee

Professor Jorge Sahade (Argentina)
Dr. Jean-Pierre Swings (Belgium)

The President informed the Assembly that formal election would take place at the Final Session of the General Assembly.

6. **Announcement of**

Official Representatives of Adhering Organizations

and

Representatives to vote on the Nominating Committee

As requested by the President, the General Secretary announced the following names:

Country	National Representative	Nominating Committee Representative
Argentina	R. Mendez	H. Levato
Australia	K.C. Freeman/ C.S.L. Keay	K.C. Freeman/ C.S.L. Keay
Austria	H.F. Haupt	H.F. Haupt
Belgium	P. Smeyers/ L. Houziaux	P. Smeyers

Brazil	J. Lépine	J. Lépine
Bulgaria	Z. Kovachev	M.K. Tsvetkov
Canada	C.R. Purton/ J.E. Hesser	J.E. Hesser
Chile	F. Noël	M. Rubio
China (Nanjing)	S.G. Wang	F. Tong
China (Taipei)	H.H. Wu	H.H. Wu
Colombia	------	------
Cuba	O. Alvarez	O. Alvarez
Czechoslovakia	V. Bumba	V. Bumba
Denmark	L.K. Christensen	L.K. Christensen
Egypt, AR	K. Aly	K. Aly
Finland	M. Valtonen/ K. Mattila	K. Lumme
France	R. Cayrel	G. Wlérick
Germany, DR	G. Ruben	H. Lorenz
Germany, FR	R. Kippenhahn	K. de Boer
Greece	J. Seiradakis	J. Seiradakis
Hungary	B. Szeidl	B. Szeidl
India	J.V. Narlikar	K.R. Sivaraman
Indonesia	B. Hidayat	B. Hidayat
Iran	Y. Sobouti	Y. Sobouti
Iraq	------	------
Ireland	P.A. Wayman	T.P. Ray
Israel	G. Shaviv	G. Shaviv
Italy	V. Castellani	G. Setti
Japan	D. Sugimoto/ K. Kodaira	D. Sugimoto
Korea DPR	------	------
Korea, RP	Nha Il Seong	Woo Jong Ok
Mexico	A. Serrano	S. Torres-Peimbert
Netherlands	H. van Woerden	H. van Woerden
New Zealand	E. Budding	E. Budding
Nigeria	S. Okoye	S. Okoye
Norway	------	------
Poland	J. Smak	B. Kolaczek
Portugal	J.P. Osorio	J.P. Osorio
Rumania	------	------

South Africa	M.W. Feast/ B. Warner	B. Warner
Spain	P.U.M. Catalan	C.A. Gimenez
Sweden	A. Sandqvist/ B. Gustafsson	A. Winnberg/ G. Lynga
Switzerland	B. Hauck	J.O. Stenflo
Turkey	D. Eryurt	Z. Aslan
United Kingdom	F.G. Smith	M. Smyth
Uruguay	G. Vicino/ G. Tancredi	G. Vicino/ G. Tancredi
USA	F. Drake	V. Trimble
USSR	A.A. Boyarchuk	V.K. Abalakin
Vatican State City	M. McCarthy	M. McCarthy
Venezuela	------	------
Yugoslavia	------	------

Acting Presidents of Commissions

The General Secretary announced that the Executive Committee had appointed the following persons to act for Presidents of Commissions unable to attend the General Assembly:

Commission	Acting President
15	J. Rahe
16	M. Davies & A. Brahic
49	L. Burlaga

7. **Appointment of the Finance Committee**

In accord with Statute 18(a), the General Assembly appointed the following Finance Committee consisting of one representative from each Adhering Body:

Country	Category/ Units	Finance Representative	Deputy
Argentina	III/4	H. Levato	
Australia	III/4	C.S.L. Keay	K.C. Freeman
Austria	I/1	H.F. Haupt	
Belgium	IV/6	P. Smeyers	
Brazil	II/2	J. Lépine	
Bulgaria	I/1	B.J. Kovachev	
Canada	VI/14	V. Gaizauskas	

Chile	I/1	G. Carrasco	
China (Nanjing)	V/10	Qu Qin-Yue	
China (Taipei)	I/1	Wu H.H.	
Colombia	I/1	------	
Cuba	I/1	O. Alvarez	
Czechoslovakia	III/4	V. Vanysek	
Denmark	II/2	J.V. Clausen	
Egypt, AR	III/4	K. Aly	
Finland	I/1	K. Mattila	
France	VII/20	G. Wlérick	
Germany, DR	II/2	G. Ruben	
Germany, FR	VII/20	M. Grewing	
Greece	II/2	J. Seiradakis	
Hungary	II/2	B. Szeidl	
India	III/4	M.B.K. Sarma	
Indonesia	I/1	M. Wiramihardja	
Iran	I/1	Y. Sobouti	
Iraq	I/1	------	
Ireland	I/1	P.A. Wayman	
Israel	II/2	G. Shaviv	
Italy	V/10	E. Proverbio	
Japan	VI/14	D. Sugimoto	K. Kodaira
Korea, DPR	I/1	------	
Korea, RP	I/1	Yun Hong Sik	
Mexico	II/2	L.C. Bazuas	A. Serrano
Netherlands	IV/6	W.B. Burton	
New Zealand	I/1	E. Budding	
Nigeria	I/1	S. Okoye	
Norway	I/1	------	
Poland	III/4	J. Smak	
Portugal	II/2	J.P. Osorio	
Rumania	II/2	------	
South Africa	III/4	M.W. Feast	I.S. Glass
Spain	II/2	C.A. Gimenez	
Sweden	III/4	D. Dravins	
Switzerland	III/4	M. Mayor	
Turkey	I/1	------	
United Kingdom	VII/20	F.G. Smith	
Uruguay	I/1	G. Vicino	G. Tancredi
USA	VIII/30	P. Boyce	

USSR	V/10	Y.S. Yatskiv
Vatican State City	I/1	J. Casanovas
Venezuela	I/1	------
Yugoslavia	II/2	------

8. Appointment of the Resolutions Committee

The President informed the Assembly that the Executive Committee proposed the establishment of a Resolutions Committee under the chairmanship of Dr. M. McCarthy, with Drs. A. H. Batten (Executive Committee Representative), L. Houziaux and H. Quintana as members. The General Assembly unanimously agreed to this composition of the Resolutions Committee.

9. Revisions of the Statutes and By-Laws

The changes adopted during the Extraordinary General Assembly are summarized in Chapter II, p. 19.

10. Resolutions submitted by Adhering Bodies

No resolutions were proposed to the XXth General Assembly by Adhering Bodies (a proposal from Sweden was withdrawn).

11. Resolutions submitted by Commissions or by Associated Inter-Union Commissions

See § 16, pp. 53-62.

The President then formally adjourned the meeting to 1988 August 11 at 10.00 and closed the Session with a word of thanks to the participants.

CHAPTER III

SECOND SESSION

Held in the Baltimore Convention Center

August 11, 1988, 10:00 to 12:00

Prof. J. Sahade, President in the Chair

Before passing to the Agenda, the General Secretary, called upon by the President, established the quorum for voting on financial matters and found eight Adhering Countries not represented. The General Assembly then appointed R. Henry, J. Narlikar & V. Trimble as Tellers.

12. Financial matters

Report of the Finance Committee

The General Secretary mentioned that copies of the financial report were in the hands of the official representatives and he thus invited the Chairman of the Finance Committee, Dr. P. Boyce, to read the report. Dr. Boyce reported as follows:

"The task of the Finance Committee is to inspect the accounts of the Union for the past three years in sufficient detail to assure that the accounts, indeed, are in order and to report its findings to the General Assembly. We compared the actual expenditures with the amounts budgeted and ascertained the cause of any major discrepancies. The Committee is also expected to look over the budget for the coming triennium, as proposed by the Executive Committee, and comment upon the fiscal implications of the projections. Much of the detailed work of the Committee was delegated to a Finance subcommittee which was appointed on 2 August, 1988. This report was modified and approved at the final meeting of the Finance Committee on 6 August, 1988.

The Finance Committee has inspected the accounts of the IAU for the period 1985-1987 and found everything to be in order. It also examined the accounting summary for the first half of 1988 and found no indication of any problems in the accounts for the current year. Certified external auditors have performed yearly audits of the IAU accounts. The auditors have found no irregularities and have certified that the accounts are an accurate representation of the state of the IAU finances. The Finance Committee, in making its examination of the accounts, has relied upon the auditors' reports for its information. Our examination leads us to confirm that the General Secretary has managed the finances in a prudent and fiscally responsible manner. In view of the complete change of personnel in the Secretariat during 1987 the General Secretary is to be congratulated for keeping the affairs of the Union running smoothly.

Operations of the Secretariat

The operations of the IAU Secretariat have benefitted greatly from the continuation of the modernization that had already started under the previous General Secretary. The membership files are up-to-date, and the ease of communicating with the Secretariat by means of Telex has been of significant help in making the meeting arrangements for this General Assembly. The IAU Secretariat soon will be moved from the gate house of the Paris Observatory into the building of the INSU/IAP. Access to better photocopy equipment, a fax machine and a tie-in to the electronic networks will help to improve further the daily operations of the Secretariat. The Finance Committee applauds this move.

In order to keep the operating costs low, our Union depends heavily upon volunteers to take on the jobs of General Secretary and Assistant General Secretary. We are grateful, not only to our busy colleagues who agree to provide their time and energy, but also to the institutions and governments who make such arrangements possible. In particular, the last two General Secretaries have enjoyed a high level of support that has contributed to keeping the operational costs of the Union at a very low level.

Comments on individual items

In looking over the accounts, the Finance Committee wishes to comment upon a few specific items. We are concerned that some of the Member countries are behind in the payment of their contributions and urge all countries to make their payment as promptly as possible.

We note that, owing to the financial problems within UNESCO, the level of support from ICSU has been cut in half during the past three years. We understand that the support is not expected to increase in the future. The proposed budget for the next triennium assumes the recent level of support from ICSU.

Since a six year contract with D. Reidel was concluded in 1985, the favorable level of income from IAU publications has been maintained and is expected to continue for the next three years. The Committee wishes to call attention to the decline in the market for the traditional astronomical publications in recent years. All publishers are feeling financial pressure that is expected to increase in the future and that may put this source of income at risk after the contract expires in 1991.

We feel it is appropriate to remind the General Assembly and the members of the Union that most of the funds that the IAU provides to the General Assembly are used to support the attendance of young astronomers. In this way, some of the contributions from member countries are also used to support young astronomers.

The Finance Committee notes that only about one-half of the funds budgeted for the exchange of astronomers in the present triennium have been used. The Finance Committee believes this indicates either that there is not sufficient awareness of this programme around the world or that there is now less need for a programme which can only provide travel support. We urge the Executive Committee to examine this programme to see how it might better fulfill the goal of promoting long-term visits of astronomers to institutions in other countries.

Operating surplus and account balances

The Committee notes that certain temporary circumstances have caused a positive net income for the past two years. These include a reduced salary expenditure in the Secretariat during 1987 and a fortuitous reduction in Executive Committee expenditures. This has resulted in an operating surplus. While this is not the deliberate operating policy of the Executive Committee, it is, nevertheless, accepted as prudent.

The accumulated balance in the IAU accounts acts as a reserve fund which now stands at somewhat more than one year of operations. The Finance Committee feels this adequate and, in view of the somewhat uncertain economic situation in the world, is very appropriate.

Proposed budget

The Committee has examined the proposed budget for the triennium 1989-1991. The income estimates are appropriately conservative. However, the expenses will grow by 12.6 percent, or slightly more than four percent per year. The members of the Executive Committee are more widely scattered, the General Assembly travel costs will be higher, and the increased IAU membership means increased costs. In view of the average inflation around the world, we find the proposed increases to be reasonable.

In forming our opinion about the costs to be expected for the support of the General Secretary and Assistant General Secretary we noted that the IAU is only covering a small fraction of the actual costs which the job entails. We note that the support of travel to IAU Symposia and Colloquia is proposed by the Executive Committee to be kept at the same level.

In order to cover the operating expenses of the Union, the Executive Committee has proposed an increase in the unit of contribution of slightly more than three percent per year. We understand the burden that these contributions impose upon some countries with large inflation rates. Nevertheless, the expenses of the Union have to be paid and the Committee strongly supports the proposed unit of contribution.

Fiscal management

The budget of the IAU is of sufficient size and complexity that we urge the Executive Committee to consider possible methods to provide appropriate assistance and advice to the General Secretary. We believe it may be possible to draw upon the expertise which exists within the membership of the IAU to accomplish this without raising costs.

Cost of General Assembly

The cost of holding a General Assembly was the subject of discussion within the Finance Committee. The current meeting has cost approximately $ 1,500,000, about 1.5 times the entire last three years of the IAU operations. The Committee notes that more than 90 percent of the cost is borne by the local hosts, including an amount for the support of young astronomers that slightly exceeds the IAU contribution.

The Finance subcommittee is concerned that the cost of holding a General Assembly may make it impossible for many countries to act as host. We feel this would be an unfortunate situation. We are also concerned about the smaller than expected attendance at the last three General Assemblies. In fact, the fraction of IAU members attending the General Assembly has declined since 1970. Since the attendance severely impacts the finances of a meeting, we find this trend to be particularly disturbing. We recommend that the Executive Committee reconsider the functions of the General Assembly, analyze the costs and the factors that drive them, and consider ways in which the functions can be accomplished more effectively and cheaply."

Votes on proposed budgets for 1988 & 1989-1991

The President thanked Dr. Boyce and put the report of the Finance Committee to the vote, taking each item separately.

- The proposed budget for 1988 was accepted unanimously.
- The increase of the unit of contribution was approved by 163 votes: there were 4 abstentions.
- The proposed budget 1989-1991 was accepted unanimously.

13. Financial Resolutions of the Executive Committee

None.

14. Resolutions submitted by the Executive Committee

The following resolutions were submitted by the Executive Committee:

Resolution A1: Amateur-Professional Cooperation in Astronomy

The XXth General Assembly of the International Astronomical Union,

recognising
the long-standing tradition of excellent and practical collaboration which has existed between amateur and professional astronomers, particularly during the first seven decades of our Union's existence;

noting
that additional communication for common projects is needed today between amateurs and professionals;

recommends
that a Working Group be established to foster this cooperation

and instructs
the General Secretary to communicate this proposal to the Executive Committee and to arrange for publication of this proposal by national and international organisations both amateur and professional.

Résolution A1 : Coopération entre les astronomes amateurs et professionnels en astronomie

La XXème Assemblée Générale de l'Union Astronomique Internationale,

reconnaissant
la tradition constante de collaboration excellente et fructueuse qui a existé entre les astronomes amateurs et professionnels, particulièrement pendant les sept premières décades de l'existence de l'Union,

notant
qu'une interaction plus poussée pour des projets communs est nécessaire aujourd'hui entre amateurs et professionnels,

recommande
l'établissement d'un Groupe de Travail pour favoriser cette coopération

et donne instruction
au Secrétaire Général de communiquer cette proposition au Comité Exécutif et de prendre les dispositions nécessaires pour la publication de cette proposition par les organisations nationales et internationales d'astronomes amateurs et d'astronomes professionnels.

Resolution A2: Adverse Environmental Impacts on Astronomy

The XXth General Assembly of the International Astronomical Union,

noting with grave concern
the increasing impact of light pollution, radio interference, space debris, and other environmental factors that adversely affect observing conditions from the ground and in space;

reaffirms
the special importance of the resolutions adopted by previous General Assemblies that relate to the protection of observatories (ground-based and in space) and of observing conditions including:

(1961) Resolutions	No. 1 & 2,	Transactions IAU XIB
(1964) Resolutions	No. 3 & 5,	Transactions IAU XIIB
(1967) Resolution	No. 2,	Transactions IAU XIIIB
(1970) Resolution	No. 10,	Transactions IAU XIVB
(1976) Resolutions	No. 8 & 9,	Transactions IAU XVIB
(1979) Resolution	No. 3,	Transactions IAU XVIIB
(1982) Resolution	No. R9,	Transactions IAU XVIIIB
(1985) Resolutions	No. B4, B5 & B7,	Transactions IAU XIXB;

strongly urges

a. that all astronomers request civil authorities and others in their countries to implement solutions to preserve the quality of observing conditions,
b. that all national organisations bring these concerns to the notice of adhering organisations, space agencies, and others in their countries;

notes with special appreciation

those agencies, communities, organisations, and individuals who have become aware of the issues and have begun to help; and

encourages

all others, everywhere, to become aware of the need to minimize the impact on the environment of light pollution, radio frequency interference, and space debris, which are causing increasingly severe impact on observing conditions for astronomy and which will compromise mankind's view of Universe;

and requests

through ICSU that SCOPE should study the nature and extent of this threat and advise the IAU of its findings.

ICSU
International Council of Scientific Unions
SCOPE
Scientific Committee on Problems of the Environment

Résolution A2 : Impacts sur l'environnement nuisibles pour l'Astronomie

La XXème Assemblée Générale de l'Union Astronomique Internationale,

constatant avec une grave inquiétude

l'impact croissant de la pollution lumineuse, des interférences dans le domaine des fréquences radio, des débris spatiaux et autres composantes de l'environnement qui affectent gravement les conditions d'observation au sol et dans l'espace,

réaffirme

l'importance spéciale des résolutions adoptées par les assemblées générales précédentes (et d'autres concernant des questions en rapport avec la protection des observatoires -au sol ou spatiaux- et des conditions d'observations passées à cette Assemblée Générale) comprenant la ou les :

(1961) Résolutions	No. 1 & 2,	Transactions UAI XIB
(1964) Résolutions	No. 3 & 5,	Transactions UAI XIIB
(1967) Résolution	No. 2,	Transactions UAI XIIIB
(1970) Résolution	No. 10,	Transactions UAI XIVB
(1976) Résolutions	No. 8 & 9,	Transactions UAI XVIB
(1979) Résolution	No. 3,	Transactions UAI XVIIB
(1982) Résolution	No. R9,	Transactions UAI XVIIIB
(1985) Résolutions	No. B4, B5 & B7,	Transactions UAI XIXB ;

insiste vigoureusement

a. pour que tous les astronomes demandent aux autorités civiles et non-civiles de mettre en oeuvre des solutions pour préserver la qualité des conditions d'observation dans leur pays,
b. pour que tous les organismes nationaux fassent connaître ces inquiétudes aux organismes adhérents, aux agences spatiales et autres, dans leur pays.

note et apprécie

le rôle joué par les individus, les communautés, les organisations et les agences qui ont pris conscience de ces problèmes et ont déjà apporté leur aide ; et, encourage

tous, en tous lieux, à prendre conscience de la nécessité de minimiser les conséquences sur l'environnement de la pollution lumineuse, des interférences hertziennes et des débris spatiaux qui dégradent de façon croissante les conditions d'observation de l'astronomie et qui sont de nature à compromettre la vision de l'Univers par l'humanité.

et demande

via ICSU, que le SCOPE étudie la nature et l'étendue de cette menace et informe l'UAI de ses conclusions.

Resolution A3: Improvement of Publications

The XXth General Assembly of the International Astronomical Union

recognising

- the need to develop clear lines of communication between the various branches of astronomy and other related scientific disciplines;
- the desirability of promoting ease of access to information contained in the astronomical literature;
- the advantages that would follow from a reduction in the variety of the editorial requirements for the submission of papers and reports; and
- the importance of identifying astronomical objects by clear and unambiguous designations; and

noting

- the growth in the cadre of young scientists trained in the use of the International System (SI) of units and widespread adoption of SI in other scientific and technical areas; and
- the substantial measure of agreement that has been reached during the drafting of the new IAU Style Manual for the preparation of astronomical papers, reports and books;

recommends

that the authors and the editors of the astronomical literature adopt the recommendations in the IAU Style Manual, which is to be published in the Transactions of the Union and reprinted for wide distribution and greater convenience;

in particular, it urges authors and editors:

1. to use only the standard SI units and those additional units that are recognised for use in astronomy, as recommended by Commission 5;
2. to adopt the conventions for citations and references that are given in the IAU Style Manual and that are exemplified in Astronomy and Astrophysics Abstracts; and
3. to ensure that all astronomical objects referred to in the literature are designated clearly and unambiguously in accordance with the recommendations of the Union.

Note:

The Executive Committee recognises that the replacement of CGS by SI units will require an adjustment of practice on the part of many astronomers; this will no doubt take time. Consequently, we urge that the total conversion from CGS to SI units by all organs of communication shall be accomplished by the time of the next General Assembly (1991).

In the meantime we request that the major journals should publish, once a year, a table of conversions between CGS and SI units, as provided by Commission 5.

SI
International System (units)
CGS
Centimeter, Gramme, Second (units)

Résolution A3 : Amélioration des Publications

La XXème Assemblée Générale de l'Union Astronomique Internationale,

reconnaissant

- le besoin de développer de bonnes voies de communication entre les différentes branches de l'astronomie et les autres disciplines scientifiques concernées;
- la désirabilité de promouvoir un accès facile à l'information contenue dans la littérature astronomique ;
- les avantages qui découleraient d'une réduction dans la variété des exigences éditoriales pour la soumission d'articles ou de rapports ; et
- l'importance de l'identification des sources astronomiques par des désignations claires non ambiguëes, et :

notant

- la croissance de l'emploi du Système International d'unités SI par les jeunes scientifiques éduqués dans ce système et l'adoption très répandue du système SI dans d'autres domaines scientifiques ou techniques ; et
- la convergence de vues atteinte pendant l'élaboration du nouveau Manuel de Style" de l'UAI pour la préparation des articles, rapports et ouvrages astronomiques ;

recommande

que les auteurs et les éditeurs adoptent pour la littérature astronomique les recommandations contenues dans le Manuel de Style de l'UAI qui va être publié dans les Transactions de l'UAI et fourni sous forme de tiré à part, pour une large distribution et pour plus de commodité ;

en particulier, elle insiste pour que les auteurs et les éditeurs :

1. utilisent seulement les unités standards SI et les unités supplémentaires qui sont acceptées pour l'astronomie, telles que recommandées par la Commission 5,
2. adoptent les conventions pour les citations et les références qui sont données dans le Manuel de Style de l'UAI et dont des exemples sont disponibles dans Astronomy and Astrophysics Abstracts ; et
3. s'assurent que toutes les sources astronomiques auxquelles référence est faite dans la littérature soient désignées clairement et sans ambiguïté en accord avec les recommandations de l'Union.

Note :

Le Comité Exécutif se rend compte que le remplacement des unités CGS par les unités SI exigera une adaptation de la part de beaucoup d'astronomes; cela prendra du temps sans doute. En conséquence, il recommande avec insistance que la conversion totale des unités CGS vers les unités SI soit effective dans tous les organes de communication à l'époque de la prochaine Assemblée Générale (1991).

Dans l'intervalle, il recommande que les revues principales publient une fois par an une table de conversion entre les unités CGS et SI, fournie par la Commission 5.

Resolution A4: International Space Year (ISY) 1992

The XXth General Assembly of the International Astronomical Union,

considering
that the International Space Year (1992) will provide a great opportunity to further international cooperation within many areas of science and technology which are closely related to astronomy and astrophysics and also that the related educational and public information efforts may make important contributions to the dissemination of knowledge, also in countries which do not normally engage in space activities, and

noting
with satisfaction the interest shown by ICSU, COSPAR, IAF and other organisations in the International Space Year,

recommends
that all IAU Adhering Bodies, IAU Commissions and individual members actively participate in International Space Year activities, also during the preparatory phases.

ISY
International Space Year
ICSU
International Council of Scientific Unions
COSPAR
Committee on Space Research
IAF
International Astronautical Federation

Résolution A4: Année Internationale de l'Espace 1992

La XXème Assemblée Générale de l'Union Astronomique Internationale,

considérant
que l'Année Internationale de l'Espace (1992) fournira une remarquable occasion pour un accroissement de la coopération internationale dans de nombreux domaines de la Science et de la Technologie en étroite relation avec l'Astronomie et l'Astrophysique, et considérant aussi que les efforts correspondants dans le domaine de l'information du public ou celui de l'éducation peuvent conduire à une contribution importante dans la dissémination des connaissances, y compris dans des pays qui ne sont pas engagés régulièrement dans des activités spatiales, et

notant
avec satisfaction l'intérêt montré par l'ICSU, le COSPAR, l'IAF et d'autres organisations pour l'Année Internationale de l'Espace,

recommande
que les organismes adhérant à l'UAI, les Commissions de l'UAI et les membres de l'UAI à titre personnel, participent activement aux activités associées à l'Année Internationale de l'Espace, y compris pendant les phases préparatoires.

Resolution A5: Cooperation to Save Hydroxyl Bands

The XXth General Assembly of the International Astronomical Union,

noting

a. the long standing concern of the International Astronomical Union for protecting radio astronomy from interference, particularly through resolutions passed at the General Assemblies in 1979, 1982 and 1985;
b. the increasing levels of harmful interference to radio astronomy, particularly from space and airborne transmitters, which diminish the advantages of locating observatories at remote sites;
c. the particularly high levels of harmful interference experienced consistently in the sub-band 1610.6-1613.8 MHz from navigation satellites which make observations of an astrophysically important hydroxyl line increasingly difficult;
d. that the 1612 MHz hydroxyl line has assumed greatly increased importance since the 1979 World Administrative Radio Conference due particularly to the discovery of numerous OH/IR stars which have been used for absolute distance determination in the Galaxy and for understanding stellar evolution;
e. that the World Administrative Radio Conference for the Mobile Services (WARC MOB-87) has also allocated the band 1610-1626.5 MHz to the Radio-Determination Satellite Service (RDSS), subject of footnote 743E of the Radio Regulations, which states that in Regions 1 and 3 harmful interference shall not be caused to the Radio Astronomy Service (RAS), and that in Region 2 several administrations have agreed to limited protection for the RAS;
f. that the WARC MOB-87 in Resolution PLEN/1 has invited the CCIR to continue its studies in order to obtain more precise results concerning the conditions of sharing in the bands 1610-1625.5 MHz and 2483.5-2500-2516.5 MHz between the RDSS on the one hand and the RAS, among other services, on the other;

urges

1. that national administrations cooperate with IUCAF to examine means to prevent harmful interference to observations in the band 1610.6-1613.8 MHz from global navigation satellite systems, particularly in designing changes to existing systems and planning new systems;
2. that administrations adhering to the International Astronomical Union and the International Telecommunication Union strive for improved protection of the RAS in the 1610.6-1613.8 MHz band by upgrading the allocation status of the RAS to that of primary service in this sub-band at the next competent World Administrative Radio Conference;
3. that IUCAF, representing the IAU, respond rapidly to the invitation to continue studying in Study Group 2 of the CCIR the conditions for successfully sharing the band 1610-1626.5 MHz and examine the problems of second harmonic emission from RDSS transmitters in the band 2483.5-2500 MHz which could affect the RAS in the band 4800-5000 MHz;
4. that administrations operating satellites or satellite systems in the aeronautical navigation satellite service at 1.5/1.6 GHz frequencies protect the RAS from harmful interference by appropriately filtering unwanted emissions;

and instructs the President

to bring this Resolution to the attention of the Secretary General of the International Telecommunication Union.

WARC
World Administrative Radio Conference
RDSS
Radio Determination Satellite Service
RAS
Radio Astronomy Service

IUCAF
Inter-Union Commission on Frequency Allocations for Radio Astronomy and Space Science

Résolution A5 : Coopération pour la sauvegarde des bandes de OH

La XXème Assemblée Générale de l'Union Astronomique Internationale,

<u>considérant</u>

a. la préoccupation constante de l'Union Astronomique Internationale en ce qui concerne la protection de la radioastronomie contre les parasites, préoccupation qui s'est manifestée en particulier par des résolutions des Assemblées Générales de 1979, 1982 et 1985 ;
b. le niveau accru des parasites nuisibles à la radioastronomie provenant en particulier d'émetteurs spatiaux ou aéroportés, qui réduit l'avantage qu'il y a à placer les observatoires dans des sites isolés ;
c. le niveau particulièrement élevé des parasites nuisibles produits régulièrement dans la sous-bande 1610.6-1613.8 MHz par des satellites de navigation, ce qui rend de plus en plus difficile les observations dans une raie de la molécule hydroxyle d'importance astrophysique ;
d. que cette raie de la molécule hydroxyle à 1612 MHz a une importance beaucoup plus grande qu'à l'époque de la Conférence WARC de 1979, en raison notamment de la découverte de nombreuses étoiles OH/IR qui sont utiles à la détermination absolue des distances dans la Galaxie et à la compréhension de l'évolution stellaire ;
e. que la Conférence Radio Administrative Mondiale pour les Services Mobiles (WARC MOB-87) a également alloué la bande 1610-1626.5 MHz au service des satellites de radio-détermination (RDSS), qui est l'objet de la note 743E des Règlements Radio, laquelle indique que dans les Régions 1 et 3 aucun parasite nuisible au service de radioastronomie (RAS) ne doit être produit et que dans la Région 2 plusieurs administrations se sont mises d'accord pour une protection limitée du RAS ;
f. que la résolution PLEN/1 de la WARC MOB-87 a invité le CCIR à continuer ses études en vue d'obtenir des résultats plus précis concernant les conditions de partage des bandes 1610-1625.5 MHz et 2483.5-2500-2516.5 MHz entre le RDSS d'une part et d'autre par le RAS,

<u>demande instamment</u>

1. que les administrations nationales coopèrent avec l'IUCAF pour examiner les moyens d'éviter des parasites nuisibles aux observations provenant des systèmes globaux de navigation par satellites dans la bande 1610.6-1613.8 MHz, en particulier lors de la préparation de changements futurs aux systèmes existants ou de projets nouveaux ;
2. que les administrations adhérant à l'Union Astronomique Internationale et à l'Union Internationale des Télécommunications s'efforcent d'augmenter la protection du RAS dans la bande 1610.6-1613.8 MHz en améliorant le statut des allocations du RAS par rapport à celles du service primaire dans cette sous-bande lors de la prochaine Conférence Radio Administrative Mondiale ;
3. que l'IUCAF, représentant l'Union Astronomique Internationale, réponde rapidement à la demande de continuer à étudier au sein du Groupe d'Etude 2 du CCIR les conditions pour un partage satisfaisant de la bande 1610-1626.5 MHz et examine le problème de l'émission dans l'harmonique 2 des émetteurs du RDSS dans la bande 2483.5-2500 MHz qui pourrait indirectement affecter le RAS dans la bande 4800-5000 MHz ;
4. que les administrations responsables des satellites et des systèmes de satellite dans le service de navigation aéronautique par satellite émettant aux fréquences comprises entre 1.5 et 1.6 GHz protègent le RAS des parasites nuisibles grâce à un filtrage approprié des émissions indésirables ;

et charge le Président

de porter la présente résolution à la connaissance du Secrétaire Général de l'Union Internationale des Télécommunications.

Resolution A6: Sharing Hydroxyl Band with Land Mobile Satellite Services

The XXth General Assembly of the Union,

considering

a. that the 1660-1660.5 MHz band is allocated to the Radio Astronomy Service on a shared, primary basis, and is used to observe hydroxyl lines, which are of the highest astrophysical importance, in many galaxies in the nearby Universe;
b. that the World Administrative Radio Conference for the Mobile Services (WARC MOB-87) has also allocated the 1660-1660.5 MHz band to the land mobile satellite service;
c. that WARC MOB-87 has added Footnote 730A to the Radio Regulations, allowing administrations to authorize aircraft stations and ship stations to communicate with space stations in the land mobile satellite service in the 1660-1660.5 MHz band;
d. that CCIR Study Group 8 has established Interim Working Party 8/14 to study, among other characteristics of mobile satellite systems, the necessary criteria for frequency sharing between the various mobile satellite systems and the other services allocated the same bands;

urges

1. that IUCAF, in representation of the International Astronomical Union, interact, as a matter of urgency, with the Interim Working Party of CCIR Study Group 8 and with Study Group 2 to work out the necessary criteria under which the Radio Astronomy Service, and the land mobile satellite service and services authorized under Footnote 730A, may share the 1660-1660.5 MHz band;
2. that administrations adhering to the International Astronomical Union and to the International Telecommunications Union bear in mind at the next competent WARC the importance of the primary allocation to the Radio Astronomy Service in the band 1660.0-1660.5 MHz;

and instructs the President

to request the Director of the CCIR to bring this Resolution to the attention of the Chairman of Interim Working Party 8/14.

WARC
: World Administrative Radio Conference

CCIR
: Comité Consultatif International des Radiocommunications

IUCAF
: Inter-Union Commission on Frequency Allocations for Radio Astronomy and Space Science

RAS
: Radio Astronomy Service

Résolution A6 : Partage de la bande Hydroxyle avec les services mobiles au sol

La XXème Assemblée Générale de l'Union Astronomique Internationale,

considérant

a. que la bande 1660-1660.5 MHz est allouée au service de la radioastronomie sur la base d'une allocation primaire partagée et est utilisée pour l'observation des raies hydroxyles qui sont de très grande importance astrophysique, dans de nombreuses galaxies proches ;

b. que la Conférence Administrative Mondiale Radio pour les Services Mobiles (WARC MOB-87) a aussi alloué la bande 1660-1660.5 MHz au service mobile terrestre de liaison avec les satellites ;
c. que WARC MOB-87 a ajouté la note 730 A aux réglementations radio permettant aux administrations d'autoriser les stations sur avion ou sur bateau de communiquer avec les stations spatiales par les services mobiles dans la bande 1660-1660.5 MHz ;
d. que le Groupe 8 d'étude du CCIR a établi un groupe de travail intérimaire 8/14 pour étudier, parmi d'autres caractéristiques des systèmes mobiles, les critères nécessaires au partage de fréquence entre les différents systèmes mobiles et les autres services auxquels certaines des bandes allouées au service des émetteurs mobiles sont aussi allouées.

recommande de façon pressante :

1. que l'IUCAF, en tant que représentant de l'UAI, agisse, pour cause d'urgence, avec le Groupe de Travail par interim du Groupe d'Etude 8 du CCIR, et avec le Groupe d'Etude 2, afin de dégager les critères nécessaires pour que le service de radioastronomie, le service de satellite mobile et les services autorisés par la note 730A puissent partager l'allocation de la bande 1660-1660.5 MHz ;
2. que les administrations adhérant à l'UAI et à l'UIT aient présente à l'esprit, lors de la prochaine WARC compétente, l'importance de l'allocation primaire au Service de Radioastronomie de la bande 1660.0-1660.5 MHz ;

demande au Président

de prier le Directeur du CCIR de porter cette résolution à l'attention du Président du Groupe de Travail par intérim 8/14.

Resolution A7: Revision of Frequency Bands for Astrophysically Significant Lines

The XXth General Assembly of the International Astronomical Union,

recalling

a. resolutions passed by the International Astronomical Union in 1979 and 1982 recommending the provision by national administrations of frequency bands for the astrophysically most important spectral lines;
b. the need expressed in those resolutions to protect these frequency bands from in-band, band-edge and sub-harmonic emissions, especially from space-borne transmitters;
c. the documentation of Study Group 2 of the CCIR in Recommendation 314 and Reports 224 and 697 concerning harmful interference to the Radio Astronomy Service;

and considering

the careful reviews by the International Astronomical Union in the period 1983-1988 of the astrophysically most important spectral lines;

recommends

that the International Astronomical Union take note of the revision of the frequencies of the astrophysically most important spectral lines listed in Tables 1 and 2 below;

and instructs the President

to bring the resolution to the attention of the Secretary General of the International Telecommunication Union.

CCIR
Comité Consultatif International des Radiocommunications

TABLE 1

Radio frenquency lines of the greatest importance to radio astronomy at frequencies below 275 GHz.

Substance	Rest Frequency	
Deuterium (DI)	327.384	MHz
Hydrogen (HI)	1420.406	MHz
Hydroxyl radical (OH)	1612.231	MHz
Hydroxyl radical (OH)	1665.402	MHz
Hydroxyl radical (OH)	1667.359	MHz
Hydroxyl radical (OH)	1720.530	MHz
Methyladyne (CH)	3263.794	MHz
Methyladyne (CH)	3335.481	MHz
Methyladyne (CH)	3349.193	MHz
Formaldehyde (H_2CO)	4829.660	MHz
Methanol (CH_3OH)	12.178	GHz
Formaldehyde (H_2CO)	14.488	GHz
Cyclopropenylidene (C_3H_2)	18.343	GHz
Water vapour (H_2O)	22.235	GHz
Ammonia (NH_3)	23.694	GHz
Ammonia (NH_3)	23.723	GHz
Ammonia (NH_3)	23.870	GHz
Silicon monoxide (SiO)	42.821	GHz
Silicon monoxide (SiO)	43.122	GHz
Carbon monosulphide (CS)	48.991	GHz
Deuterated formylium (DCO^+)	72.039	GHz
Silicon monoxide (SiO)	86.243	GHz
Formylium ($H^{13}CO^+$)	86.754	GHz
Ethynyl radical (C_2H)	87.3	GHz
Hydrogen cyanide (HCN)	88.632	GHz
Formylium (HCO^+)	89.189	GHz
Hydrogen isocyanide (HNC)	90.664	GHz
Diazenylium (N_2H^+)	93.17	GHz
Carbon monosulphide (CS)	97.981	GHz
Carbon monoxide ($C^{18}O$)	109.782	GHz
Carbon monoxide (^{13}CO)	110.201	GHz
Carbon monoxide ($C^{17}O$)	112.359	GHz
Carbon monoxide (CO)	115.271	GHz
Formaldehyde ($H_2{}^{13}CO$)	137.450	GHz
Formaldehyde (H_2CO)	140.840	GHz
Carbon monosulphide (CS)	146.969	GHz
Water vapour (H_2O)	183.310	GHz
Carbon monoxide ($C^{18}O$)	219.560	GHz
Carbon monoxide (^{13}CO)	220.399	GHz
Carbon monoxide (CO)	230.538	GHz
Carbon monosulphide (CS)	244.953	GHz
Hydrogen cyanide (HCN)	265.886	GHz
Formylium (HCO^+)	267.557	GHz
Hydrogen isocyanide (HNC)	271.981	GHz

TABLE 2

Radio frequency lines of the greatest importance
to radio astronomy at frequencies between 275 GHz & 900 GHz
(not allocated to Radio Astronomy in the Radio Regulations)
(* Central line of a group of three)

Diazenylium (N_2H^+)	279.511
Carbon monoxide ($C^{18}O$)	329.330
Carbon monoxide (^{13}CO)	330.587
Carbon monosulphide (CS)	342.883
Carbon monoxide (CO)	345.796
Hydrogen cyanide (HCN)	354.484
Formyl ion (HCO^+)	356.734
Diazenylium (N_2H^+)	372.672
Water vapour (H_2O)	380.197
Carbon monoxide ($C^{18}O$)	439.088
Carbon monoxide (^{13}CO)	440.765
Carbon monoxide (CO)	461.041
Heavy water (HDO)	464.925
Carbon (CI)	492.162
Water vapour ($H_2\,^{18}O$)	547.676
Water vapour (H_2O)	556.936
Ammonia ($^{15}NH_3$)	572.113
Ammonia (NH_3)	572.498
Hydrochloric acid (HCl)	625.918 *
Carbon monoxide (CO)	691.473
Hydrogen cyanide (HCN)	797.433
Formyl ion (HCO^+)	802.653
Carbon monoxide (CO)	806.652
Carbon (CI)	809.350

Résolution A7 : Révision des bandes de fréquences pour les raies d'intérêt astrophysique

La XXème Assemblée Générale de l'Union Astronomique Internationale,

rappelant

a. les résolutions de l'UAI en 1979 et 1982 recommandant la mise à disposition par les administrations nationales de bandes de fréquences pour les raies spectrales les plus importantes pour l'astrophysique ;

b. la nécessité exprimée dans ces résolutions de protéger ces bandes de fréquences des émissions dans la bande, en bordure de bande et sous harmoniques, spécialement issues des émetteurs à bord d'engins spatiaux ;

c. la documentation du Groupe d'étude 2 du CCIR dans la recommandation 314 et les rapports 224 et 697 concernant les parasites nuisibles au service de la radioastronomie ;

et considérant

les soigneuses révisions de la part de l'UAI dans la période 1983-1988 des raies spectrales les plus importantes pour l'astrophysique ;

recommande

que l'Exécutif de l'UAI prenne note de la révision des fréquences des raies spectrales les plus importantes en astrophysique répertoriées dans les tableaux I et II (cf. pp. 49 & 50),

et demande au Président

de porter cette résolution à l'attention du Secrétaire Général de l'Union Internationale des Télécommunications.

15. Resolutions proposed by the Resolutions Committee

At the request of the President, Dr. M. McCarthy, Chairman of the Resolutions Committee, reported on the work of the Committee. Ten resolutions were put forward by the Commissions (French translations are also included, see § 16, pp. 53-62).

Prior to reading the resolutions, Dr. McCarthy made the following announcement:

"Dear Colleagues,

In drafting proposals of the XXth General Assembly of the IAU, two trends can be noticed:

1. There were fewer specific proposals for common action requested by the Commissions of the Union. This we believe reflects an increased frequency of communications at the international level of Members assembled at symposia or colloquia sponsored by the Union, as well as other meetings for Members through national groups.

2. Exceptions to this trend were noted in an increased number of resolutions and requests in the field of astronomical ecology, concerned with the preservation of dark skies for earth based obervations and those concerned with safeguarding bands in the radio frequency spectrum required for the advance of astronomy and those concerned with alarminq restrictions for all science through the presence of an increasing amount of debris in space.

We thank the Executive Committee for their concerned care in guiding proposals concerning actions of astronomers outside and beyond the scope of strict commission matters: these are of concern to all scientists, but exceed the scope and competence of the Union's appointed Committee on Resolutions. I thank the Members of the Committee on Resolutions for their help and devoted service: Professor Dr. Leo Houziaux, Professor Dr. Alan H. Batten, Professor Dr. H. Quintana and all of Professor Swing's excellent secretarial staff."

Resolution B1: Extensions to FITS

The XXth General Assembly of the International Astronomical Union,

considering
the present situation of the transfer of catalogue and table data in digital form among astronomical institutes; and

noting
that significant improvements in portability can be made;

recommends
that all astronomical computer facilities recognise and support the rules for general extensions to the Flexible Image Transport System (FITS) including the extension for the exchange of catalogue and table data as described in Astronomy and Astrophysics Supplement Series 73, pp 359-364 and pp 365-372 (1988).

Résolution B1 : Extensions à FITS

La XXème Assemblée Générale de l'Union Astronomique Internationale,

considérant
la situation actuelle du transfert de données numériques sous forme de tables ou de catalogues entre établissements astronomiques,

notant
que des améliorations significatives de leur portabilité peuvent être accomplies,

recommande
que tous les centres de calcul astronomiques reconnaissent et appliquent les règles d'extension générales au Système Flexible de Transport d'Images (FITS), y compris l'extension pour l'échange de catalogues et de tables décrite dans Astronomy and Astrophysics Supplement Series 73, pp. 359 à 364 et pp. 365 à 372 (1988).

Resolution B2: Working Group on FITS

The XXth General Assembly of the International Astronomical Union,

considering
the high importance of the Flexible Image Transport System (FITS) for the exchange of digital data between astronomical institutes and for astronomical archives;

decides
to form a Working Group on FITS to maintain the existing FITS standards and to review, approve and maintain future extensions to FITS, recommended practices for FITS implementations, and the thesaurus of approved FITS keywords.

Résolution B2 : Groupe de travail sur FITS

La XXème Assemblée Générale de l'Union Astronomique Internationale,

considérant
la grande importance du Système Flexible de Transport d'Images (FITS) pour l'échange de données numériques entre les établissements astronomiques et pour les archives astronomiques,

décide
de former un groupe de travail sur FITS pour maintenir les normes existantes de FITS et pour examiner, approuver et maintenir les extensions futures à FITS, ainsi que les pratiques recommandées dans la mise en oeuvre de FITS, et le thesaurus des mots-clés approuvés de FITS.

Resolution B3: Endorsement of Commission Resolutions

The XXth General Assembly of the International Astronomical Union,

having
full confidence in its Commissions,

endorses
the Resolutions submitted by them to the Resolutions Committee for publication in the official languages of the Union, French and English, in Transactions IAU XXB.

Resolution B3: Soutien des Résolutions des Commissions

La XXème Assemblée Générale de l'Union Astronomique Internationale,

accordant
son entière confiance à ses Commissions,

souscrit
aux autres résolutions qu'elles ont soumises au Comité des Résolutions, pour être publiées dans les deux langues officielles de l'Union, le français et l'anglais dans les Transactions de l'UAI, volume XXB.

16. Resolutions Proposed by Commissions

The following resolutions were proposed by Commissions:

Resolution C1: Working Group on Reference Frames (WGRF) Resolutions

Commissions 4, 7, 8 & 24

recommend
the adoption of the following resolutions after their joint meeting to discuss the progress and needs of the Working Group on Reference Frames:

1. in order to avoid a confusing proliferation of reference frames, the FK5 should be retained as the IAU reference at optical wavelengths for the present and immediate future;
2. in order to derive the maximum possible information from the accumulated classical observations, and most especially from the fundamental observations, ab initio discussions of these latter observations should be encouraged and supported;
3. the International Astronomical Union should adopt a celestial reference based upon a consistent set of coordinates for a sufficient number of suitable extragalactic objects when the required observational data have been successfully obtained and appropriately analyzed. This reference frame should be based upon a common, simultaneous discussion of the observations using agreed upon conventions. This reference frame is likely to be based, initially at least, exclusively upon radio astrometry, and transformations between this reference frame and the conventional celestial and terrestrial reference systems as well as the dynamical frame should be defined. The reference frame should be updated as required;
4. the determination of the positions of radio sources at all possible wavelengths should be continued and accelerated so as to achieve the best possible all sky coverage and overall accuracy, while testing the suitability of candidate sources; the International Astronomical Union should encourage institutions to provide adequate time on appropriate instruments to ensure that the necessary astrometric observations are obtained;
5. the detection of radio stars and the determination of their positions and proper motions should be a major goal of astrometry; the determination of optical positions and proper motions of stars with respect to extragalactic objects should be encouraged. All applicable methods, particularly astrometry on large reflectors, should be used;

6. optical and infrared astrometric interferometry should be developed vigorously for use on the ground and possibly later in space. The related efforts in imaging interferometry have astrometric implications and these developments should also be supported. In all cases, the direct determination of the positions of extragalactic objects at optical/infrared wavelengths must be a major goal.

WGRF
Working Group on Reference Frames

Résolution C1 : Résolutions du Groupe de Travail sur les Repères de Référence (WGRF)

Les Commissions 4,7,8 & 24

recommandent
l'adoption des résolutions suivantes après leur discussion commune au sujet des progrès accomplis par le Groupe de Travail sur les repères de référence et sur les besoins de ce dernier :

1. que, afin d'éviter une prolifération gênante des repères de référence, le FK 5 soit conservé comme repère de référence de l'UAI aux longueurs d'onde optiques pour le moment et dans un avenir immédiat ;
2. que, afin de déduire l'information maximale possible des observations classiques cumulées, et plus particulièrement des observations fondamentales, des discussions ab initio de ces dernières soient encouragées et soutenues ;
3. que l'Union Astronomique Internationale adopte un repère de référence céleste fondé sur un ensemble cohérent de coordonnées, pour un nombre suffisant de sources extragalactiques convenables, après que les données observationnelles requises aient été obtenues avec succès et aient été convenablement analysées. Ce repère de référence devrait être fondé sur une discussion commune des observations et en utilisant des conventions ayant fait l'objet d'un accord préalable. Ce système de référence sera vraisemblablement, du moins au début, fondé exclusivement sur l'astrométrie dans le domaine des radiofréquences. Des transformations entre ce système de référence et les systèmes traditionnels terrestres et célestes devraient être définies, de même qu'avec le repère de référence de la dynamique. Le repère de référence devrait être révisé en fonction des besoins.
4. que soit continuée et accélérée la détermination de la position des radio-sources à toutes les longueurs d'onde possibles, de manière à réaliser la meilleure couverture du ciel et la meilleure précision globale, tout en vérifiant l'adéquation des sources candidates. L'Union Astronomique Internationale devrait encourager les organismes compétents à fournir suffisamment de temps sur les instruments appropriés pour assurer l'acquisition des observations astrométriques nécessaires ;
5. que la détection des radio-étoiles et la détermination de leur position et de leurs mouvements propres deviennent un but majeur de l'astrométrie ; que la détermination des positions optiques et des mouvements propres des étoiles par rapport aux sources extra-galactiques soit encouragée. Toutes les méthodes applicables, en particulier l'astrométrie au moyen des grands télescopes devraient être utilisées ;
6. que l'interférométrie optique et infrarouge soit développée vigoureusement au sol dans un premier temps, et plus tard sans doute dans l'espace. Les efforts associés en imagerie par interférométrie ont des implications astrométriques et ces développements devraient aussi être soutenus. Dans tous les cas, la détermination directe de la position des sources extragalactiques aux longueurs d'onde optique et infrarouge doit être un objectif majeur.

Resolution C2: WGRS - A Continuing Intercommission Project

Commissions 4,7,8,19,20,24,31,33 & 40

noting
the proliferation of Working and Study Groups which deal with various matters of concern to these Commissions;

recognising
the necessity of considering such matters carefully along with the inevitability of scientific interrelationships among them;

thank
the Chairpersons and Members of the Working Groups on Nutation and Astronomical Constants for their efforts; and

recommend
1. that the Working Group on Reference Systems (WGRS) be continued as an intercommission project and that it concern itself with Nutation, Astronomical Constants, Origins, Reference Frames and Time;
2. that appropriate Study Groups be formed as required and that the current chairman continue in office, and that Commissions 4,7,8,19,20,24,31,33 and 40 and the IAG be invited to appoint members;
3. that the International Astronomical Union support the efforts of the Intercommission Project by providing funds for travel of members to attend the Working Group meetings;
4. that the WGRS produce a draft report with specific recommendations at least six months before the General Assembly;
5. that close ties be maintained between the International Astronomical Union, as represented by the WGRS, and the Geodesic Community, as represented by the IAG/IUGG;
6. that a close liaison with the IERS be continued.

IAG
International Association of Geodesy
IUGG
International Union of Geodesy & Geophysics
IERS
International Earth Rotation Service

Résolution C2 : Reconduction du WGRS

Les Commissions 4,7,8,19,20,24,31,33 & 40

constatant
la prolifération de groupes d'étude et de travail s'occupant de différents sujets concernant ces commissions ;

reconnaissant
la nécessité de prendre en considération ces sujets soigneusement, ainsi que les interconnexions scientifiques inévitables entre eux ;

remercient
de leurs efforts les Président(e)s et les membres des groupes de travail sur la nutation et sur les constantes astronomiques ; et

recommandent

1. que le Groupe de Travail sur les Systèmes de Référence (WGRS) soit reconduit en tant que groupe de travail intercommission et qu'il s'occupe de la nutation, des constantes astronomiques, des origines, des repères de référence et du temps ;
2. que soient créés des groupes d'études appropriés si nécessaire, que le Président actuel poursuive son mandat et que les commissions 4, 7, 8, 19, 20, 24, 31, 33 et 40 et l'IAG soient invitées à y nommer des membres ;
3. que l'Union Astronomique Internationale soutienne les efforts de ce groupe intercommission en fournissant des crédits pour les voyages des membres se rendant aux réunions du Groupe de Travail ;
4. que le Groupe de Travail sur les Systèmes de Référence fournisse un rapport provisoire avec des recommandations spécifiques au moins six mois avant l'Assemblée Générale ;
5. que des liens étroits soient maintenus entre l'Union Astronomique Internationale, représentée par ce Groupe de Travail, et la communauté géodésique représentée par l'IAG/IUGG ;
6. qu'une liaison suivie avec l'IERS soit maintenue.

Resolution C3: Flare 22 Programme

Commission 10

recognising

the value of a coordinated scientific programme for the study of the physical processes and mechanisms in flares and solar active phenomena, which are common to a large variety of astrophysical objects and are responsible for planetary phenomena that affect the terrestrial environment;

noting

that a comprehensive study of all these components of solar activity is beyond the capabilities of anyone country, and that several countries -including the USA, the USSR and China (Nanjing)- already have detailed plans for coordinated studies during the next solar maximum;

and noting

that these countries have expressed willingness to cooperate in a coordinated international campaign;

proposes

that during the maximum of cycle 22, the International Astronomical Union co-sponsors the Flare 22 programme, under the auspices of the Solar Terrestrial Energy Programme (STEP) of SCOSTEP; and

recommends

that a member of Commission 10 be appointed to the Flare 22 Steering Committee.

STEP
: Solar Terrestrial Energy Programme

SCOSTEP
: Scientific Committee on Solar-Terrestrial Physics

Résolution C3: Programme d'étude des éruptions du cycle 22

La Commission 10

reconnaissant
la valeur d'un programme scientifique coordonné pour l'étude des processus physiques et des mécanismes produisant les éruptions chromosphériques et les phénomènes actifs solaires, lesquels sont communs à une grande variété d'objets astrophysiques et qui sont responsables des phénomènes interplanétaires affectant l'environnement terrestre,

prenant note
qu'une étude exhaustive de toutes les composantes de l'activité solaire dépasse les capacités d'un seul pays, et que, d'autre part, plusieurs pays -dont les Etats-Unis d'Amérique, l'URSS et la Chine (Nanjing)- ont des projets d'études coordonnées pour le prochain maximum solaire,

et prenant note
que ces pays ont exprimé leur volonté de coopérer dans une campagne internationale,

propose
que, durant le maximum du cycle 22, l'Union Astronomique Internationale parraine le programme Flare 22 sous les auspices du Solar Terrestrial Energy Programme (STEP) du SCOSTEP,

et recommande
qu'un membre de la Commission 10 soit désigné comme membre du comité de direction de Flare 22.

Resolution C4: Continuous Survey of Solar Phenomena

Commissions 10 & 12

considering
the scientific importance of ensuring a continuous long term survey of solar phenomena as observed in photospheric, chromospheric and coronal layers, in the solar wind and in solar-terrestrial relations;

recommends
1. the pursuit of these types of continuous observations wherever they are already conducted;
2. that other observatories should undertake similar observations, in order to ensure broad longitude coverage;
3. that financial assistance be provided by the International Astronomical Union for particular programmes which have been previously recommended by Commissions 10 and 12 and in particular the essential Debrecen photoheliographic surveys which face a most difficult financial situation.

Résolution C4 : Surveillance continue des phénomènes solaires

Les Commissions 10 & 12

considérant
l'importance scientifique d'assurer une surveillance continue à long terme des phénomènes solaires observés dans les couches photosphériques chromosphérique et coronale, dans le vent solaire et dans les relations Soleil-Terre ;

recommandent

1. la poursuite de ce genre d'observations continues là où elles sont déjà conduites ;
2. que d'autres observatoires entreprennent des observations similaires afin d'assurer une large couverture en longitude ;
3. qu'une aide financière soit fournie par l'Union Astronomique Internationale pour des programmes particuliers qui ont été recommandés par les Commissions 10 et 12, en particulier les surveillances essentielles photohéliographiques Debrecen qui sont confrontées à une situation financière très difficile.

Resolution C5: Databases on Minor Planets

Commission 15

endorses

the continued maintenance of a database on minor planets; and

recommends

the establishment of a comparable database on comets. The Working Groups on Minor Planets and on Comets are charged with the responsibility of defining and monitoring the compilation, updating, and dissemination of the respective databases.

Résolution C5 : Bases de données sur les petites planètes

La Commission 15

appuie

la continuation de la maintenance d'une base de données sur les petites planètes ; et

recommande

l'établissement d'une base de données comparable pour les comètes. Les groupes de travail sur les petites planètes et les comètes sont investis de la responsabilité de définir et de diriger la compilation, la mise à jour et la diffusion de ces bases de données, respectivement.

Resolution C6: Extension of Global Position Catalogues

Commission 24

emphasizes

the urgent need to extend global position catalogue work to substantially fainter magnitudes (fainter than m_V= 15) by all available and newly developed instrumentation.

Résolution C6 : Extension des Catalogues Globaux de Positions

La Commission 24

insiste

sur l'urgence d'étendre les travaux sur les catalogues de position couvrant tout le ciel à des magnitudes beaucoup plus élevées (m_V = 15 et bien au-delà) tant à l'aide des instruments disponibles que ceux en cours de développement.

Resolution C7: Availability of the Global Positioning System (GPS)

Commission 31

considering
a. that the Global Positioning System (GPS) has provided an invaluable service to astrometry and to international timing; and
b. that millisecond pulsar timing has progressed to unprecedented precision and yields new astrophysical insights in our Galaxy, while promising more; and
c. that the precision of the Database, which dates from October 1984 is of high quality due to the availability of GPS; and
d. that the continuity of this Database is very important for millisecond pulsar metrology and the international timing community;

thanks
those responsible for the invaluable service GPS has provided, and urges its full scale continuance.

Résolution C7 : Disponibilité du GPS

La Commission 31

considérant
a. que le Global Positional System (GPS) a rendu des services inestimables en astrométrie et dans la mesure internationale du temps ;
b. que le chronométrage des pulsars millisecondes a progressé jusqu'à une précision inégalée et fournit des vues nouvelles sur l'astrophysique de la galaxie, tout en promettant encore davantage ; et
c. que la précision de la base de données, débutant en octobre 1984 est de grande qualité grâce à la disponibilité du GPS ; et
d. que la continuité de cette base de données est très importante pour la métrologie des pulsars millisecondes et la communauté internationale de la mesure du temps ;

remercie
les responsables qui ont permis que le GPS ait pu rendre ces services inestimables et recommande vivement la continuation de ces services dans leur intégrité.

Resolution C8: Need for Accurate Time

Commission 31

considering
a. that there is a scientific need for accurate time and frequency comparisons between the national time scales and the new frequency standards under development; and
b. that at the moment, it is not possible to compare accurately the best available atomic frequency standards; and
c. that methods for time and frequency comparisons are now in the process of being evaluated, for example, VLBI, one-way and two-way pseudo-random noise signals, laser techniques on ground and in satellites, and TV signals;

recommends
1. that investigations on all the proposed or new time comparison methods should be actively pursued;

2. that simultaneous campaigns of mutual comparison should be performed; and
3. that the relevant activities be coordinated by, and results be published under the auspices of, the BIPM.

BIPM
Bureau International des Poids et Mesures

Résolution C8 : Besoin d'un temps exact

La Commission 31

considérant
a. qu'il existe un besoin scientifique pour des comparaisons exactes du temps et des fréquences des échelles nationales avec les nouveaux étalons de fréquence en cours de développement ; et,
b. qu'en ce moment, il n'est pas possible de comparer avec exactitude les meilleurs étalons atomiques de fréquence disponible et
c. que plusieurs méthodes de comparaison de temps et de fréquences sont maintenant en cours d'évaluation, par exemple, VLBI, signaux pseudo-aléatoires à un sens ou aller et retour, techniques laser au sol ou embarquées et signaux de télévision ;

recommande
1. que l'étude de toutes les méthodes déjà proposées ou nouvelles de comparaison de temps soit activement poursuivie ;
2. que des campagnes simultanées de comparaisons mutuelles soient exécutées, et
3. que les activités correspondantes soient coordonnées par le BIPM, et les résultats publiés sous les auspices de ce dernier.

Resolution C9: IAU Contribution to FAGS

Commissions 19 & 31

considering
the importance of the development of the International Earth Rotation Service (IERS) for many fields in astronomy, geodynamics and astrophysics, and the necessity for this service, as well as for the other astronomical services of the Federation of Astronomical and Geophysical Data Analysis Services (FAGS) which foster the participation of new countries, coordinate international and inter-technique activity in observation and analysis, and disseminate worlwide high precision data; and

recognising
that the support of FAGS provides a unique possibility for developing such activities;

recommend
that the Presidents of Commissions 19 and 31, in collaboration with the Presidents of other IAU Commissions involved in services supported by FAGS, propose a report to the IAU Executive Committee by June 1989, recommending an increased contribution to FAGS.

Résolution C9 : Contribution de l'UAI à FAGS

Les Commissions 19 & 31

considérant
l'importance du développement du Service International de la Rotation Terrestre (IERS) pour de nombreux domaines de l'astronomie, de la géodynamique et de l'astrophysique et la nécessité de disposer de ce service ainsi que les autres services astronomiques de la Fédération des Services d'Analyse de Données en Astronomie et en Géophysique (FAGS), qui encourage la participation de nouveaux pays, coordonne les activités internationales et intertechniques pour les observations et leur analyse, et diffuse dans le monde entier des données de haute précision ; et

reconnaissant
que le soutien de FAGS fournit une possibilité unique pour le développement de telles activités ;

recommandent
que les Présidents des Commissions 19 et 31, en collaboration avec les Présidents des autres Commissions de l'UAI concernés par des services relevant de FAGS, proposent un rapport au Comité Exécutif de l'UAI d'ici juin 1989, recommandant un accroissement de la contribution de l'UAI à FAGS.

Resolution C10: Needed History on BIH and IPMS

Commissions 19 & 31

noting
the significant impact and contribution that the BIH and IPMS have made to astronomy and geodesy over the years of their existence; and

recognising
that a suitable history of their activities does not exist;

ask
that the Presidents of the IAU, the IAG, and URSI request B. Guinot and S. Yumi to write a history of the two Services.

BIH
Bureau International de l'Heure
IPMS
International Polar Motion Service
IAG
International Association of Geodesy
URSI
Union Radio-Scientifique Internationale

Résolution C10 : Besoin de l'histoire du BIH et du IPMS

Les Commissions 19 & 31

notant
l'impact significatif que le BIH et l'ISPM ont eu sur l'astronomie et la géodésie pendant leurs années d'existence et leur contribution à ces deux disciplines, ;
reconnaissant
qu'il n'existe pas une histoire adéquate de leurs activités ;

demandent

que les Président de l'UAI, de l'IAG et de l'URSI sollicitent de B. Guinot et S. Yumi d'écrire l'histoire de ces deux services.

17. Appointment of the Special Nominating Committee (SNC)

The President asked the General Secretary to report on the names of the Members proposed for appointment by the General Assembly to the Special Nominating Committee 1988-1991. These persons will be convened by the President of the IAU for the purpose of proposing names to the XXIst General Assembly (1991) for the IAU Executive Committee membership (1991-1994). The four persons appointed are:

M. McCarthy	(Vatican)
G. Field	(USA)
R. Sunyaev	(USSR)
L. Woltjer	(Netherlands/ESO)

The Member of the SNC appointed by the Executive Committee is:

J. Rahe	(Germany, FR)

These appointments were unanimously confirmed by the General Assembly.

Note:

The President and Past President are Members of the SNC. The General Secretary and Assistant General Secretary are consultants to the SNC.

18. Nomination of new Members of the Union

The General Secretary announced that the Executive Committee had, on the proposal of the Adhering Bodies and with the advice of the Nominating Committee, admitted 792 new Members to the Union. The names of the new Members were displayed in a prominent place during the course of the General Assembly. The names will be incorporated in the alphabetical list of IAU Members to appear in Transactions Vol. XXB.

19. Applications for IAU Membership

Application from Algeria

The President invited Dr. H. Benhallou, representative of Algeria, to present an application for full membership from Algeria:

"Chers Collègues,

L'Algérie, pays en voie de développement, a pleinement pris conscience de l'intérêt de la recherche scientifique, aussi bien du point de vue culturel que du point de vue moyens de développement. Concernant l'astronomie, nous n'avons pas oublié que pendant plus d'un siècle au cours de la présence française, l'Observatoire Astronomique de Bouzoniah a connu une intense activité de recherche, particulièrement en participant au programme international de la Carte du Ciel. Par ailleurs, la lunette méridienne, l'astrolabe, et le coudé avaient contribué à de nombreuses observations.

Au lendemain de l'indépendance, chez nous, la priorité a été donnée à la formation du potentiel scientifique humain, ainsi qu'à la sauvegarde des possibilités de relance de toute activité de recherche en général.

Puis, les choses ont favorablement évolué. C'est ainsi que l'Algérie s'est dotée d'un centre de recherche en astronomie, astrophysique et géophysique et est déjà membre depuis quelques années de l'IUGG.

Aujourd'hui, je suis bien content d'être parmi vous pour vous confirmer officiellement notre demande d'adhésion à l'Union Astronomique Internationale. En effet, notre potentiel humain en chercheurs en astronomie est consistant (20 magisters et docteurs), un début de coopération est lancé avec la France et nous estimons que nous pouvons honorablement faire face à nos responsabilités vis à vis de l'UAI tout en espérant sa contribution au développement de l'astronomie en Algérie.

Je vous remercie."

Application from Iceland

The President invited Professor T. Saemundsson, representative of Iceland, to present an application for full membership from Iceland:

"Mr. President, Members of the Union, Ladies and Gentlemen,

Some time ago, my good friend, Dr. Derek McNally, Assistant General Secretary of the IAU, wrote me a letter suggesting that Iceland apply for membership of the IAU. Being an honest fellow, Derek did not dwell on the supposed benefits to astronomy of the proposed membership. Instead, he made it quite plain that the IAU needed more money and that the Executive Commitee was searching every corner of the globe -or should I say the Solar System- for possible contributors.

I must admit that my first reaction to Derek's suggestion was one of scepticism. Iceland is a small country, with no astronomical observatory. Opportunities for professional work in astronomy are extremely limited. The University of Iceland does not offer a degree course in astronomy. Few Icelanders have gone abroad to study astronomy or astrophysics, but those who have returned to Iceland have seldom been able to continue working in their area of specialization. Instead, in order to make a living, they have had to seek employment in other fields.

However, there are signs of progress. Thirty years ago, when I was working for my degree, there was only one person in Iceland with a degree in astronomy. Today we have half a dozen people who are knowledgeable in fields as diverse as solar-terrestrial relations, cosmic rays, neutron stars, plasma physics and the history of astronomy. Astronomy courses at the University of Iceland are being expanded, and there is serious talk about forming a professional astronomical society. Public interest in astronomy is high and there is an amateur astronomical association with a membership of 100 or so.

For a small country like Iceland, cooperation with other countries in areas like astronomy is essential. Without such cooperation, each person is bound to remain isolated in his own field, with no one to talk to and no access to advanced observational facilities. Fortunately, technical developments are fast reducing the geographic isolation of astronomers everywhere. The advent of electronic mail, for instance, has made it incredibly easy for individuals to correspond with colleagues in distant countries. This was brought home to me vividly last year when a supernova in the Large Magellanic Cloud created a steady stream of news which could be read almost without delay at a computer terminal. Such a breakthrough in communications is likely to strengthen the ties between astronomers worldwide. For places like Iceland the results might well be dramatic.

In the light of these developments, Iceland's application to join the IAU is seen to be, not just an empty gesture, but a logical step in international cooperation. I see reason to hope that this form of cooperation will stimulate and support the growth of an astronomical community in Iceland.

Thank you."

Application from Malaysia

The President invited Dr. M. Othman, representative of Malaysia, to present an application for associate membership from Malaysia:

"Although Malaysia has taken significant strides in other fields of science, the science of astronomy and space has largely been neglected. Recent interest, however, has resulted in a level of activity which is by no means negligible.

The teaching of astronomy at the undergraduate level is done at one college and three universities, namely the National University, the University of Malaysia and University of Technology. The students taking these courses range from non-science majors, surveyors to final year physics students. A postgraduate programme is presently available at only one university, namely the National University. Research in astronomy is carried out by astronomers in three universities, namely University of Science Penang (calendrical astronomy), University of Malaysia and the National University (M-stars and cosmology).

There are currently four scientists actively involved in astronomy, that is, they are teaching or publishing papers. Three of these have Ph.D's and one an MSc. Three more are working towards postgraduate degrees in fields that are related to astronomy. Most of the astrophysical research done to date has been carried out overseas.

In terms of popularization of astronomy, there have been several exhibitions for the public in different areas in Malaysia. There are two amateur societies, namely the Astronomical Society of Malaysia and the Islamic Astronomical Society. In June 1988, an international conference on the Islamic Calendar was held in Penang.

As far as the future is concerned, University of Science in Penang, in association with the Islamic Religious Department, will set up an Astronomy Center. An observatory is being planned by the National University which will strengthen the existing postgraduate programme and where we hope to carry out collaborative astrophysical work with other countries.

In education, two more universities are planning to offer introductory astronomy courses. The Ministry of Education has decided to include astronomy as a separate subject in the new secondary school curriculum. The old one did not have any astronomy in it. A Planetarium in Kuala Lumpur is currently being considered by the Malaysian government.

The interest in astronomy in Malaysia is high and the greatest demand is in education. The number of qualified astronomers who are able to teach in the Malaysian language is very small indeed and we hope to increase this number very soon.

We join the IAU in the hope of strengthening international contact and, more importantly, to establish international collaborative programmes.

Ladies and Gentlemen, on behalf of the Institute of Physics, Malaysia, I hereby submit for your consideration an application for Malaysia to be an Associate Member of the International Astronomical Union.

Thank you."

Application from Morocco

The President invited Dr. S. Kadiri, representative of Morocco, to present an application for full membership from Morocco:

"Mr. President, Ladies and Gentlemen,

When we learned that Professor Sahade was to visit Morocco in November 1987, we were quite concerned. What would the President of the IAU think of Moroccan astronomy?

To be sure, we have a young group of University professors and a young group of research workers at the National Research Center, but our resources are modest. On this latter point, Professor Sahade showed us that the hope for the development of astronomy in Morocco lies in us; and for that, we are extremely grateful to him.

Allow me to take advantage of this occasion that brings us together to thank Dr. Swings, the Members of his secretariat, and you, the Members of the IAU , for having made it possible for us be here.

Mr. President, Members of the Executive Committee, Ladies and Gentlemen, we shall be happy to become part of your family. Thus the Moroccan delegation states its wish to become a Member of the IAU.

Thank you."

Application from Peru

The President invited Dr. M.L. Aguilar, representative of Peru, to present an application for associate membership from Peru:

"Mr. President, Members of the Executive Committee of the International Astronomical Union, Astronomers of all countries, Ladies and Gentlemen,

I address you as Member of the National Committee of Astronomy and Astrophysics and, on behalf of Doctor Carlos del Rio, chairman of the National Council of Science and Technology, CONCYTEC, from Peru.

Mr. President, I wish to formally request the admission of my country, Peru, as an Associate Member of the International Astronomical Union.

The National Council of Science and Technology, CONCYTEC, is carrying out an important scientific, economic and social development under the highest human values and for this reason it has decided to support astronomical and astrophysical sciences. Peru is a country with an ancient culture which has earlier cultivated Astronomy and other sciences and arts. Our culture is now ready to meet the sciences of today.

The Visiting Lecturers' Programme is currently running at the Universidad Nacional Mayor de San Marcos since 1984. This University is the oldest in America and has an important position in the history of my country.

We are sure that the entry of Peru into the International Astronomical Union will be of great benefit not only to Astronomy in our country, but it will also provide many future opportunities for increased international collaboration in related sciences.

We are ready to join the international community and hope for your positive reaction.

Thanks."

Application from Saudi Arabia

The President invited Dr. Fadhl A.N. Mohammed, representative of Saudi Arabia, to present an application for full membership from Saudi Arabia.

"Mr. President, Members of the Executive Committee of the International Astronomical Union, Fellow Astronomers, Ladies and Gentlemen,

I am glad to address you on behalf of our President of King Abdulaziz City for Science and Technology (KACST), Riyadh, Saudi Arabia. This is the Government organization for promoting large-scale Astronomical Projects in the Kingdom along with other applied researches.

Regarding astronomical activities in Saudi Arabia, there are seven universities, out of which two universities have full-fledged Departments of Astronomy supported by their respective observatories. The other universities offer optional courses in Astronomy to the undergraduate students of Mathematics and Physics.

King Abdulaziz City for Science and Technology, which has membership of a large number of International Organizations in various fields, is actively engaged in preparing astronomical facilities in the Kingdom. It has carried out extensive site-testing surveys at different locations in our country. It came out that astronomical seeing is uniform over Saudi Arabia and the best sites are in the south-west region of the country. Hence a final site for a national observatory of the Kingdom has been selected which will contain reasonably large telescopes and other astronomical instruments such as meridian circle, radio telescopes, etc...

King Abdulaziz City for Science and Technology has established a few observatories, spread over the Kingdom, to sight the young crescent for Arabic months. A Landsat project for remote-sensing is already in operation and one of its two 10-meter antennas has been modified for VLBI observations. A contract has been signed with Australia to provide laser and lunar-ranging. The 75 cm laser-ranging telescope has been modified to have an additional Nasmith focus for optical astronomy. KACST is also planning to enter into the project of GONG. There is a number of planetaria being built to promote astronomical awareness in schools and public media.

Mr. President, remembering the past contribution of Arab astronomers and rich cultural heritage in Astronomy, Saudi Arabia with its above-mentioned large-scale astronomical programmes has determined to go ahead side-by-side with other nations to promote the cause of Astronomy. Regular contacts with the IAU in General Assemblies at Montreal and New Delhi convinced us to approach the IAU for the membership of our Country. Hence, on behalf of the President of the King Abdulaziz City for Science and Technology, I am applying for its membership.

Mr. President, please accept our best wishes for the IAU in promoting international cooperation among its Member countries.

Thanking all of you and best regards."

The General Secretary informed the General Assembly that all 4 applications for full membership and both applications for associate membership had been carefully examined by the Executive Committee and had agreed that the criteria for membership of the status applied for had been met in each case. The General Secretary therefore moved that the General Assembly admit Algeria, Iceland, Morocco and Saudi Arabia as new adhering Countries of the Union and Malaysia and Peru as the first new adhering associate Countries. The motion was accepted unanimously.

20. Changes in Commissions

The General Secretary read the proposals of the Executive Committee:

Commission Presidents and Vice-Presidents

	Commission	**Presidents**, Vice-President(s)
4	Ephemerides	**K. Seidelmann** (USA) B. Yallop (UK)
5	Documentation and Astronomical Data	**G. Wilkins** (UK) B. Hauck (Switzerland)
6	Astronomical Telegrams	**E. Roemer** (USA) J. Grindlay (USA)
7	Celestial Mechanics	**J. Henrard** (Belgium) A. Deprit (USA)
8	Positional Astronomy	**M. Miyamoto** (Japan) L.V. Morrison (UK)
9	Instruments and Techniques	**J. Davis** (Australia) J.C. Bhattacharrya (India)
10	Solar Activity	**E.R. Priest** (UK) V. Gaizauskas (Canada)
12	Radiation and Structure of the Solar Atmosphere	**J.W. Harvey** (USA) J.O. Stenflo (Switzerland)
14	Atomic and Molecular Data	**S. Sahal-Brechot** (France) W.L. Wiese (USA)
15	Physical Study of Comets, Minor Planets and Meteorites	**J. Rahe** (Germany, FR) A.W. Harris (USA)
16	Physical Study of Planets and Satellites	**A. Brahic** (France) M. Marov (USSR) & D. Morrison (USA)
19	Rotation of the Earth	**M. Feissel** (France) B. Kolaczek (Poland)
20	Positions and Motions of Minor Planets, Comets and Satellites	**R. West** (Germany, FR) A. Carusi (Italy)
21	Light of the Night Sky	**A.-C. Levasseur-Regourd** (France) M.S. Hanner (USA)
22	Meteors and Interplanetary Dust	**C.S.L. Keay** (Australia) J. Stohl (Czechoslovakia)
24	Photographic Astrometry	**W.F. van Altena** (USA) Ch. de Vegt (Germany, FR)

25	Stellar Photometry and Polarimetry	**I.S. McLean** (UK) A.T. Young (USA)
26	Double and Multiple Stars	**H. McAlister** (USA) H.A. Abt (USA)
27	Variable Stars	**M. Breger** (Austria) J. Percy (Canada)
28	Galaxies	**G.A. Tammann** (Switzerland) E.Ye Khachikian (USSR)
29	Stellar Spectra	**P. Conti** (USA) D.L. Lambert (USA)
30	Radial Velocities	**D. Latham** (USA) G. Burki (Switzerland)
31	Time	**P. Pâquet** (Belgium) E. Proverbio (Italy)
33	Structure and Dynamics of the Galactic System	**M. Mayor** (Switzerland) L. Blitz (USA)
34	Interstellar Matter	**J.S. Mathis** (USA) H. Habing (Netherlands)
35	Stellar Constitution	**A. Maeder** (Switzerland) P. Demarque (USA)
36	Theory of Stellar Atmospheres	**D.F. Gray** (Canada) W. Kalkofen (USA)
37	Star Clusters and Associations	**G.L.H. Harris** (Canada) C. Pilachowski (USA)
38	Exchange of Astronomers	**F.G. Smith** (UK) J. Sahade (Argentina)
40	Radio Astronomy	**P. Mezger** (Germany, FR) M. Morimoto (Japan)
41	History of Astronomy	**J.D. North** (Netherlands) S. Débarbat (France)
42	Close Binary Stars	**R.H. Koch** (USA) Y. Kondo (USA)
44	Astronomy from Space	**E. Jenkins** (USA) J. Trümper (Germany, FR)
45	Stellar Classification	**M. Golay** (Switzerland) D. MacConnell (USA)
46	Teaching of Astronomy	**A. Sandqvist** (Sweden) L. Gougenheim (France)
47	Cosmology	**K. Sato** (Japan) R.B. Partridge (USA)
48	High Energy Astrophysics	**R.A. Sunyaev** (USSR) J. Ostriker (USA)
49	The Interplanetary Plasma and the Heliosphere	**L.F. Burlaga** (USA) B. Buti (India)
50	Protection of Existing and Potential Observatory Sites	**D.L. Crawford** (USA) P.D. Murdin (UK)
51	Bioastronomy: Search for Extraterrestrial Life	**G. Marx** (Hungary) R. Brown (Australia)
	Working Group for Planetary System Nomenclature	**H. Masursky** (USA)

This proposal was received by acclamation by the General Assembly.

Organising Committees of Commissions

The General Secretary informed that, in order to save time, it had been decided not to present the lists of Members in Organising Committees of Commissions, but that the lists were available at the IAU Secretariat for inspection. They will be printed in Transactions XXB, Chapter VII.

21. IAU Representatives to other Organisations

	Organisation	Representative 1988-1991
ICSU	International Council of Scientific Unions General Committee	D. McNally
BIPM/CCDS	Bureau International des Poids & Mesures Working Group on the Temps Atomique International of the Consultative Committee for the Definition of the Second	G. Winkler
CCIR	International Radio Consultative Committee Study Group 2	J. Whiteoak A.R. Thompson
	Study Group 7	S. Leschiutta
CIE	Compagnie Internationale de l'Eclairage	D. Crawford
CODATA	Committee for Data for Science & Technology	G. Westerhout
COSPAR	Committee on Space Research	D. McNally
	COSPAR ISC B	J. Rahe
	COSPAR ISC D	S. Grzedzielski
	COSPAR ISC E	Y. Kondo
	COSPAR Sub. Committee E1	R. Sunyaev
	COSPAR Sub. Committee E2	M. Pick
COSTED	Committee on Science & Technology in Developing Countries	D. McNally
CTS	Committee on the Teaching of Science	L. Gouguenheim
EPS	European Physical Society Conference Committee	J. Bergeron
FAGS	Federation of Astronomical & Geophysical Services	E. Tandberg-Hanssen
IAF	International Astronautical Federation	Y. Kondo
IERS	International Earth Rotation Service	Ya. Yatskiv
ISY	International Space Year	L. Gouguenheim
IUCAF	Inter-Union Commission on Frequency Allocation for Radio Astronomy & Space Science	B.A. Doubinsky N. Kaifu V.L. Pankonin G. Swarup

IUPAP	International Union of Pure & Applied Physics	V. Trimble
IUWDS	International Ursigram & World Day Service	H. Coffey
QBSA	Quarterly Bulletin on Solar Activity	E. Hiei
SCOPE	Scientific Committee on Problems of Environment	R. Cayrel
SCOSTEP	Scientific Committee on Solar-Terrestrial Physics	S.T. Wu
URSI	Union Radio-Scientifique Internationale	J. Baldwin

22. Place and Date of the XXst General Assembly.

The President called upon Dr. R. Mendez to present the invitation of Argentina for the 21st General Assembly to be held in Buenos Aires. He reported that the work has already begun in Argentina. The Ministry of Education has declared the XXIst General Assembly as a project of national interest. They have also received the support of the National Research Council and of the University of La Plata. The Local Organizing Committee, composed of 12 persons, has been established. The National Organizing Committee will take major decisions before the end of the year.

The General Assembly accepted this invitation with acclamation and the President asked Prof. R. Mendez to convey the acceptance and the gratitude of the Union to the Ministry of Education of Argentina.

The proposed dates for the XXIst General Assembly are 23rd July-2nd August 1991, in order to allow astronomers involved in the observation of the total eclipse of the Sun on July 11, 1991, to attend.

23. Election to the Union of a President, a President-Elect, six Vice-Presidents, a General Secretary, and an Assistant General Secretary

The General Assembly approved by acclamation the proposal of the President that Prof. Y. Kozai be elected the new President of the Union, for the term 1989-1991.

The General Assembly also approved by acclamation the proposal that Prof. Acad. A.A. Boyarchuk be elected the President-Elect of the Union for the term 1989-1991.

The President then moved that Profs. V. Radhakrishnan, Dr. M.S. Roberts, Dr. Ye Shu-Hua be elected Vice-Presidents for the term 1989-1991. This motion was approved by acclamation.

The President finally proposed that Dr. D. McNally be elected General Secretary of the Union, and Dr. J. Bergeron Assistant General Secretary of the Union, for the term 1988-1991. This proposal was approved by acclamation.

The President then invited Prof. Y. Kozai, Prof. V. Radhakrishnan, Dr. M.S. Roberts, Dr. Ye Shu Hua and Dr. J. Bergeron to join the Executive Committee on the platform.

Following these elections, the IAU Executive Committee for the period 1988-1991 will thus be as follows:

EXECUTIVE COMMITTEE

1989-1991

President:

Prof. Y. Kozai	National Astronomical Observatory, Mitaka, Tokyo 181, Japan Telephone: 422-41-3650 Telex: 02822307 TAOMTK

General Secretary:

Dr. D. McNally	IAU Secretariat, Rm 318/319, 98bis, bd Arago, 75014 Paris, France Telephone: (33-1) 43 25 83 58 Telex: 205671 IAU F Fax: (33-1) 40 51 21 00 Home Institute: University London Observatory Mill Hill Park London NW7 2QS, UK Telephone: 01-959 0421 Telex: 28722 UCPHYS G Fax: 01-380 7145

Assistant General Secretary:

Dr. J. Bergeron	Institut d'Astrophysique, 98bis, bd Arago, F-75014 Paris, France Telephone: (33-1) 43 20 14 25 Telex: 205671 IAU F Fax: (33-1) 40 51 21 00

Vice-Presidents:

Dr. A.H. Batten	Herzberg Institute of Astrophysics, Dominion Astrophysical Observatory, 5071 West Saanich Road, Victoria BC, V8X 4M6, Canada
Prof. R. Kippenhahn	Max-Planck-Institut für Physik & Astrophysik, Karl-Schwarzschild-Strasse 1, D-8046 Garching bei München, Germany, FR
Prof. P.O. Lindblad	Stockholm Observatory, S6133 00 Saltsjöbaden, Sweden
Prof. V. Radhakrishnan	Raman Research Institute, Sadashivanagar, Bangalore 560 080, India
Dr. M.S. Roberts	NRAO, Edgemont Road, Charlottesville VA 22903, USA
Dr. Ye Shu Hua	Shangai Observatory, Shangai, China, PR

President-Elect:

Dr. A.A. Boyarchuk	Astronomical Council, USSR Academy of Sciences, Pyatnitskaya Ul. 48, 109017 Moscow, USSR

Advisers:

Prof. J. Sahade	C.C. No. 677, Observatorio Astronomico, Universidad Nacional de La Plata, 1900 La Plata (Bs. As.), Argentina
Dr. J.-P. Swings	Institut d'Astrophysique, 5, avenue de Cointe, B-4200 Cointe-Ougrée, Belgium

24. Address by the President 1985-1988

Prof. Sahade spoke as follows:

"My Dear Colleagues and Friends,

Now it is time to listen to the President of the Union for the triennium 1988-1991, who you have just elected by acclamation. But before, let me say a few final words.

In the first place, I would say that I have felt very honored and privileged with my term as President of the IAU and that I have enjoyed my task tremendously. I have been kept very busy but happy.

I have tried to devote as much time as possible to the Union and I have been rewarded with the pleasure of meeting many colleagues in a number of countries and, in some cases, of helping them acquire a new interest in our organization and in the development of Astronomy. In every circumstance, the contacts have been most useful for the two sides, I believe.

Three years is a very short time to analyze deliberately all the changes that may be needed in an Union such as ours. During the past term we proposed, as you know, a couple of modifications to the structure of the IAU that were endorsed by the Executive Committee and adopted at the Extraordinary General Assembly held on August 2nd.

I am convinced that other changes may still be necessary, particularly in regard to the role of the President and perhaps that of the Officers. I did not feel like bringing them up during my term, for obvious reasons, but I am mentioning this here to make it easier for my successors to start a discussion if they should feel the same way as I do.

My three years in office have given me the wonderful experience of dealing with an earnest, friendly and cooperative Executive Committee, and I think we should be very grateful to its membership for the way they have served the Union. In particular, our thanks should go to the retiring Vice-Presidents Bob Kraft, Manuel Peimbert and Yaroslav Yatskiv, and to the retiring Advisers, Robert Hanbury Brown and Richard West, whose calm and experienced opinions were always so valuable and so welcome.

The General Secretaries actually deserve a special paragraph, truly special thanks, because I think that, from now on, when we wish to give an example of real team work we should bring up the example of Jean-Pierre and Derek who have worked very hard, have offered all they could to our Union and have made it naturally possible for the Officers to work as really one single person. With Jean-Pierre I have constantly and pleasantly been in touch by mail, by telex or by telephone, and I think the symbiosis between the two of us he talked about in Delhi, did materialize very nicely and very effectively for the benefit of the affairs of the Union. I will certainly miss the continuous, friendly and warm exchange of views and reflections, and certainly hope that the strong ties of friendship that did develop between Jean-Pierre, Derek and myself will continue throughout the years.

The Secretariat in Paris, now in the very able hands of Monique Orine, the very dynamic administrative assistant, and her very efficient helper, Huguette Gigan, the part-time bilingual Secretary, has been a great help whenever needed and they have also responded efficiently to my requests and to my many messages by telex. Many, many thanks. I should also thank Denise Fraipont, Jean-Pierre's Secretary, and Valerie Peerless, Derek's Secretary, who were always ready to show their goodwill and their eagerness to help.

If you permit me, I would like to place on record my great indebtness to my Secretary, Viviana Soler, who served the Union with loyalty, hard work and warm enthusiasm. When she accepted the job, I am sure she did not expect to have to work so hard, and she did it with an ever present nice smile and the best of goodwill. My task would have been very difficult without her assistance, and I feel indeed most grateful to Viviana.

Last but not least, I should point out that my work as President of the Union was encouraged by the support received spontaneously from the University of La Plata, Argentina. The University provided me with a fully equipped house for office, telex and telephone services and took care of the salary for Viviana Soler. Thanks to Rector Pesack, grants were assigned by the Secretary of Science and Technlogy, by CONICET (Consejo Nacional de Investigaciones Cientificas y Técnicas) and by the University of La Plata. Thanks to Rector Plastino, a grant was assigned to us by the Research Commission of the Province of Buenos Aires. Continuous support also from the Faculty of Astronomical Sciences of the La Plata University through Dean Mondinalli, and from the Argentine Institute of Radioastronomy, through Director Colomb, is gratefully acknowledged. All such help, in infrastructure and in funds, made it possible for me to carry my duties as President without charging the Union for our operational expenses.

Although the local organizers will be thanked very warmly in a few minutes, I cannot refrain from expressing once again my admiration of the work of the National Organizing Committee and the Local Organizing Committee who have done a really wonderful, careful and perfect job. Congratulations and warm thanks! I am particularly happy that I have ended my Presidency with such magnificently organized General Assembly! Last night's Closing Banquet really deserves a separate paragraph, but it has been so wonderfully done that it is hard for me to find the right words to praise it!

Finally, thanks to all of you for honoring me with the opportunity of serving the Union and for making this General Assembly a very successful one.

Now, I have the privilege of turning the floor over to my successor, for 1988-1991, Professor Yoshihide Kozai."

25. Address by the President 1988-1991

Prof. Y. Kozai addressed the General Assembly as follows:

"Professor Sahade, Distinguished Guests, Ladies and Gentlemen,

It is indeed a great honor for me to be nominated and to serve as the President of the International Astronomical Union.

Many years ago when I was young only two or three delegates could attend the General Assembly from my country and after they came back home they reported to us on the activities of the General Assembly, which was recognized as the most important astronomy meeting. In fact there were only a few meetings in those days. By having heard such reports I dreamed that someday I could become a Member of the Union and attend a General Assembly; however, at that time I did not think it would be possible for me because of our financial situation. Still, very fortunately, I was admitted as a Member of the Union in 1961 and could attend the General Assembly at Berkeley as I was then working in the United States.

As an astronomer who was educated and trained in an isolated country in the Far East I have appreciated the importance of international cooperation and meetings and even visiting astronomers abroad. I still remember how I was excited when I could see many eminent astronomers, whom I had known only by name, i.e. by reading textbooks and scientific journals. After having had such experiences, I have had an idea that only after I could see authors I could understand correctly what they wanted to say in their papers.

Since then the world has been changed for everybody. Now several astronomers complain that there are too many meetings of interest other than General Assemblies, whereas also I know many people who desired to attend the General Assembly at least, but could not find enough funds to do so. In fact we must recognize that the IAU is a Union with many astronomers under very different conditions.

I am afraid that a more difficult problem is now facing many astronomers at their home institutes. In fact in almost every country we have faced financial difficulties, namely budget cuts, staff cuts and so on, and even some observatories are under threat of partial or complete closure. Moreover, for many observatories environmental problems have become serious because of artificial lighting and radio noise. Indeed we have many problems to solve in front of us.

Therefore, I feel that I am nominated as the President of the Union at a very severe time for astronomers and the IAU. In spite of many difficulties I believe that there are very many exciting problems we desire to understand in the world of astronomy, some of them were discussed during this General Assembly. Therefore, I am sure that despite such difficulties, astronomical researches will be developed rapidly in coming years.

In the coming three years as the President I will make every effort to cooperate with you and to try to make the Union and the General Assembly more attractive, particularly for young astronomers. To do this I would appreciate having your advice and suggestions, particularly, from the outgoing President, Professor Sahade, and the General Secretary, Dr. Swings, who have contributed very much to promote the activities of the IAU and to develop astronomy in the past years. I would like also to thank all the Members of the IAU and of the Executive Committee who have made every effort in the organization of the IAU.

Meantime, I will try to improve my accent.

Thank you"

26. Closing Ceremonies

The President proposed a vote of thanks to all the organizers, who so efficiently contributed to the great success of the XXth General Assembly, on behalf of all participants.

Address by Dr. G. Cayrel de Strobel, on behalf of the participants

"President Sahade, Professor Drake, Professors Davidsen and Giacconi, all Members of the National US Organising and Local Organising Committees and of the Executive Committee of the IAU, dear Friends and Colleagues,

I have the great honor to have been chosen for expressing our gratitutde and admiration for the perfect organisation of the XXth General Assembly of the IAU.

Both the Science and the material organisation have been outstanding.

In spite of the absence of results from the Space Telescope everyone of us has felt in Baltimore the spirit of the Space Institute and I felt myself for this, a real difference between this General Assembly and the 12 former General Assemblies I have attended in my long astronomical carreer.

Each of us was feeling, here in Baltimore that soon something extraordinary will happen in Astronomy and was preparing himself for this event.

Is it not great also, that to compensate for the absence of real Space Telescope observations the obscure star Sanduleak -69°202 turned into a Supernova and that Comet Halley just appeared to be observed for the first time from space, to supply alternate scientific highlights for this General Assembly?

We enjoyed everything from humble single Commission meetings to the "Rise and Fall of Quasars".

We are returning back to our countries, the head full of new things and happy to have discovered the charms of a superbly renovated Baltimore.

A further return of IAU General Assemblies is the periodic encounters with so many old friends, that we meet unfortunately only every three years.

At the Rome General Assembly in 1952 I was proud to be thanked for my services as local hostess and now, on the other side of the gate, I am proud to be able to thank, on behalf of my colleagues, the IAU Executives and Secretariat, all the Local Organizers and the US National Organising Committee who did so much for the success of the General Assembly.

Thank you."

Address by Mrs. A.H. Batten
on behalf of the Guest Participants

"Mr. President, Members of the IAU and other Ladies and Gentlemen,

On behalf of the guests who have accompanied IAU members to Baltimore, I would like to express our appreciation to the many people involved, as official organisers or as volunteers, in making our days here full of interest.

From the planning of group tours, to the answering of unforeseen and awkward questions, your thoughtful and enthusiastic hospitality was always apparent; and the trajectory of world-wide understanding has been made more firm and beautiful, as we met old friends and made new ones, in gardens, museums, markets, boats and buses.

Nous, les accompagnateurs inscrits, qui venons de pays différents et parlons des langues diverses, tenons à remercier le Comité Local d'Organisation et les Volontaires de nous avoir donné l'occasion de connaître divers aspects de la vie à Baltimore et dans ses environs.

A vous, nos amis, merci beaucoup.

Au revoir."

Address by the retiring General Secretary, Dr. J.-P. Swings

"Ladies and Gentlemen,

La plupart des remerciements que je tiens à formuler ont, en fait, été exprimés lors de la première session de l'Assemblée Générale : je me contenterai de les reprendre ici très brièvement.

Since most of my acknowledgements were delivered at the first session of the General Assembly, let me simply summarize them here, without repeating any name, but by outlining the following:

- the excellent team work I enjoyed with the other two Officers, as well as the very positive support from the members of the Executive Committee;

- the outstanding secretarial help I had in Paris and in Liège;

- the beautiful collaboration in which I was involved with the Local Organizing Committee in the preparation of this General Assembly;

- the financial support I received from the Ministry of Education in my country.

What happened between August 2nd and today concerns some of the topics I had mentioned in my report at the first session of the General Assembly. I will take a few items, one by one:

- the help I received in the organization of this General Assembly: I let you judge the result. The only thing I can say is that I heard very few complaints... so I take that as a positive judgement... and as a nice reward, for the Local Organizing Committee as well as for the IAU;

- the contacts with the press: I believe they have been numerous and fruitful. Talking about the press, the very local one this time, I was really impressed by the excellence of IAU Today. Here again, the collaboration between the newspaper staff and the IAU was very good, and I did appreciate it a lot. I said I would not mention names, so I won't, but I wish to thank all those who contributed so well to the success of IAU Today;

- the involvement of the IAU in the series of TV broadcasts on the observatories around the world: you just saw an un-edited preview of an excerpt. I think we can look forward to a beautiful result in the end that will contribute to the popularization of our science;

- the anti-pollution activities: they have been numerous, from press conferences to contributions at meetings, and they will continue in two days with IAU Colloquium 112 in Washington;

- the entry of new countries: four full members and two associate members were welcomed in the Union a few minutes ago: this is of course the result of contacts, sometimes personal, sometimes through colleagues knowing particular situations that, in some cases, started before "my" triennnium, and I thank those individuals for their collaboration;

- the promotion and development of astronomy, and the cooperation between amateur and professional astronomers: two Working Groups have been created to that effect.

These are all past events; let me say a few words about the future, and in particular to my successor Derek McNally. As you know, Derek, in the secretariat in Paris, we used the phone, telex and telefax quite extensively; in the new offices where you will be going, you will have the opportunity to use new links toward the astronomical community, e.g. via electronic mail (I just hope you don't get swamped with too many messages !). In a moment I will give you two items; but before I do so, I cannot resist using the sentence from Erasmus Darwin you brought to my attention a few months ago. It reads "A fool... is a man who never tried an experiment in his life".

Well, I believe being an IAU General Secretary, and organizing a General Assembly is some sort of an experiment (a rewarding one even)... so, maybe, I can now feel relieved. Anyhow, as Maarten Schmidt kindly pointed out while showing slides of the cloverleaf quasar the other night, a General Secretary can still participate in some scientific research (I suppose this implies that he has a few graduate students or postdocs or collaborators doing the work, but never mind), and I sincerely hope, Derek, that you will also be able to continue doing some research as well. In any case, the task you accepted is a challenge, and it does have rewards. Now let me end by transferring to you two symbols of our Union:

- the IAU paper knife(*) that dates back to Donald Sadler, when he was IAU General Secretary (1958-1964);

- the keys of the present secretariat in Paris, and I do this while wishing you and the new Executive Committee the best of luck for the coming triennium.

Bonne chance, et meilleurs voeux de succès à la nouvelle équipe à la tête de l'UAI."

Address by the newly elected General Secretary, Dr. D. McNally

"Monsieur le Président, Membres de l'Union, Mesdames et Messieurs,

Vous m'avez fait l'honneur de m'élire aujourd'hui Secrétaire Général de l'UAI. Je me ferai un plaisir de servir et l'Union et l'Astronomie. Ce sera l'occasion pour moi de m'acquitter partiellement de la dette que j'ai envers une science qui m'a tant donné.

J'espère avoir le plaisir de rencontrer beaucoup d'entre vous au cours de ces trois prochaines années ou, du moins, avoir l'occasion d'être en relation avec le plus grand nombre.

Encore une fois, je vous remercie de votre confiance.

Mr. President, Members of the Union, Ladies and Gentlemen,

In 1979, the Union elected as General Secretary, an Englishman from Ireland. That proved a most felicitous choice. Today you have done me the honour of electing me as General Secretary -an Irishman from England. I hope that this election will prove less asymmetrical than my favourite definition of a Ski Resort, namely, a place where daughters seek husbands and husbands seek daughters. I look forward to this opportunity to serve the Union and the science of Astronomy. It will be an opportunity to discharge some part of the debt that I owe to a science which has given me so much. I also look forward to seeing as many of you as I can in the next three years or at the very least, to being in communication.

Thank you for your confidence."

(*) Unfortunately, this item was stolen from the General Secretary's office the night before the second session of the XXth General Assembly!

CHAPTER IV

REPORT OF THE EXECUTIVE COMMITTEE

1985-1988

EXECUTIVE COMMITTEE

1985-1988

President:

Prof. J. Sahade	C.C. No. 677, Observatorio Astronomico, Universidad Nacional de La Plata, 1900 La Plata (Bs. As.), Argentina

Vice-Presidents:

Dr. A.H. Batten	Herzberg Institute of Astrophysics, Dominion Astrophysical Observatory, 5071 West Saanich Road, Victoria BC, V8X 4M6, Canada
Prof. R. Kippenhahn	MPI für Physik & Astrophysik, Karl-Schwarzschild-Strasse 1, D-8046 Garching bei München, Germany, FR
Prof. R.P. Kraft	Lick Observatory, University of California, Santa Cruz, CA 95064, USA
Prof. P.O. Lindblad	Stockholm Observatory, S6133 00 Saltsjöbaden, Sweden
Prof. M. Peimbert	Instituto de Astronomia, Apartado Postal 70264, Mexico 04510 DF, Mexico
Prof. Ya. S. Yatskiv	Main Astronomical Observatory, Ukrainian Academy of Sciences, SU-252127 Kiev, USSR

General Secretary:

Dr. J.-P. Swings	Institut d'Astrophysique, 5, avenue de Cointe, B-4200 Cointe-Ougrée, Belgium

Assistant General Secretary:

Dr. D. McNally	University of London Observatory, Mill Hill Park, London NW7 2QS, UK

Advisers:

Emeritus Prof. R. Hanbury Brown	37 Beatty Street, Balgowlah, NSW 2093, Australia
Dr. R.W. West	European South Observatory, Karl-Schwarzschild Strasse 2, D-8046, Garching bei München, Germany, FR

1. Introduction

The present report covers the period from the end of the XIXth General Assembly up to mid 1988.

The report by the General Secretary at the XXth General Assembly for the period 1st January-31 July, 1988 is to be found pp. 26-28.

A summarized financial report of the Union audited accounts for the calendar years 1985, 1986 & 1987 and a brief resume of the Union's accounts for 1988 are also included.

2. Administration

Executive Committee Meetings

The Executive Committee held the following meetings:

53rd Meeting:	Delhi, India, November 17,18 & 26, 1985
54th Meeting:	Delhi, India, November 28, 1986
55th Meeting:	Liège, Belgium, September 19-22, 1986
56th Meeting:	Santa Cruz, USA, September 9-12, 1987

The 53rd Meeting was chaired by Professor R. Hanbury Brown while the 54th was chaired by the newly elected President, Professor J. Sahade; the new Members of the Executive Committee were all present, as well as the two advisers. The 55th and 56th Meetings were also chaired by Professor J. Sahade.

Officers' Meetings

The President, the General Secretary and the Assistant General Secretary met as follows:

November 16, 1985	Delhi, India
May 14, 1986	Paris, France
September 18, 1986	Liège, Belgium
March 11-12, 1987	Paris, France
September 8, 1987	Santa Cruz, USA
November 6, 1987	Paris, France
March 10-11, 1988	Baltimore, USA

IAU Secretariat

The secretariat is in the process of moving to the following address:

98bis, boulevard Arago
F-75014 - PARIS, France

The phone and telex numbers remain unchanged.

IAU Staff

Miss D. Lours and Mrs. B. Manning left the secretariat on April 27, and June 25, 1987, respectively. They have been replaced by Mrs. Monique Orine, administrative assistant, and Mrs. Huguette Gigan, part-time bilingual secretary, as of September 1, and October 15, 1987, respectively.

Adhering Countries

Nigerian membership of IAU was ratified by the XIXth General Assembly (motion adopted unanimously, see Transactions XIXB, p. 54).

Members of the IAU

The XIXth General Assembly welcomed 935 new members: the total number of IAU members thus became 6027, on 28 November 1985. The number, as of June 1988, is 5940.

Consultants to IAU Commissions

The 40 IAU Commissions had a total of approximately 200 consultants.

3. Commissions of the IAU

The name of **Commission 51** has become "Bioastronomy: Search for Extraterrestrial Life", as of the end of the XIXth General Assembly.

Commission Reports, covering the period July 1, 1984, to June 30, 1987 have been published (Reports on Astronomy; IAU Transactions XXA, 708 pp.).

An updated list of the IAU Working Groups, belonging to one or more Commissions, has been published (cf. IB 59, pp. 11-13).

A few specific points about some Commission activities, especially those providing services to the astronomical community, are listed below:

Commission 5

Work has continued on the new IAU Style Manual, largely due to the active role of G. Wilkins (cf. IB 60, p. 26): the Manual is expected to be presented at the XXth General Assembly. Contacts have also been made with editors of the main astronomical journals to attempt to coordinate the rules of publication.

Commission 6

The IAU Central Bureau for Astronomical Telegrams (Director: B.G. Marsden) issued the following number of circulars and "telegram books" respectively in 1985, 1986, and 1987: 132, 35; 135, 41; 230, 55; i.e. a total of 497 circulars and of 131 "telegram books".

Commissions 19 and 31

The new International Earth Rotation Service (IERS) established in 1987 by the IAU and the IUGG, has become operational on January 1, 1988 (cf. IB 59, p. 15).

Commission 20

Together with Commission 6 and the IAU Working Group on Planetary System Nomenclature (WGPSN), procedures were established for assigning names and designations to various classes of newly discovered bodies in the solar system. The IAU Minor Planet Center (Director: B.G. Marsden) issued 3330 MPCs and numbered no less than 554 new minor planets during 1985-1987.

Commission 22

The IAU Meteor Data Center, created in Lund, Sweden (director: B.A. Lindblad) has computed and archived an impressive number of orbits. A report by B.A. Lindblad has been published for the period 1985-1987 (cf. IB 59, p. 17 & IB 60, p. 27).

Commission 38

Under the guidance of its President, E.A. Müller, this Commission continued to give assistance to (young) astronomers under the rules published in IB 55, pp. 25-27. The names of the 23 grant recipients for the period February 1985-November 1987 have been published in IB 59, pp. 18-19.

Commission 46: Visiting Lecturers' Programmes (VLP)

a. VLP in Lima (Peru): see IB 56, p. 14; 57, p. 16 & 58, p. 16. A limited extension of this VLP will start shortly.

b. VLP in Asuncion (Paraguay): a description appeared in IB 58, pp. 16-17 & 60, p. 27. The present contract is only for one year, starting in 1988. The possibility of the extension of the contract will be examined by the next Executive Committee.

Commission 46

New rules concerning the membership and activities of Commission 46 are under preparation. Two IAU-UNESCO International Schools for Young Astronomers were succesfully held in Beijing (10-30 August 1986) and Porto (15-27 September 1986).

4. Scientific meetings

Symposia (22)

114 Relativity in Celestial Mechanics and Astrometry - High Precision Dynamical Theories and Observational Verifications
Leningrad, USSR, May 28-31, 1985

115 Star Forming Regions
Tokyo, Japan , November 11-15, 1985

116 Luminous Stars and Associations in Galaxies
Porto Hell, Greece, May 26-31, 1985

117 Dark Matter in the Universe
Princeton, USA, June 24-28, 1985

118 Instrumentation and Research Programs for Small Telescopes
Christchurch, New Zealand, December 2-6, 1985

119 Quasars
Bangalore, India, December 2-6, 1985

120 Astrochemistry
Goa, India, December 3-7, 1985

121 Observational Evidence of Activity in Galaxies
Byurakan, Armenia, USSR, June 3-7, 1986

122 Circumstellar Matter
Heidelberg, Germany, FR, June 23-27, 1986

123 Advances in Helio- and Asteroseismology
Aarhus, Denmark, July 7-11, 1986

124 Observational Cosmology
Shanghai, China, PR, August 25-29, 1986

125 The Origin and Evolution of Neutron Stars
Nanjing, China, PR, May 26-29, 1986

126 Globular Cluster Systems in Galaxies
Cambridge, MA, USA, August 25-29, 1986

127 Structure and Dynamics of Elliptical Galaxies
Princeton, USA, May 28-31, 1986

128 Earth's Rotation and Reference Frames for Geodesy and Geodynamics
Washington, USA, October 20-24, 1986

129 The Impact of VLBI on Astrophysics, Astrometry and Geophysics
Cambridge, MA, USA, May 31 - June 3, 1987

130 Evolution of Large Scale Structures in the Universe
Balatonfured, Hungary, June 15-20, 1987

131 Planetary Nebulae
Mexico City, Mexico, October 5-9, 1987

132 The Impact of Very High S/N Spectroscopy on Stellar Physics
Paris, France, June 29-July 3, 1987

133 Mapping the Sky - Past Heritage and Future Directions
Paris, France, June 1-5, 1987

135 Interstellar Dust
Mountain View, CA, USA, July 26-30, 1988

136 The Galactic Center
Los Angeles, CA, USA, July 25-29, 1988

Colloquia (21)

87 Hydrogen Deficient Stars and Related Objects
Mysore, India, November 10-15, 1985

88 Stellar Radial Velocities
Schenectady, NY, USA, October 24-27, 1984

89 Radiation Hydrodynamics in Stars and Compact Objects
Copenhagen, Denmark, June 11-20, 1985

90 Upper Main Sequence Stars with Anomalous Abundances
Nauchny, Crimea, May 14-17, 1985

91 History of Oriental Astronomy
New Delhi, India, November 14-17, 1985

92 Physics of Be Stars
Boulder, CO, USA, August 18-22, 1986

93 Cataclysmic Variables
Bamberg, Germany, FR, June 16-20, 1986

94 Physics of Formation of FeII Lines Outside LTE
Capri, Italy, July 4-8, 1986

95 Second Conference on Faint Blue Stars
Tucson, AZ, USA, May 31 - June 3, 1987

96 The Few Body Problem
Turku, Finland, June 14-19, 1987

97 Wide Components in Double and Multiple Stars: Problems of Observation and Interpretation
Brussels, Belgium, June 8-13, 1987

98 The Contribution of Amateur Astronomers to Astronomy
Paris, France, June 20-24, 1987

99 Bioastronomy - The Next Steps
Lake Balaton, Hungary, June 22-27, 1987

100 Fundamentals of Astrometry
Belgrade, Yugoslavia, September 8-11, 1987

101 Interaction of Supernova Remnants with the Interstellar Medium
Penticton, BC, Canada, June 9-12, 1987

102 UV and X-Ray Spectroscopy of Astrophysical and Laboratory Plasmas
Beaulieu-sur-Mer, France, September 9-11, 1987

103 The Symbiotic Phenomenon
Torun, Poland, 18-21 August, 1987

105 The Teaching of Astronomy
Williamstown, NY, USA, July 27-30, 1988

106 Evolution of Peculiar Red Giant Stars
Bloomington, IN, USA, July 27-29, 1988

109 Applications of Computer Technology to Dynamical Astronomy
Gaithersburg, MD, USA, July 27-29, 1988

110 Library and Information Services in Astronomy
Washington, DC, USA, July 28-30, August 1, 1988

Co-sponsored Meetings

CNES/IAU Meeting on "Comparative Study of Magnetospheric Systems"
La Londe des Maures, France September 9-14, 1985

IAG/IAU/COSPAR Symposium on "Figure and Dynamics of the Earth, Moon, and Planets"
Praha, Czechoslovakia, September 15-20, 1986

IAU/ESLAB Symposium on "Exploration of Halley's Comet"
Heidelberg, Germany, FR, October 27-31, 1986

COSPAR/IAU Symposium on "UV Space Astronomy: Physical Processes in the Local Interstellar Medium"
Toulouse, France, June 30 - July 12, 1986

COSPAR/IAU Symposium on "Latest Results on Venus and Uranus Missions"
Toulouse, France, June 30-July 12, 1986

COSPAR/IAU Symposium on "Solar and Stellar Activity"
Toulouse, France, June 30-July 12, 1986

COSPAR/IAU Symposium on "Comets Halley and Giacobini-Zinner"
Toulouse, France, June 30-July 12, 1986

SCOSTEP/IAU 6th International Symposium on "Solar-Terrestrial Physics"
Toulouse, France, June 30-July 5, 1986

SCOSTEP/IAU Symposium "Synopsis of the Solar Maximum Analysis"
Toulouse, France, July 2-5, 1986

Symposium on "Image Detection and Quality", with Commission Internationale d'Optique,
Paris, July 16-18, 1986

COSPAR/IAU Symposium on "Physics of Compact Objects: Theory versus Observations"
Sofia, Bulgaria, July 13-19, 1987

COSPAR/IAU/IUGG/IUPAP/SCOSTEP/URSI/WMO Symposium on "The Middle Atmosphere after MAP"
Espoo, Finland, 19-23 July, 1988

COSPAR/IAU/SCOSTEP Symposium on "International Heliospheric Study"
Espoo, Finland, 20-23 July, 1988

COSPAR/IAU/IUPAP Symposium on "Cosmic Ray Studies in Space"
Espoo, Finland, 25-26 July, 1988

COSPAR/IAGA/IAU Symposium on "Studies of Cometary Environments: Modelling and Space Missions"
Espoo, Finland, 25-26 July, 1988

COSPAR/IAU/IUPAP Symposium on "Relativistic Gravitation"
Espoo, Finland, 18-19 July, 1988

COSPAR/IAA/IAU/IUGG/IUGS Symposium on "The Outer Planets: Current Knowledge, Future Prospects"
Espoo, Finland, 19-20 July, 1988

COSPAR/IAA/IAU Symposium on "Advances and Perspectives in X-Ray and Gamma-Ray Astronomy"
Espoo, Finland, 19-20 July, 1988

COSPAR/IAA/IAU/IUGG/IUGS Symposium on "Reappraisal of Moon and Mars/Phobos/Deimos: Preparation for Renewed Exploration"
Espoo, Finland, 21-22 July, 1988

COSPAR/IAA/IAU Symposium on "Magnetic Energy Conversion in Astrophysical and Laboratory Plasmas"
Espoo, Finland, 21-23 July, 1988

COSPAR/IAU Symposium on "Space Observations of Solar Variability"
Espoo, Finland, 27-28 July, 1988

COSPAR/IAU Workshop on "Future Planetary Missions"
Toulouse, France, June 30 -July 12, 1986

COSPAR/IAU Workshop on "Contamination of Environment of Space Shuttle for Astronomical Observations,
Toulouse, France, June 30 -July 12, 1986

COSPAR/IAU Workshop on "Chemical Evolution of Outer Planets, Satellites, and Small Bodies"
Toulouse, France, June 30 - July 12, 1986

COSPAR/IAA/IAU/IUGG/IUGS Workshop on "Future Planetary Missions"
Espoo, Finland, 20 July 1988

COSPAR/IAGA/IAU Workshop on "The Middle and Upper Atmosphere of Venus"
Espoo, Finland, 25-27 July, 1988

COSPAR/IAU/SCOSTEP Workshop on "Scientific Planning for the Solar Maximum and Beyond"
Espoo, Finland, 25-27 July, 1988

COSPAR/IAU/IUGG Workshop on "Origin and Evolution of Planetary and Satellite Systems"
Espoo, Finland, 29-30 July, 1988

Regional Meetings (4)

Ninth European Regional Astronomy Meeting on "Activity in Stars and Galaxies" (with EPS)
Leicester, UK, September 2-5, 1986

Fifth Latin-American Regional Astronomy Meeting,
Merida, Yucatan, Mexico,
October 6-10, 1986

Tenth European Regional Astronomy Meeting (with EPS),
Praha, Czechoslovakia, August 24-29, 1987

Fourth Asian-Pacific Regional Astronomy Meeting (with EPS),
Beijing, China, PR, October 5-9, 1987

IAU/UNESCO Young Astronomers' Schools (2)

XIVth IAU-UNESCO International School for Young Astronomers
Beijing, China, PR, 10-30 August, 1986

XVth IAU-UNESCO International School for Young Astronomers
Porto, Portugal, 8-20 September 1986

XIXth General Assembly, New Delhi, India, 19-28 November, 1985

IAU Transactions, Volume XIXB (1986), contain a report of the Proceedings of the General Assembly of the Union, including the resolutions passed at the Assembly. It also contains the reports of Commission meetings during the period of the General Assembly.

5. Publications

Publisher

In November 1984, the contract with D. Reidel Publishing Co. was renewed for a six-year period, from 1985-1991. The contract was continued when Reidel became a part of Kluwer Academic Publishers.

Sales

Number of copies sold (**hard**, soft) by D. Reidel Publishing Company in the period 1985-1987:

Transactions:

XIIIA (**2**), XIIIB (**2**), XIVA (**2**), XIVB (**2**), XVA (**1**), XVB (**5**), XVIA (**3**), XVIB (**5**), XVIIA part 1 (**7**), XVIIA part 2 (**6**), XVIIA part 3 (**7**), XVIIB (**13**), XVIIIA (**30**), XVIIIB (**40**), XIXA (**431**), XIXB (**404**).

Highlights:

I (**10**), II (**13**), III (**17**), IV 1 (**8**, 9), IV 2 (**8**, 9), V (**21**, 8), VI (**20**, 36), VII (**198**, 208).

Symposia Proceedings:

32 (**20**), 33 (**1**), 34 (**4**), 35 (**4**), 37 (**4**), 38 (**20**), 39 (**8**), 40 (**9**), 43 (**18**), 44 (**14**), 45 (**32**), 46 (**5**), 47 (**8**), 48 (**25**), 49 (**4**, 22), 50 (**15**, 16), 51 (**5**, 20), 52 (**3**, 15), 53 (**2**,9), 54 (**14**, 1), 55 (**6**, 16), 56 (**3**, 11), 57 (**4**, 11), 58 (**6**, 9), 59 (**12**, 23), 60 (**8**, 8), 61 (**9**, 17), 62 (**5**, 25), 63 (**22**, 0), 64 (**6**, 22), 65 (**6**, 3), 66 (**12**, 29), 67 (**52**, 22), 68 (**5**, 2), 69 (**4**, 24), 70 (**5**, 14), 71 (**13**, 14), 72 (**9**, 12), 73 (**12**, 20), 74 (**6**, 32), 75 (**15**, 54), 76 (**15**, 7), 77 (**10**, 34), 78 (**12**, 9), 79 (**15**, 30), 80 (**28**, 8), 81 (**7**, 40), 82 (**12**, 24), 83 (**14**, 33), 84 (**22**, 4), 85 (**8**, 19), 86 (**14**, 13), 87 (**21**, 11), 88 (**9**, 17), 89 (**9**, 14), 90 (**19**, 7), 91 (**18**, 9), 92 (**22**, 10), 93 (**14**, 14), 94 (**11**, 25), 95 (**18**, 7), 96 (**25**, 16), 97 (**21**, 13), 98 (**5**, 8), 99 (**12**, 18), 100 (**28**, 37), 101 (**38**,26), 102 (**33**, 41), 103 (**28**, 34), 104 (**49**, 50), 105 (**58**, 107), 106 (**405**, 352),107 (**276**, 197), 108 (**69**, 60), 109 (**322**, 208), 110 (**34**, 93), 111 (**373**, 253), 112 (**446**, 438), 113 (**373**, 287), 114 (**364**, 195), 115 (**302**, 213), 116 (**306**, 112), 117 (**290**, 273), 118 (**309**, 217), 119 (**331**, 311), 120 (**306**, 115).

Information Bulletin

Six issues of the Information Bulletin (n° 55 through 60) have been sent, free of charge, to IAU Members, consultants, scientific institutions, and selected international organizations. Each print-run was 7500 copies.

6. Relations to other organisations

IAU was represented by the General Secretary at the following meetings of the International Council of Scientific Unions (ICSU): 20th meeting of the General Committee (München, Germany, FR, 1985), 21st General Assembly and 21st/22nd meetings of the General Committee (Bern, Switzerland, 1986) and 23rd meeting of the General Committee (Rome, Italy, 1987). The highlights concern future programmes such as the International Space Year (1992), and the large International Geosphere-Biosphere Programme (IGBP).

IAU was also represented at meetings of the ICSU Committee for Space Research (COSPAR): XXVIth Plenary meeting of COSPAR (Toulouse, France, 1986): J.-P. Swings and R. Wilson; XXVIIth Plenaary meeting of COSPAR (Espoo, Finland, 1988): J.-P. Swings.

The following IAU Representatives to other ICSU and international bodies were active during the period 1985-1988: it is to be noted that one minor change occurred: the present IAU representative to IAF (International Astronautical Federation) is Dr. Y. Kondo, President, IAU Commission 44:

	Organisation	Representative 1985-1988
ICSU	International Council of Scientific Unions General Committee	R.M. West (until Sept. 86) J.-P. Swings (after)
BIH (*)	Bureau International de l'Heure	P. Pâquet G. Winkler
CCDS	Consultative Committee for the Definition of the Second	J. Benavente
	Working Group on the Temps Atomique International	G. Winkler
CCIR	International Radio Consultative Committee Study Group 2	J. Whiteoak A.R. Thompson
	Study Group 7	J. Pilkington
CODATA	Committee for Data for Science & Technology	G. Westerhout
COSPAR	Committee on Space Research	J.-P. Swings
COSTED	Committee on Science & Technology in Developing Countries	J.-P. Swings
CTS	Committee on the Teaching of Science	L. Gouguenheim
EPS	European Physical Society Conference Committee	D. McNally
FAGS	Federation of Astronomical & Geophysical Services	J. Kovalevsky E. Tandberg-Hanssen
IAF	International Astronautical Federation	M. Papagiannis/ Y. Kondo

ICSTI (**)	International Council for Scientific & Technical Information	G. Wilkins
IPMS (*)	International Polar Motion Service	Ye Shu-Hua
IUCAF	Inter-Union Commission on Frequency Allocation for Radio Astronomy & Space Science	R. Schilizzi G. Swarup
IUPAP	International Union of Pure & Applied Physics	V. Trimble
IUWDS	International Ursigram & World Day Service	H. Coffey
QBSA	Quarterly Bulletin on Solar Activity	E. Hiei
SCOPE	Scientific Committee on Problems of Environment	R. Cayrel
SCOSTEP	Scientific Committee on Solar-Terrestrial Physics	S.T. Wu
URSI	Union Radio Scientifique Internationale	J. Baldwin
CIE	Compagnie Internationale de l'Eclairage	D. Crawford
IERS (*)	International Earth Rotation Service	G. Wilkins

(*) IERS subsummed BIH & IPMS with effect from 1988 January 01, 1988.
(**) IAU Membership in ICSTI was discontinued in 1987.

7. Financial matters

Summarized accounts

The 3 year (1985-1987) summarized accounts are given in the following table together with the budgetary estimates. The data is extracted from the "Vérification des Comptes" for 1985, 1986 & 1987 as certified by the IAU Auditors, R. Bâcle, HEC (1985, 1986) and P. Meyssonnier EC (1987), Paris. The unit is Swiss Francs throughout.

The ICSU mean conversion rates for Swiss Francs for 1985-1987 are:

	1 US $	1Dfl.	1FF	1DM	1 Rupee
1985	2.4600	0.7423	0.2730	0.8378	0.2018
1986	1.8175	0.7323	0.2602		0.1466
1987	1.5050	0.7386	0.2492		

TABLE I

Summary 1985-1987

	Income	BUDGET	1985	1986	1987	TOTAL
1.	Contributions	1244580,00	356459,32	475002,10	399178,41	1230639,83
2.	Publications	127000,00	12247,79	73235,08	59100,25	144583,12
3.	Interest	102000,00	47698,59	22688,98	24782,38	95169,95
4.	UNESCO/ICSU	120000,00	68852,94	47358,60	30981,93	147193,47
5.	Miscellaneous	0,00	861,00	4571,98	1150,72	6583,70
		1593580,00	486119,64	622856,74	515193,69	1624170,07
	Expenditure					
1.	Administration	492000,00	175854,45	196049,48	161138,53	533042,46
2.	ICSU Contribution	32700,00	11586,60	6582,99	9833,25	28002,84
3.	Commission expenses	23600,00		5717,95	327.45	6045,40
4.	Commission projects	128000,00				
.	Exchange of Astronomers		31151,54	18500,53	8327.97	57980,04
.	Teaching		23747,00			23747,00
.	Other projects		19819,50	8202,56	1000,00	29022,06
5.	General Assembly	213000,00	199633,06	22332,30	3980,00	225945,36
6.	IAU Publications	50000,00	23722,40	16592,45	32814,48	73129,33
7.	Publication for Executive Committee & Developing Countries	34000,00	2623,47	7619,91	4526,45	14769,83
8.	Executive Committee Meeting	80000,00	1360,92	19564,00	23146,38	44071.30
9.	Officers meeting	23000,00	3670,30	5274,69	24396,85	33341,84
10.	Symposia & Colloquia	335000,00	96616,28	116160,03	151104,83	363881.14
11.	Inter Union Commission	33000,00	13899,00	4816,38	4484,42	23199,80
12.	Projects of Executive Committee	11000,00				
13.	Representation	23000,00	8948,70	7904,27		16852,97
14.	Bank charges	3500,00	1983,53	1144,71	3341.63	6469.87
15.	International School for Young Astronomers	50000,00		31543,75		31543,75
16.	Visiting Lecturer's Programmes	48000,00		12784,96	8833,46	21618.42
17.	Regional Meetings	65000,00		32000,00	16556,58	48556,58
18.	Special Nominating Committee	5000,00				
19.	Miscellaneous	0,00	119,90	159,75 88,19		367,84
	Loss on transfers	0,00	2110,68	512,41	652,02	3275,11
		1649800,00	616847,33	513551,31	454464,30	1584862,94

Comments on summarized accounts

Income:

Total income was in excess of budget.

1. Shortfall in contribution with respect to budget is largely due to late or non-payment.

4. The ICSU contribution shows a steady decline which is continuing.

Expenditure:

1. Administrative costs were seriously underestimated in the budget. The savings in 1987 were a consequence of a change of staff and the savings of 1987 will not carry forward.

4. Commission Projects were over budget.

9. The Officers Meeting expenses were over budget in 1987 -this was a direct consequence of the staff changes.

15. There was a considerable underspend on the VLP programme.

18. 1986 includes an unsupported miscellaneous expenditure of 88.19 not detected until 1987.

Balance 1985-1987

The balance for the years 1985-1987 is set out below:

TABLE 2

	1985	1986	1987
Income	486119,64	622856,74	515193,69
Expenditure	- 616847,33	- 513551,29	- 454464,50
Re/devaluation	23672,68	- 93529,04	- 36317,23
Balance of the preceding year	826143,03	719088,03	735134,44
Balance	719088,03	735134,44	759546,40 (*)

(*) The accounts for Symposium 130 have affected this balance as follows:

Income	55950,01
Expenditure	33533,39
Balance	22416,62

Total balance as of 31.12.87 781963,02

The balance for 1987 must be increased by 22416,62 as a result of IAU Symposium 130 (income +55950.01; expenditure -33533.39) giving a reserve of 781963.02 as of 1987 December 31. The losses on valuation result from changes in major currencies -in particular the US dollar which suffered heavy devaluation during the period of accounts. The present balance is about 20 % in excess of one year's working. It should be noted that ICSU is currently reviewing the balances held by member Unions.

Residual budget 1988

The residual budget for 1988 is set out in Table 3. The figures are provisional, have not been audited and do not include loss on exchange.

TABLE 3

		EC Budget	Residual Inc/Exp	Difference (B-R)
1.	Contributions	453130	461105	- 7975
2.	Publications	55000	55551	- 551
3.	Interest	35000	22060	12940
4.	UNESCO/ICSU Contribution	35000	25122	9878
5.	Miscellaneous	-	19370	- 19370
	Total income	578130	583208	- 5078
Expenditure				
1.	Administration	183500	177719	5781
2.	ICSU Subvention	11300	6521	4779
3.	Commission expenses	4000	2702	1283
4.	Commission projects			
4.1.	Exchange of astronomers	24000	15077	8923
4.2.	Other projects	14500	2595	11905
5.	General Assembly	240000	337500	- 97500
6.	IAU Publications	15000	15662	- 662
7.	Publications for EC and developing countries	12000	2702	9298
8.	EC Meetings	-	623	- 623
9.	Officers Meetings	8000	2549	- 5451
10.	Symposia & Colloquia	125000	105228	19772
11.	Inter-Union Commissions	9000	6159	2841
12.	Projects of EC	5000	-	5000
13.	Representation	8000	4028	3972
14.	Bank charges	1000	2438	- 1438
15.1.	Young Astr. Schools	-	-	-
15.2.	VLP	24000	2894	21106
16.	Regional Meetings	-	-	-
17.	SNC	-	-	-
18.	Miscellaneous	-	-	-
	Total expenditure:	684300	684397	- 97
Excess of expenditure over income:		- 106170	- 101189	- 4981

Adopted Budget for 1989-1991

The budget for 1989-1991 was approved at the XXth General Assembly and is given below:

TABLE 4

BUDGET (1989-1991)

All amounts in Swiss Francs

INCOME		Triennial (1989-91)	Annual 1989	1990	1991
1.	CONTRIBUTIONS				
1.1	Adhering Organisations (238 units)	1 532 720	493 850	510 510	528 360
1.2.	ICSU/UNESCO	78 000	26 000	26 000	26 000
1.2.1.	ICSU Allocations				
1.2.2.	ICSU Grants				
1.2.3.	UNESCO Contracts				
2.	PUBLICATIONS	165 000	55 000	55 000	55 000
3.	INTERESTS, etc.	75 000	25 000	25 000	25 000
4.	OTHER RECEIPTS	-	-	-	-
	TOTAL INCOME	1 850 720	599 850	616 510	634 360

Notes:

1.1.

a. Changes of category:
. Argentina: II to III: now 4 units
. China: III to V: now 10 units

b. Possibility of around 4 new Members in category I. Total:
1986-1988: 226 units 1988-1991: 238 units

c. Same inflation considered for 1989-1991 as for previous triennium; very difficult to estimate in present crisis period.
Proposed unit of contribution:
2075 SF for 1989
2145 SF for 1990
2220 SF for 1991

1.2. The income from ICSU/UNESCO has decreased in the last years. It is 17 335 US $ for 1988 (in November 87 dollars !). Therefore the sum indicated in the draft can only be realistically introduced as 26 000 SF.

2. 6 year contract with Reidel Publ.: same royalties as for 1986-1988 triennium.

EXPENSES		Triennial (1989-91)	Annual 1989	1990	1991
1.	EXECUTIVE COMMITTEE, etc.				
1.1.	EC Meetings	80 000	40 000	40 000	-
1.2.	Officers' Meetings	24 000	8 000	8 000	8 000
1.3.	SNC Expenses	5 000	-	5 000	-
1.4.	Other	-	-	-	-
	TOTAL 1.	109 000	48 000	53 000	8 000
2.	PUBLICATIONS				
2.1.	Information Bulletin	63 000	20 000	21 000	22 000
2.2.	Other Publications	-	-	-	-
2.3.	Distribution to Developing Countries	30 000	10 000	10 000	10 000
	TOTAL 2.	93 000	30 000	31 000	32 000
3.	SCIENTIFIC AND RELATED ACTIVITIES				
3.1.	Meetings				
3.1.1.	General Assemblies	240 000	-	-	240 000
3.1.2.	Symposia and Colloquia	365 000	120 000	120 000	125 000
3.1.3.	Regional Astronomy Meetings	65 000	32 000	33 000	-
3.1.4.	Young Astronomers' Schools	50 000	25 000	25 000	-
3.1.5.	Visiting Lecturers' Programmes and follow-ups	72 000	24 000	24 000	24 000
3.1.6.	Other	-	-	-	-
	Total 3.1.	792 000	201 000	202 000	389 000
3.2.	Commissions Activities				
3.2.1.	Commission Expenses	15 000	5 000	5 000	5 000
3.2.2.	Commission Projects				
3.2.2.1.	Exchange of Astronomers	70 000	23 000	23 000	24 000
3.2.2.2.	IAU Telegram Bureau	10 500	3 500	3 500	3 500
3.2.2.3.	IAU Minor Planet Center	10 500	3 500	3 500	3 500
3.2.2.4.	Variable Star Catalogue	10 500	3 500	3 500	3 500
3.2.2.5.	IAU Meteor Data Center	3 000	1 000	1 000	1 000
3.2.2.6.	Other (incl. IERS)	6 000	2 000	2 000	2 000
	Total 3.2.	125 500	41 500	41 500	42 500
3.3.	Relations With Other Organisations				
3.3.1.	Dues to ICSU	38 500	12 500	12 800	13 200
3.3.2.	Support of Inter-Union Commissions	18 000	6 000	6 000	6 000
3.3.3.	Representation to Other Organisations (including liaison members)	24 000	8 000	8 000	8 000
	Total 3.3.	80 500	26 500	26 800	27 200
3.4.	Other Activities				
3.4.1.	Executive Committee Projects	15 000	5 000	5 000	5 000
3.4.2.	Other Projects	-	-	-	-
	Total 3.4.	15 000	5 000	5 000	5 000
	TOTAL 3.	1 013 000	274 000	275 300	461 700

4.	ADMINISTRATION				
4.1.	Secretariat				
4.1.1.	Salaries and Social Charges, etc.	375 000	120 000	125 000	130 000
4.1.2.	Office Expenses	165 000	50 000	55 000	60 000
4.1.2.1.	Rent, Heating, Light, Cleaning, etc.				
4.1.2.2.	Communication (PTT)				
4.1.2.3.	Equipment (Major)				
4.1.2.4.	Office Supplies				
4.1.3.	General Secretary's Expenses	57 000	18 000	19 000	20 000
4.1.3.1.	Travel, Subsistence				
4.1.3.2.	Representation				
4.1.4.	President's Expenses	18 000	6 000	6 000	6 000
4.1.5.	Ass. General Secretary's Expenses +	12 000	4 000	4 000	4 000
	Local help (AGS)	3 000	1 000	1 000	1 000
	Total 4.1.	630 000	199 000	210 000	221 000
4.2.	Other				
4.2.1.	Bank Charges	3 000	1 000	1 000	1 000
4.2.2.	Audit Fees, Legal Advice	4 500	1 500	1 500	1 500
4.2.3.	Loss on Exchange	-	-	-	-
4.2.4.	Miscellaneous	-	-	-	-
	Total 4.2.	7 500	2 500	2 500	2 500
	TOTAL 4.	637 500	201 500	212 500	223 500
	TOTAL EXPENDITURE	1 852 500	553 500	571 800	727 200
	INCOME - EXPENDITURE	- 1 780	+ 46 350	+ 44 710	- 92 840

Notes:

1.1. EC meetings: less expensive than expected during 1986-1988, because of locations and absences at EC 56.
For future: include President-Elect.

2.1. Mailing of Information Bulletin had been underestimated for previous triennium; one has also to take into account hundreds of new members after the XXth General Assembly.

3. No major change foreseen for next triennium.

4.1. Secretariat

4.1.1. Salaries: had been strongly underestimated for 1986-88 triennium; however, because of new personnel (i.e. one full-time administrative assistant, and one part-time (2/3) bilingual secretary), costs will not increase much during 1989-1991.

4.1.2. The location of the Paris secretariat may change in 1988, so that an estimate of the expenses is hard to make. A very slight increase over the figure of the previous triennium was made.

4.1.3. G.S. expenses: here also, underestimated for 1986-88.

4.1.4. President's expenses: one has to include trips for useful contacts in several countries: increase this item.

8. **List of Adhering Organizations**

(Not reproduced here, see Chapter VII).

9. **List of deceased Members**

(See updated list on pp. 25-26).

10. **List of IAU publications**

Transactions and Highlights

Transactions XIXA: (Delhi, 1985) pp. viii + 736, D. Reidel Publ. Co., 1985

Transactions XIXB: (Delhi, 1985) pp. xii + 656, D. Reidel Publ. Co., 1986

Transactions XXA: (Baltimore, 1988) pp. viii + 708, Kluwer Academic Publ., 1988

Highlights of Astronomy, **7**, as presented at the XIXth General Assembly of the IAU, 1985, pp. xvi + 900, D. Reidel Publ. Co., 1986

IAU Symposia Volumes (edited by Reidel Publ. Co. & by Kluwer Academic Publ.)

106 The Milky Way Galaxy
H. van Woerden, R.D.J. Allen & W.B. Burton, pp. xxiv + 660, 1985

109 Astrometric Techniques
H.K. Eichhorn & R.J. Leacock, pp. xxi + 838, 1986

111 Calibration of Fundamental Stellar Quantities
D.S. Hayes et al., pp. xxiv + 646, 1985

112 The Search for Extraterrestrial Life: Recent Developments
M. Papagiannis, pp. xxvi + 580, 1985

113 Dynamics of Star Clusters
J. Goodman & P. Hut, 644 p., 1985

114 Relativity in Celestial Mechanics and Astrometry, High Precision Dynamical Theories and Observational Verifications
J. Kovalevsky & V.A. Brumberg, pp. xvii + 426, 1986

115 Star Forming Regions
M. Peimbert & J. Jugaku, pp. xxxiv + 734, 1985

116 Luminous Stars and Associations in Galaxies
C.W.H. de Loore, A.J. Willis & P. Laskarides, pp. xx + 530, 1986

117 Dark Matter in the Universe
J. Kormendy & G.R. Knapp, pp. xxix + 596, 1986

118 Instrumentation and Research Programs for Small Telescopes
P.L. Cottrell & J.B. Hearnshaw, pp. xxii + 482, 1986

119 Quasars
G. Swarup & V.K. Kapahi, pp. xxii + 606, 1986

120 Astrochemistry
M.S. Vardya & S.P. Tarafdar, pp. xxviii + 640, 1986

121 Observational Evidence of Activity in Galaxies
E. Ye Khachikian, K.J. Fricke & J. Melnick, pp. xxii + 601, 1987

122 Circumstellar Matter
I. Appenzeller & C. Jordan, pp. xxiv + 607, 1987

123 Advances in Helio- and Astroseismology
J. Christensen-Dalsgaard & S. Frandsen, pp. xxi + 604, 1988

124 Observational Cosmology
A. Hewitt, G. Burbidge & Li Zhi-Fang, pp. xxiii + 854, 1987

125 The Origin and Evolution of Neutron Stars
D.J. Helfand & J.-H. Huang, pp. xx + 572, 1987

126 Globular Cluster Systems in Galaxies
J.E. Grindlay & A.G. Davis Philip, pp. xxvii+ 750, 1987

127 Structure and Dynamics of Elliptical Galaxies
T. de Zeeuw, pp. xxv + 579, 1987

128 The Earth's Rotation and Reference Frames for Geodesy
A.K. Babcock & G.A. Wilkins, pp. xviii + 470, 1988

129 The Impact of VLBI on Astrophysics and Geophysics
M.J. Reid & J.M. Moran, pp. xxiv + 599, 1988

132 The Impact of Very High S/N Spectroscopy on Stellar Physics
G. Cayrel de Strobel & M. Spite, pp. xxxiii + 626, 1988

IAU Colloquia Volumes

81 Local Interstellar Medium
Y. Kondo, F.C. Bruhweiler & B.D. Savage, NASA Conference Publication 2345, 356 p., 1984

82 Cepheids, Theory and Observations
B.F. Madore, Cambridge University Press, pp. ix + 300, 1985

83 Dynamics of Comets: Their Origin and Evolution
A. Carusi & G.B. Valsecchi, D. Reidel Publ. Co., pp. xii + 400, 1985

84 Longitude Zero Symposium 1884-1984
S.R. Malin, A.E. Roy & P. Beer, Pergamon Press, in the series Vistas in Astronomy, **28**, parts 1/2, 407 p., 1985

85 Properties and Interactions of Interplanetary Dust
R.H. Giese & P. Lamy, D. Reidel Publ. Co., pp. xxxvi + 444, 1985

86 Eighth International Colloquium on Ultraviolet and X-ray Spectroscopy of Astrophysical Laboratory Plasmas
G.A. Doschek, published by the Naval Research Laboratory, 1986

87 Hydrogen Deficient Stars and Related Objects
H. Hunger, D. Schönberner & N. Kameswara Rao, D. Reidel Publ. Co,. in the series Astrophysics and Space Science Library **128**, pp. xx + 506, 1986

88 Stellar Radial Velocities
A.G. Davis Philip & D.W. Latham, L. Davis Press, pp. xi + 455, 1985

89 Radiation Hydrodynamics in Stars and Compact Objects
D. Mihalas & K.-H.A. Winkler, Springer-Verlag in the series Lecture Notes in Physics, **255**, pp. ix + 454, 1986

90 Upper Main Sequence Stars with Anomalous Abundances
C.R. Cowley, M.M. Dworetsky & C. Mégessier, D. Reidel Publ. Co., in the series Astrophysics and Space Science Library, **125**, pp. xiv + 489, 1986

91 History of Oriental Astronomy
G. Swarup, Cambridge University Press, pp. xiv + 289, 1987

92 Physics of Be Stars
A. Slettebak & T.P. Snow, Cambridge University Press, pp. xvii + 557, 1987

93 Cataclysmic Variables
H. Drechsel, J. Rahe & Y. Kondo, D. Reidel Publ. Co., reprinted from Astrophysics and Space Science, **130** n° 1 & 2, **131**, n° 1 & 2, 1987

94 Physics of Formation of FeII Lines Outside LTE
R. Viotti, A. Vittone & M. Friedjung, D. Reidel Publ. Co., in the series Astrophysics and Space Science Library, **138**, 1988

95 The Second Conference on Faint Blue Stars
A.G. Davis Philip, D.S. Hayes & J.W. Liebert, L. Davis Press, Inc., pp. xxix + 778, 1987

96 The Few Body Problem
M.J. Valtonen, D. Reidel Publ. Co., 1988

97 Wide Components in Double and Multiple Stars
J. Dommanget, E.L. van Dessel & Z. Kopal, D. Reidel Publ. Co., reprinted from Astrophysics and Space Science, **142**, 1988

98 Stargazers, The Contribution of Amateurs to Astronomy
S. Dunlop & M. Gerbaldi, Springer-Verlag, pp. xvii + 237, 1988
&
La contribution des Astronomes Amateurs à l'astronomie
Observations & Travaux, Hors série N° 1, 2, 3 & 4
Société Astronomique de France

99 Bioastronomy, Next Steps
G. Marx, D. Reidel Publ. Co., 1988

101 Supernova Remnants and the Interstellar Medium
R.S. Roger & T.L. Landecker, Cambridge University Press, pp. xi + 540, 1988

102 UV and X-ray Spectroscopy of Astrophysical and Laboratory Plasmas
F. Bely-Dubau & P. Faucher, Journal de Physique, **49**, pp. xvii + 400, 1988

103 The Symbiotic Phenomenon
J. Mikolajewska, M. Friedjung, S.J. Kenyon & R. Viotti, eds., Kluwer Academic Publishers, in the Series Astrophysics and Space Science Library, **145**, pp. xvi + 365, 1988

Proceedings of Regional Astronomy Meetings

Seventh European
Florence, Italy, December 12-16, 1983
Frontiers of Astronomy & Astrophysics
R. Pallavicini, Italian Astronomical Society, Florence, Italy

Eighth European
Toulouse, France, September 17-21, 1984
New Aspects of Galaxy Photometry
J.-L. Nieto, Springer-Verlag in the series Lecture Notes in Physics, **232**, 350 p., 1985
High Resolution in Solar Physics
R. Muller, Springer-Verlag in the series Lecture Notes in Physics, **233**, 320 p., 1985
Nearby Molecular Clouds
G. Serra, Springer-Verlag in the series Lecture Notes in Physics, **237**, 242 p., 1985

Third Asian-Pacific
Kyoto, Japan, September 30-October 5, 1984
Ed. M. Kitamura & E. Budding, D. Reidel Publ. Co, part I: 552 p.; part II: 264 p., 1986

Fourth Latin-American
Rio de Janeiro, Brazil, November 18-23, 1984
Ed. P. Pismis & S. Torres-Peimbert, special issue of Revista Mexicana de Astronomia y Astrofisica, **12**, 452 p., February 1986

Fourth Asian-Pacific
Beijing, China, PR, October 5-9, 1987
Vistas in Astronomy, **31**, 1988, Science Press (Pergamon Press), Beijing

Miscellaneous Publications

Image Detection and Quality
SFO & ANRT meeting, co-sponsored by the IAU, SPIE, eds., 1987

Commission Publications

Commission **10:** Quarterly Bulletin on Solar Activity.
Published at the Tokyo Astronomical Observatory, F. Moriyama, Mitaka, Tokyo 181, Japan.

Commission **10:** Sunspot Bulletin.
Published by the Sunspot Index Data Center (SIDC), Dr. A. Koeckelenbergh, 3 Avenue Circulaire, B-1180 Bruxelles, Belgium.

Commission **19:** Circulaires du Bureau International de l'Heure.
A monthly publication of Bureau International de l'Heure, 61, Avenue de l'Observatoire, 75014 Paris, France (now terminated).

Commission **19:** Monthly Notes of the International Polar Motion Service.
Prepared and distributed by the Central Bureau of the International Polar Motion Service, International Latitude Observatory of Mizusawa-shi, Iwate-ken, Japan.

Commission **20:** Minor Planet Circulars.
Issued by the Minor Planet Center, B.G. Marsden, Center for Astrophysics, 60 Garden Street, Cambridge, MA 02138, USA

Commission **27:** Information Bulletin on Variable Stars.
Prepared and distributed by the Konkoly Observatory of the Hungarian Academy of Sciences , 1525 Budapest XII, Box 67, Hungary.

Commission **27:** Catalogue of Variable Stars
Catalogue of Suspected Variable Stars.
Name-lists of Variable Stars.
Distributed by Publishing House Nauka, Moscow, USSR

Commission **29:** Nucleosynthesis in the Galaxy from the Study of Low-Mass Stars
Proceedings ot the IAU Commission 29 Meeting, held in New-Delhi, India, 1985
G. Cayrel de Strobel & M. Parthasarathy, Eds., Journal of Astrophysics & Astronomy, **8**, No. 2, 1987

Commission **46:** Newsletter on the Teaching of Astronomy.

Commission **46:** Astronomy Education Materials.

CHAPTER V

REPORTS OF MEETINGS OF COMMISSIONS

COMMISSION No. 4

EPHEMERIDES (EPHEMERIDES)

Report of Meeting 1988 August 4

PRESIDENT : B. Morando SECRETARY : B.D. Yallop

It was the President's sad duty to report the deaths of two former past Presidents of the Commission, W. Fricke who had also been a Vice-President of the IAU and a President of Commission 8, and D. H. Sadler who had been Superintendent of H.M. Nautical Almanac Office from 1936 to 1969 and was once General Secretary of the IAU.

1 . Organisation and Membership

The President's proposals for the Officers of the Commission for the next years were adopted as follows :

President : P. K. Seidelmann
Vice-President : B. D. Yallop

Organising Committee : V.K. Abalakin, J. Chapront, R. L. Duncombe, H. Kinoshita, Y. Kubo, J.H. Lieske, B. Morando, H. Schwan, Fu Tong.

New members of the Commission : A. K. Bhatnagar, M. Chapront-Touzé, J. Coma-Samartin, M. E. Davies, XX Newhall, A. Shiriyaiv, R.Wielen.

Consultants : J. Meeus, H. J. Felber, K.H. Steinert, T. Fukushima.

Retiring member : G.E. Taylor.

2 . Commission 4 Circular

The President announced that he has published the first issue of IAU Commission 4 Circular and hoped to publish another issue but he did not get any response from the members of the commission. P.K. Seidelmann said that it was a useful way of communicating to those members of the Commission who were unable to attend the IAU General Assembly, and it should be issued at least once a year.

3 . Resolutions of the Working Group on Reference Frames

The resolution of the Working Group on Reference Frames, which is a joint project of Commissions 4, 7, 8, 19, 20, 24, 31, 33 and 40 were discussed.
Commission 19 and 31 had already changed the first resolution to leave the formation of a study group open and names had been omitted. The changed resolution was agreed. R. L. Duncombe said that now there was no need to discuss

resolutions 7, 8 and 9 of the Report of the Working Group as they had been taken care of. The remainder of the resolutions were accepted with a few abstentions. E.M. Standish expressed a misgiving about resolution 5 because he felt that the observing requirements could not be achieved for another ten years. E. M. Gaposchkin pointed out that the observers needed encouragements which was an important aim of the resolution.

Report of Meeting 1988 August 6

PRESIDENT : B. Morando SECRETARY : B.D. Yallop

1. New Astronomical Centre in Penang

M. Ilyas spoke about the new astronomical centre which is under construction at the equatorial location of Penang in Malaysia. The centre is part of the Malaysian Science University. Initial facilities will include a telescope and photoelectric equipment. Any assistance from other observatories would be welcome. They intend to publish an ephemeris eventually.

2. Latest observed values for precession and nutation

J. O. Dickey read a paper by the Lunar Laser Ranger Working Group giving the latest values for the constants of precession and nutation obtained from 5629 points in 19 years. The value of the constant of precession obtained by Fricke is now down from 1."10 to 0." 86 per century.

3. Reports from Nautical Almanac Offices

P. K. Seidelmann discussed the work of the U.S. Naval Observatory. All the publications had continued except for the Ephemeris for land surveyors which terminated at the 1988 issue. The Floppy Almanac has now been produced for each year up to 1999. The Almanac for Computers is also available on a floppy disc for current years. Sight reduction procedures have been developed and a concise table for sight reduction will be published in the Nautical Almanac from 1989 onwards. Research has been carried out on planetary satellites and the Chinese observations of the satellites of Jupiter have been used. A "Moonwatch" was organised for the evening of 28 April 1987 to test different formulae for predicting the first visibility of new moon across North America.

B.D. Yallop reported on the activities of H. M. Nautical Almanac Office. The staff complement had been reduced to 6. There was a redivision of responsabilities between USNO and NAO for various sections of the Astronomical Almanac and the publication date has been brought forward . Compact Data for Navigation and Astronomy for the years 1986-1990 was published by HMSO in 1985. A new section on Sight Reduction Procedures has been added to Nautical Almanac 1989 which includes a method for direct computation using a programmable calculator. Four papers concerned with mean and apparent place reductions have been written in collaboration with the USNO.

B. Morando discussed the work of the Bureau des Longitudes. J.-E. Arlot had been studying the satellites of planets. It was found that a mixture of functions and polynomials required fewer coefficients for fast moving satellites. This information was now available on floppy discs. There was also a floppy disc which contained the Moon ephemeris (ELP) from -1500 to 2000 and new ephemerides of the satellites of Uranus called GUST 86. Apart from Connaissance des Temps, Ephémérides Astronomiques (or Annuaire du Bureau des Longitudes) and Ephémérides Nautiques, Bureau des Longitudes publishes supplements with ephemerides and configurations for the main satellites of Jupiter and Saturn and Notes Scientifiques et Techniques devoted to various topics (comets, minor planets, etc.)

Y. Kubo described some of the work of the Japanese Hydrographic Department. A supplement to the Nautical Almanac for 1989 has been produced containing Chebyshev coefficients of the Sun, Moon, planets and 45 stars. A preliminary study on the numerical solution to precession and nutation of a rigid Earth has been made.

The report of the International Lunar Occultation Centre listed the numbers of timings in the period 1981 to 1987, which reached a peak in 1983 and showed an increase again in 1987. Corrections to the orbital longitude and latitude of the Moon deduced from the lunar occultations observations received by the ILOC, based on the IAU (1976) system of astronomical coordinates have been published for 1971 to 1986.

V. K. Abalakin reviewed the work of the Almanac Office in Leningrad. About 2300 copies of the Astronomical Yearbook are sold each year. The volume for 1990 has been submitted for publication, whilst copy for 1991 is in an advanced stage. An English version of the contents and explanation is being prepared. Software has been supplied to observatories in the USSR. Numerical integration has been carried out for the Phobos mission. Orbital libration of the Moon has been studied theoretically.

H. Schwan talked briefly about the work at Heidelberg. The apparent places in the APFS 1988 include both the systematic and individual corrections FK5-FK4 which are now available. The preface has been translated into five languages, and V. K. Abalakin was responsible for the Russian version. Work has been done on the Chinese calendar and it is now possible to convert precisely to Julian Date.

The meeting then discussed items of general interest. It was suggested that a collection of original volumes of all almanacs should be made. M. Standish drew attention to he four outer satellites of Jupiter which are drifting off from the ephemeris predictions by a few tenths of an arc second. J. Lieske mentioned that JPL had a dial in ephemeris.

Report of Meeting 1988 August 10

1. A dictionary of Almanacs

Tong Fu requested that a document be issued which explains what is published in each almanac. P. K. Seidelmann asked that Almanac Offices send him information so that he could start the work.

2. Resolution of the Working Group of Reference Frames

B. Morando reported that Commissions 19 and 31 had recommended a change of wording. P. K. Seidelmann suggested that the Chairman of the Working Group should be J. Hughes.

3. Changeover to J2000

P. K. Seidelmann reported that Commission 20 were the last to change to J2000. They were undecided whether to convert the individual observations or the orbital elements to J2000.

4. IAU Working Group on Astronomical Software and Computer Communications

The President read the proposals for a working group on astronomical software and computers to the Commission. G.A. Wilkins said that it was premature since Commission 5 had already set up a working group on computer software, and there was no need to set up another Commission when Commission 5 could handle the problems.

5. Future Developments in Ephemerides

P. K. Seidelman discussed the way he thought ephemeris offices would change in the future. He could see a continuing need for published volumes of almanacs and tables. There will also be a growing demand for data in alternative forms. For example, almanacs on floppy discs, such as the Almanac for Computers, and interactive almanacs, which can be interrogated by telephone. There would be a greater demand for software packages such as apparent place routines. It was important to standardise software distribution, to continuously improve the data and find more compact ways of representation. It was also important to monitor user needs, which were changing. There was a continuing increase in the non-scientific use of the data especially by religious groups.

P. Kammeyer gave a detailed talk about a 250 year compressed lunar and planetary ephemeris based on DE200. The floating point Chebyshev coefficients have been expressed in integer form which compresses the data to 900K-bytes. One of the first uses of this package will be to insert it into the multi-year floppy.

L.E. Dogget said that the Floppy Almanac will now be called the Interactive Computer Ephemeris which is being developed at USNO by himself, T.S. Carroll, P. Kammeyer, W. Tangren and W. Harris. The purpose is to make the data published in the Astronomical Almanac, the Nautical Almanac and the Air Almanac available on a micro-computer. The users will be Astronomers, Navigators, Surveyors, Amateur Astronomers and others.The data will be stored in the form of Chebyshev series and will contain heliocentric, geocentric and physical data except for eclipses, occultations and minor planet data. There will be two versions, a multi-year floppy suitable for IBM Pc's and an interactive MacAlmanac suitable for MacIntosh computers.

H.Schwan explained the status of the FK5. Data was available on a magnetic tape which included the extension of 2000 faint stars down tomagnitude 9. 5 and 980 bright stars from the FK4 Supplement.

V. K. Abalakin said that the demand for the Astronomical Yearbook was about the same as ever. The Moon's hourly had been replaced by Chebyshev coefficients. The apparent places of 777 stars selected for time and latitude services and geodesy had been produced.

B. Morando mentioned some recent development at the Bureau des Longitudes. Millisecond pulsars require a precise theory of the motion of the Earth especially over a short time scale. J. L. Barnier had produced a new theory in his theses which produced good results for the radio telescope at Nançay. Improvements to the theory of the motion of satellites were being made and a new theory of Hyperion should appear soon.

In France they have a system called Minitel where the user had a keyboard and a monitor screen. Information such as train time tables, and personal bank balance is available. The Paris Observatory and the Bureau des Longitudes have been asked to prepare a package of astronomical information to be included in this services. The latitude and longitude for topocentric information is deduced from the post codes.

COMMISSION No. 5

DOCUMENTATION AND ASTRONOMICAL DATA

(DOCUMENTATION ET DONNEES ASTRONOMIQUES)

Report of Meetings: 4, 5, and 10 August 1988

PRESIDENT: G. A. Wilkins

Session 1. 4 August 1988 09.00-10.30

BUSINESS MEETING (1) Chairman: G. A. Wilkins Secretary: G. Lyngå

1. Introduction. The President opened the meeting by paying tribute to the memory of Prof. Dr. Walter Fricke and Prof. Dr. Albert G. Velghe, who had been active members of the Commission for many years. He drew attention to the extensive programme of Commission meetings and to the need for secretaries to assist in the preparation of the reports; some items of business would be finalised on 10 August after the technical meetings. About 20 members of the Commission were present.

2. Report for 1985-88. The President reviewed briefly the activities of the Commission since the meeting in Delhi. He had prepared three Newsletters giving announcements and reports, but he had been disappointed at the small number of contributions received from members. Preprints of the formal report for 1985-87 (Trans. IAU 20A, 7-12) had been distributed with the Newsletter to all members and to about 50 other persons who were known to be interested in the activities. Astronomical data continued to be the major concern of the Commission, but there had also been a considerable revival of activity in the fields of documentation. There had been a meeting of the editors of the principal astronomical journals, but it had not been possible to reach full agreement on some of the major items in the new IAU Style Manual, which is to be published in Trans. IAU 20B. The Union continued to be represented on CODATA by G. Westerhout, but had terminated its membership of ICSTI since, at present, very few of its activities are of direct relevance to astronomy.

IAU Colloquium 110 on Library and Information Services in Astronomy had been very successful. It had just been held in Washington, D.C., on 27-30 July and at the Goddard Space Flight Center on 1 August. The attendance, by 125 persons from 26 countries, had been much larger than originally expected; the Local Organising Committee had obtained financial support, from both organisations and individuals, for many persons who would otherwise not have been able to attend. The Joint Discussion on New Developments in Documentation and Data Services for Astronomers had been held the previous day (3 August). It had provided a very useful forum for the transfer of information and ideas between astronomers and others who provide library and other services. These two meetings had served to stimulate a greater awareness of, and interest in, the activities of the Commission.

3. Officers. It was noted that, in accordance with past practice, the President (G. A. Wilkins) and Vice-President (B. Hauck) had been nominated to serve for a second triennium. [These nominations were accepted by the IAU Executive Committee.] Final consideration of the nominations for members of the Scientific Organising Committee and for chairmen of the Working Groups of the Commission was deferred to the second business meeting on 10 August.

4. Membership. The response to questionnaires, which had been distributed with Newsletter No. 3 and at the Colloquium and Joint Discussion, had shown that many persons wished to become members or consultants of the Commission. It was agreed to consider nominations at the second business meeting after the President had consulted the General Secretary about the policy on the admission of additional members and consultants. The President stated that G. R. Riegler had offered to

liaise between Commissions 5 and 44 (Astronomy from Space). It was agreed that it would be useful to establish a list of persons who would each liaise with one other Commission with which they were actively associated.

5. Activities. There was a general discussion about the current programme of activities of the Commission. It was agreed that the Working Groups (WGs) on Astronomical Data and Designations should continue; the former includes a task group on FITS (see below). The WG on Classification, under the chairmanship of P. Lantos, had produced a vocabulary of astronomical terms, which would be discussed during session 5; Lantos had suggested that a new WG on Information Retrieval be set up. It was agreed that this WG should cover the revision of UDC 52 and the compilation of a thesaurus of astronomical terms.

The President drew attention to the increased interest in editorial matters that had been stimulated by the preparation of the new IAU Style Manual, and it was agreed that the WG on Editorial Policy should be revived.

The discussions at IAU Colloquium 110 had demonstrated the need to extend the cooperation that already existed between astronomical librarians; the Physics, Astronomy and Mathematics Division (PAM) of the Special Libraries Association (SLA) acted as a useful forum and it was agreed to follow up the suggestion of setting up a joint WG on Astronomical Libraries between PAM and Commission 5.

C. R. Benn drew attention to the need for a working group that would encourage and facilitate the exchange of computer software through the use of the new computer networks. Concern was expressed that some other Commissions already had such WGs, but eventually it was agreed that it would be appropriate for Commission 5 to take on this activity, especially in respect of software of general application.

6. Working Groups. A provisional list of Working Groups and Task Groups, and of their chairmen, for 1988-91 was drawn up. P. Grosbol requested that the Task Group on FITS (Flexible Image Transport System) should be a separate permanent Working Group (and not part of the WG on astronomical Data) so that its importance would be more apparent to national agencies which should be encouraged to adopt the standards prepared by the Group; he described the recent work on the extension of the standards to cover catalogues and other tabular data. The decision on the list was deferred until the second business meeting to allow for further discussions during the technical sessions.

7. Resolutions.

P. Grosbol introduced two resolutions on FITS: the first concerned the extension of the standards and the second the setting up of a permanent working group., After discussion it was agreed that the President should forward these resolutions to the Resolutions Committee of the Union. [They were adopted by the General Assembly and are given as resolutions B1 and B2.]

The President draw attention to the draft text of a resolution on the preparation of astronomical papers to be proposed by the Executive Committee of the Union concerning the form of references and the use of SI units. He explained that he had been invited to speak to the Executive Committee at its meeting on 31 July about the IAU Style Manual and he had identified a few points of particular concern. After discussion it was agreed that he should suggest to the Resolutions Committee that the resolution by extended to refer explicitly to the Style Manual and to the designation of objects. [The matter was discussed again at session 6 on editorial policy and by the Executive Committee on 9 August. The revised resolution was adopted by the General Assembly, and is given as resolution A3 on the Improvement of Publications.]

8. Representatives. It was agreed that G. Westerhout should be nominated to the Executive Committee as the IAU representative on CODATA.

The business meeting was adjourned until 10 August.

Session 2. 4 August 1988 11.00-12.30

TECHNICAL AIDS FOR DATA Chairman: G. Westerhout Secretary: P. J. Hanisch

The meeting of the Working Group on Astronomical Data was opened by the Chairman, who pointed out that, in spite of three WG newsletters, there had not been much activity, other than by the FITS Committee. He requested that WG members communicate regularly with the Chairman.

1. Networks. D. Wells reviewed the NSFNET, the U.S. National Research Network. It connects the supercomputer centres and MSF regional networks, using TCP/IP protocol. Backbone network speeds will be upgraded in steps from 56 to 1544 Kb/s. IBM routers are expected to reach 45 Mb/s by 1991, contingent on cost. SPAN and MSFNET are full-function, remote login networks, not just mail and file transfer. Through the various networks, virtual-image displays (IRAF, AIPS, IDI/Trieste), database servers, and archive servers are accommodated, as well as remote observing.

2 STARLINK. R. J. Dickens reported on the UK national interactive data-analysis facility STARLINK, started in 1979 and based on a network of VAX 780 computers. There are now 11 major nodes, 7 minor nodes, 12 remote nodes, and 1000 users. All nodes are linked through JANET. About 500K lines of code, contributed from all STARLINK nodes, are available to users.

3. ASTRONET. M. Pucillo discussed the Italian ASTRONET, connecting 70 CPUs. The primary effort is the standardization of software, based on MIDAS, including portable graphics software. The speed of the network connections (TCP/IP and DECNET) is being improved and a "distributed computing system" over a LAN is being implemented.

4. SPAN. M. VanSteenberg reported on the NASA network SPAN, which now has 2000 registered nodes and is managed by the National Space Science Data Center. Primary routers are located at NASA centers GSFC, MSFC, JSC and JPL. The backbone speed is 56 Kb/s, and there are connections to Europe (ESOC at Darmstadt) and Chile (SN). SPAN has supported several NASA missions, and it supports use of the IUE archive at GSFC (400 images per month requested) and IUE remote observing.

5. New storage technologies. M. Rushton presented an overview of the WORM (write once read many) optical disks, CD-ROM (5¼-inch optical disk) and helical-scan high-volume cassette tape technologies. Key issues are the existence of media standards (international organizations or de facto standards), cost, and lifetime. WORM optical disks (12-inch and 5¼-inch) have a top capacity of 2-3 GB (12-inch double sided), error rates (uncorrected) of 1 in 1 Tb, data transfer rates of 100-200 Kb/s, and lifetimes of 30 years. Unfortunately, the media are not interchangeable. A 12-inch drive costs about $13000, host computer adapter $2000, software device driver up to $6000, 12-inch disk $350. The next generation is due in under 2 years, and considerable capacity and transfer rate increases are expected. The CD-ROM technology is standardized. CD-ROM disks store up to 650 Mb. A master is produced first from tape, disk, network, etc. ($1500-300 per disk), after which copies are made ($2-10 per disk). A drive costs $500-1000. The helical-scan recording technique, using 8-mm videocassette tapes, stores 2 GB on a $10 cassette. A complete subsystem costs $5000-7000. This very informative review ended with a short report on juke boxes, costing anywhere from $15K to over $100K.

6. FITS. P. Grosbol reported on the activities of the FITS task group and pointed out that details of the FITS Tables Extension, including blocking factors, printable arrays, etc. are given in Astron. Astrophys. Suppl. Ser. 73, 359-372, 1988. He presented two resolutions on FITS and the Working Group agreed that: (1) the use of the Tables Extension should be an IAU recommended practice; and (2) a new FITS WG should maintain and review FITS standards.

Session 3. 4 August 1988 14.00-15.30

DATA CATALOGUES Chairman: G. Westerhout Secretary: G. Westerhout

1. Data centre reports. P. Dubois reported that the Strasbourg Data Centre (CDS) now has 5000 catalogues, and the SIMBAD database contains data and bibliography on 600 000 stars and 100 000 nonstellar objects. SIMBAD is heavily used in the planning of space projects. A colloquium on catalogue precision is foreseen for 1989. W. H. Warren reported that the NASA Astronomical Data Center (ADC) has processed 1830 requests, distributing 1097 catalogues in digital form since the New Delhi General Assembly. An ADC Online Information System has been developed to allow users to obtain information and submit requests via SPAN and Internet by logging into the National Space Science Data Center VAX 8650 Computer. Smaller data sets are now being disseminated electronically via SPAN and BITNET. S. Nishimura reported that the Japanese data centre will be moved from the Kanazawa Institute of Technology to Tokyo, where it will be operated within the new National Observatory.

2. Database systems. S. Lubow highlighted the differences between commercial and special purpose database systems. He pointed out that relational systems have great flexibility, allowing a very large set of queries to be expressed, the use of nonprocedural query languages, and easily reorganizable and highly portable (tabular) data. There are two major uses in astronomy: ground-support systems and data catalogues. Both Europe and the U.S. have formed working groups to develop efficient remote-access, highly-portable data and software, and powerful user interfaces. An important question is whether to use commercial rather than special-purpose database/file systems. The commercial systems have a very powerful background, standard query language, run on general hardware, but are expensive (from $2000 to over $100 000). Special-purpose systems are expensive to develop but software is in the public domain. A major issue is that major commercial systems currently do not provide good capabilities for multi-dimensional searches in, for example, the HST guide-star catalogue. More powerful general purpose, commercial systems are, however, under active development.

3. STARCAT. F. Ochsenbein described the STARCAT (Space Telescope ARchive and CATalog) online system, implemented on a VAX computer and an IDM 500 database machine, giving users access to a wide variety of astronomical catalogues and including query capabilities using OMNIBASE, Although the system was designed primarily for the archiving and retrieval of HST data, its interfaces to MIDAS (Munich Image Data Analysis System) and SIMBAD and to complete astronomical catalogues make it a valuable tool for data analysis. The system is available at ESO, the Canadian Space Astronomy Data Centre (Victoria), The Space Telescope Science Institute, and the U.S. National Space Science Data Center. Future developments include an enhanced and modified query system, graphics capability, an IRAS interface, and the implementation of additional astronomical catalogues.

4. Catalogue precision. G. Lyngå draw attention to the problems that are caused when catalogues do not include statements of the precision of the data, thus leading non-experts to assume precisions far beyond what the catalogue can provide; he proposed that a resolution urging the inclusion of estimates of precision in all catalogues. G. A. Wilkins pointed out that this subject is addressed in the CODATA guide for the presentation of data; journal editors are, however, reluctant to exercise detailed control. W. H. Warren commented that in principle the data centres could exercise control by refusing to accept catalogues without proper documentation. The problem must be solved and some action by the Working Group is needed in the next triennium.

5. Where are the data? W. H. Warren described the problems associated with locating particular astronomical catalogues and data when they are needed. The development of an astrophysics master directory (MD) is in progress in connection with NASA's Era of the Great Observatories. The MD will be an online system through which a user can discover or verify the existence of desired data and find out where the data are located. For catalogue data, the Astronomical Data Center at NASA has developed an online system that gives access to the large list

of catalogues available for distribution. Catalogues can be located by category (type of data), short titles, and keywords that have been assigned to each catalogue. Brief descriptions of the catalogues are currently being entered into the system.

6. Archiving of astronomical data. Carlos Jaschek sent a review on this subject which was summarized by Gart Westerhout in his absence. He pointed out that we have recently re-discovered the importance of archiving: modern methods can improve the interpretation of old observations, and many phenomena turn out to be variable in time. Transactions of a meeting in Montpellier, France, were published in CDS Bulletin No. 31 (1986), and they show that the importance of archiving observations has not yet become apparent to many colleagues. In CDS Bulletin No 35 (1988) an analysis will be published of a questionnaire sent to all observatories; 139 replies were received from 33 countries. A short summary follows: 73% report that an observing file exists, but only 47% of these (i.e. 35% of all archives) are computer readable. Only 45% of institutions have a well-defined archiving policy. This is surprising, because most of the other things concerning observations are ruled by well-defined policies. Only 67% of all observations are fully documented! And finally, 67% would like to see the publication of a set of guidelines. Jaschek concluded that guidelines should be drawn up by a small committee, mostly in the form of broad recommendations (such as "start from now rather than from 1910" and "make observing files computer readable") which should be widely distributed and might, hopefully, lead to an IAU resolution at the next General Assembly.

Session 4. 5 August 1988 09.00-10.30

DESIGNATIONS Chairman: F. Spite Secretary: K. S. de Boer

1. Introduction. C. Jaschek, the Chairman of the Working Group on Designations, was unable to travel to Baltimore, but his introductory remarks were read by the Chairman of the session, F. Spite. He drew attention to the importance of preparing clear recommendations to authors for publication in the IAU Style Manual and in journals.

2. Membership. It was agreed that the members of the WG for 1988-91 should be:

K. S. de Boer, H. Bond, H. Dickel, O. B. Dluzhnevskaya, P. Dubois, P. Hodge, M.C. Lortet, J. M. Read, F. Ochsenbein, F. Spite (Chairman) and J. Young.

3. General problems. P. Dubois spoke about the problem of choosing appropriate designations for use in the SIMBAD database, which allows retrieval of bibliographic and other information for specific objects. Many examples of inadequate notations have been found, and it was sometimes not possible to use the notations given in the First Directory of Nomenclature. C. R. Benn drew attention to the fact that 10% of the objects observed with the Isaac Newton Telescope in 1984-87 could not be identified from the designations entered in the log by the observer, thus making impossible the retrieval of the data from the archive. M.-C. Lortet spoke about erroneous and careless practices in the literature, and she reiterated the need for precision in observing logs, especially for stars in small groups.

4. Galaxies. K. S. de Boer reported that Commission 28 had revived its WG on Nomenclature. New systems are needed for designating faint galaxies and objects within galaxies.

5. Planetary nebulae. H. Dickel reported on proposals by a WG of Commission 34 for a new system for the designation of planetary nebulae. This led to a general discussion on the use of special symbols to show that the type of a new object is not yet certain or that the type indicated by an existing designation is doubtful or definitely wrong. The general view was that such symbols should not be part of the designation.

6. Memorandum on designations. The Chairman introduced a revised memorandum on designations that had been prepared for discussion with the intention that editors of journals would be asked to publish it regularly. Amongst others, H. Abt (editor of the Astrophysical Journal) commented that he found the document long and confusing, and he would expect authors and copyeditors to ignore it. It was eventually agreed that a small group should be formed to redraft the memorandum. [The new version was circulated for comment by de Boer and the substance is to be given in the new Style Manual.]

7. Other points. It was agreed that the 'clearing house' for enquiries about designations should be continued even though only a few enquiries had been received. The chairman emphasised the need for the use of typefaces that would clearly differentiate between the numeral zero and the letter O and between the numeral one and the lower-case letter L. The managers of databases should refuse to include catalogues and lists that do not conform with the basic recommendations on designations. H. Dickel had provided many short contributions about designations for IAU Today (the daily newspaper of the General Assembly); the use of SIMBAD had been demonstrated in the exhibition area; these actions have drawn attention to both the problems and the possible solutions.

Session 5 5 August 1988 11.00-12.30

CLASSIFICATION Chairman: B. Hauck Secretary, W. H. Warren, Jr

1. Introduction. The Chairman, P. Lantos, of the Working Group on Classification and Information Retrieval, was unable to attend the Assembly and so G. A. Wilkins introduced the three main topics of the agenda. There was then a general discussion about the problems of the use of keywords and classification codes.

2. Revision of UDC 52. Wilkins drew attention to the desirability of making another revision of the astronomical schedule (52) in the Universal Decimal Classification (UDC) in order to take account of the major developments in astronomy since the last major revision in 1975. The UDC schedules are widely used throughout the world (although not in North America where the Library of Congress scheme is dominant) in libraries and for information retrieval; it can be used to provide a very precise, language-independent description of the subject matter of a paper or book. At present he is the link between the WG and the International Federation for Information and Documentation (FID), which has the overall responsibility for UDC. He asked for volunteers by astronomers who would be prepared to examine and amend the schedule for their areas of expertise.

3. Vocabulary. J.-C. Pecker introduced, on behalf of Lantos, the latest version of the 'IAU Vocabulary' of keywords, which had been distributed before and during the meeting. The list contains about 1600 terms and is divided into seven chapters that cover broad areas. He suggested that the list be included in the IAU Style Manual so that it would readily be available for the indexing of astronomical papers. H. Abt considered that authors should be asked to suggest subject-headings that would be appropriate for use in the index of the journal concerned. After a lengthy discussion, it was agreed that the list should be published, but it first needed further detailed study and revision by the WG. Hauck suggested that it could be published in, for example, the information bulletin of the Strasbourg Data Centre.

4. Thesaurus. R. M. Shobbrook reviewed briefly the project to develop an IAU Thesaurus, which she had discussed in greater detail during Joint Discussion 1. The Thesaurus would supersede the Vocabulary since it would give more guidance to the indexer about the use of preferred terms and to the searcher about other related terms. It would also indicate the corresponding classification codes in other systems. The preliminary listing of the Thesaurus had been largely developed by librarians and she now needed the assistance of astronomers in checking the listings for corrections and completeness in their areas of expertise, She also saw the need for financial support for the purchase of

appropriate software for developing the cross-references and maintaining the thesaurus. W. Lück offered to make available the facilities of the Fachinformationszentrum (FIZ) at Karlsruhe for this purpose.

Session 6. 5 August 1988 14.00-15.30

EDITORIAL POLICY Chairman: G. A. Wilkins Secretary: B. Corbin

1. Introduction. In introducing the session the Chairman pointed out that the Commission did not have a working group on editorial policy, but he recommended that a small group be formed under the chairmanship of P. A. Wayman, a former General Secretary of the Union. Its major task during the coming triennium would be to review the new IAU Style Manual, but it should also consider other aspects of IAU publications.

2. Style Manual. Wilkins reviewed the stages in the development of the third draft of the IAU Style Manual, which would replace the 1971 edition of the IAU Style Book. It had unfortunately not been possible to arrange a meeting with the editors of the principal journals until May 1988; he had been disappointed to find that the editors did not consider it desirable that astronomers should conform to internationally agreed recommendations in respect of such matters as the use of SI units and the abbreviations of the titles of journals in lists of references. He was pleased, however, that the IAU Executive Committee had decided to propose a positive resolution on these points. (See paragraph 7 of the report on Session 1.) It is intended to publish the Manual in volume 20B of the Transactions of the IAU, even though it has not been possible to obtain agreement on some points.

H. Abt, the editor of the Astrophysical Journal (Ap.J.), commented that there had been a wide measure of agreement on other points at the meeting of editors. He considered that editors could not force astronomers to use SI units and he claimed that the use of the standard abbreviations for the titles of the principal astronomical journals, instead of the very short abbreviations now used in Ap.J., would lead to significant increases in the cost and price of the journal. A lively discussion ensued on these and other topics. It appeared to be generally felt that if the very short abbreviations are to be used, they should be restricted to the very commonly cited journals and their meanings should be published in the journal for the benefit of persons who are not familiar with them.

It was agreed that the Style Manual and other such reference documents should be published as separate reprints. A brief summary of the principal recommendations of the Style Manual should also be made widely available.

3. Code of Practice. Wilkins drew attention to the statement on 'Guidelines for Publication' which had been posted on the notice-board of the Commission. This had been prepared by the American Geophysical Union, but appeared to be equally applicable to astronomers as to geophysicists. There was general agreement to the proposal that this statement be included in the Style Manual, and some considered that such a statement should be adopted by the IAU.

Session 7. 5 August 1988 16.00-17.30

ELECTRONIC MAIL Chairman: B. Hauck Secretary: A. Fiala

The Chairman explained that the meeting had been called to allow an opportunity for the further exchange of information and ideas about the use of computer networks for the transmission of messages (as in electronic mail) and the publication of information (as in electronic bulletin boards), These topics had been presented and discussed during session 3 of Joint Discussion 1 on 3 August. C. R. Benn gave a brief review and there was then a general discussion amongst the small number present. It was agreed that it would be useful for Commission 5 to act as forum for further consideration of the ways in which such techniques could be better used by astronomers; for example, to encourage the use of a simple system for usernames for electronic mail.

Session 8 | 10 August 1988 | 11.00-12.30

BUSINESS MEETING (2) Chairman: G. A. Wilkins Secretary: C. R. Benn

9. Review of activities. The President reviewed briefly the actions being taken as a result of the discussions during the technical sessions. He drew attention to the revised text that he had suggested for the resolution by the Executive Committee on the improvement of publications; none of those present raised any objection to the revised wording. The joint meeting with Commission 46 on the problems of developing countries had been well attended and he hoped it would prove to have been very productive (see the following report).

10. Working Groups. The following list of Working Groups and Chairmen for 1988-91 was endorsed.

W.G.	Astronomical data	G. Westerhout
W.G.	Information retrieval	L. D. Schmadel
T.G.	UDC 52	G. A. Wilkins
T.G.	Thesaurus	R. M. Shobbrook
W.G.	Designations	F. Spite
W.G.	Editorial policy	P. A. Wayman
W.G.	FITS Standards	P. Grosbol
W.G.	Computer software	C. R. Benn
J.W.G.	Astronomical libraries	W. H. Warren, Jr

The Task Groups on UDC 52 and the Thesaurus are to be regarded as activities of the Working Group on Information Retrieval. The WG on Astronomical libraries is to be set up jointly with the Special Libraries Association. The membership and terms of reference of each Group are to be determined by the Chairman in consultation with the President of the Commission.

11. Organising Committee. It was agreed that the following persons would be members of the Organising Committee for 1988-91:

O. B. Dluzhnevskaya, C. O. R. Jaschek, J. M. Mead, L. D. Schmadel, F. Spite, W. H. Warren, Jr, P. A. Wayman and G. Westerhout.

12. Membership. It was agreed that all the 25 members of the Union, including 7 new members, who had indicated their wish to join the Commission, should be accepted. The President stated that the General Secretary had indicated that it would be appropriate for the Commission to have not more than about 10 persons as consultants, and so it was necessary to make a rather arbitrary selection from amongst those who were likely to contribute to the work of the Commission during the coming triennium. It was agreed that the President should write to the Presidents of other Commissions to enquire if they wished to nominate anyone to liaise with Commission 5 on matters of common concern.

13. Other business. There being no other business the President closed the session by expressing his thanks to all who contributed to the work of the Commission during the past triennium, and especially to those who had contributed to Colloquium 110 and to the meetings during the General Assembly.

COMMISSION 5: DOCUMENTATION AND ASTRONOMICAL DATA
COMMISSION 46: THE TEACHING OF ASTRONOMY

Report of Joint Commission Meeting on 9 August 1988 1400-1700

PROBLEMS OF DEVELOPING COUNTRIES

Chairman: G. A. Wilkins
Secretary: C. Iwaniszewska

1. Introduction. The President of Commission 5 stated that the aim of the meeting was to review the problems of developing countries in respect of documentation, data retrieval and teaching and to consider what action should be taken to alleviate these problems. Three papers were presented and there was an extended discussion that related mainly to the acquisition of publications for teaching and research. It soon became clear that the discussions at the preceding IAU Colloquia No. 105 on the Teaching of Astronomy and No. 110 on Library and Information Services in Astronomy had drawn attention to similar problems and had led to similar proposals, which are outlined below. The meeting was attended by at least 55 persons from 32 countries. A. H. Batten, the Chairman of the new IAU Working Group on the Promotion of Astronomy, sent apologies for his absence as he was required to attend a meeting of the IAU Executive Committee; he considered that the new Group must operate in close collaboration with Commissions 5, 46 and 38 (Exchange of Astronomers).

2. Educational material for astronomy in developing countries. J. R. Percy (Canada) summarised the discussions at Colloquium 105 under six headings:

(a) Acquiring current material from abroad. Current monographs and journals are essential for graduate teaching and research and for the professional development of the instructors. The major problems are due to the shortage of funds and currency restrictions. Commission 46 produces a triennial listing of educational material in all major languages to help institutions to select the most appropriate material. Donations by authors, publishers and astronomical institutions should be encouraged, but appropriate arrangements are required.

(b) Acquiring less-current material from abroad. Several organisations are now engaged in collecting and distributing surplus books and journals, but better coordination is required to avoid waste of effort.

(c) Producing translations of important material. Translations are useful but they require good translators (who are poorly rewarded), as well as the cooperation of authors and publishers. School curriculums differ from country to country and it could be helpful to develop a standard curriculum and appropriate resource material.

(d) Reprinting material locally. Low-cost editions of books and journals could be produced locally and the IAU could help in obtaining the cooperation of authors and publishers.

(e) Reprinting with local modifications. The new desk-top publishing methods could allow 'reprints' to reflect the local cultural and scientific environment.

(f) Local authorship and publication. The ultimate aim might be to produce all educational material locally, but potential authors are few and usually overloaded, while the publishers are reluctant to produce books for a limited market. The IAU might be able to help authors to find free photographs and diagrams and to underwrite some publications.

J.-C. Pecker (France) drew attention to the possibility of getting support from the educational as well as the scientific division of UNESCO.

3. Resource sharing between libraries. A. Ratnakar (India) drew attention to the importance of information about the availability of publications and to some of the suggestions put forward at Colloquium 110. (a) Preprints are now very valuable, but are not sent to all observatories; there should be an international centre for preprints. (b) Union lists of journals are available for some western countries; it would be useful to have an international list to help librarians to locate nearby copies of journals. (c) There should be a centre that collects and distributes disposal lists of surplus books and journals; this might be the Third

World Academy in Trieste. (d) In each country there should be one person or library that acts as a contact point for information about its resources. (e) Arrangements should be made to allow astronomers in countries without computer networks to obtain information by telex or post from the principal databases. (f) More journals should give information about papers that will be published.

Some of the comments on these points were as follows. (a) The Space Telescope Science Institute has offered to extend the distribution of the list of preprints that it receives; astronomers should be encouraged to send a copy of each preprint to STScI. Others wondered whether the costs of compilation and distribution could be justified. (c) Publications written in the same language group as that of the developing country are particularly valuable. (d) Consideration should be given to the possibility of using compact discs (CD-ROM) for personal computers to make databases more readily available. (e) "Current Contents" was considered to be more useful than preprint lists, but some observatories cannot afford to buy it.

4. International Space Year. L. Gougenheim (France) drew attention to the proposals for the International Space Year (ISY), which is to be held in 1992, and especially to the activities in education and public information in countries which do not normally engage in space activities. She considered that the opportunity should be taken to produce for public education: carefully selected sets of slides; appropriate written materials, including sets for students; planetarium programmes; and an ISY guidebook. These materials should have a worldwide copyright-free distribution. Scholarships to the International Space University for students from the Third World are proposed. The IAU should endeavour to ensure that full support is given to these activities by the Space Agencies. (The IAU General Assembly adopted a resolution recommending that IAU organisations and individual members should participate actively in ISY activities.)

5. Surveys. The Chairman drew attention to the questionnaire about astronomical libraries that had been distributed with Commission 5 Newsletter No. 3, and asked for further responses. S. Torres-Piembert (Mexico) considered that it would be useful to have a separate survey for library and teaching resources in developing countries. A form was hurriedly prepared during the interval and responses from delegates from eight countries were received; they revealed the small number of astronomical institutions in each country and that some do not receive even the principal astronomical journals and some do not have adequate access to personal computers and telex.

6. Costs of publication and page charges. It was suggested that publishers should be encouraged to allow libraries in the developing countries to buy journals at the reduced rate for individuals; it was pointed out that in some cases individuals could not obtain copies because there was no library subscription. Some agents (in India) refuse to allow discounts to IAU members. Attention was drawn to the problems posed by page charges and it was stressed that any request to waive charges should be made at the time of submission. A representative of UNESCO said that it is possible for individuals to buy (with local currency) coupons that can be used to purchase books or equipment in hard currency; applications must be made through the National Committee.

7. Links between institutes. There was general agreement that arrangements should be made to develop links at a personal level between institutes in developed and developing countries. An interchange of information about activities and resources could lead to the provision of assistance in a variety of forms, including information and advice as well as of publications and other materials. Such a system would help to avoid the situation where surplus publications are supplied but are then found to be unsuitable. It is important to ensure that free material is not subject to heavy postage and customs charges; they may often by avoided by sending the material via the embassy.

8. Teaching of astronomy. Finally, E. Kennedy (Canada) suggested that the history of astronomy, and especially its applications to surveying and timekeeping, is a very appropriate topic for education in developing countries, and M. L. Aguilar (Peru) stressed the value of the visiting lecturers programme.

COMMISSION No. 6

ASTRONOMICAL TELEGRAMS (TELEGRAMMES ASTRONOMIQUES)

Report of Meeting, 5 August 1988

President: A. Mrkos Secretaries: E. Roemer, B. G. Marsden

President Mrkos called the meeting to order and requested members to stand in memory of Carlos Ulrrico Cesco and Karl-August Thernöe, each of whom had died during 1987. Though long retired from the Commission, the latter had assisted in the work of the Central Bureau for Astronomical Telegrams since the 1940s, serving as Director from 1960 until the Bureau moved from Copenhagen to Cambridge at the end of 1964. Mrkos also remarked on the Commission's cosponsorship of IAU Colloquium No. 98 on the role of amateurs in astronomy.

Reporting on the recent activities of the Bureau, Director Marsden remarked that the past triennium had seen substantial changes. Although the Computer Service, which allows users to log in to computers at the Smithsonian Astrophysical Observatory and read the IAU *Circulars* as soon as they are issued, has been operating since early 1984, the appearance of SN 1987A caused the number of these subscribers suddenly to triple—to more than 100. Fortunately, SAO's system of VAX computers was just at the time in the process of being incorporated into SPAN, the Space Physics Analysis Network, and the necessary arrangements could be quickly completed. Over the course of only a few days in February 1987 the IAU *Circulars* were changed from what was essentially a printed publication with an electronic version to an electronic publication with a printed version. From then on, *Circulars* were entered into the Computer Service whenever needed, at all hours of the day and on any day of the week, while the printed copies were issued and mailed in batches once or sometimes twice each week. NASA had very generously made it possible for legitimate subscribers to use free or very inexpensive local computer networks to access the NSSDC computer, whence they were switched to SAO. Finally, in July 1988, the Bureau began routinely sending the *Circulars* to the 50-percent or so of the Computer Service subscribers who could be reached by SPAN or BITNET. While *Circulars* can sometimes be received within minutes, these networks do sometimes introduce extensive delays, but users still have the option of logging in themselves to see if more recent *Circulars* are available.

Marsden emphasized that the printed *Circulars* had not been rendered completely obsolete. Unfortunately, electronic communication was currently restricted to the nations of the first world. In the case of disagreement, the printed version should still be regarded as the "official" one, and the computer dissemination does not allow the use of subscripts and Greek letters, for example. Beginning in April 1988 the appearance of the printed *Circulars* was greatly improved with the use of TEX: to some extent they now resemble the old typeset Copenhagen *Circulars*.

With the electronic *Circulars*, the traditional coded telegraphic distribution of information had now become rather obsolete, at least in the first world, and Marsden was trying to discourage subscribers from using it. He recognized, however, that telexes and cablegrams—though expensive by modern standards—were still very necessary for communication to many other countries. The Bureau was in the process of eliminating its old-fashioned punched-paper-tape telex machines and would in the very near future be transferring telex activity to computers and Western Union's EASYLINK system.

In answer to a question from R. M. West, Marsden noted that there were still about 70 or 80 telegram (including mailgram) subscribers in North America, and perhaps 100 more in other parts of the world. A. Ratnakar, an information scientist from India, urged the continued use of telegrams, noting that the printed *Circulars* with information on the millisecond pulsar reached his country too late to be of much use. Since the telegrams have heretofore been restricted to information on transient objects, use of them for objects such as pulsars would be a considerable break with tradition and potentially extremely expensive. There was some discussion of the possibility of telefaxing selected *Circulars*, but both Marsden and West commented adversely on this very limited technology. It seemed more promising to encourage the use of electronic mail in the second and third worlds, and in this connection Marsden was asked to prepare a note for the daily *IAU Today* newspaper in the hope of finding appropriate routes and contacts. While on the general subject of telegrams, Marsden mentioned a letter from a U.S. amateur astronomer proposing changes in the five categories currently used. Traditionally, the great majority of the telegram subscribers had received just single messages announcing the discoveries of bright comets and novae. Follow-up information was sent to more specialized categories that also covered discovery announcements of faint objects. It seemed reasonable now to modify the categories so that those interested in ephemerides of bright comets would not also have to pay for messages about comets that were too faint for them to observe.

In a discussion on the Central Bureau's finances, Marsden noted that the introduction in late 1985 of a MicroVAX for the Bureau's use (also in conjunction with the Minor Planet Center), as well as NASA's generosity in response to SN 1987A by providing for a backup MicroVAX, had considerably reduced the direct charges made to SAO for usage of its central computers. The strictly calendrical accounting system, introduced in 1986, for subscriptions to the *Circulars* had also been basically beneficial, in spite of the tremendous increase in the rate of issue of the *Circulars*—more than 200 per year since SN 1987A. The salaries of assistant directors Green and Bardwell must be paid entirely from subscriptions to the IAU *Circulars* and *Minor Planet Circulars*. Although brought up as a sore point at the 1985 meeting, the 20-percent overhead charged by the Smithsonian Institution on all the expenditures of the Central Bureau and Minor Planet Center was now considered reasonable. The IAU subventions represent a small part of the total budget, but they are necessary if the service is to be maintained to important observatories in parts of the world from which it is often difficult to receive timely payments. Roemer wondered about the possible financial impact of secondary electronic dissemination of the IAU *Circulars*. Judging by the continuing steady number of primary subscribers, it did not seem that this was much of a problem, although it could clearly become one in the future. It was obviously impossible, and rather unreasonable, to clamp down on secondary distribution entirely, and a better approach was simply to ask secondary distributors to encourage their customers to take out direct subscriptions.

Mention was made of the fiasco in November 1987 following the unfortunate and widespread announcement of the unconfirmed discovery of a supposed supernova in M31. That no other erroneous information disseminated by the Central Bureau in more than 65 years of operation was accorded such a response was testament to the Bureau's generally remarkable reliability. Y. Kozai noted that the Executive Committee had received a complaint about the Bureau's publication of "predictions in retrospect". Marsden was aware of the particular action that led to this complaint; it had come at the height of the activity on SN 1987A, and he felt that it was an isolated occurrence.

The new officers of the Commission were confirmed as Roemer, President, and J. Grindlay, Vice President; and West joined Marsden as a member of the Organizing Committee. In addition to West, the following were confirmed as new members of the Commission: K. Aksnes, A. V. Filippenko, A. C. Gilmore, S. Isobe and A. S. Sharov, the last-named replacing D. Ya. Martynov as representative for the Sternberg Astronomical Institute.

COMMISSION No. 7

CELESTIAL MECHANICS (MECANIQUE CELESTE)

Reports of meetings on August 4 and 8

PRESIDENT: V.Brumberg
SECRETARY: S. Ferraz-Mello

BUSINESS MEETING

A business session was held on August 4. A second brief meeting was held on August 8. Items addressed were:

1. Election of Organizing Committee The commission elected the following officers and members of the Organizing Committee for the term 1988 to 1991:

President:	J.Henrard
Vice President:	A.Deprit
Members:	K.B.Bhatnagar
	V.A.Brumberg (Past President)
	J.Chapront
	S.Ferraz-Mello
	Cl.Froeschlé
	J.D.Hadjidemetriou
	K.Kholshevnikov
	H.Kinoshita
	He Miaofu
	A.Milani
	A.E.Roy
	P.K.Seidelmann

2. Election of New Members of the Commission

The National Committees have requested that the following new members of IAU become members of Commission 7 and they were approved after a short presentation: A.J.Abad (Spain), E.L.Akim (USSR), L.E.Doggett (USA), G.Dourneau (France), L.Duriez (France), R.S.Gomes (Brazil), Cheng Huang (China), P.Kammeyer (USA), J.Klokocnik (Czechoslovakia), J.Laskar (France), J.J.Lissauer (USA), T.Pauwels (Belgium), S.Segan (Yugoslavia), V.I.Skripnichenko (USSR), M.L.Sein-Echaluce (Spain), J.R.Taborda (Portugal), A.L.Whipple (USA), T.Yokoyama (Brazil) and B.Zafiropoulos (Greece).

3. Consultants

The Commission elected the following consultants for the term 1988-1991: V.I.Arnold (USSR), K.Meyer (USA), J.Moser (Switzerland), D.Saari (USA), S.K.Shrivastava (India), C. Simo (Spain), M. Soffel (German F.R.), J.Waldvogel (Switzerland) and J.Wisdom (USA).

4. Deceased Members

The commission payed tribute to the memory of Professors G.N.Duboshin, P.E.Elyasberg, O.C.Mohler and A.A. Orlov, members of the commission deceased since the past General Assembly.

5. Recommendations of Working Groups.
Dr. J.Kovalevsky presented to Commission 7 the motion of the Working Group on the theory of Nutation to correct the 1980 nutation series. He also presented the package of motions of the Working Group on Reference Frame proposing the adoption of a celestial reference frame made of a consistent set of coordinates of a sufficient number of extragalactic objects.

6. Recommendations of the joint meeting of Commissions 4, 7, 8, 19, 24, 31 and 40.
Dr. J.Kovalevsky presented the recommendation issued of the joint meeting For Milliarcsecond or Better Accuracy of fusion of the Working Groups on Nutation, Reference Frame and Astronomical Constants in just one working group, the Working Group on Reference Systems, whose work would be divided in 4 study groups: Nutation, Astronomical Constants Origin and Time.

7. The journal CELESTIAL MECHANICS
The commission has been advised that the Celestial Mechanics Institute has elected Professor Jacques Henrard as the next Executive Editor of the journal. Prof. Henrard will be in charge after January 1st., 1989, when other editorial changes will be announced. The commission expressed its gratitude to Professor Morris Davis for the excellent work done during his term as executive editor.

SCIENTIFIC SESSIONS

Six invited review papers were presented in the sessions held on August 8:

A. The Few Body Problem (M.Valtonen)
B. Stability of the Solar System (A.Milani)
C. Chaotic Behaviour (J.Wisdom)
D. Evolutionary Problems (S.J.Peale)
E. Ring Dynamics (N.Borderies)
F. Secular Resonances in the asteroid belt (Ch.Froeschlé).

The following abstracts were provided by the lecturers. Full length papers will be published in *Celestial Mechanics*.

STABILITY OF THE SOLAR SYSTEM (A.Milani)

A significant improvement in our understanding of the problem of the stability of the solar system resulted from the notion of regular conditionnally periodic orbits inextricably mixed with chaotic and escape orbits. A realistic perspective of solution of the problem must therefore rely on the estimate of the difference between a KAM-type orbit and the real motion over a finite time span. The achievement of such goal meets huge difficulties in all the computational procedures. In the development of analytical theories it is not any more possible to ignore the small divisor problem and the divergence of the perturbative series as a practical - not only theoretical - computational difficulty, while KAM theories do not provide a constructive algorithm. Numerical integration cannot yet directly answer questions over the stability of a realistically modelled solar system over a timescale comparable to its real age; extrapolation beyond the computed interval are not possible within a rigorous framework. Nevertheless, long term integrations by numerical and seminumerical methods of the outer planets over $10^8 - 10^9$ years (LONGSTOP, digital orrery) and of the inner planets over $10^6 - 10^7$ years (Laskar, Richardson) have greatly extended the timespan over which the dynamical behaviour of the planets is understood. The interpretation of these results is not easy because indications of a non regular behaviour have been detected, e.g., in the spectrum and in the proper modes of both, the LONGSTOP and the Laskar integrations, but the extent of the macroscopic changes which might be associated with the non regular behaviour is not known.

SECULAR RESONANCES IN THE ASTEROID BELT (Ch.Froeschlé)

According to analytic theories a secular resonance occurs if the orbital precessional rate of an asteroid is equal to one of the eigenfrequencies of the system of planetary orbits in the frame of secular perturbations theory. The old linear theory, valid only for small eccentricities and inclinations, was improved in 1969 by Williams whose theory is valid for high eccentricities and inclinations and also yields the location of secular resonances in the proper elements space. New theories by Nakai and Kinoshita (1985) and by Yoshikawa (1987), for the resonance ν_{16} and ν_6, as well as recent numerical experiments give quantitative results on the variations in eccentricity and inclination. They also allow us to investigate the topological structure of secular resonance and the possible chaotic character of secular resonant motions.

RING DYNAMICS (N.Borderies)

1. *Observational discoveries by Voyager concerning the rings of Uranus*: Voyager 2 discovered a tenth narrow ring between the rings ϵ and δ, a broad faint inner ring located approximatively 1500 km inside the innermost narrow 6 ring, a continuous distribution of small particles throughout the Uranus ring system and 10 new satellites orbiting near the rings. 2. *Resonances in the Uranian rings*: 1986 U7 and 1986 U8 are the shepherd satellites of the ϵ ring; 1986 U7 is the outer shepherd of the δ ring, and is a shepherd of the tenth ring discovered by Voyager, 1986 U1R; 1986 U8 is the outer shepherd of the γ ring (Porco amd Goldreich, *Astron.J.*93, 724, 1987 and French *et al.*, *Icarus*73, 349, 1988). 3. *Shapes of the Uranian rings*: Most of the rings of Uranus are best fitted by inclined elliptical ring models corresponding to a normal mode m=1. However, the γ ring contains, in addition to the mode $m = 1$, the mode $m = 0$, which means that it undergoes radial oscillations with a frequency equal to the epicyclic frequency, and the delta ring is best fitted by the mode $m = 2$, which means that it has the shape of an ellipse centered at the planet (French *et al.*, *Icarus*73, 349, 1988). 4. *Problems with the self-gravity model in the Uranus rings*: The self-gravity model allows us to explain how elliptical narrow rings can maintain their shape in terms of a balance between effects due to the oblateness of the planet and effects due to the self-gravity (Goldreich and Tremaine, *Astron.J.*84, 1638, 1979 and Borderies, Goldreich and Tremaine, *Astron.J.*88, 1560, 1983). This model encounters serious difficulties with the rings α and β (Marouf, Gresh and Tyler, *Bull.AAS*19, 883, 1987). 5. *Arcs of rings around Neptune*: A natural explanation for incomplete rings around Neptune, as revealed by recent occultation observations (Hubbard *et al.*, *Nature*319, 636, 1986), is that they consist of arcs of particles librating around a corotation resonance. Two models were presented (Lissauer, *Nature*318, 544, 1985 and Goldreich, Tremaine and Borderies, *Astron.J.*92, 490, 1986). 6. *Possibility of polar rings around Neptune*: Because of the presence of Triton, a satellite of Neptune with an unusual orbit in that sense that it is very inclined although close to the planet, rings with high inclinations are possible around Neptune (Dobrovolskis, *Icarus* 43, 222 1980 and Borderies, to appear in *Icarus*). Preliminary results indicate that such rings would be stable (Dobrovolskis, Steiman-Cameron and Borderies, research in progress). 7. *Formation of sharp edges by shepherd satellites*: A recent model of the shape of perturbed streamlines near an edge maintained by a close satellite faithfully reproduces the sharp edges which bound the Encke division. It is found that this striking feature is related to the local reversal of the viscous transport of angular momentum in the most strongly perturbed regions (Borderies, Goldreich and Tremaine, submitted to *Icarus*).

COMMISSION No. 8

POSITIONAL ASTRONOMY (ASTRONOMIE DE POSITION)

PRESIDENT : Y. Requième VICE-PRESIDENT : M. Miyamoto

ORGANIZING COMMITTEE : P. Benevides-Soares, D.P. Duma, L. Helmer, J. Hughes, L. Lindegren, Luo Ding-Jiang, F. Noël, G.I. Pinigin, L. Quijano, S. Sadzakov, H. Schwan, C.A. Smith, M. Yoshizawa.

Prior to Commission 8 meetings the Joint Meeting "For milliarcsecond or better accuracy" was held on August 3 including Commissions 4, 7, 8, 19, 24, 31 and 40. (Chairman : K. Seidelmann). The proceedings will be published elsewhere.

Business meeting, 4 August 1988
Chairman : Y. Requième

The President welcomed members in attendance and asked for a moment of silence in memory of four Commission members : W. Fricke, J. Larink, W.H. Robertson, G. Teleki who passed away since the last General Assembly.

The Commission unanimously approved the election of the new President, M. Miyamoto, and Vice-President, L.V. Morrison. The results of the election of the new Organizing Committee were also accepted. The members of the O.C. are : P. Benevides, D.P. Duma, L. Helmer, J. Hughes, Hu Ning-Sheng, J. Kovalevsky, L. Lindegren, F. Noël, G. Pinigin, Y. Requième (ex-officio), H. Schwan, C.A. Smith, M. Yoschizawa.

The following were approved as new members of the Commission : M. Lattanzi, J. Muiños, K. Nakajima, I. Pakvor, M. Soma, C. Turon-Lacarrieu, F. Van Leeuwen. In the following days, M. Catalan, M. Lehman, G. Pizzichini, T.A. Spoelstra and G. Westerhout applied for membership and were announced to the General Secretary.

Since the last General Assembly S. Lautsen, W. Gliese and T. Dambara have resigned.

The list of consultants was reviewed and approved : C. Cole, M. Dachich, K.R. Joschi, G. Kaplan, T. Rafferty and K. Steinert.

SRS Committee

The final report of the SRS Committee prepared by J. Hughes, D.D. Polojentsev, C.A. Smith and L. Yagudin was presented by C.A. Smith. The work on the compilation of the Southern Reference Star catalog undertaken jointly by the Pulkovo and the U.S. Naval Observatory is completed. A catalog of 20 488 positions, referred without proper motions to the FK4 system and the equinox and equator of B1950.0 at the mean epoch of observation, has been produced. This catalog has also been transformed to the FK5-J2000 basis and is available with approximate, preliminary proper motions for the convenience of the user.

The final SRS catalog was compiled from a combination of the Pulkovo and Washington preliminary catalogs with equal weights. However regional deviations between these catalogs led to bring the Pulkovo right ascension system closer to the FK4 system by reducing it to the Washington right ascension system. The declination systems of the preliminary SRS catalogs were in excellent systematic agreement.

The average mean error of the right ascensions at the mean epoch of observation is 0.07 arcsec. The corresponding value for the declinations is 0.10 arcsec. Preparation of a full description of the compilation procedures is in progress. It will be included with the publication of the S.R.S. catalog data.

On behalf of the astrometric community the President thanked the members of the S.R.S. Committee for efficient work and declared the activities of this Committee closed.

Star List Working Group

Members are G. Carrasco, T.E. Corbin, L. Helmer, M. Miyamoto, D.D. Polojentsev and Y. Requième. Two new members were proposed : S. Roeser and Luo Ding-Jiang. T.E. Corbin was proposed and accepted to act as chairman of this W.G.

During the last triennum the Pulkovo Observatory proposed the observation of 6 000 bright stars, 2 500 double stars and 1 500 reference stars near radiosources.

The urgent need of a re-observation of the I.R.S. was once more stressed. It was noted that 97 % of the I.R.S. stars have been accepted in the Hipparcos Input Catalog (including the supplement proposed by T.E. Corbin).

L. Helmer and L.V. Morrison have prepared a candidate list for a faint extension of the fundamental reference frame, to be observed by automatic meridian circles (or possibly astrolabes).

Horizontal Meridian Circle Study Group

No member of this group was participating in the General Assembly. So the President reviewed briefly progress reported by the members : axial automatic meridian circles under development at Nikolaev and Kiev (Pinigin et al., Lazarenko et al.), regular observations of the Pulkovo automatic horizontal meridian circle (Naumov, Gumerov and Pinigin), construction of a new automatic horizontal instrument at Pulkovo from an agreement between Pulkovo, Kiev, Nikolaev and Kazan. Furthermore the construction of the Danish-Chinese horizontal glass meridian circle is progressing satisfactorily.

Working Group on Astronomical Refraction

J. Hughes was confirmed as chairman. The important event of this triennum was the International Workshop on Astronomical Refraction held at Belgrade on September 1987 in memoriam to George Teleki where met 75 participants. (Publication of the Belgrade Astronomical Observatory.) The main topics were : refraction corrections in radioastronomy, ray tracing, meteorological information, refraction tables, measurements of refraction and flexure, corrections for ionosphere on VLBI observations, and terrestrial measurements. J. Hughes also stressed development of the two colour methods, studies on chromatic refraction and clarification of the concept of local pure refraction. At last he presented some figures showing the importance of the local refraction around the U.S. Naval Observatory.

Astrolabe Working Group

The report on activities was presented by S. Débarbat, interim chairman, on behalf of G. Billaud.

Most of the existing astrolabes in the world have actively participated in the MERIT campaign for the astronomical determination of the Earth rotation parameters by star observations relative to the local vertical. They were shown to be among the best classical instruments for such determinations.

Due to the changes which occured at the beginning of this year for ERP, a large number of astrolabes have moved their activities to catalog observations (Argentina, Brasil, Chile, China, Spain). The results of these individual catalogs will have to be compared with the Hipparcos Output Catalog. Such a comparison will allow the determination of the systematic errors before a new "Catalogue Général d'Astrolabes" (CGA) is established. This new CGA will be of higher quality ; it will be possible to use it to improve the proper motions and ensure the durability of the Hipparcos reference system.

Furthermore the combination of results from solar and stellar observations will, in the future, permit absolute catalogs to be formed in right ascension.

In order to extend the benefit of the Hipparcos results and, perhaps, in some cases, to improve them, the quality of the astrolabes must be increased as much as possible and this will be done if they are completely automated. Two large photoelectric astrolabes (26 cm) are in the process of final adjustment in China and in France, the CERGA photoelectric astrolabe is now in full automatic operation.

Important projects were stressed :

a) Close cooperation of the different astrolabes in order to obtain the best from their zonal observations (including Sun, planets and radiostar measurements).
b) Organization of one or two chains of measures to connect the northern and southern hemispheres (for exemple China, Latin America, Europe).

After the official report S. Débarbat read a few personal words of G. Billaud : "In reason of my imminent retirement, I decided to resign from chairmanship of the Astrolabe Working Group. But I would not forget to express thanks to all those who, during the last fifteen years, have collaborated with the W.G. For my part I have had the uncommon pleasure of performing a task which I greatly enjoyed with persons whom I highly esteemed".

H. Schwan emphasized that the astrolabe catalogs have been integrated in the FK5 with a very high weight.

In the following day members of the W.G. participated in a meeting initiated by S. Débarbat. F. Chollet (France) was designated as the new chairman and the general organization of the W.G. was discussed.

Discussion of Resolutions

The recommendations included in the report of the W.G. on Reference Frames (given during the Joint Meeting JCM1 on August 3) were proposed as resolutions by J. Hughes and discussed one after the other. The different resolutions were adopted with modifications. A. Murray and Y. Requième met afterwards with K. Seidelmann in order to reconcile the modifications suggested by the involved commissions before final adoption by the Resolution Committee.

A proposal of joint resolution with Commission 12, recommending that long-term ground-based astrometric observations of the Sun by transit instruments and astrolabes should be associated with programs of solar diameter monitoring, was accepted by our Commission but finally rejected by Commission 12.

Y. Requième and C. Turon reported on the Hipparcos Input Catalog which will be published by 1990 with in addition a complementary catalog containing the proposed stars not accepted for the Hipparcos mission for different technical reasons. Concerning the Input Catalog it was stressed that the number of accepted faint stars $11 < V < 13$ is only 1 600, showing the interest of the faint reference frame proposed by L. Helmer for ground-based observations. As to the complementary catalog, about 20 000 stars have an high astrometric interest (IRS, planetary and lunar occultations, minor planet crossing points) and 32 000 are high priority stars of astrophysical interest (variable stars, O-B stars, F stars for galactic structure, red dwarfs, halo stars). After calibration of the instrumental errors of the automatic meridian circles and astrolabes by using the Hipparcos Output Catalog, an accuracy of 0.02"-0.04" is expected for these ground-based instruments which could be used for observation of this complementary catalog. Using old catalogs including Astrographic Catalog proper motions could be derived with an accuracy of 0.003"/year.

Astrolabes, 5 August 1988
Chairman : H. Schwan

P. Benevides reviewed the method of Krejnin for obtaining absolute declinations at two different zenith distances at a given site. In collaboration with L.B.F. Clauzet he applied that method to 3 years of observations made with the Valinhos astrolabe for zenith distances 30° and 45°. Absolute corrections to the declinations of 28 FK4 stars were given with a mean error of 0.30". Uncertainties are due, to a large extent, to remaining magnitude and color effects. A full pupil photoelectric astrolabe could yield in principle 3 to 4 times better accuracy.

F. Chollet described the modified astrolabe of Paris Observatory and its current program. The stability and the possibilities of the instrument have been improved, especially by using a Zerodur prism ; it is possible to observe stars (limiting magnitude 6.5), planets and the Sun in a declination zone up to 120°. The same type of instrument is under development at Santiago, but could be also installed in the long term in Turkey, Algeria and possibly Morocco.

Some last results of the Danjon astrolabe at Santiago were given by F. Noël. Observations of Mars, Jupiter, Saturn and Uranus are in progress. For Uranus an offset between the astrolabe and the ephemerides of roughly -0.3" was found in right ascension. (A quite similar value was obtained at Bordeaux.) Observations of α SCO-A and 9 SGR were also presented. The Santiago astrolabe will be modified : the quartz prism will be replaced by cervit angles in order to stabilize the instrumental zenith distance and to make possible observations at $z = 30°$ and $z = 60°$.

A. Poma reviewed the Cagliari and Merate astrolabe catalogs. Both programs consist of 11 groups with 237 FK4 stars at Cagliari and 198 at Merate. The average error of the residuals (after applying the group corrections) is about 0.05"-0.08". Preliminary results show no significant magnitude and spectral type correction.

S. Débarbat presented a paper of F. Laclare on the astrolabe observations of the Sun. Using different reflecting prisms on a more elaborated astrolabe, observations at 10 different zenith distances are now possible. The mean error is 0.3" to 0.4" according to the zenith distance. In addition the diameter of the Sun is regularly measured (up to 40 determinations per day) ; the mean value obtained by Laclare from 2 700 measurements (1975-1987) is 959.45". Journet obtained with the same instrument from 1 200 measurements 959.03" (1984-1987). Variations of the diameter were clearly detected with periodicity of 10 years, 970 days, 330, 220, 75 and 50 days. A new prototype of automatic solar astrolabe is in service at CERGA, with CCD measurements and Zerodur reflecting prism of variable angle. Astrolabe observers are encouraged to modify theirs instruments to monitor the solar diameter variations.

Galactic reference frame and catalogs, 5 August 1988
Chairman : M. Miyamoto

H. Schwan reported on the "basic FK5" and its extension to new bright and faint fundamental stars. A tape version of the basic FK5, giving mean positions, proper motions, mean epochs and mean errors for the classical 1 535 fundamental stars, has been generated and is distributed on request. The printed version of the catalog is in preparation. Systematic and individual corrections to the FK4 have been derived.

Mean positions and proper motions for the bright FK4 Sup. stars (5.5 to 7.0) were derived at Heidelberg on the basis of about 125 catalogs of observations made since 1900. The precision of the proper motion (0.15"/cy and 0.18"/cy in right ascension and declination) is comparable to the FK4 one. A tape version for this "Bright extension" (980 FK4 Sup. stars in the FK5 system) has been generated and is also available on request.

The extension of the FK5 to fainter magnitudes is being made jointly with the U.S. Naval Observatory. Within his work of determining proper motions for all the IRS stars, T. Corbin is selecting about 2 000 new faint fundamental stars in the magnitude range 6.5 to 9.5. The selection in the AGK3R stars has been completed, the work on the southern part (SRS) is in progress.

T.E. Corbin presented a progress report on the compilation of the SRS proper motion catalog (after the completion of the final SRS position catalog). Since the poor observational histories of these stars made it impossible to compile a system south of -30° from catalogs referred directly to the FK4, the Greenwich transit circle series was used in the north to study the behavior of magnitude equation with declination. No significant variation was found for the traveling wire micrometers whereas this was not true for the fixed wire variety. Next a system of mean positions and motions, with good results north of -30°, was formed by combining data from catalogs observed with screens. The magnitude equation study was then used to bring the Cape observations made after 1915 into this system to extend south of -30° with a total of 3 890 stars. The mean errors of the proper motions in this extended system are ± 0.35"/cy and ± 0.38"/cy in right ascension and declination. Currently this system is being used to reduce the Cordoba and La Plata series, and finally all other southern catalogs that can contribute to the SRS proper motions.

L. Helmer presented a candidate catalog for faint extension to the fundamental reference frame, which could be most useful for photographic work. The faint stars (m_V = 11.5 - 12.0) have been selected at Copenhagen University Observatory with a mean density of one per square degree from three zones of the Astrographic Catalog : 1 488 in the Greenwich zone, 9 180 in the French zone and 1 630 in the Vatican zone. In the Northern remaining zones, selection is in progress. In addition 6 452 faint AGK3 stars have been selected. The positions are brought on the FK5 system and proper motions are derived where possible. The candidate list is proposed for reobserving on automatic meridian circles.

R. Wielen described the work started at Heidelberg (A.R.I.) to collect all available astrometric data on positions and proper motions of stars in a computerized data bank, called ARIGFH. All relevant astrometric observations since Bradley (1750) will be put into machine-readable form and reduced to a common system (FK5, then Hipparcos). From this unprecedented data base the A.R.I. intends to derive positions and proper motions in a way which makes best use of all the observational information obtained over more than 200 years. The final catalog, called ARIGC, should have utmost accuracy for all sufficiently well-observed stars.

M. Yoshizawa presented the second annual catalog Tokyo PMC 86 (3 873 stars observed at least two times) based on the FK5 system. The construction of the catalog was made in the classical night-to-night reduction mode and also in a global adjustment mode. The systematic trends obtained from the observed (O-C)'s for the FK5 stars confirm that a systematic error $\Delta\delta_\delta$ of about 0.1 arcsec still exists in the new basic FK5 (as already found at Bordeaux and La Palma) in the zone $+40° < \delta < +60°$.

P. Benevides described an iterative solution of singular least squares problems. With R. Teixeira he applied the overlapping method to the differential reduction of right ascensions obtained by meridian circles. The algorithm was shown to display linear convergence, even if the original coefficient matrix is rank deficient, when the final limit will be dependent on the starting iterate. The corresponding procedure for any general least-squares problem was outlined.

Common interests of Commissions 8 and 24, 6 August 1988
Chairmen : A. Upgren and Y. Requième

Y. Requième introduced a general discussion regarding the names and objectives of the Commissions 8 and 24. The present name of Commission 8 "Positional Astronomy" appears somewhat vague and could be depreciated by other colleagues insofar as it does not reflect clearly the main scientific objective of our commission which is in fact the construction of a galactic reference system. Taking into account the importance given to the problems of frames during the last general assemblies, it was suggested to change for "Stellar (or galactic) reference frame". It was also agreed that Commission 24 should change its name, but the only proposal was "Astrometry", which appeared too strongly similar to "Positional Astronomy". At this point, A. Murray raised the question of a possible merging of our both commissions. After a long discussion Y. Requième called for an informal vote, and the proposal was rejected by 19 votes to 15 and 7 abstentions (many other members did not express their opinion). Finally a consensus was not found about renaming of the commissions.

Furthermore it was pointed out that Commissions 19 and 31 were very active in the domain of the reference frames and could integrate it in their own objectives. M. Feissel confirmed that the subject had been discussed in their last meeting. As a matter of fact the extragalactic reference frame is become one of the main objectives of the Commission 19 but the whole domain of Reference systems is not well covered by any commission. The idea of a permanent Joint Working Group appears as the best solution. Finally it was recommended that" the Working Group on Reference Systems (WGRS) be continued as an intercommission project and that it concern itself with nutation, astronomical constants, origins, reference frames and time" (see Resolution C2).

Seven scientific papers were presented after that discussion. As they rather related to the Commission 24 objectives, the abstracts are to be found in the report of that commission.

Link between reference frames, 8 August 1988
Chairman : J. Hughes

K. Johnston presented a joint five year program to establish a radio-optical reference frame based upon extragalactic radiosources. The primary catalog includes 400 well distributed objects with no extended radio and optical structure ($z > 0.8$), observations were started in 1987. The radiopositions are obtained by VLBI Mark III S/X measurements. The expected accuracy is 1-3 mas for $\delta > -30°$ and 10 mas for $\delta < -30°$. The optical positions will be obtained from astrographs (Hamburg, USNO) and prime focus plates on 4 m class telescopes. In addition J. Russell detailed the recent radio observations of the N.R.L. aimed at establishing a global radio reference frame.

C. Ma reported on the GSFC catalog, including 182 extragalactic radiosources. Positions were obtained with formal errors less than 1 mas from 237681 pairs of Mark III VLBI group delay and phase delay rate observations divided into 600 data sets from 1979 to 1988. The sources are distributed fairly evenly above -30° with 68 sources in the Southern hemisphere. Analysis under different conditions and external compa-

of La Palma (latitude N 28°.75). Since then 3 annual catalogs containing 35 000 stars and 5 000 observations of solar system objects have been published. The continuing program is divided into 5 parts :

a) Optical reference frame
FK5, IRS (+90 to -45), faint extension (V = 11-12, 1 star per square degree, see report of L. Helmer above), 1 500 Hipparcos stars, PZT zones. About 65 % of the IRS has been completed and observation of the faint extension has just started.

b) Linking optical and radio reference frames
Faint reference stars in extragalactic radio fields (about half complete), radio-stars and Hipparcos link stars in extragalactic fields.

c) Stellar kinematics
Six lists of about 20 000 stars, selected by spectral type and luminosity class for improvement of proper motions.

d) Solar system
Mars, Callisto (+ Europa in future), Titan (+ Iapetus), Uranus, Neptune (+ Triton), (Pluto), 59 minor planets in Hipparcos list.

e) Monitoring variable stars
RS CVn, etc.

G. Kaplan presented the program to be observed by the Mount Wilson optical interferometer (see below J. Hughes et al.).

M. Miyamoto (with M. Soma) talked on a proposal of faint minor planets for improving the fundamental reference system. In a problem to link the stellar reference frame (FK5) with the dynamical one, they have simulated the accuracy in deriving the equinox correction ΔE, the equator correction ΔD, the obliquity correction $\Delta\varepsilon$, and the longitude correction $\Delta\lambda$ from the observations of some minor planets. The orbital calculations of these minor planets are based on the planetary ephemeris DE200.

If they rely upon the meridian observation for the brightest four minor planets, it takes over 15 years to determine the equinox correction ΔE, for example, with the accuracy of ±0.05" level. On the other hand, it is shown that if they can observe the faint three minor planets (Nos. 1620, 2100 and 3103) with the orbital period shorter than 2 years by a combination of the meridian observation with the photographic astrometry, then the accuracy of the determination of ΔE attains ±0.04" level within the observational period of 3 years. Thus, the combination of the meridian observation with the photographic astrometry for faint minor planets with short orbital periods would increase remarkably the observational efficiency.

A paper of D.P. Duma and Yu. N. Ivaschenko was read by Ya. Yatskiv, describing a method for determining the relative orientation between radio and optical reference frames on the basis of non-synchronous observations of artificial satellites in both radio and optical wavelength ranges. For its realization it is sufficient to observe an artificial satellite with respect to both systems during the same period of time. If the instrumental errors of the observations are negligible, it is possible, after having calculated the orbital elements of the satellite, to obtain the three angles P, Q and R of the relative orientation between the radio and optical systems. Positions of a geostationary satellite with a special photo-cassette mounted on a Zeiss double wide-angle astrograph (F = 2 m, d = 40 cm) and time registration (±2.5 ms) have been obtained with a precision of ±0.3". Taking that error on the satellite positions and a number of positions equal to 72, the corresponding errors on P, Q, R are respectively 0.07", 0.06" and 0.10".

Another paper was read on proper motions of double stars determined at Belgrade. S. Sadzakov and M. Dacich have observed with the Askania meridian circle 1 611 double stars not suitable for any photographic or photoelectric observations (small separation). Using the AGK2 as first epoch, proper motions were derived.

risons indicate that the reference frame defined by the radiosource catalog should be reliable at the 1 mas level.

A preliminary comparison of the optical and radio reference frames based upon VLA and optical positions of 40 radiostars (mainly in the Northern hemisphere) was given by De Vegt. Optical positions are photographic and based upon AGK3RN and SRS catalogs which have been transformed to FK5 using the tables as provided by ARI. Most positions agree within 0.1 arcsec, regional deviations of the same amplitude are indicated.

A paper of Luo Ding-Jiang and Peng Yizhi was read reporting the measurements of 24 radiostars by the Beijing photoelectric astrolabe Mark II with a mean precision of 0.05". This observing program will be continued, not only at Beijing but also at Shaanxi, Shangaï and Yunnan.

L.V. Morrison showed that the optical reference systems of the Carlsberg automatic meridian circle and the Bordeaux meridian circle were very similar and were in fact a smoothed version of the FK5. Comparison of 19 CAMC and 17 Bordeaux optical positions of radiostars (obtained in these systems) with VLA and VLBI radio positions showed clearly that 54 CAM has an offset in R.A. of 0.5 arcsec. There is no clear evidence of any other offsets greater than 0.2 arcsec.

Observing programs, 8 August 1988
Chairman : P. Benevides

T.E. Corbin presented the Pole-to-pole observing program involving the USNO's Six-Inch transit circle (Washington) and Seven-Inch transit circle (Black Birch, NZ) commenced in the Northern hemisphere in September 1985 and in the Southern hemisphere in June 1987. The program is being conducted in both fundamental and differential modes. The differential observing involves IRS (AGK3R + SRS stars) each night in zones 15° to 25° wide with FK5 stars extending 5° beyond each zone's boundaries. The fundamental observations involve observing each night : FK5 stars in Kustner lists, radio stars, Solar system objects and a fundamental azimuth during the period from the autumnal equinox to the vernal equinox. The day observations include : the Sun, Mercury, Venus, day stars to m=3.1 with the Six-Inch and to m=5.2 with the Seven-Inch and daytime azimuth stars with the Seven-Inch. Most Seven-Inch day tours have sufficient data to be reduced independently thus giving an unprecedented direct link between the daytime Solar system objects and the stellar frame. Mars is observed day or night with both instruments. Thus far 45 000 observations of IRS have been made with the Six-Inch and 12 000 observations with the Seven-Inch. The two instruments together have also made the following : 64 000 observations of FK5 stars, 6 500 observations of day stars and 4 700 observations of Solar system objects. The mean errors of a single fundamental observation are : ±0.21" in RA and ±0.22" in Dec for the Six-Inch and ±0.20" in RA and ±0.25" in Dec for the Seven-Inch. The higher value for the Seven-Inch declinations reflects the fact that the instrument is at a new site and some atmospheric parameters have not yet been modeled. In addition, a definitive value of the flexure has not been adopted. The differential observations of the IRS show external mean errors of ±0.15" in RA and ±0.16" in Dec for the Six-Inch. The Seven-Inch does not yet have sufficient IRS data to determine this quantity. The Eight Inch transit circle in Flagstaff has for the past five years participated in special differential programs such as the Halley Reference Stars, the Radio Source Reference Stars, Radio Stars and requests for observations of objects of astrophysical interest. This instrument will cease operation on September 1, 1988 and the USNO's participation in the observation of the primary RSRS will be terminated.

L.V. Morrison (with L. Helmer and M. Catalan) reported about the observational program of the Carlsberg automatic meridian circle commenced in May 1984 on the island

Space and ground-based instrumentation, 8 August 1988
Chairman : L.V. Morrison

P. Hemenway reviewed the principles of the interferometric fine guidance sensors of the Hubble Space Telescope. The standard astrometric modes of position, moving target and transfer position were presented. The overall verification plan, from the engineering "orbital verification" (2 months) through "science verification" (calibration of parameters such as field distorsions) and the initial "guaranteed time observations" (5 months) as well as the planned astrometric tests,were described.

A. and M. Meinel reported on the project "Thousand Astronomical Unit mission" (TAU). The JPL study of an astrometric spacecraft mission that would provide a baseline of 1 000 AU (TAU) after a cruise of 50 years is continuing under JPL and NASA support. This study has identified 1) the need for earlier astrometric satellites, and 2) the need for a QSO reference system. The earlier missions would be desirable both to extend the current range of parallaxes and to demonstrate the several orders of magnitude improved precision of new techniques. If TAU were limited to current astrometric precision of 1 mas, a baseline of 1 000 AU would yield precision parallaxes out to 50 to 100 kpc. Improvement to 1.0 arcsec (potentially possible with a space interferometer) would permit parallax measurements from Earth-orbiting satellites to be increased to distances of 100 to 200 kpc. With this same precision TAU could yield parallaxes to distances of 50 to 100 Mpc. The JPL study emphasizes the need for increased funding for development and demonstration of advanced astrometric techniques.

J. Kovalevsky presented a progress report on the Hipparcos European satellite to be launched by June 1989. Environmental tests of the flight model satellite are now achieved and have confirmed that the initial accuracy specifications will be kept. The satellite is currently in storage configuration until reactivation commences by the end of this year. The different steps of the data processing and sphere reconstitution were detailed. The results obtained from simulations by both consortia FAST and NDAC are convergent. Furthermore work of the TYCHO Consortium is progressing satisfactorily.

J. Hughes described progress in the design of the USNO Astrometric optical interforemeter. The main objectives are limiting magnitude at least equal to 10 (brighter quasars as an ultimate goal), multiple baselines (at least two, orthogonal, 15-20 m), real-time metrology, two-colour work, sky coverage 120°. The preliminary design includes :

a) 1.25 m siderostats with air-bearings and laser metrology on pier invar reference ;
b) Beam collapsers (1 m, F/2 parabolas) tilted, giving output beams 7-10 cm ;
c) Vacuum pipe from piers to delay lines and for delay line themselves ;
d) Dispersed fringe detector : fiber optic lines from the combined beams to the spectrometer associated with a PAPA photon camera.

L. Helmer presented the first results obtained with the new moving slit micrometer mounted on the Carlsberg automatic meridian circle at La Palma in March 1988. The micrometer has been used during three months on an enhanced observing program including some very faint stars of magnitude 13.0 - 15.0. The preliminary results show that the positional accuracy has been improved by about 0.06 second of arc, giving a zenith mean error of ±0.12" in both right ascension and declination. The limiting magnitude of the instrument was proved to be 15.0.

G. Carrasco described the modernization of the Repsold meridian circle at Cerro-Calan. The construction of an electronic chronograph, a new system of contact and the automatic drive of the observations, for the improvement of the RA observations, are already finished. For the Dec. observations, the automatic reading of the micrometer screw was completed and the installation of eight Reticon RL512G for the circle reading is in progress. The registration of the meteorological parameters and the setting of the telescope will be automatized, using a microcomputer. The re-observation of the Santiago 67 catalog are started.

A paper of G.I. Pinigin was presented by V. Abalakin on the capabilities of a meridian circle with the horizontal design (HMC). The Pulkovo HMC, in regular operation since 1986, is equipped with automatic setting (accuracy 2" obtained in 15 seconds) and automatic divided circle readings (0.02" with 4 microscopes in 12 seconds). The star transits are measured by active slit micrometers mounted on the fixed collimators. The two-side metallic mirror is monolithic with the rotation axis, giving very small flexure and collimation variation (pivot error 0.1 μ). The vertical temperature gradient in the horizontal tubes was eliminated by forced ventilation between the inner and outer tubes. Seasonal and other azimuth variations did not exceed 0.4" for two years. The standard errors are $\pm 0\overset{s}{.}011 \sec\delta$ in RA and $\pm 0\overset{''}{.}20 \sec z$ in Dec. up to m = 11. The second generation HMC will have an improved mirror and horizontal tubes with increased focal length, aiming to obtain star coordinates with an accuracy of 0.02"- 0.03" up to m = 12-13.

COMMISSION No. 9

INSTRUMENTS AND TECHNIQUES (INSTRUMENTS ET TECHNIQUES)

PRESIDENT: C.M. Humphries VICE PRESIDENT: J. Davis

Given here are reports on the scientific and business meetings held in Baltimore involving Commission 9 only. Reports on the following Joint Commission Meetings co-sponsored by Commission 9 (JCM 3 and JCM 6), and on three additional meetings organised jointly with other Commissions, will be found elsewhere in this Volume or in Highlights of Astronomy (Volume 8):

- High Angular Resolution Imaging from the Ground (JCM 3)
- Stellar Photometry with Array Detectors (JCM 6)
- Problems of IR Extinction and Standardization (with Commission 25)
- Future Space Programs (with Commission 44)
- Gamma-ray, X-ray, Extreme and Far UV, IR and Radio Astronomy from Space (with Commissions 40 and 44)

BUSINESS SESSION: 9 August 1988

Reporting on the period 1985-88, the President said that the areas of interest of Commission 9 had been enlarged by including infrared and sub-millimetre wave telescope technology and instrumentation. Although no specialist symposia had been organised by the Commission in the period under review (since these had been catered for by a wide range of topical meetings arranged by other bodies), it was felt that an important function of the Commission should continue to be the provision of a full programme of scientific sessions at each General Assembly. To this end, at the present General Assembly in Baltimore, Commission 9 had arranged or co-sponsored thirteen such sessions. Attendances at these (audiences of 75 - 200 were obtained consistently, and at one session - the Saturday morning tutorial on adaptive optics - only standing room remained) suggested that this policy was a fruitful one. The active Working Groups of the Commission continued to be those on Photoelectronic Detectors, Astronomical Photography, and High Angular Resolution Interferometry.

The following officers of the Commission were nominated and elected for the period 1988-91: President: J. Davis; Vice-President: J.C. Bhattacharyya; Organising Committee: E. Becklin (subject to confirmation), M. Cullum, J.-L. Heudier (Chairman, Astronomical Photography W.G.), C.M. Humphries, G. Lelièvre (Chairman, Photoelectronic Detector W.G.), E.H. Richardson, L.I. Snezhko, W.J. Tango (Chairman, High Angular Resolution Interferometry W.G.), R.G. Tull.

The names of the following new members were presented to the meeting and accepted: Guoxiang Ai, N.M. Ashok, M.T. Bridgeland, J. Guibert, B.W. Hadley, P.V. Kulkarni, I.S. McLean, M.C. Morris, G. Pizzichini, M. Pucillo, J. Rowntree, M. Valtonen, W.W. Weiss, R.A. Windhorst.

A discussion followed on the role of IAU symposia or colloquia on astronomical optics and instrumentation, particularly in view of the increasing tendency for such meetings to be held regularly and frequently (at least in the U.S.A. and Europe) by The Society of Photo-Optical Instrumentation Engineers (SPIE) and by some of the larger observatories. The view was expressed that contact between the Commission and SPIE on the scheduling of meetings and their content could be beneficial to each side; it was agreed that this should be pursued further by the incoming President and his organising committee.

SCIENTIFIC SESSIONS AND WORKING GROUP MEETINGS

Progress on Telescopes (Visible, IR And Sub-mm): 3 August 1988

Chairman: C.M. Humphries

J. Nelson (U. California, Berkeley) opened the session by describing the recent history and current status of the 10 metre (f/1.75 Ritchey-Chrêtien) Keck telescope project. With the enclosure for the telescope already erected at Mauna Kea, the telescope itself is scheduled for installation there in 1989, with the testing phase starting in 1990. In the current programme of stress-polishing the optics, the focal lengths of individual segments are required to be matched to within 300 microns.

SEST, the only large sub-millimetre telescope in the southern hemisphere, was constructed as a collaborative venture by Sweden and by ESO, and has already begun operation at La Silla (first light in March 1987; scheduled observations started in April 1988). Described in a presentation by R.S. Booth (Onsala Space Observatory), SEST is an open-air telescope making extensive use of carbon fibre in its construction. The low temperature coefficient of this material, combined with its high tensile strength, reduces thermal distortion and allows the telescope to operate under wind loads of 14 m s^{-1}. The reflector profile has been set to 65 microns rms from the best-fit paraboloid and further improvements will be made in the next few months to reach the specified accuracy of 50 microns rms. At present it has dual polarization receivers covering the frequency ranges 80 - 117 GHz and 220 - 240 GHz, and a 350 GHz system is under construction.

A new infrared telescope has been completed and is now in operation at the Rothney Astrophysical Observatory in the Canadian Rockies (E.F. Milone and T.A. Clark, U. Calgary). The telescope has an alt-alt mounting and is driven in two axes by computer controlled friction disc drives. Present instrumentation includes broadband and CVF photometry. A contract between the University of Calgary and the Astrophysical Research Consortium for the polishing of a 1.8m honeycomb mirror will enable the current 1.5m metal mirror to be replaced.

L.P. Bautz of the National Science Foundation gave a review of recent developments and plans for new observing facilities in the U.S.A. Included in these are: the Caltech 10 metre sub-millimetre dish being commissioned at Mauna Kea; the 10 antenna x 25 metre VLBA network due for completion in 1992; the 3.5m Apache Point telescope at Sacramento Peak, New Mexico being built by the Astrophysical Research Consortium, the honeycomb mirror for which has already been manufactured; the Keck telescope (referred to above); the conversion of the MMT to use a 6.5m monolithic primary mirror; the Magellan 8m telescope project; the 2 x 8m Columbus project with a single mount but equivalent to an aperture of 11.3m; proposed NOAO projects for 8m telescopes; and the segmented Spectroscopic Survey Telescope (see below).

A status report on the 6m SAO telescope in the USSR was provided by L.I. Snezhko. Replacement of the primary mirror in 1983-5 improved imaging quality to give 90 per cent encircled energy diameters of 0.9 arcsec on-axis at the prime focus and 1.1 arcsec at 4.5 arcmin off-axis. Improvements have also been made to the setting accuracy (+/- 3 arcsec) of the telescope and to the environment in the dome (by installing an air ventilation system). Most observations at present fall into the following categories: extragalactic spectroscopy using a TV scanner; high and moderate resolution stellar spectroscopy; speckle interferometry; superhigh time resolution photometric studies; stellar magnetic field measurements. A 2-dimensional detection system with digital storage and processing (the QUANT system) has been installed recently and this is used in an echelle spectrograph operated with a long slit, in direct imaging, and in multi-object spectroscopy. Approximately 30% of the observational time is allocated to the Special Astrophysical Observatory by the Time Allocation Committee (Chairman: Dr. V.Yu. Terebizh), and the remaining time goes to external programs including 15% to foreign users.

Harlan Smith (U. Texas, Austin) described the Spectroscopic Survey Telescope project (at a session on 9 August). This is a joint program between the Universities of Texas and Penn State, with the former responsible for the mechanical design and development and the latter for the optical work. The telescope has a constant elevation tilt (altitude 60 degrees), and thus behaves like a zenith telescope, but rotates on an azimuth bearing so that most of the northern hemisphere becomes accessible except for a small cap overhead. The primary mirror is segmented and uses 85 separate 1 metre diameter spherical mirrors on a common mount to give a total surface area larger than that of a 9m mirror, though an average of about 8m is normally used. When the telescope is in operation at fixed azimuth, the object being observed will move through the 12° field of view in approx. 1 hour and this is tracked by a moving (x-y-z) spherical aberration corrector and detector system. The focal surface is linked to one of several spectrographs in the basement of the building by an optical fibre link. Work on manufacturing the mirror segments (using 5cm thick Pyrex, with students providing the manpower) and the invar tetrahedral support structure, is well under way.

Additional presentations were given by: K. Kodaira "The Japanese National Large Telescope"; F. Merkle "Recent Developments at ESO"; P.V. Kulkarni "The Mt. Abu Telescope"; Li Depei "8-Metre Mirror Fabrication"; D. Downes "IRAM"; A. Boksenberg "Recent Developments at RGO and La Palma".

Active and Adaptive Optics: 3 August and 6 August 1988

Chairmen: F. Merkle, 3 August
L. Goad, 6 August

J. Hardy (Itek Corp.) reviewed methods of wavefront sensing and proceeded to consider the properties of (1) the atmospheric structure function as a function of projected actuator spacing for an adaptive mirror, and (2) the combined optical transfer function of the telescope and atmosphere in terms of the fitting errors when the actuator spacing (L) is not matched with r_o. In the latter case when photon noise, subaperture averaging, timing and other experimental errors are included, the SNR falls off rapidly as $L > r_o$ and partial compensation soon becomes less effective than full compensation (though still capable of yielding improved imaging). The conclusion was that, since full compensation is the better prospect, the need is to reduce the cost of adaptive mirror actuator systems (currently approx. $1000 - 2000 per actuator channel including drive electronics).

A status report on the 37 element adaptive system being developed and tested at Kitt Peak for infrared use was given by L. Goad (NOAO, Tucson). Wavefront sensing is accomplished with Hartmann-Shack optics feeding a GaAs image intensifier and Reticon detector, and the servo loop is operated at a frequency of 100Hz with a correction bandwidth of approx. 30Hz. For completing the loop,

Goad distinguished between modal control (in which Nth order Zernike polynomials are calculated) and zonal control (which is slower than modal control).

Fred Forbes (NOAO, Tucson) described the piezoelectric bimorph mirrors that are being developed by NOAO. Two thin piezoelectric wafers are cemented together with a metal electrode in between. One of the faces is chosen as the reflecting surface and is polished flat, while the other face, also polished, has a pattern of hexagonal electrodes deposited on it. Voltages applied to the electrodes produce bending moments that give local curvature on the mirror. These mirrors are small and lightweight and have excellent frequency response (up to about 10kHz).

The status of the ESO 19-element adaptive system for infrared applications and the future goals for adaptive optics with the VLT were described by F. Merkle (ESO). A system in which three adaptive mirrors would be used sequentially was considered for visible wavelengths: the first of these would be a standard tip-tilt mirror, whereas the other two would be optimally designed so that separate corrections could be made for low and high spatial frequency components. The low frequency one would have fewer than 50 sub-aperture elements, large amplitude (10 microns) and slow response (30 ms); the high frequency one would have up to 1000 elements, relatively small amplitude (1 micron) but fast response (3 ms). Also mentioned by Merkle was the group at Chengdu in China (Institute of Optics and Electronics, Academica Sinica) that has developed a prototype adaptive system and has achieved closed-loop operation in laser applications.

The use of adaptive optics in long baseline interferometry was discussed by P. Léna (Observatoire de Paris) including SNR considerations at infrared wavelengths, partial atmospheric compensation and deconvolution techniques.

O. von der Lühe (NOAO, Sacramento Peak) described the 19 hexagon segment system developed by Lockheed Palo Alto and tested jointly with NOAO on a 1m solar telescope at Sacramento Peak. Results shown on a video recording demonstrated the improved resolution that had been obtained for imaging solar granulation.

An example of a 512 element segmented adaptive system that can be purchased commercially was given by Bill Hulburd (Thermo Electron Technology Corp., formerly Western Research Corp.). The gaps between the segments occupy less than 2% of the area. For 3-axis control of the segments and piston movements up to 10 microns such a system would cost approx. $300 per segment, i.e. $100 per degree of freedom.

A. Wirth (Adaptive Optics Associates) reported on the lenslet arrays for Hartmann-Shack wavefront sensors, and processor boards with parallel 10MHz channels, currently available from AOA for adaptive systems. The parent company, United Technologies, also supplies adaptive mirrors using electrostrictive actuators.

At the Adaptive Optics 'tutorial' on August 6, the design of systems and their realisation was considered in more detail and from a practical point of view. Presentations were given by Larry Goad, Allan Wirth, Fritz Merkle and Renaud Foy; among the topics covered were wavefront sensors and sampling, types of adaptive mirror, signal processing, loop control, and laser-generated artificial guide stars.

CCD and Array Imagers for the Visible and Infrared: 8 August 1988

Chairman: M. Cullum

This session was concerned principally with the status of various array detector development projects, and with the experience of some of the groups that have been testing and using them.

Optical Detectors

Martin Cullum (ESO) reviewed the situation regarding the two major European manufacturers of optical CCDs: EEV(GEC) in England and Thomson-CSF in France. Both of these firms have been actively working in the last year towards achieving CCDs that are larger in area than those that have been widely available in the past, as well as the manufacture of thinned devices.

Thomson have recently completed a development contract, funded jointly by ESO and INSU, to produce a modified version of their standard TH7882 CCD that can be butted on three sides. The individual chips have 400 x 579 pixels of 23 microns across. This project has yielded CCDs of excellent cosmetic quality, $6e^-$ readout noise and an improved CTE compared to some previous Thomson CCDs. To complement this work, ESO and the Toulouse Observatory have developed a machine to align and mount a 2 x N array of individual CCDs on a common substrate with a precision of 2 microns for any pixel on the combined array. This machine is currently undergoing final tests in Toulouse and will shortly be installed at ESO in Garching.

Thomson have also started to deliver samples of their new 1024 x 1024 CCD with 19 micron pixels, and some of the first test results from this chip are described below.

With support from CNES, the French Space Agency, Thomson have also produced some thinned versions of their 576 x 384 pixel CCD. These have been evaluated by Thomson and by Laurent Vigroux and his colleagues at CEN-Saclay. Although the prototype versions are still not fully optimized and exhibit a few problems such as non-uniform annealing, the RQE in the visible and especially in the soft X-ray and UV region shortward of about 280nm is significantly better than the standard CCDs. In the visible, a peak RQE of about 53% has been measured, going down to about 10% at 320nm.

EEV have also produced the first wafers for their new family of 'large' CCDs. These will all have 22.5 micron pixels with the formats: 298 x 1152, 770 x 1152 and 1242 x 1152. The first tests of these devices at EEV are expected in October 1988.

Under contract from the Royal Greenwich Observatory and the Anglo-Australian Observatory, EEV have also recently delivered the first batch of thinned P8603 CCDs for evaluation. These have been tested by David Thorne at the RGO who has measured peak quantum efficiencies at 650nm of between 65 and 75%. The sensitivity of these devices at 340nm was somewhat lower than expected, with RQE figures averaging about 5%, probably again due to non-optimized backside surface treatment. The prototype chips were of an unsupported pellicle design that exhibited flatness deviations of up to 150 microns peak-to-peak for some chips. Future thinned chips will be mounted on a glass supporting plate on the electrode side to overcome this problem.

One of the prototype thinned EEV CCDs from the RGO has been given a fluorescent coating at ESO to see whether it is mechanically feasible to coat an unsupported pellicle using a spinning technique, and to see whether any significant improvement in the UV RQE could be realised. The results of this experiment were encouraging, yielding an efficiency in excess of 35% right through the atmospheric UV region, and peaking at about 80% in the visible.

Patrick Waddell (CFHT) reported on aspects of the CCD implementation at the CFHT. One aspect felt to be of considerable importance for an observatory like the CFH, where instrument change-overs occur frequently, was that of the adoption of observatory and international standards for the mechanical and optical design of dewars, and for the electrical and data transfer interfaces. This was necessary to assure maximum versatility and reliability of the different systems installed.

Considerable attention has also been paid to the 'observer interface' between the observer and the instrument and host computer. To a large extent, modern visual interaction methods with the control computer have been employed to avoid the need for visiting astronomers having to remember command strings. The data acquisition system at the CFHT currently employs an HP9000 series computer with a Sun pre-processing workstation and IRAF software.

As well as RCA and Thomson CCDs, one of the new Ford/Photometrics 512 x 512 CCDs has also recently been tested. This was of very good cosmetic quality and had an excellent CTE even at very low signal levels. The readout noise has been measured to be about 7 e^-.

Lloyd Robinson (Lick Observatory) presented information about the status of the new Reticon 1200 x 400 CCD. This project, funded jointly by EG&G-Reticon and the National Science Foundation, aimed at producing a large area CCD of good quality suitable, in particular, for spectrometric applications.

The project was started by Reticon in July 1987, and the first sample CCDs, at present unthinned and front illuminated, were received at the Lick Observatory in April 1988. The initial tests have revealed that the chip is, indeed, very good: almost perfect cosmetically with virtually no charge pockets, a full-well capacity of about 450,000 electrons, and a readout noise of 4 e^-. The CTE, revealed by cosmic-ray events, is also excellent. Another interesting characteristic of the device is the possibility of being able to hold the clock levels low during an integration without charge spreading along the columns. This 'multi-phase-pinned' mode of operation gives a significantly reduced dark current level compared to conventional CCDs, thus enabling the detector to be operated at a higher temperature. The dark signal measured in this mode of operation was less than 20 electrons pixel^{-1} hour^{-1} at -110 C.

Reticon will deliver the first thinned versions of the chip, with hopefully a good UV response, before the end of 1988. One final, but important, point made was that the unthinned version of the chip is already available for sale !

Bill Schempp (Photometrics Ltd.) gave some brief results from the first tests of the new 1024 x 1024 pixel CCD from Thomson which had recently been received by Photometrics. Although this was a low grade prototype device with a number of cosmetic defects, the performance was encouragingly good. The readout noise was 8 e^- and the full-well capacity about 250,000 electrons. At a temperature of -37 C, at which all the initial tests were carried out, the dark current was 13 e^- pixel^{-1} sec^{-1}. Further tests at lower temperatures were under way.

In a general discussion on the progress of the Tektronix 512 x 512 and 2048 x 2048 CCDs, several participants were able to support the rumours that Tektronix has, at last, managed to produce CCDs which, although not perfect, were virtually free of the charge pocket problem that has long plagued them.

Altogether, the scene for optical CCDs looked much more rosy at this IAU General Assembly than at the last one. It is clear from the data presented during the session that excellent chips have been produced with characteristics that were only dreamed of several years ago, not only by the traditional manufacturers of these devices but also by 'silicon foundries' having little or no previous CCD experience. It is also interesting to note how many of the recent CCD development projects have been funded by astronomical institutes or produced primarily for astronomical applications. This is probably indicative of the maturity of CCD technology that such developments have become a realistic and affordable possibilty for the astronomical community.

Infrared Array Detectors

Zoran Ninkov (U. of Rochester) gave a review of infrared array detectors that are currently available from commercial vendors, and described both the technological differences as well as the achieved or anticipated performance figures. The IR detectors that are generally preferred for astronomy at the present time are those of a hybrid construction, where the detective element and readout stage are fabricated from different materials. The detector materials traditionally used at infrared wavelengths have been either intrinsic or extrinsic semi-conductors. A newer class of very heavily doped materials - 'Impurity Band Conductors' (or IBC) - have recently become available that show considerable promise. The potential advantages of IBC detectors include a more uniform response and radiation hardness. To control the potentially large dark current with such detectors, a blocking layer of undoped material is grown between the active layer and the electrode, whence the acronym BIB detectors (blocked impurity band).

The method of manufacturing the chip and the internal architecture also have an important effect on the eventual performance of the detector. Epitaxially grown detector materials, although not yet so widely available as the bulk materials commonly used at the present time, as well as the use of Mesa architectures, hold the promise of improvements in the response uniformity and dark current as these become available.

Of the three most common technologies employed to read out infrared arrays - CCD, Reticon and Direct Readout or DRO - it was pointed out that DRO arrays have several advantages. They are usable at liquid Helium temperatures and allow selective and non-destructive readout modes, although with the possible disadvantage of non-linearity. There would also appear to be advantages in changing from the N-MOS FETs, found in many of today's DRO arrays, to P-MOS because of their lower ultimate 1/f noise at low temperatures.

IR arrays with pixel formats up to about 60 x 60 pixels are already in use at most of the world's major observatories. Devices with 128 x 128 and 256 x 256 already are on the horizon with a readout noise of less than 100 e^- and very low dark current levels.

Photoelectronic Detectors Other than CCDs: 9 August 1988

Chairman: G. Lelièvre

The session started with a rescheduled presentation of the Spectroscopic Survey Telescope by Harlan J. Smith (University of Texas). A summary of this contribution can be found in the session devoted to telescopes. The rest of the session dealt with detectors other than CCDs, mostly photon-counting detectors. The status of various development projects was described as well as some experience in using them.

Mark Clampin (STScI) discussed the design and development of Ranicon detectors for optical photon-counting imaging with ground-based telescopes. The significant factors that determine the performance of these detectors are found to be the proximity focussing stage, the microchannel plate stack (MCP) and the signal processing electronics. The low photon-counting efficiency typically found with optical MCP-based detectors (due to ion barrier films) presents an additional consideration. A new approach to the signal processing electronics reduces non-linearity, while achieving increased processing speeds and a position error corresponding to less than 21 microns FWHM.

An advanced Ranicon incorporating a reduced proximity-focussing gap and an unfilmed input plate in the MCP stack shows significiant improvement in detector performance. For the advanced Ranicon, the spatial resolution is shown to be typically 40 microns FWHM at 650nm and the point spread function to be stationary over the active imaging area, principally due to the removal of non-linearities and noise sources from the signal processing electronics.

Gerard Wlérick (Observatoire de Paris) presented the results of deep soundings using the Lallemand electronographic camera at the Canada-France-Hawaii Telescope. With its 81mm diameter photocathode and a resolution better than 70 lp/mm, the equivalent number of pixels is 2×10^7, making it particularly suitable for large field photometry. Special care is devoted to high accuracy photometry using fiducial marks engraved on the photocathode for registration, flat fielding at twilight, and screens to evaluate the background noise. The data handling, after digitization, is similar to the procedure used for CCD data. At the CFHT Cassegrain focus (F/8), the sampling (camera + PDS) is about 0.12 arcsec over a field of 9.5 arcmin; this large field is essential for large surveys requiring photoelectric standards in the field. It also provides a better knowledge of telescope imperfections (such as scattered light) which are important for a correct data reduction. Many results were presented including a UBV Catalogue of all objects in a selected field of SA 57, up to B=25.5.

Martin Cullum (ESO) in the absence of J Gethyn Timothy, presented the development status of projects related to the Multi-Anode Microchannel Array (MAMA) as well as results from evaluations at ESO (point spread function, linearity, count rates). Large format cameras (2048 x 2048) are being developed for use in the Hubble Space Telescope and are scheduled to be built by early 1989. Other MAMA applications include rocket flights, speckle imaging and astrometry. The flat field response of the bialkali detector is found to vary with exposure time, attributable solely to problems with the photocathode. Measurements show that centroid positions may be computed from MAMA data to an accuracy of at least 0.01 pixel. The DQE is 0.22 of the cathode RQE. With an unfilmed MCP, the DQE would be expected to be significantly higher.

Renaud Foy (Observatoire de Paris and CERGA) presented CP40, the Photon-Counting Camera developed at CERGA and Observatoire d'Haute Provence under Alain Blazit's supervision (CERGA). Four Thomson CCDs are fed through a fibre optics reducer and splitter, with a common stack of 40mm diameter Varo image tubes coupled with a RTC microchannel plate intensifier. The hardware photon-centroiding processor gives an accuracy of 1/4 pixel, providing 3072 x 2304 logical pixels. Due to the Varo first stage, the quantum efficiency remains satisfactory (about 12%) and does not lose primary photo-events, unlike MCP intensifiers which lose about half of the photon-events. The CCDs are used in video mode. Further developments include duplications of the CP40 at Paris Observatory, an improvement of the Direct Memory Access board and an increase of the readout speed. Many applications are already supported by the CP40 prototype: speckle imaging, multi-telescope interferometry, multi-slit spectroscopy, segmented pupil imaging etc.

As an application of the large format of the CP40, Gérard Lelièvre (Observatoire de Paris) showed the recent results of the Segmented Pupil Imagers at CFHT and Pic du Midi. As a result of recording the data at video rate, the eight sub-pupil images can be recentered, selected and recombined by computer at the site immediately after the observations are made. Sub-pupil images with resolution between 0.25" and 0.4" were presented while the full pupil images were typically between 0.6" and 0.8".

Alex Boksenberg (RGO) gave an update on IPCS developments and presented some new ideas related to detectors for the Herschel telescope. During the discussion. Trung Hua (Laboratoire d'Astronomie Spatiale) demonstrated that old photon-counting cameras are well alive and can still produce good quality science.

Clearly photon-counting detectors play a dominant role in key fields such as low-light level spectroscopy; in recording fast events to achieve very high time resolution; in diffraction limited applications (speckle interferometry); in high spectral resolution; and in atmospheric turbulence limited cases (direct imaging with a-posteriori fast guiding). Electronography remains a unique method for deep surveys in large and crowded fields as long as large format CCDs or arrays of CCDs are not available. All contributors emphasised the necessity of having large-area detectors, particularly for large telescopes under design or forseen within five to ten years. At the present time, it may be better to consider using photon-counting detectors than to anticipate rapid developments in CCD technology.

Photographic Working Group (PWG): 4 August 1988

Chairmen: D. Malin/J.-L. Heudier

The following scientific and technical papers were presented:
Performance of 2 New Image Processing Algorithms to Improve the Image Visibility of Astronomical Photographic Pictures (S. Koutchmy).
Astronomical Results from Electronography at the CFH Telescope (G. Wlérick).
An Ultradeep Photographic Survey with the 100" DuPont Direct Camera (R.A. Windhorst).
Hypersensitization of Infrared Plates and Vacuum Hypersensitization (A. Maury).
Plate Addition and Batch-to-batch Variation in Hypering Properties (D. Malin).

Review of Plate Manufacturing Facilities of Eastman Kodak Co. (J. Burdsall) - open discussion on Burdsall's paper and on the future availability of astronomical emulsions.
Review of poster papers (J.-L. Heudier).

Poster papers exhibited were:

Astronomical application of Konica SRV 3200 Film (K. Tomita).
A new tuning fork? (D. Block and B. Tully).
Detection of new Herbig-Haro objects on deep Schmidt plates (B. Reipurth and C. Madsen).
Total eclipse of the sun of 18 March, 1988 (F.J. Heyden).
Report on hypersensitization at the Bosscha Observatory (Hidayat, Wiramihardja, Raharto and Kogure).
Second Palomar Sky Survey: First Results (A. Maury).
The light echo of SN 1987A (D. Malin).
Hypersensitization of Kodak IV-N photoemulsions (V. Tsintsarov, M. Panov and Ts. Georgiev).
Semi-automated identification of IRAS point sources using UKST plates and the Cosmos measuring machine (A. Savage et al).
Hypersensitization Facilities at the Indian Insitute of Astrophysics (A. Rajamohan, J.C. Bhattacharyya and K.R. Sivaraman).

Apart from a brief description of plate addition by Malin, the second session consisted mainly of topics in photographic technology, especially concerning plate production. Malin had provided a list of IIIa J and F emulsion batch numbers extending over 10 years to David Jeanmarie (Al Millikan's successor) and John Burdsall of the Eastman Kodak Company with an indication of the response of these batches to nitrogen hydrogen hypersensitization. The practical response of these emulsions agreed well with the sensitometric tests conducted after manufacture and it was agreed that plate speeds could generally be increased somewhat, perhaps on average by 30% without affecting those sensitometric properties that are essential for astronomy. Burdsall agreed to produce, and Malin agreed to test, experimental batches of a variety of emulsion types with improved response to hypersensitization and ultimate long-exposure speed.

The presentation by Burdsall of the Eastman Kodak Company was a particularly valuable contribution. It gave details of plate manufacture that were not previously available and the talk was frequently interrupted by pressing questions. Mr. Burdsall was very forthcoming in his answers and discussion was continued for a considerable time after the conclusion of his paper. This was joined by Gordon Brown, Product Planner for Scientific Photography at Eastman Kodak who explained that the spectroscopic plates used by astronomers represented only a small fraction (less than 10%) of total plate production by the Company. The largest users of plates are in the graphic arts field.

Mr Brown explained that plate production was extremely labour-intensive (a point amply confirmed by Burdsall's pictures of the plant) and that economic factors had forced a major review of product lines during the previous 12 months. Many emulsion types coated on glass had been discontinued, including types IIaD and 098-04. However, as a result of strong representations from members of the Working Group, these had now been reinstated though with some special ordering conditions that were established during later discussion in a subsequent meeting of the Organising Committee of the PWG. The scientific meeting closed after two extremely useful and productive sessions.

Business Meeting of the PWG

Committee members present were: D. Malin (Chairman); J.-L. Heudier (Chairman-elect); J. Davis; B. Hidayat; C. Humphries; K. Ishida; R. Sivaraman; M Tsvetkov; R. West (present for first part only) and as observers: G. Brown and J. Burdsall (Eastman Kodak); C. Madsen; A. Maury.

1. Relations with Eastman Kodak

Several incidents had contributed to a deterioration of relationships between the PWG and the Eastman Kodak Company. Almost all of these could be traced to communication problems and all have now been very satisfactorily rectified. We were impressed with the enthusiasm with which new ideas for collaboration between the astronomical community and the Company were discussed.

Emulsion types IIaD and 098-04 will continue to be manufactured, but with 100 ft^2 (about 9.5 m^2) minimum order quantities. These will now have special order numbers (to be advised) and will be manufactured when accumulated orders approach 100 ft^2 or twice a year, the ordering windows closing at the end of December and end of June each year. While orders will be accumulated by Eastman Kodak it is strongly recommended that small users of these products use the services of the officers of the PWG to consolidate and share orders with other observatories.

2. A Journal for Astronomical Photography

The meeting expressed its regret and disappointment at the sudden demise of the AAS Photobulletin, formerly the official organ of the PWG and the American Astronomical Society. It was felt that there was still a need for such a journal and two alternatives were discussed. The Chairman had received a letter from the editor of the Journal of Photographic Science, published in England by the Royal Photographic Society (RPS), offering that journal as an alternative for Working Group publications. Although this Journal does not levy page charges, few if any participants in the PWG were members of the RPS and most of the other papers in the journal would not be of direct interest.

After considerable discussion, which revealed that Eastman Kodak Company were prepared to donate $10,000 a year to the project, it was decided to consider further an offer from the Sky Publishing Corporation to produce a new journal, under the editorial control of a panel selected by the PWG, to publish refereed papers on astronomical photography. The journal would be quite separate and distinct from Sky and Telescope magazine and would be of a quality similar to that of the AAS Photobulletin. Approval for this course of action has been given by Commission 9 and requested from the IAU Executive Committee.

3. Election of New Officers and Consultants

The meeting considered that while the Organising Committee gave good international representation, the custom of having only one delegate from each country restricted the availability of experts in our speciality. It was decided therefore to appoint initially three prominent specialists as consultants to the Committee. The new committee and its consultants are listed below. An asterisk indicates new members.

4. Future of the Working Group

Despite the steady decline in users of photography, it was the opinion of the meeting that the Group should continue as an independent part of IAU Commission 9 and not be absorbed into the Working Group that covers electronic 2-dimensional detectors (as has occurred with the AAS Photographic Working Group). It is our belief that photography will continue as a technique for the specialist in the forseeable future; general practitioners are now using CCD's with great success for many problems that were previously investigated photographically. To stay competitive in the areas where it is especially valuable (sky surveys, panoramic detection of faint objects), more sophisticated techniques and a greater level of specialist skills are required. The Working Group is an important repository of those skills. This point has been discussed more fully in 'The Age of the Specialist' in the procedings of the Working Group meeting held in Jena in April, 1987 (published as Astrophotography 87, ed. S. Marx, Springer-Verlag, February 1987).

5. Election of Organising Committee

The following agreed to serve on the Organising Committee for 1988-1991:

Chairman: Jean-Louis Heudier (France); Secretary: Jorge Schumann (FRG).

Members: John Davis (ex-officio, Australia); Olga Dokuchaeva (USSR); Brian Hadley (UK); Bambang Hidayat (Indonesia); Kei-ishi Ishida (Japan); Hideo Maehara (Japan); David Malin (Australia); Siegfried Marx (DDR); R. Rajamohan (India); William Schoening (USA); Alex Smith (to be confirmed, USA); Milcho Tsvetkov (Bulgaria); Richard West (Denmark); Olga Zichova (Czechoslovakia).

Consultants: John Burdsall (Eastman Kodak Co., USA); Claus Madsen (ESO, Germany FRG); Alain Maury (France).

WG on High Angular Resolution Interferometry: 10 August 1988

Chairman: J Davis

A total of 12 reports were presented followed by a short business session.

J. Davis (University of Sydney), "SUSI - A Progress Report". Construction of the Sydney University Stellar Interferometer (SUSI) commenced in 1987 at the Paul Wild Observatory in northern New South Wales. Initially the new instrument will have a North-South array of siderostat stations to give baselines covering the range 5-640m. Starlight will be directed by the 0.20m diameter siderostats via fixed mirrors and an evacuated pipe system to a central laboratory containing a T 420m path equalisation system and beam combining optics. The design is based on a successful prototype instrument and features active wavefront tilt correction and a fringe-tracking optical path equalisation system. The limiting magnitude is expected to be +8 and first light is planned for early 1990.

William J. Tango (University of Sydney), "Dispersion in Long Baseline Stellar Interferometry". The path compensating system used to match the paths in a long baseline interferometer should strictly be in vacuo, but for practical reasons it may be convenient to operate the compensator in a stable air-conditioned environment. This arrangement introduces a large differential air path and the resulting dispersion will seriously affect the measurement of fringe visibility, particularly when wide optical bandwidths are used. Dispersion compensators consisting of variable thicknesses of one or more glasses can be used to correct for the air dispersion. A single glass is not sufficient if the excess air path exceeds ~10m, but two glasses can correct the dispersion for air paths up to 500m when the optical bandwidth is less than ~100nm.

J. A. Hughes (USNO), "The USNO Astrometric Optical Interferometer". Preliminary planning for the USNO astrometric, optical interferometer continues. The instrument will operate as a multi-r_0 system using a dispersed fringe technique. Large (1 to 1.5 meters) siderostats will feed one-meter beam compressors in the present plan. Evacuated pipes will carry the light to the central beam combiners. The possibility of using air-bearings for the siderostats is under investigation. As much of the beam combining optics and delay lines as possible will be underground. The layout of the baselines is under discussion, but two independent baselines approximately twenty meters long will be used. At least two passbands will be utilized to reduce atmospheric effects. Preliminary simulations indicate that it should be possible to reach the brighter quasars and thus provide an inertial reference system. The goal is to have a working instrument installed by 1993.

Yves Rabbia (CERGA), "Michelson Interferometry with "small" Telescopes at CERGA". "Small" implies "working with few speckles" and includes observations at 2.2 and 10 microns with the 2 x 1m telescopes and at 0.4 to 0.7 microns with the 2 x 26cm telescopes of the I2T. The 2 x 1m telescopes lie on an E-W baseline providing a 3.5m to 15m continuously varying separation. Observations are made in wide band (~ K and N filters). A double cat's eye device on a step-by-step moving carriage forms a discrete delay-line. At each station of the carriage, a Fourier interferogram is recorded when fringes pass the zero path difference. A signal to noise ratio of ~100 has been obtained for Alpha Orionis at 10 microns. Programs in astrophysics and astrometry are under way and improvements are planned for optimized operation and increased sensitivity.

The N-S baseline of the I2T has been extended to 140m. Fringes are observed in dispersed light in a spectral window of 30nm. The limiting magnitude is V ~ 4.5. Efforts are in progress to increase sensitivity and accuracy through upgrading of hardware and software (telescope mechanics, automatic pointing, image tracking, 2 dim. detector, fringe detection, data acquisition and reduction). Future plans include the development of the I2T into InT (n telescopes) for imaging using phase closure etc. A third telescope is under construction.

W.A. Traub (CFA), "Current Status of IOTA". The Infrared-Optical Telescope Array (IOTA) is a collaborative project, involving the SAO, Harvard, the Universities of Massachusetts and Wyoming, and Lincoln Laboratories (MIT), to build an interferometer on Mt Hopkins, Arizona. Two 0.45m diameter siderostats are being built with a third planned for the near future. IOTA will have two arms, one north-east ~40m long and one south-east ~20m long. Two long discrete delay lines, and two short continuous delay lines will be housed in a vacuum. Dispersed fringe detection will be used in the visible, with a matched-filter fringe pattern recognition technique. Delay line control will be fine-tuned using the measured fringe pattern. Phase closure and multi-wavelength-baseline information will be used in a hybrid mapping scheme for bright objects, but amplitude measurements only will be possible for faint objects. The system is intended to serve as a test bed and prototype for a larger system.

John Baldwin (Cavendish laboratories), "COAST - A Progress Report". COAST is planned to be a four-telescope interferometer using measurements of 6 visibilities and 3 closure phases for reconstructing images in the red and near infrared. The instrument, with maximum baselines of ~100m, is to be sited at Cambridge (UK). Two telescopes have been funded and construction of the first siderostat is nearing completion using a 50cm flat followed by fixed horizontal 40cm Cassegrain optics. The housing for the path compensators, combining optics and detectors, now half completed, is a 30m long corrugated steel tunnel with a 1m covering of earth. Work on avalanche photodiodes for photon counting detectors is progressing; counting rates up to 350 kHz have been achieved and 2-4 MHz can be expected. We hope to have first fringes in about a year.

H.A. McAlister (Georgia State University), "The CHARA Interferometer Project". The CHARA interferometer array is envisioned as consisting of seven 1m telescopes in a Y-configuration contained within a circle of 400 meters diameter. The limiting resolution of 0.1 mas would be applied to fundamental problems in stellar astrophysics. Simultaneous access to all seven beams is planned with delay lines for each arm, to allow the development and use of imaging algorithms. Correcting only for overall tilt is expected to provide a limiting magnitude of V = +11 while higher order wavefront conditioning may extend this to V = +14. Anderson Mesa, near Flagstaff, Arizona, is being tested as the site for the array.

G.P. di Benedetto (IFC Milan), "Saturation of the Variance of Optical Path Fluctuations in a Long Baseline Michelson Interferometer". The prototype I2T interferometer at CERGA was operated at 2.2 microns to measure optical path difference fluctuations as a function of the baselength. Data for the star Arcturus at 8.8m, 13.8m and 17.3m, in good seeing conditions (: 1 arcsec), show saturation of the RMS phase errors at a level smaller (up to a factor 4) than that expected from the Kolmogorov model. By assuming that such discrepancy is due to outer-scale effects, an external length of turbulence of a few tens of meters may be derived.

R. Foy (CERGA) "The VISIR Project". VISIR is a 3 x 1.50m telescope interferometer project intended for astrophysical programs and for preparation of the interferometric mode of the VLT. A Phase-A study of a dedicated telescope, funded by INSU/CNRS and carried out at IRAM, has resulted in the very compact skew-axis mount proposed by Plathner. This design could be useful for the VLT's Auxiliary Telescopes. A study of a transporter, to allow 2D continuous telescope movement for path matching or for hypersynthesis, will start in the fall. It is suggested that VISIR should be an international facility and that linked operation with the VLT's Auxiliary Telescopes would be productive.

R.V. Stachnik (NASA), "Interferometry and NASA". As a consequence of a series of ESA and NASA-sponsored conferences on space optical interferometry, NASA Headquarters requested its JPL center to undertake a study of supporting Space-Specific Technologies. The aim was to assist the Astrophysics Division in guiding development of needed technologies by the technololgy development arm of NASA. At present, limited "astrophysics" funding is being put into optical interferometry, but signfiicant long term "technology" funding is possible. Factors influencing NASA interest in space interferometry are the appearance of several very highly rated optical interferometers in the recent "explorer" scientific satellite evaluation and the recent release of a National Academy of Sciences/Space Science Board document recommending optical interferometers as part of NASA's long-range plan.

L. Mertz "Optical Beam Combining". An inexpensive lenticular receiver provides complex visibility information from individual photons. The information can be averaged directly in polar coordinates with digital low-pass filters for improving the precision while tracking fringes. These instruments could function with a pair of telescopes to provide single-baseline stellar interferometry, or with more telescopes to get multiple baselines for phase closure renditions of optical aperture synthesis.

F. Merkle (ESO), "NOAO-ESO Workshops". Two workshops on ground-based interferometric imaging have been held, one at Oracle, Arizona, USA, in January 1987 and the other at Garching, FRG, in March 1988. The proceedings of the first meeting were published by NOAO and publication by ESO of the proceedings of the second is imminent.

Business: It was agreed to continue the annual publication list and to circulate the WG mailing list with electronic mail addresses. A WG newsletter was proposed and will be investigated by the incoming chairman of the WG. Future meetings were discussed and it was agreed that an international symposium should be organized for circa 1990. The WG Organizing Committee for 1988-1991 is J Baldwin, Y Balega, R Foy, J Gay, F Merkle, M Shao, R V Stachnik, W J Tango (Chairman), C H Townes and P Venkatakrishnan.

COMMISSION No. 10

SOLAR ACTIVITY (ACTIVITE SOLAIRE)

COMMISSION No. 12

RADIATION AND STRUCTURE OF THE SOLAR ATMOSPHERE

(RADIATION ET STRUCTURE DE L'ATMOSPHERE SOLAIRE)

Report of Meetings on 3, 4, 5, 6 and 8 August 1988

PRESIDENT (10): M. Pick SECRETARY (10): A. Benz
PRESIDENT (12): M. Kuperus SECRETARY (12): J. Harvey

JOINT BUSINESS MEETINC (5 August; 09:00-12:50, 17:40-18:05)

The President of Commission 12, M. Kuperus, opened the meeting by explaining that due to the close cooperation which exists between Commissions 10 and 12 and the strong overlap in membership, all activities of the two Commissions at the present General Assembly were organized jointly. He expressed his thanks to the Organizing Committee and vice president of Commission 12 for their aid in conducting the work of the Commission and to the President of Commisson 10, M. Pick, for her cooperation during the past three years. The following officers and organizing committees were elected for Commission 12: President: J.W. Harvey; Vice President: J.O. Stenflo; Organizing Committees: H. Ando, R. Falciani, Ai Guoxiang, E.A. Gurtovenko, M. Kuperus (past president), R. Muller, T. Roca Cortes, M. Schuessler, K.R. Sivaraman, N.O. Weiss. Applications for membership in Commission 12 by 31 individuals were accepted by acclamation. The Chairman of the Working Group on Eclipses presented the report found below.

The President of Commission 10, M. Pick, expressed her thanks to the Organizing Committee and vice president of Commission 10 and to M. Kuperus for their aid and cooperation. She reported that 72 applications for membership in Commission 10 were received of which 52 were from new IAU members. The following officers and organizing committee were elected for Commission 10: President: E.R. Priest; Vice President: V. Gaizauskas; Organizing Committee: E. Antonucci, A.O. Benz (secretary), S. Enome, O. Engvold, Fang Chen, E.A. Gurtovenko, M. Machado, M. Pick (past president), E.A. Tandberg-Hanssen.

M. Pick, IAU representative to COSPAR Subcommission E.2 reported on the recent COSPAR meeting and presented a list of symposia and workshops proposed for the COSPAR meeting in 1990. Some of these meetings have been proposed for IAU co-sponsorship. She also presented a list of proposed symposia and colloquia which have been recommended to the IAU Executive Committee for IAU sponsorship. M. Pick announced that nine future IAU colloquia and symposia have been approved by the executive committee on recommendation of Commissions 10 and 12.

The first volume of the continuation of the Greenwich Photoheliograph Results now produced by the Debrecen Observatory has appeared. L. Gesztelyi reported that serious budget cuts threaten this effort. Debrecen needs help from the IAU in obtaining film and a small computer. A resolution addressing this issue was prepared (see below).

E. Hiei, the present editor of the Quarterly Bulletin on Solar Activity, reported on the history and current status of this IAU-assisted publication: the Quarterly Bulletin on Solar Activity contains five parts: I Sunspots, II Synoptic Charts of Solar Magnetic Fields, III Eruptions Chromospheriques Brilliantes, IV Intensité de la Couronne Solaire, and V Solar Radio Emission.

Part I is edited from the data of 74 observatories and institutes at the Sunspot Index Data Center, Bruxelles. Part II is reproduced from the daily solar magnetograms obtained at the Mount Wilson Observatory. Hα flare data from 25 observatories are collected and compiled in Part III at the Meudon Observatory. Coronal intensities of 5303 line and 6374 line measured at 4 coronal observatories and the synoptic map of the coronal intensity are prepared in Part IV at the Kislovodsk Observatory. Part V is compiled from the radio data of 39 observatories at the Research Institute of Atmospherics (Toyokawa).

Each part of QBSA is published when it is ready for printing. Part I, II, III, IV and V are now published until 1985, 1985, June 1982, 1985, and 1984, respectively.

H. Coffey reported on IAU Commission 10 standing committees:

IUWDS Report

The IUWDS is a permanent service of URSI, IAU and IUGG which "aims to provide information rapidly to the world scientific community to assist in the planning, coordination, and conduct of scientific work in relevant disciplines". The current Chairman is Dr. Richard Thompson, IPS, Australia. Secretary is Gary Heckman, NOAA, ERL, Boulder. Secretary for World Days is Helen Coffey, NOAA, NESDIS, Boulder. The Steering Committee is made up of representatives from the worldwide network of Regional Warning Centers and Auxiliary Regional Warning Centers. Publications include the International Geophysical Calendar which recommends dates for solar and geophysical observations which cannot be carried out continuously. The Spacewarn Bulletin is issued montly by WDC-A for Rockets and Satellites in response to recommendations by COSPAR. Geoalerts and Ursigram messages are sent in real time by telex to Regional Warning Centers around the globe. This is the lifeblood of the IUWDS network. Expenses are borne by the host institutions. In addition, IUWDS holds workshops on solar terrestrial predictions. The 1984 Workshop in Meudon resulted in a 636 page Proceedings and a special edition of Artificial Satellites in 1987 by the Polish Academy of Sciences Space Research Centre. The next Workshop is planned for October 1989 in Leura, Australia (near Sydney). IUWDS provides a number of very useful services to the solar community, and is always open to suggestions on how to better fit the needs of the scientific community.

Solar Interplanetary Variability (SIV)

The SIV program, 1988-1990, sponsored by SCOSTEP, is international, interdisciplinary and includes spacecraft and ground-based observations. It is an effort to look at the transition between solar minimum and solar maximum. At minimum, the large scale solar wind structure consists principally of a few recurrent streams which corotate with the Sun. At maximum, changes in the solar wind are dominated by transient events that occur in intervals much less than one solar rotation. SIV looks at how it evolves from one to the other. More computer networking is encouraged to exchange and analyze scientific data. Workshops will be held to study specific events and/or the evolution of solar-interplanetary structures.

Sunspot Index Data Center (SIDC), Belgium

The SIDC produces the daily International Sunspot Number Ri. It broadcasts a monthly provisional number within 24 hours following the end of each month. It also forecasts Ri 6, 9 and 12 months ahead as requested by CCIR. Quarterly, it issues the final International Sunspot Number Ri. For the IAU QBSA it provides Ri, Rc (an index for the central zone of the disk), and also an estimate of the daily sunspot area. 90 stations currently submit data for the Ri. The SIDC monthly "Sunspot Bulletin" is mailed to 350 addresses before the sixth of the month. The latest forecast of sunspot maximum is 170 ± 25 in September 1989. The work of the engaged data collectors and their great service for the community is acknowledged by the President.

S.T. Wu reported on planned SCOSTEP activities, including STEP (Solar-Terrestrial Energy Program) planned for 1990-5, and the next Solar Terrestrial Physics Symposium to be held in 1990 at the next COSPAR meeting. He also described organizational changes of SCOSTEP including a change of its constitution, the proposed organization of STEP and the relation of SCOSTEP to the International Geosphere Biosphere Program. Wu is reelected to represent the IAU in SCOSTEP.

The activity and budget of FAGS (Federation of Astronomical and Geophysical Data Services) are highlighted by E.A. Tandberg-Hanssen.

Report from FAGS

The Federation of Astronomical and Geophysical Data Analysis Services (FAGS) was established by ICSU to oversee the activities of the following services.

International Earth Rotation Service, IERS
Quarterly Bulletin of Solar Activity, QBSA
Intl. Service for Geomagn. Indices, ISGI
Permanent Service for mean sea level, PSMSL
Bureau Gravimétrique international, BGI
Intl. Centre for Earth Tides, ICET
Intl. Ursigram and World Days Service, IUWDS
World Glacier Monitoring Service, WGMS
Sunspot Index Data Centre, SIDC
Centre de Données Stellaires, CDS

Of these series, the QBSI, IUWDS and SIDC are of direct interest to Commission 10, and they receive a modest financial support from FAGS.

The IAU is represented on the FAGS council by Drs. Kovalevsky and Tandberg-Hanssen. At the Commission 10 and 12 joint business meeting on 5/8/88, Tandberg-Hanssen gave a report on the activities of FAGS since the 19th General Assembly. After the nomination by the President of commission 10 for Tandberg-Hanssen to continue to represent IAU on the FAGS council for the following 3 years, Tandberg-Hanssen was so elected.

J.C. Pecker presented two resolutions for consideration by Commissions 10 and 12. The first resolution was accepted:

Resolution of Commissions 10 and 12

Considering
the scientific importance of insuring a continuous survey, over a long period of time, of solar phenomena as observed in photospheric, chromospheric and coronal layers, in the solar wind and in solar-terrestrial relations;
recommends
a) the pursuit of these types of continuous observations wherever they are already conducted,
b) their undertaking in other observatories especially in consideration of broad longitude coverage to insure continuity,
c) financial assistance by the IAU for the particular programs which have previously been recommended by Commissions 10 and 12, and in particular the essential Debrecen photoheliograph surveys which face a difficult financial situation.

The second resolution concerning long term astrometric measurement of the solar diameter did not receive enough support for passage.

B. Dennis reported on the Flares 22 program sponsored by SCOSTEP under the umbrella of STEP. The Chairman is M. Machado and a steering committee will be determined. A resolution to have this program recognized by the IAU is unanimously accepted:

Resolution of Commission 10

Recognizing
the value of a coordinated scientific programme for the study of physical processes and mechanisms heading to solar active phenomena and flares, and recognizing further that most of these processes and mechanisms are common to a large variety of astrophysical objects and are responsible for interplanetary phenomena which affect the terrestrial environment.

Noting:
that a comprehensive study of all these components of solar activity exceeds the capabilities of any single country.
Further noting that several countries - the USA, USSR and People's Republic of China - already have written plans for coordinated studies during the next solar maximum, and that they have expressed their willingness to participate in a coordinated international campaign.

Commission 10
proposes that the IAU cosponsors the Flare 22 programme, to be in effect during the maximum of solar cycle 22 under the umbrella of SCOSTEP's Solar Terrestrial Energy Programme (STEP)

also recommends that a member of Commission 10 be appointed to the Flare 22 steering committee.

The Presidents of Commissions 10 and 12 then described concerns voiced at the IAU Commission President's meeting. A declining attendance and interest in the general assembly, particulary among young astronomers, were of particular concern. A lively discussion followed with several proposals offered from the floor. Among these were suggestions to have more one day meetings, allow contributed papers to be presented, publish papers from the meetings in the manner of COSPAR, reduce meeting competition with COSPAR, and improve information about the content of meetings prior to the meeting.

A discussion of a merger of Commissions 10 and 12 into a single commission on the Sun occupied the remainder of the business meeting. The Presidents of Commissions 10 and 12 stated that the division of solar research into different domains is not justifiable on scientific grounds. The activity of the solar commissions could be improved if forces were joined. Thus, they propose a merger. A vigorous discussion followed with both supporting and dissenting views expressed. The prevailing view was support for the merger on scientific grounds but concern about weakening the political role of solar research in respect to other astronomical research presented by IAU commissions. A straw vote favoured merger by a ratio of more than two to one, however since less than 5% of the membership of the Commissions was present, the following resolution was adopted:

The joint meeting of Commissions 10 and 12 supports in principle the merger of the two Commissions. The meeting further requests the organizing committees of the two Commissions to produce within the next twelve months a specific proposal for such a merger, that this proposal be put to a postal ballot of the members of the two commissions and if supported by a majority of those voting in each commission then the proposal be forwarded to the IAU Executive Committee as a formal proposal for the merger of Commissions 10 and 12.

SCIENTIFIC MEETINGS

1. Solar and Stellar Coronae (4 August; 09:00-12:30, 14:00-17:30)

This Joint Commission Meeting (Commissions 10, 12, 29, 35, 36, 44) was organized by R. Falciani and E. Priest. The meeting was dedicated to the memory of the late Gordon Newkirk, Jr. and consisted of invited papers and contributed posters. Publication will be in the Transactions of the IAU.

2. Large-Scale Computer Simulation of Solar Active Phenomena (5 August; 14:00-17:30)

This joint meeting sponsored by Commissions 10 and 12 was organized by G. van Hoven and S.T. Wu. A summary will be given in the transactions of the IAU.

3. Results of Systematic Observations of the Sun (8 August; 09:00-12:30, 14:00-17:30.

This joint meeting sponsored by Commissions 10 and 12 was organized by J.C. Pecker and P. Wilson. The meeting was dedicated to Prof. Helen Dodson Prince who was unfortunately unable to attend. Both invited papers and contributed posters were presented. A summary will be given in the transactions of the IAU.

WORKING GROUP ON ECLIPSES (1985-1988)

Chairman: E. Hiei

Several eclipses occurred during 1985-88, but the total eclipse of 13 March 1988 in Indonesia, the Philippines, and the north-west Pacific, was scientifically attractive. There was a Philippine National Committee for the eclipse.

Before the next General Assembly in 1991, two total eclipses will occur: one on 22 July 1990 in Finland and U.S.S.R., the other on 11 July 1991 in Hawaii, Mexico, Central America, Colombia, and Brazil.

The Working Group held meetings on 3 and 6 August to discuss the astronomical information concerning the eclipses, scientific objectives, and a Coordinator who may assist members in obtaining non-astronomical information as well as logistic assistance, and to exchange information on the possible observing sites.

LARGE-SCALE COMPUTER SIMULATION OF SOLAR ACTIVE PHENOMENA: SUMMARY OF A SESSION AT THE IAU GENERAL ASSEMBLY ON AUGUST 2 - 11, 1988, BALTIMORE, MARYLAND, U. S. A.

G. Van Hoven
University of California, Irvine
and
S. T. Wu
University of Alabama in Huntsville

ABSTRACT. The speakers at this session have demonstrated, by argument and example, that large-scale computer simulation represents a powerful method for understanding the nonlinear physics of solar active phenomena. The past decade, in particular, has seen the advent of ever more powerful large-scale computers and simulation algorithms, which have accelerated this trend. Thus, a number of complex problems concerning the nonlinear dynamics and energetics of astrophysical activity have now become solvable and have begun to reveal new and interesting observable phenomena. To illustrate this scenario we have organized this session to present current results which are obtained by computer simulation, along with their theoretical backgrounds. Some of the highlights are described in the following.

1. ACTIVE REGION EVOLUTION AND FILAMENT CONDENSATION (D. D. SCHNACK)

Among the important unsolved problems of coronal physics are the formation and evolution of prominences, the dynamics of active regions, and convective-input mechanisms for coronal heating. These long-wavelength, low-frequency, nonlinear, dynamical phenomena are, to an excellent approximation, well described by the equations of resistive magnetohydrodynamics (MHD), and are generally characterized by motions that occur on widely separated timescales. For example, dynamical events that evolve on an intermediate timescale may occur as a result of slow changes of the global configuration. Such a scenario may lead to an eruptive prominence or a solar flare, while also exciting Alfven waves.

Historically, attempts at understanding these dynamical events have followed three courses: The conception of imaginative cartoons; the calculation of sequences of static equilibria; and the numerical solution of the time-dependent equations of motion as they evolve in response to driven boundary conditions. The limitations of the first approach are obvious. The second approach can describe adiabatic evolution, but is often restricted by the assumptions made for analytic simplification, and cannot address equations of stability or disruption. The third approach is potentially the most fruitful, but has been limited by insufficient computer resources and/or inefficient algorithms. Recently, however, advances in computational techniques and the general accessibility of supercomputers have made solutions by numerical simulation feasible. Three recent examples of such efforts are described briefly in the following.

Two-dimensional, nonlinear simulations have demonstrated for the first time the self-consistent formation of a cool, dense filament in a magnetized coronal plasma (Van Hoven et al., 1987). These filaments are thermally insulated from the rest of the corona by a sheared magnetic field, $[\vec{B}(\vec{r})]$ which has been found to be crucial to their formation. The evolution proceeds dynamically in two steps with radiative cooling followed by parallel (to $\vec{B}$) condensation, as described in Section 2. The dynamics of active regions have been studied by two- and three-dimensional simulations of the evolution of a coronal arcade in response to an imposed photospheric shear flow (Wu et al., 1983; Mikic et al., 1988a). The evolution proceeds through a slow increase in coronal magnetic energy. The work of Mikic et al. shows that, when a critical value of magnetic shear is exceeded, the configuration becomes unstable and then evolves on a fast timescale, releasing a significant amount of the accumulated magnetic energy and ejecting a plasmoid. These calculations demonstrate that a significant amount of magnetic energy can be slowly stored in the corona and then quickly released; the resulting self-consistent dynamics have many features in common with two-ribbon flares.

Three-dimensional simulations are attempting to address questions concerning the heating of the corona by the dissipation of electric currents arising from small scale fields (Parker, 1983). Ideal MHD simulations have considered the dynamical response of the corona to sequences of long-wavelength braidings at the photosphere. It has been found that the quadratic nonlinearities in the MHD equations can lead to current structures in the corona whose steadily decreasing scale widths are much smaller than those imposed at the boundaries (Mikic et al., 1988b). Exact equilibria are obtained without distinct current sheets (i.e., true discontinuities). Further studies are required to determine the dissipation and heating rates in these growing current structures.

2. RADIATIVE CONDENSATION INSTABILITIES (L. SPARKS AND G. VAN HOVEN)

Prominences are relatively dense, gaseous structures which form in a few hours and extend thousands of kilometers into the sun's atmosphere. It is believed that prominences result from a radiative-cooling instability (Field, 1965) that occurs when heat flow is inhibited by solar magnetic fields (Chiuderi and Van Hoven, 1979). This instability is driven by optically-thin radiation, mainly from heavy-ion lines, which increases as the temperature falls from coronal values. The resulting local pressure drop drives mass inflow along the magnetic field, producing runaway cooling and condensation. The UCI (University of California, Irvine) group has been engaged in a long-term study of the condensation instability in the presence of magnetic shear. Linear studies (Van Hoven et al., 1986, and references cited therein) have discovered a new kinematic condensation mode in which energy balance (esp. anisotropic thermal conduction) is paramount in setting the characteristic scale of spatial variation. For such short-wavelength modes (Drake, Sparks and Van Hoven, 1988), which are of greatest interest in explaining prominence formation, plasma flow is primarily parallel to the magnetic field. Hence, the field does not impede plasma compression, and significant density enhancement can occur.

To study the nonlinear evolution of the condensation process, we have conducted numerical simulations using a two-dimensional sheared-field model, with the results shown in Figure 1. The process occurs in two stages, starting from a low-level noise excitation as shown in Figure 1.(a). First the temperature drops to a relative level of ~ 0.03 in the central portion of the magnetic shear layer, where electron heat conduction is inhibited by the field-line tilt, as shown in Figure 1.(b). At this point, the density has only increased by a factor of ~ 2, as in Van Hoven, Sparks, and Schnack (1987), but significant parallel (to $\vec{B}$) pressure gradients have arisen away from the center of the shear layer. In the second stage, the plasma flows along (off-center) field lines, further increasing the radiation, leading to the narrow (knife-blade-like) dense and cool filament shown in Figure 1.(c), which has n,T values agreeing with prominence observations.

3. MAGNETIC RECONNECTION: PARALLEL ELECTRIC FIELDS AND HELICITY (K. SCHINDLER)

Large-scale numerical studies in cosmic plasma dynamics profit from tools that allow one to optimize the choice of parameters and to provide a suitable and effective framework for analyzing the results.

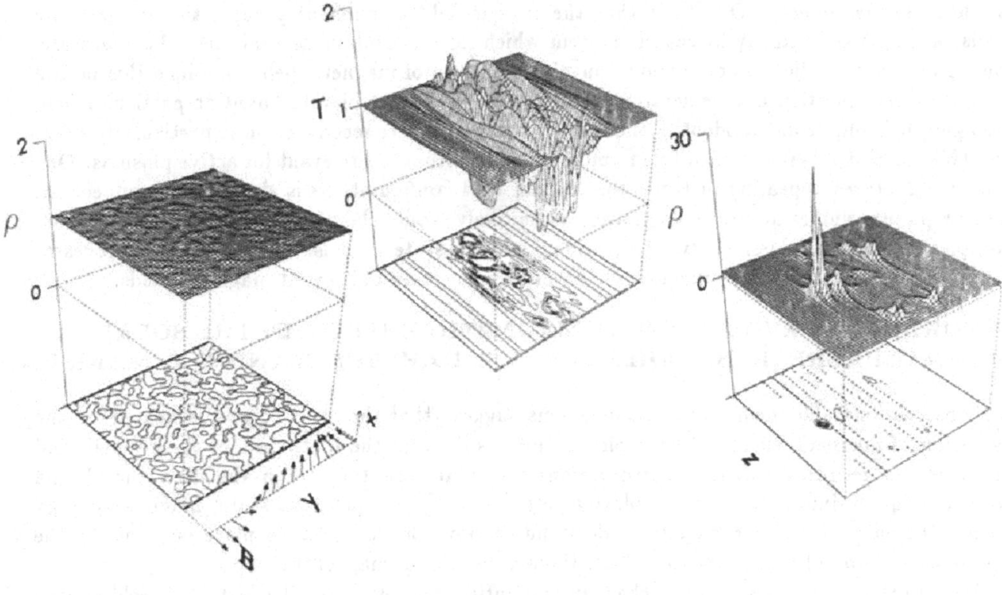

Figure 1. Filament condensation in a sheared magnetic field, as indicated in part (a). The left-hand figure shows the initial, random, density input, in a coordinate system in which the y axis is proportional to sinh(y), and the z axis is uniform. Part (b) displays the temperature profile (and contours) after the first nonlinear radiative-instability phase which is nearly isochoric. Part (c) shows the density profile late in the second, nearly isobaric phase, along with the parallel to (to $\vec{B}$) flow vectors. The sharp prominence condensation has width (in y) and height (in z) of $\sim 10^{-1}$ and 1 magnetic shear scales, respectively. The minimum temperature is $\sim 10^{-2}T_c$ and maximum density $\sim 10^{1.4}$, n_c, in agreement with filament/prominence observations.

For numerical stability studies, an investigation of the bifurcation properties of the underlying equilibrium problem proves particularly useful. Clearly, a linear instability associated with a forward bifurcation (with the control parameter increasing for decreasing stability) will not lead to dramatic dynamic effects such as required for solar flares. It is suggested that the evolution on branches that are stable with respect to dissipationless theory but unstable with respect to dissipative perturbations might play an important role in cosmic plasma activity. The onset point would be determined by mechanisms such as microinstabilities that rapidly enhance dissipation. An important example is the spontaneous formation and ejection of plasmoids. In particular, this process raises the question of whether collisional losses are sufficiently strong to provide the required dissipation (taking enhancement by current concentration into account) or whether collective loss mechanisms are required.

The majority of theoretical models of stellar and magnetospheric activity involve magnetic reconnection in one way or another. In fact, it seems reasonable to expect that progress in cosmic plasma activity depends on sufficient knowledge about reconnection. Up to now, mostly symmetric systems (e. g., those with translational in variance) have been modified by two-dimensional, analytical and computational techniques. The step to three-dimensional descriptions poses a number of conceptual problems, since the notion of magnetic reconnection in terms of separatrices and separators of the magnetic field seems to be inadequate in the general case. A concept that avoids such difficulties is based on the breakdown of magnetic-line conservation together with a localized non-ideal plasma process. One finds that the electric-field component parallel to the magnetic field is the essential quantity in generic system which do not contain neutral lines. Furthermore, spontaneous reconnection processes must involve a change of magnetic helicity. Since this notion of magnetic reconnection is considerably broader than the earlier picture based on particular field topologies, it is important to identify those configurations where reconnection is particularly effective. This concept is being applied to a number of model situations relevant for active plasmas. One of the new features appearing in three-dimensional field configurations is stochastic reconnection. The new picture unifies all processes that one intuitively would classify as magnetic reconnection. Examples include the classical, two-dimensional, steady-state models, time-dependent processes such as the tearing instability, and the reconnection relaxation of sheared magnetic fields.

4. TURBULENT EVOLUTION OF MAGNETIC ENERGY IN THE SOLAR ATMOSPHERE (R. B. DAHLBURG, J. P. DAHLBURG, AND J. T. MARISKA)

Both observations and theoretical considerations suggest that the solar corona is heated by the dissipation of magnetic energy. Photospheric motions convect the footpoints of the coronal field structures; the resulting magnetic perturbations propagate into the corona and drive the plasma into a non-equilibrium state. As the plasma settles into a new quiescent state, magnetic energy is converted into heat. One step to understanding how this might take place is to study the viscoresistive decay of a helically turbulent, three-dimensional magnetofluid.

We find that, when there is no background potential magnetic field, the perturbed field relaxes through a series of force-free (but non-constant $\alpha = J/B$) states to a new equilibrium characterized by larger-scale structures. During this relaxation almost as much heat is released through viscous dissipation as through current sheets. When the same calculation is performed with a background potential field present, we find that heat production is inhibited and structure tends to develop parallel to the background field. We also see evidence for the generation of Alfven waves, which might be observable in the solar corona. The detailed results are given by Dahlburg et al. (1988).

5. NUMERICAL TECHNIQUES FOR THE MODELING OF MAGNETIC FIELDS (G. A. GARY AND S. T. WU)

The upward extrapolation of photospheric fields employing the nonlinear force-free magnetic-fields equations is being pursued by a number of methods and investigators. The general process has been classified by Aly (1988) either as proceeding either by iterative methods or by the progressive extension method, which are generally variational or numerical-difference techniques respectively. The variational methods being investigated include those introduced by T. Sakurai (Univ. of Tokyo/Japan), D. Pridmore-Brown (Aerospace Corp./USA), P. A. Sturrock (Stanford Univ./USA), A. van Ballegooijen (Center for Astrophysics/ USA), M. Semel (Meudon Obs./France), and R. Kress (Univ. of Gottin gen/FDR). The numerical-difference method is being investigated by S. T. Wu (UAH/USA).

As a result of the studies of J. J. Aly, H. Grad, M. M. Molodensky, B. C. Low and others, the important questions of convergence, stability, uniqueness, and existence are being answered. There are upper bounds to the strength of the possible electric currents in an active region for force-free fields. The existence of singularities remains an open question of analytic vs. discontinuous solutions. However the comparison of the two methods of extrapolation, in the near future, will provide a crucial test of our general ability to extrapolate the photospheric field into the corona. The iterative method insures that the force-free fields are stable against variations of the boundary conditions, while the progressive-extension method provides necessary-and-sufficient conditions for the unique determination of all the first derivatives, and the uniqueness of the solution in some domain.

Using the numerical-difference method, Wu et al. (1984) have shown the convergence of an algorithm both for analytic models and for data from the Marshall Space Flight Center vector magnetograph. The analytic model shows that the numerical solution has errors less than a few percent at the present time. Further testing will provide the solar community with an additional procedure independent of T. Sakurai's code (1981), allowing for the verification of the extrapolated fields.

6. CONCLUDING REMARKS

Other interesting results were presented concering the shocks produced by impulsively driven reconnection (Forbes, 1988), nonlinear reconnection in the solar corona (Steinolfson) and on coronal-loop thermal instability (Einaudi and Demoulin, 1988). In short, the presentations of this session showed the importance of large-scale numerical simulations and the role they will play in the understanding of the nonlinear dynamics and energetics of solar active phenomena.

ACKNOWLEDGEMENT

The contributions of GVH to this report were supported by NSF and NASA; and those of STW by NASA grant NAGW-9 and Air Force grant AFSOR- 88-0013.

REFERENCES

Aly, J. J., (1988) "On the construction of the nonlinear-force-free coronal magnetic field from boundary data", Solar Phys. (submitted).

Chiuderi, C., and Van Hoven, G., (1979) "The dynamics of filament formation: the thermal instability in a sheared magnetic field", Astrophys. J. (Letts.), 232, L69.

Drake, J. F., Sparks, L., and Van Hoven, G., (1988) "Radiative instabilities in a sheared magnetic field", Phys. Fluids, 31, 813.

Dahlburg, R. B., Dahlburg, J. P. and Mariska, J. T., (1988) "Helical Magnetohydrodynamic Turbulence and the Coronal Heating Problem" Astron. Astrophys., 198, 300.

Einaudi, G., and Demoulin, P., (1988), in preparation.

Field, G. B., (1965), "Thermal Instability", Astrophys. J., 142, 531.

Forbes, T. G., (1988) "Shocks Produced by Impulsively Driven Reconnection" Solar Physics (in press).

Mikic, Z., Barnes, D. C, and Schnack, D. D., (1988a) "Dynamical evolution of a solar coronal magnetic field arcade", Astrophys. J., (to appear).

Mikic, Z., Schnack, D. D, and Van Hoven, G., (1988b) "Creation of current filaments in the solar corona" Astrophys. J., (to appear).

Parker, E. N., (1983) "Magnetic neutral sheets in evolving fields. I. General theory", Astrophys. J., 264, 635.

Sakurai, T., (1981), "Calculation of force-free magnetic field with non-constant α), Solar Phys., 69, 343.

Van Hoven, G., Sparks, L. and Schnack, D. D., (1987) "Nonlinear radiative condensation in a sheared magnetic field", Astrophys. J. (Letts), 317, L91.

Van Hoven, G., Sparks, L, and Tachi, T., (1986) "Ideal condensations due to perpendicular thermal conduction in a sheared magnetic field", Astrophys., J., 300, 249.

Wu, S. T., Hu, Y. Q., Nakagawa, Y. and Tandberg-Hanssen, E., (1983) "Induced Mass and Wave Motions in the Lower Solar Atmosphere. I. Effects of Shear Motion of Flux Tubes", Astrophys. J., 266, 866.

Wu, S. T., Chang, H. M. and Hagyard, M. J., (1984), "On the Numerical Computation of Nonlinear Force-Free Magnetic Fields" in Measurements of Solar Vector Magnetic Fields (M. J. Hagyard, ed.) NASA Conference Publ. 2374, pp. 17-26.

COMMISSION No. 14

ATOMIC AND MOLECULAR DATA

(DONNEES ATOMIQUES ET MOLECULAIRES)

Report of Meetings 3 and 9 August 1988

PRESIDENT: R.W. Nicholls

Business Session, 3 August 1988

1. Activities and Scope of the Commission

A brief review was given of the substance of the discussion of the previous business meeting of Nov. 21 1985 on the subject matter of the commission. There was general endorsement of the view of that meeting that the purpose of the Commission was to keep a watching brief on the energy exchange processes of atomic and molecular physics, and of related spectral and structure data of atomic and molecular species. Such information is essential to the use of astronomical observations for astrophysical diagnosis and modelling.

In the discussion which followed it was emphasised that many research communities (including chemical physics, laser physics, plasma physics, atmospheric physics and chemistry and astrophysics) produced data of importance to the commission, some of whose members were also members of the commission.

The research areas of particular interest to the commission include:

Photon phenomena (spectral wavelengths, intensities and transition probability data)
Particle phenomena (atomic, molecular, ionic and electronic collision phenomena and cross sections)
Line broadening phenomena

Accordingly the following working groups were established for 1985-88:

Atomic Spectra and Wavelength Standards (W.C. Martin)
Atomic Transition Probabilities (W.L. Wiese)
Collision Processes (A. Dalgarno)
Line Broadening (N. Feautrier)
Molecular Structure and Transition Data (W.H. Parkinson)

2. Appointment of Officers for the period 1988-91

The following committee was approved:-

President: S. Sahal-Brechot
Vice President: W.L. Wiese
Organizing Committee:
A.H. Gabriel
T. Kato
F.J. Lovas
S.L. Mandel'shtam
R.W. Nicholls
H. Nussbaumer
W.H. Parkinson
Z.R. Rudzigas
W.L. Wiese

The commission expressed its gratitude to A. Dalgarno for his outstanding services for a number of years as chairman of the working group on collision processes.

Working Groups and their Chairmen for 1988-91 were approved as follows:

1: Atomic Spectra and Wavelength Standards (W.C. Martin)
2: Atomic Transition Probabilities (W.L. Wiese)
3: Collision Processes (S. Sahl-Brechot and W.L. Wiese to discuss appointment of new Chairman)
4: Line Broadening (N. Feautrier)
5: Molecular Structure and Transition Data: (W.H. Parkinson)

Scientific Session, 9 August 1988

SPECTROSCOPIC CONSTANTS AND FRANCK CONDON FACTORS FOR THE MOLECULES OBSERVED IN THE COMETARY SPECTRUM OF HALLEY

B. Petropoulos
Research Center for Astronomy and Applied Mathematics
Athens Academy

Spectroscopic constants have been reviewed for the molecules observed recently in the spectrum of comet Halley. Franck-Condon factors have been calculated for these molecules.

SURVEY OF THE NEEDS OF ASTRONOMERS FOR NEW OR IMPROVED ATOMIC AND MOLECULAR PARAMETERS

Peter L. Smith
Harvard-Smithsonian Center for Astrophysics
60 Garden St., Cambridge MA 02138 U.S.A.

In 1984, the "Committee on Line Spectra of the Elements - Atomic Spectroscopy" [CLSE-AS] of the U.S. National Research Council, began some informal surveys of the requirements for new and improved atomic data in a number of fields. As a member of that committee, I agreed to enquire about such needs, as well as for molecular parameters for "optical" wavelength (30nm to 10μ) astronomy. Notice of the survey was mailed to 440 astronomers, who had demonstrated an interest in optical wavelength astronomical spectroscopy, distributed to all registrants at the June 1988 meeting of the American Astronomical Society, and published as letters to the editors of Newsletters of the American and Canadian Astronomical Societies. A poster paper that described the survey was presented at the January 1988 meeting of the American Astronomical Society [see Bull. Am. Astron. Soc. 19, p.1064, 1987]. Those who responded to the notices were asked to document their data needs in a particular format that was intended to enable producers of such data rapidly to understand the requirements and to be stimulated into providing them.

About 35 requests for atomic and molecular data were received. Some were well considered, documented, multipage summaries. Others were brief electronic mail statements. A number of the requests were from workshop chairs, working groups, satellite instrument teams, groups of collaborators etc., and therefore represented the collected opinions of a number of researchers. All requests will be collated, reproduced, and distributed, especially to producers of atomic and molecular data, both theorists and experimenters.

In this brief summary of the survey, it is impossible to provide a comprehensive review of the many types of data that are needed. However, one type of request stood out. A significant fraction of the responses requested greatly improved wavelength and energy level data, accurate to one part in 10^6 to 10^7 for common species, such as singly ort doubly ionized iron peak and rare-earth elements, as well as Li-like and Be-like ions.

There have been some encouraging improvements in laboratory capabilities for this sort of work in recent years, in particular, the development of the Imperial College (UK) and the Los Alamos National Laboratory (USA) Fourier transform spectrometers. It is likely that these instruments will be frequently employed in obtaining better atomic and (molecular) wavelength and energy-level data for astronomical spectroscopy. Much of the progress in astronomical research depends upon knowledge or expertise from other scientific fields; atomic and molecular data are typical. Astronomers who need such information must support activities by making their needs known, by collaborating in obtaining funds for research, by participating in the collection and analysis of the data, and by encouraging publication of the results in the astronomical literature.

Those who wish to obtain a copy of the compilation of requests for new atomic and molecular data should contact the author at the above address. Scientists with an interest in encouraging the acquisition of more and better atomic data for use in all areas of research, development and manufacturing, are referred to the proceedings of a related workshop on atomic data needs for analytical chemistry: "Needs for Fundamental Reference Data for Analytic Atomic Spectroscopy", edited by P. Boumans and A. Scheeline [Spectrochimica Acta, 43B pp 1-127, 1988]. Consideration of this report will show the considerable overlap between the atomic data needs of astronomers and those of analytical chemists.

I thank W.H. Parkinson, A. Dalgarno and R.E. Stencel for their advice and comments when this survey was being planned, and J.W. Linsky, S.J. Shawl and C.D. Scarfe for their assistance in publicizing it.

The work was supported in part by NASA Grant NSG-7304 to Harvard University.

ATOMIC SPECTRA - AN UPDATE

W.C. Martin
National Bureau of Standards
Gaithersburg
MD 20899
USA

A few forthcoming papers can be cited here to update the 1988 report of Working Group 1. The issue of J. Opt. Soc. Am. B for October 1988 (Vol. 5, No. 10) features a group of papers submitted as a tribute to Charlotte E. Moore-Sitterly on the occasion of her ninetieth birthday. Included are a review by S. Johansson and C.R. Cowley of research on the first three spectra of the iron-group elements, with special emphasis on the completeness of the line lists and analyses in view of astrophysical needs; a paper by C.M. Brown *et al.* on Fe I absorption in the 1550-3215 Å region that includes some 3000 lines; new high-accuracy measurements by R.C.M. Learner and A.P. Thorne of 300 Fe I lines in the 3830-5760 Å range; and accurate wavelengths for 558 Pt II lines over the range 1032-5223 Å as determined by J. Reader *et al.* The Pt I and Pt II spectra are of special interest for wavelength standards and will be used for calibration of spectra obtained with the High-Resolution Spectrograph of the Hubble Space Telescope.

The Atomic Energy Levels Data Center at the National Bureau of Standards has begun a program of compilations of wavelengths and energy-level classifications for spectral lines of elements in all ionization stages. A paper by V. Kaufman and J. Sugar giving these data for all the Sc spectra (Sc I - Sc XXI) will appear in J. Phys. Chem. Ref. Data 17, No. 4 (1988), and a similar compilation for the Si spectra is underway. Reference (87) of the Working-Group 1 report was cited incorrectly; the correct reference for the compilation of spectral data and Grotrian diagrams for Ni IX - Ni XXVIII is T. Shirai *et al.*: 1987, At. Data Nucl. Data Tables 37, p.235.

ATOMIC TRANSITION PROBABILITY MEASUREMENTS WITH THE NSO 1-m FTS

J.W. Brault
NOAO/NSO
Tucson, Arizona
and
W. Whaling
California Institute of Technology
Pasadena, California

The simplest method for measuring atomic transition probabilities is a two step process requiring (1) the radiative lifetime of the upper level of the transition and (2) the branching fraction, the probability that the upper level decays via the particular transition of interest. This latter probability is found by measuring the relative intensity of all of the decay channels that contribute significantly to the total decay strength of the upper level.

The first step has been greatly advanced by modern pulsed tunable UV lasers combined with nanosecond electronics. Atomic lifetimes are now measured routinely with 5% precision, and much better in special cases. This report deals with efforts to achieve similar precision and reliability in the spectrophotometry of branching ratios and with the exploration of a technique that would avoid the necessity to measure lifetimes for all levels.

Fourier transform spectroscopy has proven to be the best way to measure branching ratios. It has long been noted for its high accuracy, resolution and throughput, but for branching measurements the most important characteristic is that all of the decay channels can be measured at the same time. Any change in the brightness of the source while the spectrum is being recorded affects all branches equally, so branching ratios are unaffected. The 1-m. FTS at the McMath Solar Telescope on Kitt Peak is the best instrument currently available for these measurements, with a broad spectral range (250-1100 nm in a single spectrum; to 5500 nm in two) that extends far enough into the UV to encompass the great majority of atomic transitions and many lines of the singly-charged ion.

Recently we have explored a scheme that would take advantage of another feature of the FTS that has not been exploited in the work mentioned above: its awesome data-collection capability. This work used an inductively coupled argon plasma (ICP) source containing a trace amount of the atom to be studied. The ICP source is widely used by analytical chemists, and many papers in the spectrochemical literature have reported that excitation conditions in the ICP closely approximate thermodynamic equilibrium. This claim was tested by measuring the relative intensity in the ICP spectrum of 659 emission lines of Mo I of known transition probability (from lifetimes/branching fractions) to find the relative population of 67 levels in the ICP source. The measured level populations fit a Boltzmann distribution with an RMS deviation of 7.5%, which is consistent with the uncertainty of the transition probabilities used in evaluating the populations. From this exponential distribution of measured level populations, the population of 204 levels for which the lifetime is not known could be interpolated. Once the population of a level is determined, the relative intensity of any emission line from that level determines the transition probability for that line. Transition probabilities for all other decay channels from the same upper level are then found from the emission branching ratios; it is not necessary to measure complete branching fractions.

Transition probabilities for 2176 lines were determined in this way with a precision limited by the 8% uncertainty in the level population, whereas for strong branches the branching fraction/lifetime method can achieve 3%, the uncertainty of the best lifetime values. However, the population method gives access to many more lines - virtually all of the classified lines in the 250 -1150 nm. range of these spectra.

In practice, the branching ratios were measured in the hollow cathode (HC) spectrum as well as in the ICP spectrum; both spectra were recorded with the D 1-m FTS. Comparison of the branching ratios in the two spectra helps eliminate difficulties with blends (in the ICP) and self-absorption (in the HC). This technique is now being applied to the MO II lines in the same spectra in collaboration with J.E. Lawler (U. Wis.), who is measuring Mo II level lifetimes needed to investigate the population distribution of ionic levels in the ICP source.

SEMIEMPIRICAL CALCULATION OF gf VALUES FOR THE IRON GROUP

Robert L. Kurucz
Harvard-Smithsonian Center for Astrophysics
Cambridge, Massachusetts, USA

ABSTRACT: Lines of iron group elements that go to excited configurations that have not yet been studied in the laboratory produce considerable opacity in stars. I have computed new line lists for the first nine spectra of the iron group elements, Ca, Sc, Ti, V, Cr, Mn, Fe, Co, and Ni including as many configurations as practical at the present time. Next I will compute opacities for the temperature range 2000K to 200000K, for abundances ranging from 0.0001 solar to 10 times solar, for microturbulent velocities 0, 1, 2, 4, and 8 km/s. Then I will compute corresponding grids of models, fluxes, colors, and spectra.

In 1970 Kurucz and Peytremann (1975) computed gf values for 1.7 million atomic lines for sequences up through nickel using scaled-Thomas-Fermi-Dirac wavefunctions and eigenvectors determined from least squares Slater parameter fits to the observed energy levels. That line list provided basic data for computing opacities (Kurucz 1979a), model stellar atmospheres (Kurucz 1979a;b), and spectra (Kurucz and Avrett 1981), and produced tremendous improvement in the compaison between theory and observation (Relyea and Kurucz 1978; Buser and Kurucz 1978).

However, it failed for cool stars and the sun because it does not include molecules.

In 1983 working with Lucio Rossi from Frascati and with John Dragon and Rod Whitaker at Los Alamos I finally completed line lists for all diatomic molecules that produce important opacity in G and K stars. Those data were combined with all existing theoretical and laboratory data, a newer calculation for Fe II (Kurucz 1981), and series extrapolations for the light elements, to produce input for computing new distribution function opacity tables. The calculations involved 17,000 000 atomic and molecular lines. 3,500,000 wavelength points, 50 temperatures and 20 pressures, and took a large amount of computer time.

As a test the opacities were used to compute a theoretical solar model and to predict solar fluxes and intensities from empirical models. There are several regions between 200 and 350 nm where the predicted solar intensities are several times higher than observed, say 85% blocking instead of the 95% observed. The integrated flux error of these regions is several per cent of the total. In a flux constant model this error is balanced by a flux error in the red. The model predicts the wrong colors. In detailed ultraviolet spectrum calculations half the intermediate strength and weak lines are missing. After many experiments, I determined that this discrepancy is caused by missing iron group atomic lines that go to excited configurations that have not been observed in the laboratory. Most laboratory work has been done with emission sources that cannot strongly populate these configurations. Stars, however, show lines in absorption without difficulty. Including these additional lines will produce a dramatic increase in opacity, both in the sun and in hotter stars. A stars have the same lines as the sun but more flux in the ultraviolet to block. In B stars and in O stars there will be large effects from third, fourth and fifth iron group ions. Envelope opacities that are used in interior and pulsation models will also be strongly affected.

I am fortunate to have been granted a large amount of computer time at the San Diego Supercomputer Center by NSF (AST-8518900) to carry out new calculations. To compute the iron group line lists I determine eigenvectors by combining least squares fits for levels that have been observed with computed Hartree-Fock integrals (scaled) for higher configurations including as many configurations as I can fit into a Cray. My computer programs have evolved from Cowan's (1968) programs. All configuration interactions are included. The following table is an example for Fe II,

even					odd			
d6 4s	d6 4d		d5 4s2	d5 4s4d	d6 4p	d6 4f	d5 4s4p	d5 4s4f
5s	5d	d6 5g	4s5s	4s5d	5p	5f	4s5p	
6s	6d	6g	4s6s	4s6d	6p	6f	4s6p	
7s	7d	7g			7p		4s7p	
8s	8d		d5 4p2		8p		4s8p	
9s			d7		9p			d4s2 4p

	even	odd
22 Configurations	5723 levels	16 Configurations 5198 levels
largest Hamiltonian	1102 x 1102	1049 x 1049
Slater parameters	729	541
(many fixed at scaled Hartree-Fock)		

The laboratory data are from the computer tapes that NBS uses to print its energy level compilations (Sugar and Corliss 1985). Transition integrals are computed with scaled-Thomas-Fermi-Dirac wavefunctions and the whole transition array is produced for each ion. The forbidden transitions are computed as well. Radiative, Stark, and van der Waals damping constants and Lande g values are automatically produced for each line.

The least squares fits to determine the energy levels are now complete for the first 10 ions of the Fe group. The most complex spectra were done first before moving toward the simpler Ca end of the iron group. Fe II has been redone several times in collaboration with Sveneric Johansson from Lund who was been visiting Goddard (Johansson and Baschek 1988). We have added many more known levels, so the line lists will include many more Fe II lines with accurate positions. Johansson has just given me revised levels for Fe I so I will also redo those calculations. I will continue to revise these calculations for the iron group as new laboratory analyses become available, and I will continue on to other elements as time permits.

The following table shows the line lists completed at the present time, with the number of electric dipole lines saved for each ion,

	I	II	III	IV	V	VI	VII	VIII	IX
Ca	48573	4227	11740	113121	330004	217929	125560	30156	22803
Sc	191253	49811	1578	16985	130563	456400	227121	136916	30587
Ti	867399	264867	23742	5079	37610	155919	356808	230705	139356
V	1156790	925330	284003	61630	8427	39525	160652	443343	231753
Cr	434773	1304043	990951	366851	73222	10886	39668	164228	454312
Mn	327741	878996	1589314	1033926	450293	79068	14024	39770	147442
Fe	789176	1264969	1604934	1776984	1008385	475750	90250	14561	39346
Co	546130	1048188	2198940	1569347	2032402	1089039	562192	88976	15185
Ni	149926	404556	1309729	1918070	1971819	2211919	967466	602486	79627

The forbidden lines have not yet been tabulated. I have sorted all the lines and they will fill approximately thirty tapes by the time I complete writing them out. Most of these lines have uncertain wavelengths because they go to predicted levels. I have produced a single tape edition of these data that has all the lines with reliable wavelengths between laboratory determined energy levels.

I have not yet made exhaustive comparisons with laboratory measurements. In general the calculations are greatly improved over my earlier work and show considerably less scatter. Some of the calculated lifetimes agree perfectly with the best measurements. Fe II lifetimes are about 15% shorter than observed. There can still be considerable scatter for lines that occur only because of configuration interactions. I found a few typographical errors in the input energy data because the output line list had lines in the wrong positions. Those spectra were recomputed.

I am now combining all these new data with my earlier atomic and molecular data to prepare the input files for my opacity calculations. I will compute opacities for the temperature range 2000K to 200000K, for abundances ranging from 0.0001 solar to 10 times solar, for microturbulent velocities 0, 1, 2, 4, and 8 km/s. I hope to finish the solar abundance models. Then I will continue with lower abundance Population II opacities and models in the coming year, and I also hope to get to enhanced abundances as well. I estimate that it will take one hundred Cray hours per abundance. I will compute grids of models, fluxes, colors, and spectra for each abundance.

REFERENCES

Buser, R. and Kurucz, R.L. 1978. Theoretical UBV colors and the temperature scale for early-type stars. Astron. Astrophys., vol. 70, 555-563.

Cowan, R.D. 1968. Theoretical calculation of atomic spectra using digital computers. Journ. Opt. Soc. Amer., vol. 58, pp.808-818.

Johansoon, Se. and Baschek, B. 1988. Term analysis of complex spectra: new experimental data for Fe II. Nucl. Instr. and Methods in Phys. Res., vol. B31, pp.222-232.

Kurucz, R.L. 1979a. Model atmospheres for G, F, A, B, and O stars. Astrophys. Journ. Suppl., vol. 40, 1-340.

____________ 1979b. A progress report on theoretical photometry. Presented at the Workshop on Problems of Calibration of Multi-Color Photometric Systems, Dudley Observatory, March 17-18; published as Dudley Observatory Report No. 14, ed. A.G. Davis Philip, 363-382.

____________ 1981. Semiempirical calculation of gf values, IV: Fe II, Smithsonian Astrophys. Obs. Spec. Rep. No. 390, 317 pp.

Kurucz, R.L. and Avrett, E.H. 1981. Solar spectrum synthesis. I. A sample atlas from 224 to 300 nm. Smithsonian Astrophys. Obs. Spec. Rep. No. 391, 139 pp.

Relyea, L.J. and Kurucz, R.L. 1978. A theoretical analysis of uvby photometry. Astrophys. Journ. Suppl., vol. 37, 45-69.

Sugar, J. and Corliss C. 1985. Atomic Energy Levels of the Iron-Period Elements: Potassium through Nickel., Journal of Physical and

Chemical Reference Data, vol. 14, Supplement 2.

IDENTIFICATION ATLASES OF MOLECULAR SPECTRA

Ralph W. Nicholls
Centre for Research in Experimental Space Science
York University
4700 Keele Street
North York, Ontario
Canada M3J 1P3

In astrophysics, as in all research fields in which molecular spectra play important roles, there has been a continuing need for compendia of information on individual band systems. Firstly there is a requirement by observers and experimenters for aids to definitive identification of observed spectral features. Secondly there are needs by observers, by those who make diagnostic interpretations of spectral profiles, and also by theorists, for definitive critical compilations of appropriate molecular data with reference to the relevant literature. The continued popularity of various editions to Pearse and Gaydon's Identification of Molecular Spectra (1) and to Huber and Herzberg's Constants of Diatomic Molecules (2) are clear evidence of this.

In recognition of the need to augment the information in Pearse and Gaydon's compilation, the author and his colleagues established in the decade starting with 1964 an Identification Atlas of Molecular (3) series. Each of these roughly 20 page documents were devoted to one (diatomic) band system. It included a general description of the band system, its appearance and occurrence, an historical survey, tables of critically compiled molecular data, and plates of vibrationally identified spectra at low, medium and high resolution. These atlases were professionally printed, and because of cost the print runs were not large.

The more than a decade hiatus in Phase 1 of this project due to lack of funds (granting agencies do not look on such scholarly compilations as research !) has recently ended and Phase 2 has been initiated. This is a direct result of the formation of the Province of Ontario Centre of Excellence, The Institute for Space and Terrestrial Science and its support.

Phase 2 atlases have the same raison d'etre as those of phase 1. However the spectra will be presented as intensity plots rather than as photographic plates. The plots will be numerically derived using our extensive realistic spectral synthesis facility (4). The atlasses will be published "in house" inexpensively using desktop published facilities. The early atlases in the Phase 2 series will be devoted to atmospheric and astrophysical band systems.

REFERENCES

Pearse, R.W.B. and Gaydon, A.G. (1976). Identification of Molecular Spectra (Fourth Edition) (First Edn. 1941, Second Edn. 1950, Third Edn. 1963) Chapman and Hall London.

Huber, K.P. and Herzberg, G. (1979). Constants of Diatomic Molecules. Van Nostrand-Reinhold New York.

Nicholls, R.W. et al. (1964-1972). Identification Atlases of Molecular Spectra
1: AlO (A-X); 2: N_2(C-B); 3: N_2^+ (B-X)
4: O_2 (B-X); 5: C_2(A-X'); 6: O_2 (A-X)
7: VO (C-X); 8: CN (A-X), CN (B-X)
Centre for Research in Experimental Space Science York University.

Nicholls, R.W. and Cann, M.W.P. (1985). Realistic Numerical Synthesis of Molecular Spectra Trans IAU XIXB 146-152 1985.

COMMISSION No. 15

PHYSICAL STUDY OF COMETS, MINOR PLANETS AND METEORITES

(ETUDE PHYSIQUE DES COMETES, DES PETITES PLANETES ET DES METEORITES)

Report of Meetings 4, 6, and 9 August 1988

PRESIDENT: L. KRESAK VICE PRESIDENT: J. RAHE

Prof. L. Kresak was unable to attend the IAU General Assembly due to a sudden illness and had asked J. Rahe to be Acting President for the duration of the General Assembly. M.F. A'Hearn agreed to serve as secretary.

Tribute is paid to those members of Commission 15 who have passed away since the last General Assembly: Dr. Ludwig BIERMANN, Dr. Nicholas BOBROVNIKOFF, Dr. Richard Heinrich GIESE, Dr. Marie-Therese MARTEL, Dr. Reginald WATERFIELD.

The Commission had one Business Session, two Science Sessions on "News in Cometary Research" (organized by C. Arpigny), one Science Session on "News in Minor Planet Research" (organized by A.W. Harris), a Joint Discussion with Commission 20 on "Major Observing Programs and Data Bases" (organized by Y. Kozai and L. Kresak), and a Joint Discussion with Commission 22 on "Interrelation of Meteor Streams, Comets and Asteroids" (organized by P. Babadzhanov and L. Kresak)

I. BUSINESS SESSION

The triennial report, previously distributed to all members of the commission, was outlined by Rahe and accepted by the Commission.

Publication of the Atlas of Comet Halley 1910 II, sponsored by the Commission and published by NASA as SP-488, was noted, as was the availability of copies from B. Donn.

Following long tradition, members of the Organizing Committee who have served two terms will step down; these include C. Chapman, H. Fechtig, and L. Shul'man. Although J. Wasson has served two terms, he has only served one term as chairman of the Working Group on Meteorites and will continue as chairman and thus on the Organizing Committee for one more term. L. Kresak, as past President, will automatically be a member of the Organizing Committee.

The following new officers, members of the Organizing Committee, members of Working Groups, new Commission members and consultants were elected:

PRESIDENT: J. Rahe (FRG) VICE PRESIDENT: A.W. Harris (USA)
SECRETARY: M.F. A'Hearn (USA)

ORGANIZING COMMITTEE: C. Arpigny (Belgium), J. C. Brandt (USA), M. J. S. Belton (USA), O. V. Dobrovol'skij (USSR), E. Gruen (FRG), H. F. Haupt (Austria), D. Hughes (UK), L. Kresak (Czechoslovkia), D. F. Lupishko (USSR), H. Rickman (Sweden), E. Tedesco (USA), J. T. Wasson (USA), S. Wyckoff (USA), V. Zappala (Italy)

A list of proposed new members and consultants of Commission 15 was presented and additional nominations were solicited. The list was approved by the Organizing Committee and subsequently by the IAU.

MEMBERS OF COMMISSION 15

M.F. A'Hearn, C. Allegre, E. Anders, D.A. Andrienko, J. Arnold, C. Arpigny, W.I. Axford, P.B. Babadzhanov, G.S.D. Babu, M.E. Bailey, E.S. Barker, D.B. Beard, M.J.S. Belton, P. Birch, J.E. Blamont, D. Bockelee-Morvan, H. Bohnhardt, E. B. Boswell, J. Bouska, E.L.G Bowell, J.C. Brandt, A. Brecher, R.H. Brown, D.E. Brownlee, W.E. Brunk, B. Buratti, L.F. Burlaga, J.A. Burns, M.P. Candy, A. Carusi, G.R. Carruther A. Cellino, Z. Ceplecha, P. Cerroni, T. Chandrasekhar, C.R. Chapman, R.D. Chapman, D.-H. Chen, V.I. Cherednichenko, J. Clairemedi, G.C. Clayton, R.N. Clayton, S.V.M. Clube, A. Cochran, W.D. Cochran, M. Combi, B.C. Cosmovici, C.G. Cristescu, J. Crovisier, D.P. Cruikshank, J. Cuypers, P.A. Daniels, A.C. Danks, H. Debehogne, J. Degewij, A.H. Delsemme, I. DePater, S.F. Dermott, D. De Sanctis, W.A. Deutschman, M. Di Martino, O.V. Dobrovolsky, B.D. Donn, F. Dossin, A.V. Douglas, M.J. Drake, M. Dryer, V.P. Dzhaspiashvili, T. Encrenaz, A. Ershkovich, E. Everhart, A. Eviatar, P. Farinella, H. Fechtig, P.D. Feldman, J.A. Fernandez, J.C. Fernandez, I. Ferrin, M. Festou, C. Froeschle, A. Fujiwara, M. Fulchignoni, T. Gehrels, J. Geiss, E. Gerard, D.M. Gibson, J. Gibson, F. Giovane, J.C. Gradie, S. Green, J.M. Greenberg, R. Greenberg, L. Grossman, S. Grudzinska, E. Grun, B.A.S. Gustafson, G. Hahn, I. Halliday, M.S. Hanner, A.W. Harris, B.W. Hapke, W.K. Hartmann, M. Harwit, I. Hasegawa, L.N.K. Haser, H.F. Haupt, E.F. Helin, G. Herzberg, Z.W. Hu, W.F. Huebner, D.W. Hughes, K. Ibadinov, W.H. Ip, W.M. Irvine, S. Isobe, V. Ivanova, W.M. Jackson, K. Jockers, T.V. Johnson, K. Keil, H.U. Keller, R.F. Knacke, Z. Knezevic, Ch. Koeberl, L. Kohoutek, V.P. Konopleva, C.T. Kowal, L. Kresak, K.S. Krishna Swamy, L.K. Kristensen, C.I. Lagerkvist, P. Lamy, P. Lancaster Brown, H.P. Larson, S.M. Larson, L.A. Lebofsky, A.C. Levasseur-Regourd, B.Yu. Levin, W. Liller, C.F. Lillie, C.A. Lindsey, J.J. Lissauer, L.Z. Liu, R. Luest, M.E. Lipschutz, K.A. Lumme, D.F. Lupishko, B.A. Lutz, R.A. Lyttleton, D.J. Malaise, S.P. Maran, B.G. Marsden, D.L. Matson, O.T. Matsuura, T.B. McCord, R.E. McCrosky, J.A.M. McDonnell, S. McKenna-Lawlor, D.D. Meisel, D.A. Mendis, B.L. Milet, F.D. Miller, R.L. Millis, P.M. Millman, D. Moehlmann, E.P. Moore, V.I. Moroz, D. Morrison, A. Mrkos, T. Mukai, M.J. Mumma, T. Nakamura, W.M. Napier, G. Nazarchuk, J.S. Neff, G. Neukum, R.L. Newburn Jr., C.R. O'Dell, J.A. O'Keefe, D. Olsson-Steel,

P. Paolicchi, J.P. Parisot, P. Pellas, H.A. Perez de Tejada, C.B. Pilcher, E.M. Pittich, P.E. Proisy, D. Prialnik-Kovetz, J. Rahe, H.J. Reitsema, L.G.A. Remy-Battiau, D.O. ReVelle, H. Rickman, E. Roemer, R.Z. Sagdeev, F. Scaltriti, D.G. Schleicher, F.P. Schloerb, H.U. Schmidt, M. Schmidt, H.J. Schober, C. Sharp, Z. Sekanina, M. Shimizu, V.G. Shkodrov, E.M. Shoemaker, L.M. Shul'man, K.R. Sivaraman, B.A. Smith, R. Smoluchowski, L.E. Snyder, M. Solc, H. Spinrad, J. Stohl, J.M.G. Surdej, J. Svoren, K. Szego, H. Takeda, H. Tanabe, J. Tatum, E.F. Tedesco, A.K. Terentjeva, D. Tholen, K. Tomita, L. Typhoon, G.B. Valsecchi, V. Vanysek, G.J. Veeder, J. Veverka, M.K. Wallis, S.Ch. Wang, J.T. Wasson, T. Wdowiak, H. Weaver, P.A. Wehinger, S. Weidenschilling, P.R. Weissman, G.W. Wetherill, F.L. Whipple, L.L. Wilkening, I. Williams, W.Z. Wisniewski, J.A. Wood, M.M. Woolfson, A. Woszczyk, S. Wyckoff, S. Yabushita, A.A. Yavnel, D.K. Yeomans, V. Zappala, J.C. Zarnecki, B.H. Zellner.

CONSULTANTS:

M.A. Barucci, R.P. Binzel, H.C. Brinton, H. Campins, P.A. Daniels, R. Decher, M.J. Drake, T.I. Gombosi, D.W.E. Green, G. Hahn, H.L.F. Houpis, D. Jewitt, J. Klinger, V. Krasnopolskij, Y. Langevin, P. Magnusson, E.P. Mazets, R.P. McCoy, L.A. McFadden, G. Nazarchuk,

M. Neugebauer, M. Niedner, S. Ostro, R.O. Pepin, X.Y. Yang.

WORKING GROUP ON COMETS: C. Arpigny (Belgium) - chairman, M. F. A'Hearn (USA), J. C. Brandt (USA), O. V. Dobrovol'skij (USSR), P. D. Feldman (USA), E. Gerard (France), W.-H. Ip (FRG), A. C. Levasseur-Regourd (France), T. Mukai (Japan), H. Rickman (Sweden), Z. Sekanina (USA), L. M. Shul'man (USSR)

WORKING GROUP ON MINOR PLANETS: V. Zappala (Italy) - chairman, P. V. Birch (Australia), S. F. Dermott (USA), P. Farrinella (Italy), J. C. Gradie (USA), H. F. Haupt (Austria), D. F. Lupishko (USSR), K.-I. Lagerkvist (Sweden), X.-H. Zhou (PRC)

WORKING GROUP ON METEORITES: J. T. Wasson (USA) - chairman, M. Fulchignoni (Italy), A. Yavnel (USSR)

It is expected that these Working Groups coordinate the corresponding Commission related activities and assist the Commission President in submitting his report to the IAU General Assembly.

WORKING GROUP REPORTS

1. Comets - C. Arpigny: The principal activity of this committee was writing the triennial report which was much longer than usual because of the wealth of new results from comet P/Halley.

2. Minor Planets - A. Harris: The largest activity was writing the triennial report. Additional activities included preparing a list of

rotational characteristics of minor planets, as directed at the Delhi General Assembly, for inclusion in the Ephemeridi Malikh Planet.

3. Meteorites: no report.

Joint Working Group on Fluffy Structures: J. M. Greenberg reported on the establishment of this working group by Commission 21 and the hope for co-sponsorship by other commissions. B. Donn was nominated for membership and accepted.

RESOLUTIONS

1. A proposal by M. Belton for an observational campaign to determine the rotational state of P/Halley's nucleus after the coma has faded was endorsed by the Commission.

2. The following proposal to endorse the maintenance of data bases on asteroids and comets by A.W. Harris was adopted by the Commission and subsequently by the IAU.:

"Commission 15 endorses the continued maintenance of a database on minor planets, and recommends the establishment of a comparable database on comets. The Working Groups on Minor Planets and on Comets are charged with the responsibility of defining the contents of the respective databases and of monitoring their compilation, updating, and dissemination."

"La Commission 15 recommande le maintien d'une banque de donnees relative aux petites planetes et preconise l'etablissement d'une telle banque de donnees pour les cometes. Les groupes de travail "Petites Planetes" et "Cometes" sont charges de definir le contenu de leurs banques de donnees respectives et d'en surveiller la compilation, la mise a jour et la diffusion."

II. SUMMARY OF SCIENTIFIC SESSIONS

1. JOINT SESSION OF COMMISSIONS 15 AND 20 ON DATABASES AND LARGE PROJECTS (ORGANIZED BY L. KRESAK AND Y. KOZAI)

INTERNATIONAL HALLEY WATCH (IHW): R. Newburn emphasized the archival work of the IHW. Noting the submission of nearly all data on Giacobini-Zinner, he discussed the progress with the data on Halley. Data are due at the JPL IHW Lead Center (after having been processed by the Discipline Specialists) by 1 May 1989. He estimated that 40% of known data was at the Lead Center with another 30% in the hands of the Discipline Specialists. The primary archive will consist of 22 Gbytes of data on Halley and 220 Mbytes of data on Giacobini-Zinner. It will be published on approximately 21 compact discs; the question whether a printed archive will also be produced, will be decided by the IHW Steering Group in mid-1990.

ASTEROIDAL DATABASE: E. Tedesco described the implementation of the two parameter (H,G) asteroidal magnitude system adopted at the 1985 IAU GA. The system is in place and values of the magnitude parameters are given in the Ephemeridi Malikh Planet beginning with the 1988 volume. Formal solutions for (H,G) which yield G<0 or G>0.5 are replaced with G=0.15 and 0.25, respectively with H derived from the forced fit.

ASTEROIDAL TAXONOMY: D. Tholen described the currently widely accepted taxonomic system which is defined in terms of the 8-color asteroidal photometric system. The taxonomic classes were defined by a cluster analysis of the data, cutting the longest branches of the minimal tree in color-color plots. Albedos derived outside the 8-color survey are used to subdivide some classes yielding a total of 14 classes. Many of the classes are similar to those defined earlier on more subjective grounds but several new classes are now recognized. A fifteenth class may be justified distinguishing members of the Eos family from other S-class asteroids.

2. JOINT SESSION OF COMMISSIONS 15 AND 22 ON METEOR STREAMS, COMETS, AND ASTEROIDS (ORGANIZED BY P. BABADZHANOV AND L. KRESAK)

D. Olsson-Steel reported on a radar search for streams associated with Apollo-type asteroids. Convincing evidence exists for streams associated with 3200 Phaethon (previously known), 1566 Icarus, 5025 P-L, 1982 TA, 984 KB, 2201 Oljato, and several others. It is impossible to determine whether they come from extinct cometary nuclei or collisional debris.

D. Olsson-Steel reported also recent observations aimed at deriving the height distribution at frequencies of 2, 6, and 54 MHz. The data show that previous measurements using VHF radars have detected only a few percent of the total influx of microgram-milligram particles.

P. Stohl and V. Porubcan discussed the Taurid meteor complex and its cometary and asteroidal associations. The analysis of all precise photographic orbits and radiants of meteors and PN-fireballs of the Taurid meteor complex associated with P/Encke has enabled them to conclude that: (1) the activity of the complex extends over 4 months (170-300 degree of solar longitude); (2) both, the continuous variations of the mean orbital elements and of the mean position of the radiant of the complex confirm that several minor showers are in fact regular berths of the complex. Their names are derived from the position of the radiant at the corresponding period of observations: Arietids, Piscids, CH-Orionids, Ro-Geminids; (3) Asteroids 2201 Oljato, 1982 TA and very probably also 5025 P-L and 1984 KB are closely associated with the southern branch of the Taurid meteor complex if the derived varied mean orbit is taken into account.

Otsukha studied Monocerotids in photographic meteor catalogues and compared the radiant of 15 meteors with that expected from Comet Mellish; good agreement is found, except that the peak occurred two days earlier than expected.

Hasegawa summarized recent Japanese studies of meteors. In addition to an extensive program of visual observations there are photographic programs which have given roughly 600 double-station observations over about 10 years and an FM radar program that has shown, e.g., that the hourly rate of eta Aquarids is constant from year to year.

Taguchi showed 4-color photographs of a persistent train from a bright Orionid meteor with strong emission in the 6000-7000 Angstrom region.

3. SCIENTIFIC SESSIONS ON "NEWS IN COMETARY RESEARCH" (ORGANIZED BY C. ARPIGNY)

B. Donn summarized our understanding of cometary nuclei separating facts that are definitely known (the nucleus is a single body, that activity occurs in isolated areas, the low albedo, the predominance of water among the volatiles, etc.) from phenomena that are still not well understood (scale of surface irregularities, whether the mantle is intrinsic or evolutionary, the composition-thickness-structure of the mantle, the cause of outbursts and active areas, and the nature of the interior). He also reviewed models of Donn, Yamamoto and others, which demonstrate the the nucleus is formed by accretion of grains.

H. Rickman summarized his techniques for modelling the non-gravitational forces and how these techniques lead to estimates of the density of cometary nuclei. He stressed that if P/Halley precesses as much as implied by Sekanina's studies, then the average insolation is peaked at the subsolar latitude thus simplifying an otherwise intractable problem. He concludes that the $0.1 < \text{density} < 0.5$ g/cc.

U. Keller discussed the outgassing as observed from Giotto. The principal results include the observations that most outgassing occurs from three main sources covering perhaps 10-20% of the surface, the dust in the jets is reddish (9% per 1000 Angstroms redder than Sun), the strongest jets can be correlated with ground-based observations, filaments of width 0.5-1.0 km are common and criss-cross each other, the fact that a breeze must carry dust past the terminator, and the significance of the absence of activity on the dark side of the nucleus. He stressed that our present picture of cometary nuclei is different than that previously held by many investigators.

Z. Sekanina described his model of active vents and noted that the floor of the vents on P/Halley and P/Encke must recede by of order 10 meters per revolution. Thus individual vents can not remain active over millennia. Encke apparently has gone through two different phases of activity, one current and one producing the Taurid meteor stream, with a long inert period between these events.

R. West showed recent (April 1988) photographs of P/Halley taken with a ccd system at the Danish 1.5-m telescope at La Silla. Combination of images for an effective 12-hour exposure and subtraction of the circularly symmetric component allows one to see a definite tail with a surface brightness near 29 mag per square arcsec. The nucleus varies with an amplitude near 1 magnitude.

M. Hanner surveyed infrared results obtained from the ground which characterize the dust. Although some emission peaks are associated with specific species, others are not. The structure in the feature at 3.36 microns is still not definitively understood whereas the structure in the 10-micron silicate feature has been associated with crystalline olivine. The similarity between infrared spectra of Halley and those of IDPs was noted.

A.-C. Levasseur-Regourd described the in-situ results on the dust, noting that differences were found in the size distributions of 'new' and 'old' dust. She also summarized the important results on the composition of the dust, particularly the discovery of the three types of particles - chondritic, CHON, and mixed. She discussed tentative results suggesting that the CHON material might come from mantles on silicate particles and the fact that the CH particles were seen further from the nucleus than most other types suggesting that either they are break-down or longer-lived products. She concluded with a brief discussion of possible polymers.

D. G. Schleicher summarized results from Earth-based observations of the neutral gas. He stressed the importance of the new species detected from Earth and the relatively high abundance of CO. He noted also that for the first time there were numerous determinations of the velocities of several species from Doppler shifts using radio and infrared FTS techniques. These results agree with measurements using traditional techniques (expansion of features) and in situ results. He also discussed relative abundances noting that P/Giacobini-Zinner was highly anomalous while P/Halley was very 'normal'. Finally he noted the extreme variability.

W.-H. Ip surveyed the in situ studies of gas, particularly ions, in P/Halley. He pointed out particularly the outstanding questions - the high abundance of sulfur ions at large distances, the discrepancy between the spherical symmetry of H2O as seen by the NMS on Giotto and the jet of H2O seen spectroscopically from Vega. He noted the direct measurement of the contact surface and the dramatic change in the ionic kinetic temperature across this surface. He also noted the controversy on the nature of the polymers and the question of the source of S and C ions.

V. Vanysek summarized recent measurements of isotopic ratios in comets noting the decreasing error bars as measurement techniques improve. He briefly discussed the implications of isotopic ratios similar to those on Earth and of a reduced 12C/13C ratio.

P. Wehinger described their determination of the 12C/13C ratio in P/Halley with the result 65 +/- 9.

D. Bockelee-Morvan summarized recent work by the group at Meudon. She compared the detection of HCN in P/Halley with the upper limits for comets Wilson and P/Giacobini-Zinner noting that the results are consistent with constant fractional abundance. She also summarized their studies of OH which show the 7.4-day periodicity and yield good estimates of the lifetimes and velocities for both H2O and OH. The magnetic field in P/Halley was determined as 50nT.

M. Niedner surveyed results from the study of photographs of plasma tails and their correlation with variations in the solar wind. In P/Halley, the first tail seems to have formed between Nov. 10 and Nov. 13, a time which may be correlated with a high speed stream in the solar wind but which may also be associated with an increase in outgassing, thus leaving the mechanism for onset indeterminate. The outstanding questions are the initial velocity of the ions, whether DEs are due to front-side or tail-side reconnection, and the nature of tail rays.

A. H. Delsemme led a brief discussion about the question which properties of P/Halley were typical of all comets and which were unique to P/Halley. A consensus was not reached on most of these questions. He also discussed some of his own work on the chemical contents of cometary nuclei and on distinguishing Oort-cloud comets from Kuiper-belt comets based on their somewhat different temperatures of formation. He concluded that P/Halley came from the Kuiper belt.

4. SCIENTIFIC SESSION ON "NEWS IN MINOR PLANET RESEARCH" (ORGANIZED BY A.W. HARRIS)

S. Ostro summarized the properties of asteroids as seen with radar. He noted that some classes such as M, seem distinguishable from others while many other classes show similar properties at radar wavelengths. He pointed out numerous unusual asteroids such as Kleopatra, noting the forthcoming occultation of this asteroid which may answer questions about its shape. He discussed recent work modelling the shape of asteroids showing the distinctly different shapes of Ivar (significantly elongated) and Apollo (more nearly circular).

V. Zappala summarized the distribution of rotational properties. He described the variation of the rotation rate with size (minimum at sizes near 125 km) and the variation of the amplitude (axial ratio) with size (decreasing from about 25 km to 125 km, with possibly a minimum there or possibly a flat distribution). He also showed that the statistical properties could be described by a bi-modal distribution of velocity-size relations. He noted the need for binary asteroids to explain the longest periods in the light curves.

Z. Knezevich described several approaches to integrating the proper elements of asteroids over very long intervals. He has used a 6-body model and compared it with a 3-body model. For some cases it gives a significant improvement, but not for all. He concluded that no single approach will work well for all asteroids.

C. Chapman presented a new bias analysis of the distribution of asteroidal taxonomic types. He noted that in a simple plot of U-B vs. B-V, more than 50% of the space defines unique types. The size distributions differ dramatically from type to type with the distribution for type C deviating most drastically from a classical collisional distribution. The subdivisions of the S type seem to be at least weakly correlated with size and albedo. The question of whether S-types correspond to stony-iron or to chondritic meteorites is still open. Other questions include: why do families not have a sensible set of taxonomic types? How can Vesta have a thin crust on a metal core when Psyche has been stripped of its crust? Where did the olivine from mantles go?

A. W. Harris discussed the comparison of observed light curves near opposition with current scattering theories. Surprisingly, some asteroids with very steep phase curves, such as Aurelia, show virtually no opposition effect whereas other asteroids such as Nysa show a strong opposition spike inside 2 degrees even though theory predicts that they should not. In general, the dark asteroids show less opposition effects than expected from current theories.

M. Fulchignoni described very briefly VIEW (Vesta International Earth Watch), a proposed effort to support the Vesta mission.

M. A'Hearn
L. Kresak
J. Rahe

COMMISSION No. 16

PHYSICAL STUDY OF PLANETS AND SATELLITES

(ETUDE PHYSIQUE DES PLANETES ET DES SATELLITES)

Report of Meeting, 10 August 1988

PRESIDENT : A. Brahic

At the 20th General Assembly of the IAU, Commission 16 held its business meeting , which opened at 9:00 a.m. on 10 August 1988, under the chairmanship of the incoming President, Dr. A. Brahic (France).

Oral minutes of the previous Business Meeting of Commission 16 were presented by Dr. D. Morrison (USA), who had been present at the meeting in India and had at that time been elected 2nd Vice President. Dr. Morrison noted that the President, Dr. G. Hunt (UK) had not been active in Commission business during his term, and that the organization of a session on planets at the 20th General Assembly had been largely due to the efforts of Dr. Brahic, to whom the commission members were grateful. Dr. Morrison also noted that attendance at the India commission meeting was very low (only about 5-6 people).

Dr. Morrison noted that at the India meeting, President V. G. Tejfel (USSR) had proposed the compilation of a volume of basic data on solar system bodies. Since that time there has been no information on progress on this volume, but after some general discussion, it was noted that commission 16 endorses Dr. Tejfel's proposed project.

Dr. Morrison noted that an informal meeting of several Commission 16 members had been recently held during the COSPAR meeting in Helsinki in July, at which time a proposed slate of new officers was drawn up with the concurrence of all those in attendance.

Dr. Brahic made a brief report on his activities on behalf of the commission.

Dr. Harold Masursky (USA) reported on the activities of the committee on nomenclature of planetary bodies. After some discussion, it was reaffirmed that the committee should retain its status as a committee rather than seek status as a commission of the IAU.

On the question of new officers, it was noted that the IAU General Secretary Dr. Swings had in earlier action rejected the slate of officers proposed following an informal business session of several Commission members at the COSPAR in Helsinki in July, 1988. The slate agreed upon in Helsinki was:

A. Brahic (France)	President
D. Morrison (USA)	1st Vice President
D. P. Cruikshank (USA)	2nd Vice President

The grounds for the rejection were the perceived large number of US scientists on the slate. Considerable discussion followed, after which it was decided by Dr. Brahic that the following revised slate should be presented:

A. Brahic (France)	President
D. Morrison (USA)	1st Vice President
M. Ya. Marov (USSR)	2nd Vice President

Dr. Brahic noted his plan to make a protest to the IAU Executive Committee on this matter, and to seek clarification on the question of whether the Commission 16 can really elect its own officers, or is the commission controlled by the IAU Executive Committee.

The Organizing Committee of Commission 16 was then discussed. The following individuals were nominated and approved by those in attendance:

J. L. Bertaux
J. Burns
D. H. Chen
D. P. Cruikshank Secretary.
M. Davies
C. de Bergh
T. Encrenaz
D. Gautier
D. Matson
H. Masursky
V. Moroz
T. Owen
V. Shevshenko
B. Smith
V. Tejfel

On the question of new and continuing members of Commission 16 Dr. Brahic noted that a large number (about 40-50) individuals had been proposed in earlier correspondence, though not all of those individuals were members of the IAU. Dr. Brahic noted that he would write to all of the proposed new members of Commission 16 to ascertain their interest in joining the Commission, and then would take steps to see that they are added to the membership list if interested.

Future meetings seeking sponsorship or endorsement by Comm. 16 include the following:

1. A meeting on Bioastronomy, SETI, and organic matter anywhere in the universe, to be held in France in 1990.

2. A conference on planetary astronomy to be held in the People's Republic of China in June 1989.

3. A meeting about Neptune to result in a book in the University of Arizona Press series on planets, to be held in Paris about two years after the Voyager Neptune encounter (which occurs in August 1989).

4. A meeting in October or November of 1989 to be held in France to discuss early results from the USSR's Phobos Mission to Mars.

Commission 16 endorses these Meetings.

D. Cruikshank (USA) noted that there is growing interest in a meeting on planets in the USSR which could be heavily attended by US scientists to discuss early results of the Phobos mission and the results of the Voyager Neptune encounter. The meeting is being proposed for late 1990. No endorsement was sought or given by the commission.

It was noted that at the scientific sessions of Commission 16 held on the previous day (9 August), attendance was exceptionally large (120-150 people). This large attendance was attributed to the high quality of the program plus the fact that Dr. Brahic had undertaken at his own initiative (and doing all the work himself) to print announcements of the session and distribute one to each member of the IAU in attendance at this General Assembly. All those present at this Business Meeting murmured their approval of Dr. Brahic's action.

In discussions of possible special topics for Commission 16 scientific sessions at the Argentinian General Assembly, T. C. Owen (USA) suggested that inner planets may be suitable, with special reports on the expected Magellan mission results. The results of Hubble Space Telescope observations of planetary bodies could also be reported. W. Irvine (USA) suggested that astrochemistry and the study of carbonaceous material would be suitable topics.

The future activities of Commission 16 were discussed, and the following items were noted, with Dr. Brahic leading the discussions:

1. An update of the list of members is needed, and new members who wish to join should be added.

2. There is a need to assess the state of planetary astronomy worldwide, and Dr. Brahic suggested that this information could be gleaned from a widely circulated letter of inquiry to planetary astronomers.

3. Dr. Brahic proposes to publish and circulate a newsletter of Commission 16 activities and interests. He seeks input to that newsletter, the first issue of which he hopes to publish in about six months.

4. Acting on a suggestion by Dr. C. Keay that Commission 16 appoint some members to his working party (organized in Commission 22 - meteors and interplanetary dust) which seeks to frame recommendations on the composition and terms of reference of a high-level panel on interplanetary pollution, Commission 16 shelved the matter until a later date.

5. Dr. R. West (USA) proposed that a working group on time-variable phenomena in the Jovian System be formed to act as a liaison with the International Jupiter Watch. Dr. West proposed the following text of a resolution:

   "Commission 16 proposes the formation of a special working group on Time-Variable Phenomena in the Jovian System to coordinate ground-based observations of variable phenomena on the surfaces of Jovian satellites, the Io atmosphere and plasma Torus, and the Jovian magnetosphere and atmosphere. This working group will form liaisons with the International Jupiter Watch Program and with a special COSPAR working group to promote long-term coordinated studies of the Jovian environment using a wide variety of ground based and space-based instruments."

   Dr. Brahic noted that it is too late to have a formal resolution passed by the IAU Executive Committee at this General Assembly, but following discussion, Commission 16 gave its informal endorsement of Dr. West's resolution.

6. Dr. B. A. Smith (USA) proposed that the scope of Commission 16 be broadened to encompass the study of protoplanetary disk phenomena (of the Beta Pictoris and related types), in a move that would be consistent with the new policy at the US National Aeronautics and Space Administration to include the study of these phenomena within the discipline of planetary science. Dr. Brahic noted that it is too late to make this a formal change at the current General Assembly, but urged that the members discussed Dr. Smith's proposal among themselves and with other astronomers, so that something formal can be undertaken at the next General Assembly.

When the meeting adjourned at about 11: 00 a.m., approximately 30 people were in attendance.

Dale P. Cruikshank

COMMISSION No. 19

ROTATION OF THE EARTH (ROTATION DE LA TERRE)

Report of meetings: 3-10 August 1988

President: W.J. Klepczynski Vice Presidents : M. Feissel
B. Kołaczek

3 August : Joint Commission Meeting 1. JCM-1

The JCM-1 was a meeting on the topic "For milliarcsecond or better accuracy", organized by P. Seidelman, jointly for commissions 4,8,19,24,31,40. The meeting included 11 invited papers concerning observational accuracies, theoretical developments and computational considerations, and the reports of four working groups, on The use of millisecond pulsars, Nutation, Astronomical constants, and Reference Systems. The report of JCM-1 will appear in Highlights of Astronomy.

4 August, 9:00 Commission 19 - Business meeting I

Chair : W.J. Klepczynski Secretary : A. Babcock

Announcements

- The IAU Executive Committee appointed Dr. Yatskiv as the IAU representative to the IERS Directing Board.
- The Commission 19 Organizing Committee will meet (on Friday, 5th) to discuss: nominations to the Organizing Committee, proposed new members (5 proposed so far), and consultants

Remarks by the President : W. Klepczynski reviewed the accomplishments on the last three years, especially (1) the establishment of the IERS and (2) the studies of atmospheric driven irregularities in the Earth rotation. He observed that much time has been spent during Commission 19 meetings on reports from individual members on their own work, but reminded that the function of the IAU is to coordinate joint efforts between astronomers at different institutions or between scientists in different disciplines (i.e. IERS, MERIT).

Election of officers : The folloging nominations had been made: President: Martine Feissel, Vice President: Ye Shu-hua. Concern was expressed by B. Kołaczek that M. Feissel should not head both IERS Central Bureau and Commission 19 since it was established tradition that those actively involved in an IAU service would not be eligible to hold and office in which they would make policy recommendations regarding the service. W. Klepczynski responded that (1) IERS responsability is shared between the Central Bureau and the Technique Coordinators, so that neither officer has full control and (2) suitable candidates for Commission office are very scarce and should not be disqualified simply because they are active observers or analysts. This opinion was strongly endorsed by I. Mueller. The vote was taken and the nominees elected with 16 for, 3 abstaining.

Review of resolution on Reference Systems The report of the WG on Reference Systems had been presented at the JCM-1 by its president J.A. Hughes. A resolution had been prepared, which recommended the continuation of the WG, and the formation of Study Groups under it, on specialized topics. The discussion on this resolution was open. M. Feissel remarked that D. McCarthy had suggested a study group (SG) be formed on Links between Reference Frames. D. McCarthy said this would avoid having SG's on individual reference frames having to cover too much territory. B. Guinot considered that this was really the job of IERS, but it was explained that IERS only concerns itself with the reference frames of the techniques in its solution, whereas the optical reference frame is also of interest. Ye suggested the creation of a new commission on reference systems to replace Commission 31. This suggestion had been made at New Delhi. D. McCarthy recalled, but the IAU had not responded to the recommendation. After some discussions on the various possibilities of new organization, it was decided to add an SG on Reference Frames to the list of SG's in the resolution of JCM-1. It was also recommended that SG's on Time, Nutation, Origins, and Astronomical constants be included in the Working Group on Reference Systems.

Recommendation on IAU contribution to FAGS There was a brief discussion of the recommendation that the IAU increase their support to the Federation of Astronomical and Geophysical Data Analysis Services (FAGS). This recommendation requests that the Presidents of Commissions 19 and 31 coordinate an action with the Presidents of IAU commissions concerned with the FAGS Services.

New astronomical organization in Japan : There was a discussion of the new National Astronomical Observatory in Japan, which encompasses the functions of ILOM as well as many other astronomical institutions.

History of IPMS and BIH : D. McCarthy suggested that now that both the IPMS and BIH have been discontinued, their histories should be written as soon as possible. It was suggested that a symposium be held on the history of these organizations, but this was not strongly endorsed. I. Mueller suggested a formal resolution to direct the heads of the BIH and IPMS to write the history of their organizations. B. Wackernagel pointed out that Commission 41 (History of Astronomy) should be involved in this effort. Ya. Yatskiv said that a session on the history of the ILS, IPMS, and BIH could be included in a meeting scheduled for next year in Leningrad. I. Mueller thought a session at the next IAU General Assembly would be preferable. In any case, the resolution should be written, and addressed to B. Guinot and S. Yumi.

<u>4 August, 11:00</u> Commission 19 - Reports of Services

Chair : W.J. Klepczynski Secretary : J. Popelar

Bureau International de l'Heure (BIH)

M. Feissel presented the report on the earth rotation activities of the Bureau International de l'Heure between 1985 and 1988 which included the reorganization into the new International Earth Rotation Service on January 1, 1988. She reviewed the activity highlights as follows.

- Earth Rotation Parameter (ERP) solutions computed on a weekly, monthly and annual basis have incorporated all high precision results provided by VLBI, lunar and satellite laser ranging (LLR, SLR) measurements.
- A homogeneous combined ERP series have been compiled for the period 1962-1987 at 5-day intervals; its precision reached 0.002" for polar motion and 0.2ms for UT1 during 1984-1987, which represents an order of magnitude improvement for the annual and shorter term variations and a fivefold improvement for longer term changes over the 25 years.

- An ERP solution for optical astrometry based on the IAU 1980 theory of nutation has been obtained for all data since 1978; A deterioration of long term stability has been detected after 1984 which is reflected in annual term fluctuations and a drift in they coordinate of the pole.
- BIH Terrestrial System (BTS) has been re-defined in 1984 and is now based on a set of the cartesian coordinates of 64 sites where at least two of 13 available space geodesy networks overlap; the French Institut Géographique National (IGN) carried out the global adjustment which also produced the transformation parameters between the input networks and BTS; the BTS(1987) makes use of the Minster and Jordan plate motion model AMO-2.
- An extragalactic Celestial Reference Frame consistent with the BTS(1987) has been derived using four radio source catalogs; the coordinates of 23 primary sources and 205 additional sources, as well as the relative orientation between the original frames and the combined frame have been determined.
- Series of pseudo Universal time based on Atmospheric Angular Momentum (AAM) have been established to facilitate studies of the seasonal and higher frequency variations in the length of day and for possible use in the quick look ERP solutions in future.
- The last BIH Annual Report for 1987 has been produced and distributed as have been hard copies of the Circulars A to F; at the same time greater use of the computer networks (GE Mark III, EARN/BITNET and SPAN) has been encouraged for data dissemination.
- In 1988 the Earth rotation section of the BIH, associated with groups in IGN and Bureau des Longitudes has been transformed into the Central Bureau of IERS, continuing to maintain the data bases and assuring system continuity.

In the discussion following the BIH report the commission president and G. Winkler expressed concern regarding the deterioration of the optical astrometry results in recent years which would jeopardize any comparison between the classical and the new space techniques. In her reply Dr. Feissel attributed this fact mainly to reduced number of observations as many observatories have been phasing out their optical astrometry programs. Prof. Ye pointed out that more than 50 instruments are still in operation and the Shanghai Observatory has assumed the responsibility of the global analysis center which will continue to facilitate comparisons with the optical technique in the future.

International Polar Motion Service (IPMS)

Dr. Yokoyama presented the activity report of the International Polar Motion Service for 1985-1988. He reviewed the organization of the new National Astronomical Observatory (NAO) of Japan which was established on July 1, 1988 and incorporates among others the former International Latitude Observatory of Mizusawa, the host organization of the IPMS between 1962 and 1987. The Earth Rotation Division and the Mizusawa Astrogeodynamics Observatory (NAOM) will continue the activities related to the Earth rotation, geodynamics and gravity.

During the last period the IPMS continued as the operational and data analysis center for the optical astrometry as part of the extended MERIT campaign. An extensive computer data base has been compiled and includes :

- all ILS observational records and results between 1899 and 1979,

- all IPMS optical astrometry data, ERP results and related information between 1962 and 1987,

- complete set of computer programs for the reduction and analysis of optical astrometry observations, star catalog adjustment and global ERP estimation including evaluation of individual station biases.

The data base is available on request and it is ready for re-processing after the Hipparcos star catalog becomes available.

Since 1985 K. Yokoyama and the IPMS staff have been preparing for the transition to the newly organized IERS. The NAOM has taken the initiative and participates in the IERS activities by organizing :

- the VLBI center for IRIS-P network while actively promoting the VERA Project, the Japanese VLBI for Earth Rotation and Astrometry; the IRIS-P network has carried out monthly experiments using Kashima, Fairbanks, Ft. Davis and Richmond stations since April 1987,
- the VLBI data analysis center which developed data reduction and ERP estimation software; IRIS-P and IRIS-A experiments have been processed since the beginning of 1987 and detailed comparison of the results with NGS has been carried out.
- the computing center for evaluating atmospheric excitation functions using the Japanese Meteorological Agency data.

The NAOM also carries out research into earth tides, ocean loading and other geodynamic effects for which it operates a transportable absolute gravimeter, a superconducting gravimeter and tiltmeters.

During the discussion Dr. Djurovic asked if effects of ground water level changes are taken into account in data reduction. Dr. Yatskiv pointed out that these effects at the Poltava Observatory have been smaller than 0.001" which is below the accuracy of current measurements. D. McCarthy noted that information on ground water levels is difficult to obtain and some countries would not release it even if available. M. Gaposhkin inquired about use of laser ranging techniques in Japan. K. Yokoyama responded that there are no plans of NAO to use LLR or SLR for earth rotation studies but he referred to the Japanese Hydrographic Institute which has investigated use of SLR.

<u>4 August, 14:00</u> Commission 19 - Time series analysis

Chair : D.D. McCarthy

The following contributions were presented

- Time series analysis as a tool for accurate access to reference systems (M. Feissel and D. Gambis).
- Time series analysis and studies of Earth rotation (Ye. Shu-hua).
- Earth rotation studies at Jet Propulsion Laboratory (J.O. Dickey).
- A note on the usage of maximum entropy spectral analysis in the determination of Earth Rotation Parameters (R. Vicente and H. Lenhardt)
- Time series analyses of Earth rotation at Kiev Observatory (Ya. Yatskiv).
- Non-linear spectra and the Earth's rotation (T.H. Eubanks).
- Analysis of the zonal tides of UT1 from 1962 to 1988 (J. Hefty and N. Capitaine).
- On the solar origin of the fifty-day fluctuation of the Earth's rotation and atmospheric circulation (Djurovic and P. Pâquet).
- Time series analyses of Lunar Laser Ranging (Ch. Veillet).

5 August, 14:00 Commission 19 - Reference frames

Chair N. Capitaine Secretary C. Veillet

Ten contributions for the sessions devoted to "Reference Frames" permitted to overview the major topics, from the concepts to the realisations of the various systems.

The basic concepts on the definitions and realisations of the reference systems have been described in details by I. Mueller. SLR, LLR or VLBI ground-based observatories are the basis of a conventional terrestrial system (CTS) realised through a combination of sites and techniques. This combination has to take into account all the information available (colocation, local survey, ...) and to include all the unknowns related to the motion of the sites, scale factors for ERP series, etc.

The intercomparison of the various celestial reference frames, as pointed out by Mrs Ye, has to be carefuly made. The differences in formulation and definition can be clear, but their implementation in the various techniques as well as in the data analysis is not straightforward. For the next decades, a four-dimensional reference system is needed and could be the goal of a new commission.

Links between VLBI and dynamical reference frames made at JPL have been described by J. Dickey. Differential VLBI observations of quasars and Mars orbiters are a possible technique for such a link, as well as VLA observations of Galilean satellites, Titan, Uranus and Neptune. None of these observations were included in DE125, but they would improve the ephemeris. VLBI and optical frame agree at the 0.02" level for the northern hemisphere, at the 0.1" level for the southern. This link should improve in a near future through VLBI observations of millisecond pulsars, differential VLBI observations of Phobos and Galileo, VLA observations of outer planets, and with the use of Hipparcos and space astrometry.

Four extragalactic radio sources catalogues (GSFC, JPL, NGS and extended GSFC) have been compared and combined after a selection of primary sources in a work made at the IERS Central Bureau presented by F. Arias. The various origins are in a 0.002" radius circle, but there are systematic deformations between JPL and NGS or GSFC catalogs, as a function of source declination. The resulting catalog has a total of 228 sources, including 23 primary ones. It has only 8 primary sources in the southern hemisphere, but should be extended in a near future.

On the Earth, the IRIS-A and IRIS-P networks are both operational. K. Yokoyama presented briefly IRIS-P. the discrepancies between the procedures used in these networks have been minimized.

Using the VLBI observations, T. Herring presented his last results on nutation corrections to the IAU (1980) nutation series. It is very clear now that the observed semi-annual and annual terms are different from the conventional values, as well as the 13.7 days term. V. Dehant presented some possible explanations of these discrepancies. She showed how the input of the Earth interior model can be changed in an attempt to explain this disagreement. In fact, an improved external tidal potential doesn't explain it. Taking into account the mantle anelasticity gives still larger differences. Variations of the core flattening added to the previous modifications permit to obtain a better agreement between the theory and the observations, but are not sufficient. Variations of the pressure at the core-mantle boundary have to be included in a near future.

Recent results on precession corrections from LLR have been presented by J. Dickey. The uncertainty is still large, but should rapidly decrease as the modern accurate data will have a higher weight.

On a more general basis, D. McCarthy discussed the use of standards for reference systems. As they are both a basis for the comparisons and a tool for the observations, there is a natural conflict on their nature. Errors in standards are in fact compensated by errors in Earth Orientation Parameters (EOP), and very often EOP become a standard for other users. Finally, the analysis leads to improve the standards continuously. If we use different standards, do we understand really the influence of these differences? Many points and questions which make the standards an open problem.

Ya. Yatskiv reported on the effect of the pole tide in the various techniques contributing to EOP determination, and proposed to add these effects to the standards.

<u>8 August, 14:00</u> Commissions 19 and 31 - International Earth Rotation Servic

Chair : Ya. Yatskiv (session 1) Secretary : G.A. Wilkins
G.A. Wilkins (session 2)
I.I. Mueller (session 3)

The first two sessions of the joint meeting of Commissions 19 and 31 on the International Earth Rotation Service (IERS) were devoted to the presentation of papers on various aspects of the development and operation of IERS, which replaced the Earth-Rotation Section of the Bureau International de l'Heure (BIH) and the International Polar Motion Service (IPMS) on 1988 January 1. The afternoon session was mainly devoted to an open discussion on matters relating to the current and future operation of IERS. In addition, W. Markowitz presented a short paper on the secular motion of the pole.

History. P. Pâquet reviewed the development of the international services for time and polar motion. The International Latitude Service (ILS) was founded in 1891 on the basis of clear evidence of variations of astronomical latitude, but regular observations from 5 stations at a common latitude did not start until 1899. The introduction of wireless (radio) time signals in 1910 led eventually to the setting up in 1919 of BIH in association with the Time Commission of the newly founded IAU. The improved precision in the determination of time led in 1955 to the establishment by BIH of a rapid service for latitudes so that corrections for polar motion could be applied in the determination of UT (and hence in the determination of longitude). The ILS was replaced by IPMS in 1962. The introduction of atomic time and the use of space geodesy data to determine the Earth rotation parameters led to considerable changes in the work of BIH.

G. A. Wilkins reviewed very briefly the programme of Project MERIT (to Monitor Earth Rotation and Intercompare the Techniques of observation and analysis) and the associated COTES programme for the establishment of a new Conventional Terrestrial System for defining positions accurately on the Earth's surface. The results had shown that the techniques of VLBI and of laser ranging provided precise and complementary ways of monitoring the changes in the rotation-vector of the Earth and of establishing celestial and terrestrial reference frames. These are now the primary techniques used in IERS, but the Service will take into account all available data of high precision.

VLBI. D. S. Robertson reviewed the status of the VLBI operations that contribute to IERS. There are now 12 stations in regular operation in the northern hemisphere, but only one in the southern hemisphere. Other stations contribute occasionally and more are being developed or planned. The radio-source catalogue that defines the celestial reference frame is limited to 23 primary sources between latitudes 80 N and 50 S. Many very small effects must be modelled in order to match the precision (0.001") of the observations; corrections to the standard values of nutation in longitude and obliquity are determined from the observations. There is good agreement with the SLR results for the coordinates x,y of the pole. Daily determinations of UT are made using transatlantic baselines; the results confirm Yoder's earth-tide model with errors of order of 0.1 ms. T. A. Herring gave further details of the determinations from the VLBI data of the corrections to the IAU (1980) nutation series. Terms with periods less than one year are determined well, but as yet it is not possible to separate fully the effects of precession from those of the long-period terms.

Laser Ranging. B. E. Schutz presented a summary of his report as Coordinator for satellite laser ranging; this technique contributes to the IERS determination of x,y, UT and length of day (LOD), and also to the determinations of the temporal variations of the gravity field and tectonic motions of the stations; moreover, the station coordinates are determined in a frame whose origin is at the centre of mass of the Earth. About 20 stations contribute regularly to IERS; it would be useful to have more stations in the centre of the Eurasian plate and in the southern hemisphere. The operational centres at Austin (CSR) and Delft (DUT) make independent solutions using 3-day and 5-day arcs, respectively. The results differ by about 0.001" in x, y and 0.2 ms in UT; similar differences are found between SLR and VLBI results. The launching of more satellites with retroreflectors may reduce the coverage of LAGEOS and so require a new approach to the determination of the earth-orientation parameters; the precision of measurement should be reduced to about 1 mm, but it may not be possible to model the atmospheric transmissions to this accuracy.

C. Veillet reported that there are only three lunar laser ranging stations in regular operation at present, but several others should be provided in a year or two. The accuracy of measurement of the range is about 3 cm. Estimates of UT can, however, be determined from observations at one station with very little computation. The values of UT are model dependent and the choice of lunar ephemeris has a significant effect on the scatter of the results. The LLR and VLBI results differ at the level of about 0.05ms after fitting. The values of the nutation corrections determined from the LLR data are about 50% larger than those from VLBI, but the estimates of the correction to the precession constant are similar. The analyses continue to confirm Einstein's General Theory of Relativity. Better results are to be expected when the new stations also contribute data regularly.

Standards. D.D. McCarthy, the chairman of the IERS Working Group on Standards, said that IERS is now formally operating under the MERIT Standards, but it is clear that some changes are needed. Unfortunately, the MERIT standards are not always being used, and some of the members of the group have been very slow in producing revised drafts of the chapters for which they have responsibility. He questioned the value of standards in such circumstances. G.A. Wilkins and I.I. Mueller both stressed the value of having clearly specified up-to-date standards to simplify the combination of results obtained by different techniques of observation and the comparison of results obtained using different models.

Central Bureau. M. Feissel described the work of the Central Bureau of IERS. It is based at the Paris Observatory, but has the support of the Institut Geographique National and the Bureau des Longitudes in respect of the work on terrestrial and celestial reference frames. Two Sub-Bureaux contribute to the general services of IERS (see following reports). In addition IERS depends on the work of 3 coordinating centres (in 2 countries), 20 analysis centres (in 10 countries) and 50 observing stations (in 30 countries). She also stressed that standardisation is necessary if the data and results from these many sources are to be combined to give useful results. The initial realizations of the IERS reference frames will be based on the BIH combined solution for 1987; there are 64 primary stations in the terrestrial frame and 23 primary sources in the celestial frame; more primary points are required, and secondary points are required to make these frames more readily accessible. The comparisons of the individual solutions for the earth-orientation parameters show that the errors are of the order of 0.002" ; the combined solution is published at an interval of 0.05 years, but values are also available at intervals of 1, 3 and 5 days. Several series of circulars are issued (weekly, monthly, and as appropriate) and the Annual Report will give the final results from the Coordinating Centres and the combined solution; Technical Reports will give more information about the methods that are used; the first will contain the new IERS Standards. The IERS results are also available on-line.

D.D. McCarthy described the activities of the Rapid Service Sub-Bureau, which is operated jointly by the U.S. Naval Observatory and the National Geodetic Survey in Washington, D.C. It issues the Weekly Bulletin A, which contains estimates of the earth-rotation parameters that are based on quick-look data received up to two days before publication and on prediction models. The differences between the quick-look estimates and the values published later in the Monthly Bulletin B, issued by the Central Bureau, are gradually being reduced and are now about 0.002-3" in x,y and 0.3 ms in UT. The errors in the predictions are about 0.015" in x,y and 30 ms in UT after 80 days. The relative contributions of the different techniques differ between polar motion and UT, and also differ from those for Bulletin B. At present the forecasts of the angular momentum of the atmosphere are not used in the predictions of the earth-rotation parameters. In the later discussions, J. O. Dickey and M. Feissel said that the use of forecast values of AAM does improve the predictions of UT.

J. O. Dickey apologised for the absence of J.A. Miller of the (U.S.) National Meteorological Centre, which has agreed to operate a Sub-Bureau for Atmospheric Angular Momentum (AAM) for two years from the spring of 1989. (At present, the U.K. Meteorological Office and the European Centre for Medium-Range Weather Forecasting send their data directly to the Central Bureau.) The main purpose of the Sub-Bureau will be to consolidate the synoptic and forecast values of AAM received from the different agencies, including those of the Japanese Meteorological Agency. The close correlation between the variation in LOD and AAM is of great scientific interest and it is hoped that forecasts of AAM will prove useful in the prediction of UT.

General discussion. The Chairman of the third session, I.I. Mueller, suggested that the first topic of discussion should be the use of standards in IERS. He pointed out that in the MERIT project it had been agreed that all analysis centres should provide results based on the MERIT standards, but they were invited to also make analyses using other constants or models and to explain the differences in the results. W.E. Carter considered that it is sufficient if the analysis groups specify how the constants and models used to obtain their results differed from those in the standards; he and T.A. Herring considered that the errors introduced by using poor standards are unacceptable. This led to a consideration of the requirements of different types of users of the IERS services. In general, the operational community does not need results of the highest-possible precision, but it does require reliable, consistent stable series based on well-documented published standards and procedures. On the other hand, the research community is striving to obtain results of the highest-possible accuracy so that it can investigate the physical causes and implications of the variations in the earth-rotation vector; to this end it wishes to use the best available constants and models, but it must also be prepared to reprocess old observational data, or, in appropriate circumstances, to apply differential corrections to old results.

T. A. Herring pointed out that the changes would not be noticed by the operational community. B. Guinot stressed that in order to make possible the combination of results from different techniques it is essential that certain standards be used in common. The general conclusion of the discussion appeared to be that new IERS Standards are required as soon as possible, and that the Directing Board should take into account the wide range of views expressed during the discussion.

W. Markowitz presented an analysis of the motion of the mean pole as obtained from independent measurements based on optical astrometry, VLBI and SLR. He showed that the secular motion of the pole over 1976-1987 derived from the different series are in agreement, which confirms the reality of this motion.

In the remaining part of the session, J.O. Dickey drew attention to the desirability of obtaining earth-rotation parameters at an interval of less than one day for comparison with the variations in AAM ; she suggested that a coordinated campaign of observations should be organized during the TOPEX mission. C. Veillet considered that LLR quick-look estimates could be useful for the rapid service, especially when more stations are in operation to reduce the gaps in the data due to poor weather.

8 August, 16:00 Commission 19 - Report of the WG on Optical Astrometry

Chair : Ye Shu-hua Secretary : B. Kołaczek

At this session the report of the Working Group on classical astrometric observations were presented by its chairperson, Ye Shu-hua. She reviewed the activity of the Working Group which consisted of meetings and Circular letters in which the future role of the classical astrometric observations in determinations of the ERP as well as in variations of latitude and UT0 was discussed.

The following conclusions were presented :

- ERP determination based on the classical astrometric observations will be carried out starting with January 1, 1988 for several years by Shanghai Observatory. At present, about 55 instruments participate in this ERP determination.
- A Working Group on re-reduction of the optical astrometric ERP series based on the HIPPARCOS system has to be established.
- Monitoring of the local vertical variations in the region of seismic activity by the classical astrometric observations of time and latitude are being carried on in some observatories. The Beijing Observatory in China would like to act as the analysis center for these effects.
- Linkage of stellar reference systems connected with re-observations of star catalogues in order to compare them with the HIPPARCOS system ought to be organized by the IAU Commission 8.

Jin Wen jing, the chief of the First Division of the Shanghai Observatory, who is responsible for the ERP determinations from classical astrometric observations presented the first report with the ERP results determined in the first three months of 1988.

Tian Jing presented the results of the first successful prediction of the earthquake in China on the basis of the classical astrometric observations of latitude and UT0 variations.

K. Yokoyama presented a proposal of a resolution for establishing a Working Group on re-reduction of classical astrometric series of the ERP. The resolution was accepted. The Commission 19 will establish membership of this Working Group.

The resolution reads as follows.

COMMISSION 19

RECOGNIZING

1. the importance of optical astrometry observations for Earth rotation studies accumulated since the last century;

2. the inconsistency of celestial reference frames and methods of reduction hitherto used by independent stations;

3. the significance of the forthcoming HIPPARCOS catalog in forming a consistent optical reference frame tied to a radio reference system; and

4. the necessity of recomputing optical astrometry Earth orientation parameters using a consistent method based on the HIPPARCOS catalog,

RECOMMENDS

setting up a Working Group to investigate which observatories have observations available in a suitable form for rereducing them, under the chairmanship of J. Vondrak, which will report to the XXIst General Assembly with recommendations for:

a. the method of computing Earth orientation parameters, and
b. ways of implementation.

10 August, 14:00 Commission 19 - Business II

Chair : W.J. Klepczynski Secretary : M. Feissel

Ye Shu-hua had informed the commission president that, having been appointed as vice-president of the IAU, she wished to withdraw from the function of Vice-president of Commission 19. B. Kolaczek was nominated by the commission president, and elected by the members as the new vice-president of the commission.

W. Klepczynski presented the list of the new Organizing Committee, members, consultants and commission members.

Organizing Committee 1988-1991 : M. Feissel (President), B. Kolaczek, (Vice-president),

P. Brosche, W.E. Carter, J.O. Dickey, D.M. Djurovic, Ji, Wen-jing, N. Mironov, D.D. McCarthy, M.G. Rochester, T. Sasao, B.E. Schutz, J. Vondrak, G.A. Wilkins.

Consultants 1988-1991 : T. Herring, H. Schuh, V.I. Sergienko, D.E. Smith.

New Members : E.F. Arias, A. Babcock, C. Boucher, M. Lehmann, X X Newhall.

Working Group on Reference Systems : The joint resolution of commissions 4,7,8,19,20,24,31,33,40 concerning the continuation of the Working Group on Reference Systems was adopted (it became later IAU resolution C2 : WGRS - A continuing intercommission project).

Proposed future meetings : A colloquium on Reference Systems will be organized in 1990 by the US Naval Observatory, to review the different topics studied by the Working Group on Reference Systems. A Symposium on Reference Systems for Astrometry and Geodesy will be held in San Juan (Argentina) in 1991, in conjunction with the 21st IAU General Assembly.

International Earth Rotation Service : K. Yokoyama informed the commission that he had been elected as Chairman of the IERS Directing Board at the meeting held on August 9. The next meetings of the IERS Directing Board are scheduled for March and August 1989. The IERS Standards should be finalized by the end of 1988.

10 August, 16:00 Members reports

Chair : W.J. Klepczynski Secretary : Jin Wen-jing

B. Kolaczek presented a report on the prediction of short periodical variations of polar motion, based on the consideration of the main components in the 20-100d period range.

The topic of second presentation is the contribution of Meteorological Factors in Chinese Area to the Excitation of the Earth's Rotation reported by Sun Yong-xiang, the Institute of Geodesy and Geophysics, Academia Sinica, China. The excitation function ψ_1 and ψ_2 of air mass redistribution during 1970-1980 and axial component χ_3 of AAM from April 1981 to December 1983 have been estimated by using the global observed meteorological data, especially including Chinese mainland data. Although the area of Chinese mainland covers only 1.9% of Earth surface area, the contributions of the Chinese area to the excitation function of air mass distribution are 14% in ψ_1 and 20% in ψ_2 and for axial component χ_3 of AAM, the contribution of Chinese meteorological data may reach 4% of the global value. It is very important to utilize data of Chinese meteorological stations for calculating ψ_1 and ψ_2, and χ_3.

Summary of the resolutions and decisions adopted by Commission 19

Commission resolution on the reduction of past optical astrometry observations in the HIPPARCOS celestial frame, and setting up a Working Group.

IAU resolutions on

- WGRS - A continuing Intercommission Project (C2)
- IAU Contribution to FAGS (C9)
- Needed History on BIH and IPMS (C10).

COMMISSION No. 20

POSITIONS AND MOTIONS OF MINOR PLANETS, COMETS AND SATELLITES (POSITIONS ET MOUVEMENTS DES PETITES PLANETES, DES COMETES ET DES SATELLITES)

Report of Meetings on 4th, 8th and 10th of August, 1988

PRESIDENT: Y. Kozai SECRETARY: D.K. Yeomans

ADMINISTRATION AND SCIENTIFIC SESSION ON OCCULTATION(Aug. 4)

The President welcomed members of the Commission to Baltimore. Noting that the Commission's Vice-President, Yu.V. Batrakov, requested that his name not be put in nomination for President he requested and received Commission's confirmation of R.M. West as President and A. Carusi as Vice-President for the coming triennium. D.K. Yeomans was confirmed as the secretary without objection. All present stood in silence for a few moments in remembrance of the Commission members who had died during the triennium: Y.-Z. Zhang and W.H. Robertson.

E. Bowell requested the establishment of a Working Group for Minor Planet Citations. This group would be responsible for preparing and maintaining a machine readable file of minor planet names that would include the citations and references. Bowell also suggested that the Group look into the possibility of awarding a suitable diploma to accompany future citations. Members of the Organizing Committee, new Commission members and consultants, as well as the members of various Working Groups were confirmed during the second administrative session on August 10. J.D. Mulholland renewed his plea that Commission members consider the advantages of converting from the current sexagesimal system to one based entirely upon decimal fractions.

As chairman of the Occultation Working Group, R. Millis gave a report of activities during the triennium. Predictions for sixteen occultations and their ground tracks were distributed to the network of observers in 1987 with the corresponding number for 1988 being thirteen. Two notable successes were the December 8, 1987 occultation by 324 Bamberga and the June 9, 1988 occultation by Pluto. Chords from some thirteen sites in China, Japan and in the USA allowed an accurate size and shape to be determined for Bamberga. Evidence for the presence of atmosphere on Pluto was one of the preliminary results from observations at seven sites in Australia, New Zealand and from the Kuiper Airborne Observatory. Ten upcoming occultation events were then reviewed. Following the December 29, 1988 occultation of minor planet 6 Hebe, the 1989 events include 87 Sylvia(Feb. 28), 324 Bamberga(Mar. 18), 4 Vesta(April 2), Saturn(July 3), 9 Metis(Aug. 6), 216 Kleopatra(Aug. 14), 4 Vesta (Aug. 19), 521 Brixia(Oct. 23), and 1 Ceres(Nov. 11).

A discussion then followed with Millis noting a very limited number of observers outside the U.S. with portable telescopes suitable for occultation work and E. Roemer noting the relatively few observatories(i.e. Lick and Lowell) with plate scale and field of view to provide reliable last minute refinements to occultation predictions. D. Dunham noted that Bordeaux Observatory was also capable of making accurate position updates, and with the northern hemisphere PPM catalogue now available, observatories having instruments with narrow fields of view can also participate in position improvements. Dunham and L. Kristensen both stressed that the occultation community should single out the most important events and distribufe the predictions as soon as possible. Millis resigned as chairman of the Working Group and after some discussions on the role of the Group, Commission members took straw votes to confirm that the Working Group on Occultations should remain in operation and that L. Wasserman should become the chairman.

SCIENTIFIC SESSION: MINOR PLANET(August 4)

The director of the Minor Planet Center, B.G. Marsden, noted that the proposal, introduced at the time of the "Asteroid II" colloquim in March 1988, concerning the processing of discovery observations of main-belt minor planets, is already having a beneficial effect. This means that *real time* discoveries are credited only when accurate positions are provided on two or more nights. Processing of single-night reports, which have typically involved 60 percent of all discoveries of minor planets, is now deliberately being delayed for several months. This change is already tending to make the astrometric database more useful for orbit and identification work and to credit discoveries to the more useful and systematic observers. While exceptions can be made, the same procedure is also recommended for non-visual discoveries of comets and Earth-approaching minor planets.

There is also value in obtaining two positions(rather than one or three) on each night and in also applying the two-night rule with regard to observations of known minor planets, particularly numbered objects or those on the critical list. Observers are strongly encouraged to provide magnitude for new discoveries and objects about whose identification they made be uncertain. Particular care should be taken in recording the UT time of observation since an extraordinarily large number of observations are reported to the Minor Planet Center with the wrong date! After the first two nights of observations, the observer needs to decide whether further follow-up is necessary. The possibility of rapid communication by electronic mail now means that the Minor Planet Center can advise the observer whether a discovery is really new. Observers should try to arrange follow-up on two further nights such that the total arc spans two dark runs, or at least ten days during a single dark run. What is *not* wanted are observations on just three nights over less than a five-day span. The Minor Planet Center would also find it helpful if an observer would send all the observations of a particular object during one dark run in no more than two communications. If identifications do not show up when the arc approaches one month, further observations will be needed to secure recovery at another opposition. Extended coverage of this type, yielding arcs of up to 90 days and more, is particularly desirable if any kind of follow-up is to be given to very faint objects, for which past identifications are rather unlikely.

V.K. Abalakin briefly reported on activities at the Institute for Theoretical Astronomy. A few copies of the 1989 edition of *Efemeridy Malyky Planet* were available at the meeting. Beginning with this edition, there are some substantial improvements in the standard opposition ephemerides. Although the total amount of space occupied is unchanged, the ephemerides are now given with greater precision, there is more specific information on accuracy, and there are more detailed data on heliocentric and geocentric distances and phase angle.

E. Bowell described the Lowell Observatory, U.K. Schmidt Telescope Asteroid Survey(LUKAS). This survey is designed to provide astrometric data for minor planets(spanning more than 60 days) in an effort to reduce the biases in the distribution of orbital elements introduced by observational selection effects on discoveries. With a magnitude limit of approximately 22, this survey will utilize automatic scanning and measurement procedures to obtain minor planet positions on wide field photographic plates.

T. Gehrels described the continuing SPACEWATCH program at Steward Observatory whereby a CCD detector on a 0.9m telescope is used in a scanning mode to efficiently collect positional data on minor planets and comets. The 320x512 element CCD detector is currently being replaced by a much larger 2048x2048 one.

E. Helin discussed the plans for a recently acquired 1.2m telescope at the Jet Propulsion Laboratory's Table Mountain Observatory. This instrument will be used to make observations on targets of opportunity and to follow up on discoveries made with over-subscribed telescopes at Palomar Mountain and elsewhere. This facility, which should be operational in approximately six months, will augument the minor planet and comet astrometric programs that are active at Oak Ridge, Steward, Palomar, Lowell and elsewhere.

Brief communications were also presented by H. Debehogne regarding the use of the 0.4m GPO telescope at La Silla for astrometric observations and by R. Rajamohan regarding the recent plans to use a 0.46m Schmidt to develop the skills required to make and reduce minor planet and comet observations. Rajamohan outlined future plans that include the use of a larger instrument to search for minor planets, comets and planet X. This initiative was welcomed by Marsden as the first astrometric program in India.

A substantial part of the session was devoted to presentations related to the proposed transfer of Commission 20 activities from B1950.0 to J2000.0. On introducing this part, Marsden reminded members that the Commission had in 1983 specifically deferred action on this matter until suitable star catalogues and charts became widely available. Following the publication of the FK5, it appeared that significant progress in this area was now being made.

C. Smith described the recent completion of a SAO type catalogue that has been put onto the FK5 system at the U.S. Naval Observatory. Schwan's regional corrections have been applied and the Aoki/Standish conversion procedures have been used to transform the positions to the epoch of J2000.0.

S. Roeser described the Positions and Proper Motions(PPM) reference star catalogue for the northern hemisphere. A total of 181 581 stars are included on the FK5 system and the entire catalogue is available on the J2000 epoch. Approximately six published positions were used to determine proper motions, and the stellar RMS position error, at present epoch, is approximately 0.3 arcsec. A preliminary version is already available on tape and the final version will be ready by the end of 1989; it will then be delivered to the National Space Science Data Center at the Goddard Space Flight Center. With the completion of the CPC2 a recent epoch star catalogue is available for the southern hemisphere. Roesner expressed the desire to complete a PPM catalogue for the southern hemisphere if the Astrographic Catalogue could be made available in machine readable form.

J. Russel provided information about the completed Space Telescope Guide Star Catalogue. Containing 20 million stars, this catalogue covers the entire sky down to mag. 15 at the galactic poles and mag. 13 at the galactic equator. It is on the FK5-J2000 system and has a mean plate epoch of 1983. However, the catalogue position errors can reach 2-3 arc seconds when positions were measured in the corners of the Schmidt plates. No proper motion information is given. The primary mode of distribution of this catalogue will be on 3-4 compact optical disks.

It was necessary also to consider procedures for the conversion of existing B1950.0 observations and orbital elements of minor planets and comets to J2000.0 and the availability of appropriate representations of the solar and planetary coordinates. Marsden pointed out that, while the formulations on page B42 and B43 of the current *Astronomical Almanac* might be appropriate for conversions between B1950.0 and J2000.0 in the case of stars, the artificially induced proper motion(arising mainly from the change in the precession constant) created a potential problem with positions of minor planets and comets.

E.M. Standish discussed the numerical DE200 planetary ephemeris and P.K. Seidelmann remarked that the U.S. Naval Observatory was preparing a floppy disk planetary ephemeris over the interval 1800-2050 based upon JPL's DE200. The precision of the available data will be at the milli-arcsecond level. Standish recommended that the existing angular elements for comets and minor planets that are approximately referred to the FK4-B1950 system be transformed to the FK5-J2000 system by the application of a simple coordinate rotation such as that given on page S37 of the *Astronomical Almanac* for 1984. This would presumably be a stopgap procedure until new orbital elements could be diretly determined using observations in the J2000 system. More sophisticated conversion procedures should be applied for transforming FK4-B1950 observations to FK5-J2000. Marsden showed an example, in which the different treatment of observations and orbits(including the solar coordinates) yielded residuals in the two systems that typically differed by 0.5 arcsec. It was proposed that an ad-hoc committee be formed to examine further the requirements and to specify schedules and procedures for the transformation.

E. Tedesco outlined the new procedures for predicting minor planet magnitude data. He noted that the H and G values given in the IRAS documentation refer to blue magnitudes while those published in *MPC* 11096 refer to visual magnitudes, the latter values for G being constrained to be in the range 0.0-0.5.

Lack of time prompted the postponement of Z. Knežević's paper on proper elements and families to the minor planet session of Commission 15. L.K. Kristensen presented a brief report on the lost minor planet, 719 Albert, noting that a reinvestigation of available data shows that they are consistent if due account is taken of the uncertainties in the times of the end-points of the photographic trails. The asteroid's orbital position is currently indeterminate so that search ephemerides cannot be computed to aid recovery.

SCIENTIFIC SESSION: COMETS(August 8)

E. Roemer remarked, in taking the chair, that the past triennium was an extremely successful one that included the flyby of six spacecrafts past comet Halley in March 1986 and a record number of comet discoveries. C. Shoemaker was credited with her 14th discovery and W. Bradfield with his 13th.

S. Nakano reported upon the activities of Japanese amateur astronomers in observing and identifying minor planets. Over three thousand astrometric observations of minor planets have been reported by Japanese amateur astronomers from 30 separate observatories. Kobayashi identified 40 recently observed minor planets with known objects, while Nakano has 37 such identifications to his credit. Marsden praised the several contributions of Japanese amateur astronomers to minor planet research and noted a parellel record with respect to observations of comets.

On behalf of E.M. Shoemaker, E. Bowell presented a request for comet position and magnitude observations at large heliocentric distances. Shoemaker's recent results for the flux of long period comets at Jupiter's orbit is 3-4 times his previous estimates. This result suggests that estimates for the number of comets in the Oort cloud should be increased and that this increased flux will affect the deduced comet cratering rates for the planets and their satellites. The revised numbers may also be in better accord with the observed population of short period comets.

B.G. Marsden gave a brief summary of the status of astrometric observing program at Oak Ridge. Although considerable thought had been given a few years ago to utilizing a mosaic of small CCD detectors and inclinometers to perform the astrometry, lack of a suitable computer interface had made it impossible to put the system into practice. With a large CCD now due in the near future, it seems appropriate to make use of it directly, without the need for inclinometers. In the meantime, the observing is still done photographically. When the program with the 1.5m reflector started in 1972 it was sometimes possible to reach objects of mag. 20, but the steadily brightening sky now makes it difficult to reach fainter than mag. 18. Nevertheless, even if recoveries of comets rarely take place at Oak Ridge, final observations and other critical positions are still frequently obtained and rapidly reduced. In the area of minor planets, it is noteworthy that 30 percent of the minor planets are newly numbered on the basis of Oak Ridge observations alone at their latest oppositions.

Roemer then noted that comets with poor observation histories are a problem to recover using the small fields of CCD-equipped telescopes. A request was made for retaining the capabilities of some wide field instruments for recovering poorly observed objects.

S. Ostro discussed the use of radar Doppler and range observations as astrometric data types for comets and minor planets to be used to refine their mathematical models and observatory positions. Radar data exist for 52 minor planets and 5 comets. Radar data are powerful astrometrically because they have fractional precision between 10^{-5} and 10^{-8} and they are made along the line-of-sight, and hence are orthogonal to optical, angular position measurements. Line-of-sight velocities can be measured to the sub mm/s level. The power of the radar data, when used in conjunction with optical data, is particularly evident when the optical data arcs are small. Ostro suggested that when the radar data are published, the transmit frequency, received time, round trip time delay and Doppler frequency shift be given rather than derived quantities such as distance and velocity.

Z. Sekania outlined his recent work on the dynamics of comets. The momentum transfered to a comet by gas outflowing from the discrete centers of activity on its nucleus depends upon their outgassing pattern and surface distribution, and on the comet's spin vector. As a rule, seasonal effects are dominant and the momentum distribution is highly asymmetrical with respect to perihelion. This results in a major contribution from the transferred momentum's radial component to the perturbation of the comet's mean motion. On the other hand, the role of a rotation lag(A.M. vs. P.M. activity) is reduced, especially for high-obliquity objects. For comets like Encke with a high equatorial obliquity, there may be no correlation between nucleus spin direction and orbital acceleration or decceleration.

D.K. Yeomans showed recent results in modelling nongravitational effects in the orbit determination studies for the comets, d'Arrest, Giacobini-Zinner, Halley and Kopff. By allowing the peak in the water vaporization curve to move to either side of perihelion in the nongravitational force model, the orbital fit is improved and the optimum offset of the water vaporization curve corresponds to the similar offset of greatest intrinsic brightness in the comet's light curve. This work also supports the notion that nongravitational effects are due primarily to a radial thrust acting asymmetrically with respect to perihelion rather than to the traverse thrust component.

Marsden remarked on the matter of availability of ephemerides. A quick poll showed that, in spite of the existence of the ephemeris-computation feature in the Central Bureau for Astronomical Telegrams/Minor Planet Center computer service and the widespread distribution of suitable computer programs, more than one-third of those present still required printed ephemerides. While the frequent publication of the *IAU Circulars* and *MPC*s is clearly appropriate for ephemerides for newly-discovered comets, the need for observers to plan ever further into the future made it desirable to enquire whether annual sources of ephemerides, such as the BAA, OAA, and ICQ Handbooks, which principally include data for the predicted returns of periodic comets, could be published farther in advance. In response to a request from the Nautical Almanac Offices, orbital elements for periodic comets are now provided in the *MPC*s several years ahead, with those for all the comets returning in 1991 already published in *MPC*s dated May 1, 1988.

M.E. Baily presented a short paper by himself and C.R. Stagg entitled "Cratering constraints on the Oort cloud inner core." A re-analysis of the relative proportion of craters produced by different types of Earth crossing bodies shows that active comets probably account for less than 10 percent. This considerably relaxes cratering constraints on models of the inner Oort cloud. The absence of evidence for large surges of comet showers suggests that the inner Oort cloud, if it exists at all, does not have to be as dense as once thought.

A. Carusi gave an update on the Long Term Evolution Project. The original long-term numerical integrations for 132 comets were complete for comets through December 1984. The first update of the catalogue will include orbits for 14 new comets in addition to 23 orbits that have been recently improved. These results should be completed by the end of 1988 and released by mid-1989.

Yeomans then reported on the successful campaign of the Astrometry Network of the International Halley Watch. For comet Giacobini-Zinner a total of 1 300 astrometric observations(Jan. 28, 1984 - March 31, 1987) have been received from 81 observatories. The corresponding numbers for comet Halley are 6 631 observations(Oct. 16, 1982 - Feb. 14, 1988) from 143 observatories. For Giacobini-Zinner, 83 percent of all data were less than three times the RMS residual and hence used in orbital solutions; for Halley 90 percent of the received data were used. All data for comets Giacobini-Zinner and Halley will be archived and made available on compact disks and, if resources allow, as printed volumes.

P. Weissman brought the Commission up-to-date on NASA's proposed rendezvous mission to a comet with an enroute flyby of a main-belt minor planet. The mission currently under consideration would launch in August 1995, fly by minor planet 449 Hamburga in January 1998 and effect a rendezvous with comet Kopff in August 2000. The spacecraft would investigate the comet throughout its perihelion passage in December 2002 with cameras and other instruments that include a nucleus penetrator to investigate its subsurface ices.

R. Farquhar summarized plans to use existing and future spacecrafts to investigate comets using a low budget approach. The Japanese spacecrafts, Sakigake and Suisei, are scheduled for Earch swingbys in 1992 with the former craft then being targeted for comet Honda-Mrkos-Pajdušáková in 1996 and the latter being targeted for comet Giacobini-Zinner in 1998. Using an Earth swingby, ESA's Giotto spacecraft can be targeted for comets Grigg-Skjellerup or Hartley 2 if upcoming tests indicate that sacecraft's camera is operational. Farquhar also mentioned studies that are underway by NASA and ISAS(Japan) to assess the possibilities of a comet coma sample return mission in the late 1990's.

SCIENTIFIC SESSION: SATELLITES(August 8)

J.-E. Arlot, chairman of the Working Group on Satellites, began the session by proposing two resolutions to make more accessible the satellite astrometric data taken by spacecrafts. The complete text of these internal resolutions follows:

(1) The Working Group on Natural Satellites of Commission 20 recommends that NASA make available to all investigators the positional observations of the satellites of the planets that have been made from spacecraft. Preferably, these data should be made available in the form of right ascensions and declinations together with necessary auxiliary data such as the position of the spacecraft at the observation time. The Viking and Mariner observations of Phobos and Deimos that have been prepared in support of the Soviet Phobos mission are excellent examples of what should be done to make this type of data available.

(2) Commission 20 recommends that its Working Group on Natural Satellites study the creation of an international center of astrometric data on the natural satellites. Under the auspices of the IAU, this center would work to make these data widely available to researchers in this area of study.

Arlot then read an abstract of a paper by V. Shor on the refinements of the orbits of Phobos and Deimos using ground-based and spacecraft observations. Together with the Viking spacecraft data, optical data over the interval 1877-1976 were used to determine the satellites' orbits.

J.R. Rohde reported that an almanac for many of the natural satellites of all the planets from Mars through Pluto was nearly complete. Rectangular coordinates and differential positions with respect to the planet will be available on floppy disks over the interval 1988-2000 from the U.S. Naval Observatory within a few months. Arlot then reported that M. Chapron-Touzé and he had produced compact ephemerides of the Galilean satellites of Jupiter on floppy disks. Rohde described his work on the outer satellites of Jupiter noting that there are systematic positive biases for the early right ascension observations of JVI and JVII.

K. Aksnes outlined the recent work of the Satellite Nomenclature Liaison Committee with regard to the Uranian satellites that were discovered by the Voyager 2 spacecraft. These ten faint satellites, all located inward of Miranda, were named after Shakespearean characters and numbered from the outermost inward. The agreed upon names and designations have been published on the *IAU Circular* 4609 dated June 8, 1988.

B. Morando delivered a short paper by J. Laskar and R.A. Jacobson on the five large Uranian satellites. Laskar's analytic theory agrees well with numerical integration results with differences of only 12-100km over the ten year interval. The actual uncertainty of their theory is of the order of 100-250km in position. Their work has recently appeared in the *Astronomy and Astrophysics*, and an ephemeris is available on floppy disk.

D. Pascu reported upon the U.S. Naval Observatory's program of CCD observations of faint satellites using the 1.5m astrometric reflector at Flagstaff, Arizona. They employ a 2 to 1 re-imaging coronagraph and a 800x800 CCD. Observations of Miranda over the interval 1981-1985 have already been published. Photometry(*B*, *V*) of the faint satellites is a useful by-product of the astrometric positions.

Morando then reported upon a program to observe the mutual eclipses of the Galilean satellites of Jupiter. He noted the important role of amateur astronomers who have been making observations with photoelectric detectors as well as visual observations. Two-dimensional photometry has been done with vido cameras. Future infrared observations would be useful to detect hot spots on satellite surface.

ADMINISTRATION II(August 10)

Y. Kozai prepared the lists of the Commission's proposed new members and consultants and of the members and chairmen of the various Working Groups. These groups, listed below, were first reviewed and then approved by the Commission members.

New members: M. Chapront-Touzé, J.R. Donnison, G. Dourneau, J.A. Fernandez, W. Ferreri, Miao-fu He, H. Kinoshita, Z. Knežević, D. Lazzaro, P.J. Message, A. Milani, T. Pauwels, D.A. Ramsay,

E.M. Shoemaker, M. Soma, E.M. Standish, D. Taylor, D.J. Tholen, Z. Vavrova, A. Whipple, M. Yuasa

Consultants: C.M. Bardwell, C. Blanco, S.J. Bus, K.I. Churyumov, R.W. Farquhar, R.A. Robinson, E. Kazimirchak-Polonskaya, H. Kosai, R.H. McNaught, S.M. Milbourn, Z.M. Pereyra, V. Protitch-Benishek, N. Samojlova-Yakhontova, T. Seki, C.S. Shoemaker, J.B. Tatum

Working Group on Comets: M.E. Bailey, N.A.Belyaev, M.P. Candy, A. Carusi, A. Gilmore, L. Kresák, B.G. Marsden, S. Nakano, H. Rickman(ch.), E. Roemer, G. Sitarski, R.M. West, P. Wild, D.K. Yeomans

Working Group on Occultations: J.C. Bhattacharyya, C. Blanco, G.L. Blow, D.W. Dunham, Miao-fu He, A.R. Klemola, L.K. Kristensen, R.L. Millis, M.D. Overbeek, V. Shor, M. Soma, G.E. Taylor, L. Wasserman(ch.)

Working Group on Satellites: K. Aksnes, J.-E. Arlot(ch.), S. Ferraz-Mello, P.A. Ianna, R.A. Jacobson, J.H. Lieske, B. Morando, J.D. Mulholland, T. Nakamura, D. Pascu, M. Rapaport, P.K. Seidelmann, V. Shor, D.B. Taylor, R. Vieira-Martins

Satellite Nomenclature Liaison Committee: K. Aksnes(chairman & delegate to WGPS), J.-E. Arlot, A. Carusi, R.M. West, P.K. Seidelman(Vice-chairman & alternate delegate to WGPS)

Minor Planet Names Committee: A. Carusi, B.G. Marsden, R.M. West

Comet Nomenclature Committee: B.G. Marsden, H. Rickman, R.M. West

Standing Committee to Oversee Publication of Photometric Data for Minor Planets: E.L.G. Bowell, A. Harris, B.G. Marsden, R.M. West

Representative to Working Group on Reference Systems: D.K. Yeomans

Study Group on the Origins of Minor Planet Names: V.K. Abalakin, E.L.G. Bowell(ch.), F. Edmondson, H. Haupt, L.K. Kristensen, B.G. Marsden, P. Millman, J.D. Mulholland, E. Roemer, L. Schmadel, K. Tomita, I. Van Houten-Groeneveld, J.X. Zhang

The Commission then considered an appeal by T. Gehrels concerning a minor planet name proposed by him but rejected by the Minor Planet Names Committee as having current political implications. It was noted that the appeal had been properly filed in accord with the resolution adopted by the Commission at the XIXth General Assembly(*Trans. IAU* XIXB, p. 185, 1986). A secret ballot by Commission members was requested, and the decision of the Minor Planet Names Committee was sustained by the vote of 31 to 6. The straw poll of the Commission on their desire to convert from the sexagesimal to the decimal system showed that most were opposed to the change at this time.

Marsden then gave a report on the ad hoc committee on coordinate conversions. The committee introduced the following internal resolution:

> Noting that progress is being made toward the general availability of necessary star catalogues, Commission 20 recommends that steps be taken to convert from B1950.0 to J2000.0 in the near future. For this purpose, an ad hoc committee of Commission 20 is hereby established to prepare for the transition, with appropriate procedures to be presented for endorsement at the next General Assembly. The target date for the change to the system of FK5 on the standard equinox J2000.0 is 1992 January 1. This requires that the work of the Committee be completed by 1990 June 30.
>
> For the Committee to discharge its duties, the Commission notes that adequate star catalogues must be available by the above target date. In this context, an *adequate* catalogue means one whose stellar density is at least as great as that of the SAOC. The Commission recognizes that machine-readable versions will satisfy many users, but notes that printed versions and associated charts are still desirable. Special effort is clearly needed to achieve the stated goal for the southern hemisphere. Commission 20 urges that groups engaged in the compilation of such catalogues be supported adequately to ensure that these tasks come to fruition in a timely manner.
>
> Additionally, the transition requires protocols for the conversion of both observations and orbital elements between the existing and the new systems, as well as for the determination of orbits and the calculation of ephemerides in the new system. The committee shall be responsible for formulating these procedures, giving attention to their internal coherence and reversibility, insofar as this be possible.
>
> The ad hoc Committee is to be known as the System Transition Committee, and its chairman shall be the Commission's representative to the IAU Working Group on Reference Systems.

This resolution and the following members of the System Transition Committee were reviewed and approved by the Commission members present:

> T.E. Corbin, P.D. Hemenway, L.K. Kristensen, K. Hurukawa,
> B.G. Marsden, J.D. Mulholland, S. Roesr, P.K. Seidelmann, V. Shor,
> C.A. Smith, E.M. Standish, R.M. West, D.K. Yeomans(ch.)

C.S. Kay, representing Commission 22, read a proposal to set up a working party to gather opinion and formulate recommended policy concerning the pollution of interplanetary space by spacecraft exhausts and solid debris. Commission 20 voiced its support of the initiative and nominated Y. Kozai and R.M. West as its representatives on the working party.

The Commission adjourned its final session to reconvene in three years at the next General Assembly in Buenos Aires, Argentina.

COMMISSION No. 21

LIGHT OF THE NIGHT SKY (LUMIERE DU CIEL NOCTURNE)

Report of Meetings 3, 4, 8 and 10 August 1988

PRESIDENT: K. Mattila SECRETARY: A.C. Levasseur-Regourd

4 August 1988

BUSINESS SESSION

The President welcomed with pleasure the large number of members attending the Commission 21 sessions in Baltimore.

1. COMMISSION MEMBERSHIP AND OFFICERS

The President reported the sad news about the death of two members of the Commission, Dr. Carlos Sánchez-Magro and Professor Richard H. Giese, and reminded briefly on their scientific work related to Commission 21. Prof. Giese was a central and beloved figure in Commission 21 for many years. He served in the Organizing Committee since 1973 and as President of Commission 21 during the triennium 1982-1985. The president invited the participants to stand up in silence in memory of these two outstanding scientists and friendly colleagues. (Obituaries have been published in Commission 21 Newsletters Nos. 9 and 11).

Concerning the officers and membership the following list was approved:
President: A.C. Levasseur-Regourd
Vice-President: M.S. Hanner
Organizing Committee: S. Bowyer, R. Dumont, Yu.I. Galperin, J. Houck, Ph. Lamy, Ch. Leinert, K. Mattila, T. Mukai
New Members: J. Clairemidi, L.B. d'Hendercourt, S.F. Dermott, A. Fujiwara, H. Hasegawa, M.G. Hauser, A. Leger, K. Lumme, T. Maihara, J.-M. Perrin, G. Toller, K. Yamashi, T. Yamamoto
Consultants 1988-91: C. Classen, M. Dubin, G. Eichorn, A. Frey, M. Ilyas, D.J. Kessler, B. Kneissel, W. Kokott, G. Lopez, J. Maucherat, J.A.M. McDonnell, R.D. Mercer, J. Michael, A. Mujica, A.W. Peterson, H. Radoski, E.T. Rusk, R. Schaefer, G.H. Schwem, W.-X. Wang, K. Weiss-Wrana, H.A. Zook

2. REPORTS ON ASTRONOMY 1988

The President thanked the members for sending information on their activities in 1985-88 and acknowledged the help provided especially by A.C. Levasseur-Regourd in the preparation of the report of the Commission.

3. THE NEWSLETTER OF THE COMMISSION

The President edited three newsletters, Nos. 9, 10 and 11 during this triennium.

4. OTHER COMMISSION ACTIVITIES IN 1985-88

The President, on behalf of the Commission, has expressed, in a letter addressed to the General Secretary, his concern about the increasing threat of light pollution especially through the planned commercial uses of near - Earth environment, e.g. the French Eiffel Tower in Space and the US Celestis & Space Services Inc. projects. Commission 21 is co-sponsor of the IAU Colloquium No. 112 "Light Pollution, Radio Interference and Space Debris", 13-16 August 1988.

Commission 21 co-sponsored the IAU symposium No. 135 "Interstellar Dust", 26-30 July 1988. It was the main sponsor of the Joint Discussion IV on August 5 1988 during the GA in Baltimore on "The Cosmic Dust Connection in Interplanetary Space: Comets, Interstellar Dust and Families of Minor Planets"; it was co-sponsored by Commissions 7, 15, 20, 22, 34 and 49.

5. FORTHCOMING ACTIVITIES

The President described the preparations under way for IAU Symposium No. 139 "Galactic and Extragalactic Background Radiation - Optical, Ultraviolet and Infrared Components", to be arranged on 12-16 June 1989 in Heidelberg, FRG. The organization of a meeting on this topic has been discussed and planned during many Commission 21 sessions in the past. Therefore it is natural that Commission 21 is the main sponsor of the Symposium, which is being co-sponsored by IAU Commissions 28, 33, 34, 44 and 47 and by COSPAR Subcommission E.1. The Chairman of the Scientific Organizing Committee is K. Mattila, and the Chairman of the Local Organizing Committee is Ch. Leinert. The host institute is the Max-Planck-Institut für Astronomie.

Dr. T. Mukai described the preparations for the proposed IAU Colloquium (now approved as IAU Coll. No. 126 by the EC) "Origin and Evolution of Interplanetary Dust", which is to be held on 27-30 August 1990 in Kyoto, Japan. This meeting will continue the series of highly successful meetings on interplanetary dust, the first of which was in 1967 in Honolulu, followed by the meetings in Heidelberg (1975), Ottawa (1979) and Marseille (1984). Commission 21 is the main sponsor and Commissions 15 and 22 as well as COSPAR are the co-sponsors of this colloquium. The chairperson of the SOC is A.C. Levasseur-Regourd and the Local Organizing Committee is chaired by H. Hasegawa.

6. WORKING GROUPS

"Working Party" on Interplanetary Pollution : Commission 21 supported the establishment by Comm. 22 of a Working Party with the task to recommend an IAU/COSPAR Panel on Interplanetary Pollution. Suitable Commission 21 representatives for the Working Party will be nominated later by the Organizing committee.

Working Group on Fluffy Structures : On the initiative of J.M. Greenberg Commission 21 decided to set up a WG on this topic. The Chairman is J.M. Greenberg, and the other members T. Mukai, R. Zerull and W.-X. Wang.

Working Group on "Standard Tables of the Light of the Night Sky". This WG is described in the report of the scientific session on 8 August 1988.

7. CLOSING REMARKS

The President expressed his thanks to the members and to the Organizing Committee of Commission 21 for their cooperation. He thanked especially the two resigning members of the OC, Dr. Tanabe and Dr. J. Weinberg for their long and valuable work in the Organizing Committee and as Commission Presidents. The incoming President expressed the members' thanks to the retiring President.

SCIENTIFIC SESSIONS

3 August 1988

THREE-DIMENSIONAL MODELLING AND INVERSION TECHNIQUES RELATED TO THE ZODIACAL AND GALACTIC LIGHT
(Chair: A.C. Levasseur-Regourd)

This session was devoted to the memory of Richard H. Giese. J. Weinberg presented some personal memories of Richard H. Giese. The following papers were presented during the session, in order of their presentation:

W.B. Burton: *Modelling the galactic infrared emission* (invited review) — no abstract available

<u>B. Kneissel</u>: *The Three-Dimensional Distribution of Zodiacal Dust and its Dynamical Structure* (invited review) — Global models of the 3-D distribution in the zodiacal dust cloud taken from visual observations prefer large number densities above the solar poles (classical fan, sombreros and bulge fan). But also multilobe structures are found. Infrared observations mostly demand for density nodes above the solar poles. Despite of this apparent divergences, many models give equal densities close to the Earth. The brightness given by each model has been evaluated with use of a scattering function similar to that for Allende-meteorites and compared to the brightness data of Levasseur-Regourd and Dumont. The convergence of models close to the Earth is confirmed. Dynamical analysis reveals that there is a least number condition for the helioecliptic dependent part of the density function, which is not satisfied by the sinks of multilobes. Furthermore the number of orbits with particles in retrograde motion according to the optimum sombreros exceeds by far the number given by other observations.

<u>J.C. Good</u>: *Modelling of Zodiacal Infrared Emisstion* (invited) — A physical model for the interplanetary dust cloud has been fit to the IRAS data. This model consists of spatial distributions for the dust volumetric emissivity and temperature, an inclination and line of nodes for the cloud, and a simple parametrization of the dust emissivity as a function of wavelength. The volumetric emissivity was found to vary as $r^{-1.8}$ [-5 $(\sin\beta_0)^{-1.3}$] and the temperature as $r^{-0.35}$. The IRAS data are limited to solar elongation angles between 60° and 120° and consequently are not sensitive to material closer to the Sun than 0.9 AU. However, comparison of the predicted model flux and the AFGL Zodiacal Infrared Project (ZIP) data (which looked to within 22° of the Sun) at 10 and 20 μm shows excellent agreement and implies that the $r^{-1.8}$ power law is good to 0.4 AU. The inclination and line of nodes are 1.7° and 69° respectively.

<u>A.C. Levasseur-Regourd and R. Dumont</u>: *Out-of-Ecliptic Nodes of Uncertainly Method, Inversion Results* (invited) — Both zodiacal light and thermal emission measurements provide integrals along a line of sight of local brightnesses. Various attempts have been made towards an inversion of the integral in the ecliptic plane. The nodes of lesser uncertainty method is now extended to the out-of-ecliptic case. When coherent sets of measurements performed towards the ecliptic pole and in the ecliptic are available, the local brightness J(β) along the Earth-ecliptic pole line is found to obey to 5 constraints. If all the possible J(βo) functions are assumed to be monotonous, then they have to constrict in a nodal region, where they can be determined with less uncertainty than elsewhere. Typically, a node of lesser uncertainty is found for βo ≈ 20°, i.e. for a distance above the ecliptic ≈ 0.36 AU or a distance to the Sun ≈ 1.07 AU. From this method, the local polarization is found to decrease with increasing βo for a given distance to the Sun, while the local bulk albedo may simultaneously increase. These results are to be compared with the in-ecliptic variation with heliocentric distance of local polarization and albedo.

<u>S.S. Hong and S.M. Kwon</u>: *Connection Between the Infrared Zodiacal Emission and the Visible Zodiacal Light* (invited) — By inverting the brightness distribution of the zodiacal thermal emission along the ecliptic plane, we have examined how the volume absorption cross section $n(r)\sigma_{abs}(r;\lambda)$ of the interplanetary dust particles varies with the heliocentric distance r for infrared wavelengths. The usual inverse power law of r is found to be inadequate for describing the spatial distribution of $n(r)\sigma_{abs}(r;\lambda)$ in the infrared. The resulting $n(r)\sigma_{abs}(r;\lambda)$'s at $\lambda = 10.87$ μm and $\lambda = 20.87$ μm decrease with r less steeply than the $r^{-1.3}$ relation, which was known for the spatial distribution of the volume scattering cross section, $n(r)\sigma_{sca}(r)$, of the dust in the visible. It is further found that the ratio of $n(r)\sigma_{abs}(r;\lambda)$'s at the two IR wavelengths varies with r. This is a strong indication that the zodiacal dust cloud contains more than one dust species and their mixing ratios vary with r. Comparison of $n(r)\sigma_{abs}(r;\lambda)$ in the infrared with $n(r)\sigma_{sca}(r)$ in the visible suggests that the $r^{-1.3}$ relation is too steep a decline for the spatial distribution of the dust number density n(r) in the ecliptic plane.

<u>M.G. Hauser and J.M. Vrtilek</u>: *IRAS Measurements of the Solar Elongation Dependence of the Zodiacal Emission and the Inversion of the Infrared Brightness Integral* — We examine the infrared brightness of interplanetary dust in the ecliptic and its variation with solar elongation angle, using IRAS data in the 12, 25, and 60 micrometer bands. Over the entire elongation range covered (60 to 120 degrees), a quadratic function is an excellent fit to the peak brightness of the zodiacal

emission as a function of solar elongation in these three IRAS bands. Using our function fits to the elongation-dependent brightness, we have applied an inversion technique related to that of Hong and Um (1987, Astrophys. J., Vol. 320, 928) to the problem of determining the value at 1 A.U. and the heliocentric distance dependence of the volumetric absorption cross-section of the interplanetary dust. We will discuss the results of the inversion approach and their sensitivity to several key assumptions.

S.F. Dermott, P.D. Nicholson, R.S. Gomes and R. Malhotra: *Modelling of the IRAS Solar System Bands* (invited) — Shortly after the discovery of the solar dust bands by the IRAS spacecraft it was suggested that the bands that extend uniformly round the sky arise from the gradual comminution of asteroids in the most prominent Hirayama asteroid families (Dermott et al.: Nature 312, 505-509 (1984)). Stringent observational tests have been devised to test this hypothesis. These tests are robust in that they are geometrical in nature and can be made with observations in a single waveband. Thus, they exploit the high resolution and pointing stability of the IRAS spacecraft, while side-stepping the problem of absolute calibration. Rigorous simulation of the observational data has been achieved by a numerical model that allows us to calculate, for any distribution of particle orbits in the solar system, the brightness scans that would have been seen by IRAS in any one observing period at any elongation angle. If we allow (and this is essential) that the orbits of the particles have decayed due to Poynting-Robertson light drag and that the orbital elements of the particles are now dispersed about those of the asteroidal family members, then the observations that we have analysed todate appear to be in close agreement with the predictions of the Hirayama family model.

S. Mukai and T. Mukai: *Backscattering by a Large Particle with Rough Surface* — A similarity of backscattering features (negative polarization and backward enhancement of intensity) exist among the scattering lights by Moon, asteroid, cometary dust and interplanetary dust. We show that multiple reflections by a micro-structured rough surface including shadowing effect cause the above features. Application to P/Halley gives a significant fit to the observed results.

8 August 1988

STANDARD TABLES OF THE COMPONENTS OF THE LIGHT OF THE NIGHT SKY

(Chair: J. Weinberg, Secretary: G. Toller)

Some of the considerations which had motivated the Organizing Committee to arrange this session were the following: Several of the IAU Commissions have used their expertise to put together recommendations for some astronomical constants, names etc. in their field. We would like to ask ourselves and discuss whether we could (and should) try to combine the existing knowledge on the different components of the light of the night sky into something like the "Standard tables for the LONS". Such tables could be used e.g. by astronomers from other fields who need to know the sky brightness at a given time and a given position on the sky. These considerations have been related so far mainly to the optical domain but are certainly also timely for the infrared.

The following papers were presented at the session, in order of their presentation:

A.C. Levasseur-Regourd and R. Dumont : *Available tables for zodiacal light intensity distribution* (invited review) — Reliable evaluations of zodiacal light intensity, polarization or emission should be available from any point of observation, at any wavelength, at any time, in any viewing direction, not only to obtain a better knowledge of the interplanetary dust complex, but also to obtain an accurate estimation of the zodiacal "foreground". Such a foreground is indeed most significant for HST, IRAS, COBE, ISO, etc., observations. A review of various tables, as given during the last two decades by Smith et al., Dumont, Frey et al., Dumont and Sanchez, Classen, Torr et al., Levasseur-Regourd and Dumont, Pfleiderer and Leuprecht, etc., is presented. The problems of the wavelength dependence (solarlike for zodiacal light) and the time dependence (no significant correlation with solar activity) have been solved. However, the tables may need to be corrected for some small scale local enhancements (dust trails,...) and for the changing position of the observer with respect to the warped symmetry surface of the zodiacal cloud.

M.G. Hauser: *Distribution of the Infrared Zodiacal Emission (IRAS)* (Invited review) — The data from the IRAS survey show that the large-scale brightness of most of the sky at wavelengths from 12 μm to 100 μm is dominated by thermal emission from interplanetary dust. A variety of techniques have been utilized to segregate and model this 'zodiacal emission'. Analyses of the annual variation of ecliptic polar brightness and latitude of peak brightness provide evidence of a tilted, warped symmetry surface of the dust cloud. Available representations of the zodiacal emission now include averages of IRAS scan data, empirical fits to the scan data, and parameterized physical models. a new version of the IRAS data with improved calibration will be released in 1988. NASA's COBE mission will extend the wavelength coverage and photometric quality of zodiacal emission measurements.

J. Weinberg: *Standard Tables of the Background Starlight Distribution* (invited review) — A brief discussion was presented on post-war efforts to provide observational data on the background starlight. Examples of background starlight derived from Pioneer 10 and from Helios A were used to illustrate some of the more recent (space) data that are now available, precautions that should be taken in their use, and general considerations required in any effort to provide such a "standard" data set.

T.N. Gautier: *Distribution of the Galactic Infrared Background Starlight* (invited review) — no abstract available

H. Tanabe: *Report on star counts at the Tokyo astronomical observatory* — Since 1968, star counts on the Palomar Sky Survey Atlas have been made by using a then newly designed instrument in order to estimate background brightness in zodiacal light and airglow observations. As the results, blue and red brightnesses of the integrated starlight are presented for 24 sky regions (total measured area: 54.73 sq.deg.) including 4 polar (NCP, NEP, NGP and SGP) and 10 ecliptic regions.

These presentations were followed by a general open discussion with the title "**Standard Tables: Why, How, When and Who?**" Some of the considerations coming up in this discussion were

Units: Papers should supply conversion factors to energy units, i.e. both to I_λ [erg cm^{-2} s^{-1} $sterad^{-1}$ $Å^{-1}$] and to I_ν [erg cm^{-2} s^{-1} $sterad^{-1}$ Hz^{-1}]. The frequently used night-sky unit $S_{10}(V)_{G2V}$ is acceptable for zodiacal light photometry but conversions should be supplied. In the photometry of the galactic and extragalactic components S_{10} unit is mostly not referred to the solar spectrum. Also the bandpasses of the filters used in the photometry should be published. A table of standard conversion factors used in the night sky photometry should be published in a widely distributed reference work, such as Allen: Astrophysical Quantities.

Caution in use of "Standard Tables" : When using zodiacal light tables one should be aware of the time-dependence of the brightness distribution due to the changing position of the observer relative to the symmetry surface of the zodiacal cloud. Caution was urged in the use of the term "standard", since there is currently no single standard or "best" data set available for any of the components of the light of the night sky (LONS). It was urged that more detail be provided on LONS reduction methodology, calibration, and units when reporting such data.

Working Group on "Standard Tables" : It was decided to form a Working Group within Commission 21 which will have the task to review the existing light of the night sky tables, and provide the "best" compilations for publication in a widely distributed reference book, such as Allen: Astrophysical Quantities. It was agreed that the members of the WG should represent the optical, IR and UV domains and equally well the different LONS components, zodiacal light, background starlight and star counts. The following persons were elected to the WG: A.C. Levasseur-Regourd

(Convener), S. Bowyer, T.N. Gautier, M.G. Hauser, J. Houck, C. Lillie, H. Tanabe and J. Weinberg, and it was agreed that the WG can invite new members.

10 August 1988

OPTICAL, IR AND UV LIGHT OF THE NIGHT SKY -- RECENT OBSERVATIONAL TOPICS AND FUTURE PLANS
(Chair: K. Mattila, M.S. Hanner)

The following papers were presented during the session, in order of their presentation:

a. Galactic and Extragalactic Components

F. Low: *Infrared Cirrus* (invited review) — no abstract available

T. Matsumoto: *Infrared Cosmic Background Radiation* (invited review) — Diffuse sky brightness was observed both in near-infrared and submillimeter regions with rocket-borne radiometers cooled by liq.N_2 and He, respectively. Around 2 µm, a considerable amount of isotropic emission was observed which can possibly be attributed to the extragalactic origin. In submillimeter region, besides excess brightness over 2.74K CBR, a significant isotropic emission at 137 µm was detected. However, its origin and nature is not certain.

C. Heiles and W.T. Reach: *Separating the Gaiactic and Zodiacal IR Emission* (invited) — We are developing a physical model of the Zodiacal dust consisting of two components, a finite disk and rings that produce the 'Zodiacal bands'. We determine the location and shape of the components from the variation of their distances with solar elongation and time of year, a kind of 'parallax'. The method involves least-square fits to parameterized models and includes Gaussians to represent the 'bands' and a term in Galactic HI to represent the Galactic emission. We have determined the local emissivity, the orbital elements of the planes of the Zodiacal disk and rings, and the height of the rings above their mean plane, all to high accuracy.

S. Bowyer: *Galactic and Extragalactic UV Background Radiation* (invited review) — Our knowledge of the night sky background in the 100-4000 Å band was reviewed. Sources of diffuse emission in the band include resonantly scattered solar radiation from HeII in the Geocorona (304 Å) and from the local interstellar medium flowing through the heliosphere (HeI 584 Å and HI 1216 Å). The local hot interstellar medium will produce line emission in the 100-912 Å band; at longer wavelengths sources of diffuse emission include hot gas in the Galaxy, scattering by dust, zodiacal light and possibly an extragalactic component.

Although our knowledge of most of these compnents is extremely limited, ongoing studies of the resonantly scattered hydrogen and helium in the heliosphere have led to an excellent characterization of the local interstellar medium. Recent studies of other components have identified emission from a hot (10^5 K) gas associated with a dynamic galactic halo. An extragalactic component has been identified associated with the UV output of distant galaxies and the intensity of the radiation puts severe constraints on the star formation rate over the last one third of a Hubble time. Finally, studies of scattered light have yielded a characterization of UV scattering dust grains.

R.C. Henry, J. Murthy and P.D. Feldman: *UVX Observations of the Cosmic Ultraviolet Background* — Observations of the cosmic background light in the wavelength range 1250 Å were made with the Johns Hopkins UVX experiment flown on the Space Shuttle Columbia in January 1986. Eight targets, at galactic latitudes from 10° to 90° were observed. It is found that the cosmic background is patchy, ranging in birghtness from 300 to 900 photons (cm^2 s sr Å$)^{-1}$, with no correlation between hydrogen column density and cosmic background brightness.

J. Murthy: *Probing the Infrared Cirrus with Ultraviolet Radiation* — We are using the IRAS data in an effort to probe the global distribution of interstellar dust in the galaxy. Our method is to search the sky survey plates for enhancements of the cirrus near hot, UV-emitting stars, which dominate the ISRF up to several parsecs from the star. In a preliminary search of 25 hot stars, we have found such enhancements near five of them, implying that the filling factor of the dust is 25 %. This work is continuing in more detail.

R.A. Kimble, R.C. Henry and F. Paresce: *The Hopkins Ultraviolet Background Explorer* — We describe the scientific goals and conceptual design of a proposed satellite mission for studying the far ultraviolet background in the wavelength range from 1230 to 1800 Å. By performing a complete sky survey as well as deep pointings with both a broadband camera and imaging long-slit spectrograph, we hope to investigate a variety of diffuse emission processes arising in the interstellar medium of our own galaxy and possibly in the intergalactic medium as well.

K. Mattila: *Prospects for Measuring the Extragalactic Background in the Optical* — There have been several attempts to measure the EBL but no generally accepted values are available so far. Because the EBL is expected to be only a minor fraction of the total night sky brightness a direct separation technique is entangled with great difficulties (see e.g. Dube, Wickes and Wilkinson, Ap.J. 232, 333, 1979). In the talk a new measurement of the EBL by Mattila and Schnur was presented in the area of the dark nebula L1642 in the anticentre direction. The method is basically the same as used by Mattila (Astron. Astrophys. 47, 77, 1976) in the L134 observations, but the observing and data analysis methods have been cosiderably improved.

b. Solar System and Atmospheric Components

A.C. Levasseur-Regourd and R. Dumont: *Interplanetary Dust: Post-IRAS View* (invited review) — Thermal emission from the interplanetary dust cloud has been confirmed by IRAS to be the most prominent component of the sky brightness in the 10-25 μm range. It is indeed a severe foreground for astronomical observations. It is also a major source of information on the local properties of the dust grains. Unlike the solar scattered light, the thermal emission of the grains is isotropic; even more progress can therefore be expected from the inversion of IRAS integrated birghtnesses.
The pre-IRAS understanding of the interplanetary dust cloud is reviewed. Recent results obtained through inversion methods are presented. In the ecliptic near 1 AU, the heliocentric gradient of temperature is about -0.33, i.e. conflicts with a greybody assumption, while the gradient of albedo is about -0.7, i.e. conflicts with a homogeneous cloud assumption. Out of the ecliptic, the dust distribution is still open to dispute. It is however neither smooth (dust bands and trails, warped symmetry surface), nor homogeneous (location dependence of local polarization or albedo). Such results imply that the mixing ratios of the various interplanetary dust sources (comets, asteroids, β meteoroids) strongly depend upon location in the cloud.

P. Temi, P. De Bernardis, S. Masi, G. Moreno, A. Salama: *Balloon Observations of the Zodiacal Thermal Emission* — Results from an IR sky survey performed at wavelengths 11, 19, 50, 108 and 225 μm during a balloon flight (ARGO) on July 30, 1984 are presented. A new physical model of the interplanetary dust thermal emission is then discussed. Based on very simple assumptions (blackbody, silicate or graphite grains, spherical and homogeneous), the model agrees with all available observations of the dust IR emission (AFGL, ZIP, ARGO, IRAS). The decrease of the volumetric absorption cross section with increasing heliocentric distance is found to follow a power law with exponent ~ -1.

J.M. Vrtilek and M.G. Hauser: *IRAS measurements of diffuse solar system radiation* — Using IRAS data, we present an overview of the properties of infrared emission from the interplanetary dust cloud. To assist in separating the smooth component of the interplanetary emission from Galactic emission, discrete sources, and the zodiacal bands, and to characterize the properties of the interplanetary emission in compact from, we introduce an empirical function whose adjustable parameters have simple geometric interpretations. This function is fitted in a lower envelope sense to individual IRAS scans; the function represents the data well, with rms residuals in the 25 micrometer band of only 0.3 MJy/sr (less than one half of one percent of the peak emission in this band). We take advantage of the nearly 1 yr duration of the IRAS observations to study the time variation of the ecliptic latitude of peak brightness in the zodiacal emission, obtaining an inclination to the ecliptic $i=1.6\pm0.1^{\circ}$ and a line of ascending nodes $\Omega = 41.1\pm0.6^{\circ}$ (1 sigma) for the symmetry surface of the interplanetary dust cloud. Disagreement in Ω between our measurements and previous determinations that sample different parts of the dust cloud supports the suggestion that the cloud lacks a true symmetry plane.

M.V. Sykes: *Infrared Emission from Zodiacal Structures Arising from Asteroidal Collisions* (invited) — The zodiacal dust bands arise from asteroid collisions. Several are now associated with the major Hirayama asteroid families, which were formed by the catastrophic breakup of larger asteroids some time in the past. Not all asteroid families, however, have dust bands, and not all dust bands are associated with asteroid families. This suggests that some bands may be the products of the present ($\leq$ 10 million years) collisional disruption of smaller asteroids. Four bands have now been resolved within a few degrees of the ecliptic - one pair associated with the Themis family, the other with the Koronis family. Comparison of dust band models with the observations indicate that the dust band particles have the same mean orbital elements as their associated asteroid family members, but have a greater dispersion in those elements. The dust band particles are found not to have experienced significant orbital decay by Poynting-Robertson drag, indicating that their lifetimes are collision dominated. Dust must eventually spiral in under radiation forces, but their surface brightness relative to the source region is small. It is predicted that long-duration spacebased dust collection experiments will detect a four-peak modulation in the interplanetary dust flux as a consequence of dust originating in the Koronis and Themis families.

S. Koutchmy: *The Inner Zodiacal Dust Cloud* (invited) — The analysis of the inner zodiacal cloud is still in a state of infancy due to the lack of a proper method for observing the faint halo of the F-corona around the solar disc. Total solar eclipses offer a rare and short opportunity to perform polarimetric, doppler shift and thermal emission measurements which are much needed. A weak evidence of an IR signature of olivine particles of 100 microns size has been presented by Mizutani et al. thanks to their balloon experiment flown during the 1983 eclipse. Spaceborne experiments have collected few exciting observations in white-light. Interplanetary travelling clouds are seen on the zodiacal cloud background monitored with the Helios Spacecraft photometers (B. Jackson) and the sudden disappearance of several sun-grazing comets were observed with the P-78 NRL coronagraph, producing a huge cloud of evaporating particles in the solar neighbourhood (D. Michels et al.). Many results are now anticipated in the frame of the SOHO mission with several coronagraphs.

P.D. Feldman, J. Murthy and R.C. Henry: *UVX Observations of Ultraviolet Zodiacal Light Emission* — Observations of the zodiacal light in the wavelength range 2000-3100 Å were made with the Johns Hopkins UVX experiment flown on the Space Shuttle Columbia in January 1986. Eight targets, at elongations from 101 to 133° were observed. It is found that the uv color (relative to the visible) increases nearly linearly with ecliptic latitude implying that the small grains responsible for the uv scattering have a larger scale height than do the larger grains responsible for the visible scattering.

P.V. Kulkarni: *Night Time All-Sky Temperature Measurement at the F-Region Height in the Earth's Atmosphere* — A new instrument pointing to the zenith with a fish eye lens, prefilter, Fabry-Perot Spectrometer and an image intensifier is described to photograph, several times during a night, the temperature broadened interference fringes from the dome of sky emitting [OI] 6300 Å. When fringes are radially scanned temperatures at several points in the sky would be obtained simultaneously. A design of the system is given.

R.H. Garstang: *Modelling the Night Sky Glow from Cities* — The author described his continuing calculations of the contribution of man-made light pollution to night sky brightness. His model has been improved by the inclusion of earth curvature and by extension to B magnitudes. Applications have been made to may U.S. Observatories and possible observatory sites. Population projections from federal and state agencies are being used to predict the future growth of light pollution.

COMMISSION No. 24

PHOTOGRAPHIC ASTROMETRY (ASTROMETRIE PHOTOGRAPHIQUE)

Report of Meetings on 3,5 and 6 August 1988

PRESIDENT: A. Upgren SECRETARY: J. Stein

Activities of Commission 24 at the XX General Assembly consisted of a business meeting and two scientific sessions, and a joint session with Commission 8, on matters of interest to both commissions.

Business Meeting (5 August 1988)

The President welcomed the members of the Commission and announced the appointment of J. Stein as secretary. He called for a moment of silence to honour the memory of J. Meurers and S. Vasilevskis, members of the Commission who have died since the previous General Assembly at New Delhi.

He next announced the commission election, and as a result of it the Officers and Organizing Committee members for the next triennium will be:

President: W.F. van Altena (USA)
Vice President: Ch. de Vegt (FRG)
Organizing Committee: A.N. Argue (UK), P. Brosche (FRG), R.B. Hanson (USA), P.A. Ianna (USA), D.D. Polozhentsev (USSR), C. Turon (FRANCE), A.R. Upgren (USA)

The following were confirmed as new members of the Commission admitted since the previous General Assembly: G.M. Ballabh, P.S. Bunclark, F. Crifo, J. Guibert, Hartkopf, M.J. Irwin, C. Lopez, B. McLean, A.B. Meinel, L. Nachado, J. Nunez, D. Pascu, G. Pizzichini, E. Schilbach, E.U. Vilkki, C.A. Williams and C.E. Worley. With the election of the above new members, the membership of Commission 24 is increased to 135.

After some discussion the following report of the Working Group on Parallax Standards was approved by the commission. The Commission expressed its thanks to the Working Group.

REPORT OF THE WORKING GROUP ON PARALLAX STANDARDS TO I.A.U. COMMISSION 24

T.E. Lutz, Chairman

The Working Group on Parallax Standard Stars has begun to collect information from all observatories with active parallax programs. All programs of which we are aware are observing stars from the standards list, and at least some of the three standard fields centered on three open clusters (The Pleiades, Praesepe and IC 4756).

The working group proposed that the standard stars be observed by HIPPARCOS. All standard stars brighter than V= 12.0 are in the HIPPARCOS input catalog at the present time.

During the discussion of the report, C. Dahn expressed a desire for more guidance from the Working Group on exposure levels for standard clusters and as to how long standard regions should be left on individual observing programs. Furthermore, he requested that the Working Group develop a coordinated plan for processing some of the accumulating data.

J. Russell reminded the commission that the development of her Praesepe Catalog encouraged the Working Group to establish certain clusters as standard regions. She also reported that the publication of the Praesepe Catalog has been delayed due to the discovery of a magnitude (or perhaps color) error.

K. Strand asked if his list of subdwarfs had been added to the list of parallax standards. A. Upgren and C. Dahn advised us that many but not all of the stars on the list were being observed.

The Commission next heard and approved a brief report of the Working Group on Optical and Radio Positions followed by a report on the Working Group on Reference frames which is published elsewhere in these proceedings since it covers several commissions. The Commission expressed its thanks to all three Working Groups.

The Commission approved a resolution introduced by the sponsoring commissions of the Joint Commission Meeting (JCM-1) entitled "Toward Millisecond or Better Accuracy" held on 3 August. The resolution and the proceedings of that meeting appear elsewhere in the transactions.

Following the reports of the working groups, the Commission turned to other business. The following resolution introduced by C.H. de Vegt was approved after discussion: "Commission 24 emphasizes the urgent need to extend global position catalog work to substantially fainter magnitudes (V= 15 and beyond) by all available and newly developed instrumentation".

The Commission next raised the problem posed by its current name, photographic astrometry. Many members have recently expressed dissatisfaction with it, in view of CCD's and other non-photographic devices now coming into widespread astrometric use. The President read a letter from the late S. Vasilevskis urging the present name of Commission 24 be changed from "Photographic Astrometry" back to the name, "Parallaxes and Proper Motions" which applied prior to 1970. After considerable discussion, a non-binding vote for any of several possible names was taken with the result that the majority favored renaming the commission "Astrometry". No further action or implementation was taken with regard to this issue.

Scientific Meetings: (5 August)

The scientific meetings of 5 August covered two consecutive sessions and dealt with a number of basic and fundamental star catalogues. It was opened by the President with a tribute to Alexander N. Vyssotsky on the occasion of the centennial of his birth. The meeting was dedicated to his memory, as will be its proceedings.

The first session was chaired by W. van Altena and the second by C.A. Murray. The following papers were presented:

W. Gliese and H. Jahreiss: "The Third Catalog of Nearby Stars. I. General View and Content"
H. Jahreiss and W. Gliese: "The Third Catalog of Nearby Stars. II. Applied Methods and Use"
A. Upgren and E. Weis: "The Vyssotsky Stars - A Centennial Summary"

S. Roeser and U. Bastian: "A New Star Catalog of SAO Type"

J. Stock and C. Abad: "The Unification of Astrometric Catalogs"

Ch. de Vegt, M. Zacharias, A. Murray and M. Penston: "CPC2 - A Progress Report"
Ch. de Vegt: "A New Type of Astrometric Telescope for the Construction of a Global High Precision Faint Star Catalog"
T. Corbin and S. Urban: "The Astrographic Catalog Reference Stars - the Northern Portion"
J. Russell: "The Astrometry of the Guide Star Catalog"

C. Turon: "The HIPPARCOS Input Catalog"

W. van Altena: "The Yale Parallax Catalog"

J. Dommanget: "The CCDM, an Astrometric Catalog for Double and Multiple Stars"

These papers will be edited by A. Upgren and D. Philip and published in full in a volume of proceedings dedicated to Dr. Vyssotsky and forming the eighth in the series of Contributions of the Van Vleck Observatory.

Joint Meeting of Commissions 8 and 24:
"Common Interests" (6 August)

This meeting dealt with matters of common interest to the two commissions and was chaired by Y. Requieme, President of Commission 8.

A discussion was held concerning the desirability of changing the names of both commissions. The majority present appeared to favor names more in keeping with scientific goals of the commissions, techniques used etc. No decision was made at this time. A discussion was also held on the desirability of merging the two commissions. A non-binding vote was taken among those present and showed the majority to be in opposition to a merger.

A proposal introduced by T. Corbin, chairman of Commission 8 Working Group on Star Lists, was made and duly approved that the membership of this working group be expanded to include one member of Commission 24. G. Roser was then appointed as the Commission 24 representative.

J.D. Mulholland made an appeal for the abandonment of the sexagesimal system for angular measure stressing its awkwardness in scientific computing. It was agreed that the formal position of the two commissions on this matter would be in favor of the replacement of the sexagesimal system with a system of decimal degrees.

The following scientific papers were presented at this joint meeting:

P. Brosche: "Plate Reduction with Orthagonal Functions"

J. Stein: "Recent Results from the Multichannel Astrometric Photometer (MAP) at Allegheny Observatory"
J. Guibert: "Photographic Astrometry with MAMA"

P. Ianna: "The McCormick CCD Parallax Programs"

C. Dahn: "The HR Diagram for 57 Intrinsically Faint Stars Derived from CCD Parallaxes and Photometry"
D. Monet: "The USNO/CalTech Program to Measure the Palomar Sky Surveys"

Other meetings involving the two commissions are discussed in the report of Commission 8 and elsewhere.

COMMISSION No. 25

STELLAR PHOTOGRAPHY AND POLARIMETRY

(PHOTOMETRIE ET POLARIMETRIE STELLAIRE)

Report of Meetings held on 5 August 1988

PRESIDENT: F. Rufener SECRETARIES: I.S. McLean
A.T. Young

The Commission 25 activities during Baltimore assembly consisted of 2 joint commission meetings, reported in Highlights of astronomy, and 2 commission meetings reported here. The joint commission meetings were (chairmen/organisers in brackets):

Aug. 4: Problems of IR Extinction and Standardization (E.F. Milone)
Aug. 8: Stellar photometry with modern array detectors (F. Rufener)

These meetings originated in our commission and were co-sponsored. The two commission meetings were held on Aug. 5.

First Scientific Session (A.T. Young, Secretary)

Absolute Calibration of the Geneva Photometric System
(B. Nicolet, presented by N. Cramer)

New absolute (expressed in SI units) spectrophotometric data have allowed us to obtain a new determination of the Geneva passbands as well as conversion formulae relating the photometric data to absolute fluxes (Rufener & Nicolet, Astron. Astrophys., in press).

The calibrations are based on the data of Hayes & Latham (1975) for α Lyrae and on those of Oke & Gunn (1983) for subdwarfs. Comparison between Geneva photometry and some other sources of spectrophotometry were carried out. A slight systematic deviation between the data of Gunn & Stryker (1983) and ours appeared for the whole range of spectral types. A similar comparison was made between the absolute fluxes of Kiehling (1986) and Geneva photometry; the agreement is good.

As an illustration, the fluxes of the Supernova 1987A, systematically monitored in the Geneva system, were showed.

Photometric Data Collections (J.C. Mermilliod, presented by B. Hauck)

L'Institut d'Astronomie de l'Université de Lausanne collaborates with the Strasbourg Data Centre since its foundation in 1972 as an associate Centre and collects photoelectric photometric data. The final aim is to prepare catalogues of photometric data and check the data before they are introduced in the SIMBAD Database. Some 75 photometric systems have been recognized, but about 10 are really active. The catalogues produced contain both the original data and weighted mean values. The latter are usually computed after assessment of the data quality by comparing them with standard values.

Mean values in the UBV system have been computed and a catalogue is prepared for publication. One of the severest problem encountered during this work was the inclusion of UBV observations of visual binaries: not because our procedure is not adapted, but because of the lack of precise information on which components have been separated or observed together. Finally the improvement of the catalogue is facing the lack of co-ordinates for several thousand anonymous stars, which delayed the achievement of the whole work.

The photometric data and references can be found in the Strasbourg Database SIMBAD by interactive interrogation or can be obtained on magnetic tapes (or microfiches) from the Strasbourg and NASA Data Centres.

The specific realisation of the Institute is an Index file which does not contain individual data but tells us in which photometric systems a star has been observed. It contains at present information on 101.500 stars. We are expanding this file to include co-ordinates, V magnitudes and duplicity and variability flags. We also plan to include information from UV and IR satellites. Our disk storage possibilities recently increased and we shall develop a photometric database offering all facilities for data compilation, analysis and retrieval and hope to be able to offer a public service.

All these data have been extensively used for the preparation of the Hipparcos satellite Input Catalogue. The comparisons of the data coming from several systems proved to be an excellent test of the data quality and revealed a number of yet undetected errors.

The most recently up-dated catalogues are:
uvby beta system: 1985, Astron. Astrophys. Suppl. 60, 61
Johnson UBVRI : 1986, Astron. Astrophys. Suppl. 65, 195
UBV system : 1987, Astron. Astrophys. Suppl. 71, 413.

Question: What limiting magnitude was used in the catalogue? Answer: No limit was imposed; all the data in the literature were used, to be as extensive as possible.

A comment was made about the reproducibility of the systems. In some observatories, the Strömgren y filter may have a center wavelength as short as 5460 Å or as long as 5550 Å, a difference of 90 Å. This is considerable if you consider the width of the filter, which is about 200 Å. If everything were linear in this region, interpolation would be OK; but the 5200 Å feature and emission lines occur, so some action should be taken against such strong variations of the filter passbands. F. Rufener was glad to hear this, because the passband definition is one of the first things to consider when doing photometry. As important as counting photons is to define what photons you are counting. We can use linear transformations, but must know what we are transforming. A.T. Young added that it is in principle impossible to make these transformations without more information, because all these systems are undersampled. "So", concluded Rufener, "the best thing is to have a good passband".

The Photometric Aspects of the Hipparcos and Tycho Mission (M. Grenon)

Besides mapping the sky and determining the attitude of the satellite, a by-product of the main channel will be photometry in a very wide band, Hp (an unfiltered S-20 photocathode). Dielectric filters plus glass filters in adjacent fields will also define a narrower-band system, called Tycho. This BT, VT system is not exactly like the Johnson system. The Tycho B is similar to the Geneva B band; the Tycho V extends a little more to the blue than Johnson V, and almost transforms linearly to it. Tycho B differs from Johnson B, showing effects from Balmer lines and reddening; but the Tycho band can be accurately reduced to the Geneva band.

The precision of 1 observation will be better than 0.01 mag for Hp $<$7, increasing to 0.12 mag at Hp = 12. Because about 100 transits will be observed per star, the means should be quite satisfactory. The sampling is irregular in time, with 5 or 6 observations in 6-hour intervals separated by gaps of 3 to 6 weeks, depending on galactic latitude. This should detect variable stars very well, provided that the long-term changes in instrumental sensitivity are monitored carefully. This is expected to be due mainly to cosmic-ray and solar-flare protons; the loss in sensitivity is about 10%/year, or 30% over the life of the mission. The goal is to have systematic errors less than 0.5%; 16.000 stars will be used to monitor the instrument continuously.

To obtain good tangential velocities from Hipparcos's proper motions, good photometric distances will be needed for stars beyond one or two hundred parsecs (where the parallax errors become too big); the absolute magnitudes must be known to 0.5 mag. This means that good ground-based multicolor photometry is needed for the 58.000 survey stars, as well as many others that will be observed. A multi-observatory campaign has produced useful data on about 15.000 stars, but some 60.000 stars remain to be done. The ESO countries have paid more than $3000 per star to get Hipparcos data; for about 1% of this, ground-based data should be obtained to make the spacecraft data useful, preferably before the end of 1993.

Question: Absorption is very unevenly distributed. In the third galactic quadrant, you could go out to 500 pc. Did you take this into account? Answer: For most of the early-type stars it is possible to get the interstellar extinction, but for the K or M giants it is almost impossible. So the survey is extended to the north. Up to 200 pc it is almost clean.

<u>Progress Report on Polarimetry Matters and VLT</u>
(J. Tinbergen, presented by I.S. McLean)

The question of polarization in the new large telescopes has been examined. There will be serious polarization effects that will affect photometry and spectrophotometry. Beam combiners cause polarization: dielectric coatings, dichroic mirrors, even reflections from aluminized surfaces, and of course diffraction gratings. So the instrument is, by accident, a crude polarimeter. Faint objects tend to be rather strongly polarized, and will give apparent intensity variations if observed with a partially-polarized telescope, especially on an alt-az mounting. To obtain photometric accuracy (not just precision -- i.e., lots of photons) requires polarization measurements, not just photometric ones. (See the Liège Colloquium in 1987, and Tinbergen's paper at the VLT meeting held by ESO at Garching, 1988, in press). Someone has to model these instruments to understand the effects of instrumental polarization. Funding agencies should support the development of polarization modulators before the beam-combining optics, to measure the instrumental effects.

<u>CCD observations of 3000 binary stars with separations less than 5 arc sec</u> (A.N. Argue, P.S. Bunclark, M.J. Irwin)

Astrometric and photometric observations of close binary stars have been made in 1986 and 1987, and are still underway, on the 1 m Jacobus Kapteyn Telescope at La Palma in order to contribute to HIPPARCOS. The objects had been selected by J. Dommanget, Coordinator of the WG on double stars responsible for that part of the Input Catalogue (40.118.013 p 153).

The CCDs used were an RCA chip (pixel size 0.3 x 0.3 arc sec) and a GEC chip (0.4 x 0.4) with field sizes each approximately 3 x 2 arc min. Used with Kitt Peak V and R filters (J. Mould System) the instrumental scales are very similar to Landolt's scales (33.113.003) with a negligible colour term (Fig. 1). Flatfielding was on the twilight sky. Night-sky emission lines were never a problem.

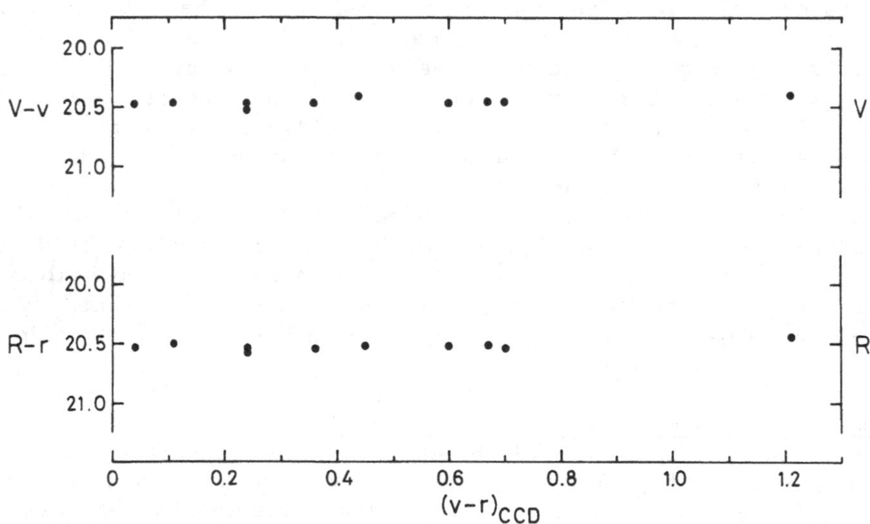

Fig. 1. Colour dependence of v and r (instrumental scales) for one night for RCA-CCD and Kitt Peak (J. Mould System) filters. Ordinates: differences from Landolt's Catalogue (33.113.003); abscissae: the instrumental colour index (v-r).

Reductions are done using a program for automatic analysis of crowded fields written by Irwin (39.036.153). The computed accuracy is a function of the separation and magnitude difference between the binary star components (Table 1). In practical terms, repeated observations on a single extinction star (e.g. Landolt 205556, magnitude $8^m.30$) on a good night gave rms residuals about the linear regression (in air mass) of $0^m.005$ in v and $0^m.008$ in r. The reduction program resolves binaries down to separations of one-half the rms seeing spread when the components are of nearly equal brightness: in practice this means that resolutions of 0.5 arc sec are fairly often achieved.

Table 1. Theoretical accuracy for GEC-CCD on 1 m Jacobus Kapteyn Telescope at La Palma (pixel size 0.3 x 0.3 arc sec) as function of separation and magnitude difference between binary star components.

	Magnitude difference in V				
Separation	0^m	1^m	2^m	3^m	4^m
1".3	< 0.01	0.02	0.03	0.05	0.10
2".6	< 0.01	0.01	0.01	0.02	0.06
3".9	< 0.01	0.01	0.01	0.02	0.06

12-Channel Photometer-Polarimeter
(J. Tinbergen, presented by I.S. McLean)

At the Observatorio del Roque de los Muchachos, on the Canary Island of La Palma, there exists a 12-photomultiplier photometer/linear polarimeter, which has several unusual modes:

a) Line photometry in up to 6 lines simultaneously (multi-line "H-beta")
b) Quasi-simultaneous photometry and linear polarimetry in up to 12 channels
c) 10-msec photometry in up to 12 simultaneous channels (still being implemented).

Information about this system is available in the Observers' Guide to the Observatory, in a Users' Manual, and in Be Star Newsletter no 16. Telescope time is allocated through committees in Britain/Netherlands and in Spain. Enquire via Royal Greenwich Observatory or Instituto de Astrofisica de Canarias.

Polarimetry of Seyfert Galaxies (U.C. Joshi)

This program aims to determine the mechanisms and energy sources of Seyferts, and to determine whether single or multiple mechanisms are required, by studying the time variations and wavelength dependence of the polarization, measured through different aperture sizes. The 1-meter telescope at Kavalur is used. The PRL polarimeter is similar to Gehrels's MINIPOL. Half-wave plates rotating (at 100 Hz) and fixed modulate the light; a computer gives on-line polarization values. The increase of polarization with decreasing aperture size, and rapid time variations, indicate a non-termal source rather than dust scattering in NGC 2992 and 3081. In NGC 3227, a large wavelength dependence suggests a large dust component. IC 4329A shows polarization concentrated to the nucleus, but also a wavelength dependence; photometry is needed, to subtract the galaxy from the central source.

Business and Second Scientific Session (I.S. McLean, Secretary)

a) Business Part

The president of the commission welcomed all new members, and in particular welcomed back to this commission Dr. Ivan King. David Crawford recommended that Mr. R. Genet becomes a member rather than a consultant, and two attendees at the meeting asked to become members also. Fifteen new members were therefore expected.

The committee then stood in silence for a short period in tribute to three deceased members (J.C. Kemp, V.B. Nikonov, A.C. Velghe).

A few resignations were reported because those members had other pursuits; the president wished them well in their new activities. A letter was received from South Africa from Dr. Cousins who would have had his 85th birthday had he been able to attend the IAU in Baltimore. All those present signed a greeting card to be forwarded on behalf of the commission.

Next the elections of the new president and vice-president took place. Following tradition, a polarimetrist (Ian McLean) and current vice-president was proposed as the new president, and Andy Young (photometrist) was put forward for vice-president. There were no other nominations. With the candidates not present votes were taken and I. McLean and A. Young were both elected by unanimous decision.

The president reported that 6 retiring members of the organising committee (OC) needed replacements and that candidates had been suggested and circulated among the OC. He presented their names to the full committee and asked for amendments or additions. None were forthcoming and the new OC members were duly elected (I.K. Knude, J.D. Landstreet, J. Lub, J. Menzies, V. Straizys, P. Wesselius).

A report was given to the Commission by R. Buser on the matter of creating a working group on synthetic photometry. It was suggested that this would be jointly with Commission 36; Drs. Gustafsson and Kurucz have agreed to serve for Comm. 36. Major activities of the working group would be as follows:

- system response functions
- stellar spectrophotometry libraries
- model stellar atmospheres flux distributions
- standard star data
- actual use of synthetic photometry: limitations & precautions
- computer time request support
- reference document.

It would be important to address the issues of calibrations, transformations and standardisation, and especially the use of synthetic photometry as a tool for linking ground-based to space-based systems.

Following some discussion, it was agreed that R. Buser and I. King would represent Commission 25 and that the president should advise the IAU secretariat that a permanent working group is to be set up.

Out of the document circulated for this purpose, we notice: "The designated working group's members are in charge of coordinating the activities and exchanges. Interested members of Commission 25 and 36, or of any other Commission, who would like to collaborate should contact the president of the Working Group. The president will report on the Working Group's activities and projects to both parent Commissions (25 and 36) during their meeting at the General Assemblies of the IAU. He will also prepare a report which will be included as part of the tri-annual report of Commission 25. The four designated members of the Working Group convened for a first meeting on August 9, discussing the major tasks to be taken up as well as the priorities with which these are to be followed. They agreed that the president (R. Buser) will establish a comprehensive list of items and projects already under way, along with a proposed strategy for the work in the group. In an iterative process, this list will be edited by the individual members of the Working Group, and adoption of a final plan of action will ensue."

Andy Young reported briefly on a photometry meeting held in Washington last year, indicating that the proceedings were to be published by NASA and circulated to all members.

The president on behalf of the commission had agreed to sponsor several IAU Colloquia and Symposia. Two such future events were discussed, an IAU Symposium on the "Magellanic Clouds" and an IAU Colloquium on "Galactic Astrophysics with Precision Photometry", both for 1990. The members agreed to co-sponsor these also.

On opening the business session for wider discussion, John Landstreet asked if it would be possible for Commission 25 to join the "peculiar A-stars" Working Group. Everyone agreed and the president undertook to pursue this item.

b) Scientific Part

A series of papers were presented to complete the scientific session. Two presentations by C. Sterken and J. Manfroid deal with the consequences of passband mismatches.

A Practical Experience (C. Sterken and J. Manfroid)

Photometrists know that a linear transformation from one instrumental system into another is only possible when certain conditions are met. There is a widespread misconception that serious problems only arise when one observes exotic objects using widely divergent instrumental passbands.

The authors carried out "multi-site" photometric measurements by simultaneously using five different quasi-standard uvby filtersets at two ESO-telescopes which are separated by only 30 meters.

Surprisingly large systematic differences show up in the Strömgren m1 and c1 indices of early B stars. This effect may systematically alter the observational H-R diagrams of reddened young clusters. (Astron. Astrophys. Supp. 71, 539)

Preservation of Authentic Photometric Systems: The Problem of Photometry on a Long Time-Baseline (J. Manfroid and C. Sterken)

In 1982 we started at ESO a programme of variable star monitoring on a long-term basis. A 50-cm telescope is committed for several months a year to carry out differential uvby measurements of about 100 variable stars of different types.

Over 5 active years we had 40 months of observations which yielded remarkable new results. The biggest problem is the preservation of the homogeneity of the dataset.

The problems are caused by the fact that different telescopes and filtersets are used, the photometers are occasionally replaced by newly designed instruments, the filters age, and the photomultipliers are sometimes replaced by ones which have different response curves.

Our conclusions are that, in spite of a consistent reduction procedure and a central data handling, it is impossible to correct some results by using analytical transformations. Moreover, systematic differences between the systems do not only show up in absolute photometry, but also in the differential data.

These talks generated considerable interest and discussion. Many members expressed concern. Most urged major observatories to standardise and D.L. Crawford urged for a "certification".

Faint Standard Stars for UBVRI (Arlo U. Landolt)

Nearly six hundred stars have been examined as possible candidates for broad-band UBVRI standard stars on the Johnson-Kron-Cousins photometric system. These new data were obtained at the 1.5-m telescope of the Cerro Tololo Inter-American Observatory over the past ten years. The detector was a RCA 31034 photomultiplier. The filters have been described in the literature (Astron. J., 88, 439, 1983).

A series of observations has established a 143-star subset from the initial six hundred stars as new standard stars in the approximate magnitude range $11 < V < 15$ and colour index range $-0.3 < (B-V) < +2.3$. These stars were observed on an average of 11 different nights and more than 20 times each. The average mean error of the mean is better than 0.005 magnitude for the magnitude and colour indices, except for (U-B) where the error is perhaps twice as large. All data have been tied into the standard stars defined in the above cited reference.

Another hundred stars, or so, almost are of standard star quality. Data acquisition is continuing; hopefully this last list of stars will be added to the above 143 in the near future.

Observational work has begun on stars reaching beyond V = 21st magnitude to provide very faint standard stars for the Hubble Space Telescope.

Geneva Photometric Catalogue (F. Rufener)

With 29.400 entries, the catalogue has been worked out from all the observations made until summer 1987. About 200.000 measures of the seven colours have been gathered in both hemispheres. Actually, the majority came from our La Silla Station (ESO, Chile). The catalogue gives informations on identification, normalized colours, usual indices and parameters, magnitude V. For each star, an evaluation of the internal precision is given and a general accuracy estimation of the data is proposed. A mag-tape version will be released by the Centre de Données Stellaires at Strasbourg. A paper-bound edition (340 pages, format 210 x 297 mm) photocomposed with Times font will be solded by Geneva Observatory (foreseen price including air-mail fee: 60 Swiss francs or 40 US$).

Dust Shell around Supernova SN1987a (U.C. Joshi et al.)

Polarimetric measurements of SN1987a in BVRI bands show that there are at least two intrinsic sources which contribute to the observed polarization; one source being the dust shell around SN1987a as is indicated by the wavelength dependence of linear and circular polarization. From the time variation of wavelength dependence of linear polarization, we have deduced the radius of dust shell ~ 0.05 pc and optical depth $\tau \sim 0.6$ at visual wavelength.

The meeting ended with a round of applause to thank the retiring president of Commission 25, F. Rufener, for his efforts over the past 3 years.

I.S. McLean
A.T. Young
F. Rufener

PROBLEMS OF INFRARED EXTINCTION AND STANDARDIZATION

Joint Commission Meeting of Commissions 25 and 9

held on August 4, 1988

Chairman and editor: *E.F. Milone*

Scientific Organizing Committee: *R. Bell, M.S. Bessell, T.A. Clark, I.S. Glass, R.L. Kurucz, I.S. McLean, F. Rufener, A.T. Young.*

A short report by *E.F. Milone*

An abstract of A.T. Young's paper

PROBLEMS OF INFRARED EXTINCTION AND STANDARDIZATION: A Short Report

E.F. Milone
University of Calgary
Calgary AB T2N 1N4 / Canada

The effects of the Earth's atmosphere on infrared radiation from astronomical sources and the nature of problems in recovering and standardizing such data were the topics discussed at this joint meeting.

A major extinction difficulty in the infrared is the failure of a linear extinction law (or any other law based on Bouguer data taken at air masses ≥ 1) to properly predict the outside-the-air magnitude in the broadband infrared. Various strategies have been suggested to overcome this difficulty, but the complications it causes for transformations may contribute to problems for IR standardization, which arise when attempts are made to combine data obtained from different sites with different photometry systems.

The wavelength interval considered was the near infrared and most speakers treated only the JHKLM passbands in the 1-5 μm region. The papers explored the sources and potential solutions of the problems created by the nature of atmospheric extinction in the near IR and with the difficulties of standardization. They therefore represented a reasonably balanced picture both of how the problems are currently dealt with and with how they might be improved. The program was opened by the chairman, who introduced the topic and the papers which followed. The first major address was entitled *Extinction and Transformation* by Andrew T. Young (San Diego State University). This talk discussed the difficulties inherent in obtaining appropriate extinction corrections for the broad-band JHKLM passbands in current use and suggested remedies. An abstract of Dr. Young's paper follows this report. The other papers given at this meeting were: *Models of Infrared Atmospheric Extinction* by K. Volk (NASA Ames), T.A. Clark and E.F. Milone (University of Calgary); *Atmospheric Extinction in the Infrared* by Ronald J. Angione (San Diego State University); *Infrared Extinction at Sutherland* by I.S. Glass and B.S. Carter SAAO); *Near-Infrared Extinction Measurements at the Indian Observatory Sites* by N.M. Ashok (Physical Research Laboratory, Ahmedabad); *Determining Solar and Stellar Spectra Above the Atmosphere by Computing Atmospheric Transmission* by Robert L. Kurucz (Harvard-Smithonian Center for Astrophysics); *JHKLM Photometry: Standard Systems, Passbands and Intrinsic Colors* by M.S. Bessell and J.M. Brett (Mt. Stromlo and Siding Spring Observatories); and *Standardization with Infrared Arrays* by Ian S. McLean (UKIRT). The latter paper explored the difficulties faced by infrared image processing work. Several papers argued for the importance of modelling the atmosphere at the time of observation, making water vapour measurements a necessity. The meeting concluded with a final *Summary* by Roger Bell (Univ. of Maryland).

Prior to Dr. Bell's summary, there was a 10-minute general discussion. A repeated theme, starting with the undersampling theorem of Dr. Young, and reprised in the concluding remarks of Dr. Bell, emphasized the importance of reexamining the placement and width of the current JHKLM system of broad-band filters which are not confined to the cleanest portions of the atmospheric infrared windows. In light of the difficulties certain to be faced by any attempt to proclaim any existing standard system as the preferred one for JHKLM work, the possibility of redefining such a system based on better passbands, seemed to have wide appeal among the participants.

The full texts of all the papers submitted questions and their responses, and other relevant material will be published elsewhere[2].

---

[2]Springer Verlag / Publishers, New York Inc., USA

EXTINCTION AND TRANSFORMATION: An Abstract

Andrew T. Young
Astronomy Department, San Diego State University
San Diego, California 92182-0334

The basic principles of heterochromatic extinction show that the approach used in the visible should not work well in the infrared, where molecular line absorption rather than continuous scattering dominates the extinction. Not only does this extinction change very rapidly with wavelength (so that stellar color becomes only weakly correlated with effective extinction), but also many of the lines are saturated (so that Forbe's curve-of-growth effect is much more severe in the IR.) Furthermore, broadband IR colors are more undersampled than those in the visible, so aliasing errors make them correlate even less with extinction.

Reduction to outside the atmosphere is difficult, but the rational approximation

$$\Delta m = \frac{(AM^2 + BM + C)}{(M + M_0)}$$

models the extinction quite well as a function of air mass, M. As M_0 is nearly indeterminate from observations over the accessible range of air masses, it can be guessed better than it can be found from observations. Good guesses for M_0, based on the synthetic photometry of Manduca and Bell (1979) from $M = 0$ to 3, give extra-atmospheric magnitudes extrapolated from the range $M = 1.0$ to 2.5 that are accurate to a few hundredths of a magnitude. Even if one assumes $M_0 = 1.0$ in every case, the largest extrapolation errors are still about three times smaller than those from linear fits (see Table III of Manduca and Bell, 1979).

Extinction can be well determined with a few stars per hour if the observations are carefully planned (Young, 1974). M should not exceed 3, beyond which its values are uncertain owing to the variable scale height of water vapor. As in the visible, the determination of nightly extinction can be strengthened enormously by reducing several nights together (Young and Irvine, 1967; Manfroid and Heck, 1983).

To prevent aliasing time-dependent extinction into the fitted parameters, one must observe both rising and setting stars and solve for time-dependent parameters, as in Rufener's (1964) "M and D"method. This is more urgent in the infrared than in the visible, because changes in water vapor can cause large effects on the IR extinction without producing an obvious visible effect. Furthermore, because the water vapor that dominates the IR extinction has a smaller scale height than the atmosphere as a whole, one must use an air mass formula that takes this into account (as well as allowing for the distinction between true and refracted zenith distances).

The only solution to the transformation problem is to satisfy the sampling theorem, which may be difficult in the IR because of gaps due to saturated telluric absorptions.

REFERENCES

Forbes, J.D. 1842, *Phil. Trans.* **132**, 225-273

Manduca, A., and Bell, R.A. 1979, *Pub. A.S.P.* **91**, 848

Manfroid, J., and Heck, A. 1983, Astron. Astrophys. **120**, 302-306

Rufener, F. 1964, *Pub. Obs. Genève*, Série A, Fasc. 66

Young, A.T., and Irvine, W.M. 1967, *Astron. J.* **72**, 945-950

Young, A.T. 1974, in *Methods of Experimental Physics*, Vol. **12** (Astrophysics, Part. A: Optical and Infrared), ed. by N. Carleton (Academic Press, New York) pp. 123-192

COMMISSION No. 26

DOUBLE AND MULTIPLE STARS (ETOILES DOUBLES ET MULTIPLES)

PRESIDENT: Karl D. Rakos SECRETARY: Harold A. McAlister

The program for Commission 26 at the Baltimore General Assembly consisted of an opening business session held on 4 August 1988 followed by three scientific sessions during the remainder of that same day. A second business session was held on the morning of 10 August 1988. The Commission also participated in the Joint Discussion on *Formation and Evolution of Stars in Binary Systems* held on 3 August 1988 and in the Joint Commission Meetings on *Stellar Photometry With Modern Array Detectors* and *Formation of Binaries*, held on 8 and 9 August respectively.

The scientific sessions held by Commission 26 and their chairmen were:

1. *HST/Hipparcos Missions and Binary Star Research* – H. McAlister and J. Dommanget
2. *Improvements in Classical Research of Double and Multiple Stars* – J. Dommanget
3. *Present and Future Cooperative Projects in Binary Star Research* – K. Rakos

FIRST BUSINESS SESSION

The President served as Session Chairman and opened the meeting by welcoming the 14 members who were present. The Chairman reviewed the consideration undertaken by the IAU Executive Committee for the abolition of Commission 26 and reported that the Commission was approved for an additional five years, at which time the appropriateness of maintaining the Commission would be reviewed. During the extended period in which this question was before the IAU Executive, the Chairman poled the 67 members of the Commission to ask their individual opinions as to the abolishment of Commission 26. Of the 41 members who replied, 37 stated that the Commission should be continued and 4 felt that it should be abolished. A similarly positive opinion had been expressed by the participants in IAU Colloquium No. 97 on *Wide Components in Double and Multiple Stars* held in Brussels, Belgium during 8–13 June 1987, by passing a resolution supporting continuation. The Chairman stated that the success of Colloquium No. 97 (a gathering that included a diversity of scientists and which was dedicated to Professor Willem J. Luyten), the opinions directly expressed to the General Secretary by commission members, the advent of powerful new techniques for the study of binary star systems, and the potential for attracting an expanded membership into the Commission were all considered by the Executive as justification for preservation of Commission 26. The great success of the Joint Discussion held on 3 August was clearly indicative of the diverse and widespread interest in binary and multiple stars.

Because of the relatively small attendance at the business meetings of the Commission during the New Delhi General Assembly, the election of new Commission officers was carried out by mail balloting during the year prior to the Baltimore General Assembly. This procedure provided for participation in elections by all members who chose to respond to the ballot. The Organizing Committee for the first time served as the Nominating Committee, and the ballot provided for writing in the names of additional candidates. Twenty-six members replied to the mail ballots and the results of the election were:

President: H. McAlister *Vice President:* H. Abt
Organizing Committee: P. Couteau, R. Harrington, A. Kiselyov, K. Rakos, E. Van Dessel, P. Bernacca
Nominating Committee: A. Poveda (Chairman), A. Batten, J. Dommanget, O. Franz, G. Salukvadze

Following a discussion of the matter of future selection of a nominating committee, it was decided that nominations for this committee would be taken from the floor during the first business session at a General Assembly, with the provision that commission members not attending the General Assembly can propose in advance a name to the President. The chairman of the nominating committee will be the person receiving the most votes with a tie being decided by the President. The work of the nominating committee will be initiated no later than one year prior to the next General Assembly. It is expected that these procedures will continue to ensure the broadest participation in the election of officers for Commission 26.

The following new members were added to Commission 26: W. van Altena (USA), A. Argue (UK), W. Bagnuolo (USA), Y. Balega (USSR), W. Beardsley (USA), W. Beavers (USA), D. Bonneau (France), P. Broche (FRG), B. Campbell (USA), L. Fredrick (USA), I. Furenlid (USA), G. Gatewood (USA), W. Gliese (FRG), J. Halbwachs (France), W. Hartkopf (USA), G. Hill (Canada), P. Ianna (USA), C. Jaschek (France), Z. Kopal (UK), D. Latham (USA), M. Lattanzi (Italy), J. Ling (Spain), L. Loden (Sweden), E. Oblak (France), T. Oswalt (USA), L. Sabados (Hungary), P. Schmidtke (USA), J. Smak (Poland), J. Stein (USA), A. Tokovinin (USSR), V. Trimble (USA), A. Upgren (USA), and M. Valtonen (Finland). This very significant increase in the membership of Commission 26 was welcomed as a major step in revitalizing the activities of the Commission. C. Hernandez (Argentina) and M. Tapia (Mexico) resigned as members of the Commission.

A letter written by P. Couteau, who could not be present in Baltimore, was presented by J. Dommanget. Dr. Couteau asked for a discussion of the future of the Commission's *Circulaire d'Information*, a publication begun by P. Muller in 1954 who continued to edit it until 1983, at which time it was turned over to Dr. Couteau. The *Circulaire* has provided a rapid means for distributing new orbits and discoveries among those active in double star research, but as pointed out by Dr. Couteau, relatively few individuals have contributed to it in recent years. In response to a question by L. Loden, it was pointed out that this is a publication that can be referenced in the literature. W. Bagnuolo asked if distribution via a computer network might enhance usefulness. The discussion concluded that the *Circulaire* should continue to be issued in fulfillment of its original goal and that its use should be publicized and encouraged. Upon the recommendation of H. Abt, the thirty years of service by Dr. Muller and the continuation of the *Circulaire* by Dr. Couteau were applauded. The new President will forward these sentiments to Dr. Couteau.

An issue related to the above discussion is the reluctance of the editors of some journals to publish measures of double stars that are not accompanied by some kind of analysis, in spite of the fact that supplement series to journals were initiated to serve in such a capacity. Such demands by editors were felt to be inappropriate and would actually tend to dampen activities in this field. Although these measurements could be directly inserted into data bases without being published in the normal means, it was agreed that this made it difficult for the users of these data to keep up with the field and would discourage young astronomers from pursuing activities in which their efforts do not receive the normal recognition from publication. It was agreed that the President would write letters to the editors of relevant journals expressing the concerns of the Commission.

SCIENTIFIC SESSIONS

Approximately 60 people were present during the Commission's scientifc sessions. For the first session, reports on the application of the Hubble Space Telescope and the HIPPARCOS astrometry satellite to the study of double and multipe stars were given by L. Fredrick (HST) and by J. Dommanget and P. Bernacca (HIPPARCOS). HST will detect binaries through the fine guidance sensors due to the high sensitivity of the FGS to non-single sources. Because the FGS will inspect very large numbers of guide stars, these sensors have the capability of a very extensive survey for duplicity. Ground-based observations using high resolution techniques can then follow up by confirming these discoveries. Efforts have been completed to minimize the impact of the FGS' inability to lock onto binaries as guide stars.

A detailed discussion of the application of HIPPARCOS to binary star astronomy is presented in the triennial report of Commission 26. A great deal of effort has gone into the preparation of the *Catalogue of the Components of Double and Multiple Stars* as a part of the HIPPARCOS *Input Catalogue.* Eighteen months following the launch of HIPPARCOS, a list of double stars will be available for circulation among the community for confirmation and subsequent measurement. The responsibility for the preparation and circulation of this first list of suspected binaries belongs to the Italian representatives to the *Input Catalogue Consortium*, and suggestions for the most effective distribution of these first results are solicited.

The second scientific session consisted of presentations by individuals whose contributions are summarized in the following abstracts:

A New Catalog of Interferometric Measurements of Binary Stars - H.A. McAlister and W.I. Hartkopf, Georgia State University. A second catalog of interferometric measurements of binary stars will be issued in the fall of 1988. The first such catalog, issued in January 1984, contained 3.363 measures of 1,123 systems along with an additional 1,863 inspections by interferometry that failed to resolve known or suspect binary systems. The second catalog, with a cutoff date of 1 July 1988, will contain 8,976 measures of 2,995 systems with an additional 3,068 negative inspections for duplicity. There has thus been a substantial increase in interferometrically obtained data for binary stars in the 4.5 years between the two catalogs. The number of papers in which these data are presented has increased from 46 to 69. The mean separation in the latest data sample is 0.347 arcsec compared with 0.32 arcsec from the first catalog. Although the number of measures and, perhaps more significantly, the time base over which these observations are being made is increasing at a useful rate, it should be emphasized that the observations are being done primarily by only four groups in the USSR, France, and the USA. There is great potential in extending interferometric programs to telescopes that have moderate apertures in the range of 0.6 to 2.5 meters. Copies of the new catalog will be distributed to members of Commission 26.

Speckle Interferometric Orbits of Binary Stars - W.I. Hartkopf, H.A. McAlister, and I. Furenlid, Georgia State University. With nearly two decades of speckle observations now available, a growing number of binary systems have completed one or more revolutions and are ripe for orbital analyses based solely or primarily upon speckle data. Orbits have been published for a number of interesting objects, including the important Hyades binary 70 Tau (Fin 342) (*A.J.*, Oct 1988), γ Per (*A.J.*, **94**, 700, 1987), and χ Dra (*A.J.*, **93**, 1236, 1987). New or improved orbits for twenty binaries are being prepared for publication. The increased overlap between the visual and spectroscopic regimes resulting from speckle is especially important and should lead to a substantial increase in the number of accurately known stellar masses. A long-term collaboration between CHARA and INAOE for the precision measurement of radial velocities is expected to eventually permit the combination of the two types of data to yield masses with accuracies at the 1% level.

Binary Star Speckle Photometry - W.G. Bagnuolo, Georgia State University. A major goal of the GSU/CHARA program of binary star speckle interferometry has been to develop methods for accurately determining the magnitudes and colors of the individual components of binary stars with separations down to the diffraction limit. A new method of analyzing speckle images, called the *fork* algorithm appears to be a useful method for such *speckle photometry*, and simulations suggest that it can attain a SNR roughly 10× that of other algorithms such as triple correlation and shift-and-add. A simple application of speckle photometry is resolving the usual 180° position angle ambiguity resulting from standard autocorrelation analyses of speckle images. In the case of the Hyades binary Finsen 342, resolving this ambiguity is not just a curiosity, because we can test the suggestion by Peterson and Solensky (1987) that the system has a short period eccentric orbit rather than a longer period circular orbit commonly assumed in the literature. Indeed, we show definitively that the period of Finsen 342 is 6.264 years and that the system has masses that are normal in comparison with other Hyades binaries. An initial estimate of the magnitude difference has also been determined. In a second application of the *fork* method, sets of speckle images of Capella taken in the Strömgren v, b, and y filters have been analyzed. These results show that the magnitude differences in v, b, and y are $m_{Aa} - m_{Ab}$ = 0.54, 0.23, and 0.09 magnitudes respectively. Thus, contrary to accepted beliefs, the more luminous star in these wavelengths is the hotter "F-star", Capella Ab, the spectroscopic secondary and the less massive component. The photometric indices are consistent with spectral types of G0 III for the secondary and G8/K0 III for the primary "G-star".

A New Program of High Precision Radial Velocity Measurements - I. Furenlid, Georgia State University, and O. Cardona, INAOE. A new international collaboration in instrument development and research has been established between INAOE and CHARA based upon spectroscopic programs to be initiated with the new 2.14–m telescope built by INAOE at the Cananea Observatory located in the state of Sonora in northern Mexico. The spectrograph, which is of the off–plane Ebert–Fastie type, uses a CCD detector system, and is fed starlight with an optical fiber bundle coupled to the telescope. For stability, the spectrograph is mounted on a platform away from the telescope. A general goal of the design has been to generate an instrument of good optical efficiency and very high stability with all major observing functions under remote control for warm–room operation. The spectrograph uses low spectral orders and only one stretch of spectrum is observed per CCD frame. The optical fiber bundle serves as an image slicer by being arranged in a close packed array at the input end and in a one–dimensional slit at the output. One major application of the new spectrograph will be to provide radial velocities of visual and interferometric binaries that are being followed by the CHARA speckle interferometry program. Improved sensitivity and signal–to–noise offer the promise of significantly improved velocities and the detection of secondaries that have eluded previous spectroscopic efforts.

Spectroscopic Observations of Close Multiple Stars - F.C. Fekel, Vanderbilt University. For over ten years, high dispersion spectroscopic observations have been obtained for a number of spectroscopic/visual multiple stars. Preliminary long–period orbits have been computed for some of these systems including ADS 14839 and ψ Sgr while nearly definitive orbits are available now for others such as HR 6469, μ Ori = ADS 4617, 13 Cet = ADS 490, and 63 Gem. When combined with visual and/or speckle data, fundamental parameters can be obtained as in the case of the double–lined spectroscopic binary χ Dra that has also been resolved by speckle interferometry. Continued collaboration with speckle observers will be of great importance for most of these systems.

Occultation of SAO 79241: A Double Star? - C. Meyer, Y. Rabbia, M. Froeschle, G. Helmer, and G. Amieux, CERGA. The disappearance behind the moon of the star SAO 79241 (HD 56176) was observed at CERGA on 27 February 1988 at a wavelength of 410 nm using the lunar lasar ranging 1.5 m telescope. The light curve clearly shows a double event, and analysis of the tracing yields the vector angular separation of 0.0994±0.0009 arcsec directed towards position angle 129.2 degrees and a magnitude difference of 0.8 magnitudes. No indication of duplicity for this star were found in the Sky Catalog 2000.0 and the SIMBAD data base at Strasbourg, and it is likely that this is a newly discovered double star.

Wide Binaries in a Sample of Common Proper Motion Stars - J.L. Halbwachs, Observatoire de Strasbourg. Common proper–motion stars were detected using data from the AGK 2/3 Catalogue (Halbwachs 1986). The probability of a pair being optical was evaluated from the apparent separation, ρ, and from the proper motion, μ. Among 113 pairs with ρ/μ between 1,000 and 3,500 years, 40% should be optical. To date, the radial velocities of both components of 50 pairs have been measured at least twice, using the CORAVEL radial velocity scanner. Nine pairs contained spectroscopic binaries, but among the 41 remaining pairs, 22 had components with identical velocities. This indicates a proportion of 54%, in agreement with the prediction. These 22 binaries have separations around 10^4 AU, and confirm that the frequency of wide binaries is not negligible.

The final scientific session was a discussion of the needs and means for cooperative undertakings in binary star research. Such cooperation will become increasingly important with the advent of HST and HIPPARCOS and with the overlap in accessibility of many systems to observation by different techniques practiced by specialists in different fields. Dommanget gave as an example of cooperation the compilation of the *HIPPARCOS Input Catalogue* by a consortium of twelve different groups, including amateur astronomers. Commission 26 can serve as a vehicle for furthering such cooperation.

FINAL BUSINESS SESSION

H. McAlister served as session chairman. The excellent service of Prof. Rakos during his term of President of Commission 26 was acknowledged with applause by the members present. Prof. Rakos has seen the Commission through an important transition period, and his efforts have ensured that double and multiple star studies will play a continuing role in the IAU.

As a brief continuation of the scientific sessions, C. Worley described his adopted nomenclature in the *Washington Double Star Catalog.* The Commission expressed its thanks to him for this substantial and continuing contribution to the field of binary star astronomy.

As the last matter of business before the Commission, the Chairman initiated a discussion concerning the desirability of a colloquium to be sponsored by Commission 26. It was concluded that such a colloquium is desirable considering the advent and extensive use of techniques at the frontier of our field.

COMMISSION No. 27

VARIABLE STARS (ETOILES VARIABLES)

Report of Meetings: August 5,9, 1988

PRESIDENT: Bela Szeidl SECRETARY: John R. Percy

BUSINESS MEETING (August 5, 1988)

1. Dr. B.G. Marsden (Central Bureau for Astronomical Telegrams) reported on the problem of the designation of extra-galactic novae. The system presently used in the IAU Circulars is ambiguous. Dr. Marsden recommended interim designations of the form "Nova LMC 1988 #2" to indicate the second nova discovered in the LMC in 1988.
   A discussion followed, in which several comments were made: the system should be (i) conveniently applicable to galaxies such as M31, M33 and the galaxies in the Virgo cluster, in which many novae may be discovered (ii) open-ended, for use in the future (iii) flexible and (iv) acceptable to the editors of the IAU Circulars. The meeting agreed to accept Dr. Marsden's recommendation.
2. Dr. M. Breger, supervisor of the IAU Archives for Unpublished Photoelectric Observations of Variable Stars, described the current status of these archives. Copies of printed files are maintained at (and can be obtained writing to) the following institutions: Centre de Données Stellaires (CDS: Strasbourg, France), the Library of the Royal Astronomical Society (London, UK) and at the Odessa Astronomical Observatory (Odessa, USSR). Electronic files (if any) are maintained in Strasbourg. The supervisor commented that he prefers printed files. Copies of data files should be sent to: M. Breger, Universitätssternwarte, Türkenschanzstrasse 17, A-1180 Wien, Austria, along with a cover sheet (see IBVS 2853 (1986) for instructions). Lists of files are published regularly in IBVS (2853 (1986)), Astron. Tsirk (1517 (1987)), and Publ. Astron. Soc. Pacific (100, 751 (1988)). At present, 196 file numbers have been assigned (45 since 1984), and 180 files have been filled. These files are now cross-referenced in the bibliographic data bases of CDS. The supervisor thanked the institutions involved in the maintenance of the Archives for their help.
3. Dr. A.M. Cherepashchuk reported on the General Catalogue of Variable Stars, which is compiled at the Sternberg Astronomical Institute in the USSR. Until now, the GCVS has been compiled from hand-written records, of which the total number is now about 500,000 and is increasing by 20,000 per annum. Less than 1 per cent of the information in the records is actually included in the GCVS, and the compilers are seeking ways to improve this situation by converting the records to machine readable form. Each GCVS file will then consist of a series of descriptive keywords, parameters and comments. The compilers are aware of the need to proceed carefully at this stage of the planning, and they would be grateful to receive suggestions from users of the catalogue.
4. Dr. B. Szeidl, co-editor of the Information Bulletin on Variable Stars, reported on the publication and distribution of the IBVS. Approximately 500 bulletins have been published in the last three years, and these have been mailed to 350 institutional addresses and 200 private addresses. The number of addresses is constantly increasing. The costs involved are very great, and the editors are seeking ways of reducing them. Various solutions (and their drawbacks) were discussed: (i) cease sending copies to "private" addresses if the recipient's institution already received a copy (private copies are more convenient and therefore more thoroughly read) (ii) charge a subscription fee (for technical reasons, it is not possible for the editors to collect subscription fees directly, though it might be done indirectly -

through the IAU secretariat, for instance) (iii) trim the mailing list periodically (this is already done) (iv) approach the IAU for financial support (the IAU would be unlikely to begin such a costly precedent) (v) keep IBVS issues to 4 pages or less, such as by using smaller print (the print is often too small already) (vi) reject papers which could or should be published in regular journals, or which are of insufficient quality (this could increase the time delay in publishing papers, and would certainly increase the work load of the editors - which is already great). Commission Vice-President M. Breger commented that the Commission should support the editors of the IBVS in finding solutions to the current problems. The Commission also affirms its gratitude to the editors for their excellent work.

5. Dr. J.A. Mattei, Director of the American Association of Variable Star Observers, discussed the current work and future plans of the association. There are 3,600 stars on the AAVSO visual program. In the past year, 260,000 observations were made by 550 observers - more than half from outside the USA - bringing the total number of observations to more than 6,000,000. All of those made since 1960 are edited and archived in machine-readable form. Earlier observations are now being computerized: this project is 60 per cent complete. The AAVSO receives about 150 requests for data each year: "real-time" information on the state of unpredictable stars, simultaneous optical observations of stars being observed at other wavelenghts, current optical data for correlating with other data, and archival data for analysis. A small fee is charged to cover the cost of compiling and sending the data. The AAVSO is presently engaged in several collaborative research projects, including one to provide improved predictions of Mira star magnitudes for the HIPPARCOS satellite input catalogue. The AAVSO is planning a meeting in 1990 in Belgium - its first outside North America - to which all variable star observers are cordially invited.
6. The following meetings have been proposed: some of them may be approved as IAU Symposia or Colloquia: in 1988: a workshop on "Astroseismology" in Vienna in December: in 1989: a meeting on "The Physics of Classical Novae" in Madrid in late June, a meeting on "Rotation and Angular Momentum Evolution of Low Mass Stars" in Catania, and a meeting on "Frontiers of Stellar Evolution" at the University of Texas, marking the 50th anniversary of McDonald Observatory: in 1990: a meeting on "Surface Inhomogeneities in Stars" in Armagh, a meeting on "The Magellanic Clouds" in Australia, a meeting as part of the "Los Alamos" series of meetings on stellar pulsation, possibly in Hungary, and the AAVSO meeting mentioned above.
7. The following slate of officers for Commission 27 was proposed and accepted: President: M. Breger, Vice-President: J.R. Percy, Organizing Committee: T.G. Barnes, J. Christensen-Dalsgaard, R.E. Gershberg, M. Jerzykiewicz, L.N. Mavridis, M. Rodono, M.A. Smith, B. Szeidl, A.M. van Genderen, and B. Warner. Several new members of the Commission were proposed prior to the General Assembly: these applicants were accepted. Further applications for membership will be considered by the officers of the Commission. The total membership is now approximately 350.
8. There was a brief discussion about whether Commission 27 should establish any Working Groups. A WG on Flare Stars existed in the past. M. Breger announced that he plans to publish an informal newsletter on Delta Scuti stars, to help to co-ordinate research in this field. Those interested in receiving such a newsletter should write to him at the address given above.
9. S. Dunlop (British Astronomical Association) raised the question of how significant research results based on visual observations (such as revised ephemerides of eclipsing binaries) could be published quickly. The IBVS does not publish results based on visual observations. No satisfactory solution to the question was proposed.

FIRST SCIENTIFIC SESSION (August 5, 1988)
Recent developments in variable star research
Chairmen: A.N. Cox and M. Breger

Eleven excellent talks were presented followed by discussions. The attendance is estimated to be 100 people with 70 people counted at one time. The program is listed below:

1. Effect of time-dependent convection on white dwarf radial pulsations
   A.N. Cox, S.G. Starrfield
2. High-overtone p-mode spectrum of the rapidly oscillating Ap star HR 1217
   D.W. Kurtz, J.M. Matthews, P. Martinez, J. Seeman, M.S. Cropper, C. Clemens, T.J. Kreidl, C. Sterken, H. Schneider, W.W. Weiss, S. Kawaler, S.O. Kepler, A. van der Peet, D. Sullivan, H.J. Wood
3. Multiple Close Frequencies of the δ Scuti Star Θ^2 Tau
   M. Breger, R. Garrido, L. Huang, S.-y. Jiang, Z.-h. Guo, M. Frueh, M. Paparo
4. Photometry of δ Scuti stars
   E. Antonello, E. Poretti
5. Short-Period Variability in Be stars
   L. Balona, J. Cuypers, J. Egan
6. High-galactic latitude F supergiants
   M. Parthasarathy
7. Variability of Herbig Ae shell stars HR 5999 and HD 163296
   D. Baade, O. Stahl
8. Observing the Function P(t)
   E.P. Belserene
9. A Period-Luminosity Relation for Anomalous Cepheids
   J.N. Nemec, A. Wehlau, C. Mendes de Oliveira
10. The RR Lyrae variables in ω Centauri
    R.J. Dickens
11. The Nature of Old Stars in the LMC
    S.M. Hughes, P.R. Wood, N. Reid

SECOND SCIENTIFIC SESSION (August 9, 1988)
Jointly with Commission 30
The Baade-Wesselink method - recent achievements and future goals
Chairman: G. Burki

In the past 12 years, the Baade-Wesselink (BW) method was revised by many authors, mainly because the recent instrumental improvements allow to obtain simultaneous high quality data in both, radial velocity and photometry.

The surface brightness technique and the future improvements have been reviewed by T.G. Barnes III and T.J. Moffett. The main advantage of this formulation, which makes use of the useful relation between V-R and the visual surface brightness (Barnes-Evans relation), is the possibility to derive the distance to the star in addition to the mean stellar radius. B.W. Carney on the one hand, and T. Lin and K. Janes on the other hand, have presented their results obtained by applying the surface brightness technique to RR Lyrae stars. They use the theoretical model atmospheres of Kurucz to derive metallicity dependent surface brightness vs. color index relations.

J.A. Fernley, A.E. Lynas-Gray, I.S. Skillen, R.F. Jameson and A.J. Longmore have applied the Infrared Flux method to determine the radius and effective temperature of the RR Lyrae star X Ari. This method is capable of estimating angular radii with an accuracy comparable to the interferometry measurements. The estimated realistic (not internal) error is +/- 7% for the radius of X Ari.

D.D. Sasselov, J.B. Lester and M.S. Fieldus have studied the velocity structure in cepheid atmospheres from infrared spectroscopy. The radial velocity curves for X Sgr and η Aql have larger amplitudes and are phase shifted with respect to the optical radial velocity curves. Of course, this fact must have an effect on the BW radii.

G. Burki, A. Arelano Ferro, L. Balona, T.G. Barnes III, C. Cacciari, S.L. Hawley, M. Imbert, T.G. Moffett and D. Sasselov have compared the results obtained by various BW programs analysing exactly the same data in photometry and radial velocity. The star chosen is W Sgr with data in radial velocity (CORAVEL) and Geneva photometry by Babel et al. (1988) and in UBVRI photometry by Moffett and Barnes (1980). The programs given similar results (at the 10% level) when the same magnitude and color index are used (ex: V and B-V, or V and B_2-V_1). On the contrary, there are systematic differences between the results obtained by using "red" magnitudes and indexes (ex: V and V-I, or G and B_2-G), which give R=65 $R_{\odot}$, and the results obtained by using V and B-V, for which R=$55_{\odot}$.

N. Simon has presented his inversion of BW technique. It is based on the fact that, while we are still not capable of selecting phases of equal surface brightness, modern observational techniques provide the means for determining, with high accuracy, phases of equal radius. According to Simon, given highly accurate data and extensive coverage of the cycle, the BW inversion technique can determine radii with only a few per cent of uncertainties.

The application of the BW method to the supernovae was revised by D. Branch. This method only applies to the early phases of supernova's evolution, when it is optically thick and has a photosphere. A distance is derived by matching a distance-independent photometric angular radius of the photosphere to a distance-dependent spectroscopic angular radius. SN 1987 A is a particularly interesting object in this context, and a distance of 55 +/- 5 kpc was derived. This result gives weight to a supernova-based extragalactic distance scale H $\cong$ 60 kms^{-1} Mpc^{-1}.

COMMISSION No. 28

GALAXIES (GALAXIES)

Report of meetings, 3, 5, 6, 8 and 9 August 1988

PRESIDENT: P.C. van der Kruit SECRETARY: G.A. Tammann

Commission 28 was involved in three all-day Joint Discussions, of which the proceedings will be published in Highlights of Astronomy, Volume 8:
JD III: **Supernova 1987A in the Large Magellanic Cloud** (Commissions 27, 28, 34, 35, 40, 44, 47, 48), 4 August,
JD VI: **Disks and Jets on Various Scales in the Universe** (Commissions 28, 34, 40, 47, 48), 9 August,
JD VII: **The Hubble Space Telescope – Status and Perspectives** (Commissions 9, 28, 44, 47), 10 August.
Furthermore, Commission 28 was involved in one full-day and two half-day Joint Commission Meetings, of which the proceedings will also be published in Highlights of Astronomy or in the sections of other Commissions:
JCM 4: **Molecules in External Galaxies** (Commissions 28, 34, 40), 5 August,
JCM 6: **Stellar Photometry with Modern Array Detectors** (Commissions 9, 25, 26, 28, 37), 8 August,
JCM 7: **Star Clusters in the Magellanic Clouds** (Commissions 28, 35, 37), 9 August.
Two further sessions were held jointly with one other Commission: **Dark Matter in our and Nearby Galaxies** (with Commission 33) and **Galaxy Redshift Determinations: Better Techniques, Better Standards** (with Commission 30). One business session and one scientific session on "New Results" were held, while the four Working Groups each organized a half-day session. Programs of sessions and, where available, short abstracts of the papers presented follow.

3 August: Business

CHAIR: P.C. van der Kruit

The following items were covered in the business session:
Draft report. The commission adopted the report and endorsed the current procedure that the President includes a copy in a circular to all Commission members soon after the report's completion.
Officers. The Commission unanimously elected G.A. Tammann (Switzerland) as President for the following three years and E.Ye. Khachikian (USSR) as Vice-president. Outgoing members of the Organizing Committee are S. d'Odorico (Italy), J. Einasto (USSR), I.D. Karachentsev (USSR), D. Lynden-Bell (UK), V.C. Rubin (USA), A. Toomre (USA) and K.-I. Wakamatsu (Japan) and these were thanked for their contributions. Continuing members are J. Lequeux (France), Li Qi-Bin (China), H. Quintana (Chili) and P.C. van der Kruit (Netherlands). The Commission elected as new members F. Bertola (Italy), R. Ellis (UK), K.C. Freeman (Australia), J.S. Gallagher (USA), S. Okamura (Japan) and V. Trimble (USA).
As a result of a review of the membership of the Commission by the President nine members voluntarily resigned. Sixty-nine members had not replied on three occasions to circulars and reply sheets by the President and were not proposed by other members for continuation of their membership; these persons were consequently voted to be removed from the membership list. The President will inform them of this by letter.
A list of 87 proposed new members was approved by the Commission.
Working Groups. The Working Groups on "Magellanic Clouds", "Galaxy Photometry and Spectrophotometry", "Internal Motions in Galaxies" and "Supernovae" were continued.
The Working Groups were asked to arrange their chairs during their sessions.
As proposed by F. Bertola, it was decided that the Working Group on "Space Schmidt Surveys" would

be transferred to Commission 44 and H.C. Arp was elected as chairman. In case Commission 44 would not accept the Working Group (as later turned out to be the case), it would continue to reside under Commission 28.
A new Working Group on "Redshifts of Galaxies" was established and J. Huchra was elected the first chairman.
The Working Group on "Nomenclature" was re-established with K.S. de Boer as chairman.
Other business. The Commission was informed of those past, future and proposed IAU Symposia and Colloquia, for which Commission 28 acted or will act as sponsoring or co-sponsoring Commission.
The Vice-president thanked the President for his work in the last three years.

3 August: Dark Matter in Our and Nearby Galaxies

CHAIR: L. Blitz

The proceedings if this joint session are contained in the report of Commission 33.

5 August: Molecules in external Galaxies

CHAIR: F. Combes

The proceedings of this Joint Commission Meeting are contained in "Highlights of Astronomy".

5 August: Working Group on Magellanic Clouds

CHAIR: M.W. Feast

The following Organizing Committee was elected: Suntzeff (USA), Wood (Australia), de Boer (FRG), Walborn (USA), Feast (South-Africa; chairman).

The proposal for a symposium on the Magellanic Clouds in Australia in 1990 was supported. It was agreed that the Working Group would attempt to co-ordinate observational work on SN1987A.

The following papers were read:
E. Hardy (Canada): **Velocity Field of the SMC (C stars)**
J. Lequeux (France): (1) **Structure of the SMC**
(2) **Emission-line and C Stars in the SMC**
T. Lloyd Evans (South-Africa): **C Stars in the Clouds**
I.S. Glass (South-Africa): **IR Observations of Cloud Miras**
S.M. Hughes (Australia): **Long-period Variables in the LMC**
D.J. Helfand and Q. Wang (USA): (1) **Hot ISM of the LMC**
(2) **Associations, Bubbles and Superbubbles in LMC**
J.V. Feitzinger (FRG): **Turbulence in the LMC**
M.-C. Lortet (France): **Current Star Formation in the Clouds**
J.C. Blades (USA): **ISM towards SN1987A**
K.S. de Boer (FRG): **ISM near SN1987A**
P. Shull (USA): **Interaction of SNe in the LMC with the ISM**
R. Wielebinski (FRG): **Radio Continuum of the Clouds**
K.S. de Boer (FRG): **Plan for a Durchmusterung of the Clouds**

Informal discussion on the last paper continued outside the meeting.

5 August: Working Group on Internal Motions in Galaxies

CHAIR: C.J. Peterson

The following papers were presented:

E. Athanassoula (France): **Gas Flows in Barred Galaxies**
I study the flow of gas and the major families of periodic orbits in and around the bars or ovals of model barred galaxies. I show that the loci of the shocks (or density enhancements) depend on a number of parameters characterising the bar and disk potentials. Thus a comparison to the dust lanes observed in real galaxies gives indications on the underlying potentials and corresponding stellar density distributions. In particular the existence of an inner Lindblad resonance is absolutely necessary for the formation of shocks at the regions where the dust lanes along the bars are observed. The extent of the x_2 and x_3 families as well as the location of the ultra harmonic resonance greatly influence the existence and shape of the shocks.

D. Bettoni and G. Galletta (Italy): **Kinematics and Photometry of SB0 Galaxies**
The preliminary results of a long term project on kinematics and photometry of SB0 galaxies are presented. the more interesting findings are:
a) In two galaxies, the edge-on system NGC 4546 and the face-on system NGC 2717, the gas and stars rotate with similar but opposite streams.
b) A typical "double wave pattern" of the bar rotation curve has been revealed in the galaxies with intermediate inclination.
c) Stellar and gas streamings perpendicular to the galaxy plane were probably detected in 6 almost face-on galaxies.
d) The observed irregularities in the gas distribution, as well as the presence of retrograde motions in two systems, should be indications of an external origin of most of the gas in SB0 galaxies.

S. Jörsäter, P.O. Lindblad (Sweden) and G.A. van Moorsel (FRG): **Gas and Dust in the Barred Galaxy NGC 1365**
In this presentation we will present results obtained from CCD photometry, optical spectra and HI observations of the barred galaxy NGC 1365. We will discuss the near-nuclear properties of the velocity field and its relation to gas and dust distribution as well as star formation. We will also discuss the properties of the global velocity field.

C.J. Peterson and M.A. Harper (USA): **Kinematics of the Ionized Gas in M83**
The velocity field of the barred galaxy M83 = NGC 5236 [type SAB(s)c] has been mapped by use of the emission lines of Hα, N[II], and S[I] on ling-slit image-tube spectrograms from the CTIO 4m telescope. Our global velocity pattern is in good agreement with the Fabry-Perot velocity field obtained by de Vaucouleurs, Pence, and Davoust (1983, Ap. J. Suppl. *53*, 17), although small systematic differences do exist between the two studies and produce a rotation curve here that is 15 percent higher. The long-slit velocities further reveal significant small scale structure in the velocity field and a complex pattern of motion in the nuclear region.

R. Buta (USA): **Kinematics and Dynamics of Inner Rings in Spiral Galaxies**
The inner rings found in early-to intermediate type barred and weakly barred spirals are usually regions where H-α emission is concentrated within a narrow range of radii. Often, there is little emission from inside or outside the ring, thereby making the global kinematics difficult to assess. The concentration of emission within the rings affords us the opportunity to explore ring kinematics in detail, however, and can provide useful information needed for comparisons with theoretical models of ringed galaxies.
The four spirals, NGC 1512, 3351, 4725, and 4736, are excellent, nearby examples of ringed galaxies whose kinematic properties have been studied using H-α Fabry-Perot interferometry. The results show that the kinematic line of nodes for circular rotation does not agree with the position angle of the major axis of the inner ring for any of these objects, suggesting either that the rings are intrinsically elliptical in shape, or that expanding/contracting motions areprevalent. Only for the NGC 4725 does the velocity-position

angle diagram of the inner ring show an unmistakable asymmetry indicative of significant non-circular motions. A possible interpretation is that this ring is expanding at 52 km/sec and rotating at 219 km/sec. However, for all four galaxies, models of stable elliptical orbits provide fits to the kinematics of the rings which are just as acceptable as pure rotation or expanding/contracting ring fits. In no case is a unique kinematic model of the ring achievable, owing to inadequate knowledge of the parent galaxy orientation parameters.

B. Jarvis (Switzerland): **Kinematical Evidence for a Central Mass Concentration in the Sombrero Galaxy (NGC 4594)**
We present new medium resolution kinematic data for the major axis of the Sombrero galaxy (NGC 4594). These data reveal a sharp mean increase of about 59 km s^{-1} in the velocity dispersion within 4.2″ of the center. The velocity gradient across the nucleus within this same distance is measured at 38 km s^{-1} $arcsec^{-1}$, insufficient to produce the observed peak in velocity dispersion by rotational broadening alone. We conclude that there is strong evidence that NGC 4594 contains a supermassive object, possibly a black hole or massive star cluster. In the simplest case of spherical symmetry we estimate a total mass $M = 1.3 \times 10^9\ M_\odot$ inside a region of radius $r = 3.5''$ (= 309 pc, D = 18.2 Mpc) of the center of the galaxy.

P.T. de Zeeuw (USA), L.S. Sparke (Netherlands), E.M. Sadler (Australia), J. Danziger (FRG), D. Bettoni and F. Bertola (Italy): **Kinematics of Gas in NGC 5077**
Hα images of the E3 galaxy NGC 5077 reveal a gaseous structure along the minor axis. A spiral patten and three central knots are present in the disk, whole a faint gaseous structure on the western side runs almost parallel to the major axis of the stellar body. Based on observations at seven different position angles, four of which are our own, we find that the velocity field of the gas is consistent with that of an inclined circular disk, but the rotation axis is significantly misaligned with both axes of the light distribution. The stellar component shows little or no rotation, and the velocity dispersion is roughly constant at ~250 km/s out to 15–20 arcsec.
The morphology and kinematics of the gas disk are used to probe the gravitational potential of the stellar body. Spherical, axisymmetric and triaxal models are considered. The intrinsic shape of the galaxy may well be nearly oblate, with the gas in a disk over the pole. The radial behaviour of the mass-to-light ratio M/L is derived.

F. Bertola, D. Bettoni, L.M. Buson and W.W. Zeilinger (Italy): **Counter-rotation in Dust-Lane Ellipticals**
Counter-rotation of gas and stars has been discovered in three elliptical galaxies which have dust lanes along their major axes. This dynamical decoupling of the two components together with that observed in minor-axis dust-lane ellipticals, where the two angular momenta are orthogonal, is the best proof of a second event, namely the acquisition of external material, in the history of dust-lane ellipticals. It is suggested that similar accretion events occur in SO galaxies giving rise to counter-rotating (and co-rotating) gas in the equatorial plane or to polar rings.

B. Whitmore (USA): **Rotation Curves for Spiral Galaxies in Clusters**
Rotation curves are presented for 21 galaxies in four large spiral-rich clusters: Cancer, Hercules, Peg I, and DC 1842–63. Although the global properties (luminosity, radius, maximum rotation velocity, and mass) of the cluster sample are similar to those of the field, there is some evidence that the amplitude of the rotation curves in the cluster galaxies are lower than those of their field counterparts. A good correlation is found between the outer gradient of the rotation curve and the galaxy's distance from the center of the cluster, in the sense that the inner galaxies tend to have falling rotation curves while the outer galaxies, and field galaxies, tend to have flat or rising rotation curves. A strong correlation is also found between the M/L gradient across a galaxy and the galaxy's position in the cluster, with the outer galaxies having steeper M/L gradients. These correlations indicate that the inner cluster environment can strip away some fraction of the mass in the outer halo of a spiral galaxy, or alternatively, may not allow the halo to form.

V.C. Rubin (USA): **Rotation Curves for Spiral Galaxies in Compact Groups**
Long-slit CCD spectra and CCD images have been obtained for 50 spiral and E galaxies in 14 Hickson groups believed to be true associations. Many galaxies exhibit tidal peculiarities, both in their images and in their rotational properties. Some characteristic peculiarities will be discussed.

E.M. Sadler (Australia) and S.M. Simkin (USA): **Kinematics of Gas Disks in Radio Galaxies**
Spectroscopic velocity measurements (Ha, [NII], and [OII]) have been obtained for the extended gas in several strong, Southern radio galaxies. these display a variety of non-circular motions superimposed on fairly regular circular velocity fields. In some cases, the peak, deprojected circular velocities exceed those for the brighter Sa galaxies.

G.G. Byrd (USA): **Triggering of Active Spiral Galaxies and QSOs in Rich Clusters**
In previous simulations of self gravitating spiral galaxies perturbed by nearby companions (Byrd et al. 1986, 1987, Astron. Astrophys.), we found that the perturbation levels in Seyferts observed by Dahari were also physically sufficient to create nuclear inflows of disk gas activity. We extend our simulations to include the effect of the gravitational field of a rich cluster on a galaxy just entering it. Disk and nuclear activity should be triggered in spirals within $\sim$700 kpc of the center of rich clusters like Coma. Tidal perturbation should be much more effective than ram pressure in triggering activity and changing spirals to SO's in rich clusters.

P. Grosbol (FRG) and G. Contopoulos (Greece): **Near Self-consistent Models of Spiral Galaxies**
The surface density response in a set of model spiral galaxies was compared with the imposed density to estimate the degree of self-consistency. It was confirmed that strong spiral structure can exist between the Inner Lindblad and the 4/1 resonance while the response outside this resonance for large amplitude generally is not well aligned with the pattern. The exact characteristics of the response is sensitive to the size of the bulge. The models show an arm segment which turns back to the center close to the 4/1 resonance. Further, the response outside the 4/1 resonance has a very patchy appearance. Similar features can be observed in galaxies like NGC 4535.

W.W. Roberts and D.S. Adler (USA): **Global Spiral Structure in Cloudy Gaseous Galactic Disks: the Need for and the Plague of Gravitational Softening in Computational Studies**
New questions are motivated through computations studies of cloudy gaseous galactic disks. Gravitational softening, although greatly needed on the one hand, is also a plague in disguise on the other. With too little softening, artificial instabilities can arise to drive unrealistic growth of random motions. With too much softening, important "nonlinear effects" of the self-gravitational gas can be suppressed. A delicate balance is necessary to capture the "multifold" role of the cloudy gaseous component in galactic disks. (This work was supported in part by NSF under Grants AST-82-04256 and AST-87-12084 and NASA under Grant NAGW-929.)

S.T. Gottesman (USA): **Constraints on the Properties of Spiral Galaxy Halos, as determined by the Kinematics of Small Satellite Systems**
We have observed a carefully selected sample of large spiral galaxies orbited by small, Magellanic-cloud like satellites. We have compared the orbital properties of the satellites with the rotational properties of the spirals. The ratio of these two independent mass estimates gives a disturbation which is significantly different than that found for binary galaxies. N-body modelling of our observation leads us to believe that our spirals have holes which are about four times the radius of their disks. Consequently, for our typical system, the isothermal halos are about four times more massive than the disks. The conclusion appears to be at variance with several strongly held beliefs concerning the structure of spirals.

6 August: Working Group on Supernovae

CHAIR: V. Trimble

Of the 24 papers proposed for this session, 17 were presented, most dealing with 1987A. K. Sato and P.J. Schinder agreed that the neutrino data can be fit with a range of equations of stat and neutron star masses. The collapsing star could have been as early as BO.7Ia (with high N) to B31A (N normal), according to N.R. Walborn's re-examination of objective prism spectra. The possible early type and higher luminosity with require rethinking of light curve and spectrum interpretation.
Meanwhile, the spectrum synthesizers (R. Wehrse, J.C. Wheeler, and W. Schmuts) concurred that we need additional UV opacity and an envelope with both steep and flat parts in its density profile to account for the continuum and lines in early spectra. Given atmospheric models, these spectra imply a Baade distance of 42±2 kpc (R.V. Wagoner) vs. 50±5 kpc from other methods. Disturbingly, the implied distance drops with time.
Lots of molecular lines have turned up in the late time IR spectra. The CO requires the emitting region to have been enriched with C and O made in the progenitor (C.M. Sharp), but we so far have no evidence for dust, either circumstellar or made in the ejecta (E. Dwek). The mixing is only one sort of deviation from layered spherical structure that must have occurred early in 1987A. Infrared speckle data (A. Chalabaev) revealed complex extended emission from June 1987 onward. In addition, faint satellite emission components on many hydrogen lines and Na I were evident during days 20 to 100 and must somehow have been associated with lumpiness in the photosphere, because the velocity splitting of the inner edges of red and blue components tracked the contemporaneous photospheric speed (N. Suntzeff). We do not understand these lines.
Also still defying explanation is the soft X-ray flux, which, if attributed to ejecta colliding with circumstellar gas (H. Hanami), requires dense, red supergiant material to be closer to the star by a factor of 100 than the 0.3 pc implied by the narrow UV lines (P. Lundqvist). The non-detection of TeV gamma rays (T. Gaisser) is not precisely mysterious, but does limit cosmic ray acceleration to less than needed if supernovae are to contribute above 10^{15} eV.
There are still supernovae other than 1987a in the universe, though perhaps not so many as we thought. H.E. Jorgensen reported first results from a search in clusters with $z = 0.2$ that was expected to have found, so far, several SNI's, but was identified only one (probable) DNII, implying an SNI rate near the low end of the possible range of 0.2–0.8 SNe per century per 10^{10} L_0^B. M. Nagasawa has been carrying out 3-d hydro-dynamic simulations of SNII's and finds that envelope is already shock break-out. Finally, N. Panagia reviewed the evidence that SN Ib's require a compact, low mass progenitor, surrounded by circumstellar material too dense easily to belong to it, and thus suggestive of a red giant - white dwarf binary.
There are two pieces of Working Group news: William Liller is now chairman, and R. Barbon and his colleagues at Asagio are updating their permanent SN catalogue to December 1988 and invite suggestions, comments, and corrections of earlier listings.

8 August: Working Group on Galaxy Photometry and Spectrophotometry

CHAIR: J.-L. Nieto

Dr. R. Buta (USA) was elected chairman of the Working Group for the period 1988 to 1991.

The following papers were presented:

K. Kodaira (Japan): **A New Parameter for Phase-space Density**
Using a homogeneous data set of 16E and 26s galaxies in the Virgo Cluster, we found a new parameter, $f = (GVD^2)^{-1}$, which has the dimension of phase-space density and shows an extremely tight correlation to the galaxy luminosity.

K. Rakos (Austria): **A Low-resolution (150 Å) Spectrophotometer for Galaxies**
In a joint project, J. Schombert, Palomar Obs., W.W. Weiss, Vienna Observatory and K.D. Rakos, Vienna Observatory, a special spectrophotometer is now under construction. Two field lenses with focal ratio of f/2 concentrate the light of the galaxy and of the local sky as two images of the main mirror on the silica fiber light guides. The fibers are formed to a single slit and a holographic concave grating corrected for spherical abberation and flat image with a focal ratio of f/2 formes an image of the spectrum on a CCD camera the spectrum from 3500 Å to 6000 Å can be folded by any filter transmission curve to get the photometry in the selected photometric system. The calibration of the spectrum can be easy made observing spectrophotometric standards. The proper sky subtraction is very simple because the sky spectrum is recorded simultaneously.

H. Lorenz (DRG): **Adaptive Digital Filters applied to Surface Photometry**
An adaptive filtering technique to the two-dimensional photometric mapping of extended objects has been presented. The key feature of this system is the "H-Transform". By comparing the results to recent studies of NGC 3379 (Capaccioli et al., 1988; Astron. Nachr. *309*, 69) no systematic errors in the reduction procedure could be found. The mean residuals are smaller than ±0.05 at a threshold of 1′1 of the night sky level.

J.-L. Prieur (France), D. Carter, A. Wilkinson (UK), W.B. Sparks (USA) and D.F. Malin (Australia): **A Statistical Study of Shell Galaxies**
Preliminary results of an extensive CCD Survey of Malin and Carter's catalogue (1983, Ap. J. *274*, 534) have been presented. Shell photometry and morphology are in agreement with the expectations of the merging models. A correlation have been found between the morphology of the shell system and the apparent flattening of the parent galaxy. Significant color differences have been measured *along* some of the shells. Recent massive nuclear star formation which has probably been triggered by the collision with the companion galaxy, has been discovered in about 20% of these objects.

E. Sadler (Australia) and J.S. Gallagher (USA): **Star-forming Regions in Small Elliptical Galaxies**
Many small ($M_B > -17$) elliptical and SO galaxies show evidence for current star formation. The star-forming regions are blue clumps with HII-region-like spectra and usually lie in the central 1–2 kpc of the galaxy. Broad-band (B, R) profiles show that these are true E/SO galaxies, not misclassified spiral or irregular systems. Thus the recent star formation history of small ellipticals appears to differ from that of more luminous early-type galaxies.

F. Bertola (Italy), M. Vietri (Italy) and W. Zeilinger (Italy): **A Search for Triaxial Bulges**
A wide sample of spiral and SO galaxies has been investigated photometrically, using CCD images obtained with the ESO telescopes in order to detect possible misalignments between the major axes of the disk and of the bulge. This phenomenon, once the effects of dust are removed, is an evidence that the bulge has a triaxial shape, as it is the ease of elliptical galaxies. A fair fraction of our galaxies shows such a twisting. The photometric data, when coupled with the kinematical ones, describing the distortions in the velocity field of the gas in the disk induced by the triaxial potential of the bulge, allow to establish with an high degree of certainty, the axial ratios of the bulge.

P. Prugniel, E. Davoust and J.-L. Nieto (France): **Eight Hierarchical Pairs of Elliptical Galaxies**
We investigated the photometric and spectroscopic proportions of a sample of eight closely interacting pools of elliptical galaxies whose mass ratio is of the order of 10 or higher. A special algorithms allows us to disantangle the contributions of the two galaxies on the usages, so that their photometric properties can be known. From (mean surface brightness, absolute magnitude) and (velocity dispersion, absolute magnitude) diagrams, we discuss the evidence for tidal heating in these pairs.

W. Jaffe (Netherlands): **The Formation of the Structure of Elliptical Galaxies**
In continuation of previous work we show that it is easy to reproduce the form of elliptical galaxies by any process that scatters the energies of individual stars by large amounts. Both the envelopes and the bodies are well reproduced. We also show that even small scattering processes with $\Delta E \propto E^{1/2}$ ma/u

nice ellipticals, and tidal heating in clusters satisfies this condition. We consider some of the consequences of this model to the statistics and evolution of galaxies and dark halos in clusters.

Hong Bae Ann (Korea): **Luminosity Distributions in Barred Galaxies**
The luminosity distribution of 3ρ barred galaxies were investigated by a new method of profile decomposition, assuming exponential function and Gaussian function along and perpendicular to bar. The fractional luminosity of bar (B/L) increase with spheroid-to-disk value (S/D) for galaxies with S/D less than 1. About 75% of the observed galaxies have B/L less than 0.1.

D. Bettoni and G. Galletta (Italy): **Peculiarities in Barred S0s**
Some peculiarities have been found analysing CCD images of barred S0 galaxies. They are:
a) in FNO systems, NGC (6684 and NGC 2983, the bar is shifted by few (~2) arcseconds with respect to the disc isophotos.
b) rings appear to be elliptic, like in the case of NGC 6684.
c) Pseudo spiral arms (composed of stars) appears in some systems.

E. Anathassoula, S. Morin, D. Puy, H. Wozniak, A. Bosma, M. Pierce, J. Lombard (France): **The Shape of Bars in Early-type Barred Galaxies**
We have examined the shape of bars in a small sample of SB0 galaxies with strong bars by fitting generalised ellipses $[(\frac{x}{a})^c + (\frac{y}{b})^c = 1]$ to be isophotes. These represent good fits and show that these bars are more rectangular like than elliptical like. The shape parameter c has a maximum near the end of the bar and the values of the maximum range between 2.7 and 5.3. The ellipticity decreases with radius starting from a maximum value around 0.8 which corresponds to an a/b of the order of 5.

R. Buta (USA): **The Nature of Rings in Early-type Disk Galaxies**
Early-type disk galaxies some of the most spectacular examples of ring phenomena among the population of normal, mostly non-interacting galaxies. In S0 galaxies these rings are most prominent at stage $S0^+$, and some of the brightest examples occur in systems possessing very weak bars or no obvious bar at all. In the absence of strong bars and significant amounts of gas, the theory implies that it would be difficult for rings to have formed in these kinds of galaxies. Therefore, ringed S0 galaxies pose a dilemma for ring formation theories that makes their properties worth exploring more carefully.
For this purpose, the surface brightness and color properties of six early-type ringed galaxies have been explored using multi-color CCD surface photometry. The sample includes two "normal" barred spirals, NGC 1350 and 139, a weakly-barred S0/a system, NGC 3081, and three "non-barred" $S0^+$ systems, NGC 7020, 7187 and 7702. The results show (1) that the rings in all of these objects are zones of enhanced blue colors regardless of whether a bar is present or not; (2) that the best defined rings have old stellar components, while the most-spiral-like ring (in NGC 1398) has no very significant stellar component; and (3) that rings and lenses are intimately connected, rather than distinct phenomena. In general, the rings of early-type disk galaxies share much in common with those of later-type barred spirals, and are probably related to the same resonances.

8 August: Stellar Photometry with Modern Array Detectors

CHAIR: F. Rufener

The proceedings of the Joint Commission Meeting are contained in the report of Commission 25.

8 August: Galaxy Redshift Determinations: Better Techniques, Better Standards

CHAIR: J. Huchra

9 August: New Results

CHAIR: G.A. Tammann

P. das Gupta and J.V. Narlikar (India): **Counts of Radio Galaxies and Evolution**
This work follows the earlier investigations of Das Gupta, Narlikar and Burbidge (1988, A. J. Jou. p. 5) in which a non-evolving radio-luminosity function (RLF) was constructed from the observed number-redshift diagram for a complete sample of radiogalaxies in a given friedman universe. Using the χ^2 and Kolmogorov-Smirnov (KS) tests it was shown that the non-evolving RLF cannot be ruled out on the basis of redshift (z) – flux density (s) plot.
Here a similar analysis is carried out for the steady state model for the 3C–R and 2.7 GHz complete samples of radio galaxies. The χ^2 and KS-test are shown to give a consistent result between the theoretical and observed $z - s$ plots. These tests and the V/v_m test does not rule out this model.

Y. Sobouti (Iran): **Symmetries of Liouville's Equation**
Liouville's and linearized Liouville-Poisson's equations have O(3) symmetry. There exists an angular momentum operator, J, in phase space which commutes with operators of these equations. J is the sum of two angular momenta in configuration and momentum spaces. This enables one (a) to classify the normal modes by the eigennumbers of J^2, and J_z, and (b) to extract the dependence of the normal modes on direction angels of the position and momentum vectors, analytically.

T.K. Chatterjee (Mexico): **Frequency Determinations of Peculiar Galaxies on a Dynamical Basis**
A study of the expected frequency of all sorts of peculiar galaxies is conducted by studying galactic collisions with varying collision parameters. Results indicate that the expected frequency of such galaxies is compatible with the observational value only if it is measured with respect to densely populated regions, using the average values of the parameters corresponding to such regions, while it falls by several orders of magnitude if we consider regions of normal density for its determination. This leads to the conclusion that most of these peculiar galaxies must have already been bound doubles, whose orbits are such as to have brought about only now the close proximity characteristic of such systems.

C.-K. Chou (China): **The Properties of Spiral Density Waves in a Magnetoactive Disk**
The propagation of spiral density waves in a differentially rotating, self-gravitating, magnetoactive and highly flattened disk is investigated by using the asymptotic theory for tightly wound spirals developed by Lin and his collaborators. We adopt the continuum fluid model as the primary basis, and out treatment will be largely analytical.
A new asymptotic dispersion relation for spiral magnetoacoustic waves with magnetic fields along the spiral arms $B_\theta(\gamma)$ is derives in the frozen field approximation. We also present a more exact local dispersion relation by using the WKB approximation and study the effects of magnetic fields on the growth rate. The stabilizing effect of the magnetic fields is spiral galaxies ($V_A^2/a^2 \simeq 0.06$, V_A = Alfven speed, a = speed of sound) is rather small. However, the field strength in solar nebula or HII regions may be considerably stronger so that $V_A^2/a^2 \simeq 0.1$ to 1. It is then expected that larger effects may result.

M. Tagger, J.F. Sygnet and R. Pellat (France): **Bars and Barred Galaxies: Linear Theory Revisited**
We reconsider the shearing sheet model of a flat disk galaxy. We show by asymptotic expansions that the global solution contains not only the usual tightly wound waves of Lin and Shu, but also a distinct, bar-like feature. Its presence, related to the Swing amplification mechanism, is due to the thin-disk nature of the galaxy. The bar feature is dominant when and where the galaxy is hot.

M. Kaufman (USA): **M81: Tests of Common Notions about Grand Design Spirals**
Although M81 is a classic density wave galaxy and the HI velocity contours show a spiral shock front, M81 does not have certain properties often attributed to grand design spirals. For example, the nonthermal

radio arms are generally centered on the gravitational potential minimum, not on the velocity shock front. The dust lanes near the velocity shock front have neither greater length nor greater extinction than those farther downstream. The giant radio HII regions are located either near or downstream from the potential minimum. The observed cross-arm distributions of dust, etc. can be understood within the context of density wave models: certain properties of spiral arms depend sensitively on the amplitude of the density wave compression or on the cloudy nature of the interstellar medium.

I.A. Issa (Egypt): **A Geometric Method to Determine the Distances of Galaxies**

T. Storchi-Bergmann and M.G. Pastoriza (Brazil): **Low Activity Nuclei with Strong [NII] Lines**

We present results from the study of the spectra of four low activity Seyfert 2 nuclei, a LINER and one of intermediate type between the two. All of them present strong [NII]$\lambda\lambda$6548+6584 emission lines. The analyzed of the absorption spectrum showed that more than 80% of the light comes from old population stars (10^{10} years) with average metallicity two times solar. We have compared the relative intensities of the emission lines with values obtained from model nebulae photoionized by a non-thermal continuum, having solar and two times solar abundances. We have concluded that the models can reproduce the line intensities for the LINER galaxy, but for the Seyfert 2 galaxies, the only way to reproduce the [NII] lines is considering an overabundance of nitrogen relative to the other elements of about 3 times solar.

U.C. Joshi (India): **Polarimetry of Seyfert Galaxies NGC 2992, 3081, 3227 and IC 4329**

J.J. Steyaert (Belgium): **Energy Generation in Quasars**

A new scheme, including new physics is proposed that could power quasars. It is based on a new class of nuclear reactions of cross-section about 10^{-33} cm^2. A mass of 10^{11} $M_\odot$ ^{3}He gas is transformed in 10^8 years into D and ^{4}He with luminosity of $8 \cdot 10^{46}$ erg s^{-1}. A variability within 1 mounth is obtained if the density is 10^9 atoms/cm^3, a value consistent with CIII] data. The scheme could be used also for active galaxies and suggests an elemental evolution of pre-quasars, quasars and galaxies with time. Data from SN 1987A absorption lines were deconvolved using the same scheme.

I.F. Mirabel (USA): **Arecibo HI Survey of the 100 most Luminous IRAS Galaxies of the Local Universe**

M.F. Struble (USA): **The "Starpile" in Abell 545**

I report the first probable visual discovery of intracluster matter (from POSS): a low surface density feature in the second richest Abell cluster, A545. Galaxy counts show it is located at the cluster center. The feature has $B - V$ ~1.3 to 1.5, the same as an elliptical galaxy at cluster z of 0.15, is ~70 kpc × 40 kpc and has at least two faint nuclei. Several galaxies, some binary, are located at its northern periphery. Its color and central location argue against an interstellar feature superimposed on the cluster. It is most similar to a cD galaxy's envelope (no nucleus).

S.P. Bhavsar (USA): **First Ranked Cluster Galaxies: A Two Population Model**

The small dispersion in the absolute magnitudes (~0.35 mag) of first-ranked galaxies in rich clusters has been the cause for much debate between the "special" and "statistical" hypotheses for these galaxies. We present statistical evidence that it is not the one or the other but a combination. The distribution in magnitudes of the brightest galaxies is best explained if they consist of two distinct populations of objects; a population of special galaxies having a Gaussian distribution of magnitudes with a small dispersion (0.21 mag), and a population of extremes of a statistical luminosity function. the best fit model requires that 63% of the clusters have a special galaxy that is on a average 0.48 magnitudes brighter than the brightest normal galaxy. The model also requires the luminosity function of galaxies in clusters to be much steeper at the very tail end, than conventionally described.

R.A. Windhorst (USA): **Discovery of a Nearly Normal Elliptical Galaxy at z = 2.39**

Data is presented on a faint galaxy that is probably a nearly normal elliptical at a redshift of 2.39. The galaxy was found on a Palomar 200 inch Four-shooter image, and is the $V \sim 23.5$ mag OPTICAL counterpart of the steep special compact, 50 mJy radio source Herc 202 from the Leiden-Berkeley Deep Survey. The object is barely extended (FWHM $\simeq$ 1.″6, total extent $\simeq$2.′6) and has fairly blue colors (Gunn $g - r = 0.03$, or $B - V \sim 0.5$ mag). It was detected in one night of integration time with the

200 inch IR systems at H (1.6μ) = 20.8 mag. I spent one night at the MMT and two nights with the MAYALL 4 meter CRYOCAM to measure its spectrum at 15 Å resolution, and found very weak ($\lesssim 2 \times$ continuum), unresolved emission lines (Lyα, CIV, CIII, and perhaps CII o HeII), all consistent with $z = 2.390 \pm 0.003$. The optical and optical-IR colors, as well as its optical morphology, are consistent with a passively evolving (giant elliptical galaxy with age $\simeq$1–2 Gyr at $z = 2.39$ (or current age of $\sim$14 Gyr if $H_o = 50$, $q_0 = 0.1$). Although the galaxy was selected as radio source, it is a 20–100 $\times$ weaker radio source than the extremely low surface brightness protogalaxy candidates in the 3CR and 1 Jy samples, discovered recently at similar redshifts. While those objects are possibly protogalaxies, whose formation might be triggered by their extremely powerful radio jets, Herc 202 has fairly compact optical morphology (visible size of 20–30 kpc), *no* strong emission lines, and a continuum consistent with that of a normal elliptical galaxy redshifted to $z = 2.4$.

I believe that Herc 202 is the closest one can get with the currently available detector technology and telescopes to find (nearly) normal elliptical galaxies at $z > 2$. The current radio source sample has 451 radio sources with 21 cm fluxes down to 50 μJy, and imaged with Fourshooter down to $V > 23$ mag. In this sample, there are another 200 radio galaxies with $V > 23$ mag, and several dozen more candidate like Herc 202.

R.W. Hunstead (Australia), M. Pettini (Australia), A. Boksenberg (UK) and A.B. Fletcher (Australia):
A Possible Primeval Galaxy at z = 2.5
QSO absorption systems in which the Lyman α profile shows damping wings have been interpreted as the HI disks of young intervening galaxies. Studies of heavy element enrichment in one such system ($z_{\rm abs}$ = 2.309 towards PHL 957) indicates an abundance of only 1/20 solar with very little evidence for dust. In another system ($z_{\rm abs} = 2.465$ towards Q0836+113) we find naarow ($\leq$50 km/s FWHM) Lyman α emission in the base of the damped Lyman α absorption line. The star formation rate inferred from the Lyman α luminosity may be as low as 1 $M_\odot$/yr.

9 August: Star Clusters in the Magellanic Clouds

CHAIR: P. Demarque

The proceedings of this Joint Commission Meeting are contained in the report of Commission 37.

COMMISSION No. 29

STELLAR SPECTRA (SPECTRES STELLAIRES)

Report of the Business and the Scientific Meetings 3-10 August 1988.

Président : G.Cayrel de Strobel Secretary : M. Spite

1. ELECTION OF VICE-PRESIDENT AND ORGANIZING COMMITTEE

Before proceeding to the election, the President recalls that there exist for Commission 29 RULES OF OPERATION (BY-LAWS), which have been prepared during the presidency of Jun Jugaku and his Organizing Committee (1982-1985). These By-Laws have been unanimously accepted in the Business meeting of Commission 29 during the XIX General Assembly in New-Delhi. They can be obtained on request from G. Cayrel de Strobel.

Essentially the By-Laws of Com 29 state that :

....."The Vice President shall be President-elect, shall assist the President in carrying out the functions of the Commission, and shall serve as Acting President if the President is temporarily unable to carry out his/her duties. The Vice President shall normally serve for a three-year term.".....

....."The Vice President shall be elected by majority vote of the members of the Commission who participate in the election....."

....."The OC will consist of the President, Vice President, and past President in addition to eight elected members by the procedures as described below. OC members will normally be elected for periods of two terms (six years). Unless elected Vice President, the OC member will then be ineligible for re-election. Past OC members will again become eligible for election after one term (three years). Four OC members will retire, and four new members will be elected, at the Commission business meeting held at each General Assembly....."

....."The election of OC members shall be conducted by secret ballot at the business meeting, with all members of the Commission present eligible to vote.....".

....."Any vacant three-year terms will be filled by those remaining candidates who have received the highest number of votes....."

....."The OC shall give highest priority to ensuring that the membership of the new OC is reasonably balanced in terms of geography and research speciality. The appointee shall be chosen from the slate of candidates considered in the election....."

After this preamble the election of new Officers of the Organizing Committee was carried out, and the following names for the new Organizing Committee were approved :

President : P.S. Conti
Vice President : D.L. Lambert

Organizing Committee : D. Baade, M.S. Bessell, A.M. Boesgaard, A.Cassatella, G. Cayrel de Strobel, B. Gustafsson, J. Smolinski, C. Pilachowski, M.Spite

2. MEMBERSHIP

2.1 Deceased :L. Goldberg, G. Haro,.E.R. Mustel.

2.2 Voluntary Resignations : J. Boulon, D.C. Morton, J. Smak., H.H. Voigt.

2.3 Admission of New Members : the membership of the following new members was approved :

M.L. AGUILAR
G. ALECIAN
M.C. ARTRU
T. ATAC
S.L. BALIUNAS
B. BARBUY
G.B. BASRI
J.E BECKMAN
S. BENSAMMAR
C. BERTOUT
H.E. BOND
J. BOUVIER
J BREYSACHER
K. BUTLER
C. CATALA
P.L. COTTRELL
L CRIVELLARI
D.N. DAWANAS
M. DENNEFELD
P. FELENBOK
E.L. FITZPATRICK
B. FOING
P. FRANCOIS
S. FRANDSEN
E.D. FRIEL
I. FURENLID
C.D. GARMANY
M.S. GIAMPAPA
D. GILLET
R. GRATTON
D. HANDLIROVA
R. HANUSCHIK
L.W. HARTMAN
U. HEBER
H. HENRICHS
H. HUBERT
K. HUNGER
JIANG SHI-YANG
C. JORDAN
S.C. JOSHI
M.T. LAGO
H. LAMERS
R. LAMONTAGNE
J.D. LANDSTREET
T. LANZ
J.M. LECONTEL
K.G. LIBBRECHT
J.W. LIEBERT
R.E. LUCK
P. MAGAIN
J.P. MAILLARD
G.W MARCY
P.L. MASSEY
Z. MIKULASEK
P. MOLARO
R. MUNDT
T. NECKEL
H. NETZER
M. NIEUWENHUIJZEN
P.E. NISSEN
J. NORRIS
T. NUGIS
M. PENSTON
F. PRADERIE
R. PRINJA
B.S. RAUTELA
R. REBOLO
R.M. RICH
B. ROBINSON
J.A. ROSE
R. RUTTEN
S.H. SAAR
J.P. SAREYAN
T. SIMON
V.V. SMITH
C. SNEDEN
D.R. SODERBLOM
G. SONNEBORN
M. STEFFEN
A. TALAVERA
F. THEVENIN
J.C. VALTIER
M.S. VASU
G. VLADILO
N. VOGT
G. WALKER
J. ZOREC

3. SYMPOSIA AND COLLOQUIA

The Commission has sponsored or cosponsored 13 IAU Conferences in 1985-1988. For details see Transactions IAU XXA p. (1988). Another Conference, "Evolution of Peculiar Red Giant Stars", (IAU Coll. 106) held in Bloomington in July 1988, was sponsored by Com 29.
The Organizing Committee has also approved sponsorship of 6 more IAU Meetings : IAU Symp. 138 " Solar Photosphere ; Structure, Convection and Magnetic Fields ", Kiev May 15-20 1989 (S.O. Stenflo). IAU Symp. "Wolf-Rayet Stars and interrelations with other massive stars in Galaxies", Bali, Indonesia, June 18-22 1990 (K.A. Van der Hucht). IAU Symp. or Coll. :" Physik of classic Novae", late spring or early summer 1989 (M. Friedjung). IAU Coll. :" The Sun and Cool Stars , activity , magnetism, dynamos", Helsinki 1989(I. Tuominen, O. Vilhu). IAU Coll. "Surface inhomogeneities in late-type-stars", Armagh N Ireland late 1990 (J.G. Doyle, P.B. Byrne). IAU Symp. "Evolution of Stars : the surface chemistry connection" Bulgaria 1990 (G.Michaud and A. Tutukov).

4. WORKING GROUPS

Working Groups (WG's) have no official status within the IAU. Hollis Johnson suggested that they exist only to (1) help the commissions in their work ; (2) encourage research and interest in their special area ; (3) cause things to happen ; (4) get people involved.

The Commission sponsored four WG's :

1. Be-Stars (D. Baade) together with Com. 45

2. Ap/Cp Stars (C. Cowley) together with Com 36 and Com 45.

3. Peculiar Red Giants (C. Jaschek and P.C. Keenan) together with Com 36 and Com 45

4. Standard Stars (A. Batten) together with Com 30 and Com 45

During the business meeting of each of the four WG's new Organizing Committees were nominated and elected. They are as follows:

Be Stars : OC (1988-1991) :

D. Baade (chairman), L.A. Barlowe, J. Dachs, V. Doazan, J.M. Marlborough, J.R. Percy, G.J. Peters.

Ap/Cp Stars : OC (1988-1991) :

I. Hubeny, D. Kurts, T. Lanz, G. Michaud, K. Sadakane (Chairman), K. Stephen.

Peculiar Red Giants : OC (1988-1991):
R. De La Reza, R. Foy, R. Garrison, H.R. Johnson (Chairman), A. Renzini, V Straizys, T. Tsuji.

Standard Stars : OC (1988-1991) :

S.J. Adelman, J. Andersen, A.H. Batten (Chairman), M. Gerbaldi, I.N. Glusheva, H. Neckel, L.S. Pasinetti.

5. INTERNATIONAL REGISTER OF STELLAR SPECTROSCOPISTS (IRSS)

The IRSS is a list of stellar spectroscopists giving names, adresses and research specialities, without regard to IAU membership. It was edited until now by W. Bonsack and by A. Settlebak. The IRSS is cosponsored by Com 45. Prof. Bonsack whos new address is :

Prof. Walter BONSACK
1016 E. El Camino Real
SUNNYVALE, CA 94087

will be happy to send computer diskettes (for IBM-PC - type computers) containing the data base for the International Register of Stellar Spectroscopists to anyone who might wish to continue the work of compiling and distributing this directory.

6. SCIENTIFIC MEETINGS

The scientific meetings inside each WG have been all very interesting and have attracted many people.

6.1 WG on Be Stars

D. Baade gave a report on the scientific activity of his WG concerning mainly the "Be-Stars New Letters" (Editor G.J. Peters) received twice a year by about 200 Institutes and Individuals. D. Baade summarized also the IAU Coll 92 on : Physics of Be Stars.Boulder, August 1986. If an other IAU Symposium or Colloquium was still to early to plan, there was broad consensus between the WG members on Be stars that a workshop to be held during the 1988-1991 period appears highly desiderable.

J.R. Percy, C.A. Grady, and G.J. Peters gave progress reports on photometric, UV spectroscopy and multifrequency observing campaigns, respectively.

Other five talks were clustered around "Empirical correlations of observational quantities in Be stars" :

R. Hanuschik : Photospheric *v sini* and characteristics of circumstellar emission lines
C. Grady : Wind and stellar *v sini*
V. Doazan : V/R ratio of optical emission lines and strength of UV resonance lines

D. McDavid : Linear continuum polarization and Hα emission strength
D. Baade : "Be-like" phenomena observed in early-type non-Be-stars

The speakers had been asked to put the emphasis on well-structured presentations of the rich observing material to resolve confusions which result from the large variety of phenomena but also from the occasionally obliterating effects of the models and interpretations applied to the data.

6.2 WG on Ap/Cp stars

C.R. Cowley reported on Ap/Cp stars and their "Peculiar Newsletter", (Editors: H. Hensberge and W. Van Rensbergen).

S.Adelman reported on the Lausanne-Workshop ("Elemental abundance analysis", editors: S.Adelman and T. Lanz, Sept 1987). G. Michaud presented and discussed a future IAU meeting, which will be chaired by him and A. Tutukov, and held in Bulgaria in late summer 1990. Michaud chose as topic for this conference: "Evolution of stars : the surface chemistry connection".

The scientific session of the WG on Ap/Cp stars consisted of invited reviews by :

J. Landstreet: Modeling magnetic and abundance patches
J. Matthews: Rapid light variations
S. Vauclair: Lithium abundances in F stars
C. Proffitt: Formation of abundance anomalies through accretion of nuclear-processed material
H. Shibahashi: Non radial pulsations of chemically peculiar stars
M. Dworetsky: Temperatures and gravities of Am stars

Brief scientific contributions were presented by R. Kurucz, P. Didelon, H. Maitzen, M. Gerbaldi, I. Savanov, K. Kumar, W. Wehlau and W. Weiss.

6.3 WG on Peculiar Red Giant Stars

Suggested and informally organized at the Strasbourg meeting ("Cool Stars with Excesses of Heavy Elements") in 1984, the WG was formally constituted at the General Assembly in India (1985) under IAU Commission 29 (Stellar Spectra) and 45 (Stellar Classification).

A twice-yearly newsletter edited and mailed by Carlos Jaschek (Strasbourg) and the impetus for the successful IAU Colloquium 106 : Evolution of Peculiar Red Giant Stars (Bloomington, Indiana, U.S.A., 27-29 July 1988) were the principal activities of the WG over the past 3 years.

The meeting, a joint session of Commissions 29, 36, and 45, was attended by approximately 60 persons. Neither of the Co-Chairpersons of the Working Group (Carlos O. Jaschek and Philip C. Keenan) were able to be present, and Hollis R. Johnson conducted this meeting.

Continuation of the newsletter on PRG Stars was discussed. It will not be continued unless a volunteer editor steps forward. One issue will, however, be edited and mailed by the new Chairman.

Future activities for the WG were discussed. Two suggested activities received strong support. (1) Because IAU Commission 42 already has plans to propose an IAU Symposium on Binary Stars near the time and the place of the General Assembly in 1991, the WG may join in organizing this meeting if it includes Red Giant in Symbiotic and Binary Systems. (2) The WG will try to promote a joint discussion or joint Commission Meeting on "Outer Atmospheres of Red Giant Stars " at the next General Assembly. The Officers will pursue both projects.

Since IAU Colloquium 106 was one of the principal recent activities of the WG, reports of that meeting were given by Hollis R. Johnson and T. Lloyd Evans.

Scientific reports were given as follows :

R. Foy (France) : Angular Diameter of χ Cyg

U.G. Jorgensen (Denmark) : Effects of Polyatomic Molecules on Carbon Star Atmospheres.

6.4 WG on Standard Stars.

The scientific part of the meeting attracted enough people to fill the room allocated and vigorous discussions were kept going through the whole meeting.

A.H. Batten and his WG once again expressed their appreciation of the work done by Laura Pasinetti in editing the "Standard Stars Newsletter". She has agreed to continue ; please support her by providing contributions.

A.H. Batten and L. Pasinetti recommended to the WG members to draw attention to areas in which standard stars are needed or should be further studied.

The following scientific reports were presented.

S.J. Adelman : The abundances of normal late B and early A type stars.
J. Andersen : The IAU radial velocity standard stars : Status 1988
M. Gerbaldi : The *v sini* of AO dwarf Stars
W. Gliese : The third edition of the Catalogue of nearby stars
T. Neckel : The solar Analogs
G. Cayrel de Strobel : Presentation of "A Catalogue of [Fe/H] determinations", 1988 edition.

At the end of the meeting a written note of P.C. Keenan was read by the chairman. This note informed that the final edition of "The revised MK standard for stars GO and later" was completed.

6.5 Joint Commission Meetings (JCM's)

Two JCM's have been sponsored by Comm. 29: JCM2 and JCM5.

JCM2 was organized jointly with Commissions: 10, 12, 29, 36, 44 and was chaired by R. FALCIANI and E. PRIEST, and centered on "Solar and Stellar Coronae". A full scientific account of this meeting will be found in the reports of Commissions 10 and 12.

JCM5 was organized jointly with Commissions: 35, 36, 37, 47 and was chaired by G. Cayrel de Strobel and centered on "Spectroscopy of individual stars in Globular Clusters and the early chemical evolution of the Galaxy". Instead of being a conventional conference with formal contributions it was organized as a scientific discussion with interventions from the floor.

Four panel discussions addressed the general subject of the spectroscopy of individual stars and the confrontation of theory and observation.

The program of the panel discussions was as follows:

I - The spectroscopic metallicities in Galactic and Magellanic globular clusters: individual chemical abundances in metal poor and metal rich clusters.

Panel leaders: M.S. Bessell, V. Castellani, R. Gratton, C. Pilachowski, M. Spite, F. Spite.

II - Metallicity of globular clusters versus metallicity of field halo stars. Do cluster and field halo stars have a common history?

Panel leaders: M.S. Bessell, J.B. Laird, D.W. Latham, R.C. Peterson, N.B. Suntzeff, J.W. Truran.

III - The chemical inhomogeneities within globular clusters: example ω Centauri.

Panel leaders: G.S. Da Costa, K.C. Freeman, R. Kraft, V.V. Smith, M. Spite.

IV - Is there an age spread among the galactic clusters?

Panel leaders: R.A. Bell, B.W. Carney, P. Demarque, R.J. Dickens.

The proceedings of JCM5 are being edited by G. Cayrel de Strobel and M. Spite and will be published by the Printing Office of Paris Observatory.

COMMISSION No. 30

RADIAL VELOCITIES (VITESSES RADIALES)

Report of Meetings 1-10 August 1988

PRESIDENT: J. Andersen SECRETARY: B. Nordström

I. Business Meetings

Two business meetings were held, on August 1 (session 1) and August 9 (session 4). The following items were on the agenda:

I.1. MEMBERSHIP

The Commission voted to welcome the following new members of Commission 30, noting with particular satisfaction the improved representation of extragalactic research:

L.A.N. da Costa (Brazil)
M. Davis (USA)
A.P. Fairall (South Africa)
F.C. Fekel (USA)
C.B. Foltz (USA)
K.C. Freeman (Australia)
G.F. Gilmore (UK)
R. Giovanelli (USA)
L. Gouguenheim (France)
J.-L. Halbwachs (France)
P. Hewett (UK)
R.W. Hilditch (UK)
J. P. Huchra (USA)
H. Levato (Argentina)
B.M. Lewis (USA)
H. Lindgren (Sweden/ESO)
L.A. Marschall (USA)
R.D. Mathieu (USA)
T. Mazeh (Israel)
R.S. McMillan (USA)
J. Melnick (Chile/ESO)
G. Meylan (Switzerland/ESO)
R.C. Peterson (USA)
H. Quintana (Chile)
K. Ratnatunga (Sri Lanka)
R.P. Stefanik (USA)
G. Wegner (USA)

C.T. Bolton (Canada) has resigned in order to respect the three-Commission limit.

I.2. COMMISSION OFFICERS

The Commission voted to approve the following slate of Commission Officers for 1988-1991:

President: D.W. Latham Vice-President: G. Burki
Organizing Committee: J. Andersen, B. Campbell, L.A.N. da Costa, A. Florsch, K.C. Freeman, R.D. McClure, and L. Prévot.

I.3. COMMISSION NEWSLETTER

At its meeting in New Delhi in 1985, the Commission accepted with thanks the offer by the incoming OC member A. Florsch to edit and publish a Commission Newsletter to inform Commission members and others interested about programmes in progress and hence to encourage coordination and collaboration where desirable. Two issues of the Newsletter have appeared in the triennium. The Editor, A. Florsch, reported that, while there is considerable interest in receiving the Newsletter, it is not matched by a corresponding willingness to contribute to it; it appears impossible to get a sufficient number of articles for the Newsletter to continue as foreseen. The Commission, therefore, resolved to discontinue the Newsletter and expressed its warm appreciation to Dr. Florsch for his efforts as Editor of the Newsletter, and to Observatoire de Strasbourg for the generous financial support of its publication.

I.4. WORKING GROUP (WG) ON RADIAL VELOCITY STANDARD STARS

As Chairman of the WG, J. Andersen presented a report on its activities, which included the conclusions of discussions of the WG held before the second business meeting. The report, which is reproduced below as Section III, concluded that while significant progress has been made in the past three years, it is still too early to propose a new set of IAU standard stars, and that work should continue at least until 1991.

The Commission approved the report and recommendations of the WG and expressed its thanks to the members, especially M. Mayor, R.D. McClure, and R.P. Stefanik, who have provided the largest amount of material on the standard stars in 1985-1988. The Commission reappointed the WG for the three-year period 1988-1991, with the following composition: J. Andersen (Chair), G. Burki, B. Campbell, D.W. Latham, M. Mayor, R.D. McClure, and R.P. Stefanik.

II. Scientific Meetings

Commission 30 held a total of six formal scientific sessions. Most of these were organized in collaboration with other Commissions, reflecting the impact of modern radial-velocity data on many fields of astronomy. The programmes of these meetings are summarized below or in the reports of the collaborating Commissions, as indicated. Moreover, members of Commission 30 made significant contributions to Joint Discussion II, "Formation and Evolution of Stars in Binary Systems"; these papers will be published in *Highlights of Astronomy, Vol.* 8 (1989, in press).

In addition, the following scientific contributions were presented at the end of the business meetings:

M. Barbier: Les Catalogues de Vitesses Radiales à l'Observatoire de Marseille (Poster).

D. Dravins: Spectral Line Asymmetries and Radial-Velocity Zero-Points.

F.C. Fekel: Radial-Velocity Observations of Close Visual Binary and Multiple Systems.

G. Isaak: Application of Optical Resonance Spectroscopy to High Precision Radial Velocity Measurements of the Sun and of Bright Stars: A Tribute to R.W. Wood of the Johns Hopkins University.

M. Mayor: Highlights of Recent Results from programmes with the CORAVELs.

R.P. Stefanik: Radial-Velocity Observing Techniques and Zero-Points with the CfA Echelle System at Oak Ridge Observatory.

II.1. RADIAL VELOCITIES OF HIGH PRECISION: STATUS AND VISTAS.

The meeting was organized and chaired by B. Campbell. It took place on August 4 (session 2), and attendance taxed the capacity of the room. The purpose of the meeting was to summarize the current status of the various operational techniques for precise radial-velocity determinations, point out the main features of what has been learned so far, and discuss directions for the future. Discussion was very lively and showed that several systems now in routine operation are reaching precisions of the order of ± 10 m s^{-1} and - equally importantly - give consistent results between the groups. It also became clear that all K giants observed, with Arcturus as the prime example, are variable at the ± 100-200 m s^{-1} level and with a variety of periods ranging from several hours to several months. Hence, the question of the general suitability of K giants as radial-velocity standard stars to the level of precision required in the future needs serious reconsideration.

The following scientific contributions were presented:

Alan W. Irwin: Long Period Radial Velocity Variations of Arcturus.

W.D. Cochran and A.P. Hatzes: The McDonald Observatory High Precision Radial Velocity Survey.

Robert S. McMillan: Accurate Radial Velocity Studies of K Giants and Solar-Type Stars.

Geoffrey W. Marcy: Precise Radial Velocities with an Absorption Cell and an Echelle Spectrograph.

Myron A. Smith: Precision Radial Velocities in Arcturus: One More Round.

Bruce Campbell: Precision Radial Velocities: And Now For Some Astrophysics.

II.2. GALAXY REDSHIFT DETERMINATIONS: BETTER TECHNIQUES, BETTER STANDARDS.

This joint meeting between Commissions 28 (*Galaxies*) and 30 took place on August 8 (sessions 3 and 4). It was also the scientific meeting of the "Working Group on Redshift Determinations" of Commission 28 and was organized by the Chairman of the WG, J.P. Huchra, on behalf of both Commissions. J.P. Huchra also served as Chairman of the session. A main purpose of the meeting was to discuss the accuracy of current techniques for redshift determinations and the consistency of the zero-points between determinations in the radio and optical regions, and in the latter case, between emission- and absorption-line redshifts. The possible adoption of a set of standard redshift galaxies, in analogy with the well-known radial-velocity standard stars utilized in stellar work, was considered in some detail. The full programme for the meeting will appear in the report of IAU Commission 28 in this volume.

II.3. KINEMATICS OF GALACTIC POPULATIONS.

This joint meeting between Commissions 33 (*Structure and Dynamics of the Galactic System*) and 30 was held on August 9 (sessions 1 and 2). It was organized on behalf of both Commissions by J. Andersen, in collaboration with B.W. Carney, K.C. Freeman, and G.F. Gilmore; J. Andersen also served as Chairman of the meeting. Its purpose was to review the intense ongoing activity in the field, try to identify areas of consensus or controversy, and hence to point to directions for the future. Although several colleagues were, regretfully, unable to attend, the programme was nevertheless a very full one. The impressive contributions of new and exciting work by several young astronomers were especially gratifying, and the large attendance and very lively discussion gave a vivid impression of a field which is indeed dynamic in more than one sense.

The scientific programme was the following:

Session 1: The Galactic Halo and Bulge.

B.W. Carney: Introduction.

T. Armandroff: Kinematics of Disk and Halo Populations of Globular Clusters and Comparison with Field Stars.

R.M. Rich: Abundances and Kinematics of K Giants in the Nuclear Bulge of our Galaxy.

A. Spaenhauer, B.F. Jones, A. Whitford, and D.M. Terndrup: Kinematics of Bulge Giants from Proper Motions.

K. Yoss: Early G Giants in the Galactic Halo.

B.W. Carney and D.W. Latham: Results from the Proper-Motion Survey.

K. Ratnatunga: Kinematic Modeling of the Stellar Components.

J. Norris and K.C. Freeman: Rotation vs. [Fe/H] for Halo Stars.

R.C. Peterson and D.W. Latham: Pal 15 and the Mass of the Outer Galaxy.

Session 2: The Thin and Thick Disks.

R. Wielen: Introduction: On Thin Disks and Flat Halos in Galaxies.

L. Blitz: Rotation of the Outer Disk.

J. Lewis: Kinematics and Chemical Properties of the Old Disk of the Galaxy.

K.C. Freeman: Kinematics of the Thick Disk and Halo.

J. Norris and E.M. Green: The Transition from Halo to Thin Disk.

C. Flynn: Kinematics in a Galactic Rotation Field.

E. Friel: Kinematics of Old Open Clusters and the Old Disk Field.

J. Laird, D.W. Latham and B.W. Carney: Evidence for a Thick Disk Population in a Proper Motion Sample.

J. Andersen and B. Nordström: Progress Report on a Kinematically Unbiased Sample of Dwarf F Stars.

II.4. THE BAADE-WESSELINK METHOD: RECENT ACHIEVEMENTS AND FUTURE GOALS.

The meeting was organized on behalf of Commissions 27 (*Variable Stars*) and 30 by G. Burki (chair), T.G. Barnes III, and B.W. Carney. It was held on August 9 (session 3), chaired by G. Burki. Its purpose was to review the recent upsurge in both the quantity and quality of B-W analyses for RR Lyraes, Cepheids, and other types of pulsating stars. This is due primarily to the systematic application of modern techniques for precise radial-velocity determinations for faint stars, but also to improvements in the theoretical tools for the analysis. In order to highlight the status of the latter, the organizers had made efforts to have several different groups analyze the same set of observations for a given star (W Sgr). Discussion in the standing-room-only audience was quite vigorous and no doubt boded well for the success of IAU Colloquium No. 111 ("The Use of Pulsating Stars in Fundamental Problems of Astronomy"), which took place the week after the General Assembly. The full programme of the meeting will be published in the report of IAU Commission 27 in this volume.

II.5. SCIENTIFIC REPORT OF THE IAU WORKING GROUP ON STANDARD STARS.

The Working Group on Standard Stars is established by Commission 45 (*Stellar Classification*), with Commissions 29 (*Stellar Spectra*) and 30 as co-sponsors (not to be confused with the Working Group on <u>Radial-Velocity</u> Standard Stars of Commission 30 itself). Its meeting, on August 10 (sessions 1 and 2), was organized and chaired by the Chairman of the WG, A.H. Batten.

On behalf of Commission 30, J. Andersen gave a summary of the work on radial-velocity standard stars which is reported in Section III below. The full programme of the meeting appears in the report of IAU Commission 45 elsewhere in this volume.

II.6. PROGRESS IN THE UNDERSTANDING OF THE DYNAMICS OF STAR CLUSTERS.

This joint meeting of Commissions 37 (*Star Clusters and Associations*) and 30 was organized on behalf of both Commissions by C.P. Pryor, and took place on August 10 (sessions 3 and 4). Its purpose was to promote cooperation between theorists and observers, formulated in the questions: "What observations are needed to check recent theoretical results?", and "What theoretical work is needed to interpret the new observations?". In order to allow an in-depth discussion, the subject was deliberately limited to <u>globular</u> clusters only, which have also seen the most intense recent theoretical and observational activity. In the same spirit, the programme was concentrated on a few, thorough reviews with ample time for discussion. The meeting attracted a large and interested audience, with front-line research well represented on both sides of the podium, and the lively debate showed that the stated goal of the meeting had been attained very successfully.

The scientific programme of the meeting was the following:

<u>Session 1: Theory</u> (Chair: C.P. Pryor).

D.C. Heggie: Core Collapse and Equipartition.
J.P. Ostriker: Binary Stars and the Evolution of Globular Clusters.
S.M. Fall: Rotation of Globular Clusters.

<u>Session 2: Observations and General Discussion</u> (Chair: I.R. King).

C.P. Pryor: Mass Functions, Velocity Anisotropy, and Binary Stars in Globular Clusters.
G. Meylan: Rotation of Globular Clusters.
P. Seitzer: Radial Velocity Observations of Cusp Clusters.
General Discussion.

<u>III. Report of the Working Group on Radial-Velocity Standard Stars</u>

SUMMARY

Recently, low-amplitude (~500 m s^{-1}) orbital motions have been shown to exist among solar-type dwarfs, presumably due to low-mass companions. Also, more irregular variations of somewhat smaller amplitude appear to be ubiquitous among late-type giants. As a consequence, no new set of primary radial-velocity standard stars can be proposed at this time, but observations will continue with the aim of reaching definite recommendations in 1991. About 300 observations of asteroids indicate that the absolute zero-points of the three main systems (CORAVEL, CfA, and DAO), and hence of the new standard system, will eventually be established to ±100 m s^{-1} or so. Recommendations for stars to be monitored 1988-1991 are made (Table I).

III.1. INTRODUCTION

By 1984, it had become apparent that the system of radial-velocity standard stars adopted by the IAU in 1955 was rapidly becoming obsolete and would be unable to fulfill its intended purpose in a future dominated by new instruments capable of yielding accuracies in the range 100-500 m s^{-1}, even on very faint stars. The issue was discussed during IAU Colloquium No. 88 (*Stellar Radial Velocities*, Ed. A.G.D. Philip and D.W. Latham, L. Davis Press, Schenectady, N.Y., 1985; referred to below as C88), see in particular the contributions by Batten (p. 325) and Mayor and Maurice (p. 299). It was agreed that a new system of standard stars was needed, in which both the velocities of individual stars and the absolute zero-point of the system were known to about ±100 m s^{-1}. At its meeting in New Delhi in 1985, Commission 30 appointed a Working Group (WG) with the task of examining the existing IAU system of radial-velocity standard stars, organizing such observations as were deemed necessary to select candidates for a new system, and presenting a preliminary list of stars to the Commission at its meeting in Baltimore in 1988 (*Trans. IAU. Vol.* XIXB, p. 237, 1986). The following have served on the WG in 1985-1988: J. Andersen (Chair), W.I. Beavers, B. Campbell, D.W. Latham, M. Mayor, R.D. McClure, and R.P Stefanik.

An account of the strategy for the selection of new radial-velocity standard stars and of the observational work carried out until November, 1987, is given in the triennial report of Commission 30 for the years 1984-1987 (*Trans. IAU. Vol.* XXA, p. 362, 1988) and will not be repeated here. The following report summarizes results and developments over the past year, the results obtained in discussions within the WG in Baltimore, and the recommendations of the WG for the future directions of the work.

III.2. INDIVIDUAL STANDARD STARS

Already Batten et al. (Publ. Dominion Astrophys. Obs. 16, p. 143, 1983) and Mayor and Maurice (C88, p. 299) identified a number of current IAU standard stars with definitely variable velocities and/or with mean velocities differing appreciably from the nominal values. In order to select future standard stars with the longest possible history of precise observations without detected variability, it was decided to base the search on those 25 stars listed by Mayor and Maurice (their Table VI) which had $|\delta|<20°$, so that all future standards would be observable from both hemispheres. These stars are listed in Table I below, and intensive observations of these stars during the period 1985-1988 were recommended. Many observations have been made, so that the material available for discussion now includes about 1000 observations of a smaller number (~12) of standards with the Center for Astrophysics (CfA) echelle system at Oak Ridge Observatory (Latham, C88, p. 21), about 2500 observations of a much larger sample (including stars south of those in Table I) from CORAVEL II at ESO in Chile (Mayor, C88, p. 35), and about 600 observations of ~35 stars from Victoria, mostly from the DAO scanner (McClure et al., C88, p. 49), but also some 2.4 Å mm^{-1} photographic observations by A.H. Batten.

There is not yet a large and homogeneous overlap of well-observed stars between these lists, but there are clear suspicions that some of them are variable at the level of ~500 m s^{-1} and with periods from a few months to years. The confirmed cases of the giant HR 152 (McClure et al., Publ. Astron. Soc. Pacific 97, p. 740, 1985) and the dwarf HD 114762 (Latham, paper at Joint Discussion II, *Highlights of Astronomy, Vol.* 8, 1989, in press), with a period of only 84 days and a 500 m s^{-1} amplitude are probably just the tip of an iceberg of yet unknown dimensions.

Continued close scrutiny is clearly needed before any of the stars can be declared constant at a level of ±100 m s^{-1}. Concerning the giant stars - the majority of the present standards (!) - the available data of high precision (±10 m s^{-1}) indicate that most or all late-type giants vary by ±100 m s^{-1} or more with a range of periods (see II.1 above) and may not at all, therefore, be suitable as standards at the desired level of precision. More dwarf candidate stars are evidently desirable, and several exist which have now been monitored for about a decade with the requisite precision. These will be added to the list in Table I and monitored over the next three years with the best possible precision. It would clearly be premature to make definite selections for a new system of radial-velocity standard stars at the present time.

III.3. ZERO-POINTS

One conclusion by Mayor and Maurice (C88, p. 299) was that the zero-points of the previous lists of "bright" ($V < 4.3$) and "faint" ($V > 4.3$) IAU standards differed by about 800 m s^{-1}. A key element in the programme of the WG was the observation of asteroids, which present starlike images, and the velocities of which can be computed to ±10 m s^{-1} or better. About 200 such observations have been made at CfA, about 35 with CORAVEL, and a similar number at DAO. In addition, sky exposures are taken frequently at both CfA and DAO. These data show that the CfA and DAO instrumental zero-points are within about 100 m s^{-1} of that defined by the asteroids, while a preliminary correction of about +400 m s^{-1} is indicated for CORAVEL. This again indicates that the absolute (asteroid) zero-point is roughly midway between those of the "bright" and "faint" IAU lists. Comparison of the observations of the standard stars (dwarfs and giants) after application of these zero-point corrections would then be expected to show agreement between all three systems.

In fact, excellent agreement is found between CORAVEL and DAO, but a significant mean difference of about 400 m s^{-1} surprisingly appears between those and the CfA velocities. Closer examination of the data shows that agreement is, in fact, obtained for the (only four) <u>dwarfs</u> in common, while the discrepancy for the <u>giants</u> alone is some 500 m s^{-1}, larger for later spectral types and/or higher luminosity. Hence, the asteroid zero-point correction is satisfactory for the (solar-type) dwarf stars, but not for the (later-type) giants. This is most likely due to a mismatch between the line spectra of the giants and that of the solar (or sky) template spectrum used in the cross-correlation, an effect which cancels out much less accurately in the 50-Å spectral range of the CfA echelle than in the 1000-1500 Å ranges covered by the two other instruments. Further studies by means of synthetic solar-type and giant spectra in the relevant range, and by observations of objects (clusters, visual binaries) containing both types of star, will be necessary in order to demonstrate the validity of this hypothesis. In order to help consolidate the comparisons between the major systems, several stars with long observational histories in at least two of these systems have been added to the list of stars to be monitored carefully over the next three years (Table I). However, prospects appear good that, eventually, the three independent, zero-points can be shown to be consistent to well within the desired ±100 m s^{-1}. The material on asteroids already appears to be about adequate to establish an absolute zero-point for the common system to that precision.

III.4. HIGH-PRECISION STANDARDS OF RELATIVE RADIAL VELOCITY

The only instrument for high-precision radial-velocity observations which has been in routine production for several years is that of the Canadian group (Campbell et al., Astrophys. J. 331, 902, 1988). While their results indicate that some <u>dwarf</u> stars may have constant radial velocities to within their precision of ±15 m s^{-1}, independent confirmation by at least one other group would be required before a set of standards of <u>relative</u> radial velocity could be proposed. An even longer span of observations of each candidate star is also desirable. As several other groups have now reached the stage of routine production of radial velocities of this kind of precision, the situation may be better in three years.

III.5. SECONDARY STANDARDS

The WG has considered the possible need for secondary standards, especially for work on faint stars with large telescopes and with detectors of limited dynamic range. It was agreed that, once a satisfactory system of primary radial-velocity standard stars has been established, secondary or regional standards can be easily established in each case, close to the faint-star targets, with one of the existing systems. Therefore, the WG does not at this time recommend that the IAU establish a system of secondary standards.

III.6. EARLY-TYPE STANDARDS

No satisfactory set of radial-velocity standard stars exists for early-type stars; the few existing candidates are inadequate with respect to both the evidence on their constancy, the accuracy of the velocity zero-point, and the coverage of spectral types (O-B-A), luminosity classes, and - especially - rotations. Better techniques for precise radial-velocity determination for such stars are needed and are being developed; once the necessary <u>precision</u> has been demonstrated, what will be needed is a set of early-type stars (visual binaries, clusters) with <u>known</u> and <u>constant</u> velocities. The selection of such stars, and the observation of solar-type calibration objects, can begin already now and are strongly encouraged.

III.7. CONCLUSIONS AND RECOMMENDATIONS

The main conclusions and recommendations of the WG may be summarized as follows:

<u>1.</u> As more time is needed to establish the constancy (or eventually determine the orbits!) of the candidates for new <u>primary standards</u>, no definitive list can be proposed at this time. In order to avoid confusion caused by intermediate systems, improved, but still preliminary velocity data will also not be given; the previous IAU system remains in force. As an interim measure, it will be proposed to the *Astronomical Almanac* that those standard stars found to vary by several km s^{-1} be omitted from future editions of the *Almanac*.

<u>2.</u> Meanwhile, observations should continue, with the highest possible accuracy, of the stars in Table I below, and of additional late-type dwarfs to be selected shortly by the WG. These should continue until 1991, with a view to establishing a definitive set of new primary standards, accurate to ±100 m s^{-1}, by then. Any observers able and willing to contribute radial-velocity observations of precision in the range 100-500 m s^{-1} are urged to make such observations of the stars in Table I and contact the Chairman or any other member of the WG so that these data may be incorporated in the new system.

<u>3.</u> A few more minor planet observations are needed, mostly to base the zero-point on a somewhat larger number of asteroids. It appears probable that the absolute zero-point of the new standard velocity system can be secured to $\leq \pm 100$ m s^{-1}.

<u>4.</u> Establishing a satsifactory system of standard stars of early spectral types (O-B-A) will take several years of effort. However, preparatory work such as development of observational techniques, and selection and observation of suitable calibration objects, can be initiated now and is strongly encouraged.

TABLE I. List of future primary standard star candidates, and other important reference stars, to be intensively observed in 1988-1991. RV_{IAU} is given in km s^{-1}.

HD No.	Name	α (2000.0)	δ	V	Sp. type	RV_{IAU}
IAU primary standard star candidates:						
693	6 Cet	$00^h 11^m 15^s.8$	-15°28'05"	4.89	F6 V	+14.7
4128	β Cet	00 43 35.3	-17 59 12	2.04	K0 III	+13.1
8779		01 26 27.2	-00 23 55	6.41	K0 IV	-5.0
18884	α Cet	03 02 16.7	+04 05 23	2.53	M1.5 III	-25.8
22484	10 Tau	03 36 52.3	+00 24 06	4.28	F9 V	+27.9
26162	43 Tau	04 09 09.9	+19 36 33	5.50	K2 III	+23.9
29139	α Tau	04 35 55.2	+16 30 33	0.85	K5 III	+54.1
36079	β Lep	05 28 14.7	-20 45 34	2.84	G5 II	-13.5
66141		08 02 15.9	+02 20 04	4.39	K2 III	+70.9
81797	α Hya	09 27 35.2	-08 39 31	1.98	K3 II-III	-4.4
89449	40 Leo	10 19 44.1	+19 28 15	4.79	F6 IV	+6.5
92588	33 Sex	10 41 24.1	-01 44 29	6.26	K1 IV	+42.8
107328	16 Vir	12 20 20.9	+03 18 45	4.96	K0 III	+35.7
114762		13 12 20.5	+17 31 01	7.31	F7 V	+49.9
124897	α Boo	14 15 39.6	+19 10 57	-0.04	K1 III	-5.3
136202	5 Ser	15 19 18.7	+01 45 55	5.06	F8 IV-V	+53.5
146051	δ Oph	16 14 20.6	-03 41 40	2.74	M0.5 III	-19.8
161096	β Oph	17 43 28.3	+04 34 02	2.77	K2 III	-12.0
171391		18 35 02.3	-10 58 38	5.14	G8 III	+6.9
182572	31 Aql	19 24 58.1	+11 56 40	5.16	G8 IV	-100.5
187691	o Aql	19 51 01.5	+10 24 56	5.11	F8 V	+0.1
203638	33 Cap	21 24 09.6	-20 51 07	5.77	K0 III	+21.9
204867	β Aqr	21 31 33.4	-05 34 16	2.91	G0 Ib	+6.7
212943	35 Peg	22 27 51.5	+04 41 44	4.79	K0 III	+54.3
213014		22 28 11.4	+17 15 48	7.70	G8 III	-39.7
Other reference stars:						
28099		04 26 39.7	+16 44 50	8.10	G0 V	+39.6
54716	63 Aur	07 11 39.3	+39 19 14	4.90	K4 III	-27.1
	M67-978	08 51 17.4	+11 45 24	9.72	K4: III	+34.7
103095		11 52 58.7	+37 43 08	6.45	G8 Vp	-99.1
113996	41 Com	13 07 10.6	+27 37 29	4.80	K5 III	-14.7
176670	λ Lyr	19 00 00.8	+32 08 44	1.47	K2.5 III	-16.9

COMMISSION No. 31

TIME (L'HEURE)

Report of Meetings: 5, 9, and 10 August 1988

PRESIDENT: D. D. McCarthy SECRETARY: H. F. Fliegel

5 August 1988

REPORTS AND ADMINISTRATIVE MATTERS:

The Report of the Commission President was presented by D. McCarthy. For lack of space, a list of references to journal articles such as published in previous Reports was omitted, but reports were included by B. Guinot on the work of the Bureau International de l'Heure and Bureau International des Poids et Mesures (BIH/ BIPM), by Y. R. Miao concerning six institutes in China, by G. Hemmleb on the Zentral Institut fuer Physik der Erde (ZIPE), by S. Starker on the Deutsche Forshungs- und Versuchsanstalt fuer Luft- und Raumfahrt (DFVLR), by H. Enslin on the Deutsches Hydrographisches Institut (DHI), by E. Proverbio on the Time Service of the Cagliari Observatory, by P. Galliano on the Istituto Elletrotecnico Nazionale (IEN), by C. Kakuta on the International Latitude Service (ILS) Observatory at Mizusawa, by Yo Kubo on the Hydrographic Department of Japan, by S. Aoki on the Tokyo Astronomical Observatory, by J. Dickey on the Jet Propulsion Laboratory (JPL), and by W. Klepczynski on the US Naval Observatory. D. McCarthy noted that a questionnaire had been mailed concerning members' activities in the field of time, and perceived needs. On the basis of this questionnaire, the list of Commission members has been revised, and two working groups have been formed: "The Use of Millisecond Pulsars and Timing of Pulsars" (chair: D. Allan); and "Time Transfer with Modern Techniques" (chair: H. Fliegel).

Draft Resolutions I thru V were discussed and amended.

SCIENTIFIC PRESENTATIONS:

A session on "The Use of Millisecond Pulsars and Timing of Pulsars" was chaired by D. Allan.

D. Backer (University of California at Berkeley) reviewed the contributions of pulsar studies to the determination of time. There are direct applications to astrometry, solar system dynamics, tests of general relativity, and also to knowledge of the interstellar medium, of matter at high energy, and of the collapse of massive stars. All these disciplines assist in the use of pulsars to form a stable time standard valid over months and years. There are 8 pulsars known with periods 11 milliseconds or less, and five radio observatories with active programs to study them: at Arecibo, Green Bank, Jodrell Bank, Nancay, and Parkes. J. Taylor (Princeton University) discussed the observability of effects of general relativity on pulsars. Three effects should in principle be observable: the Shapiro time delay due to ray bending within the solar system, advance of perihelion and change of period in binary pulsars, and gravity wave events. B. Guinot (BIPM) presented the status of International Atomic Time (TAI), against which pulsars are measured. Errors in TAI frequency affect determinations of pulsar periods, frequency drift affects determination of period rate of change, and annual effects produce an incorrect Earth reference orbit. The annual variation in laboratory time is about 3 parts in 10 E14, and is at least partly an effect of varying humidity. In the formation of TAI from about 170 participating clocks, such systematic effects are largely removed by use of an averaging algorithm, forming TAI by applying steering to the free atomic time scale EAL. E. M. Standish (JPL) analyzed the errors in the determination of the Earth ephemeris, and showed that they cannot account for the reported annual variations in determinations of pulsar periods. The session was summarized by D. Allan, who noted that the limit to the accuracy of reductions of pulsar data is clearly set by the international time scale; and he suggested that stored Hg- ion devices may extend this limit for time scales of 10 E6 seconds and greater.

9 August 1988

SCIENTIFIC PRESENTATIONS:

A session on "The Relativistic Aspects of Time" was chaired by P. Paquet.

K. Seidelmann summarized the activity of the working group on reference systems, which was tasked to define Barycentric Dynamical Time (TDB) and Terrestrial Dynamical Time (TDT) and their relationship to TAI. Specific problems were identified: the specification of the units of TDB and TDT; the use of the word "dynamical"; and the question whether TAI is "coordinate" or "proper" time. Seidelmann presented the following recommendations. Let the word "dynamical" be dropped, and replace the existing terms by TT, Terrestrial Time, defined in terms of the SI second on the rotating geoid, and TB, [Solar] Barycentric Time, defined as TT plus periodic terms only. A different proposal was made by V. Brumberg, who recommended adopting three distinct reference systems: BRS, GRS, and TRS -- Barycentric, Geocentric, and Topocentric Reference Systems -- related to one another by rigorous relativistic transformations between

JOINT MEETING: COMMISSIONS 19 AND 31:

A joint meeting for scientific presentations was shared between Commissions 19 and 31. V. Zharov presented the design of an experiment to use a laser gyroscope to determine Earth rotation. Z. Shi reviewed the first use of GPS common view time transfer and synchronization in China. G. Winkler showed the design of the prototype Hg- ion frequency standard built by Hewlett- Packard and in initial operation at the US Naval Observatory. B. Xu discussed the concepts underlying the definitions of UT1, reviewing the proposals of Aoki et al. (1982), Aoki and Kinoshita (1983), and Aoki (1987). A paper by P. Silva on timekeeping in Brazil was read by D. McCarthy.

10 August 1988

SCIENTIFIC PRESENTATIONS:

A session on "Modern Techniques of Time Transfer" was chaired by H. Fliegel.

B. Guinot detailed the use of the GPS to form TAI, and the elimination of systematic errors from the procedure. The common view schedules developed by Allan at the US National Bureau of Standards (NBS) -- renamed the National Institute of Standards and Technology (NIST) -- are employed, in which the same sequence of Navstar observations are maintained for six months or more. The BIPM takes data for 10 days, and forms normal points centered on MJD's ending in the digit 9. Errors in station coordinates are eliminated using long spans of data, including observations not in the common view schedules. The outstanding error sources still remaining are due to multipath and the ionosphere. J. McK. Luck reported on the use of TV from the Australian National Communications Satellite System (AUSSAT), taking quasi- simultaneous GPS and AUSSAT measurements at several stations to correct the AUSSAT orbit differentially. W. Klepczynski presented a report on the Laser Ranging from Stationary Orbit experiment (LASSO), which is managed thru the LASSO Operations Coordinating Group (LOCG). Since the METEOSAT - P2 satellite on which the experiment is hosted rotates at about 90 rpm, initial efforts have been focussed on scheduling ranges when the event timers are in Earth station view, and on data reduction. The system is reported to be almost operational. D. Allan described the Civil GPS User Steering Committee currently chaired by the US Department of Transportation (DOT). H. Fliegel concluded the session by leading an informal discussion on work yet to be done to make GPS more useful. It was reported that the BIPM will lead an international program to produce accurate, uniform values for antenna geographic coordinates and receiver delays.

the systems, including non- periodic changes of scale. N. Ashby (University of Colorado) showed what such transformations would entail. B. Yu recommended adopting a conventional nonrotating geocentric coordinate system as the basis for TAI. Turning to the subject of basic theory, H. P. K. Yilmaz sketched his alternative formulation of the Einstein field equations, recommending it as more responsive to the problems of celestial mechanics than the standard formulation. C. Alley reviewed the status of the University of Maryland experiment to measure the one- way speed of light, using endpoints at the Goddard Optical Research Facility and at the US Naval Observatory. M. K. Fujimoto and S. Aoki recommended adopting for the astronomical constant a value such that the speed of light will be 173.1446333 AU / day.

REPORTS:

B. Guinot presented the Report from the BIPM. He noted progress in the algorithms used to form TAI, and in time comparisons, especially by use of the Global Positioning System (GPS). G. Winkler announced the forthcoming meeting of the Comite Consultatif pour la Definition de la Seconde (CCDS), scheduled for 17- 18 April 1989 in Paris. The meeting will have three objectives: "to identify the problems arising in the daily operation of international coordination for time; to exchange ideas for possible improvements; [and] to reach a consensus on matters which require better coordination". J. Kovalevsky presented Resolutions 3 thru 5 which were adopted at the 18th assembly of the General Conference on Weights and Measures (CGPM).

The members approved the following nominations for 1988 - 1991:

President: P. Paquet Vice President: E. Proverbio

It was voted to include the chairs of the two working groups (Pulsars, Time Transfer) as members of the Organizing Committee:

Organizing Committee: D. Allan, N. Blinov, H. Fliegel, M. Fujimoto, M.Granveaud, B. Guinot, W. Klepczynski, J. Kovalevsky, Y. Miao, I. Mueller, J. Pilkington, Y. Shu Hua.

Representatives: to CCDS and BIPM: G. Winkler;
to the Federation of Astronomical and Geophysical data analysis Services (FAGS): J. Kovalevsky;
to Study Group 7 of the International Radio Consultative Committee (CCIR), S. Leschiutta.

The members approved unanimously the proposed list of new members and consultants to the Commission.

COMMISSION No. 34

INTERSTELLAR MATTER (MATIERE INTERSTELLAIRE)

Report of Meeting, 4-10 August 1988

PRESIDENT : M.J.Lequeux
VICE-PRESIDENT : J.Mathis

Business session, 10 August

The President summarized the activity of the Commission between the General Assemblies and presented the report for the period 1985-88. The subject of Intergalactic Medium has been covered for the first time in the report. The President thanks the members of the Organizing Committee as well as two other persons, T.Landecker and Y.Terzian, for their participation in the writing of the report. Some discussion took place about the format and usefulness of the report. Various opinions were expressed , but the accents was over the need to emphasize more the most interesting recent developments.

Membership and Structure

Commission 34 is the second largest of the IAU with 532 members after the previous 1985 General Assembly, compared to 470 before. During this triennium about 140 persons have expressed their desire to join the Commission and are going to be accepted at the end of this General Assembly. 2 members deceased to the knowledge of the President (E.Dibay and M.T.Martel) and 10 resigned. Thus our Commission is increasing very fast, showing the growing interest in the astronomical community for interstellar matter.

A new working group is added to the two existing ones (Planetary Nebulae and Interstellar Medium Nomenclature) : the working group on Astrochemistry chaired by A.Dalgarno.

Organizing Committee

Based on recommendations that he received from members and past officers of the commission, the President proposed H.Habing as the new Vice President and M.Dopita, R.Genzel, N.Kaifu, C.Lada and L.Rodriguez as new members of the Organizing Committee. As has been the convention, the present Vice-President J.Mathis will take over as President at the end of the XXth General Assembly. The continuing OC members are K.de Boer, D.Flower, J.Lequeux, B.Shustov, D.York. S.D'Odorico, B.Elmegreen, M.Peimbert, P.Shaver and P.Wannier will retire from the OC at the end of the current General Assembly.

Symposia, Colloquia and Scientific Sessions

Commission 34 has sponsored or co-sponsored during the last 3 years the following IAU Symposia : No 122, Circumstellar Matter (Heidelberg, 23-27 June 1987); No 131, Planetary Nebulae (Mexico City, 5-9 October 1987); No 135, Interstellar Dust (Santa Clara, Calif., 26-30 July 1988) and No 136, The Galactic Center (Los Angeles, 25-29 July 1988). In addition, the Commission has co-sponsored during the General Assembly 5 Joint Discussion

(Formation and Evolution of Stars in Binary Systems ; Supernova 1987A in the Large Magellanic Cloud; The Cosmic Dust Connection in Interplanetary Space : Comets, Interstellar Dust, and Families of Minor Planets; Atomic and Molecular Data for Astrochemistry; Disks and Jets on Various Scales in the Universe) and one Joint Commission Meeting (Molecules in External Galaxies). There have been also meetings of the three working groups of the Commission, including a full-day meeting on Planetary Nebulae. In view of all this activity, there has been no other scientific meeting of Commission 34.

The future symposia and colloquia sponsored and co-sponsored by the Commission and already approved by the Executive Committee are : Colloquium on Structure and Dynamics of the Interstellar Medium (Granada, 17-21 April 1989); Symposium 139 on Galactic and Extragalactic Background Radiation (Heidelberg, 12-16 June 1989); Symposium 140 on Galactic and Extragalactic Magnetic Fields (Heidelberg, 19-23 June 1989); a Symposium on Extragalactic Molecular Clouds : Dynamics and Evolution of Galaxies (Paris, 4-9 June 1990); a Symposium on Fragmentation of Molecular Clouds and Star Formation (Grenoble, presumably 11-15 June 1990); a Symposium on the Interstellar Medium away from the Galactic Plane (Leiden, 18-23 June 1990). There might also be a Colloquium on Radio Recombination Lines to be held in USSR in 1989.

The President mentioned a few other meetings which might be of interest to members of the Commission, in particular in 1989 : From Diffuse Matter to Stars and Planets (Institut d'Astrophysique, Paris, and June or early July); Chemistry and Spectroscopy of Interstellar Molecules (PACIFICHEM, Honolulu, Dec. 17-22).

Working groups

The Working Group on Planetary Nebulae chaired by Y.Terzian organized a one-day meeting whose program is reported later. A new and extensive catalogue of planetary nebulae prepared by A.Acker (Strasbourg Observatory and Stellar Data Centre) is almost completed and should be available soon. The working group will propose another Symposium on Planetary Nebulae in August or September 1992, in Innsbrück (these Symposia occur at 5 years intervals).

The report of the Working Group on Interstellar Medium Nomenclature chaired by H.Dickel has been published by H.Dickel, M.C. Lortet and K.de Boer in Astronomy and Astrophysics Supplement Series,68, 75 (1987). The designation system conforms to the IAU resolutions (see Vol.XIX B of the Transactions of the IAU General Assembly, Resolutions C3 and C12, pp.40 and 49) and has been the basis of the advices given on matters of designation of diffuse objects by a "Clearing House" formed by H.R.Dickel, C.Jaschek, M.C.Lortet, J.M.Mead and W.H.Warren Jr.The Working Group will continue for the next three years as a resource group for the Clearing House and in general to assist IAU in resolving related problems. Its present members are T.Chester, K.de Boer, H.Dickel, J.Dickey, M.Felli, L.Higgs, L.Kohoutek, M.Kutner, M.C.Lortet, R.Manchester, J.Mead, J.Moran and N.Panagia.

The new Working Group on Astrochemistry is preparing a request for a Symposium on Astrochemistry in 1991 in South America, 6 years after the one in Goa.

Commission 34 Meeting

The proceedings of the Joint Discussions co-sponsored by Commission 34 will be published in Highlights of Astronomy, Vol.8. We give only here the programme of the other meetings.

Planetary Nebulae

B.Balick : Morphology and Structure of Planetary Nebulae
You-Hua Chu : Extended Haloes of Planetary Nebulae
M.Perinotto : Winds from Central Stars of Planetary Nebulae
P.Van der Veen,H.Habing,T.Geballe: Evolution of Proto-Planetary Nebulae
B.Murray Lewis : Radio Frequency Observations of OH/IR Stars
J.H.Cahn : The Distance Scale of Planetary Nebulae
M.Azzopardi , N.Meyssonnier, J.Lequeux : Planetary Nebulae in Nearby Galaxies
H.Nussbaumer : From Symbiotic Stars to Planetary Nebulae
W.J.Maciel : Galactic Vertical Gradients and Planetary Nebulae
D.C.V.Mallik, M.Peimbert : Filling Factor Determinations and their Effect of Planetary Nebulae
N.Rowlands,J.R.Houck,T.Herter,G.E.Gull,M.F.Strutskie: Electron Temperatures in the Hight Excitation Zones of Planetary Nebulae
D.C.V.Mallik, S.K.Jain, B.G.Anandarao, D.P.K.Banerjee: Internal Kinematics of NGC 2440
H.E.Bond, R.Ciardullo : CCD Photometric Monitoring of Planetary Nuclei
A.Acker, B.Stenholm, J.Köppen, G.Jasniewicz : First Results of the Spectrophotometric Survey of Planetary Nebulae
G.Jasniewicz, A.Acker : Periodic Variations of the Nucleus of Abell 35
D.Middlemass, R.E.S.Clegg : On Spectroscopic Detection of Planetary Nebulae Faint Haloes

Molecules in External Galaxies (Joint Commission Meeting with Commissions 28 and 40)

M.Guélin : CO(2-1) and (1-0) in M51 and edge-on galaxies
A.Hjalmarson : Molecular-cloud spiral arms in M51 and IC342, and results from tidal interaction modeling
K.Y.Lo : Interferometric maps in external galaxies
S.Vogel :Interferometric CO map of M51
N.Nakai : Molecular clouds in Maffei2
R.Kawabe : Nobeyama interferometer results on Maffei2
F.Combes :CO in NG4438 and tidal stripping in the Virgo Cluster
Y.Sofue : Molecular gas in the nuclear regions of spiral galaxies
P.Solomon : The molecular content of isolated and interacting spiral galaxies
F.Verter : Systematic properties of CO emission from galaxies
J.Young : The molecular gas content and star formation efficiency in nearby spiral galaxies
D.Adler, W.Roberts : Can galactic GMCs be identified from l-v

diagrams

R.Genzel, A.Eckart : Far-infrared and sub-mm spectra of extragalactic sources

A.Sargent : Recent CO(2-1) observations of galaxies from CSO

A.Stark : Molecules in Galaxies:results from Bell Laboratories

R.Wielebinski : M82: ^{12}CO and ^{13}CO(2-1) data

C.Henkel : Molecular clouds in dwarf irregulars

M.Rubio : Molecular clouds in the LMC and SMC

T.Wiklind : CO emission in early-type galaxies

F.Mirabel : The M(H_2) /M(HI) ratio and OH emission in luminous infrared galaxies

D.Sanders : Molecular gas in ultra luminous infrared galaxies and quasars

N.Scoville : Molecular gas in the nuclei of luminous IRAS galaxies

COMMISSION No. 35

STELLAR CONSTITUTION (CONSTITUTION DES ETOILES)

Report of Meetings: 5 and 6 August 1988

PRESIDENT: D. Sugimoto SECRETARY: A. Maeder and J. C. Wheeler

5 August 1988

I. BUSINESS MEETING

It was started by a moment of silence with the members standing in respect for a member, Max Krook, that had passed away since the last meeting in 1985.

New officers were voted for in an election held in 1987, and the results were reported by the President. The President, Vice-President and the Organizing Committee members for 1988-91 were endorsed during the General Assembly. The new President is Andre Maeder (Switzerland) and the new Vice-President is Pierre Demarque (USA). During the last triennium the number of Organizing Committee members exceeded the usual number by one, but it was corrected at this meeting. The new Organizing Committee consists of the six carryover people and the three new ones: D.O. Gough (UK), I. Iben, Jr. (USA), R. Kippenhahn (FRG), K. Nomoto (Japan), Y. Osaki (Japan), D. Sugimoto (Japan), J.W. Truran (USA), A.V. Tutukov (USSR), and J.C. Wheeler (USA).

From the membership list four names were deleted; one for the deceased, another for having left from astronomy, and the two others for resignation from Commission 35 because their interests are now in other topics of astronomy. Twenty new members were accepted of which five had been the members of the Union and fifteen were new members accepted at this General Assembly. The total membership is now 308. This implies that the interest is being kept at high level in this field of Stellar Constitution.

One of the members proposed that Commission 35 should recommend an astronomer from a country for a new membership of the Union. There was much discussion, because the National Committe of his country had just forgotten to observe the deadline to propse it to the Union, and because the proposed astronomer was as yet a Ph.D. candidate. However, considering the importance of getting into communication with countries having a small number of astronomers, Commission 35 accepted it and proposed him for the membership of the Union.

Four proposed IAU sponsored meetings were discussed. Their titles were 1) Physics of Classical Novae, 2) Inside the Sun, 3) Rotation and Angular Momentum Evolution of Low-Mass Stars, 4) The Sun and Cool Stars; activity, magnetism and dynamos, and 5) Evolution of Stars; the photospheric abundance connection. They were considered to be convened in 1989 or 1990. Some of them got enthusiastic support, which had already or have been recommended to the Exective Committee. Others got some comments concerning the scope of the topics which are now being taken into consideration by the proposers.

During the last triennium four circular letters were mailed to all members of Commission 35. One of the members expressed his opinion that communication among the members was nevertheless not enough, that the Commission should set up a research project to concentrate the coming triennium in specified topics, and that the Organizing Committee should meet somewhere to secure it at least once during the triennium. However, considering that Commission 35 is oriented more to theoretical than observational researches, it was stressed as more important to exchange free ideas than to concentrate in specified topics. In this relation use of electronic mail was strongly encouraged and the Astronomical E-Mail Directory, that was prepared and being expanded by Dr. Chris Benn at the Royal Greenwich Observatory, will be most helpful.

II. SCIENTIFIC MEETINGS

There were six scientific meetings of Commission 35; two concerned directly with its main topics of stellar structure and evolution, three as the Joint Commission Meetings with Commissions 10, 12, 29, 36, and 44 on Solar and stellar coronae, with 29, 36, 37, and 47 on Spectroscopy of individual stars in globular clusters and the early chemical evolution of the galaxy, and with 28 and 37 on Star clusters in the Magellanic Clouds, and the last one as a brainstorming session on Early type pulsating stars. Only the first two of these meetings are reported in what follows. The Joint Commission Meetings are reported elsewhere. The brainstorming session was organized by M. Aizenman but no report will be given here because of its informal nature.

Commission 35 also supported two Joint Discussions: Formation and evolution of stars in binary systems, and Supernova 1987A in the Large Magellanic Cloud. Their Proceedings will be found in Hilights of Astronomy.

5 August 1988

Final Stages of Stellar Evolution

Chaired by J. Craig Wheeler

J. Liebert Luminosity Function of White Dwarfs
E. Robinson Observational Constraints on Binary White Dwarfs
I. Iben, Jr. Theory of Binary White Dwarfs
Z. Barkat Carbon Ignition in White Dwarfs

With the attention SN 1987A has brought to massive stars, the topic of a special Joint Discussion, the decision was made to focus this Comission meeting on white dwarfs. The study of white dwarfs has been one of the most exciting topics in the field of stellar evolution in the last few years yielding new insights into binary evolution, supernovae, and cosmology. The program was chosen to reflect these developments.

Liebert of the University of Arizona discussed the determination and implication of the luminosity function of white dwarfs. White dwarfs are not observed at magnitude limits beyond which fainter main sequence dwarfs are still detected. This implies a real cutoff in the white dwarf luminosity function. Theoretical cooling curves suggest that white dwarfs should live longer than a Hubble time and hence the paucity of dim white dwarfs sets a limit to the age of the white dwarfs. This age, presumably the age of the Galactic disk is about 9 billion years, a very low number by most expectations. Liebert reported that improved parallaxes have moved some stars to lower L and reduced this figure to about 8 billion years. Comments from the audience suggested that re-examination of opacities, etc., have resulted in little change in independent cooling calculations, with the exception of the effects of crystallization if it comes in at just the right epoch. Liebert remarked that some white dwarfs may be missing, hidden by binary companions, by having small tangential velocities or by scale height inflection, reducing the local space density, but none of these seem adequate to shift the observed downturn in the luminosity function. Liebert finally pointed out that a small sample of white dwarfs with high proper motion signifying halo stars may represent a first step toward a luminosity function for Population II stars.

Robinson of the University of Texas summarized the situation regarding the incidence of binary white dwarfs. Naive estimates suggest that as many as 1 in 10 white dwarfs should be in a close binary system. This is an important question because such systems can be an important source of gravitational radiation background, supernova progenitors, and testing grounds for theories of binary evolution. Of 13 published white dwarf pairs (plus one or two added from the audience), 6 are visual binaries of large separation evolving as separate stars, 1 is astrometric, 2 are spectroscopic, still with periods too long to be subject to gravitational radiation, and only four are interacting. Among the latter, none are candidates for supernova explosions because of the low total mass. Several surveys have been conducted for binary white dwarfs. Spectral classification schemes have found four, but the technique is subject to strong selection effects, making it less systematic. Radial velocity surveys have turned up 1 candidate in 67 systems surveyed. For a local supernova rate of $(4 \pm 2.5) \times 10^{-4}$ yr^{-1} pc^{-3} the required density of binary white dwarfs with orbital period shorter than 3 hrs is $(1.6 \pm 1) \times 10^{-5}$ pc^{-3}. Robinson and Shafter set an observational limit on the total number of binary white dwarfs 3×10^{-5} pc^{-3} of which only about 1/10 could be supernova progenitors. Thus the derived upper limit is two to eight times less than the required rate for SN Ia progenitors.

There are several problems that need to be studied further. One is the rate of supernova which is still strongly subject to small number statistics. The work of Robinson and Shafter only specifically constrains systems born with orbital period in excess of 3 hours. If they are born at ultrashort periods, the current constraints may not apply. Finally Robinson cautioned that a selectively high proportion of high mass white dwarfs could undergo gravitational settling to become DC white dwarfs. The lack of lines in these systems would render them invisible to any radial velocity search technique. Thus it remains possible that SN Ia arise selectively from massive DC-DC white dwarf pairs.

Iben of the University of Illinois discussed current theoretical ideas on the origin and evolution of binary white dwarfs. He cited GP Com as an established He-He pair and HZ 29 as a possible CO-He pair. He also argued that the sdO and sdB stars could arise in the fusing of a CO and a He dwarf and hence be indirect evidence for binary white dwarf evolution. Related arguments suggest an origin of white dwarfs like 40 Eri B with only 0.43 solar masses which are unexplained by standard single star theory.

Iben has constructed linear series of helium shell burning models which could result from binary dwarf coalescence and identified a "super horizontal branch" where tracks converge in the H-R diagram. This feature intersects the location of R CrB stars and hot helium stars given a plausible adjustment in their absolute magnitudes. This may provide evidence that R CrB stars result from the merger of CO and He dwarfs.

Finally, Iben discussed problems involved in the formation and subsequent common envelope evolution of thermally pulsing AGB stars undergoing Case C mass exchange. He noted that the common envelope could lead to friction and the reduction of the separation of the two cores, but that ejection of the envelope by the released gravitational energy of the settling cores was problematical. The basic planetary nebula ejection mechanism may play a role.

Barkat of Hebrew University discussed an important issue in the process of degenerate carbon ignition and thermonuclear explosion, the role of the convective Urca process. Degenerate carbon burning leads to thermonuclear runaway when the burning rate exceeds the neutrino loss rate. The runaway burning generates a growing convective core, but the situation changes when the convective core extends beyond a critical density where the Fermi energy corresponds to the threshold for electron capture for a particular constituent. The convection then circulates mother and daughter nuclei past the critical threshold, so that alternate electron capture and decay occur with no net change in composition but the loss of a neutrino or anti-neutrino on each half of the cycle. This cyclic process is thus a copious source of cooling neutrinos in principle. The net effect has been uncertain, however, because the capture leaves a hole in the Fermi sea to be filled by a cascading electron and the decay deposits an electron with finite kinetic energy and thus the process also is a potential source of heating. Barkat argued that the convection will rapidly produce a steady state in which there is no net change in composition with time, but a gradient in composition of mothers, daughters and electrons, not a uniform composition. The currents in these species implied by the gradient can be shown to exactly balance the heating terms so that in steady state the heating is identically zero and the convective Urca process is a strong source of net cooling as originally proposed by Paczynski.

Barkat further argued that the convective Urca cooling is unstable. The cooling is a very steep function of temperature since the process depends so sensitively on the extension of the convective core beyond the Urca shell at the critical threshold density. A positive perturbation in the core entropy will lead to much stronger neutrino losses and regulation. A negative perturbation, however, will lead to a smaller convective core and Urca losses reduced to balance the decreased nuclear energy production. Such a state will be susceptible to negative entropy perturbations due to ordinary neutrino losses and the only stable state is one in which the Urca process shuts off both itself and the thermonuclear burning to the point where the latter is balanced by the ordinary neutrino losses.

This process of central carbon ignition, thermal runaway, growing convective core, initiation of convective Urca cooling, and shutting down the burning by cooling the core will happen quickly in the core. The resulting configuration will have a strong step function downward in the inner temperature distribution in the convective core bounded by the Urca shell. The subsequent evolution will be dictated by the slow growth of the core mass and central density due to accretion. Further modeling should show whether the result is collapse if the central density reaches the critical threshold for dynamical electron capture/general relativistic instability, or explosion due to off-center carbon ignition in the hotter material which falls just beyond the Urca shell.

6 August 1988

General Scientific Session

Chaired by A. Maeder

A.N. Cox   Solar Structure, Oscillations and Neutrinos
G. Michaud   Li Abundance as a Constraint on Particle Transport
J. Truran   Non-Spherical Effects: Case Study of Novae
M. Kato, I. Hachisu, and H. Saio
    Mass Loss during Nova and Helium-Nova Outbursts
S. Sofia, M. Pinsonneault, S. Kawaler, and R. Demarque
    New Results on the Evolution of the Rotating Sun
C. Chiosi   NGC 1866, an Observational Test for Convective Overshooting

This meeting was essentially devoted to hydrodynamical effects intervening in the course of stellar evolution.

Cox of Los Alamos National Laboratory constructed two precision solar models with the Iben evolution program; one with no-diffusion of the internal atomic nuclei, and the other with the effects of gravitational settling, thermal diffusion, and concentration gradient due to diffusion. Equation of state and opacity were fitted to the latest theoretical data. Then the opacity at the bottom of convection zone was revised to increase by 15 - 20 percent which allowed a better agreement with the observed solar p-mode frequencies. His theoretical p-mode frequencies are now lower only by a few microhertz than those observed. The helium concentrations of the initial mixture were Y = 0.291 and 0.289 for the no-diffusion and diffusion models, respectively. The diffusion model evolved to a surface helium concentration of Y = 0.256 at the solar age, and the heavy element concentration, which was initially Z = 0.0200, decreased to Z = 0.0179 by diffusive settling. Calculation of g-mode solutions showed that they did not have equal period spacings until high radial order. Nonadiabatic solutions for these g-modes enabled him to predict their relative surface visibility. The high helium concentration which was necessitated by his equation of state, however, resulted in high central temperatures with 9 SNUs from the ^8B and 1.5 SNUs from the ^7Be reactions. Neither a model with iron condensed-out deeper than the convection zone, nor a model with the presumed WIMPs to cool the central regions did reproduce observed p-mode separations.

Michaud of the University of Montreal gave an alternate explanation for the Li abundance gap (at T_{eff} about 6700 K) which is observed in clusters with ages between 4 and 20 x 10^8 years but not in young clusters. According to usual understanding, the T_{eff}, the width in T_{eff} and the maximum underabundance factor were thought as well explained in terms of gravitational settling at the age of Hyades. However, Boesgaard noticed that the Li abundance gap occurred where the equatorial rotational velocity increased rapidly with T_{eff}, and she suggested that these two are related each other. Michaud and Charbonneau expressed the envelope of observed rotational velocities in Hyades as a function of T_{eff}. Then they calculated the transport of Li-free matter by meridional circulation from the region of temperature higher than 2.5 x 10^6 K, and they were able to reproduce Li abundance gap with the observed width and depth.

Truran of the University of Illinois made a recall of the general properties of classical novae. In particular, various phases in evolution of classical novae were identified; accretion phase, thermonuclear runaway to maximum, post-maximum evolution in outburst, and return to minimum. Among many possible non-spherical effects, he discussed two of them. Firstly, as for the effects of magnetic field, he discussed the disruption of the accretion disk, the alteration of the accretion flow, the absence of shear mixing, and the possibility of polar thermonuclear ignition. Secondly, as for the common envelope evolution, he pointed out that it may result in observable effects on the visual light curve during the outburst.

Kato (Keio University) et al. followed evolution of shell flashes by a steady state approach describing it with stellar-wind mass-loss solutions. The mass loss rate was obtained uniquely: It is high while the star has extended envelope but it decreases as photospheric radius shrinks. The wind mass-loss ceases when the envelope mass reduces to a critical value. The time-dependent models were also calculated in helium shell flashes. The mass loss due to the Roche lobe overflow was taken into account. The mass accumulation ratio, i.e., the ratio of the mass remaining on the white dwarf after nova explosion to the mass having accumulated during the preceding accretion phase, was also calculated. It is less than about 1/3 for high accretion rate and much less than 0.1 for low accretion rate for the case of hydrogen-flash nova. The Roche lobe overflow reduces this value. In the case of helium-flash nova this ratio is less than 0.5 for the accretion rate lower than 1 x 10^{-7} solar mass per year and about unity for accretion rate higher than about 1 x 10^{-6} solar mass per year.

Sofia et al. of Yale University developed a new code for evolution of rotating stars and applied it to the sun. Although the basic philosophy of Endal and Sofia (1976, 1978, 1981) was followed in the initial conditions and transport of angular momentum, the coding was totally independent and many details were reformulated: Accounts were taken for angular momentum loss via a magnetic wind and for angular momentum redistribution by rotationally induced instabilities. The resulting models have an oblateness in agreement with observed upper limits. The rotation curves of the models of the present-day sun must show the following features: The layers outer than 0.6 solar radius exhibit minimal differential rotation, while the central core inner than 0.2 solar radius preserves its initial rapid rotation. These basic features persist through a wide range of model parameters. Transport of angular momentum leads to rotationally induced mixing which explains the observed Li depletion in the sun. The rotation curves in their models of the present-day sun are in qualitative agreement with the estimates from current analysis of solar oscillation for the layers outer than 0.6 solar radius.

Chiosi of Padova University studied the color-magnitude (CM) diagram and the main sequence luminosity function of the young LMC globular cluster NGC 1866 with the aim at discriminating among possible scenarios for evolution of intermediate mass stars. To this purpose, Johnson B and V CCD photometry of 1517 stars was obtained in the central region of NGC 1866, and of 640 stars in the nearby field. The CM diagram of this region was corrected for photometric incompleteness and for contamination with field stars. Comparison with the observational integrated luminosity function clearly demonstrated that models with convective overshoot ought to be preferred. In other words, the star counts in the cluster suggested that the ratio of core H- to He-burning lifetimes and the range of luminosity spanned by main sequence stars must be lower and wider, respectively, than those given by classical computations without the convective overshoot. This is possible only if such efficient mixing processes with convective overshoot really take place not only in the latest stages but also over all the preceding stages of the stellar evolution. It should be noticed that only such known mixing processes as semi-convection and/or breathing convection would not be able to satisfy the observational demand.

COMMISSION No. 36

THEORY OF STELLAR ATMOSPHERES

(THEORIE DES ATMOSPHERES STELLAIRES)

Report of Meetings, 3, 4 and 9 August 1988

PRESIDENT: K. Kodaira. SECRETARY: D. F. Gray.

Business Meeting, 3 Aug., 11:00 am

I. COMMISSION ACTIVITIES

The President outlined the activities of the Commission in the period following the 19th General Assembly. The items included were (1) 1 IAU Symposium and 5 Colloquia sponsored or cosponsored, (2) 2 acting Working Groups associated with Commission 36, (3) Commission Report for the IAU Transaction A, and (4) 5 Commission Circulars and 1 supplement issued.

II. MEMBERSHIP

a) DEATH: Cannon Christopher

b) VOLUNTARY RESIGNATIONS: Muller Edith, Dumont Simone, Sobouti Yosev

c) NEW MEMBERS: The following names were approved as new commission members.

(i) Present IAU Members: Brown Alexander, Freire Ferrero Rubens, Landstreet John D.

(ii) New IAU Members: Catala Claude, Krikorian Ralph, Steffen Matthias, Takeda Yoichi, Narasimha Delampady, Bodo Gianluige, Carlson Mats, Cugier Henryk, Bowen George M., Fitzpatrick Edmund L., Friend David B., Hartkopf William I., Liebert James W., Musielak Zozislaw E., Chugaj Nikolai N., Lyubimkov Leonid S., Nikogosian Artur G.

(iii) New IAU Member proposed by Commission 36: Abbott David C.

III. ELECTION OF PRESIDENT AND VICE PRESIDENT

Basing upon the opinion surveys by means of Commission Circulars and upon the recomendation by the IAU Executive Committee, Commission President made a proposal, which was endorsed by the attendants.

Dr. D. F. Gray was approved as President. From the three nominees for Vice President, W. Kalkofen, A. G. Hearn, and R. Kudritzki, Dr. W. Kalkofen was approved as Vice President.

IV. ELECTION OF ORGANIZING COMMITTEE

Basing upon the opinion surveys by means of Commission Circulars, Commission President made a proposal, which was approved by the attendants. The new OC members are: D. F. Gray, W. Kalkofen, J. Cassinelli, L. Cram, A. G. Hearn, M. Seaton, J. L. Linsky, A. Peraiah, F. Praderie, A. Sapar, T. Tsuji, and R. Wehrse.

V. FUTURE ACTIVITIES

(i) IAU Meetings: 5 symposia and 2 colloquia asking for the sponsorship or cosponsorship of Commission 36 were planned for 1989 and 1990. A brief discussion developed about the similarity of the topics among the meetings concerning the magnetohydrodynamic properties of atmospheres, especially in common with the meetings supported by the commissions related to the solar physics.

(ii) Working Groups: 2 new WG were proposed to which Commission 36 should be associated; WG on Astrochemistry and WG on Synthetic Photometry.

VI. ON OTHER MEETINGS

The president introduced the scientific meetings related to Commission 36 which took place during and just after the General Assembly.

Scientific Meeting on "Accuracy of Abundances", 4 Aug., 2:00 pm

CHAIRPERSON: R. Wehrse
(1) Introduction (R. Wehrse)
(2) Transition Probability (W. Wiese)
(3) Excitation- and Ionization Crosssections (K. Butler)
(4) Line-Broadening Constants (M. Dimitrievic, read by B. Baschek)
(5) OB Mainsequence Stars (R. Mendez)
(6) B-F Mainsequence Stars (K. Sadakane)
(7) Solar-type Stars (F. Spite)
(8) Cool Mainsequence Stars (M. Bessell)
(9) Extragalactic Stars (B. Baschek)
Butler pointed out in particular to the "Opacity Project" initiated by M. Seaton to recalculate all relevant crossections according to the best method available, the close-coupling method. K. Sadakane appealed the necessity to promote the intercalibration among various institutions engaged in the abundance analyses. About 100 persons attended.

Scientific Meeting on "Stellar Surface Features", 9 Aug., 9:00 am

CHAIRPERSON: D. F. Gray
(1) Photometry of Starspots (M. Radono)
(2) Theoretical Aspects of Spots (P. Fox)
(3) Ap-star Mapping (W. Wehlau)
(4) Spectroscopy of Starspots (F. Fekel)
(5) Starpatches and Bisector Mapping (D. Gray)
(6) Techniques of Starspot Mapping (I. Tuominen)
Gray introduced in particular the new concept of starpatches whose main property is an enhanced velocity field. About 100 persons attended.

The reports of the following meetings associated with Commission 36 are given in the reports of the commissions in the parentheses:
WG on Ap/Cp Stars, 3 Aug., 2:00 pm (Commission 29)
WG on Peculiar Red Giants, 10 Aug., 2:00 pm (Commission 29)
Joint Commission Meeting on "Spectroscopy of Individual Stars in Globular Clusters and the Early Chemical Evolution of the Galaxy", 8 Aug., 9:00 am (Commission 29)
Joint Commission Meeting on "Solar and Stellar Coronae", 4 Aug., 9:00 am (Commission 10+12)

COMMISSION No. 37

STAR CLUSTERS AND ASSOCIATIONS

(AMAS STELLAIRES ET ASSOCIATIONS)

PRESIDENT: D.C. Heggie

In addition to the Commission sessions described below, the Commission also participated in the following:

1. Joint Discussion II: Formation and Evolution of Stars in Binary Systems (organiser: R.C. Smith)

2. Joint Commission Meeting 5: Spectroscopy of Individual Stars in Globular Clusters and the Early Chemical Evolution of the Galaxy (organiser: G. Cayrel de Strobel)

3. Joint Commission Meeting 6: Stellar Photometry with Modern Array Detectors (organiser: F. Rufener)

4. Session 37/4 = 26/6 = 42/2: Formation of Close and Wide Binaries (organiser: V.L. Trimble)

5. Session 37/5 = 30/4: Progress in the Understanding of the Dynamics of Star Clusters (organiser: C. Pryor)

Reports of the Joint Discussion and of JCM 6 will be found in *Highlights of Astronomy*. It is intended that the proceedings of JCM 5 will appear in due course as a publication of the Paris Observatory. Reports on sessions 37/4 and 37/5 will be found in this volume under the reports of Commissions 42 and 30, respectively.

<u>5 August 1988</u>

BUSINESS SESSION SECRETARY: A.G.D. Philip

1. Scientific Organising Committee, 1988-91

It was agreed to propose the following:

G.L.H. Harris (President)
C. Pilachowski (Vice-President)
D.C. Heggie (Past President)
J.E. Hesser
J-C. Mermilliod
R.R. Shobbrook
J.L. Zhao
K.A. Janes

D.A. Vandenberg
J. Claria

2. Commission membership

The President gave the names of 22 IAU members, and 14 proposed members of the IAU, who had applied to join Commission 37. There had been one resignation, and one non-member of the Union was continuing as a consultant of the Commission. These changes, which were approved by the meeting, bring the membership of the Commission to 201.

3. Stars in open clusters

J.-C. Mermilliod described the development of his database on observational data for stars in open clusters, and recommended that future observations should be better coordinated, in order to identify the clusters and the kinds of data where improvements would be most useful. There was much discussion on these issues, which extended also to globular clusters and to associations, and on the future structure of the Commission Triennial Report. In conclusion it was agreed to set up a Working Group on the Acquisition and Compilation of Data on Clusters and Associations, with the following members : J.-C. Mermilliod (Convener), K. Janes, C. Pilachowski, J.E. Hesser, R.D. Mathieu, R.E. White, J.-L. Heudier, G. Lyngå and G.L.H. Harris. A suggestion that a review paper should be written, to summarise the issues involved, was also approved.

4. Atlas of cluster photographs

Good progress on this project, which had been set in motion at the 18th General Assembly, was reported by J.-C. Mermilliod and J.-L. Heudier. The greatest remaining need is for large scale plates of the smallest clusters.

5. Other Business

G.L.H. Harris reported on suggestions, received from Commission members, on a system of nomenclature for clusters in extragalactic cluster systems. Discussion concentrated on the possibility of names based on radial coordinates relative to the centre of the parent galaxy. C. Pilachowski reported on relevant items from Joint Discussion I (Documentation, Data Services and Astronomers), especially archiving of observational data.

5 and 6 August 1988

SCIENTIFIC SESSION: POSTER SESSION

SECRETARY: D.C. Heggie

The following papers were displayed, and a discussion session, chaired by K.C. Freeman, was held in the afternoon of 5 August.

K. Akiyama & D. Sugimoto: Evolution of self-gravitating many-body system with rotation.

E.J. Alfaro & J. Cabrera-Cano: NGC 752 revisited - membership study from Ladovski's proper motion study.

H.C. Bhatt & R. Sagar: Distances to open star clusters - the kinematical method.

E. Brocato, R. Buonanno, V. Castellani & A. Walker: LMC cluster NGC 1866: a new investigation.

I.R. Brodie, R.D. Cannon, R.J. Dickens, W.K. Griffiths & R.G. Noble: Cluster membership determination of stars at the MS turnoff of ω Centauri.

J. Cabrera-Cano & E.J. Alfaro: A non-parametric approach to the membership problem in open clusters.

C.S. Chiosi, A. Bertelli, G. Meylan & S. Ortolani: Globular clusters in the LMC: NGC 1866, a test for convective overshoot.

C.A. Christian, J.N. Heasley, E.D. Friel & K.A. Janes: The giant branch of Mayall II: a globular cluster in M31.

M.S. Chun: Radial colour gradient in 47 Tuc.

A. Dapergolas, E. Kontizas, F. Pasiani, M. Pucillo & P. Santin: An age estimate for the NGC 456, 460 and 465 SMC constellation.

A.J. Delgado & E.J. Alfaro: On the calibration of intrinsic colours in *uvby* photometry

A.J. Delgado, E.J. Alfaro & A. Aparicio: CCD differential photometry of stars in the young open cluster NGC 7128. Preliminary results.

R.J. Dickens, I.R. Brodie, E.A. Bingham & S.P. Caldwell: A catalogue of magnitudes and colours in the globular cluster omega Centauri.

T. Lloyd Evans & L.G. Underhill: Correlated abundance variations in ω Centauri.

E.M. Green, M.S. Bessell, P.Demarque & C.R. King: An age spread among galactic globular clusters using the revised Yale isochrones and a new semi-empirical UBVRI calibration.

W.E. Harris, J.W. Allwright, C.J. Pritchet & S. van den Bergh: Globular Cluster Systems in Virgo: New observations from CFHT.

D.C. Heggie, J.Goodman & P. Hut: On the exponential divergence of N-body systems.

E. Kontizas, M. Kontizas & M. Metaxa: Small faint clusters in the LMC.

E. Kontizas, M. Kontizas, G. Sedmak & R. Smareglia: Ellipticities at R_h of the LMC star clusters.

E. Kontizas, M. Kontizas & E. Xiradaki: The stellar content of binary star clusters in the LMC.

P. Mazzei & L. Pigatto: The Pleiades age and the sequential star formation.

J.-C. Mermilliod: BDA - a database for stars in open clusters.

M. Metaxa, M. Kontizas, E. Kontizas: 8 star clusters of the LMC periphery with evidence of halo region.

J.L. Zhao: Current and future studies of open clusters at Shanghai Observatory.

At the poster discussion session two issues in particular received much attention: (i) problems of the determination of cluster ages by model isochrone fitting, and (ii) the complexities of abundances in stars in globular clusters.

6 August 1988

SCIENTIFIC SESSION: THE ABUNDANCE SPREAD WITHIN GLOBULAR CLUSTERS CHAIRMAN: T. Lloyd Evans

The following papers were presented:

N. Suntzeff: The metal-poor clusters.
G.H. Smith: The metal-rich clusters.
R.J. Dickens: ω Centauri.
C.A. Pilachowski: High resolution spectroscopy - quantitative results and critical tests.
A.V. Sweigart: Theoretical considerations.

These presentations were followed by a panel discussion involving R.A. Bell, K.C. Freeman, R. Gratton, J.E. Hesser, and R.P. Kraft. It is intended that the invited review papers will be included in a publication of the Paris observatory.

9 August 1988

SCIENTIFIC SESSION: STAR CLUSTERS IN THE MAGELLANIC CLOUDS (JCM7) CHAIRMAN: P. Demarque

There were three main speakers:

G. Meylan: Structure and Dynamics

Abstract. Because of lack of good observations, the structure and dynamics of the Magellanic clouds' star clusters are still largely unknown. The fact that globular clusters of all ages exist in the clouds gives a unique opportunity to check the theoretical prediction concerning a relation between age and ellipticity (the youngest clusters being the flattest). Surface brightness profiles (from surface photometry and star counts) and velocity dispersions (from integrated light spectra and individual stars) will allow a systematic use of King-Michie models to establish reliable comparison with clusters in the Galaxy. Collapsed clusters exist in the LMC (two such clusters have been observed near the bar) and surveys undertaken will soon give a general census.

P.R. Wood: Advanced Stages of Stellar Evolution in Magellanic Cloud Clusters

Abstract. The flurry of activity involving the discovery, infrared photometry, and low-dispersion spectroscopy of AGB stars in intermediate age clusters in the Magellanic Clouds has now slowed down after providing a major set of constraints which theoretical models must satisfy. The generally predicted evolutionary sequence in spectral type up the AGB from M-S-C has been verified (Lloyd Evans 1984, MNRAS, **208**, 447). The luminosity at which this transition M-S-C occurs can be derived as a function of cluster age/turnoff mass using the infrared data and the growing number of accurate cluster ages now in the literature (see compilation in Mould and Da Costa 1987, CTIO 25th Anniversary Symposium). The theoreticians are struggling with limited success to produce carbon stars at the observed luminosities (Lattanzio 1988, in IAU Colloquium 106, *Evolution of Peculiar Red Giants*). Another parameter that has been determined as a function of cluster age/turnoff mass is the luminosity at the AGB tip (Mould and Aaronson 1986, Ap.J., **303**, 10). The limit luminosity is determined by mass loss, probably of two separate kinds. Note that the tip luminosity is well above the luminosity for onset of shell flashes (Lattanzio 1988). At luminosities below $M_{bol} \sim -6$, a strong stellar wind ('superwind') of the type operating typically around OH/IR stars and dust-enshrouded carbon stars should operate; a Reimers-type mass loss law seems to (qualitatively) predict the AGB tip luminosity in this case (Mould and Aaronson 1986). However, recent derivations of mass loss rates indicate that real red giant stars do not follow the Reimers mass loss law (Wood 1987, in *Stellar Pulsation*, eds. A.N. Cox *et al*) so that any agreement between the observed tip luminosities and those derived from a Reimers mass loss law must be considered fortuitious. A second mechanism, radiation pressure ejection, may come into play in stars more luminous than $M_{bol} \sim -6$ (Wood and Faulkner 1986, Ap.J., **307**, 659). AGB stars more luminous than $M_{bol} \sim -6$ do exist on the AGB (Hughes and Wood 1987, Proc. Astr. Soc. Australia, **7**, 147); these stars are probably the most massive AGB stars.

The pre-AGB evolutionary phases of stars in Magellanic Cloud Clusters have also been receiving recent attention. Searches for variable stars are yielding new candidates (new Cepheids in NGC1866 have been identified by Storm *et al* 1988, Astr. Ap., **190**, L18). Constraints on the distance moduli to the SMC and LMC have been derived by Seidel, Da Costa and Demarque 1987, Ap.J., **313**, 192 from clump star luminosities in intermediate age clusters. An area of greatly increasing activity is the determination of accurate HR diagrams for Magellanic Cloud clusters. From the stellar evolution point of view, these results will be invaluable in determining the extent of overshoot during the main sequence phase of evolution. The data and analyses obtained so far are contradictory: for example, compare Chiosi and Pigatto 1986, Ap.J., **308**, 1 and Mateo and Hodge 1987, Ap.J., **320**, 626; and Brocato, Buonanno, Castellani and Walker 1988 (preprint) and Chiosi, Bertelli, Meylan and Ortolani (preprint). Determination of accurate HR diagrams for Magellanic Cloud clusters with ground-based CCDs and with the HST will be an active area of research over the next few years.

G.S. DaCosta: The Clusters as Signposts for the Chemical Evolution of the Magellanic Clouds

Abstract. Based on a compilation of the most recently available data for the star clusters of the Magellanic Clouds, the relation between age and abundance in the Large and Small Magellanic Clouds has been investigated. In the LMC, there is a distinct lack of well studied clusters with ages between 3 billion years and the age, taken as 15 billion years, of the small number of cluster analogues to the galactic halo globular clusters. The lack of clusters in this age range is ascribed to a combination of the effect of fading as clusters age, which encourages the selection of younger clusters in magnitude limited surveys, and the disruption of clusters, for which Hodge has recently determined a timescale of the order of 1 billion years. The lack of clusters between 3 and 15 Gyrs means the age-abundance relation for the LMC is only loosely constrained by the star clusters: the mean abundance appears to have risen slowly from [Fe/H] $\simeq -0.6$ at 2 Gyrs to its present value of [Fe/H]

$\simeq -0.3$ dex. Prior to 2 Gyrs, the abundance must have risen from approximately [Fe/H] $= -2.0$ dex, the mean abundance of the oldest clusters, but there is little information on the rate of this enrichment.

In contrast to the LMC, the SMC does contain bright clusters in the 3 - 12 Gyr age interval, with Kron 3 being the archetypal example. On the other hand, and again in contrast to the LMC, there are no clusters in the SMC that are as old as the Galactic halo globular clusters. Using these clusters as a guide, it appears that the abundance of the SMC rose slowly but steadily to [Fe/H] $\simeq -0.7$ dex at an age of approximately 1 Gyr. Since that epoch, the abundance has remained essentially constant.

A comparison has also been made between the ages and abundances determined from integrated techniques, both spectroscopic and photometric, and those inferred from direct c-m diagram studies. It is concluded that while most photometric techniques have good age sensitivity up to ages of approximately 2 - 3 Gyr, they fail to discriminate between clusters of older age. Age discrimination for these older clusters can be achieved with spectroscopic data using the hydrogen-line strength versus metal-line strength diagram; better sensitivity occurs for high S/N and higher resolution observations. Ultimately, when additional clusters have been observed spectroscopically, it should be possible to provide an accurately calibrated version of this diagram that can be used to study the clusters of other galaxies to which only integrated techniques can be applied.

In addition a number of shorter contributions were included:

S.M. Fall, R.A.W. Elson and K.C. Freeman: The Internal Structure and Kinematics of Rich Young Star Clusters in the Large Magellanic Cloud

Abstract. We have determined the surface brightness profiles of ten young star clusters in the LMC, using aperture photometry and star counts. The ages of the clusters range from 8 million years to 300 million years. The profiles extend over 8-10 mag in surface brightness and to radii of about 4 arcmin, much farther than in previous studies. Most of the clusters in our sample do not appear to be tidally limited. At large radii the projected density falls off as $r^{-\gamma}$ with $2.2 < \gamma < 3.2$, and the median value of γ equals 2.6. With one possible exception, we find no evidence for mass segregation.

In a second study, we have measured the velocities of individual stars within three of the young LMC clusters. The internal velocity dispersions are small, about 1km/s, requiring accurate determinations of the observational errors. From the surface brightness profiles and the velocity dispersions, we estimate the masses and mass-to-light ratios of the clusters. With these results and our estimates of the gravitational field of the LMC, we can determine whether the clusters are in fact tidally limited. It appears that they are not. This confirms our previous suggestion, which was based on the surface brightness profiles and stellar population models to estimate the masses. We find that up to 50% of the masses in the clusters currently extends beyond eventual tidal radii as unbound halos. The most likely explanation is that the clusters or protoclusters lost much of their mass, either in the form of stellar ejecta or in gas that was expelled during star formation.

P. Linde, G. Lyngå and B. Westerlund: A Comparison of the LMC Clusters LW 55 and HS 96

Abstract. The aim of a project now under way at the Lund and Uppsala Observatories is to compare open clusters and field stars in two regions on either side of the bar in the LMC. The discussion of the clusters in these regions aims at relating the star formation history of the areas to the present stellar content of the open clusters. As a first step we now present a photometry of the two clusters LW 55 and HS 96. Their ages and luminosity functions will be compared and discussed.

Finally, *M. Kontizas* described recent work on several issues: mass segregation, binary clusters, extended cluster halos, and cluster ellipticity; *J.-C. Mermilliod* discussed field-star contamination in simulated clusters in the Clouds; *B. Baschek* outlined work at Heidelberg on near-main sequence B stars in cloud clusters; and *C. Chiosi* discussed integrated properties of the clusters, especially the bimodal distribution of B-V.

COMMISSION No. 38

EXCHANGE OF ASTRONOMERS (ECHANGE DES ASTRONOMES)

Chairman: Edith A. Muller (President). Secretary: Kam-Ching Leung.

1. The Commission agreed the following proposals for 1988-91.

1.1 Proposals for Officers

President:	Professor Sir Francis Graham-Smith (UK)
Vice President:	Professor Jorge Sahade (Argentina)

1.2 Organising Committee

A.A. Boyarchuk	(USSR)
A. Florsch	(France
H. Jørgensen	(Denmark)
Y. Kozai	(Japan)
K.Ch. Leung	(USA)
E.A. Müller	(Switzerland), ex. off.
G. Swarup	(India)
Wang Shou Gua	(P.R. China)

1.3 Ordinary Membership

A.A. Al Sabti	(Iraq)
M.K. Aly	(Eqypt)
B. Caccin	(Italy)
H. Haupt	(Austria)
D. MacRae	(Canada)
M. Marik	(Hungary)
Il-Seong Nha	(Rep.Korea)
S. Ninkovic	(Yugoslavia)
S. Okoye	(Nigeria)
Paul M. Routley	(USA)
G. Ruben	(DRG)
C.R. Tolbert	(USA)
E. van den Heuvel	(Netherlands)
F. Bradshaw Wood	(USA)
Shu-Hua Ye	(China)

1.4 Consultant

S. Ruither	(UNESCO)

2. Guidelines

It was agreed to continue on the same principles, and to ask for the Guidelines to appear every year in the January issue of the IAU Bulletin. Minor additions and clarifications have been incorporated in the Application Procedure, which is printed below.

3. Joint Session of Commissions 5 and 46 on Developing Countries (Aug 9)

The President will attend this meeting as Observer.

F. Graham-Smith
President
October 1988

COMMISSION No. 40

RADIO ASTRONOMY (RADIOASTRONOMIE)

Report of Meetings, 3, 4, 5, 8, 9, 10 August, 1988

PRESIDENT: J.E. Baldwin VICE-PRESIDENT: P.G. Mezger

Business Meetings SECRETARY: T.L. Wilson

I NEW OFFICERS

Peter Mezger (FRG) was elected President and M. Morimoto (Japan) Vice-President for the period 1988-1991. Continuing members of the Organising Committee are J.E. Baldwin (UK), A. Baudry (France), R.S. Booth (Sweden), D.L. Jauncey (Australia), N. Kaifu (Japan), V.K. Kapahi (India), L. Matveyenko (USSR), G.D. Nicolson (South Africa), E.R. Seaquist (Canada). The following were elected as new members of the SOC: D.C. Backer (USA), R. Fanti (Italy), R. Guesten (FRG), J.M. Moran (USA), J.M. van der Hulst (Netherlands), Q.F. Yin (China).

II NEW MEMBERS

75 new members were admitted by the Commission bringing the total membership to over 750. The Commission continued its policy of not restricting further growth in its numbers. The advantages of a very large Commission with common interests in technique applied to a wide range of astronomical interests were thought to outweigh the difficulties of communication with the members. The full programme of Commission meetings together with its involvement in many Joint Discussions and Joint Commission Meetings supported this view.

III IUCAF

John Findlay, the retiring Chairman of IUCAF, presented a wide-ranging report of IUCAF activities. This explanatory review, which covers the reasons for the formation of IUCAF, its mandate, its work in recent years in coordinating efforts to protect frequency bands used for radioastronomy from man-made interference and its prospects for the future, is reprinted at the end of this report on Commission 40 activities. The Commission thanked John Findlay for the difficult and long-sustained work he had undertaken on their behalf.

Brian Robinson, the new Chairman of IUCAF, reported that the increase in membership of IUCAF, sought by the previous Chairman, had been approved by the IAU. The IAU representation is increased from two to four; G. Swarup (India), V.L. Pankonin (USA), B.A. Doubinski (USSR) and N. Kaifu (Japan) were unanimously approved as members. Both IAU and URSI have agreed to increased funding for IUCAF. Thanks were given to the retiring member R. Schilizzi (Netherlands) and also to the retiring Secretary, A.R. Thompson (USA), who gave a report of CCIR activities in the period 1986-90. This raised many issues concerning present and potential future radio interference problems which were later addressed by the Working Group on the Protection of Spectral Lines (see section IV below) and were the subject of the IAU Resolutions discussed in section V. It was felt that it would be most valuable if a short list of CCIR reports could be distributed to Commission 40 members by its next President.

IV REPORT OF WORKING GROUPS

(a) Nomenclature

Burke (Chairman) reminded the Commission that this WG had presented its final report at the previous General Assembly. However, this occasion provided an opportunity to calm fears expressed by commission members that the use of J2000 coordinates would involve changing the names of well known objects. This was emphatically not the case. The Commission were cheered by this, thanked the Chairman and the WG and agreed to its dissolution.

(b) Protection of molecular line frequencies

Robinson (Chairman), Baudry, Cohen, Hjalmarson, Kahlmann, Kaufmann, Morimoto, Pankonin, Radhakrishnan, Slysh, Spoelstra, Tiuri, Turner, Webster, Wilson.

The work of the Group continues to expand due both to the continual reassessment of the importance of molecular lines in the existing lists and new candidates and to the growing number of actual and potential sources of radio interference. Discussions covered revision of the lists of important molecular lines, interference from GLONASS at 1612 MHz, frequency allocations at 1612 MHz for the radio-determination satellite service (RDSS), the Land Mobile Satellite allocation at 1660 - 1660.5 MHz, spread-spectrum interference from satellites and long term strategy for protection of passive radioastronomy bands at future WARC meetings. These topics formed the basis for Resolutions 5 and 6 listed in section V.

The revised lists of radio frequency lines of the greatest importance to astronomy are published with Resolution 7 elsewhere in these Transactions.

In view of the growing need for the attendance by radioastronomers at meetings concerned with frequency allocations and radio interference, the support of all Observatory Directors was requested by the Commission for supplying travel funds to appropriate meetings in their Region. Many of those present agreed in principle to contribute to a common fund or to support members of their own staff for this purpose.

(c) VLBI

Johnston (Chairman), Ananthakrishnan, Backer, Biraud, Booth, Legg, Cohen, Grueff, Manchester, Kaufmann, Kellermann, Kus, Matveyenko, Morimoto, Nicolson, Pauliny-Toth, Schilizzi, Wilkinson, Yeh.

The meetings of the Working Group provide a focus for exchange of information on the present status of VLBI observatories throughout the world and reports on Playback Systems. The important issue of the availability of recording systems in the USA and Europe was also discussed.

V RESOLUTIONS

The problem of radio interference in radioastronomical observations, in particular due to transmissions from satellite systems, both actual and planned for the future, was a matter requiring the urgent attention of the Commission. The Working Group on the Protection of Molecular Line Frequencies under its Chairman Brian Robinson was instrumental in drafting Resolutions on these topics which had the unanimous support of the Commission. They were accepted as Resolutions of the Executive Committee and are printed elsewhere in these Transactions as Resolutions 5: Cooperation to save Hydroxyl Bands, 6: Sharing Hydroxyl Band with Land Mobile Satellite Services and 7: Revision of Frequency Bands for Astrophysically Significant Lines. The concerns of Commission 40 amongst others are also expressed in Resolution 2: Adverse Environmental Impacts on Astronomy.

VI SYMPOSIA

Commision 40 has taken part in planning several Symposia and colloquia during the three years since the last General Assembly (Symposia 124, 129, 134, 136; Colloquia 101, 112).

Symposia in various states of planning which involve the Commission included the following topics (contact person in brackets): Recombination lines (Kardashev), Paired and interacting galaxies (Sulentic), Extragalactic molecular clouds: dynamics and evolution of galaxies (Combes), Radioastronomical seeing (Wang), Magellanic Clouds and interaction with the Milky Way (Haynes), Magnetic fields in galaxies (Wielebinski), Magnetospheric structure and the emission mechanisms of pulsars (Rankin), Reference systems (J.A. Hughes), Structure of Molecular Clouds (Falgarone), Radio interferometry (Cornwell).

VII RADIO CATALOGUES

The attention of members was drawn by E.C. Campbell (Canada) to a list of all radio surveys from 1955 to the present which is to be published soon in Astronomy and Astrophysics Supplement.

Scientific Sessions

I. COMMISSION MEETINGS

Instrumentation, August 3, 1988 Chairman: R.S. Booth

Recent developments at Westerbork	E. Raimond
The upgrade of the Dominion Radio Astrophysical Observatory synthesis telescope	L.A. Higgs
The proposed Canadian multi-element synthesis telescope	L.A. Higgs
Recent developments at Jodrell Bank and the extension of MERLIN	P. Thomasson
The Very Long Baseline Array	J.D. Romney
The NRAO 12 m Telescope	D.T. Emerson
Improvements of the Nobeyama 45 m telescope	N. Kaifu
The Swedish-ESO Submillimetre Telescope; some results	R.S. Booth
The Nobeyama Millimetre-wave Array	M. Ishiguro
The NRAO millimetre array project	R.L. Brown
The Australia Telescope	R.D. Ekers

Solar System and Galactic Research, August 4, 1988 Chairman: N. Kaifu

Solar radio research at Clark Lake Radio Observatory	M.R. Kundu
HCN and OH observations of recent comets	D. Bockelée-Morvan
A model of core emission from pulsars	J. Rankin
A new flaring radio source in Cygnus	L.A. Higgs
OH/IR stars at the Galactic Centre	A. Winnberg
Mm-wave recombination lines	M. Gordon
Observations of NH_3 emission with the Nobeyama mm-wave array	S. Okumura-Kawabe
Evolution of the local galactic magnetic field	S.J. Goldstein
New continuum surveys with the 100-m Telescope	R. Wielebinski
All sky survey at 34.5 MHz	K.S. Dwarakanath
Extreme scattering events	R. Feidler
Evolution of the morphology of supernova remnants with pulsars	D. Bhattacharja
Various HII regions in Cep A	V. Hughes
The case of the optical identification of the 6.1 msec pulsar P1953+29	V. Boriakoff

Extragalactic Research, August 4, 1988 Chairman:

Observations of the Coma cluster at 327 MHz	L. Feretti
The end of the jet in 3C 33 north	L. Rudnick
Development of high-frequency radio outbursts in AGN	E. Valtaoja
Polarisation asymmetry in double sources	R.G. Conway

Update on fluctuations in the cosmic microwave background	R.D. Davies
VLA observations of a new sample of Molonglo quasars: aspect dependence of the optical continuum	V.K. Kapahi
Moderately compact steep-spectrum sources	C. Akuja
Magnetic fields in external galaxies	R. Wielebinski
Recent results from Westerbork	C. O'Dea
OJ287 as a precessing eclipsing binary supermassive black hole system	M. Valtonen
OH Megamasers in galaxies - new results from Nancay	L. Gougenheim
Extragalactic supernovae as radio sources	K. Weiler
An unusually aligned radio source	K. Menon
Recent observations of gravitational lenses	D. Walsh

<u>High Angular Resolution for Radio Astronomy in Space, August 10, 1988 (with Commission 44)</u> <u>Chairmen: B.F. Burke, Y. Kondo</u>

TDRSS experiment	R. Linfield
RADIOASTRON	N. Kardashev
VSOP	N. Kaifu
QUASAT status	R. Schilizzi
Orbital studies	R. Schilizzi
Prospects for low frequency space VLBI	K. Weiler
Two theorists' views	Paczynski, A. Ferrari
Speculations for the future	R.S. Booth, N. Kardashev

II JOINT COMMISSION MEETINGS

For Milliarcsec or Better Accuracy (4,7,8,19,24,31,40)	August 3
High Angular Resolution Imaging from the ground (9,40)	August 5
Molecules in External Galaxies (28,34,40)	August 5

Reports of these meetings are published in "Highlights of Astronomy" Vol 8

III JOINT DISCUSSIONS

SN1987A in the LMC	August 4
Atomic and Molecular Data for Astrochemistry	August 8
Disks and Jets on Various Scales in the Universe	August 9

IUCAF Report to the International Astronomical Union

by John W. Findlay

1. Introduction

Since I shall formally resign as Chairman of the Commission at the time of the forthcoming 20th General Assembly of the IAU it seems appropriate that the IUCAF Report to the IAU should be made to serve as a review of the work of IUCAF from its inception and also to give a short look into the future. In this form it may be of interest to all the bodies which the Commission has tried to serve, so copies will be sent to ICSU, URSI and COSPAR.

2. The formation of IUCAF

The need to have bands of radio frequencies available for scientists to use at various parts of the spectrum was discussed first by URSI at the IXth General Assembly in Zurich in 1950. It was the fairly new science of radio astronomy which made the matter urgent, and between 1950 and 1957 the subject was studied by URSI, the IAU and, as soon as it was formed, by COSPAR. When the XIIth General Assembly of URSI convened at Boulder in 1957 with Lloyd V. Berkner as President, Sub-Commission Ve of the radio astronomy Commission V was formed to prepare for the forthcoming International Telecommunications Union (ITU) World Administrative Radio Conference (WARC) to be held in late 1959.

The tasks for this sub-Commission (of which the author became chairman in the early summer of 1958) were to establish the scientific requirements for the protection of bands of radio frequencies throughout the spectrum and then to take action to get these requirements presented to the WARC and, if possible, to get them approved. International agreement was reached on the scientific requirements for radio astronomy and this was formalized in a CCIR Recommendation No. 314 approved by the CCIR Plenary Meeting held in Los Angeles in April 1959. By this time the IAU had held a General Assembly in Moscow and in August 1958 has passed a resolution supporting the work of Commission Ve and calling for bands to be cleared not only for hydrogen-line observations, but also for a series of bands at octave intervals throughout the radio spectrum. COSPAR came into being in 1958 to carry on the rocket and satellite work started in the International Geophysical Year, and so also had an obvious interest in getting space science and space operations included in the 1959 WARC.

A WARC is a major meeting called by the ITU to regulate the use of the radio frequency spectrum. A WARC is essentially a treaty-making occasion and so its pattern of work is formal. Only member countries may make submissions to the WARC and these are usually worked out well in advance and given some limited diplomatic exposure before the WARC starts. Thus the second task given to Ve required that at least one administration would enter the needs of the scientists, and, if possible, several administrations would support them. As it turned out it was Netherlands which made the vital submission; this was largely due to the leadership of Professor Jan Oort. Since it was clear that the IAU and COSPAR were also concerned with the work of the WARC, a joint respresentation was arranged, with at least one person acting for all as an observer at the Conference. This co-operation can be regarded as the unofficial start of IUCAF.

The official formation of IUCAF was largely due to Lloyd Berkner. It was obvious when the work of Ve was reported at the XIIIth General Assembly of URSI at London in 1960 that the task of working to get frequencies allocated for radio astronomy and space science should be done by an Inter-Union body. Berkner arranged a series of meetings in London which led directly to the formation of IUCAF as an ICSU Inter-Union Commission with URSI as parent Union and IAU and COSPAR as the other founding bodies.

During its first years IUCAF was composed of twelve members, four being appointed by each of the founding bodies. The Director of the CCIR and the Chairman of the International Frequency Registration Board (IFRB) act as advisers to IUCAF and there is a secretary. After some years the appointed membership was reduced to six.

3. The mandate of IUCAF

At its formation IUCAF was charged with the following tasks:

(a) To study and coordinate the requirements for radio frequency allocations for radio astronomy and space science, and to make these requirements known to the national and international bodies responsible for frequency allocations;

(b) To take action aimed at insuring that harmful interference is not caused to radio astronomy or space science, operating within the allocated bands, by other radio services.

The ITU usually holds a full WARC every twenty years but the growth of activities in space made it necessary to hold a WARC specially devoted to space frequencies in 1963. Radio astronomers were fortunate in having their service also included. IUCAF was deeply involved in this WARC, and the outcome was a much improved allocation table. The next full WARC, held in 1979, was again an opportunity to improve and extend the frequency protection for the IUCAF sciences. One important change made by this WARC was the inclusion of radio astronomy and space research together in a number of "passive" bands where no radio frequency power may be emitted at all by any service.

It will be clear from what has been written so far that the task in (a) above becomes most important when a WARC, either full scale or specialised, is planned for the near future. IUCAF attempts to keep up-to-date with the changes or growth of the scientific requirements of radio astronomy and space science by close contact with the relevant Commissions of URSI, IAU and COSPAR. This proves to be easier for radio astronomy than for space science, chiefly because of differences in the ways in which experimental work is planned. For example, a set of radio astronomical observations of the lines emitted by molecules in a particular region of sky determines at once the frequencies to be used. If the molecular lines are not judged to be important, they will not have got any form of "protection" and the observer will have to do his best to find a place and time suitable for his observations. But the lines may have been judged as important (and IUCAF, the IAU and the CCIR make and up-date such judgements) and then they are likely to have some degree of protection. And if the science appears to justify an attempt to change the degree of protection, the proposal can be made to the ITU for such a change. In some space science there can also be a clear picture of the frequencies which are scientifically most important. Obervations made from a satellite of the earth's atmosphere are an example. So also are some earth observation experiments, and studies of the ionosphere where frequencies in exact harmonic relation are needed. But quite a lot of space science does not call for exact frequency requirements although it may well need specific bandwidths to be available. In order to be informed on as wide a base as possible of the requirements of space science, IUCAF has become associated with the work of the Space Frequency Co-ordinating Group (SFCG), in addition to keeping in touch with the work of the CCIR Study Group 2 and the relevant COSPAR Commissions.

The SFCG consists of representatives from several space agencies and meets usually once a year. It is intended to provide a forum for multi-lateral discussion and coordination of spectrum and orbit matters of mutual interest concerning a number of space radio communication services. The group is particularly interested in getting an early understanding of future plans for space systems and in making suggestions with regard to current and future frequency needs. Members of the group also have an interest in radio astronomy and radar astronomy when these are relevant to spacecraft missions. For some years two members of IUCAF have been attending SFCG meetings as observers, and the informal interchanges have proved very useful. Despite the slightly informality, Dr. F. Horner of IUCAF acted as SFCG chairman at their meeting in Paris at the end of 1987.

4. IUCAF in recent years

(a) World Administrative Radio Conferences

The most recent full-scale WARC was held in 1979 and has already been referred to in paragraph 3 above. Two more specialised WARC's have taken place since that time, one on the use of the geostationary orbit met first in 1985 (ORB-85) and will convene for a second session at the end of August 1988. A WARC on the mobile services took place in Geneva in September 1987. This showed the pressure to introduce various services using satellite systems. The suggested frequencies lie close to the OH line frequencies. There was one serious outcome from this WARC as far as radioastronomy is concerned, and that was the re-allocation of the 1660 to 1660.5 MHz band to the Land Mobile Satellite Service, thus making it an equal partner in that band with radioastronomy. This situation is very unsatisfactory and will need action in the near future. The CCIR Study Group 8 has formed an International Working Party to study mobile satellite systems and it may be possible that this group will also consider the problems of sharing with radioastronomy.

It seems likely that the ITU will hold a WARC in the next few years to review the usage of the band of frequencies from one to three GHz. This band includes the hydrogen and hydroxyl lines and thus is of paramount importance to radioastronomy. If such a WARC is to take place it will require preparation by IUCAF and radioastronomers world-wide.

(b) The "passive" frequency bands

Since about 1983 IUCAF has tried to act more energetically to look at the problems of harmful interference being radiated into the "passive" bands. These are the bands which at the 1979 WARC became allocated to passive services, such as Radioastronomy and Space Research (Passive). There have been cases of space-borne systems being planned or implemented in the 1600-1700 MHz part of the spectrum which appeared to be threatening. One such system, the launch of two spacecraft to Venus and then on to rendezvous with comet Halley used a spacecraft frequency of 1667.8 MHz which fell within a passive radio astronomy and space research band. There were good practical reasons for the choice of this frequency and the power levels radiated were low. The mission was very successful; IUCAF helped to inform radioastronomers of the transmission schedules of the spacecraft, and no interference was reported. (No problems were anticipated in space science since the power levels at earth were very low.)

The concerns that space-borne transmissions would impinge on the passive bands was expressed in resolutions passed within the past four years by URSI, and IAU and COSPAR. These asked administrations planning experiments requiring radio transmissions from space to use IUCAF to help protect sensitive passive radio observations.

This agreement on the need for further action by IUCAF on this difficult subject is certainly welcome; but the format for that action is far from clear. IUCAF has encouraged the international watch for potential or actual cases of improper usage of the protected bands. At present there are cases of damaging interference from satellites in the 1610-1614 MHz band, which is protected for radio observations of the OH line. Such cases can be studied and the characteristics of the interference can be made known to scientists in countries around the world. IUCAF documents on such interference go as a matter of course to the IFRB as an adviser to IUCAF. The relationship between IUCAF and the SFCG described in paragraph 3 also opens a channel by which planners of activities in space can be kept alert to the need to protect the protected bands. In fact, the SFCG at this meeting in Tokyo in April 1986 adopted a Resolution very similar to the statements from URSI, the IAU and COSPAR.

(c) Meetings of European Radio astronomers

Meetings of radio astronomers from most European observatories have been held in the spring of 1987 and again in 1988. These were organised by H.C. Kahlmann and T.A. Th. Spoelstra of the Netherlands Foundation for Radioastronomy. The purpose of the meetings was to recognise the real and growing threat to radio astronomy from interference generated by other spectrum users and then to identify means by which this threat could be countered. The objectives of these meetings were clearly parallel to those of IUCAF and so members of the Commission attended each meeting. At the first meeting the following recommendation was agreed:

An informal and representative international group of radio astronomers, meeting in Paris on March 31st and April 1st 1987

CONSIDERING

1. the increasing threat to the survival of radio astronomy from transmissions of active services in some frequency bands, especially when air-borne or space stations are used by these services,

2. the general lack of awareness among astronomers that the radio windows to the universe are in danger of being at least partially closed by man-made interference,

3. that IUCAF exists for conveying to the ITU and to national communication authorities information on the frequency requirements of radio astronomers and space scientists

4. that nevertheless, it is necessary for firm proposals to the ITU be made through one or more national administrations

RECOMMENDS THAT IUCAF

1. continue efforts, both directly and through its parent unions, to raise the level of interest on matters of frequency allocation among astronomers in Europe and elsewhere,

2. request IAU to inform the managements of all radio astronomy observatories of the need to improve contacts with the users of active radio services in their countries, with the object of increasing general awareness of the damaging effects which those services could have on the radio astronomy service,

3. ensure that national scientific academies are made aware of the increasing threats to the survival of radio astronomy in many frequency bands and are requested to use their influence in obtaining satisfactory solutions to the problems of interference to observations.

(d) IUCAF and other meetings

IUCAF met together during the URSI Assembly in Tel Aviv in August 1987. Since the IAU Assembly in New Delhi the Commission has had representatives at the following meetings:

(a) The 6th meeting of the SFCG, Tokyo, April 21-25 1986.

(b) The XXVIth meeting of COSPAR, Toulouse, June 30 - July 11 1986.

(c) The meetings of European Radio astronomers, Paris, April 31 - May

1st 1987 and Bonn, February 23 - 24th 1988.

(d) The 7th meeting of the SFCG, Paris November 16th - 20th 1987.

Throughout the period F. Horner has been the chairman of CCIR Study Group 2 and has thus been present at meetings of that group and also at the Joint Working Party for the ITU Orbit Conference 1988.

5. IUCAF in the next few years

It will be clear that the main tasks facing IUCAF in the coming years will be to maintain, and if possible to improve, the status of the IUCAF sciences in the degree of protection that the ITU Radio Regulations afford. Not only will changes in the frequency allocations be proposed by other services, but also as other services increase their usage of bands throughout the spectrum so will the likelihood of interference to passive services increase.

As has already been explained, the most serious immediate difficulties will face radioastronomy, but space research will also need attention.

The means by which scientific and technical information can be formulated and supplied to the ITU is through the work of the CCIR. This is a regularly scheduled and ongoing task carried out by study groups, working parties etc., and the output is presented at CCIR Plenary meetings. Input to the CCIR is from the various Administrations which join in the work. To make this input effective in, for example, radioastronomy it is essential that in countries with strong radioastronomy programs there is direct interchange from scientists to the part of the administration working with the CCIR.

In a parallel manner, the discussions which take place within an administration while a WARC is being planned must allow for input in a fairly formal way from the scientists to the administration. Once a WARC is in progress, if matters of serious scientific importance arise, well-managed administrations will see to it that their delegates at the WARC are assisted in making correct choices.

But it will be obvious that these tasks of laying good scientific and technical ground-work with the CCIR and of making both careful preparation before a WARC and having a well-briefed delegation at a WARC are serious and difficult. The IUCAF experience shows that only a few administrations are able to fulfil these objectives.

It would seem that IUCAF should be able to lead in simplifying these processes. Again, using radioastronomy as an example, IUCAF has been able in the past to generate well before a WARC a position paper which carries an international agreement on what radioastronomers hope for at the WARC's end. This process worked reasonably well up to the conclusion of the 1979 WARC. But, for example, at the WARC MOB-87 questions arose on sharing an allocation between a mobile service and radioastronomy. An unsatisfactory conclusion was reached (see para. 4(a); this suggests that the IUCAF mechanism may be unable to help in the much faster-moving world.

If ways could be found to improve the links within countries between scientists and their administrations it might be that some of the agreed international opinions could be taken into consideration at an earlier stage. One very welcome development is the radioastronomers meetings in Europe (para. 4(c). if one or more such regional associations were formed the co-ordination of opinions could be much improved.

Lastly, IUCAF has concluded that the number of members appointed by URSI and the IAU should be increased from two to four. This would make the membership eight for radioastronomy and two for space research. At present this change has been presented to URSI and if approval comes from the IAU and COSPAR it is likely to come about. It recognises the need for more help from radioastronomers around the world.

COMMISSION No. 41

HISTORY OF ASTRONOMY (HISTOIRE DE L'ASTRONOMIE)

PRESIDENT J. A. EDDY

The 20th General Assembly of the IAU was held in Baltimore from 2 to 11 August 1988, and commission 41 held six sessions in all. The retiring President, J. A. Eddy, conducted a business meeting at which new officers for 1988-91 were installed:

President:	J. D. North
Vice-president:	S. Débarbat
Organizing Committee:	A. A. Gurshtein
	D. H. De Vorkin
	J. A. Eddy
	Xi Ze-Zong

A report on the present state of the General History of Astronomy was presented by O. Gingerich: there was one volume (4A) published, a second (2A) was in press, and a third (1A) was likely to be ready to go to press in 1989. Volume 3 should be ready by the end of the same year.

S. Débarbat raised the question of library and archive conservation, and noted that despite our lamentations in 1967, and similar regrets expressed in 1977 by Commission 5 (Documentation and Astronomical Data), personal papers of great importance to the history of astronomy continue to be dispersed or destroyed. It was resolved that she be invited to act on behalf of the Commission in setting up a working party to explore the problem, jointly with Commission 5.

Those present at the business meeting learned with regret of the deaths of two valued members of Commission 41, Z. Horsky and E. V. Douglas. Short addresses commemorating their work were given, respectively by O. Gingerich and E. S. Kennedy.

Themes addressed in papers presented at the sessions that followed included: the solar eclipse of July 8 1842, as described in Austrian scientific literature and novels (M.Firneis); coordinates of the stars as extracted from the Hipparchus commentary (J. Brunet); the Ford Foundation and the European Southern Observatory (F. K. Edmonson); the guest star of A.D. 185 (Y. Huang); a chronology of Chinese astronomical records (T. Kiang); pre-telescope sunspot sightings (address by retiring President J. A. Eddy); astronomical alignments at Newgrange (T. Ray); the observatory of Jai Singh (J. Saad-Cook); the great Copernicus chase (O. Gingerich); the U.S.Navy corps of Professors of Mathematics (C. J. Peterson); use of eclipse records of ancient China, Japan and Korea in the study of the Earth's palaeorotation (L. Zhisen); the history of astronomy as a means of disseminating astronomical knowledge (S.Débarbat); a French-Canadian text of the sphere (E. S. Kennedy); and the globe from Matelica (A. Carusi).

The two sessions held on 6 August were given over to a series of formal papers followed by a round-table disscussion on issues in the history of space astronomy, organized by D.H. de Vorkin. An examination was made of the state of modern astronomy and how it has been influenced by access to space, whether from balloons, sounding rockets, satellites, space probes, or manned spacecraft. The historians taking part were R. Doel, B. Hevly, R. Smith, J. Tatarewicz, and C. Waff, while W. E. Brunk, K. G. Henize and H. Friedman provided a critical commentary based on personal involvement and experience. The moderator of the sessions was M. Harwit. These sessions were highly successful: not only were the papers of high quality, but they were tightly knit and formed a coherent whole, and well illustrated the virtues of organizing meetings along thematic lines.

It is perhaps worth recording that all sessions were usually well attended, with occasionally as many as 75 in the auditorium, and rarely fewer than 50.

In closing the final session, the new President, J. D. North, obtained the approval of members present for the Commission's support of a motion by Commissions 19 and 31 that a history be prepared of the Bureau International de l'Heure and the International Polar Motion Service. The rider was added, that the two authors proposed (B. Guinot and S. Yumi) to obtain the services of a professional historian of astronomy, to advise and assist where possible.

The names of four historians were announced as new members of the Union at the closing assembly: D. King, M. Yano, E. S. Kennedy, and J. L. Lankford.

COMMISSION No. 42

CLOSE BINARY STARS (ETOILES DOUBLES SERREES)

Report of Meeting, 4 August 1988

PRESIDENT: J. Smak SECRETARY: B. J. Hrivnak

The business meeting of the Commission was held in Room 103 of the Convention Center on August 4 from 9:00-10:30, and was followed by a scientific meeting from 11:00-12:30.

The President extended a welcome to all present, and introduced a tentative agenda which was accepted. Dr. Bruce J. Hrivnak was then elected as Secretary for the meeting.

The President read the names of Commission members who had died during the previous three years: M. Fracassini, J. C. Kemp, and M. W. Ovenden and asked the attending members for a moment of silence to pay tribute to their memory.

The President announced that the Organizing Committee had recommended to the IAU Executive Committee that the new Commission President should be Robert H. Koch and the new Vice-President - Yoji Kondo. These recommendations were endorsed by the Commission. Continuing members of the Organizing Committee for 1988-91 are: E. Budding, A. M. Cherepashchuk, K.-C. Leung, J. Rahe, M. Rodono, J. Smak (as past President), and G. Shaviv. Vacancies on the Organizing Committee were then filled by the nomination and approval of A. Gimenez, R. W. Hilditch, R. F. Webbink, and A. Yamasaki, with the intention that a radio astronomer be added as soon as possible. The President expressed his appreciation to the Vice-President R. H. Koch and to the Organizing Committee for their support and cooperation over the past three years. Concerning the new Commission members, the President informed that all those who were nominated were approved by the Organizing Committee, and a final check is being performed to see that they have been elected as Union members based upon nomination by their National Committees.

Tibor Herzeg discussed the status of the *Bibliography and Program Notes*, especially with regard to present difficult problems of printing and disseminating the pages. He raised for discussion the question whether the *BPN* should continue and, if so, whether he should continue as Editor. Following discussion, the continuation of both the *Bibliography* and Dr. Herzeg in his post was endorsed. The following resolution was adopted:

> Recognizing the great value of the *Bibliography and Program Notes* to the work of IAU Commission 42, the Commission expresses its gratitude to Dr. T. Herzeg and strongly encourages his continued work in editing this publication.

Alan H. Batten reported on the status of the *Eighth Catalogue of Orbital Elements of Spectroscopic Binary Systems*, noting some of the financial problems and difficult editorial decisions. A discussion ensued on the most appropriate means to publish the *Catalogue*, its detailed format, and an editorial successor to Dr. Batten for future editions. The following resolution was adopted:

> Recognizing the great value of the *Eighth Catalogue of Orbital Elements of Spectroscopic Binary Systems* to the work of IAU Commission 42, the Commission expresses its gratitude to Dr. A. H. Batten for his editorial and scholarly service.

Robert H. Koch briefly addressed the subject of a new *Finding List of Close Binary Systems*, noting a similarity with the problems raised by Dr. Batten. Shortly a query will be mailed to users in order to gather information regarding the desirability of a new edition. The following resolution was adopted:

> Recognizing the great value of the *Finding List of Close Binary Systems* to the work of IAU Commission 42, the Commission expresses its gratitude to Dr. R. H. Koch and Dr. F. B. Wood.

R. H. Koch drew attention to two principles conveyed to the Commission by the IAU President J. Sahade. These are: (1) to note the circumstances of astronomers in disadvantaged locations and bring them into active ground-based and space work, and (2) to draw attention to coordinated observational programs of specific objects or classes of objects for specific intervals of time. He reported that the Organizing Committee was sympathetic to the principle of the first of these and drew attention to a Working Group concerned with Multi-Wavelength and Multi-Facility Programs.

A. H. Batten reported on the program of the IAU Colloquium No. 107, *Algols*, to be held in Victoria following the General Assembly.

The Commission recognized with appreciation the leadership by J. Smak as President during the past triennium.

During the scientific session the following papers were presented:

S. J. Schiller - The Hyades Binary HD 27130
M. B. K. Sarma - Distortion Wave Characteristics and Spot Modeling of the RS CVn Eclipsing Binary SV Cam
J. V. Clausen - Recent Results from the Copenhagen Binary Project
A. Gimenez and A. Claret - On the Internal Structure of Main-Sequence Stars
G. Djurasevic - Analysis of Close Binaries by Applying Inverse-Problem Method
A. Cherepashchuk - Photometry of SS 433 during 1979-87
G. Bakos - AW UMa.

FORMATION OF CLOSE AND WIDE BINARIES (Session 42/4 = 37/4 = 26/6) 9 August 1988

reported by V. Trimble

Alan Boss reviewed the four traditional mechanisms for forming binary systems -- fragmentation, fission, capture, and separate nucleation, as well as the statistical properties of known binaries, which the formation mechanisms must account for. He concluded that we now understand both the data and the physical processes well enough to decide that the vast majority of field binaries must have originated from fragmentation. Additional fragmentation results were presented by Shinji Narita and Michel Bossi, the former suggesting that it may be very difficult to get successive stages of fragmentation (because the first will already leave gas clouds that are too centrally condensed to fragment further), and the latter reporting that a thermodynamic consideration of the process indicates the possibility of producing a bimodal distribution of mass ratios.

Fission simulations were reported by Harold Williams and Richard Durisen (in absentia). They concur that fission can make transient spiral arms, rings, and discs, but, so far, no binaries. Most of the mass, however, remains in a triaxial central bar, whose fate under further collapse is still to be determined.

Stellar encounters and captures are important in dense environments like cluster cores. Willy Benz reported that, when two-body encounters are close enough for something interesting to happen, disc formation and mergers are more likely that binary capture. Douglas Heggie presented results for three-body encounters and captures. What happens as a function of total energy, and the rate of formation of permanent bound pairs are now solved problems for clusters consisting of stars of a single mass. A clever way of generalizing to multi-mass systems is badly needed.

Observers were under-represented relative to theorists, four of the originally scheduled speakers having developed assorted conflicts. Ramiro de la Reza discussed the properties of several pre-main-sequence binaries, including the isolated T Tau V 4046 Sgr. He noted that perhaps the most serious problem associated with such systems is understanding how they form, apparently in isolation from either gas or other young stars. Jean-Louis Halbwachs showed results of his analysis of samples of visual and spectroscopic binaries with main sequenc primaries. After allowance for selection effects, he concludes that the distribution of mass ratios in both samples is the same and consistent with random selection of components from a standard initial mass function. Discussion participants noted that it is possible to choose binary samples with different statistical properties, the BY Dra stars favoring large mass ratios and the K giants studied by R. Griffin small ones.

Finally, three talks addressed topics related to young binaries. Mikio Nagasawa has carried out a variety of three-dimensional, hydrodynamic simulations of tidal disruptions and related processes. David Bradstreet's investigation of the kinematics of W UMa stars finds that they have a relatively large average space motion of 66 km/s and also a velocity dispersion characteristic of old disc populations. This favors a model for their formation in which systems gradually evolve into contact from initial periods of 2-5 days. Tsevi Mazeh presented three samples of binaries of known ages in which the shortest orbit period that has not yet been circularized is correlated with age, from 10.7 days in M67 at 4-5 X 10^9 yr, through 5.7 days in clusters at 10^8 yr, to 4 days in pre-MS systems with 10^6 yr ages. These numbers imply a dependence of circularization time or orbit period much less steep than the 16/3 power law given by simple theory.

COMMISSION No. 44

ASTRONOMY FROM SPACE (L'ASTRONOMIE A PARTIR DE L'ESPACE)

Report of Meetings

PRESIDENT: Y. Kondo SECRETARY: A. Michalitsianos

I. BUSINESS SESSION

(A) New Officers. At the business session on August 4, the following new officers were elected for the next triennial period.

President - E. B. Jenkins
Vice President - J. Trümper
Secretary - A. Michalitsianos

Organizing Committee: B.F. Burke, G.W. Clark, G.G. Fazio, J.B. Hutchings, S.D. Jordan, Y. Kondo (Past President), K.A. Pounds (Past Vice President), J. Rahe, B.D. Savage, G.B. Sholomitsky, R.A. Sunyaev, Y. Tanaka, W. Wamsteker.

(B) New Members: The following individuals have been elected as new members of of our Commission: I.D. Ahmad, C.L. Bennett, A. Brown, J. Butler, P.S. Butterworth, K.G. Carpenter, F. Codova, M. Elvis, U.O. Fisk, B.H. Foing, M.G. Hauser, C.L. Imhoff, K.V.K. Iyengar, K.S. Long, J.C. Mather, J. Mead, F. Melia, C.A. Norman, G.K.H. Oertel, P.M. Perry, J. Pipher, G. Pizzichini, R.Z. Sagdeev, J. Sahade, W. Sanders, K.-H Schmidt, C. Sheffield, P.L. Smith, G. Sonneborn, J. E. Steiner, M. Valtonen, O. Vilhu.

(C) Working Groups. Three proposals for working groups were submitted to Commission 44. Two were accepted and the third was declined.

(1) Working Group on multi-wavelength astrophysics was established as a multi-Commission Working Group. The members of this working group are J. Bregman, J. Butler, F. Cordova, T. Courvoisier (Chairman), B. Foing, M. Malkan, C. Sterken and F. Verbunt. J. Bregsacher, G. Riegler and N. White are advisors to the Working Group.

(2) A proposal by Harlan Smith for "Lunar-Based Astronomy" Working Group was accepted. H. Smith was designated as its chairman.

(3) A proposal to transfer a Working Group on "Space Schmidt Survey" from Commission 28 was turned down. If modifications are made to the name and the charter of the Working Group so that they will become consistent with the basic policies of Commission 44, which require avoiding involvement in a lobbying effort for a space experiment, they may submit another proposal at the next meeting of Commission 44 in 1991.

(D) Sponsorship of meetings.

(1) Commission 44 is the principal sponsor of IAU Colloquium No. 123 on "Observatories in Earth Orbit and Beyond"; Commissions 9 and 48 are cosponsors. It is planned for the autumn of 1989 at Goddard Space Flight Center.

(2) Commission 44 is cosponsoring IAU Colloquium No. 139 on "Galactic and Extragalactic Radiation", which is to be held in Heidelberg, Germany 12-16 June 1989.

(3) Commission 44 is also sponsoring a colloquium on "Extreme UV Astronomy", to be held at Berkeley on 19-20 January 1989.

(II) Science Sessions.

Commission 44 cosponsored several Joint Discussions and Joint Meetings at the XXth IAU General Assembly in Baltimore. Since they are to be reported elsewhere in this volume, we will describe those science sessions for which our Commission was the primary sponsor.

(A) Future Space Programs. This joint meeting with Commission 9 was organized and chaired by Y. Kondo and was held on August 8. The invited speakers were C. Pellerin of NASA, R. Bonnet of ESA, Y. Tanaka of Japan and N. Kardashev of the U.S.S.R., who described the future space programs of their respective country or organization.

(1) Pellerin discussed NASA's plans for supporting astrophysical data analysis program using the results from the ongoing IUE (ultraviolet) and the projected COBE (infrared), Hubble Space Telescope (ultraviolet, visible and infrared), ROSAT (X-ray), GRO (gamma-ray), EUVE (extreme ultraviolet), IRTS (infrared), ISO (infrared), Astro-D (X-ray), XTE (X-ray) and AXAF (X-ray) astronomical satellites.

(2) Bonnet gave a summary of ESA's plans for astronomical satellites, including the so-called "Horizon 2000" program. Among the approved near-term programs are IUE, Giotto Extended Mission (comets), Hipparcos (astrometry), HST (with NASA), Ulysses (solar polar mission with NASA), and ISO (possibly with NASA, Japan and Australia). The long-term programs, known also as the "Horizon 2000" programs, consist of SOHO (solar heliosphere observatory with NASA), Cluster (space plasma physics mission with NASA), XMM (X-ray multi-mirror), FIRST (sub-millimeter spectroscopy), and CNSR (comet nucleus sample return mission with NASA). Medium-size projects under consideration include GRASP (gamma-ray), Lyman (far and extreme ultraviolet satellite probably with NASA and Canada), Quasat (radio interferometry mission with NASA and Canadian participation), Cassini (interplanetary mission to Saturn with NASA), and Vesta (interplanetary mission to small bodies - USSR, CNES and NASA collaboration).

(3) Tanaka described the ongoing X-ray satellite Ginga and discussed Japan's plans for Solar A (X-ray), SFU (multi-purpose space flyer unit), IRTS (infrared), ASTRO-D (X-ray), and Space VLBI (long base-line interferometry) missions.

(4) Kardashev reviewed U.S.S.R.'s future astronomical satellite programs consisting of Radioastron, RELICT 2 (infrared), AELITA (infrared), SUVT (ultraviolet), GRANAT (X-ray and gamma-ray), Spectrum-X (X-ray and gamma-ray), Gamma-1 (gamma-ray) and XG100 (a 100 square meter telescope planned for the year 2000).

(B) Highlights of the IUE Satellite Observatory. This science session, which was held to commemorate the first 10 years of the successful guest observer program of the IUE, was organized by Y. Kondo and W. Wamsteker and was held on August 3. The meeting was chaired by Kondo and B. Burke. The invited speakers and their topics were: W. Moos on Io Torus; R. Wagener on Upper Atmospheres of the Planets; C. la Dous on Cataclysmic Variables; E. Guinan on interacting binaries; P. Judge on stellar chromospheres; W. R. Hamman on W-R star atmospheres; M. Elvis on active galactic nuclei; and R. Chevalier on galactic haloes.

(C) Infrared Astronomy. This joint science session with Commission 9 was organized by J. Pipher, co-chaired by Pipher and M. Werner, and was held on August 6. The invited speakers and their topics were G. Neugebauer - introduction; E. Dwek, P. Solomon and F. Boulanger on IRAS; M. Harwitt on ISO; G. Fazio on SIRTF; J. Mather on COBE; H. Okuda on IRTS; N. Kardashev on RELICT 2 and AELITA; D. Downes on FIRST/LDR; and M. Longair - summary.

(D) Far and Extreme Ultraviolet Astronomy. This joint science session with Commission 9 was organized and chaired by A. Michalitsianos and was held on August 9. Invited speakers and their topics were: H. Nussbaumer on stellar spectroscopy in the far and extreme UV; D. York on the physics of the interstellar and intergalactic medium; and J. Krolik on extragalactic astrophysics. The meeting was closed with a panel discussion, chaired by F. Bruhweiler, on the importance of extreme and far UV observations to astrophysics.

(E) Gamma-ray and X-ray astronomy. This joint science session with Commission 9 was organized and chaired by Y. Tanaka and was held on August 9. The invited speakers and their topics were: R. Remillard on the nature of the hard X-ray sources from HEAO-1 survey; G. Fabiano on X-ray observations of normal galaxies; Y. Tanaka on the results from Ginga observations; N. V. D. Klis on guasi-periodic oscillations in compact X-ray sources; R. Burg on cosmic X-ray background; R. Pallavicini on X-ray study of stars; and K. Hurley on gamma-ray bursts.

(F) Radio interferometry from space. This joint science session with Commission 40 was organized and chaired by B. Burker and Y. Kondo. The invited speakers and their topics were: R. Linfield on TDRSS experiment; N. Kardashev on Radioastron; N. Kaifu on VSOP; R. Schilizzi on Quasat status; D. Murphy on orbital studies; K. Weiler on prospects for low-frequency space VLBI; A. Ferrari - a theorist's view; A. Baudry and N. Kardashev - speculations for the future.

COMMISSION No. 45

STELLAR CLASSIFICATION (CLASSIFICATION STELLAIRE)

Report of meetings in Baltimore, Maryland, USA, August 1988

President: R.F. Garrison Secretary: D.J. MacConnell

Business Meeting, 8 August 1988

I. REPORT OF THE PRESIDENT:

A. IAU Reports on Astronomy:

There were new headings for the IAU reports this time, reflecting the increase in activity in the fields of automatic classification and classification of extra-atmospheric spectra.

It was decided that a given person should be asked to write a report for at least two terms, since it takes at least one term to learn how to do it efficiently and properly, but not much more than two, since it is a somewhat onerous task.

The IAU reports are very valuable, since not all of the references are accessible in the Astronomy and Astrophysics Abstracts under the obvious headings. The personal bibliographic files of an expert in the field usually contain much more. Of course, if the reporter uses only the abstracts, it is not worth the effort. Thus a compilation drawing on the two is more valuable than either separately.

B. Support of Colloquia, Symposia, etc. by Commission 45:

Past:

Symposium #118 "Instrumentation and Research Programmes for Small Telescopes"
Colloquium #93 "Physics of Be Stars"
Colloquium #95 "Faint Blue Stars"
Workshop "Astronomy with Large Databases"
Colloquium #106 "Evolution of Peculiar Red Giant Stars"
Colloquium #110 "Library and Information Services in Astronomy"

Proposed:

"Evolution of Stars: The Photospheric-Abundance Connection" Bulgaria 1990 summer.
"Astrophysics of the Galaxy through Precision Photometry" New York 1990 October.
"WR and Luminous Blue Stars" Indonesia, 1990 June.
"The Magellanic Clouds" Australia

II. MEMBERSHIP:

A. New Members: G.S.D. Babu (India), R. Coluzzi (Italy), G. Pizzichini (Italy), and Yamashita (Japan).

B. Resigning Members: D. Evans (USA), Osawa (Japan), B. Warner (South Africa), R. West (Germany), P. Hill (UK), R. Herman (France).

C. Deceased Members: Fracassini, Stromgren.

A suggestion was made that the list should be cleaned up by asking members in the next newsletter if they wish to remain on the list. At the same time, it was suggested that we co-opt new members by inviting people who are writing papers in the field, but who are not members. New IAU members can be proposed by the commissions as well as by member countries. In that way, we can remain an active commission of interested people.

III. OFFICERS:

A. President: M. Golay (Switzerland)

B. Vice President: D.J. MacConnell (USA) with E.H. Olsen (Denmark) as second

C. Scientific Organizing Committee:

1. Remaining:
   R.F. Garrison (Canada) as past president
   N. Houk (USA)
   T. Lloyd-Evans (South Africa)
   E.H. Olsen (Denmark)

2. Retiring:
   Claria (Argentina) - requested
   Heck (France) - after 2 terms
   Slettebak (USA) - after 4 terms!
   Straizys (USSR) - as past-past president

3. Proposed (and adopted) New Members of the S.O.C.:
   C. Corbally (Vatican) H. Levato (Argentina)
   N.R. Walborn (USA) R. Wing (USA)
   Zdanavicius (USSR)

IV. REPORT OF THE WORKING GROUPS:

After considerable correspondence with various people regarding guidelines for Working Groups, I was surprised to find that:

1) There is no "official status" for working groups; i.e. they don't exist as far as the IAU is concerned.
2) There are NO guidelines.
3) Any or no commissions are needed for "sponsorship," which in any case is not defined.
4) There is no IAU financial support to WG's or their newsletters.
5) There is no need to name a WG before proposing a meeting.

So, we set them up as we wish; they are merely a convenience for large commissions. However, the IAU Secretariat and the Executive Committee wish to be informed about the creation of a Working Group and the name of the chair-person.

A. Ap Stars:

It was suggested that the name "Ap" stars continue to be used to describe the class, even though the class includes Bp and Fp stars, since "CP" implies a somewhat premature interpretation.

Two half-day sessions on Ap stars were held during the IAU, one on 3 August and the other on 5 August. The new Scientific Organizing Committee includes: K. Sadakane (Japan) (Chair), G. Michaud (Canada), D.W. Kurtz (South Africa), Lanz, and I. Hubeny (USA).

The working group was formed in 1979 jointly with Commissions 29, 36 and 45. Since Commission 29 is the "parent," it is assumed that the main report of the scientific sessions will be reported there.

B. Be Stars:

D. Baade reported that the Be Star Newsletter is published twice a year and 200 copies are mailed. There have been 6 issues totalling 160 pages. Gerry Peters will continue as editor and the newsletter is supported financially by ESO. There is regular updating of the bibliography, providing a good resource for workers in the field.

A scientific session was held on the morning of 6 August at the I.A.U. General Assembly in Baltimore.

A meeting is planned for 1991-2 as well as a workshop at an unspecified time. It was suggested that other stars, such as B stars with extended atmospheres, be included.

The Scientific Organizing Committee consists of Baade (Chair), Balona, Dachs, Doazan, Percy, Peters, Marlborough.

C. Standard Stars:

A. H. Batten reported that the working group on standard stars (Commissions 29, 30, and 45) has continued to operate during the last triennium, its most important function being the production, twice a year, of the Standard-Star Information Bulletin, under the capable editorship of L. Pasinetti.

Some members of the group have been involved in Commission 30's re-evaluation of the standard-velocity stars, which was discussed intensively at this Assembly. The Working Group met on 10 August, both for organizational matters and for scientific communications from Adelman, Andersen, Gerbaldi and Neckel.

The Scientific Organizing Committee is composed as follows: S. Adelman, J. Andersen, A. Batten (CHAIR) M. Gerbaldi, I. Glushneva, H. Neckel, and L. Pasinetti.

D. Peculiar Red Giants:

An IAU Colloquium on the "Evolution of Peculiar Red Giant Stars" (#106) was held in Bloomington, Indiana from 26-30 July 1988.

Suggested and informally organized at the Strasbourg meeting ("Cool Stars with Excesses of Heavy Elements") in 1984, the Working Group was formally constituted at the General Assembly in India (1985) under IAU Commissions 29 (Stellar Spectra) and 45 (Stellar Classification.

A twice-yearly newsletter is edited and mailed by C. Jaschek (Strasbourg). A new editor is needed.

Scientific reports were given by R. Foy ("Angular Diameter of Chi Cygni") and U. Jorgensen ("Effects of Polyatomic Molecules on Carbon Star Atmospheres").

The Scientific Organizing committee is composed of R. Foy, R. Garrison, H. Johnson (CHAIR), A. Renzini, R. de la Reza, V. Straizys, T. Tsuji, and R. Wing.

E. Photometric and Spectroscopic Data:

A.G.D. Philip reported that he and D. Egret are making a hard copy version of the standard star data available from the Strasbourg Data Center.

It was suggested during the meeting that Buscombe's catalogue of MK types would be useful if references were included; however, as is (without references), it is actually a disservice to the field of classification.

The Scientific Organizing Committee includes D. Hayes, C. Jaschek, Nandy, A.G.D. Philip (CHAIR) and W. Warren.

V. FUTURE MEETINGS:

I.A.U. General Assembly, Argentina in 1991; others in section I.

VI. RESOLUTIONS: (no resolutions were proposed)

VII. BY-LAWS:

The possibility of developing by-laws for our Commission was considered and rejected. Members present felt that by-laws are not necessary.

VIII. FUTURE SCIENTIFIC DIRECTIONS FOR COMMISSION 45:

Spectral classification is being done with digital detectors and the new methodology must be calibrated. One problem with tracings is that the eye is drawn to central depths in forming line ratios, but the bottom of a line is where the signal-to-noise ratio is the worst. There are several similar problems to be considered if we are to retain the level of discrimination that has been achieved by photographic techniques. (See report below on scientific session.)

N. Houk has just published volume 4 of the Michigan Spectral Catalogue. Work is in progress on volume 5, which will straddle the equator. Three-quarters of the northern hemisphere plates have been taken. The northern survey spectra go farther into the UV, so classifications may be slightly improved over the southern plates.

Automatic classification will be needed for any future projects involving extensions to the HD - e.g. the next 10 million stars. Work is underway by Kurtz and LaSala to develop sophisticated, pattern-recognition algorithms. See report below on scientific session.

The AAT multi-object spectrograph will provide lots of new digital data. Perhaps "quick-look" types could be provided automatically, followed by a more careful study of the most interesting stars.

Houk pointed out that good photometry is needed for the HD stars. The coverage is especially poor in the Southern Hemisphere, where the spectroscopic classifications are now complete. SAAO will probably not do it. Grenon suggested that Hipparcos will provide much new data, but it won't be of the highest precision because of the transformation necessary to obtain UBV data. The ST guide star program will also provide data, but of low precision. MacConnell suggested that qualified amateurs might be drafted to the project. Another possibility is a dedicated automatic photometric telescope. It is ironic that the difficult task of classification by eye is more than half finished, but the purportedly easier task of carrying out photometry is far from complete.

The IAU Traveling Telescope was mentioned. It has a small, uncooled photometer and a basic low-dispersion spectrograph.

Scientific Sessions, 9 August 1988: Baltimore

I. CLASSIFICATION USING DIGITAL DETECTORS:

Five papers were given during this session. A very brief summary of the main points is given hereinafter.

A. "Some Recipes for Classifying Digital Stellar Spectra" by C.J. Corbally

In the process of comparing digital spectroscopic specimens by eye, it is important to have a consistent "recipe" in order to get repeatable results. After catching the photons, the response of each pixel must be "flattened." It is then useful to convert the pixel numbers to wavelength, remove the effects of atmospheric extinction, and calibrate the counts/sec in terms of flux. The effect should be neutral for the same telescope/detector combination. After folding in the desired smoothing of pixels, the effect is a gain of information for a small number of pixels and a loss for a large number. Rectification may be desirable; while it tends to lose information on the continuum profile, it can gain information accessibility by standardization and readability.

The display of the spectrum on the screen (or tracing) can have a profound effect on the classification process. For example, the wavelength range covered for a given line depth affects whether it is the central depth of the line or the equivalent width which is most noticeable. A problem arises if there is too large a wavelength range along with too large a line depth, because the eye will notice the ratio of line depths more than the ratio of equivalent widths. The problem is that the S/N is poorest at the bottom of strong lines and can lead to serious discrepancies in classifications if the central depth is used.

The choice of signal-to-noise ratio is important. While the canonical value of S/N for IIa-0 plates is said to be about 30, that is for an unwidened spectrum. Widening can increase the S/N to the equivalent of well over 100. A S/N of 300 or better is needed to carry out classifications which are equivalent in discrimination to the best photographic work.

Not mentioned above are spectrum subtraction, fourier noise reduction, automatic classification, and other manipulations. The advantages of digital spectroscopy are many and the disadvantages are few, but the pitfalls are many and some of them are very dangerous. It is sobering to remember the look of raw data. However, the appropriate consistent recipe can and will lead to classifiable spectrograms appropriate to the problem. The novice should beware the pitfalls.

B. "UV Spectral Classification: Why Do It?"
by C.D. Garmany

Is there a direct UV-optical correlation, and are there exceptions to the correlation? The answer is a definite yes to both questions.

It is neither cost effective nor time effective to use IUE strictly for spectral classification, but there are many uses for existing archived data.

The UV has certain very definite advantages. In particular, it is important to compare UV data with optical data to look for anomalies. One can examine different layers of a star's atmosphere in the UV. As a bonus, it is possible to observe extragalactic hot stars with this tiny 0.5-m telescope in about the same time as with a 4-m optical telescope.

The IUE gives digital spectra with resolution of either 6A or 0.25A covering 1170-2000A (SWP) or 2000-3200A (LWP). The S/N is only about 30, though spectra can be co-added. The archives are available to all astronomers and it is possible to carry out on-line searches of the catalogue of observations.

The archives contain a total of 34,000 SWP images and 25,000 LWR and LWP images. These figures can be broken down as follows:

TYPES	HI-DISP	LO-DISP
O V,I	1117	932
B0-B5 V	1710	1581
B0-B5 III-I	1049	766
B5-B9 V	438	362
B5-B9 III-I	208	213
A0-A9 V	305	491

(Multiple observations of the same star are included in these figures, but WR, Oe, Be, Ae, sdO, and sdB are not included.) Wu, et al. (1983 NASA-IUE Newsletter #22) published an IUE UV Spectral atlas of standard stars; Heck, et al. (1984 ESA SP-1052) produced an atlas of "normal" stars. Additional Atlases of hot stars have been published by Dean and Bruhweiler (1985 Ap.J.Suppl. vol.57, p.133), by Walborn, Nichols-Bohlin, and Panek (1985 NASA Ref. Publ. 1155) and by Faraggiana et al. (1986 Ap.J.Suppl. vol.61, p.719).

Walborn and Panek have published a series of papers on the correlation of UV spectra with optical types for O stars. Abbott (1978) showed a correlation of stellar escape velocity with wind terminal velocity. Hutchings (1982), and Bruhweiler, Parsons, and Wray (1982) noted a correlation of wind-line strength with the metallicity derived from the optical.

(This preliminary discussion was followed by the showing of a large number of IUE images to illustrate several interesting lines and correlations).

C. "Near-IR CCD Spectroscopy of Red Stars"
by M.S. Bessell

D. "Digital Spectral Classification of OB Stars"
by N.R. Walborn

Digital spectroscopy has come of age from the viewpoint of spectral classification. The quality is entirely comparable to that previously obtained photographically at 60 A/mm and 1.2 mm widening, permitting extension of the MK methodology to large numbers of faint objects. Examples of digital data from three different instruments compared with photographic spectrograms illustrate the point:

1) B supergiants observed by E. Fitzpatrick with the Shectman system at the CTIO 1-m telescope. Resolution 1.2A, 3800-5000A, S/N 50-60.

2) O3 and previously unobserved stars to 15th magnitude in the 30 Doradus central cluster observed by N.R.W. and C. Blades at the AAT with the IPCS and multiple-object, fiber feed. A system under development for the UK Schmidt will provide 400 objects in a single exposure.

3) Ofpe/WN9 spectra observed with the CTIO 4-m SIT vidicon.

An extensive OB digital atlas is planned by N.R.W. and E. Fitzpatrick.

E. "Comments on Classification of B Stars with IUE"
by J. Rountree

The purpose of the project is to examine the applicability of the MK System in the ultraviolet region of the spectrum, to define a set of standard stars, and to establish UV classification criteria.

The sample includes 100 stars with "normal" MK types (no e, p, or n) in the range B0-B8, III-V. Most have also been classified at Strasbourg.

The data consist of high-dispersion IUE spectra taken with the SWP camera, rebinned to 0.25A and normalized. The spectra are plotted on a uniform scale and individually mounted.

The methodology is to group the spectra by existing MK spectral type, to define luminosity classes within the spectral range group, to select standards, and to move discordant stars into an earlier or later group. The process is then repeated until no further changes are needed.

Potential problems include: rotation, interstellar lines, stellar winds and variability.

The status of the project is that the material has been prepared, trends have been identified, and the B2 group has been classified. The target date for completion is December 1988 and publication by NASA of an Atlas will be sometime in 1989.

II. AUTOMATIC CLASSIFICATION

There were 4 papers originally planned for this session, but 2 of the authors were unable, at the last minute, to attend the IAU.

A. "Pattern Recognition Techniques
Applied to a Homogeneous Data Set"
by Jerry LaSala

1. Introduction

Houk, using traditional visual classification techniques, will take about 30 years to reclassify the 225,000 HD stars. Imagine trying to classify all the stars in the BD, CD, and CPD. The next 10 million will be impossible to do by traditional techniques.

The mean errors in Houk's subjective work are:

Houk-Houk	0.44	subtypes (for repeated classifications)
Houk-Garrison	0.63	subtypes
Houk-Cowley	1.3	subtypes

The goals of automatic spectral classification (see West 1973 in IAU Symposium #50) are: high speed, great endurance; homogeneity (elimination of personal errors); accuracy, detection of variability, and possibility of classification of higher dimension.

There are two approaches: criterion evaluation and pattern recognition. In the former, specified criteria (e.g. widths or strengths of specific lines) are automatically measured, then calibrated in terms of desired quantities (e.g. spectral type and luminosity class).

Pattern recognition, on the other hand, involves the comparison of the "total appearance" of the spectrum with spectra of standard stars and can include cross correlation and multivariate statistical methods.

Most of the previous work was reviewed by Kurtz in "The MK Process and Stellar Classification" (edited by Garrison, 1984).

The criterion evaluation approach has been used with up to 450 specified criteria, and gives mean errors comparable to the best visual classifiers at MK dispersion, but this approach is not practical for objective-prism spectra because of its dependence on absolute measurements.

Cross-correlation techniques give mean errors comparable to visual classifiers, but the metric distance method can give improved results. (See Kurtz in "The MK Process"). The latter also gives as bonuses: the resolution of the classification scheme it has developed, insight into possible higher dimensions, and a criterion for quantitative determination of peculiar features.

2. Application to a homogeneous data set

Plates from the Michigan Spectral Survey were borrowed from N. Houk. These were IIa-0 plates with spectra widened to 0.8mm and a dispersion of 108 A/mm. A total of 3000 spectra have been scanned. (Previous studies have involved a maximum of 150 spectra).

The spectra were digitized on an APM, using a raster scan of each spectrum (offset to avoid overlap if necessary) of 128x1024 15-micron pixels. A median filter was used perpendicular to the dispersion to minimize plate defects; this gave 1024-element spectra. The spectra were rectified.

Cross correlation was performed to obtain initial classifications. These were compared with known Houk classifications. Then the metric distance algorithm was applied to classify the stars. This process flags as peculiar any spectrum farther than some maximum distance from <u>any</u> standard. It also flags as peculiar any feature which differs from the standard by more than 3 sigma. A measure of the quality of the classification is obtained from the metric distance.

Illustrations of spectra at all stages were presented for B stars. The beauty of this technique is that it "discovers" the criteria, based on the experience of previous classifications of standards, in much the same way as the visual classifier does. In fact, it is likely that the full development of the method will involve using Houk's 225,000 classifications to teach the machine to classify so it can automatically perform the next several million. In other words, it builds on and extends the traditional techniques rather than serving as a poor substitute.

B. "The Technique of Automatic Quantitative Stellar Spectral Classification Using Stepwise Linear Regression"
by V. Malyuto and T Shvelidze

Using objective-prism spectra with a dispersion of 166 A/mm, the authors have classified standard F-G-K stars. Criteria were selected from previous work at Abastumani and a linear regression model was adopted. A stepping procedure was introduced to choose the best subsets for independent variables.

Standard stars were classified using this mathematical technique and were compared with their standard values. The r.m.s. differences were found to be +/- 0.11 for spectral classes, +/- 0.5 for luminosity classes, and +/- 0.28 for [Fe/H]. Internally, for those spectra which were repeated, the r.m.s. differences were essentially zero. The only significant deviation was in [Fe/H] for the star HD 201251.

The conclusion is that there is general consistency between the calculated and the standard values of the main physical parameters for standard stars in the sample. The technique is well suited for large-scaled galactic structure investigations. A program is underway to study the main meridional cross-section of the Galaxy.

COMMISSION No. 46

TEACHING OF ASTRONOMY

(L'ENSEIGNEMENT DE L'ASTRONOMIE)

Report of Meetings held in Baltimore

PRESIDENT: C.Iwaniszewska SECRETARIES: M.Gerbaldi
L.Gouguenheim

Sessions 1 and 2, August 6,1988

BUSINESS SESSIONS

1. Report of Commission, National Reports

The President reminded the structure of the Commission, which is a Committee of the IAU Executive Committee, and has therefore different rules from other IAU Commissions. Its objective is to further development and improvement of astronomical education at all levels throughout the world. These aims are achieved by:

a. various projects, initiated, maintained and to be developed, like the preparing of National Reports, organizing International Schools for Young Astronomers, maintaining Visiting Lecturers' Programmes in developing countries, printing the Astronomy Education Material, developing the new project of Travelling Telescope,

b. dissemination of information on astronomy teaching at various levels, through publication of the Commission Newsletter, preparing teachers' local meetings on the occasion of IAU General Assemblies, and finally, organizing special meetings like the IAU Colloquium No.105, which took place in July 1988.

The President's Report for 1985-1988 has been approved. Twenty eight National Reports on the state of Astronomy Education in individual countries prepared by the National Representatives have been already published in Newsletter No.24, a few late ones will appear in 1989. The National Committees for Astronomy in 18 countries appointed new National Representatives to Commission 46. There are however still a few countries where either no representatives could be found,or the proposed persons were still no IAU members. The list of the present 41 National Representatives is given below.

2. New Rules for Commission Membership

According to the Rules prepared in 1973 by President Edith A.Müller and Vice President Derek McNally, the Commission was composed mainly of National Representatives, a few regular members, and consultants (not IAU members). In view however of the growing interests in the Commission activity expressed by many IAU members wishing to join the Commission, the Assistant General Secretary Derek McNally elaborated new Membership Rules, accepted by the IAU Executive Committee. Three categories of members are introduced:

Category 1 - IAU members nominated by National Committees or Adhering Bodies of each country, one person per country, who maintain a liaison between the Commission and the country, prepare triennial reports; in fact they are National Representatives, and they can vote on Commission and financial matters;

Category 2 - individual IAU members showing an active interest in the development and improvement of astronomy education, having a right to vote on Commission matters except finance matters;

Category 3 - non-IAU members invited by the Commission to serve for one term, with no voting rights; in fact they are former consultants of the Commission.

It ought to be very strongly pointed out here that the condition of remaining a Commission 46 member <u>of every category for another term is to contribute to the work of the Commission</u> during the triennium between the General Assemblies. Well before every General Assembly the President or Vice President should ask all members what they have done for the Commission 46, and then suggest to withdraw persons who are no longer interested. If then, even the National Representatives fail to make a contribution to the work of the Commission, they will not be accepted by the President as continuing in Category 1, even though renominated by their National Committees. Further, if a given country is not an IAU member, then astronomers from that country can become either Category 2 members (in case of individual IAU membership) or Category 3 - in case of non-IAU members. Finally, if a country becomes an IAU member, then there must exist an Adhering Body, who will have to nominate a National Representative from among IAU individual members, as Category 1 member. After some discussion the new rules have been voted and approved by the Commission.

The President then presented a list of persons, who applied for Commission membership according to the new rules. The list has been approved, so that from now on Commission 46 has a total of 134 members in 3 Categories. New Organizing Committee members have been proposed and approved.

3. Commission 46 Members for 1988-1991

President: Aage Sandqvist, Sweden,
Vice President: Lucienne Gouguenheim, France,
Organizing Committee: Leo Houziaux, Belgium; Syuzo Isobe, Japan; Cecylia Iwaniszewska, Poland; Josip Kleczek, Czechoslovakia; Jay Pasachoff, USA; John Percy, Canada; Robert R.Robbins, USA; Donat G.Wentzel,USA; Richard West, Denmark; William Zealey, Australia.

Category 1 Members (National Representatives)

1. Australia: Alex Rodgers
2. Austria: Hermann Haupt
3. Belgium: Arlette Noels
4. Brazil: Silvio Ferraz-Mello
5. Bulgaria: Nikolai S.Nikolov
6. Canada: Richard Bochonko
7. Chile: Jose Maza
8. China, Nanjing: Ke-jia Feng
9. China, Taipei: Chun-Shan Shen
10. Colombia: Eduardo Brieva
11. Czechoslovakia: Josip Kleczek
12. Denmark: Hans J.Fogh Olsen
13. Egypt: A.Aiad
14. Finland: Heikki Oja
15. France: Lucienne Gouguenheim
16. Germany D.R.: Helmut Zimmermann
17. Germany F.R: Wolfhard Schlosser
18. Greece: L.N.Mavridis
19. Hungary: Gabor Szecsenyi-Nagy
20. India: Souriraja Ramadurai
21. Indonesia: Bambang Hidajat
22. Italy: Eduardo Proverbio
23. Japan: Syuzo Isobe
24. Korea Rep.: Jong-Ok Woo
25. Mexico: Silvia Torres-Peimbert
26. Netherlands: L.L.E.Braes
27. New Zealand: Edward Budding
28. Nigeria: Samuel Okoye
29. Norway: Jan-Erik Solheim
30. Poland: Cecylia Iwaniszewska
31. Portugal: Jose Osorio
32. South Africa: Anthony Fairall
33. Spain: Maria Catala-Poch
34. Sweden: Aage Sandqvist
35. Switzerland: Louis Martinet
36. United Kingdom: David Clarke
37. Uruguay: Julio Fernandez
38. USA: Jay Pasachoff
39. USSR: Edward Kononovich
40. Vatican City State: M.F.McCarthy
41. Venezuela: Nuria Calvet

Category 2 Members (Regular Members)

1. Agnes Acker, France
2. Henri Andrillat, France
3. S.M.Razaullah Ansari, India
4. Mihail Bacalov, Bulgaria
5. Priscilla Benson, USA
6. Elvira Botez, Roumania
7. Lucette Bottinelli, France
8. Noah Brosch, Israel
9. William Buscombe, USA
10. Joseph Chamberlain, USA
11. S.J.Codina-Landaberry, Brazil
12. Heather Couper, UK
13. Zhen-Hua Cui, China, Nanjing
14. Jon Darius, UK
15. David Dupuy, USA
16. Marie-France Duval, France
17. J.P.Emerson, UK
18. M.J.Fernandez-Figuerroa, Spain
19. Richard T.Fienberg, USA
20. Julieta Fierro, Mexico
21. Roberto Gallino, Italy
22. Michele Gerbaldi, France
23. Owen Gingerich, USA
24. Hardev Gurm, India
25. Jean-Louis Heudier, France
26. Darrel Hoff, USA
27. Leo Houziaux, Belgium
28. Mohammad Ilyas, Malaysia
29. C.D.Impey, USA
30. Allan Jarrett, South Africa
31. Hans-Ulrich Keller, Germany F.R.
32. John E.Kennedy, Canada
33. Christopher Kitchin, UK
34. Vladimir Kourganoff, France
35. Jerzy Kreiner, Poland
36. Edwin C.Krupp, USA
37. Maria Teresa Lago, Portugal
38. Liu Lin, China, Nanjing
39. Irene B.Little-Marenin, UK
40. Nicholas Lomb, Australia
41. Xing-Yuan Ma, China, Nanjing
42. Walter Maciel, Brazil
43. Ronald Madison, UK
44. Julian Marsh, UK
45. Derek McNally, UK
46. Gospodin Momchev, Bulgaria
47. Guy Moreels, France
48. Juan Muzzio, Argentina
49. Wayne Osborn, USA
50. Terry D.Oswalt, USA
51. Mazlan Othman, Malaysia
52. Naoaki Owaki, Japan

53.Jean-Paul Parisot, France
54.John Percy, Canada
55.Mario Rigutti, Italy
56.Robert R.Robbins, USA
57.Leif Robinson, USA
58.Curt Roslund, Sweden
59.Archie Roy, UK
60.John Safko, USA
61.Blas Sanahuja, Spain
62.P.P.Saxena, India
63.Natcheva Sbirkova, Bulgaria
64.David G.Schleicher, USA
65.Thomas Schmidt, Germany F.R.
66.Edward Schmitter, Nigeria
67.Daniel Schroeder, USA
68.Henry Shipman, USA
69.Jaromir Siroky, Czechoslovakia
70.Vladimir Stefl, Czechoslovakia
71.Bjorn Stenholm, Sweden
72.Alexei Stoev, Bulgaria
73.Jiri Svestka, Czechoslovakia
74.Jose Rosa Taborda, Portugal
75.Alexis Troche-Boggino,Paraguay
76.Sylvie Vauclair, France
77.Simeon Vladimirov, Bulgaria
78.Vladis Vujnovic, Yugoslavia
79.Donat G.Wentzel, USA
80.Richard M.West, Denmark
81.Richard M.Williamon, USA
82.William Zealey, Australia
83.Michael Zeilik, USA

Category 3 Members (Consultants)

1.Maria Luiza Aguilar, Peru
2.Oscar Alvarez, Cuba
3.Robert Duker, USA
4.Mary Kay Hemenway, USA
5.Reza Hayavi M.Khajehpour, Iran
6.Kevin P.Marshall, UK
7.Nidia Morrell, Argentina
8.Andrei Serban, Israel
9.Yupa Vanichai, Thailand
10.Gonzalo Vicino, Uruguay

4. Astronomy Education Material (AEM)

Since 1970 every three years Astronomy Education Materials have been published in three language groups: part A - English, part B - Slavic, part C - all other languages, offering in some cases very extensive lists of astronomical publications. In 1988 part C has been already published in Commission 46 Newsletter No.25, part A and B will appear later. A long discussion followed the report of Michele Gerbaldi on the opinions of National Representatives about the advisability of having the AEM printed. Finally, it was proposed to prepare in future only shorter, more selective lists in 4 languages:part A - English, part B - Russian, part C - French and Spanish. The selection criteria for AEM will have to be decided by the future Organizing Committee.

5. Visiting Lecturers' Programme VLP

There are now two VLP in progress, that of Peru which is still continued, and that of Paraguay which started in spring 1988. The VLP for Nigeria is delayed until the radio telescope installed there will be put into operation, and China's application for VLP is being studied by the Executive Committee. Donat G.Wentzel has been nominated as a permament coordinator of this Commission project.

6. Planetarium Group

As he considered that some planetarium staff members are much isolated from observatories and astronomical news, Hans J.F.Olsen proposed to start getting more information about help that could be offered by the Commission. If it will be found advisable, then a Working Group on Planetarium will be formed at the next IAU General Assembly.

7. International Schools for Young Astronomers (ISYA)

The XIVth ISYA has been held in August 1986 in China, the XVth - in September 1986 in Portugal. The XVIth ISYA in Cuba for Central America announced for 1988 has been postponed till August 1989. The next applicants are: Malaysia - for South-East Asia, and Morocco - for Arabic countries, but no final decision has been taken until now. Donat Wentzel commented on the higher prestige gained by a given country when either the ISYA or the VLP have been organized there by the IAU.

8. Travelling Telescope Project (TT)

Initiated by Derek McNally and Richard West, the project of lending a small telescope to developing countries in conjunction with VLP or ISYA has been finally concluded when John Percy obtained the necessary funds through the Canadian Commission for UNESCO. The instrument has been specially brought by Percy from Canada to Baltimore, and shown to the audience. TT is a small Celestron-8 with a photometer, spectrograph, 35 mm camera, equipped with necessary manuals and instructions for carrying out simple research projects. It belongs to the IAU, applications for use should be sent to the President of Commission 46. John Percy will act as coordinator of this project.

9. IAU Colloquium No.105 "The Teaching of Astronomy"

Commission 46 had been the only IAU Commission sponsoring IAU Colloquium No.105 "The Teaching of Astronomy" held in Williamstown, Mass., on July 26-30, 1988.It was attended by 162 persons coming from 31 countries. The Scientific Committee was chaired by John Percy, the Local Committee - by Jay Pasachoff. The colloquium session programmes varied from teaching at different levels and interdisciplinary approaches, to video-films, computer programmes and planetarium activities. A special feature of the meeting has been the comparatively large attendance of participants from developing countries, hence many discussions centred on their specific difficulties, from the training of astronomers to the publishing of textbooks. The colloquium proceedings edited by J.Percy and J.Pasachoff will be published by Cambridge University Press.

10. Newsletter

John Percy has been editor and Leo Houziaux - printer and distributer of the Commission Newsletter during the last term, publishing 7 issues of about 200 pages in total. Both have been thanked for their work, which they wish to continue. The editor asked members for more cooperation, for sending educational material, especially as every Commission member has been reminded of his duty to contribute to the work of the Commission.

11. Relations with ICSU and ISY

Lucienne Gouguenheim reported on her contacts with the International Council of Scientific Unions - Committee for Teaching Science (ICSU - CTS), where she met representants of other teaching commissions of scientific unions and exchanged valuable information. She is also acting as ICSU-CTS representative in the ICSU Committee for the International Space Year (ISY), scheduled for 1992. The major themes for ISY activities are connected with science, applications, education and public information, hence they fall within Commission 46 interests.

12. Contributed Papers

Two short contributed papers have been presented at the session. One, by Titus A.Th.Spoelstra on interference problems in radio astronomy, which should be presented also when teaching astronomy. Another, by Syuzo Isobe, on the existence of a Working Group on Teaching Astronomy in the Asian-Pacific Region, with members from Australia, China, Egypt, Indonesia, Korea Rep., Malaysia, and S.Isobe from Japan as chairman. A future exchange of teaching materials as well as astronomy teachers is planned.

Session 3, August 9, 1988
(jointly with Com.5)

DEVELOPING COUNTRIES

George A.Wilkins of Commission 5 presided the joint meeting of Commissions 5 and 46 devoted to reviewing problems of developing countries in respect to documentation and teaching, and to considering what help might be offered. John Percy gave an account of the discussion on textbooks held during IAU Colloquium No.105. He spoke about the difficulties in obtaining funds for buying books from abroad, the ways of preparing local publications, better suited to specific education conditions in a given country. A. Ratnakar brought into focus some ideas discussed at the recent IAU Colloquium No. 110 "Library and Information Services in Astronomy", related to preprints and publications distribution. Participants commented on the impor -

tance of having "Current Contents" in astronomical institutions (J. Muzzio), on a possibility of obtaining reduced rates for main astronomical journals sold to libraries (A.Sandqvist, H.J.F.Olsen), or to individual astronomers (H.Gurm, S.Okoye), on help in sending journals through twinning some countries (G.A.Wilkins), on the values of publications sent from neighbouring countries of the same language group (M.L.Aguilar).

Lucienne Gouguenheim told about the ways by which less developed countries can benefit through the ISY, which intends to produce astronomical materials for public education as well as to offer scholarships. However all these activities will be made possible only if the Space Agencies will give the necessary financial support. A discussion brought further points of immediate interest for researchers, as information on the possibility of obtaining a waiver of printing charges (I.Onuora, S.Torres-Peimbert), of obtaining free-of-charge books from retiring professionals (A.Sandqvist), of getting help at embassies when sending books to developing countries (A.Aiad), of UNESCO coupons for foreign books (E.A.Müller).

Sessions 4 and 5, August 10, 1988

A COMPARATIVE STUDY OF UNIVERSITY DEGREES

Aage Sandqvist introduced the session on astronomy teaching programmes at university level - by asking what are the benefits of the diversity in the world's educational systems? What is being done to increase the mobility of students in world regional areas ? What can Commission 46 do to help students overcome barriers ? Leo Houziaux mentioned the system now reigning in Western Europe, where there is a mutual recognition of Astronomy Degrees, and the programme for stimulating doctoral theses, ERASMUS, is connecting 15 institutions in 11 countries. In 1976 the number of IAU members per million inhabitants has been 4.1 for Europe and 5.1 for USA. In 1987 the corresponding numbers have been 6.2 and 7.5, respectively. David Clarke's paper dealt with the general situation in obtaining astronomy degrees in UK universities, while Michael Dworetsky told about the programme of astronomy studies at the University College in London.

The requirements for obtaining a degree in USSR have been presented in a paper of Edward Kononovich. There are three degrees in astronomy: graduation at the end of university studies, the degree of candidate of physical-mathematical sciences, and the degree of doctor of sciences. Each degree requires a preparation of thesis, presented at special meetings with oponents and referees. Alfonso Serrano spoke about the situation in Mexico, where the National Autonomous University in Mexico City is chosen by 300 000 students. A first Ph.D. programme in Astronomy is just being introduced in the country. Jay Pasachoff introduced the complex situation in USA, where the majority of students attend for four years colleges, and then only 10 % go for graduate studies at universities. Although some universities are rated very high, no correlation has been found between a successful career and the place of study.

The situation in Uruguay has been reviewed by Gonzalo Tancredi, with an emphasis on the revival of research and teaching that took place in that country in 1985. Hardev Gurm told about the new programme of university studies now being introduced in India. Only two Indian universities, that in Patiala and Osmania University in Hyderabad, conduct now graduate studies in astronomy. However the situatio will improve in the near future, as it has been mentioned by Souriraja Ramadurai, with the opening in 1989 of an Inter-University Centre for Astronomy and Astrophysics on the campus of the University of Poona, where M.Sc and Ph.D. programmes as well as other advanced worshops will be conducted. A discussion then centred on the question of diversity of curriculae, which ought to be related to the specific situation in a given country. Uruguay and USSR proposed to put more emphasis on the importance of including astronomy in secondary school programmes, special attention being paid to the possibilities of introducing astronomy as a separate subject. Such a situation exists now only in 4 countries in the world: Germany D.R., Greece, USSR and Uruguay.

Since this had been the last session of Commission 46 during the XXth IAU General Assembly, the President thanked warmly all members of the Organizing Committee, all organizers of the IAU Colloquium No.105, and finally, the director and staff of the Institute of Astronomy in Torun (Poland) ,where she is working, for their help with the sending of her numerous letters and countless telexes during the past three years.

Special Session, August 1, 1988

TEACHERS' MEETING

About 80 schoolteachers, planetarium instructors, science-centre specialists together with 20 astronomers attended the traditional meeting with local teachers on the occasion of IAU General Assemblies. The meeting has been organized by Harry Shipman with the cooperation of the Division of Continuing Education of the University of Delaware. Information on project STAR ("Science Teaching through its Astronomical Roots") - a new programme for highschool astronomy carried on at the Harvard-Smithsonian Centre for Astrophysics met with a keen interest. Other lectures have been devoted to the Supernova 1987, new conceptions of the Solar System, observations of solar eclipses, and finally, about the available astronomy teaching resources.

COMMISSION No. 47

COSMOLOGY (COSMOLOGIE)

Report of Meetings on 3 and 8 August 1988

PRESIDENT: G. Setti. VICE PRESIDENT: K. Sato.

I. BUSINESS MEETING

At the beginning of the meeting the president asked for a minute of silence to commemorate the recent deaths of two eminent members of the Commission, Professor G.C. McVittie and Professor Ya.B. Zel'dovich, also past Commission president, whose contributions to science, and to the advancement of cosmology in particular, have been outstanding.

The president informed the members that the great majority of the Commission members who answered the questionnaire circulated on March 19, 1987 expressed the opinion that the vice president should take over the office of president at the end of the 3-year term. Accordingly, he has proposed to the IAU Executive the nomination of Prof. Katsuhiko Sato (Japan) as president of Comm. 47 for the period 1988-1991. He also informed the members that in consultation with members of the Organizing Committee he has proposed Prof. R. Bruce Partridge (U.S.A.) as vice president. These nominations were endorsed by the participating members of the Commission, who also approved the new composition of the Commission's Organizing Committee as follows: A. Dressler (U.S.A.), L.Z. Fang (China), J.V. Narlikar (India), M.J. Rees (U.K.), H. Reeves (France), G. Setti (Italy), S. Shandarin (USSR), P. Shaver (Canada) and V. Trimble (U.S.A.). All new members have been contacted and have accepted to be "active" members of the O.C.

Membership of the Commission. The president indicated that according to the resolution approved at the time of the Gen. Assembly in New Delhi he had asked the Commission members to confirm explicitly their interest in remaining members of Comm. 47. 126 affirmative answers and one negative out of 279 members were received. With the recent additions, the Commission's membership has now grown to a total of just above 300 members. In view of the steadily increasing size of the Commission membership the incoming president should make a second and final enquiry, and the members who will persistently not answer will have to be automatically erased from the Commission membership.

II. SCIENTIFIC SESSIONS

Seven sessions were held on different topics, each of them introduced by a review of the main observational and theoretical developments obtained in recent years. Each session was chaired by the corresponding reviewer. The content of the sessions was as follows:

1) Microwave Background (rev., G. De Zotti)
Sub-mm Spectrum of the CMB (T. Matsumoto)
New Limits on CMB Anisotropies (R. Windhorst)

2) Very Early Universe (rev., K. Sato)
Optimistic Cosmological Model (N.S. Kardashev)
On Dirac's Large Number Hypothesis (F.A.N. Mohammed)
A Temporally Homogeneous Approach to the Early Universe (I.E. Segal)

3) Primordial Nucleosynthesis (rev. H. Reeves)
The Effects of Quark-Hadron Phase Transition on BBN (K. Sato)
Lithium Abundances in Pop. II Stars (D.K. Duncan)

4) Large Scale Distribution of Matter (rev., J.E. Peebles)
Filamentary Structure in 3-D Surveys and 2-D Wedges (S. Bhavsar)
Components of the Large Scale Velocity Field (A. Szalay)
Statistical Studies of Complete Galaxy Samples (J.F. Nicoll)

5) Distribution of Matter at Large Redshifts ($z \gtrsim 1$) (rev., A. Szalay)
Cosmological Constraints on the Clustering of XRB Sources (P. Mészaros)
Evolutionary Status of Galaxies at $z \sim 2.5$ (R.W. Hunstead)
4C41.17: a Radio Galaxy at $z = 3.8$ (K. Chambers, G. Miley)

6) Formation of Structures and Evolution (rev., S. Shandarin)
Primeval and "Rejuvenated" Galaxies: A New Search for Them (L. Ozernoy)
First-Ranked Cluster Galaxies: A Two Population Model (S. Bhavsar)

7) Quasar Evolution and Clustering Properties (rev., P. Shaver)
Limits on Dust in Damped Lyα Systems and the Obscuration of Quasars (M. Fall)
Quantization in Lyα Absorption Lines in QSOs: Determining q_o (W.G. Tifft)
Gravitational Micro-Lensing from Distant Masses (S. Refsdal)

III. ABSTRACTS

Following are the abstracts of some of the contributed papers. The two review papers by G. De Zotti and H. Reeves and the contribution by T. Matsumoto will be published in the "Highlights" of the IAU Gen. Assembly.

1) New Limits to Fluctuations in the CMB (R.A. Windhorst, J. Kristian, E.B. Fomalont and K.I. Kellermann):

We present the first part of a long term, 250 hour, deep survey with the VLA at 6 cm. The field is centered at the Palomar Ultradeep Survey Area, PUDS-2 ($14^h\ 16^m + 53°$), for which we collected ~ 30 hours of direct CCD data with the Palomar 200 inch four-shooter. This field will also be observed for ~ 40 hours with Space Telescope by the WFPC-team in their GTO time.
Currently, we have completed the observation, calibration, and reduction of 50 hours in the D (1987), and 50 hours in the C-array (1988). The full synthesis will be achieved through an additional 50 hours in the B-array (1989) and A-array (1990), followed by a final D-array integration to increase sensitivity on MWBG fluctuations. The combined 100 hours D+C array map has an rms noise of 4.5 μJy. This extends the normalized, differential 6 cm source counts down to ~ 20 μJy directly, and down to ~ 4 μJy through a P(D) analysis, both with a slope of -2.1 ($N(>S) \alpha S^{-1.1}$). This confirms the steep slope of the upturn discovered in the 21 cm counts. We subtracted this discrete source population from the noise distribution and performed a careful variance analysis of the remainder, in comparison with a previous deep 50 hour D-array integration. The resulting 2σ upper limits to fluctuations in the MWBG are now lowered to 12×10^{-5} at 18", and 6×10^{-5} at 60" scales.

2) Very Early Universe (review by K. Sato):

The research on inflation in the inhomogeneous universe has been reviewed. The inflation, which was proposed for the explanation of flatness and horizon problem, paradoxically, has been investigated usually in the context of R-W (homogeneous, isotropic) models. It has been concluded that: a) Strict cosmological "no hair theorem" (all the inhomogeneities are damped by inflation) does not hold, but "weak no hair theorem" works. b) Inhomogeneities evolve to wormholes and become child universes (causally disconnected region) if suitable conditions are satisfied. In this scenario, the universe we are now observing is one of the locally flat regions.

3) Optimistic Cosmological Model (N.S. Kardashev):

It was demonstrated that for a certain type of hidden mass (negative vacuum density and flat domain walls) in positive curvature models a regime of periodic oscillations of the Universe without approaching the singularity, and even a regime of steady-state, can be realized. A possibility of evolution from inflation and phase-transition stage to periodic oscillation and steady-state was considered. An opportunity for observational testing and some unresolved problems for the model were also considered.

4) On Dirac's Large Number Hypothesis (F.A.N. Mohammed):

The expression for redshifts is a dimensionless number containing Hubble's constant, hence inverse of Hubble's constant is inserted in Planck's relation to get a dimensionless number dependent on the frequency of a photon. For photons available in (p,p) annihilation this number approaches numerical coincidences.

5) A Temporally Homogeneous Approach to the Early Universe (I.E. Segal):

The chronometric theory explains many observed phenomena without any adjustable parameters such as q_0 or Λ, or evolution, and is not in disagreement with any directly observed complete samples of galaxies, quasars, or radio sources. In particular, it directly predicts the isotropy of the background radiation and the Planck-law spectrum at low frequencies; the redness of the observed spectral shift; the relation to redshift of flux, angular diameter, superluminal proper motion, and number counts, for specified classes of sources, in complete samples; etc.

With the adjunction of further, testable, simple and natural hypotheses, it is explanatory of galaxy formation without observable effect on the background radiation (via infrequent but stochastically inevitable development of regions of extreme temperature and density); the relative flatness of the spectrum of the X-ray background (by implying the capacity for repeated circuits of the universe by sufficiently energetic photons); the missing mass in clusters (by compatibility with a partially distance-dependent attribution of the redshifts of apparent cluster members). Among other determinations, it provides an explicit estimate for the extragalactic distance scale (or radius of the universe) from observations on a random sample of superluminal proper motions, which in conjunction with recent data provides effective reconciliation of disparate estimates of the Hubble parameter.

On an objective scientific basis it therefore appears quite preferable to primeval explosion scenarios and the Doppler theory of the redshift.

6) Filamentary Structure in 3-D Surveys and 2-D "Wedges" (S.P. Bhavsar):

We test the reality of linear features apparent to the eye in the distribution of galaxies in the CfA redshift survey. A filament finding algorithm - the Minimal Spanning Tree (MST) - is used to identify the filaments, and a data permuting technique, that leaves intact small scale structure but randomizes it on large scales, is used to determine the statistical significance of filamentary features. We find that the filaments are real. This is the first objective, statistical evidence of their physical existence, in a general 3-D survey.

Tests are also performed on numerical simulations (by Mellott) with various initial conditions having hot and cold dark matter. Comparisons are made among both 3-D cubes extracted from the numerical models, and 2-D "wedges" of data. Ways of identifying and quantifying the significance of filamentary and other non-specific patterns are discussed.

7) Components of the Large Scale Velocity Field (A. Szalay):

The multipole expansion of the peculiar velocities in the elliptical galaxy sample of Davies et al. (1988) was discussed, with special emphasis on the monopole, dipole and quadrupole components. It was found that the subtraction of the Hubble flow has removed a major part of the monopole, but it remained there in the subsamples. Also, image processing of velocity pictures revealed alignment of the dipole and quadrupole anisotropies.

8) Statistical Studies of Complete Galaxy Samples (J.F. Nicoll and I.E. Segal):

Objective, reproducible, statistically efficient studies of complete samples of magnitude-redshift and angular diameter-redshift relations in complete galaxy samples in part invalidate (samples of Nilson and of Visvanathan) and in part contraindicate (RSA and CfA samples) the Hubble law, in the absence of further hypotheses postulating local irregularities responsible for the deviations between prediction and observation.
Computer simulations inclusive of major known types of local irregularities provide no indication that they can reconcile the Hubble law with observation, and indicate rather conclusively that peculiar velocities, superclustering, and motion of the Galaxy cannot do so. In contrast the Lundmark law (quadratic redshift-distance relation) is in excellent agreement with observations, optimal among power laws, and its relative superiority of fit is unaffected by any of the assumed local irregularities introduced into the computer simulation.
On an objective scientific basis, the Lundmark law therefore appears quite preferable to the Hubble law.

9) Cosmological Constraints on the Clustering of XRB Sources (Z. Bagoly, H. Mészaros and P. Mészaros):

We discuss limits on the large scale clustering of the sources of the X-ray background at redshifts $z > 0.5$. It is found that for structures of mean present separations around 30 Mpc, the comoving mean density of structures can be constant or may at most grow as the first power of the universal lengthscale. One can set an upper limit of about a factor 35 on the luminosity dispersion at a given redshift, if the density of sources is the present estimated lower limit $n = 5\ 10^3$/sq.deg., or a factor 10^2 if it is $n = 4\ 10^4$/sq.deg. If at $z > 0.5$ there exists a constant comoving number of structures of present mean separations about 60 Mpc and overdensity exceeding a factor two, then the closure parameter is limited to $\Omega_o < 0.4$. Clusterings on larger scales comparable to recently reported complexes or great attractors may not exist at such redshifts unless their filling factor is below about 0.03%.

10) Evolutionary Status of Galaxies at $z \sim 2.5$ (R.W. Hunstead, M. Pettini, A. Boksenberg and A.B. Fletcher):

QSO absorption systems in which the Lyman alpha profile shows damping wings have been interpreted as the H I disks of young intervening galaxies. Studies of heavy element enrichment in one such system [z(abs) = 2.309 towards PHL 957] indicates an abundance of only 1/20 solar with very little evidence for dust. In another system [z(abs) = 2.465 towards Q0836+113] we find narrow (≤ 50 km/s FWHM) Lyman alpha emission in the base of the damped Lyman alpha absorption line. The star formation rate inferred from the Lyman alpha luminosity may be as low as 1 $M_\odot$/year.

11) 4C41.17: a Radio Galaxy at $z = 3.8$ (K.C. Chambers, G.K. Miley and W. van Breugel):

Galaxies associated with powerful radio sources are among the most frequently used cosmological probes, because their enormous radio luminosities enable them to be easily pinpointed out to large distances. Based on groundwork laid with the Westerbork telescope during the mid-seventies (Tielens, Miley and Willis

1979; Blumenthal and Miley 1979), some of us have recently developed a unique technique for finding galaxies with extremely high redshifts. This method uses the fact that ultra-steep spectrum radio sources are systematically much more luminous than normal-spectrum sources. During the last two years we have conducted a multispectral investigation of a northern 4C sample of ultra-steep spectrum radio sources and their counterparts.
Of our 33 4C objects, 93% were identified with faint, presumably high-redshift galaxies. The spectroscopy has shown that about half of these objects have strong emission lines and are definitely at high redshifts. Preliminary analysis has resulted in 8 candidates for galaxies having z > 2, of which 8 are confirmed identifications with optically extended objects having more than one emission line. The confirmed redshifts are z = 2.2, 2.3, 2.5, 2.6, 2.9, 3.8, and 3.8. The largest redshift is associated with the radio galaxy 4C41.17. The lines and continuum in 4C41.17 both extend over several arcseconds proving the nature of the object beyond doubt. These redshifts should be compared with z = 1.8 for 3C326.1, the largest galaxy redshift known until about one year ago (McCarthy et al. 1987a).
This opens up an exciting opportunity for studying the properties of the early Universe. For the first time we can measure the spatial distribution of brightness and velocity for gas in objects out to at least a redshift of 4. This will provide unique information about young and forming galaxies. In addition statistics of the space density of these objects and its redshift dependence are capable of providing considerable leverage on the geometry and evolution of the Universe.
Our 4C sample covered less than a third of the sky and included only the brightest radio sources. The obvious next step is to extend our search technique to larger and fainter samples of radio sources. The spatial density distribution of these objects and its dependence on redshift is a topic of fundamental interest. The Lyman alpha emission and radio emission from 4C41.17 could easily be seen if it were at a redshift of 6, beyond which Lyman alpha would be shifted into the near-infrared.
We have therefore begun multiwavelength observations of several samples of objects carefully chosen so that dependence of subsequent results on the various radio properties can be disentangled and to find more z > 2 galaxies. Preliminary samples have been selected from the Parkes, and Texas surveys and comprise 431 objects. In addition we are planning similar selections of subsamples of Molonglo and Westerbork sources. Radio images are already being obtained with the VLA, or the Molonglo Synthesis Radio Telescope (MOST) in collaboration with Richard Hunstead (Univ. of Sydney). Eventually, it is intended to obtain higher resolution images of a subsample of the most interesting southern sources with the Australia Telescope (AT).

12) Primeval and "Rejuvenated" Galaxies: a New Search for Them (L.M. Ozernoy):

Current ideas on primeval galaxies are usually based on an implicit assumption that the environmental effects are not of crucial importance for both the formation and the early evolution of those galaxies. However, galaxies mostly belong to multiple systems and loose groups whose formation was presumably quite close in time to the formation of the galaxies themselves. Since distances between galaxies in such aggregates are of the order of galaxy sizes, and the velocity dispersions of galaxies are comparable to the intragalactic velocity dispersions, close interactions between the primeval galaxies, including merging, could play a decisive role in their early history.
An immediate result of this line of attack is that primeval galaxies should have some common features with ultraluminous starburst galaxies. In particular, their thermal continuum spectrum, after an initial short stage of dust formation, is expected to peak in the far infrared (at $\lambda_0 \sim 100$ μm, which is roughly independent of whether the grain temperatures are determined by radiation of O-B stars or by dissipation of kinetic energy in collisions). Such a peak redshifted to submillimeter wave range is a signature of primeval galaxies; it can be used for a crude dating of the epoch of galaxy formation.

The subsequent evolution of young galaxies could be accompanied by mutual encounters with the companions and, as a result, by recurrent bursts of star formation. Such an environmentally regulated evolution results in a non-monotonous character of both the star formation in, and luminosity evolution of, young galaxies. Obviously, the nearby starburst galaxies demonstrate one of these latest burst phases. Therefore, the starburst galaxies (including those which are to be found at larger redshifts) might be representatives of "rejuvenated" galaxies, i.e., of rather old galaxies with "fresh" stars formed. Some recent candidates for young galaxies seem to be rejuvenated, rather than genuine primeval, galaxies.

13) First-Ranked Cluster Galaxies: a Two Population Model (S.P. Bhavsar):

The brightest galaxies in rich clusters have become the classic standard yard-sticks in observational Cosmology because of their uniform luminosities. The dispersion in the absolute magnitudes is only 0.35 mag. It has been debated whether these galaxies belong to a special class of objects (Peach, Sandage, Tremaine and Richstone) or are just the tail end of a statistical luminosity function (Peebles, Geller). Bhavsar and Barrow have used results from extreme value theory to test the statistical hypothesis, and found it to be incompatible with the current observations. What then is the explanation for the distribution of M_1 for these galaxies?
The M_1 distribution of these galaxies can best be explained by a model in which they are drawn from two distinct populations - a population of "special" objects distributed normally with a small dispersion in magnitude; and a population of extremes of a statistical distribution, also with a small dispersion in magnitude. The maximum likelihood fit of this model with data gives valuable information on the statistics of these objects, with implications for cD galaxies. It is also an independent and unique way to probe the luminosity function of galaxies at the bright end, and suggests it to be much steeper than conventionally described.

14) Limits on Dust in Damped Lyα Systems and the Obscuration of Quasars (S.M. Fall and Y. Pei):

The damped Lyα systems discovered in the spectra of quasars at high redshifts are natural places to search for dust. They have column densities of neutral hydrogen greater than 10^{20} cm^{-2} and may be protogalaxies or galactic disks in an early, gas-rich phase. We compare the spectra of quasars in the Wolfe et al. survey that have damped Lyα with those that do not have damped Lyα to obtain statistical information about the reddening by dust. Our results are given in terms of the dimensionless dust-to-gas ratio $k \equiv 10^{21} (\tau_B/N_H)$ cm^{-2}, where τ_B is the optical depth in the B band in the rest frame of an absorber and N_H is the column density of neutral hydrogen. Using non-parametric tests, we find, at the 95% confidence level, $k \leq 0.41$ (Gal), $k \leq 0.29$ (LMC) and $k \leq 0.19$ (SMC), depending on whether the extinction curve is assumed to have the same shape as that in the Milky Way or the Large or Small Magellanic Clouds. Our upper limit on the dust-to-gas ratio in damped Lyα systems are half the observed value in the Milky Way but are several times larger than the observed values in the Magellanic Clouds. We also develop a new method to set limits on the mean and the variance of the optical depth along random lines of sight. This includes a correction for the effect, pointed out by Ostriker and Heisler, that highly obscured quasars are less likely to be included in optically-selected samples than quasars with little dust in the foreground. Our results for the mean optical depth in the B band of the observer can be approximated by the formula $\bar{\tau}_B(z) = 0.4\tau_*[1+z)^{5/2}-1]$; using the upper limits on the dust-to-gas ratio in the damped Lyα systems, we obtain $\tau_* \leq 0.06$ (LMC) and $\tau_* \leq 0.05$ (SMC). For comparison, the first models by Ostriker and Heisler predict $\tau_* = 0.4$ or 0.8 and their more recent models predict $\tau_* = 0.16$. We conclude that neither set of models is consistent with our limits. As a consequence, the apparent cutoff in the counts of quasars at $z \simeq 3$ is probably not caused by dust in the damped Lyα systems. All the limits derived in this paper could be reduced significantly or a positive detection could be made by determining more accurately the spectral indices of the quasars in the Wolfe et al. survey.

15) Quantization in Ly-α Absorption Lines in QSOs: Determining q_o (W.G. Tifft and W.J. Cocke):

Current versions of the theory of the quantization of extragalactic redshifts (e.g., Cocke 1985, Ap.J. 288, 22) all state that the basic quantization interval should - locally - be proportional to $H(t)^{1/2}$ where H(t) is the Hubble "constant" at time t. QSO Lyman-α forests are rich enough in absorption lines that one might hope to find quantization in them and thus be able to say something about H(t) at different epochs.
Standard cosmology gives the relation $H(t) = H_o(1+z)\ (2q_oz+1)^{1/2}$, where q_o is the present value of the deceleration parameter and H_o is the present value of the Hubble constant. Determining H(t) at any non-zero redshift therefore provides a value of q_o.
We have analyzed four independent sets of Ly-α forest data covering a wide range of redshifts ($1.8 \leq z \leq 3.7$). Velocity differences of $c\delta z$ between neighboring absorption lines were evaluated in the local rest frame by dividing by 1+z, where z is the averaged z of the line pair. We then normalized them back to the present epoch by multiplying by the expression $(H_o/H(t))^{1/2}$, in accord with the theoretical results quoted at the top of the panel. The result is a normalized velocity difference

$$\tilde{\delta}v = (H_o/H(t))^{1/2}\ c\delta z/(1+z) = (1+z)^{-3/2}(2q_oz+1)^{-1/4}c\delta z.$$

Then a histogram of the $\tilde{\delta}$vs should show the quantization at 24.15, 36.22, 48.30, and/or 72.45 km/s at some preferred value of q_o, since the dependence on H(t) has been normalized out. In fact, all four data sets show quantization at multiples of 24.15 km/s, like the dwarf galaxy samples analyzed in 1984. In all four data sets, one of the strongly preferred values of q_o is 1/2.

16) Gravitational Micro-Lensing From Distant Masses (S. Refsdal):

Recent observations of QSO 2237+030 show 4 images with identical spectra (Yee, 1988; Schneider et al., 1988; Roberts and Yee, 1988). This is convincing evidence for a gravitational lens system with the lensing galaxy unusually close to us ($z_G = 0.039$). For this system the expected effects from micro-lensing are larger, and can be more accurately estimated than for any other system presently known. For a quasar radius of ≈ 0.1 l.y., one tenth of the mass in stars and an effective transverse velocity in the observer plane of 600 km/s, we do expect for each of the 4 images a typical change of 0.05 magnitude per year due to micro-lensing. This estimate is based on theoretical calculations by Kayser, Refsdal and Stabell (1986) which cover the relevant parameter values. Since the time delay Δt between the images is only about 1 day or less, micro-lensing will produce a change in luminosity ratios, and an observed such change will be a proof of micro-lensing, since intrinsic variability will show up "simultaneously" in all images.

COMMISSION No. 48

HIGH-ENERGY ASTROPHYSICS

(ASTROPHYSIQUE DES HAUTES ENERGIES)

Report of Meetings, 8,9 and 10 August 1988

PRESIDENT : Catherine J. Cesarsky

INTRODUCTION

The Scientific sessions of Commission 48 at the XX General Assembly of the International Astronomical Union in Baltimore, USA, were focussed, on the one hand, on the most exciting recent event in the field : Supernova 1987A, and on the other hand on novel theoretical ideas. The first two sessions, held on August 8 and 9, were devoted to SN 1987A : the first to high energy observations, the second to theory ; the second session ended with an animated general discussion of this fascinating object. The third-session, on August 10, was fully occupied by four extensive reviews, dealing with some recent developments in theoretical high energy astrophysics. Here is a list of the presentations given :

Session 1 : High energy observations of SN 1987A (chairman : F. Pacini)

Satellite results :

Y. Tanaka : (Institute of Space and Astronautical Science, Tokyo)
Results of the Ginga satellite.

J. Trümper : (Max-Planck Institut für Extraterrestrische Physik, Garching)
The hard X-ray emission of SN 1987A (MIR-KVANT)

S.M. Matz : (N.R.L. Washington)
SMM gamma-ray observations of SN 1987A.

Balloon results :

W.S. Paciesas : (NASA/MSFC, Huntsville)
Balloon-borne hard X ray experiment

P. Ubertini : (Istituto di Astrofisica Spaziale, Frascati)
Hard X-ray behaviour of SN 1987A

W.R. Cook : (Caltech, Pasadena)
An imaging observation of SN 1987A at gamma-ray energies.

B.J. Teegarden : (G.S.F.C., Washington)
GRIS observations of gamma radiation from SN 1987A : a preliminary report.

A.D. Zych : (University of California, Riverside)
Gamma-ray observation of SN 1987A above 1 MeV.

Ground observations :

E. Budding : (Carter National Observatory of New-Zealand, Wellington)
TeV gamma-rays from SN 1987A.

Session 2 : Supernova 1987A at high energies : theory
(Chairman : C. Cesarsky)

R. Mc Cray (JILA, Boulder)
X-rays from supernova 1987A.

M. Salvati : (Osservatorio Astrofisico di Arcetri, Florence)
X-rays from supernova 1987A : beneath the radio active layers.

T. Stanev : (Bartol Research Institute, Newark)
Acceleration of cosmic rays and production of high energy gamma-ray signals at SN 1987A

R.A. Chevalier : (University of Virginia, Charlottesville)
High energy phenomena related to SN 1987A.

General discussion, led by C. Cesarsky (SAp, Saclay).

Session 3 : Theoretical problems in contemporary high energy astrophysics. (Chairman : M. Rees).

R. Terlevitch : (Royal Greenwich Observatory, Hertsmonceux)
Star formation in Seyfert nuclei.

P. Meszaros : (Pennsylvania State University, University Park)
General relativistic effects in neutron stars.

J.P. Ostriker : (Princeton University, Princeton)
Superconducting strings and the high energy background.

R. Svensson : (Nordita, Copenhagen)
Pair plasmas in astrophysics.

A summary of the invited papers presented in all three sessions is presented below.

Session 1 : High energy observations of SN 1987A

Experiments on board of three satellites have been following up the supernova since it first appeared at high energies in the summer of 1987 : the japanese Ginga satellite in X rays from 6 to 28 keV, the soviet-european complex of hard X-ray instruments on the MIR-KVANT module, and the american SMM satellite in gamma rays. Detailed accounts of the recent results of these three satellites were presented in Baltimore.

Y. Tanaka recalled that X-ray emission from SN 1987A has been observed continuously by the Ginga satellite since July 1987, in the energy range 6-28 keV, with a set of proportional counters of effective area 4000 cm^2. The shape of the hard X-ray spectrum, above 15 keV, is flat, perfectly compatible with the Compton tail expected from the degradation of ^{56}Co-decay gamma rays. The spectrum of the soft component, at energies below 16 keV, is as expected from thermal bremsstrahlung from a gas of temperature in the range 10-12 keV. Apart from a burst in January 88, the X-ray flux appears to have been declining very slowly between September 87 and July 88 ; the possibility of a steady flux cannot be rejected.

In January 1988, a burst of X rays was observed ; the intensity increased over two weeks, then declined over one month. The emissivity of the soft component rose by a factor of ~ 3 ; the spectrum changed as well. During the flare, an emission line was visible at 6.8±0.2 keV ; it is interpreted as due to hydrogen-like or helium-like iron ions ; fluorescent emission would have occurred at lower energy (6.4 keV).

J. Trümper reported the results of the MIR-KVANT observations. According to the author : It was a lucky coïncidence that the KVANT module carrying the X-ray observatory "Röntgen" was launched a few weeks after SN 1987A was first observed. The docking to the MIR station was successfully completed in early April and the astronomical observations commenced on ~ 3 June 1988. The set of X-ray detectors includes a gas scintillation proportional counter (3-20 keV), a coded mask camera TTM (1 arcmin, 2-25 keV) and two Phoswich detectors (HEXE, 15-200 keV and PULSAR X-1 45-1000 keV).

Unfortunately, during June and July SN 1987A was observed only for rather short intervals (in total 3.25h), i.e. with limited sensitivity. The first long set of observations took place mid August 1987. During this period the hard X-ray flux of SN 1987A was first detected with the MIR-HEXE in the collimator rocking mode. In this mode source and background measurements are made simultaneously with two parts of the detector array. Almost all later measurements were performed by rocking the whole MIR station in order to allow the operation of the Pulsar X-1 instrument which has fixed collimators.

Between August 1987 and April 1988 a total of ~ 96 hours has been spent on SN 1987A. Figures 1 and 2 summarize the data obtained with the MIR-HEXE during this time.

After the failure of the TTM star tracker in April 1988 the Supernova has been observed for another ~ 6 hours (until mid August 1988). For this data only preliminary attitude information is available which comes from the MIR station. We prefer to derive source fluxes only after this attitude data has been calibrated. This calibration is in progress.

J. Trümper also commented on the comparison of MIR-KVANT data with other data : The Ginga data obtained at energies between 6 and 27.6 keV is rather complementary to those of the MIR-KVANT in terms of energy coverage. The high energy part of the Ginga spectrum agrees reasonably well with the MIR-HEXE data in the overlapping energy range. The low energy component (6-16 keV) seen by Ginga depicts rather large intensity fluctuations, and must be of different origin. Several balloon observations of SN 1987A have been made in fall 1987 and spring 1988 by the MSFC, the Frascati and the Caltech groups. The resulting hard X-ray fluxes confirm the data obtained with MIR-HEXE within the error limits which are quite large for the balloon experiments. Finally, J. Trümper reminded us that the hard X-ray emission of the SN 1987A had been predicted as a result of Comptonization of Cobalt 56 gamma rays in the expanding shell. However, the early GINGA and MIR-KVANT measurements showed a faster increase than the predicted one, suggesting a considerable mixing of the Cobalt 56 in the inner part of the shell. The rather flat light curve observed requires that the effect of the Cobalt 56 decay be largely compensated by a change in the Compton optical depth. However, this compensation cannot continue for a long time and one may expect that the hard X-ray flux will decrease later in 1988.

The gamma ray spectrometer on-board of the SMM satellite has been monitoring the ^{56}Co decay lines from SN1987A. The results were summarized by **S.M. Matz** : The Gamma-Ray Spectrometer (GRS) on NASA's Solar Maximum Mission satellite detected gamma ray line emission from the decay of ^{56}Co in SN 1987A, beginning in 1987 August. Since the emission from the supernova has been continuously mlonitored with the GRS. The average fluxes from 1987 August 1 to 1988 April 6 are $(6.6\pm2.0)\times10^{-4}$ and $(6.4\pm1.7)\times10^{-4}$ photons $\text{-cm}^{-2}\text{-s}^{-1}$ for the 847 and 1238 keV lines, respectively. The quoted errors represent a combination of statistical and systematic uncertainties. The best fit energies for the two lines during this period are 840±6 and 1239±10 keV. There has been no marked increase in the intensity of either line since the end of October, and the data are consistent with a constant flux since August 1.

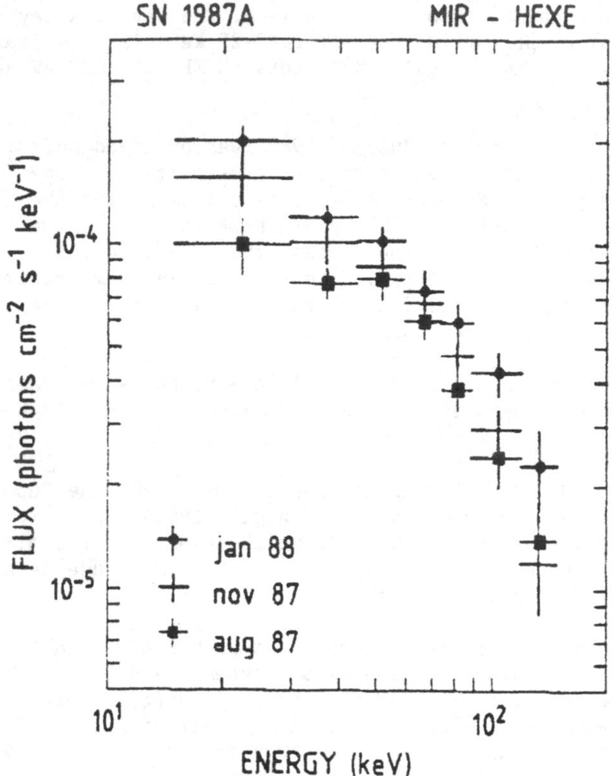

This was followed by a series of short presentations of balloon results. **W. Paciesas** spoke on behalf of the NASA-Lockheed collaboration : the region of the Large Magellanic Cloud containing SN 1987A was observed during a balloon-flight of a hard X-ray telescope on 8-10 April 1988. Significant continuum emission was detected in the 45-400 keV range which is attributed to SN 1987A. Compared to an observation with the same instrument in October 1987, the source intensity decreased by approximately 50% while the spectral shape remained qualitatively the same. This may represent the first clear indication that the hard X-ray emission is entering the declining phase expected as the ejecta become optically thin to the ^{56}Co gamma-rays.

If there is a luminous (~ 10^{38} ergs/s) X-ray source inside SN 1987A, it should cause the bolometric light curve to become flat by t=2.5 years. Spectral evidence (at ~ 20 keV) for a central compact source might also show up at roughly the same time. A pulsar will most likely be detected first (t>4-5 years) at optical wavelengths, or possibly at hard (>40 keV) X-rays if the envelope is fractured.

P. Ubertini presented the results of a hard X-ray experiment developed by an international consortium including groups from Italy, Australia, Germany, UK and USA :

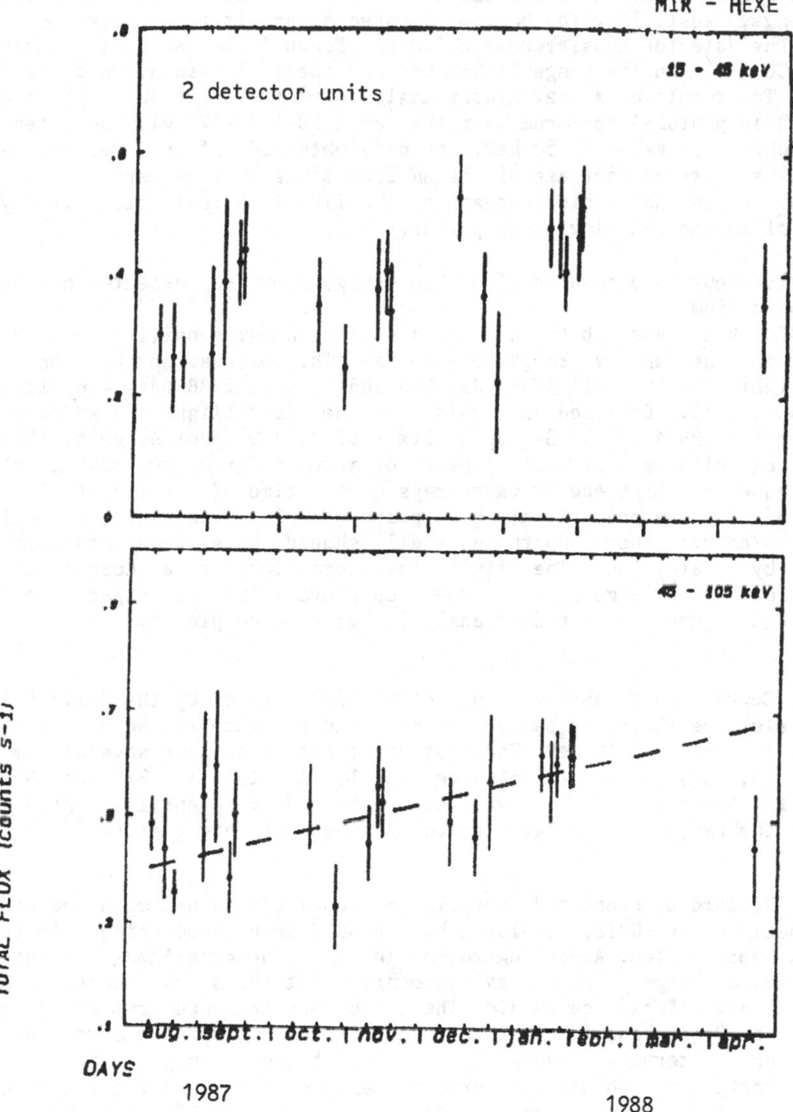
MIR - HEXE
2 detector units
15 - 45 keV
45 - 105 keV
TOTAL FLUX (counts s-1)
aug. sept. oct. nov. dec. jan. febr. mar. apr.
DAYS
1987
1988

On April 5th 1988, the Supernova 1987A was observed between 00.00 and 09.55 UT with a hard X-ray detector flown on a stratospheric balloon launched from Alice Springs, Australia. The balloon floated at an altitude corresponding to 3 to 4 mbar. The detector consisted of a 2.0 bar Xenon-filled MWPC with a sensitive area of 500 cm^2 in the range 15-180 keV and spectral resolution of 8% FWHM at 130 keV . The result of a preliminary analysis of the X-ray data, shows a quite hard ($\alpha \sim 1.3$ in photons) spectrum over the range 20-160 keV, with an intensity of about 10^{-4} ph/cm^2 s keV at 50 keV. The data obtained, if compared with earlier observations suggest an increase of the SN flux since last January.
P. Ubertini noted that, this increase of the flux is a real effect it may be an indication of an emerging low power pulsar.

He also mentioned results of a high energy gamma ray detector developed by the same consortium :

SN1987A has been observed with a spark chamber sensitive to high energy gamma-rays in the energy range 50-500 MeV. The observations were made during balloon flights on 19 April 1987 (day 55) and 5 April 1988 (day 406) from Alice Spring (Australia). Detailed analysis of the first flight showed no positive signal from SN 1987A at the 3-σ upper limit of 2.9×10^{-5} photons/cm^2s. This limit is consistent with an expanding mass of a few solar masses making the shell totally opaque to high energy gamma-rays at the time of the flight. Quick-look analysis of the second flight has placed a 3-σ limit to the flux at 9×10^{-4} photons/cm^2s. The supernova shell should have been transparent to gamma-rays by that time. The limit therefore implies a cosmic-ray proton luminosity of $<5 \times 10^{41}$ ergs/s from the supernova. The team expects to further constrain this figure when the data analysis has been completed.

W.R. Cook showed observations of SN 1987A, taken by the Caltech imaging gamma-ray telescope (GRIP), a balloon-borne coded-aperture telescope operating in the range 10 keV to 10 MeV. The instrument has been flown several times from Alice Springs, Australia, to observe SN 1987A. In May 87, the SN was not detected. In November 87, it was seen at 4.9σ ; in April 88, at 5.4σ. The spectrum in the range 40-1300 keV (including lines) is very flat ($\alpha E^{-0.9}$).

B.J. Teegarden presented surprising results obtained with the Gamma Ray Imaging Spectrometer (GRIS) developed by the GSFC in collaboration with the Bell and Sandia Laboratories. According to the author : the experiment consists of an array of seven large high-purity germanium detectors surrounded by a thick massive NaI anticoïncidence shield. The flight was launched from Alice Springs, Australia. On 30 Apr. and 1 May, 1988 11.3 hours of SN 1987A data were obtained. The experiment alternated every 20 minutes between target and background pointings. Background pointings were normally done by rotating the experiment azimuth by 180° while holding elevation constant. The ^{56}Co supernova line at 1238 keV was detected at a flux level of $(8.1 \pm 1.7) \times 10^{-4}$ photons/cm^2-sec. (The error quoted is statistical only ; systematic errors remain to be evaluated). The statistical significance of this result is 4.9σ. The 1238 keV line width is significantly larger than the instrumental line width. A preliminary estimate is 15-25 keV FWHM corresponding to a velocity range of 3600-6000km/sec. Such large broadening may indicate that a substantial amount of the freshly synthetized material has mixed with faster moving material in the overlying mantle. The data also indicate an excess in the vicinity of the 847 keV ^{56}Co line. Analysis is complicated by the presence of a strong instrumental background line at 844 keV. In addition a strong continuum signal was detected in the 120-700 keV band. Further results will be forthcoming.

Finally, **D. Zych** described the University of California Compton Double

Scatter Telescope, which features two arrays of scintillators of area $1m^2$, one in plastic and the other in NaI. This imaging telescope is sensitive to gamma-rays from 1 to 30 MeV. The instrument was flown from Alice Springs on April 15, 1988. The data analysis was on going at the time of the Baltimore IAU.

E. Budding made a brief presentation of the results of the JANZOS collaboration, which includes laboratories from Australia, New Zealand and Japan. The experiment, installed in the Black Birch Range in New Zealand, consists in one array of scintillators for the detection of UHE (>100TeV) gamma rays, and a system of mirrors to detect Cerenkov light from VHE (>1TeV) gamma rays. The observations started in October 1987, and until now only upper limits are available , except for an interesting event : on January 14 and 15, at a time coïncident with the X-ray flare observed by GINGA, JANZOS detected an excess flux, of order 2.10^{-11} $cm^{-2}s^{-1}$, of TeV gamma rays. The total energy emitted in this burst is comparable to that in the GINGA X ray burst (~ 10^{43} ergs). The probability of this excess to be a chance occurrence was estimated by the speaker at 10^{-4}. This would be the first indication of acceleration of high energy cosmic rays by the compact object at the center of SN 1987A.

- 0 -

Session 2 : Supernova 1987A at high energies : theory.

The subject was introduced by **R. Mc Cray**, who gave a vivid summary of our current understanding of this problem. The interpretation of the hard (>15 keV) X-rays from SN 1987A is straightforward : gamma rays from ^{56}Co decay are degraded by Comptonization to a typical emergent photon energy

$$h\nu \sim mec^2/\tau^2_c$$

where τ_c is the Compton optical depth from the Co to the surface. Photoelectric absorption produces a sharp low energy cut off in the emergent spectrum at approximately 20 $\zeta^{1/4}$ keV, where ζ is the metallicity in solar units of the outer envelope. Detailed Monte-Carlo calculations show that the X-ray spectrum is a power-law, with photon spectral index α ~ 1.5 when the X-rays first emerge. The spectrum becomes harder ($\alpha \rightarrow 1.0$) with time. This general behavior of the spectrum is generic to the gamma ray down-Comptonization mechanism and is insensitive to the details of the supernova model. However, the time of first emergence of the X-rays is sensitive to the model. Calculations based on models with stratified internal composition predicted that the X-rays would be detected initially at t ~ 8 months after outburst. It is possible to account for the observed emergence of hard X-rays at t ~ 4 months by revising the interior model in such a way that the ^{56}Co is close to the surface. "Mixed" models, in which the ^{56}Co is stirred throughout the inner $5M_o$, can fit the X-ray and gamma ray observations fairly well ; so can "fractured" models, in which there are deep holes exposing part of the interior ^{56}Co.

In contrast, there is no satisfactory interpretation of the soft (<15 keV) X-rays observed by the Ginga satellite. They cannot be produced by down-Comptonization of gamma rays. Models have been constructed to fit the observed spectrum in which the X-rays come from shocks formed by the interaction of supernova ejecta with circumstellar matter. However, these models require circumstellar densities that are much greater than would be expected near the supernova progenitor star and that are in conflict with the early observations of radio emission from SN1987A. It has also been suggested that the soft X-rays are produced intrinsically by a neutron or synchrotron nebula at the center of the supernova. This interpretation would require that the supernova envelope has holes that are much more transparent (by factors $>10^3$) than the average envelope. If this explanation is correct, it is easier to account for the observed soft

X-ray variability and spectrum with an accretion source than with a Crab Nebula-like pulsar.

M. Salvati presented work done with R. Bandiera and F. Pacini on the possible contribution of a non-thermal nebula (plerion), produced by a rotating neutron star, to the high energy emission of SN1987A. Granted that the bolometric light curve of the SN and the hard X ray spectrum are well accounted for by the presence of $0.07M_o$ of ^{56}Co after the explosion, there are still the problems of the early appearance of the hard radiation, and of the radiation below the photoelectric cut off. The authors argue that these characteristics of SN 1987A can be understood if two mechanisms are simultaneously at work. They should be relevant respectively, above and, below a photon energy ~ 20 keV ; besides being the location of the expected cut-off, this energy also corresponds to an observed spectral feature, indeed suggestive of a physical discontinuity.

A good candidate for the soft X-ray emission is synchrotron radiation from a non-thermal nebula, powered by a central neutron star. This can be detected if the envelope has undergone a process of early fragmentation, leading to the opening of transparent lines of sight towards the central region. Most of the hard X-ray flux can be provided by radioactivity, although an important contribution from the pulsar nebula cannot be ruled out. A possible fit to the spectrum observed at day 190 with these two components, is shown in figure 3 (where the dotted line represents the maximum, unperturbed plerion emission).

M. Salvati also emphasized the fact that, if the plerion is absent, the hard X ray flux of SN 1987A will have decreased considerably by the end of 1988.

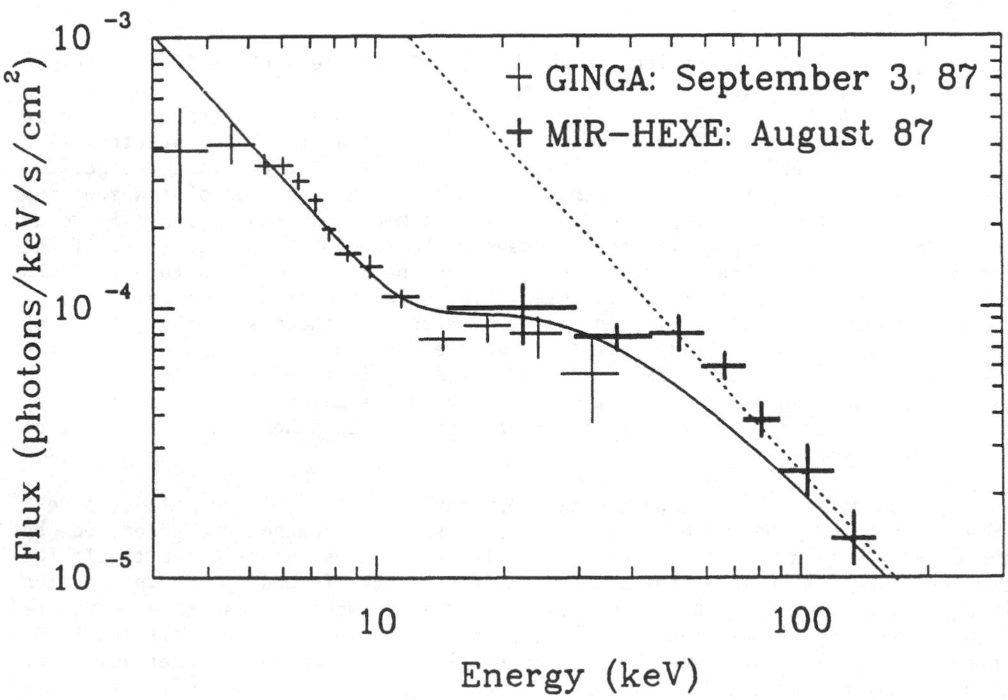

T. Stanev presented a study, made in collaboration with T. Gaisser, of the acceleration of cosmic rays and the production of high energy gamma-ray signals at SN 1987A. They developed a model which assumes that the pulsar radiation drives a MHD wind with luminosity close to the total pulsar luminosity :

$L_p=4.10^{39}(B^2_{12}/P^4_{10})$erg/s, where B_{12} is the pulsar surface magnetic field in units of 10^{12} gauss and P_{10} the period in units of 10ms. The wind is confined inside the ejecta, a shock occurs at the radius where the wind ram pressure is balanced by the accumulated energy density of the wind, and charged particles are accelerated by first order Fermi acceleration at the shock. The acceleration of protons is only limited by the diffusion time away from the shock and for the minimum value of the diffusion coefficient the maximum achievable proton energy is 10^{17} eV for B_{12} and P_{10}=1.

Gamma rays are produced in the inelastic collisions of the protons with stellar material ; the exact amount energy to go into γ-rays depends on the degree of mixing η of shell material with the shocked wind containing the accelerated cosmic rays.

An observable photon signal will occur between characteristic times t_1 when the shell becomes optically thin to >TeV photons and t_2 when the shell becomes too dispersed for significant collision loss. t_1 occurs when the column density of the shell is roughly one radiation length, ~1 year after the explosion. One can imagine two extremes that bracket t_2 :

(a) If the accelerated protons are not contained in the shell the production of secondaries will cease when the column density of the shell becomes much less than one interaction length. This case was studied by Sato and his collaborators.

(b) If the accelerated particles are confined indefinitely in the expanding remnant, t_2 will occur significantly later, after several years, when the expansion rate exceeds the interaction rate. In this case both the average time dependence and the magnitude of the signal will also depend on the nature and extent of mixing of the high energy particles with the expanding shell.

Light curves expected for confinement and for free propagation are shown in figure 4. The maximum γ-ray flux at Earth, corresponding to η=1, and full confinement of cosmic rays, was estimated through numerical calculations to be well above the observational limit - $3.5x10^{-11}$ $cm^{-2}s^{-1}$ for cosmic ray luminosity $L_{CR}=10^{39}$ erg/s and integral proton spectral index γ_p=1.

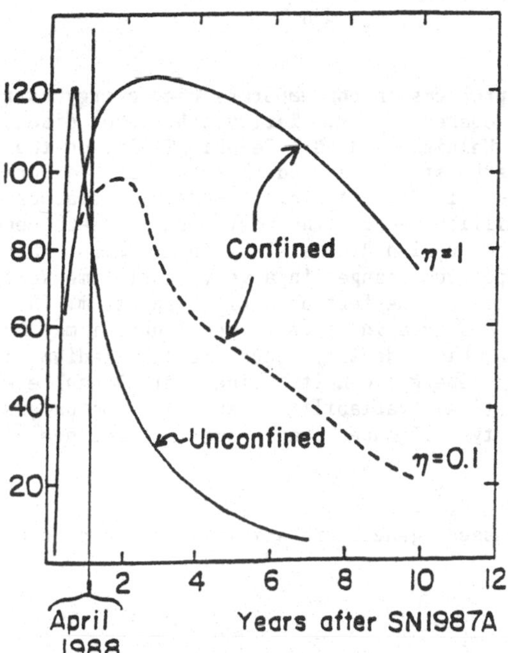

R. Chevalier dwelt on other high-energy phenomena related to the supernova :

1. When the shock front first breaks out of the atmosphere of the progenitor star, the photosphere is heated to a temperature >300,000K. Because the opacity is dominated by electron scattering, the emitted photons originate deep within the star and a significant fraction of the radiation may be at soft X-ray energies. This pulse, lasting only minutes, was missed since it occured before the first optical observation. Scattering by interstellar dust (light echo effects) gives another chance to observe this radiation. However, the strong forward scattering of dust at X-ray energies implies that the echo would only be observable with current techniques during the first few months. Another effect of the hard photospheric radiation is ionization of the circumstellar medium. IUE observations of NV in a circumstellar shell suggest an effective temperature >500,000K.

2. As the supernova shock front moves out into the surrounding medium, it can give rise to X-ray emission and non thermal radio emission. The radio emission that was observed from SN 1987A during the first month was probably due to the shock interaction with the B3 I supergiant wind. The shock front energy X-rays have been detected with the Ginga Observatory, but the observed flux variability is a difficulty for a circumstellar interaction model. The shock front will eventually reach the dense matter that has been observed with the IUE, but this could take about 30 years. An estimate of the 1 GHz radio flux at that time is 10Jy.

3. The neutrinos observed from SN 1987A suggest the formation of a neutron star and the ejection of $0.07M_{o}$ of ^{56}Ni indicates that there was not a large fall-back of matter onto the neutron star. The total luminosity of the supernova shows that an energetic pulsar is not present and suggests that neutron star accretion may become the dominant energy source. Estimates of the accretion rate in a supernova show that it is several orders of magnitudes larger than the Eddington limit rate. Under these conditions, an envelope can build up around the neutron star that prevents the escape of X-rays from close to the neutron star.

- o -

Session 3 : Theoretical problems in contemporary high energy astrophysics.

The session was opened by **R. Terlevitch**, who discussed his work (in collaboration with J. Melnick and G. Tenorio Tagle) on the relation between Seyfert nuclei and starbursts. He argued that nuclear starbursts can reproduce the observed properties of Seyfert nuclei. Models of starburst evolution show that after about 3 million years, the most massive stars become very hot and luminous ; they have been called "warmers" by the authors. When "warmers" begin to appear, the nebular spectrum changes in a very short time scale from a normal, low-excitation HII region, to a Seyfert or a Liner spectrum. The so called "broad line emission" regions observed in Seyfert type 1 nuclei may be young supernova remnants evolving in a high density interstellar medium : their observed properties (size, total energy, density, line widths) can be matched by such a model. The observed optical variability can also be understood in terms of sporadic explosions of type II supernovae. Thus the presence of a massive back hole may not be required.

P. Meszaros discussed general relativistic effects in neutron stars. The study of General Relativistic effects in neutron stars can lead to a determination of the radius of these objects, which when coupled with dynamic mass determinations allows one to reach conclusions concerning the equation of state, as well as concerning the energetics and spectral peculiarities of the emission. Observations and models of X-ray bursters lead to radii in the range 1.5 to 2.5 R_s (Schwarzschild units), the lower value coming from gravitational redshifts interpretations of absorption lines. For QPO sources, the magnetospheric beat frequency models can explain the very low upper limit to coherent pulsations if the radius is in the range 1.7-2.0 R_s. In accreting X-ray pulsars, accretion torque determinations in two objects showing cyclotron lines again suggest a value <2-2.5 R_s, while values <2 R_s would help explain why systems which are suspected to be fan beam emittors mimic the spectral signatures of pencil beam emittors. Finally, gamma-ray bursters, if they are accreting magnetized neutron stars, require radii <2.5R_s in order to explain the high incidence of 5-10 MeV bursts seen by SMM in a simple model where the low energy -absorption features are interpreted as straight cyclotron absorption. In the last three types of objects, it is the light bending which plays the crucial role, in QPOs suppressing the coherent pulsations from a low magnetic field broad cap, in X-ray pulsars in shifting the pulse phase of the emission from a small magnetic accretion colum, and in gamma-ray bursters in broadening the escape cone of high energy photons which avoid magnetic pair formation.

J.P. Ostriker presented recent work done at Princeton on superconducting strings and the high energy background. Reviews dealing with the properties of superconducting strings by Vilenkin (1985) and Thompson and Witten (1988) describe the physical basis for believing that superconducting strings may exist, the physical properties of such material, and the cosmological consequences of their existence.

For the present purpose it suffices to say that two parameters will determine most of the astrophysical consequences : the mass per unit length (of order 10^{22} gm/cm)μ expressed as the dimensionless number $G\mu/c^2 \sim 10^{-6}$ and the ratio of the electromagnetic output to the gravitational wave output, designated by f.

For cases of interest f is of the order unity. One cosmic volume $(c/H)^3$ will typically have one long string passing through it with a fractal distributions of irregularities down to very small scales, so that given the very high tenion, relativistic chaotic motions are induced which lead to self intersections and the generation of a network of smaller and smaller loops. These ultimately find equilibria stable against further fragmentation and then they oscillate losing energy via gravitational and electromagnetic radiation.

The dying old loops of length L inject an energy into their surroundings equal to $f\mu Lc^2$ in a manner quite analogous of a rapidly rotating pulsar. That is, the primary loss mechanism is via a Goldreich-Julian type relativistic MHD wind with a significant fraction of the output converted to high energy photons. Thus surrounding each radiation loop will be a giant version of the Crab nebula with quite analogous processes occuring within it for generation of X-rays, γ-rays and cosmic rays. In addition the loops will excavate large bubbles in the baryon distribution capable of identification with observed aspects of large scale cosmic structure. Radio frequency emission is relatively suppressed due to synchrotro self absorption but Compton processes are significantlty enhanced due to strong interaction with the Cosmic Background (CBR).

At late epochs, the energy of the string will produce a high frequency distortion of the CBR through the Sunyaev Zeldovich effect which is significant : the induced y parameter is approximately $y=2\times10^{-2}f$ for $(G\mu/c^2)=10^{-6}$. Since most of this distortion is produced during a period $Z>10^2$, when the universe is optically thick to electron scattering, the accompanying angular distortions are on scales greater than an arc minute and of amount $\Delta y_{rms}=10^{-5}(f=1;G\mu/c^2=10^{-6})$, low enough to be under current obervational limits.

Finally, R. Svensson gave an able summary of studies on pair plasmas in astrophysics developped over the last six years. He distinguished two processes :

a) Thermal pair production

When the electron temperature of plasma reaches approximately $m_e c^2/k \simeq 10^9-10^{10}$K, i.e. when the kinetic energies becomes of the same order as the electron rest mass energy then pair production may become important.

Detailed studies showed that for plasmas in pair equilibrium there exists a maximum temperature above which the annihilation of pairs cannot keep up with the pair production. The compactness :

$$l = \frac{L}{R}\frac{\sigma T}{m_e c^3} = 3000\left(\frac{L}{L_{Edd}}\right)\left(\frac{3R_s}{R}\right)$$

is the crucial parameter, and in pair dominated plasmas the maximum temperature is a unique monotonically but slowly decreasing function of the compactness. For compactnesses $1<l<1000$, expected in compact objects, the maximum temperature is of the order 100-500 keV. Another important result is that an annihilation line is not expected from plasmas in pair equilibrium. At high temperatures other emission processes dominate the broad line and at smaller temperatures the Thomson depth is large enough to wipe out the line.

b) Non thermal Pair Production

Four main results emerge from detailed studies .

1. When the acceleration process puts most of the power into the highest energy particles, then in general most of the "primary" synchrotron and Compton radiation will be at the highest photon energies. If this energy exceeds $m_e c^2$ and the compactness is greater than unity, then photon-photon interactions will initiate a pair cascade that redistributes the gamma-ray luminosity into the UV-X-ray range thereby changing the spectral shape.

2. Characteristic spectral breaks are introduced with the main one occurring around $m_e c^2$.

3. The pair yield, PY, i.e. the fractin of the injectd power that ends up as pair rest mass, can reach values of 10%.

4. The Thomson depth of cooled nonrelativistic pairs can reach values of $\tau_{pair} \simeq (PY\ l)^{1/2} \sim 20$.

He discussed then the application of these results to accretion flows.

A simple estimate of the maximum number of nonthermal pairs per accreted proton onto a black hole is obtained as follows : with an accretion efficiency of 10% there is about $0.1m_p c^2=100$ MeV per proton available to make radiation and ultimately at most 100 pairs. A fraction, say 50% , may end up as nonthermal radiation and a fraction PY<10% of this becomes pairs, i.e. 5 pairs or 10 particles per proton.

In two-temperature accretion flows (ion-torus) the presence of pairs affect the onset of Coulomb-cooling by the ions and the consequent collapse of the torus.

In spherical accretion flows the amount of pairs for sub-Eddington accretion rates is negligible (as the PY<<1). At Eddington rates the maximum pair-to-proton ratio derived above is reached, while at super-Eddington accretion rates the density is large enough for the pairs to annihilate on a inflow time scale reducing their number to negligible amounts.

Examples of application of these results to active galactic nuclei, to the cosmic X-ray background, and to gamma-ray bursts were given.

COMMISSION No. 49

THE INTERPLANETARY PLASMA AND THE HELIOSPHERE

(LE PLASMA INTERPLANETAIRE ET L'HELIOSPHERE)

Report of Meeting held in conjunction with the 20th General Assembly

President: Stanislaw Grzedzielski

Vice President: Leonard F. Burlaga

I. Report of Business Meeting

A business meeting chaired by L. Burlaga was held on August 3, 1988.

1. Election of Organizing Committee. The Commission elected the following officers and members of the Organizing Committee for the term 1988 - 1991:

President: Dr. Leonard F. Burlaga (USA)

Vice President: Prof. Bimla Buti (India)

Organizing Committee:

P. Bochsler (Switzerland)

A. Eviatar (Israel)

R. Jokipii (USA)

H. Ripkin (Germany, FR)

I. W. Roxburg (England)

E. Sarris (Greece)

T. Watanabe (Japan)

2. New Members.

The following new members were unanimously approved: J. Buchner, E. Chassifiere, J. Joselyn, H. Lundstedt, R. Sagdeev, Y. Ming Wang, and Tyan Yeh.

3. Working Group.

It was agreed to establish a Working Group on Plasma Astrophysics in Commission 49, and Prof. B. Buti was elected to chair this group. One objective of the Working Group will be to encourage and facilitate the interaction of members of Commission 49 with members of other Commissions who are concerned with problems of Plasma Astrophysics.

COMMISSION No. 50

PROTECTION OF EXISTING AND POTENTIAL OBSERVATORY SITES

(PROTECTION DES SITES D'OBSERVATOIRES EXISTANTS ET POTENTIELS)

PRESIDENT: S. van den Bergh
VICE-PRESIDENT: D. Crawford
ORGANIZING COMMITTEE: C. Blanco, V. Blanco, G.V. Coyne, A.A. Hoag, Y. Kozai, V. Pankonin, C.A. Torrez, J. Tremko, M.F. Walker

Due to the fact that IAU Colloquium No. 112 on "Light Pollution, Radio Interference and Space Debris" was scheduled immediately following the General Assembly the activities of Commission 50 at the Baltimore meeting were somewhat restricted. Reports on site surveys in Tibet and Anatolia were presented by Drs. Shou-Guan Wang and Zeki Aslan, respectively. Furthermore two panel discussions focussing on problems in radio astronomy were held. The first of these, which was moderated by V. Pankonin, dealt with interference in radio astronomy. Panelists included B. Robinson (Australia), J. Galt (Canada), L. Bottinelli (France), G. Swarup (India), T. Spoelstra (The Netherlands), R. Booth (Sweden), J. Ponsenby (United Kingdom) and M. Davis (U.S.A.). Each participant discussed the major interference problems in their countries. From a global perspective problems caused by transmissions from the GLONASS Satellite System seemed to be the most serious. A second panel discussion dealt with radio astronomy and CCIR, WARCs and the ITU. This panel discussion was moderated by V. Pankonin. A.R. Thompson discussed CCIR, T. Gergely WARCs, and J. Findlay and B. Robinson spoke about IUCAF. Each panelist discussed the interaction of their organization/country with CCIR and WARCs.

At the Baltimore meeting Dr. David Crawford (U.S.A.) was elected as the new President, and Dr. Paul Murdin (United Kingdom) was elected Vice-President of the Commission.

COMMISSION No. 51

BIOASTRONOMY : SEARCH FOR EXTRATERRESTRIAL LIFE

(BIOASTRONOMIE : RECHERCHE DE LA VIE DANS L'UNIVERS)

Report of Meeting ofAugust 9, 1988

President: F.D. Drake

BUSINESS MEETING

I. NEW OFFICERS

George Marx (Hungary) was elected President and Ronald Brown (Australia) Vice-President for the period 1988-1990. M. Papagiannis (USA) continued as Secretary and will serve as such over the period 1988-1990. The members of the scientific organizing committee for the period 1988-1990 are:

F. Drake (USA)
G. Gatewood (USA)
S. Gulkis (USA)
J. Heidman (France)
J. Jugaku (Japan)
N. Kardashev (USSR)
V. Slyath (USSR)
J. Tarter (USA).

II. NEW MEMBERS

Approximately 40 new members were added to the Commission, bringing the total membership to about 280. The Commission also has some 40 consultants, an unusually large number because some of the prime interests of the Commission are in scientific areas not ordinarily addressed by the IAU. Thus the total number of scientists active in the Commission is about 320. This number again illustrates the widespread interest in the work of the Commission, a phenomenon which has been evident since the founding of the Commission at the XVIII General Assembly at Patras.

III.I. SCIENTIFIC MEETINGS

The Commission held its second scientific meeting, IAU Colloquium 99, at Lake Balaton, Hungary, in June 1987. Some 200 scientists participated in this one-week meeting, which was held at excellent facilities on the shores of Lake Balaton. Arrangements for the meeting were excellent. The officers of the Commission were especially pleased to see the active participation, including many scientific presentations of people who have not been active in bioastronomy before. The proceedings of the Colloquium have now been published in the volume "Bioastronomy.... the Next Steps", edited by George Marx.

At Baltimore, the Commission received an invitation from its French colleagues to hold the next scientific meeting in the summer of 1990 at an attractive site in the French Alps. The Commission was delighted to accept this invitation. The Chairman of the local organizing committee will be Jean Heidman. It is anticipated that this will be a very popular meeting, particularly because of the rapidly growing activity in the areas of science relevant to Commission 51. The necessary approvals and support from the IAU administration are being sought.

IV. SCIENTIFIC SESSIONS

The Commission greatly increased the number of scientific sessions which is sponsored at the General Assembly. Although only one scientific session was held at the XIX General Assembly, four were held at Baltimore. Attendance was excellent. Again, the Commission officers and the session organizers were pleased to see participation by people who had not previously been active in the affairs of Commission 51. The titles of the scientific sessions were:

1) The Search for Other Planetary Systems;
2) Ongoing Searches for Extraterrestrial Radio Signals;
3) Proposed Radio Searches for Extraterrestrial Radio Signals;
4) Proposal for New Approaches to the Search for Extraterrestrial Life.

WORKING GROUP FOR PLANETARY SYSTEM NOMENCLATURE

(GROUPE DE TRAVAIL POUR LA NOMENCLATURE DU SYSTEME PLANETAIRE)

PRESIDENT	H. Masursky
MEMBERS	K. Aksnes, M. Fulchignoni, G.E. Hunt, M.Ya Marov P.M. Millman, D. Morrison, T.C. Owen, V.V. Shevchenko, B.A. Smith, V.G. Tejfel
CONSULTANTS	J.M. Boyce, G.E. Burba, A,M. Komkov,

Notice of the meetings held during the first two years of the triennium are included in volume XXA, p. 804-807 of this series. The Sixteenth meeting of the Working Group was held in Helsinki, Finland in July 1988, and the seventeenth meeting was held at Baltimore in August 1988 in connection with the General Assembly. At these meetings the following resolutions were proposed:

1. In proposing names for newly discovered satellites of the solar system it is recommended that first consideration be given to procedures already established within the IAU and that, where possible, confusing duplications with asteroid names be avoided.
2. An aquatic theme will be used for features identified during Voyager's encounter with Neptune and its satellites. Names from Greek or Roman mythology associated with neptune or Poseidon or the oceans will be used for new satellites. Surface features on Triton will be named for water creatures or sacred lakes, rivers, or seas (excluding Greek and Roman names). large ringed basins, if present, will be named for the primordial sea.
3. Additional features identified on Phobos by the Soviet mission will be named for characters and places in American and French Science fiction books about Mars.
4. The Small Bodies Task Group was activated under the chairmanship of Marciello Fulchignoni; other members include A. Brahic, Y.C. Chang, T. Gombosi, S. Isobe, L. Kasantfomoliti, D.F. Lupishko, D. Morrison, J. Veverka. This Task Group will propose names for features on the nuclei of Halleys and other comets, and additional features recognized on the surface of Phobos. We thank D. Morrison for serving as chairman pro tempore of this Task Group.
5. The name Colombo was proposed for a Saturnian ring feature.
6. Three names proposed for features on Mars and listed in volume XXA of the IAU Proceedings volume were dropped: Amet, Musmar, and Jeki. Other corrections to the lists of names printed in volume XXA of the IAU Proceedings are:

	Incorrect	Correct		Incorrect	Correct
Mars					
	Ecus Chasma	Echus Chasma		Cavi Frigores	Cavi Frigorēs
Jupiter,					
Callisto:					
	Holdr	Höldr			
Uranian Satellites:					
Miranda			Ariel		
	Silicia Regio	Sicilia Regio		Onagh	Oonagh
Titania			Oberon		
	Messima Chasma	Messina Chasma		Nommur chasma	Mommur Chasma

Names for 144 features were proposed in Volume 20A of the IAU Proceedings: 32 names for Martian features, 20 names for features on Jovian satellites, 82 names for features on Uranian satellites, and 10 names for small Uranian satellites. In addition, we propose for adoption the following 124 names: 100 names for Martian features, 1 feature on the Martian satellite Phobos, 22 features on Callisto, and 1 feature on the Saturn's satellite Rhea:

MARS

Crater:			
Aban	16.2	249.1	Town in USSR
Ajon	16.8	257.9	Town in USSR
Aktaj	20.7	46.5	Town in USSR
Angu	20.3	254.5	Town in Zaire
Argas	23.5	50.1	Town in USSR
Arta	21.6	54.2	Town in USSR
Bada	20.5	50.7	Town in USSR
Bak	18.3	256.3	Town in Hungary
Balvicar	16.5	53.1	Town in Scotland
Beltra	18.3	257.7	Town in Ireland
Bira	25.3	45.9	Town in USSR
Bland	18.5	251.5	Town in Missouri, USA
Calamar	18.4	54.9	Town in Colombia
Camargo	17.9	250.6	Town in Bolivia
Cantoura	15.0	51.5	Town in Venezuela
Chafe	15.4	257.9	Town in Nigeria
Clova	21.7	52.0	Town in Quebec, Canada
Cue	16.7	246.2	Town in Australia
Darvel	17.9	51.0	Town in Scotland
Dersu	25.8	51.9	Town in USSR
Dixie	17.9	55.9	Town in Georgia, USA
Doon	23.8	250.8	Town in Ontario, Canada
Dush	22.6	54.0	Town in Egypt
Floq	15.2	253.0	Town in Albania
Glendore	18.5	51.7	Town in Ireland
Herculaneum	19.2	58.9	Town in Italy
Ikej	21.1	247.6	Town near USSR
Imgr	19.3	249.0	Town in USSR
Ins	24.8	251.2	Town in Switzerland
Jama	21.6	53.1	Town in Tunisia
Kachug	18.4	252.5	Town near Irkutsk, USSR
Kasra	22.2	256.4	Town in Tunisia
Khanpur	20.8	257.5	Town in Pakistan
Korph	19.6	254.7	Town in USSR
Koy	21.6	50.4	Town in USSR
Kular	16.6	252.0	Town in USSR
Lapri	20.6	252.6	Town in USSR
Leuk	24.1	55.0	Town in Switzerland
Littleton	15.9	253.0	Town in Maine, USA
Mago	16.2	254.8	Town in USSR
Mandora	12.4	53.6	Town in Australia
Medrissa	18.8	56.6	Town in Algeria
Meget	19.1	252.9	Town near Irkutsk, USSR
Mirtos	22.3	51.7	Town in Crete (Greece)
Montevallo	15.3	54.2	Town in Alabama, USA
Neive	23.5	253.0	Town in Italy
Never	23.8	254.4	Town in USSR
Northport	18.6	54.2	Town in Alabama, USA
Olenek	20.1	54.2	Town in Yakut, USSR
Pica	20.0	53.2	Town in Chile
Olom	23.2	57.7	Town in USSR
Ōme	20.9	256.1	Town in Japan
Onon	16.4	257.6	Town in Mongolia
Palana	21.3	257.5	Town in Kamchatka, USSR
Pâros	22.0°N	98.0°W	Modern Greek village on island of same name
Pina	18.6	248.5	Town in Panama
Platte	16.3	247.1	Town in South Dakota, USA

(Mars craters, cont.)

Pompeii	19.0	59.1	Ruined town in Italy
Poynting	8.5°N	113.0°W	J.H. Poynting; first to theorize that small objects orbiting sun were drawn into it.
Qibā	17.4	257.0	Town in Saudi Arabia
Sabo	25.4	48.9	Town in USSR
Sevi	19.1	257.1	Town in USSR
Sian	20.1	48.0	Town in USSR
Sinda	16.0	248.9	Town in USSR
Souris	19.7	246.9	Town in Manitoba, Canada
Tarma	16.8	250.3	Town in Peru
Tejn	15.7	253.8	Town in Denmark
Telz	21.4	249.1	Town in East Germany
Tepko	15.5	256.5	Town in Australia
Thermia	19.9	250.9	Town in Greece
Tibrikot	12.7	54.9	Town in Nepal
Tokko	22.7	250.8	Town in russian USSR
Tokma	21.6	251.6	Town in USSR
Tolon	18.5	255.1	Town in Russian USSR
Tomari	20.1	246.4	Town on Sakhalin Is., USSR
Torbay	18.1	246.1	Town in Australia
Tuapi	17.3	255.8	Town in Nicaragua
Tumul	15.0	255.5	Town in USSR
Turma	17.5	252.1	Town in USSR
Ulu	22.8	252.9	Town in USSR
Urk	23.4	248.7	Town in Netherlands
Utan	24.4	246.3	Town in USSR
Wum	16.1	246.2	Town in Camaroon
Voza	23.5	53.5	Town in Solomon Islands
Xui	15.3	247.6	Town in Brazil
Yalata	22.1	254.0	Town in Australia
Yebra	21.1	254.4	Town in Spain
Colles			
Colles Nili	37/40 N	291/309W	Classical albedo feature, Portus Nili at 38°N, 295°W; also on 40292, 35297 35302
Dorsum			
Iamuna Dorsa	20.4/21.5	50.1/50.4	For albedo feature Iamuna
Fossa			
Gigas Fossae	2.5/5.0°N	128.0/130°W 140°W	Albedo feature Gigas at 0°,
Mensa:			
Labeatis Mensa	25/26°N	73.5/75°W	Albedo feature Labeatis Lacus, at 30°N, 75°W
Lunae Mensa	24/24.5N	61.5/63.5°W	Albedo feature, Lunae Palus, at 17°N, 65°W
Nilokeras Mensae	28.5/34.5N	47/54.5°W	Albedo feature Nilokeras, at 30°N, 55°W
Sacra Mensa	21/27.5°N	63/73.5°W	Albedo feature Sacra Insula, at 20°N, 57°W
Tempe Mensa	27.5/28.5°N	63/64°W	Albedo feature Tempe at 40°N, 70°W

(Mars, cont.)

Scopulus			
Xanthe Scop.	18.8/20.5	51.8/52.4	For albedo feature Xanthe
Vallis:			
Anio Vallis	37.8/38.4N	304/304.5°W	Classical river in Italy; modern Aniene and Teverone rivers
Clanis Valles	33.9/34.1°N	301/302°W	Classical river in Etruria; present Chiana river, Italy
Clasia Vallis	33.3/34.6°N	302/303.4	Classical river in Umbria, Italy
Hypsas Vallis	33.9/34.1°N	301.7/302.4	Classical river in Sicily

JUPITER

CALLISTO			
Crater:			
Ahti	42.0!N	103.5°W	Finnish water god
Ajleke	23.0°N	101.5°W	Saami god of holidays
Aziren	34.0°N	178.0°W	Estonian spirit of death
Egres	41.5°N	175.5°W	Karelian deity of bean harvest
Hijsi	61.5°N	169.5°W	Karelian deity of hunting
Jųmal	59.0°N	120.0°W	Estonian sky god
Kari	48.0°N	118.5°W	Ottar's ancestor
Kul'	63.0°N	124.0°W	Komi wood spirit
Ljekio	48.0°N	162.0°W	Finnish god of grass
Maderatcha	31.0°N	95.5°W	Saami sky god
Nirkes	30.0°N	163.5°W hunt	Karelian patron of squirrel
Norov-Ava	55.0°N	115.0°W	Mordovian mistress of field
Omol'	42.5°N	118.5°W	Komi wood spirit
Reginn	40.5°W	90.5°W	Norse dwarf
Rongoteus	54.0°N	107.0°W	Karelian deity of rye harvest
Rota	28.0°N	110.0°W	Deity of underground world
Tapio	31.0°N	110.0°W	Finnish wood deity
Tontu	28.0N	100.5°W	Finnish god of housekeeping
Tyll	43.5°N	165.5°W	Estonian epic hero
Vanapagan	38.5°N	158.5°W	Estonian wicked giant
Vu-murt	22.0°N	171.5°W	Estonian water spirit
Vutash	32.0°N	103.0°W	Estonian water spirit

SATURN

Rhea			
Crater			
Tirawa	35°N	150°W	Great spirit of Pawnee tribe, Norh America; created first men.

SMALL BODIES

Phobos

Dorsum			
Kepler Dorsum			For Johannes Kepler, 1571-1630; German astronomer

CHAPTER VI

FUNCTIONS AND STATUTES OF THE UNION

1. **Useful addresses**

The address of the Secretariat and members of the Executive Committee are noted, page 71.

2. **History of the Union**

A brief guide to the former Presidents, General Secretaries, etc... of the Union can be found in IB 55, pp. 6-7 & 61, p. 2.

3. **IAU Representatives to other Organisations**

The list of the triennium 1988-1991 is given on pp. 69 & 70 of this volume.

4. **Services of the Union**

A summary of the activities of the Union Services may be found in Transactions XVIIB, pp. 387-393.

Rules for scientific meetings

General

One of the essential tasks of the Union is to encourage circulation of ideas and to disseminate information internationally on current scientific results by organizing scientific meetings. Scientific Meetings are:

a. Symposia
b. Colloquia
c. Regional Astronomy Meetings
d. Commission Meetings and/or Joint Commission Meetings
e. Joint Discussions during General Assemblies

Except in the case (c), each meeting has to be sponsored by one or more Commissions. For Symposia there should always be more than one Commission involved, a principal sponsoring Commission with other co-sponsoring Commissions. In the case (d), the Commission Organising Committee is the Organising Committee of the Meeting. In all other cases, a Scientific Organising Committee is (SOC) appointed by the Executive Committee (EC), with the advice of the proposer of the meeting.

The Chairman of the SOC is appointed by the EC. It may be the case that the President of the sponsoring Commission is proposed as Chairman of the SOC, but it is not necessarily so. A proposer may propose himself as Chairman of the SOC. It is the responsibility of the Chairman of the SOC to actively involve the Members of the SOC (who would normally be Members of the IAU - in multidisciplinary meetings such restriction may not be desirable) in defining the range of topics to be considered and in suggesting a wide range of possible contributors - in particular younger astronomers.

Symposia, Regional Meetings and Joint Discussions cover broader fields and have larger attendance, in general, than Colloquia. Proceedings of Symposia are published by the IAU in conjunction with a Publishing House. Whether or not Proceedings of Colloquia and Regional Meetings are published, and in what form, is decided by the SOC of each Colloquium and Regional Meeting. Colloquia and Commission Meetings deal with narrower topics and are usually shorter. Immediately following the end of an IAU Symposium, Colloquium or Regional Meeting, the Chairman of the SOC supplies a short summary of the Meeting for publication in the Information Bulletin.

Regional Meetings are proposed by a host organisation within the group of countries concerned. They should be confined to years when no IAU General Assembly is held. The EC can designate a meeting organised by a specific national or regional organisation to be a Regional Astronomy Meeting. Regional Meetings should embrace as wide a spectrum of astronomical disciplines as possible.

Proposals in respect of Symposia, Colloquia and Regional Meetings are addressed on the appropriate form to the EC through the Assistant General Secretary. Such proposals should be preceded by informal enquiries made at the earliest possible opportunity. Enquiries in respect of Symposia and Colloquia should be directed to the Assistant General Secretary in the first instance; in all other cases, the first approach should be to the General Secretary. The formal proposal should be accompanied or preceded by letters of support from the Presidents of sponsoring and co-sponsoring Commissions. The following information is essential to a proposal:

1. the title;
2. date and duration;
3. location;
4. sponsoring and co-sponsoring Commissions, other co-sponsoring scientific Unions;
5. suggested composition of the SOC. No more than ten names should be included, one of the names being proposed as Chairman. There should be several countries represented and one institution should not be represented by more than one person unless some other representation (e.g. a co-sponsoring Commission or relevant working group, etc.) is involved thereby. For Regional Meetings, the Assistant General Secretary is an ex-officio Member of the SOC and in these cases more than ten names can be included, if desired;
6. suggested Chairman and composition (or total number) of the proposed Local Organising Committee (LOC);
7. name and address for maintenance of contact;
8. the estimated number of participants;
9. the financial support expected from sources other than the IAU;
10. whether or not the maximum IAU financial support is requested;
11. the names of the proposed Editor, or Editors, of Proceedings, if these are to be published and their addresses;
12. the outline scientific programme;
13. a detailed account of why the proposed meeting is useful and necessary at the time proposed and its relationship with other meetings.

Participation in Symposia and Colloquia is by invitation of the Chairman of the SOC. Invitations may be sought by suitably qualified scientists who are Members of the IAU or who are resident in an Adhering Country. While Symposia and Colloquia must be open to all qualified to participate, it is recognised that there may be occasions where logistical limitation of total number attending may need to be imposed by the SOC.

It is essential that there be no restriction based on sex, race, colour, nationality, religious or political affiliation, imposed on bona fide scientists attending internationally supported scientific meetings, either by the organisers of Symposia and Colloquia, or, by authorities in the host country. Assurance that the ICSU Rules will be adhered to must be honoured. Failure to observe those Rules will normally result in withdrawal of IAU scientific sponsorship and financial support.

In the case of Regional Astronomy Meetings, invitation to attend is open to all interested astronomers of a particular geographical region (but not excluding other interested participants) for which invitations are primarily intended. The region will be outlined by the organisers in agreement with the EC.

Contents of meetings

Commission Meetings held in the course of General Assemblies are of two kinds:

- Business Meetings, including the organisation of specific projects, and

- Scientific Meetings on specific subjects. Commission Meetings can be organised jointly by two or more Commissions in order to deal with a topic of mutual interest; alternatively, a Joint Discussion, with an agreed SOC, can cover a topic of wider interest. Proposals for Meetings of Commissions and Joint Discussions during a General Assembly are addressed to the General Secretary, who decides the time-table for such proposals.

The agenda of Scientific Meetings may consist of any or all of the following:

- Invited Review Papers,
- Invited Papers,
- Contributed Papers,
- Contributed Poster Papers,
- Video, Film,
- Discussions, including Panel Discussions.

Invited Review Papers

They should aim to be comprehensive reviews of major areas of interest to the Symposium, Colloquium, etc. Chairmen of the SOC have a responsibility to ensure that Invited Review Papers are truly comprehensive and only be published as such if adequate standards of breadth and depth are met.

Invited Papers

They are less comprehensive reviews than Invited Review Papers of a topic of recent and/or immediate scientific interest. In issuing an invitation to present an invited paper, the Chairman of the SOC should outline the scope of the work to be addressed. The SOC can include proposed contributed and poster papers in this category for extended publication if the scientific interest of the papers so warrant.

Contributed Papers

They are papers on a specific topic requested or accepted by the SOC, through the Chairman, for presentation in the main meeting.

Contributed Poster Papers

These are papers on a specific topic requested or accepted by the SOC, through the Chairman, for presentation in special poster sessions scheduled as a regular part of the meeting.

Film and Video

Facilities for the presentation of video material as well as film should be made available where possible.

Discussion

Discussion from the floor following papers is regarded as an essential form of communication. The use of Panel Discussions is not frequently exploited though it is recognised that such format has limited effective applicability. While discretion needs to be exercised, it would be reasonable to devote at least 1/3 of total meeting time to discussion.

The SOC, having decided on the subject(s) of the meeting, must then decide the general layout of the programme and the mode of, and balance between, Invited Review Papers, Invited Papers, Contributed Papers, including Posters, and Discussion, in the light of suggestions from membership of the SOC or EC. It can then proceed to decisions on the Invited Speakers, paying due regard to the distinctions between Invited Review and Invited Papers; the acceptance of a limited number of contributed papers for presentation within the meeting or as posters; the requirements to be expected for video or film presentation followed by the requirements for translation, recording of discussion, organisation of the secretariat, etc. in agreement with the LOC.

The SOC should pay particular regard to adequate provision for poster sessions if a feature of the proposed meeting. It is important that a limited number of posters are on display at any time (e.g. change of posters daily) and that presenters have adequate room in the neighbourhood of their poster stand. Each presenter should have a display board whose area should not be less than 1m x 0.7m. Adequate time to view posters should be set aside (posters are not the equivalent of an evening's entertainment) within the timetable of the meeting. Arrangements should be considered for a review of posters to be presented during the meeting.

Timetable and other commitments

A timely presentation of proposals to the EC, taking into account the annual interval between EC meetings, is necessary if a satisfactory procedure of prior review is to be attained. The following types of notice are required:

i. Preliminary notice must be received at the earliest possible date, so that the proposers can be informed of any similar proposals being made elsewhere. In the case of Regional Meetings, where the General Secretary receives proposals in the first instance, the outline proposal should normally be available 2-3 years in advance and contact maintained thereafter; in the case of all meetings ability to place adequate prior notice in the IAU Information Bulletin is a necesary service for IAU Members generally. This requirement implies at least one year between the time the meeting is proposed on the official proposal form and the time of the meeting.

ii. A firm proposal, still subject to approval and minor alteration by the EC, must be received in time for details to be investigated by the Assistant General Secretary in respect of inter alia,

. composition of SOC,
. availibility of IAU financial support,
. suitable nature and non-duplication of scientific content,

prior to presentation for approval by the EC.

Fully documented proposals are submitted by the Assistant General Secretary to the EC. The EC either accepts or rejects the proposals. Since the EC normally meets annually in August/September and since it is desirable that discussion of critical points takes place at a formal meeting, rather than by correspondence, the actual dates for preparing proposals should take into account the dates of the EC meetings. March-April is a suitable time for making firm proposals for symposia and colloquia. The Assistant General Secretary will not normally submit proposals received after July 1 to the next meeting of the EC.

If the EC accepts a proposal it decides the title, place, date, Chairman and Members of SOC, Chairman of LOC, financial support, Editor (in case of symposia) and the scientific topics. Normally all these decisions are in accord with the firm proposals made by the organisers, but is is sometimes necessary to make changes in order to ensure the international character and proper scientific coverage and to avoid duplication with other meetings. If a meeting is proposed with less than one year's notice, it may be rejected because there is no time available for such changes to be made. However, in the case of Colloquia only, a proposal can be accepted by the EC at between eight and twelve months' notice if it is deemed to meet the normal IAU requirements without alteration, provided always that one announcement in the IAU Information Bulletin (prepared in May and November, and published in June and January), will give five months' notice to IAU Members.

Presidents of related Commissions should be informed of the proposals by the organisers and invited to comment in their capacity as President.

The Chairman of the SOC,

1. invites participants to a Symposium or Colloquium after the meeting is accepted by the EC;
2. accepts or rejects requested invitations;
3. invites contributions and sets a deadline for submission of abstracts;
4. informs the Assistant General Secretary of the following:

   at least five months before the meeting:
   recommendations for all travel grants to participants whose countries of residence impose currency restrictions;

   at least three months before the meeting:
   recommendations for all other travel grants;

   at least two months before the meeting:
   detailed programme, choice of participants, local organisation, plus list of IAU grantees, plus sums, plus receipts;

   within one month of the end of the meeting:
   the final programme, list of actual participants; principal abstracts and supplies a short report of the meeting for inclusion in the Information Bulletin; final list of those awarded travel grants, the account of each award and the receipients signed receipts should also be forwarded.

The Assistant General Secretary should also be informed of any changes in the arrangements for, or programme of, a meeting.

Participation of the IAU in Joint Symposia

The IAU is willing to participate in Symposia, jointly with other Members of the ICSU family, i.e. Unions or Inter-Union Committees, Working Groups, etc. either as a principal sponsor or as a co-sponsor, or in a minor way.

As a principal sponsor, the IAU organises a Symposium in accordance with its normal procedures, the proposal being made in consultation with other Unions. It invites the co-operation and participation with other Unions in the business of the SOC through representatives appointed by those Unions to the SOC. The IAU invites other Unions to discuss matters affecting the publication of Proceedings, but would normally arrange for publication of the Proceedings as one of the current IAU Symposium volumes.

The IAU would normally expect the other Unions to contribute financially towards the expenses of the Symposium.

As a co-sponsor, the IAU would expect to be invited to appoint a representative on the appropriate Organising Committee, to be consulted through that representative on all important questions, and to be kept informed on decisions on other matters; the IAU would accept a leading role where called upon to do so, and would, in principle, contribute towards the expenses. The IAU would expect to be consulted in regard to publication of the Proceedings.

The IAU suggests that Symposia organised jointly by one or more Unions should always be announced in the following terms e.g.

Organised by the IAU in co-operation with (e.g.) IUGG and COSPAR,

or

Organised by (e.g.) IUHPS in co-operation with the IAU.

In case of Colloquia or Regional Meetings, co-sponsorship can be proposed to the EC, who will decide upon acceptance or otherwise according to current resources and requirements, without specific restriction.

Financial considerations

The EC allocates to each Symposium, Colloquium and Regional Meeting a limited sum of money which is meant as a "catalyst" to encourage other international organisations or national bodies to provide similar support. The maximum amount is specified in Swiss Francs, converted into other currencies, if necessary, according to current ICSU exchange rates. Colloquia receive smaller amounts than Symposia.

The IAU financial allocation is to be used principally for travel expenses of some of the participants. It can be used for partially supporting subsistence expenses of participants but in these cases should be limited to the nominal hotel or hostel charges rather than a per diem figure. If, following the conclusion of the meeting, some funds have not been allocated or have not been used, and the local organisation requires further funds to meet its expenses, the remaining IAU allocation can at the discretion of the General Secretary, be used to contribute towards local expenses.

While participants at IAU meetings are encouraged to use their best endeavours to obtain local, national or other international support for travel and subsistence, application can be made to the Chairman of the SOC for consideration for an IAU Travel Grant. An IAU Travel Grant application form is available to the SOC, but if different forms of application are in use to allocate other money, these forms can effectively replace the IAU travel grant forms. Having assessed the financial support available to applicants from other sources, the Chairman suggests to the Assistant General Secretary how the limited sum available could be distributed. This recommendation should reach the Assistant General Secretary not later than three months (five months for participants whose countries of residence impose currency restrictions) before the meeting, so that prospective recipients can suitably arrange their affairs. The Assistant General Secretary will arrange payment to recipients either direct, or in the form of bank drafts to the Chairman, LOC on an IAU account; arrangements to cash such drafts at a convenient place are the responsibility of the LOC.

IAU funds are intended to meet only a part of the travel and subsistence expenses of some participants; the Chairman of the SOC has to decide whether to help a moderate number of people with a small amount each, or to assist one or two principal participants appreciably. Either solution, or a compromise, is acceptable to the EC.

Expenses of the Secretariat of the Symposium, Colloquium or Regional Meeting are customarily met by organisations in the host country. Expenses of the LOC for receptions, excursions, buses, etc., are normally met either by the host municipality, national organisation, or institute, or by means of participants' contributions in the form of registration fees or itemized charges, or by a combination of these sources. For all IAU sponsored meetings, strenuous efforts should be made to keep registration fees and other charges to an absolute minimum.

Local organisation

The original proposal for a Symposium, Colloquium or Regional Meeting includes a proposed name for approval by the EC for the Chairman of the LOC. Even before such approval is finalized, the proposed Chairman normally forms a LOC and informs the Chairman of the SOC of its composition.

Under the personal guidance of the Chairman, the LOC takes care of the smooth running of the meeting. It does not receive financial help from the IAU, the necessary expenses being met by the local funds or by contributions from the participants.

The requirements of local organisation are generally as follows:

1. Meeting Rooms suitable for the expected number of participants and for the presentation of scientific papers should be reserved.

   Adequate space for poster sessions should be reserved. (Adequate space means sufficient separation of adjacent posters, determination and notification to presenters of display space, provision of a chair for presenter). It is important that the LOC make provision for a supply of pins, sticky tape, etc. for mounting poster material and for notification of participants of the time and venue of poster displays.

   Arrangements should be made for the display of visual materials: overheads, slides, films and videos. Participants should be advised of the film and video standard(s) available at the meeting venue.

2. Arrangements should be made for the reproduction of participants' documents.

3. Sufficient secretarial and technical assistance should be secured, with careful attention to the requirement for projection equipment, microphones, tape recorders, etc.

4. In conjunction with the requirements of the SOC, arrangements should be made to record verbal discussion. Reliance on tape recordings is often unsatisfactory and providing each contributor with a sheet of paper on which to record or summarize his remarks is advisable.

5. Information on accommodation should be agreed with the Chairman of SOC and sent
   . to the EC for acceptance, and
   . to prospective participants in good time.
   Block reservations are often advisable.

6. All participants should be asked to send their wishes as regards accommodation, excursions and social events to the LOC.

7. Receptions and excursions can be organised during a free period within the meeting, or just before or after the meeting. A Guest Programme is usually welcome.

8. Participants should be informed of the reservations made for them and how to reach their hotel or the meeting rooms on arrival.

9. The LOC should provide a Preliminary and a Final Programme, including useful auxiliary information, to be distributed to each participant at the appropriate time. A list of Participants, produced on about the second day of the meeting, is also extremely valuable if it corresponds closely with those actually present.

Publications of Symposium proceedings

The IAU believes that the Proceedings of Symposia are of general interest for a considerable period of time and that early publication in uniform style to a high standard is desirable. Publication and distribution have therefore been entrusted to a commercial publishing house.

The main responsibility of the IAU as joint publisher is the maintenance of a high standard of scientific value, originality and accuracy. The commercial publishing house has been contracted to ensure early publication and thereafter to take financial responsibility.

The EC, by approving the choice of Editor or Editors, places the main burden of maintaining the required scientific standard on one or two IAU Members on the understanding that they are familiar with the scientific matter of the Symposium and are persons with some experience in editorial tasks. The Editors receive no financial renumeration for their service to the Union.

It is essential that the Editor, or one of two Joint Editors, should have an excellent knowledge of the English language.

The Editor is responsible for the scientific value, the appearance and rapid delivery to the Publisher (usually within three months of the end of the Symposium) of the Proceedings.

The main editorial tasks are:

1. To inform participants in ample time before the meeting in what general form their contributions should be submitted and what arrangements have been made with the publisher for receipt of camera-ready copy. The number of printed pages available to each contributor should be determined in good time.

2. To inform the participants about IAU rules for publication of IAU Proceedings and to emphasize that any contributed papers must be refereed before acceptance for publication.

3. In advance of the Symposium, in close consultation with the SOC and LOC, to agree and arrange the precise details for recording and reporting the scientific discussion that takes place at the meeting. Difficulties in this respect must not cause undue delay in preparing for publication - it is better to sacrifice discussion rather than hold up publication.

4. To arrange with Members of the SOC for the refereeing of any contributed paper if an Editor is unable to do so.

5. To reduce the length of papers and discussion, to avoid duplication and to improve presentation where necessary.

6. To check whether IAU rules have been followed in each contribution and to arrange for re-typing if necessary.

7. To write the Introduction, Table of Contents, and obtain a Final Summary of the Symposium, maintaining uniformity with recent IAU Symposium Proceedings.

8. To maintain all necessary contact with the Publishing House, in accordance with current "Instructions for Editors" available from the Assistant General Secretary.

9. To maintain close contact with the Assistant General Secretary on all matters affecting progress of publication arrangements, especially keeping him informed of the material sent to the publishers and of any unexpected delays or alterations. In general, the Editor will not be able to allow time for substantial revision.

The length of IAU Symposium volumes should not exceed 500 printed pages. It would be the responsibility of the Symposium editor(s) to ensure that the page allocation for authors was planned before the Symposium and that authors are kept to their respective allocations.

The SOC will decide what (significant) fraction of the Symposium should be devoted to invited review and invited papers. The remainder would be devoted to contributed (including poster) papers. Recent experience shows that large numbers of contributed papers are being received and the SOC must make a decision on whether or not to accept a contributed paper for publication; if a SOC decides to publish contributed papers they should be published either in not more than two printed pages <u>or</u> as a one page abstract <u>or</u> a 1/2 page abstract <u>or</u> by title only.

Editors should refuse any tabular material, exceeding half a page. Tabular and graphical material must be contained within the total page allowance for all authors, in extenso algebraic derivations, lists of observations or other extended tabular or graphical material are inappropriate in a Symposium volume.

The SOC should accept the role of scientific referees in respect of invited review and invited papers. The editor should review each contributed and poster paper for relevance to the subject of the Symposium consulting SOC members where necessary.

While it is policy of the IAU that discussion should form an important part of all Symposia, it is up to each SOC to decide whether or not such discussion is published in the Symposium volume. If discussion is to be published it should be edited into a compact form and not be allowed to generate pages of low information content.

Editors must take steps to ensure the minimum number of blank pages for Symposium volumes (Attractive cartoons on pages devoid of science are not a remedy).

It is the policy of the IAU and that of their publishers to publish in camera-ready form. It is found that such a policy gives reasonable uniformity of appearance combined with speed of production. Authors will be sent camera-ready sheets, instructions on their use and the total amount of space available to them, by the publishers prior to the Symposium. The final camera-ready manuscripts should be sent to the Editor(s) either before the beginning of the Symposium or handed over at the Symposium. Papers not available to the Editor(s) in camera-ready form by the end of the Symposium will be deemed to have been withdrawn from publication.

In order to obtain a presentation that will indicate that the volumes of IAU Symposia form a series, it is requested that Editors adhere to the following recommendations:

The Title Page should include explicitly the words:

1. International Astronomical Union
   Union Astronomique Internationale

2. **"The Symposium Title"**

3. Proceedings of the **"No."** Symposium of the International Astronomical Union held in **"place"**, **"country"**, **"date"**

   followed by, if appropriate, the words

   "Organised by the IAU in cooperation with ....",

   in which the list of organisations is limited to the Scientific Unions, the Scientific Committees and Inter-Union Commissions of ICSU. Participation of UNESCO will be acknowledged by the following wording at the foot of the Title Page:

   "Published for the International Council of Scientific Unions with financial assistance from UNESCO".

4. **"Editor(s)"**
   "affiliation", **"place"**, **"country"**.

5. **"Publisher"**

Introduction

The Editor's introduction should mention circumstances of the organisation of the Symposium, and should list the supporting organisations and the Members of the Scientific and LOCs. It should express appreciation to those to whom it is due. The support of the IAU and other Unions, etc., should be recognised as well as that of other international, national, or local organisations.

Alphabetical Index of names and subject headings, with reference to the pagination are to be arranged in the camera-ready copy by the publisher(s).

Symposium volumes should be published 6 - 8 months after the Symposium.

Participants obtain some free reprints of their contributions and other copies can be ordered. All IAU Members can purchase Symposium volumes at reduced prices.

Publication of Colloquium Proceedings

The publication of IAU Colloquia should follow the same guidelines as for IAU Symposia. However, unless produced by the IAU publisher, the relevant manuscripts need not be in camera-ready form and some variation is allowed. Manuscript length is at the discretion of the SOC who have responsibility for the decision on whether or not to publish and the format of the publication. However, in order to facilitate archival retrieval all published IAU Colloquia proceedings must adopt the same form of Title Page. The Title Page should follow the same format as for Symposia replacing "Symposium" by "Colloquium" as appropriate i.e.

The Title Page should include explicitely

1. International Astronomical Union
   Union Astronomique Internationale

2. **"Title of Colloquium"**

3. Proceedings of the "**No.**" Colloquium of the International Astronomical Union, held in "**place**", "**country**", "**date**"

   followed by the words, if appropriate,

   "Organised by the IAU in cooperation with .....",

   in which the list of organisations is limited to the Scientific Unions, the Scientific Committees and Inter-Union Commissions of ICSU. Participation of UNESCO will be acknowledged by the following wording at the foot of the Title Page:

   "Published for the International Astronomical Council of Scientific Unions with financial assistance from UNESCO".

4. "**Editor(s)**"
   "**affiliation**", "**place**", "**country**".

5. "**Publisher**"

Publication of Regional Meetings Proceedings

If it is decided to publish the proceedings of a Regional Meeting, the same guidelines as for an IAU Symposium or Colloquium should be followed as far as is practicable given the format of the meeting and the method of publication adopted.

In order to facilitate archival retrieval the Title Page should indicate explicitly below any other title of the meeting

"Proceedings of the .... Regional Meeting of the International Astronomical Union, held in "**place**", "**country**", "**date**"

followed, if appropriate, by the words:

"Organised by the IAU in cooperation with ....."

6. **Statutes, By-Laws & Working Rules**

STATUTES

I. Denomination, Objects and Domicile

1. The International Astronomical Union (referred to as the Union) is a non-governmental organization, whose objects are :
   (a) to develop astronomy through international co-operation,
   (b) to promote the study and development of astronomy in all aspects,
   (c) to further and safeguard the interests of astronomy.

2. The legal domicile of the Union is Brussels.

II. Adherence to the Union

3. The Union adheres to the International Council of Scientific Unions.

III. Composition of the Union

4. The Union is composed of :
   (a) full members (Adhering Countries)
   (b) associate members (Associate Countries)
   (c) individual members (Members).

IV. Affiliated Organizations

5. The Union may admit the affiliation of international non-governmental organizations which contribute to the development of astronomy.

V. Adhering Countries

6. Countries adhere to the Union

   either :

   (a) through the organization by which they adhere to the International Council of Scientific Unions, or through a National Committee of Astronomy approved by that organization,

   or

   (b) if they do not adhere to the International Council of Scientific Unions, through a National Committee of Astronomy recognized by the Executive Committee of the Union.

   (c) The Adhering Organizations and National Committees of Astronomy are referred to as adhering bodies.

7. Adherence of a country to the Union is approved, on the proposal of the Executive Committee, by the General Assembly ; it terminates if the country withdraws from the Union or if the country has not paid its dues for five years.

8. Adhering Countries are classified in categories. The number of categories shall be specified in the By-laws. A country requesting adherence shall specify the category in which it desires to be classed. The specification may be declined by the Executive Committee if the category proposed is manifestly inadequate.

VI. Associate Countries

9. Countries that would like to join the Union while developing Astronomy in their territory may do so as Associate Members.

10. The Adhering Body for Associate Members may be the organization by which the country adheres to the International Council of Scientific Unions or through an institution of higher learning or a National Research Council.

11. Countries are accepted as Associate Members by the General Assembly, on the proposal of the Executive Committee, for a maximum interval of nine years, at the end of which they either become a full Member, or they resign from the Union.

12. During the probationary period, the Union, if asked by the Adhering Organization, may agree to help in the development of Astronomy in that country through the Visiting Lecturers' Programme and/or any other appropriate programme.

VII. Members

13. Members are admitted to the Union by the Executive Committee, on the proposal of an adhering body referred to in article 6, with regard to their achievements in some branch of astronomy.

VIII. General Assembly

14. (a) The work of the Union is directed by the General Assembly of representatives of Adhering Countries and of Members. Each Adhering Country appoints a representative authorized to vote in its name.

    (b) The General Assembly draws up By-laws governing the application of the Statutes.

    (c) It appoints an Executive Committee to implement the decisions of the General Assembly, and to direct the affairs of the Union in the interval between meetings of two successive ordinary General Assemblies. The Executive Committee reports to the General Assembly. The General Assembly, in accepting the report of the Executive Committee, discharges it of liability.

15. (a) On questions concerning the administration of the Union, not involving its budget, voting at the General Assembly is by Adhering Country, each country having one vote. Adhering Countries which have not paid their annual contributions up to 31 December of the year preceding the General Assembly may not participate in the voting.

    (b) On questions involving the budget of the Union, voting is similarly by Adhering Country, under the same conditions and with the same reservations as in article 15(a), the number of votes for each Adhering Country being one greater than the number of its category, as defined in article 8.

    (c) Adhering Countries may vote by correspondence on questions on the agenda for the General Assembly.

    (d) A vote is valid only if at least two thirds of the Adhering Countries having the right to vote by virtue of article 15(a) participate in it.

    (e) Associate Countries have the right to vote only on questions concerning associate membership.

16. On scientific questions not involving the budget of the Union the Members of the Union each have one vote.

17. On all questions in articles 15 and 16, decisions are taken by an absolute majority of the votes cast. However, a decision to change the Statutes is only valid if taken with the approval of at least two thirds of the votes of the Adhering Countries having the right to vote by virtue of article 15(a).

18. A motion to change the Statutes can only be discussed if it appears, in specific terms, on the agenda for the General Assembly.

IX. Executive Committee

19. The Executive Committee consists of the President of the Union, the President-elect, six Vice-Presidents, the General Secretary and the Assistant General Secretary elected by the General Assembly on the proposal of a Special Nominating Committee. The President-elect will normally become President of the succeeding Executive Committee.

X. Commissions of the Union

20. The General Assembly forms Commissions for such purposes as it may decide.

XI. Legal Representation of the Union

21. The General Secretary is the legal representative of the Union.

XII. Budget and Dues

22. (a) For each ordinary General Assembly the Executive Committee prepares a budget proposal covering the period to the next ordinary General Assembly, together with the accounts of the Union for the preceding period. It submits these to the Finance Committee for consideration ; this Finance Committee consists of one member nominated by each adhering body and approved by the General Assembly. At its first meeting during the General Assembly, the Finance Committee elects a Chairman from among its members.

(b) The Finance Committee examines the accounts of the Union from the point of view of responsible expenditure within the intent of the previous General Assembly, and it considers whether the proposed budget is adequate to implement the policy of the General Assembly, as interpreted by the Executive Committee. It submits reports on these matters to the General Assembly for approval of the account and decision on the budget.

(c) Each Adhering Country pays annually to the Union a number of units of contribution according to its category. The number of units of contribution for each category shall be specified in the By-laws.

(d) Associate Countries pay annually one unit of contribution.

(e) The amount of the unit of contribution is determined by the General Assembly, on the proposal of the Executive Committee and with the advice of the Finance Committee.

(f) The payment of contributions is the responsibility of the adhering bodies. The liability of each Adhering Country in respect of the Union is limited to the amount of that country's dues to the Union.

(g) An Adhering Country that ceases to adhere to the Union resigns at the same time its rights to a share in the assets of the Union.

XIII. Dissolution of the Union

23. The decision to dissolve the Union is only valid if taken with the approval of three quarters of the votes of the Adhering Countries having the right to vote by virtue of article 15(a).

XIV. Emergency Powers

24. If, through events outside the control of the Union, circumstances arise in which it is impracticable to comply with the provisions of these Statutes and of the By-laws drawn up by the General Assembly, the organs and officers of the Union, in the order specified below, shall take such actions as they deem necessary for the continued operation of the Union. Such action shall be reported to a higher authority immediately this becomes practicable until such time as an extraordinary General Assembly can be convened. The following is the order of authority :

    The General Assembly ; an extraordinary General Assembly ; the Executive Committee in meeting or by correspondence ; the President of the Union ; the General Secretary ; or failing the practicability or availability of any of the above, one of the Vice-Presidents.

XV. Final Clauses

25. These Statutes enter into force on 2 August 1988.

26. The present Statutes are being published in French and English versions. In case of doubt, the French version is the only authority.

BY - LAWS

I. Membership

1. Applications of countries for adherence to the International Astronomical Union (referred to as the Union) are examined by the Executive Committee and submitted to the General Assembly for approval.

2. Proposed changes in the list of Members are, with due regard to the suggestions of the Presidents of Commissions, submitted for advice to the Nominating Committee, consisting of one representative of each Adhering Country designated by the appropriate adhering body, before decision by the Executive Committee.

3. Commissions may, with the approval of the Executive Committee, co-opt consultants whom they consider may contribute to their work. The adherence of consultants expires on the last day of the ordinary General Assembly next following their admission, unless renewed.

4. An affiliated organization may participate in the work of the Union as mutually agreed between the organization and the Executive Committee.

II. General Assembly

5. The Union meets in ordinary General Assembly, as a rule, once every three years. The place and date of the ordinary General Assembly unless determined by the General Assembly at its previous meeting, shall be fixed by the Executive Committee and communicated to the adhering bodies at least six months beforehand.

6. The President, with the consent of the Executive Committee, may summon an extraordinary General Assembly. He must do so at the request of one third of the Adhering Countries.

7. The agenda of business for each ordinary General Assembly is determined by the Executive Committee and is communicated to the adhering bodies at least four months before the first day of the meeting. It shall include the proposal of the Executive Committee in regard to the unit of contribution as called for in article 24.

8. (a) Any motion or proposal received by the General Secretary at least five months before the first day of an ordinary General Assembly, whether from an adhering body, from a Commission of the Union, or from an Inter-Union Commission on which the Union is represented, must be placed on the agenda.

   (b) A motion or proposal concerning the administration or budget of the Union which does not appear on the agenda prepared by the Executive Committee, or any amendment to a motion that appears on the agenda, shall only be discussed with the prior approval of at least two thirds of the votes of Adhering Countries represented at the General Assembly and having the right to vote by virtue of Statute 15(a).

9. If there is doubt as to the administrative or scientific character of a question giving rise to a vote, the President determines the issue.

10. Where there is an equal division of votes, the President determines the issue.

11. The President may invite representatives of other organizations, scientists and young astronomers to participate in the General Assembly. Subject to the agreement of the

Executive Committee he may delegate this privilege concerning representatives of other organizations to the General Secretary, and concerning scientists and young astronomers to the adhering bodies.

III. Special Nominating Committee

12. (a) Proposals for elections to the President of the Union, a President-elect, six Vice-Presidents, the General Secretary and the Assistant General Secretary are submitted to the General Assembly by the Special Nominating Committee. This consists of the President and past President of the Union, a member proposed by the retiring Executive Committee, and four members elected by the Nominating Committee from among twelve Members proposed by Presidents of Commissions. Other than the President and immediate past President, present and former members of the Executive Committee shall not serve on the Special Nominating Committee. No two members of the Special Nominating Committee shall belong to the same country.

(b) The General Secretary and the Assistant General Secretary participate in the work of the Special Nominating Committee in an advisory capacity.

(c) The Special Nominating Committee is appointed by the General Assembly to which it reports direct. It remains in office until the end to the ordinary General Assembly next following that of its appointment, and it may fill any vacancy occurring among its members.

IV. Officers and Executive Committee

13. (a) The President of the Union remains in office until the end of the ordinary General Assembly next following that of his election ; the Vice-Presidents remain in office until the end of the second ordinary General Assembly following that of their election. They may not be re-elected immediately to the same offices.

(b) The General Secretary and the Assistant General Secretary remain in office until the end of the ordinary General Assembly next following that of their election. Normally the Assistant General Secretary succeeds the General Secretary though both officers may be re-elected for another term.

(c) The election takes place at the last session of the General Assembly, the names of the candidates proposed having been announced at a previous session.

14. The retiring President and the retiring General Secretary become advisers to the Executive Committee until the end of the ordinary General Assembly next following that of their retirement. They participate in the work of the Executive Committee and attend its meetings without voting right.

15. The Executive Committee may fill any vacancy occurring among its members. Any person so appointed remains in office until the next ordinary General Assembly.

16. The Executive Committee may draw up and publish Working Rules to implement the Statutes and By-laws.

17. The Executive Committee appoints the Union's representative to the International Council of Scientific Unions ; if not already an elected member of the Executive Committee, this representative will become its adviser.

18. (a) The General Secretary is responsible to the Executive Committee for not incurring expenditure in excess of the funds at his disposal.

(b) An administrative office, under the direction of the General Secretary, conducts the correspondence, administers the funds, and preserves the archives of the Union.

V. Commissions

19. (a) The Commissions of the Union shall pursue the scientific objects of the Union by activities such as the study of special branches of astronomy, the encouragement of collective investigations, and the discussion of questions relating to international agreements or to standardization.

    (b) The Commissions of the Union shall prepare reports on the work with which they are concerned.

20. Each Commission consists of :

    (a) a President and at least one Vice-President elected by the General Assembly on the proposal of the Executive Committee. They remain in office until the end of the ordinary General Assembly next following that of their election. They are not normally re-eligible,

    (b) an Organizing Committee, whose members are appointed by the Commission subject to the approval by the Executive Committee. The Organizing Committee assists the President and Vice-President(s) in their duties. A Commission may decide that it needs no Organizing Committee,

    (c) Members of the Union, appointed by the President, Vice-President(s) and the Organizing Committee, in consideration of their special interests ; their appointment is subject to the confirmation by the Executive Committee.

21. Between two ordinary General Assemblies, Presidents of Commissions may co-opt, from among Members of the Union, new members to the Organizing Committees and to the Commissions themselves.

22. Commissions draw up their own rules. Decisions within Commissions are taken according to the vote of their members, and they become effective once they are approved by the Executive Committee.

VI. Adhering Bodies

23. The functions of the Adhering Bodies are to promote and co-ordinate, in their respective territories, the study of the various branches of astronomy, more especially in relation to their international requirements. They are entitled to submit to the Executive Committee motions for discussions by the General Assembly.

VII. Finances

24. Each Adhering Country pays annually to the Union a number of units of contribution according to its category as follows :

Category as defined in Statute 8 :	1	2	3	4	5	6	7	8
Number of units of contribution :	1	2	4	6	10	14	20	30

25. The income of the Union is to be devoted to its objects, including

    (a) costs of publication and expenses of administration ;

    (b) the promotion of astronomical enterprises requiring international co-operation ;

    (c) the contribution due from the Union to the International Council of Scientific Unions.

26. Funds derived from donations are used by the Union in accordance with the wishes expressed by the donors.

VIII. Publications

27. The Union has the copyright to all materials printed in its publications, unless otherwise arranged.

28. Members of the Union are entitled to receive the publications of the Union free of charge or at reduced prices at the discretion of the Executive Committee taking due regard of the financial situation of the Union.

IX. Final Clauses

29. These By-laws enter into force on 2 August 1988. They can be changed with the approval of an absolute majority of the votes of the Adhering Countries having the right to vote by virtue of Statute 15(a).

30. The present By-laws are being published in French and English versions. In case of doubt, the French version is the only authority.

WORKING RULES

I. Publications

1. The publications of the International Astronomical Union, approved in the budget by the General Assembly, are prepared by the Administrative Office of the Union.

2. Commissions of the Union may, with the approval of the Executive Committee, issue their publications independently.

3. The distribution of publications of the Union is decided, on the proposal of the General Secretary, by the Executive Committee.

4. Members may purchase the publications of the Union at reduced prices.

II. Membership

A. Adhering Countries

5. Applications of countries for adherence to the Union are examined by the Executive Committee for
   (a) the adequacy of the category in which the country wishes to be classified ;
   (b) the present state and expected development of astronomy in the applying country ;
   (c) the degree to which the prospective adhering body is representative of its country's astronomical activity.

6. Applications proposing an adequate annual contribution to the Union shall, with the recommendation of the Executive Committee, be submitted to the General Assembly for decision.

B. Members

7. Individuals proposed for Union membership should, as a rule, be chosen from among astronomers and scientists, whose activity is closely linked with astronomy taking into account
   (a) the standard of their scientific achievement
   (b) the extent to which their scientific activity involves research in astronomy
   (c) their desire to assist in the fulfilment of the aims of the Union.

8. Young astronomers should be considered eligible for membership after they have shown their capability (as a rule Ph.D. or equivalent) of and experience (some years of successful activity) in conducting original research.

9. For full time professional astronomers the achievement in astronomy may consist either of original research or of substantial contributions to major observational programmes.

10. Others are eligible for membership only if they are making original contributions closely linked with astronomical research.

11. Eight months before an ordinary General Assembly, adhering bodies will be asked to propose new Members. The proposals should reach the General Secretary not later than five months before the first session of the General Assembly. Proposals received after the closing date will only be taken into consideration if the delay is justified by exceptional circumstances.

12. Each proposal shall be written separately. It should include the name, first names and postal address of the candidate, preferably that of his/her Institute or Observatory, his/her place and date of birth, the University and the year of his/her Ph.D. or equivalent title, his/her present occupation, titles and bibliographic data of two or three of his/her more important papers or publications, and details, if any, worthy to be considered by the Nominating Committee.

13. (a) Presidents of Union Commissions wishing to suggest new Members for admission should address their suggestions to the General Secretary five months before the first session of an ordinary General Assembly. The proposals should contain particulars as in article 12.

(b) The General Secretary notifies the adhering bodies in questions of such suggestions.

14. The General Secretary shall prepare two lists for the Nominating Committee.

(a) One containing the candidates proposed by the adhering bodies,

(b) the other containing those suggested by Presidents of Commissions, but not included among the proposals of adhering bodies.

15. The Nominating Committee prepares the final proposals for Union membership from the two lists as mentioned in article 14.

16. Adhering bodies should propose cancellation of Members who have left the field of astronomy for other interests, unless they continue to contribute to astronomy. Such proposals should be announced to the Member concerned and to the General Secretary.

17. The alphabetical list of Union Members will be published by the General Secretary in the Transactions of each ordinary General Assembly.

III. Commission Membership

18. Members of Union Commissions are co-opted by Commissions. The rules governing the procedure of such co-option are drawn up by the Commissions themselves.

19. Commissions should choose, or approve of, Commission members taking into account their special interests, in particular their scientific activity in the appropriate fields of research and their contribution to the work of the Commission. They may,

(a) invite Members to become members of their Commission,

(b) remove members who have not contributed to the work of the Commission,

(c) accept or reject applications for membership from existing or proposed Members,

(d) suggest non-Members for election as Members, thus enabling them to become members of the Commission.

20. Members may not, as a rule, be members of more than three Commissions.

21. Members may apply for Commission membership by writing to the President of the Commission concerned. Such applications should only be made if the Member is actively engaged in the appropriate field of research and is prepared to contribute to the work of the Commission.

22. Members of Commissions may resign from a Commission by writing to its President.

23. Adhering bodies, in sending in their proposals for new Members, may also suggest one Commission for each candidate.

24. The General Secretary will record and analyse the list of members of Commissions ; if necessary he will try to resolve any outstanding anomalies.

25. The list of Commission members will be published by the General Secretary in the Transactions of each ordinary General Assembly.

IV. Consultants

26. Eligible as Consultants are non-astronomers in a position to further the interest in astronomy.

27. Proposals of Commissons for the approval of consultants should, as a rule, reach the General Secretary not later than five months before the first session of an ordinary General Assembly.

28. The General Secretary shall prepare a list of those proposed for admission as consultants and submit it to the Executive Committee for approval.

29. The Administrative Office will maintain an alphabetical list of consultants.

30. Consultants may participate in the meetings of the Union. They may have the right to vote in the respective Commission. They receive, free of charge, the Information Bulletin of the Union.

V. Scientific Meetings

31. The General Secretary shall publish rules for scientific meetings organized or sponsored by the Union.

VI. External Contacts

32. No dealings with third parties, attributable to the Union, shall be undertaken by any Member of the Union except on the authority of the General Secretary.

33. Representatives of the Union in other bodies, especially ICSU Committees and ICSU Inter-Union Committees, shall be appointed by the Executive Committee. Nominations are sought from Presidents of appropriate Commissions.

34. Expenses incurred by Representatives of the Union in other bodies will be reimbursed at the discretion of the General Secretary, within the provisions of the Budget Estimate adopted by the General Assembly. Representatives are required to obtain prior approval of the General Secretary before incurring such expenses.

VII. General Assemblies

35. The General Secretary distributes the budget prepared by the Executive Committee to National Committes of Astronomy and/or Adhering Organizations for comments eight months before the General Assembly.

Statuts, règlements & directives

STATUTS

I. Dénomination, Buts et Domicile

1. L'Union Astronomique Internationale (ci-après dénommée l'Union) est une organisation non-gouvernementale, qui a pour buts de :

   (a) développer l'astronomie par la coopération internationale,
   (b) encourager l'étude et le développement de l'astronomie sous tous ses aspects,
   (c) servir et sauvegarder les intérêts de l'astronomie.

2. L'Union a son siège légal à Bruxelles.

II. Affiliation de l'Union

3. L'Union adhère au Conseil International des Unions Scientifiques.

III. Membres de l'Union

4. L'Union a pour membres :

   (a) des personnes morales (Pays adhérents)
   (b) des personnes morales associées (Pays associés)
   (c) des membres individuels (Membres).

IV. Organisations Affiliées

5. L'Union peut accepter l'affiliation d'organisations internationales non-gouvernementales qui contribuent au développement de l'astronomie.

V. Pays Adhérents

6. Les pays adhèrent à l'Union

   soit :

   (a) par l'intermédiaire de l'organisation par laquelle ils adhèrent au Conseil International des Unions Scientifiques, ou par l'intermédiaire d'un Comité National d'Astronomie approuvé par cette organisation,

   soit :

   (b) s'ils n'adhèrent pas au Conseil International des Unions Scientifiques, par l'intermédiaire d'un Comité National d'Astronomie reconnu par le Comité Exécutif de l'Union.

   (c) Les Organisations ou Comités mentionnés à l'article 6(a) et les Comités Nationaux d'Astronomie mentionnés à l'article 6(b) sont dénommés ci-après organismes adhérents.

7. L'adhésion d'un pays à l'Union est proposée par le Comité Exécutif et approuvée par l'Assemblée Générale : elle prend fin si le pays se retire de l'Union ou si le pays n'a pas payé sa contribution durant cinq ans.

8. Les Pays Adhérents sont répartis en catégories. Le nombre des catégories est fixé par le Règlement. Un pays qui sollicite son adhésion indique la catégorie dans laquelle il désire

être classé. La proposition peut être refusée par le Comité Exécutif si la catégorie est manifestement inadéquate.

VI. Pays Associés

9. Les pays souhaitant faire partie de l'Union tout en développant l'astronomie dans leur territoire peuvent le faire à titre de Membres Associés.

10. L'organisme adhérent d'un pays associé peut être soit l'organisation par l'intermédiaire de laquelle le pays adhère au Conseil International des Unions Scientifiques, soit une institution d'éducation supérieure ou un conseil scientifique national.

11. Les pays sont acceptés en qualité de Membres Associés par l'Assemblée Générale, sur proposition du Comité Exécutif, pour une période maximale de neuf ans au terme de laquelle ils deviennent membres à part entière, ou se retirent de l'Union.

12. Durant la période probatoire, l'Union peut accepter, à la requête de l'organisation adhérente, d'aider au développement de l'astronomie dans ce pays via le Programme de Professeurs Visiteurs et/ou de tout autre programme adéquat.

VII. Membres

13. Les Membres sont admis dans l'Union par le Comité Exécutif, sur proposition de l'un des organismes adhérents mentionnés à l'article 6, en considération de leur activité dans une branche de l'astronomie.

VIII. Assemblée Générale

14. (a) L'activité de l'Union est dirigée par l'Assemblée Générale des représentants des Pays Adhérents et des Membres. Chaque Pays Adhérent nomme un représentant autorisé à voter en son nom.

(b) L'Assemblée Générale rédige un Règlement qui précise les modalités d'application des Statuts.

(c) Elle nomme un Comité Exécutif chargé d'exécuter les décisions de l'Assemblée Générale, et d'administrer l'Union pendant la periode séparant les réunions de deux Assemblées Générales ordinaires successives. Le Comité Exécutif rend compte de sa gestion à l'Assemblée Générale. L'Assemblée Générale, en acceptant le rapport du Comité Exécutif, le décharge de sa responsabilité.

15. (a) Sur les questions concernant l'administration de l'Union, sans implication budgétaire, le vote à l'Assemblée Générale a lieu par Pays Adhérent, chaque pays disposant d'une voix. Les Pays Adhérents qui ne sont pas à jour de leurs cotisations annuelles au 31 décembre de l'année précédant l'Assemblée Générale ne peuvent pas participer aux votes.

(b) Sur les questions engageant le budget de l'Union, le vote a lieu de même par Pays Adhérent, dans les conditions et avec les réserves prévues à l'article 15(a), le nombre de voix de chaque Pays Adhérent étant égal à l'indice de sa catégorie, définie conformément à l'article 8, augmenté d'une unité.

(c) Les Pays Adhérents peuvent voter par correspondance sur les questions figurant à l'ordre du jour de l'Assemblée Générale.

(d) Un scrutin n'est valable que si au moins deux tiers des Pays Adhérents disposant du droit de vote en vertu de l'article 15(a) y prennent part.

(e) Les Pays Associés ne peuvent voter que sur des questions concernant les Membres Associés.

16. Sur les questions scientifiques n'engageant pas le budget de l'Union, les Membres de l'Union disposent chacun d'une voix.

17. Sur toutes les questions prévues aux articles 15 et 16, les décisions sont prises à la majorité absolue des suffrages. Cependant, une décision de modification des Statuts n'est valable que si elle a été prise à la majorité des deux tiers des voix des Pays Adhérents qui disposent du droit de vote en vertu de l'article 15(a).

18. Une proposition de modification des Statuts ne peut être discutée que si elle figure, en tant que telle, à l'ordre du jour de l'Assemblée Générale.

IX. Comité Exécutif

19. Le Comité Exécutif se compose du Président de l'Union, du "Président-elect", de six Vice-Présidents, du Secrétaire Général et du Secrétaire Général Adjoint, élus par l'Assemblée Générale sur la proposition du Comité Spécial des Nominations. Le "Président-elect" deviendra normalement le Président du prochain Comité Exécutif.

X. Commissions de l'Union

20. L'Assemblée Générale crée des Commissions en vue d'assurer la réalisation des buts qu'elle se propose.

XI. Représentation Légale de l'Union

21. Le Secrétaire Général est le représentant légal de l'Union.

XII. Budget et Cotisations

22. (a) Pour chaque Assemblée Générale ordinaire, le Comité Exécutif prépare un projet de budget pour la période à courir jusqu'à l'Assemblée Générale ordinaire suivante, ainsi que les comptes de l'Union pour la période précédente. Il les soumet au Comité des Finances pour examen ; ce Comité des Finances est composé de membres nommés par les organismes adhérents, à raison d'un membre par organisme, et il est approuvé par l'Assemblée Générale. Lors de sa première séance pendant l'Assemblée Générale, le Comité des Finances élit un Président parmi ses membres.

(b) Le Comité des Finances examine les comptes de l'Union pour voir si les dépenses engagées ont été conformes aux vœux émis lors de la précédente réunion de l'Assemblée Générale et il s'assure que le budget proposé vise à la poursuite de la politique de l'Assemblée Générale, telle qu'elle est interprétée par le Comité Exécutif. Il présente des rapports sur ces questions qu'il soumet à l'Assemblée Générale pour approbation des comptes, et pour décision sur le budget.

(c) Chaque Pays Adhérent verse annuellement à l'Union un nombre d'unités de cotisation qui est fonction de sa catégorie. Le nombre d'unités de cotisation pour chaque catégorie est fixé par le Règlement.

(d) La cotisation annuelle des Pays Associés s'élève à une unité de contribution.

(e) Le montant de l'unité de cotisation est fixé par l'Assemblée Générale, sur la proposition du Comité Exécutif et avec l'avis du Comité des Finances.

(f) Le paiement des cotisations est à la charge des organismes adhérents. La responsabilité de chaque Pays Adhérent envers l'Union est limitée au montant des cotisations dues par ce pays à l'Union.

(g) Un Pays Adhérent qui cesse d'adhérer à l'Union renonce de ce fait à ses droits sur l'actif de l'Union.

XIII. Dissolution de l'Union

23. La décision de dissoudre l'Union n'est valable que si elle est prise à la majorité des trois quarts des voix des Pays Adhérents qui disposent du droit de vote en vertu de l'article 15(a).

XIV. Dévolution de l'Autorité en Cas de Force Majeure.

24. Si, par suite d'événements indépendants de la volonté de l'Union, des circonstances apparaissent qui rendent impossible le respect des clauses de ces Statuts et du Règlement établi par l'Assemblée Générale, les organes et membres du Comité Exécutif de l'Union, dans l'ordre fixé ci-dessous, prendront toutes dispositions qu'ils jugeront nécessaires pour la continuation du fonctionnement de l'Union. Ces dispositions devront être soumises à une autorité supérieure dès que cela deviendra possible, jusqu'à ce qu'une Assemblée Générale extraordinaire puisse être réunie. L'autorité est dévolue dans l'ordre ci-dessous :

    l'Assemblée Générale ; une Assemblée Générale extraordinaire ; le Comité Exécutif, réuni ou par correspondance ; Le Président de l'Union ; Le Secrétaire Général ; ou, à défaut de la possibilité de recourir à l'une de ces autorités ou de leur disponibilité, un des Vice-Présidents.

XV. Clauses Finales

25. Ces Statuts entrent en vigueur le 2 Août 1988.

26. Les présents Statuts sont publiés en versions française et anglaise. En cas d'incertitude, la version française fait seule autorité.

REGLEMENT

I. Les Membres de l'Union

1. Les demandes d'adhésion des pays à l'Union Astronomique Internationale (ci-après dénommée l'Union) sont examinées par le Comité Exécutif et soumises à l'approbation de l'Assemblée Générale.

2. Les propositions de modifications de la liste des Membres sont, après examen attentif des suggestions des Présidents de Commissions, soumises pour avis au Comité des Nominations, composé d'un représentant de chaque Pays Adhérent désigné par l'organisme adhérent habilité, avant la décision du Comité Exécutif.

3. Les Commissions peuvent, avec l'approbation du Comité Exécutif, coopter des consultants qu'elles jugent en mesure d'apporter une contribution utile à leur travail. L'adhésion des consultants a pour terme le dernier jour de la première Assemblée Générale ordinaire qui suit leur admission, à moins qu'elle ne soit renouvelée.

4. Une organisation affiliée peut participer au travail de l'Union dans les conditions fixées par accord entre l'organisation et le Comité Exécutif.

II. L'Assemblée Générale

5. L'Union se réunit en Assemblée Générale ordinaire régulièrement une fois tous les trois ans. Si le lieu et la date de l'Assemblée Générale ordinaire n'ont pas été décidés lors de la précédente Assemblée Générale, ils sont fixés par le Comité Exécutif et communiqués aux organismes adhérents au moins six mois à l'avance.

6. Le Président peut convoquer, avec l'accord du Comité Exécutif, une Assemblée Générale extraordinaire. Il est tenu de le faire à la demande du tiers des Pays Adhérents.

7. L'Ordre du Jour de chaque Assemblée Générale ordinaire est arrêté par le Comité Exécutif et communiqué aux Organismes Adhérents au moins quatre mois avant le premier jour de la réunion. Il devra inclure la proposition du Comité Exécutif concernant le montant de l'unité de cotisation qui permet l'application de l'article 24.

8. (a) L'Ordre du Jour doit inclure toute motion ou proposition reçue par le Secrétaire Général au moins cinq mois avant le premier jour d'une Assemblée Générale ordinaire, qu'elle émane d'un organisme adhérent, d'une Commission de l'Union, ou d'une Commission mixte dans laquelle l'Union est représentée.

   (b) Une motion ou proposition concernant l'administration ou le budget de l'Union qui ne figure pas à l'Ordre du Jour, préparé par le Comité Exécutif, ou tout amendement à une motion qui figure à l'Ordre du Jour, ne peut être discuté qu'avec l'accord préalable des deux tiers au moins des voix des Pays Adhérents représentés à l'Assemblée Générale et disposant du droit de vote en vertu de l'article 15(a) des Statuts.

9. S'il y a doute sur le caractère administratif ou scientifique d'une question donnant lieu à un vote, l'avis du Président est prépondérant.

10. En cas de partage égal des voix, le Président a voix prépondérante.

11. Le Président peut inviter des représentants d'autres organisations, des scientifiques et de jeunes astronomes à participer à l'Assemblée Générale. Avec l'accord du Comité Exécutif, il peut déléguer ce privilège au Secrétaire Général en ce qui concerne les représentants d'autres organisations, aux organismes adhérents en ce qui concerne les scientifiques et les jeunes astronomes.

III. Le Comité Spécial des Nominations

12. (a) Les propositions pour les élections du Président de l'Union, du Président-elect, des six Vice-Présidents, du Secrétaire Général et du Secrétaire Général Adjoint sont soumises à l'Assemblée Générale par le Comité Spécial des Nominations. Ce Comité se compose du Président en fonction et du Président sortant, d'un membre proposé par le Comité Exécutif sortant et n'appartenant ni au Comité Exécutif actuel ni au Comité Exécutif précédent, et de quatre membres élus par le Comité des Nominations parmi douze membres proposés par les Présidents de Commissions. A l'exception du Président en fonction et du Président sortant, les membres actuels et les anciens membres du Comité Exécutif ne doivent pas faire partie du Comité Spécial des Nominations. Les membres du Comité Spécial des Nominations doivent tous appartenir à des pays différents.

(b) Le Secrétaire Général et le Secrétaire Général Adjoint participent au travail du Comité Spécial des Nominations à titre consultatif.

(c) Le Comité Spécial des Nominations est nommé par l'Assemblée Générale et est responsable directement devant elle. Il reste en fonction jusqu'à la fin de l'Assemblée Générale ordinaire qui suit immédiatement sa nomination, et il peut combler toute vacance survenant parmi ses membres.

IV. Le Comité Exécutif et ses Membres

13. (a) Le Président de l'Union reste en fonction jusqu'à la fin de l'Assemblée Générale ordinaire qui suit immédiatement celle de son élection ; les Vice-Présidents restent en fonction jusqu'à la fin de la deuxième Assemblée Générale ordinaire qui suit celle de leur élection. Ils ne sont pas rééligibles immédiatement pour les mêmes fonctions.

(b) Le Secrétaire Général et le Secrétaire Général Adjoint restent en fonction jusqu'à la fin de l'Assemblée Générale ordinaire qui suit immédiatement celle de leur élection. Normalement, le Secrétaire Général Adjoint succède au Secrétaire Général, mais l'un et l'autre peuvent être réélus aux mêmes fonctions pour une seconde période consécutive.

(c) Les élections ont lieu au cours de la dernière réunion de l'Assemblée Générale, les noms des candidats proposés ayant été annoncés au cours d'une réunion antérieure.

14. Le Président sortant et le Secrétaire Général sortant deviennent conseillers du Comité Exécutif jusqu'à la fin de l'Assemblée Générale ordinaire qui suit immédiatement celle de la fin de leur mandat. Ils participent au travail du Comité Exécutif et assistent à ses réunions sans droit de vote.

15. Le Comité Exécutif peut combler toute vacance survenant en son sein. Toute personne ainsi nommée reste en fonction jusqu'à l'Assemblée Générale ordinaire suivante.

16. Le Comité Exécutif peut rédiger et publier des Directives pour expliciter les Statuts et le Règlement.

17. Le Comité Exécutif nomme le représentant de l'Union qui doit siéger au sein du Conseil International des Unions Scientifiques ; si ce représentant n'est pas déjà un membre élu du Comité Exécutif, il devient conseiller.

18. (a) Le Secrétaire Général est responsable auprès du Comité Exécutif des dépenses qu'il engage, qui ne doivent pas dépasser le montant des fonds mis à sa disposition.

(b) Un bureau administratif, sous la direction du Secrétaire Général, est chargé de la correspondance, de la gestion des fonds de l'Union, et de la conservation des archives.

V. Commissions

19. (a) Les Commissions de l'Union poursuivent les buts scientifiques de l'Union par des moyens tels que l'étude de domaines particuliers de l'Astronomie, l'encouragement de recherches collectives et la discussion de questions relatives aux accords internationaux et à la standardisation.

(b) Les Commissions de l'Union établissent des rapports sur les sujets qui leur ont été confiés.

20. Chaque Commission se compose de :

(a) un Président et au moins un Vice-Président élus par l'Assemblée Générale sur la proposition du Comité Exécutif. Ils demeurent en fonction jusqu'à la fin de l'Assemblée Générale ordinaire qui suit immédiatement celle de leur élection. Ils ne sont pas normalement rééligibles,

(b) un Comité d'Organisation, dont les membres sont désignés par la Commission sous réserve de l'approbation du Comité Exécutif. Le Comité d'Organisation assiste le Président et le(s) Vice-Président(s) dans leur tâche. Une Commission peut décider qu'elle n'a pas besoin de Comité d'Organisation,

(c) des membres de l'Union, nommés par les Présidents, Vice-Président(s) et Comité d'Organisation, en considération de leurs spécialités ; leur désignation est soumise à confirmation par le Comité Exécutif.

21. Entre deux Assemblées ordinaires, les Présidents de Commissions peuvent coopter, parmi les Membres de l'Union, de nouveaux membres des Comités d'Organisation et des Commissions elles-mêmes.

22. Les Commissions rédigent leur propre règlement. Les décisions sont prises, à l'intérieur des Commissions, par un vote de leurs membres et elles deviennent d'application après approbation par le Comité Exécutif.

VI. Organismes Adhérents

23. Le rôle des organismes adhérents est d'encourager et de coordonner, sur leurs territoires respectifs, l'étude des diverses branches de l'astronomie, particulièrement en ce qui concerne leurs besoins sur le plan international. Ils ont le droit de soumettre au Comité Exécutif des propositions pour discussion par l'Assemblée Générale.

VII. Finances

24. Chaque Pays Adhérent verse à l'Union une cotisation annuelle, qui est un multiple de l'unité de cotisation en fonction de sa catégorie, comme suit :

Catégories définies conformément à l'article 8 des Statuts	:	1	2	3	4	5	6	7	8
Nombre respectif d'unités de cotisations	:	1	2	4	6	10	14	20	30

25. Les ressources de l'Union sont consacrées à la poursuite de ses buts, y compris :

(a) les frais de publication et les dépenses administratives ;

(b) l'encouragement des activités astronomiques qui nécessitent la coopération internationale ;

(c) la cotisation due par l'Union au Conseil International des Unions Scientifiques.

26. Les ressources provenant de dons sont utilisées par l'Union en tenant compte des vœux exprimés par les donateurs.

VIII. Publications

27. L'Union a la propriété littéraire de tous les textes imprimés dans ses publications, sauf accord différent.

28. Les Membres de l'Union ont le droit de recevoir les publications de l'Union gratuitement ou à prix réduit, à la discrétion du Comité Exécutif qui décide en fonction de la situation financière de l'Union.

IX. Clauses Finales

29. Ce règlement entre en vigueur le 2 Août 1988. Il peut être modifié avec l'approbation de la majorité absolue des voix des Pays Adhérents qui disposent du droit de vote en vertu de l'article 15(a) des Statuts.

30. Le présent règlement est publié en versions française et anglaise. En cas d'incertitude, la version française fait seule autorité.

DIRECTIVES

I. Publications

1. Les publications de l'Union Astronomique Internationale, approuvées dans le budget par l'Assemblée Générale, sont préparées par le Bureau Administratif de l'Union.

2. Les Commissions de l'Union peuvent, avec l'approbation du Comité Exécutif, avoir leurs propres publications.

3. Le Comité Exécutif décide, sur la proposition du Secrétaire Général, des modalités de distribution des publications de l'Union.

4. Les Membres de l'Union peuvent acquérir les publications de l'Union à un prix réduit.

II. Appartenance à l'Union

A. Pays Adhérents

5. Les demandes d'adhésion à l'Union formulées par les pays sont examinées par le Comité Exécutif compte tenu des points suivants :
   (a) justesse du choix de la catégorie dans laquelle le pays souhaite être classé ;
   (b) situation actuelle de l'Astronomie dans le pays formulant la demande, et ses possibilités de développement ;
   (c) mesure dans laquelle le futur organisme adhérent est représentatif de l'activité astronomique de son pays.

6. Les demandes proposant une contribution annuelle appropriée seront soumises pour décision à l'Assemblée Générale, avec la recommandation du Comité Exécutif.

B. Membres

7. Les personnes proposées pour devenir Membres de l'Union doivent en principe être choisies parmi des astronomes et des chercheurs dont les activités sont liées à l'astronomie, compte tenu de :
   (a) la qualité de leur œuvre scientifique ;
   (b) la mesure dans laquelle leur activité scientifique implique des recherches astronomiques ;
   (c) leur désir de contribuer à la poursuite des buts de l'Union.

8. Les jeunes astronomes doivent être considérés comme pouvant devenir Membres de l'Union dès qu'ils ont fait la preuve de leur capacité (en principe par une thèse de doctorat ou son équivalent) et de leur aptitude (quelques années d'activité fructueuse) à mener une recherche personnelle.

9. Pour les astronomes professionnels, leur contribution à l'astronomie peut consister soit en des recherches personnelles, soit en une collaboration assidue à des programmes importants d'observations.

10. Les autres personnes ne peuvent devenir Membres de l'Union que si certains de leurs travaux originaux concernent étroitement la recherche astronomique.

11. Huit mois avant une Assemblée Générale ordinaire, il sera demandé aux organismes adhérents de proposer de nouveaux Membres. Les propositions devront parvenir au Secrétaire Général au moins cinq mois avant la première session de l'Assemblée Générale. Les propositions reçues après cette date limite ne seront prises en considération que si des circonstances exceptionnelles justifient le retard.

12. Chaque proposition du nouveau Membre doit être présentée séparément et indiquer le nom, les prénoms et l'adresse postale du candidat (de préférence celle de son Institut ou Observatoire), ses date et lieu de naissance, l'Université devant laquelle il a soutenu sa thèse ou le diplôme équivalent, la date de soutenance, la situation actuelle du candidat, les titres et renseignements bibliographiques de deux ou trois de ses articles ou publications les plus significatifs et, s'il y a lieu, tous les renseignements susceptibles d'être pris en considération par le Comité des Nominations.

13. (a) Les Présidents de Commissions qui désirent suggérer de nouveaux membres doivent adresser leurs suggestions au Secrétaire Général au moins cinq mois avant la première session d'une Assemblée Générale ordinaire. Les propositions devront fournir les mêmes renseignements que ceux mentionnés à l'article 12.

    (b) Le Secrétaire Général fait part de ces suggestions aux organismes adhérents intéressés.

14. Le Secrétaire Général préparera deux listes pour le Comité des Nominations

    (a) l'une contenant les noms des candidats proposés par les organismes adhérents,

    (b) l'autre contenant les noms des candidats proposés par les Présidents de Commissions, mais qui ne sont pas déjà inclus dans les propositions des organismes adhérents.

15. A partir des deux listes mentionnées à l'article 14, le Comité des Nominations prépare les propositions définitives de nouveaux membres de l'Union.

16. Les organismes adhérents peuvent proposer la radiation de Membres ayant abandonné le domaine de l'astronomie pour d'autres activités, à moins qu'ils ne continuent à apporter une contribution à l'astronomie. Ces propositions doivent être portées à la connaissance du Secrétaire Général et du Membre concerné.

17. Le Secrétaire Général publiera la liste alphabétique des Membres de l'Union dans les Transactions de chaque Assemblée Générale ordinaire.

III. Membres des Commissions

18. Les membres des Commissions de l'Union sont cooptés par les Commissions. Cette procédure est régie par des règles établies par les Commissions elle-mêmes.

19. Les Commissions devraient choisir, ou approuver, la liste des membres de leurs commissions compte tenu de la spécialité de ces personnes, en particulier de leur activité scientifique dans le domaine de recherche de la Commission, et de leur contribution au travail de la Commission. Elles peuvent

    (a) inviter les Membres de l'Union à devenir membres de la Commission,

    (b) radier les membres de la Commission qui n'ont pas contribué à son activité,

    (c) accepter ou refuser les demandes présentées par des Membres de l'Union, ou par des personnes proposées comme tels, en vue d'appartenir à la Commission,

    (d) suggérer l'élection comme Membres de l'Union de personnes n'y appartenant pas, ce qui leur permettrait alors de devenir membres de la Commission.

20. Les Membres de l'Union ne peuvent pas, en règle générale, appartenir à plus de trois Commissions.

21. Les membres de l'Union peuvent demander à être admis dans une Commission en écrivant au Président de cette Commission. Ils ne devraient faire cette demande que si leur propre activité rentre dans le cadre des recherches de la Commission et s'ils sont décidés à contribuer au travail de la Commission.

22. Les membres des Commissions peuvent se retirer d'une Commission en écrivant à son Président.

23. En envoyant leurs propositions de nouveaux Membres, les organismes adhérents peuvent également suggérer le choix d'une Commission pour chaque candidat.

24. Le Secrétaire Général enregistrera et analysera la liste des membres des Commissions ; si cela est nécessaire, il tentera de trouver une solution aux anomalies évidentes.

25. Le Secrétaire Général publiera la liste des membres des Commissions dans les Transactions de chaque Assemblée Générale ordinaire.

IV. Consultants

26. Peuvent être élus Consultants des personnes qui ne sont pas astronomes, mais qui sont susceptibles de servir les intérêts de l'astronomie.

27. Les Commissions doivent en principe envoyer, pour approbation, leurs propositions de consultants au Secrétaire Général au moins cinq mois avant la première session d'une Assemblée Générale ordinaire.

28. Le Secrétaire Général préparera une liste des personnes proposées comme consultants et la soumettra pour approbation au Comité Exécutif.

29. Le Bureau Administratif établira une liste alphabétique des consultants.

30. Les consultants peuvent participer aux réunions de l'Union. Ils peuvent avoir droit de vote dans leurs Commissions respectives. Ils reçoivent gratuitement le Bulletin d'Information de l'Union.

V. Réunions Scientifiques

31. Le Secrétaire Général publiera un règlement pour les réunions scientifiques organisées ou parrainées par l'Union.

VI. Contacts Extérieurs

32. Aucune relation avec des tiers, imputable à l'Union, ne sera entreprise par quiconque membre de l'Union, si ce n'est sous l'autorité du Secrétaire Général.

33. Les représentants de l'Union dans d'autres organisations, en particulier les Comités de l'ICSU et les Commissions Inter-Unions, seront désignés par le Comité Exécutif. Les noms sont proposés par les Présidents des Commissions concernées.

34. Les dépenses encourues par les représentants de l'Union dans d'autres organisations seront remboursées à la discrétion du Secrétaire Général, dans les limites du Budget adopté par l'Assemblée Générale. Les représentants sont priés d'obtenir l'accord préalable du Secrétaire Général avant d'engager ces dépenses.

VII. Assemblées Générales

35. Huit mois avant l'Assemblée Générale, le Secrétaire Général envoie aux Comités Nationaux d'Astronomie et aux Organisations Adhérentes le budget préparé par le Comité Exécutif, pour commentaires.

CHAPTER VII

MEMBERSHIP

Expression of Intent

In the application of Statute 5, it is the current intention of the Executive Committee that, Individual Members of the Union in countries that cease participation retain full individual membership of the Union.

This Chapter is composed as follows:

1. List of Adhering Countries
2. Membership of Commissions
3. Geographical Repartition of Members
   - within Adhering Countries
   - within Non Adhering Countries
4. Alphabetical List of Members

1. List of adhering countries (57)

The year of adherence and approximate number of IAU Members residing in the different adhering countries are indicated (as of May 1989). The asterisk indicates that these countries are Associate National Members as opposed to Full National Members (see Chapter II, p. 19).

Country	Adhering Organization	Year	Members
Algeria	Centre de Recherche en Astronomie, Astrophysique & Géophysique B.P. 63 Bouzareah Alger	1988	-
Argentina	Consejo Nacional de Investigaciones Cientificas y Técnicas Rivadavia 1917 1033 Buenos Aires	1927	52
Australia	Australian Academy of Sciences PO Box 783 Canberra City, ACT 2601	1939	145
Austria	Bundesministerium für Wissehschaften & Forschung Minoritenplatz 5 A-1010 Wien	1955	28
Belgium	Administration des Affaires Communes et des Etablissements Scientifiques de l'Etat Boulevard Pachéco 34 6e étage B-1000 Bruxelles	1920	78
Brazil	Conselho Nacional de Desenvolvimento Cientifico e Tecnologico - CNPq Av. W3 Norte, Quadra 507-B Caixa Postal 11-1142 70740 Brasilia DF	1961	66
Bulgaria	Bulgarian Academy of Sciences 7th November Street 1 1000 Sofia	1957	43
Canada	National Research Council of Canada International Relations Ottawa, Ontario K1A OR6	1957	203
Chile	Universidad de Chile Observatorio Astronomico Nacional Casilla 36-D Santiago	1947	40

China	Chinese Astronomical Society Purple Mountain Observatory Academia Sinica Nanking	1935	307
	The Astronomical Societey of the Republic of China Academia Sinica Taipei 115	1959	14
Colombia	Universidad Nacional de Colombia Facultad de Ciencias Apartado Aereo No. 5997 Bogota	1967	2
Cuba	Academia de Ciencias de Cuba Callee 212-2906 e/29 y 31 La Lisa, La Coronela La Habana 2	1970	1
Czechoslovakia	Czechoslovak Narodni Komitet Astronomicky Budecska 6 12023 Praha 2	1922	80
Denmark	Kongelige Danske Videnskabernes Selskab H.C. Andersen Boulevard 35 DK-1553 Kobenhavn V	1922	44
Egypt (AR)	Academy of Scientific Research & Technology Dept. of Scientific Societies and International Unions 101 Kasr El-Einy Street Cairo	1925	38
Finland	Académie des Sciences et Lettres Snellmaninkatu 9-11 Helsinki 17	1948	26
France	Académie des Sciences COFUSI 23, quai Conti F-75006 Paris	1920	509
Germany, DR	Akademie der Wissenschaften der DDR Nationalkomitee für Astronomie Rosa-Luxemburg Strasse 17A DDR-1502 Postdam	1951	47
Germany, FR	National Coucil of Observatories Universitäts-Sternwarte Geismatrlandstrasse 11 D-3400 Göttingen	1951	334

			MS
Greece	Academy of Athens Panepistimiou Str. Athens	1920	75
Hungary	Hungarian Academy of Sciences Roosevelt Tér 9 Budapest V	1947	33
Iceland	Ministry of Education Science Institute University of Iceland Dunhaga 3 IS-107 Reykjavik	1988	2
India	Indian National Science Academy Bahadur Shah Zafar Marg New Delhi 110002	1964	188
Indonesia	Indonesian Institute of Sciences (LIPI) Gedung Widya Graha Jl. Jend. Gatot Subroto 10 Jakarta Selatan	1979	7
Iran	University of Tehran Office of International Relations Tehran	1969	8
Iraq	Council for Scientific Research Astronomy and Space Research Center PO Box 2441 Jadiryah Baghdad	1976	8
Ireland	The Royal Irish Academy Academy House 19 Dawson Street Dublin 2	1947	24
Israel	The Israel Academy of Sciences and Humanities Albert Einstein Square, Talbieh Jerusalem 91040	1954	42
Italy	Consiglio Nazionale delle Ricerche Piazzale delle Scienze 7 I-00100 Roma	1920	325
Japan	Science Council of Japan 22-34 Roppongi 7 chome Minato-Ku Tokyo 106	1920	305
Korea, DPR	Academy of Sciences Pyongyang	1961	21

Korea (RP)	Korean Astronomical Society Dpt of Astronomy Seoul National University Seoul 151	1973	19
Malaysia (*)	University Kebangsaan Malaysia Dpt of Physics 43600 Ukm, Bangi, Selangor	1988	2
Mexico	Instituto de Astronomia, UNAM Apartado Postal 70-264 Ciudad Universitaria 04510 DF Mexico	1921	50
Morocco	Centre National de Coordination & de Planification de la Recherche Scientifique & Technique 52 Charia Omar Ibn Khattab B.P. 1346 RP Agdal	1988	-
Netherlands	Koninklijke Nederlandse Akademie van Wetenschappen Kloveniersburgwal 29 NL-1011 JV Amsterdam	1922	140
New Zealand	The Royal Society of New Zealand PO Box 598 Wellington	1964	24
Nigeria	Nigerian Academy of Sciences Faculty of Science, University of Lagos P.M.B. 1004 University of Lagos Post Office Lagos	1985	4
Norway	Det Norske Videnskaps-Akademi i Oslo Drammensveien 78 Oslo 2	1922	19
Peru (*)	National Council of Science & Technology Camilo Carrillo 118-9 Piso Lima 11	1988	1
Poland	Polskiej Akademii Nauk Palac Kultury i Nauki Skrytka pocztowa 24 00-901 Warszawa	1922	81
Portugal	Secçao Portuguesa das Unioes Internacionais Astronomica e Geodesica e Geofisica Praça de Estrela Lisboa 1200	1924	16

Rumania	Rumanian National Committee of Astronomy Astronomical Observatory Cutitul de Argint 5 PO Box 28 75212 Bucarest	1928	16
Saudi Arabia	King Abdulaziz City for Science & Technology Directorate of Technology P.O. Box 6086 Riyadh 11442	1988	4
South Africa	The South African ICSU Secretariat Foundation for Research Development PO Box 395 0001 Pretoria	1938	35
Spain	Comision Nacional de Astronomia Instituto Geografico y Cadastral General Ibanez 3 Madrid (3)	1922	122
Sweden	Kungl. Vetenskapsakademien PO Box 50005 S-104 05 Stockholm 50	1925	76
Switzerland	Schweizerische Naturforschende Gesellschaft Zentralsekretariat Postfach 2535 Hirschengraben 11 CH-3001 Bern	1923	52
Turkey	Astronomi Denergi Baskani Faculty of Arts & Sciences Middle East Technical University 07631 Ankara	1961	40
United Kingdom	The Royal Society 6, Carlton House Terrace London SW1Y 5AG	1920	456
Uruguay	Ministerio de Educacion y Cultura Comite Nacional de Astronomia Reconquista 535 Montevideo	1970	1
USA	National Academy of Sciences Office of International Affairs 2101 Constitution Avenue N.W. Washington, DC 20418	1920	1924
USSR	Academy o Sciences of the USSR Foreign Relations Department Leninskij Prospekt 14 Moscow 71	1935	421

Vatican City	Pontificia Academia delle Scienze I-00120 Citta del Vaticano	1932	5
Venezuela	Centro de Investigaciones de Astronomia Apartado 264 Mérida 5101-A	1953	8
Yugoslavia	Savez Drustava Matematicara, Fizicara i Astronoma Jugoslosiavije Institut za Matematiku i Fiziku Cetinjski put bb 81000 Titograd	1935	38

A total of 57 countries now adhere to the Union -55 as Full Members and 2 as Associate Members.

The total individual membership of the adhering countries is 6649, 12 members reside in countries which do not yet adhere to the Union (Lebanon, Lybia, Pakistan, Paraguay, Philipines, Singapore, Sri Lanka, Thailand & Venezuela), giving a total membership of 6661.

2. Membership of Commissions

COMMISSION No. 4

EPHEMERIDES (EPHEMERIDES)

President : SEIDELMANN P KENNETH DR

Vice-President(s) : YALLOP BERNARD D DR

Organizing Committee: ABALAKIN VICTOR K DR
CHAPRONT JEAN DR
DUNCOMBE RAYNOR L DR
KINOSHITA HIROSHI DR
KUBO YOSHIO
LIESKE JAY H DR
MORANDO BRUNO L DR
SCHWAN HEINER DR
TONG FU

Members:

AOKI SHINKO PROF
ARIAS DE GREIFF J PROF
ARLOT JEAN-EUDES
BANDYOPADHYAY A DR
BEC-BORSENBERGER ANNICK
BHATNAGAR ASHOK KUMAR
BRETAGNON PIERRE DR
BRUMBERG VICTOR A DR
CAPITAINE NICOLE
CATALAN MANUEL DR
CHAPRONT-TOUZE MICHELLE
CHOLLET FERNAND DR
COMA JUAN CARLOS
DAVIES MERTON E MR
DE CASTRO ANGEL DR
DEPRIT ANDRE PROF
DI XIAO-HUA
DICKEY JEAN O'BRIEN
DOGGETT LEROY E DR
DUNHAM DAVID W
FIALA ALAN D DR
FOMINOV ALEXANDR M DR
FURSENKO M A DR
GLEBOVA NINA I DR
GONDOLATSCH FRIEDRICH PRF
HAUPT RALPH F
HENRARD JACQUES PROF
ILYAS MOHAMMAD DR
JANICZEK PAUL M DR
JOHNSTON KENNETH J
KING ROBERT WILSON JR DR
KLEPCZYNSKI WILLIAM J DR
KOLACZEK BARBARA DR
KRASINSKY GEORGE A DR
LAHIRI N C
LASKAR JACQUES DR
LEDERLE TRUDPERT DR
LEHMANN MAREK DR
LI GI MAN
LI HYOK HO
LI NENG-YAO
LIU BAO-LIN
MORRISON LESLIE V
MUELLER IVAN I PROF
NEWHALL X X DR
O'HANDLEY DOUGLAS A DR
OESTERWINTER CLAUS
REASENBERG ROBERT D DR
ROMERO PEREZ M PILAR
ROSSELLO GASPAR
SHAPIRO IRWIN I PROF
SHIRYAEV ALEXANDER A DR
SIMON JEAN-LOUIS MR
SINZI AKIRA M DR
SOCHILINA ALLA S DR
STANDISH E MYLES DR
TING YEOU-TSWEN
VAN FLANDERN THOMAS DR
WACKERNAGEL H BEAT DR
WIELEN ROLAND PROF DR
WILKINS GEORGE A DR
WILLIAMS JAMES G DR
WINKLER GERNOT M R DR
XIAN DING-ZHANG
YAMAZAKI AKIRA DR
ZHAO XIAN-ZI

COMMISSION No. 5

DOCUMENTATION AND ASTRONOMICAL DATA

(DOCUMENTATION ET DONNEES ASTRONOMIQUES)

President : WILKINS GEORGE A DR

Vice-President(s) : HAUCK BERNARD PROF

Organizing Committee:
DLUZHNEVSKAYA O B DR
JASCHEK CARLOS O R PROF
MEAD JAYLEE MONTAGUE DR
SCHMADEL LUTZ D DR
SPITE FRANCOIS M DR
WARREN WAYNE H JR DR
WAYMAN PATRICK A PROF
WESTERHOUT GART DR

Members:
ABALAKIN VICTOR K DR
ABT HELMUT A DR
BAKER NORMAN H PROF
BENACCHIO LEOPOLDO
BENN CHRIS R DR
BESSELL MICHAEL S DR
BIDELMAN WILLIAM P PROF
BOUSKA JIRI DR
COLUZZI REGINA DR
DAVIS MORRIS S PROF
DAVIS ROBERT J DR
DE BOER KLAAS SJOERDS DR
DEWHIRST DAVID W DR
DICKEL HELENE R DR
DIXON ROBERT S DR
DUBOIS PASCAL DR
DUCATI JORGE RICARDO DR
DUNCOMBE RAYNOR L DR
EGRET DANIEL DR
GARSTANG ROY H PROF
GRIFFIN ROGER F DR
GROSBOL PREBEN JOHNSON DR
GUIBERT JEAN DR
HANISCH ROBERT J DR
HARVEL CHRISTOPHER ALVIN
HECK ANDRE DR
HEFELE HERBERT PH D
HEINRICH INGE
HEINTZ WULFF D DR
HUANG BI-KUN
JENKNER HELMUT DR
KADLA ZDENKA I DR
KALBERLA PETER
LANTOS PIERRE DR
LEDERLE TRUDPERT DR
LEQUEUX JAMES DR
LIU JINMING
LONSDALE CAROL J DR
LORTET MARIE CLAIRE
LYNGA GOSTA DR
MARTYNOV D YA PROF DR
MCLEAN BRIAN JOHN
MCNALLY DEREK DR
MCNAMARA DELBERT H DR
MEADOWS A JACK PROF
MEIN PIERRE
MERMILLIOD JEAN-CLAUDE DR
MITTON SIMON DR
NISHIMURA SHIRO DR
OCHSENBEIN FRANCOIS DR
OGORODNIKOV KYRILL P PROF
PAMYATNIKH A A DR
PASINETTI LAURA E PROF
PECKER JEAN-CLAUDE PROF
PHILIP A G DAVIS
POLECHOVA PAVLA DR
PUCILLO MAURO DR
QUINTANA HERNAN DR
RAIMOND ERNST DR
RATNATUNGA KAVAN U.
REMY BATTIAU LILIANE G A
RENSON P F M DR
ROMAN NANCY G DR
RUSSO GUIDO DR
SCHILBACH ELENA DR
SCHLUETER A PROF DR
SCHMIDT K H DR
SEDMAK GIORGIO PROF
SHAKESHAFT JOHN R DR
SHCHERBINA-SAMOJLOVA I DR
TERASHITA YOICHI PROF
TRITTON SUSAN BARBARA
UESUGI AKIRA DR
WALLACE PATRICK T MR
WEIDEMANN VOLKER PROF
WELLS DONALD C III DR
WORLEY CHARLES E
WRIGHT ALAN E DR

COMMISSION No. 6

ASTRONOMICAL TELEGRAMS (TELEGRAMMES ASTRONOMIQUES)

President : ROEMER ELIZABETH PROF

Vice-President(s) : GRINDLAY JONATHAN E DR

Organizing Committee: MARSDEN BRIAN G DR
WEST RICHARD M DR

Members:

AKSNES KAARE DR
BIRAUD FRANCOIS DR
CANDY MICHAEL P MR
CUNNINGHAM LELAND E PROF
EVERHART EDGAR DR
FILIPPENKO ALEXEI V DR
GILMORE ALAN C MR
HERS JAN MR
ISOBE SYUZO DR
KOZAI YOSHIHIDE PROF
MRKOS ANTONIN DR
NAKANO SYUICHI
POUNDS KENNETH A PROF
ROSINO LEONIDA PROF
SHAROV A S DR

COMMISSION No. 7

CELESTIAL MECHANICS (MECANIQUE CELESTE)

President : HENRARD JACQUES PROF

Vice-President(s) : DEPRIT ANDRE PROF

Organizing Committee: BHATNAGAR K B DR
BRUMBERG VICTOR A DR
CHAPRONT JEAN DR
FERRAZ-MELLO S PROF DR
FROESCHLE CLAUDE DR
HADJIDEMETRIOU JOHN D
HE MIAO-FU
KHOLSHEVNIKOV K V DR
KINOSHITA HIROSHI DR
MILANI ANDREA
ROY ARCHIE E PROF
SEIDELMANN P KENNETH DR

Members:

ABAD ALBERTO J DR
ABALAKIN VICTOR K DR
AHMED MOSTAFA
AKIM EFRAIM L DR
AKSENOV E P PROF DR
AKSNES KAARE DR
ALTAVISTA CARLOS A DR
ANTONACOPOULOS GREG PROF
AOKI SHINKO PROF
BAGHOS BALEGH B DR
BALMINO GEORGES G DR
BARBERIS BRUNO
BATRAKOV YU V DR
BEC-BORSENBERGER ANNICK
BENEST DANIEL DR
BETTIS DALE G PROF
BOIGEY FRANCOISE
BORDERIES NICOLE
BOZIS GEORGE PROF
BRETAGNON PIERRE DR
BRIEVA EDUARDO PROF
BROOKES CLIVE J DR
BROUCKE ROGER DR
CALAME ODILE DR
CANDY MICHAEL P MR
CARANICOLAS NICHOLAS DR.
CEFOLA PAUL J DR
CHAPRONT-TOUZE MICHELLE
CHEN ZHEN
CHOI KYU-HONG
CID PALACIOS RAFAEL PROF
CONTOPOULOS GEORGE PROF
COOK ALAN H PROF
COUNSELMAN CHARLES C PROF
CUI CHUNFANG
CUI DOU-XING
CUNNINGHAM LELAND E PROF
DANBY J M ANTHONY DR
DAVIS MORRIS S PROF
DEMIN V G PROF DR
DIN HUA
DOGGETT LEROY E DR
DORMAND JOHN RICHARD DR
DOURNEAU GERARD DR
DROZYNER ANDRZEJ
DUNCOMBE RAYNOR L DR
DURIEZ LUC DR
DVORAK RUDOLF DR
EICHHORN HEINRICH K DR
ELIPE SANCHEZ ANTONIO
EMELIANOV NIKOLAJ V DR
ERDI B DR
EVERHART EDGAR DR
FABRE HERVE DR
FARINELLA PAOLO DR
FERNANDEZ SILVIA M. DR
FERRER MARTINEZ SEBASTIAN
FIALA ALAN D DR
FONG CHU-GANG
GALIBINA I V DR
GALLETTO DIONIGI
GAPOSCHKIN EDWARD M DR
GARFINKEL BORIS DR
GASKA STANISLAW DR
GIACAGLIA GIORGIO E PROF
GOLDREICH P DR
GOMES RODNEY D S DR
GONZALEZ CAMACHO ANTONIO
GOUDAS CONSTANTINE L PROF
GREBENIKOV E A PROF DR
GREENBERG RICHARD DR
GROUSHINSKY N P PROF DR
HAMID S EL DIN DR
HANSLMEIER ARNOLD
HEGGIE DOUGLAS C DR
HELALI YHYA E DR
HENON MICHEL C DR
HORI GENICHIRO PROF
HUANG CHENG DR
HUANG TIANYI
IVANOVA VIOLETA DR
IZVEKOV V A DR
JANICZEK PAUL M DR
JEFFERYS WILLIAM H DR
JOURNET ALAIN
JOVANOVIC BOZIDAR
JUPP ALAN H DR
KAMMEYER PETER C DR
KATSIS DEMETRIUS DR
KAULA WILLIAM M PROF
KING-HELE DESMOND G DR
KLOKOCNIK JAROSLAV DR
KOVALEVSKY JEAN DR

KOZAI YOSHIHIDE PROF
KRASINSKY GEORGE A DR
KUSTAANHEIMO PAUL E PROF
LALA PETR DR
LASKAR JACQUES DR
LAZOVIC JOVAN P PROF
LEMAITRE ANNE DR
LIESKE JAY H DR
LISSAUER JACK J DR
LU BEN-KUI
LUNDQUIST CHARLES A DR
MAGNARADZE NINA G DR
MARCHAL CHRISTIAN DR
MARKELLOS VASSILIS V DR
MARSDEN BRIAN G DR
MATAS VLADIMIR R DR
MAVRAGANIS A G PROF
MEIRE RAPHAEL
MELBOURNE WILLIAM G DR
MERMAN G A DR
MESSAGE PHILIP J DR
MIGNARD FRANCOIS DR
MIKKOLA SEPPO DR
MOONS MICHELE B M M
MORANDO BRUNO L DR
MULHOLLAND J DERRAL DR
MUSEN PETER DR
MYACHIN VLADIMIR F DR
NACOZY PAUL E DR
NAHON FERNAND PROF
NOBILI ANNA M
NOSKOV BORIS N DR
NOVOSELOV V S PROF DR
O'HANDLEY DOUGLAS A DR
OESTERWINTER CLAUS
OMAROV TUKEN B PROF
ORUS JUAN J PROF
OSORIO JOSE J S P PROF
PAL ARPAD PROF DR
PAUWELS T DR
PEALE STANTON J PROF
PETROVSKAYA M S DR
PIERCE A KEITH DR
POPOVIC BOZIDAR PROF DR
ROBINSON WILLIAM J DR
RYABOV YU A PROF DR
SAGNIER JEAN-LOUIS DR
SCHOLL HANS DR
SCHUBART JOACHIM DR
SCONZO PASQUALE DR
SEGAN STEVO
SEHNAL LADISLAV DR
SEIN-ECHALUCE M LUISA DR
SESSIN WAGNER DR
SHAPIRO IRWIN I PROF
SHARAF SH G DR
SIDLICHOVSKY MILOS DR
SIMA ZDISLAV DR
SIMON JEAN-LOUIS MR
SINCLAIR ANDREW T DR
SIRY JOSEPH W
SKRIPNICHENKO VLADIMIR DR
STANDISH E MYLES DR
STELLMACHER IRENE DR
SULTANOV G F ACAD
SUN YI-SUI
SZEBEHELY VICTOR G PROF
TABORDA JOSE ROSA DR
TATEVYAN S K DR
TAWADROS MAHET JACOUB DR
TAYLOR DONALD BOGGIA DR
THIRY YVES R PROF
TONG FU
VALTONEN MAURI J PROF
VARVOGLIS H DR
VASHKOV'YAK SOF'YA N DR
VEILLET CHRISTIAN
VILHENA DE MORAES R DR
VINTI JOHN P DR
WALCH JEAN-JACQUES
WALKER IAN WALTER
WHIPPLE ARTHUR L DR
WILLIAMS CAROL A
WNUK EDWIN
WU LIAN-DA
XU PINXIN
YAROV-YAROVOJ M S DR
YI ZHAO-HUA
YOKOYAMA TADASHI DR
YOSHIDA HARUO
YOSHIDA JUNZO PROF
YUASA MANABU DR
ZAFIROPOULOS BASIL DR
ZARE KHALIL DR
ZHANG SHENG-PAN
ZHAO XIAN-ZI
ZHENG JIA-QING
ZHENG XUE-TANG
ZHOU HONG-NAN
ZHU WEN-YAO

COMMISSION No. 8

POSITIONAL ASTRONOMY (ASTRONOMIE DE POSITION)

President : MIYAMOTO MASANORI DR

Vice-President(s) : MORRISON LESLIE V

Organizing Committee: BENEVIDES SOARES P DR
DUMA DMITRIJ P DR
HELMER LEIF
HU NING-SHENG
HUGHES JAMES A DR
KOVALEVSKY JEAN DR
LINDEGREN LENNART DR
NOEL FERNANDO
PINIGIN GENNADIJ I DR
REQUIEME YVES DR
SCHWAN HEINER DR
SMITH CLAYTON A JR DR
YOSHIZAWA MASANORI DR

Members:

ANGUITA CLAUDIO A DR
ARGYRAKOS JEAN PROF DR
BACCHUS PIERRE PROF
BACKER DONALD CH DR
BAGILDINSKIJ BRONISLAV K
BEM JERZY DR
BIEN REINHOLD DR
BILLAUD GERARD J
BRANHAM RICHARD L JR
BROUW W N DR
BYKOV MIKLE F DR
CARESTIA REINALDO A DR
CARRASCO GUILLERMO DR
CATALAN MANUEL DR
CHA DU JIN
CHAMBERLAIN JOSEPH M DR
CHERNEGA N A A DR
CHIUMIENTO GIUSEPPE
CHLISTOVSKY FRANCA DR
CHOLLET FERNAND DR
CLAUZET LUIZ B FERREIRA
CORBIN THOMAS ELBERT DR
COSTA EDGARDO DR
COUNSELMAN CHARLES C PROF
CRIFO FRANCOISE DR
DE VEGT CH PROF DR
DEBARBAT SUZANNE V DR
DEJAIFFE RENE J DR
DICK STEVEN J
DJUROVIC DRAGUTIN M DR
DRAVSKIKH A F DR
DUNCOMBE RAYNOR L DR
EICHHORN HEINRICH K DR
FABRICIUS CLAUS V
FEDOROVA RIMMA T DR
FEISSEL MARTINE DR
FOMIN VALERY A DR
FURUKAWA KIICHIRO DR
GAUSS F STEPHEN
GRUDLER PIERRE
GUBANOV VADIM S DR
GULYAEV A P DR
HARWOOD DENNIS MR
HEINTZ WULFF D DR
HEMENWAY PAUL D
HOEG ERIK DR
HUA YING-MIN
JACKSON PAUL DR
JIANG CHONG-GUO
JOHNSTON KENNETH J
JOURNET ALAIN
KHARIN A S DR
KLOCK B L DR
KOKURIN YURIJ L DR
KONIN V V DR
KOSIN GENNADIJ S DR
LACLARE F MR
LACROUTE PIERRE A PROF
LATTANZI MARIO G
LEDERLE TRUDPERT DR
LEHMANN MAREK DR
LI DONG-MING
LI NENG-YAO
LI ZHI-FANG
LI ZHIGANG
LOPEZ JOSE A ING
LOYOLA PATRICIO DR
LU CHUN-LIN
LUO DING-JIANG
MANRIQUE WALTER T PROF
MAO WEI
MAVRIDIS L N PROF
MELCHIOR PAUL J PROF DIR
MIAO YONG-KUAN
MITIC LJUBISA A DR
MUINOS JOSE L DR
NAKAJIMA KOICHI DR
NEFEDEVA ANTONINA I PROF
NEMIRO ANDREJ A DR PROF
NIKOLOFF IVAN DR
OLSEN FOGH H J
OSORIO JOSE J S P PROF
PAKVOR IVAN
PERRYMAN MICHAEL A C
PETROV G M DR
PHAM-VAN JACQUELINE MME
PILOWSKI K PROF DR
PODOBED V V DR
POLNITZKY GERHARD DR
POLOZHENTSEV DIMITRIJ DR

POMA ANGELO DR
PROVERBIO EDOARDO PROF
PUGLIANO ANTONIO PROF.
QIAN ZHI-HAN DR
QUIJANO LUIS
RAIMOND ERNST DR
REIZ ANDERS PROF
ROESER SIEGFRIED DR
ROUSSEAU JEAN-MICHEL MR
RUSSELL JANE L DR
RUSU I DR
SADZAKOV SOFIJA DR
SALETIC DUSAN
SANCHEZ MANUEL
SATO KOICHI DR
SCHMEIDLER F PROF DR
SEVARLIC BRANISLAV M PROF
SHEN KAIXIAN
SHI GUANG-CHEN
SIMS KENNETH P DR
SOEDERHJELM STAFFAN DR
SOLARIC NIKOLA
SOMA MITSURU DR
SPOELSTRA T A TH DR
STANGE LOTHAR
STONE RONALD CECIL
THOBURN CHRISTINE
THOMAS DAVID V DR
TURON-LACARRIEU C DR
VAN LEEUWEN FLOOR DR
WALLACE PATRICK T MR
WALTER HANS G DR
WESTERHOUT GART DR
WIELEN ROLAND PROF DR
XIA YI-FEI
XIE LIANGYUN
XU BANG-XIN
XU TONG-QI
YAMAZAKI AKIRA DR
YASUDA HARUO PROF DR
YATSKIV YA S DR
YE SHU-HUA
YU KYUNG-LOH PROF
ZHANG HUI
ZVEREV MITROFAN S PROF DR

COMMISSION No. 9

INSTRUMENTS AND TECHNIQUES (INSTRUMENTS ET TECHNIQUES)

President : DAVIS JOHN DR

Vice-President(s) : BHATTACHARYYA J C PROF

Organizing Committee:
BECKLIN ERIC E DR
CULLUM MARTIN DR
HEUDIER JEAN-LOUIS DR
HUMPHRIES COLIN M DR
LELIEVRE GERARD DR
MORRIS MICHAEL C
RICHARDSON E HARVEY DR
SNEZHKO LEONID I
TANGO WILLIAM J. DR
TULL ROBERT G

Members:

ABLES HAROLD D DR
AI GUOXIANG
AIME C DR
ALBRECHT RUDOLF DR
APARICI JUAN DR
ARNAUD JEAN PAUL
ASHOK N M DR
ATHERTON PAUL DAVID
BABCOCK HORACE W DR
BAO KEREN
BARANNE A DR
BARROSO JR JAIR
BARWIG HEINZ
BAUM WILLIAM A DR
BEER REINHARD DR
BENSAMMAR SLIMANE DR
BINGHAM RICHARD G DR
BLITZSTEIN WILLIAM DR
BONNEAU DANIEL
BORGNINO JULIEN DR
BOYCE PETER B DR
BRAULT JAMES W DR
BRECKINRIDGE JAMES B DR
BREJDO IZABELLA I DR
BRIDGELAND MICHAEL DR
BURTON W BUTLER DR
CAMPBELL BRUCE DR
CAO CHANGXIN
CHARVIN PIERRE PR
CHELLI ALAIN
CHRISTY JAMES WALTER DR
CLARKE DAVID DR
COHEN RICHARD S
COOKE JOHN ALAN
CORNEJO ALEJANDRO A DR
CRAWFORD DAVID L DR
CURRIE DOUGLAS G DR
DAN XHI-XIANG
DESAI JYOTINDRA N
DOBRONRAVIN PETER DR
DOKUCHAEVA OLGA D DR
DRAVINS DAINIS PROF
DREHER JOHN W
DUCHESNE MAURICE DR
DUNKELMAN LAWRENCE
EDWIN ROGER P
ENGVOLD ODDBJOERN DR
FABRICANT DANIEL G
FEHRENBACH CHARLES PROF
FELLGETT PETER PROF
FLETCHER J MURRAY
FOMENKO ALEXANDR F DR
FORD W KENT JR DR
FORT BERNARD P DR
FOY RENAUD DR
FU DELIAN
GALAN MAXIMINO J
GAO BILIE
GAUSS F STEPHEN
GAY JEAN DR
GILLINGHAM PETER MR
GLASS IAN STEWART DR
GONG SHOU-SHEN
GRAY PETER MURRAY
GRIFFITHS RICHARD E DR
GRIGORJEV VICTOR M DR
GRUNDMANN WALTER
GUIBERT JEAN DR
GUTCKE DIETRICH
HADLEY BRIAN W
HALLAM KENNETH L DR
HAMMERSCHLAG ROBERT H DR
HANISCH ROBERT J DR
HAO YUN-XIANG
HARMER CHARLES F W MR
HARMER DIANNE L MRS
HECKATHORN HARRY M
HEWITT ANTHONY V DR
HILLIARD R DR
HONEYCUTT R KENT PROF
HOOGHOUDT B G IR
HU JING-YAO
HU NING-SHENG
HUANG TIE-QIN
HYSOM EDMUND J
ILYAS MOHAMMAD DR
IOANNISIANI B K DR
ISMAILOV TOFIK K
JAYARAJAN A P MR
JEFFERS STANLEY DR
JELLEY JOHN V PHD
JENKNER HELMUT DR
JIANG SHI-YANG
JONES BARBARA
KARACHENTSEV I D DR
KARPINSKIJ VADIM N DR
KIPPER TONU DR
KISSELL KENNETH E DR
KLOCK B L DR

KOEHLER H PROF DR
KOEHLER PETER
KOPYLOV I M DR
KOROVYAKOVSKIJ YURIJ P DR
KOVACHEV B J DR
KREIDL TOBIAS J N
KUEHNE CHRISTOPH F
KULKARNI PRABHAKAR V PROF
LABEYRIE ANTOINE DR
LAQUES PIERRE DR
LASKER BARRY M DR
LEMAITRE GERARD R DR
LI DEPEI
LI TING
LIVINGSTON WILLIAM C
LOCHMAN JAN
LYNCH DAVID K
MACK PETER DR
MAHRA H S DR
MAILLARD JEAN-PIERRE DR
MALIN DAVID F MR
MARTINS DONALD HENRY DR
MCLEAN IAN S DR
MCMULLAN DENNIS DR
MEINEL ADEN B PROF
MENG XINMIN
MERKLE FRITZ DR
MERTZ LAWRENCE N DR
MIKHELSON NIKOLAJ N DR
MILLIKAN ALLAN G MR
MINAROVJECH MILAN
MORGAN BRIAN LEALAN
MORTON DONALD C DR
MURRAY STEPHEN S DR
NAKAI YOSHIHIRO
NELSON JERRY
NIEMI AIMO
NISHIMURA SHIRO DR
NUNES ROGERIO S DE SOUSA
O'DELL CHARLES R DR
ODGERS GRAHAM J DR
PASIAN FABIO
PENNY ALAN JOHN DR
PERRYMAN MICHAEL A C
PETFORD A DAVID DR
PETROV PETER P DR
PICAT JEAN-PIERRE DR
PRITCHET CHRISTOPHER J DR
PROKOF'EVA VALENTINA V DR
PUCILLO MAURO DR
RACINE RENE DR
RAKOS KARL D PROF
RAMSEY LAWRENCE W DR
REAY NEWRICK K DR
REDFERN MICHAEL R DR
RING JAMES PROF
ROBINSON LLOYD B DR
RODDIER CLAUDE DR
RODDIER FRANCOIS PROF
ROSCH JEAN PROF
ROUNTREE JANET DR
RUDER HANNS
RUSCONI LUIGIA DR
RYLOV VALERIJ S DR
SAXENA A K DR
SCHROEDER DANIEL J PROF
SCHULTZ G V DR
SCHUMANN JOERG DIETER DR
SCHUSTER HANS-EMIL
SEDMAK GIORGIO PROF
SERVAN BERNARD
SHAKHBAZYAN YURIJ L DR
SHCHEGLOV P V DR
SHEN CHANGJUN
SHEN PARN-AN
SHIVANANDAN KANDIAH DR
SIM MARY E MISS
SMITH CHARLES DITTO
SMYTH MICHAEL J DR
STESHENKO N V DR
STOREY JOHN W V DR
SU DING-QIANG
SWINGS JEAN-PIERRE DR
TRAUB WESLEY ARTHUR
TUEG HELMUT DR
ULICH BOBBY LEE
VALNICEK BORIS DR
VALTONEN MAURI J PROF
VAN CITTERS GORDON W DR
VELKOV KIRIL
VLADIMIROV SIMEON
VRBA FREDERICK J DR
WALKER ALISTAIR ROBIN DR
WALKER GORDON A H PROF
WALKER MERLE F PROF
WALLACE PATRICK T MR
WAMPLER E JOSEPH PROF
WANG LAN-JUAN
WANG YANAN
WANG YIMING
WATSON FREDERICK GARNETT
WEISS WERNER W DR
WEST RICHARD M DR
WESTPHAL JAMES A PROF
WILCOCK WILLIAM L PROF
WINDHORST ROGIER A DR
WLERICK GERARD DR
WOEHL HUBERTUS DR
WORDEN SIMON P DR
WORSWICK SUSAN
WU LIN-XIANG
WYLLER ARNE A PROF
WYNNE CHARLES G PROF
YANG SHI JIE
YAO ZHENG-QIU
YE BINXUN
ZACHAROV IGOR DR
ZAMBON GIULIO DR.
ZEALEY WILLIAM J DR
ZHANG YOUYI
ZHU NENGHONG

COMMISSION No. 10

SOLAR ACTIVITY (ACTIVITE SOLAIRE)

President : PRIEST ERIC R PROF

Vice-President(s) : GAIZAUSKAS VICTOR DR

Organizing Committee: ANTONUCCI ESTER DR
BENZ ARNOLD DR
ENGVOLD ODDBJOERN DR
ENOME SHINZO PROF
FANG CHENG
GURTOVENKO E A DR
MACHADO MARCOS
PICK MONIQUE DR
TANDBERG-HANSSEN EINAR A

Members:

ABBASOV ALIK R DR
ABRAMI ALBERTO PROF
AHLUWALIA HARJIT SINGH DR
AI GUOXIANG
ALISSANDRAKIS C PH D
ALTROCK RICHARD C DR
ALTSCHULER MARTIN D PROF
ALY M KHAIRY PROF
AMBROZ PAVEL DR
ANDERSON KINSEY A PROF
ANTALOVA ANNA
ANTIOCHOS SPIRO KOSTA
ASCHWANDEN MARKUS DR
ATAC TAMER
ATHAY R GRANT DR
AURASS HENRY DR
AVIGNON YVETTE DR
BAGARE S P DR
BALLI EDIBE PROF
BANIN V G DR
BARROW COLIN H DR
BATCHELOR DAVID ALLEN DR
BECKERS JACQUES M DR
BEEBE HERBERT A
BELL BARBARA DR
BELVEDERE GAETANO DR
BERGER MITCHELL DR
BHATNAGAR ARVIND DR
BOCCHIA ROMEO DR
BOHN HORST-ULRICH
BOMMIER VERONIQUE DR
BOUGERET J L DR
BOYER RENE
BRANDT PETER N
BRAY ROBERT J DR
BROWN JOHN C PROF
BROWNING PHILIPPA DR
BRUECKNER GUENTER E DR
BRUNER MARILYN E DR
BRUZEK ANTON DR
BUCHNER JORG DR
BUMBA VACLAV DR
CADEZ VLADIMIR
CALLY PAUL S DR
CANE HILARY VIVIEN
CARLQVIST PER A DR
CHAMBE GILBERT
CHANDRA SURESH DR
CHAPMAN GARY A DR
CHEN BIAO
CHEN ZHENCHENG
CHENG CHUNG-CHIEH DR
CHERTOPRUD V E DR
CHIUDERI-DRAGO FRANCA PR
CHUPP EDWARD L DR
CIMINO MASSIMO A PROF
CLIVER EDWARD W
COFFEY HELEN E MS
COLLADOS MANUEL DR
COOK JOHN W
COUTREZ RAYMOND A J PROF
COVINGTON ARTHUR E
CRANNELL CAROL JO DR
CSADA IMRE K DR
CULHANE LEONARD PROF
DATLOWE DAYTON DR
DE GROOT T DR
DE JAGER CORNELIS PROF
DEL TORO INIESTA JOSE DR
DENNIS BRIAN ROY DR
DERE KENNETH PAUL
DERMENDJIEV VLADIMIR DR
DEUBNER FRANZ-LUDWIG DR
DEZSO LORANT PROF
DIALETIS DIMITRIS DR
DING YOU-JI
DIZER MUAMMER PROF
DOLLFUS AUDOUIN PROF
DRYER MURRAY DR
DUBOIS MARC A
DUBOV EMIL E PROF
DULK GEORGE A PROF
DUNCAN ROBERT A PROF
DUNN RICHARD B DR
DWIVEDI BHOLA NATH DR
EDDY JOHN A DR
ELSTE GUNTHER H DR
ELWERT GERHARD PROF
EMSLIE A. GORDON
ERICKSON WILLIAM C DR
FALCHI AMBRETTA
FALCIANI ROBERTO DR
FAN DAXIONG
FARNIK FRANTISEK
FEIBELMAN WALTER A DR
FENG KE-JIA
FOING BERNARD H DR
FORBES TERRY G DR
FORTINI TERESA DR
FOSSAT ERIC DR
FRIEDMAN HERBERT DR
FRITZOVA-SVESTKA L DR

FU QI JUN
GABRIEL ALAN H
GALLOWAY DAVID DR
GARCIA DE LA ROSA JOSE I
GELFREIKH GEORGIJ B DR
GERGELY TOMAS ESTEBAN DR
GESZTELYI LIDIA
GILLILAND RONALD LYNN
GILMAN PETER A DR
GLATZMAIER GARY A
GLEISSBERG WOLFGANG PROF
GNEVYSHEVA RAISA S DR
GODOLI GIOVANNI PROF
GOKHALE MORESHWAR HARI PR
GOOSSENS MARCEL DR
GOPASYUK S I DR
GU XIAO-MA
GURMAN JOSEPH B DR
HAGEN JOHN P
HAGYARD MONA JUNE
HAMMER REINER
HANASZ JAN DR
HANSEN RICHARD T MR
HANSLMEIER ARNOLD
HARVEY JOHN W DR
HASAN SAIYID STRAJUL
HATHAWAY DAVID H DR
HAUG EBERHARD DR
HAYWARD JOHN
HEINZEL PETR DR
HENOUX JEAN-CLAUDE DR
HEYDEN FRANCIS J SJ DR
HIEI EIJIRO DR
HILDNER ERNEST DR
HIRAYAMA TADASHI PROF
HOLMAN GORDON D
HOLZER THOMAS EDWARD DR
HONG HYON IK
HOOD ALAN
HOWARD ROBERT F DR
HOYNG PETER DR
HUANG YOU-RAN
HUDSON HUGH S DR
HURFORD GORDON JAMES
HYDER C L DR
IOSHPA B A DR
IVANCHUK VICTOR I DR
JAIN RAJMAL DR
JAKIMIEC JERZY PROF
JENSEN EBERHART PROF
JIANG YAO-TIAO
JOCKERS KLAUS DR
JONES HARRISON PRICE DR
JORDAN STUART D DR
JOSELYN JO ANN C DR
JOVANOVIC BOZIDAR
KAHLER STEPHEN W DR
KAI KEIZO DR
KALMAN BELA DR
KANE SHARAD R DR
KANG JIN SOK
KARLICKY MARIAN
KARPEN JUDITH T
KAUFMANN PIERRE PROF
KIPLINGER ALAN L DR
KJELDSETH-MOE OLAV DR
KLECZEK JOSIP DR
KLEIN KARL LUDWIG DR
KLVANA MIROSLAV
KNOSKA STEFAN
KOECKELENBERGH ANDRE DR
KOPECKY MILOSLAV DR
KOSTIK ROMAN I
KOTRC PAVEL
KOUTCHMY SERGE DR
KOVACS AGNES DR
KRAUSE F DR
KRIVSKY LADISLAV DR
KRUEGER ALBRECHT DR
KUBOTA JUN DR
KUENZEL HORST
KUKLIN G V DR
KUNDU MUKUL R DR
KUPERUS MAX PROF DR
KUROCHKA L N DR
KUROKAWA HIROKI DR
LANDECKER PETER BRUCE DR
LANDMAN DONALD ALAN
LANG KENNETH R ASST PROF
LANTOS PIERRE DR
LEIBACHER JOHN DR
LEROY BERNARD DR
LEROY JEAN-LOUIS
LI CHUN-SHENG
LI SON JAE
LI WEI BAO
LIN YUANZHANG
LIU XINPING PROF.
LIVSHITS M A DR
LOUGHHEAD RALPH E DR
LOW BOON CHYE
LUNDSTEDT HENRIK DR
LUO BAO-RONG
LUO XIANHAN
LUSTIG GUENTER DR
MACHADO JOSE M A B DR
MACKINNON ALEXANDER L
MACQUEEN ROBERT M DR
MACRIS CONSTANTIN J PROF
MAKAROV VALENTINE I
MAKITA MITSUGU DR
MALHERBE JEAN MARIE DR
MALITSON HARRIET H MS
MALTBY PER PROF
MALVILLE J MCKIM PROF
MANDELSTAM S L PROF
MARTRES MARIE-JOSEPHE
MASON GLENN M
MATSUURA OSCAR T DR
MATTIG W PROF DR
MAXWELL ALAN DR
MCCABE MARIE K MS
MCINTOSH PATRICK S
MCKENNA LAWLOR SUSAN
MCLEAN DONALD J DR
MEIN PIERRE
MELROSE DONALD B PROF
MERGENTALER JAN PROF
MICHALITSIANOS ANDREW
MICHARD RAYMOND DR
MOGILEVSKIJ EH I DR
MOISEEV I G DR
MORENO-INSERTIS FERNANDO
MORETON G E
MORIYAMA FUMIO PROF
MOROZHENKO N N DR
MOTTA SANTO DR
MULLER RICHARD DR
MUSIELAK ZDZISLAW E DR
NAGASAWA SHINGO PROF
NAKAGAWA YOSHINARI DR
NAKAJIMA HIROSHI
NAMBA OSAMU DR
NEIDIG DONALD F DR
NELSON GRAHAM JOHN DR
NEUPERT WERNER M DR
NISHI KEIZO DR
NOCERA LUIGI DR
NOYES ROBERT W PROF
NUSSBAUMER HARRY PROF
OBRIDKO VLADIMIR N DR
OHKI KENICHIRO DR
ORRALL FRANK Q PROF
OZGUC ATILA
PALLAVICINI ROBERTO DR
PALLE PERE-LLUIS DR
PAN LIANDE
PAP JUDIT
PARKINSON JOHN H DR
PARKINSON WILLIAM H
PATERNO LUCIO PROF
PEDERSEN BENT M DR
PETROSIAN VAHE PROF
PFLUG KLAUS DR
PHILLIPS KENNETH J H
PIDDINGTON JACK H RES FEL
PNEUMAN GERALD W
POLAND ARTHUR I DR
POLETTO GIANNINA PROF
POLUPAN P N DR
POQUERUSSE MICHEL
PREKA-PAPADEMA P DR
PROKAKIS THEODORE J DR
QIAN JING-KUI
RAADU MICHAEL A DR
RABIN DOUGLAS MARK
RAO A PRAMESH DR

RAOULT ANTOINETTE DR
REES DAVID ELWYN DR
REGULO CLARA DR
ROCA CORTES TEODORO
ROMPOLT BOGDAN DR
ROXBURGH IAN W PROF
RUSIN VOJTECH
RUZICKOVA-TOPOLOVA B DR
SAITO KUNIJI PROF
SAWYER CONSTANCE B DR
SCHLUETER A PROF DR
SCHMIEDER BRIGITTE DR
SEMEL MEIR DR
SHEELEY NEIL R DR
SHINE RICHARD A DR
SIMON GUY
SMALDONE LUIGI ANTONIO
SOLANKI SAMI K DR
SPRUIT HENK C DR
STEPANIAN N N DR
STEWART RONALD T MR
STRONG KEITH T DR
SUEMOTO ZENZABURO PROF DR
SYKORA JULIUS DR
TAKAKURA TATSUO PROF EMER
TANAKA KATSUO DR
TAPPING KENNETH F
THOMAS ROGER J DR
TRELLIS MICHEL DR
TROTTET GERARD DR
UNDERWOOD JAMES H DR
VAN ALLEN JAMES A PROF
VAUGHAN ARTHUR H DR
VENKATESAN DORASWAMY DR
VILMER NICOLE DR
VITINSKIJ YURIJ I DR
WANG JIA-LONG
WENTZEL DONAT G DR
WILSON PETER R PROF
WOLTJER LODEWIJK PROF
XANTHAKIS JOHN N PROF
YAKOVKIN N A DR
YE SHI-HUI
YOU JIAN-QI
ZAPPALA ROSARIO ALDO DR
ZHANG HE-QI
ZHELYAZKOV IVAN DR
ZIRIN HAROLD DR
ZWAAN CORNELIS PROF DR

RAY CHOUDHURI ARNAB DR
REEVES EDMOND M DR
ROBERTS BERNARD DR
ROEMER MAX PROF
ROSCH JEAN PROF
ROZELOT JEAN P
RUST DAVID M DR
RYBANSKY MILAN
SAKURAI KUNITOMO PROF
SCHATTEN KENNETH H DR
SCHMAHL EDWARD J DR
SCHOBER HANS J DR
SHAPLEY ALAN H
SHI ZHONG-XIAN
SILBERBERG REIN DR
SITNIK G F PROF
SMITH DEAN F DR
SOTIROVSKI PASCAL DR
STELLMACHER GOETZ
STEPANOV ALEXANDER V DR
STIX MICHAEL DR
STURROCK PETER A PROF
SUN KAI
SYLWESTER BARBARA DR
TALON RAOUL DR
TANDON JAGDISH NARAIN DR
TESKE RICHARD G PROF
TIFREA EMILIA DR
TREUMANN RUDOLF A. DR
TSUBAKI TOKIO PROF
URPO SEPPO I
VAN HOVEN GERARD DR
VECK NICHOLAS
VERHEEST FRANK PROF
VINLUAN RENATO
VYALSHIN GENNADIJ F DR
WANG JING-XIU
WIEHR EBERHARD DR
WITTMANN AXEL D. PH D
WU FEI
XU AO-AO
YANG HAI SHOU
YEH TYAN DR
YUN HONG-SIK PROF
ZELENKA ANTOINE DR
ZHANG ZHEN-DA
ZHOU DAOQI
ZLOBEC PAOLO DR

RAYROLE JEAN R DR
REEVES HUBERT PROF
ROBINSON JR RICHARD D DR
ROMANCHUK PAVEL R DR
ROUDIER THIERRY DR
RUBASHEV BORIS M DR
RUZDJAK VLADIMIR DR
SAEMUNDSON THORSTEINN
SAKURAI TAKASHI DR
SCHINDLER KARL PROF DR
SCHMIDT H U DR
SCHROETER EGON H PROF
SHEA MARGARET A DR
SHIBASAKI KIYOTO
SIMNETT GEORGE M
SLONIM E M DR
SMOL'KOV GENNADIJ YA DR
SPICER DANIEL SHIELDS DR
STENFLO JAN O DR
STESHENKO N V DR
STOKER PIETER H
SUDA JAN
SVESTKA ZDENEK DR
SYLWESTER JANUSZ
TAMENAGA TATSUO DR
TANG YU-HUA
THOMAS JOHN H PROF
TLAMICHA ANTONIN DR
TRITAKIS BASIL P DR
TSUNETA SAKU DR
VALNICEK BORIS DR
VAN'T VEER FRANS DR
VELKOV KIRIL
VIAL JEAN-CLAUDE
VINOD S KRISHAN MRS DR
WALDMEIER MAX PROF DR
WANG YI-MING DR
WILD JOHN PAUL DR
WOEHL HUBERTUS DR
WU SHI TSAN DR
XU ZHENTAO
YAO JIN-XING
YOSHIMURA HIROKAZU DR
ZACHARIADIS THEODOSIOS DR
ZHANG BAI-RONG
ZHAO REN-YANG
ZHUGZHDA YUZEF D DR
ZOU YI-XIN

COMMISSION No. 12

RADIATION AND STRUCTURE OF THE SOLAR ATMOSPHERE

(RADIATION ET STRUCTURE DE L'ATMOSPHERE SOLAIRE)

President : HARVEY JOHN W DR

Vice-President(s) : STENFLO JAN O DR

Organizing Committee:
AI GUOXIANG
ANDO HIROYASU DR
FALCIANI ROBERTO DR
GURTOVENKO E A DR
KUPERUS MAX PROF DR
MULLER RICHARD DR
ROCA CORTES TEODORO
SCHUESSLER MANFRED DR
SIVARAMAN K R DR
WEISS NIGEL O DR

Members:

ACTON LOREN W DR
ADAM MADGE G DR
AIME C DR
ALISSANDRAKIS C PH D
ALTROCK RICHARD C DR
ALTSCHULER MARTIN D PROF
ANDERSEN BO NYBORG DR
ANSARI S M RAZAULLAH PROF
ANTIA H M DR
ARNAUD JEAN PAUL
ATHAY R GRANT DR
AYRES THOMAS R
BALIUNAS SALLIE L
BALTHASAR HORST DR
BEARD DAVID B DR
BECKERS JACQUES M DR
BECKMAN JOHN E PROF
BEEBE HERBERT A
BEL NICOLE J DR
BENFORD GREGORY DR
BHATNAGAR ARVIND DR
BHATTACHARYYA J C PROF
BILLINGS DONALD E PROF
BLACKWELL DONALD E PROF
BLAMONT JACQUES E PROF
BOCCHIA ROMEO DR
BOEHM KARL-HEINZ PROF
BOEHM-VITENSE ERIKA PROF
BOHN HORST-ULRICH
BOMMIER VERONIQUE DR
BONNET ROGER M DR
BOOK DAVID L
BORNMANN PATRICIA L DR
BOUGERET J L DR
BRANDT PETER N
BRAULT JAMES W DR
BRAY ROBERT J DR
BRECKINRIDGE JAMES B DR
BRUECKNER GUENTER E DR
BRUNER MARILYN E DR
BRUZEK ANTON DR
BUMBA VACLAV DR
CADEZ VLADIMIR
CAVALLINI FABIO
CEPPATELLI GUIDO DR
CHAMBE GILBERT
CHAN KWING LAM
CHAPMAN GARY A DR
CHEN BIAO
CHENG CHUNG-CHIEH DR
CHISTYAKOV VLADIMIR E DR
CHRISTENSEN-DALSGAARD J
CHVOJKOVA WOYK E DR
CLARK THOMAS ALAN DR
COLLADOS MANUEL DR
COOK JOHN W
COX ARTHUR N DR
CRAIG IAN JONATHAN D DR
CRAM LAWRENCE EDWARD DR
CUI LIAN-SHU
DARA HELEN DR
DE JAGER CORNELIS PROF
DEL TORO INIESTA JOSE DR
DELACHE PHILIPPE J DR
DELBOUILLE LUC PROF
DEMARQUE P PROF
DEUBNER FRANZ-LUDWIG DR
DEZSO LORANT PROF
DOGAN NADIR PROF
DRAVINS DAINIS PROF
DUBOV EMIL E PROF
DUMONT SIMONE DR
DUNKELMAN LAWRENCE
DUNN RICHARD B DR
DUVALL THOMAS L JR
EDMONDS FRANK N JR DR
EINAUDI GIORGIO
ELLIOTT IAN DR
ELSTE GUNTHER H DR
EPSTEIN GABRIEL LEO DR
EVANS J V DR
FANG CHENG
FELDMAN URI
FIALA ALAN D DR
FONTENLA JUAN MANUEL DR
FOSSAT ERIC DR
FOUKAL PETER V DR
FRAZIER EDWARD N DR
FRIEDMAN HERBERT DR
FROEHLICH CLAUS
GABRIEL ALAN H
GAIZAUSKAS VICTOR DR
GAUR V P
GLATZMAIER GARY A
GNEVYSHEV MSTISLAV N DR
GODOLI GIOVANNI PROF
GOKDOGAN NUZHET PROF
GOLDMAN MARTIN V
GOMEZ MARIA THERESA DR

GOPALSWAMY N DR
GOPASYUK S I DR
GORDON CHARLOTTE PROF
GREVESSE N DR
GU XIAO-MA
HAGYARD MONA JUNE
HAMMER REINER
HAUPT RALPH F
HEINZEL PETR DR
HIEI EIJIRO DR
HILDNER ERNEST DR
HIRAYAMA TADASHI PROF
HOLWEGER HARTMUT PROF
HORTON BRIAN H DR
HOTINLI METIN DR
HOUSE LEWIS L DR
HOWARD ROBERT F DR
HOYNG PETER DR
ILLING RAINER M E
JABBAR SABEH RHAMAN
JEFFERIES JOHN T DR
JONES HARRISON PRICE DR
JORDAN CAROLE DR
JORDAN STUART D DR
JOSHI G C DR
KALKOFEN WOLFGANG DR
KALMAN BELA DR
KARPEN JUDITH T
KARPINSKIJ VADIM N DR
KATO SHOJI PROF
KAUFMANN PIERRE PROF
KAWAGUCHI ICHIRO PROF
KEIL STEPHEN L
KHETSURIANI TSIALA S DR
KLEIN KARL LUDWIG DR
KNEER FRANZ DR
KNOLKER MICHAEL DR
KONONOVICH EDWARD V DR
KOPECKY MILOSLAV DR
KOSTIK ROMAN I
KOTOV VALERY DR
KOTRC PAVEL
KOUTCHMY SERGE DR
KOYAMA SHIN PROF DR
KRAEMER GERHARD DR
KUBICELA ALEKSANDAR DR
KUKLIN G V DR
KUNDU MUKUL R DR
KUROCHKA L N DR
LA BONTE BARRY JAMES
LABS DIETRICH PROF
LANDI DEGL'INNOCENTI E PR
LANDI DEGL'INNOCENTI M
LANDMAN DONALD ALAN
LANDOLFI MARCO
LANTOS PIERRE DR
LEIBACHER JOHN DR
LEIGHTON R B PROF
LEROY JEAN-LOUIS
LIN YUANZHANG
LINSKY JEFFREY L DR
LIVINGSTON WILLIAM C
LOCKE JACK L DR
LOPEZ-ARROYO M
LOUGHHEAD RALPH E DR
LUEST REIMAR PROF
LUSTIG GUENTER DR
MAKAROV VALENTINE I
MAKAROVA ELENA A DR
MAKITA MITSUGU DR
MARIK MIKLOS DR.
MARILLI ETTORE DR
MARISKA JOHN THOMAS
MARMOLINO CIRO
MATSUSHIMA SATOSHI DR
MATTIG W PROF DR
MCKENNA LAWLOR SUSAN
MEIN PIERRE
MERGENTALER JAN PROF
MEWE R DR
MEYER FRIEDRICH DR
MICHARD RAYMOND DR
MIGEOTTE MARCEL V PROF
MIHALAS DIMITRI DR
MILKEY ROBERT W DR
MORENO-INSERTIS FERNANDO
MORIYAMA FUMIO PROF
MOURADIAN ZADIG M DR
MUELLER EDITH A PROF
MUNRO RICHARD H DR
NAMBA OSAMU DR
NECKEL HEINZ DR
NESIS ANASTASIOS DR
NICOLAS KENNETH ROBERT
NICOLET MARCEL PROF
NISHI KEIZO DR
NORDLUND AKE DR
NOYES ROBERT W PROF
ORRALL FRANK Q PROF
OSTER LUDWIG F PROF DR
PALLE PERE-LLUIS DR
PANDE MAHESH CHANDRA DR
PAPATHANASOGLOU D DR
PARKINSON WILLIAM H
PASACHOFF JAY M PROF
PECKER JEAN-CLAUDE PROF
PEYTURAUX ROGER H PROF
PFLUG KLAUS DR
PHILLIPS KENNETH J H
PIERCE A KEITH DR
POQUERUSSE MICHEL
PRIEST ERIC R PROF
PROKAKIS THEODORE J DR
RABIN DOUGLAS MARK
RADICK RICHARD R DR
RAOULT ANTOINETTE DR
REES DAVID ELWYN DR
REEVES EDMOND M DR
REGULO CLARA DR
RIGHINI-COHEN GIOVANNA DR
RIGUTTI MARIO PROF
ROBERTI GIUSEPPE DR
ROBERTS BERNARD DR
RODDIER FRANCOIS PROF
ROLAND GINETTE DR
ROUDIER THIERRY DR
ROVIRA MARTA GRACIELA
RUSIN VOJTECH
RUTTEN ROBERT J. DR
RYBANSKY MILAN
SAKAI JUNICHI
SAKURAI TAKASHI DR
SAMAIN DENYS DR
SAUVAL A JACQUES DR
SCHMAHL EDWARD J DR
SCHMIDT WOLFGANG DR
SCHMITT DIETER DR
SCHOBER HANS J DR
SCHWARTZ STEVEN JAY
SEATON MICHAEL J PROF
SEMEL MEIR DR
SEVERINO GIUSEPPE
SHALLIS MICHAEL J DR
SHEELEY NEIL R DR
SHEN LONG-XIANG
SHINE RICHARD A DR
SIMON GEORGE W DR
SIMON GUY
SINGH JAGDEV DR
SINHA K DR
SITNIK G F PROF
SITTERLY CHARLOTTE M DR
SKUMANICH ANDRE PROF
SMITH PETER L DR
SOBOLEV VLADISLAV M DR
SOLANKI SAMI K DR
SONG MU-TAO
SOTIROVSKI PASCAL DR
SOUFFRIN PIERRE B DR
SPICER DANIEL SHIELDS DR
STAUDE JUERGEN DR
STEBBINS ROBIN
STEFFEN MATTHIAS DR
STIX MICHAEL DR
SUEMOTO ZENZABURO PROF DR
SVESTKA ZDENEK DR
SWENSSON JOHN W DR
TANAKA KATSUO DR
TANDBERG-HANSSEN EINAR A
TEPLITSKAYA R B DR
THOMAS JOHN H PROF
THOMAS RICHARD N DR
TORELLI M DR
TOUSEY RICHARD DR
TRIPATHI B M DR
TSAP T T DR
TSIROPOULA GEORGIA DR

TSUBAKI TOKIO PROF
UCHIDA YUTAKA PROF
UNNO WASABURO PROF
UUS UNDO DR
VAN HOVEN GERARD DR
VASILEVA GALINA J DR
VAUGHAN ARTHUR H DR
VENKATAKRISHNAN P DR
VIAL JEAN-CLAUDE
VILMER NICOLE DR
VITINSKIJ YURIJ I DR
VOLONTE SERGE DR
VUKICEVIC K M PROF DR
WALDMEIER MAX PROF DR
WANG JING-XIU
WANG ZHEN-YI
WARWICK JAMES W DR
WENTZEL DONAT G DR
WILSON PETER R PROF
WITTMANN AXEL D. PH D
WOEHL HUBERTUS DR
WORDEN SIMON P DR
WU FEI
WU LIN-XIANG
WYLLER ARNE A PROF
YOSHIMURA HIROKAZU DR
YOU JIAN-QI
YOUSSEF NAHED H DR
YUN HONG-SIK PROF
ZARRO DOMINIC M DR
ZELENKA ANTOINE DR
ZHOU DAOQI
ZHUGZHDA YUZEF D DR
ZIRIN HAROLD DR
ZIRKER JACK B DR
ZWAAN CORNELIS PROF DR

COMMISSION No. 14

ATOMIC AND MOLECULAR DATA

(DONNEES ATOMIQUES ET MOLECULAIRES)

President : SAHAL-BRECHOT SYLVIE DR

Vice-President(s) : WIESE WOLFGANG L DR

Organizing Committee: GABRIEL ALAN H
KATO TAKAKO DR
LOVAS FRANCIS JOHN DR
MANDELSTAM S L PROF
NICHOLLS RALPH W PROF
NUSSBAUMER HARRY PROF
PARKINSON WILLIAM H
RUDZIKAS ZENONAS B

Members:

ADELMAN SAUL J DR	ANDREW KENNETH L PROF	ARDUINI-MALINOVSKY M. DR
ARTRU MARIE-CHRISTINE DR	BAIRD KENNETH M DR	BARNARD HANNES A J DR
BARROW RICHARD F DR	BATES DAVID R PROF	BELY OLEG DR
BELY-DUBAU FRANCOISE	BERRINGTON KEITH ADRIAN	BIEMONT EMILE DR
BLACK JOHN HARRY DR	BLAHA MILAN DR	BOMMIER VERONIQUE DR
BRANSCOMB L M DR	BRAULT JAMES W DR	BROMAGE GORDON E DR
BURGESS ALAN DR	CARBON DUANE F DR	CARROLL P KEVIN PROF
CARVER JOHN H PROF	COOK ALAN H PROF	CORLISS C H DR
CORNILLE MARGUERITE DR	CZYZAK STANLEY J DR	DALGARNO ALEXANDER PROF
DAVIS SUMNER P DR	DE FREES DOUGLAS J DR	DELSEMME ARMAND H PROF DR
DESESQUELLES JEAN DR	DIERCKSEN GEERD H F PH D	DRESSLER KURT PROF
DUBAU JACQUES DR	DUFAY MAURICE PROF	EDLEN BENGT PROF
ENGELHARD E J G PROF DR	EPSTEIN GABRIEL LEO DR	FAUCHER PAUL DR
FEAUTRIER NICOLE DR	FELENBOK PAUL DR	FINK UWE DR
FLOWER DAVID R DR	GALLAGHER JEAN DR	GARGAUD MURIEL DR
GARSTANG ROY H PROF	GARTON W R S PROF	GLAGOLEVSKIJ JU V DR
GOLDBACH CLAUDINE MME	GRANT IAN P DR	GREEN LOUIS C PROF
HEDDLE DOUGLAS W O PROF	HEFFERLIN RAY A PROF	HEROLD HEINZ
HERZBERG GERHARD DR	HESSER JAMES E DR	HOUSE LEWIS L DR
HUBER MARTIN C E DR	HUEBNER WALTER F DR	ILIEV ILIAN
IRWIN ALAN W DR	JACQUINOT PIERRE DR	JOHNSON DONALD R DR
JOHNSON FRED M PROF DR	JOLY FRANCOIS DR	JORDAN CAROLE DR
JORDAN H L DR DIREKTOR	KENNEDY EUGENE T	KESSLER KARL G DR
KIELKOPF JOHN F DR	KIM ZONG DOK	KING R B DR
KINGSTON ARTHUR E PROF	KIPPER TONU DR	KIRBY KATE P DR
KOHL JOHN L DR	KROTO HAROLD PROF.	LAGERQVIST ALBIN PROF
LANDMAN DONALD ALAN	LANG JAMES DR	LAUNAY JEAN-MICHEL DR
LAWRENCE G M DR	LAYZER DAVID PROF	LE DOURNEUF MARYVONNE
LEGER ALAIN DR	LOULERGUE MICHELLE DR	LUTZ BARRY L DR
MAILLARD JEAN-PIERRE DR	MARTIN WILLIAM C DR	MASON HELEN E DR
MCWHIRTER R W PETER DR	MEWE R DR	MIGEOTTE MARCEL V PROF
MONFILS ANDRE G PROF	MUMMA MICHAEL JON	NEWSOM GERALD H PROF
NOLLEZ GERARD DR	OBI SHINYA PROF	OETKEN L DR
OKA TAKESHI DR	OMONT ALAIN PROF	ORTON GLENN S DR
PEACH GILLIAN DR	PETRINI DANIEL DR	PETROPOULOS BASIL CH DR

PETTINI MARCO	PFENNIG HANS H DR	PHILLIPS JOHN G PROF
PROKOF'EV VLADIMIR K PROF	QUERCI FRANCOIS R DR	RAO K NARAHARI
RICHTER JOHANNES PROF	ROSS JOHN E R DR	ROUEFF EVELYNE M A DR
RUDER HANNS	SCHADEE AERT DR	SCHRIJVER JOHANNES DR
SEATON MICHAEL J PROF	SHARP CHRISTOPHER DR	SHORE BRUCE W
SITTERLY CHARLOTTE M DR	SMITH GEOFFREY DR	SMITH PETER L DR
SMITH WM HAYDEN PROF	SOMERVILLE WILLIAM B DR	SORENSEN GUNNAR DR
STEENMAN-CLARK LOIS DR	STEHLE CHANTAL DR	STREL'NITSKIJ VLADIMIR DR
SUMMERS HUGH P DR	SWINGS JEAN-PIERRE DR	TAKAYANAGI KAZUO PROF
TATUM JEREMY B DR	TOUSEY RICHARD DR	TOZZI GIAN PAOLO
TREFFTZ ELEONORE E DR	VAN REGEMORTER HENRI DR	VAN RENSBERGEN WALTER DR
VARSHALOVICH DIMITRIJ PR	VOELK HEINRICH J PROF	VOLONTE SERGE DR
VUJNOVIC VLADIS DR	WARES GORDON W DR	WENIGER SCHAME DR
WILSON ROBERT PROF SIR	WINNEWISSER GISBERT DR	WUNNER GUENTER
YOUNG LOUISE GRAY DR	ZEIPPEN CLAUDE DR	ZENG QIN DR
ZIRIN HAROLD DR		

COMMISSION No. 15

PHYSICAL STUDY OF COMETS, MINOR PLANETS AND METEORITES

(ETUDE PHYSIQUE DES COMETES, DES PETITES PLANETES ET DES METEORITES)

President : RAHE JURGEN PROF

Vice-President(s) : HARRIS ALAN WILLIAM DR

Organizing Committee: A'HEARN MICHAEL F DR
ARPIGNY CLAUDE PROF
BELTON MICHAEL J S DR
BRANDT JOHN C DR
DOBROVOLSKY OLEG V PROF
GRUEN EBERHARD DR
HAUPT HERMANN F PROF
HUGHES DAVID W DR
KRESAK LUBOR DR
LUPISHKO DMITRIJ F
RICKMAN HANS DR
TEDESCO EDWARD F
WASSON JOHN T
WYCKOFF SUSAN DR
ZAPPALA VINCENZO PROF

Members:
ALLEGRE CLAUDE PROF
ANDERS EDWARD PROF
ANDRIENKO DMITRY A DR
ARNOLD JAMES R DR
AXFORD W IAN PROF
BABADZHANOV PULAT B DR
BAILEY MARK EDWARD
BARKER EDWIN S DR
BEARD DAVID B DR
BIRCH PETER MR
BLAMONT JACQUES E PROF
BOCKELEE-MORVAN DOMINIQUE
BOEHNHARDT HERMANN DR
BOUSKA JIRI DR
BOWELL EDWARD L G DR
BRECHER AVIVA DR PROF
BROWN ROBERT HAMILTON
BROWNLEE DONALD E PROF
BRUNK WILLIAM E DR
BURATTI BONNIE J DR
BURLAGA LEONARD F DR
BURNS JOSEPH A PROF
CAMPINS HUMBERTO DR
CANDY MICHAEL P MR
CARRUTHERS GEORGE R DR
CARUSI ANDREA
CELLINO ALBERTO DR
CEPLECHA ZDENEK DR
CERRONI PRISCILLA DR
CHANDRASEKHAR T DR
CHAPMAN CLARK R DR
CHAPMAN ROBERT D DR
CHEN DAO-HAN
CHEREDNICHENKO V I DR
CLAIREMIDI JACQUES DR
CLAYTON GEOFFREY C DR
CLAYTON ROBERT N DR
CLUBE S V M DR
COCHRAN ANITA L DR
COCHRAN WILLIAM DAVID DR
COMBI MICHAEL R DR
COSMOVICI BATALLI C DR
CRISTESCU CORNELIA G DR
CROVISIER JACQUES
CRUIKSHANK DALE P DR
CUYPERS JAN DR
DANKS ANTHONY C DR
DE PATER IMKE
DE SANCTIS GIOVANNI
DEBEHOGNE HENRI DR SC
DEGEWIJ JOHAN DR
DELSEMME ARMAND H PROF DR
DERMOTT STANLEY F
DEUTSCHMAN WILLIAM A DR
DI MARTINO MARIO
DONN BERTRAM D
DOSSIN F DR
DRYER MURRAY DR
DZHAPIASHVILI VICTOR P DR
ENCRENAZ THERESE DR
ERSHKOVICH ALEXANDER PROF
EVERHART EDGAR DR
EVIATAR AHARON PROF
FARINELLA PAOLO DR
FECHTIG HUGO DR
FELDMAN PAUL DONALD DR
FERNANDEZ JEAN-CLAUDE DR
FERNANDEZ JULIO A DR
FERRIN IGNACIO
FESTOU MICHEL C DR
FROESCHLE CHRISTIANE D DR
FUJIWARA AKIRA DR
FULCHIGNONI MARCELLO PROF
GEHRELS TOM PROF
GEISS JOHANNES PROF
GERARD ERIC DR
GIBSON DAVID MICHAEL DR
GIBSON JAMES
GIOVANE FRANK
GRADIE JONATHAN CAREY
GREEN SIMON F
GREENBERG J MAYO DR
GREENBERG RICHARD DR
GROSSMAN LAWRENCE PROF

GRUDZINSKA STEFANIA DR
HANNER MARTHA S DR
HARTMANN WILLIAM K
HASER LEO N K DR
HU ZHONG-WEI
IP WING-HUEN
IVANOVA VIOLETA DR
JOHNSON TORRENCE V DR
KNACKE ROGER F DR
KOHOUTEK LUBOS DR
KRISHNA SWAMY K S DR
LAMY PHILIPPE DR
LARSON STEPHEN M
LILLER WILLIAM DR
LIPSCHUTZ MICHAEL E DR
LIU ZONGLI
LUMME KARI A DR
MALAISE DANIEL J DR
MATSON DENNIS L DR
MCCROSKY RICHARD E DR
MEISEL DAVID D DR
MILLER FREEMAN D PROF
MOEHLMANN DIEDRICH
MORRISON DAVID PROF
MUMMA MICHAEL JON
NEFF JOHN S
NIEDNER MALCOLM B DR
OLSSON-STEEL DUNCAN I DR
PELLAS PAUL DR
PITTICH EDUARD M DR
REITSEMA HAROLD J
ROEMER ELIZABETH PROF
SCHLEICHER DAVID G DR
SCHMIDT MAARTEN PROF
SHARP CHRISTOPHER DR
SHOEMAKER EUGENE M
SMITH BRADFORD A PROF
SOLC MARTIN
SURDEJ JEAN M G
TAKEDA HIDENORI DR
TERENTJEVA ALEXANDRA K DR
TYPHOON LEE
VEEDER GLENN J DR
WANG SI-CHAO
WEHINGER PETER A DR
WETHERILL GEORGE W
WILLIAMS IWAN P DR
WOOLFSON MICHAEL M PROF
YAVNEL ALEXANDER A DR
ZELLNER BENJAMIN H DR

GUSTAFSON BO A S
HAPKE BRUCE W DR
HARWIT MARTIN PROF
HELIN ELEANOR FRANCIS
HUEBNER WALTER F DR
IRVINE WILLIAM M PROF
JACKSON WILLIAM M DR
KEIL KLAUS DR
KNEZEVIC ZORAN
KONOPLEVA VARVARA P DR
KRISTENSEN LEIF KAHL DR
LANCASTER BROWN PETER
LEBOFSKY LARRY ALLEN
LILLIE CHARLES F DR
LISSAUER JACK J DR
LOPES ROSALY DR
LUTZ BARRY L DR
MARAN STEPHEN P DR
MATSUURA OSCAR T DR
MCDONNELL J A M PROF
MENDIS DEVAMITTA ASOKA DR
MILLIS ROBERT L DR
MOORE ELLIOTT P PROF
MRKOS ANTONIN DR
NAKAMURA TSUKO DR
NEUKUM G DR
O'DELL CHARLES R DR
PAOLICCHI PAOLO DR
PEREZ-DE-TEJADA H A DR
PRIALNIK-KOVETZ DINA DR
REMY BATTIAU LILIANE G A
SAGDEEV ROALD Z DR
SCHLOERB F. PETER
SCHOBER HANS J DR
SHIMIZU MIKIO PROF
SHUL'MAN L M DR
SMOLUCHOWSKI ROMAN PROF
SPINRAD HYRON PROF
SVOREN JAN
TANABE HIROYOSHI DR
THOLEN DAVID J DR
VALSECCHI GIOVANNI B DR
VEVERKA JOSEPH DR
WDOWIAK THOMAS J DR
WEIDENSCHILLING S J DR
WHIPPLE FRED L DR
WISNIEWSKI WIESLAW Z
WOSZCZYK ANDRZEJ PROF
YEOMANS DONALD K DR
ZHOU XING-HAI

HALLIDAY IAN DR
HARRIS ALAN WILLIAM
HASEGAWA ICHIRO DR
HERZBERG GERHARD DR
IBADINOV KHURSANDKUL DR
ISOBE SYUZO DR
JOCKERS KLAUS DR
KELLER HANS ULRICH DR
KOEBERL CHRISTIAN DR
KOWAL CHARLES THOMAS
LAGERKVIST CLAES-INGVAR
LARSON HAROLD P DR
LEVASSEUR-REGOURD A.C. PR
LINDSEY CHARLES ALLAN
LIU LIN-ZHONG
LUEST RHEA DR
LYTTLETON RAYMOND A PROF
MARSDEN BRIAN G DR
MCCORD THOMAS B DR
MCKENNA LAWLOR SUSAN
MILET BERNARD L DR
MILLMAN PETER M DR
MOROZ V I PROF DR
MUKAI TADASHI DR
NAPIER WILLIAM M DR
NEWBURN RAY L JR
O'KEEFE JOHN A DR
PARISOT JEAN-PAUL
PILCHER CARL BERNARD DR
PROISY PAUL E DR
REVELLE DOUGLAS ORSON DR
SCALTRITI FRANCO DR
SCHMIDT H U DR
SEKANINA ZDENEK DR
SHKODROV V G DR
SIVARAMAN K R DR
SNYDER LEWIS E
STOHL JAN DR
SZEGO KAROLY DR
TATUM JEREMY B DR
TOMITA KOICHIRO MR
VANYSEK VLADIMIR PROF
WALLIS MAX K DR
WEAVER HAROLD F PROF
WEISSMAN PAUL ROBERT
WILKENING LAUREL L DR
WOOD JOHN A DR
YABUSHITA SHIN A PROF
ZARNECKI JAN CHARLES DR

COMMISSION No. 16

PHYSICAL STUDY OF PLANETS AND SATELLITES

(ETUDE PHYSIQUE DES PLANETES ET DES SATELLITES)

President : BRAHIC ANDRE DR

Vice-President(s) : MAROV MIKHAIL YA PROF
MORRISON DAVID PROF

Organizing Committee: BERTAUX J L DR
BURNS JOSEPH A PROF
CHEN DAO-HAN
DAVIES MERTON E MR
DE BERGH CATHERINE DR
ENCRENAZ THERESE DR
GAUTIER DANIEL
MASURSKY HAROLD DR
MATSON DENNIS L DR
MOROZ V I PROF DR
OWEN TOBIAS C PROF
SHEVCHENKO VLADISLAV V DR
SMITH BRADFORD A PROF
TEJFEL VIKTOR G DR

Members:

AKABANE TOKUHIDE DR	APPLEBY JOHN F	ARTHUR DAVID W G
ATREYA SUSHIL K	BARROW COLIN H DR	BATTANER EDUARDO DR
BAUM WILLIAM A DR	BAZILEVSKY ALEXANDR T	BEEBE RETA FAYE DR
BEER REINHARD DR	BELTON MICHAEL J S DR	BENDER PETER L DR
BERGE GLENN L DR	BERGSTRALH JAY T DR	BHATIA R K DR
BLAMONT JACQUES E PROF	BOBROV M S DR	BONDARENKO L N DR
BOSMA PIETER B DR	BOSS ALAN P DR	BOYCE PETER B DR
BOYER CHARLES	BRECHER AVIVA DR PROF	BROADFOOT A LYLE DR
BROWN ROBERT HAMILTON	BRUNK WILLIAM E DR	BURATTI BONNIE J DR
BYSTROV NIKOLAI F DR	CALAME ODILE DR	CALDWELL JOHN JAMES
CAMERON WINIFRED S MRS	CAMICHEL HENRI DR	CAMPBELL DONALD B
CATALANO SANTO DR	CHAMBERLAIN JOSEPH W PROF	CHAPMAN CLARK R DR
CLAIREMIDI JACQUES DR	COCHRAN ANITA L DR	COLLINSON EDWARD H
COLOMBO G PROF DR	COMBI MICHAEL R DR	CONNES JANINE DR
COUNSELMAN CHARLES C PROF	CRUIKSHANK DALE P DR	DE PATER IMKE
DEGEWIJ JOHAN DR	DERMOTT STANLEY F	DICKEL JOHN R
DICKEY JEAN O'BRIEN	DOLLFUS AUDOUIN PROF	DRAKE FRANK D PROF
DURRANCE SAMUEL T DR	DZHAPIASHVILI VICTOR P DR	EL-BAZ FAROUK DR
ELLIOT JAMES L DR	ELSTON WOLFGANG E PROF	ESHLEMAN VON R PROF
ESPOSITO LARRY W	FARINELLA PAOLO DR	FIELDER GILBERT DR
FINK UWE DR	FOX KENNETH DR	FOX W E MR
FUJIWARA AKIRA DR	GEAKE JOHN E DR	GEHRELS TOM PROF
GEISS JOHANNES PROF	GICLAS HENRY L MR	GIERASCH PETER J DR
GOLD THOMAS PROF	GOLDREICH P DR	GOLDSTEIN RICHARD M DR
GOODY R M	GORENSTEIN PAUL DR	GOUDAS CONSTANTINE L PROF
GREEN JACK PROF	GROSSMAN LAWRENCE PROF	GUERIN PIERRE DR
GUEST JOHN E DR	GULKIS SAMUEL DR	GURSHTEIN ALEXANDER A DR
HABIBULLIN SH T PROF DR	HAGFORS T DR	HALL JOHN S DR

HALLIDAY IAN DR
HERZBERG GERHARD DR
HIDE RAYMOND PROF
HOLBERG JAY B
HOREDT GEORG PAUL DR
HOVENIER J W DR
HU ZHONG-WEI
HUBBARD WILLIAM B PROF
HUNTEN DONALD M PROF
IRVINE WILLIAM M PROF
IWASAKI KYOSUKE DR
JEFFREYS HAROLD PROF SIR
JOHNSON TORRENCE V DR
JURGENS RAYMOND F
KARANDIKAR R V PROF
KAULA WILLIAM M PROF
KILADZE R I DR
KISLYUK VITALIJ S DR
KOPAL ZDENEK PROF
KOWAL CHARLES THOMAS
KSANFOMALITI L V DR
KUMAR SHIV S PROF
KURT V G DR
KUZMIN ARKADII D PROF DR
LANE ARTHUR LONNE DR
LARSON HAROLD P DR
LARSON STEPHEN M
LEIKIN G A DR
LEWIS J S
LISSAUER JACK J DR
LOCKWOOD G WESLEY DR
LOPES ROSALY DR
LOPEZ-MORENO JOSE JUAN
LOPEZ-PUERTAS MANUEL
LUMME KARI A DR
LUTZ BARRY L DR
MAHRA H S DR
MARTYNOV D YA PROF DR
MAYER CORNELL H
MCCORD THOMAS B DR
MCELROY M B DR
MEADOWS A JACK PROF
MIDDLEHURST BARBARA M MS
MIKHAL JOSEPH SIDKY PROF
MILLIS ROBERT L DR
MILLMAN PETER M DR
MIYAMOTO SIGENORI PROF
MOEHLMANN DIEDRICH
MOLINA ANTONIO
MOORE PATRICK DR
MOROZHENKO A V DR
MOUTSOULAS MICHAEL PROF
MULHOLLAND J DERRAL DR
MUMMA MICHAEL JON
MURPHY ROBERT E DR
NAKAGAWA YOSHITSUGU DR
NESS NORMAN F DR
NEUKUM G DR
O'KEEFE JOHN A DR
OTTELET I J DR
PANG KEVIN
PAOLICCHI PAOLO DR
PETROPOULOS BASIL CH DR
PETTENGILL GORDON H PROF
POLLACK JAMES B DR
RAO M N DR
RODRIGO RAFAEL
ROSCH JEAN PROF
RUNCORN S K PROF
RUSKOL EUGENIA L DR
SAFRONOV VICTOR S DR
SAGAN CARL DR
SAISSAC JOSEPH DR
SCHLEICHER DAVID G DR
SCHLOERB F. PETER
SHAPIRO IRWIN I PROF
SHIMIZU MIKIO PROF
SHIMIZU TSUTOMU PROF EMER
SHOEMAKER EUGENE M
SINTON WILLIAM M
SJOGREN WILLIAM L MR
SMITH HARLAN J PROF
SMOLUCHOWSKI ROMAN PROF
SODERBLOM LARRY DR
SONETT CHARLES P PROF
STOEV ALEXEI
STONE EDWARD C DR
STROBEL DARRELL F
STROM ROBERT G PROF
STRONG JOHN D PROF
SYNNOTT STEPHEN P
TERRILE RICHARD JOHN
THOLEN DAVID J DR
THOMPSON THOMAS WILLIAM
TOMBAUGH CLYDE W PROF
TRAFTON LAURENCE M DR
TRAN-MINH FRANCOISE DR
TROITSKY V S PROF DR
TYLER JR G LEONARD DR
VAN ALLEN JAMES A PROF
VAN FLANDERN THOMAS DR
VEVERKA JOSEPH DR
WALKER ROBERT M A PROF
WALLACE LLOYD V DR
WASSERMAN LAWRENCE H DR
WASSON JOHN T
WEIDENSCHILLING S J DR
WEIMER THEOPHILE P F DR
WEST ROBERT ALAN
WETHERILL GEORGE W
WHITAKER EWEN A
WILDEY ROBERT L PROF DR
WILLIAMS IWAN P DR
WILLIAMS JAMES G DR
WOOD JOHN A DR
WOOLFSON MICHAEL M PROF
WOSZCZYK ANDRZEJ PROF
YODER CHARLES F
YOUNG ANDREW T DR
YOUNG LOUISE GRAY DR
ZHANG MING-CHANG

COMMISSION No. 19

ROTATION OF THE EARTH (ROTATION DE LA TERRE)

President : FEISSEL MARTINE DR

Vice-President(s) : KOLACZEK BARBARA DR

Organizing Committee: BROSCHE PETER PROF
CARTER WILLIAM EUGENE
DICKEY JEAN O'BRIEN
DJUROVIC DRAGUTIN M DR
JIN WEN-JING
MCCARTHY DENNIS D DR
MIRONOV NIKOLAY T
ROCHESTER MICHAEL G PROF
SASAO TETSUO DR
SCHUTZ BOB EWALD
VONDRAK JAN DR
WILKINS GEORGE A DR

Members:

ARABELOS DIMITRIOS DR
ARIAS ELISA FELICITAS
BABCOCK ALICE K DR
BANG YONG GOL
BARLIER FRANCOIS E DR
BARRETO LUIZ MUNIZ PROF
BENDER PETER L DR
BILLAUD GERARD J
BLINOV N S DR
BONANOMI JACQUES DR
BOUCHER CLAUDE DR
CALAME ODILE DR
CANNON WAYNE H DR
CAPITAINE NICOLE
CHEN XING
CHIUMIENTO GIUSEPPE
CURRIE DOUGLAS G DR
DAVIES JOHN G DR
DEBARBAT SUZANNE V DR
DEJAIFFE RENE J DR
DICKMAN STEVEN R
DRAMBA C PROF
ELSMORE BRUCE DR
ENSLIN HEINZ DR
FANSELOW JOHN LYMAN
FLIEGEL HENRY F
FONG CHU-GANG
FUJISHITA MITSUMI DR
FURUKAWA KIICHIRO DR
GAIGNEBET JEAN DR
GAO BUXI
GAPOSCHKIN EDWARD M DR
GROTEN ERWIN PROF
GUINOT BERNARD R PROF
HALL R GLENN DR
HAN TIANQI
HELLWIG HELMUT WILHELM DR
HEMMLEB GERHARD DR
HIDE RAYMOND PROF.
HOSOYAMA KENNOSHUKE DR
HUA YING-MIN
IIJIMA SHIGETAKA PROF
JAKS WALDEMAR DR
JEFFREYS HAROLD PROF SIR
JI HONG-QING
KAKUTA CHUICHI DR
KALMYKOV A M DR
KING ROBERT WILSON JR DR
KLEPCZYNSKI WILLIAM J DR
KNOWLES STEPHEN H DR
KOKURIN YURIJ L DR
KOSTINA LIDIJA D DR
LAMBECK KURT PROF
LEDERLE TRUDPERT DR
LEFEBVRE MICHEL DR
LEHMANN MAREK DR
LI ZHENG-XIN DR
LIESKE JAY H DR
LUO DING-JIANG
LUO SHI-FANG
MANABE SEIJI DR
MARKOWITZ WILLIAM DR
MATSAKIS DEMETRIOS N
MEINIG MANFRED DR
MELBOURNE WILLIAM G DR
MELCHIOR PAUL J PROF DIR
MERRIAM JAMES B
MIETELSKI JAN S DR
MILOVANOVIC VLADETA DR
MOCZKO JANUSZ DR
MORGAN PETER DR
MORRISON LESLIE V
MUELLER IVAN I PROF
NAUMOV VITALIJ A DR
NEWHALL X X DR
NIEMI AIMO
NOBILI ANNA M
O'HORA NATHY P J
OKAMOTO ISAO DR
OKAZAKI SEICHI DR
OOE MASATSUGU DR
ORTE ALBERTO
OTERMA LIISI PROF
PAN XIAO-PEI
PAQUET PAUL EG DR
PARIJSKIJ N N PROF
PERDOMO RAUL
PILKINGTON JOHN D H DR
POMA ANGELO DR
POPELAR JOSEF DR
PRODAN Y I DR
PROVERBIO EDOARDO PROF
RANDIC LEO PROF DR

REN JIANG-PING
ROBERTSON DOUGLAS S
RUDER HANNS
RUNCORN S K PROF
RUSU I DR
RYKHLOVA LIDIJA V DR
SADZAKOV SOFIJA DR
SAKHAROV VLADIMIR I DR
SANCHEZ MANUEL
SATO KOICHI DR
SEKIGUCHI NAOSUKE PROF
SEVARLIC BRANISLAV M PROF
SEVILLA MIGUEL J DR
SHAPIRO IRWIN I PROF
SHI GUANG-CHEN
SIDORENKOV NIKOLAY S
SILVERBERG ERIC C DR
SLADE MARTIN A III DR
SMITH F GRAHAM PROF
SMITH HUMPHRY M
SMYLIE DOUGLAS E DR
STANILA GEORGE DR
STEPHENSON F RICHARD DR
STOYKO ANNA
SUGAWA CHIKARA DR
SUN YONGXIANG
TAKAGI SHIGETSUGU DR
TAPLEY BYRON D DR
TARADY VLADIMIR K DR
THOMAS DAVID V DR
TORAO MASAHISA
TSAO MO PROF
TSUBOKAWA IETSUNE DR
VEILLET CHRISTIAN
VEIS GEORGE PH D
VICENTE RAIMUNDO O PROF
WAKO KOJIRO DR
WAN TONG-SHAN
WANG ZHENG MING
WARD WILLIAM R DR
WILLIAMS JAMES G DR
WILSON P DR
WINKLER GERNOT M R DR
WU SHOU-XIAN
XIA JIONGYU
XIAO NAI-YUAN
XU TONG-QI
YANG FUMIN
YATSKIV YA S DR
YE SHU-HUA
YOKOYAMA KOICHI DR
YUMI SHIGERU PROF DR
ZHANG GUO-DONG
ZHAO MING
ZHENG DA-WEI
ZHU YONG-HE

COMMISSION No. 20

POSITIONS AND MOTIONS OF MINOR PLANETS, COMETS AND SATELLITES

(POSITIONS ET MOUVEMENTS DES PETITES PLANETES, DES COMETES ET DES SATELLITES)

President : WEST RICHARD M DR

Vice-President(s) : CARUSI ANDREA

Organizing Committee:
AKSNES KAARE DR
ARLOT JEAN-EUDES
KOZAI YOSHIHIDE PROF
KRESAK LUBOR DR
MARSDEN BRIAN G DR
RICKMAN HANS DR
SHOR VIKTOR A DR
WASSERMAN LAWRENCE H DR
YEOMANS DONALD K DR

Members:

A'HEARN MICHAEL F DR
ABALAKIN VICTOR K DR
AREND S DR
BABADZHANOV PULAT B DR
BAILEY MARK EDWARD
BATRAKOV YU V DR
BEC-BORSENBERGER ANNICK
BELYAEV NIKOLAJ A DR
BENEST DANIEL DR
BENNETT JOHN CAISTER MR
BIEN REINHOLD DR
BOERNGEN FREIMUT DR PH
BOWELL EDWARD L G DR
BRANHAM RICHARD L JR
BURNS JOSEPH A PROF
CALAME ODILE DR
CANDY MICHAEL P MR
CHAPRONT-TOUZE MICHELLE
CHERNYKH N S DR
CHIO CHOL ZONG
CHURMS JOSEPH
CRISTESCU CORNELIA G DR
CUNNINGHAM LELAND E PROF
DE PASCUAL MARTINEZ M DR
DE SANCTIS GIOVANNI
DEBEHOGNE HENRI DR SC
DELSEMME ARMAND H PROF DR
DIRIKIS M A DR
DOLLFUS AUDOUIN PROF
DONNISON JOHN RICHARD DR
DOURNEAU GERARD DR
DUNHAM DAVID W
DVORAK RUDOLF DR
EDMONDSON FRANK K PROF
ELLIOT JAMES L DR
EVDOKIMOV YU V DR
EVERHART EDGAR DR
FERNANDEZ JULIO A DR
FERRAZ-MELLO S PROF DR
FERRERI WALTER
FORTI GIUSEPPE DR
FRANKLIN FRED A DR
FREITAS MOURAO R R DR
FROESCHLE CLAUDE DR
FURUKAWA KIICHIRO DR
GALIBINA I V DR
GARFINKEL BORIS DR
GEHRELS TOM PROF
GIBSON JAMES
GICLAS HENRY L MR
GILMORE ALAN C MR
GREENBERG RICHARD DR
HARRINGTON ROBERT S DR
HARRIS ALAN WILLIAM DR
HASEGAWA ICHIRO DR
HAUPT HERMANN F PROF
HE MIAO-FU
HELIN ELEANOR FRANCIS
HEMENWAY PAUL D
HENRARD JACQUES PROF
HERS JAN MR
HEUDIER JEAN-LOUIS DR
HURNIK HIERONIM PROF
IANNA PHILIP A
IVANOVA VIOLETA DR
IZVEKOV V A DR
KHATISASHVILI ALFEZ SH DR
KIANG TAO PROF
KINOSHITA HIROSHI DR
KISSELEVA TAMARA P
KLEMOLA ARNOLD R DR
KNEZEVIC ZORAN
KOHOUTEK LUBOS DR
KOWAL CHARLES THOMAS
KRISTENSEN LEIF KAHL DR
LAGERKVIST CLAES-INGVAR
LAZZARO DANIELA DR
LIESKE JAY H DR
LINDBLAD BERTIL A DR
LOMB NICHOLAS RALPH DR
LOVAS MIKLOS
MACHADO LUIZ E. DA SILVA
MAHRA H S DR
MCCROSKY RICHARD E DR
MESSAGE PHILIP J DR
MILANI ANDREA
MILET BERNARD L DR
MILLIS ROBERT L DR
MINTZ BLANCO BETTY MRS
MORANDO BRUNO L DR
MRKOS ANTONIN DR
MULHOLLAND J DERRAL DR
MURRAY CARL D
NACOZY PAUL E DR
NAKAMURA TSUKO DR
NAKANO SYUICHI
NOBILI ANNA M
OTERMA LIISI PROF
PASCU DAN DR
PAUWELS T DR
PIERCE DAVID ALLEN
PITTICH EDUARD M DR

POPOVIC BOZIDAR PROF DR
RAMSAY DONALD A DR
ROEMER ELIZABETH PROF
SCHOBER HANS J DR
SCHUBART JOACHIM DR
SEKANINA ZDENEK DR
SHOEMAKER EUGENE M
SOMA MITSURU DR
SULTANOV G F ACAD
TAYLOR DONALD BOGGIA DR
TORRES CARLOS DR
VAN FLANDERN THOMAS DR
VAVROVA ZDENKA DR
VU DUONG TUYEN DR
WHIPPLE FRED L DR
WILLIAMS JAMES G DR
YUASA MANABU DR
ZHANG JIA-XIANG

PROTICH MILORAD B
RAPAPORT MICHEL DR
SAGNIER JEAN-LOUIS DR
SCHOLL HANS DR
SCHUSTER HANS-EMIL
SHELUS PETER J DR
SINCLAIR ANDREW T DR
STANDISH E MYLES DR
SVOREN JAN
THOLEN DAVID J DR
VAGHI SERGIO DR
VAN HOUTEN C J DR
VEILLET CHRISTIAN
WEISSMAN PAUL ROBERT
WILD PAUL PROF
WROBLEWSKI HERBERT DR
ZADUNAISKY PEDRO E PROF
ZIOLKOWSKI KRZYSZTOF DR

QUIJANO LUIS
REITSEMA HAROLD J
SCHMADEL LUTZ D DR
SCHRUTKA-RECHTENSTAMM PR.
SEIDELMANN P KENNETH DR
SHKODROV V G DR
SITARSKI GRZEGORZ PROF
STELLMACHER IRENE DR
SYNNOTT STEPHEN P
TOMITA KOICHIRO MR
VALSECCHI GIOVANNI B DR
VAN HOUTEN-GROENEVELD I
VIEIRA MARTINS ROBERTO DR
WHIPPLE ARTHUR L DR
WILLIAMS IWAN P DR
YABUSHITA SHIN A PROF
ZAPPALA VINCENZO PROF

COMMISSION No. 21

LIGHT OF THE NIGHT SKY (LUMIERE DU CIEL NOCTURNE)

President : LEVASSEUR-REGOURD A.C. PR

Vice-President(s) : HANNER MARTHA S DR

Organizing Committee: BOWYER C STUART PROF
DUMONT RENE DR
GALPERIN YU I PROF
HOUCK JAMES R
LAMY PHILIPPE DR
LEINERT CHRISTOPH DR
MATTILA KALEVI DR
MUKAI TADASHI DR

Members:

ALVAREZ P	ANDERSON KINSEY A PROF	ANGIONE RONALD J DR
BAGGALEY WILLIAM J PROF	BANOS COSMAS J DR	BATES DAVID R PROF
BELKOVICH O I DR	BLACKWELL DONALD E PROF	BLAMONT JACQUES E PROF
BROADFOOT A LYLE DR	CHAMBERLAIN JOSEPH W PROF	CLAIREMIDI JACQUES DR
DACHS JOACHIM PROF DR	DERMOTT STANLEY F	DIVARI N B DR
DUFAY MAURICE PROF	DUNKELMAN LAWRENCE	ELSAESSER HANS PROF
FECHTIG HUGO DR	FELDMAN PAUL DONALD DR	FISHKOVA LUISA M PROF
FUJIWARA AKIRA DR	GIOVANE FRANK	GREENBERG J MAYO DR
GRUEN EBERHARD DR	HALLIDAY IAN DR	HARWIT MARTIN PROF
HASEGAWA HIROICHI DR	HAUG ULRICH PROF	HAUSER MICHAEL G DR
HENDECOURT D' LOUIS DR	HENRY RICHARD C. PROF.	HOFMANN WILFRIED DR
HONG SEUNG SOO DR	IVANOV-KHOLODNY G S DR	JAMES JOHN F MR
JARRETT ALAN H PROF	JOUBERT MARTINE	KAPLAN J DR
KARANDIKAR R V PROF	KARYGINA ZOYA V DR	KOUTCHMY SERGE DR
KULKARNI PRABHAKAR V PROF	LEGER ALAIN DR	LILLIE CHARLES F DR
LOPEZ-MORENO JOSE JUAN	LOPEZ-PUERTAS MANUEL	LUMME KARI A DR
MAIHARA TOSHINORI DR	MATSUMOTO TOSHIO DR	MEGRELISHVILI T G PROF
MISCONI NEBIL YOUSIF DR	MOLINA ANTONIO	MORGAN DAVID H DR
MUKAI SONOYO DR	NAWAR SAMIR DR	NEUZIL LUDEK DR
NEY EDWARD P PROF	NICOLET MARCEL PROF	NISHIMURA TETSUO DR
PARESCE FRANCESCO DR	PERRIN JEAN MARIE DR	PFLEIDERER JORG PROF
PITZ ECKHART DR	RAPAPORT MICHEL DR	RIPKEN HARTMUT W DR
ROACH FRANKLIN E	ROBLEY R DR	RODRIGO RAFAEL
ROOSEN ROBERT G DR	ROZHKOVSKIJ DIMITRIJ A	SANCHEZ FRANCISCO PROF
SANCHEZ-SAAVEDRA M LUISA	SAXENA P P DR	SHAROV A S DR
SHEFOV NICOLAI N	SOBERMAN ROBERT K DR	SPARROW JAMES G DR
STAUDE HANS JAKOB PH D	TANABE HIROYOSHI DR	TOLLER GARY N DR
TOROSHLIDZE TEIMURAZ I DR	TRUTSE YU L DR	TYSON JOHN A DR
VAN ALLEN JAMES A PROF	VAN DE HULST H C PROF DR	WALLACE LLOYD V DR
WEILL GILBERT M DR	WEINBERG J L DR	WENIGER SCHAME DR
WITT ADOLF N DR	WOLSTENCROFT RAMON D DR	WOOLFSON MICHAEL M PROF
YAMAKOSHI KAZUO	YAMAMOTO TETSUO DR	YAMASHITA KOJUN DR
ZERULL REINER H DR		

COMMISSION No. 22

METEORS AND INTERPLANETARY DUST

(METEORES ET LA POUSSIERE INTERPLANETAIRE)

President : KEAY COLIN S L PROF

Vice-President(s) : STOHL JAN DR

Organizing Committee:
BABADZHANOV PULAT B DR
BAGGALEY WILLIAM J PROF
BELKOVICH O I DR
CEPLECHA ZDENEK DR
GRUEN EBERHARD DR
HASEGAWA ICHIRO DR
JONES JAMES DR
KOEBERL CHRISTIAN DR
REVELLE DOUGLAS ORSON DR
WILLIAMS IWAN P DR

Members:

ABBOTT WILLIAM N DR	BEARD DAVID B DR	BHANDARI N DR
BIBARSOV RAVIL'SH DR	BLACKWELL ALAN TREVOR	BROWNLEE DONALD E PROF
CARUSI ANDREA	CEVOLANI GIORDANO	CLIFTON KENNETH ST
CLUBE S V M DR	DAVIES JOHN G DR	DJORGOVSKI STANISLAV DR
ELFORD WILLIAM GRAHAM DR	FECHTIG HUGO DR	FIREMAN EDWARD L
FORTI GIUSEPPE DR	GLASS BILLY PRICE DR	GOSWAMI J N DR
HAJDUK ANTON DR	HAJDUKOVA MARIA	HALLIDAY IAN DR
HANNER MARTHA S DR	HARVEY GALE A DR	HASEGAWA HIROICHI DR
HAWKES ROBERT LEWIS DR	HAWKINS GERALD S DR	HEY JAMES STANLEY DR
HODGE PAUL W PROF	HONG SEUNG SOO DR	HUGHES DAVID W DR
JACCHIA LUIGI G DR	JENNISON ROGER C PROF	KAISER THOMAS R PROF
KAPISINSKY IGOR	KASHSCHEEV B L PROF DR	KOSTYLEV K V DR
KRAMER KH N DR	KRESAK LUBOR DR	KRESAKOVA MARGITA DR
KRUCHINENKO VITALIY G	KVIZ ZDENEK DR	LAMY PHILIPPE DR
LEBEDINETS VLADIMIR N DR	LEVASSEUR-REGOURD A.C. PR	LINDBLAD BERTIL A DR
LOVELL SIR BERNARD PROF	MARVIN URSULA B DR	MCCROSKY RICHARD E DR
MCDONNELL J A M PROF	MCINTOSH BRUCE A DR	MEISEL DAVID D DR
MILES HOWARD G MR	MILLMAN PETER M DR	MISCONI NEBIL YOUSIF DR
NAKAZAWA KIYOSHI DR	NAPIER WILLIAM M DR	NEWBURN RAY L JR
NUTH JOSEPH A III	O'KEEFE JOHN A DR	OLSSON-STEEL DUNCAN I DR
PADEVET VLADIMIR DR	PECINA PETR	PLAVEC ZDENKA DR
POLNITZKY GERHARD DR	PORUBCAN VLADIMIR DR	RAJCHL JAROSLAV DR
RIPKEN HARTMUT W DR	ROOSEN ROBERT G DR	RUSSELL JOHN A PROF
SEKANINA ZDENEK DR	SHAO CHENG-YUAN	SHESTAKA IVAN S DR
SIMEK MILOS DR	SOBERMAN ROBERT K DR	SVESTKA JIRI DR
TEDESCO EDWARD F	TERENTJEVA ALEXANDRA K DR	TOMITA KOICHIRO MR
VERNIANI FRANCO PROF	WANG DE-CHANG	WEINBERG J L DR
WETHERILL GEORGE W	WHIPPLE FRED L DR	WOOD JOHN A DR
WOOLFSON MICHAEL M PROF	YAMAKOSHI KAZUO	YAVNEL ALEXANDER A DR
YEOMANS DONALD K DR	ZVOLANKOVA JUDITA	

COMMISSION No. 24

PHOTOGRAPHIC ASTROMETRY (ASTROMETRIE PHOTOGRAPHIQUE)

President : VAN ALTENA WILLIAM F PROF

Vice-President(s) : DE VEGT CH PROF DR

Organizing Committee: ARGUE A NOEL MR
BROSCHE PETER PROF
HANSON ROBERT B DR
IANNA PHILIP A
POLOZHENTSEV DIMITRIJ DR
TURON-LACARRIEU C DR
UPGREN ARTHUR R DR

Members:

ABHYANKAR KRISHNA D PROF
BALLABH G M DR
BASTIAN ULRICH
BENEDICT GEORGE F DR
BLAAUW ADRIAAN PROF DR
BRANHAM RICHARD L JR
BRONNIKOVA NINA M
BUNCLARK PETER STEPHEN DR
CHIU LIANG-TAI GEORGE
CHRISTY JAMES WALTER DR
CHURMS JOSEPH
CLUBE S V M DR
CONNES PIERRE DR
CORBIN THOMAS ELBERT DR
CREZE MICHEL DR
CRIFO FRANCOISE DR
CUDWORTH KYLE MCCABE DR
DAHN CONARD CURTIS DR
DELHAYE JEAN PROF
DOMMANGET J DR
DOUGLASS GEOFFREY G
EICHHORN HEINRICH K DR
ELSMORE BRUCE DR
FALLON FREDERICK W DR
FANSELOW JOHN LYMAN
FIRNEIS FRIEDRICH J DR
FIRNEIS MARIA G DR
FRACASTORO MARIO G PROF
FRANZ OTTO G DR
FREDRICK LAURENCE W PROF
FRESNEAU ALAIN DR
GALLOUET LOUIS DR
GATEWOOD GEORGE DIRECTOR
GICLAS HENRY L MR
GOYAL A N DR
GUIBERT JEAN DR
HARRINGTON ROBERT S DR
HARTKOPF WILLIAM I DR
HARWOOD DENNIS MR
HEINTZ WULFF D DR
HEMENWAY PAUL D
HERSHEY JOHN L DR
HILL GRAHAM DR
HOFFLEIT E DORRIT DR
HUGHES JAMES A DR
IRWIN MICHAEL JOHN DR
JAHREISS HARTMUT DR
JEFFERYS WILLIAM H DR
JOHNSTON KENNETH J
JONES BURTON DR
JONES DEREK H P DR
KANAEV IVAN I DR
KISLYUK VITALIJ S DR
KLEMOLA ARNOLD R DR
KLOCK B L DR
KOLCHINSKIJ I G DR
KOVALEVSKY JEAN DR
LACROUTE PIERRE A PROF
LAPUSHKA K K DR
LATYPOV A A DR
LE POOLE RUDOLF S DR
LIPPINCOTT SARAH LEE DR
LOPEZ CARLOS LIC
LOZINSKIJ A M DR
LU PHILLIP K DR
LUTZ THOMAS E DR
LUYTEN WILLEM J PROF
MACHADO LUIZ E. DA SILVA
MARSCHALL LAURENCE A
MCALISTER HAROLD A DR
MCLEAN BRIAN JOHN
MEINEL ADEN B PROF
MENNESSIER MARIE-ODILE DR
MONET DAVID G
NICHOLSON WILLIAM
NUNEZ JORGE DR
OJA TARMO PROF
ONEGINA A B DR
PAN RONG-SHI
PASCU DAN DR
PERRYMAN MICHAEL A C
PODOBED V V DR
POTTER HEINO I DR
PROCHAZKA FRANZ V DR
QIN DAO
QUIJANO LUIS
REQUIEME YVES DR
RIZVANOV NAUFAL G DR
ROEMER ELIZABETH PROF
ROESER SIEGFRIED DR
RUDER HANNS
RUSSELL JANE L DR
SANDERS W L PROF
SCHILBACH ELENA DR
SHI GUANG-CHEN
SIMS KENNETH P DR
SMITH CLAYTON A JR DR
STANGE LOTHAR
STEIN JOHN WILLIAM
STOCK JURGEN D
STONE RONALD CECIL
STRAND KAJ AA DR
THOMAS DAVID V DR
VALBOUSQUET ARMAND DR
VAN DE KAMP PETER
VILKKI ERKKI U
WALTER HANS G DR
WAN LAI

WANG JIA-JI
WASSERMAN LAWRENCE H DR
WESSELINK ADRIAAN J DR
WESTERHOUT GART DR
WHITE GRAEME LINDSAY DR
WILLIAMS CAROL A
WORLEY CHARLES E
WROBLEWSKI HERBERT DR
YOUNIS SAAD M
ZHOU XING-HAI

COMMISSION No. 25

STELLAR PHOTOGRAPHY AND POLARIMETRY

(PHOTOMETRIE ET POLARIMETRIE STELLAIRE)

President : MCLEAN IAN S DR

Vice-President(s) : YOUNG ANDREW T DR

Organizing Committee:
BUSER ROLAND DR
KNUDE JENS KIRKESKOV DR
LANDSTREET JOHN D PROF
LUB JAN DR
MENZIES JOHN W DR
MILLER JOSEPH S PROF
MILONE EUGENE F PROF
PENNY ALAN JOHN DR
RUFENER FREDY G PROF
STRAIZYS V PROF DR
VRBA FREDERICK J DR
WESSELIUS PAUL R DR
WIELEBINSKI RICHARD PROF

Members:

ABLES HAROLD D DR
ANGEL J ROGER P PROF
ARGUE A NOEL MR
ASHOK N M DR
BALDINELLI LUIGI DR
BECKER WILHELM PROF
BLANCO VICTOR M DR
BORGMAN JAN DR PROF
BROWN DOUGLAS NASON
CELIS LEOPOLDO DR
COYNE GEORGE V DR
DAHN CONARD CURTIS DR
DOLAN JOSEPH F DR
EDWARDS PAUL J DR
FERNIE J DONALD PROF
GEHRELS TOM PROF
GERBALDI MICHELE DR
GOLAY MARCEL PROF
GRAUER ALBERT D
GUETTER HARRY HENDRIK
HARDIE R PROF
HAYES DONALD S DR
HILDITCH RONALD W DR
HOLMBERG ERIK B PROF
HYLAND A R HARRY DR
JERZYKIEWICZ MIKOLAJ DR
KAWARA KIMIAKI
KING IVAN R PROF
KUNKEL WILLIAM E DR
LANDOLT ARLO U PROF

ADELMAN SAUL J DR
ANGIONE RONALD J DR
ARNAUD JEAN PAUL
AXON DAVID
BARNES III THOMAS G DR
BEHR ALFRED PROF EMERITUS
BLECHA ANDRE BORIS G DR
BORRA ERMANNO F DR
BRUCK HERMANN A PROF
CHUGAJNOV P F DR
CRAWFORD DAVID L DR
DENOYELLE JOZEF KIC
DUBOUT RENEE
EELSALU HEINO DR
FORTE JUAN CARLOS DR
GEHRZ ROBERT DOUGLAS DR
GHOSH S K DR
GOY GERALD PROF
GRENON MICHEL DR
GUTIERREZ-MORENO A DR MRS
HARWOOD DENNIS MR
HECK ANDRE DR
HILL PHILIP W DR
HU JING-YAO
IRWIN ALAN W DR
JOSHI SURESH CHANDRA DR
KEPLER S O
KOCH ROBERT H DR
KVIZ ZDENEK DR
LASKARIDES PAUL G ASSPROF

ALBRECHT RUDOLF DR
ANTHONY-TWAROG BARBARA J
ARSENIJEVIC JELISAVETA
BAHNG JOHN D R PROF
BECK RAINER
BESSELL MICHAEL S DR
BOOKMYER BEVERLY B DR
BREGER MICHEL DR
CARNEY BRUCE WILLIAM
COUSINS A W J DR
DACHS JOACHIM PROF DR
DESHPANDE M R DR
DUCATI JORGE RICARDO DR
FEINSTEIN ALEJANDRO DR
GALLOUET LOUIS DR
GENET R M DR
GLASS IAN STEWART DR
GRAHAM JOHN A DR
GREWING MICHAEL PROF
HALL DOUGLAS S DR
HAUCK BERNARD PROF
HENSBERGE HERMAN
HILTNER W ALBERT PROF
HUANG LIN
IYENGAR K V K PROF
JOSHI U C DR
KILKENNY DAVID DR
KULKARNI PRABHAKAR V PROF
LABHARDT LUKAS
LASKER BARRY M DR

LENZEN RAINER DR
LI SIN HYONG
LOCKWOOD G WESLEY DR
LUNA HOMERO G. DR
MAITZEN HANS M DR
MANFROID JEAN DR
MARKKANEN TAPIO DR
MARRACO HUGO G DR
MASANI A PROF
MAYER PAVEL DR
MCCARTHY MARTIN F DR
MENDOZA V EUGENIO E DR
MIANES PIERRE DR
MINTZ BLANCO BETTY MRS
MITCHELL RICHARD MR
MOFFETT THOMAS J PROF
MORENO HUGO PROF
MORRIS STEPHEN C DR
MULLER A B DR
MUMFORD GEORGE S PROF
NICOLET BERNARD
NOTNI P DR
OBLAK EDOUARD
OESTREICHER ROLAND
PAGE ARTHUR MR
PEDREROS MARIO DR
PEL JAN WILLEM DR
PERRY CHARLES L DR
PFAU WERNER
PFEIFFER RAYMOND J
PHILIP A G DAVIS
PIIROLA VILPPU E DR
RAO P VIVEKANANDA DR
ROBINSON EDWARD LEWIS DR
ROSLUND CURT DR
RYDGREN ALFRED ERIC JR DR
SARMA M B K PROF
SCHMIDT EDWARD G
SCHOENEICH W DR
SHAKHOVSKOJ NIKOLAY M DR
SHAWL STEPHEN J DR
SMITH CHARLES DITTO
SMYTH MICHAEL J DR
STEINLIN ULI PROF
STERKEN CHRISTIAAN LEO DR
STOCK JURGEN D
STOCKMAN HERVEY S JR DR
STONE REMINGTON P S DR
STROHMEIER WOLFGANG PROF
SULLIVAN DENIS JOHN DR
SZKODY PAULA DR
TANDON S N PROF
TAPIA-PEREZ SANTIAGO
TINBERGEN JAAP DR
TODORAN IOAN DR
TOLBERT CHARLES R DR
TRODAHL HARRY JOSEPH DR
ULRICH BRUCE T PROF
URECHE VASILE DR
VARDANIAN R A DR
VAUGHAN ARTHUR H DR
VERMA R P DR
VISVANATHAN NATARAJAN DR
WALKER ALISTAIR ROBIN DR
WALLENQUIST AAKE A E PROF
WALRAVEN TH DR
WANG CHUAN-JIN
WARREN WAYNE H JR DR
WEISTROP DONNA DR
WESSELINK ADRIAAN J DR
WHITE NATHANIEL M DR
WILLSTROP RODERICK V DR
WINIARSKI MACIEJ
WISNIEWSKI WIESLAW Z
WOO JONG OK
WRAMDEMARK STIG S O DR
YAMASHITA YASUMASA PROF
YIN JI-SHENG

COMMISSION No. 26

DOUBLE AND MULTIPLE STARS (ETOILES DOUBLES ET MULTIPLES)

President : MCALISTER HAROLD A DR

Vice-President(s) : ABT HELMUT A DR

Organizing Committee:
BERNACCA P L PROF
COUTEAU PAUL PROF
HARRINGTON ROBERT S DR
KISELYOV ALEXEJ A DR
RAKOS KARL D PROF
VAN DESSEL EDWIN LUDO DR

Members:

ALLEN CHRISTINE	AREND S DR	ARGUE A NOEL MR
BACCHUS PIERRE PROF	BAGNUOLO WILLIAM G JR DR	BAIZE PAUL DR
BALEGA YURI YU.	BATTEN ALAN H DR	BEARDSLEY WALLACE R DR
BEAVERS WILLET I DR	BONNEAU DANIEL	BROSCHE PETER PROF
CABRITA EZEQUIEL DR	CAMPBELL ALISON DR	CAMPBELL BRUCE DR
CESTER BRUNO PROF	CHEN ZHEN	CULVER ROGER BRUCE DR
DADAEV ALEKSANDR N DR	DOCOBO DURANTEZ JOSE A	DOMMANGET J DR
DUNHAM DAVID W	EICHHORN HEINRICH K DR	FEKEL FRANCIS C
FERRER OSVALDO EDUARDO DR	FLETCHER J MURRAY	FRACASTORO MARIO G PROF
FRANZ OTTO G DR	FREDRICK LAURENCE W PROF	FREITAS MOURAO R R DR
FURENLID INGEMAR K DR	GATEWOOD GEORGE DIRECTOR	GEYER EDWARD H PROF DR
HALBWACHS JEAN LOUIS DR	HARTKOPF WILLIAM I DR	HERSHEY JOHN L DR
HIDAJAT BAMBANG PROF DR	HILL GRAHAM DR	HOLDEN FRANK
IANNA PHILIP A	ISHIDA GORO DR	JASCHEK CARLOS O R PROF
KOPAL ZDENEK PROF	KUMSISHVILI J I DR	LATHAM DAVID W DR
LATTANZI MARIO G	LING J DR	LIPPINCOTT SARAH LEE DR
LODEN KERSTIN R DR	LODEN LARS OLOF PROF	LUYTEN WILLEM J PROF
MAGALASHVILI N L DR	MEYER CLAUDE DR	MIKKOLA SEPPO DR
MORBEY CHRISTOPHER L	MOREL PIERRE JACQUES DR	MULLER PAUL
OBLAK EDOUARD	OSWALT TERRY D DR	PANNUNZIO RENATO
POPOVIC GEORGIJE DR	POVEDA ARCADIO DR	RUSSELL JANE L DR
SALUKVADZE G N DR	SCARDIA MARCO	SCARFE COLIN D DR
SCHMIDTKE PAUL C DR	SHUL'BERG A M DR	SMAK JOSEPH I PROF
STEIN JOHN WILLIAM	STRAND KAJ AA DR	SZABADOS LASZLO PH D
TOKOVININ ANDREJ A DR	TRIMBLE VIRGINIA L DR	UPGREN ARTHUR R DR
VALBOUSQUET ARMAND DR	VALTONEN MAURI J PROF	VAN ALTENA WILLIAM F PROF
VAN DE KAMP PETER	VAN DER HUCHT KAREL A DR	WALKER RICHARD L
WEIS EDWARD W DR	WIETH-KNUDSEN NIELS P DR	WILSON RAYMOND H DR
WORLEY CHARLES E	YAN LIN-SHAN	

COMMISSION No. 27

VARIABLE STARS (ETOILES VARIABLES)

President : BREGER MICHEL DR

Vice-President(s) : PERCY JOHN R PROF

Organizing Committee: BARNES III THOMAS G DR
CHRISTENSEN-DALSGAARD J
GERSHBERG R E DR
JERZYKIEWICZ MIKOLAJ DR
MAVRIDIS L N PROF
RODONO MARCELLO DR
SMITH MYRON A ASST PROF
SZEIDL BELA DR
VAN GENDEREN A M DR
WARNER BRIAN PROF

Members:

AIZENMAN MORRIS L DR
ALANIA I F DR
ALBINSON JAMES DR
ALFARO EMILIO JAVIER
ANDO HIROYASU DR
ANTIPOVA LYUDMILA DR
ANTONELLO ELIO
ARELLANO FERRO ARMANDO
ARKHIPOVA V P DR
ARSENIJEVIC JELISAVETA
ASTERIADIS GEORGIOS DR
AVGOLOUPIS STAVROS DR
BAADE DIETRICH DR
BAGLIN ANNIE DR
BAKER NORMAN H PROF
BAKOS GUSTAV A PROF
BALONA LUIS ANTERO DR
BARTOLINI CORRADO
BARWIG HEINZ
BASTIEN PIERRE DR
BATESON FRANK M OBE DR
BATH GEOFFREY T DR
BAUER WENDY HAGEN
BEDOGNI ROBERTO
BELSERENE EMILIA P
BELVEDERE GAETANO DR
BENSON PRISCILLA J DR
BERTHOMIEU GABRIELLE DR
BESSELL MICHAEL S DR
BIANCHINI ANTONIO DR
BOCHONKO D RICHARD DR
BOLTON C THOMAS PROF
BOND HOWARD E DR
BOPP BERNARD W DR
BOULON JACQUES J DR
BOWEN GEORGE H DR
BOYARCHUK A A DR
BOYARCHUK MARGARITA E DR
BROWN DOUGLAS NASON
BURKI GILBERT DR
BUSKO IVO C DR
BUTLER C JOHN DR
BUTLER DENNIS DR
BYRNE PATRICK B DR
CAMERON ANDREW COLLIER DR
CATCHPOLE ROBIN M DR
CHAVIRA ENRIQUE SR
CHEREPASHCHUK A M PROF
CHRISTY ROBERT F DR
CHUGAJNOV P F DR
COGAN BRUCE C DR
COHEN MARTIN DR
CONNOLLY LEO PAUL
CONTADAKIS MICHAEL E DR
COULSON IAIN M DR
COUTTS-CLEMENT CHRISTINE
COX ARTHUR N DR
CUYPERS JAN DR
DE GROOT MART DR
DELGADO ANTONIO JESUS
DEMERS SERGE DR
DEUPREE ROBERT G DR
DICKENS ROBERT J DR
DJORGOVSKI STANISLAV DR
DOWNES RONALD A DR
DUNLOP STORM
DUPUY DAVID L DR
DZIEMBOWSKI WOJCIECH PROF
EDWARDS PAUL J DR
EFREMOV YURY N DR
EL-BASSUNY ALAWY A A
ESKIOGLU A NIHAT
EVANS ANEURIN
EVANS NANCY REMAGE DR
FADEYEV YURI A
FEAST MICHAEL W PROF
FEIBELMAN WALTER A DR
FERLAND GARY JOSEPH
FERNIE J DONALD PROF
FITCH WALTER S DR
FRIEDJUNG MICHAEL DR
FROLOV M S DR
GAHM GOESTA F DR
GALLAGHER III JOHN S DR
GARRIDO RAFAEL
GASCOIGNE S C B DR
GENET R M DR
GEYER EDWARD H PROF DR
GIBSON DAVID MICHAEL DR
GIEREN WOLFGANG P DR
GIES DOUGLAS R DR
GLAGOLEVSKIJ JU V DR
GODOLI GIOVANNI PROF
GORBATSKY VITALIJ G PROF
GOUGH DOUGLAS O DR
GOUPIL MARIE JOSE
GRAHAM JOHN A DR
GRASDALEN GARY L DR
GRYGAR JIRI DR

GUERRERO GIANANTONIO DR
GUINAN EDWARD FRANCIS DR.
GURM HARDEV S PROF
GURSKY HERBERT DR
HACKWELL JOHN A DR
HAEFNER REINHOLD DR
HAISCH BERNHARD MICHAEL
HALL DOUGLAS S DR
HAMDY M A M DR
HANSEN CARL J PROF
HARMANEC PETR DR
HEISER ARNOLD M DR
HERBIG GEORGE H DR
HERR RICHARD B DR
HERS JAN MR
HESSER JAMES E DR
HEYDEN FRANCIS J SJ DR
HILL HENRY ALLEN DR
HILL PHILIP W DR
HOFFLEIT E DORRIT DR
HOUK NANCY DR
HUENEMOERDER DAVID P DR
HUTCHINGS JOHN B DR
IBEN ICKO JR PROF
JARZEBOWSKI TADEUSZ DR
JEFFERY CHRISTOPHER S DR
JEWELL PHILIP R DR
JIANG SHI-YANG
JONES ALBERT F MR
KADOURI TALIB HADI
KANYO SANDOR DR
KARP ALAN HERSH DR
KEPLER S O
KIM CHUL HEE DR
KIM TU HWAN
KIPLINGER ALAN L DR
KIPPENHAHN RUDOLF PROF
KJURKCHIEVA DIANA DR
KOPYLOV I M DR
KRAFT ROBERT P PROF
KRAUTTER JOACHIM DR
KREINER JERZY MAREK DR
KRZEMINSKI WOJCIECH DR
KUBIAK MARCIN A DR
KUHI LEONARD V PROF
KUMSISHVILI J I DR
KUNKEL WILLIAM E DR
KURTZ DONALD WAYNE DR
KWEE K K DR
LAGO MARIA TERESA V T PR.
LANDOLT ARLO U PROF
LANEY CLIFTON D DR
LASKARIDES PAUL G ASSPROF
LAZARO CARLOS DR
LEITE SCHEID PAULO DR
LEUNG KAM CHING PROF
LITTLE-MARENIN IRENE R DR
LIU ZONGLI
LOCKWOOD G WESLEY DR
LOPEZ DE COCA M D P DR
LUB JAN DR
MADORE BARRY FRANCIS DR
MAEDER ANDRE PROF
MAFFEI PAOLO PROF
MAHDY HAMED A DR
MAHMOUD FAROUK M A B DR
MAHRA H S DR
MAKARENKO EKATERINA N DR
MANNINO GIUSEPPE PROF
MANTEGAZZA LUCIANO
MARGRAVE THOMAS EWING JR
MARTIN WILLIAM L DR
MASANI A PROF
MATTEI JANET AKYUZ DR
MAYALL MARGARET W
MCGRAW JOHN T DR
MCNAMARA DELBERT H DR
MENNESSIER MARIE-ODILE DR
METZ KLAUS DR
MILONE EUGENE F PROF
MILONE LUIS A DR
MIRZOYAN L V DR PROF
MOFFETT THOMAS J PROF
MORGULEFF NINA ING
MORRISON NANCY DUNLAP DR
MUMFORD GEORGE S PROF
MURDIN PAUL G DR
NATHER R EDWARD
NEFF JOHN S
NIARCHOS PANAYIOTIS PH D
NIKOLOV ANDREJ DR
NUGIS TIIT
O'DONOGHUE DARRAGH DR
ODGERS GRAHAM J DR
OLAH KATALIN DR
OPOLSKI ANTONI PROF
OSAWA KIYOTERU DR
OSWALT TERRY D DR
PAPALOIZOU JOHN C B DR
PAPARO MARGIT DR
PAPOUSEK JIRI
PARSAMYAN ELMA S DR
PARTHASARATHY M DR
PATERNO LUCIO PROF
PETERSEN J O DR
PETROV PETER P DR
PETTERSEN BJOERN RAGNVALD
PIIROLA VILPPU E DR
POPOVA MALINA D PROF DR
PRINGLE JAMES E DR
PROVOST JANINE DR
PSKOVSKIJ JU P DR
PUGACH ALEXANDER F DR
RAKOS KARL D PROF
RAO N KAMESWARA
RENSON P F M DR
RICHTER G A DR
ROBINSON EDWARD LEWIS DR
RODGERS ALEX W DR
ROMANO GIULIANO PROF
ROMANOV YURI S DR
ROSINO LEONIDA PROF
ROUNTREE JANET DR
RUSSEV RUSCHO DR
SADIK AZIZ R DR
SAMUS NIKOLAI N DR
SANDMANN WILLIAM HENRY
SANYAL ASHIT DR
SAREYAN JEAN-PIERRE DR
SARMA M B K PROF
SATO NAONOBU PROF
SAWYER-HOGG HELEN B DR
SCHAEFER BRADLEY E DR
SCHMIDT EDWARD G
SCHOEMBS ROLF DR
SCHWARTZ PHILIP R DR
SCHWARZENBERG-CZERNY A
SCUFLAIRE RICHARD DR
SHARA MICHAEL DR
SHERWOOD WILLIAM A DR
SHOBBROOK ROBERT R DR
SINVHAL SHAMBHU DAYAL DR.
SMAK JOSEPH I PROF
SMEYERS PAUL PROF
SMITH HARLAN J PROF
SOLIMAN MOHAMED AHMED
SRIVASTAVA RAM KUMAR DR
STARRFIELD SUMNER PROF
STELLINGWERF ROBERT F DR
STEPIEN KAZIMIERZ DR
STERKEN CHRISTIAAN LEO DR
STOBIE ROBERT S DR
STROHMEIER WOLFGANG PROF
STROM KAREN M
STROM STEPHEN E
SZABADOS LASZLO PH D
SZECSENYI-NAGY GABOR DR
SZKODY PAULA DR
TAKEUTI MINE DR
TAMMANN G ANDREAS PROF DR
TEMPESTI PIERO PROF
TERZAN AGOP DR
THOMPSON KEITH DR
TJIN-A-DJIE HERMAN R E DR
TORRES CARLOS ALBERTO DR
TREMKO JOZEF DR
TSIOUMIS ALEXANDROS DR
TSVETKOV MILCHO K DR
TURNER DAVID G DR
TUTUKOV A V DR
TYLENDA ROMUALD DR
USHER PETER D DR
VALTIER JEAN-CLAUDE DR
VAN AGT S L TH J DR
VERHEEST FRANK PROF
VIOTTI ROBERTO DR
VOGT NIKOLAUS DR

WACHMANN A A PROF DR
WAELKENS CHRISTOFFEL
WALKER EDWARD N MR
WALKER MERLE F PROF
WALKER WILLIAM S G
WALLERSTEIN GEORGE PROF
WALRAVEN TH DR
WEBBINK RONALD F DR
WEHLAU AMELIA DR
WEHLAU WILLIAM H PROF
WEISS WERNER W DR
WENZEL W DR
WESSELINK ADRIAAN J DR
WHITELOCK PATRICIA ANN DR
WILLIAMON RICHARD M
WILLSON LEE ANNE DR
WILSON LIONEL DR
WING ROBERT F PROF
WISNIEWSKI WIESLAW Z
WOOD PETER R DR
WRIGHT FRANCES W DR
XIONG DA-RUN
YAO BAO-AN
ZUCKERMAN BEN M DR

COMMISSION No. 28

GALAXIES (GALAXIES)

President : TAMMANN G ANDREAS PROF DR

Vice-President(s) : KHACHIKIAN E YE PROF

Organizing Committee: BERTOLA FRANCESCO PROF
ELLIS RICHARD S
FREEMAN KENNETH C PROF
GALLAGHER III JOHN S DR
LEQUEUX JAMES DR
LI QI-BIN
OKAMURA SADANORI DR
QUINTANA HERNAN DR
TRIMBLE VIRGINIA L DR
VAN DER KRUIT PIETER C DR

Members:

ABLES HAROLD D DR	AFANAS'EV VIKTOR L DR	AGUERO ESTELA L DR
AHMAD FAROOQ DR	ALCAINO GONZALO DR	ALLADIN SALEH MOHAMED DR
ALLEN RONALD J DR	ALLOIN DANIELLE DR	AMBARTSUMIAN V A PROF DR
ANDRILLAT YVETTE DR	ARDEBERG ARNE L PROF	ARKHIPOVA V P DR
ARP HALTON DR	ATHANASSOULA EVANGELIE DR	AZZOPARDI MARC DR
BAHCALL JOHN N PROF	BAILEY MARK EDWARD	BAJAJA E DR
BALDWIN JACK A DR	BALKOWSKI-MAUGER CH DR	BALLABH G M DR
BARBON ROBERTO PROF	BARTHEL PETER DR	BASU BAIDYANATH PROF
BAUM WILLIAM A DR	BECK RAINER	BENDINELLI ORAZIO
BENEDICT GEORGE F DR	BERGERON JACQUELINE A DR	BERGVALL NILS AKE SIGVARD
BERKHUIJSEN ELLY M DR	BETTONI DANIELA DR	BHATTACHARYYA TARA DR
BICA EDUARDO L D DR	BIERMANN PETER L DR	BIJAOUI ALBERT DR
BINETTE LUC	BINGGELI BRUNO	BINNEY JAMES J DR
BIRKINSHAW MARK	BLITZ LEO	BLUMENTHAL GEORGE R DR
BOERNGEN FREIMUT DR PH	BOESHAAR GREGORY ORTH DR	BOKSENBERG ALEC PROF
BORCHKHADZE TENGIZ M DR	BOSMA ALBERT DR	BOTTINELLI LUCETTE DR
BRACCESI ALESSANDRO PROF	BRECHER KENNETH PROF	BRINKMANN WOLFGANG
BRINKS ELIAS DR	BRODIE JEAN P	BROSCH NOAH DR
BROSCHE PETER PROF	BURBIDGE E MARGARET PROF	BURBIDGE GEOFFREY R PROF
BURNS JACK O'NEAL JR	BURSTEIN DAVID	BUTA RONALD J DR
BUTCHER HARVEY R PROF DR	BYRD GENE G DR	CAMERON LUZIUS MARTIN
CAMPUSANO LUIS E	CANNON RUSSELL D DR	CAPACCIOLI MASSIMO DR
CARIGNAN CLAUDE DR	CARRANZA GUSTAVO J DR	CARSWELL ROBERT F DR
CARTER DAVID DR	CASERTANO STEFANO DR	CHALABAEV ALMAS DR
CHAMARAUX PIERRE DR	CHEN JIAN-SHENG	CHEN ZHENCHENG
CHINCARINI GUIDO L DR	CHOU CHIH-KANG DR	CHU YAOQUAN
CHUGAIJ NIKOLAI N DR	CHUVAEV K K DR	CLAVEL JEAN
COHEN ROSS D DR	COLIN JACQUES DR	COMTE GEORGES DR
CONTOPOULOS GEORGE PROF	CORWIN HAROLD G JR	COUCH WARRICK DR
COURTES GEORGES PROF	COWSIK RAMANATH	D'ODORICO SANDRO DR
DA COSTA NICOLAI L.-A	DANKS ANTHONY C DR	DAVIDSEN ARTHUR FALNES DR
DAVIES RODNEY D PROF	DAVIS MARC DR	DE BOER KLAAS SJOERDS DR
DE BRUYN A. GER DR	DE LA NOE JEROME DR	DE ROBERTIS M M DR

DE SILVA L.N.K. DR
DE VAUCOULEURS GERARD PR
DE ZEEUW PIETER T DR
DEJONGHE HERWIG BERT DR
DEKEL AVISHAI
DI FAZIO ALBERTO
DI SEREGO ALIGHIERI S DR
DIAZ ANGELES ISABEL DR
DICKENS ROBERT J DR
DICKEY JOHN M
DONAS JOSE DR
DONNER KARL JOHAN
DOTTORI HORACIO A DR
DRESSEL LINDA L
DRESSLER ALAN
DUBOIS PASCAL DR
DUFOUR REGINALD JAMES
DULTZIN-HACYAN D. DR
DUVAL MARIE-FRANCE
EDMUNDS MICHAEL GEOFFREY
EFSTATHIOU GEORGE
EINASTO JAAN DR
EKERS RONALD D DR
ELMEGREEN DEBRA MELOY
ELVIS MARTIN S DR
ELVIUS AINA M PROF
EMERSON DAVID
EVANS ROBERT REV
EVANS ROGER G DR
FABBIANO GIUSEPPINA
FABER SANDRA M PROF
FABRICANT DANIEL G
FAIRALL ANTHONY P PROF
FALL S MICHAEL DR
FEAST MICHAEL W PROF
FEITZINGER JOHANNES PROF
FERLAND GARY JOSEPH
FERRINI FEDERICO
FIELD GEORGE B PROF
FILIPPENKO ALEXEI V DR
FLIN PIOTR
FLORSCH ALPHONSE DR
FOLTZ CRAIG B.
FORD HOLLAND C RES PROF
FORD W KENT JR DR
FORTE JUAN CARLOS DR
FOUQUE PASCAL DR
FREEDMAN WENDY L DR
FRICKE KLAUS DR
FRIED JOSEF WILHELM DR
FROGEL JAY ALBERT DR
FTACLAS CHRIST
FUCHS BURKHARD DR
FUJIMOTO MASAYUKI DR
FUKUGITA MASATAKA DR
GALLETTA GIUSEPPE PROF
GAMALELDIN ABDULLA I DR
GARCIA LAMBAS DIEGO DR
GASCOIGNE S C B DR
GELLER MARGARET JOAN
GEORGIEV TSVETAN DR
GERHARD ORTWIN
GHIGO FRANCIS D DR
GHOSH P DR
GIOVANARDI CARLO
GIOVANELLI RICCARDO DR
GLASS IAN STEWART DR
GORGAS GARCIA JAVIER DR
GOSS W MILLER PROF
GOTTESMAN STEPHEN T DR
GOUGUENHEIM LUCIENNE
GRAHAM JOHN A DR
GRANDI STEVEN ALDRIDGE DR
GRASDALEN GARY L DR
GRIFFITHS RICHARD E DR
GUNN JAMES E PROF
GURZADIAN G A PROF DR
HAGEN-THORN VLADIMIR A DR
HAMABE MASARU DR
HARA TETSUYA DR
HARDY EDUARDO
HARMS RICHARD JAMES DR
HE XIANG-TAO
HECKMAN TIMOTHY M
HEESCHEN DAVID S DR
HEIDMANN JEAN DR
HELOU GEORGE DR
HENIZE KARL G ASTRONAUT
HENRY RICHARD B C DR
HEWITT ADELAIDE
HEWITT ANTHONY V DR
HICKSON PAUL DR
HINTZEN PAUL MICHAEL N DR
HJALMARSON AKE G DR
HODGE PAUL W PROF
HOLMBERG ERIK B PROF
HOYLE FRED SIR
HU FU-XING
HUA CHON TRUNG DR
HUANG JIE-HAO
HUANG KE-LIANG
HUANG SONG-NIAN DR
HUANG YONGWEI
HUCHRA JOHN PETER DR
HUCHTMEIER WALTER K DR
HUMMEL EDSHO
HUMPHREYS ROBERTA M PROF
HUNSTEAD RICHARD W DR
HUNTER CHRISTOPHER PROF
HUNTER JAMES H PROF
ICHIKAWA TAKASHI
ILLINGWORTH GARTH D DR
IMPEY CHRISTOPHER D DR
IRWIN JUDITH DR
ISRAEL FRANK P DR
ISSA ALI DR
JAFFE WALTER JOSEPH DR
JOG CHANDA J DR
JONES THOMAS WALTER DR
JOSHI MOHAN N PROF
JOSHI U C DR
JUGAKU JUN DR
JUNKKARINEN VESA T DR
KALAFI MANOUCHER
KALINKOV MARIN P DR
KALLOGLIAN ARSEN T DR
KANEKO NOBORU DR
KAPAHI V K DR
KAPAHI VIJAY, K.
KARACHENTSEV I D DR
KATGERT PETER DR
KAUFMAN MICHELE DR
KEEL WILLIAM C
KELLERMANN KENNETH I DR
KENNICUTT ROBERT C JR
KERR FRANK J DR
KING IVAN R PROF
KINMAN THOMAS D DR
KIRSHNER ROBERT PAUL DR
KLEIN ULRICH
KNAPP GILLIAN R DR
KOCHHAR R K DR
KODAIRA KEIICHI PROF
KOGOSHVILI NATELA G
KOLLATSCHNY WOLFRAM DR
KOO DAVID C-Y DR
KORMENDY JOHN DR
KRAAN-KORTEWEG RENEE C DR
KRISHNA GOPAL
KRON RICHARD G
KRUMM NATHAN ALLYN
KUNTH DANIEL
KUSTAANHEIMO PAUL E PROF
LAFON JEAN-PIERRE J DR
LARSON RICHARD B PROF
LASKER BARRY M DR
LAUBERTS ANDRIS DR
LAUSBERG ANDRE DR
LAWRENCE ANDREW DR
LE FEVRE OLIVIER DR
LEACOCK ROBERT JAY
LELIEVRE GERARD DR
LI JING
LI XIAO-QING
LILLER WILLIAM DR
LILLY SIMON J DR
LIN CHIA C PROF
LINDBLAD PER OLOF PROF
LIPOVETSKY V A
LIU RU-LIANG
LIU YONG-ZHEN
LO KWOK-YUNG DR
LONSDALE CAROL J DR
LOOSE HANS-HERMANN DR
LOPEZ ROSARIO DR
LORENZ HILMAR
LORTET MARIE CLAIRE

LOW FRANK J DR
LUCEY JOHN DR
LUGGER PHYLLIS M
LUMINET JEAN-PIERRE
LYNDEN-BELL DONALD PROF
LYNDS BEVERLY T DR
LYNDS ROGER C DR
MA ER
MACALPINE GORDON M
MACCHETTO FERDINANDO DR
MADORE BARRY FRANCIS DR
MARCELIN MICHEL
MARQUES DOS SANTOS P PROF
MARTIN MARIA CRISTINA DR
MARTINET LOUIS PROF
MATHEWSON DONALD S PROF
MAURICE ERIC N
MAVRIDES STAMATIA DR
MCBREEN BRIAN PHILIP DR
MCCREA WILLIAM SIR
MEDIAVILLA EVENCIO DR
MEIKLE WILLIAM P S
MENON T K PROF
MILET BERNARD L DR
MILEY G K DR
MILLER HUGH R PROF
MILLER JOSEPH S PROF
MILLER RICHARD H DR
MILLS BERNARD Y PROF
MINEVA VENETA DR
MIRABEL IGOR FELIX DR
MOLES MARIANO J DR
MOORWOOD ALAN F M
MORGAN WILLIAM W PROF
MOSS CHRISTOPHER DR
MOULD JEREMY R
MUNOZ-TUNON CASIANA
MURRAY STEPHEN S DR
MUZZIO JUAN C PROF
NARLIKAR JAYANT V PROF
NIETO JEAN-LUC
NITYANANDA R DR
NOGUCHI MASAFUMI DR
NOONAN THOMAS W PROF
NOREAU LOUIS DR
NORMAN COLIN A PROF
NULSEN PAUL DR
O'CONNELL ROBERT WEST DR
O'DEA CHRISTOPHER P DR
OEMLER AUGUSTUS JR DR
OKE J BEVERLEY PROF
OLEAK H DR
OMER GUY C JR PROF
OORT JAN H PROF
OSTERBROCK DONALD E PROF
OWEN FRAZER NELSON DR
PACHOLCZYK ANDRZEJ G PROF
PAGE THORNTON L DR
PALMER PHILIP DR
PALUMBO GIORGIO G C DR
PAN RONG-SHI
PAPAYANNOPOULOS TH DR
PASTORIZA MIRIANI G DR
PATUREL GEORGES
PEIMBERT MANUEL DR
PENG QIU-HE
PEREA-DUARTE JAIME D DR
PETERS WILLIAM L III DR
PETERSON BRADLEY MICHAEL
PETERSON CHARLES JOHN DR
PETROV GEORGY TRENDAFILOV
PFENNIGER DANIEL DR
PHILLIPS MARK M DR
PICKLES ANDREW JOHN DR
PISMIS DE RECILLAS PARIS
POPOV VASIL NIKOLOV
POVEDA ARCADIO DR
PRABHU TUSHAR P
PRENDERGAST KEVIN H PROF
PRESS WILLIAM H DR
PREVOT-BURNICHON M.L. DR
PRITCHET CHRISTOPHER J DR
PRONIK I I DR
PRONIK V I DR
PROUST DOMINIQUE
RAFANELLI PIERO DR
RANA NARAYAN CHANDRA DR
REAVES GIBSON PROF
REBEIROT EDITH DR
REPHAELI YOEL DR
RICHER HARVEY B DR
RICHSTONE DOUGLAS O DR
ROBERTS MORTON S DR
ROBERTS WILLIAM W JR PROF
ROBERTSON JAMES GORDON DR
RODRIGUEZ-ESPINOSA JOSE
ROESER HERMANN-JOSEF DR
ROOD HERBERT J
ROOS NICOLAAS DR
ROSA MICHAEL RICHARD DR
ROSADO MARGARITA DR
ROSE JAMES ANTHONY
ROTS ARNOLD H DR
RUBIN VERA C DR
RUDNICKI KONRAD PROF
RYDBECK GUSTAF H B DR
SADLER ELAINE MARGARET
SALVADOR-SOLE EDUARDO
SANAHUJA BLAS
SANCISI RENZO DR
SANDERS DAVID B DR
SANDERS ROBERT DR
SANROMA MANUEL DR
SAPRE A K DR
SARAZIN CRAIG L DR
SARGENT WALLACE L W DR
SASAKI TOSHIYUKI DR
SASLAW WILLIAM C PROF
SASTRY SHANKARA K
SAVAGE ANN DR
SCHMIDT K H DR
SCHMIDT MAARTEN PROF
SCHUECKING E L DR
SCHULTZ G V DR
SCHULZ HARTMUT DR
SCHUSTER HANS-EMIL
SCHWARZ ULRICH J DR
SCHWEIZER FRANCOIS DR
SCIAMA DENNIS W DR
SCOVILLE NICHOLAS Z
SEARLE LEONARD DR
SEIDEN PHILIP E
SELLWOOD JEREMY ARTHUR
SERSIC J L DR
SETTI GIANCARLO PROF
SHAHBAZIAN ROMELIA K DR
SHAKESHAFT JOHN R DR
SHARPLES RAY DR
SHAVER PETER A DR
SHAYA EDWARD J DR
SHERWOOD WILLIAM A DR
SHIELDS GREGORY A DR
SHORE STEVEN N
SHOSTAK G SETH DR
SIMIEN FRANCOIS DR
SIMKIN SUSAN M DR
SITKO MICHAEL L
SKILLMAN EVAN D DR
SMITH BRUCE F DR
SMITH HARDING E JR DR
SMITH HAYWOOD C DR
SMITH MALCOLM G DR
SOBOUTI YOUSEF PROF
SOLTAN ANDRZEJ MARIA DR
SONG GUO-XUAN
SPARKS WILLIAM BRIAN
SPINRAD HYRON PROF
STAVELEY-SMITH LISTER DR
STEPANIAN A A DR
STONE REMINGTON P S DR
STORCHI-BERGMAN THAISA DR
STROM RICHARD G DR
STROM ROBERT G PROF
SU HONG-JUN
SUBRAHMANYAM P V DR
SULENTIC JACK W DR
SULLIVAN WOODRUFF T III
TAKASE BUNSHIRO PROF
TALBOT RAYMOND J JR DR
TANAKA YUTAKA D DR
TANIGUCHI YOSHIAKI DR
TELESCO CHARLES M DR
TERLEVICH ROBERTO JUAN
TERZIAN YERVANT PROF
THAKUR RATNA KUMAR DR
THOMPSON LAIRD A DR
THONNARD NORBERT DR

THUAN TRINH XUAN DR
TIFFT WILLIAM G PROF
TONG YI
TOOMRE ALAR DR
TOVMASSIAN H M DR
TREMAINE SCOTT DUNCAN
TRINCHIERI GINEVRA
TULLY RICHARD BRENT DR
TURNER EDWIN L DR
TYSON JOHN A DR
ULRICH MARIE-HELENE D DR
UNGER STEPHEN DR
URBANIK MAREK DR
VALENTIJN EDWIN A DR
VALTONEN MAURI J PROF
VAN ALBADA TJEERD S DR
VAN DEN BERGH SIDNEY PROF
VAN DER HULST JAN M DR
VAN DER LAAN H PROF DR
VAN GENDEREN A M DR
VAN GORKOM JACQUELINE H
VAN MOORSEL GUSTAAF DR
VAN WOERDEN HUGO PROF DR
VARMA RAM KUMAR PROF
VERON MARIE-PAULE DR
VERON PHILIPPE DR
VISVANATHAN NATARAJAN DR
VOGLIS NIKOS DR
VORONTSOV-VEL'YAMINOV B A
VRTILEK JAN M DR
WAKAMATSU KEN-ICHI DR
WALTERBOS RENE A M DR
WARD MARTIN JOHN
WARNER JOHN W DR
WEEDMAN DANIEL W PROF
WEHINGER PETER A DR
WELCH GARY A DR
WESTERLUND BENGT E PROF
WHITE SIMON DAVID MANION
WHITFORD ALBERT E PROF
WHITMORE BRADLEY C
WIELEBINSKI RICHARD PROF
WIELEN ROLAND PROF DR
WIITA PAUL JOSEPH
WILD PAUL PROF
WILKINSON ALTHEA
WILLIAMS BARBARA A
WILLIAMS ROBERT E DR
WILLIAMS THEODORE B DR
WILLS BEVERLEY J DR
WILLS DEREK DR
WILSON ALBERT G DR
WINDHORST ROGIER A DR
WLERICK GERARD DR
WOOSLEY S E PROF
WORRALL DIANA MARY
WYNN-WILLIAMS C G DR
YAMAGATA TOMOHIKO DR
YOUNG JUDITH SHARN
ZAMORANO JAIME DR
ZASOV ANATOLE V DR
ZAVATTI FRANCO
ZHOU YOU-YUAN
ZINN ROBERT J DR

COMMISSION No. 29

STELLAR SPECTRA (SPECTRES STELLAIRES)

President : CONTI PETER S DR

Vice-President(s) : LAMBERT DAVID L PROF

Organizing Committee: BAADE DIETRICH DR
BESSELL MICHAEL S DR
BOESGAARD ANN M PROF
CASSATELLA ANGELO DR
CAYREL DE STROBEL GIUSA
GUSTAFSSON BENGT DR
PILACHOWSKI CATHERINE DR
SMOLINSKI JAN DR
SPITE MONIQUE DR

Members:

ABHYANKAR KRISHNA D PROF
ABT HELMUT A DR
ADELMAN SAUL J DR
AGUILAR MARIA LUISA
AIKMAN G CHRIS L
ALECIAN GEORGES DR
ALLER LAWRENCE HUGH
ANDRILLAT HENRI L PROF
ANDRILLAT YVETTE DR
APPENZELLER IMMO PROF
ARTRU MARIE-CHRISTINE DR
ASLANOV I A DR
ATAC TAMER
BALIUNAS SALLIE L
BARATTA GIOVANNI BATTISTA
BARBUY BEATRIZ DR
BARRY DON C DR
BASRI GIBOR B
BAUER WENDY HAGEN
BECKMAN JOHN E PROF
BENSAMMAR SLIMANE DR
BERGER JACQUES G DR
BERTOUT CLAUDE
BIDELMAN WILLIAM P PROF
BOGGESS ALBERT DR
BOND HOWARD E DR
BONSACK WALTER K PROF
BOUVIER JEROME
BOUVIER PIERRE PROF
BOYARCHUK A A DR
BRANDI ELISANDE ESTELA DR
BREYSACHER JACQUES
BROWN DOUGLAS NASON
BRUHWEILER FRED C JR
BUES IRMELA D DR
BURKHART CLAUDE DR
BUSCOMBE WILLIAM PROF
BUTCHER HARVEY R PROF DR
BUTLER KEITH DR
CAMPBELL BRUCE DR
CARNEY BRUCE WILLIAM
CARPENTER KENNETH G DR
CASTELLI FIORELLA DR
CATALA CLAUDE DR
CATALANO SANTO DR
CATCHPOLE ROBIN M DR
CAYREL ROGER DR
CLIMENHAGA JOHN L PROF
CODE ARTHUR D
COTTRELL PETER LEDSAM
COWLEY ANNE P DR
COWLEY CHARLES R PROF
CRIVELLARI LUCIO
DAWANAS DJONI N DR
DE GROOT MART DR
DENNEFELD MICHEL
DIVAN LUCIENNE DR
DOAZAN VERA DR
DOBRONRAVIN PETER DR
DOLIDZE MADONA V DR
DRAVINS DAINIS PROF
DUNCAN DOUGLAS KEVIN DR
DWORETSKY MICHAEL M DR
EDMONDS FRANK N JR DR
FARAGGIANA ROSANNA PROF
FEAST MICHAEL W PROF
FELENBOK PAUL DR
FERNANDEZ-FIGUEROA M J DR
FITZPATRICK EDWARD L DR
FLOQUET MICHELE DR
FOING BERNARD H DR
FOY RENAUD DR
FRANCOIS PATRICK DR
FRANDSEN SOEREN PROF
FREIRE FERRERO RUBENS G
FRIEDJUNG MICHAEL DR
FRIEL EILEEN D DR
FRINGANT ANNE-MARIE DR
FUJITA YOSHIO PROF
FURENLID INGEMAR K DR
GARMANY CATHERINE D DR
GARRISON ROBERT F PROF
GEHREN THOMAS PH D
GERBALDI MICHELE DR
GERSHBERG R E DR
GIAMPAPA MARK S
GILLET D DR
GILRA DAYA P DR
GLAGOLEVSKIJ JU V DR
GLUSHNEVA I N DR
GOEBEL JOHN H DR
GRATTON LIVIO PROF
GRATTON R G DR
GRAY DAVID F PROF
GREENSTEIN J L PROF
GRIFFIN RITA E M DR
GRIFFIN ROGER F DR
GROTH HANS G PROF DR
GUTHRIE BRUCE N G DR
HACK MARGHERITA PROF
HANDLIROVA DAGMAR DR
HANUSCHIK REINHARD DR

HARMANEC PETR DR
HARMER CHARLES F W MR
HARMER DIANNE L MRS
HARTMANN LEE WILLIAM
HEARNSHAW JOHN B DR
HEBER ULRICH
HEINTZE J R W DR
HENIZE KARL G ASTRONAUT
HENRICHS HUBERTUS F DR
HERBIG GEORGE H DR
HIRAI MASANORI DR
HIRATA RYUKO
HOUZIAUX L PROF
HUANG CHANG-CHUN
HUBENY IVAN
HUBERT HENRI DR
HUBERT-DELPLACE A.-M. DR
HUENEMOERDER DAVID P DR
HUNGER KURT PROF
HYLAND A R HARRY DR
JASCHEK CARLOS O R PROF
JASCHEK MERCEDES DR
JIANG SHI-YANG
JOHNSON HOLLIS R PROF
JORDAN CAROLE DR
JOSHI SURESH CHANDRA DR
JUGAKU JUN DR
KEENAN PHILIP C PROF EMER
KHARITONOV ANDREJ V DR
KHOKHLOVA V L DR
KING R B DR
KIPPER TONU DR
KITCHIN CHRISTOPHER R DR
KODAIRA KEIICHI PROF
KOGURE TOMOKAZU DR
KOMAROV N S DR
KOPYLOV I M DR
KOUBSKY PAVEL
KOVACHEV B J DR
KRAFT ROBERT P PROF
KREMPEC-KRYGIER JANINA DR
KUMAJGORODSKAYA RAISA DR
LABS DIETRICH PROF
LAGO MARIA TERESA V T PR.
LAMERS H J G L M DR
LAMLA ERICH E DR
LAMONTAGNE ROBERT DR
LANDSTREET JOHN D PROF
LANGER GEORGE EDWARD DR
LANZ THIERRY DR
LARSSON-LEANDER G PROF
LE CONTEL JEAN-MICHEL
LECKRONE DAVID S DR
LESTER JOHN B DR
LEVATO ORLANDO HUGO DR
LIBBRECHT K G DR
LIEBERT JAMES W DR
LITTLE-MARENIN IRENE R DR
LOCANTHI DOROTHY DAVIS DR
LUCK R EARLE DR
LUNDSTROM INGEMAR DR
LYNAS-GRAY ANTHONY E
MAGAIN PIERRE DR
MAILLARD JEAN-PIERRE DR
MAITZEN HANS M DR
MALARODA STELLA M DR
MASSEY PHILIP L
MATHYS GAUTIER DR
MCNAMARA DELBERT H DR
MEGESSIER CLAUDE DR
MIKULASEK ZDENEK DR
MILLIGAN J E
MOFFAT ANTHONY F J DR
MOLARO PAOLO DR
MOOS HENRY WARREN DR
MORGULEFF NINA ING
MOROSSI CARLO
MORRISON NANCY DUNLAP DR
MUNDT REINHARD DR
NECKEL HEINZ DR
NETZER HAGAI DR
NICHOLLS RALPH W PROF
NIEUWENHUIJZEN HANS DR
NIKITIN A A DR
NISHIMURA SHIRO DR
NISSEN POUL E PROF
NORRIS JOHN DR
NUGIS TIIT
OETKEN L DR
OKE J BEVERLEY PROF
ORLOV MIKHAIL DR
OSAWA KIYOTERU DR
PAGEL BERNARD E J PROF
PARSONS SIDNEY B DR
PARTHASARATHY M DR
PASINETTI LAURA E PROF
PATERSON-BEECKMANS F
PEDOUSSAUT ANDRE
PEERY BENJAMIN F PROF
PENSTON MARGARET
PERRIN MARIE-NOEL DR
PETERS GERALDINE JOAN DR
PETERSON RUTH CAROL DR
PLAVEC MIREK J PROF
POECKERT ROLAND H DR
PRADERIE FRANCOISE DR
PRESTON GEORGE W DR
PRINJA RAMAN DR
QUERCI FRANCOIS R DR
QUERCI MONIQUE DR
RAMELLA MASSIMO
RAO N KAMESWARA
RAUTELA B S DR
REBOLO RAFAEL DR
REGO FERNANDEZ M DR
REIMERS DIETER PROF
RINGUELET ADELA E DR
ROBINSON BRIAN J DR
RODGERS ALEX W DR
ROSE JAMES ANTHONY
ROSSI LUCIO
RUTTEN ROBERT J. DR
SAAR ENN DR
SADAKANE KOZO DR
SAHADE JORGE PROF
SAREYAN JEAN-PIERRE DR
SCHILD RUDOLPH E DR
SCHOLZ GERHARD DR
SEGGEWISS WILHELM PROF
SHCHEGOLEV DIMITRIJ E DR
SHORE STEVEN N
SIMON THEODORE
SINNERSTAD ULF E PROF
SLETTEBAK ARNE PROF
SMITH MYRON A ASST PROF
SMITH VERNE V DR
SNEDEN CHRISTOPHER A
SNOW THEODORE P PROF
SODERBLOM DAVID R
SONNEBORN GEORGE DR
SPITE FRANCOIS M DR
STALIO ROBERTO DR
STAWIKOWSKI ANTONI DR
STECHER THEODORE P
STEFFEN MATTHIAS DR
STENCEL ROBERT EDWARD
SUNTZEFF NICHOLAS B
SVOLOPOULOS SOTIRIOS PROF
SWENSSON JOHN W DR
SWINGS JEAN-PIERRE DR
TAFFARA SALVATORE PROF
TAKADA-HIDAI MASAHIDE DR
TALAVERA A DR
THEVENIN FREDERIC DR
THOMPSON G I DR
TSUJI TAKASHI
TUOMINEN ILKKA V DR
UNDERHILL ANNE B DR
UTSUMI KAZUHIKO DR
VALTIER JEAN-CLAUDE DR
VAN DER HUCHT KAREL A DR
VAN'T VEER-MENNERET CL DR
VASU-MALLIK SUSHMA DR
VILHU OSMI DR
VIOTTI ROBERTO DR
VLADILO GIOVANNI DR
VOGT NIKOLAUS DR
VOGT STEVEN SCOTT
VREUX JEAN MARIE DR
WALKER GORDON A H PROF
WALLERSTEIN GEORGE PROF
WATERWORTH MICHAEL DR
WEGNER GARY ALAN
WEHINGER PETER A DR
WEHLAU AMELIA DR
WEHLAU WILLIAM H PROF
WEISS WERNER W DR

WELLMANN PETER PROF DR	WENIGER SCHAME DR	WILLIAMS PEREDUR M DR
WILSON ROBERT PROF SIR	WING ROBERT F PROF	WOLF BERNHARD PH D
WOLFF SIDNEY C DR	WOOD III H J DR	WRIGHT KENNETH O DR
WYCKOFF SUSAN DR	WYLLER ARNE A PROF	YAMASHITA YASUMASA PROF
ZOREC JEAN DR		

COMMISSION No. 30

RADIAL VELOCITIES (VITESSES RADIALES)

President : LATHAM DAVID W DR

Vice-President(s) : BURKI GILBERT DR

Organizing Committee:
ANDERSEN JOHANNES
CAMPBELL BRUCE DR
DA COSTA NICOLAI L.-A
FLORSCH ALPHONSE DR
FREEMAN KENNETH C PROF
MCCLURE ROBERT D PROF
PREVOT LOUIS DR

Members:

ABT HELMUT A DR	AZZOPARDI MARC DR	BALONA LUIS ANTERO DR
BARBIER-BROSSAT M DR	BATTEN ALAN H DR	BEARDSLEY WALLACE R DR
BEAVERS WILLET I DR	BERTIAU FLOR C PROF	BOULON JACQUES J DR
BREGER MICHEL DR	BURNAGE ROBERT	CARNEY BRUCE WILLIAM
CARQUILLAT JEAN-MICHEL	COCHRAN WILLIAM DAVID DR	CRAMPTON DAVID DR
DAVIS MARC DR	DE JONGE J K DR	DE VAUCOULEURS GERARD PR
DUFLOT MARCELLE DR	EDMONDSON FRANK K PROF	EELSALU HEINO DR
FAIRALL ANTHONY P PROF	FEHRENBACH CHARLES PROF	FEKEL FRANCIS C
FLETCHER J MURRAY	FOLTZ CRAIG B.	GEORGELIN YVON P DR
GIESEKING FRANK DR	GILMORE GERARD FRANCIS	GIOVANELLI RICCARDO DR
GOUGUENHEIM LUCIENNE	GRIFFIN ROGER F DR	HALBWACHS JEAN LOUIS DR
HEINTZE J R W DR	HEWETT PAUL	HILDITCH RONALD W DR
HILL GRAHAM DR	HRIVNAK BRUCE J	HUANG CHANG-CHUN
HUBE DOUGLAS P DR	HUCHRA JOHN PETER DR	IMBERT MAURICE DR
KADOURI TALIB HADI	KARACHENTSEV I D DR	KRAFT ROBERT P PROF
LEVATO ORLANDO HUGO DR	LEWIS BRIAN MURRAY DR	LINDGREN HARRI
MARSCHALL LAURENCE A	MARTIN NICOLE DR	MATHIEU ROBERT D DR
MAURICE ERIC N	MAYOR MICHEL DR	MAZEH TSEVI DR
MCMILLAN ROBERT S DR	MELNICK GARY J	MEYLAN GEORGES DR
MORBEY CHRISTOPHER L	NORDSTROM BIRGITTA DR	OETKEN L DR
PEDOUSSAUT ANDRE	PERRY CHARLES L DR	PETERSON RUTH CAROL DR
PHILIP A G DAVIS	POPOV VICTOR S DR	PRESTON GEORGE W DR
QUINTANA HERNAN DR	RATNATUNGA KAVAN U.	REBEIROT EDITH DR
ROMANOV YURI S DR	RUBIN VERA C DR	SANWAL N B DR
SCARFE COLIN D DR	SMITH MYRON A ASST PROF	STEFANIK ROBERT DR
STOCK JURGEN D	VAN DESSEL EDWIN LUDO DR	WEGNER GARY ALAN
WILLSTROP RODERICK V DR	YOSS KENNETH M DR	

COMMISSION No. 31

TIME (L'HEURE)

President : PAQUET PAUL EG DR

Vice-President(s) : PROVERBIO EDOARDO PROF

Organizing Committee:
ALLAN DAVID W MR
BLINOV N S DR
FLIEGEL HENRY F
FUJIMOTO MASA-KATSU DR
GUINOT BERNARD R PROF
HEMMLEB GERHARD DR
KLEPCZYNSKI WILLIAM J DR
KOVALEVSKY JEAN DR
MIAO YONG-RUI
MUELLER IVAN I PROF
PILKINGTON JOHN D H DR
YE SHU-HUA

Members:

ABELE MARIS K DR	AFANASJEVA PRASKOVYA M DR	AOKI SHINKO PROF
BABCOCK ALICE K DR	BELOTSERKOVSKIJ DAVID J	BENAVENTE JOSE
BENDER PETER L DR	BOLOIX RAFAEL DR	BONANOMI JACQUES DR
BRUMBERG VICTOR A DR	CAPRIOLI GIUSEPPE PROF	CARTER WILLIAM EUGENE
CATALAN MANUEL DR	CHAMBERLAIN JOSEPH M DR	DICKEY JEAN O'BRIEN
DOMINSKI IRENEUSZ DR	ENSLIN HEINZ DR	FALLON FREDERICK W DR
FEISSEL MARTINE DR	GAIGNEBET JEAN DR	GRUDLER PIERRE
HALL R GLENN DR	HAN TIANQI	HARA KEN NOSUKE DR
HELLWIG HELMUT WILHELM DR	HERS JAN MR	IIJIMA SHIGETAKA PROF
JIN WEN-JING	KAKUTA CHUICHI DR	KESSLER KARL G DR
KOBAYASHI YUKISAYU	KOLACZEK BARBARA DR	LESCHIUTTA S PROF
LIANG ZHONG-HUAN	LIESKE JAY H DR	LU BEN-KUI
LUO DINGCHANG	LUO SHI-FANG	MARKOWITZ WILLIAM DR
MATHUR B S DR	MATSAKIS DEMETRIOS N	MCCARTHY DENNIS D DR
MEINIG MANFRED DR	MELBOURNE WILLIAM G DR	MELCHIOR PAUL J PROF DIR
MORGAN PETER DR	NAUMOV VITALIJ A DR	NEWHALL X X DR
NIIMI YUKIO	NOEL FERNANDO	ORTE ALBERTO
PARCELIER PIERRE DR	POPELAR JOSEF DR	PUSHKIN SERGEY B DR
RANDIC LEO PROF DR	ROBERTSON DOUGLAS S	SCHULER WALTER DR
SMITH HUMPHRY M	SMYLIE DOUGLAS E DR	SONG JIN-AN
STANILA GEORGE DR	STOYKO ANNA	TSUCHIYA ATSUSHI DR PROF
VICENTE RAIMUNDO O PROF	WACKERNAGEL H BEAT DR	WEBROVA LUDMILA DR
WIETH-KNUDSEN NIELS P DR	WILKINS GEORGE A DR	WINKLER GERNOT M R DR
WU SHOU-XIAN	YANG KE-JUN	YUMI SHIGERU PROF DR
ZHAI ZAOCHENG	ZHANG JINTONG	ZHAO GANG
ZHENG YING	ZHUANG QIXIANG	

COMMISSION No. 33

STRUCTURE AND DYNAMICS OF THE GALACTIC SYSTEM
(STRUCTURE ET DYNAMIQUE DU SYSTEME GALACTIQUE)

President : MAYOR MICHEL DR

Vice-President(s) : BLITZ LEO

Organizing Committee: BAHCALL JOHN N PROF
BALAZS LAJOS G DR
BINNEY JAMES J DR
EINASTO JAAN DR
LYNGA GOSTA DR
TOSA MAKOTO DR
WIELEN ROLAND PROF DR

Members:

AARSETH SVERRE J DR
ADAMSON ANDREW DR
AFANAS'EV VIKTOR L DR
AGEKJAN TATEOS A PROF
ALTENHOFF WILHELM J DR
AMBARTSUMIAN V A PROF DR
AMBASTHA A K DR
ANDRLE PAVEL DR
ANTONOV VADIM A DR
AOKI SHINKO PROF
ARDEBERG ARNE L PROF
ASTERIADIS GEORGIOS DR
ATHANASSOULA EVANGELIE DR
BALBUS STEVEN A DR
BALDWIN JOHN E DR
BARBANIS BASIL PROF
BARBERIS BRUNO
BASH FRANK N PROF
BASU BAIDYANATH PROF
BAUD BOUDEWIJN DR
BECKER WILHELM PROF
BERKHUIJSEN ELLY M DR
BLAAUW ADRIAAN PROF DR
BLANCO VICTOR M DR
BLOEMEN JOHANNES B G M DR
BOULON JACQUES J DR
BRONFMAN LEONARDO DR
BURKE BERNARD F DR
BURTON W BUTLER DR
CALDWELL JOHN A R
CANE HILARY VIVIEN
CARRASCO LUIS DR
CASWELL JAMES L DR
CHEN ZHEN
CHRISTODOULOU DMITRIS DR
CHURCHWELL EDWARD B DR
CIURLA TADEUSZ
CLEMENS DAN P DR
CLUBE S V M DR
COHEN RICHARD S
COLIN JACQUES DR
COMINS NEIL FRANCIS
CONTOPOULOS GEORGE PROF
COSTA EDGARDO DR
COURTES GEORGES PROF
CRAMPTON DAVID DR
CRAWFORD DAVID L DR
CREZE MICHEL DR
CUDWORTH KYLE MCCABE DR
CUPERMAN SAMI PROF
DAVIES RODNEY D PROF
DE JONG TEIJE DR
DEKEL AVISHAI
DELHAYE JEAN PROF
DENOYELLE JOZEF KIC
DICKEL HELENE R DR
DICKEL JOHN R
DICKMAN ROBERT L DR
DIETER NANNIELOU H DR
DOWNES DENNIS DR
DRILLING JOHN S
DUCATI JORGE RICARDO DR
DZIGVASHVILI R M DR
EDMONDSON FRANK K PROF
EFREMOV YURY N DR
EGRET DANIEL DR
ELMEGREEN DEBRA MELOY
ELSAESSER HANS PROF
ELVIUS TORD PROF EMERITUS
EVANGELIDIS E DR
FABER SANDRA M PROF
FALL S MICHAEL DR
FEAST MICHAEL W PROF
FEHRENBACH CHARLES PROF
FEITZINGER JOHANNES PROF
FENKART ROLF P PROF DR
FIGUERAS FRANCESCA DR
FITZGERALD M PIM PROF
FREEMAN KENNETH C PROF
FUCHS BURKHARD DR
FUJIMOTO MASA-KATSU DR
FUKUNAGA MASATAKA DR
GALLETTO DIONIGI
GARZON FRANCISCO DR
GENKIN IGOR L PROF DR
GEORGELIN YVON P DR
GEORGELIN YVONNE M DR
GILMORE GERARD FRANCIS
GOLDREICH P DR
GORDON MARK A DR
GRAYZECK EDWIN J DR
GYLDENKERNE KJELD DR
HABE ASAO
HABING H J DR
HAMAJIMA KIYOTOSHI DR
HARTKOPF WILLIAM I DR
HAUG ULRICH PROF
HAWKINS MICHAEL R S
HAYLI AVRAM PROF
HEILES CARL PROF
HENON MICHEL C DR
HERBST WILLIAM DR
HERMAN JACOBUS DR
HOBBS ROBERT W DR
HORI GENICHIRO PROF
HRON JOSEF DR
HUANG SONG-NIAN DR
HUGHES VICTOR A PROF

HULSBOSCH A N M DR
HUMPHREYS ROBERTA M PROF
HUNTER CHRISTOPHER PROF
IKEUCHI SATORU DR
INAGAKI SHOGO DR
INNANEN KIMMO A PROF
IRWIN JOHN B PROF
ISOBE SYUZO DR
ISRAEL FRANK P DR
IWANISZEWSKA CECYLIA DR
IWANOWSKA WILHELMINA PROF
IYE MASANORI DR
JACKSON PETER DOUGLAS DR
JAHREISS HARTMUT DR
JASCHEK CARLOS O R PROF
JASNIEWICZ GERARD DR
JIANG DONG-RONG
JOG CHANDA J DR
JOHNSON HUGH M DR
JONAS JUSTIN LEONARD
JONES DEREK H P DR
KABURAKI MASAKI PROF
KALANDADZE N B DR
KALNAJS AGRIS J DR
KASUMOV FIKRET K O DR
KATO SHOJI PROF
KERR FRANK J DR
KHARADZE E K PROF
KING IVAN R PROF
KINMAN THOMAS D DR
KLARE GERHARD DR
KNAPP GILLIAN R DR
KOLESNIK IGOR G DR
KOLESNIK L N DR
KORMENDY JOHN DR
KULSRUD RUSSELL M DR
KUTUZOV S A DR
LAFON JEAN-PIERRE J DR
LARSON RICHARD B PROF
LECAR MYRON DR
LEE SANG GAK
LI JING
LIN CHIA C PROF
LINDBLAD PER OLOF PROF
LOCKMAN FELIX J
LODEN KERSTIN R DR
LODEN LARS OLOF PROF
LUNEL MADELEINE DR
LUYTEN WILLEM J PROF
LYNDEN-BELL DONALD PROF
MACCONNELL DARRELL J DR
MACRAE DONALD A PROF
MANCHESTER RICHARD N DR
MARK JAMES WAI-KEE DR
MAROCHNIK L S PROF DR
MARTINET LOUIS PROF
MATHEWSON DONALD S PROF
MAVRIDIS L N PROF
MCCARTHY MARTIN F DR
MCGREGOR PETER JOHN DR
MEATHERINGHAM STEPHEN DR
MENNESSIER MARIE-ODILE DR
MEYLAN GEORGES DR
MEZGER PETER G PROF
MIKKOLA SEPPO DR
MILLER RICHARD H DR
MIRABEL IGOR FELIX DR
MIRZOYAN L V DR PROF
MIYAMOTO MASANORI DR
MOFFAT ANTHONY F J DR
MONET DAVID G
MONNET GUY J DR
MORRIS MARK ROOT DR
MUENCH GUIDO PROF
MUZZIO JUAN C PROF
NAHON FERNAND PROF
NECKEL TH DR
NELSON ALISTAIR H DR
NINKOVIC SLOBODAN
NISHIDA MINORU PROF
NISHIDA MITSUGU
OGORODNIKOV KYRILL P PROF
OJA TARMO PROF
OKUDA HARUYUKI DR PROF
OLANO CARLOS ALBERTO DR
OLLONGREN A PROF DR
OORT JAN H PROF
OSTRIKER JEREMIAH P PROF
PALMER PATRICK E PROF
PALOUS JAN DR
PAPAYANNOPOULOS TH DR
PAULS THOMAS ALBERT DR
PAVLOVSKAYA E D DR
PEIMBERT MANUEL DR
PEREK LUBOS DR
PERRY CHARLES L DR
PESCH PETER DR
PHILIP A G DAVIS
PIER JEFFREY R DR
PILOWSKI K PROF DR
PISMIS DE RECILLAS PARIS
PRICE R MARCUS DR
PRIESTER WOLFGANG PROF
QIAN ZHONG-YU
RABOLLI MONICA DR
RAMBERG JOERAN M PROF
REID NEILL
REIF KLAUS DR
RIEGEL KURT W DR
ROBERTS MORTON S DR
ROBERTS WILLIAM W JR PROF
ROBINSON BRIAN J DR
ROHLFS K PROF DR
RONG JIAN-XIANG
RUBIN VERA C DR
RUIZ MARIA TERESA DR
RYBICKI GEORGE B DR
SAAR ENN DR
SALA FERRAN DR
SANCHEZ-SAAVEDRA M LUISA
SANDQVIST AAGE DR
SANDULEAK NICHOLAS DR
SCHMIDT HANS PROF
SCHMIDT K H DR
SCHMIDT MAARTEN PROF
SCHMIDT-KALER TH PROF
SEGGEWISS WILHELM PROF
SELLWOOD JEREMY ARTHUR
SHANE WILLIAM W DR
SHAROV A S DR
SHER DAVID DR
SHIMIZU TSUTOMU PROF EMER
SHU FRANK H PROF
SHUTER WILLIAM L H DR
SIMONSON S CHRISTIAN DR
SLETTEBAK ARNE PROF
SOLOMON PHILIP M DR
SONG GUO-XUAN
SPARKE LINDA
SPIEGEL E DR
STECKER FLOYD W DR
STEFANOVITCH-GOMEZ A E DR
STEINLIN ULI PROF
STEPHENSON C BRUCE PROF
STIBBS DOUGLAS W N PROF
STROBEL ANDRZEJ DR
STURCH CONRAD R DR
SVOLOPOULOS SOTIRIOS PROF
SZEBEHELY VICTOR G PROF
TAMMANN G ANDREAS PROF DR
TERZIDES CHARALAMBOS DR
THE PIK-SIN PROF
THIELHEIM KLAUS O DR
TOBIN WILLIAM
TOMISAKA KOHJI DR
TONG YI
TOOMRE ALAR DR
TOOMRE JURI
TORRA JORDI DR
TREFZGER CHARLES F DR
TSIOUMIS ALEXANDROS DR
TURON-LACARRIEU C DR
UPGREN ARTHUR R DR
VAN DER KRUIT PIETER C DR
VAN WOERDEN HUGO PROF DR
VANDERVOORT PETER O DR
VEGA E. IRENE DR
VENUGOPAL V R DR
VERSCHUUR GERRIT L PROF
VETESNIK MIROSLAV DR
VOROSHILOV V I DR
WAYMAN PATRICK A PROF
WEAVER HAROLD F PROF
WEISTROP DONNA DR
WESTERHOUT GART DR
WESTERLUND BENGT E PROF
WHITE RAYMOND E DR

WHITEOAK J B DR	WHITTET DOUGLAS C B DR	WIELEBINSKI RICHARD PROF
WILSON THOMAS L DR	WOLTJER LODEWIJK PROF	WOODWARD PAUL R DR
WRAMDEMARK STIG S O DR	YOSHII YUZURU DR	YOUNIS SAAD M
YUAN CHI PROF	ZHANG BIN	ZHAO JUN-LIANG

COMMISSION No. 34

INTERSTELLAR MATTER (MATIERE INTERSTELLAIRE)

President : MATHIS JOHN S PROF

Vice-President(s) : HABING H J DR

Organizing Committee: DE BOER KLAAS SJOERDS DR
DOPITA MICHAEL ANDREW DR
FLOWER DAVID R DR
GENZEL REINHARD DR
KAIFU NORIO DR
LADA CHARLES JOSEPH DR
LEQUEUX JAMES DR
RODRIGUEZ LUIS F
SHUSTOV BORIS M DR
YORK DONALD G DR

Members:

AANNESTAD PER ARNE DR
ACKER AGNES PROF DR
AIAD A PROF.
AITKEN DAVID K DR
AKABANE KENJI A PROF
ALDROVANDI S M VIEGAS DR
ALLER LAWRENCE HUGH
ALTENHOFF WILHELM J DR
ANANTHARAMAIAH K R DR
ANDERS EDWARD PROF
ANDREW BRYAN H DR
ANDRIESSE CORNELIS D DR
ANDRILLAT HENRI L PROF
ANDRILLAT YVETTE DR
ARKHIPOVA V P DR
ARNY THOMAS T DR
AVERY LORNE W DR
AXFORD W IAN PROF
BAARS JACOB W M DR
BAART EDWARD E PROF
BALDWIN JOHN E DR
BALUTEAU JEAN-PAUL
BANIA THOMAS MICHAEL
BARLOW MICHAEL J DR
BARNES AARON DR
BARRETT ALAN H PROF
BASH FRANK N PROF
BAUDRY ALAIN DR
BECKLIN ERIC E DR
BECKMAN JOHN E PROF
BECKWITH STEVEN V W
BEDOGNI ROBERTO
BEL NICOLE J DR
BERGERON JACQUELINE A DR
BERKHUIJSEN ELLY M DR
BERNAT ANDREW PLOUS DR
BERTOUT CLAUDE
BHATT H C DR
BIANCHI LUCIANA
BIEGING JOHN HAROLD DR
BIGNELL R CARL DR
BINETTE LUC
BIRKLE KURT PH D
BLACK JOHN HARRY DR
BLADES JOHN CHRIS DR
BLAIR GUY NORMAN DR
BLESS ROBERT C PROF
BLITZ LEO
BOCHKAREV NIKOLAY G DR
BODE MICHAEL F
BODENHEIMER PETER PROF
BOESHAAR GREGORY ORTH DR
BOGGESS ALBERT DR
BOHLIN RALPH C DR
BOLAND WILFRIED
BORGMAN JAN DR PROF
BOULANGER FRANCOIS
BOUVIER JEROME
BRAND PETER W J L DR
BRAUNSFURTH EDWARD PH D
BRINKMANN WOLFGANG
BROMAGE GORDON E DR
BROWN RONALD D PROF
BRUHWEILER FRED C JR
BURGESS ALAN DR
BURKE BERNARD F DR
BURTON W BUTLER DR
BYSTROVA NATALIJA V DR
CANTO JORGE DR
CAPLAN JAMES
CAPPA DE NICOLAU CRISTINA
CAPRIOTTI EUGENE R DR
CAPUZZO DOLCETTA ROBERTO
CARRUTHERS GEORGE R DR
CASWELL JAMES L DR
CERRUTI-SOLA MONICA
CERSOSIMO JUAN CARLOS DR
CESARSKY CATHERINE J DR
CESARSKY DIEGO A DR
CHEVALIER ROGER A DR
CHINI ROLF
CHOPINET MARGUERITE DR
CHU YOU-HUA
CHURCHWELL EDWARD B DR
CLARK FRANK OLIVER DR
CLEGG ROBIN E S DR
CODE ARTHUR D
COHEN MARSHALL H PROF
COLLIN-SOUFFRIN SUZY DR
COLOMB FERNANDO R DR
COSTERO RAFAEL
COURTES GEORGES PROF
COWIE LENNOX LAUCHLAN DR
COX DONALD P PROF
COYNE GEORGE V DR
CRANE PHILIPPE
CROVISIER JACQUES
CRUVELLIER PAUL E DR
CUDABACK DAVID D DR

CUGNON PIERRE DR
CZYZAK STANLEY J DR
D'ODORICO SANDRO DR
DAHN CONARD CURTIS DR
DALGARNO ALEXANDER PROF
DANKS ANTHONY C DR
DAVIES RODNEY D PROF
DE JONG TEIJE DR
DE LA NOE JEROME DR
DEGUCHI SHUJI DR
DEHARVENG LISE DR
DEWDNEY PETER E F DR
DI FAZIO ALBERTO
DICKEL HELENE R DR
DICKEL JOHN R
DICKEY JOHN M
DINERSTEIN HARRIET L
DISNEY MICHAEL J PROF
DOKUCHAEVA OLGA D DR
DONN BERTRAM D
DORSCHNER JOHANN DR
DOTTORI HORACIO A DR
DOWNES DENNIS DR
DRAINE BRUCE T
DRAPATZ SIEGFRIED W DR
DREHER JOHN W
DUBNER GLORIA DR
DUBOUT RENEE
DUFOUR REGINALD JAMES
DULEY WALTER W PROF
DUPREE ANDREA K DR
DWEK ELI
DYSON JOHN E DR
EL SHALABY MOHAMED
ELITZUR MOSHE
ELLIOTT KENNETH H DR
ELMEGREEN BRUCE GORDON DR
ELMEGREEN DEBRA MELOY
ELVIUS AINA M PROF
EMERSON JAMES P
ENCRENAZ PIERRE J DR
ESIPOV VALENTIN F DR
EVANS ANEURIN
EVANS NEAL J II ASS PROF
FALGARONE EDITH
FALK SYDNEY W JR DR
FALLE SAMUEL A DR
FAN YING
FAULKNER DONALD J DR
FEDERMAN STEVEN ROBERT
FEIBELMAN WALTER A DR
FEITZINGER JOHANNES PROF
FELLI MARCELLO DR
FELTEN JAMES E DR
FERLET ROGER DR
FERRINI FEDERICO
FIELD DAVID
FIELD GEORGE B PROF
FIERRO JULIETA
FISCHER JACQUELINE
FLANNERY BRIAN PAUL DR
FORD HOLLAND C RES PROF
FORSTER JAMES RICHARD DR
FRIEDEMANN CHRISTIAN DR
FRISCH PRISCILLA
FUKUI YASUO DR
FURNISS IAN
GARDNER FRANCIS F DR
GAUSTAD JOHN E PROF
GAY JEAN DR
GEBALLE THOMAS R DR
GEHRELS TOM PROF
GEORGELIN YVON P DR
GERARD ERIC DR
GEROLA HUMBERTO DR
GEZARI DANIEL YSA DR
GILRA DAYA P DR
GIOVANELLI RICCARDO DR
GODFREY PETER DOUGLAS DR
GOEBEL JOHN H DR
GOLDREICH P DR
GOLDSMITH DONALD W. DR.
GOLDSMITH PAUL F DR
GOLDSTEIN SAMUEL J PROF
GOLDSWORTHY FREDERICK A
GORDON COURTNEY P PROF
GORDON MARK A DR
GOSACHINSKIJ I V DR
GOSS W MILLER PROF
GRAHAM DAVID A
GRASDALEN GARY L DR
GREENBERG J MAYO DR
GREWING MICHAEL PROF
GUELIN MICHEL DR
GUESTEN ROLF
GULL THEODORE R DR
GURZADIAN G A PROF DR
GUSEINOV O H PROF
HACKWELL JOHN A DR
HALL JOHN S DR
HARDEBECK ELLEN G DR
HARRINGTON J PATRICK DR
HARRIS ALAN WILLIAM
HARRIS STELLA
HARTEN RONALD H DR
HARTL HERBERT DR
HARTQUIST THOMAS WILBUR
HARVEY PAUL MICHAEL DR
HAYNES RAYMOND F PROF
HECHT JAMES H DR
HEILES CARL PROF
HELFER H LAWRENCE PROF
HENDECOURT D' LOUIS DR
HENIZE KARL G ASTRONAUT
HENKEL CHRISTIAN
HERZBERG GERHARD DR
HIDAJAT BAMBANG PROF DR
HIGGS LLOYD A DR
HILDEBRAND ROGER H
HIPPELEIN HANS H DR
HJALMARSON AKE G DR
HJELLMING ROBERT M DR
HOBBS LEWIS M DR
HOEGLUND BERTIL PROF
HOLLENBACH DAVID JOHN DR
HOLLIS JAN MICHAEL DR
HONG SEUNG SOO DR
HOUZIAUX L PROF
HUA CHON TRUNG DR
HUGHES VICTOR A PROF
HULSBOSCH A N M DR
HUMMER DAVID G DR
HUTCHINGS JOHN B DR
IRVINE WILLIAM M PROF
ISOBE SYUZO DR
ISRAEL FRANK P DR
ISSA ALI DR
ITOH HIROSHI DR
IYENGAR K V K PROF
JABIR NIAMA LAFTA
JACOBY GEORGE H
JAFFE DANIEL T
JENKINS L F MS
JENNINGS R E PROF
JOHNSON FRED M PROF DR
JOHNSON HUGH M DR
JOHNSTON KENNETH J
JONES FRANK CULVER DR
JURA MICHAEL DR
KAFATOS MINAS DR
KAFTAN MAY A DR
KAHN FRANZ D PROF
KALER JAMES B PROF
KAMIJO FUMIO PROF DR
KAZES ILYA DR
KEGEL WILHELM H PROF
KENNICUTT ROBERT C JR
KERR FRANK J DR
KHARADZE E K PROF
KHROMOV G S DR
KIMURA HIROSHI DR
KIRKPATRICK RONALD C DR
KIRSHNER ROBERT PAUL DR
KNACKE ROGER F DR
KNAPP GILLIAN R DR
KNUDE JENS KIRKESKOV DR
KOEPPEN JOACHIM DR
KOHOUTEK LUBOS DR
KOLESNIK IGOR G DR
KONDO YOJI DR
KOORNNEEF JAN DR
KOSTYAKOVA ELENA B DR
KRAUTTER JOACHIM DR
KREYSA ERNST
KRISHNA SWAMY K S DR
KUIPER THOMAS B H DR
KUMAR C KRISHNA DR

KUNDU MUKUL R DR
KUNTH DANIEL
KUTNER MARC LESLIE DR
KWITTER KAREN BETH DR
KWOK SUN DR
KYLAFIS NIKOLAOS D DR
LAFON JEAN-PIERRE J DR
LANGER WILLIAM DAVID DR
LASKER BARRY M DR
LAURENT CLAUDINE DR
LE SQUEREN ANNE-MARIE DR
LEE TERENCE J DR
LEGER ALAIN DR
LEPINE JACQUES R D DR
LEUNG CHUN MING DR
LILLER WILLIAM DR
LIN CHIA C PROF
LINKE RICHARD ALAN DR
LISZT HARVEY STEVEN
LO KWOK-YUNG DR
LOCKMAN FELIX J
LOREN ROBERT BRUCE DR
LORTET MARIE CLAIRE
LOUISE RAYMOND PROF
LOVAS FRANCIS JOHN DR
LOW FRANK J DR
LOZINSKAYA TAT'YANA A DR
LUCAS ROBERT DR
LYNDS BEVERLY T DR
MACIEL WALTER J DR
MACLEOD JOHN M DR
MAIHARA TOSHINORI DR
MALLIK D C V DR
MANCHESTER RICHARD N DR
MANFROID JEAN DR
MARTIN ROBERT N DR
MARTIN-PINTADO JESUS
MASSON COLIN R
MATHEWS WILLIAM G PROF
MATHEWSON DONALD S PROF
MATTILA KALEVI DR
MCCALL MARSHALL LESTER DR
MCCRAY RICHARD DR
MCCREA J DERMOTT
MCGEE RICHARD X DR
MCKEE CHRISTOPHER F PROF
MCKEITH CONAL D DR
MCNALLY DEREK DR
MEABURN J DR
MEBOLD ULRICH DR PROF
MEIER ROBERT R
MELNICK GARY J
MENDEZ ROBERTO H DR
MENON T K PROF
MENZIES JOHN W DR
MESZAROS PETER DR
MEZGER PETER G PROF
MICHALITSIANOS ANDREW
MILLAR THOMAS J DR
MILLER JOSEPH S PROF
MILNE DOUGLAS K DR
MININ I N PROF
MINN YOUNG KEY DR
MITCHELL GEORGE F DR
MIYAMA SYOKEN
MIZUNO SHUN
MO JING-ER
MORGAN DAVID H DR
MORIMOTO MASAKI DR
MORRIS MARK ROOT DR
MORTON DONALD C DR
MOUSCHOVIAS TELEMACHOS CH
MUENCH GUIDO PROF
MUFSON STUART LEE DR
MYERS PHILIP C
NAKADA YOSHIKAZU DR
NANDY KASHINATH DR
NEUGEBAUER GERRY DR
NGUYEN-QUANG RIEU DR
NORDH H LENNART DR
NORMAN COLIN A PROF
NULSEN PAUL DR
NUSSBAUMER HARRY PROF
NUTH JOSEPH A III
O'DELL CHARLES R DR
O'DELL STEPHEN L
OHTANI HIROSHI DR
OKUDA HARUYUKI DR PROF
OLOFSSON HANS
OMONT ALAIN PROF
ONAKA TAKASHI
OSAKI TORU DR
OSBORNE JOHN L DR
OSTERBROCK DONALD E PROF
OZERNOY LEONID M PROF
PAGEL BERNARD E J PROF
PALLA FRANCESCO
PALMER PATRICK E PROF
PANAGIA NINO DR
PANKONIN VERNON LEE DR
PARKER EUGENE N
PAULS THOMAS ALBERT DR
PECKER JEAN-CLAUDE PROF
PEIMBERT MANUEL DR
PENSTON MICHAEL V DR
PENZIAS ARNO A DR
PEQUIGNOT DANIEL
PERAULT MICHEL
PERINOTTO MARIO PROF
PERSI PAOLO
PETERS WILLIAM L III DR
PETROSIAN VAHE PROF
PHILLIPS JOHN PETER
PHILLIPS THOMAS GOULD DR
PISMIS DE RECILLAS PARIS
POEPPEL WOLFGANG G L DR
POTTASCH STUART R PROF
PRASAD SHEO S
PREITE-MARTINEZ ANDREA DR
PRICE R MARCUS DR
PRONIK I I DR
PSKOVSKIJ JU P DR
PUGET JEAN-LOUP DR
QIN ZHI-HAI
RADHAKRISHNAN V PROF
RAIMOND ERNST DR
RAYMOND JOHN CHARLES
REIPURTH BO
RENGARAJAN T N DR
REYNOLDS RONALD J DR
RICKARD LEE J DR
RIGHINI-COHEN GIOVANNA DR
ROBBINS R ROBERT PROF
ROBERTS WILLIAM W JR PROF
ROBINSON BRIAN J DR
ROESER HANS-PETER
ROGER ROBERT S DR
ROGERS ALAN E E DR
ROHLFS K PROF DR
ROSA MICHAEL RICHARD DR
ROSADO MARGARITA DR
ROSE WILLIAM K DR
ROSINO LEONIDA PROF
ROXBURGH IAN W PROF
ROZHKOVSKIJ DIMITRIJ A
ROZYCZKA MICHAL
SABANO YUTAKA DR
SABBADIN FRANCO DR
SALINARI PIERO
SALPETER EDWIN E PROF
SANCHEZ-SAAVEDRA M LUISA
SANCISI RENZO DR
SANDELL GORAN HANS L DR
SANDQVIST AAGE DR
SARAZIN CRAIG L DR
SARGENT ANNEILA I
SARMA N V G PROF
SATO FUMIO DR
SATO SHUJI DR
SAVAGE BLAIR D DR
SAVEDOFF MALCOLM P PROF
SCALO JOHN MICHAEL
SCARROTT STANLEY M DR
SCHALEN CARL PROF
SCHATZMAN EVRY PROF
SCHERB FRANK PROF
SCHEUER PETER A G DR
SCHMID-BURGK J DR PROF
SCHMIDT THOMAS DR
SCHMIDT-KALER TH PROF
SCHULTZ G V DR
SCHULZ ROLF ANDREAS
SCHWARTZ PHILIP R DR
SCHWARTZ RICHARD D
SCHWARZ ULRICH J DR
SCOTT EUGENE HOWARD
SCOVILLE NICHOLAS Z

SEATON MICHAEL J PROF
SEKI MUNEZO DR
SHAH GHANSHYAM A DR
SHANE WILLIAM W DR
SHAO CHENG-YUAN
SHAPIRO STUART L
SHARPLESS STEWART PROF
SHAVER PETER A DR
SHAWL STEPHEN J DR
SHCHEGLOV P V DR
SHERWOOD WILLIAM A DR
SHIELDS GREGORY A DR
SHU FRANK H PROF
SHULL JOHN MICHAEL
SHUTER WILLIAM L H DR
SILBERBERG REIN DR
SILK JOSEPH I PROF
SILVESTRO GIOVANNI
SIMONS STUART DR
SINGH PATAN DEEN DR
SITKO MICHAEL L
SIVAN JEAN-PIERRE DR
SKILLING JOHN DR
SMITH BARHAM W DR
SMITH HOWARD ALAN
SMITH PETER L DR
SNELL RONALD L
SNOW THEODORE P PROF
SOBOLEV V V DR
SOFIA SABATINO PROF
SOFUE YOSHIAKI PROF
SOLC MARTIN
SOLOMON PHILIP M DR
SOMERVILLE WILLIAM B DR
SPITZER LYMAN JR DR
STANGA RUGGERO
STECHER THEODORE P
STENHOLM BJOERN DR
STROM RICHARD G DR
SU BUMEI
SUN JIN
SUZUKI YOSHIMASA PROF
TAKAKUBO KEIYA PROF
TAMURA SHINICHI DR
TARAFDAR SHANKAR P DR
TAYLOR KENNETH N R PROF
TENORIO-TAGLE G DR
TERZIAN YERVANT PROF
THADDEUS PATRICK PROF
THE PIK-SIN PROF
THOMPSON A RICHARD DR
THONNARD NORBERT DR
THRONSON HARLEY ANDREW JR
TORRES-PEIMBERT SILVIA DR
TOSI MONICA
TOWNES CHARLES HARD DR
TREFFERS RICHARD R
TURNER BARRY E DR
TURNER KENNETH C DR
ULRICH MARIE-HELENE D DR
URASIN LIRIK A DR
VAN DE HULST H C PROF DR
VAN DER LAAN H PROF DR
VAN DISHOECK EWINE F DR
VAN GORKOM JACQUELINE H
VAN WOERDEN HUGO PROF DR
VANDEN BOUT PAUL A
VANYSEK VLADIMIR PROF
VARSHALOVICH DIMITRIJ PR
VERSCHUUR GERRIT L PROF
VIALA YVES
VIALLEFOND FRANCOIS
VIDAL JEAN-LOUIS DR
VIDAL-MADJAR ALFRED DR
VINER MELVYN R DR
VISVANATHAN NATARAJAN DR
VORONTSOV-VEL'YAMINOV B A
VRBA FREDERICK J DR
WALKER GORDON A H PROF
WALMSLEY C MALCOLM DR
WANNIER PETER GREGORY DR
WATT GRAEME DAVID
WEAVER HAROLD F PROF
WEBSTER B LOUISE DR
WEILER KURT W DR
WEISHEIT JON C DR
WENDKER HEINRICH J PROF
WESSELIUS PAUL R DR
WEYMANN RAY J PROF
WHITE GLENN J
WHITE RICHARD L
WHITELOCK PATRICIA ANN DR
WHITEOAK J B DR
WHITTET DOUGLAS C B DR
WHITWORTH ANTHONY PETER
WICKRAMASINGHE N C PROF
WILLIAMS DAVID A PROF
WILLIAMS ROBERT E DR
WILLIS ALLAN J DR
WILLNER STEVEN PAUL DR
WILSON ROBERT W DR
WILSON THOMAS L DR
WINNBERG ANDERS DR
WINNEWISSER GISBERT DR
WITT ADOLF N DR
WOLSTENCROFT RAMON D DR
WOLSZCZAN ALEXANDER DR
WOLTJER LODEWIJK PROF
WOODWARD PAUL R DR
WOOLF NEVILLE J
WOOTTEN HENRY ALWYN
WRIGHT EDWARD L DR
WU CHI CHAO DR
WYNN-WILLIAMS C G DR
XIANG DELIN
XING JUN
YABUSHITA SHIN A PROF
YORKE HAROLD W DR
YOUNIS SAAD M
ZEALEY WILLIAM J DR
ZEILIK MICHAEL II DR
ZENG QIN DR
ZHANG CHENG-YUE
ZHOU ZHEN-PU
ZIMMERMANN HELMUT DR
ZUCKERMAN BEN M DR

COMMISSION No. 35

STELLAR CONSTITUTION (CONSTITUTION DES ETOILES)

President : MAEDER ANDRE PROF

Vice-President(s) : DEMARQUE P PROF

Organizing Committee: GOUGH DOUGLAS O DR
IBEN ICKO JR PROF
KIPPENHAHN RUDOLF PROF
NOMOTO KEN'ICHI DR
OSAKI YOJI DR
SUGIMOTO DAIICHIRO PROF
TRURAN JAMES W JR
TUTUKOV A V DR
WHEELER J CRAIG PROF

Members:
AIAD A PROF.
ANGELOV TRAJKO
ARAI KENZO DR
ARNOULD MARCEL L DR
BAKER NORMAN H PROF
BECKER STEPHEN A
BISNOVATYI-KOGAN G S DR
BODENHEIMER PETER PROF
BOSS ALAN P DR
BURBIDGE GEOFFREY R PROF
CAMERON ALASTAIR G W PROF
CARSON T R DR
CAUGHLAN GEORGEANNE R
CHECHETKIN VALERIJ M DR
CHIU HONG-YEE DR
CHRISTY ROBERT F DR
COWAN JOHN J DR
DAS MRINAL KANTI
DE JAGER CORNELIS PROF
DEINZER W PROF DR
DINGENS P PROF DR
DZIEMBOWSKI WOJCIECH PROF
EGGLETON PETER P DR
EPSTEIN ISADORE PROF
EZER-ERYURT DILHAN PROF
FAULKNER JOHN PROF
FORBES J E DR
FRANTSMAN YU L DR
GALLINO ROBERTO
GIANNONE PIETRO PROF
GLATZMAIER GARY A
GRAHAM ERIC DR
HACHISU IZUMI DR
HERNANZ MARGARITA DR
AIZENMAN MORRIS L DR
ANTIA H M DR
ARIMOTO NOBUO DR
AUDOUZE JEAN PROF
BAYM GORDON ALAN DR
BENZ WILLY
BLUDMAN SIDNEY A PROF
BOEHM KARL-HEINZ PROF
BROWNLEE ROBERT R DR
CALLEBAUT DIRK K DR
CANAL RAMON M DR
CASTELLANI VITTORIO PROF
CHAN KWING LAM
CHEVALIER CLAUDE DR
CHKHIKVADZE IAKOB N
COHEN JEFFREY M DR
COWLING THOMAS G PROF
DAVIS CECIL G JR
DE LOORE CAMIEL PROF
DESPAIN KEITH HOWARD DR
DLUZHNEVSKAYA O B DR
EDWARDS ALAN CH DR
EMINZADE T A DR
ERGMA E V DR
FADEYEV YURI A
FLANNERY BRIAN PAUL DR
FOSSAT ERIC DR
FUJIMOTO MASAYUKI DR
GEROYANNIS VASSILIS S DR
GINGOLD ROBERT ARTHUR DR
GONG SHU-MO
GREGGIO LAURA DR
HAYASHI CHUSHIRO PROF
HILF EBERHARD R H PH D
ANAND S P S DR
APPENZELLER IMMO PROF
ARNETT W DAVID PROF
BAGLIN ANNIE DR
BEAUDET GILLES DR
BERTHOMIEU GABRIELLE DR
BOCCHIA ROMEO DR
BONDI HERMANN PROF SIR
BUCHLER J ROBERT PROF
CALOI VITTORIA DR
CAPUTO FILIPPINA DR
CASTOR JOHN I DR
CHANDRASEKHAR S PROF
CHITRE SHASHIKUMAR M DR
CHRISTENSEN-DALSGAARD J
CONNOLLY LEO PAUL
D'ANTONA FRANCESCA DR
DE GREVE JEAN-PIERRE DR
DEARBORN DAVID PAUL S DR
DEUPREE ROBERT G DR
DURISEN RICHARD H DR
EDWARDS TERRY W
ENDAL ANDREW S DR
ERIGUCHI YOSHIHARU DR
FAULKNER DONALD J DR
FONTAINE GILLES DR
FOWLER WILLIAM A PROF
GABRIEL MAURICE R DR
GHEIDARI S NASSIRI DR
GIRIDHAR SUNETRA DR
GOUPIL MARIE JOSE
GURM HARDEV S PROF
HENRY RICHARD B C DR
HITOTSUYANAGI JUICHI PROF

HOSHI REIUN DR
HOYLE FRED SIR
HUANG RUN-QIAN
HUMPHREYS ROBERTA M PROF
ILIEV ILIAN
IMSHENNIK V S DR
ISAAK GEORGE R PROF
ISERN JORGE DR
ISHIZUKA TOSHIHISA DR
ITOH NAOKI DR
JAMES RICHARD A DR
KAEHLER HELMUTH DR
KAMINISHI KEISUKE PROF
KATO MARIKO
KHOZOV GENNADIJ V
KIGUCHI MASAYOSHI DR
KING DAVID S PROF
KNOLKER MICHAEL DR
KOCHHAR R K DR
KOESTER DETLEV DR
KOTHARI D S DR
KOVETZ ATTAY PROF
KOZLOWSKI MACIEJ DR
KUMAR SHIV S PROF
KUSHWAHA R S PROF
LABAY JAVIER
LAMB DONALD QUINCY JR DR
LAMB SUSAN ANN DR
LARSON RICHARD B PROF
LASKARIDES PAUL G ASSPROF
LASOTA JEAN-PIERRE DR
LATOUR JEAN J
LEBOVITZ NORMAN R PROF
LEPINE JACQUES R D DR
LI HEN
LI ZONG-WEI
LIEBERT JAMES W DR
LINNELL ALBERT P PROF
LITTLETON JOHN E
LIVIO MARIO
MAHESWARAN MURUGESAPILLAI
MALLIK D C V DR
MARX GYORGY PROF.
MASANI A PROF
MASSEVICH ALLA G DR
MATTEUCCI FRANCESCA
MAZUREK THADDEUS JOHN DR
MAZZITELLI ITALO DR
MCCREA J DERMOTT
MELIK-ALAVERDIAN YU DR
MESTEL LEON PROF
MEYER-HOFMEISTER E DR
MICHAUD GEORGES J DR
MITALAS ROMAS ASSOC PROF
MIYAJI SHIGEKI DR
MOELLENHOFF CLAUS DR
MONAGHAN JOSEPH J DR
MOORE DANIEL R DR
MORGAN JOHN ADRIAN
MORRIS STEPHEN C DR
MOSS DAVID L DR
MUELLER EWALD
NADYOZHIN D K DR
NAKAMURA TAKASHI DR
NAKANO TAKENORI DR
NAKAZAWA KIYOSHI DR
NARASIMHA DELAMPADY DR
NARIAI HIDEKAZU PROF
NARITA SHINJI DR
NEWMAN MICHAEL JOHN DR
NISHIDA MINORU PROF
NOELS ARLETTE DR
ODELL ANDREW P
OHYAMA NOBORU PROF
OKAMOTO ISAO DR
OSTRIKER JEREMIAH P PROF
OSWALT TERRY D DR
PACZYNSKI BOHDAN PROF
PAMYATNIKH A A DR
PANDE GIRISH CHANDRA PROF
PAPALOIZOU JOHN C B DR
PEARCE GILLIAN DR
PHILLIPS MARK M DR
PINES DAVID PROF
PINOTSIS ANTONIS D DR
PLAVEC MIREK J PROF
PORFIR'EV V V DR
POVEDA ARCADIO DR
PRENTICE ANDREW J R DR
PRIALNIK-KOVETZ DINA DR
PROVOST JANINE DR
QU QIN-YUE
RAEDLER K H DR
RAMADURAI SOURIRAJA DR
REEVES HUBERT PROF
REIZ ANDERS PROF
RENZINI ALVIO PROF
ROOD ROBERT T DR
ROUSE CARL A DR
ROXBURGH IAN W PROF
RUBEN G PROF DR
SACKMANN I JULIANA DR
SAIO HIDEYUKI DR
SAKASHITA SHIRO PROF
SALPETER EDWIN E PROF
SATO KATSUHIKO PROF
SAVEDOFF MALCOLM P PROF
SAVONIJE GERRIT JAN DR
SCALO JOHN MICHAEL
SCHATTEN KENNETH H DR
SCHATZMAN EVRY PROF
SCHILD HANSRUEDI
SCHOENBERNER DETLEF PROF
SCHRAMM DAVID N PROF
SCHUTZ BERNARD F PROF
SCHWARZSCHILD MARTIN PROF
SCUFLAIRE RICHARD DR
SEARS RICHARD LANGLEY DR
SEIDOV ZAKIR F DR
SENGBUSCH KURT V DR
SHAVIV GIORA DR
SHIBAHASHI HIROMOTO DR
SHIBATA YUKIO DR
SHUSTOV BORIS M DR
SIENKIEWICZ RYSZARD DR
SIGNORE MONIQUE DR
SILVESTRO GIOVANNI
SION EDWARD MICHAEL
SMEYERS PAUL PROF
SMITH ROBERT CONNON DR
SOBOUTI YOUSEF PROF
SOFIA SABATINO PROF
SOUFFRIN PIERRE B DR
SPARKS WARREN M DR
SPIEGEL E DR
SREENIVASAN S RANGA PROF
STARRFIELD SUMNER PROF
STELLINGWERF ROBERT F DR
STIBBS DOUGLAS W N PROF
STRITTMATTER PETER A PROF
SUDA KAZUO PROF
SWEET PETER A PROF
SWEIGART ALLEN V DR
TAAM RONALD EVERETT DR
TAKAHARA MARIKO
TASSOUL MONIQUE DR
TAYLER ROGER J PROF
THIELEMANN FRIEDRICH-KARL
THOMAS HANS-CHRISTOPH DR
TJIN-A-DJIE HERMAN R E DR
TOHLINE JOEL EDWARD
TOOMRE JURI
TRIMBLE VIRGINIA L DR
TUOMINEN ILKKA V DR
TYPHOON LEE
UCHIDA JUICHI DR
ULRICH ROGER K PROF
UNNO WASABURO PROF
UUS UNDO DR
VAN DEN HEUVEL EDWARD P J
VAN DER BORGHT RENE PROF
VAN DER RAAY HERMAN B
VAN HORN HUGH M PROF
VAN RIPER KENNETH A DR
VANDENBERG DON DR
VARDYA M S DR
VAUCLAIR GERARD P DR
VILA SAMUEL C PROF
VILHU OSMI DR
WARD RICHARD A DR
WEAVER THOMAS A DR
WEBBINK RONALD F DR
WEIGERT ALFRED PROF
WEISS NIGEL O DR
WILLSON LEE ANNE DR
WILSON ROBERT E PROF
WINKLER KARL-HEINZ A DR
WOOD PETER R DR
WOOSLEY S E PROF
XIONG DA-RUN

YORKE HAROLD W DR
YUNGELSON LEV R
ZAHN JEAN-PAUL DR
ZHEVAKIN S A PROF DR
ZIOLKOWSKI JANUSZ DR

COMMISSION No. 36

THEORY OF STELLAR ATMOSPHERES

(THEORIE DES ATMOSPHERES STELLAIRES)

President : GRAY DAVID F PROF

Vice-President(s) : KALKOFEN WOLFGANG DR

Organizing Committee: CASSINELLI JOSEPH P DR
CRAM LAWRENCE EDWARD DR
HEARN ANTHONY G DR
LINSKY JEFFREY L DR
PERAIAH ANNAMANENI DR
PRADERIE FRANCOISE DR
SAPAR ARVED DR
SEATON MICHAEL J PROF
TSUJI TAKASHI
WEHRSE RAINER DR

Members:

ABBOTT DAVID C DR
ABHYANKAR KRISHNA D PROF
ALLER LAWRENCE HUGH
ALTROCK RICHARD C DR
ARPIGNY CLAUDE PROF
ATHAY R GRANT DR
AUER LAWRENCE H DR
AUMAN JASON R PROF
AVRETT EUGENE H DR
BAIRD SCOTT R
BASCHEK BODO PROF
BELL ROGER A DR
BERNAT ANDREW PLOUS DR
BERTOUT CLAUDE
BLANCO CARLO DR
BLESS ROBERT C PROF
BODO GIANLUIGI DR
BOEHM KARL-HEINZ PROF
BOEHM-VITENSE ERIKA PROF
BOEHME SIEGFRIED DR
BOESGAARD ANN M PROF
BOWEN GEORGE H DR
BROWN ALEXANDER
BROWN DOUGLAS NASON
BUES IRMELA D DR
CARBON DUANE F DR
CARLSSON MATS DR
CARSON T R DR
CASTOR JOHN I DR
CATALA CLAUDE DR
CAYREL DE STROBEL GIUSA
CAYREL ROGER DR
CHAN KWING LAM
CHEN PEISHENG
CHUGAIJ NIKOLAI N DR
CONTI PETER S DR
COWLEY CHARLES R PROF
CRIVELLARI LUCIO
CUGIER HENRYK DR
CUNY YVETTE J DR
DAVIS CECIL G JR
DELACHE PHILIPPE J DR
DOMKE HELMUT PH D
DOYLE JOHN GERARD
DRAKE STEPHEN A
DRAVINS DAINIS PROF
DUFTON PHILIP L DR
DUPREE ANDREA K DR
EDMONDS FRANK N JR DR
ELSTE GUNTHER H DR
ERIKSSON KJELL DR
EVANGELIDIS E DR
FARAGGIANA ROSANNA PROF
FINN G D DR
FITZPATRICK EDWARD L DR
FONTENLA JUAN MANUEL DR
FOY RENAUD DR
FREIRE FERRERO RUBENS G
FRIEND DAVID B DR
FRISCH HELENE DR
FRISCH URIEL DR
FROESCHLE CHRISTIANE D DR
GAIL HANS-PETER DR
GEBBIE KATHARINE B DR
GOKDOGAN NUZHET PROF
GORDON CHARLOTTE PROF
GRANT IAN P DR
GREENSTEIN J L PROF
GREVESSE N DR
GRININ VLADIMIR P DR
GROTH HANS G PROF DR
GUSSMANN E A DR
GUSTAFSSON BENGT DR
HACK MARGHERITA PROF
HAISCH BERNHARD MICHAEL
HAMMAN WOLF-RAINER
HARTMANN LEE WILLIAM
HARUTYUNIAN HAIK A DR
HEASLEY JAMES NORTON
HEBER ULRICH
HEKELA JAN DR
HEROLD HEINZ
HITOTSUYANAGI JUICHI PROF
HOLWEGER HARTMUT PROF
HOLZER THOMAS EDWARD DR
HOTINLI METIN DR
HOUSE LEWIS L DR
HUBENY IVAN
HUMMER DAVID G DR
HUNGER KURT PROF
HUTCHINGS JOHN B DR
IVANOV VSEVOLOD V DR PROF
JEFFERIES JOHN T DR
JOHNSON HOLLIS R PROF
KADOURI TALIB HADI
KAMP LUCAS WILLEM DR
KANDEL ROBERT S DR
KARP ALAN HERSH DR
KHOKHLOVA V L DR

KLEIN RICHARD I DR
KOLESOV A K DR
KRISHNA SWAMY K S DR
KUMAR SHIV S PROF
LAMBERT DAVID L PROF
LIEBERT JAMES W DR
LYUBIMKOV LEONID S DR
MASSAGLIA SILVANO
MICHAUD GEORGES J DR
MNATSAKANIAN MAMIKON A DR
MUSIELAK ZDZISLAW E DR
NARASIMHA DELAMPADY DR
NIKOGHOSSIAN ARTHUR G DR
ORRALL FRANK Q PROF
PALLAVICINI ROBERTO DR
PECKER JEAN-CLAUDE PROF
POTTASCH STUART R PROF
RACHKOVSKY D N DR
ROSS JOHN E R DR
RYBICKI GEORGE B DR
SCHARMER GOERAN BJARNE
SCHMUTZ WERNER
SEDLMAYER ERWIN DR
SIMON KLAUS PETER
SITNIK G F PROF
SNIJDERS MATTHEUS A J DR
SPIEGEL E DR
SPRUIT HENK C DR
STEIN ROBERT F ASSOC PROF
STROM STEPHEN E
TARAFDAR SHANKAR P DR
TRAVING GERHARD PROF
ULMSCHNEIDER PETER PROF
VAN REGEMORTER HENRI DR
VARDAVAS ILIAS MIHAIL
WATANABE TETSUYA
WELLMANN PETER PROF DR
WILLSON LEE ANNE DR
WOEHL HUBERTUS DR
YANOVITSKIJ EDGARD G DR
ZWAAN CORNELIS PROF DR

KODAIRA KEIICHI PROF
KONTIZAS EVANGELOS DR
KUDRITZKI ROLF-PETER PH D
KURUCZ ROBERT L DR
LANDSTREET JOHN D PROF
LINNELL ALBERT P PROF
MADEJ JERZY
MATSUMOTO MASAMICHI PROF
MIHALAS DIMITRI DR
MUENCH GUIDO PROF
MUTSCHLECNER J PAUL DR
NARIAI KYOJI DR
NORDLUND AKE DR
OXENIUS JOACHIM DR
PANEK ROBERT J DR
PETERS GERALDINE JOAN DR
QUERCI FRANCOIS R DR
RAMSEY LAWRENCE W DR
ROVIRA MARTA GRACIELA
SAITO KUNIJI PROF
SCHMALBERGER DONALD C DR
SCHOENBERNER DETLEF PROF
SHINE RICHARD A DR
SIMON THEODORE
SKUMANICH ANDRE PROF
SOBOLEV V V DR
SPITE FRANCOIS M DR
STALIO ROBERTO DR
STEPIEN KAZIMIERZ DR
SWIHART THOMAS L DR
THOMAS RICHARD N DR
UENO SUEO PROF
UNDERHILL ANNE B DR
VAN'T VEER FRANS DR
VARDYA M S DR
WEBER STEPHEN VANCE
WHITE RICHARD L
WILSON PETER R PROF
WRIGHT KENNETH O DR
YORKE HAROLD W DR

KOESTER DETLEV DR
KRIKORIAN RALPH DR
KUHI LEONARD V PROF
KUSHWAHA R S PROF
LEIBACHER JOHN DR
LIU CAIPIN
MARLBOROUGH J M PROF
MATSUSHIMA SATOSHI DR
MIYAMOTO SIGENORI PROF
MUKAI SONOYO DR
NAGIRNER DMITRIJ I DR
NEFF JOHN S
O'MARA BERNARD J PROF
PAGEL BERNARD E J PROF
PASINETTI LAURA E PROF
PHILLIPS JOHN G PROF
QUERCI MONIQUE DR
REIMERS DIETER PROF
RUTTEN ROBERT J. DR
SAKHIBULLIN NAIL A DR
SCHMID-BURGK J DR PROF
SCHOLZ M PROF
SHIPMAN HENRY L DR
SIMONNEAU EDUARDO DR
SNEZHKO LEONID I
SOUFFRIN PIERRE B DR
SPITE MONIQUE DR
STEFFEN MATTHIAS DR
STIBBS DOUGLAS W N PROF
TAKEDA YOICHI DR
TOOMRE JURI
UESUGI AKIRA DR
UNNO WASABURO PROF
VAN'T VEER-MENNERET CL DR
VIIK TONU DR
WEIDEMANN VOLKER PROF
WICKRAMASINGHE N C PROF
WILSON S J
WYLLER ARNE A PROF
ZAHN JEAN-PAUL DR

COMMISSION No. 37

STAR CLUSTERS AND ASSOCIATIONS

(AMAS STELLAIRES ET ASSOCIATIONS)

President : HARRIS GRETCHEN L H DR

Vice-President(s) : PILACHOWSKI CATHERINE DR

Organizing Committee: CLARIA JUAN DR
HEGGIE DOUGLAS C DR
HESSER JAMES E DR
JANES KENNETH A DR
MERMILLIOD JEAN-CLAUDE DR
SHOBBROOK ROBERT R DR
VANDENBERG DON DR
ZHAO JUN-LIANG

Members:

AARSETH SVERRE J DR
AIAD A PROF.
ALKSNIS ANDREJS DR
BALAZS BELA A DR
BELL ROGER A DR
BOUVIER PIERRE PROF
BYRD GENE G DR
CANNON RUSSELL D DR
CARNEY BRUCE WILLIAM
CHIOSI CESARE S DR
COLIN JACQUES DR
D'ANTONA FRANCESCA DR
DEJONGHE HERWIG BERT DR
DICKENS ROBERT J DR
EINASTO JAAN DR
FALL S MICHAEL DR
FITZGERALD M PIM PROF
FUSI PECCI FLAVIO
GRATTON R G DR
GRINDLAY JONATHAN E DR
HARRIS HUGH C
HASSAN S M DR
HENON MICHEL C DR
HILLS JACK G DR
INAGAKI SHOGO DR
JOSHI U C DR
KILAMBI G C DR
KONTIZAS MARY DR
KUN MARIA DR
LAPASSET EMILIO DR
LLOYD EVANS THOMAS DR
LYNDEN-BELL DONALD PROF
MARRACO HUGO G DR
MAYOR MICHEL DR
MEYLAN GEORGES DR
ABOU-EL-ELLA MOHAMED S DR
ALCAINO GONZALO DR
ALLEN CHRISTINE
BARKHATOVA KLAUDIA PROF
BIJAOUI ALBERT DR
BURKHEAD MARTIN S
CALLEBAUT DIRK K DR
CAPUTO FILIPPINA DR
CASTELLANI VITTORIO PROF
CHRISTIAN CAROL ANN
CUDWORTH KYLE MCCABE DR
DA COSTA GARY STEWART DR
DEMERS SERGE DR
DLUZHNEVSKAYA O B DR
EL-BASSUNY ALAWY A A
FEAST MICHAEL W PROF
FORTE JUAN CARLOS DR
GASCOIGNE S C B DR
GREEN ELIZABETH M. DR
GRUBISSICH C PROF DR
HARRIS WILLIAM E DR
HAWARDEN TIMOTHY G DR
HERBST WILLIAM DR
HUT PIET
ISHIDA KEIICHI PROF
KADLA ZDENKA I DR
KING IVAN R PROF
KRAFT ROBERT P PROF
LADA CHARLES JOSEPH DR
LARSSON-LEANDER G PROF
LODEN LARS OLOF PROF
LYNGA GOSTA DR
MARTINS DONALD HENRY DR
MENON T K PROF
MIKKOLA SEPPO DR
AGEKJAN TATEOS A PROF
ALFARO EMILIO JAVIER
AURIERE MICHEL
BECKER WILHELM PROF
BLAAUW ADRIAAN PROF DR
BUTLER DENNIS DR
CALOI VITTORIA DR
CAPUZZO DOLCETTA ROBERTO
CHAVARRIA-K. CARLOS
CHUN MUN-SUK DR
CUFFEY J MR
DAUBE-KURZEMNIECE I A DR
DI FAZIO ALBERTO
EFREMOV YURY N DR
ELMEGREEN BRUCE GORDON DR
FEINSTEIN ALEJANDRO DR
FREEMAN KENNETH C PROF
GOLAY MARCEL PROF
GRIFFITHS WILLIAM K
HANES DAVID A DR
HARVEL CHRISTOPHER ALVIN
HAZEN MARTHA L DR
HEUDIER JEAN-LOUIS DR
ILLINGWORTH GARTH D DR
JONES DEREK H P DR
KAMP LUCAS WILLEM DR
KONTIZAS EVANGELOS DR
KRON GERALD E DR
LANDOLT ARLO U PROF
LAVAL ANNIE DR
LU PHILLIP K DR
MARKKANEN TAPIO DR
MATTEUCCI FRANCESCA
MENZIES JOHN W DR
MOFFAT ANTHONY F J DR

MOULD JEREMY R
MUZZIO JUAN C PROF
NEMEC JAMES
NESCI ROBERTO
NEWELL EDWARD B DR
NISSEN POUL E PROF
OGURA KATSUO DR
ORTOLANI SERGIO
OSBORN WAYNE DR
OSMAN ANAS MOHAMED DR
PARSAMYAN ELMA S DR
PEDREROS MARIO DR
PENNY ALAN JOHN DR
PETERSON CHARLES JOHN DR
PETROVSKAYA M S DR
PHILIP A G DAVIS
PISKUNOV ANATOLY E
PLATAIS IMANT K DR
POPOVA MALINA D PROF DR
POVEDA ARCADIO DR
PRITCHET CHRISTOPHER J DR
QIAN BO-CHEN
RAJAMOHAN R DR
RAM SAGAR DR
RENZINI ALVIO PROF
RICHER HARVEY B DR
ROSINO LEONIDA PROF
ROTH-HOPPNER MARIA LUISE
ROUNTREE JANET DR
RUPRECHT JAROSLAV DR
RUSSEVA TATJANA
SALUKVADZE G N DR
SAMUS NIKOLAI N DR
SANDERS W L PROF
SAWYER-HOGG HELEN B DR
SCARIA K K DR
SCHILD HANSRUEDI
SCHUSTER HANS-EMIL
SHAROV A S DR
SHAWL STEPHEN J DR
SHER DAVID DR
SIMODA MAHIRO PROF
STETSON PETER B. DR
SZECSENYI-NAGY GABOR DR
TERZAN AGOP DR
THE PIK-SIN PROF
TORNAMBE AMEDEO
TSVETKOVA KATIA
TURNER DAVID G DR
TWAROG BRUCE A
UPGREN ARTHUR R DR
VAN ALTENA WILLIAM F PROF
VAN DEN BERGH SIDNEY PROF
WALKER GORDON A H PROF
WALKER MERLE F PROF
WALLENQUIST AAKE A E PROF
WAN LAI
WARREN WAYNE H JR DR
WEAVER HAROLD F PROF
WEHLAU AMELIA DR
WHITE RAYMOND E DR
WIELEN ROLAND PROF DR
WRAMDEMARK STIG S O DR
WU HSIN-HENG DR
ZINN ROBERT J DR

COMMISSION No. 38

EXCHANGE OF ASTRONOMERS (ECHANGE DES ASTRONOMES)

President : SMITH F GRAHAM PROF

Vice-President(s) : SAHADE JORGE PROF

Organizing Committee:
BOYARCHUK A A DR
FLORSCH ALPHONSE DR
JORGENSEN HENNING E PROF
KOZAI YOSHIHIDE PROF
LEUNG KAM CHING PROF
MUELLER EDITH A PROF
SWARUP GOVIND PROF
WANG SHOU-GUAN

Members:

AL-SABTI ABDUL ADIM DR	ALY M KHAIRY PROF	CACCIN BRUNO
HAUPT HERMANN F PROF	MACRAE DONALD A PROF	MARIK MIKLOS DR.
NHA IL-SEONG DR	NINKOVIC SLOBODAN	OKOYE SAMUEL E PROF
ROUTLY PAUL M DR	RUBEN G PROF DR	TOLBERT CHARLES R DR
VAN DEN HEUVEL EDWARD P J	WOOD F BRADSHAW PROF	YE SHU-HUA

COMMISSION No. 40

RADIO ASTRONOMY (RADIOASTRONOMIE)

President : MEZGER PETER G PROF

Vice-President(s) : MORIMOTO MASAKI DR

Organizing Committee:
BACKER DONALD CH DR
BALDWIN JOHN E DR
BAUDRY ALAIN DR
BOOTH ROY S PROF
FANTI ROBERTO
GUESTEN ROLF
JAUNCEY DAVID L DR
KAIFU NORIO DR
KAPAHI V K DR
MATVEYENKO L I DR
MORAN JAMES M DR
NICOLSON GEORGE D DR
SEAQUIST ERNEST R PROF
VAN DER HULST JAN M DR
YIN QI-FENG

Members:

ABDULLA SHAKER ABDUL AZIZ	ABLES JOHN G DR	ABRAMI ALBERTO PROF
ADE PETER A R DR	AKABANE KENJI A PROF	AKUJOR CHIDI E.
ALEXANDER JOSEPH K	ALEXANDER PAUL DR	ALLEN RONALD J DR
ALLER HUGH D DR	ALLER MARGO F DR	ALTENHOFF WILHELM J DR
ANANTHARAMAIAH K R DR	ANDREW BRYAN H DR	APARICI JUAN DR
ARNAL MARCELO EDMUNDO DR	ASCHWANDEN MARKUS DR	ASSOUSA GEORGE ELIAS DR
AURASS HENRY DR	AVERY LORNE W DR	AVIGNON YVETTE DR
AXON DAVID	BAARS JACOB W M DR	BAART EDWARD E PROF
BAATH LARS B DR	BAGRI DURGADAS S	BALKLAVS A E DR
BALONEK THOMAS J DR	BANHATTI D G DR	BARRETT ALAN H PROF
BARROW COLIN H DR	BARTEL NORBERT HARALD DR	BARTHEL PETER DR
BARVAINIS RICHARD DR	BASH FRANK N PROF	BASU DIPAK DR
BATES RICHARD HEATON T DR	BATTY MICHAEL DR	BECK RAINER
BENN CHRIS R DR	BENNETT CHARLES L DR	BENSON PRISCILLA J DR
BENZ ARNOLD DR	BERGE GLENN L DR	BERKHUIJSEN ELLY M DR
BHANDARI RAJENDRA DR	BHONSLE RAJARAM V PROF	BIEGING JOHN HAROLD DR
BIERMANN PETER L DR	BIGNELL R CARL DR	BIRAUD FRANCOIS DR
BIRKINSHAW MARK	BLAIR DAVID GERALD	BLANDFORD ROGER DAVID DR
BLOEMHOF ERIC E DR	BOCKELEE-MORVAN DOMINIQUE	BOISCHOT ANDRE DR
BOLTON JOHN G	BORIAKOFF VALENTIN	BOTTINELLI LUCETTE DR
BOWERS PHILLIP F	BRACEWELL RONALD N PROF	BRAUDE SEMION YA PROF AG
BREGMAN JACOB D IR	BRIDLE ALAN H PROF	BRINKS ELIAS DR
BRODERICK JOHN DR	BROTEN NORMAN W	BROUW W N DR
BROWNE IAN W A DR	BURBIDGE GEOFFREY R PROF	BURKE BERNARD F DR
CAROUBALOS C A PROF	CARR THOMAS D PROF	CASTETS ALAIN DR
CASWELL JAMES L DR	CATARZI MARCO DR	CERNICHARO JOSE DR
CHAN KWING LAM	CHEN HONGSHENG	CHIKADA YOSHIHIRO DR
CHINI ROLF	CHRISTIANSEN WAYNE A	CHRISTIANSEN WILBUR PROF

CHU HAN-SHU PROF.
CLARK FRANK OLIVER DR
COHEN RAYMOND J DR
COLEMAN PAUL HENRY DR
CONWAY ROBIN G DR
COTTON WILLIAM D Jr
CRANE PATRICK C
CUDABACK DAVID D DR
DAINTREE EDWARD J DR
DAVIES RODNEY D PROF
DE GROOT T DR
DE RUITER HANS RUDOLF
DELANNOY JEAN DR
DESHPANDE AVINASH
DICKEL HELENE R DR
DICKMAN ROBERT L DR
DOUBINSKIJ B A DR
DOWNS GEORGE S DR
DRAVSKIKH A F DR
DULK GEORGE A PROF
EKERS RONALD D DR
ELSMORE BRUCE DR
ENOME SHINZO PROF
ERIKSEN GUNNAR PROF
EWING MARTIN S
FELDMAN PAUL A DR
FERETTI LUIGINA
FINDLAY JOHN W DR
FOMALONT EDWARD B DR
FRATER ROBERT H DR
FRISK URBAN DR
GALT JOHN A DR
GAYLARD MICHAEL JOHN
GELFREIKH GEORGIJ B DR
GERGELY TOMAS ESTEBAN DR
GINZBURG VITALY L PROF
GOLD THOMAS PROF
GOLDWIRE HENRY C JR
GORDON MARK A DR
GOSS W MILLER PROF
GRAHAM DAVID A
GREWING MICHAEL PROF
GULKIS SAMUEL DR
HADDOCK FRED T DR
HAN FU
HANBURY BROWN ROBERT PROF
HARDEE PHILIP
HARTZ THEODORE R DR
HASLAM C GLYN T DR
HAYNES RAYMOND F PROF
HEIDMANN JEAN DR
HENKEL CHRISTIAN
HIGGS LLOYD A DR
HIRABAYASHI HISASHI DR
HO PAUL T P
HOEGBOM JAN A DR
HOLLIS JAN MICHAEL DR
HUCHTMEIER WALTER K DR
HULSBOSCH A N M DR

CLARK BARRY G DR
CLEMENS DAN P DR
COHEN RICHARD S
CONDON JAMES J DR
CORDES JAMES M
COUTREZ RAYMOND A J PROF
CROOM DAVID L DR
CUI ZHENXING
DAISHIDO TSUNEAKI PROF
DAVIS MICHAEL M DR
DE JAGER CORNELIS PROF
DE YOUNG DAVID S DR
DENISSE JEAN-FRANCOIS DR
DESPOIS DIDIER DR
DICKEL JOHN R
DIETER NANNIELOU H DR
DOUGLAS JAMES N PROF
DRAKE FRANK D PROF
DREHER JOHN W
DWARAKANATH K S
ELGAROY OYSTEIN PROF
ELWERT GERHARD PROF
EPSTEIN EUGENE E DR
ESHLEMAN VON R PROF
FEIGELSON ERIC D DR
FELLI MARCELLO DR
FERRARI ATTILIO DR
FLEISCHER ROBERT DR
FORT DAVID NORMAN DR
FRIBERG PER
FUERST ERNST DR
GARAY GUIDO DR
GEBLER KARL-HEINZ DR
GENT HUBERT MR
GHIGO FRANCIS D DR
GIOIA ISABELLA M DR
GOLDSMITH PAUL F DR
GONZE ROGER F J IR
GORGOLEWSKI STANISLAW PR
GOTTESMAN STEPHEN T DR
GREGORINI LORETTA
GUELIN MICHEL DR
GULL STEPHEN F DR
HAGEN JOHN P
HAN WENJUN
HANISCH ROBERT J DR
HARRIS DANIEL E DR
HASCHICK AUBREY
HAYASHI MASAHIKO DR
HAZARD CYRIL DR
HEILES CARL PROF
HEWISH ANTONY PROF
HILF EBERHARD R H PH D
HJALMARSON AKE G DR
HOANG BINH DY DR
HOEGLUND BERTIL PROF
HOOGHOUDT B G IR
HUGHES PHILIP
HUNSTEAD RICHARD W DR

CLARK DAVID H DR
COHEN MARSHALL H PROF
COLE TREVOR WILLIAM PROF
CONKLIN EDWARD K
COSTAIN CARMAN H DR
COVINGTON ARTHUR E
CROVISIER JACQUES
DAGKESAMANSKY RUSTAM D DR
DAVIES JOHN G DR
DAVIS ROBERT J DR
DE LA NOE JEROME DR
DEGAONKAR S S DR
DENT WILLIAM A PROF
DEWDNEY PETER E F DR
DICKEY JOHN M
DIXON ROBERT S DR
DOWNES DENNIS DR
DRAKE STEPHEN A
DUFFETT-SMITH PETER JAMES
DYSON F J DR
ELLIS G R A PROF
EMERSON DAVID
ERICKSON WILLIAM C DR
EVANS KENTON DOWER DR
FEIX GERHARD DR
FELTEN JAMES E DR
FIELD GEORGE B PROF
FLETT ALISTAIR M
FOUQUE PASCAL DR
FRIEDMAN HERBERT DR
FUKUI YASUO DR
GARDNER FRANCIS F DR
GELDZAHLER BERNARD J
GENZEL REINHARD DR
GIBSON DAVID MICHAEL DR
GIOVANNINI GABRIELE
GOLDSTEIN SAMUEL J PROF
GOPALSWAMY N DR
GOSACHINSKIJ I V DR
GOWER J F R DR
GREGORY PHILIP C DR
GUIDICE DONALD A DR
GWINN CARL R DR
HAMILTON P A DR
HANASZ JAN DR
HANKINS TIMOTHY HAMILTON
HARTEN RONALD H DR
HASEGAWA TETSUO DR
HAYNES MARTHA P
HEESCHEN DAVID S DR
HELOU GEORGE DR
HEY JAMES STANLEY DR
HILLS RICHARD E DR
HJELLMING ROBERT M DR
HOBBS ROBERT W DR
HOGG DAVID E DR
HOWARD WILLIAM E III DR
HUGHES VICTOR A PROF
IKHSANOV ROBERT N DR

IKHSANOVA VERA N DR
INATANI JUNJI
INOUE MAKOTO DR
ISHIGURO MASATO PROF.
JAFFE WALTER JOSEPH DR
JANSSEN MICHAEL ALLEN
JENKINS CHARLES R
JENNISON ROGER C PROF
JEWELL PHILIP R DR
JI SHUCHEN DR
JIN SHEN-ZENG
JOHANSSON LARS ERIK B DR
JOHNSON DONALD R DR
JOHNSTON KENNETH J
JOLY FRANCOIS DR
JONAS JUSTIN LEONARD
JONES DAYTON L
JOSHI MOHAN N PROF
KAFTAN MAY A DR
KAHLMANN H C DR
KAHN FRANZ D PROF
KAI KEIZO DR
KAKINUMA TAKAKIYO T PROF
KALBERLA PETER
KAMEYA OSAMU DR
KANG GON IK
KAPAHI VIJAY, K.
KARDASHEV N S DR
KASUGA TAKASHI
KAUFMANN PIERRE PROF
KAWABATA KINAKI PROF
KAZES ILYA DR
KELLERMANN KENNETH I DR
KENDERDINE SIDNEY DR
KERR FRANK J DR
KESTEVEN MICHAEL J L DR
KISLYAKOV ALBERT G DR
KLEIN KARL LUDWIG DR
KLEIN ULRICH
KO HSIEN C PROF
KOJOIAN GABRIEL DR
KOTELNIKOV V A ACAD
KRAUS JOHN D PROF
KREYSA ERNST
KRISHNA GOPAL
KRISHNAMOHAN S DR
KRISHNAN THIRUVENKATA MR
KRONBERG PHILIPP DR
KRUEGEL ENDRIK DR
KUIJPERS H. JAN M.E. DR
KUIPER THOMAS B H DR
KULKARNI PRABHAKAR V PROF
KULKARNI SHRINIVAS R DR
KULKARNI V K DR
KUNDT WOLFGANG PROF DR
KUNDU MUKUL R DR
KURIL-CHIK V N DR
KUS ANDRZEJ JAN DR
KUTNER MARC LESLIE DR
KUZMIN ARKADII D PROF DR
KWOK SUN DR
LADA CHARLES JOSEPH DR
LAING ROBERT
LANG KENNETH R ASST PROF
LANGER WILLIAM DAVID DR
LANTOS PIERRE DR
LARGE MICHAEL I DR
LASENBY ANTHONY
LAWRENCE CHARLES R DR
LE SQUEREN ANNE-MARIE DR
LEBLANC YOLANDE DR
LEGG THOMAS H DR
LEPINE JACQUES R D DR
LEQUEUX JAMES DR
LESTRADE JEAN FRANCOIS DR
LEUNG CHUN MING DR
LEVREAULT RUSSELL M DR
LI CHUN-SHENG
LI GYONG WON
LI HONG-WEI
LIANG SHI-GUANG
LILLEY EDWARD A PROF
LINKE RICHARD ALAN DR
LITTLE LESLIE T DR
LO KWOK-YUNG DR
LOCKE JACK L DR
LOCKMAN FELIX J
LONGAIR M S PROF
LOREN ROBERT BRUCE DR
LORENZ HILMAR
LOVELL SIR BERNARD PROF
LOZINSKAYA TAT'YANA A DR
LU YANG
LUO XIANHAN
LYNE ANDREW G DR
MACCHETTO FERDINANDO DR
MACDONALD GEOFFREY H DR
MACDONALD JAMES
MACHALSKI JERZY DR
MACLEOD JOHN M DR
MACRAE DONALD A PROF
MANCHESTER RICHARD N DR
MANDOLESI NAZZARENO
MARAN STEPHEN P DR
MARCAIDE JUAN-MARIA DR
MARQUES DOS SANTOS P PROF
MARSCHER ALAN PATRICK
MARTIN ROBERT N DR
MASLOWSKI JOZEF DR
MASSON COLIN R
MATHESON DAVID NICHOLAS
MATSAKIS DEMETRIOS N
MATTHEWS HENRY E DR
MATTILA KALEVI DR
MAXWELL ALAN DR
MAYER CORNELL H
MCADAM W BRUCE DR
MCCULLOCH PETER M DR
MCKENNA LAWLOR SUSAN
MCLEAN DONALD J DR
MEBOLD ULRICH DR PROF
MEEKS M LITTLETON DR
MEIER DAVID L
MENON T K PROF
MICHALEC ADAM
MILEY G K DR
MILLS BERNARD Y PROF
MILNE DOUGLAS K DR
MILOGRADOV-TURIN JELENA
MIRABEL IGOR FELIX DR
MITCHELL KENNETH J DR
MOISEEV I G DR
MOLCHANOV A P PROF
MORISON IAN MR
MORITA KAZUHIKO
MORIYAMA FUMIO PROF
MORRAS RICARDO DR.
MORRIS DAVID DR
MORRIS MARK ROOT DR
MUNDY LEE G DR
MURDOCH HUGH S DR
MUTEL ROBERT LUCIEN
MUXLOW THOMAS
MYERS PHILIP C
NADEAU DANIEL DR
NAGNIBEDA VALERY G DR
NAN REN-DONG
NGUYEN-QUANG RIEU DR
O'DEA CHRISTOPHER P DR
O'SULLIVAN JOHN DAVID DR
OKOYE SAMUEL E PROF
ONUORA LESLEY IRENE DR
OORT JAN H PROF
OSTERBROCK DONALD E PROF
OWEN FRAZER NELSON DR
PACHOLCZYK ANDRZEJ G PROF
PADMAN RACHAEL
PADRIELLI LUCIA
PALMER PATRICK E PROF
PANKONIN VERNON LEE DR
PAPAGIANNIS MICHAEL D PRO
PAREDES JOSE MARIA DR
PARIJSKIJ YU N DR
PARKER EDWARD A DR
PARMA PAOLA
PARRISH ALLAN DR.
PASACHOFF JAY M PROF
PAULINY TOTH IVAN K K DR
PAULS THOMAS ALBERT DR
PAYNE DAVID G
PEARSON TIMOTHY J
PEDLAR ALAN DR
PENG YUN-LOU
PENZIAS ARNO A DR
PERLEY RICHARD ALAN
PETERS WILLIAM L III DR
PETTENGILL GORDON H PROF

PHILLIPS THOMAS GOULD DR	PICK MONIQUE DR	PONSONBY JOHN E B DR
POOLEY GUY DR	PORCAS RICHARD DR	PRESTON ROBERT ARTHUR
PREUSS EUGEN DR	PRICE R MARCUS DR	PRIESTER WOLFGANG PROF
PUSCHELL JEFFERY JOHN	QIAN SHAN-JIE	QIU YU-HAI
RADHAKRISHNAN V PROF	RAIMOND ERNST DR	RAMATY REUVEN DR
RAO A PRAMESH DR	RAOULT ANTOINETTE DR	RAY THOMAS P
RAZIN V A DR	READHEAD ANTHONY C S DR	REBER GROTE DR
REICH WOLFGANG	REID MARK JONATHAN DR	REIF KLAUS DR
REYES FRANCISCO DR	RIBES JEAN-CLAUDE DR	RICKARD LEE J DR
RICKETT BARNABY JAMES DR	RIIHIMAA JORMA J DR	RILEY JULIA M DR
ROBERTS DAVID HALL DR	ROBERTS MORTON S DR	ROBERTSON DOUGLAS S
ROBERTSON JAMES GORDON DR	ROBINSON BRIAN J DR	ROBINSON JR RICHARD D DR
RODRIGUEZ LUIS F	ROEDER ROBERT C PROF	ROENNAENG BERNT O DR
ROESER HANS-PETER	ROGER ROBERT S DR	ROGERS ALAN E E DR
ROGSTAD DAVID H DR	ROHLFS K PROF DR	ROMNEY JONATHAN D DR
ROWSON BARRIE DR	RUBIN ROBERT HOWARD	RUBIO MONICA DR
RUDNICK LAWRENCE DR	RYDBECK GUSTAF H B DR	RYDBECK OLOF E H PROF
RYZHKOV NIKOLAI F DR	SAIKIA D J DR	SALPETER EDWIN E PROF
SANAMIAN V A DR	SANDELL GORAN HANS L DR	SANDERS DAVID B DR
SARGENT ANNEILA I	SARMA N V G PROF	SASTRY CH V
SATO FUMIO DR	SAUNDERS RICHARD D.E.	SAVAGE ANN DR
SAWANT HANUMANT S DR	SCALISE JR EUGENIO DR	SCHEUER PETER A G DR
SCHILIZZI RICHARD T DR	SCHLICKEISER REINHARD DR	SCHMIDT MAARTEN PROF
SCHULTZ G V DR	SCHULZ ROLF ANDREAS	SCHWARTZ PHILIP R DR
SCHWARZ ULRICH J DR	SCOTT JOHN S DR	SCOTT PAUL F DR
SEIELSTAD GEORGE A	SEIRADAKIS JOHN HUGH DR	SETTI GIANCARLO PROF
SHAFFER DAVID B DR	SHAKESHAFT JOHN R DR	SHAVER PETER A DR
SHERIDAN K V DR	SHIMMINS ALBERT JOHN	SHOLOMITSKY G B DR
SHUTER WILLIAM L H DR	SIEBER WOLFGANG PH D	SIMON PAUL A DR
SINHA RAMESHWAR P	SKILLMAN EVAN D DR	SLADE MARTIN A III DR
SLEE O B DR	SLYSH VJACHOSLAV I DR	SMITH ALEX G PROF
SMITH DEAN F DR	SMITH F GRAHAM PROF	SMOL'KOV GENNADIJ YA DR
SOBOLEVA N S DR	SOFUE YOSHIAKI PROF	SOROCHENKO R L DR
SPENCER JOHN HOWARD	SPENCER RALPH E DR	SPOELSTRA T A TH DR
SRAMEK RICHARD A DR	STAHR-CARPENTER M DR	STANLEY G J
STANNARD DAVID DR	STEFFEN MATTHIAS DR	STEINBERG JEAN-LOUIS DR
STEWART PAUL DR	STEWART RONALD T MR	STONE R G DR
STROM RICHARD G DR	STRUKOV IGOR A DR	SUKUMAR SUNDARAJAN DR
SULLIVAN WOODRUFF T III	SWARUP GOVIND PROF	SWENSON GEORGE W JR PROF
TABARA HIROTO DR	TAKAGI KOJIRO PROF	TAKAKUBO KEIYA PROF
TAKAKURA TATSUO PROF EMER	TAKANO TOSHIAKI DR	TANAKA RIICHIRO PROF
TARTER JILL C DR	TAYLOR A R DR	TERZIAN YERVANT PROF
THOMASSON PETER DR	THOMPSON A RICHARD DR	THUM CLEMENS DR
TLAMICHA ANTONIN DR	TOFANI GIANNI PROF	TOLBERT CHARLES R DR
TOMASI PAOLO DR	TOVMASSIAN H M DR	TOWNES CHARLES HARD DR
TRITTON KEITH P DR	TROITSKY V S PROF DR	TROLAND THOMAS HUGH
TURLO ZYGMUNT DR	TURNER BARRY E DR	TURNER KENNETH C DR
TURTLE A J DR	UDAL'TSOV V A DR	UKITA NOBUHARU
ULRICH BRUCE T PROF	ULRICH MARIE-HELENE D DR	UNGER STEPHEN DR
UNWIN STEPHEN C	URPO SEPPO I	USON JUAN M DR
VALLEE JACQUES P DR	VALTAOJA ESKO	VALTONEN MAURI J PROF
VAN DE HULST H C PROF DR	VAN DER KRUIT PIETER C DR	VAN DER LAAN H PROF DR
VAN GORKOM JACQUELINE H	VAN NIEUWKOOP J DR IR	VAN WOERDEN HUGO PROF DR
VANDEN BOUT PAUL A	VELUSAMY T DR	VENUGOPAL V R DR
VERON PHILIPPE DR	VERSCHUUR GERRIT L PROF	VERTER FRANCES DR
VINER MELVYN R DR	VIVEKANAND M DR	VOGEL STUART NEWCOMBE DR
WADE CAMPBELL M DR	WALKER ROBERT C DR	WALL JASPER V DR
WALMSLEY C MALCOLM DR	WALSH DENNIS DR	WAN TONG-SHAN
WANG JING-SHENG	WANG SHOU-GUAN	WANNIER PETER GREGORY DR

WARDLE JOHN F C PROF
WARNER PETER J DR
WARWICK JAMES W DR
WEI MINGZHI
WEILER KURT W DR
WELCH WILLIAM J PROF
WELLINGTON KELVIN DR
WENDKER HEINRICH J PROF
WESTERHOUT GART DR
WESTFOLD KEVIN C PROF
WHITEOAK J B DR
WICKRAMASINGHE N C PROF
WIELEBINSKI RICHARD PROF
WILD JOHN PAUL DR
WILKINSON PETER N DR
WILLIS ANTHONY GORDON DR
WILLS BEVERLEY J DR
WILLS DEREK DR
WILLSON ROBERT FREDERICK
WILSON ANDREW S DR
WILSON ROBERT W DR
WILSON THOMAS L DR
WILSON WILLIAM J DR
WINDHORST ROGIER A DR
WINK JOERN ERHARD DR
WINNBERG ANDERS DR
WINNEWISSER GISBERT DR
WITZEL ARNO DR
WOLSZCZAN ALEXANDER DR
WOLTJER LODEWIJK PROF
WOODSWORTH ANDREW W.DR
WOOTTEN HENRY ALWYN
WRIGHT ALAN E DR
WU HUAI-WEI
WU SHENGYIN
WU XINJI
XIA ZHIGUO DR
XU PEI-YUAN
XU ZHI-CAI
YANG JIAN
YOUNIS SAAD M
ZAITSEV VALERII V DR
ZENSUS J.-ANTON DR
ZHANG FU JUN
ZHELEZNIAKOV VLADIMIR V
ZHENG YI-JIA
ZHOU TI-JIAN
ZIEBA STANISLAW DR
ZLOBEC PAOLO DR
ZUCKERMAN BEN M DR

COMMISSION No. 41

HISTORY OF ASTRONOMY (HISTOIRE DE L'ASTRONOMIE)

President : NORTH JOHN DAVID PROF

Vice-President(s) : DEBARBAT SUZANNE V DR

Organizing Committee: DEVORKIN DAVID H
EDDY JOHN A DR
GURSHTEIN ALEXANDER A DR
XI ZE-ZONG

Members:

ANSARI S M RAZAULLAH PROF
ARGYRAKOS JEAN PROF DR
BADOLATI ENNIO
BANDYOPADHYAY A DR
BENSON PRISCILLA J DR
BERENDZEN RICHARD DR
BISHOP ROY L DR
BO SHU-REN
BRUNET JEAN-PIERRE DR
CARLSON JOHN B
CHEN ZUN-GUI
CIMINO MASSIMO A PROF
CORNEJO ALEJANDRO A DR
CUI ZHEN-HUA
DADIC ZARKO DR
DARIUS JON DR
DEEMING TERENCE J DR
DEKKER E DR
DEWHIRST DAVID W DR
DICK STEVEN J
DOBRZYCKI JERZY PROF
EDMONDSON FRANK K PROF
EELSALU HEINO DR
ERPYLEV N P DR
FERNIE J DONALD PROF
FERRARI D'OCCHIEPPO K DR
FIRNEIS MARIA G DR
FLORIDES PETROS S PROF
FREIESLEBEN H C DR
FREITAS MOURAO R R DR
GINGERICH OWEN PROF
HAWKINS GERALD S DR
HAYLI AVRAM PROF
HEGGIE DOUGLAS C DR
HERRMANN DIETER PROF DR
HOSKIN MICHAEL A DR
HOWSE H DEREK
HYSOM EDMUND J
IDLIS G M DR
JACKISCH GERHARD DR
KENNEDY JOHN E PROF
KHROMOV G S DR
KIANG TAO PROF
KING DAVID S PROF
KING HENRY C DR
KRUPP EDWIN C DR
KUNITZSCH PAUL PROF.
LANG KENNETH R ASST PROF
LEVY JACQUES R DR
LI ZHI-SEN
LIU JINYI
LOPES ROSALY DR
MALIN STUART
MCKENNA LAWLOR SUSAN
MEADOWS A JACK PROF
MERLEAU-PONTY J PROF
MOESGAARD KRISTIAN P
NADAL ROBERT
NAKAYAMA SHIGERU DR
NICOLAIDIS EFTHYMIOS DR
OMER GUY C JR PROF
ORCHISTON WAYNE DR
OSTERBROCK DONALD E PROF
PEDERSEN OLAF PROF
PETERSON CHARLES JOHN DR
PETRI WINFRIED PROF DR
PIGATTO LUISA DR
PINGREE DAVID PROF
POGO ALEXANDER DR
PORTER NEIL A PROF
POULLE EMMANUEL PROF
PROKAKIS THEODORE J DR
PROVERBIO EDOARDO PROF
QUAN HEJUN
RONAN COLIN A
RYBKA PRZEMYSLAW DR
SATO NAONOBU PROF
SBIRKOVA-NATCHEVA T
SCHAEFER BRADLEY E DR
SHUKLA K
SIGNORE MONIQUE DR
SOLC MARTIN
STEPHENSON F RICHARD DR
STOEV ALEXEI
SULLIVAN WOODRUFF T III
SUNDMAN ANITA DR
SVOLOPOULOS SOTIRIOS PROF
SWERDLOW NOEL PROF
TATON RENE PROF
THOREN VICTOR E PROF
VERDET JEAN-PIERRE DR
WANG DE-CHANG
WATTENBERG D PROF
WHITAKER EWEN A
WHITE GRAEME LINDSAY DR
WHITROW GERALD JAMES PROF
WILSON CURTIS A
WRIGHT HELEN
XU ZHENTAO
YABUUTI KIYOSHI PROF
YEOMANS DONALD K DR
ZHANF SHOUZHONG DR
ZHANG PEIYU
ZHUANG WEIFENG
ZOSIMOVICH IRINA D

COMMISSION No. 42

CLOSE BINARY STARS (ETOILES DOUBLES SERREES)

President : KOCH ROBERT H DR

Vice-President(s) : KONDO YOJI DR

Organizing Committee:
BUDDING EDWIN DR
CHEREPASHCHUK A M PROF
GIMENEZ ALVARO
HILDITCH RONALD W DR
LEUNG KAM CHING PROF
RAHE JURGEN PROF
RODONO MARCELLO DR
SHAVIV GIORA DR
SMAK JOSEPH I PROF
WEBBINK RONALD F DR
YAMASAKI ATSUMA DR

Members:

ABHYANKAR KRISHNA D PROF
AL-NAIMY HAMID M K DR
ANDERSEN JOHANNES
ANTIPOVA LYUDMILA DR
ANTONOPOULOU E DR
AWADALLA NABIL SHOUKRY DR
BARTOLINI CORRADO
BATESON FRANK M OBE DR
BATH GEOFFREY T DR
BATTEN ALAN H DR
BLAIR WILLIAM P DR
BLITZSTEIN WILLIAM DR
BOLTON C THOMAS PROF
BONAZZOLA SILVANO DR
BOOKMYER BEVERLY B DR
BOPP BERNARD W DR
BRADSTREET DAVID H DR
BRANDI ELISANDE ESTELA DR
BREINHORST ROBERT A DR
BROGLIA PIETRO DR
BROWNLEE ROBERT R DR
BRUHWEILER FRED C JR
BUNNER ALAN N DR
BUSSO MAURIZIO
CALLANAN PAUL DR
CATALANO SANTO DR
CESTER BRUNO PROF
CHAMBLISS CARLSON R DR
CHANMUGAM GANESAR PROF
CHAPMAN ROBERT D DR
CHEN KWAN-YU PROF
CHOCHOL DRAHOMIR
CHOI KYU-HONG
CILLIE G G PROF
CLARIA JUAN DR
CLAUSEN JENS VIGGO LEKTOR
COLLINS GEORGE W II PROF
COWLEY ANNE P DR
CRISTALDI SALVATORE DR
DADAEV ALEKSANDR N DR
DE GREVE JEAN-PIERRE DR
DE GROOT MART DR
DE KORT JULES J DR
DE LOORE CAMIEL PROF
DELGADO ANTONIO JESUS
DEMIRCAN OSMAN DR
DORFI ERNST ANTON DR
DOUGHTY NOEL A DR
DRECHSEL HORST DR
DUERBECK HILMAR W DR
DURISEN RICHARD H DR
DUSCHL WOLFGANG J DR
EATON JOEL A DR
ECHEVARRIA JUAN DR
EGGLETON PETER P DR
FAULKNER JOHN PROF
FEKEL FRANCIS C
FERLUGA STENO DR
FERRARI D'OCCHIEPPO K DR
FERRER OSVALDO EDUARDO DR
FIRMANI CLAUDIO A PROF
FLANNERY BRIAN PAUL DR
FRACASTORO MARIO G PROF
FRANK JUHAN
FRANTSMAN YU L DR
FREDRICK LAURENCE W PROF
FRIEDJUNG MICHAEL DR
GARMANY CATHERINE D DR
GEYER EDWARD H PROF DR
GIANNONE PIETRO PROF
GIBSON DAVID MICHAEL DR
GIOVANNELLI FRANCO DR
GIURICIN GIULIANO
GOLDMAN ITZHAK DR
GRYGAR JIRI DR
GUINAN EDWARD FRANCIS DR
GULLIVER AUSTIN FRASER DR
GURSKY HERBERT DR
GUSEINOV O H PROF
GYLDENKERNE KJELD DR
HADRAVA PETR
HALL DOUGLAS S DR
HAMMERSCHLAG-HENSBERGE G
HANAWA TOMOYUKI DR
HARMANEC PETR DR
HASSALL BARBARA J M DR
HAZLEHURST JOHN DR
HEINTZ WULFF D DR
HENSLER GERHARD
HERCZEG TIBOR J PROF DR
HILL GRAHAM DR
HJELLMING ROBERT M DR
HOFFMANN MARTIN DR
HOLT STEPHEN S
HONEYCUTT R KENT PROF
HORAK TOMAS B DR

HORIUCHI RITOKU DR	HRIVNAK BRUCE J	HUANG RUN-QIAN
HUBE DOUGLAS P DR	HUTCHINGS JOHN B DR	IBANOGLU C DR
IMAMURA JAMES DR	IMBERT MAURICE DR	IRWIN JOHN B PROF
JABBAR SABEH RHAMAN	JASCHEK CARLOS O R PROF	JASNIEWICZ GERARD DR
JOSS PAUL CHRISTOPHER DR	JURKEVICH IGOR DR	KADOURI TALIB HADI
KALUZNY JANUSZ DR	KANDPAL CHANDRA D	KARETNIKOV VALENTIN G R
KAWABATA SHUSAKU PROF	KENYON SCOTT J DR	KITAMURA M PROF
KJURKCHIEVA DIANA DR	KOPAL ZDENEK PROF	KOUBSKY PAVEL
KRAFT ROBERT P PROF	KRAICHEVA ZDRAVSKA DR	KRAUTTER JOACHIM DR
KREINER JERZY MAREK DR	KRIZ SVATOPLUK DR	KRON KATHERINE GORDON
KRUCHINENKO VITALIY G	KRUSZEWSKI ANDRZEJ PROF	KRZEMINSKI WOJCIECH DR
KUMSIASHVILY MZIA I DR	KURPINSKA-WINIARSKA M DR	KVIZ ZDENEK DR
KWEE K K DR	LA DOUS CONSTANZE A DR	LACY CLAUD H DR
LAMB DONALD QUINCY JR DR	LANDOLT ARLO U PROF	LAPASSET EMILIO DR
LARSSON-LEANDER G PROF	LAVROV M I PROF	LI ZHONGYUAN
LINNELL ALBERT P PROF	LIU QINGYAO DR	LIU XUEFU
LUCY LEON B PROF	LYUTY VICTOR M DR	MACDONALD JAMES
MACERONI CARLA	MAGALASHVILI N L DR	MAMMANO AUGUSTO DR
MARDIROSSIAN FABIO	MARILLI ETTORE DR	MARINO BRIAN F ENG
MARTYNOV D YA PROF DR	MATTEI JANET AKYUZ DR	MAUDER HORST PROF DR
MAYER PAVEL DR	MAZEH TSEVI DR	MCCLUSKEY GEORGE E JR DR
MELIA FULVIO DR	MERRILL JOHN E DR	MEYER-HOFMEISTER E DR
MEZZETTI MARINO	MIKOLAJEWSKA JOANNA DR	MIKULASEK ZDENEK DR
MILONE EUGENE F PROF	MINESHIGE SHIN DR	MIYAJI SHIGEKI DR
MOCHNACKI STEPHAN W DR	MORGAN THOMAS H DR	MUMFORD GEORGE S PROF
NAKAMURA YASUHISA	NARIAI KYOJI DR	NATHER R EDWARD
NELSON BURT DR	NEWSOM GERALD H PROF	NHA IL-SEONG DR
NORDSTROM BIRGITTA DR	OH KYU DONG DR	OKAZAKI AKIRA DR
OLIVER JOHN PARKER DR	OLSON EDWARD C PROF	OLSSON-STEEL DUNCAN I DR
OSAKI YOJI DR	PACZYNSKI BOHDAN PROF	PADALIA T D DR
PARK HONG SUH DR	PARTHASARATHY M DR	PATKOS LASZLO DR
PETERS GERALDINE JOAN DR	PICCIONI ADALBERTO	PIIROLA VILPPU E DR
PLAVEC MIREK J PROF	POLIDAN RONALD S	POPPER DANIEL M PROF
PRINGLE JAMES E DR	PUSTYL'NIK IZOLD B DR	QIAO GUOJUN
RAFERT JAMES BRUCE	RAHUNEN TIMO	RAKOS KARL D PROF
REFSDAL S PROF DR	REUNING ERNEST G DR	RITTER HANS DR
ROBB RUSSEL M	ROBERTSON JOHN ALISTAIR	ROBINSON EDWARD LEWIS DR
ROVITHIS PETER DR	ROVITHIS-LIVANIOU HELEN	ROXBURGH IAN W PROF
RUCINSKI SLAWOMIR M DR	SADIK AZIZ R DR	SAHADE JORGE PROF
SAIJO KEIICHI	SANWAL N B DR	SANYAL ASHIT DR
SAVONIJE GERRIT JAN DR	SCALTRITI FRANCO DR	SCARFE COLIN D DR
SCHILLER STEPHEN	SCHMIDT HANS PROF	SCHMIDTKE PAUL C DR
SCHOBER HANS J DR	SCHOEFFEL EBERHARD F DR	SEGGEWISS WILHELM PROF
SEMENIUK IRENA DR	SHAKURA NICHOLAJ I DR	SHEN LIANG-ZHAO
SHU FRANK H PROF	SHUL'BERG A M DR	SIMA ZDISLAV DR
SIMMONS JOHN FRANCIS L	SINVHAL SHAMBHU DAYAL DR.	SISTERO ROBERTO F DR
SMITH ROBERT CONNON DR	SOBIESKI STANLEY DR	SOEDERHJELM STAFFAN DR
SOLHEIM JAN ERIK	SPARKS WARREN M DR	SRIVASTAVA J B DR
SRIVASTAVA RAM KUMAR DR	STARRFIELD SUMNER PROF	STEIMAN-CAMERON THOMAS DR
STEINER JOAO E DR	STENCEL ROBERT EDWARD	STROHMEIER WOLFGANG PROF
SUGIMOTO DAIICHIRO PROF	SUNDMAN ANITA DR	SVECHNIKOVA MARIA A DR
SZAFRANIEC ROZALIA DR	SZKODY PAULA DR	TAAM RONALD EVERETT DR
TAN HUISONG	THOMPSON KEITH DR	TODORAN IOAN DR
TREMKO JOZEF DR	TRIMBLE VIRGINIA L DR	TUTUKOV A V DR
URECHE VASILE DR	VAN DEN HEUVEL EDWARD P J	VAN HAMME WALTER
VAN PARADIJS JOHANNES DR	VAN'T VEER FRANS DR	VAZ LUIZ PAULO RIBEIRO
VETESNIK MIROSLAV DR	VILHU OSMI DR	WADE RICHARD ALAN DR
WALKER RICHARD L	WALKER WILLIAM S G	WALTER KURT PROF DR
WARD MARTIN JOHN	WARGAU WALTER F DR	WARNER BRIAN PROF

WEHLAU WILLIAM H PROF
WEIGERT ALFRED PROF
WEILER EDWARD J DR
WELLMANN PETER PROF DR
WESSELINK ADRIAAN J DR
WILLIAMON RICHARD M
WILLIAMS ROBERT E DR
WILSON ROBERT E PROF
WOOD DAVID B DR
WOOD F BRADSHAW PROF
WRIGHT KENNETH O DR
ZHAI DI-SHENG
ZHANG ER-HE DR
ZHANG JINTONG
ZHOU DAOQI
ZHOU HONG-NAN
ZHU CI-SHENG
ZIOLKOWSKI JANUSZ DR
ZUIDERWIJK EDWARDUS J

COMMISSION No. 44

ASTRONOMY FROM SPACE (L'ASTRONOMIE A PARTIR DE L'ESPACE)

President : JENKINS EDWARD B DR

Vice-President(s) : TRUEMPER JOACHIM PROF

Organizing Committee: BURKE BERNARD F DR
CLARK GEORGE W PROF
FAZIO GIOVANNI G DR
HUTCHINGS JOHN B DR
JORDAN STUART D DR
KONDO YOJI DR
MICHALITSIANOS ANDREW
POUNDS KENNETH A PROF
RAHE JURGEN PROF
SAVAGE BLAIR D DR
SHOLOMITSKY G B DR
SUNYAEV RASHID A DR
TANAKA YASUO PROF
WAMSTEKER WILLEM DR

Members:

ACTON LOREN W DR
AGRAWAL P C DR
AHMAD IMAD ALDEAN DR
ALEXANDER JOSEPH K
ASCHENBACH BERND PH D
AYRES THOMAS R
BALIUNAS SALLIE L
BENEDICT GEORGE F DR
BENNETT CHARLES L DR
BERGERON JACQUELINE A DR
BERNACCA P L PROF
BIANCHI LUCIANA
BLAMONT JACQUES E PROF
BLEEKER JOHAN A M DR IR
BLESS ROBERT C PROF
BOGGESS ALBERT DR
BOGGESS NANCY W DR
BOHLIN RALPH C DR
BOKSENBERG ALEC PROF
BONNET ROGER M DR
BOUGERET J L DR
BOWYER C STUART PROF
BOYARCHUK A A DR
BOYD ROBERT L F PROF SIR
BRANDT JOHN C DR
BRINKMAN BERT C DR
BROWN ALEXANDER
BRUECKNER GUENTER E DR
BRUHWEILER FRED C JR
BRUNER MARILYN E DR
BUMBA VACLAV DR
BUNNER ALAN N DR
BURGER MARIJKE DR
BURTON WILLIAM M
BUTLER C JOHN DR
BUTTERWORTH PAUL
CAMPBELL MURRAY F
CARPENTER KENNETH G DR
CARROLL P KEVIN PROF
CARVER JOHN H PROF
CATURA RICHARD C DR
CHAPMAN ROBERT D DR
CHARLES PHILIP ALLAN
CHOCHOL DRAHOMIR
CHUBB TALBOT A DR
CLARK THOMAS ALAN DR
CODE ARTHUR D
CORDOVA FRANCE A D
COURTES GEORGES PROF
COURVOISIER THIERRY J.-L.
CRANNELL CAROL JO DR
CULHANE LEONARD PROF
DAVIDSEN ARTHUR FALNES DR
DAVIS ROBERT J DR
DE JAGER CORNELIS PROF
DENNIS BRIAN ROY DR
DI COCCO GUIDO
DOLAN JOSEPH F DR
DUNKELMAN LAWRENCE
DUPREE ANDREA K DR
EL-RAEY MOHAMED E DR
ELVIS MARTIN S DR
FABRICANT DANIEL G
FARAGGIANA ROSANNA PROF
FELDMAN PAUL DONALD DR
FERRARI TONIOLO MARCO
FICHTEL CARL E DR
FISHER PHILIP C
FISHMAN GERALD J
FITTON BRIAN DR
FOING BERNARD H DR
FREDGA KERSTIN PROF
FRIEDMAN HERBERT DR
FRISK URBAN DR
FU CHENG-QI
FURNISS IAN
GABRIEL ALAN H
GEZARI DANIEL YSA DR
GIACCONI RICCARDO PROF
GILRA DAYA P DR
GLASER HAROLD DR
GOLD THOMAS PROF
GONDHALEKAR PRABHAKAR DR
GREWING MICHAEL PROF
GREYBER HOWARD D DR
GRIFFITHS RICHARD E DR
GULL THEODORE R DR

GURSKY HERBERT DR
GUSEINOV O H PROF
HACK MARGHERITA PROF
HADDOCK FRED T DR
HALLAM KENNETH L DR
HAN ZHENG-ZHONG
HANG HENG-RONG
HARMS RICHARD JAMES DR
HARTZ THEODORE R DR
HARVEY CHRISTOPHER C DR
HARVEY PAUL MICHAEL DR
HAUSER MICHAEL G DR
HAYAKAWA SATIO PROF
HEARN ANTHONY G DR
HECKATHORN HARRY M
HEISE JOHN DR
HELMKEN HENRY F DR
HENIZE KARL G ASTRONAUT
HENOUX JEAN-CLAUDE DR
HENSBERGE HERMAN
HINTEREGGER HANS E DR
HOFFMAN JEFFREY ALAN DR
HOLBERG JAY B
HOLT STEPHEN S
HOUZIAUX L PROF
HOWARTH IAN DONALD
HOYNG PETER DR
HU WEN-RUI
HUBER MARTIN C E DR
IMHOFF CATHERINE L DR
INOUE HAJIME DR
IYENGAR K V K PROF
JAMAR CLAUDE A J DR
JORDAN CAROLE DR
KAFATOS MINAS DR
KARPINSKIJ VADIM N DR
KASTURIRANGAN K DR
KIMBLE RANDY A DR
KOCH-MIRAMOND LYDIE DR
KRAEMER GERHARD DR
KRAUSHAAR WILLIAM L PROF
KURT V G DR
LAMERS H J G L M DR
LECKRONE DAVID S DR
LEMAIRE PHILIPPE DR
LEWIN WALTER H G PROF
LI TIPEI
LI ZHONGYUAN
LINDBLAD BERTIL A DR
LINSKY JEFFREY L DR
LINSLEY JOHN
LONG KNOX S DR
LOVELL SIR BERNARD PROF
LUEST REIMAR PROF
MA YU-QIAN
MACCHETTO FERDINANDO DR
MALAISE DANIEL J DR
MALITSON HARRIET H MS
MANARA ALESSANDRO A DR
MANDELSTAM S L PROF
MANDOLESI NAZZARENO
MARAN STEPHEN P DR
MARAR T M K
MAROV MIKHAIL YA PROF
MATHER JOHN CROMWELL
MATSUOKA MASARU DR
MCCLUSKEY GEORGE E JR DR
MCCRACKEN KENNETH G DR
MCWHIRTER R W PETER DR
MEAD JAYLEE MONTAGUE DR
MELIA FULVIO DR
MELNICK GARY J
MEWE R DR
MIYAMOTO SIGENORI PROF
MODISETTE JERRY L PROF
MONET DAVID G
MONFILS ANDRE G PROF
MOOS HENRY WARREN DR
MORGAN THOMAS H DR
MORTON DONALD C DR
MUELLER EDITH A PROF
MURDOCK THOMAS LEE
NESS NORMAN F DR
NEUPERT WERNER M DR
NORDH H LENNART DR
NORMAN COLIN A PROF
NOVICK ROBERT
NOVOTNY VACLAV
NOYES ROBERT W PROF
O'MONGAIN EON
ODA MINORU PROF
ODA NAOKI
OERTEL GOETZ K DR
OGAWARA YOSHIAKI
OKUDA TORU
OLTHOF HINDERICUS DR
OWEN TOBIAS C PROF
PACIESAS WILLIAM S DR
PACINI FRANCO PROF
PAPAGIANNIS MICHAEL D PRO
PARKINSON JOHN H DR
PARKINSON WILLIAM H
PERRY PETER M DR
PETERS GERALDINE JOAN DR
PETERSON LAURENCE E PROF
PHILLIPS KENNETH J H
PINKAU K PROF
PIPHER JUDITH L
POLIDAN RONALD S
PRICE STEPHAN DONALD
PROKOF'EV VLADIMIR K PROF
PROSZYNSKI MIECZYSLAW
RAO RAMACHANDRA V PROF
REES MARTIN J PROF
REEVES EDMOND M DR
RENSE WILLIAM A DR
RIGHINI-COHEN GIOVANNA DR
ROMAN NANCY G DR
ROSENDHAL JEFFREY D DR
RUBEN G PROF DR
RUDER HANNS
SAGDEEV ROALD Z DR
SAHADE JORGE PROF
SANDERS WILTON TURNER III
SATO KATSUHIKO PROF
SCHMIDT K H DR
SCHOENEICH W DR
SCHULTZ G V DR
SCHWARTZ DANIEL A DR
SCHWARTZ STEVEN JAY
SELVELLI PIERLUIGI DR
SHEFFIELD CHARLES DR
SHIVANANDAN KANDIAH DR
SILVESTRO GIOVANNI
SIMON PAUL A DR
SIMON PAUL C DR
SMITH BRADFORD A PROF
SMITH HARLAN J PROF
SMITH HOWARD ALAN
SMITH LINDA J
SMITH PETER L DR
SNOW THEODORE P PROF
SOFIA SABATINO PROF
SONNEBORN GEORGE DR
SPADA GIANFRANCO DR
SPEER R J DR
SPITZER LYMAN JR DR
STACHNIK ROBERT V
STAUBERT RUDIGER PROF DR
STECHER THEODORE P
STEINBERG JEAN-LOUIS DR
STEINER JOAO E DR
STENCEL ROBERT EDWARD
STERN ROBERT ALLAN
STIER MARK T
STOCKMAN HERVEY S JR DR
STONE R G DR
SU WAN-ZHEN
TAKAKURA TATSUO PROF EMER
THOMAS ROGER J DR
TOVMASSIAN H M DR
TRAUB WESLEY ARTHUR
TSUNEMI HIROSHI DR
UNDERHILL ANNE B DR
UNDERWOOD JAMES H DR
UPSON WALTER L II DR
VALNICEK BORIS DR
VALTONEN MAURI J PROF
VAN BEEK FRANK PROF DR
VAN DE HULST H C PROF DR
VAN DER HUCHT KAREL A DR
VAN DUINEN R J DR
VAN SPEYBROECK LEON P DR
VIAL JEAN-CLAUDE
VIDAL-MADJAR ALFRED DR
VILHU OSMI DR
VIOTTI ROBERTO DR

WALSH DENNIS DR
WANG SHUI
WARNER JOHN W DR
WEILER EDWARD J DR
WEINBERG J L DR
WESSELIUS PAUL R DR
WESTPHAL JAMES A PROF
WILLIS ALLAN J DR
WILLNER STEVEN PAUL DR
WILSON ROBERT PROF SIR
WRAY JAMES D DR
WU CHI CHAO DR
WUNNER GUENTER
YAMASHITA KOJUN DR
ZARNECKI JAN CHARLES DR
ZOMBECK MARTIN V DR
ZOU HUI-CHENG

COMMISSION No. 45

STELLAR CLASSIFICATION (CLASSIFICATION STELLAIRE)

President : GOLAY MARCEL PROF

Vice-President(s) : MACCONNELL DARRELL J DR

Organizing Committee:
CORBALLY CHRISTOPHER
GARRISON ROBERT F PROF
HOUK NANCY DR
LEVATO ORLANDO HUGO DR
LLOYD EVANS THOMAS DR
OLSEN ERIK H
WALBORN NOLAN R DR
WING ROBERT F PROF
ZDANAVICIUS KAZIMERAS DR

Members:
ALBERS HENRY PROF
ARDEBERG ARNE L PROF
BABU G S D
BAHNG JOHN D R PROF
BARBIER-BROSSAT M DR
BARRY DON C DR
BARTAYA R A DR
BELL ROGER A DR
BIDELMAN WILLIAM P PROF
BLANCO VICTOR M DR
BUSCOMBE WILLIAM PROF
BUSER ROLAND DR
CELIS LEOPOLDO DR
CESTER BRUNO PROF
CHEREPASHCHUK A M PROF
CHRISTY JAMES WALTER DR
CLARIA JUAN DR
COLUZZI REGINA DR
COWLEY ANNE P DR
CRAMPTON DAVID DR
CRAWFORD DAVID L DR
DIVAN LUCIENNE DR
DUFLOT MARCELLE DR
EGRET DANIEL DR
ELVIUS TORD PROF EMERITUS
FEAST MICHAEL W PROF
FEHRENBACH CHARLES PROF
FUKUDA ICHIRO
GERBALDI MICHELE DR
GEYER EDWARD H PROF DR
GLAGOLEVSKIJ JU V DR
GUETTER HARRY HENDRIK
GURZADIAN G A PROF DR
HACK MARGHERITA PROF
HALLAM KENNETH L DR
HAUCK BERNARD PROF
HAYES DONALD S DR
HECK ANDRE DR
HENIZE KARL G ASTRONAUT
HILTNER W ALBERT PROF
HUANG LIN
HUMPHREYS ROBERTA M PROF
JASCHEK CARLOS O R PROF
JASCHEK MERCEDES DR
KEENAN PHILIP C PROF EMER
KHARADZE E K PROF
KRON GERALD E DR
LABHARDT LUKAS
LEE SANG GAK
LODEN KERSTIN R DR
LOW FRANK J DR
LUTZ JULIE H DR
LYNGA GOSTA DR
MAEHARA HIDEO DR
MALARODA STELLA M DR
MCCARTHY MARTIN F DR
MCCLURE ROBERT D PROF
MCNAMARA DELBERT H DR
MEAD JAYLEE MONTAGUE DR
MENDOZA V EUGENIO E DR
MORGAN WILLIAM W PROF
MORGULEFF NINA ING
MOROSSI CARLO
NANDY KASHINATH DR
NICOLET BERNARD
NORTH PIERRE
NOTNI P DR
OJA TARMO PROF
OSBORN WAYNE DR
PARSONS SIDNEY B DR
PASINETTI LAURA E PROF
PERRY CHARLES L DR
PHILIP A G DAVIS
PRESTON GEORGE W DR
RAUTELA B S DR
ROMAN NANCY G DR
ROUNTREE JANET DR
RUDKJOBING MOGENS PROF
SANDULEAK NICHOLAS DR
SANWAL N B DR
SCHILD RUDOLPH E DR
SCHMIDT-KALER TH PROF
SEITTER WALTRAUT C PROF
SHARPLESS STEWART PROF
SINNERSTAD ULF E PROF
SLETTEBAK ARNE PROF
STEINLIN ULI PROF
STEPHENSON C BRUCE PROF
STOCK JURGEN D
STRAIZYS V PROF DR
STROBEL ANDRZEJ DR
UPGREN ARTHUR R DR
WALKER GORDON A H PROF
WARREN WAYNE H JR DR
WESSELIUS PAUL R DR
WESTERLUND BENGT E PROF
WILLIAMS JOHN A DR
WU HSIN-HENG DR
WYCKOFF SUSAN DR
YAMASHITA YASUMASA PROF
YOSS KENNETH M DR

COMMISSION No. 46

TEACHING OF ASTRONOMY

(L'ENSEIGNEMENT DE L'ASTRONOMIE)

President : SANDQVIST AAGE DR

Vice-President(s) : GOUGUENHEIM LUCIENNE

Organizing Committee: HOUZIAUX L PROF
ISOBE SYUZO DR
IWANISZEWSKA CECYLIA DR
KLECZEK JOSIP DR
PASACHOFF JAY M PROF
PERCY JOHN R PROF
ROBBINS R ROBERT PROF
WENTZEL DONAT G DR
WEST RICHARD M DR
ZEALEY WILLIAM J DR

Members:

ACKER AGNES PROF DR	AIAD A PROF.	ANDRILLAT HENRI L PROF
ANSARI S M RAZAULLAH PROF	BACALOV MIHAIL	BENSON PRISCILLA J DR
BOCHONKO D RICHARD DR	BOTEZ ELVIRA DR	BOTTINELLI LUCETTE DR
BRAES L L E DR	BRIEVA EDUARDO PROF	BROSCH NOAH DR
BUDDING EDWIN DR	BUSCOMBE WILLIAM PROF	CALVET NURIA DR
CATALA MARIA ASUNCION DR	CHAMBERLAIN JOSEPH M DR	CODINA LANDABERRY SAYD J
COUPER HEATHER MISS	CUI ZHEN-HUA	DARIUS JON DR
DUPUY DAVID L DR	DUVAL MARIE-FRANCE	EMERSON DAVID
FAIRALL ANTHONY P PROF	FENG KE-JIA	FERNANDEZ JULIO A DR
FERNANDEZ-FIGUEROA M J DR	FERRAZ-MELLO S PROF DR	FIENBERG RICHARD T DR
FIERRO JULIETA	GALLINO ROBERTO	GERBALDI MICHELE DR
GINGERICH OWEN PROF	GURM HARDEV S PROF	HAUPT HERMANN F PROF
HEUDIER JEAN-LOUIS DR	HIDAJAT BAMBANG PROF DR	HOFF DARREL BARTON
ILYAS MOHAMMAD DR	IMPEY CHRISTOPHER D DR	JARRETT ALAN H PROF
KELLER HANS ULRICH DR	KENNEDY JOHN E PROF	KITCHIN CHRISTOPHER R DR
KONONOVICH EDWARD V DR	KOURGANOFF VLADIMIR PROF	KREINER JERZY MAREK DR
KRUPP EDWIN C DR	LAGO MARIA TERESA V T PR.	LITTLE-MARENIN IRENE R DR
LOMB NICHOLAS RALPH DR	MA XING-YUAN	MACIEL WALTER J DR
MADDISON RONALD CH DR	MARSH JULIAN C D	MARTINET LOUIS PROF
MAVRIDIS L N PROF	MAZA JOSE	MCCARTHY MARTIN F DR
MCNALLY DEREK DR	MOMCHEV GOSPODIN	MOREELS GUY DR
MUZZIO JUAN C PROF	NICOLOV NIKOLAI S DR	NOELS ARLETTE DR
OJA HEIKKI DR	OKOYE SAMUEL E PROF	OLSEN FOGH H J
OSBORN WAYNE DR	OSORIO JOSE J S P PROF	OTHMAN MAZLAN
OWAKI NAOAKI DR	PARISOT JEAN-PAUL	PROVERBIO EDOARDO PROF
RAMADURAI SOURIRAJA DR	RIGUTTI MARIO PROF	ROBINSON LEIF J
RODGERS ALEX W DR	ROSLUND CURT DR	ROY ARCHIE E PROF
SAFKO JOHN L	SANAHUJA BLAS	SAXENA P P DR
SBIRKOVA-NATCHEVA T	SCHLEICHER DAVID G DR	SCHLOSSER WOLFHARD PROF
SCHMIDT THOMAS DR	SCHMITTER EDWARD F DR	SCHROEDER DANIEL J PROF
SHEN CHUN-SHAN	SHIPMAN HENRY L DR	SIROKY JAROMIR DR
SOLHEIM JAN ERIK	STEFL VLADIMIR	STENHOLM LARS
STOEV ALEXEI	SVESTKA JIRI DR	SZECSENYI-NAGY GABOR DR

TABORDA JOSE ROSA DR
TORRES-PEIMBERT SILVIA DR
TROCHE-BOGGINO A E DR
VAUCLAIR SYLVIE D DR
VLADIMIROV SIMEON
VUJNOVIC VLADIS DR
WILLIAMON RICHARD M
WOO JONG OK
ZEILIK MICHAEL II DR
ZIMMERMANN HELMUT DR

COMMISSION No. 47

COSMOLOGY (COSMOLOGIE)

President : SATO KATSUHIKO PROF

Vice-President(s) : PARTRIDGE ROBERT B PROF

Organizing Committee: DRESSLER ALAN
FANG LI-ZHI
NARLIKAR JAYANT V PROF
REES MARTIN J PROF
REEVES HUBERT PROF
SETTI GIANCARLO PROF
SHANDARIN SERGEI F DR
SHAVER PETER A DR
TRIMBLE VIRGINIA L DR

Members:

ALFVEN HANNES PROF	ALLAN PETER M	ANDRILLAT HENRI L PROF
AUDOUZE JEAN PROF	AULUCK FAQIR CHAND PROF	AZUMA TAKAHIRO DR
BALDWIN JOHN E DR	BARBERIS BRUNO	BARDEEN JAMES M PROF
BARNOTHY JENO DR PROF	BARROW JOHN DAVID	BASU DIPAK DR
BECKMAN JOHN E PROF	BEL NICOLE J DR	BELINSKY VLADIMIR DR
BENNETT CHARLES L DR	BERGERON JACQUELINE A DR	BERTOLA FRANCESCO PROF
BHAVSAR SUKETU P	BICKNELL GEOFFREY V DR	BIGNAMI GIOVANNI F
BIRKINSHAW MARK	BLUDMAN SIDNEY A PROF	BOKSENBERG ALEC PROF
BOND JOHN RICHARD	BONDI HERMANN PROF SIR	BONNOR W B PROF
BOYLE BRIAN DR	BRECHER KENNETH PROF	BURBIDGE GEOFFREY R PROF
CALVANI MASSIMO DR	CARR BERNARD JOHN	CAVALIERE ALFONSO G PROF
CHANG KYONGAE DR	CHEN JIAN-SHENG	CHENG FU-HUA
CHENG FU-ZHEN	CHINCARINI GUIDO L DR	CHITRE DATTAKUMAR M DR
CHU YAOQUAN	CLARIA JUAN DR	COCKE WILLIAM JOHN PROF
COHEN JEFFREY M DR	COHEN ROSS D DR	CONDON JAMES J DR
DADHICH NARESH DR	DANESE LUIGI DR	DAS P K DR
DATTA BHASKAR DR	DAVIDSON WILLIAM PROF	DAVIES PAUL CHARLES W
DAVIES ROGER L DR	DAVIS MARC DR	DAVIS MICHAEL M DR
DE RUITER HANS RUDOLF	DE VAUCOULEURS GERARD PR	DE ZOTTI GIANFRANCO DR
DEKEL AVISHAI	DEMARET JACQUES DR	DICKE ROBERT H PROF
DIONYSIOU DEMETRIOS PROF	DOROSHKEVICH A G DR	DULTZIN-HACYAN D. DR
DYER CHARLES CHESTER DR	EFSTATHIOU GEORGE	EHLERS JURGEN PROF
EINASTO JAAN DR	ELLIS GEORGE F R PROF	ELLIS RICHARD S
ELVIS MARTIN S DR	FABER SANDRA M PROF	FALK SYDNEY W JR DR
FALL S MICHAEL DR	FELTEN JAMES E DR	FIELD GEORGE B PROF
FILIPPENKO ALEXEI V DR	FLORIDES PETROS S PROF	FOCARDI PAOLA DR
FONG RICHARD	FORD HOLLAND C RES PROF	FORMAN WILLIAM RICHARD DR
FOUQUE PASCAL DR	FRANCESCHINI ALBERTO	FRENK CARLOS S
FUJIMOTO MITSUAKI DR	FUKUGITA MASATAKA DR	FUKUI TAKAO DR
GALLETTO DIONIGI	GELLER MARGARET JOAN	GIALLONGO EMANUELE DR
GIURICIN GIULIANO	GODART ODON PROF	GOLD THOMAS PROF
GOLDSMITH DONALD W. DR.	GORET PHILIPPE DR	GRATTON LIVIO PROF
GREGORY STEPHEN ALBERT DR	GREYBER HOWARD D DR	GRISHCHUK L P DR
GUNN JAMES E PROF	HACYAN SHAHEN DR.	HARA KEN NOSUKE DR

HARDY EDUARDO	HARMS RICHARD JAMES DR	HARRISON EDWARD R PROF
HAWKING STEPHEN W PROF	HAYAKAWA SATIO PROF	HAYASHI CHUSHIRO PROF
HE XIANG-TAO	HEAVENS ALAN DR	HEIDMANN JEAN DR
HELLER MICHAEL PROF	HEWETT PAUL	HEWITT ADELAIDE
HOYLE FRED SIR	HU ESTHER M DR	HUCHRA JOHN PETER DR
ICKE VINCENT DR	IKEUCHI SATORU DR	IMPEY CHRISTOPHER D DR
IYER B R DR	JAROSZYNSKI MICHAL	JAUNCEY DAVID L DR
JIANG SHUDING	JONES BERNARD J T DR	JOSHI MOHAN N PROF
JUNKKARINEN VESA T DR	JUSZKIEWICZ ROMAN	KAPOOR RAMESH CHANDER
KARACHENTSEV I D DR	KASPER U DR	KATO SHOJI PROF
KAWABATA KINAKI PROF	KELLERMANN KENNETH I DR	KEMBHAVI AJIT K
KIM JIK SU	KODAMA HIDEO	KOLB EDWARD W DR
KOO DAVID C-Y DR	KORMENDY JOHN DR	KOVETZ ATTAY PROF
KOZLOVSKY B Z DR	KRASINSKI ANDRZEJ PROF.	KRISS GERARD A DR
KUNTH DANIEL	KUSTAANHEIMO PAUL E PROF	LACHIEZE-REY MARC
LAKE KAYLL WILLIAM DR	LASOTA JEAN-PIERRE DR	LAUSBERG ANDRE DR
LAYZER DAVID PROF	LEQUEUX JAMES DR	LI ZHI-FANG
LIEBSCHER DIERCK-E DR	LILLY SIMON J DR	LIU LIAO
LIU YONG-ZHEN	LONGAIR M S PROF	LU TAN
LUMINET JEAN-PIERRE	LYNDEN-BELL DONALD PROF	MACCALLUM MALCOLM A H
MAEDA KEI-ICHI DR	MANDOLESI NAZZARENO	MARANO BRUNO
MARDIROSSIAN FABIO	MAREK JOHN	MATERNE JUERGEN DR
MATHER JOHN CROMWELL	MATSUMOTO TOSHIO DR	MATZNER RICHARD A PROF
MAVRIDES STAMATIA DR	MCCREA J DERMOTT	MCCREA WILLIAM SIR
MELOTT ADRIAN L PROF.	MERAT PARVIZ	MERIGHI ROBERTO DR
MESZAROS ATTILA DR	MESZAROS PETER DR	MEYER DAVID M DR
MEZZETTI MARINO	MISNER CHARLES W PROF	MORRISON PHILIP PROF
MULLER RICHARD A	NARASIMHA DELAMPADY DR	NARIAI HIDEKAZU PROF
NEEMAN YUVAL PROF	NISHIDA MINORU PROF	NOERDLINGER PETER D PROF
NOONAN THOMAS W PROF	NORMAN COLIN A PROF	NOTTALE LAURENT
NOVELLO MARIO DR	NOVIKOV I D DR	O'CONNELL ROBERT WEST DR
OEMLER AUGUSTUS JR DR	OKOYE SAMUEL E PROF	OMER GUY C JR PROF
OMNES ROLAND PROF	OORT JAN H PROF	OZERNOY LEONID M PROF
OZSVATH I PROF	PAAL GYORGY DR	PACHNER JAROSLAV PROF
PADMANABHAN T DR	PADRIELLI LUCIA	PAGE DON NELSON
PAN RONG-SHI	PEACOCK JOHN ANDREW	PECKER JEAN-CLAUDE PROF
PEEBLES P JAMES E	PENZIAS ARNO A DR	PERRYMAN MICHAEL A C
PERSIDES SOTIRIOS C	PETERSON BRUCE A DR	PETROSIAN VAHE PROF
PRESS WILLIAM H DR	PUGET JEAN-LOUP DR	QU QIN-YUE
RAMELLA MASSIMO	RAYCHAUDHURI AMALKUMAR DR	RINDLER WOLFGANG PROF
RIVOLO ARTHUR REX	ROBERTS DAVID HALL DR	ROBINSON I PROF
ROEDER ROBERT C PROF	ROSQUIST KJELL	ROWAN-ROBINSON MICHAEL DR
ROXBURGH IAN W PROF	RUBIN VERA C DR	RUDDY VINCENT P DR
RUDNICK LAWRENCE DR	SAAR ENN DR	SALVADOR-SOLE EDUARDO
SANZ JOSE L DR	SAPAR ARVED DR	SARGENT WALLACE L W DR
SASAKI MISAO	SATO HUMITAKA PROF	SAVAGE ANN DR
SCHATZMAN EVRY PROF	SCHEUER PETER A G DR	SCHMIDT MAARTEN PROF
SCHNEIDER JEAN	SCHNEIDER PETER DR	SCHRAMM DAVID N PROF
SCHUECKING E L DR	SCHULTZ G V DR	SCIAMA DENNIS W DR
SCOTT ELIZABETH L PROF	SEGAL IRVING E DR	SEIDEN PHILIP E
SEIELSTAD GEORGE A	SERSIC J L DR	SHAVIV GIORA DR
SHAYA EDWARD J DR	SHIVANANDAN KANDIAH DR	SIGNORE MONIQUE DR
SILK JOSEPH I PROF	SIMON RENE L E PROF	SISTERO ROBERTO F DR
SMITH HARDING E JR DR	SMITH RODNEY M DR	SMOOT III GEORGE F.
SOKOLOWSKI LECH	SPYROU NICOLAOS PROF	STECKER FLOYD W DR
STEIGMAN GARY PROF	STEWART JOHN MALCOLM DR	STOEGER WILLIAM R DR
STRUBLE MITCHELL F	STRUKOV IGOR A DR	SUBRAHMANYA C R
SUNYAEV RASHID A DR	SURDEJ JEAN M G	SZALAY ALEX DR
TAMMANN G ANDREAS PROF DR	TANABE KENJI DR	TARTER JILL C DR

TAUBER GERALD E PROF
TAYLER ROGER J PROF
THOMPSON LAIRD A DR
THUAN TRINH XUAN DR
TIFFT WILLIAM G PROF
TIPLER FRANK JENNINGS DR
TOMIMATSU AKIRA DR
TOMITA KENJI PROF
TREDER H J PROF DR
TREMAINE SCOTT DUNCAN
TREVESE DARIO
TULLY RICHARD BRENT DR
TURNER EDWIN L DR
TURNER MICHAEL S
TYSON JOHN A DR
TYTLER DAVID DR
USON JUAN M DR
VAGNETTI FAUSTO
VAIDYA P C PROF
VAN DER LAAN H PROF DR
VANYSEK VLADIMIR PROF
VETTOLANI GIAMPAOLO
VISHNIAC ETHAN T
VISHVESHWARA C V PROF
VOGLIS NIKOS DR
VON BORZESZKOWSKI H H DR
WAGONER ROBERT V PROF
WAINWRIGHT JOHN DR
WANG RENCHUAN
WEBSTER ADRIAN S DR
WEBSTER RACHEL
WEINBERG STEVEN DR
WESSON PAUL S DR
WHEELER JOHN A DR
WHITE SIMON DAVID MANION
WHITROW GERALD JAMES PROF
WILKINSON DAVID T
WILL CLIFFORD M DR
WILSON ALBERT G DR
WINDHORST ROGIER A DR
WOLTJER LODEWIJK PROF
WRIGHT EDWARD L DR
XANTHOPOULOS B C DR
XIANG SHOUPING
XIAO XING HUA
YANG LAN-TIAN
ZAMORANI GIOVANNI
ZEL'MANOV A L DR
ZHANG JIA-LU
ZHANG ZHEN-JIU
ZHOU YOU-YUAN
ZHU SHI-CHANG
ZHU XINGFENG
ZIEBA STANISLAW DR
ZOU ZHEN-LONG
ZUIDERWIJK EDWARDUS J

COMMISSION No. 48

HIGH-ENERGY ASTROPHYSICS

(ASTROPHYSIQUE DES HAUTES ENERGIES)

President : SUNYAEV RASHID A DR

Vice-President(s) : OSTRIKER JEREMIAH P PROF

Organizing Committee:
CESARSKY CATHERINE J DR
CLARK GEORGE W PROF
GIACCONI RICCARDO PROF
PACINI FRANCO PROF
QU QIN-YUE
SALPETER EDWIN E PROF
SCHEUER PETER A G DR
SCHRAMM DAVID N PROF
TRIMBLE VIRGINIA L DR
TRUEMPER JOACHIM PROF
WOLFENDALE ARNOLD W PROF
WOLTJER LODEWIJK PROF

Members:

ABRAMOWICZ MAREK DR	ADAMS DAVID J DR	AGRAWAL P C DR
AHLUWALIA HARJIT SINGH DR	ALFVEN HANNES PROF	APPARAO K M V DR
ARNAUD MONIQUE	ARNOULD MARCEL L DR	ARONS JONATHAN
ASCHENBACH BERND PH D	ASSEO ESTELLE DR	AUDOUZE JEAN PROF
AXFORD W IAN PROF	BAAN WILLEM A	BARNOTHY JENO DR PROF
BASU DIPAK DR	BAYM GORDON ALAN DR	BECKER ROBERT HOWARD
BEGELMAN MITCHELL CRAIG	BENFORD GREGORY DR	BERGERON JACQUELINE A DR
BICKNELL GEOFFREY V DR	BIERMANN PETER L DR	BIGNAMI GIOVANNI F
BISWAS SUKUMAR DR	BLANDFORD ROGER DAVID DR	BLEEKER JOHAN A M DR IR
BLUDMAN SIDNEY A PROF	BONAZZOLA SILVANO DR	BONOMETTO SILVIO A DR
BOYD ROBERT L F PROF SIR	BRECHER KENNETH PROF	BUNNER ALAN N DR
BURBIDGE GEOFFREY R PROF	BURROWS ADAM SETH	CAMERON ALASTAIR G W PROF
CASH WEBSTER C JR	CASSE MICHEL DR	CATURA RICHARD C DR
CAUGHLAN GEORGEANNE R	CAVALIERE ALFONSO G PROF	CHANDRASEKHAR S PROF
CHECHETKIN VALERIJ M DR	CHIAN ABRAHAM CHIAN-LONG	CHITRE SHASHIKUMAR M DR
CHUBB TALBOT A DR	CHUPP EDWARD L DR	COHEN JEFFREY M DR
COLLIN-SOUFFRIN SUZY DR	CONDON JAMES J DR	COWIE LENNOX LAUCHLAN DR
COWSIK RAMANATH	CRUISE ADRIAN MICHAEL DR	CULHANE LEONARD PROF
CURIR ANNA	DA COSTA JOSE MARQUES DR	DADHICH NARESH DR
DAHLE S V DR	DAUTCOURT G DR	DAVIDSEN ARTHUR FALNES DR
DAVIDSON WILLIAM PROF	DAVIS LEVERETT JR PROF	DAVIS MICHAEL M DR
DE FELICE FERNANDO DR	DE GRAAF T DR	DE YOUNG DAVID S DR
DEBRUNNER HERMANN DR	DENNIS BRIAN ROY DR	DEWITT BRYCE S DR
DICKE ROBERT H PROF	DISNEY MICHAEL J PROF	DOLAN JOSEPH F DR
DRAKE FRANK D PROF	DRURY LUKE O'CONNOR DR	DUORAH HIRA LAL DR
DUROUCHOUX PHILIPPE	DUTHIE JOSEPH G PROF	EDWARDS PAUL J DR
EICHLER DAVID DR	EILEK JEAN	ELVIS MARTIN S DR
EVANS W DOYLE	FABIAN ANDREW C DR	FANG LI-ZHI
FAZIO GIOVANNI G DR	FELTEN JAMES E DR	FENTON K B DR
FERRARI ATTILIO DR	FICHTEL CARL E DR	FIELD GEORGE B PROF
FISHER PHILIP C	FORMAN WILLIAM RICHARD DR	FOWLER WILLIAM A PROF

FRANCESCHINI ALBERTO FRANDSEN SOEREN PROF FRANK JUHAN
FRANSSON CLAES FRIEDMAN HERBERT DR GAISSER THOMAS K
GALEOTTI PIERO PROF GARMIRE GORDON P PROF GINZBURG VITALY L PROF
GOLD THOMAS PROF GOLDSMITH DONALD W. DR. GONZALES-A WALTER D DR
GREISEN KENNETH I PROF GREWING MICHAEL PROF GREYBER HOWARD D DR
GRIFFITHS RICHARD E DR GRINDLAY JONATHAN E DR GUNN JAMES E PROF
GURSKY HERBERT DR GUSEINOV O H PROF HALL ANDREW NORMAN
HANG HENG-RONG HARWIT MARTIN PROF HAUBOLD HANS JOACHIM
HAWKING STEPHEN W PROF HAYAKAWA SATIO PROF HAYMES ROBERT C PROF
HEISE JOHN DR HELFAND DAVID JOHN HENRIKSEN RICHARD N DR
HENRY RICHARD C. PROF. HOANG BINH DY DR HOFFMAN JEFFREY ALAN DR
HOLLOWAY NIGEL J DR HOYLE FRED SIR HUANG KE-LIANG
ICHIMARU SETSUO DR INOUE HAJIME DR IPSER JAMES R PROF
ISRAEL WERNER PROF ITO KENSAI A PROF JACKSON JOHN CHARLES DR
JAFFE WALTER JOSEPH DR JELLEY JOHN V PHD JOKIPII J R PROF
JONES FRANK CULVER DR JONES THOMAS WALTER DR JOSS PAUL CHRISTOPHER DR
KAFKA PETER KAHN FRANZ D PROF KAPOOR RAMESH CHANDER
KATZ JONATHAN I KELLERMANN KENNETH I DR KELLOGG EDWIN M DR
KEMBHAVI AJIT K KIRK JOHN DR KLINKHAMER FRANS DR
KOCHAROV GRANT E PROF KOCH-MIRAMOND LYDIE DR KOLB EDWARD W DR
KONDO MASAAKI DR KOYAMA KATSUJI KOZLOWSKI MACIEJ DR
KREISEL E PROF KRISTIANSSON KRISTER PROF KULSRUD RUSSELL M DR
KUNDT WOLFGANG PROF DR KURT V G DR LAMB DONALD QUINCY JR DR
LAMB FREDERICK K PROF LAMB SUSAN ANN DR LAMPTON MICHAEL
LASHER GORDON JEWETT DR LATTIMER JAMES M DR LEA SUSAN MAUREEN DR
LI QI-BIN LI TIPEI LI YUAN-JIE
LI ZONG-WEI LINSLEY JOHN LIU JINYI
LIU RU-LIANG LONGAIR M S PROF LOVELACE RICHARD V E DR
LU TAN LUEST REIMAR PROF LUMINET JEAN-PIERRE
LYNDEN-BELL DONALD PROF MA YU-QIAN MACCACARO TOMMASO DR
MACCAGNI DARIO MACCHETTO FERDINANDO DR MARTIN INACIO MALMONGE DR
MASON GLENN M MASON KEITH OWEN MATSUOKA MASARU DR
MAZUREK THADDEUS JOHN DR MCBREEN BRIAN PHILIP DR MCCRAY RICHARD DR
MELROSE DONALD B PROF MESTEL LEON PROF MESZAROS PETER DR
MEYER FRIEDRICH DR MEYER JEAN-PAUL DR MICHEL F CURTIS PROF
MILLER JOHN C DR MIYAJI SHIGEKI DR MIYAMOTO SIGENORI PROF
MORRISON PHILIP PROF MURAKAMI TOSHIO NAIDENOV VICTOR O
NEEMAN YUVAL PROF NITYANANDA R DR NOMOTO KEN'ICHI DR
NORMAN COLIN A PROF NOVICK ROBERT NULSEN PAUL DR
ODA MINORU PROF OGAWARA YOSHIAKI OKOYE SAMUEL E PROF
OZERNOY LEONID M PROF O'CONNELL ROBERT F PROF O'SULLIVAN DENIS F
PACHOLCZYK ANDRZEJ G PROF PAGE CLIVE G DR PALUMBO GIORGIO G C DR
PARKER EUGENE N PARKINSON JOHN H DR PAULINY TOTH IVAN K K DR
PENG QIU-HE PEROLA GIUSEPPE C DR PETERSON BRUCE A DR
PETERSON LAURENCE E PROF PETROSIAN VAHE PROF PIDDINGTON JACK H RES FEL
PINKAU K PROF PORTER NEIL A PROF POUNDS KENNETH A PROF
PRASANNA A R DR PREUSS EUGEN DR PROTHEROE RAYMOND J DR
QUINTANA HERNAN DR RADHAKRISHNAN V PROF RAMADURAI SOURIRAJA DR
RAUBENHEIMER BAREND C PR RAZDAN HIRALAL REES MARTIN J PROF
REEVES HUBERT PROF RENGARAJAN T N DR ROSNER ROBERT
ROSSI BRUNO B PROF RUFFINI REMO SALVATI MARCO
SANDERS WILTON TURNER III SARTORI LEO PROF SASLAW WILLIAM C PROF
SAVEDOFF MALCOLM P PROF SCARGLE JEFFREY D DR SCHATTEN KENNETH H DR
SCHATZMAN EVRY PROF SCHILIZZI RICHARD T DR SCHNOPPER HERBERT W DR
SCHREIER ETHAN J DR SCHWARTZ DANIEL A DR SCIAMA DENNIS W DR
SCOTT JOHN S DR SEIELSTAD GEORGE A SETTI GIANCARLO PROF
SEWARD FREDERICK D SHAHAM JACOB PROF SHAKURA NICHOLAJ I DR
SHAPIRO MAURICE M PROF SHAVER PETER A DR SHAVIV GIORA DR
SHIELDS GREGORY A DR SHUKRE C S DR SIGNORE MONIQUE DR

SIKORA MAREK
SILBERBERG REIN DR
SKILLING JOHN DR
SMITH BARHAM W DR
SMITH F GRAHAM PROF
SOFIA SABATINO PROF
SPADA GIANFRANCO DR
STAUBERT RUDIGER PROF DR
STECKER FLOYD W DR
STEIGMAN GARY PROF
STEINER JOAO E DR
STEPANIAN A A DR
STEPHENS S A DR
STOCKMAN HERVEY S JR DR
STRONG IAN B DR
STURROCK PETER A PROF
SVENSSON ROLAND
SWANK JEAN HEBB
TAKAHARA FUMIO DR
TANAKA YASUO DR
TAYLER ROGER J PROF
TERRELL NELSON JAMES JR
THORNE KIP S PROF
TOMIMATSU AKIRA DR
TRURAN JAMES W JR
TRUSSONI EDOARDO
TSURUTA SACHIKO DR
VALTONEN MAURI J PROF
VAN DEN HEUVEL EDWARD P J
VAN RIPER KENNETH A DR
VIDAL NISSIM V DR
VOELK HEINRICH J PROF
WANAS M I DR
WANG DEYU
WANG SHOU-GUAN
WANG YI-MING DR
WANG ZHEN-RU
WEAVER THOMAS A DR
WEBSTER ADRIAN S DR
WEISHEIT JON C DR
WEISSKOPF MARTIN CH DR
WENTZEL DONAT G DR
WESTFOLD KEVIN C PROF
WHEELER JOHN A DR
WILL CLIFFORD M DR
WILSON JAMES R DR
WOLSTENCROFT RAMON D DR
WORRALL DIANA MARY
YANG HAI SHOU
YANG LAN-TIAN
YOU JUNHAN
ZAMORANI GIOVANNI
ZHANG HE-QI
ZHANG JIA-LU
ZHANG ZHEN-JIU
ZOMBECK MARTIN V DR

COMMISSION No. 49

THE INTERPLANETARY PLASMA AND THE HELIOSPHERE

(LE PLASMA INTERPLANETAIRE ET L'HELIOSPHERE)

President : BURLAGA LEONARD F DR

Vice-President(s) : BUTI BIMLA PROF

Organizing Committee:
BOCHSLER PETER
EVIATAR AHARON PROF
JOKIPII J R PROF
RIPKEN HARTMUT W DR
ROXBURGH IAN W PROF
SARRIS EMMANUEL T PH D
WATANABE TAKASHI DR

Members:

AHLUWALIA HARJIT SINGH DR
ANANTHAKRISHNAN S
ANDERSON KINSEY A PROF
BARNES AARON DR
BARROW COLIN H DR
BARTH CHARLES A PROF
BERTAUX J L DR
BLACKWELL DONALD E PROF
BLUM PETER PROF
BONNET ROGER M DR
BRANDT JOHN C DR
BUCHNER JORG DR
CHAMBERLAIN JOSEPH W PROF
CHASSEFIERE ERIC
CHEN BIAO
CUPERMAN SAMI PROF
DE JAGER CORNELIS PROF
DELACHE PHILIPPE J DR
DOLGINOV ARKADY Z PROF DR
DRYER MURRAY DR
DURNEY BERNARD DR
DYSON JOHN E DR
ERGMA E V DR
ESHLEMAN VON R PROF
FAHR HANS JOERG PROF DR
FEYNMAN JOAN DR
FIELD GEORGE B PROF
GOSLING JOHN T DR
GRZEDZIELSKI STANISLAW PR
HABBAL SHADIA RIFAI
HARVEY CHRISTOPHER C DR
HEYVAERTS JEAN DR
HOLLWEG JOSEPH V
HOLZER THOMAS EDWARD DR
IONSON JAMES ALBERT
JOSELYN JO ANN C DR
KAKINUMA TAKAKIYO T PROF
KELLER HORST UWE DR
LAFON JEAN-PIERRE J DR
LEVY EUGENE H DR
LI XIAO-QING
LOTOVA N A DR
LUEST REIMAR PROF
LUNDSTEDT HENRIK DR
MACQUEEN ROBERT M DR
MANGENEY ANDRE DR
MASON GLENN M
MATSUURA OSCAR T DR
MAVROMICHALAKI HELEN DR
MENDIS DEVAMITTA ASOKA DR
MESTEL LEON PROF
MICHEL F CURTIS PROF
MOUSSAS XENOPHON PH D
NAKAGAWA YOSHINARI DR
PARESCE FRANCESCO DR
PARKER EUGENE N
PERKINS FRANCIS W DR
PFLUG KLAUS DR
PNEUMAN GERALD W
RAADU MICHAEL A DR
READHEAD ANTHONY C S DR
REAY NEWRICK K DR
RICKETT BARNABY JAMES DR
RIDDLE ANTHONY C DR
ROACH FRANKLIN E
ROSNER ROBERT
RUSSELL CHRISTOPHER T
SAGDEEV ROALD Z DR
SASTRI HANUMATH J DR
SAWYER CONSTANCE B DR
SCHATZMAN EVRY PROF
SCHERB FRANK PROF
SCHMIDT H U DR
SCHREIBER ROMAN
SCHWARTZ STEVEN JAY
SETTI GIANCARLO PROF
SHAWHAN STANLEY D DR
SHEA MARGARET A DR
SMITH DEAN F DR
SONETT CHARLES P PROF
STONE R G DR
STURROCK PETER A PROF
SUESS STEVEN T DR
TRITAKIS BASIL P DR
VAINSTEIN L A DR
VAN ALLEN JAMES A PROF
VINOD S KRISHAN MRS DR
WALLIS MAX K DR
WANG YI-MING DR
WELLER CHARLES S DR
WILD JOHN PAUL DR
WU SHI TSAN DR
YEH TYAN DR

COMMISSION No. 50

PROTECTION OF EXISTING AND POTENTIAL OBSERVATORY SITES

(PROTECTION DES SITES D'OBSERVATOIRES EXISTANTS ET POTENTIELS)

President : CRAWFORD DAVID L DR

Vice-President(s) : MURDIN PAUL G DR

Organizing Committee: BLANCO CARLO DR
BLANCO VICTOR M DR
COYNE GEORGE V DR
HOAG ARTHUR A DR
KOZAI YOSHIHIDE PROF
PANKONIN VERNON LEE DR
TORRES CARLOS ALBERTO DR
TREMKO JOZEF DR
VAN DEN BERGH SIDNEY PROF
WALKER MERLE F PROF

Members:

ALY M KHAIRY PROF
BARRETO LUIZ MUNIZ PROF
BURSTEIN DAVID
DOMMANGET J DR
GALAN MAXIMINO J
HIDAJAT BAMBANG PROF DR
KUBICELA ALEKSANDAR DR
MARKKANEN TAPIO DR
MAVRIDIS L N PROF
NELSON BURT DR
SCHILIZZI RICHARD T DR
TORRES CARLOS DR
WOSZCZYK ANDRZEJ PROF
ARDEBERG ARNE L PROF
BENSAMMAR SLIMANE DR
CAYREL ROGER DR
DUNKELMAN LAWRENCE
GOEBEL ERNST DR
HUANG YIN-LIANG
LEIBOWITZ ELIA M DR
MARX SIEGFRIED DR
MCCARTHY MARTIN F DR
OSORIO JOSE J S P PROF
SHCHEGLOV P V DR
WAYMAN PATRICK A PROF
WU MING-CHAN
ARIAS DE GREIFF J PROF
BHATTACHARYYA J C PROF
DAVIS JOHN DR
EDWARDS PAUL J DR
HELMER LEIF
JIANG SHI-YANG
MAHRA H S DR
MATTIG W PROF DR
MENZIES JOHN W DR
SANCHEZ FRANCISCO PROF
SMITH F GRAHAM PROF
WOOLF NEVILLE J
ZHANG BAI-RONG

COMMISSION No. 51

BIOASTRONOMY : SEARCH FOR EXTRATERRESTRIAL LIFE

(BIOASTRONOMIE : RECHERCHE DE LA VIE DANS L'UNIVERS)

President : MARX GYORGY PROF.

Vice-President(s) : BROWN RONALD D PROF
KARDASHEV N S DR

Organizing Committee: CONNES PIERRE DR
GATEWOOD GEORGE DIRECTOR
JUGAKU JUN DR
PACINI FRANCO PROF
REES MARTIN J PROF
TROITSKY V S PROF DR

Members:

AL-NAIMY HAMID M K DR	AL-SABTI ABDUL ADIM DR	ALMAR IVAN PROF
AMBARTSUMIAN V A PROF DR	ANDO HIROYASU DR	BAKOS GUSTAV A PROF
BALAZS BELA A DR	BALL JOHN A DR	BANIA THOMAS MICHAEL
BARBIERI CESARE PROF	BASU BAIDYANATH PROF	BASU DIPAK DR
BAUM WILLIAM A DR	BEAUDET GILLES DR	BECKMAN JOHN E PROF
BECKWITH STEVEN V W	BEEBE RETA FAYE DR	BENEST DANIEL DR
BERENDZEN RICHARD DR	BERNACCA P L PROF	BILLINGHAM JOHN
BIRAUD FRANCOIS DR	BLESS ROBERT C PROF	BOWYER C STUART PROF
BOYCE PETER B DR	BRACEWELL RONALD N PROF	BRODERICK JOHN DR
BURKE BERNARD F DR	CAMPBELL BRUCE DR	CAMPUSANO LUIS E
CARLSON JOHN B	CARR THOMAS D PROF	CHAISSON ERIC J PROF
CHOU KYONG CHOL PROF	CLARK THOMAS A DR	COUPER HEATHER MISS
CURRIE DOUGLAS G DR	DAIGNE GERARD	DARIUS JON DR
DAVIS MICHAEL M DR	DAWE JOHN ALAN DR	DE GRAAFF W DR
DE JAGER CORNELIS PROF	DE JONGE J K DR	DE LOORE CAMIEL PROF
DE VINCENZI DONALD DR	DELSEMME ARMAND H PROF DR	DICK STEVEN J
DIXON ROBERT S DR	DJORGOVSKI STANISLAV DR	DORSCHNER JOHANN DR
DOWNS GEORGE S DR	DRAKE FRANK D PROF	DYSON F J DR
ECCLES MICHAEL J DR	ELLIS GEORGE F R PROF	EPSTEIN EUGENE E DR
EVANS NEAL J II ASS PROF	FAZIO GIOVANNI G DR	FEJES ISTVAN DR
FELDMAN PAUL A DR	FIELD GEORGE B PROF	FIRNEIS FRIEDRICH J DR
FIRNEIS MARIA G DR	FISHER PHILIP C	FREDRICK LAURENCE W PROF
FUJIMOTO MASA-KATSU DR	FUJIMOTO MITSUAKI DR	GEHRELS TOM PROF
GHIGO FRANCIS D DR	GINZBURG VITALY L PROF	GIOVANNELLI FRANCO DR
GODOLI GIOVANNI PROF	GOLDSMITH DONALD W. DR.	GOTT III J RICHARD
GOUDIS CHRISTOS D PROF	GREENBERG J MAYO DR	GREENSTEIN J L PROF
GREGORY PHILIP C DR	GULKIS SAMUEL DR	GUNN JAMES E PROF
GURM HARDEV S PROF	HADDOCK FRED T DR	HAISCH BERNHARD MICHAEL
HAJDUK ANTON DR	HARRINGTON ROBERT S DR	HARRISON EDWARD R PROF
HART MICHAEL H DR	HECK ANDRE DR	HEESCHEN DAVID S DR
HEIDMANN JEAN DR	HERCZEG TIBOR J PROF DR	HERSHEY JOHN L DR
HEUDIER JEAN-LOUIS DR	HIRABAYASHI HISASHI DR	HOANG BINH DY DR
HOLLIS JAN MICHAEL DR	HOROWITZ PAUL PROF	HUNTEN DONALD M PROF
HUNTER JAMES H PROF	HYSOM EDMUND J	IDLIS G M DR
IRVINE WILLIAM M PROF	ISRAEL FRANK P DR	JASTROW ROBERT
JEFFERS STANLEY DR	JENNISON ROGER C PROF	JONES ERIC M

KAFATOS MINAS DR	KAFKA PETER	KAUFMANN PIERRE PROF
KELLER HANS ULRICH DR	KELLERMANN KENNETH I DR	KLEIN MICHAEL J DR
KNOWLES STEPHEN H DR	KOCER D DR	KOCH ROBERT H DR
KOEBERL CHRISTIAN DR	KRAUS JOHN D PROF	KSANFOMALITI L V DR
KUIPER THOMAS B H DR	KUZMIN ARKADII D PROF DR	LAFON JEAN-PIERRE J DR
LAQUES PIERRE DR	LEE SANG GAK	LILLEY EDWARD A PROF
LIPPINCOTT SARAH LEE DR	LODEN LARS OLOF PROF	LOVELL SIR BERNARD PROF
MAFFEI PAOLO PROF	MARGRAVE THOMAS EWING JR	MAROV MIKHAIL YA PROF
MARTIN ANTHONY R DR	MARTIN MARIA CRISTINA DR	MATSAKIS DEMETRIOS N
MATSUDA TAKUYA PROF	MAVRIDIS L N PROF	MCALISTER HAROLD A DR
MCDONOUGH THOMAS R DR	MENDOZA V EUGENIO E DR	MILET BERNARD L DR
MINN YOUNG KEY DR	MIRABEL IGOR FELIX DR	MORIMOTO MASAKI DR
MOROZ V I PROF DR	MORRIS MARK ROOT DR	MORRISON PHILIP PROF
MOUTSOULAS MICHAEL PROF	MULLER RICHARD A	NAKAGAWA YOSHINARI DR
NIARCHOS PANAYIOTIS PH D	ODA MINORU PROF	OLIVER BERNARD M DR
OLIVER BERNARD M PROF	OLLONGREN A PROF DR	OSTRIKER JEREMIAH P PROF
OWEN TOBIAS C PROF	PAGE THORNTON L DR	PAPAGIANNIS MICHAEL D PRO
PARIJSKIJ YU N DR	PASINETTI LAURA E PROF	PEREK LUBOS DR
PESEK RUDOLPH PROF	POLLACK JAMES B DR	PONNAMPERUMA CYRIL PROF
PONSONBY JOHN E B DR	PROCHAZKA FRANZ V DR	PURCELL EDWARD M PROF
QIU PUZHANG, ASS. PROF.	QUINTANA HERNAN DR	QUINTANA JOSE M DR
RAJAMOHAN R DR	REAY NEWRICK K DR	RIIHIMAA JORMA J DR
ROBINSON LEIF J	ROOD ROBERT T DR	ROWAN-ROBINSON MICHAEL DR
RUBIN ROBERT HOWARD	RUSSELL JANE L DR	SAGAN CARL DR
SAKURAI KUNITOMO PROF	SANCISI RENZO DR	SCARGLE JEFFREY D DR
SCHATZMAN EVRY PROF	SCHILD RUDOLPH E DR	SCHNEIDER JEAN
SCHOBER HANS J DR	SEEGER CHARLES LOUIS III	SEIELSTAD GEORGE A
SEIRADAKIS JOHN HUGH DR	SHAPIRO MAURICE M PROF	SHEN CHUN-SHAN
SHIMIZU MIKIO PROF	SHOSTAK G SETH DR	SHUTER WILLIAM L H DR
SINGH H P	SIVARAM C DR	SLYSH VJACHOSLAV I DR
SMITH F GRAHAM PROF	SMITH HARLAN J PROF	SMITH HOWARD ALAN
SNYDER LEWIS E	SOFUE YOSHIAKI PROF	STALIO ROBERTO DR
STEIN JOHN WILLIAM	STRAIZYS V PROF DR	STURROCK PETER A PROF
SULLIVAN WOODRUFF T III	TAKADA-HIDAI MASAHIDE DR	TARTER JILL C DR
TAVAKOL REZA	TEDESCO EDWARD F	TEJFEL VIKTOR G DR
TERZIAN YERVANT PROF	THADDEUS PATRICK PROF	TOLBERT CHARLES R DR
TOVMASSIAN H M DR	TOWNES CHARLES HARD DR	TRIMBLE VIRGINIA L DR
TURNER EDWIN L DR	VALBOUSQUET ARMAND DR	VALLEE JACQUES P DR
VAN DE KAMP PETER	VAN FLANDERN THOMAS DR	VARSHALOVICH DIMITRIJ PR
VAZQUEZ MANUEL DR	VENUGOPAL V R DR	VERSCHUUR GERRIT L PROF
VOGT NIKOLAUS DR	VON HOERNER SEBASTIAN DR	WALLIS MAX K DR
WATSON FREDERICK GARNETT	WELCH WILLIAM J PROF	WETHERILL GEORGE W
WIELEBINSKI RICHARD PROF	WILLIAMS IWAN P DR	WILLSON ROBERT FREDERICK
WOLSTENCROFT RAMON D DR	ZUCKERMAN BEN M DR	

WORKING GROUP FOR PLANETARY SYSTEM NOMENCLATURE

(GROUPE DE TRAVAIL POUR LA NOMENCLATURE DU SYSTEME PLANETAIRE)

President : H. Masursky

Members :

K. Aksnes, M. Fulchignoni, M. Ya. Marov, P.M. Millman, D. Morrison, T.C. Owen, V.V. Shevchenko, B.A. Smith, V.G. Tejfel

Task Groups :

1) Lunar Nomenclature :

V.V. Shevchenko (Chairman)
A. Dollfus, F. El-Baz, H. Masursky, P.M. Millman, S.K. Runcorn, E.A. Whitaker

2) Mercury Nomenclature :

D. Morrison (Chairman)
D.P. Campbell, M.E. Davies, A. Dollfus, N.P. Erpylev, J.E. Guest

3) Venus Nomenclature :

M. Ya. Marov (Chairman)
A.T. Basilevsky, D.B. Campbell, R.M. Goldstein, R.F. Jurgens, H. Masursky, G.H. Pettengill, Y.F. Tjuflin

4) Mars Nomenclature :

B.A. Smith (Chairman)
A. Dollfus, M. Ya. Marov, Ya. Martynov, H. Masursky, S. Miyamoto, C. Sagan

5) Outer Solar System Momenclature

T.C. Owen (Chairman)
K. Aksnes, A.T. Basilevsky, R. Beebe, M.S. Bobrov, A. Brahic, M.E. Davies, N.P. Erpylev, H. Masursky, B.A. Smith, V.G. Tejfel

6) Surface on Asteroids and Comets

D. Morrison
J. Veverka, A. Brahic, M. Fulchignoni, T. Gombosi, L. Ksantfomaliti, Y. Yatskiv, Y.C. Chang, S. Isobe

7) Small Bodies

D. Morrison (Chairman pro tempore)
A. Brahic, Y.C. Chang, T. Gombosi, S. Isobe, L. Ksanfomoliti, D. Lupishko, D. Morrison, J. Verveka

3. Geographical repartition of Members

Adhering Countries

Country : ARGENTINA

Members :

- AGUERO ESTELA L DR
- ALTAVISTA CARLOS A DR
- ARIAS ELISA FELICITAS
- ARNAL MARCELO EDMUNDO DR
- BAJAJA E DR
- BRANDI ELISANDE ESTELA DR
- BRANHAM RICHARD L JR
- CAPPA DE NICOLAU CRISTINA
- CARESTIA REINALDO A DR
- CARRANZA GUSTAVO J DR
- CLARIA JUAN DR
- COLOMB FERNANDO R DR
- DUBNER GLORIA DR
- FEINSTEIN ALEJANDRO DR
- FERNANDEZ SILVIA M. DR
- FERRER OSVALDO EDUARDO DR
- FILLOY EMILIO MANUEL E.E.
- FORTE JUAN CARLOS DR
- GARCIA LAMBAS DIEGO DR
- HERNANDEZ CARLOS ALBERTO
- IANNINI GUALBERTO DR
- LANDI-DESSY J DR
- LAPASSET EMILIO DR
- LEVATO ORLANDO HUGO DR
- LOPEZ CARLOS LIC
- LOPEZ GARCIA ZULEMA L DR
- LOPEZ JOSE A ING
- LOPEZ-GARCIA FRANCISCO DR
- LUNA HOMERO G. DR
- MACHADO MARCOS
- MALARODA STELLA M DR
- MANRIQUE WALTER T PROF
- MARABINI RODOLFO JOSE
- MARRACO HUGO G DR
- MARRACO HUGO G DR
- MARTIN MARIA CRISTINA DR
- MENDEZ ROBERTO H DR
- MILONE LUIS A DR
- MIRABEL IGOR FELIX DR
- MORRAS RICARDO DR.
- NIEMELA VIRPI S DR
- OLANO CARLOS ALBERTO DR
- PERDOMO RAUL
- POEPPEL WOLFGANG G L DR
- RABOLLI MONICA DR
- RINGUELET ADELA E DR
- ROVIRA MARTA GRACIELA
- SAHADE JORGE PROF
- SERSIC J L DR
- SISTERO ROBERTO F DR
- VEGA E. IRENE DR
- ZADUNAISKY PEDRO E PROF

Country : AUSTRALIA

Members :

ABLES JOHN G DR
AITKEN DAVID K DR
ALLEN DAVID A DR
BAILEY JEREMY A
BATTY MICHAEL DR
BESSELL MICHAEL S DR
BICKNELL GEOFFREY V DR
BIRCH PETER MR
BLAIR DAVID GERALD
BOLTON JOHN G
BOOTH ANDREW J
BOWEN EDWARD G DR
BOYLE BRIAN DR
BRAY ROBERT J DR
BROWN RONALD D PROF
CALLY PAUL S DR
CANDY MICHAEL P MR
CANE HILARY VIVIEN
CANNON RUSSELL D DR
CARTER DAVID DR
CARVER JOHN H PROF
CASWELL JAMES L DR
CHRISTIANSEN WILBUR PROF
COGAN BRUCE C DR
COLE TREVOR WILLIAM PROF
COUCH WARRICK DR
CRAM LAWRENCE EDWARD DR
DAVIS JOHN DR
DAWE JOHN ALAN DR
DOPITA MICHAEL ANDREW DR
DUNCAN ROBERT A PROF
DURRANT CHRISTOPHER J DR
EDWARDS PAUL J DR
EKERS RONALD D DR
ELFORD WILLIAM GRAHAM DR
ELLIS G R A PROF
ERICKSON WILLIAM C DR
EVANS ROBERT REV
FAULKNER DONALD J DR
FENTON K B DR
FORSTER JAMES RICHARD DR
FRATER ROBERT H DR
FREEMAN KENNETH C PROF
GALLOWAY DAVID DR
GARDNER FRANCIS F DR
GASCOIGNE S C B DR
GILLINGHAM PETER MR
GINGOLD ROBERT ARTHUR DR
GODFREY PETER DOUGLAS DR
GOLLNOW H DR
GOTTLIEB KURT
GRAY PETER MURRAY
GREEN ELIZABETH M. DR
HAMILTON P A DR
HARWOOD DENNIS MR
HAYNES RAYMOND F PROF
HORTON BRIAN H DR
HOSKING ROGER J PROF
HUGHES SHAUN
HUNSTEAD RICHARD W DR
HYLAND A R HARRY DR
JAUNCEY DAVID L DR
KALNAJS AGRIS J DR
KEAY COLIN S L PROF
KESTEVEN MICHAEL J L DR
KVIZ ZDENEK DR
LAMBECK KURT PROF
LARGE MICHAEL I DR
LOMB NICHOLAS RALPH DR
LOUGHHEAD RALPH E DR
MALIN DAVID F MR
MANCHESTER RICHARD N DR
MATHEWSON DONALD S PROF
MCADAM W BRUCE DR
MCCRACKEN KENNETH G DR
MCCULLOCH PETER M DR
MCGEE RICHARD X DR
MCGREGOR PETER JOHN DR
MCLEAN DONALD J DR
MELROSE DONALD B PROF
MILLS BERNARD Y PROF
MILNE DOUGLAS K DR
MINNET HARRY C MR
MONAGHAN JOSEPH J DR
MORETON G E
MORGAN PETER DR
MULLALY RICHARD F DR
MURDOCH HUGH S DR
NELSON GRAHAM JOHN DR
NEWELL EDWARD B DR
NIKOLOFF IVAN DR
NORRIS JOHN DR
NORRIS RAYMOND PAUL
NULSEN PAUL DR
O'MARA BERNARD J PROF
O'SULLIVAN JOHN DAVID DR
OLSSON-STEEL DUNCAN I DR
PAGE ARTHUR MR
PETERSON BRUCE A DR
PIDDINGTON JACK H RES FEL
PRENTICE ANDREW J R DR
PROTHEROE RAYMOND J DR
REBER GROTE DR
REES DAVID ELWYN DR
ROBERTSON JAMES GORDON DR
ROBINSON BRIAN J DR
ROBINSON JR RICHARD D DR
RODGERS ALEX W DR
ROSS JOHN E R DR
SADLER ELAINE MARGARET
SAVAGE ANN DR
SHARPLES RAY DR
SHERIDAN K V DR
SHIMMINS ALBERT JOHN
SHOBBROOK ROBERT R DR
SIMS KENNETH P DR
SLEE O B DR
SMITH LINDSEY F DR
SPARROW JAMES G DR
STAVELEY-SMITH LISTER DR
STEWART RONALD T MR
STOREY JOHN W V DR
TANGO WILLIAM J. DR
TAYLOR KENNETH N R PROF
THOMPSON KEITH DR
TUOHY IAN R DR
TURTLE A J DR
VAN DER BORGHT RENE PROF
VARDAVAS ILIAS MIHAIL
VISVANATHAN NATARAJAN DR
WATERWORTH MICHAEL DR
WATSON FREDERICK GARNETT
WEBSTER B LOUISE DR
WELLINGTON KELVIN DR
WESTFOLD KEVIN C PROF
WHITE GRAEME LINDSAY DR
WHITEOAK J B DR
WICKRAMASINGHE D T DR
WILD JOHN PAUL DR
WILSON BRIAN G PROF
WILSON PETER R PROF
WOOD PETER R DR
WRIGHT ALAN E DR
ZAMBON GIULIO DR.
ZEALEY WILLIAM J DR

Country : AUSTRIA

Members :

AUNER GERHARD DR
DORFI ERNST ANTON DR
FIRNEIS FRIEDRICH J DR
HANSLMEIER ARNOLD
HRON JOSEF DR
LUSTIG GUENTER DR
POLNITZKY GERHARD DR
SCHNELL ANNELIESE DR
SCHRUTKA-RECHTENSTAMM PR.
WEISS WERNER W DR
BALAZS BELA A DR
DVORAK RUDOLF DR
FIRNEIS MARIA G DR
HARTL HERBERT DR
JACKSON PAUL DR
MAITZEN HANS M DR
PROCHAZKA FRANZ V DR
SCHOBER HANS J DR
STIFT MARTIN JOHANNES DR
BREGER MICHEL DR
FERRARI D'OCCHIEPPO K DR
GOEBEL ERNST DR
HAUPT HERMANN F PROF
KOEBERL CHRISTIAN DR
PFLEIDERER JORG PROF
RAKOS KARL D PROF
SCHROLL ALFRED DR
WEINBERGER RONALD DR

Country : BELGIUM

Members :

AREND S DR
BAECK NICOLE A L DR
BOSMAN-CRESPIN DENISE
CALLEBAUT DIRK K DR
CUYPERS JAN DR
DEBEHOGNE HENRI DR SC
DEJONGHE HERWIG BERT DR
DEMARET JACQUES DR
DINGENS P PROF DR
ELST ERIC WALTER DR
GONZE ROGER F J IR
HENRARD JACQUES PROF
JAMAR CLAUDE A J DR
LEMAITRE ANNE DR
MANFROID JEAN DR
MIGEOTTE MARCEL V PROF
MOONS MICHELE B M M
OTTELET I J DR
PAUWELS T DR
RENSON P F M DR
SAUVAL A JACQUES DR
SIMON RENE L E PROF
STEYAERT HERMAN PROF DR
SWINGS JEAN-PIERRE DR
VERBEEK PAUL DR
WAELKENS CHRISTOFFEL
ARNOULD MARCEL L DR
BERTIAU FLOR C PROF
BRIHAYE CHARLES C A DR
COUTREZ RAYMOND A J PROF
DE GREVE JEAN-PIERRE DR
DEHANT VERONIQUE DR
DELBOUILLE LUC PROF
DENIS CARLO DR
DOMMANGET J DR
GABRIEL MAURICE R DR
GOOSSENS MARCEL DR
HENSBERGE HERMAN
KOECKELENBERGH ANDRE DR
MAGAIN PIERRE DR
MEIRE RAPHAEL
MOERDIJK WILLY G DR
NICOLET MARCEL PROF
OXENIUS JOACHIM DR
PERDANG JEAN M DR
ROBE H A G DR
SCUFLAIRE RICHARD DR
SMEYERS PAUL PROF
SURDEJ JEAN M G
VAN DESSEL EDWIN LUDO DR
VERHEEST FRANK PROF
WARZEE J DR
ARPIGNY CLAUDE PROF
BIEMONT EMILE DR
BURGER MARIJKE DR
CUGNON PIERRE DR
DE LOORE CAMIEL PROF
DEJAIFFE RENE J DR
DELCROIX ANDRE J S DR
DENOYELLE JOZEF KIC
DOSSIN F DR
GODART ODON PROF
GREVESSE N DR
HOUZIAUX L PROF
LAUSBERG ANDRE DR
MALAISE DANIEL J DR
MELCHIOR PAUL J PROF DIR
MONFILS ANDRE G PROF
NOELS ARLETTE DR
PAQUET PAUL EG DR
REMY BATTIAU LILIANE G A
ROLAND GINETTE DR
SIMON PAUL C DR
STERKEN CHRISTIAAN LEO DR
SVALGAARD LEIF DR
VAN RENSBERGEN WALTER DR
VREUX JEAN MARIE DR
ZANDER RODOLPHE DR

Country : BRAZIL

Members :

ABRAHAM ZULEMA DR
ALDROVANDI RUBEN DR
BARBUY BEATRIZ DR
BARRETO LUIZ MUNIZ PROF
BARROSO JR JAIR
BENEVIDES SOARES P DR
BICA EDUARDO L D DR
BUSKO IVO C DR
CAPELATO HUGO VICENTE DR
CHIAN ABRAHAM CHIAN-LONG
CLAUZET LUIZ B FERREIRA
CODINA LANDABERRY SAYD J
DA COSTA JOSE MARQUES DR
DA COSTA NICOLAI L.-A
DA ROCHA VIEIRA E DR
DA SILVA LICIO DR
DE FREITAS PACHECO J A DR
DE LA REZA RAMIRO DR
DE SOUZA RONALDO DR
DOTTORI HORACIO A DR
DUCATI JORGE RICARDO DR
FAUNDEZ-ABANS M DR
FERRAZ-MELLO S PROF DR
FREITAS MOURAO R R DR
GIACAGLIA GIORGIO E PROF
GOMES ALERCIO M DR
GOMIDE FERNANDO DE MELLO
GONZALES-A WALTER D DR
GRIJO DE OLIVEIRA A K DR
GRUENWALD RUTH DR
JANOT-PACHECO EDUARDO DR
KAUFMANN PIERRE PROF
KEPLER S O
LAZZARO DANIELA DR
LEITE SCHEID PAULO DR
LEPINE JACQUES R D DR
LOISEAU NORA DR
MACHADO LUIZ E. DA SILVA
MACIEL WALTER J DR
MAGALHAES ANTONIO MARIO
MARQUES DOS SANTOS P PROF
MARTIN INACIO MALMONGE DR
MATSUURA OSCAR T DR
NICOLACI DA COSTA LUIZ-A.
NOVELLO MARIO DR
OPHER REUVEN PROF
PALMEIRA RICARDO A R DR
PASTORIZA MIRIANI G DR
PIAZZA LILIANA RIZZO
QUARTA MARIA LUCIA
QUAST GERMANO RODRIGO
RAO K RAMANUJA DR
SAWANT HANUMANT S DR
SCALISE JR EUGENIO DR
SCHUCH NELSON JORGE
SESSIN WAGNER DR
SINGH PATAN DEEN DR
STEINER JOAO E DR
STORCHI-BERGMAN THAISA DR
TAKAGI SHIGETSUGU DR
TORRES CARLOS ALBERTO DR
VAZ LUIZ PAULO RIBEIRO
VIEIRA MARTINS ROBERTO DR
VILHENA DE MORAES R DR
VILLELA THYRSO V. DR.
YOKOYAMA TADASHI DR

Country : BULGARIA

Members :

BACALOV MIHAIL
BONEV BONU K MR
DERMENDJIEV VLADIMIR DR
DOBRITSCHEV V M MR
FILIPOV LATCHEZAR
GEORGIEV TSVETAN DR
ILIEV ILIAN
IVANOV GEORGI R DR
IVANOVA VIOLETA DR
KALINKOV MARIN P DR
KJURKCHIEVA DIANA DR
KOLEV DIMITAR ZDRAVKOV
KOVACHEV B J DR
KRAICHEVA ZDRAVSKA DR
KUNCHEV PETER DR
MINEVA VENETA DR
MOMCHEV GOSPODIN
NICOLOV NIKOLAI S DR
NIKOLOV ANDREJ DR
PANOV KIRIL DR
PETROV GEORGY TRENDAFILOV
PETROV NIKOLAI
POPOV VASIL NIKOLOV
POPOVA MALINA D PROF DR
RADOSLAVOVA TSVETANKA
RAIKOVA DONKA DR
RUSSEV RUSCHO DR
RUSSEVA TATJANA
SBIRKOVA-NATCHEVA T
SHKODROV V G DR
SPASOVA NEDKA MARINOVA
STOEV ALEXEI
TOMOV ALEXANDER NIKOLOV
TSVETANOV ZLATAN IVANOV
TSVETKOV MILCHO K DR
TSVETKOV TSVETAN DR
TSVETKOVA KATIA
UMLENSKI VASIL
VELKOV KIRIL
VLADIMIROV SIMEON
YANKULOVA IVANKA DR
ZHELYAZKOV IVAN DR
ZLATEV SLAVEY

Country : CANADA

Members :

AIKMAN G CHRIS L
ANDREW BRYAN H DR
ARGYLE P E DR
AUMAN JASON R PROF
AVERY LORNE W DR
BAIRD KENNETH M DR
BAKOS GUSTAV A PROF
BARKER PAUL K DR
BARNARD HANNES A J DR
BASTIEN PIERRE DR
BATTEN ALAN H DR
BEAUDET GILLES DR
BELL MORLEY B
BINETTE LUC
BISHOP ROY L DR
BLACKWELL ALAN TREVOR
BOCHONKO D RICHARD DR
BOLTON C THOMAS PROF
BORRA ERMANNO F DR
BRANDIE GEORGE W DR
BROTEN NORMAN W
BURKE J ANTHONY DR
CALDWELL JOHN JAMES
CAMPBELL BRUCE DR
CANNON WAYNE H DR
CARIGNAN CLAUDE DR
CARLBERG RAYMOND GARY DR
CHAU WAI Y PROF
CLARK THOMAS ALAN DR
CLARKE THOMAS R DR
CLEMENT MAURICE J PROF
CLIMENHAGA JOHN L PROF
CLUTTON-BROCK MARTIN DR
COSTAIN CARMAN H DR
COSTAIN CECIL C DR
COUTTS-CLEMENT CHRISTINE
COVINGTON ARTHUR E
CRABTREE DENNIS DR
CRAMPTON DAVID DR
DE ROBERTIS M M DR
DEMERS SERGE DR
DEWDNEY PETER E F DR
DOHERTY LORNE H DR
DULEY WALTER W PROF
DYER CHARLES CHESTER DR
EVANS NANCY REMAGE DR
FAHLMAN GREGORY G DR
FELDMAN PAUL A DR
FERNIE J DONALD PROF
FICH MICHEL DR
FITZGERALD M PIM PROF
FLETCHER J MURRAY
FONTAINE GILLES DR
FORT DAVID NORMAN DR
FRIEL EILEEN D DR
GAETZ TERRANCE J DR
GAIZAUSKAS VICTOR DR
GALT JOHN A DR
GARRISON ROBERT F PROF
GOWER ANN C DR
GOWER J F R DR
GRAY DAVID F PROF
GREGORY PHILIP C DR
GRIFFITH JOHN S PROF
GRUNDMANN WALTER
GULLIVER AUSTIN FRASER DR
HALLIDAY IAN DR
HANES DAVID A DR
HARDY EDUARDO
HARRIS GRETCHEN L H DR
HARRIS WILLIAM E DR
HARROWER GEORGE A DR
HARTWICK F DAVID A DR
HARTZ THEODORE R DR
HASEGAWA TATSUHIKO DR
HAWKES ROBERT LEWIS DR
HENRIKSEN RICHARD N DR
HERZBERG GERHARD DR
HESSER JAMES E DR
HICKSON PAUL DR
HIGGS LLOYD A DR
HILL GRAHAM DR
HUBE DOUGLAS P DR
HUGHES VICTOR A PROF
HUTCHINGS JOHN B DR
INNANEN KIMMO A PROF
IRWIN ALAN W DR
IRWIN JUDITH DR
ISRAEL WERNER PROF
JEFFERS STANLEY DR
JONCAS GILLES DR
JONES JAMES DR
KAMPER KARL W DR
KENNEDY JOHN E PROF
KOEHLER JAMES A PROF
KORMENDY JOHN DR
KRONBERG PHILIPP DR
KWOK SUN DR
LAKE KAYLL WILLIAM DR
LAMONTAGNE ROBERT DR
LANDECKER THOMAS L DR
LANDSTREET JOHN D PROF
LAPOINTE S M DR
LEAHY DENIS A DR
LEGG THOMAS H DR
LESTER JOHN B DR
LOCKE JACK L DR
LOWE ROBERT P DR
MACLEOD JOHN M DR
MACRAE DONALD A PROF
MADORE BARRY FRANCIS DR
MANN PATRICK J DR
MARLBOROUGH J M PROF
MARTIN PETER G PROF
MCCALL MARSHALL LESTER DR
MCCLURE ROBERT D PROF
MCCUTCHEON WILLIAM H PROF
MCDONALD J K PETRIE DR
MCINTOSH BRUCE A DR
MENON T K PROF
MERRIAM JAMES B
MICHAUD GEORGES J DR
MILLMAN PETER M DR
MILONE EUGENE F PROF
MITALAS ROMAS ASSOC PROF
MITCHELL GEORGE F DR
MOCHNACKI STEPHAN W DR
MOFFAT ANTHONY F J DR
MOFFAT JOHN W. DR
MOORHEAD JAMES M DR
MORBEY CHRISTOPHER L
MORRIS STEPHEN C DR
MORTON DONALD C DR
NADEAU DANIEL DR
NAQVI S I H PROF
NEMEC JAMES
NICHOLLS RALPH W PROF
NOREAU LOUIS DR
ODGERS GRAHAM J DR
PACHNER JAROSLAV PROF
PATHRIA RAJ K PROF
PEDREROS MARIO DR
PERCY JOHN R PROF
PINEAULT SERGE DR
POECKERT ROLAND H DR
POPELAR JOSEF DR
PRITCHET CHRISTOPHER J DR
PRYCE MAURICE H L DR
PURTON CHRISTOPHER R DR
RACINE RENE DR
RAMSAY DONALD A DR
RATNATUNGA KAVAN U.
REED B CAMERON DR
RICE JOHN B DR
RICHARDSON E HARVEY DR
RICHER HARVEY B DR
ROBB RUSSEL M
ROCHESTER MICHAEL G PROF
ROGER ROBERT S DR

Country : CANADA

Members :

ROGERS CHRISTOPHER DR
ROY JEAN-RENE
SCARFE COLIN D DR
SHUTER WILLIAM L H DR
SREENIVASAN S RANGA PROF
TAPPING KENNETH F
TATUM JEREMY B DR
TURNER DAVID G DR
VAN DEN BERGH SIDNEY PROF
VINER MELVYN R DR
WEBSTER RACHEL
WELCH GARY A DR
WILLIS ANTHONY GORDON DR
WRIGHT KENNETH O DR
ZHANG CHENG-YUE
ROTTENBERG J A DR
RUCINSKI SLAWOMIR M DR
SCRIMGER J. NORMAN DR
SMITH G H DR
STETSON PETER B. DR
TASSOUL JEAN-LOUIS PROF
TAYLOR A R DR
UNDERHILL ANNE B DR
VANDENBERG DON DR
WAINWRIGHT JOHN DR
WEHLAU AMELIA DR
WESEMAEL FRANCOIS DR
WOODSWORTH ANDREW W.DR
YEE HOWARD K.C. DR
ZHANG SHENG-PAN
ROUTLEDGE DAVID DR
SAWYER-HOGG HELEN B DR
SEAQUIST ERNEST R PROF
SMYLIE DOUGLAS E DR
SUTHERLAND PETER G DR
TASSOUL MONIQUE DR
TREMAINE SCOTT DUNCAN
VALLEE JACQUES P DR
VENKATESAN DORASWAMY DR
WALKER GORDON A H PROF
WEHLAU WILLIAM H PROF
WESSON PAUL S DR
WOOLSEY E G
YEN JUI-LIN PROF

Country : CHILE

Members :

ALCAINO GONZALO DR
APARICI JUAN DR
BRONFMAN LEONARDO DR
CELIS LEOPOLDO DR
INGERSON THOMAS DR
KUNKEL WILLIAM E DR
LOYOLA PATRICIO DR
MELNICK JORGE
NOEL FERNANDO
QUINTANA HERNAN DR
RUBIO MONICA DR
SCHWARZ HUGO E
VOGT NIKOLAUS DR
WROBLEWSKI HERBERT DR
ALVAREZ HECTOR DR
BALDWIN JACK A DR
CAMPUSANO LUIS E
COSTA EDGARDO DR
KAWARA KIMIAKI
LILLER WILLIAM DR
MAY J
MINTZ BLANCO BETTY MRS
PEDERSEN HOLGER DR
REIPURTH BO
RUIZ MARIA TERESA DR
SUNTZEFF NICHOLAS B
WALKER ALISTAIR ROBIN DR
ANGUITA CLAUDIO A DR
BLANCO VICTOR M DR
CARRASCO GUILLERMO DR
GUTIERREZ-MORENO A DR MRS
KRZEMINSKI WOJCIECH DR
LINDGREN HARRI
MAZA JOSE
MORENO HUGO PROF
PHILLIPS MARK M DR
REYES FRANCISCO DR
SCHUSTER HANS-EMIL
TORRES CARLOS DR
WILLIAMS ROBERT E DR

Country : CHINA, PEOPLE'S REP.

Members :

AI GUOXIANG
BAO KEREN
BIAN YU-LIN
BO SHU-REN
CAO CHANGXIN
CAO LIHONG
CAO SHENGLIN
CHEN BIAO
CHEN CHUAN-LE
CHEN DAO-HAN
CHEN HONGSHENG
CHEN JIAN-SHENG
CHEN PEISHENG
CHEN XIAO-ZHONG
CHEN XING
CHEN ZHEN
CHEN ZHENCHENG
CHEN ZUN-GUI
CHENG FU-HUA
CHENG FU-ZHEN
CHU HAN-SHU PROF.
CHU YAOQUAN
CUI CHUNFANG
CUI DOU-XING
CUI LIAN-SHU
CUI ZHEN-HUA
CUI ZHENXING
DAN XHI-XIANG
DENG ZUGAN DR
DI XIAO-HUA
DIN HUA
DING YOU-JI
FAN DAXIONG
FAN YING
FANG CHENG
FANG LI-ZHI
FENG HESHENG
FENG KE-JIA
FONG CHU-GANG
FU CHENG-QI
FU DELIAN
FU QI JUN
GAO BILIE
GAO BUXI
GONG HUI-REN
GONG SHOU-SHEN
GONG SHU-MO
GU XIAO-MA
GUO NEI-SHU DR
GUO QUAN SHI
HAN FU
HAN TIANQI
HAN WENJUN
HAN ZHENG-ZHONG
HANG HENG-RONG
HAO YUN-XIANG
HE MIAO-FU
HE XIANG-TAO
HSIANG YAN-YU
HU FU-XING
HU JING-YAO
HU NING-SHENG
HU WEN-RUI
HU ZHONG-WEI
HUA YING-MIN
HUANG BI-KUN
HUANG CHANG-CHUN
HUANG CHENG DR
HUANG JIE-HAO
HUANG KE-LIANG
HUANG KUN-YI
HUANG LIN
HUANG RUN-QIAN
HUANG SONG-NIAN DR
HUANG TIANYI
HUANG TIE-QIN
HUANG YIN-LIANG
HUANG YONGWEI
HUANG YOU-RAN
JI HONG-QING
JI SHUCHEN DR
JIANG CHONG-GUO
JIANG DONG-RONG
JIANG SHI-YANG
JIANG SHUDING
JIANG YAO-TIAO
JIANG ZHAOJI
JIN BIAOREN DR
JIN SHEN-ZENG
JIN WEN-JING
KIMURA HIROSHI DR
LI CHUN-SHENG
LI DEPEI
LI DONG-MING
LI HEN
LI HONG-WEI
LI JING
LI NENG-YAO
LI QI-BIN
LI TING
LI TIPEI
LI WEI BAO
LI XIAO-QING
LI YUAN-JIE
LI ZHENG-XIN DR
LI ZHI-FANG
LI ZHI-SEN
LI ZHIGANG
LI ZHONGYUAN
LI ZONG-WEI
LI ZONG-YUN
LIANG SHI-GUANG
LIANG ZHONG-HUAN
LIN YUANZHANG
LIU BAO-LIN
LIU CAIPIN
LIU JINMING
LIU JINYI
LIU LIAO
LIU LIN
LIU LIN-ZHONG
LIU QINGYAO DR
LIU RU-LIANG
LIU XINPING PROF.
LIU XUEFU
LIU YONG-ZHEN
LIU ZONGLI
LU BEN-KUI
LU CHUN-LIN
LU TAN
LU YANG
LUO BAO-RONG
LUO DING-JIANG
LUO DINGCHANG
LUO SHI-FANG
LUO XIANHAN
MA ER
MA XING-YUAN
MA YU-QIAN
MAO WEI
MENG XINMIN
MIAO YONG-KUAN
MIAO YONG-RUI
MO JING-ER
NAN REN-DONG
PAN JUN-HUA
PAN LIANDE
PAN NING-BAO
PAN RONG-SHI
PAN XIAO-PEI
PENG QIU-HE
PENG YUN-LOU
QI GUANRONG
QIAN BO-CHEN
QIAN JING-KUI
QIAN SHAN-JIE
QIAN ZHI-HAN DR
QIAN ZHONG-YU
QIAO GUOJUN

Country : CHINA, PEOPLE'S REP.

Members :

QIN DAO	QIN SONG-NIAN	QIN ZHI-HAI
QIU PUZHANG, ASS. PROF.	QIU YU-HAI	QU QIN-YUE
QUAN HEJUN	REN JIANG-PING	RONG JIAN-XIANG
SHEN CHANGJUN	SHEN KAIXIAN	SHEN LIANG-ZHAO
SHEN LONG-XIANG	SHEN PARN-AN	SHI GUANG-CHEN
SHI ZHONG-XIAN	SONG GUO-XUAN	SONG JIN-AN
SONG MU-TAO	SU BUMEI	SU DING-QIANG
SU HONG-JUN	SU WAN-ZHEN	SUN JIN
SUN KAI	SUN YI-SUI	SUN YONGXIANG
TAN HUISONG	TANG YU-HUA	TONG FU
TONG YI	WAN LAI	WAN TONG-SHAN
WANG CHUAN-JIN	WANG DE-CHANG	WANG DEYU
WANG JIA-JI	WANG JIA-LONG	WANG JING-SHENG
WANG JING-XIU	WANG LAN-JUAN	WANG RENCHUAN
WANG SHOU-GUAN	WANG SHUI	WANG SHUNDE DR
WANG SI-CHAO	WANG YANAN	WANG YIMING
WANG ZHEN-RU	WANG ZHEN-YI	WANG ZHENG MING
WEI MINGZHI	WU FEI	WU HUAI-WEI
WU LIAN-DA	WU LIN-XIANG	WU MING-CHAN
WU SHENGYIN	WU SHOU-XIAN	WU XINJI
WU XUE-LIN	XI ZE-ZONG	XIA JIONGYU
XIA YI-FEI	XIAN DING-ZHANG	XIANG DELIN
XIANG SHOUPING	XIAO NAI-YUAN	XIAO XING HUA
XIE GUANG-ZHONG	XIE LIANGYUN	XING JUN
XIONG DA-RUN	XU AO-AO	XU BANG-XIN
XU JIA-YAN	XU PEI-YUAN	XU PINXIN
XU TONG-QI	XU ZHENTAO	XU ZHI-CAI
YAN LIN-SHAN	YANG FUMIN	YANG HAI SHOU
YANG JIAN	YANG LAN-TIAN	YANG SHI JIE
YAO BAO-AN	YAO JIN-XING	YAO ZHENG-QIU
YE BINXUN	YE SHI-HUI	YE SHU-HUA
YE WENWEI	YI ZHAO-HUA	YIN JI-SHENG
YIN QI-FENG	YOU JIAN-QI	YOU JUNHAN
YU XIN ALFRED DR	YUE ZENG-YUAN	ZENG QIN DR
ZHAI DI-SHENG	ZHAI ZAOCHENG	ZHANG BAI-RONG
ZHANG BIN	ZHANG ER-HE DR	ZHANG FU JUN
ZHANG GUO-DONG	ZHANG HE-QI	ZHANG HUI
ZHANG JIA-LU	ZHANG JIA-XIANG	ZHANG JINTONG
ZHANG MING-CHANG	ZHANG PEIYU	ZHANG XIU ZHONG
ZHANG YOUYI	ZHANG ZHEN-DA	ZHANG ZHEN-JIU
ZHAO GANG	ZHAO JUN-LIANG	ZHAO MING
ZHAO REN-YANG	ZHAO XIAN-ZI	ZHENG DA-WEI
ZHENG JIA-QING	ZHENG WENGUANG	ZHENG XUE-TANG
ZHENG YI-JIA	ZHENG YING	ZHOU DAOQI
ZHOU HONG-NAN	ZHOU TI-JIAN	ZHOU XING-HAI
ZHOU YOU-YUAN	ZHOU ZHEN-PU	ZHU CI-SHENG
ZHU NENGHONG	ZHU SHI-CHANG	ZHU WEN-YAO
ZHU XINGFENG	ZHU YONG-HE	ZHUANG QIXIANG
ZHUANG WEIFENG	ZOU HUI-CHENG	ZOU YI-XIN
ZOU ZHEN-LONG		

Country : CHINA, TAIWAN

Members :

CHOU CHIH-KANG DR
CHOU DEAN-YI DR
FU-SHONG KUO
HSIANG-KUANG TSENG
HUANG YI-LONG DR
HUANG YINN-NIEN DR
LING CHIH-BING DR
NEE TSU-WEI DR
SHEN CHUN-SHAN
TING YEOU-TSWEN
TSAI CHANG-HSIEN DIRECTOR
TSAO MO PROF
TYPHOON LEE
WU HSIN-HENG DR

Country : COLOMBIA

Members :

ARIAS DE GREIFF J PROF
BRIEVA EDUARDO PROF

Country : CUBA

Members :

ALVAREZ OSCAR

Country : CZECHOSLOVAKIA

Members :

AMBROZ PAVEL DR
ANDRLE PAVEL DR
ANTALOVA ANNA
BICAK JIRI DR
BOUSKA JIRI DR
BUMBA VACLAV DR
BURSA MILAN DR
CEPLECHA ZDENEK DR
CHOCHOL DRAHOMIR
CHVOJKOVA WOYK E DR
FARNIK FRANTISEK
FISCHER STANISLAV DR
GRYGAR JIRI DR
HADRAVA PETR
HAJDUK ANTON DR
HAJDUKOVA MARIA
HANDLIROVA DAGMAR DR
HARMANEC PETR DR
HEINZEL PETR DR
HEKELA JAN DR
HORAK TOMAS B DR
HORAK ZDENEK PROF DR
HORSKY JAN PROF
HUDEC RENE DR
KAPISINSKY IGOR
KARLICKY MARIAN
KLECZEK JOSIP DR
KLOKOCNIK JAROSLAV DR
KLVANA MIROSLAV
KNOSKA STEFAN
KOPECKY MILOSLAV DR
KOTRC PAVEL
KOUBSKY PAVEL
KRESAK LUBOR DR
KRESAKOVA MARGITA DR
KRIVSKY LADISLAV DR
KRIZ SVATOPLUK DR
LETFUS VOJTECH DR
LOCHMAN JAN
MAYER PAVEL DR
MESZAROS ATTILA DR
MIKULASEK ZDENEK DR
MINAROVJECH MILAN
MRKOS ANTONIN DR
NEUZIL LUDEK DR
NOVOTNY VACLAV
ONDERLICKA BEDRICH DR
PADEVET VLADIMIR DR
PALOUS JAN DR
PAPOUSEK JIRI
PECINA PETR
PEREK LUBOS DR
PESEK RUDOLPH PROF
PITTICH EDUARD M DR
POLECHOVA PAVLA DR
PORUBCAN VLADIMIR DR
RAJCHL JAROSLAV DR
RUPRECHT JAROSLAV DR
RUSIN VOJTECH
RUZICKOVA-TOPOLOVA B DR
RYBANSKY MILAN
SEHNAL LADISLAV DR
SIDLICHOVSKY MILOS DR
SIMA ZDISLAV DR
SIMEK MILOS DR
SIROKY JAROMIR DR
SOLC IVAN DR
SOLC MARTIN
STEFL VLADIMIR
STOHL JAN DR
SUDA JAN
SVESTKA JIRI DR
SVOREN JAN
SYKORA JULIUS DR
TLAMICHA ANTONIN DR
TREMKO JOZEF DR
VALNICEK BORIS DR
VANYSEK VLADIMIR PROF
VAVROVA ZDENKA DR
VONDRAK JAN DR
WEBROVA LUDMILA DR
ZACHAROV IGOR DR
ZVOLANKOVA JUDITA

Country : DENMARK

Members :

ANDERSEN JOHANNES
BAERENTZEN JORN
CHRISTENSEN PER R DR
CHRISTENSEN-DALSGAARD J
CLAUSEN JENS VIGGO LEKTOR
EINICKE O H DR
FABRICIUS CLAUS V
FLORENTIN-NIELSEN RALPH
FRANDSEN SOEREN PROF
GAMMELGAARD PETER MAG SCI
GYLDENKERNE KJELD DR
HANSEN LEIF LECTURER
HELMER LEIF
HELT BODIL E
HOEG ERIK DR
JOHANSEN KAREN T LEKTOR
JONES BERNARD J T DR
JONES JANET E DR
JORGENSEN HENNING E PROF
JORGENSEN UFFE GRAE DR
KJAERGAARD PER DR
KNUDE JENS KIRKESKOV DR
KRISTENSEN LEIF KAHL DR
KUSTAANHEIMO PAUL E PROF
LUND NIELS
MADSEN JES
MOESGAARD KRISTIAN P
NISSEN POUL E PROF
NORDLUND AKE DR
NORGAARD-NIELSEN HANS U
OLSEN ERIK H
OLSEN FOGH H J
PEDERSEN OLAF PROF
PETERSEN J O DR
REIZ ANDERS PROF
RUDKJOBING MOGENS PROF
SCHNOPPER HERBERT W DR
SOMMER-LARSEN JESPER DR
SORENSEN GUNNAR DR
SVENSSON ROLAND
THOMSEN BJARNE B LECT
ULFBECK OLE DR
WESTERGAARD NIELS J DR
WIETH-KNUDSEN NIELS P DR

Country : EGYPT

Members :

ABOU-EL-ELLA MOHAMED S DR
ABULAZM MOHAMED SAMIR
AHMED IMAM IBRAHIM PROF
AHMED MOSTAFA
AIAD A PROF.
ALY M KHAIRY PROF
AWAD MERVAT EL-SAID DR
AWADALLA NABIL SHOUKRY DR
BAGHOS BALEGH B DR
EL NAWAWAY MOHAMED SALEH
EL SHALABY MOHAMED
EL-BASSUNY ALAWY A A
EL-RAEY MOHAMED E DR
EL-SHAARAWY M B DR
GALAL A A DR
GAMALELDIN ABDULLA I DR
GHOBROS ROSHDY AZER DR
HAMDY M A M DR
HAMID S EL DIN DR
HASSAN S M DR
HELALI YHYA E DR
ISSA ALI DR
KAMEL OSMAN M DR
MAHDY HAMED A DR
MAHMOUD ABUBAKR ABDELKAWI
MAHMOUD FAROUK M A B DR
MARIE M A DR
MIKHAIL FAHMY I PROF DR
MIKHAL JOSEPH SIDKY PROF
NAWAR SAMIR DR
OSMAN ANAS MOHAMED DR
SHALTOUT MESALAM A M DR
SHARAF MOHAMED ADEL DR
SOLIMAN MOHAMED AHMED
TAWADROS MAHET JACOUB DR
WANAS M I DR
YOUSEF SHAHINAZ M DR
YOUSSEF NAHED H DR

Country : FINLAND

Members :

DONNER KARL JOHAN
HAEMEEN ANTTILA KAARLE A
HAIKALA LAURI K
JAAKKOLA TOIVO S
JARNEFELT GUSTAF J PROF
KULTIMA JOHANNES
LUMME KARI A DR
MARKKANEN TAPIO DR
MATTILA KALEVI DR
MIKKOLA SEPPO DR
NIEMI AIMO
OJA HEIKKI DR
OTERMA LIISI PROF
PIIROLA VILPPU E DR
QVIST BERTIL PROF
RAHUNEN TIMO
RAITALA JOUKO T
RIIHIMAA JORMA J DR
SANDELL GORAN HANS L DR
TEERIKORPI VELI PEKKA DR
TIURI MARTTI PROF
TUOMINEN ILKKA V DR
URPO SEPPO I
VALTAOJA ESKO
VALTONEN MAURI J PROF
VILHU OSMI DR

Country : FRANCE

Members :

ABU EL ATA NABIL DR
AIME C DR
ALLOIN DANIELLE DR
ARDUINI-MALINOVSKY M. DR
ARNAUD MONIQUE
ASSEO ESTELLE DR
AUDOUZE JEAN PROF
AUVERGNE MICHEL
BACCHUS PIERRE PROF
BALKOWSKI-MAUGER CH DR
BARANNE A DR
BAUDRY ALAIN DR
BELY OLEG DR
BENSAMMAR SLIMANE DR
BERGER JACQUES G DR
BERRUYER-DESIROTTE N DR
BERTOUT CLAUDE
BIRAUD FRANCOIS DR
BOCCHIA ROMEO DR
BOISCHOT ANDRE DR
BONNEAU DANIEL
BOSMA ALBERT DR
BOUGERET J L DR
BOULON JACQUES J DR
BOYER RENE
BRIOT DANIELLE DR
BRYANT JOHN DR
CALAME ODILE DR
CAPITAINE NICOLE
CASOLI FABIENNE DR
CATALA CLAUDE DR
CAZENAVE ANNY DR
CESARSKY DIEGO A DR
CHAMBE GILBERT
CHARVIN PIERRE PR
CHOLLET FERNAND DR
CLAIREMIDI JACQUES DR
COMBES FRANCOISE DR
CONNES JANINE DR
COUPINOT GERARD DR
CREZE MICHEL DR
CRUVELLIER PAUL E DR
DAPPEN WERNER
DE LA NOE JEROME DR
DEHARVENG LISE DR
DELANNOY JEAN DR
DENISSE JEAN-FRANCOIS DR
DESPOIS DIDIER DR
DOLEZ NOEL DR
DOURNEAU GERARD DR
DUBOIS MARC A
DUCHESNE MAURICE DR
DUMONT RENE DR

ACKER AGNES PROF DR
ALECIAN GEORGES DR
ANDRILLAT HENRI L PROF
ARIMOTO NOBUO DR
ARTRU MARIE-CHRISTINE DR
ATHANASSOULA EVANGELIE DR
AUGARDE RENEE DR
AVIGNON YVETTE DR
BAGLIN ANNIE DR
BALMINO GEORGES G DR
BARBIER-BROSSAT M DR
BEC-BORSENBERGER ANNICK
BELY-DUBAU FRANCOISE
BERGEAT JACQUES G DR
BERGER XAVIER DR
BERTAUX J L DR
BIJAOUI ALBERT DR
BLAMONT JACQUES E PROF
BOCKELEE-MORVAN DOMINIQUE
BOMMIER VERONIQUE DR
BONNET ROGER M DR
BOTTINELLI LUCETTE DR
BOULANGER FRANCOIS
BOUVIER JEROME
BRAHIC ANDRE DR
BRUNET JEAN-PIERRE DR
BURKHART CLAUDE DR
CAMICHEL HENRI DR
CAPLAN JAMES
CASSE MICHEL DR
CAYREL DE STROBEL GIUSA
CELNIKIER LUDWIK DR
CHALABAEV ALMAS DR
CHAPRONT JEAN DR
CHASSEFIERE ERIC
CHOPINET MARGUERITE DR
COLIN JACQUES DR
COMBES MICHEL
CONNES PIERRE DR
COURTES GEORGES PROF
CRIFO FRANCOISE DR
CUNY YVETTE J DR
DAVOUST EMMANUEL
DEBARBAT SUZANNE V DR
DELABOUDINIERE J.-P.
DELHAYE JEAN PROF
DENNEFELD MICHEL
DIVAN LUCIENNE DR
DOLLFUS AUDOUIN PROF
DOWNES DENNIS DR
DUBOIS PASCAL DR
DUFAY MAURICE PROF
DUMONT SIMONE DR

AGRINIER BERNARD L MR
ALLEGRE CLAUDE PROF
ANDRILLAT YVETTE DR
ARLOT JEAN-EUDES
ARTZNER GUY
AUBIER MONIQUE G DR
AURIERE MICHEL
AZZOPARDI MARC DR
BATZE PAUL DR
BALUTEAU JEAN-PAUL
BARLIER FRANCOIS E DR
BEL NICOLE J DR
BENEST DANIEL DR
BERGER CHRISTIANE DR
BERGERON JACQUELINE A DR
BERTHOMIEU GABRIELLE DR
BILLAUD GERARD J
BLAZIT ALAIN DR
BOIGEY FRANCOISE
BONAZZOLA SILVANO DR
BORGNINO JULIEN DR
BOUCHER CLAUDE DR
BOULESTEIX JACQUES
BOYER CHARLES
BRETAGNON PIERRE DR
BRUSTON PAUL DR
BURNAGE ROBERT
CANAVAGGIA RENEE DR
CARQUILLAT JEAN-MICHEL
CASTETS ALAIN DR
CAYREL ROGER DR
CESARSKY CATHERINE J DR
CHAMARAUX PIERRE DR
CHAPRONT-TOUZE MICHELLE
CHEVALIER CLAUDE DR
CHRISTOPHE-GLAUME J DR
COLLIN-SOUFFRIN SUZY DR
COMTE GEORGES DR
CORNILLE MARGUERITE DR
COUTEAU PAUL PROF
CROVISIER JACQUES
DAIGNE GERARD
DE BERGH CATHERINE DR
DEHARVENG JEAN-MICHEL DR
DELACHE PHILIPPE J DR
DEMARCQ JEAN ING
DESESQUELLES JEAN DR
DOAZAN VERA DR
DONAS JOSE DR
DUBAU JACQUES DR
DUBOUT RENEE
DUFLOT MARCELLE DR
DURIEZ LUC DR

Country : FRANCE

Members :

DUROUCHOUX PHILIPPE
DUVAL MARIE-FRANCE
DUVERT GILLES DR
EDELMAN COLETTE DR
EGRET DANIEL DR
ENCRENAZ PIERRE J DR
ENCRENAZ THERESE DR
FABRE HERVE DR
FAUCHER PAUL DR
FEAUTRIER NICOLE DR
FEHRENBACH CHARLES PROF
FEISSEL MARTINE DR
FELENBOK PAUL DR
FERLET ROGER DR
FERNANDEZ JEAN-CLAUDE DR
FERRANDO PHILIPPE DR
FESTOU MICHEL C DR
FLOQUET MICHELE DR
FLORSCH ALPHONSE DR
FOING BERNARD H DR
FORT BERNARD P DR
FOSSAT ERIC DR
FOUQUE PASCAL DR
FOY RENAUD DR
FRANCOIS PATRICK DR
FREIRE FERRERO RUBENS G
FRESNEAU ALAIN DR
FRIEDJUNG MICHAEL DR
FRINGANT ANNE-MARIE DR
FRISCH HELENE DR
FRISCH URIEL DR
FROESCHLE CHRISTIANE D DR
FROESCHLE CLAUDE DR
FROESCHLE MICHEL DR
GABRIEL ALAN H
GAIGNEBET JEAN DR
GALLOUET LOUIS DR
GAMBIS DANIEL DR
GARGAUD MURIEL DR
GARNIER ROBERT ING
GAUTIER DANIEL
GAY JEAN DR
GEORGELIN YVON P DR
GEORGELIN YVONNE M DR
GERARD ERIC DR
GERBAL DANIEL DR
GERBALDI MICHELE DR
GERIN MARYVONNE DR
GILLET D DR
GIRAUD EDMOND
GOLDBACH CLAUDINE MME
GONCZI GEORGES
GORDON CHARLOTTE PROF
GORET PHILIPPE DR
GOUGUENHEIM LUCIENNE
GOUPIL MARIE JOSE
GOUTTEBROZE PIERRE DR
GREC GERARD
GRENIER SUZANNE
GREVE ALBERT DR
GRUDLER PIERRE
GRY CECILE DR
GUELIN MICHEL DR
GUERIN PIERRE DR
GUIBERT JEAN DR
GUINOT BERNARD R PROF
HALBWACHS JEAN LOUIS DR
HARVEY CHRISTOPHER C DR
HAYLI AVRAM PROF
HECK ANDRE DR
HECQUET JOSETTE DR
HEIDMANN JEAN DR
HENDECOURT D' LOUIS DR
HENON MICHEL C DR
HENOUX JEAN-CLAUDE DR
HEUDIER JEAN-LOUIS DR
HEYVAERTS JEAN DR
HOANG BINH DY DR
HUA CHON TRUNG DR
HUBERT HENRI DR
HUBERT-DELPLACE A.-M. DR
IMBERT MAURICE DR
IRIGOYEN MAYLIS
ISRAEL GUY MARCEL DR
JACQUINOT PIERRE DR
JASCHEK CARLOS O R PROF
JASCHEK MERCEDES DR
JASNIEWICZ GERARD DR
JOLY FRANCOIS DR
JOLY MONIQUE
JOUBERT MARTINE
JOURNET ALAIN
JUNG JEAN DR
KAHANE CLAUDINE DR
KANDEL ROBERT S DR
KAZES ILYA DR
KLEIN KARL LUDWIG DR
KOCH-MIRAMOND LYDIE DR
KOURGANOFF VLADIMIR PROF
KOUTCHMY SERGE DR
KOVALEVSKY JEAN DR
KRIKORIAN RALPH DR
KUNTH DANIEL
LABEYRIE ANTOINE DR
LABEYRIE JACQUES DR
LACHIEZE-REY MARC
LACLARE F MR
LACROUTE PIERRE A PROF
LAFFINEUR MARIUS MR
LAFON JEAN-PIERRE J DR
LALLEMENT ROSINE DR
LAMY PHILIPPE DR
LANNES ANDRE DR
LANTOS PIERRE DR
LAQUES PIERRE DR
LASKAR JACQUES DR
LASOTA JEAN-PIERRE DR
LATOUR JEAN J
LAUNAY JEAN-MICHEL DR
LAURENT CLAUDINE DR
LAVAL ANNIE DR
LAZAREFF BERNARD DR
LE BORGNE JEAN FRANCOIS
LE CONTEL JEAN-MICHEL
LE DOURNEUF MARYVONNE
LE FEVRE OLIVIER DR
LE SQUEREN ANNE-MARIE DR
LEBLANC YOLANDE DR
LEBRETON YVELINE DR
LEFEBVRE MICHEL DR
LEFEVRE JEAN DR
LEGER ALAIN DR
LELIEVRE GERARD DR
LEMAIRE PHILIPPE DR
LEMAITRE GERARD R DR
LENA PIERRE J PROF
LEORAT JACQUES DR
LEQUEUX JAMES DR
LEROY BERNARD DR
LEROY JEAN-LOUIS
LESTRADE JEAN FRANCOIS DR
LEVASSEUR-REGOURD A.C. PR
LEVY JACQUES R DR
LORTET MARIE CLAIRE
LOSCO LUCETTE DR
LOUISE RAYMOND PROF
LOULERGUE MICHELLE DR
LUCAS ROBERT DR
LUEST REIMAR PROF
LUMINET JEAN-PIERRE
LUNEL MADELEINE DR
MAGNAN CHRISTIAN DR
MAILLARD JEAN-PIERRE DR
MALHERBE JEAN MARIE DR
MANGENEY ANDRE DR
MARCELIN MICHEL
MARCHAL CHRISTIAN DR
MARIOTTI JEAN MARIE DR
MARTIN FRANCOIS DR

Country : FRANCE

Members :

MARTIN NICOLE DR
MARTRES MARIE-JOSEPHE
MASNOU FRANCOISE DR
MASNOU J L DR
MATHEZ GUY
MAURICE ERIC N
MAURON NICOLAS DR
MAVRIDES STAMATIA DR
MAZURE ALAIN DR
MCCARROLL RONALD PROF
MEGESSIER CLAUDE DR
MEIN NICOLE DR
MEIN PIERRE
MELLIER YANNICK DR
MENEGUZZI MAURICE M DR
MENNESSIER MARIE-ODILE DR
MERAT PARVIZ
MERAT PARVIZ DR
MERCIER CLAUDE DR
MERLEAU-PONTY J PROF
MEYER CLAUDE DR
MEYER JEAN-PAUL DR
MIANES PIERRE DR
MICHARD RAYMOND DR
MIGNARD FRANCOIS DR
MILET BERNARD L DR
MILLET JEAN DR
MILLIARD BRUNO
MOCHKOVITCH ROBERT DR
MONNET GUY J DR
MONTES CARLOS DR
MONTMERLE THIERRY DR
MORANDO BRUNO L DR
MOREELS GUY DR
MOREL PIERRE JACQUES DR
MORGULEFF NINA ING
MORRIS DAVID DR
MOURADIAN ZADIG M DR
MULLER PAUL
MULLER RICHARD DR
NADAL ROBERT
NAHON FERNAND PROF
NGUYEN-QUANG RIEU DR
NIETO JEAN-LUC
NITTMAN JOHANN
NOLLEZ GERARD DR
NOTTALE LAURENT
OBLAK EDOUARD
OMNES ROLAND PROF
OMONT ALAIN PROF
PARCELIER PIERRE DR
PARISOT JEAN-PAUL
PATUREL GEORGES
PECKER JEAN-CLAUDE PROF
PEDERSEN BENT M DR
PEDOUSSAUT ANDRE
PELLAS PAUL DR
PELLET ANDRE
PELLETIER GUY DR
PEQUIGNOT DANIEL
PERAULT MICHEL
PERRIER CHRISTIAN DR
PERRIN JEAN MARIE DR
PERRIN MARIE-NOEL DR
PETON ALAIN DR
PETRINI DANIEL DR
PEYTURAUX ROGER H PROF
PHAM-VAN JACQUELINE MME
PICAT JEAN-PIERRE DR
PICK MONIQUE DR
POQUERUSSE MICHEL
POULLE EMMANUEL PROF
POUMEYROL FERNAND MR
POUQUET ANNICK DR
POYET JEAN-PIERRE DR
PRADERIE FRANCOISE DR
PREVOT LOUIS DR
PREVOT-BURNICHON M.L. DR
PROISY PAUL E DR
PROUST DOMINIQUE
PROVOST JANINE DR
PUGET JEAN-LOUP DR
QUERCI FRANCOIS R DR
QUERCI MONIQUE DR
RABBIA YVES DR
RAOULT ANTOINETTE DR
RAPAPORT MICHEL DR
RAYROLE JEAN R DR
REBEIROT EDITH DR
REEVES HUBERT PROF
REINISCH GILBERT DR
REQUIEME YVES DR
RIBES ELIZABETH DR
RIBES JEAN-CLAUDE DR
RICORT GILBERT DR
ROBLEY R DR
ROCCA-VOLMERANGE BRIGITTE
RODDIER CLAUDE DR
ROQUES SYLVIE DR
ROSCH JEAN PROF
ROSTAS FRANCOIS DR
ROTHENFLUG ROBERT DR
ROUDIER THIERRY DR
ROUEFF EVELYNE M A DR
ROUSSEAU JEAN-MICHEL MR
ROUSSEAU JEANINE DR
ROZELOT JEAN P
RYTER CHARLES E DR
SAGNIER JEAN-LOUIS DR
SAHAL-BRECHOT SYLVIE DR
SAISSAC JOSEPH DR
SAMAIN DENYS DR
SAREYAN JEAN-PIERRE DR
SCHAEFFER RICHARD DR
SCHATZMAN EVRY PROF
SCHEIDECKER JEAN-PAUL DR
SCHMIEDER BRIGITTE DR
SCHNEIDER JEAN
SCHUMACHER GERARD DR
SEMEL MEIR DR
SERVAN BERNARD
SIBILLE FRANCOIS
SIGNORE MONIQUE DR
SIMIEN FRANCOIS DR
SIMON GUY
SIMON JEAN-LOUIS MR
SIMON PAUL A DR
SIMONNEAU EDUARDO DR
SIVAN JEAN-PIERRE DR
SOL HELENE DR
SORU-ESCAUT IRINA MRS
SOTIROVSKI PASCAL DR
SOUFFRIN PIERRE B DR
SOULIE GUY
SPITE FRANCOIS M DR
SPITE MONIQUE DR
STASINSKA GRAZYNA DR
STEENMAN-CLARK LOIS DR
STEFANOVITCH-GOMEZ A E DR
STEHLE CHANTAL DR
STEINBERG JEAN-LOUIS DR
STELLMACHER GOETZ
STELLMACHER IRENE DR
STOYKO ANNA
TALON RAOUL DR
TARRAB IRENE
TATON RENE PROF
TERRIEN JEAN
TERZAN AGOP DR
TEXEREAU JEAN M
THEVENIN FREDERIC DR
THIRY YVES R PROF
TRAN MINH NGUYET DR
TRAN-MINH FRANCOISE DR
TRELLIS MICHEL DR
TROTTET GERARD DR
TULLY JOHN A DR
TURON PIERRE
TURON-LACARRIEU C DR

Country : FRANCE

Members :

VALBOUSQUET ARMAND DR
VAN REGEMORTER HENRI DR
VAPILLON LOIC J DR
VEILLET CHRISTIAN
VERON PHILIPPE DR
VIALLEFOND FRANCOIS
VIGIER JEAN-PIERRE DR
VOLONTE SERGE DR
WALCH JEAN-JACQUES
WLERICK GERARD DR
ZEIPPEN CLAUDE DR
VALIRON PIERRE DR
VAN'T VEER FRANS DR
VAUCLAIR GERARD P DR
VERDET JEAN-PIERRE DR
VIAL JEAN-CLAUDE
VIDAL JEAN-LOUIS DR
VILMER NICOLE DR
VU DUONG TUYEN DR
WEIMER THEOPHILE P F DR
WOLTJER LODEWIJK PROF
ZOREC JEAN DR
VALTIER JEAN-CLAUDE DR
VAN'T VEER-MENNERET CL DR
VAUCLAIR SYLVIE D DR
VERON MARIE-PAULE DR
VIALA YVES
VIDAL-MADJAR ALFRED DR
VITON MAURICE DR
VUILLEMIN ANDRE DR
WENIGER SCHAME DR
ZAHN JEAN-PAUL DR

Country : GERMANY, D.R.

Members :

AURASS HENRY DR
BOERNGEN FREIMUT DR PH
DOMKE HELMUT PH D
GUSSMANN E A DR
HERRMANN DIETER PROF DR
JENSCH A
KRAUSE F DR
KUENZEL HORST
MARX SIEGFRIED DR
NOTNI P DR
PFAU WERNER
RICHTER G A DR
SCHMIDT K H DR
STANGE LOTHAR
VON BORZESZKOWSKI H H DR
WENZEL W DR
BECK H G
BUCHNER JORG DR
DORSCHNER JOHANN DR
GUTCKE DIETRICH
JACKISCH GERHARD DR
KASPER U DR
KREISEL E PROF
LIEBSCHER DIERCK-E DR
MEINIG MANFRED DR
OETKEN L DR
PFLUG KLAUS DR
RUBEN G PROF DR
SCHOENEICH W DR
STAUDE JUERGEN DR
WAGNER CHRISTIAN U DR
ZIMMERMANN HELMUT DR
BOEHME ANNELIES DR
DAUTCOURT G DR
FRIEDEMANN CHRISTIAN DR
HEMMLEB GERHARD DR
JAEGER FRIEDRICH W PROF
KOEHLER PETER
KRUEGER ALBRECHT DR
LORENZ HILMAR
MOEHLMANN DIEDRICH
OLEAK H DR
RAEDLER K H DR
SCHILBACH ELENA DR
SCHOLZ GERHARD DR
TREDER H J PROF DR
WATTENBERG D PROF

Country : GERMANY, F.R.

Members :

ALBRECHT RUDOLF DR
ALTENHOFF WILHELM J DR
ANZER ULRICH DR
APPENZELLER IMMO PROF
ASCHENBACH BERND PH D
AXFORD W IAN PROF
BAADE DIETRICH DR
BAARS JACOB W M DR
BAHNER KLAUS DR
BALTHASAR HORST DR
BARWIG HEINZ
BASCHEK BODO PROF
BASTIAN ULRICH
BECK RAINER
BECKERS JACQUES M DR
BEHR ALFRED PROF EMERITUS
BENVENUTI PIERO DR
BERKHUIJSEN ELLY M DR
BEUERMANN KLAUS P PROF
BIEN REINHOLD DR
BIERMANN PETER L DR
BIRKLE KURT PH D
BLUM PETER PROF
BOEHME SIEGFRIED DR
BOEHNHARDT HERMANN DR
BOERNER GERHARD DR
BOHN HORST-ULRICH
BOHRMANN ALFRED PROF
BRANDT PETER N
BRAUNINGER HEINRICH DR
BRAUNSFURTH EDWARD PH D
BREINHORST ROBERT A DR
BREYSACHER JACQUES
BRINKMANN WOLFGANG
BROSCHE PETER PROF
BRUCH ALBERT
BRUZEK ANTON DR
BUES IRMELA D DR
BUTLER KEITH DR
CHE-BOHNENSTENGEL ANNE
CHINI ROLF
CRANE PHILIPPE
CULLUM MARTIN DR
D'ODORICO SANDRO DR
DACHS JOACHIM PROF DR
DANZIGER I JOHN DR
DE BOER KLAAS SJOERDS DR
DE VEGT CH PROF DR
DEINZER W PROF DR
DEUBNER FRANZ-LUDWIG DR
DI SEREGO ALIGHIERI S DR
DIERCKSEN GEERD H F PH D
DORENWENDT KLAUS DR
DRAPATZ SIEGFRIED W DR
DRECHSEL HORST DR
DROZYNER ANDRZEJ
DUERBECK HILMAR W DR
DUSCHL WOLFGANG J DR
EHLERS JURGEN PROF
EL EID MOUNIB DR
ELSAESSER HANS PROF
ELWERT GERHARD PROF
ENARD DANIEL DR
ENGELHARD E J G PROF DR
ENSLIN HEINZ DR
FAHR HANS JOERG PROF DR
FECHTIG HUGO DR
FEITZINGER JOHANNES PROF
FEIX GERHARD DR
FOSBURY ROBERT A E DR
FRANK JUHAN
FREIESLEBEN H C DR
FRICKE KLAUS DR
FRIED JOSEF WILHELM DR
FUCHS BURKHARD DR
FUERST ERNST DR
GAIL HANS-PETER DR
GARAY GUIDO DR
GEBLER KARL-HEINZ DR
GEFFERT MICHAEL DR
GEHREN THOMAS PH D
GENZEL REINHARD DR
GEYER EDWARD H PROF DR
GIEREN WOLFGANG P DR
GIESEKING FRANK DR
GLATZEL WOLFGANG DR
GLEISSBERG WOLFGANG PROF
GONDOLATSCH FRIEDRICH PRF
GRAHAM DAVID A
GRAHL BERND H DR
GREWING MICHAEL PROF
GROOTE DETLEF
GROSBOL PREBEN JOHNSON DR
GROSSMANN-DOERTH U DR
GROTEN ERWIN PROF
GROTH HANS G PROF DR
GRUEN EBERHARD DR
GUESTEN ROLF
HACHENBERG OTTO PROF DR
HAEFNER REINHOLD DR
HAERENDEL G DR
HAMMAN WOLF-RAINER
HAMMER REINER
HANUSCHIK REINHARD DR
HARTQUIST THOMAS WILBUR
HARVEY GALE A DR
HASER LEO N K DR
HASLAM C GLYN T DR
HAUG EBERHARD DR
HAUG ULRICH PROF
HAUPT WOLFGANG DR
HAZLEHURST JOHN DR
HEBER ULRICH
HEFELE HERBERT PH D
HEINRICH INGE
HEMPE KLAUS
HENKEL CHRISTIAN
HENSLER GERHARD
HERMAN JACOBUS DR
HEROLD HEINZ
HILF EBERHARD R H PH D
HILLEBRANDT WOLFGANG PH D
HIPPELEIN HANS H DR
HIRTH WOLFGANG ERNST PH D
HOFFMANN MARTIN DR
HOFMANN WILFRIED DR
HOLWEGER HARTMUT PROF
HOREDT GEORG PAUL DR
HOUSE FRANKLIN C DR
HUCHTMEIER WALTER K DR
HUNGER KURT PROF
IP WING-HUEN
ISSERSTEDT JOERG DR
JAHREISS HARTMUT DR
JOCKERS KLAUS DR
JOERSAETER STEVEN DR
JORDAN H L DR DIREKTOR
KAEHLER HELMUTH DR
KAFKA PETER
KALBERLA PETER
KANBACH GOTTFRIED DR
KAUFMANN JENS PETER DR
KEGEL WILHELM H PROF
KELLER HANS ULRICH DR
KELLER HORST UWE DR
KIPPENHAHN RUDOLF PROF
KIRK JOHN DR
KLARE GERHARD DR
KLEIN ULRICH
KNEER FRANZ DR
KNOLKER MICHAEL DR
KOEHLER H PROF DR
KOEPPEN JOACHIM DR
KOHOUTEK LUBOS DR
KOLLATSCHNY WOLFRAM DR
KRAEMER GERHARD DR
KRAUTTER JOACHIM DR
KREYSA ERNST
KRUEGEL ENDRIK DR

Country : GERMANY, F.R.

Members :

KUDRITZKI ROLF-PETER PH D
KUEHNE CHRISTOPH F
KUEHR HELMUT
KUNDT WOLFGANG PROF DR
KUNITZSCH PAUL PROF.
LABS DIETRICH PROF
LAMLA ERICH E DR
LAUBERTS ANDRIS DR
LEDERLE TRUDPERT DR
LEINERT CHRISTOPH DR
LEMKE DIETRICH DR
LENZEN RAINER DR
LOOSE HANS-HERMANN DR
LUCY LEON B PROF
LUEST RHEA DR
MATAS VLADIMIR R DR
MATERNE JUERGEN DR
MATTIG W PROF DR
MAUDER HORST PROF DR
MEBOLD ULRICH DR PROF
MERKLE FRITZ DR
METZ KLAUS DR
MEURS EVERT DR
MEYER FRIEDRICH DR
MEYER-HOFMEISTER E DR
MEZGER PETER G PROF
MOELLENHOFF CLAUS DR
MOORWOOD ALAN F M
MUELLER EWALD
MUENCH GUIDO PROF
MUNDT REINHARD DR
NECKEL HEINZ DR
NECKEL TH DR
NESIS ANASTASIOS DR
NEUKUM G DR
OCHSENBEIN FRANCOIS DR
OESTREICHER ROLAND
PAULINY TOTH IVAN K K DR
PETRI WINFRIED PROF DR
PFENNIG HANS H DR
PIERRE MARGUERITE DR
PILOWSKI K PROF DR
PINKAU K PROF
PITZ ECKHART DR
POHL ECKHARD DR
PORCAS RICHARD DR
PREUSS EUGEN DR
PRIESTER WOLFGANG PROF
REFSDAL S PROF DR
REICH WOLFGANG
REIF KLAUS DR
REIMERS DIETER PROF
RICHTER JOHANNES PROF
RITTER HANS DR
ROEMER MAX PROF
ROESER HANS-PETER
ROESER HERMANN-JOSEF DR
ROESER SIEGFRIED DR
ROHLFS K PROF DR
ROSA DOROTHEA DR
ROSA MICHAEL RICHARD DR
ROTH-HOPPNER MARIA LUISE
RUDER HANNS
SCHAIFERS KARL DR
SCHEFFLER HELMUT PROF
SCHILLER KARL PROF DR
SCHINDLER KARL PROF DR
SCHLICKEISER REINHARD DR
SCHLOSSER WOLFHARD PROF
SCHLUETER A PROF DR
SCHLUETER DIETER PROF
SCHMADEL LUTZ D DR
SCHMAHL GUENTER PROF
SCHMEIDLER F PROF DR
SCHMID-BURGK J DR PROF
SCHMIDT H U DR
SCHMIDT HANS PROF
SCHMIDT THOMAS DR
SCHMIDT WOLFGANG DR
SCHMIDT-KALER TH PROF
SCHMITT DIETER DR
SCHNEIDER PETER DR
SCHNUR GERHARD F O
SCHOEFFEL EBERHARD F DR
SCHOEMBS ROLF DR
SCHOENBERNER DETLEF PROF
SCHOENFELDER VOLKER DR
SCHOLL HANS DR
SCHOLZ M PROF
SCHROEDER ROLF DR
SCHROETER EGON H PROF
SCHRUEFER EBERHARD DR
SCHUBART JOACHIM DR
SCHUESSLER MANFRED DR
SCHULTZ G V DR
SCHULZ HARTMUT DR
SCHULZ ROLF ANDREAS
SCHUMANN JOERG DIETER DR
SCHWAN HEINER DR
SCHWARTZ ROLF PH D
SEDLMAYER ERWIN DR
SEGGEWISS WILHELM PROF
SEITTER WALTRAUT C PROF
SENGBUSCH KURT V DR
SETTI GIANCARLO PROF
SHARP CHRISTOPHER DR
SHAVER PETER A DR
SHERWOOD WILLIAM A DR
SIEBER WOLFGANG PH D
SIMON KLAUS PETER
SNIJDERS MATTHEUS A J DR
SOLF JOSEF DR
SOLLAZZO CLAUDIO
STAHL OTMAR RICHARD DR
STAUBERT RUDIGER PROF DR
STAUDE HANS JAKOB PH D
STEFFEN MATTHIAS DR
STEINLE HELMUT DR
STIX MICHAEL DR
STRASSL HANS L PROF
STROHMEIER WOLFGANG PROF
STUMPFF PETER PROF DR
TAKEDA YOICHI DR
TARENGHI MASSIMO DR
TENORIO-TAGLE G DR
THIELEMANN FRIEDRICH-KARL
THIELHEIM KLAUS O DR
THOMAS HANS-CHRISTOPH DR
TRAVING GERHARD PROF
TREFFTZ ELEONORE E DR
TREUMANN RUDOLF A. DR
TRUEMPER JOACHIM PROF
TSCHARNUTER WERNER M DR
TUEG HELMUT DR
ULMSCHNEIDER PETER PROF
ULRICH BRUCE T PROF
ULRICH MARIE-HELENE D DR
UNSOELD ALBRECHT PROF
URBARZ H DR
VAN DER LAAN H PROF DR
VAN MOORSEL GUSTAAF DR
VOELK HEINRICH J PROF
VOIGT HANS H PROF
VOLLAND H DR
VON DER HEIDE JOHANN DR
VON HOERNER SEBASTIAN DR
VON WEIZSAECKER C F PROF
WACHMANN A A PROF DR
WALMSLEY C MALCOLM DR
WALTER HANS G DR
WALTER KURT PROF DR
WAMPLER E JOSEPH PROF
WEHRSE RAINER DR
WEIDEMANN VOLKER PROF
WEIGERT ALFRED PROF
WELLMANN PETER PROF DR
WENDKER HEINRICH J PROF
WEST RICHARD M DR
WIEHR EBERHARD DR

Country : GERMANY, F.R.

Members :

WIELEBINSKI RICHARD PROF
WIELEN ROLAND PROF DR
WILSON P DR
WILSON RAYMOND N DR
WILSON THOMAS L DR
WINK JOERN ERHARD DR
WINNEWISSER GISBERT DR
WITTMANN AXEL D. PH D
WITZEL ARNO DR
WOEHL HUBERTUS DR
WOLF BERNHARD PH D
WOLF RAINER E A DR
WUNNER GUENTER
YORKE HAROLD W DR
ZEKL HANS WILHELM
ZERULL REINER H DR

Country : GREECE

Members :

ABBOTT WILLIAM N DR
ALISSANDRAKIS C PH D
ANTONACOPOULOS GREG PROF
ANTONOPOULOU E DR
ARABELOS DIMITRIOS DR
ARGYRAKOS JEAN PROF DR
ASTERIADIS GEORGIOS DR
AVGOLOUPIS STAVROS DR
BANOS COSMAS J DR
BANOS GEORGE J PROF
BARBANIS BASIL PROF
BOZIS GEORGE PROF
CARANICOLAS NICHOLAS DR.
CAROUBALOS C A PROF
CONTADAKIS MICHAEL E DR
CONTOPOULOS GEORGE PROF
DANEZIS EMMANUEL DR
DARA HELEN DR
DIALETIS DIMITRIS DR
DIONYSIOU DEMETRIOS PROF
GEROYANNIS VASSILIS S DR
GOUDAS CONSTANTINE L PROF
GOUDIS CHRISTOS D PROF
HADJIDEMETRIOU JOHN D
KATSIS DEMETRIUS DR
KONTIZAS EVANGELOS DR
KONTIZAS MARY DR
KOUVELIOTOU CHRYSSA DR
KYLAFIS NIKOLAOS D DR
LASKARIDES PAUL G ASSPROF
MACRIS CONSTANTIN J PROF
MARKELLOS VASSILIS V DR
MAVRAGANIS A G PROF
MAVRIDIS L N PROF
MAVROMICHALAKI HELEN DR
MERZANIDES CONSTANTINOS
MOUSSAS XENOPHON PH D
MOUTSOULAS MICHAEL PROF
NIARCHOS PANAYIOTIS PH D
NICOLAIDIS EFTHYMIOS DR
PAPAELIAS PHILIP DR
PAPATHANASOGLOU D DR
PAPAYANNOPOULOS TH DR
PERSIDES SOTIRIOS C
PETROPOULOS BASIL CH DR
PINOTSIS ANTONIS D DR
PLAKIDIS STAVROS PROF
POULAKOS CONSTANTINE DR
PREKA-PAPADEMA P DR
PROKAKIS THEODORE J DR
ROVITHIS PETER DR
ROVITHIS-LIVANIOU HELEN
SARRIS ELEFTHERIOS PH D
SARRIS EMMANUEL T PH D
SEIRADAKIS JOHN HUGH DR
SPITHAS ELEFTERIOS N DR
SPYROU NICOLAOS PROF
SVOLOPOULOS SOTIRIOS PROF
TERZIDES CHARALAMBOS DR
THEODOSSIOU EFSTRATIOS DR
TRITAKIS BASIL P DR
TSIKOUDI VASSILIKI PH D
TSIOUMIS ALEXANDROS DR
TSIROPOULA GEORGIA DR
VARVOGLIS H DR
VEIS GEORGE PH D
VENTURA JOSEPH DR
VLACHOS DEMETRIUS G PROF
VLAHOS LOUKAS DR
VOGLIS NIKOS DR
XANTHAKIS JOHN N PROF
XANTHOPOULOS B C DR
ZACHARIADIS THEODOSIOS DR
ZAFIROPOULOS BASIL DR
ZIKIDES MICHAEL C DR

Country : HUNGARY

Members :

ALMAR IVAN PROF
BALAZS LAJOS G DR
BARCZA SZABOLCS DR
BARLAI KATALIN DR
CSADA IMRE K DR
DEZSO LORANT PROF
ERDI B DR
FEJES ISTVAN DR
GERLEI OTTO
GESZTELYI LIDIA
HORVATH ANDRAS DR
ILL MARTON J DR
ILLES ALMAR ERZSEBET DR
JANKOVICS ISTVAN DR
KALMAN BELA DR
KANYO SANDOR DR
KELEMEN JANOS
KOVACS AGNES DR
KOVACS GEZA DR
KUN MARIA DR
LOVAS MIKLOS
MARIK MIKLOS DR.
MARX GYORGY PROF.
OLAH KATALIN DR
PAAL GYORGY DR
PAPARO MARGIT DR
PATKOS LASZLO DR
SZABADOS LASZLO PH D
SZALAY ALEX DR
SZECSENYI-NAGY GABOR DR
SZEGO KAROLY DR
SZEIDL BELA DR
VERES FERENC

Country : ICELAND

Members :

GUDMUNDSSON EINAR H
SAEMUNDSON THORSTEINN

Country : INDIA

Members :

ABHYANKAR KRISHNA D PROF
AGRAWAL P C DR
AHMAD FAROOQ DR
ALLADIN SALEH MOHAMED DR
ALURKAR S K DR
AMBASTHA A K DR
ANANTHAKRISHNAN S
ANSARI S M RAZAULLAH PROF
ANTIA H M DR
APPARAO K M V DR
ASHOK N M DR
AULUCK FAQIR CHAND PROF
BABU G S D
BAGARE S P DR
BALLABH G M DR
BANDYOPADHYAY A DR
BANERJI SRIRANJAN DR
BANHATTI D G DR
BASU BAIDYANATH PROF
BASU DIPAK DR
BHANDARI N DR
BHANDARI RAJENDRA DR
BHAT CHAMAN LAL DR
BHAT NARAYANA P DR
BHATIA PREM K DR
BHATIA R K DR
BHATIA V B DR
BHATNAGAR ARVIND DR
BHATNAGAR ASHOK KUMAR
BHATNAGAR K B DR
BHATT H C DR
BHATTACHARYA DIPANKAR
BHATTACHARYYA J C PROF
BHATTACHARYYA TARA DR
BHONSLE RAJARAM V PROF
BISWAS SUKUMAR DR
BODDAPATI G ANANDARAO DR
BUTI BIMLA PROF
CHANDRA SURESH DR
CHANDRASEKHAR T DR
CHITRE SHASHIKUMAR M DR
COWSIK RAMANATH
DADHICH NARESH DR
DAMLE S V DR
DAS MRINAL KANTI
DAS P K DR
DATTA BHASKAR DR
DEGAONKAR S S DR
DESAI JYOTINDRA N
DESHPANDE AVINASH
DESHPANDE M R DR
DUORAH HIRA LAL DR
DURGAPRASAD N DR
DWARAKANATH K S
DWIVEDI BHOLA NATH DR
GAUR V P
GHOSH P DR
GHOSH S K DR
GIRIDHAR SUNETRA DR
GOKHALE MORESHWAR HARI PR
GOPALA RAO U V MR
GOSWAMI J N DR
GOYAL A N DR
GURM HARDEV S PROF
HASAN SAIYID SIRAJUL
IYENGAR K V K PROF
IYER B R DR
JAIN RAJMAL DR
JAIN SURENDRA DR
JAYARAJAN A P MR
JOG CHANDA J DR
JOSHI G C DR
JOSHI MOHAN N PROF
JOSHI SURESH CHANDRA DR
JOSHI U C DR
KANDPAL CHANDRA D
KAPOOR RAMESH CHANDER
KARANDIKAR R V PROF
KASTURIRANGAN K DR
KEMBHAVI AJIT K
KILAMBI G C DR
KOCHHAR R K DR
KOTHARI D S DR
KRISHNA GOPAL
KRISHNA SWAMY K S DR
KRISHNAMOHAN S DR
KRISHNAN THIRUVENKATA MR
KULKARNI PRABHAKAR V PROF
KULKARNI V K DR
KUSHWAHA R S PROF
LAHIRI N C
LAL DEVENDRA
MAHRA H S DR
MALLIK D C V DR
MANCHANDA R K DR
MARAR T M K
MATHUR B S DR
MITRA A P DR
MOHAN CHANDER DR
NARANAN S PROF
NARASIMHA DELAMPADY DR
NARAYANA J V
NARLIKAR JAYANT V PROF
NITYANANDA R DR
PADALIA T D DR
PADMANABHAN T DR
PANDE GIRISH CHANDRA PROF
PANDE MAHESH CHANDRA DR
PANDEY A K
PANDEY S K
PARTHASARATHY M DR
PATI A K
PERAIAH ANNAMANENI DR
PRABHU TUSHAR P
PRASANNA A R DR
PRATAP R DR
PUNETHA LALIT MOHAN DR
RADHAKRISHNAN V PROF
RAGHAVAN NIRUPAMA DR
RAJAMOHAN R DR
RAJU P K DR
RAKSHIT H PROF
RAM SAGAR DR
RAMADURAI SOURIRAJA DR
RAMAMURTHY SWAMINATHAN
RAMANA MURTHY P V DR
RANA NARAYAN CHANDRA DR
RAO M N DR
RAO N KAMESWARA
RAO P VIVEKANANDA DR
RAO RAMACHANDRA V PROF
RAUTELA B S DR
RAY ALAK DR
RAY CHOUDHURI ARNAB DR
RAYCHAUDHURI AMALKUMAR DR
RAZDAN HIRALAL
RENGARAJAN T N DR
SAHA SWAPAN KUMAR DR
SAIKIA D J DR
SANWAL N B DR
SAPRE A K DR

Country : INDIA

Members :

SARMA M B K PROF
SASTRY CH V
SAXENA P P DR
SHAH GHANSHYAM A DR
SINGH H P
SINHA K DR
SIVARAMAN K R DR
SRIVASTAVA J B DR
SUBRAHMANYA C R
TALWAR SATYA P DR
TARAFDAR SHANKAR P DR
TREHAN SURINDAR K PROF
VARDYA M S DR
VELUSAMY T DR
VERMA SATYA DEV DR
VIVEKANAND M DR
SARMA N V G PROF
SASTRY SHANKARA K
SCARIA K K DR
SHUKLA K
SINGH JAGDEV DR
SINVHAL SHAMBHU DAYAL DR.
SREEKANTAN B V DR
SRIVASTAVA RAM KUMAR DR
SUBRAHMANYAM P V DR
TANDON JAGDISH NARAIN DR
THAKUR RATNA KUMAR DR
TRIPATHI B M DR
VARMA RAM KUMAR PROF
VENUGOPAL V R DR
VINOD S KRISHAN MRS DR
VIVEKANANDA RAO
SASTRI HANUMATH J DR
SAXENA A K DR
SEN S N DR
SHUKRE C S DR
SINGH KULINDER PAL DR
SIVARAM C DR
SRINIVASAN G
STEPHENS S A DR
SWARUP GOVIND PROF
TANDON S N PROF
TONWAR SURESH C PROF.
VAIDYA P C PROF
VASU-MALLIK SUSHMA DR
VERMA R P DR
VISHVESHWARA C V PROF

Country : INDONESIA

Members :

DAWANAS DJONI N DR
RADIMAN IRATIUS
WIRAMIHARDJA SUHARDJA DR
HIDAJAT BAMBANG PROF DR
SIREGAR SURYADI DR
IBRAHIM JORGA
SUTANTYO WINARDI

Country : IRAN

Members :

ADJABSHIRIZADEH ALI
KIASATPOOR AHMAD PROF
SOBOUTI YOUSEF PROF
GHEIDARI S NASSIRI DR
MALAKPUR IRADJ DR
TEHERANY D
KALAFI MANOUCHER
MOVAHED REZA DR

Country : IRAQ

Members :

ABDULLA SHAKER ABDUL AZIZ
JABBAR SABEH RHAMAN
SADIK AZIZ R DR
AL-NAIMY HAMID M K DR
JABIR NIAMA LAFTA
YOUNIS SAAD M
AL-SABTI ABDUL ADIM DR
KADOURI TALIB HADI

Country : IRELAND

Members :

CARROLL P KEVIN PROF
CAWLEY MICHAEL DR
DRURY LUKE O'CONNOR DR
ELLIOTT IAN DR
FAHY EDWARD F PROF
FEGAN DAVID J DR
FLORIDES PETROS S PROF
GRIMLEY PETER DR
HOEY MICHAEL J DR
KENNEDY EUGENE T
KIANG TAO PROF
MCBREEN BRIAN PHILIP DR
MCCREA J DERMOTT
MCKEITH NIALL ENDA DR
MCKENNA LAWLOR SUSAN
O'CONNOR SEAMUS L DR
O'MONGAIN EON
O'SULLIVAN DENIS F
PORTER NEIL A PROF
RAY THOMAS P
REDFERN MICHAEL R DR
RUDDY VINCENT P DR
WAYMAN PATRICK A PROF
WRIXON GERARD T DR

Country : ISRAEL

Members :

BARKAT ZALMAN PROF
BEKENSTEIN JACOB D DR
BRAUN ARIE
BROSCH NOAH DR
CUPERMAN SAMI PROF
DEKEL AVISHAI
ERSHKOVICH ALEXANDER PROF
EVIATAR AHARON PROF
FINZI ARRIGO DR
GLASNER SHIMON AMI
GOLDMAN ITZHAK DR
GOLDSMITH S DR
GRADSZTAJN ELI DR
HARPAZ AMOS DR
HORWITZ GERALD PROF
IBBETSON PETER AARON DR
JOSEPH J H DR
KATZ JOSEPH DR
KOVETZ ATTAY PROF
KOZLOVSKY B Z DR
LEIBOWITZ ELIA M DR
LIVIO MARIO
MAZEH TSEVI DR
MEKLER YURI PROF
NEEMAN YUVAL PROF
NEMIROFF ROBERT DR
NETZER HAGAI DR
OHRING GEORGE PROF
PEKERIS CHAIM LEIB PROF
PRIALNIK-KOVETZ DINA DR
RAKAVY GIDEON PROF
REPHAELI YOEL DR
SACK NOAM DR
SADEH D DR
SEGALUVITZ ALEXANDER DR
SHAVIV GIORA DR
STEINITZ RAPHAEL PROF
TAUBER GERALD E PROF
TUCHMAN YITZHAK
VAGER ZEEV DR
VIDAL NISSIM V DR
YEIVIN Y PROF

Country : ITALY

Members :

ABRAMI ALBERTO PROF
ABRAMOWICZ MAREK DR
AIELLO SANTI DR
ALTAMORE ALDO
ANGELETTI LUCIO DR
ANILE ANGELO M
ANTONELLO ELIO
ANTONUCCI ESTER DR
AURIEMMA GIULIO DR
BADOLATI ENNIO
BAFFA CARLO DR
BALDINELLI LUIGI DR
BALLARIO M C PROF
BANDIERA RINO DR
BARATTA GIOVANNI BATTISTA
BARBARO G DR
BARBERIS BRUNO
BARBIERI CESARE PROF
BARBON ROBERTO PROF
BARLETTI RAFFAELE ENG
BARTOLINI CORRADO
BATTISTINI PIERLUIGI DR
BEDOGNI ROBERTO
BELVEDERE GAETANO DR
BENACCHIO LEOPOLDO
BENDINELLI ORAZIO
BERNACCA P L PROF
BERTELLI GIANPAOLO DR
BERTIN GIUSEPPE PROF
BERTOLA FRANCESCO PROF
BETTONI DANIELA DR
BIANCHI LUCIANA
BIANCHINI ANTONIO DR
BIGNAMI GIOVANNI F
BLANCO CARLO DR
BODO GIANLUIGI DR
BONIFAZI ANGELO DR
BONOLI FABRIZIO
BONOMETTO SILVIO A DR
BRACCESI ALESSANDRO PROF
BRINI DOMENICO PROF
BROGLIA PIETRO DR
BUONANNO ROBERTO
BUSON LUCIO M DR
BUSSO MAURIZIO
CACCIANI ALESSANDRO PROF
CACCIARI CARLA DR
CACCIN BRUNO
CALOI VITTORIA DR
CALVANI MASSIMO DR
CANTU ALBERTO M DR
CAPACCIOLI MASSIMO DR
CAPPELLARO ENRICO DR
CAPRIOLI GIUSEPPE PROF
CAPUTO FILIPPINA DR
CAPUZZO DOLCETTA ROBERTO
CARUSI ANDREA
CASSATELLA ANGELO DR
CASTELLANI VITTORIO PROF
CASTELLI FIORELLA DR
CATALANO FRANCESCO A DR
CATALANO SANTO DR
CATARZI MARCO DR
CAVALIERE ALFONSO G PROF
CAVALLINI FABIO
CAZZOLA PAOLO DR
CELLINO ALBERTO DR
CEPPATELLI GUIDO DR
CERRONI PRISCILLA DR
CERRUTI-SOLA MONICA
CESTER BRUNO PROF
CEVOLANI GIORDANO
CHINCARINI GUIDO L DR
CHIOSI CESARE S DR
CHIUDERI CLAUDIO PROF
CHIUDERI-DRAGO FRANCA PR
CHIUMIENTO GIUSEPPE
CHLISTOVSKY FRANCA DR
CIATTI FRANCO DR
CIMINO MASSIMO A PROF
COLOMBO G PROF DR
COLUZZI REGINA DR
COMORETTO GIOVANNI
CONCONI PAOLO DR
CORADINI ANGIOLETTA
COSMOVICI BATALLI C DR
COSTA ENRICO
CRISTALDI SALVATORE DR
CRISTIANI STEFANO DR
CRIVELLARI LUCIO
CUGUSI LEONINO DR
CURIR ANNA
D'ANTONA FRANCESCA DR
DALLAPORTA N PROF
DANESE LUIGI DR
DE BIASE GIUSEPPE A DR
DE FELICE FERNANDO DR
DE RUITER HANS RUDOLF
DE SABBATA V PROF DR
DE SANCTIS GIOVANNI
DE ZOTTI GIANFRANCO DR
DELLA VALLE MASSIMO DR
DELLI SANTI SAVERIO
DI COCCO GUIDO
DI FAZIO ALBERTO
DI MARTINO MARIO
DI TULLIO GRAZIELLA DR
EINAUDI GIORGIO
ELLIS GEORGE F R PROF
FALCHI AMBRETTA
FALCIANI ROBERTO DR
FALOMO RENATO DR
FANTI CARLA GIOVANNINI
FANTI ROBERTO
FARAGGIANA ROSANNA PROF
FARINELLA PAOLO DR
FEDERICI LUCIANA
FELLI MARCELLO DR
FERETTI LUIGINA
FERLUGA STENO DR
FERRARI ATTILIO DR
FERRARI TONIOLO MARCO
FERRERI WALTER
FERRINI FEDERICO
FICARRA ANTONINO DR
FOCARDI PAOLA DR
FORTI GIUSEPPE DR
FORTINI TERESA DR
FRACASTORO MARIO G PROF
FRANCESCHINI ALBERTO
FULCHIGNONI MARCELLO PROF
FUSCO-FEMIANO ROBERTO
FUSI PECCI FLAVIO
GALEOTTI PIERO PROF
GALLETTA GIUSEPPE PROF
GALLETTO DIONIGI
GALLINO ROBERTO
GIACHETTI RICCARDO PROF
GIALLONGO EMANUELE DR
GIANNONE PIETRO PROF
GIANNUZZI MARIA A DR
GIOVANARDI CARLO
GIOVANNELLI FRANCO DR
GIOVANNINI GABRIELE
GIURICIN GIULIANO
GODOLI GIOVANNI PROF
GOMEZ MARIA THERESA DR
GRATTON LIVIO PROF
GRATTON R G DR
GREGGIO LAURA DR
GREGORINI LORETTA
GRUBISSICH C PROF DR
GRUEFF GAVRIL DR
GUARNIERI ADRIANO DR
GUERRERO GIANANTONIO DR
HACK MARGHERITA PROF
KRANJC ALDO DR
LA PADULA CESARE
LAI SEBASTIANA

Country : ITALY

Members :

LANDI DEGL'INNOCENTI E PR
LANDI DEGL'INNOCENTI M
LANDINI MASSIMO PROF
LANDOLFI MARCO
LARI CARLO DR
LATTANZI MARIO G
LESCHIUTTA S PROF
LISEAU RENE DR
LIST FRANCO DR
LUCCHIN FRANCESCO
MACCAGNI DARIO
MACERONI CARLA
MAFFEI PAOLO PROF
MAGNI GIANFRANCO
MALAGNINI MARIA LUCIA
MAMMANO AUGUSTO DR
MANARA ALESSANDRO A DR
MANCUSO SANTI PROF
MANDOLESI NAZZARENO
MANNINO GIUSEPPE PROF
MANTEGAZZA LUCIANO
MANTOVANI FRANCO
MARANO BRUNO
MARASCHI LAURA DR
MARDIROSSIAN FABIO
MARGONI RINO
MARILLI ETTORE DR
MARMOLINO CIRO
MARTINI ALDO DR
MASANI A PROF
MASSAGLIA SILVANO
MATTEUCCI FRANCESCA
MAZZITELLI ITALO DR
MAZZONI MASSIMO DR
MAZZUCCONI FABRIZIO DR
MERIGHI ROBERTO DR
MESSINA ANTONIO
MEZZETTI MARINO
MILANI ANDREA
MILANO LEOPOLDO DR
MISSANA MARCO DR
MISSANA NATALE PROF
MOLARO PAOLO DR
MONSIGNORI FOSSI BRUNA DR
MOROSSI CARLO
MOTTA SANTO DR
MUREDDU LEONARDO DR
NATALI GIULIANO DR
NATTA ANTONELLA DR
NESCI ROBERTO
NOBILI ANNA M
NOBILI L DR
NOCERA LUIGI DR
NOCI GIANCARLO PROF
OCCHIONERO FRANCO PROF
OLIVA ERNESTO DR
ORTOLANI SERGIO
PACINI FRANCO PROF
PADRIELLI LUCIA
PALAGI FRANCESCO
PALLA FRANCESCO
PALLAVICINI ROBERTO DR
PALUMBO GIORGIO G C DR
PANNUNZIO RENATO
PAOLICCHI PAOLO DR
PARMA PAOLA
PASIAN FABIO
PASINETTI LAURA E PROF
PASTORI LIVIO
PATERNO LUCIO PROF
PATRIARCHI PATRIZIO DR
PERINOTTO MARIO PROF
PEROLA GIUSEPPE C DR
PERSI PAOLO
PETTINI MARCO
PICCIONI ADALBERTO
PIGATTO LUISA DR
PINTO GIROLAMO PROF
PIOTTO GIAMPAOLLO
PIRRONELLO VALERIO
PIZZELLA G DR
PIZZICHINI GRAZIELLA
POLCARO V F
POLETTO GIANNINA PROF
POMA ANGELO DR
PORETTI ENNIO
PREITE-MARTINEZ ANDREA DR
PROVERBIO EDOARDO PROF
PUCILLO MAURO DR
PUGLIANO ANTONIO PROF
RAFANELLI PIERO DR
RAMELLA MASSIMO
RANIERI MARCELLO
RENZINI ALVIO PROF
RIGHINI ALBERTO PROF
RIGUTTI MARIO PROF
ROBERTI GIUSEPPE DR
RODONO MARCELLO DR
ROMANO GIULIANO PROF
ROSINO LEONIDA PROF
ROSSI LUCIO
RUFFINI REMO
RUSCONI LUIGIA DR
RUSSO GUIDO DR
SABBADIN FRANCO DR
SAGGION ANTONIO PROF
SALINARI PIERO
SALVATI MARCO
SANTIN PAOLO DR
SCALTRITTI FRANCO DR
SCARDIA MARCO
SCIAMA DENNIS W DR
SECCO LUIGI DR
SEDMAK GIORGIO PROF
SELVELLI PIERLUIGI DR
SEMENZATO ROBERTO
SERIO SALVATORE DR
SEVERINO GIUSEPPE
SILVESTRO GIOVANNI
SMALDONE LUIGI ANTONIO
SMRIGLIO FILIPPO PROF
SPADA GIANFRANCO DR
STAGNI RUGGERO
STALIO ROBERTO DR
STANGA RUGGERO
STRAFELLA FRANCESCO
STRAZZULLA GIOVANNI
TAFFARA SALVATORE PROF
TAGLIAFERRI GIUSEPPE PROF
TANZELLA-NITTI GIUSEPPE
TANZI ENRICO G
TEMPESTI PIERO PROF
TOFANI GIANNI PROF
TOMASI PAOLO DR
TORELLI M DR
TORNAMBE AMEDEO
TORRICELLI GUIDETTA DR
TOSI MONICA
TOZZI GIAN PAOLO
TREVESE DARIO
TRINCHIERI GINEVRA
TRUSSONI EDOARDO
TURATTO MASSIMO DR
UBERTINI PIETRO
URAS SILVANO DR
VAGNETTI FAUSTO
VAIANA GIUSEPPE S DR
VALSECCHI GIOVANNI B DR
VERGNANO A PROF
VERNIANI FRANCO PROF
VETTOLANI GIAMPAOLO
VIETRI MARIO DR
VIGOTTI MARIO
VIOTTI ROBERTO DR
VIRGOPIA NICOLA PROF
VITTONE ALBERTO ANGELO
VITTORIO NICOLA
VLADILO GIOVANNI DR
ZAMORANI GIOVANNI

Country : ITALY

Members :

ZANINETTI LORENZO
ZAPPALA ROSARIO ALDO DR
ZAPPALA VINCENZO PROF
ZAVATTI FRANCO
ZITELLI VALENTINA DR
ZLOBEC PAOLO DR
ZUCCARELLO FRANCESCA

Country : JAPAN

Members :

AIKAWA TOSHIKI
AKABANE KENJI A PROF
AKABANE TOKUHIDE DR
ANDO HIROYASU DR
AOKI SHINKO PROF
ARAI KENZO DR
AZUMA TAKAHIRO DR
CHIKADA YOSHIHIRO DR
DAISHIDO TSUNEAKI PROF
DEGUCHI SHUJI DR
ENOME SHINZO PROF
ERIGUCHI YOSHIHARU DR
FUJIMOTO MASA-KATSU DR
FUJIMOTO MASAYUKI DR
FUJIMOTO MITSUAKI DR
FUJISHITA MITSUMI DR
FUJITA YOSHIO PROF
FUJIWARA AKIRA DR
FUKUDA ICHIRO
FUKUE JUN DR
FUKUGITA MASATAKA DR
FUKUI TAKAO DR
FUKUI YASUO DR
FUKUNAGA MASATAKA DR
FUKUSHIMA TOSHIO DR
FURUKAWA KIICHIRO DR
HABE ASAO
HACHISU IZUMI DR
HAMABE MASARU DR
HAMADA TETSUO PROF
HAMAJIMA KIYOTOSHI DR
HANAWA TOMOYUKI DR
HARA KEN NOSUKE DR
HARA TADAYOSHI DR
HARA TETSUYA DR
HASEGAWA HIROICHI DR
HASEGAWA ICHIRO DR
HASEGAWA TETSUO DR
HAYAKAWA SATIO PROF
HAYASHI CHUSHIRO PROF
HAYASHI MASAHIKO DR
HIEI EIJIRO DR
HIRABAYASHI HISASHI DR
HIRAI MASANORI DR
HIRATA RYUKO
HIRAYAMA TADASHI PROF
HITOTSUYANAGI JUICHI PROF
HORI GENICHIRO PROF
HORIUCHI RITOKU DR
HOSHI REIUN DR
HOSOKAWA YOSHIMASA H PROF
HOSOYAMA KENNOSHUKE DR
ICHIKAWA TAKASHI
ICHIMARU SETSUO DR
IIJIMA SHIGETAKA PROF
IKEUCHI SATORU DR
INAGAKI SHOGO DR
INATANI JUNJI
INOUE HAJIME DR
INOUE MAKOTO DR
INOUE TAKESHI PROF
IRIYAMA JUN DR
ISHIDA GORO DR
ISHIDA KEIICHI PROF
ISHIGURO MASATO PROF.
ISHIZAWA TOSHIAKI A PROF
ISHIZUKA TOSHIHISA DR
ISOBE SYUZO DR
ITO KENSAI A PROF
ITOH HIROSHI DR
ITOH NAOKI DR
IWASAKI KYOSUKE DR
IYE MASANORI DR
JUGAKU JUN DR
KABURAKI MASAKI PROF
KAI KEIZO DR
KAIFU NORIO DR
KAKINUMA TAKAKIYO T PROF
KAKUTA CHUICHI DR
KAMEYA OSAMU DR
KAMIJO FUMIO PROF DR
KAMINISHI KEISUKE PROF
KANEKO NOBORU DR
KASUGA TAKASHI
KATO MARIKO
KATO SHOJI PROF
KATO TAKAKO DR
KAWABATA KINAKI PROF
KAWABATA KIYOSHI
KAWABATA SHUSAKU PROF
KAWAGUCHI ICHIRO PROF
KAWATA YOSHIYUKI DR
KIGUCHI MASAYOSHI DR
KIKUCHI SADAEMON PROF
KINOSHITA HIROSHI DR
KITAMOTO SHUNJI DR
KITAMURA M PROF
KITAMURA SEIICHI DR
KOBAYASHI ETSUKE DR
KOBAYASHI YUKISAYU
KODAIRA KEIICHI PROF
KODAMA HIDEO
KOGURE TOMOKAZU DR
KONDO MASAAKI DR
KONDO MASAYUKI DR
KOSUGI TAKEO
KOYAMA KATSUJI
KOYAMA SHIN PROF DR
KOZAI YOSHIHIDE PROF
KUBO YOSHIO
KUBOTA JUN DR
KUNIEDA HIDEYO DR
KUROKAWA HIROKI DR
MAEDA KEI-ICHI DR
MAEDA KOITIRO
MAEHARA HIDEO DR
MAIHARA TOSHINORI DR
MAKINO FUMIYOSHI DR
MAKISHIMA KAZUO
MAKITA MITSUGU DR
MANABE SEIJI DR
MATSUDA TAKUYA PROF
MATSUMOTO MASAMICHI PROF
MATSUMOTO TOSHIO DR
MATSUOKA MASARU DR
MIKAMI TAKAO DR
MITSUDA KAZUHISA DR
MIYAJI SHIGEKI DR
MIYAMA SYOKEN
MIYAMOTO MASANORI DR
MIYAMOTO SIGENORI PROF
MIYAMOTO SYOTARO PROF DR
MIZUNO SHUN
MORIMOTO MASAKI DR
MORITA KAZUHIKO
MORIYAMA FUMIO PROF
MUKAI SONOYO DR
MUKAI TADASHI DR
MURAKAMI TOSHIO
NAGASAWA SHINGO PROF
NAGASE FUMIAKI DR
NAKADA YOSHIKAZU DR
NAKAGAWA NAOYA DR
NAKAGAWA YOSHINARI DR
NAKAGAWA YOSHITSUGU DR
NAKAI YOSHIHIRO
NAKAJIMA HIROSHI
NAKAJIMA KOICHI DR
NAKAMURA TAKASHI DR
NAKAMURA TSUKO DR
NAKAMURA YASUHISA
NAKANO SABURO DR
NAKANO TAKENORI DR
NAKAYAMA SHIGERU DR
NAKAZAWA KIYOSHI DR
NARIAI HIDEKAZU PROF
NARIAI KYOJI DR
NARITA SHINJI DR
NIIMI YUKIO

Country : JAPAN

Members :

NISHI KEIZO DR
NISHIDA MINORU PROF
NISHIDA MITSUGU
NISHIMURA JUN DR
NISHIMURA MASAKI
NISHIMURA SHIRO DR
NOGUCHI MASAFUMI DR
NOMOTO KEN'ICHI DR
OBI SHINYA PROF
ODA MINORU PROF
ODA NAOKI
OGAWARA YOSHIAKI
OGURA KATSUO DR
OHASHI TAKAYA DR
OHKI KENICHIRO DR
OHTANI HIROSHI DR
OHYAMA NOBORU PROF
OKAMOTO ISAO DR
OKAMURA SADANORI DR
OKAZAKI AKIRA DR
OKAZAKI SEIICHI DR
OKI TOSIO PROF DR
OKUDA HARUYUKI DR PROF
OKUDA TORU
ONAKA TAKASHI
ONO YORO PROF
OOE MASATSUGU DR
OSAKI TORU DR
OSAKI YOJI DR
OSAWA KIYOTERU DR
OWAKI NAOAKI DR
SABANO YUTAKA DR
SADAKANE KOZO DR
SAIJO KEIICHI
SAIO HIDEYUKI DR
SAITO KUNIJI PROF
SAITO MAMORU DR
SAITO SUMISABURO DR
SAKAI JUNICHI
SAKASHITA SHIRO PROF
SAKURAI KUNITOMO PROF
SAKURAI TAKASHI DR
SAKURAI TAKEO T PROF
SASAKI MISAO
SASAKI TOSHIYUKI DR
SASAO TETSUO DR
SATO FUMIO DR
SATO HUMITAKA PROF
SATO KATSUHIKO PROF
SATO KOICHI DR
SATO NAONOBU PROF
SATO SHUJI DR
SATO YUZO DR
SEKI MUNEZO DR
SEKIGUCHI NAOSUKE PROF
SHIBAHASHI HIROMOTO DR
SHIBASAKI KIYOTO
SHIBATA KAZUNARI DR
SHIBATA SHINPEI DR
SHIBATA YUKIO DR
SHIMIZU MIKIO PROF
SHIMIZU TSUTOMU PROF EMER
SIMODA MAHIRO PROF
SINZI AKIRA M DR
SOFUE YOSHIAKI PROF
SOMA MITSURU DR
SUDA KAZUO PROF
SUEMOTO ZENZABURO PROF DR
SUGAWA CHIKARA DR
SUGIMOTO DAIICHIRO PROF
SUZUKI YOSHIMASA PROF
TABARA HIROTO DR
TAKADA-HIDAI MASAHIDE DR
TAKAGI KOJIRO PROF
TAKAHARA FUMIO DR
TAKAHARA MARIKO
TAKAKUBO KEIYA PROF
TAKAKURA TATSUO PROF EMER
TAKANO TOSHIAKI DR
TAKARADA KATSUO DR
TAKASE BUNSHIRO PROF
TAKAYANAGI KAZUO PROF
TAKEDA HIDENORI DR
TAKENOUCHI TADAO DR
TAKEUTI MINE DR
TAMENAGA TATSUO DR
TAMURA SHINICHI DR
TANABE HIROYOSHI DR
TANABE KENJI DR
TANABE TOSHIHIKO DR
TANAKA KATSUO DR
TANAKA RIICHIRO PROF
TANAKA WATARU DR
TANAKA YASUO DR
TANAKA YASUO PROF
TANAKA YUTAKA D DR
TANIGUCHI YOSHIAKI DR
TAWARA YUZURU DR
TERASHITA YOICHI PROF
TOMIMATSU AKIRA DR
TOMISAKA KOHJI DR
TOMITA KENJI PROF
TOMITA KOICHIRO MR
TORAO MASAHISA
TOSA MAKOTO DR
TOSHIKI DR
TSUBAKI TOKIO PROF
TSUBOKAWA IETSUNE DR
TSUCHIYA ATSUSHI DR PROF
TSUJI TAKASHI
TSUNEMI HIROSHI DR
TSUNETA SAKU DR
TSURUTA SACHIKO DR
UCHIDA JUICHI DR
UCHIDA YUTAKA PROF
UENO SUEO PROF
UESUGI AKIRA DR
UKITA NOBUHARU
UNNO WASABURO PROF
UTSUMI KAZUHIKO DR
WAKAMATSU KEN-ICHI DR
WAKO KOJIRO DR
WASHIMI HARUICHI DR
WATANABE TAKASHI DR
WATANABE TETSUYA
YABUSHITA SHIN A PROF
YABUUTI KIYOSHI PROF
YAMAGATA TOMOHIKO DR
YAMAGUCHI SHICHIRO
YAMAKOSHI KAZUO
YAMAMOTO TETSUO DR
YAMASAKI ATSUMA DR
YAMASHITA KOJUN DR
YAMASHITA YASUMASA PROF
YAMAZAKI AKIRA DR
YASUDA HARUO PROF DR
YOKOSAWA MASAYOSHI DR
YOKOYAMA KOICHI DR
YONEYAMA TADAOKI DR
YOSHIDA HARUO
YOSHIDA JUNZO PROF
YOSHII YUZURU DR
YOSHIMURA HIROKAZU DR
YOSHIZAWA MASANORI DR
YUASA MANABU DR
YUMI SHIGERU PROF DR

Country : KOREA DPR

Members :

BAEK CHANG RYONG
BANG YONG GOL
CHA DU JIN
CHA GI UNG
CHIO CHOL ZONG
CHOI WON CHOL
DONG IL ZUN
HONG HYON IK
KANG GON IK
KANG JIN SOK
KIM JIK SU
KIM YONG HYOK DR
KIM YONG UK
KIM YUL
KIM ZONG DOK
LI GI MAN
LI GYONG WON
LI HYOK HO
LI J Y
LI SIN HYONG
LI SON JAE

Country : KOREA, REPUBLIC

Members :

CHANG KYONGAE DR
CHOE SEUNG URN DR
CHOI KYU-HONG
CHOU KYONG CHOL PROF
CHUN MUN-SUK DR
HONG SEUNG SOO DR
HYUN JONG-JUNE PROF
KIM CHUL HEE DR
KIM TU HWAN
LEE SANG GAK
LEE SEE-WOO DR
MINN YOUNG KEY DR
NHA IL-SEONG DR
OH KYU DONG DR
PARK HONG SUH DR
SHIM WOON-TAIK PROF
WOO JONG OK
YU KYUNG-LOH PROF
YUN HONG-SIK PROF

Country : MALAYSIA

Members :

ILYAS MOHAMMAD DR
OTHMAN MAZLAN

Country : MEXICO

Members :

ALLEN CHRISTINE
ARELLANO FERRO ARMANDO
BISIACCHI GIANFRANCO DR
CANTO JORGE DR
CARDONA OCTAVIO DR
CARRASCO LUIS DR
CHAVARRIA-K. CARLOS
CHAVIRA ENRIQUE SR
CHELLI ALAIN
CORNEJO ALEJANDRO A DR
COSTERO RAFAEL
CRUZ-GONZALEZ IRENE
DALTABUIT ENRIQUE DR
DE LA HERRAN V JOSE ENG
ECHEVARRIA JUAN DR
ECHEVERRIA ROMAN JUAN M.
FIERRO JULIETA
FIRMANI CLAUDIO A PROF
FRANCO JOSE DR
GALINDO TREJO JESUS DR
GARCIA-BARRETO JOSE A
GONZALEZ G
HACYAN SHAHEN DR.
HERRERA MIGUEL ANGEL DR
KLAPP JAIME DR
KOENIGSBERGER GLORIA
LOPEZ JOSE ALBERTO DR
MALACARA DANIEL
MARTINEZ MARIO DR
MENDEZ MANUEL DR
MENDOZA V EUGENIO E DR
OBREGON DIAZ OCTAVIO J DR
PEIMBERT MANUEL DR
PENA JOSE
PENICHE ROSARIO DR
PEREZ-DE-TEJADA H A DR
PEREZ-PERAZA JORGE DR
PISMIS DE RECILLAS PARIS
POVEDA ARCADIO DR
RECILLAS-CRUZ ELSA DR
RODRIGUEZ LUIS F
ROSADO MARGARITA DR
ROTH MIGUEL R DR
SARMIENTO-GALAN A F DR
SCHUSTER WILLIAM JOHN DR
SEKIGUCHI KAZUHIRO DR
SERRANO ALFONSO DR
TAPIA MAURICIO DR
TORRES-PEIMBERT SILVIA DR
WARMAN JOSEF DR

Country : NETHERLANDS

Members :

ACHTERBERG ABRAHAM DR
ANDERSEN BO NYBORG DR
ANDRIESSE CORNELIS D DR
ATANASIJEVIC IVAN DR
BARTHEL PETER DR
BAUD BOUDEWIJN DR
BEINTEMA DOUWE A DR
BLAAUW ADRIAAN PROF DR
BLEEKER JOHAN A M DR IR
BLOEMEN JOHANNES B G M DR
BOLAND WILFRIED
BORGMAN JAN DR PROF
BOSMA PIETER B DR
BRAES L L E DR
BREGMAN JACOB D IR
BREUKERS R J L H DR
BRINKMAN BERT C DR
BROUW W N DR
BURGER J J DR IR
BURTON W BUTLER DR
BUTCHER HARVEY R PROF DR
BYLEVELD WILLEM DR
CASERTANO STEFANO DR
COLEMAN PAUL HENRY DR
DE BRUYN A. GER DR
DE GRAAF T DR
DE GRAAFF W DR
DE GRAAUW TH DR
DE GROOT T DR
DE JAGER CORNELIS PROF
DE JONG TEIJE DR
DE KORT JULES J DR
DE KORTE PIETER A J DR
DEERENBERG A.J.M. DR
DEGEWIJ JOHAN DR
DEKKER E DR
FITTON BRIAN DR
FRISK URBAN DR
FRITZOVA-SVESTKA L DR
GREENBERG J MAYO DR
HABING H J DR
HAMMERSCHLAG ROBERT H DR
HAMMERSCHLAG-HENSBERGE G
HEARN ANTHONY G DR
HEINTZE J R W DR
HEISE JOHN DR
HENRICHS HUBERTUS F DR
HERMSEN WILLEM DR
HOEKSTRA ROEL DR
HOOGHOUDT B G IR
HOVENIER J W DR
HOYNG PETER DR
HUBENET HENRI DR
HUBER MARTIN C E DR
HULSBOSCH A N M DR
ICKE VINCENT DR
ISRAEL FRANK P DR
IVES JOHN CHRISTOPHER MR
JAKOBSEN PETER
KAASTRA JELLE S DR
KAHLMANN H C DR
KATGERT PETER DR
KATGERT-MERKELIJN J K DR
KUIJPERS H. JAN M.E. DR
KUPERUS MAX PROF DR
KWEE K K DR
LAMERS H J G L M DR
LE POOLE RUDOLF S DR
LUB JAN DR
METCALFE LEO DR
MEWE R DR
MILEY G K DR
MULLER A B DR
MULLER C A PROF JR
NAMBA OSAMU DR
NIEUWENHUIJZEN HANS DR
NORTH JOHN DAVID PROF
O'DEA CHRISTOPHER P DR
OLLONGREN A PROF DR
OLNON FRISO
OLTHOF HINDERICUS DR
OORT JAN H PROF
PATERSON-BEECKMANS F
PEL JAN WILLEM DR
PERRYMAN MICHAEL A C
POTTASCH STUART R PROF
RAIMOND ERNST DR
RAO A PRAMESH DR
RIPKEN HARTMUT W DR
ROOS NICOLAAS DR
ROSENBERG J DR
RUTTEN ROBERT J. DR
SANCISI RENZO DR
SANDERS ROBERT DR
SAVONIJE GERRIT JAN DR
SCHADEE AERT DR
SCHEEPMAKER ANTON DR
SCHILIZZI RICHARD T DR
SCHRIJVER JOHANNES DR
SCHWARZ ULRICH J DR
SHANE WILLIAM W DR
SHOSTAK G SETH DR
SMIT J A PROF
SPARKE LINDA
SPOELSTRA T A TH DR
STEVENS GERARD A DR
STROM RICHARD G DR
SVESTKA ZDENEK DR
SWANENBURG B N DR
TAKENS ROELF JAN DR
THE PIK-SIN PROF
TINBERGEN JAAP DR
TJIN-A-DJIE HERMAN R E DR
VAGHI SERGIO DR
VALENTIJN EDWIN A DR
VAN AGT S L TH J DR
VAN ALBADA TJEERD S DR
VAN BEEK FRANK PROF DR
VAN BUEREN HENDRIK G PROF
VAN DE HULST H C PROF DR
VAN DE KAMP PETER
VAN DE STADT HERMAN DR
VAN DEN HEUVEL EDWARD P J
VAN DER HUCHT KAREL A DR
VAN DER HULST JAN M DR
VAN DER KLIS MICHIEL DR
VAN DER KRUIT PIETER C DR
VAN DIGGELEN J DR
VAN DUINEN R J DR
VAN GENDEREN A M DR
VAN HERK G
VAN HOUTEN C J DR
VAN HOUTEN-GROENEVELD I
VAN NIEUWKOOP J DR IR
VAN PARADIJS JOHANNES DR
VAN WOERDEN HUGO PROF DR
VERBUNT FRANCISCUS DR
WESSELIUS PAUL R DR
WIJNBERGEN JAN DR
ZWAAN CORNELIS PROF DR

Country : NEW ZEALAND

Members :

ALLEN WILLIAM
BAGGALEY WILLIAM J PROF
BATES RICHARD HEATON T DR
BATESON FRANK M OBE DR
BLOW GRAHAM L
BUDDING EDWIN DR
COTTRELL PETER LEDSAM
CRAIG IAN JONATHAN D DR
DODD RICHARD J DR
DOUGHTY NOEL A DR
DOUGLASS GEOFFREY G
GILMORE ALAN C MR
HEARNSHAW JOHN B DR
JONES ALBERT F MR
KERR ROY P PROF
MARINO BRIAN F ENG
ORCHISTON WAYNE DR
RUMSEY NORMAN J
SCHATTEN KENNETH H DR
STONE RONALD CECIL
SULLIVAN DENIS JOHN DR
TOBIN WILLIAM
TRODAHL HARRY JOSEPH DR
WALKER WILLIAM S G

Country : NIGERIA

Members :

AKUJOR CHIDI E.
OKOYE SAMUEL E PROF
ONUORA LESLEY IRENE DR
SCHMITTER EDWARD F DR

Country : NORWAY

Members :

AKSNES KAARE DR
BRAHDE ROLF
CARLSSON MATS DR
ELGAROY OYSTEIN PROF
ENGVOLD ODDBJOERN DR
ERIKSEN GUNNAR PROF
HAUGE OIVIND DR
HAVNES OVE DR
JENSEN EBERHART PROF
KJELDSETH-MOE OLAV DR
LEER EGIL PROF
MALTBY PER PROF
OESTGAARD ERLEND
OWREN LEIF DR
PETTERSEN BJOERN RAGNVALD
RINGNES TRULS S DR
SOLHEIM JAN ERIK
STABELL ROLF DR
TRULSEN JAN K PROF

Country : PERU

Members :

AGUILAR MARIA LUISA

Country : POLAND

Members :

BEM JERZY DR
CIURLA TADEUSZ
CUGIER HENRYK DR
CZERNY BOZENA DR
DOBRZYCKI JERZY PROF
DOMINSKI IRENEUSZ DR
DZIEMBOWSKI WOJCIECH PROF
FLIN PIOTR
GASKA STANISLAW DR
GLEBOCKI ROBERT PROF
GORGOLEWSKI STANISLAW PR
GRABOWSKI BOLESLAW DR
GRUDZINSKA STEFANIA DR
GRZEDZIELSKI STANISLAW PR
HAENSEL PAWEL DR
HANASZ JAN DR
HELLER MICHAEL PROF
HURNIK HIERONIM PROF
IWANISZEWSKA CECYLIA DR
IWANOWSKA WILHELMINA PROF
JAKIMIEC JERZY PROF
JAKS WALDEMAR DR
JAROSZYNSKI MICHAL
JARZEBOWSKI TADEUSZ DR
JERZYKIEWICZ MIKOLAJ DR
JUSZKIEWICZ ROMAN
KALUZNY JANUSZ DR
KOLACZEK BARBARA DR
KOZIEL KAROL PROF DR
KOZLOWSKI MACIEJ DR
KRASINSKI ANDRZEJ PROF.
KREINER JERZY MAREK DR
KRELOWSKI JACEK DR
KREMPEC-KRYGIER JANINA DR
KRUSZEWSKI ANDRZEJ PROF
KUBIAK MARCIN A DR
KURPINSKA-WINIARSKA M DR
KUS ANDRZEJ JAN DR
LEHMANN MAREK DR
MACHALSKI JERZY DR
MADEJ JERZY
MASLOWSKI JOZEF DR
MERGENTALER JAN PROF
MICHALEC ADAM
MIETELSKI JAN S DR
MIKOLAJEWSKA JOANNA DR
MOCZKO JANUSZ DR
OPOLSKI ANTONI PROF
PACZYNSKI BOHDAN PROF
PROSZYNSKI MIECZYSLAW
ROMPOLT BOGDAN DR
ROZYCZKA MICHAL
RUDAK BRONISLAW
RYBKA PRZEMYSLAW DR
SCHREIBER ROMAN
SCHWARZENBERG-CZERNY A
SEMENIUK IRENA DR
SERAFIN RICHARD AUGUST
SIENKIEWICZ RYSZARD DR
SIKORA MAREK
SIKORSKI JERZY DR
SITARSKI GRZEGORZ PROF
SMAK JOSEPH I PROF
SMOLINSKI JAN DR
SOKOLOWSKI LECH
SOLTAN ANDRZEJ MARIA DR
STAWIKOWSKI ANTONI DR
STEPIEN KAZIMIERZ DR
STROBEL ANDRZEJ DR
SYLWESTER BARBARA DR
SYLWESTER JANUSZ
SZAFRANIEC ROZALIA DR
TURLO ZYGMUNT DR
TYLENDA ROMUALD DR
URBANIK MAREK DR
WINIARSKI MACIEJ
WNUK EDWIN
WOSZCZYK ANDRZEJ PROF
ZIEBA STANISLAW DR
ZIOLKOWSKI JANUSZ DR
ZIOLKOWSKI KRZYSZTOF DR

Country : PORTUGAL

Members :

CABRITA EZEQUIEL DR
CAMPOS L M BRAGA DA COSTA
COELHO BALSA MARIO C DR
DA SILVA A V C S
DE BRITO e ABREU J C DR
DOS REIS M PROF
LAGO MARIA TERESA V T PR.
MACHADO JOSE M A B DR
MAGALHAES ANTONIO A S ENG
MARQUES MANUEL N DR
NUNES ROGERIO S DE SOUSA
OSORIO JOSE J S P PROF
PASCOAL ANTONIO J B SCI
TABORDA JOSE ROSA DR
TAVARES J T L DR
VICENTE RAIMUNDO O PROF

Country : RUMANIA

Members :

BOTEZ ELVIRA DR
CRISTESCU CORNELIA G DR
DINESCU A DR
DINULESCU NICOLAE I PROF
DRAMBA C PROF
LUNGU NICOLAIE DR
MIHAILA IERONIM PROF
NADOLSCHI V PROF DR
OPROTU TIBERIU DR
PAL ARPAD PROF DR
RUSU I DR
RUSU L DR
STANILA GEORGE DR
TIFREA EMILIA DR
TODORAN IOAN DR
URECHE VASILE DR

Country : SAUDI ARABIA

Members :

BROSTERHUS E B F DR
HAMZAOGLU ESAT E H DR
TOPAKTAS LATIF A DR
TUFEKCIOGLU ZEKI DR

Country : SOUTH AFRICA

Members :

BAART EDWARD E PROF
BALONA LUIS ANTERO DR
BARRETT PAUL EVERETT DR
BENNETT JOHN CAISTER MR
CALDWELL JOHN A R
CATCHPOLE ROBIN M DR
CHURMS JOSEPH
CILLIE G G PROF
COUSINS A W J DR
DE JAGER GERHARD PROF
DE JAGER OCKER C DR
ENGELBRECHT CHRISTIAN DR
EVANGELIDIS E DR
FAIRALL ANTHONY P PROF
FEAST MICHAEL W PROF
GAYLARD MICHAEL JOHN
GLASS IAN STEWART DR
HERS JAN MR
HIRST WILLIAM P
JARRETT ALAN H PROF
JONAS JUSTIN LEONARD
KILKENNY DAVID DR
KURTZ DONALD WAYNE DR
LANEY CLIFTON D DR
LLOYD EVANS THOMAS DR
MENZIES JOHN W DR
NICOLSON GEORGE D DR
O'DONOGHUE DARRAGH DR
OVERBEEK MICHIEL DANIEL
RAUBENHEIMER BAREND C PR
STOKER PIETER H
WALRAVEN TH DR
WARGAU WALTER F DR
WARNER BRIAN PROF
WHITELOCK PATRICIA ANN DR

Country : SPAIN

Members :

ABAD ALBERTO J DR	ALFARO EMILIO JAVIER	ALVAREZ P
ARRIBAS SANTIAGO DR	BACHILLER RAFAEL DR	BATTANER EDUARDO DR
BECKMAN JOHN E PROF	BENAVENTE JOSE	BOLOIX RAFAEL DR
BONET JOSE A	BUITRAGO JESUS	BUJARRABAL VALENTIN
CALVO MANUEL	CAMARENA BADIA VICENTE PR	CANAL RAMON M DR
CARDUS ALMEDA J O MR	CATALA MARIA ASUNCION DR	CATALAN MANUEL DR
CERNICHARO JOSE DR	CID PALACIOS RAFAEL PROF	CLAVEL JEAN
CODINA VIDAL J M DR	COLLADOS MANUEL DR	COMA JUAN CARLOS
CORNIDE MANUEL	COSTA VICTOR DR	DE CASTRO ANGEL DR
DE CASTRO ELISA	DE PASCUAL MARTINEZ M DR	DEL RIO GERARDO DR
DEL TORO INIESTA JOSE DR	DELGADO ANTONIO JESUS	DIAZ ANGELES ISABEL DR
DOCOBO DURANTEZ JOSE A	DULTZIN-HACYAN D. DR	EIROA DE SAN FRANCISCO C
ELIPE SANCHEZ ANTONIO	ESTALELLA ROBERT	FERNANDEZ-FIGUEROA M J DR
FERRER MARTINEZ SEBASTIAN	FIGUERAS FRANCESCA DR	FUENSALIDA JIMENEZ J DR
GALAN MAXIMINO J	GARCIA DE LA ROSA JOSE I	GARCIA-PELAYO JOSE DR
GARRIDO RAFAEL	GARZON FRANCISCO DR	GILMOZZI ROBERTO
GIMENEZ ALVARO	GOMEZ GONZALEZ JESUS DR	GONZALEZ CAMACHO ANTONIO
GONZALEZ-RIESTRA R DR	GORGAS GARCIA JAVIER DR	HERNANZ MARGARITA DR
HERRERO DAVO ARTEMIO DR	HIDALGO MIGUEL A DR	ISERN JORGE DR
JORDI NEBOT CARME DR	LABAY JAVIER	LAHULLA J FORNIES DR
LAZARO CARLOS DR	LING J DR	LOPEZ DE COCA M D P DR
LOPEZ ROSARIO DR	LOPEZ-ARROYO M	LOPEZ-MORENO JOSE JUAN
LOPEZ-PUERTAS MANUEL	MARCAIDE JUAN-MARIA DR	MARTIN-DIAZ CARLOS DR
MARTIN-LORON M DR	MARTIN-PINTADO JESUS	MARTINEZ ROGER CARLOS DR
MEDIAVILLA EVENCIO DR	MOLES MARIANO J DR	MOLINA ANTONIO
MORALES-DURAN CARMEN	MORENO-INSERTIS FERNANDO	MUINOS JOSE L DR
MUNOZ-TUNON CASIANA	NUNEZ JORGE DR	ORTE ALBERTO
ORUS JUAN J PROF	PALLE PERE-LLUIS DR	PAREDES JOSE MARIA DR
PENSADO JOSE DR	PEREA-DUARTE JAIME D DR	PLANESAS PERE
PRIETO MERCEDES	QUIJANO LUIS	QUINTANA JOSE M DR
REBOLO RAFAEL DR	REGLERO-VELASCO VICTOR DR	REGO FERNANDEZ M DR
REGULO CLARA DR	ROCA CORTES TEODORO	RODRIGO RAFAEL
RODRIGUEZ-ESPINOSA JOSE	ROLLAND ANGEL DR	ROMERO PEREZ M PILAR
ROSSELLO GASPAR	SALA FERRAN DR	SALVADOR-SOLE EDUARDO
SANAHUJA BLAS	SANCHEZ FRANCISCO PROF	SANCHEZ MANUEL
SANCHEZ-SAAVEDRA M LUISA	SANROMA MANUEL DR	SANZ JOSE L DR
SEIN-ECHALUCE M LUISA DR	SEVILLA MIGUEL J DR	SIMO CHARLES DR
STEPPE HANS DR	TALAVERA A DR	THUM CLEMENS DR
TORRA JORDI DR	TORROJA J PROF	VAZQUEZ MANUEL DR
VILCHEZ MEDINA JOSE M DR	VIVES TEODORO JOSE DR	WAMSTEKER WILLEM DR
WYLLER ARNE A PROF	ZAMORANO JAIME DR	

Country : SWEDEN

Members :

ADOLFSSON TORD DR
ALFVEN HANNES PROF
ARDEBERG ARNE L PROF
BAATH LARS B DR
BERGVALL NILS AKE SIGVARD
BJORNSSON CLAES-INGVAR
BOOTH ROY S PROF
CARLQVIST PER A DR
CATO B TORGNY DR
DRAVINS DAINIS PROF
EDLEN BENGT PROF
ELLDER JOEL DR
ELVIUS AINA M PROF
ELVIUS TORD PROF EMERITUS
ERIKSSON KJELL DR
FAELTHAMMAR CARL GUNNE PR
FRANSSON CLAES
FREDGA KERSTIN PROF
FRIBERG PER
GAHM GOESTA F DR
GUSTAFSON BO A S
GUSTAFSSON BENGT DR
HANSSON NILS DR
HJALMARSON AKE G DR
HOEGBOM JAN A DR
HOEGLUND BERTIL PROF
HOLMBERG ERIK B PROF
JOHANSSON LARS ERIK B DR
KOLLBERG ERIK L PROF
KRISTENSON HENRIK DR
KRISTIANSSON KRISTER PROF
LAGERKVIST CLAES-INGVAR
LAGERQVIST ALBIN PROF
LARSSON-LEANDER G PROF
LAURENT BERTEL E PROF
LEHNERT B P PROF
LINDBLAD BERTIL A DR
LINDBLAD PER OLOF PROF
LINDEGREN LENNART DR
LODEN KERSTIN R DR
LODEN LARS OLOF PROF
LUNDSTEDT HENRIK DR
LUNDSTROM INGEMAR DR
LYNGA GOSTA DR
LYTTKENS EJNAR DR
NILSON PETER DR
NORDH H LENNART DR
NYMAN LARS-AKE DR
OEHMAN YNGVE PROF
OJA TARMO PROF
OLOFSSON HANS
OLOFSSON S GOERAN DR
RAADU MICHAEL A DR
RAMBERG JOERAN M PROF
RICKMAN HANS DR
ROENNAENG BERNT O DR
ROSLUND CURT DR
ROSQUIST KJELL
RYDBECK GUSTAF H B DR
RYDBECK OLOF E H PROF
SAHAI RAGHVENDRA DR
SANDQVIST AAGE DR
SCHALEN CARL PROF
SCHARMER GOERAN BJARNE
SINNERSTAD ULF E PROF
SOEDERHJELM STAFFAN DR
STENHOLM BJOERN DR
STENHOLM LARS
SUNDMAN ANITA DR
SWENSSON JOHN W DR
VAN GRONINGEN ERNST DR
WALLENQUIST AAKE A E PROF
WESTERLUND BENGT E PROF
WIEDLING TOR DR
WINNBERG ANDERS DR
WRAMDEMARK STIG S O DR

Country : SWITZERLAND

Members :

BARTHOLDI PAUL DR
BECKER WILHELM PROF
BENZ ARNOLD DR
BINGGELI BRUNO
BLECHA ANDRE BORIS G DR
BOCHSLER PETER
BONANOMI JACQUES DR
BOUVIER PIERRE PROF
BURKI GILBERT DR
BUSER ROLAND DR
CAMERON LUZIUS MARTIN
CHMIELEWSKI YVES DR
COURVOISIER THIERRY J.-L.
DEBRUNNER HERMANN DR
DRESSLER KURT PROF
DUERST JOHANNES DR
FENKART ROLF P PROF DR
FROEHLICH CLAUS
GEISS JOHANNES PROF
GOLAY MARCEL PROF
GOY GERALD PROF
GRENON MICHEL DR
HAUCK BERNARD PROF
JAVET PIERRE PROF
KRAAN-KORTEWEG RENEE C DR
LABHARDT LUKAS
LANZ THIERRY DR
MAEDER ANDRE PROF
MAETZLER CHRISTIAN DR
MAGUN ANDREAS DR
MARTINET LOUIS PROF
MATHYS GAUTIER DR
MAYOR MICHEL DR
MERMILLIOD JEAN-CLAUDE DR
MUELLER EDITH A PROF
MUELLER HELMUT O PROF DR
NICOLET BERNARD
NORTH PIERRE
NUSSBAUMER HARRY PROF
PFENNIGER DANIEL DR
RUFENER FREDY G PROF
SCHANDA ERWIN PROF
SCHULER WALTER DR
SPAENHAUER ANDREAS MARTIN
STEINLIN ULI PROF
STENFLO JAN O DR
TAMMANN G ANDREAS PROF DR
TREFZGER CHARLES F DR
WALDMEIER MAX PROF DR
WILD PAUL PROF
XIA ZHIGUO DR
ZELENKA ANTOINE DR

Country : TURKEY

Members :

AKCAYLI MELEK M A DR
AKYOL MUSTAFA UNAL PROF
ASLAN ZEKI DR
ATAC TAMER
AVCIOGLU KAMURAN PROF DR
AYDIN CEMAL PROF DR
BALLI EDIBE PROF
BOLCAL CETIN DR
BOYDAG-YILDIZDOGDU F S
BOZKURT SUKRU DR
DEMIRCAN OSMAN DR
DERMAN I ETHEM DR
DIZER MUAMMER PROF
DOGAN NADIR PROF
ENGIN SEMANUR PROF
ERCAN E. NIHAL
ERTAN A YENER DR
ESKIOGLU A NIHAT
EZER-ERYURT DILHAN PROF
GOELBASI ORHAN DR
GOKDOGAN NUZHET PROF
GUDUR N DR
GULMEN OMUR DR
HAZER S DR
HOTINLI METIN DR
IBANOGLU C DR
KANDEMIR GUELCIN
KARAALI SALIH DR
KIRAL ADNAN PROF
KIRBIYIK HALIL DR
KOCER D DR
MARSOGLU A DR
MENTESE HUSEYIN DR
OZGUC ATILA
PEKUENLUE E RENNAN DR
SEZER CENGIZ DR
TEKTUNALI H GOKMEN DR
TUNCA ZEYNEL DR
YILMAZ FATMA DR
YILMAZ NIHAL DR

Country : URUGUAY

Members :

FERNANDEZ JULIO A DR

Country : U.K.

Members :

AARSETH SVERRE J DR
ADAM MADGE G DR
ADAMS DAVID J DR
ADAMSON ANDREW DR
ADE PETER A R DR
ALBINSON JAMES DR
ALEXANDER JOHN B
ALEXANDER PAUL DR
ALLAN PETER M
ALLEN ANTHONY JOHN DR
ANDERSON BRYAN DR
ANDREWS DAVID A DR
ANDREWS PETER J DR
ARDAVAN HOUSHANG DR
ARGUE A NOEL MR
ASSOUSA GEORGE ELIAS DR
ATHERTON PAUL DAVID
AXON DAVID
BAILEY MARK EDWARD
BALDWIN JOHN E DR
BARLOW MICHAEL J DR
BAROCAS VINICIO PROF
BARROW COLIN H DR
BARROW JOHN DAVID
BARROW RICHARD F DR
BASTIN JOHN A PROF
BATES BRIAN DR
BATES DAVID R PROF
BATH GEOFFREY T DR
BEALE JOHN S DR
BEGGS DENIS W MR
BELL BURNELL S JOCELYN DR
BELL KENNETH LLOYD DR
BENN CHRIS R DR
BERGER MITCHELL DR
BERRINGTON KEITH ADRIAN
BINGHAM RICHARD G DR
BINNEY JAMES J DR
BLACKMAN CLINTON PAUL DR
BLACKWELL DONALD E PROF
BODE MICHAEL F
BOKSENBERG ALEC PROF
BONDI HERMANN PROF SIR
BONNOR W B PROF
BOYD ROBERT L F PROF SIR
BRAND PETER W J L DR
BRANDUARDI-RAYMONT G
BRANSON NICHOLAS J B A DR
BRIDGELAND MICHAEL DR
BROMAGE GORDON E DR
BROOKES CLIVE J DR
BROWN JOHN C PROF
BROWNE IAN W A DR
BROWNING PHILIPPA DR
BRUCK HERMANN A PROF
BRUCK MARY T DR
BUNCLARK PETER STEPHEN DR
BURGESS ALAN DR
BURGESS DAVID D PROF
BURTON WILLIAM M
BUTCHINS SYDNEY ADAIR
BUTLER C JOHN DR
BYRNE PATRICK B DR
CALLANAN PAUL DR
CAMERON ANDREW COLLIER DR
CAMPBELL JAMES W
CARR BERNARD JOHN
CARSON T R DR
CARSWELL ROBERT F DR
CHARLES PHILIP ALLAN
CLARK DAVID H DR
CLARKE DAVID DR
CLEGG PETER E DR
CLEGG ROBIN E S DR
CLUBE S V M DR
COHEN RAYMOND J DR
COLLINSON EDWARD H
CONWAY ROBIN G DR
COOK ALAN H PROF
COOKE B A DR
COOKE JOHN ALAN
COUPER HEATHER MISS
COWLING THOMAS G PROF
CROOM DAVID L DR
CRUISE ADRIAN MICHAEL DR
CULHANE LEONARD PROF
CZERNY MICHAL DR
DAINTREE EDWARD J DR
DARIUS JON DR
DAVIDSON WILLIAM PROF
DAVIES JOHN G DR
DAVIES PAUL CHARLES W
DAVIES RODNEY D PROF
DAVIS RICHARD J DR
DE GROOT MART DR
DENNISON P A DR
DEWHIRST DAVID W DR
DICKENS ROBERT J DR
DISNEY MICHAEL J PROF
DONNISON JOHN RICHARD DR
DORMAND JOHN RICHARD DR
DOWNES ANN JULIET B
DOYLE JOHN GERARD
DREVER RONALD W P DR
DREW JANET
DUFFETT-SMITH PETER JAMES
DUFTON PHILIP L DR
DUNLOP STORM
DWORETSKY MICHAEL M DR
DYSON JOHN E DR
ECCLES MICHAEL J DR
EDMUNDS MICHAEL GEOFFREY
EDWIN ROGER P
EFSTATHIOU GEORGE
EGGLETON PETER P DR
ELLIOTT KENNETH H DR
ELLIS RICHARD S
ELSMORE BRUCE DR
EMERSON DAVID
EMERSON JAMES P
EVANS ANEURIN
EVANS KENTON DOWER DR
EVANS ROGER G DR
FABIAN ANDREW C DR
FALLE SAMUEL A DR
FAWELL DEREK R DR
FELLGETT PETER PROF
FIELD DAVID
FIELDER GILBERT DR
FLETT ALISTAIR M
FLOWER DAVID R DR
FONG RICHARD
FOX W E MR
FRENK CARLOS S
FURNISS IAN
GARTON W R S PROF
GEAKE JOHN E DR
GENT HUBERT MR
GERHARD ORTWIN
GIETZEN JOSEPH W
GILMORE GERARD FRANCIS
GLENCROSS WILLIAM M DR
GODWIN JON GUNNAR DR
GOLDSWORTHY FREDERICK A
GONDHALEKAR PRABHAKAR DR
GOUGH DOUGLAS O DR
GRAINGER JOHN F DR
GRANT IAN P DR
GREEN ROBIN M DR
GREEN SIMON F
GRIFFIN MATTHEW J DR
GRIFFIN RITA E M DR
GRIFFIN ROGER F DR
GRIFFITHS WILLIAM K
GUEST JOHN E DR
GULL STEPHEN F DR
GUTHRIE BRUCE N G DR
HADLEY BRIAN W
HALL ANDREW NORMAN

Country : U.K.

Members :

HANBURY BROWN ROBERT PROF
HARMER DIANNE L MRS
HARRIS ALAN WILLIAM
HARRIS STELLA
HARRISON RICHARD A DR
HARTLEY KENNETH F DR
HASSALL BARBARA J M DR
HAWARDEN TIMOTHY G DR
HAWKING STEPHEN W PROF
HAWKINS MICHAEL R S
HAYWARD JOHN
HAZARD CYRIL DR
HEAVENS ALAN DR
HEDDLE DOUGLAS W O PROF
HEGGIE DOUGLAS C DR
HEWETT PAUL
HEWISH ANTONY PROF
HEY JAMES STANLEY DR
HIDE RAYMOND PROF
HILDITCH RONALD W DR
HILL PHILIP W DR
HILLS RICHARD E DR
HILTON JOHN DR
HOLDEN FRANK
HOLLOWAY NIGEL J DR
HOOD ALAN
HOOD ALAN DR
HOSKIN MICHAEL A DR
HOWARTH IAN DONALD
HOWSE H DEREK
HOYLE FRED SIR
HUGHES DAVID W DR
HUMMEL EDSHO
HUMPHRIES COLIN M DR
HUTCHEON RICHARD J DR
HYSOM EDMUND J
IRELAND JOHN G DR
IRWIN MICHAEL JOHN DR
ISAAK GEORGE R PROF
JACKSON JOHN CHARLES DR
JAMES JOHN F MR
JAMES RICHARD A DR
JAMESON RICHARD F DR
JEFFERY CHRISTOPHER S DR
JEFFREYS HAROLD PROF SIR
JELLEY JOHN V PHD
JENKINS CHARLES R
JENNINGS R E PROF
JENNISON ROGER C PROF
JONES DEREK H P DR
JORDAN CAROLE DR
JORDEN PAUL RICHARD
JOSEPH ROBERT D DR
JUPP ALAN H DR
KAHN FRANZ D PROF
KAISER THOMAS R PROF
KAZANTZIS PANAYOTIS DR
KENDERDINE SIDNEY DR
KIBBLEWHITE EDWARD J DR
KING ANDREW R DR
KING HENRY C DR
KING-HELE DESMOND G DR
KINGSTON ARTHUR E PROF
KITCHIN CHRISTOPHER R DR
KOPAL ZDENEK PROF
KROTO HAROLD PROF.
LA DOUS CONSTANZE A DR
LAING ROBERT
LANCASTER BROWN PETER
LANG JAMES DR
LASENBY ANTHONY
LAWRENCE ANDREW DR
LEE TERENCE J DR
LITTLE LESLIE T DR
LONGAIR M S PROF
LOVELL SIR BERNARD PROF
LUCEY JOHN DR
LYNAS-GRAY ANTHONY E
LYNDEN-BELL DONALD PROF
LYNE ANDREW G DR
LYTTLETON RAYMOND A PROF
MACCALLUM MALCOLM A H
MACDONALD GEOFFREY H DR
MACGILLIVRAY HARVEY T DR
MACKAY CRAIG D DR
MACKINNON ALEXANDER L
MADDISON RONALD CH DR
MALIN STUART
MALLIA EDWARD A DR
MAREK JOHN
MARSDEN PHILIP L PROF
MARSH JULIAN C D
MARSHALL KEVIN
MARTIN ANTHONY R DR
MARTIN DEREK H PROF
MARTIN WILLIAM L DR
MASON HELEN E DR
MASON KEITH OWEN
MATHESON DAVID NICHOLAS
MCCREA WILLIAM SIR
MCDONNELL J A M PROF
MCHARDY IAN MICHAEL DR
MCKEITH CONAL D DR
MCMULLAN DENNIS DR
MCNALLY DEREK DR
MCWHIRTER R W PETER DR
MEABURN J DR
MEADOWS A JACK PROF
MEIKLE WILLIAM P S
MESSAGE PHILIP J DR
MESTEL LEON PROF
MILES HOWARD G MR
MILLAR THOMAS J DR
MILLER JOHN C DR
MILLS ALLAN A DR
MITTON JACQUELINE
MITTON SIMON DR
MOFFATT HENRY KEITH PROF
MONTEIRO TANIA DR
MOORE DANIEL R DR
MOORE PATRICK DR
MORGAN BRIAN LEALAN
MORGAN DAVID H DR
MORISON IAN MR
MORRIS MICHAEL C
MORRISON LESLIE V
MOSS CHRISTOPHER DR
MOSS DAVID L DR
MURDIN PAUL G DR
MURRAY CARL D
MURRAY JOHN B DR
MUXLOW THOMAS
NANDY KASHINATH DR
NAPIER WILLIAM M DR
NELSON ALISTAIR H DR
NICHOLSON WILLIAM
O'HORA NATHY P J
OSBORNE JOHN L DR
PADMAN RACHAEL
PAGE CLIVE G DR
PAGEL BERNARD E J PROF
PALMER PHILIP DR
PAPALOIZOU JOHN C B DR
PARKER EDWARD A DR
PARKINSON JOHN H DR
PAXTON HAROLD J B R
PEACH GILLIAN DR
PEACH JOHN V DR
PEACOCK JOHN ANDREW
PEARCE GILLIAN DR
PENNY ALAN JOHN DR
PENSTON MARGARET
PENSTON MICHAEL V DR
PERRY JUDITH J DR
PETFORD A DAVID DR
PETROU MARIA DR
PETTINI MAX
PHILLIPS JOHN PETER
PHILLIPS KENNETH J H

Country : U.K.

Members :

PIKE CHRISTOPHER DAVID
PILKINGTON JOHN D H DR
PONMAN TREVOR DR
PONSONBY JOHN E B DR
POOLEY GUY DR
POUNDS KENNETH A PROF
PRIEST ERIC R PROF
PRINGLE JAMES E DR
PRINJA RAMAN DR
PYE JOHN P DR
QUENBY JOHN J DR
RACKHAM THOMAS W DR
RAINE DEREK J DR
RAPLEY CHRISTOPHER G DR
REAY NEWRICK K DR
REES MARTIN J PROF
RICHARDSON KEVIN J
RILEY JULIA M DR
RING JAMES PROF
ROBERTS BERNARD DR
ROBERTSON JOHN ALISTAIR
ROBINSON WILLIAM J DR
ROBSON IAN E DR
RONAN COLIN A
ROWAN-ROBINSON MICHAEL DR
ROWSON BARRIE DR
ROXBURGH IAN W PROF
ROY ARCHIE E PROF
RUNCORN S K PROF
SANFORD PETER WILLIAM MR
SAUNDERS RICHARD D.E.
SCARROTT STANLEY M DR
SCHEUER PETER A G DR
SCHILD HANSRUEDI
SCHUTZ BERNARD F PROF
SCHWARTZ STEVEN JAY
SCOTT PAUL F DR
SEATON MICHAEL J PROF
SELLWOOD JEREMY ARTHUR
SEYMOUR P A H
SHAKESHAFT JOHN R DR
SHALLIS MICHAEL J DR
SIM MARY E MISS
SIMMONS JOHN FRANCIS L
SIMNETT GEORGE M
SIMONS STUART DR
SINCLAIR ANDREW T DR
SISSON GEORGE M MR
SKILLEN IAN DR
SKILLING JOHN DR
SKINNER GERALD DR
SMITH F GRAHAM PROF
SMITH GEOFFREY DR
SMITH HUMPHRY M
SMITH LINDA J
SMITH ROBERT CONNON DR
SMITH RODNEY M DR
SMYTH MICHAEL J DR
SOLANKI SAMI K DR
SOMERVILLE WILLIAM B DR
SORENSEN SOREN-AKSEL DR
SPARKS WILLIAM BRIAN
SPEER R J DR
SPENCER RALPH E DR
STANNARD DAVID DR
STEPHENSON F RICHARD DR
STEWART JOHN MALCOLM DR
STEWART PAUL DR
STIBBS DOUGLAS W N PROF
STICKLAND DAVID J DR
STOBIE ROBERT S DR
SUBRAMANIAN KANDASWAMY DR
SUMMERS HUGH P DR
SWEET PETER A PROF
SYKES-HART AVRIL B DR
TAVAKOL REZA
TAYLER ROGER J PROF
TAYLOR DONALD BOGGIA DR
TAYLOR KEITH DR
TER HAAR DIRK
TERLEVICH ELENA DR
TERLEVICH ROBERTO JUAN
THOBURN CHRISTINE
THOMAS DAVID V DR
THOMASSON PETER DR
THOMPSON G I DR
TOZER DAVID C DR
TRITTON KEITH P DR
TRITTON SUSAN BARBARA
TURNER MARTIN J L DR
TWISS R Q DR
TWORKOWSKI ANDRZEJ S
UNGER STEPHEN DR
VAN BREDA IAN G DR
VAN DER RAAY HERMAN B
VAN LEEUWEN FLOOR DR
VECK NICHOLAS
WALKER EDWARD N MR
WALKER IAN WALTER
WALL JASPER V DR
WALLACE PATRICK T MR
WALLIS MAX K DR
WALSH DENNIS DR
WARD MARTIN JOHN
WARNER PETER J DR
WARWICK ROBERT S DR
WEISS NIGEL O DR
WELLGATE G BERNARD MR
WHITE GLENN J
WHITROW GERALD JAMES PROF
WHITTET DOUGLAS C B DR
WHITWORTH ANTHONY PETER
WICKRAMASINGHE N C PROF
WILCOCK WILLIAM L PROF
WILKINS GEORGE A DR
WILKINSON ALTHEA
WILKINSON PETER N DR
WILLIAMS DAVID A PROF
WILLIAMS IWAN P DR
WILLIAMS PEREDUR M DR
WILLIS ALLAN J DR
WILLMORE A PETER PROF
WILLSTROP RODERICK V DR
WILSON LIONEL DR
WILSON MICHAEL JOHN DR
WILSON ROBERT PROF SIR
WOLFENDALE ARNOLD W PROF
WOLSTENCROFT RAMON D DR
WOOD ROGER DR
WOOLFSON MICHAEL M PROF
WORRALL GORDON DR
WORSWICK SUSAN
WYNNE CHARLES G PROF
YALLOP BERNARD D DR
YULIN BIAN
ZARNECKI JAN CHARLES DR
ZINNECKER HANS
ZUIDERWIJK EDWARDUS J

Country : U.S.A.

Members :

A'HEARN MICHAEL F DR
AANNESTAD PER ARNE DR
ABBOTT DAVID C DR
ABLES HAROLD D DR
ABT HELMUT A DR
ACTON LOREN W DR
ADAMS A N MR
ADAMS JAMES H JR DR
ADAMS THOMAS F DR
ADEL ARTHUR F PROF EMER
ADELMAN SAUL J DR
AHLUWALIA HARJIT SINGH DR
AHMAD IMAD ALDEAN DR
AIZENMAN MORRIS L DR
ALBERS HENRY PROF
ALDROVANDI S M VIEGAS DR
ALEXANDER JOSEPH K
ALLAN DAVID W MR
ALLEN RONALD J DR
ALLER HUGH D DR
ALLER LAWRENCE HUGH
ALLER MARGO F DR
ALTROCK RICHARD C DR
ALTSCHULER MARTIN D PROF
ANAND S P S DR
ANANTHARAMAIAH K R DR
ANDERS EDWARD PROF
ANDERSEN TORBEN BRENDER
ANDERSON CHRISTOPHER M DR
ANDERSON KINSEY A PROF
ANDERSON KURT S
ANDREW KENNETH L PROF
ANGEL J ROGER P PROF
ANGIONE RONALD J DR
ANTHONY-TWAROG BARBARA J
ANTIOCHOS SPIRO KOSTA
APPLEBY JOHN F
APPLETON PHILIP NOEL DR
ARNAUD JEAN PAUL
ARNETT W DAVID PROF
ARNOLD JAMES R DR
ARNQUIST WARREN N DR
ARNY THOMAS T DR
ARONS JONATHAN
ARP HALTON DR
ARTHUR DAVID W G
ASCHWANDEN MARKUS DR
ATHAY R GRANT DR
ATKINSON ROBERT D'E DR
ATREYA SUSHIL K
AUER LAWRENCE H DR
AUGASON GORDON C DR
AVRETT EUGENE H DR
AYRES THOMAS R
BAAN WILLEM A
BABCOCK ALICE K DR
BABCOCK HORACE W DR
BACKER DONALD CH DR
BAGNUOLO WILLIAM G JR DR
BAGRI DURGADAS S
BAHCALL JOHN N PROF
BAHNG JOHN D R PROF
BAIRD GEORGE A DR
BAIRD SCOTT R
BAKER JAMES GILBERT DR
BAKER NORMAN H PROF
BALBUS STEVEN A DR
BALDWIN RALPH B
BALICK BRUCE PROF
BALIUNAS SALLIE L
BALL JOHN A DR
BALLY JOHN DR
BALONEK THOMAS J DR
BANDERMANN L W DR
BANTA THOMAS MICHAEL
BARDEEN JAMES M PROF
BARKER EDWIN S DR
BARKER TIMOTHY DR
BARNES AARON DR
BARNES III THOMAS G DR
BARNOTHY JENO DR PROF
BARRETT ALAN H PROF
BARRY DON C DR
BARTEL NORBERT HARALD DR
BARTH CHARLES A PROF
BARVAINIS RICHARD DR
BASART JOHN P
BASH FRANK N PROF
BASRI GIBOR B
BATCHELOR DAVID ALLEN DR
BAUER CARL A DR
BAUER WENDY HAGEN
BAUM WILLIAM A DR
BAUSTIAN W W MR
BAUTZ LAURA P DR
BAYM GORDON ALAN DR
BEARD DAVID B DR
BEARDSLEY WALLACE R DR
BEAVERS WILLET I DR
BECKER ROBERT A DR
BECKER ROBERT HOWARD
BECKER STEPHEN A
BECKLIN ERIC E DR
BECKWITH STEVEN V W
BEEBE HERBERT A
BEEBE RETA FAYE DR
BEER REINHARD DR
BEGELMAN MITCHELL CRAIG
BELL BARBARA DR
BELL ROGER A DR
BELSERENE EMILIA P
BELTON MICHAEL J S DR
BENDER PETER L DR
BENEDICT GEORGE F DR
BENFORD GREGORY DR
BENNETT CHARLES L DR
BENSON PRISCILLA J DR
BENZ WILLY
BERENDZEN RICHARD DR
BERG RICHARD A DR
BERGE GLENN L DR
BERGSTRALH JAY T DR
BERMAN ROBERT HIRAM DR
BERNAT ANDREW PLOUS DR
BETTIS DALE G PROF
BHAVSAR SUKETU P
BIDELMAN WILLIAM P PROF
BIEGING JOHN HAROLD DR
BIGNELL R CARL DR
BILLINGHAM JOHN
BILLINGS DONALD E PROF
BIRKINSHAW MARK
BLACK JOHN HARRY DR
BLADES JOHN CHRIS DR
BLAHA MILAN DR
BLAIR GUY NORMAN DR
BLAIR WILLIAM P DR
BLANDFORD ROGER DAVID DR
BLASIUS KARL RICHARD DR
BLESS ROBERT C PROF
BLITZ LEO
BLITZSTEIN WILLIAM DR
BLOEMHOF ERIC E DR
BLUDMAN SIDNEY A PROF
BLUMENTHAL GEORGE R DR
BODENHEIMER PETER PROF
BOEHM KARL-HEINZ PROF
BOEHM-VITENSE ERIKA PROF
BOESGAARD ANN M PROF
BOESHAAR GREGORY ORTH DR
BOGGESS ALBERT DR
BOGGESS NANCY W DR
BOHANNAN BRUCE EDWARD
BOHLIN J DAVID DR
BOHLIN RALPH C DR
BOLDT ELIHU DR
BOLEY FORREST I
BOND HOWARD E DR
BOND JOHN RICHARD

Country : U.S.A.

Members :

BONSACK WALTER K PROF
BOOK DAVID L
BOOKMYER BEVERLY B DR
BOPP BERNARD W DR
BORD DONALD JOHN
BORDERIES NICOLE
BORIAKOFF VALENTIN
BORNMANN PATRICIA L DR
BOSS ALAN P DR
BOWELL EDWARD L G DR
BOWEN GEORGE H DR
BOWERS PHILLIP F
BOWYER C STUART PROF
BOYCE PETER B DR
BOYNTON PAUL EDWARD DR
BRACEWELL RONALD N PROF
BRADSTREET DAVID H DR
BRANCH DAVID R DR
BRANDT JOHN C DR
BRANSCOMB L M DR
BRAULT JAMES W DR
BRECHER AVIVA DR PROF
BRECHER KENNETH PROF
BRECKINRIDGE JAMES B DR
BREGMAN JOEL N
BRIDLE ALAN H PROF
BRINKS ELIAS DR
BROADFOOT A LYLE DR
BRODERICK JOHN DR
BRODIE JEAN P
BROUCKE ROGER DR
BROWN ALEXANDER
BROWN DOUGLAS NASON
BROWN HARRISON DR
BROWN ROBERT HAMILTON
BROWN ROBERT L DR
BROWNLEE DONALD E PROF
BROWNLEE ROBERT R DR
BRUCATO ROBERT J.
BRUECKNER GUENTER E DR
BRUHWEILER FRED C JR
BRUNER MARILYN E DR
BRUNING DAVID H DR
BRUNK WILLIAM E DR
BUCHLER J ROBERT PROF
BUFF JAMES S DR
BUHL DAVID DR
BUNNER ALAN N DR
BURATTI BONNIE J DR
BURBIDGE E MARGARET PROF
BURBIDGE GEOFFREY R PROF
BURKE BERNARD F DR
BURKHEAD MARTIN S
BURLAGA LEONARD F DR
BURNS JACK O'NEAL JR
BURNS JOSEPH A PROF
BURROWS ADAM SETH
BURSTEIN DAVID
BUSCOMBE WILLIAM PROF
BUTA RONALD J DR
BUTLER DENNIS DR
BUTTERWORTH PAUL
BYARD PAUL L DR
BYRD GENE G DR
CAHN JULIUS H PROF
CAILLAULT JEAN PIERRE DR
CAMERON ALASTAIR G W PROF
CAMERON WINIFRED S MRS
CAMPBELL ALISON DR
CAMPBELL BELVA G S DR
CAMPBELL DONALD B
CAMPBELL MURRAY F
CAMPINS HUMBERTO DR
CANFIELD RICHARD C DR
CANIZARES CLAUDE R PROF
CAPEN CHARLES F
CAPRIOTTI EUGENE R DR
CARBON DUANE F DR
CARLETON NATHANIEL P DR
CARLSON JOHN B
CARNEY BRUCE WILLIAM
CAROFF LAWRENCE J
CARPENTER KENNETH G DR
CARPENTER LLOYD DR
CARR THOMAS D PROF
CARRUTHERS GEORGE R DR
CARSENTY URI DR
CARTER WILLIAM EUGENE
CASH WEBSTER C JR
CASSINELLI JOSEPH P DR
CASTELLI JOHN P
CASTOR JOHN I DR
CATON DANIEL B DR
CATURA RICHARD C DR
CAUGHLAN GEORGEANNE R
CEFOLA PAUL J DR
CENTRELLA JOAN M DR
CERSOSIMO JUAN CARLOS DR
CHAFFEE FREDERIC H DR
CHAISSON ERIC J PROF
CHAMBERLAIN JOSEPH M DR
CHAMBERLAIN JOSEPH W PROF
CHAMBLISS CARLSON R DR
CHAN KWING LAM
CHANDRA SUBHASH
CHANDRASEKHAR S PROF
CHANMUGAM GANESAR PROF
CHAPMAN CLARK R DR
CHAPMAN GARY A DR
CHAPMAN ROBERT D DR
CHEN KWAN-YU PROF
CHENG CHUNG-CHIEH DR
CHEVALIER ROGER A DR
CHITRE DATTAKUMAR M DR
CHIU HONG-YEE DR
CHIU LIANG-TAI GEORGE
CHRISTIAN CAROL ANN
CHRISTIANSEN WAYNE A
CHRISTODOULOU DMITRIS DR
CHRISTY JAMES WALTER DR
CHRISTY ROBERT F DR
CHU YOU-HUA
CHUBB TALBOT A DR
CHUPP EDWARD L DR
CHURCHWELL EDWARD B DR
CLARK ALFRED JR PROF
CLARK BARRY G DR
CLARK FRANK OLIVER DR
CLARK GEORGE W PROF
CLARK THOMAS A DR
CLARKE JOHN T
CLAYTON DONALD D PROF
CLAYTON GEOFFREY C DR
CLAYTON ROBERT N DR
CLEMENS DAN P DR
CLIFTON KENNETH ST
CLINE THOMAS L DR
CLIVER EDWARD W
COCHRAN ANITA L DR
COCHRAN WILLIAM DAVID DR
COCKE WILLIAM JOHN PROF
CODE ARTHUR D
COFFEEN DAVID L DR
COFFEY HELEN E MS
COHEN JEFFREY M DR
COHEN JUDITH DR
COHEN LEON PROF
COHEN MARSHALL H PROF
COHEN MARTIN DR
COHEN RICHARD S
COHEN ROSS D DR
COHN HALDAN N
COLBURN DAVID S DR
COLGATE STIRLING A DR
COLLINS GEORGE W II PROF
COMBI MICHAEL R DR
COMINS NEIL FRANCIS
CONDON JAMES J DR
CONKLIN EDWARD K

Country : U.S.A.

Members :

CONNOLLY LEO PAUL
CONTI PETER S DR
COOK JOHN W
CORBALLY CHRISTOPHER
CORBIN THOMAS ELBERT DR
CORDES JAMES M
CORDOVA FRANCE A D
CORLISS C H DR
CORWIN HAROLD G JR
COTTON WILLIAM D Jr
COULSON IAIN M DR
COUNSELMAN CHARLES C PROF
COWAN JOHN J DR
COWIE LENNOX LAUCHLAN DR
COWLEY ANNE P DR
COWLEY CHARLES R PROF
COX ARTHUR N DR
COX DONALD P PROF
CRAINE ERIC RICHARD DR
CRANE PATRICK C
CRANNELL CAROL JO DR
CRAWFORD DAVID L DR
CRUIKSHANK DALE P DR
CRUTCHER RICHARD M DR
CUDABACK DAVID D DR
CUDWORTH KYLE MCCABE DR
CUFFEY J MR
CULVER ROGER BRUCE DR
CUNNINGHAM LELAND E PROF
CURRIE DOUGLAS G DR
CZYZAK STANLEY J DR
DA COSTA GARY STEWART DR
DAHN CONARD CURTIS DR
DALGARNO ALEXANDER PROF
DANBY J M ANTHONY DR
DANKS ANTHONY C DR
DATLOWE DAYTON DR
DAVIDSEN ARTHUR FALNES DR
DAVIDSON KRIS DR
DAVIES MERTON E MR
DAVIES ROGER L DR
DAVIS CECIL G JR
DAVIS LEVERETT JR PROF
DAVIS MARC DR
DAVIS MICHAEL M DR
DAVIS MORRIS S PROF
DAVIS ROBERT J DR
DAVIS SUMNER P DR
DE FREES DOUGLAS J DR
DE JONGE J K DR
DE PATER IMKE
DE VAUCOULEURS GERARD PR
DE VINCENZI DONALD DR
DE YOUNG DAVID S DR
DE ZEEUW PIETER T DR
DEARBORN DAVID PAUL S DR
DEEMING TERENCE J DR
DELSEMME ARMAND H PROF DR
DEMARQUE P PROF
DENNIS BRIAN ROY DR
DENNISON EDWIN W DR
DENT WILLIAM A PROF
DEPRIT ANDRE PROF
DERE KENNETH PAUL
DERMOTT STANLEY F
DESPAIN KEITH HOWARD DR
DEUPREE ROBERT G DR
DEUTSCHMAN WILLIAM A DR
DEVINNEY EDWARD J DR
DEVORKIN DAVID H
DEWITT BRYCE S DR
DEWITT JOHN H JR
DEWITT-MORETTE CECILE PR
DICK STEVEN J
DICKE ROBERT H PROF
DICKEL HELENE R DR
DICKEL JOHN R
DICKEY JEAN O'BRIEN
DICKEY JOHN M
DICKINSON DALE F DR
DICKMAN ROBERT L DR
DICKMAN STEVEN R
DIETER NANNIELOU H DR
DINERSTEIN HARRIET L
DIXON ROBERT S DR
DJORGOVSKI STANISLAV DR
DOGGETT LEROY E DR
DOHERTY LOWELL R PROF
DOLAN JOSEPH F DR
DONN BERTRAM D
DOSCHEK GEORGE A DR
DOUGLAS JAMES N PROF
DOWNES RONALD A DR
DOWNS GEORGE S DR
DRAINE BRUCE T
DRAKE FRANK D PROF
DRAKE STEPHEN A
DREHER JOHN W
DRESSEL LINDA L
DRESSLER ALAN
DRILLING JOHN S
DRYER MURRAY DR
DUFOUR REGINALD JAMES
DULK GEORGE A PROF
DUNCAN DOUGLAS KEVIN DR
DUNCOMBE RAYNOR L DR
DUNHAM DAVID W
DUNKELMAN LAWRENCE
DUNN RICHARD B DR
DUPREE ANDREA K DR
DUPUY DAVID L DR
DURISEN RICHARD H DR
DURNEY BERNARD DR
DURRANCE SAMUEL T DR
DUTHIE JOSEPH G PROF
DUVALL THOMAS L JR
DWEK ELI
DYCK M DR
DYER EDWARD R DR
DYSON F J DR
EATON JOEL A DR
EDDY JOHN A DR
EDMONDS FRANK N JR DR
EDMONDSON FRANK K PROF
EDWARDS ALAN CH DR
EDWARDS TERRY W
EICHHORN HEINRICH K DR
EICHLER DAVID DR
EILEK JEAN
EL-BAZ FAROUK DR
ELITZUR MOSHE
ELLIOT JAMES L DR
ELMEGREEN BRUCE GORDON DR
ELMEGREEN DEBRA MELOY
ELSTE GUNTHER H DR
ELSTON WOLFGANG E PROF
ELVIS MARTIN S DR
EMSLIE A. GORDON
ENDAL ANDREW S DR
EPPS HARLAND WARREN PROF
EPSTEIN EUGENE E DR
EPSTEIN GABRIEL LEO DR
EPSTEIN ISADORE PROF
EPSTEIN RICHARD I DR
ESHLEMAN VON R PROF
ESPOSITO F PAUL PROF
ESPOSITO LARRY W
EUBANKS THOMAS M DR
EVANS J V DR
EVANS JOHN W DR
EVANS NEAL J II ASS PROF
EVANS W DOYLE
EVERHART EDGAR DR
EWEN HAROLD I DR
EWING MARTIN S
FABBIANO GIUSEPPINA
FABER SANDRA M PROF
FABRICANT DANIEL G
FALGARONE EDITH

Country : U.S.A.

Members :

FALK SYDNEY W JR DR
FALLON FREDERICK W DR
FAY THEODORE D DR
FEIBELMAN WALTER A DR
FELDMAN PAUL DONALD DR
FELTEN JAMES E DR
FIALA ALAN D DR
FIENBERG RICHARD T DR
FINK UWE DR
FIROR JOHN W DR
FISHER J RICHARD
FISHMAN GERALD J
FIX JOHN D DR
FLIEGEL HENRY F
FOMALONT EDWARD B DR
FORBES TERRY G DR
FORMAN WILLIAM RICHARD DR
FOWLER WILLIAM A PROF
FRANZ OTTO G DR
FREEDMAN WENDY L DR
FRIEDMAN HERBERT DR
FROGEL JAY ALBERT DR
FTACLAS CHRIST
GALLAGHER III JOHN S DR
GAPOSCHKIN EDWARD M DR
GARMANY CATHERINE D DR
GARY DALE E
GAUSS F STEPHEN
GEBBIE KATHARINE B DR
GELDZAHLER BERNARD J
GERGELY TOMAS ESTEBAN DR
GHIGO FRANCIS D DR
GIBSON DAVID MICHAEL DR
GIERASCH PETER J DR
GILMAN PETER A DR
GIOIA ISABELLA M DR
GLASER HAROLD DR
GLASSGOLD ALFRED E PROF
GOLD THOMAS PROF
GOLDSMITH DONALD W. DR.
GOLDSTEIN RICHARD M DR
GOLUB LEON DR
GOODY R M
GORDON KURTISS J PROF
GORENSTEIN PAUL DR
GOTT III J RICHARD
GOULD ROBERT J PROF
GRAHAM ERIC DR
GRASDALEN GARY L DR
GREEN JACK PROF
GREENBERG RICHARD DR
GREGORY STEPHEN ALBERT DR
GRIFFITHS RICHARD E DR

FALL S MICHAEL DR
FANSELOW JOHN LYMAN
FAZIO GIOVANNI G DR
FEIGELSON ERIC D DR
FELDMAN URI
FERLAND GARY JOSEPH
FICHTEL CARL E DR
FILIPPENKO ALEXEI V DR
FINN G D DR
FISCHEL DAVID DR
FISHER PHILIP C
FITCH WALTER S DR
FLANNERY BRIAN PAUL DR
FOGARTY WILLIAM G DR
FONTENLA JUAN MANUEL DR
FORD HOLLAND C RES PROF
FORREST WILLIAM JOHN
FOX KENNETH DR
FRAZIER EDWARD N DR
FRENCH RICHARD G
FRIEND DAVID B DR
FROST KENNETH J DR
FURENLID INGEMAR K DR
GALLAGHER JEAN DR
GARFINKEL BORIS DR
GARMIRE GORDON P PROF
GATEWOOD GEORGE DIRECTOR
GAUSTAD JOHN E PROF
GEHRELS TOM PROF
GELLER MARGARET JOAN
GEROLA HUMBERTO DR
GIACCONI RICCARDO PROF
GIBSON JAMES
GIES DOUGLAS R DR
GILRA DAYA P DR
GIOVANE FRANK
GLASPEY JOHN W DR
GLATZMAIER GARY A
GOLDMAN MARTIN V
GOLDSMITH PAUL F DR
GOLDSTEIN SAMUEL J PROF
GOMES RODNEY D S DR
GOPALSWAMY N DR
GORDON MARK A DR
GOSLING JOHN T DR
GOTTESMAN STEPHEN T DR
GRABOSKE HAROLD C JR
GRAHAM JOHN A DR
GRAUER ALBERT D
GREEN LOUIS C PROF
GREENSTEIN GEORGE PROF
GREISEN KENNETH I PROF
GRINDLAY JONATHAN E DR

FALLER JAMES E PROF
FAULKNER JOHN PROF
FEDERMAN STEVEN ROBERT
FEKEL FRANCIS C
FELDMAN URI DR
FEYNMAN JOAN DR
FIELD GEORGE B PROF
FINDLAY JOHN W DR
FIREMAN EDWARD L
FISCHER JACQUELINE
FISHER RICHARD R DR
FITZPATRICK EDWARD L DR
FLEISCHER ROBERT DR
FOLTZ CRAIG B.
FORBES J E DR
FORD W KENT JR DR
FOUKAL PETER V DR
FRANKLIN FRED A DR
FREDRICK LAURENCE W PROF
FRIEDLANDER MICHAEL PROF
FRISCH PRISCILLA
FRYE GLENN M PROF
GAISSER THOMAS K
GALLET ROGER M
GARLICK GEORGE F DR
GARSTANG ROY H PROF
GATLEY IAN
GEBALLE THOMAS R DR
GEHRZ ROBERT DOUGLAS DR
GENET R M DR
GEZARI DANIEL YSA DR
GIAMPAPA MARK S
GICLAS HENRY L MR
GILLILAND RONALD LYNN
GINGERICH OWEN PROF
GIOVANELLI RICCARDO DR
GLASS BILLY PRICE DR
GOEBEL JOHN H DR
GOLDREICH P DR
GOLDSMITH PAUL F DR
GOLDWIRE HENRY C JR
GOODE PHILIP R
GORDON COURTNEY P PROF
GORENSTEIN MARC V
GOSS W MILLER PROF
GOTTLIEB CARL A DR
GRADIE JONATHAN CAREY
GRANDI STEVEN ALDRIDGE DR
GRAYZECK EDWIN J DR
GREEN RICHARD F DR
GREENSTEIN J L PROF
GREYBER HOWARD D DR
GROSS PETER G PROF

Country : U.S.A.

Members :

GROSSMAN ALLEN S PROF
GROSSMAN LAWRENCE PROF
GROTH EDWARD J III
GUETTER HARRY HENDRIK
GUIDICE DONALD A DR
GUINAN EDWARD FRANCIS DR
GULKIS SAMUEL DR
GULL THEODORE R DR
GUNN JAMES E PROF
GURMAN JOSEPH B DR
GURSKY HERBERT DR
GWINN CARL R DR
HABBAL SHADIA RIFAI
HACKWELL JOHN A DR
HADDOCK FRED T DR
HAGEN JOHN P
HAGFORS T DR
HAGYARD MONA JUNE
HAISCH BERNHARD MICHAEL
HALL DONALD N DR
HALL DOUGLAS S DR
HALL JOHN S DR
HALL R GLENN DR
HALLAM KENNETH L DR
HANISCH ROBERT J DR
HANKINS TIMOTHY HAMILTON
HANNER MARTHA S DR
HANSEN CARL J PROF
HANSEN RICHARD T MR
HANSON ROBERT B DR
HAPKE BRUCE W DR
HARDEBECK ELLEN G DR
HARDEE PHILIP
HARDIE R PROF
HARMER CHARLES F W MR
HARMS RICHARD JAMES DR
HARNDEN FRANK R Jr
HARRINGTON J PATRICK DR
HARRINGTON ROBERT S DR
HARRIS ALAN WILLIAM DR
HARRIS DANIEL E DR
HARRIS HUGH C
HARRISON EDWARD R PROF
HART MICHAEL H DR
HARTEN RONALD H DR
HARTKOPF WILLIAM I DR
HARTMANN LEE WILLIAM
HARTMANN WILLIAM K
HARTOOG MARK RICHARD DR
HARVEL CHRISTOPHER ALVIN
HARVEY JOHN W DR
HARVEY PAUL MICHAEL DR
HARWIT MARTIN PROF
HASCHICK AUBREY
HATHAWAY DAVID H DR
HAUBOLD HANS JOACHIM
HAUPT RALPH F
HAUSER MICHAEL G DR
HAVLEN ROBERT J DR
HAWKINS GERALD S DR
HAYES DONALD S DR
HAYMES ROBERT C PROF
HAYNES MARTHA P
HAZEN MARTHA L DR
HEAP SARA R DR
HEASLEY JAMES NORTON
HECHT JAMES H DR
HECKATHORN HARRY M
HECKMAN TIMOTHY M
HEESCHEN DAVID S DR
HEFFERLIN RAY A PROF
HEGYI DENNIS J ASSOC PROF
HEILES CARL PROF
HEINTZ WULFF D DR
HEISER ARNOLD M DR
HELFAND DAVID JOHN
HELFER H LAWRENCE PROF
HELIN ELEANOR FRANCIS
HELLWIG HELMUT WILHELM DR
HELMKEN HENRY F DR
HELOU GEORGE DR
HEMENWAY PAUL D
HENIZE KARL G ASTRONAUT
HENRY RICHARD B C DR
HENRY RICHARD C. PROF.
HERBIG GEORGE H DR
HERBST ERIC DR
HERBST WILLIAM DR
HERCZEG TIBOR J PROF DR
HERR RICHARD B DR
HERSHEY JOHN L DR
HERTZ PAUL L DR
HEWITT ADELAIDE
HEWITT ANTHONY V DR
HIBBS ALBERT R MGR PLANS
HILDEBRAND ROGER H
HILDNER ERNEST DR
HILL HENRY ALLEN DR
HILLIARD R DR
HILLS JACK G DR
HILTNER W ALBERT PROF
HINKLE KENNETH H
HINTEREGGER HANS E DR
HINTZEN PAUL MICHAEL N DR
HJELLMING ROBERT M DR
HO PAUL T P
HOAG ARTHUR A DR
HOBBS LEWIS M DR
HOBBS ROBERT W DR
HODGE PAUL W PROF
HOESSEL JOHN GREG
HOFF DARREL BARTON
HOFFLEIT E DORRIT DR
HOFFMAN JEFFREY ALAN DR
HOGAN CRAIG J DR
HOGG DAVID E DR
HOLBERG JAY B
HOLLENBACH DAVID JOHN DR
HOLLIS JAN MICHAEL DR
HOLLWEG JOSEPH V
HOLMAN GORDON D
HOLT STEPHEN S
HOLZER THOMAS EDWARD DR
HONEYCUTT R KENT PROF
HOROWITZ PAUL PROF
HOUCK JAMES R
HOUK NANCY DR
HOUSE LEWIS L DR
HOWARD ROBERT F DR
HOWARD W MICHAEL DR
HOWARD WILLIAM E III DR
HRIVNAK BRUCE J
HU ESTHER M DR
HUBBARD WILLIAM B PROF
HUBENY IVAN
HUCHRA JOHN PETER DR
HUDSON HUGH S DR
HUEBNER WALTER F DR
HUENEMOERDER DAVID P DR
HUGHES JAMES A DR
HUGHES JOHN P DR
HUGHES PHILIP
HUGUENIN G RICHARD
HUMMER DAVID G DR
HUMPHREYS ROBERTA M PROF
HUNDHAUSEN ARTHUR DR
HUNTEN DONALD M PROF
HUNTER CHRISTOPHER PROF
HUNTER DEIDRE ANN
HUNTER JAMES H PROF
HURFORD GORDON JAMES
HUT PIET
HYDER C L DR
IANNA PHILIP A
IBEN ICKO JR PROF
ILLING RAINER M E
ILLINGWORTH GARTH D DR
IMAMURA JAMES DR
IMHOFF CATHERINE L DR

Country : U.S.A.

Members :

IMPEY CHRISTOPHER D DR
IRVINE WILLIAM M PROF
JACKSON PETER DOUGLAS DR
JACOBSEN THEODOR S PROF
JAFFE WALTER JOSEPH DR
JANSSEN MICHAEL ALLEN
JEFFERYS WILLIAM H DR
JENKNER HELMUT DR
JOHNSON DONALD R DR
JOHNSON HUGH M DR
JOKIPII J R PROF
JONES DAYTON L
JONES HARRISON PRICE DR
JOSELYN JO ANN C DR
JURA MICHAEL DR
KAFATOS MINAS DR
KAITCHUCK RONALD H
KAMMEYER PETER C DR
KAPAHI V K DR
KAPLAN LEWIS D DR
KATZ JONATHAN I
KAWALER STEVEN D DR
KEIL KLAUS DR
KELLER GEOFFREY
KENNICUTT ROBERT C JR
KERR FRANK J DR
KIELKOPF JOHN F DR
KING IVAN R PROF
KINMAN THOMAS D DR
KIRKPATRICK RONALD C DR
KLARMANN JOSEPH PROF
KLEINMANN DOUGLAS E DR
KLEPCZYNSKI WILLIAM J DR
KLIORE ARVYDAS JOSEPH DR
KNAPP GILLIAN R DR
KO HSIEN C PROF
KOESTER DETLEV DR
KOLB EDWARD W DR
KOO DAVID C-Y DR
KOVAR N S DR
KRAFT ROBERT P PROF
KREIDL TOBIAS J N
KRISTIAN JEROME DR
KRON GERALD E DR
KRUMM NATHAN ALLYN
KUIPER THOMAS B H DR
KUMAR C KRISHNA DR
KUNDU MUKUL R DR
KUTNER MARC LESLIE DR
LA BONTE BARRY JAMES
LADA CHARLES JOSEPH DR
LAMB FREDERICK K PROF
LAMBERT DAVID L PROF

IONSON JAMES ALBERT
IRWIN JOHN B PROF
JACKSON WILLIAM M DR
JACOBY GEORGE H
JAMES KENNETH A DR
JASTROW ROBERT
JENKINS EDWARD B DR
JENNER DAVID C DR
JOHNSON FRED M PROF DR
JOHNSON TORRENCE V DR
JONES BARBARA
JONES ERIC M
JONES THOMAS WALTER DR
JOSS PAUL CHRISTOPHER DR
JURGENS RAYMOND F
KAFTAN MAY A DR
KALER JAMES B PROF
KAMP LUCAS WILLEM DR
KAPAHI VIJAY, K.
KARP ALAN HERSH DR
KAUFMAN MICHELE DR
KEEL WILLIAM C
KEIL STEPHEN L
KELLERMANN KENNETH I DR
KENT STEPHEN M
KESSLER KARL G DR
KIMBLE RANDY A DR
KING R B DR
KIPLINGER ALAN L DR
KIRSHNER ROBERT PAUL DR
KLEIN MICHAEL J DR
KLEMOLA ARNOLD R DR
KLINGLESMITH DANIEL A DR
KLOCK B L DR
KNIFFEN DONALD A DR
KOCH DAVID G
KOHL JOHN L DR
KONDO YOJI DR
KOORNNEEF JAN DR
KOVAR ROBERT P DR
KRAUS JOHN D PROF
KRIEGER ALLEN S DR
KROGDAHL W S DR
KRON KATHERINE GORDON
KRUPP EDWIN C DR
KULKARNI SHRINIVAS R DR
KUMAR SHAILENDRA
KURFESS JAMES D
KUTTER G SIEGFRIED DR
LACY CLAUD H DR
LALA PETR DR
LAMB RICHARD C DR
LAMPTON MICHAEL

IPSER JAMES R PROF
JACCHIA LUIGI G DR
JACOBS KENNETH C DR
JAFFE DANIEL T
JANICZEK PAUL M DR
JEFFERIES JOHN T DR
JENKINS L F MS
JEWELL PHILIP R DR
JOHNSON HOLLIS R PROF
JOHNSTON KENNETH J
JONES BURTON DR
JONES FRANK CULVER DR
JORDAN STUART D DR
JUNKKARINEN VESA T DR
JURKEVICH IGOR DR
KAHLER STEPHEN W DR
KALKOFEN WOLFGANG DR
KANE SHARAD R DR
KAPLAN J DR
KARPEN JUDITH T
KAULA WILLIAM M PROF
KEENAN PHILIP C PROF EMER
KELLER CHARLES F
KELLOGG EDWIN M DR
KENYON SCOTT J DR
KHARE BISHUN N DR
KING DAVID S PROF
KING ROBERT WILSON JR DR
KIRBY KATE P DR
KISSELL KENNETH E DR
KLEIN RICHARD I DR
KLEMPERER W K DR
KLINKHAMER FRANS DR
KNACKE ROGER F DR
KNOWLES STEPHEN H DR
KOCH ROBERT H DR
KOJOIAN GABRIEL DR
KONIGL ARIEH DR
KOPP ROGER A DR
KOWAL CHARLES THOMAS
KRAUSHAAR WILLIAM L PROF
KRISS GERARD A DR
KROLIK JULIAN H
KRON RICHARD G
KUHI LEONARD V PROF
KULSRUD RUSSELL M DR
KUMAR SHIV S PROF
KURUCZ ROBERT L DR
KWITTER KAREN BETH DR
LACY JOHN H DR
LAMB DONALD QUINCY JR DR
LAMB SUSAN ANN DR
LANDE KENNETH PROF

Country : U.S.A.

Members :

LANDECKER PETER BRUCE DR
LANE ADAIR P
LANGER GEORGE EDWARD DR
LARSON RICHARD B PROF
LASKER BARRY M DR
LAUTMAN D A DR
LAWRIE DAVID G
LEACOCK ROBERT JAY
LECAR MYRON DR
LEIBACHER JOHN DR
LESTER DANIEL F DR
LEVINE RANDOLPH H DR
LEWIN WALTER H G PROF
LI NED C DR
LIEBERT JAMES W DR
LILLIE CHARLES F DR
LIN DOUGLAS N. C. DR
LINGENFELTER RICHARD E
LINSKY JEFFREY L DR
LIPSCHUTZ MICHAEL E DR
LITTLE-MARENIN IRENE R DR
LIU SOU-YANG DR
LOCANTHI DOROTHY DAVIS DR
LONG KNOX S DR
LOPES ROSALY DR
LOVELACE RICHARD V E DR
LU PHILLIP K DR
LUGGER PHYLLIS M
LUTZ JULIE H DR
LYNCH DAVID K
MACALPINE GORDON M
MACCONNELL DARRELL J DR
MACQUEEN ROBERT M DR
MALVILLE J MCKIM PROF
MARGON BRUCE H PROF
MARK JAMES WAI-KEE DR
MARSCHER ALAN PATRICK
MARTIN WILLIAM C DR
MASON GLENN M
MASURSKY HAROLD DR
MATHIEU ROBERT D DR
MATSON DENNIS L DR
MATTHEWS HENRY E DR
MAX CLAIRE E DR
MAYALL NICHOLAS U ASTRON
MAZUREK THADDEUS JOHN DR
MCCAMMON DAN
MCCLINTOCK JEFFREY E DR
MCCRAY RICHARD DR
MCDONOUGH THOMAS R DR
MCGRAW JOHN T DR
MCLAREN ROBERT A DR
MCMILLAN ROBERT S DR

LANDMAN DONALD ALAN
LANE ARTHUR LONNE DR
LANGER WILLIAM DAVID DR
LARSON STEPHEN M
LATHAM DAVID W DR
LAWRENCE CHARLES R DR
LAYZER DAVID PROF
LEBOFSKY LARRY ALLEN
LECKRONE DAVID S DR
LEIGHTON R B PROF
LEUNG CHUN MING DR
LEVREAULT RUSSELL M DR
LEWIS BRIAN MURRAY DR
LIBBRECHT K G DR
LIESKE JAY H DR
LILLY SIMON J DR
LINCOLN J VIRGINIA MISS
LINKE RICHARD ALAN DR
LINSLEY JOHN
LISSAUER JACK J DR
LITTLETON JOHN E
LIVINGSTON WILLIAM C
LOCKMAN FELIX J
LONGMORE ANDREW J
LOREN ROBERT BRUCE DR
LOW BOON CHYE
LUCK R EARLE DR
LUNDQUIST CHARLES A DR
LUTZ THOMAS E DR
LYNDS BEVERLY T DR
MACCACARO TOMMASO DR
MACDONALD JAMES
MACY WILLIAM WRAY DR
MANSFIELD VICTOR N PROF
MARGRAVE THOMAS EWING JR
MARKOWITZ WILLIAM DR
MARSDEN BRIAN G DR
MARTINS DONALD HENRY DR
MASSEY PHILIP L
MATHER JOHN CROMWELL
MATHIS JOHN S PROF
MATSUSHIMA SATOSHI DR
MATTHEWS THOMAS A DR
MAXWELL ALAN DR
MAYER CORNELL H
MCALISTER HAROLD A DR
MCCARTHY DENNIS D DR
MCCLUSKEY GEORGE E JR DR
MCCROSKY RICHARD E DR
MCELROY M B DR
MCINTOSH PATRICK S
MCLEAN BRIAN JOHN
MCNAMARA DELBERT H DR

LANDOLT ARLO U PROF
LANG KENNETH R ASST PROF
LARSON HAROLD P DR
LASHER GORDON JEWETT DR
LATTIMER JAMES M DR
LAWRENCE G M DR
LEA SUSAN MAUREEN DR
LEBOVITZ NORMAN R PROF
LEE PAUL D DR
LEPP STEPHEN H DR
LEUNG KAM CHING PROF
LEVY EUGENE H DR
LEWIS J S
LIDDELL U MR
LILLEY EDWARD A PROF
LIN CHIA C PROF
LINDSEY CHARLES ALLAN
LINNELL ALBERT P PROF
LIPPINCOTT SARAH LEE DR
LISZT HARVEY STEVEN
LITVAK MARVIN M DR
LO KWOK-YUNG DR
LOCKWOOD G WESLEY DR
LONSDALE CAROL J DR
LOVAS FRANCIS JOHN DR
LOW FRANK J DR
LUCKE PETER B DR
LUTZ BARRY L DR
LUYTEN WILLEM J PROF
LYNDS ROGER C DR
MACCHETTO FERDINANDO DR
MACK PETER DR
MALITSON HARRIET H MS
MARAN STEPHEN P DR
MARISKA JOHN THOMAS
MARSCHALL LAURENCE A
MARTIN ROBERT N DR
MARVIN URSULA B DR
MASSON COLIN R
MATHEWS WILLIAM G PROF
MATSAKIS DEMETRIOS N
MATTEI JANET AKYUZ DR
MATZNER RICHARD A PROF
MAYALL MARGARET W
MAYFIELD EARLE B DR
MCCABE MARIE K MS
MCCLAIN EDWARD F
MCCORD THOMAS B DR
MCDONALD FRANK B DR
MCGIMSEY BEN Q JR DR
MCKEE CHRISTOPHER F PROF
MCLEAN IAN S DR
MEAD JAYLEE MONTAGUE DR

Country : U.S.A.

Members :

MEATHERINGHAM STEPHEN DR
MEIER ROBERT R
MELBOURNE WILLIAM G DR
MELOTT ADRIAN L PROF.
MERTZ LAWRENCE N DR
MEYERS KARIE ANN
MICHEL F CURTIS PROF
MIHALAS BARBARA R WEIBEL
MILKEY ROBERT W DR
MILLER JOSEPH S PROF
MILLIKAN ALLAN G MR
MISCONI NEBIL YOUSIF DR
MITCHELL RICHARD MR
MODISETTE JERRY L PROF
MONET DAVID G
MOORE RONALD L DR
MORGAN JOHN ADRIAN
MORRIS CHARLES S
MORRISON DAVID PROF
MORTON G A DR
MOUSCHOVIAS TELEMACHOS CH
MUKHERJEE KRISHNA
MULLAN DERMOTT J DR
MUMMA MICHAEL JON
MURDOCK THOMAS LEE
MUSEN PETER DR
MUTEL ROBERT LUCIEN
MYERS PHILIP C
NARAYAN RAMESH DR
NEIDIG DONALD F DR
NELSON ROBERT M
NEUPERT WERNER M DR
NEWMAN MICHAEL JOHN DR
NEY EDWARD P PROF
NIELL ARTHUR E DR
NOERDLINGER PETER D PROF
NORMAN COLIN A PROF
NUTH JOSEPH A III
O'DELL CHARLES R DR
O'KEEFE JOHN A DR
ODENWALD STEN F DR
OERTEL GOETZ K DR
OKE J BEVERLEY PROF
OLIVER JOHN PARKER DR
OMER GUY C JR PROF
ORTON GLENN S DR
OSTER LUDWIG F PROF DR
OSWALT TERRY D DR
OZERNOY LEONID M PROF
PACIESAS WILLIAM S DR
PALMER PATRICK E PROF
PANG KEVIN
PAPAGIANNIS MICHAEL D PR

MEEKS M LITTLETON DR
MEINEL ADEN B PROF
MELIA FULVIO DR
MENDIS DEVAMITTA ASOKA DR
MESZAROS PETER DR
MEYLAN GEORGES DR
MICZAIKA G R DR
MIHALAS DIMITRI DR
MILLER FREEMAN D PROF
MILLER RICHARD H DR
MILLIS ROBERT L DR
MISNER CHARLES W PROF
MITCHELL WALTER E JR
MOFFETT THOMAS J PROF
MOOK DELO E PROF
MOOS HENRY WARREN DR
MORGAN THOMAS H DR
MORRIS MARK ROOT DR
MORRISON NANCY DUNLAP DR
MOTZ LLOYD PROF
MUELLER IVAN I PROF
MULDERS GERARD F W
MULLER RICHARD A
MUNDY LEE G DR
MURPHY ROBERT E DR
MUSIELAK ZDZISLAW E DR
MUTSCHLECNER J PAUL DR
NACOZY PAUL E DR
NATHER R EDWARD
NELSON BURT DR
NESS NORMAN F DR
NEWBURN RAY L JR
NEWSOM GERALD H PROF
NICOLAS KENNETH ROBERT
NILSSON CARL DR
NOONAN THOMAS W PROF
NOVICK ROBERT
O'CONNELL ROBERT F PROF
O'DELL STEPHEN L
O'LEARY BRIAN T
OEGERLE WILLIAM R
OESTERWINTER CLAUS
OLIVER BERNARD M DR
OLSEN KENNETH H DR
ORLIN HYMAN DR
OSBORN WAYNE DR
OSTERBROCK DONALD E PROF
OWEN FRAZER NELSON DR
OZSVATH I PROF
PAGE DON NELSON
PANAGIA NINO DR
PANKONIN VERNON LEE DR
PAPALIOLIOS COSTAS DR

MEIER DAVID L
MEISEL DAVID D DR
MELNICK GARY J
MERRILL JOHN E DR
MEYER DAVID M DR
MICHALITSIANOS ANDREW
MIDDLEHURST BARBARA M MS
MIKESELL ALFRED H MR
MILLER HUGH R PROF
MILLIGAN J E
MINESHIGE SHIN DR
MITCHELL KENNETH J DR
MODALI SARMA B DR
MOLNAR MICHAEL R PROF
MOORE ELLIOTT P PROF
MORAN JAMES M DR
MORGAN WILLIAM W PROF
MORRIS STEVEN DR
MORRISON PHILIP PROF
MOULD JEREMY R
MUFSON STUART LEE DR
MULHOLLAND J DERRAL DR
MUMFORD GEORGE S PROF
MUNRO RICHARD H DR
MURRAY STEPHEN S DR
MUSMAN STEVEN DR
MUZZIO JUAN C PROF
NAKANO SYUICHI
NEFF JOHN S
NELSON JERRY
NEUGEBAUER GERRY DR
NEWHALL X X DR
NEWTON ROBERT R DR
NIEDNER MALCOLM B DR
NISHIMURA TETSUO DR
NORDSTROM BIRGITTA DR
NOYES ROBERT W PROF
O'CONNELL ROBERT WEST DR
O'HANDLEY DOUGLAS A DR
ODELL ANDREW P
OEMLER AUGUSTUS JR DR
OKA TAKESHI DR
OLIVER BERNARD M PROF
OLSON EDWARD C PROF
ORRALL FRANK Q PROF
OSMER PATRICK S DR
OSTRIKER JEREMIAH P PROF
OWEN TOBIAS C PROF
PACHOLCZYK ANDRZEJ G PROF
PAGE THORNTON L DR
PANEK ROBERT J DR
PAP JUDIT
PARESCE FRANCESCO DR

Country : U.S.A.

Members :

PARISE RONALD A DR
PARKER EUGENE N
PARKER ROBERT A R
PARKINSON TRUMAN DR
PARKINSON WILLIAM H
PARRISH ALLAN DR.
PARSONS SIDNEY B DR
PARTRIDGE ROBERT B PROF
PASACHOFF JAY M PROF
PASCU DAN DR
PAULS THOMAS ALBERT DR
PAYNE DAVID G
PEALE STANTON J PROF
PEARSON TIMOTHY J
PEDLAR ALAN DR
PEEBLES P JAMES E
PEERY BENJAMIN F PROF
PELLERIN JR CHARLES J DR
PENZIAS ARNO A DR
PERKINS FRANCIS W DR
PERLEY RICHARD ALAN
PERRY CHARLES L DR
PERRY PETER M DR
PESCH PETER DR
PETERS GERALDINE JOAN DR
PETERS WILLIAM L III DR
PETERSON BRADLEY MICHAEL
PETERSON CHARLES JOHN DR
PETERSON LAURENCE E PROF
PETERSON RUTH CAROL DR
PETRO LARRY DAVID
PETROSIAN VAHE PROF
PETTENGILL GORDON H PROF
PFEIFFER RAYMOND J
PHILIP A G DAVIS
PHILLIPS JOHN G PROF
PHILLIPS THOMAS GOULD DR
PICKLES ANDREW JOHN DR
PIER JEFFREY R DR
PIERCE A KEITH DR
PIERCE DAVID ALLEN
PILACHOWSKI CATHERINE DR
PILCHER CARL BERNARD DR
PINES DAVID PROF
PINGREE DAVID PROF
PIPHER JUDITH L
PLAVEC MIREK J PROF
PLAVEC ZDENKA DR
PNEUMAN GERALD W
POGO ALEXANDER DR
POLAND ARTHUR I DR
POLIDAN RONALD S
POLLACK JAMES B DR
PONNAMPERUMA CYRIL PROF
POPPER DANIEL M PROF
PRADHAN DR
PRASAD SHEO S
PRAVDO STEVEN H
PRENDERGAST KEVIN H PROF
PRESS WILLIAM H DR
PRESTON GEORGE W DR
PRESTON ROBERT ARTHUR
PRICE MICHAEL J. DR.
PRICE R MARCUS DR
PRICE STEPHAN DONALD
PRINCE HELEN DODSON PROF
PROBSTEIN R F DR
PROTHEROE WILLIAM M PROF
PUETTER RICHARD C DR
PURCELL EDWARD M PROF
PUSCHELL JEFFERY JOHN
PYPER SMITH DIANE M DR
QUIRK WILLIAM J DR
RABIN DOUGLAS MARK
RADICK RICHARD R DR
RADOSKI HENRY R DR
RAFERT JAMES BRUCE
RAHE JURGEN PROF
RAMATY REUVEN DR
RAMSEY LAWRENCE W DR
RANK DAVID M PROF
RANKIN JOANNA M DR
RAO K NARAHARI
RAYMOND JOHN CHARLES
READHEAD ANTHONY C S DR
REASENBERG ROBERT D DR
REAVES GIBSON PROF
REEVES EDMOND M DR
REID MARK JONATHAN DR
REID NEILL
REITSEMA HAROLD J
RENSE WILLIAM A DR
REUNING ERNEST G DR
REVELLE DOUGLAS ORSON DR
REYNOLDS JOHN H PROF
REYNOLDS RONALD J DR
REYNOLDS STEPHEN P
RHODES EDWARD J JR
RICHARDSON R S
RICHSTONE DOUGLAS O DR
RICKARD JAMES JOSEPH DR
RICKARD LEE J DR
RICKER GEORGE R DR
RICKETT BARNABY JAMES DR
RIDDLE ANTHONY C DR
RIEGEL KURT W DR
RIGHINI-COHEN GIOVANNA DR
RINDLER WOLFGANG PROF
RIVOLO ARTHUR REX
ROACH FRANKLIN E
ROARK TERRY P PROF
ROBBINS R ROBERT PROF
ROBERTS DAVID HALL DR
ROBERTS MORTON S DR
ROBERTS WALTER ORR DR
ROBERTS WILLIAM W JR PROF
ROBERTSON DOUGLAS S
ROBINSON EDWARD LEWIS DR
ROBINSON I PROF
ROBINSON LEIF J
ROBINSON LLOYD B DR
RODDIER FRANCOIS PROF
RODMAN RICHARD B DR
ROEDER ROBERT C PROF
ROEMER ELIZABETH PROF
ROGERS ALAN E E DR
ROGERSON JOHN B PROF
ROGSTAD DAVID H DR
ROMAN NANCY G DR
ROMANISHIN WILLIAM DR
ROMNEY JONATHAN D DR
ROOD HERBERT J
ROOD ROBERT T DR
ROOSEN ROBERT G DR
ROSE JAMES ANTHONY
ROSE WILLIAM K DR
ROSEN EDWARD DR
ROSENDHAL JEFFREY D DR
ROSNER ROBERT
ROSS DENNIS K PROF
ROSSI BRUNO B PROF
ROTS ARNOLD H DR
ROUNTREE JANET DR
ROUSE CARL A DR
ROUTLY PAUL M DR
RUBIN ROBERT HOWARD
RUBIN VERA C DR
RUDERMAN MALVIN A
RUDNICK LAWRENCE DR
RUDNICKI KONRAD PROF
RUGGE HUGO R DR
RULE BRUCE H
RUSSELL CHRISTOPHER T
RUSSELL JANE L DR
RUSSELL JOHN A PROF
RUST DAVID M DR
RYBICKI GEORGE B DR
RYDGREN ALFRED ERIC JR DR
SACKMANN I JULIANA DR

Country : U.S.A.

Members :

SAFKO JOHN L
SALO HEIKKI
SANDAGE ALLAN
SANDERS WILTON TURNER III
SANDULEAK NICHOLAS DR
SARGENT ANNEILA I
SASLAW WILLIAM C PROF
SAWYER CONSTANCE B DR
SCHAEFER BRADLEY E DR
SCHILD RUDOLPH E DR
SCHLESINGER BARRY M DR
SCHMALBERGER DONALD C DR
SCHMIDTKE PAUL Ç DR
SCHOOLMAN STEPHEN A DR
SCHROEDER DANIEL J PROF
SCHUTZ BOB EWALD
SCHWARTZ RICHARD D
SCONZO PASQUALE DR
SCOTT JOHN S DR
SEARS RICHARD LANGLEY DR
SEGAL IRVING E DR
SEIELSTAD GEORGE A
SHAFFER DAVID B DR
SHAPERO DONALD C DR
SHAPIRO STUART L
SHARPLESS STEWART PROF
SHAW R WILLIAM PROF
SHAYA EDWARD J DR
SHEFFIELD CHARLES DR
SHER DAVID DR
SHIPMAN HENRY L DR
SHORE BRUCE W
SHULL JOHN MICHAEL
SILVERBERG ERIC C DR
SIMON MICHAL PROF
SIMONSON S CHRISTIAN DR
SION EDWARD MICHAEL
SITTERLY CHARLOTTE M DR
SKILLMAN EVAN D DR
SLETTEBAK ARNE PROF
SMITH ANDREW M DR
SMITH BRUCE F DR
SMITH DEAN F DR
SMITH HARLAN J PROF
SMITH HOWARD ALAN
SMITH PETER L DR
SMOLUCHOWSKI ROMAN PROF
SNELL RONALD L
SOBERMAN ROBERT K DR
SODERBLOM LARRY DR
SOLOMON PHILIP M DR
SPARKS WARREN M DR
SPIEGEL E DR

SAGAN CARL DR
SALPETER EDWIN E PROF
SANDERS DAVID B DR
SANDFORD MAXWELL T II
SANYAL ASHIT DR
SARGENT WALLACE L W DR
SAVAGE BLAIR D DR
SCALO JOHN MICHAEL
SCHERB FRANK PROF
SCHILLER STEPHEN
SCHLOERB F. PETER
SCHMIDT EDWARD G
SCHMUTZ WERNER
SCHRAMM DAVID N PROF
SCHUECKING E L DR
SCHWARTZ DANIEL A DR
SCHWARZSCHILD MARTIN PROF
SCOTT ELIZABETH L PROF
SCOVILLE NICHOLAS Z
SEEGER CHARLES LOUIS III
SEIDELMANN P KENNETH DR
SEKANINA ZDENEK DR
SHAHAM JACOB PROF
SHAPIRO IRWIN I PROF
SHAPLEY ALAN H
SHAW JAMES SCOTT DR
SHAWHAN STANLEY D DR
SHEA MARGARET A DR
SHELUS PETER J DR
SHIELDS GREGORY A DR
SHIVANANDAN KANDIAH DR
SHORE STEVEN N
SILBERBERG REIN DR
SIMKIN SUSAN M DR
SIMON NORMAN R PROF
SINHA RAMESHWAR P
SIRY JOSEPH W
SJOGREN WILLIAM L MR
SKUMANICH ANDRE PROF
SLOVAK MARK HAINES DR
SMITH BARHAM W DR
SMITH CHARLES DITTO
SMITH ELSKE V P DR
SMITH HAYWOOD C DR
SMITH MALCOLM G DR
SMITH VERNE V DR
SMOOT III GEORGE F.
SNOW THEODORE P PROF
SOBIESKI STANLEY DR
SOFIA SABATINO PROF
SONETT CHARLES P PROF
SPENCER JOHN HOWARD
SPINRAD HYRON PROF

SALISBURY J W DR
SAMPSON DOUGLAS H PROF
SANDERS W L PROF
SANDMANN WILLIAM HENRY
SARAZIN CRAIG L DR
SARTORI LEO PROF
SAVEDOFF MALCOLM P PROF
SCARGLE JEFFREY D DR
SCHERRER PHILIP H DR
SCHLEICHER DAVID G DR
SCHMAHL EDWARD J DR
SCHMIDT MAARTEN PROF
SCHNEPS MATTHEW H
SCHREIER ETHAN J DR
SCHULTE D H DR
SCHWARTZ PHILIP R DR
SCHWEIZER FRANCOIS DR
SCOTT EUGENE HOWARD
SEARLE LEONARD DR
SEEGER PHILIP A DR
SEIDEN PHILIP E
SEWARD FREDERICK D
SHAO CHENG-YUAN
SHAPIRO MAURICE M PROF
SHARA MICHAEL DR
SHAW JOHN H PROF
SHAWL STEPHEN J DR
SHEELEY NEIL R DR
SHEN BENJAMIN S P PROF
SHINE RICHARD A DR
SHOEMAKER EUGENE M
SHU FRANK H PROF
SILK JOSEPH I PROF
SIMON GEORGE W DR
SIMON THEODORE
SINTON WILLIAM M
SITKO MICHAEL L
SKALAFURIS ANGELO J
SLADE MARTIN A III DR
SMITH ALEX G PROF
SMITH BRADFORD A PROF
SMITH CLAYTON A JR DR
SMITH HARDING E JR DR
SMITH HORACE A
SMITH MYRON A ASST PROF
SMITH WM HAYDEN PROF
SNEDEN CHRISTOPHER A
SNYDER LEWIS E
SODERBLOM DAVID R
SOIFER BARUCH T DR
SONNEBORN GEORGE DR
SPICER DANIEL SHIELDS DR
SPITZER LYMAN JR DR

Country : U.S.A.

Members :

SPRUIT HENK C DR
SRAMEK RICHARD A DR
STACEY GORDON J DR
STACHNIK ROBERT V
STAHR-CARPENTER M DR
STANDISH E MYLES DR
STANFORD SPENCER A
STANLEY G J
STARK ANTONY A
STARRFIELD SUMNER PROF
STEBBINS ROBIN
STECHER THEODORE P
STECKER FLOYD W DR
STEFANIK ROBERT DR
STEIGER W R PROF
STEIGMAN GARY PROF
STEIMAN-CAMERON THOMAS DR
STEIN JOHN WILLIAM
STEIN ROBERT F ASSOC PROF
STEIN WAYNE A PROF
STELLINGWERF ROBERT F DR
STENCEL ROBERT EDWARD
STEPHENSON C BRUCE PROF
STERN ROBERT ALLAN
STIER MARK T
STINEBRING DANIEL R
STOCKMAN HERVEY S JR DR
STOCKTON ALAN N DR
STONE EDWARD C DR
STONE R G DR
STONE REMINGTON P S DR
STRAND KAJ AA DR
STRITTMATTER PETER A PROF
STROBEL DARRELL F
STROM KAREN M
STROM ROBERT G PROF
STROM STEPHEN E
STRONG IAN B DR
STRONG JOHN D PROF
STRONG KEITH T DR
STRUBLE MITCHELL F
STRUCK-MARCELL CURTIS J
STRYKER LINDA L
STURCH CONRAD R DR
STURROCK PETER A PROF
SUESS STEVEN T DR
SUKUMAR SUNDARAJAN DR
SULENTIC JACK W DR
SULLIVAN WOODRUFF T III
SWANK JEAN HEBB
SWEIGART ALLEN V DR
SWENSON GEORGE W JR PROF
SWERDLOW NOEL PROF
SWIHART THOMAS L DR
SYNNOTT STEPHEN P
SZEBEHELY VICTOR G PROF
SZKODY PAULA DR
TAAM RONALD EVERETT DR
TADEMARU EUGENE DR
TALBOT RAYMOND J JR DR
TANDBERG-HANSSEN EINAR A
TAPIA-PEREZ SANTIAGO
TAPLEY BYRON D DR
TARNSTROM GUY DR
TARTER C BRUCE DR
TARTER JILL C DR
TAYLOR DONALD J DR
TAYLOR JOSEPH H PROF
TEDESCO EDWARD F
TELESCO CHARLES M DR
TERRELL NELSON JAMES JR
TERRILE RICHARD JOHN
TERZIAN YERVANT PROF
TESKE RICHARD G PROF
THADDEUS PATRICK PROF
THOLEN DAVID J DR
THOMAS JOHN H PROF
THOMAS RICHARD N DR
THOMAS ROGER J DR
THOMPSON A RICHARD DR
THOMPSON LAIRD A DR
THOMPSON RODGER I PROF
THOMPSON THOMAS WILLIAM
THONNARD NORBERT DR
THOREN VICTOR E PROF
THORNE KIP S PROF
THORSTENSEN JOHN R
THRONSON HARLEY ANDREW JR
THUAN TRINH XUAN DR
TIFFT WILLIAM G PROF
TIMOTHY J GETHYN DR
TIPLER FRANK JENNINGS DR
TOHLINE JOEL EDWARD
TOLBERT CHARLES R DR
TOLLER GARY N DR
TOMASKO MARTIN G DR
TOMBAUGH CLYDE W PROF
TOOMRE ALAR DR
TOOMRE JURI
TOUSEY RICHARD DR
TOWNES CHARLES HARD DR
TRAFTON LAURENCE M DR
TRAUB WESLEY ARTHUR
TREFFERS RICHARD R
TREXLER JAMES H MR
TRIMBLE VIRGINIA L DR
TROLAND THOMAS HUGH
TRURAN JAMES W JR
TUCKER WALLACE H DR
TULL ROBERT G
TULLY RICHARD BRENT DR
TURNER BARRY E DR
TURNER EDWIN L DR
TURNER KENNETH C DR
TURNER MICHAEL S
TWAROG BRUCE A
TYLER JR G LEONARD DR
TYSON JOHN A DR
TYTLER DAVID DR
ULICH BOBBY LEE
ULMER MELVILLE P PROF
ULRICH ROGER K PROF
UNDERWOOD JAMES H DR
UNWIN STEPHEN C
UOMOTO ALAN K DR
UPGREN ARTHUR R DR
UPSON WALTER L II DR
UPTON E K L DR
USHER PETER D DR
USON JUAN M DR
VAN ALLEN JAMES A PROF
VAN ALTENA WILLIAM F PROF
VAN BLERKOM DAVID J PROF
VAN BREUGEL WIL
VAN CITTERS GORDON W DR
VAN DISHOECK EWINE F DR
VAN DORN BRADT HALE DR
VAN FLANDERN THOMAS DR
VAN GORKOM JACQUELINE H
VAN HAMME WALTER
VAN HORN HUGH M PROF
VAN HOVEN GERARD DR
VAN RIPER KENNETH A DR
VAN SPEYBROECK LEON P DR
VANDEN BOUT PAUL A
VANDERVOORT PETER O DR
VAUGHAN ARTHUR H DR
VEEDER GLENN J DR
VENKATAKRISHNAN P DR
VERSCHUUR GERRIT L PROF
VERTER FRANCES DR
VESECKY J F DR
VEVERKA JOSEPH DR
VILA SAMUEL C PROF
VILKKI ERKKI U
VINTI JOHN P DR
VISHNIAC ETHAN T
VOGEL STUART NEWCOMBE DR
VOGT STEVEN SCOTT

Country : U.S.A.

Members :

VORPAHL JOAN A DR
VRBA FREDERICK J DR
VRTILEK JAN M DR
WACKERNAGEL H BEAT DR
WADDINGTON C JAKE PROF
WADE CAMPBELL M DR
WADE RICHARD ALAN DR
WAGNER RAYMOND L DR
WAGNER ROBERT M DR
WAGNER WILLIAM J DR
WAGONER ROBERT V PROF
WALBORN NOLAN R DR
WALKER ARTHUR B C JR PROF
WALKER HELEN J.
WALKER MERLE F PROF
WALKER RICHARD L
WALKER ROBERT C DR
WALKER ROBERT M A PROF
WALLACE LLOYD V DR
WALLACE RICHARD K
WALLERSTEIN GEORGE PROF
WALTER FREDERICK M
WALTERBOS RENE A M DR
WANG YI-MING DR
WANNIER PETER GREGORY DR
WARD RICHARD A DR
WARD WILLIAM R DR
WARDLE JOHN F C PROF
WARES GORDON W DR
WARNER JOHN W DR
WARREN WAYNE H JR DR
WARWICK JAMES W DR
WASSERMAN LAWRENCE H DR
WASSON JOHN T
WATSON WILLIAM D PROF
WATT GRAEME DAVID
WDOWIAK THOMAS J DR
WEAVER HAROLD F PROF
WEAVER THOMAS A DR
WEBBER JOHN C DR
WEBBINK RONALD F DR
WEBER STEPHEN VANCE
WEBSTER ADRIAN S DR
WEEDMAN DANIEL W PROF
WEEKES TREVOR C DR
WEGNER GARY ALAN
WEHINGER PETER A DR
WEIDENSCHILLING S J DR
WEILER EDWARD J DR
WEILER KURT W DR
WEILL GILBERT M DR
WEINBERG J L DR
WEINBERG STEVEN DR
WEIS EDWARD W DR
WEISBERG JOEL MARK
WEISHEIT JON C DR
WEISSKOPF MARTIN CH DR
WEISSMAN PAUL ROBERT
WEISTROP DONNA DR
WELCH WILLIAM J PROF
WELLER CHARLES S DR
WELLS DONALD C III DR
WENTZEL DONAT G DR
WESSELINK ADRIAAN J DR
WEST ROBERT ALAN
WESTERHOUT GART DR
WESTPHAL JAMES A PROF
WETHERILL GEORGE W
WEYMANN RAY J PROF
WHEELER J CRAIG PROF
WHEELER JOHN A DR
WHIPPLE ARTHUR L DR
WHIPPLE FRED L DR
WHITAKER EWEN A
WHITE NATHANIEL M DR
WHITE ORAN R DR
WHITE R STEPHEN PROF
WHITE RAYMOND E DR
WHITE RICHARD E
WHITE RICHARD L
WHITE SIMON DAVID MANION
WHITFORD ALBERT E PROF
WHITMORE BRADLEY C
WHITNEY BALFOUR S
WHITNEY CHARLES A PROF
WHITTLE D MARK DR
WIDING KENNETH G DR
WIESE WOLFGANG L DR
WIITA PAUL JOSEPH
WILDEY ROBERT L PROF DR
WILKENING LAUREL L DR
WILKES BELINDA J
WILKINSON DAVID T
WILL CLIFFORD M DR
WILLIAMON RICHARD M
WILLIAMS BARBARA A
WILLIAMS CAROL A
WILLIAMS JAMES G DR
WILLIAMS JOHN A DR
WILLIAMS THEODORE B DR
WILLNER STEVEN PAUL DR
WILLS BEVERLEY J DR
WILLS DEREK DR
WILLSON LEE ANNE DR
WILLSON ROBERT FREDERICK
WILSON ALBERT G DR
WILSON ANDREW S DR
WILSON CURTIS A
WILSON JAMES R DR
WILSON RAYMOND H DR
WILSON ROBERT E PROF
WILSON ROBERT W DR
WILSON WILLIAM J DR
WINCKLER JOHN R PROF
WINDHORST ROGIER A DR
WING ROBERT F PROF
WINGET DONALD E
WINKLER GERNOT M R DR
WINKLER KARL-HEINZ A DR
WINKLER PAUL FRANK DR
WISNIEWSKI WIESLAW Z
WITHBROE GEORGE L DR
WITT ADOLF N DR
WITTEN LOUIS PROF
WOLFE ARTHUR M PROF
WOLFF SIDNEY C DR
WOLFSON C JACOB
WOLSZCZAN ALEKSANDER DR
WOLSZCZAN ALEXANDER DR
WOOD DAVID B DR
WOOD F BRADSHAW PROF
WOOD III H J DR
WOOD JOHN A DR
WOODWARD PAUL R DR
WOOLF NEVILLE J
WOOSLEY S E PROF
WOOTTEN HENRY ALWYN
WORDEN SIMON P DR
WORLEY CHARLES E
WORRALL DIANA MARY
WRAY JAMES D DR
WRIGHT EDWARD L DR
WRIGHT FRANCES W DR
WRIGHT HELEN
WRIGHT JAMES P DR
WRIGHT MELVYN C H DR
WU CHI CHAO DR
WU SHI TSAN DR
WYCKOFF SUSAN DR
WYNN-WILLIAMS C G DR
YAHIL AMOS DR.
YANG KE-JUN
YAPLEE B S
YEH TYAN DR
YEOMANS DONALD K DR
YODER CHARLES F
YORK DONALD G DR
YOSS KENNETH M DR
YOUNG ANDREW T DR

Country : U.S.A.

Members :

YOUNG ARTHUR DR.
YOUNG JUDITH SHARN
YOUNG LOUISE GRAY DR
YUAN CHI PROF
ZABRISKIE F R PROF
ZARE KHALIL DR
ZARRO DOMINIC M DR
ZEILIK MICHAEL II DR
ZELLNER BENJAMIN H DR
ZENSUS J.-ANTON DR
ZHANF SHOUZHONG DR
ZINN ROBERT J DR
ZIRIN HAROLD DR
ZIRKER JACK B DR
ZOMBECK MARTIN V DR
ZUCKERMAN BEN M DR

Country : U.S.S.R.

Members :

ABALAKIN VICTOR K DR
ABBASOV ALIK R DR
ABELE MARTS K DR
AFANAS'EV VIKTOR L DR
AFANASJEVA PRASKOVYA M DR
AGEKJAN TATEOS A PROF
AKIM EFRAIM L DR
AKSENOV E P PROF DR
ALANTA T F DR
ALKSNIS ANDREJS DR
AMBARTSUMIAN V A PROF DR
ANDRIENKO DMITRY A DR
ANTIPOVA LYUDMILA DR
ANTONOV VADIM A DR
ARKHIPOVA V P DR
ASLANOV I A DR
BABADZHANIANC MICHAIL DR
BABADZHANOV PULAT B DR
BAGILDINSKIJ BRONISLAV K
BALEGA YURI YU.
BALKLAVS A E DR
BANIN V G DR
BARKHATOVA KLAUDIA PROF
BARTAYA R A DR
BATRAKOV YU V DR
BAZILEVSKY ALEXANDR T
BELINSKY VLADIMIR DR
BELKOVICH O I DR
BELOTSERKOVSKIJ DAVID J
BELYAEV NIKOLAJ A DR
BIBARSOV RAVIL'SH DR
BISNOVATYI-KOGAN G S DR
BLINOV N S DR
BOBROV M S DR
BOCHKAREV NIKOLAY G DR
BONDARENKO L N DR
BORCHKHADZE TENGIZ M DR
BOYARCHUK A A DR
BOYARCHUK MARGARITA E DR
BRATIJCHUK MATRONA V
BRAUDE SEMION YA PROF AG
BREJDO IZABELLA I DR
BRONNIKOVA NINA M
BRUMBERG VICTOR A DR
BYKOV MIKLE F DR
BYSTROV NIKOLAI F DR
BYSTROVA NATALIJA V DR
CHECHETKIN VALERIJ M DR
CHEREDNICHENKO V I DR
CHEREPASHCHUK A M PROF
CHERNEGA N A A DR
CHERNYKH N S DR
CHERTOPRUD V E DR
CHISTYAKOV VLADIMIR E DR
CHKHIKVADZE IAKOB N
CHUGAIJ NIKOLAI N DR
CHUGAJNOV P F DR
CHUVAEV K K DR
DADAEV ALEKSANDR N DR
DAGKESAMANSKY RUSTAM D DR
DANILOV VLADIMIR M DR
DAUBE-KURZEMNIECE I A DR
DEMIN V G PROF DR
DENISYUK EDVARD K DR
DIRIKIS M A DR
DIVARI N B DR
DLUZHNEVSKAYA O B DR
DOBRONRAVIN PETER DR
DOBROVOLSKY OLEG V PROF
DOKUCHAEVA OLGA D DR
DOLGINOV ARKADY Z PROF DR
DOLIDZE MADONA V DR
DOROSHKEVICH A G DR
DOUBINSKIJ B A DR
DRAVSKIKH A F DR
DROFA VASILIY K DR
DUBOV EMIL E PROF
DUMA DMITRIJ P DR
DZHAPIASHVILI VICTOR P DR
DZIGVASHVILI R M DR
EELSALU HEINO DR
EFREMOV YU I DR
EFREMOV YURY N DR
EINASTO JAAN DR
EMELIANOV NIKOLAJ V DR
EMINZADE T A DR
ERGMA E V DR
ERPYLEV N P DR
ESIPOV VALENTIN F DR
EVDOKIMOV YU V DR
FADEYEV YURI A
FEDOROVA RIMMA T DR
FISHKOVA LUTSA M PROF
FOMENKO ALEXANDR F DR
FOMIN VALERY A DR
FOMINOV ALEXANDR M DR
FRANTSMAN YU L DR
FRIDMAN ALEKSEY M
FROLOV M S DR
FURSENKO M A DR
GALIBINA I V DR
GALPERIN YU I PROF
GELFREIKH GEORGIJ B DR
GENKIN IGOR L PROF DR
GERSHBERG R E DR
GINZBURG VITALY L PROF
GLAGOLEVSKIJ JU V DR
GLEBOVA NINA I DR
GLUSHNEVA I N DR
GNEDIN YURIJ N DR
GNEVYSHEV MSTISLAV N DR
GNEVYSHEVA RAISA S DR
GOPASYUK S I DR
GORBATSKY VITALIJ G PROF
GORDON ISAAC M DR
GOSACHINSKIJ I V DR
GREBENIKOV E A PROF DR
GREGUL A YA DR
GRIGORJEV VICTOR M DR
GRININ VLADIMIR P DR
GRISHCHUK L P DR
GROUSHINSKY N P PROF DR
GUBANOV VADIM S DR
GULYAEV A P DR
GULYAEV RUDOLF A DR
GURSHTEIN ALEXANDER A DR
GURTOVENKO E A DR
GURZADIAN G A PROF DR
GUSEINOV O H PROF
GUSEJNOV RAGIM EH DR
HABIBULLIN SH T PROF DR
HAGEN-THORN VLADIMIR A DR
HARUTYUNIAN HAIK A DR
IBADINOV KHURSANDKUL DR
IDLIS G M DR
IKHSANOV ROBERT N DR
IKHSANOVA VERA N DR
IMSHENNIK V S DR
IOANNISIANI B K DR
IOSHPA B A DR
ISMAILOV TOFIK K
IVANCHUK VICTOR I DR
IVANOV VSEVOLOD V DR PROF
IVANOV-KHOLODNY G S DR
IZVEKOV V A DR
KADLA ZDENKA I DR
KALANDADZE N B DR
KALLOGLIAN ARSEN T DR
KALMYKOV A M DR
KANAEV IVAN I DR
KARACHENTSEV I D DR
KARDASHEV N S DR
KARETNIKOV VALENTIN G R
KARPINSKIJ VADIM N DR
KARYGINA ZOYA V DR
KASHSCHEEV B L PROF DR
KASUMOV FIKRET K O DR
KHACHIKIAN E YE PROF
KHARADZE E K PROF

Country : U.S.S.R.

Members :

KHARIN A S DR
KHARITONOV ANDREJ V DR
KHATISASHVILI ALFEZ SH DR
KHETSURIANI TSIALA S DR
KHOKHLOVA V L DR
KHOLSHEVNIKOV K V DR
KHOZOV GENNADIJ V
KHROMOV G S DR
KILADZE R I DR
KIPPER TONU DR
KISELYOV ALEXEJ A DR
KISLYAKOV ALBERT G DR
KISLYUK VITALIJ S DR
KISSELEVA TAMARA P
KLIMISHIN I A PROF
KOCHAROV GRANT E PROF
KOGOSHVILI NATELA G
KOKURIN YURIJ L DR
KOLCHINSKIJ I G DR
KOLESNIK IGOR G DR
KOLESNIK L N DR
KOLESOV A K DR
KOMAROV N S DR
KONIN V V DR
KONONOVICH EDWARD V DR
KONOPLEVA VARVARA P DR
KOPYLOV I M DR
KORCHAK A A DR
KOROVYAKOVSKIJ YURIJ P DR
KOSIN GENNADIJ S DR
KOSTIK ROMAN I
KOSTINA LIDIJA D DR
KOSTYAKOVA ELENA B DR
KOSTYLEV K V DR
KOTELNIKOV V A ACAD
KOTOV VALERY DR
KOVAL I K DR
KRAMER KH N DR
KRASINSKY GEORGE A DR
KRASSOVSKY V I DR
KRUCHINENKO VITALIY G
KSANFOMALITI L V DR
KUKLIN G V DR
KUMAJGORODSKAYA RAISA DR
KUMSIASHVILY MZIA I DR
KUMSISHVILI J I DR
KURIL-CHIK V N DR
KUROCHKA L N DR
KURT V G DR
KUTUZOV S A DR
KUZMIN ARKADIJ D PROF DR
LAPUSHKA K K DR
LATYPOV A A DR
LAVROV M I PROF
LAVRUKHINA A K PROF DR
LEBEDINETS VLADIMIR N DR
LEIKIN G A DR
LINNIK V P PROF DR
LIPOVETSKY V A
LIVSHITS M A DR
LOTOVA N A DR
LOZINSKAYA TAT'YANA A DR
LOZINSKIJ A M DR
LUPISHKO DMITRIJ F
LYUBIMKOV LEONID S DR
LYUTY VICTOR M DR
MAGAKIAN TIGRAN Y DR
MAGALASHVILI N L DR
MAGNARADZE NINA G DR
MAKARENKO EKATERINA N DR
MAKAROV VALENTINE I
MAKAROVA ELENA A DR
MANDELSTAM S L PROF
MAROCHNIK L S PROF DR
MAROV MIKHAIL YA PROF
MARTYNOV D YA PROF DR
MASSEVICH ALLA G DR
MATVEYENKO L I DR
MEDVEDEV YURI A DR
MEGRELISHVILI T G PROF
MELIK-ALAVERDIAN YU DR
MEN' A V DR
MERMAN G A DR
MERMAN NATALIA V DR
MIKHELSON NIKOLAJ N DR
MININ I N PROF
MIRONOV NIKOLAY T
MIRZOYAN L V DR PROF
MITROFANOVA LYUDMILA A DR
MNATSAKANIAN MAMIKON A DR
MOGILEVSKIJ EH I DR
MOISEEV I G DR
MOLCHANOV A P PROF
MOROZ V I PROF DR
MOROZHENKO A V DR
MOROZHENKO N N DR
MYACHIN VLADIMIR F DR
NADYOZHIN D K DR
NAGIRNER DMITRIJ I DR
NAGNIBEDA VALERY G DR
NAIDENOV VICTOR O
NAUMOV VITALIJ A DR
NEFEDEVA ANTONINA I PROF
NEMIRO ANDREJ A DR PROF
NIKITIN A A DR
NIKOGHOSSIAN ARTHUR G DR
NOSKOV BORIS N DR
NOVIKOV I D DR
NOVIKOV SERGEJ B DR
NOVOSELOV V S PROF DR
NUGIS TIIT
OBASHEV SAKEN O DR
OBRIDKO VLADIMIR N DR
OGORODNIKOV KYRILL P PROF
OMAROV TUKEN B PROF
ONEGINA A B DR
ORLOV MIKHAIL DR
PAMYATNIKH A A DR
PARIJSKIJ N N PROF
PARIJSKIJ YU N DR
PARSAMYAN ELMA S DR
PAVLOVSKAYA E D DR
PETROV G M DR
PETROV GENNADIJ M
PETROV GEORGIJ I PROF DR
PETROV PETER P DR
PETROVSKAYA M S DR
PINIGIN GENNADIJ I DR
PISKUNOV ANATOLY E
PLATAIS IMANT K DR
PODOBED V V DR
POLOZHENTSEV DIMITRIJ DR
POLUPAN P N DR
POPOV VICTOR S DR
PORFIR'EV V V DR
POTTER HEINO I DR
PRODAN Y I DR
PROKOF'EV VLADIMIR K PROF
PROKOF'EVA IRINA A DR
PROKOF'EVA VALENTINA V DR
PRONIK I I DR
PRONIK V I DR
PSKOVSKIJ JU P DR
PUGACH ALEXANDER F DR
PUSHKIN SERGEY B DR
PUSTYL'NIK IZOLD B DR
RACHKOVSKY D N DR
RAZIN V A DR
RIZVANOV NAUFAL G DR
ROMANCHUK PAVEL R DR
ROMANOV YURI S DR
ROZHKOVSKIJ DIMITRIJ A
RUBASHEV BORIS M DR
RUDZIKAS ZENONAS B
RUSKOL EUGENIA L DR
RYABOV YU A PROF DR
RYKHLOVA LIDIJA V DR
RYLOV VALERIJ S DR
RYZHKOV NIKOLAI F DR

Country : U.S.S.R.

Members :

RZHIGA OLEG N DR
SAGDEEV ROALD Z DR
SALUKVADZE G N DR
SANDAKOVA E V DR
SHAHBAZIAN ROMELIA K DR
SHAKURA NICHOLAJ I DR
SHAROV A S DR
SHCHERBINA-SAMOJLOVA I DR
SHESTAKA IVAN S DR
SHIRYAEV ALEXANDER A DR
SHUL'BERG A M DR
SHUSTOV BORIS M DR
SKRIPNICHENKO VLADIMIR DR
SMOL'KOV GENNADIJ YA DR
SOBOLEV VLADISLAV M DR
SOROCHENKO R L DR
STEPANIAN N N DR
STRAIZYS V PROF DR
SULTANOV G F ACAD
TARADY VLADIMIR K DR
TEPLITSKAYA R B DR
TOKOVININ ANDREJ A DR
TROITSKY V S PROF DR
TSEYTLIN NAUM M
URASIN LIRIK A DR
VARDANIAN R A DR
VASILEVA GALINA J DR
VITINSKIJ YURIJ I DR
VOROSHILOV V I DR
YANOVITSKIJ EDGARD G DR
YAVNEL ALEXANDER A DR
ZAITSEV VALERII V DR
ZEL'MANOV A L DR
ZHEVAKIN S A PROF DR
ZVEREV MITROFAN S PROF DR

SAAR ENN DR
SAKHAROV VLADIMIR I DR
SAMUS NIKOLAI N DR
SAPAR ARVED DR
SHAKHBAZYAN YURIJ L DR
SHANDARIN SERGEI F DR
SHCHEGLOV P V DR
SHEFFER EUGENE K DR
SHEVCHENKO VLADISLAV V DR
SHOLOMITSKY G B DR
SHUL'MAN L M DR
SIDORENKOV NIKOLAY S
SLONIM E M DR
SNEZHKO LEONID I
SOBOLEVA N S DR
STANKEVICH KAZIMIR S DR
STEPANOV ALEXANDER V DR
STREL'NITSKIJ VLADIMIR DR
SUNYAEV RASHID A DR
TATEVYAN S K DR
TEREBIZH VALERY YU DR
TOROSHLIDZE TEIMURAZ I DR
TRUTSE YU L DR
TUTUKOV A V DR
UUS UNDO DR
VARSHALOVICH DIMITRIJ PR
VEISMANN UNO DR
VITYAZEV VENEAMIN V DR
VYALSHIN GENNADIJ F DR
YAROV-YAROVOJ M S DR
YULDASHBAEV TAIMAS S
ZASOV ANATOLE V DR
ZHAGAR YOURI H DR
ZHUGZHDA YUZEF D DR

SAFRONOV VICTOR S DR
SAKHIBULLIN NAIL A DR
SANAMIAN V A DR
SEIDOV ZAKIR F DR
SHAKHOVSKOJ NIKOLAY M DR
SHARAF SH G DR
SHCHEGOLEV DIMITRIJ E DR
SHEFOV NICOLAI N
SHIRYAEV A V DR
SHOR VIKTOR A DR
SHULOV OLEG S DR
SITNIK G F PROF
SLYSH VJACHOSLAV I DR
SOBOLEV V V DR
SOCHILINA ALLA S DR
STEPANIAN A A DR
STESHENKO N V DR
STRUKOV IGOR A DR
SVECHNIKOVA MARTA A DR
TEJFEL VIKTOR G DR
TERENTJEVA ALEXANDRA K DR
TOVMASSIAN H M DR
TSAP T T DR
UDAL'TSOV V A DR
VAINSTEIN L A DR
VASHKOV'YAK SOF'YA N DR
VIIK TONU DR
VORONTSOV-VEL'YAMINOV B A
YAKOVKIN N A DR
YATSKIV YA S DR
YUNGELSON LEV R
ZDANAVICIUS KAZIMERAS DR
ZHELEZNIAKOV VLADIMIR V
ZOSIMOVICH IRINA D

Country : VATICAN CITY

Members :

BOYLE RICHARD P. DR
CASANOVAS JUAN DR
COYNE GEORGE V DR
MCCARTHY MARTIN F DR
STOEGER WILLIAM R DR

Country : VENEZUELA

Members :

ABDALA JOSE DR
BRUZUAL GUSTAVO
CALVET NURIA DR
FERRIN IGNACIO
FUENMAYOR FRANCISCO J DR
IBANEZ S. MIGUEL H. DR
MENDOZA CLAUDIO
STOCK JURGEN D

Country : YUGOSLAVIA

Members :

ANGELOV TRAJKO
ARSENIJEVIC JELISAVETA
CADEZ ANDREJ DR
CADEZ VLADIMIR
DADIC ZARKO DR
DIMITRIJEVIC MILAN
DINTINJANA BOJAN DR
DJURASEVIC GOJKO
DJUROVIC DRAGUTIN M DR
DOMINKO FRAN PROF DR
JOVANOVIC BOZIDAR
KILAR BOGDAN DR
KNEZEVIC ZORAN
KUBICELA ALEKSANDAR DR
KUZMANOSKI MIKE
LAZOVIC JOVAN P PROF
LUKACEVIC ILIJA S DR
MILOGRADOV-TURIN JELENA
MILOVANOVIC VLADETA DR
MITIC LJUBISA A DR
NINKOVIC SLOBODAN
PAKVOR IVAN
PAVLOVSKI KRESIMIR
POPOVIC BOZIDAR PROF DR
POPOVIC GEORGIJE DR
PROTICH MILORAD B
RANDIC LEO PROF DR
RUZDJAK VLADIMIR DR
SADZAKOV SOFIJA DR
SALETIC DUSAN
SEGAN STEVO
SEVARLIC BRANISLAV M PROF
SIMOVLJEVITCH JOVAN L DR
SOLARIC NIKOLA
VINCE ISTVAN
VUJNOVIC VLADIS DR
VUKICEVIC K M PROF DR
ZWITTER TOMAZ

Country : LEBANON

Members :

PLASSARD J DR

Country : LIBYA

Members :

VETESNIK MIROSLAV DR

Country : PAKISTAN

Members :

QUAMAR JAWAID

Country : PARAGUAY

Members :

TROCHE-BOGGINO A E DR

Country : PHILIPPINES

Members :

HEYDEN FRANCIS J SJ DR VINLUAN RENATO

Country : SINGAPORE

Members :

WAN FOOK SUN WILSON S J

Country : SRI LANKA

Members :

DE SILVA L.N.K. DR MAHESWARAN MURUGESAPILLAI

Country : THAILAND

Members :

PORNCHAI P.-TANAKUN SONGSATHAPORN RUANGSAK DR

4. Alphabetical list of Members

Note:

In this list, the letter adjacent to the Commission's number, has the following meaning:

P for President
V for Vice-President
C for Member of the Organizing Committee of the Commission

Reprints of this list are available from the IAU Publisher:

Kluwer Academic Publishers
P.O. Box 17
3300 A DORDRECHT, The Netherlands

Telephone (0)78-334911, Telex 29245, Telefax (0)78-334254

A'HEARN MICHAEL F DR
ASTRONOMY PROGRAM
UNIVERSITY OF MARYLAND
COLLEGE PARK MD 20742
U.S.A.
TEL: 301-454-6076
TLX: 710-826-0352
TLF:
EML:
COM: 15C,20

AANNESTAD PER ARNE DR
PHYSICS DEPT
ARIZONA STATE UNIVERSITY
TEMPE AZ 85287
U.S.A.
TEL: 602-965-3644
TLX:
TLF:
EML:
COM: 34

AARSETH SVERRE J DR
INSTITUTE OF ASTRONOMY
MADINGLEY ROAD
CAMBRIDGE CB3 OHA
U.K.
TEL: 62204
TLX: 817297 ASTRON G
TLF:
EML:
COM: 33.37

ABAD ALBERTO J DR
DPTO FISICA TEORICA
UNIVERSIDAD DE ZARAGOZA
ED. MATEMATICAS
50009 ZARAGOZA
SPAIN
TEL: (34)76-357011
TLX: 58198
TLF:
EML:
COM: 07

ABALAKIN VICTOR K DR
USSR ACADEMY OF SCIENCES
CENTR. ASTR. OBSERVATORY
PULKOVO
196140 LENINGRAD M-140
U.S.S.R.
TEL: 298-2242
TLX: 12261 PENIKS
TLF:
EML:
COM: 04C.05.07.20

ABBASOV ALIK R DR
SCIENT. & INDUSTRIAL ASS.
OF COSMIC RESEARCH
159 LENIN PROSPECT
370106 BAKU
U.S.S.R.
TEL:
TLX:
TLF:
EML:
COM: 10

ABBOTT DAVID C DR
JILA
UNIVERSITY OF COLORADO
BOULDER CO 80309
U.S.A.
TEL:
TLX:
TLF:
EML:
COM: 36

ABBOTT WILLIAM N DR
UNIVERSITY OF ATHENS
MICHALACOPOULOU 42
GR-11528 ATHENS
GREECE
TEL: 7213352
TLX:
TLF:
EML:
COM: 22.

ABDALA JOSE DR
AVENIDA SOLANO P.B.No. 1
EDIFICIO ARAGUANEY
CHACAITO CARACAS 1050
VENEZUELA
TEL:
TLX:
TLF:
EML:
COM:

ABDULLA SHAKER ABDUL AZIZ
ASTRONOMY & SPACE RES CTR
COUNCIL FOR SCI RESEARCH
P O BOX 2441
JADIRIYAH. BAGHDAD
IRAQ
TEL: 7765127
TLX: 2187 BATHILMI IK
TLF:
EML:
COM: 40

ABELE MARIS K DR
LATVIAN STATE UNIVERSITY
ASTRONOMICAL OBSERVATORY
226098 RIGA
U.S.S.R.
TEL:
TLX:
TLF:
EML:
COM: 31

ABHYANKAR KRISHNA D PROF
DEPT OF ASTRONOMY
OSMANIA UNIVERSITY
HYDERABAD 500 007
INDIA
TEL: 851672
TLX:
TLF:
EML:
COM: 24.29.36.42

ABLES HAROLD D DR
FLAGSTAFF STATION
US NAVAL OBSERVATORY
P O BOX 1149
FLAGSTAFF AZ 86002
U.S.A.
TEL: 602-779-5132
TLX: 26230 ASTRO
TLF:
EML:
COM: 09.25.28

ABLES JOHN G DR
CSIRO
DIVISION OF RADIOPHYSICS
P.O.BOX 76
EPPING NSW 2121
AUSTRALIA
TEL:
TLX:
TLF:
EML:
COM: 40

ABOU-EL-ELLA MOHAMED S DR
HELWAN OBSERVATORY (HIAG)
HELWAN-CAIRO
EGYPT
TEL: 780645
TLX: 93070 HIAG UN
TLF:
EML:
COM: 37

ABRAHAM ZULEMA DR
INST PESQUISAS ESPACIAIS
CRAAM RADIO OBSERVATORIO
CAIXA POSTAL 515
12200 S. JOSE DOS CAMPOS
BRAZIL
TEL: 011-826-6588
TLX: 01134061
TLF:
EML:
COM:

ABRAMI ALBERTO PROF
OSSERVATORIO ASTRONOMICO
DI TRIESTE
I-34131 TRIESTE
ITALY
TEL:
TLX:
TLF:
EML:
COM: 10.40

ABRAMOWICZ MAREK DR
INTERNATIONAL SCHOOL FOR
ADVANCED STUDIES
I-34014 TRIESTE
ITALY
TEL: 040-224281
TLX: 460392 ICTP
TLF:
EML:
COM: 48.

ABT HELMUT A DR
KITT PEAK NATIONAL OBS
BOX 26732
TUCSON AZ 85726
U.S.A.
TEL: 602-325-9215
TLX: 0666-484 AURA NOAO
TLF:
EML:
COM: 05.26V.29.30

ABU EL ATA NABIL DR
BUREAU DES LONGITUDES
77 AVE DENFERT-ROCHEREAU
F-75014 PARIS
FRANCE
TEL: 1-43-20-13-30
TLX: 270776 F
TLF:
EML:
COM:

ABULAZM MOHAMED SAMIR
NATIONAL INST OF ASTRON
AND GEOPHYS. RESEARCHES
HELWAN OBSERVATORY
HELWAN
EGYPT
TEL: 78-2683/0645
TLX: 93070 HIAG UN
TLF:
EML:
COM:

ACHTERBERG ABRAHAM DR
STERREKUNDIG INSTITUUT
POSTBUS 80000
PRINCETONPLEIN 5
3508 TA UTRECHT
NETHERLANDS
TEL: 0-30-535200
TLX: 40048 FYLUT
TLF:
EML: BITNET:WWMACHTOMUTRUUO
COM:

ACKER AGNES PROF DR
OBSERVATOIRE
UNIVERSITE DE STRASBOURG
11 RUE DE L UNIVERSITE
F-67000 STRASBOURG
FRANCE
TEL: (88)35-43-00
TLX: 890506 STAROBS
TLF:
EML:
COM: 34.46

ACTON LOREN W DR
91-01 255 LOCKHEED PALO
ALTO RESEARCH LAB.
3251 HANOVER ST
PALO ALTO CA 94303
U.S.A.
TEL: 415-424-3267
TLX: 346409 LMSC SUVL
TLF:
EML:
COM: 12.44

ADAM MADGE G DR
DEPT OF ASTROPHYSICS
SOUTH PARKS ROAD
OXFORD OX1 3RQ
U.K.
TEL:
TLX:
TLF:
EML:
COM: 12

ADAMS A W MR
6549 N. 35TH ROAD
ARLINGTON VA 22213
U.S.A.
TEL: 703-532-8246
TLX:
TLF:
EML:
COM:

ADAMS DAVID J DR
ASTRONOMY DEPT
THE UNIVERSITY
LEICESTER LE1 7RH
U.K.
TEL: 0533-554455
TLX: 341198
TLF:
EML:
COM: 48

ADAMS JAMES F JR DR
CODE 4154.2
NAVAL RESEARCH LABORATORY
WASHINGTON DC 20375-5000
U.S.A.
TEL: 202 767 2747
TLX:
TLF:
EML: SPAN::11335::ADAMS
COM:

ADAMS THOMAS F DR
LOS ALAMOS SCIENTIFIC LAB
G-7 MS329
P O BOX 1663
LOS ALAMOS NM 87545
U.S.A.
TEL: 505-667-6384
TLX:
TLF:
EML:
COM:

ADAMSON ANDREW DR
SCHOOL OF PHYSICS & ASTRO
LANCASHIRE POLYTECHNIC
CORPORATION STREET
PRESTON PR1 2TQ
U.K.
TEL:
TLX:
TLF:
EML:
COM: 33

ADE PETER A R DR
PHYSICS DEPT
QUEEN MARY COLLEGE
MILE END ROAD
LONDON E1 4NS
U.K.
TEL: 01-980-4811
TLX: 893750
TLF:
EML:
COM: 40

ADEL ARTHUR F PROF EMER
N. ARIZONA UNIVERSITY
BOX 5679
FLAGSTAFF AZ 86001
U.S.A.
TEL: 602-774-6597
TLX:
TLF:
EML:
COM:

ADELMAN SAUL J DR
DEPARTMENT OF PHYSICS
THE CITADEL
CHARLESTON SC 29409
U.S.A.
TEL: 803-792-6943
TLX:
TLF:
EML:
COM: 14.25.29

ADJABSHIRIZADEH ALI
CTR FOR ASTRON RESEARCH
KHADJEH NASSIR ALODIN OBS
UNIVERSITY OF TABRIZ
TABRIZ 51664
IRAN
TEL: 0098-041-32564
TLX:
TLF:
EML:
COM:

ADOLFSSON TORD DR
KRAGEHOLMSGATAN 12
S-216 19 MALMOE
SWEDEN
TEL: 040-157586
TLX:
TLF:
EML:
COM:

AFANAS'EV VIKTOR L DR
SPECIAL ASTROPHYS. OBS.
NIZHNIJ ARKHYS
357140 STAVROPOLSKY KRAI
U.S.S.R.
TEL:
TLX:
TLF:
EML:
COM: 28.33

AFANASJEVA PRASKOVYA M DR
PULKOVO OBSERVATORY
196140 LENINGRAD
U.S.S.R.
TEL: 298-22-42
TLX:
TLF:
EML:
COM: 31

AGEKJAN TATEOS A PROF
UNIVERSITY OF LENINGRAD
OBSERVATORY
199178 LENINGRAD
U.S.S.R.
TEL:
TLX:
TLF:
EML:
COM: 33.37

AGRAWAL P C DR
TATA INST. OF FUNDAMENTAL
RESEARCH
HOMI BHABHA RD
COLABA. BOMBAY 400 005
INDIA
TEL: 4952311 EXT.393
TLX: 011 3009 TIFR IN
TLF:
EML:
COM: 44.48

AGRINIER BERNARD L MR
C E N SACLAY
BP NO 2
F-91190 GIF-YVETTE
FRANCE
TEL:
TLX:
TLF:
EML:
COM:

AGUERO ESTELA L DR
OBSERVATORIO ASTRONOMICO
LAPRIDA NO 854
5000 CORDOBA
ARGENTINA
TEL: 36876 40613
TLX: 51822 BUCOR
TLF:
EML:
COM: 28

AGUILAR MARIA LUISA
UNSM
FAC. CIENCIAS FISICAS
AV ARICA 830
LIMA 5
PERU
TEL: 24 3961/52 1343
TLX:
TLF:
EML:
COM: 29

AHLUWALIA HARJIT SINGH DR
DEPT PHYSICS & ASTRONOMY
UNIVERSITY OF NEW MEXICO
800 YALE BLVD N E
ALBUQUERQUE NM 87131
U.S.A.
TEL: 505-277-2941
TLX: 660461
TLF:
EML:
COM: 10.48.49

AHMAD FAROOQ DR
DEPT OF PHYSICS
UNIVERSITY OF KASHMIR
SRINAGAR 190 006 KASHMIR
INDIA
TEL: 71559
TLX:
TLF:
EML:
COM: 28

AHMAD IMAD ALDEAN DR
IMAD-AD DEAN INC.
4323 ROSEDALE AVENUE
BETHESDA MD 20814
U.S.A.
TEL: 301 565 4714
TLX:
TLF:
EML:
COM: 44

AHMED IMAM IBRAHIM PROF
DEPT OF ASTRONOMY
FACULTY OF SCIENCE
CAIRO UNIVERSITY
GIZA CAIRO
EGYPT
TEL:
TLX:
TLF:
EML:
COM:

AHMED MOSTAFA
ASTRONOMY & METEOROL DEPT
FACULTY OF SCIENCE
CAIRO UNIVERSITY
CAIRO
EGYPT
TEL:
TLX:
TLF:
EML:
COM: 07

AI GUOXIANG
BEIJING ASTRONOMICAL OBS
BEIJING
CHINA, PEOPLE'S REP.
TEL:
TLX:
TLF:
EML:
COM: 09.10.12C

AIAD A PROF.
DEPT OF ASTRONOMY
FACULTY OF SCIENCE
CAIRO UNIVERSITY
GEZA ORMAN
EGYPT
TEL: 869538
TLX:
TLF:
EML:
COM: 34.35.37.46

AIELLO SANTI DR
DEPARTMENT OF PHYSICS
VIA L PANCALDO 3/45
FIRENZE
ITALY
TEL: 055 435939
TLX:
TLF:
EML:
COM:

AIKMAN G CHRIS L
DOMINION ASTROPHYS OBS
5071 W SAANICH ROAD
RR 5
VICTORIA BC V8X 4M6
CANADA
TEL: 604-388-1975
TLX: 049-7295
TLF:
EML:
COM: 29.

AIME C DR
DEPT D'ASTROPHYSIQUE
UNIVERSITE DE NICE
PARC VALROSE
F-06034 NICE CEDEX
FRANCE
TEL: 93-51-91-00
TLX:
TLF:
EML:
COM: 09.12.

AITKEN DAVID K DR
PHYSICS DEPT (RAAF)
UNIV OF MELBOURNE
PARKVILLE VIC 3052
AUSTRALIA
TEL: 03-341-6818
TLX: 35185 UNIMEL
TLF:
EML:
COM: 34

AIZENMAN MORRIS L DR
DIV ASTRONOMICAL SC
NATL SCI FOUNDATION
RM 615. 1800 G ST NW
WASHINGTON DC 20550
U.S.A.
TEL: 202-357-7643
TLX:
TLF:
EML:
COM: 27.35.

AKABANE KENJI A PROF
TOKYO ASTRONOMICAL
OBSERVATORY
OSAWA MITAKA
TOKYO 181
JAPAN
TEL: 0267-98-2831
TLX: 3329005 TAO NRO J
TLF:
EML:
COM: 34.40.

AKABANE TOKUHIDE DR
HIDA OBSERVATORY
KAMITAKARA
GIFU 506-13
JAPAN
TEL: 0578-6-2311
TLX:
TLF:
EML:
COM: 16

AKCAYLI MELEK M A DR
EGE UNIVERSITY
FEN FAKULTESI
GOK BILIMLERI ENSTITUSU
BORNOVA-IZMIR
TURKEY
TEL:
TLX:
TLF:
EML:
COM:

AKIN EFRAIM L DR
INSTITUTE OF APPLIED MATH
USSR ACADEMY OF SCIENCES
MIUSSAKAYA SQ4
125047 MOSCOW
U.S.S.R.
TEL: 251 37 39
TLX:
TLF:
EML:
COM: 07

AKSENOV E P PROF DR
STERNBERG STATE
ASTRONOMICAL INSTITUTE
UNIVERSITETSKIJ PROSP 13
119899 MOSCOW
U.S.S.R.
TEL: 139-28-58
TLX:
TLF:
EML:
COM: 07

AKSNES KAARE DR
NORWEGIAN DEFENCE
RESEARCH ESTABLISHMENT
P O BOX 25
N-2007 KJELLER
NORWAY
TEL: 2-737650
TLX: 76528
TLF:
EML:
COM: 06.07.20C

AKUJOR CHIDI E.
UNIVERSITY OF NIGERIA
DPT PHYSICS & ASTRONOMY
NSUKKA, ANAMBRA STATE
NIGERIA
TEL: 042-771532
TLX: 51496N LION NG
TLF:
EML:
CON: 40

AKYOL MUSTAFA UNAL PROF
SELCUK UNIVERSITY
FACULTY OF EDUCATION
42090 KONYA
TURKEY
TEL:
TLX:
TLF:
EML:
CON:

AL-NAIMY HAMID M K DR
ASTRONOMY & SPACE RES CTR
COUNCIL FOR SCI RESEARCH
P O BOX 2441
JADIRIYAH. BAGHDAD
IRAQ
TEL: 7765127
TLX: 212187
TLF:
EML:
CON: 42.51.

AL-SABTI ABDUL ADIM DR
PHYSICS DEPT
SCIENCE COLLEGE
BAGHDAD UNIVERSITY
JADIRIYAH. BAGHDAD
IRAQ
TEL: 5552340
TLX:
TLF:
EML:
CON: 38.51

ALANIA I F DR
ABASTUMANI ASTROPHYSICAL
OBSERVATORY
383762 ABASTUMANI.GEORGIA
U.S.S.R.
TEL: 225/244
TLX:
TLF:
EML:
CON: 27.

ALBERS HENRY PROF
VASSAR COLLEGE OBS
POUGHKEEPSIE NY 12601
U.S.A.
TEL: 914-452-7000
TLX:
TLF:
EML:
CON: 45.

ALBINSON JAMES DR
DPT OF PHYSICS
UNIVERSITY OF KEELE
STAFFS ST5 5BG
U.K.
TEL: 0782 621111
TLX: 36113 UNVLIB G
TLF:
EML: JANET:JSA@UK.AC.KL.PH.STAR
CON: 27

ALBRECHT RUDOLF DR
SPACE TELESCOPE EUROPEAN
COORDINATING FACILITY
KARL-SCHWARZSCHILD-STR 2
D-8046 GARCHING B MUNCHEN
GERMANY. F.R.
TEL: 89-320-06-287
TLX: 528 282 22 EO D
TLF:
EML:
CON: 09.25.

ALCAINO GONZALO DR
INSTITUTO ISAAC NEWTON
CASILLA 8-9
CORREO 9
SANTIAGO
CHILE
TEL: 472-013
TLX: c/o ESO 240853 ESOGO
TLF:
EML:
CON: 28.37

ALDROVANDI RUBEN DR
INST DI FISICA TEORICA
RUA PAMPLONA 145
01405 SAO PAULO SP
BRAZIL
TEL: 288-5643
TLX:
TLF:
EML:
CON:

ALDROVANDI S M VIEGAS DR
OHIO STATE UNIVERSITY
DEPT OF PHYSICS
174 WEST 18TH AVE.
COLUMBUS OH 43210
U.S.A.
TEL:
TLX:
TLF:
EML:
CON: 34

ALECIAN GEORGES DR
OBSERVATOIRE DE PARIS
SECTION DE MEUDON
DAF
F-92195 MEUDON PL CEDEX
FRANCE
TEL: 1-45-34-74-20
TLX: 201571 LAM
TLF:
EML:
CON: 29

ALEXANDER JOHN B
ROYAL GREENWICH OBS
HERSTMONCEUX CASTLE
HAILSHAM BN27 1RP
U.K.
TEL: 0323-833171
TLX: 87451
TLF:
EML:
CON:

ALEXANDER JOSEPH K
OFFICE OF CHIEF SCIENTIST
CODE P
NASA HEADQUARTERS
WASHINGTON DC 20546
U.S.A.
TEL:
TLX:
TLF:
EML:
CON: 40.44

ALEXANDER PAUL DR
MULLARD RADIO AST. OBS.
CAVENDISH LABORATORY
MADINGLEY ROAD
CAMBRIDGE CB3 0HE
U.K.
TEL: 223-66477
TLX: 81292 CAVLAB G
TLF:
EML: PA25@UK.AC.CAM.PHX
CON: 40

ALFARO EMILIO JAVIER
INST. ASTROFISICA DE
ANDALUCIA. APDO 2144
PROFESOR ALBAREDA 1
18080 GRANADA
SPAIN
TEL: 12 13 11
TLX: 78573 IAAG E
TLF:
EML:
CON: 27.37

ALFVEN HANNES PROF
DEPT OF PLASMA PHYSICS
ROYAL INST OF TECHNOLOGY
S-100 44 STOCKHOLM
SWEDEN
TEL: 46-08-787 70 00
TLX: 10389 KTHB
TLF:
EML:
CON: 47.48.

ALISSANDRAKIS C PH D
LAB OF ASTROPHYSICS
UNIVERSITY OF ATHENS
PANEPISTIMIOPOLIS
GR-15771 ATHENS
GREECE
TEL: 1-735122
TLX:
TLF:
EML:
CON: 10.12

ALKSNIS ANDREJS DR
RADIOASTROPHYSICAL
OBSERVATORY
226524 RIGA LATVIA
U.S.S.R.
TEL: 226796 RIGQ
TLX:
TLF:
EML:
CON: 37

ALLADIN SALEH MOHAMED DR
DEPT OF ASTRONOMY
OSMANIA UNIVERSITY
HYDERABAD 500 007
INDIA
TEL: 71116
TLX:
TLF:
EML:
CON: 28

ALLAN DAVID W MR
BUREAU OF STANDARDS
TIME & FREQUENCY DIV.
CODE 524
BOULDER CO 80302
U.S.A.
TEL: 303-497-5637
TLX: 910-940-5906
TLF:
EML:
COM: 31C

ALLAN PETER M
ASTRONOMY DEPT
UNIVERSITY OF MANCHESTER
MANCHESTER M13 9PL
U.K.
TEL: 061-273-7121
TLX:
TLF:
EML:
COM: 47

ALLEGRE CLAUDE PROF
INST PHYSIQUE DU GLOBE
4 PLACE JUSSIEU
F-75005 PARIS
FRANCE
TEL:
TLX:
TLF:
EML:
COM: 15

ALLEN ANTHONY JOHN DR
ASTRONOMY UNIT SCHOOL OF
MATHEM. SCIENCES QUEEN
MARY COL. MILE END ROAD
LONDON E1 4NS
U.K.
TEL: 1-980-4811
TLX: 893750 QMCUOL G
TLF:
EML: allen@UK.AC.QMC.MATHS
COM:

ALLEN CHRISTINE
INSTITUTO DE ASTRONOMIA
UNAM
APDO POSTAL 70-264
04510 MEXICO DF
MEXICO
TEL:
TLX:
TLF:
EML:
COM: 26.37.

ALLEN DAVID A DR
ANGLO AUSTRALIAN
OBSERVATORY
PO BOX 296
EPPING NSW 2121
AUSTRALIA
TEL: 02-868-1666
TLX: 23999 OSTD AA
TLF:
EML:
COM:

ALLEN RONALD J DR
ASTRONOMY DEPARTMENT
UNIVERSITY OF ILLINOIS
1011 W. SPRINGFIELD AVE
URBANA IL 61801
U.S.A.
TEL: 217-333-3090
TLX: 510-011-969
TLF:
EML:
COM: 28.40

ALLEN WILLIAM
ADAMS LANE OBSERVATORY
46 ADAMS LANE
BLENHEIM
NEW ZEALAND
TEL: (057)87258
TLX:
TLF:
EML:
COM:

ALLER HUGH D DR
DEPT OF ASTRONOMY
UNIVERSITY OF MICHIGAN
810 DENNISON BUILDING
ANN ARBOR MI 48109-1090
U.S.A.
TEL: 313-764-3466
TLX:
TLF:
EML:
COM: 40

ALLER LAWRENCE HUGH
ASTRONOMY DEPT
UNIVERSITY OF CALIFORNIA
MATH-SCIENCES BLDG
LOS ANGELES CA 90024
U.S.A.
TEL: 213-825-3515
TLX: 910-342-7597
TLF:
EML:
COM: 29.34.36

ALLER MARGO F DR
ASTRONOMY DEPT
UNIVERSITY MICHIGAN
ANN ARBOR MI 48109-1090
U.S.A.
TEL: 313-764-3465
TLX: 810-223-6056
TLF:
EML:
COM: 40

ALLOIN DANIELLE DR
OBSERVATOIRE DE PARIS
SECTION DE MEUDON
F-92195 MEUDON PL CEDEX
FRANCE
TEL: 1-45-34-74-04
TLX: 201571 F
TLF:
EML:
COM: 28

ALMAR IVAN PROF
KONKOLY OBSERVATORY
BOX 67
H-1525 BUDAPEST
HUNGARY
TEL: 366-621
TLX: 227460
TLF:
EML:
COM: 51

ALTAMORE ALDO
ISTITUTO ASTRONOMICO
UNIVERSITA DI ROMA
VIA G.M. LANCISI 29
I-00161 ROMA
ITALY
TEL: 06-8442977
TLX: 613255 INFRO
TLF:
EML:
COM:

ALTAVISTA CARLOS A DR
OBSERVATORIO ASTRONOMICO
PASEO DEL BOSQUE
1900 LA PLATA
ARGENTINA
TEL: 21-7308
TLX: 31151 BULAP
TLF:
EML:
COM: 07

ALTENHOFF WILHELM J DR
MPI FUER RADIOASTRONOMIE
AUF DEM HUEGEL 69
D-5300 BONN
GERMANY. F.R.
TEL: 0228-525293
TLX: 0886440 MPIFR D
TLF:
EML:
COM: 33.34.40

ALTROCK RICHARD C DR
AIR FORCE GEOPHYSICS LAB
NATIONAL SOLAR OBS
SUNSPOT NM 88349
U.S.A.
TEL: 505-434-1390
TLX: 066-484
TLF:
EML:
COM: 10.12.36

ALTSCHULER MARTIN D PROF
DEPT RAD THERAPY.BOX 522
HOSP UNIV OF PENNSYLVANIA
3400 SPRUCE ST
PHILADELPHIA PA 19104
U.S.A.
TEL: 215-662-6472
TLX:
TLF:
EML:
COM: 10.12

ALURKAR S K DR
PHYSICAL RESEARCH LAB
NAVRANGPURA
AHMEDABAD 380 009
INDIA
TEL: 462129
TLX: 121-397 PRL IN
TLF:
EML:
COM:

ALVAREZ HECTOR DR
DEPTO DE ASTRONOMIA
UNIVERSIDAD DE CHILE
CASILLA 36-D
SANTIAGO
CHILE
TEL: 2294101
TLX:
TLF:
EML:
COM:

ALVAREZ OSCAR
INSTITUTE OF GEOPHYSICS
AND ASTRONOMY
212 ST NUMBER 2906
LISA HAVANA CITY
CUBA
TEL: 218416
TLX: 0511240
TLF:
EML:
COM:

ALVAREZ P
INST DE ASTROFISICA DE
CANARIAS
LA LAGUNA
38071 TENERIFE
SPAIN
TEL:
TLX: 92640 IACE E
TLF:
EML:
COM: 21

ALY M KHAIRY PROF
HELWAN OBSERVATORY
24 MONTAZAH STR.
HELIOPOLIS. CAIRO
EGYPT
TEL: 448832
TLX:
TLF:
EML:
COM: 10,38.50

AMBARTSUMIAN V A PROF DR
BYURAKAN ASTROPHYSICAL
OBSERVATORY
378433 ARMENIA
U.S.S.R.
TEL: 52-45-80
TLX: 412623
TLF:
EML:
COM: 28.33.51

AMBASTHA A K DR
UDAIPUR SOLAR OBSERVATORY
11 VIDYA MARG
UDAIPUR 313 001
INDIA
TEL: 25626
TLX:
TLF:
EML:
COM: 33

AMBROZ PAVEL DR
ASTRONOMICAL INSTITUTE
CZECH. ACAD. OF SCIENCES
251 65 ONDREJOV
CZECHOSLOVAKIA
TEL: 72-45-25
TLX: 121579 ASTR C
TLF:
EML:
COM: 10

ANAND S P S DR
APPLIED RESEARCH CORP.
8201 CORPORATE DRIVE
SUITE 920
LANDOVER MD 20785
U.S.A.
TEL: 301-459-8442
TLX:
TLF:
EML:
COM: 35

ANANTHAKRISHNAN S
RADIOASTRONOMY CENTER
TATA INSTITUTE OF
FUNDAMENTAL RESEARCH
OOTACAMUND 643 001
INDIA
TEL: 2032
TLX: 853241
TLF:
EML:
COM: 49

ANANTHARAMAIAH K R DR
NTL RADIO ASTRONOMY OBS
VLA SITE
PO BOX 6
SOCORRO NM 87801
U.S.A.
TEL: 505-772-4306
TLX: 9109881710
TLF:
EML:
COM: 34.40

ANDERS EDWARD PROF
ENRICO FERMI INSTITUTE
UNIVERSITY OF CHICAGO
5640 S. ELLIS AVE
CHICAGO IL 60637
U.S.A.
TEL: 312-962-7108
TLX: 6871133
TLF:
EML:
COM: 15.34

ANDERSEN BO NYBORG DR
TREUBSTRAAT 13
2221 AL KATWIJK
NETHERLANDS
TEL:
TLX:
TLF:
EML:
COM: 12

ANDERSEN JOHANNES
COPENHAGEN UNIVER. OBSER.
BRORFELDEVEJ 23
DK-4340 TOLLOSE
DENMARK
TEL: 45 03 48 81 95
TLX: 44155 DANAST DK
TLF:
EML:
COM: 30C.42

ANDERSEN TORBEN BRENDER
OPTICAL DESIGN GR. D/254E
LOCKHEED RESEARCH LABS
3251 HANOVER STREET
PALO ALTO CA 94304
U.S.A.
TEL:
TLX:
TLF:
EML:
COM:

ANDERSON BRYAN DR
UNIVERSITY OF MANCHESTER
JODRELL BANK
MACCLESFIELD SK11 9DL
U.K.
TEL:
TLX:
TLF:
EML:
COM:

ANDERSON CHRISTOPHER M DR
WASHBURN OBSERVATORY
UNIVERSITY OF WISCONSIN
MADISON WI 53706
U.S.A.
TEL: 608-262-0492
TLX:
TLF:
EML:
COM:

ANDERSON KINSEY A PROF
SPACE SCIENCES LAB
UNIVERSITY OF CALIFORNIA
BERKELEY CA 94720
U.S.A.
TEL: 415-642-1313
TLX: 910-3667945 UC SPACE
TLF:
EML:
COM: 10.21.49

ANDERSON KURT S
NEW MEXICO STATE UNIV
DEPT OF ASTRONOMY
LAS CRUCES NM 88003
U.S.A.
TEL: 505-646-1032
TLX: 210-983-0549 NMSUCI
TLF:
EML:
COM:

ANDO HIROYASU DR
TOKYO ASTRONOMICAL
OBSERVATORY
OSAWA 2-21-1. MITAKA
TOKYO 181
JAPAN
TEL: 0422-32-5111
TLX: 2822307 TAOMK J
TLF:
EML:
COM: 12C.27.51

ANDREW BRYAN H DR
CHIEF. PROGRAM SERVICES
NATL RESEARCH COUNCIL
OTTAWA ONT K1A 0R6
CANADA
TEL: 613-993-3731
TLX: 0533145
TLF:
EML:
COM: 34.40

ANDREW KENNETH L PROF
DEPT OF PHYSICS
PURDUE UNIVERSITY
W LAFAYETTE IN 47907
U.S.A.
TEL: 317-494-5540
TLX:
TLF:
EML:
COM: 14

ANDREWS DAVID A DR
ARMAGH OBSERVATORY
ARMAGH BT61 9DG
U.K.
TEL:
TLX:
TLF:
EML:
COM:

ANDREWS PETER J DR
ROYAL GREENWICH OBS
HERSTMONCEUX CASTLE
HAILSHAM BN27 1RP
U.K.
TEL: 0323-83-3171
TLX: 87451
TLF:
EML:
COM:

ANDRIENKO DMITRY A DR
VLADIMIRSKAYA 51 53
FLAT 13
252003 KIEV
U.S.S.R.
TEL: 25-07-75
TLX: 132201
TLF:
EML:
COM: 15

ANDRIESSE CORNELIS D DR
VALERIUSLAAN 15
NL-6865 JA DOORWERTH
NETHERLANDS
TEL: 085-332794
TLX:
TLF:
EML:
COM: 34

ANDRILLAT HENRI L PROF
LABORATOIRE D'ASTRONOMIE
UNIV DES SCIENCES & TECH-
NIQUES DU LANGUEDOC
F-34060 MONTPELLIER
FRANCE
TEL: 67-63-25-37
TLX: 490944 USTMONT F
TLF:
EML:
COM: 29.34.46.47

ANDRILLAT YVETTE DR
LABORATOIRE D'ASTRONOMIE
UNIV DES SCIENCES & TECH-
NIQUES DU LANGUEDOC
F-34060 MONTPELLIER
FRANCE
TEL:
TLX:
TLF:
EML:
COM: 28.29.34

ANDRLE PAVEL DR
ASTRONOMICAL INSTITUTE
CZECH. ACAD. OF SCIENCES
BUDECSKA 6
120 23 PRAHA 2
CZECHOSLOVAKIA
TEL: 25-87-57
TLX: 122486
TLF:
EML:
COM: 33

ANGEL J ROGER P PROF
STEWARD OBSERVATORY
UNIVERSITY OF ARIZONA
TUCSON AZ 85721
U.S.A.
TEL: 602-621-6541
TLX: 467175
TLF:
EML:
COM: 25

ANGELETTI LUCIO DR
OSSERVATORIO ASTRONOMICO
VIA DEL PARCO MELLINI 84
I-00136 ROMA
ITALY
TEL: 06-34-70-56
TLX:
TLF:
EML:
COM:

ANGELOV TRAJKO
INSTITUTE OF ASTRONOMY
UNIVERSITY OF BELGRADE
STUDENTSKI TRG 16
YU-11000 BELGRADE
YUGOSLAVIA
TEL: 011-638-715
TLX:
TLF:
EML:
COM: 35

ANGIONE RONALD J DR
ASTRONOMY DEPT
SAN DIEGO STATE UNIV
SAN DIEGO CA 92182
U.S.A.
TEL: 619-265-6183
TLX:
TLF:
EML:
COM: 21.25

ANGUITA CLAUDIO A DR
OBS ASTRONOMICO NACIONAL
UNIVERSIDAD DE CHILE
CASILLA 36-D
SANTIAGO
CHILE
TEL: 229-4101
TLX:
TLF:
EML:
COM: 08

ANILE ANGELO M
DIPARTIMENTO MATEMATICA
CITTA UNIVERSITARIA
I-95123 CATANIA
ITALY
TEL: 095-330533
TLX:
TLF:
EML:
COM:

ANSARI S M RAZAULLAH PROF
PHYSICS DEPT
ALIGARH MUSLIM UNIVERSITY
ALIGARH UP 202 001
INDIA
TEL: 4568
TLX:
TLF:
EML:
COM: 12,41.46

ANTALOVA ANNA
ASTRONOMICAL INSTITUTE
CZECH. ACAD. OF SCIENCES
05 960 TATRANSKA LOMNICA
CZECHOSLOVAKIA
TEL:
TLX:
TLF:
EML:
COM: 10

ANTHONY-TWAROG BARBARA J
DEPT PHYSICS & ASTRONOMY
UNIVERSITY OF KANSAS
LAWRENCE KS 66045
U.S.A.
TEL: 913-864-4933
TLX:
TLF:
EML:
COM: 25

ANTIA H M DR
TATA INSTITUTE OF
FUNDAMENTAL RESEARCH
HOMI BHABHA ROAD
BOMBAY 400005
INDIA
TEL: 4952311
TLX: 011 3009 TIFR IN
TLF:
EML:
COM: 12.35

ANTIOCHOS SPIRO KOSTA
NAVAL RESEARCH LABORATORY
CODE 4170 SA
4555 OVERLOOK AVE. S.W.
WASHINGTON DC 20375
U.S.A.
TEL: 202-767-6199
TLX:
TLF:
EML:
COM: 10

ANTIPOVA LYUDMILA DR
ASTRONOMICAL COUNCIL OF
THE USSR ACAD. OF SCI.
PYATNITSKAYA 48
MOSCOW 109017 USSR
U.S.S.R.
TEL: 231 06 80
TLX: 411 576 ASCON SU
TLF:
EML:
COM: 27.42

ANTONACOPOULOS GREG PROF
DEPT OF ASTRONOMY
UNIVERSITY OF PATRAS
GR-26110 PATRAS
GREECE
TEL: 991-145
TLX:
TLF:
EML:
COM: 07

ANTONELLO ELIO
OSSERVATORIO ASTRONOMICO
VIA E. BIANCHI 46
I-22055 MERATE
ITALY
TEL: 039-592035
TLX:
TLF:
EML:
COM: 27

ANTONOPOULOU E DR
DEPT OF ASTRONOMY
UNIVERSITY OF ATHENS
PANEPISTIMIOPOLIS
GR-15771 ATHENS
GREECE
TEL:
TLX:
TLF:
EML:
COM: 42

ANTONOV VADIM A DR
LENINGRAD STATE UNIV
ASTRONOMICAL OBSERVATORY
199178 LENINGRAD
U.S.S.R.
TEL:
TLX:
TLF:
EML:
COM: 33

ANTONUCCI ESTER DR
ISTITUTO DI FISICA
UNIVERSITA DI TORINO
CORSO D'AZEGLIO 46
I-10125 TORINO
ITALY
TEL: 011-657694
TLX: 211041 INFNTO I
TLF:
EML:
COM: 10C

ANZER ULRICH DR
MPI FUER PHYSIK UND
ASTROPHYSIK
KARL-SCHWARZSCHILD-STR 1
D-8046 GARCHING B MUNCHEN
GERMANY. F.R.
TEL: 089-32990
TLX: 524629 ASTRO D
TLF:
EML:
COM:

AOKI SHINKO PROF
TOKYO ASTRONOMICAL
OBSERVATORY
MITAKA-SHI
TOKYO 181
JAPAN
TEL: 0422-32-5111
TLX: 2822307 TAOMTK J
TLF:
EML:
COM: 04.07.31.33

APARICI JUAN DR
DEPTO DE ASTRONOMIA
UNIVERSIDAD DE CHILE
CASILLA 36-D
SANTIAGO
CHILE
TEL: 2294101
TLX: 440005 ATTN OBSERVAL
TLF:
EML:
COM: 09.40

APPARAO K M V DR
TATA INSTITUTE OF
FUNDAMENTAL RESEARCH
HOMI BHABHA RD
BOMBAY 400 005
INDIA
TEL: 219111x341
TLX: 011-3009 TIFR IN
TLF:
EML:
COM: 48

APPENZELLER IMMO PROF
LANDESSTERNWARTE
KOENIGSTUHL
D-6900 HEIDELBERG 1
GERMANY. F.R.
TEL: 06221-10036
TLX: 461789 MPIA D
TLF:
EML:
COM: 29.35

APPLEBY JOHN F
JET PROPULSION LAB.
CALTECH MS 183-301
4800 OAK GROVE DIRVE
PASADENA CA 91109
U.S.A.
TEL: 818-354-3943
TLX:
TLF:
EML:
COM: 16

APPLETON PHILIP NOEL DR
DEPT. OF PHYSICS
12PHYSICS BUILDING
IOWA STATE UNIVERSITY
AMES IA 50011
U.S.A.
TEL: 515-294 3667
TLX:
TLF:
EML: BITNET:S1.PNA@ISUMVS
COM:

ARABELOS DIMITRIOS DR
DEPT GEODESY & SURVEYING
UNIVERSITY THESSALONIKI
GR-54006 THESSALONIKI
GREECE
TEL: 003031-99-2693
TLX: 412181 AUTH GR
TLF:
EML:
COM: 19

ARAI KENZO DR
DEPT OF PHYSICS
KUMAMOTO UNIVERSITY
KUMAMOTO 860
JAPAN
TEL: 096-344-2111
TLX:
TLF:
EML:
COM: 35

ARDAVAN HOUSHANG DR
INSTITUTE OF ASTRONOMY
MADINGLEY ROAD
CAMBRIDGE CB3 0HA
U.K.
TEL: 0223-3375
TLX: 817297 ASTRON G
TLF:
EML:
COM:

ARDEBERG ARNE L PROF
LUND OBSERVATORY
BOX 43
S-221 00 LUND
SWEDEN
TEL: 046-10-72-90
TLX: 33199 OBSMOT S
TLF:
EML:
COM: 28.33.45.50

ARDUINI-MALINOVSKY M. DR
CNES
2 PLACE MAURICE QUENTIN
F-75039 PARIS CEDEX 01
FRANCE
TEL:
TLX:
TLF:
EML:
COM: 14

ARELLANO FERRO ARMANDO
INSTITUTO DE ASTRONOMIA
APDO POSTAL 70-264
CIUDAD UNIVERSITARIA
04510 MEXICO DF
MEXICO
TEL: 905-548-5305
TLX: 01760155 CICME
TLF:
EML:
COM: 27

ARENDT S DR
AVENUE DE SATURNE 11
B-1180 BRUXELLES
BELGIUM
TEL:
TLX:
TLF:
EML:
COM: 20.26

ARGUE A NOEL DR
INSTITUTE OF ASTRONOMY
THE OBSERVATORIES
MADINGLEY ROAD
CAMBRIDGE CB3 0HA
U.K.
TEL: 0223-62204-63
TLX: 817297 ASTRON G
TLF:
EML:
COM: 24C.25.26

ARGYLE P E DR
DOMINION RADIO ASTROPHY-
SICAL OBSERVATORY
BOX 248
PENTICTON BC V2A 6K3
CANADA
TEL:
TLX:
TLF:
EML:
COM:

ARGYRAKOS JEAN PROF DR
193 PATISSION ST
GR-11253 ATHENS
GREECE
TEL: 8677000
TLX:
TLF:
EML:
COM: 08.41

ARIAS DE GREIFF J PROF
OBSERVATORIO NACIONAL
APARTADO 2584
BOGOTA 1. D.E.
COLOMBIA
TEL:
TLX:
TLF:
EML:
COM: 04.50

ARIAS ELISA FELICITAS
FACULTAD DE CIENCIAS
UNIV NACIONAL DE LA PLATA
PASEO DEL BOSQUE
1900 LA PLATA (BS.AS.)
ARGENTINA
TEL:
TLX:
TLF:
EML:
COM: 19

ARIMOTO NOBUO DR
OBSERVATOIRE DE PARIS
SECTION DE MEUDON
LAN
F-92195 MEUDON PL CEDEX
FRANCE
TEL: 1-45-34-75-70
TLX: 207571 F
TLF:
EML:
COM: 35

ARKHIPOVA V P DR
STERNBERG STATE ASTRO-
NOMICAL INSTITUTE
119899 MOSCOW V-234
U.S.S.R.
TEL: 139-26-57
TLX:
TLF:
EML:
COM: 27.28.34

ARLOT JEAN-EUDES
BUREAU DES LONGITUDES
77 AVE DENFERT-ROCHEREAU
F-75014 PARIS
FRANCE
TEL: 1-40-51-22-67
TLX:
TLF:
EML:
COM: 04.20C

ARNAL MARCELO EDMUNDO DR
INSTITUTO ARGENTINO
DE RADIOASTRONOMIA
CASILLA DE CORREO 5
1894 VILLA ELISA (Bs.As.)
ARGENTINA
TEL: 21-43793
TLX:
TLF:
EML:
COM: 40

ARNAUD JEAN PAUL
CANADA-FRANCE-HAWAII
TELESCOPE
PO BOX 1597
KAMUELA HI 96743
U.S.A.
TEL: 808-885-7944
TLX: 633147 CFHT
TLF:
EML:
COM: 09.12.25

ARNAUD MONIQUE
SERVICE D'ASTROPHYSIQUE.
CEN SACLAY
91191 GIF SUR YVETTE
FRANCE
TEL: 1-69087017
TLX: 604860 PHYSPAC F
TLF:
EML: SPAN:32779::ARNAUD
COM: 48

ARNETT W DAVID PROF
ENRICO FERMI INSTITUTE
933 E. 56TH AVE
CHICAGO IL 60637
U.S.A.
TEL: 312-962-8208
TLX: 910-221-5617
TLF:
EML:
COM: 35

ARNOLD JAMES R DR
UNIVERSITY OF CALIFORNIA
DEPT OF CHEMISTRY. B-017
LA JOLLA CA 92093
U.S.A.
TEL: 619 534 2908
TLX: 9103371271
TLF:
EML: Bitnet:jarnold@ucsd
COM: 15

ARNOULD MARCEL L DR
INSTITUT D'ASTRONOMIE
UNIV LIBRE DE BRUXELLES
B-1050 BRUXELLES
BELGIUM
TEL: 2-649-00-30
TLX: 23069 UNILIB
TLF:
EML:
COM: 35.48

ARNQUIST WARREN N DR
8127 DELGANY AVE
PLAYA DEL REY CA 90291
U.S.A.
TEL: 213-821-2724
TLX:
TLF:
EML:
COM:

ARNY THOMAS T DR
DEPT PHYSICS & ASTRONOMY
UNIV OF MASSACHUSETTS
GRC TOWER B
AMHERST MA 01003
U.S.A.
TEL: 413-545-2194
TLX:
TLF:
EML:
COM: 34

ARONS JONATHAN
DEPT OF ASTRONOMY
UNIVERSITY OF CALIFORNIA
601 CAMPBELL HALL
BERKELEY CA 94720
U.S.A.
TEL: 415-642-4730
TLX: 820181 UCB AST RAL
TLF:
EML:
COM: 48

ARP HALTON DR
MT WILSON & LAS CAMPANAS
OBSERVATORIES
813 SANTA BARBARA ST
PASADENA CA 91101
U.S.A.
TEL: 213-577-1122
TLX:
TLF:
EML:
COM: 28

ARPIGNY CLAUDE PROF
INSTITUT D'ASTROPHYSIQUE
UNIVERSITE DE LIEGE
AVENUE DE COINTE 5
B-4200 COINTE-OUGREE
BELGIUM
TEL: 041-529980-263
TLX:
TLF:
EML:
COM: 15C.36

ARRIBAS SANTIAGO DR
INST DE ASTROFISICA
DE CANARIAS
UNIVER. DE LA LAGUNA
LA LAGUNA TENERIFE 38200
SPAIN
TEL: 34 22 26 22 11
TLX: 92640
TLF:
EML: SPAN:IAC::SAN
COM:

ARSENIJEVIC JELISAVETA
ASTRONOMICAL OBSERVATORY
VOLGINA 7
YU-11050 BEOGRAD
YUGOSLAVIA
TEL:
TLX:
TLF:
EML:
COM: 25.27

ARTHUR DAVID W G
US GEOLOGICAL SURVEY
FLAGSTAFF AZ 86001
U.S.A.
TEL:
TLX:
TLF:
EML:
COM: 16

ARTRU MARIE-CHRISTINE DR
OBSERVATOIRE DE PARIS
SECTION DE MEUDON
ASTROPHYS FONDAMENTALE
F-92195 MEUDON PL CEDEX
FRANCE
TEL: 1-45-34-75-76
TLX: 201571 LAM
TLF:
EML:
COM: 14.29

ARTZNER GUY
IAS
BP 10
F-91371 VERRIERES-LE-B.
FRANCE
TEL: 1-64-47-43-09
TLX: 600252
TLF:
EML:
COM:

ASCHENBACH BERND PE D
MPI F. PHYSIK & ASTROPHYS
INST F. EXTRATERR PHYSIK
KARL-SCHWARZSCHILD-STR 1
D-8046 GARCHING B MUNCHEN
GERMANY, F.R.
TEL:
TLX:
TLF:
EML:
COM: 44.48

ASCHWANDEN MARKUS DR
GODDARD SPACE FLIGHT CEN.
NASA CODE 602.6
GREENBELT MD 20771
U.S.A.
TEL: (301)286 5424
TLX:
TLF:
EML: SPAN:ISIS::MARKUS
COM: 10.40

ASHOK N M DR
PHYS. RESEARCH LABORATORY
NAVRANGPURA
AHMEDABAD 380 009
INDIA
TEL: 462129
TLX: 121397
TLF:
EML:
COM: 09.25

ASLAN ZEKI DR
INONU UNIVERSITY
FACULTY OF SCIENCES
MALATYA
TURKEY
TEL:
TLX:
TLF:
EML:
COM:

ASLANOV I A DR
SHEMAKHA ASTROPHYSICAL
OBSERVATORY
373243 AZERBAIDZAN
U.S.S.R.
TEL:
TLX:
TLF:
EML:
COM: 29

ASSEO ESTELLE DR
CENTRE PHYSIQUE THEORIQUE
ECOLE POLYTECHNIQUE
F-91128 PALAISEAU
FRANCE
TEL: 1-69-41-82-00
TLX: 691596
TLF:
EML:
COM: 48

ASSOUSA GEORGE ELIAS DR
545 BOYLSTON STREET
SUITE 901
BOSTON MA 02116
U.K.
TEL:
TLX:
TLF:
EML:
COM: 40

ASTERIADIS GEORGIOS DR
DEPT GEODESY & SURVEYING
UNIVERSITY THESSALONIKI
GR-54006 THESSALONIKI
GREECE
TEL: 003-31-992693
TLX: 412181 AUTH GR
TLF:
EML:
COM: 27.33

ATAC TAMER
KANDILLI OBSERVATORY
BOSPHORUS UNIVERSITY
CENGELKOY
ISTANBUL
TURKEY
TEL: 3320240
TLX: 26411 BOUN TR
TLF:
EML:
COM: 10.29

ATANASIJEVIC IVAN DR
FACULTY OF SCIENCES
NL-6500 GL NIJMEGEN
NETHERLANDS
TEL:
TLX:
TLF:
EML:
COM:

ATHANASSOULA EVANGELIE DR
OBSERVATOIRE DE MARSEILLE
2 PLACE LE VERRIER
F-13248 MARSEILLE CEDEX 4
FRANCE
TEL: 91-95-90-88
TLX: 420241 F
TLF:
EML:
COM: 28.33

ATHAY R GRANT DR
HIGH ALTITUDE OBSERVATORY
P O BOX 3000
BOULDER CO 80307
U.S.A.
TEL: 303-497-1556
TLX:
TLF:
EML:
COM: 10.12.36

ATHERTON PAUL DAVID
ASTRONOMY GROUP
BLACKETT LABORATORY
IMPERIAL COLLEGE
LONDON SW7
U.K.
TEL:
TLX:
TLF:
EML:
COM: 09

ATKINSON ROBERT D'E DR
SWAIN HALL WEST 319
ASTRONOMY DEPT
INDIANA UNIVERSITY
BLOOMINGTON IN 47401
U.S.A.
TEL:
TLX:
TLF:
EML:
COM:

ATREYA SUSHIL K
UNIVERSITY OF MICHIGAN
DEPT ATM. & OCEANIC SCI.
SPACE RESEARCH BLDG
ANN ARBOR MI 48109-2143
U.S.A.
TEL: 313-764-3335
TLX: 8102236056
TLF:
EML:
COM: 16

AUDIER MONIQUE G DR
OBSERVATOIRE DE PARIS
SECTION DE MEUDON
F-92195 MEUDON PL CEDEX
FRANCE
TEL: 1-45-34-77-55
TLX: 270912
TLF:
EML:
COM:

AUDOUZE JEAN PROF
INSTITUT D'ASTROPHYSIQUE
98 BIS BOULEVARD ARAGO
F-75014 PARIS
FRANCE
TEL: 1-43-20-14-25
TLX:
TLF:
EML:
COM: 35.47.48

AUER LAWRENCE H DR
LOS ALAMOS NATL LAB
ESS-5, MS F665
LOS ALAMOS NM 87545
U.S.A.
TEL: 505-667-5824
TLX:
TLF:
EML:
COM: 36

AUGARDE RENEE DR
OBSERVATOIRE DE MARSEILLE
2 PLACE LE VERRIER
F-13248 MARSEILLE CEDEX 4
FRANCE
TEL: 91-95-90-88
TLX: 620241
TLF:
EML:
COM:

AUGASON GORDON C DR
M 245 6 NASA AMES
RESEARCH CENTER
MOFFETT FIELD CA 94035
U.S.A.
TEL: 415 694 4156
TLX:
TLF:
EML:
COM:

AULUCK FAQIR CHAND PROF
DEPT OF PHYSICS AND
ASTROPHYSICS
UNIVERSITY OF DELHI
DELHI 110 007
INDIA
TEL: 2918993
TLX:
TLF:
EML:
COM: 47

AUMAN JASON R PROF
DEPT GEOPHYS & ASTRONOMY
UNIV OF BRITISH COLUMBIA
VANCOUVER BC V6T 1W5
CANADA
TEL: 604-228-2892
TLX:
TLF:
EML:
COM: 36

AUNER GERHARD DR
INSTITUT FUER ASTRONOMIE
TUERKENSCHANZSTR 17
A-1180 WIEN
AUSTRIA
TEL:
TLX:
TLF:
EML:
COM:

AURASS HENRY DR
AKADEMIE DER WISSENSCHA.
DER DDR ZENTRALINSTITUT
FUR ASTROPHYSIK
DDR 1501 TREMSDORF
GERMANY. D.R.
TEL: MICHENDORF 2261
TLX: 15420
TLF:
EML:
COM: 10.40

AURIEMMA GIULIO DR
DIPARTIMENTO DI FISICA
UNIVERSITA DI ROMA
PIAZZALE A. MORO 2
I-00187 ROMA
ITALY
TEL: 396-4976336
TLX: 613255 INFNRO
TLF:
EML:
COM:

AURIERE MICHEL
OBSERVATOIRE PIC-DU-MIDI
ET TOULOUSE
F-65200 BAGNERES-DE-B.
FRANCE
TEL: 62-95-19-69
TLX:
TLF:
EML:
COM: 37

AUVERGNE MICHEL
OBSERVATOIRE DE NICE
BP 139
F-06003 NICE CEDEX
FRANCE
TEL: 93-89-04-20
TLX:
TLF:
EML:
COM:

AVCIOGLU KAMURAN PROF DR
UNIVERSITY OBSERVATORY
UNIVERSITE
ISTANBUL
TURKEY
TEL: 90-1-522-35-97
TLX:
TLF:
EML:
COM:

AVERY LORNE W DR
HERZBERG INST ASTROPHYS
NATIONAL RESEARCH COUNCIL
OTTAWA ONT K1A 0R6
CANADA
TEL: 613-993-6060
TLX:
TLF:
EML:
COM: 34.40

AVGOLOUPIS STAVROS DR
SECTION OF ASTROPHYSICS
DPT OF PHYSICS
UNIV. OF THESSALONIKI
GR 54006 THESSALONIKI
GREECE
TEL: 31-991357
TLX: 0412181 AUTH GR
TLF:
EML:
COM: 27

AVIGNON YVETTE DR
OBSERVATOIRE DE PARIS
SECTION DE MEUDON
F-92195 MEUDON PL CEDEX
FRANCE
TEL: 1-45-34-77-71
TLX: 200590
TLF:
EML:
COM: 10.40

AVRETT EUGENE H DR
CENTER FOR ASTROPHYSICS
60 GARDEN STREET
CAMBRIDGE MA 02138
U.S.A.
TEL: 617-495-7423
TLX: 921428 SATELLITE CAM
TLF:
EML:
COM: 36

AWAD MERVAT EL-SAID DR
DEPT OF ASTRONOMY
FACULTY OF SCIENCE
CAIRO UNIVERSITY
CAIRO
EGYPT
TEL:
TLX:
TLF:
EML:
COM:

AWADALLA NABIL SHOUKRY DR
HELWAN OBSERVATORY (NIAG)
HELWAN-CAIRO
EGYPT
TEL: 780645
TLX: 93070 NIAG UN
TLF:
EML:
COM: 42

AXFORD W IAN PROF
MPI FUER AERONOMIE
POSTFACH 20
D-3411 KATLENBURG-LINDAU
GERMANY. F.R.
TEL: 05556-41-414
TLX: 965527
TLF:
EML:
COM: 15.34.48

AXON DAVID
UNIVERSITY OF MANCHESTER
NUFFIELD RADIO ASTRON LAB
JODRELL BANK
MACCLESFIELD SK11 9DL
U.K.
TEL: 0477-71321
TLX: 36149
TLF:
EML:
COM: 25.40

AYDIN CEMAL PROF DR
FEN FAKULTESI
ASTRONOMI BOLUMU
BESEVLER. ANKARA
TURKEY
TEL: 232105-94
TLX:
TLF:
EML:
COM:

AYRES THOMAS R
UNIVERSITY OF COLORADO
CTR/ASTROPHYS & SPACE AST
CAMPUS BOX 391
BOULDER CO 80309
U.S.A.
TEL: 303-492-5320
TLX: 755842 JILA
TLF:
EML:
COM: 12.44

AZUMA TAKAHIRO DR
DOKKYO UNIVERSITY
SOKA
SAITAMA 340
JAPAN
TEL: 0489 42 1111
TLX: 2972005
TLF:
EML:
COM: 47

AZZOPARDI MARC DR
OBSERVATOIRE DE MARSEILLE
2 PLACE LE VERRIER
F-13248 MARSEILLE CEDEX 4
FRANCE
TEL: 91 95 90 88
TLX: 420241 F
TLF:
EML:
COM: 28.30

BAADE DIETRICH DR
ST/ECF
C/O ESO
KARL-SCHWARZSCHILD-STR 2
D-8046 GARCHING B MUNCHEN
GERMANY. F.R.
TEL: 49-89-32006388
TLX: 05 282820 EO D
TLF:
EML:
COM: 27.29C

BAAN WILLEM A
ARECIBO OBSERVATORY
PO BOX 995
ARECIBO PR 00613
U.S.A.
TEL: 809-878-2612
TLX: 385638
TLF:
EML:
COM: 48

BAARS JACOB W M DR
MPI FUER RADIOASTRONOMIE
AUF DEM HUEGEL 69
D-5300 BONN
GERMANY. F.R.
TEL: 0228-525310
TLX: 886440
TLF:
EML:
COM: 34.40

BAART EDWARD E PROF
DEPT OF PHYSICS
RHODES UNIVERSITY
P O BOX 94
GRAHAMSTOWN 6140
SOUTH AFRICA
TEL: 0461-7128
TLX: 244226
TLF:
EML:
COM: 34.40

BAATH LARS B DR
PL 8481
S-439 00 ONSALA
SWEDEN
TEL:
TLX: 2400 ONSPACE S
TLF:
EML:
COM: 40

BABADZHANIANC MICHAIL DR
ASTRONOMICAL OBSERVATORY
LENINGRAD UNIVERSITY
LENINGRAD
U.S.S.R.
TEL:
TLX:
TLF:
EML:
COM:

BABADZHANOV PULAT B DR
ASTROPHYSICAL INSTITUTE
TADJIK ACAD OF SCIENCES
734670 DUSHANBE
U.S.S.R.
TEL:
TLX:
TLF:
EML:
COM: 15.20.22C

BABCOCK ALICE K DR
NAVAL OBSERVATORY
TIME SERV. ALTERN. STAT.
11820 S W 166TH STREET
MIAMI FL 33177
U.S.A.
TEL: 305 235 0515
TLX:
TLF:
EML: TELEM:RICH.VLBI/OMNET MAIL/USA
COM: 19.31

BABCOCK HORACE W DR
MT WILSON & LAS CAMPANAS
OBSERVATORIES
813 SANTA BARBARA ST
PASADENA CA 91101
U.S.A.
TEL: 818-577-1122
TLX:
TLF:
EML:
COM: 09

BABU G S D
INDIAN INST. ASTROPHYSICS
KORAMANGALA
BANGALORE 560034
INDIA
TEL: 569179
TLX: 8452763 IIAB IN
TLF:
EML:
COM: 45

BACALOV MIHAIL
PEOPLE'S ASTRONOMICAL
OBSERVATORY AND PLANET
DIMITROVGRAD
BULGARIA
TEL: 0035903913797
TLX:
TLF:
EML:
COM: 46

BACCHUS PIERRE PROF
LABORATOIRE D'ASTRONOMIE
UNIVERSITE DE LILLE I
1 IMP. DE L'OBSERVATOIRE
F-59000 LILLE
FRANCE
TEL: 20-52-44-24
TLX:
TLF:
EML:
COM: 08.26

BACHILLER RAFAEL DR
CENTRO ASTRONOMICO DE
YEBES AP 148
E-19080 GUADALAJARA
SPAIN
TEL: (34)29 03 11
TLX: 23465 IGC E
TLF:
EML:
COM:

BACKER DONALD CH DR
RADIO ASTRONOMY LAB
UNIVERSITY OF CALIFORNIA
601 CAMPBELL HALL
BERKELEY CA 94720
U.S.A.
TEL: 415-NGC-5118
TLX: 820181 UCB AST RAL
TLF:
EML:
COM: 08.40C

BADOLATI ENNIO
VIA GIUSEPPE COTRONEI 11
I-80129 NAPOLI
ITALY
TEL: 081-243245
TLX:
TLF:
EML:
COM: 41

BAECK NICOLE A L DR
191 E-3 PLEIN
B-9218 GENT
BELGIUM
TEL:
TLX:
TLF:
EML:
COM:

BAEK CHANG RYONG
PHYSICS DEPT
KIM IL SUNG UNIVERSITY
TAESONG DISTRICT
PYONGYANG
KOREA DPR
TEL:
TLX:
TLF:
EML:
COM:

BAERENTZEN JORN
ELMEHOJVEJ 66
DK-8270 HOJBJERG
DENMARK
TEL: 45-6-272428
TLX:
TLF:
EML:
COM:

BAFFA CARLO DR
OSSERVATORIO ASTROFISICO
DI ARCETRI.
LARGO E.FERMI 5
FIRENZE 50125
ITALY
TEL: 55 27521
TLX: 572268 ARCETR I
TLF:
EML:
COM:

BAGARE S P DR
INDIAN INSTITUTE OF
ASTROPHYSICS
BANGALORE 560 034
INDIA
TEL: 566585/566497
TLX: 845763 IIAB IN
TLF:
EML:
COM: 10

BAGGALEY WILLIAM J PROF
PHYSICS DEPT
UNIVERSITY OF CANTERBURY
CHRISTCHURCH
NEW ZEALAND
TEL: 482-009 x767
TLX: 4144 UNICANT NZ
TLF:
EML:
COM: 21.22C

BAGHOS BALEGH B DR
HELWAN OBSERVATORY (HIAG)
HELWAN-CAIRO
EGYPT
TEL: 780645 - 934948
TLX: 93070 HIAG
TLF:
EML:
COM: 07

BAGILDINSKIJ BRONISLAV K
PULKOVO OBSERVATORY
196140 LENINGRAD
U.S.S.R.
TEL:
TLX:
TLF:
EML:
COM: 08

BAGLIN ANNIE DR
OBSERVATOIRE DE NICE
BP 139
F-06003 NICE CEDEX
FRANCE
TEL: 93-89-04-20
TLX: 460004
TLF:
EML:
COM: 27.35

BAGNUOLO WILLIAM G JR DR
CENTER FOR HIGH ANGULAR
RESOL ASTRONOMY
GEORGIA STATE UNIVERSITY
ATLANTA GA 30303
U.S.A.
TEL: 404 651 2932
TLX:
TLF:
EML:
COM: 26

BAGRI DURGADAS S
N.R.A.O.
P.O. BOX 0
SOCORRO NM 87801-0387
U.S.A.
TEL: 505-772-4011
TLX: 910-988-1710
TLF:
EML:
COM: 40

BAHCALL JOHN N PROF
INST FOR ADVANCED STUDY
OLDEN LN.. BLDG E
PRINCETON NJ 08540
U.S.A.
TEL: 609-734-8054
TLX: 837680
TLF:
EML:
COM: 28.33C

BAHNER KLAUS DR
MPI FUER ASTRONOMIE
ADOLF-KOLPING-STR 5
D-6903 NECKARGEMUEND
GERMANY. F.R.
TEL: 06223-3735
TLX:
TLF:
EML:
COM:

BAHNG JOHN D R PROF
DEARBORN OBSERVATORY
NORTHWESTERN UNIVERSITY
EVANSTON IL 60201
U.S.A.
TEL: 312-491-8645
TLX:
TLF:
EML:
COM: 25.45

BAILEY JEREMY A
ANGLO-AUSTRALIAN OBS
P.O. BOX 296
EPPING NSW 2121
AUSTRALIA
TEL: 02-868-1666
TLX: 23999 AAOSYD AA
TLF:
EML:
COM:

BAILEY MARK EDWARD
DEPT OF ASTRONOMY
THE UNIVERSITY
MANCHESTER M13 9PL
U.K.
TEL: 061-273-7121
TLX:
TLF:
EML:
COM: 15.20.28

BAIRD GEORGE A DR
DPT PHYSICS & ASTRONOMY
BENEDICTINE COLLEGE
BELFIELD
ATCHISON KS 66002-1499DUB
U.S.A.
TEL:
TLX:
TLF:
EML:
COM:

BAIRD KENNETH M DR
DIVISION OF PHYSICS
NATL RES COUNCIL CANADA
MONTREAL ROAD
OTTAWA ONT K1A OR6
CANADA
TEL:
TLX:
TLF:
EML:
COM: 14

BAIRD SCOTT R
DEPT PHYSICS & ASTRONOMY
BENEDICTINE COLLEGE
K-14
ATCHISON KS 66002-1499
U.S.A.
TEL: 913-367-5340
TLX:
TLF:
EML:
COM: 36

BAIZE PAUL DR
6 RUE DAUBIGNY
F-75017 PARIS
FRANCE
TEL:
TLX:
TLF:
EML:
COM: 26

BAJAJA E DR
INST. ARG. DE RADIOAST.
CC N 5
1894 VILLA ELISA
ARGENTINA
TEL: 2143793/870230
TLX:
TLF:
EML:
COM: 28

BAKER JAMES GILBERT DR
14 FRENCH DRIVE
BEDFORD NH 03102
U.S.A.
TEL: 603-472-5860
TLX:
TLF:
EML:
COM:

BAKER NORMAN H PROF
ASTRONOMY DEPARTMENT
COLUMBIA UNIVERSITY
PUPIN HALL 538 W 120
NEW YORK NY 10027
U.S.A.
TEL: 212-280-3280
TLX: 220094 COLU UR
TLF:
EML:
COM: 05.27.35

BAKOS GUSTAV A PROF
DEPT OF PHYSICS
UNIVERSITY WATERLOO
WATERLOO ONT N2L 3G1
CANADA
TEL: 519-885-1211
TLX:
TLF:
EML:
COM: 27.51

BALAZS BELA A DR
INSTITUT FUER ASTRONOMIE
TUERKENSCHANZSTR 17
A-1180 WIEN
AUSTRIA
TEL: 141019
TLX:
TLF:
EML:
COM: 37.51

BALAZS LAJOS G DR
KONKOLY OBSERVATORY
BOX 67
H-1525 BUDAPEST
HUNGARY
TEL: 166-506.166-426
TLX: 227460 KONOB H
TLF:
EML:
COM: 33C

BALBUS STEVEN A DR
UNIVERSITY OF VIRGINIA
PO BOX 3818 UNIV. STATION
CHARLOTTESVILLE VA 22903
U.S.A.
TEL: 804 924 4897
TLX:
TLF:
EML: BITNET:sb@virginia
COM: 33

BALDINELLI LUIGI DR
C P 1630
I-40100 BOLOGNA
ITALY
TEL: 051-227002
TLX:
TLF:
EML:
COM: 25

BALDWIN JACK A DR
CERRO TOLOLO
INTERAMERICAN OBSERVATORY
CASILLA 603
LA SERENA
CHILE
TEL: 213352
TLX:
TLF:
EML:
COM: 28

BALDWIN JOHN E DR
CAVENDISH LABORATORY
MADINGLEY ROAD
CAMBRIDGE CB3 OHE
U.K.
TEL: 223-66477
TLX: 81292 CAVLAB G
TLF:
EML:
COM: 33.34.40C.47

BALDWIN RALPH B
6190 GATEHOUSE DR S.E.
GRAND RAPIDS MI49506
U.S.A.
TEL: 619-949-6190
TLX:
TLF:
EML:
COM:

BALEGA YURI YU.
SPECIAL ASTROPH. OBSERV.
NIZHNI ARKHYS
35714 STAVROPOLSKY KRAJ
U.S.S.R.
TEL:
TLX:
TLF:
EML:
COM: 26

BALICK BRUCE PROF
UNIVERSITY OF WASHINGTON
ASTRONOMY DEPT FM-20
SEATTLE WA 98195
U.S.A.
TEL: 206-543-7683
TLX: 4740096 UW UI
TLF:
EML:
COM:

BALIUNAS SALLIE L
CENTER FOR ASTROPHYSICS
60 GARDEN STREET
CAMBRIDGE MA 02138
U.S.A.
TEL: 617-495-7415
TLX:
TLF:
EML:
COM: 12.29.44

BALKLAVS A E DR
RADIOPHYSICAL OBSERVATORY
LATVIAN ACADEMY OF SCI.
TURGENEVA 19
226524 RIGA
U.S.S.R.
TEL:
TLX:
TLF:
EML:
COM: 40

BALKOWSKI-MAUGER CH DR
OBSERVATOIRE DE PARIS
SECTION DE MEUDON
DAF
F-92195 MEUDON PL CEDEX
FRANCE
TEL: 1-45-34-75-56
TLX: 201571 LAM
TLF:
EML:
COM: 28

BALL JOHN A DR
NEROC HAYSTACK OBS
OFF ROUTE 40
WESTFORD MA 01886
U.S.A.
TEL: 617-692-4764
TLX:
TLF:
EML:
COM: 51

BALLABH G M DR
DEPT OF ASTRONOMY
OSMANIA UNIVERSITY
HYDERABAD 500 007
INDIA
TEL: 71951 x 247
TLX:
TLF:
EML:
COM: 24.28

BALLARIO M C PROF
OSSERVATORIO ASTRONOMICO
DI ARCETRI
VIA S LEONARDO
I-50100 FIRENZE
ITALY
TEL:
TLX:
TLF:
EML:
COM:

BALLI EDIBE PROF
UNIVERSITE KASATHANESI
ISTANBUL
TURKEY
TEL:
TLX:
TLF:
EML:
COM: 10

BALLY JOHN DR
HOH L 245
AT&T BELL LABORATORIES
HOLMDEL NJ 07733
U.S.A.
TEL: 201 888 7124
TLX: 219879 BTLH UR
TLF:
EML: jb@hohn-2.att.com
COM:

BALMINO GEORGES G DR
CNES/GRGS/BGI
18 AVE EDOUARD BELIN
F-31055 TOULOUSE CEDEX
FRANCE
TEL: 61-27-44-27
TLX: 531081 CNEST B F
TLF:
EML:
COM: 07

BALONA LUIS ANTERO DR
S A A O
P O BOX 9
OBSERVATORY 7935
SOUTH AFRICA
TEL: 47-0025
TLX: 20309
TLF:
EML:
COM: 27.30

BALONEK THOMAS J DR
DPT OF PHYS. & ASTRONOMY
COLGATE UNIVERSITY
HAMILTON NY 13346
U.S.A.
TEL: 315 824 1000
TLX:
TLF:
EML: BITNET:TBALONEK@COLGATEU
COM: 40

BALTHASAR HORST DR
UNIVERSTAETS STERNWARTE
GEISMARLANDSTRASSE 11
D- GOETTINGEN 3400
GERMANY. F.R.
TEL: 0551-39-50-48
TLX: 96753
TLF:
EML: BITNET:HBALTHA@DGOGWDG1
COM: 12

BALUTEAU JEAN-PAUL
OBS DE HAUTE-PROVENCE
F-04870 ST-MICHEL-L'OBS.
FRANCE
TEL: 92-76-63-68
TLX: 410690 F
TLF:
EML:
COM: 34

BANDERMANN L W DR
21131 GRENOLA DRIVE
CUPERTINO CA 95014
U.S.A.
TEL:
TLX:
TLF:
EML:
COM:

BANDIERA RINO DR
OSSERVATORIO ASTROFISICO
DI ARCETRI
LARGO E FERMI 5
FIRENZE 50125
ITALY
TEL: 055 2752 249
TLX: 572268
TLF:
EML: BANDIERA@ASTRFI.INFN.IT
COM:

BANDYOPADHYAY A DR
POSITIONAL ASTRONOMY CTR
P-546. BLOCK N
NEW ALIPORE
CALCUTTA 700 053
INDIA
TEL: 450321
TLX:
TLF:
EML:
COM: 04.41

BANERJI SRIRANJAN DR
DEPARTMENT OF PHYSICS
THE UNIVERSITY OF BURDWAN
BURDWAN 713104
INDIA
TEL:
TLX:
TLF:
EML:
COM:

BANG YONG GOL
PYONGYANG ASTRON OBS
ACADEMY OF SCIENCES DPRK
TAESONG DISTRICT
PYONGYANG
KOREA DPR
TEL:
TLX:
TLF:
EML:
COM: 19

BANHATTI D G DR
SCHOOL OF PHYSICS
MADURAI KAMARAJ UNIVERS.
PALKALAINAGAR
MADURAI 625021
INDIA
TEL: 85257
TLX: 445308 MKU IN
TLF:
EML:
COM: 40

BANIA THOMAS MICHAEL
DEPT OF ASTRONOMY
BOSTON UNIVERSITY
725 COMMONWEALTH AVE
BOSTON MA 02215
U.S.A.
TEL: 617-353-3652
TLX: 95129 BOS UNIV BSN
TLF:
EML:
COM: 34.51

BANIN V G DR
SIBIZMIR
P B 4
664697 IRKUTSK 33
U.S.S.R.
TEL: 6-02-65
TLX:
TLF:
EML:
COM: 10

BANOS COSMAS J DR
ASTRONOMICAL INSTITUTE
NATIONAL OBSERVATORY
P O 20048
GR-11810 ATHENS
GREECE
TEL: 3461-191
TLX: 215530 OBSA GR
TLF:
EML:
COM: 27

BANOS GEORGE J PROF
UNIVERSITY OF IOANNINA
PHYSICS DEPT
DIV OF ASTRO-GEOPHYSICS
GR-45332 IOANNINA
GREECE
TEL: 0651-91697
TLX: 322160
TLF:
EML:
COM:

BAO KEREN
NANJING ASTRONOMICAL
INSTRUMENT FACTORY
NANJING
CHINA. PEOPLE'S REP.
TEL: 46191
TLX: 34136 GLYNJ c/o NAIF
TLF:
EML:
COM: 09

BARANNE A DR
OBSERVATOIRE DE MARSEILLE
2 PLACE LE VERRIER
F-13248 MARSEILLE CEDEX 4
FRANCE
TEL: 91-95-90-88
TLX:
TLF:
EML:
COM: 09

BARATTA GIOVANNI BATTISTA
OSSERVATORIO ASTRONOMICO
DI ROMA
V. DEL PARCO MELLINI 84
I-00136 ROMA
ITALY
TEL: 06-347-056
TLX:
TLF:
EML:
COM: 29

BARBANIS BASIL PROF
DEPT OF ASTRONOMY
UNIVERSITY THESSALONIKI
GR-54006 THESSALONIKI
GREECE
TEL: 0030-31-991357
TLX: 412181
TLF:
EML:
COM: 33

BARBARO G DR
OSSERVATORIO ASTRONOMICO
VICOLO DELL'OSSERVATORIO
I-35100 PADOVA
ITALY
TEL: 049-661499
TLX: 430176 UNPADU I
TLF:
EML:
COM:

BARBERIS BRUNO
IST. DI FISICA MATEMATICA
UNIVERSITA DI TORINO
VIA CARLO ALBERTO 10
I-10123 TORINO
ITALY
TEL: 011-539214
TLX:
TLF:
EML:
COM: 07.33.47

BARBIER-BROSSAT M DR
OBSERVATOIRE DE MARSEILLE
2 PLACE LE VERRIER
F-13248 MARSEILLE CX 04
FRANCE
TEL: 91-95-90-88
TLX: 420241 F
TLF:
EML:
COM: 30.45

BARBIERI CESARE PROF
INSTITUTO DI ASTRONOMIA
UNIVERSITA DI PADOVA
VIC. DELL'OSSERVATORIO 5
I-35100 PADOVA
ITALY
TEL: 049-661499
TLX: 430176 UNPADU I
TLF:
EML:
COM: 51

BARBON ROBERTO PROF
OSSERVATORIO ASTROFISICO
I-36012 ASIAGO VI
ITALY
TEL: 0424-62665
TLX: 430110 SETURIST
TLF:
EML:
COM: 28

BARBUY BEATRIZ DR
UNIVERSIDADE DE SAO PAULO
DEPT DE ASTRONOMIA
C P 30627
SAO PAULO 01051
BRAZIL
TEL: 11 577 8599
TLX: 1156735 IAGK BR
TLF:
EML:
COM: 29

BARCZA SZABOLCS DR
KONKOLY OBSERVATORY
THEGE UT 13/17
BOX 67
H-1525 BUDAPEST
HUNGARY
TEL: 1-166-426
TLX: 227460 KONOB H
TLF:
EML:
COM:

BARDEEN JAMES M PROF
PHYSICS DEPT FM-15
UNIVERSITY OF WASHINGTON
SEATTLE WA 98195
U.S.A.
TEL: 206-545-2394
TLX: 4740096 UW UI
TLF:
EML:
COM: 47

BARKAT ZALMAN PROF
RACAH INST. OF PHYSICS
HEBREW UNIV. OF JERUSALEM
JERUSALEM 91904
ISRAEL
TEL: 02-584490
TLX: 25391
TLF:
EML:
COM:

BARKER EDWIN S DR
ASTRONOMY DEPT
UNIVERSITY OF TEXAS
RLM 15.308
AUSTIN TX 78712
U.S.A.
TEL: 512-471-4461
TLX: 910-874-1351
TLF:
EML:
COM: 15

BARKER PAUL K DR
PHYSICS DEPT.
YORK UNIVERSITY
4700 KEELE STREET
NORTH YORK ONTARIO M3J1P3
CANADA
TEL:
TLX:
TLF:
EML:
COM:

BARKER TIMOTHY DR
DPT OF PHYS. & ASTRONOMY
WHEATON COLLEGE
NORTON MA 02766
U.S.A.
TEL: 508 285 7722
TLX:
TLF:
EML:
COM:

BARKHATOVA KLAUDIA PROF
STATE UNIVERSITY
LENIN ST 51
620083 SVERDLOVSK
U.S.S.R.
TEL:
TLX:
TLF:
EML:
COM: 37

BARLAI KATALIN DR
KONKOLY OBSERVATORY
THEGE UT 13/17
BOX 67
H-1525 BUDAPEST
HUNGARY
TEL: 36-1-366-621
TLX: 227460 KONOB H
TLF:
EML:
COM:

BARLETTI RAFFAELE ENG
OSSERVATORIO ARCETRI
LARGO E. FERMI 5
I-50125 FIRENZE
ITALY
TEL: 055-2752253
TLX: 572268
TLF:
EML:
COM:

BARLIER FRANCOIS E DR
CERGA
AVENUE COPERNIC
F-06130 GRASSE
FRANCE
TEL: 93-36-58-49
TLX: 470865
TLF:
EML:
COM: 19

BARLOW MICHAEL J DR
DEPT PHYSICS & ASTRONOMY
UNIVERSITY COLLEGE LONDON
GOWER STREET
LONDON WC1E 6BT
U.K.
TEL: 01-387-7050
TLX: 28722 UCPHYS G
TLF:
EML:
COM: 34

BARNARD HANNES A J DR
PHYSICS DEPT
UNIV OF BRITISH COLUMBIA
6224 AGRICULTURE ROAD
VANCOUVER BC V6T 2A6
CANADA
TEL: 604-228-2894
TLX: 04-508576
TLF:
EML:
COM: 14

BARNES AARON DR
NASA AMES RESEARCH CENTER
CODE 245-3
MOFFETT FIELD CA 94035
U.S.A.
TEL: 415-694-5506
TLX:
TLF:
EML:
COM: 34.49

BARNES III THOMAS G DR
DEPT OF ASTRONOMY
UNIVERSITY OF TEXAS
MC DONALD OBSERVATORY
AUSTIN TX 78712
U.S.A.
TEL: 512-471-4461
TLX: 910-874-1351
TLF:
EML:
COM: 25.27C

BARNOTHY JENO DR PROF
833 LINCOLN STREET
EVANSTON IL 60201
U.S.A. -
TEL: 312-328-5729
TLX:
TLF:
EML:
COM: 47.48

BAROCAS VINICIO PROF
11 FEWLANDS AVENUE
FULWOOD
PRESTON PR2 4QR
U.K.
TEL: 0772-719249
TLX:
TLF:
EML:
COM:

BARRETO LUIZ MUNIZ PROF
OBSERVATORIO NACIONAL
RUA GENERAL BRUCE 586
20921 RIO DE JANEIRO
BRAZIL
TEL: 021-580-7313
TLX: 02121288 OBSN
TLF:
EML:
COM: 19.50

BARRETT ALAN H PROF
15 TREASURE RD.
MARATHON FL 33050
U.S.A.
TEL:
TLX:
TLF:
EML:
COM: 34.40

BARRETT PAUL EVERETT DR
SOUTH AFRICAN ASTRON OBS
PO BOX 9
OBSERVATORY 7935
SOUTH AFRICA
TEL: 021 470025 125
TLX: 520309 SAAO SA
TLF:
EML:
COM:

BARROSO JR JAIR
CNPQ - OBSERVATORIO NACL
LAB. NACL. ASTROFISICA
CAIXA POSTAL 21
37500 ITAJUBA MG
BRAZIL
TEL: 035-6220788
TLX: 031-2603
TLF:
EML:
COM: 09

BARROW COLIN H DR
DPT OF PHYSICS
HAILEY COLLEGE
HERTFORD SG13 7WU
U.K.
TEL:
TLX:
TLF:
EML:
COM: 10.16.40.49

BARROW JOHN DAVID
ASTRONOMY CENTRE
UNIVERSITY OF SUSSEX
FALMER
BRIGHTON BN1 9QH
U.K.
TEL: 0273-606755
TLX: 877159 UNISEX G
TLF:
EML:
COM: 47

BARROW RICHARD F DR
PHYSICAL CHEMISTRY
LABORATORY
SOUTH PARKS RD
OXFORD OX1 3QZ
U.K.
TEL: 0865-53322
TLX:
TLF:
EML:
COM: 14

BARRY DON C DR
DEPT OF ASTRONOMY
UNIV. OF SOUTHERN CALIF.
LOS ANGELES CA 90089
U.S.A.
TEL: 213-743-2764
TLX:
TLF:
EML:
COM: 29.45.

BARTAYA R A DR
ABASTUMANI ASTROPHYSICAL
OBSERVATORY
383762 ABASTUMANI.GEORGIA
U.S.S.R.
TEL: 237 ABASTUMANI
TLX: 327409
TLF:
EML:
COM: 45

BARTEL NORBERT HARALD DR
CENTER FOR ASTROPHYSICS
60 GARDEN STREET
CAMBRIDGE MA 02138
U.S.A.
TEL: 617-495-9278
TLX: 921428
TLF:
EML:
COM: 40

BARTH CHARLES A PROF
LASP
UNIVERSITY OF COLORADO
BOX 392
BOULDER CO 80309
U.S.A.
TEL: 303-492-7502
TLX:
TLF:
EML:
COM: 49

BARTHEL PETER DR
KAPTEYN ASTRO. INSTITUTE
UNIV OF GRONINGEN
PO BOX 800
9700 AV GRONINGEN
NETHERLANDS
TEL: 50 634073
TLX: 53572 STARS NL
TLF:
EML:
COM: 28.40

BARTHOLDI PAUL DR
OBSERVATOIRE DE GENEVE
CH-1290 SAUVERNY
SWITZERLAND
TEL: 22-55-26-11
TLX: 27720 OBSG CH
TLF:
EML:
COM:

BARTOLINI CORRADO
DIPARTIMENTO ASTRONOMIA
UNIVERSITA DI BOLOGNA
VIA ZAMBONI 33
I-40126 BOLOGNA
ITALY
TEL: 051-226677
TLX: 211664
TLF:
EML:
COM: 27.42

BARVAINIS RICHARD DR
HAYSTACK OBSERVATORY
WESTFORD MA 01886
U.S.A.
TEL: 617 692 4764
TLX: 948149
TLF:
EML:
COM: 40

BARWIG HEINZ
INST F ASTRON & ASTROPHYS
UNIVERSITAET MUNCHEN
SCHEINERSTR. 1
D-8000 MUNCHEN 80
GERMANY, F.R.
TEL: 089-98-90-21
TLX: 529815 UNIVM D
TLF:
EML:
COM: 09.27

BASART JOHN P
COOVER HALL
IOWA STATE UNIVERSITY
AMES IA 50011
U.S.A.
TEL: 515-294-2663
TLX:
TLF:
EML:
COM:

BASCHEK BODO PROF
INSTITUT F THEORETISCHE
ASTROPHYSIK
IM NEUENHEIMER F 561
D-6900 HEIDELBERG
GERMANY, F.R.
TEL: 49-6221-562837
TLX: 461515 UNIHD D
TLF:
EML:
COM: 36

BASH FRANK N PROF
ASTRONOMY DEPT
UNIVERSITY OF TEXAS
R.L. MOORE BLDG
AUSTIN TX 78712
U.S.A.
TEL: 512-471-4461
TLX: 910-874-1351
TLF:
EML:
COM: 33.34.40

BASRI GIBOR B
ASTRONOMY DEPARTMENT
UNIVERSITY OF CALIFORNIA
BERKELEY CA 94720
U.S.A.
TEL: 415-642-8198
TLX: 820181 UCB ASTRAL UD
TLF:
EML:
COM: 29

BASTIAN ULRICH
ASTRONOMISCHES-RECHEN
INSTITUT
MOENCHHOFSTR. 12-14
D-6900 HEIDELBERG
GERMANY, F.R.
TEL: 06221-49026
TLX: 461336 ARI HD D
TLF:
EML:
COM: 24

BASTIEN PIERRE DR
DEPT DE PHYSIQUE
UNIVERSITE DE MONTREAL
C P 6128 SUCC A
MONTREAL PQ H3C 3J7
CANADA
TEL: 514-343-7355
TLX: 055-62425 UDEMPHYSAS
TLF:
EML:
CON: 27

BASTIN JOHN A PROF
PHYSICS DEPT
QUEEN MARY COLLEGE
MILE END ROAD
LONDON E1 4NS
U.K.
TEL:
TLX:
TLF:
EML:
CON:

BASU BAIDYANATH PROF
APPLIED MATHEMATICS DEPT
CALCUTTA UNIVERSITY
92 A P C ROAD
CALCUTTA 700 009
INDIA
TEL:
TLX:
TLF:
EML:
CON: 28.33.51

BASU DIPAK DR
DEPT. OF PHYSICS
UNIVERSITY OF WEST INDIES
ST AUGUSTINE
TRINIDAD WEST INDIES
INDIA
TEL:
TLX:
TLF:
EML:
CON: 40.47.48.51

BATCHELOR DAVID ALLEN DR
MAIL CODE 633
NASA GODDARD SPACE FLIGHT
CENTER
GREENBELT MD 20771
U.S.A.
TEL: 301 286 2988
TLX: 892339
TLF:
EML:
CON: 10

BATES BRIAN DR
DEPT OF PURE & APPL PHYS
QUEEN'S UNIVERSITY
BELFAST BT7 1NN
U.K.
TEL: 245133
TLX: 74487
TLF:
EML:
CON:

BATES DAVID R PROF
DEPT OF APPLIED MATHS
QUEEN'S UNIVERSITY
BELFAST BT7 1NN
U.K.
TEL:
TLX:
TLF:
EML:
CON: 14.21

BATES RICHARD HEATON T DR
ELECTRICAL & ELECTRONIC
ENG. DEPT
UNIVERSITY OF CANTERBURY
CHRISTCHURCH
NEW ZEALAND
TEL: 03-482-009 x336
TLX:
TLF:
EML:
CON: 40

BATESON FRANK M OBE DR
ASTRONOMICAL RESEARCH LTD
P O BOX 3093
GREERTON TAURANGA
NEW ZEALAND
TEL: 64-075-410-216
TLX: 2880 CPO TG NZ
TLF:
EML:
CON: 27.42

BATH GEOFFREY T DR
DEPT OF ASTROPHYSICS
UNIVERSITY OF OXFORD
OXFORD OX1 3RQ
U.K.
TEL: 511336
TLX: 83295
TLF:
EML:
CON: 27.42

BATRAKOV YU V DR
INSTITUTE OF THEORETICAL
ASTRONOMY
NABEREZHNAYA KUTUZOVA 10
191187 LENINGRAD
U.S.S.R.
TEL: 272-40-23
TLX: 121578 ITA SU
TLF:
EML:
CON: 07.20

BATTANER EDUARDO DR
DPTO FIS TIERRA & COSMOS
FACULTAD DE CIENCIAS
AVDA FUENTENUEVA
GRANADA
SPAIN
TEL: 958-202212-306
TLX:
TLF:
EML:
CON: 16

BATTEN ALAN H DR
HERZBERG INST ASTROPHYS
DOMINION ASTROPHYS OBS
5071 W SAANICH RD
VICTORIA BC V8X 4M6
CANADA
TEL: 604-388-0009
TLX: 0497295
TLF:
EML:
CON: 26.30.42

BATTISTINI PIERLUIGI DR
OSSERVATORIO ASTRONOMICO
VIA ZAMBONI 33
I-40126 BOLOGNA
ITALY
TEL:
TLX:
TLF:
EML:
CON:

BATTY MICHAEL DR
CSIRO
DIVISION OF RADIOPHYSICS
P.O. BOX 76
EPPING NSW 2121
AUSTRALIA
TEL: 02-868-0222
TLX: 26230 ASTRO AA
TLF:
EML:
CON: 40

BAUD BOUDEWIJN DR
FOKKER B.V.
SPACE DIVISION
POSTBUS 7600
NL-1117 ZJ SHIPHOL
NETHERLANDS
TEL: 020-5449111
TLX:
TLF:
EML:
CON: 33

BAUDRY ALAIN DR
OBSERVATOIRE DE BORDEAUX
F-33270 FLOIRAC
FRANCE
TEL: 56-86-43-30
TLX:
TLF:
EML:
CON: 34.40C

BAUER CARL A DR
PENNA STATE UNIVERSITY
506 DAVEY
UNIVERSITY PARK PA 16802
U.S.A.
TEL:
TLX:
TLF:
EML:
CON:

BAUER WENDY HAGEN
WHITIN OBSERVATORY
WELLESLEY COLLEGE
WELLESLEY MA 02181
U.S.A.
TEL: 617-235-0320
TLX:
TLF:
EML:
CON: 27,29

BAUM WILLIAM A DR
LOWELL OBSERVATORY
1400 W. MARS HILL RD
FLAGSTAFF AZ 86001
U.S.A.
TEL: 602-774-3358
TLX:
TLF:
EML:
CON: 09.16.28.51

BAUSTIAN W W MR
KITT PEAK NAT OBSERVATORY
PO BOX 26732
950 N. CHERRY AVE
TUCSON AZ 85726
U.S.A.
TEL:
TLX:
TLF:
EML:
COM:

BANTZ LAURA P DR
DIV ASTRONOMICAL SCIENCES
NATL SCIENCE FOUNDATION
WASHINGTON DC 20550
U.S.A.
TEL: 202-357-9488
TLX:
TLF:
EML:
COM:

BATH GORDON ALAN DR
DEPT OF PHYSICS
UNIVERSITY OF ILLINOIS
URBANA IL 61801
U.S.A.
TEL: 217-333-4363
TLX: 910-830-6599 PHYSICS
TLF:
EML:
COM: 35.48

BAZILEVSKY ALEXANDR T
VERNADSKY INST GEOCHEM &
ANALYTICAL CHEMISTRY
KOSYGIN STR 19
117334 MOSCOW
U.S.S.R.
TEL:
TLX:
TLF:
EML:
COM: 16

BEALE JOHN S DR
231 MARLBOROUGH ROAD
SWINDON SN3 1NN
U.K.
TEL: 0793-34725
TLX:
TLF:
EML:
COM:

BEARD DAVID B DR
DEPT PHYSICS & ASTRONOMY
UNIVERSITY OF KANSAS
LAWRENCE KS 66045
U.S.A.
TEL: 913-864-3752
TLX:
TLF:
EML:
COM: 12.15.22

BEARDSLEY WALLACE R DR
PO BOX 531
NEWARK CA 94560
U.S.A.
TEL: 415-792-2786
TLX:
TLF:
EML:
COM: 26.30

BEAUDET GILLES DR
DEPT DE PHYSIQUE
UNIVERSITE DE MONTREAL
CP 6128
MONTREAL PQ H3C 3J7
CANADA
TEL: 514-343-6669
TLX: 055-62425
TLF:
EML:
COM: 35.51

BEAVERS WILLET I DR
ERWIN W FICK OBSERVATORY
IOWA STATE UNIVERSITY
AMES IA 50011
U.S.A.
TEL: 515-294-3667
TLX:
TLF:
EML:
COM: 26,30

BEC-BORSENBERGER ANNICK
BUREAU DES LONGITUDES
77 AVE DENFERT-ROCHEREAU
F-75014 PARIS
FRANCE
TEL: 1-43-20-12-10
TLX:
TLF:
EML:
COM: 04.07.20

BECK H G
VEB CARL ZEISS
FORSCHUNGSZENTRUM
CARL-ZEISS STR 1
DDR-6900 JENA
GERMANY. D.R.
TEL:
TLX:
TLF:
EML:
COM:

BECK RAINER
MPI FUER RADIOASTRONOMIE
AUF DEM HUEGEL 69
D-5300 BONN 1
GERMANY. F.R.
TEL: 0228-525-320
TLX: 886440 MPIFR D
TLF:
EML:
COM: 25.28.40

BECKER ROBERT A DR
PO BOX 4609
CARMEL CA 93921
U.S.A.
TEL:
TLX:
TLF:
EML:
COM:

BECKER ROBERT HOWARD
PHYSICS DEPT
UNIVERSITY OF CALIFORNIA
DAVIS CA 95616
U.S.A.
TEL: 916-752-6921
TLX: 910-531-0785 UC DAVS
TLF:
EML:
COM: 48

BECKER STEPHEN A
LOS ALAMOS NATIONAL LAB.
APPL. THEORET. PHYS. DIV.
PO BOX 1663, MS B220
LOS ALAMOS NM 87545
U.S.A.
TEL: 505-667-8931
TLX: 660495
TLF:
EML:
COM: 35

BECKER WILHELM PROF
ASTRONOMISCHES INSTITUT
UNIVERSITAET BASEL
VENUSSTRASSE 7
CH-4102 BINNINGEN
SWITZERLAND
TEL: 061-22-77-11
TLX:
TLF:
EML:
COM: 25.33.37

BECKERS JACQUES M DR
E.S.O.
KARL SCHWARZSCHILD STR.2
D-8046 GARCHING B.MUENCHEN
GERMANY. F.R.
TEL: 49 89 320 06 0
TLX: 05 28 282 22 EO D
TLF:
EML:
COM: 10.12

BECKLIN ERIC E DR
INSTITUTE FOR ASTRONOMY
2680 WOODLAWN DRIVE
HONOLULU HI 96822
U.S.A.
TEL: 808-948-6666
TLX: 723 8459 UHAST HR
TLF:
EML:
COM: 09C.34

BECKMAN JOHN E PROF
INSTITUTO DE FISICA
DE CANARIAS
LA LAGUNA
38071 TENERIFE
SPAIN
TEL:
TLX:
TLF:
EML:
COM: 12.29.34.47.51

BECKWITH STEVEN V W
DEPT OF ASTRONOMY
SPACE SCIENCES BLDG
CORNELL UNIVERSITY
ITHACA NY 14853
U.S.A.
TEL: 607-256-4805
TLX:
TLF:
EML:
COM: 34.51

BEDOGNI ROBERTO
OSERVATORIO DI ASTRONOMIA
UNIVERSITA DEGLI STUDI
C P 596
I-40100 BOLOGNA
ITALY
TEL: 051-222956
TLX: 211664 INFNBO I
TLF:
EML:
COM: 27.34

BEEBE HERBERT A
NEW MEXICO STATE UNIV.
DEPT OF ASTRONOMY
LAS CRUCES NM 88003
U.S.A.
TEL: 505-646-4438
TLX: 910-983-0549 NMSUC
TLF:
EML:
COM: 10.12

BEEBE RETA FAYE DR
NEW MEXICO STATE UNIV
DEPT OF ASTRONOMY
BOX 4500
LAS CRUCES NM 88003
U.S.A.
TEL: 505-646-1938
TLX:
TLF:
EML:
COM: 16.51

BEER REINHARD DR
183-301
JET PROPULSION LAB
4800 OAK GROVE DR
PASADENA CA 91109
U.S.A.
TEL: 818-354-4748
TLX:
TLF:
EML:
COM: 09.16

BEGELMAN MITCHELL CRAIG
JILA
UNIVERSITY OF COLORADO
CAMPUS BOX 440
BOULDER CO 80309
U.S.A.
TEL: 303-492-7856
TLX: 755842 JILA
TLF:
EML:
COM: 48

BEGGS DENIS W DR
INSTITUTE OF ASTRONOMY
MADINGLEY ROAD
CAMBRIDGE CB3 0HA
U.K.
TEL: 0223-62204
TLX: 817297 ASTRON G
TLF:
EML:
COM:

BEHR ALFRED PROF EMERITUS
ESCHENWEG 3
D-3406 BOVENDEN
GERMANY, F.R.
TEL: 0551-8897
TLX:
TLF:
EML:
COM: 25

BEINTEMA DOUWE A DR
SPACE RESEARCH DEPT
UNIVERSITY OF GRONINGEN
P O BOX 800
NL-9700 AV GRONINGEN
NETHERLANDS
TEL: 050-116631
TLX: 53572
TLF:
EML:
COM:

BEKENSTEIN JACOB D DR
PHYSICS DEPT
BEN GURION UNIVERSITY
P O B 653
BEERSHEVA 84105
ISRAEL
TEL: 057-664271
TLX: 5253 UNASI IL
TLF:
EML:
COM:

BEL NICOLE J DR
OBSERVATOIRE DE PARIS
SECTION DE MEUDON
F-92195 MEUDON PL CEDEX
FRANCE
TEL: 1-45-34-74-12
TLX: 201571 LAM
TLF:
EML:
COM: 12.34.47

BELINSKY VLADIMIR DR
LANDAU INST.THEOR.PHYSICS
KOSYGIN STR.2
117940 MOSCOW
U.S.S.R.
TEL: 137 32 44
TLX:
TLF:
EML:
COM: 47

BELKOVICH O I DR
ENGELHARDT ASTRONOMICAL
OBSERVATORY
OBSERVATORY STATION
422526 KAZAN
U.S.S.R.
TEL: 324827
TLX:
TLF:
EML:
COM: 21.22C

BELL BARBARA DR
CENTER FOR ASTROPHYSICS
60 GARDEN STREET
CAMBRIDGE MA 02138
U.S.A.
TEL: 617-495-2688
TLX:
TLF:
EML:
COM: 10

BELL BURNELL S JOCELYN DR
ROYAL OBSERVATORY
BLACKFORD HILL
EDINBURGH EH9 3HJ
U.K.
TEL: 031-667-3321
TLX: 72383 ROEDIN G
TLF:
EML:
COM:

BELL KENNETH LLOYD DR
DEPT OF APPLIED MATH
QUEEN'S UNIVERSITY
BELFAST BT7 1NN
U.K.
TEL: 245133
TLX: 74487 QUBADM
TLF:
EML:
COM:

BELL MORLEY B
HERZBERG INST ASTROPHYS
NATL RESEARCH COUNCIL
OTTAWA ONT K1A 0R6
CANADA
TEL: 613-993-6060
TLX: 0533715
TLF:
EML:
COM:

BELL ROGER A DR
ASTRONOMY PROGRAM
UNIVERSITY OF MARYLAND
COLLEGE PARK MD 20742
U.S.A.
TEL: 301-454-6282
TLX: 887294
TLF:
EML:
COM: 36.37.45

BELOTSERKOVSKIJ DAVID J
VNIIFTRI
GOSSTANDART USSR
LENINSKY PROSPECT 9
117049 MOSCOW
U.S.S.R.
TEL: 236-40-44
TLX: 411378 GOST
TLF:
EML:
COM: 31

BELSERENE EMILIA P
MARIA MITCHELL OBS
3 VESTAL STREET
NANTUCKET MA 02554.
U.S.A.
TEL: 617-228-9273
TLX:
TLF:
EML:
COM: 27

BELTON MICHAEL J S DR
SOLAR SYSTEM PROGRAM
NATL OPTICAL ASTRON OBS
950 N. CHERRY AVE
TUCSON AZ 85726
U.S.A.
TEL: 602-327-5511
TLX: 666-484 AURA KNPO TU
TLF:
EML:
COM: 15C.16

BELVEDERE GAETANO DR
ISTITUTO DI ASTRONOMIA
CITTA UNIVERSITARIA
I-95125 CATANIA
ITALY
TEL: 095-330533
TLX: 970359 ASTRCT I
TLF:
EML:
COM: 10.27

BELY OLEG DR
OBSERVATOIRE DE NICE
BP 139
F-06003 NICE CEDEX
FRANCE
TEL: 93-89-04-20
TLX:
TLF:
EML:
COM: 14

BELY-DUBAU FRANCOISE
OBSERVATOIRE DE NICE
BP 139
F-06003 NICE CEDEX
FRANCE
TEL: 93-89-04-20
TLX:
TLF:
EML:
COM: 14

BELYAEV NIKOLAJ A DR
INST FOR THEORETICAL
ASTRONOMY
10 KUTUZOV QUAY
191187 LENINGRAD
U.S.S.R.
TEL: 279-06-67
TLX: 121578 ITA SU
TLF:
EML:
COM: 20

BEN JERZY DR
ASTRONOMICAL OBSERVATORY
WROCLAW UNIVERSITY
UL. KOPERNIKA 11
51-622 WROCLAW
POLAND
TEL:
TLX:
TLF:
EML:
COM: 08

BENACCHIO LEOPOLDO
OSSERVATORIO ASTRONOMICO
VICOLO DELL'OSSERVATORIO
I-35122 PADOVA
ITALY
TEL: 049-661499
TLX: 430176 UNPADU I
TLF:
EML:
COM: 05

BENAVENTE JOSE
INSTITUTO Y OBSERVATORIO
DE MARINA
CECILIO PUJAZON S/N
11110 S. FERNANDO (CADIZ)
SPAIN
TEL: 956-883548
TLX: 76108 IOM E
TLF:
EML:
COM: 31

BENDER PETER L DR
JILA
UNIVERSITY OF COLORADO
BOX 440
BOULDER CO 80309
U.S.A.
TEL: 303-492-6793
TLX: 755842 JILA
TLF:
EML:
COM: 16.19.31

BENDINELLI ORAZIO
DIPT. DI ASTRONOMIA
VIA ZAMBONI 33
I-40126 BOLOGNA
ITALY
TEL: 051-226677/956
TLX: 211664 INFNBO I
TLF:
EML:
COM: 28

BENEDICT GEORGE F DR
DEPT OF ASTRONOMY
UNIVERSITY OF TEXAS
AUSTIN TX 78712
U.S.A.
TEL: 512-471-4461
TLX:
TLF:
EML:
COM: 24.28.44

BENEST DANIEL DR
OBSERVATOIRE DE NICE
BP 139
F-06003 NICE CEDEX
FRANCE
TEL: 93-89-04-20
TLX: 460004
TLF:
EML:
COM: 07.20.51

BENEVIDES SOARES P DR
INST ASTRON. E GEOFISICO
CAIXA POSTAL 30627
01051 SAO PAULO SP
BRAZIL
TEL: 11-275-3720
TLX: 1136221 IAGM BR
TLF:
EML:
COM: 08C

BENFORD GREGORY DR
PHYSICS DEPT
UNIVERSITY OF CALIFORNIA
IRVINE CA 92717
U.S.A.
TEL: 714-856-5147
TLX:
TLF:
EML:
COM: 12.48

BENN CHRIS R DR
ROYAL GREENWICH OBSERV.
HAILSHAM
EAST SUSSEX BN27 1RP
U.K.
TEL: 323 833171
TLX: 87451 RGOBS GB
TLF:
EML: JANET:CRB@UK.AC.RGO.STAR
COM: 05.40

BENNETT CHARLES L DR
NASA
GSFC
CODE 685
GREENBELT MD 20771
U.S.A.
TEL: 301 286 3902
TLX:
TLF:
EML: SPAN:CHAMP::BENNETT
COM: 40.44.47

BENNETT JOHN CAISTER MR
90 MALAN STREET
RIVIERA
PRETORIA 0084
SOUTH AFRICA
TEL: 012-704895
TLX:
TLF:
EML:
COM: 20

BENSAMMAR SLIMANE DR
OBSERVATOIRE DE PARIS
SECTION DE MEUDON
F-92195 MEUDON PL CEDEX
FRANCE
TEL: 1-45-34-78-35
TLX: 270912 OBSASTR F
TLF:
EML:
COM: 09.29.50

BENSON PRISCILLA J DR
WHITIN OBSERVATORY
WELLESLEY COLLEGE
WELLESLEY MA 02181
U.S.A.
TEL: 617 235 0320
TLX:
TLF:
EML: PBENSON@LUCY.WELLESLEY.EDU
COM: 27.40.41.46

BENVENUTI PIERO DR
ST/ECF
C/O ESO
KARL-SCHWARZSCHILD-STR 2
D-8046 GARCHING B MUNCHEN
GERMANY, F.R.
TEL: 49-89-32006291
TLX: 52828222 EO D
TLF:
EML:
COM:

BENZ ARNOLD DR
GRUPPE F RADIOASTRONOMIE
INSTITUT FUR ASTRONOMIE
ETH-ZENTRUM
CH-8092 ZUERICH
SWITZERLAND
TEL: 1-256-42-23
TLX: 53178 ETHBI CH
TLF:
EML:
COM: 10C.40

BENZ WILLY
CENTER FOR ASTROPHYSICS
HARVARD COLLEGE OBS.
60 GARDEN STREET
CAMBRIDGE MA 02138
U.S.A.
TEL: 617-495-9889
TLX:
TLF:
EML:
COM: 35

BERENDZEN RICHARD DR
PRESIDENT'S OFFICE
THE AMERICAN UNIVERSITY
WASHINGTON DC 20016
U.S.A.
TEL: 202-885-2121
TLX:
TLF:
EML:
COM: 41,51

BERG RICHARD A DR
HQ DEFENCE MAPPING AGENCY
BUILDING 56
US NAVAL OBSERVATORY
WASHINGTON DC 20305
U.S.A.
TEL:
TLX:
TLF:
EML:
COM:

BERGE GLENN L DR
OWENS VALLEY RADIO OBS
CALTECH 170-25
PASADENA CA 91125
U.S.A.
TEL: 818-356-6969
TLX: 675425
TLF:
EML:
COM: 16.40

BERGEAT JACQUES G DR
OBSERVATOIRE DE LYON
F-69230 ST-GENIS-LAVAL
FRANCE
TEL: 78-56-07-05
TLX:
TLF:
EML:
COM:

BERGER CHRISTIANE DR
CERGA
AVENUE COPERNIC
F-06130 GRASSE
FRANCE
TEL: 93-36-58-49
TLX: 470865 F
TLF:
EML:
COM:

BERGER JACQUES G DR
OBSERVATOIRE DE PARIS
61 AVE DE L OBSERVATOIRE
F-75014 PARIS
FRANCE
TEL: 1-40-51-22-47
TLX: 270776 OBSPARIS
TLF:
EML:
COM: 29

BERGER MITCHELL DR
DPT. OF MATHEMATICAL
SCIENCES
THE UNIVERSITY
ST ANDREWS KY16 9SS FIFE
U.K.
TEL: 334 76161
TLX: 76213 SAULIB GB
TLF:
EML: EARN%SOLAR::MBERGER
COM: 10

BERGER XAVIER DR
LAB ECOTHERMIQUE SOLAIRE
SOPHIA ANTIPOLIS
BP 21
F-06562 VALBONNE CEDEX
FRANCE
TEL: 93-65-34-00
TLX: 970134 F
TLF:
EML:
COM:

BERGERON JACQUELINE A DR
INSTITUT D'ASTROPHYSIQUE
98 BIS BOULEVARD ARAGO
F-75014 PARIS
FRANCE
TEL: 1-43-20-14-25
TLX: 270070 c/o IAP
TLF:
EML:
COM: 28,34,44,47.48

BERGSTRALH JAY T DR
JPL
M/X 183-301
4800 OAK GROVE DRIVE
PASADENA CA 91109
U.S.A.
TEL: 818-354-2296
TLX:
TLF:
EML:
COM: 16

BERGVALL NILS AKE SIGVARD
ASTRONOMISKA OBSERVATORIE
BOX 515
S-751 20 UPPSALA
SWEDEN
TEL:
TLX:
TLF:
EML:
COM: 28

BERKHUIJSEN ELLY M DR
MPI FUER RADIOASTRONOMIE
AUF DEM HUEGEL 69
D-5300 BONN 1
GERMANY. F.R.
TEL:
TLX: 886440 MPIFR D
TLF:
EML:
COM: 28,33,34,40

BERMAN ROBERT HIRAM DR
MIT
RM 36-227
77 MASSACHUSETTS AVE
CAMBRIDGE MA 02139
U.S.A.
TEL: 617-253-1000
TLX:
TLF:
EML:
COM:

BERNACCA P L PROF
OSSERVATORIO ASTROFISICO
DELL'UNIVERSITA'
I-36012 ASIAGO (VICENZA)
ITALY
TEL: 0424-62505
TLX: 430110 SETOUR
TLF:
EML:
COM: 26C,44.51

BERNAT ANDREW PLOUS DR
COMPUTER SCIENCE DEPT
UNIVERSITY OF TEXAS
AT EL PASO
EL PASO TX 79968
U.S.A.
TEL: 915-747-5494
TLX:
TLF:
EML:
COM: 34.36

BERRINGTON KEITH ADRIAN
DEPT OF APPLIED MATHS
QUEEN'S UNIVERSITY
BELFAST BT7 1NN
U.K.
TEL:
TLX:
TLF:
EML:
COM: 14

BERRUYER-DESIROTTE N DR
OBSERVATOIRE DE NICE
BP 139
06003 NICE CEDEX
FRANCE
TEL: 92 00 30 11
TLX: 460 004
TLF:
EML:
COM:

BERTAUX J L DR
SERVICE D'AERONOMIE
BP NO 3
F-91370 VERRIERES-LE-B.
FRANCE
TEL: 1-69-20-31-16
TLX: 692400 F
TLF:
EML:
COM: 16C.49

BERTELLI GIANPAOLO DR
DIPARTIMENTO DI ASTRONOMI
VICOLO DELL OSSERV. 5
PADOVA 35122
ITALY
TEL: 049 661499
TLX: 432071 ASTROS I
TLF:
EML:
COM:

BERTHOMIEU GABRIELLE DR
OBSERVATOIRE DE NICE
BP 139
F-06003 NICE CEDEX
FRANCE
TEL: 93-89-04-20
TLX: 46004 F
TLF:
EML:
COM: 27.35

BERTIAU FLOR C PROF
WAVERSEBAAN 220
B-3030 HEVERLEE
BELGIUM
TEL:
TLX:
TLF:
EML:
COM: 30

BERTIN GIUSEPPE PROF
SCUOLA NORMALE SUPERIORE
PIAZZA DEI CAVALIERI
I-56100 PISA
ITALY
TEL: 050-597265
TLX: 590548 SNSPI I
TLF:
EML:
COM:

BERTOLA FRANCESCO PROF
OSSERVATORIO ASTRONOMICO
VICOLO DELL'OSSERVATORIO
I-35100 PADOVA
ITALY
TEL: 049-661499
TLX: 430176 UNPADU I
TLF:
EML:
COM: 28C.47

BERTOUT CLAUDE
INSTITUT D'ASTROPHYSIQUE
98 BIS BOULEVARD ARAGO
F-75014 PARIS
FRANCE
TEL: 1-43-20-14-25
TLX:
TLF:
EML:
COM: 29.34.36

BESSELL MICHAEL S DR
MT STROMLO OBSERVATORY
WODEN P.O. ACT 2606
AUSTRALIA
TEL: 062-881111
TLX: 62270 CANOPUS AA
TLF:
EML:
COM: 05.25.27.29C

BETTIS DALE G PROF
TICOM
UNIVERSITY OF TEXAS
AT AUSTIN
AUSTIN TX 78712
U.S.A.
TEL:
TLX:
TLF:
EML:
COM: 07

BETTONI DANIELA DR
OSSERVATORIO ASTRONOMICO
VICOLO DELL'OSSERVATORIO
I-35122 PADOVA
ITALY
TEL: 049-661499
TLX: 430176 UNPADU I
TLF:
EML:
COM: 28

BEUERMANN KLAUS P PROF
INSTITUT FUR ASTRONOMIE &
ASTROPHYSIK ZU BERLIN
ERNST-REUTER-PL 7
D-1000 BERLIN
GERMANY, F.R.
TEL:
TLX:
TLF:
EML:
COM:

BHANDARI N DR
PHYSICAL RESEARCH LAB
NAVRANGPURA
AHMEDABAD 380 009
INDIA
TEL: 462129
TLX: 0121397
TLF:
EML:
COM: 22

BHANDARI RAJENDRA DR
RAMAN RESEARCH INSTITUTE
BANGALORE 560 080
INDIA
TEL: 812-360122
TLX: 8452671 RRI IN
TLF:
EML:
COM: 40

BHAT CHAMAN LAL DR
BHABHA ATOMIC RESEARCH
CENTRE NRL
SRINAGAR 190 006
KASHMIR
INDIA
TEL: 74965
TLX:
TLF:
EML:
COM:

BHAT NARAYANA P DR
TATA INSTITUTE OF
FUNDAMENTAL RESEARCH
HOMI BHABHA ROAD
BOMBAY 400005
INDIA
TEL: 4952311
TLX: 0113009 TIFR IN
TLF:
EML:
COM:

BHATIA PREM K DR
DEPT OF MATHEMATICS
UNIVERSITY OF JODHPUR
JODHPUR 342 001
INDIA
TEL:
TLX:
TLF:
EML:
COM:

BHATIA R K DR
DEPT OF ASTRONOMY
OSMANIA UNIVERSITY
HYDERABAD 500 007
INDIA
TEL:
TLX:
TLF:
EML:
COM: 16

BHATIA V B DR
DEPT PHYSICS & ASTROPHYS
DELHI UNIVERSITY
DELHI-110 007
INDIA
TEL: 2918993
TLX:
TLF:
EML:
COM:

BHATNAGAR ARVIND DR
UDAIPUR SOLAR OBSERVATORY
11 VIDYA MARG
UDAIPUR RAJAS 313 001
INDIA
TEL: 25626-23861
TLX:
TLF:
EML:
COM: 10.12

BHATNAGAR ASHOK KUMAR
POSITIONAL ASTR. CENTER
P 546 BLOCK N 1ST FL
NEW ALIPORE
CALCUTTA 700053
INDIA
TEL: 450321. 493541
TLX:
TLF:
EML:
COM: 04

BHATNAGAR K B DR
ZAKIR HUSSAIN COLLEGE
UNIVERSITY OF DELHI
AJMERI GATE
NEW DELHI 110 006
INDIA
TEL: 522802
TLX:
TLF:
EML:
COM: 07C

BHATT H C DR
INDIAN INSTITUTE OF
ASTROPHYSICS
SARJAPUR ROAD
BANGALORE 560 034
INDIA
TEL: 566585
TLX: 845763 IIAB IN
TLF:
EML:
COM: 34

BHATTACHARYA DIPANKAR
RAMAN RESEARCH INSTITUTE
BANGALORE 560080
INDIA
TEL: (812)340122
TLX: 845-2671
TLF:
EML:
COM:

BHATTACHARYYA J C PROF
INDIAN INSTITUTE OF
ASTROPHYSICS
BANGALORE 560 034
INDIA
TEL: 566583/566585
TLX: 845763 IIAB IN
TLF:
EML:
COM: 09V.12.50

BHATTACHARYYA TARA DR
JOGAMAYA DEVI COLLEGE
92 SYAMAPRADAD MUKERJEE
CALCUTTA 700 026
INDIA
TEL:
TLX:
TLF:
EML:
COM: 28

BHAVSAR SUKETU P
DEPT PHYSICS & ASTRONOMY
UNIVERSITY OF KENTUCKY
LEXINGTON KY 40506-0055
U.S.A.
TEL: 606-257-6722
TLX:
TLF:
EML:
COM: 47

BHONSLE RAJARAM V PROF
PHYSICAL RESEARCH LAB
NAVRANGPURA
AHMEDABAD 380 009
INDIA
TEL: 462129
TLX: 121397 PRL IN
TLF:
EML:
COM: 40

BIAN YU-LIN
BEIJING ASTRONOMICAL OBS
BEIJING 100080
CHINA. PEOPLE'S REP.
TEL:
TLX: 22040 BAOAS CN
TLF:
EML:
COM:

BIANCHI LUCIANA
OSSERVATORIO ASTRONOMICO
DI TORINO
I-10025 PINO TORINESE
ITALY
TEL: 011-842040
TLX: 213236 TO ASTR I
TLF:
EML:
COM: 34.44

BIANCHINI ANTONIO DR
OSSERVATORIO ASTROFISICO
I-36012 ASIAGO
ITALY
TEL: 0424-62665
TLX:
TLF:
EML:
COM: 27

BIBARSOV RAVIL'SH DR
ASTROPHYSICAL INSTITUTE
TADJIK ACAD OF SCIENCES
734670 DUSHANBE
U.S.S.R.
TEL:
TLX:
TLF:
EML:
COM: 22

BICA EDUARDO L D DR
INSTITUTO DE FISICA UFRGS
AV BENTO GONCALVES 9500
PORTO ALEGRE RS
90000
BRAZIL
TEL: 512 364677
TLX: 515730 CCUF BR
TLF:
EML:
COM: 28

BICAK JIRI DR
DEPT OF MATH PHYSICS
CHARLES UNIVERSITY
HOLESOVICKACH 2
180 00 PRAHA 8
CZECHOSLOVAKIA
TEL: 849951
TLX:
TLF:
EML:
COM:

BICKNELL GEOFFREY V DR
MOUNT STROMLO & SIDING
SPRING OBSERVATORIES
PRIVATE BAG
WODEN P.O. ACT 2606
AUSTRALIA
TEL: 61-62-88-1111
TLX: 62270 AA
TLF:
EML:
COM: 47.48

BIDELMAN WILLIAM P PROF
WARNER & SWASEY OBS
CASE WESTERN RESERVE UNIV
CLEVELAND OH 44106
U.S.A.
TEL: 216-368-6699
TLX:
TLF:
EML:
COM: 05.29.45

BIEGING JOHN HAROLD DR
RADIO ASTRONOMY LAB
CAMPBELL HALL
UNIVERSITY OF CALIFORNIA
BERKELEY CA 94720
U.S.A.
TEL: 415-642-6931
TLX:
TLF:
EML:
COM: 34.40

BIEMONT EMILE DR
INSTITUT D'ASTROPHYSIQUE
UNIVERSITE DE LIEGE
AVENUE DE COINTE 5
B-4200 COINTE-OUGREE
BELGIUM
TEL: 41-52-99-80
TLX:
TLF:
EML:
COM: 14

BIEN REINHOLD DR
ASTRONOMISCHES RECHEN-
INSTITUT
MOENCHHOFSTR 12-14
D-6900 HEIDELBERG
GERMANY. F.R.
TEL: 06221/49026
TLX: 461336 ARIHD D
TLF:
EML:
COM: 08.20

BIERMANN PETER L DR
MPI FUER RADIOASTRONOMIE
AUF DEM HUEGEL 69
D-5300 BONN 1
GERMANY. F.R.
TEL: 228-525279
TLX: 886440 MPIFR D
TLF:
EML:
COM: 28.40.48

BIGNAMI GIOVANNI F
INSTITUTO FISICA COSMICA
CONSIGLIO NAZION. RICERCH
15/A VIA BASSINI
I 20133 MILANO
ITALY
TEL: 39-2-2367587
TLX: 313839 MUACNR I
TLF:
EML:
COM: 47.48

BIGNELL R CARL DR
NRAO-VLA
PO BOX 0
SOCORRO NM 87801
U.S.A.
TEL: 505-772-4242
TLX: 910-988-1710
TLF:
EML:
COM: 34.40

BIJAOUI ALBERT DR
OBSERVATOIRE DE NICE
BP 139
F-06003 NICE CEDEX
FRANCE
TEL: 93-89-04-20
TLX: 460004
TLF:
EML:
COM: 28.37

BILLAUD GERARD J
CERGA
AVENUE COPERNIC
F-06130 GRASSE
FRANCE
TEL: 93-36-58-49
TLX: 470865
TLF:
EML:
COM: 08.19

BILLINGHAM JOHN
LIFE SCIENCE DIVISION
NASA AMES RESEARCH CTR
MOFFETT FIELD CA 94035
U.S.A.
TEL: 415-694-5181
TLX: 348408 NASA AMES MOF
TLF:
EML:
COM: 51

BILLINGS DONALD E PROF
UNIVERSITY OF COLORADO
DEPT OF ASTROGEOPHYSICS
BOULDER CO 80309
U.S.A.
TEL:
TLX:
TLF:
EML:
COM: 12

BINETTE LUC
CITA
60 ST. GEORGE STREET
UNIV. OF TORONTO
TORONTO M5S 1A1
CANADA
TEL: 416 978 8497
TLX: 06-218915 UT ENG TOR
TLF:
EML:
COM: 28.34

BINGGELI BRUNO
ASTRONOMISCHES INSTITUT
UNIVERSITAET BASEL
VENUSSTRASSE 7
CH-4102 BINNINGEN
SWITZERLAND
TEL:
TLX:
TLF:
EML:
COM: 28

BINGHAM RICHARD G DR
ROYAL GREENWICH OBS
HERSTMONCEUX CASTLE
HAILSHAM BN27 1RP
U.K.
TEL: 0323-833171
TLX: 87451
TLF:
EML:
COM: 09

BINNEY JAMES J DR
DEPT THEORETICAL PHYSICS
1 KEBLE ROAD
OXFORD OX1 3NP
U.K.
TEL: 865-53281
TLX: 83295 NUCLOX G
TLF:
EML:
COM: 28.33C

BIRAUD FRANCOIS DR
OBSERVATOIRE DE PARIS
SECTION DE MEUDON
F-92195 MEUDON PL CEDEX
FRANCE
TEL: 1-45-07-76-02
TLX: 270912
TLF:
EML:
COM: 06.40.51

BIRCH PETER MR
PERTH OBSERVATORY
BICKLEY, W. AUSTR. 6076
AUSTRALIA
TEL: 09-2938-255
TLX:
TLF:
EML:
COM: 15

BIRKINSHAW MARK
HARVARD UNIVERSITY
DEPT OF ASTRONOMY
60 GARDEN STREET
CAMBRIDGE MA 02138
U.S.A.
TEL: 617-495-9092
TLX: 921428
TLF:
EML:
COM: 28.40.47

BIRKLE KURT PH D
MPI FUER ASTRONOMIE
KOENIGSTUHL
D-6900 HEIDELBERG 1
GERMANY, F.R.
TEL:
TLX:
TLF:
EML:
COM: 34

BISHOP ROY L DR
DEPT OF PHYSICS
ACADIA UNIVERSITY
WOLFVILLE NS BOP 1X0
CANADA
TEL: 902-542-2201
TLX:
TLF:
EML:
COM: 41

BISIACCHI GIANFRANCO DR
INSTITUTO DE ASTRONOMIA
UNAM
APDO POSTAL 70-264
04510 MEXICO DF
MEXICO
TEL: 548-4537
TLX: 1760155 CICME
TLF:
EML:
COM:

BISNOVATYI-KOGAN G S DR
SPACE RESEARCH INSTITUTE
USSR ACADEMY OF SCIENCES
PROFSOYUZNAYA 84/32
117810 MOSCOW
U.S.S.R.
TEL: 333-31-22
TLX: 411498 STARSU
TLF:
EML:
COM: 35

BISWAS SUKUMAR DR
COSMIC RAY GROUP
TATA INST FUND RESEARCH
HOMI BHABHA RD
BOMBAY 400 005
INDIA
TEL: 91-22-219111
TLX: 113009 TIFR IN
TLF:
EML:
COM: 48

BJORNSSON CLAES-INGVAR
STOCKHOLM OBSERVATORY
S-133 00 SALTSJOBADEN
SWEDEN
TEL: 08-7170195
TLX: 12972 SOBSERV S
TLF:
EML:
COM:

BLAAUW ADRIAAN PROF DR
KAPTEYN LABORATORY
P O BOX 800
NL-9700 AV GRONINGEN
NETHERLANDS
TEL: 050-634084
TLX: 53572 STARS NL
TLF:
EML:
COM: 24.33 37

BLACK JOHN HARRY DR
STEWARD OBSERVATORY
UNIVERSITY OF ARIZONA
TUCSON AZ 85721
U.S.A.
TEL: 602-621-6531
TLX: 467175
TLF:
EML:
COM: 14.34

BLACKMAN CLINTON PAUL DR
CARLSTON LODGE
CAMPSIE RD. TORRANCE
GLASGOW G64 4HD
U.K.
TEL:
TLX:
TLF:
EML:
COM:

BLACKWELL ALAN TREVOR
METEORITE PROJECT
BOX 464
SUB POST OFFICE 6
SASKATOON S7N OWO
CANADA
TEL:
TLX:
TLF:
EML:
COM: 22

BLACKWELL DONALD E PROF
DEPT OF ASTROPHYSICS
SOUTH PARKS ROAD
OXFORD OX1 3RQ
U.K.
TEL: 0865-511336
TLX:
TLF:
EML:
COM: 12.21.49

BLADES JOHN CHRIS DR
SPACE TELESCOPE SCI INST
HOMEWOOD CAMPUS
3700 SAN MARTIN DR
BALTIMORE MD 21218
U.S.A.
TEL: 301-338-4805
TLX: 6849101 STSCI UW
TLF:
EML:
COM: 34

BLAHA MILAN DR
NAVAL RESEARCH LABORATORY
CODE 4720
WASHINGTON DC 20375
U.S.A.
TEL:
TLX:
TLF:
EML:
COM: 14

BLAIR DAVID GERALD
PHYSICS DEPARTMENT
UNIVERSITY OF W.AUSTRALIA
NEDLANDS WA 6009
AUSTRALIA
TEL:
TLX: 92992 AA
TLF:
EML:
COM: 40

BLAIR GUY NORMAN DR
9460 SW CHERAW COURT
TUALATIN. OR 97062
U.S.A.
TEL:
TLX:
TLF:
EML:
COM: 34

BLAIR WILLIAM P DR
DPT OF PHYS. & ASTRONOMY
THE JOHNS HOPKINS UNIV.
CHARLES AND 34TH STREETS
BALTIMORE MD 21218
U.S.A.
TEL: 301 338 8447
TLX: 9102400225
TLF:
EML: SPAN:SCIVAX::WBLAIR
COM: 42

BLAMONT JACQUES E PROF
C N E S
2 PLACE MAURICE QUENTIN
F-75039 PARIS CEDEX 01
FRANCE
TEL: 1-45-08-76-12
TLX: 214674
TLF:
EML:
COM: 12.15.16.21.44

BLANCO CARLO DR
UNIVERSITA DI CATANIA
ISTITUTO DI ASTRONOMIA
VIALE A. DORIA 6
I-95125 CATANIA
ITALY
TEL: 095-330533
TLX: 970359 ASTRCT I
TLF:
EML:
COM: 36.50C

BLANCO VICTOR M DR
CERRO TOLOLO
INTERAMERICAN OBSERVATORY
CASILLA 603
LA SERENA
CHILE
TEL: 213352
TLX: 34-645227 AURA CT
TLF:
EML:
COM: 25.33.45.50C

BLANDFORD ROGER DAVID DR
THEORETICAL ASTROPHYSICS
CALTECH 130-33
PASADENA CA 91125
U.S.A.
TEL: 213-356-4200
TLX: 675429
TLF:
EML:
COM: 40.48

BLASIUS KARL RICHARD DR
3839 MYRTLE
LONG BEACH CA 90807
U.S.A.
TEL:
TLX:
TLF:
EML:
COM:

BLAZIT ALAIN DR
OBSERV. DE LA COTE D'AZUR
CAUSSOLS
06460-ST VALLIER DE THIEY
FRANCE
TEL: 93 42 62 70
TLX: 461 402
TLF:
EML:
COM:

BLECHA ANDRE BORIS G DR
16 RUE ET. DUMONT
CH-1204 GENEVE
SWITZERLAND
TEL:
TLX:
TLF:
EML:
COM: 25

BLEEKER JOHAN A M DR IR
SPACE RESEARCH LABORATORY
BENELUXLAAN 21
NL-3527 HS UTRECHT
NETHERLANDS
TEL: 030-937145
TLX: 47224 ASTRO NL
TLF:
EML:
COM: 44.48

BLESS ROBERT C PROF
ASTRONOMY DEPT
UNIVERSITY OF WISCONSIN
475 N. CHARTER ST
MADISON WI 53706
U.S.A.
TEL: 608-262-1715
TLX:
TLF:
EML:
COM: 34.36.44.51

BLINOV N S DR
STERNBERG STATE
ASTRONOMICAL INSTITUTE
UNIVERSITETSTKIJ PROSP.13
119899 MOSCOW
U.S.S.R.
TEL: 139-10-49
TLX:
TLF:
EML:
COM: 19.31C

BLITZ LEO
ASTRONOMY PROGRAM
UNIVERSITY OF MARYLAND
COLLEGE PARK MD 20742
U.S.A.
TEL: 301-454-3001
TLX: 710-826-0352
TLF:
EML:
COM: 28,33V,34

BLITZSTEIN WILLIAM DR
DEPT ASTRON & ASTROPHYS
UNIV OF PENNSYLVANIA
DAVID RITTENHOUSE LAB E1
PHILADELPHIA PA 19104
U.S.A.
TEL: 215-898-7899
TLX: 834621
TLF:
EML:
COM: 09.42

BLOEMEN JOHANNES B G M DR
STERREWACHT
P.O. BOX 9513
2300 RA LEIDEN
NETHERLANDS
TEL: (31)71 275818
TLX: 39058 ASTRO NL
TLF:
EML:
COM: 33

BLOEMHOF ERIC E DR
HARVARD-SMITHSONIAN
CENTER FOR ASTROPHYSICS
60 GARDEN STREET
CAMBRIDGE MA 02138
U.S.A.
TEL: 617 495 7314
TLX:
TLF:
EML: BLOEMHOF@CFA
COM: 40

BLOW GRAHAM L
CARTER OBSERVATORY
PO BOX 2909
WELLINGTON
NEW ZEALAND
TEL: 4 728 167
TLX: NZ 30172 NATOBS
TLF:
EML:
COM:

BLUDMAN SIDNEY A PROF
DEPT OF PHYSICS
UNIV OF PENNSYLVANIA
PHILADELPHIA PA 19104
U.S.A.
TEL: 215-898-8151
TLX: 831908
TLF:
EML:
COM: 35.47.48

BLUM PETER PROF
INSTITUT F. ASTROPHYSIK
UNIVERSITAET BONN
AUF DEM HUEGEL 71
D-5300 BONN
GERMANY. F.R.
TEL: 0228-73-36-65
TLX: 0886440 MPIFR
TLF:
EML:
COM: 49

BLUMENTHAL GEORGE R DR
LICK OBSERVATORY
UNIVERSITY OF CALIFORNIA
SANTA CRUZ CA 95064
U.S.A.
TEL: 408 429 2005
TLX:
TLF:
EML: BITNET:george@portal
COM: 28

BO SHU-REN
INST F.HISTORY OF NAT SCI
1 GONG YUAN WEST ROAD
BEIJING
CHINA. PEOPLE'S REP.
TEL:
TLX:
TLF:
EML:
COM: 41

BOBROV M S DR
ASTRONOMICAL COUNCIL
USSR ACADEMY OF SCIENCES
PYATNITSKAYA UL 48
109017 MOSCOW
U.S.S.R.
TEL: 231-39-80
TLX: 412623 SCSTP SU
TLF:
EML:
COM: 16

BOCCHIA ROMEO DR
OBSERVATOIRE DE BORDEAUX
AVENUE P SEMIROT
F-33270 FLOIRAC
FRANCE
TEL: 56-86-43-30
TLX:
TLF:
EML:
COM: 10.12.35

BOCHKAREV NIKOLAY G DR
STERNBERG STATE ASTR INST
1179899 MOSCOW
U.S.S.R.
TEL:
TLX:
TLF:
EML:
COM: 34

BOCHONKO D RICHARD DR
DEPT MATH & ASTRONOMY
UNIVERSITY OF MANITOBA
WINNIPEG MB R3T 2N8
CANADA
TEL: 204-474-9501
TLX:
TLF:
EML:
COM: 27.46

BOCHSLER PETER
PHYSIKALISCHES INSTITUT
UNIVERSITAET BERN
SIDLERSTRASSE 5
CH-3012 BERN
SWITZERLAND
TEL: 0041-31-65-4419
TLX: 32320 PHYBE CH
TLF:
EML:
COM: 49C

BOCKELEE-MORVAN DOMINIQUE
OBSERVATOIRE DE PARIS
SECTION MEUDON
5 PLACE JULES JANSSEN
92195 MEUDON CEDEX
FRANCE
TEL: 1-45 07 76 05
TLX: 270912 OBSASTR
TLF:
EML:
COM: 15.40

BODDAPATI G ANANDARAO DR
PHYSICAL RESEARCH LAB.
ROOM 760
AHMEDABAD 380 009
INDIA
TEL: 462129
TLX: 121397
TLF:
EML:
COM:

BODE MICHAEL F
SCHOOL OF PHYSICS & ASTR
LANCASHIRE POLYTECHNIC
PRESTON PR1 2TQ
U.K.
TEL:
TLX:
TLF:
EML:
COM: 34

BODENHEIMER PETER PROF
LICK OBSERVATORY
UNIVERSITY OF CALIFORNIA
SANTA CRUZ CA 95064
U.S.A.
TEL: 408-429-2064
TLX:
TLF:
EML:
COM: 34.35

BODO GIANLUIGI DR
OSSERVATORIO ASTRONOMICO
DI TORINO
PINO TORINESE 10025
ITALY
TEL: 11 841067
TLX: 213236 TO ASTRI
TLF:
EML:
COM: 36.39

BOEHM KARL-HEINZ PROF
ASTRONOMY DEPT
UNIVERSITY OF WASHINGTON
SEATTLE WA 98195
U.S.A.
TEL: 206-543-2888
TLX: 4740096
TLF:
EML:
COM: 12.35.36

BOEHM-VITENSE ERIKA PROF
ASTRONOMY DEPT
UNIVERSITY OF WASHINGTON
FM 20
SEATTLE WA 98195
U.S.A.
TEL: 206-543-4858
TLX:
TLF:
EML:
COM: 12.36

BOEHME ANNELIES DR
HEINRICH HERTZ INSTITUTE
SOLAR TERRESTR PHYSICS
TELEGRAFENBERG
DDR-1500 POTSDAM
GERMANY. D.R.
TEL:
TLX:
TLF:
EML:
COM:

BOEHME SIEGFRIED DR
ASTRONOMISCHES
RECHEN INSTITUT
MOENCHHOFSTR 12-14
D-6900 HEIDELBERG
GERMANY. F.R.
TEL: 06221-49026
TLX:
TLF:
EML:
COM: 36

BOEHNHARDT HERMANN DR
DR. REMEIS OBSERVATORY
ERLANGEN NUREMBERG UNIV.
STERNWARTSTRASSE 7
D 8600 BAMBERG
GERMANY. F.R.
TEL: 0951 57708
TLX: 629830 UNIER D
TLF:
EML: EARN/BITNET:HPA122@DEERRZEEO
COM: 15

BOERNER GERHARD DR
MPI F PHYSIK & ASTROPHYS
FOEHRINGER RING 6
D-8000 MUENCHEN
GERMANY. F.R.
TEL:
TLX:
TLF:
EML:
COM:

BOERNGEN FREIMUT DR PH
ZNTRLINST. F. ASTROPHYSIK
KARL-SCHWARZSCHILD-OBS
DDR-6901 TAUTENBURG
GERMANY. D.R.
TEL: JENA 23530
TLX:
TLF:
EML:
COM: 20.28

BOESGAARD ANN M PROF
INSTITUTE FOR ASTRONOMY
2680 WOODLAWN DR
HONOLULU HI 96822
U.S.A.
TEL: 808-948-8756
TLX: 723-8459
TLF:
EML:
COM: 29C.36

BOESHAAR GREGORY ORTH DR
SPACE TELESCOPE INSTITUTE
HOMEWOOD CAMPUS
3700 SAN MARTIN DRIVE
BALTIMORE MD 21218
U.S.A.
TEL:
TLX:
TLF:
EML:
COM: 28.34

BOGGESS ALBERT DR
NASA/GSFC
CODE 680
GREENBELT MD 20771
U.S.A.
TEL: 301-286-5975
TLX:
TLF:
EML:
COM: 29.34.44

BOGGESS NANCY W DR
NASA HEADQUARTERS
CODE EZ
WASHINGTON DC 20546
U.S.A.
TEL: 202-453-1469
TLX:
TLF:
EML:
COM: 44

BOHANNAN BRUCE EDWARD
SOMMERS-BAUSCH OBS
UNIVERSITY OF COLORADO
BOX 391
BOULDER CO 80309
U.S.A.
TEL: 303-492-8782
TLX:
TLF:
EML:
COM:

BOHLIN J DAVID DR
NASA HEADQUARTERS
CODE EZ
WASHINGTON DC 20546
U.S.A.
TEL: 202-453-1466
TLX: 89530
TLF:
EML:
COM:

BOHLIN RALPH C DR
SPACE TELESCOPE SCI INST
HOMEWOOD CAMPUS
3700 SAN MARTIN DRIVE
BALTIMORE MD 21218
U.S.A.
TEL: 301-338-4804
TLX: 6849101 STSCI UW
TLF:
EML:
COM: 34.44

BOHN HORST-ULRICH
FRAUNHOFER GES. FHAK
WALDPARKSTR. 41
D-8012 OTTOBRUNN
GERMANY. F.R.
TEL: 089-6013086
TLX:
TLF:
EML:
COM: 10.12

BOHRMANN ALFRED PROF
SCHAEFSTR 23
D-2050 HAMBURG 80
GERMANY. F.R.
TEL: 7399800
TLX:
TLF:
EML:
COM:

BOIGEY FRANCOISE
INTA. LAB MECAN. CELESTE
UNIVERSITE P & M CURIE
4 PLACE JUSSIEU. TOUR 66
F-75230 PARIS CEDEX 05
FRANCE
TEL:
TLX:
TLF:
EML:
COM: 07

BOISCHOT ANDRE DR
OBSERVATOIRE DE PARIS
SECTION DE MEUDON
F-92195 MEUDON PL CEDEX
FRANCE
TEL: 1-45-07-77-74
TLX: 200590 CNET
TLF:
EML:
COM: 40

BOKSENBERG ALEC PROF
ROYAL GREENWICH OBS
HERSTMONCEUX CASTLE
HAILSHAM BN27 1RP
U.K.
TEL: 323-833171
TLX: 87451 RGOBSY G
TLF:
EML:
COM: 28.44.47

BOLAND WILFRIED
RADIO OBSERVATORY
POSTBUS 2
7990 AA DWINGELOO
NETHERLANDS
TEL: 31-5219 7244
TLX: 42043
TLF:
EML:
COM: 34

BOLCAL CETIN DR
PHYSICS DEPARTMENT
ISTANBUL UNIVERSITY
34459 VEZNECILER
ISTANBUL
TURKEY
TEL: [illegible]3320740
TLX: 26401 BOUNTR
TLF:
EML:
COM:

BOLDT ELIHU DR
NASA/GSFC
CODE 661
GREENBELT MD 20771
U.S.A.
TEL: 301-286-5853
TLX: 89675 NASCOM-GBLT
TLF:
EML:
COM:

BOLEY FORREST I
WILDER LABORATORY
DARTMOUTH COLLEGE
HANOVER NH 03755
U.S.A.
TEL: 603-646-2966
TLX:
TLF:
EML:
COM:

BOLOIX RAFAEL DR
REAL INSTITUTO Y
OBSERVATORIO DE LA ARMADA
11110 SAN FERNANDO CADIZ
SPAIN
TEL: 56 883548
TLX: 76108 ION E
TLF:
EML:
COM: 31

BOLTON C THOMAS PROF
DAVID DUNLAP OBSERVATORY
P O BOX 360
RICHMOND HILL ONT L4C 4Y6
CANADA
TEL: 416-884-9652
TLX: 698-6766 TOR
TLF:
EML:
COM: 27.42

BOLTON JOHN G
39 PANORAMA CRESCENT
BUDERIM QLD 4556
AUSTRALIA
TEL: 071-453374
TLX:
TLF:
EML:
COM: 40

BONNIER VERONIQUE DR
UA 812 DAMAP
OBSERV. DE PARIS MEUDON
PLACE JULES JANSSEN
92195 MEUDON CEDEX
FRANCE
TEL: 1-45 07 74 54
TLX: 201571
TLF:
EML: 28726::BONNIER
COM: 10.12.14

BONANONI JACQUES DR
OBSERVATOIRE CANTONAL
CH-2000 NEUCHATEL
SWITZERLAND
TEL: 038-24 18 61
TLX:
TLF:
EML:
COM: 19.31

BONAZZOLA SILVANO DR
OBSERVATOIRE DE PARIS
SECTION DE MEUDON
F-92195 MEUDON PL CEDEX
FRANCE
TEL: 1-45-07-74-29
TLX: 201571 LAM
TLF:
EML:
COM: 42.48

BOND HOWARD E DR
SPACE TELESCOPE SCI INST
3700 SAN MARTIN DRIVE
BALTIMORE MD 21218
U.S.A.
TEL: 301-338-4718
TLX: 6849101
TLF:
EML:
COM: 27.29

BOND JOHN RICHARD
DEPT OF PHYSICS
STANFORD UNIVERSITY
STANFORD CA 94305
U.S.A.
TEL: 415-497-1775
TLX:
TLF:
EML:
COM: 47

BONDARENKO L N DR
STERNBERG STATE ASTR INST
PROSPECT 13
UNIVERSITETSKY PROSP. 13
119899 MOSCOW V-234
U.S.S.R.
TEL: 139-3721
TLX:
TLF:
EML:
COM: 16

BONDI HERMANN PROF SIR
CHURCHILL COLLEGE
CAMBRIDGE CB3 ODS
U.K.
TEL:
TLX:
TLF:
EML:
COM: 35.47

BONET JOSE A
INSTITUTO DE ASTROFISICA
DE CANARIAS
38071 TENERIFE
SPAIN
TEL:
TLX:
TLF:
EML:
COM:

BONEV BONU K MR
PEOPLE'S ASTR.OBSERVATORY
ST. AVGUSTA TRATANA 29/8
6000 STARA ZAGORA
BULGARIA
TEL:
TLX:
TLF:
EML:
COM:

BONIFAZI ANGELO DR
OSSERVATORIO ASTRONOMICO
I-40100 BOLOGNA
ITALY
TEL:
TLX:
TLF:
EML:
COM:

BONNEAU DANIEL
CERGA
OBSERVATOIRE DU CALERN
F-06460 ST VALLIER DE T.
FRANCE
TEL: 93-42-62-70
TLX: 461402
TLF:
EML:
COM: 09.26

BONNET ROGER M DR
ESA
8-10 RUE MARIO NIKIS
F-75738 PARIS CEDEX 15
FRANCE
TEL: 1-42-73-71-07
TLX: ESA 202746
TLF:
EML:
COM: 12.44.49

BONNOR W B PROF
1 SOUTH BANK TERRACE
SURBITON. SURREY KT6 6DG
U.K.
TEL: 1-399-1103
TLX:
TLF:
EML:
COM: 47

BONOLI FABRIZIO
OSSERVATORIO ASTRONOMICO
UNIVERSITARIO
C P 596
I-40100 BOLOGNA
ITALY
TEL: 051-222956
TLX: 211664 INFNBO I
TLF:
EML:
COM:

BONOMETTO SILVIO A DR
ISTITUTO DI FISICA
G GALILEI 8
VIA MARZOLO
I-35100 PADOVA
ITALY
TEL:
TLX:
TLF:
EML:
COM: 48

BONSACK WALTER K PROF
INSTITUTE FOR ASTRONOMY
2690 WOODLAWN DRIVE
HONOLULU HI 96822
U.S.A.
TEL:
TLX:
TLF:
EML:
COM: 29

BOOK DAVID L
USE NAVAL RESEARCH LAB
CODE 4040
WASHINGTON DC 20375
U.S.A.
TEL:
TLX:
TLF:
EML:
COM: 12

BOOKMYER BEVERLY B DR
DEPT OF PHYS & ASTRONOMY
CLEMSON UNIVERSITY
CLEMSON SC 29631
U.S.A.
TEL: 803-656-3417
TLX:
TLF:
EML:
COM: 25.42

BOOTH ANDREW J
CHATTERTON ASTRONOMY DEPT
SCHOOL OF PHYSICS
UNIVERSITY OF SYDNEY
SYDNEY
AUSTRALIA
TEL:
TLX:
TLF:
EML:
COM:

BOOTH ROY S PROF
ONSALA SPACE OBSERVATORY
S-439 00 ONSALA
SWEDEN
TEL: 46-300-62590
TLX: 2400 ONSPACE S
TLF:
EML:
COM: 40C

BOPP BERNARD W DR
DEPT PHYSICS & ASTRONOMY
UNIVERSITY OF TOLEDO
TOLEDO OH 43606
U.S.A.
TEL: 419-537-2274
TLX:
TLF:
EML:
COM: 27.42

BORCHKHADZE TENGIZ M DR
ABASTUMANI ASTROPHYSICAL
OBSERVATORY
383762 ABASTUMANI.GEORGIA
U.S.S.R.
TEL:
TLX:
TLF:
EML:
COM: 28

BORD DONALD JOHN
DEPT OF NATURAL SCIENCES
UNIVERSITY OF MICHIGAN
DEARBORN
DEARBORN MI 48128
U.S.A.
TEL: 313-593-5483
TLX:
TLF:
EML:
COM:

BORDERIES NICOLE
JPL 301-150
4800 OAK GROVE DRIVE
PASADENA CA 91109
U.S.A.
TEL: 818-354-8211
TLX: 675429
TLF:
EML:
COM: 07

BORGMAN JAN DR PROF
KAPTEYN OBSERVATORY
MENSINGHEWEG 20
NL-9301 KA RODEN DR
NETHERLANDS
TEL:
TLX:
TLF:
EML:
COM: 25.34

BORGNINO JULIEN DR
DEPT D'ASTROPHYSIQUE
UNIVERSITE DE NICE
PARC VALROSE
F-06034 NICE CEDEX
FRANCE
TEL:
TLX:
TLF:
EML:
COM: 09

BORIAKOFF VALENTIN
NAIC
420 SPACE SCIENCES BLDG
CORNELL UNIVERSITY
ITHACA NY 14853
U.S.A.
TEL: 607-256-3734
TLX: 932454
TLF:
EML:
COM: 40

BORNMANN PATRICIA L DR
NOAA ERL (CIRES)
SPACE ENVIRONMENT LAB.
325 BROADWAY R/E/SE
BOULDER CO 80303
U.S.A.
TEL: 303 497 3532
TLX: 45897 SOLTERWARN BDR
TLF:
EML: SPAN:SELVAX:pbornmann
COM: 12

BORRA ERMANNO F DR
DEPT DE PHYSIQUE
UNIVERSITE LAVAL
STE FOY PQ G1K 7P4
CANADA
TEL: 418-656-7405
TLX: 051-31621
TLF:
EML:
COM: 25

BOSMA ALBERT DR
OBSERVATOIRE DE MARSEILLE
2 PLACE LE VERRIER
F-13248 MARSEILLE CEDEX 4
FRANCE
TEL: 91-95-90-88
TLX: 420241
TLF:
EML:
COM: 28

BOSMA PIETER B DR
DEPT PHYSICS & ASTRONOMY
FREE UNIVERSITY
DE BOELELAAN 1081
NL-1081 HV AMSTERDAM
NETHERLANDS
TEL: 020-5485318
TLX:
TLF:
EML:
COM: 16

BOSMAN-CRESPIN DENISE
BLVD D AVROY 68
BTE 093
B-4000 LIEGE
BELGIUM
TEL: 0032-41-237486
TLX:
TLF:
EML:
COM:

BOSS ALAN P DR
CARNEGIE INST.OF WASHIN.
DEPT OF TERR. MAGNETISM
5241 BROAD BR. RD. NW
WASHINGTON DC 20015
U.S.A.
TEL: (202)686 4402
TLX: 440427
TLF:
EML:
COM: 16.35

BOTEZ ELVIRA DR
INSTITUT D'ENSEIGNEMENT
SUPERIEUR
13 RUE EM. BODNARAS
5800 SUCEAVA
RUMANIA
TEL: 98716147
TLX:
TLF:
EML:
COM: 46

BOTTINELLI LUCETTE DR
OBSERVATOIRE DE PARIS
SECTION DE MEUDON
RADIOASTRONOMIE
F-92195 MEUDON PL CEDEX
FRANCE
TEL: 1-45-07-76-04
TLX: 270912 OBSASTR F
TLF:
EML:
COM: 28.40.46

BOUCHER CLAUDE DR
INSTITUT GEOGRAPHIQUE NTL
2 AV PASTEUR
94160 SAINT MANDE
FRANCE
TEL: 43 74 12 15
TLX:
TLF:
EML:
COM: 19

BOUGERET J L DR
OBSERVATOIRE DE PARIS
SECTION DE MEUDON
DESPA
F-92195 MEUDON PL CEDEX
FRANCE
TEL: 1-45-07-77-04
TLX: 204464
TLF:
EML:
COM: 10.12.44

BOULANGER FRANCOIS
RADIOASTRONOMIE ENS
24 RUE LHOMOND
75231 PARIS CEDEX 05
FRANCE
TEL:
TLX:
TLF:
EML:
COM: 34

BOULESTEIX JACQUES
OBSERVATOIRE DE MARSEILLE
2 PLACE LE VERRIER
F-13248 MARSEILLE
FRANCE
TEL: 91-95-90-88
TLX: 420241F
TLF:
EML:
COM:

BOULON JACQUES J DR
OBSERVATOIRE PARIS
61 AVE DE L'OBSERVATOIRE
F-75014 PARIS
FRANCE
TEL: 1-40-51-22-53
TLX: 270776 OBSPARIS
TLF:
EML:
COM: 27.30.33

BOUSKA JIRI DR
DEPT OF ASTRONOMY
CHARLES UNIVERSITY
SVEDSKA 8
150 00 PRAHA
CZECHOSLOVAKIA
TEL: 42-2-540395
TLX: 121673 MFF
TLF:
EML:
COM: 05.15

BOUVIER JEROME
INSTITUT D'ASTROPHYSIQUE
98 BIS BD ARAGO
75014 PARIS
FRANCE
TEL: 43 20 14 25
TLX: 205671 IAU F
TLF:
EML: BOUVIER@FRIAP51
COM: 29.34

BOUVIER PIERRE PROF
OBSERVATOIRE DE GENEVE
CH-1290 SAUVERNY
SWITZERLAND
TEL:
TLX:
TLF:
EML:
COM: 29.37

BOWELL EDWARD L G DR
LOWELL OBSERVATORY
1400 W. MARS HILL RD
FLAGSTAFF AZ 86001
U.S.A.
TEL: 602-774-3358
TLX:
TLF:
EML:
COM: 15.20

BOWEN EDWARD G DR
1/39 CLARKE STREET
NARRABEEN NSW 2101
AUSTRALIA
TEL: 98 8565
TLX:
TLF:
EML:
COM:

BOWEN GEORGE H DR
PHYSICS DEPARTMENT
IOWA STATE UNIVERSITY
AMES IA 50011
U.S.A.
TEL: 515 294 7659
TLX:
TLF:
EML: BITNET:S1.GHB@ISUMVS
COM: 27.36

BOWERS PHILLIP F
NAVAL RESEARCH LABORATORY
CODE 4134
WASHINGTON DC 20375
U.S.A.
TEL: 202-767-2495
TLX:
TLF:
EML:
COM: 40

BOWYER C STUART PROF
UNIVERSITY OF CALIFORNIA
ASTRONOMY DEPT
BERKELEY CA 94720
U.S.A.
TEL: 415-642-1648
TLX: 910-366 7945
TLF:
EML:
COM: 21C.44.51

BOYARCHUK A A DR
ASTRONMICAL COUNCIL
USSR ACADEMY OF SCIENCES
PYATNITSKAYA UL. 48
109017 MOSCOW
U.S.S.R.
TEL:
TLX: 411576 ASCON SU
TLF:
EML:
COM: 27.29.38C.44

BOYARCHUK MARGARITA E DR
ASTRONOMICAL COUNCIL
USSR ACADEMY OF SCIENCES
PYATNITSKAYA UL. 48
109017 MOSCOW
U.S.S.R.
TEL:
TLX:
TLF:
EML:
COM: 27

BOYCE PETER B DR
AMERICAN ASTRON SOCIETY
2000 FLORIDA AVE N.W.
SUITE 300
WASHINGTON DC 20009
U.S.A.
TEL: 202-328-2010
TLX: 257588 AASW UR
TLF:
EML:
COM: 09.16.51

BOYD ROBERT L F PROF SIR
41 CHURCH STREET
LITTLEHAMPTON BN17 5PU
U.K.
TEL:
TLX:
TLF:
EML:
COM: 44.48

BOYDAG-YILDIZDOGDU F S
ACADEMY OF ISTANBUL
ENG & ARCHITECTURE
DEPT OF PHYSICS
ISTANBUL
TURKEY
TEL:
TLX:
TLF:
EML:
COM:

BOYER CHARLES
26 RUE ANDRE DELICIEUX
F-31400 TOULOUSE
FRANCE
TEL:
TLX:
TLF:
EML:
COM: 16

BOYER RENE
OBSERVATOIRE DE PARIS
SECTION DE MEUDON
DASOP
F-92195 MEUDON PL CEDEX
FRANCE
TEL: 1-45-07-77-41
TLX: 201571 LAM
TLF:
EML:
COM: 10

BOYLE BRIAN DR
ANGLO AUSTRALIAN OBS.
PO BOX 296
EPPING NSW 2121
AUSTRALIA
TEL: 02 868 1666
TLX: 23999 AAOSYD AA
TLF:
EML:
COM: 47

BOYLE RICHARD P. DR
VATICAN OBSERVATORY
I 00120 CITTA DEL VATICANO
VATICAN CITY
TEL: 39 6 698 5266
TLX: 504 2020 VATOBS VA
TLF:
EML: SPECOLA_VAT@ASTROM.SPAN
COM:

BOYNTON PAUL EDWARD DR
ASTRONOMY DEPT
UNIVERSITY OF WASHINGTON
SEATTLE WA 98195
U.S.A.
TEL:
TLX:
TLF:
EML:
COM:

BOZIS GEORGE PROF
DEPT THEORET MECHANICS
UNIVERSITY THESSALONIKI
GR-54006 THESSALONIKI
GREECE
TEL: 031-992845
TLX:
TLF:
EML:
COM: 07

BOZKURT SUKRU DR
EGE UNIVERSITY
OBSERVATORY
P K 21
BORNOVA-IZMIR
TURKEY
TEL: 180 306 (HOME)
TLX:
TLF:
EML:
COM:

BRACCESI ALESSANDRO PROF
DIPTO DI ASTRONOMIA
VIA ZAMBONI 33
I-40126 BOLOGNA
ITALY
TEL: 051-222956
TLX: 211664 INFNBOI
TLF:
EML:
COM: 28

BRACEWELL RONALD N PROF
STANFORD UNIVERSITY
DURAND 329 A
STANFORD CA 94305
U.S.A.
TEL: 415-497-3545
TLX:
TLF:
EML:
COM: 40.51

BRADSTREET DAVID H DR
DPT OF PHYSICAL SCIENCE
EASTERN COLLEGE
ST DAVIDS PA 19087
U.S.A.
TEL: 215 341 5945
TLX:
TLF:
EML:
COM: 42

BRAES L L E DR
STERREWACHT
POSTBUS 9513
NL-2300 RA LEIDEN
NETHERLANDS
TEL: 071-272727
TLX:
TLF:
EML:
COM: 46

BRAHDE ROLF
INST THEORET ASTROPHYSICS
UNIVERSITY OF OSLO
N-0315 BLINDERN. OSLO 3
NORWAY
TEL: 2-456508
TLX:
TLF:
EML:
COM:

BRAHIC ANDRE DR
OBSERVATOIRE DE PARIS
SECTION DE MEUDON
F-92195 MEUDON PL CEDEX
FRANCE
TEL: 1-45-07-74-02
TLX: 201571 LAM
TLF:
EML:
COM: 16P

BRANCH DAVID R DR
DEPT PHYSICS & ASTRONOMY
UNIVERSITY OF OKLAHOMA
NORMAN OK 73019
U.S.A.
TEL: 405-325-3961
TLX: 9108306521
TLF:
EML:
COM:

BRAND PETER W J L DR
DEPT OF ASTRONOMY
UNIVERSITY OF EDINBURGH
ROYAL OBSERVATORY
EDINBURGH EH9 3HJ
U.K.
TEL: 031-667-3321
TLX: 72383 ROE EDIN G
TLF:
EML:
COM: 34

BRANDI ELISANDE ESTELA DR
OBSERVATORIO ASTRONOMICO
UNIVERSIDAD NACIONAL
PASEO DEL BOSQUE
1900 LA PLATA
ARGENTINA
TEL: 21-1761
TLX:
TLF:
EML:
COM: 29.42

BRANDIE GEORGE W DR
ENVIRONMENTAL ENGINEERING
DUPUIS HALL
QUEEN'S UNIVERSITY
KINGSTON ONT K7L 3N6
CANADA
TEL:
TLX:
TLF:
EML:
COM:

BRANDT JOHN C DR
LAB. ATMOS. &
SPACE PHYSICS. BOX 392
UNIVERSITY OF COLORADO
BOULDER. CO 80309-0392
U.S.A.
TEL: 303-492-3215
TLX: 9109403441
TLF:
EML:
COM: 15C.44.49

BRANDT PETER N
KIEPENHEUER INSTITUT FUER
SONNENPHYSIK
SCHOENECKSTR. 6
D-7800 FREIBURG BR.
GERMANY. F.R.
TEL: 0761-32864
TLX: 7721552
TLF:
EML:
COM: 10.12

BRANDUARDI-RAYMONT G
MULLARD SPACE SCIENCE LAB
HOLMBURY ST MARY
DORKING. SURREY RH5 6NT
U.K.
TEL: 030-670-292
TLX: 859185
TLF:
EML:
COM:

BRANHAM RICHARD L JR
CENTRO REGIONAL DE INVEST
CIENTIFICAS Y TECNOL.
CASILLA DE CORREO 131
5500 MENDOZA
ARGENTINA
TEL: 061-2411794
TLX: 55438 CYTME AR
TLF:
EML:
COM: 08.20.24

BRANSCOMB L M DR
NAT BUREAU OF STANDARDS
WASHINGTON DC 20025
U.S.A.
TEL:
TLX:
TLF:
EML:
COM: 14

BRANSON NICHOLAS J B A DR
GENERAL BOARD OFFICE
THE OLD SCHOOLS
CAMBRIDGE CB2 1TT
U.K.
TEL:
TLX:
TLF:
EML:
COM:

BRATIJCHUK MATRONA V
UZHGOROD STATE UNIVERSITY
HORKIY 46
294000 UZHGOROD
U.S.S.R.
TEL: 3-60-55
TLX: 274155 KNIGA
TLF:
EML:
COM:

BRAUDE SEMION YA PROF AG
INST RADIOPHY ELECTR
UKRAINIAN ACADEMY OF SCI
310085 KHARKOV
U.S.S.R.
TEL: 441092
TLX:
TLF:
EML:
COM: 40

BRAULT JAMES W DR
NAT'L SOLAR OBSERVATORY
PO BOX 26732
950 N. CHERRY AVE
TUCSON AZ 85726
U.S.A.
TEL: 325-9363
TLX: 666484 AURA NOAO TUC
TLF:
EML:
COM: 09.12.14

BRAUN ARIE
RACAH INST. OF PHYSICS
HEBREW UNIV. OF JERUSALEM
JERUSALEM 91904
ISRAEL
TEL: 02-584521
TLX: 25391 HU IL
TLF:
EML:
CON:

BRAUNINGER HEINRICH DR
MPI F PHYSIK & ASTROPHYS
INST F. EXTRATERR. PHYSIK
D-8046 GARCHING B MUNCHEN
GERMANY. F.R.
TEL: 089-3299-566
TLX:
TLF:
EML:
CON:

BRAUNSFURTH EDUARD PH D
IM HAARMANNSBOCH 99A
D-4630 BOCHUM
GERMANY. F.R.
TEL:
TLX:
TLF:
EML:
CON: 34

BRAY ROBERT J DR
CSIRO
DIV OF APPLIED PHYSICS
P.O.BOX 218
LINDFIELD NSW 2070
AUSTRALIA
TEL: 467-6354
TLX: 26296
TLF:
EML:
CON: 10.12

BRECHER AVIVA DR PROF
35 MADISON STREET
BELMONT MA 02178
U.S.A.
TEL: 617-489-1386
TLX:
TLF:
EML:
CON: 15.16

BRECHER KENNETH PROF
DEPT OF ASTRONOMY
BOSTON UNIVERSITY
725 COMMONWEALTH AVE
BOSTON MA 02215
U.S.A.
TEL: 617-353-3423
TLX: 95-1289 BIS UNIV BSN
TLF:
EML:
CON: 28.47.48

BRECKINRIDGE JAMES B DR
JPL/CALTECH
MS 183-301
4800 OAK GROVE DR
PASADENA CA 91103
U.S.A.
TEL: 213-354-6785
TLX: 675429
TLF:
EML:
CON: 09.12

BREGER MICHEL DR
INSTITUT FUER ASTRONOMIE
TUERKENSCHANZSTR 17
A-1180 WIEN
AUSTRIA
TEL: 222-34-53-605
TLX: 133099 VIAST A
TLF:
EML:
CON: 25.27P.30

BREGMAN JACOB D IR
NETHERLANDS FOUNDATION
RADIOASTRONOMY
POSTBUS 2
NL-7990 AA DWINGELOO
NETHERLANDS
TEL: 05219-7244
TLX: 42043
TLF:
EML:
CON: 40

BREGMAN JOEL N
NRAO
EDGEMONT ROAD
CHARLOTTESVILLE VA 22903
U.S.A.
TEL: 804-296-0235
TLX:
TLF:
EML:
CON:

BREINHORST ROBERT A DR
ASTRONOMISCHES INSTITUT
STERNWARTE
AUF DEM HUEGEL 71
D-5300 BONN 1
GERMANY. F.R.
TEL: 0228-733660
TLX:
TLF:
EML:
CON: 42

BREJDO IZABELLA I DR
PULKOVO OBSERVATORY
196140 LENINGRAD
U.S.S.R.
TEL: 297-94-59
TLX:
TLF:
EML:
CON: 09

BRETAGNON PIERRE DR
BUREAU DES LONGITUDES
77 AVE DENFERT-ROCHEREAU
F-75014 PARIS
FRANCE
TEL: 1-40-51-22-69
TLX:
TLF:
EML:
CON: 04.07

BREUKERS R J L H DR
STERREWACHT
POSTBUS 9513
2300 RA LEIDEN
NETHERLANDS
TEL:
TLX:
TLF:
EML:
CON:

BREYSACHER JACQUES
ESO
KARL-SCHWARZSCHILD STR. 2
D 8046 GARCHING B MUNCHEN
GERMANY. F.R.
TEL: (089)32006224
TLX: 5282820
TLF:
EML:
CON: 29

BRIDGELAND MICHAEL DR
INSTITUTE OF ASTRONOMY
MADINGLEY ROAD
CAMBRIDGE CB3 0HA
U.K.
TEL: 0223 337524
TLX: 817297 ASTRON G
TLF:
EML:
CON: 09

BRIDLE ALAN H PROF
NRAO
EDGEMONT ROAD
CHARLOTTESVILLE VA 22903
U.S.A.
TEL: 804-296-0375
TLX: 910-997-0174
TLF:
EML:
CON: 40

BRIEVA EDUARDO PROF
OBSERVATORIO NACIONAL
APARTADO 2584
BOGOTA 1. D.E.
COLOMBIA
TEL: 423786
TLX:
TLF:
EML:
CON: 07.46

BRIHAYE CHARLES C A DR
UNIV LIBRE DE BRUXELLES
50 AVE F.D. ROOSEVELT
B-1050 BRUXELLES
BELGIUM
TEL: 02-6876928
TLX:
TLF:
EML:
CON:

BRINI DOMENICO PROF
LABORATORIO TESRE
VIA CASTAGNOLI 1
I-40100 BOLOGNA
ITALY
TEL:
TLX:
TLF:
EML:
CON:

BRINKMAN BERT C DR
SPACE RESEARCH LABORATORY
BENELUXLAAN 21
NL-3527 HS UTRECHT
NETHERLANDS
TEL:
TLX:
TLF:
EML:
COM: 44

BRINKMANN WOLFGANG
MPI F PHYS & ASTROPHYSIK
INST F EXTRATERR PHYSIK
KARL-SCHWARZSCHILD-STR 1
D-8046 GARCHING B MUNCHEN
GERMANY. F.R.
TEL: 893299877
TLX: 05215845 XTER D
TLF:
EML:
COM: 28.34

BRINKS ELIAS DR
N R A O
P.O. BOX 0
SOCORRO NM 87801-0387
U.S.A.
TEL: 505 835 7000
TLX: 910 9881710
TLF:
EML: EBRINKS@NRAO
COM: 28.40

BRIOT DANIELLE DR
OBSERVATOIRE DE PARIS
61 AVE DE L'OBSERVATOIRE
F-75014 PARIS
FRANCE
TEL: 1-40-51-22-39
TLX: 270776
TLF:
EML:
COM:

BROADFOOT A LYLE DR
UNIVERSITY OF ARIZONA
LUNAR/PLANETARY LAB. WEST
901 GOULD-SIMPSON BUILD.
TUCSON AZ 85721
U.S.A.
TEL: 602-621-4301
TLX: 910-952-1143
TLF:
EML:
COM: 16.21

BRODERICK JOHN DR
PHYSICS DEPT
VPI & SU
BLACKSBURG VA 24061
U.S.A.
TEL: 703-961-5321
TLX: 910-3331861 VPIBKS
TLF:
EML:
COM: 40.51

BRODIE JEAN P
SPACE SCIENCES LABORATORY
UNIVERSITY OF CALIFORNIA
BERKELEY CA 94720
U.S.A.
TEL: 415-642-1579
TLX: 910-366-7945
TLF:
EML:
COM: 28

BROGLIA PIETRO DR
OSSERVATORIO ASTRONOMICO
VIA E. BIANCHI 46
I-22055 MERATE COMO
ITALY
TEL: 039-592035
TLX:
TLF:
EML:
COM: 42

BROMAGE GORDON E DR
ASTROPHYSICS GROUP
RUTHERFORD APPLETON LAB
CHILTON DIDCOT OX11 0QX
U.K.
TEL: 235-21900
TLX: 83159
TLF:
EML:
COM: 14.34

BRONFMAN LEONARDO DR
DEPARTAMENTO DE ASTRONOMI
UNIVERSIDAD DE CHILE
CASILLA 36-D
SANTIAGO
CHILE
TEL: 2281941
TLX: 340260 PBVTR CK
TLF:
EML:
COM: 33

BRONNIKOVA NINA M
PULKOVO OBSERVATORY
196140 LENINGRAD
U.S.S.R.
TEL:
TLX:
TLF:
EML:
COM: 24

BROOKES CLIVE J DR
EARTH. SATELLITE RES UNIT
DEPT OF MATHEMATICS
ASTON UNIVERSITY
BIRMINGHAM B4 7ET
U.K.
TEL: 21-359-3611
TLX: 335787
TLF:
EML:
COM: 07

BROSCH NOAH DR
WISE OBSERVATORY
TEL AVIV UNIVERSITY
RAMAT AVIV
TEL AVIV 69978
ISRAEL
TEL: 972 3 413 788
TLX: 342171 VERSY IL
TLF:
EML: BITNET:E38@TAUNOS.
COM: 28.46

BROSCHE PETER PROF
OBSERVATORIUM HOHER LIST
UNIV STERNWARTE BONN
D-5568 DAUN
GERMANY. F.R.
TEL: 06592-2150
TLX:
TLF:
EML:
COM: 19C.24C.26.28

BROSTERHUS E B F DR
C/O LOCKHEED CITY
PO BOX 6308
JEDDAH
SAUDI ARABIA
TEL: 02-656-2501x355
TLX:
TLF:
EML:
COM:

BROTEN NORMAN W
HERZBERG INST ASTROPHYS
NATL RESEARCH COUNCIL
OTTAWA ONT K1A 0R6
CANADA
TEL: 613-593-6060
TLX:
TLF:
EML:
COM: 40

BROUCKE ROGER DR
7203 RUNNING ROPE CIRCLE
AUSTIN TX 78731
U.S.A.
TEL: 512-345-6435
TLX:
TLF:
EML:
COM: 07

BROUW W N DR
RADIOSTERRENWACHT
POSTBUS 2
NL-7990 AA DWINGELOO
NETHERLANDS
TEL: 05219-7244
TLX: 42043 SRZM NL
TLF:
EML:
COM: 08.40

BROWN ALEXANDER
JILA
UNIVERSITY OF COLORADO
BOULDER CO 80309
U.S.A.
TEL: 303-492-8962
TLX: 755842 JILA
TLF:
EML:
COM: 36.44

BROWN DOUGLAS NASON
UNIVERSITY OF WASHINGTON
DEPT OF ASTRONOMY. FM-20
SEATTLE WA 98195
U.S.A.
TEL: 2065436313/2888
TLX:
TLF:
EML:
COM: 25.27.29.36

BROWN HARRISON DR
3005 LA MANCHA DRIVE
ALBUQUERQUE NM 87104
U.S.A.
TEL:
TLX:
TLF:
EML:
COM:

BROWN JOHN C PROF
UNIVERSITY OF GLASGOW
DEPT PHYSICS & ASTRONOMY
GLASGOW G12 8QQ
U.K.
TEL: 041-330-5182
TLX: 777070 UNIGLA
TLF:
EML:
COM: 10

BROWN ROBERT HAMILTON
JPL/CALTECH
MS 183-501
4800 OAK GROVE DRIVE
PASADENA CA 91109
U.S.A.
TEL: 818-354-2517
TLX:
TLF:
EML:
COM: 15.16

BROWN ROBERT L DR
NRAO
EDGEMONT ROAD
CHARLOTTESVILLE VA 22901
U.S.A.
TEL: 804-296-0232
TLX: 910-997-0174
TLF:
EML:
COM:

BROWN RONALD D PROF
CHEMISTRY DEPT
MONASH UNIVERSITY
WELLINGTON ROAD
CLAYTON VIC 3168
AUSTRALIA
TEL:
TLX:
TLF:
EML:
COM: 34.51V

BROWNE IAN W A DR
NUFFIELD RADIO ASTR LABS
JODRELL BANK
MACCLESFIELD SK119DL
U.K.
TEL: 0477-71321
TLX: 36149
TLF:
EML:
COM: 40

BROWNING PHILIPPA DR
DEPT OF PURE & APPLIED
PHYSICS
PO BOX 88
MANCHESTER M60 1QD
U.K.
TEL: 061 236 3311
TLX: 666094
TLF:
EML: MCCPPBOUK.AC.UMRCC.CMS
COM: 10

BROWNLEE DONALD E PROF
DEPT OF ASTRONOMY
UNIVERSITY OF WASHINGTON
SEATTLE WA 98195
U.S.A.
TEL: 206-543-2888
TLX:
TLF:
EML:
COM: 15.22

BROWNLEE ROBERT R DR
MS F670
LOS ALAMOS SCIENTIFIC LAB
LOS ALAMOS NM 87544
U.S.A.
TEL: 505-662-6427
TLX:
TLF:
EML:
COM: 35.42

BRUCATO ROBERT J.
PALOMA OBSERV. 105-24
CALTECH
PASADENA CA 91125
U.S.A.
TEL: 818 356-4035
TLX: 675425 OR 188192
TLF:
EML:
COM:

BRUCH ALBERT
ASTRONOMISCHES INSTITUT
DER UNIVERSITAET MUENSTER
DOMAGKSTR 75
D-4400 MUENSTER
GERMANY. F.R.
TEL:
TLX:
TLF:
EML:
COM:

BRUCK HERMANN A PROF
CRAIGOWER
PENICUIK EH26 9LA
U.K.
TEL: 968-75918
TLX:
TLF:
EML:
COM: 25

BRUCK MARY T DR
ROYAL OBSERVATORY
EDINBURGH EH9 3HJ
U.K.
TEL: 31-667-3321
TLX: 72383
TLF:
EML:
COM:

BRUECKNER GUENTER E DR
CODE 4160
NAVAL RESEARCH LABORATORY
WASHINGTON DC 20375-5000
U.S.A.
TEL: 202-767-3287
TLX:
TLF:
EML:
COM: 10.12.44

BRUHWEILER FRED C JR
10102 GARDINER AVE
SILVER SPRING MD 20902
U.S.A.
TEL:
TLX:
TLF:
EML:
COM: 29.34.42.44

BRUMBERG VICTOR A DR
INSTITUTE OF APPL.ASTRON.
USSR ACADEMY OF SCIENCES
8 ZHDANOV STREET
197042 LENINGRAD
U.S.S.R.
TEL:
TLX:
TLF:
EML:
COM: 04.07C.31

BRUNER MARILYN E DR
LOCKHEED PALO ALTO RES
LAB DEPT 91-20 BLDG 255
3251 HANOVER ST
PALO ALTO CA 94304
U.S.A.
TEL: 415-858-4023
TLX: 346409 LMSC
TLF:
EML:
COM: 10.12.44

BRUNET JEAN-PIERRE DR
OBS PIC-DU-MIDI TOULOUSE
14 AVENUE EDOUARD BELIN
F-31400 TOULOUSE
FRANCE
TEL: 61-25-21-01
TLX: 503776
TLF:
EML:
COM: 41

BRUNING DAVID H DR
DEPARTMENT OF PHYSICS
UNIVERSITY OF LOUISVILLE
LOUISVILLE KY 40292
U.S.A.
TEL: 502 588 6787
TLX:
TLF:
EML: BITNET:dhbrun01@ulkyvx
COM:

BRUNK WILLIAM E DR
OFF SPACE SC CODE SL
NASA HEADQUARTERS
400 MARYLAND AVE S W
WASHINGTON DC 20546
U.S.A.
TEL: 202-453-1596
TLX:
TLF:
EML:
COM: 15.16

BRUSTON PAUL DR
L P S P
BP NO 10
F-91371 VERRIERES-LE-B.
FRANCE
TEL: 1-69-20-10-60
TLX:
TLF:
EML:
COM:

BRUZEK ANTON DR
SCHWAIGHOFSTR 7
D-7800 FREIBURG
GERMANY. F.R.
TEL: 0761-78522
TLX:
TLF:
EML:
COM: 10.12

BRUZUAL GUSTAVO
C I D A
APARTADO POSTAL 264
MERIDA 5101-A
VENEZUELA
TEL: 58-74-639930
TLX: 74174 CIDA VC
TLF:
EML:
COM:

BRYANT JOHN DR
47 AV. FELIX FAURE
75015 PARIS
FRANCE
TEL: (1)45 57 76 47
TLX:
TLF:
EML:
COM:

BUCHLER J ROBERT PROF
DEPT OF PHYSICS
UNIVERSITY OF FLORIDA
GAINESVILLE FL 32611
U.S.A.
TEL: 904-373-9942
TLX:
TLF:
EML:
COM: 35

BUCHNER JORG DR
ZENTRALINSTITUT FUR
ASTROPHYSIK DER ADW
ROSA LUXEMBURG STR 17A
POTSDAM 1591
GERMANY. D.R.
TEL: 7620
TLX: 15305 VDEPDH
TLF:
EML:
COM: 10.49

BUDDING EDWIN DR
CARTER OBSERVATORY
P O BOX 2909
WELLINGTON 1
NEW ZEALAND
TEL: 04-728-167
TLX: 30177 NATOBS NZ
TLF:
EML:
COM: 42.46

BUES IRMELA D DR
REMEIS STERNWARTE
STERNWARTSTR 7
D-8600 BAMBERG
GERMANY. F.R.
TEL: 0951-57708
TLX:
TLF:
EML:
COM: 29.36

BUFF JAMES S DR
DEPT PHYSICS & ASTRONOMY
DARTMOUTH COLLEGE
HANOVER NH 03755
U.S.A.
TEL:
TLX:
TLF:
EML:
COM:

BUHL DAVID DR
INFRARED & RADIO ASTR BR.
NASA/GSFC. CODE 693
GREENBELT MD 20771
U.S.A.
TEL: 301-286-8810
TLX:
TLF:
EML:
COM:

BUITRAGO JESUS
INST DE ASTROFISICA DE
CANARIAS
LA LAGUNA
38071 TENERIFE
SPAIN
TEL: 922-26-22-[illegible]
TLX: 92640
TLF:
EML:
COM:

BUJARRABAL VALENTIN
CENTRO ASTRON DE YEBES
O A N
APARTADO CORREOS 148
19080 GUADALAJARA
SPAIN
TEL: 11-[illegible]23358
TLX:
TLF:
EML:
COM:

BUMBA VACLAV DR
ASTRONOMICAL INSTITUTE
CZECH. ACAD. OF SCIENCES
251 65 ONDREJOV
CZECHOSLOVAKIA
TEL: 72-45-25
TLX: 121579 ASTR C
TLF:
EML:
COM: 10.12.44

BUNCLARK PETER STEPHEN DR
INSTITUTE OF ASTRONOMY
MADINGLEY ROAD
CAMBRIDGE CB3 OHA
U.K.
TEL: 0223 337548
TLX: 817297 ASTRON G
TLF:
EML:
COM: 24

BUNNER ALAN N DR
PERKIN-ELMER CORP
M S 897
100 WOOSTER HEIGHTS RD
DANBURY CT 06810
U.S.A.
TEL: 203-797-6339
TLX:
TLF:
EML:
COM: 42.44.48

BUONANNO ROBERTO
OSS ASTRON SU MONTE MARIO
VIA DEL PARCO MELLINI 84
I-00136 ROMA
ITALY
TEL: 06-347056
TLX:
TLF:
EML:
COM:

BURATTI BONNIE J DR
JET PROPULSION LABORATORY
4800 OAK GROVE DR 183-501
PASADENA CA 91109
U.S.A.
TEL: 818 354 7427
TLX:
TLF:
EML:
COM: 15.16

BURBIDGE E MARGARET PROF
CTR ASTROPHYS & SPACE SCI
UNIVERSITY OF CALIFORNIA
SAN DIEGO MC C-011
LA JOLLA CA 92093
U.S.A.
TEL: 619-452-4477
TLX:
TLF:
EML:
COM: 28

BURBIDGE GEOFFREY R PROF
CTR ASTROPHYS & SPACE SCI
UNIVERSITY OF CALIFORNIA
SAN DIEGO MC C-011
LA JOLLA CA 92093
U.S.A.
TEL: 619-452-6626
TLX:
TLF:
EML:
COM: 28.35.40.47.48

BURGER J J DR IR
ESTEC
POSTBUS 299
NL-2200 AG NOORDWIJK
NETHERLANDS
TEL: 01719-84404
TLX: 39098
TLF:
EML:
COM:

BURGER MARIJKE DR
OBS ROYAL DE BELGIQUE
RINGLAAN 3
B-1180 BRUSSEL
BELGIUM
TEL: 02-37522484
TLX: 21565 OBSBEL B
TLF:
EML:
COM: 44

BURGESS ALAN DR
DEPT OF APPLIED MATHS
SILVER STREET
CAMBRIDGE CB3 9EW
U.K.
TEL:
TLX:
TLF:
EML:
COM: 14.34

BURGESS DAVID D PROF
BLACKETT LABORATORY
IMPERIAL COLLEGE OF
SCIENCE & TECHNOLOGY
LONDON SW7 2BZ
U.K.
TEL: 1-589-5111x6931
TLX:
TLF:
EML:
COM:

BURKE BERNARD F DR
DEPT OF PHYSICS
M I T RM 26-335
CAMBRIDGE MA 02139
U.S.A.
TEL: 617-253-2572
TLX: 92-1473
TLF:
EML:
COM: 33.34.40.44C.51

BURKE J ANTHONY DR
DEPT OF PHYSICS
UNIVERSITY OF VICTORIA
VICTORIA BC V8W 2Y2
CANADA
TEL: 604-721-7743
TLX:
TLF:
EML:
COM:

BURKHART CLAUDE DR
OBSERVATOIRE DE LYON
F-69230 ST-GENIS-LAVAL
FRANCE
TEL: 78-56-07-05
TLX: 310-926
TLF:
EML:
COM: 29

BURKHEAD MARTIN S
ASTRONOMY DEPT
INDIANA UNIVERSITY
SWAIN HALL WEST
BLOOMINGTON IN 47405
U.S.A.
TEL: 812-335-6917
TLX:
TLF:
EML:
COM: 37

BURKI GILBERT DR
OBSERVATOIRE DE GENEVE
CH-1290 SAUVERNY
SWITZERLAND
TEL: 22-55-26-11
TLX: 27720
TLF:
EML:
COM: 27.30V

BURLAGA LEONARD F DR
NASA/GSFC
CODE 692
GREENBELT MD 20771
U.S.A.
TEL:
TLX:
TLF:
EML:
COM: 15.49P

BURNAGE ROBERT
OBS DE HAUTE-PROVENCE
F-04870 ST-MICHEL-L'OBS.
FRANCE
TEL: 92-76-63-68
TLX: 410690
TLF:
EML:
COM: 30

BURNS JACK O'NEAL JR
PHYSICS & ASTRONOMY DEPT
UNIVERSITY OF NEW MEXICO
800 YALE BLVD N.E.
ALBUQUERQUE NM 87131
U.S.A.
TEL: 505-277-2705
TLX:
TLF:
EML:
COM: 28

BURNS JOSEPH A PROF
CORNELL UNIVERSITY
THURSTON HALL
ITHACA NY 14850
U.S.A.
TEL: 607-256-4875
TLX: 937478
TLF:
EML:
COM: 15.16C.20

BURROWS ADAM SETH
DEPT OF PHYSICS
SUNY
STONY BROOK NY 11794
U.S.A.
TEL: 516-246-6810
TLX:
TLF:
EML:
COM: 48

BURSA MILAN DR
ASTRONOMICAL INSTITUTE
CZECH. ACAD. OF SCIENCES
BUDECSKA 6
120 23 PRAHA 2
CZECHOSLOVAKIA
TEL: 25-05-51
TLX: 122486
TLF:
EML:
COM:

BURSTEIN DAVID
ARIZONA STATE UNIVERSITY
DEPT OF PHYSICS
TEMPE AZ 85281
U.S.A.
TEL:
TLX:
TLF:
EML:
COM: 28.50

BURTON W BUTLER DR
STERREWACHT
POSTBUS 9513
NL-2300 RA LEIDEN
NETHERLANDS
TEL: 071-272727
TLX: 39058 ASTRON
TLF:
EML:
COM: 09.33.34

BURTON WILLIAM M
SPACE & ASTROPHYS DIV
RUTHERFORD APPLETON LAB
CHILTON DIDCOT OX11 0QX
U.K.
TEL: 235-21900
TLX: 83159
TLF:
EML:
COM: 44

BUSCOMBE WILLIAM PROF
ASTRONOMY DEPT
NORTHWESTERN UNIVERSITY
EVANSTON IL 60201
U.S.A.
TEL: 312-491-7527
TLX:
TLF:
EML:
COM: 29.45.46

BUSER ROLAND DR
ASTRONOMISCHES INSTITUT
UNIVERSITAET BASEL
VENUSSTRASSE 7
CH-4102 BINNINGEN
SWITZERLAND
TEL: 061-227711
TLX:
TLF:
EML:
COM: 25C.45

BUSKO IVO C DR
MCT/INST PESQUISAS ESPEC.
AVE DOS ASTRONAUTAS 1758
JARDIM DA GRANJA
12200 SAO JOSE DOS CAMPOS
BRAZIL
TEL: 22-9977 x 397
TLX: 012-33530 INPE BR
TLF:
EML:
COM: 27

BUSON LUCIO M DR
OSSERVATORIO ASTRONOMICO
VICOLO DELL'OSSERVAT. 5
35122 PADOVA
ITALY
TEL: 049 661499
TLX: 432071 ASTROS I
TLF:
EML: BUSON@ASTRPD.INFNET
COM:

BUSSO MAURIZIO
OSSERVATORIO ASTRONOMICO
DI TORINO
I-10025 PINO TORINESE
ITALY
TEL: 011-84-1067
TLX: 213239 TO ASTR I
TLF:
EML:
COM: 42

BUTA RONALD J DR
DEPARTMENT OF ASTRONOMY
UNIVERSITY OF TEXAS
AUSTIN TX 78712
U.S.A.
TEL: 512 471 3466
TLX: 910 874 1351
TLF:
EML:
COM: 28

BUTCHER HARVEY R PROF DR
KAPTEYN ASTRONOMICAL INST
POSTBUS 800
NL-9700 AV GRONINGEN
NETHERLANDS
TEL: 05908-19631
TLX: 53767 VSWPO NL
TLF:
EML:
COM: 28.29

BUTCHINS SYDNEY ADAIR
DEPART. OF CIVIL AVIATION
STUDIES FACUL. OF MATHS
THE MINORIES
TOWER HILL EC3N 1JY
U.K.
TEL: 01 722-7344
TLX:
TLF:
EML:
COM:

BUTI BIMLA PROF
PHYSICAL RESEARCH LAB
NAVRANGPURA
AHMEDABAD 380 009
INDIA
TEL: 462129
TLX: 121-397 IN
TLF:
EML:
COM: 49V

BUTLER C JOHN DR
ARMAGH OBSERVATORY
COLLEGE HILL
ARMAGH BT61 9DG N IRELAND
U.K.
TEL: 0861-522-928
TLX: 747937 ARMOBS G
TLF:
EML:
COM: 27.44

BUTLER DENNIS DR
ASTRONOMY DEPT
YALE UNIVERSITY
BOX 2023 YALE STATION
NEW HAVEN CT 06520
U.S.A.
TEL:
TLX:
TLF:
EML:
COM: 27.37

BUTLER KEITH DR
INSTITUT FUR ASTRONOMIE
UND ASTROPHYSIK
SCHEINERSTRASSE 1
8000 MUNCHEN 80
GERMANY. F.R.
TEL: 089-98-90-21
TLX:
TLF:
EML:
COM: 29

BUTTERWORTH PAUL
NASA GODDARD SPACE
FLIGHT CENTER
CODE 633
GREENBELT MD 20771
U.S.A.
TEL: 301-286-3995
TLX:
TLF:
EML:
COM: 44

BYARD PAUL L DR
DEPT OF ASTRONOMY
OHIO STATE UNIVERSITY
174 W. 18TH AVE
COLUMBUS OH 43210
U.S.A.
TEL: 614-422-1773
TLX:
TLF:
EML:
COM:

BYKOV MIKLE F DR
ASTRONOMICAL INSTITUTE
UZBEKIAN ACADEMY OF SCI
700000 TASHKENT
U.S.S.R.
TEL:
TLX:
TLF:
EML:
COM: 08

BYLEVELD WILLEM DR
OMNIVERSUM SPACE THEATRE
PRES KENNEDYLAAN 5
2517 JK THE HAGUE
NETHERLANDS
TEL: (0)70547479
TLX:
TLF: (0)70524280
EML:
COM:

BYRD GENE G DR
DEPT PHYSICS & ASTRONOMY
UNIVERSITY OF ALABAMA
BOX 1921
UNIVERSITY AL 35486
U.S.A.
TEL: 205-348-5050
TLX:
TLF:
EML:
COM: 28.37

BYRNE PATRICK B DR
ARMAGH OBSERVATORY
ARMAGH BT61 9DG
U.K.
TEL: 44-861-522928
TLX: 747937 ARMOBS G
TLF:
EML:
COM: 27

BYSTROV NIKOLAI F DR
PULKOVO OBSERVATORY
196140 LENINGRAD
U.S.S.R.
TEL: 297-94-81
TLX:
TLF:
EML:
COM: 16

BYSTROVA NATALIJA V DR
SPECIAL ASTROPHYSICAL
OBSERVATORY
LENINGRAD BRANCH
196140 LENINGRAD
U.S.S.R.
TEL: 297-9452
TLX:
TLF:
EML:
COM: 34

CABRITA EZEQUIEL DR
OBS ASTRONOMICO DE LISBOA
TAPADA DA LISBOA
1300 LISBOA
PORTUGAL
TEL: 637351-634669
TLX:
TLF:
EML:
COM: 26

CACCIANI ALESSANDRO PROF
DIPARTIMENTO DI FISICA
UNIVERSITA LA SAPIENZA
PIAZZALE ALDO MORO 2
I-00185 ROMA
ITALY
TEL: 06-4976-265
TLX: 613255 INFNRO
TLF:
EML:
COM:

CACCIARI CARLA DR
OSSERVATORIO ASTRONOMICO
CP 596
40100 BOLOGNA
ITALY
TEL: 39 51 259301
TLX: 520634 INFNBO I
TLF:
EML:
COM:

CACCIN BRUNO
DIPARTIMENTO DI FISICA
VIA RAIMONDO SNC
UNIVERSITA TOR VERGATA
I-00173 ROMA
ITALY
TEL: 39-6-79792323
TLX: 626382 FIUNTV
TLF:
EML:
COM: 38

CADEZ ANDREJ DR
DEPT OF PHYSICS
UNIVERSITY OF LJUBLJANA
JADRANSKA 19
61000 LJUBLJANA
YUGOSLAVIA
TEL: 265-061
TLX:
TLF:
EML:
COM:

CADEZ VLADIMIR
INSTITUTE OF PHYSICS
P.O. BOX 57
YU-11001 BEOGRAD
YUGOSLAVIA
TEL: 011-212-219
TLX: 11002 INFIZ YU
TLF:
EML:
COM: 10.12

CAHN JULIUS H PROF
UNIVERSITY OF ILLINOIS
DEPARTMENT OF ASTRONOMY
1011 WEST SPRINFIELD AVE
URBANA IL 61801
U.S.A.
TEL: 217-333-3090
TLX:
TLF:
EML:
COM:

CAILLAULT JEAN PIERRE DR
DEPARTMENT OF PHYSICS &
ASTRONOMY UNIVERSITY OF
GEORGIA
ATHENS GA 30602
U.S.A.
TEL: 404 542 2883
TLX: 490 999 1619
TLF:
EML:
COM:

CALAME ODILE DR
CERGA
AVENUE COPERNIC
F-06130 GRASSE
FRANCE
TEL: 93-36-58-49
TLX: 470865
TLF:
EML:
COM: 07.16.19.20

CALDWELL JOHN A R
SAAO
P.O.BOX 9
OBSERVATORY 7935
SOUTH AFRICA
TEL: 021 47 0025
TLX: 57-20309
TLF:
EML:
COM: 33

CALDWELL JOHN JAMES
YORK UNIVERSITY
DEPT. OF PHYSICS
4700 KEELE STREET
NORTH YORK. ONT. M3J1P3
CANADA
TEL: 416-736-2100
TLX: 065224736 YORK U TOR
TLF:
EML:
COM: 16

CALLANAN PAUL DR
DEPARTMENT OF ASTROPHYS.
UNIVERSITY OF OXFORD
SOUTH PARKS ROAD
OXFORD OX1 3RQ
U.K.
TEL: 865272092
TLX: 83295 NUCLOX G
TLF:
EML:
COM: 42

CALLEBAUT DIRK K DR
DEPT OF PHYSICS UIA
UIA
UNIVERSITEITSPLEIN 1
B-2610 WILRIJK-ANTWERPEN
BELGIUM
TEL: 3-828-2528
TLX: 33646
TLF:
EML:
COM: 35.37

CALLY PAUL S DR
DPT OF MATHEMATICS
MONASH UNIVERSITY
CLAYTON
VICTORIA 3168
AUSTRALIA
TEL: 03 565 4471
TLX: MONASH AA 32691
TLF:
EML: apm150f@vaxc.cc.monash.edu.au
COM: 10

CALOI VITTORIA DR
CNR
ASTROFISICA SPAZIALE
C P 67
I-00044 FRASCATI
ITALY
TEL: 06-942-5654
TLX: 610261 CNR FRA
TLF:
EML:
COM: 35.37

CALVANI MASSIMO DR
ISTITUTO DI ASTRONOMIA
VIC. DELL'OSSERVATORIO 5
I-35122 PADOVA
ITALY
TEL: 049-661499
TLX: 430176 UNPADU I
TLF:
EML:
COM: 47

CALVET NURIA DR
C I D A
APARTADO POSTAL 264
MERIDA 5101-A
VENEZUELA
TEL: 58-74-639930
TLX: 74174 CIDA VC
TLF:
EML:
COM: 46

CALVO MANUEL
DPTO DE ASTRONOMIA
UNIVERSIDAD DE ZARAGOZA
50009 ZARAGOZA
SPAIN
TEL: 357011
TLX:
TLF:
EML:
COM:

CAMARENA BADIA VICENTE PR
DPTO MATEMATICA APLICADA
ETS INGENIEROS INDUSTR
UNIVERSIDAD DE ZARAGOZA
50009 ZARAGOZA
SPAIN
TEL:
TLX:
TLF:
EML:
COM:

CAMERON ALASTAIR G W PROF
HARVARD COLLEGE OBS
60 GARDEN STREET
CAMBRIDGE MA 02138
U.S.A.
TEL: 617-495-5374
TLX:
TLF:
EML:
COM: 35.48

CAMERON ANDREW COLLIER DR
ASTRONOMY CENTRE PHYSICS
BUILDING UNIVERSITY OF
SUSSEX
FALMER BRIGHTON BN1 9QH
U.K.
TEL: 273 678117
TLX: 877159 BHVTXS G
TLF:
EML:
COM: 27

CAMERON LUZIUS MARTIN
ASTRONOMISCHES INSTITUT
DER UNIVERSITAT BASEL
VENUSSTRASSE 7
CH 4102 BINNINGEN
SWITZERLAND
TEL: 061 22 77 11
TLX:
TLF:
EML:
COM: 28

CAMERON WINIFRED S MRS
5087 IRON SPRINGS RD
PRESCOTT AZ 86301
U.S.A.
TEL:
TLX:
TLF:
EML:
COM: 16

CAMICHEL HENRI DR
24, AVE C. FLAMMARION
F-31500 TOULOUSE
FRANCE
TEL: 61-48-96-91
TLX:
TLF:
EML:
COM: 16

CAMPBELL ALISON DR
CARNEGIE INST.OF WASHING.
DEPT. OF TERR. MAGNETISM
5241 BROAD BRANCH RD. NW
WASHINGTON DC 20015
U.S.A.
TEL: (202)686 4397
TLX:
TLF:
EML: BITNET:AWC@STSCI
COM: 26

CAMPBELL BELVA G S DR
DPT OF PHYS. & ASTRONOMY
UNIVERSITY OF NEW MEXICO
ALBUQUERQUE NM 87131
U.S.A.
TEL: 505 277 5148
TLX:
TLF:
EML: BITNET:BEL@UNMB
COM:

CAMPBELL BRUCE DR
4537 RITHETWOOD PLC.
VICTORIA BC V8X 4J9
CANADA
TEL:
TLX:
TLF:
EML:
COM: 09.26.29.30C.51

CAMPBELL DONALD B
SPACE SCIENCES BUILDING
CORNELL UNIVERSITY
ITHACA NY 14853-1247
U.S.A.
TEL: 607 255 5274
TLX: 932454
TLF:
EML:
COM: 16

CAMPBELL JAMES W
ROYAL OBSERVATORY
BLACKFORD HILL
EDINBURGH EH9 3HJ
U.K.
TEL:
TLX:
TLF:
EML:
COM:

CAMPBELL MURRAY F
DEPT PHYSICS & ASTRONOMY
COLBY COLLEGE
WATERVILLE ME 04901
U.S.A.
TEL: 207-872-3251
TLX:
TLF:
EML:
COM: 44

CAMPINS HUMBERTO DR
PLANETARY SCIENCE INSTIT.
SAIC
2030 E SPEEDWAY SUITE 201
TUCSON AZ 85719
U.S.A.
TEL: 602 881 0337
TLX:
TLF:
EML:
COM: 15

CAMPOS L M BRAGA DA COSTA
INST SUPERIOR TECNICO
AVE ROVISCO PAIS
1096 LISBOA CODEX
PORTUGAL
TEL: 800525
TLX: 63423 ISTUTL P
TLF:
EML:
COM:

CAMPUSANO LUIS E
DPTO DE ASTRONOMIA
UNIVERSIDAD DE CHILE
CASILLA 36-DRD BELIN
SANTIAGO
CHILE
TEL: 02-22941011
TLX: 44001 PBCZ
TLF:
EML:
COM: 28.51

CANAL RAMON M DR
DPTO FISICA DE ATMOSFERA
UNIVERSIDAD DE BARCELONA
AVDA. DIAGONAL 645
08028 BARCELONA
SPAIN
TEL:
TLX:
TLF:
EML:
COM: 35

CANAVAGGIA RENEE DR
OBSERVATOIRE DE PARIS
61 AVE DE L'OBSERVATOIRE
F-75014 PARIS
FRANCE
TEL: 1-43-20-12-10
TLX: 270776
TLF:
EML:
COM:

CANDY MICHAEL P MR
PERTH OBSERVATORY
WALNUT ROAD
BICKLEY 6076 WEST AUSTR
AUSTRALIA
TEL:
TLX:
TLF:
EML:
COM: 06.07.15.20

CANE HILARY VIVIEN
DEPT. OF PHYSICS
UNIVERSITY OF TASMANIA
GPO BOX 252C
HOBART, TASMANIA 7001
AUSTRALIA
TEL: 61-02-202-401
TLX: AA58150
TLF:
EML:
COM: 10.33

CANFIELD RICHARD C DR
2680 WOODLAWN DRIVE
HONOLULU HI 96822
U.S.A.
TEL:
TLX:
TLF:
EML:
COM:

CANIZARES CLAUDE R PROF
M I T
RM 37-501
CAMBRIDGE MA 02139
U.S.A.
TEL: 617-253-7500
TLX: 921473 MITCAM
TLF:
EML:
COM:

CANNON RUSSELL D DR
ANGLO-AUSTRALIAN OBS
PO BOX 296
EPPING NSW 2121
AUSTRALIA
TEL: 02-868-1666
TLX: 23999 AAOSYD
TLF:
EML:
COM: 28.37

CANNON WAYNE H DR
DEPTS/PHYS/EARTH & ATM SC
YORK UNIVERSITY
4700 KEELE STREET
DOWNSVIEW ONT M3J1P3
CANADA
TEL: 416-667-6410
TLX: 06524736
TLF:
EML:
COM: 19

CANTO JORGE DR
INSTITUTO DE ASTRONOMIA
UNAM
APDO POSTAL 70-264
04510 MEXICO DF
MEXICO
TEL: 5485305
TLX: 1760155 CIC ME
TLF:
EML:
COM: 34

CANTU ALBERTO M DR
IST CIBERNETICA/BIOFISICA
CNR
I-16032 CAMOGLI (GE)
ITALY
TEL: 0185-770646
TLX:
TLF:
EML:
COM:

CAO CHANGXIN
NANJING ASTRONOMICAL
INSTRUMENT FACTORY
NANJING
CHINA. PEOPLE'S REP.
TEL: 46191
TLX: 34136 GLYNJ CN
TLF:
EML:
COM: 09

CAO LIHONG
PURPLE MOUNTAIN OBS
ACADEMIA SINICA
NANJING
CHINA. PEOPLE'S REP.
TEL:
TLX:
TLF:
EML:
COM:

CAO SHENGLIN
DEPT OF ASTRONOMY BEIJING
NORMAL UNIVERSITY 100875
BEIJING
CHINA. PEOPLE'S REP.
TEL: 201 2255
TLX: 222701
TLF:
EML:
COM:

CAPACCIOLI MASSIMO DR
OSSERVATORIO ASTRONOMICO
UNIVERSITY OF PADOVA
VIC. DELL'OSSERVATORIO 5
I-35122 PADOVA
ITALY
TEL: 049-66-14-99
TLX: 430176 UNIPDU I
TLF:
EML:
COM: 28

CAPELATO HUGO VICENTE DR
INSTITUTO DE PESQUISAS
ESPACIAIS INPE/MCT DPT
ASTRO. CAIXA POSTAL 515
12201 SAO JOSE DOS CAMPOS
BRAZIL
TEL: 123 229977
TLX: 1233530 INPE BR
TLF:
EML:
COM:

CAPEN CHARLES F
SOLIS LACUS OBSERVATORY
RT.2. BOX 262 E
CUBA NO 65453
U.S.A.
TEL:
TLX:
TLF:
EML:
COM:

CAPITAINE NICOLE
OBSERVATOIRE DE PARIS
61 AVE DE L'OBSERVATOIRE
F-75014 PARIS
FRANCE
TEL: 1-40-51-22-31
TLX: 270776
TLF:
EML:
COM: 04.19

CAPLAN JAMES
OBSERVATOIRE DE MARSEILLE
2 PLACE LE VERRIER
F-13248 MARSEILLE CEDEX 4
FRANCE
TEL: 91-95-90-88
TLX:
TLF:
EML:
COM: 34

CAPPA DE NICOLAU CRISTINA
INSTITUTO ARGENTINO DE
RADIOASTRONOMIA
CASILLA DE CORREO No. 5
1894 VILLA ELISA (Bs.As.)
ARGENTINA
TEL:
TLX:
TLF:
EML:
COM: 34

CAPPELLARO ENRICO DR
OSSERVATORIO ASTRONOMICO
DI PADOVA
VICOLO DELL'OSSERV. 5
PADOVA 35122
ITALY
TEL: 049 661499
TLX: 432071 ASTROS I
TLF:
EML: SPAN:ASTRPD::CAPPELLARO
COM:

CAPRIOLI GIUSEPPE PROF
OSSERVATORIO ASTRONOMICO
DI ROMA
VIA TRIONFALE 204
I-00136 ROMA
ITALY
TEL: 06-347050
TLX:
TLF:
EML:
COM: 12

CAPRIOTTI EUGENE R DR
DEPT OF ASTRONOMY
OHIO STATE UNIVERSITY
5058 ALPHEUS SMITH LAB
COLUMBUS OH 43210
U.S.A.
TEL: 614-422-1773
TLX:
TLF:
EML:
COM: 34

CAPUTO FILIPPINA DR
CNR ASTROFISICA SPAZIALE
C P 67
I-00044 FRASCATI. ROMA
ITALY
TEL: 06-942-5651
TLX: 610261 CNR FRA
TLF:
EML:
COM: 35.37

CAPUZZO DOLCETTA ROBERTO
ISTITUTO ASTRONOMICO
UNIVERSITA LA SAPIENZA
VIA G.M. LANCISI 29
I-00161 ROMA
ITALY
TEL: 06-867525
TLX:
TLF:
EML:
COM: 14.37

CARANICOLAS NICHOLAS DR.
DEPT OF ASTRONOMY
UNIVERSITY THESSALONIKI
GR-54006 THESSALONIKI
GREECE
TEL: 031-991357/59
TLX:
TLF:
EML:
COM: 07

CARBON DUANE F DR
NASA AMES RESEARCH CNTR.
M.S. 245-6
MOFFETT FIELD. CA 94035
U.S.A.
TEL: 408-429-2149
TLX:
TLF:
EML:
COM: 14.36

CARDONA OCTAVIO DR
INST NAC DE ASTROFISICA
OPTICA Y ELECTRONICA
APDO POSTAL 216
72000 PUEBLA. PUE.
MEXICO
TEL: 22-470500
TLX:
TLF:
EML:
COM:

CARDUS ALMEDA J O MR
OBSERVATORIO DEL EBRO
ROQUETES (TARRAGONA)
SPAIN
TEL: 977-500511
TLX:
TLF:
EML:
COM:

CARESTIA REINALDO A DR
OBS ASTRONOMICO F AGUILAR
AV BENAVIDEZ 8175 OESTE
RIVADAVIA
5400 SAN JUAN
ARGENTINA
TEL: 0054-064-231615
TLX:
TLF:
EML:
COM: 08

CARIGNAN CLAUDE DR
DPT DE PHYSIQUE
UNIVERSITE DE MONTREAL
CP 6128 SUCC A
MONTREAL QUE H3C 3J7
CANADA
TEL: 514 343 7355
TLX:
TLF:
EML: CARIGNAN@CC.UMONTREAL.CA
COM: 28

CARLBERG RAYMOND GARY DR
PHYSICS DEPARTMENT
YORK UNIVERSITY
TORONTO ONT M3J IP3
CANADA
TEL: 416-667-3851
TLX:
TLF:
EML:
COM:

CARLETON NATHANIEL P DR
SMITHSONIAN ASTROPHYSICAL
OBSERVATORY
60 GARDEN ST
CAMBRIDGE MA 02138
U.S.A.
TEL: 617-495-7405
TLX: 921428 SATELLITE CAM
TLF:
EML:
COM:

CARLQVIST PER A DR
DEPT OF PLASMA PHYSICS
ROYAL INST OF TECHNOLOGY
S-100 44 STOCKHOLM 70
SWEDEN
TEL: 46-8-787-7697
TLX:
TLF:
EML:
COM: 10

CARLSON JOHN B
CENTER FOR ARCHAEOASTRON.
P.O. BOX X
COLLEGE PARK MD 20740
U.S.A.
TEL: 301-864-6637
TLX:
TLF:
EML:
COM: 41.51

CARLSSON MATS DR
INST. THEORE. ASTROPHYS.
P.O. BOX 1029
BLINDERN
N-0315 OSLO 3
NORWAY
TEL:
TLX:
TLF:
EML:
COM: 36

CARNEY BRUCE WILLIAM
DEPT OF PHYS & ASTRONOMY
PHILLIPS HALL 039A
UNIV OF NORTH CAROLINA
CHAPEL HILL NC 27514
U.S.A.
TEL: 919-962-3023
TLX:
TLF:
EML:
COM: 25.29.30.37

CAROFF LAWRENCE J
SPACE SCIENCE DIVISION
NASA AMES RESEARCH CTR
MS 245-6
MOFFETT FIELD CA 94035
U.S.A.
TEL: 415-694-5523
TLX:
TLF:
EML:
COM:

CAROUBALOS C A PROF
LAB ELECTRONIC PHYSICS
UNIVERSITY OF ATHENS
KYBRIA TYPA-ILISSIA
ATHENS 144
GREECE
TEL: 7244096/11119
TLX: 215530 OBSA GR
TLF:
EML:
COM: 40

CARPENTER KENNETH G DR
CODE 681
NASA
GSFC
GREENBELT MD 20771
U.S.A.
TEL: 301 286 3453
TLX: 89675 K CARPENTER
TLF:
EML: SPAN:6172::HRSCARPENTER
COM: 29.44

CARPENTER LLOYD DR
EG/G WASH ANALYT SERV CTR
5000 PHILADELPHIA WAY
LANHAM MD 20706
U.S.A.
TEL:
TLX:
TLF:
EML:
COM:

CARQUILLAT JEAN-MICHEL
OBSERVATOIRE PIC-DU-MIDI
ET TOULOUSE
14 AVENUE EDOUARD BELIN
F-31400 TOULOUSE
FRANCE
TEL: 61-25-21-01
TLX: 530776 OBSTLSE
TLF:
EML:
COM: 30

CARR BERNARD JOHN
SCHOOL OF MATHEM. SCI.
QUEEN MARY COLLEGE
MILE END ROAD
LONDON E1 4NS
U.K.
TEL: 01-980-4811
TLX:
TLF:
EML:
COM: 47

CARR THOMAS D PROF
DEPT OF ASTRONOMY
UNIVERSITY OF FLORIDA
GAINESVILLE FL 32611
U.S.A.
TEL: 904-392-2066
TLX:
TLF:
EML:
COM: 40.51

CARRANZA GUSTAVO J DR
LAPRIDA 880
5000 CORDOBA
ARGENTINA
TEL:
TLX:
TLF:
EML:
COM: 28

CARRASCO GUILLERMO DR
OBS ASTRONOMICO NACIONAL
UNIVERSIDAD DE CHILE
CASILLA 36-D
SANTIAGO
CHILE
TEL: 229-40-02
TLX:
TLF:
EML:
COM: 08

CARRASCO LUIS DR
INSTITUTO DE ASTRONOMIA
UNAM
APDO POSTAL 70-264
04510 MEXICO DF
MEXICO
TEL: 905-548-5305
TLX:
TLF:
EML:
COM: 33

CARROLL P KEVIN PROF
PHYSICS DEPT
UNIVERSITY COLLEGE
BELFIELD
DUBLIN 4
IRELAND
TEL:
TLX:
TLF:
EML:
COM: 14.44

CARRUTHERS GEORGE R DR
SPACE SCIENCE DIVISION
US NAVAL RESEARCH LAB
CODE 7123
WASHINGTON DC 20375
U.S.A.
TEL: 202-767-2764
TLX:
TLF:
EML:
COM: 15.34

CARSENTY UNI DR
UNIVERSITY OF MARYLAND
ASTRONOMY PROGRAM
COLLEGE PARK MD 20742
U.S.A.
TEL: (301)454 5850
TLX:
TLF:
EML: carsenty%Astro.umd.Edu@umd2
COM:

CARSON T R DR
UNIVERSITY OBSERVATORY
BUCHANAN GARDENS
ST ANDREWS, FIFE KY16 9SS
U.K.
TEL:
TLX:
TLF:
EML:
COM: 35.36

CARSWELL ROBERT F DR
INSTITUTE OF ASTRONOMY
MADINGLEY ROAD
CAMBRIDGE CB3 0HA
U.K.
TEL: 223-62204
TLX: 817297 ASTRON G
TLF:
EML:
CON: 28

CARTER DAVID DR
MT STROMLO OBSERVATORY
PRIVATE BAG
WODEN P.O. ACT 2606
AUSTRALIA
TEL: 062-88-1111
TLX: 62270 AA
TLF:
EML:
CON: 28

CARTER WILLIAM EUGENE
NATL GEODETIC SURVEY
N/CG 114
ADV. TECHNOL. SECT.. GRDL
ROCKVILLE MD 20852
U.S.A.
TEL: 301-443-8423
TLX:
TLF:
EML:
CON: 19C.31

CARUSI ANDREA
IST ASTROFISICA SPAZIALE
DEP PLANETOLOGIA
VIALE DELL'UNIVERSITA' 11
I-00185 ROMA
ITALY
TEL: 06-4956951
TLX: 610261 CNRFRA
TLF:
EML:
CON: 15.20V.22

CARVER JOHN H PROF
RESEARCH SCHOOL OF
PHYSICAL SCIENCES
AUSTRALIAN NAT UNIVERSITY
CANBERRA ACT 2601
AUSTRALIA
TEL: 062-492476
TLX: 62615 APHYS
TLF:
EML:
CON: 14.44

CASANOVAS JUAN DR
VATICAN OBSERVATORY
I-00120 CITTA DEL VATICANO
VATICAN CITY
TEL: 698-3411/5266
TLX: 2020 VATOBS VA
TLF:
EML:
CON:

CASERTANO STEFANO DR
KAPTEYN LABORATORIUM
POST BUS 800
9700 AV GRONINGEN
NETHERLANDS
TEL: 050 634052
TLX: 53572 STARS NL
TLF:
EML: STEFANO@HGRRUG5
CON: 28

CASH WEBSTER C JR
LASP
UNIVERSITY OF COLORADO
BOX 392
BOULDER CO 80309
U.S.A.
TEL: 303-492-3708
TLX:
TLF:
EML:
CON: 48

CASOLI FABIENNE DR
RADIOASTRONOMIE MILLIMETR
ENS. 24 RUE LHOMOND
75231 PARIS CEDEX 05
FRANCE
TEL: 43 29 12 25
TLX: 202601 F NORMSUP
TLF:
EML: CASOLI@FRULM11
CON:

CASSATELLA ANGELO DR
LAB ASTROFISICA SPAZIALE
C P 67
I-00044 FRASCATI. ROMA
ITALY
TEL:
TLX:
TLF:
EML:
CON: 29C

CASSE MICHEL DR
SECTION D'ASTROPHYSIQUE
CEN SACLAY
BP NO 2
F-91190 GIF-S-YVETTE
FRANCE
TEL:
TLX:
TLF:
EML:
CON: 48

CASSINELLI JOSEPH P DR
ASTRONOMY DEPT
UNIVERSITY OF WISCONSIN
475 N. CHARTER ST.
MADISON WI 53706
U.S.A.
TEL: 608-262-1752
TLX: 265452 UOFWISC MDS
TLF:
EML:
CON: 36C

CASTELLANI VITTORIO PROF
ISTITUTO ASTRONOMICO
UNIVERSITA DI ROMA
VIA LANCISI 29
I-00161 ROMA
ITALY
TEL: 06-86-7525
TLX:
TLF:
EML:
CON: 35.37

CASTELLI FIORELLA DR
OSSERVATORIO ASTRONOMICO
VIA TIEPOLO 11
I-34131 TRIESTE
ITALY
TEL: 040-793921
TLX: 461137 OAT I
TLF:
EML:
CON: 29

CASTELLI JOHN P
AFGL-PHP
HANSCOM AFB
BEDFORD MA 01731
U.S.A.
TEL:
TLX:
TLF:
EML:
CON:

CASTETS ALAIN DR
OBSERVATOIRE DE GRENOBLE
CERMO BP 53X
38041 GRENOBLE CEDEX
FRANCE
TEL: 76 51 47 86
TLX:
TLF:
EML: EARN:CASTETS@FRAG51
CON: 40

CASTOR JOHN I DR
LAWRENCE LIVERMORE
NATIONAL LAB L-23
PO BOX 808
LIVERMORE CA 94550
U.S.A.
TEL: 415-422-4664
TLX: 910-3868339 LLNL
TLF:
EML:
CON: 35.36

CASWELL JAMES L DR
CSIRO
DIVISION OF RADIOPHYSICS
P.O.BOX 76
EPPING NSW 2121
AUSTRALIA
TEL: 02-868-0222
TLX: 26230 ASTRO AA
TLF:
EML:
CON: 33.34.40

CATALA CLAUDE DR
OBSERVATOIRE DE MEUDON
PLACE JULES JANSSEN
92195 MEUDON PRINC.CEDEX
FRANCE
TEL: 1-45 07 76 63
TLX: 204464
TLF:
EML:
CON: 29.36

CATALA MARIA ASUNCION DR
DPTO FISICA DE ATMOSFERA
UNIVERSIDAD DE BARCELONA
AVDA. DIAGONAL 645
08028 BARCELONA
SPAIN
TEL: 330-73-11/244
TLX:
TLF:
EML:
CON: 46

CATALAN MANUEL DR
INSTITUTO Y OBSERVATORIO
DE MARINA
SAN FERNANDO (CADIZ)
SPAIN
TEL: 883548
TLX: 76108
TLF:
EML:
COM: 04.08.31

CATALANO FRANCESCO A DR
ISTITUTO DI ASTRONOMIA
CITTA UNIVERSITARIA
I-95125 CATANIA
ITALY
TEL: 095-330533
TLX: 970359 ASTRCT I
TLF:
EML:
COM:

CATALANO SANTO DR
ISTITUTO DI ASTRONOMIA
CITTA UNIVERSITARIA
I-95125 CATANIA
ITALY
TEL: 095-330533
TLX: 970359 ASTRCT I
TLF:
EML:
COM: 16.29.42

CATARZI MARCO DR
ARCETRI OBSERVATORY
L.E. FERMI 5
FLORENCE
ITALY
TEL: 55 27521
TLX: 572268 ARCETR I
TLF:
EML: 38954::CATARZI
COM: 40

CATCHPOLE ROBIN M DR
S A A O
P O BOX 9
OBSERVATORY 7935
SOUTH AFRICA
TEL: 470025
TLX: 5720309
TLF:
EML:
COM: 27.29

CATO B TORGEY DR
NORDISK TELESATELLITSTAT.
BOX 107
S-457 00 TANUMSHEDE
SWEDEN
TEL: 46-525-291-55
TLX: 20164 NORDSAT S
TLF:
EML:
COM:

CATON DANIEL B DR
ASSISTANT PROFESSOR
OF PHYSICS AND ASTRONOMY
APPALACHIAN STATE UNIV.
BOONE NC 28608
U.S.A.
TEL: (704)262 2446
TLX: 888370 OR 62671500
TLF:
EML: BITNET:CATONDB@APPSTATE
COM:

CATURA RICHARD C DR
LOCKHEED RESEARCH LAB
DEPT 91-20 BLDG 255
3251 HANOVER ST
PALO ALTO CA 94304
U.S.A.
TEL: 415-858-4066
TLX: 346409 LMSC SUVL
TLF:
EML:
COM: 44.48

CAUGHLAN GEORGEANNE R
PHYSICS DEPT
MONTANA STATE UNIVERSITY
BOZEMAN MT 59717
U.S.A.
TEL: 406-994-6170
TLX:
TLF:
EML:
COM: 35.48

CAVALIERE ALFONSO G PROF
ASTROFISICA. DIP FISICA
II UNIVERSITA DI ROMA
VIA O. RAIMONDO
I-00173 ROMA
ITALY
TEL:
TLX:
TLF:
EML:
COM: 47.48

CAVALLINI FABIO
OSSERVATORIO ASTROFISICO
DI ARCETRI
LARGO E. FERMI 5
I-50125 FIRENZE
ITALY
TEL: 055-2752200
TLX: 572268
TLF:
EML:
COM: 12

CAWLEY MICHAEL DR
PHYSICS DEPARTMENT
ST PATRICKS COLLEGE
MAYNOOTH
CO KILDARE
IRELAND
TEL: 01 285222 EX499
TLX:
TLF:
EML: BITNET:MCAWLEY@VAX1.MAY.IE
COM:

CAYREL DE STROBEL GIUSA
OBSERVATOIRE DE PARIS
SECTION DE MEUDON
F-92195 MEUDON PL CEDEX
FRANCE
TEL: 1-45-07-78-63
TLX:
TLF:
EML:
COM: 29C.36

CAYREL ROGER DR
OBSERVATOIRE DE PARIS
61 AVE DE L'OBSERVATOIRE
F-75014 PARIS
FRANCE
TEL: 1-40-51-22-51
TLX: 270776
TLF:
EML:
COM: 29.36.50

CAZENAVE ANNY DR
CNRS-GRGS
18 AVENUE EDOUARD BELIN
F-31400 TOULOUSE
FRANCE
TEL: 61-27-40-11
TLX: 531081
TLF:
EML:
COM:

CAZZOLA PAOLO DR
OSSERVATORIO ASTRONOMICO
VICOLO DELL'OSSERVATORIO
I-35100 PADOVA
ITALY
TEL:
TLX:
TLF:
EML:
COM:

CEPOLA PAUL J DR
MAIL STATION 64
C S DRAPER LAB
555 TECHNOLOGY SQ
CAMBRIDGE MA 02139
U.S.A.
TEL: 617-258-1787
TLX:
TLF:
EML:
COM: 07

CELIS LEOPOLDO DR
PONTIF. UNIV. CATOLICA
DEPTO DE ASTRONOMIA
CASILLA 6014
SANTIAGO
CHILE
TEL:
TLX:
TLF:
EML:
COM: 25.45

CELLINO ALBERTO DR
OSSERVATORIO ASTRONOMICO
DI TORINO STRADA
OSSERVATORIO
PINO TORINESE 10025
ITALY
TEL: 011 842040
TLX: 213236 TOASTR
TLF:
EML: CELLINO@ASTTOZ.INFN.IT
COM: 15

CELNIKIER LUDWIK DR
OBSERVATOIRE DE PARIS
SECTION DE MEUDON
F-92195 MEUDON PL CEDEX
FRANCE
TEL: 1-45-07-74-20
TLX: 201571 LAM
TLF:
EML:
COM:

CENTRELLA JOAN M DR
DEPARTMENT OF PHYSICS
DREXEL UNIVERSITY
PHILADELPHIA PA 19104
U.S.A.
TEL: 215 895 2715
TLX:
TLF:
EML:
COM:

CEPLECHA ZDENEK DR
ASTRONOMICAL INSTITUTE
CZECH. ACAD. OF SCIENCES
OBSERVATORY
251 65 ONDREJOV
CZECHOSLOVAKIA
TEL: 724525
TLX: 121579
TLF:
EML:
COM: 15.22C

CEPPATELLI GUIDO DR
OSSERVATORIO ASTROFISICO
DI ARCETRI
LARGO E. FERMI 5
I-50125 FIRENZE
ITALY
TEL: 055-2752200
TLX: 572268
TLF:
EML:
COM: 12

CERNICHARO JOSE DR
CENTRO ASTRONOMICO DE
YEBES AP 148
E-19080 GUADALAJARA
SPAIN
TEL: 11 290311
TLX: 58279508 IRAM
TLF:
EML:
COM: 40

CERRONI PRISCILLA DR
INST. DI ASTROF. SPAZIALE
REPARTO DI PLANETOLOGIA
VIALE DELL'UNIVERSITA' 11
00185 ROMA
ITALY
TEL: 4956951
TLX: 610261 CNR FRA
TLF: 4454969
EML: BITNET:CARUSI@IRMUNISA
COM: 15

CERRUTI-SOLA MONICA
OSSERVATORIO ASTROFISICO
DI ARCETRI
LARGO E. FERMI 5
I-50125 FIRENZE
ITALY
TEL: 055-2752258
TLX: 572268 ARCETR I
TLF:
EML:
COM: 34

CERSOSIMO JUAN CARLOS DR
ARECIBO OBSERVATORY
POST OFFICE BOX 995
ARECIBO P.R. 00613
PUERTO RICO
U.S.A.
TEL:
TLX:
TLF:
EML:
COM: 34

CESARSKY CATHERINE J DR
INST RECHERCHE FONDAMENT.
CEN SACLAY
DPHG/SAP-BAT 28
F-91191 GIF/YVETTE CEDEX
FRANCE
TEL: 1-69-08-39-12
TLX: 690860
TLF:
EML:
COM: 34.48

CESARSKY DIEGO A DR
INSTITUT D'ASTROPHYSIQUE
98 BIS BD ARAGO
F-75014 PARIS
FRANCE
TEL: 1-43-20-14-25
TLX:
TLF:
EML:
COM: 34

CESTER BRUNO PROF
OSSERVATORIO ASTRONOMICO
VIA TIEPOLO 11
I-34131 TRIESTE
ITALY
TEL: 040-793921/221
TLX: 461137 OAT I
TLF:
EML:
COM: 26.42.45

CEVOLANI GIORDANO
FISBAT-CNR
VIA DE' CASTAGNOLI 1
I-40126 BOLOGNA
ITALY
TEL: 051-9593/94
TLX: 511350
TLF:
EML:
COM: 22

CHA DU JIN
PYONGYANG ASTRON OBS
ACADEMY OF SCIENCES DPRK
TAESONG DISTRICT
PYONGYANG
KOREA DPR
TEL:
TLX:
TLF:
EML:
COM: 08

CHA GI UNG
PYONGYANG ASTRON OBS
ACADEMY OF SCIENCES DPRK
TAESONG DISTRICT
PYONGYANG
KOREA DPR
TEL:
TLX:
TLF:
EML:
COM:

CHAFFEE FREDERIC H DR
MULTIPLE MIRROR TELES OBS
UNIVERSITY OF ARIZONA
TUCSON AZ 85721
U.S.A.
TEL:
TLX:
TLF:
EML:
COM:

CHAISSON ERIC J PROF
SPACE TELESC. SCI. INST.
HOPKINS HOMEWOOD CAMPUS
BALTIMORE MD 21218
U.S.A.
TEL: 301-338-4757
TLX: 6849101
TLF:
EML:
COM: 51

CHALABAEV ALMAS DR
OBSERVATOIRE DE HAUTE
PROVENCE
04870 ST MICHEL L'OBSERV
FRANCE
TEL: 92 76 63 68
TLX: 410690 F
TLF:
EML: CHALABAEV@FRONI51
COM: 28

CHAMARAUX PIERRE DR
OBSERVATOIRE DE PARIS
SECTION DE MEUDON
F-92195 MEUDON
FRANCE
TEL: 1-45-07-75-94
TLX: 270912
TLF:
EML:
COM: 28

CHAMBE GILBERT
OBSERVATOIRE DE PARIS
SECTION DE MEUDON
DASOP
F-92195 MEUDON PL CEDEX
FRANCE
TEL: 1-45-34-77-93
TLX:
TLF:
EML:
COM: 10.12

CHAMBERLAIN JOSEPH M DR
ADLER PLANETARIUM
1300 S. LAKE SHORE DR
CHICAGO IL 60605
U.S.A.
TEL: 312-322-0325
TLX:
TLF:
EML:
COM: 08.31.46

CHAMBERLAIN JOSEPH W PROF
SPACE PHYS & ASTRON DEPT
RICE UNIVERSITY
HOUSTON TX 77001
U.S.A.
TEL: 713-527-8101
TLX: 556457
TLF:
EML:
COM: 16.21.49

CHAMBLISS CARLSON R DR
DEPT OF PHYS SCIENCES
KUTZTOWN UNIVERSITY
KUTZTOWN PA 19530
U.S.A.
TEL: 215-683-4439
TLX:
TLF:
EML:
COM: 42

CHAN KWING LAM
APPLIED RESEARCH CORP.
8201 CORPORATE DRIVE
LANDOVER MD 20785
U.S.A.
TEL: 301-459-8442
TLX:
TLF:
EML:
COM: 12.35.36.40

CHANDRA SUBHASH
NIS PHILIPS LABS
345 SCARBOROUGH RD
BRIAR CLIFF NY 10510
U.S.A.
TEL:
TLX:
TLF:
EML:
COM:

CHANDRA SURESH DR
PHYSICS DEPARTMENT
GORAKHPUR UNIVERSITY
GORAKHPUR 273 009
INDIA
TEL:
TLX:
TLF:
EML:
COM: 10

CHANDRASEKHAR S PROF
LAB ASTRON & SPACE RES
933 E 56TH STREET
CHICAGO IL 60637
U.S.A.
TEL: 312-962-7860
TLX:
TLF:
EML:
COM: 35.48

CHANDRASEKHAR T DR
PHYSICAL RESEARCH
LABORATORY
AHMEDABAD 380 009
INDIA
TEL: 462129
TLX: 121397 PRL IN
TLF:
EML:
COM: 15

CHANG KYONGAE DR
DEPT. PHYSICS & OPTICS
CHUNGJU UNIVERSITY
CHUNGJU
KOREA. REPUBLIC
TEL:
TLX:
TLF:
EML:
COM: 47

CHANMUGAM GANESAN PROF
PHYS & ASTRONOMY DEPT
LOUISIANA STATE UNIV
BATON ROUGE LA 70803-4001
U.S.A.
TEL: 504-388-6894
TLX:
TLF:
EML:
COM: 42

CHAPMAN CLARK R DR
PLANETARY SCI INSTITUTE
2030 E SPEEDWAY
SUITE 201
TUCSON AZ 85719
U.S.A.
TEL: 602-881-0332
TLX:
TLF:
EML:
COM: 15.16

CHAPMAN GARY A DR
SAN FERNANDO OBSERVATORY
DEPT PHYSICS & ASTRONOMY
CALIFORNIA STATE UNIV.
NORTHRIDGE CA 91330
U.S.A.
TEL: 818-885-2775
TLX:
TLF:
EML:
COM: 10.12

CHAPMAN ROBERT D DR
CODE PD311
LYNDON JOHNSON SPACE CTR
HOUSTON TX 77058
U.S.A.
TEL:
TLX:
TLF:
EML:
COM: 15.42.44

CHAPRONT JEAN DR
BUREAU DES LONGITUDES
77 AVE DENFERT-ROCHEREAU
F-75014 PARIS
FRANCE
TEL: 1-40-51-22-71
TLX:
TLF:
EML:
COM: 04C.07C

CHAPRONT-TOUZE MICHELLE
BUREAU DES LONGITUDES
77 AVE DENFERT-ROCHEREAU
F-75014 PARIS
FRANCE
TEL: 1-40-51-22-66
TLX:
TLF:
EML:
COM: 04.07.20

CHARLES PHILIP ALLAN
DEPT OF ASTROPHYSICS
UNIVERSITY OF OXFORD
SOUTH PARKS ROAD
OXFORD OX1 3RQ
U.K.
TEL: 0865-511336x536
TLX: 83295 NUCLOX
TLF:
EML:
COM: 44

CHARVIN PIERRE PR
OBSERVATOIRE DE PARIS
61 AVE DE L'OBSERVATOIRE
F-75014 PARIS
FRANCE
TEL: 1-40-51-21-57
TLX: OBS PARIS 270 776
TLF:
EML:
COM: 09

CHASSEFIERE ERIC
SERVICE D'AERONOMIE DU
CNRS. BP 3
91371VERRIERES LE BUISSON
FRANCE
TEL: 64 47 42 11
TLX: 602400 F
TLF:
EML:
COM: 49

CHAU WAI Y PROF
PHYSICS DEPT
QUEEN'S UNIVERSITY
KINGSTON ONT K7L 3N6
CANADA
TEL: 613-547-3526
TLX:
TLF:
EML:
COM:

CHAVARRIA-K. CARLOS
INSTITUTO DE ASTRONOMIA
UNAM
APARTADO POSTAL 70-264
04510 MEXICO DF
MEXICO
TEL:
TLX:
TLF:
EML:
COM: 37

CHAVIRA ENRIQUE SR
INAOE
AP POSTALES 216 y 51
72000 PUEBLA. PUE.
MEXICO
TEL: 47-05-00
TLX:
TLF:
EML:
COM: 27

CHE-BONNENSTENGEL ANNE
SUELZBRACKRING 39A
D-2050 HAMBURG 80
GERMANY. F.R.
TEL: 040-7738550
TLX:
TLF:
EML:
COM:

CHECHETKIN VALERIJ M DR
INSTITUTE OF APLIED MATH
USSR ACADEMY OF SCIENCES
MIUSSKAYA SQ4
125047 MOSCOW
U.S.S.R.
TEL: 251 37 39
TLX:
TLF:
EML:
COM: 35.48

CHELLI ALAIN
UNAM
APARTADO POSTAL 70-264
04510 MEXICO DF
MEXICO
TEL: 548-3712/5306
TLX: 1760155
TLF:
EML:
COM: 09

CHEN BIAO
PURPLE MOUNTAIN
OBSERVATORY
NANJING
CHINA. PEOPLE'S REP.
TEL: 46700
TLX: 34144 PMONJ CN
TLF:
EML:
COM: 10.12.49

CHEN CHUAN-LE
BEIJING ASTRONOMICAL OBS
ACADEMIA SINICA
BEIJING
CHINA. PEOPLE'S REP.
TEL: 281698 BEIJING
TLX: 22040 BAOAS CN
TLF:
EML:
COM:

CHEN DAO-HAN
PURPLE MOUNTAIN
OBSERVATORY
NANJING
CHINA. PEOPLE'S REP.
TEL: 31096
TLX: 34144 PMONJ CN
TLF:
EML:
COM: 15.16C

CHEN HONGSHENG
BEIJING ASTRONOMICAL OBS
ACADEMIA SINICA
BEIJING
CHINA. PEOPLE'S REP.
TEL:
TLX: 9053
TLF:
EML:
COM: 40

CHEN JIAN-SHENG
BEIJING ASTRONOMICAL OBS
ACADEMIA SINICA
PEKING
CHINA. PEOPLE'S REP.
TEL:
TLX: 22040 BAOAS CN
TLF:
EML:
COM: 28.47

CHEN KWAN-YU PROF
DEPT PHYSICS & ASTRONOMY
UNIVERSITY OF FLORIDA
GAINESVILLE FL 32611
U.S.A.
TEL: 904-392-2055
TLX:
TLF:
EML:
COM: 42

CHEN PEISHENG
YUNNAN OBSERVATORY
P.O. BOX 110
KUNMING. YUNNAN PROVINCE
CHINA. PEOPLE'S REP.
TEL: 72946
TLX:
TLF:
EML:
COM: 36

CHEN XIAO-ZHONG
BEIJING PLANETARIUM
BEIJING
CHINA. PEOPLE'S REP.
TEL:
TLX:
TLF:
EML:
COM:

CHEN XING
SHANGHAI OBSERVATORY
ACADEMIA SINICA
SHANGHAI
CHINA. PEOPLE'S REP.
TEL: 380696
TLX: 33164 SHAO CN
TLF:
EML:
COM: 19

CHEN ZHEN
PURPLE MOUNTAIN
OBSERVATORY
NANJING
CHINA. PEOPLE'S REP.
TEL: 46700
TLX: 34144 PMONJ CN
TLF:
EML:
COM: 07.26.33

CHEN ZHENCHENG
BEIJING ASTRONOMICAL OBS
ACADEMIA SINICA
BEIJING
CHINA. PEOPLE'S REP.
TEL: 281698
TLX: 9053
TLF:
EML:
COM: 10.28

CHEN ZUN-GUI
BEIJING PLANETARIUM
BEIJING
CHINA. PEOPLE'S REP.
TEL:
TLX:
TLF:
EML:
COM: 41

CHENG CHUNG-CHIEH DR
NAVAL RESEARCH LABORATORY
CODE 4175CC
WASHINGTON DC 20375
U.S.A.
TEL: 202-767-2350
TLX:
TLF:
EML:
COM: 10.12

CHENG FU-HUA
CENTER FOR ASTROPHYSICS
UNIV SCIENCE & TECHNOLOGY
HEFEI. ANHUI
CHINA. PEOPLE'S REP.
TEL: 63300-526 HEFEI
TLX: 90028 USTC CN
TLF:
EML:
COM: 47

CHENG FU-ZHEN
CENTER FOR ASTROPHYSICS
UNIV SCIENCE & TECHNOLOGY
HEFEI. ANHUI PROVINCE
CHINA. PEOPLE'S REP.
TEL: 63300 HEFEIx987
TLX: 90028 USTC CN
TLF:
EML:
COM: 47

CHEREDNICHENKO V I DR
KIEV POLYTECHNICAL INST
252056 KIEV
U.S.S.R.
TEL:
TLX:
TLF:
EML:
COM: 15

CHEREPASHCHUK A M PROF
STERNBERG STATE
ASTRONOMICAL INSTITUTE
119899 MOSCOW
U.S.S.R.
TEL: 139-38-38
TLX:
TLF:
EML:
COM: 27.42C.45

CHERNEGA N A DR
OBSERVATORY OF THE
KIEV UNIVERSITY
OBSERVATORNAYA 3
252053 KIEV
U.S.S.R.
TEL: 26-23-91
TLX:
TLF:
EML:
COM: 08

CHERNYKH N S DR
CRIMEAN ASTROPHYSICAL
OBSERVATORY
NAUCHNY
334413 CRIMEA
U.S.S.R.
TEL:
TLX:
TLF:
EML:
COM: 20

CHERTOPRUD V E DR
HYDROMETEOROLOGICAL CTR
OF THE USSR
123376 MOSCOW
U.S.S.R.
TEL:
TLX:
TLF:
EML:
COM: 10

CHEVALIER CLAUDE DR
OBS DE HAUTE-PROVENCE
F-04870 ST-MICHEL-L'OBS.
FRANCE
TEL: 92-76-63-68
TLX: 410690
TLF:
EML:
COM: 35

CHEVALIER ROGER A DR
DEPT OF ASTRONOMY
UNIVERSITY OF VIRGINIA
PO BOX 3818
CHARLOTTESVILLE VA 2290?
U.S.A.
TEL: 804-924-4889
TLX:
TLF:
EML:
COM: 34

CHIAN ABRAHAM CHIAN-LONG
INST PESQUISAS ESPACIAIS
CAIXA POSTAL 515
12200 S. JOSE DOS CAMPOS
BRAZIL
TEL: 0123-22-9977
TLX: 011-33530
TLF:
EML:
COM: 48

CHIKADA YOSHIHIRO DR
NOBEYAMA RADIO OBS
TOKYO ASTRONOMICAL OBS
NOBEYAMA
NAGANO PREF 384-13
JAPAN
TEL: 267-98-2831
TLX: 3329005 TAONRO
TLF:
EML:
COM: 40

CHINCARINI GUIDO L DR
OSSERVATORIO ASTRONOMICO
VIA E BIANCHI 46
I-22055 MERATE
ITALY
TEL: 039-596412
TLX:
TLF:
EML:
COM: 28.47

CHINI ROLF
MPI FUER RADIOASTRONOMIE
AUF DEM HUEGEL 69
D-5300 BONN
GERMANY F.R.
TEL:
TLX: 886440
TLF:
EML:
COM: 34.40

CHIO CHOL ZONG
PYONGYANG ASTRON OBS
ACADEMY OF SCIENCES DPRK
TAESONG DISTRICT
PYONGYANG
KOREA DPR
TEL:
TLX:
TLF:
EML:
COM: 20

CHIOSI CESARE S DR
ISTITUTO DI ASTRONOMIA
UNIVERSITA DI PADOVA
I-35100 PADOVA
ITALY
TEL: 049-66-1499
TLX: 430176 UNDAPU I
TLF:
EML:
COM: 37

CHISTYAKOV VLADIMIR E DR
USSURIISK SOLAR STATION
PRIMORSKY KRAY
692533 GORNOTAEZHNOE
U.S.S.R.
TEL: 91121 USSURIISK
TLX: 213954 SOLNZE
TLF:
EML:
COM: 12

CHITRE DATTAKUMAR M DR
COMPUTER SCIENCES CORP.
8728 COLESVILLE RD
SILVER SPRING MD 20910
U.S.A.
TEL:
TLX:
TLF:
EML:
COM: 47

CHITRE SHASHIKUMAR M DR
TATA INSTITUTE OF
FUNDAMENTAL RESEARCH
HOMI BHABHA RD
BOMBAY 400 005
INDIA
TEL: 219111
TLX: 011-3009 TIFR IN
TLF:
EML:
COM: 35.48

CHIU HONG-YEE DR
MUDD BLDG RM 828
COLUMBIA UNIVERSITY
NEW YORK NY 10027
U.S.A.
TEL:
TLX:
TLF:
EML:
COM: 35

CHIU LIANG-TAI GEORGE
T J WATSON RESEARCH CTR
BOX 218
YORKTOWN HEIGHTS NY 10598
U.S.A.
TEL: 914-945-2436
TLX:
TLF:
EML:
COM: 24

CHIUDERI CLAUDIO PROF
DIP. DI ASTRONOMIA
UNIVERSITA DEGLI STUDI
LARGO E. FERMI 5
I-50125 FIRENZE
ITALY
TEL: 055-27521
TLX: 572268 ARCETR-I
TLF:
EML:
COM:

CHIUDERI-DRAGO FRANCA PR
OSSERVATORIO ASTROFISICO
DI ARCETRI
LARGO E. FERMI 5
I-50125 FIRENZE
ITALY
TEL: 055-2752251
TLX:
TLF:
EML:
COM: 10

CHIUMIENTO GIUSEPPE
OSS ASTRONOMICO DI TORINO
STRADA OSSERVATORIO 20
I-10025 PINO TORINESE
ITALY
TEL: 011-841067
TLX: 213236 TOASTR I
TLF:
EML:
COM: 08.19

CHKHIKVADZE IAKOB N
ABASTUMANI ASTROPHYSICAL
OBSERVATORY
383762 ABASTUMANI.GEORGIA
U.S.S.R.
TEL: 2-78
TLX: 127409
TLF:
EML:
COM: 35

CHLISTOVSKY FRANCA DR
OSSERVATORIO ASTRONOMICO
DI BRERA
VIA BRERA 28
I-20121 MILANO
ITALY
TEL:
TLX:
TLF:
EML:
COM: 08

CHMIELEWSKI YVES DR
OBSERVATOIRE DE GENEVE
CHEMIN DES MAILLETTES 51
CH-1290 SAUVERNY
SWITZERLAND
TEL: 22-55 26 11
TLX: 2 77 20 OBSG CH
TLF:
EML:
COM:

CHOCHOL DRAHOMIR
ASTRONOMICAL INSTITUTE
CZECH. ACAD. OF SCIENCES
059 60 TATRANSKA LOMNICA
CZECHOSLOVAKIA
TEL: 969-967866
TLX: 78277
TLF:
EML:
COM: 42.44

CHOE SEUNG URN DR
SEOUL NATL UNIVERSITY
EARTH SCIENCE/EDUCATION
SINLIM DONG. GWANG GU
SEOUL 151742
KOREA. REPUBLIC
TEL:
TLX:
TLF:
EML:
COM:

CHOI KYU-HONG
DEPT ASTRON & METEOROLOGY
YONSEI UNIVERSITY
134 SHINCHON. SUDAEMUN
SEOUL 120
KOREA. REPUBLIC
TEL: 02-392-0131
TLX:
TLF:
EML:
COM: 07.42

CHOI WON CHOL
PYONGYANG ASTRON OBS
ACADEMY OF SCIENCES DPRK
TEASONG DISTRICT
PYONGYANG
KOREA DPR
TEL: 5-3334. 5-3239
TLX:
TLF:
EML:
COM:

CHOLLET FERNAND DR
OBSERVATOIRE DE PARIS
61 AVE DE L'OBSERVATOIRE
F-75014 PARIS
FRANCE
TEL: 1-40-51-22-05
TLX: 270776 OBS PARIS
TLF:
EML:
COM: 04.08

CHOPINET MARGUERITE DR
57. RUE THIERS
F-92100 BOULOGNE
FRANCE
TEL: 1-47-61-11-44
TLX:
TLF:
EML:
COM: 34

CHOU CHIH-KANG DR
ASTRONOMY.
NAT. CENTRAL UNIVERSITY
CHUNG-LI
TAIWAN. ROC
CHINA. TAIWAN
TEL: 886 3 4251375
TLX:
TLF:
EML:
COM: 28

CHOU DEAN-YI DR
PHYSICS DPT
TSING HUA UNIVERSITY
HSIN CHU 30043
CHINA. TAIWAN
TEL:
TLX:
TLF:
EML: BITNET:DYCHOU@TWNCTU01
COM:

CHOU KYONG CHOL PROF
KYUNG HEE UNIVERSITY
DEPT ASTRONOMY & SPACE
SCIENCE
SEOUL 131
KOREA. REPUBLIC
TEL: 966-0061/5
TLX:
TLF:
EML:
COM: 51

CHRISTENSEN PER R DR
NIELS BOHR INSTITUTE
BLEGDAMSVEJ 17
DK 2100 COPENHAGEN O
DENMARK
TEL: 31 42 16 16
TLX: 15216 NBI DK
TLF:
EML: PERRE@NBIVAX.NBI.DK
COM:

CHRISTENSEN-DALSGAARD J
INSTITUTE OF ASTRONOMY
UNIVERSITY OF AARHUS
DK-8000 AARHUS C
DENMARK
TEL: 06-12-88-99
TLX: 64767 AAUSCI DK
TLF:
EML:
COM: 12.27C.35

CHRISTIAN CAROL ANN
CANADA-FRANCE-HAWAII
TELESCOPE CORPORATION
BOX 1597
KAMUELA HI 96743
U.S.A.
TEL: 808-885-7944
TLX: 633147 CFHT
TLF:
EML:
COM: 37

CHRISTIANSEN WAYNE A
DEPT PHYSICS & ASTRONOMY
UNIVERSITY OF N. CAROLINA
CHAPEL HILL NC 27514
U.S.A.
TEL: 919-962-3011
TLX:
TLF:
EML:
COM: 40

CHRISTIANSEN WILBUR PROF
RMB 436
MAC'S REEF RD
VIA BUNGENDORE 2621
AUSTRALIA
TEL: 062-303287
TLX:
TLF:
EML:
COM: 40

CHRISTODOULOU DMITRIS DR
DPT OF PHYS. & ASTRONOMY
LOUISIANA STATE UNIV.
BATON ROUGE LA 70803-4001
U.S.A.
TEL: 504 388 2261679
TLX:
TLF:
EML: BITNET:PHCHRI@LSUMVS
COM: 33

CHRISTOPHE-GLAUME J DR
SERVICE D'AERONOMIE
BP NO 3
F-91370 VERRIERES-LE-B.
FRANCE
TEL: 1-69-20-10-60
TLX:
TLF:
EML:
COM:

CHRISTY JAMES WALTER DR
HUGHES AIRCRAFT CO.
1720 W. NIONA PL.
TUCSON AZ 85704
U.S.A.
TEL: 602-297-1377
TLX:
TLF:
EML:
COM: 09.24.45

CHRISTY ROBERT F DR
CALTECH
PASADENA CA 91125
U.S.A.
TEL: 213-795-6811
TLX:
TLF:
EML:
COM: 27.35

CHU HAN-SHU PROF.
PURPLE MOUNTAIN OBS
ACADEMIA SINICA
NANJING
CHINA. PEOPLE'S REP.
TEL: 86 25 301096
TLX: 34144 PMOMJ CN
TLF:
EML:
COM: 40

CHU YAOQUAN
CENTER FOR ASTROPHYSICS
UNIV. SCIENCE & TECHN.
HEFEI. ANHUI
CHINA. PEOPLE'S REP.
TEL:
TLX:
TLF:
EML:
COM: 28.47

CHU YOU-HUA
UNIVERSITY OF ILLINOIS
ASTRONOMY DEPT
1011 W. SPRINGFIELD AVE.
URBANA IL 61801
U.S.A.
TEL: 217-333-5535
TLX:
TLF:
EML:
COM: 34

CHUBB TALBOT A DR
5023 N. 38TH STREET
ARLINGTON VA 22207
U.S.A.
TEL:
TLX:
TLF:
EML:
COM: 44.48

CHUGAIJ NIKOLAI N DR
ASTRONOMICAL COUNCIL
USSR ACADEMY OF SCIENCES
PYATNITSKAYA 48
MOSCOW USSR 109017
U.S.S.R.
TEL: 231 21 29
TLX: 411576 ASCON SU
TLF:
EML:
COM: 28.36

CHUGAJNOV P F DR
CRIMEAN ASTROPHYSICAL OBS
USSR ACADEMY OF SCIENCES
NAUCHNIIY
334413 CRIMEA
U.S.S.R.
TEL:
TLX:
TLF:
EML:
COM: 25.27

CHUN MUN-SUK DR
DEPT ASTR & METEOROLOGY
COLLEGE OF SCIENCE
YONSEI UNIVERSITY
SEOUL
KOREA. REPUBLIC
TEL: 2-392-0131
TLX:
TLF:
EML:
COM: 37

CHUPP EDWARD L DR
PHYSICS DEPT
UNIV OF NEW HAMPSHIRE
DEMERITT HALL
DURHAM NH 03824
U.S.A.
TEL: 603-862-2750
TLX: 950030
TLF:
EML:
COM: 10.48

CHURCHWELL EDWARD B DR
WASHBURN OBSERVATORY
UNIVERSITY OF WISCONSIN
475 N. CHARTER ST
MADISON WI 53706
U.S.A.
TEL: 608-262-7857
TLX: 265452 UOFWISC MDS
TLF:
EML:
COM: 33.34

CHURMS JOSEPH
S A A O
P O BOX 9
OBSERVATORY 7935
SOUTH AFRICA
TEL: 021-47-0025
TLX:
TLF:
EML:
COM: 20.24

CHUVAEV K K DR
CRIMEAN ASTROPHYSICAL OBS
USSR ACADEMY OF SCIENCES
NAUCHNY
334413 CRIMEA
U.S.S.R.
TEL:
TLX:
TLF:
EML:
COM: 28

CHVOJKOVA WOYK E DR
ASTRONOMICAL INSTITUTE
CZECH. ACAD. OF SCIENCES
BUDECSKA 6
120 23 PRAHA 2
CZECHOSLOVAKIA
TEL:
TLX:
TLF:
EML:
COM: 12

CIATTI FRANCO DR
OSSERVATORIO ASTROFISICO
I-36012 ASIAGO VI
ITALY
TEL: 0424-62665
TLX: 430110 SETURIST
TLF:
EML:
COM:

CID PALACIOS RAFAEL PROF
FACULTAD DE CIENCIAS
DEPT DE ASTRONOMIA
CIUDAD UNIVERSITARIA
50009 ZARAGOZA
SPAIN
TEL: 357011
TLX:
TLF:
EML:
COM: 07

CILLIE G G PROF
4 MINSERIE STREET
STELLENBOSCH 7600
SOUTH AFRICA
TEL: 02231-3515
TLX:
TLF:
EML:
COM: 42

CIMINO MASSIMO A PROF
VIA A. CADLOLO 19
I-00136 ROMA
ITALY
TEL: 06-34-92-598
TLX:
TLF:
EML:
COM: 10.41

CIURLA TADEUSZ
ASTRONOMICAL INSTITUTE
WROCLAW UNIVERSITY
UL. KOPERNIKA 11
51-622 WROCLAW
POLAND
TEL:
TLX:
TLF:
EML:
COM: 33

CLAIRENIDI JACQUES DR
OBSERVATOIRE DE BESANCON
41 BIS AVENUE DE
L'OBSERVATOIRE
25000 BESANCON
FRANCE
TEL: 81 66 69 00
TLX: 361144 F
TLF:
EML: CLAIRENI@FROBES51
COM: 15.16.21

CLARIA JUAN DR
OBSERVATORIO ASTRONOMICO
LAPRIDA 854
5000 CORDOBA
ARGENTINA
TEL: 26876 OR 40613
TLX: 51822 BUCOR
TLF:
EML:
COM: 37C.42.45.47

CLARK ALFRED JR PROF
UNIVERSITY OF ROCHESTER
DEPT OF MECHANICAL
ROCHESTER NY 14627
U.S.A.
TEL:
TLX:
TLF:
EML:
COM:

CLARK BARRY G DR
NRAO
VLA PROJECT
PO BOX 0
SOCORRO NM 87801
U.S.A.
TEL: 505-772-4011
TLX: 9109881710
TLF:
EML:
COM: 40

CLARK DAVID H DR
SIENCE DIVISION. SCIENCE
NORTH STAR AVE.
SWINDON SN2 1ET
U.K.
TEL: 0793-26222
TLX: 449466
TLF:
EML:
COM: 40

CLARK FRANK OLIVER DR
DEPT PHYSICS & ASTRONOMY
UNIVERSITY OF KENTUCKY
LEXINGTON KY 40506
U.S.A.
TEL: 606-257-3376
TLX:
TLF:
EML:
COM: 34.40

CLARK GEORGE W PROF
MIT
ROOM 37-611
CAMBRIDGE MA 02139
U.S.A.
TEL: 617-253-5842
TLX:
TLF:
EML:
COM: 44C.48C

CLARK THOMAS A DR
NASA/GSFC
CODE 974
GREENBELT MD 20771
U.S.A.
TEL: 301-286-5957
TLX:
TLF:
EML:
COM: 51

CLARK THOMAS ALAN DR
UNIVERSITY OF CALGARY
PHYSICS DEPARTMENT
2500 UNIVERSITY DRIVE NW
CALGARY ATL T2N 1N4
CANADA
TEL: 403-284-5392
TLX:
TLF:
EML:
COM: 12.44

CLARKE DAVID DR
THE UNIVERSITY
DEPARTMENT OF ASTRONOMY
GLASGOW G12 8QQ
U.K.
TEL: 41-339-8855
TLX: 778421
TLF:
EML:
COM: 09

CLARKE JOHN T
NASA/GSFC
HUBBLE SPACE TELESCOPE
CODE 681
GREENBELT MD 20771
U.S.A.
TEL: 301-286-5781
TLX: 710-828-9716
TLF:
EML:
COM:

CLARKE THOMAS R DR
MCLAUGHLIN PLANETARIUM
ROYAL ONTARIO MUSEUM
100 QUEENS PARK CRESCENT
TORONTO ONT M5S 2C6
CANADA
TEL: 416-998-8551
TLX:
TLF:
EML:
COM:

CLAUSEN JENS VIGGO LEKTOR
COPENHAGEN UNIVERSITY OBS
BRORFELDEVEJ 23
DK-4340 TOLLOSE
DENMARK
TEL: 45-3-488195
TLX: 44155 DANAST
TLF:
EML:
COM: 42

CLAUZET LUIZ B FERREIRA
INST ASTRON. E GEOFISICO
UNIVERSIDADE DE SAO PAULO
CAIXA POSTAL 30627
01051 SAO PAULO
BRAZIL
TEL:
TLX:
TLF:
EML:
COM: 08

CLAVEL JEAN
IUE OBSERVATORY
ESA
APARTADO 54065
28080 MADRID
SPAIN
TEL: 34-1-401-9661
TLX: 42444 VILSE
TLF:
EML:
COM: 28

CLAYTON DONALD D PROF
DEPT SPACE PHYS & ASTRON
RICE UNIVERSITY
HOUSTON TX 77001
U.S.A.
TEL: 713-527-8101
TLX:
TLF:
EML:
COM:

CLAYTON GEOFFREY C DR
NASA HEADQUARTERS
CODE EZ
WASHINGTON DC 20546
U.S.A.
TEL: 202 453 1469
TLX:
TLF:
EML: SPAN:POLLUX::GCLAYTON
COM: 15

CLAYTON ROBERT N DR
ENRICO FERMI INSTITUTE
UNIVERSITY OF CHICAGO
CHICAGO IL 60637
U.S.A.
TEL: 312 702 7777
TLX:
TLF:
EML:
COM: 15

CLEGG PETER E DR
QUEEN MARY COLLEGE
MILE END ROAD
LONDON E1 4NS
U.K.
TEL: 01-980-4811
TLX: 893-750 QMCUOL
TLF:
EML:
COM:

CLEGG ROBIN E S DR
UNIVERSITY COLLEGE LONDON
DEPT PHYSICS & ASTRONOMY
GOWER STREET
LONDON WC1E 6BT
U.K.
TEL: 01-387-7050x382
TLX: 28722 UCPHYS G
TLF:
EML:
COM: 34

CLEMENS DAN P DR
DEPT. OF ASTRONOMY
BOSTON UNIVERSITY
725 COMMONWEALTH AVENUE
BOSTON MA 02215
U.S.A.
TEL: 617 353 6140
TLX:
TLF:
EML:
COM: 33.40

CLEMENT MAURICE J PROF
UNIVERSITY OF TORONTO
DEPARTMENT OF ASTRONOMY
TORONTO ONT M5S 1A7
CANADA
TEL: 416-978-4833
TLX: 06-986766
TLF:
EML:
COM:

CLIFTON KENNETH ST
NASA MARSHALL SPACE
FLIGHT CENTER ES 63
HUNTSVILLE AL 35812
U.S.A.
TEL: 205-453-2305
TLX: 594416
TLF:
EML:
COM: 22

CLINENHAGA JOHN L PROF
UNIVERSITY OF VICTORIA
DEPARTMENT OF PHYSICS
VICTORIA BC V8W 2Y2
CANADA
TEL: 604-721-7741
TLF: 0497222
TLF:
EML:
COM: 29

CLINE THOMAS L DR
LHEA
CODE 661
NASA GSFC
GREENBELT MD 20771
U.S.A.
TEL: 301 286 8375
TLX: 89675 NASCOM GBLT
TLF:
EML: SPAN:6197::CLINE
COM:

CLIVER EDWARD W
US AIR FORCE GEOPHYS LAB
SPACE PHYSICS DIVISION
HANSCOM AIR FORCE BASE
BEDFORD MA 01731
U.S.A.
TEL: 617-861-3975
TLX: 928123 AFGL HANSCOM
TLF:
EML:
COM: 10

CLUBE S V M DR
UNIVERSITY OF OXFORD
DEPT OF ASTROPHYSICS
SOUTH PARKS ROAD
OXFORD OX1 3RQ
U.K.
TEL: 0865-511336
TLX:
TLF:
EML:
COM: 15.22.24.33

CLUTTON-BROCK MARTIN DR
UNIVERSITY OF MANITOBA
DEPARTMENT OF MATHEMATICS
WINNIPEG MB R3T 2N2
CANADA
TEL: 204-261-9255
TLX:
TLF:
EML:
COM:

COCHRAN ANITA L DR
ASTRONOMY DEPARTMENT
THE UNIVERSITY OF TEXAS
AUSTIN TX 78712
U.S.A.
TEL: 512 471 1471
TLX: 910 874 1351
TLF:
EML: BITNET:aslj720@uta3081
COM: 15.16

COCHRAN WILLIAM DAVID DR
UNIVERSITY OF TEXAS
DEPARTMENT OF ASTRONOMY
AUSTIN TX 78712
U.S.A.
TEL: 512-471-4461
TLX:
TLF:
EML:
COM: 15.30

COCKE WILLIAM JOHN PROF
STEWARD OBSERVATORY
UNIVERSITY OF ARIZONA
TUCSON AZ 85721
U.S.A.
TEL: 602-621-6540
TLX:
TLF:
EML:
COM: 47

CODE ARTHUR D
WASHBURN OBSERVATORY
UNIVERSITY OF WISCONSIN
475 N. CHARTER ST
MADISON WI 53706
U.S.A.
TEL: 608-262-9594
TLX:
TLF:
EML:
COM: 29.34.44

CODINA LANDABERRY SAYD J
OBSERVATORIO NACIONAL
RUA GENERAL BRUCE 586
20921 RIO DE JANEIRO RJ
BRAZIL
TEL: 580-7313 x 267
TLX: 21288
TLF:
EML:
COM: 46

CODINA VIDAL J M DR
FABRA OBSERVATORY
GRAN VIA DE LOS CORTES
CATALANES 679
08013 BARCELONA
SPAIN
TEL: 34-3-2454766
TLX:
TLF:
EML:
COM:

COELHO BALSA MARIO C DR
RUA TRINDADE COELHO 21
2o DTO
3000 COIMBRA
PORTUGAL
TEL:
TLX:
TLF:
EML:
COM:

COFFEEN DAVID L DR
BOX 151
HASTINGS/HUDSON NY 10706
U.S.A.
TEL: 914-478-2594
TLX:
TLF:
EML:
COM:

COFFEY HELEN E MS
NOAA
NGDC E/GC2
325 BROADWAY
BOULDER CO 80303
U.S.A.
TEL: 303-497-6223
TLX: 592811 NOAA MASC BDR
TLF:
EML:
COM: 10

COGAN BRUCE C DR
MOUNT STROMLO OBSERVATORY
PRIVATE BAG
WODEN P.O. ACT 2606
AUSTRALIA
TEL: 062-88-1111
TLX: 62270
TLF:
EML:
COM: 27

COHEN JEFFREY M DR
UNIV OF PENNSYLVANIA
PHYSICS DEPARTMENT
PHILADELPHIA PA 19174
U.S.A.
TEL:
TLX:
TLF:
EML:
COM: 35.47.48

COHEN JUDITH DR
KITT PEAK NATIONAL OBS
P O BOX 26732
TUCSON AZ 85726
U.S.A.
TEL:
TLX:
TLF:
EML:
COM:

COHEN LEON PROF
HUNTER COLLEGE
DEPARTMENT OF PHYSICS
695 PARK AVE
NEW YORK NY 10021
U.S.A.
TEL: 212-570-5696
TLX:
TLF:
EML:
COM:

COHEN MARSHALL H PROF
CALTECH
105-24
PASADENA CA 91125
U.S.A.
TEL: 213-356-4000
TLX: 675425
TLF:
EML:
COM: 34.40

COHEN MARTIN DR
UNIVERSITY OF CALIFORNIA
RADIO ASTRONOMY LAB
601 CAMPBELL HALL
BERKELEY CA 94720
U.S.A.
TEL: 415-642-2833
TLX: 820181 UCB AST RALUD
TLF:
EML:
COM: 27

COHEN RAYMOND J DR
NUFFIELD RADIO ASTR LABS
JODRELL BANK
MACCLESFIELD SK11 9DL
U.K.
TEL: 0477-71321
TLX: 36149
TLF:
EML:
COM: 40

COHEN RICHARD S
COLUMBIA UNIVERSITY
INST FOR SPACE STUDIES
2880 BROADWAY
NEW YORK NY 10025
U.S.A.
TEL: 212-678-5611
TLX:
TLF:
EML:
COM: 09.33.40

COHEN ROSS D DR
UNIVERSITY OF CALIFORNIA
SAN DIEGO
CASS DEPT C 011
LA JOLLA CA 92093
U.S.A.
TEL: 619 534 2664
TLX:
TLF:
EML:
COM: 28.47

COHN HALDAN N
INDIANA UNIVERSITY
DEPT OF ASTRONOMY
SWAIN WEST 319
BLOOMINGTON IN 47405
U.S.A.
TEL: 812-335-4174
TLX:
TLF:
EML:
COM:

COLBURN DAVID S DR
1944 WAVERLEY STREET
PALO ALTO CA 94301
U.S.A.
TEL:
TLX:
TLF:
EML:
COM:

COLE TREVOR WILLIAM PROF
SCHOOL OF ELECTRICAL ENG
UNIVERSITY OF SYDNEY
SYDNEY NSW 2006
AUSTRALIA
TEL: 02-692-2682
TLX:
TLF:
EML:
COM: 40

COLEMAN PAUL HENRY DR
KAPTEYN ASTRON. INST.
PO BOX 800
9700 AV GRONINGEN
NETHERLANDS
TEL: 50 634064
TLX: 53572 STARS NL
TLF:
EML: BITNET:GRUFFONGRRUG5
COM: 40

COLGATE STIRLING A DR
THEORETICAL DIVISION
LOS ALAMOS SCIENTIFIC LAB
MS 275 B
LOS ALAMOS NM 87545
U.S.A.
TEL: 505-667-2897
TLX:
TLF:
EML:
COM:

COLIN JACQUES DR
OBSERVATOIRE DE BORDEAUX
B.P. 21
F-33270 FLOIRACH
FRANCE
TEL: 56-86-43-30
TLX:
TLF:
EML:
COM: 28.37.33

COLLADOS MANUEL DR
INSTITUTO DE ASTROFISICA
DE CANARIAS
38200 LA LAGUNA TENERIFE
SPAIN
TEL: 34 22 26 22 11
TLX: 92640
TLF:
EML:
COM: 10.12

COLLIN-SOUFFRIN SUZY DR
INSTITUT D'ASTROPHYSIQUE
98 BIS. BOULEVARD ARAGO
F-75014 PARIS
FRANCE
TEL: 1-43-20-14-25
TLX:
TLF:
EML:
COM: 34.48

COLLINS GEORGE W II PROF
THE OHIO STATE UNIVERSITY
174 W. 18TH AVENUE
COLUMBUS OH 43210
U.S.A.
TEL: 614-422-5467
TLX:
TLF:
EML:
COM: 42

COLLINSON EDWARD H
THE COPSE
CHURCH LANE
PLAYFORD
IPSWICH. SUFFOLK IP6 9DB
U.K.
TEL: 62-22-57 IPSWIC
TLX:
TLF:
EML:
COM: 16

COLOMB FERNANDO R DR
INSTITUTO ARGENTINO DE
RADIOASTRONOMIA
CASILLA DE CORREO NO 5
1894 VILLA ELISA (Bs.As.)
ARGENTINA
TEL: 021-43793
TLX: 18052 CICYT AR
TLF:
EML:
COM: 34

COLOMBO G PROF DR
ISTITUTO MECCANICA APPL
UNIVERSITA DI PADOVA
VIA F. MARZOLO 9
I-35100 PADOVA
ITALY
TEL:
TLX:
TLF:
EML:
COM: 16

COLUZZI REGINA DR
OSSERVATORIO ASTRONOMICO
DI ROMA. VIALE DEL PARCO
MELLINI 84
00136 ROMA
ITALY
TEL: (06) 34 70 56
TLX: 626226 OA ROMA I
TLF:
EML:
COM: 05.45

COMA JUAN CARLOS
INSTITUTO Y OBSERVATORIO
DE MARINA
SAN FERNANDO (CADIZ)
SPAIN
TEL:
TLX:
TLF:
EML:
COM: 04

COMBES FRANCOISE DR
OBSERVATOIRE DE PARIS
SECTION DE MEUDON
DEMIRM
F-92195 MEUDON PL CEDEX
FRANCE
TEL: 1-45-07-78-98
TLX: 270912 OBSASTR
TLF:
EML:
COM:

COMBES MICHEL
OBSERVATOIRE DE PARIS
SECTION DE MEUDON
F-92195 MEUDON PL CEDEX
FRANCE
TEL: 1-45-07-76-91
TLX: 201571 LAM
TLF:
EML:
COM:

COMBI MICHAEL R DR
AER INC
840 MEMORIAL DRIVE
CAMBRIDGE MA 02139
U.S.A.
TEL: 617 547 6207
TLX:
TLF:
EML:
COM: 15.16

COMINS NEIL FRANCIS
UNIVERSITY OF MAINE
DEPT PHYSICS & ASTRONOMY
BENNETT HALL
ORONO ME 04469
U.S.A.
TEL: 207-581-1037
TLX:
TLF:
EML:
COM: 33

CONORETTO GIOVANNI
OSSERVATORIO ASTROFISICO
DI ARCETRI
LARGO E FERMI 5
FIRENZE 50125
ITALY
TEL: 55 27521
TLX: 572268 ARCETR I
TLF:
EML:
COM:

CONTE GEORGES DR
OBSERVATOIRE DE MARSEILLE
2 PLACE LE VERRIER
13248 MARSEILLE CEDEX 4
FRANCE
TEL: 91 95 90 88
TLX: 420241 F
TLF:
EML: EARN::"CONTE@FROMRS51"
COM: 28

CONCONI PAOLO DR
OSSERVATORIO ASTRONOMICO
DI BRERA
I-22055 MILANO
ITALY
TEL:
TLX:
TLF:
EML:
COM:

CONDON JAMES J DR
NRAO
EDGEMONT ROAD
CHARLOTTESVILLE VA 22903
U.S.A.
TEL: 804-296-0211
TLX: 910-997-0174
TLF:
EML:
COM: 40.47.48

CONKLIN EDWARD K
FORTH INC
111 N. SEPULVEDA BLVD 300
MANHATTAN BEACH CA 90266
U.S.A.
TEL:
TLX: 275182 FORT UR
TLF:
EML:
COM: 40

CONNES JANINE DR
CIRCE
BP 63
F-91406 ORSAY CEDEX
FRANCE
TEL: 1-69-28-76-75
TLX: FACORS 692166 F
TLF:
EML:
COM: 16

CONNES PIERRE DR
SERVICE D'AERONOMIE
B P 3
F-91370 VERRIERES-LE-B.
FRANCE
TEL: 1-64-47-42-77
TLX: 692400
TLF:
EML:
COM: 24.51C

CONNOLLY LEO PAUL
DEPT. OF PHYSICS
CALIFORNIA ST. UNIVERSITY
5500 UNIVERSITY PARKWAY
SAN BERNARDINO CA 92407
U.S.A.
TEL: 714-880-5400
TLX:
TLF:
EML:
COM: 27.35

CONTADAKIS MICHAEL E DR
DEPT GEODESY & SURVEYING
UNIVERSITY THESSALONIKI
UNIV. BOX 503
GR-54006 THESSALONIKI
GREECE
TEL: 003-031-99-2693
TLX: 412181 AUTH GR
TLF:
EML:
COM: 27

CONTI PETER S DR
JILA
UNIVERSITY OF COLORADO
BOX 440
BOULDER CO 80309
U.S.A.
TEL: 303-492-8913
TLX: 755842 JILA
TLF:
EML:
COM: 29P.36

CONTOPOULOS GEORGE PROF
UNIVERSITY OF ATHENS
ASTRONOMY DEPARTMENT
PANEPISTIMIOPOLIS
GR-15771 ATHENS
GREECE
TEL: 01-7243-211
TLX:
TLF:
EML:
COM: 07.28.33

CONWAY ROBIN G DR
NUFFIELD RADIO ASTR LABS
JODRELL BANK
MACCLESFIELD SK11 9DL
U.K.
TEL: 0477-71321
TLX: 36149
TLF:
EML:
COM: 40

COOK ALAN H PROF
DEPT PHYS/UNIV CAMBRIDGE
THE MASTER'S LODGE
SELWYN COLLEGE
CAMBRIDGE CB3 9DQ
U.K.
TEL: 223-62381 Ex 29
TLX: 81292 CAVLAB
TLF:
EML:
COM: 07.14

COOK JOHN W
8032 SLEEPY VIEW LN
SPRINGFIELD VA 22153
U.S.A.
TEL: 202-767-2161
TLX:
TLF:
EML:
COM: 10.12

COOKE B A DR
LEICESTER UNIVERSITY
X-RAY ASTRONOMY GROUP
PHYSICS DEPARTMENT
LEICESTER LE1 7RH
U.K.
TEL: 533-554455 E188
TLX: 341664
TLF:
EML:
COM:

COOKE JOHN ALAN
UNIVERSITY OF EDINBURGH
DEPT OF ASTRONOMY
ROYAL OBSERVATORY
EDINBURGH EH9 3HJ
U.K.
TEL: 031-667-3221
TLX: 72383 ROEDIN G
TLF:
EML:
COM: 09

CORADINI ANGIOLETTA
IST ASTROFISICA SPAZIALE
REP PLANETOLOGIA
VIALE UNIVERSITA 11
I-00185 ROMA
ITALY
TEL: 06-495-6951
TLX: 680489 CNR FRA
TLF:
EML:
COM:

CORBALLY CHRISTOPHER
VATICAN OBS. RES. GROUP
STEWARD OBSERVATORY
UNIVERSITY OF ARIZONA
TUCSON AZ 85721
U.S.A.
TEL: 602-621-3225
TLX: 467175
TLF:
EML:
COM: 45C

CORBIN THOMAS ELBERT DR
US NAVAL OBSERVATORY
ASTROMETRY DEPT
WASHINGTON DC 20390
U.S.A.
TEL: 202-653-1557
TLX: 710-8221970
TLF:
EML:
COM: 08.24

CORDES JAMES M
CORNELL UNIVERSITY
SPACE SCIENCE BUILDING
ITHACA NY 14853
U.S.A.
TEL: 607-256-3734
TLX: 932 458
TLF:
EML:
COM: 40

CORDOVA FRANCE A B
LOS ALAMOS NATIONAL LAB
MS D436
LOS ALAMOS NM 87545
U.S.A.
TEL: 505-667-3904
TLX:
TLF:
EML:
COM: 44

CORLISS C H DR
FOREST HILLS LABORATORY
2955 ALBEMARLE STREET NW
2955 ALBEMARLE STR N.W.
WASHINGTON DC 20008
U.S.A.
TEL: 202-362-6085
TLX:
TLF:
EML:
COM: 14

CORNEJO ALEJANDRO A DR
INAOE
AP POSTALES 216 y 51
72000 PUEBLA. PUE.
MEXICO
TEL: 47-05-00
TLX:
TLF:
EML:
COM: 09.41

CORNIDE MANUEL
DEPTO DE ASTROFISICA
FACULTAD DE FISICA
UNIVERSIDAD COMPLUTENSE
28040 MADRID
SPAIN
TEL: 449-53-16
TLX: 47273 FFUC
TLF:
EML:
COM:

CORNILLE MARGUERITE DR
UA 812 OBS.DE PARIS-
SECTION DE MEUDON
5 PLACE JULES JANSSEN
92195 MEUDON PRINC.CEDEX
FRANCE
TEL: 1-45 07 74 55
TLX: 201571
TLF:
EML: BITNET:FRORS31
COM: 14

CORWIN HAROLD G JR
UNIVERSITY OF TEXAS
DEPARTMENT OF ASTRONOMY
RLM 15.308
AUSTIN TX 78712-1083
U.S.A.
TEL: 512-471-4461
TLX: 9108741351
TLF:
EML:
COM: 28

COSMOVICI BATALLI C DR
IST FISICA SPAZIO INTERPL
CNR
I-00044 FRASCATI (ROMA)
ITALY
TEL: 06-9423801
TLX: 610261 I
TLF:
EML:
COM: 15

COSTA EDGARDO DR
DEPTO DE ASTRONOMIA
UNIVERSIDAD DE CHILE
CASILLA 36-D
SANTIAGO
CHILE
TEL:
TLX:
TLF:
EML:
COM: 08.33

COSTA ENRICO
IST ASTROFISICA SPAZIALE
C P 67
I-00044 FRASCATI
ITALY
TEL: 06-942-5655
TLX: 610261 I
TLF:
EML:
COM:

COSTA VICTOR DR
INSTITUTO DE ASTROFISICA
DE ANDALUCIA
APARTADO 2144
18080 GRANADA
SPAIN
TEL: 58-121311
TLX: 78573 IAGG E
TLF:
EML:
COM:

COSTAIN CARMAN H DR
DOMINION RADIO ASTROPHY-
SICAL OBSERVATORY
BOX 248
PENTICTON BC V2A 6K3
CANADA
TEL: 604-497-5321
TLX: 048-88127
TLF:
EML:
COM: 40

COSTAIN CECIL C DR
DIVISION OF PHYSICS
NAT RES COUNCIL OF CANADA
OTTAWA K1A 0R6
CANADA
TEL:
TLX:
TLF:
EML:
COM:

COSTERO RAFAEL
INSTITUTO DE ASTRONOMIA
UNAM
APDO POSTAL 70-264
04510 MEXICO DF
MEXICO
TEL: 548-5305
TLX: 1760155 CICME
TLF:
EML:
COM: 34

COTTON WILLIAM D Jr
NRAO
EDGEMONT ROAD
CHARLOTTESVILLE VA 22901
U.S.A.
TEL: 804-296-0319
TLX: 5105875482
TLF:
EML:
COM: 40

COTTRELL PETER LEDSAM
DEPT OF PHYSICS
UNIVERSITY OF CANTERBURY
CHRISTCHURCH 1
NEW ZEALAND
TEL: 03-482-009
TLX: 4144 NZ
TLF:
EML:
COM: 29

COUCH WARRICK DR
ANGLO-AUSTRALIAN OBS
P.O. BOX 296
EPPING NSW 2121
AUSTRALIA
TEL: 02-868-1666
TLX: 23999 AAOSYD AA
TLF:
EML:
COM: 28

COULSON IAIN M DR
UNITED KINGDOM TELESCOPES
665 KOMOHANA STREET
HILO HI 96720
U.S.A.
TEL:
TLX:
TLF:
EML:
COM: 27

COUNSELMAN CHARLES C PROF
MIT
DEPT EARTH & PLANET SCI
ROOM 54-620
CAMBRIDGE MA 02139
U.S.A.
TEL: 617-253-7902
TLX: 921473 MIT CAM
TLF:
EML:
COM: 07.08.16

COUPER HEATHER MISS
55 COLOMB STREET
GREENWICH
LONDON SE10 9HZ
U.K.
TEL: 1 853 0574
TLX:
TLF:
EML:
COM: 46.51

COUPINOT GERARD DR
OBS PIC-DU-MIDI/TOULOUSE
9 RUE DE LA MOULETTE
F-65200 BAGNERES
FRANCE
TEL: 61-95-19-69
TLX:
TLF:
EML:
COM:

COURTES GEORGES PROF
OBSERVATOIRE DE MARSEILLE
2 PLACE LE VERRIER
F-13248 MARSEILLE CEDEX 4
FRANCE
TEL: 91-95-90-88
TLX: 410594
TLF:
EML:
COM: 28.33.34.44

COURVOISIER THIERRY J.-L.
OBSERVATOIRE DE GENEVE
CH 1290 SAUVERNY
SWITZERLAND
TEL: 41 22 55 26 11
TLX:
TLF:
EML:
COM: 44

COUSINS A W J DR
S A A O
P O BOX 9
OBSERVATORY 7935
SOUTH AFRICA
TEL: 021-47-0025
TLX: 5720309
TLF:
EML:
COM: 25

COUTEAU PAUL PROF
OBSERVATOIRE DE NICE
BP 139
F-06003 NICE CEDEX
FRANCE
TEL: 93-89-04-20
TLX:
TLF:
EML:
COM: 26C

COUTREZ RAYMOND A J PROF
6 RUE EGIDE BOUVIER
B-1160 BRUXELLES
BELGIUM
TEL:
TLX:
TLF:
EML:
COM: 10.40

COUTTS-CLEMENT CHRISTINE
UNIVERSITY OF TORONTO
DEPARTMENT OF ASTRONOMY
TORONTO ONT M5S 1A7
CANADA
TEL: 1-416-978-5186
TLX:
TLF:
EML:
COM: 27

COVINGTON ARTHUR E
131 COLLEGE STREET
KINGSTON ON K7L 4L7
CANADA
TEL:
TLX:
TLF:
EML:
COM: 10.40

COWAN JOHN J DR
UNIVERSITY OF OKLAHOMA
DEPT PHYSICS & ASTRONOMY
NORMAN OK 73019
U.S.A.
TEL: 405-325-3961
TLX:
TLF:
EML:
COM: 35

COWIE LENNOX LAUCHLAN DR
SPACE TELESCOPE SCI INST
HOMEWOOD CAMPUS
3700 SAN MARTIN DR
BALTIMORE MD 21218
U.S.A.
TEL:
TLX:
TLF:
EML:
COM: 34 48

COWLEY ANNE P DR
ARIZONA STATE UNIVERSITY
PHYSICS DEPARTMENT
TEMPE AZ 85287
U.S.A.
TEL: 602-965-2919
TLX:
TLF:
EML:
COM: 29.42.45

COWLEY CHARLES R PROF
UNIVERSITY OF MICHIGAN
ASTRONOMY DEPARTMENT
ANN ARBOR MI 48109-1090
U.S.A.
TEL: 313-764-3437
TLX: 810-2236056
TLF:
EML:
COM: 29.36

COWLING THOMAS G PROF
19 HOLLIN GARDENS
HEADINGLEY
LEEDS LS16 5NL
U.K.
TEL: 785-342 LEEDS
TLX:
TLF:
EML:
COM: 35

COWSIK RAMANATH
TATA INSTITUTE OF
FUNDAMENTAL RESEARCH
HOMI BHABHA RD
BOMBAY 400 005
INDIA
TEL:
TLX:
TLF:
EML:
COM: 28.48

COX ARTHUR N DR
LOS ALAMOS NATIONAL LAB
P O BOX 1663
LOS ALAMOS NM 87545
U.S.A.
TEL: 505-667-7648
TLX: 910-988-1773
TLF:
EML:
COM: 12.27

COX DONALD P PROF
UNIVERSITY OF WISCONSIN
DEPT OF ASTRONOMY
1150 UNIVERSITY AVENUE
MADISON WI 53706
U.S.A.
TEL: 608-262-5916
TLX:
TLF:
EML:
COM: 34

COYNE GEORGE V DR
SPECOLA VATICANA
I-00120 VATICANO CITY
VATICAN CITY
TEL: 06-698-3411
TLX: 2020 VAT OBS
TLF:
EML:
COM: 25.34.50C

CRABTREE DENNIS DR
DAO CSADC
5071 W SAANICH ROAD
VICTORIA BC V8X 4M6
CANADA
TEL: 604 388 0025
TLX: 049 7295
TLF:
EML:
COM:

CRAIG IAN JONATHAN D DR
DEPT APPLIED MATHEMATICS
UNIVERSITY OF WAIKATO
HAMILTON
NEW ZEALAND
TEL: 62889
TLX:
TLF:
EML:
COM: 12

CRAINE ERIC RICHARD DR
WESTERN RESEARCH CO
5061 W CAMINO DE GIRASOL
TUCSON AZ 85745
U.S.A.
TEL: 602-743-7377
TLX:
TLF:
EML:
COM:

CRAM LAWRENCE EDWARD DR
ASTROPHYSICS DEPT
UNIVERSITY OF SYDNEY
SYDNEY NSW 2006
AUSTRALIA
TEL:
TLX:
TLF:
EML:
COM: 12.36C

CRAMPTON DAVID DR
DOMINION ASTROPHYS OBS
5071 W SAANICH ROAD
RR 5
VICTORIA BC V8X 4M6
CANADA
TEL: 604-388-3900
TLX: 0497295
TLF:
EML:
COM: 30.33.45

CRANE PATRICK C
NRAO
PO BOX O
SOCORRO NM 87801
U.S.A.
TEL: 505-772-4011
TLX: 910-988-1710
TLF:
EML:
COM: 40

CRANE PHILIPPE
ESO
KARL-SCHWARZSCHILD-STR 2
D-8046 GARCHING B MUNCHEN
GERMANY. F.R.
TEL: 49-89-79-20-98
TLX: 528-28222 EO D
TLF:
EML:
COM: 34

CRANNELL CAROL JO DR
NASA/GSFC
CODE 682
GREENBELT MD 20771
U.S.A.
TEL: 301-286-5007
TLX: 89675
TLF:
EML:
COM: 10.44

CRAWFORD DAVID L DR
KITT PEAK NATIONAL OBS
BOX 26732
950 N. CHERRY AVENUE
TUCSON AZ 85726
U.S.A.
TEL: 602-325-9346
TLX: 666-484 AURA NOAO
TLF:
EML:
COM: 09.25.33.45.50P

CREZE MICHEL DR
OBSERVATOIRE DE BESANCON
41 AVE DE L'OBSERVATOIRE
F-25000 BESANCON
FRANCE
TEL: 81-80-22-66
TLX: OBSBES 361144 F
TLF:
EML:
COM: 24.33

CRIFO FRANCOISE DR
OBSERVATOIRE DE PARIS
SECTION DE MEUDON
DEPEG
F-92195 MEUDON PL CEDEX
FRANCE
TEL: 1-45-07-78-34
TLX: 201571 F
TLF:
EML:
COM: 08.24

CRISTALDI SALVATORE DR
OSSERVATORIO ASTROFISICO
CITTA UNIVERSITARIA
VIALE ARTALE ALAGONA 75
I-95126 CATANIA
ITALY
TEL: 033-07-34
TLX: 970359 ASTRCT I
TLF:
EML:
COM: 47

CRISTESCU CORNELIA G DR
ASTRONOMICAL OBSERVATORY
CUTITUL DE ARGINT 5
75212 BUCAREST
RUMANIA
TEL: 23-68-92
TLX:
TLF:
EML:
COM: 15.20

CRISTIANI STEFANO DR
INSTITUTO DI ASTRONOMIA
DELLA UNIVERSI. DI PADOVA
VICOLO DELL OSSERV. 5
PADOVA I 35122
ITALY
TEL: 661499
TLX: 432071
TLF:
EML: SPAN:39003::CRISTIANI
COM:

CRIVELLARI LUCIO
OSSERVATORIO ASTRONOMICO
DI TRIESTE
VIA G.B. TIEPOLO 11
I-34131 TRIESTE
ITALY
TEL: 040-793221
TLX: 461137 OAT I
TLF:
EML:
COM: 29.36

CROOM DAVID L DR
RUTHERFORD APPLETON LAB
CHILTON DIDCOT OX11 0QX
U.K.
TEL: 0235-21900
TLX: 83159
TLF:
EML:
COM: 40

CROVISIER JACQUES
OBSERVATOIRE DE PARIS
SECTION DE MEUDON
F-92195 MEUDON PL CEDEX
FRANCE
TEL: 1-45-07-75-99
TLX: 270912
TLF:
EML:
COM: 15.34.40

CRUIKSHANK DALE P DR
NASA AMES RESEARCH CENTER
MS 245-6
MOFFETT FIELD CA 94035
U.S.A.
TEL:
TLX:
TLF:
EML:
COM: 15.16

CRUISE ADRIAN MICHAEL DR
RUTHERFORD & APPLETON
LABORATORIES
DITTON PARK
SLOUGH SL3 9JX
U.K.
TEL:
TLX:
TLF:
EML:
COM: 44

CRUTCHER RICHARD M DR
UNIV ILLINOIS ASTRON DEPT
341 ASTRON BLDG
1011 W. SPRINGFIELD AVE
URBANA IL 61801
U.S.A.
TEL: 217-333-9581
TLX:
TLF:
EML:
COM:

CRUVELLIER PAUL E DR
LAB ASTRONOMIE SPATIALE
TRAVERSE DU SIPHON
LES TROIS LUCS
F-13012 MARSEILLE
FRANCE
TEL: 91-66-08-32
TLX: 420584
TLF:
EML:
COM: 34

CRUZ-GONZALEZ IRENE
INSTITUTO DE ASTRONOMIA
UNAM
APDO POSTAL 70-264
04510 MEXICO DF
MEXICO
TEL: 905-548-5306
TLX:
TLF:
EML:
COM:

CSADA IMRE K DR
KONKOLY OBSERVATORY
THEGE UT 13/17
BOX 67
H-1121 BUDAPEST
HUNGARY
TEL: 166-426
TLX: 227460
TLF:
EML:
COM: 10

CUDABACK DAVID D DR
UNIVERSITY OF CALIFORNIA
RADIO ASTRONOMY LAB
BERKELEY CA 94720
U.S.A.
TEL: 415-642-5724
TLX: 820181 UCB AST
TLF:
EML:
COM: 34.40

CUDWORTH KYLE MCCABE DR
YERKES OBSERVATORY
UNIVERSITY OF CHICAGO
BOX 258
WILLIAMS BAY WI 53191
U.S.A.
TEL: 414-245-5555
TLX:
TLF:
EML:
COM: 24.33.37

CUFFEY J MR
NEW MEXICO STATE UNIV
DEPT EARTH SCI & ASTRON
UNIVERSITY PARK NM 88001
U.S.A.
TEL:
TLX:
TLF:
EML:
COM: 37

CUGIER HENRYK DR
ASTRONOMICAL INSTITUTEE
OF THE WROCKAW UNIVERSITY
UL KOPERNIKA 11
51 622 WROCLAW
POLAND
TEL: 48 24 34
TLX: 712791 UWR PL
TLF:
EML:
COM: 36

CUGNON PIERRE DR
OBS ROYAL DE BELGIQUE
AVENUE CIRCULAIRE 3
B-1180 BRUXELLES
BELGIUM
TEL: 2-375-24-84
TLX: 21565 OBSBEL
TLF:
EML:
COM: 34

CUGUSI LEONINO DR
DIPT DI SCIENZE FISICHE
UNIVERSITA DI CAGLIARI
VIA OSPEDALE 72
I-09100 CAGLIARI
ITALY
TEL: 070-664770
TLX:
TLF:
EML:
COM:

CUI CHUNFANG
WUHAN TECHNICAL UNIV. OF
SURVEYING AND MAPPING
WUHAN
CHINA. PEOPLE'S REP.
TEL:
TLX:
TLF:
EML:
COM: 07

CUI DOU-XING
CHANGCHUN ARTIFICIAL
SATELLITE OBSERVATORY
P.O. BOX 1067
CHANGCHUN
CHINA. PEOPLE'S REP.
TEL: 42859
TLX: 2421 CHANGCHUN
TLF:
EML:
COM: 07

CUI LIAN-SHU
DEPT OF ASTRONOMY
NANJING UNIVERSITY
NANJING
CHINA. PEOPLE'S REP.
TEL: 37551
TLX: 0909
TLF:
EML:
COM: 12

CUI ZHEN-HUA
BEIJING PLANETARIUM
138 XI ZHI MEN WAI STR.
WEST DISTRICT
BEIJING 01
CHINA. PEOPLE'S REP.
TEL: 89 3003
TLX:
TLF:
EML:
COM: 41.46

CUI ZHENXING
BEIJING ASTRONOMICAL OBS
BEIJING 100080
CHINA. PEOPLE'S REP.
TEL: 282100 BEIJING
TLX: 22040 BAOAS CN
TLF:
EML:
COM: 40

CULHANE LEONARD PROF
MULLARD SPACE SCIENCE LAB
UNIVERSITY COLLEGE LONDON
HOLMBURY ST MARY
DORKING RH5 6NS
U.K.
TEL: 030670-292
TLX: 859185 UCMSSL G
TLF:
EML:
COM: 10.44.48

CULLUM MARTIN DR
ESO
KARL SCHWARZSCHILD STR. 2
D 8046 GARCHING
GERMANY. F.R.
TEL: (089) 320 060
TLX: 5282820 EO D
TLF:
EML: EARN:cullum@dgaeso51
COM: 09C

CULVER ROGER BRUCE DR
COLORADO STATE UNIVERSITY
DEPARTMENT OF PHYSICS
FORT COLLINS CO 80523
U.S.A.
TEL: 303-491-6206
TLX: 9109309000 ENGRCSUFT
TLF:
EML:
COM: 26

CUNNINGHAM LELAND E PROF
1520 CREST AVENUE
RICHMOND CA 94805
U.S.A.
TEL:
TLX:
TLF:
EML:
COM: 06.07.20

CUNY YVETTE J DR
OBSERVATOIRE DE PARIS
SECTION DE MEUDON
F-92195 MEUDON PL CEDEX
FRANCE
TEL: 1-45-07-78-38
TLX:
TLF:
EML:
COM: 36

CUPERMAN SAMI PROF
TEL-AVIV UNIVERSITY
DEPT PHYSICS & ASTRONOMY
RAMAT-AVIV 69978
ISRAEL
TEL: 03-420217/425697
TLX: 342171 VERSY IL
TLF:
EML:
COM: 33.49

CURIR ANNA
OSS ASTRONOMICO DI TORINO
STRADA OSSERVATORIO 20
I-10025 PINO TORINESE
ITALY
TEL: 011-84-10-67
TLX: 213236 TO ASTR I
TLF:
EML:
COM: 48

CURRIE DOUGLAS G DR
UNIVERSITY OF MARYLAND
DEPT PHYSICS & ASTRONOMY
COLLEGE PARK MD 20742
U.S.A.
TEL: 301-454-3405
TLX:
TLF:
EML:
COM: 09.19.51

CUYPERS JAN DR
KONINKLIJKE STERRENWACHT
VAN BELGIE
RINGLAAN 3
B 1180 BRUSSEL
BELGIUM
TEL: 3223752484
TLX: 21565
TLF:
EML:
COM: 15.27

CZERNY BOZENA DR
COPERNICUS ASTRONOMICAL
CENTER POLISH ACAD OF
SCIENCE BARTYCKA 18
00716 WARSAW
POLAND
TEL: WARSAW 41 00 41
TLX:
TLF:
EML:
COM:

CZERNY MICHAL DR
ASTRONOMY DEPARTMENT
LEICESTER UNIVERSITY
UNIVERSITY ROAD
LEICESTER LE1 7RH
U.K.
TEL: 533 522073
TLX: 347250 LEICUN
TLF:
EML:
COM:

CZYZAK STANLEY J DR
OHIO STATE UNIVERSITY
DEPT PHYSICS & ASTRONOMY
174 W. 18TH ST
COLUMBUS OH 43210
U.S.A.
TEL: 614-422-6543
TLX:
TLF:
EML:
COM: 14.34

D'ANTONA FRANCESCA DR
OSSERVATORIO ASTRONOMICO
DI ROMA
MONTE PORZIO
I-00040 ROMA
ITALY
TEL: 06 9449019
TLX:
TLF:
EML:
COM: 35.37

D'ODORICO SANDRO DR
ESO
KARL-SCHWARZSCHILD-STR 2
D-8046 GARCHING B MUNCHEN
GERMANY. F.R.
TEL: 89-320-06-00
TLX: 528-28-222
TLF:
EML:
COM: 28.34

DA COSTA GARY STEWART DR
YALE UNIVERSITY
DEPARTMENT OF ASTRONOMY
BOX 6666
NEW HAVEN CT 06511
U.S.A.
TEL: 203-436-3460
TLX:
TLF:
EML:
COM: 37

DA COSTA JOSE MARQUES DR
INST PESQUISAS ESPACIAIS
INPE
CAIXA POSTAL 515
12200 S. JOSE DOS CAMPOS
BRAZIL
TEL: 0123-229977
TLX: 1133530 INPEBR
TLF:
EML:
COM: 48

DA COSTA NICOLAI L.-A
OBSERVATORIO NACIONAL
RUA GENERAL BRUCE 586
SAO CRISTOVAO
20921 RIO DE JANEIRO
BRAZIL
TEL:
TLX:
TLF:
EML:
COM: 28.30C

DA ROCHA VIEIRA E DR
INSTITUTO DE FISICA
UNIVISERSIDADE FEDERAL
DO RIO GRANDE DO SUL
90000 PORTO ALEGRE RS
BRAZIL
TEL: 0512-21-7666
TLX: 0511055 UFRSBR
TLF:
EML:
COM:

DA SILVA A V C S
OBSERVATORIO ASTRONOMICO
UNIVERSIDADE SANTA CLARA
3000 COIMBRA
PORTUGAL
TEL:
TLX:
TLF:
EML:
COM:

DA SILVA LICIO DR
OBSERVATORIO NACIONAL
RUA GENERAL BRUCE 586
20921 RIO DE JANEIRO RJ
BRAZIL
TEL: 580-7313
TLX: 21288
TLF:
EML:
COM:

DACHS JOACHIM PROF DR
ASTRONOMISCHES INSTITUT
RUHR-UNIVERSITAET
POSTFACH 102148
D-4630 BOCHUM 1
GERMANY. F.R.
TEL: 0234-7003454
TLX: 0825860
TLF:
EML:
COM: 21.25

DADAEV ALEKSANDR N DR
PULKOVO OBSERVATORY
196140 LENINGRAD
U.S.S.R.
TEL:
TLX:
TLF:
EML:
COM: 26.42

DADHICH NARESH DR
DEPT OF MATHEMATICS
UNIVERSITY OF POONA
PUNE 411 007
INDIA
TEL: 56061/91
TLX:
TLF:
EML:
COM: 47.48

DADIC ZARKO DR
ZAVOD ZA POVIJEST
ZNANOSTI JAZU
ANTE KOVACICA 5
41000 ZAGREB
YUGOSLAVIA
TEL: 3841-440124
TLX:
TLF:
EML:
COM: 41

DAGKESAMANSKY RUSTAM D DR
LEBEDEV PHYSICAL INSTITU.
LENINSKY PROSPEKT 53
117924 MOSCOW
U.S.S.R.
TEL: 135 14 29
TLX: 411479 NEOD SU
TLF:
EML:
COM: 40

DAHN CONARD CURTIS DR
US NAVAL OBSERVATORY
PO BOX 1149
FLAGSTAFF AZ 86002
U.S.A.
TEL: 602-779-5132
TLX:
TLF:
EML:
COM: 24.25.34

DAIGNE GERARD
OBSERVATOIRE DE BORDEAUX
B.P. 21
F-33270 FLOIRAC
FRANCE
TEL: 1-56-86-43-30
TLX:
TLF:
EML:
COM: 51

DAINTREE EDWARD J DR
NUFFIELD RADIO ASTRO LABS
JODRELL BANK
MACCLESFIELD SK11 9DL
U.K.
TEL: 0477-71321
TLX: 36149
TLF:
EML:
COM: 40

DAISHIDO TSUNEAKI PROF
DEPARTMENT OF SCIENCE
SCHOOL OF EDUCATION
WASEDA UNIVERSITY
SHINJUKU-KU TOKYO 160
JAPAN
TEL: 2-203-4141
TLX: 2323280 WASEDA J
TLF:
EML:
COM: 40

DALGARNO ALEXANDER PROF
CENTER FOR ASTROPHYSICS
60 GARDEN STREET
CAMBRIDGE MA 02138
U.S.A.
TEL: 617-495-4403
TLX: 921428
TLF:
EML:
COM: 14.34

DALLAPORTA N PROF
ISTITUTO DI ASTRONOMIA
UNIVERSITA
VIC. DELL'OSSERVATORIO 5
I-35100 PADOVA
ITALY
TEL: 049-66-14-99
TLX:
TLF:
EML:
COM:

DALTABUIT ENRIQUE DR
INSTITUTO DE ASTRONOMIA
UNAM
APDO POSTAL 70-264
04510 MEXICO DF
MEXICO
TEL:
TLX:
TLF:
EML:
COM:

DAMLE S V DR
TATA INSTITUTE OF
FUNDAMENTAL RESEARCH
BOMBAY 400 005
INDIA
TEL: 219111
TLX: 0113009 TIFR IN
TLF:
EML:
COM: 48

DAN XHI-XIANG
SHANGHAI OBSERVATORY
ACADEMIA SINICA
SHANGHAI
CHINA PEOPLE'S REP.
TEL: 386191
TLX: 33164 SHAO CN
TLF:
EML:
COM: 09

DANBY J M ANTHONY DR
DEPARTMENT OF MATHEMATICS
N.CAROLINA STATE UNIV
RALEIGH NC 27695-8205
U.S.A.
TEL: 919-737-3210
TLX:
TLF:
EML:
COM: 07

DANESE LUIGI DR
OSSERVATORIO ASTRONOMICO
VIC. DELL'OSSERVATORIO 5
I-35100 PADOVA
ITALY
TEL: 049-66-1499
TLX: 430176 UNPADU-I
TLF:
EML:
COM: 47

DANEZIS EMMANUEL DR
SECTION OF ASTROPHYSICS
UNIVERS. OF ATHENS
GR 15783 ZOGRAFOS ATHENS
GREECE
TEL:
TLX:
TLF:
EML:
COM:

DANILOV VLADIMIR M DR
URAL STATE UNIVERSITY
LENIN PROSPECT 51
SVERDLOVSK USSR
U.S.S.R.
TEL: 22 33 86
TLX:
TLF:
EML:
COM:

DANKS ANTHONY C DR
ARC
8201 CORPORATE DRIVE
SUITE 920
LANDOVER MD 20785
U.S.A.
TEL: 301-459-8833
TLX:
TLF:
EML:
COM: 15.28.34

DANZIGER I JOHN DR
ESO
KARL-SCHWARZSCHILD-STR 2
D-8046 GARCHING B MUNCHEN
GERMANY. F.R.
TEL:
TLX:
TLF:
EML:
COM:

DAPPEN WERNER
OBSERVATOIRE DE PARIS
SECTION DE MEUDON
92195 MEUDON PL CEDEX
FRANCE
TEL: 45-34-75-30
TLX:
TLF:
EML:
COM:

DARA HELEN DR
RES.CENTER FOR ASTRONOMY
ACADEMY OF ATHENS
14 ANAGNOSTOPOULOU STR.
GR 10673 ATHENS
GREECE
TEL: 3613 589
TLX:
TLF:
EML: EXAKAZO@GRATHUN1
COM: 12

DARIUS JON DR
SCIENCE MUSEUM
LONDON SW7 2DD
U.K.
TEL: 01-589-3456x643
TLX: 21200 SCMLIB G
TLF:
EML:
COM: 41.46.51

DAS MRINAL KANTI
DEPT PHYSICS/DELHI UNIV
SRI VENKATESWARA COLLEGE
DHAULA KUAN
NEW DELHI 110 021
INDIA
TEL:
TLX:
TLF:
EML:
COM: 35

DAS P K DR
INDIAN INSTITUTE OF
ASTROPHYSICS
BANGALORE 560 034
INDIA
TEL: 566585
TLX: 845763 IIAB IN
TLF:
EML:
COM: 47

DATLOWE DAYTON DR
LOCKHEED PALO A. RES LAB
DEPT 91-20 BLDG 255
3251 HANOVER STREET
PALO ALTO CA 94304
U.S.A.
TEL: 415-858-4074
TLX:
TLF:
EML:
COM: 10

DATTA BHASKAR DR
INDIAN INSTITUTE OF
ASTROPHYSICS
BANGALORE 560 034
INDIA
TEL: 566585 / 566497
TLX: 845763 IIAB IN
TLF:
EML:
COM: 47

DAUBE-KURZEMNIECE I A DR
RADIOASTROPHYSICAL OBS
LATVIAN ACAD OF SCIENCES
TURGENEVA 19
226524 RIGA LATVIA
U.S.S.R.
TEL: 226796
TLX:
TLF:
EML:
COM: 37

DAUTCOURT G DR
ZNTRLINST. F. ASTROPHYSIK
STERNWARTE BABELSBERG
ROSA-LUXEMBURG-STR 17A
DDR-1502 POTSDAM
GERMANY. D.R.
TEL:
TLX:
TLF:
EML:
COM: 48

DAVIDSEN ARTHUR FALNES DR
DEPT PHYSICS & ASTRONOMY
JOHNS HOPKINS UNIVERSITY
CHARLES & 34TH STREETS
BALTIMORE MD 21218
U.S.A.
TEL: 301-338-7370
TLX:
TLF:
EML:
COM: 28.44.48

DAVIDSON KRIS DR
SCHOOL PHYS & ASTRONOMY
UNIVERSITY OF MINNESOTA
116 CHURCH ST S.E.
MINNEAPOLIS MN 55455
U.S.A.
TEL: 612-373-7795
TLX:
TLF:
EML:
COM:

DAVIDSON WILLIAM PROF
25 PADDOCK CLOSE
MANSFIELD NOTTS NG21 9PL
U.K.
TEL:
TLX:
TLF:
EML:
COM: 47.48

DAVIES JOHN G DR
NUFFIELD RADIO ASTRO LABS
JODRELL BANK
MACCLESFIELD SK11 9DL
U.K.
TEL: 0477-71321
TLX: 36174
TLF:
EML:
COM: 19.22.40

DAVIES MERTON E MR
THE RAND CORPORATION
1700 MAIN STREET
SANTA MONICA CA 90406
U.S.A.
TEL: 213-393-0411
TLX:
TLF:
EML:
COM: 04.16C

DAVIES PAUL CHARLES W
SCHOOL OF PHYSICS
THE UNIVERSITY
NEWCASTLE/TYNE NE1 7RU
U.K.
TEL:
TLX:
TLF:
EML:
COM: 47

DAVIES ROBERT D PROF
NUFFIELD RADIO ASTR LABS
JODRELL BANK
MACCLESFIELD SK11 9DL
U.K.
TEL: 0477-71321
TLX: 36149
TLF:
EML:
COM: 28.33.34.40

DAVIES ROGER L DR
KITT PEAK NTL OBSERVATORY
PO BOX 26732
950 N CHERRY AVENUE
TUCSON AZ 85726
U.S.A.
TEL: 602 325 9353
TLX: 0666 484 AURA NOAO
TLF:
EML:
COM: 47

DAVIS CECIL G JR
UNIVERSITY OF CALIFORNIA
LOS ALAMOS NATIONAL LAB
GROUP P-15 / MS D 406
LOS ALAMOS NM 87545
U.S.A.
TEL: 505-667-5908
TLX:
TLF:
EML:
COM: 35.36

DAVIS JOHN DR
SCHOOL OF PHYSICS
UNIVERSITY OF SYDNEY
SYDNEY NSW 2006
AUSTRALIA
TEL: 02-692-3604
TLX: 26169 UNISYD AA
TLF:
EML:
COM: 09P.50

DAVIS LEVERETT JR PROF
CALTECH 405-47
PASADENA CA 91125
U.S.A.
TEL: 818-356-4243
TLX:
TLF:
EML:
COM: 48

DAVIS MARC DR
DEPARTMENT OF ASTRONOMY
UNIVERSITY OF CALIFORNIA
601 CAMPBELL HALL
BERKELEY CA 94720
U.S.A.
TEL: 415-642-5156
TLX: 820181 UCB AST
TLF:
EML:
COM: 28.30.47

DAVIS MICHAEL M DR
ARECIBO OBSERVATORY
PO BOX 995
ARECIBO PR 00613
U.S.A.
TEL: 809-878-2612
TLX: 385638
TLF:
EML:
COM: 40.47.48.51

DAVIS MORRIS S PROF
DEPT PHYSICS & ASTRONOMY
UNIVERSITY OF N. CAROLINA
284 PHILLIPS HALL 039A
CHAPEL HILL NC 27514
U.S.A.
TEL: 919-962-3011
TLX:
TLF:
EML:
COM: 05.07

DAVIS RICHARD J DR
NUFFIELD RADIO ASTR LABS
JODRELL BANK
MACCLESFIELD SK11 9DL
U.K.
TEL:
TLX:
TLF:
EML:
COM:

DAVIS ROBERT J DR
OPTICAL/INFRARED ASTR DIV
SMITHSONIAN ASTROPHYS OBS
60 GARDEN ST. MS 20
CAMBRIDGE MA 02138
U.S.A.
TEL: 616-495-7335
TLX: 921428 SATELLITE CAM
TLF:
EML:
COM: 05.40.44

DAVIS SUMNER P DR
PHYSICS DEPARTMENT
UNIVERSITY OF CALIFORNIA
BERKELEY CA 94708
U.S.A.
TEL: 415 642 4857
TLX:
TLF:
EML:
COM: 14

DAVOUST EMMANUEL
OBSERVATOIRE PIC DU MIDI
ET TOULOUSE
14 AVENUE EDOUARD BELIN
F-31400 TOULOUSE
FRANCE
TEL: 61-25-21-01
TLX: 530776
TLF:
EML:
COM:

DAWANAS DJONI N DR
DEP OF ASTRO. BOSSCHA OBS
BANDUNG INST OF TECH
JL: GANESHA 10. BANDUNG
BANDUNG
INDONESIA
TEL:
TLX: 28324 ITB BD
TLF:
EML:
COM: 29

DAWE JOHN ALAN DR
A.N.U.
SIDING SPRING OBSERVATORY
PRIVATE BAG
COONABARABRAN 2357
AUSTRALIA
TEL: 068-426-221
TLX: 63945 AA CANOPUS
TLF:
EML:
COM: 51

DE BERGH CATHERINE DR
OBSERVATOIRE DE PARIS
SECTION DE MEUDON
LAB ASTRO INFRAROUGE
F-92195 MEUDON PL CEDEX
FRANCE
TEL: 1-45 07 76 66
TLX: 201571 LAM
TLF:
EML:
COM: 16C

DE BIASE GIUSEPPE A DR
OSSERVATORIO ASTRONOMICO
VIA DEL PARCO MELLINI 84
I-00136 ROMA
ITALY
TEL:
TLX:
TLF:
EML:
COM:

DE BOER KLAAS SJOERDS DR
ASTRONOMISCHES INSTITUT
UNIVERSITAET BONN
AUF DEM HUEGEL 71
D-5300 BONN
GERMANY. F.R.
TEL: 48-228-733656
TLX: 886440
TLF:
EML:
COM: 05.28.34C

DE BRITO e ABREU J C DR
RUA DO OLIVAL 142
1200 LISBOA
PORTUGAL
TEL: 60-72-00
TLX:
TLF:
EML:
COM:

DE BRUYN A. GER DR
RADIOSTERREWACHT
POSTBUS 2
NL-7990 AA DWINGELOO
NETHERLANDS
TEL: 05219-7244
TLX: 42043 SRZM NL
TLF:
EML:
COM: 28

DE CASTRO ANGEL DR
NTL ASTRONOMICAL OBS.
ALFONSO XII-3
APARTADO 12354
28014 MADRID
SPAIN
TEL:
TLX:
TLF:
EML:
COM: 04

DE CASTRO ELISA
DEPTO DE ASTROFISICA
FACULTAD DE FISICA
UNIVERSIDAD COMPLUTENSE
28040 MADRID
SPAIN
TEL: 449-53-16
TLX: 47273 FFUC
TLF:
EML:
COM:

DE FELICE FERNANDO DR
DIP. FISICA G. GALILEI
CITTA UNIVERSITARIA
VIA MARZOLO 8
I-35100 PADOVA
ITALY
TEL: 049-844-278
TLX: 430308 DFGGPDI
TLF:
EML:
COM: 48

DE FREES DOUGLAS J DR
MOLECULAR RESEARCH INST.
701 WELCH ROAD SUITE 213
PALO ALTO CA 94304
U.S.A.
TEL: 415 723 6039
TLX:
TLF:
EML:
COM: 14

DE FREITAS PACHECO J A DR
OBSERVATORIO DE SAO PAULO
UNIVERSIDADE DE SAO PAULO
CAIXA POSTAL 30627
01051 SAO PAULO
BRAZIL
TEL: 021-717-3518
TLX:
TLF:
EML:
COM:

DE GRAAF T DR
INSTITUUT VOOR FONETISCHE
WETENSCHAPPEN
GROTE ROZENSTRAAT 31
NL-9712 TG GRONINGEN
NETHERLANDS
TEL:
TLX:
TLF:
EML:
COM: 48

DE GRAAFF W DR
APPELGAARDE 117
NL-3992 JD HOUTEN
NETHERLANDS
TEL:
TLX:
TLF:
EML:
COM: 51

DE GRAAUW TH DR
ESA
SPACE SCIENCE DEPARTMENT
POSTBUS 299
NL-2200 AG NOORDWIJK
NETHERLANDS
TEL:
TLX:
TLF:
EML:
COM:

DE GREVE JEAN-PIERRE DR
ASTROPHYSICAL INSTITUTE
VRIJE UNIV BRUSSEL
PLEINLAAN 2
B-1050 BRUSSELS
BELGIUM
TEL: 32-2-6413498
TLX: 61051 VUBCO
TLF:
EML:
COM: 35.42

DE GROOT MART DR
ARMAGH OBSERVATORY
COLLEGE HILL
ARMAGH BT61 9DG
U.K.
TEL: 0861-522928
TLX: 747937 ARMOBS G
TLF:
EML:
COM: 27.29.42

DE GROOT T DR
STERREKUNDIG INSTITUUT
PRINCETONPLEIN 5
POSTBUS 80000
NL-3508 TA UTRECHT
NETHERLANDS
TEL: 30 535 200
TLX: 47224 ASTRO
TLF:
EML:
COM: 10.40

DE JAGER CORNELIS PROF
ASTRONOMICAL OBSERVATORY
LAB FOR SPACE RESEARCH
BENELUXLAAN 21
NL-3527 HS UTRECHT
NETHERLANDS
TEL: 030-937145
TLX: 47224
TLF:
EML:
COM: 10.12.35.40.44.49.51

DE JAGER GERHARD PROF
DEPT PHYSICS/ELECTRONICS
RHODES UNIVERSITY
P O BOX 94
GRAHAMSTOWN 6140
SOUTH AFRICA
TEL: 0461-7128
TLX: 244226
TLF:
EML:
COM:

DE JAGER OCKER C DR
DEPT OF PHYSICS
POTCHEFSTROOM UNIVERSITY
2520 POTCHEFSTROOM
SOUTH AFRICA
TEL: 148 992418
TLX: 34 6019
TLF:
EML:
COM:

DE JONG TEIJE DR
ASTRONOMICAL INSTITUTE
ROETERSSTRAAT 15
NL-1018 WB AMSTERDAM
NETHERLANDS
TEL: 20-5223004
TLX: 16460 FACWN NL
TLF:
EML:
COM: 33.34

DE JONGE J K DR
DEPARTMENT OF ASTRONOMY
UNIVERSITY OF PITTSBURGH
RIVERVIEW PARK
PITTSBURGH PA 15214
U.S.A.
TEL:
TLX:
TLF:
EML:
COM: 30.51

DE KORT JULES J DR
HOUTLAAN 4
NL-6500 GV NIJMEGEN
NETHERLANDS
TEL:
TLX:
TLF:
EML:
COM: 42

DE KORTE PIETER A J DR
LAB. FOR SPACE RESEARCH
BENELUXLAAN 21
NL-3527 HS UTRECHT
NETHERLANDS
TEL: 31 30-937145
TLX: 47224
TLF:
EML:
COM:

DE LA HERRAN V JOSE ENG
INSTITUTO DE ASTRONOMIA
MEXICO
APDO POSTAL 971
MEXICO 1 DF
MEXICO
TEL:
TLX:
TLF:
EML:
COM:

DE LA NOE JEROME DR
OBSERVATOIRE DE BORDEAUX
AVENUE PIERRE SEMIROT
BP 21
F-33270 FLOIRAC
FRANCE
TEL: 56-86-43-30
TLX:
TLF:
EML:
COM: 28.34.40

DE LA REZA RAMIRO DR
OBSERVATORIO NACIONAL
RUA GENERAL BRUCE 586
20921 RIO DE JANEIRO RJ
BRAZIL
TEL: 580-7313
TLX: 21288
TLF:
EML:
COM:

DE LOORE CAMIEL PROF
ASTROPHYSICAL INSTITUTE
VRIJE UNIVERSITEIT
PLEINLAAN 2 76
B-1050 BRUSSEL
BELGIUM
TEL: 32-2-6413496
TLX: 61051 VUBCO B
TLF:
EML:
COM: 35.42.51

DE PASCUAL MARTINEZ M DR
OBSERVATORIO ASTRONOMICO
ALFONSO XII 3 & 5
28014 MADRID
SPAIN
TEL: 2270107
TLX: 23475 IGC
TLF:
EML:
COM: 20

DE PATER IMKE
UNIVERSITY OF CALIFORNIA
ASTRONOMY DEPT
601 CAMPBELL HALL
BERKELEY CA 94720
U.S.A.
TEL: 415-642-1947
TLX:
TLF:
EML:
COM: 15.16

DE ROBERTIS M M DR
DPT OF PHYSICS
YORK UNIVERSITY
4700 KEELE STREET
NORTH YORK ON M3J 1P3
CANADA
TEL: 416 736 2100
TLX: 065 24736
TLF:
EML:
COM: 28

DE RUITER HANS RUDOLF
IST DI RADIOASTRONOMIA
VIA IRNERIO 46
I-40126 BOLOGNA
ITALY
TEL: 051-23-28-56
TLX: 211664 INFN BO I
TLF:
EML:
COM: 40.47

DE SABBATA V PROF DR
ISTITUTO DI FISICA
UNIVERSITA DI BOLOGNA
VIA IRNERIO 46
I-40100 BOLOGNA
ITALY
TEL: 260991/051
TLX:
TLF:
EML:
COM:

DE SANCTIS GIOVANNI
OSSERVATORIO ASTRONOMICO
DI TORINO
STRADA OSSERVATORIO 20
I-10025 PINO TORINESE
ITALY
TEL: 011-841067
TLX: 213236 TOASTR I
TLF:
EML:
COM: 15.20

DE SILVA L.N.K. DR
DEPT OF MATHEMATICS
UNIVERSITY OF COLOMBO
COLOMBO 03
SRI LANKA
TEL:
TLX:
TLF:
EML:
COM: 28

DE SOUZA RONALDO DR
INST ASTRONOM E GEOFISICO
UNIV DE SAO PAULO
CAIXA POSTAL 30627
01051 SAO PAULO SP
BRAZIL
TEL: 5778599
TLX: 36221 IAGM BR
TLF:
EML:
COM:

DE VAUCOULEURS GERARD PR
DEPARTMENT OF ASTRONOMY
UNIVERSITY OF TEXAS
RLM 15.212
AUSTIN TX 78712
U.S.A.
TEL: 512-471-4461
TLX:
TLF:
EML:
COM: 28.30.47

DE VEGT CH PROF DR
HAMBURGER STERNWARTE
GOJENBERGSWEG 112
D-2050 HAMBURG 80
GERMANY. F.R.
TEL: 40-7252-4128
TLX: 21788 HAMST
TLF:
EML:
COM: 08.24V

DE VINCENZI DONALD DR
MAIL CODE 239-11
NASA AMES RESEARCH CENTER
MOFFETT FIELD. CA 94035
U.S.A.
TEL:
TLX:
TLF:
EML:
COM: 51

DE YOUNG DAVID S DR
KITT PEAK NAT OBSERVATORY
PO BOX 26732
TUCSON AZ 85726
U.S.A.
TEL: 602-327-5511
TLX:
TLF:
EML:
COM: 40.48

DE ZEEUW PIETER T DR
THEORETICAL ASTROPHYSICS
130-33
CALIFORNIA INST. TECHNOL.
PASADENA CA 911250
U.S.A.
TEL: 818 356 4519
TLX: 675425 CALTECH PSD
TLF:
EML:
COM: 28

DE ZOTTI GIANFRANCO DR
ISTITUTO DI ASTRONOMIA
VIC. DELL'OSSERVATORIO 5
I-35122 PADOVA
ITALY
TEL: 049-66-1499
TLX: 430176 UNPADU I
TLF:
EML:
COM: 47

DEARBORN DAVID PAUL S DR
LAWRENCE LIVERMORE LAB
L-23
PO BOX 808
LIVERMORE CA 94550
U.S.A.
TEL:
TLX:
TLF:
EML:
COM: 35

DEBARBAT SUZANNE V DR
OBSERVATOIRE DE PARIS
61 AVE DE L'OBSERVATOIRE
F-75014 PARIS
FRANCE
TEL: 1-40-51-22-09
TLX: 270776
TLF:
EML:
COM: 08.19.41V

DEBEHOGNE HENRI DR SC
OBS ROYAL DE BELGIQUE
AVENUE CIRCULAIRE 3
B-1180 BRUXELLES
BELGIUM
TEL: 02-3743801
TLX: 21565 B
TLF:
EML:
COM: 15.20

DEBRUNNER HERMANN DR
PHYSIKALISCHES INSTITUTE
UNIVERSITAET BERN
SIDLERSTRASSE 5
CH-3000 BERN
SWITZERLAND
TEL: 31-65-40-51
TLX: 32320 CH
TLF:
EML:
COM: 48

DEEMING TERENCE J DR
ICARUS RESEARCH
P.O. BOX 540205
HOUSTON TX 77254
U.S.A.
TEL: 713-772-8414
TLX:
TLF:
EML:
COM: 41

DEERENBERG A.J.M. DR
SPACE RESEARCH LABORATORY
HUYGENS LABORATORY
WASSENAARSEWEG 78
NL-2300 RA LEIDEN
NETHERLANDS
TEL: 071-272727
TLX:
TLF:
EML:
COM:

DEGAONKAR S S DR
PHYS. RESEARCH LABORATORY
NAVRANGPURA
AHMEDABAD 380 009
INDIA
TEL: 462129
TLX: 121397 PRL IN
TLF:
EML:
COM: 40

DEGEWIJ JOHAN DR
NOBBERMANSTRAAT 66
NL-2313 GS LEIDEN
NETHERLANDS
TEL:
TLX:
TLF:
EML:
COM: 15.16

DEGUCHI SHUJI DR
NOBEYAMA RADIO OBSERVAT.
MINAMIMAKI
MINAMISAKU
NAGANO 384-13
JAPAN
TEL: 0267 98 2831
TLX: 3329005 TAONRO J
TLF:
EML:
COM: 34

DEHANT VERONIQUE DR
OBSERVATOIRE ROYAL
DE BELGIQUE
AVENUE CIRCULAIRE 3
B-1180 BRUXELLES
BELGIUM
TEL: 32 2 3752484
TLX: 21565 OBSBEL B
TLF:
EML:
COM:

DEHARVENG JEAN-MICHEL DR
LAS
TRAVERSE DU SIPHON
LES TROIS LUCS
F-13012 MARSEILLE
FRANCE
TEL: 91-66-08-32
TLX: 420 584 F
TLF:
EML:
COM:

DEHARVENG LISE DR
OBSERVATOIRE DE MARSEILLE
2 PLACE LE VERRIER
F-13248 MARSEILLE CEDEX 4
FRANCE
TEL: 91-95-90-88
TLX:
TLF:
EML:
COM: 34

DEINZER W PROF DR
UNIVERSITAETS-STERNWARTE
GEISMARLANDSTR 11
D-3400 GOETTINGEN
GERMANY. F.R.
TEL: 0551-395044
TLX: 96753 USTERN D
TLF:
EML:
COM: 35

DEJAIFFE RENE J DR
OBS ROYAL DE BELGIQUE
AVENUE CIRCULAIRE 3
B-1180 BRUXELLES
BELGIUM
TEL: 3752484
TLX: 21565 OBSBEL B
TLF:
EML:
COM: 08.19

DEJONGHE HERWIG BERT DR
R.U.G.
STERREKUNDIG OBSERVAT.
KRIJGSLAAN 281-S9
B-9000 GENT
BELGIUM
TEL:
TLX:
TLF:
EML:
COM: 28.37

DEKEL AVISHAI
THE HEBREW UNIVERSITY
DEPT OF PHYSICS
JERUSALEM
ISRAEL
TEL: 972-2-584 605
TLX:
TLF:
EML:
COM: 28.33.47

DEKKER E DR
MUSEUM BOERHAAVE
STEENSTRAAT 1A
NL-2312 BS LEIDEN
NETHERLANDS
TEL: 071-123084
TLX:
TLF:
EML:
COM: 41

DEL RIO GERARDO DR
NAT'L ASTRONOMICAL OBS
ALFONSO XII-3
28014 MADRID
SPAIN
TEL: 91-2270107/1935
TLX: 23465 IGCE
TLF:
EML:
COM:

DEL TORO INIESTA JOSE DR
INSTITUTO DE ASTROFISICA
DE CANARIAS
38200 LA LAGUNA TENERIFE
SPAIN
TEL: 34 22 26 22 11
TLX: 92640 IACE
TLF: 26 30 05
EML: SPAN:IAC::JTI
COM: 10.12

DELABOUDINIERE J.-P.
IAS
BP 10
F-91371 VERRIERES-LE-B.
FRANCE
TEL: 64 47 43 05
TLX: 600252 LPSP VB
TLF:
EML:
COM:

DELACHE PHILIPPE J DR
OBSERVATOIRE DE NICE
BP 139
F-06003 NICE CEDEX
FRANCE
TEL: 93-89-04-20
TLX: 460004
TLF:
EML:
COM: 12.36.49

DELANNOY JEAN DR
IRAM
DOMAINE UNIVERSITAIRE
VOIE 10
F-38406 ST MARTIN D'HERES
FRANCE
TEL: 76-42-33-83
TLX: 980753
TLF:
EML:
COM: 40

DELBOUILLE LUC PROF
INSTITUT D'ASTROPHYSIQUE
UNIVERSITE DE LIEGE
AVENUE DE COINTE 5
B-4200 COINTE-OUGREE
BELGIUM
TEL: 041 52 99 80
TLX: 41264 ASTRLG B
TLF:
EML:
COM: 12

DELCROIX ANDRE J S DR
19A RUE E VANDERVELDE
B-7230 FRAMERIES
BELGIUM
TEL:
TLX:
TLF:
EML:
COM:

DELGADO ANTONIO JESUS
INST DE ASTROFISICA DE
ANDALUCIA. APDO 2144
PROFESOR ALBAREDA 1
18080 GRANADA
SPAIN
TEL: 58-12-13-00
TLX: 78573 IAAG E
TLF:
EML:
COM: 27.42

DELHAYE JEAN PROF
2 RUE DE LA PLEIADE
F-94240 L'HAY-LES-ROSES
FRANCE
TEL: 1-46-64-57-71
TLX:
TLF:
EML:
COM: 24.33

DELLA VALLE MASSIMO DR
DIPARTIMENTO ASTRONOMIA
UNIVERSITA DI PADOVA
VICOLO OSSERVATORIO 5
35122 PADOVA
ITALY
TEL: 049 661499
TLX: 432071
TLF:
EML: INFNET:39003::DELLAVALLE
COM:

DELLI SANTI SAVERIO
OSSERVATORIO ASTRONOMICO
UNIVERSITARIO
C P 596
I-40100 BOLOGNA
ITALY
TEL: 051-222956
TLX: 211664 INFNBO I
TLF:
EML:
COM:

DELSEMME ARMAND H PROF DR
DEPT OF PHYS & ASTRONOMY
UNIVERSITY OF TOLEDO
2801 W BANCROFT STREET
TOLEDO OH 43606
U.S.A.
TEL: 419-537-2654
TLX:
TLF:
EML:
COM: 14.15.20.51

DEMARCQ JEAN ING
OBSERVATOIRE DE NICE
BP 139
F-06003 NICE CEDEX
FRANCE
TEL: 93-55-89-65
TLX:
TLF:
EML:
COM:

DEMARET JACQUES DR
158/033 AV L'OBSERVATOIRE
B-4000 LIEGE
BELGIUM
TEL: 041-52-72-61
TLX: 41264 ASTROLIEGE
TLF:
EML:
COM: 47

DEMARQUE P PROF
YALE UNIV OBSERVATORY
260 WHITNEY AVENUE
PO BOX 6666
NEW HAVEN CT 06511
U.S.A.
TEL: 203-436-8246
TLX:
TLF:
EML:
COM: 12.35V.37

DEMERS SERGE DR
DEPARTEMENT DE PHYSIQUE
UNIVERSITE DE MONTREAL
CP 6128 SUCC A
MONTREAL PQ H3C 3J7
CANADA
TEL: 514-343-6718
TLX: 05562425
TLF:
EML:
COM: 27.37

DEMIN V G PROF DR
STERNBERG STATE
ASTRONOMICAL INSTITUTE
UNIVERSITETSKIJ PROSP 13
119899 MOSCOW
U.S.S.R.
TEL: 139-36-81
TLX:
TLF:
EML:
COM: 07

DEMIRCAN OSMAN DR
PHYSICS DEPARTMENT
MIDDLE EAST & TECHN UNIV
ANKARA
TURKEY
TEL: 237100/3253 ANK
TLX: 42761 OOTK TR
TLF:
EML:
COM: 42

DENG ZUGAN DR
GRADUATE SCHOOL
DEPT. OF PHYSICS
P.O. BOX 3908
BEIJING 100039
CHINA. PEOPLE'S REP.
TEL: 01 289461
TLX: 22040 BAOAS CN
TLF:
EML:
COM:

DENIS CARLO DR
INSTITUT D'ASTROPHYSIQUE
UNIVERSITE DE LIEGE
AVENUE DE COINTE 5
B-4200 COINTE-OUGREE
BELGIUM
TEL: 41-52-99-80
TLX: 41264
TLF:
EML:
COM:

DENISSE JEAN-FRANCOIS DR
48 RUE MR. LE PRINCE
F-75006 PARIS
FRANCE
TEL: 1-43-29-48-74
TLX:
TLF:
EML:
COM: 40

DENISYUK EDWARD K DR
ASTROPHYSICAL INSTITUTE
480068 ALMA ATA
U.S.S.R.
TEL:
TLX:
TLF:
EML:
COM:

DENNEFELD MICHEL
INSTITUT D'ASTROPHYSIQUE
98 BIS BD ARAGO
75014 PARIS
FRANCE
TEL: 1 43 20 14 25
TLX: 205671 IAU F
TLF:
EML: SPAN:IAPOBS::DENNEFELD
COM: 29

DENNIS BRIAN ROY DR
NASA/GSFC
CODE 682
GREENBELT MD 20771
U.S.A.
TEL: 301-286-6604
TLX: 89675
TLF:
EML:
COM: 10.44.48

DENNISON EDWIN W DR
ELECTRONIC VISION CO
11526 SORRENTO
VALLEY ROAD
SAN DIEGO CA 92121
U.S.A.
TEL:
TLX:
TLF:
EML:
COM:

DENNISON P A DR
TRINITY HALL
UNIVERSITY OF CAMBRIDGE
CAMBRIDGE CB3
U.K.
TEL:
TLX:
TLF:
EML:
COM:

DENOYELLE JOZEF LIC
KON STERRENWACHT V.BELGIE
RINGLAAN 3
B-1180 BRUSSEL
BELGIUM
TEL: 02-375-2484
TLX: 21565 OBSBEL
TLF:
EML:
COM: 25.33

DENT WILLIAM A PROF
DEPT PHYSICS & ASTRONOMY
TOWER B
UNIV OF MASSACHUSETTS
AMHERST MA 01003
U.S.A.
TEL: 413-545-3665
TLX:
TLF:
EML:
COM: 40

DEPRIT ANDRE PROF
CTR FOR APPLIED MATHS
NATL BUREAU OF STANDARDS
GAITHERSBURG MD 20899
U.S.A.
TEL: 301-921-2631
TLX:
TLF:
EML:
COM: 04.07V

DERE KENNETH PAUL
NAVAL RESEARCH LABORATORY
CODE 4163
WASHINGTON DC 20375
U.S.A.
TEL: 202-767-2517
TLX:
TLF:
EML:
COM: 10

DERMAN I ETHEM DR
ANKARA UNIVERSITY
FACULTY OF SCIENCE
ANKARA
TURKEY
TEL: 41-236550/0109
TLX:
TLF:
EML:
COM:

DERMENDJIEV VLADIMIR DR
DEPT OF ASTRONOMY
BULGARIAN ACAD SCIENCES
72 LENIN BLVD
1784 SOFIA
BULGARIA
TEL: 73-41-269
TLX: 23561 ECF BAN BG
TLF:
EML:
COM: 10

DERMOTT STANLEY F
CORNELL UNIVERSITY
CTR RADIOPHYS & SPACE RES
SPACE SICENCES BLDG
ITHACA NY 14853
U.S.A.
TEL: 607-256-3727
TLX:
TLF:
EML:
COM: 15.16.21

DESAI JYOTINDRA N
PHYSICAL RESEARCH LAB
ROOM No. 763
AHMEDABAD 380 009
INDIA
TEL: 462-129
TLX: 121397
TLF:
EML:
COM: 09

DESESQUELLES JEAN DR
UNIVERSITE LYON I
CAMPUS LA DOUA
F-69622 VILLEURBANNE CDX
FRANCE
TEL: 78-89-81-24
TLX: 380273 IPN F
TLF:
EML:
COM: 14

DESHPANDE AVINASH
RAMAN RESEARCH INSTITUTE
BANGALORE 560 080
INDIA
TEL: 812 360 122
TLX: 845 2671 RRI IN
TLF:
EML:
COM: 40

DESHPANDE M R DR
PHYSICAL RESEARCH LAB
NAVRANGPURA
AHMEDABAD 380 009
INDIA
TEL: 462129
TLX: 121397 PRL IN
TLF:
EML:
COM: 25

DESPAIN KEITH HOWARD DR
LOS ALAMOS NATIONAL LAB
X-2, MS B220
PO BOX 1663
LOS ALAMOS NM 87545
U.S.A.
TEL: 505-667-2388
TLX:
TLF:
EML:
COM: 35

DESPOIS DIDIER DR
OBSERVATOIRE DE BORDEAUX
AVENUE P. SEMIROT
BP 89
33270 FLOIRAC
FRANCE
TEL: 56 86 43 30
TLX:
TLF:
EML: EARN:DESPOIS@FRBOR51
COM: 40

DEUBNER FRANZ-LUDWIG DR
INST F ASTRONOMIE &
ASTROPHYSIK
AM HUBLAND
D-8700 WUERZBURG
GERMANY. F.R.
TEL: 0931-8885030
TLX: 68671 UNIWBG D
TLF:
EML:
COM: 10.12

DEUPREE ROBERT G DR
ESS-5 MS F665
LOS ALAMOS NATIONAL LAB
PO BOX 1663
LOS ALAMOS NM 87545
U.S.A.
TEL: 505-667-8215
TLX:
TLF:
EML:
COM: 27.35

DEUTSCHMAN WILLIAM A DR
MATH. & PHYSICS DEPT
SEMON HALL
OREGON INST OF TECHNOLOGY
KLAMATH FALLS OR 97601
U.S.A.
TEL: 503-882-6321
TLX:
TLF:
EML:
COM: 15

DEVINNEY EDWARD J DR
100 UNION AVE
DELANCO NJ 08075
U.S.A.
TEL: 609-764-1250
TLX:
TLF:
EML:
COM:

DEVORKIN DAVID H
NATL AIR & SPACE MUSEUM
SMITHSONIAN INSTITUTION
WASHINGTON DC 20560
U.S.A.
TEL: 202-357-2828
TLX:
TLF:
EML:
COM: 41C

DEWDNEY PETER E F DR
DOMINION RADIO ASTROPHYS
OBSERVATORY
P O BOX 248
PENTICTON BC V2A 6K3
CANADA
TEL: 604-497-5321
TLX: 048-88127
TLF:
EML:
COM: 34.40

DEWHIRST DAVID W DR
INSTITUTE OF ASTRONOMY
THE OBSERVATORIES
MADINGLEY ROAD
CAMBRIDGE CB3 0HA
U.K.
TEL: 0223-62204
TLX: 817297 ASTRON G
TLF:
EML:
COM: 05.41

DEWITT BRYCE S DR
DEPT OF PHYSICS
UNIVERSITY OF TEXAS
AUSTIN TX 78712
U.S.A.
TEL: 512-471-5055
TLX: 9108741305
TLF:
EML:
COM: 48

DEWITT JOHN H JR
3602 WOODS HILL RD
NASHVILLE TN 37215
U.S.A.
TEL: 615-383-8272
TLX:
TLF:
EML:
COM:

DEWITT-MORETTE CECILE PR
DEPT OF PHYSICS
UNIVERSITY OF TEXAS
9.220 R.L. MOORE HALL
AUSTIN TX 78712
U.S.A.
TEL: 512-471-1052
TLX: 910-8741305
TLF:
EML:
COM:

DEZSO LORANT PROF
HELIOPHYSICAL OBSERVATORY
H-4010 DEBRECEN
HUNGARY
TEL: 52-11-015
TLX: 72517 DEOBS H
TLF:
EML:
COM: 10.12

DI COCCO GUIDO
ISTITUTO TE.S.R.E.-C.N.R.
VIA DE CASTAGNOLI 1
I-40126 BOLOGNA
ITALY
TEL: 051-519593
TLX: 511350 CNR BO I
TLF:
EML:
COM: 44

DI FAZIO ALBERTO
OSSERVATORIO ASTRONOMICO
DI ROMA
VIA DEL PARCO MELLINI 84
I-00136 ROMA
ITALY
TEL: 06-34-70-56
TLX: 613103 PPRMT I
TLF:
EML:
COM: 28.34.37

DI MARTINO MARIO
OSSERVATORIO ASTRONOMICO
DI TORINO
STRADA OSSERVATORIO 20
I-10025 PINO TORINESE
ITALY
TEL: 011-841067
TLX: 213236 TO ASTR I
TLF:
EML:
COM: 15

DI SEREGO ALIGHIERI S DR
SPACE TELESCOPE EUROPEAN
COORDINATING FACILITY
KARL-SCHWARZSCHILD-STR 2
D-8046 GARCHING B MUNCHEN
GERMANY, F.R.
TEL:
TLX:
TLF:
EML:
COM: 28

DI TULLIO GRAZIELLA DR
OSSERVATORIO ASTRONOMICO
I-35100 PADOVA
ITALY
TEL:
TLX:
TLF:
EML:
COM:

DI XIAO-HUA
PURPLE MOUNTAIN
OBSERVATORY
NANJING
CHINA, PEOPLE'S REP.
TEL: 37609
TLX: 34144 PMONJ CN
TLF:
EML:
COM: 04

DIALETIS DIMITRIS DR
NTL OBSERVATORY OF ATHENS
P.O. BOX 20048
GR 11810 ATHENS
GREECE
TEL: 3461191 8040619
TLX: 21 5530 OBSA GR
TLF:
EML:
COM: 10

DIAZ ANGELES ISABEL DR
DPTO. DE FISICA TEORICA
C-XI UNIVERSIDAD AUTONOMA
DE MADRID CANTOBLANCO
28049 MADRID
SPAIN
TEL: 34 1 3974223
TLX: 27810
TLF:
EML: BITNET:ADIAZ@EMDUAM11
COM: 28

DICK STEVEN J
DEPT. OF THE NAVY
U.S. NAVAL OBSERVATORY
34TH & MASSACH. AVE., NW
WASHINGTON DC 20392-5100
U.S.A.
TEL:
TLX:
TLF:
EML:
COM: 08.41.51

DICKE ROBERT H PROF
JOSEPH HENRY LABS
PHYSICS DEPT
PRINCETON UNIVERSITY
PRINCETON NJ 08540
U.S.A.
TEL: 609-452-4317
TLX:
TLF:
EML:
COM: 47.48

DICKEL HELENE R DR
ASTRONOMY DEPT
UNIVERSITY OF ILLINOIS
1011 W. SPRINGFIELD AVE
URBANA IL 61801-3000
U.S.A.
TEL: 217-333-5602
TLX: 910-245-2434 AST
TLF:
EML:
COM: 05.33.34.40

DICKEL JOHN R
341 ASTRONOMY BLDG
UNIVERSITY OF ILLINOIS
1011 W. SPRINGFIELD AVE
URBANA IL 61801
U.S.A.
TEL: 217-333-5532
TLX: 910-245-2434 PURCH
TLF:
EML:
COM: 16.33.34.40

DICKENS ROBERT J DR
RUTHERFORD APPLETON LAB
SPACE & ASTROPHYS DIV
CHILTON DIDCOT OX11 0QX
U.K.
TEL: 0235-21900
TLX: 83159
TLF:
EML:
COM: 27.28.37

DICKEY JEAN O'BRIEN
JPL - CALTECH
MS 138-208
4800 OAK GROVE DRIVE
PASADENA CA 91109
U.S.A.
TEL: 818-354-3235
TLX: 675429
TLF:
EML:
COM: 04.16.19C.31

DICKEY JOHN M
UNIVERSITY OF MINNESOTA
DEPT OF ASTRONOMY
116 CHURCH STREET S.E.
MINNEAPOLIS MN 55455
U.S.A.
TEL: 612-373-3308
TLX:
TLF:
EML:
COM: 28.34.40

DICKINSON DALE F DR
LOCKHEED RESEARCH LAB
92-20 205
3251 HANOVER ST
PALO ALTO CA 94304
U.S.A.
TEL: 415-424-2701
TLX:
TLF:
EML:
COM:

DICKMAN ROBERT L DR
RADIO ASTRONOMY GRC
UNIV. OF MASSACHUSETTS
AMHERST MA 01003
U.S.A.
TEL: 413 545 0925
TLX: 95 5491
TLF:
EML:
COM: 33.40

DICKMAN STEVEN R
DEPT GEOLOGICAL SCIENCES
STATE UNIV OF NEW YORK
BINGHAMTON NY 13901
U.S.A.
TEL: 607-777-4378
TLX:
TLF:
EML:
COM: 19

DIERCKSEN GEERD H F PH D
MPI F. PHYSIK & ASTROPHYS
KARL-SCHWARZSCHILD-STR 1
D-8046 GARCHING B MUNCHEN
GERMANY. F.R.
TEL: 89-32-990
TLX: 524629 ASTRO D
TLF:
EML:
COM: 14

DIETER NANNIELOU H DR
CLAY ROAD
N. THEFORD VT 05054
U.S.A.
TEL: 802-333-4079
TLX:
TLF:
EML:
COM: 33.40

DIMITRIJEVIC MILAN
ASTRONOMSKA OPSERVATORIJA
VOLGINA 7
YU-11050 BEOGRAD
YUGOSLAVIA
TEL: 419357
TLX:
TLF:
EML:
COM:

DIN HUA
DEPT OF ASTRONOMY
NANJING UNIVERSITY
NANJING
CHINA. PEOPLE'S REP.
TEL:
TLX: 34151 PRCNU CN
TLF:
EML:
COM: 07

DINERSTEIN HARRIET L
UNIV OF TEXAS AT AUSTIN
ASTRONOMY DEPT
RLM 15.308
AUSTIN TX 78712
U.S.A.
TEL: 512-471-3449
TLX: 910-874-1351
TLF:
EML:
COM: 34

DINESCU A DR
INSTITUT DE GEODESIE
PHOTOGRAMM CARTOGRAPHIE
1A BLVD DE L'EXPOSITION
78334 BUCAREST
RUMANIA
TEL:
TLX:
TLF:
EML:
COM:

DING YOU-JI
YUNNAN OBSERVATORY
P O BOX 110
KUNMING. YUNNAN PROVINCE
CHINA. PEOPLE'S REP.
TEL: 22034
TLX: 64040 YUOBS CN
TLF:
EML:
COM: 10

DINGENS P PROF DR
KORTRIJKSE STEENWEG 763
B-9000 GENT
BELGIUM
TEL: 091-221966
TLX:
TLF:
EML:
COM: 35

DINTINJANA BOJAN DR
ASTRONOMICAL OBSERVATORY
UNIVERZA EDVARDA KARELJA
JANDRANSKA 19
YU 61111 LJUBLJANA
YUGOSLAVIA
TEL:
TLX:
TLF:
EML:
COM:

DINULESCU NICOLAE I PROF
SOSEANA KISELEFF 13
SECTOR 1
72168 BUCAREST
RUMANIA
TEL:
TLX:
TLF:
EML:
COM:

DIONYSIOU DEMETRIOS PROF
HELLENIC AIR-FORCE ACAD.
DEKELIA-ATTICA
18 ANASSIAS STR
GR-11634 ATHENS
GREECE
TEL: 7238436-2466366
TLX:
TLF:
EML:
COM: 47

DIRIKIS M A DR
LATVIAN STATE UNIVERSITY
ASTRONOMICAL OBSERVATORY
226098 RIGA
U.S.S.R.
TEL:
TLX:
TLF:
EML:
COM: 20

DISNEY MICHAEL J PROF
ASTROPHYSICS GROUP
UNIV. WALES COLL. CARDIFF
PO BOX 913
CARDIFF CF1 3TH
U.K.
TEL: 222-874785
TLX: 498635
TLF:
EML:
COM: 34.4A

DIVAN LUCIENNE DR
INSTITUT D'ASTROPHYSIQUE
98 BIS BOULEVARD ARAGO
F-75014 PARIS
FRANCE
TEL: 1-43-20-14-25
TLX: IMAG 270-0-70
TLF:
EML:
COM: 29.45

DIVARI N B DR
ODESSA POLYTECHNICAL INST
270044 ODESSA
U.S.S.R.
TEL:
TLX:
TLF:
EML:
COM: 21

DIXON ROBERT S DR
OHIO STATE UNIVERSITY
RADIO OBSERVATORY
2015 NEIL AVE
COLUMBUS OH 43210
U.S.A.
TEL: 614-422-6789
TLX:
TLF:
EML:
COM: 05.40.51

DIZER MUAMMER PROF
KANDILI OBSERVATORY
BOGAZICI UNIVERSITY
CENGELKOY
ISTANBUL
TURKEY
TEL: 3320277
TLX:
TLF:
EML:
COM: 10

DJORGOVSKI STANISLAV DR
ASTRONOMY DEPARTMENT
105 24 ROBINSON LAB.
CALTECH
PASADENA CA 91125
U.S.A.
TEL: 818 356 4415
TLX: 675425
TLF:
EML: SPAN:GEORGE@6035
COM: 22.27.51

DJURASEVIC GOJKO
UNIVERSITY OF BELGRADE
ASTRONOMSKA OPSERVATORIJA
VOLGINA 7
11050 BEOGRAD
YUGOSLAVIA
TEL: 011 419 553
TLX:
TLF:
EML:
COM:

DJUROVIC DRAGUTIN M DR
DEPT OF ASTRONOMY
FACULTY OF SCIENCES
STUDENTSKI TRG 16
YU-11000 BEOGRAD
YUGOSLAVIA
TEL: 011-420-221
TLX:
TLF:
EML:
COM: 08.19C

DLUZHNEVSKAYA O B DR
ASTRONOMICAL COUNCIL
USSR ACADEMY OF SCIENCES
PYATNITSKAYA UL 48
109017 MOSCOW
U.S.S.R.
TEL: 231-54-61
TLX: 412623 SCSTP SU
TLF:
EML:
COM: 05C.35.37

DOAZAN VERA DR
OBSERVATOIRE DE PARIS
61 AVE DE L'OBSERVATOIRE
F-75014 PARIS
FRANCE
TEL: 1-40-51-22-35
TLX: OBS PARIS 270776
TLF:
EML:
COM: 29

DOBRITSCHEV V M MR
DEPT OF ASTRONOMY
BULGARIAN ACAD SCIENCES
7TH NOVEMBER STR 1
1000 SOFIA
BULGARIA
TEL: 7341
TLX: 23561 ECFBAN BG
TLF:
EML:
COM:

DOBRONRAVIN PETER DR
CRIMEAN ASTROPHYSICAL
OBSERVATORY
NAUCHNYJ 2-2
334413 CRIMEA
U.S.S.R.
TEL:
TLX:
TLF:
EML:
COM: 09.29

DOBROVOLSKY OLEG V PROF
INST OF ASTROPHYSICS
SVIRIDENKO ST 22
734670 DUSHANBE
U.S.S.R.
TEL:
TLX:
TLF:
EML:
COM: 15C

DOBRZYCKI JERZY PROF
HISTORY OF SCIENCE
POLISH ACAD OF SCIENCES
GWIAZDZISTA 27/169
01-814 WARSZAWA
POLAND
TEL: 33-22-03
TLX:
TLF:
EML:
COM: 41

DOCOBO DURANTEZ JOSE A
OBSERVATORIO ASTRONOMICO
RAMON MARIA ALLER
P O BOX 197
SANTIAGO DE COMPOSTELA
SPAIN
TEL:
TLX:
TLF:
EML:
COM: 26

DODD RICHARD J DR
CARTER OBSERVATORY
P O BOX 2909
WELLINGTON 1
NEW ZEALAND
TEL: 728-167
TLX: 30172 MATOBS NZ
TLF:
EML:
COM:

DOGAN NADIR PROF
ANKARA UNIVERSITY
FEN FAKULTESI
ANKARA
TURKEY
TEL:
TLX:
TLF:
EML:
COM: 12

DOGGETT LEROY E DR
NAUTICAL ALMANAC OFFICE
US NAVAL OBSERVATORY
WASHINGTON DC 20392
U.S.A.
TEL: 202 653 1572
TLX:
TLF:
EML:
COM: 04.07

DOHERTY LORNE H DR
HERZBERG INST ASTROPHYS
NATIONAL RESEARCH COUNCIL
OTTAWA ONT K1A 0R6
CANADA
TEL: 613-993-6060
TLX: 053-367-15
TLF:
EML:
COM:

DOHERTY LOWELL R PROF
ASTRONOMY DEPT
UNIVERSITY OF WISCONSIN
475 N. CHARTER ST
MADISON WI 53706
U.S.A.
TEL: 608-262-1249
TLX:
TLF:
EML:
COM:

DOKUCHAEVA OLGA D DR
STERNBERG ASTRONOMICAL
INSTITUTE
UNIVERSITY PROSPECT 13
119899 MOSCOW
U.S.S.R.
TEL:
TLX:
TLF:
EML:
COM: 09.34

DOLAN JOSEPH F DR
NASA/GSFC
CODE 681
GREENBELT MD 20771
U.S.A.
TEL: 301-286-5920
TLX: 89675
TLF:
EML:
COM: 25.44.48

DOLEZ NOEL DR
OBSERVATOIRE MIDI
PYRENEES
14 AVENUE EDOUARD BELIN
31400 TOULOUSE
FRANCE
TEL:
TLX:
TLF:
EML:
COM:

DOLGINOV ARKADY Z PROF DR
IOFFE PHYSICAL TECH INST
194021 LENINGRAD
U.S.S.R.
TEL:
TLX:
TLF:
EML:
COM: 44

DOLIDZE MADONA V DR
ABASTUMANI ASTROPHYSICAL
OBSERVATORY
383762 ABASTUMANI.GEORGIA
U.S.S.R.
TEL:
TLX:
TLF:
EML:
COM: 29

DOLLFUS AUDOUIN PROF
OBSERVATOIRE DE PARIS
SECTION DE MEUDON
F-92195 MEUDON PL CEDEX
FRANCE
TEL: 1-45-34-75-30
TLX:
TLF:
EML:
COM: 10.16.20

DOMINKO FRAN PROF DR
SARANOVICEVA 11
YU-61000 LJUBLJANA
YUGOSLAVIA
TEL: 061-322-210
TLX:
TLF:
EML:
COM:

DOMINSKI IRENEUSZ DR
ASTRONOMICAL LATITUDE OBS
BOROWIEC
62-035 KORNIK
POLAND
TEL: POZNAN 170187
TLX: 412623 AOS PL
TLF:
EML:
COM: 31

DONER HELMUT PH D
ZNTRLINST. F. ASTROPHYSIK
ROSA-LUXEMBURG-STR 17A
DDR-1502 POTSDAM
GERMANY. D.R.
TEL:
TLX:
TLF:
EML:
COM: 36

DOMMANGET J DR
OBS ROYAL DE BELGIQUE
AVENUE CIRCULAIRE 3
B-1180 BRUXELLES
BELGIUM
TEL: 2-375-24-84
TLX: 21565
TLF:
EML:
COM: 24.26.50

DONAS JOSE DR
L.A.S.
TRAVERSE DU SIPHON
LES TROIS LUCS
13012 MARSEILLE
FRANCE
TEL: 91 05 59 00
TLX: 420584 F
TLF:
EML: BITNET:DONAS@FRLASM51
COM: 28

DONG IL ZUN
PYONGYANG ASTRON OBS
ACADEMY OF SCIENCES DPRK
TAESONG DISTRICT
PYONGYANG
KOREA DPR
TEL:
TLX:
TLF:
EML:
COM:

DONN BERTRAM D
NASA/GSFC
CODE 691
GREENBELT MD 20771
U.S.A.
TEL: 301-286-6859
TLX: 89675
TLF:
EML:
COM: 15.34

DONNER KARL JOHAN
OBS. & ASTR. LAB
TAHTITORNINMAKI
SF-00130 HELSINKI
FINLAND
TEL:
TLX:
TLF:
EML:
COM: 28

DONNISON JOHN RICHARD DR
DEPT OF MATH SCIENCES
GOLDSMITHS' COLLEGE
NEW CROSS
LONDON SE14 6NW
U.K.
TEL: 01-6927171
TLX:
TLF:
EML:
COM: 20

DOPITA MICHAEL ANDREW DR
MOUNT STROMLO OBSERVATORY
PRIVATE BAG
WODEN P.O. ACT 2606
AUSTRALIA
TEL: 062-88-1111
TLX: 62270 CANOPUS AA
TLF:
EML:
COM: 34C

DORENWENDT KLAUS DR
PHYSIKALISCH-TECHNISCHES
BUNDESANSTALT
BUNDESALLEE 100
D-3300 BRAUNSCHWEIG
GERMANY. F.R.
TEL: 531-592-12-10
TLX: 9-52-822 PTB D
TLF:
EML:
COM:

DORFI ERNST ANTON DR
INSTITUT FUR ASTRONOMIE
UNIVERSITAET WIEN
TUERKENSCHANZSTR. 17
A-1180 WIEN
AUSTRIA
TEL: 345360-188
TLX: 133099
TLF:
EML:
COM: 42

DORMAND JOHN RICHARD DR
MATHEMATICS DEPARTMENT
TEESSIDE POLYTECHNIC
MIDDLESBROUGH
CLEVELAND TS1 3BA
U.K.
TEL: 642-218121x4365
TLX:
TLF:
EML:
COM: 07

DOROSHKEVICH A G DR
INST OF APPLIED MATHS
USSR ACADEMY OF SCIENCES
125047 MOSCOW
U.S.S.R.
TEL: 972-37-14
TLX:
TLF:
EML:
COM: 47

DORSCHNER JOHANN DR
UNIV STERNWARTE JENA
SCHILLERGAESSCHEN 2
DDR-6900 JENA
GERMANY. D.R.
TEL: 8222637
TLX: 5886134
TLF:
EML:
COM: 34.51

DOS REIS N PROF
OBSERVATORIO ASTRONOMICO
3000 COIMBRA
PORTUGAL
TEL:
TLX:
TLF:
EML:
COM:

DOSCHEK GEORGE A DR
NAVAL RESEARCH LABORATORY
CODE 4170
WASHINGTON DC 20375
U.S.A.
TEL: 202-767-6473
TLX:
TLF:
EML:
COM:

DOSSIN F DR
INSTITUT D'ASTROPHYSIAUE
UNIVERSITE DE LIEGE
AVENUE DE COINTE 5
B-4200 COINTE-OUGREE
BELGIUM
TEL: 041-52-99-80
TLX: 41264 ASTRLG
TLF:
EML:
COM: 15

DOTTORI HORACIO A DR
INSTITUTO DE FISICA
UNIV RIO GRANDE DO SUL
AV. BENTO GONCALVES 9500
90049 PORTO ALEGRE - RS
BRAZIL
TEL: 0055-512-364677
TLX: 511055 UFRS BR
TLF:
EML:
COM: 28.34

DOUBINSKIJ B A DR
INST OF RADIO & ELECTRON
USSR ACADEMY OF SCIENCES
103907 MOSCOW
U.S.S.R.
TEL:
TLX:
TLF:
EML:
COM: 40

DOUGHTY NOEL A DR
PHYSICS DEPARTMENT
UNIVERSITY OF CANTERBURY
CHRISTCHURCH 1
NEW ZEALAND
TEL:
TLX:
TLF:
EML:
COM: 42

DOUGLAS JAMES N PROF
DEPARTMENT OF ASTRONOMY
UNIVERSITY OF TEXAS
R.L. MOORE HALL
AUSTIN TX 78712-1083
U.S.A.
TEL: 512-471-4461
TLX: 910874-1351
TLF:
EML:
COM: 40

DOUGLASS GEOFFREY G
BLACK BIRCH ASTROMETRY
OBSERVATORY
PO BOX 770
BLENHEIM
NEW ZEALAND
TEL: 64-057-87-164
TLX:
TLF:
EML:
COM: 24

DOURNEAU GERARD DR
OBSERVATOIRE DE L'UNIVER.
DE BORDEAUX I
AVE. PIERRE SEMIROT
33270 FLOIRAC
FRANCE
TEL: 56 86 43 30
TLX:
TLF:
EML: EARN::DOURNEAU@FROBOR51
COM: 07.20

DOWNES ANN JULIET B
MULLARD RADIO AST OBS
CAVENDISH LABORATORY
MADINGLEY ROAD
CAMBRIDGE CB3 0HE
U.K.
TEL: 0223-66477
TLX: 81292
TLF:
EML:
COM:

DOWNES DENNIS DR
IRAM
VOIE 10
DOMAINE UNIVERSITAIRE
F-38406 ST-MARTIN-D'HERES
FRANCE
TEL: 76-42-33-83
TLX: 980753
TLF:
EML:
COM: 33.34.40

DOWNES RONALD A DR
APPLIED RES. CORP.
8201 CORPORATE DRIVE
SUITE 920
LANDOVER MD 20785
U.S.A.
TEL: 301 459 8833
TLX:
TLF:
EML: SPAN:CHAMP::FOS
COM: 27

DOWNS GEORGE S DR
MIT LINCOLN LABORATORY
ROOM B285
PO BOX 73
LEXINGTON MA 02173
U.S.A.
TEL:
TLX:
TLF:
EML:
COM: 40.51

DOYLE JOHN GERARD
ARMAGH OBSERVATORY
ARMAGH BT61 9DG
U.K.
TEL: 861-522-928
TLX: 747937
TLF:
EML:
COM: 36

DRAINE BRUCE T
PRINCETON UNIVERSITY OBS
PRINCETON UNIVERSITY
PEYTON HALL
PRINCETON NJ 08544
U.S.A.
TEL: 609-452-3574
TLX:
TLF:
EML:
COM: 34

DRAKE FRANK D PROF
DIV OF NATURAL SCIENCES
BOARD OF STUDIES ASTR &
ASTROPHYS UNIV CALIFORNIA
SANTA CRUZ CA 95064
U.S.A.
TEL: 408-429-2501
TLX:
TLF: 408-429-0146
EML:
COM: 16.40.48.51

DRAKE STEPHEN A
NASA/GSFC
CODE 602.6 UVSP
GREENBELT MD 20771
U.S.A.
TEL: 301-286-7985
TLX:
TLF:
EML:
COM: 36.40

DRAMBA C PROF
ASTRONOMICAL OBSERVATORY
P O BOX 28
CUTITUL DE ARGINT 5
75212 BUCAREST
ROMANIA
TEL: 753998; 193407
TLX:
TLF:
EML:
COM: 19

DRAPATZ SIEGFRIED W DR
MPI F EXTRATERRESTRISCHE
PHYSIK
KARL-SCHWARZSCHILD-STR 1
D-8046 GARCHING B MUNCHEN
GERMANY. F.R.
TEL: 089-3299880
TLX: 05215845
TLF:
EML:
COM: 34

DRAVINS DAINIS PROF
LUND OBSERVATORY
BOX 43
S-221 00 LUND
SWEDEN
TEL: 46-10 70 00
TLX: 33199 OBSNOT S
TLF:
EML:
COM: 09.12.29.36

DRAVSKIKH A F DR
SPECIAL ASTROPHYSICAL OBS
USSR ACADEMY OF SCIENCES
LENINGRAD BRANCH
196140 LENINGRAD
U.S.S.R.
TEL: 297-94-52
TLX:
TLF:
EML:
COM: 08.40

DRECHSEL HORST DR
DR-REMEIS-STERNWARTE
ASTR INST UNIV ERLANGEN-N
STERNWARTSTR 7
D-8600 BAMBERG
GERMANY. F.R.
TEL: 0951-57708
TLX: 629830 UNIER D
TLF:
EML:
COM: 42

DREHER JOHN W
M I T
DEPT OF PHYSICS
ROOM 26-315
CAMBRIDGE MA 02139
U.S.A.
TEL: 617-253-8519
TLX: 921473
TLF:
EML:
COM: 09.34.40

DRESSEL LINDA L
DEPT SPACE PHYS & ASTRON
RICE UNIVERSITY
PO BOX 1892
HOUSTON TX 77251
U.S.A.
TEL: 713-527-8101
TLX:
TLF:
EML:
COM: 28

DRESSLER ALAN
MT WILSON & LAS CAMPANAS
OBSERVATORIES
813 SANTA BARBARA STREET
PASADENA CA 91101-1292
U.S.A.
TEL: 818-304-0245
TLX: 675425 CALTECH PSD
TLF:
EML:
COM: 28.47C

DRESSLER KURT PROF
LAB PHYSIK CHEMIE
ETH-ZENTRUM
CH-8092 ZUERICH
SWITZERLAND
TEL: 256-4441
TLX: 53178 ETHBI CH
TLF:
EML:
COM: 14

DREVER RONALD W P DR
DEPT NATURAL PHILOSOPHY
GLASGOW UNIVERSITY
GLASGOW G12 8QQ
U.K.
TEL:
TLX:
TLF:
EML:
COM:

DREW JANET
DEPT OF ASTROPHYSICS
SOUTH PARKS ROAD
OXFORD OX1 3RQ
U.K.
TEL: 0865-511336
TLX: 851-83295
TLF:
EML:
COM:

DRILLING JOHN S
DEPT PHYSICS & ASTRONOMY
LOUISIANA STATE UNIV
BATON ROUGE LA 70803
U.S.A.
TEL: 504-388-6795
TLX:
TLF:
EML:
COM: 33

DROFA VASILIY K DR
DEPARTMENT OF ASTRONO
KIEV UNIVERSITY
252127 KIEV
U.S.S.R.
TEL:
TLX:
TLF:
EML:
COM:

DROZYNER ANDRZEJ
DGFI MARSTALLPLATZ 8
8000 MUNCHEN 22
GERMANY, F.R.
TEL:
TLX:
TLF:
EML:
COM: 07

DRURY LUKE O'CONNOR DR
DUBLIN INSTITUTE FOR
ADVANCED STUDIES
5 MERRION SQUARE
DUBLIN 2
IRELAND
TEL: 353 1 774321
TLX: 31687 DIAS
TLF:
EML: LD@DIASCP.uucp
COM: 48

DRYER MURRAY DR
SPACE ENVIRONMENT LAB
NOAA ERL (R/E/SE)
325 BROADWAY
BOULDER CO 80303
U.S.A.
TEL: 303-497-3978
TLX: 592811 NOAA MASC BDR
TLF:
EML:
COM: 10.15.49

DUBAU JACQUES DR
OBSERVATOIRE DE PARIS
SECTION DE MEUDON
F-92195 MEUDON PL CED
FRANCE
TEL: 1-45-07-74-56
TLX: 270912 OBSASTR
TLF:
EML:
COM: 14

DUBNER GLORIA DR
INSTITUTO ARGENTINO DE
RADIOASTRONOMIA
CASILLA DE CORREO No. 5
1894 VILLA ELISA (Bs.As.)
ARGENTINA
TEL: 21-4-3793
TLX: 22414 CEDOC AR
TLF:
EML:
COM: 34

DUBOIS MARC A
CEA
D R F C
BP 6
F-92260 FONTENAY-AUX-ROS.
FRANCE
TEL: 1-46-54-78-81
TLX:
TLF:
EML:
COM: 10

DUBOIS PASCAL DR
OBS DE STRASBOURG
11 RUE DE L'UNIVERSITE
F-67000 STRASBOURG
FRANCE
TEL: 88-35-43-00
TLX:
TLF:
EML:
COM: 05.28

DUBOUT RENEE
OBSERVATOIRE DE LYON
AVENUE CHARLES ANDRE
F-69230 ST-GENIS-LAVAL
FRANCE
TEL: 78-56-07-05
TLX: 310926
TLF:
EML:
COM: 25.34

DUBOV EMIL E PROF
WDC-B2
MOLODEZHNAYA 3
117296 MOSCOW
U.S.S.R.
TEL: 923-55-71
TLX: 411478 SGC SU
TLF:
EML:
COM: 10.12

DUCATI JORGE RICARDO DR
INSTITUTO DE FISICA
UFRGS
AV. BENTO GONCALVES 9500
90000 PORTO ALEGRE - RS
BRAZIL
TEL: 512-364677
TLX: 051-1055 UFRS BR
TLF:
EML:
COM: 05.25.33

DUCHESNE MAURICE DR
OBSERVATOIRE DE PARIS
61 AVE DE L'OBSERVATOIRE
F-75014 PARIS
FRANCE
TEL: 1-43-20-12-10
TLX:
TLF:
EML:
COM: 09

DUERBECK HILMAR W DR
ASTRONOMISCHES INSTITU
UNIVERSITAT MUENSTER
DOMAGKSTR. 75
D-4400 MUENSTER
GERMANY, F.R.
TEL: 251-833561
TLX:
TLF:
EML:
COM: 42

DUERST JOHANNES DR
LANGWIES
CH-8821 SCHOENENBERG
SWITZERLAND
TEL: 01-788-1785
TLX:
TLF:
EML:
COM:

DUFAY MAURICE PROF
UNIVERSITE CLAUDE-BERNARD
LYON I
43 BD DU 11 NOVEMBRE
F-69621 VILLEURBANNE
FRANCE
TEL:
TLX:
TLF:
EML:
COM: 14.21

DUFFETT-SMITH PETER JAMES
MULLARD RADIO ASTRON OBS
CAVENDISH LABORATORY
MADINGLEY ROAD
CAMBRIDGE CB3 OHE
U.K.
TEL: 0223-66477
TLX: 81292 CAVLAB G
TLF:
EML:
COM: 40

DUFLOT MARCELLE DR
OBSERVATOIRE DE MARSEILLE
2 PLACE LE VERRIER
F-13248 MARSEILLE CEDEX 4
FRANCE
TEL: 91-95-90-88
TLX: 420241
TLF:
EML:
COM: 30.45

DUFOUR REGINALD JAMES
SPACE PHYSICS DEPARTMENT
RICE UNIVERSITY
204K SPACE SCIENCE BLDG.
HOUSTON TX 77001
U.S.A.
TEL: 713-527-8101
TLX:
TLF:
EML:
COM: 28.34

DUFTON PHILIP L DR
DEPT PURE & APPLIED PHYS
QUEEN'S UNIVERSITY
BELFAST BT7 1NN
U.K.
TEL: 245133
TLX: 74487
TLF:
EML:
COM: 36

DULEY WALTER W PROF
PHYSICS DEPARTMENT
YORK UNIVERSITY
4700 KEELE STREET
DOWNSVIEW ONT M3J1P3
CANADA
TEL: 416-667-3040
TLX:
TLF:
EML:
COM: 34

DULK GEORGE A PROF
DEPT ASTROPHYS SCIENCES
UNIVERSITY OF COLORADO
CB 391
BOULDER CO 80309
U.S.A.
TEL: 303-492-8788
TLX:
TLF:
EML:
COM: 10.40

DULTZIN-HACYAN D. DR
ASTROFISICA ANDALUCIA
APDO CORREOS 2144
GRANADA 18080
SPAIN
TEL: 58-121300
TLX: 78573 IAAGE
TLF:
EML:
COM: 28.47

DUMA DMITRIJ P DR
MAIN ASTRONOMICAL OBS
UKRAINIAN ACADEMY OF SCI
252127 KIEV
U.S.S.R.
TEL: 66-31-10
TLX: 131406 SKY SU
TLF:
EML:
COM: 08C

DUMONT RENE DR
OBSERVATOIRE DE BORDEAUX
B.P. 21
F-33270 FLOIRAC
FRANCE
TEL: 56-86-43-30
TLX:
TLF:
EML:
COM: 21C

DUMONT SIMONE DR
INSTITUT D'ASTROPHYSIQUE
98 BIS BOULEVARD ARAGO
F-75014 PARIS
FRANCE
TEL: 1-43-20-14-25
TLX:
TLF:
EML:
COM: 12

DUNCAN DOUGLAS KEVIN DR
SPACE TELESCOPE SC INST
HOMEWOOD CAMPUS
3700 SAN MARTIN DR
BALTIMORE MD 21218
U.S.A.
TEL: 301-338-4935
TLX:
TLF:
EML:
COM: 29

DUNCAN ROBERT A PROF
CSIRO
DIVISION OF RADIOPHYSICS
P.O.BOX 76
EPPING NSW 2121
AUSTRALIA
TEL:
TLX:
TLF:
EML:
COM: 10

DUNCOMBE RAYNOR L DR
DEPT OF AEROSPACE ENG
UNIVERSITY OF TEXAS
AUSTIN TX 78712
U.S.A.
TEL: 512-471-4239
TLX: 704265 CSRUTX UD
TLF:
EML:
COM: 04C.05.07.08

DUNHAM DAVID W
COMPUTER SCIENCES CORP.
10110 AEROSPACE RD
LANHAM SEABROOK MD 20706
U.S.A.
TEL: 301-794-1392
TLX: 7108259636 CSCSSD
TLF:
EML:
COM: 04.20.26

DUNKELMAN LAURENCE
LUNAR & PLANETARY LAB
UNIVERSITY OF ARIZONA
PO BOX 36241
TUCSON AZ 85740
U.S.A.
TEL: 602-621-6963
TLX:
TLF:
EML:
COM: 09.12.21.44.50

DUNLOP STORM
140 STOCKS LANE
EAST WITTERING
CHICHESTER W SUS.PO20 8NT
U.K.
TEL: 0243 670354
TLX: 9312-1107-40
TLF:
EML: TELECOM GOLD72:MAG100665
COM: 27

DUNN RICHARD B DR
NATL SOLAR OBSERVATORY
SUNSPOT NM 88349
U.S.A.
TEL: 505-434-1390
TLX:
TLF:
EML:
COM: 10.12

DUORAH HIRA LAL DR
DEPARTMENT OF PHYSICS
GAUHATI UNIVERSITY
GUWAHATI (ASSAM) 781014
INDIA
TEL: GUWUHATI 88531
TLX:
TLF:
EML:
COM: 48

DUPREE ANDREA K DR
SOLAR & STELLAR DIVISION
CENTER FOR ASTROPHYSICS
60 GARDEN STREET
CAMBRIDGE MA 02138
U.S.A.
TEL: 617-495-7489
TLX: 921428 SATELLITE CAM
TLF:
EML:
COM: 34.36.44

DUPUY DAVID L DR
DEPARTMENT OF PHYSICS
VIRGINIA MILITARY INST
LEXINGTON VA 24450
U.S.A.
TEL: 703-463-6225
TLX:
TLF:
EML:
COM: 27.46

DURGAPRASAD N DR
TATA INSTITUTE OF
FUNDAMENTAL RESEARCH
BOMBAY 400 005
INDIA
TEL: 219111 x 342
TLX: 0113009
TLF:
EML:
COM:

DURIEZ LUC DR
LABORATOIRE D'ASTRONOMIE
1 IMPASSE DE
L'OBSERVATOIRE
59000 LILLE
FRANCE
TEL: 20 52 44 24
TLX:
TLF:
EML: EARN:DURIEZ@FRCITL71
COM: 07

DURISEN RICHARD H DR
DEPARTMENT OF ASTRONOMY
INDIANA UNIVERSITY
SWAIN WEST 319
BLOOMINGTON IN 47405
U.S.A.
TEL: 812-335-6921
TLX:
TLF:
EML:
COM: 35.42

DURNEY BERNARD DR
NATL SOLAR OBSERVATORY
P.O. BOX 26732
TUCSON AZ 85726
U.S.A.
TEL: 602-327-5511
TLX:
TLF:
EML:
COM: 49

DUROUCHOUX PHILIPPE
CEA CEN/SACLAY
DPHG/SAP
F-91191 GIF/YVETTE CEDEX
FRANCE
TEL: 69-08-33-76
TLX: 690860 PHYSPAC F
TLF:
EML:
COM: 48

DURRANCE SAMUEL T DR
DPT OF PHYS. & ASTRONOMY
JOHNS HOPKINS UNIVERSITY
CHARLES AND 34TH STREETS
BALTIMORE MD 21218
U.S.A.
TEL: 301 338 8707
TLX: 9102400225 JHU CAS
TLF:
EML:
COM: 16

DURRANT CHRISTOPHER J DR
DEPT APPLIED MATHEMATICS
UNIVERSITY OF SYDNEY
SYDNEY NSW 2006
AUSTRALIA
TEL: 02-692-3373
TLX: 20056 FISHLIB AA
TLF:
EML:
COM:

DUSCHL WOLFGANG J DR
INSTITUT F THEORETISCHE
ASTROPHYSIK
IM NEUENHEIMER F 561
D-6900 HEIDELBERG
GERMANY. F.R.
TEL: 49(6221)562967
TLX: 46515 UNIHD D
TLF:
EML: BITNET:CJ0@DHDURZ1
COM: 42

DUTHIE JOSEPH G PROF
UNIVERSITY OF ROCHESTER
DEPT PHYSICS & ASTRONOMY
ROCHESTER NY 14627
U.S.A.
TEL:
TLX:
TLF:
EML:
COM: 48

DUVAL MARIE-FRANCE
OBSERVATOIRE DE MARSEILLE
2 PLACE LE VERRIER
F-13248 MARSEILLE CEDEX 4
FRANCE
TEL: 91-95-90-88
TLX:
TLF:
EML:
COM: 28.46

DUVALL THOMAS L JR
NATL SOLAR OBSERVATORY
PO BOX 26732
950 N. CHERRY AVE
TUCSON AZ 85726
U.S.A.
TEL: 602-325-9338
TLX: 666-484 AURA-KPNO-
TLF:
EML:
COM: 12

DUVERT GILLES DR
GROUPE D'ASTROPHYSIQUE
OBSERVATOIRE DE GRENOBLE
BP 53X
38401 GRENOBLE CEDEX
FRANCE
TEL: 76 51 48 85
TLX:
TLF:
EML: BITNET:DUVERT@FRGAG51
COM:

DVORAK RUDOLF DR
INSTITUT FUER ASTRONOMIE
UNIVERSITAETSSTERNWARTE
TUERKENSCHANZSTRASSE 17
A-1180 VIENNA
AUSTRIA
TEL: 222-34-53-600
TLX:
TLF:
EML:
COM: 07.20

DWARAKANATH K S
RAMAN RESEARCH INST.
SADASHIVANAGAR
BANGALORE 560 080
INDIA
TEL: 812 360 122
TLX: 845 2671 RRI IN
TLF:
EML:
COM: 40

DWEK ELI
NASA/GSFC CODE 697
LAB EXTRATERRESTR.PHYSICS
GREENBELT MD 20771
U.S.A.
TEL: 301-286-6209
TLX:
TLF:
EML:
COM: 34

DWIVEDI BHOLA NATH DR
DEPT APPLIED PHYSICS.I.T.
BANARAS HINDU UNIVERSITY
VARANASI 221 005
INDIA
TEL:
TLX: 0545-208 TECH IN
TLF:
EML:
COM: 10

DWORETSKY MICHAEL M DR
DEPT PHYSICS & ASTRONOMY
UNIVERSITY COLLEGE LONDON
GOWER STREET
LONDON WC1E 6BT
U.K.
TEL: 01-387-7050
TLX: 28722
TLF:
EML:
COM: 29

DYCK M DR
KITT PEAK NATIONAL OBS
PO BOX 26732
TUCSON AZ 85726
U.S.A.
TEL:
TLX:
TLF:
EML:
COM:

DYER CHARLES CHESTER DR
PHYS SCS GR RM S-650
SCARBOROUGH COLLEGE
UNIVERSITY OF TORONTO
TORONTO ONT M1C 1A4
CANADA
TEL: 416-284-3318
TLX:
TLF:
EML:
COM: 47

DYER EDWARD R DR
3626 DAVIS STREET N.W.
WASHINGTON DC 20007
U.S.A.
TEL:
TLX:
TLF:
EML:
COM:

DYSON F J DR
INST FOR ADVANCED STUDY
PRINCETON NJ 08540
U.S.A.
TEL: 609-734-8055
TLX:
TLF:
EML:
COM: 40.51

DYSON JOHN E DR
ASTRONOMY DEPARTMENT
UNIVERSITY OF MANCHESTER
MANCHESTER M13 9PL
U.K.
TEL: 061-273-7121
TLX: 668932
TLF:
EML:
COM: 34.49

DZHAPIASHVILI VICTOR P DR
ABASTUMANI ASTROPHYSICAL
OBSERVATORY
383762 ABASTUMANI.GEORGIA
U.S.S.R.
TEL:
TLX:
TLF:
EML:
COM: 15.16

DZIEMBOWSKI WOJCIECH PROF
ASTRONOMICAL CENTER
UL. BARTYCKA 18
00-716 WARSAW
POLAND
TEL:
TLX:
TLF:
EML:
COM: 27.35

DZIGVASHVILI R M DR
ABATSUMANI ASTROPHYSICAL
OBSERVATORY
383762 ABASTUMANI
U.S.S.R.
TEL:
TLX:
TLF:
EML:
COM: 33

EATON JOEL A DR
INDIANA UNIVERSITY
ASTRONOMY DEPT
SWAIN HALL WEST 319
BLOOMINGTON IN 47405
U.S.A.
TEL: 812-335-4176
TLX:
TLF:
EML:
COM: 42

ECCLES MICHAEL J DR
SUNNYSIDE
BALLENCRIEFF TOLL
BATHGATE EH48 4LD
U.K.
TEL: 0506-53989
TLX: 727484
TLF:
EML:
COM: 51

ECHEVARRIA JUAN DR
OBSERV. ASTRONOMICO NL
APARTADO POSTAL 877
ENSENADA BC 22800
MEXICO
TEL:
TLX:
TLF:
EML:
COM: 42

ECHEVERRIA ROMAN JUAN M.
APDO POSTAL 877
22860 ENSENADA. B. CALIF.
MEXICO
TEL:
TLX:
TLF:
EML:
COM:

EDDY JOHN A DR
UNIV. CORP FOR ATMOS. RES
PO BOX 3000
BOULDER CO 80307
U.S.A.
TEL: 303-497-1150
TLX: 45694
TLF:
EML:
COM: 10.41C

EDELMAN COLETTE DR
BUREAU DES LONGITUDES
77 AV. DENFERT ROCHEREAU
75014 PARIS
FRANCE
TEL: 1 40 51 22 72
TLX:
TLF:
EML:
COM:

EDLEN BENGT PROF
DEPARTMENT OF PHYSICS
UNIVERSITY OF LUND
SOELVEGATAN 14
S-223 62 LUND
SWEDEN
TEL: 046-107730
TLX:
TLF:
EML:
COM: 14

EDMONDS FRANK N JR DR
DEPARTMENT OF ASTRONOMY
UNIVERSITY OF TEXAS
RLM 15.212
AUSTIN TX 78712
U.S.A.
TEL: 512-471-4461
TLX:
TLF:
EML:
COM: 12.29.36

EDMONDSON FRANK K PROF
GOETHE LINK OBSERVATORY
INDIANA UNIVERSITY
319 A SWAIN HALL WEST
BLOOMINGTON IN 47405
U.S.A.
TEL: 817-335-6918
TLX:
TLF:
EML:
COM: 20.30.33.41

EDMUNDS MICHAEL GEOFFREY
DEPT APPLIED MATH & AST
UNIVERSITY COLLEGE
PO BOX 78
CARDIFF CF1 1XL
U.K.
TEL: 0222-44211
TLX: 498635 ULIBCF
TLF:
EML:
COM: 28

EDWARDS ALAN CH DR
APARTMENT 4
101 HIAWATHA AVENUE
SANTA CRUZ CA 95062
U.S.A.
TEL:
TLX:
TLF:
EML:
COM: 35

EDWARDS PAUL J DR
MT STROMLO OBSERVATORY
PRIVATE BAG
WODEN P.O. ACT 2606
AUSTRALIA
TEL: 062-88-1111
TLX: 68270 AA
TLF:
EML:
COM: 25.27.48.50

EDWARDS TERRY W
DEPT PHYSICS & ASTRONOMY
UNIVERSITY OF MISSOURI
COLUMBIA MO 65211
U.S.A.
TEL: 314-882-3036
TLX:
TLF:
EML:
COM: 35

EDWIN ROGER P
UNIVERSITY OBSERVATORY
BUCHANAN GARDENS
ST ANDREWS. FIFE KY16 9LZ
U.K.
TEL:
TLX:
TLF:
EML:
COM: 09

EELSALU HEINO DR
TARTU OBSERVATORY
202444 TORAVERE. ESTONIA
U.S.S.R.
TEL: 41469 TARTU
TLX:
TLF:
EML:
COM: 25.30.41

EFREMOV YU I DR
INST FOR APPLIED MATHS
USSR ACADEMY OF SCIENCES
125047 MOSCOW
U.S.S.R.
TEL:
TLX:
TLF:
EML:
COM:

EFREMOV YURY N DR
STERNBERG ASTRON INST
UNIVERSITETSKY PROSP. 13
119899 MOSCOW
U.S.S.R.
TEL: 139 26 57
TLX:
TLF:
EML:
COM: 27.33.37

EFSTATHIOU GEORGE
INSTITUTE OF ASTRONOMY
MADINGLEY ROAD
CAMBRIDGE CB3 0HA
U.K.
TEL: 0223-62204
TLX: 817297 ASTRON G
TLF:
EML:
COM: 28.47

EGGLETON PETER P DR
INSTITUTE OF ASTRONOMY
MADINGLEY ROAD
CAMBRIDGE CB3 0HA
U.K.
TEL: 223-62204
TLX: 817297 ASTRON G
TLF:
EML:
COM: 35.42

EGRET DANIEL DR
OBS DE STRASBOURG
11 RUE DE L'UNIVERSITE
F-67000 STRASBOURG
FRANCE
TEL: 88-35-43-00
TLX: 890506 STAROBS
TLF:
EML:
COM: 05.33.45

EHLERS JURGEN PROF
MPI FUER PHYSIK UND
ASTROPHYSIK
KARL-SCHWARZSCHILD-STR 1
D-8046 GARCHING B MUNCHEN
GERMANY. F.R.
TEL: 089-3299-0444
TLX: 524629 ASTRO D
TLF:
EML:
COM: 47

EICHHORN HEINRICH K DR
DEPARTMENT OF ASTRONOMY
UNIVERSITY OF FLORIDA
211 SPACE SC RES BLDG
GAINESVILLE FL 32611
U.S.A.
TEL: 904-392-2052
TLX:
TLF:
EML:
COM: 07.08.24.26

EICHLER DAVID DR
ASTRONOMY PROGRAM
UNIVERSITY OF MARYLAND
COLLEGE PK MD 20742
U.S.A.
TEL: 301-454-6448
TLX: 710-8260352
TLF:
EML:
COM: 48

EILEK JEAN
PHYSICS DEPARTMENT
NEW MEXICO TECH
SOCORRO NM 87801
U.S.A.
TEL: 505-835-5433
TLX:
TLF:
EML:
COM: 48

EINASTO JAAN DR
TARTU ASTROPHYSICAL OBS
ESTONIAN ACAD OF SCIENCES
202444 TORAVERE. ESTONIA
U.S.S.R.
TEL:
TLX:
TLF:
EML:
COM: 28.33C.37.47

EINAUDI GIORGIO
DIP. DI ASTRONOMIA
UNIVERSITA DEGLI STUDI
LARGO E. FERMI 5
I-50125 FIRENZE
ITALY
TEL: 055 27521
TLX: 572268 ARCETR I
TLF:
EML:
COM: 12

EINICKE O H DR
UNIVERSITY OBSERVATORY
OESTER VOLDGADE 3
DK-1350 COPENHAGEN K
DENMARK
TEL: 1-14-17-90
TLX:
TLF:
EML:
COM:

EIROA DE SAN FRANCISCO C
OBSERVATORIO ASTRONOMICO
NACIONAL
CALLE ALFONSO XII. 3
28014 MADRID
SPAIN
TEL: 1 2270107
TLX: 49880 OANM
TLF:
EML:
COM:

EKERS RONALD D DR
CSIRO DIVISION OF
RADIOPHYSICS
PO BOX 76
EPPING NSW 2121
AUSTRALIA
TEL: 868-0222
TLX: 26230
TLF:
EML:
COM: 28.40

EL EID MOUNIB DR
UNIVERSITATS-STERNWARTE
GEISMARLANDSTR.11
98 BIS. BOULEVARD ARAGO
D-3400 GOTTINGEN
GERMANY. F.R.
TEL: 395042
TLX: 96753
TLF:
EML:
COM:

EL NAWAWAY MOHAMED SALEH
ACADEMY OF SCIENTIFIC
RESEARCH AND TECHNOLOGY
101 KASR EL EINI STREET
CAIRO
EGYPT
TEL:
TLX:
TLF:
EML:
COM:

EL SHALABY MOHAMED
ASTRONOMY & METEOROL DEPT
CAIRO UNIVERSITY
CAIRO
EGYPT
TEL:
TLX:
TLF:
EML:
COM: 34

EL-BASSUNY ALAWY A A
HIAG
HELWAN CAIRO
EGYPT
TEL: 782683
TLX: 9703 HIAG UN
TLF:
EML:
COM: 27.37

EL-BAZ FAROUK DR
ITEK OPTICAL SYSTEMS
10 MAGUIRE ROAD
LEXINGTON MA 02173
U.S.A.
TEL: 617-276-2532
TLX: 923456
TLF:
EML:
COM: 16

EL-RABY MOHAMED E DR
DEPT ENVIRONMENT. STUDIES
INST GRADUATE STUD.& RES.
UNIVERSITY OF ALEXANDRIA
ALEXANDRIA
EGYPT
TEL:
TLX:
TLF:
EML:
COM: 44

EL-SHAARAWY M B DR
HELWAN OBSERVATORY
HELWAN-CAIRO
EGYPT
TEL:
TLX:
TLF:
EML:
COM:

ELFORD WILLIAM GRAHAM DR
DEPARTMENT OF PHYSICS
UNIVERSITY OF ADELAIDE
GPO BOX 498
ADELAIDE 5001
AUSTRALIA
TEL: 02-228-5371
TLX: 89141 UNIVQD AA
TLF:
EML:
COM: 22

ELGAROY OYSTEIN PROF
INST THEORET ASTROPHYSICS
UNIVERSITY OF OSLO
P O BOX 1029
N-0315 BLINDERN. OSLO 3
NORWAY
TEL: 02-456-504
TLX:
TLF:
EML:
COM: 40

ELIPE SANCHEZ ANTONIO
DPTO FIS TIERRA & COSMOS
UNIVERSIDAD DE ZARAGOZA
50009 ZARAGOZA
SPAIN
TEL: 976-357011
TLX: 58198
TLF:
EML:
COM: 07

ELITZUR MOSHE
DEPT PHYSICS & ASTRONOMY
UNIVERSITY OF KENTUCKY
LEXINGTON KY 40506-0055
U.S.A.
TEL: 606-257-4720
TLX:
TLF:
EML:
COM: 34

ELLDER JOEL DR
ONSALA SPACE OBSERVATORY
S-439 34 ONSALA
SWEDEN
TEL:
TLX:
TLF:
EML:
COM:

ELLIOT JAMES L DR
DEPT EARTH & PLANET SCI
MIT
BLDG 54-422A
CAMBRIDGE MA 02139
U.S.A.
TEL: 617-253-6308
TLX: 921473 MIT CAM
TLF:
EML:
COM: 16.20

ELLIOTT IAN DR
DUNSINK OBSERVATORY
DUBLIN 15
IRELAND
TEL: 1-387-959
TLX: 31687 DIAS EI
TLF:
EML:
COM: 12

ELLIOTT KENNETH H DR
DEPT OF SPACE RESEARCH
UNIVERSITY OF BIRMINGHAM
PO BOX 363
BIRMINGHAM B15 2TT
U.K.
TEL: 021-472-1301
TLX: 338938
TLF:
EML:
COM: 34

ELLIS G R A PROF
UNIVERSITY OF TASMANIA
P.O.BOX 252C
HOBART. TASMANIA
AUSTRALIA
TEL:
TLX: 58150
TLF:
EML:
COM: 40

ELLIS GEORGE F R PROF
SISSA
STRADA COSTIERA 11
MIRAMARE
TRIESTE 34 014
ITALY
TEL: 40 224118
TLX: 460 392 ICTP
TLF:
EML:
COM: 47.51

ELLIS RICHARD S
PHYSICS DEPARTMENT
DURHAM UNIVERSITY
SOUTH ROAD
DURHAM DH1 3LE
U.K.
TEL:
TLX:
TLF:
EML:
COM: 28C.47

ELMEGREEN BRUCE GORDON DR
IBM
THOMAS J. WATSON RES CTR
PO BOX 218
YORKTOWN HEIGHTS NY 10598
U.S.A.
TEL: 914-945-2448
TLX: 137456
TLF:
EML:
COM: 34.37

ELMEGREEN DEBRA MELOY
IBM
THOMAS J. WATSON RES CTR
PO BOX 218
YORKTOWN HEIGHTS NY 10598
U.S.A.
TEL: 914-945-2448
TLX: 137456
TLF:
EML:
COM: 28.33.34

ELSAESSER HANS PROF
MPI FUR ASTRONOMIE
KOENIGSTUHL
D-6900 HEIDELBERG
GERMANY. F.R.
TEL: 62-21-528-200
TLX:
TLF:
EML:
COM: 21.33

ELSMORE BRUCE DR
CAVENDISH LABORATORY
MADINGLEY ROAD
CAMBRIDGE CB3 OHE
U.K.
TEL: 0223-66477
TLX: 81292
TLF:
EML:
COM: 19.24.40

ELST ERIC WALTER DR
KONINKLIJKE STERRENWACHT
VAN BELGIE
RINGLAAN 3
B-1180 BRUSSEL
BELGIUM
TEL:
TLX:
TLF:
EML:
COM:

ELSTE GUNTHER H DR
DEPARTMENT OF ASTRONOMY
UNIVERSITY OF MICHIGAN
DENNISON BUILDING
ANN ARBOR MI 48109
U.S.A.
TEL: 313-764-3444
TLX:
TLF:
EML:
COM: 10.12.36

ELSTON WOLFGANG E PROF
DEPARTMENT OF GEOLOGY
UNIVERSITY OF NEW MEXICO
ALBUQUERQUE NM 87131
U.S.A.
TEL: 505-277-5339
TLX: 660461
TLF:
EML:
COM: 16

ELVIS MARTIN S DR
HARVARD SMITHSONIAN CTR
FOR ASTROPHYSICS
60 GARDEN STREET
CAMBRIDGE MA 02138
U.S.A.
TEL: 617-495-7442
TLX: 921428
TLF:
EML:
COM: 28.44.47.48

ELVIUS AINA M PROF
STOCKHOLM OBSERVATORY
S-133 00 SALTSJOBADEN
SWEDEN
TEL: 08-7170195
TLX: 12972 SODDSERV S
TLF:
EML:
COM: 28.34

ELVIUS TORD PROF EMERITUS
NORRLANDSGATAN 34F
S-752 29 UPPSALA
SWEDEN
TEL: 018-100857
TLX:
TLF:
EML:
COM: 33.45

ELVERT GERHARD PROF
LEHRSTUHL F THEORETISCHE
ASTROPHYSIK
UNIVERSITAET TUEBINGEN
D-7400 TUEBINGEN
GERMANY. F.R.
TEL: 07071/296483
TLX: 7-262714 ALT D
TLF:
EML:
COM: 10.40

EMELIANOV NIKOLAJ V DR
STERNBERG STATE
ASTRONOMICAL INSTITUTE
UNIVERSITETSKIJ PROSP 13
119899 MOSCOW
U.S.S.R.
TEL: 139-37-64
TLX:
TLF:
EML:
COM: 07

EMERSON DAVID
DEPT OF ASTRONOMY
UNIVERSITY OF EDINBURGH
ROYAL OBSERVATORY
EDINBURGH EH9 3HJ
U.K.
TEL:
TLX:
TLF:
EML:
COM: 28.40.46

EMERSON JAMES P
DEPT OF PHYSICS
QUEEN MARY COLLEGE
MILE END ROAD
LONDON E1 4NS
U.K.
TEL: 01-980-4811
TLX: 893750 QMCUOL G
TLF:
EML:
COM: 34

EMINZADE T A DR
SHEMAKHA ASTROPHYSICAL
OBSERVATORY
373243 AZERBAIDZAN
U.S.S.R.
TEL:
TLX:
TLF:
EML:
COM: 35

EMSLIE A. GORDON
UNIVERSITY OF ALABAMA
DEPT OF PHYSICS
HUNTSVILLE AL 35899
U.S.A.
TEL: 205-895-6167
TLX:
TLF:
EML:
COM: 10

EMARD DANIEL DR
ESO
KARL-SCHARZSCHILDSTR. 2
D-8046 GARCHING B MUNCHEN
GERMANY. F.R.
TEL: 89-320 06 251
TLX:
TLF:
EML:
COM:

ENCRENAZ PIERRE J DR
ECOLE NORMALE SUPERIEURE
24 RUE LHOMOND
F-75005 PARIS
FRANCE
TEL: 1-43-29-12-35
TLX:
TLF:
EML:
COM: 34

ENCRENAZ THERESE DR
OBSERVATOIRE DE PARIS
SECTION DE MEUDON
GROUPE PLANETES
F-92195 MEUDON PL CEDEX
FRANCE
TEL: 1-45-07-76-60
TLX: 204464
TLF:
EML:
COM: 15.16C

ENDAL ANDREW S DR
APPLIED RESEARCH CO
8201 CORPORATE DRIVE
LANDOVER MD 20785
U.S.A.
TEL: 301-459-8442
TLX:
TLF:
EML:
COM: 35

ENGELBRECHT CHRISTIAN DR
UNIVERSITY OF STELLENBOSC
PRETORIA
SOUTH AFRICA
TEL:
TLX:
TLF:
EML:
COM:

ENGELHARD E J G PROF DR
PHYS-TECHN-BUNDESANSTALT
BRAUNSCHWEIG
SACKRING 34
D-3300 BRAUNSCHWEIG
GERMANY. F.R.
TEL: 0531-56365
TLX:
TLF:
EML:
COM: 14

ENGIN SEMANUR PROF
DEPARTMENT OF ASTRONOMY
UNIVERSITY OF ANKARA
FEN FAKULTESI
ANKARA
TURKEY
TEL:
TLX:
TLF:
EML:
COM:

ENGVOLD ODDBJOERN DR
INST THEORET ASTROPHYSICS
UNIVERSITY OF OSLO
P O BOX 1029
N-0315 BLINDERN. OSLO 3
NORWAY
TEL:
TLX:
TLF:
EML:
COM: 09 10C

ENOME SHINZO PROF
TOYOKAWA OBSERVATORY
NAGOYA UNIVERSITY
13 HONOHARA 3-CHOME
TOYOKAWA 442
JAPAN
TEL: 5338-6-3154
TLX: 4322-310 TYKW J
TLF:
EML:
COM: 10C.40

ENSLIN HEINZ DR
TIERLAENDER WEG 5
2057 REINBECK
GERMANY. F.R.
TEL: 040-7223616194
TLX:
TLF:
EML:
COM: 19.31

EPPS HARLAND WARREN PROF
DEPARTMENT OF ASTRONOMY
UNIVERSITY OF CALIFORNIA
MATH SCI RM 8983
LOS ANGELES CA 90024
U.S.A.
TEL: 213-825-3025
TLX: 910-3427597
TLF:
EML:
COM:

EPSTEIN EUGENE E DR
THE AEROSPACE CORPORATION
2118 PATRICIA AVE
LOS ANGELES CA 90025
U.S.A.
TEL: 213-648-6798
TLX: 664460
TLF:
EML:
COM: 40.51

EPSTEIN GABRIEL LEO DR
NASA/GSFC
CODE 682
GREENBELT MD 20771
U.S.A.
TEL:
TLX:
TLF:
EML:
COM: 12.14

EPSTEIN ISADORE PROF
ASTRONOMY DEPARTMENT
COLUMBIA UNIVERSITY
PUPIN PHYSICAL LABS
NEW YORK NY 10027
U.S.A.
TEL: 212-280-3280
TLX: 125953 COLUMBIA
TLF:
EML:
COM: 35

EPSTEIN RICHARD I DR
LOS ALAMOS NATIONAL LAB
MS 436
LOS ALAMOS NM 87545
U.S.A.
TEL: 505-667-9595
TLX:
TLF:
EML:
COM:

ERCAN E. NIHAL
KANDILLI OBSERVATORY
BOSPHORUS UNIVERSITY
CENGELKOY
ISTANBUL 81220
TURKEY
TEL: 3320240/41
TLX: 26411 BOUN TR
TLF:
EML:
COM:

ERDI B DR
ASTRONOMICAL DEPARTMENT
LORAND EOTVOS UNIVERSITY
KUN BELA TER 2
H-1083 BUDAPEST
HUNGARY
TEL: 141019
TLX:
TLF:
EML:
COM: 07

ERGMA E V DR
ASTRONOMICAL COUNCIL
USSR ACADEMY OF SCIENCES
PYATNITSKAYA UL 48
109017 MOSCOW
U.S.S.R.
TEL: 231-54-61
TLX: 412623 SCSTP SU
TLF:
EML:
COM: 35.49

ERICKSON WILLIAM C DR
DEPT. OF PHYSICSM
UNIVERSITY OF TASMANIA
GPO BOX 252C
HOBART, TASMANIA 7001
AUSTRALIA
TEL: 61-02-202-401
TLX: AA58150
TLF:
EML:
COM: 10.40

ERIGUCHI YOSHIHARU DR
DEPT EARTH SC & ASTRONOMY
COLL ARTS & SC/UNIV TOKYO
KOMABA MEGURO
TOKYO 153
JAPAN
TEL: 03-467-1171x439
TLX: 25510 UNITOKYO
TLF:
EML:
COM: 35

ERIKSEN GUNNAR PROF
INST THEORET ASTROPHYSICS
UNIVERSITY OF OSLO
P O BOX 1029
N-0315 BLINDERN, OSLO 3
NORWAY
TEL: 02-45-65-15
TLX:
TLF:
EML:
COM: 40

ERIKSSON KJELL DR
ASTRONOMISKA
OBSERVATORIET
BOX 515
S-751 20 UPPSALA
SWEDEN
TEL: 18-11-24-88
TLX: 76024 UNIVUPS S
TLF:
EML:
COM: 36

ERPYLEV N P DR
ASTRONOMICAL COUNCIL
USSR ACADEMY OF SCIENCES
PYATNITSKAYA UL 48
109017 MOSCOW
U.S.S.R.
TEL: 231-54-61
TLX: 412623 SCSTP SU
TLF:
EML:
COM: 41

ERSHKOVICH ALEXANDER PROF
DEPT GEOPHYS & PLANET SCI
TEL-AVIV UNIVERSITY
TEL-AVIV 69978
ISRAEL
TEL: 03-413505
TLX: 342171 VERSY IL
TLF:
EML:
COM: 15

ERTAN A YENER DR
EGE UNIVERSITY
FEN FAKULTESI
ASTRONOMI BOLUMU
BORNOVA-IZMIR
TURKEY
TEL:
TLX:
TLF:
EML:
COM:

ESHLEMAN VON R PROF
DURAND 221
STANFORD UNIVERSITY
STANFORD CA 94305
U.S.A.
TEL: 415-497-3531
TLX:
TLF:
EML:
COM: 16.40.49

ESIPOV VALENTIN F DR
STERNBERG STATE ASTR INST
117234 MOSCOW
U.S.S.R.
TEL:
TLX:
TLF:
EML:
COM: 34

ESKIOGLU A NIHAT
DEVLET MUHENDISLIK
MIMARLIK AKADEMISI
ADAPAZARI, SAKARYA
TURKEY
TEL:
TLX:
TLF:
EML:
COM: 27

ESPOSITO F PAUL PROF
DEPT OF PHYSICS 11
UNIVERSITY OF CINCINNATI
CINCINNATI OH 45221
U.S.A.
TEL: 513-475-2233
TLX:
TLF:
EML:
COM:

ESPOSITO LARRY W
LASP
UNIVERSITY OF COLORADO
CAMPUS BOX 392
BOULDER CO 80309
U.S.A.
TEL: 303-492-7325
TLX:
TLF:
EML:
COM: 16

ESTALELLA ROBERT
DPTO FISICA DE ATMOSFERA
UNIVERSIDAD DE BARCELONA
AVDA. DIAGONAL 645
08028 BARCELONA
SPAIN
TEL: 330-73-11/298
TLX:
TLF:
EML:
COM:

EUBANKS THOMAS M DR
JET PROPULSION LABORATORY
4800 OAK GROVE DRIVE
PASADENA CA 91109
U.S.A.
TEL: 818 354 2900
TLX:
TLF:
EML:
COM:

EVANGELIDIS K DR
PLASMA PHYSICS DIVISION
NUCOR - PELINDABA
PRIVATE BAG X256
PRETORIA 0001
SOUTH AFRICA
TEL: 27-12-21-3311
TLX: 30253 SA
TLF:
EML:
COM: 33.36

EVANS ANEURIN
DEPT OF PHYSICS
UNIVERSITY OF KEELE
KEELE ST5 5BG
U.K.
TEL: 0782-621111
TLX: 36113 UNKLIB G
TLF:
EML:
COM: 27.34

EVANS J V DR
COMSAT LABORATORIES
22300 COMSAT DR
CLARKSBURG MD 20871
U.S.A.
TEL: 301-428-4422
TLX: 908753
TLF:
EML:
COM: 12

EVANS JOHN V DR
1 BAYA ROAD
ELDORADO
SANTA FE NM 87503
U.S.A.
TEL:
TLX:
TLF:
EML:
COM:

EVANS KENTON DOWER DR
PHYSICS DEPARTMENT
THE UNIVERSITY
LEICESTER LE1 7RH
U.K.
TEL: 0533 554455
TLX: 341664
TLF:
EML:
COM: 40

EVANS NANCY REMAGE DR
DEPT OF ASTRONOMY
UNIVERSITY OF TORONTO
TORONTO ONT M5S 1A7
CANADA
TEL:
TLX:
TLF:
EML:
COM: [illegible]

EVANS NEAL J II ASS PROF
DEPARTMENT OF ASTRONOMY
UNIVERSITY OF TEXAS
AUSTIN TX 78712
U.S.A.
TEL: 512-471-4461
TLX:
TLF:
EML:
COM: 34.51

EVANS ROBERT REV
57 TALBOT ROAD
HAZELBROOK NSW 2779
AUSTRALIA
TEL:
TLX:
TLF:
EML:
COM: 28

EVANS ROGER G DR
RUTHERFORD LABORATORY
DIDCOT OX11 0QX
U.K.
TEL: 0235-21900
TLX: 83159 RUTHLB G
TLF:
EML:
COM: 28

EVANS W DOYLE
390 EL CONEJO
LOS ALAMOS NM 87544
U.S.A.
TEL: 505-667-3644
TLX:
TLF:
EML:
COM: 48

EVDOKIMOV YU V DR
ENGELHARDT ASTR OBS
422526 KAZAN
U.S.S.R.
TEL:
TLX:
TLF:
EML:
COM: 20

EVERHART EDGAR DR
985 DICK MOUNTAIN DR.
BAILEY CO 80421
U.S.A.
TEL:
TLX:
TLF:
EML:
COM: 06.07.15.20

EVIATAR AHARON PROF
DEPT GEOPHYS & PLANET SCI
TEL-AVIV UNIVERSITY
TEL-AVIV 69978
ISRAEL
TEL: 03-420620
TLX: 342171 VERSY IL
TLF:
EML:
COM: 15.49C

EWEN HAROLD I DR
60 BEAVER ROAD
WESTON MA 02193
U.S.A.
TEL:
TLX:
TLF:
EML:
COM:

EWING MARTIN S
CALTECH 102-24
PASADENA CA 91125
U.S.A.
TEL: 818-356-4970
TLX: 675425
TLF:
EML:
COM: 40

EZER-ERYURT DILHAN PROF
MIDDLE EAST TECHN UNIV
ANKARA
TURKEY
TEL: 23-71-00 x 3255
TLX: 42761 ODTK TR
TLF:
EML:
COM: 35

FABBIANO GIUSEPPINA
HARVARD-SMITHSONIAN CTR
FOR ASTROPHYSICS
60 GARDEN STREET
CAMBRIDGE MA 02138
U.S.A.
TEL: 617-495-7204
TLX: 921428 SATELLITE CAM
TLF:
EML:
COM: 28

FABER SANDRA M PROF
LICK OBSERVATORY
UNIVERSITY OF CALIFORNIA
SANTA CRUZ CA 95064
U.S.A.
TEL: 408-429-2944
TLX:
TLF:
EML:
COM: 28.33.47

FABIAN ANDREW C DR
INSTITUTE OF ASTRONOMY
MADINGLEY ROAD
CAMBRIDGE CB3 OHE
U.K.
TEL:
TLX:
TLF:
EML:
CON: 48

FABRE HERVE DR
2 AVENUE MARECHAL FOCH
F-06310 BEAULIEU/MER
FRANCE
TEL:
TLX:
TLF:
EML:
CON: 07

FABRICANT DANIEL G
HARVARD-SMITHSONIAN CTR
FOR ASTROPHYSICS
60 GARDEN STREET
CAMBRIDGE MA 02138
U.S.A.
TEL: 617-495-7398
TLX: 921428 SATELLITE CAM
TLF:
EML:
CON: 09.28.44

FABRICIUS CLAUS V
COPENHAGEN UNIVERSITY OBS
BRORFELDE
DK-4340 TOLLOSE
DENMARK
TEL: 45-3-488-195
TLX: 44155 DANAST DK
TLF:
EML:
CON: 08

FADEYEV YURI A
ASTRONOMICAL COUNCIL
USSR ACADEMY OF SCIENCES
PYATNITSKAYA STR 48
109017 MOSCOW
U.S.S.R.
TEL: 231-54-61
TLX: 412623 SCSTP SU
TLF:
EML:
CON: 27.35

FAELTHAMMAR CARL GUNNE PR
DEPT OF PLASMA PHYSICS
ROYAL INST OF TECHNOLOGY
S-100 44 STOCKHOLM 70
SWEDEN
TEL: 0-8-687-7685
TLX: 10389 KTHB
TLF:
EML:
CON:

FAHLMAN GREGORY G DR
DEPT GEOPHYS & ASTRONOMY
UNIV OF BRITISH COLUMBIA
2075 WESBROOK PLACE
VANCOUVER BC V6T 1W5
CANADA
TEL: 604-228-4891
TLX: 04543425
TLF:
EML:
CON:

FAHR HANS JOERG PROF DR
INSTITUT FUR ASTROPHYSIK
DER UNIVERSITAET BONN
AUF DEM HUEGEL 71
D-5300 BONN
GERMANY, F.R.
TEL: 0228-733677
TLX: 886440 RFI
TLF:
EML:
CON: 49

FAHY EDWARD F PROF
PHYSICS DEPT
UNIVERSITY COLLEGE
CORK
IRELAND
TEL: 021-26871
TLX: 26050
TLF:
EML:
CON:

FAIRALL ANTHONY P PROF
DEPT OF ASTRONOMY
UNIVERSITY OF CAPE TOWN
RONDEBOSCE 7700
SOUTH AFRICA
TEL: 21-698531 x 629
TLX: 5721439
TLF:
EML:
CON: 28.30.46

FALCHI ANDRETTA
OSSERVATORIO ASTROFISICO
DI ARCETRI
LARGO E. FERMI 5
I-50125 FIRENZE
ITALY
TEL: 055-2752235
TLX: 572268 ARCETR I
TLF:
EML:
CON: 10

FALCIANI ROBERTO DR
DIP. DI ASTRONOMIA
UNIVERSITA DEGLI STUDI
LARGO E. FERMI 5
I-50125 FIRENZE
ITALY
TEL: 055-27521
TLX: 572268 ARCETR I
TLF:
EML:
CON: 10.12C

FALGARONE EDITH
CALTECH
DOWNS LAB. OF PHYSICS
405-47
PASADENA CA 91195
U.S.A.
TEL: 818-2432438
TLX:
TLF:
EML:
CON: 34

FALK SYDNEY W JR DR
DEPT OF ASTRONOMY
UNIVERSITY OF TEXAS
AUSTIN TX 78712
U.S.A.
TEL:
TLX:
TLF:
EML:
CON: 34.47

FALL S MICHAEL DR
SPACE TELESCOPE SCI INST
HOMEWOOD CAMPUS
3700 SAN MARTIN DRIVE
BALTIMORE MD 21218
U.S.A.
TEL:
TLX:
TLF:
EML:
CON: 28.33.37.47

FALLE SAMUEL A DR
DEPT APPLIED MATHEMATICS
UNIVERSITY OF LEEDS
LEEDS LS2 9JT
U.K.
TEL: 532-431-751
TLX:
TLF:
EML:
CON: 34

FALLER JAMES E PROF
JILA/NBS
UNIVERSITY OF COLORADO
BOULDER CO 80309
U.S.A.
TEL: 303-492-8509
TLX: 755842 JILA
TLF:
EML:
CON:

FALLON FREDERICK W DR
N/CG 114
NOAA/NGS
6010 EXECUTIVE BLVD
ROCKVILLE MD 20852
U.S.A.
TEL: 301-443-8474
TLX:
TLF:
EML:
CON: 24.31

FALOMO RENATO DR
OSSERV.ASTRON.DI PADOVA
VICOLO DELL'OSSERV.5
35122 PADOVA
ITALY
TEL: 39-49-661499
TLX: 432071 ASTROS I
TLF:
EML: ASTRPD::FALOMO
CON:

FAN DAXIONG
PURPLE MOUNTAIN
OBSERVATORY
NANJING
CHINA, PEOPLE'S REP.
TEL: (025)303583
TLX: 34144 PMONJ CN
TLF:
EML:
CON: 10

FAN YING
DEPT OF ASTRONOMY
BEIJING NORMAL UNIVERSITY
BEIJING 100082
CHINA. PEOPLE'S REP.
TEL: 653531-6285
TLX:
TLF:
EML:
COM: 34

FANG CHENG
DEPT OF ASTRONOMY
NANJING UNIVERSITY
NANJING
CHINA. PEOPLE'S REP.
TEL: 34651-2882
TLX: 34151 PRCNU CN
TLF:
EML:
COM: 10C.12

FANG LI-ZHI
BEIJING OBSERVATORY
ACADEMIA SINICA
BEIJING
CHINA. PEOPLE'S REP.
TEL: 281968
TLX: 22040 BAOAS CN
TLF:
EML:
COM: 47C.48

FANSELOW JOHN LYMAN
JET PROPULSION LAB
M/S 264-748
4800 OAK GROVE DR
PASADENA CA 91109
U.S.A.
TEL: 213-354-6323
TLX: 675429
TLF:
EML:
COM: 19.24

FANTI CARLA GIOVANNINI
ISTITUTO RADIOASTRONOMIA
VIA IRNERIO 46
I-40126 BOLOGNA
ITALY
TEL: 051-232856/57
TLX: 211664 INFN BO
TLF:
EML:
COM:

FANTI ROBERTO
ISTITUTO DI FISICA
UNIVERSITA DI BOLOGNA
VIA IRNERIO 46
I-40126 BOLOGNA
ITALY
TEL: 232856/57
TLX: 211664 INFN BO
TLF:
EML:
COM: 40C

FARAGGIANA ROSANNA PROF
OSSERVATORIO ASTRONOMICO
VIA TIEPOLO 11
I-34131 TRIESTE
ITALY
TEL: 040-793921
TLX: 461137 OAT I
TLF:
EML:
COM: 29.36.44

FARINELLA PAOLO DR
ISTITUTO DI MATEMATICA
UNIVERSITA' DI PISA
VIA BUONARROTI 2
I-56100 PISA
ITALY
TEL:
TLX:
TLF:
EML:
COM: 07.15.16

FARNIK FRANTISEK
ASTRONOMICAL INSTITUTE
CZECH. ACAD. OF SCIENCES
ONDREJOV OBSERVATORY
251 65 ONDREJOV
CZECHOSLOVAKIA
TEL: 204-999-201/202
TLX: 121579
TLF:
EML:
COM: 10

FAUCHER PAUL DR
OBSERVATOIRE DE NICE
BP 139
F-06003 NICE CEDEX
FRANCE
TEL: 93-89-04-20
TLX: 460004
TLF:
EML:
COM: 14

FAULKNER DONALD J DR
MT STROMLO & SIDING
SPRING OBSERVATORIES
PRIVATE BAG
WODEN P.O. ACT 2606
AUSTRALIA
TEL: 062-88-1111
TLX: 62270
TLF:
EML:
COM: 34.35

FAULKNER JOHN PROF
LICK OBSERVATORY
UNIVERSITY OF CALIFORNIA
SANTA CRUZ CA 95064
U.S.A.
TEL: 408-429-2815
TLX:
TLF:
EML:
COM: 35.42

FAUNDEZ-ABANS M DR
LABORATORIO NACIONAL DE
ASTROFISICA
CAIXA POSTAL 21
37500 ITAJUBA MG
BRAZIL
TEL: 035-6220788
TLX: 312603 OBNA BR
TLF:
EML:
COM:

FAVELL DEREK R DR
UNIVERSITY OF LONDON OBS
MILL HILL PARK
LONDON NW7 2QS
U.K.
TEL:
TLX:
TLF:
EML:
COM:

FAY THEODORE D DR
MAIL STOP 19
TELEDYNE BROWN ENG
CUMMINGS RES PARK
HUNTSVILLE AL 35807
U.S.A.
TEL:
TLX:
TLF:
EML:
COM:

FAZIO GIOVANNI G DR
CENTER FOR ASTROPHYSICS
HCO/SAO
60 GARDEN ST
CAMBRIDGE MA 02138
U.S.A.
TEL: 617-495-7458
TLX: 921428 SATELLITE CAM
TLF:
EML:
COM: 44C.48.51

FEAST MICHAEL W PROF
S A A O
P O BOX 9
OBSERVATORY 7935
SOUTH AFRICA
TEL: (27)21 47 00 25
TLX: 520309
TLF:
EML:
COM: 27.28.29.33.37.45

FEAUTRIER NICOLE DR
OBSERVATOIRE DE PARIS
SECTION DE MEUDON
F-92195 MEUDON PL CEDEX
FRANCE
TEL: 1-45-07-75-52
TLX: 201571 LAM
TLF:
EML:
COM: 14

FECHTIG HUGO DR
SANSENWEG 3
D-6906 LEIMEN
GERMANY. F.R.
TEL:
TLX:
TLF:
EML:
COM: 15.21.22

FEDERICI LUCIANA
DIPTO DI ASTRONOMIA
UNIVERSITA DEGLI STUDI
C P 596
I-40100 BOLOGNA
ITALY
TEL:
TLX: 211664 INFN BO I
TLF:
EML:
COM:

FEDERMAN STEVEN ROBERT
DEPT. OF PHYSICS & ASTRON
THE UNIVERSITY OF TOLEDO
TOLEDO OH 43606
U.S.A.
TEL: 419-537-2652
TLX:
TLF:
EML:
COM: 34

FEDOROVA RIMMA T DR
ASTRONOMICAL OBSERVATORY
NIKOLAEV BRANCH OF THE
MAIN ASTRONOMICAL OBS
327000 NIKOLAEV
U.S.S.R.
TEL: 37-57-14
TLX:
TLF:
EML:
COM: 08

FEGAN DAVID J DR
PHYSICS DEPT
UNIVERSITY COLLEGE
BELFIELD
DUBLIN 4
IRELAND
TEL: 692244
TLX: 32693 UCD EI
TLF:
EML:
COM:

FEHRENBACH CHARLES PROF
LES MAGNANARELLES
LOURMARIN
F-84160 CADENET
FRANCE
TEL: 90-68-00-28
TLX:
TLF:
EML:
COM: 09.30.33.45

FEIBELMAN WALTER A DR
NASA/GSFC
CODE 685
GREENBELT MD 20771
U.S.A.
TEL: 301-286-5272
TLX:
TLF:
EML:
COM: 10.27.34

FEIGELSON ERIC D DR
DEPARTMENT OF ASTRONOMY
PENNSYLVANIA STATE UNIV.
UNIVERSITY PARK PA 16802
U.S.A.
TEL: 814 865 0162
TLX: 842 510
TLF:
EML: INTERNET:EDF@ASTRO.PSU.EDU
COM: 40

FEINSTEIN ALEJANDRO DR
OBSERVATORIO ASTRONOMICO
1900 LA PLATA
ARGENTINA
TEL: 021-21-7308
TLX: 31216 CESLA AR
TLF:
EML:
COM: 25.37

FEISSEL MARTINE DR
OBSERVATOIRE DE PARIS
61 AVE DE L'OBSERVATOIRE
F-75014 PARIS
FRANCE
TEL: 1-40-51-22-26
TLX:
TLF:
EML:
COM: 08.19P.31

FEITZINGER JOHANNES PROF
ASTRONOMISCHES INSTITUT
RUHR UNIVERSITAET BOCHUM
POSTFACH 102148
D-4630 BOCHUM
GERMANY. F.R.
TEL: 0234-700-3450
TLX: 825860-1 RUB D
TLF:
EML:
COM: 28.33.34

FEIX GERHARD DR
RUHR UNIVERSITAET BOCHUM
DEPT XII
POSTFACH 102148
D-4630 BOCHUM
GERMANY. F.R.
TEL: 0234-700-2051
TLX: 0825860
TLF:
EML:
COM: 40

FEJES ISTVAN DR
FOMI SATELLITE
GEODETIC OBSERVATORY
BOX 546
H-1373 BUDAPEST
HUNGARY
TEL:
TLX:
TLF:
EML:
COM: 51

FEKEL FRANCIS C
VANDERBILT UNIVERSITY
DYER OBSERVATORY
NASHVILLE TN 37235
U.S.A.
TEL: 615-322-2804
TLX:
TLF:
EML:
COM: 26.30.42

FELDMAN PAUL A DR
HERZBERG INST ASTROPHYS
NATL RESEARCH COUNCIL
100 SUSSEX DR
OTTAWA ONT K1A 0R6
CANADA
TEL: 613-993-6060
TLX: 0533715
TLF:
EML:
COM: 40.51

FELDMAN PAUL DONALD DR
DEPT PHYSICS & ASTRONOMY
JOHNS HOPKINS UNIVERSITY
BALTIMORE MD 21218
U.S.A.
TEL: 301-338-7339
TLX: 710-234-1090
TLF:
EML:
COM: 15.21.44

FELDMAN URI
HOLBURT CTR FOR SPACE RES
NAVAL RESEARCH LABORATORY
WASHINGTON DC 20375
U.S.A.
TEL: 202-767-3286
TLX:
TLF:
EML:
COM: 12

FELDMAN URI DR
NAVAL RESEARCH LABORATORY
NRL 4174
WASHINGTON DC 20375-5000
U.S.A.
TEL:
TLX:
TLF:
EML:
COM:

FELENBOK PAUL DR
OBSERVATOIRE DE PARIS
SECTION DE MEUDON
F-92195 MEUDON PL CEDEX
FRANCE
TEL: 1-45-07-75-23
TLX: 201571 LAM
TLF:
EML:
COM: 14.29

FELLGETT PETER PROF
DEPT OF CYBERNETICS
3 EARLY GATE
WHITEKNIGHTS
READING RG6 2AL
U.K.
TEL: 0734-65758
TLX: 847813
TLF:
EML:
COM: 09

FELLI MARCELLO DR
OSSERVATORIO ASTROFISICO
DI ARCETRI
LARGO E. FERMI 5
I-50125 FIRENZE
ITALY
TEL: 055-2752240
TLX: 572268
TLF:
EML:
COM: 34.40

FELTEN JAMES E DR
NASA - CODE 685
GODDARD SPACE FLIGHT CTR
GREENBELT MD 20771
U.S.A.
TEL: 301-552-1526
TLX:
TLF:
EML:
COM: 34.40.47.48

FENG HESHENG
YUNNAN OBSERVATORY
KUNMING
CHINA, PEOPLE'S REP.
TEL:
TLX:
TLF:
EML:
COM:

FENG KE-JIA
DEPT OF ASTRONOMY
BEIJING NORMAL UNIVERSITY
BEIJING 80
CHINA, PEOPLE'S REP.
TEL: 65-3531 x 6967
TLX:
TLF:
EML:
COM: 10.46

FENKART ROLF P PROF DR
ASTRONOMISCHES INSTITUT
UNIVERSITAET BASEL
VENUSSTRASSE 7
CH-4102 BINNINGEN
SWITZERLAND
TEL: 061-22-77-11
TLX:
TLF:
EML:
COM: 33

FENTON K B DR
DEPT OF PHYSICS
UNIVERSITY OF TASMANIA
P.O.BOX 252C
HOBART TASMANIA 7001
AUSTRALIA
TEL: 002-202411
TLX: 58150 AA
TLF:
EML:
COM: 48

FERETTI LUIGINA
ISTITUTO RADIOASTRONOMIA
VIA IRNERIO 46
I-40126 BOLOGNA
ITALY
TEL: 051-232856
TLX: 520634 INFNBO I
TLF:
EML:
COM: 40

FERLAND GARY JOSEPH
ASTRONOMY DEPARTMENT
OHIO STATE UNIVERSITY
COLUMBUS OH 43210
U.S.A.
TEL: 614-422-1773
TLX: 8104821715
TLF:
EML:
COM: 27.28

FERLET ROGER DR
INSTITUT D'ASTROPHYSIQUE
DE PARIS 98 BIS BD ARAGO
75014 PARIS
FRANCE
TEL: (33-1)43201425
TLX: 205671 IAU F
TLF:
EML: EARN:FERLET@FRIAP51
COM: 34

FERLUGA STENO DR
UNIVERSITA'DI TRIESTE
DIPARTIMENTO ASTRONOMIA
VIA G.B. TIEPOLO 11
TRIESTE 309390
ITALY
TEL: 040 763917
TLX: 461137 OAT I
TLF: 040-309418
EML: ASTRONET:ASTRTS::FERLUGA
COM: 42

FERNANDEZ JEAN-CLAUDE DR
OBSERVATOIRE DE NICE
BP 139
F-06003 NICE CEDEX
FRANCE
TEL: 93-89-04-20
TLX:
TLF:
EML:
COM: 15

FERNANDEZ JULIO A DR
DEPT DE ASTRONOMIA
UNIV DE REP MONTEVIDEO
TRISTAN NARVAJA 1674
MONTEVIDEO
URUGUAY
TEL: (598-2)-49 11 044
TLX: UDELAR UY 26692
TLF:
EML:
COM: 15.20.46

FERNANDEZ SILVIA M. DR
OBSERVATORIO ASTRONOMICO
LAPRIDA 854
5000 CORDOBA
ARGENTINA
TEL: 51-40613/36676
TLX: 51822 BUCOR
TLF:
EML:
COM: 07

FERNANDEZ-FIGUEROA M J DR
ASTROFIS FAC DE FISICAS
UNIVERSIDAD COMPLUTENSE
CIUDAD UNIVERSITARIA
28040 MADRID
SPAIN
TEL: 4-49-53-16
TLX: 47273 FF UC
TLF:
EML:
COM: 29.46

FERNIE J DONALD PROF
DAVID DUNLAP OBSERVATORY
P O BOX 360
RICHMOND HILL ONT L4C 4Y6
CANADA
TEL: 416-884-9562
TLX: 06-986766 TELEXPERTS
TLF:
EML:
COM: 25.27 41

FERRANDO PHILIPPE DR
SERVICE D'ASTROPHYSIQUE.
DPHG-SAP CEN SACLAY
91191 GIF SUR YVETTE
FRANCE
TEL: (33-1)69082020
TLX: 604860 PHYSPAC F
TLF:
EML: BITNET:FERRANDO@12779
COM:

FERRARI ATTILIO DR
ISTITUTO DI FISICA
GENERALE DELL'UNIVERSITA
CORSO M. D'AZEGLIO 46
I-10125 TORINO
ITALY
TEL: 011-657694
TLX: 211041 INFN TO I
TLF:
EML:
COM: 40.48

FERRARI D'OCCHIEPPO K DR
OESTERREICHISCHE AKADEMIE
DER WISSENSCHAFTEN
DR-IGNAZ-SEIPEL-PLATZ 2
A-1010 WIEN
AUSTRIA
TEL: 222-72-81991
TLX: 01-12628
TLF:
EML:
COM: 41.42

FERRARI TONIOLO MARCO
IST ASTROFISICA SPAZIALE
C P 67
I-00044 FRASCATI
ITALY
TEL: 06-9425651
TLX: 610261 CNR-FRA I
TLF:
EML:
COM: 44

FERRAZ-MELLO S PROF DR
UNIVERSIDADE DE SAO PAULO
DEPT ASTRONOMIA
CAIXA POSTAL 30627
01051 SAO PAULO SP
BRAZIL
TEL: 11-549-6709
TLX: 1136221 IAGM BR
TLF:
EML:
COM: 07C 20.46

FERRER MARTINEZ SEBASTIAN
DPTO FIS TIERRA & COSMOS
UNIVERSIDAD DE ZARAGOZA
50009 ZARAGOZA
SPAIN
TEL: 976-357011
TLX: 58198
TLF:
EML:
COM: 07

FERRER OSVALDO EDUARDO DR
UNIV NACIONAL DE LA PLATA
FACULTAD DE CIENCIAS
ASTRON Y GEOFISICAS
1900 LA PLATA
ARGENTINA
TEL:
TLX:
TLF:
EML:
COM: 16.42

FERRERI WALTER
OSSERVATORIO ASTRONOMICO
DI TORINO
STR OSSERVATORIO N.20
PINO TORINESE 10025
ITALY
TEL: 011 842040
TLX: 213236 ASTP I
TLF:
EML:
COM: 20

FERRIN IGNACIO
UNIVERSIDAD DE LOS ANDES
FACULTAD DE CIENCIAS
DEPTO DE FISICA
MERIDA 5101
VENEZUELA
TEL:
TLX:
TLF:
EML:
COM: 15

FERRINI FEDERICO
ISTITUTO DI ASTRONOMIA
UNIVERSITA DI PISA
PIAZZA TORRICELLI 2
I-56100 PISA
ITALY
TEL: 050-43343
TLX:
TLF:
EML:
COM: 28.34

FESTOU MICHEL C DR
OBSERVATOIRE DE BESANCON
41B AVE DE L'OBSERVATOIRE
F-25044 BESANCON
FRANCE
TEL: 81 80 22 66
TLX: 361144 OBSBES F
TLF:
EML:
COM: 15

FEYNMAN JOAN DR
JPL
144-218
4800 OAK GROVE DRIVE
PASADENA CA 91109
U.S.A.
TEL: 818-354-3454
TLX: 675429
TLF:
EML:
COM: 49

FIALA ALAN D DR
NAUTICAL ALMANAC OFFICE
US NAVAL OBSERVATORY
34B MASSACHUSETTS AVE NW
WASHINGTON DC 20390
U.S.A.
TEL: 202-653-1274
TLX: 710-822-1970
TLF:
EML:
COM: 04.07.12

FICARRA ANTONINO DR
IST DI RADIOASTRONOMIA
VIA IRNERIO 46
I-40126 BOLOGNA
ITALY
TEL: 051-232856
TLX: 211664 INFN BO I
TLF:
EML:
COM:

FICH MICHEL DR
PHYSICS DEPARTMENT
UNIVERSITY OF WATERLOO
WATERLOO, ONTARIO N2L 3G1
CANADA
TEL: 519 885 1577
TLX: 06 955259
TLF:
EML: BITNET:FICHOWATSCI
COM:

FICHTEL CARL E DR
NASA/GSFC
CODE 660
GREENBELT MD 20771
U.S.A.
TEL: 301-286-6281
TLX: 89675
TLF:
EML:
COM: 44.48

FIELD DAVID
SCHOOL OF CHEMISTRY
CANTOCKS CLOSE
BRISTOL BS8 1TS
U.K.
TEL: 0272-24161 x505
TLX: 444174 BUPHYS
TLF:
EML:
COM: 34

FIELD GEORGE B PROF
CENTER FOR ASTROPHYSICS
60 GARDEN ST
CAMBRIDGE MA 02138
U.S.A.
TEL: 617-495-4721
TLX: 921428 SATELLITE CAM
TLF:
EML:
COM: 28.34.40.47.48.49.51

FIELDER GILBERT DR
LUNAR AND PLANETARY UNIT
E S DEPT
LANCASTER UNIVERSITY
LANCASTER LA1 4YB
U.K.
TEL: 65201
TLX: 65111
TLF:
EML:
COM: 16

FIENBERG RICHARD T DR
SKY & TELESCOPE
49 BAY STATE ROAD
CAMBRIDGE MA 02138
U.S.A.
TEL: 617 864 7360
TLX:
TLF:
EML: SPAN:CFA::FIENBERG
COM: 46

FIERRO JULIETA
INSTITUTO DE ASTRONOMIA
UNAM
APDO POSTAL 70-264
04510 MEXICO DF
MEXICO
TEL:
TLX:
TLF:
EML:
COM: 34.46

FIGUERAS FRANCESCA DR
DEPT DE ASTRONOMIA
UNIV. DE BARCELONA
AVENIDA DIAGONAL 647
08028 BARCELONA
SPAIN
TEL: 34-3 3307311
TLX:
TLF:
EML: D3FAFFS0@EBOUB011
COM: 33

FILIPOV LATCHEZAR
CENTR LAB FOR SPACE RES
BULGARIAN ACAD SCIENCES
MOSKOVA STR 6
1000 SOFIA
BULGARIA
TEL: 87-09-78
TLX: 23351 CLSR BG
TLF:
EML:
COM:

FILIPPENKO ALEXEI V DR
DEPARTMENT OF ASTRONOMY
UNIVERSITY OF CALIFORNIA
BERKELEY CA 94720
U.S.A.
TEL: 415 642 1813
TLX: 820181
TLF:
EML: BITNET:ALEX@BKYAST
COM: 06.28.47

FILLOY EMILIO MANUEL R.E.
INSTITUTO ARGENTINO DE
RADIOASTRONOMIA
CASILLE DE CORREO No. 5
1894 VILLA ELISA (Bs.As.)
ARGENTINA
TEL: 4-3793
TLX:
TLF:
EML:
COM:

FINDLAY JOHN W DR
NRAO
EDGEMONT ROAD
CHARLOTTESVILLE VA 22901
U.S.A.
TEL:
TLX: 910-997-0174
TLF:
EML:
COM: 40

FINK UWE DR
LUNAR & PLANETARY LAB
UNIVERSITY OF ARIZONA
TUCSON AZ 85721
U.S.A.
TEL: 602-621-2736
TLX: 9109521143
TLF:
EML:
COM: 14.16

FINN G D DR

U.S.A.
TEL:
TLX:
TLF:
EML:
COM: 36

FINZI ARRIGO DR
DEPT OF MATHEMATICS
TECHNION. I.I.T.
HAIFA 32000
ISRAEL
TEL:
TLX: 46406 TECON IT
TLF:
EML:
COM:

FIREMAN EDWARD L
SMITHSONIAN ASTROPHYSICAL
OBSERVATORY
60 GARDEN ST
CAMBRIDGE MA 02138
U.S.A.
TEL: 617-495-7271
TLX:
TLF:
EML:
COM: 22

FIRMANI CLAUDIO A PROF
INSTITUTO DE ASTRONOMIA
UNAM
APDO POSTAL 70-264
04510 MEXICO DF
MEXICO
TEL: 905-548-3712
TLX: 1760155 CICME
TLF:
EML:
COM: 42

FIRNEIS FRIEDRICH J DR
INST INFO PROC/OEAW
SONNENFELSGASSE 19/2
A-1010 WIEN
AUSTRIA
TEL:
TLX:
TLF:
EML:
COM: 24.51

FIRNEIS MARIA G DR
INSTITUT FUER ASTRONOMIE
TUERKENSCHANZSTR 17
A-1180 WIEN
AUSTRIA
TEL: 0222-34-53-60
TLX:
TLF:
EML:
COM: 24.41.51

FIROR JOHN W DR
NCAR
PO BOX 3000
BOULDER CO 80307
U.S.A.
TEL: 303-497-1600
TLX: 45694
TLF:
EML:
COM:

FISCHEL DAVID DR
EARTH OBSERVATION
SATELLITE COMPANY
4300 FORBES BLVD
LANHAM MD 20706
U.S.A.
TEL: 552 0500
TLX: 277685
TLF:
EML:
COM:

FISCHER JACQUELINE
NAVAL RESEARCH LABORATORY
CODE 4138F
WASHINGTON DC 20375
U.S.A.
TEL: 202-767-3058
TLX:
TLF:
EML:
COM: 34

FISCHER STANISLAV DR
ASTRONOMICAL INSTITUTE
CZECH. ACAD. OF SCIENCES
BUDECSKA 6
120 23 PRAHA 2
CZECHOSLOVAKIA
TEL: 252438
TLX: 122486 ASTRC
TLF:
EML:
COM:

FISHER J RICHARD
NRAO
PO BOX 2
GREEN BANK WV 24944
U.S.A.
TEL: 304-456-2011
TLX: 710-938-1530
TLF:
EML:
COM:

FISHER PHILIP C
RUFFNER ASSOCIATES
PO BOX 7070
MENLO PARK CA 94026
U.S.A.
TEL:
TLX:
TLF:
EML:
COM: 44.48.51

FISHER RICHARD R DR
HIGH ALTITUDE OBSERVATORY
PO BOX 3000
BOULDER CO 80307
U.S.A.
TEL: 303-494-5151
TLX:
TLF:
EML:
COM:

FISHKOVA LUISA M PROF
ABASTUMANI ASTROPHYSICAL
OBSERVATORY
383762 ABASTUMANI.GEORGIA
U.S.S.R.
TEL:
TLX:
TLF:
EML:
COM: 21

FISHMAN GERALD J
MSFC/ASTROPHYSICS BRANCH
SPACE SCIENCES LAB ES-62
HUNTSVILLE AL 35812
U.S.A.
TEL: 205-453-0117
TLX:
TLF:
EML:
COM: 44

FITCH WALTER S DR
P.O. BOX 100
ORACLE. AZ 85623
U.S.A.
TEL: 602-896-2911
TLX: 467175
TLF:
EML:
COM: 27

FITTON BRIAN DR
ESTEC. ASTRONOMY DIVISION
POSTBUS 299
NL-2200 AG NOORDWIJK
NETHERLANDS
TEL: 02524-4635
TLX:
TLF:
EML:
COM: 44

FITZGERALD M PIM PROF
DEPT OF PHYSICS
UNIVERSITY OF WATERLOO
WATERLOO ONT N2L 3G1
CANADA
TEL: 519-885-1572
TLX:
TLF:
EML:
COM: 33.37

FITZPATRICK EDWARD L DR
PRINCETON UNIVERSITY OBS.
PEYTON HALL
PRINCETON NJ 08544
U.S.A.
TEL: 609 452 3702
TLX:
TLF: 609 243 7333
EML: FITZ@ASTROVAX.PRINCETON.EDU
COM: 29.36

FIX JOHN D DR
DEPT PHYSICS & ASTRONOMY
UNIVERSITY OF IOWA
IOWA CITY IA 52240
U.S.A.
TEL: 319-353-7064
TLX: 910-525-1398
TLF:
EML:
COM:

FLANNERY BRIAN PAUL DR
EXXON RES & ENGINEERING
ROUTE 22 EAST
ANNANDALE NJ 08801
U.S.A.
TEL: 201-730-2540
TLX: 136140 EXXONRES
TLF:
EML:
COM: 34.35.42

FLEISCHER ROBERT DR
ROUTE 1
BOX 41 A
KEEDYSVILLE MD 21756
U.S.A.
TEL: 301-432-8870
TLX:
TLF:
EML:
COM: 40

FLETCHER J MURRAY
DOMINION ASTROPHYS OBS
5071 W SAANICH ROAD
RR 5
VICTORIA BC V8X 4M6
CANADA
TEL: 604-388-3905
TLX: 049-7295
TLF:
EML:
COM: 09.26.30

FLETT ALISTAIR M
UNIVERSITY OF ABERDEEN
DEPTARTMENT OF PHYSICS
ABERDEEN AB9 2UE
U.K.
TEL: 0224-40241
TLX: 73458 UNIABN G
TLF:
EML:
COM: 40

FLIEGEL HENRY F
3730 EL MORENO AVENUE
PO BOX 8682
LA CRESCENTA CA 91214
U.S.A.
TEL: 213-648-7452
TLX:
TLF:
EML:
COM: 19.31C

FLIN PIOTR
JAGIELLONIAN UNIVERSITY
OBSERVATORY
UL. ORLA 171
30-244 KRAKOW
POLAND
TEL:
TLX: 0322297 UJ PL
TLF:
EML:
COM: 28

FLOQUET MICHELE DR
OBSERVATOIRE DE PARIS
SECTION DE MEUDON
DEPEG
F-92195 MEUDON PL CEDEX
FRANCE
TEL: 1-45-07-78-51
TLX:
TLF:
EML:
COM: 29

FLORENTIN-NIELSEN RALPH
COPENHAGEN UNIVERSITY OBS
BRORFELDEVEJ 23
DK-4340 TOLLOSE
DENMARK
TEL: 3-488195
TLX: 44155 DANAST
TLF:
EML:
COM:

FLORIDES PETROS S PROF
SCHOOL OF MATHEMATICS
TRINITY COLLEGE
DUBLIN 2
IRELAND
TEL: 772941
TLX: 25442 TCD EI
TLF:
EML:
COM: 41.47

FLORSCH ALPHONSE DR
OBS DE STRASBOURG
11 RUE DE L'UNIVERSITE
F-67000 STRASBOURG
FRANCE
TEL: 88-35-43-00
TLX: 890506 STAROBS F
TLF:
EML:
COM: 28.30C.38C

FLOWER DAVID R DR
DEPT OF PHYSICS
UNIVERSITY OF DURHAM
DURHAM DH1 3LE
U.K.
TEL: 0385-64971
TLX: 537351
TLF:
EML:
COM: 14.34C

FOCARDI PAOLA DR
DIPART. DI ASTRONOMIA
CASELLA POSTALE 596
BOLOGNA 40100
ITALY
TEL: 051 259301
TLX: 520634 INFNBO I
TLF:
EML: SPAN:ASTBO3::PAOLA
COM: 47

FOGARTY WILLIAM G DR
IBM
NCMD
411 EAST WISCONSIN AVE.
MILWAUKEE WI 53202
U.S.A.
TEL:
TLX:
TLF:
EML:
COM:

FOING BERNARD H DR
INST. D'ASTRO. SPATIALE
BP 10
91371 VERRIERES BUISSON
FRANCE
TEL: 64474328
TLX: 600252 F LPSPVB
TLF:
EML: SPAN:IAPOBS::FOING
COM: 10.29.44

FOLTZ CRAIG B.
MULT. MIRROR TEL. OBS.
UNIVERSITY OF ARIZONA
TUCSON AZ 85721
U.S.A.
TEL: 602-621-1269
TLX: 467175
TLF:
EML:
COM: 28.30

FOMALONT EDWARD B DR
NRAO
PO BOX O
SOCORRO NM 87801
U.S.A.
TEL:
TLX: 910-988-1710
TLF:
EML:
COM: 40

FOMENKO ALEXANDR F DR
SPECIAL ASTROPHYS OBS
NIZHNIJ ARKHYZ
357140 STAVROPOLSKIJ KRAJ
U.S.S.R.
TEL:
TLX:
TLF:
EML:
COM: 09

FOMIN VALERY A DR
PULKOVO OBSERVATORY
196140 LENINGRAD
U.S.S.R.
TEL:
TLX:
TLF:
EML:
COM: 08

FOMINOV ALEXANDR M DR
INST OF THEORET ASTRONOMY
USSR ACADEMY OF SCIENCES
10 KUTUZOV QUAY
191187 LENINGRAD
U.S.S.R.
TEL: 278-88-98
TLX: 121578 ITA SU
TLF:
EML:
COM: 04

FONG CHU-GANG
SHANGHAI OBSERVATORY
ACADEMIA SINICA
SHANGHAI
CHINA, PEOPLE'S REP.
TEL: 386191
TLX: 33164 SHAO CN
TLF:
EML:
COM: 07.19

FONG RICHARD
DEPT OF PHYSICS
UNIVERSITY OF DURHAM
SOUTH ROAD
DURHAM DH1 3LE
U.K.
TEL: 64971
TLX: 537351
TLF:
EML:
COM: 47

FONTAINE GILLES DR
DEPT OF PHYSICS
UNIVERSITY OF MONTREAL
P O BOX 6128
MONTREAL PQ H3C 3J7
CANADA
TEL: 514-343-6680
TLX: 05562425
TLF:
EML:
COM: 35

FONTENLA JUAN MANUEL DR
ES 52/MSFC
NASA
HUNTSVILLE AL 35812
U.S.A.
TEL:
TLX:
TLF:
EML:
COM: 12.36

FORBES J E DR
PO BOX 88120
INDIANAPOLIS IN 46208
U.S.A.
TEL:
TLX:
TLF:
EML:
COM: 35

FORBES TERRY G DR
SPACE SCIENCE CENTER EOS
SCIENCE ENGIN. RES. BLDG
UNIV. OF NEW HAMPSHIRE
DURHAM NH 03857
U.S.A.
TEL: 603 862 3872
TLX: 950030 UNH PHYS
TLF:
EML: BITNET%"T_FORBES@UNHH"
COM: 10

FORD HOLLAND C RES PROF
SPACE TELESCOPE SCI INST
JOHNS HOPKINS UNIVERSITY
HOMEWOOD CAMPUS
BALTIMORE MD 21218
U.S.A.
TEL: 301-338-4803
TLX:
TLF:
EML:
COM: 28.34.47

FORD W KENT JR DR
DEPT TERRESTR. MAGNETISM
CARNEGIE INST. WASHINGTON
5241 BROAD BRANCH RD N.W.
WASHINGTON DC 20015
U.S.A.
TEL: 202-966-0863
TLX: 440427 MAGN UI
TLF:
EML:
COM: 09.28

FORMAN WILLIAM RICHARD DR
SMITHSONIAN ASTROPHYS OBS
60 GARDEN STREET
CAMBRIDGE MA 02138
U.S.A.
TEL: 617-495-7210
TLX: 92-1428
TLF:
EML:
COM: 47.48

FORREST WILLIAM JOHN
DEPT PHYSICS & ASTRONOMY
UNIVERSITY OF ROCHESTER
ROCHESTER NY 14627
U.S.A.
TEL: 716-275-4343
TLX:
TLF:
EML:
COM:

FORSTER JAMES RICHARD DR
CSIRO
DIVISION OF RADIOPHYSICS
P.O.BOX 76
EPPING NSW 2121
AUSTRALIA
TEL: 02-868-0222
TLX: 26230 ASTRO AA
TLF:
EML:
COM: 34

FORT BERNARD P DR
OBSERVATOIRE PIC-DU-MIDI
ET TOULOUSE
14 AVENUE EDOUARD BELIN
F-31400 TOULOUSE
FRANCE
TEL: 61-25-21-01
TLX: 530776 F
TLF:
EML:
COM: 09

FORT DAVID NORMAN DR
733 LONDSDALE ROAD
OTTAWA K1K 0J9
CANADA
TEL:
TLX:
TLF:
EML:
COM: 40

FORTE JUAN CARLOS DR
UNIV NACIONAL DE LA PLATA
FAC DE CIENCIAS ASTRON Y
GEOFISICAS
1900 LA PLATA
ARGENTINA
TEL:
TLX:
TLF:
EML:
COM: 25.28.37

FORTI GIUSEPPE DR
OSSERVATORIO ASTROFISICO
DI ARCETRI
LARGO E. FERMI 5
I-50125 FIRENZE
ITALY
TEL: 055-2752236
TLX: 572268 ARCETR
TLF:
EML:
COM: 20.22

FORTINI TERESA DR
VIA F.D. GUERRAZZI 19
I-00152 ROMA
ITALY
TEL:
TLX:
TLF:
EML:
COM: 10

FOSBURY ROBERT A E DR
ST/ECF
C/O ESO
KARL-SCHWARZSCHILD-STR 2
D-8046 GARCHING B MUNCHEN
GERMANY. F.R.
TEL: 49-89-32006235
TLX: 52828222 EO D
TLF:
EML:
COM:

FOSSAT ERIC DR
OBSERVATOIRE DE NICE
BP 139
F-06003 NICE CEDEX
FRANCE
TEL: 93-89-04-20
TLX: 460004
TLF:
EML:
COM: 10.12.35

FOUKAL PETER V DR
CAMBRIDGE RESEARCH &
INSTRUMENTATIONN INC.
21 ERIE STREET
CAMBRIDGE MA 02139
U.S.A.
TEL: 617-491-2627
TLX:
TLF:
EML:
COM: 12

FOUQUE PASCAL DR
OBSERVATOIRE DE PARIS
SECTION MEUDON
5 PLACE JULES JANSSEN
92195 MEUDON CEDEX
FRANCE
TEL: 45 07 76 08
TLX: 270 912 OBSASTR
TLF:
EML: SPAN:28726::FOUQUE
COM: 28.40.47

FOWLER WILLIAM A PROF
CALTECH 106-38
PASADENA CA 91125
U.S.A.
TEL: 818-356-4272
TLX:
TLF:
EML:
COM: 35.46

FOX KENNETH DR
DEPT PHYSICS & ASTRONOMY
UNIVERSITY OF TENNESSEE
503 PHYSICS
KNOXVILLE TN 37996-1200
U.S.A.
TEL: 615-974-2288
TLX:
TLF:
EML:
COM: 16

FOX W E MR
BRITISH ASTRONOMICAL ASS
40 WINDSOR ROAD
NEWARK NOTTINGHAMS
U.K.
TEL: 0636-704-932
TLX:
TLF:
EML:
COM: 16

FOY RENAUD DR
CERGA
OBSERVATOIRE DE CALERN
CAUSSOLS
F-06460 ST VALLIER DE T.
FRANCE
TEL:
TLX: 461402 CERGLOBS
TLF:
EML:
COM: 09.29.36

FRACASTORO MARIO G PROF
VIA MONVISO 3
I-10025 PINO TORINESE
ITALY
TEL: 011-840493
TLX:
TLF:
EML:
COM: 24.26.42

FRANCESCHINI ALBERTO
DIPATIMENTO DI FISICA
UNIVERSITA DI PADOVA
VIA MARZOLO 8
I-35100 PADOVA
ITALY
TEL:
TLX:
TLF:
EML:
COM: 47.48

FRANCO JOSE DR
INSTITUTO DE ASTRONOMIA
APDO POSTAL 70 264
04510 MEXICO D.F.
MEXICO
TEL:
TLX:
TLF:
EML:
COM:

FRANCOIS PATRICK DR
OBSERVATOIRE DE PARIS
DASGAL
92195 MEUDON PPL CEDEX
FRANCE
TEL: 45 07 78 67
TLX: 270912
TLF: 45 07 74 72
EML:
COM: 29

FRANDSEN SOEREN PROF
INSTITUTE OF ASTRONOMY
UNIVERSITY OF AARHUS
DK-8000 AARHUS C
DENMARK
TEL: 6-128899
TLX: 64767 AAUSCI DK
TLF:
EML:
COM: 29.48

FRANK JUHAN
MAX PLANCK INSTITUT
KARL SCHWARZSCHILDSTRAS.1
8046 GARCHING B MUNCHEN
GERMANY. F.R.
TEL:
TLX:
TLF:
EML:
COM: 42.48

FRANKLIN FRED A DR
PLANETARY SCIENCE DIV
CENTER FOR ASTROPHYSICS
60 GARDEN STREET
CAMBRIDGE MA 02138
U.S.A.
TEL: 617-495-7230
TLX:
TLF:
EML:
COM: 20

FRANSSON CLAES
STOCKHOLM OBSERVATORY
S-133 00 SALTSJOBADEN
SWEDEN
TEL: 46-871-70195
TLX: 12972 SOBSERV S
TLF:
EML:
COM: 48

FRANTSMAN YU L DR
RADIOASTROPHYSICAL OBS
LATVIAN ACAD OF SCIENCES
226524 RIGA
U.S.S.R.
TEL: 226006
TLX:
TLF:
EML:
COM: 15.42

FRANZ OTTO G DR
LOWELL OBSERVATORY
1400 W. MARS HILL RD
FLAGSTAFF AZ 86001
U.S.A.
TEL: 602-774-3358
TLX:
TLF:
EML:
COM: 24.26

FRATER ROBERT H DR
CHIEF CSIRO DIV RADIOPHYS
P.O.BOX 76
EPPING NSW 2121
AUSTRALIA
TEL: 02-868-0222
TLX: 26230 ASTRO
TLF:
EML:
COM: 40

FRAZIER EDWARD N DR
TRW
1 SPACE PARK
REDONDO BEACH CA 90278
U.S.A.
TEL: 213-535-4723
TLX:
TLF:
EML:
COM: 12

FREDGA KERSTIN PROF
SWEDISH BOARD F SPACE ACT
BOX 4006
S-171 54 SOLNA
SWEDEN
TEL: 08-733-6486
TLX: 17128 SPACECO S
TLF:
EML:
COM: 44

FREDRICK LAURENCE W PROF
LEANDER MCCORMICK OBS
UNIV. OF VIRGINIA
BOX 3818 UNIV STATION
CHARLOTTESVILLE VA 22903
U.S.A.
TEL: 804-924-4905
TLX: 510-587-5453 (TWX)
TLF:
EML:
COM: 24.26.42.51

FREEDMAN WENDY L DR
THE OBSERVATORIES OF THE
CARNEGIE INSTI.OF WASHIN.
813 SANTA BARBARA STREET
PASADENA CA 91101
U.S.A.
TEL: 818 577 1122
TLX:
TLF:
EML: wendy@mwlco.caltech.edu
COM: 28

FREEMAN KENNETH C PROF
MT STROMLO OBSERVATORY
PRIVATE BAG
WODEN PO
CANBERRA ACT 2606
AUSTRALIA
TEL: 062-881111
TLX: 62270 CANOPUS AA
TLF:
EML:
COM: 28C.30C.33.37

FREISLEBEN H C DR
APP 6120
FLORENTINER STR 20
D-7000 STUTTGART 75
GERMANY. F.R.
TEL: 0711-4702-6120
TLX:
TLF:
EML:
COM: 41

FREIRE FERRERO RUBENS G
LPSP (IAS)
B.P.10
F-91371 VERRIERES-LE-BUISSON
FRANCE
TEL: 64-47-42-45
TLX: 606 252
TLF:
EML:
COM: 29.36

FREITAS MOURAO R R DR
MUSEU ASTR E CIENCIAS
AFINS/CNPQ RUA GEN BRUCE
SAN CRISTOVAO
20921 RIO DE JANEIRO
BRAZIL
TEL: 580-7154/7204
TLX: 22653
TLF:
EML:
COM: 20.26.41

FRENCH RICHARD G
M I T
DEPT OF EARTH. ATMOSPH.
& PLANET. SCI.. 54-422
CAMBRIDGE MA 02139
U.S.A.
TEL: 617-253-3392
TLX:
TLF:
EML:
COM:

FRENK CARLOS S
PHYSICS DEPARTMENT
UNIVERSITY OF DURHAM
SOUTH ROAD
DURHAM DH1 3LE
U.K.
TEL: 0385-64971
TLX: 537351 DURLIB G
TLF:
EML:
COM: 47

FRESNEAU ALAIN DR
OBSERVATOIRE ASTRONOMIQUE
11. RUE DE L'UNIVERSITE
F-67000 STRASBOURG
FRANCE
TEL: 88-35-82-00
TLX: 890506 STAROBS F
TLF:
EML:
COM: 24

FRIBERG PER
ONSALA SPACE OBSERVATORY
CHALMERS UNIV TECHNOLOGY
S-439 00 ONSALA
SWEDEN
TEL: 0300-60650
TLX: 8542400 ONSPACE
TLF:
EML:
COM: 40

FRICKE KLAUS DR
UNIVERSITAETSSTERNWARTE
UNIVERSITAET GOETTINGEN
GEISMARLANDSTR 11
D-3400 GOETTINGEN
GERMANY. F.R.
TEL: 1149551-395051
TLX: 96753 USTERN D
TLF:
EML:
COM: 26

FRIDMAN ALEKSEY M
ASTRONOMICAL COUNCIL
USSR ACADEMY OF SCIENCES
PYATNITSKAYA STR 48
109017 MOSCOW
U.S.S.R.
TEL: 231-54-61
TLX: 412623 SCSTP SU
TLF:
EML:
COM:

FRIED JOSEF WILHELM DR
MPI FUER ASTRONOMIE
KOENIGSTUHL
D-6900 HEIDELBERG 1
GERMANY. F.R.
TEL: 06221-5281
TLX: 461789 MPIA D
TLF:
EML:
COM: 28

FRIEDEMANN CHRISTIAN DR
UNIV STERNWARTE JENA
SCHILLERGAESSCHEN 2
DDR-6900 JENA
GERMANY. D.R.
TEL: 8222637/27122
TLX: 05886134
TLF:
EML:
COM: 34

FRIEDJUNG MICHAEL DR
INSTITUT D'ASTROPHYSIQUE
98 BIS BOULEVARD ARAGO
F-75014 PARIS
FRANCE
TEL: 1-43-20-14-25
TLX: 270776 OBS
TLF:
EML:
COM: 27.29.42

FRIEDLANDER MICHAEL PROF
DEPARTMENT OF PHYSICS
WASHINGTON UNIVERSITY
ST LOUIS MO 63130
U.S.A.
TEL: 314-889-6279
TLX:
TLF:
EML:
COM:

FRIEDMAN HERBERT DR
US NAVAL RESEARCH LAB
CODE 7100
WASHINGTON DC 20375
U.S.A.
TEL:
TLX:
TLF:
EML:
COM: 10.12.40.44.48

FRIEL EILEEN D DR
DOMINION ASTROPH. OBSERV.
5071 WEST SAANICH ROAD
VICTORIA BC V8X 4M6
CANADA
TEL: 604 388-0062
TLX:
TLF:
EML:
COM: 29

FRIEND DAVID B DR
DPT OF PHYS. & ASTRONOMY
WILLIAMS COLLEGE
WILLIAMSTOWN MA 01267
U.S.A.
TEL: 413 597 2817
TLX:
TLF:
EML: BITNET:DBFRIEND@WILLIAMS
COM: 36

FRINGANT ANNE-MARIE DR
OBSERVATOIRE DE PARIS
61 AVE DE L'OBSERVATOIRE
F-75014 PARIS
FRANCE
TEL: 1-40-51-22-48
TLX:
TLF:
EML:
COM: 29

FRISCH HELENE DR
OBSERVATOIRE DE NICE
BP 139
F-06003 NICE CEDEX
FRANCE
TEL: 93-89-04-20
TLX: 460004
TLF:
EML:
COM: 36

FRISCH PRISCILLA
UNIVERSITY OF CHICAGO
ASTRONOMY & ASTROPHYS CTR
5640 S. ELLIS AVE
CHICAGO IL 60637
U.S.A.
TEL: 312-962-8211
TLX: 910-221-5617
TLF:
EML:
COM: 34

FRISCH URIEL DR
OBSERVATOIRE DE NICE
BP 139
F-06003 NICE CEDEX
FRANCE
TEL: 93-89-04-20
TLX: 460004
TLF:
EML:
COM: 36

FRISK URBAN DR
SPACE SCIENCE DPT
ESTEC
POSTBUS 299
NL-2200 AG NOORDWIJK
NETHERLANDS
TEL:
TLX:
TLF:
EML:
COM: 40.44

FRITZOVA-SVESTKA L DR
DOPPERSTRAAT 147
NL-3752 JC BUNSCHOTEN
NETHERLANDS
TEL: 03499-84403
TLX:
TLF:
EML:
COM: 10

FROEHLICH CLAUS
WORLD RADIATION CENTER
PHYSIKALISCH-METEOROL OBS
POSFACH 173
CH-7260 DAVOS-DORF
SWITZERLAND
TEL: 41-083-521-31
TLX: 74732 PMOD CH
TLF:
EML:
COM: 12

FROESCHLE CHRISTIANE D DR
OBSERVATOIRE DE NICE
BP 139
F-06003 NICE CEDEX
FRANCE
TEL:
TLX:
TLF:
EML:
COM: 15.36

FROESCHLE CLAUDE DR
OBSERVATOIRE DE NICE
BP 139
F-06003 NICE CEDEX
FRANCE
TEL: 93-89-04-20
TLX:
TLF:
EML:
COM: 07C.20

FROESCHLE MICHEL DR
CERGA
AVENUE COPERNIC
F-06130 GRASSE
FRANCE
TEL:
TLX:
TLF:
EML:
COM:

FROGEL JAY ALBERT DR
NAT OPTICAL ASTR OBS
PO BOX 26732
950 N. CHERRY AVENUE
TUCSON AZ 85726
U.S.A.
TEL: 602-327-5511
TLX:
TLF:
EML:
COM: 28

FROLOV M S DR
ASTRONOMICAL COUNCIL
USSR ACADEMY OF SCIENCES
PYATNITSKAYA UL 48
109017 MOSCOW
U.S.S.R.
TEL: 231-54-61
TLX: 412623 SCSTP SU
TLF:
EML:
COM: 27

FROST KENNETH J DR
NASA/GSFC. SPACE STATION
OFFICE. CODE 400.7
GREENBELT RD. BLDG 16
GREENBELT MD 20771
U.S.A.
TEL: 301-286-8874
TLX:
TLF:
EML:
COM:

FRYE GLENN M PROF
PHYSICS DEPARTMENT
CASE WESTERN RESERVE UNIV
ROCK BUILDING
CLEVELAND OH 44106
U.S.A.
TEL: 216-368-2997
TLX:
TLF:
EML:
COM:

FTACLAS CHRIST
SPACE SCIENCE DIVISION
PERKIN-ELMER CORP.
100 WOOSTER HEIGHTS RD
DANBURY CT 06810-7589
U.S.A.
TEL: 203-797-6448
TLX:
TLF:
EML:
COM: 28

FU CHENG-QI
SHANGHAI OBSERVATORY
ACADEMIA SINICA
SHANGHAI
CHINA. PEOPLE'S REP.
TEL: 386191
TLX: 33164 SHAO CN
TLF:
EML:
COM: 44

FU DELIAN
BEIJING ASTRONOMICAL OBS
ACADEMIA SINICA
BEIJING
CHINA. PEOPLE'S REP.
TEL: 282070
TLX: 22040 BAOAS CN
TLF:
EML:
COM: 09

FU QI JUN
BEIJING OBSERVATORY
ACADEMIA SINICA
BEIJING
CHINA. PEOPLE'S REP.
TEL: 275580
TLX: 22040 BAOAS CN
TLF:
EML:
COM: 10

FUCHS BURKHARD DR
ASTRONOMISCHES-RECHEN
INSTITUT
MOENCHHOFSTR. 12-14
D-6900 HEIDELBERG 1
GERMANY. F.R.
TEL: 06221-49026
TLX: 461336 ARIHD D
TLF:
EML:
COM: 28.33

FUENMAYOR FRANCISCO J DR
UNIVERSIDAD DE LOS ANDES
FACULTAD DE CIENCIAS
DEPARTAMENTO DE FISICA
MERIDA 5101
VENEZUELA
TEL: 074-63-99-30
TLX: 74173 CDCH-ULA
TLF:
EML:
COM:

FUENSALIDA JIMENEZ J DR
INSTITUTO DE ASTROFISICA
DE CANARIAS
38200 LA LAGUNA TENERIFE
SPAIN
TEL: 34 22 26 22 11
TLX: 92640
TLF:
EML:
COM:

FUERST ERNST DR
MPI FUER RADIOASTRONOMIE
AUF DEM HUEGEL 69
D-5300 BONN
GERMANY. F.R.
TEL:
TLX: 886440 MPIFR D
TLF:
EML:
COM: 40

FUJIMOTO MASAYUKI DR
NIIGATA UNIVERSITY
FACULTY OF EDUCATION
8050 IKARASHI-2
NIIGATA 950-21
JAPAN
TEL:
TLX:
TLF:
EML:
COM: 28.35

FUJIMOTO MASA-KATSU DR
TOKYO ASTR OBSERVATORY
MITAKA
TOKYO 181
JAPAN
TEL: 0422-32-5111
TLX: 2822307 TAOMTK J
TLF:
EML:
COM: 31C.33.51

FUJIMOTO MITSUAKI DR
DEPT OF PHYSICS
NAGOYA UNIVERSITY
NAGOYA 464
JAPAN
TEL:
TLX:
TLF:
EML:
COM: 47.51

FUJISHITA MITSUMI DR
DIV. OF EARTH ROTATION
NAT. ASTRONO. OBSERVATORY
2-12 HOSHIGAOKA
MIZUSAWA. IWATE 023
JAPAN
TEL: 197 24 7111
TLX: 837628 ILSMIZ J
TLF:
EML: TELEMAIL ID:MIZUSAWA
COM: 19

FUJITA YOSHIO PROF
DEPARTMENT OF ASTRONOMY
UNIVERSITY OF TOKYO
BUNKYO KU
TOKYO 113
JAPAN
TEL: 423-74-4186
TLX:
TLF:
EML:
COM: 29

FUJIWARA AKIRA DR
DEPARTMENT OF PHYSICS
KYOTO UNIVERSITY
KITASHIRAKAWA SAKYOKU
KYOTO 606
JAPAN
TEL: 075-751-2111
TLX: 5422693 LIBKYU J
TLF:
EML:
COM: 15.16.21

FUKUDA ICHIRO
DEPT OF PHYSICS
KANAZAWA INST OF TECHNOL.
7-1 OGIGAOKA. NONOICHI
ISHIKAWA 921
JAPAN
TEL: 0762-48-1100
TLX: 5122456 KIT LC J
TLF:
EML:
COM: 45

FUKUE JUN DR
ASTRONOMICAL INSTITUTE
OSAKA KYOIKU UNIVERSITY
4-88 MINAMIKAWAHORICHO
OSAKA 543
JAPAN
TEL:
TLX:
TLF:
EML:
COM:

FUKUGITA MASATAKA DR
RESEARCH INSTITUTE FOR
FUNDAMENTAL PHYSICS
KYOTO UNIVERSITY
KYOTO 606
JAPAN
TEL: 075 711 1381
TLX: 5423179 RIFPK
TLF:
EML:
COM: 28.47

FUKUI TAKAO DR
DEPT OF LIBERAL ARTS
DOKKYO UNIVERSITY
SAKAE-MACHI 600
SOKA SAITAMA
JAPAN
TEL: 0489-42-1111
TLX:
TLF:
EML:
COM: 47

FUKUI YASUO DR
DEPARTMENT OF PHYSICS
NAGOYA UNIVERSITY
FUROCHO CHIKUSAKU
NAGOYA 464
JAPAN
TEL: 052-781-5111
TLX: 4477323 SCUNAG J
TLF:
EML:
COM: 34.40

FUKUNAGA MASATAKA DR
DEPARTMENT OF ASTRONOMY
TOHOKU UNIVERSITY
AOBA ARAMAKI SENDAI MIYAG
JAPAN
TEL: 22 222 1800
TLX:
TLF:
EML:
COM: 33

FUKUSHIMA TOSHIO DR
SATELLITE GEODESY OFFICE
HYDROGRAPHIC DEPT
5 3 1 TSUKIJI CHUO KU
TOKYO 104
JAPAN
TEL: 3 541 3811
TLX: 2522452 HDJODC J
TLF:
EML: GE MarkIII NODE-RC28 GGDHDJ
COM:

FULCHIGNONI MARCELLO PROF
IST. ASTROFISICA SPAZIALE
C.N.R.-C.N.R.
I-00044 FRASCATI
ITALY
TEL: 6-4956951
TLX: 680489 CNR FRA
TLF:
EML:
COM: 15

FURENLID INGEMAR K DR
DEPT PHYSICS & ASTRONOMY
GEORGIA STATE UNIVERSITY
ATLANTA GA 30303
U.S.A.
TEL: 404-658-2932
TLX:
TLF:
EML:
COM: 26.29

FURNISS IAN
DEPT PHYSICS & ASTRONOMY
UNIVERSITY COLLEGE LONDON
GOWER STREET
LONDON WC1E 6BT
U.K.
TEL: 01-387-7050
TLX: 28722 UCPHYS
TLF:
EML:
COM: 34.44

FURSENKO M A DR
INST OF THEORET ASTRONOMY
USSR ACADEMY OF SCIENCES
10 KUTUZOV QUAY
191187 LENINGRAD
U.S.S.R.
TEL: 278-88-98
TLX: 121578 ITA SU
TLF:
EML:
COM: 04

FURUKAWA KIICHIRO DR
TOKYO ASTRONOMICAL OBS
OSAWA MITAKA
181 TOKYO
JAPAN
TEL:
TLX:
TLF:
EML:
COM: 08.19.20

FUSCO-FEMIANO ROBERTO
IST. ASTROFISICA SPAZIALE
C P 67
I-00044 FRASCATI
ITALY
TEL: 9425655
TLX: 610261
TLF:
EML:
COM:

FUSI PECCI FLAVIO
OSSERVATORIO ASTRONOMICO
UNIVERSITARIO
C P 596
I-40100 BOLOGNA
ITALY
TEL: 051-222956
TLX: 211664 INFNBO-I
TLF:
EML:
COM: 37

FU-SHONG KUO
DEPARTMENT OF PHYSICS
INST PHYSICS & ASTRONOMY
NAT CENTRAL UNIVERSITY
CHUNG LI
CHINA. TAIWAN
TEL:
TLX:
TLF:
EML:
COM:

GABRIEL ALAN H
INSTITUT D'ASTROPHYSIQUE
SPATIALE
P.O. BOX 10
F-91371 VERRIERES LE B
FRANCE
TEL: 1-64-47-43-15
TLX: 600252
TLF:
EML:
COM: 10.12.14C.44

GABRIEL MAURICE R DR
INSTITUT D'ASTROPHYSIQUE
UNIVERSITY OF LIEGE
AVENUE DE COINTE 5
B-4200 COINTE-OUGREE
BELGIUM
TEL: 041-52-99-80
TLX: 41264 ASTRLG
TLF:
EML:
COM: 35

GAETZ TERRANCE J DR
DPT OF ASTRONOMY
UNIV. WESTERN ONTARIO
LONDON ONTARIO N6A 3K7
CANADA
TEL: 519 661 3183
TLX: 0647134
TLF:
EML: BITNET:GAETZ@UWOVAX
COM:

GAHM GOESTA F DR
STOCKHOLM OBSERVATORY
S-133 00 SALTSJOEBADEN
SWEDEN
TEL: 08-717-0637
TLX: 12972
TLF:
EML:
COM: 27

GAIGNEBET JEAN DR
CERGA
AVENUE COPERNIC
F-06130 GRASSE
FRANCE
TEL: 93-36-58-49
TLX: 470865 F
TLF:
EML:
COM: 19.31

GAIL HANS-PETER DR
INST THEORET ASTROPHYSIK
DER UNIVERSITAET
IM NEUENHEIMER FELD 294
D-6900 HEIDELBERG 1
GERMANY. F.R.
TEL:
TLX:
TLF:
EML:
COM: 16

GAISSER THOMAS K
BARTOL RESEARCH FOUND.
UNIVERSITY OF DELAWARE
NEWARK DE 19716
U.S.A.
TEL: 302-451-8111
TLX: 510-666-0805 BARTOL
TLF:
EML:
COM: 48

GAIZAUSKAS VICTOR DR
HERZBERG INST ASTROPHYS
NATL RESEARCH COUNCIL
OTTAWA ONT K1A 0R6
CANADA
TEL: 613-993-7395
TLX: 0533715 NRCOTT
TLF:
EML:
COM: 10V.12

GALAL A A DR
HELWAN OBSERVATORY
HELWAN-CAIRO
EGYPT
TEL: 780645. 782683
TLX:
TLF:
EML:
COM:

GALAN MAXIMINO J
M & G ENGS.
S. MARTIN DE PORRES 45
28035 MADRID
SPAIN
TEL: 341-216-0995
TLX:
TLF:
EML:
COM: 09.50

GALEOTTI PIERO PROF
IST. DI COSMO-GEOFISICA
CORSO FIUME 4
I-10133 TORINO
ITALY
TEL: 011-658979
TLX: 224379 COSMOT I
TLF:
EML:
COM: 48

GALIBINA I V DR
INSTITUTE OF THEORETICAL
ASTRONOMY
10 KUTUZOV QUAY
191187 LENINGRAD
U.S.S.R.
TEL: 186-19-74
TLX: 121578 ITA SU
TLF:
EML:
COM: 07.20

GALINDO TREJO JESUS DR
INST. ASTRONOMIA. UNAM
APARTADO POSTAL 70-264
04510 MEXICO: D.F.
MEXICO
TEL: 905 5485305
TLX: 1760155 CICME
TLF:
EML: BITNET:JGAL@UNAMVM1
COM:

GALLAGHER III JOHN S DR
LOWELL OBSERVATORY
MARS HILL RD
1400 WEST
FLAGSTAFF AZ 86001
U.S.A.
TEL: 602-774-3358
TLX:
TLF:
EML:
COM: 27.28C

GALLAGHER JEAN DR
NAT. INS. OF STANDARDS
& TECHNO. PHYSICS A 323
GAITHERSBURG MD 20899
U.S.A.
TEL: 301-975 2204
TLX:
TLF:
EML: BITNET:JWG00001@NBS
COM: 14

GALLET ROGER M
964 7TH STREET
BOULDER CO 80302
U.S.A.
TEL:
TLX:
TLF:
EML:
COM:

GALLETTA GIUSEPPE PROF
ISTITUTO DI ASTRONOMIA
UNIVERSITA DI PADOVA
VIC. DELL'OSSERVATORIO 5
I-35122 PADOVA
ITALY
TEL: 049-66-1499
TLX: 430176 UNPADU I
TLF:
EML:
COM: 28

GALLETTO DIONIGI
IST. DI FISICA MATEMATICA
UNIVERSITA DI TORINO
VIA CARLO ALBERTO 10
I-10123 TORINO
ITALY
TEL: 011-539214
TLX:
TLF:
EML:
COM: 07.33.47

GALLINO ROBERTO
IST. DI FISICA GENERALE
DELL'UNIVERSITA'
CORSO M. D'AZEGLIO 46
I-10125 TORINO
ITALY
TEL: 011-655103
TLX: 21104 INFN TO
TLF:
EML:
COM: 35.46

GALLOUET LOUIS DR
OBSERVATOIRE DE PARIS
61 AVE DE L'OBSERVATOIRE
F-75014 PARIS
FRANCE
TEL: 1-40-51-22-07
TLX:
TLF:
EML:
COM: 24.25

GALLOWAY DAVID DR
DEPT OF APPLIED MATHS
UNIVERSITY OF SYDNEY
SYDNEY N.S.W. 2006
AUSTRALIA
TEL: 692-2222
TLX: 26169 UNISYD AA
TLF:
EML:
COM: 10

GALPERIN YU I PROF
SPACE RESEARCH INSTITUTE
USSR ACADEMY OF SCIENCES
PROFSOJUSNAYA 84/32
117810 MOSCOW GSP-7
U.S.S.R.
TEL: 333-31-22
TLX: 411498 STAR SU
TLF:
EML:
COM: 21C

GALT JOHN A DR
DOMINION RADIO ASTROPHY-
SICAL OBSERVATORY
P O BOX 248
PENTICTON BC V2A 6K3
CANADA
TEL: 604-497-5321
TLX: 048-88127
TLF:
EML:
CON: 40

GAMALELDIN ABDULLA I DR
HELWAN OBSERVATORY
HELWAN-CAIRO
EGYPT
TEL: 780645
TLX:
TLF:
EML:
CON: 28

GAMBIS DANIEL DR
OBSERVATOIRE DE PARIS
61 AV. DE L'OBSERVATOIRE
75014 PARIS
FRANCE
TEL: 33-1-40512233
TLX: 270776 OBS F
TLF:
EML: GAMBIS@FRIAP51
CON:

GAMMELGAARD PETER MAG SCI
INSTITUTE OF ASTRONOMY
UNIVERSITY OF AARHUS
LANGELANDSGADE
DK-8000 AARHUS C
DENMARK
TEL: 06-128899
TLX: 64767 AAUSCI DK
TLF:
EML:
CON:

GAO BILIE
NANJING ASTRONOMICAL
INSTRUMENT FACTORY
NANJING
CHINA, PEOPLE'S REP.
TEL:
TLX:
TLF:
EML:
CON: 09

GAO BUXI
INSTITUTE OF GEODESY &
GEOPHYSICS
XU DONG LU
WUCHAN, HUBEI
CHINA, PEOPLE'S REP.
TEL: 813805
TLX:
TLF:
EML:
CON: 19

GAPOSCHKIN EDWARD M DR
55 FARMCREST AVE
LEXINGTON MA 02173
U.S.A.
TEL: 617-862-2538
TLX:
TLF:
EML:
CON: 07.19

GARAY GUIDO DR
ESO
KARL-SCHWARZSCHILD-STR 2
D-8046 GARCHING B MUNCHEN
GERMANY, F.R.
TEL:
TLX:
TLF:
EML:
CON: 40

GARCIA DE LA ROSA JOSE I
INST DE ASTROFISICA DE
CANARIAS
LA LAGUNA
38071 TENERIFE
SPAIN
TEL: 922-26-22-11
TLX: 92640
TLF:
EML:
CON: 10

GARCIA LAMBAS DIEGO DR
FACULDAD DE CIENCIAS
UNIV. NACIONAL LA PLATA
PASEO DEL BOSQUET
1900 LA PLATA (BS.AS.)
ARGENTINA
TEL: 54 051 40613
TLX:
TLF:
EML:
CON: 28

GARCIA-BARRETO JOSE A
INST. DE ASTRONOMIA UNAM
OBS ASTRONOMICO NACIONAL
APDO POSTAL 877
22860 ENSENADA, B. CALIF.
MEXICO
TEL: 667-830-93
TLX:
TLF:
EML:
CON:

GARCIA-PELAYO JOSE DR
INST DE ASTROFISICA DE
ANDALUCIA, APDO 2144
PROFESOR ALBAREDA 1
18080 GRANADA
SPAIN
TEL: 25-61-03
TLX:
TLF:
EML:
CON:

GARDNER FRANCIS F DR
CSIRO
DIVISION OF RADIOPHYSICS
P.O.76
EPPING NSW 2121
AUSTRALIA
TEL: 02-868-0222
TLX: 26230 AA
TLF:
EML:
CON: 34.40

GARFINKEL BORIS DR
YALE UNIVERSITY OBS
NEW HAVEN CT 06520
U.S.A.
TEL: 203-436-3460
TLX:
TLF:
EML:
CON: 07.20

GARGAUD MURIEL DR
OBSERVATOIRE DE BORDEAUX
BP 89
33270 FLOIRAC
FRANCE
TEL: 56 86 43 30
TLX:
TLF:
EML: MURIEL@FRBOBO51
CON: 14

GARLICK GEORGE F DR
267 SOUTH BELOIT AVE
LOS ANGELES CA 90049
U.S.A.
TEL: 213-472-3512
TLX:
TLF:
EML:
CON:

GARMANY CATHERINE D DR
JILA
UNIVERSITY OF COLORADO
BOULDER CO 80309
U.S.A.
TEL: 303-492-7836
TLX:
TLF:
EML:
CON: 29.42

GARMIRE GORDON P PROF
PENNA STATE UNIVERSITY
525 DAVEY LAB
UNIVERSITY PARK PA 16802
U.S.A.
TEL: 814-865-0418
TLX: 842510 PENNSTBSTR SC
TLF:
EML:
CON: 48

GARNIER ROBERT ING
OBSERVATOIRE DE LYON
F-69230 ST-GENIS-LAVAL
FRANCE
TEL: 78-56-07-05
TLX:
TLF:
EML:
CON:

GARRIDO RAFAEL
INST DE ASTROFISICA DE
ANDALUCIA, APDO 2144
PROFESOR ALBAREDA 1
18080 GRANADA
SPAIN
TEL: 58-121311
TLX: 78753
TLF:
EML:
CON: 27

GARRISON ROBERT F PROF
DAVID DUNLAP OBSERVATORY
P O BOX 360
RICHMOND HILL ONT L4C 4Y6
CANADA
TEL: 416-884-9562
TLX: 06-986766
TLF:
EML:
COM: 29.45C

GARSTANG ROY H PROF
JILA
UNIVERSITY OF COLORADO
BOULDER CO 80309
U.S.A.
TEL: 303-492-7795
TLX: 755842 JILA
TLF:
EML:
COM: 05.14

GARTON W R S PROF
BLACKETT LABORATORY
IMPERIAL COLLEGE
LONDON SW7 2BZ
U.K.
TEL: 0233-21657
TLX: 261503
TLF:
EML:
COM: 14

GARY DALE E
CALTECH
SOLAR ASTRONOMY 264-33
PASADENA CA 91125
U.S.A.
TEL: 818-356-3863
TLX: 675425 CALTECH PSD
TLF:
EML:
COM:

GARZON FRANCISCO DR
INSTITUTO DE ASTROFISICA
DE CANARIAS
38200 LA LAGUNA TENERIFE
SPAIN
TEL: 34 22 26 22 11
TLX: 92640
TLF:
EML: SPAN:IAC::FGL
COM: 33

GASCOIGNE S C B DR
MT STROMLO OBSERVATORY
WODEN P.O. ACT 2606
AUSTRALIA
TEL:
TLX:
TLF:
EML:
COM: 27.28.37

GASKA STANISLAW DR
INSTITUTE ASTRONOMY
UL CHOPINA 12-18
87-100 TORUN
POLAND
TEL:
TLX:
TLF:
EML:
COM: 07

GATEWOOD GEORGE DIRECTOR
ALLEGHENY OBSERVATORY
OBSERVATORY STATION
PITTSBURG PA 15214
U.S.A.
TEL: 412-321-2400
TLX:
TLF:
EML:
COM: 24.26.51C

GATLEY IAN
UK INFRARED TELESCOPE
665 KOMOHANA STREET
HILO HI 96720
U.S.A.
TEL: 808-961-3756
TLX: 633135
TLF:
EML:
COM:

GAUR V P
UTTAR PRADESH STATE OBS
MANORA PEAK
NAINITAL (UP) 263 129
INDIA
TEL: 2136
TLX:
TLF:
EML:
COM: 12

GAUSS F STEPHEN
US NAVAL OBSERVATORY
WASHINGTON DC 20390
U.S.A.
TEL: 202-653-1510
TLX:
TLF:
EML:
COM: 08.09

GAUSTAD JOHN E PROF
DEPT OF ASTRONOMY
SWARTHMORE COLLEGE
SWARTHMORE PA 19081
U.S.A.
TEL: 215-447-7271
TLX:
TLF:
EML:
COM: 34

GAUTIER DANIEL
OBSERVATOIRE DE PARIS
SECTION DE MEUDON
F-92195 MEUDON PL CEDEX
FRANCE
TEL: 1-45-07-77-07
TLX: 201571 LAM
TLF:
EML:
COM: 16C

GAY JEAN DR
CERGA
AVENUE COPERNIC
F-06130 GRASSE
FRANCE
TEL: 93-36-58-49
TLX: 470865
TLF:
EML:
COM: 09.34

GAYLARD MICHAEL JOHN
HRAO
PO BOX 443
KRUGERSDORP 1740
SOUTH AFRICA
TEL: 011-6424692
TLX: 321006
TLF:
EML:
COM: 40

GEAKE JOHN E DR
PHYSICS DEPT
UMIST
MANCHESTER M60 1QD
U.K.
TEL: 61-236-3311
TLX: 666094
TLF:
EML:
COM: 16

GEBALLE THOMAS R DR
UK TELESCOPES
665 KOMOHANA STREET
HILO HI 96720
U.S.A.
TEL: 808 961 3756
TLX: 633135
TLF:
EML: NSSDCA::PSI%UKTH::TOM
COM: 34

GEBBIE KATHARINE B DR
JILA
UNIVERSITY OF COLORADO
BOULDER CO 80309
U.S.A.
TEL: 303-492-7825
TLX: 755842 JILA
TLF:
EML:
COM: 36

GEBLER KARL-HEINZ DR
RADIOASTRONOMISCHES INST
DER UNIVERSITAET BONN
AUF DEM HUEGEL 71
D-5300 BONN 1
GERMANY. F.R.
TEL: 0228-733662
TLX:
TLF:
EML:
COM: 40

GEFFERT MICHAEL DR
OBSERVATORIUM HOHER LIST
D 5568 DAUN
GERMANY. F.R.
TEL: 06592 2150
TLX:
TLF:
EML:
COM:

GEHRELS TOM PROF
LUNAR LABORATORY
UNIVERSITY OF ARIZONA
TUCSON AZ 85721
U.S.A.
TEL: 602-621-6970
TLX:
TLF:
EML:
COM: 15.16.20.25.34.51

GEHREN THOMAS PH D
INST ASTRON & ASTROPHYSIK
UNIVERSITAETS STERNWARTE
SCHEINERSTRASSE 1
D-8000 MUENCHEN 80
GERMANY. F.R.
TEL: 89-98-90-21
TLX: 529815 UNIVERS D
TLF:
EML:
COM: 29

GEHRZ ROBERT DOUGLAS DR
DEPT PHYSICS & ASTRONOMY
UNIVERSITY OF WYOMING
UNIV STATION BOX 3905
LARAMIE WY 82071
U.S.A.
TEL: 307-766-6176
TLX:
TLF:
EML:
COM: 25

GEISS JOHANNES PROF
PHYSIK INSTITUT
UNIVERSITAET BERN
SIDLERSTRASSE 5
CH-3012 BERN
SWITZERLAND
TEL: 31-65-44-02
TLX: 32320
TLF:
EML:
COM: 15.16

GELDZAHLER BERNARD J
NAVAL RESEARCH LABORATORY
CODE 4121.6
WASHINGTON DC 20375
U.S.A.
TEL:
TLX:
TLF:
EML:
COM: 40

GELFREIKH GEORGIJ B DR
PULKOVO OBSERVATORY
196140 LENINGRAD
U.S.S.R.
TEL:
TLX:
TLF:
EML:
COM: 10.40

GELLER MARGARET JOAN
CENTER FOR ASTROPHYSICS
60 GARDEN STREET
CAMBRIDGE MA 02138
U.S.A.
TEL: 617-495-7409
TLX: 921428 SATELLITE CAM
TLF:
EML:
COM: 28.47

GENET R M DR
FAIRBORN OBSERVATORY
1750 S. PRICE RD.ROOM 127
TEMPE AZ 85281
U.S.A.
TEL: 602-968 3899
TLX:
TLF:
EML:
COM: 25.27

GENKIN IGOR L PROF DR
PHYSICS FACULTY
KAZAKH STATE UNIVERSITY
KOMSOMOLSKAYA 96
480012 ALMA ATA
U.S.S.R.
TEL: 67-70-18
TLX:
TLF:
EML:
COM: 33

GENT HUBERT MR
PROSPECT HOUSE
SHERWOOD LANE
WORCESTER WR2 4NX
U.K.
TEL: 422 186
TLX:
TLF:
EML:
COM: 40

GENZEL REINHARD DR
MAX PLANCK INSTITUT FUER
PHYSIK UND ASTROPHYSIK
D 8046 GARCHING B MUNCHEN
GERMANY F.R.
TEL:
TLX:
TLF:
EML:
COM: 34C.40

GEORGELIN YVON P DR
OBSERVATOIRE DE MARSEILLE
2 PLACE LE VERRIER
F-13248 MARSEILLE CEDEX
FRANCE
TEL: 91-95-90-88
TLX: 420241 F
TLF:
EML:
COM: 30.33.34

GEORGELIN YVONNE M DR
OBSERVATOIRE DE MARSEILLE
2 PLACE LE VERRIER
F-13248 MARSEILLE CEDEX 4
FRANCE
TEL: 91-95-90-88
TLX: 420241 F
TLF:
EML:
COM: 33

GEORGIEV TSVETAN DR
DEPT OF ASTRONOMY AND
NATIONAL ASTRONO. OBSERV.
72 LENIN BLVD
1184 SOFIA
BULGARIA
TEL: 758 927
TLX: 23561
TLF:
EML:
COM: 28

GERARD ERIC DR
OBSERVATOIRE DE PARIS
SECTION DE MEUDON
F-92195 MEUDON PL CEDEX
FRANCE
TEL: 1-45-07-76-07
TLX: 270912 OBSASTR.
TLF:
EML:
COM: 15.34

GERBAL DANIEL DR
OBSERVATOIRE DE PARIS
SECTION DE MEUDON
F-92195 MEUDON PL CEDEX
FRANCE
TEL: 1-45-07-74-19
TLX: 201571 LAM
TLF:
EML:
COM:

GERBALDI MICHELE DR
INSTITUT D'ASTROPHYSIQUE
98 BIS BD ARAGO
F-75014 PARIS
FRANCE
TEL: 1-43-20-14-25
TLX:
TLF:
EML:
COM: 25.29.45.46

GERGELY TOMAS ESTEBAN DR
DIV ASTRONOMICAL SCIENCES
NATL SCIENCE FOUNDATION
1800 G STREET N.W.
WASHINGTON DC 20550
U.S.A.
TEL: 202-357-9696
TLX:
TLF:
EML:
COM: 10.40

GERHARD ORTWIN
INSTITUTE OF ASTRONOMY
MADINGLEY ROAD
CAMBRIDGE CB3 0HA
U.K.
TEL: 0223-337548
TLX:
TLF:
EML:
COM: 28

GERIN MARYVONNE DR
RADIO ASTRONO. MILLIME.
LAB. DE PHYSIQUE
E.N.S. 24 RUE LHOMOND
75231 PARIS CEDEX 05
FRANCE
TEL: (33-1)43291225
TLX:
TLF:
EML: GERIN@FRULM11
COM:

GERLEI OTTO
HELIOPHYSICAL OBSERVATORY
H-4010 DEBRECEN
HUNGARY
TEL:
TLX:
TLF:
EML:
COM:

GEROLA HUMBERTO DR
IBM CORPORATION
DEPT K64/282
5600 COTTLE ROAD
SAN JOSE CA 95193
U.S.A.
TEL:
TLX:
TLF:
EML:
COM: 34

GEROVANNIS VASSILIS S DR
DEPT OF ASTRONOMY
UNIVERSITY OF PATRAS
GR-26110 PATRAS
GREECE
TEL:
TLX:
TLF:
EML:
COM: 35

GERSHBERG R E DR
CRIMEAN ASTROPHYSICAL OBS
NAUCHNY
334413 CRIMEA
U.S.S.R.
TEL:
TLX:
TLF:
EML:
COM: 27C.29

GESZTELYI LIDIA
DEBRECEN HELOPHYSICAL
OBSERVATORY
BOX 30
H-4010 DEBRECEN
HUNGARY
TEL: 52-11-015
TLX: 072517 DEOBS H
TLF:
EML:
COM: 10

GEYER EDWARD H PROF DR
OBSERVATORIUM HOHER LIST
UNIVERSITAET BONN
D-5568 DAUN/EIFEL
GERMANY. F.R.
TEL: 06592-2150
TLX:
TLF:
EML:
COM: 26.27.42.45

GEZARI DANIEL YSA DR
NASA/GSFC
CODE 693
GREENBELT MD 20771
U.S.A.
TEL: 301-286-3432
TLX:
TLF:
EML:
COM: 34.44

GHEIDARI S NASSIRI DR
PHYSICS DPT
& BIRUNI OBSERVATORY
SHIRAZ UNIVERSITY
SHIRAZ
IRAN
TEL:
TLX:
TLF:
EML:
COM: 35

GHIGO FRANCIS D DR
N.R.A.O.
PO BOX 2
GREENBANK. WV 24944
U.S.A.
TEL:
TLX:
TLF:
EML:
COM: 28.40.51

GHOBROS ROSHDY AZER DR
HELWAN OBSERVATORY
HELWAN-CAIRO
EGYPT
TEL:
TLX:
TLF:
EML:
COM:

GHOSH P DR
TATA INSTITUTE OF
FUNDAMENTAL RESEARCH
BOMBAY 400 005
INDIA
TEL: 21-9111 EXT 260
TLX: 011-3009
TLF:
EML:
COM: 28

GHOSH S K DR
TATA INSTITUTE OF
FUNDAMENTAL RESEARCH
HOMI BHABHA RD
BOMBAY 400 005
INDIA
TEL: 219111
TLX: 011-3009 TIFR IN
TLF:
EML:
COM: 25

GIACAGLIA GIORGIO E PROF
ESCOLA POLITECHNICA
UNIVERSIDADE DE SAO PAULO
CAIXA POSTAL 8174
05508 SAO PAULO
BRAZIL
TEL: 55-11-32237
TLX:
TLF:
EML:
COM: 07

GIACCONI RICCARDO PROF
SPACE TELESCOPE SCI INST
HOMEWOOD CAMPUS
3700 SAN MARTIN DRIVE
BALTIMORE MD 21218
U.S.A.
TEL: 301-338-4711
TLX: 6849101 ST SCI
TLF:
EML:
COM: 44.48C

GIACHETTI RICCARDO PROF
UNIVERSITY OF FLORENCE
DEPARTMENT OF PHYSICS
LARGO E. FERMI 2
I-50100 FIRENZE
ITALY
TEL: 229-8141
TLX: 572570
TLF:
EML:
COM:

GIALLONGO EMANUELE DR
OSSERVATORIO ASTRONOMICO
CAPODIMONTE
NAPOLI
ITALY
TEL:
TLX:
TLF:
EML:
COM: 47

GIAMPAPA MARK S
NATL SOLAR OBSERVATORY
PO BOX 26732
950 N. CHERRY AVE
TUCSON AZ 85726-6732
U.S.A.
TEL: 602-327-5511
TLX: 0666484 AURA NOAOTUC
TLF:
EML:
COM: 29

GIANNONE PIETRO PROF
OSSERVATORIO ASTRONOMICO
VIALE DEL PARCO MELLINI 8
I-00136 ROMA
ITALY
TEL: 3452794
TLX:
TLF:
EML:
COM: 35.42

GIANNUZZI MARIA A DR
DIPT. DI MATEMATICA
UNIV DI ROMA LA SAPIENZA
PIAZZA GRAMSCI 5
I-00041 ALBANO/LAZIALE
ITALY
TEL: 06-932-11-01
TLX:
TLF:
EML:
COM:

GIBSON DAVID MICHAEL DR
PHYSICS DEPT
NM INST MINING TECHN
CAMPUS STATION
SOCORRO NM 87801
U.S.A.
TEL: 505-835-5340
TLX:
TLF:
EML:
COM: 15.27.40.42

GIBSON JAMES
ITT/FEC & JPL. CALTECH
MAIL STOP 238-322
4800 OAK GROVE DRIVE
PASADENA CA 91103
U.S.A.
TEL: 818-354-2900
TLX: 675429
TLF:
EML:
COM: 15.20

GICLAS HENRY L MR
120 E. ELM AVE
FLAGSTAFF AZ 86001
U.S.A.
TEL: 602-774-4769
TLX:
TLF:
EML:
COM: 16.20.24

GIERASCH PETER J DR
DEPARTMENT OF ASTRONOMY
CORNELL UNIVERSITY
ITHACA NY 14853
U.S.A.
TEL: 607-256-3507
TLX:
TLF:
EML:
COM: 16

GIEREN WOLFGANG P DR
OBSERVATORIUM HOHER LIST
DER UNIV.STERNWARTE BONN
D-5568 DAUN/EIFEL
GERMANY. F.R.
TEL:
TLX:
TLF:
EML:
COM: 27

GIES DOUGLAS R DR
DEPT. OF PHYS.& ASTRONOMY
GEORGIA STATE UNIVERSITY
ATLANTA GA 30303
U.S.A
TEL: 404 651 2932
TLX:
TLF:
EML: BITNET:PHYDDG@GSUVM1
COM: 27

GIESEKING FRANK DR
OBSERVATORIUM HOHER LIST
UNIVERSITAET BONN
D-5568 DAUN
GERMANY. F.R.
TEL:
TLX:
TLF:
EML:
COM: 30

GIETZEN JOSEPH W
ROYAL GREENWICH OBS
HAILSHAM BN27 1RP
U.K.
TEL: 32-181-3171
TLX:
TLF:
EML:
COM:

GILLET D DR
OBS.DE HAUTE PROVENCE
F-04870 ST-MICHEL-L'OBS.
FRANCE
TEL: 92-76-63-68
TLX: 410690
TLF:
EML: BITNET:GILLET@FRONI51
COM: 29

GILLILAND RONALD LYNN
SPACE TELESCOPE SCIENCE
INSTITUTE
3700 SAN MARTIN DRIVE
BALTIMORE MD 21218
U.S.A.
TEL:
TLX:
TLF:
EML:
COM: 10

GILLINGHAM PETER MR
ANGLO AUSTRALIAN OBS
PRIVATE BAG
COONABARABRAN NSW 2357
AUSTRALIA
TEL: 068-42-1122
TLX: 63945AA
TLF:
EML:
COM: 09

GILMAN PETER A DR
HIGH ALTITUDE OBSERVATORY
NCAR
PO BOX 3000
BOULDER CO 80307
U.S.A.
TEL: 303-497-1560
TLX:
TLF:
EML:
COM: 10

GILMORE ALAN C MR
MT JOHN OBSERVATORY
P O BOX 57
LAKE TEKAPO
NEW ZEALAND
TEL: 64-5-056-813
TLX:
TLF:
EML:
COM: 06.20

GILMORE GERARD FRANCIS
INST OF ASTRONOMY
MADINGLEY ROAD
CAMBRIDGE CB3 OHA
U.K.
TEL: 0223-62204
TLX: 81797 ASTRON G
TLF:
EML:
COM: 30.33

GILMOZZI ROBERTO
ESA IUE OBSERVATORY
APARTADO 54065
28080 MADRID
SPAIN
TEL: 34-1-4019661
TLX: 42555
TLF:
EML:
COM:

GILRA DAYA P DR
SM SYSTEMS & RESEARCH CO.
8401 CORPORATE DRIVE
SUITE 450
LANDOVER MD 20785
U.S.A.
TEL: 301-763-4483
TLX:
TLF:
EML:
COM: 29.34.44

GIMENEZ ALVARO
INSTITUTO DE ASTROFISICA
DE ANDALUCIA
APARTADO 2144
18080 GRANADA
SPAIN
TEL: 58-12131116
TLX: 78573 IAAG E
TLF:
EML:
COM: 42C

GINGERICH OWEN PROF
CENTER FOR ASTROPHYSICS
60 GARDEN STREET
CAMBRIDGE MA 02138
U.S.A.
TEL: 617-495-7216
TLX: 921428 SATELLITE CAM
TLF:
EML:
COM: 41.46

GINGOLD ROBERT ARTHUR DR
ANU SUPERCOMPUTER FACILIT
CSC AUSTRAL.NAT.UNIVERS.
GPO BOX 4
CANBERRA
AUSTRALIA
TEL: 062-493437
TLX:
TLF:
EML:
COM: 35

GINZBURG VITALY L PROF
P N LEBEDEV PHYS INST
LENINSKY PROSPECT 53
117924 MOSCOW B 333
U.S.S.R.
TEL:
TLX:
TLF:
EML:
COM: 40.48.51

GIOIA ISABELLA M DR
HARVARD-SMITHSONIAN CTR
FOR ASTROPHYSICS
60 GARDEN STREET
CAMBRIDGE MA 02138
U.S.A.
TEL: 617-495-7138
TLX: 921428 SATELLITE CAM
TLF:
EML:
COM: 40

GIOVANARDI CARLO
OSSERVATORIO ASTROFISICO
DI ARCETRI
LARGO E. FERMI 5
I-50125 FIRENZE
ITALY
TEL: 055-2752239
TLX: 572268 ARCETR I
TLF:
EML:
COM: 28

GIOVANE FRANK
NSF ASTRON. SCIENCES
1800 G STREET NW
WASHINGTON DC 20550
U.S.A.
TEL:
TLX:
TLF:
EML:
COM: 15.21

GIOVANELLI RICCARDO DR
NAIC. CORNELL UNIVERSITY
ARECIBO OBSERVATORY
PO BOX 995
ARECIBO PR 00613
U.S.A.
TEL: 809-878-2612
TLX: 385638
TLF:
EML:
COM: 28.30.34

GIOVANNELLI FRANCO DR
IST. ASTROFISICA SPAZIALE
CP 67
I-00044 FRASCATI
ITALY
TEL: 6-942-565155
TLX: 610261 CNRFRA I
TLF:
EML:
COM: 42.51

GIOVANNINI GABRIELE
IST. DI RADIOASTRONOMIA
VIA IRNERIO 46
I-40126 BOLOGNA
ITALY
TEL: 051-232-856
TLX: 211664 INFN BO I
TLF:
EML:
COM: 40

GIRAUD EDMOND
LAB ASTROPHYS. THEORIQUE
COLLEGE DE FRANCE
98 BIS BOULEVARD ARAGO
F-75014 PARIS
FRANCE
TEL: 1-43-20-14-25
TLX:
TLF:
EML:
COM:

GIRIDHAR SUNETRA DR
INDIAN INSTITUTE OF
ASTROPHYSICS
BANGALORE 560 034
INDIA
TEL:
TLX: 845763 IIAB IN
TLF:
EML:
COM: 35

GIURICIN GIULIANO
DIPT. DI ASTRONOMIA
UNIVERSITA DI TRIESTE
VIA G.B. TIEPOLO 11
I-34131 TRIESTE
ITALY
TEL: 040-768005
TLX: 461137 OAT I
TLF:
EML:
COM: 42.47

GLAGOLEVSKIJ JU V DR
SPEC ASTROPHYSICAL OBS
USSR ACADEMY OF SCIENCES
NIZHNIJ ARKHYZ
357147 STAVROPOLSKIJ KRAJ
U.S.S.R.
TEL: 93-577
TLX:
TLF:
EML:
COM: 14.27.29.45

GLASER HAROLD DR
1346 BONITA ST
BERKELEY CA 94709
U.S.A.
TEL: 415-527-1860
TLX:
TLF:
EML:
COM: 44

GLASNER SHIMON AMI
RACAH INST. OF PHYSICS
HEBREW UNIV. OF JERUSALEM
JERUSALEM 91904
ISRAEL
TEL: 02-58-4521
TLX: 25391 HUIL
TLF:
EML:
COM:

GLASPEY JOHN W DR
CANADA FRANCE HAWAII
TELESCOPE CORPORATION
PO BOX 1597
KAMUELA HI 96743
U.S.A.
TEL:
TLX:
TLF:
EML:
COM:

GLASS BILLY PRICE DR
DEPT OF GEOLOGY
UNIVERSITY OF DELAWARE
NEWARK DE 19716
U.S.A.
TEL: 302-451-8458
TLX:
TLF:
EML:
COM: 22

GLASS IAN STEWART DR
S A A O
P O BOX 9
OBSERVATORY 7935
SOUTH AFRICA
TEL: 021-47-00-25
TLX: 57-20309 SA
TLF:
EML:
COM: 09.25.28

GLASSGOLD ALFRED E PROF
NEW YORK UNIVERSITY
PHYSICS DEPARTMENT
4 WASHINGTON PLACE
NEW YORK NY 10003
U.S.A.
TEL: 212-598-2020
TLX: 235128 NYU UR
TLF:
EML:
COM:

GLATZEL WOLFGANG DR
MAX PLANCK INSTITUT FUER
PHYSIK UND ASTROPHYSIK
KARL-SCHWARZSCHILDSTR. 1
8046-GARCHING B. MUNCHEN
GERMANY. F.R.
TEL: 089-32990
TLX: 524629
TLF:
EML: BITNET:WOG @ DGAIPP1S
COM:

GLATZMAIER GARY A
LOS ALAMOS NATIONAL LAB
ESS-5 MSF665
LOS ALAMOS NM 87545
U.S.A.
TEL: 505-667-7647
TLX:
TLF:
EML:
COM: 10.12.35

GLEBOCKI ROBERT PROF
INSTITUTE OF THEORETICAL
PHYSICS & ASTROPHYSICS
UL. WITA STWOSZA 57
80-952 GDANSK
POLAND
TEL: 41-87-00
TLX: 0512706 IFAS
TLF:
EML:
COM:

GLEBOVA NINA I DR
INST THEORET ASTRONOMY
USSR ACADEMY OF SCIENCES
10 KUTUZOV QUAY
191187 LENINGRAD
U.S.S.R.
TEL: 278-88-98
TLX: 121578 ITA SU
TLF:
EML:
COM: 04

GLEISSBERG WOLFGANG PROF
BUCHENWEG 12
D-6374 OBERURSEL
GERMANY. F.R.
TEL:
TLX:
TLF:
EML:
COM: 10

GLENCROSS WILLIAM M DR
DEPT PHYSICS & ASTRONOMY
UNIVERSITY COLLEGE LONDON
GOWER STREET
LONDON WC1E 6BT
U.K.
TEL: 01-387-7050
TLX: 28722 UCPHYS G
TLF:
EML:
CON:

GLUSHNEVA I N DR
STERNBERG STATE ASTR INST
UNIVERSITETSKIJ PROSP. 13
119899 MOSCOW
U.S.S.R.
TEL: 139-20-46
TLX:
TLF:
EML:
CON: 29

GNEDIN YURIJ N DR
LENINGRAD PHYS TECH INST
ACADEMY OF SCIENCES
194021 LENINGRAD
U.S.S.R.
TEL:
TLX:
TLF:
EML:
CON:

GNEVYSHEV MSTISLAV N DR
PULKOVO OBSERVATORY
196140 LENINGRAD
U.S.S.R.
TEL:
TLX:
TLF:
EML:
CON: 12

GNEVYSHEVA RAISA S DR
PULKOVO OBSERVATORY
196140 LENINGRAD
U.S.S.R.
TEL:
TLX:
TLF:
EML:
CON: 10

GODART ODON PROF
RUE DE CHATEAU 96
B-1488 BOUSVAL
BELGIUM
TEL: 010-613-817
TLX:
TLF:
EML:
CON: 47

GODFREY PETER DOUGLAS DR
CHEMISTRY DEPARTMENT
MONASH UNIVERSITY
WELLINGTON ROAD
CLAYTON VIC 3168
AUSTRALIA
TEL: 03-541-0811
TLX: 32691 AA
TLF:
EML:
CON: 34

GODOLI GIOVANNI PROF
DIPART. DI ASTRONOMIA
UNIVERSITA DEGLI STUDI
LARGO E. FERMI 5
I-50125 FIRENZE
ITALY
TEL: 055-27521
TLX: 572268 ARCETR I
TLF:
EML:
CON: 10.12.27.51

GODWIN JON GUNNAR DR
UNIVERSITY OBSERVATORY
SOUTH PARKS ROAD
OXFORD OX1 3RQ
U.K.
TEL: 0865-511336/507
TLX: 83295 NUCLOX G
TLF:
EML:
CON:

GOEBEL ERNST DR
INSTITUT FUER ASTRONOMIE
UNIVERSITAET WIEN
TUERKENSCHANZSTR 17
A-1180 WIEN
AUSTRIA
TEL: 0222-345360186
TLX:
TLF:
EML:
CON: 50

GOEBEL JOHN H DR
SPACE SC DIVISION 244/7
SPACE TECHNOL RES BRANCH
NASA-AMES RESEARCH CTR
MOFFETT FIELD CA 94035
U.S.A.
TEL: 415-694-6525
TLX:
TLF:
EML:
CON: 29 34

GOELBASI ORHAN DR
B.U. KANDILLI OBSERVATORY
CENGELKOY
ISTANBUL
TURKEY
TEL: 90-1332-02-41/2
TLX: 26411 BOUNTR
TLF:
EML:
CON:

GOKDOGAN NUZHET PROF
UNIVERSITY OBSERVATORY
UNIVERSITY OF ISTANBUL
ISTANBUL
TURKEY
TEL:
TLX:
TLF:
EML:
CON: 12.36

GOKHALE MORESHWAR HARI PR
INDIAN INSTITUTE OF
ASTROPHYSICS
BANGALORE 560 034
INDIA
TEL: 566585
TLX: 845763 IIAB IN
TLF:
EML:
CON: 10

GOLAY MARCEL PROF
OBSERVATOIRE DE GENEVE
CHEMIN DES MAILLETTES 51
CH-1290 SAUVERNY
SWITZERLAND
TEL: 022-55-26-11
TLX: 45419209 OBSG CH
TLF:
EML: SPAN ELVAD::APN::PSR
CON: 25.37.45P

GOLD THOMAS PROF
CTR F/RADIOPHYS & SP RES
SPACE SCIENCE BLDG
CORNELL UNIVERSITY
ITHACA NY 14853
U.S.A.
TEL: 607-256-5284
TLX: 937478
TLF:
EML:
CON: 16.40.44.47.48

GOLDBACH CLAUDINE MME
INSTITUT D'ASTROPHYSIQUE
98 BIS BOULEVARD ARAGO
F-75014 PARIS
FRANCE
TEL: 1-43-20-14-25
TLX:
TLF:
EML:
CON: 14

GOLDMAN ITZHAK DR
SCHOOL OF PHYS.& ASTRON.
TEL AVIV UNIVERSITY
TEL AVIV 69978
ISRAEL
TEL: 972-3-5450303
TLX: 342171 VERSY IL
TLF:
EML: BITNET:E17@TAUNOS
CON: 42

GOLDMAN MARTIN V
DEPARTMENT ASTRO-GEOPHYS
UNIVERSITY OF COLORADO
CAMPUS BOX 391
BOULDER CO 80309
U.S.A.
TEL: 303-492-8896
TLX:
TLF:
EML:
CON: 12

GOLDREICH P DR
CALTECH
PASADENA CA 91125
U.S.A.
TEL: 213-356-6193
TLX:
TLF:
EML:
CON: 07.16.33.34

GOLDSMITH DONALD W. DR.
INTERSTELLAR MEDIA
2153 RUSSELL STREET
BERKELEY CA 94705
U.S.A.
TEL: 415-848-1989
TLX:
TLF:
EML:
COM: 34.47.48.51

GOLDSMITH PAUL F DR
DEPT PHYSICS & ASTRONOMY
GRC TOWER B. ROOM 626
UNIV OF MASSACHUSETTS
AMHERST MA 01003
U.S.A.
TEL:
TLX:
TLF:
EML:
COM: 34.40

GOLDSMITH PAUL F DR
LEADMINE HILL ROAD 5
AMHERST MA 01002
U.S.A.
TEL: 413 545 0925
TLX: 95 5491
TLF:
EML:
COM:

GOLDSMITH S DR
DEPT PHYSICS & ASTRONOMY
TEL-AVIV UNIVERSITY
TEL-AVIV 69978
ISRAEL
TEL: 03-420-303
TLX: 342171 VRRSY
TLF:
EML:
COM:

GOLDSTEIN RICHARD M DR
JPL - CALTECH
MS 183-701
4800 OAK GROVE DRIVE
PASADENA CA 91011
U.S.A.
TEL: 818-354-6999
TLX:
TLF:
EML:
COM: 16

GOLDSTEIN SAMUEL J PROF
UNIVERSITY OF VIRGINIA
PO BOX 3818
CHARLOTTESVILLE VA 22903
U.S.A.
TEL:
TLX:
TLF:
EML:
COM: 34.40

GOLDSWORTHY FREDERICK A
SCHOOL OF MATHEMATICS
UNIVERSITY OF LEEDS
LEEDS LS2 9JT
U.K.
TEL: 0532-431751
TLX:
TLF:
EML:
COM: 34

GOLDWIRE HENRY C JR
UNIVERSITY OF CALIFORNIA
LANL
PO BOX 808 L-451
LIVERMORE CA 94550
U.S.A.
TEL: 415-423-0160
TLX:
TLF:
EML:
COM: 40

GOLLNOW H DR
MOUNT STROMLO OBSERVATORY
CANBERRA ACT
AUSTRALIA
TEL:
TLX:
TLF:
EML:
COM:

GOLUB LEON DR
HARVARD COLLEGE OBS
60 GARDEN ST
CAMBRIDGE MA 02138
U.S.A.
TEL: 617-495-7177
TLX:
TLF:
EML:
COM:

GOMES ALERCIO M DR
R GAVIAO PEIXOTO 13
AP 1401
ICARAI 24000
NITEROI RJ
BRAZIL
TEL:
TLX:
TLF:
EML:
COM: 97

GOMES RODNEY B S DR
CORNELL UNIVERSITY
SPACE SCIENCES BLD.
ITACA NY 14853
U.S.A.
TEL: 607 255 4709
TLX:
TLF:
EML: Gomes@astrosun.TN.CORNELL.EDU
COM: 07

GOMEZ GONZALEZ JESUS DR
PASEO IMPERIAL 29 6H
MADRID 5
SPAIN
TEL:
TLX:
TLF:
EML:
COM:

GOMEZ MARIA THERESA DR
OSSERVATORIO ASTRONOMICO
I-80131 NAPOLI
ITALY
TEL:
TLX:
TLF:
EML:
COM: 12

GOMIDE FERNANDO DE MELLO
DEPARTAMENTO DE FISICA
INST TECN. DE AERONAUTICA
12225 S. JOSE DOS CAMPOS
BRAZIL
TEL: 0123-22-9088
TLX: 0113393 CTAE BR
TLF:
EML:
COM:

GONCZI GEORGES
OBSERVATOIRE DE NICE
BP 139
F-06003 NICE CEDEX
FRANCE
TEL: 93-89-04-20
TLX:
TLF:
EML:
COM:

GONDHALEKAR PRABHAKAR DR
RUTHERFORD & APPLETON LAB
CHILTON DIDCOT X11 0QX
U.K.
TEL: 0235-21900
TLX: 83159
TLF:
EML:
COM: 44

GONDOLATSCH FRIEDRICH PRF
ASTRONOMISCHES RECHEN
INSTITUT
MOENCHHOFSTR 12-14
D-6900 HEIDELBERG
GERMANY. F.R.
TEL: 06221-49026
TLX:
TLF:
EML:
COM: 04

GONG HUI-REN
SHANGHAI OBSERVATORY
ACADEMIA SINICA
SHANGHAI
CHINA. PEOPLE'S REP.
TEL: 386191
TLX: 33164 SHAO CN
TLF:
EML:
COM:

GONG SHOU-SHEN
SHANGHAI OBSERVATORY
ACADEMIA SINICA
SHANGHAI
CHINA. PEOPLE'S REP.
TEL: 386191
TLX: 33164 SHAO CN
TLF:
EML:
COM: 09

GONG SHU-MO
PURPLE MOUNTAIN
OBSERVATORY
NANJING
CHINA. PEOPLE'S REP.
TEL: 46700
TLX: 34144 PMO CN
TLF:
EML:
COM: 35

GONZALES-A WALTER D DR
INST PESQUISAS ESPACIAIS
INPE
CAIXA POSTAL 515
12200 S. JOSE DOS CAMPOS
BRAZIL
TEL: 0623-229977
TLX: 011-33530 INPE BR
TLF:
EML:
COM: 48

GONZALEZ CAMACHO ANTONIO
INST DE ASTRON & GEODESIA
FAC DE CIENCIAS MATEMAT.
UNIVERSIDAD COMPLUTENSE
28040 MADRID
SPAIN
TEL: 91-2442501
TLX:
TLF:
EML:
COM: 07

GONZALEZ G
INAOE
AP POSTALES 216 y 51
72000 PUEBLA. PUE.
MEXICO
TEL:
TLX:
TLF:
EML:
COM:

GONZALEZ-RIESTRA R DR
IUE OBSEERVATORY
VILSPA
P.O. BOX 54065
28080 MADRID
SPAIN
TEL: 4019661
TLX: 42555 VILS E
TLF:
EML: EARN:IUEHOT@DDAESA10
COM:

GONZE ROGER F J IR
OBS ROYAL DE BELGIQUE
AVE CIRCULAIRE 3
B-1180 BRUXELLES
BELGIUM
TEL: 375-24-84
TLX: 21565
TLF:
EML:
COM: 40

GOODE PHILIP R
NJ INST. OF TECHNOLOGY
DEPT OF PHYSICS
323 HIGH STREET
NEWARK NJ 07102
U.S.A.
TEL: 201-596-3562
TLX:
TLF:
EML:
COM:

GOODY R M
CEPP
PIERCE HALL
29 OXFORD STREET
CAMBRIDGE MA 02138
U.S.A.
TEL: 617-495-4517
TLX:
TLF:
EML:
COM: 16

GOOSSENS MARCEL DR
ASTRONOMISCH INSTITUUT
KATHOLIEKE UNIV LEUVEN
CELESTIJNENLAAN 200 B
B-3030 HEVERLEE
BELGIUM
TEL:
TLX:
TLF:
EML:
COM: 10

GOPALA RAO U V MR
SATELLITE METEOROLOGY
INDIAN METEO. DEPT.
LODI ROAD / MAUSAM BHAVAN
NEW DELHI 110 003
INDIA
TEL:
TLX:
TLF:
EML:
COM:

GOPALSWAMY N DR
ASTRONOMY PROGRAM
UNIVERSITY OF MARYLAND
COLLEGE PARK MD 20742
U.S.A.
TEL: 301-454-6649
TLX: 62891478
TLF:
EML:
COM: 12.40

GOPASYUK S I DR
CRIMEAN ASTROPHYS OBS
USSR ACADEMY OF SCIENCES
334413 NAUCHNIY CRIMEA
U.S.S.R.
TEL:
TLX:
TLF:
EML:
COM: 10.12

GORBATSKY VITALIJ G PROF
LENINGRAD UNIVERSITY
ASTRONOMICAL OBSERVATORY
BIBLIOTECHNAJA PL 2
198904 LENINGRAD
U.S.S.R.
TEL: 257-94-91
TLX:
TLF:
EML:
COM: 27

GORDON CHARLOTTE PROF
11 RUE TOURNEFORT
F-75005 PARIS
FRANCE
TEL:
TLX:
TLF:
EML:
COM: 12.36

GORDON COURTNEY P PROF
HAMPSHIRE COLLEGE
AMHERST MA 01002
U.S.A.
TEL:
TLX:
TLF:
EML:
COM: 34

GORDON ISAAC M DR
INST OF RADIO PHYS & ELEC
310085 KHARKOV
U.S.S.R.
TEL:
TLX:
TLF:
EML:
COM:

GORDON KURTISS J PROF
FIVE COLL ASTR DEPT
HAMPSHIRE COLLEGE
AMHERST MA 01002
U.S.A.
TEL: 413-549-4600
TLX:
TLF:
EML:
COM:

GORDON MARK A DR
NRAO
949 N. CHERRY AVE.
CAMPUS BLDG 65
TUCSON AZ 85721-0655
U.S.A.
TEL: 602-882-8250
TLX:
TLF:
EML:
COM: 33.34.40

GORENSTEIN MARC V
HARVARD-SMITHSONIAN CTR
FOR ASTROPHYSICS
60 GARDEN STREET / MS-42
CAMBRIDGE MA 02138
U.S.A.
TEL: 617-495-9296
TLX:
TLF:
EML:
COM:

GORENSTEIN PAUL DR
CENTER FOR ASTROPHYSICS
60 GARDEN STREET
CAMBRIDGE MA 02138
U.S.A.
TEL: 617-495-7250
TLX: 921428 SATELLITE CAM
TLF:
EML:
COM: 16

GORET PHILIPPE DR
SECTION D'ASTROPHYSIQUE
CEN SACLAY
F-91190 GIF/YVETTE
FRANCE
TEL: 1-69-08-44-63
TLX: 690860 PHYSPAC F
TLF:
EML:
COM: 47

GORGAS GARCIA JAVIER DR
DEPT DE ASTROFISICA
FAC DE FISICAS
UNIVERSIDAD COMPLUTENSE
28040 MADRID
SPAIN
TEL: 5495316
TLX: 47273
TLF:
EML: EARN/BITNET:W062@EMDUCM11
COM: 28

GORGOLEWSKI STANISLAW PR
KATEDRA RADIOASTR/CHAIR
RADIOASTR/COPERNICUS UNIV
UL. CHOPINA 12/18
87-100 TORUN
POLAND
TEL: 20651 TORUN
TLX: 0552324 TRAO PL
TLF:
EML:
COM: 40

GOSACHINSKIJ I V DR
LENINGRAD BRANCH OF SAO
PULKOVO
196140 LENINGRAD
U.S.S.R.
TEL: 2979452
TLX: 321262
TLF:
EML:
COM: 34.40

GOSLING JOHN T DR
LOS ALAMOS NATIONAL LAB
ESS 8 - MS D 438
LOS ALAMOS NM 87545
U.S.A.
TEL: 505-667-5389
TLX: 660495
TLF:
EML:
COM: 49

GOSS W MILLER PROF
NRAO/VLA
P.O. BOX O
SOCORRO NM 87801
U.S.A.
TEL: 505-772-4011
TLX: 910-988-1710
TLF:
EML:
COM: 28.34.40

GOSWAMI J N DR
PHYS. RESEARCH LABORATORY
NAVRANGPURA
AHMEDABAD 380 009
INDIA
TEL: 462129
TLX: 0121397 PRL IN
TLF:
EML:
COM: 22

GOTT III J RICHARD
DEPT ASTROPHYSICAL SC
PRINCETON UNIVERSITY
PRINCETON NJ 08540
U.S.A.
TEL: 609-452-3813
TLX:
TLF:
EML:
COM: 51

GOTTESMAN STEPHEN T DR
DEPARTMENT OF ASTRONOMY
UNIVERSITY OF FLORIDA
GAINESVILLE FL 32611
U.S.A.
TEL: 904-392-2050/52
TLX: 8108252308
TLF:
EML:
COM: 28.40

GOTTLIEB CARL A DR
GODDARD INST/SPACE STUD
2880 BROADWAY
NEW YORK NY 10025
U.S.A.
TEL: 212-678-5566
TLX:
TLF:
EML:
COM:

GOTTLIEB KURT
46 JENNINGS STREET
CURTIN ACT 2605
AUSTRALIA
TEL: 062-814166
TLX:
TLF:
EML:
COM:

GOUDAS CONSTANTINE L PROF
DEPT OF MATHEMATICS
UNIVERSITY OF PATRAS
GR-26110 PATRAS
GREECE
TEL: 991-889
TLX: 312739 EPAP GR
TLF:
EML:
COM: 07.16

GOUDIS CHRISTOS D PROF
UNIVERSITY OF PATRAS
DEPT OF ASTRONOMY
GR-26110 PATRAS
GREECE
TEL:
TLX:
TLF:
EML:
COM: 51

GOUGH DOUGLAS O DR
INSTITUTE OF ASTRONOMY
MADINGLEY ROAD
CAMBRIDGE CB3 OHA
U.K.
TEL: 223-62204
TLX: 817297
TLF:
EML:
COM: 27.35C.36

GOUGUENHEIM LUCIENNE
OBSERVATOIRE DE PARIS
SECTION DE MEUDON
RADIOASTRONOMIE
F-92195 MEUDON PL CEDEX
FRANCE
TEL: 1-45-07-76-04
TLX:
TLF:
EML:
COM: 28.30.46V

GOULD ROBERT J PROF
PHYSICS DEPARTMENT B-019
UNIV/CALIF AT SAN DIEGO
LA JOLLA CA 92093
U.S.A.
TEL: 619-452-3649
TLX:
TLF:
EML:
COM:

GOUPIL MARIE JOSE
OBSERVATOIRE DE PARIS
SECTION MEUDON. DASGAL
92195 MEUDON PR. CEDEX
FRANCE
TEL: 45 07 78 80
TLX: 270912 OBSASTR
TLF:
EML: EARN:GOUPIL@FRMEU51
COM: 27.35

GOUTTEBROZE PIERRE DR
INSTITUT D'ASTROPHYSIQUE
SPATIALE
BP NO 10
F-91371 VERRIERES-LE B.
FRANCE
TEL: 1-64-47-42-04
TLX: 600252 LPSP VB
TLF:
EML:
COM:

GOWER ANN C DR
UNIV OF VICTORIA OBS
VICTORIA BC V8W 2Y2
CANADA
TEL:
TLX:
TLF:
EML:
COM:

GOWER J F R DR
1615 MCTAVISH ROAD
R R 2
SIDNEY BC V8L 3S1
CANADA
TEL: 604-656-5457
TLX:
TLF:
EML:
COM: 40

GOY GERALD PROF
OBSERVATOIRE DE GENEVE
CH-1290 SAUVERNY
SWITZERLAND
TEL: 22-552-611
TLX: 27720 OBSG CH
TLF:
EML:
COM: 25

GOYAL A N DR
DEPT OF MATHEMATICS
UNIVERSITY OF RAJASTHAN
JAIPUR 302 004
INDIA
TEL: 74060 (HOME)
TLX:
TLF:
EML:
COM: 24

GRABOSKE HAROLD C JR
LAWRENCE LIVERMORE LAB
PO BOX 808
LIVERMORE CA 94550
U.S.A.
TEL: 415-422-7262
TLX:
TLF:
EML:
COM:

GRABOWSKI BOLESLAW DR
INSTITUTE OF PHYSICS
UL OLESKA 48
UL. OLESKA 48
45-951 OPOLE
POLAND
TEL: 358-41
TLX: 0732230 WSP PL
TLF:
EML:
COM:

GRADIE JONATHAN CAREY
HAWAII INST OF GEOPHYSICS
DIV PLANETARY GEOSCIENCES
UNIVERSITY OF HAWAII
HONOLULU HI 96822
U.S.A.
TEL: 808-948-6488
TLX:
TLF:
EML:
COM: 15

GRADSZTAJN ELI DR
DEPT PHYSICS & ASTRONOMY
TEL-AVIV UNIVERSITY
RAMAT-AVIV 69978
ISRAEL
TEL:
TLX:
TLF:
EML:
COM:

GRAHAM DAVID A
MPI FUER RADIOASTRONOMIE
AUF DEM HUEGEL 69
D-5300 BONN 1
GERMANY, F.R.
TEL: 228-525282
TLX: 886440 MPIFR D
TLF:
EML:
COM: 34.40

GRAHAM ERIC DR
PO BOX 8298
DURANGO CO 81301
U.S.A.
TEL:
TLX:
TLF:
EML:
COM: 35

GRAHAM JOHN A DR
DEPT TERRESTR. MAGNETISM
CARNEGIE INST. WASHINGTON
5241 BROAD BRANCH RD N.W.
WASHINGTON DC 20015
U.S.A.
TEL: 202-966-0863
TLX: 440427 MAGN UI
TLF:
EML:
COM: 25.27.28

GRAHL BERND H DR
MPI FUER RADIOASTRONOMIE
AUF DEM HUEGEL 69
D-5300 BONN
GERMANY, F.R.
TEL: 02257-3112
TLX: 8869114 MPIR D
TLF:
EML:
COM:

GRAINGER JOHN F DR
PHYSICS DEPARTMENT
UMIST
MANCHESTER M60 1QD
U.K.
TEL: 061-236-3311
TLX: 666094
TLF:
EML:
COM:

GRANDI STEVEN ALDRIDGE DR
NATL OPTICAL ASTRONOMY
OBSERVATORY
PO BOX 26732
TUCSON AZ 85726
U.S.A.
TEL: 602-327-5511
TLX:
TLF:
EML:
COM: 28

GRANT IAN P DR
PEMBROKE COLLEGE
OXFORD OX1 1DW
U.K.
TEL: 0865-242-271
TLX:
TLF:
EML:
COM: 14.36

GRASDALEN GARY L DR
DEPT PHYSICS & ASTRONOMY
UNIVERSITY OF WYOMING
PO BOX 3905 UN STA
LARAMIE WY 82071
U.S.A.
TEL: 307-766-4385
TLX:
TLF:
EML:
COM: 27.28.34

GRATTON LIVIO PROF
IST. ASTROFISICA SPAZIALE
C P 67
I-00044 FRASCATI
ITALY
TEL:
TLX:
TLF:
EML:
COM: 29.47

GRATTON R G DR
OSSERVATORIO ASTRONOMICO
DI ROMA
VIA DELL'OSSERVATORIO
0040 MONTE PORZIO (ROMA)
ITALY
TEL: 0039-6-9449019
TLX:
TLF:
EML: SPAN:17468::RAFFAELE
COM: 29.37

GRAUER ALBERT D
DEPT PHYSICS & ASTRONOMY
UALR
33RD & UNIVERSITY
LITTLE ROCK AR 72204
U.S.A.
TEL: 501-569-3275
TLX:
TLF:
EML:
COM: 25

GRAY DAVID F PROF
DEPT OF ASTRONOMY
UNIV OF WESTERN ONTARIO
PHYSICS-ASTRONOMY BLDG
LONDON ONT N6A 3K7
CANADA
TEL: 519-679-3184
TLX:
TLF:
EML:
COM: 29.36P

GRAY PETER MURRAY
ANGLO-AUSTRALIAN OBS.
P.O. BOX 296
EPPING NSW 2121
AUSTRALIA
TEL: 02-868-1666
TLX: 23999 AAOSYD AA
TLF:
EML:
COM: 09

GRAYZECK EDWIN J DR
PHYSICS DEPARTMENT
UNIVERSITY OF NEVADA
4505 S MARYLAND PARKWAY
LAS VEGAS NV 89154
U.S.A.
TEL: 702-739-3507
TLX:
TLF:
EML:
COM: 33

GREBENIKOV E A PROF DR
INST THEOR & EXPER PHYS
117259 MOSCOW
U.S.S.R.
TEL:
TLX:
TLF:
EML:
COM: 07

GREC GERARD
DEPT D'ASTROPHYSIQUE
UNIVERSITE DE NICE
PARC VALROSE
F-06034 NICE
FRANCE
TEL:
TLX:
TLF:
EML:
COM:

GREEN ELIZABETH M. DR
MT STROMLO OBSERVATORY
AUSTRALIAN NAT UNIVERSITY
PRIVATE BAG
WODEN P.O. ACT 2606
AUSTRALIA
TEL: 062-88-1111
TLX: 62270 AA
TLF:
EML:
COM: 37

GREEN JACK PROF
DEPT OF GEOLOGY
CALIF STATE UNIVERSITY
LONG BEACH CA 90840
U.S.A.
TEL: 213-498-4809
TLX:
TLF:
EML:
COM: 16

GREEN LOUIS C PROF
HAVERFORD COLLEGE
7901 COLLEGE AVENUE
HAVERFORD PA 19041
U.S.A.
TEL: 215-649-0265
TLX:
TLF:
EML:
COM: 14

GREEN RICHARD F DR
KITT PEAK NTL OBSERVATORY
PO BOX 26732
950 N CHERRY STREET
TUCSON AZ 85726
U.S.A.
TEL: 602 325 9299
TLX: 0666 484 AURA NOAO
TLF:
EML:
COM:

GREEN ROBIN M DR
DEPT OF ASTRONOMY
GLASGOW UNIVERSITY
GLASGOW G12 8QQ
U.K.
TEL: 041-339-8855
TLX: 778421
TLF:
EML:
COM:

GREEN SIMON F
PHYSICS LABORATORY
UNIVERSITY OF KENT
CANTERBURY CT2 7NR
U.K.
TEL: 0227 764000
TLX: 965449
TLF:
EML:
COM: 15

GREENBERG J MAYO DR
HUYGENS LABORATORY
UNIVERSITY OF LEIDEN
WASSENAARSEWEG 78
NL-2300 RA LEIDEN
NETHERLANDS
TEL: 071-275700
TLX: 39058 ASTRO NL
TLF:
EML:
COM: 15.21.34.51

GREENBERG RICHARD DR
LUNAR & PLANETARY LAB
UNIVERSITY OF ARIZONA
TUCSON AZ 85721
U.S.A.
TEL: 602-621-6940
TLX:
TLF:
EML:
COM: 07.15.20

GREENSTEIN GEORGE PROF
ASTRONOMY DEPARTMENT
AMHERST COLLEGE
AMHERST MA 01002
U.S.A.
TEL: 413-542-2075
TLX:
TLF:
EML:
COM:

GREENSTEIN J L PROF
PALOMAR OBSERVATORY
CALIF INST OF TECHNOLOGY
1201 E CALIFORNIA ST
PASADENA CA 91125
U.S.A.
TEL: 818-356-4006
TLX:
TLF:
EML:
COM: 29.36.51

GREGGIO LAURA DR
DIPARTMENTO DI ASTRONOMIA
UNIVERSITA DI BOLOGNO
CP 596
40125 BOLOGNA
ITALY
TEL: 051 259413
TLX: 520634 INFNBO 1
TLF:
EML: SPAN/DECNET:37928::LAURA
COM: 35

GREGORINI LORETTA
IST. DI RADIOASTRONOMIA
VIA IRNERIO 46
I-40126 BOLOGNA
ITALY
TEL: 051-232-856
TLX: 211664 INFN BOT
TLF:
EML:
COM: 40

GREGORY PHILIP C DR
PHYSICS DEPT
UNIV OF BRITISH COLUMBIA
6224 AGRICULTURAL RD
VANCOUVER BC V6T 1W5
CANADA
TEL: 604-228-6417
TLX: 04-508576
TLF:
EML:
COM: 40.51

GREGORY STEPHEN ALBERT DR
PHYSICS DEPT
BOWLING GREEN STATE UNIV
BOWLING GREEN OH 43403
U.S.A.
TEL:
TLX:
TLF:
EML:
COM: 47

GREGUL A YA DR
OBSERVATORY OF THE KIEV
UNIVERSITY
OBSERVATORNAYA 3
252053 KIEV
U.S.S.R.
TEL: 262391
TLX:
TLF:
EML:
COM:

GREISEN KENNETH I PROF
336 FOREST HOME DR
ITHACA NY 14850
U.S.A.
TEL: 607-257-1650
TLX:
TLF:
EML:
COM: 48

GRENIER SUZANNE
OBSERVATOIRE DE PARIS
SECTION DE MEUDON
F-92195 MEUDON PL CEDEX
FRANCE
TEL: 1-45-07-78-41
TLX: 201571 LAM
TLF:
EML:
COM:

GRENON MICHEL DR
OBSERVATOIRE DE GENEVE
CH-1290 SAUVERNY
SWITZERLAND
TEL: 22-55-26-11
TLX: 27720 OBSG CH
TLF:
EML:
COM: 25

GREVE ALBERT DR
IRAM
300 RUE DE LA PISCINE
DOMAINE UNIVERSITAIRE
38406 ST MARTIN D'HERES
FRANCE
TEL: 76 82 49 31
TLX:
TLF:
EML:
COM:

GREVESSE N DR
INSTITUT D'ASTROPHYSIQUE
UNIVERSITE DE LIEGE
AVENUE DE COINTE 5
B-4200 COINTE-OUGREE
BELGIUM
TEL: 41-52-99-80
TLX: 41264
TLF:
EML:
COM: 12.36

GREWING MICHAEL PROF
ASTRON INST UNIVERSITAET
WALDHAUSERSTR 64
D-7400 TUEBINGEN
GERMANY. F.R.
TEL: 7071-292486
TLX: 07262714 AIT D
TLF:
EML:
COM: 25.34.40.44.48

GREYBER HOWARD D DR
10125 FALLS ROAD
POTOMAC MD 20854
U.S.A.
TEL:
TLX:
TLF:
EML:
COM: 44.47.48

GRIFFIN MATTHEW J DR
PHYSICS DPT
QUEEN MARY COLLEGE
MILE END ROAD
LONDON E1 4NS
U.K.
TEL: (01)980 4811
TLX: 893750
TLF:
EML: MJG@UK.AC.QMC.STAR
COM:

GRIFFIN RITA E M DR
THE OBSERVATORIES
MADINGLEY ROAD
CAMBRIDGE CB3 0HA
U.K.
TEL: 223-62204
TLX: 817297 ASTRON G
TLF:
EML:
COM: 29

GRIFFIN ROGER F DR
THE OBSERVATORIES
MADINGLEY ROAD
CAMBRIDGE CB3 0HA
U.K.
TEL: 44-223-62204
TLX: 817297 ASTRON G
TLF:
EML:
COM: 05.29.30

GRIFFITH JOHN S PROF
DEPT OF MATH SCIENCE
LAKEHEAD UNIVERSITY
THUNDER BAY ONT P7B 5E1
CANADA
TEL: 807-345-2121
TLX:
TLF:
EML:
COM:

GRIFFITHS RICHARD E DR
SPACE TELESCOPE SCI INST
3700 SAN MARTIN DRIVE
BALTIMORE MD 21218
U.S.A.
TEL: LEEDS 431751
TLX: 556473 UNILDS G
TLF:
EML:
COM: 09.28.44.48

GRIFFITHS WILLIAM K
DEPT OF PHYSICS
THE UNIVERSITY
LEEDS LS2 9JT
U.K.
TEL:
TLX:
TLF:
EML:
COM: 37

GRIGORJEV VICTOR M DR
SIBERIAN INST/TERR MAGN
IONOSPH RADIO WAVE PROP
P BOX 4
664697 IRKUTSK
U.S.S.R.
TEL:
TLX:
TLF:
EML:
COM: 09

GRIJO DE OLIVEIRA A K DR
OBSERVATORIO NACIONAL CNP
RUA GENERAL BRUCE 586
20921 RIO DE JANEIRO RJ
BRAZIL
TEL:
TLX:
TLF:
EML:
COM:

GRIMLEY PETER DR
PHYSICS DEPT
ST PATRICK COLLEGE
MAYNOOTH
CO KILDARE
IRELAND
TEL: 3513 1 285222
TLX: 31493
TLF:
EML:
COM:

GRINDLAY JONATHAN E DR
HARVARD OBSERVATORY
CENTER FOR ASTROPHYSICS
60 GARDEN STREET
CAMBRIDGE MA 02138
U.S.A.
TEL: 617-495-7204
TLX: 921428 SATELLITE CAM
TLF:
EML:
COM: 06VP.37.48

GRININ VLADIMIR P DR
CRIMEAN ASTROPHYS OBS
USSR ACADEMY OF SCIENCES
334413 NAUCHNY CRIMEA
U.S.S.R.
TEL:
TLX:
TLF:
EML:
COM: 36

GRISCHUK L P DR
STERNBERG STATE ASTR INST
119899 MOSCOW V-234
U.S.S.R.
TEL: 139-50-06
TLX:
TLF:
EML:
COM: 47

GROOTE DETLEF
HAMBURGER STERNWARTE
GOJENSBERGWEG 112
D-2050 HAMBURG 80
GERMANY. F.R.
TEL: 040-72524112
TLX:
TLF:
EML:
COM:

GROSBOL PREBEN JOHNSON DR
ESO
KARL-SCHWARZSCHILD-STR 2
D-8046 GARCHING B MUNCHEN
GERMANY. F.R.
TEL: 089-320-06-237
TLX: 52828222 EOD
TLF:
EML:
COM: 05

GROSS PETER G PROF
714 OXFORD ROAD
BALA CYNWYD PA 19004
U.S.A.
TEL:
TLX:
TLF:
EML:
COM:

GROSSMAN ALLEN S PROF
ERWIN FICK OBSERVATORY
IOWA STATE UNIVERSITY
AMES IA 50011
U.S.A.
TEL: 515-294-3666
TLX:
TLF:
EML:
COM:

GROSSMAN LAWRENCE PROF
DEPT GEOPHYSICAL SCIENCES
UNIVERSITY OF CHICAGO
5734 SOUTH ELLIS AVE
CHICAGO IL 60637
U.S.A.
TEL: 312-962-8153
TLX:
TLF:
EML:
COM: 15.16

GROSSMANN-DOERTH U DR
KIEPENHEUER INSTITUT
FUER SONNENPHYSIK
SCHOENECKSTR 6
D-7800 FREIBURG
GERMANY. F.R.
TEL: 0761-32864
TLX: 7721552 KIS D
TLF:
EML:
COM:

GROTEN ERWIN PROF
INST/PHYSIKALISCHE GEOD
PETERSENSTR 13
D-6100 DARMSTADT
GERMANY. F.R.
TEL: 0-6151-16-3109
TLX: 419579 TH D
TLF:
EML:
COM: 19

GROTH EDWARD J III
PHYSICS DEPT
PRINCETON UNIVERSITY
JADWIN HALL
PRINCETON NJ 08544
U.S.A.
TEL: 609-452-4361
TLX:
TLF:
EML:
COM:

GROTH HANS G PROF DR
INST ASTRON & ASTROPHYS
UNIVERSITAT MUENCHEN
SCHEINERSTRASSE 1
D-8000 MUENCHEN 80
GERMANY. F.R.
TEL: 089-989021
TLX:
TLF:
EML:
COM: 29.36

GROUSHINSKY N P PROF DR
STERNBERG STATE
ASTRONOMICAL INSTITUTE
UNIVERSITETSKIJ PROSP. 13
119899 MOSCOW
U.S.S.R.
TEL:
TLX:
TLF:
EML:
COM: 07

GRUBISSICH C PROF DR
VIA AOSTA 34/5
I-35142 PADOVA
ITALY
TEL: 049-38301
TLX:
TLF:
EML:
COM: 37

GRUDLER PIERRE
C E R G A
AVE COPERNIC
F-06130 GRASSE
FRANCE
TEL: 93-36-58-49
TLX:
TLF:
EML:
COM: 08.31

GRUDZINSKA STEFANIA DR
ASTRONOMICAL INSTITUTE
UL. CHOPINA 12/18
87-100 TORUN
POLAND
TEL: 20655
TLX: 0552234 ASTR PL
TLF:
EML:
COM: 15

GRUEFF GAVRIL DR
LAB. DI RADIOASTRONOMIA
CNR
VIA IRNERIO 46
I-40126 BOLOGNA
ITALY
TEL:
TLX:
TLF:
EML:
COM:

GRUEN EBERHARD DR
MPI FUER KERNPHYSIK
POSTFACH 103 980
D-6900 HEIDELBERG
GERMANY. F.R.
TEL: 6621-516478
TLX: 461666 MPIHD D
TLF:
EML:
COM: 15C.21.22C

GRUENWALD RUTH DR
INSTITUTO ASTRONOMICO
UNIVERSIDADE DE SAO PAULO
CAIXA POSTAL 30627
01051 SAO PAULO SP
BRAZIL
TEL: 011 5778599
TLX: 36221 IAGM BR
TLF:
EML:
COM:

GRUNDMANN WALTER
DOMINION ASTROPHYS OBS
5071 W SAANICH ROAD
VICTORIA BC V8X 4M6
CANADA
TEL: 604-388-3157
TLX: 0497295
TLF:
EML:
COM: 09

GRY CECILE DR
LAB. D'ASTRON. SPATIALE
TRAVERSE DU SIPHON
LES TROIS LUCS
13012 MARSEILLE
FRANCE
TEL: 33 91 05 59 00
TLX: 420 584
TLF:
EML: EARN:CECILE@FRLASM51
COM:

GRYGAR JIRI DR
INSTITUTE OF PHYSICS
CZECH. ACAD. OF SCIENCES
250 68 REZ
CZECHOSLOVAKIA
TEL: 84-42-41
TLX: 122626 CS
TLF:
EML:
COM: 27.42

GRZEDZIELSKI STANISLAW PR
SPACE RESEARCH CENTER
POLISH ACAD OF SCIENCES
UL. ORDONA 21
01-237 WARSZAWA
POLAND
TEL:
TLX:
TLF:
EML:
COM: 49

GU XIAO-MA
YUNNAN OBSERVATORY
ACADEMIA SINICA
KUNMING
CHINA. PEOPLE'S REP.
TEL: 72946 KUNMING
TLX: 64040 YUOBS CN
TLF:
EML:
COM: 10.12

GUARNIERI ADRIANO DR
OSSERVATORIO ASTRONOMICO
VIA ZAMBONI 33
I-40126 BOLOGNA
ITALY
TEL:
TLX:
TLF:
EML:
COM:

GUBANOV VADIM S DR
PULKOVO OBSERVATORY
196140 LENINGRAD
U.S.S.R.
TEL: 297-94-81
TLX:
TLF:
EML:
COM: 08

GUDMUNDSSON EINAR H
RAUNVISINDASTOFNUN
HASKOLANS
DUNHAGA 3
IS-107 REYKJAVIK
ICELAND
TEL: 21340
TLX: 2307 ISINFO
TLF:
EML:
COM:

GUDUR N DR
EGE UNIVERSITY
OBSERVATORY
P K 21
BORNOVA-IZMIR
TURKEY
TEL: 9051180110/2326
TLX:
TLF:
EML:
COM:

GUELIN MICHEL DR
INST RADIOASTR MILLIMETR
VOIE 10
DOMAINE UNIV DE GRENOBLE
F-38406 ST MARTIN D'HERES
FRANCE
TEL: 76-42-33-83
TLX: 980753 IRAM
TLF:
EML:
COM: 34.40

GUERIN PIERRE DR
INSTITUT D'ASTROPHYSIQUE
98 BIS BOULEVARD ARAGO
F-75014 PARIS
FRANCE
TEL: 1-43-20-14-25
TLX: 270070
TLF:
EML:
COM: 16

GUERRERO GIANANTONIO DR
OSSERVATORIO ASTRONOMICO
VIA E. BIANCHI 46
I-22055 MERATE (COMO)
ITALY
TEL: 592035
TLX:
TLF:
EML:
COM: 27

GUEST JOHN E DR
UNIVERSITY OF LONDON OBS
MILL HILL PARK
LONDON NW7 2QS
U.K.
TEL: 01-959-7367
TLX: 28722 UCPHYS
TLF:
EML:
COM: 16

GUESTEN ROLF
MPI FUER RADIOASTRONOMIE
AUF DEM HUEGEL 69
D-5300 BONN 1
GERMANY. F.R.
TEL: 49-228-525-379
TLX:
TLF:
EML:
COM: 34.40C

GUETTER HARRY HENDRIK
US NAVAL OBSERVATORY
PO BOX 1149
FLAGSTAFF AZ 86002
U.S.A.
TEL: 602-779-5132
TLX:
TLF:
EML:
COM: 25.45

GUIBERT JEAN DR
OBSERVATOIRE DE PARIS
61 AVE DE L'OBSERVATOIRE
F-75014 PARIS
FRANCE
TEL: 1-40-51-20-98
TLX:
TLF:
EML:
COM: 05.09.24

GUIDICE DONALD A DR
A F GEOPHYSICS LABORATORY
HANSCOM AFB
BEDFORD MA 01731
U.S.A.
TEL: 617-861-3989
TLX:
TLF:
EML:
COM: 40

GUINAN EDWARD FRANCIS DR
DEPT OF ASTRONOMY
VILLANOVA UNIVERSITY
VILLANOVA PA 19085
U.S.A.
TEL: 215-527-2100
TLX:
TLF:
EML:
COM: 27.42

GUINOT BERNARD R PROF
B.I.P.M.
PAVILLON DE BRETEUIL
F-92312 SEVRES CEDEX
FRANCE
TEL: 1-45-34-00-51
TLX: 201067 BIPM F
TLF:
EML:
COM: 19.31C

GULKIS SAMUEL DR
JET PROPULSION LABORATORY
4800 OAK GROVE DRIVE
PASADENA CA 91109
U.S.A.
TEL: 213-354-5708
TLX:
TLF:
EML:
COM: 16.40.51

GULL STEPHEN F DR
CAVENDISH LABORATORY
MADINGLEY ROAD
CAMBRIDGE CB3 OHE
U.K.
TEL: 223-66477
TLX: 81292
TLF:
EML:
COM: 40

GULL THEODORE R DR
LAB. ASTRON & SOLAR PHYS
NASA/GSFC. CODE 680
GREENBELT MD 20771
U.S.A.
TEL: 301-286-8060
TLX: 710-8289716
TLF:
EML:
COM: 34.44

GULLIVER AUSTIN FRASER DR
DEPT OF PHYSICS
UNIVERSITY OF BRANDON
BRANDON MANITOBA R7A 6A9
CANADA
TEL: 204-728-9520
TLX:
TLF:
EML:
COM: 42

GULMEN OMUR DR
EGE UNIVERSITY
OBSERVATORY
P K 21
BORNOVA-IZMIR
TURKEY
TEL: 90-51-180110
TLX:
TLF:
EML:
COM:

GULYAEV A P DR
STERNBERG ASTR INSTITUT
UNIVERSITETSKIJ PROSP. 13
119899 MOSCOW
U.S.S.R.
TEL: 139-19-70
TLX:
TLF:
EML:
COM: 08

GULYAEV RUDOLF A DR
IZMIRAN
AKADEMGORODOK
142092 MOSCOW REGION
U.S.S.R.
TEL:
TLX:
TLF:
EML:
COM:

GUNN JAMES E PROF
DEPT ASTROPHYSICAL SCI
PRINCETON UNIVERSITY
PEYTON HALL
PRINCETON NJ 08544
U.S.A.
TEL: 609-452-3802
TLX:
TLF:
EML:
COM: 28.47.48.51

GUO WEI-SHU DR
NANJING ASTRONOMICAL
INSTRUMENT FACTORY
ACADEMIA SINICA PO BOX846
NANJING
CHINA. PEOPLE'S REP.
TEL: 646191
TLX: 34136
TLF:
EML:
COM:

GUO QUAN SHI
PURPLE MOUNTAIN
OBSERVATORY
NANJING
CHINA. PEOPLE'S REP.
TEL:
TLX:
TLF:
EML:
COM:

GURM HARDEV S PROF
DEPT/ASTR & SPACE SCI
PANJABI UNIVERSITY
PATIALA 147 002
INDIA
TEL: 73262 x 96
TLX:
TLF:
EML:
COM: 27.35.46.51

GURMAN JOSEPH B DR
CODE 602.6
NASA
GSFC
GREENBELT MD 20771
U.S.A.
TEL: 301 286 7599
TLX: 89675
TLF:
EML: SPAN:SOLMAX::GURMAN
COM: 10

GURSHTEIN ALEXANDER A DR
INST HIST OF SCI & TECHN
USSR ACADEMY OF SCIENCES
STAROPANSKY 1/5
103012 MOSCOW
U.S.S.R.
TEL:
TLX:
TLF:
EML:
COM: 16.41C

GURSKY HERBERT DR
NAVAL RESEARCH LABORATORY
CODE 4100
WASHINGTON DC 20375
U.S.A.
TEL: 202-767-6343
TLX:
TLF:
EML:
COM: 27.42.44.48

GURTOVENKO E A DR
MAIN ASTRONOMICAL OBS
UKRAINIAN ACAD OF SCI
252127 KIEV
U.S.S.R.
TEL: 66-10-65
TLX: 131406
TLF:
EML:
COM: 10C.12C

GURZADIAN G A PROF DR
BYURAKAN ASTROPHYS OBS
378433 ARMENIA
U.S.S.R.
TEL:
TLX:
TLF:
EML:
COM: 28.34.45

GUSEINOV O H PROF
INSTITUTE OF PHYSICS
NARIMANOV AVENUE 33
370143 BAKU
U.S.S.R.
TEL: 39-39-51
TLX:
TLF:
EML:
COM: 34.42.44.48

GUSEJNOV RAGIM EH DR
SHEMAKHA ASTROPHYS OBS
373243 AZERBAIDZAN
U.S.S.R.
TEL:
TLX:
TLF:
EML:
COM:

GUSSMANN E A DR
ZNTRLINST. F. ASTROPHYSIK
ROSA-LUXEMBURG-STR 17A
DDR-1502 POTSDAM
GERMANY. D.R.
TEL:
TLX:
TLF:
EML:
COM: 36

GUSTAFSON BO A S
LUND OBSERVATORY
BOX 43
S-221 00 LUND
SWEDEN
TEL:
TLX:
TLF:
EML:
COM: 15

GUSTAFSSON BENGT DR
ASTRONOMICAL OBSERVATORY
BOX 515
S-751 20 UPPSALA
SWEDEN
TEL:
TLX:
TLF:
EML:
COM: 29C.36

GUTCKE DIETRICH
CARL-ZEISS STR 1
DDR-6900 JENA
GERMANY. D.R.
TEL:
TLX:
TLF:
EML:
COM: 09

GUTHRIE BRUCE N G DR
ROYAL OBSERVATORY
BLACKFORD HILL
EDINBURGH EH9 3HJ
U.K.
TEL: 031-667-3321
TLX: 72383 ROEDIN G
TLF:
EML:
COM: 29

GUTIERREZ-MORENO A DR MRS
DEPTO DE ASTRONOMIA
UNIVERSIDAD DE CHILE
CASILLA 36-D
SANTIAGO
CHILE
TEL: 2294101/2294002
TLX:
TLF:
EML:
COM: 25

GUINN CARL R DR
HARVARD-SMITHSONIAN CENT.
FOR ASTROPHYSICS - MS 42
60 GARDEN STREET
CAMBRIDGE MA 02138
U.S.A.
TEL: 617 495 7329
TLX: 92 1428
TLF:
EML: BITNET:GUINNOCFA
COM: 40

GYLDENKERNE KJELD DR
COPENHAGEN UNIVERSITY OBS
BRORFELDEVEJ 23
DK-4340 TOLLOSE
DENMARK
TEL: 3-488-195
TLX: 44155
TLF:
EML:
COM: 33.42

HABBAL SHADIA RIFAI
HARVARD SMITHSONIAN CTR
FOR ASTROPHYSICS
60 GARDEN STREET
CAMBRIDGE MA 02138
U.S.A.
TEL: 617-495-7348
TLX: 921428
TLF:
EML:
COM: 49

HABE ASAO
DEPT OF PHYSICS
HOKKAIDO UNIVERSITY
SAPPORO
HOKKAIDO 060
JAPAN
TEL: 11-711-2111
TLX:
TLF:
EML:
COM: 33

HABIBULLIN SH T PROF DR
KAZAN UNIV OBSERVATORY
LENIN STREET 18
420008 KAZAN
U.S.S.R.
TEL: 323641
TLX:
TLF:
EML:
COM: 16

HABING H J DR
STERREWACHT
POSTBUS 9513
NL-2300 RA LEIDEN
NETHERLANDS
TEL: 071-272727
TLX: 39058
TLF:
EML:
COM: 33.34V

HACHENBERG OTTO PROF DR
RADIOASTRONOMISCHES INST
UNIVERSITAET BONN
AUF DEM HUEGEL 71
D-5300 BONN
GERMANY. F.R.
TEL:
TLX:
TLF:
EML:
COM:

HACHISU IZUMI DR
DEPT OF AERONAUTICAL
ENGINEERING
KYOTO UNIVERSITY
KYOTO 606
JAPAN
TEL: 075-751-2111
TLX: 05422693 LIBKYU J
TLF:
EML:
COM: 35

HACK MARGHERITA PROF
OSSERVATORIO ASTRONOMICO
VIA TIEPOLO 11
I-34131 TRIESTE
ITALY
TEL: 040-793921
TLX: 461137 OAT I
TLF:
EML:
COM: 29.36.44.45

HACKWELL JOHN A DR
DEPT PHYSICS & ASTRONOMY
UNIVERSITY OF WYOMING
LARAMIE WY 82070
U.S.A.
TEL: 307-766-6296
TLX:
TLF:
EML:
COM: 27.34

HACYAN SHAHEN DR.
INSTITUTO DE ASTRONOMIA
UNAM
APDO POSTAL 70-264
04510 MEXICO DF
MEXICO
TEL: 905-548-5305
TLX: 1760155 CICME
TLF:
EML:
COM: 47

HADDOCK FRED T DR
UNIVERSITY OF MICHIGAN
937 PHYSICS-ASTRONOMY BLG
ANN ARBOR MI 48104
U.S.A.
TEL: 313-764-3430
TLX:
TLF:
EML:
COM: 40.44.51

HADJIDEMETRIOU JOHN D
DEPT THEORET MECHANICS
UNIVERSITY THESSALONIKI
GR-54006 THESSALONIKI
GREECE
TEL: 031-99-14-10
TLX:
TLF:
EML:
COM: 07C

HADLEY BRIAN W
ROYAL OBSERVATORY
BLACKFORD HILL
EDINBURGH EH9 3HJ
U.K.
TEL: 031 668 8296
TLX: 72383 ROEDIN G
TLF:
EML: BUHOUX.AC.ROE.STAR.
COM: 09

HADRAVA PETR
ASTRONOMICAL INSTITUTE
OF THE CZECHOSLOVAK
ACADEMY OF SCIENCES
251 65 ONDREJOV
CZECHOSLOVAKIA
TEL:
TLX:
TLF:
EML:
COM: 42

HAEFNER REINHOLD DR
UNIVERSITAETS STERNWARTE
SCHEINERSTRASSE 1
D-8000 MUENCHEN 80
GERMANY. F.R.
TEL: 089-98-9021
TLX:
TLF:
EML:
COM: 27

HAEMEEN ANTTILA KAARLE A
ASTRONOMY DEPT
UNIVERSITY OF OULU
SF-90100 OULU 10
FINLAND
TEL:
TLX:
TLF:
EML:
COM:

HAENSEL PAWEL DR
N COPERNICUS ASTRONOMICAL
CENTER
UL. BARTYCKA 18
00-716 WARSAW
POLAND
TEL: 410828
TLX: 81 3978 ZAPLAN PL
TLF:
EML:
COM:

HAERENDEL G DR
MPI F. PHYSIK & ASTROPHYS
INST F. EXTRATERR. PHYSIK
D-8046 GARCHING B MUNCHEN
GERMANY. F.R.
TEL: 089-3299-516
TLX: 05215845 XTERD
TLF:
EML:
COM:

HAGEN JOHN P
613 W. PARK AVE
STATE COLLEGE PA 16803
U.S.A.
TEL: 1-814-237-3031
TLX:
TLF:
EML:
COM: 10.40

HAGEN-THORN VLADIMIR A DR
ASTRONOMICAL OBSERVATORY
BIBLIOTECHNAJA PL.2
198904 LENINGRAD
U.S.S.R.
TEL: 257-94-91
TLX:
TLF:
EML:
COM: 28

HAGFORS T DR
NATL ASTRON & IONOSPHERE
CTR.. SPACE SCIENCES BLDG
CORNELL UNIVERSITY
ITHACA NY 14853
U.S.A.
TEL: 607-256-3734
TLX: 932454
TLF:
EML:
COM: 16

HAGYARD MONA JUNE
NASA MARSHALL SFC
CODE ES52
HUNTSVILLE AL 35812
U.S.A.
TEL: 205-453-5687
TLX: 594416 NASA/MSFC HTV
TLF:
EML:
COM: 10.12

HAIKALA LAURI K
OBS & ASTROPHYSICS LAB
UNIVERSITY OF HELSINKI
TAHTITORNINMAKI
SF-00130 HELSINKI 13
FINLAND
TEL: 1912948 HELSINK
TLX: 124690 UNIH SF
TLF:
EML:
COM:

HAISCH BERNHARD MICHAEL
LOCKHEED PALO ALTO
RESEARCH LABORATORY
DIV 91-20 BLDG 255
PALO ALTO CA 94304
U.S.A.
TEL: 415-858-4073
TLX: 346409
TLF:
EML:
COM: 27.36.51

HAJDUK ANTON DR
ASTRONOMICAL INSTITUTE
SLOVAK ACAD. OF SCIENCES
842 28 BRATISLAVA
CZECHOSLOVAKIA
TEL: 427-375157
TLX: 93373 SEIS
TLF:
EML:
COM: 22.51

HAJDUKOVA MARIA
DEPARTMENT OF ASTRONOMY
COMENIUS UNIVERSITY
MLYNSKA DOLINA
842 15 BRATISLAVA
CZECHOSLOVAKIA
TEL: 427-320-003
TLX:
TLF:
EML:
COM: 22

HALBWACHS JEAN LOUIS DR
OBSERV. DE STRASBOURG
11 RUE DE L'UNIVERSITE
67000 STRASBOURG
FRANCE
TEL: 88358200 441
TLX: 890506
TLF:
EML: EARN:U011030@FRCCSC21
COM: 26.30

HALL ANDREW NORMAN
DEPT OF ASTROPHYSICS
SOUTH PARKS ROAD
OXFORD OX1 3RQ
U.K.
TEL:
TLX:
TLF:
EML:
COM: 48

HALL DONALD N DR
INSTITUTE FOR ASTRONOMY
2680 WOODLAWN DRIVE
HONOLULU HI 96822
U.S.A.
TEL: 808-948-8312
TLX: 723-8459
TLF:
EML:
COM:

HALL DOUGLAS S DR
DYER OBSERVATORY
VANDERBILT UNIVERSITY
NASHVILLE TN 37235
U.S.A.
TEL: 615-373-4897
TLX: 554323
TLF:
EML:
COM: 25.27.42

HALL JOHN S DR
110 RED BUTTE DRIVE
SEDONA AZ 86336
U.S.A.
TEL: 602-284-1738
TLX:
TLF:
EML:
COM: 16.34

HALL R GLENN DR
3612 SPRING STREET
CHEVY CHASE MD 20815
U.S.A.
TEL: 301-652-7221
TLX:
TLF:
EML:
COM: 19.31

HALLAM KENNETH L DR
LAB. ASTRON & SOLAR PHYS
NASA/GSFC
GREENBELT MD 20771
U.S.A.
TEL: 301-286-6081
TLX:
TLF:
EML:
COM: 09.44.45

HALLIDAY IAN DR
HERZBERG INST ASTROPHYS
NATL RESEARCH COUNCIL
OTTAWA ONT K1A 0R6
CANADA
TEL: 613-990-0704
TLX: 0533715
TLF:
EML:
COM: 15.16.21.22

HAMABE MASARU DR
KISO OBS OF TOKYO ASTRON
OBSERVATORY
MITAKE-MURA. KISO
NAGANO 397-01
JAPAN
TEL: 26452-3360
TLX: 3347577 KSOOBS J
TLF:
EML:
COM: 28

HAMADA TETSUO PROF
DEPT OF PHYSICS
IBARAKI UNIVERSITY
310 MITO
JAPAN
TEL: 0292-252-35-24
TLX:
TLF:
EML:
COM:

HAMAJIMA KIYOTOSHI DR
KISO BRANCH OF THE
TOKYO ASTRON OBSERVATORY
MITAKEMURA KISOGUN
NAGANOKEN 397-01
JAPAN
TEL:
TLX:
TLF:
EML:
COM: 33

HAMDY M A M DR
HELWAN OBSERVATORY
HELWAN-CAIRO
EGYPT
TEL: 780645-782683
TLX: 93070 HIAG UN
TLF:
EML:
COM: 27

HAMID S EL DIN DR
DEPT OF ASTRONOMY
FACULTY OF SCIENCE
UNIVERSITY FOUAD
GIZA CAIRO
EGYPT
TEL:
TLX:
TLF:
EML:
COM: 07

HAMILTON P A DR
UNIVERSITY OF TASMANIA
PHYSICS DEPARTMENT
BOX 252C G.P.O.
HOBART. TASMANIA
AUSTRALIA
TEL: 002-20-2419
TLX: 58150
TLF:
EML:
COM: 40

HAMMAN WOLF-RAINER
INST THEOR PHYS & STERNW
UNIVERSITAET KIEL
OLSHAUSENSTR
D-2300 KIEL
GERMANY. F.R.
TEL: 0431-8804101
TLX: 292706 IAPKI D
TLF:
EML:
COM: 36

HAMMER REINER
KIEPENHEUER INSTITUT FUER
SONNENPHYSIK
SCHOENECKSTR 6
D-7800 FREIBURG
GERMANY. F.R.
TEL: 0761-32864
TLX: 7721552 KIS D
TLF:
EML:
COM: 10.12

HAMMERSCHLAG ROBERT H DR
STERREKUNDIG INSTITUUT
POSTBUS 80 000
PRINCETONPLEIN 5
NL-3508 TA UTRECHT
NETHERLANDS
TEL: 030-535200
TLX: 40048 FYLUT NL
TLF:
EML: BITNET:WHHMAIL@HUTRUU0
COM: 09

HAMMERSCHLAG-HENSBERGE G
ASTRONOMICAL INSTITUTE
UNIVERSITY OF AMSTERDAM
ROETERSSTRAAT 15
NL-1018 WB AMSTERDAM
NETHERLANDS
TEL: 020-522-3004
TLX: 16460
TLF:
EML:
COM: 42

HANZAOGLU ESAT E H DR
KING SAUD UNIVERSITY
COLLEGE OF SCIENCE
P.O. BOX 2455
RIYADH 11453
SAUDI ARABIA
TEL:
TLX:
TLF:
EML:
COM:

HAN FU
PURPLE MOUNTAIN
OBSERVATORY
NANJING
CHINA. PEOPLE'S REP.
TEL: 33738
TLX: 34144 PMONJ CN
TLF:
EML:
COM: 40

HAN TIANQI
INST OF GEODESY & GEOPHYS
ACADEMIA SINICA
WUCHANG. HUBEI
CHINA. PEOPLE'S REP.
TEL: 813712-570
TLX:
TLF:
EML:
COM: 19.31

HAN WENJUN
BEIJING ASTRONOMICAL OBS
ACADEMIA SINICA
BEIJING
CHINA. PEOPLE'S REP.
TEL: 281698
TLX:
TLF:
EML:
COM: 40

HAN ZHENG-ZHONG
PURPLE MOUNTAIN
OBSERVATORY
NANJING
CHINA. PEOPLE'S REP.
TEL: 33583
TLX: 34144
TLF:
EML:
COM: 44

HANASZ JAN DR
ASTRONOMICAL INSTITUTE
POLISH ACAD OF SCIENCES
UL. CHOPINA 12/18
87-100-TORUN
POLAND
TEL: 260-37
TLX: 0552234 ASTR PL
TLF:
EML:
COM: 10.40

HANAWA TOMOYUKI DR
DEPT OF ASTROPHYSICS
NAGOYA UNIVERSITY
CHIKUSA KU NAGOYA 464-01
JAPAN
TEL: 52 781 6769
TLX: 4477323
TLF:
EML: BITNET:B42287@JPNKUDPC
COM: 42

HANBURY BROWN ROBERT PROF
WHITE COTTAGE
PENTON MEWSEY
HANTS SP11 ORQ
U.K.
TEL: (026477) 2334
TLX:
TLF:
EML:
COM: 40

HANDLIROVA DAGMAR DR
N. COPERNICUS OBSERVATORY
AND PLANETARIUM
KRAVI HORA
61600 BRNO
CZECHOSLOVAKIA
TEL: 425744374
TLX:
TLF:
EML:
COM: 29

HANES DAVID A DR
QUEEN'S UNIVERSITY
PHYSICS DEPT
ASTRONOMY GROUP
KINGSTON ONT K7L 3N6
CANADA
TEL: 613-547-5750
TLX:
TLF:
EML:
COM: 37

HANG HENG-RONG
PURPLE MOUNTAIN
OBSERVATORY
NANJING
CHINA. PEOPLE'S REP.
TEL: 33583
TLX: 34144 PMONJ CN
TLF:
EML:
COM: 44.48

HANISCH ROBERT J DR
SPACE TELESCOPE
SCIENCE INSTITUTE
3700 SAN MARTIN DRIVE
BALTIMORE MD 21218
U.S.A.
TEL: 301 338 4910
TLX: 6849101
TLF:
EML: SPAN:SCIVAX::HANISCH
COM: 05.09.40

HANKINS TIMOTHY HAMILTON
DEPT. OF PHYS. & ASTRON.
NEW MEXICO INSTITUTE OF
MINING & TECHNOLOGY
SOCORO NM 87801
U.S.A.
TEL: 505 476 8011
TLX:
TLF:
EML:
COM: 40

HANNER MARTHA S DR
JPL
MS T1166
4800 OAK GROVE DR.
PASADENA CA 91109
U.S.A.
TEL: 818-354-4100
TLX: 675429
TLF:
EML:
COM: 15.21V.22

HANSEN CARL J PROF
JILA
UNIVERSITY OF COLORADO
BOX 440
BOULDER CO 80309
U.S.A.
TEL: 303-492-7811
TLX: 755842 JILA
TLF:
EML:
COM: 27

HANSEN LEIF LECTURER
UNIVERSITY OBSERVATORY
OESTER VOLDGADE 3
DK-1350 COPENHAGEN K
DENMARK
TEL: 1-14 17 90
TLX: 44155 DANAST DK
TLF:
EML:
COM:

HANSEN RICHARD T MR
ENGINEERING 138. VANC
150 S. HUNTINGTON AVE
BOSTON MA 02130
U.S.A.
TEL: 6177342534-HOME
TLX:
TLF:
EML:
COM: 10

HANSLMEIER ARNOLD
INSTITUT FUER ASTRONOMIE
KARL-FRANZENS-UNIVERSITAT
UNIVERSITAETSPLATZ 5
A-8010 GRAZ
AUSTRIA
TEL: 0316-380-5275
TLX:
TLF:
EML:
COM: 07.10

HANSON ROBERT B DR
LICK OBSERVATORY
UNIVERSITY OF CALIFORNIA
SANTA CRUZ CA 95064
U.S.A.
TEL: 408-429-2755
TLX:
TLF:
EML:
COM: 24C

HANSSON NILS DR
LUND OBSERVATORY
BOX 43
S-221 00 LUND
SWEDEN
TEL: 46-107000
TLX: 33533 LUNIVER S
TLF:
EML:
COM:

HANUSCHIK REINHARD DR
ASTRONOMISCHES INSTITUT
RUHR UNIVERSITAT BOCHUM
POSTFACH 102148
D 4630 BOCHUM 1
GERMANY. F.R.
TEL: 0234 7003450
TLX: 0825860
TLF:
EML:
COM: 29

HAO YUN-XIANG
DEPT OF ASTRONOMY
BEIJING NORMAL UNIVERSITY
BEIJING
CHINA. PEOPLE'S REP.
TEL: 656531-6285
TLX:
TLF:
EML:
COM: 09

HAPKE BRUCE W DR
DEPT GEOL & PLANETARY SCI
UNIVERSITY OF PITTSBURGH
321 OLD ENGINEERING HALL
PITTSBURGH PA 15235
U.S.A.
TEL: 412-624-4719
TLX:
TLF:
EML:
COM: 15

HARA KEN NOSUKE DR
SENDAI-SHIRITSU-JOSHI
SENIOR HIGH SCHOOL
KASHIWAGI 1 3 1
SENDAI 980
JAPAN
TEL:
TLX:
TLF:
EML:
COM: 31.47

HARA TADAYOSHI DR
DIV. OF EARTH ROTATION
NAT. ASTRON. OBSERVATORY
2-12 HOSHIGAOKA MIZUSA
IWATE 023
JAPAN
TEL: 197 24 7111
TLX: 837628 ILSMIZ J
TLF:
EML:
COM:

HARA TETSUYA DR
DEPT OF PHYSICS
KYOTO SANGYO UNIVERSITY
KITAKU KAMIGAMO
KYOTO 603
JAPAN
TEL: 075-701-2151
TLX: 5422661 KSU J
TLF:
EML:
COM: 28

HARDEBECK ELLEN G DR
3106 TUMBLEWEED RD
BISHOP CA 93514
U.S.A.
TEL:
TLX:
TLF:
EML:
COM: 34

HARDEE PHILIP
DEPT PHYSICS & ASTRONOMY
UNIVERSITY OF ALABAMA
BOX 1921
UNIVERSITY AL 35486
U.S.A.
TEL: 205-348-5050
TLX:
TLF:
EML:
COM: 40

HARDIE R PROF
DYER OBSERVATORY
VANDERBILT UNIVERSITY
NASHVILLE TN 37235
U.S.A.
TEL:
TLX:
TLF:
EML:
COM: 25

HARDY EDUARDO
DEPT DE PHYSIQUE
UNIVERSITE LAVAL
FAC DES SCS & DE GENIE
QUEBEC G1K 7P4
CANADA
TEL: 418-656-2960
TLX: 369-5131621
TLF:
EML:
COM: 28.47

HARMANEC PETR DR
ASTRONOMICAL INSTITUTE
CZECH. ACAD. OF SCIENCES
251 65 ONDREJOV
CZECHOSLOVAKIA
TEL: 724525
TLX: 121579
TLF:
EML:
COM: 27.29.42

HARMER CHARLES F W MR
NOAO
950 N.CHERRY AVENUE
P.O. BOX 26732
TUCSON AZ 85726-6732
U.S.A.
TEL: 602-327-5511
TLX: 0666-484 AURA NOAO
TLF:
EML:
COM: 09.29

HARMER DIANNE L MRS
RARDE CA4
BUILDING Q10
FORT HALSTEAD
SEVENOAKS. KENT
U.K.
TEL: 0959-32222X2395
TLX:
TLF:
EML:
COM: 09.29

HARMS RICHARD JAMES DR
APPLIED RESEARCH CORP.
8201 CORPORATE DRIVE
SUITE 920
LANDOVER MD 20785
U.S.A.
TEL: 301-459-8442
TLX:
TLF:
EML:
COM: 28.44.47

HARNDEN FRANK R Jr
HARVARD-SMITHSONIAN CTR
FOR ASTROPHYSICS
60 GARDEN STREET
CAMBRIDGE MA 02138
U.S.A.
TEL: 617-495-7143
TLX: 921428 SATELLITE CAM
TLF:
EML:
COM:

HARPAZ AMOS DR
DEPT PHYS & SPACE RESEARC
THE TECHNION
HAIFA 32000
ISRAEL
TEL: 972 4293521
TLX:
TLF:
EML: BITNET:PHR89AH@TECHNION
COM:

HARRINGTON J PATRICK DR
ASTRONOMY PROGRAM
UNIVERSITY OF MARYLAND
COLLEGE PK MD 20742
U.S.A.
TEL: 301-454-5944
TLX: 7108260352
TLF:
EML:
COM: 34

HARRINGTON ROBERT S DR
US NAVAL OBSERVATORY
WASHINGTON DC 20392
U.S.A.
TEL: 202-653-1533
TLX:
TLF:
EML:
COM: 20.24.26C.51

HARRIS ALAN WILLIAM
SERC
RUTHERFORD APPLETON LAB
CHILTON. DIDCOT OX11 0QX
U.K.
TEL: 0235-21900
TLX: 83159 RUTHLB G
TLF:
EML:
COM: 15.34

HARRIS ALAN WILLIAM DR
JPL
MS 183-501
4800 OAK GROVE DRIVE
PASADENA CA 91109
U.S.A.
TEL: 818-354-6741
TLX: 675429/9105883294/69
TLF:
EML:
COM: 15V.20

HARRIS DANIEL E DR
CENTER FOR ASTROPHYSICS
60 GARDEN STREET
CAMBRIDGE MA 02138
U.S.A.
TEL: 617-495-7148
TLX: 921428
TLF:
EML:
COM: 40

HARRIS GRETCHEN L H DR
DEPT OF PHYSICS
UNIVERSITY OF WATERLOO
WATERLOO ON N2L 3G1
CANADA
TEL: 519-885-1211
TLX: 069-55259
TLF:
EML:
COM: 37P

HARRIS HUGH C
US NAVAL OBSERVATORY
PO BOX 1149
FLAGSTAFF AZ 86002
U.S.A.
TEL: 602-779-5132
TLX:
TLF:
EML:
COM: 37

HARRIS STELLA
DEPT OF PHYSICS
QUEEN MARY COLLEGE
MILE END ROAD
LONDON E1 4NS
U.K.
TEL: 1-980-4811x4050
TLX: 893750
TLF:
EML:
COM: 34

HARRIS WILLIAM E DR
DEPT OF PHYSICS
MCMASTER UNIVERSITY
HAMILTON ONT L8S 4M1
CANADA
TEL: 416-525-9140
TLX: 369-0618347
TLF:
EML:
COM: 37

HARRISON EDWARD R PROF
DEPT PHYSICS & ASTRONOMY
UNIV OF MASSACHUSETTS
AMHERST MA 01003
U.S.A.
TEL: 413-545-2194
TLX:
TLF:
EML:
COM: 47.51

HARRISON RICHARD A DR
ASTROPHYS DIVISION
RUTHERFORD APPLETON LAB
CHILTON DIDCOT
OXON OX11 0QX
U.K.
TEL: 0235 446497
TLX: 83159 RUTHLB G
TLF:
EML: SPAN:19457::RAH
COM:

HARROWER GEORGE A DR
LAKEHEAD UNIVERSITY
THUNDER BAY ONT P7B 5E1
CANADA
TEL: 807-345-2121
TLX:
TLF:
EML:
COM:

HART MICHAEL H DR
7301 MASONVILLE DRIVE
ANNANDALE VA 22003
U.S.A.
TEL:
TLX:
TLF:
EML:
COM: 51

HARTEN RONALD H DR
RCA ASTRO ELECTRONIC
TB-1
PO BOX 800
PRINCETON NJ 08540
U.S.A.
TEL: 609-426-3551
TLX:
TLF:
EML:
COM: 34.40

HARTKOPF WILLIAM I DR
CENTER FOR HIGH ANGULAR
RESOLUTION ASTRONOMY
GEORGIA STATE UNIVERSITY
ATLANTA GA 30303 3083
U.S.A.
TEL: 404 651 2932
TLX:
TLF:
EML:
COM: 24.26.33

HARTL HERBERT DR
INSTITUT FUER ASTRONOMIE
TECHNIKERSTRASSE 15
A-6020 INNSBRUCK
AUSTRIA
TEL: 5522-748-5251
TLX:
TLF:
EML:
COM: 34

HARTLEY KENNETH F DR
RUTHERFORD APPLETON LAB
CHILTON. DIDCOT OX11 0QX
U.K.
TEL: 0235-21900
TLX: 83159
TLF:
EML:
COM:

HARTMANN LEE WILLIAM
CENTER FOR ASTROPHYSICS
60 GARDEN STREET
CAMBRIDGE MA
U.S.A.
TEL: 617-495-7487
TLX:
TLF:
EML:
COM: 29.36

HARTMANN WILLIAM K
PLANETARY SCIENCE INST
2030 E SPEEDWAY
SUITE 201
TUCSON AZ 85719
U.S.A.
TEL: 602-881-0332
TLX:
TLF:
EML:
COM: 15

HARTOOG MARK RICHARD DR
LICK OBSERVATORY
UNIVERSITY OF CALIFORNIA
SANTA CRUZ CA 95064
U.S.A.
TEL:
TLX:
TLF:
EML:
COM:

HARTQUIST THOMAS WILBUR
MPI FUR PHYS & ASTROPHYS
KARL-SCHWARZSCHILD-STR 1
D-8046 GARCHING B MUNCHEN
GERMANY. F.R.
TEL: 089-3299-838
TLX: 05215845 XTERR D
TLF:
EML:
COM: 34

HARTWICK F DAVID A DR
UNIVERSITY OF VICTORIA
DEPT OF PHYSICS
VICTORIA BC V8W 2Y2
CANADA
TEL: 604-721-7742
TLX: 049-7222
TLF:
EML:
COM:

HARTZ THEODORE R DR
915 MOUNTAINVIEW AVENUE
OTTAWA ONT K2B 5G3
CANADA
TEL: 613-596-1211
TLX:
TLF:
EML:
COM: 40.44

HARUTYUNIAN HAIK A DR
BYURAKAN ASTROPHYSICAL
OBSERVATORY
378433 BYURAKAN ARMENIA
U.S.S.R.
TEL: 28-3453/4142
TLX: 411576 AS CON SU
TLF:
EML:
COM: 36

HARVEL CHRISTOPHER ALVIN
6161 STEVEN'S FOREST RD
COLUMBIA MD 21045
U.S.A.
TEL: 301-964-0211
TLX:
TLF:
EML:
COM: 05.37

HARVEY CHRISTOPHER C DR
OBSERVATOIRE DE PARIS
SECTION DE MEUDON
F-92195 MEUDON PL CEDEX
FRANCE
TEL: 1-45-07-76-69
TLX: 204464
TLF:
EML:
COM: 44.49

HARVEY GALE A DR
INST F GESCHICHTE D
NATURWISSENSCHAFTEN
GOETHE UNIVERSITAET
D-6000 FRANKFURT
GERMANY. F.R.
TEL:
TLX:
TLF:
EML:
COM: 22

HARVEY JOHN W DR
NATL SOLAR OBSERVATORY
PO BOX 26732
950 N. CHERRY AVE
TUCSON AZ 85726
U.S.A.
TEL: 602-327-5511
TLX:
TLF:
EML:
COM: 10.12P

HARVEY PAUL MICHAEL DR
DEPT OF ASTRONOMY
UNIVERSITY OF TEXAS
AUSTIN TX 78712
U.S.A.
TEL: 512-471-4461
TLX: 910-874-1351
TLF:
EML:
COM: 34.44

HARWIT MARTIN PROF
ASTRONOMY DEPT.
SPACE SCIENCE BLDG
CORNELL UNIVERSITY
ITHACA NY 14853
U.S.A.
TEL: 607-256-4805
TLX:
TLF:
EML:
COM: 15.21.48

HARWOOD DENNIS MR
PERTH OBSERVATORY
BICKLEY. W. AUSTR. 6076
AUSTRALIA
TEL: 09-2938-255
TLX:
TLF:
EML:
COM: 08.24.25

HASAN SAIYID SIRAJUL
INDIAN INSTITUTE OF
ASTROPHYSICS
BANGALORE 560 034
INDIA
TEL:
TLX:
TLF:
EML:
COM: 10

HASCHICK AUBREY
HAYSTACK OBSERVATORY
WESTFORD MA 01886
U.S.A.
TEL: 617-692-4764
TLX:
TLF:
EML:
COM: 40

HASEGAWA HIROICHI DR
DEPT OF PHYSICS
KYOTO UNIVERSITY
SAKYO-KU
KYOTO 606
JAPAN
TEL: 0757512111x3833
TLX:
TLF:
EML:
COM: 21.22

HASEGAWA ICHIRO DR
4-8-5 FUJIWARADAI-KITA
KITA-KU
KOBE 651-13
JAPAN
TEL: 81-78-982-5255
TLX:
TLF:
EML:
COM: 15.20.22C

HASEGAWA TATSUHIKO DR
DEPT. OF ASTRONOMY
ST MARY'S UNIVERSITY
923 ROBIE STREET
HALIFAX NOVA SCOT.B3H 3C3
CANADA
TEL: 902 420 5823
TLX:
TLF:
EML: BITNET:HASEGAWA@STMARYS
COM:

HASEGAWA TETSUO DR
NOBEYAMA RADIO OBSERVAT.
TOKYO ASTRONO OBSERVATORY
NOBEYAMA. MINAMIMAKIMURA
NAGANO 384 13
JAPAN
TEL: 267 98 2831
TLX: 3329005
TLF:
EML:
COM: 40

HASER LEO N K DR
MPI F. EXTRATERR. PHYSIK
D-8046 GARCHING B MUNCHEN
GERMANY. F.R.
TEL: 89-329-98-03
TLX: 5215845 XTER D
TLF:
EML:
COM: 15

HASLAM C GLYN T DR
MPI FUER RADIOASTRONOMIE
AUF DEM HUEGEL 69
D-5300 BONN
GERMANY. F.R.
TEL:
TLX: 886440 MPIFR D
TLF:
EML:
COM: 40

HASSALL BARBARA J M DR
ROYAL GREENWICH OBSERVA.
MADINGLEY ROAD
CAMBRIDGE CB3 0EZ
U.K.
TEL: 0223 337548
TLX: 817297 ASTRON G
TLF:
EML: JANET:BJMH@UK.AC.CAM.AST-STAR
COM: 42

HASSAN S M DR
HIAG
HELWAN-CAIRO
EGYPT
TEL: 780645. 782683
TLX: 93070
TLF:
EML:
COM: 37

HATHAWAY DAVID H DR
CODE ES52
NASA
MARSHALL SPACE FLIGHT CTR
HUNTSVILLE AL 35812
U.S.A.
TEL: 205 544 7610
TLX:
TLF:
EML:
COM: 10

HAUBOLD HANS JOACHIM
OUTER SPACE AFFAIRS DIV.
ROOM S-3260B
UNITED NATIONS
NEW YORK NY 10017.
U.S.A.
TEL:
TLX:
TLF:
EML:
COM: 48

HAUCK BERNARD PROF
INSTITUT D'ASTRONOMIE
UNIVERSITE DE LAUSANNE
CH-1290 CHAVANNES-DES-B.
SWITZERLAND
TEL: 022-55-26-11
TLX: 27720 OBSG CH
TLF:
EML:
COM: 05V.25.45

HAUG EBERHARD DR
MOZARTSTRASSE 20
D-7430 METZINGEN
GERMANY, F.R.
TEL: 07071/296483
TLX:
TLF:
EML:
COM: 10

HAUG ULRICH PROF
HAMBURGER STERNWARTE
GOJENSBERGWEG 112
D-2050 HAMBURG 80
GERMANY, F.R.
TEL: 040-7252-4131
TLX: 217884 HAMST D
TLF:
EML:
COM: 21.33

HAUGE OIVIND DR
INST THEORET ASTROPHYSICS
UNIVERSITY OF OSLO
P O BOX 1029
N-0315 BLINDERN, OSLO 3
NORWAY
TEL: 245-65-06
TLX:
TLF:
EML:
COM:

HAUPT HERMANN F PROF
INSTITUT FUER ASTRONOMIE
DER UNIVERSITAET
UNIVERSITAETSPLATZ 5
A-8010 GRAZ
AUSTRIA
TEL: 0316-380-5271
TLX: 31078A
TLF:
EML:
COM: 15C.20.38.46

HAUPT RALPH F
3701 DULWICK DRIVE
SILVER SPRING MD 20906
U.S.A.
TEL: 301-598-7868
TLX:
TLF:
EML:
COM: 04.12

HAUPT WOLFGANG DR
GIRONDELLE 105
D-4630 BOCHUM
GERMANY, F.R.
TEL:
TLX:
TLF:
EML:
COM:

HAUSER MICHAEL G DR
LAB FOR ASTRONOMY AND
SOLAR PHYSICS, CODE 680
NASA GSFC
GREENBELT MD 20771
U.S.A.
TEL: 301 286 8701
TLX:
TLF:
EML: MHAUSER:NASA NASAMAIL
COM: 21.44

HAVLEN ROBERT J DR
NRAO
EDGEMONT ROAD
CHARLOTTESVILLE VA 22901
U.S.A.
TEL: 804-296-0223
TLX: 9109970174
TLF:
EML:
COM:

HAVNES OVE DR
AURORAL OBSERVATORY
UNIVERSITY OF TROMSO
P O 953
N-9001 TROMSO
NORWAY
TEL: 83-86060
TLX: 64124 AUROB N
TLF:
EML:
COM:

HAWARDEN TIMOTHY G DR
ROYAL OBSERVATORY
BLACKFORD HILL
EDINBURGH EH9 3HJ
U.K.
TEL: 031-667-3321
TLX: 72383 ROEDIN G
TLF:
EML:
COM: 37

HAWKES ROBERT LEWIS DR
PHYSICS DEPARTMENT
MOUNT ALLISON UNIVERSITY
SACKVILLE
NEW BRUNSWICK E0A 3C0
CANADA
TEL: 506 364 2580
TLX:
TLF:
EML: BITNET:A014@MTAH
COM: 22

HAWKING STEPHEN W PROF
DEPT OF APPLIED MATHS
AND THEORETICAL PHYSICS
SILVER STREET
CAMBRIDGE CB3 9EW
U.K.
TEL: 223-351645
TLX: 81240 CAMSPL G
TLF:
EML:
COM: 47.48

HAWKINS GERALD S DR
CONSUL 906
2400 VIRGINIA QVE NW
WASHINGTON DC 20037
U.S.A.
TEL: 202-485-2050
TLX:
TLF:
EML:
COM: 22.41

HAWKINS MICHAEL R S
ROYAL OBSERVATORY
BLACKFORD HILL
EDINBURGH EH9 3HJ
U.K.
TEL:
TLX:
TLF:
EML:
COM: 33

HAYAKAWA SATIO PROF
DEPT PHYSICS/ASTROPHYSICS
NAGOYA UNIVERSITY
FUROKO CHIKUSAKU
NAGOYA 464
JAPAN
TEL: 052-781-5111
TLX: 4477323 SCUNAG J
TLF:
EML:
COM: 44.47.48

HAYASHI CHUSHIRO PROF
MOMOYAMA YOGORO-CHO 1
FUSHIMI-KU
KYOTO 612
JAPAN
TEL: 075-611-1062
TLX:
TLF:
EML:
COM: 35.47

HAYASHI MASAHIKO DR
DEPT OF ASTRONOMY
UNIVERSITY OF TOKYO
BUNKYO KU
TOKYO 113
JAPAN
TEL: 3 812 2111
TLX: 2722126 UTGAB J
TLF:
EML:
COM: 40

HAYES DONALD S DR
PO BOX 1907
SCOTTSDALE AZ 85252
U.S.A.
TEL: 602-947-3572
TLX:
TLF:
EML:
COM: 25.45

HAYLI AVRAM PROF
OBSERVATOIRE DE LYON
F-69230 ST-GENIS-LAVAL
FRANCE
TEL: 78-56-07-05
TLX: 310916
TLF:
EML:
COM: 33.41

HAYNES ROBERT C PROF
DEPT SPACE PHYS & ASTRON
RICE UNIVERSITY
HOUSTON TX 77001
U.S.A.
TEL: 713-527-4045
TLX: 556457
TLF:
EML:
COM: 48

HAYNES MARTHA P
ASTRONOMY DEPT
CORNELL UNIVERSITY
SPACE SCIENCES BUILDING
ITHACA NY 14853
U.S.A.
TEL: 607-256-3734
TLX: 932454
TLF:
EML:
COM: 40

HAYNES RAYMOND F PROF
CSIRO
DIVISION OF RADIOPHYSICS
P.O.BOX 76
EPPING NSW 2121
AUSTRALIA
TEL:
TLX:
TLF:
EML:
COM: 34.40

HAYWARD JOHN
DEPT OF MATHEMATICS
NAPIER COLLEGE
COLINTON RD.
EDINBURGH EH10 5DT
U.K.
TEL:
TLX:
TLF:
EML:
COM: 10

HAZARD CYRIL DR
INSTITUTE OF ASTRONOMY
MADINGLEY ROAD
CAMBRIDGE CB3 OHE
U.K.
TEL:
TLX:
TLF:
EML:
COM: 40

HAZEN MARTHA L DR
HARVARD COLLEGE OBS
60 GARDEN STREET
CAMBRIDGE MA 02138
U.S.A.
TEL: 617-495-3362
TLX:
TLF:
EML:
COM: 37

HAZER S DR
FACULTY OF SCIENCE
DEPT OF ASTRONOMY
P K 21
BORNOVA-IZMIR
TURKEY
TEL:
TLX:
TLF:
EML:
COM:

HAZLEHURST JOHN DR
HAMBURGER STERNWARTE
GOJENSBERGSWEG 112
D-2050 HAMBURG 80
GERMANY. F.R.
TEL:
TLX:
TLF:
EML:
COM: 42

HE MIAO-FU
SHANGHAI OBSERVATORY
ACADEMIA SINICA
SHANGHAI
CHINA. PEOPLE'S REP.
TEL: 386191
TLX: 33164 SHAO CN
TLF:
EML:
COM: 07C.20

HE XIANG-TAO
DEPT OF ASTRONOMY
BEIJING NORMAL UNIVERSITY
BEIJING
CHINA. PEOPLE'S REP.
TEL: 656531-6285
TLX:
TLF:
EML:
COM: 28.47

HEAP SARA R DR
NASA GSFC
CODE 672
GREENBELT MD 20771
U.S.A.
TEL:
TLX:
TLF:
EML:
COM:

HEARN ANTHONY G DR
STERREKUNDIG INSTITUUT
PRINCETONPLEIN 5
POSTBUS 80 000
NL-3508 TA UTRECHT
NETHERLANDS
TEL: 030-535202
TLX: 40048 FYLUT NL
TLF:
EML: BITNET:HEARNTONY@HUTRUUC
COM: 36C.44

HEARNSHAW JOHN B DR
DEPT OF PHYSICS
UNIVERSITY OF CANTERBURY
PRIVATE BAG
CHRISTCHURCH
NEW ZEALAND
TEL: 03-482009 x 771
TLX: 4144 UNICANT NZ
TLF:
EML:
COM: 29

HEASLEY JAMES NORTON
INSTITUTE FOR ASTRONOMY
2680 WOODLAWN DRIVE
HONOLULU HI 96822
U.S.A.
TEL: 808-948-6826
TLX: 7238459 UHAST UR
TLF:
EML:
COM: 36

HEAVENS ALAN DR
DEPARTMENT OF ASTRONOMY
UNIV OF EDINBURGH. ROYAL
OBSERV. BLACKFORD HILL
EDINBURGH EH9 3HJ
U.K.
TEL: 44 31 668 8352
TLX: 72383 ROEDIN G
TLF:
EML: AFH@UK.AC.ROE.STAR
COM: 47

HEBER ULRICH
INST THEOR PHYS & STERNW
UNIVERSITAET KIEL
LEIBNIZSTR
D-2300 KIEL 1
GERMANY. F.R.
TEL: 0431-880-4103
TLX: 292706 IAPKI D
TLF:
EML:
COM: 29.36

HECHT JAMES H DR
THE AEROSPACE CORP
M2-255
PO BOX 92957
LOS ANGELES CA 90009
U.S.A.
TEL: 213 3367017
TLX:
TLF:
EML:
COM: 34

HECK ANDRE DR
OBS DE STRASBOURG
11 RUE DE L'UNIVERSITE
F-67000 STRASBOURG
FRANCE
TEL: 88-35-43-00
TLX: 890506 STAVOBS F
TLF:
EML:
COM: 05.25.45.51

HECKATHORN HARRY M
US NAVAL RESEARCH LAB
CODE 4143-2
WASHINGTON DC 20375
U.S.A.
TEL: 202-767-2764
TLX:
TLF:
EML:
COM: 09.44

HECKMAN TIMOTHY M
UNIVERSITY OF MARYLAND
ASTRONOMY PROGRAM
COLLEGE PARK MD 20742
U.S.A.
TEL: 301-454-3001
TLX: 7108260352 ASTR CORP
TLF:
EML:
COM: 28

HECQUET JOSETTE DR
OBSERVATOIRE MIDI
PYRENEES 14 AVENUE E.
BELIN
31400 TOULOUSE
FRANCE
TEL: 61 25 21 01
TLX: 530776 F
TLF:
EML:
COM:

HEDDLE DOUGLAS W O PROF
DEPT OF PHYSICS
ROYAL HOLLOWAY & BEDFORD
NEW COLLEGE
EGHAM. SURREY TW20 0EX
U.K.
TEL: 0784-35351
TLX: 935504
TLF:
EML:
COM: 14

HEESCHEN DAVID S DR
NRAO
EDGEMONT ROAD
CHARLOTTESVILLE VA 22901
U.S.A.
TEL:
TLX: 910-997-0174
TLF:
EML:
COM: 28.40.51

HEFELE HERBERT PH D
MPI FUER ASTRONOMIE
KONIGSTUHL
D-6900 HEIDELBERG 1
GERMANY. F.R.
TEL:
TLX:
TLF:
EML:
COM: 05

HEFFERLIN RAY A PROF
PHYSICS DEPT
SOUTHERN COLLEGE
DRAWER H
COLLEGEDALE TN 37315-0370
U.S.A.
TEL: 615-238-2869
TLX:
TLF:
EML:
COM: 14

HEGGIE DOUGLAS C DR
UNIVERSITY OF EDINBURGH
DEPARTMENT OF MATHEMATICS
KING'S BUILDINGS
EDINBURGH EH9 3JZ
U.K.
TEL: 31-667-1081
TLX: 727442 UNIVEDG
TLF:
EML:
COM: 07.37C.41

HEGYI DENNIS J ASSOC PROF
RANDALL LABORATORY
UNIVERSITY OF MICHIGAN
ANN ARBOR MI 48109
U.S.A.
TEL: 313-764-5448
TLX: 810-2236056
TLF:
EML:
COM:

HEIDMANN JEAN DR
OBSERVATOIRE DE PARIS
SECTION DE MEUDON
F-92195 MEUDON PL CEDEX
FRANCE
TEL: 1-45-07-75-98
TLX: 270912 F
TLF:
EML:
COM: 28.40.47.51

HEILES CARL PROF
ASTRONOMY DEPT
UNIVERSITY OF CALIFORNIA
BERKELEY CA 94720
U.S.A.
TEL: 415-642-4510
TLX: 820181 UCB AST RAL
TLF:
EML:
COM: 33.34.40

HEINRICH INGE
ASTRONOMISCHES RECHEN-
INSTITUT
MOENCHHOFSTR 12-14
D-6900 HEIDELBERG 1
GERMANY. F.R.
TEL:
TLX:
TLF:
EML:
COM: 05

HEINTZ WULFF D DR
DEPT OF ASTRONOMY
SWARTHMORE COLLEGE
SWARTHMORE PA 19081
U.S.A.
TEL: 215-447-7265
TLX:
TLF:
EML:
COM: 05.08.24.42

HEINTZE J R W DR
STERREKUNDIG INSTITUUT
PRINCETONPLEIN 5
POSTBUS 80 000
NL-3508 TA UTRECHT
NETHERLANDS
TEL: (0)30-535200
TLX: 40048 FYLUT NL
TLF:
EML: BITNET:WNNMAIL@HUTRUUO
COM: 29.30

HEINZEL PETR DR
ASTRONOMICAL INSTITUTE OF
THE CZECHOSLOVAK ACADEMY
OF SCIENCES
251 65 ONDREJOV
CZECHOSLOVAKIA
TEL: 422 724525
TLX: 121579 ASTR C
TLF:
EML:
COM: 10.12

HEISE JOHN DR
SPACE RESEARCH LABORATORY
BENELUXLAAN 21
NL-3527 HS UTRECHT
NETHERLANDS
TEL: 030-937145
TLX: 47224 ASTRO NL
TLF:
EML:
COM: 44.48

HEISER ARNOLD M DR
DYER OBSERVATORY
VANDERBILT UNIVERSITY
BOX 1803-STA B
NASHVILLE TN 37235
U.S.A.
TEL: 615-373-4897
TLX:
TLF:
EML:
COM: 27

HEKELA JAN DR
ASTRONOMICAL INSTITUTE
CZECH. ACAD. OF SCIENCES
251 65 ONDREJOV
CZECHOSLOVAKIA
TEL:
TLX:
TLF:
EML:
COM: 36

HELALI YHYA E DR
HELWAN OBSERVATORY
HELWAN-CAIRO
EGYPT
TEL:
TLX:
TLF:
EML:
COM: 07

HELFAND DAVID JOHN
COLUMBIA ASTROPHYSICS LAB
538 WEST 120TH ST
NEW YORK NY 10027
U.S.A.
TEL: 212-280-2150
TLX:
TLF:
EML:
COM: 48

HELFER H LAWRENCE PROF
DEPT PHYSICS & ASTRONOMY
UNIVERSITY OF ROCHESTER
ROCHESTER NY 14627
U.S.A.
TEL: 716-275-4377
TLX:
TLF:
EML:
COM: 34

HELIN ELEANOR FRANCIS
JET PROPULSION LAB
MS 183-501
4800 OAK GROVE DRIVE
PASADENA CA 91109
U.S.A.
TEL: 818-354-4606
TLX: 67-5429
TLF:
EML:
COM: 15.20

HELLER MICHAEL PROF
POWSTANCOW WARSAWY 13/94
33-110 TARNOW
POLAND
TEL:
TLX:
TLF:
EML:
COM: 47

HELLWIG HELMUT WILHELM DR
FREQUENCY & TIME SYSTEMS
34 TOZER ROAD
BEVERLY MA 01915
U.S.A.
TEL: 617-927-8220
TLX: 940518
TLF:
EML:
COM: 19.31

HELMER LEIF
COPENHAGEN UNIVERSITY OBS
BRORFELDE
DK-4340 TOLLOSE
DENMARK
TEL: 3-488195
TLX: 44155
TLF:
EML:
COM: 08C.50

HELMKEN HENRY F DR
SMITHSONIAN ASTROPHYS OBS
60 GARDEN ST
CAMBRIDGE MA 02138
U.S.A.
TEL:
TLX:
TLF:
EML:
COM: 44

HELOU GEORGE DR
IPAC 100-22
CALTECH
PASADENA CA 91125
U.S.A.
TEL: 818 584 2928
TLX: 584 9945
TLF:
EML: BITNET:HELOU@IPACOHAMLET.
COM: 28.40

HELT BODIL E
UNIVERSITY OBSERVATORY
OESTER VOLDGADE 3
DK-1350 COPENHAGEN K
DENMARK
TEL: 1-14 17 90
TLX: 44155 DANAST DK
TLF:
EML:
COM:

HEMENWAY PAUL D
ASTRONOMY DEPT
UNIVERSITY OF TEXAS
R.L. MOORE HALL 15.308
AUSTIN TX 78712
U.S.A.
TEL: 512-471-4461
TLX:
TLF:
EML:
COM: 08.20.24

HEMMLEB GERHARD DR
ZENTRALINSTITUT FUR
PHYSIK DER ERDE
TELEGRAFENBERG A 17
DDR-1500 POTSDAM
GERMANY. D.R.
TEL: 4551
TLX: 15305 VDE PDM DD
TLF:
EML:
COM: 19.31C

HEMPE KLAUS
GRENZWEG 24 B
D-2057 REINBEK
GERMANY. F.R.
TEL: 040-710-56-28
TLX:
TLF:
EML:
COM:

HENDECOURT D' LOUIS DR
GR. PHYSIQUE DES SOLIDES
T23. UNIVERSITE PARIS VII
4 PLACE JUSSIEU
F-75251 PARIS CEDEX 05
FRANCE
TEL:
TLX:
TLF:
EML:
COM: 21.34

HENIZE KARL G ASTRONAUT
CODE CB
NASA/JOHNSON SPACE CENTER
HOUSTON TX 77058
U.S.A.
TEL: 713-483-2411
TLX:
TLF:
EML:
COM: 28.29.34.44.45

HENKEL CHRISTIAN
MPI FUER RADIOASTRONOMIE
AUF DEM HUEGEL 69
D-5300 BONN 1
GERMANY. F.R.
TEL:
TLX: 886440 MPIFR D
TLF:
EML:
COM: 34.40

HENON MICHEL C DR
OBSERVATOIRE DE NICE
B P 139
F-06003 NICE CEDEX
FRANCE
TEL: 93-89-04-20
TLX: 460004
TLF:
EML:
COM: 07.33.37

HENOUX JEAN-CLAUDE DR
OBSERVATOIRE DE PARIS
SECTION DE MEUDON
F-92195 MEUDON PL CEDEX
FRANCE
TEL: 1-45-07-78-03
TLX:
TLF:
EML:
COM: 10.44

HENRARD JACQUES PROF
FACULTES UNIV DE NAMUR
RUE DE BRUXELLES 61
B-5000 NAMUR
BELGIUM
TEL: 81-22-90-61
TLX: 59222
TLF:
EML:
COM: 04.07P.20

HENRICHS HUBERTUS F DR
ASTRONOMICAL INSTITUTE
UNIVERSITY OF AMSTERDAM
ROETERSSTRAAT 15
1018WB AMSTERDAM
NETHERLANDS
TEL: 31205256038
TLX: 16460
TLF:
EML:
COM: 29

HENRIKSEN RICHARD N DR
ASTRONOMY GROUP
DEPT OF PHYSICS
QUEEN'S UNIVERSITY
KINGSTON ONT K7L 3N6
CANADA
TEL: 613-547-5536
TLX:
TLF:
EML:
COM: 48

HENRY RICHARD B C DR
DPT OF PHYS. & ASTRONOMY
UNIVERSITY OF OKLAHOMA
NORMAN OK 73019
U.S.A.
TEL: 405 325 3961
TLX:
TLF:
EML:
COM: 28.35

HENRY RICHARD C. PROF.
DEPT PHYSICS & ASTRONOMY
JOHNS HOPKINS UNIVERSITY
BALTIMORE MD 21218
U.S.A.
TEL: 301-338-7350
TLX:
TLF:
EML:
COM: 21.48

HENSBERGE HERMAN
ASTROPHYSICAL INSTITUTE
VRIJE UNIV BRUSSEL
PLEINLAAN 2
B-1050 BRUSSEL
BELGIUM
TEL: 02-641-3468
TLX: 61051 VUBCO
TLF:
EML:
COM: 25.44

HENSLER GERHARD
INST F ASTRON & ASTROPHYS
UNIVERSITAET MUENCHEN
SCHEINERSTR. 1
D-8000 MUNCHEN 80
GERMANY. F.R.
TEL: 089-98-90-71
TLX: 529815 UNIVM D
TLF:
EML:
COM: 42

HERBIG GEORGE H DR
INSTITUTE FOR ASTRONOMY
UNIVERSITY OF HAWAII
2680 WOODLAWN DRIVE
HONOLULU. HI 96822
U.S.A.
TEL: 808-948-8312
TLX: 723-8459
TLF:
EML:
COM: 27.29

HERBST ERIC DR
DEPT OF PHYSICS
DUKE UNIVERSITY
DURHAM NC 27706
U.S.A.
TEL: 919-684-8180
TLX: 802829 DUKTELECOM-DU
TLF:
EML:
COM:

HERBST WILLIAM DR
ASTRONOMY DEPT
WESLEYAN UNIVERSITY
MIDDLETOWN CT 06457
U.S.A.
TEL: 203-347-9411
TLX:
TLF:
EML:
COM: 33.37

HERCZEG TIBOR J PROF DR
DEPT PHYSICS & ASTRONOMY
UNIVERSITY OF OKLAHOMA
NORMAN OK 73019
U.S.A.
TEL: 405-325-3961
TLX:
TLF:
EML:
COM: 42.51

HERMAN JACOBUS DR
MAX PLANCK INSTITUT FUR
EXTRATERRESTRISCE PHYSIK
D-8046
GARCHING BEI MUNCHEN
GERMANY. F.R.
TEL:
TLX:
TLF:
EML:
COM: 33

HERMSEN WILLEM DR
LAB FOR SPACE RESEARCH
NIELS BOHRWEG 2
PO BOX 9504
2300 RA LEIDEN
NETHERLANDS
TEL: 071 275810
TLX: 39058 ASTRO NL
TLF:
EML:
COM:

HERNANDEZ CARLOS ALBERTO
OBSERVATORIO ASTRONOMICO
PASEO DEL BOSQUE
1900 LA PLATA
ARGENTINA
TEL:
TLX:
TLF:
EML:
COM:

HERNANZ MARGARITA DR
DEP. DE FISICA (ETSEIB)
UNIV. POLITEC. CATALUNA
AVDA DIAGONAL 6479
08028 BARCELONA
SPAIN
TEL: 34 3 2495800
TLX:
TLF:
EML: BITNET:EDRUPC51@HERNANZ
COM: 35

HEROLD HEINZ
THEORETISCHE ASTROPHYSIK
UNIVERSITY TUEBINGEN
AUF DER MORGENSTELLE 12.C
D-7400 TUEBINGEN
GERMANY. F.R.
TEL: 07071/292041
TLX:
TLF:
EML:
COM: 14.36

HERR RICHARD B DR
PHYSICS DEPT
UNIVERSITY OF DELAWARE
NEWARK DE 19716
U.S.A.
TEL: 302-451-2673
TLX:
TLF:
EML:
COM: 27

HERRERA MIGUEL ANGEL DR
INSTITUTO ASTRONOMIA
UNAM
APARTADO POSTAL 70-264
MEXICO 04510 DF
MEXICO
TEL: 5 48 53 05
TLX: 1760155 CICME
TLF:
EML: BITNET:MAHERROUNAMVM1
COM:

HERRERO DAVO ARTEMIO DR
INSTITUTO DE ASTROFISICA
DE CANARIAS
38200 LA LAGUNA TENERIFE
SPAIN
TEL: 34 22 26 22 11
TLX: 92640
TLF:
EML:
COM:

HERRMANN DIETER PROF DR
ARCHENHOLD STERNWARTE
ALT TREPTOW 1
DDR-1193 BERLIN
GERMANY. D.R.
TEL: 272-8871 x 494
TLX:
TLF:
EML:
COM: 41

HERS JAN MR
P O BOX 48
SEDGEFIELD 6573
SOUTH AFRICA
TEL: 04455-736
TLX:
TLF:
EML:
COM: 06.20.27.31

HERSHEY JOHN L DR
US NAVAL OBSERVATORY
34TH & MASSACHUSETTS AVE
WASHINGTON DC 20390
U.S.A.
TEL: 202-653-1554
TLX: 710-822-1970
TLF:
EML:
COM: 24.26.51

HERTZ PAUL L DR
CODE 4121.5
NAVAL RESEARCH LABORATORY
WASHINGTON DC 20375-5000
U.S.A.
TEL: 202 767 2438
TLX:
TLF:
EML:
COM:

HERZBERG GERHARD DR
HERZBERG INST ASTROPHYS
NATL RESEARCH COUNCIL
OTTAWA ONT K1A 0R6
CANADA
TEL: 613-990-0917
TLX: 0533715
TLF:
EML:
COM: 14.15.16.34

HESSER JAMES E DR
DOMINION ASTROPHYS OBS
5071 W SAANICH ROAD
RR 5
VICTORIA BC V8X 4M6
CANADA
TEL: 604-388-3974
TLX: 0497295
TLF:
EML:
COM: 14.27.37C

HEUDIER JEAN-LOUIS DR
CERGA
CAUSSOLS
06460 ST VALLIER-DE-THIEY
FRANCE
TEL: 93-42-62-70
TLX: 461402 CERGOBS F
TLF:
EML:
COM: 09C.20.24.37.46.51

HEWETT PAUL
INSTITUTE OF ASTRONOMY
UNIVERSITY OF CAMBRIDGE
MADINGLEY ROAD
CAMBRIDGE CB3 0HA
U.K.
TEL: 223-62204
TLX: 817297 ASTRON G
TLF:
EML:
COM: 30.47

HEWISH ANTONY PROF
CAVENDISH LABORATORY
MADINGLEY RD
CAMBRIDGE CB3 OHE
U.K.
TEL: 0223-66477
TLX: 81292
TLF:
EML:
COM: 40

HEWITT ADELAIDE
CASS MAIL CODE C 011
UNIV OF CALIFORNIA
SAN DIEGO
LA JOLLA CA 92093
U.S.A.
TEL: 619 534 6627
TLX:
TLF:
EML: INT:hewitt%cass.span@ucsd.edu
COM: 28.47

HEWITT ANTHONY V DR
G.E. SPACE SYSTEMS
8080 GRAINGER COURT
SPRINGFIELD VA 22153
U.S.A.
TEL: 703-569-8800
TLX:
TLF:
EML:
COM: 09.28

HEY JAMES STANLEY DR
4 SHORTLANDS CLOSE
EASTBOURNE BN22 OJE
U.K.
TEL:
TLX:
TLF:
EML:
COM: 22.40

HEYDEN FRANCIS J SJ DR
MANILA OBSERVATORY
P O BOX 1231
MANILA
PHILIPPINES
TEL: 999-417
TLX:
TLF:
EML:
COM: 10.27

HEYVAERTS JEAN DR
OBSERVATOIRE DE PARIS
SECTION DE MEUDON
F-92195 MEUDON PL CEDEX
FRANCE
TEL: 1-45-07-74-05
TLX: 201571
TLF:
EML:
COM: 49

HIBBS ALBERT R MGR PLANS
JPL
4800 OAK GROVE DRIVE
PASADENA CA 91103
U.S.A.
TEL:
TLX:
TLF:
EML:
COM:

HICKSON PAUL DR
DEPT GEOPHYS & ASTRONOMY
UNIV OF BRITISH COLUMBIA
2219 MAIN MALL
VANCOUVER BC V6T 1W5
CANADA
TEL: 604-228-2267
TLX: 045-4245
TLF:
EML:
COM: 28

HIDAJAT BAMBANG PROF DR
BOSSCHA OBSERVATORY
LEMBANG. JAVA
INDONESIA
TEL: 6001 LEMBANG
TLX: 28234 BD ITB
TLF:
EML:
COM: 26.34.46.50

HIDALGO MIGUEL A DR
FACULTAD DE CIENCIAS
FISICAS
CIUDAD UNIVERSITARIA
50009 ZARAGOZA
SPAIN
TEL:
TLX:
TLF:
EML:
COM:

HIDE RAYMOND PROF
GEOPHYSICAL FLUID
DYNAMICS LABORATORY
METEOROLOGICAL OFFICE
BRACKNELL. BERKS RG12 2SZ
U.K.
TEL: 0344-420242
TLX: 849801
TLF:
EML:
COM: 16.19

HIEI EIJIRO DR
TOKYO ASTRONOMICAL OBS
OSAWA MITAKA
TOKYO 181
JAPAN
TEL: 0422-32-5111
TLX: 2822307
TLF:
EML:
COM: 10.12

HIGGS LLOYD A DR
DOMINION RADIO ASTROPHY-
SICAL OBSERVATORY
DRC. P O BOX 248
PENTICTON BC V2A 6K3
CANADA
TEL: 604-497-5321
TLX: 048-88127
TLF:
EML:
COM: 34.40

HILDEBRAND ROGER H
ENRICO FERMI INSTITUTE
UNIVERSITY OF CHICAGO
5640 S ELLIS AVE
CHICAGO IL 60637
U.S.A.
TEL: 312-962-7581
TLX:
TLF:
EML:
COM: 34

HILDITCH RONALD W DR
UNIVERSITY OBSERVATORY
BUCHANAN GARDENS
ST ANDREWS KY16 9LZ
U.K.
TEL: 0334-76161
TLX: 76213 SAULIB G
TLF:
EML:
COM: 25.30.42C

HILDNER ERNEST DR
DIRECTOR
SPACE ENVIRON. LABORATORY
325 BROADWAY (R/S/SE)
BOULDER CO 80303-3328
U.S.A.
TEL:
TLX:
TLF:
EML:
COM: 10.12

HILF EBERHARD R H PH D
PAUL-WAGNER-STR 56
D-6100 DARMSTADT
GERMANY. F.R.
TEL:
TLX:
TLF:
EML:
COM: 35.40

HILL GRAHAM DR
DOMINION ASTROPHYS OBS
5071 W SAANICH ROAD
RR 5
VICTORIA BC V8X 4M6
CANADA
TEL: 602-388-3935
TLX: 049-7295
TLF:
EML:
COM: 24.26.30.42

HILL HENRY ALLEN DR
DEPT OF PHYSICS
UNIVERSITY OF ARIZONA
BLDG 81
TUCSON AZ 85721
U.S.A.
TEL: 602-621-6784
TLX: 910-9521143
TLF:
EML:
COM: 27

HILL PHILIP W DR
UNIVERSITY OBSERVATORY
BUCHANAN GARDENS
ST ANDREWS KY16 9LZ
U.K.
TEL: 0334-76161
TLX: 76213 SAULIB G
TLF:
EML:
COM: 25.27

HILLEBRANDT WOLFGANG PH D
MPI F. PHYSIK & ASTROPHYS
KARL-SCHWARZSCHILD-STR 1
D-8046 GARCHING B MUNCHEN
GERMANY. F.R.
TEL: 49-89-32999409
TLX:
TLF:
EML:
COM:

HILLIARD R DR

U.S.A.
TEL:
TLX:
TLF:
EML:
COM: 09

HILLS JACK G DR
LOS ALAMOS NATL LAB
THEORETICAL DIVISION T6
MS B-288
LOS ALAMOS NM 87545
U.S.A.
TEL: 505-667-9157
TLX:
TLF:
EML:
COM: 37

HILLS RICHARD E DR
CAVENDISH LABORATORY
MADINGLEY ROAD
CAMBRIDGE CB3 OHE
U.K.
TEL: 0223-66477
TLX: 81292
TLF:
EML:
COM: 40

HILTNER W ALBERT PROF
DEPT OF ASTRONOMY
UNIVERSITY OF MICHIGAN
ANN ARBOR MI 48109
U.S.A.
TEL: 313-764-3452
TLX:
TLF:
EML:
COM: 25.45

HILTON JOHN DR
DEPART OF MATH SCIENCES
GOLDSMITHS' COLLEGE
NEW CROSS LONDON SE14 6NW
U.K.
TEL:
TLX:
TLF:
EML:
COM:

HINKLE KENNETH H
KPNO. NOAO
PO BOX 26732
950 N. CHERRY AVE
TUCSON AZ 85726
U.S.A.
TEL: 602-327-5511
TLX: 0666-484 AURA NOAO T
TLF:
EML:
COM:

HINTEREGGER HANS E DR
AIR FORCE GEOPHYSICS LAB.
HANSCOM FIELD
BEDFORD MA 01731
U.S.A.
TEL:
TLX:
TLF:
EML:
COM: 44

HINTZEN PAUL MICHAEL N DR
NASA/GSFC
CODE 681
GREENBELT MD 20771
U.S.A.
TEL: 301-286-5101
TLX:
TLF:
EML:
COM: 28

HIPPELEIN HANS H DR
MPI FUER ASTRONOMIE
KOENIGSTUHL
D-6900 HEIDELBERG
GERMANY. F.R.
TEL:
TLX:
TLF:
EML:
COM: 34

HIRABAYASHI HISASHI DR
NOBEYAMA RADIO OBS
NOBEYAMA
MINAMIMAKI
NAGANO 384-13
JAPAN
TEL: 2679-8-2831
TLX: 3329005
TLF:
EML:
COM: 40.51

HIRAI MASANORI DR
FUKUOKA UNIVERSITY
OF EDUCATION
729 MUNAKATA
FUKUOKA 811-41
JAPAN
TEL: 094-032-2381
TLX:
TLF:
EML:
COM: 29

HIRATA RYUKO
DEPT OF ASTRONOMY
KYOTO UNIVERSITY
SAKYO-KU
KYOTO 606
JAPAN
TEL:
TLX:
TLF:
EML:
COM: 29

HIRAYAMA TADASHI PROF
TOKYO ASTRONOMICAL OBS
OSAWA MITAKA
TOKYO 181
JAPAN
TEL: 0422-32-5111
TLX: 2822307 TAOMTK-J
TLF:
EML:
COM: 10.12

HIRST WILLIAM P
1 CLIFFORD CRESCENT
BERGVLIET 7945
SOUTH AFRICA
TEL:
TLX:
TLF:
EML:
COM:

HIRTH WOLFGANG ERNST PH D
THEODOR-HEUS-STR 18
D-5354 WEILERSWIST
GERMANY. F.R.
TEL:
TLX:
TLF:
EML:
COM:

HITOTSUYANAGI JUICHI PROF
KATAHIRA 1-CHOME
4-6-402
SENDAI 980
JAPAN
TEL: 0222-27-9351
TLX:
TLF:
EML:
COM: 35.36

HJALMARSON AKE G DR
ONSALA SPACE OBSERVATORY
S-439 00 ONSALA
SWEDEN
TEL: 300-60651
TLX: 2400 ONSPACE
TLF:
EML:
COM: 28.34.40

HJELLMING ROBERT M DR
NRAO
PO BOX 0
SOCORRO NM 87801-0387
U.S.A.
TEL: 505-835-7273
TLX: 910-988-1710
TLF:
EML: BITNET:RHJELLMI@NRAO.edu
COM: 34.40.42

HO PAUL T P
HARVARD UNIVERSITY
60 GARDEN STREET
CAMBRIDGE MA 02138
U.S.A.
TEL: 617-495-3627
TLX: 921428
TLF:
EML:
COM: 40

HOAG ARTHUR A DR
4410 E. 14TH STREET
TUCSON AZ 85711
U.S.A.
TEL: 602-795-8644
TLX:
TLF:
EML:
COM: 50C

HOANG BINH DY DR
OBSERVATOIRE DE PARIS
SECTION DE MEUDON
LAN
F-92195 MEUDON PL CEDEX
FRANCE
TEL: 1-45-07-74-45
TLX: 201571
TLF:
EML:
COM: 40.48.51

HOBBS LEWIS M DR
YERKES OBSERVATORY
UNIVERSITY OF CHICAGO
BOX 258
WILLIAMS BAY WI 53191
U.S.A.
TEL: 414-245-5555
TLX:
TLF:
EML:
COM: 34

HOBBS ROBERT W DR
COMPUTER TECHN ASSOCIATES
1 MARYLAND CORPORATE CTR
7501 FORBES BLVD/S. 201
LANHAM MD 20706
U.S.A.
TEL: 301-464-5300
TLX:
TLF:
EML:
COM: 33.40

HODGE PAUL W PROF
ASTRONOMY FM20
UNIVERSITY OF WASHINGTON
SEATTLE WA 98195
U.S.A.
TEL: 206-543-2888
TLX: 9104740096
TLF:
EML:
COM: 22.28

HOEG ERIK DR
UNIVERSITY OBSERVATORY
OESTER VOLDGADE 3
DK-1350 COPENHAGEN K
DENMARK
TEL: 1-14-17-90
TLX: 44155 DANAST
TLF:
EML:
COM: 08

HOEGBOM JAN A DR
STOCKHOLM OBSERVATORY
S-133 00 SALTSJOEBADEN
SWEDEN
TEL: 08-7170195
TLX: 12972 SOBSERV S
TLF:
EML:
COM: 40

HOEGLUND BERTIL PROF
ONSALA SPACE OBSERVATORY
S-439 00 ONSALA
SWEDEN
TEL: 0300-60652
TLX: 2400
TLF:
EML:
COM: 34.40

HOEKSTRA ROEL DR
TPD/TNO/TH
POSTBUS 155
NL-2600 AD DELFT
NETHERLANDS
TEL:
TLX:
TLF:
EML:
COM:

HOESSEL JOHN GREG
WASHBURN OBSERVATORY
UNIV OF WISCONSIN-MADISON
475 N. CHARTER STREET
MADISON WI 53706
U.S.A.
TEL: 608-262-1752
TLX:
TLF:
EML:
COM:

HOEY MICHAEL J DR
PHYSICS DEPT
UNIVERSITY COLLEGE
BELFIELD
DUBLIN 4
IRELAND
TEL:
TLX:
TLF:
EML:
COM:

HOFF DARREL BARTON
PROJECT STAR
CENTER FOR ASTROPHYSICS
60 GARDEN STREET
CAMBRIDGE MA 02138
U.S.A.
TEL: 617-495-9798
TLX:
TLF:
EML:
COM: 46

HOFFLEIT E DORRIT DR
DEPT OF ASTRONOMY
YALE UNIVERSITY
BOX 6666
NEW HAVEN CT 06511
U.S.A.
TEL:
TLX:
TLF:
EML:
COM: 24.27

HOFFMAN JEFFREY ALAN DR
NASA-JSC
CODE CB-4
HOUSTON TX 77058
U.S.A.
TEL: 713-483-2411
TLX:
TLF:
EML:
COM: 44.48

HOFFMANN MARTIN DR
OBSERVATORIUM HOHER LIST
STERNWARTE DER
UNIVERSITAET BONN
D-5568 DAUN
GERMANY. F.R.
TEL:
TLX:
TLF:
EML:
COM: 42

HOFMANN WILFRIED DR
ASTRONOMISCHES RECHEN-
INSTITUT
MOENCHHOFSTR 12-14
D-6900 HEIDELBERG 1
GERMANY. F.R.
TEL: 06221-49026
TLX: 461336 ARIHD D
TLF:
EML:
COM: 21

HOGAN CRAIG J DR
STEWARD OBSERVATORY
UNIVERSITY OF ARIZONA
TUCSON AZ 85721
U.S.A.
TEL: 602 621 6533
TLX: 467 175
TLF:
EML:
COM:

HOGG DAVID E DR
NRAO
EDGEMONT RD
CHARLOTTESVILLE VA 22903
U.S.A.
TEL: 804-296-0220
TLX: 910-997-0174
TLF:
EML:
COM: 40

HOLBERG JAY B
LUNAR/PLANETARY LAB. WEST
UNIVERSITY OF ARIZONA
901 GOULD-SIMPSON BUILD.
TUCSON AZ 85721
U.S.A.
TEL: 602-621-4303
TLX: 9109521143
TLF:
EML:
COM: 16.44

HOLDEN FRANK
2 COLWICH CRESCENT
KINGSTON HILL
STAFFORD ST16 3XP
U.K.
TEL: 0785-53120
TLX:
TLF:
EML:
COM: 26

HOLLENBACH DAVID JOHN DR
NASA/AMES RESEARCH CENTER
MS 245-6
MOFFETT FIELD CA 94035
U.S.A.
TEL: 415-997-6426
TLX:
TLF:
EML:
COM: 34

HOLLIS JAN MICHAEL DR
NASA/GSFC
SCIENCE OPERATIONS BRANCH
CODE 684
GREENBELT MD 20771
U.S.A.
TEL: 301-286-7591
TLX:
TLF:
EML:
COM: 34.40.51

HOLLOWAY NIGEL J DR
SAFETY & RELIABILITY DIR.
WIGSHAW LANE
CULCHETH
WARRINGTON WA3 4NE
U.K.
TEL: 095-31244
TLX: 629301
TLF:
EML:
COM: 48

HOLLWEG JOSEPH V
DEPT OF PHYSICS
UNIV OF NEW HAMPSHIRE
DEMERITT HALL
DURHAM NH 03824
U.S.A.
TEL: 603-862-3869
TLX:
TLF:
EML:
COM: 49

HOLMAN GORDON D
NASA/GSFC
CODE 682
GREENBELT MD 20771
U.S.A.
TEL: 301-286-7921
TLX:
TLF:
EML:
COM: 10

HOLMBERG ERIK B PROF
EKELIDEN 2
S-433 00 PARTILLE
SWEDEN
TEL: 031-265842
TLX:
TLF:
EML:
COM: 25.28

HOLT STEPHEN S
NASA/GSFC
CODE 660
GREENBELT MD 20771
U.S.A.
TEL: 301-286-8801
TLX:
TLF:
EML:
COM: 42.44

HOLWEGER HARTMUT PROF
INST THEOR PHYS & STERNW
UNIVERSITAET KIEL
OLSHAUSENSTR
D-2300 KIEL
GERMANY. F.R.
TEL: 8804107 KIEL
TLX: 292706 IAPKI D
TLF:
EML:
COM: 12.36

HOLZER THOMAS EDWARD DR
HIGH ALTITUDE OBSERVATORY
NCAR
PO BOX 3000
BOULDER CO 80307
U.S.A.
TEL: 303-497-1536
TLX: 45694
TLF:
EML:
COM: 10.36.49

HONEYCUTT R KENT PROF
ASTRONOMY DEPT
INDIANA UNIVERSITY
SWAIN HALL WEST
BLOOMINGTON IN 47405
U.S.A.
TEL: 812-335-6916
TLX:
TLF:
EML:
COM: 09.42

HONG HYON IK
FACULTY OF PHYSICS
KIM IL SUNG UNIVERSITY
TAESONG DISTRICT
PYONGYANG
KOREA DPR
TEL:
TLX:
TLF:
EML:
COM: 10

HONG SEUNG SOO DR
DEPT OF ASTRONOMY
COLLEGE NATURAL SCIENCES
SEOUL NATIONAL UNIVERSITY
SEOUL 151
KOREA. REPUBLIC
TEL: 877-2131
TLX: 29664
TLF:
EML:
COM: 21.22.34

HOOD ALAN
APPLIED MATHS DEPT
THE UNIVERSITY
ST ANDREWS. FIFE
U.K.
TEL: 0334-76161
TLX: 76213
TLF:
EML:
COM: 10

HOOD ALAN DR
DEPT MATHEMATICAL SCIENCE
UNIVERSITY OF ST ANDREWS
ST ANDREWS
FIFE KY16 9SS
U.K.
TEL: 0334 76213
TLX: 932110846 SA G
TLF:
EML: AHSAH@SAVB.ST-AND.AC.UK.
COM:

HOOGHOUDT B G IR
PRINSENLAAN 10
NL-2341 KT OEGSTGEEST
NETHERLANDS
TEL: 071-172524
TLX:
TLF:
EML:
COM: 09.40

HORAK TOMAS B DR
VIRK
BIECNA 1
815 58 BRATISLAVA
CZECHOSLOVAKIA
TEL: 07-338451
TLX:
TLF:
EML:
COM: 42

HORAK ZDENEK PROF DR
VIETNAMSKA 2
160 00 PRAHA 6
CZECHOSLOVAKIA
TEL:
TLX:
TLF:
EML:
COM:

HOREDT GEORG PAUL DR
DFVLR
D-8031 WESSLING
GERMANY. F.R.
TEL:
TLX:
TLF:
EML:
COM: 16

HORI GENICHIRO PROF
DEPT OF ASTRONOMY
UNIVERSITY OF TOKYO
BUNKYO
TOKYO 113
JAPAN
TEL: 03-8122111x4251
TLX: 33659 UTYOSCI J
TLF:
EML:
COM: 07.33

HORIUCHI RITOKU DR
INST. FOR FUSION THEORY
HIROSHIMA UNIVERSITY
HIGASHISENDA-MACHI.NAKAKU
HIROSHIMA 730
JAPAN
TEL: 82 241 1221
TLX: 652712 HIFT J
TLF:
EML:
COM: 42

HOROWITZ PAUL PROF
DEPT OF PHYSICS
HARVARD UNIVERSITY
CAMBRIDGE MA 02138
U.S.A.
TEL: 617-495-3265
TLX: 4992111
TLF:
EML:
COM: 51

HORSKY JAN PROF
DEPT OF THEORETICAL PHYS
PURKYNE UNIVERSITY
KOTLARSKA 2
611 37 BRNO
CZECHOSLOVAKIA
TEL: 51112
TLX:
TLF:
EML:
COM:

HORTON BRIAN H DR
JACABRI ENT.
PO BOX 309
GOOLWA S.A. 5214
AUSTRALIA
TEL: 085 553376
TLX:
TLF:
EML:
COM: 12

HORVATH ANDRAS DR
TIT PLANETARIUM &
URANIA OBSERVATORY
BOX 46
H-1476 BUDAPEST
HUNGARY
TEL: 334-525
TLX:
TLF:
EML:
COM:

HORWITZ GERALD PROF
RACAH INST. OF PHYSICS
HEBREW UNIV. OF JERUSALEM
JERUSALEM 91904
ISRAEL
TEL: 584592
TLX:
TLF:
EML:
COM:

HOSHI REIUN DR
DEPT OF PHYSICS
RIKKYO UNIVERSITY
NISHI-IKEBUKURO 3-CH
TOSHIMA-KU TOKYO 171
JAPAN
TEL: 03-985-2414
TLX:
TLF:
EML:
COM: 35

HOSKIN MICHAEL A DR
CHURCHILL COLLEGE
CAMBRIDGE CB3 ODS
U.K.
TEL: 0223-358381
TLX:
TLF:
EML:
COM: 41

HOSKING ROGER J PROF
JAMES COOK UNIVERSITY
OF NORTH QUEENSLAND
TOWNSVILLE Q 4811
AUSTRALIA
TEL: 077 81 41113
TLX: AA47009
TLF:
EML:
COM:

HOSOKAWA YOSHIMASA H PROF
SAKIGAOKA 3-4-9
FUNABASHI-CITY
CHIBA PREFECTURE 274
JAPAN
TEL: 0474-48-6679
TLX:
TLF:
EML:
COM:

HOSOYAMA KENNOSHUKE DR
INTL LATITUDE OBSERVATORY
HOSHIGAOKA 2-CHOME
MIZUSAWA IWATE 023
JAPAN
TEL:
TLX:
TLF:
EML:
COM: 19

HOTINLI METIN DR
UNIVERSITE
RASATHANESI
BEYAZIT. ISTANBUL
TURKEY
TEL:
TLX:
TLF:
EML:
COM: 12.36

HOUCK JAMES R
ASTRONOMY DEPT
CORNELL UNIVERSITY
220 SPACE SCIENCE BLDG
ITHACA NY 14853
U.S.A.
TEL: 607-256-4806
TLX: 937478 ITCA
TLF:
EML:
COM: 21C

HOUK NANCY DR
DEPT OF ASTRONOMY
UNIVERSITY OF MICHIGAN
1045 PHYS-ASTRO BLDG
ANN ARBOR MI 48109
U.S.A.
TEL: 313-764-3436
TLX:
TLF:
EML:
COM: 27.45C

HOUSE FRANKLIN C DR
HEIDENREICHSTR 42
D-6100 DARMSTADT
GERMANY. F.R.
TEL: 06151-422412
TLX:
TLF:
EML:
COM:

HOUSE LEWIS L DR
HIGH ALTITUDE OBSERVATORY
NCAR
PO BOX 3000
BOULDER CO 80303
U.S.A.
TEL: 303-494-5151
TLX:
TLF:
EML:
COM: 12.14.36

HOUZIAUX L PROF
INSTITUT D'ASTROPHYSIQUE
UNIVERSITE DE LIEGE
AVENUE DE COINTE 5
B-4200 COINTE-OUGREE
BELGIUM
TEL: 31-41570180x494
TLX: 41264 ASTRLG B
TLF:
EML:
COM: 29.34.44.46C

HOVENIER J W DR
FREE UNIVERSITY
DEPT PHYSICS & ASTRONOMY
DE BOELELAAN 1081
NL-1081 HV AMSTERDAM
NETHERLANDS
TEL: 020-540-2414
TLX:
TLF:
EML:
COM: 16

HOWARD ROBERT F DR
NATL SOLAR OBSERVATORY
PO BOX 26732
TUCSON AZ 85726-6732
U.S.A.
TEL: 602-327-5511
TLX: 0666484 AURA NOAOTUC
TLF:
EML:
COM: 10.12

HOWARD W MICHAEL DR
L. LIVERMORE NATIONAL LAB
L-297
LIVERMORE CA 94550
U.S.A.
TEL: 415-422-4138
TLX:
TLF:
EML:
COM:

HOWARD WILLIAM E III DR
US NAVAL SPACE COMMAND
31 WOODLAWN TERRACE
FREDERICKSBURG
VA 22405-3358
U.S.A.
TEL: 703-663-7841
TLX:
TLF:
EML:
COM: 40

HOWARTH IAN DONALD
DEPT PHYSICS & ASTRONOMY
UNIVERSITY COLLEGE LONDON
GOWER STREET
LONDON WC1E 6BT
U.K.
TEL: 01-387-7050
TLX: 28722
TLF:
EML:
COM: 44

HOWSE M DEREK
12 BARNFIELD ROAD
RIVERHEAD
SEVENOAKS. KENT TN13 2AY
U.K.
TEL: 0732-454366
TLX:
TLF:
EML:
COM: 41

HOYLE FRED SIR
102 ADMIRALS WALK
WEST CLIFF ROAD
BOURNEMOUTH
DORSET BH2 5HF
U.K.
TEL:
TLX:
TLF:
EML:
COM: 28.35.47.48

HOYNG PETER DR
SPACE RESEARCH LABORATORY
BENELUXLAAN 21
NL-3527 HS UTRECHT
NETHERLANDS
TEL: 030-937-145
TLX: 47224 ASTRO NL
TLF:
EML:
COM: 10.12.44

HRIVNAK BRUCE J
VALPARAISO UNIVERSITY
PHYSICS DEPT
VALPARAISO IN 46383
U.S.A.
TEL: 219-464-5379
TLX:
TLF:
EML:
COM: 30.42

HRON JOSEF DR
INSTITUT FUR ASTRONOMIE
UNIVERSITAET WIEN
TUERKENSCHANZSTRASSE 17
A-1180 WIEN
AUSTRIA
TEL: (0222)345360
TLX: 133099 viast a
TLF: USERID:A8201DAB
EML: EARN:AWIUNI11@A8201DAB
COM: 33

HSIANG YAN-YU
BEIJING ASTRONOMICAL OBS
ACADEMIA SINICA
BEIJING
CHINA. PEOPLE'S REP.
TEL: 281698
TLX: 22040 BAOAS CN
TLF:
EML:
COM:

HSIANG-KUANG TSENG
INST PHYSICS & ASTRONOMY
NATL CENTRAL UNIVERSITY
CHUNG LI
CHINA. TAIWAN
TEL:
TLX:
TLF:
EML:
COM:

HU ESTHER M DR
INSTITUTE FOR ASTRONOMY
2680 WOODLAWN DRIVE
HONOLULU HI 96822
U.S.A.
TEL: 808 948 7190
TLX: 723 8459 UHAST HR
TLF:
EML: INT:hu@uhifa.ifa.hawaii.edu
COM: 47

HU FU-XING
PURPLE MOUNTAIN
OBSERVATORY ACADEMIA
SINICA
NANKING
CHINA. PEOPLE'S REP.
TEL:
TLX:
TLF:
EML:
COM: 28

HU JING-YAO
BEIJING OBSERVATORY
BEIJING
CHINA. PEOPLE'S REP.
TEL: 281698
TLX: 22040 BAOAS CN
TLF:
EML:
COM: 09.25

HU NING-SHENG
NANJING ASTRONOMICAL
INSTRUMENT FACTORY
NANJING JIANGSU PROVINCE
CHINA. PEOPLE'S REP.
TEL: 46191
TLX: 34136 GLYNJ CN :NAIF
TLF:
EML:
COM: 08C.09

HU WEN-RUI
INSTITUTE OF MECHANICS
ACADEMIA SINICA
BEIJING
CHINA. PEOPLE'S REP.
TEL: 28-4185
TLX: 22474 ASCHI CN
TLF:
EML:
COM: 44

HU ZHONG-WEI
DEPT OF ASTRONOMY
NANJING UNIVERSITY
NANJING
CHINA. PEOPLE'S REP.
TEL: 37651
TLX: 0909
TLF:
EML:
COM: 15.16

HUA CHON TRUNG DR
L.A.S.
LES TROIS LUCS
IMPASSE DU SIPHON
13012 MARSEILLE
FRANCE
TEL: 91 055932
TLX: 420584 F
TLF:
EML:
COM: 28.34

HUA YING-MIN
P.O. BOX 18
LINTONG
XIAN
CHINA. PEOPLE'S REP.
TEL:
TLX:
TLF:
EML:
COM: 08.19

HUANG BI-KUN
LIBRARY
PURPLE MOUNTAIN OBS
ACADEMIA SINICA
NANJING
CHINA. PEOPLE'S REP.
TEL: 307521
TLX: 34144 PMONJ CN
TLF:
EML:
COM: 05

HUANG CHANG-CHUN
PURPLE MOUNTAIN OBS
NANJING
CHINA. PEOPLE'S REP.
TEL: 46700/42817
TLX: 34144 PMONJ CN
TLF:
EML:
COM: 29.30

HUANG CHENG DR
SHANGHAI OBSERVATORY
ACADEMIA SINICA
SHANGHAI
CHINA. PEOPLE'S REP.
TEL:
TLX:
TLF:
EML:
COM: 07

HUANG JIE-HAO
ASTROPHYSICS INSTITUTE
NANJING UNIVERSITY
NANJING
CHINA. PEOPLE'S REP.
TEL:
TLX: 34151 PRCNU CN
TLF:
EML:
COM: 28

HUANG KE-LIANG
ASTRONOMY DEPARTMENT
NANJING UNIVERSITY
NANJING
CHINA. PEOPLE'S REP.
TEL: 34651 EXT 2882
TLX: 34151 PRCNU CN
TLF:
EML:
COM: 28.48

HUANG KUN-YI
PURPLE MOUNTAIN
OBSERVATORY
NANJING
CHINA. PEOPLE'S REP.
TEL: 32893
TLX: 34144 PMONJ CN
TLF:
EML:
COM:

HUANG LIN
BEIJING ASTR OBSERVATORY
ACADEMIA SINICA
BEIJING
CHINA. PEOPLE'S REP.
TEL: 28-16-98
TLX: 22040 BAOAS CN
TLF:
EML:
COM: 25.45

HUANG RUN-QIAN
YUNNAN OBSERVATORY
KUNMING. YUNNAN PROVINCE
CHINA. PEOPLE'S REP.
TEL:
TLX: 64040 YUOBS CN
TLF:
EML:
COM: 35.42

HUANG SONG-NIAN DR
SHANGHAI OBSERVATORY
80 NAN DAN ROAD
SHANGHAI
CHINA. PEOPLE'S REP.
TEL:
TLX:
TLF:
EML:
COM: 28.33

HUANG TIANYI
DEPT OF ASTRONOMY
NANJING UNIVERSITY
NANJING
CHINA. PEOPLE'S REP.
TEL:
TLX: 34151 PRCNU CN
TLF:
EML:
COM: 07

HUANG YIN-QIN
NANJING ASTRONOMICAL
INSTRUMENT FACTORY
NANJING
CHINA. PEOPLE'S REP.
TEL: 46191
TLX: 34136 GLYNJ CN
TLF:
EML:
COM: 09

HUANG YINN-NIEN DR
TELECOM TRAINING INST
MINISTRY OF COMMUNICATION
168 MINCHU RD.. PANCHIAO
TAIPEI 22077
CHINA. TAIWAN
TEL: 02 9639260
TLX: 31202 TELTRAINS
TLF:
EML:
COM:

HUANG YIN-LIANG
TIANJIN INSTITUTE OF
TECHNOLOGY
123 MACHANG ROAD
TIANJIN
CHINA. PEOPLE'S REP.
TEL:
TLX:
TLF:
EML:
COM: 50

HUANG YI-LONG DR
INST OF HISTORY
NAT. TSING HUA UNIVERSITY
HSINCHU 30043
CHINA. TAIWAN
TEL: 035 716780
TLX:
TLF:
EML:
COM:

HUANG YONGWEI
BEIJING ASTRONOMICAL OBS.
ACADEMIA SINICA
BEIJING
CHINA. PEOPLE'S REP.
TEL: 281698
TLX: 9053
TLF:
EML:
COM: 28

HUANG YOU-RAN
DEPT OF ASTRONOMY
NANJING UNIVERSITY
NANJING
CHINA. PEOPLE'S REP.
TEL:
TLX: 34151
TLF:
EML:
COM: 10

HUBBARD WILLIAM B PROF
PLANETARY SCIENCES DEPT
UNIVERSITY OF ARIZONA
TUCSON AZ 85721
U.S.A.
TEL: 602-621-6942
TLX: 9109521143
TLF:
EML:
COM: 16

HUBE DOUGLAS P DR
DEPT OF PHYSICS
UNIVERSITY OF ALBERTA
EDMONTON ALB T6G 2J1
CANADA
TEL: 403-432-5410
TLX:
TLF:
EML:
COM: 30.42

HUBENET HENRI DR
STERREKUNDIG INSTITUUT
PRINCETONPLEIN 5
POSTBUS 80 000
NL-3508 TA UTRECHT
NETHERLANDS
TEL: 030-535200
TLX: 40048 FYLUT NL
TLF:
EML:
COM:

HUBENY IVAN
JILA
UNIVERSITY OF COLORADO
CAMPUS BOX 440
BOULDER CO 80309-0440
U.S.A.
TEL: 303-492 7838
TLX: 755842 JILA
TLF:
EML:
COM: 29.36

HUBER MARTIN C E DR
ESA - SPACE SCIENCE DEPT
P.O. BOX 299
2200 AG NOORDWIJK
NETHERLANDS
TEL: 31-1719-83552
TLX: 39098
TLF:
EML:
COM: 14.44

HUBERT HENRI DR
DASGAL
OBSERVATOIRE DE MEUDON
92195 MEUDON PR. CEDEX
FRANCE
TEL: 1-45 07 78 50
TLX:
TLF:
EML:
COM: 29

HUBERT-DELPLACE A.-M. DR
OBSERVATOIRE DE PARIS
SECTION DE MEUDON
F-92195 MEUDON PL CEDEX
FRANCE
TEL: 1-45-34-78-56
TLX: 270912 OBSASTR
TLF:
EML:
COM: 29

HUCHRA JOHN PETER DR
CENTER FOR ASTROPHYSICS
60 GARDEN STREET
CAMBRIDGE MA 02138
U.S.A.
TEL: 617-495-7375
TLX: 921428 SATELLITE CAM
TLF:
EML:
COM: 28.30.47

HUCHTMEIER WALTER K DR
MPI FUER RADIOASTRONOMIE
AUF DEM HUEGEL 69
D-5300 BONN 1
GERMANY. F.R.
TEL: 228-525-215
TLX: 886440 MPIFR D
TLF:
EML:
COM: 28.40

HUDEC RENE DR
ASTRONOMICAL INSTITUTE
251 65 ONDREJOV
CZECHOSLOVAKIA
TEL: (204)85201
TLX: 121579
TLF:
EML:
COM:

HUDSON HUGH S DR
PHYSICS C-011
UCSD
LA JOLLA CA 92093
U.S.A.
TEL: 619-452-4476
TLX:
TLF:
EML:
COM: 10

HUEBNER WALTER F DR
SOUTHWEST RESEARCH INST.
6220 CULEBRA ROAD
P.O.DRAWER 28510
SAN ANTONIO TX 78284
U.S.A.
TEL: 512-522-2730
TLX: 244846
TLF:
EML:
COM: 14.15

HUENEMOERDER DAVID P DR
DEPARTMENT OF ASTRONOMY
525 DAVEY LABORATORY
PENNSYLVANIA STATE UNIV.
UNIVERSITY PARK PA 16802
U.S.A.
TEL: 814 865 6601
TLX:
TLF:
EML: INTERNET:dph@astro.psu.edu
COM: 27.29

HUGHES DAVID W DR
DEPT OF PHYSICS
THE UNIVERSITY
SHEFFIELD S3 7RH
U.K.
TEL: 0742-78555
TLX: 54348 ULSHEF G
TLF:
EML:
COM: 15C.22

HUGHES JAMES A DR
US NAVAL OBSERVATORY
WASHINGTON DC 20390
U.S.A.
TEL:
TLX:
TLF:
EML:
COM: 08C.24

HUGHES JOHN P DR
HARVARD-SMITHSONIAN CTR
FOR ASTROPHYSICS
60 GARDEN STREET
CAMBRIDGE MA 02138
U.S.A.
TEL: 617 495 7142
TLX: 921428 SATELLITE CAM
TLF:
EML: BITNET:HUGHES@CFA.
COM:

HUGHES PHILIP
DEPT OF ASTRONOMY
UNIVERSITY OF MICHIGAN
ANN ARBOR MI 48109-1090
U.S.A.
TEL: 313-764-3430
TLX:
TLF:
EML:
COM: 40

HUGHES SHAUN
U.K. SCHMIDT TELESCOPE
ANGLO AUSTRAL. OBSERV.
PRIVATE BAG
COONABARABRAN NSW 2357
AUSTRALIA
TEL: 068 421622
TLX:
TLF:
EML: PSI%AAOCBN::SMH
COM:

HUGHES VICTOR A PROF
DEPT OF PHYSICS
QUEEN'S UNIVERSITY
KINGSTON ONT K7L 3N6
CANADA
TEL: 613-547-6633
TLX:
TLF:
EML:
COM: 33.34.40

HUGUENIN G RICHARD
MULTITECH CORPORATION
PO BOX 109
SOUTH DEERFIELD RES PARK
SOUTH DEERFIELD MA 01373
U.S.A.
TEL: 413-665-8551
TLX: 3719862 TRUB
TLF:
EML:
COM:

HULSBOSCH A N M DR
STERRENKUNDIG INSTITUT
KATHOLIEKE UNIVERSITEIT
TOERNOOIVELD
NL-6525 ED NIJMEGEN
NETHERLANDS
TEL: 080-558833
TLX: 48228
TLF:
EML:
COM: 33.34.40

HUMMEL EDSKO
NRAL JODRELL BANK
MACCLESFIELD
CHESHIRE SK11 9DL
U.K.
TEL:
TLX:
TLF:
EML:
COM: 28

HUMMER DAVID G DR
JILA
UNIVERSITY OF COLORADO
BOX 440
BOULDER CO 80309
U.S.A.
TEL: 303-492-7837
TLX: 755842 JILA
TLF:
EML:
COM: 34.36

HUMPHREYS ROBERTA M PROF
ASTRONOMY DEPT
UNIVERSITY OF MINNESOTA
116 CHURCH STREET S.E.
MINNEAPOLIS MN 55455
U.S.A.
TEL: 612-373-9747
TLX:
TLF:
EML:
COM: 28.33.35.45

HUMPHRIES COLIN M DR
RORAL OBSERVATORY
BLACKFORD HILL
EDINBURG EH9 3HJ
U.K.
TEL: 316673321
TLX: 72383
TLF:
EML:
COM: 09C

HUNDHAUSEN ARTHUR DR
HIGH ALTITUDE OBSERVATORY
PO BOX 3000
BOULDER CO 80302
U.S.A.
TEL:
TLX:
TLF:
EML:
COM:

HUNGER KURT PROF
INST THEOR PHYS & STERNW
NEUE UNIV PHYSIK ZENTRUM
OLSHAUSENST 40 N61C
D-2300 KIEL 1
GERMANY. F.R.
TEL: 0431-880-4110
TLX: 292706IAOKI
TLF:
EML:
COM: 29.36

HUNSTEAD RICHARD W DR
SCHOOL OF PHYSICS
UNIVERSITY OF SYDNEY
SYDNEY NSW 2006
AUSTRALIA
TEL: 02-692-3871
TLX: 26169 UNISYD AA
TLF:
EML:
COM: 28.40

HUNTEN DONALD M PROF
LUNAR AND PLANETARY LAB
UNIVERSITY OF ARIZONA
TUCSON AZ 85721
U.S.A.
TEL: 602-621-4002
TLX:
TLF:
EML:
COM: 16.51

HUNTER CHRISTOPHER PROF
MATHEMATICS DEPT
FLORIDA STATE UNIVERSITY
TALLAHASSEE FL 32306
U.S.A.
TEL: 904-644-2488
TLX:
TLF:
EML:
COM: 28.33

HUNTER DEIDRE ANN
DEPT TERRESTR. MAGNETISM
CARNEGIE INST. WASHINGTON
5241 BROAD BRANCH RD N.W.
WASHINGTON DC 20015
U.S.A.
TEL: 202-966-0863
TLX: 440427 MAGN UI
TLF:
EML:
COM:

HUNTER JAMES H PROF
DEPT OF ASTRONOMY
T W BRYANT SPACE SCI BLDG
UNIVERSITY OF FLORIDA
GAINESVILLE FL 32611
U.S.A.
TEL: 904-392-1078
TLX:
TLF:
EML:
COM: 28.51

HURFORD GORDON JAMES
CALTECH 264-33
PASADENA CA 91125
U.S.A.
TEL: 818-356-3866
TLX: 675425
TLF:
EML:
COM: 10

HURNIK HIERONIM PROF
ASTRONOMICAL OBSERVATORY
A MICKIEWICZ UNIVERSITY
SLONECZNA 36
60-286 POZNAN
POLAND
TEL: 679-670
TLX:
TLF:
EML:
COM: 20

HUT PIET
INST. FOR ADVANCED STUDY
PRINCETON NJ 08540
U.S.A.
TEL: 609-734-8075
TLX: 229734 IAS UR
TLF:
EML:
COM: 37

HUTCHEON RICHARD J DR
X-RAY ASTRONOMY GROUP
PHYSICS DEPT
UNIVERSITY OF LEICESTER
LEICESTER LE1 7RH
U.K.
TEL:
TLX:
TLF:
EML:
COM:

HUTCHINGS JOHN B DR
DOMINION ASTROPHYS OBS
5071 W SAANICH ROAD
VICTORIA BC V8X 4M6
CANADA
TEL: 604-388-3909
TLX: 049-7295
TLF:
EML:
COM: 27.34.36.42.44C

HYDER C L DR
HIGH ALTITUDE OBSERVATORY
PO BOX 3000
BOULDER CO 80307
U.S.A.
TEL:
TLX:
TLF:
EML:
COM: 10

HYLAND A R HARRY DR
MT STROMLO OBSERVATORY
WODEN P.O. ACT 2606
AUSTRALIA
TEL: 062-881111
TLX: 62270 CANOPUS AA
TLF:
EML:
COM: 25.29

HYSON EDMUND J
8 EAST DRIVE
CALDECOTE
CAMBRIDGE CB3 7NZ
U.K.
TEL: 954-211137
TLX:
TLF:
EML:
COM: 09.41.51

HYUN JONG-JUNE PROF
SEOUL NATIONAL UNIVERSITY
SINLIM-DONG
KWANAK-KU
SEOUL 151
KOREA. REPUBLIC
TEL: 877-3010/2542
TLX:
TLF:
EML:
COM:

IANNA PHILIP A
LEANDER MCCORMIC OBS.
UNIV. OF VIRGINIA
PO BOX 3818 UNIV STATION
CHARLOTTESVILLE VA 22903
U.S.A.
TEL: 804-924-4898
TLX:
TLF:
EML:
COM: 20.24C.26

IANNINI GUALBERTO DR
OBSERVATORIO ASTRONOMICO
LAPRIDA 854
5000 CORDOBA
ARGENTINA
TEL:
TLX:
TLF:
EML:
COM:

IBADINOV KHURSANDKUL DR
INSTITUTE OF ASTROPHYSICS
TADJIK ACAD OF SCIENCES
734670 DUSHANBE
U.S.S.R.
TEL:
TLX:
TLF:
EML:
COM: 15

IBANEZ S. MIGUEL H. DR
UNIVERSIDAD DE LOS ANDES
FACULDAD DE CIENCIAS
DEPTO DE FISICA
MERIDA
VENEZUELA
TEL: 639930/637477
TLX: 74174 CIDA
TLF:
EML:
COM:

IBANOGLU C DR
EGE UNIVERSITY
FACULTY OF SCIENCE
BORNOVA-IZMIR
TURKEY
TEL: 180110-2332
TLX:
TLF:
EML:
COM: 42

IBBETSON PETER AARON DR
WISE OBSERVATORY
TEL-AVIV UNIVERSITY
RAMAT AVIV
TEL-AVIV 69978
ISRAEL
TEL: 972-3-413788
TLX: 342171 VERSY IL
TLF:
EML:
COM:

IBEN ICKO JR PROF
ASTR DEPT/UNIV ILLINOIS
349 ASTRONOMY BLDG
1011 W SPRINGFIELD AVE
URBANA IL 61801
U.S.A.
TEL: 217-333-3090
TLX: 9102452434 AST
TLF:
EML:
COM: 27.35C.37

IBRAHIM JORGA
DEPT OF ASTRONOMY
INSTITUTE OF TECHNOLOGY
JALAN TAMANSARI 64
BANDUNG
INDONESIA
TEL:
TLX:
TLF:
EML:
COM:

ICHIKAWA TAKASHI
DEPT OF ASTRONOMY
UNIVERSITY OF KYOTO
KITASHIRAKAWA.SAKYOKU
KYOTO 606
JAPAN
TEL: 75 751 3890
TLX:
TLF:
EML:
COM: 28

ICHIMARU SETSUO DR
DEPT OF PHYSICS
UNIVERSITY OF TOKYO
BUNKYO-KU
TOKYO 113
JAPAN
TEL: 03-812-2111
TLX: UTPHYSIC J23472
TLF:
EML:
COM: 48

ICKE VINCENT DR
STERREWACHT LEIDEN
POSTBUS 9513
NL-2300 RA LEIDEN
NETHERLANDS
TEL: 071-272727
TLX: 39058 ASTRO NL
TLF:
EML:
COM: 47

IDLIS G M DR
INSTITUTE FOR HISTORY OF
SCIENCES AND TECHNOLOGY
USSR ACADEMY OF SCIENCES
103012 MOSCOW
U.S.S.R.
TEL: 2281969
TLX:
TLF:
EML:
COM: 41.51

IIJIMA SHIGETAKA PROF
MUSASHI INSTITUTE OF
TECHNOLOGY
TAMAZUTSUMI. SETAGAYA-KU
TOKYO 158
JAPAN
TEL: 03-703-3111
TLX:
TLF:
EML:
COM: 19.31

IKEUCHI SATORU DR
TOKYO ASTRONOMICAL OBS
UNIVERSITY OF TOKYO
MITAKA TOKYO 181
JAPAN
TEL: 0422-32-5111
TLX: 2822307
TLF:
EML:
COM: 33.47

IKHSANOV ROBERT N DR
PULKOVO OBSERVATORY
196140 LENINGRAD
U.S.S.R.
TEL:
TLX:
TLF:
EML:
COM: 40

IKHSANOVA VERA N DR
PULKOVO OBSERVATORY
196140 LENINGRAD
U.S.S.R.
TEL:
TLX:
TLF:
EML:
COM: 40

ILIEV ILIAN
NATL ASTRON OBSERVATORY
P.O. BOX 136
BG-4700 SMOLYAN
BULGARIA
TEL: 73-41-559
TLX: 23561
TLF:
EML:
COM: 14.35

ILL MARTON J DR
KONKOLY OBSERVATORY
TOTH KALMAN U 19
H-6501 BAJA
HUNGARY
TEL: 11-064
TLX: 281303
TLF:
EML:
COM:

ILLES ALMAR ERZSEBET DR
KONKOLY OBSERVATORY
BOX 67
H-1525 BUDAPEST
HUNGARY
TEL: 366-621
TLX: 227460
TLF:
EML:
COM:

ILLING RAINER M E
BALL AEROSP. SYSTEMS DIV.
PO BOX 1062
BOULDER CO 80306
U.S.A.
TEL: 303-939-5888
TLX:
TLF:
EML:
COM: 12

ILLINGWORTH GARTH D DR
LICK OBSERVATORY
UNIVERSITY OF CALIFORNIA
SANTA CRUZ CA 95064
U.S.A.
TEL:
TLX:
TLF:
EML:
COM: 28.37

ILYAS MOHAMMAD DR
SCHOOL OF PHYSICS
UNIVERSITI SAINS MALAYSIA
11800 USM
PENANG
MALAYSIA
TEL: 883822
TLX: 40254 MA
TLF:
EML:
COM: 04.09.46

IMAMURA JAMES DR
DPT OF PHYSICS
UNIVERSITY OF OREGON
EUGENE OR 97403
U.S.A.
TEL: 503 686 5212
TLX:
TLF:
EML: Imamura@astro.UOREGON.EDU
COM: 42

IMBERT MAURICE DR
OBSERVATOIRE DE MARSEILLE
2 PLACE LE VERRIER
F-13248 MARSEILLE CEDEX 4
FRANCE
TEL: 91-95-90-88
TLX: 420241
TLF:
EML:
COM: 30.42

IMHOFF CATHERINE L DR
IUE OBSERVATORY
CODE 684.9
NASA GSFC
GREENBELT MD 20771
U.S.A.
TEL: 301 286 5749
TLX: 89675
TLF:
EML:
COM: 44

IMPEY CHRISTOPHER D DR
STEWARD OBSERVATORY
UNIVERSITY OF ARIZONA
TUCSON AZ 85721
U.S.A.
TEL: 602 621 6522
TLX: 467175
TLF:
EML: impey@soipl.as.arizona.edu
COM: 28.46.47

IMSHENNIK V S DR
INSTITUTE OF THEORETICAL
AND EXPERIMENTAL PHYSICS
B. CHEREMUSHKINSKAYA 25
117259 MOSCOW
U.S.S.R.
TEL: 123-02-92
TLX: 411059 CERII SU
TLF:
EML:
COM: 35

INAGAKI SHOGO DR
DEPT OF ASTRONOMY
FACULTY OF SCIENCE
UNIVERSITY OF KYOTO
KYOTO 606
JAPAN
TEL: 075-751-2111
TLX: 5422693 LIBKYU J
TLF:
EML:
COM: 33.37

INATANI JUNJI
NOBEYAMA RADIO OBS
TOKYO ASTRON OBSERVATORY
NOBEYAMA. MINAMISAKU
NAGANO 384-13
JAPAN
TEL: 267-98-2831
TLX: 3329005 TAO NRO J
TLF:
EML:
COM: 40

INGERSON THOMAS DR
CERRO TOLOLO INTER-
AMERICAN OBSERVATORY
CASILLA 603
LA SERENA CHILE
CHILE
TEL: 213 352
TLX: 620 301 AURA CT
TLF:
EML: tingerson@noao.edu
COM:

INNANEN KIMMO A PROF
CRESS PHYSICS DEPT
YORK UNIVERSITY
4700 KEELE STREET
NORTH YORK ON M3J1P3
CANADA
TEL: 416-667-3837
TLX: 06524736
TLF:
EML:
COM: 33

INOUE MAKOTO DR
INSTITUTE OF SPACE &
ASTRONAUTICAL SCIENCE
KOMABA. MEGURO-KU
TOKYO 153
JAPAN
TEL: 03-467-1111
TLX: 24550 J
TLF:
EML:
COM: 44.48

INOUE MAKOTO DR
NOBEYAMA RADIO OBS
MINAMI-MAKIMURA
MINAMI-SAKU
NAGANO 384-13
JAPAN
TEL: 0267-98-2831
TLX: 3329005 TAONRO J
TLF:
EML:
COM: 40

INOUE TAKESHI PROF
KYOTO SANGYO UNIVERSITY
KAMIGAMO
KYOTO 603
JAPAN
TEL: 075-701-2151
TLX: 5422661 KSU J
TLF:
EML:
COM:

IOANNISIANI B K DR
MAIN ASTRONOMICAL OBS
PULKOVO
196140 LENINGRAD
U.S.S.R.
TEL:
TLX:
TLF:
EML:
COM: 09

IONSON JAMES ALBERT
NASA/GSFC
CODE 682
GREENBELT MD 20771
U.S.A.
TEL: 301-286-6184
TLX:
TLF:
EML:
COM: 49

IOSEPA B A DR
INST OF TERR MAGNETITISM
AND IONOSPHERE
142092 TROITSK
U.S.S.R.
TEL: 3321921
TLX: 412623 SCP
TLF:
EML:
COM: 10

IP WING-HUEN
MPI FUER ASTRONOMIE
D-3411 KATLENBURG-LINDAU
GERMANY. F.R.
TEL: 0049-555-6416
TLX: 0965527
TLF:
EML:
COM: 15

IPSER JAMES R PROF
DEPT OF PHYSICS
UNIVERSITY OF FLORIDA
WILLIAMSON HALL
GAINESVILLE FL 32611
U.S.A.
TEL: 904-392-0521
TLX:
TLF:
EML:
COM: 46

IRELAND JOHN G DR
C/O 13 GORDON ROAD
BELVEDERE. KENT DA17 6EA
U.K.
TEL:
TLX:
TLF:
EML:
COM:

IRIGOYEN MAYLIS
UNIVERSITE DE PARIS II
12 PLACE DU PANTHEON
75005 PARIS
FRANCE
TEL:
TLX:
TLF:
EML:
COM:

IRIYAMA JUN DR
FACULTY OF ENGINEERING
CHUBU UNIVERSITY
1200 MATSUMOTO
KASUGAI-SHI.AICHI 487
JAPAN
TEL: 568 51 [illegible]
TLX:
TLF:
EML:
COM: 17

IRVINE WILLIAM M PROF
RADIO ASTRONOMY
UNIV OF MASSACHUSETTS
619 GRC TOWER B
AMHERST MA 01003
U.S.A.
TEL: 413-545-0733
TLX: 955491 UNIV MASS AMS
TLF:
EML:
COM: 15.16.34.51

IRWIN ALAN W DR
DEPT OF PHYSICS
UNIVERSITY OF VICTORIA
P O BOX 1700
VICTORIA V8W 2Y2
CANADA
TEL: 604-721-7700
TLX: 049-7222
TLF:
EML:
COM: 14.25

IRWIN JOHN B PROF
2744 N. TYNDALL AVE
TUCSON AZ 85719
U.S.A.
TEL: 602-623-7423
TLX:
TLF:
EML:
COM: 33.42

IRWIN JUDITH DR
DEPARTMENT OF ASTRONOMY
UNIVERSITY OF TORONTO
60 ST GEORGES STREET
TORONTO M5S 1A1
CANADA
TEL: 416 978 5558
TLX: 06 218915
TLF:
EML:
COM: 28

IRWIN MICHAEL JOHN DR
INSTITUTE OF ASTRONOMY
MADINGLEY ROAD
CAMBRIDGE CB3 0HA
U.K.
TEL: 0223 337548
TLX:
TLF:
EML: MJI@UK.AC.CAM.AST-STAR
COM: 24

ISAAK GEORGE R PROF
DEPT OF PHYSICS
UNIVERSITY OF BIRMINGHAM
P O BOX 363
BIRMINGHAM B15 2TT
U.K.
TEL: 021-472-1301
TLX: 338938 SPAPHY G
TLF:
EML:
COM: 35

ISERN JORGE DR
C/SEPULVEDA 83-6-3A
08015 BARCELONA
SPAIN
TEL:
TLX:
TLF:
EML:
COM: 35

ISHIDA GORO DR
BROADCAST UNIVERSITY
23-11 AKABANE-NISHI
1 CHOME. KITA-KU
TOKYO 115
JAPAN
TEL: 03-909-3871
TLX:
TLF:
EML:
COM: 26

ISHIDA KEIICHI PROF
TOKYO ASTRONOMICAL OBS
2-21-1 OSAWA MITAKA
TOKYO 181
JAPAN
TEL: 04-22-32-5211
TLX: 2822307 TAOMTK J
TLF:
EML:
COM: 37

ISHIGURO MASATO PROF.
NOBEYAMA RADIO OBSERV.
NOBEYAMA MINAMIMAKI
MINAMISAKU
NAGANO 384 13
JAPAN
TEL: 0267 98 2831
TLX: 3329005 NAONRO J
TLF:
EML:
COM: 40

ISHIZAWA TOSHIAKI A PROF
DEPT OF ASTRONOMY
UNIVERSITY OF KYOTO
KYOTO 606
JAPAN
TEL: 075-751-2111
TLX: 5422693 LIBKYU J
TLF:
EML:
COM:

ISHIZUKA TOSHIHISA DR
DEPT OF PHYSICS
IBARAKI UNIVERSITY
2-1-1 BUNKYO
MITO 310
JAPAN
TEL: 0292-26-1621
TLX:
TLF:
EML:
COM: 35

ISMAILOV TOFIK K
SPACE RES SCI INDUSTR ENT
AZERBAIJAN SSR ACAD SCI
PROSPECT LENINA 159
370106 BAKU
U.S.S.R.
TEL: 62-93-88
TLX: 142407
TLF:
EML:
COM: 09

ISOBE SYUZO DR
TOKYO ASTRONOMICAL OBS
MITAKA
TOKYO 181
JAPAN
TEL: 0422-32-5211
TLX: 02822307 TAOMTK J
TLF:
EML:
COM: 06.15.33.34.46C

ISRAEL FRANK P DR
STERREWACHT
POSTBUS 9513
NL-2300 RA LEIDEN
NETHERLANDS
TEL:
TLX:
TLF:
EML:
COM: 28.33.34.51

ISRAEL GUY MARCEL DR
SERVICE D'AERONOMIE CNRS
BP 3
F-91370 VERRIERES-LE-B.
FRANCE
TEL: 1-64-47-42-89
TLX:
TLF:
EML:
COM:

ISRAEL WERNER PROF
PHYSICS DEPT
UNIVERSITY OF ALBERTA
ALBERTA
EDMONTON AL T6G 2J1
CANADA
TEL: 403-432-3552
TLX: 0372979
TLF:
EML:
COM: 48

ISSA ALI DR
HELWAN OBSERVATORY
HELWAN-CAIRO
EGYPT
TEL: 780645. 782683
TLX: 93070
TLF:
EML:
COM: 28.34

ISSERSTEDT JOERG DR
INSTITUT FUER ASTRONOMIE
UND ASTROPHYSIK
AM HUBLAND
D-8700 WUERZBURG
GERMANY. F.R.
TEL:
TLX:
TLF:
EML:
COM:

ITO KENSAI A PROF
RIKKYO UNIVERSITY
DEPT OF PHYSICS
NISHI-IKEBUKURO
TOKYO 171
JAPAN
TEL: 03-985-2384
TLX:
TLF:
EML:
COM: 48

ITOH HIROSHI DR
DEPT OF ASTRONOMY
UNIVERSITY OF KYOTO
KYOTO 606
JAPAN
TEL: 075-751-2111
TLX:
TLF:
EML:
COM: 34

ITOH NAOKI DR
DEPT OF PHYSICS
SOPHIA UNIVERSITY
7-1 KIOI-CHO CHIYODA-KU
TOKYO 102
JAPAN
TEL: 03-238-3431
TLX:
TLF:
EML:
COM: 35

IVANCHUK VICTOR I DR
KIEV UNIVERSITY
OBSERVATORNAYA 3
252053 KIEV
U.S.S.R.
TEL:
TLX:
TLF:
EML:
COM: 10

IVANOV GEORGI R DR
UNIVERSITY OF SOFIA
DEPT OF ASTRONOMY
ANTON IVANOV STR 5
1126 SOFIA
BULGARIA
TEL:
TLX:
TLF:
EML:
COM:

IVANOV VSEVOLOD V DR PROF
ASTRONOMICAL OBSERVATORY
LENINGRAD UNIVERSITY
BIBLIOTECHNAJA PL 2
198904 LENINGRAD
U.S.S.R.
TEL: 257-94-91
TLX:
TLF:
EML:
COM: 36

IVANOVA VIOLETA DR
DEPT OF ASTRONOMY AND
NATL ASTRON OBSERVATORY
72 LENIN BLVD
1784 SOFIA
BULGARIA
TEL: 7341-559
TLX: 23561 ECF BAN BG
TLF:
EML:
COM: 07.15.20

IVANOV-KHOLODNY G S DR
IZMIRAN
USSR ACADEMY OF SCIENCES
142092 TROITSK
MOSCOW REGION
U.S.S.R.
TEL:
TLX:
TLF:
EML:
COM: 21

IVES JOHN CHRISTOPHER MR
ESTEC
POSTBUS 299
NL-2200 AG NOORDWIJK
NETHERLANDS
TEL: 01719-83629
TLX: 39098
TLF:
EML:
COM:

IWANISZEWSKA CECYLIA DR
INSTITUTE OF ASTRONOMY
N COPERNICUS UNIVERSITY
UL. CHOPINA 12/18
87-100 TORUN
POLAND
TEL: 2-60-18
TLX: 0552234 ASTR PL
TLF:
EML:
COM: 33.46C

IWANOWSKA WILHELMINA PROF
INSTITUTE OF ASTRONOMY
UL. CHOPINA 12/18
87-100 TORUN
POLAND
TEL: 260-18
TLX: 86412 PL
TLF:
EML:
COM: 33

IWASAKI KYOSUKE DR
KWASAN OBSERVATORY
YAMASHINA
KYOTO 607
JAPAN
TEL: 075-581-1235
TLX: 5422693 LIBKYUJ
TLF:
EML:
COM: 16

IYE MASANORI DR
TOKYO ASTRONOMICAL OBS
UNIVERSITY OF TOKYO
MITAKA 181
JAPAN
TEL: 0422-32-511x313
TLX: 2822307 TAOMTK J
TLF:
EML:
COM: 33

IYENGAR K V K PROF
TATA INSTITUTE OF
FUNDAMENTAL RESEARCH
BOMBAY 400 005
INDIA
TEL: 219111 x339
TLX: 113009 TIFR IN
TLF:
EML:
COM: 25.34.44

IYER B R DR
RAMAN RESEARCH INSTITUTE
BANGALORE 560 080
INDIA
TEL: 812-360122
TLX: 8452671 RRI IN
TLF:
EML:
COM: 47

IZVEKOV V A DR
INSTITUTE OF THEORETICAL
ASTRONOMY
10 KUTUZOV QUAY
191187 LENINGRAD
U.S.S.R.
TEL: 272-40-23
TLX: 121578 ITA SU
TLF:
EML:
COM: 07.20

JAAKKOLA TOIVO S
OBSERVATORY
TAHTITORNINMAKI
SF-00130 HELSINKI 13
FINLAND
TEL: 35-801-912907
TLX: 124690 UNIH SF
TLF:
EML:
COM:

JABBAR SABEH RHAMAN
ASTRONOMY & SPACE RES CTR
COUNCIL FOR SCI RESEARCH
P O BOX 2441
JADIRIYAH. BAGHDAD
IRAQ
TEL: 7765127
TLX: 213976 SRC IK
TLF:
EML:
COM: 12.42

JABIR NIAMA LAFTA
ASTRONOMY & SPACE RES CTR
COUNCIL FOR SCI RESEARCH
P O BOX 2441
JADIRIYAH. BAGHDAD
IRAQ
TEL: 7765127
TLX: 213976 SRC IK
TLF:
EML:
COM: 34

JACCHIA LUIGI G DR
CENTER FOR ASTROPHYSICS
60 GARDEN ST
CAMBRIDGE MA 02138
U.S.A.
TEL: 617-495-7213
TLX:
TLF:
EML:
COM: 22

JACKISCH GERHARD DR
ZNTRLINST. F. ASTROPHYSIK
STERNWARTE BABELSBERG
DDR-6400 SONNEBERG
GERMANY. D.R.
TEL:
TLX:
TLF:
EML:
COM: 41

JACKSON JOHN CHARLES DR
16 THE PARK
NEWARK NG24 1SG
U.K.
TEL:
TLX:
TLF:
EML:
COM: 48

JACKSON PAUL DR
INSTITUT FUER ASTRONOMIE
DER UNIVERSITAET WIEN
TUERKENSCHANZSTR 17
A-1180 WIEN
AUSTRIA
TEL:
TLX:
TLF:
EML:
COM: 08

JACKSON PETER DOUGLAS DR
ASTRONOMY PROGRAM
UNIVERSITY OF MARYLAND
COLLEGE PK MD 20742
U.S.A.
TEL: 301-454-6302
TLX:
TLF:
EML:
COM: 33

JACKSON WILLIAM M DR
DEPT OF CHEMISTRY
UNIVERSITY OF CALIFORNIA
ROOM 214
DAVIS CA 95616
U.S.A.
TEL: 916-752-0503
TLX:
TLF:
EML:
COM: 15

JACOBS KENNETH C DR
PHYSICS DEPT
HOLLINS COLLEGE
BOX 9661
ROANOKE VA 24020
U.S.A.
TEL: 703-362-6472
TLX:
TLF:
EML:
COM:

JACOBSEN THEODOR S PROF
6205 17TH AVE N.E.
SEATTLE WA 98115
U.S.A.
TEL: 206-523-5245
TLX:
TLF:
EML:
COM:

JACOBY GEORGE H
KITT PEAK NAT OBSERVATORY
PO BOX 26732
TUCSON AZ 85726
U.S.A.
TEL: 602-325-9292
TLX:
TLF:
EML:
COM: 34

JACQUINOT PIERRE DR
LABORATOIRE AIME COTTON
BAT 505
UNIVERSITE PARIS SUD
F-91405 ORSAY CEDEX
FRANCE
TEL:
TLX:
TLF:
EML:
COM: 14

JAEGER FRIEDRICH W PROF
TELEGRAFENBERG A 33
DDR-1500 POTSDAM
GERMANY. D.R.
TEL: 4551
TLX:
TLF:
EML:
COM:

JAFFE DANIEL T
UNIVERSITY OF CALIFORNIA
SPACE SCIENCES LABORATORY
BERKELEY CA 94720
U.S.A.
TEL: 415-642-1930
TLX:
TLF:
EML:
COM: 34

JAFFE WALTER JOSEPH DR
SPACE TELESCOPE SCI INST
3700 SAN MARTIN DRIVE
BALTIMORE MD 21218
U.S.A.
TEL: 301-338-4762
TLX: 684 9101 STSCI
TLF:
EML:
COM: 28.40.48

JAHREISS HARTMUT DR
ASTRONOMISCHES RECHEN-
INSTITUT
MOENCHHOFSTR 12-14
D-6900 HEIDELBERG 1
GERMANY. F.R.
TEL: 06221/49026
TLX: 461 336 ARIHD D
TLF:
EML:
COM: 24.33

JAIN RAJMAL DR
UDAIPUR SOLAR OBSERVATORY
11 VIDYA MARG
UDAIPUR-313001
INDIA
TEL: 25626 - 27457
TLX:
TLF:
EML:
COM: 10

JAIN SURENDRA DR
INDIAN INSTITUTE OF
ASTROPHYSICS
SARJAPUR ROAD
BANGALORE 560034
INDIA
TEL: 569702
TLX: 8452763 IIAB IN
TLF:
EML:
COM:

JAKIMIEC JERZY PROF
ASTRONOMICAL INSTITUTE
UL. KOPERNIKA 11
51-622 WROCLAW
POLAND
TEL: 482434
TLX: 0712791 UWRPL
TLF:
EML:
COM: 10

JAKOBSEN PETER
ASTROPHYSICS DEPARTMENT
ESA SPACE SCI DEPT/ESTEC
POSTBUS 299
NL-2200 AG NOORDWIJK
NETHERLANDS
TEL: 0171-983-3614
TLX: 39098
TLF:
EML:
COM:

JAKS WALDEMAR DR
ASTRONOMICAL LATITUDE OBS
SPACE RESEARCH CENTRE PAS
BOROWIEC
62-035 KORNIK
POLAND
TEL: POZNAN 170187
TLX: 0412623 AOS PL
TLF:
EML:
COM: 19

JAMAR CLAUDE A J DR
IAL SPACE/UNIV DE LIEGE
AVENUE DU PRE-AILY
B-4900 ANGLEUR-LIEGE
BELGIUM
TEL: 41676760
TLX: 41320 IAL SF
TLF:
EML:
COM: 44

JAMES JOHN F MR
SCHUSTER LABORATORY
THE UNIVERSITY
MANCHESTER M13 9PL
U.K.
TEL: 061-273-7121
TLX:
TLF:
EML:
COM: 21

JAMES RICHARD A DR
DEPARTMENT OF ASTRONOMY
THE UNIVERSITY
MANCHESTER M13 9PL
U.K.
TEL:
TLX:
TLF:
EML:
COM: 35

JAMESON RICHARD F DR
ASTRONOMY DEPARTMENT
THE UNIVERSITY
LEICESTER LE1 7RH
U.K.
TEL: 0533-554455
TLX: 341198
TLF:
EML:
COM:

JAMES KENNETH A DR
ASTRONOMY DEPT
BOSTON UNIVERSITY
725 COMMONWEALTH AVE
BOSTON MA 02215
U.S.A.
TEL: 617-353-2627
TLX: 95-1289 BOS UNIV BSN
TLF:
EML:
COM: 370

JANICZEK PAUL M DR
US NAVAL OBSERVATORY
348 MASSACHUSETTS AVE NW
348 MASSACHUSETTS AVE N.W
WASHINGTON DC 20390-5100
U.S.A.
TEL: 202-653-1569
TLX: 710-822-1970
TLF:
EML:
COM: 04.07

JANKOVICS ISTVAN DR
KONKOLY OBSERVATORY
BOX 67
H-1525 BUDAPEST
HUNGARY
TEL: 166-426
TLX: 227460
TLF:
EML:
COM:

JANOT-PACHECO EDUARDO DR
INSTITUTO ASTRONOMICO
CAIXA POSTAL 30627
01051 SAO PAULO SP
BRAZIL
TEL: 011 577 8599
TLX: 36221 IAGM BR
TLF:
EML:
COM:

JANSSEN MICHAEL ALLEN
JET PROPULSION LAB
MAIL STOP 183-301
4800 OAK GROVE DRIVE
PASADENA CA 91109
U.S.A.
TEL: 213-354-7247
TLX:
TLF:
EML:
COM: 40

JARNEFELT GUSTAF J PROF
LAAJASUONTIE 27
SF-00320 HELSINKI 32
FINLAND
TEL:
TLX:
TLF:
EML:
COM:

JAROSZYNSKI MICHAL
WARSAW UNIVERSITY
OBSERVATORY
AL. UJAZDOWSKIE 4
00-478 WARSAW
POLAND
TEL: 29 40 11
TLX: 813978 ZAPAN PL
TLF:
EML:
COM: 47

JARRETT ALAN H PROF
BOYDEN OBSERVATORY
P O BOX 334
BLOEMFONTEIN 9300
SOUTH AFRICA
TEL: 051-37605
TLX: 767666 SA
TLF:
EML:
COM: 21.46

JARZEBOWSKI TADEUSZ DR
ASTRONOMICAL INSTITUTE
KOPERNIKA 11
51-622 WROCLAW
POLAND
TEL:
TLX:
TLF:
EML:
COM: 27

JASCHEK CARLOS O R PROF
OBSERVATOIRE
11 RUE DE L'UNIVERSITE
F-67000 STRASBOURG
FRANCE
TEL: 88-35-43-00
TLX: 890506 STAROBS
TLF:
EML:
COM: 05C.26.29.33.42.45

JASCHEK MERCEDES DR
OBSERVATOIRE
11 RUE DE L'UNIVERSITE
F-67000 STRASBOURG
FRANCE
TEL: 88-35-43-00
TLX: 890506 STAROBS F
TLF:
EML:
COM: 29.45

JASNIEWICZ GERARD DR
OBSERVATOIRE DE
STRASBOURG
11 RUE DE L'UNIVERSITE
67000 STRASBOURG
FRANCE
TEL: 88 35 82 00
TLX: 890 506 STAROBS
TLF:
EML: EARN:U011090FRCCSC21
COM: 33.42

JASTROW ROBERT
INST FOR SPACE STUDIES
2880 BROADWAY
NEW YORK NY 10025
U.S.A.
TEL:
TLX:
TLF:
EML:
COM: 51

JAUNCEY DAVID L DR
CSIRO
DIVISION OF RADIOPHYSICS
P.O.BOX 76
EPPING NSW 2121
AUSTRALIA
TEL: 062-46-5558
TLX: 26230 ASTRO
TLF:
EML:
COM: 40C.47

JAVET PIERRE PROF
AVENUE DE BEAUMONT 36
CH-1012 LAUSANNE
SWITZERLAND
TEL:
TLX:
TLF:
EML:
COM:

JAYARAJAN A P MR
INDIAN INSTITUTE OF
ASTROPHYSICS
KORAMANGALA
BANGALORE 560 034
INDIA
TEL: 566585
TLX: 845763 IIAB IN
TLF:
EML:
COM: 09

JEFFERIES JOHN T DR
NATL OPTICAL ASTR OBS
950 NORTH CHERRY AVE
TUCSON AZ 85719
U.S.A.
TEL: 602-881-1950
TLX: 0666484 AURANOAGTUC
TLF:
EML:
COM: 12.36

JEFFERS STANLEY DR
CRESS PHYSICS DEPT
YORK UNIVERSITY
4700 KEELE ST
DOWNSVIEW ONT M3J 1P3
CANADA
TEL: 416-667-3851
TLX:
TLF:
EML:
COM: 09.51

JEFFERY CHRISTOPHER S DR
UNIVERSITY OBSERVATORY
BUCHANAN GARDENS
ST ANDREWS FIFE KY16 9LZ
SCOTLAND
U.K.
TEL:
TLX:
TLF:
EML:
COM: 27

JEFFERYS WILLIAM H DR
ASTRONOMY DEPARTMENT
UNIVERSITY OF TEXAS
AUSTIN TX 78712
U.S.A.
TEL: 512-471-4461
TLX:
TLF:
EML:
COM: 07.24

JEFFREYS HAROLD PROF SIR
160 HUNTINGDON ROAD
CAMBRIDGE CB3 0LB
U.K.
TEL: 356153
TLX:
TLF:
EML:
COM: 16.19

JELLEY JOHN V PHD
29 ABBOTT ROAD
ABINGDON OX14 1DT
U.K.
TEL: 0235-21040
TLX:
TLF:
EML:
COM: 09.44

JENKINS CHARLES R
ROYAL GREENWICH OBS
HERSTMONCEUX CASTLE
HAILSHAM BN27 1RP
U.K.
TEL: 0323-833171
TLX: 87451
TLF:
EML:
COM: 40

JENKINS EDWARD B DR
PRINCETON UNIVERSITY OBS
PRINCETON NJ 08544
U.S.A.
TEL: 609-452-3826
TLX: 322409 ASTRO PRIN
TLF:
EML:
COM: 44P

JENKINS L F MS
YALE UNIV OBSERVATORY
BOX 2023 YALE STATION
NEW HAVEN CT 06520
U.S.A.
TEL:
TLX:
TLF:
EML:
COM: 34

JENKNER HELMUT DR
SPACE TELESCOPE SCI INST
3700 SAN MARTIN DRIVE
BALTIMORE MD 21218
U.S.A.
TEL: 301-338-4842
TLX: 6849101 STSCI
TLF:
EML:
COM: 05.09

JENNER DAVID C DR
DEPT OF ASTRONOMY
UNIVERSITY OF WASHINGTON
FM-20
SEATTLE WA 98195
U.S.A.
TEL: 206-543-6182
TLX:
TLF:
EML:
COM:

JENNINGS R E PROF
DEPT PHYSICS & ASTRONOMY
UNIVERSITY COLLEGE LONDON
GOWER STREET
LONDON WC1E 6BT
U.K.
TEL: 01-387-7050
TLX: 28722
TLF:
EML:
COM: 34

JENNISON ROGER C PROF
ELECTRONICS LABORATORY
UNIVERSITY OF KENT
CANTERBURY CT2 7NT
U.K.
TEL:
TLX: 965449
TLF:
EML:
COM: 22.40.51

JENSCH A
PESTALOZZISTR 9
DDR-6900 JENA
GERMANY. D.R.
TEL:
TLX:
TLF:
EML:
COM:

JENSEN EBERHART PROF
INST THEORET ASTROPHYSICS
UNIVERSITY OF OSLO
P O BOX 1029
N-0315 BLINDERN. OSLO 3
NORWAY
TEL: 02-456502
TLX: 72425N UNIOS
TLF:
EML:
COM: 10

JERZYKIEWICZ MIKOLAJ DR
ASTRONOMICAL INSTITUTE
WROCLAW UNIVERSITY
KOPERNIKA 11
51-622 WROCLAW
POLAND
TEL: 48-24-34
TLX: 0712791 UWR PL
TLF:
EML:
COM: 25.27C

JEWELL PHILIP R DR
NRAO
CAMPUS BLDG 65
949 N. CHERRY AVE
TUCSON AZ 85721-0655
U.S.A.
TEL: 602 882 8250
TLX: 9102409524 NRAO TUC
TLF:
EML: BITNET:PJEWELL@NRAO
COM: 27.40

JI HONG-QING
INTNATL LATITUTDE STATION
TIANJIN
CHINA. PEOPLE'S REP.
TEL:
TLX:
TLF:
EML:
COM: 19

JI SHUCHEN DR
PO BOX 110 KUNMING
YUNNAN PROVINCE
YUNNAN
CHINA. PEOPLE'S REP.
TEL: 72946
TLX: 64040 YUOBS CN
TLF:
EML:
COM: 40

JIANG CHONG-GUO
YUNNAN OBSERVATORY
P.O. BOX 110
KUNMING
CHINA. PEOPLE'S REP.
TEL: 72946
TLX:
TLF:
EML:
COM: 38

JIANG DONG-RONG
SHANGHAI OBSERVATORY
ACADEMIA SINICA
SHANGHAI
CHINA. PEOPLE'S REP.
TEL: 386191
TLX: 33164 SHAO CN
TLF:
EML:
COM: 33

JIANG SHI-YANG
BEIJING ASTRONOMICAL OBS
BEIJING
CHINA. PEOPLE'S REP.
TEL: 28-1698
TLX: 22040 BAOAS CN
TLF:
EML:
COM: 09.29.27.50

JIANG SHUDING
GRADUATE SCHOOL
UNIV SCIENCE & TECHNOLOGY
P.O. BOX 3908
BEIJING
CHINA. PEOPLE'S REP.
TEL: 817031-253
TLX:
TLF:
EML:
COM: 47

JIANG YAO-TIAO
DEPARTMENT OF ASTRONOMY
NANJING UNIVERSITY
NANJING
CHINA. PEOPLE'S REP.
TEL: 34151 NANJING
TLX:
TLF:
EML:
COM: 10

JIANG ZHAOJI
BEIJING ASTRONOMICAL OBS
BEIJING
CHINA. PEOPLE'S REP.
TEL:
TLX: 22040 BAOAS CN
TLF:
EML:
COM:

JIN BIAOREN DR
WUHAN TECHNICAL UNIVERS
DEPT. OF GEODESY
39 LO-YU ROAD
WUHAN 430070
CHINA. PEOPLE'S REP.
TEL: 875571
TLX: 40210 WTUSM CN
TLF:
EML:
COM:

JIN SHEN-ZENG
BEIJING ASTRONOMICAL OBS
ACADEMIA SINICA
BEIJING
CHINA. PEOPLE'S REP.
TEL: 28 1698
TLX: 22040 BAOBS CN
TLF:
EML:
COM: 40

JIN WEN-JING
SHANGHAI OBSERVATORY
ACADEMIA SINICA
80 NAN DAN ROAD
SHANGHAI
CHINA. PEOPLE'S REP.
TEL: 386191
TLX: 33164 SHAO CN
TLF:
EML:
COM: 19C.31

JOCKERS KLAUS DR
MPI FUER AERONOMIE
POSTFACH 20
D-3411 KATLENBURG-LINDAU
GERMANY. F.R.
TEL: 05556-411
TLX: 965527 AERLI D
TLF:
EML:
COM: 10.15

JOERSAETER STEVEN DR
ST-ECF
KARL SCHWARSCHILD STR 2
8046 GARCHING BEI MUNCHEN
GERMANY. F.R.
TEL: 49 89 32006286
TLX: 52828222 EO D
TLF:
EML: BITNET:STEVEN@DGAESI51
COM:

JOG CHANDA J DR
DEPT OF PHYSICS
INDIAN INST. OF SCIENCES
BANGALORE 560 012
INDIA
TEL: 364411315
TLX: 845 8349 IISC IN
TLF:
EML:
COM: 28.33

JOHANSEN KAREN T LEKTOR
COPENHAGEN UNIVERSITY OBS
BRORFELDEVEJ 23
DK-4340 TOLLOSE
DENMARK
TEL: 03-488195
TLX:
TLF:
EML:
COM:

JOHANSSON LARS ERIK B DR
ARVESGAERDE 18
S-417 44 GOETEBORG
SWEDEN
TEL:
TLX:
TLF:
EML:
COM: 40

JOHNSON DONALD R DR
NATL BUREAU OF STANDARDS
BLDG 221. ROOM A363
GAITHERSBURG MD 20899
U.S.A.
TEL: 301-921-2828
TLX:
TLF:
EML:
COM: 14.40

JOHNSON FRED M PROF DR
DEPT PHYSICS & ASTRONOMY
CALIFORNIA STATE UNIV
FULLERTON CA 92634
U.S.A.
TEL: 714-773-3366
TLX:
TLF:
EML:
COM: 14.34

JOHNSON HOLLIS R PROF
ASTRONOMY DEPT
INDIANA UNIVERSITY
SWAIN WEST 319
BLOOMINGTON IN 47405
U.S.A.
TEL: 812-335-4172
TLX: 272279
TLF:
EML:
COM: 29.36

JOHNSON HUGH M DR
1017 NEWELL ROAD
PALO ALTO CA 94303
U.S.A.
TEL: 415-326-7223
TLX:
TLF:
EML:
COM: 33.34

JOHNSON TORRENCE V DR
JET PROPULSION LABORATORY
MAILSTOP 183-301
4800 OAK GROVE DRIVE
PASADENA CA 91109
U.S.A.
TEL: 818-354-2761
TLX: 67-5429
TLF:
EML:
COM: 15.16

JOHNSTON KENNETH J
NAVAL RESEARCH LABORATORY
CODE 7134
WASHINGTON DC 20375
U.S.A.
TEL: 202-767-2351
TLX:
TLF:
EML:
COM: 04.08.24.34.40

JOKIPII J R PROF
DEPT OF PLANET. SCIENCES
UNIVERSITY OF ARIZONA
TUCSON AZ 85721
U.S.A.
TEL: 602-621-4256
TLX:
TLF:
EML:
COM: 48.49C

JOLY FRANCOIS DR
UNIVERSITE DE BORDEAUX 1
123 RUE LAMARTINE
F-33400 TALENCE
FRANCE
TEL:
TLX:
TLF:
EML:
COM: 14.40

JOLY MONIQUE
OBSERVATOIRE DE PARIS
SECTION DE MEUDON
F-92195 MEUDON PL CEDEX
FRANCE
TEL: 1-45-34-75-70
TLX: 201571 F
TLF:
EML:
COM:

JONAS JUSTIN LEONARD
DEPT. OF PHYS. & ELECTR.
RHODES UNIVERSITY
GRAHAMSTOWN 6140
SOUTH AFRICA
TEL: (461)22023
TLX: 244226 RUANT SA
TLF:
EML:
COM: 33.40

JONCAS GILLES DR
DEPARTMENT DE PHYSIQUE
FAC. SCIENCES ET GENIE
UNIVERSITY LAVAL
STE FOY QUEBEC G1K 7P4
CANADA
TEL: 418 656 2652
TLX:
TLF:
EML: BITNET:11500410LAVALVX1
COM:

JONES ALBERT F MR
31 RANUI RD. STOKE
NELSON
NEW ZEALAND
TEL: 054-73-905
TLX:
TLF:
EML:
COM: 27

JONES BARBARA
UNIV OF CALIFORNIA AT
SAN DIEGO
CASS/C-011
LA JOLLA CA 92093
U.S.A.
TEL: 714-452-4474
TLX:
TLF:
EML:
COM: 09

JONES BERNARD J T DR
NORDITA
BLEGDAMSVEJ 17
DK-2100 COPENHAGEN
DENMARK
TEL: 01-42-16-16
TLX: 15216 NBI DK
TLF:
EML:
COM: 47

JONES BURTON DR
LICK OBSERVATORY
UNIVERSITY OF CALIFORNIA
SANTA CRUZ CA 95064
U.S.A.
TEL: 408-429-2384
TLX:
TLF:
EML:
COM: 24

JONES DAYTON L
JET PROPULSION LABORATORY
MAIL CODE 238-700
4800 OAK GROVE DRIVE
PASADENA CA 91109
U.S.A.
TEL: 818-354-7774
TLX: 675429
TLF:
EML:
COM: 40

JONES DEREK E P DR
ROYAL GREENWICH OBS
HAILSHAM BN27 1RP
U.K.
TEL: 323-833171
TLX: 87451 RGOBSY G
TLF:
EML:
COM: 24.33.37

JONES ERIC M
LOS ALAMOS NATL LAB
MS F-665
LOS ALAMOS NM 87545
U.S.A.
TEL: 505-667-6386
TLX:
TLF:
EML:
COM: 51

JONES FRANK CULVER DR
NASA/GSFC
CODE 665
GREENBELT MD 20771
U.S.A.
TEL: 301-286-5506
TLX: 710-828-9716
TLF:
EML:
COM: 34.48

JONES HARRISON PRICE DR
KITT PEAK NATL OBS
SOLAR STATION
900 N. CHERRY AVENUE
TUCSON AZ 85726
U.S.A.
TEL: 602-325-9354
TLX:
TLF:
EML:
COM: 10.12

JONES JAMES DR
DEPT OF PHYSICS
UNIV OF WESTERN ONTARIO
LONDON ONT N6A 5B9
CANADA
TEL: 519-661-3283
TLX:
TLF:
EML:
COM: 22C

JONES JANET E DR
NORDITA
BLEGDAMSVEJ 17
DK-2100 COPENHAGEN
DENMARK
TEL: 01-42-16-16
TLX: 15216 NBI DK
TLF:
EML:
COM:

JONES THOMAS WALTER DR
DEPT OF ASTRONOMY
UNIVERSITY OF MINNESOTA
116 CHURCH ST SE
MINNEAPOLIS MN 55455
U.S.A.
TEL: 612-373-3307
TLX:
TLF:
EML:
COM: 28.48

JORDAN CAROLE DR
DEPT THEORETICAL PHYSICS
OXFORD UNIVERSITY
1 KEBLE ROAD
OXFORD OX1 3NP
U.K.
TEL: 865-53281
TLX: 83295 NUCLOX
TLF:
EML:
COM: 12.14.29.44

JORDAN H L DR DIREKTOR
INSTITUT F. PLASMAPHYSIK
KERNFORSCHUNGSANLAGE
JUELICH GMBH PF 365
D-5170 JUELICH 1
GERMANY. F.R.
TEL:
TLX:
TLF:
EML:
COM: 14

JORDAN STUART D DR
LAB ASTRON & SOLAR PHYS
NASA/GSFC. CODE 682
GREENBELT MD 20771
U.S.A.
TEL: 301-286-8811
TLX: 89675
TLF:
EML:
COM: 10.12.44C

JORDEN PAUL RICHARD
ROYAL GREENWICH OBS
HERSTMONCEUX CASTLE
HAILSHAM BN27 1RP
U.K.
TEL:
TLX:
TLF:
EML:
COM:

JORDI NEBOT CARME DR
DEPT DE ASTRONOMIA
UNIVERSITAT DE BARCELONA
AVDA DIAGONAL 647
08028 BARCELONA
SPAIN
TEL: 34 3 330 73 11
TLX:
TLF:
EML: D3FACJN0@EBOUB011
COM:

JORGENSEN HENNING E PROF
UNIVERSITY OBSERVATORY
OESTER VOLDGADE 3
DK-1350 COPENHAGEN K
DENMARK
TEL: 1-14 17 90
TLX: 44155
TLF:
EML:
COM: 38C

JORGENSEN UFFE GRAE DR
NIELS BOHR INSTITUTE
BLEGDAMSVEJ 17
2100 COPENHAGEN
DENMARK
TEL: 01421616
TLX: 15216
TLF:
EML:
COM:

JOSELYN JO ANN C DR
NOAA R/E/SE2
325 BROADWAY
BOULDER CO 80303
U.S.A.
TEL: 303 497 5147
TLX: 888776 NOAA BLDR
TLF:
EML: SPAN:SELVAX::JJOSELYN
COM: 10.49

JOSEPH J H DR
DEPT GEOPHYS & PLANET SCI
TEL-AVIV UNIVERSITY
RAMAT-AVIV 69978
ISRAEL
TEL: 3-420-633
TLX: 342171 VERSY IL
TLF:
EML:
COM:

JOSEPH ROBERT D DR
BLACKETT LABORATORY
ASTROPHYSICS GROUP
IMPERIAL COLLEGE
LONDON SW7 2BZ
U.K.
TEL: 1-589-5111x6660
TLX: 261503
TLF:
EML:
COM:

JOSHI G C DR
UTTAR PRADESH STATE OBS
MANORA PEAK
NAINITAL 263 129
INDIA
TEL:
TLX:
TLF:
EML:
COM: 12

JOSHI MOHAN N PROF
RADIO ASTRONOMY CENTER
TIFR
POST BOX 8
UDHAGAMANDALAM 643 001
INDIA
TEL: 203
TLX: 8458488 TIFR IN
TLF:
EML:
COM: 28.40.47

JOSHI SURESH CHANDRA DR
UP STATE OBSERVATORY
MANORA PEAK
NAINI TAL 263 129
INDIA
TEL: 2136
TLX:
TLF:
EML:
COM: 25.29

JOSHI U C DR
PHYSICAL RESEARCH LAB
NAVRANGPURA
AHMEDABAD 380 009
INDIA
TEL: 462-129
TLX: 121337
TLF:
EML:
COM: 25.28.37

JOSS PAUL CHRISTOPHER DR
MIT
ROOM 6-203
CAMBRIDGE MA 02139
U.S.A.
TEL: 617-243-4845
TLX:
TLF:
EML:
COM: 42.43

JOUBERT MARTINE
LAB D'ASTRONOMIE SPATIALE
TRAVERSE DU SIPHON
LES TROIS LUCS
F-13012 MARSEILLE
FRANCE
TEL: 91-66-08-32
TLX: 420584
TLF:
EML:
COM: 21

JOURNET ALAIN
CERGA
AVENUE COPERNIC
F-06130 GRASSE
FRANCE
TEL: 93-36-58-49
TLX: 470865
TLF:
EML:
COM: 07.08

JOVANOVIC BOZIDAR
FACULTY OF AGRICULTURE
INST VATERRANGING
VELJKA VLAHOVICA 2
YU-21000 NOVI SAD
YUGOSLAVIA
TEL: 009382158366
TLX:
TLF:
EML:
COM: 07.10

JUGAKU JUN DR
TOKYO ASTRONOMICAL OBS
OSAWA MITAKA
TOKYO 181
JAPAN
TEL:
TLX: 2822307 TAOMTKJ
TLF:
EML:
COM: 28.29.51C

JUNG JEAN DR
THOMSON
173 BD HAUSSMANN
F-75379 PARIS CEDEX 08
FRANCE
TEL: 1-45-61-96-00
TLX: 204780 TCSF
TLF:
EML:
COM:

JUNKKARINEN VESA T DR
CASS C 011
UNIVERSITY OF CALIFORNIA
SAN DIEGO
LA JOLLA CA 92093
U.S.A.
TEL: 619 534 0735
TLX:
TLF:
EML: SPAN:27783::VESA
COM: 28.47

JUPP ALAN H DR
DEPT APPL MATH THEOR PHYS
UNIVERSITY OF LIVERPOOL
PO BOX 147
LIVERPOOL L69 3BX
U.K.
TEL: 051-709-6022
TLX: 627095
TLF:
EML:
COM: 07

JURA MICHAEL DR
UCLA
DEPT OF ASTRONOMY
MATH SCIENCES BLDG
LOS ANGELES CA 90024
U.S.A.
TEL: 213-825-4302
TLX:
TLF:
EML:
COM: 34

JURGENS RAYMOND F
JET PROPULSION LAB
MS 238/420
4800 OAK GROVE DRIVE
PASADENA CA 91109
U.S.A.
TEL: 818-354-4974
TLX: 675429
TLF:
EML:
COM: 16

JURKEVICH IGOR DR
3130 PORT WAY
ANNAPOLIS MD 21403
U.S.A.
TEL: 202-767-2003
TLX:
TLF:
EML:
COM: 42

JUSZKIEWICZ ROMAN
COPERNICUS ASTRON CENTER
UL. BARTYCKA 18
00-716 WARSAW
POLAND
TEL:
TLX:
TLF:
EML:
COM: 47

KAASTRA JELLE S DR
SPACE RESEARCH LABORATORY
PO BOX 9504
2300 RA LEIDEN
NETHERLANDS
TEL: 071 275818
TLX: 39058 ASTRO NL
TLF:
EML:
COM:

KABURAKI MASAKI PROF
4-24-9 KICHIJYOJI
MINAMI MUSASHINO
TOKYO 180
JAPAN
TEL:
TLX:
TLF:
EML:
COM: 33

KADLA ZDENKA I DR
PULKOVO OBSERVATORY
196140 LENINGRAD
U.S.S.R.
TEL:
TLX:
TLF:
EML:
COM: 05.37

KADOURI TALIB HADI
ASTRONOMY & SPACE RES CTR
COUNCIL FOR SCI RESEARCH
P O BOX 2441
JADIRIYAH. BAGHDAD
IRAQ
TEL: 00-96417765127
TLX: 212976 SRC IK
TLF:
EML:
COM: 27.30.36 42

KAEHLER HELMUTH DR
HAMBURGER STERNWARTE
GOJENBERGSWEG 112
D-2050 HAMBURG 80
GERMANY. F.R.
TEL:
TLX:
TLF:
EML:
COM: 35

KAFATOS MINAS DR
PHYSICS DEPT
GEORGE MASON UNIVERSITY
FAIRFAX VA 22030
U.S.A.
TEL: 703-323-2303
TLX:
TLF:
EML:
COM: 34.44.51

KAFKA PETER
MPI F. PHYSIK & ASTROPHYS
INSTITUT FUR ASTROPHYSIK
KARL-SCHWARZSCHILD-STR 1
D-8046 GARCHING B MUNCHEN
GERMANY. F.R.
TEL: 89-3299-0
TLX: 524629 ASTRO D
TLF:
EML:
COM: 48.51

KAFTAN MAY A DR
NRAO
PO BOX 2
GREEN BANK VA 24944
U.S.A.
TEL:
TLX: 710-938-1530
TLF:
EML:
COM: 34.40

KAHANE CLAUDINE DR
GROUPE D'ASTROPHYSIQUE
OBSERVATOIRE DE GRENOBLE
CERMO BP 68
38402 ST MARTIN D'HERES
FRANCE
TEL: 76 51 46 00
TLX:
TLF:
EML: KAHANE@FRGAG51
COM:

KAHLER STEPHEN W DR
AIR FORCE GEOPHYSICS LAB
SPACE PHYSICS DIV (PHP)
HANSCOM AIR FORCE BASE
BEDFORD MA 01731
U.S.A.
TEL: 617-861-3975
TLX:
TLF:
EML:
COM: 10

KAHLMANN H C DR
NFRA
PO BOX 2
7990 AA DWINGELOO
NETHERLANDS
TEL: (0) 5939 421
TLX: 42043
TLF:
EML:
COM: 40

KAHN FRANZ D PROF
DEPT OF ASTRONOMY
THE UNIVERSITY
MANCHESTER M13 9PL
U.K.
TEL: 61-273-7121
TLX: 668932 MCHRUL G
TLF:
EML:
COM: 34.40.48

KAI KEIZO DR
TOKYO ASTRON OBSERVATORY
21-2-1 OSAWA. MITAKA
TOKYO 181
JAPAN
TEL: 0422-32-5111
TLX: 3329005 TAOMYO J
TLF:
EML:
COM: 10.40

KAIFU NORIO DR
NOBEYAMA RADIO OBSERV.
NOBEYAMA. MINAMISAKU
NAGANO 384 13
JAPAN
TEL:
TLX:
TLF:
EML:
COM: 34C.40C

KAISER THOMAS R PROF
DEPT OF PHYSICS
THE UNIVERSITY
SHEFFIELD S3 7RH
U.K.
TEL: 0742-78555x4277
TLX: 547216 UGSHEF G
TLF:
EML:
COM: 22

KAITCHUCK RONALD H
OHIO STATE UNIVERSITY
DEPT OF ASTRONOMY
174 WEST 18TH AVENUE
COLUMBUS OH 43210
U.S.A.
TEL: 614-422-4579
TLX:
TLF:
EML:
COM:

KAKINUMA TAKAKIYO T PROF
RESEARCH INSTITUTE
OF ATMOSPHERICS
NAGOYA UNIVERSITY
TOYOKAWA AICHI 442
JAPAN
TEL: 05338-6-3154
TLX:
TLF:
EML:
COM: 40.49

KAKUTA CHUICHI DR
INTL LATITUDE OBSERVATORY
HOSHIGAOKA 2-12
MIZUSAWA IWATE 023
JAPAN
TEL: 0197-24-7111
TLX: 837628 ILSMIZ J
TLF:
EML:
COM: 19.31

KALAFI MANOUCHER
CTR FOR ASTRON RESEARCH
DEPT OF PHYSICS
TABRIZ UNIVERSITY
TABRIZ
IRAN
TEL: 041-32564
TLX:
TLF:
EML:
COM: 28

KALANDADZE N B DR
ABASTUMANI ASTROPHYSICAL
OBSERVATORY
383762 ABASTUMANI
U.S.S.R.
TEL: 227
TLX: 327409 TERMIT
TLF:
EML:
COM: 33

KALBERLA PETER
RADIOASTRONOMISCHES INST
DER UNIVERSITAET BONN
AUF DEM HUEGEL 71
D-5300 BONN 1
GERMANY. F.R.
TEL: 0228-733645
TLX: 0886440
TLF:
EML:
COM: 05.40

KALER JAMES B PROF
UNIVERSITY OF ILLINOIS
349 ASTRONOMY BLDG
1011 W. SPRINGFIELD
URBANA IL 61801
U.S.A.
TEL: 217-333-9382
TLX: 910-2452434 AST
TLF:
EML:
COM: 34

KALINKOV MARIN P DR
DEPT OF ASTRONOMY
BULGARIAN ACAD SCIENCES
72 LENIN BLVD
1784 SOFIA
BULGARIA
TEL:
TLX: 22774 CLANP BG
TLF:
EML:
COM: 28

KALKOFEN WOLFGANG DR
SMITHSONIAN ASTROPHYSICAL
OBSERVATORY
60 GARDEN STREET
CAMBRIDGE MA 02138
U.S.A.
TEL: 617-495-7285
TLX:
TLF:
EML:
COM: 12.36V

KALLOGLIAN ARSEN T DR
BYURAKAN ASTROPHYSICAL
OBSERVATORY
378433 BYURAKAN. ARMENIA
U.S.S.R.
TEL:
TLX:
TLF:
EML:
COM: 28

KALMAN BELA DR
HELIOPHYSICAL OBSERVATORY
PO BOX 30
H-4010 DEBRECEN
HUNGARY
TEL: 52-11-015
TLX: 72517 DEOBS H
TLF:
EML:
COM: 10.12

KALMYKOV A M DR
ASTRONOMICAL INSTITUTE
UZBEK ACADEMY OF SCIENCES
700000 TASHKENT
U.S.S.R.
TEL:
TLX:
TLF:
EML:
COM: 19

KALNAJS AGRIS J DR
MT STROMLO OBSERVATORY
WODEN P.O. ACT 2606
AUSTRALIA
TEL: 062-881111-248
TLX: 62270 AA
TLF:
EML:
COM: 33

KALUZNY JANUSZ DR
ASTRONOMICAL OBSERVATORY
UNIVERSITY OF WARSAW
AL. UJAZDOWSKIE 4
00-478 WARSAW
POLAND
TEL: 294011
TLX: 817063
TLF:
EML:
COM: 42

KAMEL OSMAN M DR
FACULTY OF SCIENCES
ASTRONOMY DEPT
CAIRO UNIVERSITY
GIZA-CAIRO
EGYPT
TEL:
TLX:
TLF:
EML:
COM:

KAMEYA OSAMU DR
NOBEYAMA RADIO OBSERVAT.
NAT. ASTRON. OBSERVATORY
NOBEYAMA. MINAMISAKU
NAGANO 384 13
JAPAN
TEL: 267 98 2831
TLX: 3329005 NAONRO J
TLF:
EML:
COM: 40

KAMIJO FUMIO PROF DR
DEPT OF ASTRONOMY
UNIVERSITY OF TOKYO
YAYOI BUNKYO KU
TOKYO 113
JAPAN
TEL:
TLX:
TLF:
EML:
COM: 34

KAMINISHI KEISUKE PROF
FACULTY OF SCIENCE
KUMAMOTO UNIVERSITY
KUROKAMI 2 CHOME
860 KUMAMOTO
JAPAN
TEL: 096-344-2111
TLX:
TLF:
EML:
COM: 35

KAMMEYER PETER C DR
US NAVAL OBSERVATORY
34 & MASSACHUSETTS AVE NW
WASHINGTON DC 20392
U.S.A.
TEL: 202 653 1563
TLX:
TLF:
EML:
COM: 07

KAMP LUCAS WILLEM DR
DEPT OF ASTRONOMY
BOSTON UNIVERSITY
725 COMMONWEALTH AVE
BOSTON MA 02215
U.S.A.
TEL:
TLX:
TLF:
EML:
COM: 36.37

KAMPER KARL W DR
DAVID DUNLAP OBSERVATORY
RICHMOND HILL. ON L4C 4Y6
CANADA
TEL: 416-884-9562
TLX:
TLF:
EML:
COM:

KANAEV IVAN I DR
PULKOVO OBSERVATORY
196140 LENINGRAD
U.S.S.R.
TEL:
TLX:
TLF:
EML:
COM: 24

KANBACH GOTTFRIED DR
MAX PLANCK INSTITUT FUR
EXTRATERR.PHYSIK
8046 GARCHING
GERMANY F.R.
TEL: 89 3299 544
TLX: 5215845 XTER D
TLF:
EML: BITNET:GOK@DGAIPP1S
COM:

KANDEL ROBERT S DR
LABORATOIRE METEOROLOGIE
DYNAMIQUE
ECOLE POLYTECHNIQUE
F-91128 PALAISEAU CEDEX
FRANCE
TEL: 1-69-41-82-00
TLX: 691596 ECOLEX F
TLF:
EML:
COM: 36

KANDEMIR GUELCIN
ISTANBUL TECHNICAL UNIV
FEN FAKULTESI. FIZIK B
MASLAK
ISTANBUL
TURKEY
TEL: 1609109
TLX:
TLF:
EML:
COM:

KANDPAL CHANDRA D
UP STATE OBSERVATORY
NAINITAL U.P. 263 129
INDIA
TEL: 2136.2325
TLX:
TLF:
EML:
COM: 42

KANE SHARAD R DR
SPACE SCIENCES LAB
UNIVERSITY OF CALIFORNIA
BERKELEY CA 94720
U.S.A.
TEL: 415-642-1719
TLX: 910-366-7945
TLF:
EML:
COM: 10

KANEKO NOBORU DR
DEPT OF PHYSICS
FACULTY OF SCIENCE
HOKKAIDO UNIVERSITY
060 SAPPORO
JAPAN
TEL: 11-716-2111
TLX: 932510 HOKUSC J
TLF:
EML:
COM: 28

KANG GON IK
PYONGYANG ASTRON OBS
ACADEMY OF SCIENCES DPRK
TAESONG DISTRICT
PYONGYANG
KOREA DPR
TEL:
TLX:
TLF:
EML:
COM: 40

KANG JIN SOK
PYONGYANG ASTRON OBS
ACADEMY OF SCIENCES DPRK
TAESONG DISTRICT
PYONGYANG
KOREA DPR
TEL:
TLX:
TLF:
EML:
COM: 10

KANYO SANDOR DR
KONKOLY OBSERVATORY
BOX 67
H-1525 BUDAPEST
HUNGARY
TEL: 166-426
TLX:
TLF:
EML:
COM: 27

KAPANI V K DR
NRAO
POST BOX NO 0
SOCORRO NM 87801
U.S.A.
TEL:
TLX:
TLF:
EML:
COM: 28.40C

KAPANI VIJAY. K.
JPL (NASA)
MAIL CODE 238-700
4800 OAK GROVE DRIVE
PASADENA CA 91109
U.S.A.
TEL: 818-354 1043
TLX: 675429
TLF:
EML:
COM: 28.40

KAPISINSKY IGOR
ASTRONOMICAL INSTITUTE
SLOVAK ACAD. OF SCIENCES
842 28 BRATISLAVA
CZECHOSLOVAKIA
TEL: 427-375157
TLX: 093355
TLF:
EML:
COM: 22

KAPLAN J DR
DEPT OF PHYSICS
UNIVERSITY OF CALIFORNIA
LOS ANGELES CA 90024
U.S.A.
TEL:
TLX:
TLF:
EML:
COM: 21

KAPLAN LEWIS D DR
ATMOSPH. & ENVIRONMENTAL
RESEARCH. INC.
840 MEMORIAL DRIVE
CAMBRIDGE MA 02139
U.S.A.
TEL: 617-547-6207
TLX: 951417 AERC
TLF:
EML:
COM:

KAPOOR RAMESH CHANDER
INDIAN INSTITUTE OF
ASTROPHYSICS
BANGALORE 560 034
INDIA
TEL: 566585
TLX: 845763 IIAB IN
TLF:
EML:
COM: 47.46

KARAALI SALIH DR
ISTANBUL UNIVERSITY
FACULTY OF SCIENCE
DEPT ASTRONOMY & SPACE SC
ISTANBUL 34
TURKEY
TEL: 5224200/610
TLX:
TLF:
EML:
COM:

KARACHENTSEV I D DR
SPEC ASTROPHYS OBS
ACAD OF SC/STAVROPOLSKIJ
ZELENCHUKSKAJA
357147 N ARKHYZ
U.S.S.R.
TEL:
TLX:
TLF:
EML:
COM: 09.28.30.47

KARANDIKAR R V PROF
DEPT OF ASTRONOMY
OSMANIA UNIVERSITY
HYDERABAD 500 007
INDIA
TEL: 71251
TLX:
TLF:
EML:
COM: 16.21

KARDASHEV N S DR
SPACE RESEARCH INSTITUTE
USSR ACADEMY OF SCIENCES
117810 MOSCOW
U.S.S.R.
TEL:
TLX:
TLF:
EML:
COM: 40.51V

KARETNIKOV VALENTIN G P
ODESSA STATE UNIVERSITY
PARK SHEVCHENKO
ASTRONOMICAL OBSERVATORY
270014 ODESSA
U.S.S.R.
TEL: 25-03-56
TLX:
TLF:
EML:
COM: 42

KARLICKY MARIAN
ASTRONOMICAL INSTITUTE
CZECH. ACAD. OF SCIENCES
OBSERVATORY
251 65 ONDREJOV
CZECHOSLOVAKIA
TEL:
TLX:
TLF:
EML:
COM: 10

KARP ALAN HERSH DR
SCIENTIFIC CENTER IBM
1530 PAGE MILL RD
PALO ALTO CA 94304
U.S.A.
TEL: 415-555-[illegible]
TLX:
TLF:
EML:
COM: 27.36

KARPEN JUDITH T
NAVAL RESEARCH LABORATORY
CODE 4175 K
WASHINGTON DC 20375
U.S.A.
TEL: 202-767-3441
TLX:
TLF:
EML:
COM: 10.12

KARPINSKIJ VADIM N DR
PULKOVO OBSERVATORY
196140 LENINGRAD
U.S.S.R.
TEL:
TLX:
TLF:
EML:
COM: 09.12.44

KARYGINA ZOYA V DR
ASTROPHYSICAL INSTITUTE
480068 ALMA-ATA
U.S.S.R.
TEL:
TLX:
TLF:
EML:
COM: 21

KASHSCHEEV B L PROF DR
KHARKOV INSTITUTE FOR
RADIOELECTRONICS
310059 KHARKOV
U.S.S.R.
TEL:
TLX:
TLF:
EML:
COM: 22

KASPER U DR
ZNTRLINST. F. ASTROPHYSIK
STERNWARTE BABELSBERG
ROSA-LUXEMBURG-STR 17A
DDR-1502 POTSDAM
GERMANY. D.R.
TEL:
TLX:
TLF:
EML:
COM: 47

KASTURIRANGAN K DR
ISRO SATELLITE CENTER
AIRPORT ROAD
VIMANAPURA POST
BANGALORE 560 017
INDIA
TEL: 54779
TLX: 845325 & 769
TLF:
EML:
COM: 44

KASUGA TAKASHI
NOBEYAMA RADIO OBS
TOKYO ASTRON OBSERVATORY
NOBEYAMA. MINAMISAKU
NAGANO 384-13
JAPAN
TEL: 267-98-2831
TLX: 3329005 TAO NRO J
TLF:
EML:
COM: 40

KASUMOV FIKRET K O DR
PHYSICS INSTITUTE
NARIMANOVA PR.33
AKADEMGORODOC
370122 BAKU
U.S.S.R.
TEL: 39 67 84
TLX:
TLF:
EML:
COM: 33

KATGERT PETER DR
STERREWACHT
POSTBUS 9513
NL-2300 RA LEIDEN
NETHERLANDS
TEL: 071-272727
TLX: 39058
TLF:
EML:
COM: 28

KATGERT-MERKELIJN J K DR
STERREWACHT
HUYGENS LABORATORIUM
WASSENAARSEWEG 78
NL-2300 RA LEIDEN
NETHERLANDS
TEL:
TLX:
TLF:
EML:
COM:

KATO MARIKO
DEPT OF ASTRONOMY
KEIO UNIVERSITY
4-1-1 HIYOSHI KOULOKU-KU
YOKOHAMA-SHI 223
JAPAN
TEL: 44-63-1111
TLX:
TLF:
EML:
COM: 35

KATO SHOJI PROF
DEPT OF ASTRONOMY
UNIVERSITY OF KYOTO
KITASHIRAKAWA OIWAKE
SAKYOKU KYOTO 606
JAPAN
TEL: 075-751-2111
TLX: 5422693 LIBKYU J
TLF:
EML:
COM: 12.33.47

KATO TAKAKO DR
INST OF PLASMA PHYSICS
NAGOYA UNIVERSITY
FURO-CHO CHIKUSA-KU
NAGOYA 464
JAPAN
TEL: 052-781-5111
TLX: 0447-3691 IPPCNU J
TLF:
EML:
COM: 14C

KATSIS DEMETRIUS DR
12 RUE VARNIS
GR-17124 NEA SMYRNE
GREECE
TEL: 9336014
TLX:
TLF:
EML:
COM: 07

KATZ JONATHAN I
DEPT OF PHYSICS
WASHINGTON UNIVERSITY
ST LOUIS MI 63130
U.S.A.
TEL: 314-889-6202
TLX:
TLF:
EML:
COM: 45

KATZ JOSEPH DR
RACAH INST. OF PHYSICS
HEBREW UNIV. OF JERUSALEM
JERUSALEM 91904
ISRAEL
TEL: 58-46-04
TLX: 25391 HUIL
TLF:
EML:
COM:

KAUFMAN MICHELE DR
PHYSICS DEPT
OHIO STATE UNIVERSITY
174 W. 18TH AVE
COLUMBUS OH 43210
U.S.A.
TEL: 614-422-5 ..
TLX:
TLF:
EML:
COM: 28

KAUFMANN JENS PETER DR
INSTITUT FUR ASTRONOMIE
TECHNISCHE UNIVERSITAT
HARDENBERGSTR. 36
D-1000 BERLIN 12
GERMANY. F.R.
TEL: 030-3145462
TLX: 184262 TUBLN D
TLF:
EML:
COM:

KAUFMANN PIERRE PROF
LAB. APL. ESPACIAIS
ESCOLA POLITECNICA.
CAIXA POSTAL 8174
05508 SAO PAULO SP
BRAZIL
TEL:
TLX:
TLF:
EML:
COM: 10.[illegible].40 51

KAULA WILLIAM M PROF
DEPT. OF EARTH &
SPACE SCIENCE
UNIVERSITY OF CALIFORNIA
LOS ANGELES CA 90024
U.S.A.
TEL:
TLX:
TLF:
EML:
COM: 07.16

KAWABATA KIWAKI PROF
DEPT OF PHYSICS
NAGOYA UNIVERSITY
FUROCHO CHIKUSAKU
NAGOYA 464
JAPAN
TEL:
TLX:
TLF:
EML:
COM: 40.47

KAWABATA KIYOSHI
DEPT OF PHYSICS COLL.SCI
SCIENCE UNIV OF TOKYO
1-3 KAGURAZAKA. SHINJUKU
TOKYO
JAPAN
TEL: 3-260-4271
TLX:
TLF:
EML:
COM:

KAWABATA SHUSAKU PROF
KYOTO GAKUEN UNIVERSITY
SOGABE-MACHI KAMEOKA
KYOTO 621
JAPAN
TEL: 07712-2-2001
TLX:
TLF:
EML:
COM: 42

KAWAGUCHI ICHIRO PROF
DEPT OF ASTRONOMY
FACULTY OF SCIENCE
UNIVERSITY OF KYOTO
606 KYOTO
JAPAN
TEL:
TLX:
TLF:
EML:
COM: 12

KAWALER STEVEN D DR
CTR SOLAR & SPACE RES.
YALE UNIVERSITY
PO BOX 6666
NEW HAVEN CT 06511
U.S.A.
TEL: 203 432 3012
TLX:
TLF:
EML: BITNET:kawaler@yalastro.
COM:

KAWARA KIMIAKI
CERRO TOLOLO
INTERAMERICAN OBSERVATORY
CASILLA 603
LA SERENA
CHILE
TEL: 56-51-213352
TLX: 34-620301
TLF:
EML:
COM: 25

KAWATA YOSHIYUKI DR
KANAZAWA INSTITUTE
OF TECHNOLOGY
NONOICHO
KANAZAWA NIWANI 921
JAPAN
TEL:
TLX:
TLF:
EML:
COM:

KAZANTZIS PANAYOTIS DR
DEPT. OF ASTRONOMY
UNIVERSITY OF GLASGOW
GLASGOW G12 8QQ
U.K.
TEL:
TLX:
TLF:
EML:
COM:

KAZES ILYA DR
OBSERVATOIRE DE PARIS
SECTION D'ASTROPHYSIQUE
F-92195 MEUDON PL CEDEX
FRANCE
TEL: 1-45-07-76-06
TLX:
TLF:
EML:
COM: 34.40

KEAY COLIN S L PROF
PHYSICS DEPT
NEWCASTLE UNIVERSITY
NEWCASTLE NSW 2308
AUSTRALIA
TEL: 049-685-235
TLX: 28194 NEWUN AA
TLF:
EML:
COM: 22P

KEEL WILLIAM C
UNIVERSITY OF ALABAMA
DEPT PHYSICS & ASTRONOMY
P.O. BOX 1921
TUSCALOOSA AL 35486
U.S.A.
TEL: 205-348-5050
TLX:
TLF:
EML:
COM: 28

KEENAN PHILIP C PROF EMER
PERKINS OBSERVATORY
BOX 449
DELAWARE OH 43015
U.S.A.
TEL: 614-363-1257
TLX: 810-482-1715
TLF:
EML:
COM: 29.45

KEGEL WILHELM H PROF
INST THEORETISCHE PHYSIK
UNIVERSITAT FRANKFURT
ROBERT-MAYER-STR 8-10
D-6000 FRANKFURT/MAIN 1
GERMANY. F.R.
TEL: 069-7982357
TLX: 413932 UNIF D
TLF:
EML:
COM: 34

KEIL KLAUS DR
DPT OF GEOLOGY
INSTITUTE OF METEORITICS
UNIVERSITY OF NEW MEXICO
ALBUQUERQUE NM 87131
U.S.A.
TEL: 505 277 4204
TLX:
TLF:
EML:
COM: 15

KEIL STEPHEN L
AIR FORCE GEOPHYSICS LAB
SOLAR RESEARCH BRANCH
SACRAMENTO PEAK OBS
SUNSPOT NM 88349
U.S.A.
TEL: 505-434-1390
TLX:
TLF:
EML:
COM: 12

KELEMEN JANOS
KONKOLY OBSERVATORY
HUNGARIAN ACADEMY OF SCI
BOX 67
H-1525 BUDAPEST
HUNGARY
TEL: 754-122
TLX: 227460 KONOB H
TLF:
EML:
COM:

KELLER CHARLES F
UNIVERSITY OF CALIFORNIA
LOS ALAMOS NATIONAL LAB
BOX 1663 MS F665
LOS ALAMOS NM 87545
U.S.A.
TEL: 505-667-5648
TLX:
TLF:
EML:
COM:

KELLER GEOFFREY
DEPT OF ASTRONOMY
OHIO STATE UNIVERSITY
174 W. 18TH AVENUE
COLUMBUS OH 43210
U.S.A.
TEL: 614-422-6279
TLX:
TLF:
EML:
COM:

KELLER HANS ULRICH DR
OBSERVATORY & PLANETARIUM
NECKARST 47
D-7000 STUTTGART-1
GERMANY. F.R.
TEL: 0711-291004
TLX: 721855 STBST D
TLF:
EML:
COM: 15.46.51

KELLER HORST UWE DR
MPI FUER AERONOMIE
POSTFACH 20
D-3411 KATLENBURG-LINDAU
GERMANY. F.R.
TEL: 05556-41-419
TLX: 965527 AENLI D
TLF:
EML:
COM: 49

KELLERMANN KENNETH I DR
NRAO
EDGEMONT ROAD
CHARLOTTESVILLE VA 22903
U.S.A.
TEL: 804-296-0240
TLX: 910-997-0174
TLF:
EML:
COM: 28.40.47.48.51

KELLOGG EDWIN M DR
CENTER FOR ASTROPHYSICS
60 GARDEN STREET MS 3
CAMBRIDGE MA 02138
U.S.A.
TEL:
TLX:
TLF:
EML:
COM: 48

KEMBHAVI AJIT K
TATA INSTITUTE OF
FUNDAMENTAL RESEARCH
HOMI BHABHA RD
BOMBAY 400 005
INDIA
TEL:
TLX:
TLF:
EML:
COM: 47.48

KENDERDINE SIDNEY DR
MULLARD RADIO ASTRON LAB
MADINGLEY ROAD
CAMBRIDGE CB3 OHE
U.K.
TEL: 223-66477
TLX: 81292
TLF:
EML:
COM: 40

KENNEDY EUGENE T
SCHOOL OF PHYSICAL SCI
NATL INST F. HIGHER EDUC
GLASNEVIN
DUBLIN 9
IRELAND
TEL: DUBLIN 370071
TLX: 30690 NIHED
TLF:
EML:
COM: 14

KENNEDY JOHN E PROF
323 LAKE CRESCENT
SASKATOON SASK S7H 3A1
CANADA
TEL: 374-4614
TLX:
TLF:
EML:
COM: 41.46

KENNICUTT ROBERT C JR
DEPT OF ASTRONOMY
UNIVERSITY OF MINNESOTA
116 CHURCH ST S.E.
MINNEAPOLIS MN 55455
U.S.A.
TEL: 612-376-5224
TLX:
TLF:
EML:
COM: 28.34

KENT STEPHEN M
HARVARD-SMITHSONIAN CTR
FOR ASTROPHYSICS
60 GARDEN STREET
CAMBRIDGE MA 02138
U.S.A.
TEL: 617-495-9681
TLX:
TLF:
EML:
COM:

KENYON SCOTT J DR
CENTER FOR ASTROPHYSICS
60 GARDEN STREET
CAMBRIDGE MA 02138
U.S.A.
TEL: 617 495 7235
TLX: 92 1428
TLF:
EML: BITNET:kenyon@cfa
COM: 42

KEPLER S O
INSTITUTO DE FISICA
UFRGS
AV. BENTO GONCALVES 9500
90049 PORTO ALEGRE - RS
BRAZIL
TEL: 0512-36-4677
TLX: 051-1055 UFRS BR
TLF:
EML:
COM: 25.27

KERR FRANK J DR
ASTRONOMY PROGRAM
UNIVERSITY OF MARYLAND
COLLEGE PK MD 20742
U.S.A.
TEL: 301-454-6302
TLX: 710-826-0352
TLF:
EML:
COM: 28.33.34.40

KERR ROY P PROF
UNIVERSITY OF CANTERBURY
PRIVATE BAG
CHRISTCHURCH
NEW ZEALAND
TEL: 482-009
TLX: NZ 4144
TLF:
EML:
COM:

KESSLER KARL G DR
NATL BUREAU OF STANDARDS
A505 ADMIN.
GAITHERSBURG MD 20899
U.S.A.
TEL: 301-921-3643
TLX: 197674 TPT
TLF:
EML:
COM: 14.31

KESTEVEN MICHAEL J L DR
DIVISION OF RADIOPHYSICS
CSIRO
P.O.BOX 76
EPPING NSW 2121
AUSTRALIA
TEL: 02-868-0222
TLX: 26230 ASTRO
TLF:
EML:
COM: 40

KHACHIKIAN E YE PROF
BYURAKAN ASTROPHYSICAL
OBSERVATORY
378433 BYURAKAN. ARMENIA
U.S.S.R.
TEL: 56-3453/55-6383
TLX:
TLF:
EML:
COM: 28V

KHARADZE E K PROF
ABASTUMANI ASTROPHYSICAL
OBSERVATORY
383762 GEORGIA
U.S.S.R.
TEL: 998891. 225460
TLX: 327409 TERNIT
TLF:
EML:
COM: 33.34.45

KHARE BISHUN N DR
CTR RADIO PHYS/SPACE RES
CORNELL UNIVERSITY
306 SPACE SCIENCES BLDG
ITHACA NY 14853
U.S.A.
TEL: 607-256-3934
TLX:
TLF:
EML:
COM:

KHARIN A S DR
MAIN ASTRON OBSERVATORY
UKRAINIAN ACAD OF SCIENCE
GOLOSEEVO
252127 KIEV
U.S.S.R.
TEL: 66 47 65
TLX: 132517 NEBO
TLF:
EML:
COM: 08

KHARITONOV ANDREJ V DR
ASTROPHYSICAL INSTITUTE
480068 ALMA-ATA
U.S.S.R.
TEL:
TLX:
TLF:
EML:
COM: 29

KHATISASHVILI ALFEZ SH DR
ABASTUMANI ASTROPHYSICAL
OBSERVATORY
383762 ABASTUMANI.GEORGIA
U.S.S.R.
TEL:
TLX:
TLF:
EML:
COM: 20

KHETSURIANI TSIALA S DR
ABASTUMANI ASTROPHYSICAL
OBSERVATORY
383762 ABASTUMANI.GEORGIA
U.S.S.R.
TEL:
TLX:
TLF:
EML:
COM: 12

KHOKHLOVA V L DR
ASTRONOMICAL COUNCIL
USSR ACADEMY OF SCIENCES
PYATNITSKAYA UL. 48
109017 MOSCOW
U.S.S.R.
TEL: 231-54-61
TLX: 412623 SCSTP SU
TLF:
EML:
COM: 29.36

KHOLSHEVNIKOV K V DR
LENINGRAD STATE UNIV
ASTRONOMICAL OBSERVATORY
BIBLIOTECHNAJA PL. 2
198904 LENINGRAD
U.S.S.R.
TEL: 257-94-88
TLX:
TLF:
EML:
COM: 07C

KHOZOV GENNADIJ V
ASTRONOMICAL OBSERVATORY
LENINGRAD STATE UNIV
BIBLIOTECHNAJA PL. 2
198904 LENINGRAD
U.S.S.R.
TEL: 2-57-94-84
TLX: 12168 PHOBOS
TLF:
EML:
COM: 35

KHROMOV G S DR
ASTRONOMICAL COUNCIL
USSR ACADEMY OF SCIENCES
PYATNITSKAYA UL. 48
109017 MOSCOW
U.S.S.R.
TEL:
TLX:
TLF:
EML:
COM: 34.41

KIANG TAO PROF
DUNSINK OBSERVATORY
CASTLEKNOCK
DUBLIN 15
IRELAND
TEL: 387-911
TLX: 31687 DIAS EI
TLF:
EML:
COM: 20.41

KIASATPOOR AHMAD PROF
PHYSICS DEPT
UNIVERSITY OF ESFAHAN
DANESHGAH E
ESFAHAN
IRAN
TEL: 031-44321
TLX: 31-2295 IRE U
TLF:
EML:
COM:

KIBBLEWHITE EDWARD J DR
INSTITUTE OF ASTRONOMY
MADINGLEY ROAD
CAMBRIDGE CB3 0HE
U.K.
TEL:
TLX:
TLF:
EML:
COM:

KIELKOPF JOHN F DR
DEPARTMENT OF PHYSICS
UNIVERSITY OF LOUISVILLE
LOUISVILLE KY 40292
U.S.A.
TEL: 502 588 6787
TLX:
TLF:
EML: BITNET:JFKIEL01@ULKYVX
COM: 14

KIGUCHI MASAYOSHI DR
RES INST F SCI & TECHNOL.
KINKI UNIVERSITY
3-4-1 KOWAKAE
HIGASHI-OSAKA 577
JAPAN
TEL: 06-721-2332
TLX:
TLF:
EML:
COM: 35

KIKUCHI SADAEMON PROF
ASTRONOMICAL INSTITUTE
TOHOKU UNIVERSITY
AOBAYAMA
SENDAI 980
JAPAN
TEL:
TLX:
TLF:
EML:
COM:

KILADZE R I DR
ABASTUMANI ASTROPHYSICAL
OBSERVATORY
383762 ABASTUMANI.GEORGIA
U.S.S.R.
TEL:
TLX:
TLF:
EML:
COM: 16

KILAMBI G C DR
DEPT OF ASTRONOMY
OSMANIA UNIVERSITY
HYDERABAD 500 007
INDIA
TEL: 71-251 x 247
TLX:
TLF:
EML:
COM: 37

KILAR BOGDAN DR
FACULTY OF GEODESY
LJUBLJANA UNIVERSITY
JAMOVA 2
61000 LJUBLJANA
YUGOSLAVIA
TEL:
TLX:
TLF:
EML:
COM:

KILKENNY DAVID DR
S A A O
P O BOX 9
OBSERVATORY 7935
SOUTH AFRICA
TEL: 021-47-0025
TLX: 57-20309 SA
TLF:
EML:
COM: 25

KIM CHUL HEE DR
DEPARTMENT OF EARTH
SCIENCE EDUCATION CHONBUK
NATIONAL UNIVERSITY
JUNJU CHONBUK
KOREA. REPUBLIC
TEL:
TLX:
TLF:
EML:
COM: 27

KIM JIK SU
PYONGYANG ASTRON OBS
ACADEMY OF SCIENCES DPRK
TAESONG DISTRICT
PYONGYANG
KOREA DPR
TEL:
TLX:
TLF:
EML:
COM: 47

KIM YU HWAN
KOREA ASTR/SPACE SC INST
36-1 WHAAM-DONG JUNG-GU
TAEJEON CHUNGCHUNGNAM-DO
300-31 TAEJEON
KOREA. REPUBLIC
TEL: 042-823-1497
TLX: 45532 K
TLF:
EML:
COM: 27

KIM YONG HYOK DR
PYONGYANG ASTRON OBS
ACADEMY OF SCIENCES DPRK
TAESONG DISTRICT
PYONGYANG
KOREA DPR
TEL: 5-3134. 53239
TLX:
TLF:
EML:
COM:

KIM YONG UK
PYONGYANG ASTRON OBS
ACADEMY OF SCIENCES DPRK
TAESONG DISTRICT
PYONGYANG
KOREA DPR
TEL:
TLX:
TLF:
EML:
COM:

KIM YUL
PYONGYANG ASTRON OBS
ACADEMY OF SCIENCES DPRK
TAESONG DISTRICT
PYONGYANG
KOREA DPR
TEL:
TLX:
TLF:
EML:
COM:

KIM ZONG DOK
PYONGYANG ASTRON OBS
ACADEMY OF SCIENCES DPRK
TAESONG DISTRICT
PYONGYANG
KOREA DPR
TEL:
TLX:
TLF:
EML:
COM: 14

KIMBLE RANDY A DR
DPT OF PHYS. & ASTRONOMY
THE JOHNS HOPKINS UNIV.
CHARLES & 34TH STREETS
BALTIMORE MD 21218
U.S.A.
TEL: 301 338 8738
TLX: 9102400225 JHU CASMD
TLF:
EML: SPAN:SCIVAX::CASA::RAK
COM: 44

KIMURA HIROSHI DR
PURPLE MOUNTAIN
OBSERVATORY
NANJING
CHINA. PEOPLE'S REP.
TEL: 33921
TLX: 34144 PMONJ CN
TLF:
EML:
COM: 34

KING ANDREW R DR
ASTRONOMY DEPT
UNIVERSITY OF LEICESTER
LEICESTER
U.K.
TEL: 0533-554455
TLX: 341198
TLF:
EML:
COM:

KING DAVID S PROF
DEPT PHYSICS & ASTRONOMY
UNIVERSITY OF NEW MEXICO
ALBUQUERQUE NM 87131
U.S.A.
TEL:
TLX:
TLF:
EML:
COM: 35.41

KING HENRY C DR
TRILLIUM
206 WHITE LION ROAD
LITTLE CHALFONT
BUCKS HP7 9NU
U.K.
TEL:
TLX:
TLF:
EML:
COM: 41

KING IVAN R PROF
ASTRONOMY DEPT
UNIVERSITY OF CALIFORNIA
BERKELEY CA 94720
U.S.A.
TEL: 415-642-2206
TLX: 820181 UCB ASTRAL
TLF:
EML:
COM: 25.28.33.37

KING R B DR
PO BOX 725
MEDOCINO CA 95460
U.S.A.
TEL:
TLX:
TLF:
EML:
COM: 14.29

KING ROBERT WILSON JR DR
DEPT OF EARTH ATMOSPHERIC
& PLANETARY SCIENCES
MIT 54-620
CAMBRIDGE MA 02139
U.S.A.
TEL: 617-253-7064
TLX: 921473 MIT CAM
TLF:
EML:
COM: 04.19

KINGSTON ARTHUR E PROF
DEPT OF APPLIED MATHS
& THEORETICAL PHYSICS
QUEEN'S UNIVERSITY
BELFAST BT7 1NN
U.K.
TEL: 0232-245133
TLX: 74487 QUB AND G
TLF:
EML:
COM: 14

KING-HELE DESMOND G DR
ROYAL AIRCRAFT ESTABL.
FARNBOROUGH. HANTS
U.K.
TEL: 0252 24461
TLX:
TLF:
EML:
COM: 07

KINMAN THOMAS D DR
KITT PEAK NATIONAL OBS
PO BOX 26732
TUCSON AZ 85726
U.S.A.
TEL: 602-327-5511
TLX: 0666-484 AURA NOAO
TLF:
EML:
COM: 28.33

KINOSHITA HIROSHI DR
TOKYO ASTRONOMICAL
OBSERVATORY
OSAWA MITAKA
181 TOKYO
JAPAN
TEL: 0422-32-5111
TLX:
TLF:
EML:
COM: 04C.07C 20

KIPLINGER ALAN L DR
CODE 682.6
NASA
GSFC
GREENBELT MD 20771
U.S.A.
TEL: 301 286 5424
TLX: 89675
TLF:
EML:
COM: 10 27

KIPPENHAHN RUDOLF PROF
MPI F PHYSIK & ASTROPHYS
KARL-SCHWARZSCHILD-STR 1
D-8046 GARCHING B MUNCHEN
GERMANY. F.R.
TEL: 089-32990
TLX: 524629 ASTRO D
TLF:
EML:
COM: 27.35C

KIPPER TONU DR
TARTU ASTROPHYSICAL OBS
ESTONIAN ACAD OF SCIENCES
TORAVERE
202444 TARTU
U.S.S.R.
TEL:
TLX:
TLF:
EML:
COM: 09.14.29

KIRAL ADNAN PROF
UNIVERSITY OBSERVATORY
BEYAZIT ISTANBUL
TURKEY
TEL:
TLX:
TLF:
EML:
COM:

KIRBIYIK HALIL DR
PHYSICS DEPT
MIDDLE EAST TECHN UNIV
ANKARA 06531
TURKEY
TEL: 237100/3528
TLX: 42761 ODTK TR
TLF:
EML:
COM:

KIRBY KATE P DR
CENTER FOR ASTROPHYSICS
60 GARDEN STREET
CAMBRIDGE MA 02138
U.S.A.
TEL: 617 495 7217
TLX:
TLF:
EML:
COM: 14

KIRK JOHN DR
MAX PLANCK INSTITUT
FUR ASTROPHYSIK
KARLSCHWARZSCHILD STR.1
D-8046 GARCHING
GERMANY F.R.
TEL:
TLX:
TLF:
EML:
COM: 48

KIRKPATRICK RONALD C DR
LOS ALAMOS NATL LAB
MS 220
LOS ALAMOS NM 87545
U.S.A.
TEL: 505-667-4812
TLX:
TLF:
EML:
COM: 34

KIRSHNER ROBERT PAUL DR
DEPT OF ASTRONOMY
HARVARD UNIVERSITY
CAMBRIDGE MA 02138
U.S.A.
TEL: 617-495-7390
TLX:
TLF:
EML:
COM: 28.34

KISELYOV ALEXEJ A DR
PULKOVO OBSERVATORY
196140 LENINGRAD
U.S.S.R.
TEL:
TLX:
TLF:
EML:
COM: 26C

KISLYAKOV ALBERT G DR
APPLIED PHYSICS INSTITUTE
ACADEMY OF SCIENCES
ULYANOV STREET 46
603600 GORKY
U.S.S.R.
TEL:
TLX:
TLF:
EML:
COM: 40

KISLYUK VITALIJ S DR
MAIN ASTRON OBSERVATORY
UKRAINIAN ACAD OF SCIENCE
GOLOSEEVO
252127 KIEV
U.S.S.R.
TEL:
TLX: 131406 SKY US
TLF:
EML:
COM: 16.24

KISSELEVA TAMARA P
PULKOVO MAIN ASTRONOMICAL
OBSERVATORY
196140 LENINGRAD
U.S.S.R.
TEL:
TLX:
TLF:
EML:
COM: 20

KITAMOTO SHUNJI DR
FACULTY OF SCIENCES OSAKA UNIVERSITY
1-1 MACHIKANEYAMA
TOYONAKA. OSAKA 560
JAPAN
TEL: 6 844 1151
TLX:
TLF:
EML: BITNET:KITAMOTO@JPNOSKFM
COM:

KITAMURA M PROF
TOKYO ASTRONOMICAL OBSERVATORY
MITAKA TOKYO 181
JAPAN
TEL: 0422-32-5111
TLX: 2822307 TAOMTK J
TLF:
EML:
COM: 42

KITAMURA SEIICHI DR
SHIGA UNIVERSITY
2 CHOME 5-1
HIRATSU
OTSU SHIGA 520
JAPAN
TEL: 775-37-0081
TLX:
TLF:
EML:
COM:

KITCHIN CHRISTOPHER R DR
HATFIELD POLYTECHNIC OBSERVATORY
BAYFORDBURY
HERTFORD HERTS SG13 8LD
U.K.
TEL: 0992-558-451
TLX: 262413
TLF:
EML:
COM: 29 46

KJAERGAARD PER DR
UNIVERSITY OBSERVATORY
OESTER VOLDGADE 3
DK-1350 COPENHAGEN K
DENMARK
TEL: 1-14-17-90
TLX: 44155 DANAST DK
TLF:
EML:
COM:

KJELDSETH-MOE OLAV DR
INST THEORET ASTROPHYSICS
UNIVERSITY OF OSLO
P O BOX 1029
N-0315 BLINDERN. OSLO 3
NORWAY
TEL: 47-2-456510
TLX: 72425 UNIOS N
TLF:
EML:
COM: 10

KJURKCHIEVA DIANA DR
DEPARTMENT OF PHYSICS
HIGHER PEDAGOGICAL INST
9700 SHOUMEN
BULGARIA
TEL: 6 31 51 216
TLX:
TLF:
EML:
COM: 27.42

KLAPP JAIME DR
DPT DE FISICA UNIVERSIDAD AUTONOMA METROPOLITANA
AVE MICHOACAN Y PURISIMA
IZTAPALAPA DF 09340
MEXICO
TEL: 515 64 42
TLX: 1764186 KEBEME
TLF:
EML:
COM:

KLARE GERHARD DR
LANDESSTERNWARTE
KOENIGSTUHL
D-6900 HEIDELBERG 1
GERMANY. F.R.
TEL: 06221/10036
TLX:
TLF:
EML:
COM: 33

KLARMANN JOSEPH PROF
WASHINGTON UNIVERSITY
DEPT OF PHYSICS
ST LOUIS MO 63130
U.S.A.
TEL: 314-889-6299
TLX: 650-2557719 MCI
TLF:
EML:
COM:

KLECZEK JOSIP DR
ASTRONOMICAL INSTITUTE
251 65 ONDREJOV
CZECHOSLOVAKIA
TEL: 72-45-25
TLX: 121579 ASTR C
TLF:
EML:
COM: 35.10.46C

KLEIN KARL LUDWIG DR
UA 324 DASOP
OBSERVATOIRE DE PARIS
5 PLACE JANSSEN
92195 MEUDON PR. CEDEX
FRANCE
TEL: 45 34 77 61
TLX:
TLF:
EML: SPAN:MEUDON::KLEIN
COM: 10.12.40

KLEIN MICHAEL J DR
SPACE SCIENCES DIVISION
JET PROPULSION LAB
BLDG 264-802
PASADENA CA 91109
U.S.A.
TEL: 818-354-7132
TLX:
TLF:
EML:
COM: 51

KLEIN RICHARD I DR
LAWRENCE LIVERMORE LAB
UNIVERSITY OF CALIFORNIA
PO BOX 808:L-23
LIVERMORE CA 94550
U.S.A.
TEL: 415-422-3548
TLX:
TLF:
EML:
COM: 36

KLEIN ULRICH
RADIOASTRONOMISCHES INST
DER UNIVERSITAET BONN
AUF DEM HUEGEL 71
D-5300 BONN 1
GERMANY. F.R.
TEL: 0228-73-3644
TLX:
TLF:
EML:
COM: 28.40

KLEINMANN DOUGLAS E DR
HONEYWELL ELECTRO OPTICS OPERATION
2 FORBES RI
LEXINGTON MA 02173
U.S.A.
TEL: 617-863-3841
TLX: 92-3477
TLF:
EML:
COM:

KLEMOLA ARNOLD R DR
LICK OBSERVATORY
UCSC
SANTA CRUZ CA 95064
U.S.A.
TEL: 408-429-2907
TLX:
TLF:
EML:
COM: 20.24

KLEMPERER W K DR
ELECTROMAG. FIELDS DIV
NATL BUREAU OF STANDARDS
325 BROADWAY
BOULDER CO 80303
U.S.A.
TEL: 303-497-3757
TLX: 592811 NOAA MASC BDR
TLF:
EML:
COM:

KLEPCZYNSKI WILLIAM J DR
US NAVAL OBSERVATORY
34 & MASSACHUSETTS AVE NW
WASHINGTON DC 20390
U.S.A.
TEL: 202-653-1521
TLX: 710-822-1970
TLF:
EML:
COM: 04.19.31C

KLINISHIN I A PROF
PEDAGOGIC INSTITUTE
PUSHKIN STR. 96 APT. 66
284000 IVANOFRANKOV SK
U.S.S.R.
TEL:
TLX:
TLF:
EML:
COM:

KLINGLESMITH DANIEL A DR
NASA/GSFC
CODE 684
GREENBELT MD 20771
U.S.A.
TEL: 301-286-6541
TLX:
TLF:
EML:
COM:

KLINKHAMER FRANS DR
PHYSICS DEPARTMENT L 413
LAWRENCE LIVERMORE NL LAB
PO BOX 808
LIVERMORE CA 94550
U.S.A.
TEL:
TLX:
TLF:
EML: BITNET:FRANS@LBL
COM: 48

KLIORE ARVYDAS JOSEPH DR
JPL
4800 OAK GROVE DRIVE
PASADENA CA 91109
U.S.A.
TEL: 818-354-6164
TLX: 675429
TLF:
EML:
COM:

KLOCK B L DR
6601 S. HOMESTAKE DRIVE
BOWIE MD 20715
U.S.A.
TEL: 301-262-1506
TLX:
TLF:
EML:
COM: 08.09.24

KLOKOCNIK JAROSLAV DR
ASTRONOMY INSTITUTE
CZECH. ACAD. SCI.
251 65 ONDREJOV OBS
CZECHOSLOVAKIA
TEL: 724525
TLX: 121579 ASTR C
TLF:
EML:
COM: 07

KLVANA MIROSLAV
ASTRONOMICAL INSTITUTE
CZECH. ACAD. OF SCIENCES
ONDREJOV OBSERVATORY
251 65 ONDREJOV
CZECHOSLOVAKIA
TEL:
TLX:
TLF:
EML:
COM: 10

KNACKE ROGER F DR
DEPT EARTH & SPACE SCI
SUNY AT STONY BROOK
STONY BROOK NY 11794
U.S.A.
TEL: 516-246-7673
TLX: 5102287767
TLF:
EML:
COM: 15.34

KNAPP GILLIAN R DR
DEPT ASTROPHYSICAL SCI
PRINCETON UNIVERSITY
PRINCETON NJ 08544
U.S.A.
TEL: 609-452-3824
TLX:
TLF:
EML:
COM: 28.33.34

KNEER FRANZ DR
UNIVERSITATS STERNWARTE
GEISMARLANDSTRASSE 11
D-3400 GOETTINGEN
GERMANY. F.R.
TEL: 0551-395-042
TLX: 96753 USTERN D
TLF:
EML:
COM: 12

KNEZEVIC ZORAN
ASTRONOMICAL OBSERVATORY
VOLGINA 7
11050 BELGRADE
YUGOSLAVIA
TEL: 11 419 357
TLX: 72610 AOB YU
TLF:
EML:
COM: 15.20

KNIFFEN DONALD A DR
CODE 662 NASA GSFC
GREENBELT MD 20771
U.S.A.
TEL:
TLX:
TLF:
EML:
COM:

KNOLKER MICHAEL DR
UNIV STERNWARTE
GOTTINGEN
GEISMARLANDSTRASSE 11
3400 GOTTINGEN
GERMANY. F.R.
TEL: 551-39-5046
TLX:
TLF:
EML:
COM: 12.35

KNOSKA STEFAN
ASTRONOMICAL INSTITUTE
SLOVAK ACAD. OF SCIENCES
059 60 TATRANSKA LOMNICA
CZECHOSLOVAKIA
TEL:
TLX:
TLF:
EML:
COM: 10

KNOWLES STEPHEN H DR
NAVAL RESEARCH LABORATORY
CODE 4185
WASHINGTON DC 20375-5000
U.S.A.
TEL: 202-262-2891
TLX:
TLF:
EML:
COM: 19.51

KNUDE JENS KIRKESKOV DR
UNIVERSITY OBSERVATORY
OESTER VOLDGADE 3
DK-1350 COPENHAGEN K
DENMARK
TEL: 1-14-17-90
TLX: 44155 DANAST DK
TLF:
EML:
COM: 25C.34

KO HSIEN C PROF
DEPT OF ELECT ENGINEERING
OHIO STATE UNIVERSITY
COLUMBUS OH 43210
U.S.A.
TEL: 614-422-2571
TLX: 24-5334
TLF:
EML:
COM: 40

KOBAYASHI KISUKE DR
SCIENCE INST OF OSAKA
PREFECTURE/13-23 KARITA 4
CHOME/SUMIYOSHI-KU
OSAKA 558
JAPAN
TEL: 06-692-1882
TLX:
TLF:
EML:
COM:

KOBAYASHI YUKISATU
TOKYO ASTRON OBSERVATORY
2-21-1 OSAWA
MITAKA
TOKYO 181
JAPAN
TEL: 0422-32-5111
TLX: 2822307 TAOMTK
TLF:
EML:
COM: 31

KOCER D DR
ASTRONOMY DEPARTMENT
ISTANBUL UNIVERSITY
34452 ISTANBUL
TURKEY
TEL:
TLX:
TLF:
EML:
COM: 51

KOCH DAVID G
SMITHSONIAN ASTROPHYSICAL
OBSERVATORY
60 GARDEN STREET
CAMBRIDGE MA 02138
U.S.A.
TEL: 617-495-7479
TLX: 921428 SATELLITE CAM
TLF:
EML:
COM:

KOCH ROBERT H DR
DEPT ASTRON & ASTROPHYS
UNIV OF PENNSYLVANIA
DAVID RITTENHOUSE LAB
PHILADELPHIA PA 19104
U.S.A.
TEL: 215-898-7882
TLX: 834621
TLF:
EML:
COM: 25.42F 51

KOCHAROV GRANT E PROF
PHYSICO-TECHNICAL INST
USSR ACADEMY OF SCIENCES
194021 LENINGRAD
U.S.S.R.
TEL: 247-91-67
TLX:
TLF:
EML:
COM: 49

KOCHHAR R K DR
INDIAN INSTITUTE OF
ASTROPHYSICS
BANGALORE 560 034
INDIA
TEL: 566585
TLX: 845763 IIAP IN
TLF:
EML:
COM: 28.35

KOCH-MIRAMOND LYDIE DR
IRF/DPHG/ASTROPHYSIQUE
CEN SACLAY
F-91191 GIF-S YVETTE CDX
FRANCE
TEL: 1-69-08-41-29
TLX: 690860 PHYSPAC
TLF:
EML:
COM: 44.48

KODAIRA KEIICHI PROF
NATIONAL ASTRONOMICAL OBS
MITAKA
TOKYO 181
JAPAN
TEL: 0422-32-5111
TLX: 2822307 TAOMTK J
TLF:
EML:
COM: 28.29.36

KODAMA HIDEO
COLL.LIB.ARTS. KYOTO UNIV
YOSHIDA NIHONMATSU-CHO
SAKYO-KU
KYOTO 606
JAPAN
TEL: 075-7512111
TLX:
TLF:
EML:
COM: 47

KOEBERL CHRISTIAN DR
INSTITUTE OF GEOCHEMISTRY
UNIVERSITY OF VIENNA
DR. KARL-LUEGER-RING 1
A-1010 VIENNA
AUSTRIA
TEL: 222-4300-2360
TLX:
TLF:
EML:
COM: 15.22C.51

KOECKELENBERGH ANDRE DR
OBS ROYAL DE BELGIQUE
AVENUE CIRCULAIRE 3
B-1180 BRUXELLES
BELGIUM
TEL: 02-375-24-84
TLX: 21565 OBSBEL B
TLF:
EML:
COM: 10

KOEHLER H PROF DR
SAUERBRUCHSTR 6
7920 HEIDENHEIM a.d.BRENZ
GERMANY. F.R.
TEL: 07321-44560
TLX:
TLF:
EML:
COM: 09

KOEHLER JAMES A PROF
PHYSICS DEPT
UNIV OF SASKATCHEWAN
SASKATOON S7N OWO
CANADA
TEL: 306-966-6442
TLX:
TLF:
EML:
COM:

KOEHLER PETER
CARL-ZEISS STR 1
DDR-6900 JENA
GERMANY D.R.
TEL:
TLX:
TLF:
EML:
COM: 09

KOENIGSBERGER GLORIA
INSTITUTO DE ASTRONOMIA
UNAM
APDO POSTAL 70-264
04510 MEXICO DF
MEXICO
TEL: 905-548-5305/06
TLX:
TLF:
EML:
COM:

KOEPPEN JOACHIM DR
INST F. THEOR ASTROPHYSIK
DER UNIV HEIDELBERG
IM NEUENHEIMER FELD 561
D-6900 HEIDELBERG
GERMANY. F.R.
TEL: 06221-562988
TLX: 461515 UNIHD D
TLF:
EML:
COM: 34

KOESTER DETLEV DR
DEPT. OF PHYSICS & ASTR.
LOUISIANA STATE UNIV.
BATON ROUGE
U.S.A.
TEL: 504 388-2261
TLX: 559184
TLF:
EML:
COM: 35.36

KOGOSHVILI NATELA G
ASTROPHYSICAL OBSERVATORY
MOUNT KANOBILI
383762 ABASTUMANI.GEORGIA
U.S.S.R.
TEL: 283
TLX:
TLF:
EML:
COM: 28

KOGURE TOMOKAZU DR
DEPT OF ASTRONOMY
FACULTY OF SCIENCE
UNIVERSITY OF KYOTO
KYOTO 606
JAPAN
TEL: 075-751-2111
TLX: 5422693 LIBKYU J
TLF:
EML:
COM: 29

KOHL JOHN L DR
CENTER FOR ASTROPHYSICS
60 GARDEN ST
CAMBRIDGE MA 02138
U.S.A.
TEL: 617-495-7377
TLX: 921428
TLF:
EML:
COM: 14

KOHOUTEK LUBOS DR
HAMBURGER STERNWARTE
GOJENBERGSWEG 112
D-2050 HAMBURG 80
GERMANY. F.R.
TEL: 40-7252-4112
TLX: 217884 HAMS D
TLF:
EML:
COM: 15.20.34

KOJOIAN GABRIEL DR
DEPT OF PHYSICS
UNIVERSITY OF WISCONSIN
EAU CLAIRE WI 54701
U.S.A.
TEL: 715-836-3148
TLX:
TLF:
EML:
COM: 40

KOKURIN YURIJ L DR
LEBEDEV PHYSICAL INST
USSR ACADEMY OF SCIENCES
LENINSKY PROSPEKT 53
117924 MOSCOW
U.S.S.R.
TEL: 135-03-60
TLX: 411479 NEOD SU
TLF:
EML:
COM: 08.19

KOLACZEK BARBARA DR
PLANETARY GEODESY DEPT
POLISH ACAD OF SCIENCES
UL. BARTYCKA 18
00-716 WARSAW
POLAND
TEL: 41-00-41
TLX: 815670 CBK PL
TLF:
EML:
COM: 04.19V.31

KOLB EDWARD W DR
THEORETICAL ASTROPHYSICS
MS 209
FERMI NTL ACCELERATOR LAB
BATAVIA IL 60510
U.S.A.
TEL: 312 840 4695
TLX: 720481
TLF:
EML:
COM: 47.48

KOLCHINSKIJ I G DR
MAIN ASTRON OBSERVATORY
UKRAINIAN ACADEMY OF SCI
OF SCIENCES
252127 KIEV
U.S.S.R.
TEL:
TLX:
TLF:
EML:
COM: 24

KOLESNIK IGOR G DR
MAIN ASTRON OBSERVATORY
UKRAINIAN ACADEMY OF SCI
GOLOSEEVO
252127 KIEV
U.S.S.R.
TEL: 663110
TLX: 131406 SKY SU
TLF:
EML:
COM: 33.34

KOLESNIK L N DR
MAIN ASTRON OBSERVATORY
UKRAINIAN ACADEMY OF SCI
252127 KIEV
U.S.S.R.
TEL: 66-08-69
TLX: 131406
TLF:
EML:
COM: 33

KOLESOV A K DR
LENINGRAD STATE UNIV
ASTRONOMICAL OBSERVATORY
199178 LENINGRAD
U.S.S.R.
TEL:
TLX:
TLF:
EML:
COM: 36

KOLEV DIMITAR ZDRAVKOV
BLOCK 1 E-20
SMOLJAN
BULGARIA
TEL:
TLX:
TLF:
EML:
COM:

KOLLATSCHNY WOLFRAM DR
UNIVERSITAETS STERNWARTE
GEISMARLANDSTRASSE 11
3400 GOETTINGEN
GERMANY, F.R.
TEL: 0551-395063
TLX: 96753
TLF:
EML: BITNET:WKOLLAT@DGOGWDG1
COM: 28

KOLLBERG ERIK L PROF
ELECTRON PHYSICS I
CHALMERS UNIV TECHNOLOGY
S-412 96 GOETEBORG
SWEDEN
TEL: 31-810100
TLX: 2400 ONSPACE
TLF:
EML:
COM:

KOMAROV N S DR
ODESSA STATE UNIVERSITY
ASTRONOMICAL OBSERVATORY
SHEVCHENKO PARK
270014 ODESSA
U.S.S.R.
TEL: 220396
TLX:
TLF:
EML:
COM: 29

KONDO MASAAKI DR
COLLEGE GENERAL EDUCATION
UNIVERSITY OF TOKYO
KOMABA 3-8-1 MEGURO-KU
TOKYO 153
JAPAN
TEL:
TLX:
TLF:
EML:
COM: 48

KONDO MASAYUKI DR
TOKYO ASTRONOMICAL OBS
UNIVERSITY OF TOKYO
MITAKA
TOKYO 181
JAPAN
TEL: 0422-32-5111
TLX: 2822307 TAOMTK J
TLF:
EML:
COM:

KONDO YOJI DR
NASA/GSFC
CODE 684
GREENBELT MD 20771
U.S.A.
TEL: 301-286-6247
TLX: 710-8289716
TLF:
EML:
COM: 34.42V.44C

KONIGL ARIEH DR
ASTRON & ASTROPHYSICS CTR
THE UNIVERSITY OF CHICAGO
5640 SOUTH ELLIS AVE
CHICAGO IL 60637
U.S.A.
TEL: 312 702 7968
TLX: 282131
TLF:
EML: SPAN:lasr::oddjob::arieh
COM:

KONIN V V DR
NIKOLAEV BRANCH OF THE
MAIN ASTRONOMICAL OBS
327000 NIKOLAEV
U.S.S.R.
TEL:
TLX:
TLF:
EML:
COM: 08

KONONOVICH EDWARD V DR
STERNBERG ASTRONOMICAL
INSTITUTE
119899 MOSCOW
U.S.S.R.
TEL:
TLX:
TLF:
EML:
COM: 12.46

KONOPLEVA VARVARA P DR
MAIN ASTRON OBSERVATORY
UKRAINIAN ACADEMY OF SCI
252127 KIEV
U.S.S.R.
TEL: 66 3110
TLX: 131406 SKY SU
TLF:
EML:
COM: 15

KONTIZAS EVANGELOS DR
NATL OBSERVATORY ATHENS
ASTRONOMICAL INSTITUTE
THISSION PO BOX 20048
GR-11810 ATHENS
GREECE
TEL: 01-3461191
TLX: 215530 OBSA GR
TLF:
EML:
COM: 36.37

KONTIZAS MARY DR
DEPT OF ASTRONOMY
UNIVERSITY OF ATHENS
PANEPISTIMIOPOLIS
GR-15771 ZOGRAFOS
GREECE
TEL: 01-7235122
TLX:
TLF:
EML:
COM: 37

KOO DAVID C-Y DR
LICK OBSERVATORY
NATURAL SCIENCE II
UNIVERSITY OF CALIFORNIA
SANTA CRUZ CA 95064
U.S.A.
TEL: 408 429 2130
TLX: 910 9971741 UNICAL
TLF:
EML: BITNET:KOO@PORTAL
COM: 28.47

KOORNNEEF JAN DR
SPACE TELESCOPE SCI INST
HOMEWOOD CAMPUS
3700 SAN MARTIN DRIVE
BALTIMORE MD 21218
U.S.A.
TEL: 301-338-4802
TLX: 6849101
TLF:
EML:
COM: 34

KOPAL ZDENEK PROF
DEPT OF ASTRONOMY
UNIVERSITY OF MANCHESTER
MANCHESTER M13 9PL
U.K.
TEL: 61-273-7121
TLX: 668932 MCHRUL G
TLF:
EML:
COM: 16.26.42

KOPECKY MILOSLAV DR
ASTRONOMICAL INSTITUTE
CZECH. ACAD. OF SCIENCES
FRICOVA 1
251 65 ONDREJOV
CZECHOSLOVAKIA
TEL: 724525
TLX: 121579
TLF:
EML:
COM: 10.12

KOPP ROGER A DR
LOS ALAMOS NATIONAL LAB
MS F531
LOS ALAMOS NM 87545
U.S.A.
TEL: 505-667-4398
TLX: 660495 LOS ALAMOS
TLF:
EML:
COM:

KOPYLOV I M DR
SPECIAL ASTROPHYSICAL OBS
USSR ACADEMY OF SCIENCES
STAVROPOL TERRITORY
357140 N ARKHYZ
U.S.S.R.
TEL: 93-159
TLX:
TLF:
EML:
COM: 09.27.29

KORCHAK A A DR
INSTITUTE OF TERRESTRIAL
MAGNETISM & IONOSPHERE
142092 TROITSK
U.S.S.R.
TEL:
TLX:
TLF:
EML:
COM:

KORMENDY JOHN DR
DOMINION ASTROPHYS OBS
5071 W SAANICH ROAD
VICTORIA BC V8X 4M6
CANADA
TEL: 604-388-3944
TLX: 0497295 NRC DAO VIC
TLF:
EML:
COM: 28.33.47

KOROVYAKOVSKIJ YURIJ P DR
SPECIAL ASTROPHYS OBS
USSR ACADEMY OF SCIENCES
NIZHNIJ ARKHYZ
357140 STAVROPOLSKIJ KRAJ
U.S.S.R.
TEL:
TLX:
TLF:
EML:
COM: 09

KOSIN GENNADIJ S DR
PULKOVO OBSERVATORY
196140 LENINGRAD
U.S.S.R.
TEL:
TLX:
TLF:
EML:
COM: 08

KOSTIK ROMAN I
MAIN ASTRONOMICAL OBS
UKRAINIAN ACADEMY OF SCI
252127 KIEV
U.S.S.R.
TEL: 66-4762
TLX: 131406 SKY SU
TLF:
EML:
COM: 10.12

KOSTINA LIDIJA D DR
PULKOVO OBSERVATORY
196140 LENINGRAD
U.S.S.R.
TEL:
TLX:
TLF:
EML:
COM: 19

KOSTYAKOVA ELENA B DR
STERNBERG STATE
ASTRONOMICAL INSTITUTE
117234 MOSCOW
U.S.S.R.
TEL:
TLX:
TLF:
EML:
COM: 34

KOSTYLEV K V DR
ENGELHARDT ASTRONOMICAL
OBSERVATORY
422526 KAZAN
U.S.S.R.
TEL:
TLX:
TLF:
EML:
COM: 22

KOSUGI TAKEO
NOBEYAMA RADIO OBS OF
TOKYO ASTRON OBSERVATORY
NOBEYAMA. MINAMISAKU
NAGANO 384-13
JAPAN
TEL: 267-98-2034
TLX: 3329005 TAONRO-J
TLF:
EML:
COM:

KOTELNIKOV V A ACAD
INST OF RADIO & ELECTRON
USSR ACADEMY OF SCIENCES
103907 MOSCOW
U.S.S.R.
TEL: 203-60-78
TLX:
TLF:
EML:
COM: 40

KOTHARI D S DR
DEPT OF PHYSICS
UNIVERSITY OF DELHI
NEW DELHI 110 007
INDIA
TEL:
TLX:
TLF:
EML:
COM: 35

KOTOV VALERY DR
CRIMEAN ASTROPHYSICAL OBS
USSR ACADEMY OF SCIENCES
NAUCHNY
334413 CRIMEA
U.S.S.R.
TEL:
TLX:
TLF:
EML:
COM: 12

KOTRC PAVEL
ASTRONOMICAL INSTITUTE
CZECH. ACAD. OF SCIENCES
251 65 ONDREJOV
CZECHOSLOVAKIA
TEL: 72-45-25
TLX: 121579
TLF:
EML:
COM: 10.12

KOUBSKY PAVEL
ASTRONOMICAL INSTITUTE
CZECH. ACAD. OF SCIENCES
ONDREJOV OBSERVATORY
251 65 ONDREJOV
CZECHOSLOVAKIA
TEL:
TLX:
TLF:
EML:
COM: 29.42

KOURGANOFF VLADIMIR PROF
20 AVE PAUL APPELL
F-75014 PARIS
FRANCE
TEL: 1-45-40-50-53
TLX:
TLF:
EML:
COM: 46

KOUTCHMY SERGE DR
INSTITUT D'ASTROPHYSIQUE
98 BIS BOULEVARD ARAGO
F-75014 PARIS
FRANCE
TEL: 1-43-20-14-25
TLX: 270070 F
TLF:
EML:
COM: 10.12.21

KOUVELIOTOU CHRYSSA DR
SECTION OF ASTROPHYSICS
DEPT OF PHYSICS
UNIVERSITY OF ATHENS
GR 15783 ZOGRAFOS ATHENS
GREECE
TEL: 7235122
TLX: 223815 UNIV GR
TLF: (301)7228981
EML:
COM:

KOVACHEV B J DR
DEPT ASTRONOMY & NTL OBS
BULGARIAN ACAD SCIENCES
72 LENIN BLVD
1784 SOFIA
BULGARIA
TEL: 758827
TLX: 23561 ECF BAN BG
TLF:
EML:
COM: 09.29

KOVACS AGNES DR
HELIOPHYSICAL OBSERVATORY
HUNGARIAN ACADEMY OF SCI
BOX 30
H-4010 DEBRECEN
HUNGARY
TEL: 52-11-015
TLX: 72517 DEOBS H
TLF:
EML:
COM: 10

KOVACS GEZA DR
KONKOLY OBSERVATORY
HUNGARIAN ACADEMY OF SCI
BOX 67
H-1525 BUDAPEST
HUNGARY
TEL: 1-166-426
TLX: 227460
TLF:
EML:
COM:

KOVAL I K DR
MAIN ASTRONOMICAL OBS
UKRAINIAN ACADEMY OF SCI
GOLOSEEVO
252127 KIEV
U.S.S.R.
TEL: 660869
TLX:
TLF:
EML:
COM:

KOVALEVSKY JEAN DR
CERGA
AVENUE COPERNIC
F-06130 GRASSE
FRANCE
TEL: 93-36-58-49
TLX: 470865 CERGA F
TLF:
EML:
COM: 07.08C.24.31C

KOVAR N S DR
PHYSICS DEPT
UNIVERSITY OF HOUSTON
HOUSTON TX 77004
U.S.A.
TEL:
TLX:
TLF:
EML:
COM:

KOVAR ROBERT P DR
9666 E. ORCHARD DR
ENGLEWOOD CO 80111
U.S.A.
TEL: 303-794-4494
TLX:
TLF:
EML:
COM:

KOVETZ ATTAY PROF
DEPT PHYSICS & ASTRONOMY
TEL-AVIV UNIVERSITY
RAMAT-AVIV 69978
ISRAEL
TEL: 3-420-234
TLX: 342-171 VERSY IL
TLF:
EML:
COM: 35.47

KOWAL CHARLES THOMAS
SPACE TELESCOPE SCI INST
HOMEWOOD CAMPUS
BALTIMORE MD 21218
U.S.A.
TEL:
TLX:
TLF:
EML:
COM: 15.16.20

KOYAMA KATSUJI
INST SPACE & ASTRON SCI
4-6-1 KOMABA
MEGURO-KU
TOKYO 153
JAPAN
TEL: 3-467-1111
TLX: 34757 ISASTRO J
TLF:
EML:
COM: 48

KOYAMA SHIN PROF DR
KAGAWA UNIVERSITY
SAIWAI CHO
TAKAMATSU 760
JAPAN
TEL: 878-61-4141
TLX:
TLF:
EML:
COM: 12

KOZAI YOSHIHIDE PROF
TOKYO ASTRONOMICAL OBS
OSAWA
MITAKA
TOKYO 181
JAPAN
TEL: 422-41-3650
TLX: 2822307 TAOMTK
TLF: 422-41-3690
EML:
COM: 06.07.20C.38C.50C

KOZIEL KAROL PROF DR
ASTRON OBSERVATORY KRAKOW
UL. 22 LIPCA 16
43-460 WISLA
POLAND
TEL: 32-42
TLX:
TLF:
EML:
COM:

KOZLOVSKY B Z DR
DEPT PHYSICS & ASTRONOMY
TEL-AVIV UNIVERSITY
RAMAT-AVIV 69978
ISRAEL
TEL:
TLX:
TLF:
EML:
COM: 47

KOZLOWSKI MACIEJ DR
ASTRONOMICAL OBSERVATORY
WARSAW UNIVERSITY
AL. UJAZDOWSKIE 4
00-478 WARSZAWA
POLAND
TEL:
TLX:
TLF:
EML:
COM: 35.48

KRAAN-KORTEWEG RENEE C DR
ASTRONOMISCHES INSTITUT
DER UNIVERSITAET BASEL
VENUSSTRASSE 7
4102 BINNINGEN
SWITZERLAND
TEL: 061 22 77 11
TLX:
TLF:
EML: BITNET:KRAAN@URZ.UNIBAS.@CERN
COM: 28

KRAEMER GERHARD DR
ASTRONOMISCHES INSTITUT
DER UNIVERSITAET
PHILOSOPHENWEG 37
D-7400 TUEBINGEN
GERMANY, F.R.
TEL:
TLX:
TLF:
EML:
COM: 12.44

KRAFT ROBERT P PROF
LICK OBSERVATORY
UNIVERSITY OF CALIFORNIA
SANTA CRUZ CA 95064
U.S.A.
TEL: 408-429-2991
TLX: 910-5987174
TLF:
EML:
COM: 27.29.30.37.42

KRAICHEVA ZDRAVSKA DR
DEPT OF ASTRONOMY
BULGARIAN ACAD SCIENCES
7TH NOVEMBER STR 1
1000 SOFIA
BULGARIA
TEL:
TLX:
TLF:
EML:
COM: 42

KRAMER KH N DR
ODESSA STATE UNIVERSITY
ASTRONOMICAL OBSERVATORY
270014 ODESSA
U.S.S.R.
TEL: 22-03-96
TLX:
TLF:
EML:
COM: 22

KRANJC ALDO DR
OSSERVATORIO ASTRONOMICO
DI BRERA
VIA BRERA 28
I-20121 MILANO
ITALY
TEL:
TLX:
TLF:
EML:
COM:

KRASINSKI ANDRZEJ PROF.
N COPERNICUS ASTRO CTR
POLISH ACADEMY SCIENCES
BARTYCKA 18
00716 WARSZAWA
POLAND
TEL: 41 0828
TLX: 813978
TLF:
EML:
COM: 47

KRASINSKY GEORGE A DR
INSTITUTE OF APPL. ASTRON
USSR ACADEMY OF SCIENCES
8 ZHDANOV STREET
197042 LENINGRAD
U.S.S.R.
TEL:
TLX:
TLF:
EML:
COM: 04.07

KRASSOVSKY V I DR
INST PHYSICS OF ATMOSPH
USSR ACADEMY OF SCIENCES
PYSHEVSKY PER. 3
109017 MOSCOW
U.S.S.R.
TEL: 231-88-62
TLX:
TLF:
EML:
COM:

KRAUS JOHN D PROF
RADIO OBSERVATORY
OHIO STATE UNIVERSITY
2015 NEIL AVE
COLUMBUS OH 43210
U.S.A.
TEL: 614-548-7895
TLX:
TLF:
EML:
COM: 40.51

KRAUSE F DR
ZNTRLINST. F. ASTROPHYSIK
ASTROPHYSIKALISCHES OBS
TELEGRAFENBERG
DDR-1500 POTSDAM
GERMANY. D.R.
TEL:
TLX:
TLF:
EML:
COM: 10

KRAUSHAAR WILLIAM L PROF
DEPT OF PHYSICS
UNIVERSITY OF WISCONSIN
1150 UNIVERSITY AVE
MADISON WI 53706
U.S.A.
TEL: 608-262-5916
TLX: 265452
TLF:
EML:
COM: 44

KRAUTTER JOACHIM DR
LANDESSTERNWARTE
KOENIGSTUHL
D-6900 HEIDELBERG
GERMANY. F.R.
TEL: 06221-10036
TLX: 461789 MPIA D
TLF:
EML:
COM: 27.34.42

KREIDL TOBIAS J N
LOWELL OBSERVATORY
1400 W. MARS HILL ROAD
FLAGSTAFF AZ 86001
U.S.A.
TEL: 602-774-3358
TLX:
TLF:
EML:
COM: 09

KREINER JERZY MAREK DR
UL.SENATORSKA 27 M7
30-106 KRAKOW
POLAND
TEL: 12-23-48-37
TLX:
TLF:
EML:
COM: 27.42.46

KREISEL E PROF
EINSTEIN LABORATORIUM
ROSA-LUXEMBURG-STR 17A
DDR-1502 POTSDAM
GERMANY. D.R.
TEL: 762-225
TLX: 15471
TLF:
EML:
COM: 48

KRELOWSKI JACEK DR
INSTITUTE OF ASTRONOMY
N COPERNICUS UNIVERSITY
UL. CHOPINA 12/18
87-100 TORUN
POLAND
TEL: 856-206-55
TLX: 055-2234 ASTR PL
TLF:
EML:
COM:

KREMPEC-KRYGIER JANINA DR
N.COPERNICUS ASTRON CTR
ASTROPHYSICAL LABORATORY
UL. CHOPINA 12/18
87-100 TORUN
POLAND
TEL: 260-18
TLX: 0552234 ASTR PL
TLF:
EML:
COM: 29

KRESAK LUBOR DR
ASTRONOMICAL INSTITUTE
SLOVAK ACAD. OF SCIENCES
DUBRAVSKA CESTA 5
842 28 BRATISLAVA
CZECHOSLOVAKIA
TEL: 427-375157
TLX: 93373 SEIS
TLF:
EML:
COM: 15C.20C.22

KRESAKOVA MARGITA DR
ASTRONOMICAL INSTITUTE
SLOVAK ACAD. OF SCIENCES
DUBRAVSKA CESTA 5
842 28 BRATISLAVA
CZECHOSLOVAKIA
TEL: 427-375157
TLX: 93373 SEIS
TLF:
EML:
COM: 22

KREYSA ERNST
MPI FUER RADIOASTRONOMIE
AUF DEM HUEGEL 69
D-5300 BONN 1
GERMANY. F.R.
TEL: 0228-525269
TLX: 886440 MPIFR D
TLF:
EML:
COM: 34.40

KRIEGER ALLEN S DR
RADIATION SCIENCE. INC.
P.O. BOX 293
BELMONT MA 02178
U.S.A.
TEL: 617-494-0335
TLX:
TLF:
EML:
COM:

KRIKORIAN RALPH DR
INSTITUT D'ASTROPHYSIQUE
98 BIS BOULEVARD ARAGO
75014 PARIS
FRANCE
TEL: 43 20 14 25
TLX:
TLF:
EML:
COM: 36

KRISHNA GOPAL
RADIOASTRONOMY CENTRE
TIFR
P O BOX 1234
BANGALORE 560 012
INDIA
TEL: 362816
TLX:
TLF:
EML:
COM: 28.40

KRISHNA SWAMY K S DR
ASTROPHYSICS GROUP
TATA INSTITUTE
COLABA
BOMBAY 400 005
INDIA
TEL: 219111
TLX: 113009 TIFR IN
TLF:
EML:
COM: 15.34.36

KRISHNAMOHAN S DR
RADIOASTRONOMY CENTRE
TIFR
P.O.BOX 1234
BANGALORE 560 012
INDIA
TEL: 364067
TLX: 8458486 TIFR IN
TLF:
EML:
COM: 40

KRISHNAN THIRUVENKATA MR
HELIOS ANTENNAS/ELECTRON.
234 AVVAI SHANMUGHAM RD
GOPALAPURAM
MADRAS 600 086
INDIA
TEL: 044-472680
TLX:
TLF:
EML:
COM: 40

KRISS GERARD A DR
DEPT OF PHYS & ASTRONOMY
THE JOHNS HOPKINS UNIV
BALTIMORE MD 21218
U.S.A.
TEL: 301 338 7679
TLX: 9101300225 JHU CASHD
TLF:
EML:
COM: 47

KRISTENSEN LEIF KAHL DR
INSTITUTE OF PHYSICS
UNIVERSITY OF AARHUS
NY MUNKEGADE
DK-8000 AARHUS C
DENMARK
TEL:
TLX:
TLF:
EML:
COM: 5 15 20

KRISTENSON HENRIK DR
SLAANBAERSVAEGEN 9
S-381 00 KALMAR
SWEDEN
TEL:
TLX:
TLF:
EML:
COM:

KRISTIAN JEROME DR
MT WILSON & LAS CAMPANAS
OBSERVATORIES
813 SANTA BARBARA STREET
PASADENA CA 91101
U.S.A.
TEL: 818-577-1122
TLX:
TLF:
EML:
COM:

KRISTIANSSON KRISTER PROF
DEPT OF PHYSICS
SOELVEGATAN 14
S-223 62 LUND
SWEDEN
TEL: 046-107726
TLX:
TLF:
EML:
COM: 48

KRIVSKY LADISLAV DR
ASTRONOMICAL INSTITUTE
CZECH. ACAD. OF SCIENCES
251 65 ONDREJOV
CZECHOSLOVAKIA
TEL: 72-45-25 PRAGUE
TLX: 121579 ASTR C
TLF:
EML:
COM: 10

KRIZ SVATOPLUK DR
ASTRONOMICAL INSTITUTE
251 65 ONDREJOV
CZECHOSLOVAKIA
TEL: 204-999201
TLX:
TLF:
EML:
COM: 42

KROGDAHL W S DR
DEPT PHYSICS & ASTRONOMY
UNIVERSITY OF KENTUCKY
LEXINGTON KY 40506
U.S.A.
TEL: 606-272-2659
TLX:
TLF:
EML:
COM:

KROLIK JULIAN H
JOHNS HOPKINS UNIVERSITY
DEPT OF PHYSICS & ASTR
BALTIMORE MD 21218
U.S.A.
TEL: 301-338-7926
TLX:
TLF:
EML:
COM:

KRON GERALD E DR
PINECREST OBSERVATORY
AT QUEEN'S COURT
2929 PONI MOI ROAD
HONOLULU HI 96815
U.S.A.
TEL: 808-922-1514
TLX:
TLF:
EML:
COM: 37.45

KRON KATHERINE GORDON
PINECREST OBSERVATORY
AT QUEEN'S COURT
2929 PONI MOI ROAD
HONOLULU HI 96815
U.S.A.
TEL: 808-922-1514
TLX:
TLF:
EML:
COM: 42

KRON RICHARD G
YERKES OBSERVATORY
PO BOX 258
WILLIAMS BAY WI 53191
U.S.A.
TEL: 312-236-5468
TLX:
TLF:
EML:
COM: 28

KRONBERG PHILIPP DR
UNIVERSITY OF TORONTO
DEPT OF ASTRONOMY
60 ST GEORGE STREET
TORONTO ONT M5S 1A7
CANADA
TEL: 416-978-4971
TLX: 06-986766 TOR
TLF:
EML:
COM: 40

KROTO HAROLD PROF.
SCHOOL OF CHEMISTRY
UNIVERSITY OF SUSSEX
FALMER
BRIGHTON BN1 9QJ
U.K.
TEL: 0273 678329
TLX:
TLF:
EML: BIT:KAFE4@CLUSTER.SUSSEX.AC.UK
COM: 14

KRUCHINENKO VITALIY G
ASTRONOMICAL OBSERVATORY
KIEV STATE UNIVERSITY
OBSERVATORNAYA STR 3
252053 KIEV
U.S.S.R.
TEL:
TLX:
TLF:
EML:
COM: 22 42

KRUEGEL ENDRIK DR
MPI FUER RADIOASTRONOMIE
AUF DEM HUEGEL 69
D-5300 BONN
GERMANY, F.R.
TEL:
TLX: 886440 MPIFR D
TLF:
EML:
COM: 40

KRUEGER ALBRECHT DR
ZNTRLINST. F. ASTROPHYSIK
TELEGRAFENBERG
DDR-1500 POTSDAM
GERMANY. D.R.
TEL: 4551
TLX: 15239 ZIAP DD
TLF:
EML:
COM: 10

KRUMM NATHAN ALLYN
PHYSICS DEPT
UNIVERSITY OF CINCINNATI
CINCINNATI OH 45221
U.S.A.
TEL: 513-475-2232
TLX:
TLF:
EML:
COM: 28

KRUPP EDWIN C DR
GRIFFITH OBSERVATORY
2800 EAST OBS ROAD
LOS ANGELES CA 90027
U.S.A.
TEL: 213 664 1181
TLX:
TLF:
EML:
COM: 41.46

KRUSZEWSKI ANDRZEJ PROF
ASTRONOMICAL OBSERVATORY
AL. UJAZDOWSKIE 4
00-478 WARSZAWA
POLAND
TEL:
TLX:
TLF:
EML:
COM: 42

KRZEMINSKI WOJCIECH DR
CARNEGIE INST WASHINGTON
LAS CAMPANAS OBSERVATORY
CASILLA 601
LA SERENA
CHILE
TEL: 213032
TLX:
TLF:
EML:
COM: 27.42

KSANTOMALITI L V DR
SPACE RESEARCH INSTITUTE
USSR ACADEMY OF SCIENCES
GSP7 PROFSOYUZNAYA 84/32
117810 MOSCOW
U.S.S.R.
TEL: 333-2322/3122
TLX: 411498 STAR SU
TLF:
EML:
COM: 16.51

KUBIAK MARCIN A DR
WARSAW UNIVERSITY
OBSERVATORY
AL. UJAZDOWSKIE 4
00-478 WARSAW
POLAND
TEL: 295346/294011
TLX: 813978 ZAPAN PL
TLF:
EML:
COM: 27

KUBICELA ALEKSANDAR DR
ASTRONOMSKA OPSERVATORIJA
VOLGINA 7
11050 BEOGRAD
YUGOSLAVIA
TEL: 011-419-357
TLX:
TLF:
EML:
COM: 12.50

KUBO YOSHIO
GEODESY & GEOPHYSICS DIV
HYDROGRAPHIC DEPT
TSUKIJI-5 CHUO-KU
TOKYO 104
JAPAN
TEL: 03-541-3811
TLX: 02522222 JAHYD J
TLF:
EML:
COM: 04C

KUBOTA JUN DR
KWASAN OBSERVATORY
YAMASHINA
KYOTO 607
JAPAN
TEL: 75-581-1235
TLX: 5422693 LIBKYU J
TLF:
EML:
COM: 10

KUDRITZKI ROLF-PETER PH D
INST F ASTRON & ASTROPHYS
SCHEINERSTR 1
D-8000 MUNCHEN 80
GERMANY. F.R.
TEL: 089-989021
TLX: 529815 UNIVM D
TLF:
EML:
COM: 36

KUENNE CHRISTOPH F
KOEFLACHER STR. 36
D-7928 GIENGEN/BRENZ
GERMANY. F.R.
TEL: 07322-44-48
TLX:
TLF:
EML:
COM: 39

KUEHR HELMUT
MPI FUER ASTRONOMIE
KOENIGSTUHL
D-6900 HEIDELBERG
GERMANY. F.R.
TEL: 6221-5281
TLX: 461789 MPIA D
TLF:
EML:
COM:

KUENZEL HORST
DIESELSTRASSE 13
DDR-1502 POTSDAM-BABELSB.
GERMANY. D.R.
TEL: 77318
TLX:
TLF:
EML:
COM: 10

KUHI LEONARD V PROF
ASTRONOMY DEPT
UNIVERSITY OF CALIFORNIA
BERKELEY CA 94720
U.S.A.
TEL: 415-642-3792
TLX:
TLF:
EML:
COM: 27.36

KUIJPERS H. JAN M.E. DR
STERREKUNDIG INSTITUUT
PRINCETONPLEIN 5
POSTBUS 80 000
NL-3508 TA UTRECHT
NETHERLANDS
TEL: 030-535200
TLX: 40048 FYLUT NL
TLF:
EML: BITNET:UWNKUIJ@HUTRUU0
COM: 40

KUIPER THOMAS B H DR
JET PROPULSION LABORATORY
169-506S
PASADENA CA 91109
U.S.A.
TEL: 818-354-5479
TLX: 675429
TLF:
EML:
COM: 34.40.51

KUKLIN G V DR
SIBIZMIR
P B 4
664697 IRKUTSK 33
U.S.S.R.
TEL: 6-02-65
TLX:
TLF:
EML:
COM: 10.12

KULKARNI PRABHAKAR V PROF
PHYSICAL RESEARCH LAB
AHMEDABAD 380 009
INDIA
TEL: 448029
TLX: 8458488 TIFR IN
TLF:
EML:
COM: 09.21.25.40

KULKARNI SHRINIVAS R DR
DEPT OF ASTRONOMY 105-24
CALTECH
PASADENA CA 91125
U.S.A.
TEL: 818 356 4010
TLX: 188192 675425
TLF:
EML:
COM: 40

KULKARNI V K DR
RADIO ASTRONOMY CENTRE
TATA INSTITUTE OF
FUNDAMENTAL RESEARCH
OOTACAMUND 643 001
INDIA
TEL: 0423-2651
TLX: 853241 RAC IN
TLF:
EML:
COM: 40

KULSRUD RUSSELL M DR
ASTROPHYSICAL SCIENCES
PRINCETON UNIVERSITY
PRINCETON NJ 08540
U.S.A.
TEL: 609-683-2610
TLX:
TLF:
EML:
COM: 33.48

KULTIMA JOHANNES
GEOPHYSICAL OBSERVATORY
SF-99600 SODANKYLAE
FINLAND
TEL:
TLX:
TLF:
EML:
COM:

KUMAJGORODSKAYA RAISA DR
SPECIAL ASTROPHYSICAL
OBSERVATORY
NIZHNIJ ARKHYZ
357147 STAVROPOLSKIJ KRAJ
U.S.S.R.
TEL: 93-415
TLX:
TLF:
EML:
COM: 29

KUMAR C KRISHNA DR
DEPT PHYSICS & ASTRONOMY
HOWARD UNIVERSITY
WASHINGTON DC 20059
U.S.A.
TEL: 202-636-6245
TLX:
TLF:
EML:
COM: 34

KUMAR SHAILENDRA
LUNAR & PLANETARY LAB
UNIVERSITY OF ARIZONA
TUCSON AZ 85721
U.S.A.
TEL:
TLX:
TLF:
EML:
COM:

KUMAR SHIV S PROF
DEPT OF ASTRONOMY
UNIVERSITY OF VIRGINIA
PO BOX 3818
CHARLOTTESVILLE VA 22903
U.S.A.
TEL: 804-924-4896
TLX:
TLF:
EML:
COM: 16.35.36

KUMSIASHVILY MZIA I DR
ABASTUMANI ASTROPHYSICAL
OBSERVATORY
383762 ABASTUMANI.GEORGIA
U.S.S.R.
TEL: 2-52
TLX: 327409
TLF:
EML:
COM: 42

KUMSISHVILI J I DR
ABASTUMANI ASTROPHYSICAL
OBSERVATORY
383762 ABASTUMANI.GEORGIA
U.S.S.R.
TEL: 2-79
TLX: 327409
TLF:
EML:
COM: 26.27

KUN MARIA DR
KONKOLY OBSERVATORY
HUNGARIAN ACADEMY OF SCI
BOX 67
H-1525 BUDAPEST
HUNGARY
TEL: 166-426/506/540
TLX: 227460 KONOB H
TLF:
EML:
COM: 37

KUNCHEV PETER DR
DEPT OF ASTRONOMY
FACULTY OF PHYSICS
ANTON IVANOV STR 5
1126 SOFIA
BULGARIA
TEL:
TLX:
TLF:
EML:
COM:

KUNDT WOLFGANG PROF DR
INSTITUT F ASTROPHYSIK
& EXTRATERR FORSCHUNG
AUF DEM HUEGEL 71
D-5300 BONN 1
GERMANY. F.R.
TEL: 02226-7400
TLX: 0886440
TLF:
EML:
COM: 40.48

KUNDU MUKUL R DR
ASTRONOMY PROGRAM
UNIVERSITY OF MARYLAND
COLLEGE PARK MD 20742
U.S.A.
TEL: 301-454-3005
TLX: 710-826-0352
TLF:
EML:
COM: 10.12.34.40

KUNIEDA HIDEYO DR
DEPT OF ASTROPHYSICS
NAGOYA UNIVERSITY
FURO-CHO. CHIKUSA-KU
NAGOYA 464
JAPAN
TEL: 52 781 5111
TLX: 4477323 SCUNAG J
TLF: 52 781 3541
EML:
COM:

KUNITZSCH PAUL PROF.
DAVIDSTRASSE 17
D-8000 MUENCHEN 81
GERMANY. F.R.
TEL:
TLX: 916280
TLF:
EML:
COM: 41

KUNKEL WILLIAM E DR
LAS CAMPANAS OBSERVATORY
COLA EL PINO S/N
CASILLA 601
LA SERENA
CHILE
TEL: 51-213032
TLX: 645227 AURA CT
TLF:
EML:
COM: 25.37

KUNTH DANIEL
INSTITUT D'ASTROPHYSIQUE
98 BIS BOULEVARD ARAGO
F-75014 PARIS
FRANCE
TEL: 1-43-20-14-25
TLX: 270070 IMAG F
TLF:
EML:
COM: 28.34.47

KUPERUS MAX PROF DR
STERREKUNDIG INSTITUUT
PRINCETONPLEIN 5
POSTBUS 80 000
NL-3508 TA UTRECHT
NETHERLANDS
TEL: 030-535200
TLX: 40048 FYLUT NL
TLF:
EML: BITNET:WNNMAIL@HUTRUU0
COM: 10.12C

KURFESS JAMES D
NAVAL RESEARCH LABORATORY
CODE 4150
4555 OVERLOOK AVE S.W.
WASHINGTON DC 20375
U.S.A.
TEL: 202-767-3182
TLX:
TLF:
EML:
COM:

KURIL-CHIK V N DR
STERNBERG STATE
ASTRONOMICAL INSTITUTE
119899 MOSCOW B-234
U.S.S.R.
TEL: 139-10-30
TLX:
TLF:
EML:
COM: 40

KUROCHKA L N DR
KIEV STATE UNIVERSITY
ASTRONOMICAL OBSERVATORY
OBSERVATORNAYA STR.3
252053 KIEV
U.S.S.R.
TEL: 26-26-91
TLX:
TLF:
EML:
COM: 10.12

KUROKAWA HIROKI DR
HIDA OBSERVATORY
UNIVERSITY OF KYOTO
KAMITAKARA. YOSHIKI-GUN
GIFU 506-13
JAPAN
TEL: 0578-6-2628
TLX:
TLF:
EML:
COM: 10

KURPINSKA-WINIARSKA M DR
ASTRONOMICAL OBSERVATORY
JAGELLONIAN UNIVERSITY
UL. ORLA 171
30-244 KRAKOW
POLAND
TEL:
TLX: 0322297 UJ PL
TLF:
EML:
COM: 42

KURT V G DR
SPACE RESEARCH INSTITUTE
USSR ACADEMY OF SCIENCES
PROFSOYUZNAYA STR 84/32
117810 MOSCOW
U.S.S.R.
TEL: 333-31-22
TLX: 411498 STAR SU
TLF:
EML:
COM: 16.44.48

KURTZ DONALD WAYNE DR
ASTRONOMY DEPT
UNIVERSITY OF CAPE TOWN
RONDEBOSCH
RONDEBOSCH 7700
SOUTH AFRICA
TEL: 69-8531
TLX: 521439
TLF:
EML:
COM: 27

KURUCZ ROBERT L DR
SMITHSONIAN ASTROPHYSICAL
OBSERVATORY
60 GARDEN STREET
CAMBRIDGE MA 02138
U.S.A.
TEL: 617-495-7429
TLX: 921428
TLF:
EML:
COM: 36

KUS ANDRZEJ JAN DR
RADIOASTRONOMY OBS
COPERNICUS UNIVERSITY
UL. CHOPINA 12/18
87-100 TORUN
POLAND
TEL: 04856-20651
TLX: 0552324 TRAO PL
TLF:
EML:
COM: 40

KUSHWAHA R S PROF
DEPT OF MATHEMATICS
UNIVERSITY OF JODHPUR
JODHPUR RAJ
INDIA
TEL:
TLX:
TLF:
EML:
COM: 35.36

KUSTAANHEIMO PAUL E PROF
DANMARKS TEKN HOJSKOLE
DIA E 451
DK-2800 LYNGBY
DENMARK
TEL: 45-2-883022
TLX:
TLF:
EML:
COM: 07.28.47

KUTNER MARC LESLIE DR
PHYSICS DEPT
RENSSELAER POLYTECHN INST
TROY NY 12180
U.S.A.
TEL: 518-266-6417
TLX:
TLF:
EML:
COM: 34.40

KUTTER G SIEGFRIED DR
NASA/GSFC
LAB. FOR ASTRONOMY AND
SOLAR PHYSICS. CODE 681
GREENBELT MD 20770
U.S.A.
TEL:
TLX:
TLF:
EML:
COM:

KUTUZOV S A DR
LENINGRAD STATE UNIV
DEPT OF APPLIED MATHS &
CONTROL PROCESSES
199164 LENINGRAD
U.S.S.R.
TEL:
TLX:
TLF:
EML:
COM: 33

KUZMANOSKI MIKE
INSTITUTE OF ASTRONOMY
UNIVERSITY OF BELGRADE
STUDENTSKI TRG 16
YU-11000 BELGRADE
YUGOSLAVIA
TEL: 011-638-715
TLX:
TLF:
EML:
COM:

KUZMIN ARKADII D PROF DR
LEBEDEV PHYSICAL INST
USSR ACADEMY OF SCIENCES
117924 MOSCOW
U.S.S.R.
TEL:
TLX: 411479 NEOD SU
TLF:
EML:
COM: 16.40.54

KVIZ ZDENEK DR
SCHOOL OF PHYSICS
UNIVERSITY OF SOUTH WALES
P.O.BOX 1
KENSINGTON NSW 2033
AUSTRALIA
TEL: 697-45-78
TLX: 26054 AA
TLF:
EML:
COM: 22.25.42

KWEE K K DR
STERREWACHT
POSTBUS 9513
NL-2300 RA LEIDEN
NETHERLANDS
TEL: 071-272727
TLX: 39058 ASTRO NL
TLF:
EML:
COM: 27.42

KWITTER KAREN BETH DR
THOMPSON PHYSICS LAB
DEPT PHYSICS & ASTRONOMY
WILLIAMS COLLEGE
WILLIAMSTOWN MA 01267
U.S.A.
TEL: 413-597-2272
TLX:
TLF:
EML:
COM: 34

KWOK SUN DR
DEPT OF PHYSICS
UNIVERSITY OF CALGARY
CALGARY AB T2N 1N4
CANADA
TEL: 403-284-5414
TLX: 038-21545
TLF:
EML:
COM: 34.40

KYLAFIS NIKOLAOS D DR
DEPT OF PHYSICS
UNIVERSITY OF CRETE
PO BOX 470
GR-71110 IRAKLION. CRETE
GREECE
TEL:
TLX:
TLF:
EML:
COM: 34

LA BONTE BARRY JAMES
INSTITUTE FOR ASTRONOMY
2680 WOODLAWN DRIVE
HONOLULU HI 96822
U.S.A.
TEL: 808-948-6531
TLX: 723-8459 UHAST HR
TLF:
EML:
COM: 12

LA DOUS CONSTANZE A DR
INSTITUTE OF ASTRONOMY
UNIVERSITY OF CAMBRIDGE
MADINGLEY ROAD
CAMBRIDGE CB3 0HA
U.K.
TEL: 44 223 337548
TLX: 817297 ASTRON G
TLF:
EML: BITNET:CADOUK.AC.CAM.AST-STAR
COM: 42

LA PADULA CESARE
IST. ASTROFISICA SPAZIALE
C P 67
I-00044 FRASCATI
ITALY
TEL:
TLX:
TLF:
EML:
COM:

LABAY JAVIER
DPTO FISICA DE ATMOSFERA
UNIVERSIDAD DE BARCELONA
AVDA. DIAGONAL 645
08028 BARCELONA
SPAIN
TEL: 330-7311
TLX:
TLF:
EML:
COM: 35

LABEYRIE ANTOINE DR
CERGA
F-06460 ST VALLIER DE T.
FRANCE
TEL: 93-42-62-70
TLX: 461402
TLF:
EML:
COM: 09

LABEYRIE JACQUES DR
CENTRE DES FAIBLES
RADIOACTIVITES
LAB MIXTE CNRS-CEA
F-91190 GIF-SUR-YVETTE
FRANCE
TEL: 1-69-07-78-28
TLX: 691137 F
TLF:
EML:
COM:

LABHARDT LUKAS
ASTRONOMISCHES INSTITUT
UNIVERSITAET BASEL
VENUSSTRASSE 7
CH-4102 BINNINGEN
SWITZERLAND
TEL: 0041-61-22-7711
TLX:
TLF:
EML:
COM: 25.45

LABS DIETRICH PROF
LANDESSTERNWARTE
KOENIGSTUHL
D-6900 HEIDELBERG 1
GERMANY. F.R.
TEL: 6221-10036
TLX: 461153 LSWHD D
TLF:
EML:
COM: 12.29

LACHIEZE-REY MARC
CEN SACLAY
DPHG/SAP
F-91191 GIF/YVETTE CEDEX
FRANCE
TEL: 1-69=08-62-92
TLX: 690860
TLF:
EML:
COM: 47

LACLARE F MR
CERGA
AVENUE COPERNIC
F-06130 GRASSE
FRANCE
TEL:
TLX:
TLF:
EML:
COM: 08

LACROUTE PIERRE A PROF
2 RUE D'ALISE
F-21000 DIJON
FRANCE
TEL: 80-66-11-54
TLX:
TLF:
EML:
COM: 08.24

LACY CLAUD H DR
DEPT OF PHYSICS
UNIVERSITY OF ARKANSAS
104 PHYSICS BUILDING
FAYETTEVILLE AR 72701
U.S.A.
TEL: 501-575-2506
TLX:
TLF:
EML:
COM: 42

LACY JOHN H DR
ASTRONOMY DEPARTMENT
UNIVERSITY OF TEXAS
RLM 15.308
AUSTIN TX 78712
U.S.A.
TEL: 512 471 1469
TLX:
TLF:
EML:
COM:

LADA CHARLES JOSEPH DR
STEWARD OBSERVATORY
UNIVERSITY OF ARIZONA
TUCSON AZ 85721
U.S.A.
TEL: 602-621-4878
TLX:
TLF:
EML:
COM: 34C.37.40

LAFFINEUR MARIUS MR
21 BLVD BRUNE
F-75014 PARIS
FRANCE
TEL:
TLX:
TLF:
EML:
COM:

LAFON JEAN-PIERRE J DR
OBSERVATOIRE DE PARIS
SECTION DE MEUDON
F-92195 MEUDON PL CEDEX
FRANCE
TEL: 1-45-07-78-58
TLX: 204464 F
TLF:
EML:
COM: 28.33.34.49.51

LAGERKVIST CLAES-INGVAR
ASTRONOMICAL OBSERVATORY
BOX 515
S-751 20 UPPSALA
SWEDEN
TEL: 018-113522
TLX: 76024 UNIV UPS S
TLF:
EML:
COM: 15.20

LAGERQVIST ALBIN PROF
INSTITUTE OF PHYSICS
VANADISVAGEN 9
S-113 46 STOCKHOLM
SWEDEN
TEL: 468-16-45-00
TLX: 15433 FYSTO S
TLF:
EML:
COM: 14

LAGO MARIA TERESA V T PR.
GRUPO DE MATEM. APLICADA
UNIVERSIDADE DO PORTO
RUA DAS TAIPAS 135
4000 PORTO
PORTUGAL
TEL: 380313
TLX: 28109
TLF:
EML:
COM: 27.29.46

LAHIRI N C
INDIAN ASTRON EPHEM UNIT
INDIAN METEOROLOGIC. DEPT
P-546 BLOCK N
NEW ALIPORE CALCUTTA
INDIA
TEL:
TLX:
TLF:
EML:
COM: 04

LAHULLA J FORNIES DR
OBSERVATORIO ASTRONOMICO
ALFONSO XII-3
28014 MADRID
SPAIN
TEL: 2270107
TLX: 22465 IGC E
TLF:
EML:
COM:

LAI SEBASTIANA
ISTITUTO DI ASTRONOMIA
VIA OSPEDALE 72
I-09100 CAGLIARI
ITALY
TEL: 070 663544
TLX:
TLF:
EML:
COM:

LAING ROBERT
ROYAL GREENWICH OBS
HERSTMONCEUX CASTLE
HAILSHAM BN27 1RP
U.K.
TEL: 0323-833171
TLX: 87451 RGOBS G
TLF:
EML:
COM: 40

LAKE KAVLL WILLIAM DR
DEPT OF PHYSICS
QUEEN'S UNIVERSITY
KINGSTON ONT K7L 3N6
CANADA
TEL: 613-547-3020
TLX:
TLF:
EML:
COM: 47

LAL DEVENDRA
PHYSICAL RESEARCH LAB.
NAVRANGPURA
AHMEDABAD 380009
INDIA
TEL:
TLX:
TLF:
EML:
COM:

LALA PETR DR
OUTER SPACE AFFAIRS DIVIS
UNITED NATIONS SECRETARIA
NEW YORK N.Y. 10017
U.S.A.
TEL:
TLX:
TLF:
EML:
COM: 07

LALLEMENT ROSINE DR
SERVICE D'AERONOMIE
DU CNRS
BP 3
91370 VERRIERES CEDEX
FRANCE
TEL: (33-1)64474235
TLX: 602400 AERONO
TLF:
EML: BITNET:ROSINE@FRIAP51
COM:

LAMB DONALD QUINCY JR DR
UNIVERSITY OF CHICAGO
5801 ELLIS AVE
CHICAGO IL 60637
U.S.A.
TEL: 312-962-8203
TLX: 269266
TLF:
EML:
COM: 35.42.48

LAMB FREDERICK K PROF
PHYSICS DEPT
UNIVERSITY OF ILLINOIS
1110 W GREEN STREET
URBANA IL 61801
U.S.A.
TEL: 217-333-6363
TLX: 6502272050 MCI
TLF:
EML:
COM: 48

LAMB RICHARD C DR
PHYSICS DEPARTMENT
IOWA STATE UNIVERSITY
AMES IA 50011
U.S.A.
TEL: 515 294 3873
TLX:
TLF:
EML: BITNET:LAMB@ALISUVAX
COM:

LAMB SUSAN ANN DR
DEPARTMENT OF PHYSICS
UNIVERSITY OF MISSOURI
8001 NATURAL BRG RD
ST LOUIS MO 63121
U.S.A.
TEL:
TLX:
TLF:
EML:
COM: 35.48

LAMBECK KURT PROF
AUSTRALIA NAT UNIVERSITY
RESEARCH SCHOOL.EARTH SCI
GPO BOX 4
CANBERRA 2600
AUSTRALIA
TEL: 49-2487
TLX: 62693
TLF:
EML:
COM: 19

LAMBERT DAVID L PROF
DEPT OF ASTRONOMY
UNIVERSITY OF TEXAS
R L MOORE HALL
AUSTIN TX 78712
U.S.A.
TEL: 512-471-4461
TLX: 9108741351
TLF:
EML:
COM: 29V.36

LAMERS H J G L M DR
SPACE RESEARCH LABORATORY
BENELUXLAAN 21
NL-3527 HS UTRECHT
NETHERLANDS
TEL: 030-937145
TLX: 47224 ASTRO NL
TLF:
EML:
COM: 29.44

LAMLA ERICH E DR
ASTRONOMISCHE INSTITUTE
STERNWARTE UNIV BONN
AUF DEM HUEGEL 71
D-5300 BONN
GERMANY. F.R.
TEL: 0228-73 36 59
TLX: 088 64 40 MPIFR D
TLF:
EML:
COM: 29

LAMONTAGNE ROBERT DR
DEPT DE PHYSIQUE
UNIVERSITY OF MONTREAL
CP 6128 SUCC A
MONTREAL QUEBEC H3C 3J7
CANADA
TEL: 514 342 7273
TLX:
TLF:
EML: 5007@CC.UMONTREAL.CA
COM: 29

LAMPTON MICHAEL
SPACE SCIENCES LABORATORY
UNIVERSITY OF CALIFORNIA
BERKELEY CA 94729
U.S.A.
TEL: 415-642-3576
TLX: 9103667945
TLF:
EML:
COM: 48

LAMY PHILIPPE DR
L.A.S.
LES TROIS LUCS
13012 MARSEILLE
FRANCE
TEL: 91 055932
TLX: 420584 F
TLF:
EML:
COM: 15.21C.22

LANCASTER BROWN PETER
10A ST PETER'S ROAD
ALDEBURGH. SUFFOLK
U.K.
TEL:
TLX:
TLF:
EML:
COM: 15

LANDE KENNETH PROF
PHYSICS DEPT
UNIV OF PENNSYLVANIA
PHILADELPHIA PA 19104
U.S.A.
TEL: 215-898-8177
TLX:
TLF:
EML:
COM:

LANDECKER PETER BRUCK DR
HUGHES AIRCRAFT CO
SPACE & COMM GR/BLDG S41
MS B322. PO BOX 92919
LOS ANGELES CA 90009
U.S.A.
TEL: 213-648-0815
TLX: 664480
TLF:
EML:
COM: 10

LANDECKER THOMAS L DR
DOMINION RADIO ASTRO-
PHYSICAL OBSERVATORY
P O BOX 248
PENTICTON BC V2A 6K3
CANADA
TEL: 604-497-5321
TLX: 048-88127
TLF:
EML:
COM:

LANDI DEGL'INNOCENTI E PR
DIPART. DI ASTRONOMIA
UNIVERSITA DEGLI STUDI
LARGO E. FERMI 5
I-50125 FIRENZE
ITALY
TEL: 055-27521
TLX: 572268 ARCETR I
TLF:
EML:
COM: 12

LANDI DEGL'INNOCENTI M
OSSERVATORIO ASTROFISICO
DI ARCETRI
LARGO E. FERMI 5
I-50125 FIRENZE
ITALY
TEL: 055-2752256
TLX: 572268 ARCETR
TLF:
EML:
COM: 12

LANDINI MASSIMO PROF
OSSERVATORIO ASTROFISICO
DI ARCETRI
LARGO E. FERMI 5
I-50125 FIRENZE
ITALY
TEL: 055-2752247
TLX:
TLF:
EML:
COM:

LANDI-DESSY J DR
OBSERVATORIO NACIONAL
5000 CORDOBA
ARGENTINA
TEL:
TLX:
TLF:
EML:
COM:

LANDMAN DONALD ALAN
INSTITUTE FOR ASTRONOMY
UNIVERSITY OF HAWAII
2680 WOODLAWN DRIVE
HONOLULU HI 96822
U.S.A.
TEL:
TLX:
TLF:
EML:
COM: 10.12.14

LANDOLFI MARCO
OSSERVATORIO ASTROFISICO
DI ARCETRI
LARGO E. FERMI 5
I-50125 FIRENZE
ITALY
TEL: 055-2752256
TLX: 572268 ARCETR
TLF:
EML:
COM: 12

LANDOLT ARLO U PROF
DEPT PHYSICS & ASTRONOMY
LOUISIANA STATE UNIV
BATON ROUGE LA 70803
U.S.A.
TEL: 504-388-8276
TLX: 559184
TLF:
EML:
COM: 25.27.37.42

LANDSTREET JOHN D PROF
DEPT OF ASTRONOMY
UNIV OF WESTERN ONTARIO
LONDON ONT N6A 3K7
CANADA
TEL: 519-679-3186
TLX:
TLF:
EML:
COM: 25C.29.36

LANE ADAIR P
BOSTON UNIVERSITY
ASTRONOMY DEPT
725 COMMONWEALTH AVE.
BOSTON MA 02215
U.S.A.
TEL: 617-353-2633
TLX:
TLF:
EML:
COM:

LANE ARTHUR LONNE DR
JET PROPULSION LABORATORY
4800 OAK GROVE DRIVE
PASADENA CA 91109
U.S.A.
TEL: 818-345-2725
TLX:
TLF:
EML:
COM: 16

LANEY CLIFTON D DR
S A A O
P O BOX 9
OBSERVATORY 7935
SOUTH AFRICA
TEL: 470-025
TLX:
TLF:
EML:
COM: 27

LANG JAMES DR
RUTHERFORD APPLETON LAB
CHILTON. DIDCOT OX11 0QX
U.K.
TEL: 0235-21900
TLX: 83159
TLF:
EML:
COM: 14

LANG KENNETH R ASST PROF
DEPT OF PHYSICS
TUFTS UNIVERSITY
ROBINSON HALL
MEDFORD MA 02155
U.S.A.
TEL: 617-381-3390
TLX:
TLF:
EML:
COM: 10.40.41

LANGER GEORGE EDWARD DR
DEPT OF PHYSICS
COLORADO COLLEGE
COLORADO SPRINGS CO 80903
U.S.A.
TEL: 303-4732233x578
TLX:
TLF:
EML:
COM: 29

LANGER WILLIAM DAVID DR
PLASMA PHYSICS LAB
PRINCETON UNIVERSITY
PO BOX 451
PRINCETON NJ 08544
U.S.A.
TEL: 609-683-2262
TLX:
TLF:
EML:
COM: 34.40

LANNES ANDRE DR
OMP
14 AVENUE E. BELIN
31400 TOULOUSE
FRANCE
TEL: 61 25 21 01
TLX: 530 776 F
TLF:
EML:
COM:

LANTOS PIERRE DR
OBSERVATOIRE DE PARIS
SECTION DE MEUDON
DASOP
F-92195 MEUDON PL CEDEX
FRANCE
TEL: 1-45-07-77-67
TLX:
TLF:
EML:
COM: 05.10.12.40

LANZ THIERRY DR
ASTRONOMICAL INSTITUTE
UNIVERSITY LAUSANNE
OBSERVATOIRE
1290-CHAVANNES DES BOIS
SWITZERLAND
TEL: 22 55 26 11
TLX: 419 209 OBSG CH
TLF:
EML: LANZ@CGEUGE54
COM: 29

LAPASSET EMILIO DR
OBSERVATORIO ASTRONOMICO
LAPRIDA 854
5000 CORDOBA
ARGENTINA
TEL: 051-36876
TLX: 51822 BUCOR
TLF:
EML:
COM: 37. 42

LAPOINTE S M DR
UNIVERSITE DU QUEBEC
2875 BOUL LAURIER
STE-FOY PQ G1V 2M3
CANADA
TEL: 418-657-3551
TLX: 051-31-623
TLF:
EML:
COM:

LAPUSHKA K K DR
LATVIAN STATE UNIVERSITY
ASTRONOMICAL OBSERVATORY
226098 RIGA
U.S.S.R.
TEL: 223149/611984
TLX:
TLF:
EML:
COM: 24

LAQUES PIERRE DR
OBS DU PIC DU MIDI
8 DU PONT DE LA MOULETTE
F-65200 BAGNERES-DE-B.
FRANCE
TEL: 62-95-19-69
TLX: 531675
TLF:
EML:
COM: 09.51

LARGE MICHAEL I DR
SCHOOL OF PHYSICS
UNIVERSITY OF SYDNEY
SYDNEY NSW 2006
AUSTRALIA
TEL: 2-692-2222
TLX: 26169 UNISYD AA
TLF:
EML:
COM: 40

LARI CARLO DR
LABORATORIO NAZIONALE DI
RADIOASTRONOMIA
VIA IRNERIO 46
I-40126 BOLOGNA
ITALY
TEL:
TLX:
TLF:
EML:
COM:

LARSON HAROLD P DR
LUNAR AND PLANETARY LAB
UNIVERSITY OF ARIZONA
TUCSON AZ 85721
U.S.A.
TEL: 602-621-6943
TLX:
TLF:
EML:
COM: 15.16

LARSON RICHARD B PROF
ASTRONOMY DEPARTMENT
YALE UNIVERSITY
BOX 6666
NEW HAVEN CT 06511
U.S.A.
TEL: 203-436-3015
TLX:
TLF:
EML:
COM: 28.33.35

LARSON STEPHEN M
DEPT PLANETARY SCIENCES
UNIVERSITY OF ARIZONA
TUCSON AZ 85721
U.S.A.
TEL: 602-621-4973
TLX: 910-952-1143
TLF:
EML:
COM: 15.16

LARSSON-LEANDER G PROF
LUND OBSERVATORY
BOX 43
S-221 00 LUND
SWEDEN
TEL: 46-10-70-00
TLX: 33199OBSNOT S
TLF:
EML:
COM: 29.37.42

LASENBY ANTHONY
MULLARD RADIO ASTRON OBS
CAVENDISH LABORATORY
MADINGLEY ROAD
CAMBRIDGE CB3 OHE
U.K.
TEL: 223-66477
TLX: 81292 CAVLAB G
TLF:
EML:
COM: 40

LASHER GORDON JEWETT DR
IBM TJ WATSON RES. CENT.
YORKTOWN HEIGHTS NY10598
U.S.A.
TEL:
TLX:
TLF:
EML:
COM: 48

LASKAR JACQUES DR
BUREAU DES LONGITUDES
77 AVE DENFERT ROCHEREAU
75014 PARIS
FRANCE
TEL: 1-40 51 22 74
TLX:
TLF:
EML:
COM: 04.07

LASKARIDES PAUL G ASSPROF
DEPT OF ASTRONOMY
UNIVERSITY OF ATHENS
PANEPISTIMIOPOLIS
GR-15771 ZOGRAFOS
GREECE
TEL: 01-7243211
TLX:
TLF:
EML:
COM: 25.27.35

LASKER BARRY M DR
SPACE TELESCOPE SC INST
HOMEWOOD CAMPUS
3700 SAN MARTIN DRIVE
BALTIMORE MD 21218
U.S.A.
TEL: 301-338-4840
TLX: 6849191 STSI
TLF:
EML:
COM: 09.25.28.34

LASOTA JEAN-PIERRE DR
OBSERVATOIRE DE PARIS
SECTION DE MEUDON
F-92195 MEUDON PL CEDEX
FRANCE
TEL: 1-45-07-74-16
TLX: 201571
TLF:
EML:
COM: 35.47

LATHAM DAVID W DR
CENTER FOR ASTROPHYSICS
60 GARDEN STREET
CAMBRIDGE MA 02138
U.S.A.
TEL: 617-495-7215
TLX: 921428 SATELLITE CAM
TLF:
EML:
COM: 26.30P

LATOUR JEAN J
OBSERVATOIRE PIC-DU-MIDI
ET TOULOUSE
14 AVENUE EDOUARD BELIN
F-31400 TOULOUSE
FRANCE
TEL: 61-25-21-01
TLX:
TLF:
EML:
COM: 35

LATTANZI MARIO G
OSSERVATORIO ASTRONOMICO
DI TORINO
10025 PINO TORINESE
ITALY
TEL: 011 8410 67
TLX: 213236 TO ASTR 1
TLF:
EML: SPAN:39181::LATTANZI
COM: 08.26

LATTIMER JAMES M DR
ASTROPHYSICS PROGRAM
DEPT EARTH & SPACE SCS
SUNY AT STONY BROOK
STONY BROOK NY 11794
U.S.A.
TEL: 516-246-8223
TLX:
TLF:
EML:
COM: 48

LATYPOV A A DR
ASTRONOMICAL INSTITUTE
UZBEKTAN ACADEMY OF SCI
700052 TASHKENT
U.S.S.R.
TEL: 358102
TLX: 116012 VBFMJA
TLF:
EML:
COM: 24

LAUBERTS ANDRIS DR
E.S.O.
KARL-SCHWARZSCHILD-STR 2
D-8046 GARCHING B MUNCHEN
GERMANY. F.R.
TEL: 089-32006363
TLX: 528 282 20 EO D
TLF:
EML:
COM: 28

LAUNAY JEAN-MICHEL DR
OBSERVATOIRE DE PARIS
SECTION DE MEUDON
F-92195 MEUDON PL CEDEX
FRANCE
TEL: 1-45-07-75-54
TLX: 201571 LAM
TLF:
EML:
COM: 14

LAURENT BERTEL E PROF
INST FOR THEORETICAL PHYS
VANADISVAGEN 9
S-113 46 STOCKHOLM
SWEDEN
TEL: 468-16-45-00
TLX: 15433 FYSTO S
TLF:
EML:
COM:

LAURENT CLAUDINE DR
INSU
77 AVE DENFERT-ROCHEREAU

FRANCE
TEL: 1-40-51-21-18
TLX: 270070
TLF:
EML:
COM: 34

LAUSBERG ANDRE DR
INSTITUT D'ASTROPHYSIQUE
UNIVERSITE DE LIEGE
AVENUE DE COINTE 5
B-4200 COINTE-OUGREE
BELGIUM
TEL: 41-52-99-80
TLX:
TLF:
EML:
COM: 28-47

LAUTMAN D A DR
SMITHSONIAN ASTROPHYS CTR
60 GARDEN STREET
CAMBRIDGE MA 02138
U.S.A.
TEL:
TLX:
TLF:
EML:
COM:

LAVAL ANNIE DR
OBSERVATOIRE DE MARSEILLE
2 PLACE LE VERRIER
F-13248 MARSEILLE CDX 04
FRANCE
TEL: 91-95-90-88
TLX: 420241 F
TLF:
EML:
COM: 37

LAVROV M I PROF
ENGELHARDT ASTR OBS
UNIVERSITY
420008 KAZAN
U.S.S.R.
TEL:
TLX:
TLF:
EML:
COM: 42

LAVRUKHINA A K PROF DR
INSTITUTE OF GEOCHEMISTRY
& ANALYTICAL CHEMISTRY
USSR ACADEMY OF SCIENCES
117334 MOSCOW
U.S.S.R.
TEL: 137-75-38
TLX:
TLF:
EML:
COM:

LAWRENCE ANDREW DR
ASTRONOMY UNIT
QUEEN MARY COLLEGE
MILE END ROAD
LONDON E1 4NS
U.K.
TEL: 01-975 5481
TLX: 893750
TLF:
EML: ALOUK.AC.QMC.STAR
COM: 28

LAWRENCE CHARLES R DR
RADIO ASTRONOMY 105 24
CALTECH
PASADENA CA 91125
U.S.A.
TEL: 818 356 4976
TLX: 675429
TLF:
EML: BITNET:CRL@CITDEIMO
COM: 40

LAWRENCE G M DR
LASP
UNIVERSITY OF COLORADO
CAMPUS BOX 392
BOULDER CO 80309
U.S.A.
TEL:
TLX:
TLF:
EML:
COM: 14

LAURIE DAVID G
THE AEROSPACE CORPORATION
PO BOX 92957
MS M4/041
LOS ANGELES CA 90009
U.S.A.
TEL: 213-648-6142
TLX:
TLF:
EML:
COM:

LAYZER DAVID PROF
HARVARD COLLEGE OBS
MAIL STOP 31
60 GARDEN STREET
CAMBRIDGE MA 02138
U.S.A.
TEL:
TLX:
TLF:
EML:
COM: 14.47

LAZAREFF BERNARD DR
OBSERVATOIRE DE GRENOBLE
BP 53 X
38041 GRENOBLE CEDEX
FRANCE
TEL:
TLX: 980134 F
TLF:
EML:
COM:

LAZARO CARLOS DR
INSTITUTO DE ASTROFISICA
DE CANARIAS
38200 LA LAGUNA TENERIFE
SPAIN
TEL: 34 22 26 22 11
TLX: 92640
TLF:
EML:
COM: 27

LAZOVIC JOVAN P PROF
DEPT OF ASTRONOMY
FACULTY OF SCIENCES
STUDENTSKI TRG 16
11000 BEOGRAD
YUGOSLAVIA
TEL: 11-638-715
TLX:
TLF:
EML:
COM: 07

LAZZARO DANIELA DR
OBS NACIONAL
DEPT DE ASTRON
RUA GENERAL BRUCE 586
RIO DE JANEIRO 20921
BRAZIL
TEL: 021-5807181
TLX: 2121288 OBSN BR
TLF:
EML: DAZA@LNCCVM
COM: 20

LE BORGNE JEAN FRANCOIS
OBSERVATOIRE DE MIDI-
PYRENEES
14 AVE EDOUARD BELIN
31400 TOULOUSE
FRANCE
TEL: 61 25 21 01
TLX: 530 776 F
TLF:
EML: BITNET:LEBORGNE@FROMP51
COM:

LE CONTEL JEAN-MICHEL
OBSERVATOIRE DE NICE
B P 139
F-06003 NICE CEDEX
FRANCE
TEL: 93-89-04-20
TLX: 460004
TLF:
EML:
COM: 29

LE DOURNEUF MARYVONNE
OBSERVATOIRE DE PARIS
SECTION DE MEUDON
F-92190 MEUDON PL CEDEX
FRANCE
TEL: 1-45-07-75-55
TLX: 201571 F
TLF:
EML:
COM: 14

LE FEVRE OLIVIER DR
DARC
OBSERVATOIRE DE MEUDON
92195 MEUDON CEDEX
FRANCE
TEL:
TLX:
TLF:
EML:
COM: 28

LE POOLE RUDOLF S DR
STERREWACHT
POSTBUS 9513
NL-2300 RA LEIDEN
NETHERLANDS
TEL: 071-272727
TLX: 39058 ASTRO NL
TLF:
EML:
COM: 24

LE SQUEREN ANNE-MARIE DR
OBSERVATOIRE DE PARIS
SECTION DE MEUDON
F-92195 MEUDON PL CEDEX
FRANCE
TEL: 1-45-07-75-95
TLX:
TLF:
EML:
COM: 34.40

LEA SUSAN MAUREEN DR
PHYSICS & ASTRONOMY DEPT
SAN FRANSISCO STATE UNIV
1600 HOLLOWAY AVE
SAN FRANCISCO CA 94132
U.S.A.
TEL: 415-469-1880
TLX:
TLF:
EML:
COM: 48

LEACOCK ROBERT JAY
DEPT OF ASTRONOMY
UNIVERSITY OF FLORIDA
211 SSRB
GAINESVILLE FL 32611
U.S.A.
TEL: 904-392-2052
TLX:
TLF:
EML:
COM: 28

LEAHY DENIS A DR
UNIVERSITY OF CALGARY
DEPT. OF PHYSICS
2500 UNIVERSITY DR N.W.
CALGARY ALBERTA T2N 1N4
CANADA
TEL: 403 220 7192
TLX:
TLF:
EML: BITNET:LEAHY@UNCAMULT
COM:

LEBEDINETS VLADIMIR N DR
ASTRONOMICAL COUNCIL
USSR ACADEMY OF SCIENCES
PYATNITSKAYA UL 48
109017 MOSCOW
U.S.S.R.
TEL:
TLX:
TLF:
EML:
COM: 22

LEBLANC YOLANDE DR
OBSERVATOIRE DE PARIS
SECTION DE MEUDON
F-92195 MEUDON PL CEDEX
FRANCE
TEL: 1-45-07-77-59
TLX:
TLF:
EML:
COM: 40

LEBOFSKY LARRY ALLEN
LUNAR & PLANETARY LAB
UNIVERSITY OF ARIZONA
TUCSON AZ 85721
U.S.A.
TEL: 602-621-6947
TLX:
TLF:
EML:
COM: 15

LEBOVITZ NORMAN R PROF
MATHEMATICS DEPT
UNIVERSITY OF CHICAGO
5734 S. UNIVERSITY AVE
CHICAGO IL 60637
U.S.A.
TEL: 312-753-8074
TLX:
TLF:
EML:
COM: 35

LEBRETON YVELINE DR
OBSERVATOIRE DE MEUDON
92195 MEUDON PPL CEDEX
FRANCE
TEL: 1-45 07 78 59
TLX: 201571 LAM
TLF:
EML: LEBRETON@FRMEU51
COM:

LECAR MYRON DR
CENTER FOR ASTROPHYSICS
60 GARDEN STREET
CAMBRIDGE MA 02138
U.S.A.
TEL: 617-495-7251
TLX: 921428 SATELLITE CAM
TLF:
EML:
COM: 33

LECKRONE DAVID S DR
NASA/GSFC
CODE 681
GREENBELT MD 20771
U.S.A.
TEL: 301-286-8904
TLX:
TLF:
EML:
COM: 29.44

LEDERLE TRUDPERT DR
ASTRONOMISCHES-RECHEN
INSTITUT
MOENCHHOFSTR 12-14
D-6900 HEIDELBERG 1
GERMANY. F.R.
TEL: 6221-49026
TLX: 461336 ARIHD D
TLF:
EML:
COM: 04.05.08.19

LEE PAUL D DR
LOUISIANA STATE UNIV
BATON ROUGE LA 70803
U.S.A.
TEL:
TLX:
TLF:
EML:
COM:

LEE SANG GAK
DEPT OF ASTRONOMY
COLLEGE NATURAL SCIENCES
SEOUL NATIONAL UNIVERSITY
SEOUL 151-00
KOREA. REPUBLIC
TEL: 877-2131/2139
TLX:
TLF:
EML:
COM: 33.45.51

LEE SEE-WOO DR
DEPT OF ASTRONOMY
SEOUL NATIONAL UNIVERSITY
SEOUL CITY
KOREA. REPUBLIC
TEL: 877-2131-9x3308
TLX:
TLF:
EML:
COM:

LEE TERENCE J DR
HEAD OF TECHNOLOGY
ROYAL OBSERVATORY
BLACKFORD HILL
EDINBURGH EH9 3HJ
U.K.
TEL: 031-667-3321
TLX: 72383 ROEDING UK
TLF:
EML:
COM: 14

LEER EGIL PROF
AURORAL OBSERVATORY
UNIVERSITY OF TROMSO
P O BOX 953
N-9001 TROMSO
NORWAY
TEL: 83-86060
TLX: 64124 AUROB N
TLF:
EML:
COM:

LEFEBVRE MICHEL DR
CNES/GRGS
18 AVENUE EDOUARD BELIN
F-31055 TOULOUSE CEDEX
FRANCE
TEL:
TLX:
TLF:
EML:
COM: 19

LEFEVRE JEAN DR
OBSERVATOIRE DE NICE
BP 139
F-06003 NICE CEDEX
FRANCE
TEL: 93-89-04-20
TLX:
TLF:
EML:
COM:

LEGER ALAIN DR
GROUPE PHYSIQUE SOLIDES
TOUR 23
UNIVERSITE PARIS 7
75251 PARIS CEDEX 05
FRANCE
TEL: 43362525 P.4671
TLX:
TLF:
EML:
COM: 14.21.34

LEGG THOMAS H DR
HERZBERG INST ASTROPHYS
NATL RESEARCH COUNCIL
OTTAWA ONT K1A 0R6
CANADA
TEL: 613-593-6060
TLX: 053-3715
TLF:
EML:
COM: 40

LEHMANN MAREK DR
SPACE RES CTR OF P.A.S.
ASTRONOMICAL LATITUDE OBS
BOROWIEC
PO 62035 KORNIK N.POZNAN
POLAND
TEL: (61)170187
TLX: 412623 AOS PL
TLF:
EML:
COM: 04.08.19

LEHNERT B P PROF
DEPT PLASMA PHYS. AND
FUSION RESEARCH
ROYAL INST. TECHNOLOGY
S-100 44 STOCKHOLM 70
SWEDEN
TEL: 7877763
TLX: 10389 KTEB S
TLF:
EML:
COM:

LEIBACHER JOHN DR
NATL SOLAR OBSERVATORY
PO BOX 26732
TUCSON AZ 85726-6732
U.S.A.
TEL: 602-325-9302
TLX: 3666-484
TLF:
EML:
COM: 10.12.36

LEIBOWITZ ELIA M DR
DEPT PHYSICS & ASTRONOMY
TEL-AVIV UNIVERSITY
TEL-AVIV 69978
ISRAEL
TEL: 03-413788
TLX: 342171 VERSY IL
TLF:
EML:
COM: 50

LEIGHTON R B PROF
CALTECH
1201 E. CALIFORNIA BLVD
PASADENA CA 91125
U.S.A.
TEL: 818-356-4286
TLX:
TLF:
EML:
COM: 12

LEIKIN G A DR
ASTRONOMICAL COUNCIL
USSR ACADEMY OF SCIENCES
PYATNITSKAYA UL 48
109017 MOSCOW
U.S.S.R.
TEL: 231-54-61
TLX: 412623 SCSTP SU
TLF:
EML:
COM: 16

LEINERT CHRISTOPH DR
MPI FUER ASTRONOMIE
KOENIGSTUHL
D-6900 HEIDELBERG 1
GERMANY. F.R.
TEL: 06221-528-264
TLX: 461789 MPIAD
TLF:
EML:
COM: 21C

LEITE SCHEID PAULO DR
OBSERVATORIO NACIONAL
RUA GENERAL BRUCE 586
20000 RIO DE JANEIRO
BRAZIL
TEL:
TLX:
TLF:
EML:
COM: 27

LELIEVRE GERARD DR
OBSERVATOIRE DE PARIS
61 AV. DE L'OBSERVATOIRE
75014 PARIS
PARIS
FRANCE
TEL: 40-51-22-55
TLX: 270776 OBS PARIS
TLF:
EML:
COM: 09C.28

LEMAIRE PHILIPPE DR
INSTITUT D'ASTROPHYSIQUE
SPATIALE
BP NO 10
F-91371 VERRIERES-LE-B
FRANCE
TEL: 1-64-47-43-12
TLX: 600252
TLF:
EML:
COM: 44

LEMAITRE ANNE DR
FAC UNIV N.D. DE LA PAIX
DEPT DE MATHEMATIQUES
REMPART DE LA VIERGE 8
B-5000 NAMUR
BELGIUM
TEL: 081-22-90-61
TLX: 59222 FACNAM B
TLF:
EML:
COM: 07

LEMAITRE GERARD R DR
OBSERVATOIRE DE MARSEILLE
2 PLACE LE VERRIER
F-13004 MARSEILLE
FRANCE
TEL: 91-95-90-88
TLX:
TLF:
EML:
COM: 09

LEMKE DIETRICH DR
MPI FUER ASTRONOMIE
KOENIGSTUHL
D-6900 HEIDELBERG 1
GERMANY. F.R.
TEL: 49-6221-528259
TLX: 461789 IMPIA-D
TLF:
EML:
COM:

LENA PIERRE J PROF
OBSERVATOIRE DE PARIS
SECTION DE MEUDON
F-92195 MEUDON PL CEDEX
FRANCE
TEL: 1-45-07-77-19
TLX: 201571F
TLF:
EML:
COM:

LENZEN RAINER DR
MPI FUER ASTRONOMIE
KOENIGSTUHL
D-6900 HEIDELBERG 1
GERMANY F.R.
TEL:
TLX:
TLF:
EML:
COM: 25

LEORAT JACQUES DR
OBSERVATOIRE DE MEUDON
SECTION DE MEUDON
F-92195 MEUDON PL CEDEX
FRANCE
TEL: 1-45-07-74-21
TLX: 201-571 LAM F
TLF:
EML:
COM:

LEPINE JACQUES R D DR
DEPTO DE ASTRONOMIA
IAG/USP
AV MIGUEL STEFANO 4200
04301 SAO PAULO SP
BRAZIL
TEL: 275-37-20
TLX: 1136221
TLF:
EML:
COM: 34.35.40

LEPP STEPHEN H DR
CENTER FOR ASTROPHYSICS
60 GARDEN STREET
CAMBRIDGE MA 02138
U.S.A.
TEL: 617 495 4086
TLX:
TLF:
EML:
COM:

LEQUEUX JAMES DR
OBSERVATOIRE DE MARSEILLE
2 PLACE LE VERRIER
F-13248 MARSEILLE CEDEX 4
FRANCE
TEL: 91-95-90-88
TLX: 420241 F
TLF:
EML:
COM: 05.28C.34C.40.47

LEROY BERNARD DR
OBSERVATOIRE DE PARIS
D.A.S.O.P.
92195 MEUDON PPL CEDEX
FRANCE
TEL: 45 07 78 12
TLX:
TLF:
EML:
COM: 10

LEROY JEAN-LOUIS
OBSERVATOIRE PIC-DU-MIDI
ET TOULOUSE
14 AVENUE EDOUARD BELIN
F-31400 TOULOUSE
FRANCE
TEL: 61-25-21-01
TLX: 530776 F
TLF:
EML: EARN:LEROY@FRMEU51
COM: 10.12

LESCHIUTTA S PROF
DIPARTEMENTO ELECTRONICA
POLITECNICO
CORSO DUCA D ABRUZZI 24
10129 TURINO
ITALY
TEL: 39 11 5567235
TLX: 220646 POLITO
TLF:
EML: BITNET:LESCHIUTTA@ITOPOLI
COM: 31

LESTER DANIEL F DR
DEPT OF ASTRONOMY
RLM HALL
UNIVERSITY OF TEXAS
AUSTIN TX 78712
U.S.A.
TEL: 512 471 3442
TLX: 910 8741351
TLF:
EML: ARPA:dfl@astro.as.utexas.edu
COM:

LESTER JOHN B DR
ASTRONOMY DEPT
ERINDALE COLLEGE
UNIVERSITY OF TORONTO
MISSISSAUGA L5L 1C6
CANADA
TEL: 416-828-5356
TLX:
TLF:
EML:
COM: 29

LESTRADE JEAN FRANCOIS DR
BUREAU DES LONGITUDES
77 AVE DENFERT ROCHEREAU
75014 PARIS
FRANCE
TEL: 40 51 22 65
TLX: 270070
TLF:
EML:
COM: 40

LETFUS VOJTECH DR
ASTRONOMICAL INSTITUTE
CZECH. ACAD. OF SCIENCES
251 65 ONDREJOV
CZECHOSLOVAKIA
TEL:
TLX:
TLF:
EML:
COM:

LEUNG CHUN MING DR
DEPT OF PHYSICS
RENSSELAER POLYTECH INST
TROY NY 12180-3590
U.S.A.
TEL: 518-266-6318
TLX:
TLF:
EML:
COM: 34.40

LEUNG KAM CHING PROF
BEHLEN OBSERVATORY
DEPT PHYSICS & ASTRONOMY
UNIVERSITY OF NEBRASKA
LINCOLN NB 68588
U.S.A.
TEL: 402-472-2770
TLX: 484340 UNL
TLF:
EML:
COM: 27.38C.42C

LEVASSEUR-REGOURD A.C. PR
SERVICE AERONOMIE CNRS
BP 3
F-91370 VERRIERES-LE-B.
FRANCE
TEL: 1-64-47-42-93
TLX: 602400
TLF:
EML:
COM: 15.21P.22

LEVATO ORLANDO HUGO DR
COMPLEJO ASTRONOMICO
EL LEONCITO
CASILLA DE CORREO 467
5400 SAN JUAN
ARGENTINA
TEL: 064-22-5718
TLX: 59134 ENTOP AR
TLF:
EML:
COM: 29.30.45C

LEVINE RANDOLPH H DR
50 CARVER ROAD
NEWTON MA 02161
U.S.A.
TEL: 617-965-5953
TLX:
TLF:
EML:
COM:

LEVREAULT RUSSELL M DR
DEPT. OF ASTRONOMY
VAN VLECK OBSERVATORY
WESLEYAN UNIVERSITY
MIDDLETOWN CT 06457
U.S.A.
TEL: 203 347 9411
TLX:
TLF:
EML: BITNET:rlevreault@wesleyan
COM: 40

LEVY EUGENE H DR
DEPT PLANETARY SCIENCES
LUNAR & PLANETARY LAB
UNIVERSITY OF ARIZONA
TUCSON AZ 85721
U.S.A.
TEL: 602-621-6962
TLX: 9109521143
TLF:
EML:
COM: 49

LEVY JACQUES R DR
OBSERVATOIRE DE PARIS
61 AVE DE L'OBSERVATOIRE
F-75014 PARIS
FRANCE
TEL: 1-43-20-12-10
TLX:
TLF:
EML:
COM: 41

LEWIN WALTER H G PROF
PHYSICS DEPT
MIT 37-627
CAMBRIDGE MA 02139
U.S.A.
TEL: 617-253-4282
TLX:
TLF:
EML:
COM: 44

LEWIS BRIAN MURRAY DR
ARECIBO OBSERVATORY
NAIC
PO BOX 95
ARECIBO PR 00613
U.S.A.
TEL: 809-878-2612
TLX:
TLF:
EML:
COM: 30

LEWIS J S
DEPT PLANETARY SCIENCES
UNIVERSITY OF ARIZONA
TUCSON AZ 85721
U.S.A.
TEL: 602-621-49[illegible]
TLX:
TLF:
EML:
COM: 16

LI CHUN-SHENG
DEPT OF ASTRONOMY
NANJING UNIVERSITY
NANJING 0909
CHINA. PEOPLE'S REP.
TEL: 34651-2882
TLX: 34151 PRCNU CN
TLF:
EML:
COM: 10.40

LI DEPEI
NANJING ASTRONOMICAL
INSTRUMENT FACTORY
NANJING
CHINA. PEOPLE'S REP.
TEL: 56191
TLX: 1133
TLF:
EML:
COM: 09

LI DONG-MING
PURPLE MOUNTAIN
OBSERVATORY
NANJING
CHINA. PEOPLE'S REP.
TEL:
TLX:
TLF:
EML:
COM: 06

LI GI HAN
PYONGYANG ASTRON OBS
ACADEMY OF SCIENCES DPRK
TAESONG DISTRICT
PYONGYANG
KOREA DPR
TEL:
TLX:
TLF:
EML:
COM: 04

LI GYONG WON
PYONGYANG ASTRON OBS
ACADEMY OF SCIENCES DPRK
TAESONG DISTRICT
PYONGYANG
KOREA DPR
TEL:
TLX:
TLF:
EML:
COM: 40

LI HEN
SHANGHAI OBSERVATORY
SHANGHAI
CHINA. PEOPLE'S REP.
TEL: 386191
TLX: 33164 SHAO CN
TLF:
EML:
COM: 15

LI HONG-WEI
DEPT OF ASTRONOMY
NANJING UNIVERSITY
NANJING
CHINA. PEOPLE'S REP.
TEL: 34651. 34751
TLX: 34151 PRCNU CN
TLF:
EML:
COM: 40

LI HYOK HO
PYONGYANG ASTRON OBS
ACADEMY OF SCIENCES DPRK
TAESONG DISTRICT
PYONGYANG
KOREA DPR
TEL:
TLX:
TLF:
EML:
COM: 04

LI J Y
PYONGYANG ASTRON OBS
ACADEMY OF SCIENCES DPRK
TAESONG DISTRICT
PYONGYANG
KOREA DPR
TEL:
TLX:
TLF:
EML:
COM:

LI JING
BEIJING OBSERVATORY
ACADEMIA SINICA
BEIJING
CHINA. PEOPLE'S REP.
TEL: 22040 BAOAS CN
TLX:
TLF:
EML:
COM: 28 33

LI NED C DR
CALIFORNIA UNIVERSITY
6531 WITHWORTH RD
LOS ANGELES CA 90035
U.S.A.
TEL:
TLX:
TLF:
EML:
COM:

LI NENG-YAO
PURPLE MOUNTAIN
OBSERVATORY
NANJING
CHINA. PEOPLE'S REP.
TEL: 37609
TLX: 34144 PMONJ CN
TLF:
EML:
COM: 04.08

LI QI-BIN
BEIJING OBSERVATORY
ACADEMIA SINICA
BEIJING
CHINA PEOPLE'S REP.
TEL: 281968
TLX: 22040 BAOAS CN
TLF:
EML:
COM: 280.46

LI SIN HYONG
PYONGYANG ASTRON OBS
ACADEMY OF SCIENCES DPRK
TAESONG DISTRICT
PYONGYANG
KOREA DPR
TEL:
TLX:
TLF:
EML:
COM: 35

LI SON JAE
PYONGYANG ASTRON OBS
ACADEMY OF SCIENCES DPRK
TAESONG DISTRICT
PYONGYANG
KOREA DPR
TEL:
TLX:
TLF:
EML:
COM: 10

LI TING
NANJING ASTRONOMICAL
INSTRUMENT FACTORY
NANJING
CHINA. PEOPLE'S REP.
TEL:
TLX:
TLF:
EML:
COM: 09

LI TIPEI
INSTITUTE OF HIGH ENERGY
PHYSICS
BEIJING
CHINA. PEOPLE'S REP.
TEL: 812971-464
TLX: 22082 IHEP CN
TLF:
EML:
COM: 44.48

LI WEI BAO
YUNNAN OBSERVATORY
P.O. BOX 110
KUNMING
CHINA. PEOPLE'S REP.
TEL:
TLX:
TLF:
EML:
COM: 10

LI XIAO-QING
PURPLE MOUNTAIN
OBSERVATORY
NANJING
CHINA. PEOPLE'S REP.
TEL: 31096
TLX: 34144 PMO NJ CN
TLF:
EML:
COM: 26.49

LI YUAN-JIE
DEPARTMENT OF PHYSICS
HUAZHONG UNIVERSITY OF
SCIENCE AND TECHNOLOGY
WUHAN
CHINA. PEOPLE'S REP.
TEL: 870541
TLX: 7122
TLF:
EML:
COM: 48

LI ZHENG-XIN DR
SHANGHAI OBSERVATORY
ACADEMIA SINICA
SHANGHAI
CHINA. PEOPLE'S REP.
TEL:
TLX:
TLF:
EML:
COM: 19

LI ZHIGANG
SHAANXI ASTRONOMICAL OBS
P.O. BOX 18
LINTONG
XIAN
CHINA. PEOPLE'S REP.
TEL: 32255 XIAN
TLX: 70121 CSAO CN
TLF:
EML:
COM: 08

LI ZHI-FANG
SHANGHAI OBSERVATORY
ACADEMIA SINICA
SHANGHAI
CHINA. PEOPLE'S REP.
TEL: 386191
TLX: 33164 SHAO CN
TLF:
EML:
COM: 08.47

LI ZHI-SEN
BEIJING ASTRONOMICAL OBS
ACADEMIA SINICA
BEIJING
CHINA. PEOPLE'S REP.
TEL: 28-1698
TLX: 9053
TLF:
EML:
COM: 41

LI ZHONGYUAN
DEPT EARTH & SPACE SCI.
UNIV SCIENCE & TECHNOLOGY
HEFEI. ANHUI
CHINA. PEOPLE'S REP.
TEL: 63300
TLX: 90028 USTC CN
TLF:
EML:
COM: 42.44

LI ZONG-WEI
DEPT OF ASTRONOMY
BEIJING NORMAL UNIVERSITY
BEIJING
CHINA. PEOPLE'S REP.
TEL: 65-6531 x 683
TLX:
TLF:
EML:
COM: 35.48

LI ZONG-YUN
DEPT OF ASTRONOMY
NANJING UNIVERSITY
NANJING
CHINA. PEOPLE'S REP.
TEL:
TLX: 34151 PRCNU CN
TLF:
EML:
COM:

LIANG SHI-GUANG
SHANGHAI OBSERVATORY
ACADEMIA SINICA
SHANGHAI
CHINA. PEOPLE'S REP.
TEL: 386191
TLX: 33164 SHAO CN
TLF:
EML:
COM: 40

LIANG ZHONG-HUAN
P.O. BOX 18
LINTONG
XIAN
CHINA. PEOPLE'S REP.
TEL: XIAN 32255
TLX: 70121 CSAO CN
TLF:
EML:
COM: 31

LIBBRECHT K G DR
BIG BEAR SOLAR OBS
CALTECH
264-33 CALTECH
PASADENA CA 91125
U.S.A.
TEL: 818 356 3777
TLX: 675425 CALTECH PSD
TLF:
EML: KGL@SUNDOG.CALTECH.EDU
COM: 29

LIDDELL U MR
NASA LUNAR & PLANET PRG
OFFICE OF SPACE SCIENCES
SPACE SCI & APPLICATIONS
WASHINGTON DC 20546
U.S.A.
TEL:
TLX:
TLF:
EML:
COM:

LIEBERT JAMES W DR
STEWARD OBSERVATORY
UNIVERSITY OF ARIZONA
TUCSON AZ 85721
U.S.A.
TEL: 602 621 4513
TLX: 621 41 410
TLF:
EML: LIEBERT@ARIZRVAX
COM: 29.35.36

LIEBSCHER DIERCK-E DR
ZNTRLINST. F. ASTROPHYSIK
STERNWARTE BABELSBERG
ROSA-LUXEMBURG-STR 17A
DDR-1502 POTSDAM
GERMANY. D.R.
TEL:
TLX:
TLF:
EML:
COM: 47

LIESKE JAY H DR
JPL/CALTECH
MS 264-664
4800 OAK GROVE DRIVE
PASADENA CA 91109
U.S.A.
TEL: 818-354-3642
TLX: 675429
TLF:
EML:
COM: 04C.07 19.20.31

LILLER WILLIAM DR
INSTITUTO ISAAC NEWTON
CASILLA 437
VINA DEL MAR
CHILE
TEL: 03-970864
TLX:
TLF:
EML:
COM: 15.28.34

LILLEY EDWARD A PROF
HARVARD COLLEGE
OBSERVATORY
60 GARDEN STREET
CAMBRIDGE MA 02138
U.S.A.
TEL: 617-495-3971
TLX: 921428 SATELLITE CAM
TLF:
EML:
COM: 40.51

LILLIE CHARLES F DR
TRW ELECTRONICS & DEFENSE
R11/1059
1 SPACE PARK
REDONDO BEACH. CA 90278
U.S.A.
TEL: 213-812-2248
TLX: 910-325-6611
TLF:
EML:
COM: 15 21

LILLY SIMON J DR
INSTITUTE FOR ASTRONOMY
2680 WOODLAWN DRIVE
HONOLULU HI 96822
U.S.A.
TEL: 808 948 6196
TLX: 8459 UHAST HR
TLF:
EML:
COM: 28.47

LIN CHIA C PROF
MIT
DEPT OF MATHEMATICS
77 MASSACHUSETTS AVE
CAMBRIDGE MA 02139
U.S.A.
TEL: 617-253-1796
TLX: 921473 MIT CAM
TLF:
EML:
COM: 28.33.34

LIN DOUGLAS N. C. DR
LICK OBSERVATORY
UNIVERSITY OF CALIFORNIA
SANTA CRUZ CA 95064
U.S.A.
TEL: 408-429-2731
TLX:
TLF:
EML:
COM:

LIN YUANZHANG
BEIJING ASTRONOMICAL OBS
ACADEMIA SINICA
BEIJING
CHINA. PEOPLE'S REP.
TEL: 281698
TLX: 22040 BAOAS CN
TLF:
EML:
COM: 10.12

LINCOLN J VIRGINIA MISS
2005 ALPINE DRIVE
BOULDER CO 80302
U.S.A.
TEL: 303-442-6757
TLX:
TLF:
EML:
COM:

LINDBLAD BERTIL B DR
LUND OBSERVATORY
BOX 43
S-221 00 LUND
SWEDEN
TEL: 46-10-70-00
TLX: 33199 OBSNOT S
TLF:
EML:
COM: 20.22.44

LINDBLAD PER OLOF PROF
STOCKHOLM OBSERVATORY
S-133 00 SALTSJOEBADEN
SWEDEN
TEL: 87-170195
TLX: 12972 SOBSERV S
TLF:
EML:
COM: 28.33

LINDEGREN LENNART DR
LUND OBSERVATORY
BOX 43
S-221 00 LUND
SWEDEN
TEL: 46-10-70-00
TLX: 33199 OBSNOT S
TLF:
EML:
COM: 08C

LINDGREN HARRI
E.S.O. LA SILLA
CASILLA 19001
SANTIAGO 19
CHILE
TEL: 6988757
TLX: 240881
TLF:
EML: BITNET:LINDGRENODGAESO51
COM: 30

LINDSEY CHARLES ALLAN
UNIVERSITY OF HAWAII
AT MANOA
2680 WOODLAWN DRIVE
HONOLULU HI 96822
U.S.A.
TEL: 808-948-6526
TLX:
TLF:
EML:
COM: 15

LING CHIN-BING DR
INSTITUTE OF MATHEMATICS
ACADEMIA SINICA
PO BOX NO 143
TAIPEI
CHINA. TAIWAN
TEL:
TLX:
TLF:
EML:
COM:

LING J DR
OBS.ASTRON. R.M. ALLER
UNIV. DE SANTIAGO
AVEN. DE LAS CIENCIAS S/N
SANTIAGO DE COMPOSTELA
SPAIN
TEL:
TLX:
TLF:
EML:
COM: 26

LINGENFELTER RICHARD E
UNIVERSITY OF CALIFORNIA
CASS C-011
LA JOLLA CA 92093
U.S.A.
TEL: 619-452-2464
TLX: 9103371271 SIOCCAN
TLF:
EML:
COM:

LINKE RICHARD ALAN DR
BELL LABORATORIESS
CRAWFORD HILL LAB
HOLMDEL NJ 07733
U.S.A.
TEL:
TLX:
TLF:
EML:
COM: 34.40

LINNELL ALBERT P PROF
DEPT PHYSICS & ASTRONOMY
MICHIGAN STATE UNIVERSITY
EAST LANSING MI 48824
U.S.A.
TEL: 517-353-6670
TLX:
TLF:
EML:
COM: 35.36.42

LINNIK V P PROF DR
MAIN ASTRONOMICAL OBS
PULKOVO
196140 LENINGRAD
U.S.S.R.
TEL:
TLX:
TLF:
EML:
COM:

LINSKY JEFFREY L DR
JILA
UNIVERSITY OF COLORADO
CAMPUS BOX 440
BOULDER CO 80309
U.S.A.
TEL: 303-492-7838
TLX: 755842 JILA
TLF:
EML:
COM: 12.36C.44

LINSLEY JOHN
DEPT PHYSICS & ASTRONOMY
UNIVERSITY OF NEW MEXICO
ALBUQUERQUE NM 87131
U.S.A.
TEL: 505-243-1924
TLX: 910989
TLF:
EML:
COM: 44.48

LIPOVETSKY V A
SPECIAL ASTROPHYSICAL OBS
NIZHNIJ ARKHYZ
357147 STAVROPOLSKIJ KRAJ
U.S.S.R.
TEL: 93-2-42
TLX:
TLF:
EML:
COM: 28

LIPPINCOTT SARAH LEE DR
SPROUL OBSERVATORY
SWARTHMORE COLLEGE
507 CEDAR LANE
SWARTHMORE PA 19081
U.S.A.
TEL: 215-543-9058
TLX:
TLF:
EML:
COM: 24.26.51

LIPSCHUTZ MICHAEL E DR
WETHERILL CHEMISTRY BLDG
PURDUE UNIVERSITY
W LAFAYETTE IN 47907
U.S.A.
TEL: 317 494 5326
TLX: 272 396
TLF:
EML: BITNET:BNAAPUNL@PURCCVM
COM: 15

LISEAU RENE DR
CNR ISTITUTO DI FISICA
D SPAZIO INTERPLANETARIO
VIA G. GALILEI.CP 27
I 00044 FRASCATI
ITALY
TEL: 39-6 9423801
TLX: 610261 CNR FRA I
TLF:
EML:
COM:

LISI FRANCO DR
OSSERVATORIO ASTROFISICO
DI ARCETRI
LARGO E.FERMI 5.
50125 FIRENZE
ITALY
TEL: 39-55 2752289
TLX: 572268 I
TLF: 39-55 220039
EML: SPAN:38954::LISI
COM:

LISSAUER JACK J DR
EARTH & SPACE SCIENCES
STATE UNIVERSITY NEW YORK
STONY BROOK NY 11794
U.S.A.
TEL: 516 632 8225
TLX:
TLF:
EML:
COM: 07.15.16

LISZT HARVEY STEVEN
NRAO
EDGEMONT ROAD
CHARLOTTESVILLE VA 22901
U.S.A.
TEL: 804-296-0344
TLX: 910-997-0714
TLF:
EML:
COM: 34

LITTLE LESLIE T DR
ELECTRONICS LAB
UNIVERSITY OF KENT
CANTERBURY. KENT CT2 7NT
U.K.
TEL: 0227-66822
TLX: 965449 UKCLIB
TLF:
EML:
COM: 40

LITTLETON JOHN E
DEPT OF PHYSICS
WEST VIRGINIA UNIVERSITY
PO BOX 6023
MORGANTOWN WV 26506-6023
U.S.A.
TEL: 304-293-3498
TLX: 710-921-0309
TLF:
EML:
COM: 35

LITTLE-MARENIN IRENE R DR
WHITIN OBSERVATORY
WELLESLEY MA 02181
U.S.A.
TEL: 617 235 5303
TLX:
TLF:
EML: ilittle@lucy.wellesley.edu
COM: 27.29.46

LITVAK MARVIN M DR
TRW INC. 01/1260
ONE SPACE PARK
REDONDO BEACH. CA 90278
U.S.A.
TEL:
TLX:
TLF:
EML:
COM:

LIU BAO-LIN
PURPLE MOUNTAIN
OBSERVATORY
NANJING
CHINA. PEOPLE'S REP.
TEL: 42817 / 46700
TLX: 34144 PMONJ CN
TLF:
EML:
COM: 04

LIU CAIPIN
PURPLE MOUNTAIN
OBSERVATORY
NANJING
CHINA. PEOPLE'S REP.
TEL: 42817 NANJING
TLX: 34144 PMONJ CN
TLF:
EML:
COM: 36

LIU JINMING
SHANGHAI ASTRON. OBSERV.
80 NANDAN ROAD
SHANGHAI
CHINA. PEOPLE'S REP.
TEL: 386191
TLX: 33164 SHAO CN
TLF:
EML:
COM: 05

LIU JINYI
INST OF HISTORY OF NAT SC
GONG YUAN WEST STREET 1
BEIJING
CHINA. PEOPLE'S REP.
TEL: 557180 BEIJING
TLX:
TLF:
EML:
COM: 41.48

LIU LIAO
DEPT OF PHYSICS
BEIJING NORMAL UNIVERSITY
BEIJING
CHINA. PEOPLE'S REP.
TEL:
TLX:
TLF:
EML:
COM: 47

LIU LIN
DEPT OF ASTRONOMY
NANJING UNIVERSITY
NANJING
CHINA. PEOPLE'S REP.
TEL: 34651 x 2882
TLX: 34151 PRCNU CN
TLF:
EML:
COM:

LIU LIN-ZHONG
PURPLE MOUNTAIN
OBSERVATORY
NANJING
CHINA. PEOPLE'S REP.
TEL: 46700
TLX: 34144 PMONT CN
TLF:
EML:
COM: 15

LIU QINGYAO DR
YUNNAN OBSERVATORY
ACADEMIA SINICA
KUNMING PO BOX 110
KUNMING 1131
CHINA. PEOPLE'S REP.
TEL: KUNMING 72946
TLX: 64040 YUOBS CN
TLF:
EML:
COM: 42

LIU RU-LIANG
PURPLE MOUNTAIN
OBSERVATORY
NANJING
CHINA. PEOPLE'S REP.
TEL: 42817 / 46700
TLX: 34144 PMONJ CN
TLF:
EML:
COM: 28.48

LIU SOU-YANG DR
COMPUTER SCIENCES CORP
SYSTEM SCIENCES DIVISION
8728 COLESVILLE ROAD
SILVER SPRING MD 20910
U.S.A.
TEL: 301-589-1545
TLX:
TLF:
EML:
COM:

LIU XINPING PROF.
INSTITUTE OF MECHANICS
CHINESE ACADEMY OF SCI.
BEIJING 100080
CHINA. PEOPLE'S REP.
TEL: 284185
TLX: 222554 MEHAS CN
TLF: 86-1 2561284
EML:
COM: 10

LIU XUEFU
DEPT OF ASTRONOMY
BEIJING NORMAL UNIVERSITY
BEIJING
CHINA. PEOPLE'S REP.
TEL: 656531-6285
TLX: 8511
TLF:
EML:
COM: 42

LIU YONG-ZHEN
GRADUATE SCHOOL
UNIV SCIENCE & TECHNOLOGY
P.O. BOX 3908
BEIJING
CHINA. PEOPLE'S REP.
TEL: 817031
TLX:
TLF:
EML:
COM: 28.47

LIU ZONGLI
BEIJING ASTRONOMICAL OBS
ACADEMIA SINICA
BEIJING 100080
CHINA, PEOPLE'S REP.
TEL:
TLX: 22040
TLF:
EML:
COM: 15.27

LIVINGSTON WILLIAM C
NOAO/NSO
PO BOX 26732
TUCSON AZ 85726
U.S.A.
TEL: 602-327-5511
TLX: 0666484 AURA NOAO TU
TLF:
EML:
COM: 09.12

LIVIO MARIO
DEPARTMENT OF PHYSICS
TECHNION
HAIFA 32000
ISRAEL
TEL: 04-293549
TLX: 46650 TECLI IL
TLF:
EML:
COM: 35

LIVSHITS M A DR
INSTITUTE OF TERRESTRIAL
MAGNETISM & IONOSPHERE
IZMIRAN
142092 TROITSK MOSCOW REG
U.S.S.R.
TEL:
TLX: 412623 SCSTP SU
TLF:
EML:
COM: 10

LLOYD EVANS THOMAS DR
S A A O
P O BOX 9
OBSERVATORY 7935
SOUTH AFRICA
TEL: 021-47-0026
TLX: 5720309 SA
TLF:
EML:
COM: 37.45C

LO KWOK-YUNG DR
OWENS VALLEY RADIO OBS
DEPARTMENT OF ASTRONOMY
CALTECH 105-24
PASADENA CA 91125
U.S.A.
TEL: 818-356-4415
TLX: 675425 CALTECH PSD
TLF:
EML:
COM: 28.34.40

LOCANTHI DOROTHY DAVIS DR
2180 PINECREST DRIVE
ALTADENA CA 91001
U.S.A.
TEL: 213-797-0629
TLX:
TLF:
EML:
COM: 29

LOCHMAN JAN
ASTRONOMICAL INSTITUTE
CZECH. ACAD. OF SCIENCES
DVORAKOVA 298
511 01 TURNOV
CZECHOSLOVAKIA
TEL: 0436-22622
TLX:
TLF:
EML:
COM: 09

LOCKE JACK L DR
250 BRAESIDE AVENUE
OTTAWA ONT K1H 7J5
CANADA
TEL: 613-523-0812
TLX:
TLF:
EML:
COM: 12.40

LOCKMAN FELIX J
NRAO
EDGEMONT ROAD
CHARLOTTESVILLE VA 22903
U.S.A.
TEL: 804-296-0211
TLX: 910-997-0174
TLF:
EML:
COM: 33.34.40

LOCKWOOD G WESLEY DR
LOWELL OBSERVATORY
1400 W. MARS HILL RD
FLAGSTAFF AZ 86001
U.S.A.
TEL: 602-774-3358
TLX:
TLF:
EML:
COM: 16.25.27

LODEN KERSTIN E DR
STOCKHOLM OBSERVATORY
S-133 00 SALTSJOEBADEN
SWEDEN
TEL: 08-7170195
TLX: 12972 SOBSERV S
TLF:
EML:
COM: 26.33.45

LODEN LARS OLOF PROF
ASTRONOMICAL OBSERVATORY
BOX 515
S-751 20 UPPSALA
SWEDEN
TEL: 018-11-44-90
TLX: 76024
TLF:
EML:
COM: 26.33.37.51

LOISEAU NORA DR
INPE
DEPT.RADIOASTRONOMIA
CAIXA POSTAL 515
12201 SAO JOSE DOS CAMPOS
BRAZIL
TEL:
TLX:
TLF:
EML:
COM:

LOMB NICHOLAS RALPH DR
SYDNEY OBSERVATORY
MUSEUM APPLIED ARTS & SC
P.O.BOX K346
HAYMARKET NSW 2000
AUSTRALIA
TEL:
TLX:
TLF:
EML:
COM: 20.46

LONG KNOX S DR
CTR FOR ASTROPHYSICAL SCI
DEPT PHYSICS & ASTRONOMY
THE JOHNS HOPKINS UNIV
BALTIMORE MD 21218
U.S.A.
TEL: 301 338 7391
TLX: 9102400225 JHUCASMD
TLF:
EML:
COM: 44

LONGAIR M S PROF
ASTRONOMER ROYAL FOR
SCOTLAND
ROYAL OBSERVATORY
EDINBURGH EH9 3HJ
U.K.
TEL: 031-667-3321
TLX: 72383
TLF:
EML:
COM: 40.47.48

LONGMORE ANDREW J
UK INFRARED TELESCOPE
UNIT
665 KOMOHANA STREET
HILO HI 96720
U.S.A.
TEL: 808-961-3756
TLX: 633135
TLF:
EML:
COM:

LONSDALE CAROL J DR
INFRARED PROCESSING &
ANALYSIS CENTER
CALTECH MS 100 22
PASADENA CA 91125
U.S.A.
TEL: 818 584 2929
TLX: 67 5429
TLF:
EML: INTERNET:CJL@IPAC.CALTECH.EDU
COM: 05.28

LOOSE HANS-HERMANN DR
UNIVERSITAETS STERNWARTE
GEISMARLANDSTRASSE 11
3400 GOETTINGEN
GERMANY, F.R.
TEL: 551-395-056/953
TLX:
TLF:
EML:
COM: 28

LOPES ROSALY DR
JET PROPULSION LABORATORY
4800 OAK GROVE DRIVE
PASADENA CA 91109
U.S.A.
TEL:
TLX:
TLF:
EML:
COM: 15.16.41

LOPEZ CARLOS LIC
OBS FELIX AGUILAR
AV BENAVIDEZ 8175 OESTE
5407 SAN JUAN
ARGENTINA
TEL:
TLX:
TLF:
EML:
COM: 24

LOPEZ DE COCA M D P DR
APDO 2144
INST DE ASTROFISICA
18080 GRANADA
SPAIN
TEL: 121311
TLX: 78573 IAAGE
TLF: 58-814530
EML: PILAR@IAA.ES
COM: 27

LOPEZ GARCIA ZULEMA L DR
OBSERVATORIO ASTRONOMICO
FELIX AGUILAR
AV BENAVIDEZ 8175 OESTE
5407 MARQUESADO (S.J.)
ARGENTINA
TEL:
TLX:
TLF:
EML:
COM:

LOPEZ JOSE A ING
OBSERVATORIO ASTRONOMICO
FELIX AGUILAR
AV BENAVIDEZ 8175 OESTE
5407 MARQUESADO (S.J.)
ARGENTINA
TEL: 064-231494
TLX: 59100 UNSJA AR
TLF:
EML:
COM: 08

LOPEZ JOSE ALBERTO DR
INSTITUTO DE ASTRONOMIA
UNAM APDO POSTAL 877
ENSENADA BC 22800
MEXICO
TEL: 667- 44580
TLX: 56539 CICEME
TLF:
EML:
COM:

LOPEZ ROSARIO DR
DEPT DE FISICA
BARCELONA UNIVERSITY
DIAGONAL 647
08028 BARCELONA
SPAIN
TEL: 343-3307311
TLX:
TLF:
EML: BITNET:D3FARLE0@EB0UB011
COM: 76

LOPEZ-ARROYO M
OBSERVATORIO ASTRONOMICO
ALFONSO XII-5
28014 MADRID
SPAIN
TEL:
TLX:
TLF:
EML:
COM: 12

LOPEZ-GARCIA FRANCISCO DR
OBSERVATORIO ASTRONOMICO
FELIX AGUILAR
AV. BENAVIDEZ 8175 OESTE
5407 MARQUESADO (S.J.)
ARGENTINA
TEL:
TLX:
TLF:
EML:
COM:

LOPEZ-MORENO JOSE JUAN
INST DE ASTROFISICA DE
ANDALUCIA. APDO 2144
PROFESOR ALBAREDA 1
18080 GRANADA
SPAIN
TEL: 58-12-13-00
TLX: 78573 IAAG E
TLF:
EML:
COM: 16.21

LOPEZ-PUERTAS MANUEL
INST DE ASTROFISICA DE
ANDALUCIA. APDO 2144
PROFESOR ALBAREDA 1
18080 GRANADA
SPAIN
TEL: 58-12-13-00
TLX: 78573 IAAG E
TLF:
EML:
COM: 16.21

LOREN ROBERT BRUCE DR
PO BOX 2915
SILVER CITY N.M. 88062
U.S.A.
TEL:
TLX:
TLF:
EML:
COM: 34 40

LORENZ HILMAR
ZNTRLINST. F. ASTROPHYSIK
AKAD. WISSENSCHAFTEN DDR
ROSA-LUXEMBURG-STR 17A
DDR-1502 POTSDAM-BABELSB.
GERMANY. D.R.
TEL:
TLX:
TLF:
EML:
COM: 28.40

LORTET MARIE CLAIRE
OBSERVATOIRE DE PARIS
SECTION DE MEUDON
DAPHE
F-92195 MEUDON PL CEDEX
FRANCE
TEL: 1-45-07-74-24
TLX: 201-571 LAM F
TLF:
EML:
COM: 35.28.34

LOSCO LUCETTE DR
FACULTE DES SCIENCES
F-25030 BESANCON CEDEX
FRANCE
TEL:
TLX:
TLF:
EML:
COM:

LOTOVA N A DR
IZMIRAN
AKADEMGORODOK
142092 MOSCOW REGION
U.S.S.R.
TEL:
TLX:
TLF:
EML:
COM: 49

LOUGHHEAD RALPH E DR
CSIRO
DIV OF APPLIED PHYSICS
P.O.BOX 218
LINDFIELD NSW 2070
AUSTRALIA
TEL: 02-467-6355
TLX: 26296
TLF:
EML:
COM: 10.12

LOUISE RAYMOND PROF
FACULTE DES SCIENCES
DEPT DE PHYSIQUE
33 RUE ST-LEU
F-80039 AMIENS
FRANCE
TEL:
TLX:
TLF:
EML:
COM: 34

LOULERGUE MICHELLE DR
OBSERVATOIRE DE PARIS
SECTION DE MEUDON
F-92195 MEUDON PL CEDEX
FRANCE
TEL: 1-45-07-74-55
TLX: 270912 OBSASTR
TLF:
EML:
COM: 14

LOVAS FRANCIS JOHN DR
MOLECULAR SPECTROSCOPIC
DIV 545
NATL BUREAU OF STANDARDS
WASHINGTON DC 20234
U.S.A.
TEL: 301-921-2023
TLX: 898993
TLF:
EML:
COM: 14C.34

LOVAS MIKLOS
KONKOLY OBSERVATORY
BOX 67
H-1525 BUDAPEST
HUNGARY
TEL: 166621 BUDAPEST
TLX: 227460 KONOB
TLF:
EML:
COM: 10

LOVELACE RICHARD V E DR
SPACE SCIENCES BLDG
CORNELL UNIVERSITY
ITHACA NY 14853
U.S.A.
TEL: 607-256-3968
TLX:
TLF:
EML:
COM: 48

LOVELL SIR BERNARD PROF
NUFFIELD RADIO ASTR LABS
JODRELL BANK
MACCLESFIELD SK11 9DL
U.K.
TEL: 0477-71321
TLX: 36149
TLF:
EML:
COM: [illegible] 40.44.51

LOW BOON CHYE
HIGH ALTITUDE OBSERVATORY
NCAR
PO BOX 3000
BOULDER CO 80307
U.S.A.
TEL: 303-497-1553
TLX: 45694
TLF:
EML:
COM: [illegible]

LOW FRANK J DR
4940 CALLE BARRIL
TUCSON AZ 85718
U.S.A.
TEL: 602-621-2779
TLX:
TLF:
EML:
COM: 28.34.45

LOWE ROBERT P DR
DEPT OF PHYSICS
UNIV OF WESTERN ONTARIO
LONDON ONT N6A 3K7
CANADA
TEL: 519-679-2917
TLX:
TLF:
EML:
COM:

LOYOLA PATRICIO DR
OBS ASTRONOMICO NACIONAL
UNIVERSIDAD DE CHILE
CASILLA 36-D
SANTIAGO
CHILE
TEL:
TLX:
TLF:
EML:
COM: 38

LOZINSKAYA TAT'YANA A DR
STERNBERG STATE
ASTRONOMICAL INSTITUTE
119899 MOSCOW B-234
U.S.S.R.
TEL: 139-10-30
TLX:
TLF:
EML:
COM: 34.40

LOZINSKIJ A M DR
ASTRONOMICAL COUNCIL
USSR ACADEMY OF SCIENCES
PYATNITSKAYA UL 48
109017 MOSCOW
U.S.S.R.
TEL: 231-54-61
TLX: 412623 SCSTP SU
TLF:
EML:
COM: 34

LU BEN-KUI
PURPLE MOUNTAIN
OBSERVATORY
NANJING
CHINA, PEOPLE'S REP.
TEL: 32893
TLX: 34144 PMONJ CN
TLF:
EML:
COM: 07.31

LU CHUN-LIN
PURPLE MOUNTAIN
OBSERVATORY
NANJING
CHINA, PEOPLE'S REP.
TEL: 42700
TLX: 34144 PMONJ CN
TLF:
EML:
COM: 08

LU PHILLIP K DR
DEPT PHYSICS & ASTRONOMY
WESTERN CONN STATE UNIV
181 WHITE ST
DANBURY CT 06810
U.S.A.
TEL: 203-797-4[illegible]8
TLX:
TLF:
EML:
COM: 24.[illegible]

LU TAN
DEPT OF ASTRONOMY
NANJING UNIVERSITY
NANJING
CHINA, PEOPLE'S REP.
TEL: 34651-2882
TLX: 34151 PRCNU CN
TLF:
EML:
COM: 47.48

LU YANG
DEPT OF ASTRONOMY
NANJING UNIVERSITY
NANJING
CHINA, PEOPLE'S REP.
TEL: 34651-2882
TLX: 34151 PRCNU CN
TLF:
EML:
COM: 40

LUB JAN DR
STERREWACHT
HUYGENS LABORATORIUM
POSTBUS 9513
NL-2300 RA LEIDEN
NETHERLANDS
TEL: 071-272727
TLX: 39066 ASTRO NL
TLF:
EML:
COM: 25C.27

LUCAS ROBERT DR
GROUPE D'ASTROPHYSIQUE
UNIV SCIENT & MEDICALE
CERMO BP 68
F-38402 ST-MARTIN-D'HERES
FRANCE
TEL: 76-51-46-00
TLX:
TLF:
EML:
COM: 34

LUCCHIN FRANCESCO
ISTITUTO DI FISICA
G. GALILEI
VIA MARZOLO 8
I-35100 PADOVA
ITALY
TEL: 049-844333
TLX: 430308 DF GGPDI
TLF:
EML:
COM:

LUCEY JOHN DR
DEPARTMENT OF PHYSICS
UNIVERSITY OF DURHAM
SOUTH ROAD
DURHAM DH1 3LE
U.K.
TEL:
TLX:
TLF:
EML:
COM: 28

LUCK R EARLE DR
DEPARTMENT OF ASTRONOMY
CASE WESTERN RES UNIV
CLEVELAND OH 44106
U.S.A.
TEL: 216 368 6697
TLX:
TLF:
EML:
COM: 29

LUCKE PETER B DR
DEPT PHYSICS & ASTRONOMY
MOUNT UNION COLLEGE
ALLIANCE OH 44601
U.S.A.
TEL: 216-821-5320
TLX:
TLF:
EML:
COM:

LUCY LEON B PROF
ESO
KARL-SCHWARZSCHILD-STR 2
D-8046 GARCHING B MUNCHEN
GERMANY F.R.
TEL: 89-32006-249
TLX: 0528282-0
TLF:
EML:
COM: 42

LUEST REIMAR PROF
EUROPEAN SPACE AGENCY
8-10 RUE MARIO NIKIS
F-75738 PARIS
FRANCE
TEL: 1-42-73-74-04
TLX: 202746 ESA
TLF:
EML:
COM: 12.44.48.49

LUEST RHEA DR
MPI FUER PHYSIK UND
ASTROPHYSIK
K-SCHWARZSCHILDSTR 1
D-8046 GARCHING B MUNCHEN
GERMANY. F.R.
TEL: 89-320-32990
TLX: 524629 ASTROD
TLF:
EML:
COM: 15

LUGGER PHYLLIS M
INDIANA UNIVERSITY
DEPT OF ASTRONOMY
SWAIN WEST 319
BLOOMINGTON IN 47405
U.S.A.
TEL: 812-335-6929
TLX:
TLF:
EML:
COM: 28

LUKACEVIC ILIJA S DR
FACULTY OF SCIENCES
DEPT OF MECHANICS
STUDENTSKI TRG 16
11000 BEOGRAD
YUGOSLAVIA
TEL:
TLX:
TLF:
EML:
COM:

LUMINET JEAN-PIERRE
OBSERVATOIRE DE PARIS
SECTION DE MEUDON
F-92195 MEUDON PL CEDEX
FRANCE
TEL: 1-45-07-74-23
TLX: 201571 F
TLF:
EML:
COM: 28.47.48

LUMME KARI A DR
OBSERVATORY
TAHTITORNINMAKI
SF-00130 HELSINKI 13
FINLAND
TEL: 1912910
TLX:
TLF:
EML:
COM: 15.16.21

LUNA HOMERO G. DR
INSTITUTO ARGENTINO DE
RADIOASTRONOMIA
CASILLA DE CORREO No. 5
1894 VILLA ELISA (Bs.As.
ARGENTINA
TEL:
TLX:
TLF:
EML:
COM: 25

LUND NIELS
DANISH SPACE RES INST
LUNDTOFTEVEJ 7
DK 2800 LYNGBY
DENMARK
TEL: 45 2 882277
TLX: 37198
TLF:
EML:
COM:

LUNDQUIST CHARLES A DR
RESEARCH INSTITUTE
THE UNIVERSITY OF ALABAMA
BOX 209
HUNTSVILLE AL 35899
U.S.A.
TEL: 205-895-6100
TLX:
TLF:
EML:
COM: 07

LUNDSTEDT HENRIK DR
LUND OBSERVATORY
BOX 43
S 22100 LUND
SWEDEN
TEL: 46 046 107294
TLX: 33199 OBSNOT S
TLF: 46- 46104614
EML: henrik@astro.lu.se
COM: 10.49

LUNDSTROM INGEMAR DR
LUND OBSERVATORY
BOX 43
S 22100 LUND
SWEDEN
TEL: 46 46 107300
TLX: 33199
TLF:
EML:
COM: 29

LUNEL MADELEINE DR
OBSERVATOIRE DE LYON
AVENUE CHARLES ANDRE
F-69230 ST-GENIS-LAVAL
FRANCE
TEL: 78-56-07-05
TLX: 310-926
TLF:
EML:
COM: 33

LUNGU NICOLAIE DR
INSTITUTUL POLITEHNIC
CATEDRA DE MATEMATICA
STR EMIL ISAC 15
3400 CLUJ NAPOCA
RUMANIA
TEL: 951-17224
TLX:
TLF:
EML:
COM:

LUO BAO-RONG
YUNNAN OBSERVATORY
KUNMING
CHINA. PEOPLE'S REP.
TEL:
TLX:
TLF:
EML:
COM: 10

LUO DINGCHANG
BEIJING ASTRONOMICAL OBS
ACADEMIA SINICA
WESTERN SUBURB
BEIJING
CHINA. PEOPLE'S REP.
TEL: 275580
TLX: 22040
TLF:
EML:
COM: 31

LUO DING-JIANG
BEIJING OBSERVATORY
ZHONG-GUAN-CUN
WESTERN SUBURB
BEIJING
CHINA. PEOPLE'S REP.
TEL: 281698
TLX: 22040 BAO ASCN
TLF:
EML:
COM: 08.19

LUO SHI-FANG
SHANGHAI OBSERVATORY
ACADEMIA SINICA
SHANGHAI
CHINA. PEOPLE'S REP.
TEL: 386191
TLX: 33164 SHAO CN
TLF:
EML:
COM: 19.31

LUO XIANHAN
DEPT OF GEOPHYSICS
BEIJING UNIVERSITY
BEIJING
CHINA. PEOPLE'S REP.
TEL: 22239
TLX:
TLF:
EML:
COM: 10.40

LUPISHKO DMITRIJ F
ASTRONOMICAL OBSERVATORY
KHARKOV UNIVERSITY
SUMSKAYA STR 35
310022 KHARKOV
U.S.S.R.
TEL: 43 24 28
TLX: 125531 ICAR
TLF:
EML:
COM: 15C

LUSTIG GUENTER DR
INSTITUT FUR ASTRONOMIE
KARL-FRANZENS-UNIVERSITAT
UNIVERSITATSPLATZ 5
A-8010 GRAZ
AUSTRIA
TEL: 0316-3805272
TLX: 0311662 UBGRZ
TLF:
EML:
COM: 10.12

LUTZ BARRY L DR
LOWELL OBSERVATORY
1400 W. MARS HILL RD
FLAGSTAFF AZ 86001
U.S.A.
TEL: 602-774-3358
TLX: 6502358958 MCI
TLF:
EML:
COM: 14.15.16

LUTZ JULIE H DR
PROGRAM IN ASTRONOMY
WASHINGTON STATE UNIV
PULLMAN WA 99164-2930
U.S.A.
TEL: 509-335-3136
TLX: 5107741091 WSUOIPPMA
TLF:
EML:
COM: 45

LUTZ THOMAS E DR
PROGRAM IN ASTRONOMY
WASHINGTON STATE UNIV
PULLMAN WA 99164-2930
U.S.A.
TEL: 509-335-3141
TLX: 5107741091 WSUOIPPMA
TLF:
EML:
COM: 24

LUYTEN WILLEM J PROF
SPACE SCIENCE CENTER
UNIVERSITY OF MINNESOTA
MINNEAPOLIS MN 55455
U.S.A.
TEL: 612-373-3366
TLX:
TLF:
EML:
COM: 24.26.33

LYNAS-GRAY ANTHONY E
DEPT PHYSICS & ASTRONOMY
UNIVERSITY COLLEGE LONDON
GOWER STREET
LONDON WC1E 6BT
U.K.
TEL:
TLX:
TLF:
EML:
COM: 29

LYNCH DAVID K
AEROSPACE CORPORATION
SPACE PHYSICS LABS
PO BOX 92957, MS M2-226
LOS ANGELES CA 90009
U.S.A.
TEL: 213-648-6686
TLX: 664460
TLF:
EML:
COM: 09

LYNDEN-BELL DONALD PROF
INSTITUTE OF ASTRONOMY
MADINGLEY ROAD
CAMBRIDGE CB3 0HA
U.K.
TEL: 0223-62204
TLX: 817297 ASTRON G
TLF:
EML:
COM: 28.33.37.47.48

LYNDS BEVERLY T DR
KITT PEAK NATL OBS
PO BOX 26732
TUCSON AZ 85726
U.S.A.
TEL: 602-325-9396
TLX: 0666-484 AURA NOAO
TLF:
EML:
COM: 28.34

LYNDS ROGER C DR
KITT PEAK NATL OBS
PO BOX 26732
TUCSON AZ 85726
U.S.A.
TEL: 602-327-5511
TLX:
TLF:
EML:
COM: 28

LYNE ANDREW G DR
NRAL
JODRELL BANK
MACCLESFIELD SK11 9DL
U.K.
TEL: 0477-71321
TLX: 36149
TLF:
EML:
COM: 40

LYNGA GOSTA DR
LUND OBSERVATORY
BOX 43
S-221 00 LUND
SWEDEN
TEL: 46-10-72-98
TLX: 33199 OBSNOT S
TLF:
EML:
COM: 05.33C.37.45

LYTTKENS EJNAR DR
SKOLGATAN 33 B
S-752 21 UPPSALA
SWEDEN
TEL:
TLX:
TLF:
EML:
COM:

LYTTLETON RAYMOND A PROF
INSTITUTE OF ASTRONOMY
CAMBRIDGE
U.K.
TEL: 0223-62204
TLX: 817297 ASTRON G
TLF:
EML:
COM: 15

LYUBIMKOV LEONID S DR
CRIMEAN ASTROPHYS OBS
USSR ACADEMY OF SCIENCES
NAUCHNY
334413 CRIMEA
U.S.S.R.
TEL: 43 29 45
TLX:
TLF:
EML:
COM: 36

LYUTY VICTOR M DR
CRIMEAN STATION OF
STERNBERG INSTITUTE
NAUCHNYJ
334413 CRIMEA
U.S.S.R.
TEL:
TLX:
TLF:
EML:
COM: 42

MA ER
BEIJING ASTRONOMICAL OBS
BEIJING
CHINA. PEOPLE'S REP.
TEL: 28-16-98
TLX: 22040 BAOAS CN
TLF:
EML:
COM: 28

MA XING-YUAN
DEPARTMENT OF GEOGRAPHY
BEIJING TEACHERS COLLEGE
BALIZHUANG
BEIJING
CHINA. PEOPLE'S REP.
TEL:
TLX:
TLF:
EML:
COM: 46

MA YU-QIAN
INSTITUTE OF HIGH ENERGY
PHYSICS
PO BOX 918-3
BEIJING
CHINA. PEOPLE'S REP.
TEL: 812-971 x 464
TLX: 22082 IHEP CN
TLF:
EML:
COM: 44.48

MACALPINE GORDON M
UNIVERSITY OF MICHIGAN
DEPT OF ASTRONOMY
ANN ARBOR MI 48109
U.S.A.
TEL: 313-764-3435
TLX: 810-223-6056
TLF:
EML:
COM: 28

MACCACARO TOMMASO DR
HARVARD-SMITHSONIAN CTR
FOR ASTROPHYSICS
60 GARDEN STREET
CAMBRIDGE MA 02138
U.S.A.
TEL: 617-495-7253
TLX: 921428 SATELLITE CAM
TLF:
EML:
COM: 48

MACCAGNI DARIO
IST. DI FISICA COSMICA
CNR
VIA BASSINI 15
I-20133 MILANO
ITALY
TEL: 02-298-237
TLX: 313839 MUACNR I
TLF:
EML:
COM: 48

MACCALLUM MALCOLM A H
SCHOOL OF MATH. SCIENCES
QUEEN MARY COLLEGE
MILE END ROAD
LONDON E1 4NS
U.K.
TEL: 01-980-4811
TLX: 893750 QMCUOL
TLF:
EML:
COM: 47

MACCHETTO FERDINANDO DR
SPACE TELESCOPE SCI INST
HOMEWOOD CAMPUS
3700 SAN MARTIN DRIVE
BALTIMORE MD 21218
U.S.A.
TEL: 301-338-4790
TLX: 6849101
TLF:
EML:
COM: 28.40.44.48

MACCONNELL DARRELL J DR
COMPUTER SCIENCES CORP.
SPACE TELESCOPE SCI. INST
3700 SAN MARTIN DRIVE
BALTIMORE MD 21218
U.S.A.
TEL: 301-338-4770
TLX:
TLF:
EML:
COM: 33.45V

MACDONALD GEOFFREY H DR
ELECTRONICS LABORATORY
UNIVERSITY OF KENT
CANTERBURY. KENT CT2 7NT
U.K.
TEL: 0227-66822 X258
TLX: 965449 UKCLIB
TLF:
EML:
COM: 40

MACDONALD JAMES
DEPARTMENT OF PHYSICS
UNIVERSITY OF DELAWARE
NEWARK DE 19716
U.S.A.
TEL: 302-451-2661
TLX:
TLF:
EML:
COM: 40.42

MACERONI CARLA
OSSERVATORIO ASTRONOMICO
DI ROMA
VIA DEL PARCO MELLINI 84
I-00136 ROMA
ITALY
TEL:
TLX:
TLF:
EML:
COM: 42

MACGILLIVRAY HARVEY T DR
ROYAL OBSERVATORY
BLACKFORD HILL
EDINBURGH EH9 3HJ
U.K.
TEL: 31-667-3321
TLX: 72383 ROEDIN G
TLF:
EML:
COM:

MACHADO JOSE M A B DR
TECHNICAL UNIVERSITY
OF LISBON
AV DA IGREJA 17 1 D
1700 LISBOA
PORTUGAL
TEL: 892225
TLX:
TLF:
EML:
COM: 10

MACHADO LUIZ E. DA SILVA
OBSERVATORIO DO VALONGO
UNIV. FEDERAL RIO DE J.
LAD. PEDRO ANTONIO. 43
20080 RIO DE JANEIRO
BRAZIL
TEL: 021-263-0685
TLX: 2122924 UFRJ BR
TLF:
EML:
COM: 20.24

MACHADO MARCOS
COMISION NACIONAL DE
INVESTIGACIONES ESPACIAL
AV MITRE 3100
SAN MIGUEL (BS. AS.)
ARGENTINA
TEL: 54 1 664 8371
TLX: 17511 LAMBA AR
TLF:
EML:
COM: 10C

MACHALSKI JERZY DR
ASTRONOMICAL OBSERVATORY
JAGIELLONIAN UNIVERSITY
UL. MAZOWIECKA 36/33
30-019 KRAKOW
POLAND
TEL:
TLX: 0322297 UJ PL
TLF:
EML:
COM: 40

MACIEL WALTER J DR
UNIVERSIDADE DE SAO PAULO
INST ASTRON. E GEOFISICO
CAIXA POSTAL 30627
01051 SAO PAULO SP
BRAZIL
TEL:
TLX:
TLF:
EML:
COM: 34.46

MACK PETER DR
M D M OBSERVATORY
H C 04
BOX 7520
TUCSON AZ 85735
U.S.A.
TEL: 602 620 53 60
TLX: 910 952 1116
TLF:
EML:
COM: 09

MACKAY CRAIG D DR
INSTITUTE OF ASTRONOMY
UNIVERSITY OF CAMBRIDGE
MADINGLEY RD
CAMBRIDGE CB3 0HA
U.K.
TEL: 44-223-62204
TLX: 817297 ASTRON G
TLF:
EML:
COM:

MACKINNON ALEXANDER L
DEPT OF ASTRONOMY
UNIVERSITY OF GLASGOW
GLASGOW G12 8QW
U.K.
TEL: 41-339-8855
TLX: 777070 UNIGLA
TLF:
EML:
COM: 10

MACLEOD JOHN M DR
HERZBERG INST ASTROPHYS
NATL RESEARCH COUNCIL
OTTAWA ONT K1A 0R6
CANADA
TEL: 613-593-6060
TLX:
TLF:
EML:
COM: 34.40

MACQUEEN ROBERT M DR
NCAR
HIGH ALTITUDE OBSERVATORY
PO BOX 3000
BOULDER CO 80307
U.S.A.
TEL: 303-497-1500
TLX: 45694
TLF:
EML:
COM: 10.49

MACRAE DONALD A PROF
DAVID DUNLAP OBSERVATORY
P O BOX 360
RICHMOND HILL ONT L4C 4Y6
CANADA
TEL: 416-884-9562
TLX:
TLF:
EML:
COM: 33.38.40

MACRIS CONSTANTIN J PROF
RSAAM
ACADEMY OF ATHENS
ANAGNOSTOPOULOU 14
GR-10673 ATHENS
GREECE
TEL: 3613589
TLX:
TLF:
EML:
COM: 10

MACY WILLIAM WRAY DR
LOCKHEED RESEARCH LAB.
BLDG 202. ORG 91-30
3251 HANOVER STREET
PALO ALTO CA 94304
U.S.A.
TEL:
TLX:
TLF:
EML:
COM:

MADDISON RONALD CH DR
UNIVERSITY OF KEELE
1 CHURCH PLANTATION
KEELE PARK
KEELE, STAFFS
U.K.
TEL: 0782-621111
TLX:
TLF:
EML:
COM: 46

MADEJ JERZY
WARSAW UNIVERSITY OBS
AL. UJAZDOWSKIE 4
00-478 WARSAW
POLAND
TEL: 4822-29-40-11
TLX:
TLF:
EML:
COM: 36

MADORE BARRY FRANCIS DR
DAVID DUNLAP OBSERVATORY
UNIVERSITY OF TORONTO
RICHMOND HL ONT L4C 4Y6
CANADA
TEL: 416-884-9562
TLX:
TLF:
EML:
COM: 27.28

MADSEN JES
ASTRONOMISK INSTITUT
UNIVERSITY OF AARHUS
DK-8000 AARHUS C
DENMARK
TEL: 06-12-88-99
TLX: 64767 AAUSCI DK
TLF:
EML:
COM:

MAEDA KEI-ICHI DR
DEPT OF PHYSICS
WASEDA UNIVERSITY
OKUBO 3-4-1 SHINJUKU-KU
TOKYO 160
JAPAN
TEL: 03 203 4141
TLX: 2323280 WASEDA J
TLF:
EML: BITNET:maeda@jpnwas00
COM: 47

MAEDA KOITIRO
DEPT OF PHYSICS
HYOGO COLL OF MEDICINE
NISHINOMIYA
HYOGO 663
JAPAN
TEL: 798-45-6111
TLX:
TLF:
EML:
COM:

MAEDER ANDRE PROF
OBSERVATOIRE DE GENEVE
CH-1290 SAUVERNY
SWITZERLAND
TEL: 22-552611
TLX: 45419209 OBSG CH
TLF:
EML:
COM: 27.35P.37

MAEHARA HIDEO DR
TOKYO ASTRONOMICAL
OBSERVATORY
OSAWA MITAKA
TOKYO 181
JAPAN
TEL: 0422-32-5111
TLX: 2822307 TAOMTK J
TLF:
EML:
COM: 45

MAETZLER CHRISTIAN DR
INSTITUTE APPLIED PHYSICS
UNIVERSITY OF BERN
SIDLERSTRASSE 5
CH-3012 BERN
SWITZERLAND
TEL: 031-65-89-11
TLX: 32320 PHYBE CH
TLF:
EML:
COM:

MAFFEI PAOLO PROF
UNIVERSITA DI PERUGIA
CATTEDRA DI ASTROFISICA
VIA DELL'ELCE DI SOTTO
I-06100 PERUGIA
ITALY
TEL: 075-45647
TLX:
TLF:
EML:
COM: 27.51

MAGAIN PIERRE DR
INSTITUT D'ASTROPHYSIQUE
UNIVERSITE DE LIEGE
AVENUE DE COINTE 5
B-4200 COINTE-OUGREE
BELGIUM
TEL:
TLX:
TLF:
EML:
COM: 29

MAGAKIAN TIGRAN Y DR
BYURAKAN ASTROPHYS OBS
378433 BYURAKAN ARMENIA
U.S.S.R.
TEL: 28 41 41
TLX: 411576 ASCOM SU
TLF:
EML:
COM:

MAGALASHVILI N L DR
ABASTUMANI ASTROPHYSICAL
OBSERVATORY
383762 ABASTUMANI GEORGIA
U.S.S.R.
TEL:
TLX:
TLF:
EML:
COM: 26.42

MAGALHAES ANTONIO A S ENG
OBSERVATORIO ASTRONOMICO
UNIVERSIDADE DO PORTO
MONTE DA VIRGEM
4400 VILA NOVA GAIA
PORTUGAL
TEL: 782-0404
TLX:
TLF:
EML:
COM:

MAGALHAES ANTONIO MARIO
INST ASTRON. E GEOFISICO
UNIVERSIDADE DE SAO PAULO
CAIXA POSTAL 30627
01051 SAO PAULO
BRAZIL
TEL: 55-11-275-3720
TLX: 1136221 IAGM BR
TLF:
EML:
COM:

MAGNAN CHRISTIAN DR
LAT
98 BIS BOULEVARD ARAGO
F-75014 PARIS
FRANCE
TEL: 1-43-20-14-25
TLX:
TLF:
EML:
COM:

MAGNARADZE NINA G DR
STATE UNIVERSITY
380043 TBILISI
U.S.S.R.
TEL:
TLX:
TLF:
EML:
COM: 07

MAGNI GIANFRANCO
IST. ASTROFISICA SPAZIALE
VIALE DELL'UNIVERSITA 11
I-00185 ROMA
ITALY
TEL:
TLX:
TLF:
EML:
COM:

MAGUN ANDREAS DR
INST APPLIED PHYSICS
SIDLERSTRASSE 5
CH-3012 BERN
SWITZERLAND
TEL: 031-658923
TLX:
TLF:
EML:
COM:

MAHDY HAMED A DR
HELWAN OBSERVATORY
HELWAN-CAIRO
EGYPT
TEL:
TLX:
TLF:
EML:
COM: 27

MAHESWARAN MURUGESAPILLAI
INST OF FUNDAMENT STUDIES
380/72 BAUDDHALOKA
MAWATHA
COLOMBO 7
SRI LANKA
TEL: 01-597536
TLX: 21700 IFS CE
TLF:
EML:
COM: 35

MAHMOUD ABUBAKR ABDELWAHI
ACADEMY OF SCIENTIFIC
RESEARCH AND TECHNOLOGY
101 KASR EL-EINI STREET
CAIRO
EGYPT
TEL:
TLX:
TLF:
EML:
COM:

MAHMOUD FAROUK M A B DR
HELWAN OBSERVATORY
HELWAN-CAIRO
EGYPT
TEL:
TLX:
TLF:
EML:
COM: 27

MAHRA H S DR
MANORA PEAK
NAINITAL 263 129
INDIA
TEL: 2136. 2583
TLX:
TLF:
EML:
COM: 29.16.10.27.50

MAIHARA TOSHINORI DR
DEPT OF PHYSICS
KYOTO UNIVERSITY
SAKYOKU
KYOTO 606
JAPAN
TEL: 075-751-2111
TLX: 5422693 LIBKYU J
TLF:
EML:
COM: 21 34

MAILLARD JEAN-PIERRE DR
INSTITUT D'ASTROPHYSIQUE
98BIS. BOULEVARD ARAGO
F-75014 PARIS
FRANCE
TEL: 33-1-43201425
TLX:
TLF:
EML:
COM: 09.14.29

MAITZEN HANS M DR
INSTITUT FUER ASTRONOMIE
TUERKENSCHANZSTR 17
A-1180 WIEN
AUSTRIA
TEL: 0222-345360-94
TLX: 116222 PHYSI A
TLF:
EML:
COM: 25.29

MAKARENKO EKATERINA N DR
ASTRONOMICAL OBSERVATORY
PARK SHEVCHENKO
270014 ODESSA
U.S.S.R.
TEL:
TLX:
TLF:
EML:
COM: 27

MAKAROV VALENTINE I
KISLOVODSK STATION OF THE
PULKOVO OBSERVATORY
357741 KISLOVODSK
U.S.S.R.
TEL:
TLX:
TLF:
EML:
COM: 10.12

MAKAROVA ELENA A DR
STERNBERG ASTRONOMICAL
INSTITUTE
117234 MOSCOW
U.S.S.R.
TEL: 139-1973
TLX: 113037 JAPET
TLF:
EML:
COM: 12

MAKINO FUMIYOSHI DR
INST SPACE & ASTRONAUT SC
6-1 KOMABA 4-CHOME
MEGURO-KU
TOKYO 153
JAPAN
TEL: 03-467-1111
TLX: 24550 SPACETKY J
TLF:
EML:
COM:

MAKISHIMA KAZUO
INST SPACE & ASTRON SCI
4-6-1 KOMABA
MEGURO-KU
TOKYO 153
JAPAN
TEL: 03-467-1111x301
TLX: 34757 ISASTRO J
TLF:
EML:
COM:

MAKITA MITSUGU DR
KWASAN AND HIDA OBSERV.
YAMASHINA
KYOTO 607
JAPAN
TEL: 075-581-1235
TLX:
TLF:
EML:
COM: 10.12

MALACARA DANIEL
CENTRO DE INVESTIGACIONES
EN OPTICA
APDO POSTAL 948
37000 LEON. GTO
MEXICO
TEL: 758-23
TLX:
TLF:
EML:
COM:

MALAGNINI MARIA LUCIA
OSSERVATORIO ASTRONOMICO
VIA TIEPOLO 11
PO BOX SUCC TRIESTE 5
I-34131 TRIESTE
ITALY
TEL: 040-793921
TLX: 461137 OAT I
TLF:
EML:
COM:

MALAISE DANIEL J DR
INSTITUT D'ASTROPHYSIQUE
UNIVERSITE DE LIEGE
AVENUE DE COINTE 5
B-4200 COINTE-OUGREE
BELGIUM
TEL:
TLX:
TLF:
EML:
COM: 15.44

MALAKPUR IRADJ DR
INSTITUTE OF GEOPHYSICS
TEHRAN UNIVERSITY
KARGAR SHOMALI
TEHRAN 14394
IRAN
TEL: 631081-3
TLX: 215319 UTIG
TLF:
EML:
COM:

MALARODA STELLA M DR
COMPLEJO ASTRONOMICO
EL LEONCITO
CASILLA DE CORREO 467
5400 SAN JUAN
ARGENTINA
TEL: 64-22-5718
TLX: 59134 ENTOP AR
TLF:
EML:
COM: 29.45

MALHERBE JEAN MARIE DR
OBSERVATOIRE DE PARIS
SECTION MEUDON DASOP
92195 MEUDON PPL CEDEX
FRANCE
TEL: 33-1 45077796
TLX: 201571 LAM
TLF:
EML: BITNET:MALHERBE@FRMEU51
COM: 10

MALIN DAVID F MR
ANGLO-AUSTRALIAN OBS
P.O.BOX 296
EPPING NSW 2121
AUSTRALIA
TEL: 02-868-1666
TLX: 23999 AA
TLF:
EML:
COM: 09

MALIN STUART
NATIONAL MARITIME MUSEUM
GREENWICH
LONDON SE10 9NF
U.K.
TEL: 01-858-1167
TLX:
TLF:
EML:
COM: 41

MALITSON HARRIET H MS
13315 MAGELLAN AVE
ROCKVILLE MD 20853
U.S.A.
TEL: 301-946-0496
TLX:
TLF:
EML:
COM: 10.44

MALLIA EDWARD A DR
DEPT OF ASTROPHYSICS
SOUTH PARKS ROAD
OXFORD OX1 3RQ
U.K.
TEL:
TLX:
TLF:
EML:
COM:

MALLIK D C V DR
INDIAN INSTITUTE OF
ASTROPHYSICS
BANGALORE 560 034
INDIA
TEL: 566585/566497
TLX: 845763 IIAB IN
TLF:
EML:
COM: 34.35

MALTBY PER PROF
INST THEORET ASTROPHYSICS
UNIVERSITY OF OSLO
P O BOX 1029
N-0315 BLINDERN OSLO 3
NORWAY
TEL: 2-456509
TLX:
TLF:
EML:
COM: 10

MALVILLE J MCKIM PROF
DEPT OF ASTROGEOPHYSICS
UNIVERSITY OF COLORADO
BOULDER CO 80302
U.S.A.
TEL: 303-492-8788
TLX:
TLF:
EML:
COM: 10

MAMMANO AUGUSTO DR
OSSERVATORIO ASTROFISICO
I-36012 ASIAGO. VICENZA
ITALY
TEL: 0424-62665
TLX:
TLF:
EML:
COM: 42

MANABE SEIJI DR
INTL LATITUDE OBSERVATORY
MIZUSAWA
IWATE 023
JAPAN
TEL:
TLX:
TLF:
EML:
COM: 19

MANARA ALESSANDRO A DR
OSSERVATORIO ASTRONOMICO
DI MILANO
VIA BRERA 28
I-20121 MILANO
ITALY
TEL: 02-87-4444
TLX:
TLF:
EML:
COM: 44

MANCHANDA R K DR
TATA INSTITUTE OF
FUNDAMENTAL RESEARCH
HOMI BHABHA RD
BOMBAY 400 005
INDIA
TEL: 219-111 x 336
TLX: 113009 TIFR IN
TLF:
EML:
COM:

MANCHESTER RICHARD N DR
CSIRO
DIVISION OF RADIOPHYSICS
P.O.BOX 76
EPPING NSW 2121
AUSTRALIA
TEL: 02-868-0225
TLX: 26320 ASTRO
TLF:
EML:
COM: 33 34.40

MANCUSO SANTI PROF
CAPODIMONTE ASTR OBS
VIA MOIARIELLO 16
I-80131 NAPOLI
ITALY
TEL: 44-01-01
TLX:
TLF:
EML:
COM:

MANDELSTAM S L PROF
INST OF SPECTROSCOPY
USSR ACADEMY OF SCIENCES
142092 AKADEMGORODOK
U.S.S.R.
TEL: 334-55-79
TLX:
TLF:
EML:
COM: 10.14C.44

MANDOLESI NAZZARENO
ISTITUTO TE.S.R.E.-C.N.R.
VIA DE CASTAGNOLI 1
I-40126 BOLOGNA
ITALY
TEL: 51-[illegible]-80-22
TLX: 511350
TLF:
EML:
COM: 40.44.47

MANFROID JEAN DR
INSTITUT D'ASTROPHYSIQUE
UNIVERSITE DE LIEGE
AVENUE DE COINTE 5
B-4200 COINTE-OUGREE
BELGIUM
TEL: 41-559980
TLX: 41264 ASTRLG
TLF:
EML:
COM: 25.34

MANGENEY ANDRE DR
OBSERVATOIRE DE MEUDON
SECTION DE MEUDON
F-92195 MEUDON PL CEDEX
FRANCE
TEL: 1-45-07-76-61
TLX:
TLF:
EML:
COM: 49

MANN PATRICK J DR
DEPT OF ASTRONOMY
UNIV OF WESTERN ONTARIO
LONDON ONT N6A 3K7
CANADA
TEL: 519 661 3183
TLX: 0647134
TLF:
EML: BITNET:2014-562@UWOVAX
COM:

MANNINO GIUSEPPE PROF
ISTITUTO MATEMATICO
VIA CAMPI 183
I-41100 MODENA
ITALY
TEL:
TLX:
TLF:
EML:
COM: 27

MANRIQUE WALTER T PROF
OBSERVATORIO ASTRONOMICO
FELIX AGUILAR
AV BENAVIDEZ 3175 OESTE
5407 MARQUESADO (S.J.
ARGENTINA
TEL:
TLX:
TLF:
EML:
COM: 08

MANSFIELD VICTOR N PROF
COLGATE UNIVERSITY
HAMILTON NY 13346
U.S.A.
TEL: 315-824-1000
TLX:
TLF:
EML:
COM:

MANTEGAZZA LUCIANO
OSSERVATORIO ASTRONOMICO
DI MERATE
VIA BIANCHI 46
I-[illegible]055 MERATE
ITALY
TEL: 039-[illegible]035
TLX:
TLF:
EML:
COM: 27

MANTOVANI FRANCO
IST. DI RADIOASTRONOMIA
VIA IRNERIO 46
I-40126 BOLOGNA
ITALY
TEL: 51-23-2856
TLX: [illegible]
TLF:
EML:
COM:

MAO WEI
YUNNAN OBSERVATORY
KUNMING. YUNNAN PROVINCE
CHINA. PEOPLE'S REP.
TEL:
TLX:
TLF:
EML:
COM: 08

MARABINI RODOLFO JOSE
UNIVERSIDAD NACIONAL
FACULTAD DE CIENCIAS
ASTRON Y GEOFISICAS
1900 LA PLATA
ARGENTINA
TEL: 021-217-308
TLX:
TLF:
EML:
COM:

MARAN STEPHEN P DR
NASA/GSFC
CODE 680
GREENBELT MD 20771
U.S.A.
TEL: 301-286-8607
TLX: 89675
TLF:
EML:
COM: 15.40.44

MARANO BRUNO
DIPT. DI ASTRONOMIA
UNIVERSITA DI BOLOGNA
C P 596
I-40100 BOLOGNA
ITALY
TEL: 222956
TLX: 211664 INFNBO I
TLF:
EML:
COM: 47

MARAR T M K
TECHNICAL PHYSICS DIV
ISRO SATELLITE CENTRE
AIRPORT RD. VIMANAPURA
BANGALORE 560 017
INDIA
TEL: 566 251
TLX:
TLF:
EML:
COM: 44

MARASCHI LAURA DR
ISTITUTO DI FISICA
VIA CELORIA 16
I-20133 MILANO
ITALY
TEL:
TLX:
TLF:
EML:
COM:

MARCAIDE JUAN-MARIA DR
INSTITUTO DE ASTROFISICA
DE ANDALUCIA
APARTADO 2144
E-18080 GRANADA
SPAIN
TEL: 34 58 121311
TLX: 78573 IAAG E
TLF:
EML:
COM: 40

MARCELIN MICHEL
OBSERVATOIRE DE MARSEILLE
2 PLACE LE VERRIER
F-13248 MARSEILLE CEDEX 4
FRANCE
TEL: 91-95-90-88
TLX: 420241 F
TLF:
EML:
COM: 28

MARCHAL CHRISTIAN DR
DEPT ETUDES DE SYNTHESE
ONERA
F-92320 CHATILLON
FRANCE
TEL: 1-46-57-11-60
TLX: 260907 F
TLF:
EML:
COM: 07

MARDIROSSIAN FABIO
DIPT. DI ASTRONOMIA
UNIVERSITA DEGLI STUDI
VIA G.B. TIEPOLO 11
I-34131 TRIESTE
ITALY
TEL: 040-793921/221
TLX: 461137 OAT I
TLF:
EML:
COM: 42.47

MAREK JOHN
44 PERCY ROAD
WREXHAM CLWYD
U.K.
TEL:
TLX:
TLF:
EML:
COM: 47

MARGON BRUCE H PROF
ASTRONOMY DEPT FM-20
UNIVERSITY OF WASHINGTON
SEATTLE WA 98195
U.S.A.
TEL: 206-543-0089
TLX: 4740096
TLF:
EML:
COM:

MARGONI RINO
OSSERVATORIO ASTROFISICO
DI ASIAGO
I-36012 ASIAGO. VICENZA
ITALY
TEL: 0424-62665
TLX: 430110 SETOUR
TLF:
EML:
COM:

MARGRAVE THOMAS EWING JR
400 JOHNSON STREET
VIENNA VA 22180
U.S.A.
TEL:
TLX:
TLF:
EML:
COM: 27.51

MARIE M A DR
DEPT OF ASTRONOMY
FACULTY OF SCIENCE
CAIRO UNIVERSITY
CAIRO
EGYPT
TEL:
TLX:
TLF:
EML:
COM:

MARIK MIKLOS DR.
ASTRONOMICAL DEPT
L. EOTVOS UNIVERSITY
KUN BELA TER 2
H-[illegible] BUDAPEST
HUNGARY
TEL: 143-019
TLX:
TLF:
EML:
COM: 12.46

MARILLI ETTORE DR
OSSERVATORIO ASTROFISICO
CITTA UNIVERSITARIA
I-95125 CATANIA
ITALY
TEL: 095-33-05-33
TLX: 970359 ASTRCT I
TLF:
EML:
COM: 12.42

MARINO BRIAN F ENG
AUCKLAND OBSERVATORY
BOX 72009
NORTHCOTE
AUCKLAND 9
NEW ZEALAND
TEL: 649 486 951
TLX:
TLF:
EML:
COM: 42

MARIOTTI JEAN MARIE DR
OBSERVATOIRE DE PARIS
PLACE JULES JANSSEN
92195 MEUDON PL. CEDEX
FRANCE
TEL: 33-1 45077570
TLX: 204464
TLF:
EML: BITNET:MARIOTTI@FRMEU51
COM:

MARISKA JOHN THOMAS
NAVAL RESEARCH LABORATORY
CODE 4175M
WASHINGTON DC 20375
U.S.A.
TEL: 202-767-2605
TLX:
TLF:
EML:
COM: 12

MARK JAMES WAI-KEE DR
PHYSICS DEPT
L.LIVERMORE NATL LAB L477
UNIVERSITY OF CALIFORNIA
LIVERMORE CA 94550
U.S.A.
TEL: 415-422-5931
TLX: 910-386-8339 UCCLLL
TLF:
EML:
COM: 33

MARKELLOS VASSILIS V DR
UNIVERSITY OF PATRAS
DEPT ENGINEERING SCIENCE
260 00 RION
GREECE
TEL: 061-991-465
TLX:
TLF:
EML:
COM: 07

MARKKANEN TAPIO DR
OBSERVATORY
TAHTITORNINMAKI
SF-00130 HELSINKI 13
FINLAND
TEL: 90-8391
TLX:
TLF:
EML:
COM: 25.37.50

MARKOWITZ WILLIAM DR
APT 15-B
2800 E. SUNRISE BLVD
FORT LAUDERDALE FL 33304
U.S.A.
TEL: 305-563-2859
TLX:
TLF:
EML:
COM: 19.31

MARLBOROUGH J M PROF
DEPT OF ASTRONOMY
UNIV OF WESTERN ONTARIO
LONDON ONT N6A 3K7
CANADA
TEL: 519-679-3184
TLX: 064-7134
TLF:
EML:
COM: 36

MARMOLINO CIRO
DIPARTIMENTO DI FISICA
MOSTRA D'OLTREMARE
PAD. 19
I-80125 NAPOLI
ITALY
TEL: 081-7253428
TLX: 720320 INFNNA I
TLF:
EML:
COM: 12

MAROCHNIK L S PROF DR
USSR ACADEMY OF SCIENCES
SPACE RESEARCH INSTITUTE
PROFSOJUSNAJA 84-32
117810 MOSCOW
U.S.S.R.
TEL: 333-31-22
TLX: 411498 STAR SU
TLF:
EML:
COM: 33

MAROV MIKHAIL YA PROF
INST OF APPLIED MATHS
USSR ACADEMY OF SCIENCES
MIUSSKAYA SQ 4
125047 MOSCOW
U.S.S.R.
TEL:
TLX:
TLF:
EML:
COM: 16V.44.51

MARQUES DOS SANTOS P PROF
INST ASTRON. E GEOFISICO
UNIVERSIDADE DE SAO PAULO
CAIXA POSTAL 30627
01051 SAO PAULO SP
BRAZIL
TEL: 11-276-3941
TLX: 36221 IAGN BR
TLF:
EML:
COM: 28.40

MARQUES MANUEL N DR
OBSERVATORIO ASTRONOMICO
TAPADA DA AJUDA
1300 LISBOA 3
PORTUGAL
TEL:
TLX:
TLF:
EML:
COM:

MARRACO HUGO G DR
UNIV NACIONAL DE LA PLATA
FACULTAD DE CIENCIAS
ASTRONOMICAS Y GEOFISICAS
1900 LA PLATA
ARGENTINA
TEL: 54-21-21-7308
TLX: 31151 BULAP AR
TLF:
EML:
COM: 25.37

MARRACO HUGO G DR
OBSERVATORIO ASTRONOMICO
1900 LA PLATA
ARGENTINA
TEL:
TLX:
TLF:
EML:
COM:

MARSCHALL LAURENCE A
GETTYSBURG COLLEGE
DEPT OF PHYSICS
GETTYSBURG PA 17325
U.S.A.
TEL: 717-337-1865
TLX:
TLF:
EML:
COM: 14.30

MARSCHER ALAN PATRICK
ASTRONOMY DEPT
BOSTON UNIVERSITY
725 COMMONWEALTH AVE
BOSTON MA 02215
U.S.A.
TEL: 617-353-5029
TLX: 951289 BOS UNIV BSN
TLF:
EML:
COM: 40

MARSDEN BRIAN G DR
SMITHSONIAN ASTROPHYSICAL
OBSERVATORY
60 GARDEN STREET
CAMBRIDGE MA 02138
U.S.A.
TEL: 617-495-7244
TLX: 7103206842 ASTROGRAM
TLF:
EML:
COM: 5PC 07.15.20C

MARSDEN PHILIP L PROF
DEPT OF PHYSICS
UNIVERSITY OF LEEDS
LEEDS LS2 9JT
U.K.
TEL: 0532-431751
TLX: 556473 UNILDS
TLF:
EML:
COM:

MARSH JULIAN C D
HATFIELD POLYTECHNIC
OBSERVATORY
BAYFORDBURY
HERTFORD. HERTS SG13 8LD
U.K.
TEL: 0992-558451
TLX: 262413
TLF:
EML:
CON: 46

MARSHALL KEVIN
19 NEEDINGWORTH ROAD
ST IVES
CAMBS PE17 4JT
U.K.
TEL:
TLX:
TLF:
EML:
CON:

MARSOGLU A DR
UNIVERSITY OBSERVATORY
UNIVERSITE
ISTANBUL
TURKEY
TEL: 11-522-35-97
TLX:
TLF:
EML:
CON:

MARTIN ANTHONY R DR
UK CULHAM LABORATORY
RM F4.135
ABINGDON OX14 3DB
U.K.
TEL: 1235-21843
TLX: 83189
TLF:
EML:
CON: 51

MARTIN DEREK H PROF
QUEEN MARY COLLEGE
MILE END ROAD
LONDON E1 4NS
U.K.
TEL:
TLX:
TLF:
EML:
CON:

MARTIN FRANCOIS DR
DEPT D'ASTROPHYSIQUE
UNIVERSITE DE NICE
PARC VALROSE
F-06034 NICE CEDEX
FRANCE
TEL: 93-51-91-00
TLX: 970281
TLF:
EML:
CON:

MARTIN INACIO MALMONGE DR
UNIV DE CAMPINAS-UNICAMP
INSTITUTO DE FISICA
DEPTO RAIOS COSMICOS
13100 CAMPINAS SP
BRAZIL
TEL: 0192-391301
TLX: 019-1150
TLF:
EML:
CON: 48

MARTIN MARIA CRISTINA DR
INSTITUTO ARGENTINO DE
RADIOASTRONOMIA
CASILLA DE CORREO NO 5
VILLA ELISA (BS.AS.)1894
ARGENTINA
TEL: 4 3793 87 0230
TLX: 31216 CESLA AR
TLF:
EML:
CON: 28.51

MARTIN NICOLE DR
OBSERVATOIRE DE MARSEILLE
2 PLACE LE VERRIER
F-13248 MARSEILLE CEDEX 4
FRANCE
TEL: 91-95-90-88
TLX: 420241
TLF:
EML:
CON: 30

MARTIN PETER G PROF
INST FOR THEOR ASTROPHYS
UNIVERSITY OF TORONTO
TORONTO ONT M5S 1A7
CANADA
TEL: 416-978-6840
TLX:
TLF:
EML:
CON:

MARTIN ROBERT N DR
STEWARD OBSERVATORY
UNIVERSITY OF ARIZONA
TUCSON AZ 85721
U.S.A.
TEL: 602-621-1539
TLX: 467175
TLF:
EML:
CON: 34.40

MARTIN WILLIAM C DR
NATL BUREAU OF STANDARDS
A167 PHYSICS BLDG
GAITHERSBURG MD 20899
U.S.A.
TEL: 301-921-2011
TLX:
TLF:
EML:
CON: 14

MARTIN WILLIAM L DR
ROYAL GREENWICH OBS
HERSTMONCEUX CASTLE
HAILSHAM BN27 1RP
U.K.
TEL: 0323-833171
TLX: 87451 RGOBSY G
TLF:
EML:
CON: 27

MARTINET LOUIS PROF
OBSERVATOIRE DE GENEVE
CH-1290 SAUVERNY
SWITZERLAND
TEL: 55-26-11
TLX: 27720
TLF:
EML:
CON: 28.33.46

MARTINEZ MARIO DR
CIESE
DEPTO DE GEOFISICA
APDO POSTAL 2732
22860 ENSENADA B. CALIF.
MEXICO
TEL:
TLX:
TLF:
EML:
CON:

MARTINEZ ROGER CARLOS DR
INSTITUTO DE ASTROFISICA
DE CANARIAS
38200 LA LAGUNA TENERIFE
SPAIN
TEL: 34 22 26 22 11
TLX: 92640
TLF:
EML:
CON:

MARTINI ALDO DR
ASTROFISICA SPAZIALE
CNR
C P 67
I-00044 FRASCATI. ROMA
ITALY
TEL:
TLX:
TLF:
EML:
CON:

MARTINS DONALD HENRY DR
DEPT PHYSICS & ASTRONOMY
UNIVERSITY OF ALASKA
3221 UAA DRIVE
ANCHORAGE AK 99508
U.S.A.
TEL: 907-786-1238
TLX:
TLF:
EML:
CON: 09.37

MARTIN-DIAZ CARLOS DR
INSTITUTO DE ASTROFISICA
DE CANARIAS
38200 LA LAGUNA TENERIFE
SPAIN
TEL: 34 22 26 22 11
TLX: 92640 IACE
TLF: 26 30 05
EML: SPAN:IAC::CMR
CON:

MARTIN-LOBON M DR
HERMANOS MIRALLES 14
MADRID 1
SPAIN
TEL:
TLX:
TLF:
EML:
CON:

MARTIN-PINTADO JESUS
CENTRO ASTRON DE YEBES
O A N
APARTADO CORREOS 148
19080 GUADALAJARA
SPAIN
TEL: 911-223358
TLX:
TLF:
EML:
COM: 34

MARTRES MARIE-JOSEPHE
OBSERVATOIRE DE PARIS
SECTION DE MEUDON
F-92195 MEUDON PL CEDEX
FRANCE
TEL: 1-45-34-75-30
TLX:
TLF:
EML:
COM: 10

MARTYNOV D YA PROF DR
STERNBERG STATE
ASTRONOMICAL INSTITUTE
117234 MOSCOW
U.S.S.R.
TEL:
TLX:
TLF:
EML:
COM: 05.16.42

MARVIN URSULA B DR
CENTER FOR ASTROPHYSICS
60 GARDEN STREET
CAMBRIDGE MA 02138
U.S.A.
TEL: 617-495-7270
TLX: 921428 SATELLITE CAM
TLF:
EML:
COM: 22

MARX GYORGY PROF.
L. EOTVOS UNIVERSITY
PUSHKIN U 5-7
H-1088 BUDAPEST
HUNGARY
TEL:
TLX:
TLF:
EML:
COM: 35.51P

MARX SIEGFRIED DR
ZNTRLINST. F. ASTROPHYSIK
KARL-SCHWARZSCHILD OBS
DDR-6901 TAUTENBURG
GERMANY. D.R.
TEL: JENA 23530
TLX: 5886284 KSOT DD
TLF:
EML:
COM: 50

MASANI A PROF
OSSERVATORIO ASTRONOMICO
DI BRERA
VIA BRERA 28
I-20100 MILANO
ITALY
TEL:
TLX:
TLF:
EML:
COM: 25.27.35

MASLOWSKI JOZEF DR
ASTRONOMICAL OBSERVATORY
UL. ORLA 171
30-244 KRAKOW
POLAND
TEL: 34-10-41
TLX: 0322297 UJ PL
TLF:
EML:
COM: 40

MASNOU FRANCOISE DR
28 ALLEE GAMBAUDERIE
F-91190 GIF-S-YVETTE
FRANCE
TEL:
TLX:
TLF:
EML:
COM:

MASNOU J L DR
OBSERVATOIRE DE PARIS
SECTION DE MEUDON
ER 176 DARC
F-92195 MEUDON PL CEDEX
FRANCE
TEL: 1-45-34-75-70
TLX: 201 571 LAM
TLF:
EML:
COM:

MASON GLENN M
UNIVERSITY OF MARYLAND
DEPT PHYSICS & ASTRONOMY
COLLEGE PARK MD 20742
U.S.A.
TEL: 301-454-2616
TLX: 71-8761125
TLF:
EML:
COM: 10.48.49

MASON HELEN E DR
DEPT APPLIED MATHS
& THEORETETICAL PHYSICS
SILVER STREET
CAMBRIDGE CB3 9EW
U.K.
TEL: 0223 351645
TLX: 81240
TLF:
EML:
COM: 14

MASON KEITH OWEN
MULLARD SPACE SCIENCE LAB
HOLMBURY ST MARY
DORKING. SURREY RH5 6NT
U.K.
TEL: 0306-70292
TLX: 859185
TLF:
EML:
COM: 48

MASSAGLIA SILVANO
IST. DI FISICA GENERALE
CORSO M. D'AZEGLIO 46
I-10125 TORINO
ITALY
TEL: 011-657694
TLX: 211041
TLF:
EML:
COM: 36

MASSEVICH ALLA G DR
ASTRONOMICAL COUNCIL
USSR ACADEMY OF SCIENCES
PYATNITSKAYA UL 48
109017 MOSCOW
U.S.S.R.
TEL: 231-54-61
TLX: 412623 SCSTP SU
TLF:
EML:
COM: 35

MASSEY PHILIP L
KITT PEAK NATIONAL OBS
PO BOX 26732
TUCSON AZ 85726-6732
U.S.A.
TEL: 602-327-5511
TLX:
TLF:
EML:
COM: 29

MASSON COLIN R
CENTER FOR ASTROPHYSICS
60 GARDEN STREET
CAMBRIDGE MA 02138
U.S.A.
TEL: 617-495-7000
TLX:
TLF:
EML:
COM: 34.40

MASURSKY HAROLD DR
US GEOLOGICAL SURVEY
BRANCH OF ASTRO GEOLOGY
2255 NORTH GEMINI DR
FLAGSTAFF AZ 86001
U.S.A.
TEL: 601-527-2003
TLX: 34461 GSA FTS SFC
TLF:
EML:
COM: 16C

MATAS VLADIMIR R DR
ASTRONOMISCHES-RECHEN
INSTITUT
MOENCHHOFSTR 12-14
D-6900 HEIDELBERG
GERMANY. F.R.
TEL:
TLX:
TLF:
EML:
COM: 37

MATERNE JUERGEN DR
ARETINSTRASSE 27
D-8000 MUNCHEN 90
GERMANY. F.R.
TEL:
TLX:
TLF:
EML:
COM: 47

MATHER JOHN CROMWELL
LAB ASTRON. SOL PHYS.
NASA/GSFC. CODE 685
GREENBELT MD 20771
U.S.A.
TEL: 301-286-8720
TLX: 89675
TLF:
EML:
COM: 44.47

MATHESON DAVID NICHOLAS
SERC
RUTHERFORD APPLETON LAB
CHILTON. DIDCOT OX11 0QX
U.K.
TEL: 0235-21900
TLX: 83159
TLF:
EML:
COM: 40

MATHEWS WILLIAM G PROF
LICK OBSERVATORY
UNIVERSITY OF CALIFORNIA
SANTA CRUZ CA 95064
U.S.A.
TEL: 408-429-2074
TLX:
TLF:
EML:
COM: 34

MATHEWSON DONALD S PROF
MT STROMLO & SIDING
SPRING OBSERVATORIES
PRIVATE BAG
WODEN P.O. ACT 2606
AUSTRALIA
TEL: 062-881111
TLX: 62270 AA
TLF:
EML:
COM: 28.33.34

MATHEZ GUY
OBSERVATOIRE PIC-DU-MIDI
ET DE TOULOUSE
14 AVENUE EDOUARD BELIN
F-31400 TOULOUSE
FRANCE
TEL: 61-25-21-01
TLX: 530776 OBSTLSE F
TLF:
EML:
COM:

MATHIEU ROBERT D DR
DEPARTMENT OF ASTRONOMY
UNIVERSITY OF WISCONSIN
MADISON WI 53706
U.S.A.
TEL: 608 262 5679
TLX: 265452
TLF:
EML: BITNET:MATHIEU@WISCMAC3
COM: 30

MATHIS JOHN S PROF
DEPT OF ASTRONOMY
UNIVERSITY OF WISCONSIN
475 N. CHARTER ST
MADISON WI 53706
U.S.A.
TEL: 608-262-5994
TLX: 265452 UOFWISC MDS
TLF:
EML:
COM: 34F

MATHUR B S DR
NATL PHYSICAL LABORATORY
TIME & FREQUENCY SECTION
HILLSIDE ROAD
NEW DELHI 110 012
INDIA
TEL: 586168
TLX: 31-62454 RSD IN
TLF:
EML:
COM: 31

MATHYS GAUTIER DR
OBSERVATOIRE DE GENEVE
CHEMIN DES MAILLETTES 51
CH-1290 SAUVERNY
SWITZERLAND
TEL: 55-26-11
TLX: 27720 OBSG CH
TLF:
EML:
COM: 29

MATSAKIS DEMETRIOS N
US NAVAL OBSERVATORY
34 & MASSACHUSETTS AVE NW
WASHINGTON DC 20390
U.S.A.
TEL: 202-653-1823
TLX:
TLF:
EML:
COM: 19.31.40.51

MATSON DENNIS L DR
JPL 183-501
4800 OAK GROVE DRIVE
PASADENA CA 91103
U.S.A.
TEL: 213-354-2984
TLX:
TLF:
EML:
COM: 15.16C

MATSUDA TAKUYA PROF
DEPT AERONAUTIC ENGINEERG
KYOTO UNIVERSITY
YOSHIDAHONMACHI SAKYOKU
KYOTO 606
JAPAN
TEL: 075-751-2111
TLX: 05422693 LIBKYU J
TLF:
EML:
COM: 51

MATSUMOTO MASAMICHI PROF
FACULTY OF ENGINEERING
GIFU UNIVERSITY
501-11 GIFU
JAPAN
TEL:
TLX:
TLF:
EML:
COM: 36

MATSUMOTO TOSHIO DR
DEPT OF PHYSICS
NAGOYA UNIVERSITY
FUROCHO CHIKUSAKU
NAGOYA 464
JAPAN
TEL: 052-781-5111
TLX: 4477323 SCUNAG J
TLF:
EML:
COM: 21.47

MATSUOKA MASARU DR
INST. PHYS. & CHEM. RES.
HIROSAWA. WAKO
SAITAMA 351-01
JAPAN
TEL: 0484-62-1111
TLX: 02962818 RIKEN J
TLF:
EML:
COM: 44.48

MATSUSHIMA SATOSHI DR
DEPT ASTRONOMY
PENNSYLVANIA STATE UNIV
525 DAVEY LAB
UNIVERSITY PARK PA 16802
U.S.A.
TEL: 814-865-0418
TLX:
TLF:
EML:
COM: 12.36

MATSUURA OSCAR T DR
DEPTO DE ASTRONOMIA
IAG-USP
CAIXA POSTAL 30627
01051 SAO PAULO SP
BRAZIL
TEL: 011-275-3720
TLX: 1136221 IAGM BR
TLF:
EML:
COM: 10.15.49

MATTEI JANET AKYUZ DR
AAVSO
25 BIRCH STREET
CAMBRIDGE MA 02138
U.S.A.
TEL: 617-354-0484
TLX:
TLF:
EML:
COM: 27.42

MATTEUCCI FRANCESCA
IST. ASTROFISICA SPAZIALE
C P 67
I-00044 FRASCATI
ITALY
TEL:
TLX:
TLF:
EML:
COM: 35.37

MATTHEWS HENRY E DR
J.C.M.T.
JOINT ASTRONOMY CENTRE
665 KOMOHANA STREET
HILO HAWAII 96720
U.S.A.
TEL: 808-961 3756
TLX: 633135
TLF:
EML:
COM: 40

MATTHEWS THOMAS A DR
ASTRONOMY PROGRAM
UNIVERSITY OF MARYLAND
COLLEGE PK MD 20742
U.S.A.
TEL:
TLX:
TLF:
EML:
COM:

MATTIG W PROF DR
KIEPENHEUER INSTITUT
FUER SONNENPHYSIK
SCHOENECKSTRASSE 6
D-7800 FREIBURG-IM-BR.
GERMANY. F.R.
TEL: 761-32864
TLX: 7721552 KIS D
TLF:
EML:
COM: 10.12.50

MATTILA KALEVI DR
OBSERVATORY
TAHTITORNINMAKI
SF-00130 HELSINKI 13
FINLAND
TEL: 90-1912947
TLX: 124690 UNIH SF
TLF:
EML:
COM: 21C.34.40

MATVEYENKO L I DR
SPACE RESEARCH INSTITUTE
USSR ACADEMY OF SCIENCES
117810 MOSCOW
U.S.S.R.
TEL: 333-31-22
TLX: 411498 STAR SU
TLF:
EML:
COM: 40C

MATZNER RICHARD A PROF
PHYSICS DEPT
UNIVERSITY OF TEXAS
AUSTIN TX 78712
U.S.A.
TEL: 512-471-5062
TLX:
TLF:
EML:
COM: 47

MAUDER HORST PROF DR
ASTRONOMISCHES INSTITUT
WALDHAUSER-STR 64
D-7400 TUEBINGEN
GERMANY. F.R.
TEL:
TLX:
TLF:
EML:
COM: 42

MAURICE ERIC N
OBSERVATOIRE DE MARSEILLE
. PLACE LE VERRIER
F-13248 MARSEILLE CDX 04
FRANCE
TEL: 91-95-90-88
TLX: 420241 F
TLF:
EML:
COM: 28.30

MAURON NICOLAS DR
OBSERVATOIRE DE TOULOUSE
PIC DU MIDI
14 AVENUE E.BELIN
31400 TOULOUSE
FRANCE
TEL: 61 25 21 01
TLX: 530776 F
TLF:
EML:
COM:

MAVRAGANIS A G PROF
DEPT OF ENG SECT OF MECH
NATL TECHN UNIV/5 HEROES
POLYTECH AVE
GR-15773 ATHENS
GREECE
TEL: 6433170
TLX:
TLF:
EML:
COM: 07

MAVRIDES STAMATIA DR
OBSERVATOIRE DE PARIS
SECTION DE MEUDON
DEPT RADIOASTRONOMIE
F-92195 MEUDON PL CEDEX
FRANCE
TEL: 1-45-07-75-97
TLX:
TLF:
EML:
COM: 28.47

MAVRIDIS L N PROF
DEPT GEODETIC ASTRONOMY
UNIVERSITY THESSALONIKI
GR-54006 THESSALONIKI
GREECE
TEL:
TLX:
TLF:
EML:
COM: 08.27C.33.46.50.51

MAVROMICHALAKI HELEN DR
UNIVERSITY / PHYSICS DEPT
NUCLEAR PHYSICS SECTION
104 SOLONOS ST
GR-13680 ATHENS
GREECE
TEL: 3639439
TLX:
TLF:
EML:
COM: 49

MAX CLAIRE E DR
LAWRENCE LIVERMORE LAB
L-413
PO BOX 808
LIVERMORE CA 94550
U.S.A.
TEL: 415-422-5442
TLX: 9103868339 UCLLLLVMR
TLF:
EML:
COM:

MAXWELL ALAN DR
HARVARD SMITHSONIAN CTR
FOR ASTROPHYSICS
60 GARDEN STREET
CAMBRIDGE MA 02138
U.S.A.
TEL: 617-495-9059
TLX:
TLF:
EML:
COM: 10.40

MAY J
OBS RADIOASTR DE MAIPU
UNIVERSIDAD DE CHILE
CASILLA 68
MAIPU
CHILE
TEL: 2294101
TLX:
TLF:
EML:
COM:

MAYALL MARGARET W
5 SPARKS STREET
CAMBRIDGE MA 02138
U.S.A.
TEL: 617-876-1563
TLX:
TLF:
EML:
COM: 27

MAYALL NICHOLAS U ASTRON
7206 E. CAMINO VECINO
TUCSON AZ 85715
U.S.A.
TEL: 602-886-2423
TLX:
TLF:
EML:
COM:

MAYER CORNELL H
SPACE SCIENCE DIVISION
NAVAL RESEARCH LAB
CODE 4130M
WASHINGTON DC 20375
U.S.A.
TEL: 202-767-2495
TLX:
TLF:
EML:
COM: 16.40

MAYER PAVEL DR
DEPT OF ASTRONOMY
CHARLES UNIVERSITY
SVEDSKA 8
150 00 PRAHA 5
CZECHOSLOVAKIA
TEL: 540395
TLX:
TLF:
EML:
COM: 25.42

MAYFIELD EARLE B DR
CALIF. POLY. STATE UNIV.
1427 BAYVIEW HEIGHTS DR.
LOS OSOS CA 93403
U.S.A.
TEL: 805-528-5231
TLX:
TLF:
EML:
COM:

MAYOR MICHEL DR
OBSERVATOIRE DE GENEVE
CHEMIN DES MAILLETTES 51
CH-1290 SAUVERNY
SWITZERLAND
TEL: 22-55-26-11
TLX: 27720 OBSG CH
TLF:
EML:
COM: 30.33P.37

MAZA JOSE
DEPTO DE ASTRONOMIA
UNIVERSIDAD DE CHILE
CASILLA 36-D
SANTIAGO
CHILE
TEL:
TLX:
TLF:
EML:
COM: 46

MAZEH TSEVI DR
WISE OBSERVATORY
TEL AVIV UNIVERSITY
RAMAT AVIV
TEL AVIV 69978
ISRAEL
TEL: (3) 420208
TLX: 342171 VERSY IL
TLF:
EML:
COM: 30.42

MAZURE ALAIN DR
LAB. D'ASTRONOMIE
U.S.T.L.
F-34060 MONTPELLIER
FRANCE
TEL: 1-67-52-35-48
TLX: 490944
TLF:
EML:
COM:

MAZUREK THADDEUS JOHN DR
MISSION RESEARCH CORP
PO DRAWER 719
SANTA BARBARA CA 93102
U.S.A.
TEL: 805-963-8761
TLX:
TLF:
EML:
COM: 35.48

MAZZITELLI ITALO DR
IST. ASTROFISICA SPAZIALE
C P 67
I-00044 FRASCATI
ITALY
TEL: 06-9421483
TLX: 610261 CNRFRA
TLF:
EML:
COM: 35

MAZZONI MASSIMO DR
DIPART. DI ASTRONOMIA
E SCIENZA DELLO SPAZIO
LARGO E. FERMI 5
FIRENZE 50125
ITALY
TEL:
TLX:
TLF:
EML:
COM:

MAZZUCCONI FABRIZIO DR
OSSERVATORIO ASTROFISICO
DI ARCETRI
LARGO E. FERMI 5
I-50125 FIRENZE
ITALY
TEL: 055-2752250
TLX:
TLF:
EML:
COM:

MCADAM W BRUCE DR
SCHOOL OF PHYSICS
UNIVERSITY OF SYDNEY
SYDNEY NSW 2006
AUSTRALIA
TEL: 692-2222
TLX: 26169 UNISYD
TLF:
EML:
COM: 40

MCALISTER HAROLD A DR
DEPT PHYSICS & ASTRONOMY
GEORGIA STATE UNIVERSITY
ATLANTA GA 30303
U.S.A.
TEL: 404-658-2932
TLX:
TLF:
EML:
COM: 24.26P.51

MCBREEN BRIAN PHILIP DR
PHYSICS DEPT
UNIVERSITY COLLEGE
BELFIELD
DUBLIN 4
IRELAND
TEL: 693244 671218
TLX: 36293
TLF:
EML:
COM: 28.48

MCCABE MARIE K MS
INSTITUTE FOR ASTRONOMY
2680 WOODLAWN DRIVE
HONOLULU HI 96822
U.S.A.
TEL: 808-948-8306
TLX: 7238459 UHAST HR
TLF:
EML:
COM: 10

MCCALL MARSHALL LESTER DR
DAVID DUNLAP OBSERVATORY
UNIVERSITY OF TORONTO
P O BOX 360
RICHMOND HILL ONT L4C 4Y6
CANADA
TEL: 416-978-4165
TLX:
TLF:
EML:
COM: 34

MCCAMMON DAN
UNIVERSITY OF WISCONSIN
PHYSICS DEPT
1150 UNIVERSITY AVE
MADISON WI 53706
U.S.A.
TEL: 608-262-5916
TLX: 265452 UOFWISC MDS
TLF:
EML:
COM:

MCCARROLL RONALD PROF
LAB D'ASTROPHYSIQUE
UNIVERSITE DE BORDEAUX I
F-33405 TALENCE
FRANCE
TEL:
TLX:
TLF:
EML:
COM:

MCCARTHY DENNIS D DR
US NAVAL OBSERVATORY
34 & MASSACHUSETTS AVE NW
WASHINGTON DC 20390
U.S.A.
TEL: 202-653-0066
TLX: 710-822-1970
TLF:
EML:
COM: 19C.31

MCCARTHY MARTIN F DR
SPECOLA VATICANA
00120 CITTA DEL VATICANO
VATICAN CITY STATE
TEL: 698-3411
TLX: 5042020 VAT OBS VA
TLF:
EML:
COM: 25.33.45.46.50

MCCLAIN EDWARD F
4133 MAPLE ROAD
MORNINGSIDE MD 20746
U.S.A.
TEL: 301-736-8933
TLX:
TLF:
EML:
COM:

MCCLINTOCK JEFFREY E DR
CENTER FOR ASTROPHYSICS
SMITHSONIAN ASTROPHYS OBS
60 GARDEN STREET
CAMBRIDGE MA 02138
U.S.A.
TEL: 617-495-7136
TLX:
TLF:
EML:
COM:

MCCLURE ROBERT D PROF
DOMINION ASTROPHYS OBS
5071 W SAANICH ROAD
RR 5
VICTORIA BC V8X 4M6
CANADA
TEL: 604-388-0230
TLX: 049-7295
TLF:
EML:
COM: 30C.45

MCCLUSKEY GEORGE E JR DR
DIV OF ASTRONOMY
DEPT OF MATHEMATICS
LEHIGH UNIVERSITY
BETHLEHEM PA 18015
U.S.A.
TEL: 215-861-3721
TLX:
TLF:
EML:
COM: 42.44

MCCORD THOMAS B DR
PLANETARY GEOSCIENCES DIV
HAWAII INST OF GEOPHYSICS
2525 CORREA RD
HONOLULU HI 96822
U.S.A.
TEL: 808-948 6488
TLX:
TLF:
EML:
COM: 15.16

MCCRACKEN KENNETH G DR
CSIRO - COSSA
LIMESTONE AVENUE
P.O. BOX 225
DICKSON ACT 2602
AUSTRALIA
TEL: 062-484-595
TLX: 62003 AA
TLF:
EML:
COM: 44

MCCRAY RICHARD DR
JILA
UNIVERSITY OF COLORADO
BOULDER CO 80309
U.S.A.
TEL: 303-492-7835
TLX:
TLF:
EML:
COM: 34.48

MCCREA J DERMOTT
DEPT OF MATH PHYSICS
UNIVERSITY COLLEGE
BELFIELD
DUBLIN 4
IRELAND
TEL:
TLX:
TLF:
EML:
COM: 34.35.47

MCCREA WILLIAM SIR
ASTRONOMY CENTRE
SUSSEX UNIVERSITY
BRIGHTON BN1 9QH
U.K.
TEL: 0273-606755
TLX: 877159 UNISEX G
TLF:
EML:
COM: 28.47

MCCROSKY RICHARD E DR
CENTER FOR ASTROPHYSICS
60 GARDEN STREET
CAMBRIDGE MA 02138
U.S.A.
TEL: 617-495-7212
TLX:
TLF:
EML:
COM: 15.20.22

MCCULLOCH PETER M DR
DEPT OF PHYSICS
UNIVERSITY OF TASMANIA
HOBART, TASMANIA
AUSTRALIA
TEL: 002-20-24-20
TLX: 58150
TLF:
EML:
COM: 40

MCCUTCHEON WILLIAM H PROF
DEPT OF PHYSICS
UNIV OF BRITISH COLUMBIA
2075 WESBROOK MALL
VANCOUVER BC V6T 2A6
CANADA
TEL: 604-228-3853
TLX: 04508576 UBCPHYSICS
TLF:
EML:
COM:

MCDONALD FRANK B DR
NASA/GSFC
CODE 660
GREENBELT MD 20771
U.S.A.
TEL:
TLX:
TLF:
EML:
COM:

MCDONALD J K PETRIE DR
2195 CUBBON DRIVE
VICTORIA BC V8R 1R4
CANADA
TEL: 604-597-6880
TLX:
TLF:
EML:
COM:

MCDONNELL J A M PROF
UNIT FOR SPACE SCIENCES
UNIVERSITY OF KENT
CANTERBURY, KENT CT2 7NR
U.K.
TEL: 0227-459616
TLX: 965449 UKCLIB G
TLF:
EML:
COM: 15.22

MCDONOUGH THOMAS R DR
CALIF INST OF TECHNOLOGY
500 S. OAK KNOLL. NO.46
PASADENA CA 91101
U.S.A.
TEL: 818-795-0147
TLX:
TLF:
EML:
COM: 51

MCELROY M B DR
DEPT OF EARTH & PLANETARY
SCIENCES
HARVARD UNIVERSITY
CAMBRIDGE MA 02138
U.S.A.
TEL:
TLX:
TLF:
EML:
COM: 16

MCGEE RICHARD X DR
CSIRO
DIVISION OF RADIOPHYSICS
P.O.BOX 76
EPPING NSW 2121
AUSTRALIA
TEL: 02-868-0222
TLX: 26230 ASTRO
TLF:
EML:
COM: 34

MCGINSEY BEN Q JR DR
DEPT OF PHYSICS & ASTR
GEORGIA STATE UNIVERSITY
UNIVERSITY PLAZA
ATLANTA GA 30303
U.S.A.
TEL: 404-658-2279
TLX:
TLF:
EML:
COM:

MCGRAW JOHN T DR
STEWARD OBSERVATORY
UNIVERSITY OF ARIZONA
TUCSON AZ 85721
U.S.A.
TEL: 602 621 5391
TLX: 467175
TLF:
EML:
COM: 27

MCGREGOR PETER JOHN DR
MT STROMLO & SIDING
SPRING OBSERVATORIES
PRIVATE BAG
WODEN P.O. ACT 2606
AUSTRALIA
TEL: 062-88-1111
TLX: 62270 CANOPUS AA
TLF:
EML:
COM: 33

MCHARDY IAN MICHAEL DR
DEPT. OF ASTROPHYSICS
NUCLEAR PHYS. BUILDING
KEBLE ROAD
OXFORD OX1 3RH
U.K.
TEL: 44 865 273341
TLX: 83295 NUCLOX G
TLF: 44 865 273418
EML: IMH@UK.AC.OX.ASTRO
COM:

MCINTOSH BRUCE A DR
HERZBERG INST ASTROPHYS
NATL RESEARCH COUNCIL
OTTAWA ONT K1A 0R6
CANADA
TEL:
TLX:
TLF:
EML:
COM: 22

MCINTOSH PATRICK S
NOAA SPACE ENVIRON LAB
R/E/SE3
325 BROADWAY
BOULDER CO 80303
U.S.A.
TEL: (303)4973795
TLX:
TLF:
EML: SPAN:9555::PMCINTOSH
COM: 10

MCKEE CHRISTOPHER F PROF
PHYSICS DEPT
UNIVERSITY OF CALIFORNIA
BERKELEY CA 94720
U.S.A.
TEL: 415-642-0805
TLX: 820181 UCB AST RAL UD
TLF:
EML:
COM: 34

MCKEITH CONAL D DR
QUEEN'S UNIVERSITY
PURE AND APPLIED PHYSICS
BELFAST BT7 1NN
U.K.
TEL: 0232-245133
TLX: 74487 QUB ADM
TLF:
EML:
COM: 34

MCKEITH NIALL ENDA DR
ST PATRICK'S COLLEGE
MAYNOOTH CO KILDARE
IRELAND
TEL:
TLX:
TLF:
EML:
COM:

MCKENNA LAWLOR SUSAN
ST PATRICK'S COLLEGE
DEPT OF EXPERIMENTAL
PHYSICS
MAYNOOTH CO KILDARE
IRELAND
TEL: 285222
TLX: 31493 SPCM EI
TLF:
EML:
COM: 10.12.15.40.41

MCLAREN ROBERT A DR
CANADA-FRANCE-HAWAII
TELESCOPE CORPORATION
PO BOX 1597
KAMUELA HI 96743
U.S.A.
TEL: 808-885-7944
TLX: 633147 CFHT
TLF:
EML:
COM:

MCLEAN BRIAN JOHN
SPACE TELESCOPE SCI INST
HOMEWOOD CAMPUS
BALTIMORE MD 21218
U.S.A.
TEL: 301-338-4101
TLX: 6849101 STSCI
TLF:
EML:
COM: 05.24

MCLEAN DONALD J DR
CSIRO
DIVISION OF RADIOPHYSICS
P.O.BOX 76
EPPING NSW 2121
AUSTRALIA
TEL: 02-868-0222
TLX: 26230 ASTRO AA
TLF:
EML:
COM: 10.40

MCLEAN IAN S DR
HAWAII HEADQUARTERS
JOIN ASTRONOMY CENTER
665 KOMOHANA STREET
HILO HI 96720
U.S.A.
TEL: 808 961 3756
TLX: 633135 JAC UW
TLF:
EML:
COM: 09.25P

MCMILLAN ROBERT S DR
LUNAR & PLANETARY LAB
SPACE SCIENCES BLDG
UNIVERSITY OF ARIZONA
TUCSON AZ 85721
U.S.A.
TEL: 602 621 6968
TLX:
TLF:
EML: NASAMAIL:RSMCMILLAN
COM: 30

MCMULLAN DENNIS DR
CAVENDISH LABORATORY
MADINGLEY ROAD
CAMBRIDGE CB3 0HE
U.K.
TEL:
TLX:
TLF:
EML:
COM: 09

MCNALLY DEREK DR
UNIVERSITY OF LONDON OBS
MILL HILL PARK
LONDON NW7 2QS
U.K.
TEL: 01-959-7367
TLX: 28722 UCPHYS G
TLF:
EML:
COM: 05.34.46

MCNAMARA DELBERT H DR
DEPT PHYSICS & ASTRONOMY
BRIGHAM YOUNG UNIVERSITY
PROVO UT 84602
U.S.A.
TEL: 801-378-2298
TLX:
TLF:
EML:
COM: 05.27.29.45

MCWHIRTER R W PETER DR
SPACE & ASTROPHYSICS DIV
RUTHERFORD APPLETON LAB
CHILTON. DIDCOT OX11 OQX
U.K.
TEL: 0235-446424
TLX: 83159 RUTHLB G
TLF:
EML:
COM: 14.44

MEABURN J DR
DEPT OF ASTRONOMY
THE UNIVERSITY
MANCHESTER M13 9PL
U.K.
TEL:
TLX:
TLF:
EML:
COM: 34

MEAD JAYLEE MONTAGUE DR
NASA/GSFC
CODE 680
GREENBELT MD 20771
U.S.A.
TEL: 301-286-8543
TLX:
TLF:
EML:
COM: 05C.44.45

MEADOWS A JACK PROF
ASTRONOMY & HISTORY
OF SCIENCE DEPT.
UNIVERSITY OF LEICESTER
LEICESTER LE1 7RH
U.K.
TEL:
TLX:
TLF:
EML:
COM: 05.16.41

MEATHERINGHAM STEPHEN DR
NASA GSFC
CODE 680
GREENBELT MD 20771
U.S.A.
TEL: 301-2863019
TLX:
TLF:
EML: SPAN:CHAMP::SJM
COM: 34

MEBOLD ULRICH DR PROF
RADIOASTRONOMISCHES INST
DER UNIVERSITAT BONN
AUF DEM HUEGEL 71
D-5300 BONN 1
GERMANY. F.R.
TEL:
TLX:
TLF:
EML:
COM: 34.40

MEDIAVILLA EVENCIO DR
INSTITUTO DE ASTROFISICA
DE CANARIAS
38200 LA LAGUNA TENERIFE
SPAIN
TEL: 34 22 26 22 11
TLX: 92640
TLF:
EML:
COM: 28

MEDVEDEV YURI A DR
ASTRONOMICAL OBSERVATORY
ODESSA STATE UNIVERSITY
270014 ODESSA
U.S.S.R.
TEL: 22-84-42
TLX:
TLF:
EML:
COM:

MEEKS M LITTLETON DR
MEEKS ASSOCIATES.INC.
P O BOX 643
LINCOLN MA 01773
U.S.A.
TEL: 617-259-0093
TLX:
TLF:
EML:
COM: 40

MEGESSIER CLAUDE DR
OBSERVATOIRE DE PARIS
SECTION DE MEUDON
LAN
F-92195 MEUDON PL CEDEX
FRANCE
TEL: 1-45-07-78-63
TLX: 201571 LAM
TLF:
EML:
COM: 29

MEGRELISHVILI T G PROF
ABASTUMANI ASTROPHYSICAL
OBSERVATORY
383762 ABASTUMANI.GEORGIA
U.S.S.R.
TEL: 22-66
TLX: 327409 TERMIT
TLF:
EML:
COM: 21

MEIER DAVID L
JET PROPULSION LABORATORY
CODE 264-700
4800 OAK GROVE DRIVE
PASADENA CA 91109
U.S.A.
TEL: 213-354-5062
TLX: 675429
TLF:
EML:
COM: 40

MEIER ROBERT R
NAVAL RESEARCH LAB
CODE 4140
WASHINGTON DC 20375
U.S.A.
TEL: 202-767-2773
TLX:
TLF:
EML:
COM: 34

MEIKLE WILLIAM P S
ASTROPHYSICS GROUP
IMPERIAL COLLEGE
PRINCE CONSORT ROAD
LONDON SW7 2BZ
U.K.
TEL: 589-5111
TLX: 261503
TLF:
EML:
COM: 28

MEIN NICOLE DR
OBSERVATOIRE DE PARIS
SECTION DE MEUDON
DASOP
F-92195 MEUDON PL CEDEX
FRANCE
TEL: 1-45-07-78-01
TLX:
TLF:
EML:
COM:

MEIN PIERRE
OBSERVATOIRE DE PARIS
SECTION DE MEUDON
5 PLACE JANSSEN
F-92195 MEUDON PL CEDEX
FRANCE
TEL: 1-45-07-78-01
TLX: 270912 OBSASTR
TLF:
EML:
COM: 05.10.12

MEINEL ADEN B PROF
JPL
MS 186-134
4800 OAK GROVE DRIVE
PASADENA CA 91109
U.S.A.
TEL: 818-354-6827
TLX:
TLF:
EML:
COM: 09.24

MEINIG MANFRED DR
ZENTRALINSTITUT FUR
PHYSIK DER ERDE
TELEGRAFENBERG A17
DDR-1500 POTSDAM
GERMANY. D.R.
TEL: 4551
TLX: 15305 VDE PDM DD
TLF:
EML:
COM: 19.31

MEIRE RAPHAEL
ASTRONOMICAL OBSERVATORY
GHENT STATE UNIVERSITY
WEIDESTRAAT 11
B-9050 EVERGEM
BELGIUM
TEL: 091-53-87-55
TLX:
TLF:
EML:
COM: 07

MEISEL DAVID D DR
DEPT PHYSICS & ASTRONOMY
STATE UNIVERSITY COLLEGE
SUNY
GENESEO NY 14454
U.S.A.
TEL: 716-245-5284
TLX:
TLF:
EML:
COM: 15.22

MEKLER YURI PROF
DEPT GEOPHYS & PLANET SCI
TEL-AVIV UNIVERSITY
TEL-AVIV 69978
ISRAEL
TEL: 3-413-535
TLX:
TLF:
EML:
COM:

MELBOURNE WILLIAM G DR
JPL - MAILSTOP 238-540
4800 OAK GROVE DRIVE
PASADENA CA 91109
U.S.A.
TEL: 818-354-5071
TLX:
TLF:
EML:
COM: 07.19.31

MELCHIOR PAUL J PROF DIR
OBS ROYAL DE BELGIQUE
AVENUE CIRCULAIRE 3
B-1180 BRUXELLES
BELGIUM
TEL: 2-375-24-84
TLX: 21565 OBSBEL B
TLF:
EML:
COM: 08.19.31

MELIA FULVIO DR
DEPARTMENT OF PHYSICS
NORTHWESTERN UNIVERSITY
EVANSTON IL 60208
U.S.A.
TEL: 312 491 4568
TLX:
TLF:
EML: SPAN:nssdca::11540::melia
COM: 42.44

MELIK-ALAVERDIAN YU DR
BYURAKAN ASTROPHYSICAL
OBSERVATORY
378433 ARMENIA
U.S.S.R.
TEL:
TLX:
TLF:
EML:
COM: 35

MELLIER YANNICK DR
OBS DE MIDI-PYRENEES
14 AV. EDOUARD BELIN
31400 TOULOUSE
FRANCE
TEL: 61 25 21 01
TLX: 530776
TLF:
EML:
COM:

MELNICK GARY J
HARVARD SMITHSONIAN CTR
FOR ASTROPHYSICS
60 GARDEN STREET
CAMBRIDGE MA 02138
U.S.A.
TEL: 617-495-7388
TLX:
TLF:
EML:
COM: 30.34.44

MELNICK JORGE
DEPTO DE ASTRONOMIA
UNIVERSIDAD DE CHILE
CASILLA 36-D
SANTIAGO
CHILE
TEL: 2294101
TLX:
TLF:
EML:
COM:

MELOTT ADRIAN L PROF.
PHYSICS AND ASTRONOMY
UNIVERSITY OF KANSAS
LAWRENCE KS 66045
U.S.A.
TEL: 913 864 4626
TLX:
TLF:
EML: BITNET:melott@ukanvax
COM: 47

MELROSE DONALD B PROF
DEPT OF THEORETICAL
PHYSICS
UNIVERSITY OF SYDNEY
SYDNEY NSW 2006
AUSTRALIA
TEL:
TLX:
TLF:
EML:
COM: 10.48

MENDEZ MANUEL DR
INSTITUTO DE ASTRONOMIA
UNAM
APDO POSTAL 70-264
04510 MEXICO DF
MEXICO
TEL:
TLX:
TLF:
EML:
COM:

MENDEZ ROBERTO H DR
INST ASTRONOMIA Y FISICA
DEL ESPACIO
CASILLA 67 SUCURSAL 28
1428 BUENOS AIRES
ARGENTINA
TEL: 781-6755
TLX: 22414 CEDOC AR
TLF:
EML:
COM: 34

MENDIS DEVAMITTA ASOKA DR
EECS
UNIVERSITY OF CALIFORNIA
AT SAN DIEGO
LA JOLLA CA 92093
U.S.A.
TEL: 619-452-2719
TLX:
TLF:
EML:
COM: 15.49

MENDOZA CLAUDIO
IBM VENEZUELA SCIENT. CTR
P.O. BOX 388
CARACAS 1010A
VENEZUELA
TEL: 02-9088697
TLX: 23283 IBMVE VC
TLF:
EML:
COM:

MENDOZA V EUGENIO E DR
INSTITUTO DE ASTRONOMIA
APARTADO 20-158
01000 MEXICO D.F.
MEXICO
TEL: 683-3094
TLX:
TLF:
EML:
COM: 25.45.51

MENEGUZZI MAURICE H DR
SERVICE D'ASTROPHYSIQUE
CEN SACLAY
F-91191 GIF-S-YVETTE
FRANCE
TEL: 1-69-08-44-18
TLX: 690860
TLF:
EML:
COM:

MENG XINMIN
YUNNAN OBSERVATORY
P.O. BOX 110
KUNMING
CHINA. PEOPLE'S REP.
TEL: 72946
TLX: 64040 YUOBS CN
TLF:
EML:
COM: 09

MENNESSIER MARIE-ODILE DR
LABORATOIRE D'ASTRONOMIE
U S T L
F-34060 MONTPELLIER
FRANCE
TEL: 67-63-91-44
TLX: 490944 USTMONT F
TLF:
EML:
COM: 24.27.33

MENON T K PROF
DEPT GEOPHYS & ASTRONOMY
UNIV OF BRITISH COLUMBIA
VANCOUVER BC V6T 1W5
CANADA
TEL: 604-228-2082
TLX: 04543245
TLF:
EML:
COM: 28.34.37.40

MENTESE HUSEYIN DR
UNIVERSITY OBSERVATORY
UNIVERSITY OF ISTANBUL
ISTANBUL
TURKEY
TEL: 522-35-97
TLX:
TLF:
EML:
COM:

MENZIES JOHN W DR
S A A O
P O BOX 9
OBSERVATORY 7935
SOUTH AFRICA
TEL: 47-0025
TLX: 5720309
TLF:
EML:
COM: 25C.34.37.50

MEN' A V DR
INST RADIOPHYS & ELECTRON
UKRAINIAN ACADEMY OF SCI
310085 KHARKOV
U.S.S.R.
TEL:
TLX:
TLF:
EML:
COM:

MERAT PARVIZ
INSTITUT D'ASTROPHYSIQUE
98 BIS BOULEVARD ARAGO
F-75014 PARIS
FRANCE
TEL: 1-43-20-14-25
TLX:
TLF:
EML:
COM: 47

MERAT PARVIZ DR
INSTITUT D'ASTROPHYSIQUE
98 BIS BOULEVARD ARAGO
75014 PARIS
FRANCE
TEL: 43 20 14 25
TLX:
TLF:
EML:
COM:

MERCIER CLAUDE DR
OBSERVATOIRE DE PARIS
SECTION DE MEUDON
DASOP
F-92195 MEUDON PL CEDEX
FRANCE
TEL: 1-45-07-78-15
TLX: 201571 F
TLF:
EML:
COM:

MERGENTALER JAN PROF
ASTRONOMICAL INSTITUTE
UL. KOPERNIKA 11
51-622 WROCLAW
POLAND
TEL: 48-23-29
TLX:
TLF:
EML:
COM: 10.12

MERIGHI ROBERTO DR
OSSERVATORIO ASTRONOMICO
BOLOGNA VIA ZAMBONI 33
BOLOGNA 40100
ITALY
TEL: 05 259401
TLX: 520634 INFNBO I
TLF:
EML: SPAN:37929::MERIGHI
COM: 47

MERKLE FRITZ DR
ESO
KARLSCHWARZSCHILD-STR 2
D-8046 GARCHING B MUNCHEN
GERMANY F.R.
TEL: 89-320-060
TLX: 05828222
TLF:
EML:
COM: 09

MERLEAU-PONTY J PROF
5 R GL DE CASTELNAU
F-75015 PARIS
FRANCE
TEL:
TLX:
TLF:
EML:
COM: 41

MERMAN G A DR
INSTITUTE OF THEORETICAL
ASTRONOMY
10 KUTUZOV QUAY
192187 LENINGRAD
U.S.S.R.
TEL:
TLX:
TLF:
EML:
COM: 07

MERMAN NATALIA V DR
PULKOVO OBSERVATORY
196140 LENINGRAD
U.S.S.R.
TEL:
TLX:
TLF:
EML:
COM:

MERMILLIOD JEAN-CLAUDE DR
INSTITUT D'ASTRONOMIE
UNIVERSITE DE LAUSANNE
CH-1290 CHAVANNES-DES-B.
SWITZERLAND
TEL: 22-55-26-11
TLX: 27720 OBSG CH
TLF:
EML:
COM: 05.37C

MERRIAM JAMES B
DEPT GEOLOGICAL SCIENCES
UNIV OF SASKATCHEWAN
SASKATOON SA S7N 0W0
CANADA
TEL: 306-966-5716
TLX:
TLF:
EML:
COM: 19

MERRILL JOHN E DR
DEPT OF ASTRONOMY
UNIVERSITY OF FLORIDA
SSRB 211
GAINESVILLE FL 32611
U.S.A.
TEL: 904-392-2052
TLX:
TLF:
EML:
COM: 42

MERTZ LAWRENCE N DR
287 FAIRFIELD COURT
PALO ALTO CA 94306
U.S.A.
TEL:
TLX:
TLF:
EML:
COM: 09

MERZANIDES CONSTANTINOS
DEPT. OF ASTRONOMY
UNIVERSITY THESSALONIKI
GR-54006 THESSALONIKI
GREECE
TEL:
TLX:
TLF:
EML:
COM:

MESSAGE PHILIP J DR
DEPT APPLIED MATHS AND
THEORETICAL PHYSICS
THE UNIVERSITY
LIVERPOOL L69 3BX
U.K.
TEL: 0510709-6022
TLX: 627095
TLF:
EML:
COM: 07.20

MESSINA ANTONIO
DIPT. DI ASTRONOMIA
C P 596
I-40100 BOLOGNA
ITALY
TEL:
TLX:
TLF:
EML:
COM:

MESTEL LEON PROF
ASTRONOMY CENTRE
UNIVERSITY OF SUSSEX
BRIGHTON BN1 9QH
U.K.
TEL: 273-60-6755
TLX: 877159 UNISEX G
TLF:
EML:
COM: 35.48.49

MESZAROS ATTILA DR
DEPARTMENT OF ASTRONOMY
AND ASTROPHYSICS
CHARLES UNIVERSITY
PRAGUE 5 SVEDSKA 8
CZECHOSLOVAKIA
TEL: 42 2 540395
TLX:
TLF:
EML:
COM: 47

MESZAROS PETER DR
PENNSYLVANIA STATE UNIV
525 DAVEY LABORATORY
DEPT OF ASTRONOMY
UNIVERSITY PARK PA 16802
U.S.A.
TEL: 814-865-2418
TLX: 842510
TLF:
EML:
COM: 34.47.48

METCALFE LEO DR
ASTROPHYSICS DIVISION
SPACE SCIENCE DEPARTMENT
ESTEC
2202 NOORDWIJK
NETHERLANDS
TEL: 01719 83636
TLX: 39098
TLF:
EML: EARN:LMETCALF@ESTEC
COM:

METZ KLAUS DR
INSTITUT F ASTRONOMIE
& ASTROPHYSIK
SCHEINERSTR 1
D-8000 MUENCHEN 80
GERMANY F.R.
TEL: 089-989021
TLX:
TLF:
EML:
COM: 37

MEURS EVERT DR
ESO
KARL SCHWARZSCHILD ST. 2
8046 GARCHING BEI MUNCHEN
GERMANY, F.R.
TEL: 49 8932006247
TLX: 52828222
TLF:
EML:
COM:

MEWE R DR
LAB VOOR RUIMTEONDERZOEK
BENELUXLAAN 21
NL-3527 HS UTRECHT
NETHERLANDS
TEL: 030-937145
TLX: 47224
TLF:
EML:
COM: 12.14.44

MEYER CLAUDE DR
C E R G A
AVENUE COPERNIC
F-06130 GRASSE
FRANCE
TEL: 93-36-58-49
TLX: 470865
TLF:
EML:
COM: 26

MEYER DAVID M DR
DEPT OF PHYS & ASTRONOMY
NORTHWESTERN UNIVERSITY
EVANSTON IL 60208
U.S.A.
TEL: 312 491 4516
TLX:
TLF:
EML:
COM: 47

MEYER FRIEDRICH DR
MPI FUER PHYSIK UND
ASTROPHYSIK
KARL-SCHWARZSCHILD-ST 1
D-8046 GARCHING B MUNCHEN
GERMANY, F.R.
TEL: 89-32990
TLX: 524629 ASTRO D
TLF:
EML:
COM: 11.48

MEYER JEAN-PAUL DR
SERVICE D'ASTROPHYSIQUE
CEN SACLAY
BP NO 2
F-91191 GIF/YVETTE CDX
FRANCE
TEL: 1-69-08-50- 5
TLX: 690860 PHYSPAC
TLF:
EML:
COM: 48

MEYERS KARIE ANN
OCCIDENTAL COLLEGE
PHYSICS DEPT
1600 CAMPUS RD
LOS ANGELES CA 90041
U.S.A.
TEL: 213-254-2500
TLX:
TLF:
EML:
COM:

MEYER-HOFMEISTER E DR
MPI F PHYSIK & ASTROPHYS
KARL-SCHWARZSCHILD-STR 1
D-8046 GARCHING B MUNCHEN
GERMANY, F.R.
TEL: 089-32990
TLX: 524629 ASTRO D
TLF:
EML:
COM: 35.42

MEYLAN GEORGES DR
SPACE TEL. SCIE. INSTIT.
3700 SAN MARTIN DRIVE
HOMEWOOD CAMPUS
BALTIMORE MD 21218
U.S.A.
TEL: 089 320 060
TLX: 05282 320 ES D
TLF:
EML: BITNET:meylan@stsci
COM: 33.35.37

MEZGER PETER G PROF
MPI FUER RADIOASTRONOMIE
AUF DEM HUEGEL 69
D-5300 BONN 1
GERMANY, F.R.
TEL: 0228-525297
TLX: 0886440 MPIFRD
TLF:
EML:
COM: 33.34.40F

MEZZETTI MARINO
OSSERVATORIO ASTRONOMICO
DI TRIESTE
VIA G.B. TIEPOLO 11
I-34 1 TRIESTE
ITALY
TEL: 040-793221
TLX: 461137 OAT I
TLF:
EML:
COM: 42.47

MIANES PIERRE DR
OBSERVATOIRE DE TOULOUSE
14 AVENUE EDOUARD BELIN
F-31400 TOULOUSE
FRANCE
TEL: 61-25-21-01
TLX:
TLF:
EML:
COM: 25

MIAO YONG-KUAN
DEPT OF ASTRONOMY
NANJING UNIVERSITY
NANJING 210008
CHINA, PEOPLE'S REP.
TEL:
TLX:
TLF:
EML:
COM: 08

MIAO YONG-RUI
SHAANXI OBSERVATORY
P.O. BOX 18
LINTONG XIAN
SHAANXI
CHINA PEOPLE'S REP.
TEL: XIAN 32255
TLX: 70121 CSAO CN
TLF:
EML:
COM: 31

MICHALEC ADAM
ASTRONOMICAL OBSERVATORY
JAGIELLONIAN UNIVERSITY
UL. ORLA 171
30-244 KRAKOW
POLAND
TEL: 221817/3856
TLX: 0322297 UJ PL
TLF:
EML:
COM: 40

MICHALITSIANOS ANDREW
NASA/GSFC
CODE 684.1
GREENBELT MD 20771
U.S.A.
TEL: 301-286-6177
TLX:
TLF:
EML:
COM: 10 34.44CS

MICHARD RAYMOND DR
OBSERVATOIRE DE NICE
BP 139
F-06003 NICE CEDEX
FRANCE
TEL: 93-89-04-20
TLX: 460004
TLF:
EML:
COM: 10.12

MICHAUD GEORGES J DR
250 DU FINISTERE
ST-LAMBERT J4S 1P5
CANADA
TEL: 514-343-6672
TLX: 055-62425 UDEMPHYSAS
TLF:
EML:
COM: 35.36

MICHEL F CURTIS PROF
RICE UNIVERSITY
HOUSTON TX 77251
U.S.A.
TEL: 713-527-4925
TLX: 556457
TLF:
EML:
COM: 48.49

MICZAIKA G R DR
TRW SYSTEMS R5-2291
1 SPACE PARK
REDONDO BEACH CA 90278
U.S.A.
TEL:
TLX:
TLF:
EML:
COM:

MIDDLEHURST BARBARA M MS
LUNAR & PLANETARY INST
3303 NASA ROAD 1
HOUSTON TX 77058
U.S.A.
TEL:
TLX:
TLF:
EML:
COM: 16

MIETELSKI JAN S DR
ASTRONOMICAL OBSERVATORY
JAGIELLONIAN UNIVERSITY
UL. ORLA 171
30-244 KRAKOW
POLAND
TEL: 48-12-22-38-56
TLX: 0322297 UJ PL
TLF:
EML:
COM: 1-

MIGEOTTE MARCEL V PROF
INSTITUT D'ASTROPHYSIQUE
UNIVERSITE DE LIEGE
AVENUE DE COINTE 5
B-4200 COINTE-OUGREE
BELGIUM
TEL: 41-52-9980
TLX: 41264 ASTRLG
TLF:
EML:
COM: 1. 14

MIGNARD FRANCOIS DR
CERGA
AVENUE COPERNIC
F-06130 GRASSE
FRANCE
TEL: 93-36-58-49
TLX: 470865
TLF:
EML:
COM: 07

MIHAILA IERONIM PROF
BUCHAREST UNIVERSITY
ACADEMIEI 14
70109 BUCAREST
RUMANIA
TEL: 133619
TLX:
TLF:
EML:
COM:

MIHALAS BARBARA R WEIBEL
HIGH ALTITUDE OBSERVATORY
NCAR
PO BOX 3000
BOULDER CO 80307
U.S.A.
TEL: 304-494-5151
TLX:
TLF:
EML:
COM:

MIHALAS DIMITRI DR
DEPARTMENT OF ASTRONOMY
UNIVERSITY OF ILLINOIS
1011 W. SPRINGFIELD AVE.
URBANA IL 61801
U.S.A.
TEL: 217-333-3090
TLX:
TLF:
EML:
COM: 12.36

MIKAMI TAKAO DR
OSAKA GAKUIN UNIVERSITY
2-36-1. KISHIBE-MINAMI
SUITA-SHI
OSAKA 564
JAPAN
TEL: 06-381-8434
TLX:
TLF:
EML:
COM:

MIKESELL ALFRED H MR
2509 N. CAMPBELL #69
TUCSON AZ 85719
U.S.A.
TEL: 602-327-2381
TLX:
TLF:
EML:
COM:

MIKHAIL FAHMY I PROF DR
AIN SHAMS UNIVERSITY
FACULTY OF SCIENCE
CAIRO
EGYPT
TEL: 575887
TLX: 94070 USHMS UN
TLF:
EML:
COM:

MIKHAL JOSEPH SIDKY PROF
HELWAN INSTITUTE OF
ASTRONOMY & GEOPHYSICS
HELWAN. CAIRO
EGYPT
TEL: 780645
TLX: 93070 HIAG UN
TLF:
EML:
COM: 16

MIKHELSON NIKOLAJ N DR
MAIN ASTRON OBSERVATORY
PULKOVO
196140 LENINGRAD
U.S.S.R.
TEL: 2-979-465
TLX:
TLF:
EML:
COM: 09

MIKKOLA SEPPO DR
TURKU UNIVERSITY OBS
ITAINEN PITKAKATU 1
SF-20520 TURKU
FINLAND
TEL: 358-21-435 822
TLX: 62638 TYF SF
TLF: 358-21-433 767
EML:
COM: 07.26.33.37

MIKOLAJEWSKA JOANNA DR
INSTITUTE OF ASTRONOMY
NICOLAUS COPERNICUS UNIV
CHOPINA 12/18
PL 87100 TORUN
POLAND
TEL: 56 26017 26018
TLX: 552234 ASTR PL
TLF:
EML:
COM: 42

MIKULASEK ZDENEK DR
N COPERNICUS OBSERVATORY
AND PLANETARIUM
KRAVI HORA
61600 BRNO
CZECHOSLOVAKIA
TEL:
TLX:
TLF:
EML:
COM: 29.42

MILANI ANDREA
ISTITUTO DI MATEMATICA
UNIVERSITA' DI PISA
VIA BUONARROTI 2
I-56100 PISA
ITALY
TEL:
TLX:
TLF:
EML:
COM: 070.20

MILANO LEOPOLDO DR
DIP. SCIENZE FISICHE
PAD. 19
MOSTRA OLTREMARE
I-80125 NAPOLI
ITALY
TEL: 39-81-7253447
TLX: 720320
TLF:
EML:
COM:

MILES HOWARD G MR
LANE PARK
PITYME. ST MINVER
WADEBRIDGE PL27 6PN
U.K.
TEL: 020-886-3153
TLX:
TLF:
EML:
COM: 22

MILET BERNARD L DR
OBSERVATOIRE DE NICE
BP 139
F-06003 NICE CEDEX
FRANCE
TEL: 93-89-04-20
TLX:
TLF:
EML:
COM: 15.20.26.51

MILEY G K DR
STERREWACHT
NIELS BOHRWEG 2
POSTBUS 9513
2300 RA LEIDEN
NETHERLANDS
TEL: 31 71 275849
TLX: 39058 ASTRO NL
TLF:
EML:
COM: 28.40

MILKEY ROBERT W DR
SPACE TEL SCIENCE INST
HOMEWOOD CAMPUS
3700 SAN MARTIN DRIVE
BALTIMORE MD 21218
U.S.A.
TEL: 301-338-4720
TLX: 6849101
TLF:
EML:
COM: 12

MILLAR THOMAS J DR
MATHEMATICS DEPT
UMIST
P O BOX 88
MANCHESTER M60 1QD
U.K.
TEL: 061-236-3311
TLX: 666094
TLF:
EML:
COM: 34

MILLER FREEMAN D PROF
UNIVERSITY OF MICHIGAN
DEPT OF ASTRONOMY
DENNISON BLDG
ANN ARBOR MI 48109
U.S.A.
TEL: 313-764-3447
TLX:
TLF:
EML:
COM: 15

MILLER HUGH R PROF
DEPT OF PHYSICS
GEORGIA STATE UNIVERSITY
ATLANTA GA 30303
U.S.A.
TEL: 404-658-2279
TLX:
TLF:
EML:
COM: 28

MILLER JOHN C DR
DEPT OF ASTROPHYSICS
UNIVERSITY OF OXFORD
SOUTH PARKS ROAD
OXFORD OX1 3RQ
U.K.
TEL: 44-865-511336
TLX: 83295 NUCLOX G
TLF:
EML:
COM: 48

MILLER JOSEPH S PROF
LICK OBSERVATORY
UNIVERSITY OF CALIFORNIA
SANTA CRUZ CA 95060
U.S.A.
TEL: 408-429-2135
TLX:
TLF:
EML:
COM: 25C.28.34

MILLER RICHARD H DR
ASTRONOMY DEPT
UNIVERSITY OF CHICAGO
5640 ELLIS AVENUE
CHICAGO IL 60637
U.S.A.
TEL: 312-962-8201
TLX: 6871133
TLF:
EML:
COM: 28.33

MILLET JEAN DR
LAS
TRAVERSE DU SIPHON
LES TROIS LUCS
F-13012 MARSEILLE
FRANCE
TEL: 91-66-08-32
TLX: 420584 ASTROSP F
TLF:
EML:
COM:

MILLIARD BRUNO
LAB ASTRONOMIE SPATIALE
TRAVERSE DU SIPHON
LES TROIS LUCS
F-13012 MARSEILLE
FRANCE
TEL: 91-05-59-00
TLX: 420584 ASTROSP F
TLF:
EML:
COM:

MILLIGAN J E
ASTROPHYSICS BRANCH
GODDARD SPACE FLIGHT CTR
GREENBELT MD 20771
U.S.A.
TEL:
TLX:
TLF:
EML:
COM: 29

MILLIKAN ALLAN G MR
RESEARCH LAB. B-59
EASTMAN KODAK CO
343 STATE STREET
ROCHESTER NY 14650
U.S.A.
TEL:
TLX:
TLF:
EML:
COM: 09

MILLIS ROBERT L DR
LOWELL OBSERVATORY
1400 W. MARS HILL ROAD
FLAGSTAFF AZ 86001
U.S.A.
TEL: 602-774-3358
TLX:
TLF:
EML:
COM: 15.16.20

MILLMAN PETER M DR
HERZBERG INST ASTROPHYS
NATL RES COUNCIL CANADA
OTTAWA ONT K1A 0R6
CANADA
TEL: 613-990-0705
TLX: 053-3715
TLF:
EML:
COM: 15.16.22

MILLS ALLAN A DR
DEPT OF ASTRONOMY
THE UNIVERSITY
LEICESTER LE1 7RH
U.K.
TEL: 0533-554455
TLX:
TLF:
EML:
COM:

MILLS BERNARD Y PROF
SCHOOL OF PHYSICS
UNIVERSITY OF SYDNEY
SYDNEY NSW 2006
AUSTRALIA
TEL: 02-692-2544
TLX: 26169 UNISYD
TLF:
EML:
COM: 28.40

MILNE DOUGLAS K DR
CSIRO
DIVISION OF RADIOPHYSICS
P.O.BOX 76
EPPING NSW 2121
AUSTRALIA
TEL: 02-868-0222
TLX: 26230 ASTRO
TLF:
EML:
COM: 34.40

MILOGRADOV-TURIN JELENA
INSTITUTE OF ASTRONOMY
UNIVERSITY OF BEOGRAD
STUDENTSKI TRG 16
YU-11000 BEOGRAD
YUGOSLAVIA
TEL: 638-715
TLX:
TLF:
EML:
COM: 40

MILONE EUGENE F PROF
PHYSICS DEPT
UNIVERSITY OF CALGARY
2500 UNIVERSITY DR NW
CALGARY AB T2N 1N4
CANADA
TEL: 403-220-5412
TLX: 03821545
TLF:
EML:
COM: 25C.27.42

MILONE LUIS A DR
OBSERVATORIO ASTRONOMICO
LAPRIDA 854
5000 CORDOBA
ARGENTINA
TEL: 36876 - 40629
TLX: 51822 BUCOR
TLF:
EML:
COM: 27

MILOVANOVIC VLADETA DR
INSTITUT ZA GEODEZIJU
BULEVAR REVOLUCIJE 73
11000 BEOGRAD
YUGOSLAVIA
TEL:
TLX:
TLF:
EML:
COM: 19

MINAROVJECH MILAN
ASTRONOMICAL INSTITUTE
SLOVAK ACAD. OF SCIENCES
059 60 TATRANSKA LOMNICA
CZECHOSLOVAKIA
TEL: 967-8668x0969
TLX: 78277
TLF:
EML:
COM: 09

MINESHIGE SHIN DR
ASTRONOMY DEPARTMENT
UNIV OF TEXAS AT AUSTIN
AUSTIN TEXAS 78712 1083
U.S.A.
TEL: 512 471 3000
TLX: 910 874 1351
TLF:
EML: BITNET:ASUS611@UTCHPC
COM: 42

MINEVA VENETA DR
DEPT OF ASTRONOMY
BULGARIAN ACADEMY OF SCI
72 LENIN BLVD
1784 SOFIA
BULGARIA
TEL: 75 88 27
TLX: 23561
TLF:
EML:
COM: 28

MININ I N PROF
LENINGRAD STATE UNIV
ASTRONOMICAL OBSERVATORY
BIBLIOTECHNAJA PL. 2
198904 LENINGRAD-PETRODV.
U.S.S.R.
TEL: 257-94-89
TLX:
TLF:
EML:
COM: 34

MINN YOUNG KEY DR
DEPT ASTRON & SPACE SCI
KYUNG HEE UNIVERSITY
YONG-IN
KYUNGGI-DO 170-23
KOREA. REPUBLIC
TEL: 02-764-6131
TLX:
TLF:
EML:
COM: 34.51

MINNETT HARRY C MR
CSIRO
DIVISION OF RADIOPHYSICS
P.O.BOX 76
EPPING NSW 2121
AUSTRALIA
TEL:
TLX:
TLF:
EML:
COM:

MINTZ BLANCO BETTY MRS
CERRO TOLOLO
INTERAMERICAN OBSERVATORY
CASILLA 603
LA SERENA
CHILE
TEL: 213352
TLX:
TLF:
EML:
COM: 20.25

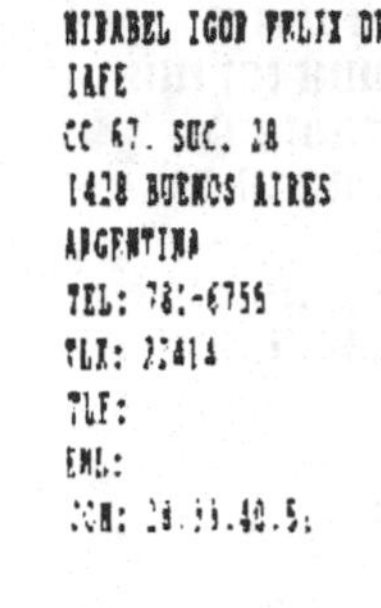

MIRABEL IGOR FELIX DR
IAFE
CC 67. SUC. 28
1428 BUENOS AIRES
ARGENTINA
TEL: 781-6755
TLX: 22414
TLF:
EML:
COM: 28.33.40.51

MIRONOV NIKOLAY T
MAIN ASTRONOMICAL OBS
UKRAINIAN ACADEMY OF SCI
252127 KIEV
U.S.S.R.
TEL: 66-47-59
TLX: 131406 SKY
TLF:
EML:
COM: 19C

MIRZOYAN L V DR PROF
BYURAKAN ASTROPHYSICAL
OBSERVATORY
378433 ARMENIA
U.S.S.R.
TEL:
TLX:
TLF:
EML:
COM: 27.33

MISCONI NEBIL YOUSIF DR
SPACE ASTRONOMY LAB
UNIVERSITY OF FLORIDA
1810 N.W. 6TH STREET
GAINESVILLE FL 32609
U.S.A.
TEL:
TLX:
TLF:
EML:
COM: 21.22

MISNER CHARLES W PROF
DEPT PHYSICS & ASTRONOMY
UNIVERSITY OF MARYLAND
COLLEGE PARK MD 20742
U.S.A.
TEL:
TLX:
TLF:
EML:
COM: 47

MISSANA MARCO DR
OSSERVATORIO ASTRONOMICO
DI BRERA
VIA CREMAGNANI 13/11
I-20059 VIMERCATE
ITALY
TEL:
TLX:
TLF:
EML:
COM:

MISSANA NATALE PROF
VIA PUCCINI 2
I-20025 PINO TORINESE
ITALY
TEL:
TLX:
TLF:
EML:
COM:

MITALAS ROMAS ASSOC PROF
DEPT OF ASTRONOMY
UNIV OF WESTERN ONTARIO
LONDON ONT N6A 5B9
CANADA
TEL: 519-679-3184
TLX:
TLF:
EML:
COM: 35

MITCHELL GEORGE F DR
DEPT OF ASTRONOMY
SAINT MARY'S UNIVERSITY
HALIFAX NS B3H 3C3
CANADA
TEL: 902-429-9780
TLX:
TLF:
EML:
COM: 34

MITCHELL KENNETH J DR
APPLIED RESEARCH CORP.
8201 CORPORATE DRIVE
SUITE 920
LANDOVER MD 20785
U.S.A.
TEL: 301 459 8442
TLX:
TLF:
EML:
COM: 40

MITCHELL RICHARD MR
12704 LA CUEVA N.E.
ALBUQUERQUE NM 87123
U.S.A.
TEL: 505-292-0309
TLX:
TLF:
EML:
COM: 25

MITCHELL WALTER E JR
ASTRONOMY DEPT
OHIO STATE UNIVERSITY
174 W. 18TH AVE
COLUMBUS OH 43210
U.S.A.
TEL: 614-422-5554
TLX:
TLF:
EML:
COM:

MITIC LJUBISA A DR
OBSERVATOIRE DE BELGRADE
VOLGINA 7
11050 BELGRADE
YUGOSLAVIA
TEL: 011-419-357
TLX:
TLF:
EML:
COM: 08

MITRA A P DR
NATIONAL PHYSICAL LAB
NEW DELHI 110 012
INDIA
TEL: 585298/581440
TLX: 3162454 RSD IN
TLF:
EML:
COM:

MITROFANOVA LYUDMILA A DR
PULKOVO OBSERVATORY
196140 LENINGRAD
U.S.S.R.
TEL:
TLX:
TLF:
EML:
COM:

MITSUDA KAZUHISA DR
INSTITUTE OF SPACE AND
ASTRONAUTICAL SCIENCES
3-1-1 YOSHINODAI
SAGAMIHARA. KANAGAWA 291
JAPAN
TEL: 427 51 3911
TLX: 134757 ISASTRC
TLF: 427 59 4253
EML:
COM:

MITTON JACQUELINE
8A CANTERBURY CLOSE
CAMBRIDGE CB4 3QQ
U.K.
TEL: 0223 355974
TLX:
TLF:
EML:
COM:

MITTON SIMON DR
CAMBRIDGE UNIV PRESS
SHAFTSBURY ROAD
CAMBRIDGE CB2 2RU
U.K.
TEL: 0-223-312-393
TLX: 817256 CUPCAM UK
TLF:
EML:
COM: 05

MIYAJI SHIGEKI DR
CHIBA UNIVERSITY
1-33, YAYOICHO
CHIBA, 260
JAPAN
TEL: 472-51-1111
TLX:
TLF:
EML:
COM: 35.42.44

MIYAMA SYOKEN
DEPT OF PHYSICS
KYOTO UNIVERSITY
KITASHIRAKAWA OIWAKECHO
KYOTO 606
JAPAN
TEL: 075-751-2111
TLX:
TLF:
EML:
COM: 34

MIYAMOTO MASANORI DR
TOKYO ASTRONOMICAL
OBSERVATORY
OSAWA MITAKA
TOKYO 181
JAPAN
TEL: 0422-32-5111
TLX: 2822307 TAOMTK J
TLF:
EML:
COM: 08P.33

MIYAMOTO SIGENORI PROF
DEPT OF PHYSICS/FAC SCI
OSAKA UNIVERSITY
MACHIKANEYAMA-CHO
TOYONAKA OSAKA 560
JAPAN
TEL: 06-844-1151
TLX:
TLF:
EML:
COM: 15.36.44.46

MIYAMOTO SYOTARO PROF DR
KASAN OBSERVATORY
YAMASHINA
KYOTO 607
JAPAN
TEL: 075-581-1235
TLX:
TLF:
EML:
COM:

MIZUNO SHUN
KANAZAWA INST TECHNOLOGY
7-1 OGIGAOKA
NONOICHIMACHI
ISHIKAWA 921
JAPAN
TEL: 0762-48-1100
TLX: 5122456 KITLC J
TLF:
EML:
COM: 34

MNATSAKANIAN MAMIKON A DR
BYURAKAN ASTROPHYSICAL
OBSERVATORY
378433 ARMENIA
U.S.S.R.
TEL: 56-34-53
TLX:
TLF:
EML:
COM: 36

MO JING-ER
PURPLE MOUNTAIN
OBSERVATORY
NANJING
CHINA, PEOPLE'S REP.
TEL: 36967
TLX: 34144 PMONJ CN
TLF:
EML:
COM: 34

MOCHKOVITCH ROBERT DR
INSTITUT D'ASTROPHYSIQUE
98 BIS BD ARAGO
75014 PARIS
FRANCE
TEL: 43 20 14 25
TLX: 270070 SU CNRS
TLF:
EML: BITNET:MOCHKOOFRIAP 51
COM:

MOCHNACKI STEPHAN W DR
DAVID DUNLAP OBSERVATORY
DEPT OF ASTRONOMY
UNIVERSITY OF TORONTO
TORONTO ONT M5S 1A7
CANADA
TEL:
TLX:
TLF:
EML:
COM: 42

MOCZKO JANUSZ DR
BOROWIEC
63-120 KORNIK
POLAND
TEL:
TLX:
TLF:
EML:
COM: 19

MODALI SARMA B DR
QSN SYSTEMS & RESEARCH
8401 CORPORATION DR.
LANDOVER MD 20785
U.S.A.
TEL: 301-459-3322
TLX:
TLF:
EML:
COM:

MODISETTE JERRY L PROF
16323 HEREFORD LN.
HOUSTON TX 77058
U.S.A.
TEL:
TLX:
TLF:
EML:
COM: 44

MOEHLMANN DIEDRICH
INST. F. KOSMOSFORSCHUNG
RUDOWER CHAUSSEE 5
DDR-1199 BERLIN
GERMANY, D.R.
TEL: 6743485
TLX: 113137 IKF DD
TLF:
EML:
COM: 15.16

MOELLENHOFF CLAUS DR
LANDESSTERNWARTE
KOENIGSTUHL
D-6900 HEIDELBERG 1
GERMANY, F.R.
TEL: 06221-10036
TLX: 461789 MPIA D
TLF:
EML:
COM: 15

MOERDIJK WILLY G DR
ASTRONOMISCH OBS R.U.G.
ST PIETERSAALSTSTRAAT 171
B-9000 GENT
BELGIUM
TEL: 091-221033
TLX:
TLF:
EML:
COM:

MOESGAARD KRISTIAN P
HISTORY OF SC DEPT
UNIVERSITY OF AARHUS
BYGADAN 1 / TORRILD
DK-8300 ODDER
DENMARK
TEL: 06-53-1004
TLX:
TLF:
EML:
COM: 41

MOFFAT ANTHONY F J DR
DEPT DE PHYSIQUE
UNIVERSITE DE MONTREAL
CP 6128 - SUCC A
MONTREAL PQ H3C 3J7
CANADA
TEL: 514-343-6682
TLX: 05562425 UDEMPHYSAS
TLF:
EML:
COM: 29.33.37

MOFFAT JOHN W. DR
DEPT OF PHYSICS
UNIVERSITY OF TORONTO
TORONTO ONT M5S 1A7
CANADA
TEL: 416-978-2949
TLX:
TLF:
EML:
COM:

MOFFATT HENRY KEITH PROF
DEPT APPLIED MATHS
& THEORETICAL PHYSICS
SILVER STREET
CAMBRIDGE CB3 9EW
U.K.
TEL: 0223-351645
TLX: 81249 CAMSPL G
TLF:
EML:
COM:

MOFFETT THOMAS J PROF
DEPT OF PHYSICS
PURDUE UNIVERSITY
WEST LAFAYETTE IN 47907
U.S.A.
TEL: 317-494-5508
TLX:
TLF:
EML:
COM: 25.27

MOGILEVSKIJ EM I DR
INSTITUTE OF TERRESTRIAL
MAGNETISM, IONOSPHERE AND
RADIO WAVE PROPAGATION
142092 TROITSK, MOSCOW R.
U.S.S.R.
TEL: 232-19-11
TLX: 412623 SCSTP SU
TLF:
EML:
COM: 10

MOHAN CHANDER DR
UNIVERSITY OF ROORKEE
DEPT. OF MATHEMATICS
ROORKEE 247 667
INDIA
TEL:
TLX:
TLF:
EML:
COM:

MOISEEV I G DR
CRIMEAN ASTROPHYS OBS
USSR ACADEMY OF SCIENCES
USSR ACADEMY OF SCIENCES
334413 CRIMEA
U.S.S.R.
TEL:
TLX:
TLF:
EML:
COM: 10.40

MOLARO PAOLO DR
OSSERVATORIO ASTRONOMICO
DI TRIESTE
VIA GB TIEPOLO 11
TRIESTE 34131
ITALY
TEL: 40 309342
TLX: 461137 OAT I
TLF:
EML:
COM: 29

MOLCHANOV A P PROF
LENINGRAD STATE UNIV
ASTRONOMICAL OBSERVATORY
199178 LENINGRAD
U.S.S.R.
TEL:
TLX:
TLF:
EML:
COM: 40

MOLES MARIANO J DR
INST DE ASTROFISICA DE
ANDALUCIA, APDO 2144
PROFESOR ALBAREDA 1
18080 GRANADA
SPAIN
TEL: 58-12-13-11
TLX: 78573 IAAG E
TLF:
EML:
COM: 28

MOLINA ANTONIO
INST DE ASTROFISICA DE
ANDALUCIA, APDO 2144
PROFESOR ALBAREDA 1
18080 GRANADA
SPAIN
TEL: 58-12-13-00
TLX: 78573 IAAG E
TLF:
EML:
COM: 16.21

MOLNAR MICHAEL R PROF
AT&T BELL LABORATORIES
ROOM 2B-422
CRAWFORDS CORNER ROAD
HOLMDEL NJ 07733
U.S.A.
TEL:
TLX:
TLF:
EML:
COM:

MONCHEV GOSPODIN
ASTRONOMICAL OBSERVATORY
P.O. BOX [illegible]
8800 SLIVEN
BULGARIA
TEL: 2-72-04
TLX:
TLF:
EML:
COM: 46

MONAGHAN JOSEPH J DR
MATHEMATICS DEPT
MONASH UNIVERSITY
CLAYTON VIC 3168
AUSTRALIA
TEL: 03-541-2563
TLX: MONASH AA 32691
TLF:
EML:
COM: 35

MONET DAVID G
US NAVAL OBSERVATORY
PO BOX 1149
FLAGSTAFF AZ 86002
U.S.A.
TEL:
TLX:
TLF:
EML:
COM: 24.33.44

MONFILS ANDRE G PROF
IAL SPACE
UNIVERSITE DE LIEGE
AVENUE DU PRE AILY
B-4900 ANGLEUR-LIEGE
BELGIUM
TEL: 041-67-66-68
TLX: 41320 IAL SP B
TLF:
EML:
COM: 14.44

MONNET GUY J DR
OBSERVATOIRE DE LYON
AVENUE CHARLES ANDRE
F-69230 ST GENIS-LAVAL
FRANCE
TEL: 78-56-07-05
TLX: 310926
TLF:
EML:
COM: 33

MONSIGNORI FOSSI BRUNA DR
OSSERVATORIO ASTROFISICO
DI ARCETRI
LARGO E. FERMI 5
I-50125 FIRENZE
ITALY
TEL: 055-2752239
TLX:
TLF:
EML:
COM:

MONTEIRO TANIA DR
DEPT. OF MATHEMATICS
ROYAL HOLLOWAY COLLEGE
UNIVERSITY OF LONDON
EGHAM SURREY TW20 OEX
U.K.
TEL: 0784-34455
TLX:
TLF:
EML: ASL0703@ULCC
COM:

MONTES CARLOS DR
OBSERVATOIRE DE NICE
BP 139
F-06003 NICE CEDEX
FRANCE
TEL: 93-89-04-20
TLX: 460004
TLF:
EML:
COM:

MONTMERLE THIERRY DR
SERVICE D'ASTROPHYSIQUE
CEN SACLAY
F-91191 GIF YVETTE CDX I
FRANCE
TEL: 1-69-08-57-22
TLX: 690860
TLF:
EML:
COM:

MOOK DELO E PROF
DEPT PHYSICS & ASTRONOMY
DARTMOUTH COLLEGE
HANOVER NH 03755
U.S.A.
TEL: 603-646-2972
TLX:
TLF:
EML:
COM:

MOONS MICHELE B M M
DEPT DE MATHEMATIQUE
FAC UNIV N.D. DE LA PAIX
REMPART DE LA VIERGE 8
B-5000 NAMUR
BELGIUM
TEL: 081-229061x2436
TLX: 59222 FACNAM B
TLF:
EML:
COM: [illegible]

MOORE DANIEL R DR
DEPT OF MATHEMATICS
IMPERIAL COLLEGE
HUXLEY BLDG. QUEEN'S GATE
LONDON SW7 2BZ
U.K.
TEL:
TLX:
TLF:
EML:
COM: 35

MOORE ELLIOTT P PROF
JOINT OBSERVATORY
FOR COMETARY RESEARCH
CAMPUS STATION
SOCORRO NM 87801
U.S.A.
TEL: 505-835-5431
TLX:
TLF:
EML:
COM: 15

MOORE PATRICK DR
FARTHINGS
39 WEST STREET
SELSEY. SUSSEX
U.K.
TEL: 0243-603-668
TLX:
TLF:
EML:
COM: 16

MOORE RONALD L DR
NASA/MARSHALL SPACE
FLIGHT CENTER ES 52
SPACE SCIENCE LABORATORY
HUNTSVILLE AL 35812
U.S.A.
TEL: 205-453-0115
TLX: 59-4416 NASA MSFC HT
TLF:
EML:
COM:

MOORHEAD JAMES M DR
ASTRONOMY DEPT
UNIVERSITY OF W. ONTARIO
LONDON ONT N6A 3K7
CANADA
TEL: 519-679-3186
TLX:
TLF:
EML:
COM:

MOORWOOD ALAN F M
ESO
KARL-SCHWARZSCHILD-STR 2
D-8046 GARCHING B MUNCHEN
GERMANY. F.R.
TEL: 089-320-06-294
TLX: 05 28 282 24 EO D
TLF:
EML:
COM: 28

MOOS HENRY WARREN DR
DEPT PHYSICS & ASTRONOMY
JOHNS HOPKINS UNIVERSITY
BALTIMORE MD 21218
U.S.A.
TEL: 301-338-7337
TLX: 7102341090
TLF:
EML:
COM: 29.44

MORALES-DURAN CARMEN
INTA
DEP. DE PROGR. ESPACIALES
TORREJON DE ARDOZ 28850
SPAIN
TEL: 91 6750700
TLX: 22026 INTA E
TLF:
EML:
COM:

MORAN JAMES M DR
CENTER FOR ASTROPHYSICS
60 GARDEN STREET
CAMBRIDGE MA 02138
U.S.A.
TEL: 617-495-7477
TLX: 921428 SATELLITE CAM
TLF:
EML:
COM: 40C

MORANDO BRUNO L DR
BUREAU DES LONGITUDES
77 AVE DENFERT-ROCHEREAU
F-75014 PARIS
FRANCE
TEL: 1-40-51-22-76
TLX:
TLF:
EML:
COM: 04C.07.20

MORBEY CHRISTOPHER L
DOMINION ASTROPHYS OBS
5071 W SAANICH ROAD
RR 5
VICTORIA BC V8X 4M6
CANADA
TEL: 604-388-0220
TLX: 0497295
TLF:
EML:
COM: 26.30

MOREELS GUY DR
OBSERVATOIRE DE BESANCON
41B AVE DE L'OBSERVATOIRE
F-25000 BESANCON
FRANCE
TEL: 81-50-22-66
TLX: 361144 F
TLF:
EML:
COM: 46

MOREL PIERRE JACQUES DR
OBSERVATOIRE DE NICE
BP 139
F-06003 NICE CEDEX
FRANCE
TEL: 93-89-04-20
TLX: 460004
TLF:
EML:
COM: 25

MORENO HUGO PROF
DEPTO DE ASTRONOMIA
UNIVERSIDAD DE CHILE
CASILLA 36-D
SANTIAGO
CHILE
TEL: 229-4101/4002
TLX: 440001 PBCZ OBSERNAL
TLF:
EML:
COM: 25

MORENO-INSERTIS FERNANDO
INST DE ASTROFISICA DE
CANARIAS
38071 TENERIFE
SPAIN
TEL: 922-262211
TLX: 92640
TLF:
EML:
COM: 10.12

MORETON G E
15-5 THE ESPLANADE
BALMORAL BEACH NSW 2088
AUSTRALIA
TEL:
TLX:
TLF:
EML:
COM: 10

MORGAN BRIAN LEALAN
BLACKETT LABORATORY
IMPERIAL COLLEGE
LONDON SW7 2BZ
U.K.
TEL: 01-589-5111
TLX: 261503
TLF:
EML:
COM: 09

MORGAN DAVID H DR
ROYAL OBSERVATORY
BLACKFORD HILL
EDINBURGH EH9 3HJ
U.K.
TEL: 031-667-3321
TLX: 72383 ROEDIN G
TLF:
EML:
COM: 21.34

MORGAN JOHN ADRIAN
THE AEROSPACE CORPORATION
MS M4/041
PO BOX 92957
LOS ANGELES CA 90009
U.S.A.
TEL:
TLX:
TLF:
EML:
COM: 35

MORGAN PETER DR
CANBERRA COLL ADV EDUC
SCHOOL OF APPLIED SCIENCE
P.O.BOX 1
BELCONNEN ACT 2616
AUSTRALIA
TEL: 062-52-2557
TLX: 62267 CANCOL AA
TLF:
EML:
COM: 19.31

MORGAN THOMAS H DR
JOHNSON SPACE CENTER
CODE SN3
HOUSTON TX 77058
U.S.A.
TEL: 713-483-5039
TLX: 762931
TLF:
EML:
COM: 42.44

MORGAN WILLIAM W PROF
YERKES OBSERVATORY
PO BOX 258
WILLIAMS WI 53191
U.S.A.
TEL: 414-245-5555
TLX:
TLF:
EML:
COM: 28.45

MORGULEFF NINA ING
INSTITUT D'ASTROPHYSIQUE
98 BIS BOULEVARD ARAGO
F-75014 PARIS
FRANCE
TEL: 1-43-20-14-25
TLX:
TLF:
EML:
COM: 27.29.45

MORIMOTO MASAKI DR
TOKYO ASTRONOMICAL
OBSERVATORY
OSAWA MITAKA
TOKYO 181
JAPAN
TEL:
TLX:
TLF:
EML:
COM: 34.40V.51

MORISON IAN MR
NRAL
JODRELL BANK
MACCLESFIELD SK11 9DL
U.K.
TEL: 0477-71321
TLX: 36149
TLF:
EML:
COM: 40

MORITA KAZUHIKO
DEPT OF PHYSICS
HOKKAIDO UNIVERSITY
NISHI 8-CHOME KITA 10-JYO
SAPPORO. HOKKAIDO 060
JAPAN
TEL: 11-711-2111
TLX:
TLF:
EML:
COM: 40

MORIYAMA FUMIO PROF
OSAKA GAKUIN UNIVERSITY
2-36-1 KISHIBE MINAMI
SUITA-SHI
OSAKA-FU 564
JAPAN
TEL: 06-381-8434
TLX:
TLF:
EML:
COM: 10.12.40

MOROSSI CARLO
OSSERVATORIO ASTRONOMICO
VIA G.B. TIEPOLO
I-34131 TRIESTE
ITALY
TEL: 40-76 85 06
TLX: 461137 OAT I
TLF:
EML:
COM: 29.45

MOROZ V I PROF DR
SPACE RESEARCH INSTITUTE
USSR ACADEMY OF SCIENCES
117810 MOSCOW
U.S.S.R.
TEL: 333-31-22
TLX: 411498 CTAP CY
TLF:
EML:
COM: 15.16C.51

MOROZHENKO A V DR
MAIN ASTRONOMICAL OBS
UKRAINIAN ACADEMY OF SCI
GOLOSEEVO
252127 KIEV
U.S.S.R.
TEL: 663110
TLX: 131406 SKY
TLF:
EML:
COM: 16

MOROZHENKO N N DR
MAIN ASTRONOMICAL OBS
UKRAINIAN ACADEMY OF SCI
GOLOSEEVO
252127 KIEV
U.S.S.R.
TEL: 663110
TLX: 131406 SKY
TLF:
EML:
COM: 10

MORRAS RICARDO DR.
INSTITUTO ARGENTINO
DE RADIOASTRONOMIA
CASILLA DE CORREO 5
1894 VILLA ELISA (Bs.As.)
ARGENTINA
TEL: 021-43793
TLX: 18052 CICYR-AR
TLF:
EML:
COM: 40

MORRIS CHARLES S
M/S 300-319
JET PROPULSION LAB
4800 OAK GROVE DRIVE
PASADENA CA 91109
U.S.A.
TEL: (818)354-8074
TLX: 675429 JPL USA
TLF:
EML: SPAN:STARS::CSM
COM:

MORRIS DAVID DR
IRAM
VOIE 10
DOMAINE UNIVERSITAIRE
F-38406 ST-MARTIN-D'HERES
FRANCE
TEL: 75-42-33-83
TLX: 980753
TLF:
EML:
COM: 40

MORRIS MARK ROOT DR
DEPT OF ASTRONOMY
MATH-SCIENCES BLDG
UCLA
LOS ANGELES CA 90024
U.S.A.
TEL: 213-825-3320
TLX:
TLF:
EML:
COM: 33.34.40.51

MORRIS MICHAEL C
ROYAL GREENWICH OBS
HERSTMONCEUX CASTLE
HAILSHAM
EAST SUSSEX BN27 1RP
U.K.
TEL: 0323-833171
TLX: 87451
TLF:
EML: JANET:UK.AC.RGO BOVEN
COM: 09C

MORRIS STEPHEN C DR
DOMINION ASTROPHYS OBS
5071 W SAANICH ROAD
RR 5
VICTORIA BC V8X 4M6
CANADA
TEL: 604-388-3976
TLX: 0497295
TLF:
EML:
COM: 25.35

MORRIS STEVEN DR
APT. 2
2860 W. 235TH STR.
TORRANCE CA 90505
U.S.A.
TEL: 213-5308708
TLX:
TLF:
EML:
COM:

MORRISON DAVID PROF
M.S. 245-1
SPACE SCIENCES DIVISION
NASA AMES RES. CENTER
MOFFETT FIELD CA 94035
U.S.A.
TEL:
TLX:
TLF:
EML:
COM: 15.16V

MORRISON LESLIE V
ROYAL GREENWICH OBS
HERSTMONCEUX CASTLE
HAILSHAM BN27 1RP
U.K.
TEL: 032-181-3171
TLX: 87451 RGOBSY G
TLF:
EML:
COM: 04.08V.19

MORRISON NANCY DUNLAP DR
DEPT PHYSICS & ASTRONOMY
UNIVERSITY OF TOLEDO
2801 W BANCROFT ST
TOLEDO OH 43606
U.S.A.
TEL: 419-537-2659
TLX:
TLF:
EML:
COM: 27.29

MORRISON PHILIP PROF
DEPT OF PHYSICS
MIT 6-205
CAMBRIDGE MA 02139
U.S.A.
TEL: 617-253-5086
TLX:
TLF:
EML:
COM: 47.48.51

MORTON DONALD C DR
HERZBERG INST ASTROPHYS
NATIONAL RESEARCH COUNCIL
100 SUSSEX DRIVE
OTTAWA ONT K1A 0R6
CANADA
TEL:
TLX:
TLF:
EML:
COM: 09.34.44

MORTON G A DR
1122 SKYCREST DR. APT 6
WALNUT CREEK CA 94595
U.S.A.
TEL: 415-933-3802
TLX:
TLF:
EML:
COM:

MOSS CHRISTOPHER DR
INSTITUTE OF ASTRONOMY
MADINGLEY ROAD
CAMBRIDGE CB3 0HA
U.K.
TEL: 0223-337548
TLX: 817297 ASTRON G
TLF:
EML:
COM: 28

MOSS DAVID L DR
MATHEMATICS DEPT
MANCHESTER UNIVERSITY
MANCHESTER M13 9PL
U.K.
TEL:
TLX:
TLF:
EML:
COM: 35

MOTTA SANTO DR
DIPT. DI MATEMATICA
CITTA UNIVERSITARIA
VIALE A. DORIA 6
I-95125 CATANIA
ITALY
TEL: 095-330533x668
TLX: 970359 ASTRCT-I
TLF:
EML:
COM: 10

MOTZ LLOYD PROF
DEPT OF ASTRONOMY
COLUMBIA UNIVERSITY
PUPIN HALL BOX 57
NEW YORK NY 10027
U.S.A.
TEL: 212-280-3279
TLX:
TLF:
EML:
COM:

MOULD JEREMY R
DIV PHYSICS & ASTRONOMY
CALTECH 105-24
PASADENA CA 91125
U.S.A.
TEL: 818-356-4168
TLX:
TLF:
EML:
COM: 28.37

MOURADIAN ZADIG M DR
OBSERVATOIRE DE PARIS
SECTION DE MEUDON
DASOP -LA326
F-92195 MEUDON PL CEDEX
FRANCE
TEL: 1-45-07-78-00
TLX:
TLF:
EML:
COM: 12

MOUSCHOVIAS TELEMACHOS CH
DEPT OF ASTRONOMY
UNIVERSITY OF ILLINOIS
1011 W. SPRINGFIELD AVE
URBANA IL 61801
U.S.A.
TEL: 217-333-3090
TLX:
TLF:
EML:
COM: 34

MOUSSAS XENOPHON PH D
NATIONAL UNIVERSITY
ASTROPHYSICS LABET
20 SKYLITSI STREET
GR-11473 ATHENS
GREECE
TEL: 7235122/8843877
TLX:
TLF:
EML:
COM: 49

MOUTSOULAS MICHAEL PROF
DEPT OF EARTH SCIENCES
UNIVERSITY OF ATHENS
GR-15784 ATHENS
GREECE
TEL: 7247569
TLX: 215255 GR
TLF:
EML:
COM: 16.51

MOVAHED REZA DR
P O BOX 6
BABOLSAR
IRAN
TEL:
TLX:
TLF:
EML:
COM:

MRKOS ANTONIN DR
DEPT OF ASTRONOMY
CHARLES UNIVERSITY
SVEDSKA 8
150 00 PRAHA 5
CZECHOSLOVAKIA
TEL:
TLX: 144307 KLET CZ
TLF:
EML:
COM: 06.15.20

MUELLER EDITH A PROF
RENNWEG 15
CH-4052 BASEL
SWITZERLAND
TEL: 061-42-31-68
TLX:
TLF:
EML:
COM: 12.38C.44

MUELLER EWALD
MPI F PHYS & ASTROPHYSIK
INSTITUT F. ASTROPHYSIK
KARL-SCHWARZSCHILD-STR. 1
D-8046 GARCHING B MUNCHEN
GERMANY. F.R.
TEL: 089-3299-0
TLX: 524629 ASTRO D
TLF:
EML:
COM: 35

MUELLER HELMUT O PROF DR
HERZOGENMUEHLESTR. 4
CH-8051 ZUERICH
SWITZERLAND
TEL: 01-41-11-47
TLX:
TLF:
EML:
COM:

MUELLER IVAN I PROF
GEODETIC SCI & SURVEYING
OHIO STATE UNIVERSITY
1958 NEIL AVENUE
COLUMBUS OH 43210-1247
U.S.A.
TEL: 614-422-2269
TLX: 245334
TLF:
EML:
COM: 04.19.31C

MUENCH GUIDO PROF
MPI FUER ASTRONOMIE
KOENIGSTUHL
D-6900 HEIDELBERG 1
GERMANY. F.R.
TEL: 06221-528210
TLX: 461 789 MPIA D
TLF:
EML:
COM: 33.34.36

MUFSON STUART LEE DR
ASTRONOMY DEPT
INDIANA UNIVERSITY
319 SWAIN WEST
BLOOMINGTON IN 47401
U.S.A.
TEL: 812-335-6917
TLX:
TLF:
EML:
COM: 34

MUINOS JOSE L DR
REAL INSTITUTO Y
OBSERVATORIO DE LA ARMADA
11110 SAN FERNANDO
CADIZ
SPAIN
TEL: 56 883548
TLX: 76108 ION E
TLF: 56 881732
EML:
COM: 08

MUKAI SONOYO DR
KANAZAWA TECHNOLOGIC INST
7-1 OGIGAOKA
NONOICHIMACHI
ISHIKAWA 921
JAPAN
TEL: 0762-48-1100
TLX: 5122456 KITLCJ
TLF:
EML:
COM: 21.36

MUKAI TADASHI DR
KANAZAWA TECHNOLOGIC INST
7-1 OGIGAOKA
NONOICHIMACHI
ISHIKAWA 921
JAPAN
TEL: 0762-48-1100
TLX: 5122456 KITLCJ
TLF:
EML:
COM: 15.21C

MUKHERJEE KRISHNA
PHYSICS & ASTRONOMY DEPT
MALOTT HALL
UNIVERSITY OF KANSAS
LAWRENCE KS 66045
U.S.A.
TEL: 913 864 4030
TLX:
TLF:
EML:
COM:

MULDERS GERARD F W
4519 EVERETT STREET
KENSINGTON MD 20895
U.S.A.
TEL: 301-564-0090
TLX:
TLF:
EML:
COM:

MULHOLLAND J DERRAL DR
SPACE ASTRONOMY LAB
UNIVERSITY OF FLORIDA
1810 NW 6TH ST
GAINESVILLE FL 32609
U.S.A.
TEL: 904-392-5450
TLX: 810-825-2308
TLF:
EML:
COM: 07.16.20

MULLALY RICHARD F DR
SCHOOL OF ELECTRICAL
ENGINEERING
UNIVERSITY OF SYDNEY
SYDNEY NSW 2006
AUSTRALIA
TEL:
TLX:
TLF:
EML:
COM:

MULLAN DERMOTT J DR
BARTOL RES FOUNDATION
UNIVERSITY OF DELAWARE
NEWARK DE 19716
U.S.A.
TEL: 301-398-1163
TLX: 5106660805
TLF:
EML:
COM:

MULLER A B DR
THOMASLAAN 40
NL-5631 GW EINDHOVEN
NETHERLANDS
TEL: 040-430322
TLX:
TLF:
EML:
COM: 25

MULLER C A PROF JR
ODINKSVELD 8
NL-7491 HB DELDEN
NETHERLANDS
TEL: 05407-2428
TLX:
TLF:
EML:
COM:

MULLER PAUL
1 RUE CHAUVAIN
F-06000 NICE
FRANCE
TEL:
TLX:
TLF:
EML:
COM: 26

MULLER RICHARD A
LAWRENCE BERKELEY LAB
BLDG 50, RM 232
BERKELEY CA 94720
U.S.A.
TEL: 415-486-5235
TLX:
TLF:
EML:
COM: 47.51

MULLER RICHARD DR
OBSERVATOIRE PIC-DU-MIDI
F-65300 BAGNERES-DE-B.
FRANCE
TEL: 62-95-00-69
TLX:
TLF:
EML:
COM: 10.12C

MUMFORD GEORGE S PROF
DEPARTMENT OF EDUCATION
TUFTS UNIVERSITY
FILENE CENTER
MEDFORD MA 02155
U.S.A.
TEL: 617-653-6923
TLX:
TLF:
EML:
COM: 25.27.42

MUMMA MICHAEL JON
NASA/GSFC
CODE 693
GREENBELT MD 20771
U.S.A.
TEL: 301-286-6994
TLX:
TLF:
EML:
COM: 14.15.16

MUNDT REINHARD DR
MAX PLANCK INSTITUT
FUR ASTRONOMIE
KONIGSTUHL
6900 HEIDELBERG
GERMANY. F.R.
TEL: 62215281/528227
TLX: 461789
TLF:
EML:
COM: 34

MUNDY LEE G DR
ASTRONOMY DEPT 105 24
CALTECH
PASADENA CA 91125
U.S.A.
TEL: 818 356 4991
TLX: 675425 CALTECH PSD
TLF:
EML:
COM: 40

MUNOZ-TUNON CASTANA
INSTITUTO DE ASTROFISICA
DE CANARIAS
38200 LA LAGUNA TENERIFE
SPAIN
TEL: 34 22 26 22 11
TLX: 92640
TLF:
EML:
COM: 28

MUNRO RICHARD E DR
2378 DENNISON LANE
BOULDER CO 80302
U.S.A.
TEL:
TLX:
TLF:
EML:
COM: 12

MURAKAMI TOSHIO
INST SPACE & ASTRON SCI
4-6-1 KOMABA
MEGURO-KU
TOKYO 153
JAPAN
TEL: 03-467-1111x303
TLX: 34757 ISASTRO J
TLF:
EML:
COM: 48

MURDIN PAUL G DR
ROYAL GREENWICH
OBSERVATORY
HERSTMONCEUX CASTLE
HAILSHAM BN27 1RP
U.K.
TEL:
TLX:
TLF:
EML:
COM: 27.50V

MURDOCH HUGH S DR
ASTROPHYSICS DEPT
UNIVERSITY OF SYDNEY
SYDNEY NSW 2006
AUSTRALIA
TEL: 02-692-2222
TLX: 26169 UNISYD
TLF:
EML:
COM: 40

MURDOCK THOMAS LEE
GENERAL RESEARCH CORP.
1891 PROFESSIONAL BLDG
LIBERTY SQUARE
DANVERS MA 01923
U.S.A.
TEL: 617-771-6584
TLX:
TLF:
EML:
COM: 44

MUREDDU LEONARDO DR
STAZIONE ASTRONOMICA
VIA OSPEDALE 72
CAGLIARI 09100
ITALY
TEL: 070 663544
TLX:
TLF:
EML:
COM:

MURPHY ROBERT E DR
NASA HEADQUARTERS
CODE EEL
WASHINGTON DC 20546
U.S.A.
TEL: 202-453-1720
TLX: 89530 NASA WSH
TLF:
EML:
COM: 15

MURRAY CARL D
SCHOOL OF MATHEMAT. SCI.
QUEEN MARY COLLEGE
MILE END ROAD
LONDON E1 4NS
U.K.
TEL: 01-980-4811
TLX: 893750
TLF:
EML:
COM: 20

MURRAY JOHN B DR
UNIVERSITY OF LONDON
OBSERVATORY
MILL HILL PARK
LONDON NW7 2QS
U.K.
TEL:
TLX:
TLF:
EML:
COM:

MURRAY STEPHEN S DR
HARVARD-SMITHSONIAN
CENTER FOR ASTROPHYSICS
60 GARDEN STREET
CAMBRIDGE MA 02138
U.S.A.
TEL: 617-495-7205
TLX: 921428 SATELLITE CAM
TLF:
EML:
COM: 09 28

MUSEN PETER DR
8804 ORBIT LANE
LANHAM MD 20801
U.S.A.
TEL: 301-552-3848
TLX:
TLF:
EML:
COM: 07

MUSIELAK ZDZISLAW E DR
SPACE SCIENCE LAB ES52
NASA/MARSHALL SPACE
FLIGHT CENTER
HUNTSVILLE AL 35812
U.S.A.
TEL: 205 544 7619
TLX: 594416
TLF:
EML: SPAN:SSL::MUSIELAK
COM: 10.36

MUSMAN STEVEN DR
NATIONAL GEODETIC SURVEY
CHARTING & GEOD. SURVEY
NOS/NOAA - N/CG112
ROCKVILLE MD 20852
U.S.A.
TEL:
TLX:
TLF:
EML:
COM:

MUTEL ROBERT LUCIEN
DEPT PHYSICS & ASTRONOMY
UNIVERSITY OF IOWA
IOWA CITY IA 52242
U.S.A.
TEL: 319-353-7205
TLX:
TLF:
EML:
COM: 40

MUTSCHLECNER J PAUL DR
ASTRONOMY DEPT
INDIANA UNIVERSITY
SWAIN HALL WEST
BLOOMINGTON IN 47405
U.S.A.
TEL:
TLX:
TLF:
EML:
COM: 36

MUXLOW THOMAS
UNIVERSITY OF MANCHESTER
NUFFIELD RADIO ASTR LABS
JODRELL BANK
MACCLESFIELD SK11 9DL
U.K.
TEL: 0477-71321
TLX: 36149
TLF:
EML:
COM: 40

MUZZIO JUAN C PROF
HARVARD SMITHSONIAN CENT.
FOR ASTROPHYSICS
60 GARDEN STREET
CAMBRIDGE MA 02138
U.S.A.
TEL:
TLX:
TLF:
EML:
COM: 28.33 37.46

MYACHIN VLADIMIR F DR
INSTITUTE OF THEORETICAL
ASTRONOMY
10 KUTUZOV QAY
192187 LENINGRAD
U.S.S.R.
TEL:
TLX:
TLF:
EML:
COM: 07

MYERS PHILIP C
HARVARD-SMITHSONIAN CTR
FOR ASTROPHYSICS, MS 42
60 GARDEN STREET
CAMBRIDGE MA 02138
U.S.A.
TEL:
TLX:
TLF:
EML:
COM: 34.40

NACOZY PAUL E DR
FEDERAL SPACE SYSTEMS
PO BOX 26712
AUSTIN TX 78755
U.S.A.
TEL: 512-467-6659
TLX:
TLF:
EML:
COM: 07.20

NADAL ROBERT
OBSERV. MIDI-PYRENEES
14 AVENUE EDOUARD BELIN
31400 TOULOUSE
FRANCE
TEL: 61 25 21 01
TLX: 530776 F
TLF:
EML:
COM: 41

NADEAU DANIEL DR
DEPARTEMENT DE PHYSIQUE
UNIVERSITE DE MONTREAL
C P 6128 SUCC A
MONTREAL PQ H3C 3J7
CANADA
TEL: 514-343-6676
TLX:
TLF:
EML:
COM: 40

NADOLSCHI V PROF DR
COM.ARDEOANI OF TESCANI
JUD
BACAU
ROMANIA
TEL:
TLX:
TLF:
EML:
COM:

NADYOZHIN D K DR
INSTITUTE OF THEORETICAL
& EXPERIMENTAL PHYSICS
CHEREMUSHKINSKAJA 25
117259 MOSCOW
U.S.S.R.
TEL: [illegible]
TLX: 411059 CERTI SU
TLF:
EML:
COM: 35

NAGASAWA SHINGO PROF
TOKYO SCIENCE UNIVERSITY
2641 YAMAZAKI
HIGASHI KAMEYAMA
NODASHI 278
JAPAN
TEL: 0471-24-1501
TLX:
TLF:
EML:
COM: 10

NAGASE FUMIAKI DR
DEPT OF ASTROPHYSICS
NAGOYA UNIVERSITY
CHIKUSA-KU, FURO-CHO
NAGOYA 464
JAPAN
TEL: 052-781-5111
TLX: 4477323 SCUNAG J
TLF:
EML:
COM:

NAGIRNER DMITRIJ I DR
LENINGRAD UNIVERSITY
ASTRONOMICAL OBSERVATORY
BIBLIOTECHNAJA PL. 2
198904 LENINGRAD-PETRODV.
U.S.S.R.
TEL: 257-94-89
TLX:
TLF:
EML:
COM: 36

NAGNIBEDA VALERY G DR
LENINGRAD UNIVERSITY
ASTRONOMICAL OBSERVATORY
BIBLIOTECHNAJA PL. 2
198904 LENINGRAD
U.S.S.R.
TEL: 257-94-91
TLX:
TLF:
EML:
COM: 40

NAHON FERNAND PROF
25 AVENUE DE L'EUROPE
F-92310 SEVRES
FRANCE
TEL: 1-45-34-18-05
TLX:
TLF:
EML:
COM: 07.33

NAIDENOV VICTOR O
A.F. IOFFE PHYS TECH INST
USSR ACADEMY OF SCIENCES
POLYTECHNICHESKAYA 26
194021 LENINGRAD
U.S.S.R.
TEL:
TLX:
TLF:
EML:
COM: 48

NAKADA YOSHIKAZU DR
DEPT OF ASTRONOMY
FACULTY OF SCIENCE
UNIV TOKYO, BUNKYO-KU
TOKYO 113
JAPAN
TEL: 03-812-2111
TLX:
TLF:
EML:
COM: 34

NAKAGAWA NAOYA DR
UNIVERSITY OF ELECTRO-
COMMUNICATIONS
CHOFU-SHI
TOKYO 182
JAPAN
TEL: 0424-83-2161
TLX: 2832446 UEC J
TLF:
EML:
COM:

NAKAGAWA YOSHINARI DR
CHIBA INST OF TECHNOLOGY
NARASHINO 275
JAPAN
TEL: 0474-75-2111
TLX:
TLF:
EML:
COM: 10.49.51

NAKAGAWA YOSHITSUGU DR
GEOPHYSICAL INSTITUTE
FACULTY OF SCIENCE
UNIVERSITY OF TOKYO
TOKYO 113
JAPAN
TEL: 3-812-2111x4311
TLX:
TLF:
EML:
COM: 16

NAKAI YOSHIHIRO
KWASAN AND HIDA
OBSERVATORIES
KYOTO UNIVERSITY
KYOTO 607
JAPAN
TEL: 075-581-1235
TLX:
TLF:
EML:
COM: 39

NAKAJIMA HIROSHI
NOBEYAMA SOLAR RADIO OBS
TOKYO ASTRON OBSERVATORY
MINAMIMAKI-MURA
NAGANO 384-13
JAPAN
TEL: 267-98-2014
TLX: 3329005 TAONRO J
TLF:
EML:
COM: 10

NAKAJIMA KOICHI DR
FACULTY OF ECONOMICS
HITOTSUBASHI UNIVERSITY
NAKA 2-1 KUNITACHI
TOKYO 186
JAPAN
TEL: 435 72 1101
TLX: 2842107
TLF:
EML:
COM: 08

NAKAMURA TAKASHI DR
DEPT OF PHYSICS
KYOTO UNIVERSITY
KYOTO 606
JAPAN
TEL:
TLX:
TLF:
EML:
COM: 35

NAKAMURA TSUKO DR
TOKYO ASTRONOMICAL OBS
OSAWA MITAKA
TOKYO 181
JAPAN
TEL: 03-812-2111
TLX:
TLF:
EML:
COM: 15 20

NAKAMURA YASUHISA
DEPT OF SCINECE EDUCATION
FACULTY OF EDUCATION
FUKUSHIMA UNIVERSITY
FUKUSHIMA 960-12
JAPAN
TEL: 0245485151/421
TLX:
TLF:
EML:
COM: 4.

NAKANO SABURO DR
KOBINATO 1-11-7
BUNKYO-KU
TOKYO 112
JAPAN
TEL:
TLX:
TLF:
EML:
COM:

NAKANO STUICHI
SMITHSONIAN OBS. M.S. 18
60 GARDEN STREET
CAMBRIDGE MA 02138
U.S.A.
TEL: 617-495-7212
TLX: 7103206842
TLF:
EML:
COM: 06.30

NAKANO TAKENORI DR
DEPT OF PHYSICS
KYOTO UNIVERSITY
KITASHIRAKAWA OIWAKE
SAKYOKU KYOTO 606
JAPAN
TEL: 075-751-2111
TLX:
TLF:
EML:
COM: 35

NAKAYAMA SHIGERU DR
FACULTY GENERAL EDUC
UNIVERSITY OF TOKYO
KOMABA MEGURO
TOKYO 153
JAPAN
TEL: 03-467-1171
TLX:
TLF:
EML:
COM: 41

NAKAZAWA KIYOSHI DR
TOKYO INST. OF TECH.
OHOKAYAMA 2-12-1
MEGUROKU TOKYO 152
JAPAN
TEL:
TLX:
TLF:
EML:
COM: 22.35

NAMBA OSAMU DR
STERREKUNDIG INSTITUUT
PRINCETONPLEIN 5
POSTBUS 80 000
NL-3508 TA UTRECHT
NETHERLANDS
TEL: 030-535200
TLX: 40048 FYLUT NL
TLF:
EML:
COM: 10 12

NAN REN-DONG
BEIJING ASTRONOMICAL OBS
ACADEMIA SINICA
BEIJING
CHINA, PEOPLE'S REP.
TEL: 28 1698
TLX: 22040 BAOBS CN
TLF:
EML:
COM: 40

NANDY KASHINATH DR
ROYAL OBSERVATORY
BLACKFORD HILL
EDINBURGH EH9 3HJ
U.K.
TEL: 31-667-3321
TLX: 72383 ROEDING
TLF:
EML:
COM: 34 45

NAPIER WILLIAM M DR
ROYAL OBSERVATORY
BLACKFORD HILL
EDINBURGH EH9 3HJ
U.K.
TEL: 031-667-3321
TLX: 72383 ROEDIN G
TLF:
EML:
COM: 15.22

NAQVI S I H PROF
DEPT PHYSICS & ASTRONOMY
UNIVERSITY OF REGINA
REGINA S4S 0A2
CANADA
TEL: 306-584-4262
TLX: 071-2683 U R REG
TLF:
EML:
COM:

NARANAN S PROF
TATA INSTITUTE OF
FUNDAMENTAL RESEARCH
HOMI BHABHA RD
BOMBAY 400 005
INDIA
TEL: 219-111
TLX: 0113009
TLF:
EML:
COM:

NARASIMHA DELAMPADY DR
TATA INST FUND RESEARCH
HOMI BHABHA ROAD
BOMBAY 400005
INDIA
TEL: 4952311
TLX: 3009 IN
TLF:
EML:
COM: 35.36.47

NARAYAN RAMESH DR
STEWARD OBSERVATORY
UNIVERSITY OF ARIZONA
TUCSON AZ 85721
U.S.A.
TEL: 602-621-2560
TLX: 467175
TLF:
EML:
COM:

NARAYANA J V
REGIONAL METEOROLOGICAL
OFFICE
4 COLLEGE ROAD
MADRAS 600 006
INDIA
TEL:
TLX:
TLF:
EML:
COM:

NARIAI HIDEKAZU PROF
NATIONAL ASTRONOMICAL OBS
MITAKA
TOKYO 181
JAPAN
TEL: 08462-2-2362
TLX:
TLF:
EML:
COM: 35.47

NARIAI KYOJI DR
TOKYO ASTRONOMICAL
OBSERVATORY
MITAKA
TOKYO 181
JAPAN
TEL: 0422-32-5111
TLX:
TLF:
EML:
COM: 36.42

NARITA SHINJI DR
DOSHISHA UNIVERSITY
KYOTO 602
JAPAN
TEL:
TLX:
TLF:
EML:
COM: 35

NARLIKAR JAYANT V PROF
TATA INSTITUTE OF
FUNDAMENTAL RESEARCH
HOMI BHABHA RD
BOMBAY 400 005
INDIA
TEL: BOMBAY 219111
TLX: 0113009 TIFR IN
TLF:
EML:
COM: 28.47C

NATALI GIULIANO DR
LAB. ASTROFISICA SPAZIALE
C P 67
I-00044 FRASCATI
ITALY
TEL:
TLX:
TLF:
EML:
COM:

NATHER R EDWARD
DEPT OF ASTRONOMY
UNIVERSITY OF TEXAS
AUSTIN TX 78712
U.S.A.
TEL:
TLX:
TLF:
EML:
COM: 27 42

NATTA ANTONELLA DR
CENTRO PER ASTRONOMIA IR
LARGO E FERMI 5
I-50125 FIRENZE
ITALY
TEL: 055-2752239
TLX: 572268 ARCETRI
TLF:
EML:
COM:

NAUKOV VITALIJ A DR
PULKOVO OBSERVATORY
196140 LENINGRAD
U.S.S.R.
TEL: 2982242
TLX:
TLF:
EML:
COM: 14 31

NAWAR SAMIR DR
HELWAN OBSERVATORY
HELWAN
EGYPT
TEL:
TLX:
TLF:
EML:
COM: [illegible]

NECKEL HEINZ DR
HAMBURGER STERNWARTE
GOJENBERGSWEG 112
D-2050 HAMBURG 80
GERMANY. F.R.
TEL: 49-40-7252-4110
TLX: 217884 HAMST D
TLF:
EML:
COM: 12.29

NECKEL TH DR
MPI FUER ASTRONOMIE
KOENIGSTUHL
D-6900 HEIDELBERG
GERMANY. F.R.
TEL: 06221-528288
TLX: 461789 MPIA D
TLF:
EML:
COM: 33

NEE TSU-WEI DR
INST PHYSICS & ASTRONOMY
NATL CENTRAL UNIVERSITY
CHUNG-LI
CHINA TAIWAN
TEL:
TLX:
TLF:
EML:
COM:

NEEMAN YUVAL PROF
DEPT PHYSICS & ASTRONOMY
TEL-AVIV UNIVERSITY
TEL-AVIV 69978
ISRAEL
TEL: 03-425411
TLX: 342171 VERSY IL
TLF:
EML:
COM: 47.48

NEFEDEVA ANTONINA I PROF
ENGELHARDT ASTRONOMICAL
OBSERVATORY
OBSERVATORY STATION
422526 KAZAN
U.S.S.R.
TEL: [illegible]4827
TLX:
TLF:
EML:
COM: 08

NEFF JOHN S
UNIVERSITY OF IOWA
605 BROOKLAND PARK DRIVE
IOWA CITY IA 52240
U.S.A.
TEL: 319-353-4340
TLX:
TLF:
EML:
COM: 15.27.36

NEIDIG DONALD F DR
AIR FORCE GEOPHYSICS LAB
NATIONAL SOLAR OBS
SUNSPOT NM 88349
U.S.A.
TEL: 505 434 7000
TLX: 0666 484 NOAC TUC
TLF:
EML: BITNET:DNEIDIG@SUNSPOT.NOAO.ED
COM: 13

NELSON ALISTAIR H DR
DEPT OF APPLIED MATHS
& ASTRONOMY
UNIVERSITY COLLEGE
CARDIFF
U.K.
TEL: 0222-44211x2668
TLX: 498635 ULIBCFG
TLF:
EML:
COM: 33

NELSON BURT DR
DEPT OF ASTRONOMY
SAN DIEGO STATE UNIV
SAN DIEGO CA 92182
U.S.A.
TEL: 619-265-6175
TLX:
TLF:
EML:
COM: 42.50

NELSON GRAHAM JOHN DR
CSIRO
DIVISION OF RADIOPHYSICS
P.O.BOX 76
EPPING NSW 2121
AUSTRALIA
TEL: 02-868-0222
TLX: 26230
TLF:
EML:
COM: 10

NELSON JERRY
LAWRENCE BERKELEY LAB
UNIVERSITY OF CALIFORNIA
BLDG 50. RM 351
BERKELEY CA 94720
U.S.A.
TEL: 415-486-5915
TLX:
TLF:
EML:
COM: 09

NELSON ROBERT M
183-501 JPL
4800 OAK GROVE DRIVE
PASADENA CA 91109
U.S.A.
TEL: [illegible]-354-68939
TLX:
TLF:
EML:
COM:

NEMEC JAMES
DEPT GEOPHYSICS & ASTRON
UNIV OF BRITISH COLUMBIA
VANCOUVER BC V6T 1W5
CANADA
TEL: 604-[illegible]-4517
TLX:
TLF:
EML:
COM: 17

NEMIRO ANDREJ A DR PROF
PULKOVO OBSERVATORY
196140 LENINGRAD
U.S.S.R.
TEL: 2982242
TLX:
TLF:
EML:
COM: 08

NEMIROFF ROBERT DR
RACAH INST OF PHYSICS
HEBREW UNIV OF JERUSALEM
JERUSALEM GIVAT RAM
ISRAEL
TEL: (02)584 928
TLX:
TLF:
EML:
COM:

NESCI ROBERTO
ISTITUTO ASTRONOMICO
UNIVERSITA DI ROMA
VIA LANCISI 29
I-00161 ROMA
ITALY
TEL: 6-867-525
TLX: 611255 INFNRO
TLF:
EML:
COM: [illegible]

NESIS ANASTASIOS DR
KIEPENHEUER INSTITUT FUR
SONNENPHYSIK
SCHOENECKSTRASSE 6
D 7800 FREIBURG
GERMANY. F.R.
TEL: 0761 [illegible]
TLX: 7721552 KISD
TLF: 761-[illegible]
EML:
COM: [illegible]

NESS NORMAN F DR
BARTOL RESEARCH INST.
UNIVERSITY OF DELAWARE
NEWARK. DE 19716
U.S.A.
TEL: 302-451-8116
TLX: 510-6665
TLF:
EML:
COM: 16.44

NETZER HAGAI DR
SCHOOL OF PHY & ASTRON
TEL AVIV UNIVERSITY
RAMAT AVIV
TEL AVIV 69978
ISRAEL
TEL: 972-3-5450108
TLX: 342171 VERST IL
TLF:
EML: BITNET:H510TAUNOS
COM: 29

NEUGEBAUER GERRY DR
PHYSICS DEPARTMENT
CALIF INST OF TECHNOLOGY
320 DOWNS
PASADENA CA 91125
U.S.A.
TEL: 818-356-4284
TLX: 675425
TLF:
EML:
COM: 34

NEUKUM G DR
D F V L R
NE-OE-PF
D-8031 WESSLING
GERMANY. F.R.
TEL: 8153-28731
TLX: 0526419 DVLOP D
TLF:
EML:
COM: 15.16

NEUPERT WERNER M DR
NASA/GSFC
CODE 680
GREENBELT MD 20771
U.S.A.
TEL: 301-286-8169
TLX:
TLF:
EML:
COM: 10.44

NEUZIL LUDEK DR
ASTRONOMICAL INSTITUTE
CZECH. ACAD. OF SCIENCES
OBSERVATORY
251 65 ONDREJOV
CZECHOSLOVAKIA
TEL:
TLX:
TLF:
EML:
COM: 22

NEWBURN RAY L JR
3226 EMERALD ISLE DRIVE
GLENDALE CA 91206
U.S.A.
TEL:
TLX:
TLF:
EML:
COM: 15.22

NEWELL EDWARD B DR
MT STROMLO & SIDING
SPRING OBSERVATORIES
PRIVATE BAG
WODEN ACT 2606
AUSTRALIA
TEL: 062-881111
TLX: AA 62270 CANOPUS
TLF:
EML:
COM: 37

NEWHALL X X DR
JPL 238 332
PASADENA CA 91109-8099
U.S.A.
TEL: 818 354 0000
TLX: 192961003
TLF:
EML: SPAN:LOGOS::XXN
COM: 04.19 31

NEWMAN MICHAEL JOHN DR
LOS ALAMOS NATIONAL LAB
X-2 MS B220
PO BOX 1663
LOS ALAMOS NM 87545
U.S.A.
TEL: 505-667-7636
TLX:
TLF:
EML:
COM: 35

NEWSOM GERALD H PROF
ASTRONOMY DEPT
OHIO STATE UNIVERSITY
174 W. 18TH AVENUE
COLUMBUS OH 43210
U.S.A.
TEL: 614-422-7082
TLX:
TLF:
EML:
COM: 14.42

NEWTON ROBERT R DR
JOHNS HOPKINS UNIVERSITY
APPLIED PHYSICS LAB
JOHNS HOPKINS RD
LAUREL MD 20707
U.S.A.
TEL: 301-953-7100
TLX:
TLF:
EML:
COM:

NEY EDWARD P PROF
DEPT OF ASTRONOMY
TATE LAB OF PHYSICS
UNIVERSITY OF MINNESOTA
MINNEAPOLIS MN 55455
U.S.A.
TEL: 612-373-4687
TLX:
TLF:
EML:
COM: 21

NGUYEN-QUANG RIEU DR
OBSERVATOIRE DE PARIS
SECTION DE MEUDON
F-92195 MEUDON PL CEDEX
FRANCE
TEL: 1-45-34-75-30
TLX: 270912
TLF:
EML:
COM: 34.40

NHA IL-SEONG DR
YONSEI UNIVERSITY OBS
134 SINCHON-DONG
SEODAEMUN-KU
SEOUL 120
KOREA. REPUBLIC
TEL: 392-0131
TLX:
TLF:
EML:
COM: 38.42

NIARCHOS PANAYIOTIS PH D
DEPT OF ASTRONOMY
UNIVERSITY OF ATHENS
PANEPISTIMIOPOLIS
GR-15771 ATHENS
GREECE
TEL:
TLX:
TLF:
EML:
COM: 27.51

NICHOLLS RALPH W PROF
CRESS. DEPT OF PHYSICS
YORK UNIVERSITY
4700 KEELE STREET
NORTH YORK ONT M3J1P3
CANADA
TEL: 416-667-383
TLX: 06524736
TLF:
EML:
COM: 14.[illegible]

NICHOLSON WILLIAM
ROYAL GREENWICH OBS
HERSTMONCEUX CASTLE
HAILSHAM BN27 1RP
U.K.
TEL:
TLX:
TLF:
EML:
COM: 24

NICOLACI DA COSTA LUIZ-A.
OBSERVATORIO NACIONAL
RUA GENERAL BRUCE 586
SAO CRISTOVAO
20921 RIO DE JANEIRO
BRAZIL
TEL: 021-580-7313
TLX: 021-21288
TLF:
EML:
COM:

NICOLAIDIS EFTHYMIOS DR
NATL RESEARCH FOUNDATION
48 VAS CONSTANTINOU AVE
11635 ATHENS
GREECE
TEL: 30 1 7210554
TLX: 224064
TLF: 7212729
EML:
COM: 41

NICOLAS KENNETH ROBERT
NAVAL RESEARCH LABORATORY
CODE 4163
OVERLOOK AVENUE
WASHINGTON DC 20375
U.S.A.
TEL: 202-767-2517
TLX:
TLF:
EML:
COM: 12

NICOLSON GEORGE D DR
HARTEBEESTHOEK RADIO ASTR
PO BOX 443
KRUGERSDORP 1740
SOUTH AFRICA
TEL: 11-642 4692
TLX: 321006 HART SA
TLF:
EML:
COM: 40C

NIEMI AIMO
TURUN YLIOPISTO
ITAINEN PITKAKATU 1
SF-20520 TURKU 52
FINLAND
TEL: 358-21-435 822
TLX: 62638 TYY SF
TLF: 358-21-433 767
EML:
COM: 09.19

NIKITIN A A DR
LENINGRAD STATE UNIV
ASTRONOMICAL OBSERVATORY
198904 LENINGRAD
U.S.S.R.
TEL: 293-22-62
TLX:
TLF:
EML:
COM: 29

NILSON PETER DR
ASTRONOMICAL OBSERVATORY
BOX 515
S-751 20 UPPSALA
SWEDEN
TEL:
TLX:
TLF:
EML:
COM:

NICOLET BERNARD
OBSERVATOIRE DE GENEVE
CH-1290 SAUVERNY
SWITZERLAND
TEL: 22-552611
TLX: 27720 OBSG CH
TLF:
EML:
COM: 25.45

NIEDNER MALCOLM B DR
LABORATORY FOR ASTRONOMY
& SOLAR PHYSICS
NASA/GSFC
GREENBELT MD 20771
U.S.A.
TEL:
TLX:
TLF:
EML:
COM: 15

NIETO JEAN-LUC
OBSERVATOIRE DE TOULOUSE
14 AVENUE EDOUARD BELIN
F-31400 TOULOUSE
FRANCE
TEL: 61-25-21-01
TLX: 530776 F
TLF:
EML:
COM: 28

NIKOGHOSSIAN ARTHUR G DR
BYURAKAN ASTROPHYSICAL
OBSERVATORY
378433 BYURAKAN ARMENIA
U.S.S.R.
TEL: 283453
TLX: 411576 ASCON SU
TLF:
EML:
COM: 36

NILSSON CARL DR
SMITHSONIAN ASTROPHYSICAL
OBSERVATORY
60 GARDEN STREET
CAMBRIDGE MA 02138
U.S.A.
TEL:
TLX:
TLF:
EML:
COM:

NICOLET MARCEL PROF
INST D'AERONOMIE SPATIALE
30 AVE DEN DOORN
B-1180 BRUXELLES
BELGIUM
TEL: 322-3742948
TLX: 21563 ESPACE B
TLF:
EML:
COM: 12.21

NIELL ARTHUR E DR
HAYSTACK OBSERVATORY
WESTFORD MA 01886
U.S.A.
TEL: 617-692-4764
TLX: 948149
TLF:
EML:
COM:

NIEUWENHUIJZEN HANS DR
STERREKUNDIG INSTITUUT
PRINCETONPLEIN 5
POSTBUS 80 000
NL-3508 TA UTRECHT
NETHERLANDS
TEL: 030-535237
TLX: 40048 FYLUT NL
TLF:
EML: BITNET:HNHANSENOHUTRUUO
COM: 29

NIKOLOFF IVAN DR
PERTH OBSERVATORY
6 PEOPLES AVENUE
GOOSEBERRY HILL 6076
AUSTRALIA
TEL: 2931865
TLX:
TLF:
EML:
COM: 08

NINKOVIC SLOBODAN
ASTRONOMICAL OBSERVATORY
VOLGINA 7
YU-11050 BEOGRAD
YUGOSLAVIA
TEL: 011-419-357
TLX:
TLF:
EML:
COM: 33.38

NICOLOV NIKOLAI S DR
UNIVERSITY OF SOFIA
DEPT OF ASTRONOMY
ANTON IVANOV STR 5
1126 SOFIA
BULGARIA
TEL: 51-24-05
TLX:
TLF:
EML:
COM: 46

NIEMELA VIRPI S DR
CALLE 51 ESQ 11
1894 VILLA ELISA (Bs.As.)
ARGENTINA
TEL:
TLX:
TLF:
EML:
COM:

NIIMI YUKIO
TOKYO ASTRONOMICAL OBS
2-21-1 OSAWA
MITAKA
TOKYO 181
JAPAN
TEL: 0422-32-5111
TLX: 2822307 TAOMTK J
TLF:
EML:
COM: 31

NIKOLOV ANDREJ DR
UNIVERSITY OF SOFIA
DEPT OF ASTRONOMY
ANTON IVANOV STR 5
1126 SOFIA
BULGARIA
TEL: 62561x375
TLX:
TLF:
EML:
COM: 27

NISHI KEIZO DR
TOKYO ASTRONOMICAL
OBSERVATORY
OSAWA MITAKA
TOKYO 181
JAPAN
TEL: 0422-32-5111
TLX: 02822307 TAOMTK J
TLF:
EML:
COM: 10.37

NISHIDA MINORU PROF
DEPT OF PHYSICS
KYOTO UNIVERSITY
KITASHIRAKAWA OIWAKE
SAKYOKU KYOTO 606
JAPAN
TEL:
TLX: 5422693 LIBKYU J
TLF:
EML:
COM: 33.35.47

NISHIDA MITSUGU
DEPT OF LITERATURE
KOBE WOMEN'S UNIVERSITY
SUMA-KU
KOBE 654
JAPAN
TEL: 078-731-4416
TLX:
TLF:
EML:
COM: 33

NISHIMURA JUN DR
INST SPACE & AERON SCI
6-1 KOMABA 4-CHOME
MEGURO-KU
TOKYO 153
JAPAN
TEL: 03-467-1111x388
TLX: J 24550 SPACETKY
TLF:
EML:
COM:

NISHIMURA MASAKI
DEPT OF PHYSICS
HOKKAIDO UNIVERSITY
NISHI 8-CHOME KITA 10-JYO
SAPPORO. HOKKAIDO 060
JAPAN
TEL: 11-71--2111
TLX:
TLF:
EML:
COM:

NISHIMURA SHIRO DR
TOKYO ASTRONOMICAL
OBSERVATORY
OSAWA MITAKA
TOKYO 181
JAPAN
TEL: 422-32-5111
TLX: 2822307 TAOMTK J
TLF:
EML:
COM: 05.09.29

NISHIMURA TETSUO DR
STEWARD OBSERVATORY
UNIVERSITY OF ARIZONA
TUCSON AZ 85721
U.S.A.
TEL: 602-621-2054
TLX: 467175
TLF:
EML:
COM: 21

NISSEN POUL E PROF
INSTITUTE OF ASTRONOMY
UNIVERSITY OF AARHUS
LANGELANDSGADE
DK-8000 AARHUS C
DENMARK
TEL: 06-128899
TLX: 64767 AAUSCI DK
TLF:
EML:
COM: 29.37

NITTMAN JOHANN
ETUDES ET FABRICATION
DOWELL SCHLUMBERGER
BP 90
F-42003 ST-ETIENNE
FRANCE
TEL: 77-37-64-23
TLX:
TLF:
EML:
COM:

NITYANANDA R DR
RAMAN RESEARCH INSTITUTE
BANGALORE 560 080
INDIA
TEL: 812-3601226
TLX: 8452671 RRI IN
TLF:
EML:
COM: 28.48

NOBILI ANNA M
ISTITUTO DI MATEMATICA
UNIVERSITA' DI PISA
VIA BUONARROTI 2
I-56100 PISA
ITALY
TEL:
TLX:
TLF:
EML:
COM: 07.19.20

NOBILI L DR
DIPT. DI FISICA
G. GALILEI
VIA MARZOLO 8
I-35131 PADOVA
ITALY
TEL: 049-844205/111
TLX: 430308 DFGGPDI
TLF:
EML:
COM:

NOCERA LUIGI DR
ISTITUTO DI FISICA
ATOMICA E MOLECOLARE
VIA GIARDINO 7
56127 PISA
ITALY
TEL: 39 50 501384
TLX:
TLF: 39 50 25175
EML: BISTAB @ ICNUCEVM
COM: 10

NOCI GIANCARLO PROF
DIPARTIM. DI ASTRONOMIA
UNIVERSITA DEGLI STUDI
LARGO E. FERMI 5
I-50125 FIRENZE
ITALY
TEL: 055-27521
TLX: 572268 ARCETR I
TLF:
EML:
COM:

NOEL FERNANDO
DEPTO DE ASTRONOMIA
UNIVERSIDAD DE CHILE
CASILLA 36-D
SANTIAGO
CHILE
TEL: 229-4101
TLX: 40853
TLF:
EML:
COM: 08C.31

NOELS ARLETTE DR
50 AVENUE DE LA PAIX
BOITE 063
B-4030 GRIVEGNEE
BELGIUM
TEL: 41-52-9980/7517
TLX:
TLF:
EML:
COM: 35.46

NOERDLINGER PETER D PROF
MICROCOSM. INC.
23720 ARLINGTON AV.
SUITE 5
TORRANCE CA 90501
U.S.A.
TEL: (213)539-9444
TLX:
TLF:
EML:
COM: 47

NOGUCHI MASAFUMI DR
NATL ASTRON. OBSERVATORY
MITAKA
TOKYO 181
JAPAN
TEL: 0422 32 5111
TLX: 2822307 TAOMTK J
TLF:
EML:
COM: 28

NOLLEZ GERARD DR
INSTITUT D'ASTROPHYSIQUE
98 BIS BOULEVARD ARAGO
F-75014 PARIS
FRANCE
TEL: 1-43-20-14-25
TLX:
TLF:
EML:
COM: 14

NOMOTO KEN'ICHI DR
DEPT EARTH SCI/ASTRONOMY
COLLEGE ARTS & SCIENCES
UNIVERSITY OF TOKYO
MEGURO-KU TOKYO 153
JAPAN
TEL: 03-467-1171
TLX: 25510 UNITOKYO
TLF:
EML:
COM: 35C.48

NOONAN THOMAS W PROF
PHYSICS DEPT
SUNY
BROCKPORT NY 14420
U.S.A.
TEL: 716-395-5581
TLX:
TLF:
EML:
COM: 28.47

NORDH H LENNART DR
STOCKHOLM OBSERVATORY
S-133 00 SALTSJOEBADEN
SWEDEN
TEL: 08-7170195
TLX: 12972
TLF:
EML:
COM: 34.44

NORDLUND AKE DR
UNIVERSITY OBSERVATORY
OESTER VOLDGADE 3
DK-1350 COPENHAGEN K
DENMARK
TEL: 1-14-17-90
TLX: 44155 DANAST
TLF:
EML:
COM: 12.36

NORDSTROM BIRGITTA DR
CENTER FOR ASTROPHYSICS
MAIL STOP 20
60 GARDEN STREET
CAMBRIDGE MA 02138
U.S.A.
TEL: 617-495-7411
TLX: 921428 SATELLITE CAM
TLF:
EML:
COM: 30.42

NORMAU LOUIS DR
DEPT. OF ASTRONOMY
UNIVERSITY OF TORONTO
TORONTO M5S 1A1
CANADA
TEL: 416 978 3146
TLX: 06986766
TLF: 416-9783921
EML: BITNET:NORMAUGUTOPPHYS
COM: 28

NORGAARD-NIELSEN HANS U
UNIVERSITY OBSERVATORY
OSTER VOLDGADE 3
DK-1350 COPENHAGEN K
DENMARK
TEL:
TLX:
TLF:
EML:
COM:

NORMAN COLIN A PROF
SPACE TELESCOPE SCI INST
HOMEWOOD CAMPUS
3700 SAN MARTIN DRIVE
BALTIMORE MD 21218
U.S.A.
TEL: 301-338-4895
TLX:
TLF:
EML:
COM: 28.34.44.47.48

NORRIS JOHN DR
MT STROMLO OBSERVATORY
PRIVATE BAG
WODEN P.O. ACT 2606
AUSTRALIA
TEL: 062-88-1111
TLX: AA 62770 CANOPUS
TLF:
EML:
COM: 29

NORRIS RAYMOND PAUL
CSIRO
DIVISION OF RADIOPHYSICS
P.O.BOX 76
EPPING NSW 2121
AUSTRALIA
TEL: 02-868 0222
TLX: 26230 ASTRO
TLF:
EML:
COM:

NORTH JOHN DAVID PROF
FILOSOFISCH INSTITUUT
RIJKSUNIVERSITEIT
GRONINGEN
NETHERLANDS
TEL: 05907-1846
TLX:
TLF:
EML:
COM: 41F

NORTH PIERRE
INSTITUT D'ASTRONOMIE
UNIVERSITE DE LAUSANNE
CH-1290 CHAVANNES-DES-B.
SWITZERLAND
TEL: 022-55-26-11
TLX: 27720 OBSG CH
TLF:
EML:
COM: 45

NOSKOV BORIS N DR
STERNBERG STATE
ASTRONOMICAL INSTITUTE
119899 MOSCOW
U.S.S.R.
TEL:
TLX:
TLF:
EML:
COM: 07

NOTNI P DR
ZNTRLINST. F. ASTROPHYSIK
STERNWARTE BABELSBERG
ROSA-LUXEMBURG-STR 17A
DDR-1502 POTSDAM
GERMANY. D.R.
TEL:
TLX:
TLF:
EML:
COM: 25.45

NOTTALE LAURENT
OBSERVATOIRE DE PARIS
SECTION DE MEUDON
DAF
F-92195 MEUDON PL CEDEX
FRANCE
TEL: 1-45-07-74-03
TLX: 201571
TLF:
EML:
COM: 47

NOVELLO MARIO DR
CTR BRAS PESQUISAS FISIC
RUA DR. XAVIER SIGAUD
150 URCA
22290 RIO DE JANEIRO
BRAZIL
TEL: (021)541 0337
TLX: 21 22563
TLF:
EML:
COM: 47

NOVICK ROBERT
DEPT OF PHYSICS
COLUMBIA UNIVERSITY
538 W. 120 STREET
NEW YORK NY 10027
U.S.A.
TEL: 212-280-3293
TLX: 22094 COLU UR
TLF:
EML:
COM: 44.48

NOVIKOV I D DR
SPACE RESEARCH INSTITUTE
USSR ACADEMY OF SCIENCES
117810 MOSCOW
U.S.S.R.
TEL:
TLX:
TLF:
EML:
COM: 47

NOVIKOV SERGEJ B DR
STERNBERG STATE
ASTRONOMICAL INSTITUTE
117234 MOSCOW
U.S.S.R.
TEL:
TLX:
TLF:
EML:
COM:

NOVOSELOV V S PROF DR
LENINGRAD STATE UNIV.
ASTRONOMICAL OBSERVATORY
BIBLIOTECHNAJA PL. 2
198904 LENINGRAD
U.S.S.R.
TEL: 257-94-91
TLX:
TLF:
EML:
COM: 07

NOVOTNY VACLAV
ASTRONOMICAL INSTITUTE
CZECH. ACAD. OF SCIENCES
ONDREJOV OBSERVATORY
251 65 ONDREJOV
CZECHOSLOVAKIA
TEL: 72-45-25
TLX: 121579
TLF:
EML:
COM: 44

NOYES ROBERT W PROF
CENTER FOR ASTROPHYSICS
60 GARDEN STREET
CAMBRIDGE MA 02138
U.S.A.
TEL: 617-495-7424
TLX: 921428 SATELLITE CAM
TLF:
EML:
COM: 10.12.44

NUGIS TIIT
W.STRUVE ASTROPHYS OBS
ESTONIAN SSR
TORAVERE
202444 TARTU
U.S.S.R.
TEL:
TLX:
TLF:
EML:
COM: 27.29

NULSEN PAUL DR
MT STROMLO & SIDING
SPRING OBSERVATORIES
PRIVATE BAG
WODEN P.O. ACT 2606
AUSTRALIA
TEL: 062-88-1111
TLX: 62270 AA
TLF:
EML:
COM: 28.34.48

NUNES ROGERIO S DE SOUSA
GRUPO DE MATEM. APLICADA
UNIVERSIDADE DO PORTO
RUA DAS TAIPAS 135
4000 PORTO
PORTUGAL
TEL: 380313/769
TLX:
TLF:
EML:
COM: 09

NUNEZ JORGE DR
OBSERVATORIO FABRA
TIBIDABO
08022 BARCELONA
SPAIN
TEL: 2475736
TLX:
TLF:
EML:
COM: 24

NUSSBAUMER HARRY PROF
INSTITUT FUER ASTRONOMIE
ETH-ZENTRUM
CH-8092 ZUERICH
SWITZERLAND
TEL: 1-256-3631
TLX: 53178 ETHBI CH
TLF:
EML:
COM: 10.14C.34

NUTH JOSEPH A III
NASA/GSFC. CODE 691
LAB EXTRATERRESTR PHYSICS
GREENBELT MD 20771
U.S.A.
TEL: 301-286-6364
TLX:
TLF:
EML:
COM: 22.34

NYMAN LARS-AKE DR
ONSALA SPACE OBSERVATORY
S 43900 ONSALA
SWEDEN
TEL: 30060651
TLX: 2400
TLF:
EML:
COM:

OBASHEV SAKEN O DR
ASTROPHYSICAL INSTITUTE
480068 ALMA-ATA
U.S.S.R.
TEL:
TLX: 275
TLF:
EML:
COM:

OBI SHINYA PROF
FAC OF GENERAL EDUCATION
UNIVERSITY OF TOKYO
KOMABA MEGURO
TOKYO 153
JAPAN
TEL:
TLX:
TLF:
EML:
COM: 14

OBLAK EDOUARD
OBSERVATOIRE DE BESANCON
41B AVE DE L'OBSERVATOIRE
F-25044 BESANCON CEDEX
FRANCE
TEL: 81-50-30-88
TLX: 361144
TLF:
EML:
COM: 25.26

OBREGON DIAZ OCTAVIO J DR
DEPART. FISICA
UNIDAD IZTAPALAPA. UAM
PO BOX 55-534
09340 MEXICO DF
MEXICO
TEL: 6860322
TLX: 1764296 UAM ME
TLF:
EML:
COM:

OBRIDKO VLADIMIR N DR
IZMIRAN
ACADEMGORODOK
142092 MOSCOW REGION
U.S.S.R.
TEL: 232-1921
TLX: 412623 SCSTP
TLF:
EML:
COM: 10

OCCHIONERO FRANCO PROF
OSSERV. ASTRONOMICO
V.LE PARCO MELLINI 84
I-00136 ROMA
ITALY
TEL: (6)3452656
TLX:
TLF:
EML:
COM:

OCHSENBEIN FRANCOIS DR
E S O
KARL-SCHWARZSCHILD-STR 2
D-8046 GARCHING B MUNCHEN
GERMANY. F.R.
TEL:
TLX:
TLF:
EML:
COM: 05

ODA MINORU PROF
WAKO-SHI
SAITAMA 351-01
JAPAN
TEL: 0484-62-1111
TLX: 02962818 RIKEN J
TLF:
EML:
COM: 44.48.51

ODA NAOKI
MICRO-ELECTRONICS RES LAB
NIPPON ELECTRIC COMPANY
4-1-1 MIYAZAKI MIYAMAE-KU
KAWASAKI. KANAGAWA 213
JAPAN
TEL:
TLX:
TLF:
EML:
COM: 44

ODELL ANDREW P
DEPT OF PHYS. & ASTRONOMY
N.AZ. UNIVERSITY
FLAGSTAFF AZ 86011
U.S.A.
TEL:
TLX:
TLF:
EML:
COM: 35

ODENWALD STEN F DR
CODE 4138.0
NAVAL RESEARCH LAB
WASHINGTON DC 20375
U.S.A.
TEL: 202 767 3010
TLX:
TLF:
EML:
COM:

ODGERS GRAHAM J DR
DOMINION ASTROPH OBS
5071 W SAANICH ROAD
RR 5
VICTORIA BC V8X 4M6
CANADA
TEL: 604-388-3977
TLX: 049-7295
TLF:
EML:
COM: 09.27

OEGERLE WILLIAM R
5924 BERWYN ROAD
COLLEGE PARK MD20740
U.S.A.
TEL:
TLX:
TLF:
EML:
COM:

OEHMAN YNGVE PROF
TRULELEN 53
S-223 67 LUND
SWEDEN
TEL: 046-143362
TLX:
TLF:
EML:
COM:

OEMLER AUGUSTUS JR DR
YALE UNIVERSITY OBS
PO BOX 6666
NEW HAVEN CT 06511
U.S.A.
TEL: 203-436-3460
TLX:
TLF:
EML:
COM: 28.47

OERTEL GOETZ K DR
AURA INC
1625 MASSACHUSETTS AVE NW
SUITE 701
WASHINGTON DC 20036
U.S.A.
TEL: 202 483 2101
TLX:
TLF:
EML: TELEMAIL: G.OERTEL
COM: 44

OESTERWINTER CLAUS
COMMANDER
NAVAL SURFACE WEAPONS CTR
K10
DAHLGREN VA 22448
U.S.A.
TEL: 703-663-7426
TLX:
TLF:
EML:
COM: 04.07

OESTGAARD ERLEND
DEPARTMENT OF PHYSICS
UNIVERSITY OF TRONDHEIM
AVH
N-7055 DRAGVOLLM
NORWAY
TEL: 07-920411 x 117
TLX:
TLF:
EML:
COM:

OESTREICHER ROLAND
LANDESSTERNWARTE
KOENIGSTUHL
D-6900 HEIDELBERG
GERMANY. F.R.
TEL: 06221-10036
TLX:
TLF:
EML:
COM: 25

OETKEN L DR
ZNTRLINST. F. ASTROPHYSIK
ASTROPHYSIK. OBSERVATOR.
TELEGRAFENBERG
DDR-1500 POTSDAM
GERMANY. D.R.
TEL:
TLX:
TLF:
EML:
COM: 14.29.30

OGAWARA YOSHIAKI
INST SPACE & ASTRON SCI
4-6-1 KOMABA
MEGURO-KU
TOKYO 153
JAPAN
TEL: 3-467-1111
TLX: 34757 ISASTRO J
TLF:
EML:
COM: 44.48

OGORODNIKOV KYRILL F PROF
DEPT OF ASTRONOMY
LENINGRAD UNIVERSITY
199164 LENINGRAD
U.S.S.R.
TEL:
TLX:
TLF:
EML:
COM: 05.33

OGURA KATSUO DR
KOKUGAKUIN UNIVERSITY
COLLEGE OF LITERATURE
HIGASHI 4-10-28
SHIBUYAKU TOKYO 150
JAPAN
TEL: 298-42-6913
TLX: 28899 SIBINBTH J
TLF:
EML:
COM: 37

OH KYU DONG DR
DEPT OF EARTH SCIENCE
CHONNAM NATL UNIVERSITY
KWANGJU. CHONNAM.
KOREA. REPUBLIC
TEL:
TLX:
TLF:
EML:
COM: 42

OHASHI TAKAYA DR
DEPARTMENT OF PHYSICS
UNIVERSITY OF TOKYO
7-3-1 HONGO BUNKYO KU
TOKYO 113
JAPAN
TEL: 3 812 2111
TLX: UTPHYSIC 23472 J
TLF: 3 817 6938
EML:
COM:

OHKI KENICHIRO DR
TOKYO ASTRONOMICAL
OBSERVATORY
OSAWA
MITAKA TOKYO 181
JAPAN
TEL: 0422-32-5111
TLX: 2822307
TLF:
EML:
COM: 10

OHRING GEORGE PROF
DEPT GEOPHYS & PLANET SCI
TEL-AVIV UNIVERSITY
TEL-AVIV 69978
ISRAEL
TEL:
TLX:
TLF:
EML:
COM:

OHTANI HIROSHI DR
DEPT OF ASTRONOMY
UNIVERSITY OF KYOTO
KYOTO 606
JAPAN
TEL: 075-751-2111
TLX: 5422693 LIBKYU J
TLF:
EML:
COM: 34

OHYAMA NOBORU PROF
FACULTY OF ENGINEERING
SHIZUOKA UNIVERSITY
3 CHOME JYOHOKU
HAMAMATSU 432
JAPAN
TEL:
TLX:
TLF:
EML:
COM: 35

OJA HEIKKI DR
OBS & ASTROPHYSICS LAB
UNIVERSITY OF HELSINKI
TAHTITORNINMAKI
SF-00130 HELSINKI 13
FINLAND
TEL: 358-0-1912942
TLX:
TLF:
EML:
COM: 46

OJA TARMO PROF
KVISTABERG OBSERVATORY
S-197 00 BRO
SWEDEN
TEL: 0758-40157
TLX:
TLF:
EML:
COM: 24.33.45

OKA TAKESHI DR
CHEMISTRY DEPT
UNIVERSITY OF CHICAGO
5735 S. ELLIS AVE
CHICAGO IL 60637
U.S.A.
TEL: 312-962-7070
TLX:
TLF:
EML:
COM: 14

OKAMOTO ISAO DR
INTERNATIONAL LATITUDE
OBSERVATORY
HOSHIGAOKA MIZUSAWA
IWATE 023
JAPAN
TEL: 0197-24-7111
TLX: 8376-28 ILSMIZ J
TLF:
EML:
COM: 19.35

OKAMURA SABAMORI DR
KISO BRANCH OF THE
TOKYO ASTRON OBSERVATORY
MITAKEMURA KISOGUN
NAGANOKEN 397-01
JAPAN
TEL: 0264-52-3360
TLX: 3347577 KSOOBS J
TLF:
EML:
COM: 28C

OKAZAKI AKIRA DR
TSUDA COLLEGE
KODAIRA
TOKYO 187
JAPAN
TEL: 0423-41-2441
TLX:
TLF:
EML:
COM: 42

OKAZAKI SEICHI DR
2-4-4 OSAWA. MITAKA
TOKYO 181
JAPAN
TEL: 0422-31-6770
TLX:
TLF:
EML:
COM: 19

OKE J BEVERLEY PROF
CALTECH 105-24
PASADENA CA 91125
U.S.A.
TEL: 818-356-4007
TLX:
TLF:
EML:
COM: 28.29

OKI TOSIO PROF DR
DEPT EARTH SC FAC OF EDUC
FUKUSHIMA UNIVERSITY
MATSUKAWA MACHI
FUKUSHIMA 960-12
JAPAN
TEL: 0245-48-5151
TLX:
TLF:
EML:
COM:

OKOYE SAMUEL E PROF
DEPT OF PHYSICS & ASTR
UNIVERSITY OF NIGERIA
NSUKKA
NIGERIA
TEL: 042-770752
TLX:
TLF:
EML:
COM: 38.40.46.47.48

OKUDA HARUYUKI DR PROF
INST F.SPACE & ASTRONAUT.
SCIENCE
4-6-1 KOMABA. MEGURO-KU
TOKYO 153
JAPAN
TEL: 03-467-1111
TLX: 24550 SPACETKY J
TLF:
EML:
COM: 33.34

OKUDA TORU
INST OF EARTH SCIENCE
HOKKAIDO UNIV OF EDUCAT
1-2 HACHIMAN-CHO
HAKODATE 040
JAPAN
TEL: 0138-41-1121
TLX:
TLF:
EML:
COM: 44

OLAH KATALIN DR
KONKOLY OBSERVATORY
HUNGARIAN ACADEMY OF SCI
BOX 67
H-1525 BUDAPEST
HUNGARY
TEL: 166-426/366-621
TLX: 227460 KONOB H
TLF:
EML:
COM: 27

OLANO CARLOS ALBERTO DR
INSTITUTO ARGENTINO DE
RADIOASTRONOMIA
CASILLA DE CORREO No. 5
1894 VILLA ELISA (Bs.As.)
ARGENTINA
TEL:
TLX:
TLF:
EML:
COM: 33

OLEAK H DR
ZNTRLINST. F. ASTROPHYSIK
STERNWARTE BABELSBERG
ROSA-LUXEMBURG-STR 17A
DDR-1502 POTSDAM
GERMANY. D.R.
TEL:
TLX:
TLF:
EML:
COM: 28

OLIVA ERNESTO DR
OSSERVATORIO ASTROFISICO
DI ARCETRI
LARGO E FERMI 5
FIRENZE 50125
ITALY
TEL: 0039 55 7752310
TLX: 572268
TLF: 0039 55 220039
EML: SPAN:38954::OLIVA
COM:

OLIVER BERNARD M DR
NASA AMES RESEARCH CTR
MS 229-8
MOFFETT FIELD CA 94035
U.S.A.
TEL: 415 694 5166
TLX: 348 408 NASA AMES
TLF:
EML:
COM: 51

OLIVER BERNARD M PROF
CHIEF SETI PROGRAM OFFICE
NASA-AMES RESEARCH CENTER
MOFFETT FIELD CA 94035
U.S.A.
TEL: 415-694-5166
TLX:
TLF:
EML:
COM: 51

OLIVER JOHN PARKER DR
DEPTARTMENT OF ASTRONOMY
UNIVERSITY OF FLORIDA
GAINESVILLE FL 32611
U.S.A.
TEL:
TLX:
TLF:
EML:
COM: 42

OLLONGREN A PROF DR
DEPT MATHS & COMPUT SCI
WASSENAARSEWEG 80
POSTBOX 9512
NL-2300 RA LEIDEN
NETHERLANDS
TEL: 071-272727-5006
TLX: 39058 ASTRO NL
TLF:
EML:
COM: 33.51

OLNON FRISO
RADIOSTERRENWACHT
STICHTING RZM
POSTBUS 2
NL-7990 AA DWINGELOO
NETHERLANDS
TEL: 05219-7244
TLX: 42043 SRZM NL
TLF:
EML:
COM:

OLOFSSON HANS
ONSALA SPACE OBSERVATORY
S-439 00 ONSALA
SWEDEN
TEL: 0300-60650
TLX: 8542400 ONSPACE
TLF:
EML:
COM: 34

OLOFSSON S GOERAN DR
STOCKHOLM OBSERVATORY
S-133 00 SALTSJOBADEN
SWEDEN
TEL: 8-7172639
TLX: 12972
TLF:
EML:
COM:

OLSEN ERIK H
COPENHAGEN UNIVERSITY OBS
BRORFELDEVEJ 23
DK-4340 TOLLOSE
DENMARK
TEL: 03-488195
TLX: 44155 DANAST DK
TLF:
EML:
COM: 45C

OLSEN FOGH H J
COPENHAGEN UNIVERSITY OBS
BRORFELDEVEJ 23
DK-4340 TOLLOSE
DENMARK
TEL: 03-488195
TLX: 44155 DQNQST
TLF:
EML:
COM: 08.46

OLSEN KENNETH H DR
LOS ALAMOS NAT LABORATORY
BOX 1663. MS C335
LOS ALAMOS NM 87545
U.S.A.
TEL: 505-667-1007
TLX:
TLF:
EML:
COM:

OLSON EDWARD C PROF
OBSERVATORY
UNIVERSITY OF ILLINOIS
1011 W. SPRINFIELD AVENUE
URBANA IL 62801
U.S.A.
TEL: 217-333-5531
TLX:
TLF:
EML:
COM: 42

OLSSON-STEEL DUNCAN I DR
DEPT OF PHYSICS
UNIVERSITY OF ADELAIDE
GPOBOX 498
ADELAIDE S A 5001
AUSTRALIA
TEL: 61 8 228 5996
TLX: UNIVAD AA 89141
TLF:
EML: dolssonsteel@f.ua.oz.au
COM: 15.22.42

OLTHOF HINDERICUS DR
ESA-ESTEC
SPACE SCIENCE DEPARTMENT
POSTBUS 299
NL-2200 AG NOORDWIJK
NETHERLANDS
TEL: 1719-86555
TLX: 39098 ESTC NL
TLF:
EML: SPAN:ESTEC 1::HOLTHOF
COM: 44

OMAROV TUKEN B PROF
ASTROPHYSICAL INSTITUTE
480068 ALMA ATA
U.S.S.R.
TEL: 64-40-40
TLX:
TLF:
EML:
COM: 07

OMER GUY C JR PROF
1080 S.W. 11TH TERRACE
GAINESVILLE FL 32601
U.S.A.
TEL: 904-378-4627
TLX:
TLF:
EML:
COM: 28.41.47

OMNES ROLAND PROF
LPTHE BAT 211
UNIVERSITE DE PARIS-SUD
F-91405 ORSAY
FRANCE
TEL: 1-69-41-77-44
TLX: 692166 FACORS
TLF:
EML:
COM: 47

OMONT ALAIN PROF
CERMO-USMG
BP 68
F-38041 ST MARTIN D'HERES
FRANCE
TEL: 76-51-47-90
TLX: 980753 F
TLF:
EML:
COM: 14.34

ONAKA TAKASHI
DEPT OF ASTRONOMY
UNIVERSITY OF TOKYO
2-11-15 YAYOI. BUNKYO-KU
TOKYO 113
JAPAN
TEL: 3-812-2111
TLX: 33659 UTYOSCI J
TLF:
EML:
COM: 34

ONDERLICKA BEDRICH DR
DEPT OF ASTRONOMY
PURKYNE UNIVERSITY
KOTLARSKA 2
611 37 BRNO
CZECHOSLOVAKIA
TEL:
TLX:
TLF:
EML:
COM:

ONEGINA A B DR
MAIN ASTRONOMICAL OBS
UKRAINIAN ACADEMY OF SCI
GOLOSEEVO
252127 KIEV
U.S.S.R.
TEL: 66-37-44
TLX:
TLF:
EML:
COM: 24

ONO TORO PROF
DEPT OF PHYSICS
UNIVERSITY HOKKAIDO
KITAHACHIJYO NISHI 8
SAPPORO HOKKAIDO 063
JAPAN
TEL:
TLX:
TLF:
EML:
COM:

ONUORA LESLEY IRENE DR
DEPT OF PHYS & ASTRONOMY
UNIV OF NIGERIA
NSUKKA
ANAMBRA STATE
NIGERIA
TEL:
TLX: 51496 ULIONS
TLF:
EML:
COM: 40

OOE MASATSUGU DR
INTERNATIONAL LATITUDE
OBSERVATORY
HOSHIGAOKA MIZUSAWA
IWATE 023
JAPAN
TEL: 0197-24-7111
TLX: 837628 MIZ J
TLF:
EML:
COM: 19

OORT JAN H PROF
PRES. KENNEDYLAAN 169
NL-2343 GZ OEGSTGEEST
NETHERLANDS
TEL:
TLX:
TLF:
EML:
COM: 28.33.40.47

OPHER REUVEN PROF
ASTRONOMY DEPT
IAG/USP
CAIXA POSTAL 30627
01051 SAO PAULO SP
BRAZIL
TEL: 275-3720
TLX: 1136221 IAGM BR
TLF:
EML:
COM:

OPOLSKI ANTONI PROF
ASTRONOMICAL OBSERVATORY
UL. KOPERNIKA 11
51-622 WROCLAW
POLAND
TEL:
TLX:
TLF:
EML:
COM: 27

OPROIU TIBERIU DR
ASTRONOMICAL OBSERVATORY
BLOC P-5 SC I 6 ET AP 10
STR BUCIUM 25
3400 CLUJ NAPOCA
RUMANIA
TEL: 951-62616
TLX:
TLF:
EML:
COM:

ORCHISTON WAYNE DR
27 MARIAN DRIVE
GISBORNE
NEW ZEALAND
TEL: N2 83 832
TLX:
TLF:
EML:
COM: 41

ORLIN HYMAN DR
NATL ACADEMY OF SCIENCES
2101 CONSTITUTION AVE NW
WASHINGTON DC 20418
U.S.A.
TEL:
TLX:
TLF:
EML:
COM:

ORLOV MIKHAIL DR
MAIN ASTRONOMICAL OBS
252127 KIEV
U.S.S.R.
TEL: 66-31-10
TLX: 131406 SKY
TLF:
EML:
COM: 29

ORRALL FRANK Q PROF
INSTITUTE FOR ASTRONOMY
2680 WOODLAWN DRIVE
HONOLULU HI 96822
U.S.A.
TEL: 808-948-8667
TLX: 723-8459
TLF:
EML:
COM: 10.12.36

ORTE ALBERTO
CECILIO PUJAZON 22-3 A
11100 S. FERNANDO (CADIZ)
SPAIN
TEL: 89-54-41
TLX:
TLF:
EML:
COM: 19.31

ORTOLANI SERGIO
OSSERVATORIO ASTRONOMICO
DI PADOVA
VIC. DELL'OSSERVATORIO 5
I-35122 PADOVA
ITALY
TEL: 049 661499
TLX: 432071 ASTROS I
TLF:
EML:
COM: 37

ORTON GLENN S DR
JPL
MS 183-301
4800 OAK GROVE DRIVE
PASADENA CA 91109
U.S.A.
TEL: 818-354-2460
TLX:
TLF:
EML:
COM: 14

ORUS JUAN J PROF
DPTO FISICA DE ATMOSFERA
UNIVERSIDAD DE BARCELONA
AVDA. DIAGONAL 645
08028 BARCELONA
SPAIN
TEL:
TLX:
TLF:
EML:
COM: 07

OSAKI TORU DR
RYUKOKU UNIVERSITY
FUKAKUSA TSUKAMOTO
FUSHIMIKU
KYOTO 612
JAPAN
TEL: 075-642-11[illegible]
TLX:
TLF:
EML:
COM: 34

OSAKI YOJI DR
DEPT OF ASTRONOMY
UNIVERSITY OF TOKYO
YAYOI BUNKYO
TOKYO 113
JAPAN
TEL: 03-812-2111
TLX: 33659 UTYOSCI J
TLF:
EML:
COM: 35C.42

OSAWA KIYOTERU DR
TOKYO ASTRONOMICAL
OBSERVATORY
OSAWA
MITAKA TOKYO 181
JAPAN
TEL: 0422-32-5111
TLX:
TLF:
EML:
COM: 27.29

OSBORN WAYNE DR
PHYSICS DEPT
CENTRAL MICHIGAN UNIV
MT PLEASANT MI 48859
U.S.A.
TEL: 517-774-3321
TLX:
TLF:
EML:
COM: 37.45.46

OSBORNE JOHN L DR
DEPT OF PHYSICS
UNIVERSITY OF DURHAM
SOUTH ROAD
DURHAM DH1 3LE
U.K.
TEL: 0385-64971
TLX: 537351 DURLIB G
TLF:
EML:
COM: 34

OSMAN ANAS MOHAMED DR
HELWAN OBSERVATORY
HELWAN-CAIRO
EGYPT
TEL: 780645 NIAG UN
TLX:
TLF:
EML:
COM: 37

OSMER PATRICK S DR
NOAO/KPNO
PO BOX 26732
TUCSON AZ 85718
U.S.A.
TEL: 602-327-5511
TLX: 666484
TLF:
EML:
COM:

OSORIO JOSE J S P PROF
OBSERVATORIO ASTRONOMICO
UNIVERSIDADE DO PORTO
MONTE DA VIRGEM
4400 VILA NOVA DE GAIA
PORTUGAL
TEL: 7820404
TLX: 22367
TLF:
EML:
COM: 07.08.46.50

OSTER LUDWIG F PROF DR
NATL SCIENCE FOUNDATION
1800 G STREET N.W.
WASHINGTON DC 20550
U.S.A.
TEL: 202-357-9857
TLX:
TLF:
EML:
COM: 12

OSTERBROCK DONALD E PROF
LICK OBSERVATORY
UNIVERSITY OF CALIFORNIA
SANTA CRUZ CA 95064
U.S.A.
TEL: 408-429-2605
TLX:
TLF:
EML:
COM: 28.34.40.41

OSTRIKER JEREMIAH P PROF
PRINCETON UNIVERSITY
OBSERVATORY
PEYTON HALL
PRINCETON NJ 08544
U.S.A.
TEL: 609-452-3800
TLX: 322409
TLF:
EML:
COM: 33.35.48V.51

OSWALT TERRY D DR
DEPT OF PHYS & SPACE SCI
FLORIDA INST TECHNOLOGY
150 W UNIVERSITY BLVD
MELBOURNE FL 32901
U.S.A.
TEL: 305 768 8098
TLX:
TLF:
EML:
COM: 26.27.35

OTERMA LIISI PROF
SIRKKALANKATU 31
SF-20700 TURKU
FINLAND
TEL: 358-21-332081
TLX:
TLF:
EML:
COM: 19.20

OTHMAN MAZLAN
DEPT OF PHYSICS
UNIVERSITI KEBANGSAAN
MALAYSIA
43600 BANGI SELANGOR
MALAYSIA
TEL: 8250001
TLX: 31496 UNIKEB MA
TLF:
EML:
COM: 46

OTTELET I J DR
INSTITUT D'ASTROPHYSIQUE
UNIVERSITE DE LIEGE
AVENUE DE COINTE 5
B-4200 COINTE-OUGREE
BELGIUM
TEL: 041-529980
TLX: 41264
TLF:
EML:
COM: 16

OVERBEEK MICHIEL DANIEL
P O BOX 212
EDENVALE 1610
SOUTH AFRICA
TEL: 11-53-5447
TLX:
TLF:
EML:
COM:

OWAKI NAOAKI DR
TOKYO GAKUGEI UNIVERSITY
NUKUIKITAMACHI
KOGANEI TOKYO 184
JAPAN
TEL:
TLX:
TLF:
EML:
COM: 46

OWEN FRAZER NELSON DR
VLA NRAO
1000 BULLOCK BLVD
PO BOX 0
SOCORRO NM 87801
U.S.A.
TEL: 505-772-4011
TLX: 910-988-1710
TLF:
EML:
COM: 28.40

OWEN TOBIAS C PROF
DEPT OF EARTH & SPACE SCI
STATE UNIVERSITY NEW YORK
STONY BROOK NY 11794
U.S.A.
TEL: 516-246-6705
TLX: 5102287767
TLF:
EML:
COM: 16C.44.51

OWREN LEIF DR
FYSISK INSTITUTT
UNIVERSITETET I BERGEN
ALLEGATEN 55
N-5000 BERGEN
NORWAY
TEL: 47-05-213050
TLX:
TLF:
EML:
COM:

OXENIUS JOACHIM DR
UNIV LIBRE BRUXELLES
CP 231
CAMPUS PLAINE ULB
B-1050 BRUSSELS
BELGIUM
TEL: 32-02-640-00-15
TLX: 23069 UNILIB B
TLF:
EML:
COM: 36

OZERNOY LEONID M PROF
CENTER FOR ASTROPHYSICS
60 GARDEN STREET
CAMBRIDGE MA 02138
U.S.A.
TEL: 617-495-4951
TLX:
TLF:
EML:
COM: 34.47.48

OZGUC ATILA
BOGAZICI UNIVERSITY
KANDILLI OBSERVATORY
CENGELKOY
ISTANBUL 81220
TURKEY
TEL: 90-1-332-02-40
TLX:
TLF:
EML:
COM: 10

OZSVATH I PROF
UNIVERSITY OF TEXAS
PROGRAMS IN MATHEMAT. SCI
PO BOX 830688
RICHARDSON TX 75083-0688
U.S.A.
TEL: 214-690-2174
TLX:
TLF:
EML:
COM: 47

O'CONNELL ROBERT F PROF
LOUISIANA STATE UNIV
BATON ROUGE LA 70803
U.S.A.
TEL: 504-388-6848
TLX:
TLF:
EML:
COM: 48

O'CONNELL ROBERT WEST DR
ASTRONOMY DEPT
UNIVERSITY OF VIRGINIA
PO BOX 3818. UNIV STATION
CHARLOTTESVILLE VA 22903
U.S.A.
TEL: 804-924-7494
TLX: 510-587-5453
TLF:
EML:
COM: 28.47

O'CONNOR SEAMUS L DR
PHYSICS DEPT
UNIVERSITY COLLEGE
BELFIELD
DUBLIN 4
IRELAND
TEL: 353-1-693244
TLX: 32693 UCD EI
TLF:
EML:
COM:

O'DEA CHRISTOPHER P DR
NETHERLANDS FOUNDATION
FOR RESEARCH IN ASTRONOMY
POSTBUS 2
7990 AA DWINGELOO
NETHERLANDS
TEL: 05219 7244
TLX: 42043 SRZM NL
TLF:
EML: BITNET:CODEA @ HGRRUG5
COM: 28.40

O'DELL CHARLES R DR
DEPT SPACE PHYS & ASTRON
RICE UNIVERSITY
PO BOX 1892
HOUSTON TX 77251
U.S.A.
TEL: 713-527-8101
TLX: 556457
TLF:
EML:
COM: 34.15.34

O'DELL STEPHEN L
SPACE SCIENCE LAB. ES-65
NASA MARSHALL SPACE
FLIGHT CENTER
HUNTSVILLE AL 35812
U.S.A.
TEL: (205)544-7708
TLX:
TLF:
EML:
COM: 34

O'DONOGHUE DARRAGH DR
DEPT OF ASTRONOMY
UNIVERSITY OF CAPE TOWN
RONDEBOSCH 7700
SOUTH AFRICA
TEL:
TLX:
TLF:
EML:
COM: 27

O'HANDLEY DOUGLAS A DR
NASA CODE Z
WASHINGTON DC 2054627A
U.S.A.
TEL: 202 4538932
TLX:
TLF:
EML:
COM: 04.07

O'HORA NATHY P J
ROYAL GREENWICH OBS
HERSTMONCEUX CASTLE
HAILSHAM BN27 1RP
U.K.
TEL:
TLX:
TLF:
EML:
COM: 19

O'KEEFE JOHN A DR
NASA/GSFC
CODE 681
GREENBELT MD 20771
U.S.A.
TEL: 301-286-8445
TLX: 89675
TLF:
EML:
COM: 15.16.22

O'LEARY BRIAN T
SAIC
2615 PACIFIC COAST HWY
SUITE 300
HERMOSA BEACH CA 90254
U.S.A.
TEL: 213-318-2611
TLX:
TLF:
EML:
COM:

O'MARA BERNARD J PROF
DEPT OF PHYSICS
UNIVERSITY OF QUEENSLAND
ST LUCIA
BRISBANE QLD 4067
AUSTRALIA
TEL:
TLX:
TLF:
EML:
COM: 36

O'MONGAIN EON
PHYSICS DEPT
UNIVERSITY COLLEGE
BELFIELD
DUBLIN 4
IRELAND
TEL: 01-693244
TLX: 32693 UCD
TLF:
EML:
COM: 44

O'SULLIVAN DENIS F
DUBLIN INSTITUTE FOR
ADVANCED STUDIES
5 MERRION SQUARE
DUBLIN 2
IRELAND
TEL: 353-1-774321
TLX: 31687 DIAS EI
TLF:
EML:
COM: 48

O'SULLIVAN JOHN DAVID DR
DIVISION OF RADIOPHYSICS
CSIRO
P.O.BOX 76
EPPING NSW 2121
AUSTRALIA
TEL:
TLX: 26230 ASTRO
TLF:
EML:
COM: 40

PAAL GYORGY DR
KONKOLY OBSERVATORY
BOX 67
H-1525 BUDAPEST
HUNGARY
TEL: 166-506
TLX: 227460 KONOB H
TLF:
EML:
COM: 47

PACHNER JAROSLAV PROF
606-55 WYNFORD HTS. CR.
TORONTO ONT M3C 1L4
CANADA
TEL: 416-447-1015
TLX:
TLF:
EML:
COM: 47

PACHOLCZYK ANDRZEJ G PROF
STEWARD OBSERVATORY
UNIVERSITY OF ARIZONA
TUCSON AZ 85721
U.S.A.
TEL: 602-621-6928
TLX: 467175
TLF:
EML:
COM: 28.40.48

PACIESAS WILLIAM S DR
DEPARTMENT OF PHYSICS
UNIVERSITY OF ALABAMA
HUNTSVILLE AL 35899
U.S.A.
TEL: 205 544 7712
TLX: 594416 ES62
TLF:
EML:
COM: 44

PACINI FRANCO PROF
DIPART. DI ASTRONOMIA
UNIVERSITA DEGLI STUDI
LARGO E. FERMI 5
I-50125 FIRENZE
ITALY
TEL: 055-27521
TLX: 572268 ARCETR-I
TLF:
EML:
COM: 44.48C.51C

PACZYNSKI BOHDAN PROF
COPERNICUS ASTRON CENTER
UL. BARTYCKA
00-716 WARSZAWA
POLAND
TEL:
TLX:
TLF:
EML:
COM: 35.42

PADALIA T D DR
UTTAR PRADESH STATE OBS
MANORA PARK
NAINITAL 263 129
INDIA
TEL: 2136
TLX:
TLF:
EML:
COM: 42

PADEVET VLADIMIR DR
ASTRONOMICAL INSTITUTE
CZECH. ACAD. OF SCIENCES
OBSERVATORY
251 65 ONDREJOV
CZECHOSLOVAKIA
TEL: 724525 PRAHA
TLX: 121579
TLF:
EML:
COM: 22

PADMAN RACHAEL
MULLARD RADIO ASTRON OBS
CAVENDISH LABORATORY
MADINGLEY ROAD
CAMBRIDGE CB3 OHE
U.K.
TEL: 223-66477
TLX: 81292
TLF:
EML:
COM: 40

PADMANABHAN T DR
ASTROPHYSICS GROUP
TATA INST.FUNDAMENTAL RES
HOMI BHABHA ROAD
BOMBAY 400 005
INDIA
TEL: 495 2311
TLX: 011-3009 TIFR IN
TLF:
EML:
COM: 47

PADRIELLI LUCIA
IST. DI RADIOASTRONOMIA
VIA IRNERIO 46
I-40126 BOLOGNA
ITALY
TEL: 051-232856
TLX: 211664 INF BO I
TLF:
EML:
COM: 40.47

PAGE ARTHUR MR
MT TAMBORINE OBSERVATORY
PO BOX 44
ASPLEY. QLD 4034
AUSTRALIA
TEL: 61-7-263-4813
TLX:
TLF:
EML:
COM: 25

PAGE CLIVE G DR
PHYSICS DEPT
UNIVERSITY OF LEICESTER
LEICESTER LE1 7RH
U.K.
TEL: 533-554455 x 23
TLX: 341664 LUXRAY G
TLF:
EML:
COM: 48

PAGE DON NELSON
104 DAVEY LABORATORY
PENNSYLVANIA STATE UNIV.
UNIVERSITY PARK PA 16802
U.S.A.
TEL: 814-863-0163
TLX: 842510
TLF:
EML:
COM: 47

PAGE THORNTON L DR
NASA JOHNSON SPACE CENTER
18639 POINT LOOKOUT DR
HOUSTON TX 77058
U.S.A.
TEL: 713-483-3728
TLX:
TLF:
EML:
COM: 28.51

PAGEL BERNARD E J PROF
ROYAL GREENWICH OBS
HERSTMONCEUX CASTLE
HAILSHAM BN27 1RP
U.K.
TEL: 0323833171
TLX: 87451 RGOBST G
TLF:
EML:
COM: 29.34.36

PAKVOR IVAN
BELGRADE OBSERVATORY
VOLGINA 7
11050 BELGRADE
YUGOSLAVIA
TEL: 38 11 419 357
TLX: 72610 AOR YU
TLF:
EML:
COM: 08

PAL ARPAD PROF DR
UNIV OF CLUJ-NAPOCA
FACULTY OF MATHEMATICS
STR KAKOCZI 72
3400 CLUJ NAPOCA
RUMANIA
TEL: 951-16101/11597
TLX:
TLF:
EML:
COM: 07

PALAGI FRANCESCO
OSSERVATORIO ASTROFISICO
DI ARCETRI
L.GO E.FERMI 5
FLORENCE 50125
ITALY
TEL: 0395527521
TLX: 572268 ARCETR I
TLF:
EML:
COM:

PALLA FRANCESCO
OSSERVATORIO ASTROFISICO
DI ARCETRI
LARGO E. FERMI 5
I-50125 FIRENZE
ITALY
TEL: 055-2752242
TLX: 572268 ARCETR I
TLF:
EML:
COM: 34

PALLAVICINI ROBERTO DR
OSSERVATORIO ASTROFISICO
DI ARCETRI
I-50125 FIRENZE
ITALY
TEL: 055-2752252
TLX: 572268 ARCETR I
TLF:
EML:
COM: 10.36

PALLE PERE-LLUIS DR
INSTITUTO DE ASTROFISICA
DE CANARIAS
38200 LA LAGUNA TENERIFE
SPAIN
TEL: 34 22 26 22 11
TLX: 92640
TLF: 34 22 26 30 05
EML: SPAN:IAC::PLP
COM: 10.12

PALMEIRA RICARDO A R DR
INST PESQUISAS ESPACIAIS
CAIXA POSTAL 515
12200 S. JOSE DOS CAMPOS
BRAZIL
TEL:
TLX:
TLF:
EML:
COM:

PALMER PATRICK E PROF
DEPT OF ASTRONOMY
UNIVERSITY OF CHICAGO
5640 S. ELLIS AVE
CHICAGO IL 60637
U.S.A.
TEL: 312-962-7972
TLX: 6871133
TLF:
EML:
COM: 33.34.40

PALMER PHILIP DR
ASTRONOMY UNIT
QUEEN MARY COLLEGE
MILE END ROAD
LONDON E1 4NS
U.K.
TEL: 44-1-975-5462
TLX: 893750
TLF:
EML: PHILIP @ QMC.MATHS.
COM: 28

PALOUS JAN DR
ASTRONOMICAL INSTITUTE
CZECH. ACAD. OF SCIENCES
BUDECSKA 6
120 23 PRAGUE 2
CZECHOSLOVAKIA
TEL: 25-87-57
TLX: 122486
TLF:
EML:
COM: 33

PALUMBO GIORGIO G C DR
TE-S R E LAB CNR
VIA DE CASTAGNOLI 1
I-40100 BOLOGNA
ITALY
TEL: 051-51-95-93
TLX: 511350 CNR-BO
TLF:
EML:
COM: 28.48

PAMYATNIKH A A DR
ASTRONOMICAL COUNCIL
USSR ACADEMY OF SCIENCES
48 PYATNITSKAYA ST
109017 MOSCOW
U.S.S.R.
TEL: 231-54-61
TLX: 412623 SCSTP SU
TLF:
EML:
COM: 05.35

PAN JUN-HUA
NANJING ASTRONOMICAL
INSTRUMENTS FACTORY
NANJING
CHINA. PEOPLE'S REP.
TEL: 46191
TLX: 34136 GLVNJ CN
TLF:
EML:
COM:

PAN LIANDE
SHAANXI OBSERVATORY
P.O. BOX 18
LINTONG
XIAN
CHINA. PEOPLE'S REP.
TEL: 3-2255
TLX: 70121 CSAO CN
TLF:
EML:
COM: 19

PAN NING-BAO
BEIJING ASTRONOMICAL OBS
ACADEMIA SINICA
BEIJING
CHINA. PEOPLE'S REP.
TEL: 281698
TLX:
TLF:
EML:
COM:

PAN RONG-SHI
SHANGHAI OBSERVATORY
ACADEMIA SINICA
SHANGHAI
CHINA. PEOPLE'S REP.
TEL: 386191
TLX: 33164 SHAO CN
TLF:
EML:
COM: 24.28.47

PAN XIAO-PEI
SHAANXI ASTRONOMICAL OBS
P.O. BOX 18
LINTONG
SHAANXI
CHINA. PEOPLE'S REP.
TEL: XIAN 32255x406
TLX: 70121 CSAO CN
TLF:
EML:
COM: 19

PANAGIA NINO DR
SPACE TELESCOPE SCI INST
3700 SAN MARTIN DRIVE
BALTIMORE MD 21218
U.S.A.
TEL: 301-338-4916
TLX: 6849101 ST SCI
TLF:
EML:
COM: 34

PANDE GIRISH CHANDRA PROF
126 ARYANAGAR
LUCKNOW 226 004
INDIA
TEL:
TLX:
TLF:
EML:
COM: 35

PANDE MAHESH CHANDRA DR
UP STATE OBSERVATORY
MANORA PEAK
NAINI TAL 263 129
INDIA
TEL: 2136
TLX:
TLF:
EML:
COM: 12

PANDEY A K
UTTAR PRADESH STATE OBS
MANORA PEAK
NAINITAL 263129
INDIA
TEL: 2136
TLX:
TLF:
EML:
COM:

PANDEY S K
DEPT OF PHYSICS
RAVISHANKAR UNIVERSITY
RAIPUR 492010
INDIA
TEL: 27064
TLX:
TLF:
EML:
COM:

PANEK ROBERT J DR
DEPT OF ASTRONOMY
525 DAVEY LAB
UNIV OF PENNSYLVANIA
UNIVERSITY PARK PA 16802
U.S.A.
TEL:
TLX:
TLF:
EML:
COM: 36

PANG KEVIN
JET PROPULSION LAB
T 1182/3
4800 OAK GROVE DRIVE
PASADENA CA 91109
U.S.A.
TEL: 818-354-5392
TLX: 675429
TLF:
EML:
COM: 16

PANKONIN VERNON LEE DR
DIV ASTRONOMICAL SCIENCES
NATL SCIENCE FOUNDATION
1800 G STREET N.W.
WASHINGTON DC 20550
U.S.A.
TEL: 202-357-9696
TLX:
TLF:
EML:
COM: 34.40.50C

PANNUNZIO RENATO
OSSERVATORIO ASTRONOMICO
STRADA OSSERVATORIO 20
I-10025 PINO TORINESE
ITALY
TEL: 011-841067
TLX: 213236 TOASTR I
TLF:
EML:
COM: 26

PANOV KIRIL DR
DEPT OF ASTRONOMY
BULGARIAN ACAD SCIENCES
7TH NOVEMBER STR 1
1000 SOFIA
BULGARIA
TEL: 7341
TLX: 23561 ECF BAN BG
TLF:
EML:
COM:

PAOLICCHI PAOLO DR
ISTITUTO DI ASTRONOMIA
UNIVERSITA' DI PISA
PIAZZA TORRICELLI 2
I-56100 PISA
ITALY
TEL: 050-43343
TLX:
TLF:
EML:
COM: 15.16

PAP JUDIT
UNIVERSITY OF COLORADO
DEPT. OF ASTROPH.SCIENCES
CAMPUS BOX 391
BOULDER CO 80309
U.S.A.
TEL:
TLX:
TLF:
EML:
COM: 10

PAPAELIAS PHILIP DR
DEPT OF PHYSICS
UNIVERSITY OF ATHENS
PANEPISTIMIOPOLIC
GR 15771 ZOGRAFOS ATHENS
GREECE
TEL: 723 51 22
TLX: 223815 UNIVGR
TLF:
EML:
COM:

PAPAGIANNIS MICHAEL D PRO
DEPT OF ASTRONOMY
BOSTON UNIVERSITY
725 COMMONWEALTH AVE
BOSTON MA 02215
U.S.A.
TEL: 617-353-2626
TLX:
TLF:
EML:
COM: 40.44.51F

PAPALIOLIOS COSTAS DR
SMITHSONIAN ASTROPHYS OBS
60 GARDEN STREET
CAMBRIDGE MA 02138
U.S.A.
TEL:
TLX:
TLF:
EML:
COM:

PAPALOIZOU JOHN C B DR
QUEEN MARY COLLEGE
DEPT. OF MATHS.
MILE END ROAD
LONDON E1 4NS
U.K.
TEL:
TLX:
TLF:
EML:
COM: 27.35

PAPARO MARGIT DR
KONKOLY OBSERVATORY
HUNGARIAN ACADEMY OF SCI
BOX 67
H-1525 BUDAPEST
HUNGARY
TEL: 166-426
TLX: 227460
TLF:
EML:
COM: 27

PAPATHANASOGLOU D DR
DEPT OF ASTRONOMY
UNIVERSITY OF ATHENS
PANEPISTIMIOPOLIS
GR-15771 ATHENS
GREECE
TEL: 7243414
TLX:
TLF:
EML:
COM: 12

PAPAYANNOPOULOS TH DR
DEPT OF ASTRONOMY
UNIVERSITY OF ATHENS
PANEPISTIMIOPOLIS
GR-15771 ZOGRAFOS
GREECE
TEL: 01 7243414
TLX:
TLF:
EML:
COM: 28.33

PAPOUSEK JIRI
DEPT OF ASTROPHYSICS KPTA
FACULTY OF SCIENCE UJEP
KOTLARSKA 2
611 37 BRNO
CZECHOSLOVAKIA
TEL: 51112
TLX:
TLF:
EML:
COM: 27

PAQUET PAUL EG DR
OBS ROYAL DE BELGIQUE
AVENUE CIRCULAIRE 3
B-1180 BRUXELLES
BELGIUM
TEL: 32-2-374-38-01
TLX: 21565 OBSBEL
TLF:
EML:
COM: 19.31F

PARCELIER PIERRE DR
OBSERVATOIRE DE PARIS
61 AVE DE L'OBSERVATOIRE
F-75014 PARIS
FRANCE
TEL: 1-43-20-12-10
TLX: 270776
TLF:
EML:
COM: 31

PAREDES JOSE MARIA DR
DEPT DE ASTRONOMIAFISICA
UNIVERSITAT DE BARCELONA
AVDA DIAGONAL 647
08028 BARCELONA
SPAIN
TEL: 34-3-3307311
TLX:
TLF:
EML: D3FAJPP0 @ EBOUB011
COM: 40

PARESCE FRANCESCO DR
SPACE TELESCOPE SCI INST
THE JOHNS HOPKINS UNIV.
HOMEWOOD CAMPUS
BALTIMORE MD 21218
U.S.A.
TEL:
TLX:
TLF:
EML:
COM: 21.49

PARIJSKIJ N N PROF
INST OF PHYSICS OF EARTH
USSR ACADEMY OF SCIENCES
123810 MOSCOW
U.S.S.R.
TEL: 252-07-21
TLX: 411196 IPZAN US
TLF:
EML:
COM: 19

PARIJSKIJ YU N DR
SPECIAL ASTROPHYSICAL OBS
USSR ACADEMY OF SCIENCES
LENINGRAD BRANCH
196140 LENINGRAD
U.S.S.R.
TEL: 2974452
TLX:
TLF:
EML:
COM: 40.51

PARISE RONALD A DR
CODE 684.9
NASA
GSFC
GREENBELT MD 20771
U.S.A.
TEL: 301 286 3896
TLX:
TLF:
EML: SPAN:UIT::PARISE
COM:

PARISOT JEAN-PAUL
OBSERVATOIRE DE BESANCON
41B AVE DE L'OBSERVATOIRE
F-25044 BESANCON CEDEX
FRANCE
TEL: 81-50-30-88
TLX: 361144 OBS
TLF:
EML:
COM: 15.46

PARK HONG SUH DR
KOREA NATL UNIVERSITY
OF EDUCATION
CHOONGBOOK 363-691
KOREA, REPUBLIC
TEL: (431) 60 3901
TLX:
TLF:
EML:
COM: 42

PARKER EDWARD A DR
ELECTRONICS LABORATORIES
THE UNIVERSITY
CANTERBURY CT2 7NT
U.K.
TEL: 0227-66822
TLX: 965449
TLF:
EML:
COM: 40

PARKER EUGENE N
LAB ASTROPHYS SPACE RES
UNIVERSITY OF CHICAGO
933 E 56TH STREET
CHICAGO IL 60637
U.S.A.
TEL: 312-962-7847
TLX: 910-221-5617
TLF:
EML:
COM: 34.48.49

PARKER ROBERT A R
NASA/JOHNSON SPACE CENTER
CODE CB
HOUSTON TX 77058
U.S.A.
TEL: 713-483-2221
TLX:
TLF:
EML:
COM:

PARKINSON JOHN H DR
MULLARD SPACE SCIENCE LAB
UNIVERSITY COLLEGE LONDON
HOLMBURY ST MARY
DORKING, SURREY RH56NS
U.K.
TEL: 030-670-292
TLX: 859185
TLF:
EML:
COM: 10 44.48

PARKINSON TRUMAN DR
KITT PEAK NATL OBS
950 N. CHERRY AVE
TUCSON AZ 85726
U.S.A.
TEL:
TLX:
TLF:
EML:
COM:

PARKINSON WILLIAM H
HARVARD COLLEGE OBS
60 GARDEN STREET
CAMBRIDGE MA 02138
U.S.A.
TEL: 617-495-4665
TLX: 921428
TLF:
EML:
COM: 10.12.14C.44

PARMA PAOLA
IST. DI RADIOASTRONOMIA
CNR
VIA IRNERIO 46
I-40126 BOLOGNA
ITALY
TEL: 051-232856
TLX:
TLF:
EML:
COM: 40

PARRISH ALLAN DR.
STATE UNIVERSITY OF N.Y.
C/O 619J GRC U MASS.
AMHERST MA 01003
U.S.A.
TEL:
TLX:
TLF:
EML:
COM: 40

PARSAMYAN ELMA S DR
BYURAKAN ASTROPHYSICAL
OBSERVATORY
378433 ARMENIA
U.S.S.R.
TEL: 56-34-53
TLX:
TLF:
EML:
COM: 27.37

PARSONS SIDNEY B DR
SPACE TELESCOPE SCI INST
HOMEWOOD CAMPUS
3700 SAN MARTIN DRIVE
BALTIMORE MD 21218
U.S.A.
TEL: 301-338-4807
TLX:
TLF:
EML:
COM: 29.45

PARTHASARATHY M DR
INDIAN INSTITUTE OF
ASTROPHYSICS
BANGALORE 560 034
INDIA
TEL: 566585/566497
TLX: 845763 IIAB IN
TLF:
EML:
COM: 27.29.42

PARTRIDGE ROBERT B PROF
HAVERFORD COLLEGE
HAVERFORD PA 19041
U.S.A.
TEL: 215-896-1144
TLX:
TLF:
EML:
COM: 47V

PASACHOFF JAY M PROF
WILLIAMS COLLEGE
HOPKINS OBSERVATORY
WILLIAMSTOWN MA 01267
U.S.A.
TEL: 413-597-2105
TLX:
TLF:
EML: BITNET:PASACHOFF@WILLIAMS
COM: 12.40.46C

PASCOAL ANTONIO J B SCI
OBSERVATORIO ASTRONOMICO
PROF MANUEL DE BARROS
MONTE DA VIRGEM
4400 VILA NOVA DE GAIA
PORTUGAL
TEL: 7820404
TLX:
TLF:
EML:
COM:

PASCU DAN DR
US NAVAL OBSERVATORY
WASHINGTON DC 20390
U.S.A.
TEL: 202-653-1178
TLX:
TLF:
EML:
COM: 20.24

PASIAN FABIO
OSSERVATORIO ASTRONOMICO
DI TRIESTE
VIA G.B. TIEPOLO 11
I-34131 TRIESTE
ITALY
TEL: 040-768005
TLX: 461137 OAT I
TLF:
EML:
COM: 09

PASINETTI LAURA E PROF
DIPARTIMENTO DI FISICA
UNIVERSITA DI MILANO
VIA CELORIA 16
I-20133 MILANO
ITALY
TEL: 2-2392275/272
TLX: 334687 INFN MI
TLF: (0039)2)2366583
EML: PASINETTI VAXMI.INFNET
COM: 05.29.36.45.51

PASTORI LIVIO
OSSERVATORIO ASTRONOMICO
VIA BIANCHI 46
I-22055 MERATE
ITALY
TEL: 592035
TLX:
TLF:
EML:
COM:

PASTORIZA MIRIANI G DR
INSTITUTO DE FISICA
UNIVERSIDADE FEDERAL
DO RIO GRANDE DO SUL
90000 PORTO ALEGRE
BRAZIL
TEL: 512-71-666
TLX: 051-1055 UFRS BR
TLF:
EML:
COM: 28

PATERNO LUCIO PROF
OSSERVATORIO ASTROFISICO
CITTA UNIVERSITARIA
I-95125 CATANIA
ITALY
TEL: 095-33-0533
TLX: 970359 ASTRCT I
TLF:
EML:
COM: 10.27

PATERSON-BEECKMANS F
VINCENT VAN GOGHLAAN 19
NL-2343 RH OEGSTGEEST
NETHERLANDS
TEL: 071-170829
TLX:
TLF:
EML:
COM: 29

PATHRIA RAJ K PROF
DEPT OF PHYSICS
UNIVERSITY OF WATERLOO
WATERLOO ONT N2L 3G1
CANADA
TEL: 519-885-1211
TLX: 069-55259
TLF:
EML:
COM:

PATI A K
INDIAN INST OF ASTROPHYS
KORAMANGALA
BANGALORE 560034
INDIA
TEL: 569702
TLX: 0845 2763
TLF:
EML:
COM:

PATKOS LASZLO DR
KONKOLY OBSERVATORY
HUNGARIAN ACADEMY OF SCI
BOX 67
H-1525 BUDAPEST
HUNGARY
TEL:
TLX: 227460
TLF:
EML:
COM: 42

PATRIARCHI PATRIZIO DR
OSSERVATORIO ASTROFISICO
DI ARCETRI
LARGO E. FERMI 5
I-50125 FIRENZE
ITALY
TEL: 055-2752282
TLX: 572268 ARCETR I
TLF:
EML:
COM:

PATUREL GEORGES
OBSERVATOIRE DE LYON
F-69230 ST-GENIS-LAVAL
FRANCE
TEL: 78-56-07-05
TLX: 310926
TLF:
EML:
COM: 28

PAULINY TOTH IVAN K K DR
MPI FUER RADIOASTRONOMIE
AUF DEM HUEGEL 69
D-5300 BONN
GERMANY. F.R.
TEL: 228-525-243
TLX: 886443
TLF:
EML:
COM: 40.48

PAULS THOMAS ALBERT DR
NAVAL RESEARCH LABORATORY
CODE 4130
WASHINGTON DC 20375-5000
U.S.A.
TEL:
TLX:
TLF:
EML:
COM: 33.34.40

PAUWELS T DR
KONINKLIJKE STERRENWACHT
VAN BELGIE
RINGLAAN 3
B 1180 BRUSSELS
BELGIUM
TEL: 32 2 3752484
TLX: 21565 OBS BEL
TLF: 32 2 374 98 22
EML:
COM: 07.20

PAVLOVSKAYA E D DR
STERNBERG STATE
ASTRONOMICAL INSTITUTE
117234 MOSCOW
U.S.S.R.
TEL:
TLX:
TLF:
EML:
COM: 33

PAVLOVSKI KRESIMIR
HVAR OBSERVATORY
FACULTY OF GEODESY
KACICEVA 26
ZAGREB 41000
YUGOSLAVIA
TEL: (41) 442600
TLX:
TLF:
EML:
COM:

PAXTON HAROLD J B R
ROYAL GREENWICH OBS
HERSTMONCEUX CASTLE
HAILSHAM BN27 1RP
U.K.
TEL:
TLX:
TLF:
EML:
COM:

PAYNE DAVID G
JET PROPULSION LAB.
MS 264-748
4800 OAK GROVE DRIVE
PASADENA CA 91109
U.S.A.
TEL:
TLX:
TLF:
EML:
COM: 40

PEACH GILLIAN DR
DEPT PHYSICS & ASTRONOMY
UNIVERSITY COLLEGE
GOWER STREET
LONDON WC1E 6BT
U.K.
TEL: 01-387-7050
TLX: 28722
TLF:
EML:
COM: 14

PEACH JOHN V DR
DEPT OF ASTROPHYSICS
SOUTH PARKS ROAD
OXFORD OX1 3RQ
U.K.
TEL: 0865-511336
TLX:
TLF:
EML:
COM:

PEACOCK JOHN ANDREW
ROYAL OBSERVATORY
BLACKFORD HILL
EDINBURGH EH9 3HJ
U.K.
TEL: 031-667-3321
TLX: 72383 ROEDIN G
TLF:
EML:
COM: 47

PEALE STANTON J PROF
DEPT OF PHYSICS
UNIVERSITY OF CALIFORNIA
SANTA BARBARA CA 93106
U.S.A.
TEL: 805-961-2977
TLX:
TLF:
EML:
COM: 07

PEARCE GILLIAN DR
DEPARTMENT OF ASTROPHYS
UNIVERSITY OF OXFORD
KEBLE ROAD
OXFORD OX1 3RH
U.K.
TEL: 273297
TLX: 83295 NUCLOX G
TLF:
EML:
COM: 55

PEARSON TIMOTHY J
OWENS VALLEY RADIO OBS
CALTECH 105-24
PASADENA CA 91125
U.S.A.
TEL: 818-356-4980
TLX: 675425
TLF:
EML:
COM: 40

PECINA PETR
ASTRONOMICAL INSTITUTE
CZECH. ACAD. OF SCIENCES
251 65 ONDREJOV
CZECHOSLOVAKIA
TEL: 25-87-57
TLX: 122486
TLF:
EML:
COM: 22

PECKER JEAN-CLAUDE PROF
COLLEGE DE FRANCE
3 RUE D'ULM
F-75231 PARIS CEDEX 05
FRANCE
TEL: (1)43291211
TLX:
TLF:
EML:
COM: 05.12.34.36.47

PEDERSEN BENT M DR
OBSERVATOIRE DE PARIS
SECTION DE MEUDON
F-92195 MEUDON PL CEDEX
FRANCE
TEL: 1-45-07-78-39
TLX:
TLF:
EML:
COM: 10

PEDERSEN HOLGER DR
E S O
CASILLA 19001
SANTIAGO 19
CHILE
TEL: 056-2-88757
TLX: 240881 ESOGO CL
TLF:
EML:
COM:

PEDERSEN OLAF PROF
HISTORY OF SCIENCE INST
UNIVERSITY OF AARHUS
NY MUNKEGADE
DK-8000 AARHUS C
DENMARK
TEL: 06-127188
TLX:
TLF:
EML:
COM: 41

PEDLAR ALAN DR
VLA PROJECT
BOX O
SOCORRO NM 87801
U.S.A.
TEL:
TLX:
TLF:
EML:
COM: 40

PEDOUSSAUT ANDRE
OBSERVATOIRE PIC-DU-MIDI
ET TOULOUSE
14 AVENUE EDOUARD BELIN
F-31400 TOULOUSE CEDEX
FRANCE
TEL: 61-25-21-01
TLX: 530776 OBSTLSE
TLF:
EML:
COM: 24.33

PEDREROS MARIO DR
ST MARY UNIVERSITY
DPT OF ASTRONOMY
HALIFAX N SCOTIA B3H 3C3
CANADA
TEL: 902 420 5640
TLX:
TLF:
EML:
COM: 25.37

PEEBLES P JAMES E
JOSEPH HENRY LABS
JADWIN HALL
PRINCETON NJ 08544
U.S.A.
TEL: 609-452-4386
TLX: 499-1517
TLF:
EML:
COM: 47

PEERY BENJAMIN F PROF
DEPT PHYSICS & ASTRONOMY
HOWARD UNIVERSITY
WASHINGTON DC 20059
U.S.A.
TEL: 202-636-6267
TLX:
TLF:
EML:
COM: 29

PEIMBERT MANUEL DR
INSTITUTO DE ASTRONOMIA
UNAM
APDO POSTAL 70-264
04510 MEXICO DF
MEXICO
TEL: 905-548-5306
TLX: 01760155 CIMCE
TLF:
EML:
COM: 28.33.34

PEKERIS CHAIM LEIB PROF
DEPT APPLIED MATHS
WEIZMANN INST OF SCIENCE
REHOVOT 76100
ISRAEL
TEL: 08-483292
TLX: 361900
TLF:
EML:
COM:

PEKUNLUE E RENNAN DR
EGE UNIVERSITY
FACULTY OF SCIENCE
BORNOVA-IZMIR
TURKEY
TEL: 222295
TLX:
TLF:
EML:
COM:

PEL JAN WILLEM DR
KAPTEIJN STERREWACHT
MENSINGHEWEG 20
NL-9301 KA RODEN
NETHERLANDS
TEL:
TLX: 53767 KSW RO NL
TLF:
EML:
COM: 25

PELLAS PAUL DR
LABORATOIRE MINERALOGIE
61 RUE BUFFON
F-75005 PARIS
FRANCE
TEL: 1-47-07-28-24
TLX:
TLF:
EML:
COM: 15

PELLERIN JR CHARLES J DR
NASA HEADQUARTERS
ASTROPHYS DIV. CODE EZ
600 INDEPENDENCE AVE SW
WASHINGTON DC 20546
U.S.A.
TEL: 202 453 1437
TLX:
TLF:
EML:
COM:

PELLET ANDRE
OBSERVATOIRE DE MARSEILLE
2 PLACE LE VERRIER
F-13248 MARSEILLE CEDEX 4
FRANCE
TEL: 91-95-90-88
TLX: 420241 F
TLF:
EML:
COM:

PELLETIER GUY DR
OBSERVATOIRE DE GRENOBLE
CERMO BP 68
38402 ST MARTIN D'HERES
FRANCE
TEL:
TLX:
TLF:
EML:
COM:

PENA JOSE
INST ASTRONOMIA. UNAM
AP. POSTAL 70-264
04510 MEXICO D.F.
MEXICO
TEL: 905-5484537
TLX:
TLF: (905)5483712
EML: BITNET:PENASOUNAMVM1
COM:

PENG QIU-HE
DEPT OF ASTRONOMY
NANJING UNIVERSITY
NANJING
CHINA. PEOPLE'S REP.
TEL: 34651 - 2882
TLX: 34151 PRCNU CN
TLF:
EML:
COM: 28.48

PENG YUN-LOU
DEPT OF ASTRONOMY
NANJING UNIVERSITY
NANJING
CHINA. PEOPLE'S REP.
TEL: 37551. 2882
TLX: 34151 PRCNU CN
TLF:
EML:
COM: 40

PENICHE ROSARIO DR
INST ASTRONOMIA
AP. POSTAL 70-264
04510 MEXICO D.F
MEXICO
TEL: 905-5484537
TLX:
TLF: (905)5483712
EML: BITNET:PENAS @ UNAMVM1
COM:

PENNY ALAN JOHN DR
RUTHERFORD APPLETON LAB
CHILTON. DIDCOT OX11 0QX
U.K.
TEL: 0235-21900
TLX: 83159 RUTHBL G
TLF:
EML:
COM: 09.25C.37

PENSADO JOSE DR
OBSERVATORIO ASTRONOMICO
ALFONSO XII-5
28014 MADRID
SPAIN
TEL: 2270107
TLX:
TLF:
EML:
COM:

PENSTON MARGARET
ROYAL GREENWICH OBS
181a HUNTINGDON ROAD
CAMBRIDGE CB3 0DJ
U.K.
TEL: 0223-277701
TLX:
TLF:
EML:
COM: 29

PENSTON MICHAEL V DR
ROYAL GREENWICH OBS
MADINGLEY ROAD
CAMBRIDGE CB3 0EZ
U.K.
TEL: 0223-337548
TLX: 817297 ASTRON G
TLF:
EML:
COM: 34

PENZIAS ARNO A DR
AT&T BELL LABORATORIES
ROOM 6A-409
600 MOUNTAIN AVENUE
MURRAY HILL NJ 07974
U.S.A.
TEL: 201-582-3361
TLX: 13-8650 OR 219348
TLF:
EML:
COM: 34.40.47

PEQUIGNOT DANIEL
OBSERVATOIRE DE PARIS
SECTION DE MEUDON
DAF
F-92195 MEUDON PL CEDEX
FRANCE
TEL: 1-45-07-74-38
TLX: 201571
TLF:
EML:
COM: 34

PERATAH ANNAMANENI DR
INDIAN INSTITUTE OF
ASTROPHYSICS
BANGALORE 560 034
INDIA
TEL: 566585. 566497
TLX: 845763 IIAB IN
TLF:
EML:
COM: 36C

PERAULT MICHEL
RADIOASTRONOMIE E.N.S.
24 RUE LHOMOND
75005 PARIS
FRANCE
TEL: 33-1-43291225
TLX: ENULM 202601
TLF: 33-1-45873489
EML: DECNET:IAPOBS::PERAULT
COM: 34

PERCY JOHN R PROF
DEPARTMENT OF ASTRONOMY
UNIVERSITY OF TORONTO
TORONTO ONT M5S 1A1
CANADA
TEL: 416-978-4971
TLX:
TLF:
EML:
COM: 27V.46C

PERDANG JEAN M DR
INSTITUT D'ASTROPHYSIQUE
UNIVERSITE DE LIEGE
AVENUE DE COINTE 5
B-4200 COINTE-OUGREE
BELGIUM
TEL: 041-52-99-80
TLX: 41264 ASTRLG B
TLF:
EML:
COM:

PERDOMO RAUL
FACULTAD DE CIENCIAS
ASTRON. y GEOFISICAS
PASEO DEL BOSQUE
1900 LA PLATA
ARGENTINA
TEL: 213-8810
TLX: 31151 BULAP
TLF:
EML:
COM: 19

PEREA-DUARTE JAIME D DR
INSTITUTO DE ASTROFISICA
DE ANDALUCIA
APDO 2144
18080 GRANADA SPAIN
SPAIN
TEL: 958-121311
TLX: 78573 IAAG E
TLF:
EML: JAIME@IAA:ES
COM: 28

PEREK LUBOS DR
ASTRONOMICAL INSTITUTE
CZECH. ACAD. OF SCIENCES
BUDECSKA 6
120 23 PRAHA 2
CZECHOSLOVAKIA
TEL: 254234
TLX: 122486
TLF:
EML:
COM: 33.51

PEREZ-DE-TEJADA H A DR
INSTITUTO DE GEOFISICA
UNAM
22860 ENSENADA. B. CALIF.
MEXICO
TEL: 706-674-0601
TLX:
TLF:
EML:
COM: 15

PEREZ-PERAZA JORGE DR
INAOE
AP POSTALES 216 y 51
72000 PUEBLA. PUE.
MEXICO
TEL: 47-04-19
TLX:
TLF:
EML:
COM:

PERINOTTO MARIO PROF
DIPART. DI ASTRONOMIA
UNIVERSITA DEGLI STUDI
LARGO E. FERMI 5
I-50125 FIRENZE
ITALY
TEL: 055-27521
TLX: 572268 ARCETR
TLF:
EML:
COM: 34

PERKINS FRANCIS W DR
PLASMA PHYSICS LAB
PRINCETON UNIVERSITY
PO BOX 451
PRINCETON NJ 08540
U.S.A.
TEL: 609-683-2603
TLX: 5106852399
TLF:
EML:
COM: 49

PERLEY RICHARD ALAN
NRAO
PO BOX 0
SOCORRO NM 87801
U.S.A.
TEL: 505-772-4011
TLX: 910-988-1710
TLF:
EML:
COM: 40

PEROLA GIUSEPPE C DR
ISTITUTO ASTRONOMICO
VIA LANCISI 29
I-00161 ROMA
ITALY
TEL: 06-867525
TLX: 613255 INFNRO
TLF:
EML:
COM: 48

PERRIER CHRISTIAN DR
OBSERVATOIRE DE LYON
RUE CHARLES ANDRE
69561 ST GENIS LAVAL CX
FRANCE
TEL: 33-78 56 07 05
TLX: 310926 OBSLYON F
TLF: 33-72 39 97 91
EML: EARN:perrier@frgag51
COM:

PERRIN JEAN MARIE DR
LABORATOIRE D'ASTRONOMIE
SPATIALE
TRAVERSE DU SIPHON
13012 MARSEILLE
FRANCE
TEL: 91055900
TLX: 420 584 E ASTROSP
TLF:
EML: BITNET:PERRIN@FRLASM51
COM: 21

PERRIN MARIE-NOEL DR
OBSERVATOIRE DE PARIS
61 AVE DE L'OBSERVATOIRE
F-75014 PARIS
FRANCE
TEL: 1-40-51-22-45
TLX:
TLF:
EML:
COM: 29

PERRY CHARLES L DR
DEPT PHYSICS & ASTRONOMY
LOUISIANA STATE UNIV
BATON ROUGE LA 70803
U.S.A.
TEL: 504-388-8287
TLX: 559184
TLF:
EML:
COM: 25.30.33.45

PERRY JUDITH J DR
INSTITUTE OF ASTRONOMY
CAMBRIDGE CB3 OHA
U.K.
TEL: 0223-62204
TLX: 817297
TLF:
EML:
COM:

PERRY PETER M DR
COMPUTER SCIENCES CORP.
8728 COLESVILLE ROAD
SILVER SPRING MD 20910
U.S.A.
TEL: 301 589 1545
TLX:
TLF:
EML:
COM: 44

PERRYMAN MICHAEL A C
ASTROPHYSICS DIVISION
SPACE SCIENCE DEPT ESA
ESTEC. POSTBUS 299
NL-2200 AG NOORDWIJK
NETHERLANDS
TEL: 01719-83615
TLX: 39098
TLF:
EML:
COM: 08.09.24.47

PERSI PAOLO
IST. ASTROFISICA SPAZIALE
C P 67
I-00044 FRASCATI
ITALY
TEL: 396-9425655
TLX: 610261 CNR FRA
TLF:
EML:
COM: 34

PERSIDES SOTIRIOS C
DEPT OF ASTRONOMY
UNIVERSITY THESSALONIKI
GR-54006 THESSALONIKI
GREECE
TEL: 991357
TLX:
TLF:
EML:
COM: 47

PESCE PETER DR
DIV ASTRONOMICAL SCIENCES
NATL SCIENCE FOUNDATION
1800 G STREET. N.W.
WASHINGTON DC 20550
U.S.A.
TEL: 202-357-7622
TLX:
TLF:
EML:
COM: 33

PESEK RUDOLPH PROF
CZECH. ACAD. OF SCIENCES
PRAGUE
CZECHOSLOVAKIA
TEL:
TLX:
TLF:
EML:
COM: 51

PETERS GERALDINE JOAN DR
SPACE SCIENCES CENTER
UNIV SOUTHERN CALIFORNIA
UNIVERSITY PARK
LOS ANGELES CA 90089-1341
U.S.A.
TEL: 213-743-6967
TLX: 4720490 USC USA
TLF:
EML:
COM: 29.36.42.44

PETERS WILLIAM L III DR
ASTRONOMY DEPT
UNIVERSITY OF TEXAS
AUSTIN TX 78712
U.S.A.
TEL:
TLX:
TLF:
EML:
COM: 28.34.40

PETERSEN J O DR
UNIVERSITY OBSERVATORY
OESTER VOLDGADE 3
DK-1350 COPENHAGEN K
DENMARK
TEL: 1-14-17-90
TLX:
TLF:
EML:
COM: 27

PETERSON BRADLEY MICHAEL
DEPT OF ASTRONOMY
OHIO STATE UNIVERSITY
174 W 18TH AVENUE
COLUMBUS OH 43210
U.S.A.
TEL: 614-422-7886
TLX:
TLF:
EML:
COM: 28

PETERSON BRUCE A DR
MT STROMLO & SIDING
SPRING OBSERVATORIES
AUSTRALIAN NAT UNIVERSITY
WODEN P.O. ACT 2606
AUSTRALIA
TEL: 61-62-88-1111
TLX: 62270
TLF:
EML:
COM: 47.48

PETERSON CHARLES JOHN DR
DEPT PHYSICS & ASTRONOMY
UNIVERSITY OF MISSOURI
223 PHYSICS BLDG
COLUMBIA MO 65211
U.S.A.
TEL: 314-882-3217
TLX:
TLF:
EML:
COM: 28.37.41

PETERSON LAURENCE E PROF
CASS C-011
UNIVERSITY OF CALIFORNIA
AT SAN DIEGO
LA JOLLA CA 92093
U.S.A.
TEL: 619-452-3461
TLX: 910-337-1271 SIOCEAN
TLF:
EML:
COM: 44.48

PETERSON RUTH CAROL DR
607 MARION PLACE
PALO ALTO CA 94301
U.S.A.
TEL: 415-321-1281
TLX:
TLF:
EML:
COM: 29.30

PETFORD A DAVID DR
DEPT OF ASTROPHYSICS
SOUTH PARKS ROAD
OXFORD OX1 3RQ
U.K.
TEL: 865-511336
TLX:
TLF:
EML:
COM: 09

PETON ALAIN DR
OBSERVATOIRE DE MARSEILLE
2 PLACE LE VERRIER
F-13248 MARSEILLE CEDEX 4
FRANCE
TEL: 91-95-90-88
TLX:
TLF:
EML:
COM:

PETRI WINFRIED PROF DR
UNTERLEITEN 2
POSTFACH 136
D-8162 SCHLIERSEE
GERMANY. F.R.
TEL: 08026-6428
TLX:
TLF:
EML:
COM: 41

PETRINI DANIEL DR
OBSERVATOIRE DE NICE
BP 139
F-06003 NICE CEDEX
FRANCE
TEL: 93-89-04-20
TLX: 460004
TLF:
EML:
COM: 14

PETRO LARRY DAVID
SPACE TELESCOPE SCI INST
HOMEWOOD CAMPUS
3700 SAN MARTIN DRIVE
BALTIMORE MD 21218
U.S.A.
TEL: 301-338-4581
TLX: 6849101
TLF:
EML:
COM:

PETROPOULOS BASIL CH DR
RES CTR ASTRON APPL MATHS
ACADEMY OF ATHENS
14 ANAGNOSTOPOULOU
GR-10673 ATHENS
GREECE
TEL: 3613589
TLX:
TLF:
EML:
COM: 14.16

PETROSIAN VAHE PROF
CTR FOR SPACE SCIENCE &
ASTROPHYSICS. ERL RM 304
STANFORD UNIVERSITY
STANFORD CA 94306
U.S.A.
TEL: 415-497-1435
TLX:
TLF:
EML:
COM: 10.34.47.48

PETROU MARIA DR
INFORMATICS. RAL
CHILTON. DIDCOT. OXON
OX11 0QX
U.K.
TEL: 0235 445498
TLX:
TLF:
EML:
COM:

PETROV G M DR
NIKOLAEV DEPT OF THE
MAIN ASTRONOMICAL OBS
327000 NIKOLAEV
U.S.S.R.
TEL: 36-18-24
TLX:
TLF:
EML:
COM: 08

PETROV GENNADIJ M
INST RADIOTECH & ELECTRON
USSR ACADEMY OF SCIENCES
MARKS AVENJU 18
103907 MOSCOW GSP-3
U.S.S.R.
TEL:
TLX:
TLF:
EML:
COM:

PETROV GEORGIJ I PROF DR
SPACE RESEARCH INSTITUTE
USSR ACADEMY OF SCIENCES
117810 MOSCOW
U.S.S.R.
TEL:
TLX:
TLF:
EML:
COM:

PETROV GEORGY TRENDAFILOV
DEPT OF ASTRONOMY
BULGARIAN ACAD SCIENCES
72 LENIN BLVD
1784 SOFIA
BULGARIA
TEL: 75-89-27
TLX: 23561 ECF BAN BG
TLF:
EML:
COM: 28

PETROV NIKOLAI
ASTRONOMICAL OBSERVATORY
PO BOX 120
9000 VARNA
BULGARIA
TEL: 22-28-90
TLX:
TLF:
EML:
COM:

PETROV PETER P DR
CRIMEAN ASTROPHYSICAL OBS
USSR ACADEMY OF SCIENCES
NAUCHNY
334413 CRIMEA
U.S.S.R.
TEL: 43 29 45
TLX:
TLF:
EML:
COM: 09.27

PETROVSKAYA M S DR
INST OF THEORET ASTRONOMY
10 KUTUZOV QUAY
191187 LENINGRAD
U.S.S.R.
TEL: 121578 ITA SU
TLX:
TLF:
EML:
COM: 07.37

PETTENGILL GORDON H PROF
MIT
RM 37-241
CAMBRIDGE MA 02139
U.S.A.
TEL: 617-253-7501
TLX: 92-1473
TLF:
EML:
COM: 16.40

PETTERSEN BJOERN RAGNVALD
INST THEORET ASTROPHYSICS
UNIVERSITY OF OSLO
P O BOX 1029
N-0315 BLINDERN OSLO 3
NORWAY
TEL: 02-45-65-01
TLX: 72705 ASTRO N
TLF:
EML:
COM: 27

PETTINI MARCO
OSSERVATORIO ASTROFISICO
DI ARCETRI
LARGO E. FERMI 5
I-50125 FIRENZE
ITALY
TEL: 055-2752282
TLX: 572268 ARCETR
TLF:
EML:
COM: 14

PETTINI MAX
ROYAL GREENWICH OBS
HERSTMONCEUX CASTLE
HAILSHAM BN27 1RP
U.K.
TEL: 44-323-833171
TLX: 87451 RGOBSY G
TLF:
EML:
COM:

PEYTURAUX ROGER H PROF
INSTITUT D'ASTROPHYSIQUE
98 BIS BOULEVARD ARAGO
F-75014 PARIS
FRANCE
TEL: 1-43-20-14-25
TLX:
TLF:
EML:
COM: 12

PFAU WERNER
UNIVERSITY OBSERVATORY
SCHILLERGAESSCHEN 2
DDR-6900 JENA
GERMANY. D.R.
TEL: 058861347
TLX:
TLF:
EML:
COM: 25

PFEIFFER RAYMOND J
6 BARBARA LANE
TITUSVILLE NJ 08560
U.S.A.
TEL: 609-883-4612
TLX:
TLF:
EML:
COM: 25

PFENNIG HANS H DR
MPI F PHYSIK & ASTROPHYS
KARL-SCHWARZSCHILD-STR 1
D-8046 GARCHING B MUNCHEN
GERMANY. F.R.
TEL: 89-32-99-94-35
TLX: 524629 ASTRO D
TLF:
EML:
COM: 14

PFENNIGER DANIEL DR
OBSERVATOIRE DE GENEVE
CH-1290 SAUVERNY
SWITZERLAND
TEL: 022-552611
TLX: 419209 OBSG CH
TLF: 41 22 553983
EML: BITNET:PFENNINGE@CGEUGE54
COM: 28

PFLEIDERER JORG PROF
INSTITUT FUER ASTRONOMIE
TECHNIKERSTRASSE 15
A-6020 INNSBRUCK
AUSTRIA
TEL: 5222-748-5251
TLX:
TLF:
EML:
COM: 21

PFLUG KLAUS DR
ZNTRLINST. F. ASTROPHYSIK
SONNENOBSERVATORIUM
EINSTEINTURM
DDR-1500 POTSDAM
GERMANY. D.R.
TEL:
TLX:
TLF:
EML:
COM: 10.12.49

PHAM-VAN JACQUELINE MME
CERGA
AVENUE COPERNIC
F-06130 GRASSE
FRANCE
TEL: 93-36-58-49
TLX: 470865
TLF:
EML:
COM: 08

PHILIP A G DAVIS
1125 OXFORD PLACE
SCHENECTADY NY 12308
U.S.A.
TEL: 518-374-5636
TLX:
TLF:
EML:
COM: 05.25.30.33.37.45

PHILLIPS JOHN G PROF
ASTRONOMY DEPT
UNIVERSITY OF CALIFORNIA
601 CAMPBELL HALL
BERKELEY CA 94720
U.S.A.
TEL: 415-642-5275
TLX:
TLF:
EML:
COM: 14.36

PHILLIPS JOHN PETER
PHYSICS DEPT
QUEEN MARY COLLEGE
MILE END ROAD
LONDON E1 4NS
U.K.
TEL: 01-980-4811
TLX: 893750 QMEUOL G
TLF:
EML:
COM: 34

PHILLIPS KENNETH J H
SPACE & ASTROPHYSICS DIV
RUTHERFORD APPLETON LAB
CHILTON. DIDCOT OX11 0QX
U.K.
TEL: 0235-21900
TLX:
TLF:
EML:
COM: 10.12.44

PHILLIPS MARK M DR
CERRO TOLOLO INTER-
AMERICAN OBSERVATORY NOAO
CASILLA 603
LA SERENA
CHILE
TEL: 51-213352
TLX: 620301 AURA CT
TLF:
EML:
COM: 28.35

PHILLIPS THOMAS GOULD DR
CALTECH
320-47
PASADENA CA 91125
U.S.A.
TEL: 818-356-4278
TLX:
TLF:
EML:
COM: 34.40

PIAZZA LILIANA RIZZO
INST. PESQUISAS ESPACIAIS
C.P. 515
12200 S. JOSE DOS CAMPOS
BRAZIL
TEL: 011-825365
TLX: 11-34061 INPE BR
TLF:
EML:
COM:

PICAT JEAN-PIERRE DR
OBSERVATOIRE PIC-DU-MIDI
& TOULOUSE
14 AVENUE EDOUARD BELIN
F-31400 TOULOUSE
FRANCE
TEL: 61-25-21-01
TLX:
TLF:
EML:
COM: 09

PICCIONI ADALBERTO
ISTITUTO ASTRONOMICO
UNIVERSITARIO
C P 596
I-40100 BOLOGNA
ITALY
TEL: 051-222956
TLX: 211664 INFNBO I
TLF:
EML:
COM: 42

PICK MONIQUE DR
OBSERVATOIRE DE PARIS
SECTION DE MEUDON
DASOP
F-92195 MEUDON PL CEDEX
FRANCE
TEL: 1-45-07-78-11
TLX: 203590
TLF:
EML:
COM: 10C.40

PICKLES ANDREW JOHN DR
INSTITUTE FOR ASTRONOMY
2680 WOODLAWN DRIVE
HONOLULU HAWAII 96822
U.S.A.
TEL: 808-948 6756
TLX: 7238459 UHAST HR
TLF: 808-9882790
EML: picles@uhifa.ifa.hawaii.edu
COM: 28

PIDDINGTON JACK H RES FEL
NATIONAL MEASUREMENT LAB
CSIRO. P O B 218
LINDFIELD
SYDNEY NSW 2070
AUSTRALIA
TEL: 467 6211
TLX: 26296 AA
TLF:
EML:
COM: 10.48

PIER JEFFREY R DR
NAVAL OBSERVATORY
FLAGSTAFF STATION
PO BOX 1149
FLAGSTAFF AZ 86002
U.S.A.
TEL: 602 779 5132
TLX:
TLF:
EML:
COM: 33

PIERCE A KEITH DR
NATL SOLAR OBSERVATORY
PO 26732
TUCSON AZ 85726
U.S.A.
TEL: 602-327-5511
TLX:
TLF:
EML:
COM: 07.12

PIERCE DAVID ALLEN
SCIENCE & MATH DIVISION
EL CAMINO COLLEGE
TORRANCE CA 90506
U.S.A.
TEL:
TLX:
TLF:
EML:
COM: 20

PIERRE MARGUERITE DR
ESO
KARL-SCHWARZSCHILD STR.2
D-8046 GARCHING B.MUNCHEN
GERMANY. F.R.
TEL: 89 320 06 293
TLX: 5282820 EO D
TLF:
EML: BITNET:PIERRE@DSONC1
COM:

PIGATTO LUISA DR
OSSERVATORIO ASTRONOMICO
VICOLO DELL OSSERV. 5
PADOVA 35122
ITALY
TEL: 049 661499
TLX: 432071
TLF:
EML:
COM: 41

PIIROLA VILPPU E DR
OBSERVATORY
TAHTITORNINMAKI
SF-00130 HELSINKI 13
FINLAND
TEL: 90-1912801
TLX: 124690 UNIH SF
TLF:
EML:
COM: 25.27.42

PIKE CHRISTOPHER DAVID
ROYAL GREENWICH OBS
HERSTMONCEUX CASTLE
HAILSHAM BN27 1RP
U.K.
TEL:
TLX:
TLF:
EML:
COM:

PILACHOWSKI CATHERINE DR
KITT PEAK NATL OBS
NTL OPTICAL ASTR OBS
PO BOX 26732
TUCSON AZ 85726
U.S.A.
TEL: 602-327-5511
TLX: 0666484 AURA NOAOTUC
TLF:
EML:
COM: 29C.37V

PILCHER CARL BERNARD DR
INSTITUTE FOR ASTRONOMY
UNIVERSITY OF HAWAII
2680 WOODLAWN DRIVE
HONOLULU HI 96822
U.S.A.
TEL: 808-948-7954
TLX: 723-8459 UHAST
TLF:
EML:
COM: 15

PILKINGTON JOHN D H DR
ROYAL GREENWICH OBS
HERSTMONCEUX CASTLE
HAILSHAM BN27 1RP
U.K.
TEL: 323 841139
TLX: 87451 RGOBSY G
TLF:
EML:
COM: 19.31C

PILOWSKI K PROF DR
GEODAETISCHES INSTITUT
TECHNISCHE UNIVERSITAET
NIENBURGER STR 1
D-3000 HANNOVER
GERMANY. F.R.
TEL:
TLX:
TLF:
EML:
COM: 08.33

PINEAULT SERGE DR
DEPT DE PHYSIQUE
UNIVERSITE LAVAL
SAINTE-FOY PQ G1K 7P4
CANADA
TEL: 418-656-3901
TLX:
TLF:
EML:
COM:

PINES DAVID PROF
DEPT OF PHYSICS
UNIVERSITY OF ILLINOIS
URBANA IL 61801
U.S.A.
TEL: 217-333-0115
TLX: 9103806599 PHYSICS S
TLF:
EML:
COM: 35

PINGREE DAVID PROF
BROWN UNIVERSITY
PO BOX 1900
PROVIDENCE RI 02912
U.S.A.
TEL: 401-863-2101
TLX:
TLF:
EML:
COM: 41

PINIGIN GENNADIJ I DR
NIKOLAYEV BRANCH
CENTRAL ASTRONOMICAL OBS
OBSERVATORNAYA 1
327001 NIKOLAYEV REGIONAL
U.S.S.R.
TEL:
TLX:
TLF:
EML:
COM: 08C

PINKAU K PROF
MPI FUER PLASMAPHYSIK
D-8046 GARCHING B MUNCHEN
GERMANY. F.R.
TEL: 89-3299-342
TLX: 05-215-808
TLF:
EML:
COM: 44.46

PINOTSIS ANTONIS D DR
DEPT OF ASTRONOMY
UNIVERSITY OF ATHENS
PANEPISTIMIOPOLIS
ATHENS 621
GREECE
TEL:
TLX:
TLF:
EML:
COM: 35

PINTO GIROLAMO PROF
OSSERVATORIO ASTRONOMICO
I-35100 PADOVA
ITALY
TEL:
TLX:
TLF:
EML:
COM:

PIOTTO GIAMPAOLLO
DIPART. DI ASTRONOMIA
VICOLO DELL'OSSERV. 5
I-35122 PADOVA
ITALY
TEL: 049-661499
TLX: 432071 ASTROS I
TLF: 049-38919
EML:
COM:

PIPHER JUDITH L
PHYSICS & ASTRONOMY DEPT
UNIVERSITY OF ROCHESTER
ROCHESTER NY 14627
U.S.A.
TEL: 716-275-4402
TLX:
TLF:
EML:
COM: 44

PIRRONELLO VALERIO
OSSERVATORIO ASTROFISICO
CITTA UNIVERSITARIA
VIALE A. DORIA
I-95125 CATANIA
ITALY
TEL: 095-330533
TLX: 970359 ASTRCT I
TLF:
EML:
COM:

PISKUNOV ANATOLY E
ASTRONOMICAL COUNCIL
USSR ACADEMY OF SCIENCES
PYATNITSKAYA 48
109017 MOSCOW
U.S.S.R.
TEL: 231-54-61
TLX: 412623 SCSTP SU
TLF:
EML:
COM: 37

PISMIS DE RECILLAS PARIS
INSTITUTO DE ASTRONOMIA
UNAM
APDO POSTAL 70-264
04510 MEXICO DF
MEXICO
TEL: 905-548-5306
TLX: 1760155 CICME
TLF:
EML:
COM: 28.33.34

PITTICH EDUARD M DR
ASTRONOMICAL INSTITUTE
SLOVAK ACAD. OF SCIENCES
DUBRAVSKA CESTA 9
842 28 BRATISLAVA
CZECHOSLOVAKIA
TEL: 427-375157
TLX: 93373 SEIS
TLF:
EML:
COM: 15.20

PITZ ECKHART DR
MPI FUER ASTRONOMIE
KOENIGSTUHL
D-6900 HEIDELBERG
GERMANY. F.R.
TEL: 06221-5281
TLX: 461789 MPIA D
TLF:
EML:
COM: 21

PIZZELLA G DR
DIPARTIMENTO DI FISICA
UNIVERSITA DI ROMA
PIAZZALE ALDO MORO 2
I-00185 ROMA
ITALY
TEL: 6-4940156
TLX: 613255 INFNRO
TLF:
EML:
COM:

PIZZICHINI GRAZIELLA
ISTITUTO TESRE/CNR
VIA DE CASTAGNOLI 1
I-40126 BOLOGNA
ITALY
TEL: 051-519593
TLX: 511350 CNR BO
TLF:
EML:
COM:

PLAKIDIS STAVROS PROF
EVRYTANIAS 16
KATO HALANDRI
GR-15231 ATHENS
GREECE
TEL: 6721770
TLX:
TLF:
EML:
COM:

PLANESAS PERE
CENTRO ASTRON DE YEBES
O A N
APARTADO CORREOS 148
19080 GUADALAJARA
SPAIN
TEL: 11-22-33-58
TLX:
TLF:
EML:
COM:

PLASSARD J DR
KSARA OBSERVATORY
KSARA
LEBANON
TEL:
TLX:
TLF:
EML:
COM:

PLATAIS IMANT K DR
RADIOASTROPHYSICAL OBS
LATVIAN ACAD OF SCIENCES
226524 RIGA
U.S.S.R.
TEL: 932088
TLX:
TLF:
EML:
COM: 37

PLAVEC MIREK J PROF
DEPT OF ASTRONOMY
UNIVERSITY OF CALIFORNIA
MS 8979
LOS ANGELES CA 90024
U.S.A.
TEL: 213-825-1672
TLX:
TLF:
EML:
COM: 29.35.42

PLAVEC ZDENKA DR
DEPT OF ASTRONOMY
UCLA
405 HILGARD AVE
LOS ANGELES CA 90024
U.S.A.
TEL: 213-206-8596
TLX:
TLF:
EML:
COM: 22

PNEUMAN GERALD W
HIGH ALTITUDE OBSERVATORY
PO BOX 3000
BOULDER CO 80302
U.S.A.
TEL: 303-497-1000
TLX: 45694
TLF:
EML:
COM: 10.49

PODOBED V V DR
STERNBERG STATE
ASTRONOMICAL INSTITUTE
117234 MOSCOW
U.S.S.R.
TEL:
TLX:
TLF:
EML:
COM: 08.24

PONCKERT ROLAND E DR
DEFENCE RESEARCH
ESTABLISHMENT PACIFIC
FMO CFB ESQUIMALT
VICTORIA BC V0S 1B0
CANADA
TEL:
TLX:
TLF:
EML:
COM: 29

POEPPEL WOLFGANG G L DR
INSTITUTO ARGENTINO
DE RADIOASTRONOMIA
CASILLA DE CORREO 5
1894 VILLA ELISA (Bs.As.)
ARGENTINA
TEL: 021-43793
TLX: 18052 CICYT-AR
TLF:
EML:
COM: 34

POGO ALEXANDER DR
MT WILSON & LAS CAMPANAS
OBSERVATORIES
813 SANTA BARBARA ST
PASADENA CA 91101
U.S.A.
TEL: 213-577-1122
TLX:
TLF:
EML:
COM: 41

POHL ECKHARD DR
STERNWARTE NUERNBERG
REGIOMONTANUSWEG 1
D-8500 NUERNBERG 20
GERMANY. F.R.
TEL: 0911-593540
TLX:
TLF:
EML:
COM:

POLAND ARTHUR I DR
NASA/GSFC
CODE 682
GREENBELT MD 20771
U.S.A.
TEL: 301-286-7334
TLX: 89675
TLF:
EML:
COM: 10

POLCARO V F
IST. ASTROFISICA SPAZIALE
C P 67
I-00044 FRASCATI
ITALY
TEL: 9425651
TLX: 610261
TLF:
EML:
COM:

POLECHOVA PAVLA DR
OBS AND PLANETARIUM
OF PRAGUE
PETRIN 205
118 46 PRAHA 1
CZECHOSLOVAKIA
TEL: 4202-5353513
TLX:
TLF:
EML:
COM: 05

POLETTO GIANNINA PROF
OSSERVATORIO ASTROFISICO
DI ARCETRI
LARGO E. FERMI 5
I-50125 FIRENZE
ITALY
TEL: 055-2752252
TLX: 572268 ARCETR I
TLF:
EML:
COM: 10

POLIDAN RONALD S
UNIVERSITY OF ARIZONA
LUNAR/PLANETARY LAB. WEST
901 GOULD-SIMPSON BUILD.
TUCSON AZ 85721
U.S.A.
TEL: 602-621-4301
TLX: 910-952-1143
TLF:
EML:
COM: 42.44

POLLACK JAMES B DR
SPACE SCIENCE DIVISION
NASA-AMES RESEARCH CTR
MS 245-3
MOFFETT FIELD CA 94035
U.S.A.
TEL: 415-694-5530
TLX:
TLF:
EML:
COM: 16.51

POLNITZKY GERHARD DR
INSTITUT FUER ASTRONOMIE
UNIVERSITAET WIEN
TUERKENSCHANZSTR 17
A-1180 WIEN
AUSTRIA
TEL: 0222-345360-90
TLX: 116222 PHYSI A
TLF:
EML:
COM: 08.22

POLOZHENTSEV DIMITRIJ DR
PULKOVO OBSERVATORY
196140 LENINGRAD
U.S.S.R.
TEL: 298-22-42
TLX:
TLF:
EML:
COM: 08.24C

POLUPAN P N DR
KIEV STATE UNIVERSITY
ASTRONOMICAL OBSERVATORY
252053 KIEV
U.S.S.R.
TEL: 26-09-08
TLX: 132201
TLF:
EML:
COM: 10

POMA ANGELO DR
INTL ASTRONOMICAL STATION
VIA OSPEDALE 72
I-09100 CAGLIARI
ITALY
TEL: 070-66-35-44
TLX: 790326 OSSAST
TLF:
EML:
COM: 08.19

PONMAN TREVOR DR
DEPT OF SPACE RESEARCH
UNIV OF BIRMINGHAM
PO BOX 363
BIRMINGHAM B15 2TT
U.K.
TEL:
TLX:
TLF:
EML:
COM:

PONNAMPERUMA CYRIL PROF
DEPT OF CHEMISTRY
UNIVERSITY OF MARYLAND
COLLEGE PARK MD 20472
U.S.A.
TEL:
TLX:
TLF:
EML:
COM: 51

PONSONBY JOHN E B DR
NRAL
JODRELL BANK
MACCLESFIELD SK11 9DL
U.K.
TEL: 0477-71321
TLX: 36149
TLF:
EML:
COM: 40.51

POOLEY GUY DR
CAVENDISH LABORATORY
MADINGLEY ROAD
CAMBRIDGE CB3 0HE
U.K.
TEL: 223-66477
TLX: 81292
TLF:
EML:
COM: 40

POPELAR JOSEF DR
DEPT ENERGY. MINES & RES.
EARTH PHYSICS BRANCH
1 OBSERVATORY CRESCENT
OTTAWA ONT K1A 0Y3
CANADA
TEL: 613-992-5419
TLX: 0533117 EMAR-OTT
TLF:
EML:
COM: 19.31

POPOV VASIL NIKOLOV
DEPT OF ASTRONOMY
BULGARIAN ACAD SCIENCES
72 LENIN BLVD
1784 SOFIA
BULGARIA
TEL: 449-477
TLX:
TLF:
EML:
COM: 28

POPOV VICTOR S DR
MAIN ASTRONOMICAL OBS
USSR ACADEMY OF SCIENCES
PULKOVO M-140
196140 LENINGRAD
U.S.S.R.
TEL:
TLX:
TLF:
EML:
COM: 30

POPOVA MALINA D PROF DR
DEPT OF ASTRONOMY
BULGARIAN ACAD SCIENCES
72 LENIN BLVD
1784 SOFIA
BULGARIA
TEL: 449-477
TLX:
TLF:
EML:
COM: 27.37

POPOVIC BOZIDAR PROF DR
OGNJENA PRICE 80
11000 BEOGRAD
YUGOSLAVIA
TEL:
TLX:
TLF:
EML:
COM: 07.20

POPOVIC GEORGIJE DR
ASTRONOMICAL OBSERVATORY
VOLGINA 7
11050 BEOGRAD
YUGOSLAVIA
TEL: 38-11-419357
TLX:
TLF:
EML:
COM: 26

POPPER DANIEL M PROF
DEPT OF ASTRONOMY
UNIVERSITY OF CALIFORNIA
LOS ANGELES CA 90024
U.S.A.
TEL: 213-825-3622
TLX: 9103427597
TLF:
EML:
COM: 42

POQUERUSSE MICHEL
OBSERVATOIRE DE PARIS
SECTION DE MEUDON
DESPA
F-92195 MEUDON PL CEDEX
FRANCE
TEL: 1-45-07-75-30
TLX: 204464
TLF:
EML:
COM: 10.12

PORCAS RICHARD DR
MPI FUER RADIOASTRONOMIE
AUF DEM HUEGEL 69
D-5300 BONN
GERMANY, F.R.
TEL: 0228-525-282
TLX: 0886440 MPIFR D
TLF:
EML:
COM: 40

PORETTI ENNIO
OSSERVATORIO ASTRONOMICO
DI BRERA
VIA E. BIANCHI 46
MERATE 22055
ITALY
TEL: 039 596412
TLX:
TLF:
EML: PORETTI@ASTMIB.INFN.IT
COM:

PORFIR'EV V V DR
KRUPSKAJA PEDAGOGOC INST
107846 MOSCOW
U.S.S.R.
TEL:
TLX:
TLF:
EML:
COM: 35

PORNCHAI P.-TANAKUN
DEPT OF PHYSICS
CHULALONGKORN UNIVERSITY
BANGKOK 10330
THAILAND
TEL: (02)252 7985
TLX: 20217 UNICHUL TH
TLF:
EML:
COM:

PORTER NEIL A PROF
PHYSICS DEPT
UNIVERSITY COLLEGE
BELFIELD
DUBLIN 4
IRELAND
TEL: 1-693-244x211
TLX: 32693 UCDEI
TLF:
EML:
COM: 41.48

PORUBCAN VLADIMIR DR
ASTRONOMICAL INSTITUTE
SLOVAK ACAD. OF SCIENCES
DUBRAVSKA 9
842 28 BRATISLAVA
CZECHOSLOVAKIA
TEL: 427-375157
TLX: 93373 SEIS
TLF:
EML:
COM: 22

POTTASCH STUART R PROF
KAPTEYN LABORATORIUM
POSTBUS 800
NL-9700 AV GRONINGEN
NETHERLANDS
TEL: 050-116641
TLX: 53572 STARS NL
TLF:
EML:
COM: 34.36

POTTER HEINO I DR
PULKOVO OBSERVATORY
196140 LENINGRAD
U.S.S.R.
TEL: 298-22-42
TLX:
TLF:
EML:
COM: 24

POULAKOS CONSTANTINE DR
RESEARCH CTR F. ASTRONOMY
AND APPLIED MATHS
ACADEMY OF ATHENS
GR-10673 ATHENS
GREECE
TEL:
TLX:
TLF:
EML:
COM:

POULLE EMMANUEL PROF
ECOLE NATLE DES CHARTES
19 RUE DE LA SORBONNE
F-75005 PARIS
FRANCE
TEL: 1-45-89-48-57
TLX:
TLF:
EML:
COM: 41

POUNEVROL FERNAND MR
OBSERVATOIRE DE BORDEAUX
AVENUE P. SEMIROT
F-33270 FLOIRAC
FRANCE
TEL: 56-86-43-30
TLX:
TLF:
EML:
COM:

POUNDS KENNETH A PROF
DEPT OF PHYSICS
THE UNIVERSITY
UNIVERSITY ROAD
LEICESTER LE1 7RH
U.K.
TEL: 0533-954455x151
TLX: 341664 LUXRAYG
TLF:
EML:
COM: 06.44C.48

POUQUET ANNICK DR
OBSERVATOIRE DE NICE
BP 139
F-06003 NICE CEDEX
FRANCE
TEL: 93-89-04-20
TLX: 460004
TLF:
EML:
COM:

POVEDA ARCADIO DR
INSTITUTO DE ASTRONOMIA
UNAM
APDO POSTAL 70-264
04510 MEXICO DF
MEXICO
TEL: 550-5805
TLX: 1760155 CICME
TLF:
EML:
COM: 26.28.35.37

POVET JEAN-PIERRE DR
OBSERVATOIRE PIC-DU-MIDI
ET DE TOULOUSE
14 AVENUE EDOUARD BELIN
F-31400 TOULOUSE
FRANCE
TEL: 61-25-21-01
TLX:
TLF:
EML:
COM:

PRABHU TUSHAR P
INDIAN INSTITUTE OF
ASTROPHYSICS
BANGALORE 560 034
INDIA
TEL:
TLX: 845763 IIAB IN
TLF:
EML:
COM: 28

PRADERIE FRANCOISE DR
OBSERVATOIRE DE PARIS
SECTION DE MEUDON
DEPT RECHERCHE SPATIALE
F-92195 MEUDON PL CEDEX
FRANCE
TEL: 1-45-07-76-51
TLX: 204464
TLF:
EML:
COM: 29.36C

PRADHAN DR
JILA QPD 525
UNIVERSITY OF COLORADO
BOULDER CO 80309
U.S.A.
TEL: 303-492-7812
TLX: 755842
TLF:
EML:
COM:

PRASAD SHEO S
JET PROPULSION LABORATORY
MS 183-601
4800 OAK GROVE DRIVE
PASADENA CA 91109
U.S.A.
TEL: 213-354-6423
TLX: 675429
TLF:
EML:
COM: 34

PRASANNA A R DR
PHYSICAL RESEARCH LAB
NAVRANGPURA
AHMEDABAD 380 009
INDIA
TEL: 462129
TLX: 021-397 PRL IN
TLF:
EML:
COM: 48

PRATAP R DR
INSTITUTE OF APPLIED
SCIENCES
COCHIN 682 317
INDIA
TEL:
TLX:
TLF:
EML:
COM:

PRAVDO STEVEN H
JET PROPULSION LABORATORY
MS 168-222
4800 OAK GROVE DRIVE
PASADENA CA 91109
U.S.A.
TEL: 818-354-4134
TLX: 910-588-3294
TLF:
EML:
COM:

PREITE-MARTINEZ ANDREA DR
IST. ASTROFISICA SPAZIALE
C P 67
I-00044 FRASCATI
ITALY
TEL:
TLX: 610261
TLF:
EML:
COM: 34

PREKA-PAPADEMA P DR
LAB OF ASTROPHYSICS
NATL UNIV OF ATHENS
PANEPISTIMIOPOLIS
GR 15783 ATHENS
GREECE
TEL: 01 7235122
TLX:
TLF:
EML: SPH75@GRATHUN1
COM: 10

PRENDERGAST KEVIN H PROF
DEPT OF ASTRONOMY
COLUMBIA UNIVERSITY
538 W. 120TH STREET
NEW YORK NY 10027
U.S.A.
TEL: 212-280-3280
TLX:
TLF:
EML:
COM: 28

PRENTICE ANDREW J R DR
DEPT OF MATHEMATICS
MONASH UNIVERSITY
CLAYTON VIC 3168
AUSTRALIA
TEL:
TLX:
TLF:
EML:
COM: 35

PRESS WILLIAM H DR
HARVARD COLLEGE OBS
60 GARDEN STREET
CAMBRIDGE MA 02138
U.S.A.
TEL: 617-495-4908
TLX: 921428 SATELLITE CAM
TLF:
EML:
COM: 28.47

PRESTON GEORGE W DR
MT WILSON & LAS CAMPANAS
OBSERVATORIES
813 SANTA BARBARA STREET
PASADENA CA 91101
U.S.A.
TEL: 818-577-1122
TLX:
TLF:
EML:
COM: 29.30.45

PRESTON ROBERT ARTHUR
138-307 JPL
4800 OAK GROVE DRIVE
PASADENA CA 91109
U.S.A.
TEL: 213-354-6895
TLX: 675429
TLF:
EML:
COM: 40

PREUSS EUGEN DR
MPI FUER RADIOASTRONOMIE
AUF DEM HUEGEL 69
D-5300 BONN 1
GERMANY. F.R.
TEL: 228-5251
TLX: 886440 MPIFR D
TLF:
EML:
COM: 40.48

PREVOT LOUIS DR
OBSERVATOIRE DE MARSEILLE
2 PLACE LE VERRIER
F-13248 MARSEILLE CDX 04
FRANCE
TEL: 91-95-90-88
TLX: 420241
TLF:
EML:
COM: 30C

PREVOT-BURNICHON M.L. DR
OBSERVATOIRE DE MARSEILLE
2 PLACE LE VERRIER
F-13248 MARSEILLE CDX 04
FRANCE
TEL: 91-95-90-88
TLX: 420241
TLF:
EML:
COM: 28

PRIALNIK-KOVETZ DINA DR
TEL AVIV UNIVERSITY
RAMAT AVIV
DEPT. OF GEOPHYSICS
TEL AVIV 69978
ISRAEL
TEL: 545 0633
TLX: 342171 VERSY IL
TLF:
EML: BITNET:B130TAUNOS
COM: 15.35

PRICE MICHAEL J. DR.
SCIENCE APPLICATIONS
5151 E BROADWAY
SUITE 1100
TUCSON AZ 85711
U.S.A.
TEL: 602-748-7400
TLX:
TLF:
EML:
COM:

PRICE R MARCUS DR
DEPT PHYSICS & ASTRONOMY
UNIVERSITY OF NEW MEXICO
ALBUQUERQUE NM 87131
U.S.A.
TEL: 505-277-2616
TLX:
TLF:
EML:
COM: 33.34.40

PRICE STEPHAN DONALD
2 POLLEY ROAD
WESTFORD MA 01886
U.S.A.
TEL: 617-861-4552
TLX:
TLF:
EML:
COM: 44

PRIEST ERIC R PROF
APPLIED MATHS DEPT
THE UNIVERSITY
ST ANDREWS. FIFE KY16 9SS
U.K.
TEL: 0334-76161
TLX: 76213 SAULIB
TLF:
EML:
COM: 10P.12

PRIESTER WOLFGANG PROF
INSTITUT F ASTROPHYSIK
AUF DEM HUEGEL 71
D-5300 BONN
GERMANY. F.R.
TEL: 0228-73-3671
TLX: 886440
TLF:
EML:
COM: 33.40

PRIETO MERCEDES
INST DE ASTROFISICA DE
CANARIAS
LA LAGUNA
38071 TENERIFE
SPAIN
TEL: 922-262211
TLX: 92640
TLF:
EML:
COM:

PRINCE HELEN DODSON PROF
4400 FILLMORE AVE
ALEXANDRIA VA 22311
U.S.A.
TEL: 703-578-1000
TLX:
TLF:
EML:
COM:

PRINGLE JAMES E DR
INSTITUTE OF ASTRONOMY
MADINGLEY ROAD
CAMBRIDGE CB3 0HA
U.K.
TEL: 0223-62204
TLX: 817297 ASTRON G
TLF:
EML:
COM: 27.42

PRINJA RAMAN DR
DEPT OF PHYSICS & ASTRON
UNIV. COLLEGE LONDON
GOWER STREET
LONDON WC1E 6BT
U.K.
TEL: 01-387 7050
TLX: 28722
TLF:
EML: BITNET:RKP@UK.AC.UCL.STARLINK
COM: 29

PRITCHET CHRISTOPHER J DR
PHYSICS DEPT
UNIVERSITY OF VICTORIA
P O BOX 1700
VICTORIA BC V8W 2Y2
CANADA
TEL: 604-721-7704
TLX: 049-7222
TLF:
EML:
COM: 09.28.37

PROBSTEIN R F DR
DEPT MECHANICAL ENGINEERG
MIT
CAMBRIDGE MA 02139
U.S.A.
TEL: 617-253-2240
TLX: 921473 MIT CAM
TLF:
EML:
COM:

PROCHAZKA FRANZ V DR
INSTITUT FUER FERSTUDIEN
UNIVERSITAET
KLAGENFURT
A-9070 KLAGENFURT
AUSTRIA
TEL: 0-472-5317
TLX:
TLF:
EML:
COM: 24.51

PRODAN Y I DR
STERNBERG STATE
ASTRONOMICAL INSTITUTE
119899 MOSCOW
U.S.S.R.
TEL: 139-55-43
TLX:
TLF:
EML:
COM: 19

PROISY PAUL E DR
OBSERVATOIRE DE LYON
F-69230 ST-GENIS-LAVAL
FRANCE
TEL: 78-56-07-05
TLX:
TLF:
EML:
COM: 15

PROKAKIS THEODORE J DR
ASTRONOMICAL INSTITUTE
NATL OBSERVATORY OF ATHEN
P O BOX 20048
GR-11810 ATHENS
GREECE
TEL: 8040619-3461191
TLX: 21 5530
TLF:
EML:
COM: 10.12.41

PROKOF'EV VLADIMIR K PROF
CRIMEAN ASTROPHYSICAL
OBSERVATORY
PO NAUCHNY
334413 CRIMEA
U.S.S.R.
TEL:
TLX:
TLF:
EML:
COM: 14.44

PROKOF'EVA IRINA A DR
PULKOVO OBSERVATORY
196140 LENINGRAD
U.S.S.R.
TEL:
TLX:
TLF:
EML:
COM:

PROKOF'EVA VALENTINA V DR
CRIMEAN ASTROPHYSICAL
OBSERVATORY
NAUCHNYJ
334413 CRIMEA
U.S.S.R.
TEL: 1-24
TLX:
TLF:
EML:
COM: 09

PRONIK I I DR
CRIMEAN ASTROPHYS OBS
USSR ACADEMY OF SCIENCES
PO NAUCHNY
334413 CRIMEA
U.S.S.R.
TEL: 569
TLX:
TLF:
EML:
COM: 28.34

PRONIK V I DR
CRIMEAN ASTROPHYS OBS
USSR ACADEMY OF SCIENCES
PO NAUCHNY
334413 CRIMEA
U.S.S.R.
TEL: 569
TLX:
TLF:
EML:
COM: 28

PROSZYNSKI MIECZYSLAW
COPERNICUS ASTRON CENTER
UL. BARTYCKA 18
00-716 WARSAW
POLAND
TEL:
TLX:
TLF:
EML:
COM: 44

PROTHEROE RAYMOND J DR
DEPT OF PHYSICS
UNIVERSITY OF ADELAIDE
ADELAIDE. S. AUSTR. 5001
AUSTRALIA
TEL: 08-228-5996
TLX: 89141 UNIVAD AA
TLF:
EML:
COM: 48

PROTHEROE WILLIAM M PROF
DEPT OF ASTRONOMY
OHIO STATE UNIVERSITY
174 W. 18TH AVENUE
COLUMBUS OH 43210
U.S.A.
TEL: 614-422-7891
TLX:
TLF:
EML:
COM:

PROTICH MILORAD B
ASTRONOMICAL OBSERVATORY
VOLGINA 7
11050 BELGRADE
YUGOSLAVIA
TEL: 011-402-365
TLX:
TLF:
EML:
COM: 20

PROUST DOMINIQUE
OBSERVATOIRE DE PARIS
SECTION DE MEUDON
DAPHE
F-92195 MEUDON PL CEDEX
FRANCE
TEL: 1-45-07-74-11
TLX: 201571 LAM
TLF:
EML:
COM: 28

PROVERBIO EDOARDO PROF
ISTITUTO DI ASTRONOMIA
VIA OSPEDALE 72
I-09100 CAGLIARI
ITALY
TEL: 070-657657
TLX: 790326 OSSAST I
TLF:
EML:
COM: 08.19.31V.46.41

PROVOST JANINE DR
OBSERVATOIRE DE NICE
BP 139
F-06003 NICE CEDEX
FRANCE
TEL: 93-89-04-20
TLX: 460004
TLF:
EML:
COM: 27.35

PRYCE MAURICE H L DR
DEPT OF PHYSICS
UNIV OF BRITISH COLUMBIA
2075 WESBROOK MALL
VANCOUVER BC V6T 1W5
CANADA
TEL:
TLX:
TLF:
EML:
COM:

PSKOVSKIJ JU P DR
STERNBERG STATE
ASTRONOMICAL INSTITUTE
119899 MOSCOW
U.S.S.R.
TEL: 139-37-21
TLX:
TLF:
EML:
COM: 27.34

PUCILLO MAURO DR
OSSERVATORIO ASTRONOMICO
VIA TIEPOLO 11
I-34131 TRIESTE
ITALY
TEL: 040-793921
TLX: 461137 OAT I
TLF:
EML:
COM: 05.09

PUETTER RICHARD C DR
C-011 CASS
UNIVERSITY OF CALIFORNIA
SAN DIEGO
LA JOLLA CA 92093
U.S.A.
TEL: 619 534 4995
TLX:
TLF:
EML:
COM:

PUGACH ALEXANDER F DR
MAIN ASTRONOMICAL OBS
UKRAINIAN ACADEMY OF
SCIENCES
252127 KIEV USSR
U.S.S.R.
TEL: 266 47 71
TLX: 131406 SKY SU
TLF:
EML:
COM: 27

PUGET JEAN-LOUP DR
RADIOASTRONOMIE
LAB. DE PHYSIQUE E.N.S.
24 RUE LHOMOND
F-75005 PARIS
FRANCE
TEL: 1-43-29-12-25
TLX: 270912
TLF:
EML:
COM: 34.47

PUGLIANO ANTONIO PROF
INST DI GEODES. E IDROGR.
ISTITUTO UNIV. NAVALE
NAPOLI VIA F. ACTON 38
NAPOLI 80133
ITALY
TEL: 081 5512330
TLX: 710417
TLF: 081 5521485
EML:
COM: 08

PUNETHA LALIT MOHAN DR
UP STATE OBSERVATORY
MANORA PEAK
NAINI TAL 263 129
INDIA
TEL: 2136
TLX:
TLF:
EML:
COM:

PURCELL EDWARD M PROF
DEPT OF PHYSICS
HARVARD UNIVERSITY
CAMBRIDGE MA 02138
U.S.A.
TEL: 617-495-2860
TLX:
TLF:
EML:
COM: 51

PURTON CHRISTOPHER R DR
DOMINION RADIO ASTROPHYS
OBSERVATORY. NRC
P O BOX 248
PENTICTON BC V2A 6K3
CANADA
TEL: 604-497-5321
TLX: 048-88127
TLF:
EML:
COM:

PUSCHELL JEFFERY JOHN
MARTIN MARIETTA
103 CHESAPEAKE PARK PLAZA
X460
BALTIMORE MD 21220
U.S.A.
TEL: 301-682-0885
TLX: 908225
TLF:
EML:
COM: 40

PUSHKIN SERGEY B DR
TIME & FREQUENCY SERVICE
GOSSTANDARD USSR
117049 MOSCOW
U.S.S.R.
TEL:
TLX:
TLF:
EML:
COM: 31

PUSTYL'NIK IZOLD B DR
TORAVERE OBSERVATORY
202444 TARTU VAJ ESTONIA
U.S.S.R.
TEL: 33439
TLX:
TLF:
EML:
COM: 42

PYE JOHN P DR
PHYSICS DEPT
UNIVERSITY OF LEICESTER
UNIVERSITY ROAD
LEICESTER LE1 7RH
U.K.
TEL: 533-554455-23
TLX: 341664 LUXRAY G
TLF:
EML:
COM:

PYPER SMITH DIANE M DR
PHYSICS DEPARTMENT
UNIVERSITY OF NEVADA
LAS VEGAS NV 89154
U.S.A.
TEL:
TLX:
TLF:
EML:
COM:

QI GUANRONG
P.O. BOX 18
LINTONG
SHAANXI
CHINA, PEOPLE'S REP.
TEL: 32255 XIAN
TLX: 70121 CSAO CN
TLF:
EML:
COM:

QIAN BO-CHEN
SHANGHAI OBSERVATORY
ACADEMIA SINICA
SHANGHAI
CHINA, PEOPLE'S REP.
TEL: 386191
TLX: 33164 SHAO CN
TLF:
EML:
COM: 37

QIAN JING-KUI
DEPT OF GEOPHYSICS
PEKING UNIVERSITY
BEIJING
CHINA, PEOPLE'S REP.
TEL: 282471-3888
TLX: 22239 PKUNI
TLF:
EML:
COM: 10

QIAN SHAN-JIE
BEIJING ASTRONOMICAL OBS
ACADEMIA SINICA
BEIJING
CHINA, PEOPLE'S REP.
TEL: 28-2194
TLX: 22040 BAOAS CN
TLF:
EML:
COM: 40

QIAN ZHI-HAN DR
SHANGHAI OBSERVATORY
ACADEMIA SINICA
SHANGHAI
CHINA, PEOPLE'S REP.
TEL: 386191
TLX: 33164 SHAO CN
TLF:
EML:
COM: 08

QIAN ZHONG-YU
BEIJING ASTRONOMICAL OBS
ACADEMIA SINICA
BEIJING
CHINA, PEOPLE'S REP.
TEL:
TLX: 22040 BAOAS CN
TLF:
EML:
COM: 33

QIAO GUOJUN
BEIJING UNIVERSITY
ROOM 214 BLDG 34
BEIJING
CHINA, PEOPLE'S REP.
TEL:
TLX:
TLF:
EML:
COM: 42

QIN DAO
PURPLE MOUNTAIN
OBSERVATORY
NANJING
CHINA, PEOPLE'S REP.
TEL: 46700
TLX: 34144 PMONJ CN
TLF:
EML:
COM: 24

QIN SONG-NIAN
YUNNAN OBSERVATORY
P.O. BOX 110
KUNMING
CHINA, PEOPLE'S REP.
TEL:
TLX:
TLF:
EML:
COM:

QIN ZHI-HAI
DEPT OF ASTRONOMY
NANJING UNIVERSITY
NANJING
CHINA, PEOPLE'S REP.
TEL: 34651-2882
TLX: 0909
TLF:
EML:
COM: 34

QIU PUZHANG, ASS. PROF.
YUNNAN OBSERVATORY
ACADEMIA SINICA
PO BOX 110
KUNMING, YUNNAN PROVINCE
CHINA, PEOPLE'S REP.
TEL: 72946 KUNMING
TLX: 64040 YUOBS CN
TLF:
EML:
COM: 51

QIU YU-HAI
BEIJING ASTRONOMICAL OBS.
ACADEMIA SINICA
BEIJING 100080
CHINA, PEOPLE'S REP.
TEL:
TLX: 22040 BAOAS CN
TLF:
EML:
COM: 40

QU QIN-YUE
DEPT OF ASTRONOMY
NANJING UNIVERSITY
NANJING
CHINA, PEOPLE'S REP.
TEL: 37551 EXT2741
TLX: 34151 PRCNU CN
TLF:
EML:
COM: 35,47,48C

QUAMAR JAVAID
D-19, STAFF TOWN
UNIVERSITY OF KARACHI
KARACHI 3201
PAKISTAN
TEL: 46 54 91
TLX:
TLF:
EML:
COM:

QUAN HEJUN
SHANGHAI OBSERVATORY
ACADEMIA SINICA
SHANGHAI
CHINA, PEOPLE'S REP.
TEL: 386191
TLX: 33164 SHAO CN
TLF:
EML:
COM: 41

QUARTA MARIA LUCIA
INSTITUTO ASTRONOMICO
UNIVERSIDADE DE SAO PAULO
CAIXA POSTAL 30627
01051 SAO PAULO SP
BRAZIL
TEL: 011 577 8599
TLX: 11 36221 IAGM BR
TLF:
EML:
COM:

QUAST GERMANO RODRIGO
OBSERVATORIO NACIONAL
RUA COLONEL RENNO 07
CAIXA POSTAL 21
37500 ITAJUBA MG
BRAZIL
TEL: 035-6220788
TLX: 031 2603
TLF:
EML:
COM:

QUENBY JOHN J DR
BLACKETT LABORATORY
IMPERIAL COLLEGE
PRINCE CONSORT ROAD
LONDON SW7 2BZ
U.K.
TEL: 1-589-5111x6661
TLX: 261503
TLF:
EML:
COM:

QUERCI FRANCOIS R DR
OBS DU PIC-DU-MIDI
ET DE TOULOUSE
14 AVENUE EDOUARD BELIN
F-31400 TOULOUSE
FRANCE
TEL: 61-25-21-01
TLX: 530776 F
TLF:
EML:
COM: 14.29.36

QUERCI MONIQUE DR
OBS DU PIC-DU-MIDI
ET DE TOULOUSE
14 AVENUE EDOUARD BELIN
F-31400 TOULOUSE
FRANCE
TEL: 61-25-21-01
TLX: 530776 F
TLF:
EML:
COM: 29.36

QUIJANO LUIS
INSTITUTO Y OBSERVATORIO
DE MARINA
SAN FERNANDO (CADIZ)
SPAIN
TEL: 956-883-548
TLX: 76108 ION E
TLF:
EML:
COM: 08.20.24

QUINTANA HERNAN DR
DEPTO DE ASTRONOMIA
UNIVERSIDAD CATOLICA
CASILLA 114-D
SANTIAGO
CHILE
TEL: 775474
TLX: 240395 PUCVA CL
TLF:
EML:
COM: 05.28C.30.48.51

QUINTANA JOSE M DR
INST DE ASTROFISICA DE
ANDALUCIA. APDO 2144
PROFESOR ALBAREDA 1
18080 GRANADA
SPAIN
TEL: 58-121300
TLX: 78573 IAAG E
TLF:
EML:
COM: 51

QUIRK WILLIAM J DR
LAWRENCE LIVERMORE NATL
LABORATORY L 35
BOX 808
LIVERMORE CA 94550
U.S.A.
TEL: 415-422-1852
TLX:
TLF:
EML:
COM:

QVIST BERTIL PROF
MATHEMATICS DEPT
ABO AKADEMI
SF-20500 ABO 50
FINLAND
TEL:
TLX:
TLF:
EML:
COM:

RAADU MICHAEL A DR
DEPT PLASMA PHYSICS
ROYAL INST OF TECHNOLOGY
S-100 44 STOCKHOLM 70
SWEDEN
TEL: 08-78-77000
TLX: 10389 KTHB STOCKHOLM
TLF:
EML:
COM: 10.49

RABBIA YVES DR
OBS. DE LA COTE D'AZUR
CERGA
AVENUE COPERNIC
06130 GRASSE
FRANCE
TEL: 93 36 58 49
TLX: 470865 CERGA
TLF:
EML: EARN::YVRABB@FRONI51
COM:

RABIN DOUGLAS MARK
NATIONAL SOLAR OBS
NATL OPTICAL ASTR OBS
PO BOX 26732
TUCSON AZ 85726-6732
U.S.A.
TEL: 602-325-9331
TLX: 0666484 AURA NOAOTUC
TLF:
EML:
COM: 10.12

RABOLLI MONICA DR
FACULTAD DE CIENCIAS
UNIVERSIDAD NAC LA PLATA
AV PASEO DEL BOSQUE
1900 LA PLATA (BS.AS.)
ARGENTINA
TEL: 54021 217308
TLX:
TLF:
EML: MRABOLLI%PSI%IAFE%SSL.SPAN
COM: 33

RACHKOVSKY D N DR
CRIMEAN ASTROPHYSICAL OBS
USSR ACADEMY OF SCIENCES
NAUCHNIY
334413 CRIMEA
U.S.S.R.
TEL: 1-03
TLX: 192
TLF:
EML:
COM: 36

RACINE RENE DR
DEPT DE PHYSIQUE
UNIVERSITE DE MONTREAL
BP 6128
MONTREAL PQ H3C 3J7
CANADA
TEL: 514-343-6718
TLX: 5561359 RZLPNUM ML
TLF:
EML:
COM: 09

RACKHAM THOMAS W DR
39 MEADOW AVENUE
GOOSTREY
CREWE CW4 8LS
U.K.
TEL: 0477-33004
TLX:
TLF:
EML:
COM:

RADHAKRISHNAN V PROF
RAMAN RESEARCH INSTITUTE
SADASHIVANAGAR
BANGALORE 560 080
INDIA
TEL: 812-360522
TLX: 8452671 RRI IN
TLF:
EML:
COM: 34.40.48

RADICK RICHARD R DR
AFGL/PHS
SUNSPOT NM 88349
U.S.A.
TEL: 505 434 1390
TLX:
TLF:
EML:
COM: 12

RADIMAN IRATIUS
BOSSCHA OBSERVATORY
LEMBANG. JAVA
INDONESIA
TEL:
TLX:
TLF:
EML:
COM:

RADOSKI HENRY R DR
AFOSR/NP
BUILDING 410
BOLLING AIR FORCE BASE
WASHINGTON DC 20332
U.S.A.
TEL: 202 767 4906
TLX:
TLF:
EML:
COM:

RADOSLAVOVA TSVETANKA
DEPT OF ASTRONOMY
BULGARIAN ACAD SCIENCES
72 LENIN BLVD
1784 SOFIA
BULGARIA
TEL:
TLX:
TLF:
EML:
COM:

RAEDLER K H DR
ZNTRLINST. F. ASTROPHYSIK
ROSA-LUXEMBURG-STR 17A
DDR-1502 POTSDAM
GERMANY. D.R.
TEL:
TLX:
TLF:
EML:
COM: 35

RAFANELLI PIERO DR
OSSERVATORIO ASTRONOMICO
VICOLO DELL'OSSERVATORIO
I-35100 PADOVA
ITALY
TEL: 49-661499
TLX:
TLF:
EML:
COM: 28

RAFERT JAMES BRUCE
DEPT PHYSICS & SPACE SCI.
FLORIDA INSTITE OF TECHNO
MELBOURNE FL 32901
U.S.A.
TEL:
TLX:
TLF:
EML:
COM: 42

RAGHAVAN NIRUPAMA DR
2133 INDIAN INSTITUTE
OF TECHNOLOGY
CAMPUS
NEW DELHI 110 029
INDIA
TEL:
TLX:
TLF:
EML:
COM:

RAHE JURGEN PROF
NASA HEADQUARTERS
CODE EL
WASHINGTON DC 20546
U.S.A.
TEL: 202-453-1590
TLX: 4974843
TLF:
EML:
COM: 15P.42C.44C

RAHUNEN TIMO
TAMPERE SAERKAENNIEMI OY
SAERKAENNIEMI
SF-33410 TAMPERE
FINLAND
TEL: 931-31333
TLX:
TLF:
EML:
COM: 42

RAIKOVA DONKA DR
DEPT OF ASTRONOMY
BULGARIAN ACAD SCIENCES
7TH NOVEMBER STR 1
1000 SOFIA
BULGARIA
TEL: 7341
TLX: 23561 ECF BAN BG
TLF:
EML:
COM:

RAIMOND ERNST DR
NETHERLAND FOUNDATION
FOR RADIOASTRONOMY
POST BUS 2
NL-7990 AA DWINGELOO
NETHERLANDS
TEL: 05219-7244
TLX: 42043
TLF:
EML:
COM: 05.08.34.40

RAINE DEREK J DR
DEPT OF ASTRONOMY
UNIVERSITY OF LEICESTER
LEICESTER LE1 7RH
U.K.
TEL: 533-554455
TLX: 341198 LEICUL
TLF:
EML:
COM:

RAITALA JOUKO T
DEPT OF ASTRONOMY
UNIVERSITY OF OULU
SF-90570 OULU 57
FINLAND
TEL: 81-35-21-06
TLX: 32375
TLF:
EML:
COM:

RAJAMOHAN R DR
INDIAN INSTITUTE OF
ASTROPHYSICS
BANGALORE 560 034
INDIA
TEL: 566497/585
TLX: 845763 IIAB IN
TLF:
EML:
COM: 37.51

RAJCHL JAROSLAV DR
ASTRONOMICAL INSTITUTE
CZECH. ACAD. OF SCIENCES
OBSERVATORY
251 65 ONDREJOV
CZECHOSLOVAKIA
TEL: 724525 PRAHA
TLX: 121579
TLF:
EML:
COM: 22

RAJU P K DR
INDIAN INSTITUTE OF
ASTROPHYSICS
BANGALORE 560 034
INDIA
TEL: 566-585
TLX: 845763 IIAB IN
TLF:
EML:
COM:

RAKAVY GIDEON PROF
EINSTEIN INST OF PHYSICS
HEBREW UNIV. OF JERUSALEM
JERUSALEM 91904
ISRAEL
TEL:
TLX:
TLF:
EML:
COM:

RAKOS KARL D PROF
INSTITUT FUER ASTRONOMIE
UNIVERSITAET WIEN
TUERKENSCHANZSTR 17
A-1180 WIEN
AUSTRIA
TEL: 0222-345360-95
TLX: 133099 VIAST A
TLF:
EML:
COM: 09.26C.27.42

RAKSHIT H PROF
BENGAL ENGINEERG COLLEGE
SIBPORE
HOWRAH
INDIA
TEL:
TLX:
TLF:
EML:
COM:

RAM SAGAR DR
INDIAN INST OF ASTROPHYS
BANGALORE 560 034
INDIA
TEL: 566585/566497
TLX: 845763 IIAB IN
TLF:
EML:
COM: 37

RAMADURAI SOURIRAJA DR
ASTRONOMY GROUP
DEPARTMENT OF PHYSICS
INDIAN INST OF SCIENCE
BANGALORE 560 012
INDIA
TEL: BGL 364411x314
TLX: 8458349 BG
TLF:
EML:
COM: 35.46.48

RAMAMURTHY SWAMINATHAN
CENTRE OF ADVANCED STUDY
IN ASTRONOMY
OSMANIA UNIVERSITY
HYDERABAD 500 007
INDIA
TEL:
TLX:
TLF:
EML:
COM:

RAMANA MURTHY P V DR
TATA INST OF FUND RES
COLABA
BOMBAY 400 005
INDIA
TEL: 495 2979
TLX: 011 3009 TIFR IN
TLF:
EML:
COM:

RAMATY REUVEN DR
LAB HIGH ENERGY ASTROPHYS
NASA/GSFC. CODE 665
GREENBELT MD 20771
U.S.A.
TEL: 301-286-8715
TLX:
TLF:
EML:
COM: 40

RAMBERG JORAN M PROF
GENVAEGEN 4
S-133 00 SALTSJOEBADEN
SWEDEN
TEL: 46-8-717-1926
TLX:
TLF:
EML:
COM: 33

RAMELLA MASSIMO
OSSERVATORIO ASTRONOMICO
VIA G.B. TIEPOLO 11
I-34131 TRIESTE
ITALY
TEL: 040-76-85-06
TLX: 461137 OAT I
TLF:
EML:
COM: 29.47

RAMSAY DONALD A DR
HERZBERG INST ASTROPHYS
NATL RES COUNCIL CANADA
100 SUSSEX DR
OTTAWA ONTARIO K1A 0R6
CANADA
TEL: 613 990 0919
TLX: 053 3715
TLF: 1 613 952 0974
EML: BITNET:DARONRCVM01
COM: 20

RAMSEY LAWRENCE W DR
DEPT OF ASTRONOMY
PENNSYLVANIA STATE UNIV
525 DAVEY LAB
UNIVERSITY PARK PA 16802
U.S.A.
TEL: 814-865-0418
TLX:
TLF:
EML:
COM: 09.36

RANA NARAYAN CHANDRA DR
THEORETICAL ASTRO GROUP
TATA INST FUND RES
HOMI BHABHA ROAD
BOMBAY 400 005
INDIA
TEL: (022) 4952311
TLX: 011 3009
TLF:
EML:
COM: 28

RANDIC LEO PROF DR
GEODETICAL FACULTY
GUNDULICEVA 54
ZAGREB 41000
YUGOSLAVIA
TEL: 041-44-66-75
TLX:
TLF:
EML:
COM: 19.31

RANIERI MARCELLO
IST. ASTROFISICA SPAZIALE
C P 67
I-00044 FRASCATI
ITALY
TEL:
TLX:
TLF:
EML:
COM:

RANK DAVID M PROF
LICK OBSERVATORY
UNIVERSITY OF CALIFORNIA
SANTA CRUZ CA 95064
U.S.A.
TEL: 408-429-2277
TLX:
TLF:
EML:
COM:

RANKIN JOANNA M DR
DEPT OF PHYSICS
UNIVERSITY OF VERMONT
A405 COOK BUILDING
BURLINGTON VT 05405
U.S.A.
TEL: 802-656-2644
TLX: 510-299-0021
TLF:
EML:
COM:

RAO A PRAMESH DR
KAPTEYN LABORATORIUM
POSTBUS 800
NL-9700 AV GRONINGEN
NETHERLANDS
TEL:
TLX:
TLF:
EML:
COM: 10.40

RAO K NARAHARI
OHIO STATE UNIVERSITY
DEPT OF PHYSICS
174 W. 18TH AVENUE
COLUMBUS OH 43210
U.S.A.
TEL: 614-422-6505
TLX:
TLF:
EML:
COM: 14

RAO K RAMANUJA DR
C/O DR. K. SURENDRA
RUA CEL. JOAO CURSINO 210
APT 92. VILA ADYANA
12200 S. JOSE DOS CAMPOS
BRAZIL
TEL:
TLX:
TLF:
EML:
COM:

RAO M N DR
PHYS. RESEARCH LABORATORY
NAVRANGPURA
AHMEDABAD 380 009
INDIA
TEL: 462129
TLX: 121397
TLF:
EML:
COM: 16

RAO N KAMESWARA
INDIAN INSTITUTE OF
ASTROPHYSICS
BANGALORE 560 034
INDIA
TEL:
TLX:
TLF:
EML:
COM: 27.29

RAO P VIVEKANANDA DR
DEPT OF ASTRONOMY
OSMANIA UNIVERSITY
HYDERABAD 500 007
INDIA
TEL:
TLX:
TLF:
EML:
COM: 25

RAO RAMACHANDRA V PROF
ISRO SATELLITE CENTER
PEENYA
BANGALORE 560 058
INDIA
TEL:
TLX:
TLF:
EML:
COM: 44

RAOULT ANTOINETTE DR
OBSERVATOIRE PARIS MEUDON
DASOP
92195 MEUDON PR. CEDEX
FRANCE
TEL: 1-45 07 77 66
TLX: 200590
TLF:
EML:
COM: 10.12.40

RAPAPORT MICHEL DR
OBSERVATOIRE DE BORDEAUX
AVENUE PIERRE SEMIROT
F-33270 FLOIRAC
FRANCE
TEL: 56-86-43-30
TLX:
TLF:
EML:
COM: 20.21

RAPLEY CHRISTOPHER G DR
MULLARD SPACE SCIENCE LAB
UNIVERSITY COLLEGE LONDON
LONDON
U.K.
TEL: 030-670-292
TLX: 859185
TLF:
EML:
COM:

RATNATUNGA KAVAN U.
DOMINION RADIO ASTR. OBS.
HERZBERG INST. OF ASTROP.
NATIONAL RESEARCH COUNCIL
PENTICTON BC V2A 6K3
CANADA
TEL:
TLX:
TLF:
EML:
COM: 05.30

RAUDENHEIMER BAREND C PR
COSMIC RAY RESEARCH UNIT
POTCHEFSTROOM UNIVERSITY
POTCHEFSTROOM 2520
SOUTH AFRICA
TEL: 01481-27511
TLX: 421363
TLF:
EML:
COM: 48

RAUTELA B S DR
UTTAR PRADESH STATE OBS
MANORA PEAK
NAINITAL 263 129
INDIA
TEL:
TLX:
TLF:
EML:
COM: 29.45

RAY ALAK DR
TATA INSTITUTE OF
FUNDAMENTAL RESEARCH
BOMBAY 400 005
INDIA
TEL:
TLX:
TLF:
EML:
COM:

RAY CHOUDHURY AHMAD DR
DEPT OF PHYSICS
INDIAN INSTITUTE OF SCI
BANGALORE 560 012
INDIA
TEL:
TLX:
TLF:
EML:
COM: 10

RAY THOMAS P
INST FOR ADVANCED STUDIES
SCHOOL OF COSMIC PHYSICS
5 MERRION SQUARE
DUBLIN 2
IRELAND
TEL: 774321
TLX: 31687 DIAS EI
TLF:
EML:
COM: 40

RAYCHAUDHURI AMALKUMAR DR
PRESIDENCY COLLEGE
COLLEGE STREET
CALCUTTA 73
INDIA
TEL:
TLX:
TLF:
EML:
COM: 47

RAYMOND JOHN CHARLES
CENTER FOR ASTROPHYSICS
60 GARDEN STREET
CAMBRIDGE MA 02138
U.S.A.
TEL:
TLX:
TLF:
EML:
COM: 34

RAYROLE JEAN R DR
OBSERVATOIRE DE PARIS
SECTION DE MEUDON
F-92195 MEUDON PL CEDEX
FRANCE
TEL: 1-45-07-77-89
TLX:
TLF:
EML:
COM: 10

RAZDAN HIRALAL
BHABHA ATOMIC RES CTR
ZAKURA SRINIGAR
KASHMIR 190 006
INDIA
TEL:
TLX:
TLF:
EML:
COM: 48

RAZIN V A DR
RADIOPHYSICAL RESEARCH
INSTITUTE
603600 GORKIJ
U.S.S.R.
TEL: 36-72-94
TLX:
TLF:
EML:
COM: 40

READHEAD ANTHONY C S DR
RADIO ASTRONOMY DEPT
CALTECH
ROBINSON BLDG
PASADENA CA 91125
U.S.A.
TEL: 213-356-4972
TLX: 675425 CALTECH PSD
TLF:
EML:
COM: 40.49

REASENBERG ROBERT D DR
CENTER FOR ASTROPHYSICS
ROOM B 217
60 GARDEN STREET
CAMBRIDGE MA 02138
U.S.A.
TEL: 617-495-7108
TLX: 921428 SATELITE CAM
TLF:
EML:
COM: 04

REAVES GIBSON PROF
DEPT OF ASTRONOMY
UNIV OF S. CALIFIFORNIA
LOS ANGELES CA 90089-1342
U.S.A.
TEL: 213-743-2039
TLX:
TLF:
EML:
COM: 28

REAY NEWRICK K DR
ASTROPHYSICS GROUP
BLACKETT LABORATORY
IMPERIAL COLLEGE
LONDON SW7 2BZ
U.K.
TEL: 1-589-5111x6669
TLX: 261503 IMPCOL
TLF:
EML:
COM: 09.49.51

REBEIROT EDITH DR
OBSERVATOIRE DE MARSEILLE
2 PLACE LE VERRIER
F-13248 MARSEILLE CDX 04
FRANCE
TEL: 91-95-90-88
TLX: 420241
TLF:
EML:
COM: 28.30

REBER GROTE DR
GENERAL DELIVERY
BOTHWELL TASM 7411
AUSTRALIA
TEL: 002-237371
TLX:
TLF:
EML:
COM: 40

REBOLO RAFAEL DR
INSTITUTO DE ASTROFISICA
DE CANARIAS
38200 LA LAGUNA TENERIFE
SPAIN
TEL: 34 22 26 22 11
TLX: 92640
TLF:
EML: SPAN:IAC::RRL
COM: 29

RECILLAS-CRUZ ELSA DR
INSTITUTO DE ASTRONOMIA
UNAM
APDO POSTAL 70-264
04510 MEXICO DF
MEXICO
TEL:
TLX:
TLF:
EML:
COM:

REDFERN MICHAEL R DR
PHYSICS DEPARTMENT
UNIVERSITY COLLEGE
GALWAY
IRELAND
TEL: 353 41 24411
TLX: 50023
TLF:
EML:
COM: 09

REED B CAMERON DR
DEPT OF PHYSICS
SAINT MARY'S UNIVERSITY
HALIFAX
NOVA SCOTIA B3H 3C3
CANADA
TEL: 902 420 5830
TLX:
TLF:
EML: BITNET:REEDOSTMARYS
COM:

REES DAVID ELWYN DR
DEPT APPLIED MATHS
UNIVERSITY OF SYDNEY
SYDNEY NSW 2006
AUSTRALIA
TEL: 02-692-3724
TLX: UNISYD AA 26169
TLF:
EML:
COM: 10.12

REES MARTIN J PROF
INSTITUTE OF ASTRONOMY
MADINGLEY RD
CAMBRIDGE CB3 0HA
U.K.
TEL: 223-62204
TLX: 817297 ASTRON G
TLF:
EML:
COM: 44.47C.48.51C

REEVES EDMOND M DR
NASA HEADQUARTERS
CODE EN
600 INDEPENDENCE AVE
WASHINGTON DC 20546
U.S.A.
TEL: 202-453-1571
TLX: 89530
TLF:
EML:
COM: 10.12.44

REEVES HUBERT PROF
SEP-SES BAT 28
CEN SACLAY
BP 2
F-91190 GIF S/YVETTE
FRANCE
TEL: 1-69-08-51-59
TLX:
TLF:
EML:
COM: 10.35.47C.48

REFSDAL S PROF DR
HAMBURGER STERNWARTE
GOJENBERGSWEG 112
D-2050 HAMBURG 80
GERMANY. F.R.
TEL: 49-40-72524124
TLX: 217884 HAMST
TLF:
EML:
COM: 42

REGLERO-VELASCO VICTOR DR
FACULTAD DE MATEMATICAS
UNIVERSIDAD DE VALENCIA
DR MOLINER 50 BURJASOT
46100 VALENCIA
SPAIN
TEL:
TLX:
TLF:
EML:
COM:

REGO FERNANDEZ M DR
ASTROFISICA
FACULTA FISICA
UNIVERSIDAD COMPLUTENSE
28040 MADRID
SPAIN
TEL: 449-53-16
TLX: 47273 FF UC
TLF:
EML:
COM: 29

REGULO CLARA DR
INSTITUTO DE ASTROFISICA
DE CANARIAS
38200 LA LAGUNA TENERIFE
SPAIN
TEL: 34 22 26 22 11
TLX: 92640
TLF:
EML: SPAN:IAC::CPD
COM: 10.12

REICH WOLFGANG
MPI FUER RADIOASTRONOMIE
AUF DEM HUEGEL 69
D-5300 BONN
GERMANY. F.R.
TEL:
TLX: 886440 MPIFR D
TLF:
EML:
COM: 40

REID MARK JONATHAN DR
CENTER FOR ASTROPHYSICS
60 GARDEN STREET
CAMBRIDGE MA 02138
U.S.A.
TEL: 617-495-7470
TLX: 921428 SATELLITE CAM
TLF:
EML:
COM: 40

REID NEILL
PALOMAR OBSERVATORY
105-24 CALTECH
PASADENA
U.S.A.
TEL: 818 356-6586
TLX:
TLF:
EML:
COM: 33

REIF KLAUS DR
RADIOASTRONOMISCHES INST
DER UNIVERSITAET BONN
AUF DEM HUEGEL 71
5300 BONN 1
GERMANY. F.R.
TEL: 228-73-36-57
TLX:
TLF:
EML: u.145ref%mpiehu%usido.uucp
COM: 33.40

REIMERS DIETER PROF
HAMBURGER STERNWARTE
UNIVERSITAET HAMBURG
GOJENBERGSWEG 112
D-2050 HAMBURG 80
GERMANY. F.R.
TEL: 040725? 4112
TLX:
TLF:
EML:
COM: 29.36

REINISCH GILBERT DR
OBSERVATOIRE DE NICE
BP 139
F-06003 NICE CEDEX
FRANCE
TEL: 93-89-04-20
TLX:
TLF:
EML:
COM:

REIPURTH BO
E.S.O.
CASILLA 19001
SANTIAGO 19
CHILE
TEL: 6988757 SANTIAG
TLX: 240881
TLF:
EML:
COM: 34

REITSEMA HAROLD J
BALL AEROSPACE SYSTEMS
DIVISION
PO BOX 1062
BOULDER CO 80306
U.S.A.
TEL: 303-441-5026
TLX:
TLF:
EML:
COM: 15.20

REIZ ANDERS PROF
LOVSPRINGSVEJ 3 B
DK-2920 CHARLOTTENLUND
DENMARK
TEL: 1-63-25-36
TLX:
TLF:
EML:
COM: 08.35

REMY BATTIAU LILIANE G A
CONSEIL DE LA RECHERCHE
UNIVERSITE DE LIEGE
7 PLACE DU XX AOUT
B-4000 LIEGE
BELGIUM
TEL: 41-42-00-80
TLX: 41397 UNIV ULG
TLF:
EML:
COM: 05.15

REN JIANG-PING
DEPT OF ASTRONOMY
NANJING UNIVERSITY
NANJING
CHINA. PEOPLE'S REP.
TEL: 34651-2882
TLX: 34151 PRCNU CN
TLF:
EML:
COM: 19

RENGARAJAN T N DR
IR ASTRONOMY
TIFR
HOMI BHABHA ROAD
BOMBAY 400 005
INDIA
TEL: 219111
TLX: 011-3009 TIFR IN
TLF:
EML:
COM: 34.48

RENSE WILLIAM A DR
DEPT PHYSICS & ASTROPHYS
UNIVERSITY OF COLORADO
DUANE PHYSICAL LABS
BOULDER CO 80302
U.S.A.
TEL: 303-492-0111
TLX:
TLF:
EML:
CON: 44

RENSON P F M DR
INSTITUT D'ASTROPHYSIQUE
UNIVERSITE DE LIEGE
AVENUE DE COINTE 5
B-4200 COINTE-OUGREE
BELGIUM
TEL: 41-52-99-80
TLX:
TLF:
EML:
CON: 05.27

RENZINI ALVIO PROF
DIPT. DI ASTRONOMIA
VIA ZAMBONI 33
I-40126 BOLOGNA
ITALY
TEL: 51-222956
TLX: 211664 INFNBO
TLF:
EML:
CON: 35.37

REPHAELI YOEL DR
SCHOOL OF PHYSICS
TEL AVIV UNIVERSITY
RAMAT AVIV
TEL AVIV 69978
ISRAEL
TEL:
TLX: 342171 VERSY IL
TLF:
EML:
CON: 28

REQUIEME YVES DR
OBSERVATOIRE DE BORDEAUX
B.P. 21
F-33270 FLOIRAC
FRANCE
TEL: 56-86-43-30
TLX:
TLF:
EML:
CON: 08C.24

REUNING ERNEST G DR
DEPT PHYSICS & ASTRONOMY
UNIVERSITY OF GEORGIA
ATHENS GA 30602
U.S.A.
TEL: 404-542-2485
TLX:
TLF:
EML:
CON: 42

REVELLE DOUGLAS ORSON DR
METEOROLOGY PROGRAM
DEPT OF GEOGRAPHY
815 DAVIS HALL
DEKALB IL 60115
U.S.A.
TEL: 815-753-0631
TLX:
TLF:
EML:
CON: 15.22C

REYES FRANCISCO DR
UNIVERSIDAD DE CHILE
DEPTO DE ASTRONOMIA
CASILLA 36-D
SANTIAGO
CHILE
TEL: 2294101
TLX: 440001 PBCZ
TLF:
EML:
CON: 40

REYNOLDS JOHN H PROF
DEPT OF PHYSICS
UNIVERSITY OF CALIFORNIA
BERKELEY CA 94720
U.S.A.
TEL: 415-642-4863
TLX: 9103667114
TLF:
EML:
CON:

REYNOLDS RONALD J DR
SPACE PHYSICS GROUP
UNIVERSITY OF WISCONSIN
1150 UNIVERSITY AVENUE
MADISON WI 53706
U.S.A.
TEL: 608-262-5916
TLX:
TLF:
EML:
CON: 34

REYNOLDS STEPHEN P
DEPARTMENT OF PHYSICS
NORTH CAROLINA STATE UNIV
BOX 8202
RALEIGH NC 27695-8202
U.S.A.
TEL: 919-737-7751
TLX:
TLF:
EML:
CON:

RHODES EDWARD J JR
11801 KILLIMORE AVE
NORTHRIDGE CA 91326
U.S.A.
TEL:
TLX:
TLF:
EML:
CON:

RIBES ELIZABETH DR
OBSERVATOIRE DE PARIS
SECTION DE MEUDON
F-92195 MEUDON PL CEDEX
FRANCE
TEL: 1-45-07-77-86
TLX:
TLF:
EML:
CON:

RIBES JEAN-CLAUDE DR
INSU
77 AVE DENFERT-ROCHEREAU
F-75014 PARIS
FRANCE
TEL: 1-43-20-13-30
TLX: 270070
TLF:
EML:
CON: 40

RICE JOHN B DR
DEPT PHYSICS & ASTRONOMY
BRANDON UNIVERSITY
BRANDON MAN R7A 6A9
CANADA
TEL: 204-727-9693
TLX: 07-502721
TLF:
EML:
CON:

RICHARDSON E HARVEY DR
HERZBERG INST ASTROPHYS
DOMINION ASTROPHYS OBS
5071 W SAANICH RD
VICTORIA BC V8X 4M6
CANADA
TEL:
TLX:
TLF:
EML:
CON: 09C

RICHARDSON KEVIN J
PHYSICS DEPARTMENT
QUEEN MARY COLLEGE
MILE END ROAD
LONDON E1 4NS
U.K.
TEL:
TLX:
TLF:
EML:
CON:

RICHARDSON R S
GRIFFITH OBSERVATORY
PO BOX 27787
LOS FELIZ STATION
LOS ANGELES CA 90027
U.S.A.
TEL:
TLX:
TLF:
EML:
CON:

RICHER HARVEY B DR
DEPT GEOPHYS & ASTRONOMY
UNIV OF BRITISH COLUMBIA
2075 WESBROOK PLACE
VANCOUVER BC V6T 1W5
CANADA
TEL: 604-228-4134
TLX:
TLF:
EML:
CON: 28.37

RICHSTONE DOUGLAS O DR
DEPARTMENT OF ASTRONOMY
UNIVERSITY OF MICHIGAN
ANN ARBOR MI 48109
U.S.A.
TEL: 313 764 3441
TLX:
TLF:
EML: D_Richstone@ub.cc.umich.edu
CON: 28

RICHTER G A DR
ZENTRALINST. F. ASTROPHYSIK
STERNWARTE BABELSBERG
ROSA-LUXEMBURG-STR 17A
DDR-6400 SONNEBERG
GERMANY. D.R.
TEL: 2287
TLX: 6288180 STEW DD
TLF:
EML:
COM: 27

RICHTER JOHANNES PROF
INST F EXPERIMENT. PHYSIK
PHYSIKZENTRUM
OLSHAUSENSTRASSE
D-2300 KIEL
GERMANY. F.R.
TEL: 0431-880-3835
TLX: 292706 IAPKID
TLF:
EML:
COM: 14

RICKARD JAMES JOSEPH DR
PO BOX 777
BORREGO SPRINGS CO 92004
U.S.A.
TEL: 714-767-5462
TLX:
TLF:
EML:
COM:

RICKARD LEE J DR
NAVAL RESEARCH LABORATORY
CODE4138RRD
WASHINGTON DC 20375-5000
U.S.A.
TEL: 202-767-2455
TLX:
TLF:
EML:
COM: 34.40

RICKER GEORGE R DR
CENTER FOR SPACE RESEARCH
MIT RM 37-527
77 MASSACHUSSETS AVENUE
CAMBRIDGE MA 02139
U.S.A.
TEL: 617-253-7532
TLX: 92-14-73
TLF:
EML:
COM:

RICKETT BARNABY JAMES DR
DEPT OF ELECTRICAL ENG
AND COMPUTER SCIENCE
UNIV CALIF AT SAN DIEGO
LA JOLLA CA 92093
U.S.A.
TEL: 619-452-2731
TLX:
TLF:
EML:
COM: 40.49

RICKMAN HANS DR
ASTRONOMISKA OBS
BOX 515
S-751 20 UPPSALA
SWEDEN
TEL: 46-18113522
TLX: 76024 UNIVUPS
TLF:
EML:
COM: 15C.20C

RICORT GILBERT DR
DEPT D'ASTROPHYSIQUE
UNIVERSITE DE NICE
PARC VALROSE
F-06034 NICE
FRANCE
TEL:
TLX:
TLF:
EML:
COM:

RIDDLE ANTHONY C DR
700 GRANT PL
BOULDER CO 80302
U.S.A.
TEL: 303-447-8127
TLX:
TLF:
EML:
COM: 49

RIEGEL KURT W DR
NATL SCIENCE FOUNDATION
1800 G STREET NW
WASHINGTON DC 20550
U.S.A.
TEL: 202-357-9450
TLX:
TLF:
EML:
COM: 33

RIGHINI ALBERTO PROF
DIPART. DI ASTRONOMIA
UNIVERSITAT DEGLI STUDI
LARGO E. FERMI 5
I-50125 FIRENZE
ITALY
TEL: 055-27521
TLX: 572268 ARCETRI I
TLF:
EML:
COM:

RIGHINI-COHEN GIOVANNA DR
DEPT EARTH & SPACE SCI
SUNY
STONY BROOK NY 11794
U.S.A.
TEL:
TLX:
TLF:
EML:
COM: 12.34.44

RIGUTTI MARIO PROF
OSSERVATORIO ASTRONOMICO
DI CAPODIMONTE
MOIARIELLO 16
I-80131 NAPOLI
ITALY
TEL: 440101
TLX:
TLF:
EML:
COM: 12.46

RIIHIMAA JORMA J DR
AARNE KARJALAINEN OBS
UNIVERSITY OF OULU
SF-90570 OULU
FINLAND
TEL:
TLX:
TLF:
EML:
COM: 40.51

RILEY JULIA M DR
MULLARD RADIO ASTRON OBS
CAVENDISH LABORATORY
MADINGLEY ROAD
CAMBRIDGE CB3 OHE
U.K.
TEL: 0223-66477
TLX: 81292
TLF:
EML:
COM: 40

RINDLER WOLFGANG PROF
UNIV OF TEXAS AT DALLAS
U.T.D.
BOX 830688
RICHARDSON TX 75083-0688
U.S.A.
TEL: 214-690 2885
TLX: 791-880
TLF:
EML:
COM: 47

RING JAMES PROF
THE BLACKETT LABORATORY
IMPERIAL COLLEGE
PRINCE CONSORT ROAD
LONDON SW7 2BZ
U.K.
TEL: 01-589-5111
TLX: 261503
TLF:
EML:
COM: 09

RINGNES TRULS S DR
INST THEORET ASTROPHYSICS
UNIVERSITY OF OSLO
P O BOX 1029
N-0315 BLINDERN. OSLO 3
NORWAY
TEL: 472-456-503
TLX: 72425 UNIOS N
TLF:
EML:
COM:

RINGUELET ADELA E DR
49 342
1900 LA PLATA
ARGENTINA
TEL: 54-21-31063
TLX: 31151 BULAP AR
TLF:
EML:
COM: 29

RIPKEN HARTMUT W DR
ESA/ESTEC
KEPLERLAAN 1
NL-2200 AG NOORDWIJK
NETHERLANDS
TEL: 01719-83163
TLX: 39098
TLF:
EML:
COM: 21.22.49C

RITTER HANS DR
UNIVERSITAETS STERNWARTE
SCHEINERSTR 1
D-8000 MUENCHEN
GERMANY. F.R.
TEL: 89-98-90-21
TLX: 529815 UNIVM D
TLF:
EML:
COM: 42

RIVOLO ARTHUR REX
UNIV OF PENNSYLVANIA
DEPT ASTRON & ASTROPHYS
PHILADELPHIA PA 19104
U.S.A.
TEL: 215-898-6250
TLX: 358300
TLF:
EML:
COM: 47

RIZVANOV NAUFAL G DR
ENGELHARDT OBSERVATORY
422526 KASAN
U.S.S.R.
TEL: 324827
TLX:
TLF:
EML:
COM: 24

ROACH FRANKLIN E
PO BOX 2065
COTTONWOOD AZ 86326
U.S.A.
TEL:
TLX:
TLF:
EML:
COM: 21.49

ROARK TERRY P PROF
KENT STATE UNIVERSITY
171 MAJORS LANE
KENT OH 44240
U.S.A.
TEL: 216-672-2220
TLX:
TLF:
EML:
COM:

ROBB RUSSEL M
DEPT. OF PHYS. & ASTR.
UNIVERSITY OF VICTORIA
VICTORIA BC V8W 2Y2
CANADA
TEL: 604 721 7750
TLX:
TLF:
EML: BITNET:ROBB@UVPHYS
COM: 42

ROBBINS R ROBERT PROF
UNIVERSITY OF TEXAS
ASTRONOMY DEPT
AUSTIN TX 78712
U.S.A.
TEL: 512-471-7312
TLX:
TLF:
EML:
COM: 34.46C

ROBE H A G DR
INSTITUT D'ASTROPHYSIQUE
UNIVERSITE DE LIEGE
AVENUE DE COINTE 5
B-4200 COINTE-OUGREE
BELGIUM
TEL: 41-52-9980
TLX: 41264 ASTRLG
TLF:
EML:
COM:

ROBERTI GIUSEPPE DR
ISTITUTO DI FISICA
PAD 19
MOSTRA D'OLTREMARE
I-80125 NAPOLI
ITALY
TEL:
TLX:
TLF:
EML:
COM: 12

ROBERTS BERNARD DR
DEPT OF APPLIED MATHS
UNIVERSITY OF ST ANDREWS
ST ANDREWS. FIFE KY16 9SS
U.K.
TEL: 0334-76161
TLX: 76213
TLF:
EML:
COM: 10.12

ROBERTS DAVID HALL DR
PHYSICS DEPARTMENT
BRANDEIS UNIVERSITY
WALTHAM MA 02254
U.S.A.
TEL: 617-647-2846
TLX: 703015
TLF:
EML:
COM: 40.47

ROBERTS MORTON S DR
NRAO
EDGEMONT ROAD
CHARLOTTESVILLE VA 22903
U.S.A.
TEL:
TLX:
TLF:
EML:
COM: 28.33.40

ROBERTS WALTER ORR DR
UNIVERSITY CORP FOR
ATMOSPHERIC RESEARCH
P O BOX 3000
BOULDER CO 80307
U.S.A.
TEL: 303-497-1610
TLX: 45694
TLF:
EML:
COM:

ROBERTS WILLIAM W JR PROF
DEPT OF APPLIED MATHS
UNIVERSITY OF VIRGINIA
THORNTON HALL
CHARLOTTESVILLE VA 22901
U.S.A.
TEL: 804-924-1038
TLX:
TLF:
EML:
COM: 28.33.34

ROBERTSON DOUGLAS S
NOAA NGS N/CG 114
11400 ROCKVILLE PIKE
ROCKVILLE MD 20852
U.S.A.
TEL: 301-443-8423
TLX:
TLF:
EML:
COM: 19.31.40

ROBERTSON JAMES GORDON DR
DEPT. OF ASTROPHYSICS
SCHOOL OF PHYSICS
UNIVERSITY OF SYDNEY
NSW 2006
AUSTRALIA
TEL:
TLX:
TLF:
EML:
COM: 28.40

ROBERTSON JOHN ALISTAIR
DEPT OF APPLIED MATHS
UNIVERSITY OF ST ANDREWS
NORTH HAUGH
ST ANDREWS. FIFE KY16 9SS
U.K.
TEL: 0334-76161
TLX: 76213
TLF:
EML:
COM: 42

ROBINSON BRIAN J DR
CSIRO
DIVISION OF RADIOPHYSICS
P.O.BOX 76
EPPING NSW 2121
AUSTRALIA
TEL: 02-868-0222
TLX: 26230 ASTRO AA
TLF:
EML:
COM: 29.33.34.40

ROBINSON EDWARD LEWIS DR
DEPT OF ASTRONOMY
UNIVERSITY OF TEXAS
AUSTIN TX 78712
U.S.A.
TEL: 512-471-3401
TLX:
TLF:
EML:
COM: 25.27.42

ROBINSON I PROF
UNIVERSITY OF TEXAS
BOX 688. MS NF 11
RICHARDSON TX 75080
U.S.A.
TEL: 214-690-2176
TLX:
TLF:
EML:
COM: 47

ROBINSON JR RICHARD D DR
ANGLO-AUSTRALIAN
OBSERVATORY
P.O.BOX 296
EPPING NSW 2121
AUSTRALIA
TEL: 02-868-1666
TLX: 23999
TLF:
EML:
COM: 10.40

ROBINSON LEIF J
SKY & TELESCOPE
49 BAY STATE RD
CAMBRIDGE MA 02238
U.S.A.
TEL: 617-864-7360
TLX:
TLF:
EML:
COM: 46.51

ROBINSON LLOYD B DR
LICK OBSERVATORY
UNIVERSITY OF CALIFORNIA
SANTA CRUZ CA 95064
U.S.A.
TEL: 408-429-2437
TLX:
TLF:
EML:
COM: 09

ROBINSON WILLIAM J DR
DEPT OF MATHEMATICS
THE UNIVERSITY
BRADFORD BD7 1DP
U.K.
TEL: 733466
TLX:
TLF:
EML:
COM: 07

ROBLEY R DR
9 ALLEES FR. VERDIER
F-31000 TOULOUSE
FRANCE
TEL: 61-52-22-73
TLX:
TLF:
EML:
COM: 21

ROBSON IAN E DR
SCHOOL OF PHYSICS & ASTR
LANCASHIRE POLYTECHNIC
PRESTON PR1 2TQ
U.K.
TEL: 772-22141x2188
TLX: 677409 LANPOL
TLF:
EML:
COM:

ROCA CORTES TEODORO
INST DE ASTROFISICA DE
CANARIAS
LA LAGUNA
38071 TENERIFE
SPAIN
TEL: 922-26-22-11
TLX: 92640
TLF:
EML:
COM: 10.12C

ROCCA-VOLMERANGE BRIGITTE
INSTITUT D'ASTROPHYSIQUE
98 BIS BOULEVARD ARAGO
F-75014 PARIS
FRANCE
TEL: 1-43-20-14-25
TLX: 270070 INSU
TLF:
EML:
COM:

ROCHESTER MICHAEL G PROF
DEPT OF EARTH SCIENCES
MEMORIAL UNIVERSITY
OF NEWFOUNDLAND
ST JOHNS. NFLD A1B 3X7
CANADA
TEL: 709-737-7565
TLX: 0164101
TLF:
EML:
COM: 19C

RODDIER CLAUDE DR
UER DE MATHS.
UNIVERSITE DE PROVENCE
PLACE VICTOR HUGO
F-13331 MARSEILLE
FRANCE
TEL:
TLX:
TLF:
EML:
COM: 09

RODDIER FRANCOIS PROF
NOAO/ADP DIVISION
PO BOX 26732
950 N. CHERRY AVENUE
TUCSON AZ 85726
U.S.A.
TEL: 602-325-9220
TLX: 0666484 AURA NOAO TU
TLF:
EML:
COM: 09.12

RODGERS ALEX W DR
MT STROMLO OBSERVATORY
WODEN P.O. ACT 2606
AUSTRALIA
TEL: 062-881111
TLX: 62270 CANOPUS AA
TLF:
EML:
COM: 27.29.46

RODMAN RICHARD B DR
65 LOCUST AVE
LEXINGTON MA 02173
U.S.A.
TEL: 617-861-8149
TLX:
TLF:
EML:
COM:

RODONO MARCELLO DR
INSTITUTE OF ASTRONOMY
UNIVERSITY OF CATANIA
VIALE ANDREA DORIA 6
I-95125 CATANIA
ITALY
TEL: 33-07-34
TLX: 970359 ASTRCT I
TLF:
EML:
COM: 27C.42C

RODRIGO RAFAEL
INST DE ASTROFISICA DE
ANDALUCIA. APDO 2144
PROFESOR ALBAREDA 1
18080 GRANADA
SPAIN
TEL: 58-12-13-00
TLX: 78573 IAAG E
TLF:
EML:
COM: 16.21

RODRIGUEZ LUIS F
INSTITUTO DE ASTRONOMIA
UNAM
APDO POSTAL 70-264
04510 MEXICO DF
MEXICO
TEL: 905-548-5306
TLX: 1760155 CICME
TLF:
EML:
COM: 34C.40

RODRIGUEZ-ESPINOSA JOSE
FACULTAD DE FISICA
UNIVERSIDAD COMPLUTENSE
28040 MADRID
SPAIN
TEL: 449 53 16
TLX: 47273
TLF:
EML:
COM: 28

ROEDER ROBERT C PROF
DEPT OF PHYSICS
SOUTHWESTERN UNIVERSITY
UNIVERSITY AVENUE
GEORGETOWN TX 78626
U.S.A.
TEL: 512-863-1633
TLX: 910-350-1677
TLF:
EML:
COM: 40.47

ROEMER ELIZABETH PROF
LUNAR AND PLANETARY LAB
UNIVERSITY OF ARIZONA
TUCSON AZ 85721
U.S.A.
TEL: 602-621-2897
TLX: 467175
TLF:
EML:
COM: 06P.15.20.24

ROEMER MAX PROF
INST F ASTROPHYSIK &
EXTRATERR FORSCHUNG
AUF DEM HUEGEL 71
D-5300 BONN 1
GERMANY. F.R.
TEL: 228-733670
TLX:
TLF:
EML:
COM: 10

ROENNAENG BERNT O DR
ONSALA SPACE OBSERVATORY
S-439 00 ONSALA
SWEDEN
TEL: 300 62637
TLX: 2400
TLF:
EML:
COM: 40

ROESER HANS-PETER
MPI FUER RADIOASTRONOMIE
AUF DEM HUEGEL 69
D-5300 BONN 1
GERMANY. F.R.
TEL: 0228/525265
TLX: 886440 MPIFR D
TLF:
EML:
COM: 34.40

ROESER HERMANN-JOSEF DR
MPI FUER ASTRONOMIE
KOENIGSTUHL
D-6900 HEIDELBERG 1
GERMANY. F.R.
TEL: 06221-528(1)206
TLX: 461789 MPIA D
TLF:
EML:
COM: 28

ROESER SIEGFRIED DR
ASTRONOMISCHES RECHEN-
INSTITUT
MOENCHHOFSTR 12-14
D-6900 HEIDELBERG 1
GERMANY. F.R.
TEL: 06221-49026
TLX: 461336 ARIHD D
TLF:
EML:
COM: 08.24

ROGER ROBERT S DR
DOMINION RADIO ASTRO-
PHYSICAL OBSERVATORY
P O BOX 248
PENTICTON BC V2A 6K3
CANADA
TEL: 604-497-5321
TLX: 048-88127
TLF:
EML:
COM: 34.40

ROGERS ALAN E E DR
HAYSTACK OBSERVATORY
WESTFORD MA 01886
U.S.A.
TEL: 617-692-4764
TLX: 948149 HAYSTACK WFRD
TLF:
EML:
COM: 34.40

ROGERS CHRISTOPHER DR
DEPT OF ASTRONOMY
UNIVERSITY OF TORONTO
TORONTO ONT M5S 1A1
CANADA
TEL: 416-978-4833
TLX: 06-986766
TLF:
EML:
COM:

ROGERSON JOHN B PROF
PRINCETON UNIVERSITY
DEPT ASTROPHYS SCIENCES
PEYTON HALL
PRINCETON NJ 08540
U.S.A.
TEL: 609-452-3806
TLX: 322409
TLF:
EML:
COM:

ROGSTAD DAVID H DR
MAIL CODE 264-748
JET PROPULSION LAB.
4800 OAK GROVE DRIVE
PASADENA CA 91109
U.S.A.
TEL:
TLX:
TLF:
EML:
COM: 40

ROHLFS K PROF DR
RUHR UNIVERSITAET BOCHUM
INSTITUT FUR ASTROPHYSIK
POSTFACH 102 148
D-4630 BOCHUM 1
GERMANY. F.R.
TEL: 0234-700-5802
TLX: 0825860
TLF:
EML:
COM: 33.34.40

ROLAND GINETTE DR
INSTITUT D'ASTROPHYSIQUE
UNIVERSITE DE LIEGE
AVENUE DE COINTE 5
B-4200 COINTE-OUGREE
BELGIUM
TEL: 41-52-99-80
TLX: 41254 ASTRLG B
TLF:
EML:
COM: 12

ROLLAND ANGEL DR
INSTITUTO DE ASTROFISICA
DE ANDALUCIA
APDO 2144
18080 GRANADA
SPAIN
TEL: 958-121-300
TLX: 78573
TLF:
EML:
COM:

ROMAN NANCY G DR
APT 306W
4260 NORTH PARK AVE
CHEVY CHASE MD 20815
U.S.A.
TEL: 301-656-6091
TLX:
TLF:
EML:
COM: 05.44.45

ROMANCHUK PAVEL R DR
KIEV STATE UNIVERSITY
ASTRONOMICAL OBSERVATORY
252053 KIEV
U.S.S.R.
TEL:
TLX:
TLF:
EML:
COM: 10

ROMANISHIN WILLIAM DR
DEPT. OF PHYS. & ASTRON.
UNIVERSITY OF OKLAHOMA
NORMAN OK 73019
U.S.A.
TEL: 405 325 3961
TLX:
TLF:
EML:
COM:

ROMANO GIULIANO PROF
V. S.ANTONIO DA PADOVA 7
I-31100 TREVISO
ITALY
TEL:
TLX:
TLF:
EML:
COM: 37

ROMANOV YURI S DR
ODESSA ASTRONOMICAL
OBSERVATORY
SHEVCHENKO PARK
270014 ODESSA
U.S.S.R.
TEL: 22-03-96
TLX:
TLF:
EML:
COM: 27.30

ROMERO PEREZ M PILAR
INST ASTRON & GEODESIA
FAC DE CIENCIAS MATEMAT.
UNIVERSIDAD COMPLUTENSE
28040 MADRID
SPAIN
TEL: 2442501
TLX:
TLF:
EML:
COM: 04

ROMNEY JONATHAN D DR
NRAO
EDGEMONT ROAD
CHARLOTTESVILLE VA 22903
U.S.A.
TEL: 804-296-0242
TLX: 910-997-0174
TLF:
EML:
COM: 40

ROMPOLT BOGDAN DR
ASTRONOMICAL INSTITUTE
UL. KOPERNIKA 11
51-622 WROCLAW
POLAND
TEL: 071-48-24-34
TLX: 0712791 UWR PL
TLF:
EML:
COM: 10

ROWAN COLIN A
13 ACORN AVENUE
BAR HILL
CAMBRIDGE CB3 8DT
U.K.
TEL: 0954-81058
TLX:
TLF:
EML:
COM: 41

RONG JIAN-XIANG
DEPT OF ASTRONOMY
NANJING UNIVERSITY
NANJING
CHINA. PEOPLE'S REP.
TEL: 34651 - 2882
TLX: 34151 PRCNU CN
TLF:
EML:
COM: 33

ROOD HERBERT J
SCHOOL OF NATURAL SCS
INST FOR ADVANCED STUDY
PRINCETON NJ 08540
U.S.A.
TEL:
TLX:
TLF:
EML:
COM: 28

ROOD ROBERT T DR
UNIVERSITY OF VIRGINIA
BOX 3818
UNIVERSITY STATION
CHARLOTTESVILLE VA 22903
U.S.A.
TEL: 804-924-4904
TLX:
TLF:
EML:
COM: 35.51

ROOS NICOLAAS DR
STERRENKUNDIG INSTITUUT
KATHOLIEKE UNIVERSITEIT
6525 ED NIJMEGEN
NETHERLANDS
TEL: 080 612080
TLX: 48228 WINAT NL
TLF:
EML: BITNET:U636003@HNYKUN11
COM: 28

ROOSEN ROBERT G DR
RAINBOW OBSERVATORY
RR1
P.O. BOX 5068
PAHOA HI 96778
U.S.A.
TEL:
TLX:
TLF:
EML:
COM: 21.22

ROQUES SYLVIE DR
OBS MIDI-PYRENEES
14 AV EDOUARD BELIN
31400 TOULOUSE
FRANCE
TEL: 61 25 21 01
TLX: 530776 OBSTLSE
TLF:
EML:
COM:

ROSA DOROTHEA DR
EMIL-KURZ-STR 4
D-8045 ISMANING
GERMANY. F.R.
TEL: 89-96-42-99
TLX:
TLF:
EML:
COM:

ROSA MICHAEL RICHARD DR
ST/ECF
C/O ESO
KARL-SCHWARZSCHILD-STR 2
D-8046 GARCHING B MUNCHEN
GERMANY. F.R.
TEL: 49-89-32006-0
TLX: 528-282-22-EO D
TLF:
EML:
COM: 28.34

ROSADO MARGARITA DR
INSTITUTO DE ASTRONOMIA
UNAM
APDO POSTAL 70-264
04510 MEXICO DF
MEXICO
TEL: 905-548-5306
TLX:
TLF:
EML:
COM: 28.34

ROSCH JEAN PROF
OBSERVATOIRES PIC-DU MIDI
ET TOULOUSE
F-65200 BAGNERES-DE-B.
FRANCE
TEL: 62-95-19-69
TLX: 531625 F
TLF:
EML:
COM: 09.10.16

ROSE JAMES ANTHONY
INSTITUTE FOR ASTRONOMY
UNIVERSITY OF HAWAII
2680 WOODLAWN DRIVE
HONOLULU HI 96822
U.S.A.
TEL:
TLX:
TLF:
EML:
COM: 28.29

ROSE WILLIAM K DR
ASTRONOMY PROGRAM
UNIVERSITY OF MARYLAND
COLLEGE PK MD 20742
U.S.A.
TEL: 301-299-2777
TLX:
TLF:
EML:
COM: 34

ROSEN EDWARD DR
DEPT OF HISTORY
CITY COLLEGE OF THE
CITY UNIVERSITY NEW YORK
NEW YORK NY 10031
U.S.A.
TEL:
TLX:
TLF:
EML:
COM:

ROSENBERG J DR
STATE UNIV OF UTRECHT
HEIDELBERLAAN 8
NL-3584 CS UTRECHT
NETHERLANDS
TEL: 030-535124
TLX:
TLF:
EML:
COM:

ROSENDHAL JEFFREY D DR
NASA HEADQUARTERS
CODE E
WASHINGTON DC 20546
U.S.A.
TEL: 202-453-1410
TLX:
TLF:
EML:
COM: 44

ROSINO LEONIDA PROF
OSSERVATORIO ASTRONOMICO
VICOLO DELL'OSSERVATORIO
I-35100 PADOVA
ITALY
TEL: 049-661499
TLX: 430176 UNPADU
TLF:
EML:
COM: 06.27.34.37

ROSLUND CURT DR
DEPT OF ASTRONOMY
CHALMERS UNIV TECHNOLOGY
S-412 96 GOTHENBURG
SWEDEN
TEL: 46-31-810-100
TLX:
TLF:
EML:
COM: 25.46

ROSNER ROBERT
UNIVERSITY OF CHICAGO
E.FERMI INST. & ASTRON.
5640 SOUTH ELLIS AVE.
CHICAGO. IL 60637
U.S.A.
TEL:
TLX:
TLF:
EML:
COM: 48 49

ROSQUIST KJELL
INST OF THEORETICAL PHYS
VANADISVAEGEN 9
S-113 46 STOCKHOLM
SWEDEN
TEL: 46-8-228160x225
TLX: 15433 FYSTO S
TLF:
EML:
COM: 47

ROSS DENNIS K PROF
PHYSICS DEPT
IOWA STATE UNIVERSITY
AMES IA 50011
U.S.A.
TEL: 515-294-6010
TLX:
TLF:
EML:
COM:

ROSS JOHN E R DR
PHYSICS DEPT
UNIVERSITY OF QUEENSLAND
ST LUCIA
BRISBANE QLD 4067
AUSTRALIA
TEL: 07-377-3429
TLX: 40315 UNIVQLD AA
TLF:
EML:
COM: 14.36

ROSSELLO GASPAR
DPTO FISICA DE ATMOSFERA
UNIVERSIDAD DE BARCELONA
AVDA. DIAGONAL 645
08028 BARCELONA
SPAIN
TEL:
TLX:
TLF:
EML:
COM: 04

ROSSI BRUNO B PROF
MIT
RM 37-667
CAMBRIDGE MA 02139
U.S.A.
TEL: 617-253-4281
TLX: 92-1473
TLF:
EML:
COM: 48

ROSSI LUCIO
IST. ASTROFISICA SPAZIALE
C P 67
I-00044 FRASCATI
ITALY
TEL: 06-9425651/2/3
TLX: 610261 CNR FRA
TLF:
EML:
COM: 29

ROSTAS FRANCOIS DR
OBSERVATOIRE DE PARIS
SECTION DE MEUDON
F-92195 MEUDON PL CEDEX
FRANCE
TEL: 1-45-07-75-65
TLX: 201571 LAM
TLF:
EML:
COM:

ROTH MIGUEL R DR
INSTITUTO DE ASTRONOMIA
UNAM
APDO POSTAL 877
22860 ENSENADA. B. CALIF.
MEXICO
TEL: 667-40887
TLX: 56539 CICEME
TLF:
EML:
COM:

ROTHENFLUG ROBERT DR
CEA SACLAY
SERVICE D'ASTROPHYSIQUE
GIF SUR YVETTE CDX 91191
FRANCE
TEL: 1 69 08 43 27
TLX: 604806
TLF:
EML: ROTHENFLUG@32779.DECNET.CERN
COM:

ROTH-KOPPNER MARIA LUISE
HAMBURGER STERNWARTE
GOJENSBERGSWEG
D-2050 HAMBURG 80
GERMANY. F.R.
TEL: 040-72524112
TLX: 217884 HAMST D
TLF:
EML:
COM: 37

ROTS ARNOLD H DR
NRAO
PO BOX 0
SOCORRO. NM 87801
U.S.A.
TEL:
TLX:
TLF:
EML:
COM: 26

ROTTENBERG J A DR
2911 BAYVIEW AVE
SUITE 110C
WILLOWDALE ONT M2K 1E8
CANADA
TEL:
TLX:
TLF:
EML:
COM:

ROUDIER THIERRY DR
OBS MIDI-PYRENEES
14 AV E. BELIN
31400 TOULOUSE
FRANCE
TEL: 61 25 21 01
TLX:
TLF:
EML:
COM: 10.12

ROUEFF EVELYNE M A DR
OBSERVATOIRE DE PARIS
SECTION DE MEUDON
DAF
F-92195 MEUDON PL CEDEX
FRANCE
TEL: 1-45-07-74-35
TLX: 201571 LAM
TLF:
EML:
COM: 14

ROUNTREE JANET DR
AFR/TECHN SERVICES STAFF
BOLLING AFB
WASHINGTON DC 20332
U.S.A.
TEL: 202-767-3968
TLX:
TLF:
EML:
COM: 09.27.37.45

ROUSE CARL A DR
627 15TH STREET
DEL MAR CA 92014
U.S.A.
TEL: 619-455-4015
TLX: 695065
TLF:
EML:
COM: 35

ROUSSEAU JEANINE DR
OBSERVATOIRE DE LYON
F-69230 ST GENIS-LAVAL
FRANCE
TEL: 78-56-07-05
TLX: 310926
TLF:
EML:
COM:

ROUSSEAU JEAN-MICHEL MR
OBSERVATOIRE DE BORDEAUX
AVENUE P. SEMIROT
F-33270 FLOIRAC
FRANCE
TEL: 56-86-43-30
TLX:
TLF:
EML:
COM: 08

ROUTLEDGE DAVID DR
ELECTRICAL ENGR DEPT
UNIVERSITY OF ALBERTA
EDMONTON AB T6G 2G7
CANADA
TEL: 403-432-5668
TLX:
TLF:
EML:
COM:

ROUTLY PAUL M DR
US NAVAL OBSERVATORY
34 & MASSACHUSETTS AVE NW
WASHINGTON DC 20390
U.S.A.
TEL: 202-653-1532
TLX:
TLF:
EML:
COM: 38

ROVIRA MARTA GRACIELA
INST. DE ASTRONOMIA Y
FISICA DEL ESPACIO
C.C. 67. SUC. 28
1428 CAPITAL
ARGENTINA
TEL:
TLX:
TLF:
EML:
COM: 12.36

ROVITHIS PETER DR
NATL OBSERVATORY ATHENS
P.O. BOX 20048
ATHENS 306
GREECE
TEL: 01-3461191
TLX: 215530 OBSA GR
TLF:
EML:
COM: 42

ROVITHIS-LIVANIOU HELEN
SECTION OF ASTROPHYSICS
ASTRONOMY AND MECHANICS
DEPT OF PHYSICS
GR-15771 ZOGRAFOS
GREECE
TEL: 01-724-3414
TLX:
TLF:
EML:
COM: 42

ROWAN-ROBINSON MICHAEL DR
DEPT OF APPLIED MATHS
QUEEN MARY COLLEGE
MILE END ROAD
LONDON E1 4NS
U.K.
TEL:
TLX:
TLF:
EML:
COM: 47.51

ROWSON BARRIE DR
NRAL
JODRELL BANK
MACCLESFIELD SK11 9DL
U.K.
TEL: 047-77-1321
TLX: 36149
TLF:
EML:
COM: 40

ROXBURGH IAN W PROF
SCHOOL OF MATHEMATICAL SC
QUEEN MARY COLLEGE
MILE END ROAD
LONDON E1 4NS
U.K.
TEL: 01-980-4811
TLX:
TLF:
EML:
COM: 10.34.35.42.47.49C

ROY ARCHIE E PROF
DEPT OF ASTRONOMY
GLASGOW UNIVERSITY
GLASGOW G12 8QQ
U.K.
TEL: 41-339-8855x502
TLX: 778421 GLASUL
TLF:
EML:
COM: 07C.46

ROY JEAN-RENE
DEPT DE PHYSIQUE
UNIVERSITE LAVAL
CITE UNIVERSITAIRE
QUEBEC G1K 7P4
CANADA
TEL: 418-656-5816
TLX: 5131621 UNILAVAL
TLF:
EML:
COM:

ROZELOT JEAN P
CERGA
AVENUE COPERNIC
F-06130 GRASSE
FRANCE
TEL: 93-36-58-49
TLX: 470865
TLF:
EML:
COM: 10

ROZHKOVSKIJ DIMITRIJ A
ASTROPHYSICAL INSTITUTE
480068 ALMA-ATA
U.S.S.R.
TEL: 62-40-40
TLX:
TLF:
EML:
COM: 21.34

ROZYCZKA MICHAL
WARSAW UNIVERSITY OBS
AL. UJAZDOWSKIE 4
00-478 WARSAW
POLAND
TEL:
TLX: 813978 ZAPAN PL
TLF:
EML:
COM: 34

RUBASHEV BORIS M DR
PULKOVO OBSERVATORY
196140 LENINGRAD
U.S.S.R.
TEL:
TLX:
TLF:
EML:
COM: 10

RUBEN G PROF DR
ZNTRLINST. F. ASTROPHYSIK
ROSA-LUXEMBURG-STR 17A
DDR-1502 POTSDAM
GERMANY. D.R.
TEL:
TLX:
TLF:
EML:
COM: 35.38.44

RUBIN ROBERT HOWARD
NASA AMES RESEARCH CENTER
MS 245-6
MOFFETT FIELD CA 94035
U.S.A.
TEL: 415-965-5528
TLX: 348408
TLF:
EML:
COM: 40.51

RUBIN VERA C DR
DEPT TERRESTR. MAGNETISM
CARNEGIE INST. WASHINGTON
5241 BROAD BRANCH RD N.W.
WASHINGTON DC 20015
U.S.A.
TEL: 202-966-0863
TLX: 440427 MAGN UI
TLF:
EML:
COM: 28.30.33.47

RUBIO MONICA DR
DEPTO DE ASTRONOMIA
UNIVERSIDAD DE CHILE
CASILLA 36-D
SANTIAGO
CHILE
TEL: 2294101
TLX: 440005 ATTN OBSERVAL
TLF:
EML:
COM: 40

RUCINSKI SLAVOMIR M DR
SPACE ASTROPHY. LABORAT.
YORK UNIVERSITY
4700 KEELE STREET
TORONTO ONTARIO M3J 1P3
CANADA
TEL: 416-665-3311
TLX: 06-986766
TLF:
EML: BITNET:FS300516@YUSOL
COM: 42

RUDAK BRONISLAW
COPERNICUS ASTRON CENTER
UL. CHOPINA 12/18
87-100 TORUN
POLAND
TEL: 26037 x 10
TLX: 813978 ZAPAN PL
TLF:
EML:
COM:

RUDDY VINCENT P DR
REGIONAL TECHNICAL COLL.
ROSSA AVENUE
CORK
IRELAND
TEL:
TLX:
TLF:
EML:
COM: 47

RUDER HANNS
LEHRSTUHL F THEORET ASTRO
PHYSIK DER UNIV TUEBINGEN
AUF DER MORGENSTELLE 12.C
D-7400 TUEBINGEN
GERMANY. F.R.
TEL: 07071/292487
TLX:
TLF:
EML:
COM: 09.14.19.24.44

RUDERMAN MALVIN A
PHYSICS DEPT
COLUMBIA UNIVERSITY
NEW YORK NY 10027
U.S.A.
TEL: 212-280 3317
TLX:
TLF:
EML:
COM:

RUDKJOBING MOGENS PROF
INSTITUTE OF ASTRONOMY
UNIVERSITY OF AARHUS
LANGELANDSGADE
DK-8000 AARHUS C
DENMARK
TEL: 06-12-88-99
TLX: 64767 AAUSCI DK
TLF:
EML:
COM: 45

RUDNICK LAWRENCE DR
UNIVERSITY OF MINNESOTA
116 CHURCH STREET S.E.
MINNEAPOLIS MN 55455
U.S.A.
TEL: 612-373-5457
TLX:
TLF:
EML:
COM: 40.47

RUDNICKI KONRAD PROF
DEPT. SPACE PHYS. & ASTRO
RICE UNIVERSITY
PO BOX 1892
HOUSTON TX 77251
U.S.A.
TEL:
TLX:
TLF:
EML:
COM: 28

RUDZIKAS ZENONAS B
INSTITUTE OF PHYSICS
LITHUANIAN SSR
K. POZELOS 54
232600 VILNIUS
U.S.S.R.
TEL: 617610
TLX:
TLF:
EML:
COM: 14C

RUFENER FREDY G PROF
OBSERVATOIRE DE GENEVE
CH-1290 SAUVERNY
SWITZERLAND
TEL: 41-22-552611
TLX: 27720 OBSG CH
TLF:
EML:
COM: 25C

RUFFINI REMO
DIPARTIMENTO DI FISICA
UNIVERSITA DI ROMA
PIAZZALE ALDO MORO 2
I-00185 ROMA
ITALY
TEL: 4976304
TLX: 613255 INFNRO I
TLF:
EML:
COM: 48

RUGGE HUGO R DR
SPACE SCIENCES LABORATORY
AEROSPACE CORPORATION
PO BOX 92957
LOS ANGELES CA 90009
U.S.A.
TEL: 213-648-7086
TLX:
TLF:
EML:
COM:

RUIZ MARIA TERESA DR
OBS ASTRONOMICO NACIONAL
UNIVERSIDAD DE CHILE
CASILLA 36-D
SANTIAGO
CHILE
TEL: 2294101
TLX:
TLF:
EML:
COM: 33

RULE BRUCE H
HALE OBSERVATORIES
2205 MONTE VISTA STREET
PASADENA CA 91107
U.S.A.
TEL: 818-794-6593
TLX:
TLF:
EML:
COM:

RUMSEY NORMAN J
21 MALONE ROAD
LOWER HUTT
NEW ZEALAND
TEL: (04)696787
TLX:
TLF:
EML:
COM:

RUNCORN S K PROF
SCHOOL OF PHYSICS
THE UNIVERSITY
NEWCASTLE/TYNE NE1 7RU
U.K.
TEL: 0632-32511
TLX: 53654 UNINEW G
TLF:
EML:
COM: 16.19

RUPRECHT JAROSLAV DR
ASTRONOMICAL INSTITUTE
CZECH. ACAD. OF SCIENCES
BUDECSKA 6
120 23 PRAHA 2
CZECHOSLOVAKIA
TEL: 258757
TLX: 122 486
TLF:
EML:
COM: 37

RUSCONI LUIGIA DR
DIPT. DI ASTRONOMIA
UNIVERSITA DI TRIESTE
VIA TIEPOLO 11
I-34131 TRIESTE
ITALY
TEL: 40-794863
TLX: 461137 OAOTI
TLF:
EML:
COM: 09

RUSIN VOJTECH
ASTRONOMICAL INSTITUTE
SLOVAK ACAD. OF SCIENCES
059 60 TATRANSKA LOMNICA
CZECHOSLOVAKIA
TEL: 0969-967866/7/8
TLX: 80-78277 AUSSAV C
TLF:
EML:
COM: 10.12

RUSKOL EUGENIA L DR
OJSCHMIDT INSTITUTE
OF PHYSICS OF THE EARTH
USSR ACADEMY OF SCIENCES
123810 MOSCOW
U.S.S.R.
TEL: 252-07-26
TLX: 411196 IPZAN SU
TLF:
EML:
COM: 16

RUSSELL CHRISTOPHER T
INST OF GEOPHYSICS
& PLANETARY PHYSICS
UNIVERSITY OF CALIFORNIA
LOS ANGELES CA 90024
U.S.A.
TEL: 213-825-3188
TLX: 910-342-6981
TLF:
EML:
COM: 49

RUSSELL JANE L DR
CODE 4130R
US NAVAL RESEARCH LAB.
WASHINGTON DC 20375
U.S.A.
TEL: 202-767-0171
TLX:
TLF:
EML:
COM: 08.24.26.51

RUSSELL JOHN A PROF
DEPT OF ASTRONOMY
UNIV SOUTHERN CALIFORNIA
UNIVERSITY PARK
LOS ANGELES CA 90089
U.S.A.
TEL: 213-743-0231
TLX:
TLF:
EML:
COM: 22

RUSSEV RUSCHO DR
UNIVERSITY OF SOFIA
DEPT OF ASTRONOMY
ANTON IVANOV STR 5
1126 SOFIA
BULGARIA
TEL: 6-25-61
TLX:
TLF:
EML:
COM: 27

RUSSEVA TATJANA
DEPT OF ASTRONOMY AND
NATL ASTRON OBSERVATORY
72 LENIN BLVD
1784 SOFIA
BULGARIA
TEL: 73-41-559
TLX: 23561 ECF BAN BG
TLF:
EML:
COM: 37

RUSSO GUIDO DR
DIP. SCIENZE FISICHE
PAD. 19
MOSTRA D'OLTREMARE
I-80125 NAPOLI
ITALY
TEL: 39-81-7253447
TLX: 720320
TLF:
EML:
COM: 05

RUST DAVID M DR
APPLIED PHYSICS LAB
JOHNS HOPKINS UNIVERSITY
JOHNS HOPKINS ROAD
LAUREL MD 20707
U.S.A.
TEL: 301-953-5414
TLX: 89-548 APL JHU LAUR
TLF:
EML:
COM: 10

RUSU I DR
ASTRONOMICAL OBSERVATORY
CUTITUL DE ARGINT 5
75212 BUCAREST 28
RUMANIA
TEL: 23-63-01
TLX:
TLF:
EML:
COM: 08.19

RUSU L DR
ASTRONOMICAL OBSERVATORY
CUTITUL DE ARGINT 5
75212 BUCAREST 28
RUMANIA
TEL: 23-63-01
TLX:
TLF:
EML:
COM:

RUTTEN ROBERT J. DR
STERREKUNDIG INSTITUUT
PRINCETONPLEIN 5
POSTBUS 80 000
NL-3508 TA UTRECHT
NETHERLANDS
TEL: 31-30-535200
TLX: 40048 FYLUT NL
TLF:
EML: BITNET:UWMMAIL@RUUTRUUG
COM: 12.29.36

RUZDJAK VLADIMIR DR
INSTITUTE OF PHYSICS
UNIVERSITY OF ZAGREB
P O BOX 304
41001 ZAGREB
YUGOSLAVIA
TEL:
TLX:
TLF:
EML:
COM: 10

RUZICKOVA-TOPOLOVA B DR
ASTRONOMICAL INSTITUTE
CZECH. ACAD. OF SCIENCES
OBSERVATORY
251 65 ONDREJOV
CZECHOSLOVAKIA
TEL: 724525 PRAHA
TLX: 121579
TLF:
EML:
COM: 10

RYABOV YU A PROF DR
MATHEMATICS DEPT OF MADI
LENINGRADSKY PROSP. 64
125319 MOSCOW
U.S.S.R.
TEL: 1550326
TLX:
TLF:
EML:
COM: 07

RYBANSKY MILAN
ASTRONOMICAL INSTITUTE
SLOVAK ACAD. OF SCIENCES
059 60 TATRANSKA LOMNICA
CZECHOSLOVAKIA
TEL: 0969-967-866
TLX: 80-78277 AUSAV C
TLF:
EML:
COM: 10.12

RYBICKI GEORGE B DR
HARVARD SMITHSONIAN
CENTER FOR ASTROPHYSICS
60 GARDEN STREET
CAMBRIDGE MA 02138
U.S.A.
TEL: 617-495-7452
TLX: 92-1428
TLF:
EML:
COM: 33.36

RYBKA PRZEMYSLAW DR
INSTITUTE OF HISTORY
OF SCIENCE
NOWY SWIAT 72
00-330 WARSZAWA
POLAND
TEL:
TLX:
TLF:
EML:
COM: 41

RYDBECK GUSTAF H B DR
ONSALA SPACE OBSERVATORY
S-439 00 ONSALA
SWEDEN
TEL: 0300-6208
TLX:
TLF:
EML:
COM: 28.40

RYDBECK OLOF E H PROF
ONSALA SPACE OBSERVATORY
S-439 00 ONSALA
SWEDEN
TEL: 0300-62081
TLX: 8542400 ONSPACE
TLF:
EML:
COM: 40

RYDGREN ALFRED ERIC JR DR
11502 SE 217 STREET
KENT WA 98031
U.S.A.
TEL:
TLX:
TLF:
EML:
COM: 25

RYKHLOVA LIDIJA V DR
ASTRONOMICAL COUNCIL
USSR ACADEMY OF SCIENCES
PYATNITSKAYA 48
109017 MOSCOW
U.S.S.R.
TEL: 231-54-61
TLX: 412623 SCSTP SU
TLF:
EML:
COM: 19

RYLOV VALERIJ S DR
SPECIAL ASTROPHYSICAL OBS
USSR ACADEMY OF SCIENCES
357140 N ARKHYZ
U.S.S.R.
TEL:
TLX:
TLF:
EML:
COM: 09

RYTER CHARLES E DR
CEN SACLAY
DPhG/SAp
BAT 28
F-91191 GIF/YVETTE CEDEX
FRANCE
TEL: 1-69-08-39-12
TLX:
TLF:
EML:
COM:

RYZHKOV NIKOLAI F DR
SPECIAL ASTROPHYSICAL OBS
LENINGRAD BRANCH
196140 LENINGRAD
U.S.S.R.
TEL:
TLX:
TLF:
EML:
COM: 40

RZHIGA OLEG N DR
INST OF RADIO & ELECTRON
USSR ACADEMY OF SCIENCES
103907 MOSCOW
U.S.S.R.
TEL:
TLX:
TLF:
EML:
COM:

SAAR ENN DR
TARTU ASTROPHYSICAL
OBSERVATORY
202444 TORAVERE. ESTONIA
U.S.S.R.
TEL:
TLX:
TLF:
EML:
COM: 29.33.47

SABANO YUTAKA DR
ASTRONOMICAL INSTITUTE
TOHOKU UNIVERSITY
ARAMAKI
SENDAI 980
JAPAN
TEL: 0222-22-1800
TLX:
TLF:
EML:
COM: 34

SABBADIN FRANCO DR
OSSERVATORIO ASTROFISICO
I-36012 ASIAGO
ITALY
TEL: 0424-62665
TLX: SETUR 430110
TLF:
EML:
COM: 34

SACK NOAH DR
DEPT OF THEORETICAL PHYS
HEBREW UNIV. OF JERUSALEM
JERUSALEM 91904
ISRAEL
TEL:
TLX:
TLF:
EML:
COM:

SACKMANN I JULIANA DR
KELLOGG RADIATION LAB
CALTECH
PASADENA CA 91125
U.S.A.
TEL: 818-356-4256
TLX:
TLF:
EML:
COM: 35

SADAKANE KOZO DR
ASTRONOMICAL INSTITUTE
OSAKA KYOIKU UNIVERSITY
TENNOJI-KU
OSAKA 543
JAPAN
TEL:
TLX:
TLF:
EML:
COM: 29

SADEH D DR
DEPT PHYSICS & ASTRONOMY
TEL-AVIV UNIVERSITY
TEL-AVIV 69978
ISRAEL
TEL: 3-420-553
TLX: 34271 VERSY
TLF:
EML:
COM:

SADIK AZIZ R DR
ASTRONOMY & SPACE RES CTR
COUNCIL FOR SCI RESEARCH
P O BOX 2441
JADIRIYAH BAGHDAD
IRAQ
TEL: 01-7765127
TLX: 213976 SRC
TLF:
EML:
COM: 27.42

SADLER ELAINE MARGARET
ANGLO-AUSTRALIAN OBS
PO BOX 296
EPPING
AUSTRALIA
TEL: 61 2 868 1666
TLX: 123999 AAOSYD AA
TLF:
EML:
COM: 28

SADZAKOV SOFIJA DR
ASTRONOMICAL OBSERVATORY
VOLGINA 7
11050 BEOGRAD
YUGOSLAVIA
TEL: 419-357/421-375
TLX:
TLF:
EML:
COM: 08.19

SAEMUNDSON THORSTEINN
RAUNVISINDASTOFNUN
HASKOLANS
DUNHAGA 3
IS-107 REYKJAVIK
ICELAND
TEL: 354-1-21340
TLX: 2307 ISINFO
TLF:
EML:
COM: 10

SAFKO JOHN L
DEPT PHYSICS & ASTRONOMY
UNIVERSITY OF S. CAROLINA
COLUMBIA SC 29208
U.S.A.
TEL: 803-777-6466
TLX: UNIVSCAROL CLB
TLF:
EML:
COM: 45

SAFRONOV VICTOR S DR
INSTITUTE OF PHYSICS
OF THE EARTH
B GRUZINSKAYA 10
123242 MOSCOW
U.S.S.R.
TEL: 252-07-26
TLX: 411196 IFZAN SU
TLF:
EML:
COM: 16

SAGAN CARL DR
CORNELL UNIVERSITY
302 SPACE SCIENCE BLDG
ITHACA NY 14853
U.S.A.
TEL: 607-256-4971
TLX: 337478
TLF:
EML:
COM: 16.51

SAGDEEV ROALD Z DR
SPACE RESEARCH INSTITUTE
USSR ACADEMY OF SCIENCES
117810 MOSCOW
U.S.S.R.
TEL: 333 14 66
TLX: 411498 STAR SU
TLF:
EML:
COM: 15.44.49

SAGGION ANTONIO PROF
ISTITUTO DI FISICA
G. GALILEI
VIA MARZOLO 8
I-35100 PADOVA
ITALY
TEL: 049-844754
TLX: 430308 DFGGPDI
TLF:
EML:
COM:

SAGNIER JEAN-LOUIS DR
BUREAU DES LONGITUDES
77 AV. DENFERT-ROCHEREAU
75014 PARIS
FRANCE
TEL: 1-40 51 22 61
TLX:
TLF:
EML:
COM: 07.20

SAHA SVAPAN KUMAR DR
INDIAN INST. OF ASTROPHY.
BANGALORE 560 034
INDIA
TEL: 569902
TLX: 845-2763 IIAB IN
TLF:
EML:
COM:

SAHADE JORGE PROF
OBSERVATORIO ASTRONOMICO
UNIV NACL DE LA PLATA
CC 677
1900 LA PLATA -Bs. As.
ARGENTINA
TEL: 54-21-249790
TLX: 31151 BULAP AR
TLF:
EML:
COM: 29.38V.42.44

SAHAI RAGHVENDRA DR
INST THEORETICAL PHYSICS
CHALMERS UNIVERSITY
UNIV OF GOTEBORG
41296 GOTEBORG
SWEDEN
TEL: (31) 721139
TLX: 2369 CHALBIB S
TLF:
EML: BITNET:TFARS@SECTHF51
COM:

SAHAL-BRECHOT SYLVIE DR
OBSERVATOIRE DE PARIS
SECTION DE MEUDON
F-92195 MEUDON PL CEDEX
FRANCE
TEL: 1-45-07-74-43
TLX: 201 571
TLF:
EML:
COM: 14P

SAIJO KEIICHI
DEPT OF PHYSICAL SCIENCES
NATIONAL SCIENCE MUSEUM
7-20 UENO PARK, TAITO-KU
TOKYO 110
JAPAN
TEL: 3-822-0111
TLX:
TLF:
EML:
COM: 42

SAIKIA D J DR
RADIO ASTRONOMY CENTRE
TIFR
POST BOX 1234
BANGALORE 560 012
INDIA
TEL: 362816 / 364062
TLX: 8458488
TLF:
EML:
COM: 40

SAIO HIDEYUKI DR
DEPT OF ASTRONOMY
UNIVERSITY OF TOKYO
BUNKYO-KU
TOKYO 113
JAPAN
TEL: 3-812-2111
TLX: UTYOSCI J33659
TLF:
EML:
COM: 35

SAISSAC JOSEPH DR
OBSERVATOIRE PIC-DU-MIDI
ET TOULOUSE
14 AVENUE EDOUARD BELIN
F 65200 BAGNERES-DE-B.
FRANCE
TEL: 62-95-19-69
TLX: 531625 S
TLF:
EML:
COM: 16

SAITO KUNIJI PROF
TOKYO ASTRONOMICAL
OBSERVATORY
OSAWA MITAKA
TOKYO 181
JAPAN
TEL:
TLX:
TLF:
EML:
COM: 10.36

SAITO MAMORU DR
DEPT OF ASTRONOMY
UNIVERSITY OF KYOTO
SAKYOKU
KYOTO 606
JAPAN
TEL: 0757512111x3904
TLX: 5422693 LIBKYU J
TLF:
EML:
COM:

SAITO SUMISABURO DR
KWASAN OBSERVATORY
YAMASHINA
KYOTO 607
JAPAN
TEL: 075-581-1235
TLX: 5422693 LIBKYU J
TLF:
EML:
COM:

SAKAI JUNICHI
FACULTY OF ENGINEERING
TOYAMA UNIVERSITY
TOYAMA, 930
JAPAN
TEL: 0764-41-1271
TLX:
TLF:
EML:
COM: 12

SAKASHITA SHIRO PROF
DEPT OF PHYSICS
HOKKAIDO UNIVERSITY
KITA 10 NISHI 8
SAPPORO HOKKAIDO 060
JAPAN
TEL: 011-716-2111
TLX:
TLF:
EML:
COM: 35

SAKHAROV VLADIMIR I DR
PULKOVO OBSERVATORY
196140 LENINGRAD
U.S.S.R.
TEL:
TLX:
TLF:
EML:
COM: 19

SAKHIBULLIN NAIL A DR
DEPT OF ASTRONOMY
KAZAN STATE UNIVERSITY
420008 KAZAN
U.S.S.R.
TEL: 32-36-41
TLX:
TLF:
EML:
COM: 36

SAKURAI KUNITOMO PROF
DEPT OF PHYSICS
KANAGAWA UNIVERSITY
KANAGAWAKU
YOKOHAMA 221
JAPAN
TEL: 045-481-5661
TLX:
TLF:
EML:
COM: 10.51

SAKURAI TAKASHI DR
TOKYO ASTRON. OBSERVATORY
2-21-1 OSAWA, MITAKA
TOKYO 181
JAPAN
TEL: 0422-32-5111
TLX: 2822307 TAOMTK J
TLF:
EML:
COM: 10.12

SAKURAI TAKEO T PROF
FACULTY OF ENGINEERING
KYOTO UNIVERSITY
SAKYO-KU
KYOTO 606
JAPAN
TEL: 75-7512111x5792
TLX: 05422693 LIBKYUJ
TLF:
EML:
COM:

SALA FERRAN DR
DEPT DE FISICA
UNIVERSITAT DE BARCELONA
AVDA DIAGONAL 647
08028 BARCELONA
SPAIN
TEL: 34 3 330 73 11
TLX:
TLF:
EML:
COM: 33

SALETIC DUSAN
ASTRONOMICAL OBSERVATORY
VOLGINA 7
11050 BEOGRAD
YUGOSLAVIA
TEL: 157-022
TLX:
TLF:
EML:
COM: 08

SALINARI PIERO
OSSERVATORIO ASTROFISICO
DI ARCETRI
LARGO E. FERMI 5
I-50125 FIRENZE
ITALY
TEL: 055-2752231
TLX: 572268
TLF:
EML:
COM: 34

SALISBURY J W DR
US GEOLOGICAL SURVEY
927 NATIONAL CENTER
RESTON VA 22092
U.S.A.
TEL: 703-860-6668
TLX: 92178
TLF:
EML:
COM:

SALO HEIKKI
JET PROPULSION LABORATORY
MAIL STOP 183-501
4800 OAK GROVE DRIVE
PASADENA CA 91109
U.S.A.
TEL: 818-354-3833
TLX:
TLF:
EML:
COM:

SALPETER EDWIN E PROF
NEWMAN LAB OF NUCLEAR STU
CORNELL UNIVERSITY
ITHACA NY 14853
U.S.A.
TEL: 607-256-3302
TLX: 937478
TLF:
EML:
COM: 34.35.40.48C

SALUKVADZE G N DR
ABASTUMANY ASTROPHYSICAL
OBSERVATORY
383762 ABASTUMANY
U.S.S.R.
TEL:
TLX:
TLF:
EML:
CON: 26.37

SALVADOR-SOLE EDUARDO
DPTO FISICA DE ATMOSFERA
UNIVERSIDAD DE BARCELONA
AVDA. DIAGONAL 645
08028 BARCELONA
SPAIN
TEL: 33007311
TLX:
TLF:
EML:
CON: 28.47

SALVATI MARCO
OSSERVATORIO DI ARCETRI
LARGO ENRICO FERMI 5
I-50125 FIRENZE
ITALY
TEL: 055-2752268
TLX: 572230 ARCETR I
TLF:
EML:
CON: 48

SAMAIN DENYS DR
INSTITUT D'ASTROPHYSIQUE
SPATIALE
BP NO 10
F-91371 VERRIERES-LE-B.
FRANCE
TEL: 1-64 47 43 04
TLX: 600 252 LPSP VE
TLF:
EML:
CON: 12

SAMPSON DOUGLAS H PROF
DEPT OF ASTRONOMY
PENNSYLVANIA STATE UNIV
525 DAVEY LAB
UNIVERSITY PARK PA 16802
U.S.A.
TEL: 814-865-0161
TLX: 842510
TLF:
EML:
CON:

SAMUS NIKOLAI N DR
ASTRONOMICAL COUNCIL
USSR ACADEMY OF SCIENCES
PYATNITSKAYA 48
109017 MOSCOW
U.S.S.R.
TEL: 231-54-61
TLX: 412623 SCSTP SU
TLF:
EML:
CON: 27.37

SANAHUJA BLAS
DPTO FISICA DE ATMOSFERA
UNIVERSIDAD DE BARCELONA
AVDA. DIAGONAL 645
08028 BARCELONA
SPAIN
TEL: 3307311 x298
TLX:
TLF:
EML:
CON: 28.46

SANAMIAN V A DR
BYURAKAN ASTROPHYSICAL
OBSERVATORY
378433 ARMENIA
U.S.S.R.
TEL: 563453
TLX:
TLF:
EML:
CON: 40

SANCHEZ FRANCISCO PROF
INSTITUTO DE ASTROFISICA
DE CANARIAS
LA LAGUNA
38071 TENERIFE
SPAIN
TEL: 922-262211
TLX: 92640 IAC E
TLF:
EML:
CON: 21.50

SANCHEZ MANUEL
INSTITUTO Y OBSERVATORIO
DE MARINA
SAN FERNANDO (CADIZ)
SPAIN
TEL: 956-883548
TLX: 76108
TLF:
EML:
CON: 08.19

SANCHEZ-SAAVEDRA M LUISA
FACULTAD DE CIENCIAS
UNIVERSIDAD DE GRANADA
18080 GRANADA
SPAIN
TEL: 958-20-22-12
TLX:
TLF:
EML:
CON: 21.33 34

SANCISI RENZO DR
KAPTEYN LABORATORIUM
POSTBUS 800
NL-9700 AV GRONINGEN
NETHERLANDS
TEL: 050-116695
TLX: 53572 STARS NL
TLF:
EML:
CON: 28.34.51

SANDAGE ALLAN
MT WILSON & LAS CAMPANAS
OBSERVATORIES
813 SANTA BARBARA ST
PASADENA CA 91101
U.S.A.
TEL: 818-577-1122
TLX:
TLF:
EML:
CON:

SANDAKOVA E V DR
KIEV STATE UNIVERSITY
ASTRONOMICAL OBSERVATORY
252053 KIEV
U.S.S.R.
TEL:
TLX:
TLF:
EML:
CON:

SANDELL GORAN HANS L DR
OBS & ASTROPHYSICS LAB
UNIVERSITY OF HELSINKI
KOPERNIKUKSENTIE 1
SF-00130 HELSINKI 13
FINLAND
TEL: 358-0-1912943
TLX:
TLF:
EML:
CON: 34.40

SANDERS DAVID B DR
INSTITUTE FOR ASTRONOMY
2680 WOODLAWN DRIVE
HONOLULU HI 96822
U.S.A.
TEL: 808 948-7399
TLX: 7238459 UHAST FF
TLF: 808 988-2790
EML: SANDERS@UHIFA.IFA.HAWAII.EDU
CON: 28.40

SANDERS ROBERT DR
KAPTEYN LABORATORY
POSTBUS 800
NL-9700 AV GRONINGEN
NETHERLANDS
TEL: 050-116695
TLX: 53572 STARS NL
TLF:
EML:
CON: 28

SANDERS W L PROF
NEW MEXICO STATE UNIV
BOX 4500
LAS CRUCES NM 88003
U.S.A.
TEL: 505-646-4914
TLX:
TLF:
EML:
CON: 24.37

SANDERS WILTON TURNER III
DEPT OF PHYSICS
UNIVERSITY OF WISCONSIN
MADISON WI 53706
U.S.A.
TEL: 608-262-5916
TLX:
TLF:
EML:
CON: 44.48

SANDFORD MAXWELL T II
LOS ALAMOS SCIENTIFIC LAB
LOS ALAMOS NM 87545
U.S.A.
TEL: 505-667-6384
TLX:
TLF:
EML:
CON:

SANDMANN WILLIAM HENRY
PHYSICS DEPT
HARVEY MUDD COLLEGE
CLAREMONT CA 91711
U.S.A.
TEL: 714-621-8024
TLX:
TLF:
EML:
COM: 27

SANDQVIST AAGE DR
STOCKHOLM OBSERVATORY
S-133 00 SALTSJOEBADEN
SWEDEN
TEL: 08-717-2149
TLX: 12972
TLF:
EML:
COM: 33.34.46P

SANDULEAK NICHOLAS DR
WARNER & SWASEY OBS
CASE WESTERN RESERVE UNIV
CLEVELAND OH 44106
U.S.A.
TEL: 216-368-6696
TLX:
TLF:
EML:
COM: 33.45

SANFORD PETER WILLIAM MR
DEPT PHYSICS & ASTRONOMY
UNIVERSITY COLLEGE LONDON
GOWER STREET
LONDON WC1E 6BT
U.K.
TEL:
TLX:
TLF:
EML:
COM:

SANROMA MANUEL DR
DEPT DE FISICA
UNIVERSITAT DE BARCELONA
AVDA DIAGONAL 647
08028 BARCELONA
SPAIN
TEL: 93 330 73 11
TLX:
TLF:
EML:
COM: 28

SANTIN PAOLO DR
OSSERVATORIO ASTRONOMICO
VIA G.B. TIEPOLO 11
CP SUCC TS 5
I-34131 TRIESTE
ITALY
TEL: 040-793921
TLX: 461137 OAT I
TLF:
EML:
COM:

SANWAL N B DR
DEPT OF ASTRONOMY
OSMANIA UNIVERSITY
HYDERABAD 500 007
INDIA
TEL: 71951 x247
TLX:
TLF:
EML:
COM: 30.42.45

SANYAL ASHIT DR
4618 OLYMPIA AVE
BELTSVILLE MD 20705
U.S.A.
TEL: 301-937-8943
TLX:
TLF:
EML:
COM: 27.42

SANZ JOSE L DR
DEPT. DE FISICA MODERNA
UNIVERSIDAD DE CANTABRIA
AVDA LOS CASTROS
39005 SANTANDER
SPAIN
TEL: 34 42 20 14 52
TLX: 35681 EDUCI E
TLF:
EML:
COM: 47

SAPAR ARVED DR
TAPEA ESTONIAN ACADEMY
OF SCIENCES
TARTU OBSERVATORY
202444 TORAVERE
U.S.S.R.
TEL:
TLX:
TLF:
EML:
COM: 36C.47

SAPRE A K DR
DEPT OF PHYSICS
RAVISHANKAR UNIVERSITY
RAIPUR 429010
INDIA
TEL: 27364
TLX:
TLF:
EML:
COM: 28

SARAZIN CRAIG L DR
DEPT OF ASTRONOMY
UNIVERSITY OF VIRGINIA
PO BOX 3818 UNIV STATION
CHARLOTTESVILLE VA 22903
U.S.A.
TEL: 804-924-4903
TLX:
TLF:
EML:
COM: 28.34

SAREYAN JEAN-PIERRE DR
OBSERVATOIRE DE NICE
BP 139
F-06003 NICE CEDEX
FRANCE
TEL: 93-89-04-20
TLX: 460004
TLF:
EML:
COM: 27.29

SARGENT ANNEILA I
CALTECH 320-47
DOWNS LAB. OF PHYSICS
PASADENA CA 91125
U.S.A.
TEL: 818-356-6622
TLX: 675425
TLF:
EML:
COM: 34.40

SARGENT WALLACE L W DR
ASTRONOMY DEPT
CALTECH 105-24
1201 E CALIFORNIA ST
PASADENA CA 91125
U.S.A.
TEL: 818-356-4055
TLX: 675425 CALTECH PSD
TLF:
EML:
COM: 28.47

SARMA M B K PROF
DEPT OF ASTRONOMY
OSMANIA UNIVERSITY
HYDERABAD 500 007
INDIA
TEL: 65228
TLX:
TLF:
EML:
COM: 25.27

SARMA N V G PROF
RAMAN RESEARCH INSTITUTE
BANGALORE 560 080
INDIA
TEL: 812-360122
TLX: 8452671 RRI IN
TLF:
EML:
COM: 34.40

SARMIENTO-GALAN A F DR
INST DE ASTRONOMIA UNAM
APARTADO POSTAL 70-264
MEXICO 04510 D.F.
MEXICO
TEL:
TLX:
TLF:
EML:
COM:

SARRIS ELEFTHERIOS PH D
NATL OBSERVATORY ATHENS
ASTRONOMICAL INSTITUTE
GR-11810 ATHENS
GREECE
TEL:
TLX: 215530 OBSA GR
TLF:
EML:
COM:

SARRIS EMMANUEL T PH D
DEPT OF ELECT ENGINEERING
DEMOCRITOS UNIV OF THRACE
GR-67100 XANTHI
GREECE
TEL: 0541-26948
TLX: 452312 POLY GR
TLF:
EML:
COM: 49C

SARTORI LEO PROF
BEHLEN LAB OF PHYSICS
UNIVERSITY OF NEBRASKA
LINCOLN NB 68588
U.S.A.
TEL:
TLX:
TLF:
EML:
COM: 48

SASAKI MISAO
RES INST FOR THEORET PHYS
HIROSHIMA UNIVERSITY
TAKEHARA-CHO
TAKEHARA 725
JAPAN
TEL: 08462-2-2362
TLX:
TLF:
EML:
COM: 47

SASAKI TOSHIYUKI DR
OKAYAMA ASTROPHYSICAL OBS
TOKYO ASTRONOMICAL OBS
HONJYO KAMOGATA
OKAYAMA 719 02
JAPAN
TEL: 86544 .156
TLX:
TLF:
EML:
COM: 28

SASAO TETSUO DR
INTERNATIONAL LATITUDE
OBSERVATORY OF MIZUSAWA
MIZUSAWA-SHI
IWATE 023
JAPAN
TEL: 197-24-7111
TLX: 837628 ILSMIZJ
TLF:
EML:
COM: 19C

SASLAW WILLIAM C PROF
ASTRONOMY DEPT
UNIVERSITY OF VIRGINIA
BOX 3818 UNIV STATION
CHARLOTTESVILLE VA 22901
U.S.A.
TEL: 804-924-4893
TLX:
TLF:
EML:
COM: 28.48

SASTRI HANUMATH J DR
INDIAN INSTITUTE OF
ASTROPHYSICS
BANGALORE 560 034
INDIA
TEL: 566-585
TLX: 845763 IIAB IN
TLF:
EML:
COM: 49

SASTRY CH V
INDIAN INSTITUTE OF
ASTROPHYSICS
BANGALORE 560 034
INDIA
TEL:
TLX:
TLF:
EML:
COM: 40

SASTRY SHANKARA K
DEPT OF ASTRONOMY
OSMANIA UNIVERSITY
HYDERABAD 500 007
INDIA
TEL:
TLX:
TLF:
EML:
COM: 26

SATO FUMIO DR
DEPT ASTRON & EARTH SCI
TOKYO GAKUGEI UNIVERSITY
KOGANEI
TOKYO 184
JAPAN
TEL: 04[illegible]
TLX:
TLF:
EML:
COM: 34.40

SATO HUMITAKA PROF
DEPARTMENT OF PHYSICS
UNIVERSITY OF KYOTO
SAKYO-KU
KYOTO 606
JAPAN
TEL:
TLX:
TLF:
EML:
COM: 47

SATO KATSUHIKO PROF
DEPT OF PHYSICS
FACULTY OF SCI UNIV TOKYO
BUNKYO-KU
TOKYO 113
JAPAN
TEL: 03-812-2111
TLX: 23472 UTPHYSIC
TLF:
EML:
COM: 35.44.47P

SATO KOICHI DR
INTERNATIONAL LATITUDE
OBSERVATORY OF MIZUSAWA
MIZUSAWA
IWATE 023
JAPAN
TEL: 0197-24-7111
TLX: 837628 ILSMIZJ
TLF:
EML:
COM: 08.19

SATO NAONOBU PROF
AKITA UNIVERSITY
1-1 TEGATA GAKUENCHO
AKITA 010
JAPAN
TEL: 0188-33-5261
TLX:
TLF:
EML:
COM: 27.41

SATO SHUJI DR
DEPT OF PHYSICS
UNIVERSITY OF KYOTO
KITASHIRAKAWA
SAKYOKU KYOTO 606
JAPAN
TEL: 075-701-5377
TLX: 5422693 LIBKYU J
TLF:
EML:
COM: 34

SATO YUZO DR
OSAWA 4-8-19, MITAKA
TOKYO 181
JAPAN
TEL:
TLX:
TLF:
EML:
COM:

SAUNDERS RICHARD D.E.
RADIO ASTRONOMY GROUP
CAVENDISH LABORATORY
MADINGLEY ROAD
CAMBRIDGE CB3 0HF
U.K.
TEL: 223-66477
TLX: 81292
TLF:
EML:
COM: 40

SAUVAL A JACQUES DR
OBS ROYAL DE BELGIQUE
AVENUE CIRCULAIRE 3
B-1180 BRUXELLES
BELGIUM
TEL: 02-375-2484
TLX: 21565 OBSBEL B
TLF:
EML:
COM: 12

SAVAGE ANN DR
UK SCHMIDT TELESCOPE
PRIVATE BAG
COONABARABRAN NSW 2357
AUSTRALIA
TEL:
TLX:
TLF:
EML:
COM: 28.40.47

SAVAGE BLAIR D DR
DEPT OF ASTRONOMY
UNIVERSITY OF WISCONSIN
475 N.CHARTER STR
MADISON WI 53706
U.S.A.
TEL: 608-262-3072
TLX: 265452 UOFWISC-MDS
TLF:
EML:
COM: 34.44C

SAVEDOFF MALCOLM P PROF
DEPT PHYSICS & ASTRONOMY
UNIVERSITY OF ROCHESTER
BAUSCH AND LOMB BLDG
ROCHESTER NY 14627
U.S.A.
TEL: 716-275-4357
TLX: 978374 UNIBOOK ROC
TLF:
EML:
COM: 34.35.48

SAVONIJE GERRIT JAN DR
ASTRONOMICAL INSTITUTE
UNIVERSITY OF AMSTERDAM
ROETERSSTRAAT 15
NL-1018 WB AMSTERDAM
NETHERLANDS
TEL: 020-5223004
TLX: 16460 FACUN NL
TLF:
EML:
COM: 35.42

SAWANT HANUMANT S DR
INPE INSTITUTO DE
PESQUISAS ESPARIAIS
CP 515
12200 SAO JOSE CAMPOS SP
BRAZIL
TEL: 0123-22 9977
TLX: 1233530
TLF: 123-218743
EML:
COM: 40

SAWYER CONSTANCE B DR
850 20TH STREET # 705
BOULDER CO 803023
U.S.A.
TEL:
TLX:
TLF:
EML:
COM: 10.49

SAWYER-HOGG HELEN B DR
DAVID DUNLAP OBSERVATORY
P O BOX 360
RICHMOND HILL ONT L4C 4Y6
CANADA
TEL:
TLX:
TLF:
EML:
COM: 27.37

SAXENA A K DR
INDIAN INSTITUTE OF
ASTROPHYSICS
BANGALORE 560 034
INDIA
TEL: 566585. 566497
TLX: 845763 IIAB IN
TLF:
EML:
COM: 09

SAXENA P P DR
DEPT OF MATHS & ASTRONOMY
LUCKNOW UNIVERSITY
LUCKNOW
INDIA
TEL:
TLX:
TLF:
EML:
COM: 21.46

SBIRKOVA-NATCHEVA T
PLANETARIUM AND PUBLIC
ASTRONOMICAL OBSERVATORY
SMOLYAN 4700 2 LENIN BLVD
POB 137
BULGARIA
TEL: 2 2953
TLX:
TLF:
EML:
COM: 41 46

SCALISE JR EUGENIO DR
INPE/CRAAM
AV DOS ASTRONAUTAS 1758
CAIXA POSTAL 515
12200 S. JOSE DOS CAMPOS
BRAZIL
TEL: 55-011-8266588
TLX: 34061 INPE BR
TLF:
EML:
COM: 40

SCALO JOHN MICHAEL
DEPT OF ASTRONOMY
UNIVERSITY OF TEXAS
AUSTIN TX 78712
U.S.A.
TEL: 512-471-4461
TLX:
TLF:
EML:
COM: 34.35

SCALTRITI FRANCO DR
OSSERVATORIO ASTRONOMICO
DI TORINO
STRADA OSSERVATORIO 20
I-10025 PINO TORINESE
ITALY
TEL: 011-841067
TLX: 213236 TOASTR I
TLF:
EML:
COM: 15.42

SCARDIA MARCO
OSSERVATORIO ASTRONOMICO
DI BRERA
VIA E. BIANCHI 46
I-22055 MERATE
ITALY
TEL:
TLX:
TLF:
EML:
COM: 26

SCARFE COLIN D DR
DEPT OF PHYSICS
UNIVERSITY OF VICTORIA
P O BOX 1700
VICTORIA BC V8W 2Y2
CANADA
TEL: 604-721-7740
TLX: 049-7222
TLF:
EML:
COM: 26.30.42

SCARGLE JEFFREY D DR
NASA-AMES RESEARCH CENTER
MS 245-3
MOFFETT FIELD CA 94035
U.S.A.
TEL: 415-694-6330
TLX:
TLF:
EML:
COM: 48.51

SCARIA K K DR
INDIAN INSTITUTE OF
ASTROPHYSICS
BANGALORE 560 034
INDIA
TEL: 566585
TLX: 845763 IIAB IN
TLF:
EML:
COM: 37

SCARROTT STANLEY M DR
PHYSICS DEPT
UNIVERSITY OF DURHAM
SOUTH ROAD
DURHAM DH1 3LE
U.K.
TEL:
TLX:
TLF:
EML:
COM: 34

SCHADEE AART DR
STERRENKUNDIG INSTITUUT
PRINCETONPLEIN 5
POSTBUS 80 000
NL-3508 TA UTRECHT
NETHERLANDS
TEL: 030-535 200
TLX: 40048 FYLUT NL
TLF:
EML: BITNET/GMMAAERSOHUTRUUO
COM: 14

SCHAEFER BRADLEY E DR
NASA GSFC
CODE 661
GREENBELT MD 20771
U.S.A.
TEL: 301 286 6955
TLX:
TLF:
EML:
COM: 27.41

SCHAEFFER RICHARD DR
PHYSIQUE THEORIQUE
CEN SACLAY
91191 GIF SUR YVETTE
FRANCE
TEL: 69 08 73 76
TLX: 690641 ENERGAT
TLF:
EML:
COM:

SCHAIFERS KARL DR
STEINBACHWEG 37
D-6900 HEIDELBERG
GERMANY. F.R.
TEL: 06221-801511
TLX:
TLF:
EML:
COM:

SCHALEN CARL PROF
LUND OBSERVATORY
BOX 43
S-221 00 LUND
SWEDEN
TEL: 46-10 70 00
TLX: 33199 OBSNOT S
TLF:
EML:
COM: 34

SCHANDA ERWIN PROF
INST OF APPLIED PHYSICS
SIDLERSTRASSE 5
CH-3012 BERN
SWITZERLAND
TEL: 031-65-89-10
TLX: 32320
TLF:
EML:
COM:

SCHARMER GOERAN BJARNE
STOCKHOLM OBSERVATORY
S-133 00 SALTSJOEBADEN
SWEDEN
TEL: 8-717-0195
TLX: 12972
TLF:
EML:
COM: 36

SCHATTEN KENNETH H DR
PHYSICS DEPT
VICTORIA UNIVERSITY
PRIVATE BAG
WELLINGTON
NEW ZEALAND
TEL:
TLX:
TLF:
EML:
COM: 10.35.46

SCHATZMAN EVRY PROF
OBSERVATOIRE DE PARIS
SECTION MEUDON. DASGAL
PLACE JULES JANSSEN
F -92195 MEUDON PPL CEDEX
FRANCE
TEL: 1-45 07 78 71
TLX: 201571
TLF:
EML:
COM: 34.35.47.48.49.51

SCHEEPMAKER ANTON DR
COSMIC RAY WORKING GROUP
HUYGENS LABORATORY
WASSENAARSEWEG 78
NL-2300 RA LEIDEN
NETHERLANDS
TEL:
TLX:
TLF:
EML:
COM:

SCHEFFLER HELMUT PROF
LANDESSTERNWARTE
KOENIGSTUHL
D-6900 HEIDELBERG 1
GERMANY. F.R.
TEL: 06221-10036
TLX:
TLF:
EML:
COM:

SCHEIBECKER JEAN-PAUL DR
OBSERVATOIRE DE NICE
BP 139
F-06003 NICE CEDEX
FRANCE
TEL: 93-89-04-20
TLX: 460004
TLF:
EML:
COM:

SCHERB FRANK PROF
PHYSICS DEPT
UNIVERSITY OF WISCONSIN
MADISON WI 53706
U.S.A.
TEL: 608-262-6879
TLX:
TLF:
EML:
COM: 34.49

SCHERRER PHILIP H DR
CTR FOR SPACE SCIENCES &
ASTROPHYSICS
STANFORD UNIVERSITY. ERL
STANFORD CA 94305
U.S.A.
TEL: 415-497-1505
TLX: 348402 STANFRD STNU
TLF:
EML:
COM:

SCHEUER PETER A G DR
CAVENDISH LABORATORY
MADINGLEY RD
CAMBRIDGE CB3 OHE
U.K.
TEL: 0223-66477x344
TLX: 81292
TLF:
EML:
COM: 34.40.47.48C

SCHILBACH ELENA DR
ZENTRALINST FUR ASTROPHYS
ROSA LUXEMBURG STR 17A
POTSDAM DDR 1591
GERMANY. D.R.
TEL:
TLX:
TLF:
EML:
COM: 05.24

SCHILD HANSRUEDI
DEPT. OF PHYS. & ASTRON.
UNIVERSITY COLLEGE LONDON
GOWER STREET
LONDON WC1E 6BT
U.K.
TEL: 01 387 7050
TLX: 28722 UCPHYS G
TLF:
EML:
COM: [illegible]

SCHILD RUDOLPH E DR
CENTER FOR ASTROPHYSICS
60 GARDEN STREET
CAMBRIDGE MA 02138
U.S.A.
TEL: 617-495-7426
TLX: 921428 SATELLITE CAM
TLF:
EML:
COM: 29.45.51

SCHILIZZI RICHARD T DR
RADIOSTERRENWACHT
POSTBUS 2
NL-7990 AA DWINGELOO
NETHERLANDS
TEL: 05219-7244
TLX: 42043 SRZM NL
TLF:
EML:
COM: 40.48.50

SCHILLER KARL PROF DR
PIRSCHWEG 6
6072 DREIEICH-BUCHSCHLAG
GERMANY. F.R.
TEL:
TLX:
TLF:
EML:
COM:

SCHILLER STEPHEN
PHYSICS DEPT
SOUTH DAKOTA STATE UNIV
BOX 2219. ROOM 310B
BROOKINGS SD 57007
U.S.A.
TEL: 605-688 4293
TLX:
TLF:
EML:
COM: 42

SCHINDLER KARL PROF DR
INST FUER THEORET PHYSIK
RUHR-UNIVERSITAET BOCHUM
D-4630 BOCHUM
GERMANY. F.R.
TEL:
TLX:
TLF:
EML:
COM: 10

SCHLEICHER DAVID G DR
LOWELL OBSERVATORY
1400 W MARS HILL ROAD
FLAGSTAFF AZ 86001
U.S.A.
TEL: 602 774 3358
TLX:
TLF:
EML:
COM: 15.16.46

SCHLESINGER BARRY M DR
SASC TECHNOLOGIES INC.
5809 ANNAPOLIS ROAD
HYATTSVILLE MD 20784
U.S.A.
TEL: 301-699-6171
TLX: 317630
TLF:
EML:
COM:

SCHLICKEISER REINHARD DR
MPI FUER RADIOASTRONOMIE
AUF DEM HUEGEL 69
D-5300 BONN
GERMANY. F.R.
TEL: 0228-5251
TLX: 886440 MPIFR D
TLF:
EML:
COM: 40

SCHLOERB F. PETER
UNIV. OF MASSACHUSETTS
DEPT PHYSICS & ASTRONOMY
AMHERST MA 01003
U.S.A.
TEL: 413-545-4303
TLX: 955491
TLF:
EML:
COM: 15.16

SCHLOSSER WOLFHARD PROF
ASTRONOMISCHES INSTITUT
POSTFACH 102148
D-4630 BOCHUM
GERMANY. F.R.
TEL: 0234-700-3454
TLX: 0825860
TLF:
EML:
COM: 46

SCHLUETER A PROF DR
MPI FUER PLASMAPHYSIK
D-8046 GARCHING B MUNCHEN
GERMANY. F.R.
TEL: 089-3299-347
TLX: 05-215808 IPP D
TLF:
EML:
COM: 05.10

SCHLUETER DIETER PROF
INST THEOR PHYS & STERNW
NEUE UNIV PHYSIK ZENTRUM
OLSHAUSENST GEB N 61C
D-2300 KIEL 1
GERMANY. F.R.
TEL: 880-4109
TLX:
TLF:
EML:
COM:

SCHMADEL LUTZ D DR
ASTRONOMISCHES RECHEN
INSTITUT
MOENCHHOFSTR 12-14
D-6900 HEIDELBERG
GERMANY. F.R.
TEL: 06221-49026
TLX: 461336 ARIHD D
TLF:
EML:
COM: 05C.20

SCHMAHL EDWARD J DR
ASTRONOMY PROGRAM
UNIVERSITY OF MARYLAND
COLLEGE PARK MD 20742
U.S.A.
TEL: 301-454-6074
TLX:
TLF:
EML:
COM: 10.12

SCHMAHL GUENTER PROF
UNIVERSITAETSSTERNWARTE
GEISMARLANDSTR 11
D-3400 GOETTINGEN
GERMANY. F.R.
TEL: 0551-395061
TLX: 96753 USTERN
TLF:
EML:
COM:

SCHMALBERGER DONALD C DR
THE ALBANY ACADEMY
ACADEMY ROAD
ALBANY NY 12208
U.S.A.
TEL: 518-465-1461
TLX:
TLF:
EML:
COM: 46

SCHMEIDLER F PROF DR
INST ASTRON & ASTROPHYSIK
SCHEINERSTR 1
D-8000 MUENCHEN 80
GERMANY. F.R.
TEL: 089-98-90-21
TLX:
TLF:
EML:
COM: 08

SCHMIDT EDWARD G
DEPT OF PHYSICS & ASTRON
UNIVERSITY OF NEBRASKA
LINCOLN NB 68588-0111
U.S.A.
TEL: 402-472-2788
TLX:
TLF:
EML:
COM: 25.27

SCHMIDT H U DR
MPI FUER PHYSIK UND
ASTROPHYSIK
KARL-SCHWARZSCHILD-STR 1
D-8046 GARCHING B MUNCHEN
GERMANY. F.R.
TEL: 89-32999413/4
TLX: 524629 ASTRO D
TLF:
EML:
COM: 10.15.49

SCHMIDT HANS PROF
UNIVERSITAETSSTERNWARTE
AUF DEM HUEGEL 71
D-5300 BONN 1
GERMANY. F.R.
TEL:
TLX:
TLF:
EML:
COM: 33.42

SCHMIDT K H DR
ZNTRLINST. F. ASTROPHYSIK
STERNWARTE BABELSBERG
ROSA-LUXEMBURG-STR 17A
DDR-1502 POTSDAM
GERMANY. D.R.
TEL:
TLX:
TLF:
EML:
COM: 05.28.33.44

SCHMIDT MAARTEN PROF
CALTECH
ASTRONOMY 105-24
PASADENA CA 91125
U.S.A.
TEL: 818-356-4204
TLX: 675425
TLF:
EML:
COM: 15.28.33.40.47

SCHMIDT THOMAS DR
RUDOLF-STEINER-SCHULE
AN DER STIFTSKIRCHE 13
D-4800 BIELEFELD 1
GERMANY. F.R.
TEL: 0521-880407
TLX:
TLF:
EML:
COM: 34.46

SCHMIDT WOLFGANG DR
KIEPENHEUER INSTITUT
FUR SONNENPHYSIK
SCHONECKSTRASSE 6
7800 FREIBURG
GERMANY. F.R.
TEL: 761/382067
TLX: 7721552 KIS D
TLF: 761-32780
EML:
COM: 12

SCHMIDTKE PAUL C DR
DEPARTMENT OF PHYSICS
ARIZONA STATE UNIVERSITY
TEMPE AZ 85287
U.S.A.
TEL: 602 965 2918
TLX:
TLF:
EML: BITNET:SCHMIDTKE@ASUCPS
COM: 26.42

SCHMIDT-KALER TH PROF
ASTRONISCHES INSTITUT
RUHR-UNIVERSITAET BOCHUM
STEINHUEGEL 105
D-5810 WITTEN
GERMANY. F.R.
TEL: 0234-7003454
TLX: 0825860
TLF:
EML:
COM: 33.34.45

SCHMID-BURGK J DR PROF
MPI FUER RADIOASTRONOMIE
AUF DEM HUEGEL 69
D-5300 BONN 1
GERMANY. F.R.
TEL: 0449-228-525271
TLX: 0886440 MPIFR D
TLF:
EML:
COM: 34.36

SCHMIEDER BRIGITTE DR
OBSERVATOIRE DE PARIS
SECTION DE MEUDON
F-92195 MEUDON PL CEDEX
FRANCE
TEL: 1-45-07-78-17
TLX:
TLF:
EML:
COM: 10

SCHMITT DIETER DR
UNIV STERNWARTE
GEISMARLANDSTRASSE 11
D 3400 GOTTINGEN
GERMANY, F.R.
TEL: 0551 395046
TLX: 96753
TLF:
EML: BITNET:dschmit@dgogwdg1
COM: 12

SCHMITTER EDWARD F DR
DEPT OF PHYSICS
UNIVERSITY OF LAGOS
AKOKA
LAGOS
NIGERIA
TEL: 01-83-78-64
TLX:
TLF:
EML:
COM: 46

SCHMUTZ WERNER
JILA
CAMPUS BOX 440
UNIVERSITY OF COLORADO
BOULDER CO 80309-0440
U.S.A.
TEL:
TLX:
TLF:
EML:
COM: 36

SCHNEIDER JEAN
OBSERVATOIRE DE PARIS
SECTION DE MEUDON
F-92195 MEUDON PL CEDEX
FRANCE
TEL: 1-45-07-74-30
TLX: 201571 LAM
TLF:
EML:
COM: 47.51

SCHNEIDER PETER DR
MAX PLANCK INSTITUT
FUR PHYSIK/ASTROPHYSIK
KARL-SCHWARZSCHILD-STR. 1
8046 GARCHING BEI MUNCHEN
GERMANY, F.R.
TEL: 89-32990
TLX: 524629
TLF:
EML:
COM: 47

SCHNELL ANNELIESE DR
INSTITUT FUER ASTRONOMIE
UNIVERSITAT WIEN
TUERKENSCHANZSTR 17
A-1180 WIEN
AUSTRIA
TEL: 222-34-51-60-91
TLX:
TLF:
EML:
COM:

SCHNEPS MATTHEW H
HARVARD-SMITHSONIAN CTR
FOR ASTROPHYSICS
60 GARDEN STREET
CAMBRIDGE MA 02138
U.S.A.
TEL: 617-495-7477
TLX: 921428 SATELLITE CAM
TLF:
EML:
COM:

SCHNOPPER HERBERT W DR
DANISH SPACE RESEARCH INS
LUNDTOFTEVEJ 7
DK-2800 LYNGBY
DENMARK
TEL: 02-88 22 77
TLX: 37198 DANBU
TLF:
EML:
COM: 48

SCHNUR GERHARD F O
ASTRONOMISCHES INSTITUT
RUHR UNIVERSITAET
POSTFACH 102148
D-4630 BOCHUM
GERMANY, F.R.
TEL:
TLX:
TLF:
EML:
COM:

SCHOBER HANS J DR
INSTITUT FUER ASTRONOMIE
UNIVERSITAETSPLATZ 5
A-8010 GRAZ
AUSTRIA
TEL: 0316-380-5273
TLX: 31078 OBSLGZ
TLF:
EML:
COM: 10.12.15.20.42.51

SCHOEFFEL EBERHARD F DR
MERIANERSTR 42
D-8600 BAMBERG
GERMANY, F.R.
TEL:
TLX:
TLF:
EML:
COM: 40

SCHOEMBS ROLF DR
INSTITUT FUER ASTRONOMIE
UND ASTROPHYSIK
SCHEINERSTR 1
D-8000 MUENCHEN 80
GERMANY, F.R.
TEL: 98-90-21
TLX:
TLF:
EML:
COM: 27

SCHOENBERNER DETLEF PROF
INST THEOR PHYS & STERNW.
OLSHAUSENSTRASSE
D-2300 KIEL
GERMANY, F.R.
TEL: 0431-8804100
TLX: 292706 IAPKI D
TLF:
EML:
COM: 35.36

SCHOENEICH W DR
ZNTRLINST. F. ASTROPHYSIK
ROSA-LUXEMBURG-STR 17A
DDR-1502 POTSDAM
GERMANY, D.R.
TEL:
TLX:
TLF:
EML:
COM: 25.44

SCHOENFELDER VOLKER DR
MPI F EXTRATERRESTRISCHE
PHYSIK
D-8046 GARCHING B MUNCHEN
GERMANY, F.R.
TEL: 49-89-3299-578
TLX: 5215845 XTER D
TLF:
EML:
COM:

SCHOLL HANS DR
ASTRONOMISCHES RECHEN
INSTITUT
MOENCHHOFSTR 12-14
D-6900 HEIDELBERG
GERMANY, F.R.
TEL:
TLX:
TLF:
EML:
COM: 07.20

SCHOLZ GERHARD DR
ZNTRLINST. F. ASTROPHYSIK
AKAD. WISSENSCHAFTEN DDR
ROSA-LUXEMBURG-STR 17A
DDR-1502 POTSDAM-BABELSB.
GERMANY, D.R.
TEL:
TLX:
TLF:
EML:
COM: 29

SCHOLZ M PROF
INST F THEORETISCHE
ASTROPHYS DER UNIVERSITAT
NEUENHEIMER FELD 561
D-6900 HEIDELBERG
GERMANY, F.R.
TEL:
TLX:
TLF:
EML:
COM: 36

SCHOOLMAN STEPHEN A DR
LOCKHEED RESEARCH LAB.
3251 HANOVER STREET
PALO ALTO CA 94304
U.S.A.
TEL:
TLX:
TLF:
EML:
COM:

SCHRAMM DAVID N PROF
UNIVERSITY CHICAGO
ASTRON & ASTROPHYS CENTER
5640 SO ELLIS AVENUE
CHICAGO IL 60637
U.S.A.
TEL: 312-962-8202
TLX: 6871133 UNCGC UW
TLF:
EML:
COM: 35.47.48C

SCHREIBER ROMAN
COPERNICUS ASTRON CENTER
ASTROPHYSICS LAB
UL. CHOPINA 12/18
87-100 TORUN
POLAND
TEL: 48-5626017
TLX: 0552234 ASTR PL
TLF:
EML:
COM: 49

SCHREIER ETHAN J DR
SPACE TELESCOPE
SCIENCE INSTITUTE
3700 SAN MARTIN DRIVE
BALTIMORE MD 21218
U.S.A.
TEL: 301 338 4740
TLX:
TLF:
EML:
COM: 48

SCHRIJVER JOHANNES DR
LABORATORIUM VOOR
RUIMTEONDERZOEK
BENELUXLAAN 21
NL-3527 HS UTRECHT
NETHERLANDS
TEL: 030-937145
TLX: 47224 ASTRO NL
TLF:
EML:
COM: 14

SCHROEDER DANIEL J PROF
DEPT PHYSICS & ASTRONOMY
BELOIT COLLEGE
BELOIT WI 53511
U.S.A.
TEL: 608-365-3391
TLX:
TLF:
EML:
COM: 09.46

SCHROEDER ROLF DR
HOSOERKENWEG 37
D-2050 HAMBURG 80
GERMANY. F.R.
TEL:
TLX:
TLF:
EML:
COM:

SCHROETER EGON H PROF
KIEPENHEUER INSTITUT
FUER SONNENPHYSIK
SCHOENECKSTRASSE 6
D-7800 FREIBURG I BR.
GERMANY. F.R.
TEL: 0761-32864
TLX: 7721552 KIS D
TLF:
EML:
COM: 10

SCHROLL ALFRED DR
SONNENOBSERVATORIUM
KANZELHOEHE
A-9521 TREFFEN
AUSTRIA
TEL: 04248-2717
TLX: 45699 SOLOBS A
TLF:
EML:
COM:

SCHRUFER EBERHARD DR
INSTITUT FUER ASTROPHYSIK
UNIVERSITAET BONN
AUF DEM HUEGEL 71
D-5300 BONN 1
GERMANY. F.R.
TEL: 228-73-33-90
TLX: 886440 MPIFR D
TLF:
EML:
COM:

SCHRUTKA-RECHTENSTAMM FR.
WILLERGASSE 27/4/7
A-1238 WIEN
AUSTRIA
TEL: 0222-8848132
TLX:
TLF:
EML:
COM: 20

SCHUBART JOACHIM DR
ASTRONOMISCHES RECHEN-
INSTITUT
MOENCHHOFSTR 12-14
D-6900 HEIDELBERG
GERMANY. F.R.
TEL: 49-6221-4-90-26
TLX: 461336 ARIHD D
TLF:
EML:
COM: 07.20

SCHUCH NELSON JORGE
OBSERVATORIO NACIONAL
UFSM/CTRO TECNOLOGIA
CIDADE UNIVERSITARIA
97100 SANTA MARIA
BRAZIL
TEL: 055-226-1616
TLX: 0552230 UFSM
TLF:
EML:
COM:

SCHUECKING E L DR
DEPT OF PHYSICS
NEW YORK UNIVERSITY
NEW YORK NY 10012
U.S.A.
TEL:
TLX:
TLF:
EML:
COM: 28.47

SCHUESSLER MANFRED DR
KIEPENHEUER-INSTITUT
FUER SONNENPHYSIK
SCHOENECKSTR 6
D-7800 FREIBURG
GERMANY. F.R.
TEL: 761-32864
TLX: 7721552 KIS D
TLF:
EML:
COM: 12C

SCHULER WALTER DR
STERNWARTE DER
KANTONSSCHULE
CH-4500 SOLOTHURN
SWITZERLAND
TEL: 065-23-20-55
TLX:
TLF:
EML:
COM: 31

SCHULTE D H DR
ITEK CORPORATION
10 MAGUIRE ROAD
LEXINGTON MA 02173
U.S.A.
TEL:
TLX:
TLF:
EML:
COM:

SCHULTZ G V DR
MPI FUER RADIOASTRONOMIE
AUF DEM HUEGEL 69
D-5300 BONN
GERMANY. F.R.
TEL: 0228-52-52-91
TLX: 0886440 ASTROD
TLF:
EML:
COM: 09.28.34.40.44.47

SCHULZ HARTMUT DR
ASTRONOMISCHES INSTITUT
UNIVERSITAT BOCHUM
POSTFACH 10 21 48
D-4630 BOCHUM 1
GERMANY. F.R.
TEL: 234-700-3454
TLX: 0825860
TLF:
EML:
COM: 28

SCHULZ ROLF ANDREAS
MPI FUER RADIOASTRONOMIE
AUF DEM HUEGEL 69
D-5300 BONN 1
GERMANY. F.R.
TEL: 228-525-232
TLX: 886440 MPIFR D
TLF:
EML:
COM: 34.40

SCHUMACHER GERARD DR
OBSERV. DE LA COTE D'AZUR
CAUSSOLS
06460 ST VALLIER DE THIEY
FRANCE
TEL: 93 42 62 70
TLX: 460004
TLF:
EML:
COM:

SCHUMANN JOERG DIETER DR
OBSERVATORIUM HOHER LIST
UNIV STERNWARTE BONN
D-5568 DAUN
GERMANY F.R.
TEL: 06592-2937
TLX:
TLF:
EML:
COM: 09

SCHUSTER HANS-EMIL
E S O
CASILLA 19001
SANTIAGO 19
CHILE
TEL: 6988757
TLX: 240881
TLF:
EML:
COM: 09.20.28.37

SCHUSTER WILLIAM JOHN DR
INSTITUTO DE ASTRONOMIA
UNAM
APDO POSTAL 877
22860 ENSENADA. B. CALIF.
MEXICO
TEL: 706-67-83093
TLX: 56539 CICE ME
TLF:
EML:
COM:

SCHUTZ BERNARD F PROF
APPLIED MATHS & ASTRONOMY
UNIVERSITY COLLEGE
CARDIFF CF1 1XL
U.K.
TEL: 0222-44211
TLX: 448635 ULIBCF
TLF:
EML:
COM: 35

SCHUTZ BOB EWALD
CENTER FOR SPACE RESEARCH
UNIVERSITY OF TEXAS
AUSTIN TX 78712
U.S.A.
TEL: 512-471-1356
TLX: 704265 CSRUTX UD
TLF:
EML:
COM: 19C

SCHWAN HEINER DR
ASTRONOMISCHES RECHEN-
INSTITUT
MOENCHHOFSTR 12-14
D-6900 HEIDELBERG
GERMANY. F.R.
TEL: 06221-49026
TLX: 461336 ARIHD D
TLF:
EML:
COM: 04C.08C

SCHWARTZ DANIEL A DR
CENTER FOR ASTROPHYSICS
60 GARDEN STREET
CAMBRIDGE MA 02138
U.S.A.
TEL: 617-495-7232
TLX:
TLF:
EML:
COM: 44.48

SCHWARTZ PHILIP R DR
NAVAL RESEARCH LABORATORY
CODE 4138
WASHINGTON DC 20375
U.S.A.
TEL: 202-767-3391
TLX:
TLF:
EML:
COM: 27.34.40

SCHWARTZ RICHARD D
PHYSICS DEPT
UNIVERSITY OF MISSOURI
8001 NATURAL BRIDGE RT
ST LOUIS MO 63121
U.S.A.
TEL: 314-553-5025
TLX: 447658 UMSL BOOKSTOR
TLF:
EML:
COM: 34

SCHWARTZ ROLF PH D
MPI FUER RADIOASTRONOMIE
AUF DEM HUEGEL 69
D-5300 BONN 1
GERMANY. F.R.
TEL: 228-525-303
TLX:
TLF:
EML:
COM:

SCHWARTZ STEVEN JAY
THEORET ASTR UNIT/MATH SC
QUEEN MARY COLLEGE
MILE END ROAD
LONDON E1 4NS
U.K.
TEL: 1-980-4811x3849
TLX:
TLF:
EML:
COM: 12.44.49

SCHWARZ HUGO E
E.S.O. LA SILLA
CASILLA 19001
SANTIAGO 19
CHILE
TEL: 056.51.213249
TLX: 240853
TLF:
EML:
COM:

SCHWARZ ULRICH J DR
KAPTEYN LABORATORIUM
POSTBUS 800
NL-9700 AV GRONINGEN
NETHERLANDS
TEL: 050-116695
TLX: 53572 STARS NL
TLF:
EML:
COM: 28.34.40

SCHWARZENBERG-CZERNY A
WARSAW UNIVERSITY OBS
AL. UJAZDOWSKIE 4
00-478 WARSAW
POLAND
TEL: 29 40 11
TLX: 813978 ZAPAN PL
TLF:
EML:
COM: 27

SCHWARZSCHILD MARTIN PROF
PRINCETON UNIVERSITY
OBSERVATORY
PEYTON HALL
PRINCETON NJ 08544
U.S.A.
TEL: 609-452-3812
TLX:
TLF:
EML:
COM: 35

SCHWEIZER FRANCOIS DR
DEPT TERREST. MAGNETISM
CARNEGIE INST. WASHINGTON
5241 BROAD BRANCH RD N.W.
WASHINGTON DC 20015
U.S.A.
TEL: 202-966-0863
TLX: 440427 MAGN UI
TLF:
EML:
COM: 28

SCIAMA DENNIS W DR
SISSA
STRADA COSTIERA 11
I-34014 TRIESTE
ITALY
TEL: 0103941-224-118
TLX: 460392
TLF:
EML:
COM: 28.47.48

SCONZO PASQUALE DR
29 OLD MYSTIC STREET
ARLINGTON MA 02174
U.S.A.
TEL: 617-646-9315
TLX:
TLF:
EML:
COM: 07

SCOTT ELIZABETH L PROF
STATISTICS
UNIVERSITY OF CALIFORNIA
367 EVANS HALL
BERKELEY CA 94720
U.S.A.
TEL: 415-642-2777
TLX: 910-366-7114 UC BERK
TLF:
EML:
COM: 47

SCOTT EUGENE HOWARD
NASA/GSFC
CODE 684.9
GREENBELT MD 20771
U.S.A.
TEL: 301-286-8746
TLX:
TLF:
EML:
COM: 34

SCOTT JOHN S DR
STEWARD OBSERVATORY
UNIVERSITY OF ARIZONA
TUCSON AZ 85721
U.S.A.
TEL:
TLX:
TLF:
EML:
COM: 40.48

SCOTT PAUL F DR
CAVENDISH LABORATORY
MADINGLEY ROAD
CAMBRIDGE CB3 OHE
U.K.
TEL: 0223-66477
TLX: 81292
TLF:
EML:
COM: 40

SCOVILLE NICHOLAS Z
ASTRONOMY & PHYSICS DEPT
UNIV OF MASSACHUSETTS
GRADUATE RESEARCH CENTER
AMHERST MA 01003
U.S.A.
TEL: 413-545-0789
TLX:
TLF:
EML:
COM: 28.34

SCRIMGER J. NORMAN DR
DEPT OF ASTRONOMY
ST MARY'S UNIVERSITY
HALIFAX NS B3H 3C3
CANADA
TEL:
TLX:
TLF:
EML:
COM:

SCUFLAIRE RICHARD DR
INSTITUT D'ASTROPHYSIQUE
UNIVERSITE DE LIEGE
AVENUE DE COINTE 5
B-4200 COINTE-OUGREE
BELGIUM
TEL: 041-52-99-80
TLX: 41264 ASTRLG B
TLF:
EML:
COM: 27.35

SEAQUIST ERNEST R PROF
DEPT OF ASTRONOMY
UNIVERSITY OF TORONTO
TORONTO ONT M5S 1A7
CANADA
TEL: 416-978-3146
TLX: 06-986766
TLF:
EML:
COM: 40C

SEARLE LEONARD DR
HALE OBSERVATORIES
813 SANTA BARBARA ST
PASADENA CA 91101
U.S.A.
TEL: 818-304-0220
TLX:
TLF:
EML:
COM: 28

SEARS RICHARD LANGLEY DR
DEPT OF ASTRONOMY
UNIVERSITY OF MICHIGAN
ANN ARBOR MI 48109
U.S.A.
TEL: 313-763-3295
TLX:
TLF:
EML:
COM: 35

SEATON MICHAEL J PROF
DEPT PHYSICS & ASTRONOMY
UNIVERSITY COLLEGE LONDON
GOWER STREET
LONDON WC1E 6BT
U.K.
TEL: 01-387-7050
TLX: 28722
TLF:
EML:
COM: 12.14.34.36C

SECCO LUIGI DR
DIPARTIMENTO DI
ASTRONOMIA
VICOLO OSSERVATORIO 5
PADOVA
ITALY
TEL: 049 661499
TLX: 430176 UNPADU I
TLF:
EML:
COM:

SEDLMAYER ERWIN DR
INST FUER ASTRONOMIE &
ASTROPHYSIK DER TECHN UNI
ERNST-REUTER-PLATZ 7
D-1000 BERLIN 10
GERMANY. F.R.
TEL:
TLX:
TLF:
EML:
COM: 36

SEDMAK GIORGIO PROF
DIPT. DI ASTRONOMIA
UNIVERSITA DI TRIESTE
VIA TIEPOLO 11
I-34131 TRIESTE
ITALY
TEL: 40-79-4863
TLX: 461137
TLF:
EML:
COM: 05.3[illegible]

SEEGER CHARLES LOUIS III
SAN FRANCISCO STATE UNIV
473 JAMES ROAD
PALO ALTO CA 94306
U.S.A.
TEL: 415-493-6005
TLX:
TLF:
EML:
COM: 51

SEEGER PHILIP A DR
LOS ALAMOS NATIONAL LAB
MS/H805
PO BOX 1663
LOS ALAMOS NM 87545
U.S.A.
TEL: 505-667-8843
TLX:
TLF:
EML:
COM:

SEGAL IRVING E DR
MIT
2-224
CAMBRIDGE MA 02139
U.S.A.
TEL: 617-253-4985
TLX:
TLF:
EML:
COM: 47

SEGALUVITZ ALEXANDER DR
11 HABANIM ST
KFAR SAVA
ISRAEL
TEL:
TLX:
TLF:
EML:
COM:

SEGAN STEVO
INSTITUTE OF ASTRONOMY
FACULTY OF SCIENCES
STUDENTSKI TRG 16
11000 BEOGRAD
YUGOSLAVIA
TEL:
TLX:
TLF:
EML:
COM: 07

SEGGEWISS WILHELM PROF
OBSERVATORIUM HOHER LIST
UNIVERSITAETS-STERNWARTE
D-5568 DAUN EIFEL
GERMANY. F.R.
TEL: 06592-2150
TLX:
TLF:
EML:
COM: 29.33.42

SEHNAL LADISLAV DR
ASTRONOMICAL INSTITUTE
CZECH. ACAD. OF SCIENCES
OBSERVATORY
251 65 ONDREJOV
CZECHOSLOVAKIA
TEL: 258757
TLX: 121579
TLF:
EML:
COM: 07

SEIDELMANN P KENNETH DR
US NAVAL OBSERVATORY
34 & MASSACHUSETTS AVE NW
WASHINGTON DC 20390
U.S.A.
TEL: 202-653-1545
TLX: 710-822-1970
TLF:
EML:
COM: 04P.07C.20

SEIDEN PHILIP E
IBM RESEARCH CENTER
PO BOX 218
YORKTOWN HEIGHTS NY 10598
U.S.A.
TEL: 914-945-1424
TLX: 137456
TLF:
EML:
COM: 28.47

SEIDOV ZAKIR F DR
SHEMAKHA ASTROPHYSICAL
OBSERVATORY
373243 SHEMAKHA. AZERB.
U.S.S.R.
TEL:
TLX:
TLF:
EML:
COM: 35

SEIELSTAD GEORGE A
NRAO
PO BOX 2
GREEN BANK WV 24944
U.S.A.
TEL: 304-456-2301
TLX: 710-938-1530
TLF:
EML:
COM: 40.47.48.51

SEIN-ECHALUCE M LUISA DR
DEPT DE MATEMATICA APLIC.
UNIVERSIDAD ZARAGOZA
AV. MARIA ZAMBRANO 50
500015 ZARAGOZA
SPAIN
TEL: (34)76 518143
TLX: 58198
TLF: 565852
EML:
COM: 07

SEIRADAKIS JOHN HUGH DR
DEPT OF ASTRONOMY
UNIVERSITY THESSALONIKI
GR-54006 THESSALONIKI
GREECE
TEL: 031-991357
TLX:
TLF:
EML:
COM: 40.51

SEITTER WALTRAUT C PROF
ASTRONOMISCHES INSTITUT
DOMAGKSTR 75
D-4400 MUENSTER/W
GERMANY. F.R.
TEL: 251-83-3561
TLX: 892529 UNI MS D
TLF:
EML:
COM: 45

SEKANINA ZDENEK DR
EARTH & SPACE SCI DIV
JPL
4800 OAK GROVE DRIVE
PASADENA CA 91103
U.S.A.
TEL: 818-354-7589
TLX:
TLF:
EML:
COM: 15.20.22

SEKI MUNEZO DR
DEPT OF EARTH SCIENCES
COLL OF GENERAL EDUCATION
TOHOKU UNIV. KAWAUCHI
SENDAI 980
JAPAN
TEL: 022-[illegible]-1800
TLX:
TLF:
EML:
COM: 34

SEKIGUCHI KAZUHIRO DR
DEPARTMENT OF ASTRONOMY
BOX 4500 NEW MEXICO STATE
UNIVERSITY
LAS CRUCES NM 88003
MEXICO
TEL: 505 646 [illegible]
TLX:
TLF:
EML:
COM:

SEKIGUCHI NAOSUKE PROF
TOKYO ASTRONOMICAL
OBSERVATORY
OSAWA. MITAKA
TOKYO 181
JAPAN
TEL: 0422-32-5111
TLX: 02822307 TAOMTK J
TLF:
EML:
COM: 19

SELLWOOD JERRY ARTHUR
THE UNIVERSITY
DEPT OF ASTRONOMY
MANCHESTER M13 9PL
U.K.
TEL: 061-273-7121
TLX:
TLF:
EML:
COM: 28.33

SELVELLI PIERLUIGI DR
OSSERVATORIO ASTRONOMICO
VIA TIEPOLO 11
I-34131 TRIESTE
ITALY
TEL: 40-793221
TLX: 461137 OAT I
TLF:
EML:
COM: 44

SEMEL MEIR DR
OBSERVATOIRE DE PARIS
SECTION DE MEUDON
F-92195 MEUDON PL CEDEX
FRANCE
TEL: 1-45-07-77-90
TLX:
TLF:
EML:
COM: 10.12

SEMENIUK IRENA DR
WARSAW UNIVERSITY OBS
AL. UJAZDOWSKIE 4
00-478 WARSZAWA
POLAND
TEL: 29-40-11/12
TLX: 815548 OAUW
TLF:
EML:
COM: 42

SEMENZATO ROBERTO
UNIVERSITA DI PADOVA
PHYSICS DEPT
VIA MARZOLO 8
I-35131 PADOVA
ITALY
TEL: 049-844-247
TLX: 430308 DF GGPD I
TLF:
EML:
COM:

SEN S N DR
INDIAN ASSOCIATION FOR
THE CULTIVATION OF SCI
JADAVPUR
INDIA
TEL:
TLX:
TLF:
EML:
COM:

SENGBUSCH KURT V DR
MPI FUER PHYSIK UND
ASTROPHYSIK
KARL-SCHWARZSCHILD-STR 1
D-8046 GARCHING B MUENCHEN
GERMANY. F.R.
TEL:
TLX:
TLF:
EML:
COM: 35

SERAFIN RICHARD AUGUST
ASTRONOMICAL OBSERVATORY
A. MICKIEWICZ UNIVERSITY
SLONECZNA 36
60-286 POZNAN
POLAND
TEL: 679 670
TLX:
TLF:
EML:
COM:

SERIO SALVATORE DR
OSSERVATORIO ASTRONOMICO
PIAZZA DEL PARLAMENTO I
PALERMO 90134
ITALY
TEL: 39 91 592451
TLX: 910402 ASTROP I
TLF:
EML: BITNET:ASTROPA@IPACUC
COM:

SERRANO ALFONSO DR
INSTITUTO DE ASTRONOMIA
UNAM
APDO POSTAL 70-264
04510 MEXICO DF
MEXICO
TEL:
TLX:
TLF:
EML:
COM:

SERSIC J L DR
OBSERVATORIO ASTRONOMICO
LAPRIDA 854
5000 CORDOBA
ARGENTINA
TEL: 051 25072
TLX: 51-822 BUCOR ORSASTR
TLF:
EML:
COM: 28.47

SERVAN BERNARD
OBSERVATOIRE DE PARIS
61 AVE DE L'OBSERVATOIRE
F-75014 PARIS
FRANCE
TEL: 1-40-51-22-36
TLX: 270776
TLF:
EML:
COM: 09

SESSIN WAGNER DR
INST TECN. DE AERONAUTICA
DEPTO DE ASTRONOMIA
12200 S. JOSE DOS CAMPOS
BRAZIL
TEL: 0123-22-9088
TLX: 011 73437 ZWG-24-73
TLF:
EML:
COM: 07

SETTI GIANCARLO PROF
ESO
KARL-SCHWARZSCHILD-STR 2
D-8046 GARCHING B MUNCHEN
GERMANY. F.R.
TEL: 049-89-320-06-0
TLX: 52828222 EO D
TLF:
EML:
COM: 28.40.47C.48.49

SEVARLIC BRANISLAV M PROF
DEPT OF ASTRONOMY
UNIVERSITY BEOGRAD
VOLGINA 7
11050 BEOGRAD
YUGOSLAVIA
TEL:
TLX:
TLF:
EML:
COM: 08.19

SEVERINO GIUSEPPE
OSSERVATORIO ASTRONOMICO
DI CAPODIMONTE
VIA MOIARIELLO 16
I-80131 NAPOLI
ITALY
TEL: 081-440101
TLX:
TLF:
EML:
COM: 12

SEVILLA MIGUEL J DR
INST DE ASTRON Y GEODESIA
ALMANSA 76
28040 MADRID
SPAIN
TEL: 2 442501
TLX:
TLF:
EML:
COM: 19

SEWARD FREDERICK D
HARVARD-SMITHSONIAN CTR
FOR ASTROPHYSICS
60 GARDEN STREET
CAMBRIDGE MA 02138
U.S.A.
TEL: 617-495-7282
TLX:
TLF:
EML:
COM: 48

SEYMOUR P A E
57 HERMITAGE ROAD
PLYMOUTH. DEVON
U.K.
TEL:
TLX:
TLF:
EML:
COM:

SEZER CENGIZ DR
EGE UNIVERSITY
FACULTY OF SCIENCE
CAMPUS P K 21
BORNOVA-IZMIR
TURKEY
TEL: 90-51-180110
TLX:
TLF:
EML:
COM:

SHAFFER DAVID B DR
NASA/GSFC
CODE 621.9
GREENBELT MD 20771
U.S.A.
TEL: 301-286-6434
TLX:
TLF:
EML:
COM: 40

SHAH GHANSHYAM A DR
INDIAN INST ASTROPHYSICS
SARJAPUR RD
KORAMANGALA
BANGALORE 560 034
INDIA
TEL: 566585 & 566497
TLX: 845-763 IIAB IN
TLF:
EML:
COM: 34

SHAHAM JACOB PROF
COLUMBIA UNIVERSITY
PHYSICS DEPARTMENT
NEW YORK NY 10027
U.S.A.
TEL: 212-280-3349
TLX: 220094 COLU UR
TLF:
EML:
COM: 48

SHAHBAZIAN ROMELLA K DR
BYURAKAN ASTROPHYSICAL
OBSERVATORY
378433 BYURAKAN. ARMENIA
U.S.S.R.
TEL: 56-34-53
TLX:
TLF:
EML:
COM: 28

SHAKESHAFT JOHN R DR
CAVENDISH LABORATORY
MADINGLEY ROAD
CAMBRIDGE CB3 OHE
U.K.
TEL: 223-66477
TLX: 81292 CAVLAB G
TLF:
EML:
COM: 05.28.40

SHAKHBAZYAN YURIJ L DR
BYURAKAN ASTROPHYSICAL
OBSERVATORY
378433 ARMENIA
U.S.S.R.
TEL:
TLX:
TLF:
EML:
COM: 09.

SHAKHOVSKOJ NIKOLAY M DR
CRIMEAN ASTROPHYSICAL OBS
USSR ACADEMY OF SCIENCES
NAUCHNIY
334413 CRIMEA
U.S.S.R.
TEL:
TLX:
TLF:
EML:
COM: 25

SHAKURA NICHOLAJ I DR
STERNBERG STATE
ASTRONOMICAL INSTITUTE
117234 MOSCOW
U.S.S.R.
TEL:
TLX:
TLF:
EML:
COM: 42.48

SHALLIS MICHAEL J DR
DEPT OF ASTROPHYSICS
SOUTH PARKS ROAD
OXFORD OX1 3RQ
U.K.
TEL:
TLX:
TLF:
EML:
COM: 12

SHALTOUT MESALAM A M DR
HELWAN OBSERVATORY
HELWAN
EGYPT
TEL:
TLX:
TLF:
EML:
COM:

SHANDARIN SERGEI F DR
INST FOR PHYSICS PROBLEMS
KOSYGIN 2
117334 MOSCOW
U.S.S.R.
TEL: 137 32 48
TLX: 113451 MAGNIT
TLF:
EML:
COM: 47C

SHANE WILLIAM W DR
STERREKUNDIG INSTITUUT
KATHOLIEKE UNIVERSITEIT
TOERNOOIVELD
NL-6525 ED NIJMEGEN
NETHERLANDS
TEL: 080-558833
TLX: 48228 WINAT NL
TLF:
EML:
COM: 33.34

SHAO CHENG-YUAN
HARVARD-SMITHSONIAN CTR
FOR ASTROPHYSICS
60 GARDEN STREET
CAMBRIDGE MA 02138
U.S.A.
TEL: 617-495-7212
TLX:
TLF:
EML:
COM: 22.34

SHAPERO DONALD C DR
BOARD ON PHYSICS & ASTR
NATIONAL RESEARCH COUNCIL
2101 CONSTITUTION AVE
WASHINGTON DC 20418
U.S.A.
TEL: 202 334 3520
TLX: 248664
TLF:
EML: BITNET:DSHAPERO@NAS
COM:

SHAPIRO IRWIN I PROF
CENTER FOR ASTROPHYSICS
ROOM P 209
60 GARDEN STREET
CAMBRIDGE MA 02138
U.S.A.
TEL: 617-495-7100
TLX: 921428 SATELLITE CAM
TLF:
EML:
COM: 04.07.16.19

SHAPIRO MAURICE M PROF
205 YOAKUM PKWY 2-1720
ALEXANDRIA VA 22304
U.S.A.
TEL:
TLX:
TLF:
EML:
COM: 48.51

SHAPIRO STUART L
CTR RADIOPHYS & SPACE RES
CORNELL UNIVERSITY
ITHACA NY 14853
U.S.A.
TEL: 607-256-4936
TLX:
TLF:
EML:
COM: 34

SHAPLEY ALAN H
NOAA
BOULDER CO 80303
U.S.A.
TEL:
TLX:
TLF:
EML:
COM: 10

SHARA MICHAEL DR
SPACE TELESCOPE SCI INST
HOMEWOOD CAMPUS
3700 SAN MARTIN DRIVE
BALTIMORE MD 21218
U.S.A.
TEL: 301-338-4743
TLX: 6849101 STSCI UW
TLF:
EML:
COM: 27

SHARAF MOHAMED ADEL DR
DEPT OF ASTRONOMY
CAIRO UNIVERSITY
CAIRO
EGYPT
TEL:
TLX:
TLF:
EML:
COM:

SHARAF SH G DR
INSTITUTE OF THEORETICAL
ASTRONOMY
10 KUTUZOV QUAY
192187 LENINGRAD
U.S.S.R.
TEL:
TLX:
TLF:
EML:
COM: 07

SHAROV A S DR
STERNBERG STATE
ASTRONOMICAL INSTITUTE
119899 MOSCOW
U.S.S.R.
TEL: 139-26-57
TLX:
TLF:
EML:
COM: 06.21.33.37

SHARP CHRISTOPHER DR
MAX PLANCK INSTITUT
FUR PHYSIK/ASTROPHYSIK
KARL-SCHWARSCHILD-STR. 1
8046 GARCHING BEI MUNCHEN
GERMANY. F.R.
TEL: 89-3299-0
TLX: 524629
TLF:
EML:
COM: 14.15

SHARPLES RAY DR
ANGLO AUSTRALIAN OBS
PO BOX 296
EPPING NSW 2121
AUSTRALIA
TEL: 02 868 1666
TLX: 123999
TLF:
EML: NSSDCASSPAN::PSI%AAOEPP::RMS
COM: 28

SHARPLESS STEWART PROF
DEPT PHYSICS & ASTRONOMY
UNIVERSITY OF ROCHESTER
ROCHESTER NY 14627
U.S.A.
TEL: 716-275-4389
TLX:
TLF:
EML:
COM: 34.45

SHAVER PETER A DR
ESO
KARL-SCHWARTZSCHILD-STR 2
D-8046 GARCHING B MUNCHEN
GERMANY. F.R.
TEL: 089-320060
TLX: 52828222 EOD
TLF:
EML:
COM: 28.34.40.47C.48

SHAVIV GIORA DR
DEPT OF PHYSICS
ISRAEL INST OF TECHNOLOGY
TECHNION
HAIFA 32000
ISRAEL
TEL:
TLX:
TLF:
EML:
COM: 35.47C.47.48

SHAW JAMES SCOTT DR
DEPT PHYSICS & ASTRONOMY
UNIVERSITY OF GEORGIA
ATHENS GA 30602
U.S.A.
TEL: 404-542-2485
TLX:
TLF:
EML:
COM:

SHAW JOHN H PROF
OHIO STATE UNIVERSITY
174 W. 18TH AVENUE
COLUMBUS OH 43210
U.S.A.
TEL: 614-422 7968
TLX:
TLF:
EML:
COM:

SHAW R WILLIAM PROF
105 HALCYON HILL
ITHACA NY 14850
U.S.A.
TEL: 607-257-1948
TLX:
TLF:
EML:
COM:

SHAWHAN STANLEY D DR
DEPT PHYSICS & ASTRONOMY
UNIVERSITY OF IOWA
IOWA CITY IA 52242
U.S.A.
TEL: 319-353-3294
TLX:
TLF:
EML:
COM: 49

SHAWL STEPHEN J DR
ASTRONOMY DEPARTMENT
UNIVERSITY OF CALIFORNIA
BERKELEY CA 94720
U.S.A.
TEL:
TLX:
TLF:
EML:
COM: 25.34.37

SHAYA EDWARD J DR
COLUMBIA UNIVERSITY
DEPT. OF ASTR., BOX 53
538 WEST 120TH STREET
NEW YORK NY 10027
U.S.A.
TEL: 212 854 6831
TLX:
TLF:
EML: BITNET:EJST@CUPHYD
COM: 28.47

SHCHEGLOV P V DR
STERNBERG STATE
ASTRONOMICAL INSTITUTE
119899 MOSCOW
U.S.S.R.
TEL: 139-19-73
TLX:
TLF:
EML:
COM: 09.34.50

SHCHERGOLEV DIMITRIJ E DR
PULKOVO OBSERVATORY
196140 LENINGRAD
U.S.S.R.
TEL:
TLX:
TLF:
EML:
COM: 24

SHCHERBINA-SAMOJLOVA I DR
INSTITUTE OF SCIENCE
AND TECHNICS INFORMATION
DEPARTMENT OF ASTRONOMY
125219 MOSCOW
U.S.S.R.
TEL: 1554237
TLX:
TLF:
EML:
COM: 05

SHEA MARGARET A DR
AIR FORCE GEOPHYSICS LAB
(PHC)
HANSCOM AFB
BEDFORD MA 01731
U.S.A.
TEL:
TLX:
TLF:
EML:
COM: 10.49

SHEELEY NEIL R DR
NAVAL RESEARCH LABORATORY
CODE 4172
WASHINGTON DC 20375
U.S.A.
TEL: 202-767-2777
TLX:
TLF:
EML:
COM: 10.12

SHEFFER EUGENE K DR
STERNBERG STATE
ASTRONOMICAL INSTITUTE
119899 MOSCOW V-234
U.S.S.R.
TEL: 1393046
TLX:
TLF:
EML:
COM:

SHEFFIELD CHARLES DR
EARTH SATELLITE CORP
7222 47TH STREET
(CHEVY CHASE)
WASHINGTON DC 20815
U.S.A.
TEL: 301 951 0104
TLX: 248618 ESCO UR
TLF:
EML:
COM: 44

SHEFOV NICOLAI N
INST OF PHYSICS OF THE
ATMOSPHERE
PYZHEVSKY 3
109017 MOSCOW
U.S.S.R.
TEL:
TLX:
TLF:
EML:
COM: 21

SHELUS PETER J DR
ASTRONOMY DEPT
UNIVERSITY OF TEXAS
RLM 15-316
AUSTIN TX 78712
U.S.A.
TEL: 512-471-3339
TLX: 910-874-1351
TLF:
EML:
COM: 20

SHEN BENJAMIN S P PROF
DEPT ASTRONOMY E1
UNIV OF PENNSYLVANIA
PHILADELPHIA PA 19104
U.S.A.
TEL: 215-898-8176
TLX:
TLF:
EML:
COM:

SHEN CHANGJUN
PURPLE MOUNTAIN
OBSERVATORY
NANJING
CHINA. PEOPLE'S REP.
TEL:
TLX: 34144 PMONJ CN
TLF:
EML:
COM: 09

SHEN CHUN-SHAN
ASTRONOMICAL STY OF CHINA
NATL TSING-HUA UNIVERSITY
HSIN-CHU 300
CHINA. TAIWAN
TEL: 886-35-719039
TLX:
TLF:
EML:
COM: 46.51

SHEN KAIXIAN
SHAANXI OBSERVATORY
P.O. BOX 18
LINTONG
XIAN
CHINA. PEOPLE'S REP.
TEL: 3-2255
TLX: 70121 CSAO CN
TLF:
EML:
COM: 08

SHEN LIANG-ZHAO
BEIJING ASTRONOMICAL OBS
BEIJING 100080
CHINA. PEOPLE'S REP.
TEL:
TLX: 22040 BAOAS CN
TLF:
EML:
COM: 42

SHEN LONG-XIANG
BEIJING OBSERVATORY
ACADEMIA SINICA
BEIJING
CHINA. PEOPLE'S REP.
TEL: 28-1968
TLX: 22040 BAOAS CN
TLF:
EML:
COM: 12

SHEN PARN-AN
NANJING ASTRONOMICAL
INSTRUMENT FACTORY
NANJING
CHINA. PEOPLE'S REP.
TEL: 46191
TLX: 34136 GLYNJ c/o NAIF
TLF:
EML:
COM: 09

SHER DAVID DR
BOX 9624
CINCINNATI OH 452098
U.S.A.
TEL: 513-871-8850
TLX:
TLF:
EML:
COM: 33.37

SHERIDAN K V DR
17B/23 THORNTON STREET
DARLING POINT NSW 2027
AUSTRALIA
TEL:
TLX:
TLF:
EML:
COM: 40

SHERWOOD WILLIAM A DR
MPI FUER RADIOASTRONOMIE
AUF DEM HUEGEL 69
D-5300 BONN
GERMANY. F.R.
TEL: 0228-525-362
TLX: 886440 ASTROD
TLF:
EML:
COM: 27.28.34

SHESTAKA IVAN S DR
ASTRONOMICAL OBSERVATORY
PARK SHEVCHENKO
270014 ODESSA
U.S.S.R.
TEL:
TLX:
TLF:
EML:
COM: 22

SHEVCHENKO VLADISLAV V DR
STERNBERG STATE
ASTRONOMICAL INSTITUTE
UNIVERSITETSKY PROSP. 13
119899 MOSCOW
U.S.S.R.
TEL:
TLX:
TLF:
EML:
COM: 16C

SHI GUANG-CHEN
PURPLE MOUNTAIN
OBSERVATORY
NANJING
CHINA. PEOPLE'S REP.
TEL: 33921
TLX: 34144 PMOAS CN
TLF:
EML:
COM: 08.19.24

SHI ZHONG-XIAN
BEIJING ASTRONOMICAL OBS
ACADEMIA SINICA
BEIJING
CHINA. PEOPLE'S REP.
TEL: 28-1698
TLX: 9053
TLF:
EML:
COM: 10

SHIBAHASHI HIROMOTO DR
DEPT OF ASTRONOMY
UNIVERSITY OF TOKYO
BUNKYO-KU
TOKYO 113
JAPAN
TEL: 03-812-2111
TLX: 33659 UTYOSCI J
TLF:
EML:
COM: 35

SHIBASAKI KIYOTO
RES INST OF ATMOSPHERICS
NAGOYA UNIVERSITY
3-13 HONOHARA. TOYOKAWA
AICHI 442
JAPAN
TEL: 5338-6-3154
TLX: 4322310 TYKV J
TLF:
EML:
COM: 10

SHIBATA KAZUNARI DR
AICHI UNIV OF EDUCATION
DEPT OF EARTH SCIENCE
1 HIROSAWA. INOGAYACHO
KARIYA AICHI 448
JAPAN
TEL: 566363111. 596
TLX:
TLF:
EML:
COM:

SHIBATA SHINPEI DR
DEPT OF PHYSICS
YAMAGATA UNIVERSITY
KOJIRAKAWA
YAMAGATA 990
JAPAN
TEL: 236 31 1421
TLX:
TLF:
EML: BITNET:B26416@JPNKUDPC
COM:

SHIBATA YUKIO DR
RESEARCH INSTITUTE FOR
SCIENTIFIC MEASUREMENTS
TOHOKU UNIVERSITY
SENDAI 980
JAPAN
TEL:
TLX:
TLF:
EML:
COM: 35

SHIELDS GREGORY A DR
DEPT OF ASTRONOMY
UNIVERSITY OF TEXAS
RLM 15.212
AUSTIN TX 78712
U.S.A.
TEL: 512-471-4461
TLX: 910-874-1351
TLF:
EML:
COM: 28.34.48

SHIN YOON-TAIK PROF
236-53 SINDAN-DONG
JOONG-KU
SEOUL 100
KOREA. REPUBLIC
TEL:
TLX:
TLF:
EML:
COM:

SHIMIZU MIKIO PROF
INST SPACE & ASTR.SCI.
1-1 YOSHINODAI 3-CHOME
SAGAMIHARA
KANAGAWA-KEN 229
JAPAN
TEL: 0427 51 3911
TLX:
TLF:
EML:
COM: 15.16.51

SHIMIZU TSUTOMU PROF EMER
ITOPIA 504
1-1 SHIOYA-CHO
TARAMIKU KOBE 655
JAPAN
TEL:
TLX:
TLF:
EML:
COM: 16.33

SHIMMINS ALBERT JOHN
18 PAGE STREET
ALBERT PARK VIC 3206
AUSTRALIA
TEL: 03-6903803
TLX:
TLF:
EML:
COM: 40

SHINE RICHARD A DR
DEPARTMENT 91-30
LOCKHEED P/ALTO RES LAB
3170 PORTER DRIVE
PALO ALTO CA 94304-1211
U.S.A.
TEL: 415-858-4135
TLX:
TLF:
EML:
COM: 10.12.36

SHIPMAN HENRY L DR
PHYSICS DEPT
UNIVERSITY OF DELAWARE
NEWARK DE 19711
U.S.A.
TEL: 302-451-2986
TLX:
TLF:
EML:
COM: 36.46

SHIRYAEV A V DR
LENINGRAD STATE UNIV
ASTRONOMICAL OBSERVATORY
199178 LENINGRAD
U.S.S.R.
TEL:
TLX:
TLF:
EML:
COM:

SHIRYAEV ALEXANDER A DR
INST OF THEOR ASTRONOMY
NAVEREZHNAYA KUTUZOVA 10
191187 LENINGRAD
U.S.S.R.
TEL: 7-812-272 40 33
TLX: 12578 ITA SU
TLF:
EML:
COM: 04

SHIVANANDAN KANDIAH DR
NAVAL RESEARCH LABORATORY
CODE 4138-S
WASHINGTON DC 20375-5000
U.S.A.
TEL: 202-767-2749
TLX: 202-767-6473
TLF:
EML:
COM: 09.44.47

SHKODROV V G DR
DEPT OF ASTRONOMY
BULGARIAN ACAD SCIENCES
72 LENIN BLVD
1784 SOFIA
BULGARIA
TEL: 7341 x 559
TLX: 23761 ECF BAN BG
TLF:
EML:
COM: 15.20

SHOBBROOK ROBERT R DR
CHATTERTON ASTRONOMY DEPT
UNIVERSITY OF SYDNEY
SYDNEY NSW 2006
AUSTRALIA
TEL: 61-2-692-3604
TLX: 26169 UNISYD AA
TLF:
EML:
COM: 27.37C

SHOEMAKER EUGENE M
BRANCH OF ASTROGEOLOGY
US GEOLOGICAL SURVEY
2255 N. GEMINI DRIVE
FLAGSTAFF AZ 86001
U.S.A.
TEL: 602-527-7181
TLX:
TLF:
EML:
COM: 15.16.20

SHOLOMITSKY G B DR
SPACE RESEARCH INSTITUTE
USSR ACADEMY OF SCIENCES
117810 MOSCOW
U.S.S.R.
TEL: 333-31-22
TLX: 411498 STAR SU
TLF:
EML:
COM: 40.44C

SHOR VIKTOR A DR
INST F. THEORET ASTRONOMY
NABEREZHNAYA KUTUZOVA 10
191187 LENINGRAD
U.S.S.R.
TEL:
TLX: 121578
TLF:
EML:
COM: 20C

SHORE BRUCE W
LAWRENCE LIVERMORE LAB
LIVERMORE CA 94550
U.S.A.
TEL: 415-447-1100
TLX:
TLF:
EML:
COM: 14

SHORE STEVEN N
ASTROPHYSICS RESEARCH CTR
NEW MEXICO TECH
SOCORRO NM 87801
U.S.A.
TEL: 505-835-5792
TLX:
TLF:
EML:
COM: 28.29

SHOSTAK G SETH DR
KAPTEYN LABORATORIUM
POSTBUS 800
NL-9700 AV GRONINGEN
NETHERLANDS
TEL: 050-116655
TLX: 53572 STARS NL
TLF:
EML:
COM: 28.51

SHU FRANK H PROF
ASTRONOMY DEPT
UNIVERSITY OF CALIFORNIA
CAMPBELL HALL
BERKELEY CA 94720
U.S.A.
TEL: 415-642-2529
TLX:
TLF:
EML:
COM: 33.34.42

SHUKLA K
DEPT MATHS & ASTRONOMY
LUCKNOW UNIVERSITY
LUCKNOW U P
INDIA
TEL:
TLX:
TLF:
EML:
COM: 41

SHUKRE C S DR
RAMAN RESEARCH INSTITUTE
SADASHIVANAGAR
BANGALORE 560 080
INDIA
TEL: 812-360122
TLX: 8425671 RRI IN
TLF:
EML:
COM: 48

SHULL JOHN MICHAEL
UNIVERSITY OF COLORADO
JILA
BOULDER CO 80309
U.S.A.
TEL: 303-492-7827
TLX:
TLF:
EML:
COM: 34

SHULOV OLEG S DR
LENINGRAD STATE UNIV
ASTRONOMICAL OBSERVATORY
199178 LENINGRAD
U.S.S.R.
TEL:
TLX:
TLF:
EML:
COM:

SHUL'BERG A M DR
ODESSA STATE UNIVERSITY
ASTRONOMICAL OBSERVATORY
270014 ODESSA
U.S.S.R.
TEL: 250356
TLX:
TLF:
EML:
COM: 26.42

SHUL'MAN L M DR
MAIN ASTRONOMICAL OBS
UKRAINIAN ACADEMY OF SCI
252127 KIEV
U.S.S.R.
TEL:
TLX: 131406 SKY SU
TLF:
EML:
COM: 15

SHUSTOV BORIS M DR
ASTRONOMICAL COUNCIL
USSR ACADEMY OF SCIENCES
PYATNITSKAYA 48
109017 MOSCOW
U.S.S.R.
TEL: 231-54-61
TLX: 412623 SCSTP SU
TLF:
EML:
COM: 34C.35

SHUTER WILLIAM L H DR
DEPT OF PHYSICS
UNIV OF BRITISH COLUMBIA
6224 AGRICULTURE ROAD
VANCOUVER BC V6T 2A6
CANADA
TEL: 604-228-4269
TLX: 04508576
TLF:
EML:
COM: 33.34.40.51

SIBILLE FRANCOIS
OBSERVATOIRE DE LYON
F-69230 ST-GENIS-LAVAL
FRANCE
TEL: 78-56-07-05
TLX: 310926
TLF:
EML:
COM:

SIDLICHOVSKY MILOS DR
ASTRONOMICAL INSTITUTE
CZECH. ACAD. OF SCIENCES
BUDECSKA 6
120 23 PRAHA 2
CZECHOSLOVAKIA
TEL: PRAHA 258757
TLX: 122486
TLF:
EML:
COM: 07

SIDORENKOV NIKOLAY S
HYDROMETCENTRE OF USSR
123376 MOSCOW
U.S.S.R.
TEL:
TLX:
TLF:
EML:
COM: 19

SIEBER WOLFGANG PH D
FACHHOCHSCHULE NIEDERRHEI
FACHBEREICH ELECTR.
REINARZSTR. 49 69
D-4150 KREFELD 1
GERMANY. F.R.
TEL: 02151-8220
TLX:
TLF:
EML:
COM: 40

SIENKIEWICZ RYSZARD DR
COPERNICUS ASTRON CENTER
UL. BARTYCKA 18
00-716 WARSAW
POLAND
TEL: 411086
TLX: 813878 ZAPAN PL
TLF:
EML:
COM: 35

SIGNORE MONIQUE DR
ECOLE NORMALE SUPERIEURE
RADIOASTRONOMIE
24 RUE LHOMOND
F-75231 PARIS CEDEX 05X
FRANCE
TEL: 1-45-29-12-25
TLX:
TLF:
EML:
COM: 35.41.47.48

SIKORA MAREK
COPERNICUS ASTR. CENTER
UL. BARTYCKA 18
00-716 WARSAW
POLAND
TEL:
TLX:
TLF:
EML:
COM: 48

SIKORSKI JERZY DR
INST OF THEOR PHYSICS
UNIV OF GDANSK
UL WITA STWOSZA 57
80 952 GDANSK
POLAND
TEL:
TLX: 0512706 IFAS PL
TLF:
EML:
COM:

SILBERBERG REIN DR
NAVAL RESEARCH LABORATORY
CODE 4154
WASHINGTON DC 20375
U.S.A.
TEL: 202-767-2803
TLX:
TLF:
EML:
COM: 10.34.48

SILK JOSEPH I PROF
ASTRONOMY DEPT
UNIVERSITY OF CALIFORNIA
BERKELEY CA 94720
U.S.A.
TEL: 415-642-2113
TLX: 820181 UCB AST
TLF:
EML:
COM: 34.47

SILVERBERG ERIC C DR
MCDONALD OBSERVATORY
UNIVERSITY OF TEXAS
PO BOX 1337
FORT DAVIS TX 79734
U.S.A.
TEL:
TLX:
TLF:
EML:
COM: 19

SILVESTRO GIOVANNI
ISTITUTO DI FISICA
UNIVERSITA DI TORINO
CORSO M. D'AZEGLIO 46
I-10125 TORINO
ITALY
TEL: 11-650-8623
TLX: 211041 INFNTO
TLF:
EML:
COM: 34.35.44

SIM MARY E MISS
ROYAL OBSERVATORY
BLACKFORD HILL
EDINBURGH EH9 3HJ
U.K.
TEL: 031-667-3321
TLX: 72383 ROEDIN G
TLF:
EML:
COM: 09

SIMA ZDISLAV DR
ASTRONOMICAL INSTITUTE
CZECH. ACAD. OF SCIENCES
BUDECSKA 6
120 23 PRAHA 2
CZECHOSLOVAKIA
TEL: 42-2-258757
TLX: 66-122486
TLF:
EML:
COM: 07.42

SIMEK MILOS DR
ASTRONOMICAL INSTITUTE
CZECH. ACAD. OF SCIENCES
251 65 ONDREJOV
CZECHOSLOVAKIA
TEL: PRAHA 724525
TLX: 121579
TLF:
EML:
COM: 22

SIMIEN FRANCOIS DR
OBSERVATOIRE DE LYON
AVENUE CHARLES ANDRE
F-69230 ST-GENIS-LAVAL
FRANCE
TEL: 78-56-07-05
TLX: 310926
TLF:
EML:
COM: 28

SIMKIN SUSAN M DR
MICHIGAN STATE UNIVERSITY
DEPT PHYSICS & ASTRONOMY
EAST LANSING MI 48824
U.S.A.
TEL: 517-353-4540
TLX:
TLF:
EML:
COM: 28

SIMMONS JOHN FRANCIS L
31 HAVELOCK STREET
GLASGOW G11 5HA
U.K.
TEL:
TLX:
TLF:
EML:
COM: 42

SIMNETT GEORGE M
DEPT OF SPACE RESEARCH
UNIVERSITY OF BIRMINGHAM
BIRMINGHAM B15 2TT
U.K.
TEL:
TLX:
TLF:
EML:
COM: 10

SIMO CHARLES DR
UNIVERSIDAD DE BARCELONA
FACULDAD DE MATEMATICAS
AV JOSE ANTONIO 585
BARCELONA 7
SPAIN
TEL:
TLX:
TLF:
EML:
COM:

SIMODA MAHIRO PROF
DEPT ASTRON/EARTH SCI
TOKYO GAKUGEI UNIVERSITY
KOGANEI
TOKYO 184
JAPAN
TEL: 0423-25-2111
TLX:
TLF:
EML:
COM: 37

SIMON GEORGE W DR
AFGL-PHS
NATL SOLAR OBSERVATORY
SUNSPOT NM 88349
U.S.A.
TEL: 505-434-1390
TLX:
TLF:
EML:
COM: 12

SIMON GUY
OBSERVATOIRE DE PARIS
SECTION DE MEUDON
F-92195 MEUDON PL CEDEX
FRANCE
TEL: 1-45-07-77-87
TLX:
TLF:
EML:
COM: 10.12

SIMON JEAN-LOUIS MR
BUREAU DES LONGITUDES
77 AVE DENFERT-ROCHEREAU
F-75014 PARIS
FRANCE
TEL: 1-43-20-12-10
TLX:
TLF:
EML:
COM: 04.07

SIMON KLAUS PETER
INST F ASTRON & ASTROPHYS
DER UNIVERSITAET MUENCHEN
SCHEINERSTR 1
D-8000 MUENCHEN 80
GERMANY. F.R.
TEL: 089-98-90-21
TLX: 529815 UNIVM D
TLF:
EML:
COM: 36

SIMON MICHAL PROF
DEPT EARTH SPACE SCIENCES
SUNY
STONY BROOK NY 11794
U.S.A.
TEL: 516-246-7672
TLX: 510-228-7767
TLF:
EML:
COM:

SIMON NORMAN R PROF
BEHLEN LAB OF PHYSICS
UNIVERSITY OF NEBRASKA
LINCOLN NE 68588-0111
U.S.A.
TEL: 402-472-2788
TLX:
TLF:
EML:
COM:

SIMON PAUL A DR
1 RUE MORTE BOUTEILLE
F-78140 VELIZY
FRANCE
TEL:
TLX:
TLF:
EML:
COM: 40.44

SIMON PAUL C DR
INSTITUT AERONOMIE SPAT.
AVENUE CIRCULAIRE 3
B-1180 BRUXELLES
BELGIUM
TEL: 2-375-15-79
TLX: 21563
TLF:
EML:
COM: 44

SIMON RENE L E PROF
INSTITUT D'ASTROPHYSIQUE
UNIVERSITE DE LIEGE
AVENUE DE COINTE 5
B-4200 COINTE-OUGREE
BELGIUM
TEL: 041-52-99-80
TLX: 41264 ASTRLG B
TLF:
EML:
COM: 47

SIMON THEODORE
INSTITUTE FOR ASTRONOMY
UNIVERSITY OF HAWAII
2680 WOODLAWN DRIVE
HONOLULU HI 96822
U.S.A.
TEL: 808-948-8968
TLX: 723-8459 UHAST HR
TLF:
EML:
COM: 29.36

SIMONNEAU EDUARDO DR
INSTITUT D'ASTROPHYSIQUE
98 BIS BOULEVARD ARAGO
F-75014 PARIS
FRANCE
TEL: 1-43-20-14-25
TLX:
TLF:
EML:
COM: 36

SIMONS STUART DR
SCHOOL OF MATHEMAT. SCI.
QUEEN MARY COLLEGE
MILE END ROAD
LONDON E1 4NS
U.K.
TEL: 01-980-4811
TLX:
TLF:
EML:
COM: 34

SIMONSON S CHRISTIAN DR
1061 RUSSELL AVENUE
LOS ALTOS CA 94022
U.S.A.
TEL: 415-968-0473
TLX:
TLF:
EML:
COM: 33

SIMOVLJEVITCH JOVAN L DR
DEPT OF ASTRONOMY
FACULTY OF SCIENCES
STUDENTSKI TRG 16
11000 BEOGRAD
YUGOSLAVIA
TEL: 011-638-715
TLX:
TLF:
EML:
COM:

SIMS KENNETH P DR
SYDNEY OBSERVATORY
OBSERVATORY PARK
SYDNEY NSW 2000
AUSTRALIA
TEL:
TLX:
TLF:
EML:
COM: 08.24

SINCLAIR ANDREW T DR
ROYAL GREENWICH OBS
HERSTMONCEUX CASTLE
HAILSHAM BN27 1RP
U.K.
TEL: 0323-833171
TLX: 87451 RGOBSY G
TLF:
EML:
COM: 07.20

SINGH H P
DEPT OF PHYSICS AND
ASTROPHYSICS
DELHI UNIVERSITY
DELHI 110 007
INDIA
TEL: 324851
TLX:
TLF:
EML:
COM: 51

SINGH JAGDEV DR
INDIAN INSTITUTE OF
ASTROPHYSICS
BANGALORE 560 034
INDIA
TEL: 566-585/566-497
TLX: 845763 IIAP IN
TLF:
EML:
COM: 12

SINGH KULINDER PAL DR
TATA INST FUNDAM RES
HOMI BHABHA ROAD
COLABA
BOMBAY 400 005
INDIA
TEL: 495 2971
TLX: 011 3009
TLF:
EML: BITNET:uunet!shakti!tifr!root
COM:

SINGH PATAN DEEN DR
UNIVERSIDADE DE SAO PAULO
INST ASTRON E GEOFISICO
CAIXA POSTAL 30627
01051 SAO PAULO SP
BRAZIL
TEL: 011-275-3720
TLX: 011-36221 IAGM BR
TLF:
EML:
COM: 34

SINHA K DR
UTTAR PRADESH STATE OBS
MANORA PEAK
NAINITAL 263 129
INDIA
TEL: 2136
TLX: CABLE : ASTRONOMY
TLF:
EML:
COM: 12

SINHA RANESHWAR P
SYSTEMS & APPL SCI CORP
4400 FORBES BLVD
LANHAM MD 20706
U.S.A.
TEL: 301-743-5203
TLX: 317630
TLF:
EML:
COM: 40

SINNERSTAD ULF E PROF
STOCKHOLM OBSERVATORY
S-133 00 SALTSJOEBADEN
SWEDEN
TEL: 08-7170195
TLX:
TLF:
EML:
COM: 29.45

SINTON WILLIAM M
UNIVERSITY OF HAWAII
INSTITUTE FOR ASTRONOMY
2680 WOODLAWN DRIVE
HONOLULU HI 96822
U.S.A.
TEL: 808-948-8307
TLX:
TLF:
EML:
COM: 16

SINVHAL SHAMBHU DAYAL DR.
4/3 SNEHALATAGANG
INDORE 452 001
INDIA
TEL:
TLX:
TLF:
EML:
COM: 27.42

SINZI AKIRA M DR
HYDROGRAPHIC DEPT
TSUKIJI 5 CHUO KU
TOKYO 104
JAPAN
TEL:
TLX:
TLF:
EML:
COM: 04

SION EDWARD MICHAEL
DEPT OF ASTRONOMY
VILLANOVA UNIVERSITY
VILLANOVA PA 19085
U.S.A.
TEL: 215-645-4822
TLX:
TLF:
EML:
COM: 35

SIREGAR SURYADI DR
BANDUNG INST TECHNOLOGY
DEPARTMENT OF ASTRONOMY
GANESHA 10
BANDUNG 40132
INDONESIA
TEL: 84254 EXT 476
TLX: 28324 ITB BANDUNG
TLF:
EML:
COM:

SIROKY JAROMIR DR
PALACKY UNIVERSITY
DEPT PHYSICS & ASTRONOMY
LENIN STR 26
771 46 OLOMOUC
CZECHOSLOVAKIA
TEL: 22451
TLX:
TLF:
EML:
COM: 46

SIRY JOSEPH W
4438 42ND STREET N.W.
WASHINGTON DC 20016
U.S.A.
TEL:
TLX:
TLF:
EML:
COM: 07

SISSON GEORGE M MR
PLANETREES
WALL
HEXHAM NE46 4EQ
U.K.
TEL: 0434-81-434
TLX:
TLF:
EML:
COM:

SISTERO ROBERTO F DR
OBSERVATORIO ASTRONOMICO
LAPRIDA 854
5000 CORDOBA
ARGENTINA
TEL: 40613 - 36876
TLX: 51822 BUCOR
TLF:
EML:
COM: 42.47

SITARSKI GRZEGORZ PROF
CENTER FOR SPACE RESEARCH
UL. BARTYCKA 18
00-716 WARSAW
POLAND
TEL: 410041
TLX: 815670 CBK PL
TLF:
EML:
COM: 20

SITKO MICHAEL L
DEPARTMENT OF PHYSICS
210 BRAUNSTEIN - ML 11
UNIVERSITY OF CINCINNATI
CINCINNATI OH 45221-0011
U.S.A.
TEL:
TLX:
TLF:
EML:
COM: 28.34

SITNIK G F PROF
STERNBERG STATE
ASTRONOMICAL INSTITUTE
UNIVERSITETSKIJ PROSP. 13
119899 MOSCOW
U.S.S.R.
TEL: 139-19-73
TLX:
TLF:
EML:
COM: 10.12.36

SITTERLY CHARLOTTE M DR
3711 BRANDYWINE ST N.W.
WASHINGTON DC 20016
U.S.A.
TEL: 202-966-9044
TLX:
TLF:
EML:
COM: 12.14

SIVAN JEAN-PIERRE DR
LAB D'ASTRONOMIE SPATIALE
TRAVERSE DU SIPHON
LES TROIS LUCS
F-13012 MARSEILLE
FRANCE
TEL: 91-66-08-32
TLX: 420584 ASTROSP
TLF:
EML:
COM: 34

SIVARAM C DR
INDIAN INSTITUTE OF
ASTROPHYSICS
BANGALORE 560 034
INDIA
TEL: 566585. 566497
TLX: 845763 IIAB IN
TLF:
EML:
COM: 51

SIVARAMAN K R DR
INDIAN INSTITUTE OF
ASTROPHYSICS
BANGALORE 560 034
INDIA
TEL: 568805
TLX: 845763 IIAB IN
TLF:
EML:
COM: 12C.15

SJOGREN WILLIAM L MR
JPL
MS 264-664
4800 OAK GROVE DRIVE
PASADENA CA 91109
U.S.A.
TEL: 818-354-4868
TLX: 675421
TLF:
EML:
COM: 16

SKALAFURIS ANGELO J
NAVAL RESEARCH LABORATORY
C-5307
WASHINGTON DC 20375-5000
U.S.A.
TEL: 302-767-3227
TLX:
TLF:
EML:
COM:

SKILLEN IAN DR
DEPARTMENT OF ASTRONOMY
UNIVERSITY OF LEICESTER
LEICESTER LE1 7RH
U.K.
TEL:
TLX:
TLF:
EML:
COM:

SKILLING JOHN DR
DEPT APPLIED MATHS
& THEORETICAL PHYSICS
SILVER STREET
CAMBRIDGE CB3 9EW
U.K.
TEL:
TLX:
TLF:
EML:
COM: 34.48

SKILLMAN EVAN D DR
DEPARTMENT OF ASTRONOMY
RLM 15 308
UNIVERSITY OF TEXAS
AUSTIN TX 78712-1083
U.S.A.
TEL: 512 471 1775
TLX: 910 8741351 TEXASTRO
TLF:
EML:
COM: 28.40

SKINNER GERALD DR
SCHOOL OF PHYS. AND RES.
UNIVERSITY OF BIRMINGHAM
PO BOX 363
BIRMINGHAM B15 2TT
U.K.
TEL: 021 414 6450
TLX: 338938
TLF:
EML: SPAN 19457::BHVAD::GKS
COM:

SKRIPNICHENKO VLADIMIR DR
INST OF THEOR ASTRONOMY
NABEREZHNAYA KUTUZOVA 10
191187 LENINGRAD
U.S.S.R.
TEL: 272 40 23
TLX:
TLF:
EML:
COM: 07

SKUMANICH ANDRE PROF
HIGH ALTITUDE OBSERVATORY
PO BOX 3000
BOULDER CO 80307
U.S.A.
TEL: 303-497-1528
TLX: 45694
TLF:
EML:
COM: 12.36

SLADE MARTIN A III DR
JPL
264-737
4800 OAK GROVE DRIVE
PASADENA CA 91109
U.S.A.
TEL: 818-354-6538
TLX:
TLF:
EML:
COM: 19.40

SLEE O B DR
CSIRO
DIVISION OF RADIOPHYSICS
P.O.BOX 76
EPPING NSW 2121
AUSTRALIA
TEL: 868-0222
TLX: 26230 ASTRO
TLF:
EML:
COM: 40

SLETTEBAK ARNE PROF
PERKINS OBSERVATORY
PO BOX 449
DELAWARE OH 43015
U.S.A.
TEL: 614-363-1257
TLX:
TLF:
EML:
COM: 29.33.45

SLONIM E M DR
ASTRONOMICAL INSTITUTE
UZBEKIAN ACADEMY OF SCI
700000 TASHKENT
U.S.S.R.
TEL:
TLX:
TLF:
EML:
COM: 10

SLOVAK MARK HAINES DR
UNIVERSITY OF WISCONSIN
DEPARTMENT OF ASTRONOMY
475 NORTH CHARTER STREET
MADISON WI 53706
U.S.A.
TEL: 608 262 7542
TLX: 265452 UOFWISC MDS
TLF:
EML: MADRAF::SLOVAK
COM:

SLYSH VJACHOSLAV I DR
SPACE RESEARCH INSTITUTE
USSR ACADEMY OF SCIENCES
117810 MOSCOW
U.S.S.R.
TEL:
TLX:
TLF:
EML:
COM: 40.51

SMAK JOSEPH I PROF
COPERNICUS ASTRON CENTER
UL. BARTYCKA 18
00-716 WARSAW
POLAND
TEL: 41-00-41
TLX: 813978 ZAPAN PL
TLF:
EML:
COM: 26.27.42C

SMALDONE LUIGI ANTONIO
DIPARTIMENTO DI FISICA
MOSTRA D'OLTREMARE
PAD 19
I-80125 NAPOLI
ITALY
TEL: 81-7253428
TLX: 720320 INFNNA I
TLF:
EML:
COM: 10

SMEYERS PAUL PROF
ASTRONOMISCH INSTITUUT
KATHOLIEKE UNIV LEUVEN
CELESTIJNENLAAN 200B
B-3000 LEUVEN
BELGIUM
TEL: 016-20-06-56
TLX: 25715 KULBI B
TLF:
EML:
COM: 27.35

SMIT J A PROF
STERREKUNDIG INSTITUUT
POSTBUS 80 000
PRINCETONPLEIN 5
NL-3508 TA UTRECHT
NETHERLANDS
TEL: 30 535 200
TLX:
TLF:
EML:
COM:

SMITH ALEX G PROF
DEPT OF ASTRONOMY
UNIVERSITY OF FLORIDA
211 SPACE SCI BLDG
GAINESVILLE FL 32611
U.S.A.
TEL: 904-392-6135
TLX:
TLF:
EML:
COM: 40

SMITH ANDREW M DR
NASA/GSFC
CODE 681
GREENBELT MD 20771
U.S.A.
TEL: 301-286-8648
TLX:
TLF:
EML:
COM:

SMITH BARHAM W DR
LOS ALAMOS NATIONAL LAB
MS D-436
LOS ALAMOS NM 87545
U.S.A.
TEL: 505-667-1585
TLX:
TLF:
EML:
COM: 34.48

SMITH BRADFORD A PROF
DEPT OF PLANETARY SCI
UNIVERSITY OF ARIZONA
TUCSON AZ 85721
U.S.A.
TEL: 602-621-6930
TLX: 910-952-1143
TLF:
EML:
COM: 15.16C.44

SMITH BRUCE F DR
THEORETICAL STUDIES BR.
NASA-AMES RESEARCH CTR
245-3
MOFFETT FIELD CA 94035
U.S.A.
TEL: 415-694-5515
TLX:
TLF:
EML:
COM: 28

SMITH CHARLES DITTO
8606 LAKE ISLE DRIVE
TAMPA FL 33617
U.S.A.
TEL:
TLX:
TLF:
EML:
COM: 09.25

SMITH CLAYTON A JR DR
US NAVAL OBSERVATORY
WASHINGTON DC 20390
U.S.A.
TEL: 202-653-1511
TLX: 710-822-1970
TLF:
EML:
COM: 08C.24

SMITH DEAN F DR
BERKELEY RESEARCH ASSOC.
290 GREEN ROCK DRIVE
BOULDER CO 80302
U.S.A.
TEL: 303-444-1922
TLX:
TLF:
EML:
COM: 10.40.49

SMITH ELSKE V P DR
COLL HUMANITIES/SCIENCES
VIRGINIA COMMONW UNIV
900 PARK AVENUE
RICHMOND VA 23284
U.S.A.
TEL: 804-257-1674
TLX:
TLF:
EML:
COM:

SMITH F GRAHAM PROF
NRAL
JODRELL BANK
MACCLESFIELD SK11 9DL
U.K.
TEL: 0477-71321
TLX: 36149
TLF:
EML:
COM: 19.38P.40.46.50.51

SMITH G H DR
DOMINION ASTROPHYSICAL
OBSERVATORY
VICTORIA B.C.
CANADA
TEL:
TLX:
TLF:
EML:
COM:

SMITH GEOFFREY DR
DEPT OF ASTROPHYSICS
SOUTH PARKS ROAD
OXFORD OX1 3RQ
U.K.
TEL: 0865-511336
TLX:
TLF:
EML:
COM: 14

SMITH HARDING E JR DR
CASS C-011
UNIVERSITY OF CALIFORNIA
LA JOLLA CA 92093
U.S.A.
TEL: 619-542-4558
TLX:
TLF:
EML:
COM: 28.47

SMITH HARLAN J PROF
ASTRONOMY DEPT
UNIVERSITY OF TEXAS
R.L. MOORE HALL 15.206
AUSTIN TX 78712
U.S.A.
TEL: 512-471-4461
TLX:
TLF:
EML:
COM: 16.27.44.51

SMITH HAYWOOD C DR
DEPT OF ASTRONOMY
211 SPACE SC RES BLDG
UNIVERSITY OF FLORIDA
GAINESVILLE FL 32611
U.S.A.
TEL: 904-392-1074
TLX:
TLF:
EML:
COM: 28

SMITH HORACE A
MICHIGAN STATE UNIVERSITY
DEPT PHYSICS & ASTRONOMY
EAST LANSING MI 48824
U.S.A.
TEL: 517-353-6784
TLX:
TLF:
EML:
COM:

SMITH HOWARD ALAN
SPACE SCIENCE DIVISION
CODE 4138SM
NAVAL RESEARCH LABORATORY
WASHINGTON DC 20375-5000
U.S.A.
TEL: 202-767-3058
TLX:
TLF:
EML:
COM: 34.44.51

SMITH HUMPHRY M
23 NORMANDALE
BEXHILL-ON-SEA TN39 3LU
U.K.
TEL: 0424-214288
TLX:
TLF:
EML:
COM: 19.31

SMITH LINDA J
DEPT PHYSICS & ASTRONOMY
UNIVERSITY COLLEGE LONDON
GOWER STREET
LONDON WC1E 6BT
U.K.
TEL: 01-387-7050x788
TLX: 28722 UCPHYS G
TLF:
EML:
COM: 44

SMITH LINDSEY F DR
7. KENNEDY RD
AUSTINMER NSW 2515
AUSTRALIA
TEL: 042-675366
TLX:
TLF:
EML:
COM:

SMITH MALCOLM G DR
UK INFRARED TELESCOPE
665 KOMOHANA STREET
HILO HI 96720
U.S.A.
TEL: 808-961-3756
TLX: 708-633-135
TLF:
EML:
COM: 28

SMITH MYRON A ASST PROF
NATL SOLAR OBSERVATORY
BOX 26732
TUCSON AZ 85726-6732
U.S.A.
TEL: 602-327-5511
TLX: 0666484 AURA/NOAO TU
TLF:
EML:
COM: 27C.29.30

SMITH PETER L DR
CENTER FOR ASTROPHYSICS
MS-50
60 GARDEN STREET
CAMBRIDGE MA 02138
U.S.A.
TEL: 617-495-4984
TLX: 921428 SATELLITE CAM
TLF:
EML:
COM: 12.14.34.44

SMITH ROBERT CONNON DR
ASTRONOMY CENTRE
UNIVERSITY OF SUSSEX
PHYSICS BLDG. FALMER
BRIGHTON BN1 9QH
U.K.
TEL: 273-606755x3101
TLX: 877159 BHVTXS G
TLF:
EML:
COM: 35.42

SMITH RODNEY M DR
DEPARTMENT OF PHYSICS
UNIVERSITY OF DURHAM
SOUTH ROAD
DURHAM DH1 3LE
U.K.
TEL:
TLX:
TLF:
EML:
COM: 47

SMITH VERNE V DR
DEPARTMENT OF ASTRONOMY
UNIVERSITY OF TEXAS
AUSTIN TX 78712
U.S.A.
TEL: 512 471 3351
TLX:
TLF:
EML:
COM: 29

SMITH WM HAYDEN PROF
MCDONNELL CENTER FOR
SPACE SCIENCES
WASHINGTON UNIVERSITY
ST LOUIS MO 63130
U.S.A.
TEL: 314-889-6574
TLX:
TLF:
EML:
COM: 14

SMOLINSKI JAN DR
COPERNICUS ASTRON CENTER
UL. CHOPINA 12/18
87-100 TORUN
POLAND
TEL:
TLX:
TLF:
EML:
COM: 29C

SMOLUCHOWSKI ROMAN PROF
ASTRONOMY DEPT
UNIVERSITY OF TEXAS
RLM HALL 15.314
AUSTIN TX 78712
U.S.A.
TEL: 512-471-1305
TLX: 910-874-1351
TLF:
EML:
COM: 15.16

SMOL'KOV GENNADIJ YA DR
SIBIZMIR
P B 4
664697 IRKUTSK 33
U.S.S.R.
TEL:
TLX:
TLF:
EML:
COM: 10.40

SMOOT III GEORGE F.
LAWRENCE BERKELEY LAB
UNIVERSITY OF CALIFORNIA
BLDG 50-230
BERKELEY CA 94720
U.S.A.
TEL: 415-486-5237
TLX:
TLF:
EML:
COM: 47

SMRIGLIO FILIPPO PROF
IST. DI ASTRONOMIA
UNIVERSITA LA SAPIENZA
VIA LANCISI 29
I-00161 ROMA
ITALY
TEL: 867525-8442977
TLX:
TLF:
EML:
COM:

SMYLIE DOUGLAS E DR
DEPT EARTH & ATMOSPH SC
YORK UNIVERSITY
4700 KEELE STREET
DOWNSVIEW ONT M3J 1P3
CANADA
TEL: 416-736-5245
TLX:
TLF:
EML:
COM: 19.31

SMYTH MICHAEL J DR
DEPT OF ASTRONOMY
ROYAL OBSERVATORY
EDINBURGH EH9 3HJ
U.K.
TEL: 031-667-3321
TLX: 72383
TLF:
EML:
COM: 09.25

SNEDEN CHRISTOPHER A
UNIVERSITY OF TEXAS
DEPT OF ASTRONOMY
AUSTIN TX 78712
U.S.A.
TEL: 512-471-4461
TLX:
TLF:
EML:
COM: 29

SNELL RONALD L
FIVE COLLEGE RADIO ASTRON
OBS. GRC TOWER B
UNIV OF MASSACHUSETTS
AMHERST MA 01003
U.S.A.
TEL: 413-545-1949
TLX: 955491
TLF:
EML:
COM: 34

SNEZHKO LEONID I
SPECIAL ASTROPHYS OBS
NIZHNIJ ARKHYZ
357147 STAVROPOLSKIJ KRAJ
U.S.S.R.
TEL: 93513
TLX: 297140 ZENIT
TLF:
EML:
COM: 09C.36

SNIJDERS MATTHEUS A J DR
ASTRONOMISCHES INSTITUT
WALDHAUSER STRASSE 64
D-7400 TUBINGEN 1
GERMANY. F.R.
TEL: 292486
TLX: 07262 714 AIT D
TLF:
EML:
COM: 36

SNOW THEODORE P PROF
CASS CB 391
UNIVERSITY OF COLORADO
BOULDER CO 80309
U.S.A.
TEL: 303-492-6857
TLX:
TLF:
EML:
COM: 29.34.44

SNYDER LEWIS E
DEPT OF ASTRONOMY
UNIVERSITY OF ILLINOIS
URBANA IL 61801
U.S.A.
TEL: 217-333-5530
TLX: 910-245-2434
TLF:
EML:
COM: 15.51

SOBERMAN ROBERT K DR
FRANKLIN RESEARCH CENTER
ARVIN-CALSPAN
20TH & RACE STREETS
PHILADELPHIA PA 19103
U.S.A.
TEL: 215-448-1058
TLX: 710-670-1889
TLF:
EML:
COM: 21.22

SOBIESKI STANLEY DR
NASA/GSFC
CODE 673
GREENBELT MD 20771
U.S.A.
TEL:
TLX:
TLF:
EML:
COM: 42

SOBOLEV V V DR
ASTRONOMICAL OBSERVATORY
LENINGRAD UNIVERSITY
199178 LENINGRAD
U.S.S.R.
TEL:
TLX:
TLF:
EML:
COM: 34.36

SOBOLEV VLADISLAV M DR
MAIN ASTRONOMICAL OBS
PULKOVO
196140 LENINGRAD
U.S.S.R.
TEL: 298-22-42
TLX:
TLF:
EML:
COM: 12

SOBOLEVA N S DR
SPECIAL ASTROPHYSICAL OBS
USSR ACADEMY OF SCIENCES
LENINGRAD BRANCH
196140 LENINGRAD
U.S.S.R.
TEL:
TLX:
TLF:
EML:
COM: 40

SOBOUTI YOUSEF PROF
DEPARTMENT OF PHYSICS
SHIRAZ UNIVERSITY
SHIRAZ
IRAN
TEL: 1198-71-57339
TLX:
TLF:
EML:
COM: 28.35

SOCHILINA ALLA S DR
INST TO THEORET ASTRONOMY
USSR ACADEMY OF SCIENCES
10 KUTUZOV QUAY
191187 LENINGRAD
U.S.S.R.
TEL: 278-88-98
TLX: 121578 ITA SU
TLF:
EML:
COM: 04

SODERBLOM DAVID R
SPACE TELESCOPE SCI INST
3700 SAN MARTIN DRIVE
BALTIMORE MD 21218
U.S.A.
TEL: 301-338-4830
TLX: 6849101 STSCI
TLF:
EML:
COM: 29

SODERBLOM LARRY DR
BRANCH OF ASTROGEOLOGIC
US GEOLOGICAL SURVEY
2555 NORTH GEMINI DRIVE
FLAGSTAFF AZ 86001
U.S.A.
TEL:
TLX:
TLF:
EML:
COM: 16

SOEDERHJELM STAFFAN DR
LUND OBSERVATORY
BOX 43
S-221 00 LUND
SWEDEN
TEL: 46-10 73 03
TLX: 33199 OBSHOT S
TLF:
EML:
COM: 08.42

SOFIA SABATINO PROF
YALE UNIVERSITY OBS
PO BOX 6666
NEW HAVEN CT 06511
U.S.A.
TEL: 203-436-3460
TLX: 710-465-3041
TLF:
EML:
COM: 34.35.44.48

SOFUE YOSHIAKI PROF
DEPT OF ASTRONOMY
UNIVERSITY OF TOKYO
YAYOI-CHO. BUNKYO-KU
113 TOKYO
JAPAN
TEL:
TLX:
TLF:
EML:
COM: 34.40.51

SOIFER BARUCH T DR
PHYSICS DEPT
CALTECH
DOWNES LAB 320-47
PASADENA CA 91125
U.S.A.
TEL: 818-356-6626
TLX: 675425
TLF:
EML:
COM:

SOKOLOWSKI LECH
ASTRONOMICAL OBSERVATORY
JAGIELLONIAN UNIVERSITY
UL. ORLA 171
30-244 KRAKOW
POLAND
TEL: 012-22-38-56
TLX: 0322723 UJ PL
TLF:
EML:
COM: 47

SOL HELENE DR
DARC
5 PLACE JULES JANSSEN
OBSERVATOIRE DE MEUDON
92195 MEUDON PPL CEDEX
FRANCE
TEL: 1-45 07 74 28
TLX: 20157 LAM
TLF:
EML: BITNET:SOL@FRMEU51
COM:

SOLANKI SAMI K DR
MATHEMATICAL INSTITUTE
UNIVERSITY OF ST ANDREWS
ST ANDREWS KY169SS
U.K.
TEL: 44-334-76161
TLX:
TLF:
EML: ANSSS@SAVA.ST-AND.AC.UK
COM: 10.12

SOLARIC NIKOLA
GEODETSKI FAKULTET
UNIVERSITY OF ZAGREB
KACICEVA 26
41000 ZAGREB
YUGOSLAVIA
TEL: 041-521-548
TLX:
TLF:
EML:
COM: 08

SOLC IVAN DR
ASTRONOMICAL INSTITUTE
GROUP OF OPTICS
DVORAKOVA 298
511 01 TURNOV
CZECHOSLOVAKIA
TEL: 42-2-540395
TLX: 121673 MFF
TLF:
EML:
COM:

SOLC MARTIN
DEPT ASTRONOMY/ASTROPHYS
CHARLES UNIVERSITY PRAGUE
SVEDSKA 8
150 00 PRAHA 5
CZECHOSLOVAKIA
TEL: 02-540395
TLX:
TLF:
EML:
COM: 15.34.42

SOLF JOSEF DR
MPI FUER ASTRONOMIE
KOENIGSTUHL
D-6900 HEIDELBERG 1
GERMANY. F.R.
TEL: 6221-528-226
TLX: 461789 MPIA D
TLF:
EML:
COM:

SOLHEIM JAN ERIK
INST F. MATEMAT REALFAG
P O BOX 953
N-9001 TROMSO
NORWAY
TEL: 083-86060
TLX: 64124
TLF:
EML:
COM: 42.46

SOLIMAN MOHAMED AHMED
HELWAN OBSERVATORY
HELWAN
EGYPT
TEL:
TLX:
TLF:
EML:
COM: 27

SOLLAZZO CLAUDIO
EUROPEAN SPACE OPERATIONS
CENTER
ROBERT-BOSCH-STR 5
D-6100 DARMSTADT
GERMANY. F.R.
TEL: 06151-8861
TLX: 419453 ESOC D
TLF:
EML:
COM:

SOLOMON PHILIP M DR
ASTROPHYSICS PROGRAM
DEPT EARTH & SPACE SCI
SUNY AT STONY BROOK
STONY BROOK NY 11794
U.S.A.
TEL: 516-246-8383
TLX: 510-228-7767
TLF:
EML:
COM: 33.34

SOLTAN ANDRZEJ MARIA DR
COPERNICUS ASTRON CENTER
UL. BARTYCKA 18
00-716 WARSAW
POLAND
TEL:
TLX:
TLF:
EML:
COM: 28

SOMA MITSURU DR
NAT. ASTRON. OBSERVATORY
OSAWA MITAKA
TOKYO 181
JAPAN
TEL: 0422 32 5111
TLX: 02822307
TLF:
EML:
COM: 08.20

SOMERVILLE WILLIAM B DR
DEPT PHYSICS & ASTRONOMY
UNIVERSITY COLLEGE LONDON
GOWER STREET
LONDON WC1E 6BT
U.K.
TEL: 01-382-7050
TLX: 28722
TLF:
EML:
COM: 14.34

SOMMER-LARSEN JESPER DR
NIELS BOHR INSTITUTE
BLEGDAMSVEJ 15
DK-2100 KBH.0
DENMARK
TEL:
TLX:
TLF:
EML:
COM:

SONETT CHARLES P PROF
DEPT PLANETARY SCIENCES
UNIVERSITY OF ARIZONA
TUCSON AZ 85721
U.S.A.
TEL: 602-621-6935
TLX: 9109521143
TLF:
EML:
COM: 16.49

SONG GUO-XUAN
SHANGHAI OBSERVATORY
ACADEMIA SINICA
SHANGHAI
CHINA. PEOPLE'S REP.
TEL: 386191
TLX: 33164 SHAO CN
TLF:
EML:
COM: 28.33

SONG JIN-AN
SHANXI ASTRON OBSERVATORY
P.O. BOX 18
LINTONG
XIAN
CHINA. PEOPLE'S REP.
TEL: XIAN 32255
TLX: 70121 CSAO CN
TLF:
EML:
COM: 31

SONG MU-TAO
PURPLE MOUNTAIN
OBSERVATORY
NANJING
CHINA. PEOPLE'S REP.
TEL:
TLX:
TLF:
EML:
COM: 12

SONGSATHAPORN RUANGSAK DR
PHYSICS DEPARTMENT
CHIANG MAI UNIVERSITY
CHIANG MAI 50002
THAILAND
TEL: 221934 X 135
TLX: 43553 UNICHIM TH
TLF:
EML:
COM:

SONNEBORN GEORGE DR
LAB. FOR ASTR. & PHYSICS
NASA GSFC
CODE 681
GREENBELT MD 20771
U.S.A.
TEL: 301 286 3665
TLX: 89675
TLF:
EML: SPAN:6471::SONNEBORN
COM: 29.44

SORENSEN GUNNAR DR
INSTITUTE OF PHYSICS
LANGELANDSGADE
DK-8000 AARHUS C
DENMARK
TEL:
TLX:
TLF:
EML:
COM: 14

SORENSEN SOREN-AKSEL DR
DEPT COMPUTER SCIENCE
UNIVERSITY COLLEGE LONDON
LONDON WC1E 6BT
U.K.
TEL:
TLX:
TLF:
EML:
COM:

SOROCHENKO R L DR
PHYSICAL INSTITUTE
USSR ACADEMY OF SCIENCES
LENINSKY PROSPECT 53
117924 MOSCOW
U.S.S.R.
TEL: 135-01-71
TLX: 411479 NEOD SU
TLF:
EML:
COM: 40

SORU-ESCAUT IRINA MRS
OBSERVATOIRE DE PARIS
SECTION DE MEUDON
F-92195 MEUDON PL CEDEX
FRANCE
TEL: 1-45-34-75-30
TLX:
TLF:
EML:
COM:

SOTIROVSKI PASCAL DR
OBSERVATOIRE DE PARIS
SECTION DE MEUDON
F-92195 MEUDON PL CEDEX
FRANCE
TEL: 1-45-07-78-02
TLX: 270912
TLF:
EML:
COM: 10.12

SOUFFRIN PIERRE B DR
OBSERVATOIRE DE NICE
BP 139
F-06003 NICE CEDEX
FRANCE
TEL: 93-89-04-20
TLX:
TLF:
EML:
COM: 12.35.36

SOULIE GUY
OBSERVATOIRE DE BORDEAUX
F-33270 FLOIRAC
FRANCE
TEL: 56-86-43-30
TLX:
TLF:
EML:
COM:

SPADA GIANFRANCO DR
T E S R E
CNR
VIA DE CASTAGNOLI 1
I-40126 BOLOGNA
ITALY
TEL: 51-95-93
TLX: 511350 CNR BO
TLF:
EML:
COM: 44.48

SPARENBAUER ANDREAS MARTIN
ASTRONOMISCHES INSTITUT
UNIVERSITAET BASEL
VENUSSTRASSE 7
CH-4102 BINNINGEN
SWITZERLAND
TEL: 061-227711
TLX:
TLF:
EML:
COM:

SPARKE LINDA
KAPTEYN LABORATORIUM
GRONINGEN UNIVERSITY
POSTBUS 800
NL-9700 AV GRONINGEN
NETHERLANDS
TEL: 050-634056
TLX: 53572 STARS NL
TLF:
EML:
COM: 33

SPARKS WARREN M DR
LOS ALAMOS NATL LAB
MS-F669
LOS ALAMOS NM 87545
U.S.A.
TEL: 505-667-4922
TLX:
TLF:
EML:
COM: 35.42

SPARKS WILLIAM BRIAN
ROYAL GREENWICH OBS
HERSTMONCEUX CASTLE
HAILSHAM BN27 1RP
U.K.
TEL: 03230833171
TLX: 87451
TLF:
EML:
CON: 28

SPARROW JAMES G DR
AERONAUTICAL RESEARCH
LABORATORIES
BOX 4331
MELBOURNE 3001
AUSTRALIA
TEL: 03-647 7623
TLX: 39391 ARL AA
TLF:
EML:
CON: 2[illegible]

SPASOVA NEDKA MARINOVA
DEPT OF ASTRONOMY
BULGARIAN ACAD SCIENCES
72 LENIN BLVD
1784 SOFIA
BULGARIA
TEL: 7341 x379
TLX:
TLF:
EML:
CON:

SPEER R J DR
DEPT OF PHYSICS
IMPERIAL COLLEGE
PRINCE CONSORT ROAD
LONDON SW7 2BZ
U.K.
TEL: 01-589-5111
TLX: 261503 IMPCOL
TLF:
EML:
CON: 44

SPENCER JOHN HOWARD
NAVAL RESEARCH LABORATORY
CODE 4134
WASHINGTON DC 20375
U.S.A.
TEL: 202-767-3050
TLX:
TLF:
EML:
CON: 40

SPENCER RALPH E DR
NUFFIELD RADIO ASTRONMY
LABORATORIES
JODRELL BANK
MACCLESFIELD SK11 9DL
U.K.
TEL: 0477-71-321
TLX: 36149 JODREL G
TLF:
EML:
CON: 40

SPICER DANIEL SHIELDS DR
NASA/GSFC
CODE 682
GREENBELT MD 20771
U.S.A.
TEL: 301-286-7334
TLX:
TLF:
EML:
CON: 10.12

SPIEGEL E DR
ASTRONOMY DEPARTMENT
COLUMBIA UNIVERSITY
NEW YORK NY 10027
U.S.A.
TEL:
TLX:
TLF:
EML:
CON: 33.35.36

SPINRAD HYRON PROF
DEPT OF ASTRONOMY
UNIVERSITY OF CALIFORNIA
BERKELEY CA 94720
U.S.A.
TEL: 415-642-2078
TLX:
TLF:
EML:
CON: 15.28

SPITE FRANCOIS M DR
OBSERVATOIRE DE PARIS
SECTION DE MEUDON
F-92195 MEUDON PL CEDEX
FRANCE
TEL: 1-45-07-78-40
TLX: 270 912
TLF:
EML:
CON: 05C.29.36

SPITE MONIQUE DR
OBSERVATOIRE DE PARIS
SECTION DE MEUDON
F-92195 MEUDON PL CEDEX
FRANCE
TEL: 1-45-07-78-39
TLX: 270 912
TLF:
EML:
CON: 29C.36

SPITHAS ELEFTERIOS N DR
DEPT OF ASTRONOMY
UNIVERSITY OF ATHENS
PANEPISTIMIOPOLIS
GR-15771 ATHENS
GREECE
TEL:
TLX:
TLF:
EML:
CON:

SPITZER LYMAN JR DR
PRINCETON UNIVERSITY
OBSERVATORY
PEYTON HALL
PRINCETON NJ 08544
U.S.A.
TEL: 609-452-3809
TLX: 322409
TLF:
EML:
CON: 34.44

SPOELSTRA T A TH DR
NETHERLANDS FOUNDATION
FOR RADIO ASTRONOMY
OUDE HOOGEVEENSEDIJK 4
NL-7991 PD DWINGELOO
NETHERLANDS
TEL: 05219-7244
TLX: 42043 SRZM NL
TLF:
EML:
CON: 08.40

SPRUIT HENK C DR
NATL SOLAR OBSERVATORY
SUNSPOT NM 88349
U.S.A.
TEL:
TLX:
TLF:
EML:
CON: 10.36

SPYROU NICOLAOS PROF
DEPT OF ASTRONOMY
UNIVERSITY THESSALONIKI
GR-54006 THESSALONIKI
GREECE
TEL: 031-992658
TLX: 412181
TLF:
EML:
CON: 47

SRAMEK RICHARD A DR
NRAO
PO BOX 0
SOCORRO NM 87801
U.S.A.
TEL: 505-772-4011
TLX: 9109881710
TLF:
EML:
CON: 40

SREEKANTAN B V DR
TATA INSTITUTE OF
FUNDAMENTAL RESEARCH
HOMI BHABHA RD
BOMBAY 400 005
INDIA
TEL: 219111
TLX: 011-3009
TLF:
EML:
CON:

SREENIVASAN S RANGA PROF
PHYSICS DEPARTMENT
UNIVERSITY OF CALGARY
2500 UNIVERSITY DR.NW
CALGARY AB T2N 1N4
CANADA
TEL: 403-284-5385
TLX:
TLF:
EML:
CON: 35

SRINIVASAN G
RAMAN RESEARCH INSTITUTE
BANGALORE 560 080
INDIA
TEL: 812-360122
TLX: 8452671 RRI IN
TLF:
EML:
CON:

SRIVASTAVA J B DR
UTTAR PRADESH STATE OBS
MANORA PEAK
NAINITAL 263 129
INDIA
TEL: 2136
TLX:
TLF:
EML:
COM: 42

SRIVASTAVA RAM KUMAR DR
UTTAR PRADESH STATE OBS
MANORA PEAK
NAINITAL
INDIA
TEL:
TLX:
TLF:
EML:
COM: 27.42

STABELL ROLF DR
INST THEORET ASTROPHYSICS
UNIVERSITY OF OSLO
P O BOX 1029
N-0315 BLINDERN. OSLO 3
NORWAY
TEL: 2-456-530
TLX: 72705 ASTRO N
TLF:
EML:
COM:

STACEY GORDON J DR
DEPARTMENT OF PHYSICS
UNIVERSITY OF CALIFORNIA
BERKELEY CA 94720
U.S.A.
TEL: 415 642 1128
TLX: 9103667114
TLF:
EML: STACEY@ISTI.SSL.BERKELEY.ED
COM:

STACHNIK ROBERT V
HARVARD-SMITHSONIAN CTR
FOR ASTROPHYSICS
60 GARDEN STREET
CAMBRIDGE MA 02138
U.S.A.
TEL: 617-495-2829
TLX: 921428 SATELLITE CAM
TLF:
EML:
COM: 44

STAGNI RUGGERO
OSSERVATORIO ASTROFISICO
DI ASIAGO
I-36012 ASIAGO
ITALY
TEL: 0424-62665
TLX: 430110 SETOUR I
TLF:
EML:
COM:

STAHL OTMAR RICHARD DR
LANDESSTERNWARTE
KOENIGSTUHL
6900 HEIDELBERG 1
GERMANY. F.R.
TEL: 6221/509 232
TLX: 461153
TLF:
EML: BITNET:BP20DHDURZ1
COM:

STAHR-CARPENTER M DR
1101 HILL TOP ROAD
CHARLOTTESVILLE VA 22903
U.S.A.
TEL: 804-293-7063
TLX:
TLF:
EML:
COM: 40

STALIO ROBERTO DR
DIPT. DI ASTRONOMIA
UNIVERSITA DI TRIESTE
VIA TIEPOLO 11
I-34131 TRIESTE
ITALY
TEL: 40-793921/221
TLX: 461137 OAT I
TLF:
EML:
COM: 29.36.51

STANDISH E MYLES DR
JET PROPULSION LAB
JPL 264-664
PASADENA CA 91109
U.S.A.
TEL: 818-354-3959
TLX:
TLF:
EML:
COM: 04.07.20

STANFORD SPENCER A
DEPTARTMENT OF ASTRONOMY
UNIVERSITY OF WISCONSIN
475 NORTH CHARTER
MADISON WI 53706
U.S.A.
TEL: 608 262 1298
TLX:
TLF:
EML:
COM:

STANGA RUGGERO
DIPART. DI ASTRONOMIA
UNIVERSITA DEGLI STUDI
LARGO E. FERMI 5
I-50125 FIRENZE
ITALY
TEL: 055 27521
TLX: 572269 ARCETR I
TLF:
EML:
COM: 34

STANGE LOTHAR
TECHNICAL UNIVERSITY
DRESDEN
MOMMSENSTR.13
DDR-8027 DRESDEN
GERMANY. D.R.
TEL: 463-4652
TLX: 02278
TLF:
EML:
COM: 08.24

STANILA GEORGE DR
ASTRONOMICAL OBSERVATORY
P.O. BOX 28
CUTITUL DE ARGINT 5
75212 BUCAREST 28
RUMANIA
TEL: 23-68-92
TLX:
TLF:
EML:
COM: 19.31

STANKEVICH KAZIMIR S DR
RADIOPHYSICAL RESEARCH
INSTITUTE
603600 GORKIJ
U.S.S.R.
TEL: 38-90-91
TLX:
TLF:
EML:
COM:

STANLEY G J
PO BOX 1348
CARMEL VALLEY CA 93924
U.S.A.
TEL: 408-659-2940
TLX:
TLF:
EML:
COM: 40

STANNARD DAVID DR
NRAL
JODRELL BANK
MACCLESFIELD SK11 9DL
U.K.
TEL: 0477-71321
TLX: 36149 JODREL G
TLF:
EML:
COM: 40

STARK ANTONY A
AT & T BELL LABORATORIES
HOH L-231
HOLMDELL NJ 07733
U.S.A.
TEL: 201-949-4842
TLX:
TLF:
EML:
COM:

STARRFIELD SUMNER PROF
DEPT OF PHYSICS
ARIZONA STATE UNIVERSITY
TEMPE AZ 85281
U.S.A.
TEL: 602-965-3561
TLX: 667391 ARIZ ST U TMP
TLF:
EML:
COM: 27.35.42

STASINSKA GRAZYNA DR
OBSERVATOIRE DE PARIS
SECTION DE MEUDON
F-92195 MEUDON PL CEDEX
FRANCE
TEL: 1-45-07-74-22
TLX: 201571 LAM
TLF:
EML:
COM:

STAUBERT RUDIGER PROF DR
ASTRONOMISCHES INSTITUT
UNIVERSITAET TUEBINGEN
WALDHAUSERSTR 64
D-7400 TUEBINGEN
GERMANY. F.R.
TEL: 7071-294980
TLX: 7262714 AIT D
TLF:
EML:
COM: 44.48

STAUDE HANS JAKOB PH D
MPI FUER ASTRONOMIE
KOENIGSTUHL
D-6900 HEIDELBERG 1
GERMANY. F.R.
TEL: 06221-528229
TLX: 461789 MPIA D
TLF:
EML:
COM: 21

STAUDE JUERGEN DR
ZNTRLINST. F. ASTROPHYSIK
SONNENOBSERVATORIUM
EINSTEINTURM
DDR-1500 POTSDAM
GERMANY. D.R.
TEL:
TLX:
TLF:
EML:
COM: 12

STAVELEY-SMITH LISTER DR
ANGLO AUSTRALIAN OBS
PO BOX 296
EPPING NSW 2121
AUSTRALIA
TEL: 02 868 1666
TLX: 123999 AAOSYD AA
TLF:
EML: ACSNET:LSS@AAOEPP.OZ
COM: 28

STANIKOWSKI ANTONI DR
ASTRONOMICAL CENTER
UL. CHOPINA 12/18
87-100 TORUN
POLAND
TEL:
TLX:
TLF:
EML:
COM: 29

STEBBINS ROBIN
JILA
UNIVERSITY OF COLORADO
BOX 440
BOULDER CA 90309-0440
U.S.A.
TEL: 303-492-6073
TLX: 755842
TLF:
EML:
COM: 12

STECHER THEODORE P
NASA/GSFC
CODE 680
GREENBELT MD 20771
U.S.A.
TEL: 301-286-8718
TLX:
TLF:
EML:
COM: 29.44.34

STECKER FLOYD W DR
HIGH ENERGY ASTROPHYS LAB
NASA/GSFC CODE 660
GREENBELT MD 20771
U.S.A.
TEL: 301-286-6057
TLX:
TLF:
EML:
COM: 33.47.48

STEENMAN-CLARK LOIS DR
OBSERVATOIRE DE NICE
BP 139
F-06003 NICE CEDEX
FRANCE
TEL: 93-89-04-20
TLX:
TLF:
EML:
COM: 14

STEFANIK ROBERT DR
OAK RIDGE OBSERVATORY
HARVARD SMITHSONIAN CTR
PINNACLE ROAD
HARVARD MASS 01451
U.S.A.
TEL: 617 495 7070
TLX:
TLF:
EML: SPAN:cfaz::stefanik
COM: 30

STEFANOVITCH-GOMEZ A E DR
OBSERVATOIRE DE PARIS
SECTION DE MEUDON
F-92195 MEUDON PL CEDEX
FRANCE
TEL: 1-45-07-78-43
TLX: 201571 F
TLF:
EML:
COM: 33

STEFFEN MATTHIAS DR
INST FUR THEOR PHYSIK
OLSHAUSENSTRASSE 40
2300 KIEL 1
GERMANY. F.R.
TEL: 0431 880 4101
TLX: 292706
TLF:
EML:
COM: 12.29.36.40

STEFL VLADIMIR
DEPT THEORET PHYS & ASTRO
FACULTY NATURAL SCIENCES
J.E. PURKYNE UNIVERSITY
611 37 BRNO
CZECHOSLOVAKIA
TEL: 51112
TLX:
TLF:
EML:
COM: 46

STEHLE CHANTAL DR
DEPT ATOMES ET MOLECULES
OBSERVATOIRE DE PARIS
92195 MEUDON PRINC CEDEX
FRANCE
TEL: 1-45 07 74 53
TLX: LAN 201571
TLF:
EML:
COM: 14

STEIGER W R PROF
CALTECH SUBMILL. OBS.
P.O. BOX 4339
HILO HI 96720
U.S.A.
TEL:
TLX:
TLF:
EML:
COM:

STEIGMAN GARY PROF
OHIO STATE UNIVERSITY
DEPT. OF PHYSICS. OSU
174 WEST 18TH AVE.LAWARE
COLUMBUS OH 43210
U.S.A.
TEL: 614-292-1999
TLX: 8104631715
TLF:
EML:
COM: 47.48

STEIMAN-CAMERON THOMAS DR
THEORETICAL STUDIES
MAIL STOP 245-3
NASA AMES RESEARCH CENTER
MOFFETT FIELD CA 94035
U.S.A.
TEL: 415 694 3120
TLX:
TLF:
EML: SPAN:GAL::TOMSC
COM: 42

STEIN JOHN WILLIAM
555 HILL STREET
SEWICKLEY PA 15143
U.S.A.
TEL: 412-741-4182
TLX:
TLF:
EML:
COM: 24.26.51

STEIN ROBERT F ASSOC PROF
PHYSICS-ASTRONOMY DEPT
MICHIGAN STATE UNIVERSITY
EAST LANSING MI 48824
U.S.A.
TEL: 517-353-8661
TLX:
TLF:
EML:
COM: 36

STEIN WAYNE A PROF
SCHOOL OF PHYS & ASTRON
UNIVERSITY OF MINNESOTA
MINNEAPOLIS MN 55455
U.S.A.
TEL: 612-373-9963
TLX:
TLF:
EML:
COM:

STEINBERG JEAN-LOUIS DR
OBSERVATOIRE DE PARIS
SECTION DE MEUDON
F-92195 MEUDON PL CEDEX
FRANCE
TEL: 1-45-07-76-96
TLX: 204464 F
TLF:
EML:
COM: 40.44

STEINER JOAO E DR
INST PESQUISAS ESPACIAIS
CAIXA POSTAL 515
AV. DOS ASTRONAUTAS 1758
12200 S. JOSE DOS CAMPOS
BRAZIL
TEL: 0123-22-9977
TLX:
TLF:
EML:
COM: 42.44.48

STEINITZ RAPHAEL PROF
PHYSICS DEPT
BEN GOURION UNIVERSITY OF
NEGEV
BEERSHEVA 84105
ISRAEL
TEL: 57-70985
TLX:
TLF:
EML:
COM:

STEINLE HELMUT DR
MAX PLANCK INSTITUT
EXTRATERRESTRISCHE PHYSIK
D 8046 GARCHING FRG
GERMANY, F.R.
TEL: 49-89 3299 9470
TLX: 215845 EXTER D
TLF: 49-89 3299 569
EML: BITNET:HCS@DGAIPP1S
COM:

STEINLIN ULI PROF
ASTRONOMISCHES INSTITUT
UNIVERSITAET BASEL
VENUSSTRASSE 7
CH-4102 BINNINGEN
SWITZERLAND
TEL: 061-227711
TLX:
TLF:
EML:
COM: 25.33.45

STELLINGWERF ROBERT F DR
MISSION RESEARCH CORP.
1720 RANDOLPH RD SE
ALBUQUERQUE NM 87106
U.S.A.
TEL: 505-843-7200
TLX:
TLF:
EML:
COM: 27.35

STELLMACHER GOETZ
INSTITUT D'ASTROPHYSIQUE
98 BIS BOULEVARD ARAGO
F-75014 PARIS
FRANCE
TEL: 1-43-20-14-25
TLX:
TLF:
EML:
COM: 10

STELLMACHER IRENE DR
BUREAU DES LONGITUDES
77 AVE DENFERT-ROCHEREAU
F-75014 PARIS
FRANCE
TEL: 1-43-20-12-10
TLX:
TLF:
EML:
COM: 07.20

STENCEL ROBERT EDWARD
CASA
UNIVERSITY OF COLORADO
CAMPUS BOX 391
BOULDER CO 80309
U.S.A.
TEL: 303-492-7178
TLX:
TLF:
EML:
COM: 29.42.44

STENFLO JAN O DR
INSTITUT FUER ASTRONOMIE
ETH ZENTRUM
CH 8092 ZURICH
SWITZERLAND
TEL: 303 497-1500
TLX: 989764
TLF:
EML:
COM: 10.12V

STENHOLM BJOERN DR
LUND OBSERVATORY
BOX 43
S 22100 LUND
SWEDEN
TEL: 046107306
TLX: 33199
TLF: 046104614
EML: bjorn@astro.lu.se
COM: 34

STENHOLM LARS
STOCKHOLM OBSERVATORY
S-133 00 SALTSJOEBADEN
SWEDEN
TEL: 08-7170195
TLX: 12972 SOBSERV S
TLF:
EML:
COM: 46

STEPANIAN A A DR
CRIMEAN ASTROPHYSICAL OBS
USSR ACADEMY OF SCIENCES
NAUCHNIY
334413 CRIMEA
U.S.S.R.
TEL:
TLX:
TLF:
EML:
COM: 28.48

STEPANIAN N N DR
CRIMEAN ASTROPHYSICAL OBS
USSR ACADEMY OF SCIENCES
NAUCHNIY
334413 CRIMEA
U.S.S.R.
TEL: 1-86. 5-55
TLX:
TLF:
EML:
COM: 10

STEPANOV ALEXANDER V DR
CRIMEAN ASTROPHYSICAL
OBSERVATORY
334247 RT-22 KATZIVELY
U.S.S.R.
TEL: YALTA 727906
TLX: 222192 VOSHOD
TLF:
EML:
COM: 10

STEPHENS S A DR
TATA INSTITUTE OF
FUNDAMENTAL RESEARCH
HOMI BHABHA RD
BOMBAY 400 005
INDIA
TEL: 219111
TLX: 011-3009 TIFR IN
TLF:
EML:
COM: 48

STEPHENSON C BRUCE PROF
WARNER & SWASEY OBS
CASE WESTERN RESERVE UNIV
CLEVELAND OH 44106
U.S.A.
TEL: 216-368-3728
TLX:
TLF:
EML:
COM: 33.45

STEPHENSON F RICHARD DR
DEPARTMENT OF PHYSICS
UNIVERSITY OF DURHAM
DURHAM DH1 3LE
U.K.
TEL: 0385-64971 x208
TLX: 537351 DURLIB G
TLF:
EML:
COM: 19.41

STEPIEN KAZIMIERZ DR
ASTRONOMICAL OBSERVATORY
AL. UJAZDOWSKIE 4
00-478 WARSAW
POLAND
TEL: 29-40-11
TLX:
TLF:
EML:
COM: 27.36

STEPPE HANS DR
IRAM
AV. DIVINA PASTORA 7
BLOQUE 6/2B
18012 GRANADA
SPAIN
TEL:
TLX:
TLF:
EML:
COM:

STERKEN CHRISTIAAN LEO DR
ASTROPHYSICAL INSTITUT
VRIJE UNIV BRUSSEL
PLEINLAAN 2
B-1050 BRUSSELS
BELGIUM
TEL: 0032-2-6413469
TLX: 61051 VUBCO
TLF:
EML:
COM: 25.27

STERN ROBERT ALLAN
LOCKHEED P. ALTO RES LAB
DEPT 91-20 / BLDG 255
3251 HANOVER ST
PALO ALTO CA 94304
U.S.A.
TEL: 415-858-4072
TLX:
TLF:
EML:
COM: 44

STESHENKO N V DR
CRIMEAN ASTROPHYSICAL OBS
USSR ACADEMY OF SCIENCES
NAUCHNIY
334413 CRIMEA
U.S.S.R.
TEL: 065-54-32-945
TLX:
TLF:
EML:
COM: 09.10

STETSON PETER B. DR
DOMINION ASTROPHYSICAL
OBSERVATORY
5071 WEST SAANICH ROAD
VICTORIA BC V8X 4M6
CANADA
TEL:
TLX:
TLF:
EML:
COM: 37

STEVENS GERARD A DR
LABORATORIUM VOOR
RUIMTEONDERZOEK
BENELUXLAAN 21
NL-3527 HS UTRECHT
NETHERLANDS
TEL:
TLX:
TLF:
EML:
COM:

STEWART JOHN MALCOLM DR
DEPT OF APPLIED MATH &
THEORETICAL PHYSICS
SILVER STREET
CAMBRIDGE CB3 9EW
U.K.
TEL: 223-351-645
TLX:
TLF:
EML:
COM: 47

STEWART PAUL DR
MATHEMATICS DEPT
THE UNIVERSITY
MANCHESTER M13 9PL
U.K.
TEL:
TLX:
TLF:
EML:
COM: 40

STEWART RONALD T MR
CSIRO
DIVISION OF RADIOPHYSICS
P.O.BOX 76
EPPING NSW 2121
AUSTRALIA
TEL: 868-0222
TLX: 26230
TLF:
EML:
COM: 10.40

STEYAERT HERMAN PROF DR
STERRENKUNDIG INSTITUUT
KRYGSLAAN 271 S9
B-9000 GENT
BELGIUM
TEL: 91-22-5715x2572
TLX:
TLF:
EML:
COM:

STIBBS DOUGLAS W N PROF
UNIVERSITY OBSERVATORY
BUCHANAN GARDENS
ST ANDREWS. FIFE KY16 9LZ
U.K.
TEL: 0334-76161
TLX: 72613 SAULIB GB
TLF:
EML:
COM: 33.35.36

STICKLAND DAVID J DR
SPACE & ASTROPHYS DIV
RUTHERFORD APPLETON LAB
CHILTON. DIDCOT OX1 0QX
U.K.
TEL: 0235-21900
TLX: 83159
TLF:
EML:
COM:

STIER MARK T
PERKIN-ELMER CORPORATION
SPACE SCIENCE DIVISION
MS-897
DANBURY CT 06810
U.S.A.
TEL: 203-797-5708
TLX: 965954
TLF:
EML:
COM: 44

STIFT MARTIN JOHANNES DR
INSTITUT FUER ASTRONOMIE
TUERKENSCHANZSTR 17
A-1180 WIEN
AUSTRIA
TEL: 0222-345360/96
TLX:
TLF:
EML:
COM:

STINEBRING DANIEL R
DEPARTMENT OF PHYSICS
PRINCETON UNIVERSITY
PRINCETON NJ 08544
U.S.A.
TEL: 609-452-5578
TLX:
TLF:
EML:
COM:

STIX MICHAEL DR
KIEPENHEUER-INSTITUT
FUER SONNENPHYSIK
SCHOENECKSTR 6
D-7800 FREIBURG IM BR.
GERMANY. F.R.
TEL:
TLX: 7721552 KIS D
TLF:
EML:
COM: 10.12

STOBIE ROBERT S DR
ROYAL OBSERVATORY
BLACKFORD HILL
EDINBURGH EH9 3HJ
U.K.
TEL: 031-667-3321
TLX: 72383 ROEDIN G
TLF:
EML:
COM: 27

STOCK JURGEN D
CENTRO DE INVESTIGACION
DE ASTRONOMIA
APARTADO 264
MERIDA
VENEZUELA
TEL: 074-639930
TLX: 74174
TLF:
EML:
COM: 24.25.30.45

STOCKMAN HERVEY S JR DR
SPACE TELESCOPE SCI INST
HOMEWOOD CAMPUS
3700 SAN MARTIN DRIVE
BALTIMORE MD 21218
U.S.A.
TEL: 301-338-4820
TLX: 6849101 STSCI UW
TLF:
EML:
COM: 25.44.48

STOCKTON ALAN N DR
INSTITUTE FOR ASTRONOMY
2680 WOODLAWN DR
HONOLULU HI 96822
U.S.A.
TEL:
TLX:
TLF:
EML:
COM:

STOEGER WILLIAM R DR
SPECOLA VATICANA
00120 CITTA DEL VATICANO
VATICAN CITY STATE
TEL: 06-698-3411
TLX: 504-2020 VATOBS VA
TLF:
EML:
COM: 47

STOEV ALEXEI
PEOPLE'S ASTRONOMICAL
OBSERVATORY AND PLANETAR
YURI GAGARIN
STARA ZAGORA
BULGARIA
TEL: 0035904243183
TLX:
TLF:
EML:
COM: 16.41.46

STOHL JAN DR
ASTRONOMICAL INSTITUTE
SLOVAK ACAD. OF SCIENCES
842 28 BRATISLAVA
CZECHOSLOVAKIA
TEL: 427-375157
TLX: 093355
TLF:
EML:
COM: 15.22V

STOKER PIETER H
COSMIC RAY RESEARCH UNIT
POTCHEFSTROOM UNIVERSITY
POTCHEFSTROOM 2520
SOUTH AFRICA
TEL: 27-1481-25360
TLX: 421363
TLF:
EML:
COM: 10

STONE EDWARD C DR
CALTECH
407-47 DOWNS LAB.
PASADENA CA 91125
U.S.A.
TEL: 818-356-8321
TLX: 188192
TLF:
EML:
COM: 16

STONE R G DR
LAB EXTRATERRESTR PHYSICS
NASA/GSFC CODE 690
GREENBELT MD 20771
U.S.A.
TEL: 301-286-8631
TLX: 710-82089716
TLF:
EML:
COM: 40.44.49

STONE REMINGTON P S DR
LICK OBSERVATORY
MOUNT HAMILTON CA 95140
U.S.A.
TEL: 408-274-1809
TLX:
TLF:
EML:
COM: 25.28

STONE RONALD CECIL
BLACK BIRCH ASTROMETRIC
OBSERVATORY
P O BOX 770
BLENHEIM
NEW ZEALAND
TEL: 64-057-87164
TLX:
TLF:
EML:
COM: 08.24

STORCHI-BERGMAN THAISA DR
INSTITUTO DE FISICA
UNIV DO RIO GRANDE DO SUL
CAIXA POSTAL 15051
PORTO ALEGRE 91500
BRAZIL
TEL: (0512)364677
TLX: 515730 CCUF BR
TLF:
EML:
COM: 28

STOREY JOHN W V DR
DEPT OF PHYSICS
UNIV NEW SOUTH WALES
P.O.BOX 1
KENSINGTON NSW 2033
AUSTRALIA
TEL: 61-2-6974591
TLX: 26054 AA
TLF:
EML:
COM: 09

STOTKO ANNA
11 RUE ERNEST CRESSON
F-75014 PARIS
FRANCE
TEL: 1-45-39-56-35
TLX:
TLF:
EML:
COM: 19.31

STRAFELLA FRANCESCO
DIPARTIMENTO DI FISICA
UNIVERSITA DI LECCE
I-73100 LECCE
ITALY
TEL: 832-627-247
TLX: 860830 UNSTLE I
TLF:
EML:
COM:

STRAIZYS V PROF DR
ASTROPHYSICAL DEPARTMENT
INSTITUTE OF PHYSICS
POZELOS 54
232600 VILNIUS, LITHUANIA
U.S.S.R.
TEL: 73-12-27
TLX:
TLF:
EML:
COM: 25C.45.51

STRAND KAJ AA DR
3200 ROWLAND PL N.W.
WASHINGTON DC 20008
U.S.A.
TEL: 202-966-0495
TLX:
TLF:
EML:
COM: 24.26

STRASSL HANS L PROF
ASTRON INST UNIV MUENSTER
DOMAGKSTR 75
D-4400 MUENSTER
GERMANY, F.R.
TEL: 0251-86-24-63
TLX:
TLF:
EML:
COM:

STRAZZULLA GIOVANNI
OSSERVATORIO ASTROFISICO
CITTA UNIVERSITARIA
I-95125 CATANIA
ITALY
TEL: 95-330533
TLX: 970359 ASTRCT I
TLF:
EML:
COM:

STREL'NITSKIJ VLADIMIR DR
ASTRONOMICAL COUNCIL
USSR ACADEMY OF SCIENCES
PYATNITSKAYA UL 48
109017 MOSCOW
U.S.S.R.
TEL: 231-54-61
TLX: 412623 SCSTP SU
TLF:
EML:
COM: 14

STRITTMATTER PETER A PROF
STEWARD OBSERVATORY
TUCSON AZ 85721
U.S.A.
TEL: 602-621-6532
TLX:
TLF:
EML:
COM: 35

STROBEL ANDRZEJ DR
INSTITUTE OF ASTRONOMY
N COPERNICUS UNIVERSITY
UL. CHOPINA 12/18
87-100 TORUN
POLAND
TEL: 260-18
TLX: 0552234 ASTR PL
TLF:
EML:
COM: 31.45

STROBEL DARRELL F
DEPT OF EARTH & PLANETARY
JOHNS HOPKINS UNIVERSITY
BALTIMORE MD 21218
U.S.A.
TEL:
TLX:
TLF:
EML:
COM: 16

STROHMEIER WOLFGANG PROF
VOLKFELDSTR 5
D-8600 BAMBERG
GERMANY, F.R.
TEL: 0951-55394
TLX:
TLF:
EML:
COM: 25.27.42

STROM KAREN M
ASTRONOMY PROGRAM
UNIV OF MASSACHUSETTS
GRC 518 B6732
AMHERST MA 07003
U.S.A.
TEL: 413-545-2290
TLX:
TLF:
EML:
COM: 27

STROM RICHARD G DR
RADIOSTERRENWACHT
POSTBUS 2
NL-7990 AA DWINGELOO
NETHERLANDS
TEL: 05219-7244
TLX: 42043 SRZM NL
TLF:
EML:
COM: 28.34.40

STROM ROBERT G PROF
DEPT OF PLANETARY SCI
UNIVERSITY OF ARIZONA
TUCSON AZ 85721
U.S.A.
TEL: 602-621-2720
TLX: 9109521143
TLF:
EML:
COM: 16.23

STROM STEPHEN E
ASTRONOMY PROGRAM
UNIV OF MASSACHUSETTS
GRC 518 B
AMHERST MA 01003
U.S.A.
TEL: 413-545-2290
TLX:
TLF:
EML:
COM: 27.36

STRONG IAN B DR
LOS ALAMOS NATIONAL LAB
MS 436
P O BOX 1663
LOS ALAMOS NM 87545
U.S.A.
TEL: 505-667-4823
TLX:
TLF:
EML:
COM: 48

STRONG JOHN D PROF
ASTRON RESEARCH FACILITY
UNIVERSITY MASSACHUSETTS
AMHERST MA 01003
U.S.A.
TEL:
TLX:
TLF:
EML:
COM: 16

STRONG KEITH T DR
DEPT 91-30 BLDG 255
LOCKHEED P.A. RES LAB
3251 HANOVER STREET
PALO ALTO CA 94304
U.S.A.
TEL: 415 354 5136
TLX: 346409 LMSC
TLF: 415 424 3333
EML:
COM: 10

STRUBLE MITCHELL F
DEPT ASTRON & ASTROPHYS
UNIV OF PENNSYLVANIA
PHILADELPHIA PA 19104
U.S.A.
TEL: 215-243-8176
TLX:
TLF:
EML:
COM: 47

STRUCK-MARCELL CURTIS J
IOWA STATE UNIVERSITY
PHYSICS DEPARTMENT
AMES IA 50011
U.S.A.
TEL: 515-294-5440
TLX:
TLF:
EML:
COM:

STRUKOV IGOR A DR
SPACE RESEARCH INSTITUTE
USSR ACADEMY OF SCIENCES
117810 MOSCOW
U.S.S.R.
TEL: 333 14 66
TLX: 411498 STAR SU
TLF:
EML:
COM: 40.47

STRYKER LINDA L
LICK OBSERVATORY
UNIVERSITY OF CALIFORNIA
SANTA CRUZ CA 95064
U.S.A.
TEL: 408-429-2844
TLX:
TLF:
EML:
COM:

STUMPFF PETER PROF DR
MPI FUER RADIOASTRONOMIE
AUF DEM HUEGEL 69
D-5300 BONN
GERMANY. F.R.
TEL: 0228-525360
TLX: 886440 MPIFR D
TLF:
EML:
COM:

STURCH CONRAD R DR
COMPUTER SCIENCES CORP
SPACE TEL SCIENCE INST
3700 SAN MARTIN DRIVE
BALTIMORE MD 21218
U.S.A.
TEL: 301-338-4856
TLX:
TLF:
EML:
COM: 33

STURROCK PETER A PROF
STANFORD UNIVERSITY
CTR FOR SPACE SCIENCE &
ASTROPHYSICS
STANFORD CA 94305
U.S.A.
TEL: 415-723-1438
TLX: 3484 STANFRD STNU
TLF:
EML:
COM: 10.48.49.51

SU BUNEI
YUNNAN OBSERVATORY
P.O. BOX 110
KUNMING
CHINA. PEOPLE'S REP.
TEL: 72946
TLX: 64040 YUOBS CN
TLF:
EML:
COM: 34

SU DING-QIANG
NANJING ASTRONOMICAL
INSTRUMENT FACTORY
JIANGSU PROVINCE
CHINA. PEOPLE'S REP.
TEL: 41193
TLX:
TLF:
EML:
COM: 09

SU HONG-JUN
PURPLE MOUNTAIN
OBSERVATORY
NANJING
CHINA. PEOPLE'S REP.
TEL: 025-36967
TLX: 34144 PMOBT CN
TLF:
EML:
COM: 28

SU WAN-ZHEN
PURPLE MOUNTAIN
OBSERVATORY
NANJING
CHINA. PEOPLE'S REP.
TEL: 33583
TLX: 34144 PMONJ CN
TLF:
EML:
COM: 44

SUBRAHMANYA C R
TATA INST. OF FUNDAMENTAL
RESEARCH
P.O.BOX 1234
BANGALORE 560 012
INDIA
TEL:
TLX:
TLF:
EML:
COM: 47

SUBRAHMANYAM P V DR
DEPT OF ASTRONOMY
OSMANIA UNIVERSITY
HYDERABAD 500 007
INDIA
TEL: 71951 x 247
TLX:
TLF:
EML:
COM: 28

SUBRAMANIAN KANDASWAMY DR
ASTRONOMY CENTRE
PHYSICS BUILDING
UNIVERSITY OF SUSSEX
FALMER BRIGHTON BN1 9QH
U.K.
TEL:
TLX:
TLF:
EML:
COM:

SUDA JAN
ASTRONOMICAL INSTITUTE
CZECH. ACAD. OF SCIENCES
ONDREJOV OBSERVATORY
251 65 ONDREJOV
CZECHOSLOVAKIA
TEL: 724525
TLX: 121579
TLF:
EML:
COM: 10

SUDA KAZUO PROF
ASTRONOMICAL INSTITUTE
TOHOKU UNIVERSITY
ARAMAKI SENDAI 980
JAPAN
TEL: 0222-22-1800
TLX:
TLF:
EML:
COM: 35

SUEMOTO ZENZABURO PROF DR
TOKYO ASTRONOMICAL
OBSERVATORY
OSAWA MITAKA
TOKYO 181
JAPAN
TEL:
TLX:
TLF:
EML:
COM: 10.12

SUESS STEVEN T DR
SPACE SCIENCES LAB
CODE ES 52
NASA/MARSHALL SFC
HUNTSVILLE AL 35812
U.S.A.
TEL: 205-453-2824
TLX:
TLF:
EML:
COM: 49

SUGAWA CHIKARA DR
HANANOI 1586-25
KASHIWA-SHI
CHIBA-KEN 277
JAPAN
TEL: 0471-33-3825
TLX:
TLF:
EML:
COM: 19

SUGIMOTO DAIICHIRO PROF
DEPT EARTH SCI & ASTRON
COLL ARTS & SCIENCES
UNIV OF TOKYO. KOMABA
MEGURO-KU TOKYO 153
JAPAN
TEL: 03-467-1171
TLX: 2426728 TODAIK J
TLF:
EML:
COM: 35C.37.42

SUKUMAR SUNDARAJAN DR
UNIVERSITY OF ILLINOIS
349 ASTRONOMY BLDG
1011 WEST SPRINGFIELD AVE
URBANA IL 61801
U.S.A.
TEL: 217 244 1187
TLX: 9102409464 ASTRODEPT
TLF:
EML: sukumar@rigel.astro.uiuc.ed
COM: 40

SULENTIC JACK W DR
DEPT PHYSICS & ASTRONOMY
UNIVERSITY OF ALABAMA
P O BOX 1921
TUSCALOOSA AL 35487
U.S.A.
TEL: 205-348-5050
TLX: 810-729-5845
TLF:
EML:
COM: 28

SULLIVAN DENIS JOHN DR
PHYSICS DEPARTMENT
VICTORIA UNIVERSITY
PRIVATE BAG
WELLINGTON
NEW ZEALAND
TEL: 721000
TLX:
TLF:
EML:
COM: 25

SULLIVAN WOODRUFF T III
DEPT. ASTRONOMY F-20
UNIVERSITY OF WASHINGTON
SEATTLE WA 98195E
U.S.A.
TEL:
TLX:
TLF:
EML:
COM: 28.40.41.51

SULTANOV G F ACAD
SHEMAKA ASTROPHYSICAL OBS
373243 AZERBAIDZAN
U.S.S.R.
TEL:
TLX:
TLF:
EML:
COM: 07.20

SUMMERS HUGH P DR
JET JOINT UNDERTAKING
CULHAM LABORATORY
ABINGDON OX14 3EA
U.K.
TEL: 0235-28822
TLX: 837505 JETEUR G
TLF:
EML:
COM: 14

SUN JIN
DEPT OF ASTRONOMY
BEIJING NORMAL UNIVERSITY
BEIJING
CHINA. PEOPLE'S REP.
TEL: 65-6531. 6285
TLX:
TLF:
EML:
COM: 34

SUN KAI
ASTROPHYSICS DIVISION
GEOPHYSICS DEPARTMENT
BEIJING UNIVERSITY
BEIJING
CHINA. PEOPLE'S REP.
TEL:
TLX: 22239 PKUNI CN
TLF:
EML:
COM: 10

SUN YI-SUI
DEPT OF ASTRONOMY
NANJING UNIVERSITY
NANJING
CHINA. PEOPLE'S REP.
TEL: 37551
TLX: 34151 PRCNU CN
TLF:
EML:
COM: 07

SUN TONGXIANG
INSTITUTE OF GEODESY &
GEOPHYSICS
XU DONG LU
WUHAN
CHINA. PEOPLE'S REP.
TEL:
TLX:
TLF:
EML:
COM: 19

SUNDMAN ANITA DR
STOCKHOLM OBSERVATORY
S-133 00 SALTSJOEBADEN
SWEDEN
TEL: 08-717-06-34
TLX: 12972 SWEDEN
TLF:
EML:
COM: 41.42

SUNTZEFF NICHOLAS B
NATL OPTICAL ASTR OBS
CTIAO
CASILLA 603
LA SERENA
CHILE
TEL: 565-121-3352
TLX: 620301 AURA CT
TLF:
EML:
COM: 29

SUNYAEV RASHID A DR
SPACE RESEARCH INSTITUTE
USSR ACADEMY OF SCIENCES
117810 MOSCOW
U.S.S.R.
TEL:
TLX:
TLF:
EML:
COM: 44C.47.48P

SURDEJ JEAN M G
INSTITUT D'ASTROPHYSIQUE
UNIVERSITE DE LIEGE
AVENUE DE COINTE 5
B-4200 COINTE-OUGREE
BELGIUM
TEL: 32-41-529980
TLX: 41264 ASTRLG B
TLF:
EML:
COM: 15.47

SUTANTYO WINARDI
BOSSCHA OBSERVATORY
LEMBANG. JAVA
INDONESIA
TEL: 6001 LEMBANG
TLX:
TLF:
EML:
COM:

SUTHERLAND PETER G DR
PHYSICS DEPT
MCMASTER UNIVERSITY
HAMILTON ONT L8S 4M1
CANADA
TEL: 416-525-9140
TLX:
TLF:
EML:
COM:

SUZUKI YOSHIMASA PROF
23-1 NAKAJIMA
HIRONOMACHI
UJI SHI 611
JAPAN
TEL:
TLX:
TLF:
EML:
COM: 34

SVALGAARD LEIF DR
BERTOGENLAAN 31
B-3202 LUBBEEK
BELGIUM
TEL:
TLX:
TLF:
EML:
COM:

SVECHNIKOVA MARIA A DR
ASTRONOMICAL DEPT OF
URALSKIJ STATE UNIV
620083 SVERDLOVSK
U.S.S.R.
TEL:
TLX:
TLF:
EML:
COM: 42

SVENSSON ROLAND
NORDITA
BLEGDAMSVEJ 17
DK-2100 COPENHAGEN
DENMARK
TEL: 00945-1-421616
TLX: 15216 NBI DK
TLF:
EML:
COM: 48

SVESTKA JIRI DR
OBSERVATORY AND
PLANETARIUM OF PRAGUE
PETRIN 205
118 46 PRAGUE 1
CZECHOSLOVAKIA
TEL:
TLX:
TLF:
EML:
COM: 27.46

SVESTKA ZDENEK DR
SPACE RESEARCH LABORATORY
BENELUXLAAN 21
NL-3527 HS UTRECHT
NETHERLANDS
TEL: 030-937145
TLX: 47224 ASTRO NL
TLF:
EML:
COM: 10.12

SVOLOPOULOS SOTIRIOS PROF
DEPT OF ASTROPHYSICS
UNIVERSITY OF ATHENS
PANEPISTIMIOPOLIS
GR-15771 ATHENS
GREECE
TEL:
TLX:
TLF:
EML:
COM: 29.33.41

SVOREN JAN
ASTRONOMICAL INSTITUTE
SLOVAK ACAD. OF SCIENCES
059 60 TATRANSKA LOMNICA
CZECHOSLOVAKIA
TEL: 42-969-967866
TLX: 78277 AU SAV CS
TLF:
EML:
COM: 15.20

SWANENBURG B N DR
SPACE RESEARCH LABORATORY
P.O. BOX 9504
WASSENAARSEWEG 78
NL-2300 RA LEIDEN
NETHERLANDS
TEL: 071-272727
TLX: 39058 ASTRO NL
TLF:
EML:
COM:

SWANK JEAN HEBB
NASA/GSFC
CODE 661
GREENBELT MD 20771
U.S.A.
TEL: 301-286-6188
TLX: 89675 NASCOM GBLT
TLF:
EML:
COM: 48

SWARUP GOVIND PROF
RADIO ASTRONOMY CENTER
TIFR - POST BOX 1234
INDIAN INST. OF SC CAMPUS
BANGALORE 560 012
INDIA
TEL: 363138. 362816
TLX: 8458438 TIFR IN
TLF:
EML:
COM: 38C.40

SWEET PETER A PROF
DEPT OF ASTRONOMY
THE UNIVERSITY
GLASGOW G12 8QW
U.K.
TEL: 041-339-8855
TLX:
TLF:
EML:
COM: 35

SWEIGART ALLEN V DR
NASA/GSFC
CODE 681
GREENBELT MD 20771
U.S.A.
TEL: 301-286-6274
TLX: 89675
TLF:
EML:
COM: 35

SWENSON GEORGE W JR PROF
ELECT & COMPUTER ENG DEPT
UNIVERSITY OF ILLINOIS
1406 WEST GREEN ST
URBANA IL 61801
U.S.A.
TEL: 217-333-4498
TLX:
TLF:
EML:
COM: 40

SWENSSON JOHN W DR
DEPT OF THEORETICAL PHYS
SOELVEGATAN 14 A
S-223 62 LUND
SWEDEN
TEL: 40-10969686
TLX:
TLF:
EML:
COM: 12.29

SWERDLOW NOEL PROF
UNIVERSITY OF CHICAGO
5640 S. ELLIS AVENUE
CHICAGO IL 60637
U.S.A.
TEL: 312-962-7969
TLX:
TLF:
EML:
COM: 41

SWIHART THOMAS L DR
STEWARD OBSERVATORY
UNIVERSITY OF ARIZONA
TUCSON AZ 85721
U.S.A.
TEL: 602-621-6525
TLX:
TLF:
EML:
COM: 36

SWINGS JEAN-PIERRE DR
INSTITUT D'ASTROPHYSIQUE
UNIVERSITE DE LIEGE
AVENUE DE COINTE 5
B-4200 COINTE-OUGREE
BELGIUM
TEL: 41-52-9980
TLX: 41264 ASTRLG B
TLF:
EML:
COM: 09.14.29

SYKES-HART AVRIL B DR
DEPT OF ASTROPHYSICS
SOUTH PARKS ROAD
OXFORD OX1 3RQ
U.K.
TEL:
TLX:
TLF:
EML:
COM:

SYKORA JULIUS DR
ASTRONOMICAL INSTITUTE
SLOVAK ACAD. OF SCIENCES
SKALNATE PLESO OBS
0969 60 TATRANSKA LOMNICA
CZECHOSLOVAKIA
TEL: 0969-967866
TLX: 78277 AUSAV CZ
TLF:
EML:
COM: 10

SYLWESTER BARBARA DR
SPACE RESEARCH CENTER
POLISH ACADEMY SCIENCES
UL. KOPERNIKA 11
51622 WROCLAW
POLAND
TEL: 483738
TLX: 0712791 UWRPL
TLF:
EML:
COM: 10

SYLWESTER JANUSZ
SPACE RESEARCH CENTER
POLISH ACAD OF SCIENCES
UL. KOPERNIKA 11
51-622 WROCLAW
POLAND
TEL: 48 18 01
TLX:
TLF:
EML:
COM: 10

SYNNOTT STEPHEN P
JET PROPULSION LABORATORY
MS 264-686
4800 OAK GROVE DRIVE
PASADENA CA 91109
U.S.A.
TEL: 818-354-5933
TLX:
TLF:
EML:
COM: 16.20

SZABADOS LASZLO PH D
KONKOLY OBSERVATORY
BOX 67
H-1525 BUDAPEST
HUNGARY
TEL: 1-166-426
TLX: 227460 KONOBH
TLF:
EML:
COM: 25.27

SZAFRANIEC ROZALIA DR
UL. KOPERNIKA 27
31-501 KRAKOW
POLAND
TEL:
TLX:
TLF:
EML:
COM: 42

SZALAY ALEX DR
DEPT ATOMIC PHYSICS
EOTVOS UNIVERSITY
PUSKIN U.5-7
1088 BUDAPEST
HUNGARY
TEL:
TLX:
TLF:
EML:
COM: 47

SZEBEHELY VICTOR G PROF
DEPT AEROSPACE ENGINEERG
UNIVERSITY OF TEXAS
WRW 414
AUSTIN TX 78712
U.S.A.
TEL: 512-471-4239
TLX: 9108741305
TLF:
EML:
COM: 07.33

SZECSENYI-NAGY GABOR DR
DEPT OF ASTRONOMY
LORAND EOTVOS UNIVERSITY
KUN BELA TER 2
H-1083 BUDAPEST
HUNGARY
TEL: 1141019
TLX:
TLF:
EML:
COM: 27.37.46

SZEGO KAROLY DR
CENTRAL RESEARCH INST
FOR PHYSICS
PO BOX 49
H 1525 BUDAPEST
HUNGARY
TEL: 36 1 551 682
TLX: 224722
TLF: 36 [illegible] 696567
EML:
COM: [illegible]5

SZEIDL BELA DR
KONKOLY OBSERVATORY
BOX 67
H-1525 BUDAPEST
HUNGARY
TEL: 1-366-621
TLX: 227460 KONOB
TLF:
EML:
COM: 27C

SZKODY PAULA DR
DEPT OF ASTRONOMY
UNIVERSITY OF WASHINGTON
SEATTLE WA 98195
U.S.A.
TEL: 206-543-1988
TLX:
TLF:
EML:
COM: 25.27.42

TAAM RONALD EVERETT DR
DEPT PHYSICS & ASTRONOMY
NORTHWESTERN UNIVERSITY
EVANSTON IL 60208
U.S.A.
TEL: 312 491 7528
TLX:
TLF:
EML:
COM: 35.42

TABARA HIROTO DR
FACULTY OF EDUCATION
UTSUNOMIYA UNIVERSITY
MINEMACHI
UTSUNOMIYA 321
JAPAN
TEL: 0286-36-1515
TLX:
TLF:
EML:
COM: 40

TABORDA JOSE ROSA DR
FACULTY OF SCIENCES
ASTRONOMICAL OBSERVATORY
R ESCOLA POLITECNICA 58
1200 LISBOA
PORTUGAL
TEL:
TLX:
TLF:
EML:
COM: 07.46

TADEMARU EUGENE DR
UNIV OF MASSACHUSETTS
AMHERST MA 01002
U.S.A.
TEL:
TLX:
TLF:
EML:
COM:

TAFFARA SALVATORE PROF
VIA CALZA 5BIS.
I-35128 PADOVA
ITALY
TEL: 049-8071-624
TLX:
TLF:
EML:
COM: 29

TAGLIAFERRI GIUSEPPE PROF
OSSERVATORIO ASTROFISICO
DI ARCETRI
LARGO E. FERMI 5
I-50125 FIRENZE
ITALY
TEL:
TLX:
TLF:
EML:
COM:

TAKADA-HIDAI MASAHIDE DR
RES INST OF CIVILIZATION
TOKAI UNIVERSITY
1117 KITAKANAME
KANAGAWA 259-12
JAPAN
TEL: 0463-58-1211
TLX: 2423402 UNITOK J
TLF:
EML:
COM: 29.51

TAKAGI KOJIRO PROF
DEPT OF PHYSICS
TOYAMA UNIVERSITY
3190 GOFUKU
TOYAMA 930
JAPAN
TEL: 0764-334716
TLX:
TLF:
EML:
COM: 40

TAKAGI SHIGETSUGU DR
DEPT. DE FISICA-CCE-UFRN
CAMPUS UNIVERSITARIO
59000 NATAL-RN
BRAZIL
TEL:
TLX:
TLF:
EML:
COM: 19

TAKAHARA FUMIO DR
DEPT. OF PHYSICS
TOKYO METROPOLITAN UNIVER
FUKAZAWA 2-1-1.SETAGAYA
TOKYO 158
JAPAN
TEL:
TLX:
TLF:
EML:
COM: 48

TAKAHARA MARIKO
DEPT OF ASTRONOMY
UNIVERSITY OF TOKYO
2-11-16 YAYOI. BUNKYO-KU
TOKYO 113
JAPAN
TEL: 03-812-2111
TLX: 33659 UTYOSCI
TLF:
EML:
COM: 35

TAKAKUBO KEIYA PROF
ASTRONOMICAL INSTITUTE
TOHOKU UNIVERSITY
ARAMAKI AZA AOBA
SENDAI 980
JAPAN
TEL: 222-22-1800
TLX: 852246 THUCOM J
TLF:
EML:
COM: 34.40

TAKAKURA TATSUO PROF EMER
DEPT OF ASTRONOMY
UNIVERSITY OF TOKYO
BUNKYO-KU
TOKYO 113
JAPAN
TEL: 03-812-2111
TLX: 33659 UTYOSCI J
TLF:
EML:
COM: 10.40.44

TAKANO TOSHIAKI DR
TOYOKAWA OBSERVATORY
NAT. ASTRON. OBSERVATORY
HONOHARA 3-13
TOYOKAWA 442
JAPAN
TEL: 5338 4 5711
TLX: 4322310 TYKW J
TLF:
EML:
COM: 40

TAKARADA KATSUO DR
KYOTO INST OF TECHNOLOGY
MATSUGASAKI SAKYOKU
KYOTO 606
JAPAN
TEL: 075-791-3211
TLX:
TLF:
EML:
COM:

TAKASE BUNSHIRO PROF
TOKYO ASTRONOMICAL
OBSERVATORY
OSAWA. MITAKA
TOKYO 181
JAPAN
TEL: 0422-32-5111
TLX: 2822307 TAOMTK
TLF:
EML:
COM: 28

TAKAYANAGI KAZUO PROF
INSTITUTE OF SPACE AND
ASTRONAUTICAL SCIENCES
1-1. YOSHINODAI 3-CHOME
SAGAMIHARA.KANAGAWA-KEN
JAPAN
TEL: 0427 51 3911
TLX: 24550 SPACETKY J
TLF:
EML:
COM: 14

TAKEDA HIDENORI DR
DEPT OF AERONAUT. ENGINRG
KYOTO UNIVERSITY
SAKYOKU
KYOTO 606
JAPAN
TEL:
TLX:
TLF:
EML:
COM: 15

TAKEDA YOICHI DR
INST THEOR ASTROPHYSICS
UNIVERSITAT HEIDELBERG
IM NEUENHEIMER FELD 561
D 6900 HEIDELBERG 1
GERMANY. F.R.
TEL: 06221 562837
TLX: 461515 UNIHD I
TLF:
EML:
COM: 36

TAKENOUCHI TADAO DR
1-28-30 KICHIJYOJI
KITA-MACHI MUSASHINO
TOKYO 180
JAPAN
TEL:
TLX:
TLF:
EML:
COM:

TAKENS ROELF JAN DR
ASTRONOMICAL INSTITUTE
ROETERSSTRAAT 15
NL-1018 WB AMSTERDAM
NETHERLANDS
TEL: 020-5223009
TLX: 16460 FAC NL
TLF:
EML:
COM:

TAKEUTI MINE DR
ASTRONOMICAL INSTITUTE
TOHOKU UNIVERSITY
ARAMAKI AZA AOBA
SENDAI 980
JAPAN
TEL: 222-22-1800
TLX: 852246 THUCOM J
TLF:
EML:
COM: 27

TALAVERA A DR
IUE OBSERVATORY
IUE VILSPA
VILLA FRANCA APDO 54065
E-28080 MADRID
SPAIN
TEL: 1 401 9661
TLX: 42555 VILSE
TLF:
EML:
COM: 29

TALBOT RAYMOND J JR DR
THE AEROSPACE CORPORATION
1927 CURTIS AVENUE
REDONDO BEACH CA 90278
U.S.A.
TEL: 213-379-9927
TLX:
TLF:
EML:
COM: 28

TALON RAOUL DR
CESR
9 AVENUE DU COLONEL ROCHE
BP 4346
F-31029 TOULOUSE CEDEX
FRANCE
TEL:
TLX:
TLF:
EML:
COM: 10

TALWAR SATYA P DR
PHYSICS DEPT
DELHI UNIVERSITY
DELHI 110 007
INDIA
TEL:
TLX:
TLF:
EML:
COM:

TAKENAGA TATSUO DR
FACULTY OF EDUCATION
MIE UNIVERSITY
TSU-SHI
MIE 514
JAPAN
TEL:
TLX:
TLF:
EML:
COM: 10

TAMMANN G ANDREAS PROF DR
ASTRONOMISCHES INSTITUT
UNIVERSITAET BASEL
VENUSSTRASSE 7
CH-4102 BINNINGEN
SWITZERLAND
TEL: 061-227711
TLX:
TLF:
EML:
COM: 27.28P.33.47

TAMURA SHINICHI DR
DEPT OF ASTRONOMY
TOHOKU UNIVERSITY
ARAMAKI
SENDAI 980
JAPAN
TEL: 222-22-1800
TLX: 852246 THUCOM J
TLF:
EML:
COM: 34

TAN HUISONG
YUNNAN OBSERVATORY
P.O. BOX 110
KUNMING
CHINA. PEOPLE'S REP.
TEL: 72946
TLX: 64040 YUOBS CN
TLF:
EML:
COM: 47

TANABE HIROYOSHI DR
TOKYO ASTRONOMICAL
OBSERVATORY
OSAWA. MITAKA
TOKYO 181
JAPAN
TEL: 0422-32-5111
TLX: 02822307 TAOMTK J
TLF:
EML:
COM: 15.21

TANABE KENJI DR
OKAYAMA UNIV. OF SCIENCE
1-1 RIDAI-CHO
OKAYAMA 700
JAPAN
TEL: 0862 52 3161
TLX:
TLF: 0862 55 3847
EML:
COM: 47

TANABE TOSHIHIKO DR
TOKYO ASTRONOMICAL OBS
MITAKA
TOKYO 181
JAPAN
TEL: 422 32 5111
TLX:
TLF:
EML:
COM:

TANAKA KATSUO DR
TOKYO ASTRONOMICAL
OBSERVATORY
OSAWA. MITAKA
TOKYO 181
JAPAN
TEL: 0422-32-5111
TLX: 2822307 TAOMTK J
TLF:
EML:
COM: 10.12

TANAKA RIICHIRO PROF
FACULTY GENERAL EDUCATION
NIIGATA UNIVERSITY
ASAHIMACHIDORI
NIIGATA 951
JAPAN
TEL:
TLX:
TLF:
EML:
COM: 40

TANAKA WATARU DR
DEPT OF ASTRONOMY
UNIVERSITY OF TOKYO
BUNKYO-KU
TOKYO 113
JAPAN
TEL: 03-812-2111
TLX:
TLF:
EML:
COM:

TANAKA YASUO DR
FACULTY OF EDUCATION
IBARAKI UNIVERSITY
BUNKYO
MITO 310
JAPAN
TEL: 292-26-1621x372
TLX:
TLF:
EML:
COM: 48

TANAKA YASUO PROF
INST SPACE & ASTRONAUT
SCIENCES
4-6-1 KOMABA.MEGURO-KU
TOKYO 153
JAPAN
TEL: 03-467-1111
TLX: J24550 SPACE TKY
TLF:
EML:
COM: 44C

TANAKA YUTAKA D DR
KOBE YAMATE WOMEN'S
JUNIOR COLLEGE
SUWAYAMA CHUO-KU
KOBE 650
JAPAN
TEL: 78 341 6060
TLX:
TLF:
EML:
COM: 28

TANDBERG-HANSSEN EINAR A
NASA/MSFC
ES01
HUNTSVILLE AL 35812
U.S.A.
TEL: 205-544-7578
TLX:
TLF:
EML:
COM: 10C.12

TANDON JAGDISH NARAIN DR
DEPT OF PHYSICS AND
ASTROPHYSICS
UNIVERSITY OF DELHI
DELHI 110 007
INDIA
TEL: 25 215 21
TLX:
TLF:
EML:
COM: 10

TANDON S N PROF
TATA INSTITUTE OF
FUNDAMENTAL RESEARCH
HOMI BHABHA RD
BOMBAY 400 005
INDIA
TEL: 219111 x 339
TLX: 0113009 TIFR IN
TLF:
EML:
COM: 25

TANG YU-HUA
DEPT OF ASTRONOMY
NANJING UNIVERSITY
NANJING
CHINA. PEOPLE'S REP.
TEL: 37651
TLX: 0909
TLF:
EML:
COM: 10

TANGO WILLIAM J. DR
SCHOOL OF PHYSICS
UNIVERSITY OF SYDNEY
SYDNEY NSW 2006
AUSTRALIA
TEL: 02-692-3953
TLX: 26169 UNISYD AA
TLF:
EML:
COM: 09C

TANIGUCHI YOSHIAKI DR
KISO OBSERVATORY
INSTITUTE OF ASTRONOMY
THE UNIVERSITY OF TOKYO
NAGANO 397-01
JAPAN
TEL: 264 52 3360
TLX: 334 7577 KSOOBS J
TLF:
EML:
COM: 28

TANZELLA-NITTI GIUSEPPE
OSSERVATORIO ASTRONOMICO
DI TORINO
I-10025 PINO TORINESE
ITALY
TEL: 011-841067
TLX: 213236 TO ASTR I
TLF:
EML:
COM:

TANZI ENRICO G
IST. DI FISICA COSMICA
CNR
VIA BASSANI 15
I-20133 MILANO
ITALY
TEL:
TLX:
TLF:
EML:
COM:

TAPIA MAURICIO DR
INSTITUTO DE ASTRONOMIA
UNAM
APDO POSTAL 877
22860 ENSENADA. B. CALIF.
MEXICO
TEL: 4-08-80/8-30-93
TLX: 56739 CICEME
TLF:
EML:
COM:

TAPIA-PEREZ SANTIAGO
STEWARD OBSERVATORY
UNIVERSITY OF ARIZONA
TUCSON AZ 85721
U.S.A.
TEL: 602-621-2876
TLX:
TLF:
EML:
COM: 25

TAPLEY BYRON D DR
DEPT AEROSPACE ENGR
AND ENGR MECHANICS
UNIV OF TEXAS. WRW 402
AUSTIN TX 78712
U.S.A.
TEL: 512-471-1356
TLX:
TLF:
EML:
COM: 19

TAPPING KENNETH F
HERZBERG INST ASTROPHYS
NATL RES COUNCIL CANADA
100 SUSSEX DRIVE
OTTAWA ONTARIO K1A OR6
CANADA
TEL: 613 991 5842
TLX: 053 3715
TLF:
EML:
COM: 10

TARADY VLADIMIR K DR
MAIN ASTRON OBSERVATORY
UKRAINIAN ACADEMY OF SCI
GOLOSEEVO
252127 KIEV
U.S.S.R.
TEL: 662286
TLX: 131406 SKY SU
TLF:
EML:
COM: 19

TARAFDAR SHANKAR P DR
TATA INSTITUTE OF
FUNDAMENTAL RESEARCH
HOMI BHABHA RD
BOMBAY 400 005
INDIA
TEL: 219111
TLX: 011-3009 TIFR IN
TLF:
EML:
COM: 34.36

TARENGHI MASSIMO DR
ESO
KARL-SCHWARZSCHILD-STR 2
D-8046 GARCHING B MUNCHEN
GERMANY. F.R.
TEL: 089-32006236
TLX: 52828222 EO D
TLF:
EML:
COM:

TARNSTROM GUY DR
MIT LINCOLN LABORATORY
PO BOX 73
LEXINGTON MA 02173
U.S.A.
TEL: 617-863-5500
TLX: 923355
TLF:
EML:
COM:

TARRAB IRENE
OBSERVATOIRE DE PARIS
61 AV. DE L'OBSERVATOIRE
F-75014 PARIS
FRANCE
TEL: 1-40-51-22-37
TLX:
TLF:
EML:
COM:

TARTER C BRUCE DR
LAWRENCE LIVERMORE LAB
L-295
UNIVERSITY OF CALIFORNIA
LIVERMORE CA 94550
U.S.A.
TEL: 415-422-4169
TLX:
TLF:
EML:
COM:

TARTER JILL C DR
STANFORD UNIVERSITY
ASTRONOMY PROGRAM
STANFORD CA 94305
U.S.A.
TEL:
TLX:
TLF:
EML:
COM: 40.47.51

TASSOUL JEAN-LOUIS PROF
DEPT DE PHYSIQUE
UNIVERSITE DE MONTREAL
CP 6128
MONTREAL PQ H3C 3J7
CANADA
TEL: 514-343-7274
TLX:
TLF:
EML:
COM:

TASSOUL MONIQUE DR
C/O DEPT DE PHYSIQUE
UNIVERSITE DE MONTREAL
C P 6128
MONTREAL P Q H3C 3J7
CANADA
TEL: 514-343-7274
TLX:
TLF:
EML:
COM: 35

TATEVYAN S K DR
ASTRONOMICAL COUNCIL
USSR ACADEMY OF SCIENCES
PYATNITSKAYA UL 48
109017 MOSCOW
U.S.S.R.
TEL: 231-54-61
TLX: 412623 SCSTP SU
TLF:
EML:
COM: 07

TATON RENE PROF
CENTRE ALEXANDRE KOYRE
12 RUE COLBERT
F-75002 PARIS
FRANCE
TEL: 1-42-97-52-85
TLX:
TLF:
EML:
COM: 41

TATUM JEREMY B DR
CLIMENHOGA OBSERVATORY
UNIVERSITY OF VICTORIA
VICTORIA BC V8W 2Y2
CANADA
TEL:
TLX:
TLF:
EML:
COM: 14.15

TAUBER GERALD E PROF
DEPT OF PHYSICS
TEL-AVIV UNIVERSITY
TEL-AVIV 69978
ISRAEL
TEL: 3-420692
TLX: 342171 VERSY IL
TLF:
EML:
COM: 47

TAVAKOL REZA
SCHOOL OF MATHEMAT. SCI.
QUEEN MARY COLLEGE
MILE END ROAD
LONDON E1 4NS
U.K.
TEL:
TLX:
TLF:
EML:
COM: 51

TAVARES J T L DR
AV DIAS DA SILVA
173 R/C ESQ
3000 COIMBRA
PORTUGAL
TEL:
TLX:
TLF:
EML:
COM:

TAWADROS NAHET JACOUB DR
HELWAN OBSERVATORY
HELWAN-CAIRO
EGYPT
TEL: 780 645
TLX: 93070 HIAG
TLF:
EML:
COM: 07

TAWARA YUZURU DR
DEPT OF ASTROPHYSICS
NAGOYA UNIVERSITY
FURO-CHO. CHIKUSA KU
NAGOYA 464
JAPAN
TEL: 52 781 5111
TLX: 4477323 SCUNGY J
TLF: 52 781 3541
EML:
COM:

TAYLER ROGER J PROF
ASTRONOMY CENTRE
UNIVERSITY OF SUSSEX
BRIGHTON BN1 9QH
U.K.
TEL: 273-606755
TLX: 877159 UNISEX G
TLF:
EML:
COM: 35.47.48

TAYLOR A R DR
DEPT OF PHYSICS
UNIVERSITY OF CALGARY
2500 UNIVERSITY DR N W
CALGARY ALBERTA T2N 1N4
CANADA
TEL: 403 220 5385
TLX: 821545
TLF: 403 289 3331
EML: BITNET:ARTAYLOR@UNCAMULT
COM: 40

TAYLOR DONALD BOGGIA DR
ROYAL GREENWICH OBS
HERSTMONCEUX CASTLE
HAILSHAM BN27 1RP
U.K.
TEL: 0323-833272
TLX: 87451 RGOBSY G
TLF:
EML:
COM: 07.23

TAYLOR DONALD J DR
DEPT PHYSICS & ASTRONOMY
UNIVERSITY OF NEBRASKA
LINCOLN NB 68588
U.S.A.
TEL: 402-472-3686
TLX:
TLF:
EML:
COM:

TAYLOR JOSEPH H PROF
PRINCETON UNIVERSITY
PHYSICS DEPARTMENT
PRINCETON NJ 08544
U.S.A.
TEL: 609-452-4368
TLX: 4993512
TLF:
EML:
COM:

TAYLOR KEITH DR
ROYAL GREENWICH OBS
HERSTMONCEUX CASTLE
HAILSHAM BN27 1RP
U.K.
TEL: 0323-833171
TLX: 87451
TLF:
EML:
COM:

TAYLOR KENNETH N R PROF
105A COPELAND ROAD
BEECROFT NSW 2119
AUSTRALIA
TEL:
TLX:
TLF:
EML:
COM: 34

TEDESCO EDWARD F
JET PROPULSION LAB
MS 183-501
4800 OAK GROVE DRIVE
PASADENA CA 91109
U.S.A.
TEL: 818-354-4739
TLX:
TLF:
EML:
COM: 15C.22.51

TEERIKORPI VELI PEKKA DR
TURKU UNIVERSITY OBS
TUORLA
SF-21500 PIIKKIO
FINLAND
TEL: 358-21-435 822
TLX: 62638 TYF SF
TLF: 358-21-433 767
EML:
COM:

TEHRANY D
65 AVENUE REY
TEHRAN
IRAN
TEL:
TLX:
TLF:
EML:
COM:

TEJFEL VIKTOR G DR
LAB OF LUNAR & PLANETARY
PHYSICS
ASTROPHYSICAL INSTITUTE
480068 ALMA-ATA
U.S.S.R.
TEL: 68-30-53
TLX:
TLF:
EML:
COM: 16C.51

TEKTUNALI H GOKMEN DR
UNIVERSITY OBSERVATORY
UNIVERSITY OF ISTANBUL
ISTANBUL
TURKEY
TEL: 9015223597
TLX:
TLF:
EML:
COM:

TELESCO CHARLES M DR
NASA MARSHALL SPACE
FLIGHT CENTER CODE ES63
SPACE SCIENCE LABORATORY
HUNTSVILLE AL 35812
U.S.A.
TEL: 205 544 7723
TLX: 594416
TLF:
EML:
COM: 28

TEMPESTI PIERO PROF
ISTITUTO DI ASTRONOMIA
DELL'UNIVERSITA
VIA G.M. LANCISI 29
I-00161 ROMA
ITALY
TEL: 06-8447977
TLX:
TLF:
EML:
COM: 27

TENORIO-TAGLE G DR
MPI F PHYSIK & ASTROPHYS
KARL-SCHWARZSCHILD-STR 1
D-8046 GARCHING B MUNCHEN
GERMANY. F.R.
TEL: 089-32990
TLX: 524629 ASTRO D
TLF:
EML:
COM: 34

TEPLITSKAYA R B DR
SIBIZMIR
P B 4
664697 IRKUTSK 33
U.S.S.R.
TEL: 6-23-65
TLX:
TLF:
EML:
COM: 12

TER HAAR DIRK
P.O. BOX 10
349 MIDDLE STREET
PETWORTH GU28 ORY
U.K.
TEL:
TLX:
TLF:
EML:
COM:

TERASHITA YOICHI PROF
KANAZAWA INST TECHNOLOGY
NONOICHI CHO
MINAMI KYOKU
KANAZAWA 921
JAPAN
TEL:
TLX:
TLF:
EML:
COM: 05

TEREBIZH VALERY YU DR
CRIMEAN STATION OF
STERNBERG ASTRON INST
NAUCHNY
334413 CRIMEA
U.S.S.R.
TEL: SIMPHEROPOL 367
TLX:
TLF:
EML:
COM:

TERENTJEVA ALEXANDRA K DR
ASTRONOMICAL COUNCIL
USSR ACADEMY OF SCIENCES
PYATNITSKAYA 48
109017 MOSCOW
U.S.S.R.
TEL: 231-54-61
TLX: 412623 SCSTP SU
TLF:
EML:
COM: 15.22

TERLEVICH ELENA DR
ROYAL GREENWICH OBSERV.
HERSTMONCEUX CASTLE
HAILSHAM BN27 1RP
U.K.
TEL: 323-833171
TLX:
TLF:
EML: ET@STARLINK.RGO.GREENWICH.AC.UK
COM:

TERLEVICH ROBERTO JUAN
ROYAL GREENWICH OBS
HERSTMONCEUX CASTLE
HAILSHAM BN27 1RP
U.K.
TEL:
TLX:
TLF:
EML:
COM: 28

TERRELL NELSON JAMES JR
ESS-9 MS/D436
LOS ALAMOS NATL LAB
BOX 1663
LOS ALAMOS NM 87545
U.S.A.
TEL: 505-667-2044
TLX:
TLF:
EML:
COM: 46

TERRIEN JEAN
103, RUE DE VERSAILLES
F-92410 VILLE D'AVRAY
FRANCE
TEL: 1-47-09-10-34
TLX:
TLF:
EML:
COM:

TERRILE RICHARD JOHN
JET PROPULSION LABORATORY
MS 183-30
PASADENA CA 91109
U.S.A.
TEL:
TLX:
TLF:
EML:
COM: 16

TERZAN AGOP DR
OBSERVATOIRE DE LYON
F-69230 ST GENIS-LAVAL
FRANCE
TEL: 78-56-07-05
TLX: 310926
TLF:
EML:
COM: 27.37

TERZIAN YERVANT PROF
CORNELL UNIVERSITY
SPACE SCIENCES BLDG
ITHACA NY 14853
U.S.A.
TEL: 607-256-4935
TLX: 932454
TLF:
EML:
COM: 28.34.40.51

TERZIDES CHARALAMBOS DR
DEPT. OF ASTRONOMY
UNIVERSITY THESSALONIKI
GR-54006 THESSALONIKI
GREECE
TEL:
TLX:
TLF:
EML:
COM: 33

TESKE RICHARD G PROF
DEPT OF ASTRONOMY
UNIVERSITY OF MICHIGAN
ANN ARBOR MI 48109
U.S.A.
TEL: 313-764-3398
TLX:
TLF:
EML:
COM: 10

TEXEREAU JEAN M
CERGA
AVENUE COPERNIC
F-06130 GRASSE
FRANCE
TEL: 93-36-58-49
TLX:
TLF:
EML:
COM:

THADDEUS PATRICK PROF
CENTER FOR ASTROPHYSICS
60 GARDEN STREET
CAMBRIDGE MA 02138
U.S.A.
TEL: 617-495-7340
TLX: 921428 SATELLITE CAM
TLF:
EML:
COM: 34.51

THAKUR RATNA KUMAR DR
DEPARTMENT OF PHYSICS
RAVISHANKAR UNIVERSITY
RAIPUR 492010
INDIA
TEL: 27064
TLX:
TLF:
EML:
COM: 28

THE PIK-SIN PROF
ASTRONOMICAL INSTITUTE
ANTON PANNEKOEK
ROETERSSTRAAT 15
NL-1018 WB AMSTERDAM
NETHERLANDS
TEL: 020-522-3004
TLX: 16460 FACUM NL
TLF:
EML:
COM: 33.34.37

THEODOSSIOU EFSTRATIOS DR
DEPARTMENT OF PHYSICS
UNIVERSITY OF ATHENS
PANEPISTIMIOPOLIS
GR 15783 ZOGRAFOS ATHENS
GREECE
TEL: 7243414
TLX:
TLF:
EML:
COM:

THEVENIN FREDERIC DR
OBS. DE LA COTE D'AZUR
BP 139
06003 NICE CEDEX
FRANCE
TEL: 92 00 30 11
TLX:
TLF:
EML: THEVENIN@FRONI51
COM: 29

THIELEMANN FRIEDRICH-KARL
MPI FUR PHYS. & ASTRON.
KARLSCHWARZSCHILD STR. 1
D-8046 GARCHING B MUNCHEN
GERMANY, F.R.
TEL:
TLX:
TLF:
EML:
COM: 35

THIELHEIM KLAUS O DR
ABTEILUNG MATHEM. PHYSIK
UNIVERSITAET KIEL
OLSHAUSENSTR 40/60
D-2300 KIEL
GERMANY, F.R.
TEL: 0431-880-3216
TLX: 292979 IFKKI
TLF:
EML:
COM: 33

THIRY YVES R PROF
UNIVERSITE DE PARIS VI
TOUR 66
4 PLACE JUSSIEU
F-75230 PARIS CEDEX 05
FRANCE
TEL: 1-43-36-25-25
TLX:
TLF:
EML:
COM: 07

THOBURN CHRISTINE
ROYAL GREENWICH OBS
HERSTMONCEUX CASTLE
HAILSHAM BN27 1RP
U.K.
TEL: 0323-833171
TLX: 87451
TLF:
EML:
COM: 08

THOLEN DAVID J DR
INSTITUTE FOR ASTRONOMY
2680 WOODLAWN DRIVE
HONOLULU HI 96822
U.S.A.
TEL: 808 948 6930
TLX: 8459 UHAST HR
TLF:
EML: tholen@uhifa.ifa.hawaii.edu
COM: 15.16.20

THOMAS DAVID V DR
SCIENCE RESEARCH COUNCIL
CENTRAL OFFICE
P O BOX 18
SWINDON SN2 1ET WILT
U.K.
TEL:
TLX:
TLF:
EML:
COM: 08.19.24

THOMAS HANS-CHRISTOPH DR
MPI FUER PHYSIK UND
ASTROPHYSIK
KARL-SCHWARZSCHILD-STR 1
D-8046 GARCHING B MUNCHEN
GERMANY. F.R.
TEL:
TLX:
TLF:
EML:
COM: 35

THOMAS JOHN H PROF
DEPT MECH & AEROSPACE SCI
UNIVERSITY OF ROCHESTER
ROCHESTER NY 14627
U.S.A.
TEL: 716-275-4083
TLX:
TLF:
EML:
COM: 10.12

THOMAS RICHARD N DR
1155 TIMBERLANE
PINEBROOK HILLS
BOULDER CO 80302
U.S.A.
TEL: 303-443-9290
TLX:
TLF:
EML:
COM: 12.36

THOMAS ROGER J DR
NASA/GSFC
CODE 682
GREENBELT MD 20771
U.S.A.
TEL: 301-286-7921
TLX:
TLF:
EML:
COM: 10.44

THOMASSON PETER DR
NRAL
JODRELL BANK
MACCLESFIELD SK11 9DL
U.K.
TEL: 0477-71321
TLX: 36149
TLF:
EML:
COM: 40

THOMPSON A RICHARD DR
NRAO
VLBA PROJECT
2015 IVY ROAD
CHARLOTTESVILLE VA 22903
U.S.A.
TEL: 804-296-0211
TLX:
TLF:
EML:
COM: 34.40

THOMPSON G I DR
ROYAL OBSERVATORY
EDINBURGH EH9 3HJ
U.K.
TEL:
TLX:
TLF:
EML:
COM: 29

THOMPSON KEITH DR
MONASH UNIVERSITY
DEPT OF PHYSICS
WELLINGTON ROAD
CLAYTON. VICT. 3168
AUSTRALIA
TEL: 03-5654090
TLX: 32691 AA
TLF:
EML:
COM: 27.42

THOMPSON LAIRD A DR
ASTR. DEPT.
UNIVERSITY OF ILLINOIS
1011 W. SPRINGFIELD
URBANA IL 61801
U.S.A.
TEL: 217 333 3090
TLX:
TLF:
EML:
COM: 28.47

THOMPSON RODGER I PROF
STEWARD OBSERVATORY
UNIVERSITY OF ARIZONA
TUCSON AZ 85721
U.S.A.
TEL: 602-621-6527
TLX: 467175
TLF:
EML:
COM:

THOMPSON THOMAS WILLIAM
PLANETARY SCI. INST.
282 S LAKE AVE..SUITE 218
PASADENA CA 91101
U.S.A.
TEL: 213 449 4955
TLX:
TLF:
EML:
COM: 16

THOMSEN BJARNE B LECT
INSTITUTE OF ASTRONOMY
UNIVERSITY OF AARHUS
DK-8000 AARHUS C
DENMARK
TEL:
TLX:
TLF:
EML:
COM:

THONNARD NORBERT DR
ATOM SCIENCES
114 RIDGEWAY CENTER
OAK RIDGE. TN 37830
U.S.A.
TEL: 615-483-1113
TLX:
TLF:
EML:
COM: 28.34

THOREN VICTOR E PROF
130 GOODBODY HALL
INDIANA UNIVERSITY
BLOOMINGTON IN 47401
U.S.A.
TEL: 812-825-5970
TLX:
TLF:
EML:
COM: 41

THORNE KIP S PROF
CALTECH 130-33
PASADENA CA 91125
U.S.A.
TEL: 213-356-4598
TLX: 675425
TLF:
EML:
COM: 48

THORSTENSEN JOHN R
DARTMOUTH COLLEGE
DEPT PHYSICS & ASTRONOMY
HANOVER NH 03755
U.S.A.
TEL: 603-646-2869
TLX:
TLF:
EML:
COM:

THRONSON HARLEY ANDREW JR
DEPT PHYSICS & ASTRONOMY
UNIVERSITY OF WYOMING
LARAMIE WY 82071
U.S.A.
TEL: 307-766-6150
TLX:
TLF:
EML:
COM: 34

THUAN TRINH XUAN DR
DEPT OF ASTRONOMY
UNIVERSITY OF VIRGINIA
BOX 3818 UNIV STATION
CHARLOTTESVILLE VA 22903
U.S.A.
TEL: 804-924-4894
TLX:
TLF:
EML:
COM: 28.47

THUN CLEMENS DR
IRAM
AV. DIVINA PASTORA 7
BLOQUE 6/2B
18012 GRANADA
SPAIN
TEL: 958-480413
TLX: 78521 IRAM E
TLF:
EML:
COM: 40

TIFFT WILLIAM G PROF
STEWARD OBSERVATORY
UNIVERSITY OF ARIZONA
TUCSON AZ 85721
U.S.A.
TEL: 602-621-6532
TLX: 467175
TLF:
EML:
COM: 28.47

TIFREA EMILIA DR
OBSERVATOIRE DE BUCAREST
CUTITUL DE ARGINT 5
75212 BUCAREST 28
RUMANIA
TEL: 23-60-10
TLX: 09-26-29
TLF:
EML:
COM: 10

TIMOTHY J GETHYN DR
CTR FOR SPACE SCIENCE &
ASTROPHYSICS
STANFORD UNIV., ERL 314
STANFORD CA 94305
U.S.A.
TEL: 415-497-3059
TLX: 348402 STANFRD STNU
TLF:
EML:
COM:

TINBERGEN JAAP DR
KAPTEYN STERREWACHT
WERKGROEP
MENSINGHEWEG 20
NL-9301 KA RODEN
NETHERLANDS
TEL: 05908-19631
TLX: 53767 KSWRO NL
TLF:
EML:
COM: 25

TING YEOU-TSWEN
ASTRONOMY SECTION
CENTRAL WEATHER BUREAU
64 KUNG YUEN ROAD
TAIPEI 100
CHINA. TAIWAN
TEL: 3713181-281
TLX:
TLF:
EML:
COM: 04

TIPLER FRANK JENNINGS DR
PHYSICS DEPT
TULANE UNIVERSITY
NEW ORLEANS LA 70118
U.S.A.
TEL:
TLX:
TLF:
EML:
COM: 47

TIURI MARTTI PROF
HELSINKI UNIV TECHNOLOGY
RÁDIO LABORATORY
OTAKAARI 5 A
SF-02150 ESPOO 15
FINLAND
TEL: 358-0-451-2545
TLX: 122771 HORTA SF
TLF:
EML:
COM:

TJIN-A-DJIE HERMAN R E DR
KOEKOELAAN 106
NL-1403 EJ BUSSUM
NETHERLANDS
TEL: 02159-17076
TLX: 16460 FACWN NL
TLF:
EML:
COM: 27.35

TLAMICHA ANTONIN DR
ASTRONOMICAL OBSERVATORY
CZECH. ACAD. OF SCIENCES
251 65 ONDREJOV
CZECHOSLOVAKIA
TEL: 72-45-25
TLX: 121579 ASTR CZ
TLF:
EML:
COM: 10.40

TOBIN WILLIAM
DEPARTMENT OF PHYSICS
UNIVERSITY OF CANTERBURY
CHRISTCHURCH 1
NEW ZEALAND
TEL:
TLX:
TLF:
EML:
COM: 33

TODORAN IOAN DR
ASTRONOMICAL OBSERVATORY
STR CIRESILOR 19
3400 CLUJ NAPOCA
RUMANIA
TEL:
TLX:
TLF:
EML:
COM: 25.42

TOFANI GIANNI PROF
OSSERVATORIO ASTROFISICO
DI ARCETRI
LARGO E. FERMI 5
I-50125 FIRENZE
ITALY
TEL: 055-2752217
TLX: 572268 ARCETRI
TLF:
EML:
COM: 40

TOHLINE JOEL EDWARD
DEPT PHYSICS & ASTRONOMY
LOUISIANA STATE UNIV.
BATON ROUGE LA 70803
U.S.A.
TEL: 504-388-6851
TLX: 559184
TLF:
EML:
COM: 35

TOKOVININ ANDREJ A DR
STERNBERG STATE
ASTRONOMICAL INSTITUTE
119899 MOSCOW
U.S.S.R.
TEL: 939 33 18
TLX:
TLF:
EML:
COM: 26

TOLBERT CHARLES R DR
LEANDER MCCORMICK OBS
BOX 3818
UNIVERSITY STATION
CHARLOTTESVILLE VA 22903
U.S.A.
TEL: 804-924-7494
TLX:
TLF:
EML:
COM: 25.38.40.51

TOLLER GARY N DR
NASA/GSFC
CODE 685.3
GREENBELT MD 20771
U.S.A.
TEL:
TLX:
TLF:
EML:
COM: 21

TOMASI PAOLO DR
LAB. DI RADIOASTRONOMIA
VIA IRNERIO 46
I-40126 BOLOGNA
ITALY
TEL:
TLX:
TLF:
EML:
COM: 40

TOMASKO MARTIN G DR
LUNAR & PLANETARY LAB
UNIVERSITY OF ARIZONA
SPACE SCIENCES BLDG
TUCSON AZ 85721
U.S.A.
TEL: 602-621-6969
TLX:
TLF:
EML:
COM:

TOMBAUGH CLYDE W PROF
DEPT OF ASTRONOMY
NEW MEXICO STATE UNIV
BOX 4500
LAS CRUCES NM 88003
U.S.A.
TEL: 505-646-2107
TLX:
TLF:
EML:
COM: 16

TOMIMATSU AKIRA DR
DEPT OF PHYSICS
NAGOYA UNIVERSITY
NAGOYA 464
JAPAN
TEL:
TLX:
TLF:
EML:
COM: 47.48

TOMISAKA KOHJI DR
FACULTY OF EDUCATION
NIIGATA UNIVERSITY
IKARASHI 2-8050
NIIGATA 950-21
JAPAN
TEL: 25 262 7269
TLX:
TLF:
EML: BITNET:c30841@jpnkudpc
COM: 33

TOMITA KENJI PROF
RES INST FOR THEORET PHYS
HIROSHIMA UNIVERSITY
TAKEHARA 725
JAPAN
TEL: 08462-2-2362
TLX:
TLF:
EML:
COM: 47

TOMITA KOICHIRO DR
11-20 4 CHOME YOGA
SETAGAYAKU
TOKYO 158
JAPAN
TEL: 037000066
TLX:
TLF:
EML:
COM: 15.20.22

TOMOV ALEXANDER NIKOLOV
DEPT OF ASTRONOMY
BULGARIAN ACAD SCIENCES
72 LENIN BLVD
1784 SOFIA
BULGARIA
TEL:
TLX:
TLF:
EML:
COM:

TONG FU
PURPLE MOUNTAIN
OBSERVATORY
NANJING
CHINA. PEOPLE'S REP.
TEL: 33921
TLX: 34144 PMOAS CN
TLF:
EML:
COM: 04C.07

TONG YI
DEPARTMENT OF ASTRONOMY
BEIJING NORMAL UNIVERSITY
19 XINJIEKOU OUT-STREET
BEIJING
CHINA. PEOPLE'S REP.
TEL: 656531-6285
TLX:
TLF:
EML:
COM: 28.33

TONWAR SURESH C PROF.
TATA INST FUND RESEARCH
HOMI BHABHA MARG
COLABA
BOMBAY 400005
INDIA
TEL: 22 495 2311
TLX: 011 3009 TIFR IN
TLF:
EML:
COM:

TOOMRE ALAR DR
MIT
ROOM 2-371
77 MASSACHUSETTS AVE
CAMBRIDGE MA 02139
U.S.A.
TEL: 617-253-4326
TLX:
TLF:
EML:
COM: 28.33

TOOMRE JURI
DEPT ASTRO-GEOPHYSICS
JILA
UNIVERSITY OF COLORADO
BOULDER CO 80309
U.S.A.
TEL: 303-492-7854
TLX:
TLF:
EML:
COM: 33.35.36

TOPAKTAS LATIF A DR
KING SAUD UNIVERSITYRY
COLLEGE OF SCIENCE
P.O. BOX 2455
RIYADH 11453
SAUDI ARABIA
TEL:
TLX:
TLF:
EML:
COM:

TORAO MASAMISA
5-7-9-401 YOYOGI
SHIBUYA
TOKYO 151
JAPAN
TEL:
TLX:
TLF:
EML:
COM: 19

TORELLI M DR
OSSERVATORIO ASTRONOMICO
VIA DEL PARCO MELLINI 84
I-00136 ROMA
ITALY
TEL: 347056
TLX:
TLF:
EML:
COM: 12

TORNAMBE AMEDEO
IST. ASTROFISICA SPAZIALE
C P 67
I-00044 FRASCATI
ITALY
TEL:
TLX:
TLF:
EML:
COM: 37

TOROSHLIDZE TEIMURAZ I DR
ABASTUMANI ASTROPHYSICAL
OBSERVATORY
383762 ABASTUMANI.GEORGIA
U.S.S.R.
TEL:
TLX:
TLF:
EML:
COM: 21

TORRA JORDI DR
DEPT DE FISICA
UNIVERSITAT DE BARCELONA
AVDA DIAGONAL 647
08028 BARCELONA
SPAIN
TEL: 34 3 330 73 11
TLX:
TLF:
EML:
COM: 33

TORRES CARLOS ALBERTO DR
OBSERVATORIO NACIONAL/LNA
RUA CORONEL RENNO 07
CAIXA POSTAL 21
37500 ITAJUBA MG
BRAZIL
TEL: 035-622-0788
TLX: 031-2603
TLF:
EML:
COM: 27.50C

TORRES CARLOS DR
OBS ASTRONOMICO NACIONAL
UNIVERSIDAD DE CHILE
CASILLA 36-D
SANTIAGO
CHILE
TEL: 56-2-229-4101
TLX: 440001 ITT BOOTH FOR
TLF:
EML:
COM: 20.50

TORRES-PEIMBERT SILVIA DR
INSTITUTO DE ASTRONOMIA
UNAM
APDO POSTAL 70-264
04510 MEXICO DF
MEXICO
TEL: 905-548-5306
TLX: 1760155 CIC ME
TLF:
EML:
COM: 34.46

TORRICELLI GUIDETTA DR
OSSERVATORIO DI ARCETRI
L.E. FERMI 5
FIRENZE 50125
ITALY
TEL: 55 2752260
TLX: 572268
TLF:
EML:
CON:

TORROJA J PROF
CATEDRA DE ASTRONOMIA
FACULTAD DE CIENCIAS
UNIVERSIDAD COMPLUTENSE
28040 MADRID
SPAIN
TEL:
TLX:
TLF:
EML:
CON:

TOSA MAKOTO DR
ASTRONOMICAL INSTITUTE
TOHOKU UNIVERSITY
SENDAI 980
JAPAN
TEL: 0222-22-1800
TLX: 852246 THUCOM J
TLF:
EML:
CON: 33C

TOSHIBI DR
ASTRONOMICAL INSTITUTE
TOHOKU UNIVERSITY
AOBAYAMA SENDAI
JAPAN
TEL:
TLX:
TLF:
EML:
CON:

TOSI MONICA
OSSERVATORIO ASTRONOMICO
UNIVERSITARIO
C P 596
I-40100 BOLOGNA
ITALY
TEL: 51-222956
TLX: 211664 INFN BO I
TLF:
EML:
CON: 34

TOUSEY RICHARD DR
NAVAL RESEARCH LABORATORY
CODE 7140
WASHINGTON DC 20375
U.S.A.
TEL: 202-767-3441
TLX:
TLF:
EML:
CON: 12.14

TOVMASSIAN H M DR
BYURAKAN ASTROPHYSICAL
OBSERVATORY
375433 ARMENIA
U.S.S.R.
TEL: 56-34-53
TLX:
TLF:
EML:
CON: 28.40.44.51

TOWNES CHARLES HARD DR
DEPT OF PHYSICS
UNIVERSITY OF CALIFORNIA
RM 557 BIRGE HALL
BERKELEY CA 94720
U.S.A.
TEL: 415-642-1128
TLX:
TLF:
EML:
CON: 34.40.51

TOZER DAVID C DR
SCHOOL OF PHYSICS
UNIVERSITY OF NEWCASTLE
NEWCASTLE/TYNE NE1 7RU
U.K.
TEL:
TLX:
TLF:
EML:
CON:

TOZZI GIAN PAOLO
OSSERVATORIO ASTROFISICO
DI ARCETRI
LARGO E. FERMI 5
I-50125 FIRENZE
ITALY
TEL: 055-2752250
TLX: 572268 ARCETR I
TLF:
EML:
CON: 14

TRAFTON LAURENCE M DR
ASTRONOMY DEPT
UNIVERSITY OF TEXAS
AT AUSTIN
AUSTIN TX 78712
U.S.A.
TEL: 512-471-1476
TLX:
TLF:
EML:
CON: 16

TRAN MINH NGUYET DR
OBSERVATOIRE DE PARIS
SECTION DE MEUDON
F-92195 MEUDON PL CEDEX
FRANCE
TEL: 1-45-07-74-47
TLX: 270912 OBSASTR
TLF:
EML:
CON:

TRAN-MINH FRANCOISE DR
DASGAL
OBSERVATOIRE PARIS MEUDON
92195 MEUDON PRINC.CEDEX
FRANCE
TEL: 33-1-45077553
TLX: 201571 LAM
TLF: 45 07 74 69
EML:
CON: 16

TRAUB WESLEY ARTHUR
CENTER FOR ASTROPHYSICS
60 GARDEN STREET
CAMBRIDGE MA 02138
U.S.A.
TEL: 617-495-7406
TLX: 921428 SATELLITE CAM
TLF:
EML:
CON: 09.44

TRAVING GERHARD PROF
INSTITUT FUER
THEORETISCHE ASTROPHYSIK
NEUENHEIMER FELD 561
D-6900 HEIDELBERG
GERMANY. F.R.
TEL: 06221-562815
TLX: 461515
TLF:
EML:
CON: 36

TREDER H J PROF DR
ZNTRLINST. F. ASTROPHYSIK
STERNWARTE BABELSBERG
ROSA-LUXEMBURG-STR 17A
DDR-1502 POTSDAM
GERMANY. D.R.
TEL: 762275
TLX: 15471 ADW RZB DD
TLF:
EML:
CON: 47

TREFFERS RICHARD R
ASTRONOMY DEPT
UNIVERSITY OF CALIFORNIA
BERKELEY CA 94720
U.S.A.
TEL: 415-642-4223
TLX:
TLF:
EML:
CON: 34

TREFFTZ ELEONORE E DR
MPI F PHYSIK UND
ASTROPHYSIK
KARL-SCHWARZSCHILD-STR 1
D-8046 GARCHING B MUNCHEN
GERMANY. F.R.
TEL: 89-32990
TLX: 524629 ASTRO D
TLF:
EML:
CON: 14

TREFZGER CHARLES F DR
ASTRONOMISCHES INSTITUT
UNIVERSITAET BASEL
VENUSSTRASSE 7
CH-4102 BINNINGEN
SWITZERLAND
TEL: 061-22-77-11
TLX:
TLF:
EML:
CON: 33

TREHAN SURINDAR K PROF
DEPT OF MATHEMATICS
PANJAB UNIVERSITY
CHANDIGARH 160 014
INDIA
TEL: 29938
TLX:
TLF:
EML:
CON:

TRELLIS MICHEL DR
OBSERVATOIRE DE NICE
BP 139
F-06003 NICE CEDEX
FRANCE
TEL: 93-89-04-20
TLX:
TLF:
EML:
COM: 10

TREMAINE SCOTT DUNCAN
CITA. MC LENNAN LABSO
UNIVERSITY OF TORONTO
60 ST GEORGE STREET
TORONTO M5S 1A1
CANADA
TEL: 416-978-6879
TLX:
TLF:
EML:
COM: 28.47

TREMKO JOZEF DR
ASTRONOMICAL INSTITUTE
SLOVAK ACAD. OF SCIENCES
SKALNATE PLESO OBS
059 60 TATRANSKA LOMNICA
CZECHOSLOVAKIA
TEL: 967866
TLX: 78277 AUSAV CZ
TLF:
EML:
COM: 27.42.50C

TREUMANN RUDOLF A. DR
MPI F. PHYS & ASTROPHYSIK
INST F EXTRATERR PHYSIK
D-8046 GARCHING B MUNCHEN
GERMANY. F.R.
TEL: 89-3299831
TLX: 5215845 XTER D
TLF:
EML:
COM: 10

TREVESE DARIO
OSSERVATORIO ASTRONOMICO
VIA DEL PARCO MELLINI 84
I-00136 ROMA
ITALY
TEL: 6-347-056
TLX:
TLF:
EML:
COM: 47

TREXLER JAMES H MR
NAVAL RESEARCH LABORATORY
OXON HILL MD 20745
U.S.A.
TEL:
TLX:
TLF:
EML:
COM:

TRIMBLE VIRGINIA L DR
DEPT OF PHYSICS
UNIVERSITY OF CALIFORNIA
IRVINE CA 92717
U.S.A.
TEL: 714-856-6948
TLX:
TLF:
EML:
COM: 26.28C.35.42.47C.48C.51

TRINCHIERI GINEVRA
OSSERVATORIO ASTROFISICO
DI ARCETRI
LARGO E. FERMI 5
I-50125 FIRENZE
ITALY
TEL: 055-2752230
TLX: 572268 ARCETR I
TLF:
EML:
COM: 28

TRIPATHI B M DR
UTTAR PRADESH STATE OBS
MANORA PEAK
NAINITAL 263 129
INDIA
TEL: 2136
TLX: CABLE : ASTRONOMY
TLF:
EML:
COM: 12

TRITAKIS BASIL P DR
ASTRONOMY & APPL MATHS
ACADEMY OF ATHENS
14 ANAGNOSTOPOULOU
GR-10673 ATHENS
GREECE
TEL: 1-3613589
TLX:
TLF:
EML:
COM: 10.49

TRITTON KEITH P DR
ROYAL GREENWICH OBS
HERSTMONCEUX CASTLE
HAILSHAM BN27 1RP
U.K.
TEL: 323 833 171
TLX:
TLF:
EML:
COM: 40

TRITTON SUSAN BARBARA
ROYAL OBSERVATORY
BLACKFORD HILL
EDINBURGH EH9 3HJ
U.K.
TEL: 031-667-3321
TLX: 72383 ROEDIN G
TLF:
EML:
COM: 05

TROCHE-BOGGINO A E DR
INST DE CIENCIAS BASICAS
UNIV NACIONAL DE ASUNCION
C.CORREO 1039-1804
ASUNCION
PARAGUAY
TEL:
TLX:
TLF:
EML:
COM: 46

TRODAHL HARRY JOSEPH DR
VICTORIA UNIVERSITY
PRIVATE BAG
WELLINGTON
NEW ZEALAND
TEL: 721-000
TLX:
TLF:
EML:
COM: 25

TROITSKY V S PROF DR
RADIOPHYSICAL RESEARCH
INSTITUTE
LYADOV STREET 25/14
603600 GORKIJ
U.S.S.R.
TEL: 36-04-40
TLX:
TLF:
EML:
COM: 16.40.51C

TROLAND THOMAS HUGH
PHYSICS-ASTRONOMY DEPT
UNIVERSITY OF KENTUCKY
LEXINGTON KY 40506
U.S.A.
TEL: 606-257-8620
TLX:
TLF:
EML:
COM: 40

TROTTET GERARD DR
OBSERVATOIRE DE PARIS
SECTION DE MEUDON
DASOP
F-92195 MEUDON PL CEDEX
FRANCE
TEL: 1-45-07-78-08
TLX:
TLF:
EML:
COM: 10

TRUEMPER JOACHIM PROF
MPI F EXTRATERRESTRISCHE
PHYSIK
D-8046 GARCHING B MUNCHEN
GERMANY. F.R.
TEL: 089-3299559
TLX: 5215845 XTER D
TLF:
EML:
COM: 44V.48C

TRULSEN JAN K PROF
UNIVERSITY OF TROMSO
P O BOX 953
N-9001 TROMSO
NORWAY
TEL:
TLX:
TLF:
EML:
COM:

TRURAN JAMES W JR
DEPT OF ASTRONOMY
UNIVERSITY OF ILLINOIS
URBANA IL 61801
U.S.A.
TEL: 217-333-3090
TLX:
TLF:
EML:
COM: 35C.48

TRUSSONI EDOARDO
IST. DI COSMOGEOFISICA
DEL CNR
CORSO FIUME 4
I-10133 TORINO
ITALY
TEL: 011-657694/8979
TLX: 211041 INFNTO
TLF:
EML:
COM: 44

TRUTSE YU L DR
INST PHYSICS OF ATMOSPH
USSR ACADEMY OF SCIENCES
109017 MOSCOW
U.S.S.R.
TEL:
TLX:
TLF:
EML:
COM: 21

TSAI CHANG-HSIEN DIRECTOR
ASTRONOMICAL STY OF CHINA
TAIPEI OBSERVATORY
TAIPEI 104
CHINA. TAIWAN
TEL:
TLX:
TLF:
EML:
COM:

TSAO MO PROF
NO 47 SEC 3
HSIN-I ROAD
TAIPEI 106
CHINA. TAIWAN
TEL: 02-7047795
TLX:
TLF:
EML:
COM: 19

TSAP T T DR
CRIMEAN ASTROPHYS OBS
USSR ACADEMY OF SCIENCES
NAUCHNIY
334413 CRIMEA
U.S.S.R.
TEL: 132
TLX:
TLF:
EML:
COM: 12

TSCHARNUTER WERNER M DR
INST. FUR THEOR. ASTROPHY
DER UNIVERSITAT
IM NEUENHEIMER FELD 561
D-6900 HEIDELBERG 1
GERMANY. F.R.
TEL:
TLX:
TLF:
EML:
COM:

TSEYTLIN NAUM M
RADIOPHYSICAL RES INST
LYADOV STR. 25/14
603600 GORKY
U.S.S.R.
TEL: 36-01-29
TLX: 1111 LUNA
TLF:
EML:
COM:

TSIKOUDI VASSILIKI PH D
DEPT OF PHYSICS
DIV OF ASTRO-GEOPHYSICS
UNIVERSITY OF IOANNINA
GR-45332 IOANNINA
GREECE
TEL: 0651-91084
TLX:
TLF:
EML:
COM:

TSIOUMIS ALEXANDROS DR
DEPT GEODETIC ASTRONOMY
UNIVERSITY THESSALONIKI
GR-54006 THESSALONIKI
GREECE
TEL:
TLX:
TLF:
EML:
COM: 27.33

TSIROPOULA GEORGIA DR
NATL OBS OF ATHENS
ASTRONOMICAL INSTITUTE
P.O. BOX 20048
GR 11810 ATHENS
GREECE
TEL: 3461191
TLX: 215530 OBS.GR
TLF:
EML:
COM: 12

TSUBAKI TOKIO PROF
DEPT OF EARTH SCIENCE
SHIGA UNIVERSITY
2-5-1 HIRATSU
OHTSU 520
JAPAN
TEL: 0775-37-0081
TLX:
TLF:
EML:
COM: 10.12

TSUBOKAWA IETSUNE DR
INTERNATIONAL LATITUDE
OBSERVATORY
MIZUSAWA
IWATE 023
JAPAN
TEL: 0197247111
TLX: 837628
TLF:
EML:
COM: 19

TSUCHITA ATSUSHI DR PROF
TOKYO ASTRONOMICAL OBS
OSAWA
MITAKA
TOKYO 181
JAPAN
TEL: 0422-32-5111
TLX: 02822307 TAOMTK J
TLF:
EML:
COM: 31

TSUJI TAKASHI
INSTITUTE OF ASTRONOMY
UNIVERSITY OF TOKYO
MITAKA
TOKYO 181
JAPAN
TEL: 0422-32-5111
TLX: 02822307 TAOMTK J
TLF:
EML:
COM: 29.36C

TSUNEMI HIROSHI DR
FACULTY OF SCIENCES
OSAKA UNIVERSITY
1-1 MACHIKANEYAMA
TOYONAKA. OSAKA 560
JAPAN
TEL: 6 844 1151
TLX:
TLF:
EML: BITNET:TSUNEMI@JPNOSKPH
COM: 44

TSUNETA SAKU DR
INSTITUTE OF ASTRONOMY
UNIVERSITY OF TOKYO
2-21-1 OSAWA MITAKA
TOKYO 181
JAPAN
TEL: 422 32 4710
TLX: 2822307 TAOMTK J
TLF:
EML: STSUNETA:NASAMAIL
COM: 10

TSURUTA SACHIKO DR
DEPT. OF PHYSICS
UNIVERSITY OF TOKYO
C/O S. ISHIMARU
TOKYO
JAPAN
TEL:
TLX:
TLF:
EML:
COM: 48

TSVETANOV ZLATAN IVANOV
DEPT OF ASTRONOMY
BULGARIAN ACAD SCIENCES
72 LENIN BLVD
1784 SOFIA
BULGARIA
TEL: 7341 x379
TLX: 23561 ECF BAN BG
TLF:
EML:
COM:

TSVETKOV MILCHO K DR
DEPT OF ASTRONOMY
BULGARIAN ACAD SCIENCES
72 LENIN BLVD
1784 SOFIA
BULGARIA
TEL: 758927
TLX: 23561 ECF BAN BG
TLF:
EML:
COM: 27

TSVETKOV TSVETAN DR
DEPT OF ASTRONOMY
FACULTY OF PHYSICS
ANTON IVANOV STR 5
1126 SOFIA
BULGARIA
TEL:
TLX:
TLF:
EML:
COM:

TSVETKOVA KATIA
DEPT OF ASTRONOMY AND
NATL ASTRON OBSERVATORY
72 LENIN BLVD
1784 SOFIA
BULGARIA
TEL: 73-41-379
TLX: 23561 ECF BAN BG
TLF:
EML:
COM: 37

TUCHMAN YTZHAK
RACAH INST. OF PHYSICS
HEBREW UNIV. OF JERUSALEM
JERUSALEM 91904
ISRAEL
TEL: 02-584417
TLX: 25391 HUIL
TLF:
EML:
COM:

TUCKER WALLACE H DR
PO BOX 266
BONSALL CA 92003
U.S.A.
TEL: 619-728-7103
TLX:
TLF:
EML:
COM:

TUEG HELMUT DR
ALFRED-WEGENER INSTITUT
FUR POLARFORSCHUNG
COLUMBUS CENTER
D-2850 BREMERHAVEN
GERMANY. F.R.
TEL:
TLX:
TLF:
EML:
COM: 09

TUFEKCIOGLU ZEKI DR
ASTRONOMY DEPT
KING ABDULAZIZ UNIV.
P.O. BOX 9028
JEDDAH 21413
SAUDI ARABIA
TEL:
TLX:
TLF:
EML:
COM:

TULL ROBERT G
DEPT OF ASTRONOMY
UNIVERSITY OF TEXAS
AT AUSTIN RLM 15 308
AUSTIN TX 78712
U.S.A.
TEL: 512-471-3337
TLX: 910-874-1351
TLF:
EML:
COM: 09C

TULLY JOHN A DR
OBSERVATOIRE DE NICE
BP 139
F-06003 NICE CEDEX
FRANCE
TEL: 93-89-04-20
TLX: 460004
TLF:
EML:
COM:

TULLY RICHARD BRENT DR
INSTITUTE FOR ASTRONOMY
UNIVERSITY OF HAWAII
2680 WOODLAWN DR
HONOLULU HI 96822
U.S.A.
TEL: 808-948-8606
TLX: 723-8459 UHAST HR
TLF:
EML:
COM: 28.47

TUNCA ZEYNEL DR
EGE UNIVERSITY
FACULTY OF SCIENCE
DEPARTMENT OF ASTRONOMY
BORNOVA-IZMIR
TURKEY
TEL: 180110-2332
TLX:
TLF:
EML:
COM:

TUOHY IAN R DR
MT STROMLO & SIDING
SPRING OBSERVATORIES
WODEN P.O. ACT 2606
AUSTRALIA
TEL: 062-88-1111
TLX: 62270 AA
TLF:
EML:
COM:

TUOMINEN ILKKA V DR
OBSERVATORY
UNIVERSITY OF HELSINKI
TAHTITORNINMAKI
SF-00130 HELSINKI 13
FINLAND
TEL: 3580-1912946
TLX:
TLF:
EML:
COM: 29.35

TURATTO MASSIMO DR
OSSERVATORIO ASTRONOMICO
DI PADOVA
VICOLO OSSERVATORIO 5
35122 PADOVA
ITALY
TEL: 049 661499
TLX: 432071
TLF:
EML: SPAN:39003::TURATTO
COM:

TURLO ZYGMUNT DR
ASTRONOMICAL CENTER
UL. CHOPINA 12/18
87-100 TORUN
POLAND
TEL:
TLX:
TLF:
EML:
COM: 40

TURNER BARRY E DR
NRAO
EDGEMONT ROAD
CHARLOTTESVILLE VA 22903
U.S.A.
TEL: 804-296-0337
TLX: 910-997-0174
TLF:
EML:
COM: 34.40

TURNER DAVID G DR
DEPT OF ASTRONOMY
ST MARY'S UNIVERSITY
HALIFAX NS B3H 3C3
CANADA
TEL: 9024299780x2254
TLX:
TLF:
EML:
COM: 27.37

TURNER EDWIN L DR
PRINCETON UNIVERSITY OBS
PEYTON HALL
PRINCETON NJ 08544
U.S.A.
TEL: 609-452-3577
TLX:
TLF:
EML:
COM: 28.47.51

TURNER KENNETH C DR
ARECIBO OBSERVATORY
PO BOX 995
ARECIBO PR 00613
U.S.A.
TEL: 809-878-2612
TLX: 385638
TLF:
EML:
COM: 34.40

TURNER MARTIN J L DR
X-RAY ASTRONOMY GROUP
PHYSICS DEPT
UNIV OF LEICESTER
LEICESTER LE1 7RH
U.K.
TEL: 533-554455
TLX: 341664 LUXRAY G
TLF:
EML:
COM:

TURNER MICHAEL S
ASTRON & ASTROPHYS CENTER
UNIVERSITY OF CHICAGO
5460 S. ELLIS AVE
CHICAGO IL 60637
U.S.A.
TEL: 312-962-7974
TLX: 6871133 UNCGO VW
TLF:
EML:
COM: 47

TURON PIERRE
40 RUE DE LUZARCHES
F-95270 SEUGY
FRANCE
TEL:
TLX:
TLF:
EML:
COM:

URECHE VASILE DR
UNIVERSITY OF CLUJ-NAPOCA
FACULTY OF MATHEMATICS
STR M KOGALNICEANU 1
3400 CLUJ NAPOCA
RUMANIA
TEL: 951-16101/11592
TLX:
TLF:
EML:
COM: 25.42

URPO SEPPO I
HELSINKI UNIV OF TECHNOL.
RADIO LABORATORY
OTAKAARI 5 A
SF-02150 ESPOO 15
FINLAND
TEL: 358-0-4512548
TLX: 122771 HORTA SF
TLF:
EML:
COM: 10.40

USHER PETER D DR
DEPT OF ASTRONOMY
PENNSYLVANIA STATE UNIV
507 DAVEY LAB
UNIVERSITY PARK PA 16802
U.S.A.
TEL: 814-865-1509
TLX: 842510 PENNSTBSTRCG
TLF:
EML:
COM: 27

USON JUAN M DR
NRAO
VLA
PO BOX O
SOCORRO NM 87801
U.S.A.
TEL: 505 835 7237
TLX: 910 988 1710
TLF:
EML:
COM: 40.47

UTSUMI KAZUHIKO DR
DEPT OF ASTRONOMY
HIROSHIMA UNIVERSITY
NAKA-KU, HIGASHI-SENDA
HIROSHIMA 730
JAPAN
TEL: 082-241-1221
TLX:
TLF:
EML:
COM: 29

UUS UNDO DR
TARTU ASTROPHYSICAL
OBSERVATORY
202444 TORAVERE, ESTONIA
U.S.S.R.
TEL:
TLX:
TLF:
EML:
COM: 12,35

VAGER ZEEV DR
DEPT OF PHYSICS
WEIZMANN INSTITUTE
REHOVOT
ISRAEL
TEL:
TLX:
TLF:
EML:
COM:

VAGHI SERGIO DR
ESTEC/PHA
POSTBUS 299
NL-2200 AG NOORDWIJK
NETHERLANDS
TEL: 01719-83453
TLX: 39098
TLF:
EML:
COM: 20

VAGNETTI FAUSTO
DIPARTIMENTO DI FISICA
II UNIVERSITA' DI ROMA
VIA ORAZIO RAIMONDO
I-00173 ROMA
ITALY
TEL: 6-7979-2323
TLX: 611462 UNIVRM
TLF:
EML:
COM: 47

VAIANA GIUSEPPE S DR
OSSERV. ASTRONOMICO
DI PALERMO
PALAZZO DEI NORMANNI
PALERMO
ITALY
TEL:
TLX:
TLF:
EML:
COM:

VAIDYA P C PROF
34 SHARDA NAGAR
PALDI
AHMEDABAD 380 007
INDIA
TEL: 413322
TLX:
TLF:
EML:
COM: 47

VAINSTEIN L A DR
PHYSICAL INSTITUTE
USSR ACADEMY OF SCIENCES
LENINSKY PROSP. 53
117924 MOSCOW
U.S.S.R.
TEL: 135-22-50
TLX:
TLF:
EML:
COM: 49

VALBOUSQUET ARMAND DR
OBS DE STRASBOURG
11 RUE DE L'UNIVERSITE
F-67000 STRASBOURG
FRANCE
TEL: 88-35-43-00
TLX:
TLF:
EML:
COM: 24.26.51

VALENTIJN EDWIN A DR
KAPTEYN LABORATORIUM
POSTBUS 800
NL-9700 AV GRONINGEN
NETHERLANDS
TEL: 050-116695
TLX:
TLF:
EML:
COM: 28

VALIRON PIERRE DR
GROUPE D'ASTROPHYSIQUE
OBSERVATOIRE DE GRENOBLE
USTMG CNRS CERMO B.P. 68
38402 ST MARTIN D'HERES C
FRANCE
TEL: 76 51 47 87
TLX: 980 753 IRAM F
TLF:
EML: BITNET:VALIRON@FRGAG51
COM:

VALLEE JACQUES P DR
HERZBERG INSTITUTE
CNRC
100 SUSSEX DRIVE
OTTAWA ONT K1A 0R6
CANADA
TEL: 613-993-6060
TLX:
TLF:
EML:
COM: 40.51

VALNICEK BORIS DR
ASTRONOMICAL INSTITUTE
CZECH. ACAD. OF SCIENCES
OBSERVATORY
251 65 ONDREJOV
CZECHOSLOVAKIA
TEL: 204-999-202
TLX: 121579
TLF:
EML:
COM: 09.10.44

VALSECCHI GIOVANNI B DR
IAS PLANETOLOGIA VIALE
DELL UNIVERSITA 11.
ROMA 00185
ITALY
TEL: 39 6 4456951
TLX: CNR FRA 610261
TLF:
EML: BITNET:GIOVANNI@IRMIAS
COM: 15.20

VALTAOJA ESKO
DEPT OF PHYSICAL SCIENCES
UNIVERSITY OF TURKU
SF-20500 TURKU 50
FINLAND
TEL: 358-21-435 822
TLX: 62643 TYF
TLF: 358-21-433 767
EML:
COM: 40

VALTIER JEAN-CLAUDE DR
OBSERVATOIRE DE NICE
BP 139
F-06003 NICE CEDEX
FRANCE
TEL: 93-89-04-20
TLX: 460004
TLF:
EML:
COM: 27.29

VALTONEN MAURI J PROF
DEPT OF PHYSICAL SCIENCES
UNIVERSITY OF TURKU
IT. PITKAKATU 1
SF-20520 TURKU 52
FINLAND
TEL: 358-21-645822
TLX: 62683 TYK SF
TLF: 358-21-633767
EML:
COM: 07.09.26.28.40.44.48

VAN AGT S L TH J DR
STERRENKUNDIG INSTITUUT
TOERNOOIVELD
NL-6525 ED NIJMEGEN
NETHERLANDS
TEL: 080-558833
TLX: 48228 WINAT NL
TLF:
EML:
COM: 27

VAN ALBADA TJEERD S DR
KAPTEYN LABORATORIUM
POSTBUS 800
NL-9700 AV GRONINGEN
NETHERLANDS
TEL: 050-116695
TLX: 53572 STARS NL
TLF:
EML:
COM: 28

VAN ALLEN JAMES A PROF
DEPT PHYSICS & ASTRONOMY
UNIVERSITY OF IOWA
IOWA CITY IA 52242
U.S.A.
TEL: 319-353-4531
TLX:
TLF:
EML:
COM: 10.16.21.49

VAN ALTENA WILLIAM F PROF
YALE UNIVERSITY OBS
PO BOX 6666
NEW HAVEN CT 06511
U.S.A.
TEL: 203-436-3318
TLX:
TLF:
EML:
COM: 24F.26.37

VAN BEEK FRANK PROF DR
TECHNICAL UNIV. OF DELFT
DEPT MECHAN ENGINEERING
MEKELWEG 2
NL-2628 CD DELFT
NETHERLANDS
TEL: 015-785396
TLX:
TLF:
EML:
COM: 44

VAN BLERKOM DAVID J PROF
ASTRONOMY DEPT
UNIV OF MASSACHUSETTS
AMHERST MA 01002
U.S.A.
TEL:
TLX:
TLF:
EML:
COM:

VAN BREDA IAN G DR
ROYAL GREENWICH OBS
HERSTMONCEUX CASTLE
HAILSHAM BN27 1RP
U.K.
TEL: 0323-833171
TLX: 87451
TLF:
EML:
COM:

VAN BREUGEL WIL
RADIOASTRONOMY LABORATORY
UNIVERSITY OF CALIFORNIA
601 CAMPBELL HALL
BERKELEY CA 94720
U.S.A.
TEL: 415-642-5275
TLX:
TLF:
EML:
COM:

VAN BUEREN HENDRIK G PROF
ADVISORY COUNCIL FOR
SCIENCE POLICY
P O BOX 18524
NL-2502 EM THE HAGUE
NETHERLANDS
TEL: 070-639922
TLX:
TLF:
EML:
COM:

VAN CITTERS GORDON W DR
DIV OF ASTRONOMICAL SCI
NTL SCIENCE FOUNDATION
WASHINGTON DC 20550
U.S.A.
TEL:
TLX:
TLF:
EML:
COM: 09

VAN DE HULST H C PROF DR
STERREWACHT
POSTBUS 9513
NL-2300 RA LEIDEN
NETHERLANDS
TEL: 071-148333
TLX: 39058
TLF:
EML:
COM: 21.34.40.44

VAN DE KAMP PETER
AMSTEL 244
NL-1017 AH AMSTERDAM
NETHERLANDS
TEL: 020-233377
TLX:
TLF:
EML:
COM: 24.26.51

VAN DE STADT HERMAN DR
KAPTEYN ASTRON.INSTITUTE
POSTBUS 800
NL-9700 AV GRONINGEN
NETHERLANDS
TEL: 50-634073
TLX: 53572
TLF:
EML:
COM:

VAN DEN BERGH SIDNEY PROF
HERZBERG INST ASTROPHYS
DOMINION ASTROPHYS OBS
5071 W SAANICH ROAD
VICTORIA BC V8X 4M6
CANADA
TEL: 604-388-3934
TLX: 0497295
TLF:
EML:
COM: 28.37.50C

VAN DEN HEUVEL EDWARD P J
ASTRONOMICAL INSTITUT
ROETERSSTRAAT 15
NL-1018 WB AMSTERDAM
NETHERLANDS
TEL: 020-5223004
TLX: 16460 FACWN NL
TLF:
EML:
COM: 35.36.42.48

VAN DER BORGHT RENE PROF
31. THE PROMENADE
ISLE OF CAPRI
SURFERS PARADISE 4217
AUSTRALIA
TEL: 385712
TLX:
TLF:
EML:
COM: 35

VAN DER HUCHT KAREL A DR
SPACE RESEARCH LABORATORY
RESEARCH UTRECHT
BENELUXLAAN 21
NL-3527 HS UTRECHT
NETHERLANDS
TEL: 030-937145
TLX: 47224
TLF:
EML:
COM: 26.29.44

VAN DER HULST JAN M DR
KAPTEYN ASTRON. INSTITUTE
POSTBUS 800
NL-9700 AV GRONINGEN
NETHERLANDS
TEL: 31-50-634054
TLX: 53572 STARS NL
TLF:
EML:
COM: 28.40C

VAN DER KLIS MICHIEL DR
ASTRONOMICAL INSTITUTE
UNIVERSITY OF AMSTERDAM
ROETERSSTRAAT 15
1018 WV AMSTERDAM
NETHERLANDS
TEL: 31 20 5256007
TLX: 16460 FACWN NL
TLF:
EML: BITNET:A416NVDKOHASARA11
COM:

VAN DER KRUIT PIETER C DR
KAPTEYN LABORATORIUM
POSTBUS 800
NL-9700 AV GRONINGEN
NETHERLANDS
TEL: 050-634073
TLX: 53572 STARS NL
TLF:
EML:
COM: 28C.33.40

VAN DER LAAN H PROF DR
E.S.O.
KARL SCHWARZSCHILDSTR.2
8046 GARCHING B. MUNCHEN
GERMANY. F.R.
TEL: 089 320 06 227
TLX: 5 282 8220 EO D
TLF:
EML:
COM: 28.34.40.47

VAN DER RAAY HERMAN B
DEPT OF PHYSICS
UNIVERSITY OF BIRMINGHAM
P O BOX 363
BIRMINGHAM B15 2TT
U.K.
TEL: 021-472-1301
TLX: 228938 SPAPHY G
TLF:
EML:
COM: 35

VAN DESSEL EDWIN LUDO DR
KONINKLIJKE STERRENWACHT
RINGLAAN 3
B-1180 BRUSSELS
BELGIUM
TEL: 32-26-735366
TLX: 21565 OBSBEL
TLF:
EML:
COM: 26C.30

VAN DIGGELEN J DR
OBSERVATORY UTRECHT
AETSVELDSELAAN 12
NL-1381 EA WEESP
NETHERLANDS
TEL:
TLX:
TLF:
EML:
COM:

VAN DISHOECK EWINE F DR
CALTECH
DIV. OF GEOL. SCI. 170-25
PASADENA CA 91125
U.S.A.
TEL: 818 356 6942
TLX: 675425
TLF: 818 568 0935
EML:
COM: 34

VAN DORN BRADT HALE DR
MIT RM 37-581
CENTER FOR SPACE RESEARCH
CAMBRIDGE MA 02139
U.S.A.
TEL: 617-253-7550
TLX: 921473 MITCAM
TLF:
EML:
COM:

VAN DUINEN R J DR
FOKKER BV
P O BOX 7600
NL-1117 ZJ SCHIPHOL
NETHERLANDS
TEL: 020-5442036
TLX: 12[illegible]
TLF:
EML:
COM: 44

VAN FLANDERN THOMAS DR
T.F. ASSOCIATES
6327 WESTERN AVE NW
WASHINGTON DC 20015
U.S.A.
TEL: 202-363-3860
TLX:
TLF:
EML:
COM: 04.16.20.51

VAN GENDEREN A M DR
STERREWACHT LEIDEN
POSTBUS 9513
NL-2300 RA LEIDEN
NETHERLANDS
TEL: 071-272727
TLX: 31476 ASTRO NL
TLF:
EML:
COM: 27C.28

VAN GORKOM JACQUELINE H
NRAO
PO BOX 0
SOCORRO NM 87801
U.S.A.
TEL: 505-772-4302
TLX: 910-497-0174
TLF:
EML:
COM: 28.34.40

VAN GRONINGEN ERNST DR
UPPSALA ASTRONOMICAL OBS
BOX 515
S-75120 UPPSALA
SWEDEN
TEL:
TLX:
TLF:
EML:
COM:

VAN HAMME WALTER
FLORIDA INTNTL UNIVERSITY
DEPT. OF PHYSICS COLLEGE
UNIVERSITY PARK
MIAMI FLORIDA 33199
U.S.A.
TEL:
TLX:
TLF:
EML:
COM: 42

VAN HERK G
STERREWACHT LEIDEN
POSTBUS 9513
NL-2300 RA LEIDEN
NETHERLANDS
TEL: 071-272727
TLX:
TLF:
EML:
COM:

VAN HORN HUGH M PROF
DEPT PHYSICS & ASTRONOMY
UNIVERSITY OF ROCHESTER
ROCHESTER NY 14627
U.S.A.
TEL: 716-275-4344
TLX:
TLF:
EML:
COM: 35

VAN HOUTEN C J DR
STERREWACHT LEIDEN
POSTBUS 9513
NL-2300 RA LEIDEN
NETHERLANDS
TEL: 071-272727
TLX: 39058 ASTRO NL
TLF:
EML:
COM: 20

VAN HOUTEN-GROENEVELD I
STERREWACHT LEIDEN
POSTBUS 9513
NL-2300 RA LEIDEN
NETHERLANDS
TEL: 071-272727
TLX: 39058 ASTRO NL
TLF:
EML:
COM: 20

VAN HOVEN GERARD DR
DEPT OF PHYSICS
UNIVERSITY OF CALIFORNIA
IRVINE CA 92717
U.S.A.
TEL: 714-856-5145
TLX: 683322 IRIN
TLF:
EML:
COM: 10.12

VAN LEEUWEN FLOOR DR
ROYAL GREENWICH OBS
HERSTMONCEUX CASTLE
HAILSHAM
EAST SUSSEX BN27 1RP
U.K.
TEL:
TLX:
TLF:
EML:
COM: 08

VAN MOORSEL GUSTAAF DR
ESO
KARLSCHWARZSCHILD STR.2
8046 GARCHING BEI MUNCHEN
GERMANY F.R.
TEL: 49 89 32006362
TLX: 52828222 EO D
TLF:
EML:
COM: 28

VAN NIEUWKOOP J DR IR
STERREKUNDIG INSTITUUT
PRINCETONPLEIN 5
POSTBUS 80 000
NL-3508 TA UTRECHT
NETHERLANDS
TEL: 030 535231
TLX: 40048 FYLUT NL
TLF:
EML: BITNET:UNNGEEYG@HUTRUU0
COM: 40

VAN PARADIJS JOHANNES DR
ASTRONOMICAL INSTITUTE
ROETERSTRAAT 15
NL-1018 WB AMSTERDAM
NETHERLANDS
TEL: 020-5223003/4
TLX: 16460 FACWN NL
TLF:
EML:
COM: 42

VAN REGEMORTER HENRI DR
OBSERVATOIRE DE PARIS
SECTION DE MEUDON
F-92195 MEUDON PL CEDEX
FRANCE
TEL: 1-45 07 74 44
TLX: 201571 F
TLF:
EML:
COM: 14.36

VAN RENSBERGEN WALTER DR
ASTROPHYSISCH INSTITUUT
VRIJE UNIVERSITEIT
PLEINLAAN 2
B-1050 BRUSSEL
BELGIUM
TEL: 02-641-34-97
TLX:
TLF:
EML:
COM: 14

VAN RIPER KENNETH A DR
LOS ALAMOS NATIONAL LAB
MS B 226 X-6
PO BOX 1663
LOS ALAMOS NM 87545
U.S.A.
TEL: 505-667-8104
TLX:
TLF:
EML:
COM: 35.48

VAN SPEYBROECK LEON P DR
CENTER FOR ASTROPHYSICS
60 GARDEN STREET
CAMBRIDGE MA 02138
U.S.A.
TEL: 617-495-7233
TLX:
TLF:
EML:
COM: 44

VAN WOERDEN HUGO PROF DR
KAPTEYN LABORATORIUM
POSTBUS 800
NL-9700 AV GRONINGEN
NETHERLANDS
TEL: 050-116695
TLX: 53572 STARS NL
TLF:
EML:
COM: 28.33.34.40

VANDEN BOUT PAUL A
NRAO
EDGEMONT ROAD
CHARLOTTESVILLE VA 22903
U.S.A.
TEL: 804-296-0241
TLX: 910-997-0174
TLF:
EML:
COM: 34.40

VANDENBERG DON DR
C/O PHYSICS DEPT
UNIVERSITY OF VICTORIA
P O BOX 1700
VICTORIA BC V8W 2Y2
CANADA
TEL: 604-721-7739
TLX: 0497222
TLF:
EML:
COM: 35.37C

VANDERVOORT PETER O DR
DEPT ASTRON & ASTROPHYS
5640 ELLIS AVENUE
CHICAGO IL 60637
U.S.A.
TEL: 312-962-8209
TLX:
TLF:
EML:
COM: 33

VANYSEK VLADIMIR PROF
DEPT OF ASTRONOMY
CHARLES UNIVERSITY
SVEDSKA 8
150 00 PRAHA 5
CZECHOSLOVAKIA
TEL: 00422-540395
TLX: 121673 MFF
TLF:
EML:
COM: 15.34.47

VAN'T VEER FRANS DR
OBSERVATOIRE DE PARIS
61 AV. DE L'OBSERVATOIRE
F-75014 PARIS
FRANCE
TEL: 1-40-51-22-21
TLX: 270776 OBSASTR
TLF:
EML:
COM: 10.36.42

VAN'T VEER-MENNERET CL DR
OBSERVATOIRE DE PARIS
61 AV. DE L'OBSERVATOIRE
F-75014 PARIS
FRANCE
TEL: 1-40 51 22 49
TLX:
TLF:
EML:
COM: 29.36

VAPILLON LOIC J DR
OBSERVATOIRE DE PARIS
SECTION DE MEUDON
F-92195 MEUDON PL CEDEX
FRANCE
TEL: 1-45-07-76-23
TLX: 201571
TLF:
EML:
COM:

VARDANYAN R A DR
BYURAKAN ASTROPHYSICAL
OBSERVATORY
378433 ARMENIA
U.S.S.R.
TEL: 28-41-42
TLX:
TLF:
EML:
COM: 25

VARDAVAS ILIAS MIHAIL
ALLIGATOR RIVERS RES INST
P.O. BOX 387
BONDI JUNCTION NSW 2022
AUSTRALIA
TEL: 02-3870697
TLX: 23984 ARRIS
TLF:
EML:
COM: 36

VARDYA M S DR
TATA INSTITUTE OF
FUNDAMENTAL RESEARCH
HOMI BHABHA RD
BOMBAY 400 005
INDIA
TEL: 219111 x221
TLX: 011-3009 TIFR IN
TLF:
EML:
COM: 35.36

VARMA RAM KUMAR PROF
PHYSICAL RESEARCH LAB
AHMEDABAD 380 009
INDIA
TEL: 272-462-129
TLX: 0121397 PRL IN
TLF:
EML:
COM: 28

VARSHALOVICH DIMITRIJ PR
PHYSIKO-TECHNICAL INST
USSR ACADEMY OF SCIENCES
194021 LENINGRAD
U.S.S.R.
TEL: 247-22-55
TLX:
TLF:
EML:
COM: 14.34.51

VARVOGLIS H DR
DEPT OF ASTRONOMY
UNIVERSITY THESSALONIKI
GR-54006 THESSALONIKI
GREECE
TEL: 30-31-991357
TLX: 412181
TLF:
EML:
COM: 07

VASHKOV'YAK SOF'YA N DR
STERNBERG STATE
ASTRONOMICAL INSTITUTE
UNIVERSITETSKIJ PROSP 13
119889 MOSCOW
U.S.S.R.
TEL: 139-37-64
TLX: 113037 JAPET
TLF:
EML:
COM: 07

VASILEVA GALINA J DR
PULKOVO OBSERVATORY
196140 LENINGRAD
U.S.S.R.
TEL:
TLX:
TLF:
EML:
COM: 12

VASU-MALLIK SUSHMA DR
INDIAN INST OF ASTROPHYS
BANGALORE 560 034
INDIA
TEL: 569179/569180
TLX: 845763 IIAB IN
TLF:
EML:
COM: 29

VAUCLAIR GERARD P DR
OBS DU PIC-DU-MIDI
ET DE TOULOUSE
14 AVENUE EDOUARD BELIN
F-31400 TOULOUSE
FRANCE
TEL: 61-25-21-01
TLX: 530776
TLF:
EML:
COM: 35

VAUCLAIR SYLVIE D DR
OBS DU PIC-DU-MIDI
ET DE TOULOUSE
14 AVENUE EDOUARD BELIN
F-31400 TOULOUSE
FRANCE
TEL: 61-25-21-01
TLX:
TLF:
EML:
COM: 46

VAUGHAN ARTHUR H DR
PERKIN-ELMER CORP
7421 ORANGEWOOD AVE
GARDEN GROVE CA 92641
U.S.A.
TEL: 714-895-1667
TLX:
TLF:
EML:
COM: 10.12.25

VAVROVA ZDENKA DR
KLET OBSERVATORY
CESKE BUDEJOVICE
BEZRUCOVA 4
CZECHOSLOVAKIA
TEL: 0337 3274
TLX:
TLF:
EML:
COM: 20

VAZ LUIZ PAULO RIBEIRO
OBSERVATORIO ASTRONOMICO
DEPTO DE FISICA-ICEX-UFMG
CAIXA POSTAL 702
30161 BELO HORIZONTE - MG
BRAZIL
TEL: 55-31-4412541
TLX: 312308 UFMG BR
TLF:
EML:
COM: 42

VAZQUEZ MANUEL DR
INSTITUTO DE ASTROFISICA
DE CANARIAS
38071 LA LAGUNA
SPAIN
TEL:
TLX: 92640 IACE E
TLF:
EML:
COM: 51

VECK NICHOLAS
MARCONI RESEARCH CENTRE
WEST HANNINGFIELD RD
GT. BADDOW
CHELMSFORD ESSEX CM2 8HN
U.K.
TEL: 0245 73331
TLX: 995016 GECRES G
TLF: 0245 75244
EML: VE08%a.gec-mrc.co.uk@uel-cs
COM: 10

VEEDER GLENN J DR
JPL
MS 183-501
4800 OAK GROVE DRIVE
PASADENA CA 91109
U.S.A.
TEL: 213-354-7388
TLX:
TLF:
EML:
COM: 15

VEGA E. IRENE DR
OBSERVATORIO ASTRONOMICO
DE LA UNLP
PASEO DEL BOSQUE s/n
1900 LA PLATA
ARGENTINA
TEL: 021-21-7308
TLX: 31151 BULAP AR
TLF:
EML:
COM: 33

VEILLET CHRISTIAN
CERGA
AVENUE COPERNIC
F-06130 GRASSE
FRANCE
TEL: 93-36-58-49
TLX: 470865
TLF:
EML:
COM: 07.19.20

VEIS GEORGE PH D
GEODESY LABORATORY
NATL TECHNICAL UNIVERSITY
ATHENS
GREECE
TEL:
TLX:
TLF:
EML:
COM: 19

VEISMANN UNO DR
TARTU OBSERVATORY
202444 ESTONIA
U.S.S.R.
TEL:
TLX:
TLF:
EML:
COM:

VELKOV KIRIL
DEPT OF ASTRONOMY AND
NATL ASTRON OBSERVATORY
72 LENIN BLVD
1784 SOFIA
BULGARIA
TEL: 73-41-614
TLX: 23561 ECF BQN BG
TLF:
EML:
COM: 09.10

VELUSAMY T DR
RADIO ASTRONOMY CTR
P O BOX NO 8
UDHAGAMANDALAM 643 001
INDIA
TEL: 2651 & 2032
TLX: 8458468 TIFR IN
TLF:
EML:
COM: 40

VENKATAKRISHNAN P DR
MARSHALL SPACE FLIGHT
CENTER. ES 52 SOLAR BR.
HUNTSVILLE. AL 35812
U.S.A.
TEL: (205)544 9404
TLX:
TLF:
EML:
COM: 12

VENKATESAN DORASWAMY DR
PHYSICS DEPT
UNIVERSITY OF CALGARY
CALGARY ALB T2N 1N4
CANADA
TEL:
TLX:
TLF:
EML:
COM: 10

VENTURA JOSEPH DR
DEPT OF PHYSICS
UNIVERSITY OF CRETE
GR 71409 HERAKLION CRETE
GREECE
TEL: 081 239757
TLX: 262728
TLF:
EML: VENTURA@GRCREARN
COM:

VENUGOPAL V R DR
RADIO ASTRONOMY
CENTRE OF TIFR
POST BOX 8
UDHAGAMANDALAM 643 001
INDIA
TEL: 2651 & 2032
TLX: 0853-241 RAC IN
TLF:
EML:
COM: 33.40.51

VERBEEK PAUL DR
GEORGE MINNELAAN 50
B-9830 S MARTENS-LATEM
BELGIUM
TEL: 09-82-61-19
TLX:
TLF:
EML:
COM:

VERBUNT FRANCISCUS DR
ASTRONOMICAL INSTITUTE
POSTBUS 80000
3508 TA UTRECHT
NETHERLANDS
TEL: 030 3299833
TLX: 05 21845 XTEB D
TLF:
EML: BITNET:FUVWDGATPPIS
COM:

VERDET JEAN-PIERRE DR
OBSERVATOIRE DE PARIS
61 AVE DE L'OBSERVATOIRE
F-75014 PARIS
FRANCE
TEL: 1-40 51 22 06
TLX:
TLF:
EML:
COM: 41

VERES FERENC
BAJA OBSERVATORY
ASTRONOMICAL INSTITUTE
BOX 110
H-6500 BAJA
HUNGARY
TEL: 11064. 12170
TLX: 281303
TLF:
EML:
COM:

VERGNANO A PROF
OSSERVATORIO ASTRONOMICO
I-10025 PINO TORINESE
ITALY
TEL:
TLX:
TLF:
EML:
COM:

VERHEEST FRANK PROF
INST TEKORET. MECHANIKA
RIJKSUNIVERSITEIT GENT
KRIJGSLAAN 281 (S9)
B-9000 GENT
BELGIUM
TEL: 091-22 57 15
TLX: 12 754 RUGENT B
TLF:
EML:
COM: 10.27

VERMA R P DR
TATA INSTITUTE OF
FUNDAMENTAL RESEARCH
HOMI BHABHA ROAD
BOMBAY 400 005
INDIA
TEL: 219111
TLX: 0113009 TIFR IN
TLF:
EML:
COM: 25

VERMA SATYA DEV DR
DEPT PHYSICS & SPACE SCI
UNIVERSITY SCHOOL OF SCI
GUJARAT UNIVERSITY
AHMEDABAD 380 009
INDIA
TEL: (0272)440920
TLX:
TLF:
EML:
COM:

VERNIANI FRANCO PROF
DIPARTIMENTO DI FISICA
VIA IRNERIO 46
I-40126 BOLOGNA
ITALY
TEL: 26-09-91
TLX: 211664
TLF:
EML:
COM: 22

VERON MARIE-PAULE DR
OBS DE HAUTE-PROVENCE
F-04870 ST-MICHEL L'OBS.
FRANCE
TEL: 92-76-63-68
TLX:
TLF:
EML:
COM: 28

VERON PHILIPPE DR
OBS DE HAUTE-PROVENCE
F-04870 ST-MICHEL L'OBS.
FRANCE
TEL: 92-76-63-68
TLX:
TLF:
EML:
COM: 28.40

VERSCHUUR GERRIT L PROF
4802 BROOKSTONE TERRACE
BOWIE. MD 2071585
U.S.A.
TEL:
TLX:
TLF:
EML:
COM: 33.34.40.51

VERTER FRANCES DR
CODE 685
NASA
GSFC
GREENBELT MD 20771
U.S.A.
TEL: 301 286 7460
TLX:
TLF:
EML:
COM: 40

VESECKY J F DR
STANFORD CTR FOR RADAR
ASTRONOMY. STANFORD UNIV.
233 DURAND
STANFORD CA 94305-4055
U.S.A.
TEL:
TLX:
TLF:
EML:
COM:

VETESNIK MIROSLAV DR
GARYOUNIS UNIVERSITY
FACULTY OF SCIENCE
POB 9480 2
BENGHAZI
LIBYA
TEL:
TLX:
TLF:
EML:
COM: 33.42

VETTOLANI GIAMPAOLO
IST. DI RADIOASTRONOMIA
CNR
VIA IRNERIO 46
I-40126 BOLOGNA
ITALY
TEL: 39-51-232856
TLX: 211664 INFN BO I
TLF:
EML:
COM: 47

VEVERKA JOSEPH DR
CORNELL UNIVERSITY
312 SPACE SCI BLDG
ITHACA NY 14853
U.S.A.
TEL: 607-256-3507
TLX: 937478
TLF:
EML:
COM: 15.16

VIAL JEAN-CLAUDE
INSTITUT D'ASTROPHYSIQUE
SPATIALE
BP 10
F-91371 VERRIERES-LE-B.
FRANCE
TEL: 1-64-47-42-17
TLX: 600252
TLF:
EML:
COM: 10.12.44

VIALA YVES
DEMIRM
OBSERVATOIRE DE MEUDON
92195 MEUDON PR CEDEX
FRANCE
TEL: 33-1-45077912
TLX: 270912 OBSASTR
TLF:
EML:
COM: 34

VIALLEFOND FRANCOIS
DEMIRM
OBSERVATOIRE DE MEUDON
92195 MEUDON PR CEDEX
FRANCE
TEL: 33-1-45077905
TLX: 270912 OBSASTR
TLF: 33-1-45077893
EML: BITNET:FVIALLEF@FRMEU51
COM: 34

VICENTE RAIMUNDO O PROF
FACULDADE CIENCIAS LISBOA
RUA MESTRE AVIZ 30 B/C
1495 LISBOA
PORTUGAL
TEL: 2112666
TLX:
TLF:
EML:
COM: 19.31

VIDAL JEAN-LOUIS DR
OBS DU PIC-DU-MIDI
F-65200 BAGNERES-DE-B.
FRANCE
TEL: 62-95-19-69
TLX: 531625
TLF:
EML:
COM: 34

VIDAL NISSIM V DR
INSTITUTE FOR SCIENCES
& TECHNOLOGY
6 UZIEL STR.
GIVAT SHMUEL 51905
ISRAEL
TEL: 03 5321189
TLX:
TLF:
EML:
COM: 48

VIDAL-MADJAR ALFRED DR
INSTITUT D'ASTROPHYSIQUE
98 BIS BOULEVARD ARAGO
F-75014 PARIS
FRANCE
TEL: 1-43-20-14-25
TLX:
TLF:
EML:
COM: 34.44

VIEIRA MARTINS ROBERTO DR
OBSERVATORIO NACIONAL
RUA GENERAL BRUCE 586
SAO CRISTOVAO
20921 RIO DE JANEIRO
BRAZIL
TEL: 021-580-7313
TLX: 021-21288
TLF:
EML:
COM: 20

VIETRI MARIO DR
OSSERVATORIO ASTROFISICO
DI ARCETRI
LARGO E.FERMI 5
FIRENZE 50125
ITALY
TEL: 055 2752249
TLX: 572268
TLF:
EML: SPAN:38954::VIETRI
COM:

VIGIER JEAN-PIERRE DR
INSTITUT H. POINCARE
11 RUE P.& M. CURIE
F-75005 PARIS
FRANCE
TEL:
TLX:
TLF:
EML:
COM:

VIGOTTI MARIO
LAB. DI RADIOASTRONOMIA
VIA IRNERIO 46
I-40126 BOLOGNA
ITALY
TEL: 51-232856
TLX:
TLF:
EML:
COM:

VIIK TONU DR
TORAVERE OBSERVATORY
202444 TARTU RAJ. ESTONIA
U.S.S.R.
TEL: 4-11-81 TARTU
TLX:
TLF:
EML:
COM: 36

VILA SAMUEL C PROF
DEPT OF ASTRONOMY
UNIV OF PENNSYLVANIA
33RD & WALNUT STREETS
PHILADELPHIA PA 19104
U.S.A.
TEL: 215-898-5994
TLX:
TLF:
EML:
COM: 35

VILCHEZ MEDINA JOSE M DR
INSTITUTO DE ASTROFISICA
DE CANARIAS
38200 LA LAGUNA TENERIFE
SPAIN
TEL: 34 22 26 22 11
TLX: 92640
TLF:
EML:
COM:

VILHENA DE MORAES R DR
DEPTO DE ASTRONOMIA
ITA-CTA
12200 S. JOSE DOS CAMPOS
BRAZIL
TEL: 55-123-229088
TLX: 01173437 ZWO-24-73
TLF:
EML:
COM: 07

VILHU OSMI DR
OBS AND ASTROPHYS LAB
UNIVERSITY OF HELSINKI
TAHTITORNINMAKI
SF-00130 HELSINKI 13
FINLAND
TEL:
TLX:
TLF:
EML:
COM: 29.35.42.44

VILKKI ERKKI U
YERKES OBSERVATORY
WILLIAMS BAY WI 53191
U.S.A.
TEL: 414 245 5555
TLX:
TLF:
EML:
COM: 24

VILLELA THYRSO V. DR.
INPE
DEPT DE ASTROFISICA
CAIXA POSTAL 515
SAO JOSE DOS CAMPOS SP
BRAZIL
TEL:
TLX:
TLF:
EML:
COM:

VILMER NICOLE DR
DASOP
SECTION D'ASTROPHYSIQUE
OBSERVATOIRE PARIS MEUDON
92195 MEUDON PR.CEDEX
FRANCE
TEL: 33-1-45077406
TLX: 200590
TLF:
EML: SPAN:MEUDON::VILMER
COM: 10.12

VINCE ISTVAN
ASTRONOMICAL OBSERVATORY
VOLGINA 7
11050 BEOGRAD
YUGOSLAVIA
TEL: (11)419357
TLX: 72610 AOB YU
TLF:
EML:
COM:

VINER MELVYN R DR
DEPT. OF METALL. ENGIN.
QUEENS'S UNIVERSITY
KINGSTON ONT K7L 3N6
CANADA
TEL:
TLX:
TLF:
EML:
COM: 34.40

VINLUAN RENATO
UNIVERSITY OF SOUTHERN
PHILIPPINES
OBRERO DAVAO CITY 9501
PHILIPPINES
TEL:
TLX:
TLF:
EML:
COM: 10

VINOD S KRISHAN MRS DR
INDIAN INSTITUTE OF
ASTROPHYSICS
BANGALORE 560 034
INDIA
TEL: 566585/566497
TLX: 845763 IIAB IN
TLF:
EML:
COM: 10.49

VINTI JOHN P DR
MIT
MEASUREMENT SYSTEMS LAB
RM W59-216
CAMBRIDGE MA 02139
U.S.A.
TEL: 417-782-2470
TLX:
TLF:
EML:
COM: 07

VIOTTI ROBERTO DR
CNR ASTROFISICA SPAZIALE
C P 67
I-00044 FRASCATI
ITALY
TEL: 06-942-5655
TLX: 610261
TLF:
EML:
COM: 27.29.44

VIRGOPIA NICOLA PROF
DIPT. DI MATEMATICA
UNIV DI ROMA LA SAPIENZA
CITTA UNIVERSITARIA
I-00185 ROMA
ITALY
TEL:
TLX:
TLF:
EML:
COM:

VISHNIAC ETHAN T
UNIVERSITY OF TEXAS
DEPT OF ASTRONOMY
AUSTIN TX 78712
U.S.A.
TEL: 512-471-1429
TLX:
TLF:
EML:
COM: 47

VISHVESHWARA C V PROF
RAMAN RESEARCH INSTITUTE
BANGALORE 560 080
INDIA
TEL: 812-360122
TLX: 8452671 RRI IN
TLF:
EML:
COM: 47

VISVANATHAN NATARAJAN DR
MT STROMLO OBSERVATORY
WODEN P.O. ACT 2606
AUSTRALIA
TEL: 062-881111
TLX: 62270 TLG CANOPUS AA
TLF:
EML:
COM: 25.28.34

VITINSKIJ YURIJ I DR
PULKOVO OBSERVATORY
196140 LENINGRAD
U.S.S.R.
TEL: 298-22-42
TLX:
TLF:
EML:
COM: 10.12

VITON MAURICE DR
L A S
TRAVERSE DU SIPHON
LES TROIS LUCS
F-13012 MARSEILLE
FRANCE
TEL:
TLX:
TLF:
EML:
COM:

VITTONE ALBERTO ANGELO
OSSERVATORIO ASTRONOMICO
DI CAPODIMONTE
VIA MOIARIELLO 16
I-80131 NAPOLI
ITALY
TEL: 81-440101
TLX:
TLF:
EML:
COM:

VITTORIO NICOLA
ISTITUTO ASTRONOMICO
UNIVERSITA DI ROMA
VIA LANCISI 29
I-00161 ROMA
ITALY
TEL:
TLX:
TLF:
EML:
COM:

VITYAZEV VENIAMIN V DR
LENINGRAD STATE UNIV
BIBLIOTECHNAYA UL2
PETRODVORETS
198904 LENINGRAD
U.S.S.R.
TEL:
TLX:
TLF:
EML:
COM:

VIVEKANAND M DR
RAMAN RESEARCH INSTITUTE
BANGALORE 560 080
INDIA
TEL: 812-360122
TLX: 8452671 RRI IN
TLF:
EML:
COM: 40

VIVEKANANDA RAO
CENTRE OF ADVANCED STUDY
IN ASTRONOMY
OSMANIA UNIVERSITY
HYDERABAD 500 007
INDIA
TEL:
TLX:
TLF:
EML:
COM:

VIVES TEODORO JOSE DR
CTR ASTRON HISPANO ALEMAN
REINA 66 9ºB
CORREOS 511
04002 ALMERIA
SPAIN
TEL: 23-09-88
TLX: 78812 DSAZ E
TLF:
EML:
COM:

VLACHOS DEMETRIOS G PROF
UNIV OF THESSALONIKI
DEPT GEODESY & SURVEYING
FACULTY OF ENGINEERING
GR-54006 THESSALONIKI
GREECE
TEL: 031-991520
TLX: 412181 AUTH GR
TLF:
EML:
COM:

VLADILO GIOVANNI DR
OSSERVATORIO ASTRONOMICO
DI TRIESTE
VIA TIEPOLO 11
TRIESTE I 34131
ITALY
TEL: 40 30 9342
TLX: 461137 OAT I
TLF:
EML:
COM: 29

VLADIMIROV SIMEON
ASTRONOMICAL OBSERVATORY
72 LENIN BLVD
POST BOX 15
SOFIA 1309
BULGARIA
TEL: 23-13-97
TLX:
TLF:
EML:
COM: 09.46

VLAHOS LOUKAS DR
SECTION OF ASTROPHYSICS
DEPT OF PHYSICS
UNIV OF THESSALONIKI
GR 54006 THESSALONIKI
GREECE
TEL: 31-991357
TLX: 0412181 AUTH GR
TLF:
EML:
COM:

VOELK HEINRICH J PROF
MPI FUER KERNPHYSIK
POSTFACH 103 980
D-6900 HEIDELBERG
GERMANY. F.R.
TEL: 6221-516-295
TLX: 461666
TLF:
EML:
COM: 14.48

VOGEL STUART NEWCOMBE DR
PHYSICS DEPARTMENT
RENSSELAER POLYTECHNIC
INSTITUTE
TROY NY 12180-3590
U.S.A.
TEL: 518 276 8415
TLX: 6716050
TLF:
EML:
COM: 40

VOGLIS NIKOS DR
SECTION OF ASTROPHYSICS
DEPT OF PHYSICS
UNIVERSITY OF ATHENS
GR 15783 ZOGRAFOS ATHENS
GREECE
TEL: 01 7243 414
TLX: 223815 UNIV GR
TLF:
EML: node:GRATHUNI.userid:SPH70
COM: 28.47

VOGT NIKOLAUS DR
DEPTO DE ASTRONOMIA
UNIVERSIDAD CATOLICA
CASILLA 6014
SANTIAGO
CHILE
TEL: 775474
TLX: 240395 PUCVA CL
TLF:
EML:
COM: 27.29.51

VOGT STEVEN SCOTT
LICK OBSERVATORY
UNIVERSITY OF CALIFORNIA
SANTA CRUZ CA 95064
U.S.A.
TEL: 408-429-2844
TLX: 910-598-4408
TLF:
EML:
COM: 29

VOIGT HANS H PROF
NIKOLAUSBERGER WEG 74.
GEISMARLANDSTR. 11
D-3400 GOETTINGEN
GERMANY. F.R.
TEL: 0551-55879
TLX:
TLF:
EML:
COM:

VOLLAND H DR
ASTRONOMISCHES INSTITUT
DER UNIVERSITAET
AUF DEM HUEGEL 71
D-5300 BONN
GERMANY. F.R.
TEL: 0228-733674
TLX: 0886440
TLF:
EML:
COM:

VOLONTE SERGE DR
ESA
8-10 RUE MARIO NIKIS
F-75007 PARIS
FRANCE
TEL:
TLX:
TLF:
EML:
COM: 12.14

VON BORZESZKOWSKI H H DR
EINSTEIN-LABORATORIUM
AKAD. WISSENSCHAFTEN DDR
ROSA-LUXEMBURG-STR 17A
DDR-1502 POTSDAM
GERMANY. D.R.
TEL: 762225
TLX:
TLF:
EML:
COM: 47

VON DER HEIDE JOHANN DR
ALARDUSSTR 12
D-2000 HAMBURG 20
GERMANY. F.R.
TEL: 40-491-4016
TLX:
TLF:
EML:
COM: 8

VON HOERNER SEBASTIAN DR
KRUMMENACKER-STR 186
D-7300 ESSLINGEN
GERMANY. F.R.
TEL:
TLX:
TLF:
EML:
COM: 51

VON WEIZSAECKER C F PROF
MAX-PLANCK INSTITUT
RIEMERSCHMID-STR 7
D-8130 STARNBERG
GERMANY. F.R.
TEL:
TLX:
TLF:
EML:
COM:

VONDRAK JAN DR
ASTRONOMICAL INSTITUTE
CZECH. ACAD. OF SCIENCES
BUDECSKA 6
120 23 PRAHA 2
CZECHOSLOVAKIA
TEL: 42-2-258757
TLX: 66-122486
TLF:
EML:
COM: 19C

VORONTSOV-VEL'YAMINOV B A
STERNBERG STATE ASTRONOM.
INSTITUTE
117234 MOSCOW
U.S.S.R.
TEL:
TLX:
TLF:
EML:
COM: 28.34

VOROSHILOV V I DR
MAIN ASTRONOMICAL OBS
UKRAINIAN ACADEMY OF SCI
252127 KIEV
U.S.S.R.
TEL: 66-31-10
TLX: 131406 SKY SU
TLF:
EML:
COM: 33

VORPAHL JOAN A DR
JET PROPULSION LAB.
CALTECH
4800 OAK GROVE DRIVE
PASADENA CA 91109
U.S.A.
TEL:
TLX:
TLF:
EML:
COM:

VRBA FREDERICK J DR
US NAVAL OBSERVATORY
PO BOX 1149
FLAGSTAFF AZ 86002
U.S.A.
TEL: 602-779-5132
TLX:
TLF:
EML:
COM: 09.25C.34

VREUX JEAN MARIE DR
INSTITUT D'ASTROPHYSIQUE
UNIVERSITE DE LIEGE
AVENUE DE COINTE 5
B-4200 COINTE-OUGREE
BELGIUM
TEL: 41-529980
TLX: 41264
TLF:
EML:
COM: 29

VRTILEK JAN M DR
HARVARD UNIVERSITY
DIV. OF APPLIED SCIENCES
29 OXFORD STREET
CAMBRIDGE MA 02138
U.S.A.
TEL: 617 495 0589
TLX:
TLF:
EML: SPAN::CFARGZ::VRTILEK
COM: 28

VU DUONG TUYEN DR
BUREAU DES LONGITUDES
77 AVE DENFERT-ROCHEREAU
F-75014 PARIS
FRANCE
TEL: 1-45-07-22-62
TLX:
TLF:
EML:
COM: 20

VUILLEMIN ANDRE DR
L.A.S.
TRAVERSE DU SIPHON
LES 3 LUCS
13012 MARSEILLE
FRANCE
TEL:
TLX:
TLF:
EML:
COM:

VUJNOVIC VLADIS DR
INSTITUTE OF PHYSICS
OF THE UNIVERSITY
P O B 304
41001 ZAGREB
YUGOSLAVIA
TEL: 041-271211
TLX: 22203 IPS YU
TLF:
EML:
COM: 14.46

VUKICEVIC K M PROF DR
DEPT OF ASTRONOMY
FACULTY OF SCIENCES
STUDENTSKI TRG 16
11000 BEOGRAD
YUGOSLAVIA
TEL:
TLX:
TLF:
EML:
COM: 12

VYALSHIN GENNADIJ F DR
MAIN ASTRONOMICAL OBS
PULKOVO
196140 LENINGRAD
U.S.S.R.
TEL:
TLX:
TLF:
EML:
COM: 10

WACHMANN A A PROF DR
SCHNIEDERSBERG 2B
D-2057 REINBEK
GERMANY. F.R.
TEL:
TLX:
TLF:
EML:
COM: 27

WACKERNAGEL H BEAT DR
51 BROADMOOR HILLS DRIVE
COLORADO SPRINGS CO 80906
U.S.A.
TEL: 303-554-3801
TLX:
TLF:
EML:
COM: 04 31

WADDINGTON C JAKE PROF
PHYSICS DEPT
UNIVERSITY OF MINNESOTA
116 CHURCH ST S.E.
MINNEAPOLIS MN 55455
U.S.A.
TEL: 612-624-2566
TLX: 910-576-2955
TLF:
EML:
COM:

WADE CAMPBELL M DR
NRAO
PO BOX O
SOCORRO NM 87801
U.S.A.
TEL: 505-835-5351
TLX: 910-988-1710
TLF:
EML:
COM: 40

WADE RICHARD ALAN DR
STEWARD OBSERVATORY
UNIVERSITY OF ARIZONA
TUCSON AZ 85721
U.S.A.
TEL: 602 621 2752
TLX: 467175
TLF:
EML: BITNET:WADE@ARIZRVAX
COM: 42

WAELKENS CHRISTOFFEL
ASTRONOMISCH INSTITUUT
KATHOLIEKE UNIV. LEUVEN
CELESTIJNENLAAN 200B
B-3030 HEVERLEE
BELGIUM
TEL: 016-20-06-56
TLX: 25715
TLF:
EML:
COM: 27

WAGNER RAYMOND L DR
FORD AEROSPACE& COMM CORP
COLORADO SPRINGS DIVISION
10440 STATE HWY. 83N
COLORADO SPRINGS CO 80908
U.S.A.
TEL: 303-594-1118
TLX:
TLF:
EML:
COM:

WAGNER ROBERT M DR
LOWELL OBSERVATORY
MARS HILL ROAD 1400 WEST
FLAGSTAFF AZ 86001
U.S.A.
TEL: 602 779 0106
TLX:
TLF:
EML:
COM:

WAGNER WILLIAM J DR
NOAA/SPACE ENVIR. LAB
325 BROADWAY
PO BOX 3000
BOULDER CO 80303
U.S.A.
TEL: 303-497-3274
TLX:
TLF:
EML:
COM:

WAGONER ROBERT V PROF
STANFORD UNIVERSITY
VARIAN PHYSICS BLDG
STANFORD CA 94305
U.S.A.
TEL: 415-723-4561
TLX: 348402
TLF:
EML:
COM: 47

WAINWRIGHT JOHN DR
DEPT OF APPLIED MATH
UNIVERSITY OF WATERLOO
WATERLOO ONTARIO N2L 3G1
CANADA
TEL: 519 885 1211
TLX: 069 55259
TLF:
EML:
COM: 47

WAKAMATSU KEN-ICHI DR
COLLEGE OF TECHNOLOGY
GIFU UNIVERSITY
GIFU 501-11
JAPAN
TEL: 582-30-1111
TLX:
TLF:
EML:
COM: 28

WAKO KOJIRO DR
INTERNATIONAL LATITUDE
OBSERVATORY
MIZUSAWA
IWATE 023
JAPAN
TEL:
TLX:
TLF:
EML:
COM: 19

WALBORN NOLAN R DR
SPACE TELESCOPE SCI INST
3700 SAN MARTIN DRIVE
BALTIMORE MD 21218
U.S.A.
TEL: 301-338-4915
TLX: 6849101 STSCI UW
TLF:
EML:
COM: 45C

WALCH JEAN-JACQUES
CERGA
AVENUE COPERNIC
F-06130 GRASSE
FRANCE
TEL: 93-36-58-49
TLX: 470865 CERGA F
TLF:
EML:
COM: 07

WALDMEIER MAX PROF DR
SWISS FEDERAL OBSERVATORY
WIRZENWEID 15
CH-8053 ZUERICH
SWITZERLAND
TEL:
TLX:
TLF:
EML:
COM: 10.12

WALKER ALISTAIR ROBIN DR
CERRO TOLOLO INTER-
AMERICAN OBSERVATORY
CASILLA 603
LA SERENA
CHILE
TEL: 09565-2-3352
TLX: 621301
TLF:
EML:
COM: 09.25

WALKER ARTHUR B C JR PROF
CENTER FOR SPACE SCIENCE
& ASTROPHYSICS
STANFORD UNIV. ERL 310
STANFORD CA 94305
U.S.A.
TEL: 415-497-1486
TLX:
TLF:
EML:
COM:

WALKER EDWARD N MR
ROYAL GREENWICH OBS
HERSTMONCEUX CASTLE
HAILSHAM BN27 1RP
U.K.
TEL: 0323-833171
TLX: 87451
TLF:
EML:
COM: 27

WALKER GORDON A H PROF
DEPT GEOPHYS & ASTRONOMY
UNIV OF BRITISH COLUMBIA
2075 WESBROOK PLACE
VANCOUVER BC V6T 1W5
CANADA
TEL: 604-228-4133
TLX: 0454245
TLF:
EML:
COM: 09.29.34.37.45

WALKER HELEN J.
AMES RESEARCH CENTER
MS 245-6
MOFFET FIELD CA 94035
U.S.A.
TEL: 415-694-4216
TLX: 348408 NASAAMES MOFF
TLF:
EML:
COM:

WALKER IAN WALTER
DEPT OF ASTRONOMY
THE UNIVERSITY
GLASGOW G12 8QQ
U.K.
TEL:
TLX:
TLF:
EML:
COM: 07

WALKER MERLE F PROF
LICK OBSERVATORY
UNIVERSITY OF CALIFORNIA
SANTA CRUZ CA 95064
U.S.A.
TEL: 408-429-2526
TLX:
TLF:
EML:
COM: 09.27.37.50C

WALKER RICHARD L
US NAVAL OBSERVATORY
FLAGSTAFF STATION
BOX 1149
FLAGSTAFF AZ 86002
U.S.A.
TEL: 602-774-6621
TLX:
TLF:
EML:
COM: 26.42

WALKER ROBERT C DR
NRAO
P O BOX O
SOCORRO NM 87801
U.S.A.
TEL: 505 8357247
TLX: 910 988 1710
TLF:
EML: BITNET:CWALKER@NRAO
COM: 40

WALKER ROBERT M A PROF
PHYSICS DEPT
WASHINGTON UNIVERSITY
BOX 1105
ST LOUIS MO 63130
U.S.A.
TEL: 314-889-6225
TLX:
TLF:
EML:
COM: 16

WALKER WILLIAM S G
14 APPLEYARD CRESC.
AUCKLAND 5
NEW ZEALAND
TEL: 09-548-736
TLX:
TLF:
EML:
COM: 27.42

WALL JASPER V DR
ROYAL GREENWICH OBS
HERSTMONCEUX CASTLE
HAILSHAM BN27 1RP
U.K.
TEL: 323-833-171
TLX: 87451
TLF:
EML:
COM: 40

WALLACE LLOYD V DR
KITT PEAK NATIONAL
OBSERVATORY
PO BOX 26732
TUCSON AZ 85726
U.S.A.
TEL: 602-327-5511
TLX:
TLF:
EML:
COM: 16.17

WALLACE PATRICK T MR
STARLINK RUTHERFORD
APPLETON LAB
CHILTON. DIDCOT OX11 0QX
U.K.
TEL: 44-235-445-472
TLX: 83159
TLF:
EML:
COM: 05.08.09

WALLACE RICHARD K
LOS ALAMOS NAT LABORATORY
X-7, MS B257
LOS ALAMOS NM 87545
U.S.A.
TEL: 505-667-5000
TLX:
TLF:
EML:
COM:

WALLENQUIST AAKE A E PROF
ASTRONOMICAL OBSERVATORY
NORRLANDSGATAN 34 B
S-752 29 UPPSALA
SWEDEN
TEL: 18-13 56 85
TLX:
TLF:
EML:
COM: 25.37

WALLERSTEIN GEORGE PROF
ASTRONOMY DEPT FM 20
UNIVERSITY OF WASHINGTON
SEATTLE WA 98195
U.S.A.
TEL: 206-543-2888
TLX:
TLF:
EML:
COM: 27.29

WALLIS MAX K DR
DEPT APPL MATHS & ASTRON
UNIVERSITY COLLEGE
P O BOX 78
CARDIFF CF1 1XL. WALES
U.K.
TEL: 222-44211
TLX: 488635
TLF:
EML:
COM: 15.49.51

WALMSLEY C MALCOLM DR
MPI FUER RADIOASTRONOMIE
AUF DEM HUEGEL 69
D-5300 BONN 1
GERMANY. F.R.
TEL: 0228-525305
TLX: 0886440 MPIFR D
TLF:
EML:
COM: 34.40

WALRAVEN TH DR
P O BOX 98
CORNELIA
ORANGE FREESTATE 9850
SOUTH AFRICA
TEL:
TLX:
TLF:
EML:
COM: 25.27

WALSH DENNIS DR
NRAL
JODRELL BANK
MACCLESFIELD SK11 9DL
U.K.
TEL: 0477-71321
TLX: 36149
TLF:
EML:
COM: 40.44

WALTER FREDERICK M
UNIVERSITY OF COLORADO
CASA
PO BOX 391
BOULDER CO 80309
U.S.A.
TEL: 303-492-7606
TLX: 9109403441
TLF:
EML:
COM:

WALTER HANS G DR
ASTRONOMISCHES RECHEN
INSTITUT
MOENCHHOFSTR 12-14
D-6900 HEIDELBERG 1
GERMANY. F.R.
TEL: 49026
TLX: 461336 ARIHD D
TLF:
EML:
COM: 08.24

WALTER KURT PROF DR
ASTRONOMISCHES INSTITUT
DER UNIVERSITAET
WALDHAEUSERSTR 64
D-7400 TUEBINGEN
GERMANY. F.R.
TEL: 07071-296126
TLX: 7262714 AIT D
TLF:
EML:
COM: 42

WALTERBOS RENE A M DR
BERKELEY ASTRONOMY DEPT
UNIVERSITY OF CALIFORNIA
BERKELEY CA 94720
U.S.A.
TEL: 415 642 8285
TLX: 820181
TLF:
EML: walterbos@bkyast.berkeley.e
COM: 28

WAMPLER E JOSEPH PROF
ESO
KARL-SCHWARZSCHILD-STR 2
D-8046 GARCHING B MUNCHEN
GERMANY. F.R.
TEL: 49-89-320-06297
TLX: 52828222 EO D
TLF:
EML:
COM: 09

WAMSTEKER WILLEM DR
ESA IUE GROUND STATION
VILLAFRANCA DE CASTILLO
P O BOX 54065
28080 MADRID
SPAIN
TEL: 34-1-401-9661
TLX: 42555
TLF:
EML:
COM: 44C

WAN FOOK SUN
DEPT OF MATHEMATICS
NATL UNIVERSITY SINGAPORE
KENT RIDGE
SINGAPORE 0511
SINGAPORE
TEL: 772-2742
TLX:
TLF:
EML:
COM:

WAN LAI
ZO-SE SECTION
SHANGHAI OBSERVATORY
SHANGHAI
CHINA. PEOPLE'S REP.
TEL: 380696
TLX: 33164 SHAO CN
TLF:
EML:
COM: 24.37

WAN TONG-SHAN
SHANGHAI OBSERVATORY
80 NAN DAN ROAD
SHANGHAI
CHINA. PEOPLE'S REP.
TEL: 386191
TLX: 33164 SHAO CN
TLF:
EML:
COM: 19.40

WANAS M I DR
DEPT OF ASTRONOMY
CAIRO UNIVERSITY
GEZA ORMAN
EGYPT
TEL:
TLX:
TLF:
EML:
COM: 48

WANG CHUAN-JIN
PURPLE MOUNTAIN
OBSERVATORY
NANJING
CHINA. PEOPLE'S REP.
TEL: 46700
TLX: 34144 PMONJ CN
TLF:
EML:
COM: 25

WANG DEYU
PURPLE MOUNTAIN
OBSERVATORY
NANJING
CHINA. PEOPLE'S REP.
TEL: 42817. 46700
TLX: 34144 PMONJ CN
TLF:
EML:
COM: 48

WANG DE-CHANG
PURPLE MOUNTAIN
OBSERVATORY
NANJING
CHINA. PEOPLE'S REP.
TEL: 646700. 644205
TLX: 34144 PMONJ CN
TLF:
EML:
COM: 22.41

WANG JIA-JI
SHANGHAI OBSERVATORY
ACADEMIA SINICA
SHANGHAI
CHINA. PEOPLE'S REP.
TEL: 386191
TLX: 33164 SHAO CN
TLF:
EML:
COM: 24

WANG JIA-LONG
BEIJING ASTRONOMICAL OBS
ACADEMIA SINICA
BEIJING
CHINA. PEOPLE'S REP.
TEL:
TLX: 22040 BAOBS CN
TLF:
EML:
COM: 10

WANG JING-SHENG
YUNNAN OBSERVATORY
P.O. BOX 110
KUNMING
CHINA. PEOPLE'S REP.
TEL: 72946
TLX: 64040 YUOBS CN
TLF:
EML:
COM: 40

WANG JING-XIU
BEIJING ASTRONOMICAL OBS
ACADEMIA SINICA
BEIJING
CHINA. PEOPLE'S REP.
TEL: 28 1698
TLX: 22040 BAOBS CN
TLF:
EML:
COM: 10.12

WANG LAN-JUAN
SHANGHAI OBSERVATORY
ACADEMIA SINICA
SHANGHAI
CHINA. PEOPLE'S REP.
TEL: 386191
TLX: 33164 SHAO CN
TLF:
EML:
COM: 09

WANG RENCHUAN
CENTER FOR ASTROPHYSICS
UNIV SCIENCE & TECHNOLOGY
HEFEI. ANHUI
CHINA. PEOPLE'S REP.
TEL:
TLX: 90028 USTC CN
TLF:
EML:
COM: 47

WANG SHOU-GUAN
BEIJING OBSERVATORY
ACADEMIA SINICA
BEIJING
CHINA. PEOPLE'S REP.
TEL: 281261
TLX: 22040 BAOAS CN
TLF:
EML:
COM: 38C.40.48

WANG SHUI
DEPT EARTH & SPACE SCI.
UNIV SCIENCE & TECHNOLOGY
HEFEI. ANHUI
CHINA. PEOPLE'S REP.
TEL: 63300-209
TLX: 4430
TLF:
EML:
COM: 44

WANG SHUNDE DR
BEIJING ASTRONOMICAL OBS
CHINESE ACAD OF SCIENCE
BEIJING 100080
CHINA. PEOPLE'S REP.
TEL: 2561264
TLX: 22040 BAOAS CN
TLF:
EML:
COM:

WANG SI-CHAO
PURPLE MOUNTAIN
OBSERVATORY
NANJING
CHINA. PEOPLE'S REP.
TEL: 44205
TLX: 34144 PMONJ CN
TLF:
EML:
COM: 15

WANG YANAN
NANJING ASTRONOMICAL
INSTRUMENT FACTORY
NANJING
CHINA. PEOPLE'S REP.
TEL: 46191
TLX: 34136 GLYNJ c/o NAIF
TLF:
EML:
COM: 09

WANG YIMING
YUNNAN OBSERVATORY
P.O. BOX 110
KUNMING. YUNNAN PROVINCE
CHINA. PEOPLE'S REP.
TEL:
TLX:
TLF:
EML:
COM: 09

WANG YI-MING DR
CODE 4172W
NAVAL RESEARCH LABORATORY
WASHINGTON DC 20375-5000
U.S.A.
TEL: 202 767 6202
TLX:
TLF:
EML:
COM: 10.48.49

WANG ZHENG MING
SHAANXI ASTRONOMICAL OBS
P.O. BOX 18
LINTONG
XIAN
CHINA. PEOPLE'S REP.
TEL: 32255 XIAN
TLX: 70121 CSAO CN
TLF:
EML:
COM: [illegible]

WANG ZHEN-RU
DEPT OF ASTRONOMY
NANJING UNIVERSITY
NANJING
CHINA. PEOPLE'S REP.
TEL: 37551 x 2685
TLX: 34151 PRCNU CN
TLF:
EML:
COM: 48

WANG ZHEN-YI
PURPLE MOUNTAIN
OBSERVATORY
NANJING
CHINA. PEOPLE'S REP.
TEL: 46700
TLX: 34144 PMONJ CN
TLF:
EML:
COM: 12

WANNIER PETER GREGORY DR
JET PROPULSION LAB 169506
CALIF INST OF TECHNOLOGY
4800 OAK GROVE DRIVE
PASADENA CA 91109
U.S.A.
TEL: 818-354-[illegible]
TLX: 67-5429
TLF:
EML:
COM: 34.40

WARD MARTIN JOHN
INSTITUTE OF ASTRONOMY
MADINGLEY ROAD
CAMBRIDGE CB3 OHA
U.K.
TEL:
TLX:
TLF:
EML:
COM: 28.42

WARD RICHARD A DR
LAWRENCE LIVERMORE LAB
L-23. A-DIVISION
PO BOX 808
LIVERMORE CA 94550
U.S.A.
TEL: 415-4[illegible]
TLX: 910-386-8339 UCLLL
TLF:
EML:
COM: 35

WARD WILLIAM R DR
JPL
4800 OAK GROVE DRIVE
PASADENA CA 91103
U.S.A.
TEL:
TLX:
TLF:
EML:
COM: 19

WARDLE JOHN F C PROF
PHYSICS DEPT
BRANDEIS UNIVERSITY
WALTHAM MA 02154
U.S.A.
TEL: 617-647-2869
TLX:
TLF:
EML:
COM: 40

WARES GORDON W DR
73 PERKINS STREET
WEST NEWTON MA 02165
U.S.A.
TEL:
TLX:
TLF:
EML:
COM: 14

WARGAU WALTER F DR
UNIV OF SOUTH AFRICA
DEPT MATHS. APPL MATHS &
ASTRONOMY. P.O. BOX 392
PRETORIA 0001
SOUTH AFRICA
TEL: 27-12-440-2133
TLX: 350068 TA UNISA [illegible]
TLF:
EML:
COM: 42

WARMAN JOSEF DR
INSTITUTO DE ASTRONOMIA
UNAM
APDO POSTAL 70-264
04510 MEXICO DF
MEXICO
TEL:
TLX:
TLF:
EML:
COM:

WARNER BRIAN PROF
ASTRONOMY DEPT
INST. OF THEOR.PHYSICS
& ASTROPHYSICS
7700 RONDEBOSCH
SOUTH AFRICA
TEL: 6502391
TLX: 521439
TLF:
EML:
COM: 27C.42

WARNER JOHN W DR
PERKIN-ELMER CORP
M/S 892
100 WOOSTER HEIGHTS RD
DANBURY CT 06810
U.S.A.
TEL: 203-796-7919
TLX:
TLF:
EML:
COM: 28.44

WARNER PETER J DR
CAVENDISH LABORATORY
MADINGLEY ROAD
CAMBRIDGE CB3 OHE
U.K.
TEL: 223-66477
TLX: 81292 CAVLAB G
TLF:
EML:
COM: 40

WARREN WAYNE H JR DR
NASA/GSFC
CODE 633.8
GREENBELT MD 20771
U.S.A.
TEL: 301-286-8310
TLX: 89675 NASCOM GBLT
TLF:
EML:
COM: 05C.25.37 45

WARWICK JAMES W DR
DEPT OF ASTROPHYS.PLANET
& ATMOSPHERIC SCIENCES
UNIVERSITY OF COLORADO
BOULDER CO 80309
U.S.A.
TEL: 303-447-9524
TLX:
TLF:
EML:
COM: 12 40

WARWICK ROBERT S DR
PHYSICS DEPT
UNIVERSITY OF LEICESTER
LEICESTER LE1 7RH
U.K.
TEL: 533-554455
TLX: 341664 LUXRAY G
TLF:
EML:
COM:

WARZEE J DR
115 RUE MATTOT
B-1410 WATERLOO
BELGIUM
TEL:
TLX:
TLF:
EML:
COM:

WASHIMI HARUICHI DR
RESEARCH INSTITUTE
OF ATMOSPHERICS
NAGOYA UNIVERSITY
TOYOKAWA AICHI 442
JAPAN
TEL: 05338-6-3154
TLX: 4323321
TLF:
EML:
COM:

WASSERMAN LAWRENCE H DR
LOWELL OBSERVATORY
1400 W. MARS HILL ROAD
FLAGSTAFF AZ 86001
U.S.A.
TEL: 602-774-3358
TLX:
TLF:
EML:
COM: 16.20C.24

WASSON JOHN T
INSTITUTE OF GEOPHYSICS
UNIVERSITY OF CALIFORNIA
LOS ANGELES CA 90024
U.S.A.
TEL: 213-825-1986
TLX:
TLF:
EML:
COM: 15C.16

WATANABE TAKASHI DR
TOKYO ASTRONOMICAL OBSERV
MITAKA
TOKY 181
JAPAN
TEL: 0532-54-3067
TLX: 2822307 TAONTK J
TLF:
EML:
COM: 49C

WATANABE TETSUYA
TOKYO ASTRONOMICAL
OBSERVATORY
2-21-1 OSAWA. MITAKA
TOKYO 181
JAPAN
TEL: 422-32-5111
TLX:
TLF:
EML:
COM: 36

WATERWORTH MICHAEL DR
UNIVERSITY OF TASMANIA
GPO BOX 252C
HOBART. TASMANIA 7001
AUSTRALIA
TEL: 61-102-202405
TLX: 58150 AA
TLF:
EML:
COM: 29

WATSON FREDERICK GARNETT
UK SCHMIDT TELESCOPE
PRIVATE BAG
COONABARABRAN NSW 2357
AUSTRALIA
TEL: 068-421622
TLX: 61[illegible]45 CANOPUS AA
TLF:
EML:
COM: 09.51

WATSON WILLIAM D PROF
PHYSICS DEPT
UNIVERSITY OF ILLINOIS
URBANA IL 61801
U.S.A.
TEL: 217-33-7240
TLX:
TLF:
EML:
COM:

WATT GRAEME DAVID
UK TELESCOPES UNIT
665 KOMOHANA ROAD
HILO HI 96720
U.S.A.
TEL: 808-961-3576
TLX: 633135
TLF:
EML:
COM: 34

WATTENBERG D PROF
LINDENHOFSTR 57
DDR-1147 BERLIN
GERMANY. D.R.
TEL: 5[illegible]7-[illegible]-[illegible]
TLX:
TLF:
EML:
COM: 41

WAYMAN PATRICK A PROF
DUNSINK OBSERVATORY
DUBLIN 15
IRELAND
TEL: 35[illegible]-1-387911
TLX: 31657 DIAS
TLF:
EML:
COM: 05C.33.50

WDOWIAK THOMAS J DR
PHYSICS DEPARTMENT
UNIVERSITY OF ALABAMA
BIRMINGHAM AL 35294
U.S.A.
TEL: 205 934 4736
TLX: 888826 UAB BHM
TLF:
EML:
COM: 15

WEAVER HAROLD F PROF
DEPT OF ASTRONOMY
UNIVERSITY OF CALIFORNIA
BERKELEY CA 94720
U.S.A.
TEL:
TLX: 820181 UCS AST
TLF:
EML:
COM: 15.33.34.37

WEAVER THOMAS A DR
PHYSICS DEPT L-17
LAWRENCE LIVERMORE LAB
PO BOX 808
LIVERMORE CA 94550
U.S.A.
TEL: 415-423-1850
TLX:
TLF:
EML:
COM: 35.48

WEBBER JOHN C DR
INTERFEROMETRICS INC.
8150 LEESBURG PIKE
VIENNA. VA 22180
U.S.A.
TEL: 703 790 8500
TLX:
TLF:
EML:
COM:

WEBBINK RONALD F DR
DEPT OF ASTRONOMY
UNIVERSITY OF ILLINOIS
1011 W. SPRINGFIELD AVE
URBANA IL 61801
U.S.A.
TEL: 217-333-9582
TLX: 910-245-2434 AST
TLF:
EML:
COM: 27.35.42C

WEBER STEPHEN VANCE
L-477
LAWRENCE LIVERMORE LAB
PO BOX 808
LIVERMORE CA 94550
U.S.A.
TEL: 415-422-5433
TLX: 910-386-8339
TLF:
EML:
COM: 36

WEBROVA LUDMILA DR
ASTRONOMICAL INSTITUTE
CZECH. ACAD. OF SCIENCES
BUDECSKA 6
120 23 PRAHA
CZECHOSLOVAKIA
TEL: 255 287
TLX: 122486
TLF:
EML:
COM: 31

WEBSTER ADRIAN S DR
UNITED KINGDOM TELESCOPES
665 KOMOHANA STREET
HILO HI 96720
U.S.A.
TEL: 808-961-3756
TLX: 633135
TLF:
EML:
COM: 47.48

WEBSTER B LOUISE DR
SCHOOL OF PHYSICS
UNIV OF NEW SOUTH WALES
P.O.BOX 1
KENSINGTON NSW 2033
AUSTRALIA
TEL: 02-697-4546
TLX: 26054 AA
TLF:
EML:
COM: 34

WEBSTER RACHEL
CANADIAN INST. FOR ASTRO.
UNIVERSITY OF TORONTO
60 ST GEORGE STREET
TORONTO ONTARIO M5S 1A7
CANADA
TEL: 416-978-8496
TLX:
TLF:
EML: BITNET:WEBSTER@UTORPHYS
COM: 47

WEEDMAN DANIEL W PROF
PENNSYLVANIA STATE UNIV
ASTRONOMY DEPARTMENT
525 DAVEY LABORATORY
UNIVERSITY PARK PA 16802
U.S.A.
TEL: 814-865-0418
TLX: 842510
TLF:
EML:
COM: 28

WEEKES TREVOR C DR
FRED LAWRENCE WHIPPLE OBS
HARVARD-SMITHSONIAN CTR
PO BOX 97
AMADO AZ 85645-0097
U.S.A.
TEL: 602-679-6741
TLX:
TLF:
EML:
COM:

WEGNER GARY ALAN
DEPT PHYSICS & ASTRONOMY
WILDER LABORATORY
DARTMOUTH COLLEGE
HANOVER NH 03755
U.S.A.
TEL: 603-646-2359
TLX:
TLF:
EML:
COM: 29.30

WEHINGER PETER A DR
PHYSICS DEPARTMENT
ASTRONOMY GROUP
ARIZONA STATE UNIVERSITY
TEMPE AZ 85287
U.S.A.
TEL: 602-965-4063
TLX: 140289 HALLEY ASU UT
TLF:
EML: BITNET:WEHINGER@ASUCPS
COM: 15.28.29

WEHLAU AMELIA DR
ASTRONOMY DEPT
UNIV OF WESTERN ONTARIO
LONDON ONT N6A 3K7
CANADA
TEL: 519-679-3186
TLX: 064-7134
TLF:
EML:
COM: 27.29.37

WEHLAU WILLIAM H PROF
ASTRONOMY DEPT
UNIV OF WESTERN ONTARIO
LONDON ONT N6A 3K7
CANADA
TEL: 519-679-3183
TLX: 064-7134
TLF:
EML:
COM: 27.29.42

WEHRSE RAINER DR
INST F THEOR ASTROPHYSIK
IM NEUENHEIMER FELD 561
D-6900 HEIDELBERG
GERMANY F.R.
TEL: 06221-562837
TLX: 461515 UNIHD [illegible]
TLF:
EML:
COM: 36C

WEI MINGZHI
BEIJING ASTRONOMICAL OBS
ACADEMIA SINICA
BEIJING
CHINA. PEOPLE'S REP.
TEL: 28 1698
TLX: 22040 BAOAS CN
TLF:
EML:
COM: 40

WEIDEMANN VOLKER PROF
INST THEOR PHYS & STERNW
NEUE UNIV PHYSIK ZENTRUM
OLSHAUSENST GEB N 61C
D-2300 KIEL 1
GERMANY. F.R.
TEL: 0431-880-4110
TLX: 292706 IAFKI D
TLF:
EML:
COM: 05.36

WEIDENSCHILLING S J DR
PLANETARY SCIENCE INST
2030 E SPEEDWAY SUITE 201
TUCSON AZ 85719
U.S.A.
TEL: 602 881 0332
TLX:
TLF:
EML:
COM: 15.16

WEIGERT ALFRED PROF
HAMBURGER STERNWARTE
GOJENSBERGWEG 112
D-2050 HAMBURG 80
GERMANY F.R.
TEL: 72244112
TLX: 217884 HAMST D
TLF:
EML:
COM: 35.42

WEILER EDWARD J DR
PRINCETON UNIVERSITY OBS
PEYTON HALL
PRINCETON NJ 08540
U.S.A.
TEL:
TLX:
TLF:
EML:
COM: 42.44

WEILER KURT W DR
NAVAL RESEARCH LABORATORY
NRL-CODE 4131
WASHINGTON DC 20375-5000
U.S.A.
TEL: 202-767-0292
TLX:
TLF:
EML: SPAN:11334::WEILER
COM: 34.40

WEILL GILBERT M DR
SPOT IMAGE CORPORATION
1897 PRESTON WHITE DRIVE
RESTON VA 22091-4326
U.S.A.
TEL: 620-22-00
TLX: 4993073
TLF:
EML:
COM: 21

WEIMER THEOPHILE P F DR
OBSERVATOIRE DE PARIS
61 AVE DE L'OBSERVATOIRE
F-75014 PARIS
FRANCE
TEL: 1-43-20-12-10
TLX:
TLF:
EML:
COM: 16

WEINBERG J L DR
SPACE ASTRONOMY LAB
UNIVERSITY OF FLORIDA
1810 N.W. 6TH STREET
GAINESVILLE FL 32609
U.S.A.
TEL: 904-392-5450
TLX: 810-825-2308 SPACELA
TLF:
EML:
COM: 21.22.44

WEINBERG STEVEN DR
DEPT OF PHYSICS
UNIVERSITY OF TEXAS
AUSTIN TX 78712
U.S.A.
TEL: 512-471-4394
TLX: 910-874-1305
TLF:
EML:
COM: 47

WEINBERGER RONALD DR
INSTITUT FUR ASTRONOMIE
TECHNIKERSTRASSE 25
A-6020 INNSBRUCK
AUSTRIA
TEL: 05222 7485251
TLX:
TLF:
EML:
COM:

WEIS EDWARD W DR
VAN VLECK OBSERVATORY
WESLEYAN UNIVERSITY
MIDDLETOWN CT 06457
U.S.A.
TEL: 203-347-9411
TLX:
TLF:
EML:
COM: 26

WEISBERG JOEL MARK
CARLETON COLLEGE
DEPT PHYSICS & ASTRONOMY
NORTHFIELD MN 55057
U.S.A.
TEL: 507-663-4367
TLX:
TLF:
EML:
COM:

WEISHEIT JON C DR
PHYSICS DEPARTMENT L-297
L. LIVERMORE NTL LAB
PO BOX 808
LIVERMORE CA 94550
U.S.A.
TEL: 415-423-4254
TLX:
TLF:
EML:
COM: 34.48

WEISS NIGEL O DR
UNIVERSITY OF CAMBRIDGE
DEPT APPL MATH/THEO PHYS
SILVER STREET
CAMBRIDGE CB3 9EW
U.K.
TEL: 0223-351645
TLX: 81240
TLF:
EML:
COM: 17C.35

WEISS WERNER W DR
INSTITUT FUER ASTRONOMIE
DER UNIVERSITAET WIEN
TUERKENSCHANZSTR 17
A-1180 WIEN
AUSTRIA
TEL: 0-22-2-34-53-60
TLX: 116222 PHYSI A
TLF:
EML:
COM: 09.27.29

WEISSKOPF MARTIN CH DR
NASA MSFC
CODE ES-65
HUNTSVILLE AL 35812
U.S.A.
TEL: 205-453-3238
TLX:
TLF:
EML:
COM: 48

WEISSMAN PAUL ROBERT
JET PROPULSION LABORATORY
MS 183-601
4800 OAK GROVE DRIVE
PASADENA CA 91109
U.S.A.
TEL: 818-354-2636
TLX: 675429
TLF:
EML:
COM: 15.20

WEISTROP DONNA DR
2191 GLENFIELD ROAD
ANNAPOLIS MD 21401
U.S.A.
TEL:
TLX:
TLF:
EML:
COM: 25.33

WELCH GARY A DR
DEPT OF ASTRONOMY
ST MARY'S UNIVERSITY
HALIFAX NS B3H 3C3
CANADA
TEL: 902-429-9780
TLX:
TLF:
EML:
COM: 28

WELCH WILLIAM J PROF
RADIO ASTRONOMY LAB
UNIVERSITY OF CALIFORNIA
601 CAMPBELL HALL
BERKELEY CA 94720
U.S.A.
TEL: 415-642-6679
TLX: 820181 UCB AST RAL
TLF:
EML:
COM: 40.51

WELLER CHARLES S DR
OBSERVABLES TECHNO.DIVIS.
CODE 1410
D.TAYLOR RESEARCH CENTER
BETHESDA MD 20084-5000
U.S.A.
TEL:
TLX:
TLF:
EML:
COM: 49

WELLGATE G BERNARD MR
CANEHEATH HOUSE
ARLINGTON
POLEGATE BN26 6SJ
U.K.
TEL:
TLX:
TLF:
EML:
COM:

WELLINGTON KELVIN DR
CSIRO
DIVISION OF RADIOPHYSICS
P.O.BOX 76
EPPING NSW 2121
AUSTRALIA
TEL: 02-8680222
TLX: 26230 AA
TLF:
EML:
COM: 40

WELLMANN PETER PROF DR
INST FUER ASTRONOMIE &
ASTROPHYSIK
SCHEINER-STR 1
D-8000 MUENCHEN 80
GERMANY. F.R.
TEL:
TLX:
TLF:
EML:
COM: 29.36.42

WELLS DONALD C III DR
NRAO
EDGEMONT ROAD
CHARLOTTESVILLE VA 22901
U.S.A.
TEL: 804-296-0211
TLX:
TLF:
EML:
COM: 05

WENDKER HEINRICH J PROF
HAMBURGER STERNWARTE
GOJENBERGSWEG 112
D-2050 HAMBURG 80
GERMANY. F.R.
TEL: 040-7252-4112
TLX: 217884 HAMST D
TLF:
EML:
COM: 34.40

WENIGER SCHANK DR
OBSERVATOIRE DE PARIS
SECTION DE MEUDON
F-92195 MEUDON PL CEDEX
FRANCE
TEL: 1-45-34-75-30
TLX:
TLF:
EML:
COM: 14.21.29

WENTZEL DONAT G DR
ASTRONOMY PROGRAM
UNIVERSITY OF MARYLAND
COLLEGE PARK MD 20742
U.S.A.
TEL:
TLX:
TLF:
EML:
COM: 10.12.46C.48

WENZEL W DR
ZNTRLINST. F. ASTROPHYSIK
STERNWARTE SONNEBERG
DDR-6400 SONNEBERG
GERMANY. D.R.
TEL:
TLX:
TLF:
EML:
COM: 27

WESEMAEL FRANCOIS DR
DEPT DE PHYSIQUE
UNIVERSITE DE MONTREAL
C P 6128 SUCC A
MONTREAL PQ H3C 3J7
CANADA
TEL: 514-343-7355
TLX: 05562425 UDEMPHYSAS
TLF:
EML:
COM:

WESSELINK ADRIAAN J DR
143 FALLS ROAD
BETHANY CT 06525
U.S.A.
TEL: 203-393-3787
TLX:
TLF:
EML:
COM: 24.25.27.42

WESSELIUS PAUL R DR
SPACE RESEARCH LABORATORY
POSTBUS 800
NL-9700 AV GRONINGEN
NETHERLANDS
TEL:
TLX:
TLF:
EML:
COM: 25C.34.44.45

WESSON PAUL S DR
DEPT OF PHYSICS
UNIVERSITY OF WATERLOO
WATERLOO ONT N2L 3G1
CANADA
TEL: 519-885-1211
TLX:
TLF:
EML:
COM: 47

WEST RICHARD M DR
ESO
KARL-SCHWARZSCHILD-STR 2
D-8046 GARCHING B MUNCHEN
GERMANY. F.R.
TEL: 89-32006275
TLX: 52828220 ESO D
TLF:
EML:
COM: 06C.09.20P.46C

WEST ROBERT ALAN
JET PROPULSION LABORATORY
MS 183-301
4800 OAK GROVE DRIVE
PASADENA CA 91109
U.S.A.
TEL: 818-354-2479
TLX: 675429
TLF:
EML:
COM: 16

WESTERGAARD NIELS J DR
DANISH SPACE RES INST
LUNDTOFTEVEJ 7
DK 2800 LYNGBY
DENMARK
TEL: 45 2 882277
TLX: 37198 DANRU DK
TLF:
EML:
COM:

WESTERHOUT GART DR
SCIENTIFIC DIRECTOR
US NAVAL OBSERVATORY
WASHINGTON DC 20390
U.S.A.
TEL: 202-653-1513
TLX: 710-822-1970
TLF:
EML:
COM: 05C.08.24.33.40

WESTERLUND BENGT E PROF
ASTRONOMICAL OBSERVATORY
BOX 515
S-751 20 UPPSALA
SWEDEN
TEL: 46-18-135157
TLX: 76024 UNIV UPPSS
TLF:
EML:
COM: 28.33.45

WESTFOLD KEVIN C PROF
MONASH UNIVERSITY
CLAYTON VIC 3168
AUSTRALIA
TEL: 03-5413080
TLX: 32691 AA
TLF:
EML:
COM: 40.48

WESTPHAL JAMES A PROF
CALTECH
170-25
1201 E. CALIFORNIA
PASADENA CA 91125
U.S.A.
TEL: 213-356-4900
TLX:
TLF:
EML:
COM: 09.44

WETHERILL GEORGE W
DEPT TERRESTR. MAGNETISM
CARNEGIE INST. WASHINGTON
5241 BROAD BRANCH RD N.W.
WASHINGTON DC 20015
U.S.A.
TEL: 202-966-0863
TLX: 440427 MAGN UI
TLF:
EML:
COM: 15.16.22.51

WEYMANN RAY J PROF
STEWARD OBSERVATORY
UNIVERSITY OF ARIZONA
TUCSON AZ 85721
U.S.A.
TEL: 602-621-2575
TLX: 467175
TLF:
EML:
COM: 34

WHEELER J CRAIG PROF
ASTRONOMY DEPT
UNIVERSITY OF TEXAS
AUSTIN TX 78712
U.S.A.
TEL: 512-471-4461
TLX: 910-874-1351
TLF:
EML:
COM: 35C

WHEELER JOHN A DR
DEPT OF PHYSICS
UNIVERSITY OF TEXAS
AUSTIN TX 78712
U.S.A.
TEL: 512-471-3751
TLX: 910-874-1351
TLF:
EML:
COM: 47.48

WHIPPLE ARTHUR L DR
MCDONALD OBSERVATORY
THE UNIVERSITY OF TEXAS
AT AUSTIN
AUSTIN. TX 78712-1351
U.S.A.
TEL: 512 4716332?
TLX:
TLF:
EML: BITNET:asag105@uta3081
COM: 07.20

WHIPPLE FRED L DR
CENTER FOR ASTROPHYSICS
60 GARDEN STREET
CAMBRIDGE MA 02138
U.S.A.
TEL: 617-495-7200
TLX:
TLF:
EML:
COM: 15.20.22

WHITAKER EWEN A
LUNAR & PLANETARY LAB
UNIVERSITY OF ARIZONA
TUCSON AZ 85711
U.S.A.
TEL: 602-621-2838
TLX: 910-952-1143
TLF:
EML:
COM: 16.41

WHITE GLENN J
PHYSICS DEPT
QUEEN MARY COLLEGE
MILE END ROAD
LONDON E1 4NS
U.K.
TEL: 980-4811 x4045
TLX: 893750
TLF:
EML:
COM: 34

WHITE GRAEME LINDSAY DR
CSIRO
DIVISION OF RADIOPHYSICS
P.O.BOX 76
EPPING NSW 2121
AUSTRALIA
TEL: 868-0222 x420
TLX: 26230 ASTRO AA
TLF:
EML:
COM: 24.41

WHITE NATHANIEL M DR
LOWELL OBSERVATORY
1400 W. MARS HILL ROAD
FLAGSTAFF AZ 86001
U.S.A.
TEL: 602-774-3358
TLX:
TLF:
EML:
COM: 25

WHITE ORAN R DR
7590 ROAD 39
MANCOS CO 80307
U.S.A.
TEL: 303-533-7318
TLX:
TLF:
EML:
COM:

WHITE R STEPHEN PROF
IGPP
UNIVERSITY OF CALIFORNIA
RIVERSIDE CA 92521
U.S.A.
TEL: 714-787-4503
TLX:
TLF:
EML:
COM:

WHITE RAYMOND E DR
STEWARD OBSERVATORY
UNIVERSITY OF ARIZONA
TUCSON AZ 85721
U.S.A.
TEL: 602-621-6526
TLX: 467175
TLF:
EML:
COM: 33.37

WHITE RICHARD E
SMITH COLLEGE
ASTRONOMY DEPARTMENT
CLARK SCIENCE CENTER
NORTHAMPTON MA 01063
U.S.A.
TEL: 413-584-2700
TLX:
TLF:
EML:
COM:

WHITE RICHARD L
SPACE TELESCOPE SCI INST
HOMEWOOD CAMPUS
3700 SAN MARTIN DRIVE
BALTIMORE MD 21218
U.S.A.
TEL: 301-338-4747
TLX:
TLF:
EML:
COM: 34.36

WHITE SIMON DAVID MANTON
STEWARD OBSERVATORY
UNIVERSITY OF ARIZONA
TUCSON AZ 85721
U.S.A.
TEL: 602-621-6530
TLX:
TLF:
EML:
COM: 28.47

WHITELOCK PATRICIA ANN DR
SAAO
P O BOX 9
OBSERVATORY
CAPE 7935
SOUTH AFRICA
TEL: 470025
TLX: 57-20309
TLF:
EML:
COM: 27.34

WHITEOAK J B DR
CSIRO
DIVISION OF RADIOPHYSICS
P.O.BOX 76
EPPING NSW 2121
AUSTRALIA
TEL: 612868-0226
TLX: 26230
TLF:
EML:
COM: 33.34.40

WHITFORD ALBERT E PROF
LICK OBSERVATORY
UNIVERSITY OF CALIFORNIA
SANTA CRUZ CA 95064
U.S.A.
TEL: 408-429-2149
TLX:
TLF:
EML:
COM: 28

WHITMORE BRADLEY C
SPACE TELESCOPE SCI INST
HOMEWOOD CAMPUS
3700 SAN MARTIN DRIVE
BALTIMORE MD 21218
U.S.A.
TEL: 301-338-4713
TLX:
TLF:
EML:
COM: 28

WHITNEY BALFOUR S
1102 E. MISSOURI
NORMAN OK 73071
U.S.A.
TEL: 405-321-3547
TLX:
TLF:
EML:
COM:

WHITNEY CHARLES A PROF
CENTER FOR ASTROPHYSICS
60 GARDEN STREET
CAMBRIDGE MA 02138
U.S.A.
TEL: 617-495-7451
TLX:
TLF:
EML:
COM:

WHITROW GERALD JAMES PROF
41 HOME PARK RD
WIMBLEDON
LONDON SW19 7HS
U.K.
TEL: 1-947-34-3467
TLX:
TLF:
EML:
COM: 41.47

WHITTET DOUGLAS C B DR
SCHOOL OF PHYSICS & ASTR
LANCASHIRE POLYTECHNIC
CORPORATION ST
PRESTON PR1 2TQ
U.K.
TEL: 772-22141
TLX:
TLF:
EML:
COM: 33.34

WHITTLE D MARK DR
DEPARTMENT OF ASTRONOMY
UNIVERSITY OF VIRGINIA
BOX 3818 UNIV STATION
CHARLOTTESVILLE VA 22903
U.S.A.
TEL: 864 924 4900
TLX: 910 997 0174 NRAO
TLF:
EML:
COM:

WHITWORTH ANTHONY PETER
DEPT OF APPLIED MATHS
& ASTRONOMY
UNIVERSITY COLLEGE
CARDIFF CF1 1XL
U.K.
TEL: 0222-44211
TLX: 498635 ULIBCFG
TLF:
EML:
COM: 34

WICKRAMASINGHE D T DR
AUSTRALIAN NAT UNIVERSITY
DEPT OF APPLIED MATHS
P.O.BOX 4
CANBERRA ACT 2600
AUSTRALIA
TEL:
TLX:
TLF:
EML:
COM:

WICKRAMASINGHE N C PROF
UNIVERSITY COLLEGE
DEPT OF APPLIED MATHS
& THEORETICAL PHYSICS
CARDIFF CF1 1XL
U.K.
TEL: 222-44211
TLX: 498635 ULIBCFG
TLF:
EML:
COM: 34.36.40

WIDING KENNETH G DR
US NAVAL RESEARCH LAB
CODE 7144
WASHINGTON DC 20375
U.S.A.
TEL: 202-767-2605
TLX:
TLF:
EML:
COM:

WIEDLING TOR DR
OSTRA VILLAVAGEN 15
S-611 36 NYKOPING
SWEDEN
TEL:
TLX:
TLF:
EML:
COM:

WIEHR EBERHARD DR
UNIVERSITAETS STERNWARTE
GEISMARLANDSTR 11
D-3400 GOETTINGEN
GERMANY. F.R.
TEL: 0551-395053
TLX:
TLF:
EML:
COM: 10

WIELEBINSKI RICHARD PROF
MPI FUER RADIOASTRONOMIE
AUF DEM HUEGEL 69
D-5300 BONN
GERMANY. F.R.
TEL: 0228-525-300
TLX: 886440 MPIFR D
TLF:
EML:
COM: 25C.28.33.40.51

WIELEN ROLAND PROF DR
ASTRONOMISCH. RECHEN-INST
MOENCHHOFSTR. 12-14
D-6900 HEIDELBERG 1
GERMANY. F.R.
TEL: 06221-490264
TLX: 461 336 ARIHD D
TLF:
EML:
COM: 04.08.28.33C.37

WIESE WOLFGANG L DR
NATL BUREAU OF STANDARDS
DIVISION 531
ROOM A267. BLDG 221
GAITHERSBURG MD 20899
U.S.A.
TEL: 301-921-2071
TLX: WU 898493
TLF:
EML:
COM: 14V

WIETH-KNUDSEN NIELS P DR
SVEND TROSTSVEJ 12
DK-1912 FREDERIKSBERG C
DENMARK
TEL: 1-249131
TLX:
TLF:
EML:
COM: 26.31

WIITA PAUL JOSEPH
DEPT PHYSICS & ASTRONOMY
GEORGIA STATE UNIVERSITY
ATLANTA GA 30303
U.S.A.
TEL: 404-658-2932
TLX:
TLF:
EML:
COM: 28

WIJNBERGEN JAN DR
LAB VOOR RUIMTEONDERZOEK
HOOGBOUW WSN
POSTBUS 800
NL-9700 AV GRONINGEN
NETHERLANDS
TEL: 050-116660
TLX: 53572 STARS NL
TLF:
EML:
COM:

WILCOCK WILLIAM L PROF
SCHOOL OF PHYSICAL AND
MOLECULAR SCIENCES
UNIV COLLEGE OF N. WALES
BANGOR GWYNEDD LL57 2UW
U.K.
TEL: 0248-351151
TLX: 61100
TLF:
EML:
COM: 09

WILD JOHN PAUL DR
CSIRO
P.O.BOX 225
DICKSON ACT 2602
AUSTRALIA
TEL:
TLX:
TLF:
EML:
COM: 10.40.49

WILD PAUL PROF
ASTRONOMISCHES INSTITUT
UNIVERSITAET BERN
SIDLERSTRASSE 5
CH-3012 BERN
SWITZERLAND
TEL: 31-65-85-96
TLX: 32320 PHYBE CH
TLF:
EML:
COM: 20.28

WILDEY ROBERT L PROF DR
NORTHERN ARIZONA UNIV
ASTROPHYSICAL OBSERVATORY
FLAGSTAFF AZ 86011
U.S.A.
TEL: 602-523-2661
TLX:
TLF:
EML:
COM: 16

WILKENING LAUREL L DR
UNIVERSITY OF ARIZONA
ADMIN BLDG 601
TUCSON AZ 85721
U.S.A.
TEL: 602-626-3513
TLX:
TLF:
EML:
COM: 15

WILKES BELINDA J
SMITHSONIAN ASTROPHYSICAL
OBSERVATORY
60 GARDEN STREET
CAMBRIDGE MA 02138
U.S.A.
TEL: 617-495-7768
TLX: 921428
TLF:
EML:
COM:

WILKINS GEORGE A DR
291 KINGS DRIVE
EASTBOURNE BN21 2YA
U.K.
TEL: 323-502686
TLX:
TLF:
EML:
COM: 04.05P.19C.31

WILKINSON ALTHEA
ASTRONOMY DEPT
UNIVERSITY OF MANCHESTER
MANCHESTER M13 9PL
U.K.
TEL: 061-273-7121
TLX:
TLF:
EML:
COM: 28

WILKINSON DAVID T
PRINCETON UNIVERSITY
JADWIN HALL
PO BOX 708
PRINCETON NJ 08540
U.S.A.
TEL: 609-452-4406
TLX:
TLF:
EML:
COM: 47

WILKINSON PETER N DR
NRAL
JODRELL BANK
MACCLESFIELD SK11 9DL
U.K.
TEL: 0-477-71321
TLX: 36149
TLF:
EML:
COM: 40

WILL CLIFFORD M DR
DEPT OF PHYSICS
WASHINGTON UNIVERSITY
ST LOUIS MO 63130
U.S.A.
TEL: 314-889-6244
TLX:
TLF:
EML:
COM: 47.48

WILLIAMON RICHARD M
FERNBANK SCIENCE CENTER
156 HEATON PARK DRIVE
ATLANTA GA 30307
U.S.A.
TEL: 404-378-4311
TLX:
TLF:
EML:
COM: 27.42.46

WILLIAMS BARBARA A
UNIVERSITY OF DELAWARE
PHYSICS DEPT
SHARP LAB.
NEWARK DE 19716
U.S.A.
TEL: 302-451-718-266
TLX:
TLF:
EML:
COM: 28

WILLIAMS CAROL A
DEPT OF MATHEMATICS
UNIVERSITY OF S. FLORIDA
TAMPA FL 33620
U.S.A.
TEL: 813-974-2643
TLX:
TLF:
EML:
COM: 07.24

WILLIAMS DAVID A PROF
MATHEMATICS DEPT
UMIST
P O BOX 88
MANCHESTER M60 1QD
U.K.
TEL: 061-236-3311
TLX: 666094
TLF:
EML:
COM: 34

WILLIAMS IWAN P DR
THEORET. ASTRONOMY UNIT
QUEEN MARY COLLEGE
MILE END ROAD
LONDON E1 4NS
U.K.
TEL: 01-980-4811
TLX: 893750
TLF:
EML:
COM: 15.16.20.22C.51

WILLIAMS JAMES G DR
JPL 264-700
4800 OAK GROVE DRIVE
PASADENA CA 91109
U.S.A.
TEL: 818-354-6466
TLX: 910-588-3269 JPL
TLF:
EML:
COM: 04.16.19.20

WILLIAMS JOHN A DR
PHYSICS DEPT
ALBION COLLEGE
ALBION MI 49224
U.S.A.
TEL: 517-629-5511
TLX:
TLF:
EML:
COM: 45

WILLIAMS PEREDUR M DR
ROYAL OBSERVATORY
BLACKFORD HILL
EDINBURGH EH9 3HJ
U.K.
TEL: 31-667-3321
TLX: 72383 ROEDIN G
TLF:
EML:
COM: 29

WILLIAMS ROBERT E DR
CERRO TOLOLO
INTERAMERICAN OBSERVATORY
CASILLA 603
LA SERENA
CHILE
TEL: 56-51-213-352
TLX: 645227 AURA CT
TLF:
EML:
COM: 28.34.42

WILLIAMS THEODORE B DR
DEPT PHYSICS & ASTRONOMY
RUTGERS UNIVERSITY
PO BOX 849
PISCATAWAY NJ 08854
U.S.A.
TEL: 201-932-2516
TLX:
TLF:
EML:
COM: 28

WILLIS ALLAN J DR
DEPT PHYSICS & ASTRONOMY
UNIVERSITY COLLEGE LONDON
GOWER STREET
LONDON WC1E 6BT
U.K.
TEL: 01-387-7050
TLX: 28722
TLF:
EML:
COM: 34.44

WILLIS ANTHONY GORDON DR
ATHABASCA UNIVERSITY
BOX 10000
ATHABASCA
ALBERTA. CANADA TOG 2RO
CANADA TOG 2RO
TEL: (403)675-6221
TLX:
TLF:
EML:
COM: 40

WILLMORE A PETER PROF
SPACE RESEARCH DEPT
UNIVERSITY OF BIRMINGHAM
PO BOX 363
BIRMINGHAM B15 2TT
U.K.
TEL: 021-472-1301
TLX: 338938 SPAPHY G
TLF:
EML:
COM:

WILLNER STEVEN PAUL DR
CENTER FOR ASTROPHYSICS
60 GARDEN STREET
CAMBRIDGE MA 02138
U.S.A.
TEL: 617-495-7123
TLX: 921428 SATELLITE CAM
TLF:
EML:
COM: 34.44

WILLS BEVERLEY J DR
ASTRONOMY DEPT RLM 15 308
UNIVERSITY OF TEXAS
AUSTIN TX 78712
U.S.A.
TEL: 512-471-3424
TLX: 910-874-1351
TLF:
EML:
COM: 28.40

WILLS DEREK DR
ASTRONOMY DEPT RLM 15 308
UNIVERSITY OF TEXAS
AUSTIN TX 78712
U.S.A.
TEL: 512-471-4461
TLX: 910-874-1351
TLF:
EML:
COM: 28.40

WILLSON LEE ANNE DR
ASTRONOMY PROGRAM
PHYSICS DEPT
IOWA STATE UNIVERSITY
AMES IA 50011
U.S.A.
TEL: 515-294-6765
TLX: 910-520-2157
TLF:
EML:
COM: 27.35.36

WILLSON ROBERT FREDERICK
DEPT OF PHYSICS
TUFTS UNIVERSITY
MEDFORD MD 02155
U.S.A.
TEL: 617-628--5000
TLX:
TLF:
EML:
COM: 40.51

WILLSTROP RODERICK V DR
INSTITUTE OF ASTRONOMY
MADINGLEY ROAD
CAMBRIDGE CB3 OHA
U.K.
TEL: 0223-62204
TLX: 817297 ASTRON G
TLF:
EML:
COM: 25.30

WILSON ALBERT G DR
RESEARCH PROGRAM STUDIES
PO BOX 113
TOPANGA CA 90290
U.S.A.
TEL: 818-716-6332
TLX:
TLF:
EML:
COM: 28.47

WILSON ANDREW S DR
ASTRONOMY PROGRAM
UNIVERSITY OF MARYLAND
COLLEGE PARK MD 20742
U.S.A.
TEL: 301-454-6361
TLX: 7108260353
TLF:
EML:
COM: 40

WILSON BRIAN G PROF
UNIVERSITY OF QUEENSLAND
55 WALCOTT STREET
ST LUCIA, QUEENSLAND 4067
AUSTRALIA
TEL: 61-7-3772200
TLX: 40315 UNIQLD AA
TLF:
EML:
COM:

WILSON CURTIS A
ST JOHN'S COLLEGE
PO BOX 1671
ANNAPOLIS MD 21404
U.S.A.
TEL: 301-263-2371
TLX:
TLF:
EML:
COM: 41

WILSON JAMES R DR
LAWRENCE LIVERMORE LAB.
L-35
LIVERMORE CA 94550
U.S.A.
TEL:
TLX:
TLF:
EML:
COM: 48

WILSON LIONEL DR
ENV. SCIENCE DEPT
LANCASTER UNIVERSITY
LANCASTER LA1 4YQ
U.K.
TEL: 524-65201x4275
TLX: 65111 LANCULG
TLF:
EML:
COM: 27

WILSON MICHAEL JOHN DR
SCHOOL OF APPLIED MATH
UNIVERSITY OF LEEDS
LEEDS LS2 9JT
U.K.
TEL:
TLX:
TLF:
EML:
COM:

WILSON P DR
INST ANGEWANDTE GEODAESIE
RICHARD-STRAUSS-ALLEE 11
D-6000 FRANKFURT/MAIN 70
GERMANY. F.R.
TEL: 49-69-6333260
TLX: 413592
TLF:
EML:
COM: 19

WILSON PETER R PROF
DEPT OF APPLIED MATHS
UNIVERSITY OF SYDNEY
SYDNEY NSW 2007
AUSTRALIA
TEL:
TLX:
TLF:
EML:
COM: 10.12.36

WILSON RAYMOND H DR
5325 GAINSBOROUGH DR
FAIRFAX VA 22032
U.S.A.
TEL: 703-978-3889
TLX:
TLF:
EML:
COM: 26

WILSON RAYMOND N DR
ESO
KARL-SCHWARZSCHILD-STR 2
D-8046 GARCHING B MUNCHEN
GERMANY. F.R.
TEL: 089-320-06-274
TLX: 528282-0 EO D
TLF:
EML:
COM:

WILSON ROBERT E PROF
DEPT OF ASTRONOMY
UNIVERSITY OF FLORIDA
GAINESVILLE FL 32611
U.S.A.
TEL: 904-392-2182
TLX:
TLF:
EML:
COM: 35.42

WILSON ROBERT PROF SIR
DEPT PHYSICS & ASTRONOMY
UNIVERSITY COLLEGE LONDON
GOWER STREET
LONDON WC1E 6BT
U.K.
TEL: 01-380-7154
TLX: 28722
TLF:
EML:
COM: 14.29.44

WILSON ROBERT W DR
AT & BELL LABORATORIES
HOH L239
BOX 400
HOLMDEL NJ 07733
U.S.A.
TEL: 201-949-3803
TLX:
TLF:
EML:
COM: 34.40

WILSON S J
DEPT OF MATHEMATICS
NATL UNIV OF SINGAPORE
KENT RIDGE
SINGAPORE 0511
SINGAPORE
TEL:
TLX:
TLF:
EML:
COM: 36

WILSON THOMAS L DR
MPI FUER RADIOASTRONOMIE
AUF DEM HUEGEL 69
D-5300 BONN
GERMANY. F.R.
TEL: 0228-525-378
TLX: 886440 MPIFR D
TLF:
EML:
COM: 33.34.40

WILSON WILLIAM J DR
JPL
BLDG 168-327
4800 OAK GROVE DRIVE
PASADENA CA 91109
U.S.A.
TEL: 818-354-5699
TLX:
TLF:
EML:
COM: 40

WINCKLER JOHN R PROF
SCHOOL OF PHYS & ASTRON
UNIVERSITY OF MINNESOTA
MINNEAPOLIS MN 55455
U.S.A.
TEL: 612-373-4688
TLX:
TLF:
EML:
COM:

WINDHORST ROGIER A DR
DEPARTMENT OF PHYSICS
ARIZONA STATE UNIVERSITY
TEMPE AZ 85287-1504
U.S.A.
TEL: 602 965 7143
TLX: 156 1058
TLF:
EML:
COM: 09.28.40.47

WING ROBERT F PROF
ASTRONOMY DEPT
OHIO STATE UNIVERSITY
174 W. 18TH AVENUE
COLUMBUS OH 43210
U.S.A.
TEL: 614-422-7876
TLX:
TLF:
EML:
COM: 27.29.45C

WINGET DONALD E
UNIV. OF TEXAS AT AUSTIN
ASTRONOMY DEPT
AUSTIN TX 78712
U.S.A.
TEL: 512-471-4461
TLX:
TLF:
EML:
COM:

WINIARSKI MACIEJ
CRACOW ASTRONOMICAL OBS
UL. ORLA 171
30-244 KRAKOW
POLAND
TEL:
TLX:
TLF:
EML:
COM: 25

WINK JOERN ERHARD DR
MPI FUER RADIOASTRONOMIE
AUF DEM HUEGEL 69
D-5300 BONN
GERMANY. F.R.
TEL:
TLX: 886440 MPIFR D
TLF:
EML:
COM: 40

WINKLER GERNOT M R DR
US NAVAL OBSERVATORY
TIME SERVICE DEPT
34TH & MASSACHUSETTS AVE
WASHINGTON DC 20390-5100
U.S.A.
TEL: 202-653-1520
TLX: 710-822-1970 NAVOBSY
TLF:
EML:
COM: 04.19.31

WINKLER KARL-HEINZ A DR
LOS ALAMOS NATIONAL LAB
X-DOT. MS-B218
PO BOX 1663
LOS ALAMOS NM 87545
U.S.A.
TEL:
TLX:
TLF:
EML:
COM: 35

WINKLER PAUL FRANK DR
DEPT OF PHYSICS
MIDDLEBURY COLLEGE
MIDDLEBURY VT 05753
U.S.A.
TEL: 802-388-3711
TLX: 353249
TLF:
EML:
COM:

WINNBERG ANDERS DR
ONSALA SPACE OBSERVATORY
S-439 00 ONSALA
SWEDEN
TEL: 46-300-60651
TLX: 2400 ONSPACE S
TLF:
EML:
COM: 34.40

WINNEWISSER GISBERT DR
UNIVERSITAT ZU KOLN
I. PHYSIKALISCHES INST
UNIVERSITATSSTRASSE 14
D-5000 KOLN 41
GERMANY. F.R.
TEL: 221-470-3567
TLX:
TLF:
EML:
COM: 14.34.40

WIRAMIHARDJA SUHARDJA DR
BOSSCHA OBSERVATORY
INST OF TECHNOLOGY
BANDUNG
LEMBANG 40391
INDONESIA
TEL:
TLX:
TLF:
EML:
COM:

WISNIEWSKI WIESLAW Z
UNIVERSITY OF ARIZONA
LUNAR & PLANETARY LAB
TUCSON AZ 85721
U.S.A.
TEL: 602-621-6956
TLX: 910-952-1143
TLF:
EML:
COM: 15.25.27

WITHBROE GEORGE L DR
CENTER FOR ASTROPHYSICS
HARVARD COLLEGE OBS
60 GARDEN STREET
CAMBRIDGE MA 02138
U.S.A.
TEL: 617-495-7438
TLX:
TLF:
EML:
COM:

WITT ADOLF N DR
RITTER ASTROPHYSICAL
RESEARCH CENTER
UNIVERSITY OF TOLEDO
TOLEDO OH 43606
U.S.A.
TEL: 419-537-2709
TLX: 810 442 1633
TLF:
EML:
COM: 21.34

WITTEN LOUIS PROF
DEPT OF PHYSICS
UNIVERSITY OF CINCINNATI
CINCINNATI OH 45221-0111
U.S.A.
TEL: 513-475-6492
TLX:
TLF:
EML:
COM:

WITTMANN AXEL D. PH D
UNIVERSITAETS-STERNWARTE
GEISMARLANDSR 11
D-3400 GOETTINGEN
GERMANY. F.R.
TEL: 0551-395042
TLX: 96753 USTERN D
TLF:
EML:
COM: 10.12

WITZEL ARNO DR
MPI FUER RADIOASTRONOMIE
AUF DEM HUEGEL 69
D-5300 BONN
GERMANY. F.R.
TEL: 525211
TLX: 886440 MPIFR D
TLF:
EML:
COM: 40

WLERICK GERARD DR
OBSERVATOIRE DE PARIS
SECTION DE MEUDON
F-92195 MEUDON PR. CEDEX
FRANCE
TEL: 1-45-07-22-40
TLX:
TLF:
EML:
COM: 09.28

WNUK EDWIN
ASTRONOMICAL OBSERVATORY
A. MICKIEWICZ UNIVERSITY
UL. SLONECZNA 36
60-286 POZNAN
POLAND
TEL:
TLX:
TLF:
EML:
COM: 07

WOEHL HUBERTUS DR
KIEPENHEUER INSTITUT
FUER SONNENPHYSIK
SCHOENECKSTRASSE 6
D-7800 FREIBURG
GERMANY. F.R.
TEL: 0049-761-32864
TLX: 7721552 KIS D
TLF:
EML:
COM: 09.10.12.36

WOLF BERNHARD PH D
LANDESSTERNWARTE
KOENIGSTUHL
D-6900 HEIDELBERG 1
GERMANY. F.R.
TEL: 06221-10036
TLX:
TLF:
EML:
COM: 29

WOLF RAINER E A DR
MPI FUER ASTRONOMIE
KOENIGSTUHL
D-6900 HEIDELBERG 1
GERMANY. F.R.
TEL: 06221-528-1
TLX: 461789 MPIA D
TLF:
EML:
COM:

WOLFE ARTHUR M PROF
DEPT PHYSICS & ASTRONOMY
UNIVERSITY OF PITTSBURGH
PITTSBURGH PA 15260
U.S.A.
TEL: 412-624-4318
TLX:
TLF:
EML:
CON:

WOLFENDALE ARNOLD W PROF
PHYSICS DEPT
THE UNIVERSITY
SOUTH ROAD
DURHAM DH1 3LE
U.K.
TEL: 0385-64971
TLX: 537351 DURLIB G
TLF:
EML:
CON: [illegible]

WOLFF SIDNEY C DR
KITT PEAK NAT OBSERVATORY
PO BOX 26732
TUCSON AZ 85726-6732
U.S.A.
TEL: 602-327-5511
TLX: 666-484 AURA NOAO
TLF:
EML:
CON: 29

WOLFSON C JACOB
LOCKHEED RESEARCH LABS
DEPT 91-30
BLDG 202
PALO ALTO CA 94304
U.S.A.
TEL: 415-424-2855
TLX:
TLF:
EML:
CON:

WOLSTENCROFT RAMON D DR
ROYAL OBSERVATORY
BLACKFORD HILL
EDINBURGH EH9 3HJ
U.K.
TEL: 031-667-3321
TLX: 72383
TLF:
EML:
CON: 21.34.45.51

WOLSZCZAN ALEKSANDER DR
ARECIBO OBSERVATORY
P.O. BOX 995
ARECIBO. PR 00613
U.S.A.
TEL:
TLX:
TLF:
EML:
CON:

WOLSZCZAN ALEXANDER DR
ARECIBO OBSERVATORY
BOX 995
ARECIBO PR 00613
U.S.A.
TEL: 809 878 2612
TLX: 385638
TLF:
EML:
CON: 34.40

WOLTJER LODEWIJK PROF
OBSERVATOIRE DE HAUTE
PROVENCE
F-04870 ST MICHEL OBSERV.
FRANCE
TEL: 92 76 63 68
TLX: 410693 ONP F
TLF:
EML:
CON: 10.33.34.40.47.48

WOO JONG OK
FAC. OF NATURAL SCIENCES
KOREA NATL UNIV OF EDUC
CHUNGWON-GUN
CHUNGBUK 320-23
KOREA. REPUBLIC
TEL: 431-40-3712
TLX:
TLF:
EML:
CON: 25.46

WOOD DAVID B DR
6 TURNING MILL ROAD
LEXINGTON MA 02173
U.S.A.
TEL:
TLX:
TLF:
EML:
CON: 42

WOOD F BRADSHAW PROF
DEPT OF ASTRONOMY
UNIVERSITY OF FLORIDA
SSRB 211
GAINESVILLE FL 32611
U.S.A.
TEL: 904-392-2059
TLX:
TLF:
EML:
CON: 38.42

WOOD III H J DR
DEPT OF ASTRONOMY
INDIANA UNIVERSITY
SWAIN HALL W 319
BLOOMINGTON IN 47405
U.S.A.
TEL:
TLX:
TLF:
EML:
CON: [illegible]

WOOD JOHN A DR
CENTER FOR ASTROPHYSICS
60 GARDEN STREET
CAMBRIDGE MA 02138
U.S.A.
TEL: 617-495-7273
TLX: 921428 SATELLITE CAM
TLF:
EML:
CON: 15.16.22

WOOD PETER R DR
MT STROMLO & SIDING
SPRING OBSERVATORIES
PRIVATE BAG
WODEN P.O. ACT 2606
AUSTRALIA
TEL: 62-881111
TLX: 62270 CANOPUS AA
TLF:
EML:
CON: 27.35

WOOD ROGER DR
ROYAL GREENWICH OBS
HERSTMONCEUX CASTLE
HAILSHAM BN27 1RP
U.K.
TEL: 323-833171x3391
TLX: 87451 RGOBSY G
TLF:
EML:
CON:

WOODSWORTH ANDREW W.DR
DOMINION ASTROPHYS. OBS.
5071 WEST SAANICH ROAD
VICTORIA V8X 4M6
CANADA
TEL: 604-388-0024
TLX: 049-7295
TLF:
EML:
CON: 40

WOODWARD PAUL R DR
DEPARTMENT OF ASTRONOMY
UNIVERSITY OF MINNESOTA
116 CHURCH STREET S.E.S
MINNEAPOLIS MS 55455
U.S.A.
TEL:
TLX:
TLF:
EML:
CON: 13.34

WOOLF NEVILLE J
STEWARD OBSERVATORY
UNIVERSITY OF ARIZONA
TUCSON AZ 85721
U.S.A.
TEL:
TLX:
TLF:
EML:
CON: 34.50

WOOLFSON MICHAEL M PROF
DEPT OF PHYSICS
UNIVERSITY OF YORK
HESLINGTON. YORK YO1 5DD
U.K.
TEL: 904-59861
TLX: 57933 YORKULG
TLF:
EML:
CON: 15.16.21.22

WOOLSEY E G
1909 LAUDER DRIVE
OTTAWA ONT K2A 1A9
CANADA
TEL:
TLX:
TLF:
EML:
CON:

WOOSLEY S E PROF
LICK OBSERVATORY
UNIVERSITY OF CALIFORNIA
SANTA CRUZ CA 95064
U.S.A.
TEL: 408-429-2976
TLX:
TLF:
EML:
COM: 28.35

WOOTTEN HENRY ALWYN
NRAO
EDGEMONT ROAD
CHARLOTTESVILLE VA 22901
U.S.A.
TEL: 804-296-0211
TLX: 510-587-5482
TLF:
EML:
COM: 34.40

WORDEN SIMON P DR
US SPACE COMMAND
1650 STONEY POINT COURT
COLORADO SPRINGS CO 80919
U.S.A.
TEL: 303 593 2469
TLX:
TLF:
EML:
COM: 09.12

WORLEY CHARLES E
ASTRON & ASTROPHYS DIV.
US NAVAL OBSERVATORY
WASHINGTON DC 20390
U.S.A.
TEL: 202-653-1588
TLX:
TLF:
EML:
COM: 05.24.26

WORRALL DIANA MARY
HARVARD-SMITHSONIAN CTR
FOR ASTROPHYSICS
60 GARDEN STREET
CAMBRIDGE MA 02138
U.S.A.
TEL: 617-495-7139
TLX: 921428 SATELLITE CAM
TLF:
EML:
COM: 28.48

WORRALL GORDON DR
BIRDSWOOD
EARDISLEY. HEREFDS
U.K.
TEL:
TLX:
TLF:
EML:
COM:

WORSWICK SUSAN
ROYAL GREENWICH OBS
HERSTMONCEUX CASTLE
HAILSHAM BN27 1RP
U.K.
TEL: 0323-833171
TLX: 87451
TLF:
EML:
COM: 09

WOSZCZYK ANDRZEJ PROF
INSTITUTE OF ASTRONOMY
UL. CHOPINA 12/18
87-100 TORUN
POLAND
TEL: 2-60-18
TLX: 00552234 ASTR PL
TLF:
EML:
COM: 15.16.50

WRAMDEMARK STIG S O DR
LUND OBSERVATORY
BOX 43
S-221 00 LUND
SWEDEN
TEL: 46-10 7303
TLX: 33199 OBSNOT S
TLF:
EML:
COM: 25.33.37

WRAY JAMES D DR
DEPT OF ASTRONOMY
UNIVERSITY OF TEXAS
RL MOORE HALL 15-212
AUSTIN TX 78712
U.S.A.
TEL:
TLX:
TLF:
EML:
COM: 44

WRIGHT ALAN E DR
AUSTRALIAN NATIONAL
RADIO ASTRONOMY OBS
P.O.BOX 276
PARKES NSW 2870
AUSTRALIA
TEL: 068-62-1677
TLX: 63999 QASAR
TLF:
EML:
COM: 05.40

WRIGHT EDWARD L DR
DEPT OF ASTRONOMY
UCLA
LOS ANGELES CA 90024
U.S.A.
TEL: 213-825-5755
TLX:
TLF:
EML:
COM: 34.47

WRIGHT FRANCES W DR
ASTRONOMY DEPT
HARVARD COLLEGE OBS
60 GARDEN ST
CAMBRIDGE MA 02138
U.S.A.
TEL: 617-495-2647
TLX:
TLF:
EML:
COM: 27

WRIGHT HELEN
THOMAS HOUSE APT. 517
1330 MASSACHUSETTS AVE.
WASHINGTON DC 20005
U.S.A.
TEL:
TLX:
TLF:
EML:
COM: 41

WRIGHT JAMES P DR
DIV ASTRONOMICAL SCIENCES
NAT SC FOUNDATION RM 615
1800 G STREET N.W.
WASHINGTON DC 20550
U.S.A.
TEL: 202-357-7639
TLX:
TLF:
EML:
COM:

WRIGHT KENNETH O DR
DOMINION ASTROPHYS OBS
5071 W SAANICH ROAD
RR 5
VICTORIA BC V8X 4M6
CANADA
TEL: 604-388-3157
TLX:
TLF:
EML:
COM: 29.36.42

WRIGHT MELVYN C H DR
RADIO ASTRONOMY LAB
UNIVERSITY OF CALIFORNIA
BERKELEY CA 94720
U.S.A.
TEL: 415-642-0420
TLX:
TLF:
EML:
COM:

WRIXON GERARD T DR
NATL MICROELECTRONICS
RESEARCH CENTER
UNIVERSITY COLLEGE
CORK
IRELAND
TEL: 353-21-508375
TLX: 26050
TLF:
EML:
COM:

WROBLEWSKI HERBERT DR
DEPTO DE ASTRONOMIA
UNIVERSIDAD DE CHILE
CASILLA 36-D
SANTIAGO
CHILE
TEL: 229-4101
TLX:
TLF:
EML:
COM: 30 24

WU CHI CHAO DR
COMPUTER SCIENCES CORP
SPACE TELESCOPE SC INST
HOMEWOOD CAMPUS
BALTIMORE MD 21218
U.S.A.
TEL: 301-338-4770
TLX: U.S.A.
TLF:
EML:
COM: 14.44

WU FEI
BEIJING OBSERVATORY
ACADEMIA SINICA
BEIJING
CHINA, PEOPLE'S REP.
TEL: 28-1698
TLX: 22040 BAOBS CN
TLF:
EML:
COM: 10.12

WU HSIN-HENG DR
DEPARTMENT OF PHYSICS
NATL CENTRAL UNIVERSITY
CHUNG-LI
CHINA, TAIWAN
TEL:
TLX:
TLF:
EML:
COM: 37.45

WU HUAI-WEI
SHANGHAI OBSERVATORY
ACADEMIA SINICA
80 NAN DAN ROAD
SHANGHAI
CHINA, PEOPLE'S REP.
TEL: 386191
TLX: 33164 SHAO CN
TLF:
EML:
COM: 40

WU LIAN-DA
PURPLE MOUNTAIN
OBSERVATORY
NANJING
CHINA, PEOPLE'S REP.
TEL: 32893
TLX: 34144 PMONJ CN
TLF:
EML:
COM: 07

WU LIN-XIANG
DEPT OF GEOPHYSICS
BEIJING UNIVERSITY
BEIJING
CHINA, PEOPLE'S REP.
TEL:
TLX:
TLF:
EML:
COM: 09.12

WU MING-CHAN
YUNNAN OBSERVATORY
P.O. BOX 110
KUNMING, YUNNAN PROVINCE
CHINA, PEOPLE'S REP.
TEL: 72946
TLX: 64040 YUOBS CN
TLF:
EML:
COM: 50

WU SHENGYIN
BEIJING ASTR OBSERVATORY
ACADEMIA SINICA
BEIJING
CHINA, PEOPLE'S REP.
TEL:
TLX: 22040 BAOAS CN
TLF:
EML:
COM: 40

WU SHI TSAN DR
SCHOOL OF ENGINEERING
UNIVERSITY OF ALABAMA
IN HUNTSVILLE
HUNTSVILLE AL 35899
U.S.A.
TEL: 205-895-6413
TLX:
TLF:
EML:
COM: 10.49

WU SHOU-XIAN
SHAANXI OBSERVATORY
P.O. BOX 18
LINTONG XIAN
SHAANXI
CHINA, PEOPLE'S REP.
TEL: 55951 XIAN
TLX: 70121 CSAO CN
TLF:
EML:
COM: 19.31

WU XINJI
DEPT OF GEOPHYSICS
PEKING UNIVERSITY
BEIJING
CHINA, PEOPLE'S REP.
TEL: 282471-3929
TLX: 22239 PKUNI CN
TLF:
EML:
COM: 40

WU XUE-LIN
NANJING ASTRONOMICAL
INSTRUMENT FACTORY
JIANGSU PROVINCE
CHINA, PEOPLE'S REP.
TEL:
TLX:
TLF:
EML:
COM:

WUNNER GUENTER
LEHRSTUHL F THEORET ASTRO
PHYSIK DER UNIV TUEBINGEN
AUF DER MORGENSTELLE 12.C
D-7400 TUEBINGEN
GERMANY, F.R.
TEL: 7071-292487
TLX: 7262714 AIT D
TLF:
EML:
COM: 14.44

WYCKOFF SUSAN DR
PHYSICS DEPT/ASTRON GROUP
ARIZONA STATE UNIVERSITY
TEMPE AZ 85287
U.S.A.
TEL: 602-965-3561
TLX: 140289 HALLEU ASU UT
TLF:
EML:
COM: 15C.29.45

WYLLER ARNE A PROF
GRUPO SUECO
APARTADO 66
SANTA CRUZ DE LA PALMA
38071 TENERIFE
SPAIN
TEL: 34-22-40-00-16
TLX:
TLF:
EML:
COM: 09.12.29.36

WYNNE CHARLES G PROF
ROYAL GREENWICH OBS
HERSTMONCEUX CASTLE
HAILSHAM BN27 1RP
U.K.
TEL: 0323-533-171
TLX: 87451
TLF:
EML:
COM: 09

WYNN-WILLIAMS C G DR
INSTITUTE FOR ASTRONOMY
UNIVERSITY OF HAWAII
2680 WOODLAWN DRIVE
HONOLULU HI 96822
U.S.A.
TEL: 808-948-8807
TLX:
TLF:
EML:
COM: 28.34

XANTHAKIS JOHN N PROF
R C A A M
ACADEMY OF ATHENS
14 ANAGNOSTOPOULOU
GR-10673 ATHENS
GREECE
TEL: 3613589
TLX:
TLF:
EML:
COM: 10

XANTHOPOULOS B C DR
DEPT OF PHYSICS
UNIVERSITY OF CRETE
GR-71110 IRAKLION
GREECE
TEL: 081-235576
TLX: 262728
TLF:
EML:
COM: 47

XI ZE-ZONG
INSTITUTE OF THE HISTORY
OF NATURAL SCIENCE
BEIJING
CHINA, PEOPLE'S REP.
TEL:
TLX:
TLF:
EML:
COM: 41C

XIA JIONGYU
INSTITUTE OF GEODESY &
GEOPHYSICS
XU DONG LU
WUCHANG, HUBEI
CHINA, PEOPLE'S REP.
TEL:
TLX:
TLF:
EML:
COM: 19

XIA YI-FEI
DEPT OF ASTRONOMY
NANJING UNIVERSITY
NANJING
CHINA. PEOPLE'S REP.
TEL: 34651-2682
TLX: 34151 PRCNU CN
TLF:
EML:
COM: 08

XIA ZHIGUO DR
INSTITUT FUER ASTRONOMIE
ETH-ZENTRUMNICA
8092 ZURICH
SWITZERLAND
TEL: 1-256-3813
TLX: 817379 EHHG CH
TLF:
EML:
COM: 40

XIAN DING-ZHANG
PURPLE MOUNTAIN
OBSERVATORY
NANJING
CHINA. PEOPLE'S REP.
TEL: 37609
TLX: 34144 PMONJ CN
TLF:
EML:
COM: 04

XIANG DELIN
PURPLE MOUNTAIN
OBSERVATORY
NANJING
CHINA. PEOPLE'S REP.
TEL: 33738
TLX: 34144 PMONJ CN
TLF:
EML:
COM: 34

XIANG SHOUPING
UNIV SCIENCE & TECHNOLOGY
HEFEI. ANHUI
CHINA. PEOPLE'S REP.
TEL:
TLX:
TLF:
EML:
COM: 47

XIAO NAI-YUAN
DEPT OF ASTRONOMY
NANJING UNIVERSITY
NANJING
CHINA. PEOPLE'S REP.
TEL: 34651-2882
TLX: 34151 PRCNU CN
TLF:
EML:
COM: 19

XIAO XING HUA
DEPT OF ASTRONOMY
BEIJING NORMAL UNIVERSITY
BEIJING
CHINA. PEOPLE'S REP.
TEL: 656531/1367
TLX:
TLF:
EML:
COM: 47

XIE GUANG-ZHONG
YUNNAN OBSERVATORY
ACADEMIA SINICA
PO BOX 110
KUNMING
CHINA. PEOPLE'S REP.
TEL: 72946
TLX: 64040 YUOBS CN
TLF:
EML:
COM:

XIE LIANGYUN
INSTITUTE OF GEODEDY &
GEOPHYSICS
XU DONG LU
WUCHANG. HUBEI
CHINA. PEOPLE'S REP.
TEL:
TLX:
TLF:
EML:
COM: 08

XING JUN
DEPT OF GEOPHYSICS
BEIJING UNIVERSITY
BEIJING
CHINA. PEOPLE'S REP.
TEL:
TLX:
TLF:
EML:
COM: 34

XIONG DA-RUN
PURPLE MOUNTAIN
OBSERVATORY
NANJING
CHINA. PEOPLE'S REP.
TEL: 42817
TLX: 34144 PMONJ CN
TLF:
EML:
COM: 27.35

XU AO-AO
DEPT OF ASTRONOMY
NANJING UNIVERSITY
NANJING
CHINA. PEOPLE'S REP.
TEL:
TLX: 34151 PRC NU CN
TLF:
EML:
COM: 10

XU BANG-XIN
DEPT OF ASTRONOMY
NANJING UNIVERSITY
NANJING
CHINA. PEOPLE'S REP.
TEL:
TLX:
TLF:
EML:
COM: 08

XU JIA-YAN
P.O. BOX 18
LINTONG
SHAANXI
CHINA. PEOPLE'S REP.
TEL:
TLX:
TLF:
EML:
COM:

XU PEI-YUAN
INST OF ELECTRON. PHYSICS
UNIV SCIENCE & TECHNOLOGY
SHANGHAI
CHINA. PEOPLE'S REP.
TEL: 951602
TLX:
TLF:
EML:
COM: 40

XU PINXIN
PURPLE MOUNTAIN
OBSERVATORY
NANJING
CHINA. PEOPLE'S REP.
TEL: 32893
TLX: 34144 PMONJ CN
TLF:
EML:
COM: 07

XU TONG-QI
SHANGHAI OBSERVATORY
80 NAN DAN ROAD
SHANGHAI
CHINA. PEOPLE'S REP.
TEL: 386191
TLX: 33164 SHAO CN
TLF:
EML:
COM: 08.19

XU ZHENTAO
PURPLE MOUNTAIN OBS.
2 WEST BEIJING ROAD
NANJING
CHINA. PEOPLE'S REP.
TEL: 31096
TLX: 34144 PMONJ CN
TLF:
EML:
COM: 10.41

XU ZHI-CAI
2 WEST BEIJING STREET
NANJING
CHINA. PEOPLE'S REP.
TEL:
TLX:
TLF:
EML:
COM: 40

YABUSHITA SHIN A PROF
DEPT APPLIED MATHS & PHYS
KYOTO UNIVERSITY
SAKYOKU
KYOTO 606
JAPAN
TEL: 075-751-2111
TLX:
TLF:
EML:
COM: 15.20.14

YABUUTI KIYOSHI PROF
20 TANAKA HIGASKI
HINOKUCH MACHI
SAKYOKU 606 KYOTO
JAPAN
TEL:
TLX:
TLF:
EML:
COM: 41

YAHIL AMOS DR.
ASTRONOMY PROGRAM
SUNY AT STONY BROOK
ESS BLDG
STONY BROOK NY 11794-2100
U.S.A.
TEL: 516-246-6545
TLX: 510-228-7767
TLF:
EML:
COM:

YAKOVKIN N A DR
KIEV STATE UNIVERSITY
ASTRONOMICAL OBSERVATORY
252053 KIEV
U.S.S.R.
TEL:
TLX:
TLF:
EML:
COM: 10

YALLOP BERNARD D DR
ROYAL GREENWICH OBS
HERSTMONCEUX CASTLE
HAILSHAM BN27 1RP
U.K.
TEL: 0323-833171
TLX: 87451
TLF:
EML:
COM: 04VP

YAMAGATA TOMOHIKO DR
NATL ASTRON. OBSERVATORY
MITAKA
TOKYO 181
JAPAN
TEL: 0422 32 5111
TLX: 02822307 TAOMTK J
TLF:
EML:
COM: 28

YAMAGUCHI SHICHIRO
DEPT OF APPLIED MATHS
FAC ENGG. GIFU UNIVERSITY
YANAGIDO
GIFU 501-11
JAPAN
TEL: 582-30-1111
TLX:
TLF:
EML:
COM:

YAMAKOSHI KAZUO
COSMIC MATTER DIVISION
COSMIC RAY RES.TOKYO UNIV
3 CHOME. 2-1 YAMASHI
TOKYO 188
JAPAN
TEL: 0424-61-4131
TLX: 02822371 ICRTU J
TLF:
EML:
COM: [illegible]

YAMAMOTO TETSUO DR
INST. SP.& ASTR. SCIENCE
1-1 YOSHINODAI 3-CHOME
SAGAMIHARA
KANAGAWA-KEN 229
JAPAN
TEL: 0427 51 3911
TLX: 24550 SPACETKY J
TLF:
EML:
COM: [illegible]

YAMASAKI ATSUMA DR
DEPT EARTH SC & ASTRONOMY
UNIVERSITY OF TOKYO
KOMABA MEGURO-KU
TOKYO 153
JAPAN
TEL:
TLX:
TLF:
EML:
COM: 42C

YAMASHITA KOJUN DR
DEPT OF PHYSICS
OSAKA UNIVERSITY
MACHIKANEYAMACHO 1-1
TOYONAKA OSAKA 560
JAPAN
TEL: 06-844-1151
TLX:
TLF:
EML:
COM: 21.44

YAMASHITA YASUMASA PROF
TOKYO ASTRONOMICAL
OBSERVATORY
OSAWA. MITAKA
TOKYO 181
JAPAN
TEL:
TLX: 2822307 TAOMTK J
TLF:
EML:
COM: 25.29 45

YAMAZAKI AKIRA DR
HYDROGRAPHIC DEPT
TSUKIJI 5 CHUO-KU
TOKYO 104
JAPAN
TEL: 03-541-3811
TLX: 0 252 2222 JHBYD J
TLF:
EML:
COM: 04.08

YAN LIN-SHAN
SHANGHAI OBSERVATORY
ACADEMIA SINICA
SHANGHAI
CHINA. PEOPLE'S REP.
TEL: 386191
TLX: 33164 SHAO CN
TLF:
EML:
COM: 26

YANG FUMIN
SHANGHAI OBSERVATORY
ACADEMIA SINICA
SHANGHAI
CHINA. PEOPLE'S REP.
TEL: 386191
TLX: 33164 SHAO CN
TLF:
EML:
COM: 19

YANG HAI SHOU
ASTROPHYSICS DIVISION
GEOPHYSICS DEPT
BEIJING UNIVERSITY
BEIJING
CHINA. PEOPLE'S REP.
TEL: 282471-3888
TLX: 22239 PKUNI CN
TLF:
EML:
COM: 10.48

YANG JIAN
PURPLE MOUNTAIN
OBSERVATORY
NANJING
CHINA. PEOPLE'S REP.
TEL:
TLX: 34144 PMONJ CN
TLF:
EML:
COM: 40

YANG KE-JUN
GEOPHYSICAL INSTITUTE
UNIVERSITY OF ALASKA
FAIRBANKS AK 99775-0800
U.S.A.
TEL:
TLX:
TLF:
EML:
COM: 31

YANG LAN-TIAN
DEPT OF PHYSICS
HUAZHONG NORMAL UNIV.
WUHAN
CHINA. PEOPLE'S REP.
TEL: 75601x300 OR401
TLX:
TLF:
EML:
COM: 47.48

YANG SHI JIE
PURPLE MOUNTAIN
OBSERVATORY
NANJING
CHINA. PEOPLE'S REP.
TEL:
TLX: 34144 PMO NJ CN
TLF:
EML:
COM: 09

YANKULOVA IVANKA DR
DEPT OF ASTRONOMY
FACULTY OF PHYSICS
ANTON IVANOV STR 5
1126 SOFIA
BULGARIA
TEL:
TLX:
TLF:
EML:
COM:

YANOVITSKIJ EDGARD G DR
MAIN ASTRONOMICAL OBS
UKRAINIAN ACADEMY OF SCI
252127 KIEV
U.S.S.R.
TEL: 66-31-10
TLX: 131406 SKY SU
TLF:
EML:
COM: 36

YAO BAO-AN
SHANGHAI OBSERVATORY
ACADEMIA SINICA
SHANGHAI
CHINA. PEOPLE'S REP.
TEL: 386191
TLX: 33164 SHAO CN
TLF:
EML:
COM: 27

YAO JIN-XING
PURPLE MOUNTAIN
OBSERVATORY
NANJING
CHINA. PEOPLE'S REP.
TEL:
TLX:
TLF:
EML:
COM: 10

YAO ZHENG-QIU
NANJING ASTRONOMICAL
INSTRUMENT FACTORY
NANJING
CHINA. PEOPLE'S REP.
TEL: 46191
TLX: 34136 GLVNJ c/o NAIF
TLF:
EML:
COM: 09

YAPLEE B S
6 CREST VIEW COURT
ROCKVILLE MD 20854
U.S.A.
TEL: 301-762-0935
TLX:
TLF:
EML:
COM:

YAROV-YAROVOJ M S DR
MATHEMATICS DEPT
MVTU
VTORAYA BAUMANSKAYA 5
107005 MOSCOW
U.S.S.R.
TEL: 267-03-92
TLX: 111572
TLF:
EML:
COM: 07

YASUDA HARUO PROF DR
TOKYO ASTRONOMICAL
OBSERVATORY
OSAWA. MITAKA
TOKYO 181
JAPAN
TEL:
TLX:
TLF:
EML:
COM: 08

YATSKIV YA S DR
MAIN ASTRONOMICAL OBS
UKRAINIAN ACADEMY OF SCI
252127 KIEV
U.S.S.R.
TEL: 663110
TLX: 131406 SKY SU
TLF:
EML:
COM: 08.19

YAVNEL ALEXANDER A DR
METEORITE COMMITTEE
USSR ACADEMY OF SCIENCES
UL M ULIANOVOJ 3 K 1
117313 MOSCOW
U.S.S.R.
TEL: 1[illegible]536
TLX:
TLF:
EML:
COM: 15.[illegible]

YE BINXUN
YUNNAN OBSERVATORY
P.O. BOX 110
KUNMING
CHINA. PEOPLE'S REP.
TEL: 72946
TLX: 64040 YUOBS CN
TLF:
EML:
COM: 09

YE SHI-HUI
PURPLE MOUNTAIN
OBSERVATORY
NANJING
CHINA. PEOPLE'S REP.
TEL: 46700
TLX: 34144 PMONTJ CN
TLF:
EML:
COM: 10

YE SHU-HUA
SHANGHAI OBSERVATORY
SHANGHAI
CHINA. PEOPLE'S REP.
TEL: 386191
TLX: 33164 SHAO CN
TLF:
EML:
COM: 08.19.31C.38

YE WENWEI
INSTITUTE OF SEISMOLOGY
STATE SEISMO BUREAU
XIAO HONG SHAN WUHAN
HUBEI
CHINA. PEOPLE'S REP.
TEL:
TLX:
TLF:
EML:
COM:

YEE HOWARD K.C. DR
DEPARTEMENT DE PHYSIQUE
UNIVERSITE DE MONTREAL
C P 6128 SUCC A
MONTREAL PQ H3C 3J7
CANADA
TEL: 514-343-7274
TLX:
TLF:
EML:
COM:

YEH TYAN DR
COOPERATIVE INST FOR
RESEARCH IN ENVIR SCI
UNIVERSITY OF COLORADO
BOULDER CO 80309
U.S.A.
TEL: 303 497 5401
TLX:
TLF:
EML:
COM: 10.49

YEIVIN Y PROF
TEL-AVIV UNIVERSITY
TEL-AVIV 69978
ISRAEL
TEL:
TLX:
TLF:
EML:
COM:

YEN JUI-LIN PROF
DEPT OF ELECTRICAL
ENGINEERING
UNIVERSITY TORONTO
TORONTO ONT M5S 1A4
CANADA
TEL: 416-978-8756
TLX:
TLF:
EML:
COM:

YEOMANS DONALD K DR
JET PROPULSION LAB
MS 301-150 G
4800 OAK GROVE DRIVE
PASADENA CA 91109
U.S.A.
TEL: 818-354-2127
TLX: 675429 JPL COMM PSD
TLF:
EML:
COM: 15.20CS.22.41

YI ZHAO-HUA
DEPT OF ASTRONOMY
NANJING UNIVERSITY
NANJING
CHINA. PEOPLE'S REP.
TEL:
TLX:
TLF:
EML:
COM: 07

YILMAZ FATMA DR
UNIVERSITY OBSERVATORY
ISTANBUL
TURKEY
TEL:
TLX:
TLF:
EML:
COM:

YILMAZ NIHAL DR
DEPT OF ASTRONOMY
FEN FAKULTESI
UNIVERSITY OF ANKARA
BESEVLER. ANKARA
TURKEY
TEL: 236-550
TLX:
TLF:
EML:
COM:

YIN JI-SHENG
BEIJING ASTRONOMICAL OBS
ACADEMIA SINICA
BEIJING
CHINA. PEOPLE'S REP.
TEL: 28-1203
TLX: 22040 BAOAS CN
TLF:
EML:
COM: 25

YIN QI-FENG
DEPT OF GEOPHYSICS
BEIJING UNIVERSITY
BEIJING
CHINA. PEOPLE'S REP.
TEL: 28-2471 x3886
TLX: 22239 PKUNI CN
TLF:
EML:
COM: 40C

YODER CHARLES F
JET PROPULSION LAB
MS 183150 1
4800 OAK GROVE DRIVE
PASADENA CA 91109
U.S.A.
TEL: 818-354-2444
TLX: 617-5429
TLF:
EML:
COM: 16

YOKOSAWA MASAYOSHI DR
DEPT OF PHYSICS
IBARAKI UNIVERSITY
2-1-1 BUNKYO MITO
IBARAKI 310
JAPAN
TEL: 292 26 1621
TLX:
TLF:
EML:
COM:

YOKOYAMA KOICHI DR
INTERNATIONAL LATITUDE
OBSERVATORY
MIZUSAWA
IWATE 023
JAPAN
TEL: 0197-24-7111
TLX: 837628
TLF:
EML:
COM: 19

YOKOYAMA TADASHI DR
UNIVERSIDADE ESTADUAL
PAULISTA
CAIXA POSTAL 178
RIO CLARO 13500
BRAZIL
TEL: 55-195340122
TLX: 011-31870
TLF:
EML:
COM: 07

YONEYAMA TADAOKI DR
2-1-16 HIBARIGAOKA-KITA
HOYA-SHI
TOKYO 202
JAPAN
TEL:
TLX:
TLF:
EML:
COM:

YORK DONALD G DR
AAC
UNIVERSITY OF CHICAGO
5640 S. ELLIS AVENUE
CHICAGO IL 60637
U.S.A.
TEL: 312-962-8930
TLX: 910-221-5617
TLF:
EML:
COM: 34C

YORKE HAROLD W DR
UNIVERSITAETS STERNWARTE
GEISMARLANDSTR 11
D-3400 GOETTINGEN
GERMANY. F.R.
TEL: 395042
TLX: 96753
TLF:
EML:
COM: 34.35.36

YOSHIDA HARUO
DEPT OF ASTRONOMY
UNIVERSITY OF TOKYO
2-11-16 YAYOI. BUNKYO-KU
TOKYO 113
JAPAN
TEL: 3-812-2111
TLX:
TLF:
EML:
COM: 07

YOSHIDA JUNZO PROF
DEPT OF PHYSICS
KYOTO SANGYO UNIVERSITY
KAMIGAMO KITA-KU
KYOTO 603
JAPAN
TEL: 075-701-2151
TLX: 5422661 KSUJ
TLF:
EML:
COM: 07

YOSHII YUZURU DR
TOKYO ASTRON OBS
UNIVERSITY OF TOKYO
MITAKA 181
JAPAN
TEL: 0422-32-5111
TLX: 2822307 TAOMTK J
TLF:
EML:
COM: 33

YOSHIMURA HIROKAZU DR
FACULTY OF SCIENCE
DEPT OF ASTRONOMY
UNIVERSITY OF TOKYO
TOKYO 113
JAPAN
TEL: 03-812-2111
TLX: 33659 UTYOSCI
TLF:
EML:
COM: 10.12

YOSHIZAWA MASANORI DR
TOKYO ASTRONOMICAL
OBSERVATORY
OSAWA 2-21-1. MITAKA
TOKYO 181
JAPAN
TEL: 0422-32-5111
TLX:
TLF:
EML:
COM: 08C

YOSS KENNETH M DR
ASTRONOMY DEPT
UNIVERSITY OF ILLINOIS
1011 W. SPRINGFIELD AVE
URBANA IL 61801
U.S.A.
TEL: 217-333-3295
TLX:
TLF:
EML:
COM: 30.45

YOU JIAN-QI
PURPLE MOUNTAIN
OBSERVATORY
NANJING
CHINA. PEOPLE'S REP.
TEL: 46700
TLX: 34144 PMONJ CN
TLF:
EML:
COM: 10.12

YOU JUNHAN
ASTROPHYS RES DIV
UNIV OF SCI & TECHNOLOGY
HEIFI. ANHUI PROVINCE
CHINA. PEOPLE'S REP.
TEL: 63300
TLX: 90028 USTC CN
TLF:
EML:
COM: 48

YOUNG ANDREW T DR
DEPT OF ASTRONOMY
SAN DIEGO STATE UNIV
SAN DIEGO CA 92182-0334
U.S.A.
TEL: 619-265-5817
TLX:
TLF:
EML:
COM: 16.25V

YOUNG ARTHUR DR
ASTRONOMY DEPT
SAN DIEGO STATE UNIV
SAN DIEGO CA 92182
U.S.A.
TEL: 619-265-6167
TLX:
TLF:
EML:
COM:

YOUNG JUDITH SHARN
FIVE COLLEGE RADIOASTR
OBSERVATORY
UNIV OF MASSACHUSETTS
AMHERST MA 01003
U.S.A.
TEL: 413-545-0789
TLX: 95-5491
TLF:
EML:
COM: 28

YOUNG LOUISE GRAY DR
DEPT OF ASTRONOMY
SAN DIEGO STATE UNIV
SAN DIEGO CA 92182
U.S.A.
TEL: 619-287-4890
TLX:
TLF:
EML:
COM: 14.16

YOUNIS SAAD M
ASTRONOMY & SPACE RES CTR
COUNCIL FOR SCI RESEARCH
P O BOX 2441
JADIRIYAH. BAGHDAD
IRAQ
TEL: 7765127
TLX: 2187 BATHILMI IK
TLF:
EML:
COM: 24.33.34.40

YOUSEF SHAHINAZ M DR
DEPT OF ASTRONOMY
FACULTY OF SCIENCE
CAIRO UNIVERSITY
CAIRO
EGYPT
TEL:
TLX:
TLF:
EML:
COM:

YOUSSEF NAHED E DR
DEPT OF ASTRONOMY
FACULTY OF SCIENCE
CAIRO UNIVERSITY
CAIRO
EGYPT
TEL: 586041
TLX:
TLF:
EML:
COM: 12

YU KYUNG-LOK PROF
SEOUL NATL UNIVERSITY
SINLIM-DONG
KWANAK-KU
SEOUL 151
KOREA. REPUBLIC
TEL:
TLX:
TLF:
EML:
COM: 08

YU XIN ALFRED DR
BEIJING ASTRONOMICAL
OBSERVATORY
ACADEMIA SINICA
BEIJING
CHINA. PEOPLE'S REP.
TEL:
TLX:
TLF:
EML:
COM:

YUAN CHI PROF
DEPT OF PHYSICS
CITY COLLEGE OF N Y
138 ST CONVENT AVE
NEW YORK NY 10031
U.S.A.
TEL: 212-690-6823
TLX:
TLF:
EML:
COM: 33

YUASA MANABU DR
DEPT OF MATH & PHYSICS
KINKI UNIVERSITY
HIGASHI-OSAKA
OSAKA 577
JAPAN
TEL:
TLX:
TLF:
EML:
COM: 07.20

YUE ZENG-YUAN
DEPT OF GEOPHYSICS
PEKING UNIVERSITY
PEKING
CHINA. PEOPLE'S REP.
TEL:
TLX:
TLF:
EML:
COM:

YULDASHBAEV TAIMAS S
ASTRONOMICAL INSTITUTE
USSR ACADEMY OF SCIENCES
700052 TASHKENT
U.S.S.R.
TEL:
TLX:
TLF:
EML:
COM:

YULIN DJAN
ROYAL OBSERVATORY
BLACKFORD HILL
EDINBURGH EH9 3HJ
U.K.
TEL: 031-667 3321
TLX: 72383
TLF:
EML:
COM:

YUMI SHIGERU PROF DR
FUJISAWA 3672-3
FUJISAWA-SHI
KANAGAWA 251
JAPAN
TEL: 0466-81-8223
TLX:
TLF:
EML:
COM: 19.31

YUN HONG-SIK PROF
SEOUL NAT UNIVERSITY
SINLIM-DONG
KWANAK-KU
SEOUL 151
KOREA. REPUBLIC
TEL: 877-2130 x3542
TLX:
TLF:
EML:
COM: 10.12

YUNGELSON LEV R
ASTRONOMICAL COUNCIL
USSR ACADEMY OF SCIENCES
PYATNITSKAYA 48
109017 MOSCOW
U.S.S.R.
TEL: 231-54-61
TLX: 412623 SCSTP SU
TLF:
EML:
COM: 35

ZABRISKIE F R PROF
RD 1
ALEXANDRIA PA 16611
U.S.A.
TEL: 814-669-4483
TLX:
TLF:
EML:
COM:

ZACHARIADIS THEODOSIOS DR
RES CENTER FOR ASTRONOMY
ACADEMY OF ATHENS
14 ANAGNOSTOPOULOU STREET
GR 10673 ATHENS
GREECE
TEL:
TLX:
TLF:
EML: EZAKA20@GRATHUN1
COM: 10

ZACHAROV IGOR DR
ASTRONOMICAL INSTITUTE
CZECH. ACAD. OF SCIENCES
OBSERVATORY
251 65 ONDREJOV
CZECHOSLOVAKIA
TEL: 724525
TLX: 121579
TLF:
EML:
COM: 39

ZADUNAISKY PEDRO E PROF
CNIE - CENTRO ESPACIAL
SAN MIGUEL
AV MITRE 3100
1663 SAN MIGUEL Bs.As.)
ARGENTINA
TEL: 667-2371
TLX:
TLF:
EML:
COM: 20

ZAFIROPOULOS BASIL DR
DEPT OF PHYSICS
UNIVERSITY OF PATRAS
GR 26110 PATRAS
GREECE
TEL: 991973
TLX: 312447 UNPA GR
TLF: 061-991909
EML:
COM: 07

ZAHN JEAN-PAUL DR
OBS DU PIC-DU-MIDI
ET DE TOULOUSE
14 AVENUE EDOUARD BELIN
F-31400 TOULOUSE
FRANCE
TEL: 61-25-21-01
TLX: 530776 F
TLF:
EML:
COM: 35.36

ZAITSEV VALERII V DR
INST OF APPLIED PHYSICS
ULYANOVA ST. 46
603600 GORKY
U.S.S.R.
TEL:
TLX:
TLF:
EML:
COM: 40

ZAMBON GIULIO DR.
10 SUMMERS PLACE
FLOREY ACT 2615
AUSTRALIA
TEL:
TLX:
TLF:
EML:
COM: 09

ZAMORANI GIOVANNI
IST. DI RADIOASTRONOMIA
CNR
VIA IRNERIO 46
I-40126 BOLOGNA
ITALY
TEL: 39-51-232856
TLX: 211664 INFN BO I
TLF:
EML:
COM: 47.48

ZAMORANO JAIME DR
DEPT ASTROFISICA
FAC C. FISICAS
UNIV. COMPLUTENSE
28040 MADRID
SPAIN
TEL: 4495316
TLX: 47272
TLF:
EML:
COM: 28

ZANDER RODOLPHE DR
INSTITUT D'ASTROPHYSIQUE
UNIVERSITE DE LIEGE
AVENUE DE COINTE 5
B-4200 COINTE-OUGREE
BELGIUM
TEL: 041-529980
TLX: 41264 ASTRLG B
TLF:
EML:
COM:

ZANINETTI LORENZO
IST. DI FISICA GENERALE
CORSO M. D'AZEGLIO 46
I-10125 TORINO
ITALY
TEL: 011-657694
TLX: 211041 INFNTO I
TLF:
EML:
COM:

ZAPPALA ROSARIO ALDO DR
ISTITUTO DI ASTRONOMIA
CITTA UNIVERSITARIA
I-95125 CATANIA
ITALY
TEL: 33-05-33 x 493
TLX: 970359 ASTRCT I
TLF:
EML:
COM: 10

ZAPPALA VINCENZO PROF
OSSERVATORIO ASTRONOMICO
I-10025 PINO TORINESE
ITALY
TEL: 11-841067
TLX: 213236 TO ASTR I
TLF:
EML:
COM: 15C.20

ZARE KHALIL DR
1180 AVALT DRIVE
MOUNTAIN VIEW CA 94040
U.S.A.
TEL: 415-940-1881
TLX:
TLF:
EML:
COM: 07

ZARNECKI JAN CHARLES DR
UNIT FOR SPACE SCIENCES
PHYSICS LABORATORIES
UNIV KENT AT CANTERBURY
CANTERBURY CT2 7NR
U.K.
TEL: 0227-66822
TLX: 965449 UKLIB
TLF:
EML:
COM: 15.44

ZARRO DOMINIC M DR
NASA GSFC
CODE 682 6
BLDG 7
GREENBELT MD 20771
U.S.A.
TEL: 301 286 2039
TLX: 89675
TLF:
EML: SOLAR::DZARRO
COM: 10

ZASOV ANATOLE V DR
STERNBERG ASTRONOMICAL
INSTITUTE
119899 MOSCOW V-234
U.S.S.R.
TEL:
TLX:
TLF:
EML:
COM: 28

ZAVATTI FRANCO
DIPT. DI ASTRONOMIA
VIA ZAMBONI 33
I-40126 BOLOGNA
ITALY
TEL: 051-222956
TLX: 211664 INFNBO I
TLF:
EML:
COM: 28

ZDANAVICIUS KAZIMERAS DR
ASTRONOMIJOS OBS
VILNIUS 31
CIURLIONIO 29. LITHUNIA
U.S.S.R.
TEL:
TLX:
TLF:
EML:
COM: 45C

ZEALEY WILLIAM J DR
UNIVERSITY OF WOLLONGONG
PHYSICS DEPT
BOX 1144
WOLLONGONG NSW 2500
AUSTRALIA
TEL: 042-270-555
TLX: 29022 AA
TLF:
EML:
COM: 09.34 48C

ZEILIK MICHAEL II DR
DEPT PHYSICS & ASTRONOMY
UNIVERSITY OF NEW MEXICO
800 YALE BLVD NE
ALBUQUERQUE NM 87131
U.S.A.
TEL: 505-277-4442
TLX:
TLF:
EML:
COM: 34.46

ZEIPPEN CLAUDE DR
OBSERVATOIRE DE PARIS
SECTION DE MEUDON
F-92195 MEUDON PL CEDEX
FRANCE
TEL: 1-45-07-74-43
TLX: 201571 LAM F
TLF:
EML:
COM: 14

ZEKL HANS WILHELM
TON BELLER GMBH
BURGSTRASSE 22
D-6140 BENSHEIM 3
GERMANY. F.R.
TEL: 06251-73001
TLX: 468352 TBELL D
TLF:
EML:
COM:

ZELENKA ANTOINE DR
DACHSLENBERGSTR. 56
CH-8180 BUELACH
SWITZERLAND
TEL:
TLX:
TLF:
EML:
COM: 10.12

ZELLNER BENJAMIN H DR
SPACE TELESCOPE INSTITUTE
HOMEWOOD CAMPUS
3700 SAN MARTIN DRIVE
BALTIMORE MD 21218
U.S.A.
TEL:
TLX:
TLF:
EML:
COM: 15

ZEL'MANOV A L DR
STERNBERG STATE
ASTRONOMICAL INSTITUTE
119899 MOSCOW
U.S.S.R.
TEL:
TLX:
TLF:
EML:
COM: 47

ZENG QIN DR
PURPLE MOUNTAIN
OBSERVATORY
NANJING
CHINA. PEOPLE'S REP.
TEL: 86 25 308516
TLX: 34144 PMONJ CN
TLF:
EML:
COM: 14.34

ZENSUS J.-ANTON DR
NRAO
P.O. BOX O
SOCORRO NM 87801
U.S.A.
TEL: 505-835-7348
TLX: 910-988-1710
TLF: 505-835-8027
EML: BITNET:AZENSUS@NRAO.EDU
COM: 40

ZERULL REINER H DR
RUHR-UNIVERSITAET BOCHUM
BEREICH EXTRATERR. PHYSIK
D-4630 BOCHUM
GERMANY. F.R.
TEL: 234-700-4576
TLX: 0825860
TLF:
EML:
COM: 21

ZHAGAR YOURI Z DR
ASTRONOMICAL OBSERVATORY
OF LATVIAN UNIVERSITY
RAINIS BUL 19
RIGA 226098
U.S.S.R.
TEL: (013-2)223149
TLX: 161171 TEMA SU
TLF:
EML:
COM:

ZHAI DI-SHENG
BEIJING ASTRONOMICAL OBS
ACADEMIA SINICA
BEIJING
CHINA. PEOPLE'S REP.
TEL: 28-1698
TLX: 22040 BAOAS CN
TLF:
EML:
COM: 42

ZHAI ZAOCHENG
SHANGHAI OBSERVATORY
ACADEMIA SINICA
80 NAN DAN ROAD
SHANGHAI
CHINA. PEOPLE'S REP.
TEL: 386191
TLX: 33164 SHAO CN
TLF:
EML:
COM: 31

ZHANF SHOUZHONG DR
414 WEST 120 ST.. APT.401
NEW YORK NY 10027
U.S.A.
TEL: 212 666 4689
TLX:
TLF:
EML:
COM: 41

ZHANG BAI-RONG
YUNNAN OBSERVATORY
KUNMING. YUNNAN PROVINCE
CHINA. PEOPLE'S REP.
TEL: 72946
TLX: 64040 YUOBS CN
TLF:
EML:
COM: 10.50

ZHANG BIN
DEPT OF GEOPHYSICS
BEIJING UNIVERSITY
BEIJING
CHINA. PEOPLE'S REP.
TEL:
TLX:
TLF:
EML:
COM: 33

ZHANG CHENG-YUE
DEPT. OF PHYSICS
UNIVERSITY OF CALGARY
2500 UNIVERSITY DRIVE NW
CALGARY. ALBERTA T2N 1N4
CANADA
TEL:
TLX:
TLF:
EML:
COM: 34

ZHANG ER-HE DR
BEIJING ASTRONOMICAL OBS
CHINESE ACAD OF SCIENCE
BEIJING 100080
CHINA. PEOPLE'S REP.
TEL: 256-6698
TLX: 22040 BAOAS CN
TLF:
EML:
COM: 42

ZHANG FU JUN
SHAGHAI OBSERVATORY
ACADEMIA SINICA
SHANGAI
CHINA. PEOPLE'S REP.
TEL:
TLX:
TLF:
EML:
COM: 40

ZHANG GUO-DONG
BEIJING OBSERVATORY
ACADEMIA SINICA
BEIJING
CHINA. PEOPLE'S REP.
TEL:
TLX:
TLF:
EML:
COM: 19

ZHANG HE-QI
PURPLE MOUNTAIN
OBSERVATORY
NANJING. JIANGSU PROVINCE
CHINA. PEOPLE'S REP.
TEL:
TLX: 34144 PMONJ CN
TLF:
EML:
COM: 10.48

ZHANG HUI
P.O. BOX 18
LINTONG
XIAN
CHINA. PEOPLE'S REP.
TEL: XIAN 32255
TLX: 70121 CSAC CN
TLF:
EML:
COM: 08

ZHANG JIA-LU
ASTROPHYSICS RESEARCH DIV
UNIV SCIENCE & TECHNOLOGY
HEFEI. ANHUI PROVINCE
CHINA. PEOPLE'S REP.
TEL: 63300 HEFEI
TLX: 90028 USTC CN
TLF:
EML:
COM: 47.48

ZHANG JIA-XIANG
PURPLE MOUNTAIN
OBSERVATORY
ACADEMIA SINICA
NANJING
CHINA. PEOPLE'S REP.
TEL:
TLX: 34144 PMONJ CN
TLF:
EML:
COM: 20

ZHANG JINTONG
INSTITUTE OF GEODESY &
GEOPHYSICS
XU DONG LU
WUHAN
CHINA. PEOPLE'S REP.
TEL:
TLX:
TLF:
EML:
COM: 31.42

ZHANG MING-CHANG
DEPT OF ASTRONOMY
NANJING UNIVERSITY
NANJING
CHINA. PEOPLE'S REP.
TEL: 37651
TLX: 0909
TLF:
EML:
COM: 16

ZHANG PEIYU
PURPLE MOUNTAIN
OBSERVATORY
NANJING
CHINA. PEOPLE'S REP.
TEL: 37521
TLX: 34144 PMONJ CN
TLF:
EML:
COM: 41

ZHANG SHENG-PAN
2 ASSINIBOINE ROAD
SUITE 720
DOWNSVIEW ONT M3J 1L1
CANADA
TEL: 736 489
TLX:
TLF:
EML:
COM: 07

ZHANG XIU ZHONG
SHAANXI ASTR. OBSERVATORY
CHINESE ACADEMY OF SCIEN.
PO BOX 18 LINTONG
SHAANXI
CHINA. PEOPLE'S REP.
TEL: 32255-336 XIAN
TLX: 70121 CSAO CN
TLF:
EML:
COM:

ZHANG YOUYI
PURPLE MOUNTAIN
OBSERVATORY
NANJING
CHINA. PEOPLE'S REP.
TEL:
TLX: 34144 PMONJ CN
TLF:
EML:
COM: 09

ZHANG ZHEN-DA
DEPT OF ASTRONOMY
NANJING UNIVERSITY
NANJING
CHINA. PEOPLE'S REP.
TEL: 34651-2882
TLX: 34151 PRCNU CN
TLF:
EML:
COM: 10

ZHANG ZHEN-JIU
DEPT OF PHYSICS
HUAZHONG NORMAL UNIV.
WUHAN
CHINA. PEOPLE'S REP.
TEL: WUHAN 75601
TLX: 6908
TLF:
EML:
COM: 47.48

ZHAO GANG
SHANGHAI OBSERVATORY
ACADEMIA SINICA
SHANGHAI
CHINA. PEOPLE'S REP.
TEL: 386191
TLX: 33164 SHAO CN
TLF:
EML:
COM: 31

ZHAO JUN-LIANG
SHANGHAI OBSERVATORY
SHANGHAI
CHINA. PEOPLE'S REP.
TEL: 386191
TLX: 33164 SHAO CN
TLF:
EML:
COM: 33.37C

ZHAO MING
SHANGHAI OBSERVATORY
ACADEMIA SINICA
SHANGHAI
CHINA. PEOPLE'S REP.
TEL: 386191
TLX: 33164 SHAO CN
TLF:
EML:
COM: 19

ZHAO REN-YANG
BEIJING ASTRONOMICAL OBS
ACADEMIA SINICA
BEIJING
CHINA. PEOPLE'S REP.
TEL:
TLX:
TLF:
EML:
COM: 10

ZHAO XIAN-ZI
PURPLE MOUNTAIN
OBSERVATORY
NANJING
CHINA. PEOPLE'S REP.
TEL: 31337
TLX: 085 34144 PMONJ CN
TLF:
EML:
COM: 04.07

ZHELEZNIAKOV VLADIMIR V
APPLIED PHYSICS INSTITUTE
USSR ACADEMY OF SCIENCES
ULYANOV STREET 46
603600 GORKIJ
U.S.S.R.
TEL:
TLX:
TLF:
EML:
COM: 40

ZHELYAZKOV IVAN DR
SOFIA UNIVERSITY
FACULTY OF PHYSICS
5 ANTON IVANOV BLVD
BG-1126 SOFIA
BULGARIA
TEL: 62561 EXT 641
TLX: 23296 SUKO R BG
TLF:
EML:
COM: 10

ZHENG DA-WEI
SHANGHAI OBSERVATORY
ACADEMIA SINICA
80 NAN DAN ROAD
SHANGHAI
CHINA. PEOPLE'S REP.
TEL: 386191
TLX: 33164 SHAO CN
TLF:
EML:
COM: 19

ZHENG JIA-QING
PURPLE MOUNTAIN
OBSERVATORY
NANJING
CHINA. PEOPLE'S REP.
TEL: 46700
TLX: 34144 PMONJ CN
TLF:
EML:
COM: 07

ZHENG WENGUANG
BEIJING ASTRONOMICAL OBS
BEIJING
CHINA. PEOPLE'S REP.
TEL:
TLX:
TLF:
EML:
COM:

ZHENG XUE-TANG
DEPT OF ASTRONOMY
BEIJING NORMAL UNIVERSITY
BEIJING
CHINA. PEOPLE'S REP.
TEL: 633531-6285
TLX: 8512
TLF:
EML:
COM: 07

ZHENG YING
PURPLE MOUNTAIN
OBSERVATORY
NANJING
CHINA. PEOPLE'S REP.
TEL:
TLX: 34144 PMONJ CN
TLF:
EML:
COM: 31

ZHENG YI-JIA
BEIJING ASTRONOMICAL OBS
BEIJING
CHINA. PEOPLE'S REP.
TEL:
TLX: 22040 BAOBS CN
TLF:
EML:
COM: 40

ZHEVAKIN S A PROF DR
RADIOPHYSICAL RESEARCH
INSTITUTE
LYADOV STREET 25/14
603600 GORKIJ
U.S.S.R.
TEL: 36-67-51
TLX:
TLF:
EML:
COM: 35

ZHOU DAOQI
DEPT OF GEOPHYSICS
PEKING UNIVERSITY
BEIJING
CHINA. PEOPLE'S REP.
TEL: 282471-3888
TLX: 22239 PKUNI
TLF:
EML:
COM: 10.12.42

ZHOU HONG-NAN
DEPT OF ASTRONOMY
NANJING UNIVERSITY
NANJING
CHINA. PEOPLE'S REP.
TEL: 34651-2882
TLX: 34151 PRCNU CN
TLF:
EML:
COM: 07.42

ZHOU YI-JIAN
DEPT OF GEOPHYSICS
PEKING UNIVERSITY
BEIJING
CHINA. PEOPLE'S REP.
TEL: 28-2471-3886
TLX: 22239 PKUNI
TLF:
EML:
COM: 40

ZHOU XING-HAI
PURPLE MOUNTAIN
OBSERVATORY
NANJING
CHINA. PEOPLE'S REP.
TEL: 46700
TLX: 34144 PMONJ CN
TLF:
EML:
COM: 15.24

ZHOU YOU-YUAN
CENTER FOR ASTROPHYSICS
UNIV SCIENCE & TECHNOLOGY
OF CHINA
HEFEI. ANHUI
CHINA. PEOPLE'S REP.
TEL: 63300 HEFEI
TLX: 90028 USTC CN
TLF:
EML:
COM: 28.47

ZHOU ZHEN-PU
PURPLE MOUNTAIN
OBSERVATORY
NANJING
CHINA. PEOPLE'S REP.
TEL: 33738
TLX: 34114 PMONTJ CN
TLF:
EML:
COM: 34

ZHU CI-SHENG
DEPT OF ASTRONOMY
NANJING UNIVERSITY
NANJING
CHINA. PEOPLE'S REP.
TEL: 37551/2882
TLX: 34151 PRCNU CN
TLF:
EML:
COM: 42

ZHU NENGHONG
SHANGHAI OBSERVATORY
ACADEMIA SINICA
SHANGHAI
CHINA. PEOPLE'S REP.
TEL: 386191
TLX: 33164 SHAO CN
TLF:
EML:
COM: 09

ZHU SHI-CHANG
DEPT OF PHYSICS
SHANGHAI TEACHERS UNIV
SHANGHAI
CHINA. PEOPLE'S REP.
TEL: 384-301
TLX: 9016
TLF:
EML:
COM: 47

ZHU WEN-YAO
SHANGHAI OBSERVATORY
ACADEMIA SINICA
SHANGHAI
CHINA. PEOPLE'S REP.
TEL: 386191
TLX: 33164 SHAO CN
TLF:
EML:
COM: 07

ZHU XINGFENG
CENTER FOR ASTROPHYSICS
UNIV SCIENCE & TECHNOLOGY
HEFEI. ANHUI
CHINA. PEOPLE'S REP.
TEL:
TLX: 90028 USTC CN
TLF:
EML:
COM: 47

ZHU YONG-HE
BEIJING ASTRONOMICAL OBS
BEIJING
CHINA. PEOPLE'S REP.
TEL: 281-1698
TLX: 22040 BAOBS CN
TLF:
EML:
COM: 19

ZHUANG QIXIANG
SHANGHAI OBSERVATORY
80 NAN DAN ROAD
SHANGHAI
CHINA. PEOPLE'S REP.
TEL: 386191
TLX: 33164 SHAO CN
TLF:
EML:
COM: 31

ZHUANG WEIFENG
BEIJING OBSERVATORY
ACADEMIA SINICA
BEIJING
CHINA. PEOPLE'S REP.
TEL:
TLX:
TLF:
EML:
COM: 41

ZHUGZHDA YUZEF D DR
INSTITUTE OF TERRESTRIAL
MAGNETISM & IONOSPHERE
142092 ACADEMGORODOK
U.S.S.R.
TEL:
TLX:
TLF:
EML:
COM: 10.12

ZIEBA STANISLAW DR
OBSERVATORIUM ASTRON.
JAGIELLONIAN UNIVERSITY
UL. ORLA 171
30-244 KRAKOW
POLAND
TEL: 223856. 221877
TLX: 0322297 UJ PL
TLF:
EML:
COM: 40.47

ZIKIDES MICHAEL C DR
DEPT OF ASTRONOMY
PANEPISTIMIOPOLIS
GR-15771 ATHENS
GREECE
TEL:
TLX:
TLF:
EML:
COM:

ZIMMERMANN HELMUT DR
UNIVERSITAETS-STERNWARTE
SCHILLERGAESSCHEN 2
DDR-6900 JENA
GERMANY. D.R.
TEL: 27122
TLX:
TLF:
EML:
COM: 34.46

ZINN ROBERT J DR
DEPT OF ASTRONOMY
YALE UNIVERSITY
260 WHITNEY AVE BOX 6666
NEW HAVEN CT 06511
U.S.A.
TEL: 203-436-3460
TLX:
TLF:
EML:
COM: 28.37

ZINNECKER HANS
ROYAL OBSERVATORY
BLACKFORD HILL
EDINBURGH EH9 3HJ
U.K.
TEL: 031-667-3321
TLX: 72383 ROEDIN G
TLF:
EML:
COM:

ZIOLKOWSKI JANUSZ DR
COPERNICUS ASTRON CENTER
UL. BARTYCKA 18
00-716 WARSAW
POLAND
TEL:
TLX:
TLF:
EML:
COM: 35.42

ZIOLKOWSKI KRZYSZTOF DR
SPACE RESEARCH CENTER
UL. BARTYCKA 18
00-716 WARSAW
POLAND
TEL: 0-22-410041
TLX:
TLF:
EML:
COM: 20

ZIRIN HAROLD DR
CALTECH 264-33
PASADENA CA 91125
U.S.A.
TEL: 818-356-3857
TLX:
TLF:
EML:
COM: 10.12.14

ZIRKER JACK B DR
NATL SOLAR OBSERVATORY
SUNSPOT NM 88349
U.S.A.
TEL: 505-434-1390
TLX:
TLF:
EML:
COM: 12

ZITELLI VALENTINA DR
DIPARTIMENTO DI ASTRON
VIA ZAMBONI 33
BOLOGNA 40100
ITALY
TEL: 051 259301
TLX: 520634 INFN I
TLF:
EML: SPAN:37929
COM:

ZLATEV SLAVEY
ASTRONOMICAL OBSERVATORY
OF KARDGALI
6600 KARDGALI
BULGARIA
TEL: 03-61-25-95
TLX: 47471
TLF:
EML:
COM:

ZLOBEC PAOLO DR
OSSERVATORIO ASTRONOMICO
G.B. TIEPOLO 11
P.O.B. SUCC. TRIESTE 5
I-34131 TRIESTE
ITALY
TEL: 40-793921
TLX: 461137 OAT I
TLF:
EML:
COM: 10.40

ZOMBECK MARTIN V DR
CENTER FOR ASTROPHYSICS
60 GARDEN STREET
CAMBRIDGE MA 02138
U.S.A.
TEL: 617-495-7227
TLX: 921428 SATELLITE CAM
TLF:
EML:
COM: 44.48

ZOREC JEAN DR
INSTITUT D'ASTROPHYSIQUE
98 BIS BD ARAGO
75014 PARIS
FRANCE
TEL: 43 20 14 25
TLX: 270070 SU CNRS F
TLF:
EML:
COM: 29

ZOSIMOVICH IRINA D
INSTITUTE OF HISTORY
ACAD SCI UKRAINIAN SSR
KIROV STR 4
252001 KIEV
U.S.S.R.
TEL: 29-02-72
TLX:
TLF:
EML:
COM: 41

ZOU HUI-CHENG
SHANGHAI OBSERVATORY
80 NAN DAN ROAD
SHANGHAI
CHINA. PEOPLE'S REP.
TEL: 386191
TLX: 33164 SHAO CNCN
TLF:
EML:
COM: 44

ZOU YI-XIN
BEIJING OBSERVATORY
CHUNGKUANTSUN
W. SUBURB
BEIJING
CHINA. PEOPLE'S REP.
TEL: 281261
TLX: 22040
TLF:
EML:
COM: 10

ZOU ZHEN-LONG
BEIJING OBSERVATORY
ACADEMIA SINICA
BEIJING
CHINA. PEOPLE'S REP.
TEL:
TLX: 22040 BAOAS CN
TLF:
EML:
COM: 47

ZUCCARELLO FRANCESCA
ISTITUTO DI ASTRONOMIA
VIALE A. DORIA 6
I-95100 CATANIA
ITALY
TEL: 330-533
TLX: 970359 ASTRCT I
TLF:
EML:
COM:

ZUCKERMAN BEN M DR
ASTRONOMY DEPT
UCLA
LOS ANGELES CA 90024
U.S.A.
TEL: 213-825-9338
TLX: 910 342 7597
TLF:
EML:
COM: 27.34.40.51

ZUIDERWIJK EDWARDUS J
ROYAL GREENWICH OBS.
HERSTMONCEUX
WINDMILL HILL
HAILSHAM BN27 1RP
U.K.
TEL: (0)323-833171
TLX:
TLF:
EML:
COM: 42.47

ZVEREV MITROFAN S PROF DR
PULKOVO OBSERVATORY
196140 LENINGRAD
U.S.S.R.
TEL:
TLX:
TLF:
EML:
COM: 08

ZVOLANKOVA JUDITA
ASTRONOMICAL INSTITUTE
SLOVAK ACAD. OF SCIENCES
842 28 BRATISLAVA
CZECHOSLOVAKIA
TEL: 427-375157
TLX: 93373 SETS
TLF:
EML:
COM: 22

ZWAAN CORNELIS PROF DR
STERREKUNDIG INSTITUUT
PRINCETONPLEIN 5
POSTBUS 80 000
NL-3508 TA UTRECHT
NETHERLANDS
TEL: 030-535223
TLX: 40048 FYLUT NL
TLF:
EML: BITNET:UNNKEES@HUTRUU0
COM: 10.12 36

M486

ZWITTER TOMAZ
ASTRONOMICAL OBSERVATORY
UNIVERSITY E. KARDELJ
JADRANSKA 19
YU 61111 LJUBLJANA
YUGOSLAVIA
TEL: (61)26506)
TLX:
TLF:
EML:
COM:

NOTES:

NOTES:

CHAPTER VIII

IAU STYLE BOOK

THE IAU STYLE MANUAL (1989)

THE PREPARATION OF ASTRONOMICAL PAPERS AND REPORTS

Prepared by

George A. Wilkins
(President of IAU Commission 5)

on behalf of the
EXECUTIVE COMMITTEE
of the
INTERNATIONAL ASTRONOMICAL UNION

Dedicated to the memory of

Donald H. Sadler

(General Secretary of the IAU, 1958–1964)

THE IAU STYLE MANUAL (1989)

CONTENTS

THE IAU STYLE MANUAL (1989)

LIST OF TABLES

PREFACE

It is a pleasure to contribute a preface to the IAU Style Manual dedicated to the memory of Donald H. Sadler. Dr Sadler, General Secretary of the International Astronomical Union (IAU) 1958-1964, prepared the first "IAU Style Book" in 1961 for use in editing and proofreading the contributions to the Transactions of the IAU. It was republished with amendments by his successor, J.-C. Pecker, in 1966 and in a shortened version by C. de Jager and A. Jappell in 1971. On the invitation of IAU Commission 5, an extensive revision was begun by S. Mitton (Cambridge University Press) but was not completed; the task of revision was passed to G. A. Wilkins just prior to the Delhi General Assembly in 1985. The first draft of the present text was distributed for comment in April 1986; it was primarily intended for use in the preparation of papers for publication in the proceedings of IAU meetings. It was then suggested that an attempt should be made to obtain an agreement with the editors of the principal astronomical journals to use a common set of recommendations in order to simplify the preparation of all astronomical papers. A meeting of editors held in May 1988 reached a wide measure of agreement and a revised draft was prepared for comment during the Baltimore General Assembly.

Since manuscripts for IAU Transactions, Symposia and Highlights are now submitted in camera-ready form, where correction after submission is almost impossible, it was decided that the new Style Manual should include general and detailed guidance on both the drafting of the text and the preparation of the typescript. The increasing use of text-processing software and of high-quality printers, which together give results that are comparable with those from commercial typesetting, has been taken into account. The present Style Manual therefore contains recommendations on style and other information for authors of papers and reports for publication by or on behalf of the IAU. 'Style' is here taken to mean the format and layout of the document together with the conventions regarding designations, references, units etc. IAU publications should maintain consistency and be of pleasing and acceptable appearance. The recommendations take into account the views of the editors of the principal astronomical journals and the recommendations of other major international organisations. It is hoped, therefore, that these recommendations will be widely adopted in all appropriate astronomical publications to the benefit of authors, editors, referees publishers, printers and, most importantly, the readers.

I would like to express the appreciation of the Union to Dr Wilkins for undertaking and bringing this onerous task to a successful completion and to Dr Mitton who made the first start. The preparation of a document of this nature requires that considerable thought be given to establishing conventions which remain unambiguous for long periods of time. Dr Wilkins is to be congratulated on the thorough manner he has brought to the production of the task. It is also a pleasure to record the Union's appreciation of the support Dr Wilkins received from Annette Hedges, Cynthia White and Sue Frizzell at the Royal Greenwich Observatory (UK) in the preparation of drafts and the final camera-ready copy of this Manual. The final text of the Manual benefited from the comments of many colleagues of whom the following deserve special mention:

H. Abt, F. Praderie, J. Shakeshaft, F. Spite and P. A. Wayman.

On behalf of the International Astronomical Union, I would like to thank all who have contributed to the preparation of the Manual and urge all members of the Union to adhere to its precepts in the preparation of their astronomical typescripts.

1988 December

D. McNally
General Secretary
International Astronomical Union

IAU-UAI Secretariat
Institut d'Astrophysique
98bis Boulevard Arago
F-75014 Paris

SECOND PREFACE

The scope of this IAU Style Manual has been extended beyond the original concept, but it has not been feasible to make all the appropriate revisions of the original arrangement and text. Even now, I am aware of some omissions as there has not been sufficient time to develop the additional material as fully as would otherwise have been desirable. I hope that readers will notify me of any errors and omissions which they notice so that a supplement or revised edition may be prepared. I would be glad to receive comments on the general recommendations and suggestions for how this manual might be made more helpful and useful. In particular I would welcome comments and suggestions from readers whose first language is not English.

By the time that this manual is published I will have retired from the Royal Greenwich Observatory, but nevertheless correspondence about it may be sent to me at the address given below.

George A. Wilkins
President, IAU Commission 5

Address until 1990 March:

Royal Greenwich Observatory
Herstmonceux Castle
Hailsham
East Sussex
U.K. BN27 1RP

Address from 1990 March:

Royal Greenwich Observatory
Madingley Road
Cambridge
U.K. CB3 0EZ

SUMMARY OF RECOMMENDATIONS BY THE INTERNATIONAL ASTRONOMICAL UNION TO AUTHORS ON THE PREPARATION OF ASTRONOMICAL PAPERS AND REPORTS

1. PURPOSE

The following recommendations apply in particular to the preparation of camera-ready copy for papers and reports for publication by, or on behalf of, the International Astronomical Union; most of them apply also to other forms of copy and to other publishers. The recommendations are intended to assist astronomers to prepare reports on their investigations and activities in a form that will result in their publication with the minimum of effort and delay and that will be effective in conveying to the readers the essential information about their methods and results. Further details and advice are given in the IAU Style Manual (1989).

2. PREPARATION OF THE DRAFT PAPER

2.1 Planning

Ensure that you are aware of the recommendations concerning content, style and layout before you start to write the manuscript or type the 'compuscript'.

Prepare an outline structure of the paper bearing in mind any particular restriction on length that may apply when the paper is to be published in the proceedings of a meeting.

In general, each paper should include an introduction giving the background and the objectives, a description of the techniques used, a statement of the results obtained, a summary of the conclusions, and a list of references.

2.2 Preliminaries

Choose a clear, informative and concise title that does not include any special symbols nor any acronyms that will be familiar only to specialists.

Write an author statement that includes all those, but only those, who have made significant contributions to the development of the paper; give affiliations and postal addresses.

Draft an abstract that states clearly and concisely the objectives, methods, and principal conclusions; keep within a length of about 200 words, or less for short papers.

Prepare a list of keywords, if required.

2.3 The text

Keep sentences short and simple in construction; avoid unusual words and long compound terms whenever possible.

Be consistent in typographical style (e.g. initial capital letters).

Use SI and other recognised units throughout. Give clear and unambiguous designations for all astronomical objects (e.g. include coordinates in standard format).

Use unambiguous notations and terminology for physical quantities, coordinates and timescales.

Give citations to other useful and relevant papers by identifying the author and year of publication as in Brown 1988; cite more than two authors as Brown et al.

Insert appropriate headings and subheadings with a decimal system of numbering. Cross-references within the text should be by section number.

Do not use footnotes for additional information or references.

2.4 Tables and illustrations

Careful consideration should be given to the choice, content and design of tables and illustrations to ensure that they are appropriate and well presented.

2.5 Checking

Prepare the manuscript and any typescript to be used as printer's copy at double spacing in order to facilitate the marking of corrections and revisions and the insertion of other instructions or comments.

Check the first typescript carefully for content (correctness, completeness and clarity) and typographical style, and mark clearly the changes to be made and other action to be taken during the preparation of the final copy.

Pay particular attention to all numerical information, including designations and references.

Whenever possible, a person other than the first author should also carry out such checking, especially if the paper is to be printed from camera-ready copy provided by the author.

3. PREPARATION OF CAMERA-READY COPY

3.1 Physical quality and format

For the typescript, use either the preprinted sheets supplied by the publisher or other good quality paper of the same size.

Use an electric typewriter with a carbon ribbon or use another type of printer that will produce sharp black characters. A dot-matrix printer should only be used if it produces letter-quality copy.

Type (print) only within the specified type area (150 x 215 mm for IAU Symposia and Highlights; 175 x 250 mm for Transactions), except that in unjustified text it is permissible for lines to extend occasionally over the right-hand margin by two or three characters so as not to break a word unnaturally. The last line may also cross the bottom limit, but letters or lines should not be squeezed together to fit them into the printed outline.

Type at a line spacing of about 4 mm (6 lines per inch) with a character width of about 2 mm (12 characters per inch). Choose a typeface that matches the one currently used in the publications and that distinguishes between the figure 0 and the letter O and between the figure 1 and the lower-case letter L. The line on the preprinted sheet for running-head and page number should be left blank. The typed sheets should be numbered in (blue) pencil in the top right-hand corner.

The typescript must be absolutely clean and any minor corrections of a few characters must be made very carefully; correcting fluid may be used, but it must be thin and dry so that the overtyped characters are of equal density. For larger corrections the whole line should be retyped on a separate sheet of paper; if additional material is to be inserted the following lines must also be retyped to the point where the corrected lines can be inserted without disturbing the remainder of the paragraph. A major omission that would seriously disturb the pagination may be inserted at the end of the paper with an appropriate heading and cross-reference.

3.2 Layout

The first page of each paper in IAU Symposia or Highlights is to include:

The title of the paper in capital letters starting at the left-hand margin and followed by 4 line spaces (blank lines).

The author's name in capital and small letters, indented 10 spaces and followed on the next line by the author's affiliation and full postal address (unless this is given elsewhere in the volume) and then by 4 line spaces.

The abstract, starting with the word ABSTRACT in capitals at the left-hand margin and followed by 2 line spaces and then the heading of the first section of the text.

All section and subsection headings are to start at the left, and have arabic (decimal) numbers, as follows:-

First-level headings in capital letters on a separate line, with two line spaces above and one below.

Second-level headings (one decimal place) in small letters on a separate line, with one line space above and below; bold type may be used.

Lower-level subheadings (two or more decimal places) in underlined small letters, followed by the text on the same line.

New paragraphs are to be preceded by a blank line; the first word is to be indented by 5 spaces (10 mm) unless it follows a heading or starts with a subheading.

All pages should normally be full, except that a new section or subsection should not be started unless there is space for at least two lines of text, and except that there should always be at least two lines at the top of a new page at the end of a paragraph.

Displayed matter, such as mathematical equations, should have one line space above and below and should be indented 5 spaces (10 mm). The numbers of an equation should normally be given in parentheses at the end of the line. Formulae should be spaced so as to avoid ambiguity and to show their structure.

3.3 Tables and illustrations

Tables should normally be typed in the same manner as the text. Columns should be separated by adequate spacing, and not by rules. The caption should precede the table, but auxiliary information may be given below the tabular matter. A narrow table should be centred; the caption should be the same width as the table. Two line spaces should be given between the table and any following or preceding text. If necessary, place a wide table sideways on the page with the top at the left of the upright page; do not include any text (other than the caption and notes) on such a page.

Illustrations should be supplied as original drawings in black ink on good quality paper or as glossy sharp photographic prints or as high-quality computer output from a laser printer. If possible, they should be prepared to a size that is appropriate for reduction with the text, but an unreduced illustration may be supplied separately. Names, numbers on axes or contours, and units should be shown clearly and unambiguously on the drawings. The size of characters and the thickness of lines should be appropriate for reproduction. The caption should be typed under the illustration, usually at the same time as the main text, and to the same width as the illustration. Clear instructions must be given about the placing of separate illustrations.

Tables and figures should be numbered in separate sequences in arabic numerals.

3.4 References

The standard form for reference to a paper in a serial publication is as follows:

Author, A. N., Year. Title of paper. Abbreviated title of serial Volume number (arabic numerals), first-last page numbers.

If there are several authors, their names should be given in the same form (up to a maximum of eight, then et. al.).

Some journals do not allow the inclusion of the title of the paper; in any case, a very long title should be shortened.

The abbreviations of the title words of serials should follow international standards as exemplified by the list in Astronomy and Astrophysics Abstracts.

The title of the serial should be given in italic type and the volume number in bold type if this facility is available; otherwise they should not be underlined.

If it is useful to give a part number, it should be given in parentheses immediately after the volume number.

The standard form for a reference to a book such as a monograph or the proceedings of a conference is as follows:

Author, A.N., Year. Title of book. Place of publication: Name of publisher.

The role of an editor is indicated by giving (Ed.), after the name.

The number of the edition may be given in parentheses after the title in the form (nth ed.).

The details of the conference, including place and date, should be given after the title if they do not form part of it.

The standard form for a reference to a paper in the proceedings of a conference is as follows:

Author, year. Title of paper. In: Editor, (Ed.), year. Title of book. Details of Conference. Place of publication: publisher. First-last page numbers.

If there are several references to the same proceedings, a separate reference may be given for the proceedings, and then the references reduce to:

Author, year. Title of paper. In: Editor year, first-last page numbers.

References are to be arranged in alphabetical order of the name of the author, and in chronological order for each author. If there is more than one reference to an author in one year, then they should be distinguished by giving a letter a, b, c, ... as a suffix to the year. Papers by a single author should be given before those in which that author is the first of two co-authors, and these should be given before these in which that author is the first of more than two co-authors. The usual rules for arranging names in alphabetical sequence should be applied in so far as they are appropriate for multiple authors.

3.5 Checking

The camera-ready copy should be checked carefully to ensure that all corrections marked on the draft have been carried through correctly without introducing any further errors.

The copy should be examined separately to verify that it conforms with the recommendations on style given above.

The abstract and the text should be read again to verify that they are clear, correct and complete.

Final corrections should be made in the manner described above, and should themselves be checked.

4. PACKING AND DESPATCH

The copy should be checked carefully to ensure that it is complete and that all pages of text and any separate illustrations, corrections or other material are clearly and correctly identified.

Camera-ready copy should be protected on both sides by cardboard and placed in a strong envelope of appropriate size or wrapped in carefully sealed paper.

The copy should be sent by airmail (unless local) letter post to the Editor or other person specified by the Editor.

THE IAU STYLE MANUAL (1989)

THE PREPARATION OF ASTRONOMICAL PAPERS, REPORTS AND BOOKS

1. GENERAL CONSIDERATIONS

1.1 Purpose of the IAU Style Manual

The report of a scientific investigation or activity should be prepared in a manner that will result in its publication with the minimum of effort and delay and in a form that will be effective in conveying to the reader the essential information about the method and results of the investigations. The recommendations in this manual are intended to assist astronomers to attain these objectives; they provide general guidance about the appropriate structure of the paper and detailed guidance about typographical style and format. Adherence to these recommendations during the writing of the paper and in the preparation of the typescript will reduce considerably the amount of further effort that will be expended by the author, editor, referees, printer and publisher before the final version of the paper is published. Moreover, their adoption will make it easier for the readers to appreciate the significance of the paper and to satisfy themselves as to the validity of the results.

The recommendations in this manual should be followed by all authors of papers and reports that are intended for publications by, or on behalf of, the International Astronomical Union (IAU). The word 'paper' is hereafter used in a general sense to include not only papers describing new discoveries or reviewing previous work, but also reports of commission meetings and working groups. Many of the recommendations apply to the preparation of books, and some specific guidance is given for the editors of the proceedings of IAU symposia and colloquia. A paper that does not conform with these recommendations may be returned for revision and retyping, and the delay may result in the paper being published late or not at all. The recommendations on content apply to all forms of printing, but the recommendations on format apply in particular to camera-ready copy that is to be used for printing, by photolithography or other similar process, in the Transactions of the IAU, in the proceedings of IAU symposia, or in other IAU publications. Guidance is also given for typescripts that will be used as printer's copy for phototypesetting.

The recommendations of other major international organisations, such as the International Council of Scientific Unions (ICSU) and the International Standards Organisation (ISO), have been followed except in a few cases where it has been considered desirable to conform to common current astronomical practices that do not conflict with practices in related fields. It is hoped that these recommendations will be acceptable to most publishers of astronomical journals so that, for example, authors and typists will not have to use different abbreviations and formats for references when submitting papers to different journals.

This style manual is mainly concerned with the final form of presentation of the paper that is submitted for publication, and so it is intended for use by typists and copyeditors, as well as by authors. In addition, it contains general guidance to authors, editors and referees about structure and content. For simplicity, pronouns of masculine gender are used for persons who may be male or female.

1.2 Responsibilities of the author, editor, referee and publisher

The publication of a scientific paper usually involves four stages:

(a) the preparation of the copy by the author;

(b) the validation of the copy by the editor and referee;

(c) the checking of proofs and the correction of errors; and

(d) the printing and distribution of the paper by the publisher.

It is important that the separate responsibilities of the author, editor, referee and publisher are properly understood. Most of the following notes are of general application, but some points apply specifically to IAU publications and do not apply to those journals that insist on the use of the house style of the publisher.

A general statement of the obligations of authors, editors and referees (or reviewers) has been issued by the American Geophysical Union (1988); it is reproduced in Appendix A since it is applicable to astronomers as well as to geophysicists. Some, but not all, of the points in it are discussed in more detail in later sections of this manual. The resolution on the improvement of publications that was adopted at the IAU General Assembly in Baltimore in August 1988 is reproduced in Appendix B.

1.21 Author. The author is responsible for submitting the 'copy' for the paper by the scheduled date and in a form that is acceptable to both the editor and the publisher. If the report is the joint work of several persons, they should agree amongst themselves on the contact author who will be responsible for submitting the copy and for any subsequent negotiations with the editor and publisher. The 'copy' will usually consist of a typescript, together with any necessary drawings, photographs and tables, in a format that is specified by the publisher. This material may take the form of camera-ready copy that will be composed, that is set in type (which is now usually on film rather than in metal). It may also, in certain circumstances, be submitted in electronic form, on a magnetic disc or via a computer network.

The author should read carefully any specific instructions provided by the editor or publisher about the length and content of the paper, the format of the copy, and any timetable for submission and subsequent stages of the publication procedure. These instructions may supplement or differ in detail from the recommendations given in this manual.

The author should obtain the agreement of the editor, in advance of submission of the copy, if for any reason he wished to depart from these instructions in any significant way. The author should inform the typist of any special requirements that are not covered by markings on the draft of the report. All authors are advised to seek critical comments from their colleagues about the content and clarity of the text before preparing the copy that will be submitted to the editor. The author is also responsible for obtaining permission to reproduce copyright material, such as Sky-Survey plates.

The author must ensure that any camera-ready copy is of an appropriate technical quality in respect of the sharpness and uniformity of the characters and of the layout of mathematical expressions and tables. Unless the author has the use of suitable equipment he should not attempt to prepare a paper that contains a high proportion of complex mathematical expressions (or otherwise requires the use of different typefaces and sizes, especially symbols, and flexible spacing). Instead, for example, he should submit only a summary of such a paper in the proceedings of a conference and submit the full paper to a journal that is printed by conventional methods.

The responsibility for the accuracy of the copy rests with the author but, if it is possible, the copy should also be checked by another person who has not been involved in its preparation; this is especially true if the information density is high (as in formulae, tables and reference lists) or if time is short. The author (or the 'contact author' for a joint paper) is also responsible for checking, and returning on schedule, any proof that is sent by the editor or publisher in order to allow the author to draw attention to any errors of composition made by the printer. This proof stage also allows opportunities for the detection and correction of errors made by the author or typist and for the improvement of the content or appearance of the paper; amendments on proof are, however, costly, time consuming and a prolific source of further errors, and so the author should endeavour to ensure that the printer's copy is correct and clearly marked to show the author's intentions.

1.22 Editor. The tasks of the editor of, say, the proceedings of a symposium are first of all to plan the structure of the volume and to inform the authors of the conditions that their contributions should satisfy. Secondly, he must decide, perhaps after seeking advice from referees, whether the contributions received meet the required standards for quality of content and presentation. If he is not satisfied he should reject the paper or return it for revision or retyping. An editor should not amend a paper in any significant respect without obtaining the agreement of the author; he should not, for example, make extensive alterations to improve the style and construction of the English without permission. On the other hand he should not accept a paper unless he is satisfied that it is appropriate, soundly based, clearly expressed and carefully prepared in accordance with the instructions given to the author. A paper that reaches unorthodox conclusions should not be rejected merely for that reason; it should be rejected if the author is unable to present the arguments clearly or is unwilling to take into account other relevant evidence.

The editor of a volume that is to be printed from camera-ready copy should not normally accept a paper that is not of an appropriate technical quality. (See the corresponding guidance for authors, above.) If necessary, he should arrange for the preparation of good-quality copy for the summary only, and encourage the author to submit the full paper to a journal printed by conventional methods.

Guidance for the editors of the proceedings of IAU symposia, colloquia and regional meetings is given in the 'Rules for Scientific Meetings' in IAU Information Bulletin 58 (June 1987), 25-31; extracts are given in Appendix C. The editor is expected to prepare an introduction, table of contents and an index to the volume (BSI 1976a).

1.23 Referee. The refereeing, or reviewing, of typescripts prior to publication is an important step in the task of ensuring that published scientific literature is of a high quality. The criteria that are to be used vary according to the purpose of the paper, which may, for example, be concerned mainly with the presentation of the results of original research, with a critical review of past research, or with a proposal for a new observational programme. The editor of the journal or other publication will usually provide a list of points for the referee to consider, but the main criteria are significance, originality and clarity.

All papers should be original in the sense that they do not largely consist of material that has already been (or is to be) published by the author or by another person. The referee should also try to verify that the author has cited relevant work by other scientists and that he has not claimed (directly or indirectly) credit for work for which he was not responsible. (Further guidance on the selection of references is given in section 2.51.)

The referee should verify that the arguments are logical, complete and presented in a manner that will be understandable by the paper's intended readership. He should also look to see that the terminology, notation, units and designations used are in accordance with the recommendations given later in this manual. He should judge the completeness and acceptability of papers that report numerical results by reference to the principles listed in section 2.32; these principles reflect the views of many scientists who have experience in the use and evaluation of published data. A referee is not, however, expected to check every detail of the paper, although he should draw attention to any errors that he notices and he should try to verify that no serious blunders have been made in algebraic or numerical derivations.

Criteria for judgement on significance are not easy to formulate, especially as the importance of an observation or other result may not be apparent until much later when it can be related to other observations or new hypotheses.

1.24 Publisher. The task of the publisher is to ensure, firstly, that the papers are printed in a clear and pleasing format and, secondly, that the complete volume is published and distributed as quickly as possible. The publisher's copyeditor is expected to mark on the copy any further instructions for the printer; these may include minor changes to bring the typescript into conformity with the standard style for the volume, but he should not alter the text in any significant respect without obtaining the agreement of the editor and the author. The arrangements for proofreading and indexing should be agreed between the editor and the publisher.

2. DRAFTING AN ASTRONOMICAL PAPER

2.1 Planning

Important factors that must be taken into account during the drafting of an astronomical paper are the context in which the paper will be published and the length that will be acceptable to the editor; these factors determine the level of knowledge and understanding by the readers that may be assumed and the style and depth of treatment that will be appropriate. In general, papers in the proceedings of conferences should give fairly broad descriptions of the background, methods and results of the investigations; such papers can be fairly brief and can be evaluated by the editor alone. The detailed report on the investigation that provides the basis for the validation of the data and the justification of the conclusions is best suited to a primary journal where it will be more rigorously refereed and where delays in the completion and acceptance of one paper do not cause delays in the publication of other papers. On the other hand, a conference may be deliberately planned so that some participants are encouraged to develop their subject in depth and so that the proceedings will provide an authoritative review of a particular field. The standard required for the contributions to such volumes will be quite different from those that contain short reports on current research.

In all cases, however, the author should develop the theme and structure of the paper before beginning detailed drafting. The initial outline should include the following items:

(a) the title of the paper;
(b) a list of contributors;
(c) an abstract that summarises the substance of the paper;
(d) a list of sections and, if appropriate, subsection headings;
(e) a list of key references;
(f) a list of figures (drawings and photographs); and
(g) a list of tables.

This outline should be updated as the drafting progresses and should be used at the end to verify that there are no accidental omissions from the final typescript. A list of keywords, or other indexing information, should be prepared if required by the journal.

In general, appendices (or annexes) should be included only if there are clear advantages either in not incorporating such material in the main structure or in not publishing it in separate papers. Any appendix that might be cited and reproduced separately from the main paper should include an informative title, an author statement, an abstract or brief introductory paragraph, the substance of the appendix, and a list of any necessary references.

2.2 The preliminaries

The title, author statement, abstract, and keywords are particularly important since they will often be reprinted in abstracts journals and stored in bibliographic databases for use in computer-based information retrieval systems; they are also used for the classification and indexing of the paper.

2.21 Title. The title should be clear, informative and concise; it should indicate both the character of the paper and the principal topic discussed in it. Short, eye-catching titles that give an inadequate or misleading impression of the subject of the paper must not be used. A title should not assume that the reader will be aware of the context in which the paper will be printed; rather, the title should be able to stand alone. (Editors of conference proceedings should examine titles of contributions with this requirement in mind.) Special characters that are not available on ordinary typewriters, and symbols for names of elements, compounds and physical quantities should normally be avoided in titles. Acronyms and abbreviations should only be given where their significance and meaning will be widely understood.

A short title for use as a headline for the right-hand pages should be suggested (or checked for suitability) by the author; it may be assumed that the context, or the names of the authors, will be indicated by the headline on the left-hand pages. These headlines are also known as 'running heads'.

An English translation of the title should be given if the original is written in another language. A French translation may be required for some IAU reports.

2.22 Author. The list of the names of the authors of a joint paper should include only those persons who have made significant contributions to its development. (See also section 2.52.) The list should be arranged in alphabetical order unless the authors consider that a different order is appropriate because of the different contributions to the work; for example, the name of a person who has played a dominant role in the work reported or in the writing of the paper may be listed first. In such cases, the editor should not change the sequence. Each name should be given in the form that the author normally uses for scientific purposes so that attributions in citation indexes will appear together. The affiliation of each author should be given in the form specified by the editor or publisher; the full postal address should be given in the proceedings of IAU symposia and colloquia, and in papers published in most journals.

If a paper has been prepared by an institution or formally constituted workshop group, it is desirable that the principal author or authors should be identified personally in the author statement so as to simplify the citation and refereeing of the paper. The role of the person concerned should be indicated briefly; full information about the status of the paper and of those who contributed to it should be given in the introduction to the paper or in the acknowledgement section, if this is more appropriate.

If a paper is not printed in a latin alphabet, the names of the authors should be transliterated into English in accordance with the appropriate standard; for example, the names of Russian authors should be transliterated from the Russian alphabet according to the scheme given in section 6.31.

2.23 Abstract. Every paper should be accompanied by an abstract in the language of the main text and by an abstract in English if the paper is in another language. For some IAU reports a translation of the abstract into French is also useful; the General Secretary should be consulted in individual cases.

An abstract should state briefly the objectives, methods and principal conclusions of the investigation or paper; it should be concise, clear and as comprehensive as is possible, preferably within a length of about 200 words, or less. (Some journals allow longer abstracts, but any abstract that is longer than this may be arbitrarily truncated in an abstracts journal or bibliographic database.) It should form a continuous text and should not take the form of a list of subject headings; rather, it should summarise in general terms the substance of the paper. Specialist jargon should not be used. Any important numerical result should be given, and the nature and extent of the numerical data given in the paper should be indicated; in general, however, the listing of numerical data in the abstract is to be avoided.

The abstract should be understandable without reference to the main text or external source of information. Bibliographic citations and explicit references to figures and tables should not be given; any necessary citation should be given in full. Mathematical formulae, special characters and abbreviations should be avoided; any abbreviation that is not in wide use, but which it is necessary to use for brevity, should be explained on its first occurrence.

2.24 Keywords. A list of keywords is useful for indexing purposes in abstracts journals (and in the volume in which the paper is printed) and for information retrieval from databases that do not include the abstract. A list of terms that are suitable for this purpose is being prepared by IAU Commission 5; a structured thesaurus is in an earlier stage of preparation. The indexes to Astronomy and Astrophysics Abstracts provide a useful source of keywords. The indexes to previous volumes of the journal in which a paper is to be published should also be consulted.

The combination of a general term with a specific term (for example, stars: chemical composition) often provide an effective way of specifying a keyword that can also serve as a subject heading for the index of the journal.

2.3 The text

2.31 General guidance. It is not the purpose of this manual to provide instruction in the art of writing English prose. It is, however, appropriate to draw attention to a few points that are particularly important for scientific papers that will be written and read by scientists who are not fluent in English.

(a) Different aspects of the paper (such as objectives, methods and results) should be treated systematically in sections with appropriate headings.

(b) Sentences should be short and simple in construction. Any long and complicated sentence should be examined, and then split into shorter sentences, even if this involves inserting a few extra words to link the sentences together.

(c) Short and familiar words should be used in preference to long and unusual words whenever possible. Long compound terms, specialist jargon and slang should be avoided.

In addition to being correct and clear, it is desirable that the text should be consistent in typographical matters such as the use of initial capital letters and the form of abbreviations. Inconsistent or unconventional typographical style, like errors in spelling and syntax, can be distracting to readers even when they are not misleading. The later sections of this manual contain much detailed guidance on such points; for example, the use of capitals for the initial letters of words is discussed in section 6.13. It must be recognised, however, that there are no absolute rules on such matters, and so the guidance given here differs in detail from that given in other publications, such as: AIP 1978; ApJ 1983; Butcher 1981; Hart 1983; Royal Society 1974; Young 1969.

Further guidance on the writing of scientific papers and reports and on English usage is given in many books, including: Barzun & Graff 1977; Booth 1985; CBE 1983; Cochran et al. 1973; Day 1989; Fowler 1965; Gowers 1986; O'Connor & Woodford 1976; Prentice-Hall 1974; Strunk & White 1979.

It is particularly important that mathematical formulae and other symbols are presented clearly and correctly, and that the terminology, notation and designations used for astronomical concepts and objects are precise and unambiguous. These matters are also discussed in some detail in sections 5 and 7 of this manual.

2.32 Data. It is important that a paper that presents numerical data obtained from astronomical observations should contain all the information that is necessary for their use and evaluation. The following principles, which have been developed by experts from many different fields of science, should be followed in the preparation of astronomical papers.

(a) The paper must describe the observational procedures used to obtain the numerical data.

(b) The paper must describe the procedure used to derive the reported results from the actual measurements.

(c) The paper must give the numerical results in a form that is as free from interpretation as possible and in such a manner that the uncertainties of the data can be re-analysed in terms of a hypothesis that is different from that considered by the author.

(d) The data must be presented in such a way that the object, system or phenomena observed and the quantities tabulated can be unambiguously identified and so that the results can be readily related to other data for the same or similar systems.

The basis and implications of these principles are discussed further in the CODATA 'Guide to the Presentation of Astronomical Data' (Wilkins 1982). The recommendations in this manual on designations (section 7.2) and units (section 5.1) are intended to ensure that principle (d) is satisfied.

2.33 Conclusions. In general, each paper should include a final section that contains a general discussion of the principal results and of any further conclusions that can be drawn from them. This section should have a separate heading so that it is clearly distinct from the main text. It should not repeat the abstract, although in many respects it may be regarded as an extended abstract.

2.4 Tables and illustrations

2.41 Tables. Careful consideration should be given to the design of tables in order to ensure that they are appropriate for their purpose. Copy for tables should be prepared in accordance with the recommendations given in section 3.4. The most important use of printed tables in astronomical papers is to provide a precise record of observational data in a form that is convenient for use by others. Tables should not normally be used when the objective is to show to low precision how, for example, the properties of a system vary with the parameters that are used to characterize it; such relationships are usually best shown in graphical form. Redundancy between tables and graphs should normally be avoided. Similarly, tables should not normally be given where the relationship between respondent and argument can be expressed mathematically in a form that can be easily evaluated by calculators or computers.

Tables that are to be printed, or reproduced on microfiche, should be designed so that they will fit conveniently on to the page (preferably in its normal orientation). The characters should be of adequate size, and there should be sufficient space between columns and rows to ensure that there is little risk of confusion in extracting data from the table. The layout should be such that rules are not required to separate adjacent columns or lines. The column headings should be chosen carefully in order to ensure that the significance of each column is clear and the unit of tabulation is unambiguous. For example, it could be given in the form Q/u, where Q is the symbol for the quantity and u is the symbol for the unit, including any appropriate power of ten. The sequence of the tabulation should be chosen to suit the most likely mode of use of the data; if necessary, each line should be numbered serially and appropriate indexes provided. The headings should be repeated if the table occupies more than one page.

Each table should have a caption that includes a title describing the content and purpose of the table concisely but explicitly, and a brief legend giving any explanation that is necessary to define the meanings of the column headings or to specify the basis of the tabulations. Notes on individual entries, including citations, may be given either on the same lines or as footnotes, depending on their length and frequency. More detailed explanatory notes may be given in the text, but ideally it should be possible to use each table without reference to the text; correspondingly, it is usually desirable that the text should be intelligible without explicit reference to the tables.

The editor should be consulted before any extensive table is prepared in final form.

2.42 Illustrations are here regarded as 'figures' or 'plates'. The term 'figure' is used to refer to a line drawing or other graphic image that can be reproduced by the same photolithographic technique as the text. The term 'plate' is used when the image must be reproduced by a special technique (e.g. colour printing) on a separate page; since the extra printing costs may be considerable the editor should be consulted in advance about their inclusion.

Figures should be carefully designed or chosen so that they are suited to their purpose. Graphs, contour maps, outline sketches of equipment, block diagrams of processes, and other line drawings should be kept as simple as possible; in particular, detailed labelling should be avoided so that the lettering on the printed page may be a reasonable size. (It should be borne in mind that such a diagram may be copied for use in a lecture, and so the heights of the letters and numerals should be not less than 5% of the height of the figure). Similarly, photographs should show clearly the object of interest and should not include a lot of confusing or irrelevant detail; colour should not be requested unless it is essential to the understanding of the points being illustrated.

Each illustration should have a caption that includes a concise, informative title and a legend that explains the significance of the symbols and labels used in it. The units in which quantities are measured, the scales of drawings and photographs, and the names of the objects illustrated should be shown clearly and unambiguously. The expressions chosen to label the axes, or to show what is represented by contour lines, should be pure numbers, such as Q/u, where Q is the symbol for the quantity and u is the symbol for the unit. Top and right-hand borders with scale marks should always be given with a graph to facilitate the extraction of numerical values from it. Ideally, the illustration, including its caption, should be useful without reference to the text, although further information may be given there. Excessive redundancy between text and caption should be avoided.

2.5 References, footnotes and acknowledgements

2.51 References. The list of references forms an important part of an astronomical paper. Its primary function in a paper that reports new results is to indicate the sources of information that have been used by the author in developing the new results or on which critical comments have been made. It can also be used to draw attention to other relevant, recent papers that are concerned with the same topic but have not directly influenced the results. The list should not be inflated by including references to all previous work; it is sufficient to draw attention to an earlier paper that contains references to the basic work in the field. In particular, editors and referees should discourage unnecessary citations of earlier papers by the author and his colleagues. Similar considerations apply to review articles; these should comment on the significance of the papers that are cited and then attempt to summarise the current status of the field. For such articles, and for the triennial reports of the presidents of IAU commissions, the aim should be to give complete coverage of useful papers over the period since the previous review was prepared; there is no value in listing papers that give incomplete preliminary accounts of work that has been published in full later or that do not contain any significant new results or insight.

It is now generally recognised that the most effective way of presenting references is by means of a consolidated list after the text of the report, rather than by giving them in the text or as footnotes. Some journals insist on the use of numbered lists in the sequence of citation in the text, but the system adopted in IAU publications is to give the author's surname and the year for the citation, and then to list the references in an alphabetical/chronological sequence as described in detail in section 4.3.

In the IAU Reports on Astronomy it has been the practice to include in the text the author's name followed by the reference number of the entry in Astronomy and Astrophysics Abstracts. This practice is much less convenient for the user and is not now recommended.

2.52 Footnotes. The use of footnotes to provide additional information or to give the author's second thoughts is strongly discouraged. In particular, information pertaining to research grants and other support should normally be included in the acknowledgements at the end of the text of the paper. Important information that is obtained after the submission of the paper and that significantly affects any of its conclusions should be given in a note at the end; attention may be drawn to this addition by the insertion of a simple reference at the appropriate place in the text.

2.53 Acknowledgements. Significant contributions by persons other than the authors to the execution of the investigation or the preparation of the paper may be acknowledged at the end of the paper. Supervision by senior staff, as well as the assistance of persons who have carried out routine or technical work under supervision, is often best acknowledged here, rather than by including their names as co-authors. This also applies where a person has provided equipment or software, but has not participated in its use for the investigation being reported. This is also the most appropriate place for recording brief details of research grants or contracts when such an acknowledgement is a condition of the award.

Acknowledgements of the sources of copyrighted material may be given with such items or they may be collected together at the end. Authors are reminded that it is their responsibility to obtain permission to reproduce material from other publications that may be subject to copyright.

3. BASIC INSTRUCTIONS FOR THE PREPARATION OF COPY

3.1 Preparation of typescripts for camera-ready copy

The following instructions apply specifically to the preparation of camera-ready copy by direct typing or with the aid of a word-processing system in which the output device is a typewriter or other printer with a limited choice of typefaces and of character and line spaces. Authors who have the use of typesetting software and laser printers should aim to produce output that corresponds closely in style to ordinary typescripts so that ordinary readers will not be distracted by glaring differences between papers. In particular, the typeface and typesizes (for text and headings) should be similar to those currently used in the series in which the paper is to be printed. Even with these restrictions, the readers will still obtain the subtle benefits of sharper, proportionally-spaced characters and of better facilities for the setting of mathematical expressions.

3.11 General instructions for the typist. The camera-ready typescript must be absolutely clean and free from creases since it will be photographically reproduced. Whenever possible use the special sheets of paper that are issued by the editor or publisher; these show in light blue the frame within which the text must be typed. There are different formats for different series of publications and only the correct style of sheet should be used. If it is necessary to use plain paper, ensure that the typescript is prepared to the correct dimensions on good-quality paper.

Use an electric typewriter or printer that gives sharp characters of even density; dot-matrix printers must not be used unless they are of exceptionally high quality. If possible use an elite typeface (12 characters per inch = 10 ch/21mm), and type the text at single spacing (6 lines per inch = 1 line/4mm) on one side only of the paper. If you have a choice, use a typeface that provides a good differentiation between the numerals zero and one and the letters O and lower case l; avoid using a sans-serif face as these are more difficult to read. If you must use a manual typewriter, use a machine that is in good condition, with clean, sharp type that strikes the paper evenly. If possible, use one-time carbon ribbons which give clean, sharp, black impressions. Normal black ribbons may be used if they are replaced regularly to avoid fading. Unavailable characters or symbols should be drawn with a thin pen and black permanent ink at the correct size; a few special characters may be formed from two standard characters.

3.12 Corrections. Do not use an eraser on the typing sheets, and do not use white correcting fluids for typing over large errors, as these methods result in bad reproduction. A correction facility on the typewriter may be used for small corrections (a few characters). The use of a fluid is permissible for correcting a single character, but it must be thin and dry so that the over-typed character is of equal density. For large corrections (one whole line or more) it is better to type the correct lines on a fresh sheet of typing paper. To correct an omission, retype the line on which it occurs and the following lines to the point where the corrected lines can be placed without disturbing the remainder of the page. Each correction line should be referred to the original typescript by quoting the page and line-number; it is not necessary to place each correction on a separate sheet. The publisher will mount all corrected lines. Type corrections with the same type-writer and typeface on the same style of paper as the original.

3.13 Layout. Type, as far as possible, to the full width and length of the blue frame on the paper supplied by the publisher or to the specified dimensions. Do not leave extra margins within the blue frame. Text should not be justified (aligned on the edge of the right-hand margin) if a typewriter or printer with fixed-width spaces is used. In unjustified text it may be permissible for lines to extend occasionally over the right-hand margin by two or three characters, so as not to break a word unnaturally. The last line may cross the bottom limit. In no circumstances should letters or lines be squeezed together to fit them into the printed outline. If possible avoid starting a new paragraph on the last line of a page or giving the last line of a paragraph at the top of a new page.

The layout of the preliminaries (title, authors, abstract) and of headings and paragraphs should conform with the instructions provided by the editor or otherwise indicated on the typing sheets. Typical instructions for papers for IAU publications are as follows:

The title should be typed in capital letters starting flush left on the fifth line. It should be followed by at least two blank lines.

The names of the authors and their affiliations and addresses should be typed in capitals and small letters with an indent of 10 spaces (20mm). Each author's initials should precede the surname (family name). A forename (given name) may be given in full only if an author always uses this style. The affiliation should, where possible, be given in English, and should comprise the name of the institution (with department if appropriate) and its location. Full stops (otherwise known as 'periods' or 'full points') should be omitted in acronyms, but should be given after initials and abbreviations; commas should not be given at the ends of the lines. If the authors have different affiliations, there should be one blank line between the information for each author (or set of authors at the same institution). The last affiliation should be followed by at least two blank lines.

The word ABSTRACT should be typed flush left in capitals and the text of the abstract should begin on the same line. The abstract should preferably be composed as a single paragraph.

The instructions on the layout of reports to be published in the Transactions of the IAU are distributed directly to the Presidents of the Commissions.

3.14 Headings and subheadings. In IAU publications the principal sections of the report should be indexed by numbered 'first-order headings' that are typed flush-left in arabic numerals and capitals. Each such heading should be preceded by two blank lines and followed by one blank line.

Subheadings should be numbered decimally (as in this manual) and typed flush-left in small letters (except for the initial capital on the first word). A second-order heading (one decimal place) should be preceded by one blank line and followed by one blank line. A third-order heading (two or more decimal places) should be preceded by one blank line, but the following text should be run on in the same line. In general, it is best to avoid more than three levels of numbered headings; two levels are often sufficient even in reports of medium length.

New paragraphs should be preceded by a blank line and should be indented 5 spaces (10 mm) from the left-hand margin. A series or list of items within a paragraph may be indicated by the use of the letters a, b, c, etc. enclosed in parentheses; the first line for each item should be indented by five spaces (10 mm). The items may be separated by blank lines if this appears to be appropriate.

The publisher will normally be responsible for the insertion of running heads (see subsection 2.21) and page numbers. The sheets (or 'folios') of the typescript should be numbered sequentially from 1 in the top right-hand corner of each sheet, using a light-blue pencil.

3.15 Numbering of tables, illustrations and equations. In general, tables, figures and equations should each be numbered sequentially in Arabic numerals from 1. Illustrations, such as colour photographs, that are to be reproduced on separate pages as 'plates' may be numbered in the same sequence as figures that are to be printed with the text; in such a case, a reference in the text should be given in this form "... in Figure 23 (Plate 3) ...". For books and long papers with many tables, figures or equations it may be more convenient to the readers if a new sequence of numbers is started with each chapter or principal section; e.g., a cross-reference of the form "Table 3-5" would refer to the fifth table in chapter or section 3.

The number of an equation should be given within parentheses at the end of the line concerned, without leaders between the equation and the number; the cross-reference to the equation should not, however, include the parentheses. Individual equations within a group of related equations may be identified by using a decimal notation; e.g., "Equation 27.3" would refer to the third equation in the group of equations identified at the end of last line of the group by (27). It is not necessary to number an equation, or displayed mathematical expression, if there are no cross-references to it or if it could be easily identified by reference to the section in which it occurs.

3.16 Illustrations. Appropriate spaces must be left on the sheets of typescript for figures that are to be reproduced with the text. All such figures must be numbered in sequence, using Arabic numerals, and each must have a caption that is complete enough for the illustration to be appreciated without reference to the text. The caption may be typed at the appropriate place in the typescript even though copy for the figure itself may be provided separately. Recommendations on the preparation of the illustrations are given in section 3.5.

3.17 References. The format of citations in the text and of bibliographic references in the list should follow the recommendations given in section 4, unless the editor has specified a different system.

3.2 Preparation of typescripts for printer's copy.

A typescript that is to be used as printer's copy should be clear and neat so as to reduce the number of errors made by the compositor. (The correction of such errors is costly, time-consuming and a prolific source of further errors.) The typescript should be typed on only one side of the paper, whose size should normally be A4 (296 x 210 mm) or the corresponding American size (279 x 216 mm). It should be double-spaced (3 lines per inch = 1 line/8mm) throughout, with adequate margins of at least 40 mm at the left and bottom. Otherwise, the instructions given in the preceding section usually apply.

Corrections must be legible and are best given at the appropriate place in the text, rather than in the margin. Instructions should be given in simple words, rather than by the use of proof-correction marks. White correcting fluid may be used freely, but heavily corrected pages should be retyped. Some publishers pass the typescripts through OCR (optical character recognition) readers to avoid the need for manual input of the text by the compositor, and so authors should aim to produce typescripts with the minimum of handwritten corrections.

Special characters and mathematical equations may be written in by hand using black ink. Greek letters and unusual symbols should be identified separately by marginal notes written in pencil. Further guidance on the presentation of mathematical formulae is given in section 5.2.

Notes for the attention of the editor should be clearly distinguished from text that is to be set in type: for example, they may be encircled and preceded by the words To Editor.

Extra copies should be supplied in accordance with the requirements of the journal.

3.3 Submission of copy in electronic or magnetic forms

Some publishers of astronomical journals and books are considering how best to utilise the new techniques for the transmission of information by electronic means (public telephone system or computer network) or in magnetic forms, such as diskettes (floppy discs) generated by desk-top computers. The former is appropriate when the 'electronic copy' is to be used to generate typescripts for consideration by the editor and referees; subsequent communications about the paper can also be by electronic-mail and the final version can be used directly by the printer either for the production of camera-ready copy or for input to a typesetting system. Diskettes may also be used to avoid the necessity for the printer to keyboard the paper from printer's copy. Diskettes are also appropriate if the author is able to use text-processing software to generate a 'compuscript' that contains all the control characters required to specify the typographical format in full detail. (Such control characters may be lost or changed if compuscript files are transmitted by electronic means.) The author must ensure that the diskette and the compuscript file are fully compatible with the system to be used by the printer. The journal editor or the publisher may be expected to supply appropriate instructions, and possibly software, for the system concerned.

3.4 Preparation of copy for tables

3.41 Tables must be carefully designed to suit the method of printing, as well as to be appropriate to their purpose as discussed in section 2.4.

Short tables may be typed directly in camera-ready copy in the appropriate places and with their captions. Tables that do not require the full width of the page should be centred between the margins of the type area. Each table caption should be typed above the table and to the same width. The caption should include the number of the table (in Arabic numerals) and a concise, informative title; further explanatory notes may be given either immediately after the title or after the table. The body of the table should be typed at single spacing (unless it includes subscripts or superscripts) with blank lines between every five lines (or other appropriate small number) to give greater legibility. At least two, preferably three, blank spaces should be allowed between successive columns. Camera-ready copy for large tables that occupy one page or more will usually be produced directly by a computer printer to save time in preparation and checking. Care must be taken to ensure that such print-outs have good contrast and are of constant density. If possible the caption and column headings should be printed at the same time; otherwise care should be taken to ensure that these are typed in a similar style and density.

Tables that are to be used as printers' copy should be typed at double spacing on separate sheets with their captions. Explicit instructions must be given for column headings, for spacings between columns and blocks of lines, and for any other special features of the table, such as the use of different founts for arguments or certain respondents. The most appropriate place for the table should be indicated in the margin of the typescript.

Large tables will usually be required in camera-ready form even if the text is to be composed by the printer. Such tables will usually be reduced in size before they are printed and this should be taken into account in their design. The reduction is usually to 70% of the original size for tables prepared on typewriters. Very wide tables should be avoided, but if necessary a table may be reproduced in landscape (broadside) format with its top along the left-hand margin of the page.

3.42 Rules. If rules are to be included on camera-ready copy (and they are best avoided), they should be drawn by an expert with the proper tools to ensure uniformity of width and density. Great care is required to ensure that the rules are parallel or perpendicular to the lines of the table.

3.5 Preparation of copy for illustrations

3.51 Line drawings must be submitted as originals, drawn in black ink on good-quality tracing paper, draughtsman's film or white matt paper, or as glossy photographic prints. Photocopies, multiliths, Verifax or Xerox copies are not normally acceptable substitutes. In computer-drawn figures it is vital to ensure that the plotter or matrix printer produces clean black lines of an appropriate width and uniform density.

Drawings for camera-ready copy must be finished to a size that is compatible with the permitted typing area. The drawings will be reduced photographically with the typescript, and the size of lettering and the width of the lines must take this into account (see also subsection 2.42). Lettering should be done with a lettering stencil or with press-on lettering; free-hand or typewritten lettering is not recommended.

Drawings for submission with printer's copy should be finished to a size that will allow for a substantial reduction to the final printed size, which may be 50% (or less) of the original size. The thickness of lines and the size of the lettering must take this into account; there should be consistency in thickness and size between all elements in the illustrations. The scale of a drawing or photograph should be indicated in a manner that is independent of the reduction factor.

Each illustration for camera-ready copy should be mounted so that the edges of the illustration (and not the edges of the paper on which it is drawn) are appropriately indented (say 20mm) with respect to the margins or adjacent text. The caption should be typed below the figure and to the same width. Large illustrations may be aligned on the left-hand margin; if it is necessary to turn it to landscape (broadside) format then the top of the illustration should be at the left-hand margin of the page. If, for any reason, it is impossible to provide artwork of the correct finished size, leave enough space in the typescript for the incorporation of the illustration and supply the artwork to the publisher, who will take care of the photographic reduction and mounting.

Illustrations for printer's copy (or unmounted illustrations for camera-ready copy) must be numbered carefully (in Arabic numerals) and the appropriate orientation indicated if there can be any doubt about it. The place where each illustration would be best inserted should be indicated by a circled note in the margin of the typescript. Captions should be typed on a separate sheet of paper.

3.52 Photographs. The reproduction of photographs necessarily involves a slight loss of quality, particularly if the picture has to be enlarged, and so photographs should be supplied as original prints that are large and glossy and that have good contrast in tone ranges and between subject and background. Photocopies of prints are not acceptable. Photographs of illustrations from printed books and journals do not reproduce well and should be avoided if possible.

Any additional lettering, arrows, or scales should be marked on transparent overlays with register marks. Be careful not to mark or score the photographs or to dent them when writing on the back. The photographs should, however, be appropriately numbered on the back by writing lightly with a soft lead pencil, never with a ballpoint pen; the appropriate orientation should be indicated if there can be any doubt about it. Captions must be typed on separate sheets of paper; scale factors should not normally be given in captions as the final reduction factor may not be the same as that expected. Paperclips, pins and staples should not be used to attach overlays or captions to the photographs, which should be protected by keeping them between sheets of cardboard.

3.6 Title pages for proceedings of IAU symposia and colloquia

Editors of the proceedings of symposia and colloquia that have been sponsored by the IAU should ensure that the title page contains all the information necessary to identify the conference by the title, or by number, or by place and date. The title page should also give the name(s) and affiliation(s) of the editor(s) and the name and place of the publisher, which may be an observatory or other non-commercial organisation. The basic recommended format (where nn represents a number in Arabic numerals) is as follows:

NAME OF UNION IN ENGLISH AND FRENCH

TITLE OF CONFERENCE
IN LARGE CAPITALS

PROCEEDINGS OF THE nnTH SYMPOSIUM OF THE
INTERNATIONAL ASTRONOMICAL UNION
HELD IN PLACE, COUNTRY
nn-nn MONTH 19nn

Details of cooperating ICSU organisations if appropriate

EDITED BY

NAME IN CAPITALS
Affiliation, Place, Country

and

SECOND EDITOR
if appropriate

NAME OF PUBLISHER
Place(s) of publication

It is desirable that full information be given on a separate title page when the papers are published in a regular or special issue of a journal or other serial; such a page is useful for classification and abstracting purposes and if the material is rebound as a separate book. This title page should also specify the name, volume and date of the journal or serial; this additional information should be given after the name of the editors.

If appropriate, the support of other organisations should be indicated by the words "Organized by the IAU in cooperation with ..." in which the list of organizations is limited to the Scientific Unions, the Scientific Committees and Inter-Union Commissions of ICSU. Participation of UNESCO will be acknowledged by the following wording at the foot of the title page, "Published for the International Council of Scientific Unions with financial assistance from UNESCO".

3.7 Checking and correcting copy

All copy should be carefully checked before it is sent to the printer, either for reproduction or composition. Any errors left in camera-ready copy will probably appear in the printed version (unless they are noticed by the editor or copyeditor) since proofs are not normally supplied to authors. Non-trivial errors on printer's copy are likely to appear in print unless an adequate time is allowed for proofreading; trivial errors may be noticed on proof but their correction may itself lead to other errors in the printed version. Checking should be carried out in separate stages; for completeness, for sense, and for detail.

3.71 Sense. A referee or a colleague who comes fresh to the paper is usually in a better position than the author to notice any error or omissions of a general character. A sentence may be ambiguous or may not convey the idea intended by the author; a figure may not actually illustrate the effect claimed in the text; a significant factor may have been overlooked either in the investigation itself or in the paper on it; the abstract may not be a proper summary of the paper; or the conclusions may not be justified in the main text. For these reasons the paper should be read critically in order to find faults of this kind before the final copy is sent to the printer.

3.72 Detail. The checking to find errors of detail should be carried out quite separately from the more general checks. If possible, separate examinations should be made for the following types of error:

(a) Errors in numerical values, including dates, in the text, tables and figures: the typed values should be checked independently of the manuscript whenever this is possible; tables should be checked systematically.

(b) Errors in, or omission of, the units associated with numerical values, especially in figures and tables; also the use of units that are not in accordance with the recommendations given in section 5.1.

(c) Errors in references: names, titles, dates, volume numbers and page numbers should be checked against the sources cited and not against other lists; the abbreviations for the titles of serials should follow the recommendations given in section 4.4. It is not unknown for authors to make errors in references to their own papers!

(d) Errors in spelling and in grammar: these may not appear to be important, but their presence is often indicative of a lack of care in other aspects of the preparation of the report, and they may be taken to indicate a lack of thoroughness in the investigation to which the report refers.

(e) Errors and ambiguities in formulae and equations: these may be detected by consistency checks or by reference to original sources.

(f) Errors and ambiguities in the designations of astronomical objects: the principles and recommendations given in section 7 should be followed.

Urgency of publication is no justification for a lack of care and attention in the preparation of a typescript for publication; rather it should be seen as a reason for taking care at all stages of the work.

3.73 Completeness. The most basic check is to ensure that the copy is physically complete; even for camera-ready copy it is necessary to ensure that no pages of text, corrections or tables have been mislaid. For printer's copy it is also necessary to check that the preliminaries, photographs, overlays, and captions are all present. For long papers it may be worthwhile to highlight on a spare copy of the text all citations and cross-references to other text, tables and figures, and then to verify systematically that all are satisfied; this will also guard against any failures to carry though any changes in the numbers of sections, tables or figures.

4. CITATIONS AND LISTS OF REFERENCES

The style of citations and bibliographic references described here must be used in all IAU publications; it is known as the Harvard system. There is, however, no system of referencing that is accepted by all international organisations, journal editors and publishers. It is hoped that the system recommended here will be adopted by editors of major astronomical journals; it is intended to simplify both typing and typesetting, and to be appropriate for computer-based information-retrieval systems. It follows in general form the standards adopted by many national and international scientific organisations (BSI 1976, 1978, 1984; ICSU AB 1978; ISO 1972, 1974).

4.1 Form of citations in the text

The citation in the text should be sufficient only to identify the appropriate entry in the list of references or to indicate the source of information that cannot be retrieved by the readers. In the former case the citation should identify the author and give the year of publication since these two items are used to determine the sequence of references in the list; the following rules should be followed:

(a) If the name of the author occurs naturally in the text, the year should be given afterwards in parentheses. Otherwise the name of the author and the year should be given in parentheses at the appropriate place in the text, without any punctuation mark between them.

(b) If there are two authors, their names should be linked by an ampersand (&) or by the word 'and' if the symbol & is not available on the typewriter.

(c) If there are three or more authors, the name of the first should be followed by the abbreviation 'et al.', meaning 'and others', even if other such citations refer to papers by the same first author with different sets of other authors.

(d) If two or more authors have the same family name and the references are to the same year, their initials should be given after the family name; otherwise, initials should be omitted in the citation.

(e) If there are two or more references with the same author (or joint author) and year, the letters, a, b, c, ... should be appended to the year to distinguish between them.

(f) If several citations for one author (or joint author) occur at the same place in the text, the years should be separated by commas; the name should not be repeated.

(g) If two or more citations for different authors occur at the same place in the text, they should be separated by semi-colons.

(h) If the citation is to a paper (or other source of information) that does not explicitly name the person (or persons) who have written or edited the paper, the citation should give a short name that is sufficient to identify the corresponding entry in the reference list where the full details are given. This short name could be, for example, the acronym for the organisation concerned or the name of the chairman of a working group; otherwise the author should be given as 'Anon'.

(i) If there are several citations to different places in the same source document, it is helpful to give in each citation the relevant page number after the year. If there are many citations to the same document it may be worthwhile to refer to it in such a way that it is not necessary to repeat the name of the author and the year on each occasion.

(j) The names of authors of papers that are not printed in a latin-type alphabet should be transliterated in accordance with the recommendations given in section 6.3.

Citations that refer to information that has not been published or that is not otherwise available for reference should give appropriate information about the source in the text, usually in the form: name of person concerned, the year, and the type of communication; examples include: 'in preparation', 'unpublished work', 'personal letter', 'oral communication'. This does not apply if a paper is 'in press' in the sense that it has been accepted and all bibliographic details except volume and page numbers can be given. A reference may be given if the meeting at which the information was presented can be specified or if an abstract of a paper has been published. If the source of information is an unpublished document that is available for reference (or copying) in an archive or library, then the identification number and place should be given in the list of references in an appropriate format.

Terms such as 'ibid.', 'idem', 'op.cit.' and 'loc. cit.' should not be used; they are often confusing and changes in the text can make them erroneous if they are not carefully checked.

4.2 Format of bibliographic references

The information to be given in a bibliographic reference depends on the type of publication concerned.

(a) A full reference to a paper in a serial publication should give the following elements: (1) the name of the author(s) and the year; (2) the title of the paper; and (3) the abbreviated title and volume number of the serial, together with the number of the first and last pages separated by a hyphen. The title of the paper can provide useful information for editors and referees as well as for readers; it is, however, the general practice in IAU publications and in many astronomical journals to omit the title of the paper. Titles of papers may be given in IAU publications if they do not cause the camera-ready copy to run over to another page. The title of the serial should be abbreviated in accordance with the recommendations in section 4.41 unless the publisher insists on the use of a different system. Some journals give only the first page number, but the inclusion of the last page number can be very helpful to readers who do not have easy direct access to the serial concerned.

(b) If the reference is to a monograph by the author cited, then it should contain the following elements: (1) the author and year; (2) the title of the monograph and the edition, if relevant; and (3) the place of publication and the name of the publisher. If the publisher operates in several places it is sufficient to give the name of the first place listed after the publisher's name on the title page of the book. The name of the town should be followed by the name of the country unless it is a well-known city. The name of the town may be omitted if it is included in the name of the publisher. The international standard book number (ISBN) may be given as an aid to identification.

(c) If the reference is to a paper within a book edited by others, then the following elements are to be given: (1) author and year; and (2) the reference for the book as in (b), followed by the page numbers concerned. If, as often occurs, there are several references to papers in the same book, then the length of the reference list should be shortened by giving in each case the second element in the form of a citation to a separate reference for the book. In both forms the second element should be preceded by 'In:'. If the journal allows the inclusion of the title of paper itself, then this should be given after the first element.

(d) If the reference is to a paper that is published in a book as part of the record of the proceedings of a conference, then the reference should include the name(s) of the editor(s), the title of the book, the identification of the conference if this is not included in the title, the place and date of the conference, and the identification of the publisher (place and name).

The names of the authors should normally be given in the list of references in exactly the same form as in the original papers, except that the initials should always be given after the surnames, which should be followed by commas when transpositions have been made. In multi-authored papers all names after the first eight should be omitted and replaced by 'et al.'.

The items within each element of the reference should be separated from each other by commas; the end of each element should be indicated by a full stop. The title of serials and books should not be underlined in camera-ready copy; they may be typed in italic if this facility is available to the author. Volume numbers should be given in arabic numerals; they should not be underlined, although they may be made bolder. Underlining on printer's copy to indicate changes of typeface should normally be left to the copyeditor.

Examples of the application of these recommendations are to be found in the list of references at the end of this manual. It is recommended that authors should always refer to the instructions provided by the editor, or to previous issues of the serial, or to other similar books by the same publisher, in order to establish the format to be used in any particular case.

4.3 Sequence of references

In IAU publications, in which the citation in the text is based on the name of the author(s) and the date, the sequence to be adopted in the list of references is based, firstly, on the alphabetic sequence of the surnames of the first authors, and secondly, on the numerical sequence of the date for each author. The standard alphabetic sequence to be used is as follows:

A B C D E F G H I J K L M N O P Q R S T U V W X Y Z

Accents and other diacritical marks should normally be ignored. Any prefix (such as 'van') should be regarded as a part of the surname; a space within the surname should be sorted before A; and any hyphen should be ignored. No distinction is to be made between upper and lower-case letters.

All references with the same first author should be collected together, giving firstly those where he is the sole author, arranged in chronological sequence, then those where there are two authors, arranged firstly in the alphabetical sequence of the names of the second author, and secondly in chronological sequence, and finally those where there are three or more authors; within the latter group the sequence may be solely by date, regardless of the names of the other authors, since they are not specified in the citation in the text. If there are two authors with the same surname then the sequence should be that of their initials. On the other hand, initials should be ignored if the surname of one author begins with the surname of another author; the shorter name should precede the longer name.

The following example illustrates the use of these rules for the sequence of references and for punctuation in the element giving the author(s) and the date.

In the list of references	Citation in the text
Brown, R., 1977.	Brown (1977) or (Brown 1977)
Brown, R., 1978.	Brown (1978)
Brown, R., Green, B. V., 1976.	Brown & Green (1976)
Brown, R., White, C., Green, B. V., 1974	Brown et al. (1974)
Brown, R., Black, A. T., White, C., Green, B. V. 1975.	Brown et al. (1975)
Browning, A., 1972.	Browning (1972)

4.4 Abbreviations to be used for the titles of serials

4.41 Standard rules. In the preparation of camera-ready copy for IAU publications, the titles of serials should be abbreviated in accordance with the list given in Astronomy and Astrophysics Abstracts, which is in turn based on the recommendations of the appropriate international organisations (e.g., see ICSU AB 1978). The following guidelines should be adopted for any words not included in this list: (a) articles, conjunctions and prepositions should usually be ignored; (b) a sufficient number of letters of other words should be included to suggest to a scientist the full words and to avoid ambiguity; and (c) the names of places, except principal capital cities, should be given in full. The initial letters of the abbreviations for an adjective should be given in lower case (small type), except where it forms part of a proper name. An acronym should not normally be used for the name of an organisation unless it is so used in the full title of the serial. If a title consists of one word it is usually given in full. A list of the principal abbreviations that are recommended for use in astronomical publications is given in Table 1A.

Table 1. Abbreviations used in references to serial publications

A. Abbreviations recommended for use in astronomical publications

Word/Root	Abbreviation	Word/Root	Abbreviation	Word/Root	Abbreviation
Abstract-	Abstr.	Interna-	Int.	Prelimin-	Prelim.
Academ-	Acad.	Interi-	Inter.	Preprint	Prepr.
Annal-	Ann.	Interplanetary	Interplanet.	Proceedings	Proc.
Annu-	Annu.	Ionos-	Ionos.	Publica-	Publ.
Astrometr-	Astrometr.	Itali-	Ital.	Publika-	Publ.
Astronom-	Astron.	Jahrbuch-	Jahrb.	Quarterly	Q.
Astrophys-	Astrophys.	Joernal-	J.	Quantit-	Quant.
Atmos-	Atmos.	Jornal-	J.	Rapport	Rapp.
Bericht-	Ber.	Journal-	J.	Record-	Rec.
Bibliot-	Bibl.	Japan	Jpn.	Relativit-	Relativ.
Bibliogra-	Bibliogr.	Laborat-	Lab.	Report	Rep.
Bolet-	Bol.	Lette, Lettr-	Lett.	Reprint	Repr.
Bulletin	Bull.	Libra-	Libr.	Research	Res.
Catalog-	Cat.	London	Lond.	Review-, Revue-	Rev.
Celestial	Celest.	Magnet-	Magn.	Royal	R.
Centr-	Cent.	Mathemat-	Math.	Satellite	Satell.
Colloqui-	Colloq.	Mechani-	Mech.	Scien-	Sci.
Comput-	Comput.	Meddel-	Medd.	Scripta, Scritt-	Scr.
Conferen-	Conf.	Memento	Mem.	Socie-	Soc.
Contributions	Contr.	Memoir	Mem.	Solar	Sol.
Current	Curr.	Memorand-	Memo.	Southern	South.
Depart-	Dep.	Monat-, Month-	Mon.	Spectroscop-	Spectrosc.
Dominion	Dom.	Natur-	Nat.	Sternwarte-	Sternw.
Deutsch	Dtsch.	National-	Natl.	Supplement	Suppl.
Edit-	Ed.	Notas	Notas	Survey	Surv.
Electroni-	Electron.	Note	Note	Sympos-, Sympoz-	Symp.
Engineer-	Eng.	Notes	Notes	System-	Syst.
Ephemeri-	Eph.	Notices	Not.	Techne-	Tech.
Experiment-	Exp.	Nouve-	Nouv.	Technolog-	Technol.
Facol-, Facul-	Fac.	Numeri-	Numer.	Telescop-	Telesc.
Fascicul-	Fasc.	Observ-	Obser.	Theoret-, Theori-	Theor.
Francais-	Fr.	Offic-	Off.	Terrestr-	Terr.
Giornale	G.	Optic-, Optik-	Opt.	Transactions	Trans.
Gazet-	Gaz.	Osserva-	Oss.	Travaux	Trav.
General	Gen.	Pacific	Pac.	Union	Union
Gesellschaft	Ges.	Paper-, Papier	Pap.	United	U.
Geschichte	Gesch.	Particle	Part.	Universi-	Univ.
Handb-	Handb.	Philosoph-	Philos.	Variable	Var.
Incorporated	Inc.	Photography-	Photogr.	Zeitschrift-	Z.
Inform-	Inf.	Photometr-	Photom.	Zentral	Zent.
Institut-	Inst.	Physi-	Phys.	Zhurnal	Zh.
Instrument-	Instrum.	Planetary	Planet.		

A full list of recommended abbreviations is given in Astronomy and Astrophysics Abstracts.

B. Short abbreviations used in some astronomical publications.

A.A.A.	Astron. Astrophys. Abstr.	Bull. A.A.S.	Bull. Am. Astron. Soc.
A.J.	Astron. J.	B.A.A.S.	Bull. Am. Astron. Soc.
A.Q.n.	Astrophys. Quant., nth ed.	M.N.R.A.S.	Mon. Not. R. Astron. Soc.
Ap.J.	Astrophys. J.	M.N.	Mon. Not. R. Astron. Soc.
Astr. Ap.	Astron. Astrophys.	Pub. A.S.P.	Publ. Astron. Soc. Pacific
A & A	Astron. Astrophys.	P.A.S.P.	Publ. Astron. Soc. Pacific

4.42 Other rules. Some journals insist on the use of other rules; some allow or require the use of very short forms for certain journals that are well-known to professional astronomers. A list of such short forms is given for information in Table 1B; these forms should not, however, be used in IAU publications. Such forms may cause difficulties for scientists and librarians who may not be familiar with them, and they may be missed or cause confusion in literature searches using computers. Their use may be justified in review articles containing many references, such as the IAU triennial reports, but the meanings of such short forms should be given at the head of the list of references.

4.5 Series of books

Series of books that may be treated as serial publications include:

Annual Reviews of Astronomy and Astrophysics. Editors: Burbidge, G. R., Layzer, D., Phillips, J. G., (from 1974,). Annual Reviews Inc., Palo Alto, California, 1963 onwards.
Abbreviation: Annu. Rev. Astron. Astrophys.

Handbuch der Physik (Encyclopedia of Physics). Editor: Flügge, S. Springer-Verlag, Berlin, 1955 onwards.
Abbreviation: Handb. Phys.

Highlights of Astronomy. Published for the International Astronomical Union. Reidel, Dordrecht-Holland, 1968 onwards at intervals of 3 years.
Abbreviation: Highlights Astron.

Landolt-Börnstein: Numerical data and functional relationships in science and technology. New series, Editor-in-chief: Hellwege, K.-H., Group VI: Astronomy, astrophysics and space research. Springer-Verlag, Berlin, 1965, 1981 onwards.
Abbreviation: Landolt-Börnstein, New Series, Group VI.

Vistas in Astronomy. Editor: Beer, P. Pergamon Press, Oxford, 1955 onwards.
Abbreviation: Vistas Astron.

5. UNITS, SYMBOLS AND FORMULAE

5.1 Units

5.11 SI Units. The international system (SI) of units, prefixes, and symbols should be used for all physical quantities except that certain special units, which are specified later, may be used in astronomy, without risk of confusion or ambiguity, in order to provide a better representation of the phenomena concerned. SI units are now used to a varying extent in all countries and disciplines, and this system is taught in almost all schools, colleges and universities. The units of the centimetre-gram-second (CGS) system and other non-SI units, which will be unfamiliar to most young scientists, should not be used even though they may be considered to have some advantages over SI units by some astronomers. The IAU Executive Committee has recommended (see Appendix B) that astronomers should complete the change-over quickly, and so, for example, reports and papers for the IAU General Assembly in 1991 should all use SI units or other recognised units.

General information about SI units can be found in the publications of national standards organisations and in many textbooks and handbooks (e.g., Anderson 1981b, Bell & Goldman 1986, Drazil 1983). There are three classes of SI units: (a) the seven base units that are regarded as dimensionally independent; (b) two supplementary, dimensionless units for plane and solid angles; and (c) derived units that are formed by combining base and supplementary units in algebraic expressions; such derived units often have special names and symbols and can be used in forming other derived units. The units of classes (a) and (b) are listed in Table 2. The units of class (c) of greatest interest to astronomers are given in Table 3 for those with simple names and symbols, and in Table 4 for those with compound names and symbols. In forming compound names division is indicated by per, while in the corresponding symbols it is permissible to use either a negative index or a solidus (oblique stroke or slash); thus the SI unit of velocity is a metre per second and the corresponding symbol is $m\ s^{-1}$ or m/s.

The space between the base units is important in such a case since ms^{-1} could be interpreted as a frequency of 1000 Hz; a space is not necessary if the preceding unit ends in a superscript; a full stop (period) may be inserted between units to remove any ambiguity; the solidus should only be used in simple expressions and must never be used twice in the same compound unit.

Table 2. The names and symbols for the SI base and supplementary units.

Quantity	SI Unit: Name	Symbol
length	metre	m
mass	kilogram	kg
time (1)	second	s
electric current	ampere	A
thermodynamic temperature	kelvin	K
amount of substance	mole	mol
luminous intensity	candela	cd
plane angle	radian	rad
solid angle	steradian	sr

(1) The abbreviation sec should not be used to denote a second of time.

Table 3. Special names and symbols for SI derived units.

Quantity	SI Unit: Name	Symbol	Expression
frequency	hertz	Hz	s^{-1}
force	newton	N	$kg\ m\ s^{-2}$
pressure, stress	pascal	Pa	$N\ m^{-2}$
energy	joule	J	N m
power	watt	W	$J\ s^{-1}$
electric charge	coulomb	C	A s
electric potential	volt	V	$J\ C^{-1}$
electric resistance	ohm	Ω	$V\ A^{-1}$
electric conductance	siemens	S	$A\ V^{-1}$
electric capacitance	farad	F	$C\ V^{-1}$
magnetic flux	weber	Wb	V s
magnetic flux density	tesla	T	$Wb\ m^{-2}$
inductance	henry	H	$Wb\ A^{-1}$
luminous flux	lumen	lm	cd sr
illuminance	lux	lx	$lm\ m^{-2}$

Table 4. Examples of SI derived units with compound names.

Quantity	SI unit: Name	Symbol
density (mass)	kilogram per cubic metre	kg m^{-3}
current density	ampere per square metre	A m^{-2}
magnetic field strength	ampere per metre	A m^{-1}
electric field strength	volt per metre	V m^{-1}
dynamic viscosity	pascal second	Pa s
heat flux density	watt per square metre	W m^{-2}
heat capacity, entropy	joule per kelvin	J K^{-1}
energy density	joule per cubic metre	J m^{-3}
permittivity	farad per metre	F m^{-1}
permeability	henry per metre	H m^{-1}
radiant intensity	watt per steradian	W sr^{-1}
radiance	watt per square metre per steradian	W m^{-2} sr^{-1}
luminance	candela per square metre	cd m^{-2}

Table 5. SI prefixes and symbols for multiples and submultiples.

Submultiple	Prefix	Symbol	Multiple	Prefix	Symbol
10^{-1}	deci	d	10	deca	da
10^{-2}	centi	c	10^{2}	hecto	h
10^{-3}	milli	m	10^{3}	kilo	k
10^{-6}	micro	µ	10^{6}	mega	M
10^{-9}	nano	n	10^{9}	giga	G
10^{-12}	pico	p	10^{12}	tera	T
10^{-15}	femto	f	10^{15}	peta	P
10^{-18}	atto	a	10^{18}	exa	E

Note: Decimal multiples and submultiples of the kilogram should be formed by attaching the appropriate SI prefix and symbol to gram and g, not to kilogram and kg.

5.12 SI prefixes. Decimal multiples and submultiples of the SI units, except the kilogram, are formed by attaching the names or symbols of the appropriate prefixes to the names or symbols of the units. The combination of the symbols for a prefix and unit is regarded as a single symbol which may be raised to a power without the use of parentheses. The recognised list of prefixes and symbols is given in Table 5. These prefixes may be attached to one or more of the unit symbols in an expression for a compound unit and to the symbol for a non-SI unit. Compound prefixes should not be used.

5.13 Non-SI units. It is recognised that some units that are not part of the international system will continue to be used in appropriate contexts. Such units are listed in Table 6; they are either defined exactly in terms of SI units or are defined in other ways and are determined by measurement. Other non-SI units, such as Imperial units and others listed in Table 7, should not normally be used.

Table 6. Non-SI units that are recognised for use in astronomy.

Quantity	Unit: Name	Symbol	Value
time (1)	minute	min or ^{m}	60 s
time (1)	hour	h or ^{h}	3600 s = 60 min
time (1)	day	d or ^{d}	86 400 s = 24 h
time	year (Julian)	a	31.5576 Ms = 365.25 d
angle (2)	second of arc	"	(π/648 000) rad
angle	minute of arc	'	(π/10 800) rad
angle	degree	°	(π/180) rad
angle (3)	revolution (cycle)	^{c} or c	2π rad
length	astronomical unit	au	0.149 598 Tm
length	parsec	pc	30.857 Pm
mass	solar mass	$M_{\odot}$	1.9891×10^{30} kg
mass	unified atomic mass unit	u	$1.660\ 540 \times 10^{-27}$ kg
energy	electron volt	eV	0.160 2177 aJ
flux density	jansky (4)	Jy	10^{-26} W m^{-2} Hz^{-1}

(1) The alternative symbol is not formally recognised in the SI system.
(2) The symbol mas is often used for a milliarcsecond (0".001).
(3) The unit and symbols are not formally recognised in the SI system.
(4) The jansky is only used in radio astronomy.
(5) The degree Celsius (°C) is used in specifying temperature for meteorological purposes, but otherwise the kelvin (K) should be used.

5.14 <u>Time and angle</u>. The units for sexagesimal measures of time and angle are included in Table 6. The names of the units of angle may be prefixed by 'arc' whenever there could be confusion with the units of time. The symbols for these measures are to be typed or printed as superscripts immediately following the numerical values; if the last sexagesimal value is divided decimally, the decimal point should be placed under, or after, the symbol for the unit; leading zeros should be inserted in sexagesimal numbers as indicated in the following examples.

$2^{d}\ 13^{h}\ 07^{m}\ 15^{s}.259 \qquad 06^{h}\ 19^{m}\ 05^{s}.18 \qquad 120^{\circ}\ 58'\ 08''.26$

These non-SI units should not normally be used for expressing intervals of time or angle that are to be used in combination with other units.

In expressing the precision or resolution of angular measurement, it is becoming common in astronomy to use the milliarcsecond as the unit, and to represent this by the symbol mas; this is preferable to other abbreviations, but its meaning should be made clear at its first occurrence. The more appropriate SI unit would be the nanoradian (1 nrad = 0.2 mas). In general, the degree with decimal subdivision is recommended for use when the radian is not suitable and when there is no requirement to use the sexagesimal subdivision. If it is more appropriate to describe an angle in terms of complete revolutions (or rotations or turns or cycles), then the most appropriate symbol appears to be a letter c; this may be used in a superior position as in 1^{c} = 360° = 2π rad = 1 rev, but it may be used as in 1 c/s = 1 Hz.

The use of units of time for the representation of angular quantities, such as hour angle, right ascension and sidereal time, is common in astronomy, but it is a source of confusion and error in some contexts, especially in formulae for numerical calculation. The symbol for a variable followed by the superscript for a unit may be used to indicate the numerical value of that variable when measured in that unit.

5.15 <u>Astronomical units</u>. The IAU System of Astronomical Constants recognises a set of astronomical units of length, mass and time for use in connection with motions in the Solar System; they are related to each other through the adopted value of the constant of gravitation when expressed in these units (IAU 1976). The symbol for the astronomical unit of length is au; the astronomical unit of time is 1 day (d) of 86 400 SI seconds (s); the astronomical unit of mass is equal to the mass of the Sun and is often denoted by $M_{\odot}$, but the special subscript makes this symbol inconvenient for general use.

An appropriate unit of length for studies of structure of the Galaxy is the parsec (pc), which is defined in terms of the astronomical unit of length (au). The unit known as the light-year is appropriate to popular expositions on astronomy and is sometimes used in scientific papers as an indicator of distance.

The IAU has used the julian century of 36 525 days in the fundamental formulae for precession, but the more appropriate basic unit for such purposes and for expressing very long periods is the year. The recognised symbol for a year is the letter a, rather than yr, which is often used in papers in English; the corresponding symbols for a century (ha and cy) should not be used. Although there are several different kinds of year (as there are several kinds of day), it is best to regard a year as a julian year of 365.25 days (31.5576 Ms) unless otherwise specified.

It should be noted that sidereal, solar and universal time are best regarded as measures of hour angle expressed in time measure; they can be used to identify instants of time, but they are not suitable for use as precise measures of intervals of time since the rate of rotation of Earth, on which they depend, is variable with respect to the SI second.

5.16 <u>Obsolete units</u>. It is strongly recommended that the non-SI units listed in Table 7 are no longer used. Some of the units listed are rarely used in current literature, but they have been included for use in the study of past literature. Imperial and other non-metric units should not be used in connection with processes or phenomena, but there are a few situations where their use may be justified (as in 'the Hale 200-inch telescope on Mount Palomar'). The equivalent value in SI units should be given in parentheses if this is likely to be helpful.

Table 7. Non-SI units and symbols whose continued use is deprecated.

Quantity	Unit: Name	Symbol	Value
length	ångström	Å	10^{-10}m = 0.1 nm
length	micron	μ	1 μm
length	fermi		1 fm
area	barn	b	10^{-28}m^{2}
volume	cubic centimetre	cc	10^{-6}m^{3}
force	dyne	dyn	10^{-5}N
energy	erg	erg	10^{-7}J
energy (2)	calorie	cal	4.1868 J
pressure	bar	bar	10^{5}Pa
pressure	standard atmosphere	atm	101 325 Pa
acceleration (gravity)	gal	Gal	10^{-2}m s^{-2}
gravity gradient	eotvos	E	10^{-9}s^{-2}
magnetic flux density	gauss	G	corresponds to 10^{-4}T
magnetic flux density	gamma	γ	corresponds to 10^{-9}T
magnetic field strength	oersted	Oe	corresponds to (1000/4π)A m^{-1}

(1) Non-metric units, such as miles, feet, inches, tons, pounds, ounces, gallons, pints, etc., should not be used except in special circumstances.
(2) There are other obsolete definitions and values for the calorie.

The definitions of the SI units and an extensive list of conversion factors for obsolete units are given by Anderson (1981b). In particular, wavelengths should be expressed in metres with the appropriate SI prefix; e.g., for wavelengths in the visual range the nanometre (nm) should be used instead of the ångström (Å), which is a source of confusion in comparisons with longer and shorter wavelengths expressed in recognised SI units. The notation of the form λ followed by a numerical value (which represents the wavelength in angstroms) should also be abandoned. The name micrometre and symbol μm should be used instead of micron and μ. In all cases, the spelling metre should be used for the unit, while the spelling meter should be used for a measuring instrument (as in micrometer). The word kilometre should be pronounced ki-lo-me-te, not kil-lom-e-ter.

If wavenumbers are used they should be based on the metre, not the centimetre; in any case the unit (m^{-1} or cm^{-1}) should be stated since they are not dimensionless quantities. The uses of frequency (in Hz) at radio wavelengths and energy (in eV) at X-ray wavelengths are appropriate for some purposes, but they serve to obscure the essential unity of the electromagnetic spectrum, and so it may be helpful to give the wavelength as well at the first occurrence; the correspondences between these units and wavelength are as follows:

wavelength in metres = $2.997\,924\,58 \times 10^{8}$/frequency in hertz

or = $1.239\,842\,4 \pm 3 \times 10^{-6}$/energy in electron-volts

5.17 Magnitude. The concept of apparent and absolute magnitude in connection with the brightness or luminosity of a star or other astronomical object will continue to be used in astronomy even though it is difficult to relate the scales of magnitude to photometric measures in the SI system. Magnitude, being the logarithm of a ratio, is to be regarded as a dimensionless quantity; the name may be abbreviated to mag without a full stop, and it should be written after the number. The use of a superscript (^{m}) is not recommended. The method of determination of a magnitude or its wavelength range may be indicated by appropriate letters in italic type as in *U, B, V*. The photometric system used should be clearly specified when precise magnitudes are given.

5.2 Symbols, formulae and technical abbreviations

5.21 Symbols. Particular attention must be paid during the preparation of both camera-ready copy and printer's copy to the presentation of symbols, especially if the typewriter or printer being used has only a normal keyboard. In preparing a manuscript for a typist, the author should be careful to distinguish between the following characters which may be confused when written carelessly or in script.

a α ∝	C c	G σ 6	g q	k κ
l I 1	N η	o ο	p ρ	Q 2
r ν	S s 5	t τ	u ν υ	U V
w ω	X χ	Z z 2	1 7	O 0

The author should mark clearly subscripts and superscripts, and distinguish between a subscript figure 1 and a comma, and between a superscript figure 1 and a prime ' (single quote).

Most typewriters and printers do not distinguish between the figure one and the lower-case letter L; or between the figure zero and the letter O; or between hyphens, dashes and minus signs. These ambiguities are of no consequence on camera-ready copy if the meaning is clear from the context, but the differences can be important in designations, and so authors should guide the reader if confusion could occur. Printer's copy should be marked appropriately so that the compositor will use the correct typefaces.

Characters or symbols that are not available on the typewriter or printer (either directly or by combining two simpler characters) should be drawn carefully on camera-ready copy with a thin pen and black permanent ink. This applies also to printer's copy that contains only simple mathematical formulae and few special symbols.

5.211 Variables. Characters for mathematical variables that are to be set in italic should be marked by underlining in pencil in the text, but it is not normally necessary to mark such characters in displayed equations since the compositor should be aware of the standard rules. Functional operators, such as sin, cos, exp, ln, are always set in roman type. Characters for variables that are to be set in bold type should be marked by a wavy underlining, or by a general instruction that is unambiguous. Underlining on camera-ready copy is best avoided, but it may be necessary in some cases to avoid ambiguity.

5.22 Formulae. Authors should lay out formulae in such a way as to simplify the typing or composition and to avoid ambiguity; for example:

xy^{-1} or x/y is usually preferable to $\frac{x}{y}$; $\exp(-t_1^2)$ is preferable to $e^{-t_1^2}$

exponents should be used in preference to root signs, as in $x^{1/n}$; the arguments of operators (or standard functions) should be placed in parentheses when they contain several terms, as in sin(a+bt); spaces of various lengths (1, 2 or 3 typewriter spaces) may be used to show the structure of mathematical expressions.

On printer's copy, complex mathematical formulae should be written out carefully by hand since it is then possible to indicate more clearly the required spacing and layout; even so the author should clarify any possible ambiguities (in subscripts, superscripts, Greek letters and other special symbols) by appropriate markings in pencil. Further guidance on the setting of mathematics is given by, for example: ApJ 1983; BSI 1961; Chaundy et al. 1954; Royal Society 1974.

5.221 Brackets. The normal sequence of use of brackets is as follows: parentheses ( ); square brackets [ ]; braces { }. These may be followed by larger sizes of these brackets in the same sequence. On camera-ready copy it is often sufficient to use only parentheses; greater care is then required in checking the copy to ensure that the numbers of left and right parentheses are equal. Angular brackets are often used to indicate an average or for some other specialised purpose.

5.222 Numbers. In order to reduce the risk of error, numbers that consist of more than 4 figures should normally be separated by spaces into groups of 3, counting from the decimal-point position. The decimal-point should be indicated by a dot at half the height of the figures, or by a full stop (or by a comma in French and some other languages); neither commas nor full stops should be used as group markers. Thus 38 932.071 17 is preferred to 38932.07117, while 38,932.071,17 and 38.932,071.17 are deprecated. It should be noticed that the integral part of a julian date is often split as a group of 3 figures followed by a group of 4 figures. Numbers in the text should not be broken at the ends of lines, and small numbers (less than say 13) should normally be written in words except when they are followed by the symbol for a unit; thus, three stars but 3 pc. The zero before the decimal point in numbers such as 0.123 should not be omitted.

5.223 Mathematical signs. The following list shows the standard signs for the most common mathematical operations and relations. Some of these signs are sometimes used with different meanings.

plus, add	$+$	smaller (less) than	$<$
minus, subtract	$-$	larger (greater) than	$>$
multiply	$\times$	smaller than or equal to	$\leq$
divide	$/$	larger than or equal to	$\geq$
equal to	$=$	much larger than	$\gg$
identically equal to	$\equiv$	tends to (approaches)	$\rightarrow$
corresponds to	$\hat{=}$	of the order of	$\sim$
approximately equal to	$\approx$	plus or minus	$\pm$
not equal to	$\neq$	proportional to	$\propto$
scalar product (of vectors)	$\cdot$	infinity	∞
vector product	$\times$		

The letter x may be used as the sign for the multiplication of two numbers, but it should not be used in symbolic formulae, when a full stop is usually preferable. A full stop should not be used in numbers such as 3×10^n. The sign ~ is often used for 'asymptotically equal to'.

5.23 Chemical and spectroscopic symbols. The symbols for chemical elements are printed in upright type; the first letter only is a capital, as in Ca, and the symbol is not followed by a full-stop. The nucleon number may be indicated as a left superscript, as in ^{198}Hg, while the atomic number may be indicated by a left subscript, as in $^{198}_{80}$Hg.

Singly-ionized atoms should be indicated by plus and minus signs as superscripts, as in H^+, Cl^-; multiple charges should be indicated by the appropriate arabic number before the sign, as in Ca^{2+}, Fe^{13+}. The use of roman numerals is still common in astronomy, but should be abandoned, except perhaps in such terms as 'HII regions'.

Spectral lines are, in general, indicated by roman capitals with perhaps a Greek suffix (which is not now written as a subscript), as in Hα. Energy levels are printed in italic, but may be typed in roman without underlining in camera-ready copy, unless this would cause confusion in the context concerned.

5.24 Particles and quanta. Upright letters are used to indicate the various types of particle and quanta. Capital Greek letters are used for hyperons and lower-case Greek letters are used for mesons. Lower-case letters are used for nucleons.

neutron	n	electron	e	pion	π	neutrino	ν
proton	p	deuteron	d	muon	μ	photon	γ

The charge of a particle is indicated by the appropriate sign as a superscript, as in π^+, π^-, π°. If no sign is shown then p and e refer to positive protons and negative electrons, respectively. Antiparticles are indicated by an overline above the symbol for the particle, as in $\bar{\nu}$ for an antineutrino.

The symbols representing the particles involved in a nuclear reaction should be specified in the following format:

initial nuclide (incoming particle(s), or quanta , outgoing particle(s) or quanta) final nuclide

For example: ^{14}N (α, p) ^{17}O ^{59}Co (n,γ) ^{60}Co

5.25 Astronomical symbols and names. In general, the special symbols for the names of planets and certain astronomical phenomena, such as conjunctions, should not be used. (A key should always be given if any such special symbols ae used.) The names of planets, bright-stars and other individual objects should be spelt with initial capital letters, but adjectival forms should begin with a small letter, as in: Jupiter, jovian satellites; the Galaxy, galactic coordinates.

In the column headings of tables, for example, the following symbols may be used for the names of the principal planets:

Me	Mercury	E	Earth	J	Jupiter	U	Uranus
V	Venus	EM	Earth-Moon	S	Saturn	N	Neptune
		Ma	Mars			P	Pluto

The Bayer designations of bright stars should be printed in upright type with lower-case Greek letters followed by the standard three-letter abbreviations for the names of the constellations that are given in Table 11. If Greek type is not available, the English name of the Greek letter should be spelt out in full. The English names for the letters of the Greek alphabet are given in section 6.32. General rules for the designation of astronomical objects are given in section 7.

The spectral classifications of stars and other similar symbols and abbreviations should be printed in upright type, as in B5 and cG2. Abbreviations for the names of catalogues are expressed in upright type without full stops, as in FK5 and SAO.

5.26 Dates and times. The instant at which an event (such as an observation) occurred is usually best represented by giving either the julian date (JD) with an appropriate number of decimal places or the calendar date followed by the time of day in conventional form. In an astronomical context the calendar date is best expressed in the sequence year-month-day; and this sequence should always be used in accordance with the international standard (ISO 2014: 1976) if the date is expressed in purely numerical form. The months and days should be represented by two-digit numbers from 01 to 12 for January to December, as in Table 8, and from 01 to 31 for the days of the month; roman numerals should not be used. This sequence is the most convenient for use in computers; for example, dates (including time of day) can be sorted into chronological sequence by treating them as if they were decimal numbers; it also avoids the ambiguity associated with the alternative systems used in Europe and the USA in which, for example, 5/3/88 could mean either 5 March or May 3, 1988. On the printed page the elements of the date (and time) should be separated by spaces or hyphens. In appropriate contexts, such as in ephemerides, the value for the day of the month may be outside the normal range; it is then understood to refer to the preceding or following month; for example, 1988 December 31 may be represented as 1989 January 0. Abbreviations for the names of the months are given in Table 8.

Table 8. Abbreviations for the names of the months.

The recommended abbreviations for the names (in English and French) of the months of the julian and gregorian calendars are given below; 3/4-letter abbreviations should be used in text, but the language-independent two-letter abbreviations may be used in tables.

	English	French			English	French	
01	Jan.	Jan.	JA	07	July	Juil.	JL
02	Feb.	Fev.	FE	08	Aug.	Aout	AU
03	Mar.	Mars	MR	09	Sep.	Sep.	SE
04	Apr.	Avr.	AR	10	Oct.	Oct.	OC
05	May	Mai	MA	11	Nov.	Nov.	NO
06	June	Juin	JN	12	Dec.	Dec.	DE

It is essential that the basis of the date and time system used for reporting observations or predictions be clearly stated since no system is free from ambiguity. The current gregorian calendar was first introduced in 1582 but the julian calendar continued to be used in some countries for several centuries. The julian calendar may be used for specifying calendar dates before it was adopted; it is then referred to as the julian proleptic calendar. References to calendar dates earlier than, say, 1800 should specify which calendarial system is being used. The calendar year 1 BC is followed by the year AD 1 in ordinary historical usage; for astronomical purposes it is convenient to denote the year 1 BC as year 0 and, more generally, the year n BC as year -(n-1).

The precision with which the time should be specified varies according to the circumstances, but it should be borne in mind that an observation may be used in many different contexts. For most purposes, current observations should be reported in the system of coordinated universal time (UTC), but the difference from universal time (UT) is less than one second and may often be ignored. Observations that are timed very precisely and that span several years may be better reported in the system of international atomic time (TAI) since this is free from step-adjustments. Current papers that deal with observations made before 1972 should specify what assumptions, if any, have been made in reducing the timescale used in the original record to that used in the new paper. For example, the adopted values of the difference between clock time and UT or between ephemeris time (ET) and UT should be stated.

The following acronyms for timescales are to be used in all languages.

UT	universal time	TAI	international atomic time
UTC	coordinated universal time	TDT	terrestrial dynamical time
ET	ephemeris time	TDB	barycentric dynamical time

In conversions from calendar date and time to julian date it must be remembered that the julian day begins at 12^hUT (noon on the Greenwich meridian), whereas current calendar days begin at 0^hUT. This difference arose from an earlier use of astronomical days which began at noon so that the same calendar date could be used throughout the night when observations were made in Europe. The scale of Greenwich mean time (GMT) was first used in connection with astronomical days but was later (especially from 1925 onwards) used in connection with civil days beginning at midnight; this name should no longer be used for reporting the times of astronomical observations.

The julian-day system may be used in conjunction with other time systems such as ephemeris time and international atomic time. The name julian ephemeris day (JED) was introduced for the former, but it is now more appropriate to use the abbreviation JD together with the abbreviation for the timescale; thus the column heading in a table could be JD (TAI) while for individual values the timescale may be indicated after the numerical value.

The modified julian date (MJD) is in widespread use for current dates to provide a shorter number for which the decimal part is zero at 0^h UTC (or TAI or UT, as specified); MJD is equal to JD minus 240 0000.5.

Certain 'epochs' (or instants of time) have a special significance in relation to the definition and use of celestial reference systems, and are indicated by adding the suffix .0 or .5 to the number for the year. They may be defined either in terms of the besselian or tropical year or in terms of the julian year, and so the appropriate letter B or J should precede the numerical value which specifies the epoch, which may be either the beginning or middle of a besselian or julian year, as in the following statements.

J1900.0 = 1900 Jan. 0.5 = JD 241 5020.0

B1950.0 = 1950 Jan. 0.923 = JD 243 3282.423

J1986.5 = 1986 July 2.625 = JD 244 6614.125

J2000.0 = 2000 Jan. 1.5 = JD 245 1545.0

These dates refer to instants of UT unless another timescale is indicated; when a precision of better than $0^d.001$ is required the timescale should be specified. Such dates may be understood to refer to besselian epochs for years before 1984 if no prefix letter is given. The letters B and J are often used to indicate that positions are referred to the reference systems for B1950.0 and J2000.0, respectively, but they have a much wider use.

5.27 Notation. In general, the notation adopted for physical quantities and mathematical variables should be that in common use in the subject area of the paper. Nevertheless the meaning of each symbol should be carefully defined when it is first used. If the notation is necessarily extensive and complex, a list of the symbols and their significance should be given at the end of the paper just before the acknowledgements.

6. LANGUAGE, SPELLING AND TRANSLITERATION

6.1 Use of various languages

6.11 Languages used by IAU. The scientific and administrative reports of the IAU are printed in English or in French, sometimes in both languages. Proper names and bibliographic references should be transliterated into the Latin alphabet where necessary. Certain items, such as speeches at official opening ceremonies, may be published in other languages and alphabets (or ideographs). American or British spelling and syntax may be used in English texts provided that one system is used consistently in each text and that no ambiguity of meaning is introduced.

6.12 Accents and diacritical marks should be given whenever this is customary; they should be marked carefully on camera-ready copy using a thin pen if they are not available on the typewriter or printer. Accents are usually omitted from capital letters in French.

6.13 Initial capitals. The use of capitals for the initial letters of words is much more common in English (and German) than in French. It is recommended that the following rules be adopted in both languages in IAU publications. The initial letter of a word should be typed or printed as a capital in the following cases: the first word of a sentence or title; names and titles of persons, but small letters are usually used in separated prefixes (as in de Sitter); individual astronomical objects (such as Earth, the Solar System, Orion, the Crab Nebula, Galactic Centre); geographical places (countries, towns); names of particular organisations (such as Commission 5), meetings (such as the General Assembly) and posts (such as the President), when referring to specific occasions or persons; titles of serial publications; names of individual objects or instruments (Voyager 2); and trade names. An initial capital letter is not required when the name of a person (or object) is used as an adjective or as the name of unit, unless it forms part of the name of an individual object (Isaac Newton Telescope). Initial capital letters should not be used for physical quantities and concepts such as right ascension; in English they are, however, normally used for languages and nationalities. Initial capitals are usually used in references to tables, figures and equations in order to highlight them.

6.14 Hyphens are used for three principal purposes: (a) to form a compound word that represents a new concept, (b) to link words that are used together as an adjective, and (c) to indicate that a word has been split at the end of a line. There are, unfortunately, no rules that are usually accepted and free from difficulties, but some general guidance may be given for each type of usage.

(a) There appears to be a growing tendency to omit the hyphen when two words are combined together to form a new concept, such as database or postcode; many common combinations continue, however, to be printed as two separate words. Hyphens are usually omitted after prefixes, even in such words as coordinate.

(b) A hyphen should normally be included in a compound adjective when one of the two words is a noun, as in second-degree harmonic and 4.2-metre telescope, even though the two words would not otherwise be joined by a hyphen.

(c) Long words at the ends of lines may be split between syllables in order to avoid excessive interword spacings in justified text; a hyphen is then inserted at the end of the line, not at the beginning of the next line. Each part of the word should include at least three letters.

6.15 Inverted commas or primes are used to indicate the beginning and end of a direct quotation from another text and to highlight a particular word or phrase. If a quotation is displayed there is no need to enclose it in inverted commas; it may be indented or printed in a smaller typeface. Attention may be drawn to a word or phrase by printing it in italics. These different usages of inverted commas may be distinguished by the use of double primes as in "This is a quotation", and by the use of single primes to indicate a 'highlight', a colloquialism, or the title of a book. It should be noticed that the punctuation of the main sentence should be outside the inverted commas. Single primes should be used for both usages if the typewriter does not have a double prime as a single character.

6.2 Names and abbreviations

6.21 Abbreviations and acronyms. In general, abbreviations of individual words or names should be followed by a full stop and a space, but the full stop is usually omitted if the last letter of the abbreviation is also the last letter of the word. Acronyms should, however, be spelt without any separation (by a space or a full stop) between the letters, and without terminating full stops. Acronyms or other abbreviations that have come into common use and are read as ordinary words may be typed or printed in lower-case letters, eccept that the initial letter may be capitalized under the rules given above (as in Lageos).

The name of an organisation to which several references are to be made should be given in full (with its usual acronym) on its first occurrence. New or specialised acronyms or abbreviations should only be introduced if it is clear that the benefits of brevity will outweigh the disadvantage of unfamiliarity to the readers.

Lists of some acronyms in common use in astronomy and related fields are given in Table 9 (organisations) and Table 10 (activities and publications). The names of constellations should be abbreviated only to the three-letter forms given in Table 11.

6.22 Names of persons. In scientific papers the names of persons are usually given without courtesy titles or initials unless the latter are required to distinguish between persons with the same surname. In reports of meetings, titles (in abbreviated forms) and initials may be given on the first occurrence of the name, but they should be omitted thereafter. The standard abbreviations for titles are as follows.

English:	Prof.	Dr	Mr	Mrs	Miss	Ms
French:	Prof.	Dr	M.	Mme	Mlle	

6.23 Names of countries and places. In the text of a paper the names or abbreviations of countries, states and cities should be expressed in the form used in the country concerned or in the form commonly used in the language of the text. In mailing addresses the local spellings should be given; postcodes or zipcodes should be included where known. Examples include:

English:	Brussels	Cologne	London	Munich	Moscow	Warsaw
French:	Bruxelles	Cologne	Londres	Munich	Moscou	Varsovie
Local form:	both and Brussel	Köln	London	München	Moscva	Warszawa

The recommended names and abbreviations for the German republics are as follows:

English:	Federal Republic of Germany	FRG
	German Democratic Republic	GDR
French:	République Fédérale d'Allemagne	RFA
	République Démocratique d'Allemagne	RDA
German	Bundesrepublik Deutschland	BRD
	Deutsche Demokratische Republik	DDR

Table 9. Acronyms for the names of organisations.

Most of the following acronyms for the names of organisations are in common use, but the meanings of these and other similar acronyms should be given on first use in any context where they may not be familiar to most readers.

AAS	American Astronomical Society
AG	Astronomische Gesellschaft (Germany)
AGU	American Geophysical Union
AIAA	American Institute of Aeronautics and Astronautics
AIG=IAG	Association Internationale de Geodesie
ASA	American Standards Association
ASA	Astronomical Society of Australia
ASP	Astronomical Society of the Pacific
ASSA	Astronomical Society of South Africa
BAA	British Astronomical Association
BIH	Bureau International de l'Heure
BIPM	Bureau International des Poids et Mesures
CCIR	Comité Consultatif International des Radiocommunications
CDS	Centre de Données Stellaires (Strasbourg, France)
CETEX	Committee on Contamination by Extraterrestrial Exploration
CNRS	Centre National de la Recherche Scientifique (France)
CODATA	Committee on Data for Science and Technology (ICSU)
COSPAR	Committee on Space Research (ICSU)
CSIRO	Commonwealth Scientific and Industrial Research Organization
EPS	European Physical Society
ESA	European Space Agency
ESO	European Southern Observatory
FAGS	Federation of Astronomical and Geophysical Data Analysis Services
GSFC	Goddard Space Flight Center (USA)
IAG	International Association of Geodesy
IAGA	International Association of Geomagnetism and Aeronomy
IAF	International Astronautical Federation
IAU	International Astronomical Union
ICSTI	International Council for Scientific & Technical Information
ICSU	International Council of Scientific Unions
ICSU-AB	Abstracting Board of ICSU; now replaced by ICSTI
IEEE	Institute of Electrical and Electronics Engineers (USA)
IERS	International Earth Rotation Service
INSPEC	Information Services for the Physical and Engineering Communities
IPMS	International Polar Motion Service; now replaced by IERS
ISO	International Standards Organization
ITU	International Telecommunication Union
IUCAF	Inter-Union Committee on Frequency Allocation for Radio and Space Science
IUCI	Inter-Union Committee on the Ionosphere
IUCS	Inter-Union Commission on Spectroscopy
IUCSTP	Inter-Union Commission on Solar-Terrestrial Physics
IUGG	International Union of Geodesy and Geophysics
IUHPS	International Union of History of Philosophy of Science
IUPAC	International Union of Pure and Applied Chemistry
IUPAP	International Union of Pure and Applied Physics
IUTAM	International Union of Theoretical and Applied Mechanics
IWDS	International Ursigram and World Days Service
JOSO	Joint Organisation for Solar Observations
JPL	Jet Propulsion Laboratory (USA)
MIT	Massachusetts Institute of Technology (USA)
MPI	Max Planck Institut (Germany)
NASA	National Aeronautics and Space Administration (USA)
NSF	National Science Foundation (USA)
RAS	Royal Astronomical Society (UK)
RASC	Royal Astronomical Society of Canada
RASNZ	Royal Astronomical Society of New Zealand

Table 9. Acronyms for the names of organisations (continued).

SAAO	South African Astronomical Observatory
SAF	Société Astronomique de France
SAI	Societa Astronomica Italiana
SAO	Smithsonian Astrophysical Observatory (USA)
SERC	Science and Engineering Research Council (UK)
SPARMO	Solar Particles and Radiation Monitoring Organization
UAI=IAU	Union Astronomique Internationale
UN	United Nations
UNESCO	United Nations Educational, Scientific and Cultural Organization
URSI	Union Radio-Scientifique Internationale
WDC	World Data Center
WMO	World Meteorological Organization

Table 10. Acronyms for astronomical activities and publications.

The following acronyms for some astronomical activities and publications are in common use, but the meanings of these and other similar acronyms should be given on first use in any context where they may not be familiar to most readers.

AAA	Astronomy and Astrophysics Abstracts
AGKn	Astronomischer Gesellschaft Katalog Nummer n
AJB	Astronomischer Jahresbericht
APFS	Apparent Places of Fundamental Stars
BD	Bonner Durchmusterung
BSC=BS	Bright Star Catalogue (Yale)
CD	Cordoba Durchmusterung
CMB	Cosmic microwave background
CPD	Cape Photographic Durchmusterung
FKn	Fundamental Katalog Nummer n
GC	General Catalogue (Washington, 1937)
GCVS	General Catalogue of Variable Stars
HD	Henry Draper Catalogue
HDE	Henry Draper (Catalogue) Extension
HEAO	High-Energy Astronomical Observatory
HR	Hertzsprung-Russell
HR	Harvard Revised Photometry Catalog
HST	Hubble Space Telescope
IC	Index Catalogue of nebulae ...
IDS	Index Catalogue of Visual Double Stars
IGY	International Geophysical Year (1957/8)
IQSY	International Quiet Sun Years (1964/5)
IRAS	Infra-Red Astronomy Satellite
ISY	International Space Year (1992)
IUE	International Ultraviolet Explorer satellite
MERIT	Monitor Earth-Rotation and Intercompare the Techniques of observation and analysis (1980/7)
NGC	New General Catalogue of nebulae ...
MHD	Magnetohydrodynamics (also known as hydromagnetics)
PZT	Photographic zenith telescope
QSO	Quasi-stellar object
SAO	Smithsonian Astrophysical Observatory Star Catalog
SI	Système International des Unités
SIMBAD	CDS database: Set of Identifications, Measurements and Bibliography for Astronomical Data
SRS	Southern Reference System
VLBI	Very-long-baseline radio interferometry
ZAMS	Zero-age main sequence

Table 11. Names and standard abbreviations of constellations.

The following list of constellation names and abbreviations is in accordance with the resolutions of the International Astronomical Union (Trans. IAU, 1, 158; 4, 221; 9, 66 and 77). The boundaries of the constellations are listed by E. Delporte, on behalf of the IAU, in, Delimitation scientifique des constellations (tables et cartes), Cambridge University Press, 1930; they lie along the meridians of right ascension and parallels of declination for the mean equator and equinox of 1875.0.

Nominative		Genitive	Nominative		Genitive
Andromeda	And	Andromedae	Lacerta	Lac	Lacertae
Antlia	Ant	Antliae	Leo	Leo	Leonis
Apus	Aps	Apodis	Leo Minor	LMi	Leonis Minoris
Aquarius	Aqr	Aquarii	Lepus	Lep	Leporis
Aquila	Aql	Aquilae	Libra	Lib	Librae
Ara	Ara	Arae	Lupus	Lup	Lupi
Argo[1]	Arg	Argus	Lynx	Lyn	Lyncis
Aries	Ari	Arietis	Lyra	Lyr	Lyrae
Auriga	Aur	Aurigae	Mensa	Men	Mensae
Bootes	Boo	Bootis	Microscopium	Mic	Microscopii
Caelum	Cae	Caeli	Monoceros	Mon	Monocerotis
Camelopardalis	Cam	Camelopardalis	Musca	Mus	Muscae
Cancer	Cnc	Cancri	Norma	Nor	Normae
Canes Venatici	CVn	Canum Venaticorum	Octans	Oct	Octantis
Canis Major	CMa	Canis Majoris	Ophiuchus	Oph	Ophiuchi
Canis Minor	CMi	Canis Minoris	Orion	Ori	Orionis
Capricornus	Cap	Capricorni	Pavo	Pav	Pavonis
Carina	Car	Carinae	Pegasus	Peg	Pegasi
Cassiopeia	Cas	Cassiopeiae	Perseus	Per	Persei
Centaurus	Cen	Centauri	Phoenix	Phe	Phoenicis
Cepheus	Cep	Cephei	Pictor	Pic	Pictoris
Cetus	Cet	Ceti	Pisces	Psc	Piscium
Chamaeleon	Cha	Chamaeleontis	Piscis Austrinus[2]	PsA	Piscis Austrini
Circinus	Cir	Circini	Puppis	Pup	Puppis
Columba	Col	Columbae	Pyxis	Pyx	Pyxidis
Coma Berenices	Com	Comae Berenices	Reticulum	Ret	Reticuli
Corona Austrina[2]	CrA	Coronae Austrinae	Sagitta	Sge	Sagittae
Corona Borealis	CrB	Coronae Borealis	Sagittarius	Sgr	Sagittarii
Corvus	Crv	Corvi	Scorpius	Sco	Scorpii
Crater	Crt	Crateris	Sculptor	Scl	Sculptoris
Crux	Cru	Crucis	Scutum	Sct	Scuti
Cygnus	Cyg	Cygni	Serpens[3]	Ser	Serpentis
Delphinus	Del	Delphini	Sextans	Sex	Sextantis
Dorado	Dor	Doradus	Taurus	Tau	Tauri
Draco	Dra	Draconis	Telescopium	Tel	Telescopii
Equuleus	Equ	Equulei	Triangulum	Tri	Trianguli
Eridanus	Eri	Eridani	Triangulum Australe	TrA	Trianguli Australis
Fornax	For	Fornacis	Tucana	Tuc	Tucanae
Gemini	Gem	Geminorum	Jrsa Major	UMa	Ursae Majoris
Grus	Gru	Gruis	Ursa Minor	UMi	Ursae Minoris
Hercules	Her	Herculis	Vela	Vel	Velorum
Horologium	Hor	Horologii	Virgo	Vir	Virginis
Hydra	Hya	Hydrae	Volans	Vol	Volantis
Hydrus	Hyi	Hydri	Vulpecula	Vul	Vulpeculae
Indus	Ind	Indi			

[1]In modern usage Argo is divided into Carina, Puppis, and Vela.
[2]Australis is sometimes used, in both nominative and genitive.
[3]Serpens may be divided into Serpens Caput and Serpens Cauda.

Note also: LMC Large Magellanic Cloud SMC Small Magellanic Cloud

6.3 Transliteration

6.31 Transliteration of Russian alphabet. The list given in Table 12, which is used in Astronomy and Astrophysics Abstracts, should be used for the transliteration of the Russian alphabet.

Table 12. The Russian alphabet.

А	а	a	П	п	p
Б	б	b	Р	р	r
В	в	v	С	с	s
Г	г	g	Т	т	t
Д	д	d	У	у	u
Е	е	e	Ф	ф	f
Ё	ё	e	Х	х	kh
Ж	ж	zh	Ц	ц	ts
З	з	z	Ч	ч	ch
И	и	i	Ш	ш	sh
Й	й	j	Щ	щ	shch
К	к	k	Ы	ы	y
Л	л	l	Ь	ь	'
М	м	m	Э	э	eh
Н	н	n	Ю	ю	yu
О	о	o	Я	я	ya

6.32 Greek alphabet. The list given in Table 13 shows the English names for the letters of the Greek alphabet and indicates some alternative forms of the letters.

Table 13. The Greek alphabet.

α	A	alpha	η	H	eta	ν	N	nu	τ	T	tau
β	B	beta	θ	Θ	theta	ξ	Ξ	xi	υ	Υ	upsilon
γ	Γ	gamma	ι	I	iota	ο	O	omicron	φ	Φ	phi
δ	Δ	delta	κ	K	kappa	π	Π	pi*	χ	X	chi
ε	E	epsilon	λ	Λ	lambda	ρ	P	rho	ψ	Ψ	psi
ζ	Z	zeta	μ	M	mu	σ	Σ	sigma#	ω	Ω	omega

* ϖ ('curly pi') is an alternative form of π for use in mathematics.

\# ς is an alternative form of σ.

7. DESIGNATION OF ASTRONOMICAL OBJECTS

7.1 General principles

7.11 Nomenclature. All authors should be careful to ensure that they use clear and unambiguous designations for the astronomical objects that they list or discuss, and referees should refer back any papers or tabulations that do not provide satisfactory designations (e.g., see Jaschek 1986). The most appropriate form of designation varies according to the type of object. Solar-system objects are usually designated by simple 'names' that may be alphabetic, numeric or alphanumeric in form; some objects have two names that are of equal standing, while others are given temporary names when they are first observed and permanent names of a different form when their properties have been established. There are standard procedures for the assignment of such names, but for galactic and extragalactic objects the situation is much more complicated and simple names are usually only appropriate for a very limited number of the brightest stars or most interesting objects of each type. Other objects are often identified by means of a sequence number in a catalogue or report, which is itself usually specified by an acronym or other code. Such identifiers when used with a cross-identification index provide access to current general catalogues, which give precise positional data and other information about stars, and also to specialised lists or catalogues, which contain data on stars and non-stellar objects that have particular characteristics, such as variable stars or globular clusters. The identifier of an object in a specialist list is often treated as the usual name of the object, but one object may appear in several such lists and the same acronym may be used for different lists. The resulting duplications and ambiguities lead to much confusion and error (e.g., see Lortet & Dickel 1984). Much effort has been devoted to gathering information about past practices and to devising better systems for the designation of stellar and non-stellar sources of radiation (e.g., see Dickel et al. 1987).

7.12 Information about the many different ways of referring to stars and non-stellar celestial objects (excluding solar-system objects) is gathered together in 'The First Dictionary of the Nomenclature of Celestial Objects' and its Supplement (Fernandez et al. 1983; Lortet & Spite 1986). These volumes should be consulted before any new abbreviation or system of designations is introduced or if help is needed in identifying objects with unknown or ambiguous names or designations. Updates of the dictionary are to be published from time to time. Cross-identification indexes are available at several data centres, and in addition the Strasbourg Data Centre (CDS) provides in the SIMBAD database a dictionary of synonyms, which have been drawn from a list of about 400 different catalogues; SIMBAD is remotely accessible through public networks.

7.13 Generalised designations. Designations by name or catalogue reference number are not satisfactory for general use for large numbers of objects or for new types of objects. It is recommended that authors use more informative designations that specify both the essential characteristics of the object (by means of an alphanumeric codename) and its position to a precision that is sufficient to distinguish it from other objects of the same type. Authors should not, however, introduce new designations for objects for which satisfactory designations already exist. The inclusion of positional information in designations will make it easier to guard against unnecessary duplication in respect of objects of the same type and to look for identities between objects of different types.

A new catalogue which provides additional information about previously known objects should give positional information and at least one earlier designation. New acronyms or abbreviations to indicate the codenames of types of objects, catalogues, authors, observatories or instruments should not be introduced unnecessarily and they should be sufficient to avoid ambiguities; a minimum of two letters should be used for each. Numerals may be included in acronyms, but special characters should be avoided. Full bibliographic information should normally be given about non-standard catalogues or reports from which data have been taken.

The following precepts should be followed in constructing the positional component of a new designation:

give truncated coordinates (not rounded);

give explicitly leading zeros and the sign of declination or latitude;

give decimal points if appropriate, but do not include spaces between the numerals;

give right ascension (in time units) before declination (in arc units) in the format hhmmss.ss±ddmmss.s, or give galactic longitude and latitude (both in degrees and decimals) with the prefix G in the format Gddd.dd±dd.dd; and

precede the right ascension with the letter J to indicate that the position is for the reference system of julian epoch J2000.0 or with B for the system of besselian epoch B1950.0.

Designations including positions should not be changed each time better positions are determined. Supplementary indicators may be added to distinguish between close objects, to indicate association with a larger body, or to specify other object parameters.

The designations of individual objects inside a larger object should specify first the larger object followed by a colon (:), and then an appropriate type or catalogue codename and the coordinates or number for the object itself.

Temporary designations may be based on identifications on a finding chart; it is important that the coordinate system, scale and orientations (N-S and E-W) of the chart be indicated clearly.

A Working group of IAU Commission 34 has developed a note on "specifications concerning names, designations, and nomenclature for astronomical radiation sources outside the solar system"; it is reproduced in appendix D. The generalised form of a designation is specified as

Origin△Sequence△(Specifier)

where △ denotes a blank space; an origin and a sequence must be given, but a specifier is optional. The origin is the codename for the catalogue or type of source; the sequence is the reference number in the catalogue or the positional information; while the specifier contains supplementary information to assist in the identification of the source. The note contains examples and other guidance.

7.2 Designation of objects in the Solar System

7.21 Planets, satellites and rings. There is an IAU Working Group for Planetary System Nomenclature that is responsible for the adoption of names for the surface features of planets and satellites and for newly discovered members of the planetary system (excluding minor planets and comets). Satellites are also designated by numbers that are assigned in the chronological sequence of discovery. Discovery claims should be sent with appropriate details to the IAU Telegram Bureau. The decisions of the Working Group are reported in the Transactions of the IAU from time to time.

7.22 Minor planets (asteroids) and comets. Numerical designations are assigned to minor planets (otherwise known as asteroids) when reliable orbital elements have been determined. The observers, or persons who determined the orbits, are allowed to propose names, usually of persons or mythological characters, that are subject to confirmation by Commission 20; proposals should be sent to the Director of the IAU Minor Planet Center.

Newly discovered comets are assigned temporary designations each consisting of the year of discovery followed by a lower-case letter that indicates the chronological sequence of the discoveries. When the discovery is confirmed the comet is also usually known by the name(s) of the discoverer(s). Subsequently, permanent numbers (in roman capitals) are assigned; they indicate for each year the chronological sequence of the passages of all of the observed comets through perihelion; periodic comets are distinguished by the code P/ in front of the year.

7.23 Meteors. Meteor showers are usually given names according to the constellations in which their radiant points occur or according to the name of the comet with which they are associated (because of an agreement in orbital elements).

Fireballs are usually identified by the dates on which they were seen. Large meteorites are usually identified by the names of the places near where they were found. Meteor craters are also identified by geographical locations.

7.3 Designation of objects outside the Solar System

7.31 Bright stars. About 1000 stars have individual proper names that are derived from early Arabic names or that have been assigned in recent times because of their peculiar characteristics or value for particular purposes (such as astronavigation). A more precise identification of the star should always be given if such a name is used, especially as the names are often different in English, French, and other languages. About 900 such names are listed in the Yale Bright Star Catalogue (BSC, Hoffleit 1982), which gives information about over 9000 stars brighter than magnitude 6.5; it also gives names based on the Bayer system (Greek letter followed by constellation name) or on Flamsteed's catalogue (number and constellation). About 1500 bright stars are listed in the Astronomical Almanac; the tabulation gives the corresponding Bayer, Flamsteed and BSC designations, as well as the mean position and other information about each star.

The 'classical' and other catalogues that are often used for the designation of stars brighter than about magnitude 10 are represented by the following abbreviations.

AGKn	BD	BSC or BS	CD	CPD	FKn
GC	HD	HDE	NZC	SAO	ZC

The full names of these catalogues are given in Table 11. These abbreviations may normally be used in IAU publications without further explanation; other abbreviations for these catalogues must not be used. These abbreviations should not be used for other catalogues, but ZC is already in use for the Zwicky catalogue of clusters of galaxies.

7.312 Faint stars. For stars not included in the catalogues listed above and for which designations have not been given in other catalogues, it is recommended that designations be assigned in the form of an acronym and position in the standard form specified in section 7.12.

7.313 Double stars. The standard form for the name of a component of a double or multiple star is IDS hhmm.mNddmmA, where the position is given by equatorial coordinates for the equinox of 1900.0, N or S is used to indicate northern (+) or southern (-) declinations, and A or B or C ... specifies a component. The IDS catalogue is updated at the U.S. Naval Observatory as WDS. The brighter double stars have alternative designations in bright-star catalogues, as well as in special lists of double or variable stars.

7.314 Variable stars. The 'classical' names of variable stars not having Bayer-type names consist of the name of the constellation preceded by (a) a code of one or two capital letters or (b) the letter V followed by a number. Harvard designations consisting of the letters HV followed by a numerical code have also been widely used for faint variables. The General Catalogue of Variable Stars and SIMBAD provide other designations and cross references. When it exists the classical name should be used in titles.

7.315 Novae. When a nova is discovered it is given a designation of the form Nova Constellation Year. The year is followed by a lower-case letter if more than one nova is discovered in a constellation in the same year. A nova may subsequently be assigned a standard designation as a variable star.

7.316 Supernovae. The IAU Central Telegram Bureau provides temporary designations for supernovae based on the year and order of discovery in the form SN1985A, SN1985B, ... SN1985Z and as required SN1985aa, SN1985ab, ... SN1985az, SN1985ba, ... This system provides continuity with the list established by Zwicky. An archival list is maintained at the California Institute of Technology.

7.317 Supernova remnants. It is recommended that designations of the form SNR followed by the position with respect to the system of J2000.0 be used for newly discovered supernova remnants.

7.318 Planetary nebulae. Planetary nebulae are usually designated by a name consisting of the code PN followed by its galactic coordinates in degrees (without a decimal point) and a serial number. The appropriate stellar designation should also be given when available.

7.319 Stellar Clusters. Some clusters have proper names. Many open clusters and globular clusters of stars are identified by their numbers in Messier's catalogue (for the brightest) or by their numbers in the New General Catalogue (NGC) and supplementary Index Catalogues (IC). There are also other lists; the current standard designation takes the form Chhmm±ddd, and should be given.

7.320 Galaxies. Some galaxies have proper names, but in general bright galaxies are also identified by their Messier, NGC or IC numbers.

7.331 Radio sources. At first, radio sources were named by giving the constellation name followed by a capital letter. Then the reference numbers in the 3rd Cambridge catalogue were commonly used to give designations such as 3C 273. The IAU adopted at Delhi (1985) a resolution concerning the nomenclature for radio sources in accordance with the general principles given in section 7.1. In particular it recommended the use of designations for newly catalogued sources in the form of a catalogue acronym followed by the right ascension and declination with respect to the reference system of J2000.0.

7.331 Pulsars. For this particular type of radio source the position should be preceded by the acronym PSR. Older names, such as CP1919 for the first pulsar that was discovered, are now obsolete.

7.332 Quasars. At present there is no special standard form for the designation of quasars.

7.34 X-ray sources. At first X-ray sources were designated by the constellation abbreviation followed by the letter X and a sequence number within the constellation, as in Sco X-1. Now designations are taken from the catalogues for the surveys by particular satellites.

7.35 Other types of objects. A wide variety of systems are in use for the designation of other types of celestial objects; many are specific to the wavelength of observation. Advice on specific problems is available from the persons listed at the end of Appendix B, where references to information about existing designations are also given.

REFERENCES

Anderson, H. L., (Ed.), 1981a. Physics Vade Mecum. New York: American Institute of Physics. ISBN 0 11 88318 289 0.

Anderson, H. L., 1981b. The international system of units (SI). In: Anderson 1981a, 5-10.

AIP 1978. Style Manual. (3rd ed.) New York: American Institute of Physics.

ApJ, 1983. A manual of style for the Astrophysical Journal and Supplement Series. (Rev. ed.) University of Chicago Press.

Barzun, J., Graff, H. F., 1977. The modern researcher. (3rd ed.) New York: Harcourt Brance Jovanovich.

Bell, R. J., Goldman, D. T., 1986. SI: The International System of Units. (5th ed.) London: Her Majesty's Stationery Officer. ISBN 0 11 887527 2.

Booth, Vernon, 1985. Communicating in science: writing and speaking. Cambridge University Press. ISBN 0 521 27771 X.

BSI, 1961. Preparation of mathematical copy and correction of proofs. BS 1219M: 1961. London: British Standards Institution.

BSI, 1976a. Recommendations for the preparation of indexes for books, periodicals and other publications. BS 3700: 1976. London: British Standards Institution.

BSI, 1976b. Recommendations for bibliographic references. BS 1629: 1976. London: British Standards Institution.

BSI, 1978. Citing publications by bibliographical references. BS 5605: 1978. London: British Standards Institution.

BSI, 1984. Specification for the abbreviation of title words and titles of periodicals. BS 4148: 1984. London: British Standards Institution.

Butcher, Judith, 1981. Copy-editing: the Cambridge handbook. (2nd ed.) Cambridge, UK: Cambridge Univ. Press. ISBN 0 521 23868 4.

CBE, 1983. CBE Editors Style Manual. (5th ed.) Bethesda, Maryland: Council of Biology Editors.

Chaundy, T. W., Barrett, P. R., Batey, C., 1954. The printing of mathematics. Oxford University Press.

Chiswall, B., Grigg, E. C. M., 1971. SI Units. Sydney: John Wiley.

Cochran, W., Fenner, P., Hill, M., 1973. Geowriting: a guide to writing, editing, and printing in Earth science. Washington, DC: American Geological Institute.

de Jager, C., & Jappell, A., 1971. Part 2 of Astronomer's handbook: Style book, Trans. Int. Astron. Union 14B, 254-264.

Day, Robert A., 1989. How to write and publish a scientific paper. (3rd ed.) Cambridge University Press. ISBN 0 521 36760 3.

Dickel, H. R., Lortet, M.-C., de Boer, K. S., 1987. Designation and nomenclature for diffuse radiating sources. Astron. Astrophys. Suppl. Ser. 68, 75-80.

Drazil, J. V., 1983. Quantities and units of measurement: a dictionary and handbook. London: Mansell Publishing Limited. ISBN 0 7201 1665 1. Wiesbaden: Oscar Brandstetter Verlag. ISBN 3 87097 117 7.

Fernandez, A., Lortet, M.-C., Spite, F., 1983. The first dictionary of nomenclature of celestial objects (Solar System excluded). Astron. Astrophys. Suppl. Ser., 52 (4), pp. 200.

Fowler, H. W., 1965. Revised by E. Gowers. A dictionary of modern English usage. (2nd ed.) Oxford: Oxford University Press.

Gowers, E., 1986. Revised by S. Greenbaum & J. Whitcut. The complete plain words (3rd ed.) London: Her Majesty's Stationery Office, ISBN 0 11701121 5; also by Penguin Books (1987).

Hart, H., 1983. Hart's rules for compositors and readers at the University Press Oxford. (39th ed.) Oxford: Oxford Univ. Press. ISBN 0 19 212983 X.

Heck, A., Manfroid, J., 1985. International directory of astronomical associations and societies (IDAAS 1986). Obser. Strasbourg: Publ. Spec. CDS 8.

Heck, A., Manfroid, J., 1986. International directory of professional astronomical institutions (IDPAI 1987). Obser. Strasbourg: Publ. Spec. CDS 9.

Hoffleit, D., 1982. The bright star catalogue. (4th ed.) New Haven, Connecticut: Yale University Observatory.

ICSU AB, 1978. International serials catalogue. Paris: International Council of Scientific Unions Abstracting Board. ISBN 92 9027 004 7.

ISO, 1972. Documentation - international code for the abbreviation of titles of periodicals. ISO 4-1972 (E). International Standards Organisation.

ISO, 1974. Documentation - international list of periodical title word abbreviations. ISO 833-1974 (E). International Standards Organisation.

Jaschek, C., 1986. Designation problems in astronomy. Q.J. Roy. Astron. Soc. 27, 60-63.

Lortet, M.-C., Dickel, H. R., 1984. Tricks and traps in astronomical nomenclature. Astron. Astrophys. Suppl. Ser. 56, 1-4.

Lortet, M.-C., Spite, E., 1986. First supplement to the first dictionary of the nomenclature of celestial objects (Solar System excluded). Astron. Astrophys. Suppl. Ser., 64(2), 329-389.

O'Connor, M., Woodford, F. P., 1976. Writing scientific papers in English. Amsterdam: Elsevier.

OUP, 1981. The Oxford dictionary for writers and editors. Oxford University Press.

Pecker, J.-C., 1966. (a) Manuel de rédaction; (b) Style book. Trans. Int. Astron. Union 12C, (a) 55-100 (in French); (b) 101-127 (in English).

Prentice-Hall, (Anon.), 1974. Words into type. Englewood Cliffs, New Jersey: Prentice-Hall.

Royal Society, 1974. General notes on the preparation of scientific papers. (3rd ed.) London: Royal Society.

Strunk, W., White, E. B., 1979. The elements of style. (3rd ed.) New York: Macmillan. ISBN 0 02 418200 1.

Young, B., (Ed.), 1969. A manual of style for authors, editors, and copywriters. (12th ed.) Chicago: University of Chicago Press. SBN 226 77008 7.

AGU GUIDELINES FOR PUBLICATION

The following 'Guidelines for Publication' have been issued by the American Geophysical Union in EOS (1988 October 11) and are appropriate to astronomy as well as to geophysics.

A. Obligations of Editors of Scientific Journals

1. An editor should give unbiased consideration to all manuscripts offered for publication, judging each on its merits without regard to race, gender, religious belief, ethnic origin, citizenship, or political philosophy of the author(s).
2. An editor should process manuscripts promptly.
3. The editor has complete responsibility and authority to accept a submitted paper for publication or to reject it. The editor may confer with associate editors or reviewers for an evaluation to use in making this decision.
4. The editor and the editorial staff should not disclose any information about a manuscript under consideration to anyone other than those from whom professional advice is sought.
5. An editor should respect the intellectual independence of authors.
6. Editorial responsibility and authority for any manuscript authored by an editor and submitted to the editor's journal should be delegated to some other qualified person, such as another editor or an associate editor of that journal. Editors should avoid situations of real or perceived conflicts of interest. If an editor chooses to participate in an ongoing scientific debate within his journal, the editor should arrange for some other qualified person to take editorial responsibility.
7. Unpublished information, arguments, or interpretations disclosed in a submitted manuscript should not be used in an editor's own research except with the consent of the author.
8. If an editor is presented with convincing evidence that the main substance or conclusions of a paper published in an editor's journal are erroneous, the editor should facilitate publication of an appropriate paper pointing out the error and, if possible, correcting it.

B. Obligations of Authors

1. An author's central obligation is to present a concise, accurate account of the research performed as well as an objective discussion of its significance.
2. A paper should contain sufficient detail and reference to public sources of information to permit the author's peers to repeat the work.
3. An author should cite those publications that have been influential in determining the nature of the reported work and that will guide the reader quickly to the earlier work that is essential for understanding the present investigation. Information obtained privately, as in conversation, correspondence, or discussion with third parties, should not be used or reported in the author's work without explicit permission from the investigator with whom the information originated. Information obtained in the course of confidential services, such as refereeing manuscripts or grant applications, should be treated similarly.
4. Fragmentation of research papers should be avoided. A scientist who has done extensive work on a system or group of related systems should organize publication so that each paper gives a complete account of a particular aspect of the general study.
5. It is inappropriate for an author to submit manuscripts describing essentially the same research to more than one journal of primary publication.
6. A criticism of a published paper may sometimes be justified; however, in no case is personal criticism considered to be appropriate.
7. To protect the integrity of authorship, only persons who have significantly contributed to the research and paper preparation should be listed as authors. The corresponding author attests to the fact that any others named as authors have seen the final version of the paper and have agreed to its submission for publication. Deceased persons who meet the criterion for co-authorship should be included, with a footnote reporting date of death. No fictitious name should be listed as an author or co-author. The author who submits a manuscript for publication accepts the responsibility of having included as co-authors all persons appropriate and none inappropriate.

C. Obligations of Reviewers of Manuscripts

1. Inasmuch as the reviewing of manuscripts is an essential step in the publication process, every scientist has an obligation to do a fair share of reviewing.
2. A chosen reviewer who feels inadequately qualified or lacks the time to judge the research reported in a manuscript should return it promptly to the editor.
3. A reviewer of a manuscript should judge objectively the quality of the manuscript and respect the intellectual independence of the authors. In no case is personal criticism appropriate.
4. A reviewer should be sensitive even to the appearance of a conflict of interest when the manuscript under review is closely related to the reviewer's work in progress or published. If in doubt, the reviewer should return the manuscript promptly without review, advising the editor of the conflict of interest or bias.
5. A reviewer should not evaluate a manuscript authored or co-authored by a person with whom the reviewer has a personal or professional connection if the relationship would bias judgment of the manuscript.
6. A reviewer should treat a manuscript sent for review as a confidential document. It should neither be shown to nor discussed with others except, in special cases, to persons from whom specific advice may be sought; in that event, the identities of those consulted should be disclosed to the editor.
7. Reviewers should explain and support their judgments adequately so that editors and authors may understand the basis of their comments. Any statement that an observation, derivation, or argument had been previously reported should be accompanied by the relevant citation.
8. A reviewer should be alert to failure of authors to cite relevant work by other scientists. A reviewer should call to the editor's attention any substantial similarity between the manuscript under consideration and any published paper or any manuscript submitted concurrently to another journal.
9. Reviewers should not use or disclose unpublished information, arguments, or interpretations contained in a manuscript under consideration, except with the consent of the author.

D. Obligations of Scientists Publishing Outside the Scientific Literature

1. A scientist publishing in the popular literature has the same basic obligation to be accurate in reporting observations and unbiased in interpreting them as when publishing in a scientific journal.
2. The scientist should strive to keep public writing, remarks, and interviews as accurate as possible consistent with effective communication.
3. A scientist should not proclaim a discovery to the public unless the support for it is of strength sufficient to warrant publication in the scientific literature. An account of the work and results that support a public pronouncement should be submitted as quickly as possible for publication in a scientific journal.

IAU RESOLUTION 3 (1988) ON THE IMPROVEMENT OF PUBLICATIONS

Resolution A3: Improvement of Publications

The XXth General Assembly of the International Astronomical Union

recognising

- the need to develop clear lines of communication between the various branches of astronomy and other related scientific disciplines;
- the desirability of promoting ease of access to information contained in the astronomical literature;
- the advantages that would follow from a reduction in the variety of the editorial requirements for the submission of papers and reports; and
- the importance of identifying astronomical objects by clear and unambiguous designations; and

noting

- the growth in the cadre of young scientists trained in the use of the International System (SI) of units and widespread adoption of SI in other scientific and technical areas; and
- the substantial measure of agreement that has been reached during the drafting of the new IAU Style Manual for the preparation of astronomical papers, reports and books;

recommends

that the authors and the editors of the astronomical literature adopt the recommendations in the IAU Style Manual, which is to be published in the Transactions of the Union and reprinted for wide distribution and greater convenience;

in particular, it urges authors and editors:

1. to use only the standard SI units and those additional units that are recognised for use in astronomy, as recommended by Commission 5;
2. to adopt the conventions for citations and references that are given in the IAU Style Manual and that are exemplified in Astronomy and Astrophysics Abstracts; and
3. to ensure that all astronomical objects referred to in the literature are designated clearly and unambiguously in accordance with the recommendations of the Union.

Note:

The Executive Committee recognises that the replacement of CGS by SI units will require an adjustment of practice on the part of many astronomers; this will no doubt take time. Consequently, we urge that the total conversion from CGS to SI units by all organs of communication shall be accomplished by the time of the next General Assembly (1991).

In the meantime we request that the major journals should publish, once a year, a table of conversions between CGS and SI units, as provided by Commission 5.

SI
: International System (units)

CGS
: Centimeter, Gramme, Second (units)

EXTRACTS FROM IAU RULES FOR SCIENTIFIC MEETINGS

The following notes on the publication of the proceedings of conferences sponsored by the IAU have been taken from the 'Rules for Scientific Meetings' given in IAU Information Bulletin 58 (June 1987). Some additional notes are given in No. 61 (Jan. 1989).

Proceedings of IAU Symposia

The IAU believes that the Proceedings of Symposia remain of general interest for a considerable period of time and that early publication in uniform style to a high standard is desirable. Publication and distribution have therefore been entrusted to a commercial publishing house.

The main responsibility of the IAU as joint publisher is the maintenance of a high standard of scientific value, originality and accuracy. The commercial publishing house has been contracted to ensure early publication and thereafter to take financial responsibility.

The Executive Committee, by approving the choice of Editor or Editors, places the main burden of maintaining the required scientific standard on one or two IAU members on the understanding that they are familiar with the scientific matter of the Symposium and are persons with some experience in editorial tasks. The Editors receive no financial renumeration for their service to the Union.

It is essential that the Editor, or one of two Joint Editors, should have an excellent knowledge of the English Language.

The Editor is responsible for the scientific value, the appearance and rapid delivery to the Publisher (usually within three months of the end of the Symposium) of the copy for the Proceedings. The main editorial tasks are:

(a) To inform participants in ample time before the meeting in what general form their contributions should be submitted and what arrangements have been made with the publisher for receipt of camera-ready copy. The number of printed pages available to each contributor should be determined in good time.

(b) To inform the participants about IAU rules for publication of IAU Proceedings and to emphasize that any contributed papers must be refereed before acceptance for publication.

(c) In advance of the Symposium, in close consultation with the SOC and LOC, to agree and arrange the precise details for recording and reporting the scientific discussion that takes place at the meeting. Difficulties in this respect must not cause undue delay in preparing for publication - it is better to sacrifice discussion rather than hold up publication.

(d) To arrange with members of the SOC for the refereeing of any contributed paper if an Editor is unable to do so.

(e) To reduce the length of papers and discussion, to avoid duplication and to improve presentation where necessary.

(f) To check whether IAU rules have been followed in each contribution and to arrange for re-typing if necessary.

(g) To write the Introduction, Table of Contents, and obtain a Final Summary of the Symposium, maintaining uniformity with recent IAU Symposium Proceedings.

(h) To maintain all necessary contact with the Publishing House, in accordance with current "Instructions for Editors" available from the Assistant General Secretary.

(i) To maintain close contact with the Assistant General Secretary on all matters affecting progress of publication arrangements, especially keeping him informed of the material sent to the publishers and of any unexpected delays or alterations. In general, the Editor will not be able to allow time for substantial revision.

In the proceedings of Symposium volumes, Invited Reviews and Invited Papers will be allowed extended publication subject to a maximum number (decided by the SOC and Editor) of camera-ready pages. Editorial discretion will be used to ensure that material previously published elsewhere is not duplicated. It should be noted that in extenso algebraic derivations, lists of observations and other tabular data are inappropriate in a Symposium volume.

The SOC has discretion to make the decision whether or not the contributed papers (oral and poster) are published in the Symposium proceedings. If a decision to publish is taken, then contributed papers (including posters), after editorial refereeing, will be published in not more than two camera-ready pages. Should the SOC decide that any contributed paper is given more extended publication, the same rules as for invited reviews and invited papers will apply.

It is the policy of the IAU and that of their publishers to publish in camera-ready form. It is found that such a policy gives reasonable uniformity of appearance combined with speed of production. Authors will be sent camera-ready sheets, instructions for their use and the total amount of space available to them, by the publishers prior to the Symposium. The final camera-ready manuscripts should be sent to the Editor(s) either before the beginning of the Symposium or handed over at the Symposium. Papers not available to the Editor(s) in camera-ready form by the end of the Symposium will be deemed to have been withdrawn from publication.

In order to obtain a presentation that will indicate that the volumes of IAU Symposia form a series, it is requested that Editors adhere to the following recommendations:

(The title page should have the format specified in section 3.6)

The Editor's introduction should mention circumstances of the organization of the Symposium, and should list the supporting organizations and the members of the Scientific and Local Organising Committees. It should express appreciation to those to whom it is due. The support of the IAU and other Unions, etc., should be recognized as well as that of other international, national, or local organizations.

(The Editors should provide an alphabetical index of names and subject headings in accordance with the instructions of the publishers.)

Symposium volumes should be published 6-8 months after the Symposium.

Participants obtain some free reprints of their contributions and other copies can be ordered. All IAU members can purchase Symposium volumes at reduced prices.

Proceedings of IAU Colloquia

The publication of IAU Colloquia should follow the same guidelines as for IAU Symposia. However, unless produced by the IAU publisher, the relevant manuscripts need not be in camera-ready form and some variation is allowed. Manuscript length is at the discretion of the SOC who have responsibility for the decision on whether or not to publish and the format of the publication. However, in order to facilitate archival retrieval all published IAU Colloquia proceedings must adopt the same form of Title Page. The title page should follow the same format as for Symposia replacing "Symposium" by "Colloquium" as appropriate.

Proceedings of IAU Regional Meetings

If it is decided to publish the proceedings of a Regional Meeting, the same guidelines as for an IAU Symposium or Colloquium should be followed as far as is practicable given the format of the meeting and the method of publication adopted. In order to facilitate archival retrieval the title page should indicate explicitly below any other title of the meeting.

Proceedings of the ... ... Regional Meeting of the International Astronomical Union, held in "place", "country", "date"

followed, if appropriate by the words "Organized by the IAU in cooperation with ...".

IAU RECOMMENDATIONS (1988) CONCERNING THE DESIGNATION OF RADIATION SOURCES OUTSIDE THE SOLAR SYSTEM

Recognizing the need for clear and unambiguous nomenclature of all astronomical sources of radiation, the prolific increase in the number of identified sources, and the requirements for data storage and information retrieval, the following set of specifications (developed and endorsed by the International Astronomical Union) is recommended for use throughout the field of astronomy for radiation sources outside the Solar System. All authors of papers and contributors to data bases of any kind are urged to adhere to these specifications, since otherwise significant data may be irretrievably lost. When existing designations are used in listings, they should never be altered. Object listings should contain a second designation and/or positional information for objects with unfamiliar names.

The *designation* of an astronomical source shall consist of the following parts:

Origin∆Sequence∆(Specifier)

Note that the triangle (∆) is used here to denote a blank space; the parentheses are required if a specifier is included. *Origin* and *sequence* are essential, *specifier* is optional; the number of blanks may be larger in machine-readable files to right justify numerical or tabular data.

The following *examples* illustrate the recommended form of astronomical designations:

NGC∆205
PKS∆1817–43
CO∆J0326.0+3041.0
H2O∆G123.4+57.6∆(VLSR=–185)

The *origin* is a "word" or acronym to specify the catalog or collection of objects. It may be constructed from catalog names (*e.g.*, NGC, BD), the names of authors (RCW), types of objects (PSR, PN), types of sources (13CO, HCN), instruments or observatories used (1E, IRAS), etc.

The following rules apply to the construction of *new* origins:

- ***Origin* shall consist of at least two characters.**
- ***Origin* shall consist of letters and/or numerals only; special characters should be avoided.**
- ***Origin* shall be unique, *i.e.*, the appropriate reference literature (see below) should be checked to avoid duplication with existing catalog designations, constellation names, abbreviations of object types, etc.**
- **The authors of a new catalog shall specify in their article which acronym is to be used in *origin*. Users shall never abbreviate *origin*.**

The *sequence* is normally a numerical field to uniquely determine the object within a catalog or collection. It may be a sequence number within a catalog (*e.g.*, HD∆224801), or it may be based on coordinates.

If coordinates in any form are used to encode an object, the following rules apply:

- Coordinates shall be preceded by a code for the reference frame, specifically G for galactic coordinates, and B for Besselian 1950 or J for Julian 2000 equatorial coordinates if confusion might be possible.
- Coordinates shall be specified as LLL.ll±BB.bb for galactic coordinates and for equatorial coordinates as HHMMSS.s±DDMMSS.s, without spaces; fewer digits may be used as appropriate.
- Coordinates shall be truncated (not rounded), thus defining a unique (small) field on the sky in which the object is located.
- Coordinates shall contain leading zeroes if necessary, and the plus or minus sign: ±BB.bb or ±DDMMSS.s.
- Coordinates used in designations shall be considered as names; therefore, they shall not be changed even if the positions become more accurately known (e.g., at a different epoch: BD -25 765 stays, even though its declination is now −26°).
- If at some stage subcomponents or multiplicity of objects is recognized, the best designation solution is to name these subcomponents with letters or numerals, which then are added to *sequence* with a colon, e.g., NGC 1818:B12.

The *specifier* is optional and allows one to indicate association with larger radiating sources (*e.g.*, M△31, W△3) or to indicate other object parameters. However, they are *not* required syntax and are enclosed in parentheses.

Examples of complete designations are:

Designation	Position	
Origin△Sequence△(Specifier)	RA (2000)	DEC (2000)
BD -3 5750	00 02 02.4	-02 45 59
H20 B0446.6+7253.7	04 46 37.3	+72 53 47
AC 211 (=1E 2127+119; in M 15)	21 30 15.54	+11 43 39.0
PN G001.2-00.3	17 49 36.9	-28 03 59
R 136:a3	05 38 42.4	-69 06 03

There exists a multitude of improper, confusing or unclear designations in the literature. General rules and advice on how to generate designations can be found in "The First Dictionary of the Nomenclature...", cited below. Examples of improper use of designations are:

BD 4° 14	declination sign is missing
N221	unclear source: NGC, or N in LMC ?
IRAS 5404-220	leading zero missing; poor position
P 43578	one-letter origin is ambiguous

THE IAU STYLE MANUAL (1989)

INDEX

Printed in the United States
By Bookmasters